Principles of

THIRD EDITION

DIGITAL **UPDATE**

Principles of Life

THIRD EDITION

DIGITAL **UPDATE**

David M. Hillis
University of Texas at Austin

Mary V. Price
Emerita,
University of California, Riverside

Richard W. Hill
Emeritus,
Michigan State University

David W. Hall

Marta J. Laskowski
Oberlin College

Austin • Boston • New York • Plymouth

Senior Vice President, STEM: Daryl Fox
Executive Program Director: Sandy Lindelof
Executive Program Manager: Lisa Lockwood
Executive Marketing Manager: Will Moore
Director of Development, STEM: Debbie Hardin
Development Editor: Marita Bley
Executive Project Manager, Content, STEM: Katrina Mangold
Director of Content: Heather Southerland
Senior Media Editor: Jennifer Compton
Executive Media Editor: Amy Thorne
Associate Editor: Nathan Livingston
Editorial Assistant: Natalie Noland
Marketing Assistant: Elizabet Cabrera
Senior Director of Content Management Enhancement: Tracey Kuehn
Senior Managing Editor: Lisa Kinne
Senior Content Project Manager: Edward Dionne
Senior Workflow Project Manager: Lisa McDowell
Director of Design, Content Management: Diana Blume
Senior Design Services Manager: Natasha Wolfe
Senior Cover Design Manager: John Callahan
Art Manager: Matthew McAdams
Director of Digital Production: Keri deManigold
Advanced Media Project Manager: Dan Comstock
Executive Permissions Editor: Cecilia Varas
Composition: Lumina Datamatics, Inc.
Printing and Binding: LSC Communications

Library of Congress Control Number: 2021947047

Student Edition Paperback:
ISBN-13: 978-1-319-45029-8
ISBN-10: 1-319-45029-6

Student Edition Loose-leaf:
ISBN-13: 978-1-319-45031-1
ISBN-10: 1-319-45031-8

International Edition:
ISBN-13: 978-1-319-49416-2
ISBN-10: 1-319-49416-1

Printed in the United States of America

1 2 3 4 5 6 27 26 25 24 23 22

Macmillan Learning
120 Broadway
New York, New York 10271
www.macmillanlearning.com

To Our Readers

Welcome to *Principles of Life*. This book is a *concept-centered* introduction to biology. When we ourselves were students, we realized that—six months after a course ended—the concepts were what we remembered. By now, educational research has confirmed that in undergraduate courses, student needs are served best by an emphasis on concepts. With its concept-centered approach, this book is written for you.

With all the rapid advances in information-sharing on the internet, simple facts are easier than ever to look up. This easy accessibility changes our relationship to facts. In particular, it places an ever-increasing premium on our ability to evaluate facts and integrate them into coherent knowledge. This reality helps explain why central organizing frameworks—concepts—are so important.

Some of you intend to go on into science or medicine. For you, *Principles of Life* recognizes that a solid foundation in concepts will be important throughout your career, helping you to assimilate and use ever-enlarging spheres of factual knowledge.

Some of you expect to move on into other interests after completing your introductory courses in science. For you, long-term knowledge will be your pay-off for your study of biology. *Principles of Life* recognizes that you will be most likely to recall concepts, not isolated facts, as years go by.

Believe it or not, scientists and educators occasionally gather together to debate how university and college courses can best serve your needs. A decade ago, a group of about 500 helped formulate a watershed report—*Vision and Change in Undergraduate Biology Education: A Call to Action*.[1]

Two questions—both of great importance to you—are stressed in *Vision and Change* (*V & C*). First, what are the core concepts that students of the twenty-first century need to understand? Second, what competencies—personal abilities—do you need to develop to succeed? *Principles of Life* is focused on helping you master both the core concepts and the competencies.

Principles of Life, as part of its conceptual approach, places central importance on the five **core concepts** pinpointed by V & C:

Principles of Life highlights the six V & C **competencies** in every chapter, helping you develop your ability to:

• evolution	• apply the process of science
• the relationship between structure and function	• use quantitative reasoning
• information flow, exchange, and storage	• use modeling and simulation
• pathways and transformations of energy and matter	• tap into the interdisciplinary nature of science
• systems biology	• communicate and collaborate with other disciplines
	• understand the relationship between science and society

[1]*Vision and Change* is published by the American Association for the Advancement of Science. To read the report, go to https://visionandchange.org/ and upload the 2011 report.

About the Authors

from left: Mary Price, David Hall, Marta Laskowski, David Hillis, Richard Hill

DAVID M. HILLIS is the Alfred W. Roark Centennial Professor in Integrative Biology at the University of Texas at Austin, where he also has directed the Center for Computational Biology and Bioinformatics, the Biodiversity Center, and the School of Biological Sciences. Dr. Hillis has taught courses in introductory biology, genetics, evolution, systematics, and biodiversity. He is a member of the National Academy of Sciences and the American Academy of Arts and Sciences. He was awarded a John D. and Catherine T. MacArthur Fellowship, and has served as President of the Society for the Study of Evolution and of the Society of Systematic Biologists. He served on the National Research Council committee that wrote the report *BIO 2010: Transforming Undergraduate Biology Education for Research Biologists* and currently serves on the Executive Committee of the National Academies Scientific Teaching Alliance.

MARY V. PRICE is Professor of Biology, Emerita, at the University of California, Riverside, and Adjunct Professor in the School of Natural Resources and the Environment at the University of Arizona. In "retirement" she continues to teach, investigate, and publish. Dr. Price has taught, mentored, and published with students at all levels and particularly enjoys leading field classes in the arid regions of North America and Australia, and the tropical forests of Central America, Africa, and Madagascar. Her research focuses on understanding not only the ecology of North American deserts and mountains but also on how science really works.

RICHARD W. HILL is Emeritus Professor in the Department of Integrative Biology at Michigan State University and a frequent Guest Investigator at Woods Hole Oceanographic Institution. He is the senior author of the leading textbook on animal physiology. Among the awards he has received are the Outstanding Faculty Award, Meritorious Faculty Award, and election as Fellow of the AAAS. His research interests include: temperature regulation and energetics in birds and mammals, especially neonates; and environmental physiology of marine tertiary sulfonium and quaternary ammonium compounds.

DAVID W. HALL taught a variety of classes at Wake Forest University, the University of Texas, and the University of Georgia during his academic career. He especially enjoyed teaching introductory biology and genetics to undergraduates and received several teaching awards for his efforts in the classroom. Ever since high school, he has been captivated by the living world but was initially overwhelmed by the enormous diversity of life. However, he soon realized that there are fundamental principles that unite all organisms, which greatly facilitates the study of biology. Helping students learn these principles was the foundation of his biological teaching.

MARTA J. LASKOWSKI is a Professor in the Biology Department at Oberlin College. Dr. Laskowski has mentored undergraduate students in research and has taught introductory biology, skills-based first year seminars (Feeding the World), plant physiology, and plant development. She heads an effort at Oberlin, funded by the HHMI Inclusive Excellence program, to enhance the climate for and success of a diverse student population in STEM. One of her numerous journal articles resulted in a *Guinness World Record* for the fastest opening flower (*Cornus canadensis*; bunchberry). A college class in developmental biology so captivated her that she decided to focus her research on discovering the intricate subcellular interactions that establish the plant root system.

All the new enhancements add not just to the learning experience of the students, but also make teaching this material that much more focused and aligned with something that is emerging as an important standard.... A welcome improvement in a biology textbook, designed for both instructors and students, which adopts key pedagogical competencies, wholly aligned with the *Vision and Change* directive."

Kamal Dulai, *University of California, Merced*

The new toolbox of active learning opportunities integrated into the Third Edition of *Principles of Life* provides numerous opportunities for students and faculty to master *Vision and Change*'s Core Competencies. If used creatively, this text contains essential tools for mastering biology."

Justen Whittall, *Santa Clara University*

Principles of Life, Third Edition Digital Update

* = New Chapter

Principles of Life—Tour of the Third Edition Digital Update

Because success as a biologist means more than just succeeding in the first biology course

If you're concerned that the practical skills of biology will be lost when you move on to the next course or take your first step into the "real world," *Principles of Life* lays a solid foundation for later courses and for your career. Expanding on its pioneering concept-driven approach, experimental data-driven exercises, and active learning focus, the Third Edition Digital Update includes features designed to help you master concepts and become skillful at solving biological problems.

Research shows that when students engage with a course, it leads to better outcomes. *Principles of Life* is a holistic solution that has been designed from the ground up to actively engage you and help develop your skills as a biologist.

With its focus on key competencies foundational to biology education and careers, self-guided adaptive learning, and online resources, *Principles of Life* is the resource you need to succeed.

THINK LIKE A **SCIENTIST**

Quantitative Reasoning

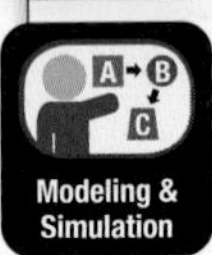

Modeling & Simulation

Changes in Earth's physical environment have affected the evolution of life

In the experiment shown in Investigation Figure 17.8, body mass of individuals in the experimental populations of *Drosophila* increased (on average) about 2 percent per generation in the high-oxygen environment (although the rate of increase was not constant over the experiment). In the Permian, giant flying insects, such as dragonflies the size of modern hawks, inhabited Earth. Is the rate of increase in body mass

A FOCUS ON SKILLS AND CORE COMPETENCIES

The AAAS *Vision and Change* report's six "core competencies," related to quantitative reasoning, simulation, and communication, are integrated both implicitly throughout the text and explicitly in a key feature, Think Like a Scientist. TLAS boxes develop these core competencies and have been designed specifically to teach you the skills you need to become a functional, practical, effective scientist.

ANALYZE THE DATA

After Kashefi and Lovley isolated Strain 121, they examined its growth at various temperatures. The table below shows generation time (time between cell divisions) at nine temperatures.

Temperature (°C)	Generation time (hr)
85	10
90	4
95	3
100	2.5
105	2
110	4
115	6
120	20

A FOCUS ON DATA

Principles of Life has always emphasized the role of research and experimentation in the introductory biology curriculum. You will learn about the scientific method and experimental design and understand how real research continues to drive our understanding of life on Earth.

Chemiosmotic Mechanism

Examining Mitochondrial Poisons: Cyanide

In your groups, predict what effect HCN would have on the inputs and outputs of these reactions. What will build up, and what will be depleted?

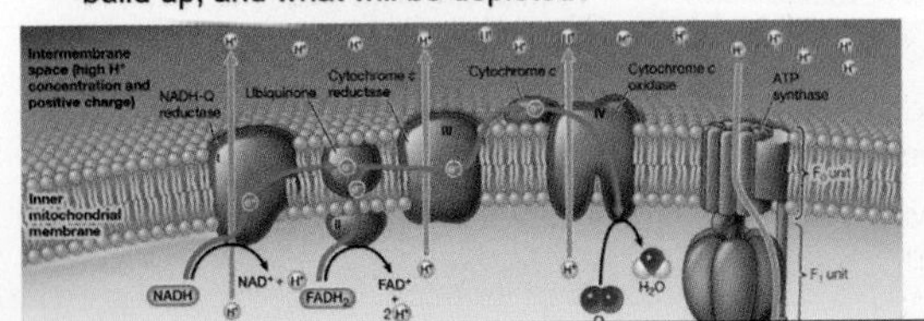

A FOCUS ON ACTIVE LEARNING

Where other texts give lip service to active learning, *Principles of Life* delivers, with an Active Learning Guide and 30 Active Learning Modules ready for classroom delivery. Built around key concepts, the ALMs provide a road map for pre-class work and in-class activities, including Apply the Data exercises, animations, videos, and quizzing directly mapped to in-text concepts and learning objectives.

Principles of Life, Third Edition Digital Update Content Updates

The Third Edition Digital Update of *Principles of Life* has not only been revised to be a more effective pedagogical tool, but it has also been updated to reflect the latest research and advances in biology.

© Nishinaga Susumu/Nature Production/Minden Pictures

CHAPTER 1

- New Section describing how science informs society using COVID-19 as an example

PART 1: CELLS

- Update to Review & Apply 2.1 to clarify terminology
- Updates to Figures throughout to improve illustrations

© Daniel Heuclin/Minden Pictures

PART 2: GENETICS

- Updates to Figures throughout to improve illustrations
- Update to Link to acknowledge in Chapter 13 students will learn that evolution would be impossible without mutation
- Revised content related to silent mutations
- Revisions to “The information for protein synthesis lies in the genetic code” subsection related to DNA mutations that are synonymous and how tRNA interacts with mRNA; throughout section silent is revised to synonymous
- Revisions to “Stability of mRNA can be regulated” subsection to discuss miRNAs and siRNAs
- Revised Table 12.2 title to: Initial Estimates of Protein-Coding Genes in a Strain of Three Species of Bacteria from Genome Sequencing; updated gene numbers in table
- Revisions to “Metagenomics reveals the diversity of viruses and prokaryotic organisms” subsection to include example about sequencing RNA from nasal samples of COVID-19 patients
- Updates to Table 12.3 to update haploid gene sizes and number of protein-coding genes
- Updated numbers in “The human genome sequence held some surprises” subsection

© Tui De Roy/Minden Pictures

PART 3: EVOLUTION

- Revision to introduction to acknowledge COVID-19
- New chapter opening art of phylogenetic trees that depict and track COVID-19 evolution
- New answer to chapter opening question about how phylogenetic methods are used to understand the origin, evolution, and spread of new diseases, like COVID-19
- Revisions in “Molecular evolution is used to study and combat diseases” subsection to acknowledge COVID-19

© Pete Oxford/Minden Pictures

PART 4: DIVERSITY

- Revisions to subsection “The great majority of prokaryotic species have never been studied” to update numbers of described bacteria and species of prokaryotic archae
- Revisions to include COVID-19 as an example of human diseases caused by positive-sense single-strand RNA viruses; update to Figure 18.23 to include description of coronaviruses
- Updates to Table 22.1 to update the approximate number of living species for rotifers and relatives and vertebrates
- Revisions to update numbers referenced to known species of amphibians, anurans, reptiles, living mammals, morphologically diverse eutherians; updates to Table 22.3 to update number of described species of marsupials and eutherians

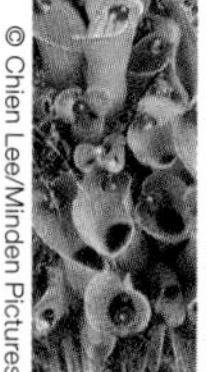

© Chien Lee/Minden Pictures

PART 5: PLANT FORM AND FUNCTION

- Updates to Figures throughout to improve illustrations
- Revisions to subsection “Cells that undergo asymmetrical cell division can produce daughter cells with an identity different from that of their parents” to clarify explanation of BASL
- Revisions to “Separation of Male and Female Gametophytes” subsection to clarify how some plant species form two separate types of flowers
- Updates to “Think Like a Scientist: Modeling Earth’s carbon cycle” to update data for mean concentration of CO_2 at Mauna Loa for the indicated month
- Review & Apply 27.1 updated by reordering questions and adding new questions #3 & #4, new graph supports question #3

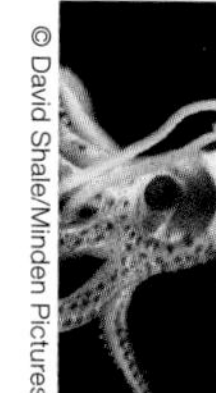

© David Shale/Minden Pictures

PART 6: ANIMAL FORM AND FUNCTION

- Updates to Figures throughout to improve illustrations
- Updates to clarify explanation in chapter opening introduction
- Revisions to “Adaptive immunity has four key features” subsection to clarify explanation of herd immunity
- Revision to Answer to chapter opening question “Do people who refuse vaccination for themselves put others at risk” to address COVID-19 vaccinations
- Added online chapter **Kidneys: Water and Salt Balance**

© John Cancalosi/Alamy Stock Photo

PART 7: ECOLOGY

- Added Media Clip to subsection “Climate is not the only factor that molds terrestrial biomes” and Figure 38.12 entitled Grasslands and Fire
- Revisions to “Life histories are diverse” subsection to update life expectancy in light of COVID-19
- Revisions to “Knowledge of metapopulation dynamics helps us conserve species and control epidemics” to acknowledge epidemics; added Activity 39.6 COVID-19 Simulation
- Revisions to placement of media Activity 40.5 to appear with Figure 40.10; renamed Evolutionary Arms Race Simulation
- Updates to climate change key term definition
- Updates to Figure 42.12 to use most recent data on greenhouse gas concentrations
- Revisions to “Recent increases in greenhouse gases are warming Earth’s surface” subsection to address latest trends and data of extreme weather events

Core Competencies

***Principles of Life* was created to ensure that you gain the knowledge you need from your introductory biology course and acquire the skills needed to succeed as a life sciences major.**

The AAAS *Vision and Change* report's six "core competencies," related to quantitative reasoning, simulation, and communication, are integrated both implicitly throughout the text and explicitly in a key feature, **THINK LIKE A SCIENTIST**.

THINK LIKE A SCIENTIST

A major goal is to align the text with the *Vision and Change* recommendations, especially as they relate to acquisition of the six core competencies. TLAS boxes explicitly develop these core competencies, and have been designed specifically to teach the skills needed to become a functional, practical, effective scientist. TLAS questions are high-level and aim to have you integrate concepts across the chapter or across chapters and ask you to *do something*.

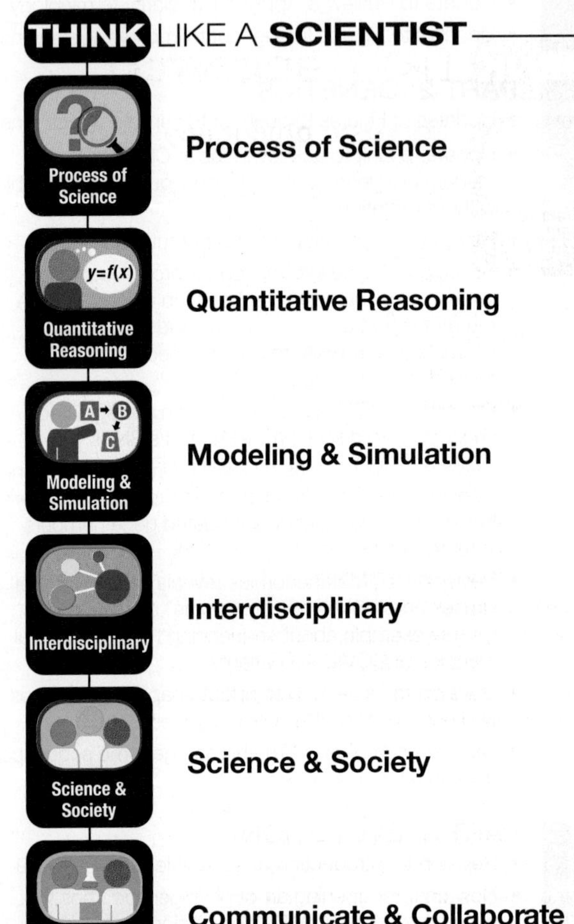

LIKE A **SCIENTIST**

Changes in Earth's physical environment have affected the evolution of life

In the experiment shown in Investigation Figure 17.8, body mass of individuals in the experimental populations of *Drosophila* increased (on average) about 2 percent per generation in the high-oxygen environment (although the rate of increase was not constant over the experiment). In the Permian, giant flying insects, such as dragonflies the size of modern hawks, inhabited Earth. Is the rate of increase in body mass seen in Investigation Figure 17.8 sufficient to account for the giant insects of the Permian? How long would it take for giant insect body size to evolve?

Here you will use quantitative reasoning and a simple model of selection to estimate how quickly insect body size could have evolved in response to higher atmospheric oxygen concentrations.

1. Suppose that the average rate of increase in dragonfly size

Very Effective—Think Like a Scientist is a great feature and I would assign this as a supplemental assignment. This feature encourages synthesis of material and development of critical-thinking skills around a relevant topic."

Sara E. Lahman, PhD, *University of Mount Olive*

Core Competencies

The **TLAS** is great. Wonderful emphasis on critical thinking and application."

Jennifer A. Metzler, *Ball State University*

THINK LIKE A SCIENTIST

Forensic phylogeny

Phylogenetic trees are used throughout biology, but only in recent years have they become important for forensic investigations. Here you will explore the relationship between science and society by applying your knowledge of phylogeny to a criminal court case.

A criminal case in Texas charged a defendant with knowingly and intentionally infecting a series of women with HIV. A phylogenetic analysis was used to demonstrate that the defendant transmitted HIV to his victims. (Other evidence was needed to prove knowledge and intent.) In this case, sequences of HIV isolated from the victims and the defendant, together with the closest sequences from an HIV database (the outgroup), were compared and used to construct a phylogenetic tree of the viruses (**FIGURE 14.10**). Viruses from each individual in the case are colored alike on the tree to the right. The labels are the codes for the individuals in the case. All of the individuals labeled CC01–CC08 are known to have engaged in sex; they represent an epidemiological cluster. (In forensic cases, samples are "blinded" to the investigators by assigning numbers to each sample, rather than using people's names. Only after the conclusions are finalized do other investigators decode the numbers to reveal the results.)

1. Which of the individuals labeled in the tree is consistent with being the source of this infection cluster? Why?
2. Why is the tree inconsistent with any of the other individuals being the source of infection within this cluster?
3. What was the purpose of including an outgroup made up of individuals who were outside the epidemiological cluster?

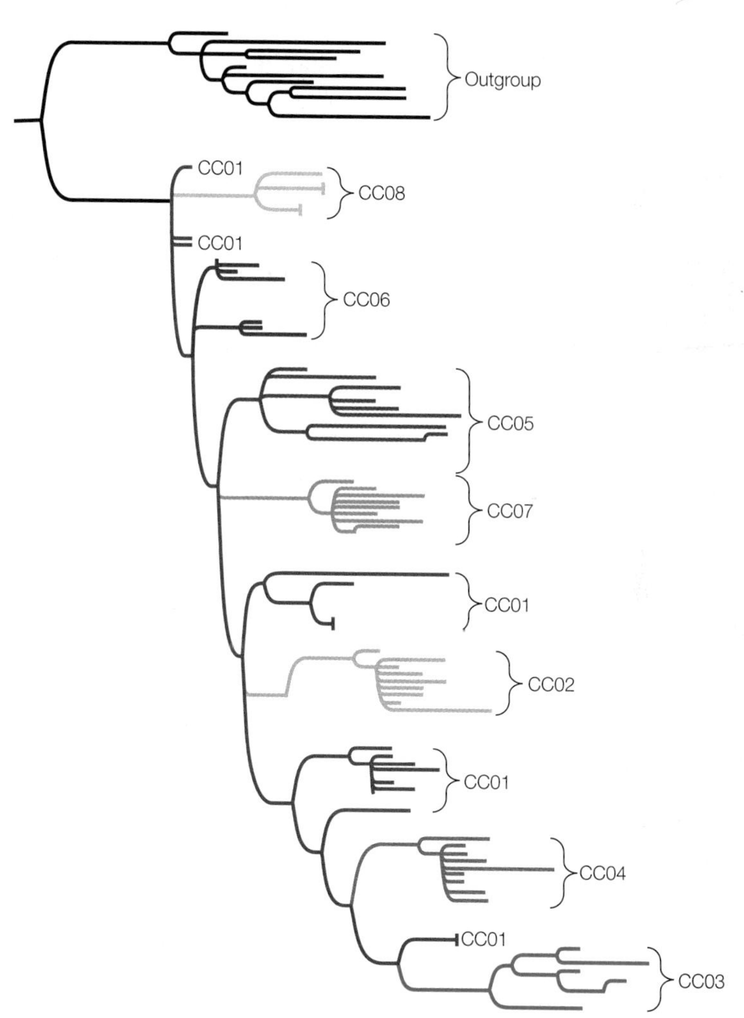

FIGURE 14.10 Forensic Phylogeny (After D. I. Scaduto et al. 2010. *Proc Natl Acad Sci USA* 107: 21242–21247.)

A complete list of the **Think Like a Scientist** boxes is shown on the following pages xii and xiii, along with their related **core competencies**.

Core Competencies

THINK LIKE A SCIENTIST

Chapter	Title	Process of Science	Quantitative Reasoning	Modeling & Simulation	Interdisciplinary	Science & Society	Communicate & Collaborate
2	Climbing the walls	●				●	
3	The origin of the molecules of life on Earth			●			
4	Advances in microscopy have resulted in greater understanding of cell structure and function		●		●		
5	The green Earth	●	●				
6	Identifying and ordering steps in signal transduction pathways	●					
7	Treating cancer	●				●	
8	Coat color inheritance in Labrador retrievers		●			●	
9	How can CODIS be used to identify suspects from a drop of blood?	●	●			●	
10	Evidence for lateral gene transfer in aphids	●					
11	Determining the regulation of the *lac* operon	●					
12	Inactivation of specific genes using CRISPR-Cas9 gene editing	●					
13	Observing and measuring phenotypic evolution	●					
14	Forensic phylogeny					●	
15	Why was the 1918–1919 influenza pandemic so severe?				●	●	●
16	Reinforcement of reproductive isolation	●					
17	Changes in Earth's physical environment have affected the evolution of life		●	●			
18	Putting bacteria to work					●	
19	Using phylogenies to make predictions				●		
20	Coevolution of plants and their pollinators	●					
21	How dependent are plants on their fungal mutualists?		●				
22	How do biologists estimate how many species are still undiscovered?			●			
23	How can one identify the anatomical parts of a plant if they appear unfamiliar?	●					
24	Testing new analytical methods: Might *Tillandsia* make useful air pollution monitors?		●			●	
25	Correlation and causation	●					
26	Impact of temperature on the bloom time of plants near Walden Pond in Concord, Massachusetts		●			●	●
27	Modeling Earth's carbon cycle		●	●		●	
28	Using quantitative reasoning to communicate with nonscientists about "burning off" extra food calories		●				●
29	Is global warming affecting animal life or not?	●				●	

Core Competencies

THINK LIKE A SCIENTIST

Chapter	Title	Process of Science	Quantitative Reasoning	Modeling & Simulation	Interdisciplinary	Science & Society	Communicate & Collaborate
30	How does a person's maximal rate of O_2 consumption vary with elevation in the mountains?		●				
31	Do some moths jam bats' echolocation mechanism?	●					
32	Commercialization of hormones: New choices for people to make	●					●
33	From the shores of ancient Rome to flashing muscle fibers: Progress in a stunning collaboration across generations of scientists, disciplines, animals, and tissues	●			●		●
34	The value of manipulative experiments	●	●		●		
35	Differentiation can be due to inhibition of transcription factors	●					
36	Avoiding incompatibilities in blood type: The immune response to the Rh factor					●	
37	How are animals reacting to global warming?	●					
38	Phylogenetic methods contribute to our understanding of biogeography	●			●		
39	Dispersal corridors can "rescue" fragmented populations from extinction	●	●				
40	Intra- and interspecific competition influence the morphology of coexisting species	●	●				
41	Additional predictions of the MacArthur-Wilson theory can be tested	●	●				
42	Computer models of Earth's climate link global warming to human activities		●	●	●	●	●

Process of Science

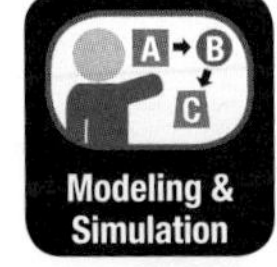
Modeling & Simulation

Science & Society

Quantitative Reasoning

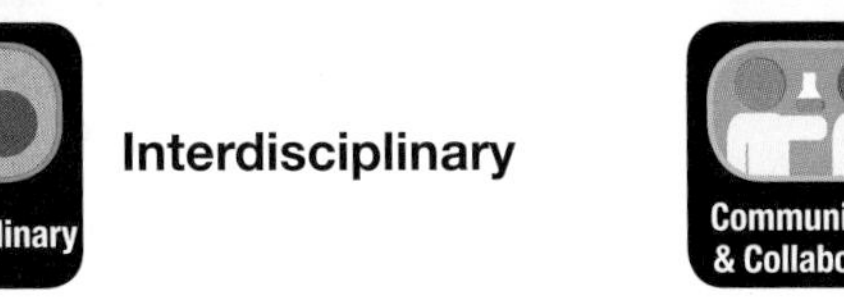
Interdisciplinary

Communicate & Collaborate

Mastering the Key Concepts

Each chapter of *Principles of Life* is built around a pedagogical framework meant to ensure a mastery of all of the important biological concepts in the introductory course.

KEY CONCEPTS

Dividing chapters into sections, every Key Concept explores a single essential concept in light of established facts and relevant experimental evidence, providing the conceptual framework for the chapter, exercises, and questions ahead.

KEY CONCEPTS

17.1 Events in Earth's History Can Be Dated

17.2 Changes in Earth's Physical Environment Have Affected the Evolution of Life

17.3 Major Events in the Evolution of Life Can Be Read in the Fossil Record

LEARNING OBJECTIVES

Learning Objectives are provided at the start of each Key Concept. The goal of Learning Objectives is to help you focus your attention as you read each section. At the end of each section, we reinforce the Learning Objectives with exercises/questions in Review & Apply. Learning Objectives encourage active learning and focus on mastering concepts and skills.

LEARNING OBJECTIVES

By the end of this key concept you should be able to:

17.1.1 Construct a geological map indicating the ages of exposed rocks and use the map to search for fossils of a given age.

17.1.2 Select appropriate methods for dating fossils and rocks from different geological time periods.

17.1.3 Place important events in biological history onto a time line of Earth's history.

REVIEW & APPLY

This feature is designed to briefly summarize the previous section and help you master concepts and competencies through questions. R&A questions are concept-specific, aligning with the Learning Objectives. With the exception of introductory concepts, R&A questions tend to be higher-level Bloom's and, when possible, ask you to engage in an activity-based answer.

REVIEW & APPLY | 17.1

R The layering of sedimentary rock strata enables geologists to determine the relative ages of fossils. Assigning actual ages to these strata requires analysis of radioactive decay, paleomagnetic dating, and fossil comparisons across strata. Geologists divide the history of life into eons, eras, and periods based on assemblages of fossil organisms found in successive layers of rocks.

A Imagine you have been assigned the job of producing a geological map of rocks that were formed between 600 and 400 million years ago (mya). You collect a sample from each of ten sites (1–10 on the map below), determine the ratio of ^{206}Pb to ^{238}U for each sample, and use these ratios to estimate the ages of the rock samples, resulting in the table on the following page.

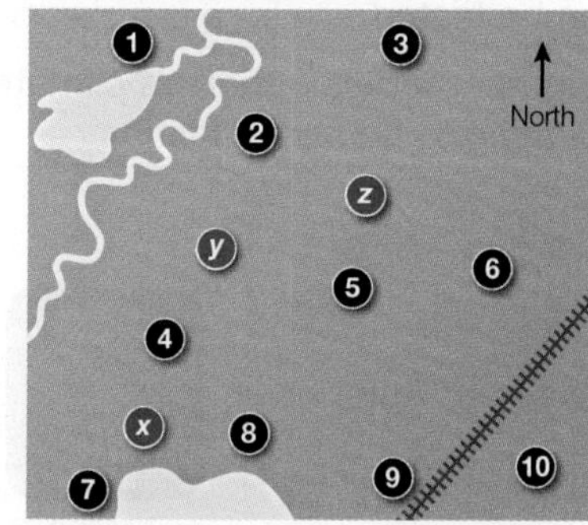

Site	$^{206}Pb/^{238}U$ ratio	Estimated age (mya)
1	0.076	474
2	0.077	479
3	0.069	431
4	0.081	505

Mastering the Key Concepts

VISUAL SUMMARIES

Visual Summaries conclude every chapter, providing a visually compelling checklist, emphasizing major chapter concepts through key figures, bullets, and lower-level Bloom's questions. The Visual Summary ensures you have mastered the major points of the chapter. The content is laid out so as to facilitate referencing back to the original chapter text and figures and directing you to relevant animations and activities.

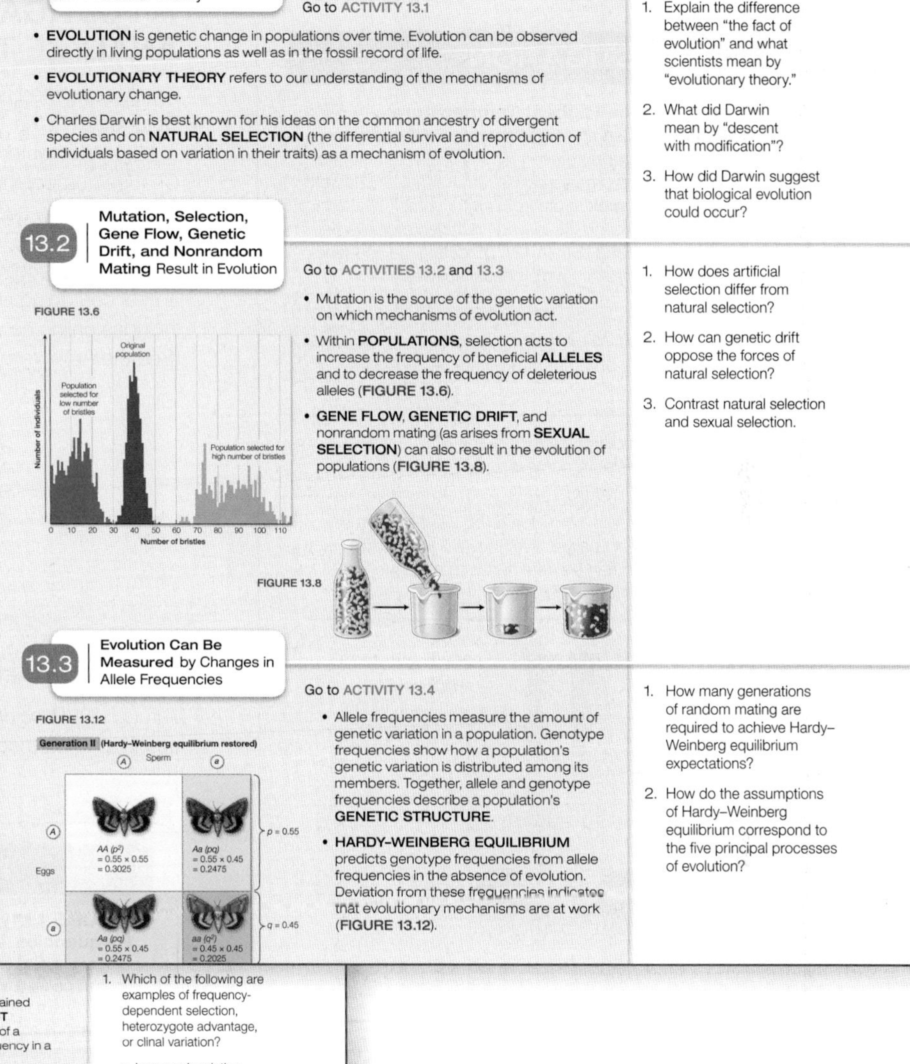

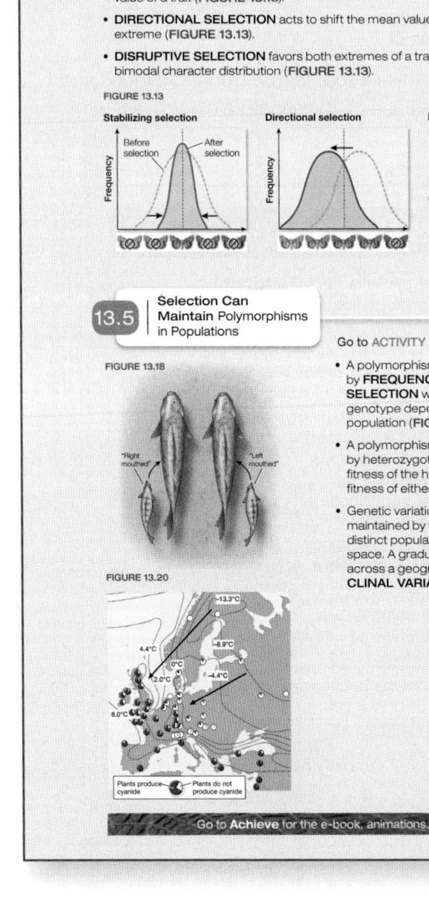

Go to ACTIVITY 13.5

• A polymorphism may be maintained by FREQUENCY-DEPENDENT SELECTION when the fitness of a genotype depends on its frequency in a population (FIGURE 13.18).
• A polymorphism may also be maintained by heterozygote advantage when the fitness of the heterozygote exceeds the fitness of either homozygote.
• Genetic variation within species may be maintained by the existence of genetically distinct populations over geographic space. A gradual change in phenotype across a geographic gradient is known as CLINAL VARIATION (FIGURE 13.20).

1. Which of the following are examples of frequency-dependent selection, heterozygote advantage, or clinal variation?

a. Increased sprinting performance of individuals with two different alleles of an actin gene, compared with individuals with two copies of just one of the alleles.

b. Increased survival of individuals with rare color patterns, compared with those with common color patterns.

c. Geographic variation in the size of deer from northern to southern latitudes.

Go to Achieve for the e-book, animations, activities, and additional resources and assignments.

> **VISUAL SUMMARY** is fantastic. It can be interpreted by itself and students can learn independently from the text. I like the questions next to the summary as well. I do like all the application questions (TLAS, R&A, and Investigations). Students always want more practice, and more application questions, so these are invaluable."
>
> Shira D. P. Rabin,
> *University of Louisville*

Developing Skills and Working with Data

Principles of Life has always been known for emphasizing the role of experimentation, data, and research in our understanding of biology. The Third Edition includes tools to help you understand how we know what we know.

INVESTIGATION

FIGURE 18.14 What Is the Highest Temperature Compatible with Life? Can any organism thrive at temperatures above 120°C? This is the temperature used for sterilization, known to destroy all previously described organisms. Kazem Kashefi and Derek Lovley isolated an unidentified prokaryote from water samples taken near a hydrothermal vent and found it survived and even multiplied at 121°C. The organism was dubbed "Strain 121," and its gene sequencing results indicate that it is a prokaryotic archaeal species.[a]

HYPOTHESIS

Some prokaryotes can survive at temperatures above 120°C.

METHOD

1. Seal samples of unidentified, iron-reducing, thermal vent prokaryotes in tubes with a medium containing Fe^{3+} as an electron acceptor. Control tubes contain Fe^{3+} but no organisms.
2. Hold both tubes in a sterilizer at 121°C for 10 hours. If the iron-reducing organisms are metabolically active, they will reduce the Fe^{3+} to Fe^{2+} (as magnetite, which can be detected with a magnet).

RESULTS

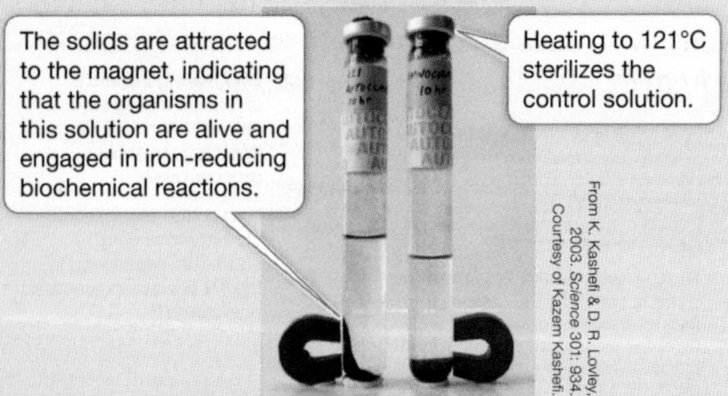

CONCLUSION

Prokaryotic archaea of Strain 121 can survive at temperatures above the previously defined sterilization limit.

ANALYZE THE DATA

After Kashefi and Lovley isolated Strain 121, they examined its growth at various temperatures. The table below shows generation time (time between cell divisions) at nine temperatures.

Temperature (°C)	Generation time (hr)
85	10
90	4
95	3
100	2.5
105	2
110	4
115	6
120	20
130	No growth, but cells not killed

INVESTIGATIONS WITH ANALYZE THE DATA QUESTIONS

Highly acclaimed by adopters, Investigations and Analyze the Data return in the Third Edition. The goal of the Investigations is to help you master both big concepts in biology and *Vision and Change* competencies. This is done by illustrating a real study and having you analyze the resulting real data. Investigations with Analyze the Data questions are higher-level Bloom's, integrating concepts within the chapter or across chapters, and encouraging activity-based answers. In addition, Achieve includes online companions to the Analyze the Data exercises. (See the Achieve section for details.)

RESEARCH TOOLS

Throughout *Principles of Life*, this feature focuses on techniques and quantitative methods scientists use to investigate biological systems.

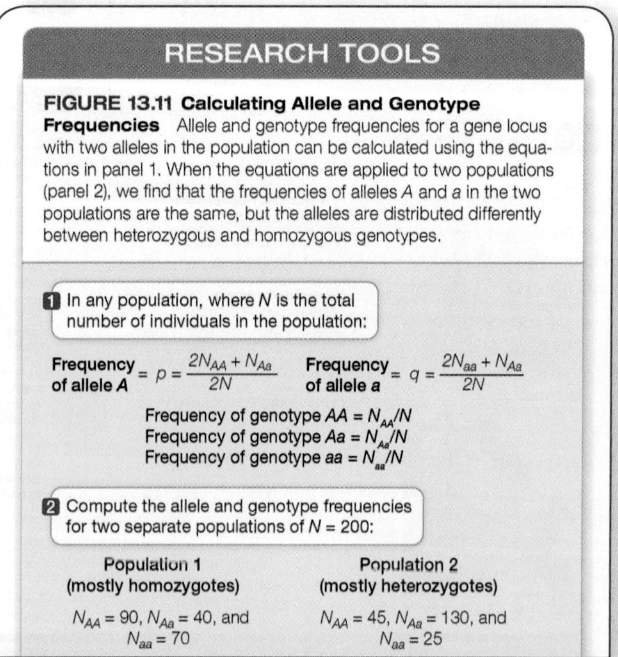

FIGURE 13.11 Calculating Allele and Genotype Frequencies Allele and genotype frequencies for a gene locus with two alleles in the population can be calculated using the equations in panel 1. When the equations are applied to two populations (panel 2), we find that the frequencies of alleles *A* and *a* in the two populations are the same, but the alleles are distributed differently between heterozygous and homozygous genotypes.

Developing Skills and Working with Data

I think this **(REVIEW & APPLY)** is a great feature. Applying what they've just read in a slightly new way will improve understanding and retention...."

Jennifer Butler, *Willamette University*

MAKING SENSE OF DATA: A STATISTICS PRIMER

This primer (an appendix in the text and also in Achieve) lays the proper groundwork for understanding statistics and data, providing helpful support for all of the quantitative exercises.

How Does Statistics Help Us Understand the Natural World?

Statistics is essential to scientific discovery. Most biological studies involve five basic steps, each of which requires statistics:

- **Step 1: Choose an Experimental Design**
  Clearly define the scientific question and the methods necessary to tackle the question.
- **Step 2: Collect Data**
  Gather information about the natural world through observations and experiments.
- **Step 3: Organize and Visualize the Data**
  Use tables, graphs, and other useful representations to gain intuition about the data.

REVIEW & APPLY | 16.3

R Allopatric speciation results from the separation of populations by geographic barriers; it is the dominant mode of speciation among most groups of organisms. Sympatric speciation may result from disruptive selection that results in ecological isolation, but polyploidy is the most common cause of sympatric speciation among plants.

A

1. Explain how speciation via polyploidy can happen in only two generations.
2. If allopatric speciation is the most prevalent mode of speciation, what do you predict about the geographic distributions of many closely related species? Does your answer differ for species that are sedentary versus highly mobile?
3. The species of Darwin's finches shown in the phylogeny in Figure 16.8 have all evolved on islands of the Galápagos archipelago within the past 3 million years. Molecular clock analysis (see Key Concept 14.3) has been used to determine the dates of the various speciation events in that phylogeny. Geological techniques for dating rock samples (see Key Concept 17.1) have been used to determine the ages of the various Galápagos islands. The table shows the number of species of Darwin's finches and the number of islands that have existed in the archipelago at several times during the past 4 million years (data from P. R. Grant. 2001. *Oikos* 92: 385–403).

Time (millions of years ago)	Number of islands	Number of finch species
0.25	18	14
0.50	18	9
0.75	9	7
1.00	6	5
2.00	4	3
3.00	4	1
4.00	3	0

   a. Plot the number of species of Darwin's finches and the number of islands in the Galápagos archipelago (dependent variables) against time (independent variable).

   b. Are the data consistent with the hypothesis that isolation of populations on newly formed islands is related to speciation in this group of birds? Why or why not?

4. If no more islands form in the Galápagos archipelago, do you think that speciation by geographic isolation will continue to occur among Darwin's finches? Why or why not? What additional data could you collect to test your hypothesis (without waiting to see if speciation occurs)?

Active Learning

Active learning continues to be central to the mission of *Principles of Life*. Features both in the text and online present you with an even more engaging experience.

CHAPTER OPENER WITH QUESTION

A short introduction with an attention-grabbing photo and compelling question gives you something to ponder while reading and studying the chapter. The chapter ends with a return to the question and some discussion of the answer.

LINKS

Links point you to additional discussion of a concept or key term elsewhere in the book, providing an opportunity for integration across chapters.

Key Concept 14.3 describes how biologists reconstruct the gene sequences of extinct organisms.

IN-FIGURE QUESTIONS

Incorporated into figures, these questions are designed to engage you and help you think about the implications of the figure/diagram. In-figure questions tend to be lower-level Bloom's, and are often amenable to in-class discussion.

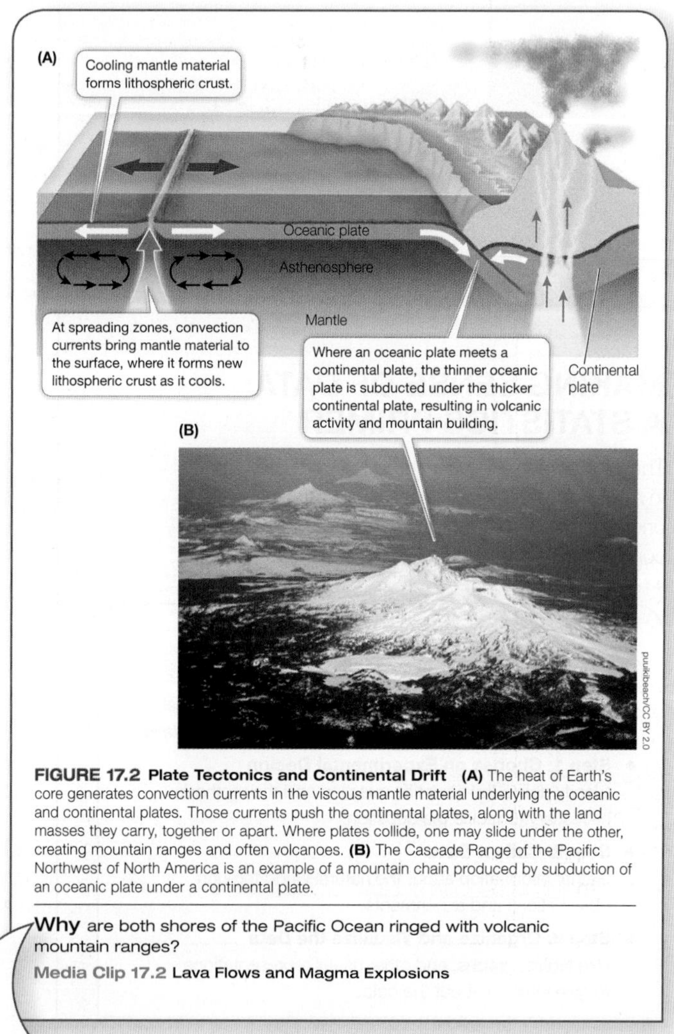

FIGURE 17.2 Plate Tectonics and Continental Drift (A) The heat of Earth's core generates convection currents in the viscous mantle material underlying the oceanic and continental plates. Those currents push the continental plates, along with the land masses they carry, together or apart. Where plates collide, one may slide under the other, creating mountain ranges and often volcanoes. **(B)** The Cascade Range of the Pacific Northwest of North America is an example of a mountain chain produced by subduction of an oceanic plate under a continental plate.

Why are both shores of the Pacific Ocean ringed with volcanic mountain ranges?

Media Clip 17.2 Lava Flows and Magma Explosions

Why are both shores of the Pacific Ocean ringed with volcanic mountain ranges?

Media Clip 17.2 Lava Flows and Magma Explosions

Active Learning

Encouraging you to be more involved while reading the textbook is just the beginning of the Active Learning approach in *Principles of Life*. For instructors who have been teaching actively for years, or those who are just beginning to use these techniques, we've created an Active Learning Guide and an accompanying set of Active Learning Modules to engage you before, during, and after class.

ACTIVE LEARNING MODULES

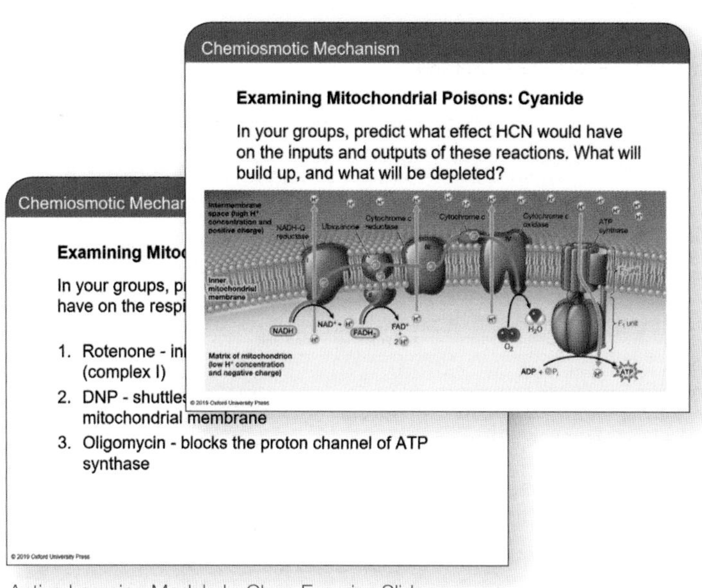

Active Learning Module In-Class Exercise Slides

The expanded Active Learning Modules provide everything instructors need to successfully implement an active approach to teaching key topics. Each module's many resources include:

- Pre-class video specifically created for the module
- Pre-quiz and post-quiz
- Handout for in-class work
- Detailed in-class exercise
- Detailed instructor's guide

These modules are easy to implement and are a great way to add more active learning to the classroom.

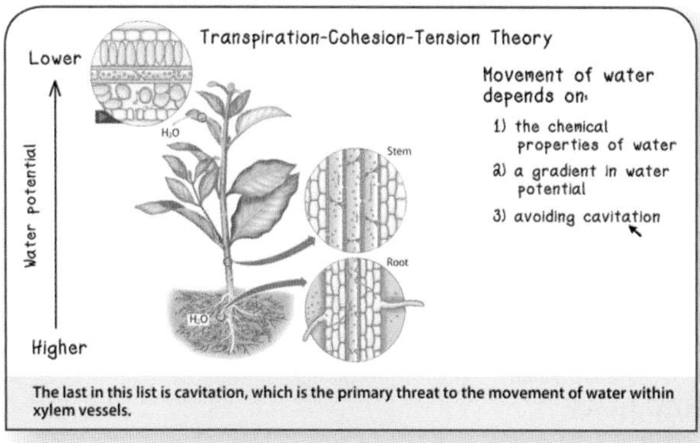

Active Learning Module In-Class Video

ACTIVE LEARNING GUIDE

The Active Learning Guide provides extensive resources and support for implementing active learning techniques in any classroom, large or small. This guide provides instructors with a thorough introduction to the concepts, techniques, and benefits of active learning. Chapter-by-chapter guidance provides strategies for how to best utilize learning resources in *Principles of Life* to teach in a more active format.

Part 1: Introduction to Active Learning

Chapter 1: What Is Active Learning?

Chapter 2: Designing Your Course for Active Learning

Chapter 3: Using Active Learning in the Classroom

Chapter 4: How to Implement *Principles of Life* Resources

Part 2: Active Learning Resources and Suggestions by Chapter

Each chapter in Part 2 of the Active Learning Guide corresponds to a textbook chapter and includes the following:

- An overview of the textbook chapter
- References to all of the student media resources, listed by Key Concept
- References to and descriptions of each Active Learning Module
- Detailed suggestions for active learning activities and exercises for each Key Concept, including "draw," "video," and "compare" exercises, think-pair-share activities, spider maps, minute papers, clicker questions, and more
- Suggestions for incorporating the in-text Links and Analyze the Data features into in-class activities

Part 3: Appendices

- Appendix A: An Overview of Bloom's Taxonomy
- Appendix B: A Guide to Using the *Principles of Life*, Third Edition Learning Objectives
- Appendix C: Learning Objectives for *Principles of Life*, Third Edition

Principles of Life, now available with 

Achieve is the culmination of years of development work put toward creating the most powerful online learning tool for biology students. It houses all of our renowned assessments, multimedia assets, e-books, and instructor resources in a powerful new platform.

Achieve supports educators and students throughout the full range of instruction, including assets suitable for pre-class preparation, in-class active learning, and post-class study and assessment. The pairing of a powerful new platform with outstanding biology content provides an unrivaled learning experience.

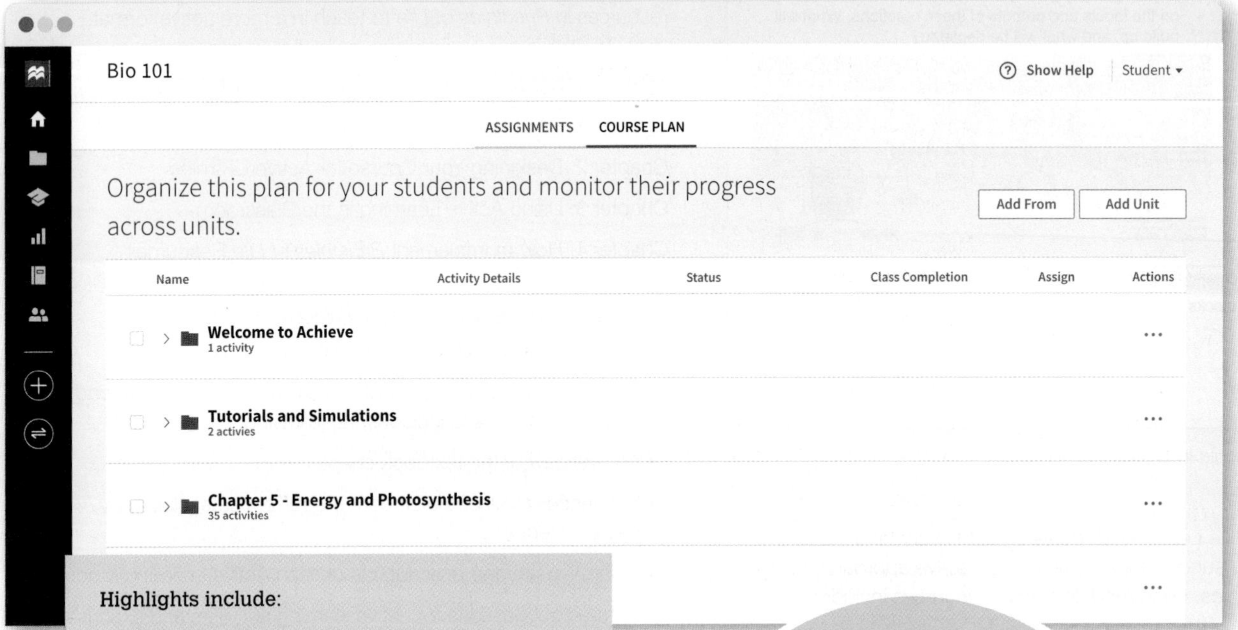

Bio 101
Show Help
Student
ASSIGNMENTS
COURSE PLAN
Organize this plan for your students and monitor their progress across units.
Add From
Add Unit
Name | Activity Details | Status | Class Completion | Assign | Actions
Welcome to Achieve
1 activity
Tutorials and Simulations
2 activies
Chapter 5 - Energy and Photosynthesis
35 activities

Highlights include:

- **A design guided by learning science research.** Co-designed through extensive collaboration and testing by both students and faculty including two levels of Institutional Review Board approval for every study of Achieve.
- **A learning path of powerful content** including pre-class, in-class, and post-class activities and assessments. A detailed Gradebook with insights for just-in-time teaching and reporting on student achievement by learning objective.
- **Easy integration and Gradebook sync** with iClicker classroom engagement solutions.
- **Simple integration** with your campus LMS and availability through **Inclusive Access** programs.

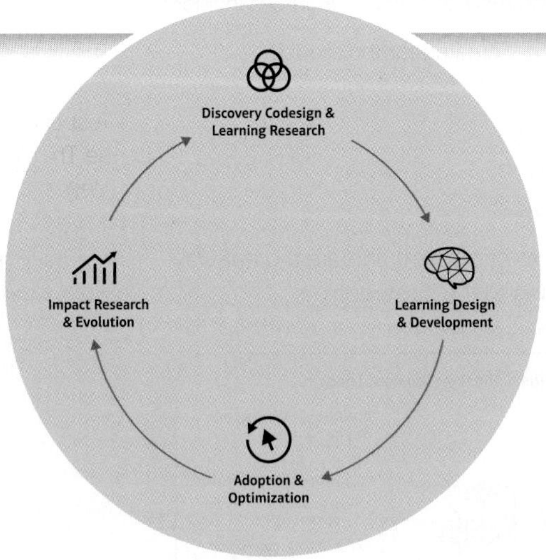

Discovery Codesign & Learning Research
Learning Design & Development
Adoption & Optimization
Impact Research & Evolution

For more information or to sign up for a demonstration of Achieve, contact your local Macmillan representative or visit **macmillanlearning.com/achieve.**

Achieve: Media and Assessment Resources

Principles of Life's assessment and media resources give instructors a range of tools for assessing student's progress before class, in class, after class, and on exams.

Assessment questions and resources are aligned to the Learning Objectives associated with each Key Concept. This provides instructors with a concrete way of assessing students on mastery of the most important material in each chapter.

FEATURE	ASSESSMENT AND MEDIA RESOURCES
Active Learning Modules	Active Learning modules include everything needed to conduct in-class activities. Resources include pre- and post-class quizzes, videos, student handouts, and instructor activity guides.
Learning Curve	LearningCurve adaptive quizzing gives students individualized question sets and feedback based on responses. All questions link back to the e-book to encourage students to read the book in preparation for class and exams. LearningCurve organizes questions by Key Concept, and instructors can easily hide questions on concepts they are not covering. New questions have been added to Learning Curve for the Third Edition Digital Update.
Activities & Simulations	Interactive activities, including simulations, are assignable and include assessment questions for students to check their understanding.
Analyze the Data	Each in-text Analyze the Data exercise is accompanied by an online companion exercise. The online companion exercise gives additional practice with the same skills addressed by the in-text exercise.
Animations	Animations with associated questions to help students visualize important concepts.
Summative Quizzes	Each chapter includes a Summative Quiz composed of 20 questions spanning the chapter's Key Concepts. Quizzes are pre-built and ready to assign. At the same time, they are completely customizable; instructors can add, revise, or remove questions to match their course content.
Test Bank	The *Principles of Life* Test Bank available in Cognero includes new questions added for the Third Edition Digital Update. Questions are offered at a variety of Bloom's levels and cover all key concepts in the text.
Achieve Item Library	The Achieve item library allows instructors to create their own assignments and author their own questions. Easily search for questions and filter by difficulty level, Bloom's level, grading, or question type or topic. *Principles of Life*, Third Edition Digital Update, includes many new questions, including kinesthetic type questions.

Preface

Principles of Life, Third Edition Digital Update

Principles of Life, Third Edition, is updated and now available with Achieve! Achieve supports students and educators throughout the full range of instruction, including assets suitable for pre-class preparation, in-class active learning, post-class study and assessment. The pairing of a powerful new platform with outstanding biology content provides an unrivaled learning experience.

Features of Achieve Include

- A design guided by learning science research. Developed with extensive collaboration and testing by students and faculty, including two levels of Institutional Review Board approval for every efficacy study of Achieve.
- A learning path of powerful content including pre-class, in-class, and post-class activities and assessments.
- A detailed Gradebook with insights for just-in-time teaching and reporting on student achievement by learning concept.
- Easy integration and Gradebook synchronization with iClicker classroom engagement solutions.
- Simple integration with your campus LMS and availability through Inclusive Access programs.

The *Principles of Life* Story

Prior to our launch of the first edition of *Principles of Life*, introductory biology textbooks for science majors presented encyclopedic summaries of biological knowledge. We believe that students who spend their time diligently memorizing myriad details and a vast terminology actually retain fewer of the concepts that are the foundation for further study in advanced courses. In *Principles of Life*, we take the opposite approach: we promote understanding over memorization. Details are important, but no modern biology textbook can begin to cover all the information biologists have learned to date, and students today have many other ways to access the details as they need them.

The conception of *Principles of Life* coincided with two major reports that supported the change to a conceptual approach; *Vision and Change in Undergraduate Biology Education: A Call to Action*, published in 2011 by the American Association for the Advancement of Science (supported by the National Science Foundation) and *BIO2010: Transforming Undergraduate Education for Future Research Biologists*, sponsored by the National Institutes of Health and the Howard Hughes Medical Institute. These reports recommend focusing on core concepts and competencies, teaching students through active learning rather than memorization, and improving the integration of statistical and computational approaches. From the first edition of *Principles of Life*, we have used our experience as authors and educators to implement these recommendations for a new approach to teaching introductory biology.

With the astute guidance of Andy Sinauer—we convened an advisory group of twenty leading biology educators and instructors in introductory biology from throughout North America. During an intensive meeting of the authors and this group, dynamic discussions led to the solidification of the core concepts we believe are essential for teaching introductory biology. The book took shape, and members of the advisory group reviewed the emerging chapters, providing feedback at every stage of the book's development.

All chapters have undergone extensive between-edition review by experts in each respective discipline, and the chapters have been revised accordingly. Active learning has always been a priority in *Principles of Life*. With the Third Edition revised, the emphasis on active learning has been dramatically enhanced—to the point that active-learning features permeate the book. We have expanded opportunities for students to apply what they have learned by using real data and examples—and have better integrated and explained the concepts of statistical analysis of data. Our coverage and application of systems concepts is expanded. With the *Think Like a Scientist* feature, we have developed opportunities for students to practice the core competencies that have become critical for modern biologists.

Principles of Life stresses the five core concepts (themes) identified in *Vision and Change* as being essential for all undergraduates to understand:

- evolution
- the relationship between structure and function
- information flow, exchange, and storage
- pathways and transformations of energy and matter
- biological systems

As we develop these concepts, we keep a steady focus on the needs of beginning students at the university level. In preparing each chapter, our central question has been, "What does a beginning student need to know?" We have then met the needs of the beginning student with a concept-centered approach that introduces facts and terms as they are needed to develop concepts, avoiding the inclusion of terms and facts for their own sake. For students who go on in biological science, *Principles of Life* provides the conceptual foundation they will need to succeed in upper-level courses. For the many students who complete their study of biology at the introductory level, *Principles of Life* recognizes that—long after a year of study—people remember concepts, not isolated facts.

Vision and Change, in addition, identified six core competencies that undergraduates must develop to succeed in science in the twenty-first century. *Principles of Life* and Achieve support students in developing these skills. *Vision and Change* argues students should be able to

- apply the process of science,
- use quantitative reasoning,

- use modeling and simulation,
- tap into the interdisciplinary nature of science,
- communicate and collaborate with other disciplines, and
- understand the relationship between science and society.

Our art program for *Principles of Life* continues to build on our success from *Life: The Science of Biology*. We pioneered the use of balloon captions to help students understand and interpret the biological processes illustrated in figures without repeatedly going back and forth between a figure, its legend, and the text. These guides help students connect critical points of figures to the concepts that are developed in the text. Conceptual diagrams are used in many places, and text-art coordination has been emphasized. When diagrams or data sets from the scientific literature are presented, readers will now be able to find those diagrams or data sets in the literature with our new referencing system.

Features of *Principles of Life*

Focus on Concepts: Each chapter is organized into a series of *Key Concepts*, each with its own *Learning Objectives*. Our focus in each Key Concept section is to identify and explain the concepts that beginning university students need to know. At the end of each Key Concept, a *Review & Apply* recaps the main points and presents questions related to the Learning Objectives for students to ponder. At the end of the chapter, the *Visual Summary* follows up with a visual and narrative review of major concepts throughout the chapter and further questions related to the Learning Objectives. Throughout, the questions we raise are deliberately designed to span the incremental levels of Bloom's Taxonomy of Cognitive Domains. Answers to all questions are included in Achieve, *Principles of Life*'s online platform.

Chapter Opener: Each chapter starts with carefully worded, professionally vetted *Learning Objectives*, which are then reinforced with each Key Concept section in the chapter, so that students can clearly see the goals they will achieve in their studies. Chapter openers have been designed with active learning in mind. Each chapter begins with a brief statement focused on major themes accompanied by a dramatic photograph and interpretive question for students to consider. These opening questions are designed so that students will be able to offer tentative answers as they start a chapter, but will be able to offer far more thorough answers as they finish. At the end of each chapter, we reprise the opening photograph with an answer of our own.

Think Like a Scientist: The *Think Like a Scientist* entry in each chapter emphasizes one or more of the six core competencies, using a system of icons to highlight the particular competencies. Topics such as manipulative experiments, proper choice of controls, meta-analysis, and communicating science to the public are presented in ways that will help each student learn more about the ways that scientists think. Questions—with answers at our online companion site, Achieve—are often provided to stimulate engagement.

Investigations: Investigations help students learn the process of science by being organized into sections on Hypothesis, Method, Results, and Conclusion. Most include a section (titled *Analyze the Data*) in which we present a subset of actual data from the published experiment. Students are asked to analyze these data and to make connections between observations, analyses, hypotheses, and conclusions. Extensive online resources are provided to expand the content on many Investigations. These resources include expanded discussions of the original research, links to the original publications, and discussion and links for any follow-up investigations that have been published. We have also included a *Making Sense of Data: A Statistics Primer* (Appendix B) to help students in developing this important skill.

Review and Apply: Each section concludes with a concise summary and a set of study questions. Many of these questions encourage students to go beyond memorization and engage more thoroughly in the process of science. As with all study questions in the book, we provide answers online in Achieve.

Visual Summary: To help students recall what they have learned, the *Visual Summary* includes both illustrations (especially helpful for visual learners) and bulleted points. The *Visual Summary* also includes additional study questions about each section in the chapter. Again, answers to all questions are provided online in Achieve.

Media Links: To help students deepen their understanding, we provide *Links* that allow students to see interconnections among such topics as molecular or cell biology, evolution, biological diversity, physiology, and ecology. The *Links* are not merely cross-references but include brief statements of pertinence, helping readers to see why they might want to follow a *Link*. We also feature *Animated Tutorials* and *Activities*, which include opportunities for students to use modeling and simulation modules to further reinforce their understanding of concepts.

Research Tools: Students need to learn about some of the major research tools that are used in biology, including major laboratory, computational, and field methods. Our *Research Tools* figures explain these tools and provide a context for how they are used by biologists. We have also included a *Working with DNA* so that students have an easy place to review and understand the major methods of molecular biology.

Active Learning: Active learning is a key component of *Principles of life*. We have implemented opportunities for active learning throughout the text and online in Achieve. In-text active learning opportunities include chapter opening questions, which give students something to ponder when reading and studying the chapter. Figure questions are designed to engage students and help them think about the implications of the figure or diagram. Active Learning modules include resources for classwork, as well as an instructor guide to support instructors in implementation of active learning.

Special Contributions

Many people contributed to the creation of the Third Edition Digital Update of *Principles of Life* (see below). However, three individuals deserve special mention for their contributions. Susan D. Hill did a masterful job in writing Chapter 35 on Animal Development. Nickolas Waser worked extensively with Mary Price on the Ecology section (Part 7) and was otherwise intimately involved in discussions of the book's planning and execution. David Sadava reprised his Chapter 36 on Immunology and provided expert editorial support on the Cells and Genetics sections (Parts 1 and 2).

Many People to Thank

In addition to the many biologists listed on the next page who provided formal reviews, each of us benefitted enormously from personal contacts with colleagues who helped us resolve issues and made critical suggestions for new material. They are: Walter Arnold, University of Veterinary Medicine (Vienna); Tobias Baskin, University of Massachusetts; Larry Gilbert, University of Texas, Austin; Harry Greene, Cornell University; Hugo Hofhuis, Wageningen University; Edward McCabe, University of Colorado and the March of Dimes Foundation; Will Petry, University of California, Irvine; Frank Price, Utica College; Thomas Ruf, University of Veterinary Medicine (Vienna); Richard Shingles, Johns Hopkins University; David Sleboda, Brown University; Viola Willemsen, Wageningen University; and Andrew Zanella, The Claremont Colleges.

Special thanks go to our editors who have guided us and the book through to the completion of the Third Edition Digital Update: Lisa Lockwood, Debbie Hardin, Marita Bley, Andy Sinauer, Danna Niedzwiecki, and Laura Green. Liz Pierson applied her outstanding copyediting skills to our manuscript. Dragonfly Studios worked with each of us to revise and create effective and beautiful line art. Mark Siddall rose to the challenge of finding new, even better photographs. We also wish to thank the entire team at Macmillan, including the Macmillan media group for their expertise in producing Achieve, the Regional Specialists, and the Regional Sales Managers.

Digital Content and Supplements

Principles of Life features a wide array of online resources to support and reinforce the material covered in the textbook. The activities, animations, and media clips referenced throughout the book are linked directly in the e-book, allowing students to instantly reference these resources from any device.

There is a wide array of instructor resources available, including multiple versions of all textbook figures, PowerPoint presentations, and a computerized test bank. PowerPoints and images are available from within Achieve, as well as the Active Learning Guide and Instructor's Manual. The computerized Test Bank can be accessed from the Instructor's Companion site.

We have enjoyed writing *Principles of Life* and wish you success. We hope that this book will serve you well.

DAVID M. HILLIS
MARY V. PRICE
RICHARD W. HILL
DAVID W. HALL
MARTA J. LASKOWSKI

Advisors and Reviewers

Scott Abella, University of Nevada, Las Vegas
Laura Altfeld, Saint Leo University
Pierette M. Appasamy, Chatham University
Kathryn Bell, Salt Lake Community College
Christine Bezotte, Elmira College
Ryan Bickel, University of Rochester
Mary Blakefield, Indiana University East
Chris Botanga, Chicago State University
Nicole Bournias-Vardiabasis, California State University, San Bernardino
Alison K. Brody, University of Vermont
Victoria Brown-Kennerly, Webster College
Winnifred M Bryant, University of Wisconsin, Eau Claire
Stephen Burnett, Clayton State University
Cheryl Burrell, Forsyth Technical Community College
Jennifer J. Butler, Willamette University
Mari Butler, Endicott College
Patrick William Cafferty, Emory University
Mickael J. Cariveau, University of Mount Olive
Billy J. Carver, Lees-McRae College
Lindsay Chaney, Snow College
Shelton Charles, Forsyth Technical Community College
Sixue Chen, University of Florida
Nicole Cintas, Northern Virginia Community College
Amanda N. Clark, Chipola College
Justin A. Compton, Springfield College
Jonna M. Coombs, Adelphi University
Andrea L. Corbett, Cleveland State University
Nancy E. Cowden, University of Lynchburg
Clayton E. Cressler, University of Nebraska, Lincoln
Timothy M. Davidson, California State University, Sacramento
C. Ainsley Davis, Bethune-Cookman University
Jill DeVito, University of Texas at Arlington
Jed H. Doelling, Salt Lake Community College
Amy L. Downing, Ohio Wesleyan University
Kamal Dulai, University of California, Merced
Jamin Eisenbach, Eastern Michigan University
Peter Ekechukwu, Horry Georgetown Technical College
W. Alex Escobar, Emory University
Cerrone R. Foster, East Tennessee State University
Melinda A. Fowler, Springfield College
Laura Francis, University of Massachusetts, Amherst
R. Adam Franssen, Longwood University
Mark Fulton, Bemidji State University
Stefanie K. Gazda, University of Florida
Marina M. Gerson, Stanislaus State University
Susan M. R. Gurney, Drexel University
Ehren F. Haderlie, Brigham Young University, Idaho
Matthew D. Halfhill, Saint Ambrose University
Valerie Haywood, Case Western Reserve University
Connie Heiman, Angelo State University
Kristy L. Henscheid, Columbia Basin College
Susan Z. Herrick, University of Connecticut
Laura H. Hill, University of Vermont
Tracie Ivy, Wofford University
Victor M. Izzo, University of Vermont
Brandon E. Jackson, Longwood University
Lance Johnson, Midland University
Kevin B. Jones, Charlestown Southern University
Douglas D. Kane, Defiance College
Joshua M. Kapfer, University of Wisconsin, Whitewater
Bretton W. Kent, University of Maryland
Moshe Khurgel, Bridgewater College
Henrik Kibak, California State University, Monterey Bay
Adam Kleinschmit, Adams State University
Richard Knapp, University of Houston
William Kristan, California State University, San Marcos
Rukmani Kuppuswami, Hill College
Sara E. Lahman, University of Mount Olive
Jennifer L. Larimore, Agnes Scott College
Tali D. Lee, University of Wisconsin, Eau Claire
Kristen A. Lennon, Hagerstown Community College
Iris I. Levin, Agnes Scott College
Kathryn L. Lipson, Western New England University
Robert E. Loeb, Penn State, Dubois
M. Wayne Mabe, Forsyth Technical Community College
Erin MacNeal Rehrig, Fitchburg State University
Chintamani S. Manish, Midland University
Jordan M. Marshall, Purdue University, Fort Wayne
Amanda J. Martino, Saint Francis University
Justin W. Merry, Saint Francis University
Jennifer A. Metzler, Ball State University
R. L. Minckley, University of Rochester
D. Blaine Moore, Kalamazoo College
Tsafrir S. Mor, Arizona State University
Mario L. Muscedere, Boston University
Barbara Musolf, Clayton State University
Vamsi J. Nalam Purdue University, Fort Wayne
Cassandra R. Nelson, Marquette University
F. A. O'Leary, Saint Edwards University
David G. Oppenheimer, University of Florida
Kate K. O'Toole, Emory University
Aditi Pai, Spelman College
Laura K. Palmer, Pennsylvania State University, Altoona
Daniel M. Pavuk, Bowling Green State University
Jay Pieczynski, Rollins College
A. A. Powolny, Spelman College
Christopher Quinn, University of Wisconsin, Milwaukee
Shira D. P. Rabin, University of Louisville
Emily S. J. Rauschert, Cleveland State University
U. G. Reinhardt, Eastern Michigan University

Leslie Ries, Georgetown University
Jessica M. Rocheleau, Western New England College
Sean M. Rollins, Fitchburg State University
Daad Saffarini, University of Wisconsin, Milwaukee
Lucia Santacruz, Bowie State University
Thomas Sasek, University of Louisiana at Monroe
Leslie J. Saucedo, University of Puget Sound
Stephanie C. Schroeder, Webster University
Paul J. Schulte, University of Nevada, Las Vegas
Leo Shapiro, University of Maryland
Richard Shingles, Johns Hopkins University
Dave Shutler, Acadia University
Robert C. Sizemore, Alcorn State University
Don Spence, Bethune-Cookman University
Shannon Stevenson, University of Minnesota, Duluth
David R. Sultemeier, University of Puget Sound
Fengjie Sun, Georgia Gwinnett College
Ken G. Sweat, Arizona State University
Casey P. terHorst, California State University, Northridge
Ximena Valderrama, Ramapo College of New Jersey
Lori Valentine Rose, Hill College
Jennifer von Reis, Columbia Basin College
Daryle Waechter-Brulla, University of Wisconsin, Whitewater
Mitchell Walkowicz, University of Massachusetts, Amherst
Suzanne Watts Gollery, Sierra Nevada College
Michael M. Webber, University of Nevada, Las Vegas
Christine L. Weilhoefer, University of Portland
Mary White, Southeastern Louisiana University
Lisa B. Whitenack, Allegheny College
Justen B. Whittall, Santa Clara University
Robert R. Wise, University of Wisconsin, Oshkosh
Irene M. Wolf, Saint Francis University
Erica B. Young, University of Wisconsin, Milwaukee

Media and Supplements Contributors

Jill DeVito, University of Texas, Arlington
Donna Francis, University of Massachusetts, Amherst
Carol Hand, Science writer
Phillip Harris, University of Alabama
Margaret Hill, Science writer
Norman Johnson, University of Massachusetts, Amherst
Carly Jordan, The George Washington University
Laurie Leonelli, New York University
Betty McGuire, Cornell University
Meredith Safford, Johns Hopkins University
John Townsend-Mehler, Montana State University
Mary Tyler, University of Maine, Orono
Robert Wise, University of Wisconsin, Oshkosh (emeritus)

Contents

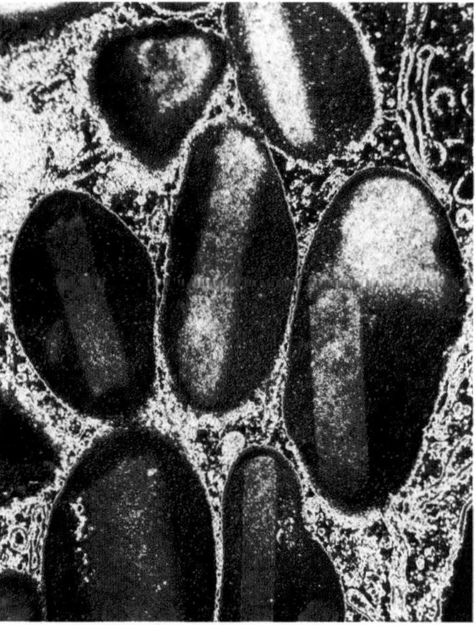

Part 2 Genetics

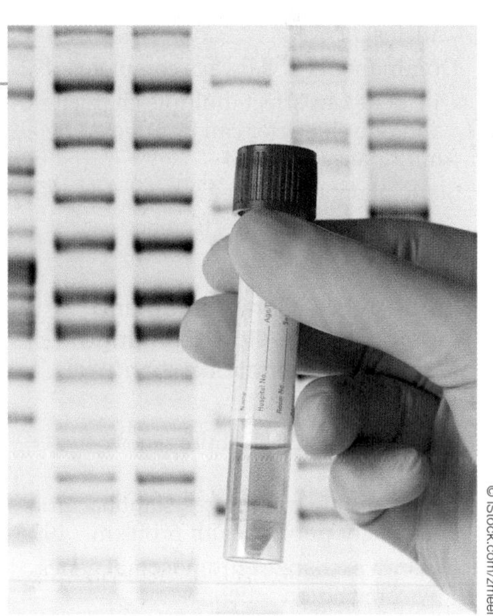

Part 3 Evolution

Part 4 Diversity

© Alex Hyde/Minden Pictures

Part 6 Animal Form and Function

Part 7 Ecology

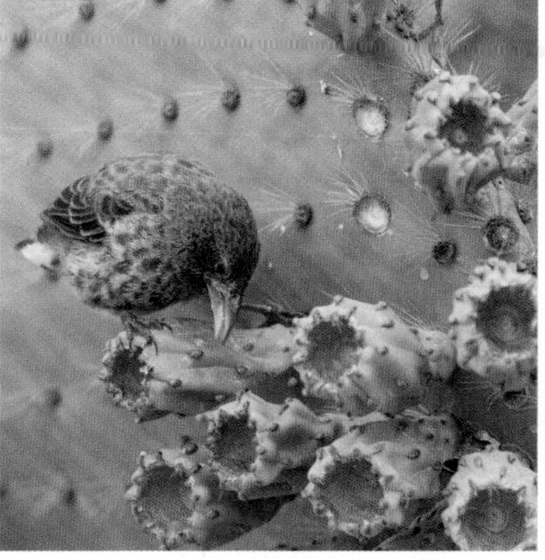

© Tui De Roy/Minden Pictures

© Yva Momatiuk and John Eastcott/Minden Pictures

Principles of Life 1

You have probably seen the African savannas in photographs or films. What did you notice? Like most people, you probably saw the vivid sunlit grasslands and the exotic animals such as giraffes. However, if you have not done so before, take a little time now to think about how these living things survive, reproduce, interact with one another, and influence their environment. With the introduction to biology in this book, we would like to inspire you to ask questions about what life is, how living systems work, and how the living world came to be as we observe it today.

Biologists have amassed a huge amount of information about the living world, and some introductory biology classes focus on memorizing details. In this book we take a different approach, focusing on the major principles of life that underlie everything in biology.

What do we mean by "principles of life"? Look at the photograph. Why is the view mostly of plants, with just a few animals? Why does the tree have the shape it does? Why have the giraffes evolved such long necks? A fundamental principle of life, namely that all living organisms require energy to grow, move, reproduce,

© iStock.com/pierivb

and maintain their bodies, can explain the prevalence of plants. Ultimately, most of the needed energy comes from the Sun. The leaves of the grasses and trees contain chlorophyll, a green pigment that absorbs the light energy from the Sun. In fact, some of the plants, such as the acacia tree in the photograph, have evolved shapes whereby they present great numbers of leaves directly to the Sun's light, maximizing solar energy absorption, while also evading the reach of leaf-eating mammals. By the process of photosynthesis, the green leaves of the plants transform water and carbon dioxide into sugar and oxygen, and in this process some of the energy from the Sun is stored in the sugar. The plants then use this sugar as a source of energy to build their tissues and grow. In turn, giraffes and other animals eat the plants to obtain energy from their tissues. A giraffe is ultimately solar-powered, as is the person who took the photograph.

The photograph also illustrates other principles of biology in addition to the flow of energy. One is that living organisms often survive and thrive by interacting with one another in complex ways. Another is that evolution has often modified organisms based on the demands of their interactions. The long neck of the giraffes, which has evolved over millions of years, permits them to reach the high leaves on the trees in their ecological community. After reading this book, you will better understand the main principles of life. You'll be able to describe how organisms capture and transform energy; pass genetic information to their offspring in reproduction; grow, develop, and behave; and interact with other organisms and with their physical environment. You will also have learned how the system of life on Earth evolved, and how it continues to change. May you always view the natural world with new insight and understanding!

KEY CONCEPTS

1.1 Living Organisms Share Common Aspects of Structure, Function, and Energy Flow

Biology is the scientific study of life, which encompasses all living things, or **organisms**. The living things we know about are all descended from a single-celled ancestor that lived on Earth almost 4 billion years ago. We can imagine that something with some similarities to life as we know it might have originated differently, perhaps on other planets. But the evidence suggests that all of life on Earth today has a single origin—a single common ancestor—and we consider all the organisms that descended from that common ancestor to be a part of life.

Life as we know it had a single origin

The overwhelming evidence for the common ancestry of life lies in the many characteristics that are shared among living organisms. Typically, living organisms

- are composed of a common set of chemical parts, such as nucleic acids (including DNA, which carries our genetic information) and amino acids (the chemical building blocks that make up proteins)
- are composed at a microscopic level of similar structures, such as cells enclosed within membranes
- depend on intricate interactions among structurally complex parts to maintain the living state
- contain genetic information that uses a nearly universal code to specify how proteins are assembled
- convert molecules obtained from their environment into new biological molecules
- extract energy from the environment and use it to carry out life functions
- replicate their genetic information in the same manner when reproducing themselves
- have a fundamental set of genes that share structural similarities
- evolve through gradual changes in their genetic information

Taken together, these shared characteristics logically lead to the conclusion that all life has a common ancestry, and that the diverse organisms that exist today originated from one life form. If life had multiple origins, there would be little reason to expect a nearly universal genetic code, or the structural similarities among many genes, or a common set of amino acids. If we were to discover something similar to life that had originated independently, such as a self-replicating system on another planet, we would expect it to be fundamentally different in these aspects. It might be similar in some ways to life on Earth, such as using genetic information to reproduce. But we would not expect the details of its genetic code, for example, to be like ours.

The simple list of shared characteristics above, however, does not describe the incredible complexity and diversity of life. Just think, for example, of the many different kinds of birds you see each month, or the many different kinds of trees. One of the major questions that biologists address today is how the great diversity of life on Earth has evolved. It is fascinating and important that all these life forms share fundamental characteristics, but the enormous diversity of life forms is also an important theme and question in biology.

A final introductory point of great significance concerns the boundaries between "living" and "nonliving." When biologists say that biology is the study of life, they generally have little difficulty distinguishing organisms that are alive from other systems that are not alive. Yet the boundaries are not always clear. One important case for debate is the viruses, which are not composed of cells and cannot carry out most functions on their own. Instead viruses use the cells they invade to perform most functions for them. Yet viruses contain genetic information, and they mutate and evolve. So even though viruses are not independent cellular organisms, their existence depends on cells, and there is strong evidence that viruses evolved from cellular life forms. For these reasons, most biologists consider viruses to be a part of life. But as viruses illustrate, the boundaries between "living" and "nonliving" are not always clear, and all biologists do not agree on exactly where we should draw the lines.

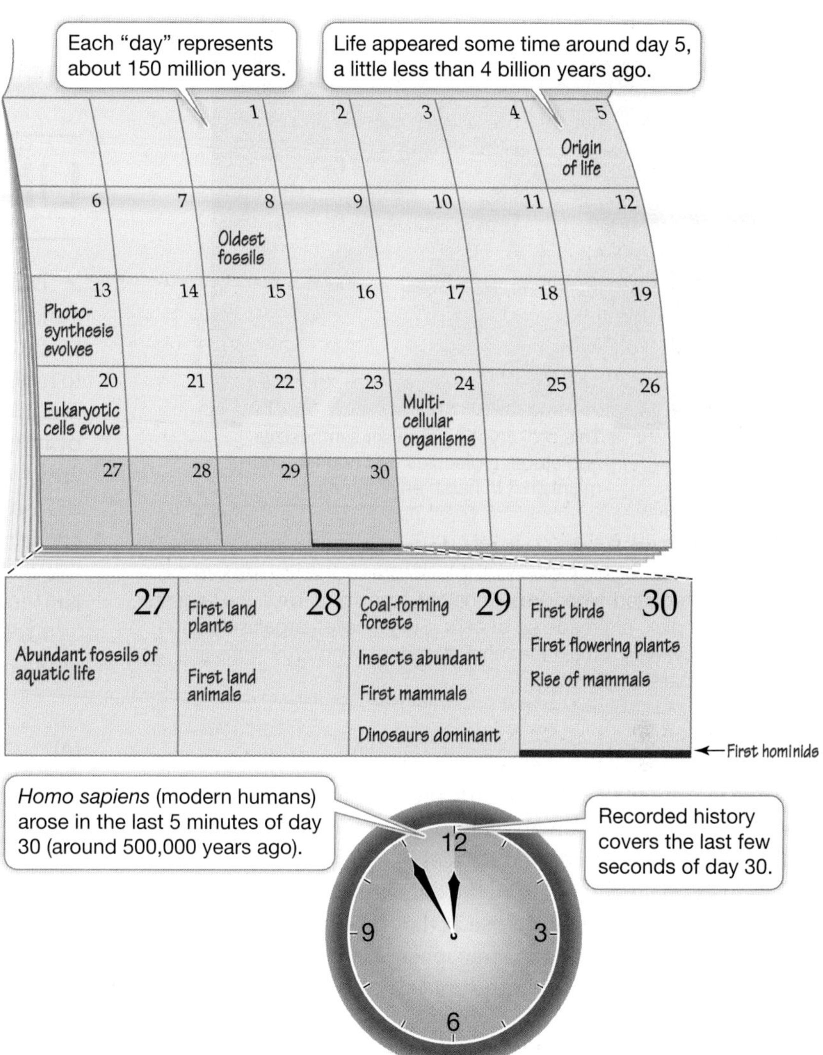

FIGURE 1.1 Life's Calendar Depicting Earth's history on the scale of a 30-day month provides a sense of the immensity of evolutionary time.

Major steps in the history of life are compatible with known physical and chemical processes

Geologists estimate that Earth formed between 4.6 and 4.5 billion years ago. At first the planet was not a very hospitable place. It was some 600 million years or more before the earliest life evolved. If we picture the history of Earth as a 30-day month, with each day representing about 150 million years, life first appeared somewhere toward the end of the first week (**FIGURE 1.1**).

How might life have arisen from nonliving matter? In thinking about this question, we must take into account that the young Earth's atmosphere, oceans, and climate all were very different than they are today. Biologists have conducted many experiments that simulate the conditions on early Earth. These experiments have confirmed that the formation of complex organic molecules under such conditions is possible, even probable.

The critical step for the evolution of life, however, was the appearance of **nucleic acids**—molecules that could reproduce themselves and also contain the information for the synthesis, or manufacture, of other large molecules with complex but stable shapes. These large, complex molecules were proteins. Their shapes varied enough to enable them to participate in increasing numbers and kinds of chemical reactions with other molecules.

CELLULAR STRUCTURE EVOLVED IN THE COMMON ANCESTOR OF LIFE In the next big step in the origin of life, a membrane surrounded and enclosed complex proteins and other biological molecules, forming a tiny **cell**. This membrane kept the enclosed components separate from the surrounding external environment. Molecules called fatty acids played a critical role because these molecules form membrane-like films instead of dissolving in water. When agitated, these films can form hollow spheres, which could have enveloped assemblages of biological molecules. The creation of a cell interior, separate from the external environment, allowed the reactants and products of chemical reactions to be concentrated, opening up the possibility that those reactions could be integrated and controlled. This natural process of membrane formation likely resulted in the first cells with the ability to reproduce—the evolution of the first cellular organisms.

For more than 2 billion years after cells originated, every organism consisted of only one cell. These first organisms were **prokaryotes**, which are made up of single cells containing genetic material and other biochemical structures enclosed in a membrane

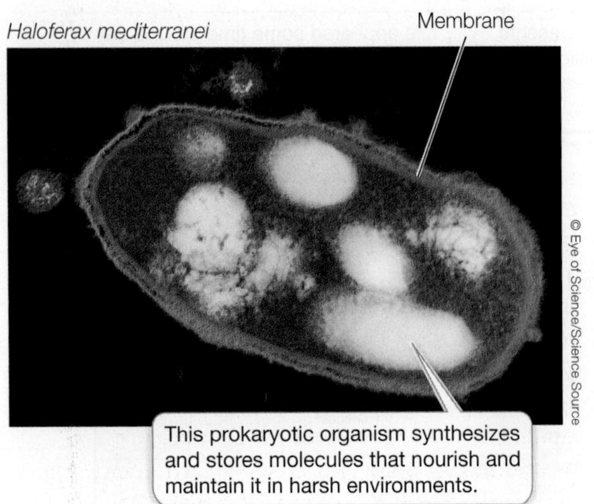

FIGURE 1.2 The Basic Unit of Life Is the Cell The concentration of reactions within the enclosing membrane of a cell allowed the evolution of integrated organisms. Today all organisms, even the largest and most complex, are made up of cells. Single-celled organisms such as this one, however, remain the most abundant living organisms (in absolute numbers) on Earth.

(**FIGURE 1.2**). Vast numbers of their descendants, such as bacteria, exist in similar form today. Early prokaryotes were confined to the oceans, which had an abundance of complex molecules they could use as raw materials and sources of energy. The water of the oceans also shielded them from the damaging effects of ultraviolet (UV) light, which was intense at that time because there was little or no oxygen (O_2) in the atmosphere, and for that reason, no protective ozone (O_3) layer in the upper atmosphere.

PHOTOSYNTHESIS ALLOWED LIVING ORGANISMS TO CAPTURE THE SUN'S ENERGY To fuel the chemical reactions inside them, the earliest prokaryotes took in molecules directly from their environment and broke down these small molecules to release and use the energy contained in their chemical bonds. Many modern prokaryotes still function this way, and very successfully.

About 2.5 billion years ago, or on day 13 of our imaginary month-long calendar of life, the emergence of photosynthesis in some prokaryotes changed the nature of life on Earth (see Figure 1.1). **Photosynthesis** is a set of chemical reactions that transforms the energy of sunlight into chemical-bond energy of the sugar glucose and other relatively small biological molecules. In turn, the chemical-bond energy of these small molecules can be tapped to power other chemical reactions inside cells, including the synthesis of large molecules, such as proteins, that are the building blocks of cells.

Photosynthesis is the basis of much of life on Earth today because its energy-capturing processes provide food not only for photosynthetic organisms themselves, but also for other organisms that eat the photosynthetic ones. Photosynthetic organisms—such as the grasses and trees in our opening photograph—use solar energy to build their tissues, and then other organisms (such as the giraffes) use those tissues as food. Early photosynthetic cells were probably similar to the present-day prokaryotes called cyanobacteria (**FIGURE 1.3**). Over time, the early photosynthetic prokaryotes became so abundant that they produced vast quantities of O_2 as a by-product of photosynthesis.

LINK The pathways that harvest chemical energy to all the kinds of biological work necessary to support metabolism are presented in **Chapter 5**.

During the early eons of life on Earth, there was no O_2 in the atmosphere. In fact, O_2 was poisonous to many of the prokaryotes that lived at that time. But organisms that tolerated O_2 were able to proliferate as O_2 slowly began to accumulate in the atmosphere. The presence of O_2 opened up vast new avenues of evolution. **Aerobic metabolism**, a set of chemical reactions that releases energy from life's molecules by using O_2, proved to be more efficient than **anaerobic metabolism**, a set of reactions that extracts energy without using O_2. For this reason, O_2 allowed organisms to live more intensely and grow larger. The majority of living organisms today use O_2 in extracting energy from molecules.

Oxygen in the atmosphere also made it possible for life to move onto land. For most of life's history, UV radiation falling

(A)

(B)

FIGURE 1.3 Photosynthetic Organisms Changed Earth's Atmosphere Cyanobacteria were the first photosynthetic organisms on Earth. **(A)** Colonies of cyanobacteria called stromatolites are known from the ancient fossil record. **(B)** Living stromatolites are still found in suitable environments on Earth today.

on Earth's surface was so intense that it destroyed any living cell that was not well shielded by water. But as a result of photosynthesis, O_2 accumulated in the atmosphere for more than 2 billion years and gradually resulted in a layer of ozone (O_3) in the upper atmosphere. By about 500 million years ago, or about day 28 on our imaginary calendar of life, the ozone layer was sufficiently dense—and absorbed enough of the Sun's UV radiation—to make it possible for organisms to leave the protection of the water and live on land (see Figure 1.1).

EUKARYOTIC CELLS AROSE THROUGH ENDOSYMBIOSIS Another important, earlier step in the history of life was the evolution of cells that are composed internally of membrane-enclosed compartments called **organelles**. Organelles were—and are—important because specialized cellular functions could be performed inside them, separated from the rest of the cell. The first organelles probably appeared about 2.5 billion years after life first appeared on Earth, or on about day 20 in Figure 1.1.

One of these organelles, the **nucleus**, came to contain the cell's genetic information. The nucleus (Latin *nux*, "nut" or "core") gives these cells their name: **eukaryotes** (Greek *eu*, "true," + *karyon*, "kernel" or "core"). The eukaryotic cell is distinct from the cells of prokaryotes (*pro*, "before"), which lack nuclei.

Some organelles are hypothesized to have originated by **endosymbiosis**, which means "living inside another" and may have occurred as a complicated consequence of the ingestion of smaller cells by larger cells. The **mitochondria** that release energy for use by a eukaryotic cell probably evolved from engulfed prokaryotic organisms. And **chloroplasts**—the organelles specialized to conduct photosynthesis in eukaryotic photosynthetic organisms—could have originated when larger eukaryotes ingested photosynthetic prokaryotes. When large cells ingested smaller ones, the two cells could then at times evolve a stable partnership (instead of the large cells always using the smaller ones as food). In this way, ingested prokaryotes could provide the large cells with sugars from photosynthesis, or perform other functions for the large cells. In return, the host cells provided a good environment for their smaller partners.

MULTICELLULARITY ALLOWED SPECIALIZATION OF TISSUES AND FUNCTIONS For the first few billion years of life, all organisms—whether prokaryotic or eukaryotic—were single-celled. At some point, the cells of some eukaryotes failed to separate after cell division and remained attached to each other. In these groupings of cells it was possible for some cells in the group to specialize in certain functions, such as reproduction, while other cells specialized in other functions, such as absorbing nutrients. **Cellular specialization** enabled multicellular eukaryotes to increase in size and become more efficient at gathering resources and living in specific environments.

Biologists can trace the evolutionary tree of life

If all the organisms on Earth today are the descendants of a single kind of unicellular organism that lived almost 4 billion years ago, how have they become so different? An organism reproduces by replicating its **genome**, which is the sum total of its genetic material, as we will discuss shortly. This replication process is not perfect, however, and changes, called **mutations**, are introduced almost every time a genome is replicated. Some mutations give rise to structural and functional changes in organisms. Within a population of organisms, as individuals mate with one another, these changes can spread while the population continues to be made up of one kind, or species, of organism. However, if something happens to isolate some members of a population from the others, structural and functional differences between the two groups will accumulate over time. The two groups may eventually differ enough that their members no longer regularly reproduce with one another. In this way the two populations become two different species.

Tens of millions of species exist on Earth today. Many times that number lived in the past but are now extinct. As biologists discover species, they give each one a scientific name called a **binomial** (because it is made up of two Latinized words). The first word identifies the species' genus—a group of species that share a recent common ancestor. The second word indicates the species. For example, the scientific name for the human species is *Homo sapiens*: *Homo* is our genus and *sapiens* our species. *Homo* is Latin for "man," and *sapiens* is from the Latin word for "wise" or "rational." These scientific names are italicized, and unlike common names, are the same in every language. This allows biologists writing in any language to refer to the same species in the same way.

Much of biology is based on comparisons among species. Our ability to make relevant comparisons has improved greatly in recent decades as a result of our relatively newfound ability to study and compare the genomes of different species. We do this by sequencing a genome (in whole or in part), which means we can determine the order of the nucleotides that serve as the building blocks of the organism's DNA. Genome sequencing and other molecular techniques have allowed biologists to add a vast array of molecular evidence to existing evolutionary knowledge based on the fossil record. The result is the ongoing compilation of **phylogenetic trees** that document and diagram evolutionary relationships as part of an overarching **tree of life**. The broadest categories of this tree are shown in **FIGURE 1.4**. (The tree is expanded in Appendix A, and you can also explore the tree interactively online.)

Although many details remain to be clarified, the broad outlines of the tree of life have been determined. Its branching patterns are based on a rich array of evidence from fossils, structures, chemical processes, behavior, and molecular analyses of genomes. Molecular data in particular have been used to separate the tree into two major branches of life: **Bacteria** and **Archaea**. One specialized group of organisms arose and diversified among the archaea, but with contributions of mitochondria and chloroplasts from endosymbiotic bacteria. We call these specialized organisms **eukaryotes** (or formally, **Eukarya**). Eukaryotes have a distinct cellular structure called a nucleus that contains most of their genetic information. Because eukaryotes have many other distinctive features as well, some of which came from bacteria and some from archaea, most biologists recognize three **domains** of life: Bacteria, Archaea, and Eukarya.

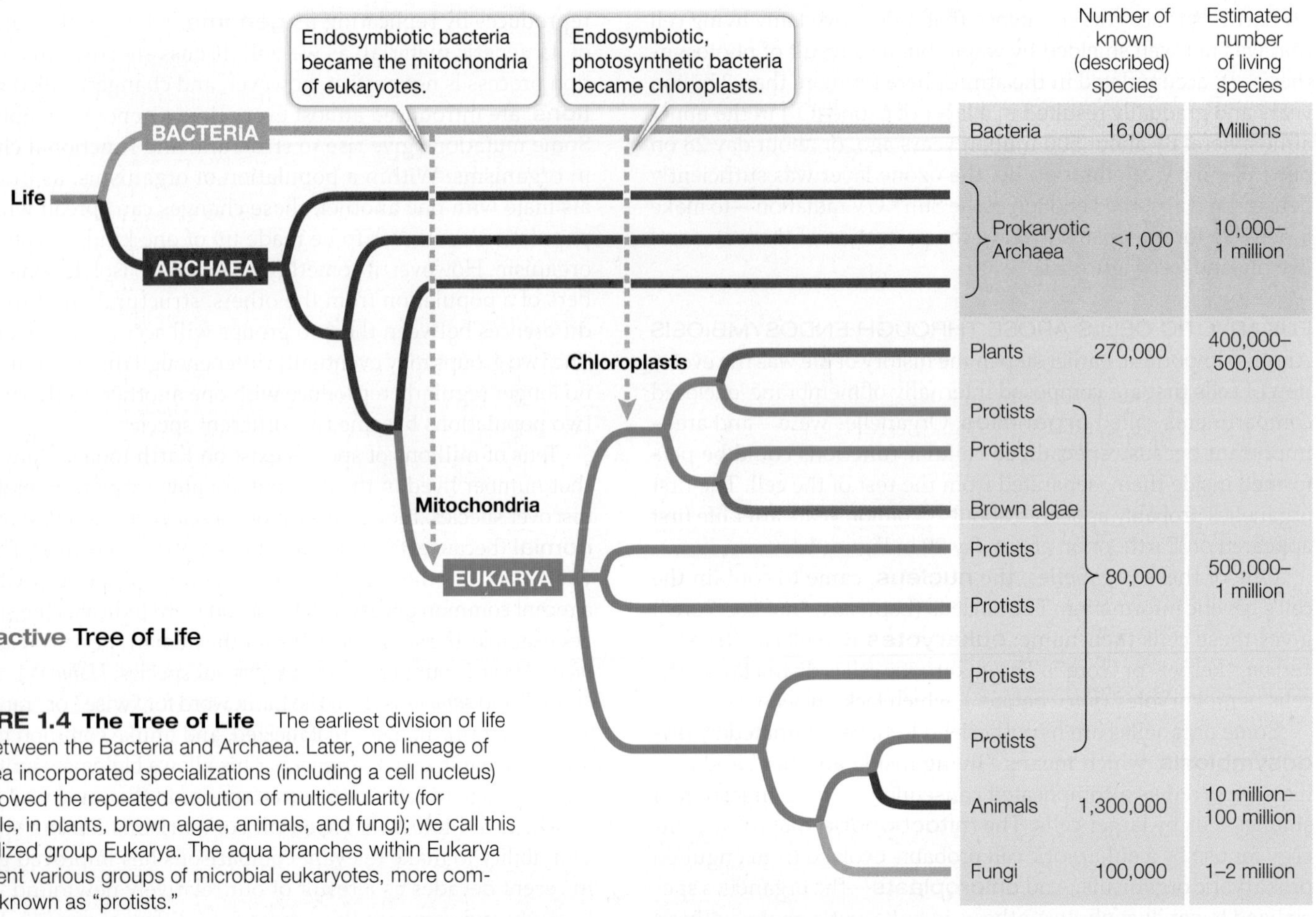

Endosymbiotic bacteria became the mitochondria of eukaryotes.

Endosymbiotic, photosynthetic bacteria became chloroplasts.

Life

BACTERIA

ARCHAEA

EUKARYA

Chloroplasts

Mitochondria

	Number of known (described) species	Estimated number of living species
Bacteria	16,000	Millions
Prokaryotic Archaea	<1,000	10,000–1 million
Plants	270,000	400,000–500,000
Protists, Protists, Brown algae, Protists, Protists, Protists, Protists	80,000	500,000–1 million
Animals	1,300,000	10 million–100 million
Fungi	100,000	1–2 million

Interactive Tree of Life

FIGURE 1.4 The Tree of Life The earliest division of life was between the Bacteria and Archaea. Later, one lineage of Archaea incorporated specializations (including a cell nucleus) that allowed the repeated evolution of multicellularity (for example, in plants, brown algae, animals, and fungi); we call this specialized group Eukarya. The aqua branches within Eukarya represent various groups of microbial eukaryotes, more commonly known as "protists."

Many (but not all) eukaryotes are multicellular. Plants, brown algae (such as kelp), fungi, and animals are examples of familiar multicellular eukaryotes. We know that multicellularity arose independently in each of these four familiar multicellular groups (as well as in several others that you will learn about) because they are each most closely related to different groups of unicellular eukaryotes (commonly called protists), as you can see from the branching pattern of Figure 1.4.

Life's unity allows discoveries in biology to be generalized

Knowledge gained from investigations of one kind of organism can, with care, be generalized to other organisms because all life is related by descent from a common ancestor, shares a genetic code, and consists of similar molecular building blocks. Biologists use certain species as **model systems** for research, knowing they can often extend their findings to other organisms, including humans.

Our basic understanding of the chemical reactions in cells came from research on bacteria but is applicable to all cells, including

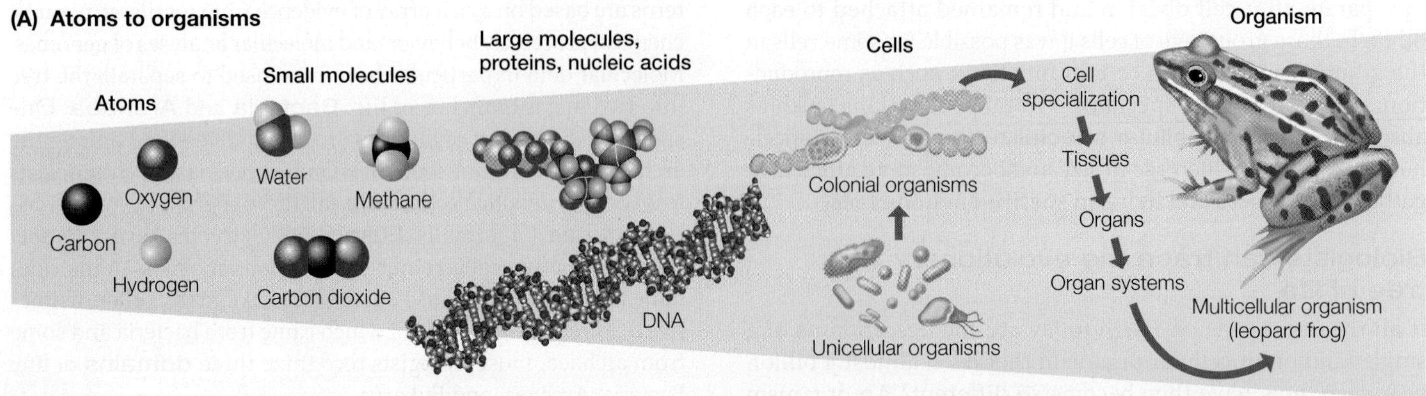

those of humans. Similarly, the biochemistry of photosynthesis—the process by which plants use sunlight to produce sugars—was largely worked out from experiments on *Chlorella*, a unicellular green alga. Much of what we know about the genes that control plant development is the result of work on *Arabidopsis thaliana*, a member of the mustard family. Knowledge about how animals develop has come from work on sea urchins, frogs, chickens, roundworms, and fruit flies. And recently, the discovery of a major gene controlling human skin color came from work on zebrafish. Being able to generalize from model systems is a powerful tool in biology.

1.2 Life Depends on Organization and Energy

All of life depends on organization. Physics gives us the second law of thermodynamics, which states that, left to themselves, organized entities tend to become more random. Any loss of organization threatens the well-being of organisms. Cells, for example, must combat the thermodynamic tendency for their molecules, structures, and systems to fall apart—to become disorganized. Energy is required to maintain organization. For this reason, cells require energy throughout their lives.

Organization is apparent in a hierarchy of levels, from molecules to ecosystems

Cells synthesize, or manufacture, proteins and other complex molecules by assembling atoms into new, highly organized configurations. Such complex molecules give cells their structure and enable them to function. For example, a fatty acid molecule that the cell synthesizes may become part of a membrane that organizes the inside of the cell by dividing it into compartments. Or a protein made by a cell may enable a specific chemical reaction to take place in the cell by helping speed up the reaction—that is, by acting as a catalyst for the reaction.

Organization is also essential for many cells to function together in a multicellular organism. As we have seen, multicellularity allows individual cells to specialize and depend on other cells for functions they themselves do not perform. But the different specialized cells also work together. For example, division of labor in a multicellular organism usually requires a circulatory system so that the functions of specialized cells in one part of the body are of use to cells in other, distant parts of the body.

Overall, a multicellular organism exhibits many hierarchical levels of organization (**FIGURE 1.5A**). Small molecules are organized into larger ones, such as DNA and proteins. Large molecules are organized into cells, and assemblages of differentiated cells are organized into **tissues**. For example, a single muscle cell cannot generate much force, but when many cells combine to form the tissue of a working muscle, considerable force and movement can be generated. Different tissue types are organized to form **organs** that accomplish specific functions. The heart, brain, and stomach are each constructed of several types of tissues, as are the roots, stems, and leaves of plants. Organs whose functions are interrelated can be grouped into **organ systems**; the esophagus, stomach, and intestines, for example, are all part of the digestive system. Because all these levels of organization are subject to the second law of thermodynamics, they all tend to degrade unless

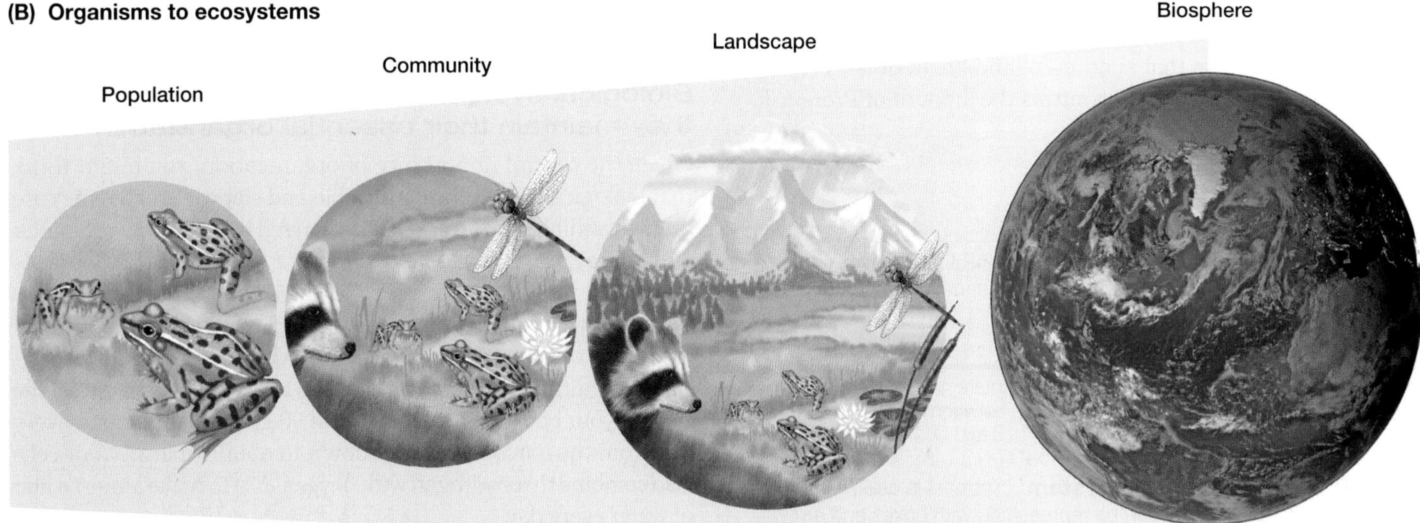

FIGURE 1.5 Life Consists of Organized Systems at a Hierarchy of Scales **(A)** The hierarchy of systems within a multicellular organism. DNA—a molecule—encodes the information for cells—a higher level of organization. Cells, in turn, are the components of still higher levels of organization: tissues, organs, and the organism itself. **(B)** Organisms interacting with their external environment form ecological systems on a hierarchy of scales. Individual organisms form the smallest ecological system. Individuals of a species form populations, which interact with other populations to form communities. Multiple communities in turn interact within landscapes at progressively larger scales until they include all the landscapes and organisms of Earth: the entire biosphere. (NASA image of Biosphere by Reto Stöckli, based on data from NASA and NOAA.)

Activity 1.1 The Hierarchy of Life

energy is applied to the system. This is why an organism must use energy to maintain its functions.

In addition to the internal hierarchy of an individual organism, there is an external hierarchy in the larger biological world where organisms interact with each other and their physical environment—forming **ecological systems**, often shortened to **ecosystems** (**FIGURE 1.5B**). Individual organisms interacting with their immediate environment form the smallest ecological system. Groups of individuals of any one species live together in **populations**, and populations of multiple species interact in ecological **communities**. Multiple communities interact within **landscapes**. The landscape of the entire Earth and all its life is known as the **biosphere**.

Now that we have recognized the internal hierarchy of organization within an individual organism and the external hierarchy in the ecosystem, we must recognize a highly important distinction between the internal and external hierarchies. All the hierarchical levels of organization within an individual organism are encoded by its single genome, so that these levels generally interact harmoniously. By contrast, the external ecological hierarchy of populations, communities, and landscapes involves interactions among multiple species with multiple genomes, so that interactions are not always harmonious. For example, individuals may prevent others of their own species from exploiting a necessary resource such as food, or they may exploit members of other species as food.

Each level of biological organization consists of systems

We have already discussed within-organism systems and ecological systems. More generally, a **system** is a set of interacting parts in which neither the parts nor the whole can be fully understood without taking into account the interactions. A simple biological system might consist of a few **components** (e.g., proteins, pools of nutrients, or organisms) and the **processes** by which the components interact (e.g., protein synthesis, nutrient metabolism, or grazing) (**FIGURE 1.6**).

Consider, for example, the system within a cell that synthesizes and controls the quantity of a particular protein, which we'll call Protein T (**FIGURE 1.7A**). The components of the system are the amino acids from which Protein T is made, Protein T, and the breakdown products of Protein T. The processes are the biochemical pathways that synthesize and break down Protein T. To understand how the cell controls the amount of Protein T, we must understand how all the components and processes in this system function.

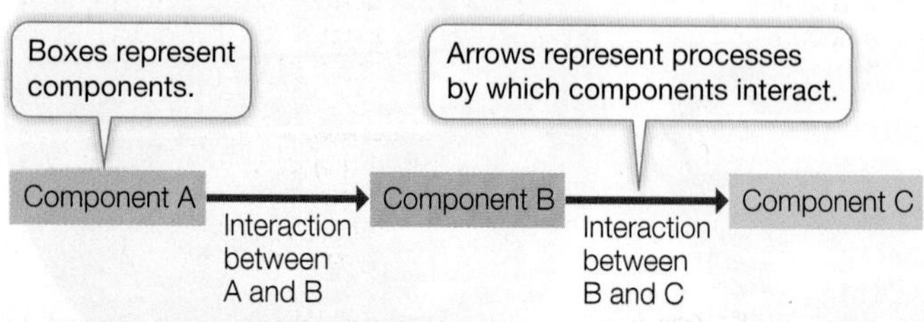

FIGURE 1.6 A Generalized System Systems in cells, whole organisms, and ecosystems can be represented with boxes and arrows.

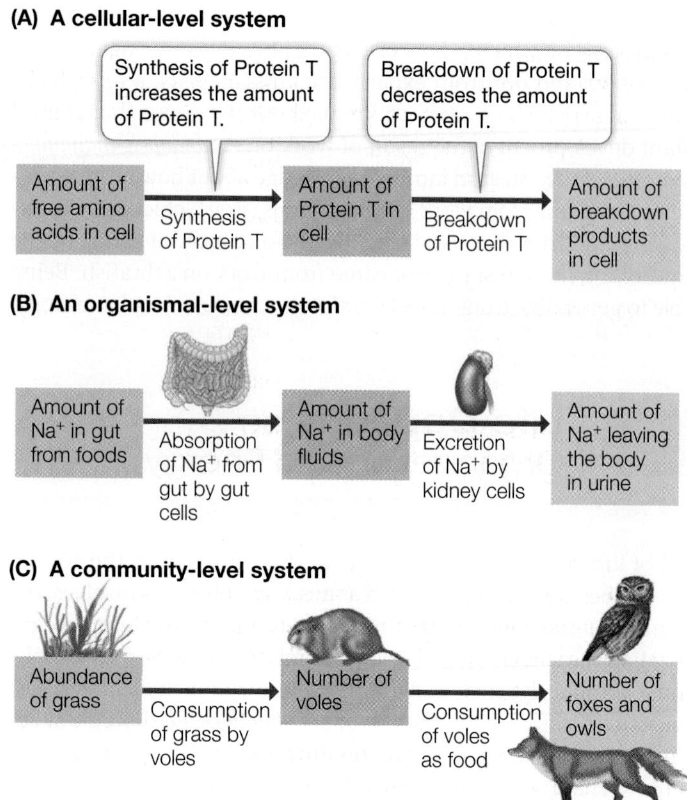

FIGURE 1.7 Organized Systems Exist at Many Levels **(A)** This cellular-level system synthesizes and breaks down a cell protein called Protein T. **(B)** This organismal-level system determines the amount (and thus the concentration) of sodium (Na^+) in the blood plasma and other extracellular body fluids of a human. **(C)** This community-level system helps determine the number of meadow voles (*Microtus pennsylvanicus*) in a field in the spring.

Systems are found at every level of biological organization. For example, our bodies have a physiological system that controls the amount of sodium (Na^+) in our body fluids (**FIGURE 1.7B**). Grass, voles, and predators (e.g., foxes and owls) are components of a community-level system (**FIGURE 1.7C**).

Biological systems are highly dynamic even as they maintain their essential organization

Given the central importance of organization, you might think that biological systems are inflexible and static. Actually, they are often incredibly **dynamic**—characterized by rapid flows of matter and energy. On average, for example, a cell in your body breaks down and rebuilds 2–3 percent of its protein molecules per day. Each day it also makes and uses more than 100,000 trillion (10^{14}) molecules of adenosine triphosphate (ATP), the molecule responsible for shuttling energy from sources to uses. Collectively, all the cells in your body liberate more than 90 grams of hydrogen every day from the foods they break down to obtain energy. Your cells also combine that hydrogen with oxygen (O_2) to make almost a liter of water every day.

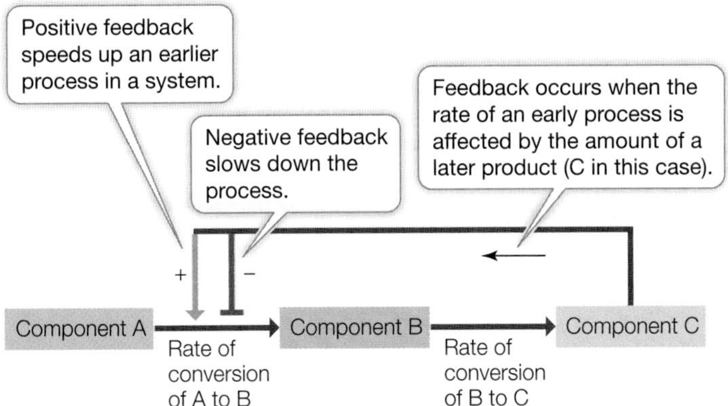

FIGURE 1.8 Feedback Can Be Positive or Negative Positive feedback tends to destabilize a system, whereas negative feedback typically stabilizes a system.

This dynamic aspect of biological systems means that they constantly exchange energy and matter with their surroundings. For example, even after a single-celled or multicellular organism has reached maturity, most of its molecules are steadily replaced as time passes. In this ceaseless, dynamic process, atoms are lost from the organism to the surrounding soil, air, or water, and they are replaced with atoms from the soil, air, or water. Yet as the atomic building blocks of any particular organism come and go, the organization of the molecules, structures, and systems in the organism persists. This fact emphasizes the central importance of organization.

Positive and negative feedback are common in biological systems

Often, the amount of one of the components of a system, such as component C in **FIGURE 1.8**, affects the rate of one of the earlier processes in the system. This effect is called **feedback** and may be described as positive or negative. Feedback is often diagrammed simply with a line and symbol, but its actual mechanism may be complex.

Positive feedback occurs in a system when a product of the system *speeds up* an earlier process. The effect of positive feedback is to cause the product to be produced faster and faster. To return to one of our earlier examples, if the breakdown products of Protein T sped up synthesis of Protein T, this would lead to more breakdown products, then even more Protein T, then even more breakdown products, and so on. Positive feedback tends to destabilize a system, but destabilization can sometimes be advantageous, provided it is ultimately brought under control.

Negative feedback occurs when a product of a system *slows down* an earlier process in the system. Often, as the product increases in amount or concentration, it exerts more and more of a slowing effect. Negative feedback stabilizes the amount of the product in this way: as a high amount of the product accumulates, that accumulation tends to reduce further production of the product. For example, if an increase in the amount of breakdown products of Protein T slowed down synthesis of Protein T, this would lead to a decreased amount of breakdown products and a return to the previous rate of Protein T synthesis. Negative feedback is very common in **regulatory systems**, which are systems that tend to stabilize amounts or concentrations.

Systems analysis is a conceptual tool for understanding all levels of organization

Biologists today employ an approach known as **systems analysis** to understand how biological systems function. In systems analysis, we identify the parts or components of a biological system and specify the processes by which the components interact (see Figure 1.6). We may also be able to specify the *rates* of these interactions and how the rates are affected by feedback. What we can do then is analyze how the system will change through time. Will the amounts of different components increase or decrease, and how quickly? Will there be any state of stable balance, or **equilibrium**, that the system eventually reaches? With a detailed systems analysis, we can analyze all these features.

To do the analysis, we write out mathematical equations that express the amounts of the different components and that include the processes and their rates. Expressed in words, such an equation for component B in Figure 1.6 has the following form:

The amount of B present at some time in the future =
the amount of B now + the amount of A converted into B
– the amount of B converted into C

We write out a similar equation for each component in the system.

We can analyze the relatively simple biological systems in Figure 1.7 by hand, but the analysis of larger systems quickly becomes very complicated and is typically carried out using computers. The approach, however, is the same: we express the rates of all processes as mathematical equations (**FIGURE 1.9**). After this analysis is done, we have a **computational model** of the biological system. If the computational model is well grounded in factual knowledge of the biological system, the model will mimic the biological system.

An important use of computational models is prediction. For instance, if atmospheric temperature affects a biological system, we can use a computational model to develop a hypothetical prediction of the future behavior of the system in a warming world by adjusting the model to take into account the expected increase in atmospheric temperature.

Animation 1.2 System Simulation

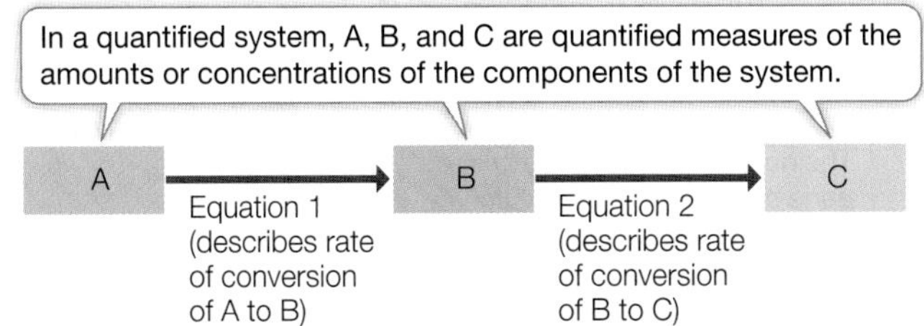

FIGURE 1.9 Mathematical Equations Allow Us to Quantify Systems The components of a system can be related through the use of mathematical equations.

1.3 | **Genetic Systems** Control the Flow, Exchange, Storage, and Use of Information

The information required to build a functional organism—the "blueprint" for its existence—is contained in the organism's genome, which as we noted earlier is the sum total of all the information encoded by its genes. The presence of genetic information and the processes by which organisms "decode" and use it to build the proteins that underlie a body's structure and function involve fundamental principles that we will discuss and expand on throughout the book.

Genomes encode the proteins that govern an organism's structure

Early in the chapter we noted the importance of self-replicating nucleic acids in the origin of life. Nucleic acid molecules contain long sequences of four subunits called **nucleotides**. The sequence of these nucleotides in **deoxyribonucleic acid**, or **DNA**, allows the organism to assemble **proteins**. Each protein-coding **gene** is a specific segment of DNA whose sequence carries the information for building one or more proteins (**FIGURE 1.10**). Proteins, in turn, are the molecules that govern the chemical reactions within cells and form much of an organism's structure. For these reasons, in biology we often say that genes "encode" proteins.

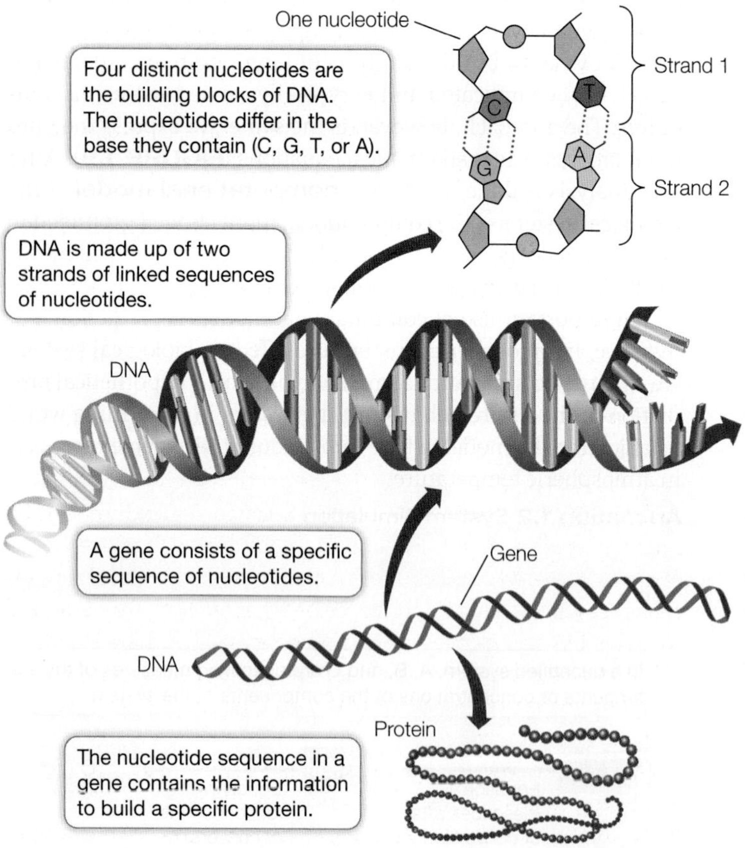

To get a sense of the amount of information encoded by your genome, let's think about an analogy between your genome and the books on a shelf at the library. If you were to write out your genome using four letters to represent the four DNA nucleotides, you would write more than 3 billion letters. Using the type size you are reading now, your genome would fill more than 1,000 books the size of this one.

All the cells of a given multicellular organism contain the same genome, yet the different cells have different functions and form different proteins. For example, oxygen-carrying hemoglobin occurs in red blood cells, gut cells produce digestive proteins, and so on. The differences among cells within an individual organism arise because different types of cells express, or use, different parts of the genome.

When mutations alter the nucleotide sequence of a gene, the protein that the gene encodes is often altered as well. Mutations may occur spontaneously, as happens when mistakes take place during replication of DNA. Mutations can also be caused by certain chemicals (such as those in cigarette smoke) and by radiation (including UV radiation from the Sun). Most mutations either are harmful or have no effect. Occasionally a mutation improves the functioning of the organism under the environmental conditions the individual encounters. Evolution by natural selection can then favor the persistence in a population of such a mutation. Mutations are the raw material of evolution.

FIGURE 1.10 DNA Is Life's Blueprint The instructions for life are contained in the sequences of nucleotides in DNA molecules. Specific DNA nucleotide sequences make up genes. The information in each gene provides the cell with the information it needs to manufacture molecules of a specific protein.

Genomes provide insights into all aspects of an organism's biology

Scientists determined the first complete DNA sequence of an organism's genome in 1976. This first sequence belonged to a virus, chosen because viral genomes are very small compared with the genomes of most cellular organisms. Two decades passed before the first bacterial genome was sequenced, in 1995. The first animal genome to be sequenced was a relatively small one—that of a roundworm—and was determined in 1998. A massive effort to sequence the complete human genome began in 1990 and finished 13 years later.

Since then, scientists have used the methods developed in those pioneering projects, as well as new DNA sequencing technologies that appear each year, to sequence genomes of thousands of species. As methods have improved, the cost and time for sequencing a complete genome have dropped dramatically. The day is rapidly approaching when the sequencing of genomes from individual organisms will be commonplace for many biological applications.

What are we learning from genome sequencing? One surprise came when some genomes turned out to contain many fewer genes than expected. For example, there are only about 20,000 different genes that encode proteins in a human genome, although most biologists had expected many times that number. Biologists can compare genomes from many species to learn how and why one species differs from another. Such comparative genomic studies allow biologists to trace the evolution of genes through time and to document how particular changes in gene sequences result in changes in structure and function, including changes responsible for inherited diseases.

The vast amount of information being collected from genome studies has led to rapid development of the field of *bioinformatics*, the study of the organization and use of biological information. In this emerging field, biologists and computer scientists work together to develop new computational tools to organize, process, and study the immense databases used in comparing genomes.

1.4 Evolution Explains the Diversity as Well as the Unity of Life

Evolution—change in the genetic makeup of biological populations through time—is a major unifying principle of biology. Any process that can lead to changes in the frequencies of genes in a population from generation to generation is an evolutionary process. A common set of evolutionary processes is at work in populations of all organisms. The constant change that occurs in these populations gives rise to all the diversity we see in life. These two themes—unity and diversity—provide a framework for organizing and thinking about the evolution of life. The similarities of all forms of life allow us to make comparisons and predictions from one species to another, as we have discussed. The differences are what make biology such a rich and exciting field for investigation and discovery.

Natural selection is an important process of evolution

Charles Darwin compiled factual evidence for the existence of evolution in his 1859 book *On the Origin of Species*. Since then, biologists have gathered massive amounts of data supporting Darwin's idea that all living organisms are descended from a common ancestor. Darwin also identified one of the most important processes that produce evolutionary change. He argued that the differing survival and reproduction among individuals in a population, which he termed **natural selection**, could account for much of the evolution of life.

LINK The story of Darwin's formulation of the evolution by natural selection is covered in **Key Concept 13.1**.

When Darwin proposed that living organisms are descended from a common ancestor and are therefore related to one another, he did not have the advantage we have today of understanding the processes of genetic inheritance. Those processes, which we will cover in depth in Chapters 7–9, were not widely understood until the early 1900s. But Darwin knew that offspring differed from their parents, even though they showed strong similarities. And he knew that any population of a plant or animal species displays individual variation.

Darwin himself bred pigeons, and he knew that if you select breeding pairs on the basis of some particular trait, then that trait is more likely to be present in their offspring than in the general population. He was well aware of how pigeon fanciers selected breeding pairs to produce offspring with unusual feather patterns, beak shapes, or body sizes. He realized that if humans could select for specific traits in organisms such as pigeons, a similar process could operate in nature. Darwin emphasized that human-imposed selection, which he called "artificial selection," has been practiced on crop plants and domesticated animals since the dawn of human civilization. In coining the term "natural selection," he argued that a similar process occurs in nature. But in nature, the "selection" occurs not by human choice but by the fact that some individuals contribute more offspring to future generations than others.

How does natural selection work? Darwin thought that differing probabilities of survival and reproductive success could account for evolutionary change. He reasoned that the reproductive capacities of plants and animals, if unchecked, would result in unlimited growth of populations. However, we do not observe such growth in nature. In most species, only a small percentage of offspring survive to reproduce. For this reason, any trait will spread in the population if that trait gives individual organisms even a small increase in the probability that they will survive and reproduce.

Because organisms with certain traits survive and reproduce best under specific sets of conditions, natural selection leads to **adaptations**: structural, physiological, or behavioral traits that increase an organism's chances of surviving and reproducing in its specific environment. For example, consider the frogs shown in **FIGURE 1.11**. Look at the frogs' feet and notice how much they differ depending on where the species lives. The greatly expanded toe pads of tree frogs are an adaptation to life in trees, whereas webbed rear feet are an adaptation to a life lived primarily in water.

Biologists often think about two different kinds of explanations for adaptations. On the one hand, we can consider the immediate genetic, physiological, neurological, and developmental processes that explain how an adaptation works. We call these **proximate explanations**. For example, a proximate explanation for the toes of tree frogs might examine the physical structure of the toe pads and explain how expansion of the toe leads to greater adhesion to a surface, such as that of a tree trunk. Such an explanation tells us how the adaptation works, but it does not explain how tree frogs came to possess such toe pads. An **ultimate explanation**, on the other hand, concerns the processes that led to the evolution of toe pads in various groups of climbing frogs. Ultimate explanations involve comparison of variation within and among species and describe how a given trait affects an organism's chances for survival and reproduction.

Although natural selection has been demonstrated in countless biological investigations, it is not the only process that results in evolution, as we will explore in Chapters 13–17. An example of another evolutionary process is genetic drift, which refers to random changes in gene frequencies in a population because of chance events. As a result of the various evolutionary processes, all biological populations evolve through time. All the evolutionary

FIGURE 1.11 Adaptations to the Environment The limbs of frogs show adaptations to the different environments of each species.
Media Clip 1.1 Wallace's Flying Frog

processes operating over the long history of Earth have led to the remarkable diversity of life on our planet today.

Evolution is a fact, as well as the basis for broader theory

The famous biologist Theodosius Dobzhansky once wrote that "Nothing in biology makes sense except in the light of evolution." Dobzhansky was emphasizing the need to include an evolutionary perspective and approach in all aspects of biological study. Everything in biology is a product of evolution, and biologists need a perspective of change and adaptation to fully understand biological systems.

You may have heard someone say that evolution is "just a theory," implying that there is some question about whether or not biological populations evolve. This is a common misunderstanding that originates in part from the different meanings of the word "theory" in everyday language and in science. In everyday speech, some people use the word "theory" to mean "hypothesis" or even—disparagingly—"a guess." In science, however, a **theory** is *a body of scientific work in which rigorously tested and well-established facts and principles are used to make predictions about the natural world.* In short, evolutionary theory is both (1) a body of knowledge supported by facts and (2) the resulting understanding of the various processes by which biological populations have changed and diversified over time, and by which Earth's populations continue to evolve.

We can observe and measure evolution directly, and many biologists conduct experiments to understand the processes that take place in evolving populations. We often observe changes in the genetic makeup of populations over relatively short time periods. For example, every year health agencies need to produce new flu vaccines, because populations of influenza viruses evolve so quickly that last year's vaccines may not be effective against this year's populations of viruses. In addition, we can directly observe a record of the history of evolution by studying fossils of organisms that died and were preserved over the almost unimaginably long periods of geological time. Exactly *how* biological populations change through time is something that is subject to testing and experimentation. The fact that biological populations evolve, however, is not disputed among biologists.

1.5 Science Is Based on Quantitative Observations, Experiments, and Reasoning

Regardless of the many different tools and methods used in research, all scientific investigations are based on quantitative observation, experimentation, and reasoning. In each of these areas, scientists are guided by an established set of scientific methodological principles.

Observing and quantifying are important skills

Many biologists are motivated by their observations of the living world. Learning *what to observe* in nature is a skill that develops with experience in biology. An intimate understanding of the **natural history** of a group of organisms—how the organisms get their food, reproduce, behave, regulate their functions, and interact with other organisms—leads to better observations and prompts biologists to ask questions about those observations. The more a biologist knows about general principles of life, the more they are likely to gain new insights from observing nature.

Although biologists have always observed the world around them, today our ability to observe is greatly extended by technologies such as electron microscopes, rapid genome sequencing, magnetic resonance imaging, and global positioning satellites. These technologies allow us to observe everything from the distribution

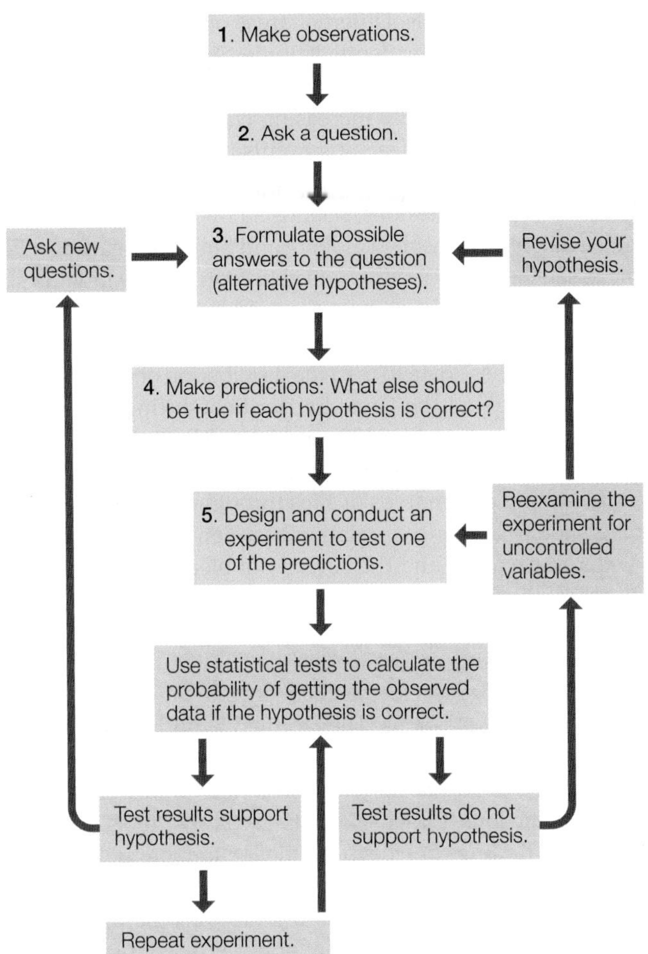

FIGURE 1.12 Scientific Methodology The process of observation, speculation and questioning, hypothesis formation, prediction, and experimentation is a cornerstone of modern science, although scientists may initiate their research at any of several different points.

Animation 1.1 Using Scientific Methodology

of molecules in the body (by using electron microscopes) to the daily movement of animals across continents and oceans (by using global positioning satellites).

As scientists we must also be able to **quantify** our observations—turn the observations into explicit counts or measures that can be used and verified by others. Whether we are testing a new drug or mapping the migrations of whales, mathematical and statistical calculations are essential. Just as there has been progress in our abilities to *observe*, there has also been progress in our abilities to *quantify*. For example, biologists once classified organisms entirely on the basis of qualitative descriptions of the physical differences among them. Today our ability to quantify the molecular and physical differences among species, combined with explicit mathematical models of the evolutionary process, enables quantitative analyses of evolutionary history. These mathematical calculations, in turn, help us compare all other aspects of the biology of different organisms.

Scientific methods combine observation, experimentation, and logic

Often, science textbooks describe "*the* scientific method," as if there is a single flow chart that all scientists follow. This view is an oversimplification. Model flow charts include much of what scientists do, but you should not conclude that scientists necessarily go through all the steps in a flow chart in one prescribed, linear order.

Observations lead to questions, and scientists make additional observations and often do experiments to answer those questions. This approach, called the hypothesis–prediction method, has five steps: (1) making observations; (2) asking questions; (3) forming hypotheses, or tentative answers to the questions; (4) making predictions based on the hypotheses; and (5) testing the predictions by making additional observations or conducting experiments. These are the steps in traditional flow charts such as the one shown in **FIGURE 1.12**.

Getting from questions to answers

Let's consider an example of how scientists start with a general question and work to find answers. To do this, let's focus on coral reefs. Notice the reef-building coral in **FIGURE 1.13**. Why is its shape so similar to that of the acacia tree in our opening photo? The answer is that reef-building corals are photosynthetic! Corals are animals and can't photosynthesize on their own; instead, photosynthesis is carried out by huge numbers of symbiotic algae that live inside the coral tissues. These microscopic algae produce sugar by photosynthesis and share the sugar with their coral hosts. Although corals also eat foods from the environment, much of their energy comes from the sugar they get from their algal symbionts.

FIGURE 1.13 Maximizing the Capture of Light This acacia tree and coral colony have similar shapes, both of which are adaptations to maximize the capture of sunlight.

Thus we are not surprised that corals have often evolved shapes that optimize the ability of their algae to collect the Sun's light striking them from above. Coral reefs have been around for a long time. Most of the world's reefs are less than 10,000 years old, but similar reefs have existed since the age of the dinosaurs.

Despite their ancient history, reef corals are in trouble today. One source of trouble is that their symbiosis with algae is often disrupted in the modern world. Starting about 70 years ago, scientists began observing episodes of widespread "bleaching" of coral reefs. Large areas of reefs were losing their usual pigmentation and becoming very pale, or even white—thus the term "bleaching." The bleached corals were weakened, and in some cases dying. Since then, bleaching episodes have become more frequent and widespread. Scientists quickly learned that during bleaching a coral loses its algal symbionts, or its algal symbionts lose their pigments. Thus bleaching represents a disruption of the coral–alga symbiosis. But why is this happening now?

To answer questions such as this, biologists begin by sifting through what is already known to arrive at possible answers, or **hypotheses**. In the case of coral reefs, biologists knew that bleaching events were associated with various environmental stresses such as pollution of coastal waters, freshwater runoff from the mainland, and a changing climate, especially increases in the temperature and acidity of the ocean. However, many of these environmental stresses occur simultaneously in nature, and so it wasn't known which of these stresses were actually causing bleaching. Biologists Ove Hoegh-Guldberg and Jason Smith wanted to know what was happening to the corals' algae and why.

Working at the Lizard Island Research Station, in the northern part of the Great Barrier Reef in Australia, Hoegh-Guldberg and Smith focused on two species of corals (*Stylophora pistillata* and *Seriatopora hystrix*) that had been exhibiting bleaching in the surrounding reefs. They collected samples of healthy corals and brought them into the laboratory, where the corals were maintained in seawater aquaria at 27°C under filtered sunlight. After the corals had adjusted to conditions in the laboratory, Hoegh-Guldberg and Smith manipulated the environment of the corals in highly defined ways. They raised the water temperature in some of the aquaria and increased the sunlight exposure in others. In both cases the corals bleached. When the scientists examined the algae inside the corals, they found that the coral–alga symbiosis was disrupted. The algae in corals exposed to high light had lost their pigmentation. The corals exposed to high temperatures had lost their algae completely. Either way, the bleached corals had been deprived of much of their food supply—algal photosynthesis—with potentially dire consequences. Fortunately, research by Cynthia Lewis, Mary Alice Coffroth, and others has shown that many species of corals have some ability to recover their algae after bleaching (see Investigation Figure 19.22).

Hoegh-Guldberg and Smith, who published their findings in 1989, demonstrated that excessively warm water temperatures alone could cause corals to bleach, but their experiments were performed in the laboratory, with just two species of corals collected from a single reef. Biologists also believed it was important to survey many coral species in the natural communities of the Great Barrier Reef. The Great Barrier Reef is 2,300 kilometers long (roughly the size of Japan or Italy) and is made up of about 600 different species of corals. Different areas of the reef vary greatly in their architecture, water depth, latitude, Sun exposure, and species composition, as well as in how well they are protected from pollution and fishing. Does increasing water temperature alone account for widespread bleaching events in this complex context in nature? Are all corals equally affected?

To answer these questions, scientists now have resources that weren't available to Hoegh-Guldberg and Smith 30 years ago. For example, satellite imagery is now used to continuously monitor sea surface temperatures throughout the world. A group of scientists led by Terry Hughes at James Cook University in Queensland, Australia, used these data to map the distribution of stressful temperatures throughout the Great Barrier Reef during three major bleaching events, in 1998, 2002, and 2016. The researchers also mapped the distribution of coral health. Using aerial surveys to assess bleaching at hundreds of sites throughout the reef, the researchers found a strong correlation between the temperature of the water at each site and the severity of bleaching, despite the many other differences that existed among the sites. Their study, and many others, have shown that ocean temperature is a reliable predictor of bleaching events worldwide.

Like other biologists, Hoegh-Guldberg and Smith made observations that led to questions. They formed hypotheses that could explain those observations. They then made predictions based on those hypotheses and designed and carried out experiments to test their predictions. Other scientists used other tools to expand on the earlier findings of Hoegh-Guldberg and Smith. It has now become clear that the world's reefs are under grave threat from ocean warming. As further warming has become more likely, biologists are trying to learn more about how corals resist and recover from heat damage.

Well-designed experiments have the potential to falsify hypotheses

Once predictions are made from a hypothesis, experiments can be designed to test those predictions. The most informative experiments have the ability to show that the prediction is wrong. If an experiment of this type is carried out and indicates the prediction is wrong, the hypothesis must be modified or rejected.

There are two general types of experiments. Both compare **data**—quantified measurements or observations—from different groups. A **controlled experiment** deliberately changes, or manipulates, one or more of the factors being tested. A **comparative experiment** compares unmanipulated data gathered from different sources.

In a *controlled experiment*, we start with groups that are as similar as possible. We predict on the basis of our hypothesis that some critical factor, or **variable**, has an effect on the phenomenon we are investigating. We devise some method to manipulate *only that variable* in an "experimental" group, and we compare the resulting data with data from an unmanipulated "control" group. **INVESTIGATION FIGURE 1.14** describes one of the controlled experiments performed by Hoegh-Guldberg and Smith to quantify the effects of elevated water temperature and light on corals.

The basis of controlled experiments is that one variable is manipulated while all others are held constant. The variable that is

INVESTIGATION

FIGURE 1.14 Controlled Experiments Manipulate a Variable Coral reefs had long been observed to bleach in response to environmental stresses, such as changes in light intensity, salinity, and water temperature. However, in natural settings such stresses often occur simultaneously, making it difficult to determine which particular stress is causing the corals to bleach. In addition, corals can lose color in two different ways. The corals may get rid of their endosymbiotic algae (zooxanthellae) that are responsible for their color. Or the corals may retain their zooxanthellae but the endosymbionts themselves may become less pigmented. It was not known which process was primarily responsible for bleaching observed in nature. To answer some of these questions, Ove Hoegh-Guldberg and G. Jason Smith performed a series of controlled laboratory experiments on two coral species collected from the Great Barrier Reef in Australia.[a]

HYPOTHESIS

Corals bleach under stress because they reduce the numbers of endosymbiotic zooxanthellae in their tissues.

METHOD

1. Collect samples of a coral species (*Stylophora pistillata* in the experiments shown here) from a reef and transfer them to aquaria in the laboratory. Maintain the coral colonies at 27°C and under 25% full sunlight.
2. For high-light treatments, expose some colonies to full sunlight, while maintaining the water temperature at 27°C. Maintain control colonies under 25% full sunlight and 27°C. After 10 days measure for each colony the chlorophyll *a* content per surface area of the colony (micrograms [μg] per cm^2), the number of zooxanthellae per area of the colony, and the amount of chlorophyll *a* in the zooxanthellae (picograms [pg] per cell) after separation from the coral host cells.
3. For high-temperature treatments, place some colonies in an aquarium in which the water is maintained at 32°C under 25% sunlight. Leave the remaining colonies (controls) at 27°C and 25% sunlight. After 4 days measure number of zooxanthellae in the colonies and the chlorophyll content of the zooxanthellae.

RESULTS

See graphs below. Error bars in all graphs represent one standard error of the mean.

CONCLUSION

The hypothesis is not supported in the case of high light–induced bleaching. However, bleaching at high water temperatures was accompanied by a decrease in the number of zooxanthellae in the corals, supporting the hypothesis.

[a]O. Hoegh-Guldberg and G. J. Smith. 1989. *J Exp Mar Biol Ecol* 129: 279–303.

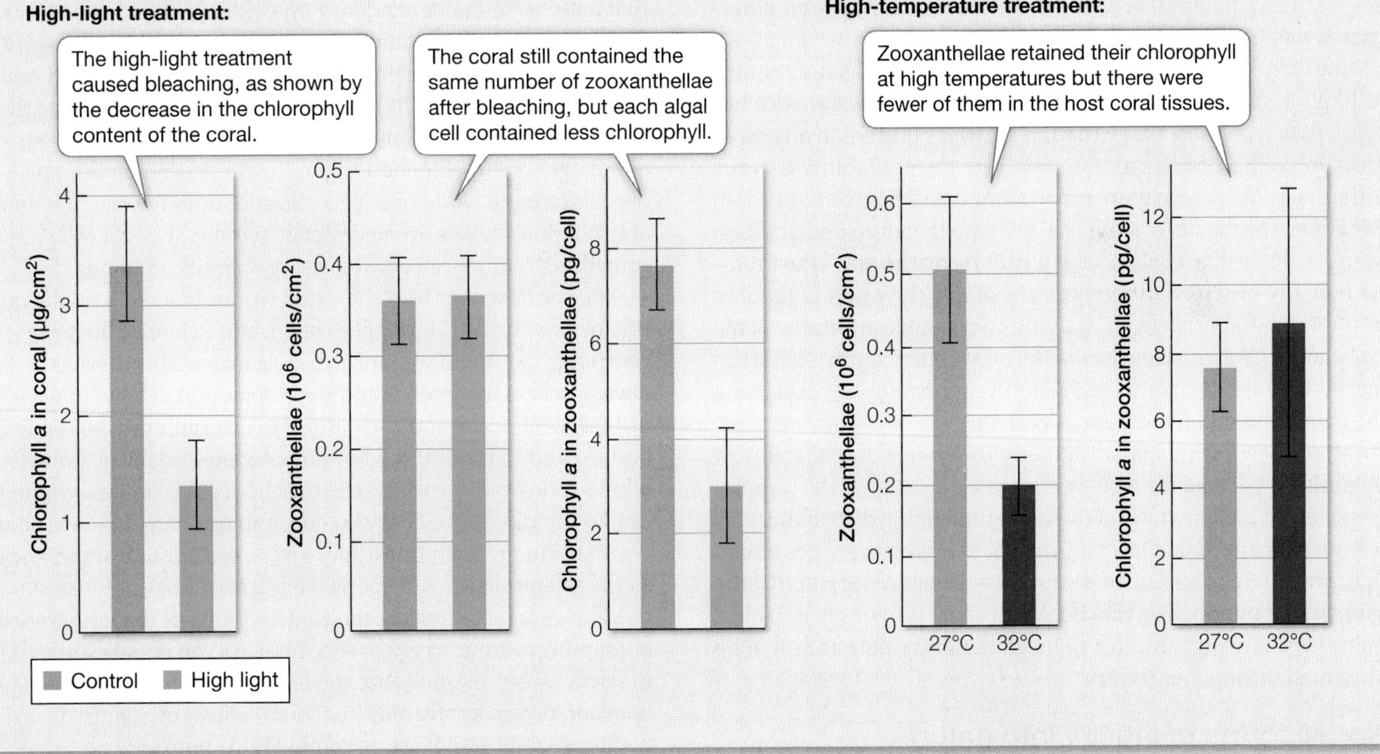

manipulated is called the *independent variable* (because an investigator can manipulate it independently of other considerations). The response that is measured is the *dependent variable* (because it is not manipulated directly by the investigator but is permitted to vary in ways that depend on the independent variable). A perfectly controlled experiment is not easy to design because biological variables are so interrelated that it is often difficult to alter just one.

A *comparative experiment* starts with the prediction that there will be a difference among naturally existing groups based on the hypothesis. In comparative experiments, we do not control or manipulate any of the variables, and often we cannot even identify all the variables that are present. We simply gather and compare data from different naturally occurring groups. Hughes and his colleagues performed a comparative experiment when they related temperature stress to coral bleaching at many sites throughout the Great Barrier Reef (**INVESTIGATION FIGURE 1.15**). Of course, the sample sites differed in many ways besides water temperature.

The results of experiments frequently reveal that the situation is more complex than the hypothesis anticipated, thus raising new questions. As a result, biologists often develop new questions, hypotheses, and experiments as they collect more data. The process of science is open-ended in this regard, and continued research leads to an ever-better understanding of the living world, with practical implications for agriculture, medicine, conservation of species, and other endeavors.

Statistical methods are essential scientific tools

Whether we do controlled or comparative experiments, at the end we have to decide whether the effect we observe—for example, a difference among groups—is a real difference. Any observed difference might instead just be a result of chance. How do we gain confidence that a measured difference is large enough or repeatable enough to represent a real effect? In other words, how do we decide in an unbiased, objective way whether a measured difference is meaningful?

Statistical significance refers to the extent to which a result is unlikely to be due to chance alone. Scientists use statistics because they recognize that variation is always present in any set of measurements. Statistical tests calculate the probability that the differences observed in an experiment could be due to random variation. The results of statistical tests are therefore probabilities. Many statistical tests start with a **null hypothesis**—the premise that any observed differences are simply the result of random differences that arise from drawing two samples from the same population. When scientists collect data, they apply statistical methods to those data to calculate the probability of observing the data if the null hypothesis is correct.

More specifically, statistical methods tell us the probability of obtaining a particular result by chance alone, even if the samples being tested are drawn from the same underlying distribution. As scientists, we need to eliminate, insofar as possible, the possibility that any differences seen are simply due to chance variation in the samples. Appendix B in this book is a short primer on statistical methods that you can refer to as you analyze data that will be presented throughout the text.

Not all forms of inquiry into nature are scientific

Science is a human endeavor that is bounded by certain standards of best practice. Other areas of scholarship share with science the practice of making observations and asking questions, but scientists are distinguished by what they do with their observations and how they answer their questions. Data, subjected to appropriate statistical analysis, are critical in testing hypotheses. Science is the most powerful approach humans have devised for learning about the world and how it works.

Scientific explanations for natural processes are objective and reliable because the hypotheses proposed *must be testable* and *must have the potential of being rejected* by direct observations and experiments. Scientists must clearly describe the methods they use to test hypotheses so that other scientists can repeat their experiments to see if they get the same results. Not all experiments are repeated, but surprising or controversial results are always subjected to independent verification. Scientists worldwide share this process of testing and rejecting hypotheses, contributing to a common body of scientific knowledge.

If you understand the methods of science, you can distinguish science from non-science. Art, music, and literature all contribute to the quality of human life, but they are not science. They do not use scientific methods to establish what is fact. Religion is not science, although religions have historically attempted to explain natural events ranging from unusual weather patterns to crop failures to human diseases. Most such phenomena that at one time were mysterious can now be explained in terms of scientific principles. Some fundamental tenets of some religious faiths, such as the existence of a supreme deity or deities, cannot be confirmed or refuted by experimentation and for this reason are outside the realm of science.

The power of science derives from the uncompromising objectivity and absolute dependence on evidence that comes from reproducible and quantifiable observations. A religious or spiritual explanation of a natural phenomenon may be coherent and satisfying for the person holding that view, but it is not testable and therefore it is not science. To invoke a supernatural explanation (such as a "creator" or "intelligent designer" with no known bounds) is to depart from the world of science. Science does not say that untestable religious beliefs are necessarily wrong, just that they are not something that we can address using scientific methods.

Science describes how the world works. It is silent on the question of how the world "ought to be." Many scientific advances that contribute to human welfare also raise major ethical issues. Recent developments in genetics and developmental biology may enable us to select the sex of our children, to use stem cells to repair our bodies, and to modify the human genome. Scientific knowledge allows us to do these things, but science cannot tell us whether or not we should do so, or if we choose to do them, how we should regulate them. Such questions are as crucial to human society as the science itself. A responsible scientist does not lose sight of these questions or neglect the contributions of the humanities in attempting to come to grips with them. As you pursue your studies in science, keep in mind that mastery of humanities may be highly relevant for you as you reason about the uses of science. Consider the big themes of biology as you read this book.

Science informs society

Science is not just about the discovery of new knowledge. In many cases, practical applications that benefit society are the primary motivation for scientific research. The COVID-19 pandemic that began to change all of our lives in 2020 is a clear example.

INVESTIGATION

FIGURE 1.15 Comparative Experiments Look for Differences among Groups Bleaching of reefs is a dramatic symptom of stress in the corals that make up reefs. Many different stresses can cause coral bleaching in local areas, including pollution, disease, excess light, and water temperatures that are too warm or too cold. But scientists wanted to know what was behind the widespread bleaching events that had been occurring more frequently worldwide. During three major bleaching events in the Great Barrier Reef—in 1998, 2002, and 2016—Terry Hughes and his colleagues measured the amount of coral bleaching at a large number of sites throughout the reef system. In this way they were able to produce a detailed map of the severity of coral bleaching and compare it with the amount of ocean warming over the same area.[a]

HYPOTHESIS

The severity of coral bleaching is correlated with the amount of ocean warming.

METHOD

1. Select a large number of sites distributed throughout the Great Barrier Reef for study: 638 sites in 1998, 631 in 2002, and 1,156 in 2016.
2. Take aerial photographs of the study sites from low-flying aircraft. Analyze the photographs to calculate the percentage of the reef's area that is bleached at each site.
3. Use sea surface temperatures detected by satellite imaging to calculate the cumulative amount of above-normal ocean temperatures (degree heating weeks, a unit that takes into account both the amount of excess heat and its duration) that occurred at each site during the 3 months prior to the assessment of reef bleaching.
4. Map the amount of ocean warming and severity of bleaching across the Great Barrier Reef. Plot warming versus coral bleaching.

RESULTS

The distribution of coral bleaching matches the distribution of heat stress. The amount of ocean heating correctly predicted the amount of bleaching at 75% of the test sites.

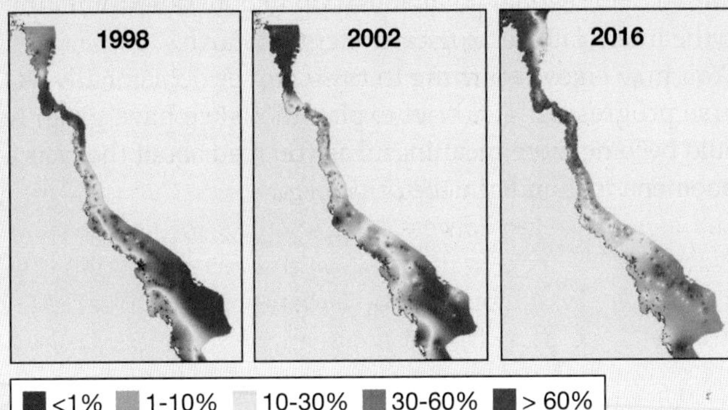

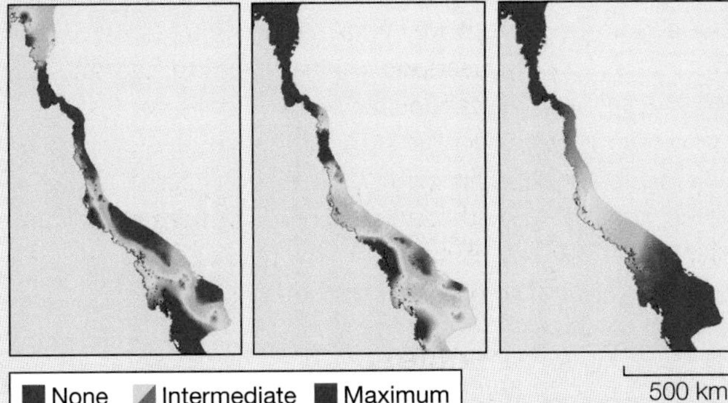

Map templates provided by Geoscience Australia, CC BY 4.0.

For these data from 2016, the amount of heat stress correlates strongly with the percent bleaching.

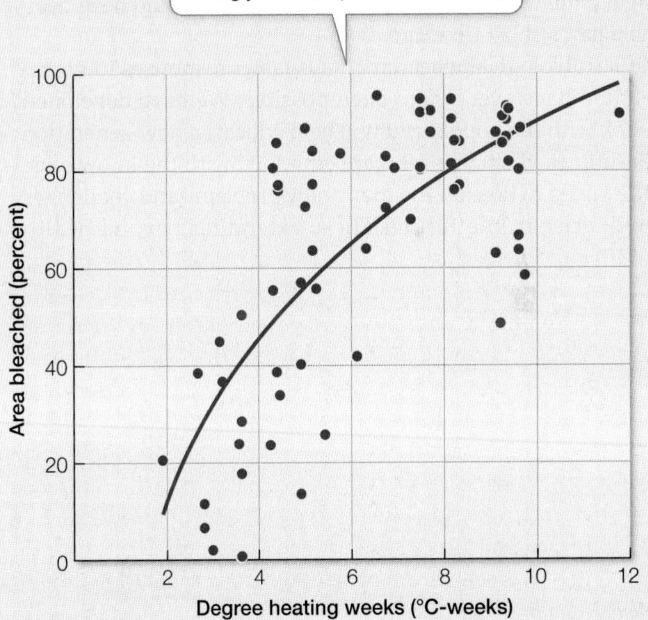

CONCLUSION

The results support the hypothesis that coral bleaching is correlated with higher water temperatures.

[a]T. P. Hughes et al. 2017. *Nature* 543: 373–377.

This pandemic had enormous social, economic, health, and political consequences for most of the world's populations, including many millions of deaths and even more cases of serious illness. Early on, biologists quickly took up the challenges of understanding this pandemic and the viral pathogen (SARS-CoV-2) that caused it. The genome of the virus was rapidly sequenced, which allowed biologists to study its structure and function (see Chapters 10–12). Viruses are especially dangerous pathogens because most viruses evolve much more rapidly than do their hosts (see Chapters 13 and 15). Once a database of genomes of the rapidly evolving virus was established, phylogenetic analyses provided rapid insights into the origin, diversification, and spread of the virus around the world (see Chapter 14). Our knowledge of other viruses and their biology provided clear insights into how the virus was transmitted and how it evolves (see Chapters 15 and 18). Biologists immediately began working on new vaccines to combat the virus, and these vaccines were developed, tested, and brought to use in record time (see Chapter 36). Population modeling of the virus allowed biologists to make predictions of its spread and develop best practices for reducing the impact of the pandemic (see Chapter 39). All of these rapid contributions by biologists greatly reduced the potential impact of the pandemic. Although deaths and illnesses from the COVID-19 pandemic were widespread, rapid study of the virus and the accompanying epidemic resulted in far less damage to the world's human populations than would have been possible just a short time ago.

Modern biological research allows faster responses to global crises than have ever before been possible. We have developed this book with the hope that it will help educate a new generation of well-informed citizens who are equipped with the knowledge that they need to take on the many new problems and challenges that will arise in the future. These extend far beyond health emergencies such as the COVID-19 pandemic. It is now clear that global climate change, the biodiversity crisis, and other environmental issues will be major challenges for our society, now and moving forward. Throughout this book, we have emphasized the importance to society of understanding and applying scientific methodology and knowledge. Our goal is to help you understand how to make the best use of scientific knowledge and practices, so that you can help make the world of tomorrow a better place for all.

This book is organized into the major themes of biology

You will see the fundamental principles of life introduced in this chapter—such as evolution—at work in each part of this book. In Part 1 you will learn about the molecular organization of life. We will discuss the origin of life, energy and matter, and how proteins and nucleic acids became the self-replicating cellular systems of life. Part 2 will describe how these self-replicating systems work and the genetic principles that explain heredity and mutation, which are the basis of evolution. In Part 3 we will go into detail about how evolution works. Part 4 will examine the products of evolution: the vast diversity of life and the many different ways organisms solve common problems such as how to reproduce, defend themselves, and obtain nutrients. Parts 5 and 6 will explore the physiological adaptations that allow plants and animals to survive and function in a wide range of physical environments. Finally, in Part 7 we will discuss these environments and the integration of individual organisms, populations, and communities into the interrelated ecological systems of Earth.

You may enjoy returning to this chapter occasionally as the course progresses. The brief explanations we have given here should become more meaningful as you read about the facts and phenomena that underlie the principles.

Part 1
Cells

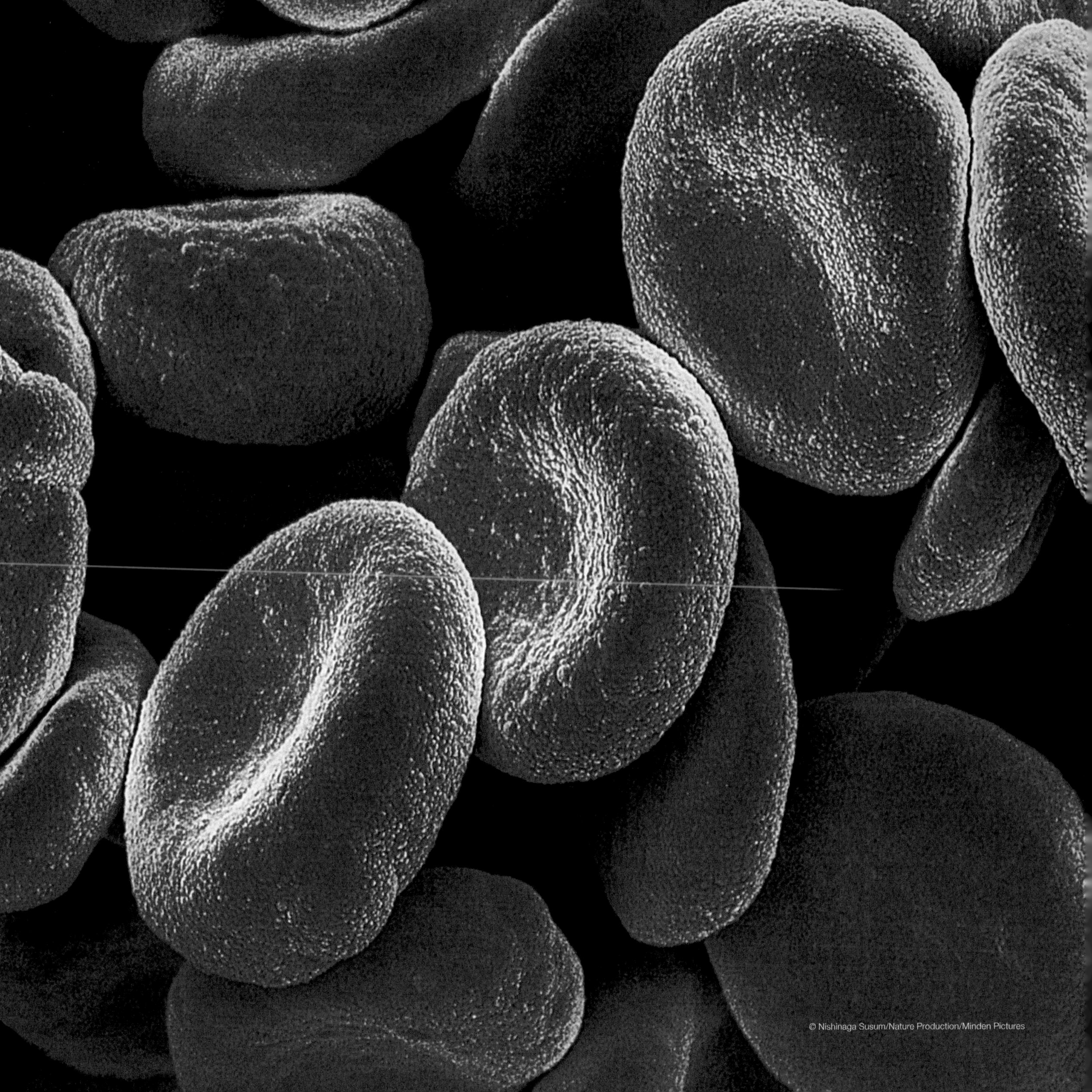

2 Life's Chemistry and the Importance of Water

A major discovery in modern biology was that living things are composed of the same chemical elements as the vast nonliving universe. All matter contained in galaxies, such as the one housing our solar system, is made up of 92 naturally occurring chemical elements. A living organism is largely composed of remarkably few of them. Just four chemical elements—hydrogen, oxygen, carbon, and nitrogen—account for more than 95 percent of the weight of all living organisms.

What properties of the four elements common in living tissues make them so important to life?

You will find the answer to this question on page 38.

KEY CONCEPTS

2.1 An Element's Atomic Structure Determines Its Properties

LEARNING OBJECTIVES

By the end of this key concept you should be able to:

2.1.1 Describe the structure of an atom.

2.1.2 Relate the atomic structure of an atom to its identity as an element.

2.1.3 Use the chemical characteristics of elements, as determined by the number of electrons in the *s* and *p* orbitals of the outermost shell, to group elements into columns on the periodic table.

All matter is composed of **atoms**. An **element** is a fundamental substance that consists of only one kind of atom. There are more than 100 different elements, though only 92 of them are naturally occurring, including the four that are most abundant in living systems—hydrogen, carbon, oxygen, and nitrogen. Elements differ in how they interact with other elements, though some tend to behave in very similar ways. These chemical behaviors are determined largely by the number and charge of three types of particles in an element's atoms.

Atoms are composed of protons, neutrons, and electrons

An atom consists of three fundamental particles: positively charged **protons**, negatively charged **electrons**, and, except for hydrogen, uncharged **neutrons**. Protons and neutrons are clustered in close proximity in the **nucleus** of an atom. Rapidly moving electrons are found outside the nucleus in **electron shells** (**FIGURE 2.1**). These shells are far from the nucleus (on the subatomic scale), such that an atom is mostly empty space. Protons and neutrons are 2,000 times larger than electrons and make up most of the mass of an atom. The unit for measuring the **atomic mass** of atoms and atomic particles is the **dalton** (**Da**). A proton and a neutron each have a mass of approximately 1 Da. In contrast, an electron has a mass of 0.0005 Da.

The number of protons in the nucleus is called the **atomic number**, and this number determines the identity of the element. In the **periodic table**, elements are presented in order of atomic number and organized into vertical columns (**FIGURE 2.2**). Elements in the same column have the same number of electrons in their outermost shell (known as the **valence shell**), which results in their having similar chemical properties. The atomic numbers of the four most common elements in living matter are 1 for hydrogen (abbreviated as H), 6 for carbon (C), 7 for nitrogen (N), and 8 for oxygen (O).

Media Clip 2.1 The Elements Song

Isotopes differ in the number of neutrons in the nucleus

The number of neutrons in the nucleus of different atoms of a particular element can vary. For example, carbon atoms usually have six neutrons but can contain seven or eight. These variants do not change the identity of the element, which is determined by the number of protons in the nucleus, but they do affect the atomic mass of the element. Atoms of an element that differ in the number of neutrons are called **isotopes**. As an example, carbon has an atomic number of 6 (so it has 6 protons), and its most common isotope has six neutrons and an atomic mass of 12. It is referred to as carbon-12 (^{12}C). Another carbon isotope has two additional neutrons in its nucleus and has an atomic mass of 14; it is referred to as carbon-14 (^{14}C).

Electrons occur in atomic orbitals in electron shells

There are many ways to depict the substructure of an atom. The three fundamental particles—protons, electrons, and neutrons—were first proposed more than a century ago, and since then many more have been discovered. Although more precise models have

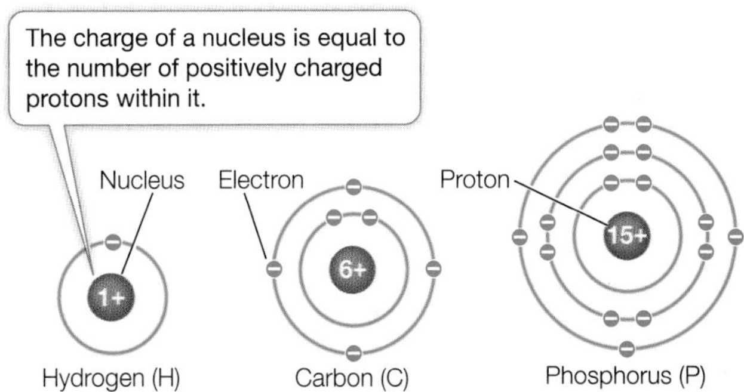

FIGURE 2.1 Bohr Models for Hydrogen, Carbon, and Phosphorus The nucleus consists of protons and neutrons. The net positive charge of the nucleus, which is equal to the number of protons, is shown for each atom. Negatively charged electrons are outside the nucleus in shells.

The element chlorine has 17 protons in its nucleus. How many electrons does it have?

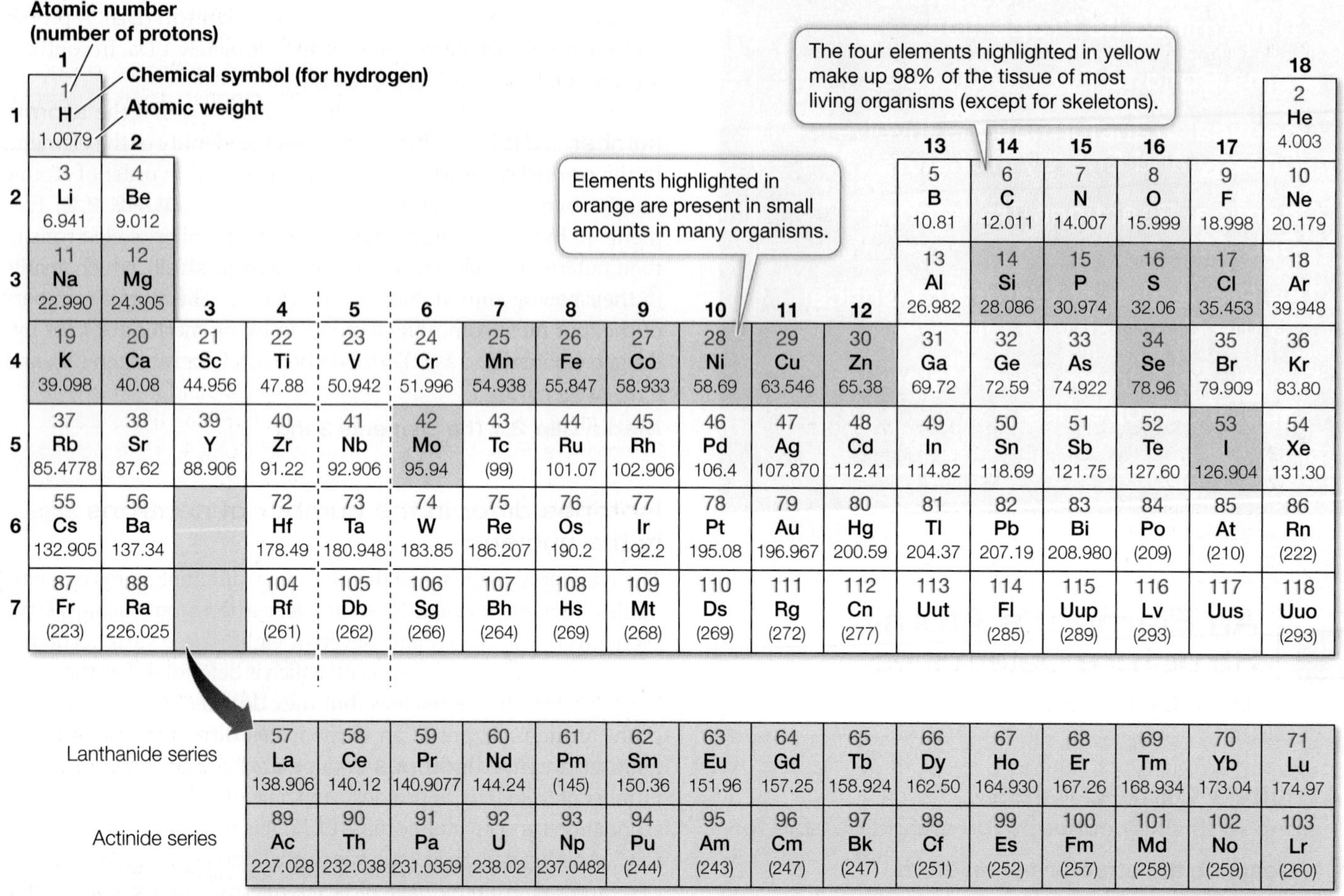

FIGURE 2.2 The Periodic Table Groups Elements According to Their Chemical Properties Elements in the same vertical column have the same number of electrons in their outermost shell. In each box, the atomic number is in red, the element symbol is in black, and the atomic weight is in gray. The atomic weight of an element is the weighted sum of the atomic masses of each of its isotopes, averaged by the frequency with which each isotope occurs in nature.

Why is this table called a "periodic" table?

been proposed, the chemistry of biological systems is commonly described by a simple model first proposed by the physicist Niels Bohr. In the Bohr model, seen in Figure 2.1, electrons orbiting the nucleus are shown in electron shells that differ in their distance from the nucleus. The innermost shell can contain at most 2 electrons, the second up to 8 electrons, and the third up to 18 electrons. More distant shells can contain even more electrons.

The Bohr model suggests that each electron orbits the nucleus somewhat like planets in our solar system orbit the Sun, but this is not what actually happens. Instead, electrons in each shell move within a defined area of space called an **atomic orbital**. Electron orbitals differ in shape, orientation, and distance from the nucleus (**FIGURE 2.3**), and each can contain one or two electrons. Every electron shell has a single, spherical *s* orbital. The second shell and subsequent shells also have three dumbbell-shaped *p* orbitals. The third shell and subsequent shells have additional orbitals. Orbitals differ in energy level; electrons in lower-shell orbitals generally have lower energy than electrons in higher-shell orbitals, and within a given shell, *s*-orbital electrons have lower energies than *p*-orbital electrons.

Activity 2.1 Electron Orbitals

Electrons in the outermost shell determine an element's chemical properties

As mentioned above, elements are organized in the periodic table (see Figure 2.2) in columns that reflect similarities in chemical properties. These similar properties in turn reflect similarities in the number of electrons found in each element's outermost shell. For example, elements in the first column of the periodic table each have one electron in the *s* orbital of the outermost shell, while each element in the second column has two electrons in that shell.

Elements tend to be most stable when the *s* and *p* orbitals in their outermost shell are full—that is, when there are two electrons in the *s* orbital and two in each of the three *p* orbitals. This tendency gives rise to the **octet rule**, which states that an atom will lose, gain, or share electrons in order to achieve a stable configuration of eight electrons in its outermost shell. The elements in the last column of the periodic table have this stable configuration

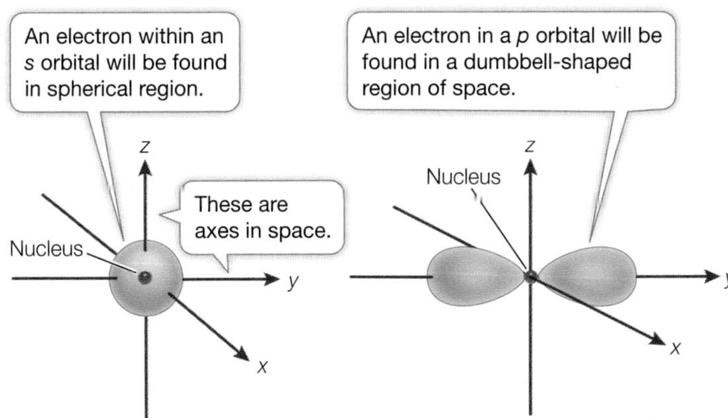

FIGURE 2.3 The Shape of the *s* and *p* Orbitals in an Electron Shell Orbitals represent the expected location of electrons around the nucleus of an atom. All electron shells have *s* orbitals. There are three *p* orbitals in the second and subsequent shells, along the *x*, *y*, and *z* geometric axes (only one is shown here, along the *y* axis). The third and subsequent shells have additional orbitals. (After OpenStax, Chemistry. OpenStax CNX. Jun 20, 2016 http://cnx.org/contents/85abf193-2bd2-4908-8563-90b8a7ac8df6@9.311. CC BY 4.0.)

Where would the two *p* orbitals that are not shown on the figure be?

and are known as the **noble gases**. They got this name because, majestically, they don't react with other elements, having full *s* and *p* orbitals in their outermost shells—that is, eight electrons in their outermost shell. As we will explain in Key Concept 2.2, elements in the other columns may interact with one another to lose or gain electrons, thereby generating a more stable electron configuration in which the *s* and *p* orbitals in their outermost shell are full. Exceptions to the octet rule do, however, exist.

An atom's tendency to attract electrons from another atom is termed its **electronegativity**. The electronegativity of an atom depends on how many electrons are in the *s* and *p* orbitals of its outermost shell (more electrons indicate higher electronegativity) and on the distance between the electrons in the outermost shell and the nucleus (the closer the electrons are to the positively charged nucleus, the greater the electronegative pull). Electronegativity thus increases from the bottom left to the top right of the periodic table; smaller atoms and those with closer to eight electrons in their outermost *s* and *p* orbitals have higher electronegativities. Note that the four elements most abundant in living systems—H, C, N, and O—are among the most electronegative atoms.

Some elements and isotopes are unstable

Some elements on the periodic table are highly unstable and as such are not found in nature. Elements with atomic numbers 95–118, for example, are made in laboratories and are highly unstable, as are some isotopes of some naturally occurring elements such as uranium. These unstable elements and isotopes can spontaneously undergo **radioactive decay**, losing atomic particles to become different isotopes of the same element, or even other elements. For example, the radioactive carbon isotope that has eight neutrons, ^{14}C, spontaneously decays to form ^{14}N (nitrogen-14), which is the common, stable isotope of that element.

LINK **Key Concept 17.1** discusses how radioactive elements or isotopes, such as ^{14}C, decay at characteristic rates, and how this property makes it possible for scientists to use them in radiometric dating of events in Earth's history.

REVIEW & APPLY | 2.1

R Living organisms are composed of the same chemical elements as the rest of the universe. An atom consists of a nucleus containing protons and neutrons, and a characteristic configuration of electrons in orbitals around the nucleus. An atom's chemical properties are determined by its atomic structure. Elements are composed of atoms with the same number of protons. Elements are grouped in the periodic table according to the number of electrons in their outermost shell.

A

1. The element oxygen has an atomic number of 8 and an isotope with an atomic mass of 16 (^{16}O). How many protons and neutrons are in the nucleus of a ^{16}O atom?
2. Using a Bohr model, sketch the locations of the protons, neutrons, and electrons in a ^{16}O oxygen atom.
3. There is another isotope of oxygen with an atomic mass of 18. With respect to the number of protons, neutrons, and electrons, how do the atoms in this ^{18}O isotope differ from ^{16}O oxygen atoms?
4. In the periodic table (see Figure 2.2), sulfur (S, atomic number 16) is just below oxygen. In terms of orbitals and electrons, describe the outermost shell of a sulfur atom.

The properties of an element are determined by the number of electrons in its outermost *s* and *p* orbitals. One important property is how atoms will interact with one another, which we consider in the next Concept.

2.2 | Atoms Bond to Form Molecules

LEARNING OBJECTIVES

By the end of this key concept you should be able to:

2.2.1 **Predict how elements will react with one another based on their electronegativities.**

2.2.2 **Describe the properties of a covalent bond when the two atoms (a) have and (b) do not have the same electronegativity.**

2.2.3 **Recognize typical polar and nonpolar covalent bonds in biological molecules.**

2.2.4 **Describe hydrogen bonds and van der Waals interactions.**

Atoms interact with one another. A **chemical bond** is an attraction between two atoms. There are several kinds of chemical bonds (**TABLE 2.1**), some of which are strong and some of which are weak. In this key concept we begin with **covalent bonds**, which are strong bonds that result from the sharing of electrons between atoms. We then examine **ionic bonds**, strong bonds that form when one atom gains and another loses one or more electrons to achieve a more stable electron configuration. Strong bonds result in the formation of molecules and compounds. A **molecule** is formed when two or more atoms join together chemically. A **compound** is a molecule formed by at least two different elements. We also discuss weaker bonds, including hydrogen bonds, which are enormously important in biology, and van der Waals interactions.

Animation 2.1 Chemical Bond Formation

Atoms share electrons in covalent bonds

A covalent bond forms when two atoms attain more stable electron configurations in their outermost (valence) shells by sharing one or more pairs of electrons. In a covalent bond, each atom contributes one member of each electron pair. This happens, for example, when two hydrogen atoms—which have just one electron each in their single electron shell—are drawn close together. When their electrons pair up, each hydrogen atom attains a stable number of electrons in its outermost shell, and the two atoms are linked in a single covalent bond, forming a hydrogen molecule (H_2) (**FIGURES 2.4** and **2.5A**).

Now consider oxygen, which has an outermost shell with a total of six electrons in its *s* and *p* orbitals. According to the octet rule, oxygen requires two more electrons to complete its valence shell. This is achieved when two oxygen atoms share two pairs of electrons, forming two covalent bonds, a so-called double bond (**FIGURE 2.5B**). Thus two oxygen atoms make up the molecule oxygen, written O_2. Covalent bonds can also form between atoms of different elements. For example, one oxygen and two hydrogen atoms make up a water molecule (H_2O) (**FIGURE 2.5C**), and

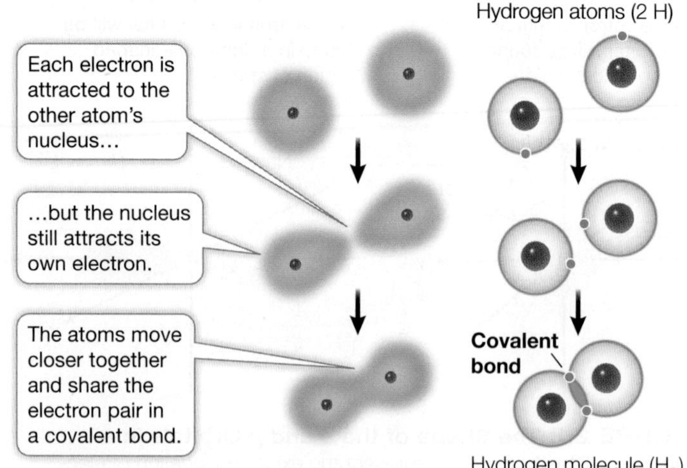

FIGURE 2.4 Electrons Shared in Covalent Bonds Two hydrogen atoms can combine to form a hydrogen molecule. A covalent bond forms when the electron orbitals of the two atoms overlap in an energetically favorable manner.

one carbon and four hydrogens make up methane (CH_4) (**FIGURE 2.5D**). Covalent bonds are conveyed in structural formulas by lines drawn between the bonded atoms, as seen in Figure 2.5. Molecules also can be visually represented by space-filling models and ball-and-stick models.

For two atoms to form a covalent bond, their electronegativities cannot be too dissimilar. The most common elements in biological molecules have similar electronegativities and thus readily form covalent bonds with one another. The properties of molecules are influenced by the characteristics of their covalent bonds. Three important aspects of covalent bonds are orientation, strength and stability, and the degree of sharing of electrons.

ORIENTATION For a given pair of elements, such as carbon bonded to hydrogen, the length of the covalent bond is always the same. And for a given atom within a molecule, the angle of

TABLE **2.1** | Chemical Bonds Important in Biological Systems

Type of bond	Strength	Notes	Structure	Energy to break bond (kilojoules/mole)
Covalent bond	Strong	Electrons shared in orbitals	H O | ‖ —N—C—	200–500 (single covalent bond) 500–700 (double covalent bond)
Ionic bond	Strong	Electrons transferred from one atom to another, resulting in electrostatic attraction between oppositely charged ions	H O | + − ‖ —N—H O—C— | H	1,100–20,000 (in ionic crystal)
Hydrogen bond	Weak	Electrostatic attraction between slightly positively charged H (in -OH or -NH) and slightly negatively charged O or N atom	H | δ+ δ− | —N—H•••••O=C—	4–50
van der Waals interactions	Very weak	Transient electrostatic attraction between two atoms, each of which is in a nonpolar covalent bond	H H H—H C H H	<4

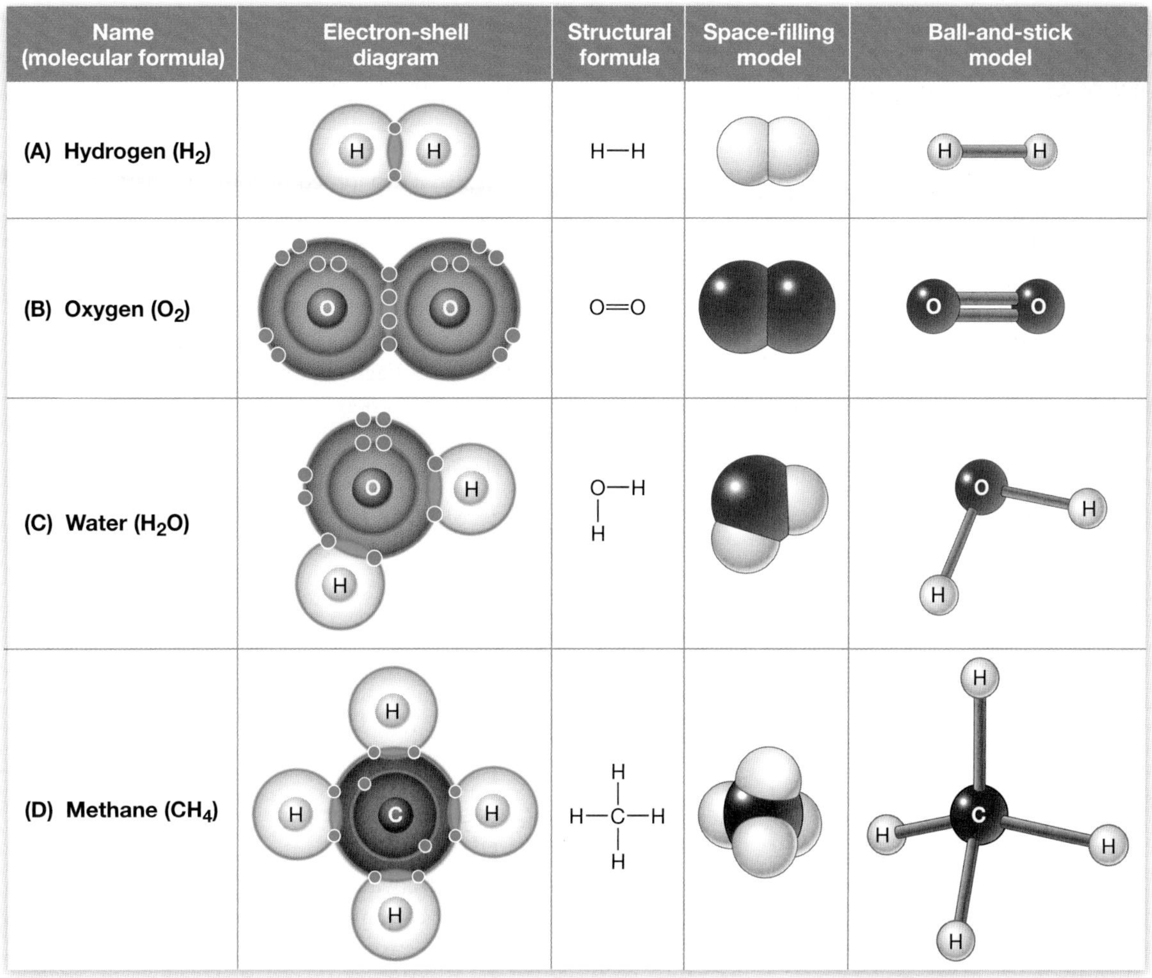

Name (molecular formula)	Electron-shell diagram	Structural formula	Space-filling model	Ball-and-stick model
(A) Hydrogen (H_2)		H—H		
(B) Oxygen (O_2)		O=O		
(C) Water (H_2O)		O—H | H		
(D) Methane (CH_4)		H | H—C—H | H		

FIGURE 2.5 Various Representations of Covalent Bonding

Draw the four representations for the molecule carbon dioxide (CO_2).

each covalent bond with respect to the others is generally the same. Furthermore, bond angles tend to be fairly consistent for particular atoms regardless of the molecule in which they occur. Look, for example, at the illustration of methane in Figure 2.5D. Because carbon has just four electrons in its outermost (valence) shell, it can form four covalent bonds to complete its valence shell. The four single covalent bonds that carbon makes with other atoms form a molecule in a tetrahedral orientation. The orientations of covalent bonds in space give molecules their three-dimensional geometry, and the shapes of molecules contribute to their biological functions.

STRENGTH AND STABILITY Covalent bonds are strong, meaning it takes a lot of energy to break them, but they vary in strength depending on the atoms involved in the bond. Differences in bond strengths are important in determining whether chemical reactions between molecules will occur.

UNEQUAL SHARING OF ELECTRONS When two atoms of the same element are covalently bonded, electrons in the bond are shared equally by both atoms. However, when the two atoms are different elements, the sharing may not be equal because of differences in electronegativity. When atoms differ little in electronegativity, electrons are shared about equally; such a covalent bond is called **nonpolar** because the "pull" of each nucleus on the electrons is about equal. In a **polar** covalent bond, by contrast, the electrons are unequally shared between two atoms because of differences in the atoms' electronegativity. The electrons are pulled closer to the nucleus of the more electronegative atom in the bonding pair. This unequal sharing results in a charge difference between the two atoms: the more electronegative atom will be slightly negatively charged (written as δ^-) and the other will be slightly positively charged (δ^+). For example, when hydrogen is covalently bound to oxygen, the shared electrons tend to be pulled closer to the oxygen nucleus, which is significantly more electronegative than hydrogen (**FIGURE 2.6**). The polarity of covalent bonds is critically important in biological systems and is the basis of hydrogen bonds, as we will see below.

Ionic bonds form between atoms by a transfer of electrons

In contrast to a covalent bond, an ionic bond forms when a highly electronegative atom pulls one or more electrons away from an atom with low electronegativity. This results in making the atom that loses its electron(s) positively charged and the one gaining

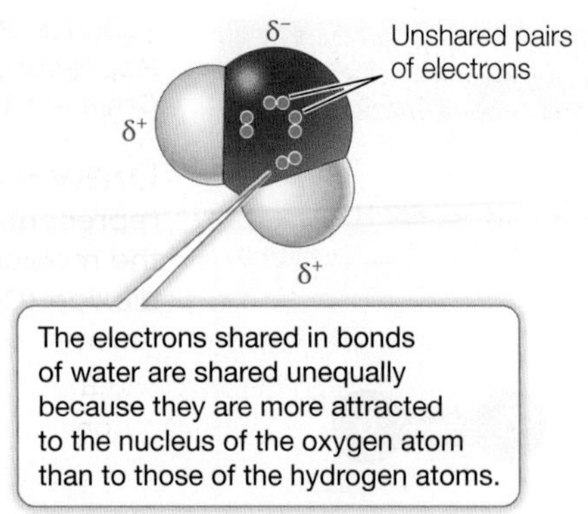

FIGURE 2.6 Polar Covalent Bonding In water, the hydrogen–oxygen bond is polar.

Indicate the polarity of the covalent bonds on the structural formula for carbon dioxide (CO_2).

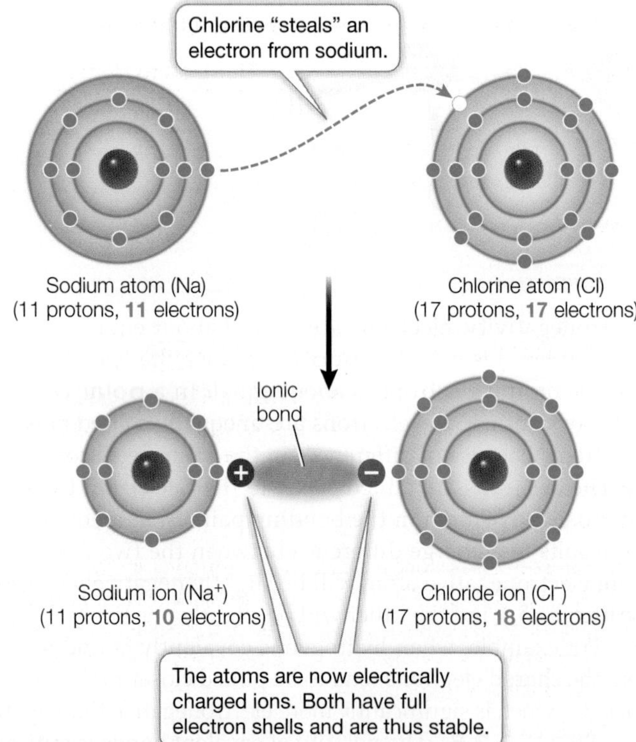

FIGURE 2.7 Ionic Bonds When a sodium (Na) atom reacts with a chlorine (Cl) atom, the more electronegative Cl fills its valence shell by gaining an electron from Na. In so doing, the Cl atom becomes a negatively charged Cl^- ion and the Na atom becomes a positively charged Na^+ ion.

Draw a similar reaction between potassium (K) and fluorine (F) (refer to the periodic table in Figure 2.2).

the electron(s) negatively charged. An atom that loses or gains one or more electrons is termed an **ion**. The positively charged ion, termed a **cation**, and the negatively charged ion, termed an **anion**, strongly attract one another because of their opposite charges. **FIGURE 2.7** illustrates this process for two elements, sodium and chloride, that are joined by ionic bonds to form sodium chloride (NaCl; table salt).

Ionic compounds tend to form crystals in which there is a regular, recurring arrangement of the anions and cations held together by the attractions of oppositely charged ions. For example, in table salt, sodium cations (Na^+) and chloride anions (Cl^-) form a regular lattice (**FIGURE 2.8**). Each cation in the crystal is attracted to all of its immediate anion neighbors, and vice versa. Because each atom in an ionic compound is being attracted to several other atoms, it makes little sense to talk about individual molecules. So when a formula for an ionic compound is given, for example NaCl for table salt, it represents the relative amount of each atom, rather than the molecular structure.

While ionic substances such as table salt are held together by strong ionic bonds, such substances often do not occur as solids in biological systems. Although there are important exceptions, ionic substances often dissolve in water and this renders the strength of the ionic bond irrelevant. Ions in solution play enormous roles in the chemistry of life, as we will see throughout this book.

Polar molecules can interact through electrostatic attractions

Atoms involved in polar covalent bonds interact through electrostatic attractions. The term **dipole** describes a separation of opposite electric charges, such as those occurring in a polar covalent bond. The slightly positively charged and negatively charged atoms in such a bond can be attracted to other slightly charged atoms in other nearby polar bonds. Such interactions are termed **dipole-dipole interactions** and they are generally weak. These interactions are often shown by the Greek letters δ^+ and δ^-, as seen in Figure 2.6, indicating weak charges, rather than stronger (+) and (–) ionic charges.

The **hydrogen bond** is an example of a dipole–dipole interaction. Covalent bonds between oxygen and hydrogen or between nitrogen and hydrogen are polar because of their large differences in electronegativity. In both cases, the hydrogen atom becomes slightly positively charged relative to the oxygen or nitrogen atom. The positively charged hydrogen atom is then electrostatically attracted to negatively charged atoms (usually nitrogen or oxygen) participating in other polar covalent bonds. Hydrogen bonds are important both within and between molecules. There are many hydrogen bonds between molecules of water, where the negatively charged (δ^-) oxygen atom of one water molecule is attracted to the positively charged (δ^+) hydrogen atoms of other water molecules (**FIGURE 2.9A**).

Although individual hydrogen bonds are weak, many can form within one molecule or between molecules. Collectively, large numbers of hydrogen bonds have considerable strength and can greatly influence the structure and properties of substances. Hydrogen bonds play important roles in determining

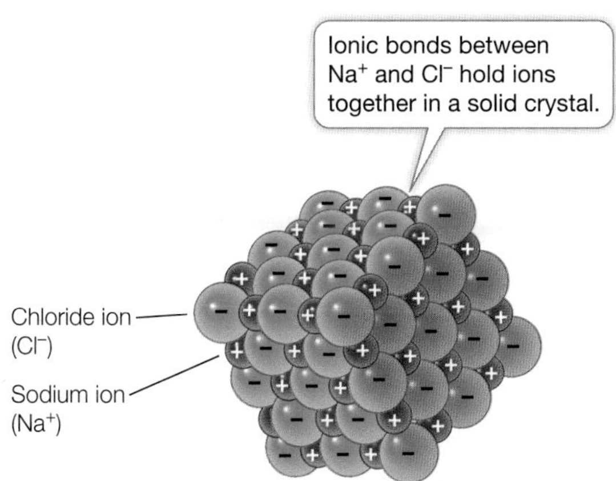

FIGURE 2.8 Lattice Structure of Sodium Chloride (NaCl) Ionic compounds can take the form of large lattices. In an ionic solid such as NaCl, oppositely charged ions are packed together into a repeating array called a crystal lattice.

Why are the Cl^- ions larger than the Na^+ ions in the lattice?

and maintaining the three-dimensional shapes of giant molecules such as DNA and proteins (**FIGURE 2.9B**), as we will discuss in Chapter 3. Hydrogen bonding also contributes to several properties of water that have great significance for life, as we will discuss in Key Concept 2.5.

Molecules can weakly attract one another through van der Waals interactions

Counterintuitively, atoms in nonpolar molecules interact with one another through electrostatic attractions between oppositely charged atoms, termed **van der Waals interactions**. Understanding van der Walls interactions hinges on recognizing that electrons are not stationary, but instead are moving in their orbitals around the nucleus. As the electrons move, their density around an atom may be asymmetrical for a brief time, resulting in a temporary dipole that is very small. This weak dipole can induce a temporary dipole in a neighboring atom if it is close enough. The temporary dipole is then electrostatically attracted to the induced dipole, similar to the dipole–dipole interactions discussed above. This electrostatic attraction is short-lived and very weak (see Table 2.1). However, there often are numerous such interactions among molecules that together can amount to a significant force. Even in polar molecules, such interactions also occur and can be significant.

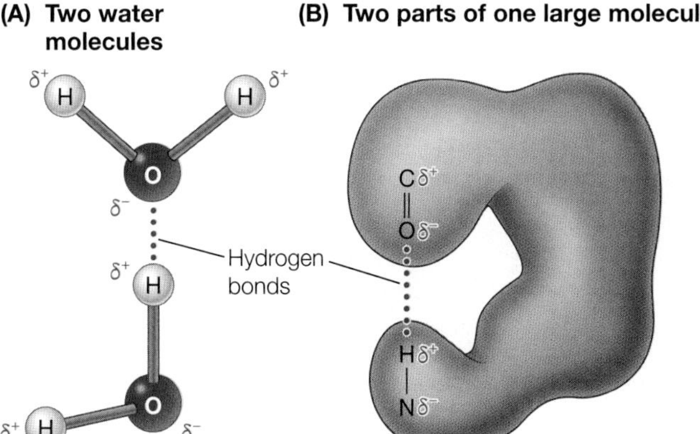

FIGURE 2.9 Hydrogen Bonds Can Occur between and within Molecules A hydrogen bond is an attraction between a slightly positively charged covalently bonded hydrogen and a slightly negatively charged covalently-bonded nitrogen or oxygen. **(A)** Hydrogen bonds can form between molecules. **(B)** Hydrogen bonds can also form between different parts of the same molecule.

REVIEW & APPLY | 2.2

R Some elements form strong covalent bonds by sharing electrons. Unequal sharing of electrons resulting from differences in the electronegativities of the participating atoms creates a polar covalent bond. Other atoms become ions by gaining or losing electrons. An ionic bond forms through the attraction of oppositely charged ions. Weaker forces such as hydrogen bonds and van der Waals interactions also attract atoms to one another within the same or different molecules.

Below is a table of elements and their electronegativities.

Element	Electronegativity
Oxygen (O)	3.4
Chlorine (Cl)	3.2
Nitrogen (N)	3.0
Carbon (C)	2.6
Phosphorus (P)	2.2
Hydrogen (H)	2.2
Sodium (Na)	0.9
Potassium (K)	0.8

If two atoms interact and the difference (if any) in electronegativity is 0.5 or less, a nonpolar covalent bond will form. At the other extreme, if the difference in electronegativity is greater than 2, electrons will not be not shared but will be transferred in an ionic bond.

A 1. For each of the following pairs of elements, indicate whether they will form an ionic bond or a covalent bond. If covalent, will the bond be polar or nonpolar? For ionic and polar covalent bonds, indicate which atom will have the positive charge.

a. K and Cl
b. C and C
c. O and P
d. N and H
e. C and O
f. H and C

Though strong, both ionic and covalent bonds can break to form new bonds, transforming some molecules into others. To understand the breaking and reforming of chemical bonds, we must first examine the role of energy in chemical transformations.

THINK LIKE A SCIENTIST

Process of Science

Science & Society

Climbing the walls

Geckos are able climbers. They can comfortably support their body weight using a single toe on a vertical surface and can climb across ceilings (**FIGURE 2.10A**). There are several possible explanations for how geckos might achieve the "foot stickiness" needed to perform these feats.

To walk on walls and ceilings, geckos could be (1) creating suction between their foot and the surface, (2) secreting a sticky adhesive, (3) using microscopic hooks to mechanically grab imperfections on the surface, (4) using electrostatic attraction to chemically bond to the surface, or (5) using van der Waals interactions to bond with the surface. Here you will apply the process of science to explore each of these explanations. As you do so, think about how an understanding of the science of gecko feet can benefit society.

The first explanation, suction, requires that the air pressure under the foot is reduced relative to the ambient air pressure. From the structure of their toe pads, it seems unlikely that geckos could create an airtight seal with a surface. Furthermore, removal of the air from a container placed briefly over a gecko on a wall does not cause the gecko to fall to the floor.

1. Why would air removal cause a gecko to fall to the floor if it were using suction to stick to the wall?

The second explanation, secreting a sticky adhesive, can be discarded because a gecko walking on a wall of glass leaves no residue, indicating it is not secreting an adhesive.

Microscopic examination shows that the hairlike structures on the toe pads do not have hooks, but instead have flattened ends (**FIGURE 2.10B** and **C**), which suggests that the third explanation, mechanical grabbing, is not correct.

2. Besides microscopic observation, what experiment could you perform to show that microscopic hooks grabbing tiny imperfections in the surface are not the explanation for the gecko's foot stickiness?

Eliminating the first three explanations, leaves us with the possibility of chemical bonding between a gecko's foot and the surface, either through electrostatic attractions or van der Waal's interactions. To test for electrostatic attractions, researchers classified different surfaces based on their polarity and then ascertained whether geckos could cling better to polar surfaces. They found that the geckos' clinging ability was unaffected by the polarity of the surface.

The last remaining explanation is that gecko feet chemically bond to the surface using van der Waals interactions. For that to be true, there must be a large area of close contact between a gecko's foot and the surface. The flattened ends of the individual microfibers on the toe pads allow this large area of close contact. Researchers have confirmed that individual fibers can bond to surfaces sufficiently strongly to explain the animal's ability to defy gravity.

Thus we deduce that geckos are able to grab surfaces tightly using only van der Waal's interactions. Mathematical models have shown exactly how the toe design functions during locomotion—in particular, how the foot is able to both grab tightly *and* release quickly from the surface. The understanding of the design of gecko toe pads has also led to a great deal of recent research attempting to engineer artificial materials with the same properties.

3. How might artificial materials with the properties of gecko toe pads be used?

(A)

(B)

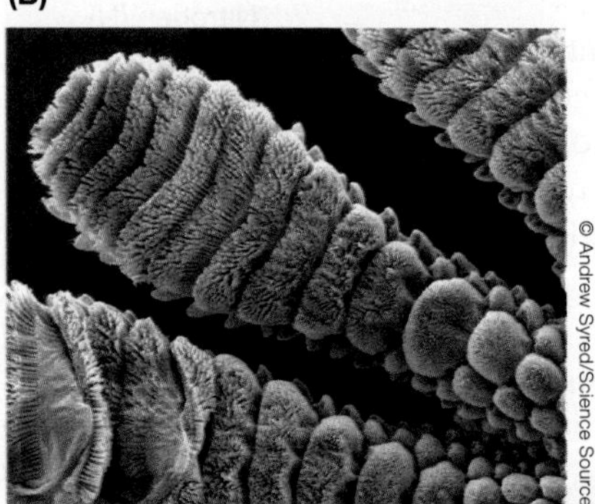

(C)

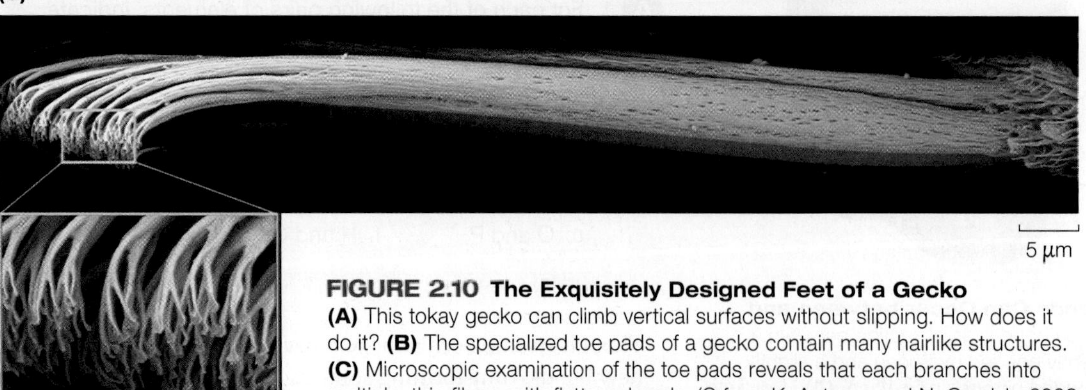

FIGURE 2.10 The Exquisitely Designed Feet of a Gecko
(A) This tokay gecko can climb vertical surfaces without slipping. How does it do it? **(B)** The specialized toe pads of a gecko contain many hairlike structures. **(C)** Microscopic examination of the toe pads reveals that each branches into multiple, thin fibers with flattened ends. (C from K. Autumn and N. Gravish. 2008. *Philos Trans A Math Phys Eng Sci* 366: 1575–1590.)

Chemical Transformations Involve Energy and Energy Transfers

LEARNING OBJECTIVES

By the end of this key concept you should be able to:

2.3.1 **Define energy based on what it can do.**

2.3.2 **Explain why energy is neither gained nor lost when it is transformed.**

2.3.3 **Explain usable energy in terms of the laws of thermodynamics.**

Typically, biological molecules are involved in biological structures such as those that form the membranes that surround cells, in tissues composed of many cells, or—at a more "macro" level—in an organ system, such as skin, composed of many tissues. Whatever their role, biological molecules are composed from simpler molecular building blocks. In some cases, molecules in biological systems are stable and do not change. However, many biological molecules are constantly changing due to interactions with other molecules (that is, because of chemical reactions) within living systems. We noted in Key Concept 2.2 that atoms bond together to achieve stability in the configuration of their electrons. The stability of a molecule is closely associated with its level of energy. Less stable molecules have higher energy and are converted during chemical reactions to more stable, lower-energy molecules. In this key concept we describe the concept of energy and the rules that govern its transformations in chemical reactions. As you will see, energy and its conversions underlie much of biology.

Change in living systems involves energy transformations

Energy is present in all matter, but it is difficult to define. **Energy** is usually understood by what it can do: it is the capacity to produce a change. All changes that have occurred, or will ever occur, anywhere in the universe involve energy transformation. That is a profound fact and tells us that energy is really important! Energy falls into two general categories, kinetic energy and potential energy, at both macro- and microscopic levels. **Kinetic energy** is the energy of movement and includes thermal, sound, and electromagnetic energy. Earth orbiting the Sun, a ball rolling across the ground, molecules moving in a liquid, electrons in orbitals, and a photon of light all have kinetic energy. **Potential energy** is stored energy and includes gravitational, elastic, chemical-bond, and nuclear energy. Examples of potential energy are a ball held above the ground, a stretched tendon, and a covalent bond.

To produce a change, energy must be transformed, either by conversion from one form of energy to another (for example, in the eye, light energy is converted to chemical-bond energy) or by being transferred from one location to another (the energy stored in food is converted to the energy used to build complex molecules), or by both. Thus when an organism moves, or warms up, or cools down, or makes noise, or produces light, energy transformations are producing those changes. For example, chemical-bond energy can be converted to motion energy when a muscle contracts and a limb moves, and electromagnetic energy can become chemical-bond energy when a plant uses light in the process of photosynthesis to synthesize sugar. Changes in energy are governed by the laws of thermodynamics. These laws apply to all matter and all energy transformations in the universe.

LINK Energy transformations will be a major subject of later chapters. We will examine, for example, how light energy is converted into chemical-bond energy in **Key Concept 5.5** and how chemical-bond energy is converted to muscle movement in **Key Concept 33.1**.

The first law of thermodynamics states that energy cannot be created or destroyed

The **first law of thermodynamics** states that *energy cannot be created or destroyed*. The first law has important implications. It means that energy has to come from somewhere and has to go somewhere when a change (such as a chemical transformation) occurs, and that during a change no energy disappears and no new energy is formed. In other words, the first law says that the total amount of energy in a closed system (one in which energy cannot escape or enter) remains constant. Thus when an organism uses energy to move, or grow, or shiver, or sweat, or make noise, or produce light, or anything else, the energy being used is being transformed from one form or place to another. Energy is always conserved (**FIGURE 2.11A**).

Many biological changes involve breaking down large, complex molecules to simpler molecules, and using the simpler ones to build other complex molecules. Think of starch in the food you eat being digested to its simple sugar building blocks, and then these simple sugars being used to build complex molecules in the body. In general, there is stored energy in complex molecules, and this is released when the molecules are broken down. By contrast, building complex molecules out of simple molecules requires energy.

The strength of a chemical bond is determined by how much energy is needed to break it. Thus a weak bond needs little energy to break it, implying it has relatively high potential energy, while a strong bond requires a lot of energy to break it and thus has low potential energy. Breaking one or more weak bonds (high potential energy) to make strong bonds (low potential energy) results in the release of chemical-bond energy, and that energy can then be used to cause change.

As we have just described, energy is released and used in biological transformations, with the total amount before and after the transformation being the same. But while the first law of thermodynamics tells us that the energy is conserved in the system, that energy is not always available in usable form for the next transformation. The second law of thermodynamics tells us that the amount of usable energy declines after each energy transformation.

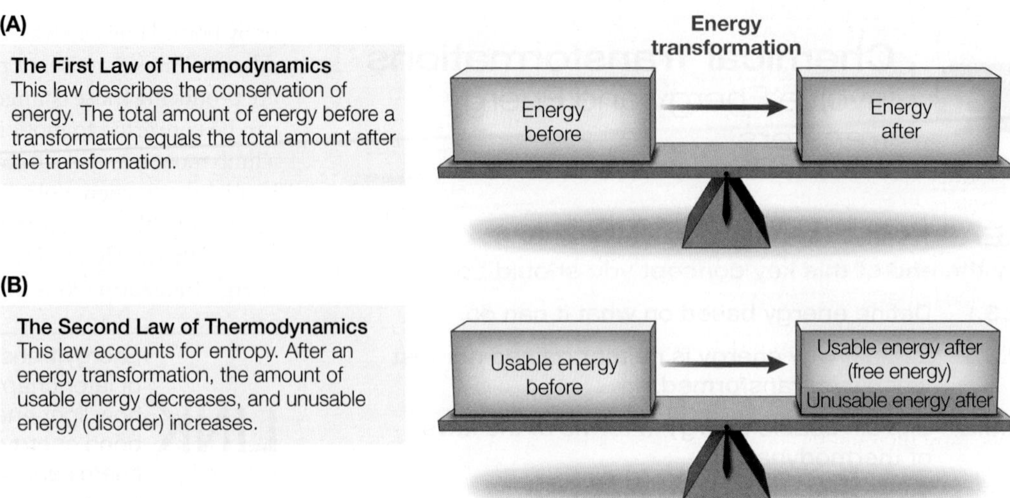

FIGURE 2.11 The Laws of Thermodynamics **(A)** The first law states that energy cannot be created or destroyed. **(B)** The second law states that after each energy transformation, some energy becomes unavailable to do work.

The second law of thermodynamics states that with each energy transformation, entropy increases

The **second law of thermodynamics** deals with entropy and states that *when an energy transformation occurs, there is an increase in entropy.* **Entropy** is often described as a measure of disorder, but it can also be considered as a measure of how spread out energy is, or how much of the energy is unusable. An increase in entropy implies that energy is becoming less concentrated, that it is becoming less usable, which is the natural tendency for all energy. Thus the second law of thermodynamics states that after each energy transformation, some of the energy in the system becomes unavailable to do work (**FIGURE 2.11B**). While we may not realize it, all of us intuitively understand the second law based on our own experiences. A hot kettle on a countertop cools down, but a room-temperature kettle does not spontaneously heat up. When a hot kettle cools, thermal energy is transferred to its surroundings: the kettle will cool to room temperature and the room will heat up an imperceptible amount. Entropy causes spreading out of the thermal energy that was initially concentrated in the kettle. Only an input of energy can warm up the kettle—that is, only an input of energy can impose order on the system.

The second law explains why some things happen (those that increase entropy) and why some don't (those that would decrease entropy). The second law tells us, for example, that an energy transfer that would increase entropy will not happen spontaneously—it can only occur if energy is added to the system (e.g., heating a kettle on the stove). The second law also tells us that in an energy transformation, some usable energy is lost, increasing entropy. In living systems, transformations often occur in steps in a pathway. In these cases there is an overall increase in entropy at each step (**FIGURE 2.12**). In the next key concept we describe the important role that energy plays in the chemical reactions that occur in living systems.

LINK The repercussions of the second law of thermodynamics can be seen in food chains, which have few steps because of the loss of usable energy at each step (see **Key Concept 41.3**).

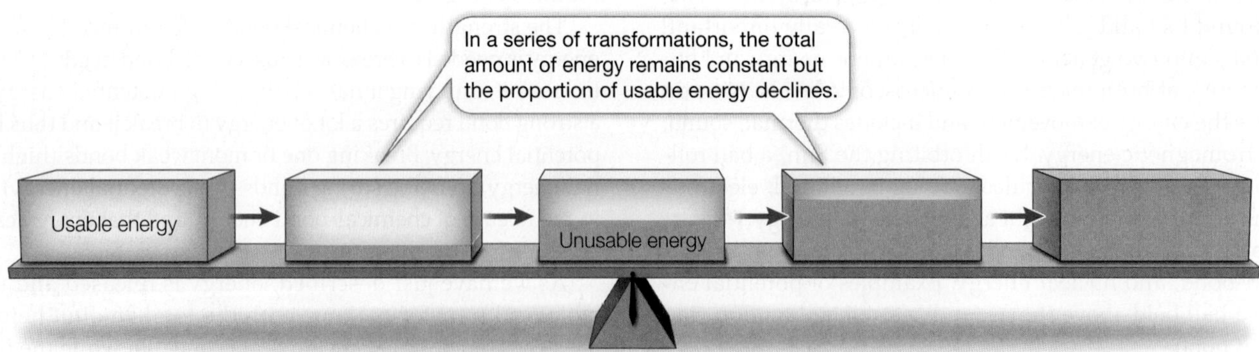

FIGURE 2.12 Usable Energy Declines in a Series of Transformations Many biological processes, ranging from biochemical pathways to ecological chains, involve several steps. Because of the second law of thermodynamics, usable energy is lost at each step.

If energy transformations resulted in 20 percent of the energy becoming unusable, how much energy would still be usable after three transformations? After five transformations?

REVIEW & APPLY | 2.3

R Energy is the capacity to produce change. Two laws of thermodynamics govern energy transformations in living (and other) systems. The first law states that in a transformation, energy is conserved, so the total amount of energy before and after the transformation is the same. The second law states that in a transformation, entropy increases. This means that the overall amount of usable energy declines after a transformation.

A
1. When you eat a granola bar and then go for a run, energy transformations occur. Beginning with the food energy in the bar, describe the forms of energy used and the changes of energy that occur.
2. After the energy transformations in Question 1, is the total energy the same? Is the total usable energy the same?

We have discussed the types of bonds that can form among atoms and described the laws of thermodynamics. We are now ready to examine how bonds break and reform during chemical reactions.

2.4 | Chemical Reactions Transform Substances

LEARNING OBJECTIVES

By the end of this key concept you should be able to:

2.4.1 **State what happens in a chemical reaction.**

2.4.2 **Describe what happens to the potential energy in an exergonic reaction and in an endergonic reaction.**

2.4.3 **Explain how activation energy, the concentration of reactants and products, and temperature are expected to alter the rate of a chemical reaction.**

A **chemical reaction** occurs when atoms combine or change their bonding partners. Consider the example of **hydrolysis**, a basic type of chemical reaction in which a molecule interacts with water, resulting in its breakdown to simpler molecules:

$$\text{Complex molecule} + H_2O \longrightarrow \text{simpler molecules}$$

The hydrolysis of the sugar sucrose produces its component molecules, glucose and fructose. We can express this reaction using a chemical equation:

$$\underset{\text{(sucrose)}}{C_{12}H_{22}O_{11}} + H_2O \longrightarrow \underset{\text{(glucose)}}{C_6H_{12}O_6} + \underset{\text{(fructose)}}{C_6H_{12}O_6}$$

On the left side of this equation are the **reactants**, sucrose and water, and on the right side are the **products**, glucose and fructose. The total numbers of atoms of each element on the left and right sides of the equation are the same (12 carbon atoms, 24 hydrogen atoms, and 12 oxygen atoms). However, the arrangement of atoms has changed (Chapter 3 will cover the specifics). According to the laws of thermodynamics, the total energy on both sides of the reaction is the same, but some energy is released that was stored in the chemical bonds of sucrose, and there is also an increase in entropy.

Chemical reactions release or require energy

A chemical reaction, such as the hydrolysis of sugar, proceeds as some bonds in the reactants are broken and new bonds form to make the products. The products of a reaction have different properties from those of the reactants. For example, glucose and fructose are different from sucrose and water. Importantly, the energy contained in the covalent bonds differs between the reactants and products, and so does the entropy. Together, the differences in bond energy and the change in entropy determine whether the total energy change, which is termed **free energy** (represented by ΔG), of the reaction is positive or negative. If it is negative, the reaction releases energy—that is, it is **exergonic** (**FIGURE 2.13A**). In other words, the chemical energy stored in the bonds of the products is less than the chemical energy stored in the bonds of reactants, there is greater entropy in the products than the reactants, or both. If the free energy is positive, the reaction is **endergonic**, meaning it requires energy to occur (**FIGURE 2.13B**). For the hydrolysis of glucose, the energy contained in the chemical bonds of sucrose and water (the reactants) is greater than the energy in the bonds of glucose and fructose (the products) and there is an increase in entropy, implying that the reaction is exergonic and will release energy when it occurs. Some of this energy can be used to do something useful; the rest will be lost as heat to the environment.

A **condensation reaction** is the reverse of a hydrolysis reaction. In a condensation reaction, two molecules combine to form a larger molecule, producing water as a by-product. We can write the condensation that forms sucrose as follows:

$$\underset{\text{(glucose)}}{C_6H_{12}O_6} + \underset{\text{(fructose)}}{C_6H_{12}O_6} \longrightarrow \underset{\text{(sucrose)}}{C_{12}H_{22}O_{11}} + H_2O$$

This reaction will have a positive free energy, meaning that it requires energy to occur and thus is endergonic (see Figure 2.13B). Condensation reactions can generate very large molecules by the sequential addition of smaller molecules. Some single molecules in living systems have millions of atoms. To synthesize such molecules in endergonic reactions requires large inputs of energy, which is obtained from exergonic reactions.

Hydrolysis and condensation reactions are common in living systems, and we will meet them repeatedly in later chapters. Hydrolysis reactions use water, are exergonic, and involve the breakdown of larger molecules to give smaller molecules with stronger bonds (less potential energy). Condensation reactions, in contrast, release water, are endergonic, and involve the production of larger molecules with weaker bonds (more potential energy). Some of the large molecules produced by condensation reactions are used to store chemical-bond energy. Stored energy can be released, and some of it is used to perform useful changes, via hydrolysis of large molecules.

(A) Exergonic reaction

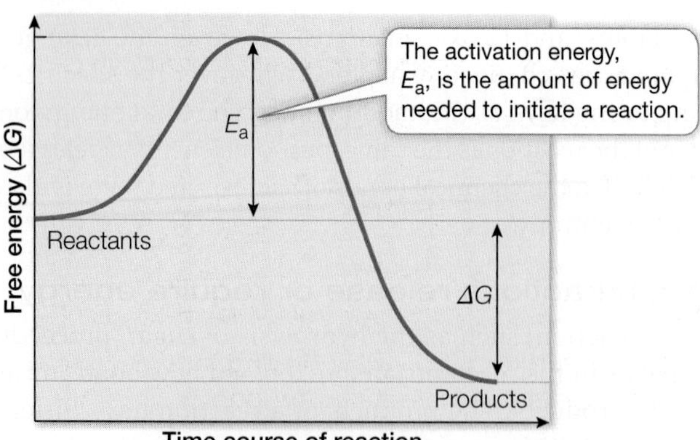

(B) Endergonic reaction

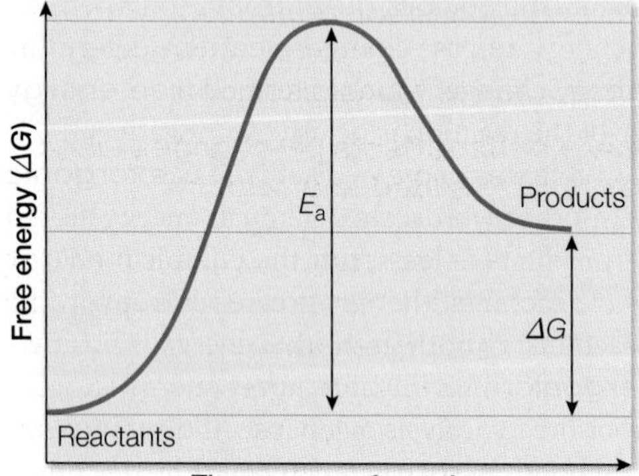

FIGURE 2.13 Energy in Reactants and Products for Exergonic and Endergonic Reactions **(A)** Exergonic reactions release energy. **(B)** Endergonic reactions require energy input.

Will either of these reactions spontaneously progress to completion?

As discussed in **Key Concepts 5.3** and **28.1** carbohydrates, lipids, and proteins are large biological molecules that can be broken down to obtain energy for biological processes.

Not all chemical reactions begin spontaneously

We might expect that the hydrolysis of sucrose would proceed spontaneously because it is exergonic. However, when sugar is added to water, only a few of the molecules of sucrose spontaneously break down into glucose and fructose. The transformation does occur, it is spontaneous, but at a very slow pace. In chemical terminology, the rate of the reaction is low. The laws of thermodynamics predict the direction of a transformation (in this case, hydrolysis rather than condensation) but not its rate. We'll say more about rate below.

Chemical transformations—both exergonic, or spontaneous (e.g., hydrolysis), and endergonic, or non-spontaneous (e.g., condensation)—require additional energy to initiate the reaction. The total amount of energy that must be supplied for a reaction to begin is termed the **activation energy (E_a)** (see Figure 2.13).

To understand why activation energy is usually required, even for exergonic reactions, consider that covalent bonds must be broken, at least partially, in reactants. The formation of covalent bonds releases energy, and thus breaking them requires energy. Once the bonds in the reactants are broken, new bonds can re-form, and if they are lower-energy (stronger) bonds than the ones that were broken, the reaction will result in an overall release of energy. This released energy may even be large enough to supply the activation energy for the next molecules to react, so that no additional energy input is required.

Consider, for example, the gasoline in a non-electric car. Clearly, the reaction of gasoline with oxygen is strongly exergonic (explosively so), but these two reactants do not react until a spark is introduced that supplies the activation energy. No further spark is needed for that particular batch of fuel after the first explosive reaction. Activation energy is an important component of all chemical reactions and is a major reason why even strongly exergonic reactions do not begin spontaneously.

If the requirement for activation energy prevents the hydrolysis of sucrose from beginning spontaneously, why do we see a few molecules of glucose and fructose when we add sucrose to water? This observation indicates that the reaction has occurred at least for some sucrose molecules. We might conclude (correctly) that the **reaction rate**, which measures how fast products (in this case glucose and fructose) are made per unit of time, is very low but not zero.

Several factors affect the rate of a reaction

Three factors strongly influence the rate of a chemical reaction: activation energy, temperature, and concentration. For a reaction to occur, the reactants have to be physically close enough to one another to interact and must have enough energy to react. Reactant molecules come into contact by colliding with one another. If a collision has sufficient energy—equal to or greater than the activation energy—then the reaction can occur. The reaction rate is increased by higher temperatures, which increase the number of collisions between molecules and the energy of those collisions. Higher concentration of reactants also increases the number of collisions.

The lower the activation energy, the more likely it is that a collision between reactants will have sufficient energy to form products. A special class of proteins called enzymes serves to lower the activation energy in chemical reactions in living systems, a topic we will discuss in depth in Chapter 3.

The reaction rate for the spontaneous hydrolysis of sucrose is quite low at room temperature. However, if we wait long enough, would all of the sucrose eventually break down into glucose and fructose? To answer this question, let's think about the reverse reaction, the condensation reaction between glucose and fructose that forms sucrose. The activation energy for this endergonic condensation reaction is higher than that for the exergonic hydrolysis reaction (see Figure 2.13) because stronger covalent bonds are being broken in the reactants (glucose and fructose). However, some collisions will have sufficient energy for the condensation reaction to occur, and so rarely, sucrose will be synthesized from a collision of glucose and fructose. That is, the hydrolysis of sucrose is a reversible reaction, in which some sucrose molecules are formed while others are being broken down.

$$\text{Sucrose} + H_2O \rightleftharpoons \text{glucose} + \text{fructose}$$

Free energy favors the forward direction (hydrolysis). However, as the forward reaction progresses, the concentration of glucose and fructose increases, which increases the rate of the reverse reaction. At some point the rates of the forward and reverse reactions will be equal and the reaction will have reached **equilibrium**, at which point the relative concentrations of reactants and products will no longer change. At equilibrium for the hydrolysis of sucrose, there will be a high concentration of glucose and fructose and very little sucrose. However, if the concentrations of glucose and fructose are raised above the equilibrium values, the reaction rate will increase in the reverse direction.

REVIEW & APPLY | 2.4

R In chemical reactions, energy is used and released when chemical bonds are formed and broken. Free energy, available to do work, is released in exergonic reactions, so the net change in chemical-bond energy in the chemical products compared with that in the reactants is negative, the entropy of products is greater than reactants, or both. A biological example is the hydrolysis of complex molecules to simpler ones. In endergonic reactions, the net change in free energy is positive, meaning that an overall energy input is required for the reaction to proceed. An example is a condensation reaction resulting in the synthesis of complex molecules from simpler ones. Both endergonic and exergonic reactions require an input of activation energy to start. Reaction rates are higher when the activation energy is lower, when the temperature is higher, or when the concentration of reactants is increased.

A

1. Predict whether these situations are endergonic or exergonic and explain your reasoning:
   a. A cell wall is formed from simpler molecules to surround a cell.
   b. Complex molecules such as proteins are digested into their constituent building blocks by hydrolysis.
2. When sucrose is dissolved in water, very little of it is hydrolyzed to glucose + fructose, even though there is a lot of water around. The following mechanisms could speed up the reaction. Explain how each one works.
   a. Adding a substance that induces sucrose molecules to be stretched so that the chemical bonds are more readily reactive with water.
   b. Heating up the solution.

In biological systems, chemical reactions occur in an aquatic environment. Understanding life's chemistry and many of its other characteristics requires an understanding of water's physical and chemical properties. In the next Concept, we consider the properties of water and some of their implications.

2.5 The Properties of Water Are Critical to the Chemistry of Life

LEARNING OBJECTIVES

By the end of this key concept you should be able to:

2.5.1 **Describe how hydrogen bonding in water affects its heat capacity, melting, evaporation, cohesion, and adhesion.**

2.5.2 **Describe the interactions of ions, polar molecules, and nonpolar molecules in water.**

2.5.3 **Describe the self-ionization of pure water to produce H_3O^+ and OH^-, and what it means to have an acidic versus a basic solution.**

2.5.4 **Explain how a buffer affects pH changes when acids or bases are added to a solution.**

It is hard to overemphasize the importance of water for living systems. The search for life on other planets usually begins with the question of whether the planet has liquid water in reasonable quantities. Planets without liquid water are generally thought to be incompatible with life. In this key concept we discuss the properties of water that make it so critical for living systems. These characteristics are related to the polarity of the covalent bonds in the water molecule and the ability to form hydrogen bonds between water molecules. **INVESTIGATION FIGURE 2.14** examines the question of where Earth's water came from.

High specific heat and high heat of vaporization enable water to moderate the temperature inside living systems

In liquid water, hydrogen bonds are constantly forming and breaking between water molecules. An individual water molecule can be involved in up to four hydrogen bonds with other water molecules. The hydrogen bonds that form between water molecules must be broken in order for water molecules to dissociate. As a consequence, water has a high **specific heat**, defined as the amount of heat that must be absorbed or lost to change the temperature of 1 gram of the substance by 1 degree Celsius. Because it takes a great deal of energy to cause water to increase in temperature, water has a high **heat of vaporization**, meaning it takes a great deal of energy to vaporize it. Water's high specific heat and high heat of vaporization have several consequences for living systems:

- Water can exist as a liquid at a broad range of temperatures. Because the boiling point of water is 100°C, it is found as a liquid in a wide variety of biological environments where organisms live.
- Because a great deal of heat is required for water to evaporate, it can be used to cool the surfaces of living things. Heat from organisms is lost through evaporative cooling at the surface.

INVESTIGATION

FIGURE 2.14 The Origin of Earth's Water Water is essential for life, and this is why the search for life beyond Earth initially often centers on finding water. About 70 percent of Earth's surface is covered with water, and the oceans provide a hospitable environment for life. The origin of Earth's water is not well established. About 4.6 billion years ago, the solar system formed from gases around the Sun. It was initially very hot when matter began to condense to form the planets (including Earth), and any water present would have been evaporated. Yet several hundred million years later, there was abundant water on Earth. Where did it come from?[a]

HYPOTHESIS

Asteroids, rocky bodies formed near the Sun that are now orbiting it in a belt between Mars and Jupiter, carried water to Earth and deposited it on impact.

METHOD

1. Identify the two isotopes of water (see Key Concept 2.1): hydrogen (H), which contains one proton in its nucleus (see Figure 2.1), and deuterium (D), which contains one proton and one neutron in its nucleus.
2. Use chemical methods such as mass spectroscopy to calculate what proportions of Earth's ocean and lake water consist of H_2O and the much rarer D_2O. This provides an isotopic "signature" for the water.
3. Obtain data from satellite images offering an estimate of the proportions of the two types of water for several celestial bodies.

RESULTS

Source	Ratio $D_2O:H_2O$
Pacific Ocean	1.6×10^{-4}
Atlantic Ocean	1.6×10^{-4}
Great Lakes	1.6×10^{-4}
Jupiter's atmosphere	2.2×10^{-5}
Saturn's atmosphere	1.8×10^{-5}
Asteroid 1	1.7×10^{-4}
Asteroid 2	1.6×10^{-4}
Asteroid 3	1.6×10^{-4}

CONCLUSION

Given the similarity of $D_2O:H_2O$ ratios in water from Earth and from the asteroids, water on Earth could have come from the impacts of asteroids. It did not come from the large planets.

ANALYZE THE DATA

It has been hypothesized that comets could be another possible source of Earth's water. Unlike asteroids that originated in the inner solar system, comets originated in the outer regions of the solar system and orbit the Sun, sometimes coming close to Earth. Satellite and ground-based chemical analyses were made noting water binding to comets. Most notable was the 2014 landing of the space probe Rosetta on Comet Churyumov-Gerasimenko.

The table shows data for water associated with comets.

Source	Ratio $D_2O:H_2O$
Comet Hale-Bopp	2.2×10^{-4}
Comet Tuttle	3.1×10^{-4}
Comet Halley	2.2×10^{-4}
Comet Churyumov-Gerasimenko	4.5×10^{-4}

1. How do the data on water from comets compare with those from Earth?
2. What can you conclude about the possibility of comets versus asteroids being a source of water on Earth?
3. What is the significance of the data from Comet Churyumov-Gerasimenko?

[a] K. Altwegg et al. 2015. *Science* 347: 1261952.

A similar **Analyze the Data** exercise may be assigned in **Achieve**.

- The large amount of water in living tissues reduces temperature fluctuations experienced because of changes in environmental temperature.

Sweating is a mechanism for cooling; perspiration helps regulate our body temperature (see **Key Concept 28.3**).

When water freezes, additional hydrogen bonds form between water molecules, causing ice to have a very organized structure. The distance between water molecules in ice is greater than that in liquid water. This means that ice is less dense than liquid water. The lower density of ice explains why lakes and oceans freeze from the top to the bottom, making it more difficult for them to freeze solid during the winter, and making it more likely that they will thaw in the summer. Imagine what would happen to aquatic life if lakes froze solid in the winter.

Hydrogen bonding accounts for the cohesion and adhesion of water

The numerous hydrogen bonds that give water a high heat capacity and high heat of vaporization also give it high cohesive and adhesive strengths. **Cohesion** is defined as the attraction of identical molecules for one another. Cohesion between water molecules is the force that generates **surface tension**, which permits a container to be filled with water above its rim without overflowing, explains why water forms droplets on leaves, and allows a spider to walk on the surface of a pond. **Adhesion** is defined as the attraction of different molecules for one another. Water can show strong adhesive attraction to various surfaces, especially those that are polar. Cohesion and adhesion permit narrow columns of liquid water to move from the roots to the leaves of plants, even very tall ones. When water evaporates from the leaves, the entire column moves upward in response to the pull of the molecules at the top (see Key Concept 24.4).

Water is a solvent for charged and polar substances

Water is often called the universal **solvent** because a variety of substances can dissolve in it. Note that dissolving a substance in a solvent does not involve the formation or breakage of covalent bonds. Rather, it entails the *distribution* of solute molecules throughout the solvent. In order for a substance to dissolve in water it must be able to form weak electrostatic attractions with water molecules via their polar nature.

Substances that dissolve easily in water are called **hydrophilic** ("water-loving"), while those that don't are known as **hydrophobic** ("water-fearing"). Why are some molecules hydrophilic and other hydrophobic? Electrostatic interactions between a polar (δ^+ and δ^-) or charged ionic solute (e.g., Na^+) and the polar water molecules (with δ^+ and δ^-) are energetically favorable. Substances that have these charged properties are hydrophilic. Ionic compounds, such as table salt (NaCl), are held together by strong ionic bonds (see Table 2.1), but they dissolve in water because the ions become surrounded by water molecules in a **hydration shell**. In a hydration shell surrounding a cation (positively charged ion, e.g., Na^+), the negatively charged polar end of the O—H bonds—that is, the oxygen "end"—interacts electrostatically with the positively charged ion in solution. For anions (negatively charged ions) in solution, it is the positively charged hydrogen "end" of the water molecule that interacts with the ion (**FIGURE 2.15**). If the sum of the electrostatic attraction forces in the hydration shells is larger than the sum of the strengths of the ionic bonds in the ionic compound and the hydrogen bonds between water molecules that were broken, then the ionic compound will dissolve. Hydration shells keep the ions relatively far apart, which implies that the ionic attraction between the cations and anions (e.g., Na^+ and Cl^-) is substantially lower than in the salt crystal. Since most chemistry in biological systems occurs in water, the high strength of ionic bonds in the dry solid (see Table 2.1) is often irrelevant.

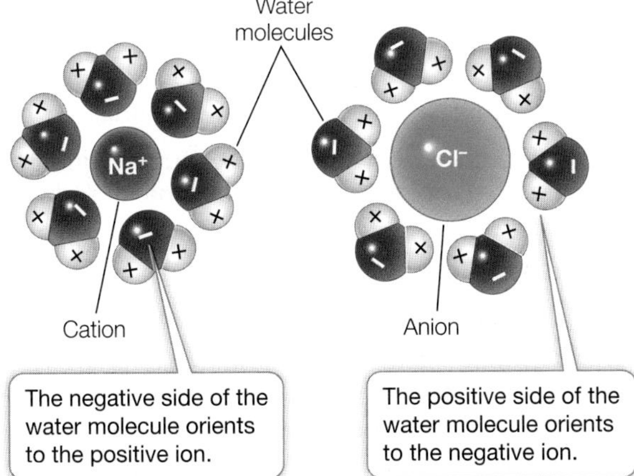

FIGURE 2.15 How Salts Dissolve in Water The ions of an ionic compound readily dissociate in water because of the electrostatic attractions of the charges in the ions with polar regions (δ^+ and δ^-) of water. Water molecules surround each ion, forming shells.

Nonpolar substances can interact with water molecules through van der Waals electrostatic interactions, but these are weak compared with the hydrogen bonds between water molecules. For this reason, the interaction of nonpolar substances with water is energetically unfavorable and so nonpolar substances are hydrophobic. For example, when a hydrophobic substance such as olive oil is mixed with water, the oil molecules rapidly aggregate, which minimizes the number of hydrogen bonds between water molecules that will be disrupted. Oil, a nonpolar substance, and water do not mix! Some molecules are **amphipathic**, meaning they have both hydrophobic and hydrophilic regions, which is critical to their function. You are familiar with one type of amphipathic substance—soaps.

Aqueous solutions may be acidic, basic, or neutral

Water is a covalent molecule. Even so, a very small fraction of water molecules can form ions in a process termed self-ionization. An H_2O molecule spontaneously ionizes into an H^+ cation (a proton) and an OH^- anion (a hydroxide ion, an example of a molecular ion). The H^+ immediately adds on to the oxygen of another water molecule to form an H_3O^+ cation. Note that H^+ (hydrogen ion, proton) and H_3O^+ (hydronium ion) are often used interchangeably.

$$2\,H_2O \leftrightharpoons H_2O + H^+ + OH^- \leftrightharpoons H_3O^+ + OH^-$$

As noted above, the amount of water that forms ions in solution is tiny. In pure water, the relative concentrations of the H_3O^+ and OH^- ions are each equal to 1×10^{-7} **moles**/liter. (A mole of water is 18 g/L, so this corresponds to 0.0000018 g.) A solution in which the concentrations of these two ions are equal is referred to as a neutral solution.

Compounds called **acids** raise the concentration of H^+ ions when dissolved in water, thereby increasing the concentration of H_3O^+ ions and raising the concentration of H_3O^+ relative to OH^-. In contrast, compounds termed **bases** produce proton acceptors such as OH^- ions when dissolved in water. OH^- ions combine with some of the H_3O^+ in the solution, forming water in the reverse reaction noted above. Adding a base to a solution *reduces* the concentration of H_3O^+ relative to OH^-. Put another way, a base is a substance that dissociates in water to give hydroxide ions (OH^-). The **pH** of an aqueous solution is defined as the negative logarithm of the concentration of H_3O^+ ions (see Appendix B for a discussion of logarithms). Pure water and neutral solutions have an H_3O^+ concentration of 10^{-7} M (moles/liter), so the pH is 7. An acidic solution has more H_3O^+ and so the pH is less than 7. A basic solution has less H_3O^+ and a pH above 7. **TABLE 2.2** shows the pH of various solutions.

TABLE **2.2** | pH of Various Solutions

Solution	pH
Hydrochloric acid	1
Stomach acid	1.5–2
Acid rain	4–5
Human urine, skin	5–6
Distilled water	7
Human blood	7.4
Seawater	8
Antacids	10
Bleach	13
Sodium hydroxide	13.9

Water exposed to the air becomes mildly acidic because some CO_2 dissolves in water to give carbonic acid, some of which then ionizes, giving bicarbonate and hydronium ions:

$$CO_2 + 2\,H_2O \rightleftharpoons H_2CO_3 + H_2O \rightleftharpoons HCO_3^- + H_3O^+$$

Thus rainwater is always mildly acidic because it contains some dissolved CO_2. The reactions shown above also act to stabilize pH. If a moderate amount of acid is added to a solution of carbonic acid, the reaction increases to the left, reducing the hydronium ion concentration and increasing the pH. Similarly, addition of a base causes the reaction to increase to the right as hydronium ions are removed, decreasing the pH. Such a solution is said to **buffer** pH changes, meaning that addition of acids and bases to the solution cause only very small pH changes. Buffers are important in living systems. As you will see in Chapter 30, the oxygen-carrying system of blood is exquisitely sensitive to changes in pH, and buffers are used there to maintain a constant level of pH at 7.35–7.45. A reduction to pH 7.00 can have severe consequences.

While rainwater is slightly acidic, seawater is quite basic, with a pH of about 8.4. Calcium carbonate ($CaCO_3$) in seawater reacts with hydronium ions, releasing CO_2 and increasing pH:

$$CaCO_3 + 2\,H_3O^+ \rightleftharpoons Ca^{2+} + CO_2 + 3\,H_2O$$

Note that increasing the acidity of seawater drives this reaction to the right, implying that calcium carbonate dissolves more easily as pH drops. This has important implications for many marine organisms, such as oysters, that use calcium carbonate as a primary component of their exterior skeletons.

Increased release of CO_2 to the atmosphere not only causes climate changes but is associated with increased acidification of the oceans, with adverse consequences for organisms there (see **Key Concept 42.4**).

REVIEW & APPLY | 2.5

R The properties of water that make it the ideal medium both inside and outside living systems include a high specific heat and high heat of vaporization, an ability to dissolve many substances important to life, and the ability to stick together in cohesion and adhesion. The properties of water are a consequence of the polar covalent bonds between its hydrogen and oxygen atoms. Polarity allows water molecules to form hydrogen bonds and thereby interact with one another. These bonds must be broken by heat energy before water heats up. In addition, hydrogen bonds make it possible for water to interact with dissolving substances, and the polar nature of water allows it to interact with polar and charged substances. Hydrophilic substances interact with and can dissolve in water. Hydrophobic substances avoid water and do not dissolve. Acids increase the hydronium ion concentration in a water solution; bases decrease the concentration.

A

1. The boiling point of water (H_2O) is 100°C, while the boiling point of methane (CH_4) is –161°C. Explain this difference in terms of hydrogen bonding between molecules.
2. CO_2 is formed in human tissues as a by-product of chemical changes and enters the bloodstream. CO_2 is removed from red blood cells by dissolving and ionizing in water:

$$CO_2 + H_2O \longrightarrow HCO_3^- + H^+$$

What is the consequence of CO_2 removal on the pH of the blood? What chemical reactions would restore the pH of the blood?

In the next concept we will see how the chemistry of biological molecules is determined by certain molecular groups that they possess.

2.6 | Functional Groups Give Molecules Specific Properties

LEARNING OBJECTIVES

By the end of this key concept you should be able to

2.6.1 **List the common functional groups found in biological molecules and their characteristics.**

2.6.2 **Describe how structural isomers differ from one another.**

Most of the molecules of living things fall into four basic categories: carbohydrates, lipids, proteins, and nucleic acids. Each of these kinds of biomolecules has its own structural and functional characteristics. We'll review the structure and function of these four kinds of molecules in Chapter 3. The properties and reactivities of these molecules are strongly influenced by the attachment of small clusters of atoms called **functional groups**. Here we describe several specific functional groups that are responsible for the characteristic chemical reactions of biomolecules.

Chemical characteristics of organic molecules depend on functional groups

The four major types of large biomolecules—carbohydrates, lipids, nucleic acids, and proteins—have carbon backbones (—C—C—C—...), and so the chemistry of life is primarily a study of carbon-based molecules. In chemistry, the study of carbon is called **organic chemistry**.

Carbon atoms form four covalent bonds with other atoms, allowing them to make structurally complex molecules. Organic molecules vary from small, containing a few atoms, to enormous, containing hundreds to many millions of atoms. The chemistry of organic molecules is determined in large part by the number and location of certain combinations of functional groups. You will repeatedly encounter several functional groups in your study of biology. Each functional group has specific chemical properties, and when attached to a larger molecule, it gives those properties to the larger molecule (**FIGURE 2.16**). The consistent chemical behavior of functional groups helps us understand the properties of the molecules that contain them, both with respect to molecular shape and reactivity. To get a sense of how functional groups can alter properties, consider the phosphate group. The phosphate group is highly polar and tends to interact with water. The addition of a phosphate group can turn a hydrophobic, nonpolar molecule into a hydrophilic, polar one, thereby affecting its interaction with the aqueous environment in a living system.

Isomers contribute to molecular diversity

While the addition of functional groups can determine how a molecule interacts with other molecules in the cell and thus results in molecular diversity, the properties of molecules can also differ due to variations in the arrangements of their atoms. Some molecules with the same chemical formula are not identical. You have seen an example of this with glucose and fructose. Both glucose and fructose have the atomic composition $C_6H_{12}O_6$, but they differ in the arrangement of their atoms and consequently in their properties. These two molecules are called **structural isomers** because they differ in how the atoms and functional groups are joined to one another. In **stereoisomers**, the atoms and functional groups are the same but the molecules differ in their three-dimensional geometry.

Functional group	Class of compounds and an example	Properties
Methyl 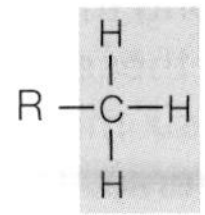 R—CH_3	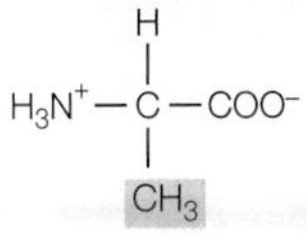 **Alkyl** H_3N^+—CH(CH_3)—COO^- Alanine	Nonpolar. Important modification of proteins and cytosine nucleotide.
Hydroxyl 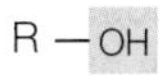 R—OH	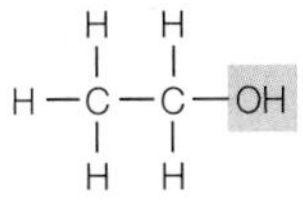 **Alcohols** H—CH_2—CH_2—OH Ethanol	Polar. Involved in hydrogen bonding. Often participates in condensation reactions. Required in side chain for phosphorylation of proteins.
Sulfhydryl 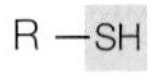 R—SH	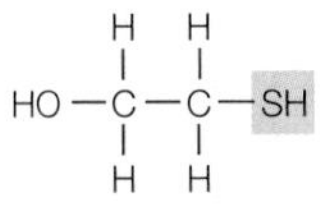 **Thiols** HO—CH_2—CH_2—SH Mercaptoethanol	Polar. Can form disulfide bridges to stabilize protein structure.
Aldehyde 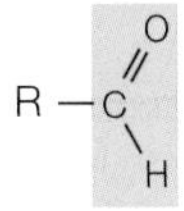 R—C(=O)—H	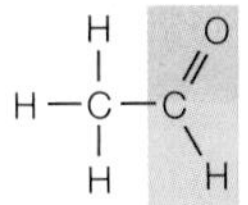 **Aldehydes** CH_3—C(=O)—H Acetaldehyde	Polar. Very reactive. Important in energy-releasing reactions.
Keto 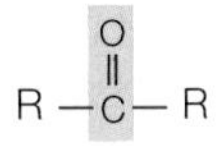 R—C(=O)—R	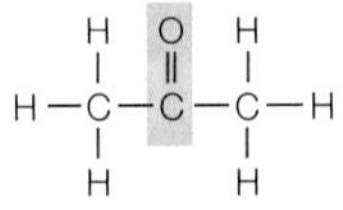 **Ketones** CH_3—C(=O)—CH_3 Acetone	Polar. Important in carbohydrates and in energy reactions.
Carboxyl 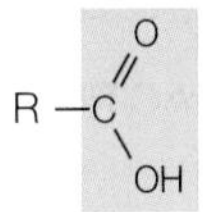 R—C(=O)—OH	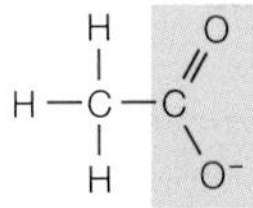 **Carboxylic acids** CH_3—C(=O)—O^- Acetic acid	Charged; acidic. Ionizes in living tissues to form $-COO^-$ (and H^+). Reacts with amino group to form peptide bond.
Amino 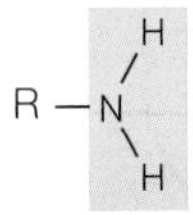 R—N(H)—H	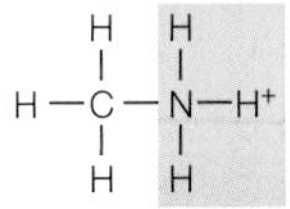 **Amines** CH_3—NH_2—H^+ Methylamine	Charged; basic. Accepts H^+ in living tissues to form $-NH_3^+$ as shown. Reacts with carboxyl group to form peptide bond.
Phosphate 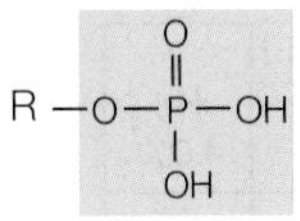 R—O—P(=O)(OH)—OH	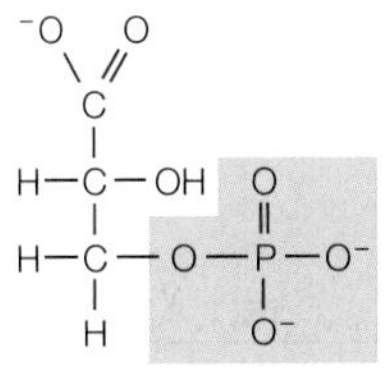 **Organic phosphates** $^-$O—C(=O)—CH(OH)—CH_2—O—P(=O)(O^-)—O^- 3-Phosphoglycerate	Charged; acidic. Ionizes in living tissues to form $-O-P-O_3^{2-}$ (and 2 H^+). Enters into condensation reactions, often with inorganic phosphate, PO_4^{3-}. When bonded to another phosphate, hydrolysis is strongly exergonic.

FIGURE 2.16 Common Functional Groups in Biological Molecules "R" refers to the rest of the organic molecule and can thus be a large variety of possibilities.

Activity 2.2 Functional Groups

REVIEW & APPLY | 2.6

R Functional groups are small groups of atoms that attach to other molecules and confer specific properties. Some functional groups are polar, and make the host molecule more polar and better able to interact with water and other polar or charged substances. Other functional groups are nonpolar, and interact with nonpolar substances either in the host molecule or in the environment. Molecular diversity is enhanced by differing arrangements of atoms in two molecules with the same overall atomic formula.

A

1. Give a few examples of the functional groups that might be found on the outside of biological molecules.
2. How do structural isomers differ?

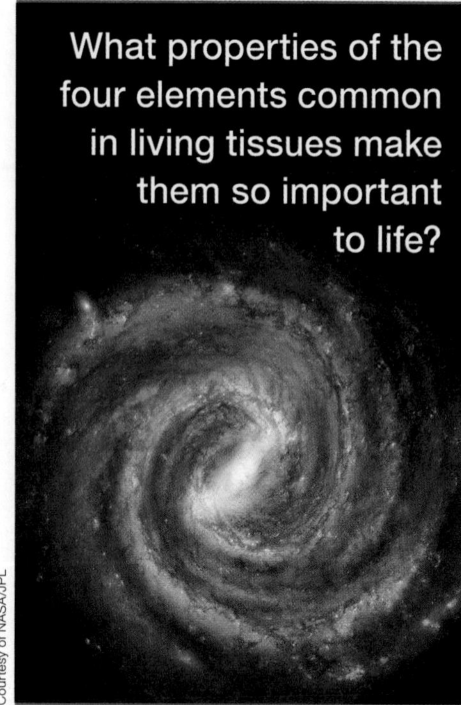

What properties of the four elements common in living tissues make them so important to life?

Courtesy of NASA/JPL

Answer: One reason why H, O, C, and N are the most common elements in living tissues is simply because they are the among the most common elements. Indeed, these four elements are among the top six most abundant elements in the universe. The other two common elements, He and Ne, are chemically inert gases and, as such, are not involved in the rich chemistry of living systems. On Earth, most of the next most common elements are ion-forming; Na, K, Ca, and Fe are examples. Ion-forming elements cannot form the large, structurally complex molecules that perform the central roles in living systems (see Chapter 3).

H and O are common in living systems because life evolved in, and is substantially composed of, water. The properties of water make it the perfect solvent for the chemical reactions necessary in living systems.

The chemistry of H, C, N, and O makes them especially suited for life. They can form both nonpolar and polar covalent bonds, participate in hydrogen bonds, and especially in functional groups, can readily form ions. Molecules with carbon backbones can be large, yielding structurally complex molecules with variable chemistry resulting from specific functional groups.

In summary, a combination of their relative abundance, coupled with specifics of their chemistry and the importance of water, explains the commonness of these four elements in living tissues. Perhaps not surprisingly, the search for life on other planets involves identifying those that have liquid water and an abundance of these four elements.

2 VISUAL SUMMARY

2.1 An Element's Atomic Structure Determines Its Properties

Go to ACTIVITY 2.1

FIGURE 2.1

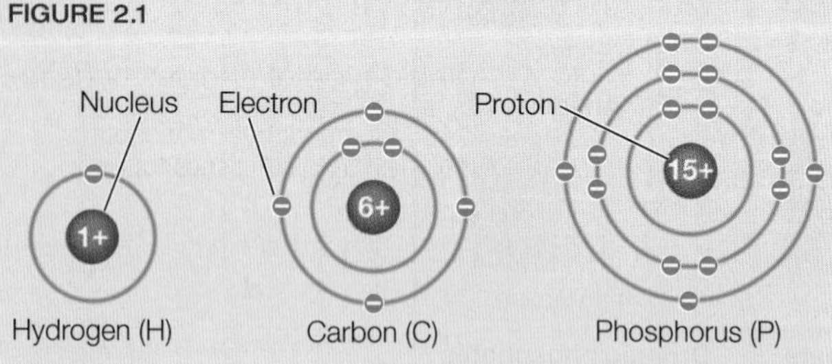

FIGURE 2.2

	1	2	13	14	15	16	17	18
1	1 H 1.0079							2 He 4.003
2	3 Li 6.941	4 Be 9.012	5 B 10.81	6 C 12.011	7 N 14.007	8 O 15.999	9 F 18.998	10 Ne 20.179
3	11 Na 22.990	12 Mg 24.305	13 Al 26.982	14 Si 28.086	15 P 30.974	16 S 32.06	17 Cl 35.453	18 Ar 39.948

- Each **ELEMENT** has a characteristic number of **PROTONS** in its **NUCLEUS** (**FIGURE 2.1**).
- The number of **ELECTRONS** in the **ATOMIC ORBITALS** of its outermost shell determines an element's chemical properties, which is the reason that groups (elements in the same column) in the **PERIODIC TABLE** behave similarly (**FIGURE 2.2**).

QUESTIONS

1. How many protons and neutrons are in the nucleus of the element aluminum (Al), with an atomic number of 13 and an atomic mass of 27?
2. How many electrons are in each of the three electron shells of Al?
3. Why do isotopes differ in atomic mass but not atomic number?

2.2 Atoms Bond to Form Molecules

Go to ANIMATION 2.1

- **ELECTRONEGATIVITY** and the number of electrons in the **VALENCE SHELL** together determine how atoms will **BOND**.
- **COVALENT BONDS** are strong bonds and may be **POLAR** (**FIGURE 2.6**).
- **IONS** are formed when electrons are transferred from one atom to another (**FIGURE 2.7**).
- Although weak, **HYDROGEN BONDS** and **VAN DER WAALS INTERACTIONS** are critically important within and between **MOLECULES**.

FIGURE 2.6

FIGURE 2.7

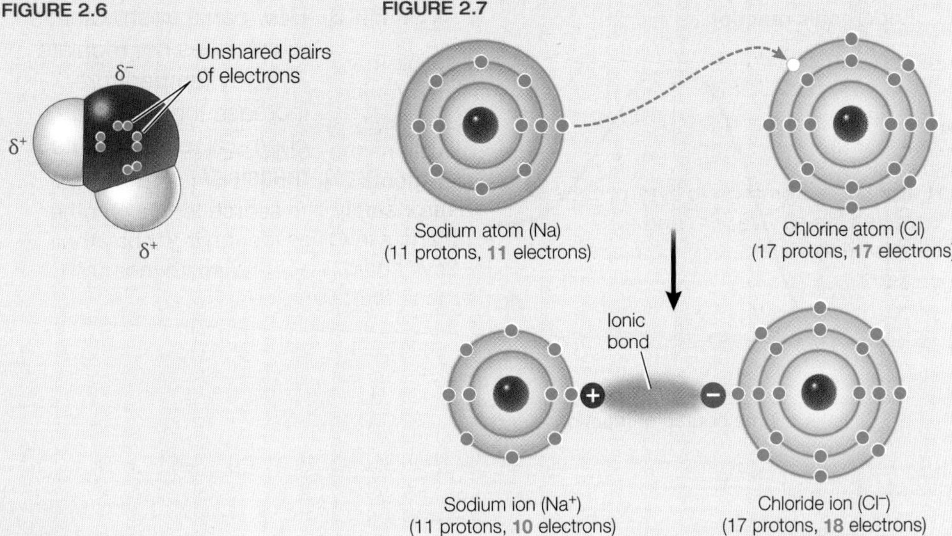

QUESTIONS

1. Why aren't covalent bonds between atoms of the same element polar?
2. Why do elements like sodium (Na) in the first group of the periodic table form ionic as opposed to covalent bonds?
3. How does the polarity of carbon dioxide (CO_2) compare with that of hexane (C_6H_{14})? Would you expect either of them to form hydrogen bonds with water?

2.3 Chemical Transformations Involve Energy and Energy Transfers

- **ENERGY** is required for any change in living systems.
- The **FIRST LAW OF THERMODYNAMICS** states that energy is conserved; none is lost and new energy is not created.
- The **SECOND LAW OF THERMODYNAMICS** states that **ENTROPY** increases when energy is converted or transferred; energy becomes more dispersed, and the amount of usable energy is reduced (**FIGURE 2.11**).

FIGURE 2.11

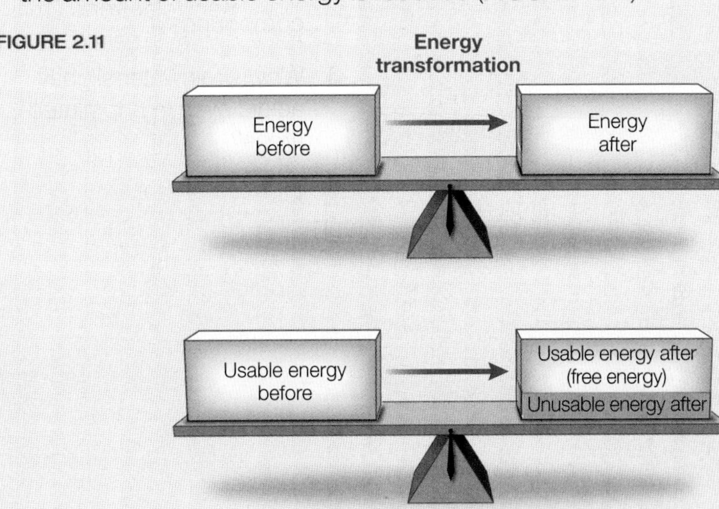

1. What is the difference between kinetic energy and potential energy?
2. Which type of energy—kinetic energy or potential energy—do the following examples represent? (a) A hamburger. (b) A diving falcon. (c) A water molecule vibrating in a hot kettle. (d) A beam of light. (e) An electric fish just before it generates current.
3. What is the relative potential energy content of strong compared with weak chemical bonds?

2.4 Chemical Reactions Transform Substances

- During a **CHEMICAL REACTION**, atoms change bonding partners.
- Chemical reactions may be **EXERGONIC** (release energy) or **ENDERGONIC** (require energy). Both types of reactions require an input of **ACTIVATION ENERGY** (**E_A**) to begin (**FIGURE 2.13**).
- The **REACTION RATE** can be altered by changes in concentration, temperature, or activation energy.

FIGURE 2.13

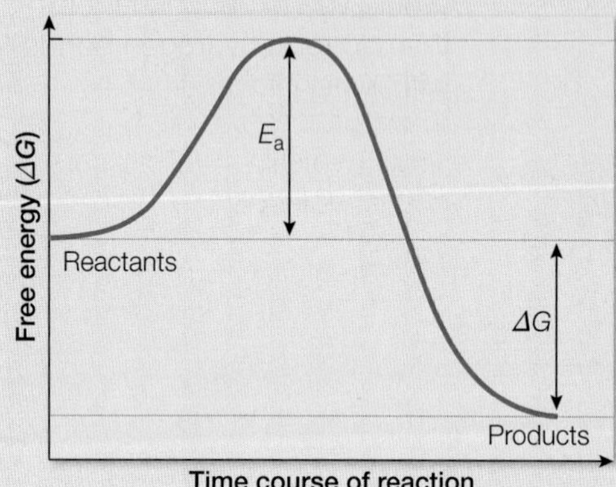

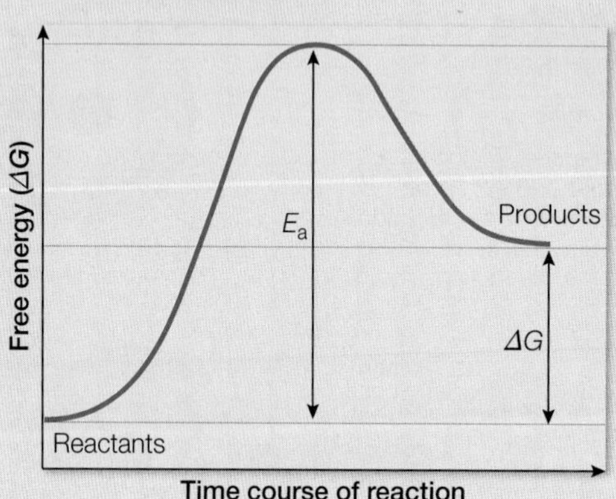

QUESTIONS

1. Can a reaction be exergonic if the products have higher chemical-bond energy than the reactants?
2. What is the difference between hydrolysis and condensation reactions? Which is usually exergonic?
3. How can a freshwater fish, which does not regulate its body temperature, increase the rate of a chemical reaction in its muscles?

2.5 The Properties of Water Are Critical to the Chemistry of Life

- Many of water's characteristics are explained by its polar O—H bonds and ability to form hydrogen bonds (**FIGURE 2.15**).
- Water is an excellent **SOLVENT** for ionic and polar compounds.
- **ACIDS** ionize in water to give H_3O^+ ions. **BASES** ionize to give OH^- ions.

FIGURE 2.15

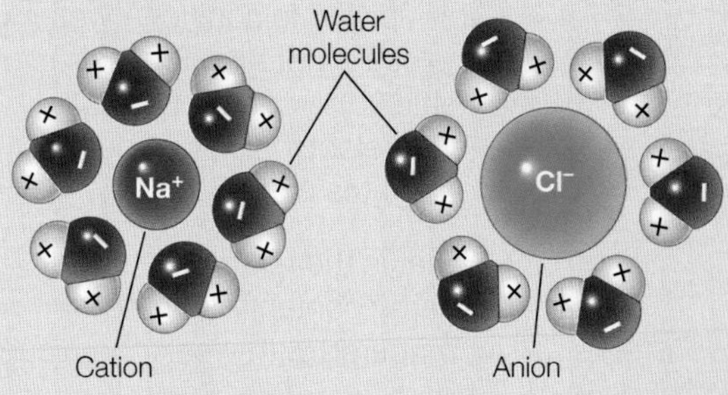

1. Why does water have a high heat capacity?
2. What is surface tension, and why does water have a high surface tension?
3. Why is water an excellent solvent for polar and ionic compounds?
4. Why is rainwater slightly acidic even in unpolluted areas?

2.6 Functional Groups Give Molecules Specific Properties

Go to ACTIVITY 2.2

- Some **FUNCTIONAL GROUPS** are common in biological molecules (**FIGURE 2.16**).
- Functional groups give molecules many of their chemical properties.
- **STRUCTURAL ISOMERS** have the same chemical formula but differ in how atoms are bonded to one another.

FIGURE 2.16

Functional group	Class of compounds and an example	Properties
Methyl 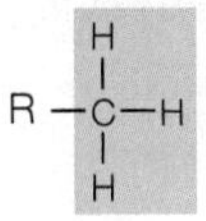 R—CH_3	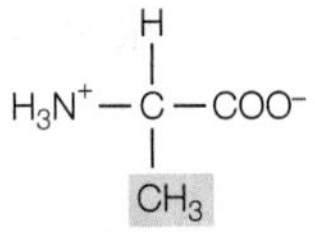 **Alkyl** H_3N^+—CH(CH_3)—COO^- Alanine	Nonpolar. Important modification of proteins and cytosine nucleotide.
Hydroxyl 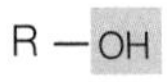 R—OH	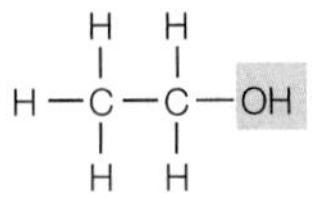 **Alcohols** H—CH_2—CH_2—OH Ethanol	Polar. Involved in hydrogen bonding. Often participates in condensation reactions. Required in side chain for phosphorylation of proteins.
Sulfhydryl 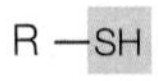 R—SH	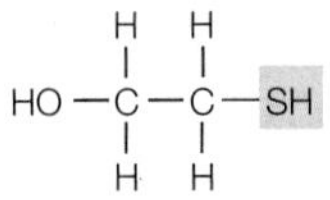 **Thiols** HO—CH_2—CH_2—SH Mercaptoethanol	Polar. Can form disulfide bridges to stabilize protein structure.

QUESTIONS

1. Why does the presence of a carboxyl group (—COOH) make a molecule more acidic?
2. The thiol functional group (—SH) is weakly polar. Sulfur is in the same group of the periodic table as oxygen, but the thiol functional group is much less polar than the hydroxyl (—OH) functional group. Why do you think this is the case?
3. Suppose a molecule has the chemical formula C_4H_{10}. Draw two structural isomers of this molecule.

Go to **Achieve** for the e-book, animations, activities, and additional resources and assignments.

3 Macromolecules

Typical cells, such as those shown here from the bacterium *Escherichia coli*, consist of about 70 percent water, 28 percent macromolecules, and 2 percent small molecules. The same four types of macromolecules appear in all cells.

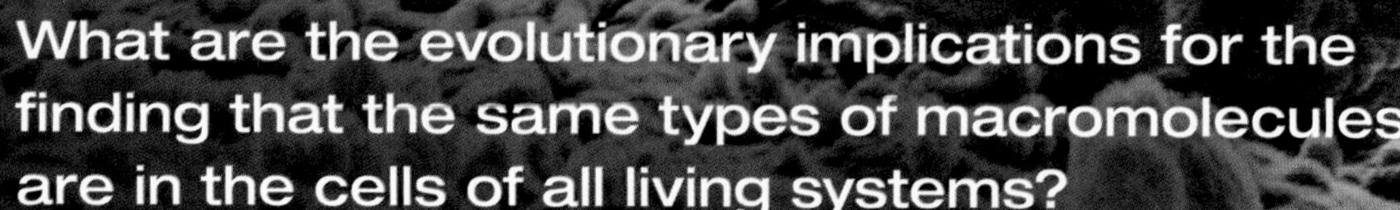

What are the evolutionary implications for the finding that the same types of macromolecules are in the cells of all living systems?

You will find the answer to this question on page 67.

USDA, ARS, EMU

KEY CONCEPTS

3.1 Lipids Are Characterized by Their Insolubility in Water

LEARNING OBJECTIVES

By the end of this key concept you should be able to:

3.1.1 **Define lipids based on their insolubility in water.**

3.1.2 **Identify triglycerides and phospholipids from their chemical structure.**

3.1.3 **Recognize the difference between saturated and unsaturated fatty acids and explain how saturation affects how closely they can pack together.**

3.1.4 **List some of the functions of lipids.**

In this chapter we consider the structure and function of the large molecules, called **macromolecules**, that are prevalent in all organisms: lipids, carbohydrates, nucleic acids, and proteins. These macromolecules are formed by endergonic reactions (see Key Concept 2.4) that form covalent bonds between smaller molecules. Carbohydrates, nucleic acids, and proteins are **polymers**, large molecules formed by covalent bonds between smaller molecules called **monomers**, which are often referred to as **residues** when in polymers. In contrast, lipids are not polymers, and are defined more by their properties than by their structure.

Lipids are structurally and functionally diverse

Lipids are a structurally and functionally diverse group of compounds defined by their insolubility in water. Lipids are insoluble in water—that is, they are hydrophobic—because of their many nonpolar covalent bonds. As we discussed in Key Concept 2.5, hydrophobic molecules (such as lipids) tend to aggregate in aqueous solutions. The structural features of several different kinds of lipids are illustrated in **FIGURE 3.1**.

Lipids contain mostly C—C and C—H nonpolar bonds, which have high chemical-bond energy compared with C—O and O—H bonds. As you will see, an important property for lipids in organisms is their melting temperature, at which they turn from a solid to a fluid. You observe lipid melting when you heat up butter, which is largely lipids. The melting points of lipids are determined by their size and by how closely the molecules pack together. Larger lipid molecules have higher melting temperatures because there are generally more van der Waals interactions between larger molecules than smaller ones. Similarly, if molecules have shapes that enable them to pack together closely, there will be more van der Waals interactions between them; this packing also results in a higher melting temperature.

Lipids play several roles in living organisms. Here we focus on the structure of lipids used in energy storage and membrane formation, but we will mention a few others to illustrate the diversity of lipid function.

Energy is stored in fats and oils

The lipids with which you are most familiar, fats and oils, are triglycerides. A **triglyceride** contains three fatty acid molecules and one glycerol molecule (see Figure 3.1B). A **fatty acid** consists of a long nonpolar hydrocarbon chain with a terminal polar carboxyl functional group (—COOH). **Glycerol** is a three-carbon molecule with three hydroxyl functional groups (—OH). These molecules are joined by condensation reactions (**FIGURE 3.2**). Triglycerides are excellent stores for chemical-bond energy.

As we will describe in **Chapter 5**, lipids yield more energy than carbohydrates or proteins during degradation.

Fatty acids can be saturated or unsaturated (see Figure 3.1A):

- In a **saturated fatty acid**, all the bonds between the carbon atoms in the hydrocarbon chain are single bonds (—C—C—C—C—). Double bonds between carbon atoms (—C═C—) do not occur in a saturated fatty acid because all the available bonds to carbon are saturated with hydrogen atoms. Saturated fatty acid molecules are relatively straight molecules and thus are able to pack together tightly, like pencils in a box. As noted above, this allows molecular interactions and a relatively high melting point. Saturated fats are usually solid at human body temperature (37°C), for example.
- In an **unsaturated fatty acid**, the hydrocarbon chain has one or more double bonds between carbon atoms (—C═C—). Linoleic acid (see Figure 3.1A), for example, has two double bonds near the middle of the hydrocarbon chain. This results in kinks to what otherwise is a linear, straight chain. Such kinks prevent the unsaturated molecules from packing together tightly, resulting in a lower melting temperature and the possibility of a fluid state at body temperature.

Phospholipids form biological membranes

The many C—C and C—H bonds in a fatty acid contribute to the molecule's hydrophobic character. However, the carboxyl functional group at the end of the molecule is not hydrophobic (C—O bonds are polar), and indeed, in living systems this functional group ionizes and forms —COO$^-$ and becomes even more hydrophilic. A fatty acid thus has two opposing chemical properties: a hydrophilic end and a long hydrophobic "tail." A molecule that

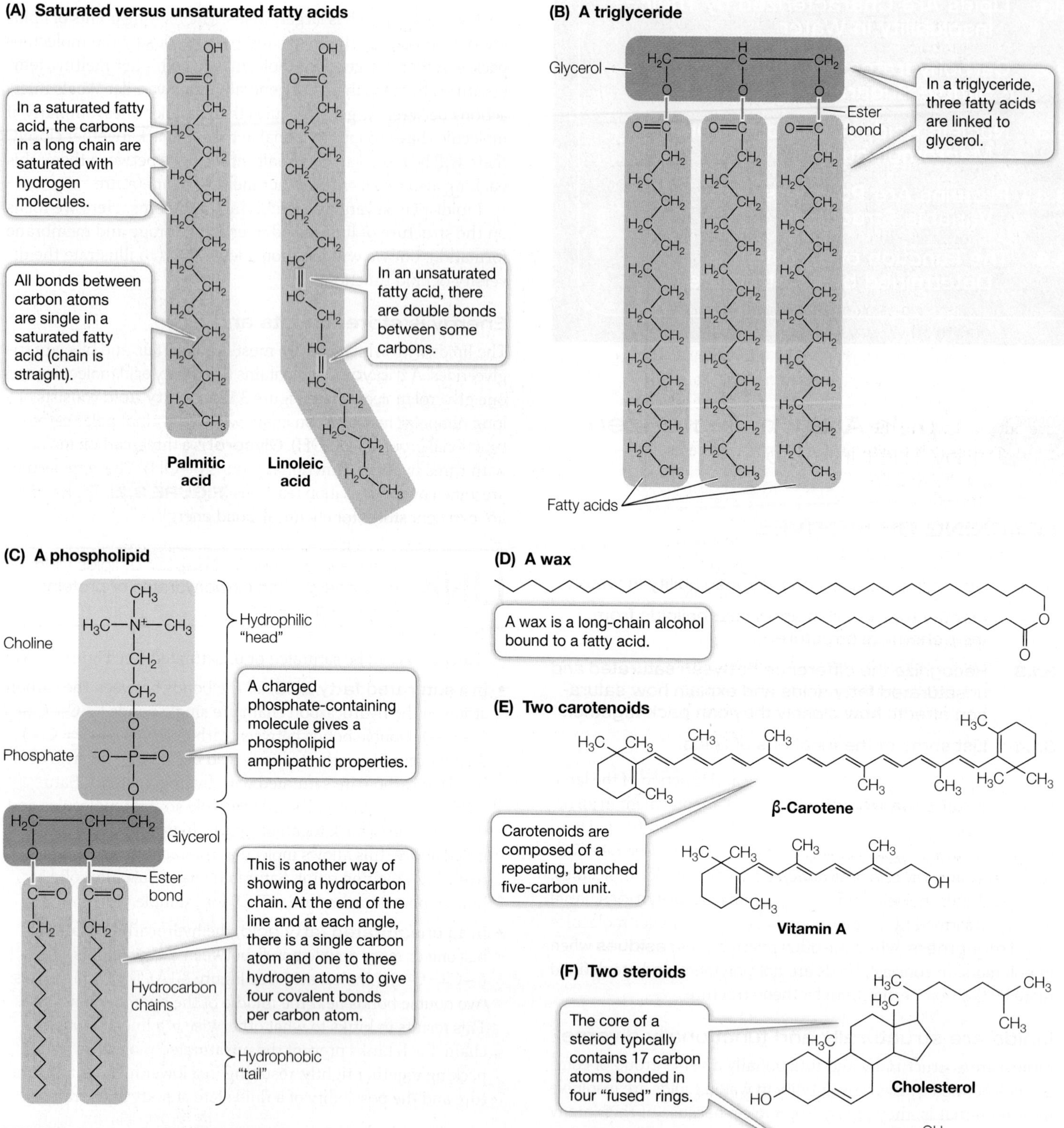

FIGURE 3.1 Basic Lipid Structures **(A)** Fatty acids are components of many lipids. Saturated fatty acids (left) are straight chains, whereas in unsaturated fatty acids (right) the double bond(s) between some carbon atoms form kinks. **(B)** Fats and oils are triglycerides. **(C)** Phospholipids are the major component of cell membranes. Their polar hydrophilic ends and nonpolar hydrophobic fatty acid tails make phospholipids amphipathic. **(D)** One of the waxes found in beeswax. The long nonpolar chains are very hydrophobic. **(E)** The alternating single and double bonds between adjacent carbons allow carotenoids such as β-carotene and vitamin A to absorb light energy. **(F)** Steroids have a distinctive ring structure. Examples shown are cholesterol and the estrogen hormone estradiol.

Why is vitamin A classified as a lipid even though it contains a polar —OH group?

is partly hydrophilic and partly hydrophobic is said to be **amphipathic**.

A triglyceride, in which a glycerol molecule is bonded to three fatty acid chains, is primarily hydrophobic. Like triglycerides, **phospholipids** also contain fatty acids bound to glycerol. However, in phospholipids a charged phosphate-containing molecule replaces one of the fatty acids, making phospholipids amphipathic (see Figure 3.1C). The charged phosphate-containing functional group (there are several different kinds in different phospholipids) attracts polar water molecules, while the two fatty acids are hydrophobic and aggregate together and with other hydrophobic substances.

In an aqueous environment, phospholipids line up in such a way that the nonpolar hydrophobic "tails" pack together and the phosphate-containing "heads" face outward, where they interact with water molecules. The result is a **bilayer**: a sheet two molecules thick, with water excluded from the core (**FIGURE 3.3A**). Although no covalent bonds link individual phospholipids in these bilayers, such stable aggregations form readily in aqueous conditions. Biological membranes are composed of this kind of **phospholipid bilayer**, along with other molecules, especially proteins (**FIGURE 3.3B**). Membranes function as barriers separating one aqueous region from another. Chapter 4 will describe biological membrane structure and function in greater detail.

In association with certain proteins, phospholipids can form single-layer spherical structures called **lipoproteins** that have

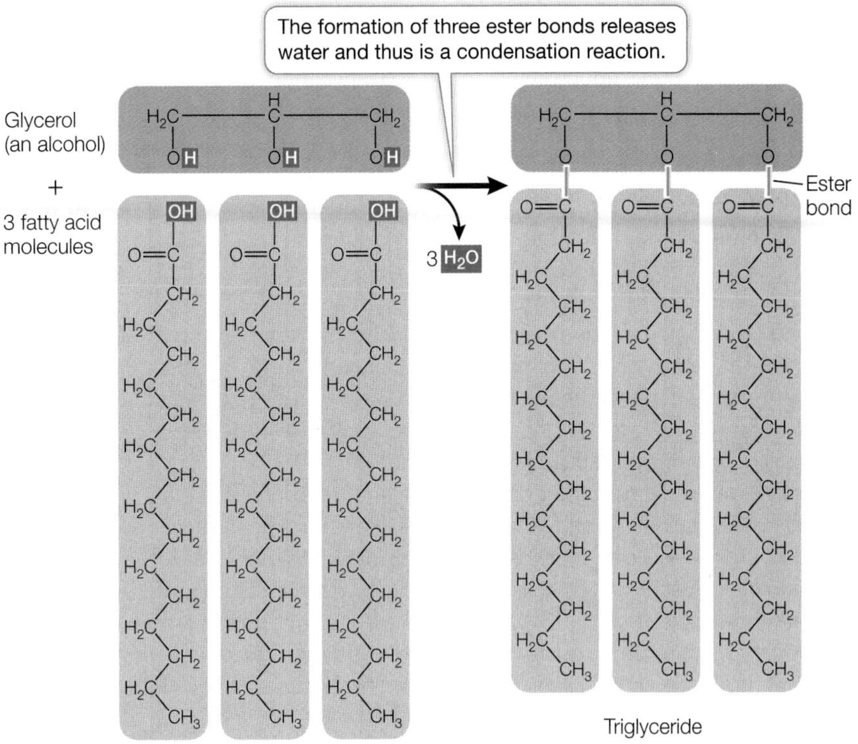

FIGURE 3.2 Synthesis of a Triglyceride A condensation reaction between glycerol and three fatty acids synthesizes a triglyceride.

Animation 3.1 Macromolecules: Lipids

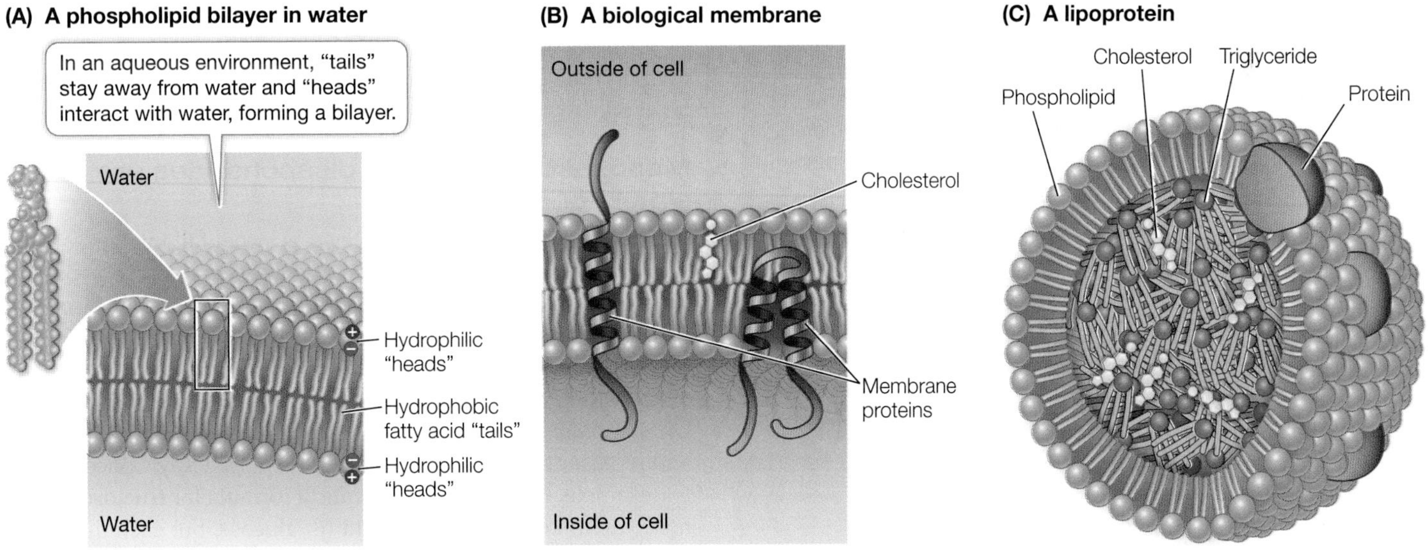

FIGURE 3.3 Phospholipids Have Hydrophilic Heads and Hydrophobic Tails (A) Phospholipids readily form a bilayer in an aqueous environment. **(B)** The general molecular structure of a biological membrane is a phospholipid bilayer that has proteins embedded in or associated with it. The phospholipid bilayer is a barrier both to the entry of substances from the environment outside the cell and to the exit of contents from inside the cell. **(C)** Phospholipids, in conjunction with certain proteins, can form lipoproteins, which transport lipids such as cholesterol in the blood.

Is a phospholipid bilayer a barrier that ions cannot cross?

hydrophobic interiors and hydrophilic exteriors and which are used for transporting lipids in aqueous solutions (**FIGURE 3.3C**). For example, lipoproteins transport triglycerides and cholesterol in the blood to and from all the tissues of the body.

Lipids have many other functions

In many vertebrates, fats are used not only for energy storage but are also important for organ protection and thermal regulation. Two types of fats, known colloquially as white fat and brown fat, are found in mammals. **White fat** serves to store energy and provide thermal insulation that helps regulate body temperature. **Brown fat** gets its color from iron-rich mitochondria and plays an important role in thermoregulation, particularly in infants. Waxes (see Figure 3.1D) coat the leaves of plants to reduce water loss and protect against pathogens (see Key Concept 27.4). Honey bees produce waxes to build their combs. Birds and mammals secrete waxes onto their hair and feathers to repel water and keep them pliable. **Carotenoids** are lipids that can absorb energy from particular wavelengths of light (e.g., β-carotene in plants and vitamin A in animals) (see Figure 3.1E). **Steroids** are an important class of lipids in plants and animals. Cholesterol is a steroid that is essential in animal membranes (see Chapter 4). Closely related to cholesterol are the steroid hormones such as estrogen (see Figure 3.1F) that function as signal molecules in both plants and animals.

LINK Some plant hormones are steroids that regulate growth and development (see **Key Concept 25.3**).

REVIEW & APPLY | 3.1

R Lipids are a diverse group of molecules made up primarily of carbon and hydrogen and that are insoluble in water. Lipids include triglycerides, phospholipids, steroids, carotenoids, and waxes. Lipids play roles in energy storage, thermoregulation, light absorption, and intracellular communication and are important elements of many biological structures. The cell membrane is composed of a phospholipid bilayer.

A
1. Fatty acids and glycerol react to form triglycerides. Why are fatty acids and triglycerides both classified as lipids while glycerol is not?
2. What is the difference between fats and oils in terms of their structure?
3. Why can phospholipids form bilayers but triglycerides cannot?

Lipids are termed macromolecules in part because they often occur in large aggregations in cells. However, their individual molecules are quite small compared to the true macromolecules, which are composed of large polymers of repeating subunits. We begin our investigation of these polymers with carbohydrates.

3.2 | Carbohydrates Are Made from Simple Sugars

LEARNING OBJECTIVES

By the end of this key concept you should be able to:

3.2.1 Draw the ring form of five- and six-carbon carbohydrates.

3.2.2 Distinguish structural isomers and stereoisomers of monosaccharides.

3.2.3 Define the difference between mono-, di-, oligo-, and polysaccharides.

3.2.4 Describe the functional consequences of linear versus branched polysaccharides, and how branching occurs in terms of which carbons are involved in the bonding.

Carbohydrates are a large group of molecules with similar atomic compositions but that differ greatly in size, chemical properties, and biological functions. **Carbohydrates** often have the general formula $C_m(H_2O)_n$, where *m* and *n* represent numbers, but this is misleading, as there are not really carbon atoms bonded to water but instead linked carbons bonded to functional groups containing O and H. Carbohydrates with 12 or fewer carbons are termed simple sugars. Most of the carbon atoms in carbohydrates have a hydroxyl (—OH) functional group. Some carbohydrates are relatively small, such as glucose, $C_6H_{12}O_6$, which is the primary energy source for many organisms. Others are composed of a large number of simple carbohydrate repeats joined together into very large polymers. An example is starch, which is a polymer of glucose and has the formula $(C_6H_{10}O_5)_k$, where *k* can be a very large number.

Carbohydrates play several roles in living organisms. We focus here on the structures of small versus large carbohydrates and how their structures relate to their function.

Monosaccharides and disaccharides are simple carbohydrates

The simple sugars that you perceive as sweet fall into two structural categories: monosaccharides and disaccharides. **Monosaccharides** (*mono*, "one," + *saccharide*, "sugar") are the simplest carbohydrates, and can exist in a linear form or, more usually, in a ring form (**FIGURE 3.4**). Monosaccharides consist of five or six carbon atoms, termed pentoses and hexoses, respectively. Different monosaccharides often have the same chemical formula and represent structural or stereoisomers (see Key Concept 2.6). For example, $C_6H_{12}O_6$ is the chemical formula for fructose, mannose, alpha (α) glucose, and beta (β) glucose, but each differs in the arrangement of their carbon, hydrogen, and oxygen atoms (**FIGURE 3.5**). Carbohydrate isomers play important roles in function—for example, in binding to and interacting with other molecules. The three-dimensional structure of polymers composed of sugars varies depending on the isomers of the simple sugars that are linked together (see next page).

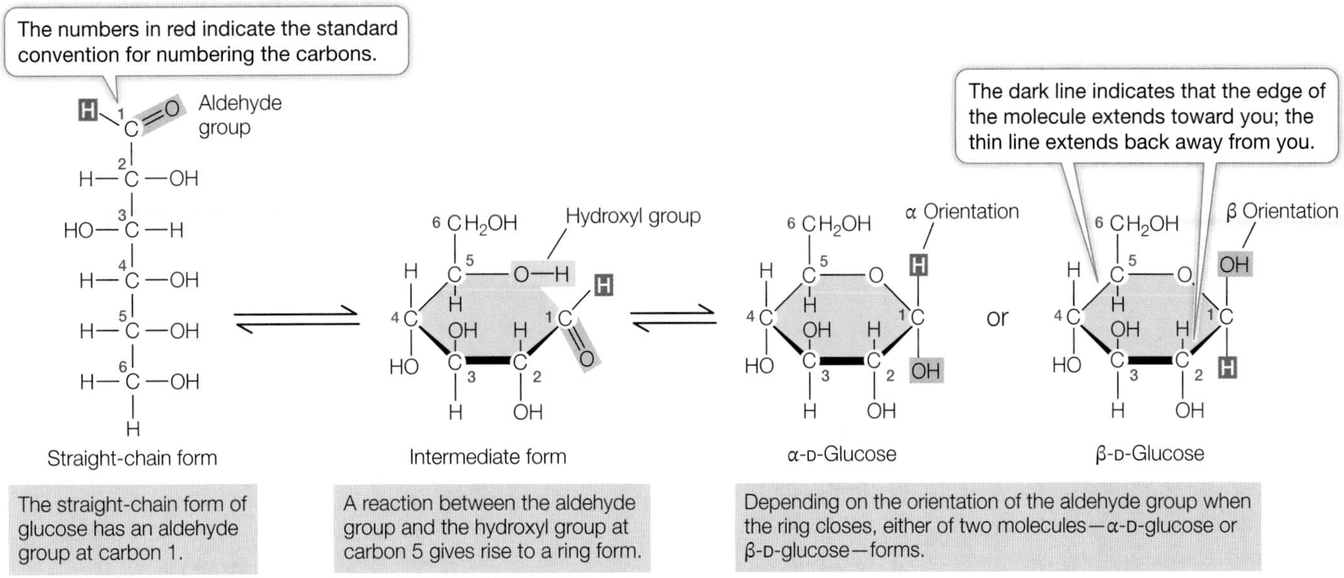

The numbers in red indicate the standard convention for numbering the carbons.

The dark line indicates that the edge of the molecule extends toward you; the thin line extends back away from you.

Straight-chain form — Intermediate form — α-D-Glucose or β-D-Glucose

The straight-chain form of glucose has an aldehyde group at carbon 1.

A reaction between the aldehyde group and the hydroxyl group at carbon 5 gives rise to a ring form.

Depending on the orientation of the aldehyde group when the ring closes, either of two molecules—α-D-glucose or β-D-glucose—forms.

FIGURE 3.4 Isomers of Glucose **(A)** Straight-chain glucose can convert to a ring form. **(B)** The alpha (α) and beta (β) forms of glucose can interconvert.

Activity 3.1 Forms of Glucose

Disaccharides are composed of two monosaccharides joined in a condensation reaction (**FIGURE 3.6**). The bond that links the sugars is called a **glycosidic bond**, or glycosidic linkage. Two hydroxyl (—OH) groups are involved in the reaction, which produces a molecule of water:

$$\text{A—OH} + \text{HO—B} \longrightarrow \text{A—O—B} + H_2O$$

where A and B are the carbon atoms that are part of two monosaccharides.

Depending on which —OH groups are involved and on their orientation, different disaccharides result. Sucrose, common table sugar, is an important disaccharide formed in plants from a glucose molecule and a fructose molecule. The bond joining the first carbon of the glucose and the second carbon of the fructose is called a 1,2 glycosidic bond (see Figure 3.6). The plant sugars maltose and cellobiose are both glucose disaccharides linked by 1,4 glycosidic bonds.

Mono- and disaccharides are the main carbohydrates in cells and can readily enter the pathways that break them down to release energy. Mono- and disaccharides can be covalently bonded to proteins or lipids, modifying their solubility and function. The monosaccharides ribose and deoxyribose (see Figure 3.5) are also a component of nucleotides that make up nucleic acid macromolecules (see Key Concept 3.3).

LINK The synthesis of glucose by plants from carbon dioxide and water using light energy is covered in **Chapter 5**.

Oligosaccharides are carbohydrates composed of a few monosaccharides

Oligosaccharides are constructed from three to ten monosaccharides joined by glycosidic bonds. Many oligosaccharides are modified by the addition of functional groups, which give them

Five-carbon sugars (pentoses)

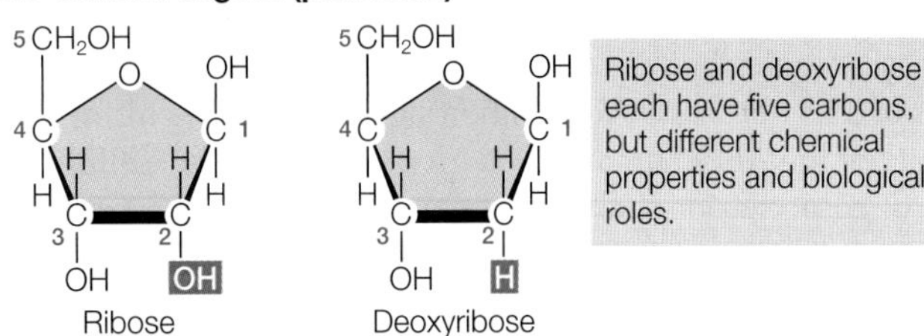

Six-carbon sugars (hexoses)

α-Mannose — α-Glucose — β-Glucose — Fructose

α-Mannose, α-glucose, and β-glucose are stereoisomers of each other. All three are structural isomers of fructose.

FIGURE 3.5 Monosaccharides Some five- and six-carbon monosaccharides.

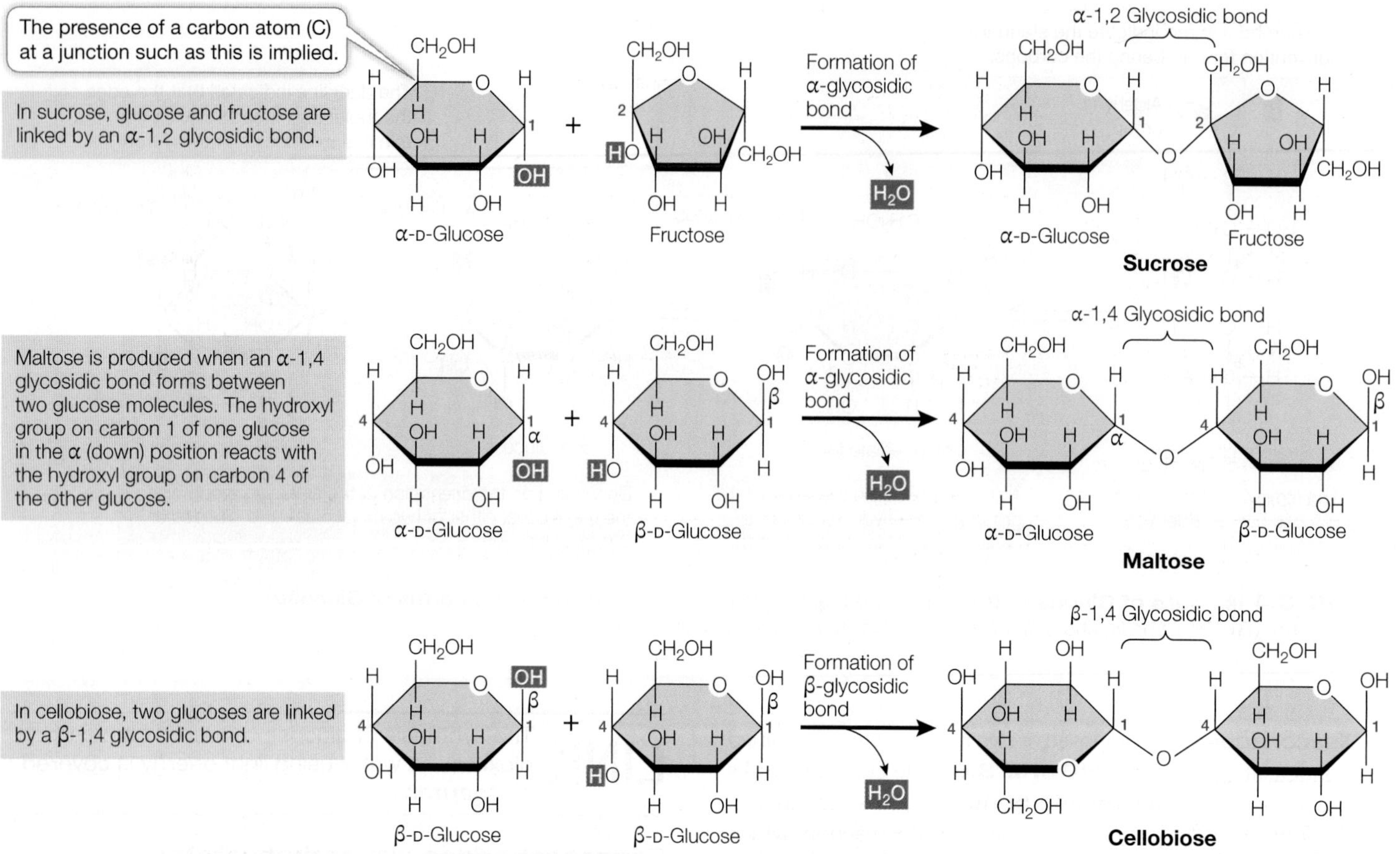

The presence of a carbon atom (C) at a junction such as this is implied.

In sucrose, glucose and fructose are linked by an α-1,2 glycosidic bond.

Maltose is produced when an α-1,4 glycosidic bond forms between two glucose molecules. The hydroxyl group on carbon 1 of one glucose in the α (down) position reacts with the hydroxyl group on carbon 4 of the other glucose.

In cellobiose, two glucoses are linked by a β-1,4 glycosidic bond.

FIGURE 3.6 Glycosidic Bonds Link Monosaccharides In a condensation reaction, two hydroxyl (—OH) groups react, liberating water, and the remaining oxygen atom covalently joins the two molecules. Depending on which —OH groups are involved, and on their orientation, different disaccharides result.

special properties. Modified and unmodified oligosaccharides are often covalently bonded to proteins or lipids, which alter their function and solubility. Oligosaccharides bound to the proteins and lipids on the outer surfaces of cells function as recognition signals. For example, the specificity of the ABO human blood group system is due to differences in oligosaccharide chains on proteins present on the surfaces of their cells.

Polysaccharides are composed of many monosaccharides

Polysaccharides are large polymers of hundreds to thousands of monosaccharides connected by glycosidic bonds (**FIGURE 3.7**). Polysaccharides can be linear chains of monomers, attached via 1,4 glycosidic bonds, or they can be branched from 1,6 glycosidic bonds. The specific monomers that make up a polymer are often referred to as residues in polysaccharides and other macromolecules.

Linear chains of polysaccharides can align in close proximity and form hydrogen bonds with neighboring chains. Parallel alignment of polysaccharide chains allows them to form dense sheets or strong fibers. The most abundant biological macromolecule on the planet is the polysaccharide cellulose, the major structural component of plant cell walls. This macromolecule is composed of β-glucose residues, attached via 1,4 glycosidic bonds (see Figure 3.7). The major molecule in arthropod skeletons, chitin, is a polymer of modified β-glucoses joined by 1,4 bonds. Cellulose and chitin are both resistant to being broken down because it difficult for chemicals or enzymes to access the bonds in these dense polysaccharides.

Key Concept 5.4 discusses why cellulose is the most abundant biological macromolecule on the planet.

Starches and glycogen are giant, branched polysaccharides of α-glucose residues, attached via 1,4 and 1,6 glycosidic bonds (see Figure 3.7). **Starches** are the principal energy storage compounds of plants, and glycogen is the principal energy storage compound in animals, fungi, and bacteria (see Figure 3.7). Starches and glycogen are water-insoluble, though the glucose residues at their borders are readily accessible and thus readily hydrolyzed into glucose monomers, which in turn can be broken down to liberate their stored energy.

(A) Molecular structure

Cellulose

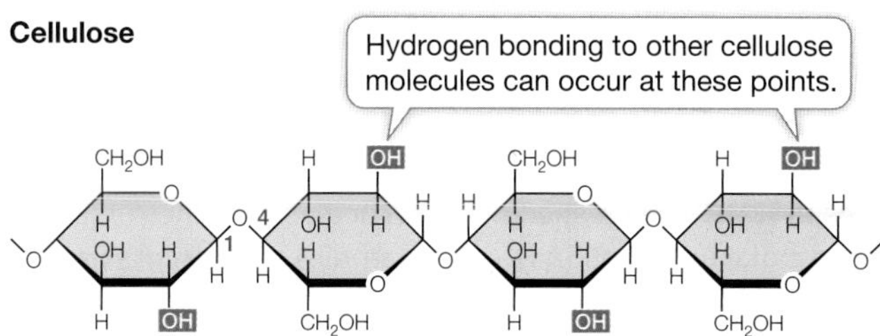

Cellulose is an unbranched polymer of glucose with β-1,4 glycosidic bonds that are chemically very stable.

Starch and glycogen

Branching occurs here.

Glycogen and starch are polymers of glucose with α-1,4 glycosidic bonds. α-1,6 Glycosidic bonds produce branching at carbon 6.

(B) Macromolecular structure

Linear (cellulose)

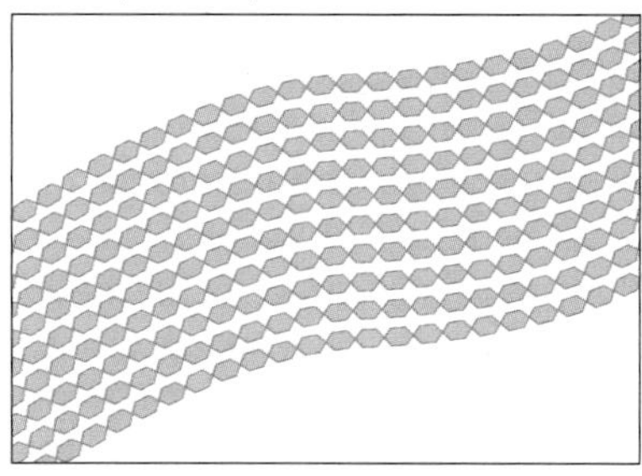

Parallel cellulose molecules form hydrogen bonds, resulting in thin fibrils.

Branched (starch)

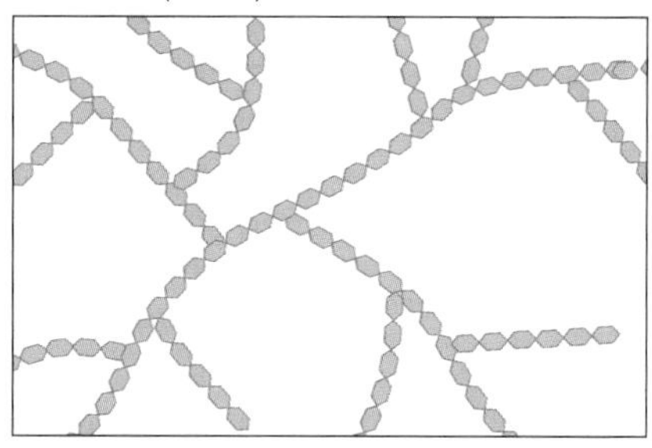

Branching limits the number of hydrogen bonds that can form in starch molecules, making starch less compact than cellulose.

Highly branched (glycogen)

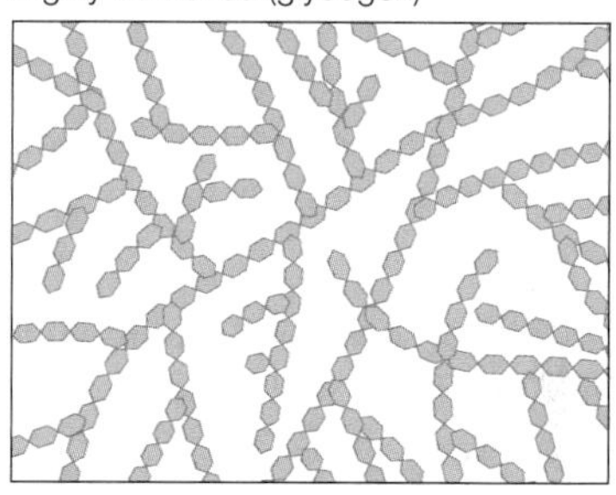

The high amount of branching in glycogen makes its solid deposits more compact than starch.

(C) Polysaccharides in cells

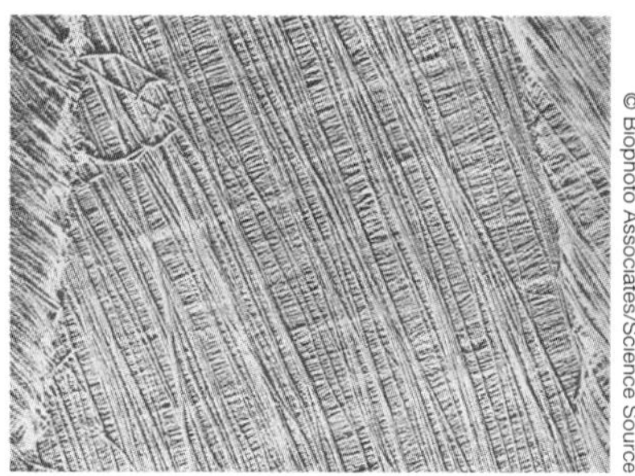

© Biophoto Associates/Science Source

Layers of cellulose fibrils, as seen in this scanning electron micrograph, give plant cell walls great strength.

© Dennis Kunkle Microscopy/Science Source

Within these potato cells, starch deposits (colored red in this scanning electron micrograph) have a granular shape.

© Don W. Fawcett/Science Source

The dark clumps in this electron micrograph are glycogen deposits in a liver cell.

FIGURE 3.7 Polysaccharides Can Be Branched or Unbranched Cellulose, starch, and glycogen have different levels of branching.

Animation 3.2 Macromolecules: Carbohydrates

Which of these three carbohydrate macromolecules—cellulose, starch, or glycogen—will have the most water molecules packed within it?

If glucose is the major source of energy in cells, why is it stored as starch or glycogen? The reason is that a high level of glucose causes water to enter cells. The movement of water is determined by osmotic pressure, which we will discuss in Key Concept 4.2. Briefly, individual glucose molecules in solution exert higher osmotic pressure than the same number of glucose molecules in fewer insoluble glycogen molecules. High osmotic pressure causes water to enter the cells. If it were not for insoluble polysaccharides, large amounts of energy would be expended to expel excess water from cells with high glucose levels.

REVIEW & APPLY | 3.2

R Carbohydrates are composed of carbon, hydrogen, and oxygen in the general formula $C_m(H_2O)_n$. Large carbohydrate polymers are composed of simple sugar monomers. Monosaccharides may have different numbers of carbon atoms; pentoses have five, and hexoses have six. Many sugars have the same chemical formula but differ in the arrangement of their atoms (that is, they are isomers). Condensation reactions join monosaccharides into disaccharides, oligosaccharides, or polysaccharides. The 1,4 glycosidic bonds result in linear polymers of glucose, whereas 1,6 glycosidic bonds result in branching. Important polysaccharides include starches, glycogen, and cellulose.

A

1. Glucose, mannose, and fructose all have the chemical formula $C_6H_{12}O_6$. Using α-glucose, α-mannose, and fructose, explain the difference between structural isomers and stereoisomers.
2. A carbohydrate consists of five monosaccharides joined by glycosidic bonds. What type of saccharide does this molecule represent?
3. Why is cellulose a structurally stronger molecule than glycogen or starch?

Carbohydrate macromolecules are relatively simple polymers, composed of large numbers of identical residues, such as α-glucose in glycogen. In the next section we will consider nucleic acid macromolecules, which are composed of four different residues and are thus substantially more diverse than carbohydrates.

3.3 Nucleic Acids Are Informational Macromolecules

LEARNING OBJECTIVES

By the end of this key concept you should be able to:

3.3.1 Recognize the structure of a nucleotide and determine whether it is a mono-, di-, or triphosphate.

3.3.2 Write down the complementary base pairing rules and know that base pairing occurs between antiparallel strands.

3.3.3 List the differences between RNA and DNA in terms of their bases, sugar, and structure.

3.3.4 Define DNA replication, transcription, and translation.

Nucleic acids are polymers that store, transmit, and express genetic (hereditary) information. There are two types of nucleic acids: **DNA (deoxyribonucleic acid)** and **RNA (ribonucleic acid)**. These macromolecules are long, unbranched polymers containing from just a few to hundreds of millions of nucleotide monomers. Although there are just four different nucleotides in both DNA and RNA, the number of possible nucleotide sequences in a DNA or RNA macromolecule is enormous.

DNA stores information, and some of that information encodes instructions for building proteins. A particular sequence of the nucleotides in DNA is expressed as a particular sequence of nucleotides in RNA, which specifies a particular sequence of amino acids that make up a protein (Key Concept 3.4 will discuss protein structure). DNA sequences also encode information for recognition by a variety of proteins that interact with DNA. Chapters 9–12 will describe in detail how DNA and RNA are involved in heredity and how their information content is expressed in biological systems. Here we focus on the chemical structures of these two nucleic acids, on which their function depends.

Nucleotides are the building blocks of nucleic acids

A **nucleotide** consists of three components: a monosaccharide (the pentose ribose or deoxyribose), a nitrogen-containing **base**, and one to three phosphate groups. The base is linked to the first carbon of the monosaccharide (termed the 1-prime or 1′ carbon), and the phosphate groups are linked to the 5′ carbon of the monosaccharide (**FIGURE 3.8**). In the absence of phosphate groups, the base and ribose or deoxyribose together make a **nucleoside**; thus, a nucleotide is a nucleoside mono-, di-, or triphosphate. The bases of the nucleic acids take one of two chemical forms: a single-ring structure called a **pyrimidine**, or a fused double-ring structure called a **purine**. Nucleotides usually contain three phosphates when they are monomers, but in nucleic acids they contain just one phosphate group.

The basic structures of DNA and RNA monomers differ in two respects:

1. The monosaccharide in DNA is deoxyribose, whereas in RNA it is ribose.
2. DNA is composed of nucleotides with bases cytosine (C), thymine (T), adenine (A), and guanine (G). RNA has A, G, and C, but its fourth base is uracil (U, not T).

During synthesis of a nucleic acid, nucleotide monomers are added one at a time. A reaction between the —OH group on the 3′ carbon atom of the ribose/deoxyribose in the last nucleotide of the existing chain and the triphosphate on the 5′ carbon of the incoming nucleotide monomer releases a pyrophosphate molecule; the resulting linkage between the two monosaccharides in the nucleotide is called a **phosphodiester bond** (**FIGURE 3.9**). Adding nucleotides in this manner forms a chain of alternating sugar and phosphate molecules called the sugar–phosphate backbone of the nucleic acid.

Nucleic acids cannot be extended using a nucleotide containing a single phosphate because the overall free energy of the reaction is positive and thus disfavored (see Key Concept 2.4). However, the free energy associated with breaking the bond between the first (α) and second (β) phosphate groups in a nucleotide containing three phosphates is sufficiently negative to make the overall synthesis reaction favorable. Because each incoming nucleotide is added to the 3′ carbon atom of the last ribose/deoxyribose in the

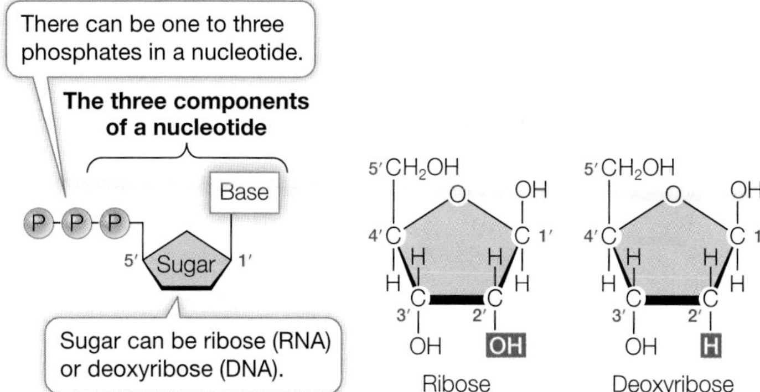

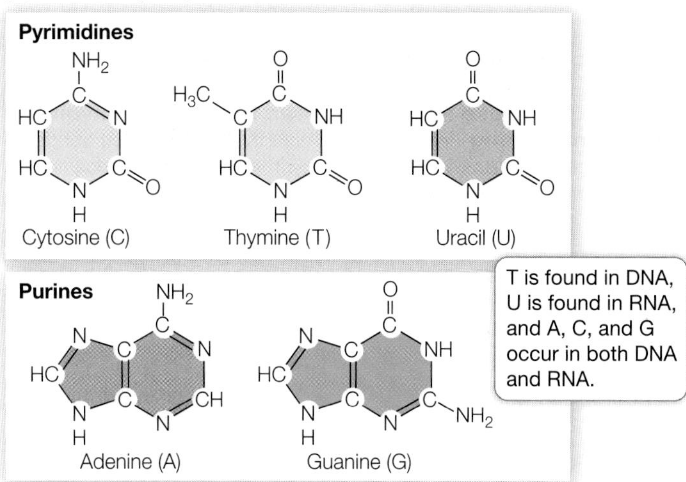

FIGURE 3.8 Nucleotide Structure **(A)** A nucleotide is a nucleoside mono-, di-, or triphosphate. The carbons on the ribose/deoxyribose are labeled with primes to distinguish them from the carbons on the bases. **(B)** The bases fall into two groups: the pyrimidines and the purines.

Activity 3.2 Nucleic Acid Building Blocks

Animation 3.3 Macromolecules: Nucleic Acids

nucleic acid chain, nucleic acids are said to *be synthesized in the 5′ to 3′ direction* (see Figure 3.9).

Nucleic acids vary in size (that is, the length of the polymer). Nucleic acids with a few to 20 monomers are called **oligonucleotides**. In living systems, oligonucleotides are primarily RNA molecules and act to regulate the synthesis of new DNA as well as help regulate the expression of information encoded in DNA. Informational nucleic acids involved in heredity are **polynucleotides**, which include the longest polymers in the living world. Some DNA molecules contain hundreds of millions of nucleotides.

LINK In the lab, synthetic DNA oligonucleotides are used in the polymerase chain reaction (PCR), which amplifies (that is, produces multiple copies of) a particular DNA sequence. See **Figure 9.13**.

FIGURE 3.9 Synthesis of a Nucleic Acid A reaction occurs between the hydroxyl group on the 3′ carbon of the growing nucleic acid chain and the first phosphate group on the 5′ carbon of the incoming nucleotide. The reaction adds a single nucleotide to the growing nucleic acid and releases a pyrophosphate molecule. The nucleic acid being synthesized here is RNA, with the sugar ribose (note —OH group on 2′ carbon).

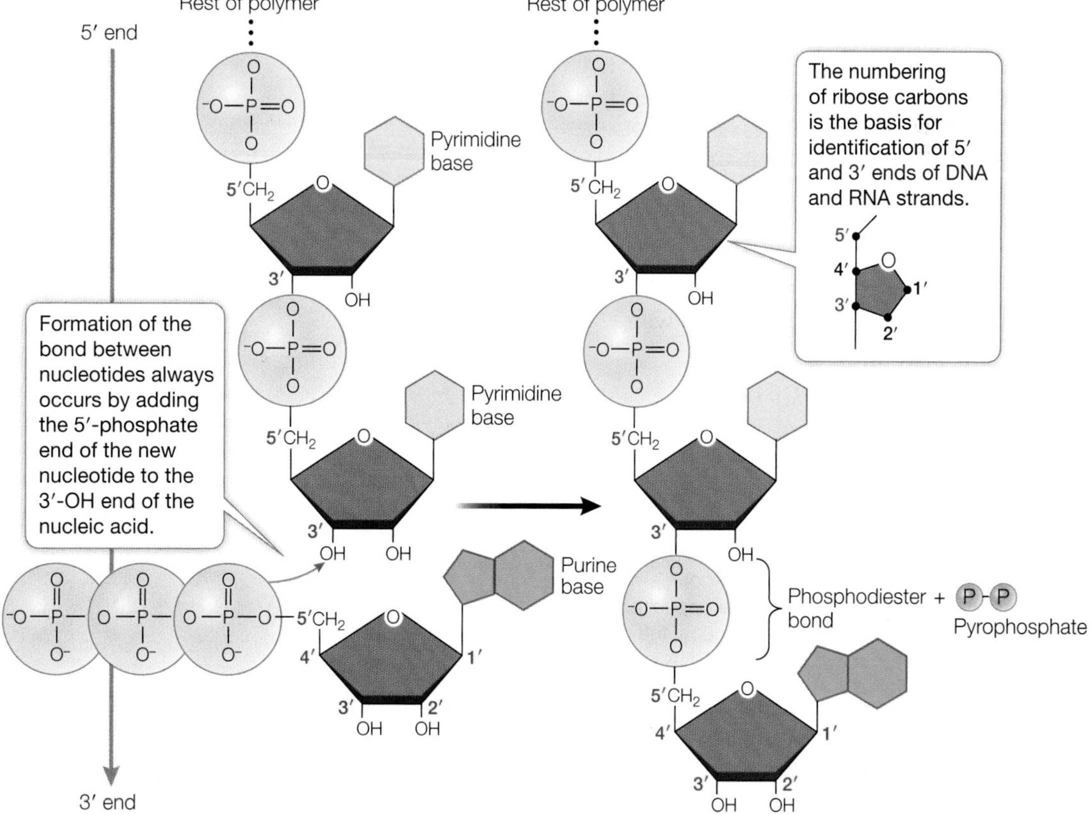

Base pairing occurs in both DNA and RNA

DNA and RNA also differ in that RNA is usually composed of a single oligonucleotide or polynucleotide strand, whereas DNA almost always contains two strands. The rules of **complementary base pairing** explain a great deal about the structure and function of both DNA and RNA. Some bases are able to "pair" with others through the formation of hydrogen bonds when two nucleic acid strands are physically close enough and in the appropriate orientation. In DNA adenine bonds with thymine, forming an A-T base pair, and cytosine bonds with guanine, forming a C-G base pair (**FIGURE 3.10**). In RNA the base pairs are A-U and C-G. Hydrogen bonds form between the partial negative charge (δ^-) on an oxygen or nitrogen atom of one base, and the partial positive charge (δ^+) on a hydrogen atom of another base (see Key Concept 2.2 to review hydrogen bonds).

Individual hydrogen bonds are relatively weak, but there are so many of them in DNA and RNA that collectively they provide a considerable force of attraction. However, this attraction is not as strong as that provided by multiple covalent bonds. This means that base pairs are relatively easy to separate with a modest input of energy. As you will see in Chapters 9 and 10, the breaking and making of hydrogen bonds in nucleic acids is vital to their roles in living systems. Let's now look in a little more detail at the structures of RNA and DNA.

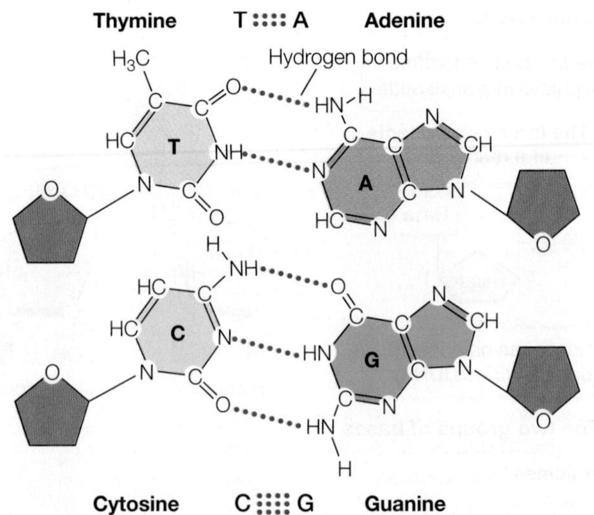

FIGURE 3.10 Base Pairing between Nucleotides Involves Hydrogen Bonding In DNA, thymine forms two hydrogen bonds with adenine, and cytosine forms three hydrogen bonds with guanine. In RNA (not shown), uracil replaces thymine and is able to form two hydrogen bonds with adenine.

Which requires more energy to pull apart, a G-C base pair or an A-T base pair?

DNA DNA is a double-stranded molecule—that is, it consists of two separate nucleic acid strands of the same length. The two strands are antiparallel: they run in opposite directions, with their 5′ ends at opposite ends of the double-stranded molecule (**FIGURE 3.11A**). DNA is a remarkably uniform molecule structurally. The A-T and G-C base pairs are about the same width (each is a double-ring purine paired with a single-ring pyrimidine), and the two polynucleotide strands form a "ladder" of constant thickness that twists into a double helix (**FIGURE 3.11B**). The sugar–phosphate groups form the sides of the ladder, and the bases with their hydrogen bonds form the rungs on the inside. The double helix is a right-handed helix (like the threads on a screw). A way of envisioning this is that if you hold DNA and look at it from one end, it twists clockwise.

Activity 3.3 DNA Structure

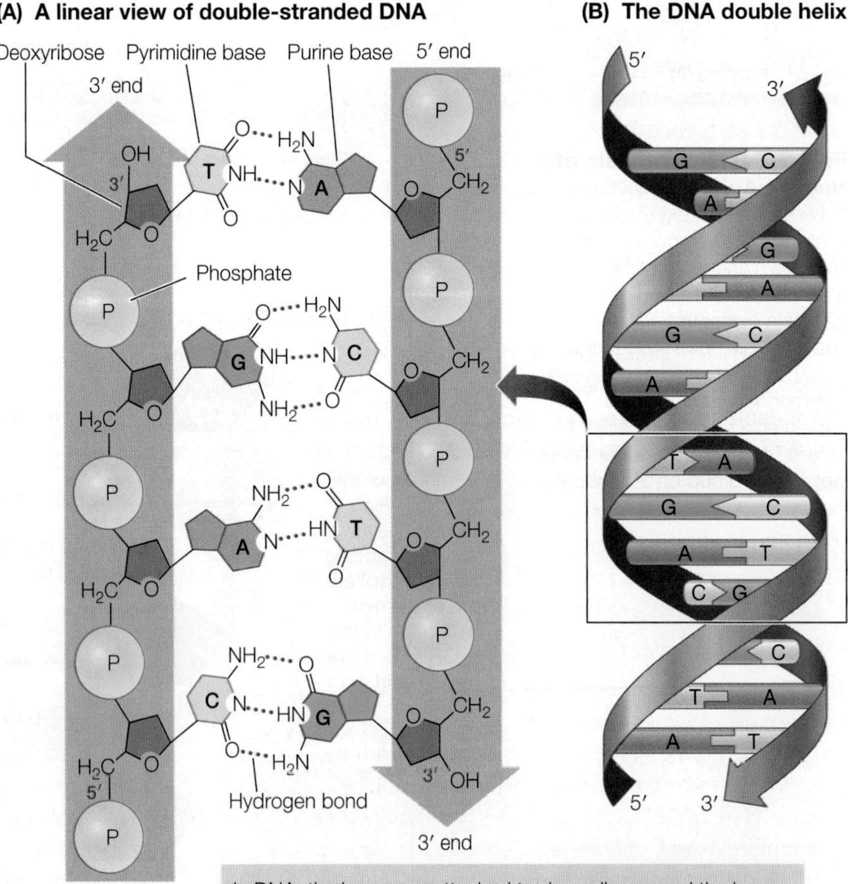

FIGURE 3.11 Structure of DNA **(A)** Antiparallel strands twist around one another and are held together by complementary base pairing. The molecule is a constant thickness because complementary base pairs always have one single-ring pyrimidine base and one double-ring purine base. Note that antiparallel strands have opposite directionality: one strand runs in the 5′-to-3′ direction, the other in the 3′-to-5′ direction. **(B)** The two strands in DNA are coiled in a right-handed double helix.

How is the information in a DNA molecule fully specified by determining the base sequence of either strand?

RNA RNA is usually a single-stranded molecule, though it often folds back on itself to form short double-stranded regions (**FIGURE 3.12**). The double-stranded regions are stabilized by complementary base pairing between nucleotides in separate parts of the molecule. The stabilized double-stranded regions give the molecule a distinctive three-dimensional shape that affects how it interacts with other molecules. It is important to realize that this folding occurs by complementary base pairing, and that the structure is thus determined by the particular order of bases in the RNA molecule. Notice that the regions that exhibit base pairing are antiparallel, as in DNA.

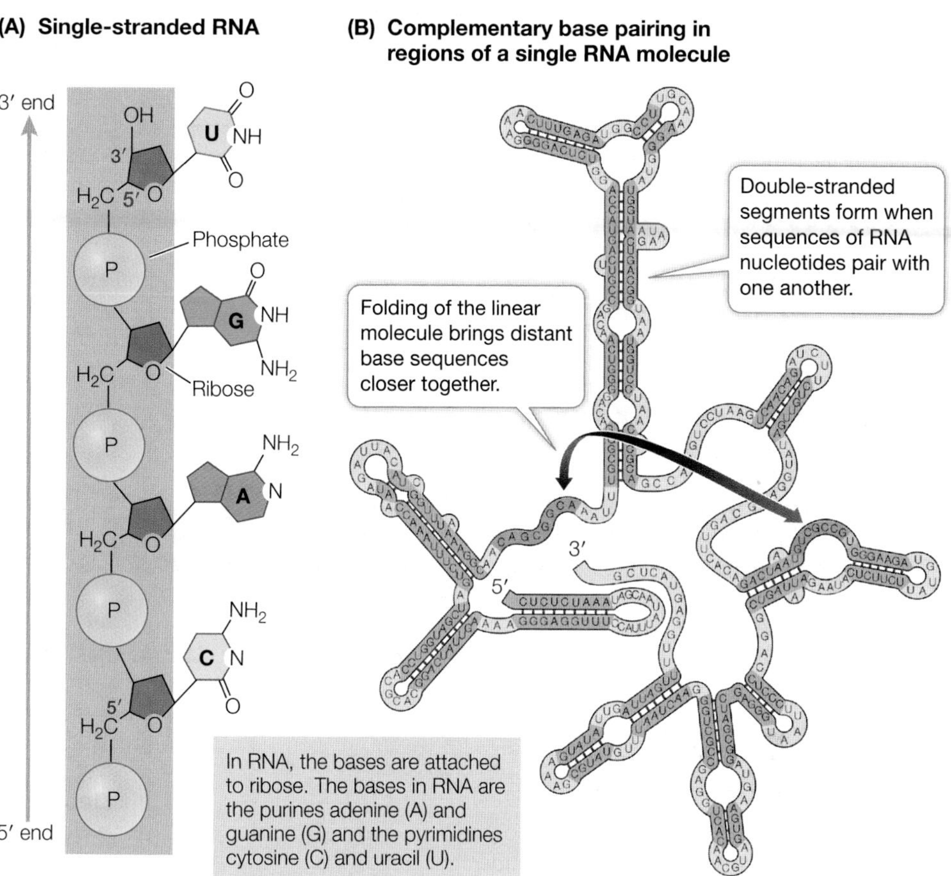

FIGURE 3.12 Structure of RNA **(A)** RNA is usually a single-stranded molecule. **(B)** Complementary base pairing can occur between sequences of nucleotides on an individual RNA molecule. A single-stranded RNA may fold back on itself, aligning complementary sequences that can be linked by hydrogen bonds, giving the molecule a distinct three-dimensional shape.

DNA carries information and is expressed through RNA

DNA is a purely informational molecule. The information is encoded in the sequence of bases carried in its strands. For example, the information encoded in the sequence 5′-TCAGCA-3′ (note that 5′ and 3′ indicate the polarity of the sequence) is different from the information in the sequence 5′-CCAGCA-3′. Because of complementary base pairing, all of the sequence information carried in one strand is also carried in the other. Thus knowing the sequence of nucleotides on one strand allows perfectly accurate reconstruction of the sequence on the other strand.

The preservation of the information in DNA is accomplished through **DNA replication**, a process by which DNA is accurately copied into a new, identical DNA molecules that can be transmitted to daughter cells and offspring. How the information in DNA is used to synthesize proteins involves two processes: transcription and translation. **Transcription** is the process by which information encoded in DNA base sequences is used to synthesize RNA. **Translation** is the process by which the information in RNA base sequences is used to synthesize proteins. The details of these processes are described in Chapters 9 and 10, but it is important to realize three points here:

1. *DNA replication and transcription depend on the base-pairing properties of the nucleotides.* In both replication and transcription, complementary base pairing to a DNA template strand is used to synthesize the new nucleic acid molecule. The resulting new DNA or RNA strand is *complementary* to the existing DNA template strand. One implication of this is that the new strand will be in the opposite orientation as the template strand; if the template strand is 3′ to 5′ reading from left to right, then the new strand will be synthesized 5′ to 3′ going from left to right.
2. *The entire DNA molecule is copied during DNA replication.* Since DNA holds essential information, it must be replicated completely so that each new cell or new offspring receives a complete set of DNA from its parent (**FIGURE 3.13A**).
3. *Gene expression is the transcription of specific DNA sequences into complementary RNA.* Sequences of DNA that are transcribed into RNAs that are themselves functional, or that represent a template for protein synthesis, are called **genes** (**FIGURE 3.13B**). The complete set of DNA in a living organism is called its **genome**. However, not all DNA in the genome encodes genes, and not all genes are expressed to make protein. For example, in humans the gene that encodes the major protein in hair (keratin) is expressed only in skin cells. The genetic information in the keratin-encoding gene is transcribed into RNA and then translated into the protein keratin in these cells. In other tissues, such as the muscles, the keratin gene is not transcribed but other genes are—for example, the genes that encode proteins present in muscles but not in skin.

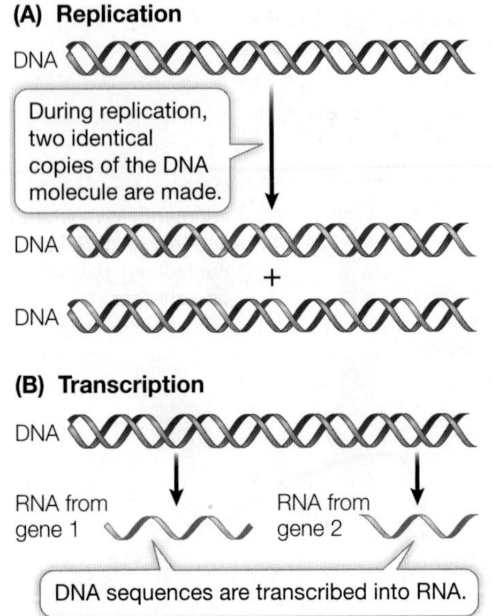

FIGURE 3.13 DNA Replication and Transcription **(A)** The entire DNA molecule is replicated. Each of the two original strands of the DNA molecule (blue) is used as a template to synthesize a new strand (red). **(B)** Only a subset of DNA sequences are transcribed at a given time. Transcription can occur from either strand of the DNA molecule.

How will the synthesized RNA molecule differ from the DNA strand used as a template?

Proteins are the most complex of the macromolecules—they are composed of different combinations of twenty amino acids, each of which has different structural and chemical properties. This diversity of their amino acid residues allows proteins to have enormous structural diversity.

REVIEW & APPLY | 3.3

R The nucleic acids DNA and RNA are polymers made up of nucleotide monomers. Each nucleotide consists of a pentose monosaccharide sugar (deoxyribose or ribose), a phosphate group, and one of five nitrogen-containing bases (C, T, A, and G in DNA; C, U, A, and G in RNA). Phosphodiester bonds join nucleotides together in strands of DNA and RNA, forming the sugar–phosphate backbone of each molecule. Complementary base pairing holds the antiparallel strands of DNA or RNA together. The sequence of nucleotides in a DNA strand carries the information that is used by RNA to encode proteins. The genetic information in DNA is passed from generation to generation in cells and organisms.

A
1. What are the components of a nucleotide found in DNA?
2. Fill in the following table.

Nucleic acid	Sugar in nucleotide	Bases present	Single or double stranded?
RNA			
DNA			

3. What are the differences among DNA replication, transcription, and translation?

3.4 | Proteins Are Polymers with Variable Structures

LEARNING OBJECTIVES

By the end of this key concept you should be able to:

3.4.1 **Describe how amino acids differ in their side chains (R groups).**

3.4.2 **Describe how amino acids are joined together in a peptide.**

3.4.3 **Distinguish primary, secondary, tertiary, and quaternary structure of proteins.**

3.4.4 **Define denaturation and list factors that can destabilize protein structure.**

Polypeptides are the fourth and final type of biological macromolecule we will discuss in this chapter. In terms of structural and functional diversity, they are at the top of the list. Compared with proteins, the lipids, carbohydrates, and nucleic acids are relatively inactive in cells. Proteins, by contrast, are the real workhorses. Their activities are largely responsible for the diverse biochemistry that makes living organisms so interesting. Here we examine the varied structures of proteins, and in the next key concept we'll discuss how their structures relate to their varied functions.

Amino acids are the building blocks of proteins

Proteins are polymers made up of tens to tens of thousands of monomers called **amino acids**. As their name suggests, the amino acids contain two functional groups: the nitrogen-containing amino group (—NH_2) and the (acidic) carboxyl group (—COOH). In most cells, ionization of amino acids causes the —COOH group to release a proton (H^+) to form —COO^-, and the —NH_2 group to gain a proton to become —NH_3^+. Thus amino acids have both acidic and basic properties (see Key Concept 2.5).

The central carbon atom of an amino acid—the α (alpha) carbon—has four available electrons for covalent bonding. In all amino acids, two of the electrons are occupied by the two functional groups noted above, and a third is occupied by a hydrogen atom. The fourth bonding electron is shared with the **R group**, or **side chain**, that differs in each amino acid. Each amino acid is identified by its R

(A) Amino acids with electrically charged hydrophilic side chains

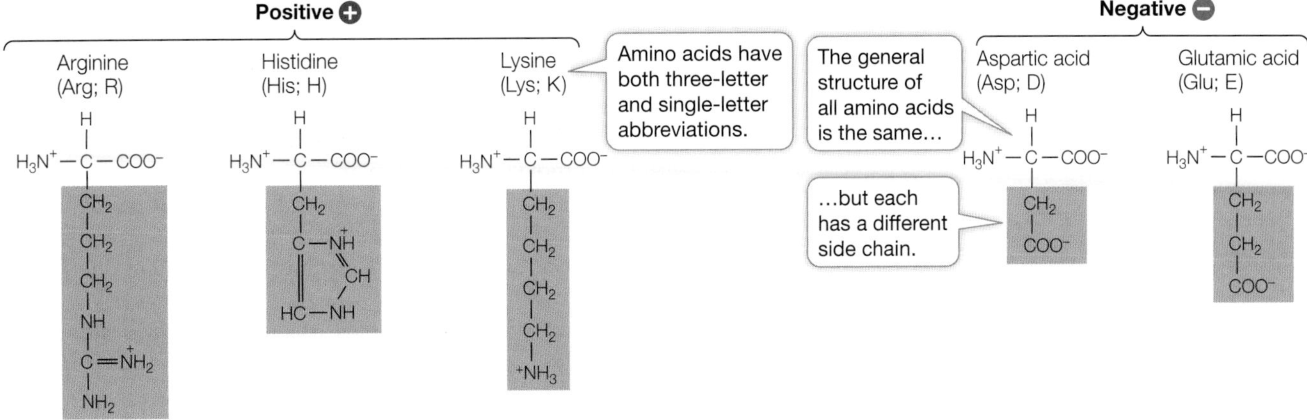

(B) Amino acids with polar but uncharged side chains (hydrophilic)

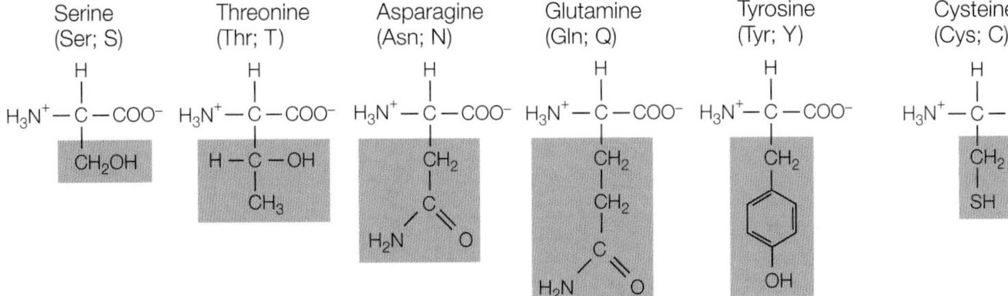

(C) Special cases

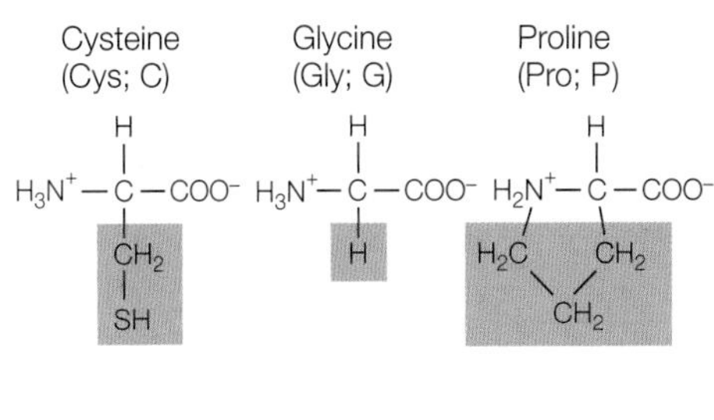

(D) Amino acids with nonpolar hydrophobic side chains

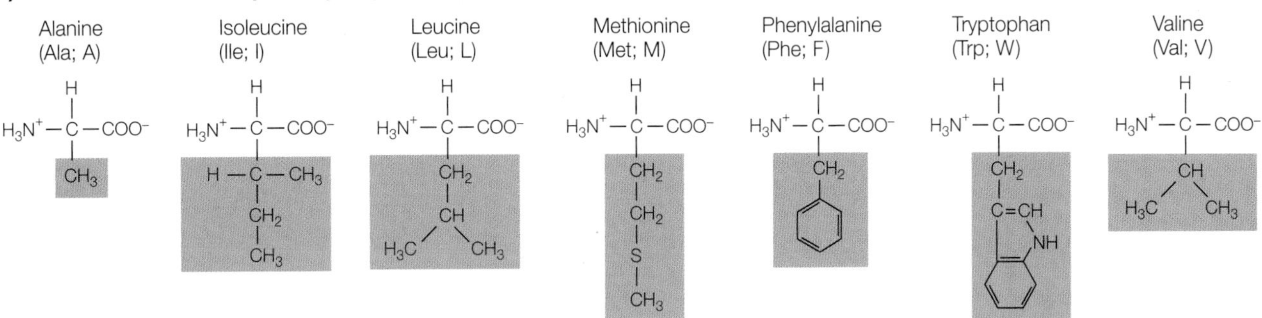

FIGURE 3.14 The Twenty Common Amino Acids
The R group for each amino acid appears in a shaded box.

Animation 3.4 Macromolecules: Proteins

group. Twenty amino acids occur extensively in the proteins of all organisms. These 20 amino acids can be grouped according to the properties conferred by their side chains (**FIGURE 3.14**):

- Five amino acids have electrically charged side chains (+1 or –1), attract water (are hydrophilic), and attract oppositely charged ions.
- Five amino acids have polar side chains (δ^+, δ^-) and tend to form hydrogen bonds with water and other polar or charged substances. These amino acids are also hydrophilic.
- Seven amino acids have side chains that are nonpolar hydrocarbons or very slightly modified hydrocarbons. In the watery environment of the cell, these hydrophobic side chains may cluster together in the interior of the protein, or interact with lipids in membranes.

Three amino acids—glycine, proline, and cysteine—are special cases, although the side chains of the former two generally are hydrophobic:

- The glycine side chain consists of a single hydrogen atom and is small enough to fit into tight corners in the interior of a protein molecule, where a larger side chain could not fit. The single hydrogen side chain allows the protein to be more flexible at sites where glycine occurs.
- Proline possesses a modified amino group that lacks a hydrogen atom and instead forms a covalent bond with the hydrocarbon side chain, resulting in a ring structure. This limits both its hydrogen-bonding ability and its ability to rotate. Thus proline often functions to stabilize bends or loops in proteins.

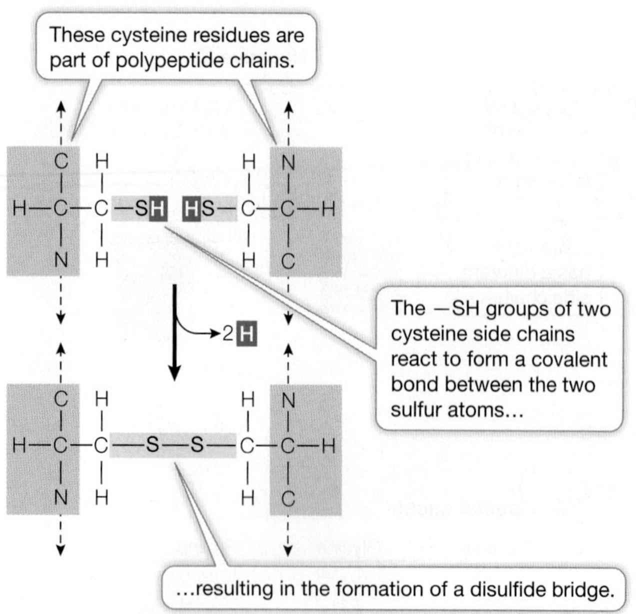

FIGURE 3.15 Formation of a Disulfide Bridge The side chains on two cysteine amino acids can form a covalent disulfide bridge, linking different parts of the same protein or different proteins together.

- The cysteine side chain, which has a terminal —SH group, can react with another cysteine side chain elsewhere in the same molecule, or on another molecule, to form a covalent bond called a disulfide bridge, or disulfide bond (—S—S—). Disulfide bridges help stabilize the three-dimensional structure of proteins (**FIGURE 3.15**). Note that most cysteines in proteins are not involved in disulfide bridges.

The 20 amino acid side chains differ in their charge, polarity, size, shape, and functional groups. Proteins thus have enormous structural diversity and specificity of interactions with other molecules and ions.

Activity 3.4 Features of Amino Acids

Amino acids are linked together by peptide bonds

Amino acids can form short polymers of 20 or fewer amino acids, called **oligopeptides** or simply **peptides**. These include some hormones and other molecules involved in signaling from one part of an organism to another. Even with their relatively short chains of amino acids, oligopeptides have distinctive three-dimensional structures.

More common are the longer polymers called **polypeptides**, each with a unique sequence of amino acids. A functional protein may be made up of one or more polypeptides. Proteins range in size from small ones such as insulin, which has 51 amino acids, to huge molecules such as the muscle protein titin, with approximately 30,000 amino acids.

Like nucleic acids, oligopeptides and polypeptides form via the sequential addition of new amino acids to the ends of existing chains. The amino group of the incoming amino acid reacts with the carboxyl group of the amino acid at the end of the peptide chain. This condensation reaction forms a **peptide bond** (**FIGURE 3.16**). Note that there is directionality in the addition of each amino acid to the growing chain, just as there is with the addition of nucleotides to a growing nucleic acid chain. In the case of oligopeptides and polypeptides, *polymerization takes place in the amino to carboxyl direction.*

Activity 3.5 Macromolecule Building Simulation

The precise sequence of amino acids is the **primary structure** of a protein. Scientists have determined the primary structures of many proteins. The theoretical number of different proteins is enormous since there are 20 different amino acids that could be present at every position, and proteins usually contain hundreds of amino acids.

LINK **Key Concept 10.3** discusses the rules of translation, which allows the amino acid sequence (primary structure) of a protein to be determined from its RNA (and thus DNA) nucleotide sequence.

Higher-level protein structure is determined by primary structure

The primary structure of a protein is established by covalent peptide bonds, but higher levels of structure are determined largely by noncovalent bonds, including hydrogen bonds, ionic bonds and other

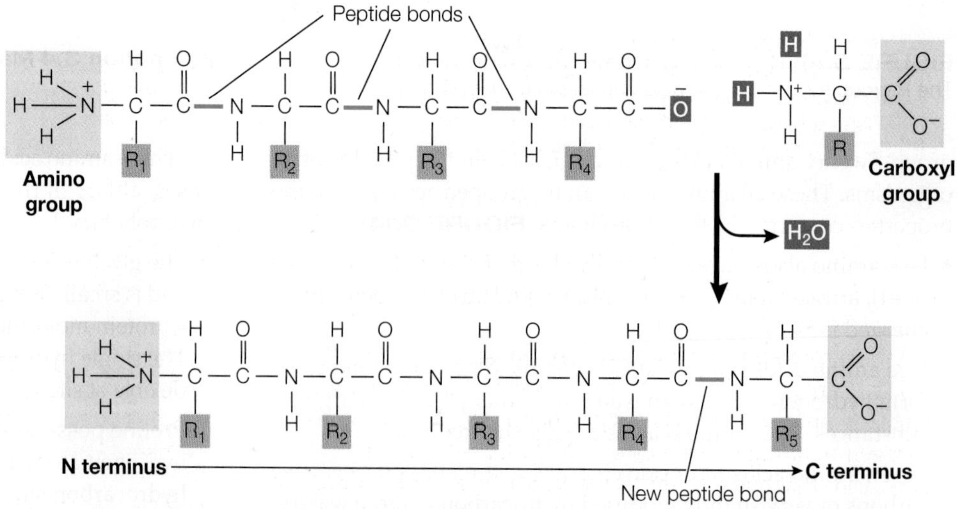

FIGURE 3.16 Peptide Bond Formation Peptide bonds are formed by condensation reactions in which the amino group of one amino acid reacts with the carboxyl group of another amino acid or peptide, releasing a molecule of water. Here a tripeptide is extended to a tetrapeptide with the formation of a peptide bond.

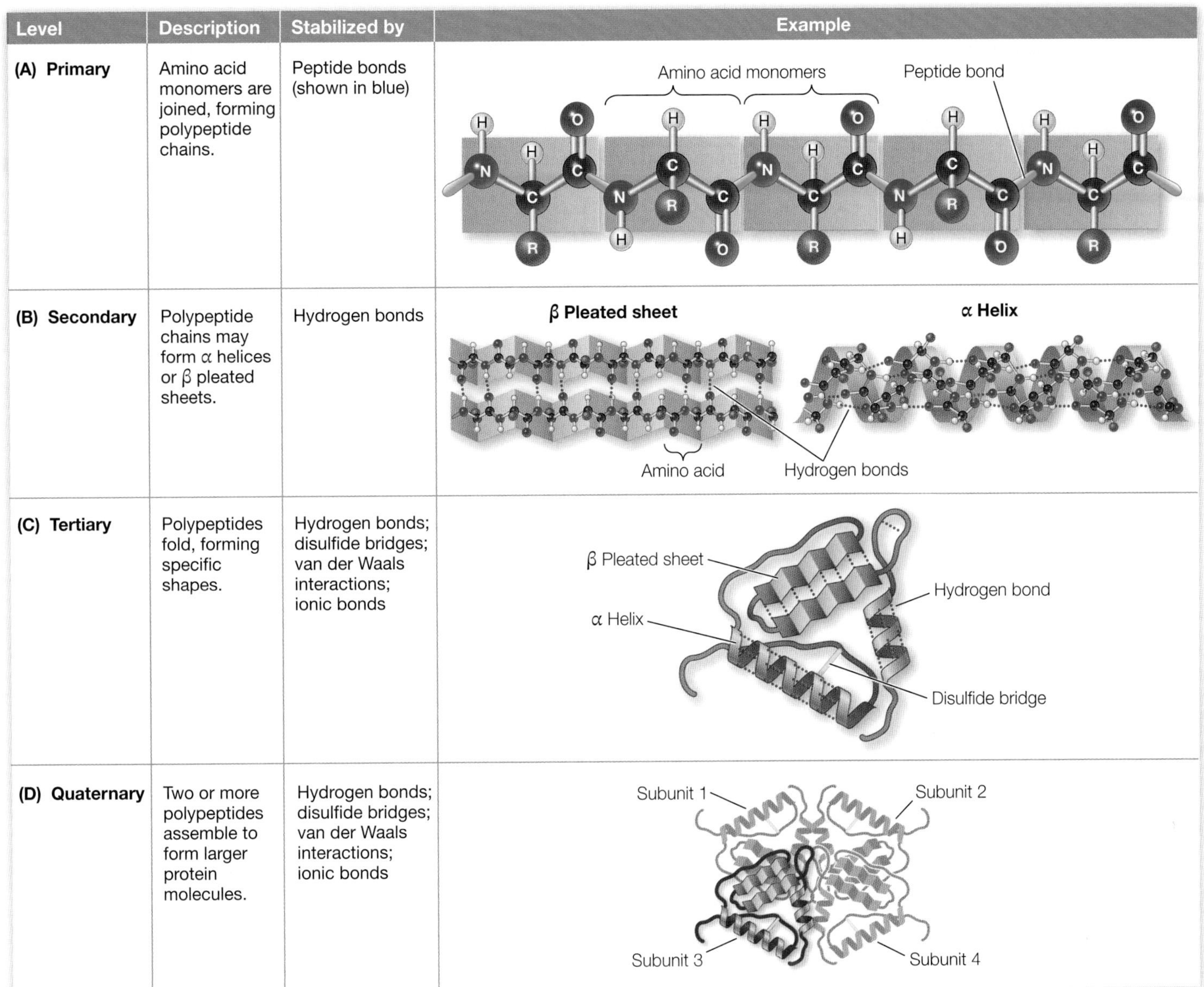

Level	Description	Stabilized by	Example
(A) Primary	Amino acid monomers are joined, forming polypeptide chains.	Peptide bonds (shown in blue)	Amino acid monomers; Peptide bond
(B) Secondary	Polypeptide chains may form α helices or β pleated sheets.	Hydrogen bonds	β Pleated sheet; α Helix; Amino acid; Hydrogen bonds
(C) Tertiary	Polypeptides fold, forming specific shapes.	Hydrogen bonds; disulfide bridges; van der Waals interactions; ionic bonds	β Pleated sheet; α Helix; Hydrogen bond; Disulfide bridge
(D) Quaternary	Two or more polypeptides assemble to form larger protein molecules.	Hydrogen bonds; disulfide bridges; van der Waals interactions; ionic bonds	Subunit 1; Subunit 2; Subunit 3; Subunit 4

FIGURE 3.17 The Four Levels of Protein Structure **(A)** The sequence of amino acids in each polypeptide is its primary structure. **(B)** The primary structure determines whether regions of the polypeptide will form α helices or β pleated sheets, which are the two common secondary structures and are stabilized by hydrogen bonds between the —NH and —CO groups in the backbone of the amino acid chain. **(C)** Additional folding creates the three-dimensional tertiary structure of a polypeptide. **(D)** Multiple identical or different polypeptides may come together to form the quaternary structure of a mature protein.

electrostatic interactions, and van der Waals interactions. Follow **FIGURE 3.17** as we describe how a protein chain becomes a three-dimensional structure.

A protein's **secondary structure** consists of regular, repeated spatial patterns in different regions of a polypeptide chain. There are two common types of secondary structure, both determined by hydrogen bonding between the —NH and —CO groups of the amino acids that make up the primary structure (see Figure 3.17B):

- The **α (alpha) helix** is right-handed like the threads on a screw. The R groups extend outward from the peptide backbone of the helix. The coiling results from hydrogen bonds that form between the —NH group on one amino acid and the —CO group on another within the same turn of the helix.
- The **β (beta) pleated sheet** is formed from two or more sequences of amino acids that are extended and aligned in the polypeptide. The sheet is stabilized by hydrogen bonds between the —NH groups and the —CO groups on the two chains, which may be in the same or opposite directions.

While secondary structure is common in proteins, with most containing α helices, β pleated sheets, or both in different regions of the same polypeptide chain, a few proteins contain neither and thus lack secondary structure. Particular amino acids tend to be absent, or common, in secondary structures. For example, proline, with its unusual ring structure and relative inflexibility, is not found in α helices, and amino acids with bulky side chains, such as tryptophan, are more likely to be in β pleated sheets.

Tertiary structure of a protein arises from bending and folding of the polypeptide chain, which results in a three-dimensional structure (see Figure 3.17C). Tertiary structure includes a buried interior as well as a surface that is exposed to the environment. The functional groups on the amino acids present on a protein's exposed outer surfaces are capable of interacting with other molecules in the cell (**FIGURE 3.18**).

Whereas hydrogen bonding between the —NH and —CO groups within and between chains is responsible for a protein's secondary structure, it is the interactions between R groups—the amino acid side chains—that determine tertiary (and quaternary) structure. These interactions include:

- *Covalent disulfide bridges* between cysteines.
- *Ionic interactions* between charged side chains. These include ionic bonds between positive and negative charges on side chains. For example, arginine (which has a positively charged R group) and glutamic acid (which has a negatively charged R group) can stabilize a particular three-dimensional structure. Conversely, repulsion of like-charged side chains can prevent incorrect three-dimensional structures from forming.
- *Hydrogen bonds* between side chains.
- *van der Waals interactions,* which stabilize associations between hydrophobic side chains.

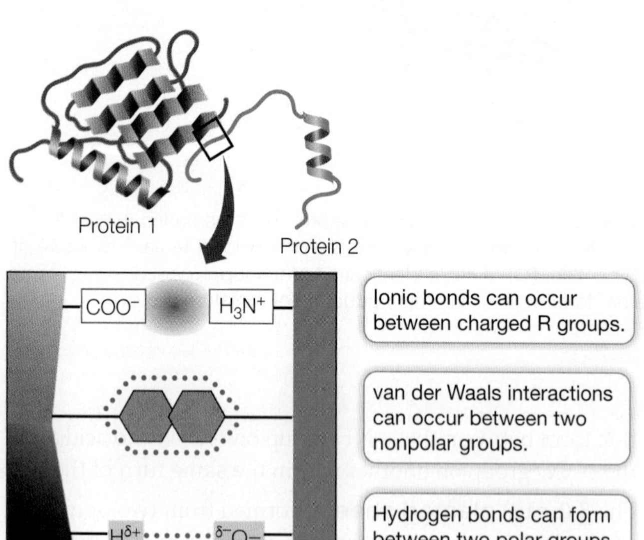

FIGURE 3.18 Noncovalent Interactions between Proteins and Other Molecules Noncovalent interactions allow a protein (brown) to bind to another protein (blue), or to another molecule, with specific properties. Noncovalent interactions also allow regions within a single protein to interact with one another.

For the hydrogen atom shown in the hydrogen bond, to what atom is it likely covalently bonded?

LINK To review the strong and weak interactions that can occur between atoms, see **Key Concept 2.2**.

Many functional proteins contain two or more polypeptide chains, called subunits, each folded into its own unique tertiary structure. The protein's **quaternary structure** results from the ways in which these subunits bind together and interact (see Figure 3.17D). All of the forces listed above can contribute to holding subunits together in the correct structural orientation. The β-pleated sheet secondary structure may form between separate polypeptide chains in the quaternary structure. Some proteins that are functional as a single polypeptide can form very large aggregations due to β sheet formation between polypeptides, which can interfere with the proper functioning of these proteins. Such misfolding and aggregation are implicated in the human nervous system abnormalities Alzheimer's disease and Parkinson's disease.

Protein structure can change depending on the environment

Both secondary and tertiary structure derive from primary structure. If a protein is heated slowly, the thermal energy input will disrupt the weaker interactions and break hydrogen bonds, causing the secondary, tertiary, and (if present) quaternary structure to break down. A protein is said to be **denatured** when its molecular conformation is so disrupted that it loses its characteristic properties. While heat will eventually break disulfide bridges, chemical treatment may be required to fully denature proteins with disulfide bridges. In many cases a denatured protein will return to its normal tertiary structure when it cools or when the denaturing chemicals are removed, demonstrating that all the information needed to specify the protein's unique shape is contained in its primary structure. **INVESTIGATION FIGURE 3.19** reviews the famous experiment by biochemist Christian Anfinsen showing that proteins will spontaneously fold into their correct three-dimensional structure as long as their primary structure is not disrupted. Denaturation can be reversible, but it can also be irreversible, especially if it allows the protein to bond to other (denatured) polypeptides to which it was not bonding before. For example, we all know that boiling an egg denatures its (albumin) proteins and is not reversible. When we boil an egg, the originally separate albumin polypeptides end up sticking to one another through β pleated sheets, other hydrogen bonding, and new disulfide bridges between molecules.

Several environmental factors can destabilize protein structure. Changes in pH alter the ionization state of amino acid side chains and thus their charge, altering their ionic interactions and thus affecting the structure of the polypeptide. High concentrations of polar solutes such as urea, $CO(NH_2)_2$, can disrupt hydrogen bonding and destabilize structure. Nonpolar substances may also denature a protein in cases where van der Waals interactions between hydrophobic groups are essential for maintaining the protein's structure.

Media Clip 3.1 Protein Structures in 3D

INVESTIGATION

FIGURE 3.19 Primary Structure Specifies Tertiary Structure Using the protein ribonuclease, Christian Anfinsen showed that proteins spontaneously fold into functionally correct three-dimensional configurations.[a,b] As long as the primary structure is not disrupted, the information for correct folding (under the right conditions) is retained.

HYPOTHESIS

Under controlled conditions that simulate the normal cellular environment, a denatured protein can refold into a functional three-dimensional structure.

METHOD

Chemically denature a functional ribonuclease (an enzyme that catalyzes the hydrolysis of RNA) so that only its primary structure (i.e., an unfolded polypeptide chain) remains. Once denaturation is complete, remove the disruptive chemicals.

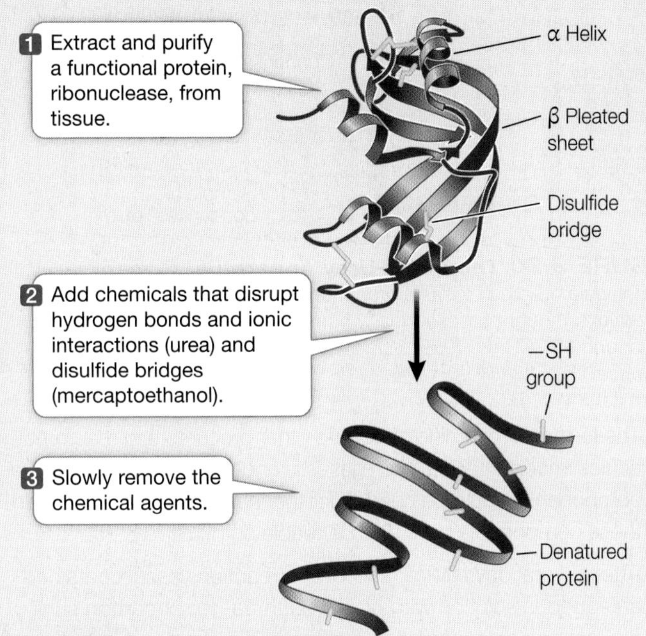

RESULTS

When the disruptive agents are removed, three-dimensional structure is restored and the protein once again is functional.

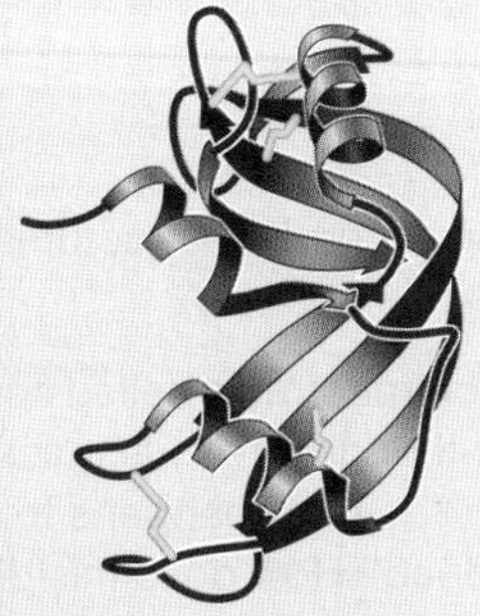

CONCLUSION

In normal cellular conditions, the primary structure of a protein specifies how it folds into a functional, three-dimensional structure.

ANALYZE THE DATA

After the tertiary structures of proteins were shown to be highly specific, the question arose as to how the order of amino acids determined the three-dimensional structure. The second protein for which structure was determined was ribonuclease A (RNase A). This enzyme was readily available from cow pancreases at slaughterhouses and, because it works in the highly acidic environment of the cow stomach, was stable compared with most proteins and easy to purify. RNase A has 124 amino acids. Among these are eight cysteine residues, which form four disulfide bridges. Were these covalent bonds between cysteines essential for the three-dimensional structure of RNase A? As outlined above, Christian Anfinsen and his colleagues set out to answer this question.

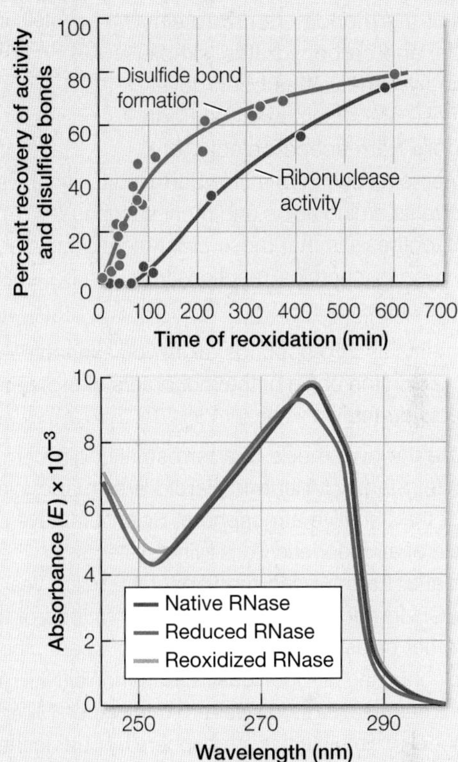

1. Initially, the disulfide bonds (S—S) in RNase A were all broken (reduced) to give sulfhydryl groups (—SH) on the cysteines, by the addition of mercaptoethanol and urea. The denatured protein was purified and placed into normal cellular (reoxidation) conditions at time zero. At various times, the amount of S—S bond re-formation and the activity of the enzyme were measured by chemical methods. The data are shown in the top graph. At what time did disulfide bonds begin to form? At what time did enzyme activity begin to appear? Explain the difference between these times.
2. The three-dimensional structure of RNase A was examined by ultraviolet spectroscopy. In this technique, the protein was exposed to different wavelengths of ultraviolet light (measured in nanometers) and the amount of light absorbed by the protein at each wavelength was measured (*E*). The results are plotted in the bottom graph. Look carefully at the plots. What are the differences between the peak absorbances of native (untreated) and reduced (denatured) RNase A? What happened when reduced RNase A was reoxidized (renatured)? What can you conclude about the structure of RNase A from these experiments?

[a] C. B. Anfinsen et al. 1961. *Proc Natl Acad Sci USA* 47: 1309–1314.

[b] F. White, Jr. 1961. *J Biol Chem* 236: 1353–1360.

A similar **Analyze the Data** exercise may be assigned in **Achieve**.

THINK LIKE A SCIENTIST

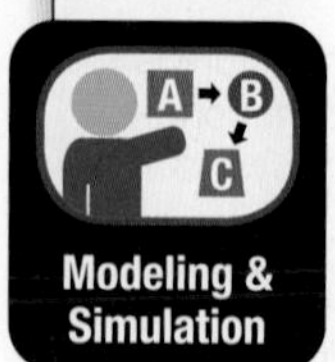

The origin of the molecules of life on Earth

In Chapter 2 we described evidence for water, a prerequisite for life, on extraterrestrial bodies such as comets and asteroids, and the possibility that these contributed to the origin of water on Earth. The same types of observations have been made for the precursors of the monomers of proteins (amino acids) and carbohydrates (simple sugars). The hypothesis has been developed that these precursors reached Earth as passengers from space objects, and then chemical reactions in primitive Earth led to what we now recognize as amino acids and monosaccharides. An alternative hypothesis is that these precursors originated on Earth, derived by chemical reactions from components in the primitive atmosphere.

1. What elements and compounds might you expect to be important in the origin of the monomers of proteins and carbohydrates?

To investigate the hypothesis of a terrestrial origin of monomers needed for life, Stanley Miller and Harold Urey made a laboratory simulation of the primitive atmosphere. Scientists have proposed that the origin of atmospheric O_2 is living systems. So the primitive atmosphere after Earth was formed about 4.5 billion years ago did not contain O_2. In addition, based on observations of the atmospheres of other celestial bodies as well as of space, the primitive atmosphere contained nitrogen gas (N_2), ammonia (NH_3), hydrogen gas (H_2), and methane (CH_4), as well as water vapor. Miller and Urey created a model to simulate this atmosphere by putting these gases into a closed glass system (**FIGURE 3.20**).

Animation 3.5 **Synthesis of Prebiotic Molecules**

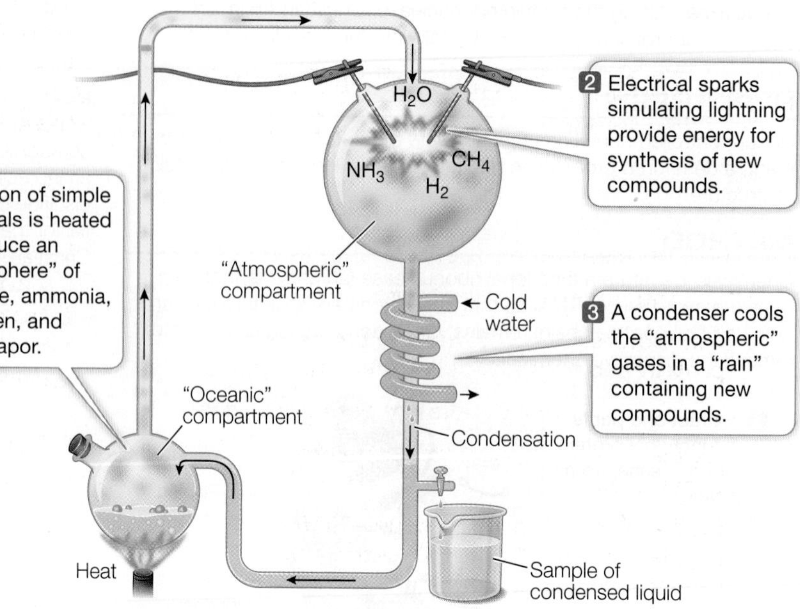

FIGURE 3.20 The Miller-Urey Apparatus Some macromolecule building blocks are spontaneously synthesized under conditions thought to reflect early conditions on Earth. (After S. L. Miller and H. C. Urey. 1959. *Science* 130: 245–251; S. L. Miller. 1953. *Science* 117: 528–519.)

2. The formation of amino acids and other precursors of macromolecules important to life from gases is endergonic. What additional components would be needed for the reactions to occur? Would these components have been available on primitive Earth?
3. After several days, Miller and Urey's reaction vessel contained an aqueous solution with dissolved three- to six-carbon sugars, amino acids, and even purines and pyrimidines. What can you conclude from the observations of the apparatus after several days?

REVIEW & APPLY | 3.4

R Protein structure can be described at four levels. The primary structure is the order of amino acids covalently linked by peptide bonds to form polypeptide chain(s). Each amino acid has the same general structure, with amino and carboxylic acid functional groups, and a variable R group that confers specific properties to that amino acid. Secondary structure involves the folding (β pleated sheet) and twisting (α helix) of the polypeptide. Tertiary structure involves the three-dimensional folding of the polypeptide into a specific shape. Most proteins also have a quaternary structure, whereby several polypeptides interact to form the functional unit. Both tertiary and quaternary structure are stabilized by hydrogen bonding, ionic bonding, van der Waals interactions, and covalent bonding (disulfide bridges) between R groups.

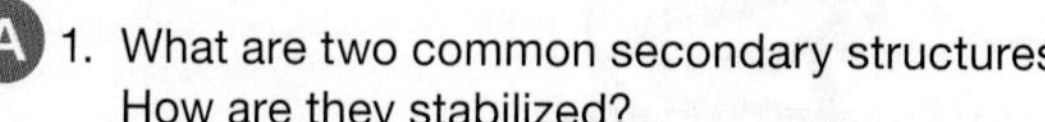

A
1. What are two common secondary structures in proteins? How are they stabilized?
2. Hair is made of a structural protein called keratin. Ironing dry hair straightens it out, though it recurls quickly in a humid environment. What do you think happens to the keratin molecules when hair is ironed? How does this differ from what happens if hair is chemically treated to make it straight?
3. Suppose the amino acid glutamic acid (see Figure 3.14A) is at a location that is critical for stabilizing tertiary structure. At pH 7, the protein is in its normal tertiary structural configuration. However, at pH 4, the protein is no longer in its normal tertiary structural configuration. Explain why at pH 4 the structure changes.
4. What factors can denature proteins?

The diversity of structures we have just discussed give proteins an enormous diversity of function as we will see in the next concept.

3.5 The Function of a Protein Is Determined by Its Structure

LEARNING OBJECTIVES

By the end of this key concept you should be able to:

3.5.1 **List the functions of proteins.**

3.5.2 **List factors that can alter the shape of a protein and thus affect its function.**

3.5.3 **Define and give examples of cofactors.**

3.5.4 **Define an enzyme and describe how an enzyme works.**

3.5.5 **Distinguish active site versus allosteric regulation, and define competitive and irreversible inhibition.**

Proteins are involved in essentially all of the processes that occur in all organisms. The list below gives you an idea of the vast array of protein functions:

- **Enzymes** are catalytic molecules that increase the rate of biochemical reactions.
- **Structural proteins** and **motor proteins** provide physical stability and enable movement.
- **Signal proteins** and **regulatory proteins** control the rates of numerous biological processes.
- **Receptor proteins** receive and respond to molecular signals from inside and outside the organism.
- **Transport proteins** carry substances within the organism and across biological membranes.
- **Defensive proteins** recognize and respond to substances or particles that invade the organism from the environment.
- **Storage proteins** store amino acids and some other substances for later use.

We will meet examples of all of these proteins in many of the chapters of this textbook. Here we discuss general properties that apply to several different classes of proteins and then go into some detail about enzymes.

A protein's function is determined by its binding characteristics

Within a living organism, a protein may bind to other proteins, other kinds of macromolecules, or a variety of smaller molecules, including ions. These interactions occur at specific sites on the protein. If a polypeptide comes into contact with another molecule, amino acid side chains on its surface may form weak bonds with functional groups on the surface of the other molecule, or with ions. The three-dimensional shape of the protein determines which amino acid side chains will interact with the substance being bound. The identity of the side chains then determines the number and kinds of bonds that can occur. *The binding is always very specific*—typically just one or a few molecules or ions can bind to a particular protein. In addition to being modified by environmental factors, as we discussed in Key Concept 3.4, the shape of a protein and its chemical properties can also be modified in several other ways.

LIGAND BINDING A **ligand** is a molecule or ion that binds to another molecule, in this case to a protein. Ligand binding is specific and can involve several weak bonds, which together make for a relatively strong interaction. The strength of the interaction is termed the **binding affinity**; higher affinity values are associated with more specific binding. Binding can cause a change in the shape of a protein, a so called **conformational change**, which can affect the protein's function (**FIGURE 3.21A**). Ligands are usually signals that bind to receptors.

SIDE CHAIN MODIFICATION The structure of a protein can be modified by the covalent bonding of a chemical group to the side chain of one or more of its amino acids. The chemical modification of just one amino acid can alter the shape and function of a protein. For example, the addition of a charged phosphate group to an uncharged R group can cause the amino acid to move to the outer surface of the protein to interact with water molecules, altering the shape of the protein in the region near the amino acid (**FIGURE 3.21B**). Side chain modification is important in enzyme regulation, as discussed below.

COFACTORS Many proteins require an additional organic, nonprotein molecule or an inorganic ion, often a metal ion, in order to function (**FIGURE 3.21C** and **TABLE 3.1**). These cofactors vary in how tightly they bind to the protein. Tightly bound cofactors include those that are covalently bound to the protein, in which case they are called prosthetic groups. Loosely bound cofactors often become chemically modified and leave the protein in order to be returned to their initial state. The adenosine triphosphate ribonucleotide (ATP) cofactor is an excellent example, often becoming converted to ADP during a reaction on one protein and then being converted back to ATP during a reaction on another protein. Loosely bound cofactors that bind to enzymes are often termed coenzymes.

PROTEOLYSIS Some proteins are nonfunctional because they are too long when first synthesized. To become functional, a part of the protein must be removed by breaking a particular peptide bond in a hydrolysis reaction—a process called **proteolysis**. Proteolysis only occurs under situations when the functional protein is required, allowing it to be synthesized but not activated until the appropriate time and location (**FIGURE 3.21D**). In some proteins, proteolysis does not result in activation but instead produces an inactive protein. Whether the result is activation or inactivation depends on the protein involved.

You will see many examples in subsequent chapters of proteins that undergo changes in structure or chemical modifications that alter their functional properties. Indeed, organisms would cease to live if they just kept their proteins as they are. Proteins are dynamic molecules, and nowhere is this better shown than in the case of enzymes.

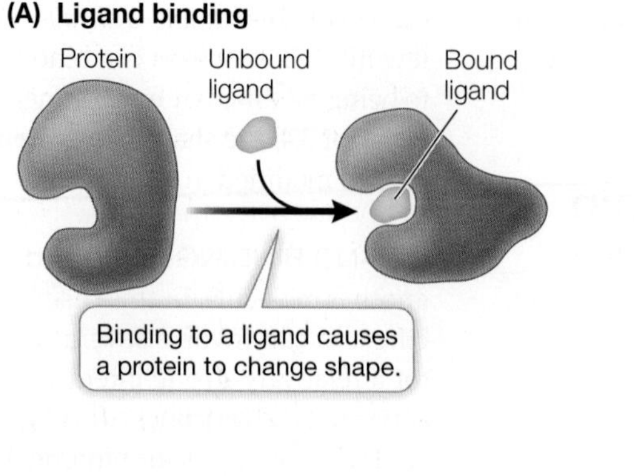

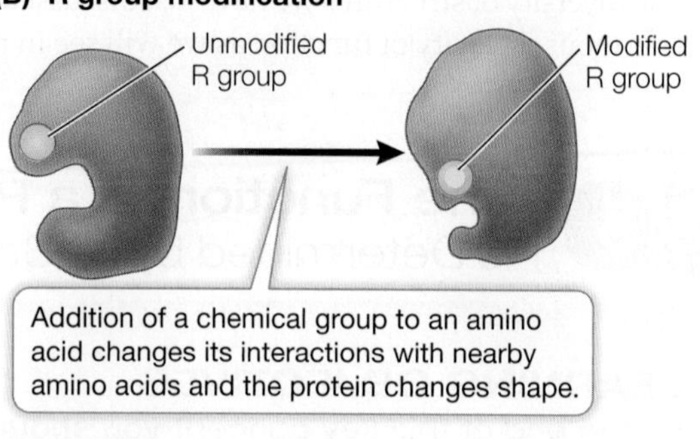

(C) Addition of cofactor

Cofactor

Substrates

Inactive protein

Active protein

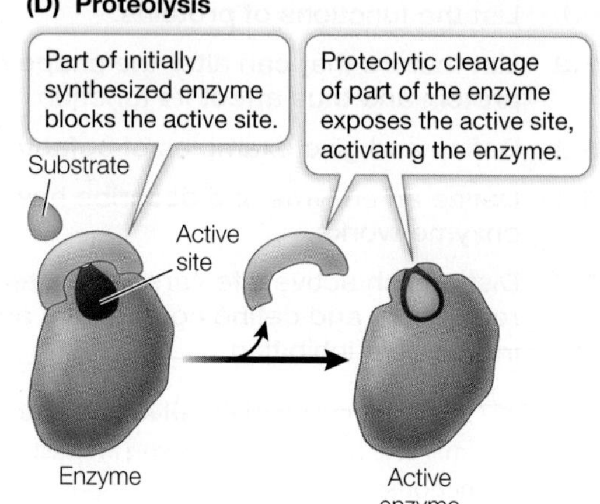

FIGURE 3.21 Changes in Protein Structure and Function Protein shape can change when a protein binds to a ligand **(A)** or when a protein's side chain is chemically modified **(B)**. **(C)** Cofactors bind to proteins, sometimes covalently, to enable protein function. **(D)** Some proteins are nonfunctional when synthesized and must be cleaved to become functional.

Are covalent bonds broken when a ligand binds to a protein and causes a conformational change?

Enzymes speed up biochemical reactions

You will recall that in Chapter 2 we discussed how the rate of a chemical reaction depends on temperature, concentration of reactants and products, free energy (endergonic versus exergonic), and activation energy. The last can be particularly important in living systems. If all the other factors such as environment, concentration, and free energy are a "go," a reaction will not occur unless a "spark" of activation energy gets it going (see Figure 2.13). In the chemistry lab, this can be done by applying heat to the reactants so they will be more likely to bump into one another in the configuration needed for the reaction to occur. Or a surface such as a metal can be added that binds the reactants and speeds up the reaction, reducing activation energy so that reactants are right next to one another, again in the right configuration.

In living systems, high heat and metals are toxic. So they need another way to lower activation energy to initiate chemical reactions. In almost all of the thousands of chemical reactions occurring in a cell at a particular time, activation energy is lowered by enzymes. These proteins are able to **catalyze** a chemical reaction, meaning they can speed it up without undergoing a permanent chemical change themselves. It is important to note that like any catalyst, an enzyme does not cause a reaction to occur, but it increases the rate of the reaction. *No catalyst makes a reaction occur that would not proceed without it.*

The vast majority of enzymes are proteins, but a few important enzymes are RNA molecules called ribozymes. An

TABLE **3.1** | Some Examples of Cofactors

Cofactor	Role in catalyzed reactions and an example
INORGANIC IONS	
Iron (Fe^{2+} or Fe^{3+})	Electron acceptor. Photosynthesis (see Key Concept 5.2)
Copper (Cu^{+} or Cu^{2+})	Oxidation–reduction. Respiratory chain in mitochondria.
Zinc (Zn^{2+})	Electron-pair acceptor. DNA synthesis.
LOOSELY BOUND ORGANIC COFACTORS (called coenzymes when protein is an enzyme)	
ATP	Provides and extracts energy; "energy currency" of cell. Membrane transport (see Key Concept 4.2).
Coenzyme A	Carries acetyl groups ($-CO-CH_3$). Cellular respiration (see Key Concept 5.2).
NAD/NADH	Carries electrons to and away from substrates. Energy metabolism (see Key Concept 5.1).
FAD/FADH	Carries electrons to and away from substrates. Energy metabolism (see Key Concept 5.1).
COVALENTLY BOUND ORGANIC COFACTORS (prosthetic groups)	
Heme	Binds ions, O_2, and electrons; contains iron. Oxygen transport in hemoglobin (see Key Concept 30.4).
Retinal	Converts light energy to chemical-bond energy. Light sensing in the eye (see Key Concept 31.4).

(A) Exergonic reaction

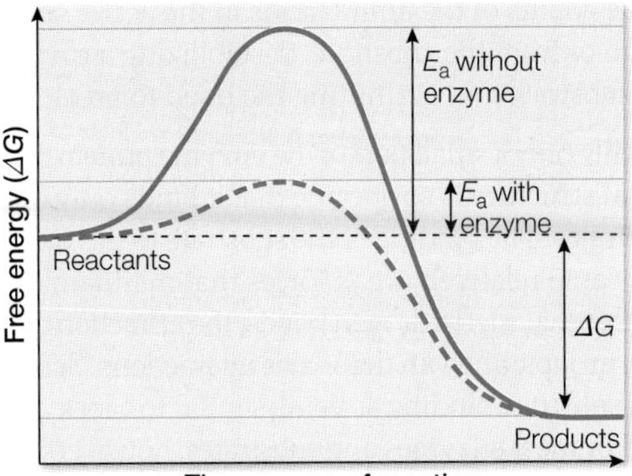

(B) Endergonic reaction

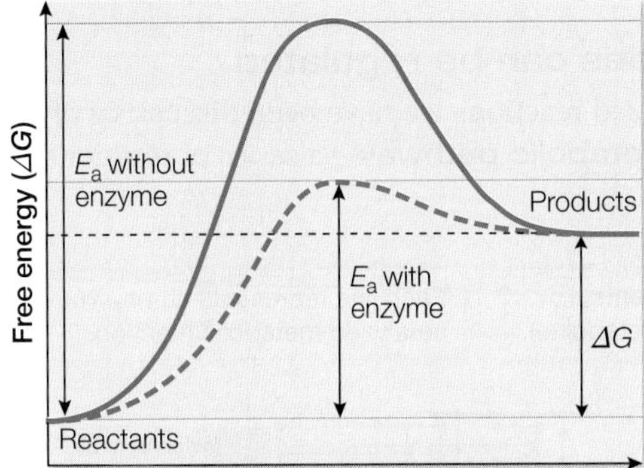

FIGURE 3.22 Enzyme Catalysis and Activation Energy When an enzyme catalyzes a reaction, it does not alter the change in free energy (endergonic or exergonic), but it lowers the activation energy (E_a).

When an exergonic reaction has an enzyme catalyst, is more useful energy produced per molecule of reactants or per unit of time?

E + S ⇌ ES ⇌ E + P

Substrates

Empty active site

Products

When the substrates bind to the active site, the two halves of the enzyme move together, changing the shape of the enzyme so that catalysis can take place.

FIGURE 3.23 Some Enzymes Change Shape When Substrate Binds to Them Shape changes result in an induced fit between enzyme (E) and substrate (S), improving the catalytic ability of the enzyme. Induced fit can be observed in the enzyme hexokinase, seen here with and without its substrates, glucose (red) and ATP (yellow), during a reaction that adds a phosphate group from ATP to glucose.

enzyme can bind the reactants in a chemical reaction and participate in the reaction itself. However, this participation does not permanently change the enzyme. At the end of the reaction, the enzyme is unchanged and available to catalyze additional reactions.

An enzyme can change the rate of a reaction substantially by reducing the activation energy (**FIGURE 3.22**). Consider, for example, the molecules of sucrose in your sweetened ice tea. Their hydrolysis to glucose and fructose (see Figure 2.13A) may take hundreds of years. However, in your lower intestine, where the protein enzyme sucrase is secreted, the same reaction occurs in 1 second! Typically, an enzyme-catalyzed reaction proceeds 10^3–10^8 times faster than the uncatalyzed reaction, and the enzyme converts 100–1,000 substrate molecules into product per second. Without enzymes, or ribozymes, it is difficult to envision how life could occur given the slow rates of most biochemical reactions.

Activity 3.6 Free Energy Changes

Enzymes bind specific reactants at their active sites

Most nonbiological catalysts are nonspecific. For example, powdered platinum metal can act as a catalyst in chemistry and catalyzes virtually any reaction in which molecular hydrogen (H_2) is a reactant. In contrast, biological catalysts are highly specific. An enzyme usually recognizes and binds to only one or a few closely related reactants, and it catalyzes only a single chemical reaction.

In an enzyme-catalyzed reaction, the reactants are called **substrates**. Substrate molecules bind to a particular site on the enzyme, called the **active site**, where catalysis takes place (**FIGURE 3.23**). The high specificity of an enzyme results from the exact three-dimensional shape (also called conformation) and chemical properties of its active site, as we described above. The names of enzymes reflect their functions and often end with the suffix "ase." For example, the enzyme sucrase catalyzes the hydrolysis of sucrose.

The binding of substrate(s) (S) to the active site of an enzyme (E) produces an **enzyme–substrate complex (ES)** which is held together by one or more means, such as hydrogen bonding, ionic bonds, van der Waals interactions, or temporary covalent bonding. The enzyme–substrate complex gives rise to product(s) (P) and enzyme (see Figure 3.23). The enzyme (E) is in the same chemical form at the end of the reaction as it was at the beginning. While bound to the substrate(s), it may change chemically, but by the end of the reaction it has been restored to its initial form and is ready to catalyze the same reaction again.

An enzyme lowers the activation energy of a reaction in one of three ways (**FIGURE 3.24**):

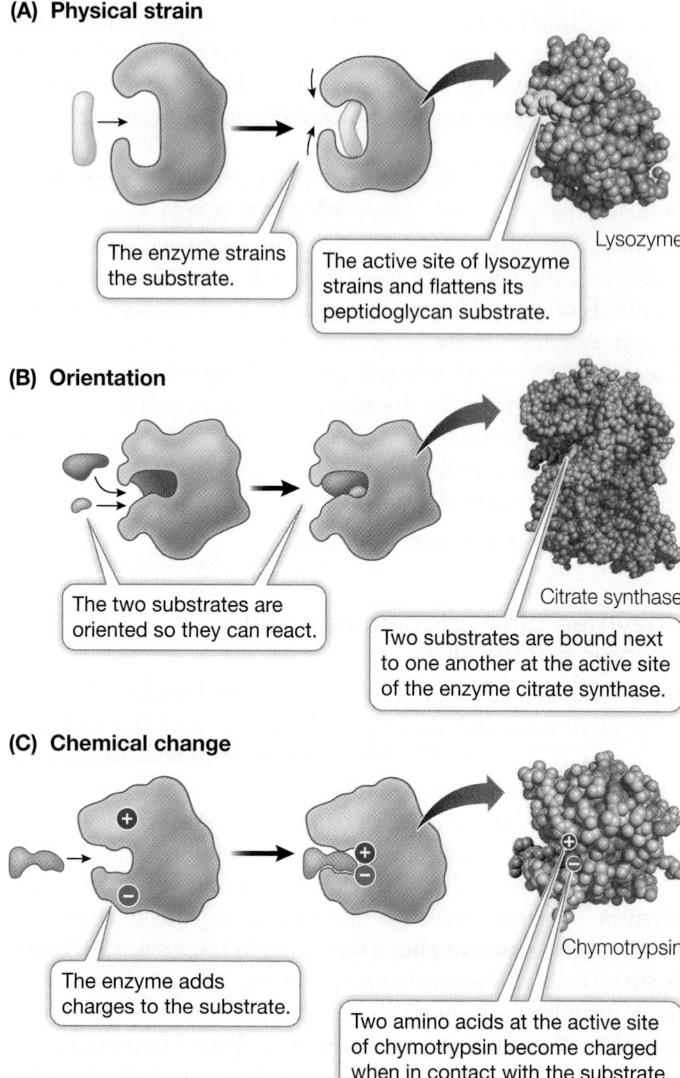

FIGURE 3.24 Chemistry at the Active Site An enzyme can lower the activation energy of a reaction in one of three ways: **(A)** inducing strain on the substrate(s), **(B)** orienting the substrate(s) appropriately for reaction, or **(C)** adding chemical groups.

1. *Inducing strain.* Once a substrate has bound to its active site, an enzyme can cause bonds in the substrate to stretch, putting it in an unstable state. This causes the strained bonds to have higher potential energy (they are weaker), making them easier to break.
2. *Substrate orientation.* When free in solution, substrates are randomly moving from place to place, vibrating, rotating, and tumbling. They only rarely have the proper orientation to react when they collide. When bound to an enzyme, however, substrates are held in the correct orientation to bring the appropriate parts of the substrates together, facilitating bond breakage or formation.
3. *Adding chemical groups.* The side chains (R groups) of an enzyme's amino acids, or of its cofactor(s), may be directly involved in the reaction. For example, in acid–base catalysis, the acidic or basic side chains of the amino acids in the active site transfer H^+ ions to or from the substrate, destabilizing a covalent bond in the substrate and permitting the bond to break.

The active site is usually only a small part of the enzyme protein. But its three-dimensional structure is so specific that it binds only one or a few related substrates. The binding of the substrate to the active site depends on the same relatively weak forces that maintain the tertiary structure of the enzyme: hydrogen bonds, the attraction and repulsion of charged groups, and van der Waals interactions. Scientists used to think of substrate binding as being similar to a lock and key fitting together. For most enzymes and substrates, however, the relationship is more like a baseball and a catcher's mitt: the substrate first binds, and then the active site changes slightly to make the binding tight. Figure 3.23 illustrates this "induced fit" phenomenon.

Enzyme activities can be regulated

The enzyme-catalyzed reactions we have been discussing often operate within a **metabolic pathway**—a series of reactions in

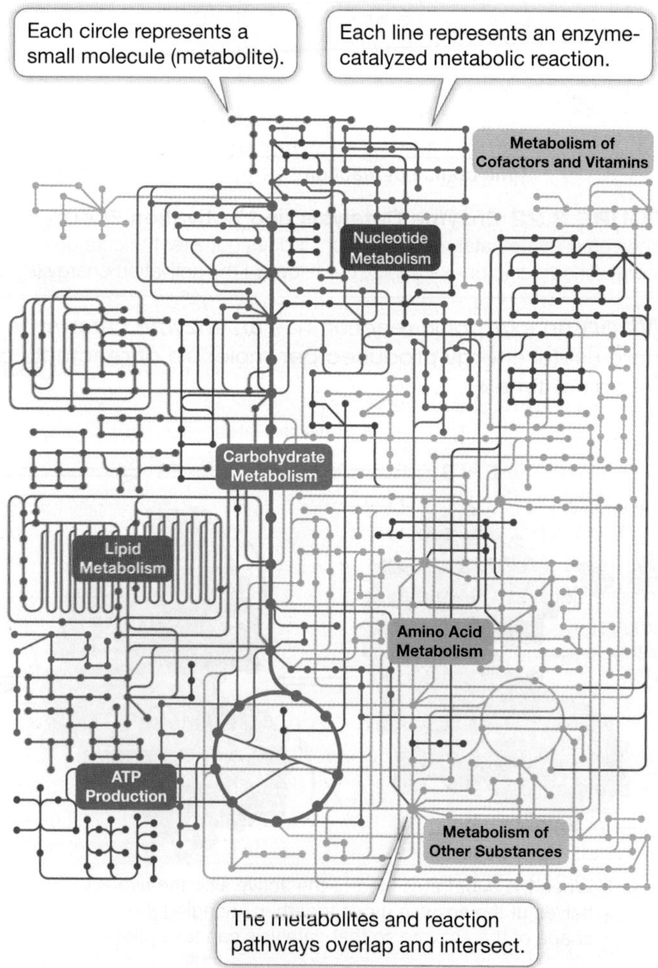

FIGURE 3.25 A Biochemical System The complex interactions of metabolic pathways can be studied using the tools of systems biology. Enzymes are a major element controlling these pathways. (After KEGG: Kyoto Encyclopedia of Genes and Genomes. Map 01100 © 2004 Kanehisa Laboratories; M. Kanehisa et al. 2017. *Nucleic Acids Res* 45: D353–D361.)

which the product of one reaction is a substrate for the next. For example, the pathway for the complete degradation of sucrose begins with sucrose and ends many reactions later with the production of CO_2 and H_2O. Energy is released along the way. As mentioned above, the first step in this particular pathway involves the enzyme sucrase, and each of the subsequent steps in this pathway are catalyzed by a specific enzyme. Pathways may be **catabolic**, breaking down molecules into smaller molecules and releasing energy, or they may be **anabolic**, synthesizing relatively complex molecules from simpler ones. A typical cell contains hundreds of enzymes that participate in many interconnecting metabolic pathways (**FIGURE 3.25**).

In multicellular organisms, different cells require different enzymes depending on their functions. In humans, for example, the enzymes involved in the breakdown of food are used in the digestive system but not in the heart. Even in single-celled organisms, different enzymes are needed in different situations. For example, at a particular time, an individual bacterial cell may live in an environment that provides lots of amino acids, such as on a forgotten steak in the back of a refrigerator. The bacterium therefore does not need to make amino acids on its own, as it can get them from the environment. To be efficient, such a cell is not likely to produce the enzymes it does not need—in this case, the enzymes involved in making amino acids.

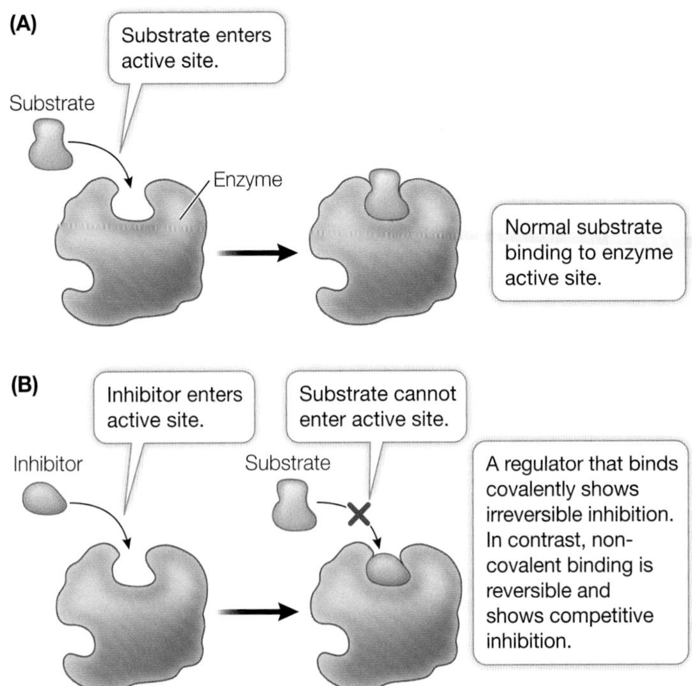

FIGURE 3.26 Active Site Inhibition of Enzyme Activity
(A) Normal binding of a substrate to an enzyme. **(B)** Binding of a regulator to the active site inhibits an enzyme's activity.

Animation 3.6 Enzyme Catalysis

LINK The fine-tuning of switching on and off the synthesis of enzymes though regulation of DNA expression is discussed in **Chapter 11**.

In addition to being able to regulate the synthesis of particular enzymes, cells also have the ability to regulate the activity of enzymes after they have been synthesized. Regulation of enzyme activity provides exquisite control of the concentration of products of metabolic pathways and prevents problems associated with having too little or too much of any metabolic substance. Using synthesis and degradation of enzymes is, in contrast, a less efficient way to control the production of a pathway, and allows larger fluctuations in concentrations. Regulation of enzyme activity involves either the binding of a regulatory molecule to the active site or the binding of a regulatory molecule to another site on the enzyme (allosteric regulation). The allosteric site is usually on a different peptide than the one containing the active site in multi-peptide proteins.

ACTIVE SITE REGULATION Regulatory molecules that bind to the active site of an enzyme are **inhibitors** that physically prevent the substrate(s) of the enzyme from entering (**FIGURE 3.26**). If an inhibitor enters the active site, covalently bonds to an amino acid side chain, and never leaves, then there is **irreversible inhibition** (Figure 3.26B). This type of inhibition is rare in normal circumstances because it permanently inactivates the enzyme. However, several drugs used in medicine, or secreted by one organism to harm another, are irreversible inhibitors. An example is the antibiotic penicillin, which is made by fungi to kill bacteria. Penicillin irreversibly binds to the active site of a transpeptidase enzyme that is responsible for building the bacterial cell wall, knocking out its function. Inability to build the cell wall leads to bacterial cell death.

In some cases an inhibitor is similar enough to a particular enzyme's natural substrate that it can bind noncovalently to the active site, yet is different enough that no chemical reaction occurs. When such a molecule is bound, it physically prevents the usual substrate from entering the active site and the enzyme is thus inactive. Such a molecule competes with the normal substrate for the enzyme and is called a competitive inhibitor. Competitive inhibition is reversible. When the concentration of the competitive inhibitor is reduced, the active site is less likely to be occupied by the inhibitor, and the enzyme regains activity.

Some pharmaceutical drugs have been developed to be competitive inhibitors of enzyme targets. For example, methotrexate is a drug designed with a structure similar to that of the metabolite dihydrofolate. The latter is converted by an enzyme to a substance that is a used in the synthesis of the bases in DNA and RNA (see Figure 3.8 for the structure of bases). Acting as a competitive inhibitor of the enzyme, methotrexate reduces base synthesis, which reduces DNA and RNA synthesis, preventing rapid division of some cancer cells.

ALLOSTERIC REGULATION Regulatory molecules that bind to an enzyme at a site distinct from the active site can cause a change in the shape (conformation) of the enzyme, altering its activity (**FIGURE 3.27**). Such control of enzyme activity is termed **allosteric regulation**, and the binding location is termed the **allosteric site**. Enzymes that show allosteric regulation usually have their allosteric and active sites on different

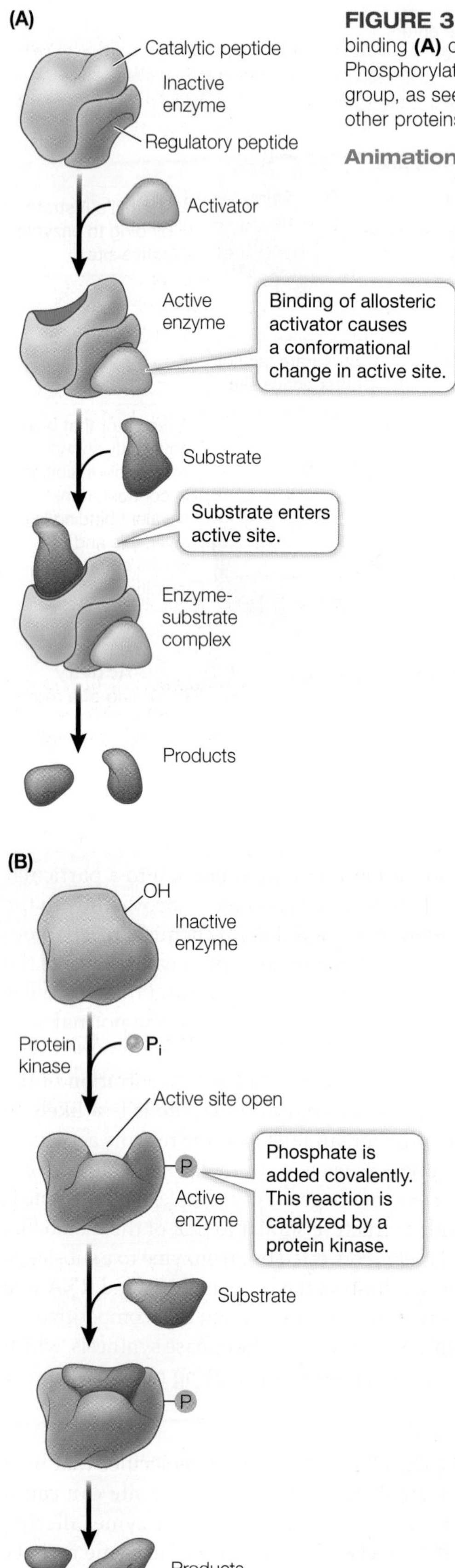

FIGURE 3.27 Allosteric Regulation of Enzyme Activity Noncovalent binding **(A)** or covalent chemical modification **(B)** can activate or inactivate an enzyme. Phosphorylation—modification of an amino acid side chain by the addition of a phosphate group, as seen in (B)—is a reversible process by which cells regulate many enzymes and other proteins.

Animation 3.7 Allosteric Regulation of Enzymes

peptides in the quaternary structure of the protein. The regulatory molecule may bind noncovalently, and reversibly, to the allosteric site. Binding may result in inhibition or, in some cases, cause an active site to become activated (see Figure 3.27A).

Some allosteric sites can be modified by the reversible, covalent binding of a molecule. A common example is the situation in which an amino acid residue (serine, threonine or tyrosine) is covalently modified by the addition of a phosphate group, in a process called phosphorylation (see Figure 3.27B). If this occurs in a hydrophobic region of the enzyme, the altered amino acid may induce a change in protein structure to interact with hydrophilic regions, because phosphate carries a negative charge. Protein phosphorylation is an extremely important mechanism by which cells regulate many different enzymes and other proteins. It is a reversible process: a class of enzymes called protein **kinases** catalyze the addition of phosphate groups to proteins, whereas protein **phosphatases** remove phosphate groups from proteins. Protein kinases and phosphatases are common across species. In humans several hundred of these enzymes are responsible for modifying about one-third of proteins. We will meet examples of these proteins many times in this book.

LINK Protein kinases are of particular importance in intracellular signaling pathways (see **Key Concepts 6.2 and 6.3**) and in the control of cell reproduction (see **Key Concept 7.5**).

REVIEW & APPLY | 3.5

R Proteins have many structural and functional roles in living systems. The diversity of protein three-dimensional structure ensures that a specific protein can bind a specific molecule (a ligand). Protein structure, and therefore function, can be altered by the binding of ligands or cofactors, by the chemical modification of amino acid side chains, and by proteolysis. Of particular importance are enzymes, proteins that catalyze (speed up) chemical transformations in living systems. Enzyme function depends on protein structure. The principles of protein structure modification, including ligand binding and the regulation of structure and therefore activity, apply to enzymes.

A
1. Why might a receptor protein's structure change when a ligand binds?
2. What does it mean to say that an enzyme is a catalyst? How does an enzyme achieve its catalytic properties?
3. How does competitive inhibition work?

What are the evolutionary implications for the finding that the same types of macromolecules are in the cells of all living systems?

USDA, ARS, EMU

Answer: As you will see in Part 3 of this book, evolution is essential for understanding biology. The fact that all organisms on Earth contain and use the same set of macromolecules—lipids, carbohydrates, nucleic acids, and proteins—implies that all organisms came from the same ancestor. While humans are certainly very different from other mammals—and certainly from bacteria—it is clear that we are related to these organisms through the same rules of genetics and DNA. Put colloquially, "human DNA is not different from mouse or bacterial DNA." Of course, the information contents of these three DNAs are very different, but how the DNAs encode information is not. This unity has allowed biologists to use organisms easily studied in the laboratory or field as stand-ins for other organisms. Many of the discoveries about inheritance (DNA) and its expression in terms of structure and function (RNA, proteins, lipids, carbohydrates) were initially made—and are still being made—in lab-grown bacteria and yeasts. These rules have been repeatedly shown to apply, albeit with some minor modifications, to humans and trees.

The unity of biological processes through macromolecules has provided a rationale for the search for extraterrestrial life. As you saw in Chapter 2, water has been found associated with the comets and asteroids that we already know impact Earth. Meteorites from these asteroids that have been found on Earth contain some of the building blocks of DNA and carbohydrates, for example. Earth-based instruments, space telescopes, and probes actually landing on other bodies in the solar system are being used to look for traces of macromolecules or their monomers, or even simpler molecules such as ammonia that could be used to make macromolecules. The assumption is that where macromolecules can exist or have existed, there is the possibility of life.

VISUAL SUMMARY 3

3.1 Lipids Are Characterized by Their Insolubility in Water

FIGURE 3.1 **A triglyceride**

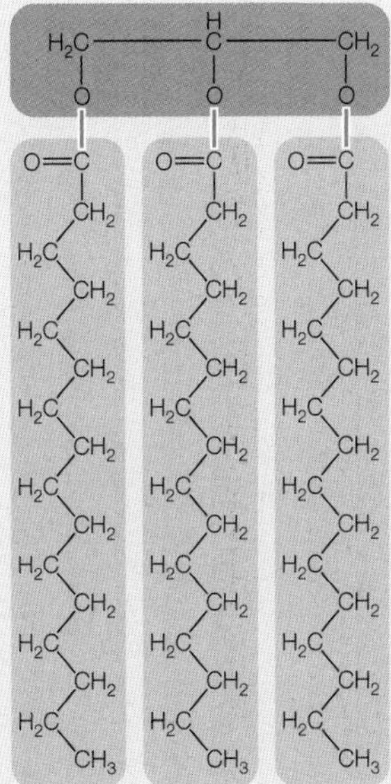

Go to ANIMATION 3.1

- **LIPIDS** are water-insoluble **MACROMOLECULES**.
- **TRIGLYCERIDES** and **PHOSPHOLIPIDS** differ in their degree of saturation (**FIGURE 3.1**).

QUESTIONS

1. Why are some triglycerides liquid and others solid at room temperature?
2. What does it mean to say that a phospholipid is amphipathic?
3. Why is a phospholipid bilayer a barrier to water-soluble molecules?

3.2 Carbohydrates Are Made from Simple Sugars

Go to ACTIVITY 3.1 and ANIMATION 3.2

FIGURE 3.5

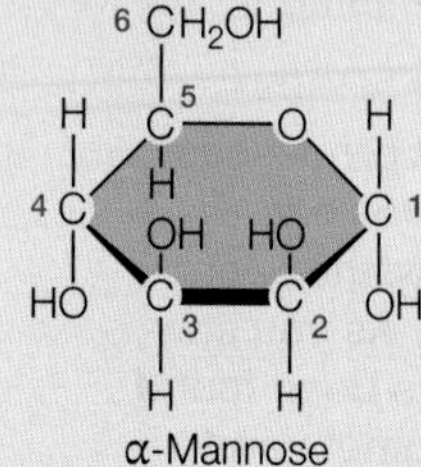

- **MONOSACCHARIDES** contain five or six carbons, form rings, and are structural isomers or stereoisomers of one another (**FIGURE 3.5**).
- **GLYCOSIDIC BONDS** connect carbohydrate monomers in **POLYSACCHARIDES**, which are used for storage and structural support (**FIGURE 3.7**).

FIGURE 3.7

Cellulose

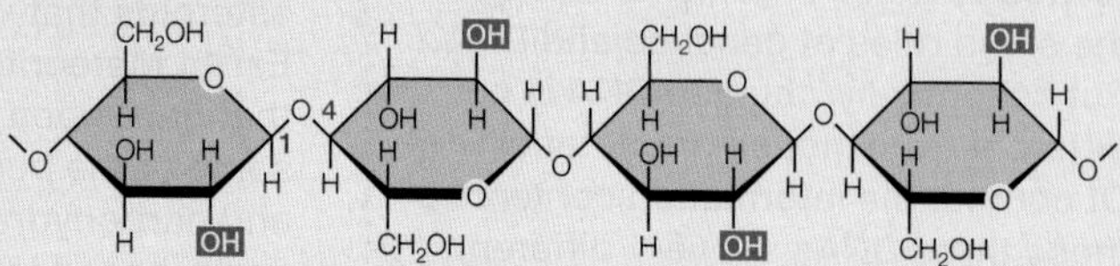

Starch and glycogen

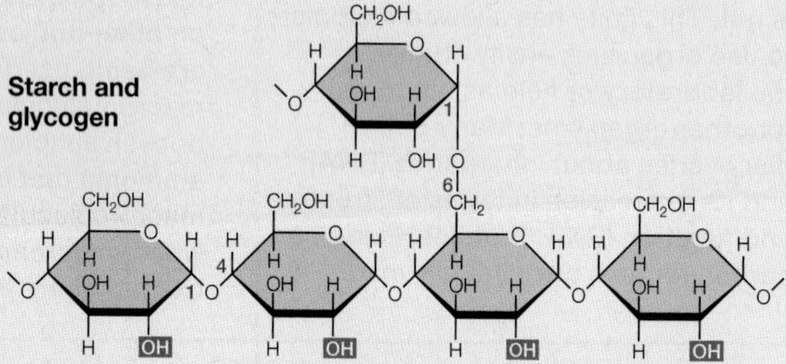

QUESTIONS

1. Why are mono-, di-, and oligosaccharides soluble in water?
2. What are the structural differences between cellulose and starch, both of which are polysaccharides found in seeds?
3. Only about 3% of a cell's contents is composed of carbohydrates, yet carbohydrates are the most common macromolecules on Earth. Why?

3.3 Nucleic Acids Are Informational Macromolecules

Go to ACTIVITIES 3.2 and 3.3 and ANIMATION 3.3

FIGURE 3.11

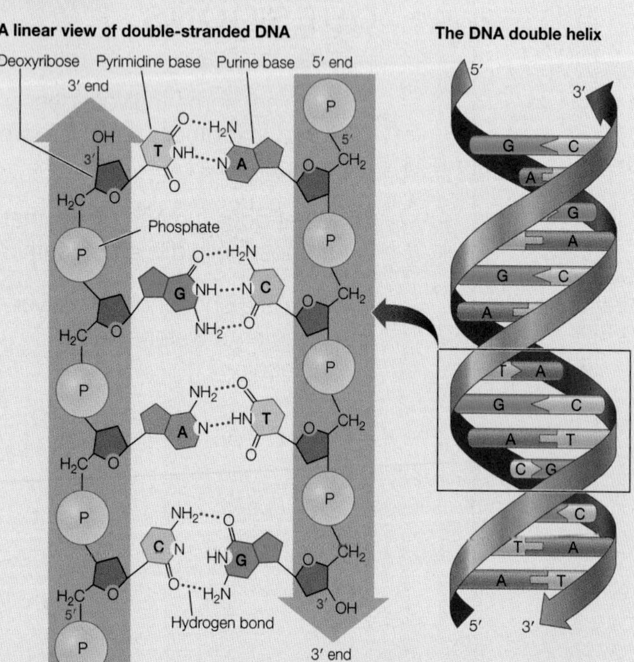

The DNA double helix

- **NUCLEIC ACIDS** have a 5′ end and a 3′ end and are composed of **NUCLEOTIDES**.
- Two antiparallel strands of nucleic acids can show **COMPLEMENTARY BASE PAIRING** (**FIGURE 3.11**).

QUESTIONS

1. How does RNA differ from DNA?
2. What is complementary base pairing?
3. What does it mean to say that all the information in DNA is fully contained on each strand?

3.4 Proteins Are Polymers with Variable Structures

Go to ACTIVITIES 3.4 and 3.5 and ANIMATIONS 3.4 and 3.5

- **AMINO ACIDS** have different chemical properties determined by their **R GROUPS** (**SIDE CHAINS**) and are joined by **PEPTIDE BONDS** to give a **PROTEIN** its **PRIMARY STRUCTURE** (**FIGURE 3.14**).
- Proteins have **SECONDARY** (**α HELICES** and **β PLEATED SHEETS**), **TERTIARY** (three-dimensional shape), and sometimes **QUATERNARY** (multiple polypeptides) **STRUCTURE** (**FIGURE 3.17**).
- **DENATURING** a protein disrupts the higher levels of structure but leaves the primary structure unaffected.

FIGURE 3.14

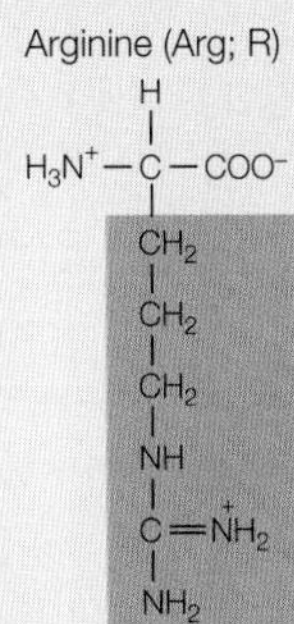

FIGURE 3.17

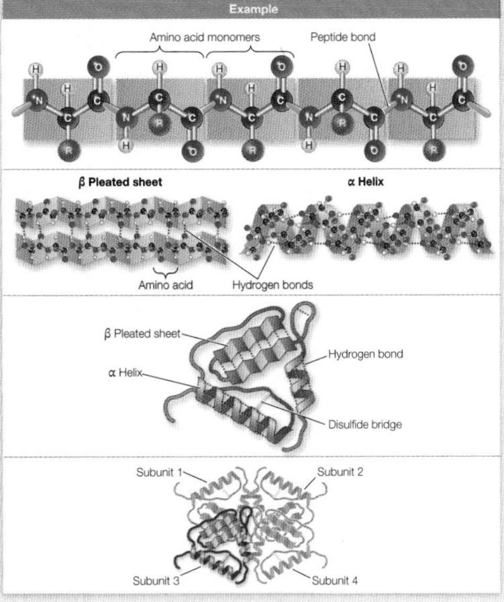

QUESTIONS

1. In a particular protein, for most of the locations where cysteines occur, replacing a cysteine with a serine has little effect on function. However, there are two cysteines that, when either is replaced, knock out function of the molecule. How might you explain both of these results?
2. Could amino acids that are far apart in the primary structure both be in the active site of an enzyme?
3. Some proteins occur in the aqueous region of the cell, and others reside in the lipid bilayer of the cell membrane. What would you expect to see with respect to hydrophilic versus hydrophobic regions of these two kinds of proteins?

3.5 The Function of a Protein Is Determined by Its Structure

Go to ACTIVITY 3.6 and ANIMATIONS 3.6 and 3.7

- Proteins show highly specific binding determined by the structure and identity of their amino acid side chains.
- **ENZYMES** are catalysts that lower the activation energy of a reaction (**FIGURE 3.22**).
- Binding of various molecules and inorganic ions and chemical modifications can alter the shape and chemical properties of proteins, thus altering their function (**FIGURE 3.27**).

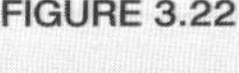

FIGURE 3.22

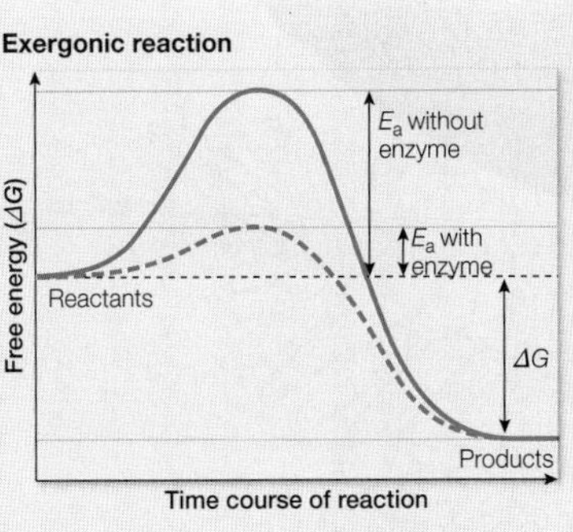

FIGURE 3.27

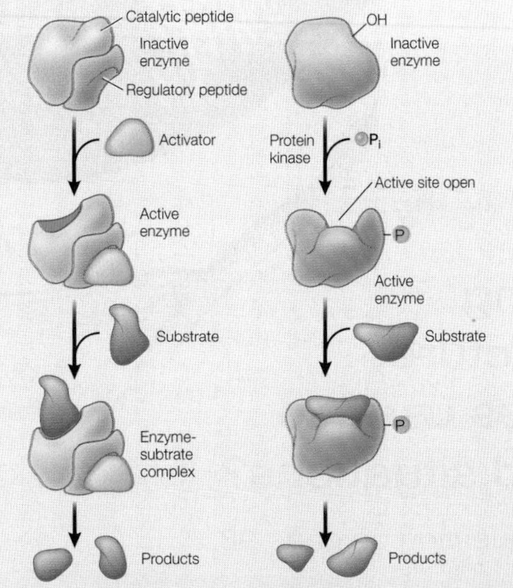

QUESTIONS

1. How can enzymes reduce the activation energy of a chemical reaction?
2. What are cofactors? Are cofactors chemically altered during an enzymatic reaction?
3. What is allosteric regulation? How might allosteric regulation be used to reduce metabolite production from a particular pathway?

4 Cell Structure and Membranes

Robert Hooke is famous in physics because he first described the law that governs elasticity. He is also famous in biology as the first person to describe cells. He applied this term to the regular, boxlike structures that he observed in slices of cork because they reminded him of monks' cells in monasteries. Cells are small, and Hooke had to use a light microscope to see them. Ever since Hooke's time, both light and electron microscopes have been essential to describe cell structure.

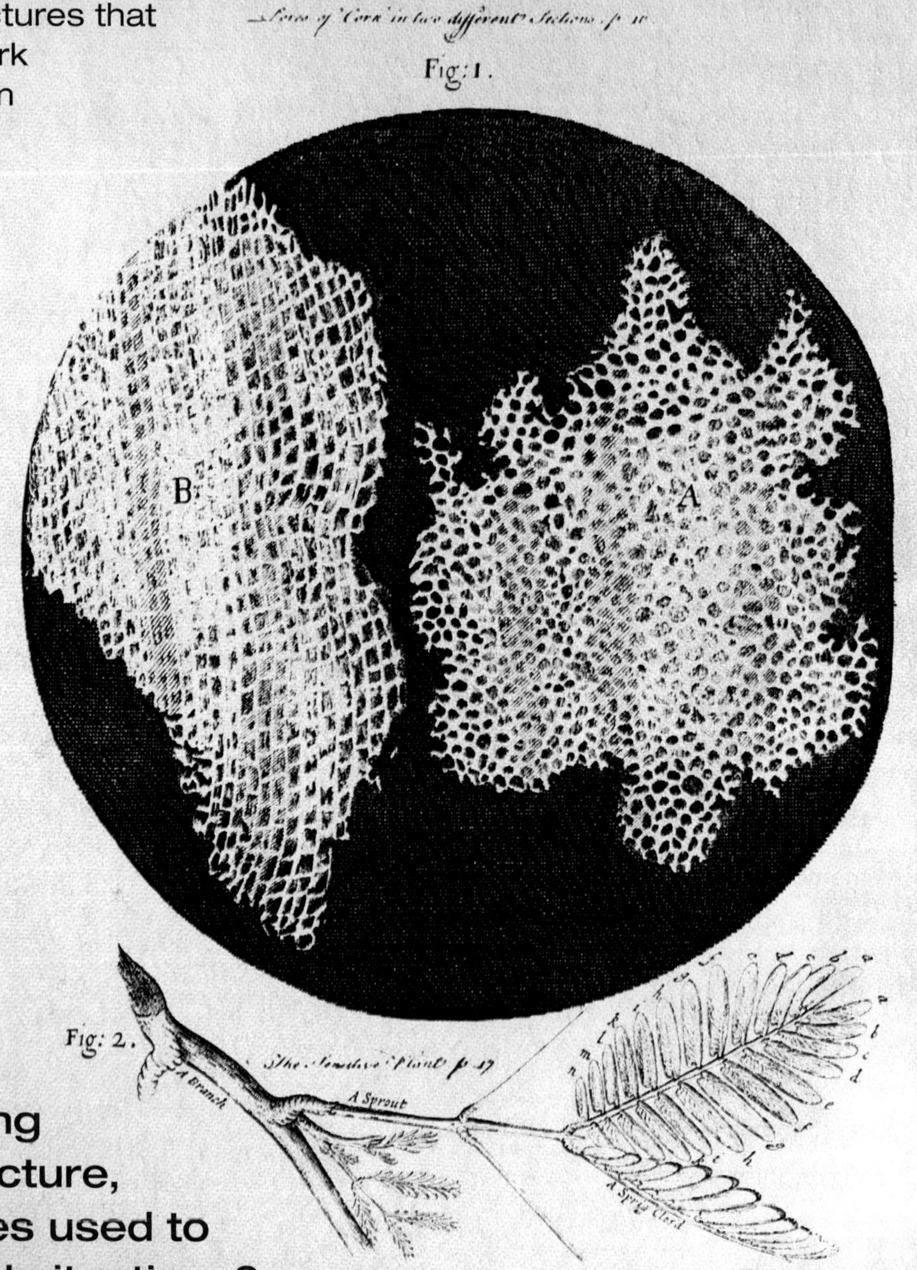

In addition to their use in describing basic biological structure, how are microscopes used to study cells in applied situations?

You will find the answer to this question on page 98.

KEY CONCEPTS

4.1 **The Cell Membrane Separates the Interior of the Cell from Its Environment**

4.2 **Passive and Active Transport Are Used by Small Molecules to Cross Membranes**

4.3 **Vesicles Are Used to Transport Large Molecules across Membranes in Eukaryotes**

4.4 **Cell Size, Shape, and Ability to Move Are Determined by Internal and External Structures**

4.5 **Compartmentalization Occurs in Prokaryotic Cells and Is Extensive in Eukaryotic Cells**

4.1 The Cell Membrane Separates the Interior of the Cell from Its Environment

LEARNING OBJECTIVES

By the end of this key concept you should be able to:

4.1.1 **Describe the structure of a typical biological membrane.**

4.1.2 **Distinguish among integral, anchored, and peripheral membrane proteins.**

4.1.3 **Describe how the inner and outer surfaces of a biological membrane can differ.**

4.1.4 **Explain how the temperature of the environment can affect the lipid composition of the cell membrane.**

The cell is the simplest structural unit of a living organism. Some organisms are single cells; other organisms, such as humans, have trillions of cell. Cells are membrane-enclosed structures that are capable of producing other cells. The **cell theory**, developed in the nineteenth century, was the first unifying principle of biology and has three critical components:

1. Cells are the fundamental units of life.
2. All living organisms are composed of cells.
3. All cells come from preexisting cells.

In the almost two centuries since the cell theory was first proposed, no organism has been discovered that is not composed of cells. The components of the cell theory are now considered to be facts about the living world. *The study of the biology of the cell is thus critical to understanding life.*

Everything inside a cell, except for the nucleus in those cells that have one, is termed the **cytoplasm**. Within the cytoplasm is water, along with thousands of substances (see Chapters 2 and 3 for a summary of molecules and ions) that differ in identity and concentration from the composition of the outside of the cell. These substances determine the cell's structure and function. The **cell membrane** performs the vital role of keeping the internal chemicals of a cell, which carry out life's functions, different from the chemical composition outside the cell. As with many of the basic processes of life at the cellular level, the essential structure and functions of membranes are similar across organisms.

A biological membrane's characteristics are determined by the chemical properties of its lipid and protein constituents. In essence, a membrane consists of a phospholipid bilayer (see Figure 3.3A) with numerous associated proteins that are embedded in it or attached to it on the outside or inside (**FIGURE 4.1**). The membrane is not a solid, unchanging structure. Rather, most of the lipids and proteins move laterally through the membrane. The term **fluid mosaic model** describes membrane structure in terms of random movement ("fluid") and separateness of the proteins and lipids ("mosaic").

Lipids form the hydrophobic core of the membrane

Phospholipids are the major lipid component of biological membranes. Phospholipids form thin bilayers, about 8 nanometers (0.008 micrometer) thick, with their charged, hydrophilic phospholipid "heads" on the external surfaces interacting with the aqueous environments, and their nonpolar hydrophobic fatty acid "tails" interacting with each other on the inside of the membrane. The hydrophobic interior acts as a barrier and prevents the diffusion of polar molecules and ions through the membrane.

Key Concept 3.1 discusses how having both hydrophobic and hydrophilic parts (that is, amphipathic properties) allows phospholipids to form a bilayer in an aqueous environment.

Although all biological membranes share a similar structure, there are many different kinds of phospholipids, and membranes from different cells and other membrane-enclosed compartments may vary greatly in their lipid composition. For example, phospholipids can differ in their fatty acid chain length (number of carbon atoms) and degree of unsaturation (number of double bonds in their fatty acids), and in the kinds of polar (phosphate-containing) groups present. The most common fatty acids in membranes have chains with 16–18 carbon atoms and 0–2 double bonds. Saturated fatty acid chains (those with no double bonds) allow close packing of phospholipids in the bilayer, whereas the "kinks" caused by double bonds in unsaturated fatty acids (see Figure 3.1A) make for less dense packing.

The steroid cholesterol (see Figure 3.1F) is an abundant component of cell membranes in animals, composing from about 20 percent to more than 50 percent of the lipid molecules. The single hydroxyl group (—OH) in cholesterol interacts with the polar heads of the phospholipids, while the nonpolar rings and hydrocarbon chain insert among the fatty acid chains of the membrane and modulate membrane fluidity (see Figure 4.1). Other lipids,

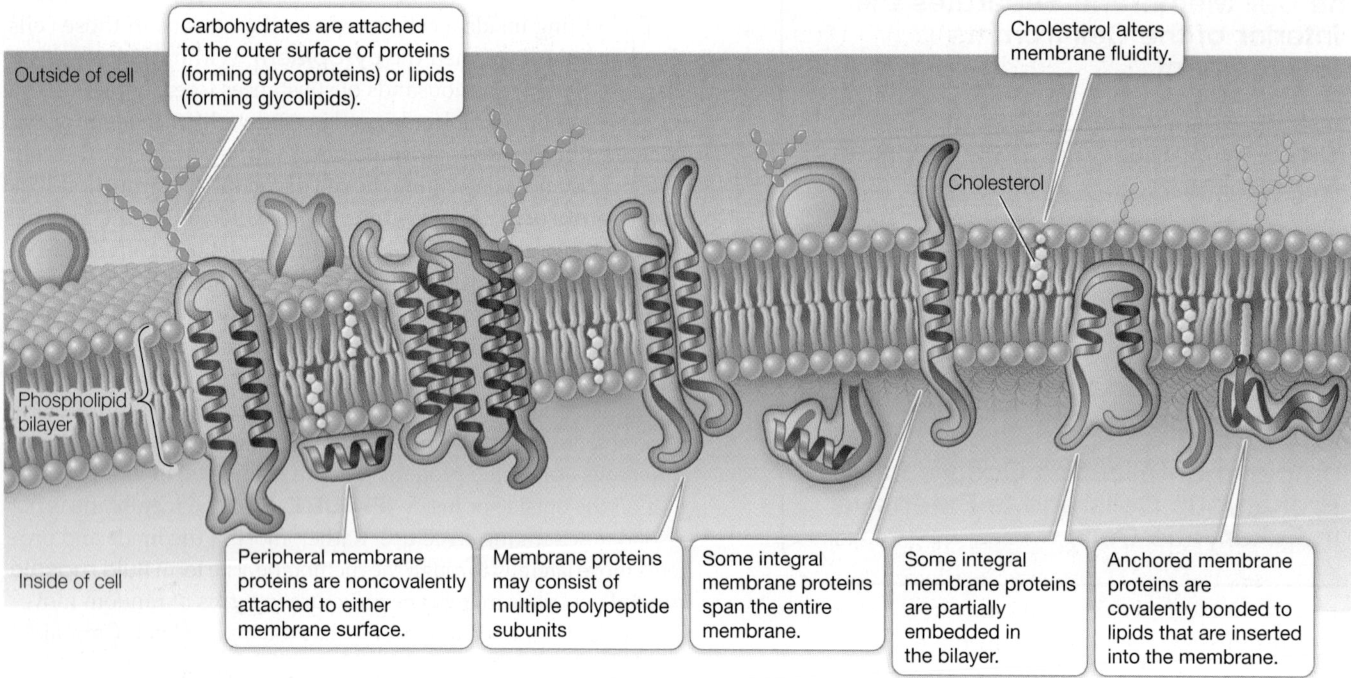

FIGURE 4.1 Fluid Mosaic Model A cell membrane consists of a phospholipid bilayer with embedded or associated proteins. Membrane proteins fall into three classes (integral, anchored, and peripheral) based on how they are associated with the lipid bilayer.

Activity 4.1 The Fluid Mosaic Model

such as the glycolipids discussed below, are also present and can act to alter membrane fluidity.

Membrane fluidity is important for function. Because phospholipids bond with one another through van der Waals interactions, they move laterally in the membrane quite easily, as these weak bonds break and re-form. A membrane composed only of phospholipids would be about as fluid as olive oil. The two most important factors influencing membrane fluidity are:

1. *Lipid composition*. Phospholipids with longer fatty acid chains have more bonds with one another than those with shorter chains and thus have lower fluidity. Saturated phospholipids pack more tightly together than those with kinked, unsaturated fatty acids, resulting in less fluid membranes. Cholesterol is able to alter interactions among the fatty acid side chains and alter membrane fluidity. At low temperatures cholesterol increases membrane fluidity, and at high temperatures it decreases fluidity. In plants, **phytosterols** are similar in structure to cholesterol and modify fluidity in plant cell membranes in a similar manner.
2. *Temperature*. Membrane fluidity declines under cold conditions because molecules move more slowly at lower temperatures. Changes in fluidity can alter the function of a membrane. A familiar example is when your fingers get numb after contact with ice, which causes a reduction in membrane fluidity in the nerve cells, which decreases nerve activity. To overcome the potentially deleterious effects of reduced fluidity at low temperature, organisms can alter the lipid composition of their membranes in cold areas of their body, or when their environment gets colder. Specifically, in cells that experience colder temperatures, phospholipids that have saturated, long fatty acid chains can be replaced with those that have unsaturated, short chains, and the amount of cholesterol can be altered appropriately. Such changes increase fluidity and are critical for the survival of plants, bacteria, and hibernating animals during winter.

Activity 4.2 Lipid Bilayer Composition Simulation

While phospholipid molecules can easily move laterally within a membrane, it is rare for a phospholipid in one half of the bilayer to spontaneously flip over to the other side. For that to happen, the polar part of the molecule would have to move through the hydrophobic interior of the membrane. Since spontaneous flip-flops are rare, the inner and outer halves of the bilayer may be quite different in the kinds of phospholipids they contain.

Proteins give a membrane special properties

All biological membranes contain proteins. Typically, cell membranes have about 1 protein molecule for every 25 phospholipid molecules. This ratio varies depending on membrane function. For example, in the inner membrane of the **mitochondrion** (plural *mitochondria*), a compartment inside the cell specialized for energy production, there is 1 protein for every 5 lipids. By contrast, the membranes found in myelin, which encloses portions of some neurons (nerve cells) and acts as an electrical insulator, have only one protein for every 70 lipids.

Recall from Figure 3.14 that some amino acids contain nonpolar hydrophobic R groups, whereas others contain polar or charged hydrophilic R groups. The arrangement of these amino acids in a membrane protein determines whether the membrane protein will insert into the nonpolar lipid bilayer and how it will be positioned. Membrane proteins and lipids generally interact noncovalently. The polar or charged regions of proteins interact with the charged heads of phospholipids, and the nonpolar hydrophobic regions of both molecules interact. There are three general types of membrane proteins (see Figure 4.1):

1. **Integral membrane proteins** are at least partly embedded in the phospholipid bilayer and thus have both exposed hydrophobic regions that interact with the interior of the membrane, and exposed hydrophilic regions that interact with the aqueous environment external to the membrane.
2. **Anchored membrane proteins** are covalently attached to fatty acids or other lipid groups but otherwise have no exposed hydrophobic regions. The hydrophobic lipid groups insert into the phospholipid bilayer and hold these proteins in association with the membrane.
3. **Peripheral membrane proteins** lack exposed hydrophobic groups and are not embedded in the bilayer. Instead, they have polar or charged regions that interact with exposed parts of integral membrane proteins, or with the charged heads of phospholipid molecules.

Proteins are asymmetrically distributed on the inner and outer surfaces of membranes, giving the two surfaces different properties. An integral membrane protein that extends all the way through the phospholipid bilayer and protrudes on both sides of a bilayer is called a **transmembrane protein**. In addition to one or more transmembrane domains (regions) that extend through the bilayer, such a protein may have domains with different specific functions on the inner and outer sides of the membrane. Some integral membrane proteins do not extend all the way through the membrane. Such partially integrated proteins, like anchored and peripheral membrane proteins, are thus located on one side of the membrane or the other. This asymmetrical arrangement gives the two surfaces of the membrane different properties. As we will see later, especially in Chapter 6, many critical functions of the cell depend on the inner and outer surfaces of the membrane having different properties.

Like phospholipids, some membrane proteins move relatively freely within the phospholipid bilayer. Cell fusion experiments illustrate this migration dramatically. When two cells fuse, a single continuous membrane forms around both cells, and some proteins from each cell distribute themselves uniformly around this membrane (**INVESTIGATION FIGURE 4.2**).

Although some proteins are free to migrate throughout the membrane, others appear to be constrained in their movement. For example, proteins involved in several signaling pathways tend to be clustered into certain regions of the cell membrane. Signaling function would be disrupted were the proteins involved in signaling to migrate. In some cases, protein movement is limited because the proteins are anchored to other

INVESTIGATION

FIGURE 4.2 Proteins Can Move Laterally in Membranes Two animal cells can be fused together in the laboratory, forming a single large hybrid cell. This phenomenon was used to test whether membrane proteins can move independently in the plane of the cell membrane.[a]

HYPOTHESIS

Proteins embedded in a membrane can move freely within the membrane, and the rate of movement will depend on the fluidity of the membrane, which will differ at different temperatures.

METHOD

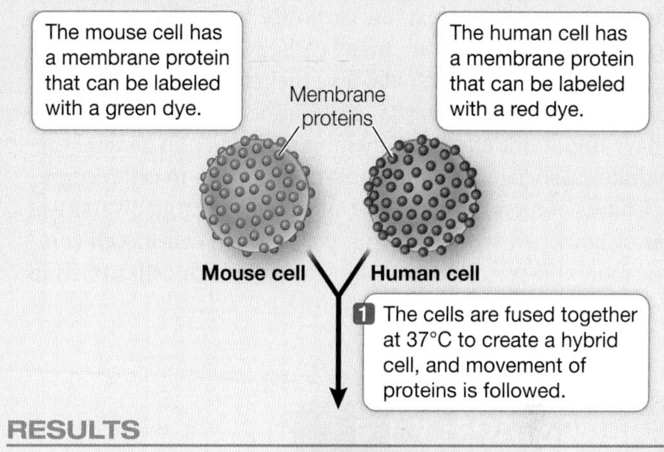

RESULTS

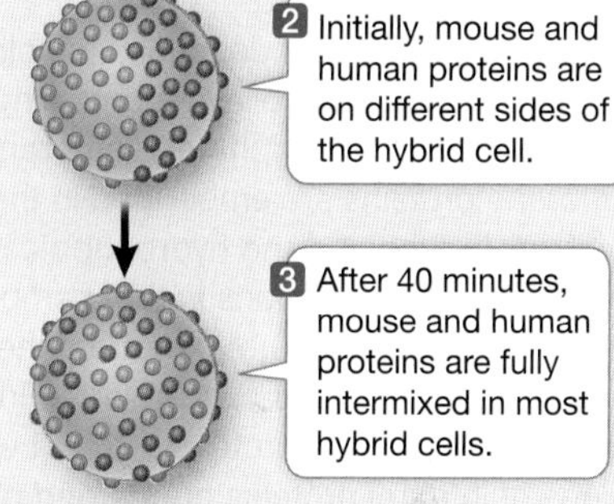

CONCLUSION

Membrane proteins can move laterally within the fluid membrane.

ANALYZE THE DATA

The experiment was repeated at various temperatures, by first allowing cells to fuse at 37°C for 10 minutes, and then changing the temperature for a further 30 minutes. The percent of hybrid cells with intermixed proteins was then measured and the following data were obtained.

Temperature (°C) after 10 min at 37°C	Hybrid cells with intermixed proteins (%)
0	0
15	8
20	42
26	77

1. Plot these data on a graph of percent hybrid cells with intermixed proteins vs. Temperature.
2. Explain these data, relating the results to membrane fluidity.
3. At 37°C most but not all hybrid cells have intermixed proteins after 40 minutes. Given that the membrane has high fluidity at 37°C, why might some cells not show complete intermixing?

[a]D. L Frye and M. Edidin. 1970. *J Cell Sci* 7: 319–335.

A similar **Analyze the Data** exercise may be assigned in **Achieve**.

elements of the cell. Some proteins appear to occur primarily in lipid rafts, which are areas of the membrane in which the lipid composition differs from the rest of the membrane. In other cases, proteins are constrained by attachment to cytoskeletal elements (see Key Concept 4.4).

Cell membrane carbohydrates are recognition sites

The carbohydrates associated with membranes are attached either to lipids, to form **glycolipids**, or to proteins, to form **glycoproteins** (when there are a few carbohydrate residues) or **proteoglycans** (when there are many carbohydrate residues). They are generally located on the outside of the cell, where they interact with substances in the external environment. While glycolipids can have important effects on membrane fluidity, in general carbohydrates associated with membranes function in cell recognition. Chains of monosaccharides can generate a large diversity of linear or branched structures that play roles in cell-to-cell communication and cell adhesion (the ways in which cells attach to one another, or to a substrate).

REVIEW & APPLY | 4.1

R The fluid mosaic model describes a biological membrane as a fluid phospholipid bilayer containing a variety of integral, anchored, and peripheral membrane proteins. Integral membrane proteins are embedded in the membrane and have both hydrophilic and hydrophobic regions that affect their position and functions in the membrane. Anchored membrane proteins do not penetrate the membrane but are attached to lipids that are inserted into the membrane. Peripheral membrane proteins interact with polar or charged regions of integral proteins or charged phospholipid heads on one side or the other of the membrane, but do not penetrate the membrane. Properties of the inner and outer sides of the cell membrane differ because of the asymmetrical distribution of membrane proteins. The fluidity of a biological membrane is affected by the length and saturation of the fatty acid side chains in the phospholipids and the amount of cholesterol (in animals) or phytosterols (in plants).

A

1. Explain why the fluidity of a cell membrane allows integral membrane proteins to move laterally but not to flip from one side of the membrane to the other.
2. For an integral membrane protein, which class of amino acids might you expect to find on the portion of the protein embedded in the interior of the membrane (see Figure 3.14)? List a few examples.
3. What is the evidence for membrane fluidity?
4. In the winter, ducks spend much of their time with their feet in very cold water. What chemical characteristics do you predict the phospholipids in the cell membranes of duck feet will have?

Membranes separate the inside of cells from the external environment. However, many substances must pass from the outside of a cell to the inside, and vice versa. In the next key concept we will see how small molecules cross biological membranes.

4.2 | Passive and Active Transport Are Used by Small Molecules to Cross Membranes

LEARNING OBJECTIVES

By the end of this key concept you should be able to:

4.2.1 **Define a selectively permeable membrane.**

4.2.2 **Distinguish between simple and facilitated diffusion.**

4.2.3 **Define osmosis and predict the direction of water movement when a cell is placed in a hypotonic, isotonic, or hypertonic solution.**

4.2.4 **Compare and contrast active and passive membrane transport mechanisms.**

4.2.5 **Predict the type of membrane transport required for polar, charged, and nonpolar molecules and atoms.**

Membranes separate an internal environment from the external environment, giving cells or cell compartments the ability to regulate their internal composition. Biological membranes allow some substances, but not others, to pass through them, a characteristic called **selective permeability**. The movement of substances across the membrane is characterized as either passive or active. **Passive transport** does not require an input of energy. For example, if a membrane is permeable to a particular substance, that substance can simply diffuse (as we describe below) across the membrane from a region of higher concentration to a region of lower concentration. **Active transport**, by contrast, requires an input of energy and the assistance of specialized membrane proteins.

Animation 4.1 Passive Transport

Simple diffusion takes place through the phospholipid bilayer

In a solution, there is a tendency for all of the components to become evenly distributed. You can see this when a drop of food coloring falls into a glass of water. Initially the pigment molecules are very concentrated, but they will move about at random, slowly spreading until the intensity of color is the same throughout the glass. A solution in which the solute molecules are uniformly distributed is said to be at equilibrium. This does not mean the molecules have stopped moving; it just means they are moving in such a way that their overall distribution does not change.

Diffusion is the process of random movement of particles toward a state of equilibrium. Diffusion is thus a net movement from regions of higher concentration to regions of lower concentration. Diffusion can be understood as a consequence of the second law of thermodynamics (see Key Concept 2.3). Recall that this law states that energy tends to become more dispersed. The thermal energy in a solution is more dispersed when the solute (dissolved) particles are more uniformly concentrated in the solution than when the same number of particles are concentrated in a particular region. Thus particles tend to move from areas of high concentration to areas of low concentration, which leads to equal concentrations throughout the solution, or equilibrium. This is what happens with a drop of food coloring in a glass of water. Initially all the food-coloring molecules are in one place, where the drop landed. This is thermodynamically unfavorable. The food-coloring molecules then diffuse so that they become randomly distributed in the water, a situation that is more energetically favorable. Now think of a cell, with a particular solute more concentrated on the outside than in the cytoplasm, inside. The solute will likewise tend to diffuse, moving from the outside with high concentration to the inside with low concentration, until there are equal concentrations on both sides of the membrane. As you will see throughout this book, cells and their membranes have numerous mechanisms to combat this tendency towards equal concentrations of solutes inside and outside of the cell. If they did not, the solute contents of cells would be the same as that of their environment.

While diffusion is quite fast in gases, it is substantially slower in liquids such as water. Diffusion is generally a very slow process in living tissues except over very short distances. For example, it would take several years for a molecule of oxygen gas (O_2) to diffuse from your lung through the tissues in your body to a cell in your fingertip! This slow rate of diffusion explains why large multicellular plants and animals evolved bulk fluid movement systems such as circulatory systems to rapidly distribute vital molecules such as O_2.

LINK Animals with high metabolic rates require high rates of O_2 transport and have evolved circulatory systems with exceptional abilities for rapid O_2 transport. The diverse circulatory systems of animals are compared in **Key Concept 30.5.**

How fast a substance diffuses, measured as the net movement of particles per unit of time, depends on several factors:

- The *diameter* of the molecules or ions: smaller molecules diffuse faster.
- The *temperature* of the solution: higher temperatures lead to faster diffusion because particles have higher thermal energy and thus move faster.
- The *concentration gradient* in the system—that is, the change in solute concentration with distance in a given direction. The greater the concentration gradient, the more rapidly a substance diffuses.
- The *area* across which a substance diffuses and the *distance* it moves. A larger surface area permits more rapid diffusion (this explains why in many cells the cell membranes are folded). As you will see in Chapter 30, keeping the distance a solute must diffuse as short as possible maximizes the diffusion rate.

How do the first three factors apply to a cell surrounded by a membrane? The cytoplasm is largely a water-based (aqueous) solution, and so is the surrounding environment. *In a complex solution (one with many different solutes), the diffusion of each solute depends only on its own concentration, not on the concentrations of other solutes.* So one might expect a substance with a higher concentration inside the cell to diffuse out, and one with a higher concentration outside the cell to diffuse in. Indeed, some small molecules can pass through the phospholipid bilayer of the membrane by **simple diffusion**, which is defined as diffusion without the need for an integral membrane protein. In general, membranes are more permeable to molecules that are small, nonpolar, uncharged, and hydrophobic; the more lipid-soluble the molecule is, the more rapidly it diffuses through the lipid bilayer. Nonpolar, small molecules such as O_2 and carbon dioxide (CO_2) readily cross membranes. Small polar molecules such as water (H_2O) and urea ($CO[NH_2]_2$) can also pass through membranes, though the permeability is lower. In contrast, electrically charged or large polar molecules, such as ions or amino acids and sugars, do not pass readily through a membrane because they are strongly hydrophilic and thus insoluble in the hydrophobic interior of the lipid bilayer. However, as we discuss below, specialized integral proteins facilitate the transport of these molecules across membranes.

Osmosis is the movement of water across membranes

Although the phospholipid bilayer is somewhat permeable to water molecules, many types of cells have specialized protein channels, called **aquaporins**, that increase water permeability. When the overall concentration of dissolved solutes differs on two sides of a water-permeable but solute-impermeable membrane, water will show net movement from the side of lower solute concentration to the side with higher solute concentration in a process termed **osmosis**. As with the diffusion of solutes, osmosis can be understood using the second law of thermodynamics. The movement of water from low to high solute concentration disperses the energy in the solute molecules, by making the solute concentrations on the two sides of the membrane more similar.

Osmotic pressure (symbolized by the Greek capital letter *pi*, Π) is defined as the pressure that needs to be applied to a solution to prevent the flow of water across a membrane by osmosis. Water will show net flow across a membrane from a solution with low osmotic pressure to one with high osmotic pressure. Importantly, every dissolved solute particle contributes equally to osmotic pressure. For example, when salt (NaCl) dissolves, both the sodium ions and the chloride ions contribute equally to osmotic pressure. We quantify osmotic pressure of a solution by measuring its **osmolarity**, which is the total number of solute particles per liter of water (the concentration of solute particles). Distilled water has an osmotic pressure equal to zero because there are no dissolved solute particles. The equation for osmotic pressure is

$$\Pi = cRT$$

where c is the osmolarity, R is the gas constant, and T is the absolute temperature (R and T may be familiar to you from your study of chemistry). Note that osmotic pressure is proportional to the osmolarity (c)—the more dissolved solutes there are, the higher the osmotic pressure.

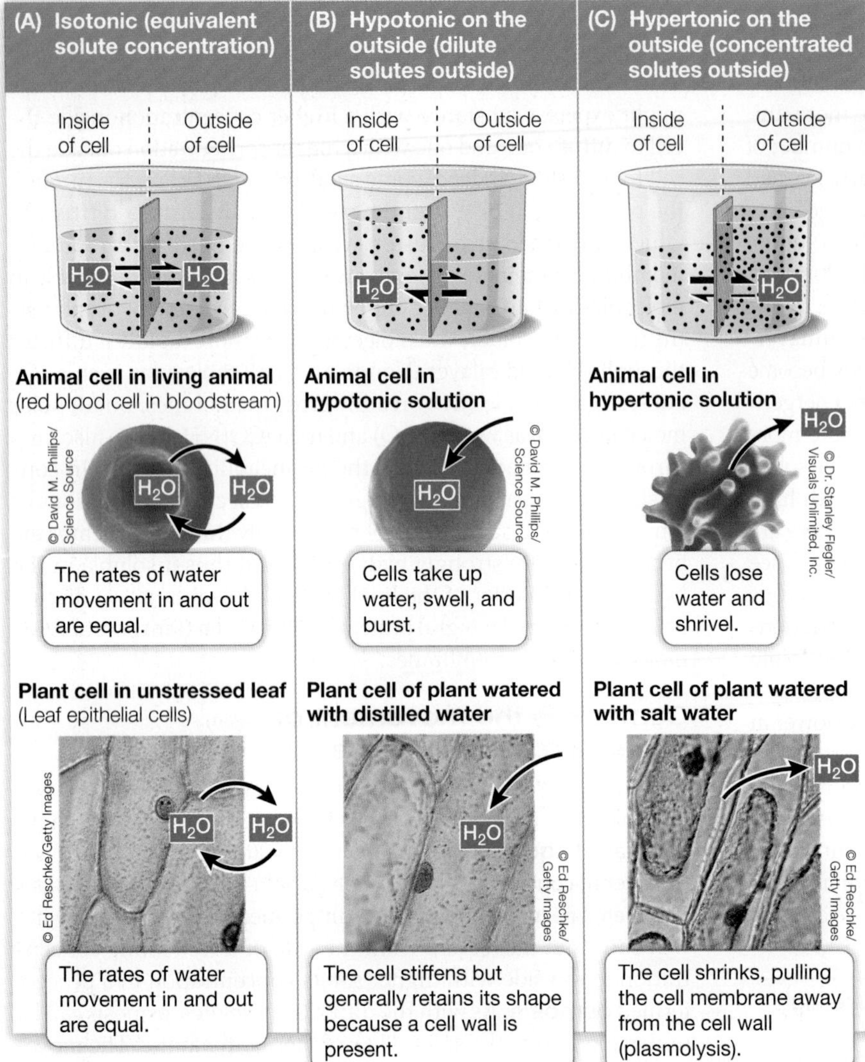

FIGURE 4.3 Net Movement of Water by Osmosis from Areas of High to Low Solute Concentration Density of particles represents the concentration of membrane-impermeable solute. Black arrows represent net movement of water across the water-permeable membrane. **(A)** In isotonic solutions there is no net water movement. **(B)** In hypotonic solutions, net water movement is into the cell. **(C)** In hypertonic solutions, net water movement is out of the cell.

At the beach you accidentally swallow a lot of salty sea water. What happens to the cells lining your gut?

LINK The concept of osmotic pressure and water movement underlies the regulation of body fluids by animal organs such as the kidney (see **Key Concept 29.2**) and the movement of water up a tree trunk (see **Key Concept 24.4**).

When determining whether osmosis will occur between two solutions separated by a membrane, one must know the concentrations of each solute in the two solutions, and the permeability of the membrane to the solutes. Solutes that cannot cross the membrane determine the tonicity of a solution. **Tonicity** is a measure of the *relative* concentration of solutes on either side of a membrane and determines the direction and extent of net water movement across a membrane. An **isotonic** solution has the same concentration of membrane-impermeable solutes as the other solution. If the solution on one side of the membrane is isotonic to the solution on the other side, there is no net water flow (**FIGURE 4.3A**). A **hypotonic** solution has a lower concentration of membrane-impermeable solutes than the other solution (**FIGURE 4.3B**). A **hypertonic** solution contains a higher concentration of membrane-impermeable solutes than the other solution (**FIGURE 4.3C**). When there is a net flow of water, it occurs from the hypotonic to the hypertonic solution. Osmolarity—the total amount of dissolved solute particles, including those that can cross the membrane—is a related, but different, measure of solute concentration. (Tonicity is affected only by those solute particles that cannot cross the membrane.) Permeable solutes will diffuse across the membrane, equalizing their concentrations on either side, and thus do not contribute to net water movement (**FIGURE 4.4**).

The concentration of membrane-impermeable solutes outside the cell determines the direction of water flow for all animal cells. In laboratory experiments, a red blood cell takes up water from a solution that is hypotonic to the cell's contents, and eventually bursts because its cell membrane cannot withstand the pressure created by the water entry (see Figure 4.3B). Conversely, the cell shrinks if the solution surrounding it is hypertonic to its contents (see Figure 4.3C). As these experiments show, the integrity of blood cells is absolutely dependent on the maintenance of a constant, and equal, concentration of total membrane-impermeable solutes in the surrounding blood plasma (see Figure 4.3A).

In contrast to animal cells, the cells of many other groups, including archaea, bacteria, plants, fungi, and some protists, have cell walls that limit their volume and keep them from bursting (see Key Concept 4.4). Cells with sturdy walls take up a limited amount of water, and in so doing they build up internal pressure against the cell wall, which prevents more water from entering. This pressure within the cell is called **turgor pressure**; it keeps the green parts of plants upright and is the driving force for enlargement of plant cells (see Key Concept 25.1).

Membrane permeability is increased by membrane proteins

Polar molecules, especially large ones, and charged substances do not readily diffuse across phospholipid bilayers. However, they

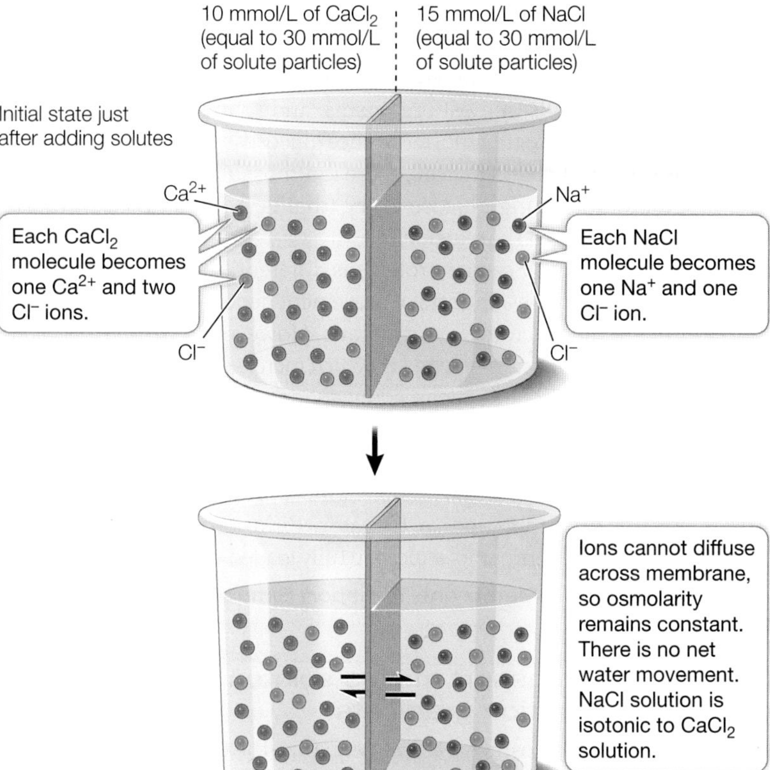

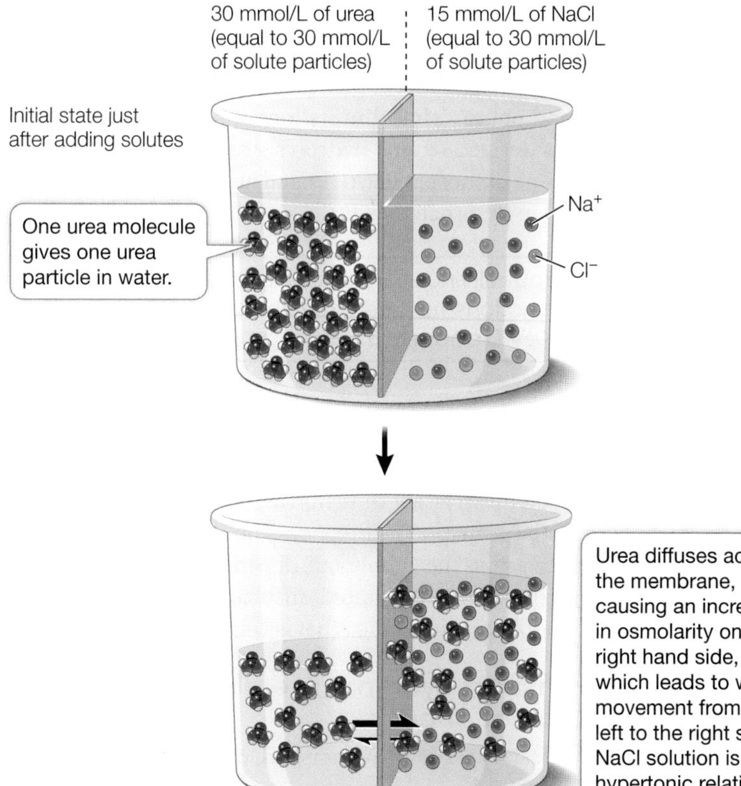

FIGURE 4.4 Osmolarity and Tonicity Osmolarity measures the total concentration of solute particles, both permeable and impermeable. Tonicity measures the relative concentration across a membrane of only those solute molecules to which the membrane is impermeable. Suppose two solutions are placed on either side of a membrane. **(A)** If the two solutions have the same osmolarity, they will be isotonic if the membrane that separates them is impermeable to all of the solute particles. Such is the case in this example, where the membrane is impermeable to Ca^{2+}, Na^+ and Cl^- ions. **(B)** In contrast, the two solutions may have the same osmolarity but not be isotonic if some solute particles can diffuse across the membrane. In this example, urea is membrane-permeable but Na^+ and Cl^- ions are not, and so urea rapidly diffuses to give equal concentrations on both sides of the membrane, causing the solution on the right side to have higher osmolarity and thus be hypertonic to the solution on the left side. This difference leads to net water flow towards the right side.

can cross the membrane passively (that is, without the input of energy) with the help of two kinds of integral transmembrane proteins, **channel proteins** and **carrier proteins.** As the name indicates, channel proteins form channels in the membrane through which certain substances can pass. Carrier proteins bind to specific substances and speed up their diffusion across the membrane.

Passive transport of solutes down their concentration gradient with the help of an integral transmembrane protein is called **facilitated diffusion**. Channel and carrier proteins that facilitate diffusion across a membrane do so in both directions. Likewise, as mentioned earlier, protein channels called aquaporins in the membranes of cells facilitate the passage of water (**FIGURE 4.5A**). Some water can also cross the membrane by "hitchhiking" with some ions, such as Na^+, as they pass through **ion channels**—channels formed by transmembrane proteins for the passage of specific ions by diffusion. (Recall that in solution, ions are surrounded by a water shell; see Figure 2.15). As you will see in later chapters, the movement of ions across membranes through channels is important in many biological processes, including ATP (adenosine triphosphate) production within the mitochondria, the electrical activity of the nervous system, and the opening of pores in plant leaves to allow gas exchange with the environment. Several types of ion channels have been identified, each of them specific for a particular ion. All of them show the same basic structure of a hydrophilic channel that allows a particular ion to pass through.

Most ion channels are "gated," meaning they can be opened or closed to ion passage. A gated channel opens when a stimulus causes a change in the three-dimensional shape of the channel (**FIGURE 4.5B**). In some cases this stimulus is the binding of a chemical signal, or ligand. Channels controlled in this way are called ligand-gated channels. In contrast, a voltage-gated channel is stimulated to open or close by a change in the voltage (electric charge) across the membrane.

(A) An aquaporin

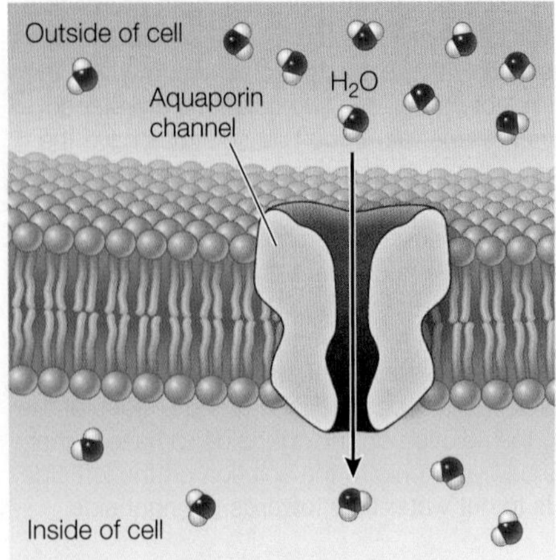

(B) A gated ion channel

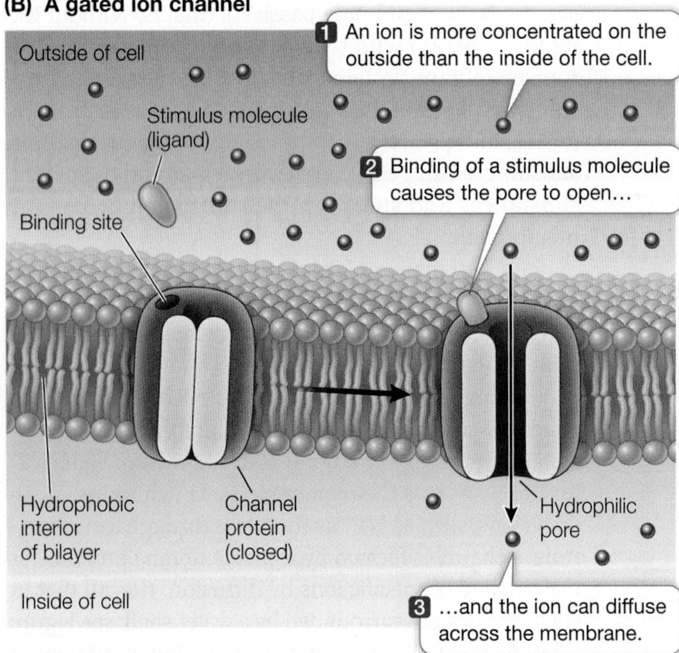

FIGURE 4.5 Channel Proteins Transport across the membrane can be facilitated by several different channel proteins. **(A)** Aquaporins increase a membrane's permeability to water. **(B)** Gated ion channel proteins increase permeability to specific ions as needed in response to a particular signal.

LINK As you will see in **Key Concept 31.2**, voltage-gated channels are especially important in the conduction of electric signals in the nervous system. Opening and closing of voltage-gated channels alters the electric charge difference across the cell membranes of nerve cells (neurons), which is critical to the transmission of information within and between cells.

Carrier proteins transport polar molecules such as sugars and amino acids. Carrier proteins facilitate diffusion by binding the transported substance on one side of the membrane and releasing it on the other side. Facilitated diffusion of the monosaccharide glucose is accomplished with carrier proteins called **glucose transporters** (**FIGURE 4.6A**). Glucose is the major energy source for cells in most organisms, and so many cell membranes contain glucose transporters. Since glucose is usually broken down as soon as it enters the cell, there is almost always a concentration gradient favoring glucose entry (that is, a higher concentration outside the cell than inside).

Transport by carrier proteins is different from simple diffusion. In both processes, the rate of movement depends on the concentration gradient across the membrane. However, in carrier-mediated transport, a point is reached at which increases in the concentration gradient are not accompanied by an increased rate of diffusion because all the carrier proteins are already occupied. At this point, the facilitated diffusion system is said to be saturated (**FIGURE 4.6B**). Because there are only a limited number of carrier protein molecules per unit of membrane area, the rate of diffusion reaches a maximum when all the carrier molecules are fully loaded with solute molecules.

Activity 4.3 Membrane Transport Simulation

Active transport moves solutes against their concentration gradients

Diffusion and osmosis tend to equalize the concentrations of solutes on either side of a membrane. However, one hallmark of cells is that their internal solute concentrations are quite different from their external environment. To achieve this, a cell must sometimes move substances *against their concentration gradients*. This process requires the input of energy and is known as active transport. Often the energy source is the hydrolysis of the nucleotide ATP. As we will discuss in Chapter 5, ATP has readily accessible chemical-bond energy stored in its terminal phosphate bond. This energy is derived from a hydrolysis reaction that converts ATP to adenosine diphosphate (ADP) and an inorganic phosphate ion (PO_4^{3-}, often designated as P_i).

LINK You will find more details about how ATP functions as an energy shuttle in cells in **Key Concept 5.1**.

The differences between diffusion and active transport of solutes are summarized in **TABLE 4.1**. In cases of simple and facilitated diffusion, ions or molecules can move down their concentration gradients in either direction across the cell membrane. In contrast, active transport is directional, and moves a substance in one direction across the membrane, depending on the transport protein's function. As in facilitated diffusion, there is a specific carrier protein for each substance that is transported.

Active transport systems are distinguished by their source of energy

There are two basic types of active transport, primary and secondary.

1. **Primary active transport** uses chemical-bond energy released during the hydrolysis of ATP or another chemical reac-

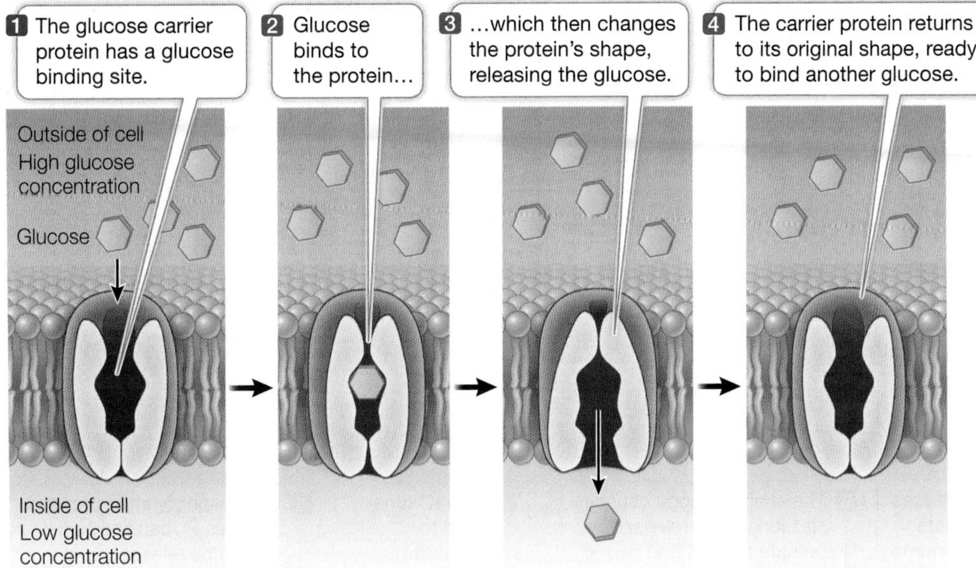

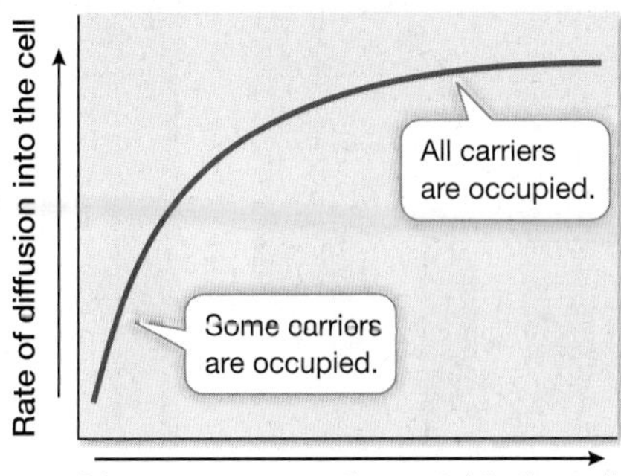

FIGURE 4.6 Facilitated Transport of Glucose by the Glucose Transporter **(A)** Transport across the membrane can be facilitated by carrier proteins, which bind to the solute. The glucose transporter binds glucose on the side of the membrane where its concentration is high and releases it on the side where its concentration is low. **(B)** The rate of glucose diffusion into the cell increases linearly with concentration when glucose concentrations outside the cell are low, but levels off and approaches a maximum at high concentrations because all glucose transporters are being used (that is, when the transporters are saturated).

tion, or light energy, to transport a solute against its concentration gradient. For example, the concentration of potassium ions (K^+) inside a cell is often much higher than the concentration in the fluid bathing the cell. In contrast, the concentration of sodium ions (Na^+) is often much higher outside the cell. These gradients are maintained by an integral protein in the cell membrane that pumps Na^+ out of the cell and K^+ into the cell. Appropriately called the **sodium–potassium (Na^+–K^+) pump**, this integral membrane-spanning protein catalyzes the hydrolysis of a molecule of ATP to ADP and a free phosphate ion (P_i) and uses the transferred energy to bring two K^+ ions into the cell and to export three Na^+ ions (**FIGURE 4.7**).

2. **Secondary active transport** uses the energy stored in an ion concentration gradient or an electrical gradient, previously generated by primary active transport, to transport the solute against its concentration gradient. Secondary transport is often termed **coupled transport** or **co-transport** because two substances are simultaneously transported across a membrane by one protein. For example, once the Na^+–K^+ pump establishes a concentration gradient of sodium ions by primary active transport, the passive diffusion of some Na^+ back into a cell can provide energy for the secondary active transport of glucose into the cell. The coupling of Na^+ movement down its concentration gradient with glucose transport up its concentration gradient is achieved by the sodium–glucose co-transporter (**FIGURE 4.8**). This co-transporter is used when glucose is absorbed from the digestive tract, where it is at low concentration, into the bloodstream, where it is at high concentration. In

TABLE **4.1** | Solute Transport across Membranes

	Simple diffusion	Facilitated diffusion (channel or carrier protein)	Active transport
Cellular energy required?	No; transport down concentration gradient	No; transport down concentration gradient	Yes; transport against concentration gradient
Driving force	Concentration gradient	Concentration gradient	ATP hydrolysis or ion gradient
Membrane protein required?	No	Yes	Yes
Specificity	No	Yes	Yes
Example	CO_2, urea	Various ions, glucose	Various ions

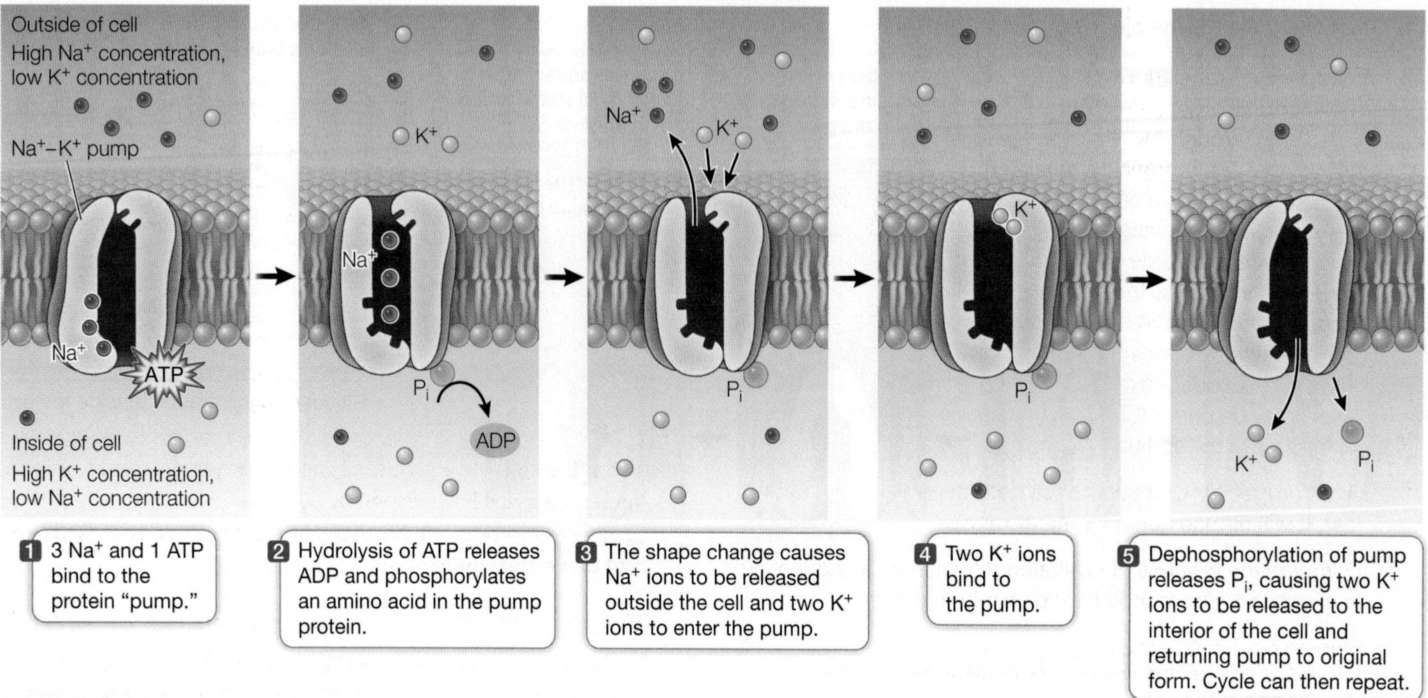

FIGURE 4.7 Primary Active Transport: The Sodium–Potassium Pump Solutes can be transported across a membrane against their concentration gradient using primary active transport. The sodium–potassium pump uses chemical-bond energy transferred during the hydrolysis of ATP to drive both Na^+ and K^+ ions against their concentration gradients.

some cases of secondary active transport, the ion and the transported molecule move in opposite directions, whereas in other cases they move in the same direction (as for glucose and Na^+ in the cells lining the digestive tract). Secondary active transport aids in the uptake of amino acids and sugars, which are essential raw materials for cell maintenance and growth.

Animation 4.2 Active Transport

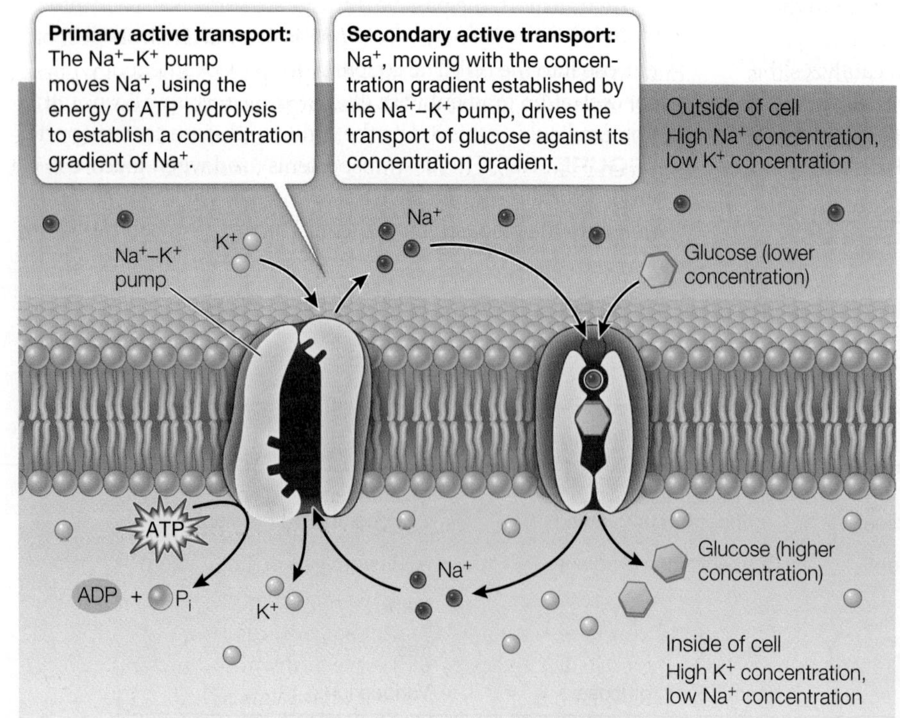

FIGURE 4.8 Secondary Active Transport: The Sodium-Glucose Co-Transporter Secondary active transport uses a concentration gradient of one solute (generated by primary active transport) to move a second solute against its concentration gradient. The sodium–glucose co-transporter uses diffusion of sodium down its concentration gradient to actively transport glucose up its concentration gradient.

Would you expect to find any cells possessing both the sodium–glucose co-transporter and the glucose transporter (see Figure 4.6) in their cell membranes?

REVIEW & APPLY | 4.2

R Biological membranes are permeable to small solutes that are nonpolar or uncharged, which can diffuse down their concentration gradients through the membrane without channel proteins or carrier proteins. Membrane permeable molecules will be found at equal concentrations on both sides of a membrane. Membranes are also permeable to water, which will show net flow from the hypotonic to hypertonic side by osmosis. Charged and larger polar molecules cannot cross the membrane without the help of integral transport proteins. Gated channel proteins and carrier proteins control diffusion by opening or closing channels. Carrier proteins can set up concentration gradients using active transport. Active transport may be primary, as in the sodium–potassium pump, or secondary, as in the sodium–glucose co–transporter.

A
1. What properties of a substance determine whether, and how fast, it will diffuse across a membrane?
2. Compare the process of facilitated diffusion by a channel protein versus a carrier protein. Which might be faster, and why?
3. Placing cut flowers in salt water causes them to rapidly wilt. In contrast, in distilled water cut flowers do not wilt. Explain.
4. Some bacteria have an integral membrane transport protein called bacteriorhodopsin. This transporter absorbs energy from light and uses it to pump protons (H^+) out of the cell, against a concentration gradient. What type of transport mechanism does this represent (passive diffusion, facilitated diffusion, primary active transport, or secondary active transport)?
5. Normally, secondary active transport couples diffusion with active transport. In brown fat, an energy store in mammals, there is a process in mitochondria in which diffusion is not coupled to active transport. This produces heat. Why?

We have examined several passive and active ways by which ions and small molecules can enter and leave cells. Next we will discuss how large molecules such as proteins, polysaccharides, and nucleic acids are transported across membranes.

4.3 Vesicles Are Used to Transport Large Molecules across Membranes in Eukaryotes

LEARNING OBJECTIVES

By the end of this key concept you should be able to:

4.3.1 **Describe the processes and purposes of endocytosis and exocytosis.**

4.3.2 **Differentiate among phagocytosis, pinocytosis, and receptor-mediated endocytosis.**

Many macromolecules are simply too large and too charged or polar to pass through biological membranes. In one sense, this is a fortunate property—cellular integrity depends on containing these macromolecules in specific locations. However, some cells must sometimes take up or **secrete** (release to the external environment) intact large molecules. Even if a membrane were permeable to such molecules, they would diffuse extremely slowly because of their large size. It takes a completely different mechanism than diffusion to move intact large molecules across membranes. In **eukaryotic cells** (those with a nucleus; see Key Concept 4.5), including plants, fungi, protists and animals, large molecule transport involves packaging the molecules and a small volume of fluid into small, membrane-enclosed structures termed **vesicles**. Transport of large molecules or particles out of the cells via vesicles is called **exocytosis**, and import of such molecules is called **endocytosis** (**FIGURE 4.9**). Both processes are possible because membranes can fuse with and bud off vesicles quite easily.

Exocytosis moves materials out of the cell

Exocytosis is the process by which materials packaged in vesicles are secreted from the cell (see Figure 4.9A). As you will see in Key Concept 4.5, proteins destined for export from the cell end up in such vesicles. After a vesicle is formed around the substance to be secreted, it migrates toward the cell membrane, with which it fuses. An opening is made to the outside of the cell. Once emptied of its contents, the vesicle membrane is incorporated into the cell membrane. Many different kinds of substances are released to the external environment via exocytosis, including waste products and secreted proteins, such as digestive enzymes from the pancreas, neurotransmitters from neurons, and materials for the construction of the plant cell wall. You will encounter these processes in later chapters.

Endocytosis moves materials into the cell

Endocytosis is a general term for a group of processes that bring small molecules, macromolecules, large particles, and even other cells into eukaryotic cells (see Figure 4.9B). In this process, the cell membrane invaginates (folds inward), forming a small pocket around materials from the environment. The pocket deepens, forming a vesicle. This vesicle separates from the cell membrane and moves with its contents to the cell's interior. The contents inside the vesicle are separated from the **cytosol** (intracellular fluid of the cell) by the vesicle membrane. There are three types of endocytosis: phagocytosis, pinocytosis, and receptor-mediated endocytosis:

1. In **phagocytosis** ("cellular eating"), a very large vesicle called the **phagosome** is formed when a cell engulfs and ingests a large particle or even an entire cell. Phagocytosis is initiated by specific cell membrane proteins called receptors that recognize a specific ligand on the surface of the particle or cell. The binding of the ligand to the receptor causes the phagocytic cell to engulf the particle or other cell. Phagocytosis is restricted to specialized cells; for example, unicellular protists use phagocytosis for feeding, and some white blood cells use phagocytosis to engulf foreign cells and substances.

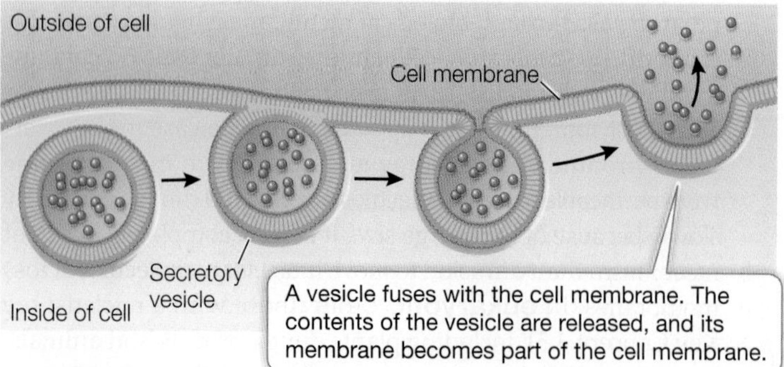

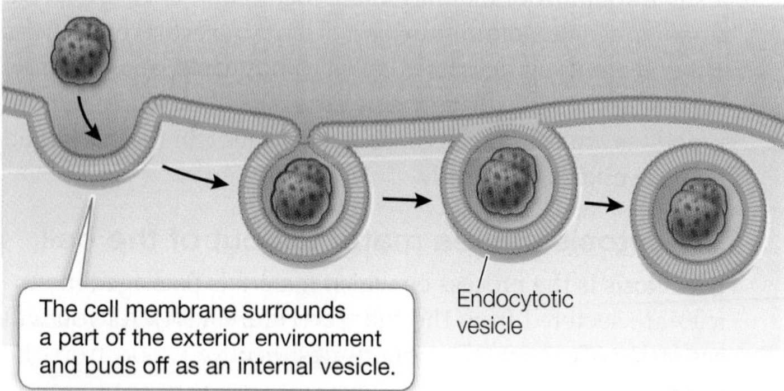

FIGURE 4.9 Exocytosis and Endocytosis Exocytosis **(A)** and endocytosis **(B)** are used, respectively, to move large particles and fluid out of and into the cell.

Animation 4.3 Endocytosis and Exocytosis

The phagosome that forms usually fuses with another vesicle in the cell called the lysosome, where the vesicle's contents are digested (see Key Concept 4.5).

Media Clip 4.1 An Amoeba Eats by Phagocytosis

2. In **pinocytosis** ("cellular drinking"), vesicles bring fluids and dissolved substances, including proteins, into the cell. Sometimes called fluid endocytosis, pinocytosis is relatively nonspecific regarding what it brings into the cell. For example, pinocytosis goes on constantly in the endothelium—the single layer of cells that separates a blood capillary from the surrounding tissue. Pinocytosis allows cells of the endothelium to rapidly acquire fluids and dissolved solutes from the blood.

3. **Receptor-mediated endocytosis** is a mechanism for bringing specific large molecules, recognized by specific receptors, into the cell. The receptors bind their specific ligand and then migrate laterally in the membrane to pits on the surface of the cell membrane. These pits are slight depressions that have their cytoplasmic surfaces coated by a protein, which is often clathrin. Clathrin (or other protein) molecules strengthen and stabilize the pit and the vesicle after it forms (**FIGURE 4.10**). The uptake process is similar to that in phagocytosis. However, in receptor-mediated endocytosis, the receptor is specific for a given macromolecule or aggregate, making it possible to control the abundance in the cell of each type of ingested molecule.

An excellent example of receptor-mediated endocytosis is the way cholesterol is taken up by most mammalian cells. Cholesterol and triglycerides, which have low solubility in water, are packaged by liver cells into lipoprotein particles (see Figure 3.3C), which are released into the bloodstream. When a particular cell requires cholesterol, it produces low-density lipoprotein (LDL) receptors, which are inserted into the cell membrane and allow LDL binding and subsequent endocytosis. In most people, the liver takes up unused LDLs for recycling. However, some people have an inherited disease called familial hypercholesterolemia. They have a defective LDL receptor in their livers, which prevents receptor-mediated endocytosis of LDLs in the liver. As a result, these people have

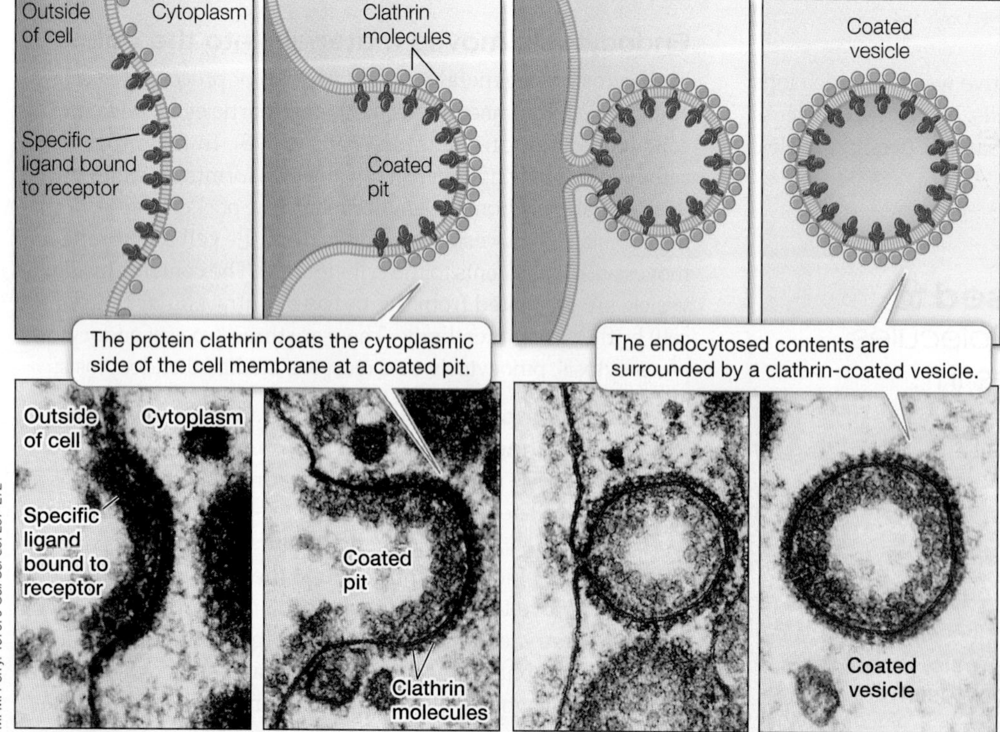

FIGURE 4.10 Receptor-Mediated Endocytosis Receptors bound to their cargo can move laterally in the membrane to coated pits, where endocytosis then occurs.

high levels of cholesterol in the blood. The cholesterol builds up on the walls of the arteries that nourish the heart, which can cause heart attacks. In extreme cases where only the defective receptor is present, children and teenagers can have heart disease typical of older people.

In recent years it has become clear that receptor-mediated endocytosis also plays an important role in cell signaling, which we will discuss in Key Concept 6.1.

REVIEW & APPLY | 4.3

R In eukaryotes, molecules that are too large, too charged, or too polar to pass through biological membranes with the help of transport proteins may instead be taken into or out of the cell in membrane-enclosed vesicles. In exocytosis, vesicles ferry large molecules out of the cell. In endocytosis, invaginations of the cell membrane form vesicles around materials from the outside environment, allowing a cell to import large molecules, fluid, or even small cells from outside.

A
1. What property of the cell membrane makes transport by vesicles feasible?
2. What is the difference between phagocytosis and pinocytosis?
3. Would a small molecule such as an amino acid enter a cell by receptor-mediated endocytosis? Why or why not?

While membranes from all organisms share numerous features, cells vary enormously in their size, shape, and ability to move. In the next key concept we will discuss the underlying causes of this variation.

4.4 Cell Size, Shape, and Ability to Move Are Determined by Internal and External Structures

LEARNING OBJECTIVES

By the end of this key concept you should be able to:

4.4.1 Explain how surface area-to-volume ratio constrains cell size.

4.4.2 List the primary functions of cytoskeletal filaments.

4.4.3 Classify the three major groups of cytoskeletal filaments based on their size and whether they show dynamic instability.

4.4.4 Compare the extracellular structures of prokaryotes, plant cells, and animal cells.

4.4.5 List the primary functions of the types of cell junctions found in animals.

If cells were simply an inside separated from an outside by a phospholipid membrane, we might expect them to be spherical, similar to an oil droplet in water. However, cells are often not round and instead come in a variety of shapes. The ability of a cell to have a shape other than spherical is due to the presence of a cell skeleton. The skeleton of cells not only allows them to have multiple shapes, but it has important implications for the size of cells and their ability to move. Before considering the effects of the skeleton on these attributes, let's consider why cells tend to be small.

Small cells have a high surface area-to-volume ratio

The vast majority of cells are tiny, with diameters ranging from about 1 to 100 micrometers. There are some exceptions, notably the relatively enormous eggs of fishes and amphibians (a few millimeters in diameter) and of reptiles and birds (a few centimeters in diameter), which store nutrients for development (see Chapter 35), and the cells of a few species of algae and bacteria that are large enough to be viewed with the unaided eye. Small cell size is a practical necessity arising from the decrease in the **surface area-to-volume ratio** for an object that increases in size but remains the same shape. For example, the surface area and volume of a sphere are proportional to the radius squared and cubed, respectively. Thus, doubling the radius of a sphere increases the surface area by fourfold and increases the volume by eightfold, resulting in the surface area-to-volume ratio declining by twofold (**FIGURE 4.11**).

The *volume* of a cell determines the amount of metabolic activity, and thus the cell's need for resources and its rate of waste production per unit of time. Since substances enter and exit the cell through the cell membrane, cell *surface area* determines the rate at which resources can enter and wastes can leave a cell. You will recall that in Key Concept 4.2 we cited surface area as a factor in the rate of diffusion of a substance across a membrane.

Radius (μm)	1	2	4
Surface area $4\pi r^2$ (μm^2)	4π	16π	64π
Volume $4/3\pi r^3$ (μm^3)	$4/3\,\pi$	$32/3\,\pi$	$256/3\,\pi$
Surface area-to-volume ratio ($\mu m^2/\mu m^3$)	3	1.5	0.75

FIGURE 4.11 Why Cells Are Small As an object increases in size, the surface area-to-volume ratio decreases if the object's shape remains the same. This fact constrains cell size. A cell cannot get so large that the surface area of the cell membrane is not adequate for the diffusion of substances into and out of the cell.

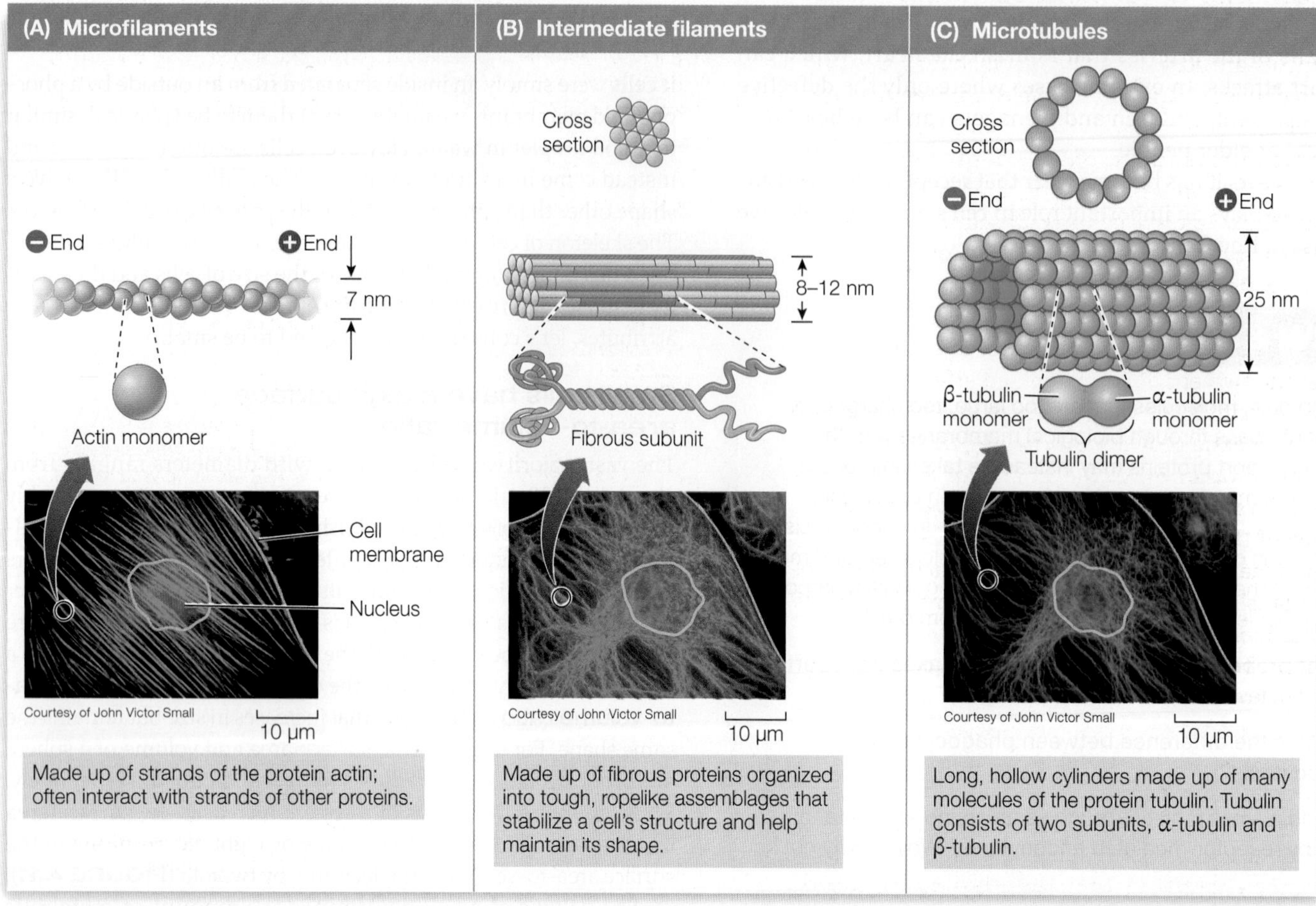

FIGURE 4.12 Cytoskeleton Filaments The three different classes of cytoskeletal filaments vary in size. Microfilaments and microtubules have directionality (a plus and a minus end).

As the surface area-to-volume ratio declines, it becomes more difficult for a cell to obtain sufficient resources and eliminate waste per of unit time, which favors small cell size. In addition, substances must move from one location to another within the cell; the smaller the cell, the more easily this is accomplished.

While surface area-to-volume ratio limits cell size, some cells change their shape as they become larger, allowing them to maintain a sufficiently high surface area-to-volume ratio. Shape change can involve increased infoldings of the cell membrane, or size increases that occur in only one dimension. For example, nerve cells in humans, which can be more than a meter in length, are extremely narrow, which keeps the surface area-to-volume ratio adequately large. For cells that grow without changes in their shape, the rate of exchange of materials per unit of area of the cell membrane must become larger to offset the reduction in surface area per unit of volume. To do this, cells can increase the number of channel and carrier proteins in the membrane.

Several kinds of protein filaments are present in cells

The interior of a cell is supported by a meshwork of protein filaments called the **cytoskeleton**, literally "cell skeleton." Each type of filament is a polymer, made up of protein monomers. The cytoskeleton has several important roles:

- It supports the cell and maintains its shape.
- It maintains the position of internal cell compartments called organelles (described in Key Concept 4.5) and other particles within the cell.
- It moves organelles and other particles around within the cell.
- It interacts with extracellular structures, helping anchor the cell in place.

Protein filaments have been best studied in eukaryotic cells, which have nuclei and membrane-surrounded organelles (in contrast to **prokaryotic cells**, such as bacteria, which do not have a nucleus and generally lack membrane-surrounded organelles). The cytoskeletal protein filaments are classified into three groups: **microfilaments**, which are the smallest in diameter and made of **actin**; **intermediate filaments**, which are made of a variety of different proteins; and **microtubules**, which are the largest in diameter and made of tubulin (**FIGURE 4.12**). While our focus will be on eukaryotic filaments, it is worth noting that prokaryotic cells also possess filaments that are equivalent to these three classes of protein filaments, with similar functions.

Microfilaments determine cell shape and function in cell movement

Microfilaments are actin polymers, each about 7 nanometers (nm) in diameter, that are assembled from actin monomers that usually add to the filament at one end (the "plus end") and are removed at the other end (the "minus end") (see Figure 4.12A). In a filament of constant length, assembly and disassembly are in equilibrium, such that the two processes occur at the same rate. The relative rates of assembly and disassembly are not the same, however, in filaments that shorten or lengthen. Assembly and disassembly may be altered by the binding of other, non-actin proteins at the two ends. The combined effects of adding monomers (assembly) and removing them (disassembly) and the roles of extra proteins that can regulate how much these processes occur make the lengths of microtubules quite variable, a situation termed **dynamic instability**. The importance of this phenomenon is shown by what happens when dynamic instability of microfilaments does not occur. The mushroom toxin phalloidin irreversibly binds to and stabilizes microfilaments so that they cannot be disassembled. This has disastrous results in cells, as they die. Usefully, however, the strong binding of phalloidin to polymerized actin makes it perfect for visualizing microfilaments in dead cells in the microscope.

Microfilaments are important in maintaining overall cell shape and causing localized shape changes in regions of the cell. For example, microfilaments are involved in the flowing movement of the cytoplasm called **cytoplasmic streaming**, and in the "pinching" contractions that divide an animal cell into two daughter cells when cells reproduce (see Key Concept 7.2). Microfilaments are also involved in the formation of cellular extensions called pseudopodia (*pseudo*, "false," + *podia*, "feet") that enable some cells, such as *Amoeba*, to move (**INVESTIGATION FIGURE 4.13**). In many types of animal cells, most notably muscle cells, actin filaments are associated with the motor protein myosin, and the interactions of the actin and myosin are responsible for cell contractions.

Intermediate filaments are diverse and stable

There are at least 50 different kinds of intermediate filaments that are grouped into six different molecular classes based on their amino acid sequence. Many occur in just a few cell types in animals. Intermediate filaments are tough, ropelike protein assemblages 8–12 nm in diameter (see Figure 4.12B). They are more permanent than the other two types of filaments and do not show dynamic instability.

Intermediate filaments anchor cell structures in place. In some cells, they help maintain the positions of the nucleus and other organelles in the cell. Intermediate filaments also resist tension. For example, one class of intermediate filament consists of fibrous keratin proteins, which are important in maintaining rigidity of body

INVESTIGATION

FIGURE 4.13 The Role of Actin Microfilaments in Cell Movement Is microfilament polymerization the mechanism responsible for cell movement in *Amoeba*? To find out, experimenters injected *Amoeba* with the drug cytochalasin B, which inhibits the polymerization of actin monomers, preventing the formation of microfilaments.[a]

HYPOTHESIS

Amoeboid cell movements are caused by microfilament polymerization.

METHOD

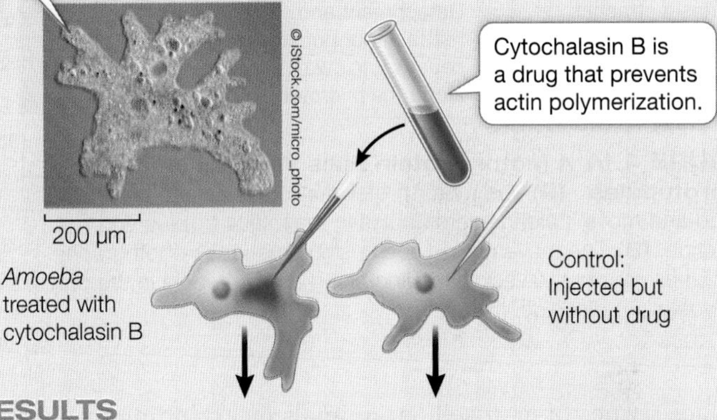

RESULTS

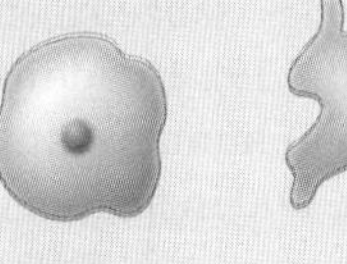

CONCLUSION

Microfilaments are essential for amoeboid cell movement.

ANALYZE THE DATA

Several important controls were done to validate the conclusions of this experiment. One was to not treat *Amoeba* at all. One was to inject *Amoeba* with a solution lacking the drug. Two other controls were to inject cycloheximide, which inhibits new protein synthesis, or colchicine, which inhibits the polymerization of tubulin. Here are all of the results:

Experimental condition	Rounded cells (%)
Untreated	4
Injected with solution lacking a drug	3
Injected with Cytochalasin B	95
Injected with Colchicine	4
Injected with Cycloheximide	3

1. Explain the reasoning behind each experimental condition. Why were the controls important?
2. Interpret the results of this experiment. What can you conclude about movements in *Amoeba* and the cytoskeleton?
3. If you wanted to demonstrate that cytochalasin B is not a general cell toxin, which could explain the lack of cell movement following injection, what experiments might you perform?

[a]T. D. Pollard and R. R Weihing. 1974. *CRC Crit Revs Biochem* 2: 1–65.

A similar **Analyze the Data** exercise may be assigned in **Achieve**.

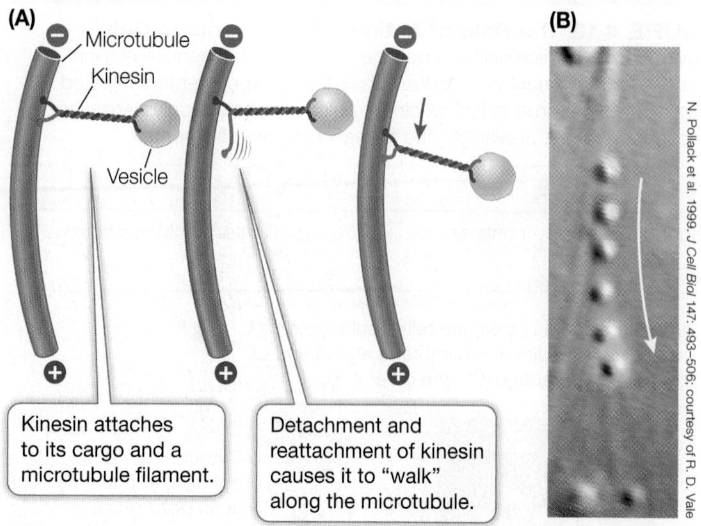

FIGURE 4.14 A Motor Protein Pulls Vesicles along Microtubules **(A)** The motor protein kinesin is able to bind to cargo and move it along microtubules in the minus-to-plus end direction. **(B)** The movement is seen using time-lapse photography of a small membrane vesicle moving along a microtubule in the protist *Dictyostelium discoideum*.

surface tissues of multicellular animals by extending through the cytoplasm and connecting to specialized membrane structures called desmosomes (see Figure 4.19). In addition, keratins are found in cells of the outer layer of human skin, making it tough and mostly water-impermeable, and in feathers, hair, claws, and fingernails.

Microtubules are the thickest elements of the cytoskeleton

Microtubules (see Figure 4.12C) are long, hollow, unbranched cylinders about 25 nm in diameter. Microtubules are polymers of the protein **tubulin**. Tubulin dimers consist of one molecule each of α-tubulin and β-tubulin. Each microtubule cylinder consists of 13 chains of tubulin dimers surrounding the hollow center. Like microfilaments, microtubules show dynamic instability, with plus and minus ends and various associated proteins. However, unlike actin microfilaments that are stable in length when rates of assembly and disassembly are equal, microtubules have a growth phase, followed by a phase where no assembly or disassembly occurs, and then a rapid disassembly phase. Both assembly and disassembly occur from the same (plus) end.

Microtubules form a rigid internal skeleton for some cells or cell regions, and they act as a framework along which motor proteins such as kinesins can move various organelles and other items within the cell. Motor proteins bind to their cargo and "walk" it along a microtubule by a repeated series of shape changes. (These shape changes are powered by energy from ATP hydrolysis.) Some motor proteins move toward the minus end of the microtubule (dyneins, for example; see below) and others move towards the plus end (kinesins, for example; **FIGURE 4.14**), allowing bidirectional movement along microtubules. Similar bidirectional movement enabled by different motor proteins occurs along actin microfilaments.

LINK Microtubules are important components of the spindle, which separates chromosomes during cell division; see **Key Concept 7.2**.

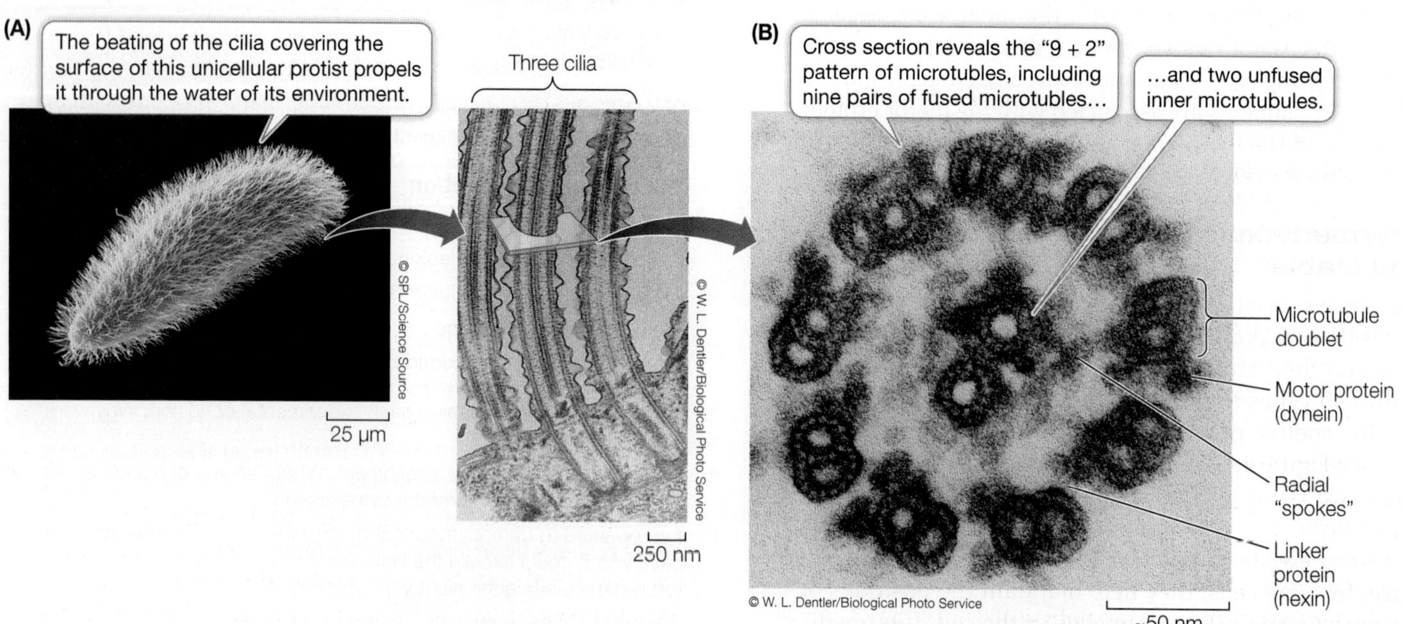

FIGURE 4.15 Cilia **(A)** Cilia are found on a variety of cells. **(B)** A cross section of a single cilium shows the underlying microtubule structure, consisting of nine microtubule doublets surrounding a pair of single microtubules, giving the 9 + 2 arrangement. The microtubule structure of a flagellum (not shown) is the same.

Some cells have projections from their cell membrane termed **flagella** (singular *flagellum*) and **cilia** (singular *cilium*) (**FIGURE 4.15**) that include a microtubule component. Cilia are short, only 0.25 micrometer in length, and can occur singly, as a **primary cilium**, or in tens to hundreds of **motile cilia**. Flagella are long, 100–200 micrometers, and occur singly, in pairs, or up to several per cell. Motile cilia and flagella are involved in movement either of the entire cell (as in protists, a group of unicellular eukaryotic organisms) or, in some ciliated cells, of fluid across the cell (as in the human respiratory system). The microtubules that are found in motile cilia and flagella not only support these structures but also cause them to bend, which moves them. (Bacterial cells have flagella that allow cell movement, but these are very different structures with no microtubules.)

In cross section, a motile cilium or eukaryotic flagellum is surrounded by the cell membrane and contains a "9 + 2" array of microtubules. As Figure 4.15B shows, nine fused pairs of microtubules—called doublets—form an outer cylinder, and one pair of unfused microtubules runs up the center. Movement of cilia and flagella is driven by a motor protein called dynein that is present on the outer nine pairs of microtubules. Dyneins cause the microtubule doublets to slide past one another in localized regions on one side of the cilium or flagellum, which causes the entire cilium or flagellum to bend in that location. The movement of dynein contractions along the cilium or flagellum results in a powerful beat that is able to propel fluid relative to the cell. Other proteins that bind to microtubules affect how the cilium or flagellum bends as it moves.

In the primary cilium, the central pair of unfused microtubules is missing and the cilium is unable to move. The primary cilium is important in signaling rather than in movement. Primary cilia have been shown to detect a variety of stimuli, including various chemical signals, light, gravity, and changes in osmolarity and temperature. They are also involved in setting up the left–right axis during vertebrate development; abnormal primary cilia can result in organs being on the wrong side of the body.

Cell walls provide support and protection

Although the cell membrane is the functional barrier between the inside and the outside of a cell, cells produce molecules and secrete them to the outside of the cell membrane. There these molecules form structures that play essential roles in protecting, supporting, or attaching cells to each other. Because they are outside the cell membrane, these structures are said to be extracellular. Several groups of living things (plants, archaea, bacteria, fungi, and some protists) build an extracellular **cell wall**, which is a relatively rigid structure exterior to the cell membrane that encloses the cell (**FIGURE 4.16**). Cell walls are tough, flexible structures that differ in their chemical makeup depending on the organism and cell type. Some are composed primarily of carbohydrates (cellulose in plants, chitin in fungi), others primarily of protein (in some archaea), and still others a mix (peptidoglycans in bacteria). The cell wall serves two primary functions. The first is structural support, and the second is protection from physical damage and pathogens. Cells with walls can be hypertonic with respect to their environment because the wall limits their volume and keeps them from bursting (see Key Concept 4.2).

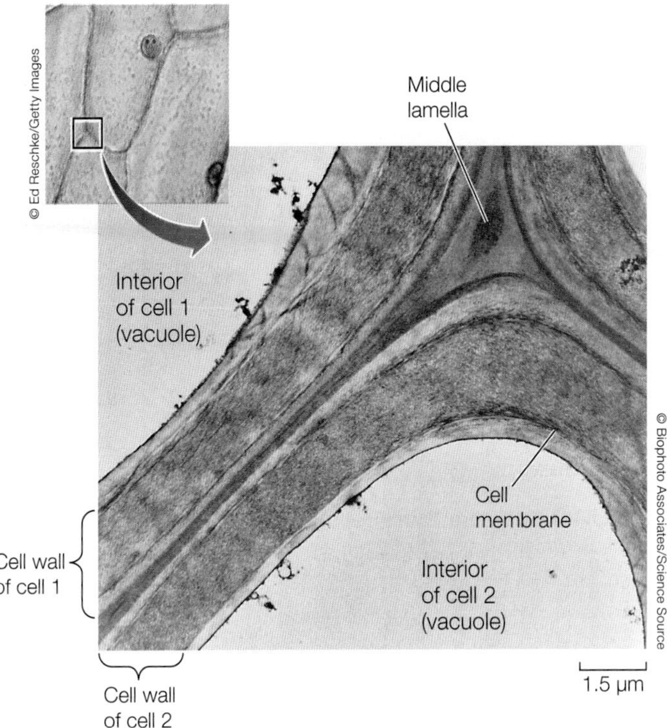

FIGURE 4.16 Cell Walls Cell walls in adjacent plant cells. An extracellular polysaccharide, pectin, forms the middle lamella that cements adjacent cell walls together.

Pectin synthesis occurs inside of cells. How might this large polysaccharide be transported to the outside of the cell?

The extracellular matrix supports various functions

Animal cells lack cell walls, but many are surrounded by, or in contact with, an **extracellular matrix** consisting of the fibrous protein **collagen** and gel-like proteoglycans (proteins with many attached carbohydrate residues). A third group of proteins links the collagen and the proteoglycan matrix together (**FIGURE 4.17**). The extracellular matrices of animal cells have several roles:

- They hold cells together in tissues.
- They contribute to the physical properties of cartilage, skin, and other tissues. For example, the mineral component of bone is laid down on an organized extracellular matrix.
- They help filter materials passing between different tissues. This is especially important in the kidney.
- They help orient cell movements during embryonic development and during tissue repair.

Proteins connect the cell membrane to the extracellular matrix. These proteins (for example, integrin) span the cell membrane and have two binding sites: one on the interior of the cell, usually to microfilaments or to intermediate filaments in the cytoplasm just below the cell surface, and the other to collagen in the extracellular matrix. These binding sites are noncovalent and reversible. When a cell moves its location in an organism, the first step is for integrin

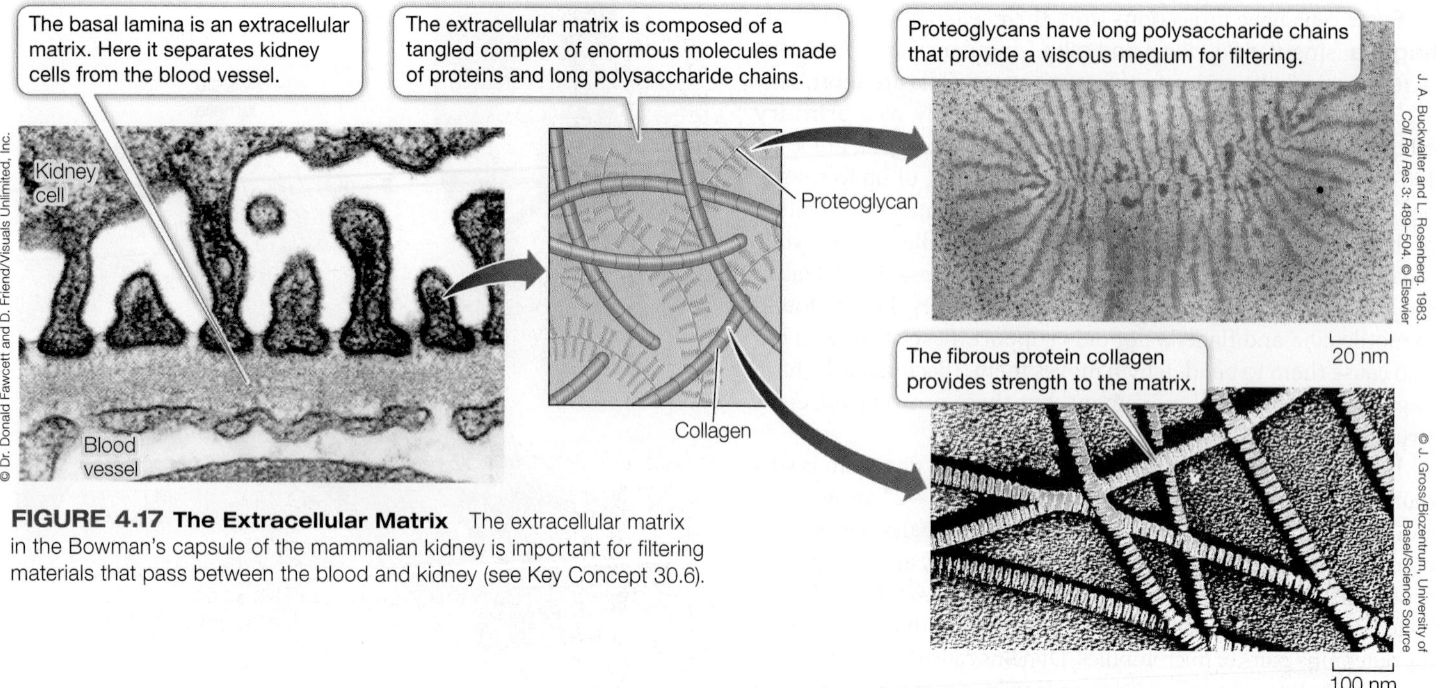

FIGURE 4.17 The Extracellular Matrix The extracellular matrix in the Bowman's capsule of the mammalian kidney is important for filtering materials that pass between the blood and kidney (see Key Concept 30.6).

to change its three-dimensional structure so that it detaches from the collagen (**FIGURE 4.18**).

Some bacterial cells also produce an extracellular matrix under some conditions. This matrix is especially important when bacteria form large aggregations of cells fixed to a surface in a so-called **biofilm**. The bacterial extracellular matrix facilitates communication among cells and the exchange of metabolites and genetic information. It thus allows an aggregation of bacterial cells to behave in a more coordinated manner, akin to cells in a single multicellular organism. Other microorganisms can also participate in biofilm formation, including several fungi. Biofilms can pose problems to human health because the extracellular matrix can prevent antibiotics from reaching bacterial cells in the interior, which makes the cells in these aggregations especially resistant to antimicrobials. Brushing your teeth to reduce plaque formation is a daily battle you wage against a microbial biofilm.

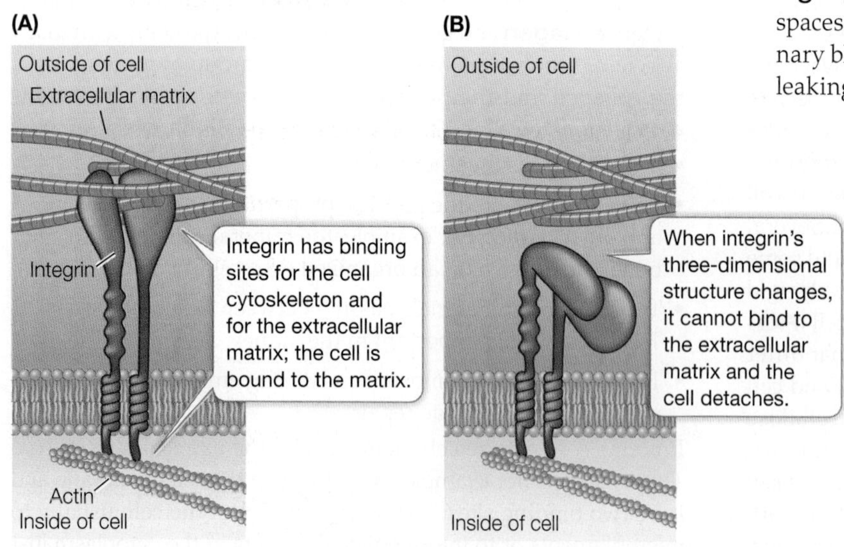

FIGURE 4.18 Cell Membrane Proteins Interact with the Extracellular Matrix Proteins such as integrin in the cell membrane enable animal cells to attach **(A)** and detach **(B)** from the extracellular matrix.

Cell junctions connect adjacent cells

In multicellular organisms, cells are joined via specialized **cell junctions**. **FIGURE 4.19** shows three types of junctions in animals:

1. **Tight junctions** prevent substances from moving through spaces between cells. For example, the epithelium of the urinary bladder contains tight junctions to prevent urine from leaking into the adjacent abdomen.
2. **Desmosomes** hold adjacent cells together with stable protein connections. This provides mechanical stability for tissues such as skin that are under stress due to movement. Desmosomes do, however, permit some materials to move around in the extracellular matrix between cells.
3. **Gap junctions** are channels that run between membrane pores in adjacent cells. Gap junctions allow substances such as ions, small molecules, and electric signals to pass between cells. In the heart, for example, gap junctions allow the rapid spread of electric current mediated by ions so the heart muscle cells can beat in unison.

While plant cells are joined together by their cell walls (see Figure 4.16B), communication between adjacent cells is made possible by numerous cell

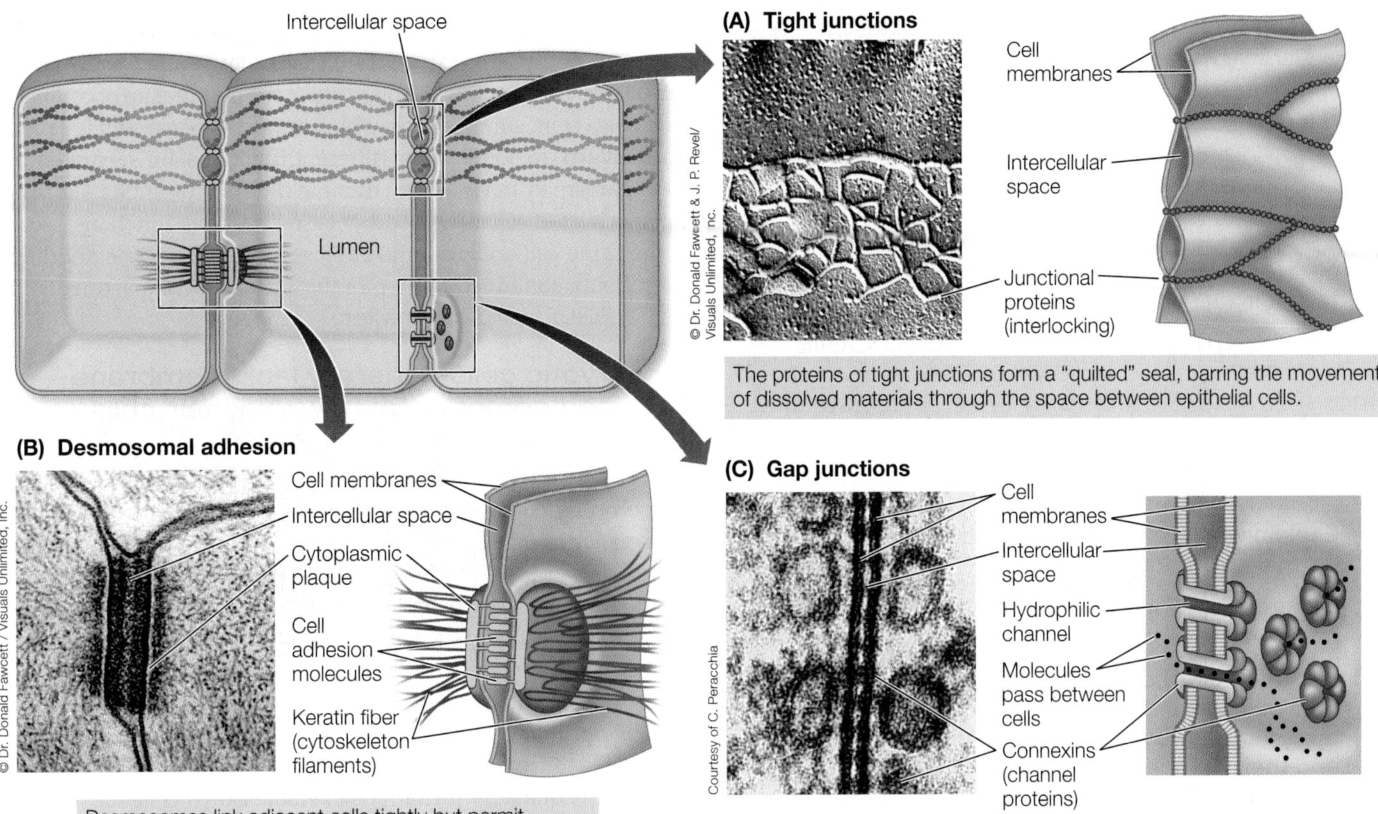

FIGURE 4.19 Three Types of Junctions in Animal Cells Tight junctions **(A)** and desmosomes **(B)** are abundant in epithelial tissues. Gap junctions **(C)** are found in some muscle and nerve tissues, in which rapid communication between cells is important. Although all three junction types are shown in the cell at the top left, all three are not necessarily seen at the same time in actual cells.

Activity 4.4 **Animal Cell Junctions**

membrane-lined channels called **plasmodesmata** (see Figure 4.20C). Plasmodesmata are similar to gap junctions between animal cells, but they allow passage of larger substances. Plasmodesmata allow water, ions, small molecules, hormones, and even some RNA and protein molecules to move between connected cells. In this way, energy-rich molecules such as sugars can be shared among plant cells, and plant hormones can affect growth at sites far from where the hormones were synthesized.

REVIEW & APPLY | 4.4

R The cytoskeleton, consisting of microfilaments, intermediate filaments, and microtubules, serves both to maintain cell structure and to provide a framework along which motor proteins can transport cargo. In addition, microtubules are a major component of cilia and flagella. Cells of some types of organisms (plants, archaea, bacteria, fungi, and some protists) have a cell wall, which serves to strengthen, support, and protect cells. In plants, plasmodesmata permit communication between cells. Animal cells, which do not have cell walls, are joined by tight junctions, desmosomes, and gap junctions.

A

1. Make a table that compares the composition, size and stability, and function of the three groups of cytoskeletal filaments.
2. Explain why motor proteins are often used to move proteins in cells, even when the movement is down their concentration gradient.
3. What kinds of cell junctions would you expect to find, and why, in the following locations?
   a. In the digestive system, where material must pass from cells lining the intestine to the blood vessels, without going through the extracellular material between cells.
   b. In a small animal, where a chemical signal passes rapidly from one cell to the next.
   c. In the heart, where cells are under constant mechanical stress caused by repeated muscular contractions.
4. When cancer spreads from its primary location to other parts of the body (a process called metastasis), tumor cells detach from their original location and then reattach at a different location. How might the integrin–collagen system be involved in this process?

Earlier in this chapter we saw how the cell membrane is a selectively permeable barrier that separates the cell from its environment. In the next key concept we'll describe how barriers are formed within cells to separate some regions from others.

4.5 Compartmentalization Occurs in Prokaryotic Cells and Is Extensive in Eukaryotic Cells

LEARNING OBJECTIVES

By the end of this key concept you should be able to:

4.5.1 Explain why compartmentalization is critical for normal cell function.

4.5.2 Describe how prokaryotic cells compartmentalize their cytoplasm.

4.5.3 List the major eukaryotic cell compartments (nucleus, endoplasmic reticulum, Golgi apparatus, lysosome, mitochondrion, plastids) and their primary functions.

Cells perform a large number of different functions. Some, the so called housekeeping functions, are shared by virtually all cells. But in most multicellular organisms, special cell types also have specific functions. Some functions are essentially the opposite of one another. For example, the breaking down of complex molecules into simpler ones, which releases energy and is called **catabolism**, is essentially the opposite of the building up of complex molecules from simpler ones, which requires energy and is called **anabolism**. In some cases, different functions are separated in the organism by having them occur in different locations—different specialized cells. But in other cases, multiple functions occur in the same cell and are physically separated from one another, often being contained in membrane-enclosed compartments in the cytoplasm called **organelles**.

Prokaryotic cells generally lack membrane-enclosed compartments within their cells

Biologists classify all living things into three domains: Archaea, Bacteria, and Eukarya. The organisms in Archaea and Bacteria are collectively called **prokaryotes**, which are generally small (0.1–5 μm in diameter), single-celled, lacking a nucleus, and usually without membrane-enclosed internal compartments (**FIGURE 4.20A**). In contrast, the cells of **eukaryotes** (which include plants, fungi, protists and animals) are much larger (10–100 μm), have a nucleus, and have numerous membrane-enclosed organelles, each of which has characteristic structure and functions (**FIGURE 4.20B** and **C**).

Since prokaryotic cells lack membrane compartments, it is reasonable to wonder whether the cytoplasm in their cells is uniform, with equal concentrations of solutes and other substances throughout the cell. Our discussion of diffusion (see Key Concept 4.2) would suggest that this would be the case. However, it has become clear in recent years that the cytoplasm of prokaryotes is not uniform.

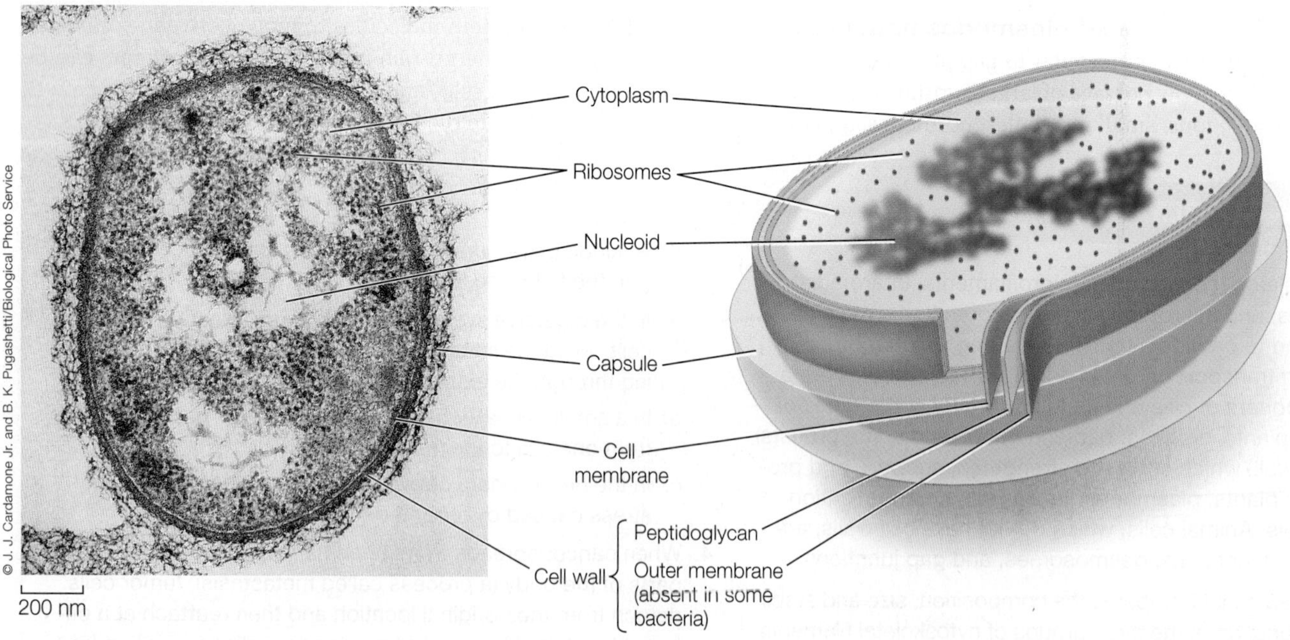

FIGURE 4.20 Cell Structure Generalized structure of a bacterial **(A)**, animal **(B)**, and plant **(C)** cell. Animal and plant cells, like all cells in the domain Eukarya, have extensive membrane compartmentalization, whereas cells in the two prokaryotic domains, Bacteria and Archaea (the latter not shown), generally have much less, if any, compartmentalization.

Media Clip 4.2 The Inner Life of a Cell

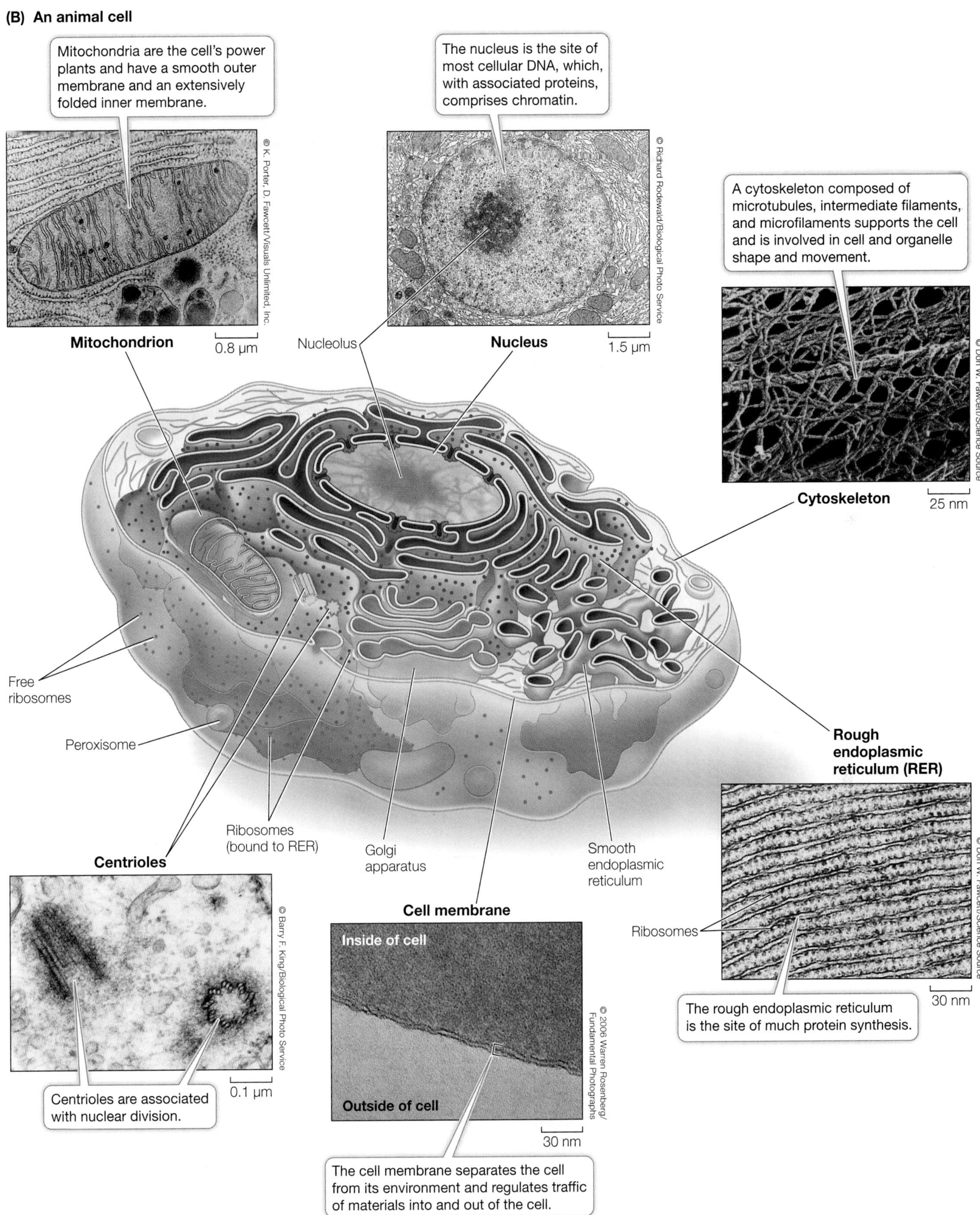
(B) An animal cell
Mitochondria are the cell's power plants and have a smooth outer membrane and an extensively folded inner membrane.
© K. Porter, D. Fawcett/Visuals Unlimited, Inc.
Mitochondrion
0.8 μm
The nucleus is the site of most cellular DNA, which, with associated proteins, comprises chromatin.
© Richard Rodewald/Biological Photo Service
Nucleolus
Nucleus
1.5 μm
A cytoskeleton composed of microtubules, intermediate filaments, and microfilaments supports the cell and is involved in cell and organelle shape and movement.
© Don W. Fawcett/Science Source
Cytoskeleton
25 nm
Free ribosomes
Peroxisome
Rough endoplasmic reticulum (RER)
Ribosomes (bound to RER)
Golgi apparatus
Smooth endoplasmic reticulum
Centrioles
© Barry F. King/Biological Photo Service
0.1 μm
Centrioles are associated with nuclear division.
Cell membrane
Inside of cell
Outside of cell
© 2006 Warren Rosenberg/ Fundamental Photographs
30 nm
The cell membrane separates the cell from its environment and regulates traffic of materials into and out of the cell.
Ribosomes
© Don W. Fawcett/Science Source
30 nm
The rough endoplasmic reticulum is the site of much protein synthesis.

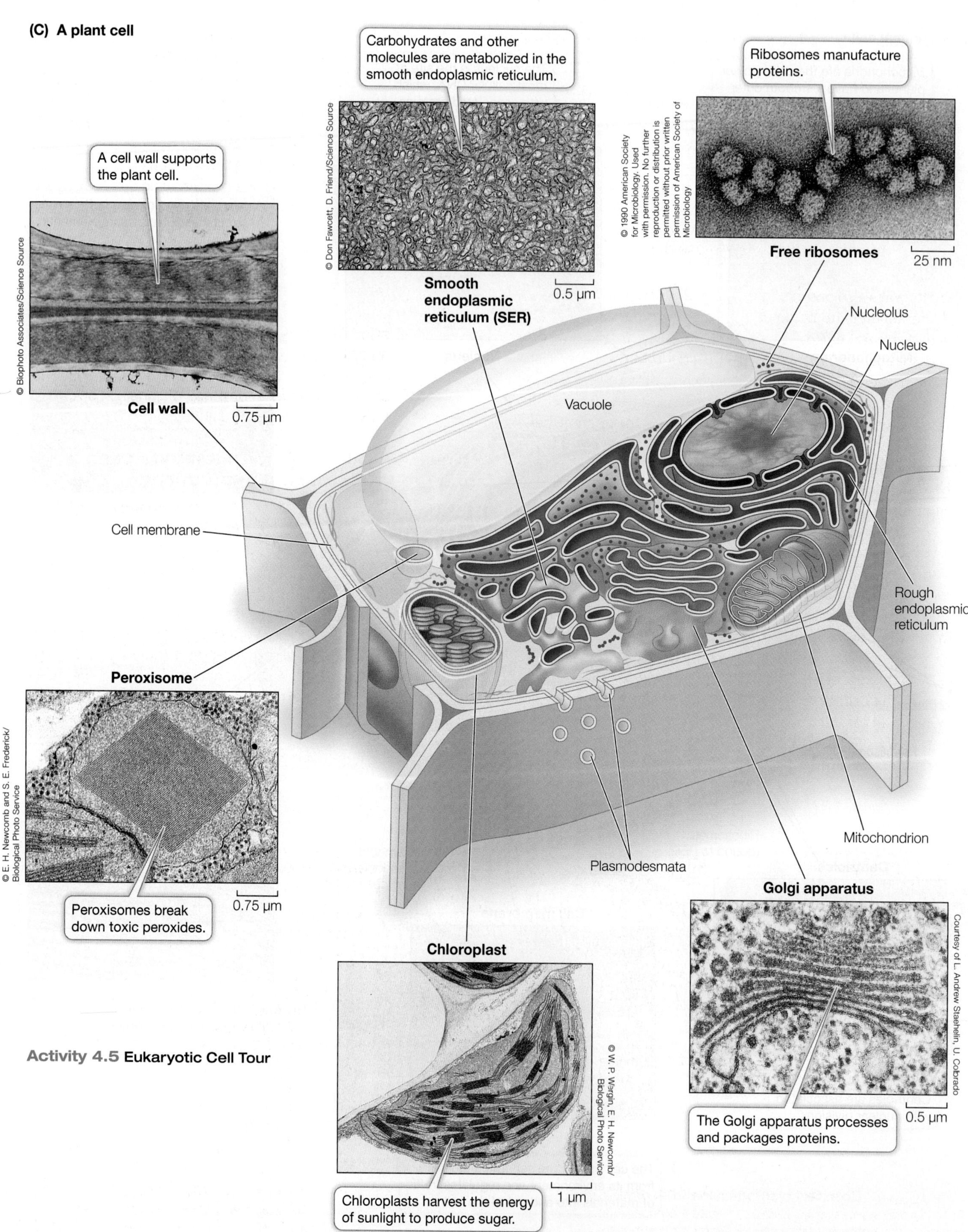

Activity 4.5 Eukaryotic Cell Tour

Prokaryotic cells contain various protein structures that are used to separate certain substances from others in the cell. These structures include carboxysomes (**FIGURE 4.21A**), microcompartments 40–200 nanometers in diameter that function to convert CO_2 into carbohydrates, and metabolosomes, which sequester toxic molecules and begin their breakdown.

LINK **Key Concept 17.2** discusses how atmospheric oxygen concentrations have changed over time, and how prokaryotes evolved not only to tolerate O_2 but to make use of it.

Prokaryotes also possess tiny encapsulin nanocompartments, 25–30 nanometers in diameter. Like microcompartments, these structures are not enclosed inside a lipid-based membrane, but instead are enclosed by a protein (encapsulin in the case of microcompartments). Inside each microcompartment is usually a single type of compartmentalized protein, often involved in breaking down toxic molecules. An example is the enzyme peroxidase, which catalyzes the breakdown of peroxides (e.g., H_2O_2) that can otherwise damage lipids and proteins. (This is also one of the functions of peroxisome organelles in eukaryotic cells; see Figure 4.20C). Another compartment, which has a different kind of function, is the gas vesicle (45 to over 100 nm in diameter), which is found in some bacteria and archaea and gives cells buoyancy in the aquatic environments in which they float. Gas vesicles are used to position the cell at an optimal depth in the water column for nutrient acquisition.

In addition to the protein compartments described above, prokaryotic cells have other ways to keep certain functions separate from the rest of the cell by having them occur in particular regions of the cell. While much is unknown, the cytoskeleton likely plays an important role, allowing motor proteins to ferry molecules into particular regions of the cell against concentration gradients. In addition, the cell's DNA is present as a single, circular molecule, called the **bacterial chromosome**, residing in a region of the cell appropriately called the **nucleoid**. Prokaryotes can also contain insoluble particles, such as lipid droplets or glycogen granules that are large enough to be visible in their cells. In addition, they house numerous **ribosomes**, which are relatively large structures (~250 nm in diameter, ten times larger than the nanocompartments noted above). Ribosomes are composed of numerous proteins and RNAs and are the site of protein synthesis (see Key Concept 10.4). While most prokaryotes lack internal membranes, they do occur in some groups. Most notably, the cyanobacteria have an extensive internal membrane system that enables them to perform photosynthesis (**FIGURE 4.21B**).

The similarity of the membranes found in photosynthetic chloroplasts to those of cyanobacteria is evidence for the endosymbiotic origins hypothesis for these organelles; see **Key Concept 19.1**.

Eukaryotic cells possess numerous kinds of membrane-enclosed compartments

Eukaryotic cells have extensive membrane compartmentalization and thus are able to effectively separate certain molecules and biochemical reactions from the rest of the cell. This impressive "division of labor" provides possibilities for regulation and efficiency that were important in the evolution of plants and animals.

Each type of organelle has a specific role in the cell. Some organelles have been characterized as factories that make specific products. Others are like power plants that take in energy in one form and convert it into a more useful form. While most organelles are found in all eukaryotic cells, there are notable exceptions. For example, plant cells differ from animal cells in having chloroplasts that are colored green by the pigment used in photosynthesis (see Figure 4.20C).

(A)

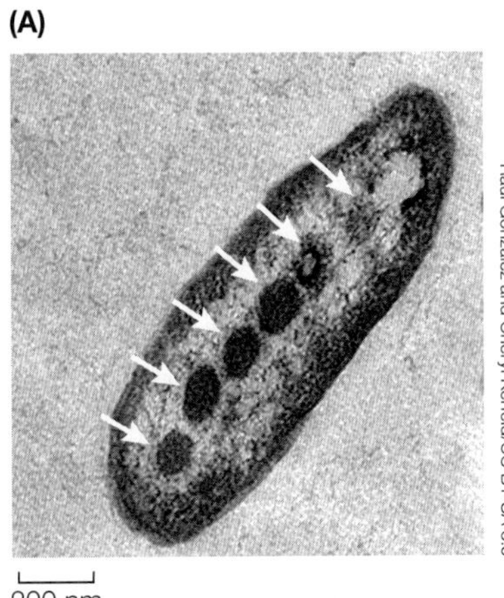

(B)

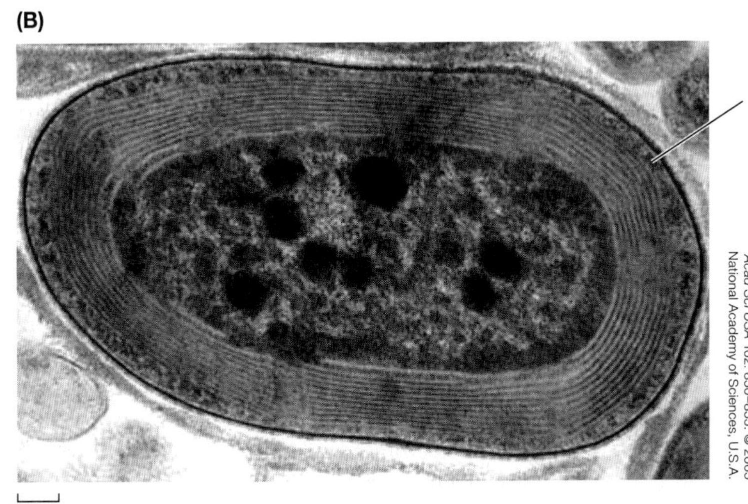

FIGURE 4.21 Compartmentalization in Prokaryotes **(A)** Many prokaryotes have specialized protein compartments. An example is the bacterial carboxysome microcompartment used to fix atmospheric CO_2 (indicated by white arrows). **(B)** Some prokaryotes, such as this cyanobacterium, have internal membrane compartments.

The nucleus contains most of the cell's DNA

The nucleus contains most of the cell's DNA, and is usually the largest organelle. At 5 micrometers in diameter, the nucleus is substantially larger than most prokaryotic cells. The nucleus is surrounded by the **nuclear envelope**, consisting of two membranes perforated by thousands of **nuclear pores** that regulate movement of substances into and out of the nucleus. The nucleus is the site of DNA replication and RNA transcription, and contains the **nucleolus**, a region where ribosomes are formed from RNA and proteins. The nucleus can be thought of as an organelle that protects DNA and creates an environment where transcription is regulated.

Inside the nucleus, each DNA molecule combines with specific proteins to form chromatin, which makes up the long, thin, linear threads called chromosomes. Different eukaryotic organisms have different numbers of chromosomes (ranging from two in one species of Australian ant to hundreds in some plants). You'll learn a great deal more about chromosomes and the transcription and translation of DNA into protein in Chapters 8 and 10.

The endomembrane system is a group of interrelated organelles

Much of the volume of many eukaryotic cells is taken up by an extensive **endomembrane system**. This interconnected system of membrane-enclosed compartments includes the nuclear envelope, endoplasmic reticulum, Golgi apparatus, and lysosomes, which are derived from the Golgi apparatus. Tiny vesicles shuttle substances between the various compartments and to (in exocytosis) and from (in endocytosis) the cell membrane. Membranes and the materials they contain are in constant motion, fusing to and detaching from one organelle to another, suggesting that all the compartments of the endomembrane system are functionally related.

ENDOPLASMIC RETICULUM The extensive networks of interconnected membranes branching throughout the cytoplasm are collectively called the **endoplasmic reticulum**, or **ER**. The interior compartment (lumen) of the ER is separate and distinct from the surrounding cytosol (see Figure 4.20B and C). The ER can enclose up to 10 percent of the interior volume of the cell, and its extensive folding results in a surface area many times greater than that of the cell membrane. There are two types of ER: rough and smooth.

The **rough endoplasmic reticulum** (**RER**) has many ribosomes attached to its outer surface, giving it a rough appearance. These ribosomes are temporarily attached to the ER membrane when they are synthesizing proteins destined for modification within the RER. Proteins destined to enter the RER include most integral membrane proteins, proteins that function inside any of the endomembrane system organelles, and secreted proteins. All of these proteins contain a specific sequence of about 20 amino acids that signals the ribosome to attach to the RER during translation (see Key Concept 10.5). Once inside the RER, proteins are chemically modified and then transported to other locations in the cell in vesicles that pinch off from the ER. Many proteins are covalently linked to carbohydrate groups, thus becoming glycoproteins. The carbohydrate groups on some secreted glycoproteins play roles in recognition and interactions between cells (see Chapter 6). Other carbohydrate groups tag proteins for delivery to specific cellular locations.

The **smooth endoplasmic reticulum** (**SER**) is connected to portions of the RER but lacks ribosomes and is more tubular (less like flattened sacs) than the RER. The SER has several important roles, among them lipid synthesis, detoxification of small molecules by chemical modification, and storage of calcium ions (Ca^{2+}) that can be rapidly released to cause a cellular response such as muscle contraction.

The amount of RER and SER in a cell varies with the cell's primary function. Secretory cells, such as those in the pancreas that secrete large quantities of digestive enzymes, are packed with RER. Liver cells, which are especially good at modifying and detoxifying molecules that enter the body from the digestive system, have abundant SER.

GOLGI APPARATUS The **Golgi apparatus** (**FIGURE 4.22**) consists of a stack of flattened membranous sacs called cisternae (singular cisterna). There can be many of these stacks in a cell. When vesicles from the RER fuse with the Golgi apparatus membrane, their proteins are released into the lumen of a Golgi apparatus cisterna, where they may be further modified. Importantly, the Golgi apparatus has an orientation such that entry of proteins from the RER occurs on one side (the *cis* face) and exit to the lysosome or the cell membrane occurs on the other side (the *trans* face). This allows sequential modification of proteins as they move from one side to the other, traveling through different cisternae, each of which has its own set of enzymes. The manner by which proteins move from one cisterna to the next has not been fully determined but may involve vesicles and small channels between cisternae. Once vesicles exit the Golgi apparatus, they travel to and fuse with the cell membrane or lysosome, or with other parts of the endomembrane system (RER, SER, Golgi apparatus, nuclear envelope).

LYSOSOMES The **primary lysosomes** originate from the Golgi apparatus and are small, densely staining organelles. They contain hydrolases (digestive enzymes) and are the sites where macromolecules—proteins, carbohydrates, nucleic acids, and lipids—are hydrolyzed into their monomers (component parts) (see Chapter 3). Some macromolecules that are hydrolyzed in lysosomes enter from outside the cell by phagocytosis (see Key Concept 4.3). The resulting vesicle moves into the cytoplasm and fuses with a primary lysosome to form a **secondary lysosome**, in which hydrolysis occurs. The products of digestion (e.g., amino acids) are transported across the membrane of the lysosome, providing monomers for other cellular processes. The "used" secondary lysosome, now containing undigested particles, then moves to the cell membrane, fuses with it, and releases the undigested contents to the environment.

Even for cells that do not perform phagocytosis, there is a continual need to break down cell components as they wear out and replace them with new ones. The programmed destruction of cell components is called autophagy, and lysosomes are where the cell

Animation 4.4
The Golgi Apparatus

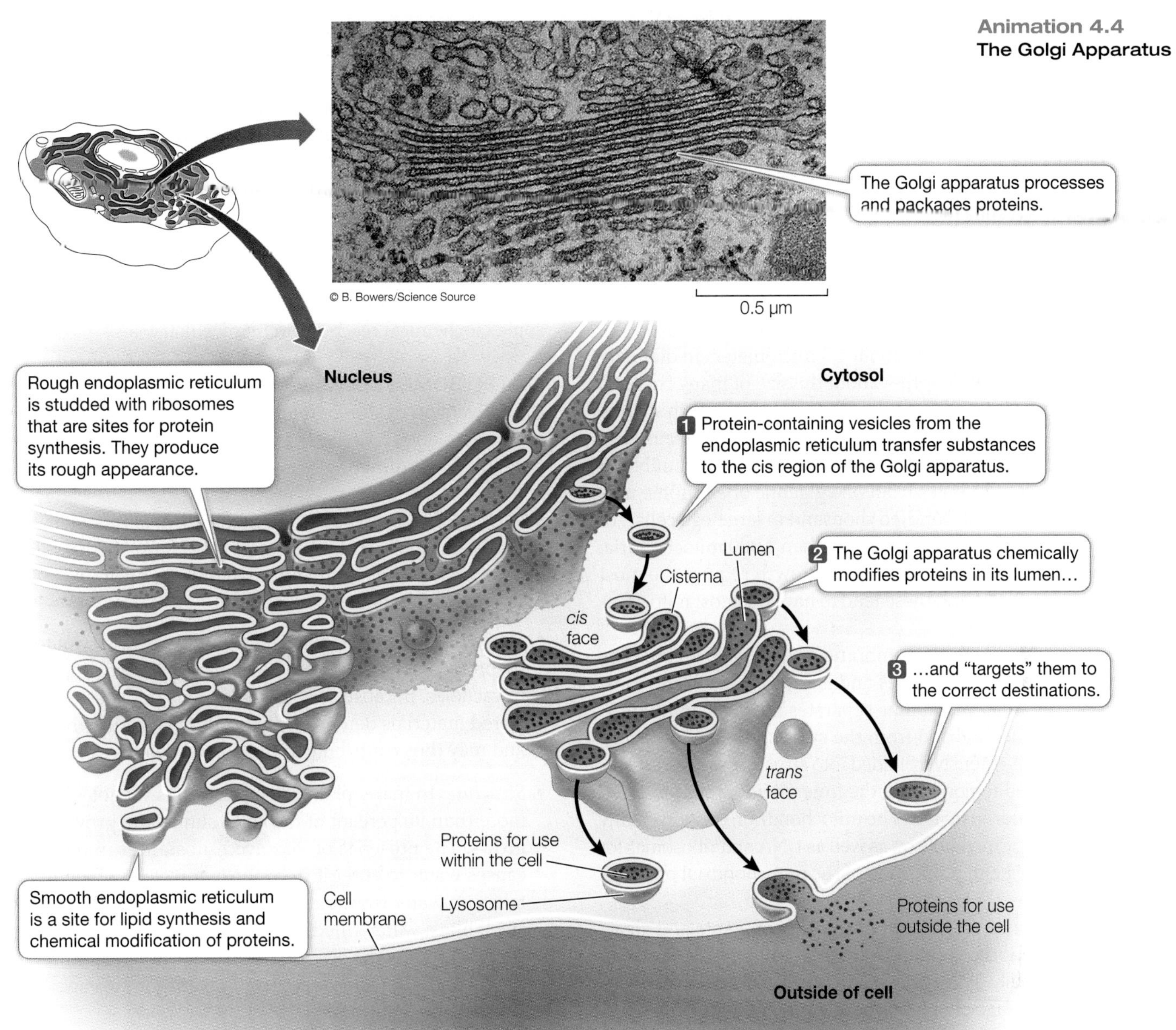

FIGURE 4.22 Processing of Proteins in the Endomembrane System Occurs in the Golgi Apparatus Proteins from the rough endoplasmic reticulum enter the Golgi apparatus via fusion to the cisterna on the *cis* face, are sequentially modified as they move through the Golgi apparatus, and then are released in vesicles from the *trans* face.

Where are proteins that are destined to end up as integral cell membrane proteins synthesized?

breaks down its own materials, even entire organelles, hydrolyzing their constituents. Failure of lysosomes to digest internal components is often harmful or even fatal. For example, in humans, Tay-Sachs disease occurs when a particular lipid called a ganglioside is not broken down in the lysosomes and instead accumulates in brain cells and damages them. In the most common form of this disease, an affected baby starts exhibiting neurological symptoms and becomes blind, deaf, and unable to swallow after 6 months of age. Death occurs before age four.

Plant cells do not appear to contain lysosomes, but the central vacuole of a plant cell (described below) may function in an equivalent capacity because it, like lysosomes, contains many digestive enzymes.

Some organelles transform energy

A cell requires energy to make the molecules it needs for activities such as growth, reproduction, responsiveness, and movement. Mitochondria (found in all eukaryotic cells) transfer chemical-bond energy from glucose to ATP, which is then used to fuel a variety of endergonic processes. Chloroplasts (found in plants and other photosynthetic cells) capture light energy from sunlight and convert it into chemical-bond energy.

MITOCHONDRIA In eukaryotic cells, the breakdown of energy-rich molecules such as the monosaccharide glucose begins in the cytosol. The molecules that result from this partial degradation enter the mitochondria, the primary function of which is to harvest the chemical-bond energy of those molecules in a form the cell can use, namely the energy-rich nucleotide ATP. We will discuss these energy-harvesting processes in Chapter 5.

A typical mitochondrion is about 1.5 micrometers in diameter and 2–8 micrometers in length—about the size of many bacteria. It contains many copies of a circular DNA molecule, much smaller than the linear DNA molecules in the eukaryotic nucleus, and can divide independently of the central nucleus. The number of mitochondria per cell ranges from one gigantic one in some unicellular protists to a few hundred thousand in large egg cells. An average human liver cell contains more than 1,000 mitochondria. Cells that are active in movement and growth require the most chemical-bond energy, and these tend to have the most mitochondria per unit of volume.

Mitochondria have two membranes (see Figure 4.20B). The outer membrane has large pores and is thus permeable to most substances. The inner membrane separates the biochemical processes of the mitochondrion from the surrounding cytosol. The inner membrane is extensively folded into structures called cristae, and the fluid-filled region inside the inner membrane is referred to as the mitochondrial matrix. The mitochondrion contains many enzymes for energy metabolism, as well as DNA and ribosomes for the synthesis of a small proportion of the mitochondrial proteins.

PLASTIDS Plastids are present in the cells of plants and algae, and like mitochondria, contain circular DNA molecules and can divide autonomously. There are several types of plastids, each specialized for the storage of polysaccharides or pigments. An important type of plastid is the **chloroplast** (see Figure 4.20C), which contains the green pigment chlorophyll and is the site of photosynthesis, a process that converts light energy into chemical-bond energy stored in carbohydrates (see Key Concept 5.5). Plastids are abundant in photosynthetic cells such as leaves, which may contain dozens of plastids. By contrast, some algae contain just one large plastid, and the size and shape of this large organelle is used to classify the organism.

Like a mitochondrion, a chloroplast is a cellular compartment enclosed within two membranes. In addition, chloroplasts contain internal membranes that look like stacks of flat, hollow discs, called **thylakoids**. As you will see in Chapter 5, light energy is converted to chemical-bond energy on the thylakoid membranes. The aqueous fluid within the chloroplast and surrounding the thylakoids is called the **stroma**, and it is there that carbohydrates are synthesized. Like the mitochondrial matrix, the chloroplast stroma contains ribosomes and DNA, which are used to synthesize some of the chloroplast proteins.

Several other membrane-enclosed organelles perform specialized functions

There are several other kinds of membrane-enclosed organelles with specialized functions.

PEROXISOMES **Peroxisomes** are small (0.2–1 μm diameter) organelles that accumulate and break down toxic peroxides, such as hydrogen peroxide (H_2O_2), which occur as the by-products of some biochemical reactions in many eukaryotes.

GLYOXYSOMES **Glyoxysomes** are organelles that carry out the conversion of stored lipids to carbohydrates. This pathway is unique to plants, so these organelles are found only in plant cells, particularly germinating seeds.

VACUOLES **Vacuoles** occur in many eukaryotic cells, but particularly those of plants and fungi (see Figure 4.20C). There can be one large vacuole or many small ones in a cell. Plant vacuoles have several functions:

- *Storage*: Like all cells, plant cells produce a variety of toxic by-products and waste products. Plants store many of these in vacuoles. Because they are poisonous or distasteful, these stored materials deter some animals from eating the plants, and may thus contribute to the plants' defenses and survival.
- *Structure*: In many plant cells, enormous vacuoles take up more than 90 percent of the cell volume and grow as the cell grows. The presence of dissolved substances in the vacuole causes water to enter it from the cytoplasm (which in turn takes up water from outside the cell), making the vacuole swell like a water-filled balloon (see Figure 4.3). The plant cell wall resists the swelling, causing the cell to stiffen from the increase in water pressure. This turgor pressure helps support the plant.
- *Reproduction*: Some pigments in the petals and fruits of flowering plants are contained in vacuoles. These pigments —the red, purple, and blue anthocyanins—are visual cues that help attract animals, which assist in pollination and seed dispersal.
- *Catabolism*: In the seeds of some plants, the vacuoles contain enzymes that hydrolyze stored seed proteins into monomers. The developing plant seedling uses these monomers as building blocks and sources of energy.

Many freshwater protists have **contractile vacuoles**, which are able to forcefully contract. Their function is to get rid of the excess water that enters the cell, which is hypertonic relative to the freshwater environment in which the cell lives. The contractile vacuole enlarges as water enters, and then abruptly contracts, forcing the water out of the cell through a special pore structure.

THINK LIKE A SCIENTIST

Interdisciplinary

Quantitative Reasoning

Advances in microscopy have resulted in greater understanding of cell structure and function

In the opening of this chapter, you learned how Robert Hooke's use of the microscope not only gave cells their name, but also led to the promulgation of the cell theory, which states that all life is based on cells. The small size of cells and cell compartments has been one of the major challenges for understanding the biology of the cell. As a result, much of what we have discovered has been closely tied to advances in techniques to view small structures. The key parameter for the ability of a microscope to allow a scientist to peer into the structure of the cell is optical resolution, a quantitative measure of how close two objects can be and still be discernible as separate. The unaided human eye has a resolution of about 0.2 millimeter (200 micrometers [μm]). Some cells are larger than this. If you shine a light on your skin, you may be able to just make out the cells on the surface. But you can't see a nucleus inside each cell—the nucleus is much smaller in diameter than 200 μm.

The invention and improvement of the light microscope (**FIGURE 4.23A**) and of the electron microscope (**FIGURE 4.23B**) have led to greatly improved resolution. By a century ago, improved lenses had led to microscopes with a resolution of 0.5 μm. This was improved over the past century to 0.2 μm, which is the practical limit for light microscopes of the type you use in the teaching laboratory.

1. A light microscope that can resolve objects that are 0.2 μm apart, such that they can be seen by a human eye, represents what level of magnification?
2. With a light microscope with 200 nm resolution, what structures in prokaryotic and eukaryotic cells could be observed?

The physical properties of light and lenses limit the resolution of microscopes using them. The electron microscope, with a beam of electrons (instead of light) focused by magnets (instead of glass lenses), improves resolution immensely. The main parameter involved in this improvement is the reduction in wavelength of the beam of electrons compared with visible light. Most electron microscopes have a resolution limit of 0.0002 μm. This is an effective magnification of 1-million-fold over the unaided human eye. (Can you calculate how this number is derived?) The main disadvantage of electron microscopes is that they cannot be used to visualize living cells.

3. What cellular structures can be visualized by electron microscopy that are not visible in light microscopy?

Recently, physicists have pushed the resolution limits of electron microscopy even lower. Some electron microscopes now have resolution as low as 50 pm (50 picometers, or 0.00005 μm). This is smaller than a carbon atom! It is now becoming possible to see life at the atomic level.

(A)

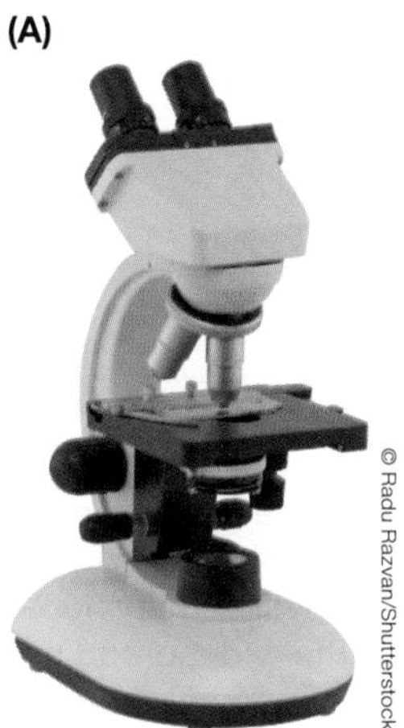

© Radu Razvan/Shutterstock

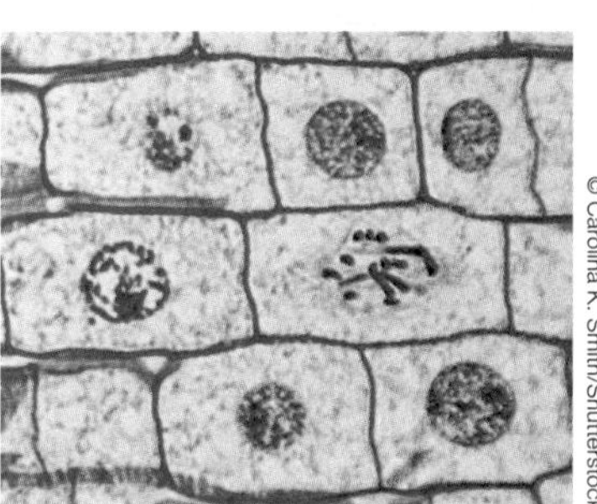

© Carolina K. Smith/Shutterstock

Onion (*Allium cepa*) root tip cells

10 μm

(B)

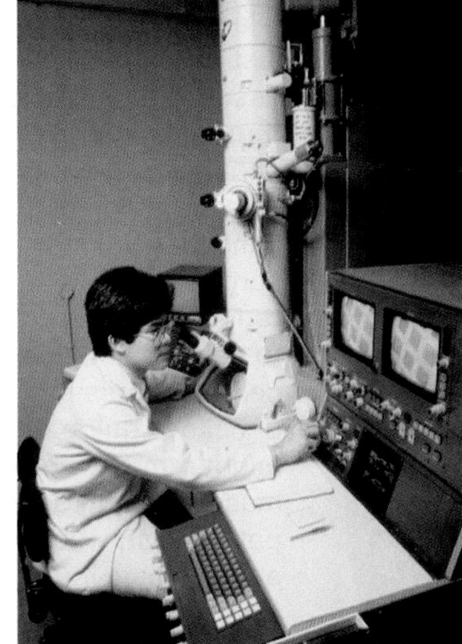

© Sinclair Stammers/Science Source

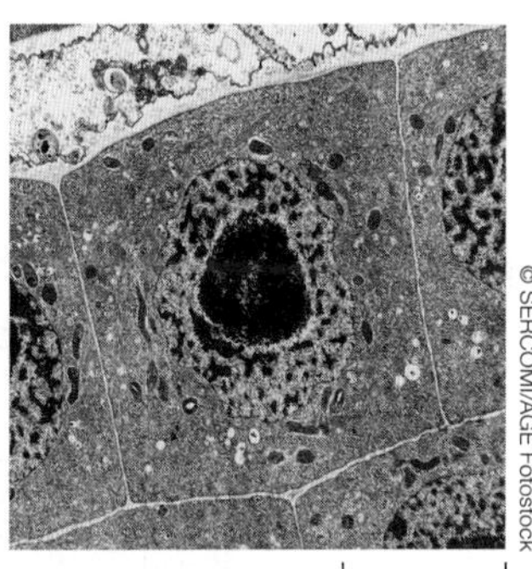

© SERCOMI/AGE Fotostock

10 μm

FIGURE 4.23 Microscopes Are Critical for Understanding the Biology of the Cell **(A)** A standard light microscope that you might find in a classroom or laboratory. **(B)** An electron microscope.

REVIEW & APPLY | 4.5

R Cell compartmentalization separates cell functions that might interfere with one another. While the cytosol in prokaryotic cells shows some organization in the form of protein-enclosed structures and unequal distribution of proteins, prokaryotic cells do not have a nucleus or other membrane-enclosed compartments. The DNA in prokaryotes generally is found as a circular chromosome in a region of the cell called the nucleoid. In contrast, all eukaryotic cells have several membrane-enclosed organelles—for example, a nucleus, mitochondria, and the components of the endomembrane system. Several organelles are interconnected in the endomembrane system, including the nuclear envelope, endoplasmic reticulum, Golgi apparatus, and various vesicles that are generated from the primary lysosomes. Plant cells have several additional distinctive features, including cell walls, chloroplasts, and vacuoles.

A

1. Why is it important to compartmentalize functions within cells?
2. How do prokaryotic cells compartmentalize cell functions?
3. List and distinguish the functions of the organelles in the endomembrane system.
4. In I-cell disease, an enzyme in the endomembrane system that normally adds phosphorylated sugar groups to proteins is lacking, and some of the proteins are not targeted to the lysosomes as they would be in normal cells. The "I" stands for inclusion bodies that appear in the cells and which are vesicles filled with nondegraded carbohydrates and lipids. What do you think happens to the proteins that are not correctly targeted to the lysosome?

In addition to their use in describing basic biological structure, how are microscopes used to study cells in applied situations?

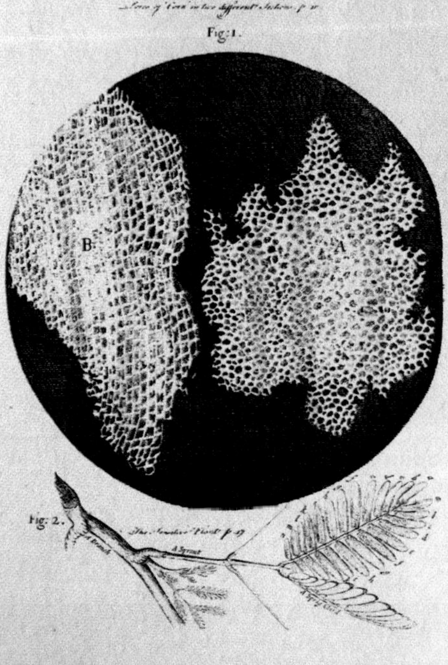

© World History Archive/Alamy Stock Photo; © roystudio/123 RF

Answer: Pathology is the study of diseases. In medicine, describing and identifying a disease is important for a physician in choosing a treatment. Until the widespread use of microscopes in medical pathology, diseases were described in general terms and treatments were nonspecific. A growth on a woman's breast could be a fluid-filled cyst (which we now know would just be drained), or a benign lump of tissue with no danger of spreading throughout the body or causing death (which now would be removed by a surgeon), or a malignant, lethal cancer (which now requires not just surgical removal but specific drug or radiation therapy). Microscopy is vital to pathologists to examine not just cell structures but also the contents of cells to guide treatment. A surgeon might remove a piece of the tissue mass in the breast and give it to the pathologist, who would slice the tissue and examine it under the microscope. Malignant dense cancer cells look very different from normal and benign cells: the tumor cells have denser and more compact nuclei, are smaller, and are arranged in clumps rather than sheets. In addition to examining the general cell structure, the pathologist would apply dyes to the tissue slices that bind specifically to proteins that are present only in malignant cancer cells, and in some cases only in certain subtypes of cancer (**FIGURE 4.24**). Thus, for example, a pathologist would use a specific dye to microscopically determine that the cells are not only breast cancer, but a certain type that has a specific treatment. The microscope made famous by Robert Hooke saves lives.

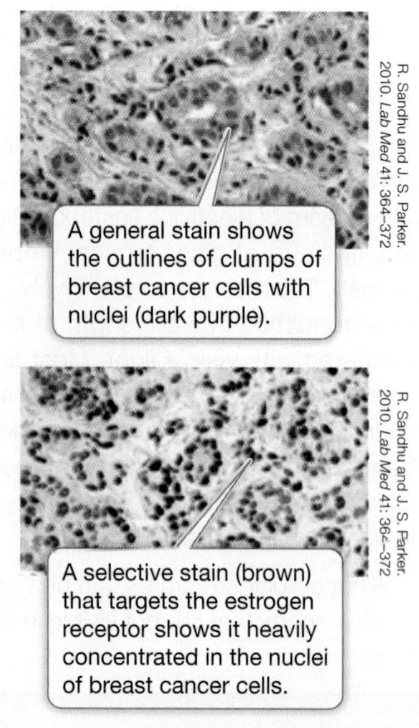

FIGURE 4.24 Staining of Cancer Cells

4.1 The Cell Membrane Separates the Interior of the Cell from Its Environment

Go to ACTIVITIES 4.1 and 4.2

- The **CELL MEMBRANE** is a fluid phospholipid bilayer with **INTEGRAL**, **ANCHORED**, and **PERIPHERAL MEMBRANE PROTEINS** (**FIGURE 4.1**).
- Membrane fluidity is affected by temperature, the saturation and length of fatty acid side chains in phospholipids, and the cholesterol content.

FIGURE 4.1

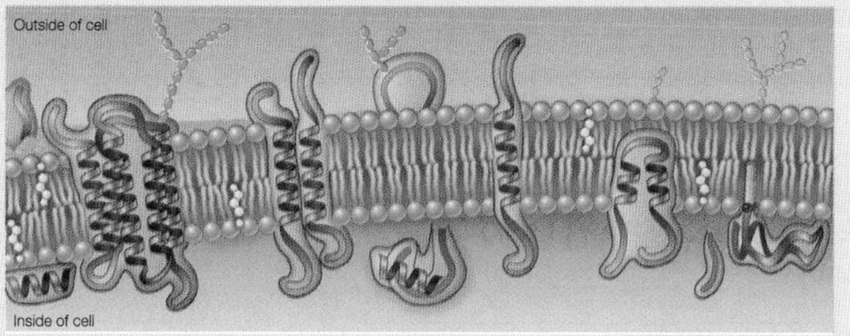

QUESTIONS

1. What are the differences between peripheral, anchored, and integral membrane proteins?
2. How might the lipid composition of a bacterial cell membrane (which lacks cholesterol) grown at 37°C differ when grown at 32°C?
3. Cells in the skin adhere to one another through specific junctions on their membranes. On which surface of the cell membrane would you expect to find proteins involved in cell recognition and adhesion?

4.2 Passive and Active Transport Are Used by Small Molecules to Cross Membranes

Go to ACTIVITY 4.3 and ANIMATIONS 4.1 and 4.2

FIGURE 4.3

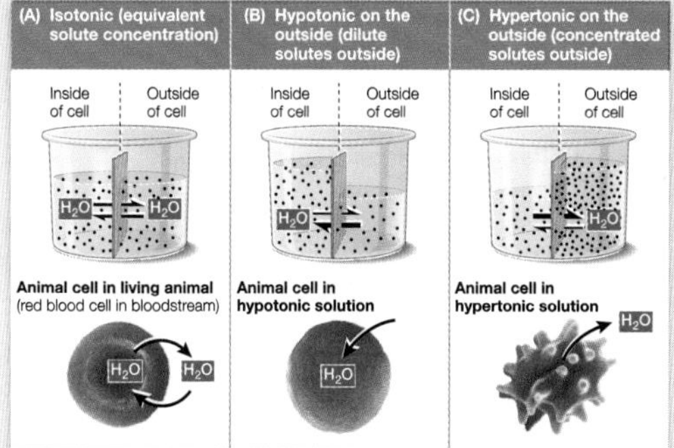

- The hydrophobic inner region of the bilayer is a barrier to many substances, but protein transporters allow the bilayer to be permeable to specific molecules and ions.
- **OSMOSIS** is the movement of water from areas of low to high solute concentration (**FIGURE 4.3**).
- **DIFFUSION** is the movement of particles down their concentration gradient and may be facilitated by **CARRIER PROTEINS** and **CHANNEL PROTEINS** including **ION CHANNELS** (**FIGURE 4.5**).
- **ACTIVE TRANSPORT**, which requires an input of energy, is required to move solutes from areas of low to high concentration (**FIGURE 4.7**).

FIGURE 4.5

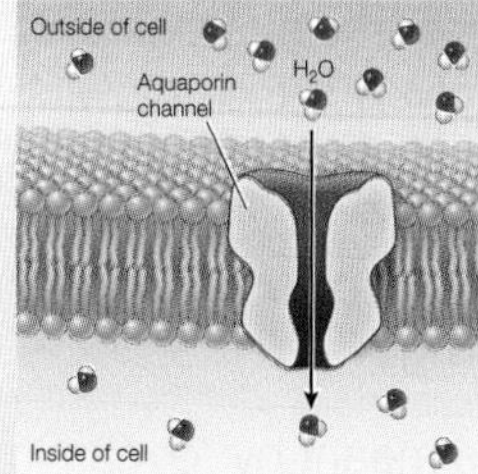

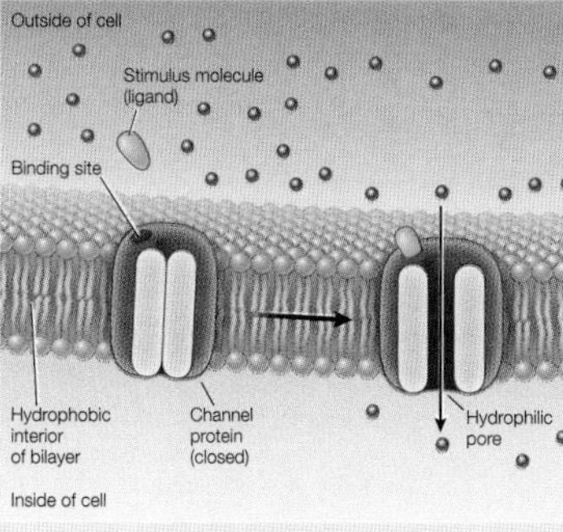

FIGURE 4.7

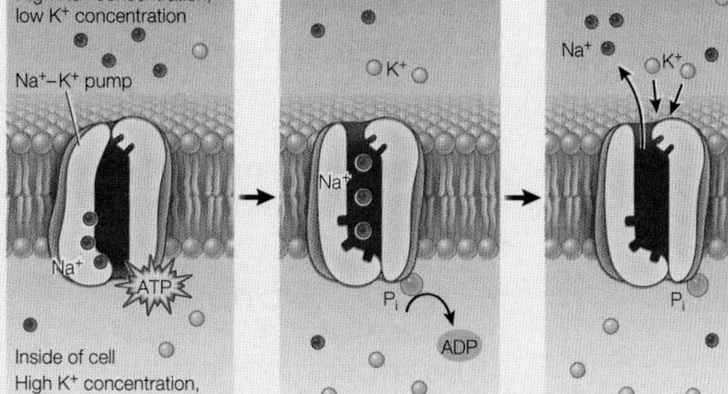

1. You treat a membrane with an enzyme that degrades proteins. Do you expect this treatment to alter the diffusion rate for CO_2? What about for sodium ions?
2. Explain why an animal cell will eventually burst when placed in a hypotonic solution. If a cell had large numbers of aquaporins in its membrane, would it be able to avoid bursting by allowing water to rapidly leave through these channels?
3. Under what conditions would each of the following two ways for glucose to enter a cell be used: (1) facilitated diffusion via a carrier protein and (2) secondary active transport?

4 VISUAL SUMMARY

4.3 Vesicles Are Used to Transport Large Molecules across Membranes in Eukaryotes

Go to ANIMATION 4.3

FIGURE 4.9

Exocytosis

Outside of cell

Cell membrane

Secretory vesicle

Inside of cell

Endocytosis

Endocytotic vesicle

- **VESICLES** transport large molecules and other substances into and out of the cell. (**FIGURE 4.9**).
- In **EXOCYTOSIS**, vesicles fuse with the cell membrane to secrete their cargo into the extracellular environment.
- In **ENDOCYTOSIS**, vesicles engulf substances from the extracellular environment. There are three types of endocytosis: **PHAGOCYTOSIS**, **PINOCYTOSIS**, and **RECEPTOR-MEDIATED ENDOCYTOSIS** (**FIGURE 4.10**).

FIGURE 4.10

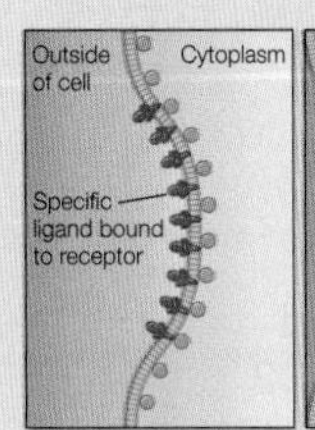

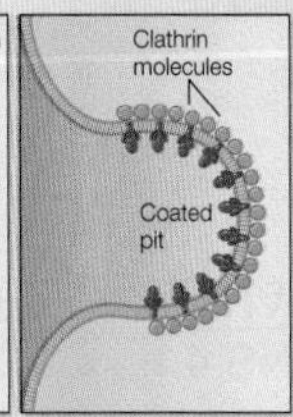

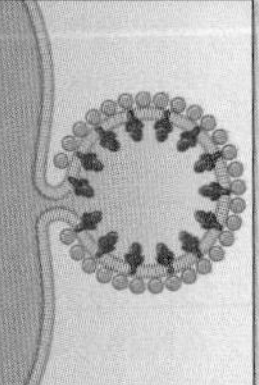

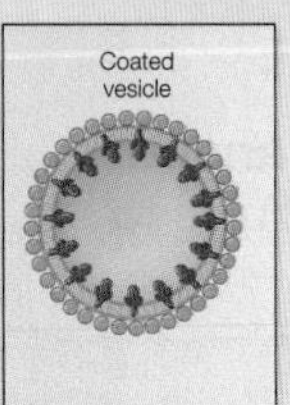

QUESTIONS

1. How can the cell be selective in what it transports across the cell membrane using endocytosis?
2. Toxic compounds are taken up from the bloodstream by endocytosis in the liver and then degraded. Why don't these compounds harm the cell when they are first engulfed?
3. Why do cells use endocytosis instead of facilitated diffusion to transport large molecules down their concentration gradient?

4.4 Cell Size, Shape, and Ability to Move Are Determined by Internal and External Structures

Go to ACTIVITY 4.4

FIGURE 4.12

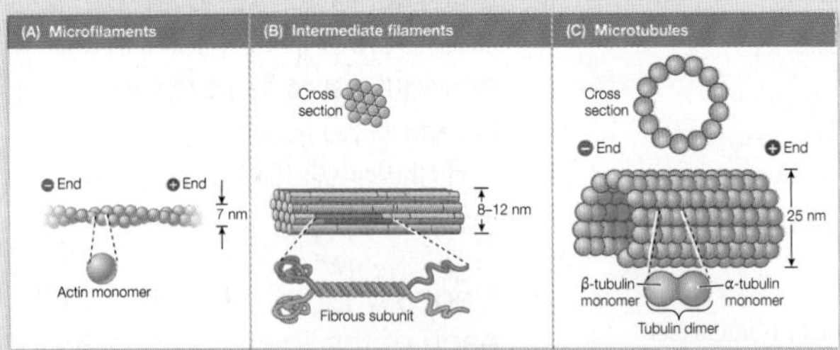

FIGURE 4.19

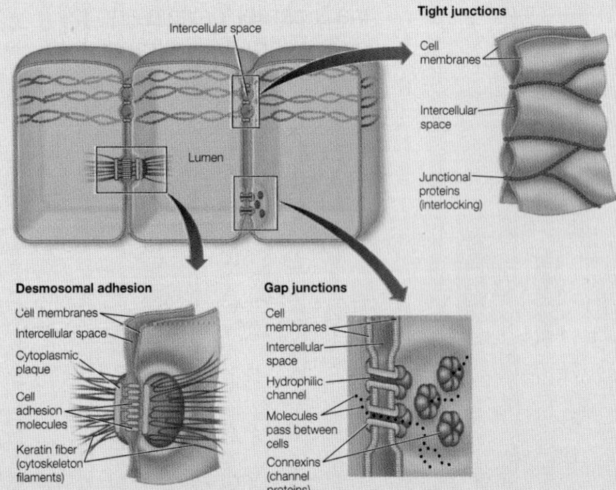

- Cell size is constrained by a cell's **SURFACE AREA-TO-VOLUME RATIO**.
- The **CYTOSKELETON** consists of **MICROFILAMENTS**, **INTERMEDIATE FILAMENTS**, and **MICROTUBULES**. These filaments maintain cell shape, hold organelles in position, promote the movement of **CILIA** and **FLAGELLA**, and help anchor the cell (**FIGURE 4.12**).
- **CELL WALLS** are important for protection and support and allow cells to survive in hypotonic external environments.
- The **EXTRACELLULAR MATRIX** is important for cell signaling, tissue support, protection from physical stress and pathogens, and various other functions.
- Animal cells can be joined by **TIGHT JUNCTIONS**, **DESMOSOMES**, and **GAP JUNCTIONS** (**FIGURE 4.19**).

QUESTIONS

1. If a cell is a perfect cube and 100 μm in length on each side, what is its surface area-to-volume ratio? If this cubic cell doubled in size, so that it was 200 μm on a side, what would its new surface area-to-volume ratio be?
2. What cytoskeletal elements are involved in the following: (a) change in cell shape, (b) movement by flagella, and (c) cytoplasmic streaming?
3. What is meant by "dynamic instability" of the cytoskeleton?

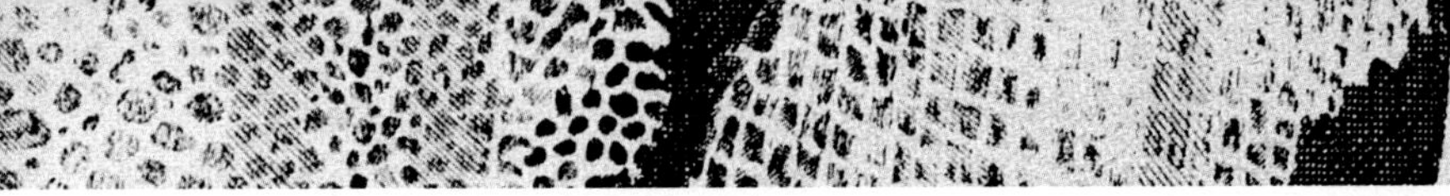

4.5 Compartmentalization Occurs in Prokaryotic Cells and Is Extensive in Eukaryotic Cells

Go to ACTIVITY 4.5 and ANIMATION 4.4

- Eukaryotic cells have numerous membrane-enclosed compartments called **ORGANELLES**, which segregate important functions away from the rest of the cell (**FIGURE 4.20**).
- The properties of the membranes enclosing organelles determine, in part, the contents of the organelle.
- Prokaryotic cells lack a nucleus but can have protein compartments (**FIGURE 4.21**). These compartments are not enclosed by a lipid-based membrane, but rather by proteins.

FIGURE 4.20

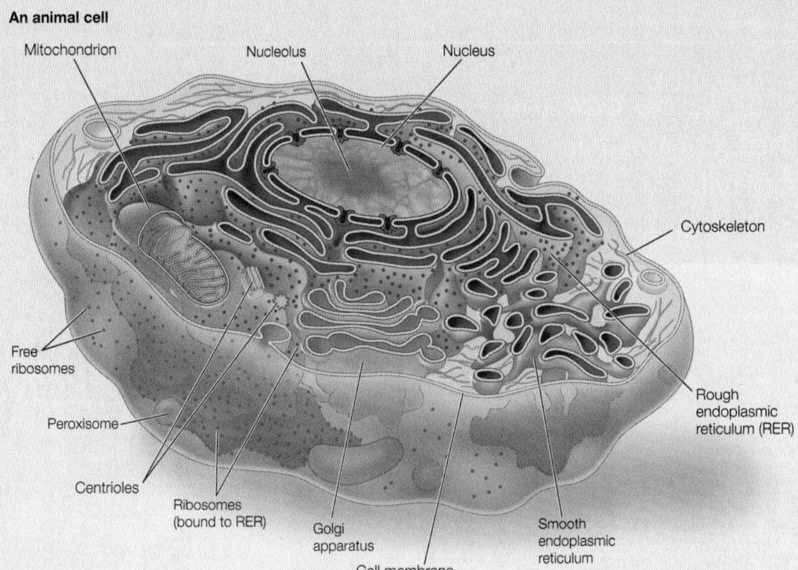

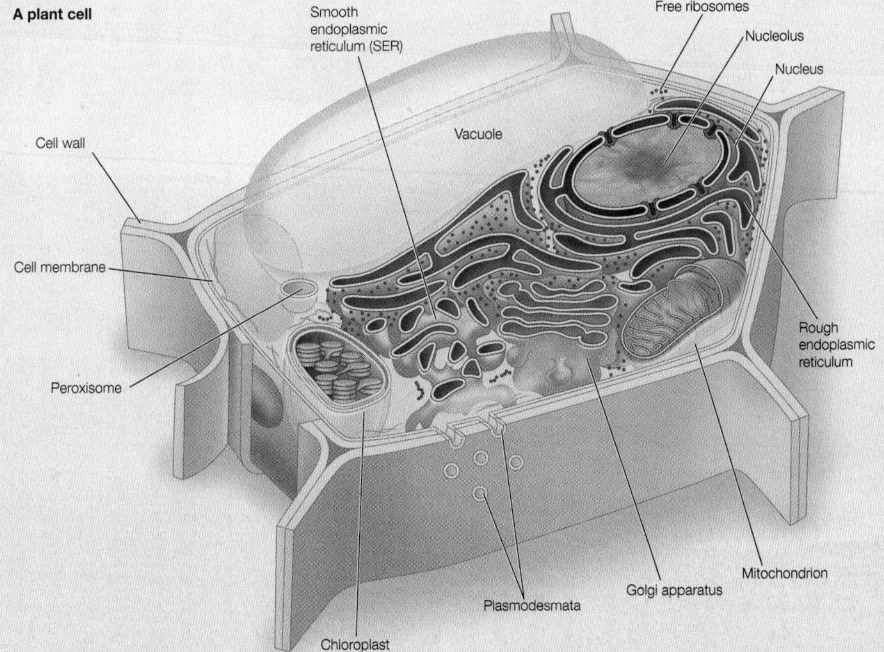

QUESTIONS

1. Even when special compartments are unavailable, the organization of cytoplasm is still observed and shows variation in concentrations of various substances in different regions of the cell. How is that possible?
2. Some bacterial cells have specialized structures in their cytoplasm that contain high concentrations of a single enzyme. What kinds of processes might take place in such structures?
3. The Golgi apparatus has been described as the "post office" organelle. Why is this description appropriate?

FIGURE 4.21

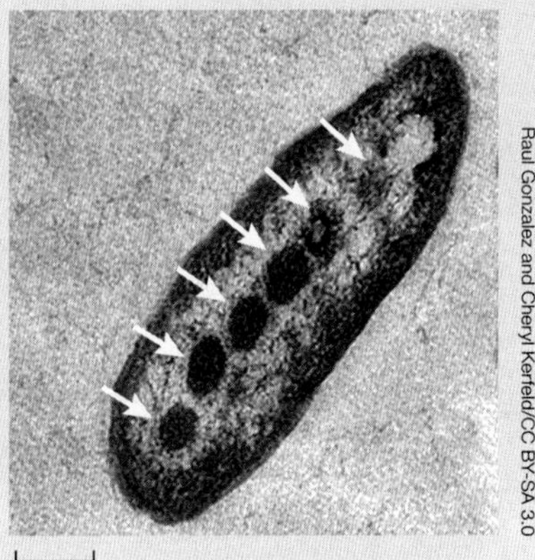

5 Cell Metabolism: Synthesis and Degradation of Biological Molecules

Earth is called the blue planet, a result of its large oceans. However, as can be seen in this image that shows the amount of a green pigment called chlorophyll, Earth should perhaps be called the green planet.

Why is Earth green?

You will find the answer to this question on page 125.

SeaWiFS Project, NASA/Goddard Space Flight Center and ORBIMAGE

KEY CONCEPTS

5.1 ATP and Reduced Coenzymes Are the Energy Currency for Biosynthesis

LEARNING OBJECTIVES

By the end of this key concept you should be able to:

5.1.1 **Explain what is meant when ATP is described as having "high-energy" phosphate bonds.**

5.1.2 **Describe how ATP is used to drive endergonic processes such as active transport.**

5.1.3 **Define a redox reaction and explain why oxidized molecules have lower chemical-bond energy than reduced molecules.**

5.1.4 **Explain how the high-energy molecules ATP and NADH are synthesized from the lower energy substrates ADP and NAD^+.**

Energy is stored in the chemical bonds of molecules, and it can be released and transformed by the metabolic pathways of living cells. A **metabolic pathway** consists of a coordinated series of biochemical reactions that convert molecules into different molecules. For example, you will see in this chapter that a certain metabolic pathway converts the six-carbon monosaccharide glucose to carbon dioxide (CO_2). This does not occur in a single step, but rather "bit by bit" in a series of chemical reactions.

Metabolic pathways have several defining characteristics:

- They consist of a series of separate, intermediate reactions.
- Each reaction is catalyzed by a specific enzyme.
- Most metabolic pathways are similar in all organisms, from prokaryotes to eukaryotes.
- Many metabolic pathways are compartmentalized, with certain steps occurring in certain regions of the cytosol or inside specific organelles (in eukaryotes) or micro- or nanocompartments (in prokaryotes).
- Each metabolic pathway is controlled by one or a few key enzymes that can be inhibited or activated, thereby determining the rate at which the pathway proceeds.

Many of the processes that a cell must perform require energy. For many organisms, this energy is obtained from the breakdown of energy-rich organic molecules that they consume. The major exceptions are cells of the numerous photosynthetic organisms, including plants, algae, and cyanobacteria, that use energy from the Sun to produce the energy-rich molecules they need. In addition, some species obtain their energy from inorganic (non-carbon-containing) sources, such as hydrogen sulfide released at hydrothermal vents in the oceans.

Recall from Key Concept 2.4 that endergonic reactions require a net input of energy to proceed, whereas exergonic reactions release energy. Within a cell, endergonic reactions can only proceed if they are coupled to exergonic reactions. How can the numerous different endergonic reactions occurring in a cell be coupled to the numerous different exergonic reactions? Energy for an endergonic reaction is required at the time the reaction occurs. This energy can be supplied directly by an exergonic reaction that occurs at the same time and in the same place. However, endergonic and exergonic reactions can also be coupled indirectly when they use the same form of chemical-bond energy—we can call it their "energy currency"—such that energy that is released from a diverse set of exergonic reactions is captured and stored. This energy is then used to drive endergonic reactions when and where they occur. The two most widely used energy currencies are the coenzymes ATP and NADH.

ATP hydrolysis releases energy

Cells use **adenosine triphosphate** (**ATP**) as their main energy currency. Some of the energy that is released in exergonic reactions is captured in chemical bonds when ATP is formed from adenosine diphosphate (ADP) and inorganic phosphate (P_i). The ATP can then be hydrolyzed at other sites in the cell, releasing free energy to drive endergonic reactions (**FIGURE 5.1**).

An active cell requires the production of millions of molecules of ATP per second to drive its biochemical machinery. You are already familiar with some of the activities in the cell that require free energy derived from the hydrolysis of ATP:

- Active transport across a membrane (see Key Concept 4.2)
- Reactions to synthesize macromolecules from their component molecules (see Chapter 3)
- Movements of vesicles and of cilia or flagella by motor proteins (see Key Concept 4.4)

The hydrolysis of a molecule of ATP yields ADP and P_i and releases energy (see Figure 5.1):

$$\text{ATP} + H_2O \rightarrow \text{ADP} + P_i + \text{free energy}$$

Under standard laboratory conditions, the change in free energy for this reaction (ΔG) is about –7.3 kcal/mol (–30 kJ/mol). The change in free energy is negative because the chemical bonds in ATP and

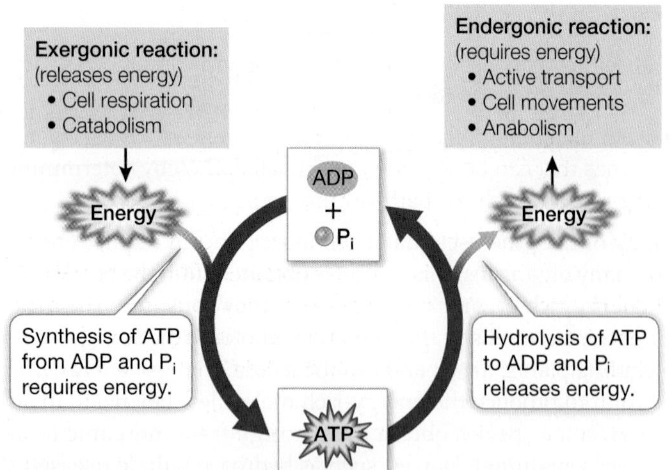

FIGURE 5.1 Coupling of Exergonic and Endergonic Reactions Using ATP and ADP An exergonic reaction releases energy that is used to drive the endergonic condensation of ADP + P_i to make ATP. At a different time and location in the cell, the energy released by the hydrolysis of ATP to ADP + P_i can then be used to drive an endergonic reaction.

In a cell, the energy released from the hydrolysis of ATP is 10 kcal/mol. If an endergonic reaction requires 15 kcal/mol, how many ATP hydrolyses would it require?

Activity 5.1 ATP and Coupled Reactions

H_2O are weaker than the chemical bonds in ADP and P_i. Weaker bonds have more chemical-bond energy, and thus ATP and H_2O together have more chemical-bond energy than ADP and P_i. Because the P—O phosphoanhydride bond has high chemical-bond energy, it is often termed a "high-energy" bond. A molecule of ATP can also be hydrolyzed to adenosine monophosphate (AMP) and a pyrophosphate ion ($P_2O_7^{4-}$, commonly abbreviated as PP_i). In this case, the phosphate bond between the first and second phosphates is broken, releasing energy. Additional energy may also be released by the subsequent hydrolysis of PP_i to two molecules of P_i.

In some reactions, ATP is formed by substrate-level phosphorylation—the enzyme-mediated direct transfer of phosphate from another molecule (the substrate) to ADP. This is the case for some reactions of the metabolic pathway for glucose breakdown (glycolysis), as we will see in Key Concept 5.2. However, the vast majority of ATP in cells is formed instead by oxidative phosphorylation, as we will also discuss in Key Concept 5.2.

Redox reactions transfer electrons and energy

Another way of transferring energy in chemical reactions is to completely or partially transfer electrons. A reaction in which one substance transfers one or more electrons to another substance is called a **reduction–oxidation**, or **redox**, **reaction**. *Oxidation is the loss* of one or more electrons by an atom, ion, or molecule, and *reduction is the gain* of one or more electrons (remember OIL RIG). Oxidation and reduction *always occur together*: as one compound is oxidized, the electrons it loses are transferred to another compound, reducing it. Compounds that attract electrons are called oxidizing agents and those that donate electrons are called reducing agents (**FIGURE 5.2A**). Note, however, that a compound might be an oxidizing agent when reacting with a strong reducing agent but could act as a reducing agent when reacting with a strong oxidizing agent.

Although oxidation and reduction are defined in terms of traffic in electrons, it is often helpful to think in terms of the gain or loss of hydrogen (H) atoms. Reactions in the cell often involve replacement of H atoms with carbon (C), nitrogen (N), oxygen (O), or sulfur (S) atoms. Hydrogen has lower electronegativity (a weaker ability to attract electrons) than these other atoms and so replacement of H by C, N, O, or S in an X—H bond (X = H, C, N, O, or S) means that electrons will be partially transferred away from X—it is oxidized because it has a lesser share of electrons in the new covalent bond. So when a molecule in a cell loses a hydrogen atom, it usually becomes oxidized. The one common atom in cells that makes this rule imperfect is P, which has lower electronegativity than H: replacement of H by P in an X—H bond reduces X because X has a *greater* share of the electrons in the new bond. Similarly, one can think of redox reactions in terms of gains or losses of oxygen atoms. Oxygen has higher electronegativity than H, C, N, P, or S, and so replacement of O by another atom in an X—O bond (X = H, C, N, O, P, or S) means electrons will be partially transferred towards X—it is reduced because it has a greater share of electrons in the new covalent bond. When a molecule in a cell loses an oxygen atom, it becomes reduced (**FIGURE 5.2B**).

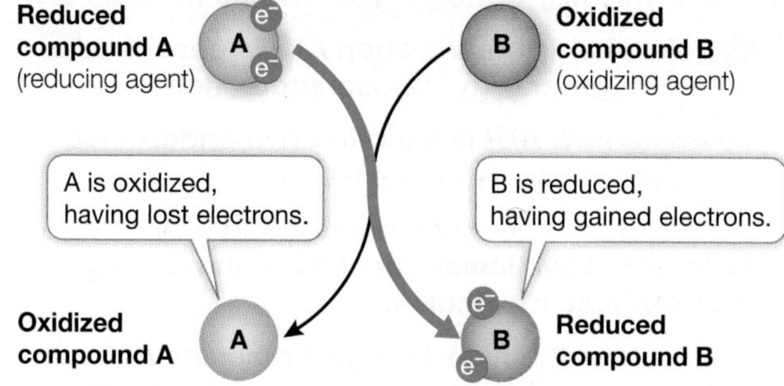

FIGURE 5.2 A Redox Reaction Involves Transfer of Electrons **(A)** Electron transfer may be complete, such that one or more electrons move from one molecule to another, resulting in the oxidized molecule becoming less negatively charged and the reduced molecule becoming more negatively charged. **(B)** Electron transfer may be partial, so that no electrons are lost or gained but an atom's share of electrons is altered. This situation is illustrated with carbon and hydrogen being reduced. The C—O and O—H bonds in the molecule on the left side of the equation are polar, with the oxygen atom having a larger share and the carbon and hydrogen atoms a smaller share of the electrons, resulting in a slight charge difference (δ^+ and δ^-). Because the C—H bond that replaces the C—O—H is nonpolar, electrons are shared almost equally in the product. Thus in this reaction carbon and hydrogen have had a partial gain of electrons and have been reduced.

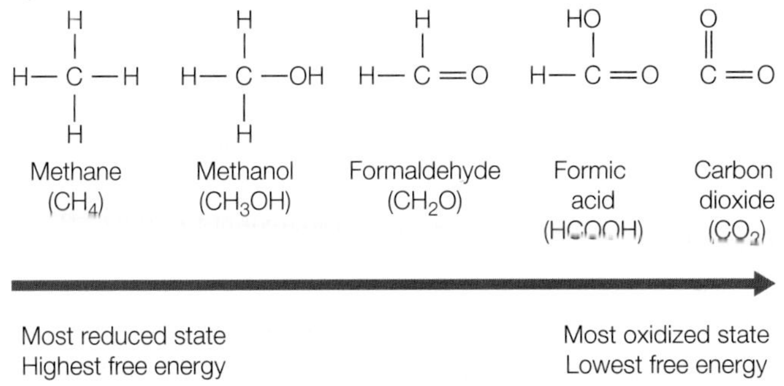

FIGURE 5.3 C—H Bonds Have Higher Free Energy (They Are Weaker) than C—O Bonds As its carbon atoms become more oxidized (less reduced), an organic molecule has lower chemical-bond energy (its bonds are stronger).

In general, the more reduced an organic molecule is, the more energy is stored in its covalent bonds. Thus molecules with more C—H bonds store more energy than molecules with C—O and O—H bonds (**FIGURE 5.3**). One implication of this is that lipids, with their high hydrocarbon content, have more chemical-bond energy per gram than do sugars or proteins (see Chapter 3 for structures). When carbon is fully oxidized, as it is in CO_2, it cannot be used as a source of energy. When organic molecules are reduced, they end up with more stored energy in their chemical bonds, and when they are oxidized, they end up with less.

When a redox reaction occurs, a reducing agent is oxidized and ends up with lower chemical-bond potential energy while an oxidizing agent is reduced and ends up with more potential energy. Overall, some chemical-bond energy is transferred from the reducing agent to the reduced product, and some is lost to entropy (see Key Concept 2.3).

Cells use the coenzyme **nicotinamide adenine dinucleotide (NAD)** as an electron carrier in redox reactions (**FIGURE 5.4**). NAD exists in two chemically distinct forms, one oxidized (NAD^+) and the other reduced (NADH). The reduction reaction

$$NAD^+ + H^+ + 2\,e^- \rightarrow NADH$$

involves the transfer of a proton (the hydrogen ion, H^+) and two electrons, which are released by the accompanying oxidation reaction of an organic molecule (R):

$$RH_2 \rightarrow R + 2H^+ + 2\,e^-$$

If the oxidation and reduction reactions are written as one coupled reaction, which is what actually happens, then we obtain:

$$RH_2 + NAD^+ \rightarrow NADH + H^+ + R$$

where R represents the rest of the molecule.

The reduced NADH molecule is readily oxidized within the cell. Oxygen is highly electronegative and readily accepts electrons from NADH. The oxidation of NADH by O_2 (which occurs in several steps):

$$NADH + H^+ + \tfrac{1}{2}\,O_2 \rightarrow NAD^+ + H_2O$$

is highly exergonic, releasing energy with a ΔG of –52.4 kcal/mol (–219 kJ/mol), which is considerably more than the energy released by the hydrolysis of ATP. Note that the oxidizing agent appears here as "½ O_2," instead of "O." This notation emphasizes that it is molecular oxygen (O_2) that acts as the oxidizing agent.

NAD is a common electron carrier in cells, but it is not the only one. Others include **flavin adenine dinucleotide (FAD/FADH$_2$)**, which also transfers electrons during glucose metabolism (see Key Concept 5.2), and **nicotinamide adenine dinucleotide phosphate (NADP$^+$/NADPH)**, which is used in photosynthesis (see Key Concept 5.5).

How do these coenzymes participate in the flow of energy within cells? The release and reuse of cellular energy can be summarized as follows:

- Organic molecules are oxidized during catabolism, releasing energy, which is mostly captured by the reduction of coenzymes such as NAD^+ (to give NADH).
- The primary energy currency of the cell is ATP; hydrolysis of ATP supplies the energy for many energy-requiring processes, including anabolism.

Because catabolism produces NADH (and similar reduced coenzymes) but most of the energy-consuming reactions require ATP, cells need a way to transfer energy from NADH to ATP; that is, oxidation of NADH (to give NAD^+) must be coupled to the synthesis of ATP from ADP and P_i. This coupling is accomplished in a process called oxidative phosphorylation, which we will discuss in Key Concept 5.2.

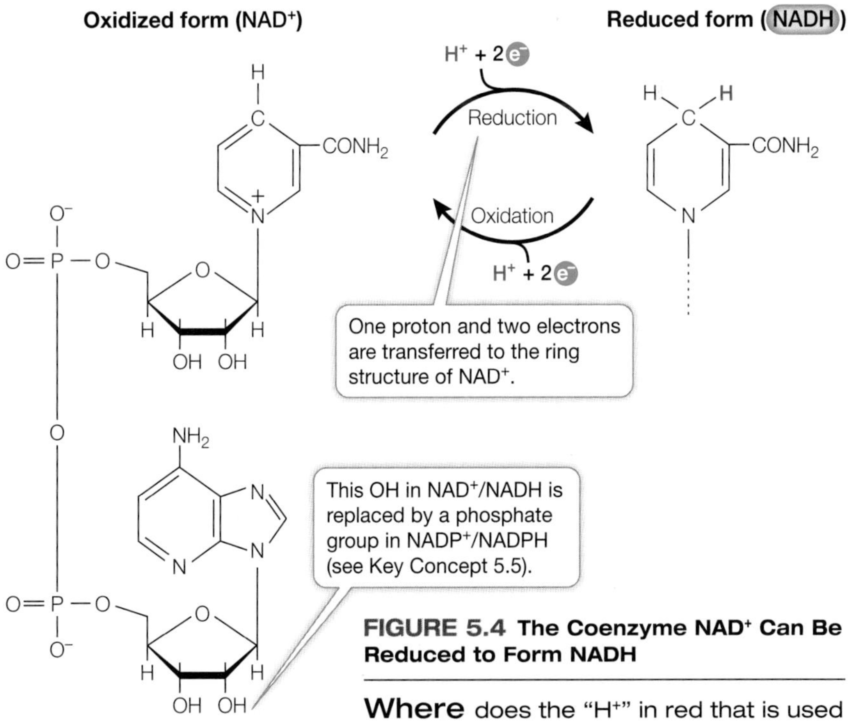

FIGURE 5.4 The Coenzyme NAD$^+$ Can Be Reduced to Form NADH

Where does the "H^+" in red that is used in the reduction of NAD^+ come from?

REVIEW & APPLY | 5.1

R When macromolecules are oxidized, they release their chemical-bond energy. Oxidation occurs in a series of steps, which allows a portion of the energy to be captured in ATP and other coenzymes such as NADH. Energy captured in these and other coenzymes is then used to synthesize additional ATP. ATP is the energy currency of the cell and is used to drive a variety of endergonic reactions by coupling them to the exergonic hydrolysis of ATP.

A 1. In Chapter 4 you learned about the sodium–potassium pump (see Figure 4.7), which transports Na^+ and K^+ ions against their concentration gradients using ATP. Describe how the high-energy bonds in ATP allow this process to occur.

2. For each of the following reactions, explain whether the carbon atoms in the reactants are being oxidized or reduced.

   a. $C_6H_{12}O_6 + 6\ O_2 \rightarrow 6\ CO_2 + 6\ H_2O$

   b. $6\ CO_2 + 6\ H_2O \rightarrow C_6H_{12}O_6 + 6\ O_2$

3. The energy required to reduce NAD^+ to form NADH is substantially more than is available from the hydrolysis of ATP. Where does the energy come from to reduce NAD^+?

We have seen that both ATP and NADH function as energy-coupling coenzymes, which are used by cells to store and transfer energy. We will now look at how cells capture energy from the catabolism of glucose to produce NADH, and then transfer this energy from NADH to ATP.

5.2 Carbohydrate Catabolism in the Presence of Oxygen Releases a Large Amount of Energy

LEARNING OBJECTIVES

By the end of this key concept you should be able to:

5.2.1 **Identify where the four stages of aerobic respiration of glucose occur in a cell and how many ATP molecules are produced in each stage.**

5.2.2 **List the differences between aerobic respiration and fermentation of glucose with respect to the products, energy released, and role of mitochondria.**

5.2.3 **Describe how carbohydrate catabolism is regulated at the enzyme level.**

Cellular respiration is the set of metabolic reactions used by cells to harvest energy from molecules such as carbohydrates. The energy can then be used to fuel endergonic reactions. Energy is released when organic molecules, with many C—C and C—H (reduced) bonds, are oxidized to CO_2. We will first consider in detail the catabolism of the carbohydrate glucose (**FIGURE 5.5**).

In the cell, the complete oxidation of glucose to CO_2 takes place in four major stages: glycolysis, pyruvate oxidation, the citric acid cycle, and oxidative phosphorylation. Several of these are multi-reaction metabolic pathways. In this section we describe each stage in turn. When the oxidation of glucose takes place in the presence of oxygen in the cell, it is called **aerobic respiration** (see Figure 5.5, left). The complete oxidation of glucose releases a considerable amount of energy:

$$\text{Glucose} + 6\ O_2 \rightarrow 6\ CO_2 + 6\ H_2O + \text{energy}$$
(686 kcal [2,872 kJ] per mole of glucose)

This equation applies whether glucose is oxidized in cells or in a chemistry lab. In the lab, this energy is released as heat, and that

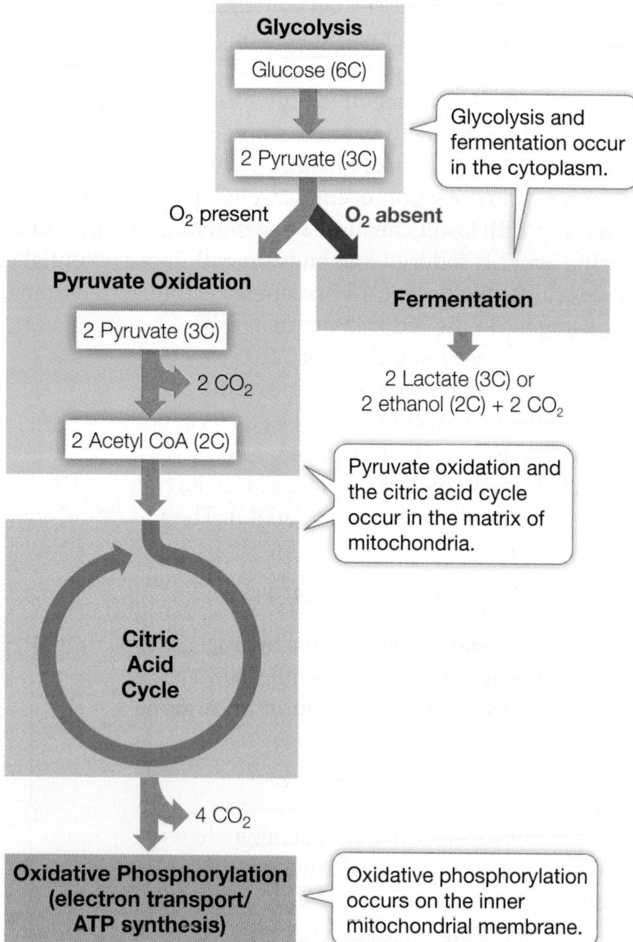

FIGURE 5.5 Pathways for Catabolizing Glucose
Respiration in the presence of oxygen involves four steps: glycolysis, pyruvate oxidation, the citric acid cycle and oxidative phosphorylation. In the absence of oxygen, respiration involves glycolysis and fermentation. "2C," "3C," and "6C" indicate the number of carbon atoms in each molecule.

Activity 5.2 Glycolysis and Fermentation

is where scientists get the exact number of kilocalories noted in the equation. In living systems, however, a significant amount of the released energy (234 kcal/mol; 34% of the total) is captured in the high-energy phosphoanhydride bonds of ATP. The efficiency of energy capture in aerobic respiration is impressive, on par with that of gasoline-powered cars, which run on the energy released from the oxidation of hydrocarbons to CO_2, and which generally have efficiencies between 25 and 50 percent. One factor that contributes to the high energy efficiency in cells is a principle of metabolism noted above: glucose is oxidized in a series of steps that release the energy in smaller amounts that can be captured relatively easily in coupled endergonic reactions (especially the reduction of NAD^+ and the phosphorylation of ADP). When molecular oxygen (O_2) is not available, glucose can be only partially oxidized, and the total energy released is substantially smaller (28 kcal [118 kJ] per mole of glucose; 2% efficiency). Breakdown of glucose in the absence of oxygen is termed **fermentation** (see Figure 5.5, right).

Aerobic respiration and fermentation of glucose both begin with glycolysis, which is the series of reactions that converts the six-carbon monosaccharide glucose into two three-carbon molecules of pyruvate (the anion of pyruvic acid). Once pyruvate is formed, it has one of two fates. In the presence of oxygen, pyruvate enters aerobic respiration, which begins with pyruvate oxidation, in which two three-carbon molecules of pyruvate are oxidized to two two-carbon molecules of acetyl CoA and two molecules of CO_2. Each of the acetyl CoA molecules then enters the citric acid cycle, which is a series of reactions that results in the complete oxidation of the acetyl group to CO_2. In the absence of oxygen, pyruvate enters the fermentation pathway, resulting in the production of either CO_2 and the two-carbon molecule ethanol (in plants and fungi), or of the three-carbon molecule lactate (in animals).

In glycolysis, glucose is partially oxidized and some energy is released

Glycolysis takes place in the cytosol and involves ten enzyme-catalyzed reactions. During glycolysis, a few of the C—H bonds in the glucose molecule are oxidized, releasing some stored energy. The final products are two molecules of pyruvate, two molecules of ATP, and two molecules of NADH. The ten steps in the metabolic pathway of glycolysis can be divided into two stages: the initial energy-investing reactions that consume chemical-bond energy stored in ATP, and the energy-harvesting reactions that produce ATP and NADH (**FIGURE 5.6**).

In the initial stage (steps 1–5 in Figure 5.6), a molecule of glucose has two phosphate groups added to it (at —OH groups on the first and sixth carbons) and undergoes a structural alteration (isomerization), producing a fructose molecule containing two phosphates. Each of these first two phosphorylations is an endergonic condensation reaction, and both the energy and phosphate groups are supplied by ATP, which is hydrolyzed to ADP. The resulting molecule is then cleaved to produce two molecules of glyceraldehyde 3-phosphate.

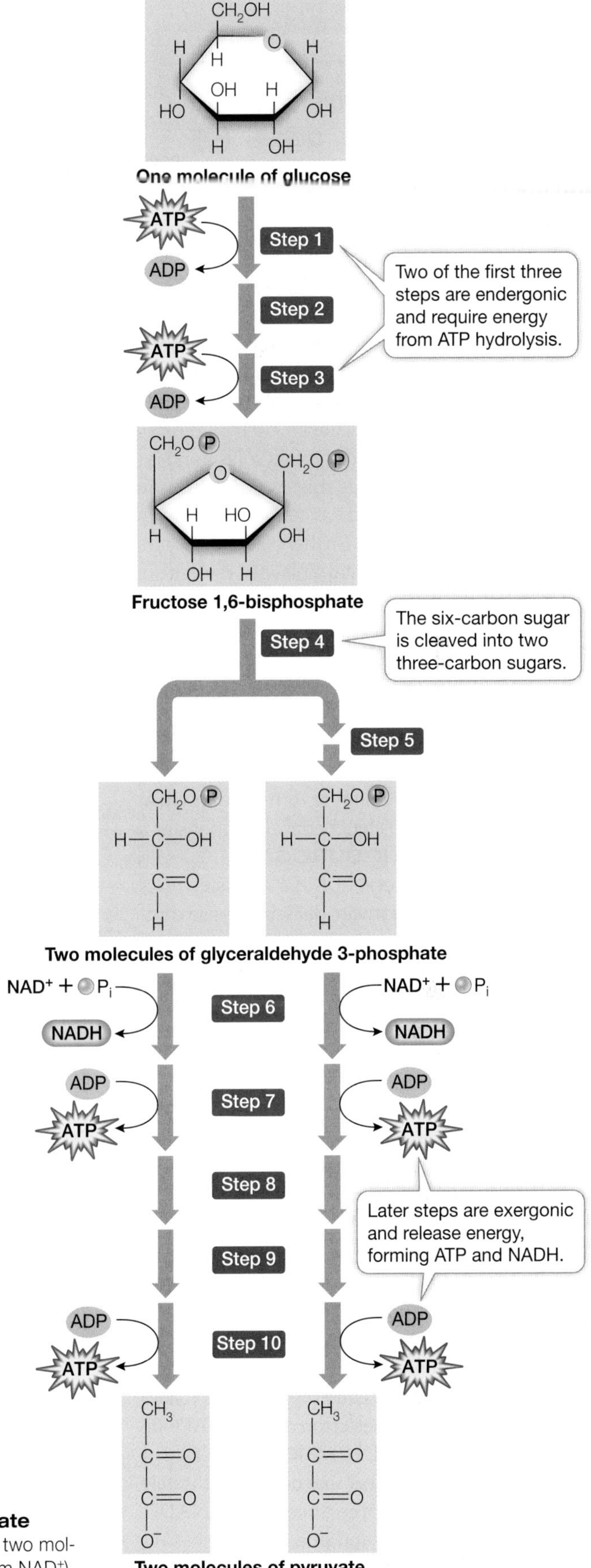

FIGURE 5.6 Glycolysis Converts Glucose to Pyruvate Overall, there is a net production of two molecules of pyruvate, two molecules of ATP (from two ADP) and two molecules of NADH (from NAD^+).

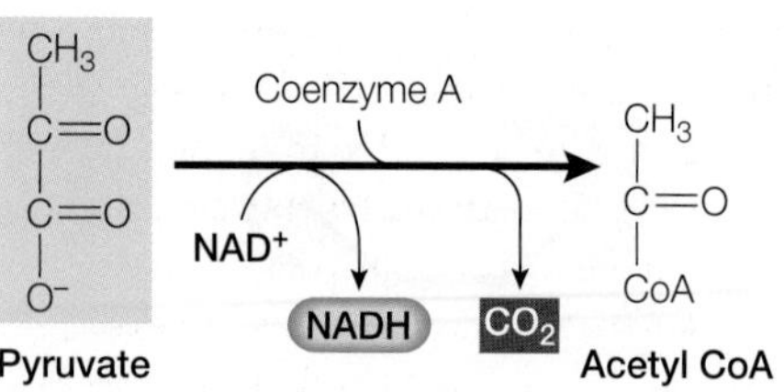

FIGURE 5.7 Oxidation of Pyruvate Pyruvate is oxidized to acetyl CoA in aerobic respiration.

Each glyceraldehyde 3-phosphate is then phosphorylated (at the —H group of the first carbon) in a redox reaction (producing NADH from NAD^+) (step 6 in Figure 5.6), and then several subsequent steps remove both phosphates, adding them to ADP to produce ATP in a process called **substrate-level phosphorylation**, resulting in the production of pyruvate (steps 7–10 in Figure 5.6).

Pyruvate is the end-product of glycolysis and is somewhat more oxidized than glucose (to see this, examine the number of covalent bonds between carbon and oxygen atoms in pyruvate versus glucose). In the presence of O_2, further oxidation occurs. In prokaryotes these subsequent reactions take place in the cytosol, but in eukaryotes they take place in the mitochondrial matrix.

Pyruvate oxidation links glycolysis and the citric acid cycle

After glycolysis, the next step in the aerobic catabolism of glucose involves the oxidation of pyruvate to a two-carbon acetate molecule and CO_2. The acetate is then bound to coenzyme A to form **acetyl CoA**, which is used in various biochemical reactions as a carrier of acetyl groups (**FIGURE 5.7**).

The formation of acetyl CoA is a multistep reaction catalyzed by the pyruvate dehydrogenase complex, which contains 60 individual proteins and 5 different coenzymes. The overall reaction is exergonic, and one molecule of NAD^+ is reduced to NADH. The main role of acetyl CoA is to donate its acetyl group to the four-carbon compound oxaloacetate, forming the six-carbon molecule citrate (the anion of citric acid). This initiates the citric acid cycle, one of life's most important energy-harvesting pathways.

The citric acid cycle completes the oxidation of glucose to CO_2

Acetyl CoA is the starting point for the citric acid cycle. This pathway of eight reactions completely oxidizes the two-carbon acetyl group to two molecules of CO_2. The free energy released from these reactions is used to synthesize guanosine triphosphate (GTP) (from guanosine diphosphate [GDP] and P_i), and to reduce the electron carriers NAD^+ (to NADH) and FAD (to $FADH_2$) (**FIGURE 5.8**). For each molecule of acetyl CoA that donates its acetate to the citric acid cycle, three molecules of NADH, one molecule of $FADH_2$, and one molecule of GTP are produced. This pathway is a cycle because the starting material, oxaloacetate, is regenerated in the last step and is ready to accept another acetate group from acetyl CoA. The citric acid cycle operates twice for each glucose molecule that enters glycolysis (once for each pyruvate).

GTP can be used directly as a source of chemical-bond energy to drive endergonic reactions in the cell, or it can phosphorylate ADP to ATP, which can then be used in this role. How are the reduced coenzymes NADH and $FADH_2$ used to generate ATP? We answer that question in the next section, and in doing so we explain how oxygen molecules (O_2) are involved in aerobic respiration. Even though glucose is fully oxidized to CO_2 in the steps we have outlined, none of these steps require oxygen molecules.

Energy is transferred from NADH to ATP by oxidative phosphorylation

As we mentioned in Key Concept 5.1, energy-consuming processes in the cell use ATP as their source of energy. In order to

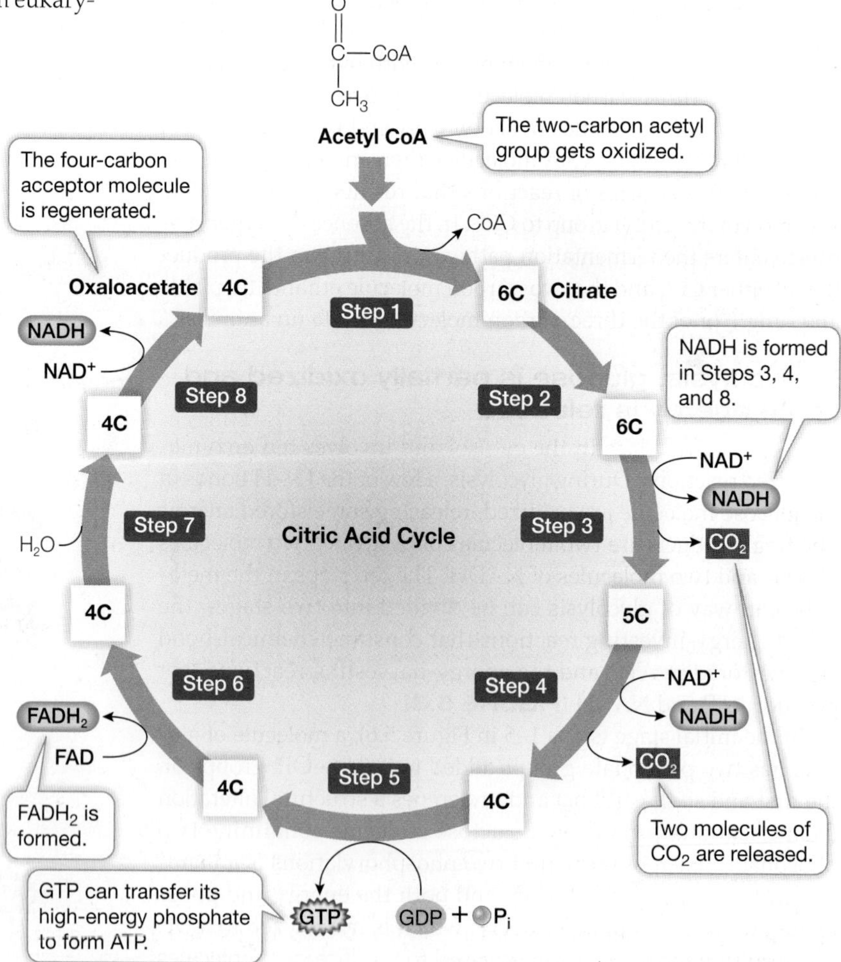

FIGURE 5.8 The Citric Acid Cycle One turn through the eight steps of the citric acid cycle results in the complete oxidation of acetyl CoA to CO_2, the production of one GTP (from GDP + P_i), and the reduction of three NAD^+ (to NADH) and one FAD (to $FADH_2$). "6C," "5C," and "4C" indicate the number of carbon atoms in each intermediate in the cycle.

Activity 5.3 The Citric Acid Cycle

fully use the energy harvested in catabolism, cells transfer energy from NADH (and $FADH_2$) to the phosphoanhydride bonds of ATP, in a process termed **oxidative phosphorylation**.

In eukaryotic mitochondria, oxidative phosphorylation is accomplished in two steps: electron transport and chemiosomosis. During electron transport, NADH (and $FADH_2$) oxidation is used to actively transport protons (H^+ ions) across the inner mitochondrial membrane, resulting in a proton gradient. During chemiosmosis, the diffusion of protons back across the membrane is then used to drive the synthesis of ATP. (In prokaryotes, oxidative phosphorylation takes place at the cell membrane.)

First let's examine how the oxidation of NADH and $FADH_2$ leads to the production of the proton gradient. For example, when NADH is oxidized to NAD^+, O_2 is reduced to H_2O:

$$NADH + H^+ + \frac{1}{2} O_2 \rightarrow NAD^+ + H_2O$$

This does not happen in a single step. Rather, there is a series of redox electron carrier proteins called the respiratory chain embedded in the inner membrane of the mitochondrion (**FIGURE 5.9**). In a process termed **electron transport**, the electrons from the oxidation of NADH and $FADH_2$ pass from one carrier to the next in the chain. The oxidation reactions are exergonic, and they release energy that is used to actively transport H^+ ions across the membrane. $FADH_2$ feeds into the chain later than NADH, and thus the electrons that $FADH_2$ delivers are able to move fewer protons across the inner mitochondrial membrane.

An important aspect of electron transport is that an oxidation reaction is always coupled with a reduction. When NADH is oxidized to NAD^+, the corresponding reduction reaction is the formation of water from O_2:

$$2\ H^+ + 2\ e^- + \frac{1}{2} O_2 \rightarrow H_2O$$

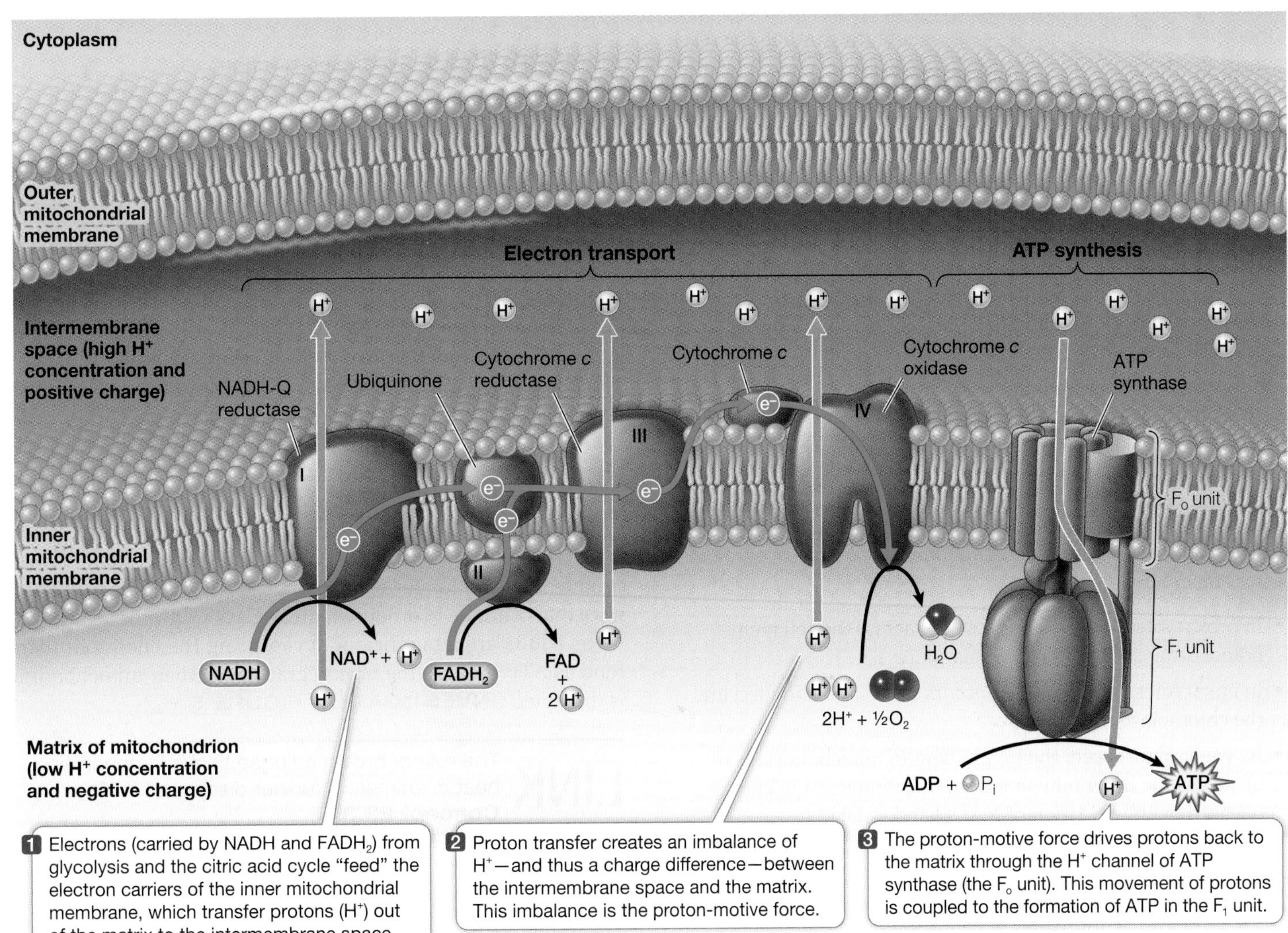

FIGURE 5.9 Oxidative Phosphorylation in the Mitochondrion The mitochondrial electron transport chain (respiratory chain) pumps protons into the intermembrane space. The resulting proton gradient is used by ATP synthase to phosphorylate ADP to ATP.

Activity 5.4 **Electron Transport Simulation**

Animation 5.1 **Electron Transport and ATP Synthesis**

So the key role of O_2 in cells—the reason we breathe and have a blood system to deliver O_2 to tissues, and the reason an enormous number of species require continuous access to molecular oxygen—is to act as an electron acceptor and become reduced, forming water. This allows the respiratory chain to continue to move electrons from one complex to the next, and to pump protons across the membrane against their concentration gradient.

Activity 5.5 The Respiratory Chain

Chemiosmosis uses the proton gradient to generate ATP

In addition to the electron transport carriers, the inner mitochondrial membrane contains an enzyme called **ATP synthase** (see Figures 5.9 and 5.10). This enzyme uses the H^+ gradient to drive the synthesis of ATP via a mechanism called **chemiosmosis**—the movement of ions across a membrane from a region of higher concentration to a region of lower concentration. Protons (H^+) cannot readily diffuse across the membrane (see Key Concept 4.2), but they can cross the membrane through the ATP synthase enzyme. ATP synthase converts the potential energy of the proton gradient (called the proton-motive force) into the chemical-bond energy in ATP.

ATP synthase is a molecular motor composed of two parts: the F_o unit, which is a transmembrane domain that functions as the H^+ channel; and the F_1 unit, which contains the active sites for ATP synthesis (**FIGURE 5.10**). The F_1 unit consists of six subunits (three each of two polypeptide chains), arranged like the segments of an orange around a central polypeptide. The potential energy set up by the proton gradient drives the passage of protons through the ring of polypeptides that make up the F_o component. This ring rotates as the protons pass through the membrane, causing part of the F_1 unit to rotate as well. ADP and P_i bind to active sites that become exposed on the F_1 unit as it rotates, and ATP is made.

The structure and function of the ATP synthase enzyme is shared by living organisms as diverse as bacteria and humans. The enzyme makes ATP at rates of up to 100 molecules per second. In all organisms, this molecular motor relies on protein gradients across membranes:

- In prokaryotes, the gradient is set up across the cell membrane, using energy from various sources.
- In eukaryotes, chemiosmosis occurs in the mitochondria and the chloroplasts.
- As we have just seen, the H^+ gradient in mitochondria is set up across the inner mitochondrial membrane, using energy released by the oxidation of NADH and FADH2.
- In chloroplasts, the H^+ gradient is set up across the thylakoid membrane using energy from light (see Key Concept 5.5). In this case, the reduced coenzyme is NADPH, a relative of NADH.

Despite these differences in detail, the mechanism of chemiosmosis is similar in almost all forms of life.

Animation 5.2 Two Experiments Demonstrate the Chemiosmotic Mechanism

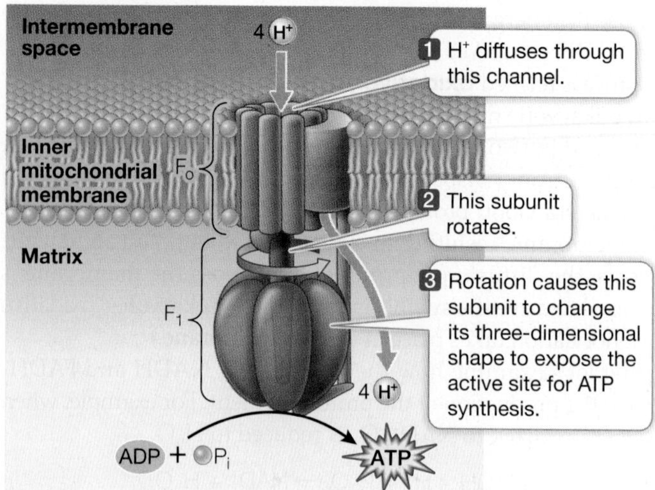

FIGURE 5.10 The Structure and Mechanism of ATP Synthase Mitochondrial ATP synthase is a rotary motor that uses the potential energy stored in the proton gradient to synthesize ATP from ADP and P_i.

Media Clip 5.1 ATP Synthase in Motion

The proton gradient can also be used to produce heat

In some situations, a membrane channel is present in the inner mitochondrial membrane that is always open to protons. The presence of this channel reduces the H^+ gradient, which implies that ATP cannot be made as efficiently. Considerable energy is released as protons move through the membrane channel down their concentration gradient. The released energy forms heat instead of being used to make ATP. In newborn human infants, a membrane protein appropriately called uncoupling protein 1 (UCP1) disrupts the H^+ gradient in the mitochondria of cells in brown adipose tissue ("brown fat"), and this results in the release of heat. Because infants lack body hair and have high surface area-to-volume ratios, they cool easily; the presence of brown adipose tissue allows them to generate heat and helps keep them warm. Once they grow larger, this method of heat production is no longer needed and UCP1 levels drop. Mutant mice that continue to produce high levels of UCP1 in their brown fat as adults are less efficient at capturing the energy in their food as ATP because the proton gradient in their mitochondria is disrupted (**INVESTIGATION FIGURE 5.11**).

The role of brown adipose tissue in generating heat in animals is further discussed in **Key Concept 28.3**.

A popular weight-loss drug in the 1930s was the synthetic uncoupling molecule dinitrophenol. There were claims of dramatic weight loss when the drug was administered to obese patients. Unfortunately, the heat that was released caused fatally high fevers, and the effective dose and fatal dose were quite close. The use of this drug was discontinued in 1938, but the general strategy of using an uncoupling molecule for weight loss remains a subject of research.

INVESTIGATION

FIGURE 5.11 Mitochondria, Genetics, and Obesity As people (and mice) get older, they tend to accumulate fat. In the course of investigating a strain of mice genetically unable to make the receptor for a hormone called ghrelin that is involved in controlling appetite, Yuxiang Sun and her team at Baylor College of Medicine compared these mice with normal mice as they aged.[a] They were surprised to find that the genetically altered (mutant) mice did not put on as much weight as their normal counterparts did. The researchers discovered that the genetic mutation in these mice caused uncoupling of oxidative phosphorylation and burning of body fat. Their experiment investigated whether levels of mitochondrial uncoupling protein 1 (UCP1) were different in the mutant and normal mice.

HYPOTHESIS

Mice that make more UCP1 burn more body fat.

METHOD

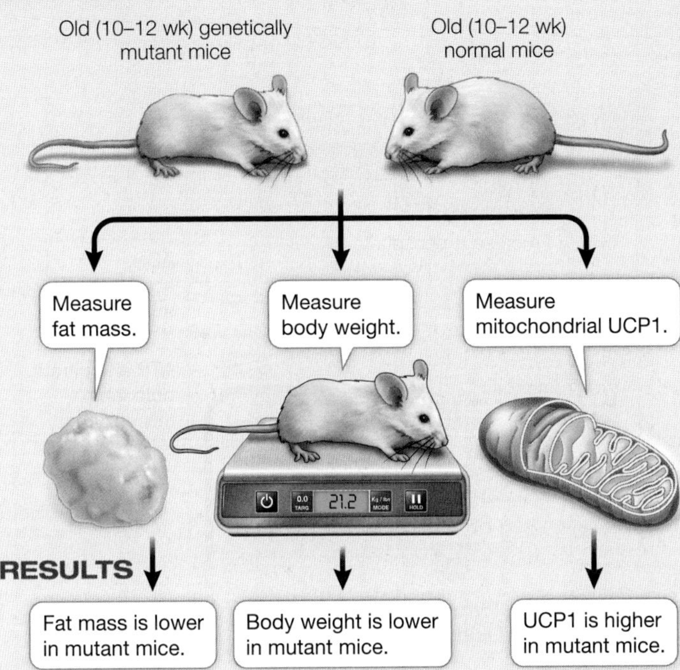

CONCLUSION

Increased UCP1 correlates with less fat and a lower body weight.

ANALYZE THE DATA

In an effort to determine whether weight gain in normal (wild-type) mice might be due to inactivity or to eating too much (two of the well-known causes of weight gain), as opposed to a biochemical abnormality (such as unusual UCP1 levels), Sun and her team examined younger (3–4 months old) and older (10–12 months old) mice of both strains. They measured weight and body composition in terms of fat and lean body tissues. The results are shown in the two graphs below. Error bars indicate ± one standard error of the mean.

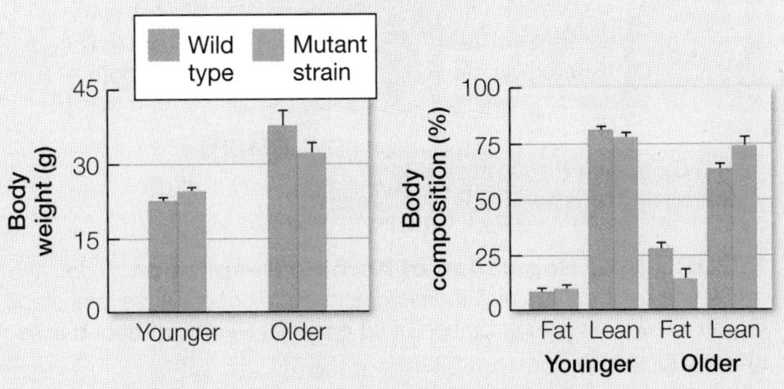

1. Did the mutant mice and wild-type mice gain weight as they aged?
2. The researchers wondered whether the difference in weight of normal and mutant older mice noted in the lower left graphs was due to changes in eating patterns or exercising. So they measured how much food the mice ate per day and measured their movements over time in a special chamber. The results are shown in the two graphs below. What can you conclude about the role of food intake and exercise on the weight differences between the two strains of mice?

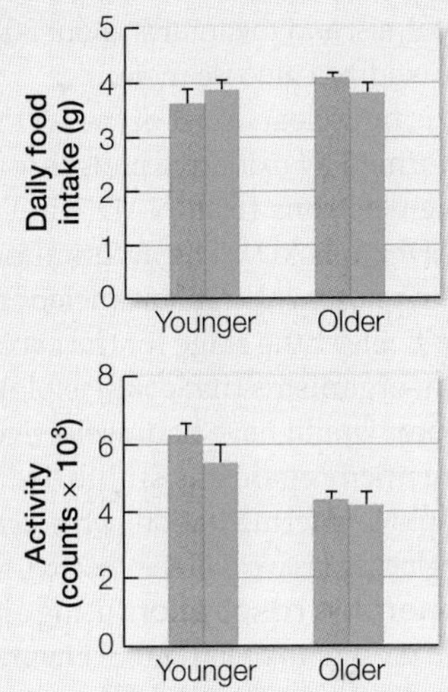

3. The UCP1 found in the inner mitochondrial membrane of brown fat cells uncouples mitochondrial electron transport (oxidation) and ATP production (phosphorylation), so that instead of being trapped as chemical energy in the formation of ATP, the energy released by oxidation is released as heat. Sun and her team measured UCP1 levels in brown fat mitochondria in older individuals of the two strains of mice. The results are shown in the graph below. What can you conclude about the role of UCP1 in the weight differences between the two strains of mice?

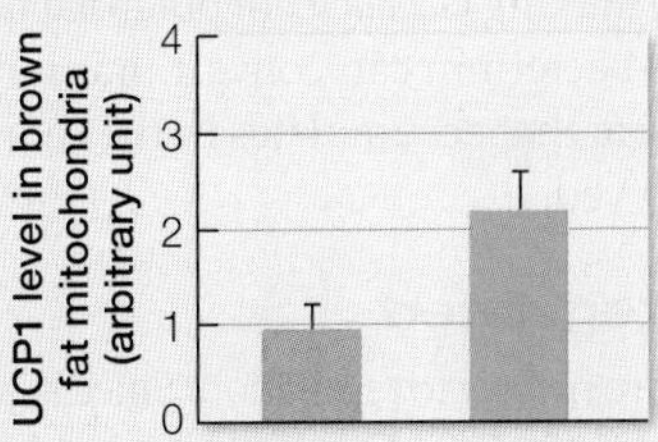

[a]X. Ma et al. 2011. *PLoS ONE* 6: e16391. © 2011 Ma et al.; CC BY 4.0.

A similar **Analyze the Data** exercise may be assigned in **Achieve**.

Oxidative phosphorylation and chemiosmosis yield a large amount of ATP

For each NADH that is produced in the mitochondrial matrix and gives electrons to the respiratory chain, about 2.5 ATP molecules are formed. For $FADH_2$, the value is about 1.5 ATP molecules. Thus the three molecules of NADH and one molecule of $FADH_2$ produced by each turn of the citric acid cycle yield about nine [(3 × 2.5) + (1 × 1.5)] molecules of ATP. Two molecules of acetyl CoA are produced from each glucose molecule, so the total is about 18 ATPs per molecule of glucose. The two NADH molecules produced by oxidation of two pyruvate molecules, which also occurs in the mitochondrial matrix, give another five ATPs. Since the two molecules of NADH produced in glycolysis have to be actively transported into the mitochondrial matrix, the net gain is less, about 1.5 molecules of ATP per molecule of NADH, giving another three ATP molecules. Add to this the two GTP molecules (which can phosphorylate ADP to ATP) formed by oxidation of two acetyl CoA molecules in the citric acid cycle and the net of two additional ATP molecules formed by substrate-level phosphorylation during glycolysis, and the total is about 30 molecules of ATP produced per fully oxidized glucose molecule.

The vital role of O_2 is now clear: most of the ATP produced in cellular respiration is formed by oxidative phosphorylation—the process of transferring electrons from NADH to O_2, resulting in the oxidation of NADH to NAD^+. The accumulation of atmospheric O_2 as a result of photosynthesis by ancient microorganisms (see Key Concept 17.2) set the stage for the evolution of oxidative phosphorylation; organisms that could exploit the O_2 as a terminal electron receptor would have had a selective advantage.

Nevertheless, many microorganisms still thrive where O_2 is scarce. Rather than employing fermentation, which captures relatively little energy from glucose (as we will see shortly), these bacteria and archaea employ **anaerobic respiration**, using alternative terminal electron acceptors present in their natural environments. Terminal electron receptors vary depending on the species and the environment, and include nitrates (NO_3^-), sulfates (SO_4^{2-}), sulfur, and various metal ions. For example, the bacterium *Geobacter metallireducens* typically lives in sediments at the bottom of streams or ponds and uses metal ions, including iron, as terminal electron acceptors:

$$Fe^{3+} + e^- \text{ (from electron transport)} \rightarrow Fe^{2+}$$

This bacterium can also use radioactive uranium ions as electron acceptors. In the process, the uranium is converted from a soluble to an insoluble form, making *Geobacter* of potential use in environmental cleanup. The bacterium can convert uranium in contaminated water into a form that accumulates in the sediment and can be more readily removed.

Respiration is regulated

Metabolic pathways are subject to regulation. Since it is enzymes that control the speed of chemical reactions in cells, regulation of a pathway usually involves control of enzyme abundance and/or activity. Regulation of enzyme activity often involves **feedback inhibition**, in which the product of a pathway binds to and inhibits an enzyme that catalyzes an early step of the pathway. Binding occurs at the active site or at an allosteric site, which is a regulatory site elsewhere on the protein (see Key Concept 3.5).

Aerobic respiration is highly regulated. For example, high concentrations of ATP, the main product of glucose catabolism, or of citrate, a compound in the citric acid cycle, inhibit an early enzyme in glycolysis. ATP also inhibits pyruvate dehydrogenase, which oxidizes pyruvate, and a few enzymes in the citric acid cycle (**FIGURE 5.12**). When levels of ATP in the cell are high, aerobic respiration slows. Sometimes feedback involves the product of one pathway speeding up reactions in another pathway. For example, citrate not only inhibits an early step in glycolysis, but also activates an enzyme that is early in the pathway for synthesizing glycogen, which is a large glucose polymer. Feedback regulation generally occurs rapidly, affecting a pathway within minutes.

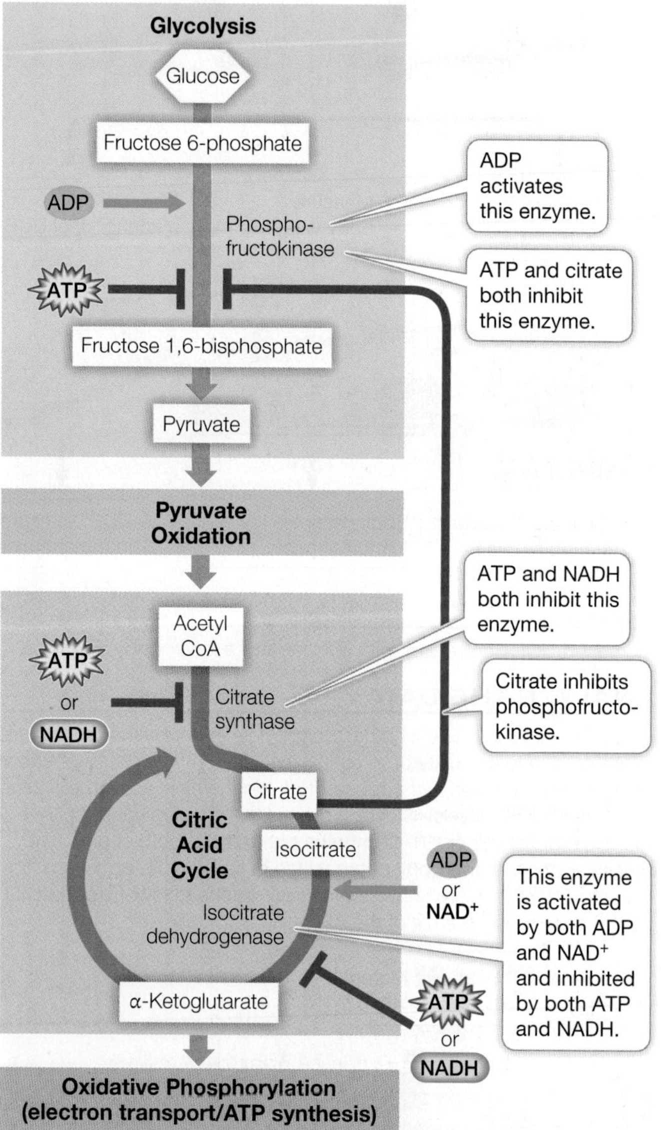

FIGURE 5.12 Regulation of Aerobic Respiration Allosteric regulation of glycolysis and the citric acid cycle occurs at several steps. Green arrows indicate activation of an enzyme. Red blunt-ended lines indicate inhibition.

Activity 5.6 Regulation of Energy Pathways

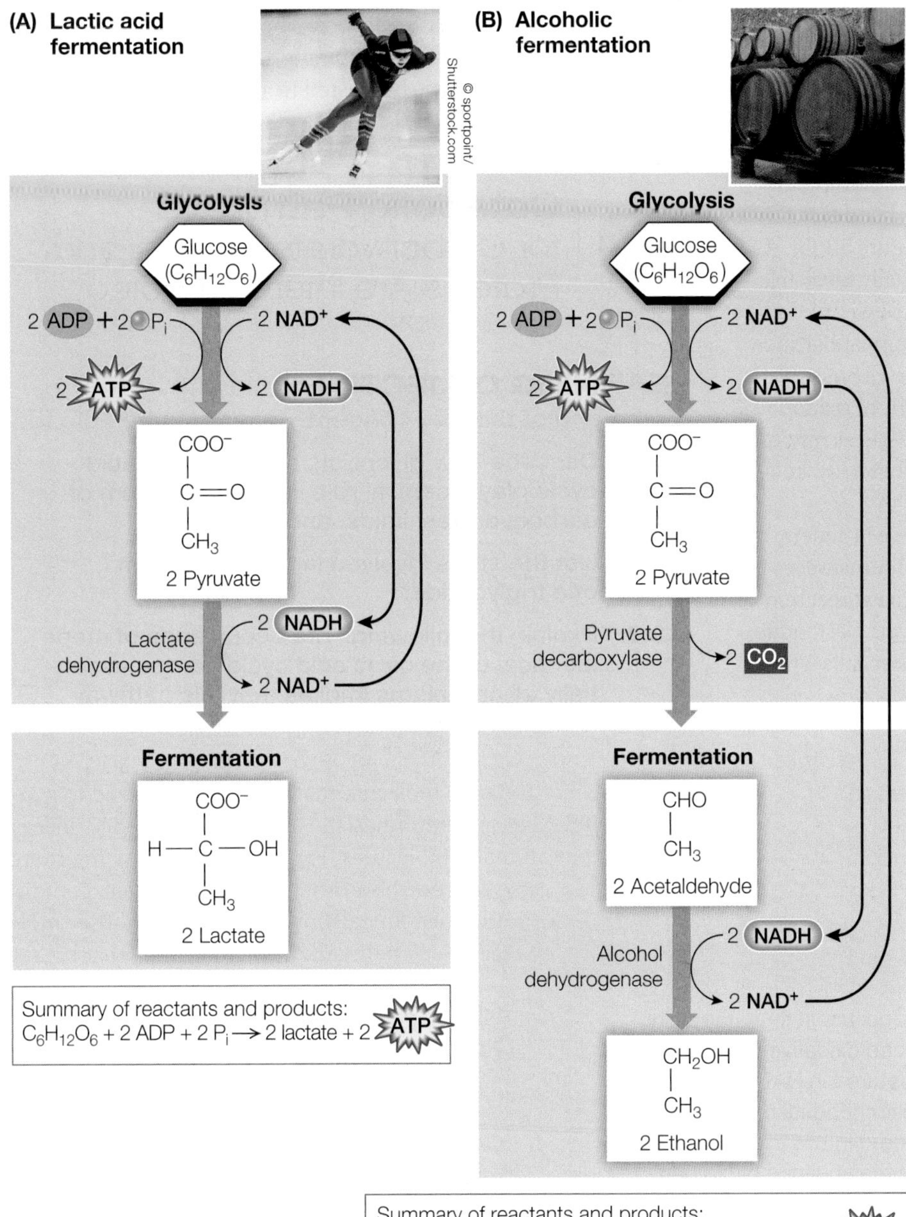

FIGURE 5.13 Fermentation (A) In lactic acid fermentation, each molecule of glucose is converted to two molecules of lactate, and two molecules of ATP are synthesized from ADP + P_i. **(B)** In alcoholic fermentation, each molecule of glucose is converted to two molecules of ethanol and two molecules of CO_2 and two molecules of ATP are synthesized from ADP + P_i. Note that many of the steps in the pathways can operate essentially in reverse.

The rate of a biochemical reaction can also be controlled by reducing or increasing the number of enzyme molecules present relative to the substrate. This can be done by altering the transcription of the genes that encode the enzymes. These events take time, and typically the effects on metabolism will take many hours to days to appear. For example, excess levels of glucose and other dietary factors can lead to increased transcription of the gene for fatty acid synthase, a key enzyme in the synthesis of fatty acids. Excess citrate produced by the citric acid cycle is broken down to acetyl CoA, which in turn is used in fatty acid synthesis. This is one reason why people accumulate fat after eating too much. The fatty acids may be catabolized later to produce acetyl CoA.

LINK The regulation of gene transcription to produce more of a protein is described in **Chapter 11**.

Fermentation of glucose releases a small amount of energy

In the absence of molecular oxygen (O_2), the respiratory chain cannot operate. (The exceptions, as we noted earlier, are the respiratory chains of anaerobic microbes adapted to use terminal electron acceptors other than oxygen.) The immediate consequence is that chemiosmosis cannot be used to generate ATP. However, there is another implication. During aerobic respiration, NADH is oxidized to NAD^+ when it donates its electrons to the respiratory chain. Without an alternative, the NADH produced by glycolysis would not be oxidized to NAD^+ and glycolysis would stop because there would be no NAD^+ to reduce, meaning that no energy could be obtained from glucose. To solve this problem, organisms use fermentation to oxidize the NADH and produce NAD^+, thus allowing glycolysis to continue (**FIGURE 5.13**).

Like glycolysis, fermentation pathways occur in the cytoplasm. There are many different types of fermentation used by different organisms, but all operate to regenerate NAD^+. During fermentation, the NADH made during glycolysis is not used by the respiratory chain to form ATP. Therefore the overall yield of ATP from fermentation is restricted to the ATP made in glycolysis (two ATP molecules per glucose molecule).

Two fermentation pathways are found in a wide variety of organisms:

1. Lactic acid fermentation, the end-product of which is lactic acid (lactate)
2. Alcoholic fermentation, the end-product of which is ethyl alcohol (ethanol)

In **lactic acid fermentation**, pyruvate serves as the electron acceptor and lactate is the product (see Figure 5.13A). This process takes place in many microorganisms and multicellular organisms, including plants and vertebrates. A notable example of lactic acid fermentation occurs in vertebrate muscle tissue. Usually, vertebrates get their energy for muscle contractions aerobically, with the circulatory system supplying O_2 to muscles. The O_2 supply is almost always adequate for small vertebrates, which explains why birds can fly long distances without resting. But in larger vertebrates such as humans, the circulatory system is unable to deliver enough O_2 when the need is great, such as during a long sprint. At this point, the muscle cells break down glycogen (a stored polysaccharide) and undergo lactic acid fermentation. The process is reversible; lactate is converted back to pyruvate once O_2 is available again.

Alcoholic fermentation takes place in certain yeasts (eukaryotic microbes) and some plant cells under anaerobic conditions. In this process, pyruvate is converted to ethanol (see Figure 5.13B). As with lactic acid fermentation, the reactions are essentially reversible.

By recycling NAD^+, fermentation allows glycolysis to continue, thus producing small amounts of ATP through substrate-level phosphorylation: a net yield of two ATPs per glucose molecule, which is much lower than the energy yield from oxidative phosphorylation. For this reason, most organisms that rely on fermentation instead of respiration are microbes that grow relatively slowly.

Although the yield of ATP per molecule of glucose is generally low with cellular anaerobic metabolism, in some circumstances it can produce an adequate supply of ATP if the enzymatic reactions in the pathway are speeded up. Indeed, this occurs in vertebrate muscle cells (see Key Concept 33.2) and in some cancer cells where O_2 is in low supply.

Activity 5.7 Energy Pathways in Cells

Activity 5.8 Energy Levels

REVIEW & APPLY | 5.2

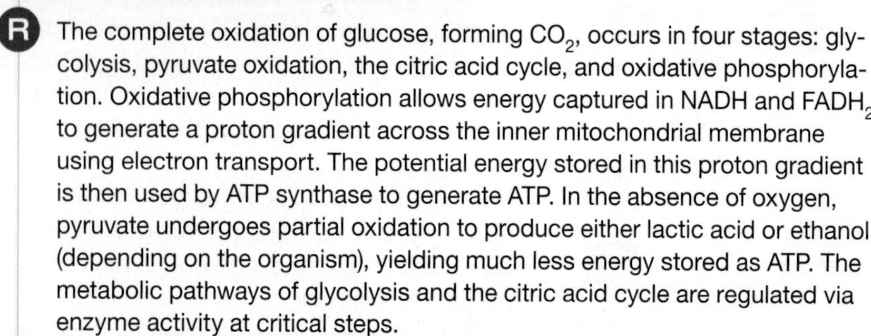

R The complete oxidation of glucose, forming CO_2, occurs in four stages: glycolysis, pyruvate oxidation, the citric acid cycle, and oxidative phosphorylation. Oxidative phosphorylation allows energy captured in NADH and $FADH_2$ to generate a proton gradient across the inner mitochondrial membrane using electron transport. The potential energy stored in this proton gradient is then used by ATP synthase to generate ATP. In the absence of oxygen, pyruvate undergoes partial oxidation to produce either lactic acid or ethanol (depending on the organism), yielding much less energy stored as ATP. The metabolic pathways of glycolysis and the citric acid cycle are regulated via enzyme activity at critical steps.

A

1. Assuming that each NADH and $FADH_2$ produced in the mitochondrial matrix can be used to synthesize about 2.5 and 1.5 molecules of ATP respectively, and that each NADH produced in the cytoplasm can be used to synthesize about 1.5 molecules of ATP, compare the energy yields (in terms of ATP per glucose molecule) from fermentation and from aerobic respiration.
2. Cyanide is a deadly toxin. It acts by binding to a heme prosthetic group that is present in the last protein complex of the respiratory chain, preventing electrons from being transferred to oxygen. What happens to aerobic respiration when cyanide is present in sufficient concentrations to block electron transport? How could a cell still generate ATP from glucose when cyanide is present?
3. Why is replenishing NAD^+ crucial to cellular metabolism? Do you think high levels of NAD^+ would activate or inhibit early glycolysis enzymes?

Now that we have discussed how cells harvest chemical-bond energy from glucose in cellular respiration, we will briefly turn our attention to how energy is harvested from other molecules.

5.3 Catabolic Pathways for Carbohydrates, Lipids, and Proteins Are Interconnected

LEARNING OBJECTIVES

By the end of this key concept you should be able to:

5.3.1 Describe how glycolysis and the citric acid cycle play a central role in the catabolism of carbohydrates, lipids, and proteins.

5.3.2 List the steps involved in the β-oxidation of a triglyceride.

5.3.3 Explain the following: The occurrence of steps identical to the citric acid cycle across essentially all organisms implies that this pathway is ancient.

Glucose is not the only molecule that can be catabolized to synthesize ATP. ATP can be synthesized using chemical-bond energy contained in all macromolecules. Fats in particular, being more reduced than other macromolecules, contain more chemical-bond energy than other molecules. A hallmark of catabolism is that breakdown products eventually enter the aerobic respiration pathways described in Key Concept 5.2. **FIGURE 5.14** outlines various metabolic pathways, including several catabolic pathways.

CARBOHYDRATES It is most straightforward to predict what happens with the breakdown of polysaccharides. These macromolecules are hydrolyzed to monosaccharides, as are disaccharides. In many cases the monosaccharides are glucose and can enter glycolysis at the first step. Other monosaccharides, such as fructose or galactose, are converted into a molecule that enters glycolysis at a slightly later step.

LIPIDS Lipids vary considerably in their chemical composition (see Key Concept 3.1), and the steps in their breakdown differ as well. The main lipids used in energy storage are triglycerides, which are first hydrolyzed in the cytoplasm to their constituents—glycerol and three fatty acids. Glycerol is converted into dihydroxyacetone phosphate, which is a three-carbon intermediate in glycolysis. Fatty acids are highly reduced molecules that are converted to acetyl CoA in a process called **β-oxidation**, catalyzed by enzymes

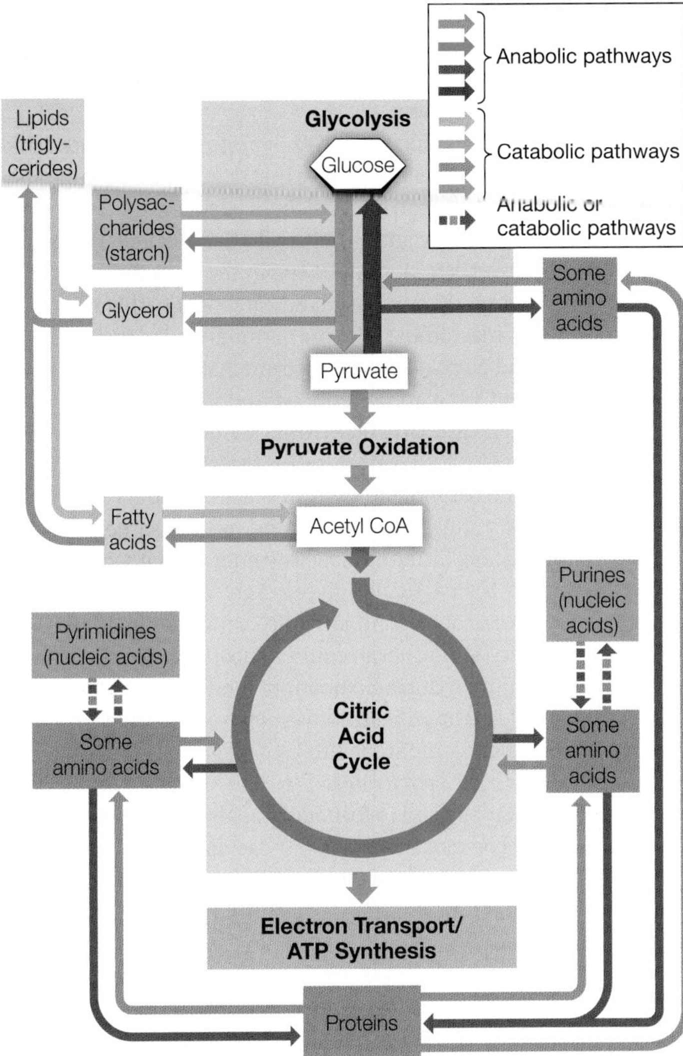

FIGURE 5.14 Metabolic Relationships The catabolic pathways of glycolysis and the citric acid cycle are central to many other metabolic pathways.

inside the mitochondrial matrix. For example, the β-oxidation of a 16-carbon (C_{16}) fatty acid occurs in several steps:

$$C_{16} \text{ fatty acid} + \text{CoA} \rightarrow C_{16} \text{ fatty acyl CoA}$$

$$C_{16} \text{ fatty acyl CoA} + \text{CoA} \rightarrow C_{14} \text{ fatty acyl CoA} + \text{acetyl CoA}$$

$$\text{(Repeat last step 6 more times)} \rightarrow 8 \text{ acetyl CoA}$$

With each removal of a two-carbon acetyl group during β-oxidation, the remaining fatty acid chain—now two carbons shorter—is left with a CoA group attached to its end (see second reaction above). The oxidation process is repeated using this fatty acyl CoA, releasing another acetyl CoA and again leaving a fatty acyl CoA, once again two carbons shorter. The acetyl CoA molecules produced by β-oxidation enter the citric acid cycle and the acetyl groups are fully oxidized to CO_2.

PROTEINS Proteins are hydrolyzed to their amino acid building blocks in a process called **proteolysis**. Proteolysis requires a small initial input of energy in the form of ATP to overcome bonds involved in protein folding, but the hydrolysis of the peptide bonds is exergonic, though no energy is captured in the form of ATP. Degradation of the resulting amino acid building blocks results in molecules that feed into glycolysis or the citric acid cycle at different points. For example, the amino acid glutamate is converted into α-ketoglutarate, an intermediate in the citric acid cycle.

LINK Animals need chemical-bond energy to live. **Key Concept 28.1** discusses how animals use the chemical-bond energy in carbohydrates, lipids, and proteins.

NUCLEIC ACIDS Nucleic acids constitute a relatively small percentage of the molecules consumed in food, and are thus seldom used as an energy source. Nevertheless, nucleic acids can be hydrolyzed to their nucleotide monomers, which can then be further catabolized into inorganic phosphate groups, bases, and ribose or deoxyribose. The latter two components of nucleic acids enter aerobic respiration in glycolysis (ribose or deoxyribose) or in the citric acid cycle (bases).

The way in which glucose and other macromolecules are oxidized is very similar across cells that respire aerobically. Even in species that live in anaerobic conditions, components of the citric acid cycle have been found. These commonalities suggest that the citric acid cycle may represent one of the earliest metabolic pathways.

REVIEW & APPLY | 5.3

R Many macromolecules can be oxidized and the released energy used to produce ATP. Catabolism of molecules other than glucose produces products that enter the metabolic pathways of aerobic respiration at different points. Carbohydrates, some amino acids, and glycerol (from triglycerides) enter at various steps in glycolysis; fatty acids enter at the acetyl CoA step, and several amino acids and nucleic acids enter at different steps in the citric acid cycle.

A

1. Proteins and triglycerides can be used as an energy source. Briefly describe where the building blocks of these macromolecules enter aerobic respiration.
2. Write the equation for the full oxidation of a triglyceride molecule that has three identical fatty acids, each containing eight carbon atoms. (*Hint*: See Key Concept 3.1.)
3. A newly discovered species of bacteria is found living in sediments in lakes. Oxygen levels in these sediments are essentially zero because aerobic bacteria that break down organic material from dead plants, algae, and insects use it up. The newly discovered species is living on the macromolecules that are not used by the aerobic bacteria. Where does the oxygen used by the aerobic bacteria end up? Since the new species lives where oxygen is absent, do you think it is likely that it is still oxidizing carbon using the steps of aerobic respiration?

You have seen how cells harvest chemical-bond energy in cellular respiration. Now we will see how that energy can be used in anabolism.

5.4 | Anabolic Pathways Use Large Amounts of ATP

LEARNING OBJECTIVES

By the end of this key concept you should be able to:

5.4.1 Explain why anabolism involves a series of small energetic steps involving coupled ATP hydrolysis.

5.4.2 Give an example of how intermediates of catabolic pathways can be diverted into anabolic pathways, and vice versa.

5.4.3 Describe how anabolic pathways can be regulated at both the enzyme and gene-expression levels.

The pathways we have seen thus far in the chapter illustrate the central role of glycolysis, pyruvate oxidation, the citric acid cycle, and oxidative phosphorylation in the aerobic metabolism of macromolecules. These pathways generate large quantities of ATP. Energy captured as ATP is readily available to perform various functions in the cell. One important function is anabolism.

Anabolism can generate macromolecules and their subunits

As mentioned in Chapter 3, macromolecules are synthesized from their building blocks. The reactions are catalyzed by various enzymes and require energy input. Glycogen, starch, and cellulose are synthesized from glucose monomers present in the cell. Triglycerides are synthesized from fatty acids and glycerol, especially in adipocytes (fat cells). Nucleic acids are synthesized from nucleotide building blocks, as we will discuss in Chapter 9. Protein synthesis from amino acids will be discussed in Chapter 10. In all cases, the synthesis of these large molecules is performed in steps that individually do not require large energy inputs. For example, while a large amount of energy is used to synthesize glycogen from glucose monomers, its stepwise synthesis, one monomer at a time, is manageable, and the energy for each step can be supplied by ATP. Likewise, the synthesis of other macromolecules involves steps that are small enough that ATP, or NADH, can supply sufficient energy.

In addition to synthesizing macromolecules from their subunits, cells often need to synthesize the subunits themselves. This need can be caused either by insufficient intake of a particular building block, or by differences in the stability or relative usage of various macromolecular monomers. All organisms can synthesize some of the monomers they need from other compounds. Those that cannot be synthesized are said to be **essential nutrients**, meaning they must be obtained from the environment. For example, in humans, 8 of the 20 amino acids cannot be synthesized from other compounds and are thus essential. These same amino acids can be synthesized by many bacteria and are thus not essential nutrients for them.

Anabolic pathways that synthesize various subunits are strikingly similar to the catabolic pathways that degrade those same subunits, and in many cases can be considered as simple reversals of the enzymatic steps. We saw such a reversal after fermentation stops, where lactate converts back to pyruvate in the presence of O_2. The relationships among the anabolic (and catabolic) pathways for the main subunits of macromolecules are summarized in Figure 5.14.

Glycolytic and citric acid cycle intermediates, instead of being oxidized to form CO_2, can be reduced and used to form glucose in a process called **gluconeogenesis** (which means "new formation of glucose"). Pyruvate is often the starting point and is obtained from the catabolism of amino acids. Some of the steps are exact reversals of glycolysis steps and use the same enzymes. However, other steps differ because the energy that would be required to reverse the particular glycolysis step is too large. As we would expect from the second law of thermodynamics (see Key Concept 2.3), glucose synthesis requires substantially more energy input than is obtained during oxidation: it takes the hydrolysis of six ATP molecules to synthesize one glucose molecule from two pyruvate molecules (compared with two ATP molecules obtained by oxidizing glucose to pyruvate).

Acetyl CoA can be used to form fatty acids in a reversal of fatty acid catabolism. The most common fatty acids have even numbers of carbons: 14, 16, or 18. These are formed by the addition of two-carbon acetyl CoA "units" one at a time until the appropriate chain length is reached.

Some intermediates in the citric acid cycle are reactants in pathways that synthesize important components of nucleic acids. For example, α-ketoglutarate and oxaloacetate are starting points for the synthesis of purines and pyrimidines, respectively. Some intermediates of glycolysis and the citric acid cycle can be used for amino acid synthesis.

Catabolism and anabolism are integrated into a system

A carbon atom from a protein in your burger can end up in DNA, fat, or CO_2, among other fates. How does the organism "decide" which metabolic pathways to follow, in which cells? With hundreds of enzymes and all of the possible interconversions, you might expect that the cellular concentrations of various biochemical molecules would fluctuate widely. Remarkably, the levels of these substances in what is called the metabolic pool—the sum total of all the biochemical molecules in a cell—are usually quite constant. This constancy implies that cells carefully regulate both the synthesis and breakdown of macromolecules and their subunits.

Consider what happens to the starch in your burger bun. In the digestive system, starch is hydrolyzed to glucose, which enters the blood. If it is needed, such as during physical activity, the glucose is distributed to and absorbed by cells in the rest of the body. But if there is already enough glucose in the blood to meet the body's needs, the excess glucose is converted into glycogen and stored in the liver. If insufficient glucose is supplied by food, glycogen is broken down, or if glycogen stores are exhausted, other molecules are used to make glucose by gluconeogenesis. The end result is that the level of glucose in the blood is remarkably constant.

LINK Regulation of glucose levels in the blood is described in **Key Concept 28.4**.

The interconnected nature of catabolism and anabolism implies that imbalances in nutrient uptake can be overcome quite effectively. Thus diets that are low in sugar will promote gluconeogenesis, whereas diets that are low in fat will promote fatty acid synthesis. This balancing can involve activation or inhibition of enzymes on short time scales, such as might occur following the consumption of a meal, or changes in enzyme expression over longer time scales, such as during prolonged changes in diet. However, in some cases, imbalances cannot be overcome by modifying enzyme activity or expression, such as when an essential amino acid is missing from the diet.

REVIEW & APPLY | 5.4

R Anabolic pathways are, to a large degree, simple reversals of catabolic pathways. Differences arise when steps of catabolism are highly endergonic, implying that it would be difficult to pay the energy cost of reversing that step using ATP or NADH. Anabolism and catabolism are integrated, and overall synthesis versus breakdown of macromolecules is often controlled by energy levels in cells: high ATP and NADH promote anabolism, resulting in the production of glycogen (or starch) and triglycerides. The manner in which anabolic and catabolic pathways are intertwined allows imbalances in nutrient uptake of different macromolecules to be readily overcome.

A

1. Gluconeogenesis uses many of the same enzymes as glycolysis. However, for a few steps of glycolysis that are highly exergonic, alternate enzymes are used. Why is this necessary?
2. Trace the biochemical pathway by which a carbon atom from a starch molecule in rice eaten today can end up in a muscle protein tomorrow.
3. The Atkins diet is high in protein and fat and very low in carbohydrates. How might this diet affect the rates of glycolysis and gluconeogenesis? Do you expect changes in enzyme expression?

We have seen how cellular respiration allows organisms to harvest chemical-bond energy from organic molecules and how this energy can be used to perform the anabolic and other energy-requiring activities of the cell. For the rest of the chapter, we'll look at how plants and other photosynthetic organisms produce organic molecules using energy from light.

5.5 Life Is Supported by the Sun: Light Energy Captured during Photosynthesis Converts Carbon Dioxide to Carbohydrates

LEARNING OBJECTIVES

By the end of this key concept you should be able to:

5.5.1 Describe the structure of the chloroplast and assign function to the various parts.

5.5.2 Explain how the Sun provides energy for carbon fixation and outline the sequence of reactions in the Calvin cycle showing where energy is used.

5.5.3 Describe the wavelengths of light energy used for photosynthesis.

5.5.4 Explain why the world is green.

The energy released by catabolic pathways in almost all organisms, including animals, plants, and prokaryotes, ultimately comes from the Sun. **Photosynthesis** (literally, "synthesis from light") is an anabolic process by which the energy of sunlight is captured and used to convert carbon dioxide (CO_2) and water (H_2O) into carbohydrates (which we represent as a six-carbon sugar, $C_6H_{12}O_6$) and oxygen gas (O_2):

$$6\ CO_2 + 12\ H_2O \rightarrow C_6H_{12}O_6 + 6\ O_2 + 6\ H_2O$$

Animation 5.3 The Source of the Oxygen Produced by Photosynthesis

Ponder for a moment this highly endergonic reaction. All that a photosynthetic organism requires is water, carbon dioxide, and sunlight and it can synthesize carbohydrates. This implies that photosynthetic organisms have an inexhaustible supply of chemical-bond energy that they capture as light energy from the Sun! The equation above is essentially the reverse of the general equation for aerobic catabolism of glucose that we discussed in Key Concept 5.2. The difference is that in the equation above there are 12 H_2O molecules on the left side, not 6 as would be expected from the equation for aerobic catabolism of glucose. The reason, as we will soon discuss in detail, is that in photosynthesis the oxygen atoms in O_2 all come from water—6 O_2 molecules thus come from 12 H_2O molecules. Some of the general molecular processes of photosynthesis are similar to those in glucose catabolism. For example, both processes involve redox reactions, electron transport, and chemiosmosis. However, the details of photosynthesis are quite different.

Photosynthesis involves two pathways (**FIGURE 5.15**):

1. The **light reactions** convert light energy into chemical-bond energy in the form of ATP and the reduced electron carrier NADPH. NADPH is similar to NADH but with an additional phosphate group attached to the sugar of its adenosine (see Figure 5.4).

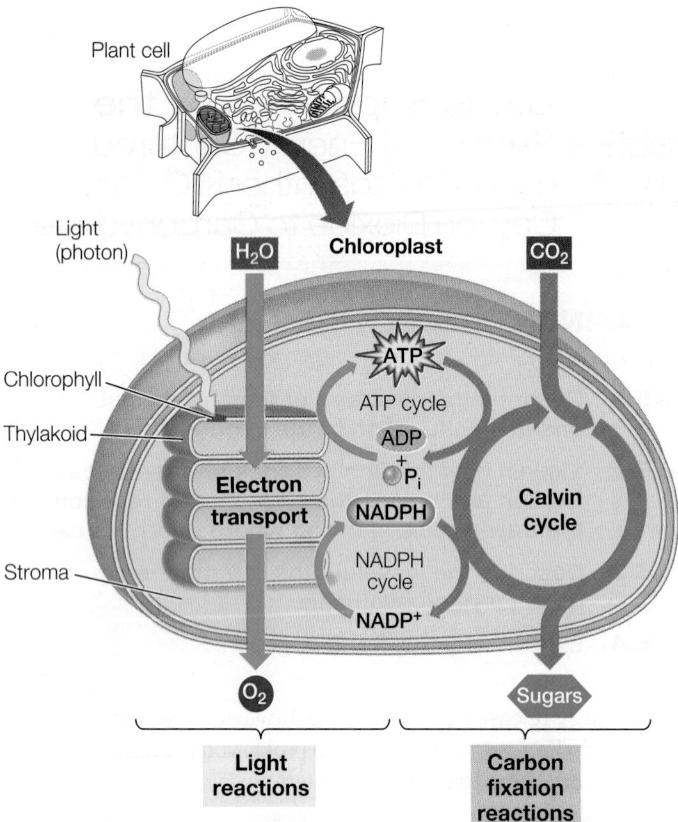

FIGURE 5.15 An Overview of Photosynthesis
Photosynthesis consists of two pathways, the light reactions and the carbon-fixation reactions. In eukaryotic plants (and algae), these pathways occur in different parts of the chloroplast—the light reactions in the thylakoids, and the carbon-fixation reactions in the stroma.

Where in the cell does the reduction of CO_2 occur, and what is the reducing agent?

2. The **carbon-fixation reactions** (also termed the light-independent, or dark, reactions) do not use light directly, but instead use the ATP and NADPH made by the light reactions, along with CO_2, to produce carbohydrates.

Both the light reactions and the carbon-fixation reactions stop in the dark because, in photosynthetic organisms, carbon-fixation requires both ATP and NADPH that are synthesized in the light reactions. In photosynthetic prokaryotes (e.g., cyanobacteria; see Figure 4.21B), the light reactions take place on internal membranes and the carbon-fixation reactions occur in the cytosol. In eukaryotic plants (and algae), which will be our focus here, both pathways proceed within the chloroplast, but they occur in different parts of that organelle (see Figure 5.15).

LINK The endosymbiotic origins hypothesis for chloroplasts is supported by their similarity to cyanobacteria and will be discussed in **Key Concept 19.1**.

Light energy is absorbed by chlorophyll and other pigments

Light is a form of energy that can be converted to other forms, such as thermal or chemical-bond energy. Earth would be a very cold place without the conversion of light energy to the thermal energy that warms the planet. Light comes in individual packets of energy termed **photons**. The amount of energy contained in a single photon depends on its electromagnetic wavelength—the shorter the wavelength, the higher the energy in the photon. Differences in wavelength are also what we perceive as color differences.

The visible portion of the electromagnetic spectrum (**FIGURE 5.16**) encompasses a fairly narrow range of wavelengths and energy levels. In order for a photon of a particular wavelength to be detected or used by an organism, that organism must have molecules that are able to absorb the photon's energy without breaking. Photons with very short wavelengths break covalent bonds when they are absorbed and thus cannot be easily captured, since molecules in the capturing system would be damaged. In contrast, long-wavelength photons have less energy, making them less useful for forming chemical-bond energy. The visible spectrum represents photons with enough energy to excite electrons to high-energy states in atoms, but not enough energy to break most organic covalent bonds.

In plants and other photosynthetic organisms, certain receptive molecules absorb photons in order to harvest their energy for biological processes. Molecules that absorb photons in the visible spectrum are termed **pigments**. Pigments absorb only specific wavelengths of light—photons with specific amounts of energy. When a photon meets a pigment molecule, one of three things can happen:

1. The photon may bounce off the molecule—it may be *scattered* or *reflected*.
2. The photon may pass through the molecule—it may be *transmitted*.
3. The photon may be *absorbed* by the molecule, adding energy to the molecule.

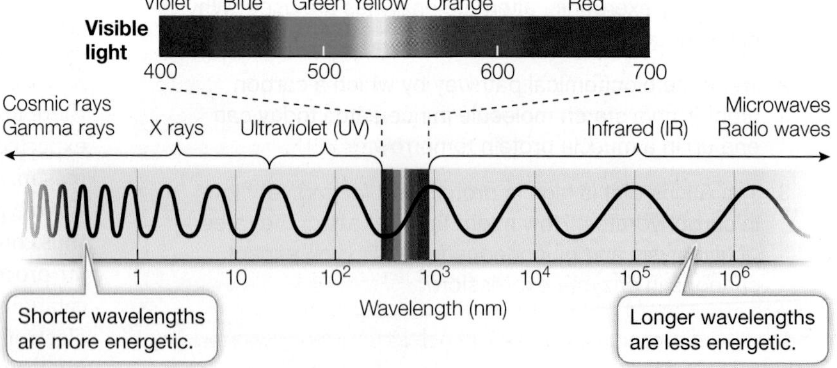

FIGURE 5.16 The Electromagnetic Spectrum The portion of the electromagnetic spectrum that is visible to humans as light is shown in detail at the top. (After S. Giannakis et al. 2016. *Appl Catal B: Environ* 199: 199–223.)

The wavelengths of photons absorbed by biological pigment molecules represent a small part of the electromagnetic spectrum. In terms of energy transfer, why don't pigment molecules absorb at wavelengths much shorter or longer than those found in the visible spectrum?

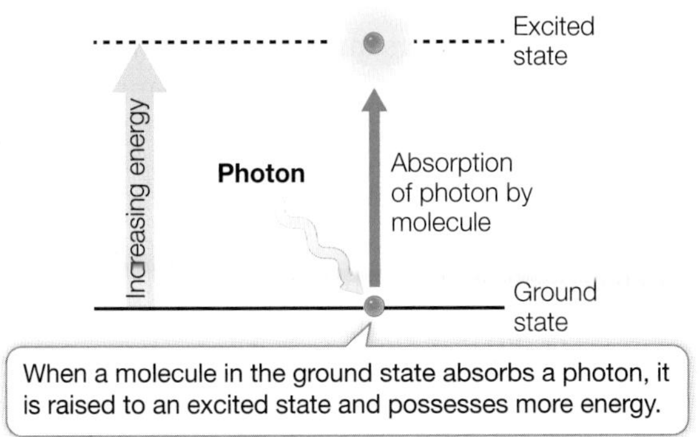

FIGURE 5.17 Excitation in a Pigment Molecule An electron in a pigment molecule absorbs a photon of light energy and is excited to a higher energy state.

Neither of the first two outcomes causes any change in the molecule. However, in the case of absorption, the photon disappears and its energy is absorbed by the molecule. The photon's *energy* cannot disappear, because according to the first law of thermodynamics, energy is neither created nor destroyed (see Key Concept 2.3). When the molecule acquires the energy of the photon, one of the electrons in one of its atoms becomes "excited" and moves outward from its normal electron shell to the next electron shell. As you may recall from Chapter 2, electrons in more distant shells have higher energy. The pigment molecule has moved from its lower energy ground state to its excited, higher energy state (**FIGURE 5.17**).

The difference in free energy between the molecule's excited state and its ground state is approximately equal to the energy of the absorbed photon (a small amount is lost to entropy). A pigment molecule is unstable and more chemically reactive when it is excited than when it is in the ground state.

Two main pigments are involved in photosynthesis: chlorophyll *a* and chlorophyll *b*. Each of these molecules is able to absorb photons of light with particular wavelengths, as shown in their **absorption spectra** (**FIGURE 5.18A**). Not surprisingly, the rate of photosynthesis as a function of supplied light, which is plotted as an **action spectrum**, has peaks that roughly correspond to the peak absorbances for these pigments (**FIGURE 5.18B**).

Both chlorophyll *a* and chlorophyll *b* have a very similar, complex ring structure, with a magnesium ion at the center (**FIGURE 5.19**). A long hydrocarbon "tail" helps anchor the chlorophyll molecule to proteins in the thylakoid membrane of the chloroplast.

The chlorophylls absorb blue and red light, which are near the two ends of the visible spectrum (see Figure 5.18). In addition, plants possess accessory pigments that absorb photons intermediate in energy between the red and the blue wavelengths, and then transfer a portion of that energy to the chlorophylls. Among these accessory pigments are carotenoids such as β-carotene, which absorb photons in the blue and blue-green wavelengths and appear deep yellow (see Figure 5.18). The phycobilins, which are found in red algae and in cyanobacteria, absorb various yellow-green, yellow, and orange wavelengths.

Some plant pigments act as sensors that regulate growth and development; see **Key Concept 25.3.**

Light absorption results in reduction of electron acceptors by chlorophyll *a* in the reaction center

The pigments in photosynthetic organisms are bound to membrane proteins that are arranged into **light-harvesting complexes**. For example, light-harvesting complex II consists of three copies of a transmembrane protein, each of which binds 12 chlorophyll and 2 carotenoid pigment molecules. Thus this single complex contains 36 chlorophyll and 6 carotenoid molecules. Several light-harvesting complexes together make up the **antenna complex**, the role of which is to capture photons of light and deliver the energy to the **reaction center**, where the energy is used to excite electrons that are then used to reduce electron acceptors (**FIGURE 5.20**). The antenna complex forms a ring around the reaction center. The antenna complex and reaction center together make up a **photosystem**, which is the structure that captures light energy and converts it into chemical-bond energy.

When chlorophyll absorbs a photon of light, it enters an excited state. This is an unstable situation, and the chlorophyll rapidly returns to its ground state, releasing most of the absorbed energy. This is an extremely rapid process—measured in picoseconds (trillionths of a second)! For most chlorophyll molecules embedded in

(A) The absorption spectra of *Anacharis*

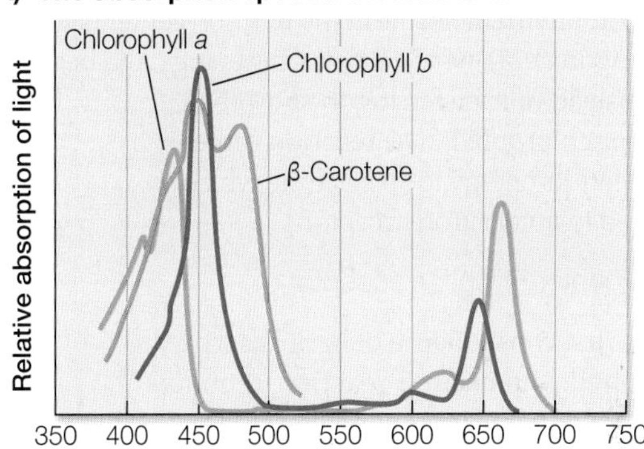

(B) The action spectrum for *Anacharis*

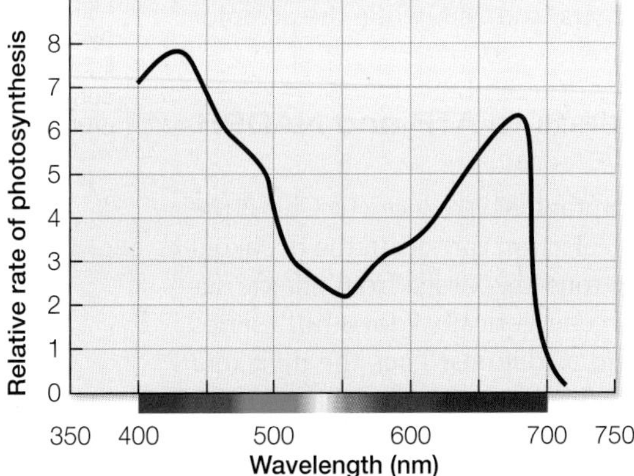

FIGURE 5.18 Absorption and Action Spectra (A) The three common photosynthetic pigments in plants have characteristic absorption spectra, which represent the wavelengths of light they will absorb. The absorption spectra shown here are for *Anacharis*, a common aquatic plant. **(B)** The action spectrum for *Anacharis*, measured by the relative photosynthetic rate of the intact plant, corresponds to the peaks in the absorption spectra of the plant's pigment molecules.

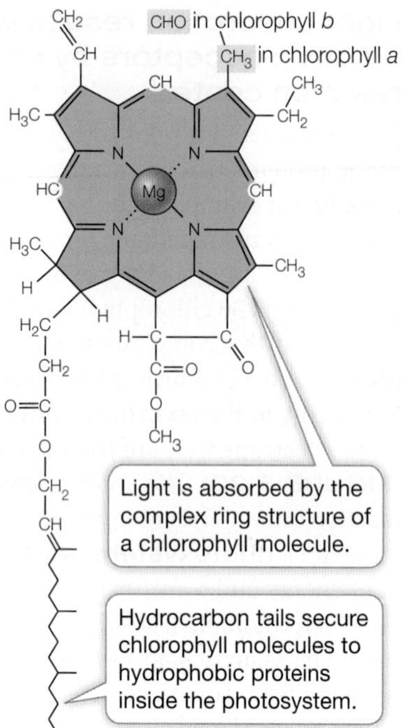

FIGURE 5.19 The Structure of Chlorophyll Pigments Both chlorophyll a and b molecules have a region that surrounds a magnesium ion and a hydrophobic "tail," which facilitates anchoring to hydrophobic regions of integral thylakoid membrane proteins.

the thylakoid membrane, the released energy is absorbed by other, adjacent chlorophyll molecules. The energy eventually arrives at a ground-state chlorophyll *a* molecule at the reaction center (symbolized by Chl; see Figure 5.21), which absorbs the energy and becomes excited (Chl*). Instead of transferring the absorbed energy to another chlorophyll, the reaction center chlorophyll transfers its energy to an electron, which is then transferred to an electron acceptor:

$$Chl^* + acceptor \rightarrow Chl^+ + acceptor^-$$

This, then, is the first consequence of light absorption by chlorophyll: the excited reaction center chlorophyll (Chl*) donates its excited electron in a redox reaction and becomes positively charged (Chl^+). As a result of this transfer of an electron, the chlorophyll gets oxidized, while the acceptor molecule is reduced.

Reduction leads to ATP and NADPH formation

The electron acceptor that is reduced by Chl* is the first of a chain of electron carriers in the thylakoid membrane. Electrons are passed from one carrier to another in an energetically "downhill" series of reductions and oxidations. Thus the thylakoid membrane has an electron transport chain similar to the respiratory chain of mitochondria that we discussed in Key Concept 5.2.

Two photosystems are used during photosynthesis:

1. **Photosystem I** (containing "P_{700}" chlorophyll at its reaction center) has reaction center chlorophyll *a* that shows highest absorbance of light at 700 nanometers (nm). This photosystem passes an excited electron to an electron transport chain, where it is used to reduce $NADP^+$, forming NADPH:

$$NADP^+ + H^+ + 2\,e^- \rightarrow NADPH$$

2. **Photosystem II** (with "P680" chlorophyll at its reaction center) has reaction center chlorophyll *a* that shows highest absorbance of light at 680 nm. This photosystem passes an excited electron to a second electron transport chain, where it is eventually used to reduce chlorophyll *a* in photosystem I. Photosystem II replenishes its electron supply by oxidizing (splitting) water, releasing oxygen:

$$H_2O \rightarrow \tfrac{1}{2}\,O_2 + 2\,H^+ + 2\,e^-$$

The energetic electrons from both photosystems are passed through a series of thylakoid membrane–bound electron carriers to a final acceptor at a lower energy level. As in the mitochondrion, a proton gradient is generated by proton transfer during electron transport, with lower pH (more H^+) in the thylakoid interior than in the stroma. The proton gradient is used by ATP synthase to phosphorylate ADP to produce ATP—a process called **photophosphorylation**. **FIGURE 5.21** shows the series of electron transport reactions that use the energy from light to generate NADPH and ATP. The pathway is noncyclic in that the energy and electrons start in one molecule (chlorophyll) and end up in another (NADPH). The term "Z scheme" is used to describe the

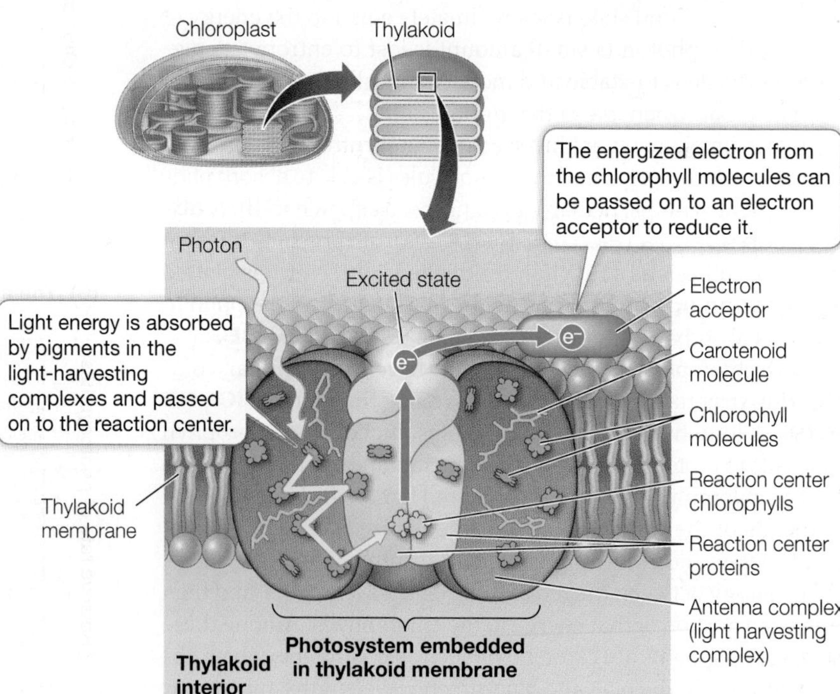

FIGURE 5.20 The Photosystem Captures Light Energy to Excite Electrons In each photosystem, multiple chlorophyll and carotenoid molecules are embedded side by side in antenna complexes that surround a specific central pair of chlorophyll *a* molecules in the reaction center. A pigment molecule absorbs a photon of light and the energy is subsequently transferred in multiple steps to a reaction center chlorophyll *a* molecule. The excited chlorophyll *a* molecule then transfers its energy to an electron, essentially converting it into chemical-bond energy, which then is used to reduce an electron acceptor.

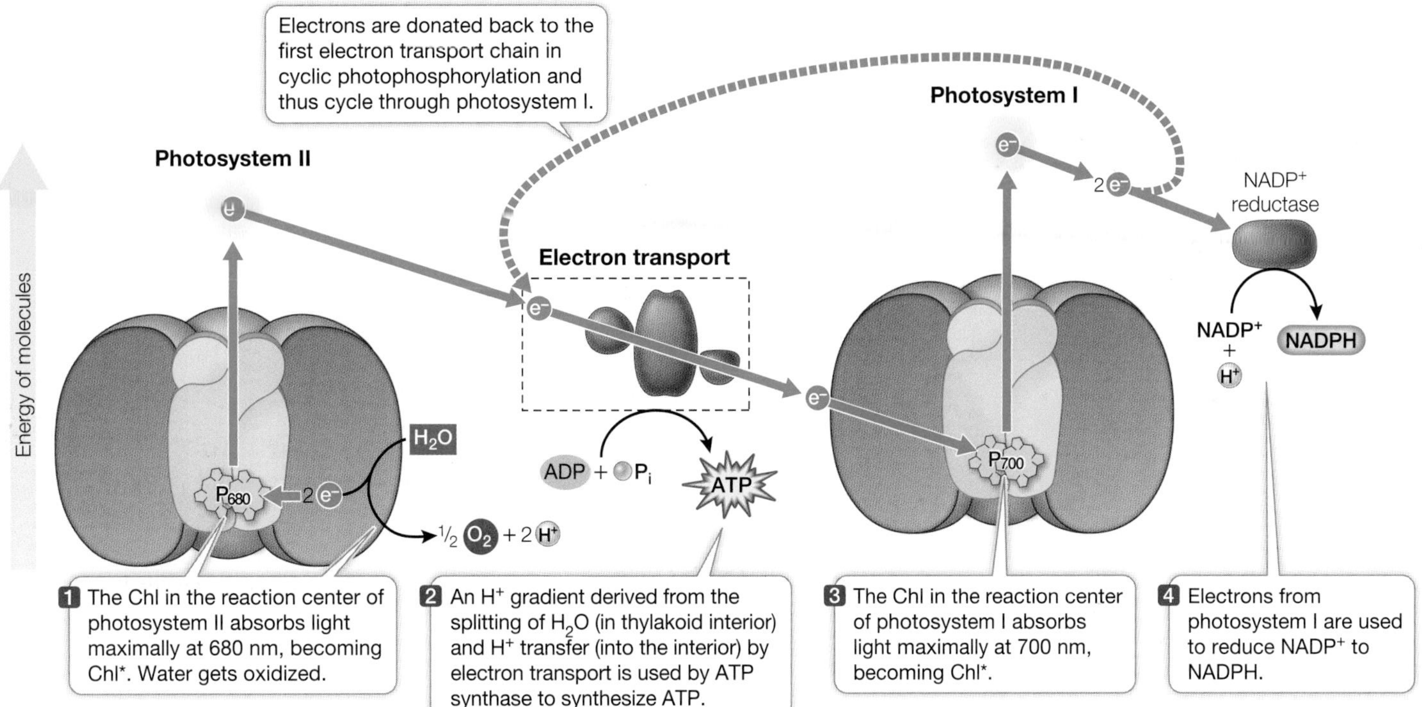

FIGURE 5.21 Electron Transport during the Light Reactions of Photosynthesis Electron transport may involve the noncyclic, or Z-scheme pathway, or the cyclic pathway. In the noncyclic pathway (solid blue lines), electrons obtained by the reaction center chlorophyll *a* from the oxidation of water are excited by photosystem II and then transferred to an electron acceptor. After passing through an electron transport chain, electrons are used to reduce chlorophyll *a* of the photosystem I reaction center. Electrons are then excited again and used to reduce another electron acceptor. They then travel down the second electron transport chain and are finally used to reduce NADP+, forming NADPH. In the cyclic pathway (dashed blue lines), electrons flow from the reaction center of photosystem I, back to the first electron transport chain and then back to photosystem I.

shape of the pathway when it is arranged as in Figure 5.21. You may have noticed that the order of the photosystems, in terms of electron movement, is photosystem II then I. This reflects the fact that photosystem I was the first one to be discovered.

Let's look in more detail at these photosystems, beginning with photosystem II.

PHOTOSYSTEM II After an excited chlorophyll *a* molecule in the reaction center (Chl*) gives up its energetic electron to reduce a chemical acceptor molecule, the chlorophyll lacks an electron and is very unstable. It has a strong tendency to "grab" an electron from another molecule to replace the one it lost—it is now a strong oxidizing agent. The replenishing electrons come from water, breaking the two O—H bonds:

$$\text{Oxidation of water: } H_2O \rightarrow \tfrac{1}{2}\,O_2 + 2\,H^+ + 2\,e^-$$

$$\text{Reduction of chlorophyll } a\text{: } 2\,e^- + 2\,Chl^+ \rightarrow 2\,Chl$$

$$\text{Overall: } 2\,Chl^+ + H_2O \rightarrow 2\,Chl + 2\,H^+ + \tfrac{1}{2}\,O_2$$

The source of the O_2 produced by photosynthesis is H_2O.

PHOTOSYSTEM I In photosystem I, an excited electron from the Chl* at the reaction center reduces an acceptor. The oxidized chlorophyll (Chl^+) now "grabs" an electron, but in this case the electron comes from photosystem II, by way of the last carrier in the electron transport system. This links the two photosystems chemically. They are also linked spatially, with both photosystems present in the thylakoid membrane. The energetic electrons from photosystem I pass through several molecules in another electron transport system and end up reducing $NADP^+$ to NADPH.

Next in the process of harvesting light energy to produce carbohydrates is the series of carbon-fixation reactions. These reactions require more ATP than NADPH. If the pathway we just described, the noncyclic or Z-scheme pathway, were the only set of light reactions operating, there might not be sufficient ATP for carbon fixation. Cyclic photophosphorylation makes up for this imbalance. This pathway uses only photosystem I and produces ATP but not NADPH; it is cyclic because the electrons flow from the reaction center of photosystem I, through the electron transport system that delivers electrons to photosystem I, and then back to photosystem I (see Figure 5.21).

Animation 5.4 Photophosphorylation

We have seen how photosystems I and II absorb light energy, which ultimately ends up as chemical-bond energy in ATP and NADPH. Let's look now at how these two energy-rich molecules are used in the carbon-fixation reactions to reduce CO_2 and thereby form carbohydrates.

The chemical-bond energy generated by the light reactions is used to convert CO_2 to carbohydrates

The energy in ATP and NADPH is used in the carbon-fixation reactions to produce carbohydrates from CO_2. During this process, the

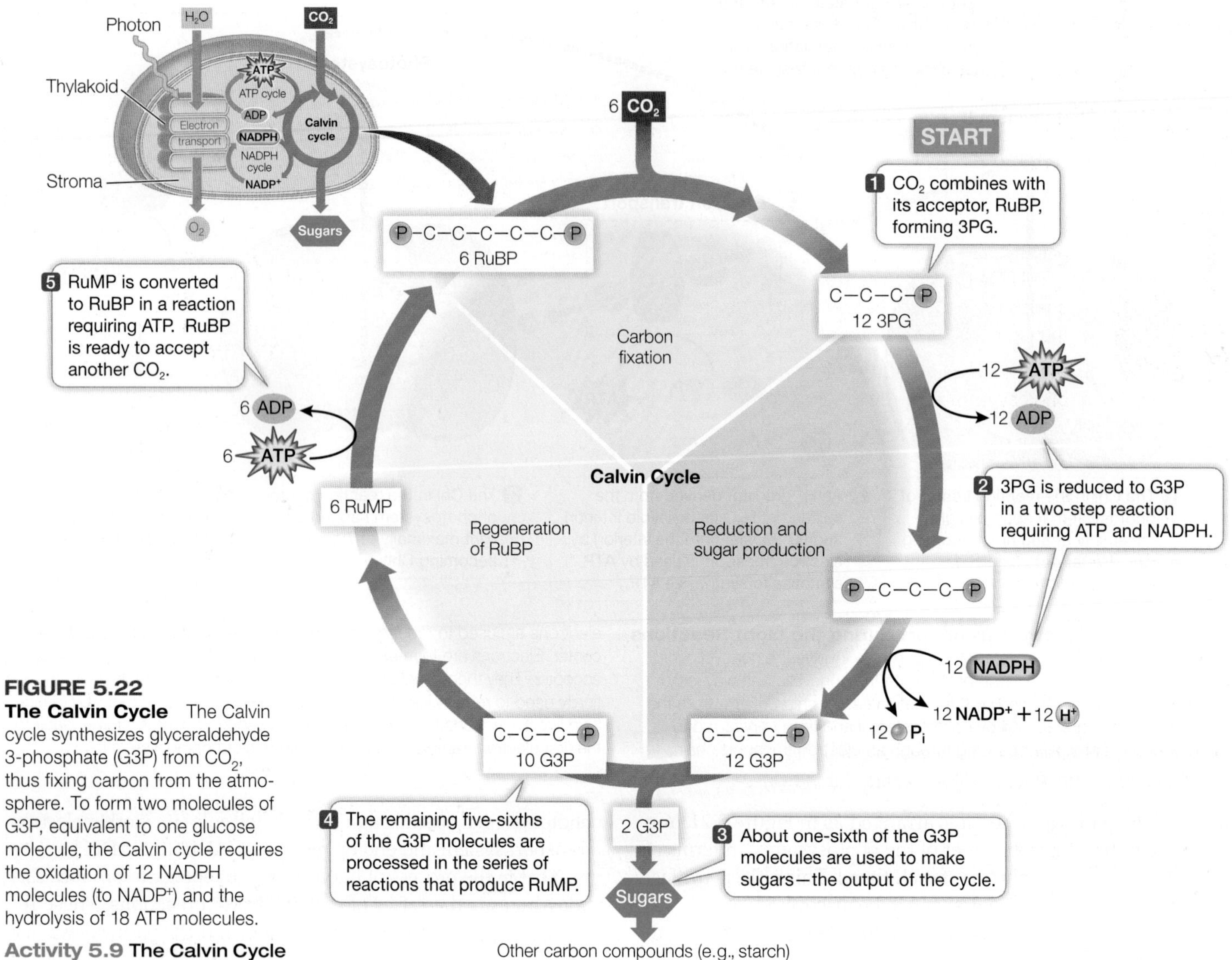

FIGURE 5.22
The Calvin Cycle The Calvin cycle synthesizes glyceraldehyde 3-phosphate (G3P) from CO_2, thus fixing carbon from the atmosphere. To form two molecules of G3P, equivalent to one glucose molecule, the Calvin cycle requires the oxidation of 12 NADPH molecules (to NADP+) and the hydrolysis of 18 ATP molecules.

Activity 5.9 The Calvin Cycle

carbon atoms are reduced. Most carbon fixation occurs only in the light, when ATP and NADPH are being generated. The metabolic pathway occurs in the stroma, or central region, of the chloroplast (see Figure 5.15) and is called the **Calvin cycle** after one of its discoverers, Melvin Calvin. As in all biochemical pathways, each reaction in the Calvin cycle is catalyzed by a specific enzyme. The cycle is composed of three distinct processes (**FIGURE 5.22**):

1. *Fixation of CO_2.* The initial reaction of the Calvin cycle adds the one-carbon CO_2 to an acceptor molecule, the five-carbon ribulose 1,5-bisphosphate (RuBP). The immediate product is a six-carbon molecule, which quickly breaks down into two three-carbon molecules called 3-phosphoglycerate (3PG; see Figure 5.22). The enzyme that catalyzes this reaction, **ribulose bisphosphate carboxylase/oxygenase (rubisco)**, is rather sluggish as enzymes go. It typically catalyzes two to three fixation reactions per second. Because of this, plants need a lot of rubisco to perform enough photosynthesis to satisfy the needs of growth and metabolism. Rubisco constitutes about half of all the protein in a leaf, and it is probably the most abundant protein on Earth.
2. *Reduction of 3PG to form glyceraldehyde 3-phosphate.* This series of reactions involves a phosphorylation (using the high-energy phosphate from an ATP made in the light reactions) and a reduction (using an NADPH made in the light reactions). The product is **glyceraldehyde 3-phosphate (G3P)**, which is a three-carbon sugar phosphate.
3. *Regeneration of the CO_2 acceptor, RuBP.* Most of the G3P ends up as ribulose monophosphate (RuMP), and ATP is used to convert this compound into RuBP. So for every "turn" of the Calvin cycle, one CO_2 is fixed and the CO_2 acceptor is regenerated.

Animation 5.5 Tracing the Pathway of CO_2

What happens to the extra G3P made by the Calvin cycle that does not go on to produce RuMP (see Figure 5.22)? About one-sixth of

THINK LIKE A SCIENTIST

Quantitative Reasoning

Process of Science

The green Earth

Why (ultimately) are photosynthetic organisms green? There are actually several other questions contained within this one:

1. Why don't plants have pigments that absorb light in the ultraviolet (UV) end of the spectrum? We already answered this question (see Key Concept 5.5): UV and shorter wavelengths have so much energy per photon that they would damage the pigments that absorbed them.
2. Why does peak absorbance, and peak photosynthetic rate, occur for photons with wavelengths around 400 nm?
3. Why don't plants have pigments that absorb photons with longer wavelengths—greater than 800 nm, for example?
4. Finally, why don't photosynthetic organisms use the middle of the visible light spectrum? In other words, why are photosynthetic organisms green and not black?

One way to make progress on answering these questions is to determine what wavelengths are available to photosynthetic organisms. **FIGURE 5.23** shows how much solar irradiance reaches the top of Earth's atmosphere, Earth's surface at sea level, and 10 meters below the ocean surface. Another useful piece of information is that the amount of energy contained in light is about 124,000/λ kJ/mol, where λ is the wavelength in nanometers.

Using Figure 5.23, construct a table that shows approximately how much energy, relative to 400-nm-wavelength light, is available for light with wavelengths of 400, 550, 650, 1,000, and 1,600 nm at Earths' surface (sea level) and 10 meters below the ocean surface. Using the values from your table, go back and answer Questions 2, 3, and 4. Do the answers to those questions suggest anything regarding whether photosynthesis evolved on land or in a marine environment?

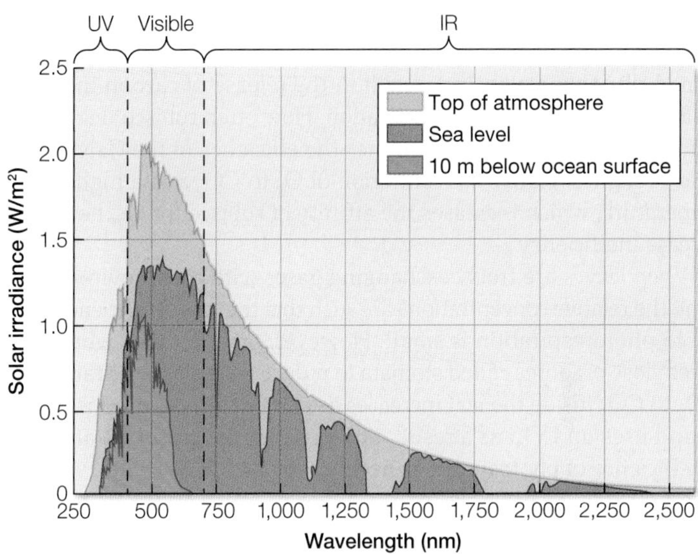

FIGURE 5.23 Solar Radiation Reaching Earth Curves show the solar irradiance, which is the amount of energy per unit time (Watts [W]) per square meter, for different wavelengths at different locations. (After R. Milo and R. Phillips. 2015. *Cell Biology by the Numbers*. Garland Science: New York. Adapted from National Renewable Energy Laboratory.)

the G3P produced in the Calvin cycle is exported out of the chloroplast to the cytosol, where it is converted to hexoses (glucose and fructose) (see step 3 of Figure 5.22). These are the hexoses that are represented as $C_6H_{12}O_6$ in the general equation for photosynthesis. These molecules may be catabolized for energy in mitochondria as part of cellular respiration; used as carbon skeletons for the synthesis of amino acids and other molecules; or converted to the disaccharide sucrose, which is transported out of the leaf to other organs in the plant.

Late in the day when glucose has accumulated inside the chloroplast, the glucose units are linked to form the polysaccharide starch. This storage carbohydrate can then be drawn on during the night so that the photosynthetic tissues can continue to export sucrose to the rest of the plant even when photosynthesis is not taking place. In addition, starch is abundant in nonphotosynthetic organs such as roots, underground stems, and seeds, where it provides a ready supply of glucose to fuel cellular activities, including plant growth.

The products of the Calvin cycle are of crucial importance to Earth's entire biosphere. The C—H covalent bonds generated by this cycle store almost all of the energy for life on Earth. Photosynthetic organisms, which are also called **autotrophs** ("self-feeders"), release most of this energy in cellular respiration, and use it to support their own growth, development, and reproduction. However, photosynthetic organisms are also the source of energy for other organisms that consume them, the **heterotrophs** ("other-feeders") that cannot photosynthesize. Heterotrophs depend on autotrophs for chemical-bond energy, which they harvest via cellular respiration. In particular, much of the energy in autotrophs resides in cellulose, the most abundant macromolecule on Earth. The glucose building blocks of cellulose are made by photosynthesis. Even when a heterotroph is consuming another heterotroph, such as when we eat a fish for dinner, the chemical-bond energy in that fish ultimately comes from the Sun: the fish could be one that fed on the cellulose (and other molecules) of photosynthetic organisms (such as algae), or it might have fed on another heterotroph that fed on photosynthetic organisms, or it might have fed on a heterotroph that fed on a heterotroph that fed on photosynthetic organisms… you get the idea!

Rubisco can use O_2 instead of CO_2 as a substrate

Rubisco fixes carbon from the atmosphere in the reaction between ribulose bisphosphate and CO_2. However, rubisco also catalyzes the reaction between ribulose bisphosphate and O_2, although at a lower rate. This reaction produces a five-carbon compound that then splits into a three-carbon compound and a two-carbon compound. The two-carbon compound is recycled in a series of steps and reenters the Calvin cycle. However, there is a cost, both in terms of energy (hydrolysis of ATP and oxidation of NADPH) and in terms of carbon: one carbon atom is fully oxidized to CO_2. Thus the reaction of rubisco with O_2, and the subsequent steps to return the product to the Calvin cycle, which are collectively termed **photorespiration**, result in the release of carbon into the atmosphere, rather than its fixation. How often rubisco reacts with O_2 rather than CO_2 determines the efficiency of the Calvin cycle. A higher relative concentration of O_2 to CO_2 and a higher temperature, which increases the affinity of rubisco for O_2, both increase inefficiency.

When leaves are freely exchanging gases with their environment, the relative concentrations are such that the loss of efficiency due to photorespiration is small. However, on a hot day, when leaves close off pores called stomata to reduce water loss, the ratio of O_2 to CO_2 inside the leaf increases as photosynthesis produces O_2 and uses up CO_2; as a result, photorespiration increases and the efficiency of photosynthesis declines.

Such is the dilemma for plants carrying out photosynthesis as we have described it so far. This type of photosynthesis and the plants that use it are called C_3 because the first detectable product of CO_2 fixation, 3-phosphoglycerate, has three carbon atoms. In C_4 and CAM (crassulacean acid metabolism) plants, however, the initial carbon-fixing reaction is different. This reaction combines a three-carbon compound and CO_2 to form a four-carbon compound as the initial product (hence "C_4"). The reaction is catalyzed by an enzyme that, unlike rubisco, does not react with oxygen. Subsequently the four-carbon compound releases CO_2, which is then fixed as usual by rubisco in the Calvin cycle.

In C_4 plants, the initial fixation of CO_2 and the second fixation of CO_2 occur in separate cells (**FIGURE 5.24A**). The release of CO_2 from the four-carbon compound maintains a locally high concentration of CO_2 around rubisco, reducing photorespiration considerably. In CAM plants, the initial fixation and second fixation of CO_2 are separated in time rather than space. In CAM plants, the initial fixation occurs at night, when leaves can more freely exchange gases with their environment without losing much water. Then, during the day, the four-carbon compound releases CO_2 near rubisco, and carbohydrate synthesis proceeds (**FIGURE 5.24B**).

LINK **Key Concept 23.1** and **Figure 23.8** provide greater detail on leaf anatomy and the different types of cells involved in photosynthesis.

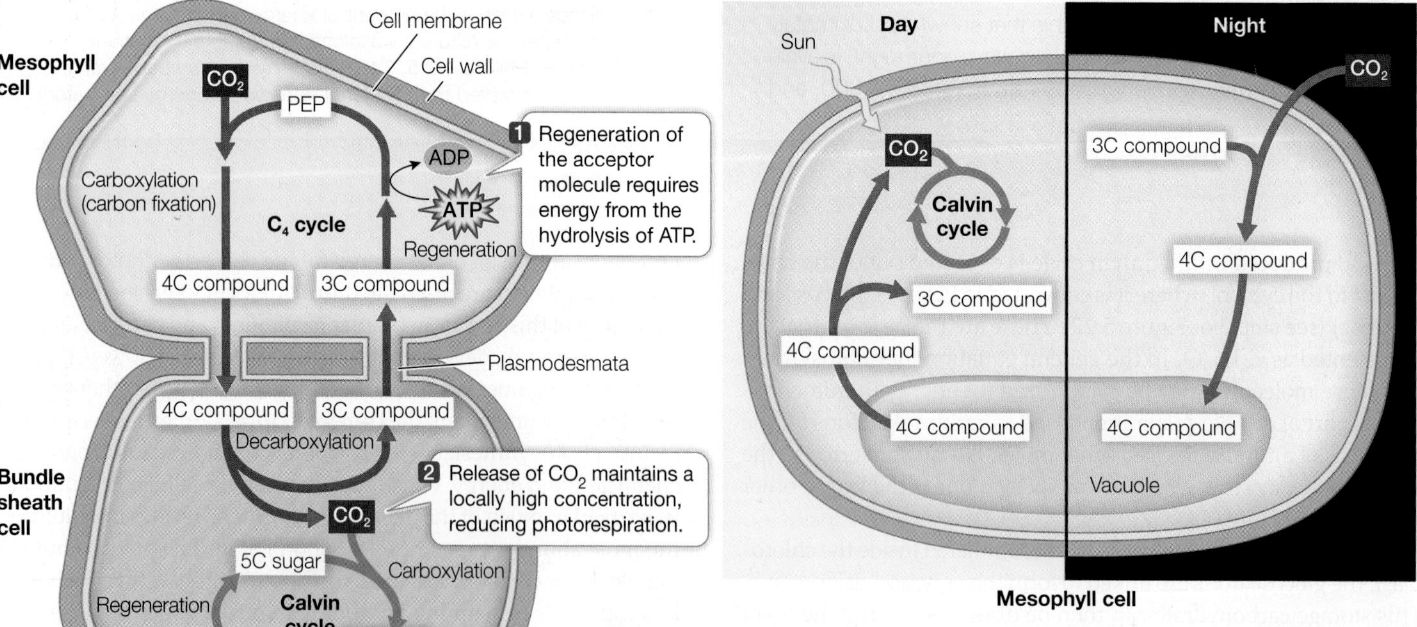

FIGURE 5.24 C_4 and CAM Plants Fix Carbon Dioxide Twice (A) C_4 plants initially fix carbon into a four-carbon (4C) compound in mesophyll cells of the leaf. This compound then releases CO_2 in the bundle sheath cells, which have high amounts of rubisco, to fix carbon in the first step of the Calvin cycle. **(B)** At night, CAM plants fix carbon into a four-carbon compound, which is then stored in the vacuole. This compound then releases CO_2 during the day to the chloroplasts, where the CO_2 is fixed by rubisco in the first step of the Calvin cycle. (B after Sabine Deviche. © Arizona Board of Regents/ASU Ask A Biologist; CC BY-SA 3.0. https://askabiologist.asu.edu/cam-plants)

We might expect that C_4 and CAM plants would have an advantage in hotter climates, where photorespiration is more likely to be a burden. That is what we find. Most plant species are C_3 plants, which initially fix CO_2 using rubisco and thus have no way to avoid photorespiration. C_3 plants, such as rice and wheat, perform best in temperate areas. C_4 plants, such as corn and sugarcane, and CAM plants, which include succulents such as cacti, perform well in hotter environments.

Media Clip 5.2 Why Are Leaves Green?

REVIEW & APPLY | 5.5

R Photosynthesis captures energy from light and uses that energy to synthesize carbohydrates from CO_2 fixed from the atmosphere. Some of the reactions require light, which is used to excite electrons. These electrons are then used in electron transport to set up a proton gradient (like mitochondrial electron transport) and to reduce $NADP^+$ to NADPH. The proton gradient is then used to synthesize ATP. The ATP and NADPH provide the energy needed by the carbon-fixation reactions that fix CO_2 in the Calvin cycle. With few exceptions, all of the chemical-bond energy that is used by living organisms can be traced back to photosynthesis.

A

1. What are the reactants and products of the light-dependent and carbon-fixation reactions of photosynthesis, and where are the reactions located in the chloroplast?
2. When a photon of light is absorbed by photosystems, how is the energy used?
3. Explain why on a hot, dry day the rate of carbon fixation by photosynthesis in a C_3 plant becomes very low but the plant continues to synthesize ATP.
4. Write equations for the production of the following in photosynthesis, and indicate whether they are oxidations, reductions, or neither: Chl^+, O_2, ATP, NADPH.
5. If green plant cells are incubated in the presence of CO_2 molecules containing radioactive carbon atoms, the fate of the carbon atoms can be followed. In an experiment, radioactive CO_2 was provided for 1 minute to plant cells, and then the cells were examined after 1, 5, 10, 20, and 30 minutes. The following molecules were labeled with radioactive carbon at some point(s): glucose, glyceraldehyde 3-phosphate (G3P), 3-phosphoglycerate (3PG), ribulose 1,5-bisphosphate (RuBP), and sucrose. List these molecules in the order in which they became labeled.

Why is Earth green?

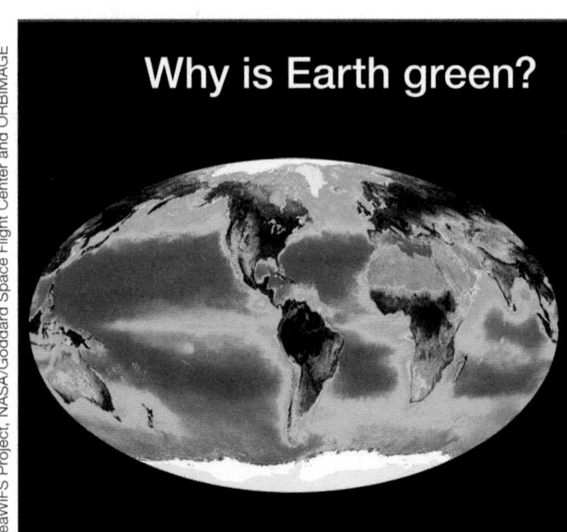

SeaWiFS Project, NASA/Goddard Space Flight Center and ORBIMAGE

Answer: One answer is, of course, that photosynthetic organisms are abundant and they use pigments, such as chlorophyll, to absorb photons of light that are blue or red. Photons in the green part of the spectrum are thus reflected off photosynthetic organisms (or pass straight through), making them appear green.

While this explanation is correct, it is a proximate explanation, meaning it explains *how* photosynthetic organisms are green. However, it is not an ultimate explanation for *why* photosynthetic pigments are green as opposed to blue or red or some other color (see THINK LIKE A SCIENTIST on p. 123 for the answer to this question). Biologists can be interested in both answers: a cell biologist might be most interested in how cells are green, while an evolutionary biologist might want to know why cells evolved to be green and not another color.

5 VISUAL SUMMARY

5.1 ATP and Reduced Coenzymes Are the Energy Currency for Biosynthesis

Go to ACTIVITY 5.1

- The high-energy phosphate bonds of **ADENOSINE TRIPHOSPHATE** (**ATP**) release energy when hydrolyzed and store energy when formed (**FIGURE 5.1**).
- The hydrolysis of ATP is an exergonic reaction that releases energy. Endergonic reactions in a cell are coupled to exergonic reactions.

FIGURE 5.1

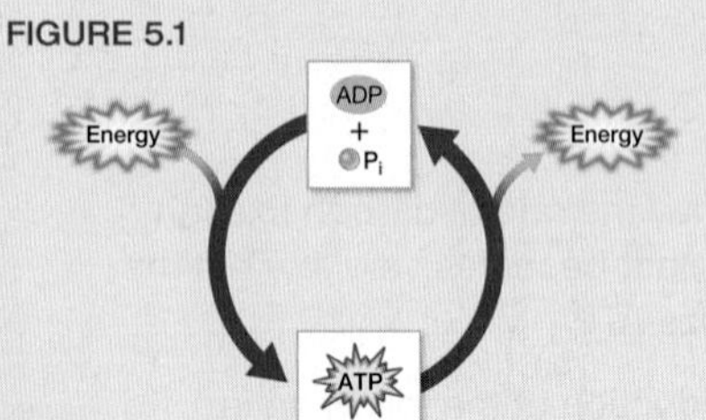

- During catabolism, molecules such as carbohydrates and lipids, which contain many C—C and C—H bonds, become oxidized, releasing energy. Oxidation is coupled to the energy-requiring reduction of coenzymes such as NAD^+, which in the process produces NADH. Such **REDOX REACTIONS** are common in metabolic pathways.

QUESTIONS

1. The formation of ATP from ADP + P_i is endergonic. How is a cell able to make this reaction occur?
2. What does it mean to say that ATP is the "energy currency" of the cell?
3. In a redox reaction, what does it mean to say that a compound is reduced or oxidized?

5.2 Carbohydrate Catabolism in the Presence of Oxygen Releases a Large Amount of Energy

Go to ACTIVITIES 5.2–5.8 and ANIMATIONS 5.1 and 5.2

- The oxidative breakdown of carbohydrates takes place in four stages with many steps, each catalyzed by an enzyme or carrier protein. Some steps release energy, which is captured by reduction of a coenzyme or phosphorylation of ADP.
- Catabolism in the presence of O_2 (**AEROBIC RESPIRATION**) releases much more energy than does catabolism in the absence of O_2 (**FERMENTATION**) (**FIGURE 5.5**).
- Some of the stages of carbohydrate catabolism, including electron transport and ATP production by chemiosmosis, occur in mitochondria (**FIGURE 5.9**).

FIGURE 5.5

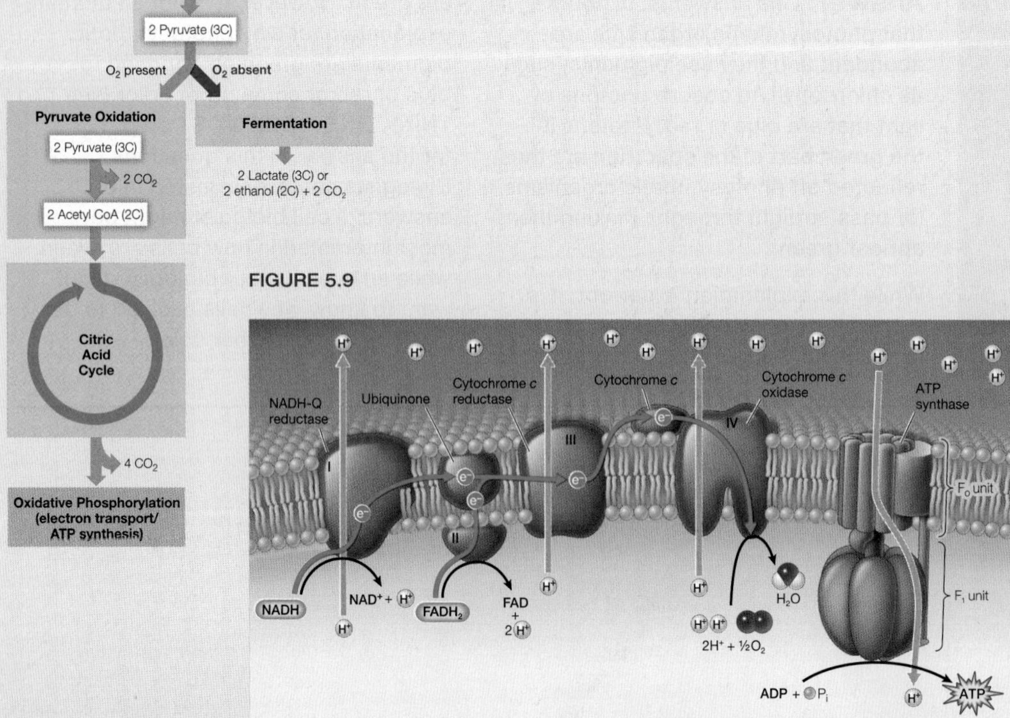

FIGURE 5.9

QUESTIONS

1. The degradation of glucose to release energy occurs in numerous steps, each of which has a reasonably small change in free energy. Why is this advantageous?
2. In the absence of oxygen, cells will switch to fermentation. The goal of respiration is to generate chemical-bond energy, but in fermentation the step from pyruvate to ethanol or lactate requires energy. Why don't cells just stop the pathway at pyruvate, or acetyl CoA, in the absence of oxygen?
3. How is the chemical-bond energy in NADH and $FADH_2$ converted to chemical-bond energy in ATP?

5.3 Catabolic Pathways for Carbohydrates, Lipids, and Proteins Are Interconnected

FIGURE 5.14

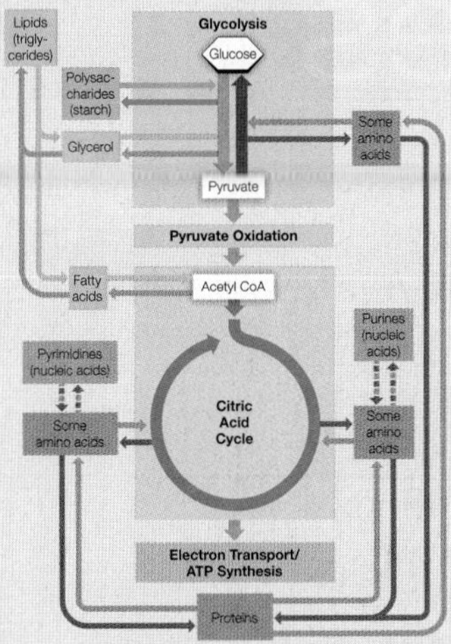

- The pathways for catabolism of carbohydrates, including glycolysis and the citric acid cycle, are central to the pathways for catabolism of lipids and proteins (**FIGURE 5.14**).
- Triglycerides are an important energy-storage molecule and are broken down to produce acetyl CoA by β-oxidation, which then enters the citric acid cycle.

QUESTIONS

1. When you go on a diet, your body switches to breaking down fat stores in adipose tissue. Describe where triglycerides enter aerobic respiration and what happens to the hydrogen and carbon atoms contained in them.
2. How many ATP molecules will be generated from one molecule of a 16-carbon fatty acid?
3. The respiratory chain in mitochondria ends up donating electrons to oxygen. However, electron transport is thought to have evolved before oxygen was present in the atmosphere. What does this indicate about early electron transport chains?

5.4 Anabolic Pathways Use Large Amounts of ATP

- Anabolic pathways involve many endergonic steps, and are thus often linked to the hydrolysis of ATP to provide energy.
- Both catabolic and anabolic pathways are regulated by enzymes.
- Anabolic pathways are often the reverse of catabolic pathways, involving the same molecules and enzymes.

1. The synthesis of macromolecule subunits, and of macromolecules themselves, occurs in numerous steps, none of which has a large change in free energy. Why are steps involving small changes in free energy advantageous?
2. When we switch to a low-carbohydrate diet, glucose levels do not drop to zero in our blood stream. Where might glucose be coming from in the short term (over the first few hours to a day or two) and long term (days to weeks)?
3. When a single-celled organism finds itself in a situation in which levels of glucose are very high for a long period of time, what kinds of metabolic enzymes might be expressed at higher levels?

5.5 Life Is Supported by the Sun: Light Energy Captured during Photosynthesis Converts Carbon Dioxide to Carbohydrates

Go to ACTIVITY 5.9 and ANIMATIONS 5.3–5.5

FIGURE 5.15

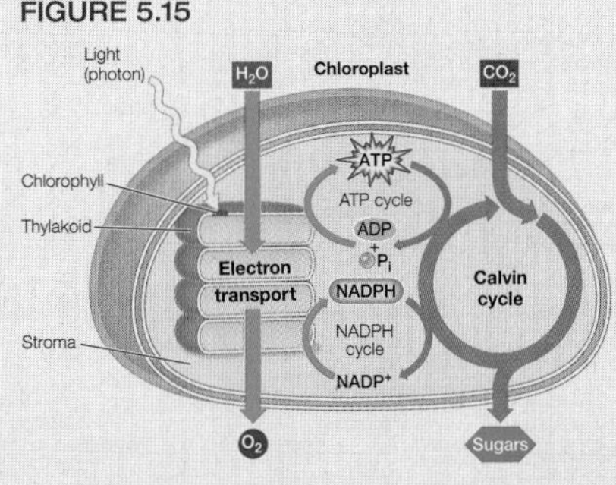

- **PHOTOSYNTHESIS** occurs in the chloroplast in eukaryotic plants and algae and on membranes in the cytosol in prokaryotes.
- There are two phases of photosynthesis: one phase captures light energy and converts it into chemical-bond energy, and the other phase uses this chemical-bond energy to synthesize carbohydrates from CO_2 (**FIGURE 5.15**).

1. In what way are chloroplasts and mitochondria similar in their function?
2. Plant cell walls are composed mostly of cellulose. Why might this be the case? (*Hint*: Examine the structure of cellulose in Key Concept 3.2.)
3. In what kinds of environments do you expect C_4 and CAM plants to do better than C_3 plants? Explain your answer.

Go to **Achieve** for the e-book, animations, activities, and additional resources and assignments.

6 Cell Signals and Responses

As evening descends, a female firefly moves up in the grass to begin watching for males flying overhead. If she observes the correct pattern of light flashes from a passing male, she will respond with a flash of her own. The male's pattern of flashes causes her brain to send a nerve impulse to her light organ to generate her response flash.

How do the cells in the light organ of a female firefly convert the nerve impulse from her brain into a flash of light?

You will find the answer to this question on page 143.

KEY CONCEPTS

6.1 Cells Detect a Variety of Signals

LEARNING OBJECTIVES

By the end of this key concept you should be able to:

6.1.1 **Explain what happens in each of the three stages of cell-to-cell signaling.**

6.1.2 **List the three ways in which a cell signal is different than an environmental signal.**

6.1.3 **Distinguish among juxtacrine, autocrine, paracrine, and endocrine signaling.**

The ability to detect environmental variables is critically important for all organisms. Once this ability evolved, the stage was set for the evolution of cell-to-cell signaling, in which one cell produces a signal to be detected by another cell. In this chapter we discuss some of the types of signals, receptors, and responses used in cell-to-cell signaling.

Organisms are able to detect environmental variables

The environment that an organism experiences varies in a multitude of ways both spatially and temporally. Not surprisingly, all organisms, from single-celled prokaryotes to multicellular eukaryotes, are able to detect and respond to some environmental variables, including light, temperature, sound, touch, magnetic fields, gravitational fields, pH, and various chemicals. For example, plants grow toward the light (phototropism), bacteria alter the rate of transcription of various genes in response to elevated temperatures, barnacle larvae attach to solid substrates, pigeons navigate using Earth's magnetic field, and carrion beetles fly toward the smell of dead animals. While responses such as these involve the whole organism, environmental variables are detected by individual cells, which then undergo changes that lead to appropriate responses. The ability of cells to sense and respond to signals is key to the maintenance of stable intracellular conditions, a theme that recurs throughout this book.

How can cells detect particular aspects of their environment? They do so by possessing various sensory receptors, including chemo-, photo-, thermo-, and mechanoreceptors. At its most fundamental level, detection of an aspect of an organism's environment involves changes in the tertiary (three-dimensional) structure of a sensory receptor. Structural changes then start a cascade of events that lead to a response by the cell (and then by the organism). The alteration in the shape of a sensory protein can occur in a variety of ways. In the last chapter we met chlorophyll and other photosynthetic pigments that are able to absorb light. In addition to chlorophyll, plants possess other pigment molecules bound to proteins that are used to detect light in order to generate an appropriate response (such as flowering or seed germination). The proteins undergo a conformational change when the attached pigment molecule absorbs a photon of light, which then leads to an appropriate cellular response.

LINK Like green plant cells, the cells in your eyes contain pigments that absorb light of particular wavelengths. These pigments are also attached to proteins. The pigment retinal (derived from vitamin A; see **Figure 3.1E**) is covalently attached to the protein opsin, forming rhodopsin. Absorption of a photon by retinal causes a change in the shape of the opsin; see **Key Concept 31.4**.

Once molecules that could detect environmental variables had evolved, the evolution of cell-to-cell signaling followed. A signal between cells is different than an environmental signal in three ways:

1. It is produced by a cell, the signaling cell.
2. It conveys specific information.
3. It has an intended recipient, the receiver or target cell.

The goal of an intercellular signal is to elicit a response in the target cell.

Response to a signal occurs in three steps

If a signal is "successful," it will result in the activation of a cellular response in the receiver cell. How does a receiver cell generate a response to a signal? There are three steps that must occur (**FIGURE 6.1**):

1. Binding of the chemical signal to its receptor
2. A cascade of events inside the cell (signal transduction)
3. The cell's response

A chemical signal, also referred to as a **ligand**, is a molecule that binds to a receptor protein. Cell signals include proteins, lipids, and other molecules. The binding of a ligand to a receptor protein is very specific, such that particular receptors have evolved to bind only to specific ligands. This means that a cell signal can be targeted to a very particular group of cells—those that have a receptor that can bind the specific signal. The same signals may be bound by different receptor molecules, which allows for responses to detection of the same signal to differ across cells. Cells without a receptor for a ligand will not be able to detect the signal.

When a receptor binds to its ligand, its three-dimensional structure changes, which alters its enzymatic activity and/or

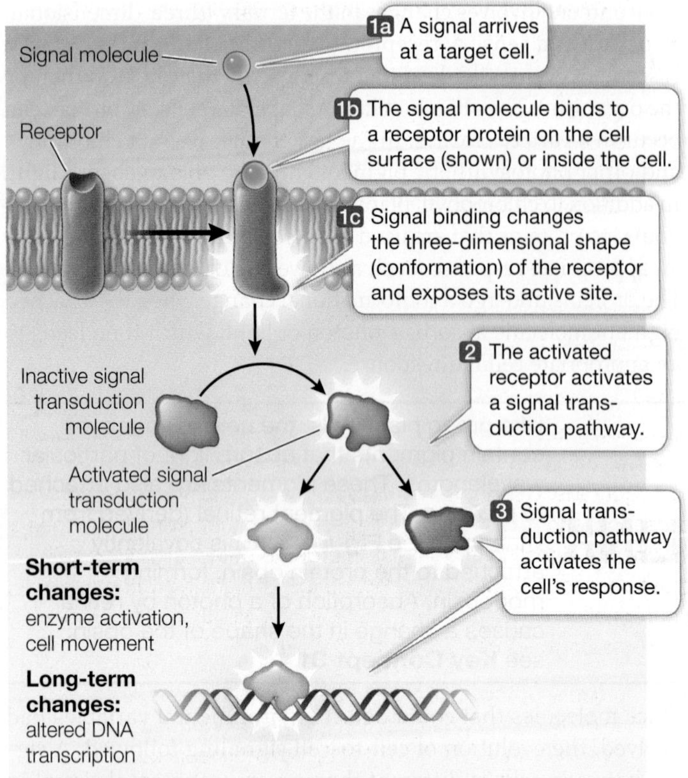

FIGURE 6.1 A Signal Transduction Pathway A signal activates a receptor, causing a change in its three-dimensional structure and setting in motion a series of events in the cell that result in short- and/or long-term responses. The steps between detection (binding of the signal) and the final response(s) are termed signal transduction.

binding properties. In some cases, the receptor itself causes the final cellular response. However, in most cases the binding of the ligand to the receptor sets in motion a series of intracellular events called **signal transduction**.

Signal transduction commonly includes a cascade (series) of events, many of which involve enzymes that exhibit allosteric regulation. Recall that allosteric regulation involves an alteration in the three-dimensional shape of a protein as a result of the binding of another molecule at a site other than the protein's active site (see Figure 3.27). In many cases, allosteric changes in signal transduction enzymes are caused by the binding of phosphate groups (phosphorylation).

The involvement of enzymes in signal transduction is very important because it can lead to amplification of the signal as it is transduced. Recall that enzymes catalyze reactions and remain unchanged, ready to catalyze again. A single activated enzyme can therefore catalyze the reaction of many substrates. Tens, hundreds, or thousands of molecules in the signal transduction pathway can be activated—all from the binding of a single chemical signal molecule to a single receptor. This makes it possible for extremely low concentrations of a signal to give a robust response.

A signal transduction pathway ends with the activation of the proteins that generate the cellular response. The response may be short-term, involving the activation of an enzyme to increase the rate of a reaction, or the opening of an ion channel to change the voltage across a membrane, or long-term, such as the modification of a regulatory protein that leads to an alteration in gene expression.

Cell signaling was an important requirement for the evolution of multicellularity. In order for differentiation to occur, in which some cells are functionally different than others, cells must communicate with one another. In single-celled organisms the number of cell signal receptors is substantially lower than in multicellular organisms. Complex multicellular organisms have a large percentage of genes that encode molecules involved in cell-to-cell signaling.

Cells are exposed to many cell signals but respond to only a subset of them

Inside a large multicellular organism, chemical signals reach target cells by local diffusion, direct contact, or bulk flow through the circulatory system in animals and the vascular system in plants. These signals are usually present in tiny concentrations (as low as 10^{-10} M) and differ in their sources and mode of delivery.

Juxtacrine signaling requires direct contact between the signaling and the receiving cell, and involves either a signal that diffuses intercellularly through gap junctions (or plasmodesmata in plants) or an interaction between signal and receptor molecules bound to the surfaces of the two cells (**FIGURE 6.2A**). In some cases the extracellular matrix may also play a role.

Gap junctions allow electric signals, ions, and chemical signals to pass between adjacent cells (see **Figure 4.19**).

Paracrine signaling occurs when signals diffuse to nearby target cells. Paracrine signaling includes synaptic signaling in which a neurotransmitter made by a nerve cell diffuses to an adjacent cell and stimulates it. In some cases a cell releases a signal for which it has a receptor. In that case the cell has an effect on itself, a process termed **autocrine signaling** (**FIGURE 6.2B**). Autocrine signaling is especially important in tumor cells, which produce growth factor signals that stimulate their own cell division, resulting in uncontrolled growth.

A signal molecule may also act from a distance. **Hormones** are chemical signals that are transported from the signaling cell to the distant receiving cell by bulk flow, such as through the circulatory system of animals or the vascular system of plants. In multicellular animals, hormones are produced by specialized cells in specific tissues, and this form of signaling is termed **endocrine signaling** (**FIGURE 6.2C**). Plant hormones are somewhat different from animal hormones in that they tend to be produced in several different locations rather than in specialized tissues (see Table 25.1).

While plant and animal cells encounter many signals, either from the environment or made by other cells, *only cells with the necessary receptors can respond*. This is very important. Every cell has in its DNA the genes for encoding all types of receptors. But it is only in certain cells that these genes are expressed. For example, liver cells and cells in the eye both have the gene that encodes photoreceptors, but it is only in certain cells in the eye that this gene is expressed.

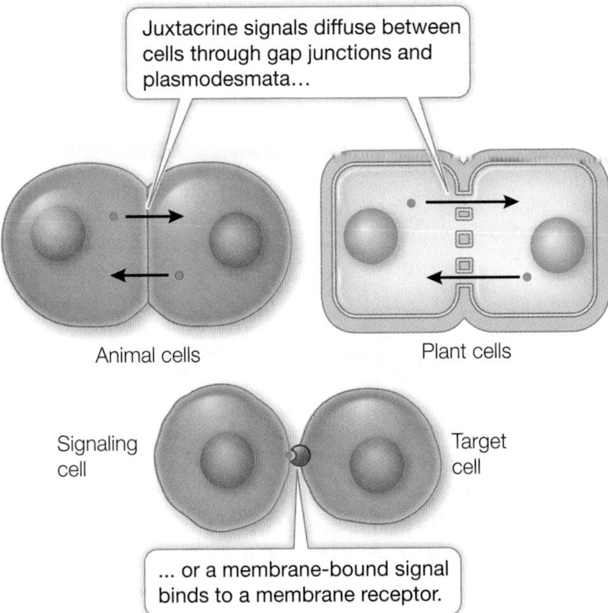

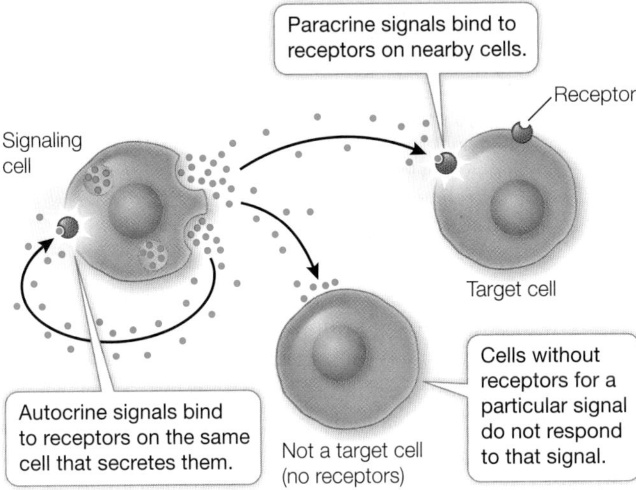

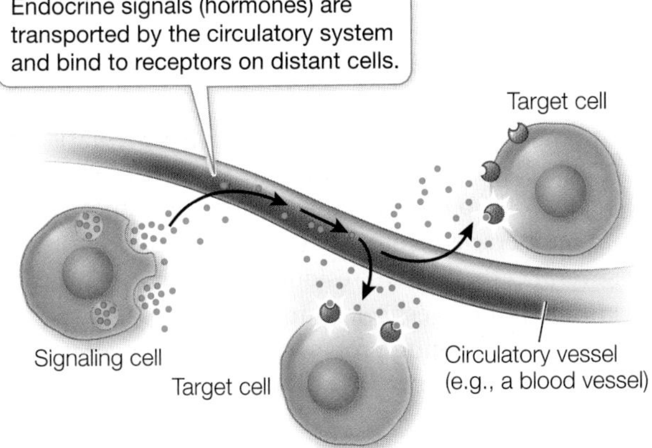

FIGURE 6.2 Chemical Signaling Systems **(A)** Signaling may require physical contact between the signaling cell and the receiving cell (juxtacrine signaling). **(B)** A signal molecule may act locally on the cell that secretes it (autocrine signaling) or on a nearby cell (paracrine signaling). **(C)** Alternatively, a signal molecule can act from a distance (endocrine signaling in animals). Long-distance signals, often termed hormones, require transport from the signaling cell to the receiving cell using bulk flow, such as found in the circulatory system.

Activity 6.1 Chemical Signaling Systems

REVIEW & APPLY | 6.1

R Signal receptors allow cells to detect a variety of environmental and cell signals. Cell-to-cell signaling involves chemical signals that are produced by one cell and carry specific information to target cells. Cell signaling may be local, including autocrine, paracrine, and juxtacrine signaling, or distant, involving hormones (endocrine signaling in animals). In order to detect signals, a cell must possess receptors, which bind to cell signals and undergo conformational changes, which eventually lead to one or more short- or long-term responses by the cell.

A

1. An organism produces a defective protein that renders it incapable of responding to a particular signal. In terms of function in signal pathways, what kind of protein might be defective?
2. When some types of cells are grown on a petri dish, they stop dividing when they come into physical contact with each other. What type of cell-to-cell signaling is involved in this response?
3. Some cells can detect pH levels in their surroundings. How might a receptor detect changes in pH?
4. Using examples, explain how light can be either an environmental signal or a cell signal.

Sensory proteins undergo conformational changes when they detect a signal. Next we will discuss some of the major classes of receptor proteins and how signal detection affects them.

6.2 Signal Molecule Receptors Can Be Classified into Several Groups

LEARNING OBJECTIVES

By the end of this key concept you should be able to:

6.2.1 **Explain how signals that bind to intracellular receptors differ from those that bind to membrane receptors.**

6.2.2 **Distinguish the three types of membrane receptors in terms of the effect of signal binding.**

There are hundreds of different signal molecule receptor proteins in a multicellular eukaryote. Each binds a particular signal

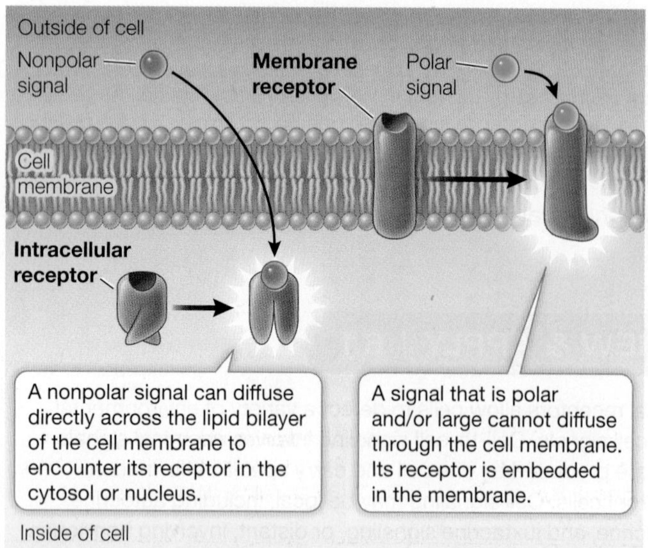

FIGURE 6.3 Receptors May Be Intracellular or Embedded in the Membrane Chemical signals that can diffuse across the cell membrane have receptors in the cytosol or nucleus. In contrast, signals that are unable to diffuse across the membrane have receptors embedded in the cell membrane.

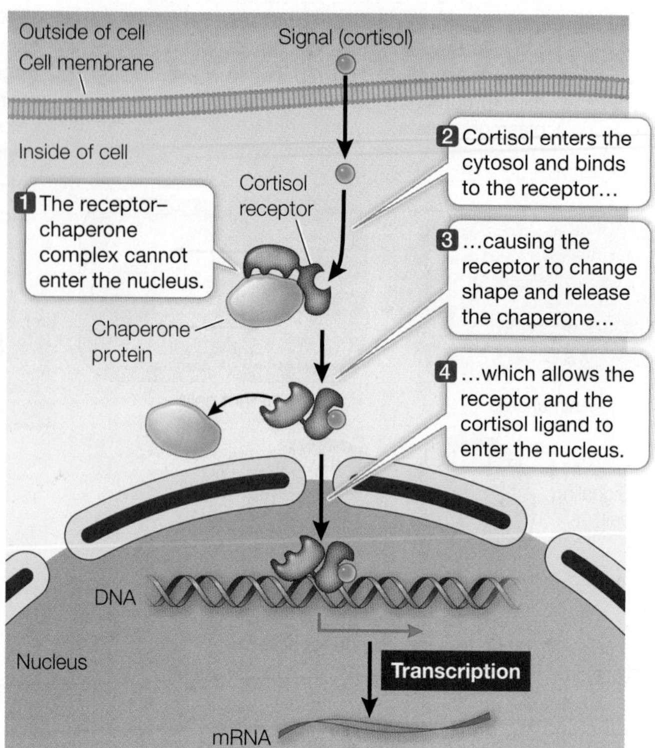

FIGURE 6.4 An Intracellular Steroid Hormone Receptor Cortisol is a steroid hormone that diffuses across the cell membrane and then binds its intracellular receptor. Binding causes the receptor to release a chaperone protein. Chaperone proteins bind a variety of other proteins, often stabilizing their three dimensional structure. In this case, chaperone binding prevents localization of the receptor to the nucleus. The hormone-bound receptor can then enter the nucleus and increase (as shown) or decrease expression of target genes.

molecule. Signal receptors can be grouped based on their location and their behavior following detection of a signal. DNA sequence analysis of various types of eukaryotic receptors indicates that closely related receptor proteins are often in gene families—sets of genes that are derived from a single parent gene and that share a similar structure and function (see Key Concept 12.3). Although we focus here on cell signals, environmental signals are detected by related receptor proteins and cause similar changes.

Receptors may be on the cell membrane or inside the cell

Although chemical signals (ligands) are highly variable, scientists can divide them into two groups based on whether or not they can diffuse through membranes. On this basis, we can classify a receptor by its location in the cell, which largely depends on the nature of its ligand (**FIGURE 6.3**). Receptors for signals that can diffuse through membranes are intracellular, found in either the cytosol or nucleus. Receptors for signals that do not cross the cell membrane are imbedded in the membrane, with their binding surface exposed to the ligand in the extracellular environment.

Recall from Chapter 4 that small or nonpolar substances can diffuse across the phospholipid bilayer of the cell membrane. If these are ligands, they bind to **intracellular receptors** (receptors located inside the cell). An important example of such signals is animal steroid hormones, such as estrogen and cortisol, which can readily diffuse across the cell membrane. Once in the cell, these hormones bind to their cytosolic receptors (**FIGURE 6.4**).

Large or polar ligands cannot cross the lipid bilayer. Insulin, for example, is a protein hormone that is too large (51 amino acids) and too hydrophilic to diffuse through the cell membrane. Ligands that are incapable of entering the cell must bind to **membrane receptors**, which are transmembrane proteins with an extracellular, ligand-binding region. In addition to the ligand-binding region, membrane receptors have a cytosolic region that is responsible for initiating signal transduction in the cell. These two sides are separated by a hydrophobic region that spans the membrane (**FIGURE 6.5**). In many cases the cytosolic region functions as an enzyme, with its active site on the cytosolic side of the membrane. The ligand binds to the extracellular region and acts as an allosteric regulator, exposing the active site of the cytosolic region. The ligand does not contribute further to the cellular response; its role is purely to "knock on the door."

Ligands bind to their receptors noncovalently and reversibly. For most ligand–receptor complexes, binding is favored, even at very low ligand concentrations, implying that whenever the signal is present, its receptors will be bound. The specificity of ligand binding is due to both to the receptor's protein sequence and to glycosylation: membrane receptors are typically glycoproteins, containing several carbohydrate residues attached to particular amino acids in the protein, especially in the extracellular region. In addition to altering binding properties, the carbohydrates affect the folding of the protein and alter the response of membrane receptors to ligand binding.

Like all biochemical processes, cell signaling can be regulated (as we will discuss in more detail in Key Concept 6.4). One way

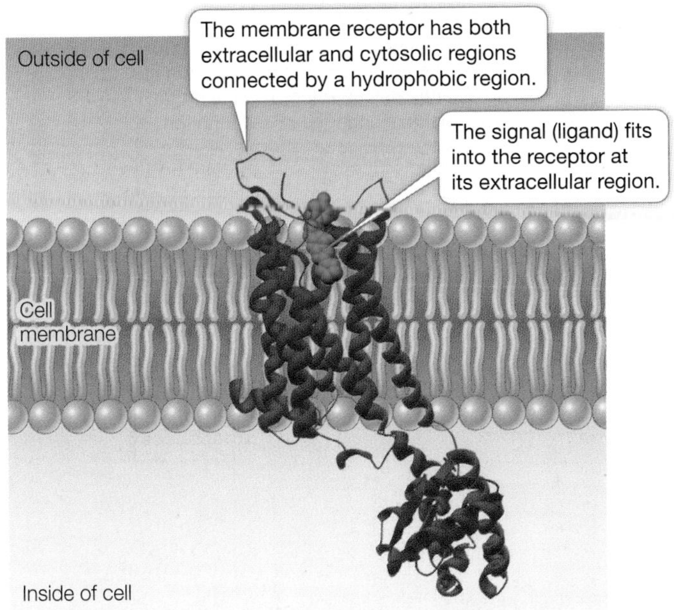

FIGURE 6.5 Binding of a Ligand (Adenosine) to Its Receptor Adenosine, when acting as a chemical signal, fits into the extracellular region of its membrane receptor, binding to it noncovalently and causing a conformational change. (Data from PDB 3EML. V.-P. Jaakola et al. 2008. *Science* 322: 1211–1217.)

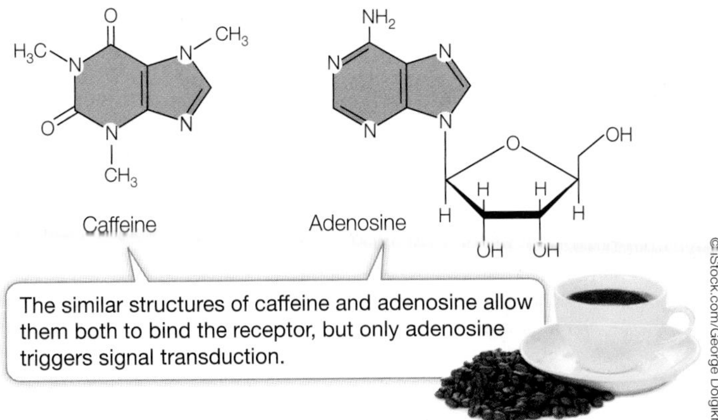

FIGURE 6.6 The Structures of Adenosine and Caffeine Caffeine is a competitive inhibitor of the adenosine receptor. Adenosine suppresses arousal, so inhibition of this signaling pathway leads to heightened activity.

Some inhibitors often have a structure similar to the normal ligand, while others do not. What might explain the difference?

cells do this is by producing inhibitors that prevent binding of the normal ligand. Such inhibitors can bind to the ligand-binding extracellular region of the receptor but not cause its activation—just as inserting the wrong key into a lock prevents the right key from being inserted. This is analogous to active site competitive inhibition of enzymes (see Key Concept 3.5). Knowing the chemistry of receptor–ligand and natural inhibitor binding allows scientists to either identify from nature or go into the chemistry lab and design molecules that can activate or inhibit signaling. For example, in the human brain there is a receptor that binds the signal adenosine (the nucleoside composed of adenine bonded to ribose). Adenosine binding leads to a signal transduction pathway that results in drowsiness. The natural molecule, caffeine, is similar enough in structure to adenosine to bind to the adenosine receptor but is not able to activate it. Binding of caffeine prevents adenosine from binding. As you know, this results in less drowsiness and increased alertness (**FIGURE 6.6**). The major drugs that affect human behavior have either been isolated from nature or designed to bind to specific receptors in the brain.

Classification of membrane receptors is based on their actions

Upon binding to an extracellular signal, a membrane receptor causes a change on the inside of the cell. Membrane receptors are categorized based on the change they initiate. In complex eukaryotes such as mammals and higher plants, there are three well-studied categories of membrane receptors: ligand-gated ion channels, G protein-coupled receptors, and protein kinase receptors. A few other receptor classes exist, but because you will see these three several times later in this book, we describe them in some detail here.

LIGAND-GATED ION CHANNELS As described in Key Concept 4.2, the cell membranes of many cells contain ligand-gated channels for ions such as Na^+, K^+, Ca^{2+}, or Cl^- (see Figure 4.5). These protein channels are receptors because their functioning depends on ligand binding. An example is the acetylcholine receptor (AChR), a ligand-gated sodium channel located in the cell membranes of skeletal muscle cells. Acetylcholine (ACh) is a neurotransmitter—a chemical signal released from nerve cells. Opening of the channel allows Na^+, which is more concentrated outside the cell than inside, to diffuse into the cell (**FIGURE 6.7**). This initiates a series of events that result in muscle contraction (see Figure 31.18).

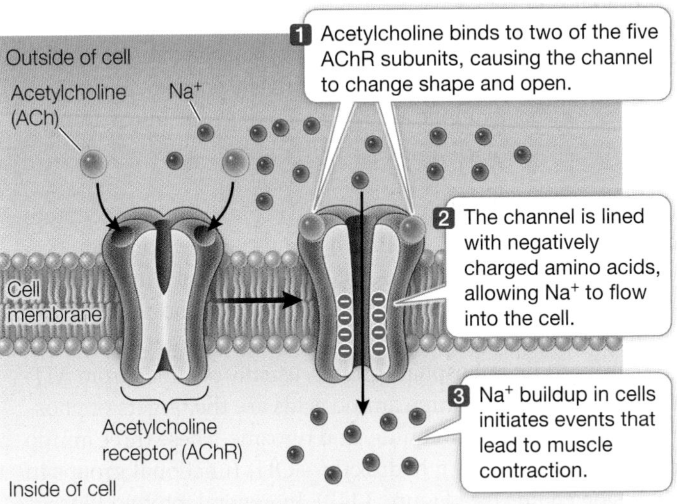

FIGURE 6.7 A Gated Ion Channel The acetylcholine receptor (AChR) is a ligand-gated ion channel for sodium ions. It is made up of five polypeptide subunits. When acetylcholine molecules (ACh) bind to two of the subunits, the gate opens and Na^+ flows into the cell.

(A) G protein-coupled receptor, G protein, and (inactive) effector protein

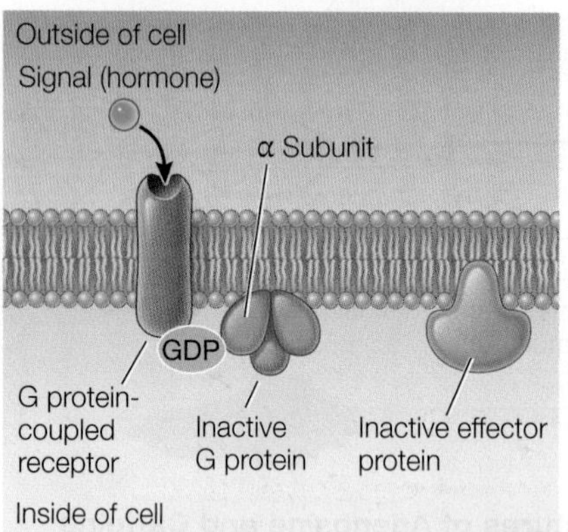

(B) Activation of the G protein

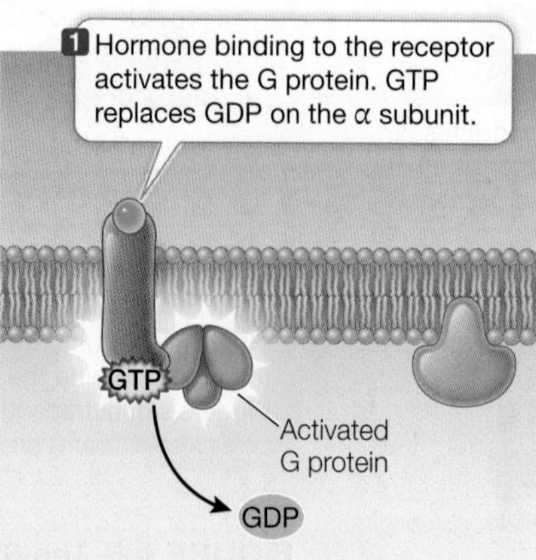

(C) Activation of the effector protein

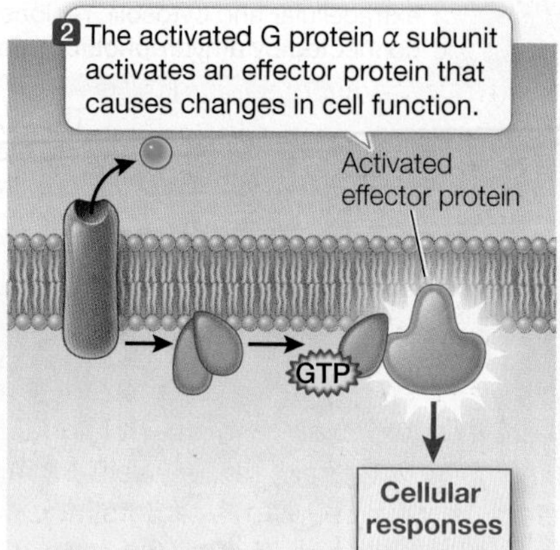

FIGURE 6.8 A G Protein-Coupled Receptor The heterotrimeric G protein is the intermediary between the receptor and an effector protein.

G PROTEIN-COUPLED RECEPTORS Another category of eukaryotic cell membrane receptor is the G protein-coupled receptors (GPCRs), which immediately activate GTP-binding proteins, or G proteins. GPCRs have seven hydrophobic membrane-spanning regions and associate with specific G proteins that are partially embedded in the cytosolic side of the cell membrane. G proteins are a diverse set of membrane proteins that can bind both GTP and GDP. The specific G proteins that interact with GPCRs are heterotrimers, meaning they are made up of three different peptide monomers. One of these subunits, termed the α subunit, is able to bind GDP or GTP. When a signal molecule is not present, such that the GPCR is not activated, the α subunit of the G protein has GDP bound. When its ligand binds, a conformational change occurs in the GPCR, which in turn causes a change in the G protein such that the α subunit exchanges GDP for GTP. The GTP-bound α subunit is then able to activate an effector protein, which is an enzyme that causes further changes in the cell, leading to downstream signal amplification and a cellular response (**FIGURE 6.8**). GPCRs are abundant in the human genome: of the approximately 20,000 protein-coding human genes, more than 800 encode GPCRs. GPCRs are especially important in the sensory systems of animals (see Key Concept 31.4).

PROTEIN KINASE RECEPTORS Protein kinase receptors change shape upon ligand binding and the new conformation exposes or activates an active site in the cytosolic region. Specifically, the cytosolic region of the ligand-bound receptor has protein kinase activity—it modifies specific target proteins in the cell by adding phosphate groups, usually obtained from ATP, to their side chains. Three amino acids are the targets of phosphorylation: serine, threonine, and tyrosine. These three amino acids are the ones with hydroxyl (—OH) functional groups in their side chains (see Figure 3.14B). In general, protein kinases catalyze the following reaction:

$$\text{ATP} + \text{protein} \xrightarrow{\text{Protein kinase}} \text{ADP} + \text{phosphorylated protein}$$

This reaction results in the covalent modification (phosphorylation) of the target protein, thereby changing its activity (see Figure 3.27B). Protein kinases are extraordinarily important in biological signaling: in humans more than 500 of the protein-coding genes are protein kinases, and there is an even higher proportion of such genes in some plants.

An example of a protein kinase receptor is the receptor for the hormone insulin (**FIGURE 6.9**). The activation of this receptor upon insulin binding results in phosphorylation of specific

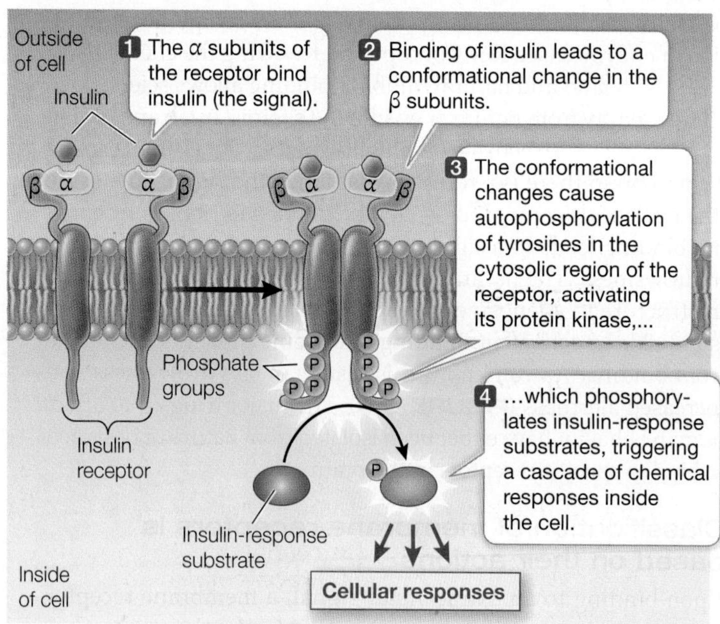

FIGURE 6.9 A Protein Kinase Receptor The mammalian hormone insulin binds to a protein kinase receptor on the outside surface of the cell and initiates a response. Upon binding of insulin, the insulin receptor undergoes a conformational change and autophosphorylates specific tyrosine residues. Once the ligand-bound receptor is phosphorylated, another conformational change allows it to bind its target proteins, which it then phosphorylates.

tyrosine residues on the receptor itself, so-called **autophosphorylation**, which in turn allows binding and subsequent phosphorylation of target proteins. Once phosphorylated, the target proteins bring about the cell's response, which includes the insertion of glucose transport proteins into the cell membrane.

REVIEW & APPLY | 6.2

R Cells possess intracellular and membrane receptors that allow them to detect membrane-soluble and membrane-insoluble chemical signals, respectively. Ligand-gated ion channels, G protein-coupled receptors, and protein kinase receptors are three types of membrane receptors grouped according to their actions.

A
1. If an intact cell is treated to remove cell surface proteins, will the cell be able to detect any signals from other cells?
2. Some cancer cells have changes in heterotrimeric G proteins that allow the α subunit to always bind to GTP, rather than having to be activated first by a GPCR. How might this GTP binding ability of a mutant G protein affect cell signaling through its associated GPCRs?
3. Tyrosine and phenylalanine have similar structures. Do you think that an insulin receptor that has a single amino acid change is more likely to be nonfunctional if a tyrosine is substituted by a phenylalanine or if a phenylalanine is substituted by a tyrosine?

When a signal activates a receptor at the cell membrane, it initiates a cascade of events in the cell that leads to a cell's response. We now turn to how signals are transduced into responses.

6.3 Signal Transduction Allows a Cell to Respond Appropriately to a Signal

LEARNING OBJECTIVES

By the end of this key concept you should be able to:

6.3.1 **List three kinds of cellular responses that may occur following signal reception and transduction.**

6.3.2 **Define a second messenger and describe the primary benefit of incorporating a second messenger in a signal transduction pathway.**

6.3.3 **Explain how a signal transduction pathway can amplify the effects of the original signal to generate a larger response.**

When a membrane receptor is activated by binding to its signal molecule, it initiates a cascade of events in the cell—a signal transduction pathway. Proteins interact with other proteins, sometimes through nonprotein intermediates, until the final responses are achieved. Through such a cascade, an initial signal can be both *amplified* and *distributed* to cause several different responses. In addition, a signaling cascade allows several different signaling pathways to be more easily integrated, as we will see in Key Concept 6.4.

Before we discuss how signals are amplified and distributed by signal transduction pathways, let's look at some of the cellular responses that can result from cell signaling.

Cell functions change in response to environmental signals

The activation of a receptor by a signal, and the subsequent transduction and amplification of the signal, ultimately leads to changes in cell function. There are many ways, both immediate and longer term, in which a cell might respond, some of which we mention here:

- *Opening of ion channels* changes the balance of ion concentrations between the outside of the cell membrane and its interior (see Figure 6.7). As you will see in Chapter 31, this results in a change in the electric potential (voltage) across the membrane, with important consequences in nerve and muscle cells. Changes in ion concentrations are extremely rapid.
- A cell response may involve the *alteration of enzyme activities*. An example is the activation of specific enzymes in liver cells exposed to the hormone epinephrine, which we discuss below. An alteration in enzyme activity promptly alters the relative rates of various metabolic pathways in a cell.
- Many signal transduction pathways lead to *alterations in gene expression*. The expression of some genes may be switched on (upregulated), whereas that of others may be switched off (downregulated). These changes in gene expression alter the abundance of proteins (often enzymes) encoded by the genes, thus altering cell function. Changes due to alterations in gene expression take longer to be realized, and are often longer lasting, than the two changes listed above. You will see many examples that highlight the importance of gene regulation throughout this book.

LINK As discussed in **Key Concept 3.5**, regulation of enzyme activity involves either the binding of a regulatory molecule to the active site to prevent binding to the substrate (an inhibitor), or the binding of a regulatory molecule to another site on the enzyme, altering its three-dimensional shape and therefore its function (allosteric regulation).

A particular signal can lead to different responses in different types of cells. For example, in heart muscle cells, the hormone epinephrine binds to its receptor and stimulates a signal transduction pathway that results in the *activation* of an enzyme that mobilizes glucose for energy and muscle contraction. However, in the smooth muscle cells that line the digestive tract, epinephrine stimulates a pathway that results in the *inhibition* of a target enzyme, allowing the muscle cells to relax. This increases the diameter of the blood vessels, allowing more nutrients to be carried from the digestive system to the rest of the body. Heart and digestive tract muscle cells respond differently to the same epinephrine signal because the signal transduction pathways stimulated by binding of this hormone are different in the different cell types. Let's take a closer look at the mechanism by which cells amplify and transduce signals to bring about these responses.

A signaling cascade can involve enzyme regulation and signal amplification

While most signal transduction pathways involve several sequential steps to result in an overall cellular response, that is not always the case. Sometimes, binding to the receptor causes the response directly. For example, binding of a steroid hormone such as estradiol to its steroid receptor in the cytosol alters the receptor's conformation, releasing a chaperone protein previously bound to the receptor. (Chaperones are proteins that bind numerous other proteins and stabilize their three-dimensional structures.) This allows the receptor to enter the nucleus, where it affects the expression of particular genes (see Figure 6.4). Such steroid hormone signaling pathways are thus as short as possible: the bound receptor directly causes the cellular response, which is a change in gene expression. Similarly, binding of a signal to ligand-gated ion channels causes them to open, immediately causing the cellular response: a change in voltage across the cell membrane.

Most signal transduction pathways involve multiple sequential steps, a so-called **signal transduction cascade**, in which particular enzymes are either activated or inhibited by other enzymes in the pathway. For example, a protein kinase adds a phosphate group to a target protein, and this covalent change alters the protein's conformation and activates or inhibits its function. In the case of activation, a previously inaccessible active site is exposed, and the target protein goes on to perform a new cellular role. In the case of inactivation, a previously accessible active site is made inaccessible, and its function ceases.

An example of a signal transduction cascade involving only proteins is the activation of mitogen-activated protein kinases, or MAPKs (**FIGURE 6.10**). Mitogens are signal molecules that simulate cell division. Many do this by causing the activation of MAPKs that then phosphorylate target proteins. These activated proteins increase the rate of expression of genes whose protein

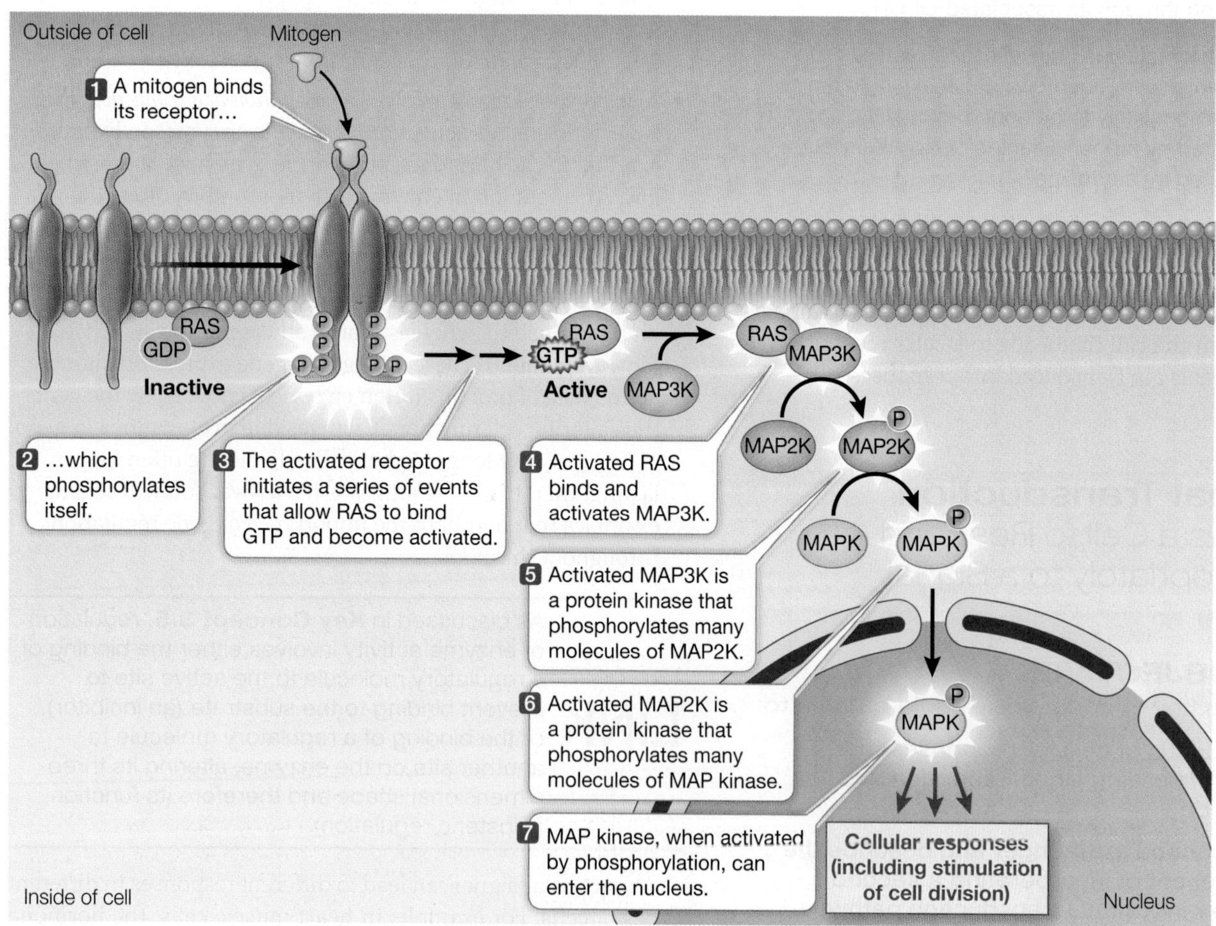

FIGURE 6.10 A Signal Transduction Cascade Binding of a mitogen to its membrane receptor sets in motion a chain of events, many of which involve phosphorylation of target proteins by protein kinases.

Animation 6.1 Signal Transduction and Cancer

THINK LIKE A SCIENTIST

Identifying and ordering steps in signal transduction pathways

Identifying the proteins that are involved in a signal transduction pathway is challenging. A genetic approach would be to try to find protein-coding genes that, when knocked out (made nonfunctional by altering the DNA), disrupt the pathway. A biochemical approach might examine changes to proteins that occur when a cell receives a stimulus. For example, what proteins are phosphorylated during signal transduction?

Suppose you use these approaches and identify many protein-coding genes, including two that encode protein kinases, X and Y, that are phosphorylated during signal transduction.

1. What amino acid(s) on these proteins might be phosphorylated?

Once you have identified protein kinases that have been phosphorylated, you would then want to determine their order in the pathway. For example, is protein kinase X involved earlier or later in the pathway than protein kinase Y? One way to answer this question is to examine the effect of a knockout of kinase X on kinase Y phosphorylation.

2. If kinase X is involved in the pathway before kinase Y, what effect would a kinase X knockout have on kinase Y phosphorylation? What about if X is involved after Y?

Once you determine the order of proteins in the pathway—for example, that X is involved before Y—you might want to determine if X phosphorylates Y directly (X → Y), meaning that Y is a substrate for X. To do this you could examine the effects of knockouts of other kinases on the phosphorylation of X and Y to see if you can find other kinases that, when knocked out, still result in phosphorylation of X but not of Y.

3. If you find such a phosphorylase, what does this tell you about whether kinase Y is a substrate for kinase X?

products promote cell division. In a nondividing cell, MAPKs are inactive and must be activated by phosphorylation, which is performed by MAPK kinases, or MAP2Ks. MAP2Ks are usually inactive and are also activated by phosphorylation, which is performed by MAP2K kinases, or MAP3Ks. MAP3Ks are activated in a variety of ways, including by binding of an active G protein called **RAS** (see Figure 6.10). There are three RAS proteins in humans, all of which are involved in activating cell division. Changes in RAS proteins that make them permanently active are among the most common differences between cancer cells, which divide uncontrollably, and normal cells. It is worth pointing out that the RAS family of G proteins are not the heterotrimeric G proteins that are associated with GPCRs (recall Figure 6.8). Instead, RAS proteins are members of the small-GTPase family of G proteins that are somewhat similar in amino acid sequence to the α subunit of the heterotrimeric G proteins.

Second messengers can stimulate signal transduction

In many signal transduction pathways, small, nonprotein molecules called **second messengers** occur at one or more steps in the signaling cascade. (First messengers are the signal molecules.) Some second messengers are hydrophilic, such as cyclic AMP (cAMP) and inositol trisphosphate (IP_3) (**FIGURE 6.11A**) and Ca^{2+} ions; some are hydrophobic, such as diacylglycerol

(A) Hydrophilic second messengers

cAMP

IP_3

(B) Hydrophobic second messengers

DAG

PIP_3

(C) A gaseous second messenger

N=O

NO

FIGURE 6.11 Examples of Second Messengers Second messengers can be **(A)** hydrophilic molecules such as cyclic AMP (cAMP) and inositol trisphosphate (IP_3), **(B)** hydrophobic molecules such as diacylglycerol (DAG) and phosphatidylinositol trisphosphate (PIP_3), and **(C)** gases such as nitric oxide (NO). Note that PIP_3 is considered a hydrophobic second messenger even though it is an amphipathic molecule because it is insoluble in water.

Where do you think hydrophobic second messengers are usually found in a cell?

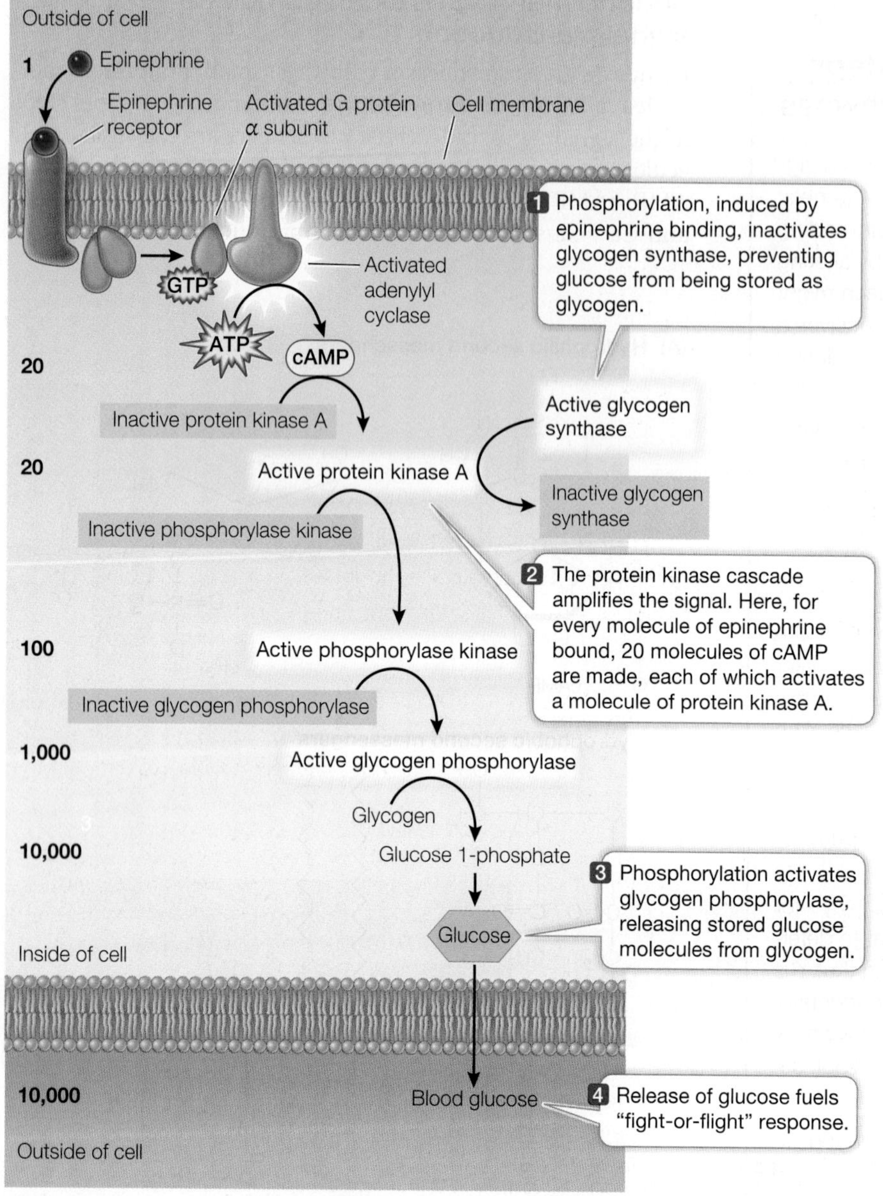

FIGURE 6.12 A Signaling Cascade Involving a Second Messenger Epinephrine activates a signaling cascade that includes the second messenger cAMP. The net result of this cascade is that each molecule of the signal, epinephrine, results in the release into the bloodstream of 10,000 molecules of glucose.

Activity 6.2 Cell Signaling and Amplification Simulation

Animation 6.2 A Signal Transduction Pathway

(DAG) and phosphatidylinositol trisphosphate (PIP_3) (**FIGURE 6.11B**); and some are gases, such as nitric oxide (NO) (**FIGURE 6.11C**). Second messengers do not have enzymatic activity themselves; rather, they act to regulate target enzymes by binding to them noncovalently (allosteric regulation). Whereas receptor binding is highly specific, second messengers allow a cell to respond to a single event at the cell membrane with *many events inside the cell*—in other words, second messengers effectively *distribute* the initial signal. Hydrophobic second messengers act on proteins associated with cell membranes, whereas hydrophilic messengers act on cytosolic proteins. Gaseous second messengers are unusual in that they can initiate responses in nearby cells because they rapidly diffuse through membranes.

A good example of a signaling cascade that uses a second messenger, and the first cascade in which a second messenger was discovered, is the G protein-coupled protein kinase pathway stimulated by epinephrine in liver cells (**FIGURE 6.12**). Binding of epinephrine to its G protein-coupled receptor results in the activation of a G protein, which in turn activates the membrane-integrated enzyme adenylyl cyclase. Adenylyl cyclase catalyzes the production of the second messenger cAMP from ATP, which then activates a key signal molecule, the enzyme protein kinase A. In turn, protein kinase A phosphorylates two other enzymes, glycogen synthase and phosphorylase kinase, with opposite effects. Phosphorylation of glycogen synthase inhibits its usual function, which is to join glucose molecules to form the energy-storing molecule glycogen, thus preventing the storage of glucose as glycogen (see Figure 6.12, STEP 1). Phosphorylation of phosphorylase kinase activates it, allowing it to phosphorylate (and activate) glycogen phosphorylase, which then catalyzes the breakdown of glycogen, releasing glucose, which can then enter the bloodstream (see Figure 6.12, STEPS 2 and 3). Thus a primary role of this signaling pathway is to mobilize glucose from glycogen stores. Glucose is then ready to fuel skeletal muscle contractions and increased heart rate, preparing the individual to meet an emergency—the so-called fight-or-flight response (see Figure 6.12, STEP 4). **INVESTIGATION FIGURE 6.13** describes the experiment that confirmed the existence of a second messenger in the epinephrine signal transduction pathway.

An important consequence of having multiple steps in a signal transduction cascade is that the signal can be amplified with each step, resulting in substantial amplification after several such steps. The amplification of the signal in the pathway illustrated in Figure 6.12 is impressive. Each molecule of epinephrine that arrives at the cell membrane ultimately results in 10,000 molecules of glucose being released into the blood.

Note that each step that involves an enzyme amplifies the signal. Steps that involve binding of allosteric regulators, such as the G protein that activates adenylyl cyclase and the cAMP second messenger that activates protein kinase A, do not amplify the signal because it takes one allosteric regulator to activate one enzyme.

INVESTIGATION

FIGURE 6.13 The Discovery of a Second Messenger Glycogen phosphorylase is activated in liver cells after epinephrine binds to a membrane receptor. Earl Sutherland and his colleagues observed that this activation could occur in a test tube only if fragments of the cell membrane were present. They designed experiments to show that a second messenger caused the activation of glycogen phosphorylase.[a]

HYPOTHESIS

Phosphorylation of liver glycogen phosphorylase in the cytosol following epinephrine binding to its cell membrane receptor involves a soluble second messenger that is in the cytosol.

METHOD

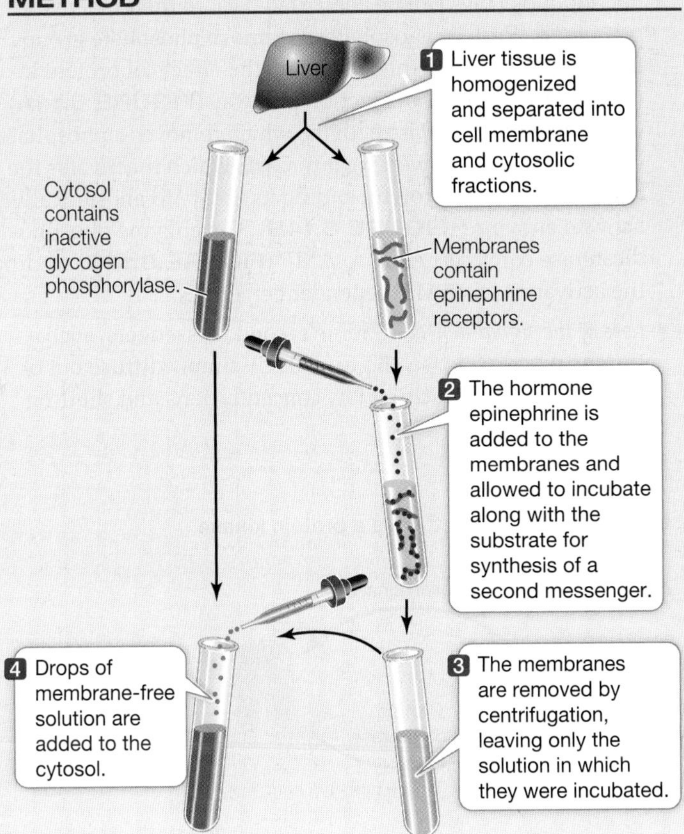

RESULTS

The level of phosphorylation of glycogen phosphorylase increases when drops from the membrane-free solution are added to the cytosol.

CONCLUSION

Since membranes themselves do not have to be present for phosphorylation to occur, a soluble second messenger, produced by hormone-activated membranes, is present in the solution and activates enzymes in the cytosol.

A similar **Analyze the Data** exercise may be assigned in **Achieve**.

ANALYZE THE DATA

The study involved two experiments. In the first experiment, cells were homogenized and then separated into cytosolic and membrane components and the following results obtained:

Condition	Phosphorylation activity (units)
Homogenized cells	0.4
Homogenized cells + epinephrine	2.5
Cytosol fraction (no cell membranes)	0.2
Cytosol fraction + epinephrine	0.4
Membrane fraction + epinephrine	0.4
Cytosol fraction + membrane fraction + epinephrine	2.0

1. What do these data indicate about the epinephrine response in liver cells?
2. Do these data indicate that a second messenger is involved in epinephrine signal transduction?

In the second experiment, cells were homogenized and then separated into cytosolic and membrane components. Membrane components were then incubated with or without epinephrine, the membrane components were removed, and the resulting membrane-free solution was added to the cytosolic fraction. The following results were obtained:

Condition	Phosphorylation activity (units)
Cytosol fraction (no membranes)	1.2
Cytosol fraction + epinephrine	1.3
Cytosol fraction + solution from membrane component incubated without epinephrine	1.4
Cytosol fraction + solution from membrane component incubated with epinephrine	2.6

3. Which condition in the table represents the one illustrated in the Method section above?
4. Propose an experiment to show that the factor that activates the enzyme is stable on heating (and therefore probably not a protein) and give predicted data.
5. Propose an experiment to show that cAMP can replace the membrane fraction + hormone treatment and give predicted data.

[a]T. W. Rall et al. 1957. *J Biol Chem* 224: 463–475.

REVIEW & APPLY | 6.3

R Signals bind to particular receptors, setting in motion one or more events that lead to a response by the cell. Common responses include movement of ions across the cell membrane, modification of enzyme activity, or changes in transcription. Signaling cascades used during transduction allows dispersion and amplification of the initial signal. In addition to proteins, especially kinases and phosphatases, signal transduction may involve nonprotein secondary messengers such as cAMP and Ca^{2+} ions.

A
1. Compare first messengers (e.g., hormones) with second messengers (e.g., cAMP) with regard to their chemical nature, where they are made, and their activity.
2. Using Figure 6.12, predict what would happen to a liver cell exposed to epinephrine and at the same time to (a) a drug that inhibits protein kinase A and (b) a drug that inhibits the hydrolysis of GTP to GDP. (Assume that both these drugs are able to cross the cell membrane.)
3. Many signals occur in very low concentrations. Using the epinephrine signal transduction pathway as an example, explain how a weak signal can cause a robust response in the cell.

So far we have discussed how signaling pathways are activated to generate responses. At some point, responses must also be turned off, often rapidly. This can be complicated when a cell's response depends on inputs from multiple pathways. How are signaling pathways integrated to allow appropriate responses?

6.4 Signal Transduction Is Highly Regulated

LEARNING OBJECTIVES

By the end of this key concept you should be able to:

6.4.1 Describe three ways in which a signal response can be turned off.

6.4.2 Describe the effects of insulin and glucagon on glycogen synthesis and breakdown.

6.4.3 Distinguish between type I and type II diabetes.

6.4.4 Identify common disorders caused by misregulation of cell signaling.

Several signals may be detected by a cell, which must then integrate the signal information in order to respond appropriately. Importantly, cells must be able to reset their signaling pathways in order to respond appropriately to new signals.

Signal pathways are actively reset

Signal transduction is a temporary event in the cell and gets turned off once the cell has responded. Turning off a signaling pathway is important as it resets the pathway, allowing the cell to respond again to the signal. A signal transduction pathway can be turned off in three ways:

- *Receptor recycling.* Membrane receptors can be endocytosed and degraded in lysosomes (see Key Concept 4.5). Intracellular receptors also turn over quite rapidly (on the order of every few hours for the estrogen receptor), which ends the response.
- *Chemical reversion of signal transduction molecules.* Signal transduction molecules can be converted back to their inactive precursors. Protein phosphatases remove phosphate groups from target proteins, thus reversing the effects of protein kinases in the signal transduction pathway (**FIGURE 6.14A**). G proteins have GTPase activity, which removes a phosphate group from GTP, converting it to GDP, which inactivates the α subunit of the G protein, ending its ability to allosterically activate enzymes (**FIGURE 6.14B**). The enzyme phosphodiesterase converts cAMP to AMP (**FIGURE 6.14C**), ending the activation of cAMP-dependent enzymes.
- *Loss of the signal molecule.* Some second messengers, such as the gas nitric oxide (see Figure 6.11C), simply diffuse out of the cells, quickly reaching low concentrations and shutting down a response.

(A) Phosphatase inactivates a protein kinase

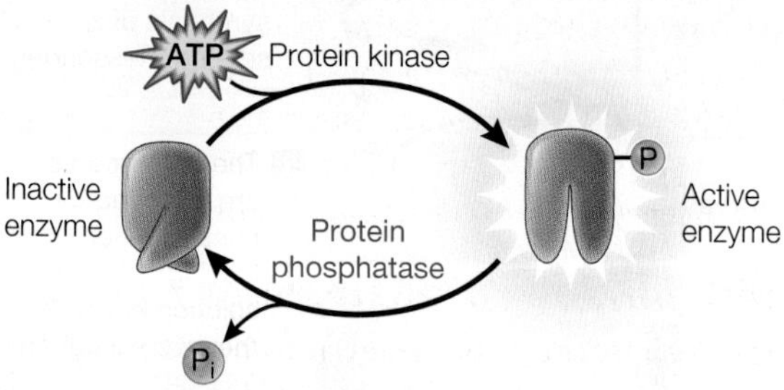

(B) GTPase inactivates a G protein

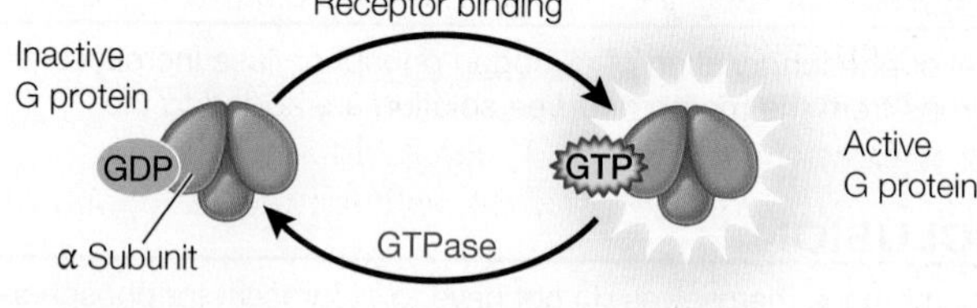

(C) Phosphodiesterase inactivates cAMP

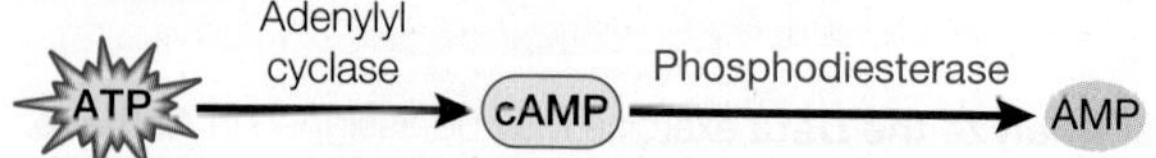

FIGURE 6.14 Signal Transduction Regulatory Mechanisms Some signals lead to the production of active signal transduction molecules such as protein kinases **(A)**, G proteins **(B)**, and cAMP **(C)**. Other enzymes (red type) inactivate or remove these active molecules.

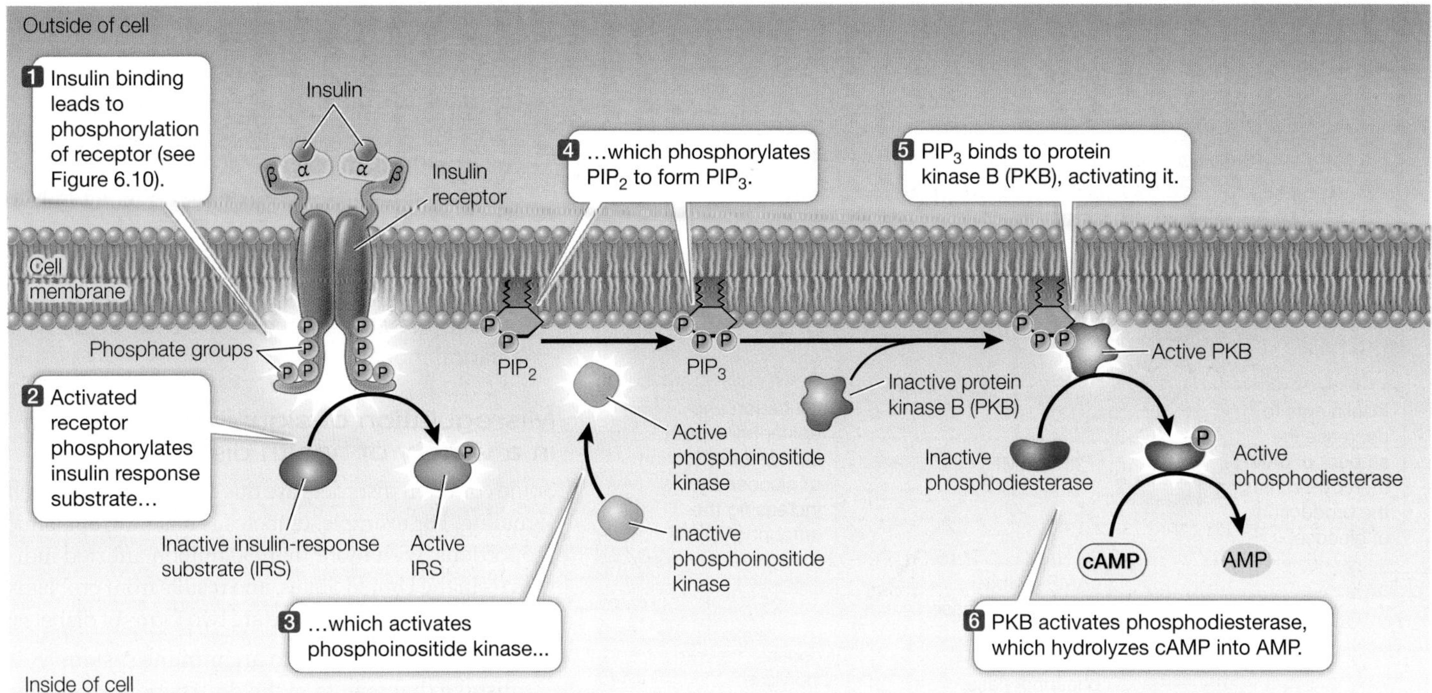

FIGURE 6.15 The Insulin Signal Transduction Pathway One result of insulin binding is an increase in cAMP breakdown. Reduced cAMP leads to an increase in glycogen synthesis and a reduction in glycogen breakdown, causing a reduction in blood glucose levels.

The balance between the processes that end a response and the magnitude of the signal determines the ultimate cellular response to a signal. Cells can alter this balance by adjusting the relative rate of synthesis versus breakdown of the enzymes involved, or by activating or inhibiting the enzymes using other molecules as allosteric effectors (see Key Concept 3.5). In addition, cell response is determined by integration across several signaling pathways.

Signaling pathways integrate multiple signals

A great deal has been learned about signal transduction pathways and cellular responses in the past two decades. As biologists teased apart specific pathways, they found that many of them are interconnected: responses are turned on by some pathways and switched off by others.

To illustrate integration of signaling pathways, we will focus on the regulation of glucose levels in blood in mammals. Glucose is a key to energy metabolism, as it is the starting material for glycolysis (see Key Concept 5.2). Many signals affect glucose levels, including the hormones epinephrine, glucagon, and insulin. We have already seen how an epinephrine signal transduction pathway affects the release of glucose stored in the polysaccharide glycogen in liver cells (see Figure 6.12). In addition to controlling the level of glucose via the synthesis versus breakdown of glycogen, the epinephrine pathway has effects on other metabolic pathways that serve to increase glucose availability: decreasing fatty acid synthesis, increasing triglyceride breakdown, and increasing gluconeogenesis (producing glucose from non-carbohydrate sources). Some of these effects can be predicted from the metabolic pathways discussed in Chapter 5 (see Figure 5.14).

The hormone glucagon is produced by the pancreas and binds to a G protein-coupled receptor in liver, muscle, and adipose (fat) cells. Binding to its GPCR activates a G protein that in turn activates adenylyl cyclase in the cell membrane. Thus the glucagon signaling pathway and the epinephrine pathway shown in Figure 6.12 produce equivalent responses, increasing glucose levels in blood.

A third signal involved in regulating glucose levels is the hormone insulin. The insulin receptor is a protein kinase receptor (see Figure 6.9), and its signaling pathway is outlined in **FIGURE 6.15**. Once stimulated, this pathway activates phosphodiesterase, which lowers cAMP levels, inhibiting the signal transduction pathway for glucagon and epinephrine, and preventing the activation of glycogen phosphorylase and the release of glucose (see Figure 6.12). Thus insulin acts to counteract the effects of glucagon and epinephrine through interacting signal transduction pathways. Yet another effect of the insulin signal transduction pathway (not shown on Figure 6.15) is to cause cytosolic vesicles to fuse with the cell membrane in some tissues such as muscle. These vesicles have numerous glucose carrier proteins in their membrane, which allow rapid uptake of glucose from the blood (see Figure 4.6). Again, the net effect is a reduction of glucose in the blood.

In most organisms, glucose is the major energy source for cells. See **Key Concept 4.2** for a description of facilitated transport of glucose in liver cells.

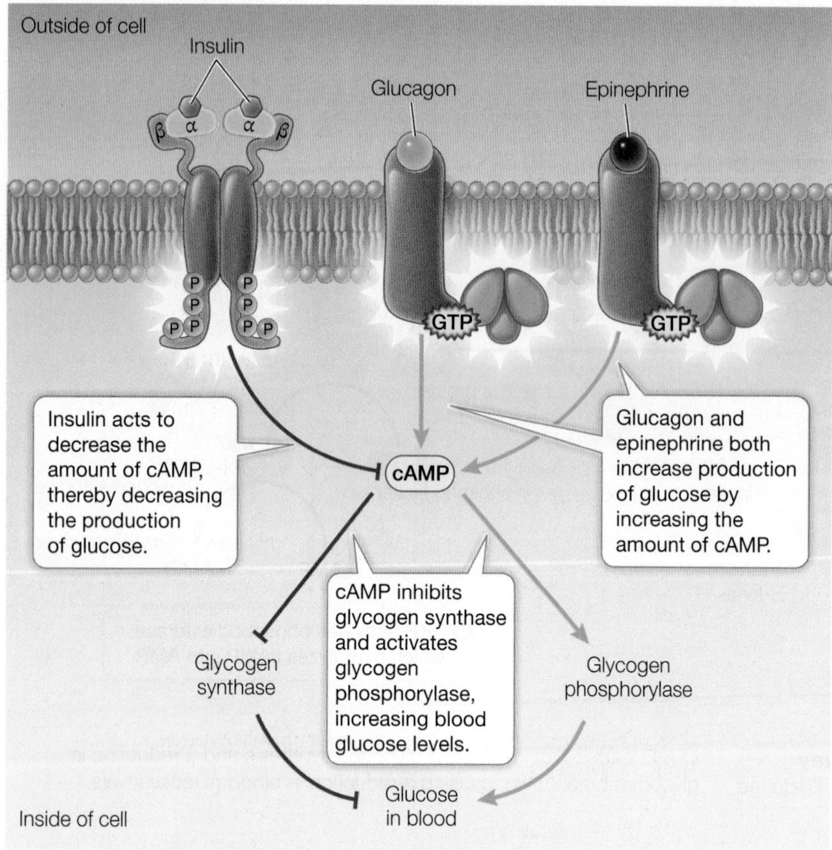

FIGURE 6.16 Three Endocrine Signals That Control Cellular Glucose Levels Green arrows indicate an effect that increases the amount of active target. Red lines indicate an effect that decreases the amount of active target. Compare this diagram with Figures 6.14 and 6.15.

In terms of glycogen synthesis and breakdown, insulin and epinephrine have opposite effects. However, they also both increase the number of glucose carriers in muscle cell membranes. Given that the main function of insulin is to reduce blood glucose levels and that of epinephrine is to prepare for muscular activity (see Chapter 31), why does increasing glucose carriers (in muscle) help with these two functions?

FIGURE 6.16 provides an integrated overview of the steps involved in blood glucose regulation stimulated by the three different signals: epinephrine, glucagon, and insulin. Considering the effects of these three signals as a single regulatory network reveals several patterns:

- *Different pathways can have the same targets.* For example, all three pathways target the enzyme glycogen synthase.
- *Signal pathways interfere with one another.* For example, the second messenger cAMP increases in concentration in response to epinephrine and glucagon but decreases in response to insulin.
- *It is challenging to predict the effects of multiple signals.* If insulin and epinephrine arrive at the same cell, what happens to glycogen breakdown and synthesis? Do blood glucose levels go up or down? Predictability is made more difficult because the relative concentrations of different receptors and signal transduction proteins vary. In addition, although you may think the picture of glucose regulation outlined in Figure 6.16 is complex, it is actually a substantially simplified view. There are several other signals that alter glycogen breakdown and synthesis. Understanding the interacting effects of multiple pathways requires the use of mathematical models.

Misregulation of signaling is involved in a variety of health disorders

Some common disorders are due to problems with cell signaling. For example, diabetes is reaching epidemic proportions, with more than 30 million affected individuals in the United States, and results from problems with insulin signaling. There are two forms of diabetes:

1. Type I diabetes is due to an immune system mistake that results in the destruction of cells in the pancreas that normally produce insulin.
2. Type II diabetes, which accounts for more than 90 percent of all diabetes, is caused primarily by obesity and is associated with insensitivity of liver cells to a low level of insulin produced by the pancreas.

In both cases the insulin signaling pathway either is not activated or the response is substantially decreased. As you can surmise from the three pathways described above, not having a functional insulin pathway allows the epinephrine and glucagon pathways to predominate, keeping glucose levels in the blood abnormally high. High blood glucose levels lead to high glucose levels in various cell types. For example, cells of the small blood vessels (capillaries) experience higher glucose levels, which alters their metabolism and surface proteins, and eventually leads to the vessels becoming leakier. Leaky vessels hinder blood flow, reducing oxygen and nutrient delivery, and result in damage to various tissues such as the retina of the eye (causing blindness) and nerves in the feet and hands.

Media Clip 6.1 Insulin and Diabetes

Another disorder with abnormal cell signaling is cancer. For example, the defining character, uncontrolled cell division, represents an issue with cell cycle regulation. We have already noted one cause: mutations in RAS proteins that make them permanently active are among the most common changes in cancer cells. These changes permanently upregulate the mitogen signaling cascade, stimulating cells to divide even though no signal is being received (see Figure 6.10).

These abnormalities in signaling pathways in disease make them targets for treatments. Type I diabetes is treated with insulin injections—adding the signal. Reduced insulin sensitivity in type II diabetes is treated with Metformin, which has a variety of effects, including inhibition of glucagon-stimulated rise in cAMP, which reduces glucose levels.

In this chapter we have focused on signaling pathways that occur during cell-to-cell signaling in animal cells. However, signal transduction pathways are important in the functioning of all living organisms. You will encounter the themes and principles presented in this chapter in many additional contexts in this text—for example, in learning how certain signals initiate the development of cells in a fertilized egg and how others hasten ripening in fruit.

REVIEW & APPLY | 6.4

R After a signal transduction pathway is activated, it is soon shut off and reset so that it is able to respond again. Signaling pathways are often integrated so that multiple signals may affect the same responses. Several common disorders are caused by errors in signaling pathways.

A

1. Why is it important for a cell response to a signal to be relatively short-lived?
2. A drug has been developed that inhibits adenylyl cyclase in liver cells. Could such a drug be used to treat diabetes?
3. Many individuals with type II diabetes have reduced insulin sensitivity. They thus have high levels of glucose in their blood but low levels in their muscle tissues. Explain this observation.

© Radim Schreiber/FireflyExperience.org

Answer: When the nerve signal arrives at the light organ, the neurotransmitter octopamine is released from the nerve cell. Octopamine binds to its G protein-coupled receptor, which activates a G protein. The active G protein initiates a signal transduction pathway that ends with the production of light (a flash). The flash is caused by the light reaction, catalyzed by the enzyme luciferase, which oxidizes its substrate luciferin:

$$\text{Luciferin} + \text{ATP} + O_2 \rightarrow \text{oxyluciferin} + \text{AMP} + PP_i + \text{light}$$

The reaction, and thus the flash, is thought to be caused by a brief pulse of oxygen supplied to the light organ that allows the reaction to proceed. Currently three hypotheses have been proposed for how this might happen:

1. Production of nitric oxide (NO) briefly shuts off mitochondrial electron transport, causing oxygen levels to rise.
2. Production of NO leads to an increase in hydrogen peroxide (from abundant peroxisomes), which is then used in place of oxygen in the light reaction.
3. Activation of ion pumps causes fluid to be pumped out of small tubes that are open to the outside, which allows oxygen to rapidly diffuse deep into the light organ.

Determining which of these hypotheses, or perhaps another, is the correct one is an area of ongoing research. The signal transduction pathway between the octopamine signal and the flash thus represents a signal transduction pathway whose details have not been worked out. Some questions in science are not yet answered!

6 VISUAL SUMMARY

6.1 Cells Detect a Variety of Signals

Go to ACTIVITY 6.1

FIGURE 6.1

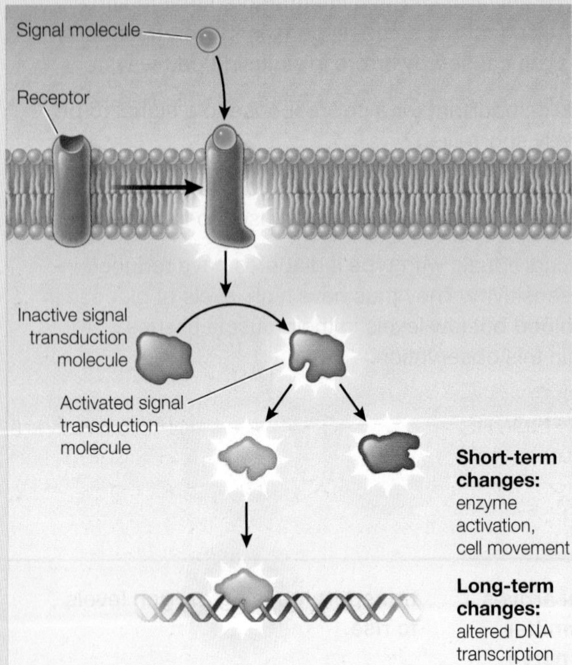

- The ability of cells to detect environmental variables allowed the evolution of cell-to-cell signaling.
- Signal detection leads to transduction and response (**FIGURE 6.1**).
- **JUXTACRINE**, **PARACRINE**, and **AUTOCRINE SIGNALING** work over short distances, whereas **ENDOCRINE SIGNALING** works over longer distances.

QUESTIONS

1. Interferons are released by cells in response to viral infection and allow neighboring cells to upregulate antiviral proteins. What type of signal is interferon?
2. How is a cell signal different than an environmental signal?
3. Why don't all cells respond to all cell signals?

6.2 Signal Molecule Receptors Can Be Classified into Several Groups

Go to ANIMATION 6.1

- Membrane-soluble signals have **INTRACELLULAR RECEPTORS**, whereas membrane-insoluble signals have **MEMBRANE RECEPTORS** (**FIGURE 6.3**).
- Common membrane receptors include ligand-gated ion channels, G protein-coupled receptors, and protein kinase receptors.
- Ligand binding can cause **AUTOPHOSPHORYLATION** of protein kinase receptors. (**FIGURE 6.9**).

FIGURE 6.3

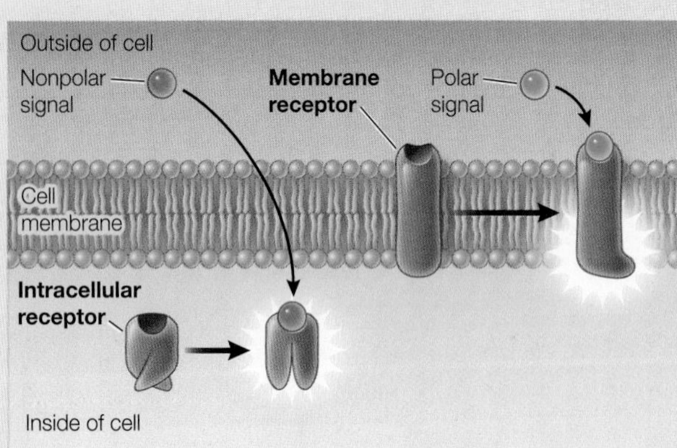

FIGURE 6.9

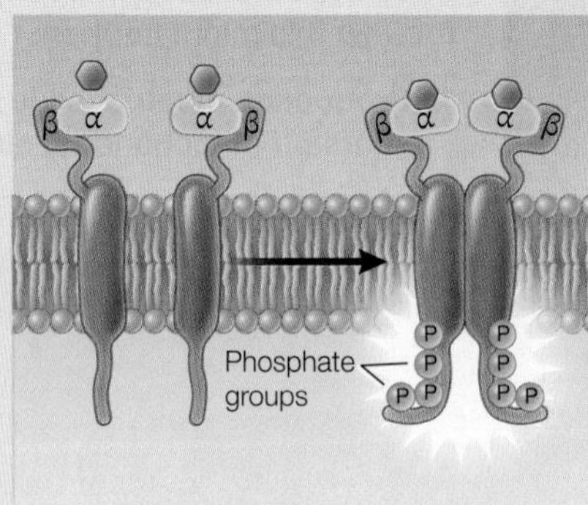

1. Why do some signals have receptors with an extracellular region?
2. How does a G protein-coupled receptor activate its associated G protein?
3. What is the effect on the relative ion concentrations on the inside and outside of a cell when a ligand-gated channel binds its signal?

6.3 Signal Transduction Allows a Cell to Respond Appropriately to a Signal

Go to ACTIVITY 6.2 and ANIMATIONS 6.1 and 6.2

FIGURE 6.10

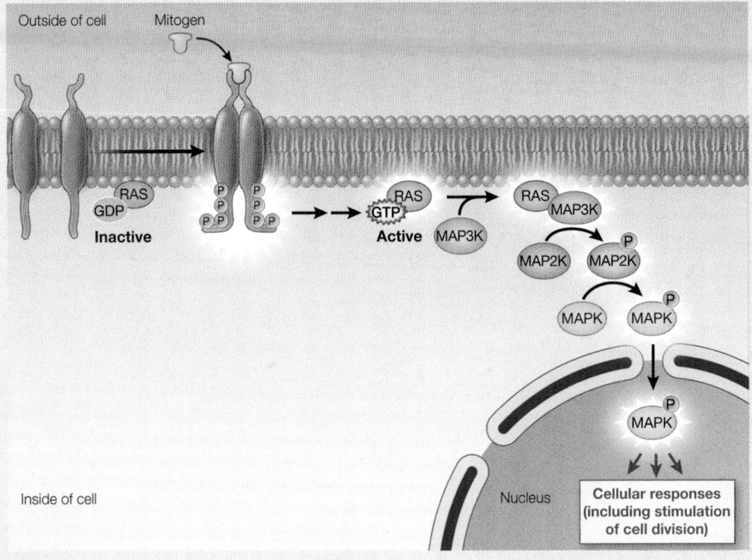

- Cell responses to signals include ion movement, modification of gene transcription rates, and alteration of enzyme activities.
- Enzymes involved in **SIGNAL TRANSDUCTION CASCADES** allow amplification of the signal (**FIGURE 6.10**).
- **SECOND MESSENGERS** are nonprotein molecules involved in **SIGNAL TRANSDUCTION**.

QUESTIONS

1. Acetylcholine is released by nerve cells where they innervate muscle cells. The response to acetylcholine binding is a rapid influx of sodium ions. What type of receptor is probably involved in this response?
2. Some cells do not modify transcription rates in response to a mitogen. Why might a cell be unable to respond?
3. When a signal is received by a cell and a signal pathway is activated, what are the three primary responses that may occur?

6.4 Signal Transduction Is Highly Regulated

- Signal responses are turned off by recycling the receptor, inactivating signal molecules, or losing a signal molecule (**FIGURE 6.14**).
- Signaling pathways are often interconnected, and responses are integrated (**FIGURE 6.16**).
- Several health disorders are caused by problems with signaling.

FIGURE 6.14

Phosphatase inactivates a protein kinase

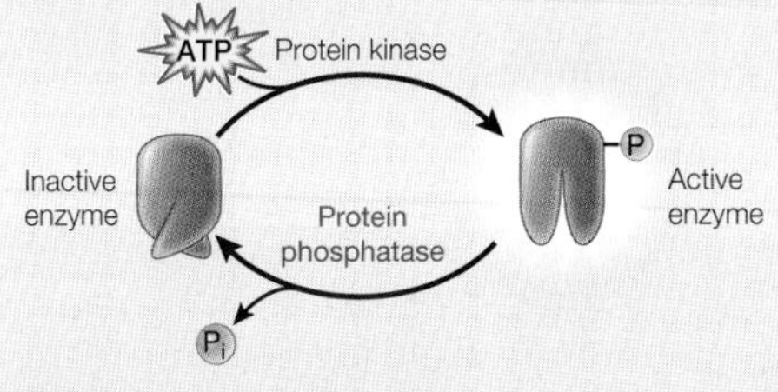

GTPase inactivates a G protein

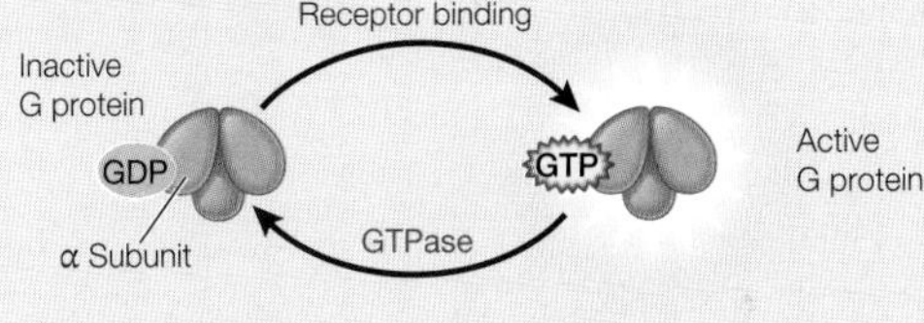

Phosphodiesterase inactivates cAMP

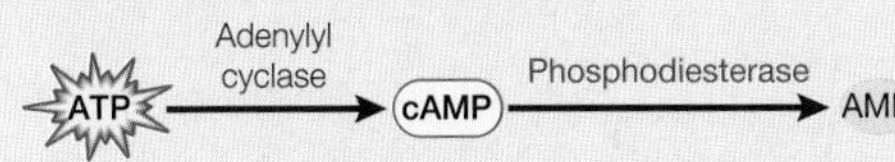

FIGURE 6.16

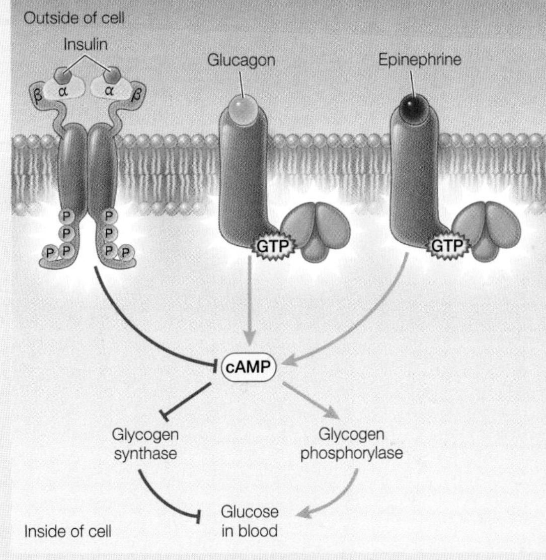

Activity 6.3 **Concept Matching: Signaling in Cells**

1. What happens if an activated signal molecule cannot be inactivated?
2. How does lack of insulin lead to high blood sugar after a meal?
3. How does the presence of the second messenger cAMP allow interference between the insulin and glucagon signaling pathways?

Go to **Achieve** for the e-book, animations, activities, and additional resources and assignments.

Part 2

Genetics

7 The Cell Cycle and Cell Division

The Pacific yew tree contains a compound in its bark, paclitaxel, that is used to treat a variety of cancers.

How does paclitaxel affect cancer cells, and how was this effect discovered?

You will find the answer to this question on page 174.

KEY CONCEPTS

- **7.1** Reproduction May Be Asexual or Sexual
- **7.2** Asexual Reproduction Results in Genetically Identical Daughter Cells
- **7.3** Sexual Reproduction by Meiosis Halves the Number of Chromosomes and Generates Genetic Diversity
- **7.4** Errors during Cell Division Can Result in Changes in Chromosome Number
- **7.5** The Cell Cycle and Cell Death Are Highly Regulated in Eukaryotes

7.1 Reproduction May Be Asexual or Sexual

LEARNING OBJECTIVES

By the end of this key concept you should be able to:

7.1.1 **Explain the difference between sexual and asexual reproduction.**

7.1.2 **Explain what is meant by binary, mitotic, and meiotic cell reproduction and state in which divisions of life they are found.**

7.1.3 **Describe and give examples of the roles of fertilization and meiosis in the diverse sexual life cycles of eukaryotes.**

In Chapter 4 we described the cell theory, which states that cells are the fundamental units of life and that all living organisms are composed of cells. In addition, the cell theory states that all cells come from preexisting cells. In this chapter we focus on how that happens—how cells give rise to other cells. The process by which one cell is able to give rise to two cells is termed **cell division**. Cell division may be asymmetrical or symmetrical. In asymmetrical cell division, mother and daughter cells can usually be distinguished from one another by size: the mother cell produces a daughter cell and remains essentially unchanged. Asymmetrical cell division is found in organisms such as yeast, which is the organism used to brew beer and leaven bread (**FIGURE 7.1A**). It is also common in more complex eukaryotes, such as plants and animals. In symmetrical division, a mother cell divides and produces two daughter cells that are essentially identical and cannot be distinguished from one another. Symmetrical cell division is seen in bacteria and several single-celled eukaryotes such as *Paramecium* (**FIGURE 7.1B**). Single-celled microbes are usually characterized by symmetrical or asymmetrical cell division. Multicellular organisms, however, usually exhibit both types of cell

(A) Asymmetrical cell division

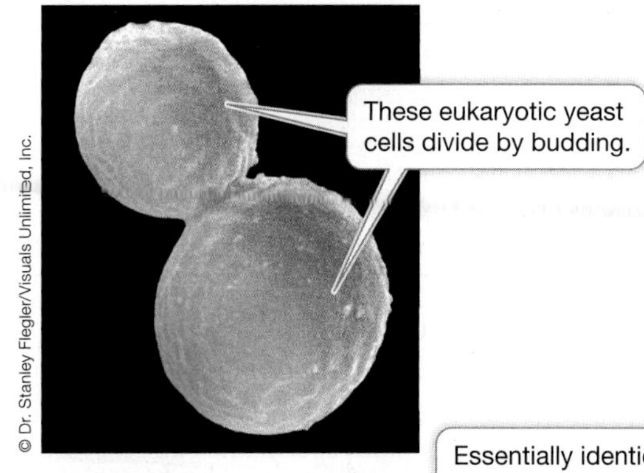

(B) Symmetrical cell division

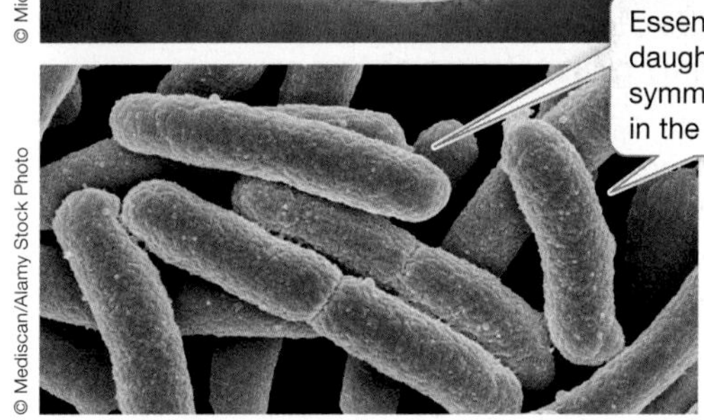

(C) Symmetrical followed by asymmetrical cell division

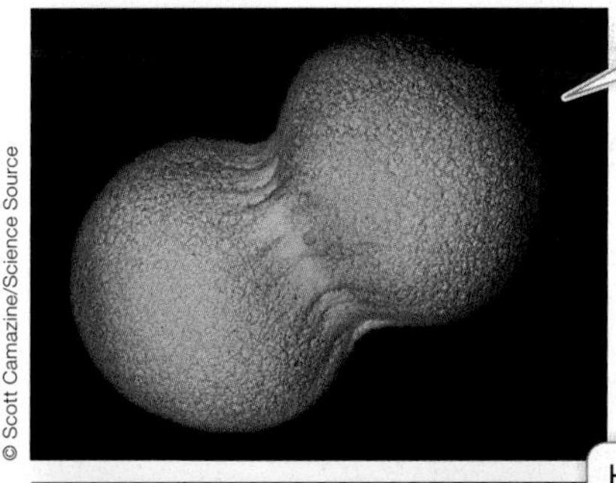

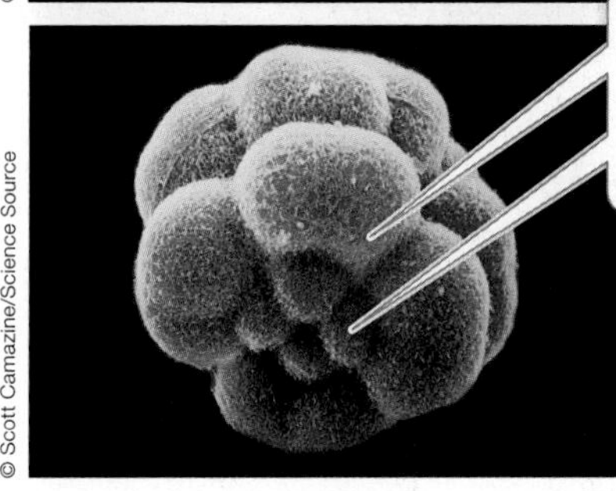

FIGURE 7.1 Cell Division Microbes undergo **(A)** asymmetrical or **(B)** symmetrical cell division. **(C)** Multicellular species often exhibit both asymmetrical and symmetrical cell divisions.

division (**FIGURE 7.1C**). While some cell divisions may appear symmetrical in terms of the size and shape of the daughter cells, there can be functional differences between the daughters, indicating that the division was actually asymmetrical.

LINK Asymmetrical cell division in terms of cell size and contents is a hallmark of early embryonic development in animals. See **Chapter 35**.

In single-celled organisms, cell division and reproduction (the production of offspring) are equivalent. However, in multicellular organisms, many rounds of cell division can occur before reproduction, and only a subset of cells (in humans, sperm and egg cells) are involved in reproduction. The multicellular adult derives from many cell divisions starting from a single cell. In addition to providing more cells for the growing organism, cell division is important for maintaining functional integrity: when cells are damaged or die, cell division of neighboring cells often provides replacements.

When the time for reproduction of an organism comes, it may be accomplished through two basic strategies: asexual reproduction or sexual reproduction. Some organisms are capable of both.

Asexual reproduction by binary fission or mitosis results in genetic constancy

Asexual reproduction is an effective means of making new individuals, and it is common in nature. The offspring resulting from asexual reproduction are **clones** of the parent organism—that is, they are genetically identical (or virtually identical) to each other and the parent. Any genetic variations among the parent and offspring of asexual reproduction are due to rare genetic changes called **mutations**, which are alterations in DNA sequence (discussed in detail in Key Concept 9.3). This small amount of variation among offspring, and between offspring and parent, contrasts with the extensive variation possible in sexually reproducing organisms, as we will see in Key Concept 7.3.

© All Canada Photos/Alamy Stock Photo

FIGURE 7.2 Asexual Reproduction in Aspen Trees All the aspen trees in this picture are essentially identical genetically, indicating that they all arose from a single tree by asexual reproduction.

Most single-celled prokaryotes reproduce by **binary fission**, a type of asexual cell division that produces two genetically identical daughter cells from a parent cell. A cell of the bacterium *Escherichia coli* is the whole organism, so when it divides to form two new cells, it is reproducing. Similarly, single-celled eukaryotes (such as yeast) can reproduce asexually through a type of cell division termed **mitosis**, which also produces two genetically identical daughter cells. In addition to being used during asexual reproduction in single-celled eukaryotes, mitosis is the type of cell division used to add new cells in a multicellular eukaryote.

Many multicellular eukaryotes, including fungi and plants and some animals, can also reproduce asexually. Perhaps the most dramatic example of this is a forest containing thousands of aspen trees (*Populus tremuloides*) in the Wasatch Mountains of Utah (**FIGURE 7.2**). DNA analyses have shown that these trees are clones—they are virtually identical genetically. Aspen can reproduce sexually, with male and female plants, but in this and many other aspen stands, all the trees are the same sex and reproduction is asexual. All the trees in this clonal colony change color in autumn at the same time because the forest actually exists as a single living organism. An extensive root system spreads through the soil, and at intervals stems form and grow into new trees.

LINK Vegetative reproduction, a type of asexual reproduction in plants, is discussed in **Key Concept 26.1**.

Sexual reproduction by meiosis results in genetic diversity

Sexual reproduction involves the production of specialized cells, called **gametes**, that fuse to produce offspring that differ genetically from one another and the parents. Because sexual reproduction involves the fusion of two cells, a sexual life cycle must also involve a type of cell division that halves the amount of DNA carried in a cell; otherwise the amount of DNA would double with each cell fusion. **Meiosis** (described in detail in Key Concept 7.3) is the type of cell division that results in the production of daughter cells (gametes) that contain only half the DNA of the original cell. Meiosis also increases genetic diversity, which is the raw material for natural selection and evolution. Sexual reproduction involving meiosis occurs only in eukaryotes. The majority of eukaryotes have a sexual (or meiotic) life cycle.

LINK **Chapter 13** discusses the importance of genetic variation for evolution.

As Chapter 4 described, the DNA in eukaryotic cells is organized into multiple structures called chromosomes. Each chromosome consists of a double-stranded molecule of DNA and associated proteins. Although all cells of multicellular organisms contain DNA, not all body cells are specialized for reproduction. A **somatic cell** is any cell in the body of a multicellular organism that is not specialized for reproduction. Somatic cells are derived by mitosis.

The somatic cells of many familiar organisms, including most vascular plants and animals, are **diploid**, meaning they each contain two *paired* sets of chromosomes. The chromosomes in each set are numbered 1, 2, 3, and so on and each chromosome differs from another in the same set in both length and the DNA sequences they contain. Because the two chromosomes in each pair are nearly identical in length and DNA sequence, they are called **homologous pairs**. One of the two sets of chromosomes comes from the organism's maternal parent (mother) and the other comes from its paternal parent (father). Thus, for each chromosome pair, one is paternal and the other is maternal. For example, in human somatic cells with 46 chromosomes, one set of 23 comes from the mother and the other set of 23 from the father. Each of us has a chromosome 1 from each parent, a chromosome 2 from each parent, and so on.

The two chromosomes in a homologous pair (called **homologs** of one another) bear corresponding, though not identical, genetic information. For example, a homologous pair of chromosomes in a pea plant may carry slightly different versions of a DNA sequence that encodes an enzyme affecting seed shape (discussed in Key Concept 8.1). One homolog may carry the version for wrinkled seeds, while the other may carry the version for round seeds.

Gametes contain only a *single set* of chromosomes—that is, the chromosomes of gametes do not have homologs. Accordingly, gametes are said to be **haploid**. During sexual reproduction, two haploid gametes fuse to form a diploid **zygote** in a process called **fertilization**. In multicellular eukaryotes, the diploid zygote then undergoes many mitotic cell divisions, resulting eventually in the mature organism. The chromosome number in the diploid zygote is denoted by $2n$, and in the haploid gamete by n, where n is the number of different (non-homologous) chromosomes.

Sexual life cycles are diverse

A life cycle is the series of stages through which an organism passes from one generation to the next. All sexual life cycles involve meiosis (the process by which diploid cells divide to produce haploid cells) and fertilization (fusion of haploid cells to produce diploid cells). In some cases, gametes develop immediately after meiosis and then fuse. In others, each haploid cell divides and develops into a haploid organism—the haploid stage of the life cycle—that eventually produces gametes by mitosis. The fusion of gametes—fertilization—results in a zygote and begins the diploid stage of the life cycle.

Since the origin of sexual reproduction, evolution has generated many different versions of the sexual life cycle. **FIGURE 7.3** shows three examples. At the two extremes, either the diploid zygote immediately undergoes meiosis and essentially all of the life cycle is haploid (the **haplontic** life cycle), or the haploid gametes immediately undergo fertilization and essentially all the life cycle is diploid (the

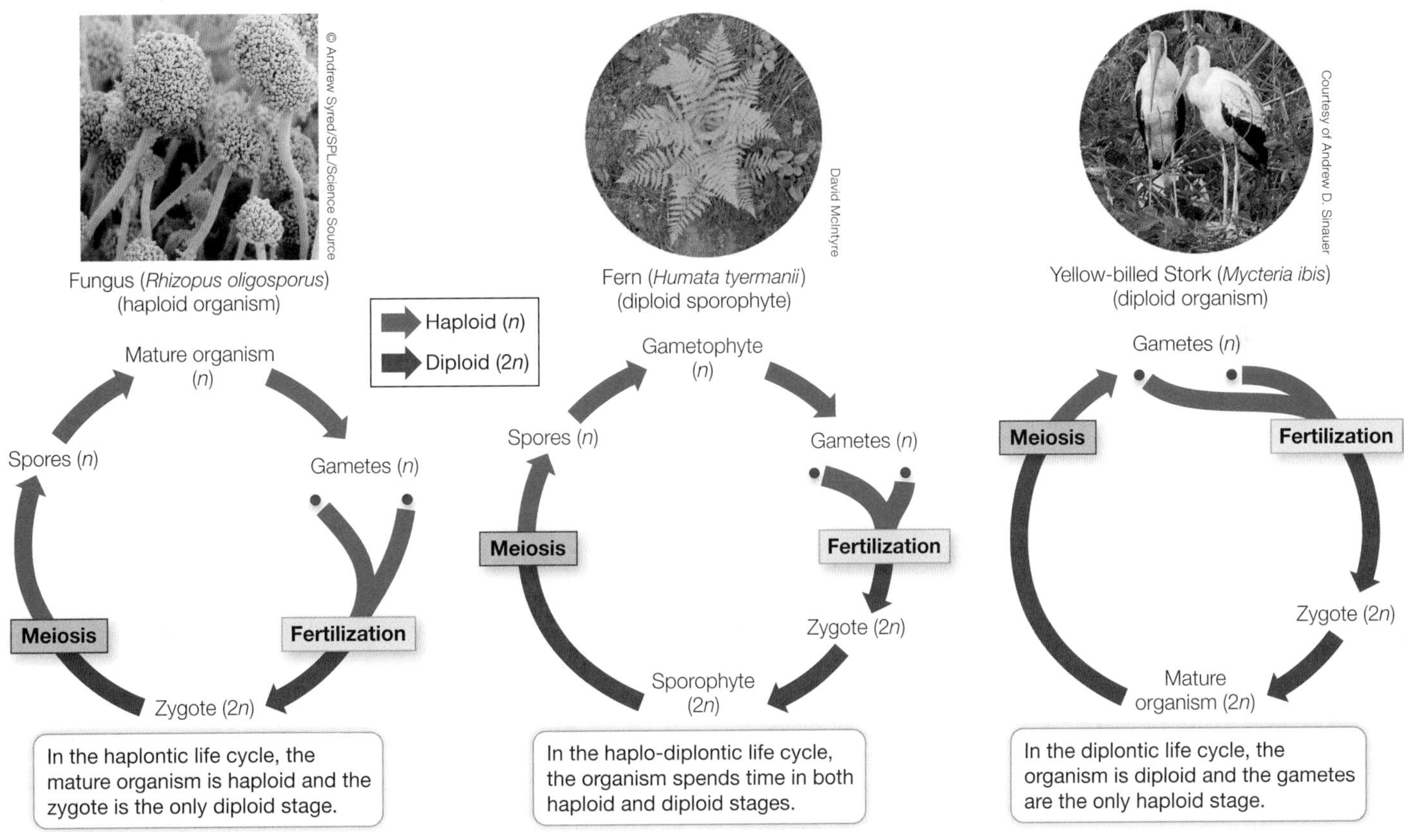

FIGURE 7.3 All Sexual Life Cycles Involve Meiosis and Fertilization Fertilization involves the fusion of two haploid gamete cells, each containing a single set of n chromosomes, to give rise to a diploid zygote cell containing two sets of n chromosomes. Meiosis involves the production of haploid cells from a diploid cell. In between these two events, there may be many mitotic cell divisions in either the diploid stage (a diplontic life cycle), the haploid stage (a haplontic life cycle), or both stages (a haplo-diplontic life cycle).

Which type of sexual life cycle do humans exhibit?

Activity 7.1 Sexual Life Cycle

diplontic life cycle). Between these extremes are **haplo-diplontic** life cycles in which the organism spends significant time in both haploid and diploid stages. We will examine the life cycles of a variety of organisms in Part 4 of this book. For now we focus on the role of sexual reproduction in generating diversity among individuals.

The essence of sexual reproduction is the *random selection of one of the two homologs of each chromosome pair of the diploid* to make a haploid gamete, followed by fusion of two haploid gametes to produce a diploid cell. As we will see in Key Concept 7.3, further diversity is introduced by events that occur during meiosis. These processes contribute to a shuffling of genetic information, so that no two individuals have exactly the same genetic constitution (unless they are clones that arise after fertilization, such as identical twins). The diversity provided by sexual reproduction opens up enormous opportunities for natural selection and evolution.

REVIEW & APPLY | 7.1

R In asexual reproduction, cells divide either by binary fission (as in single-celled prokaryotes such as bacteria) or by mitosis (in eukaryotes). Eukaryotic cells may also undergo a sexual life cycle in which the gametes form by meiosis. Sexual reproduction involves the fusion of two haploid gametes to form a diploid zygote. Sexual life cycles vary across eukaryotic organisms; they may be haplontic, diplontic, or haplo-diplontic, depending on the relative time spent in the haploid versus diploid stages.

A
1. In terms of the genetic diversity produced, what is the main difference between sexual and asexual reproduction?
2. In single-celled organisms, reproduction and cell division are equivalent. However, this is not true in multicellular organisms. Why not?
3. Describe the two events that characterize a sexual, or meiotic, life cycle.

We have briefly mentioned the different types of cell division and the roles they play in the life cycles of organisms. Now let's look in more detail at the processes of cell division, starting with binary fission and mitosis.

7.2 Asexual Reproduction Results in Genetically Identical Daughter Cells

LEARNING OBJECTIVES

By the end of this key concept you should be able to:

7.2.1 Describe how binary fission results in identical daughter cells.

7.2.2 Describe the four phases of the eukaryotic cell cycle.

7.2.3 Identify the five stages of mitosis, in order, and state what the chromosomes are doing in each stage.

In order for any cell to divide, the following four events must occur:

1. **Cell division signals**: One or more signals are required to initiate cell division. The signals may originate from either inside or outside the cell.
2. **DNA replication**: The dividing cell's genetic material (DNA) must be duplicated so that each of the two new cells will have a full complement of genetic information.
3. **DNA segregation**: The replicated DNA must be distributed appropriately to the two daughter cells, so that each receives a copy of every chromosome.
4. **Cytokinesis**: The cytoplasm must divide to form the two new cells, each surrounded by a cell membrane and a cell wall in organisms that have one.

Let's see how these events occur during the processes of binary fission in prokaryotes and mitosis in eukaryotes.

Prokaryotic cells divide by binary fission

In prokaryotes, cell division results in reproduction: one individual becomes two. A dividing cell first grows in size and replicates its DNA, and then it separates the cytoplasm and segregates its DNA into two new cells in the process called binary fission (**FIGURE 7.4**).

CELL DIVISION SIGNALS External factors such as environmental conditions and nutrient concentrations are common signals for prokaryotes. For example, the bacterium *Bacillus subtilis* can divide every 30 minutes under ideal conditions. But when nutrients in its environment are low, it stops dividing. It then resumes dividing when conditions improve.

DNA REPLICATION In most prokaryotic cells, almost all of the genetic information is carried on one single, circular chromosome. Two regions of the prokaryotic chromosome play functional roles in DNA replication:

- *ori*: the site where replication of the circular chromosome starts (the *ori*gin of replication)
- *ter*: the site where replication ends (the *ter*minus of replication)

Chromosome replication begins at the *ori* site and moves toward the *ter* site. When replication is complete, the two daughter DNA molecules separate and segregate from one another to opposite ends of the cell.

DNA SEGREGATION Replication begins near the center of the cell, and as it proceeds, the *ori* regions move toward opposite ends of the cell; segregation and replication co-occur. DNA sequences adjacent to the *ori* region bind proteins that are essential for this segregation. This is an active process involving

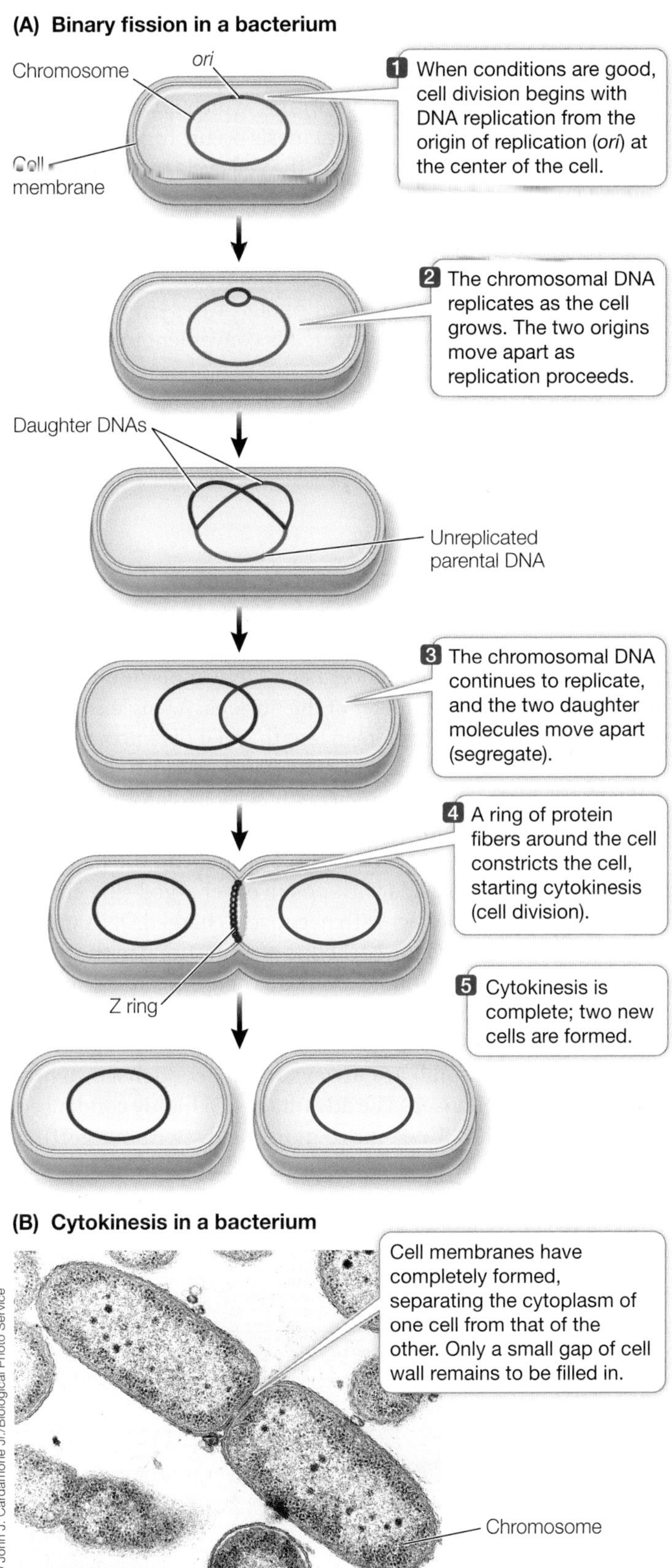

FIGURE 7.4 Binary Fission in a Bacterium **(A)** An overview of binary fission. DNA replication begins at the *ori* site, replicated DNA segregates to either end of the cell, and cytokinesis produces two cells. **(B)** Two identical daughter cells produced by binary fission. (A after K. Jonas. 2014. *Curr Opin Microbiol* 18: 54–60.)

proteins that bind specifically to DNA, and ATP is hydrolyzed, indicating this is an energy-dependent process. Components of the prokaryotic cytoskeleton (see Figure 4.12) are involved in the segregation process. In particular, a bacterial protein that is structurally related to actin but functionally related to tubulin provides a filament along which the *ori* regions and their associated proteins move.

CYTOKINESIS The actual division of a single cell and its contents into two cells begins soon after DNA segregation begins. Initially there is a pinching in of the cell membrane caused by the contraction of a ring of fibers, the **Z-ring**, on the inside surface of the membrane (similar to a drawstring on shorts being slowly tightened). The major component of these fibers is structurally similar to tubulin (the protein that makes up eukaryotic microtubules), but its function is analogous to that of actin. As DNA segregation proceeds, the membrane pinches in more, and once the daughter DNA molecules are fully segregated, new cell wall materials are deposited, which finally separate the two cells.

Eukaryotic cells divide by mitosis followed by cytokinesis

Mitotic cell division in eukaryotes, as in prokaryotes, results in identical daughter cells. However, the details are quite different. In eukaryotes, cell division occurs within the context of the **cell cycle**: the phases through which a single eukaryotic cell passes to produce daughter cells by cell division.

CELL DIVISION SIGNALS Like prokaryotes, single-celled eukaryotes often undergo cell division when environmental conditions are appropriate. For multicellular eukaryotes, the internal environmental conditions experienced by cells are often suitable for cell division. However, the fact that environmental conditions are suitable does not necessarily mean that it occurs. The reason is that cell division signals in multicellular eukaryotes are related to the function of the entire organism. For example, there is no need to produce additional liver cells when the liver has reached its optimal size. Once a cell is stimulated to begin dividing, internal signals come into play to ensure that cell division occurs correctly. We will discuss these signals in Key Concept 7.5.

DNA REPLICATION Unlike prokaryotes, eukaryotes have more than one chromosome. While the mechanics of DNA replication are essentially the same in eukaryotes and prokaryotes, there are two major differences. First, eukaryotic DNA replication starts at numerous origins of replication, not at just one as in prokaryotes. Second, DNA replication occurs during a specified part of the cell cycle and does not overlap with the segregation of DNA into daughter cells.

DNA SEGREGATION Because eukaryotes have multiple chromosomes, segregation of the two sets of DNA molecules produced by replication is somewhat more complicated than it is in prokaryotes. When a cell divides, one copy of each chromosome in the parental cell must end up in each of the two new cells—for example, each new somatic cell in a human will have all 46 chromosomes. Chromosomes become highly condensed before they segregate, which makes them easier to move into the two new nuclei. The **spindle**, a special cytoskeletal structure composed of microtubules, is involved in moving chromosomes during segregation.

CYTOKINESIS The process of cytokinesis in plant cells (which have cell walls) is different than in animal cells (which do not have cell walls). We describe both processes below.

As noted above, in eukaryotes cell division occurs within the context of the cell cycle. The eukaryotic cell cycle can be divided into four phases: G1 (or G0), S, G2, and M (**FIGURE 7.5**):

1. *G1 (or G0) phase.* Cells that are preparing to undergo cell division are in **G1 phase** of the cell cycle. G1 is the phase in which most eukaryotic cells are found, and this is where they carry out their normal functions. Cells that are not yet dividing are in an arrested state termed **G0**. Cells remain in G0 or G1 unless they receive appropriate signals to enter the next phase of the cell cycle.
2. *S phase.* Cells in **S phase** (synthesis phase) of the cell cycle are actively replicating their DNA.

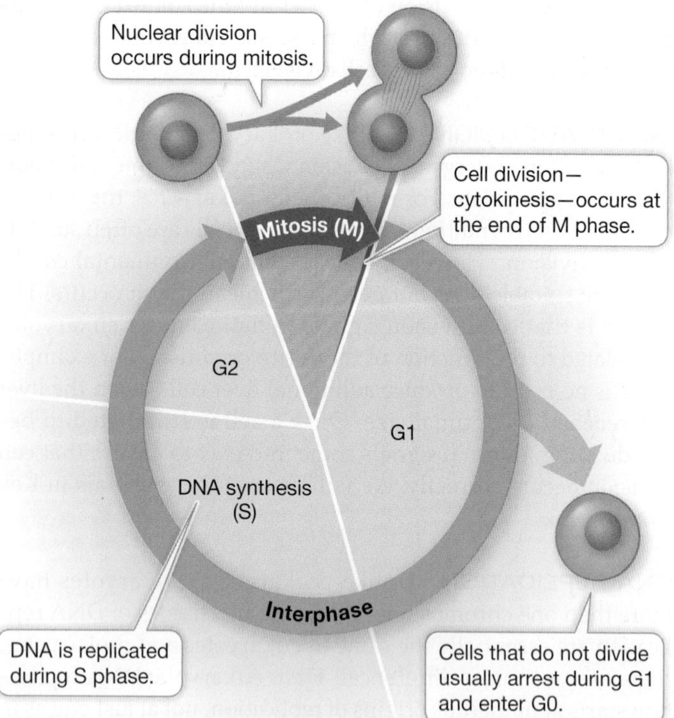

FIGURE 7.5 The Eukaryotic Cell Cycle Nondividing cells are in G1, or in an arrested state termed G0. G1, S, and G2 together make up interphase. Mitosis plus cytokinesis together make up M phase.

What is the most striking difference between the eukaryotic and prokaryotic cell cycles?

3. *G2 phase.* Once DNA synthesis is complete, cells enter **G2 phase**, where they prepare for mitosis.
4. *M phase.* Finally cells enter **M phase**, which is when DNA segregation in mitosis and cytokinesis occur. Together, G1, S, and G2 are termed **interphase**; the cell cycle can be partitioned into M phase and interphase.

In mitosis, a single nucleus gives rise to two daughter nuclei that each contain the same number of chromosomes as the parent nucleus. Although mitosis is a continuous process in which each event flows smoothly into the next, it is convenient to subdivide it into a series of stages: prophase, prometaphase, metaphase, anaphase, and telophase. We will examine these stages in more detail, but first let's briefly examine how the structure of the eukaryotic chromosome changes in preparation for mitosis.

Animation 7.1 Mitosis

Eukaryotic chromosomes are compacted into chromatin

In G1 of the cell cycle, each chromosome in the nucleus consists of a long linear DNA sequence. Each DNA molecule is many times longer than the diameter of the nucleus. A single chromosome can be many centimeters in length, yet it fits into a nucleus that is only a few micrometers in diameter (1 cm = 10,000 μm). To fit, DNA is heavily compacted by being wrapped around special proteins. This packaging is highly organized and involves many different proteins, the most common of which are the histones. Histones have amino acids with positively charged side chains on the surfaces that interact with negatively charged DNA (recall the phosphate groups in the backbone of each strand in the helix; see Figure 3.11). DNA and its associated proteins together are termed **chromatin**.

The level of packaging of DNA can vary. Some regions of DNA on a chromosome are always highly condensed, while others are less condensed. The advantage of a highly condensed chromosome is that it can move more easily, without getting entangled with other chromosomes or material in the cell. The disadvantage is that other proteins are unable to interact with it. You will see in later chapters that the interaction of proteins with DNA is key to regulating DNA's expression as genetic material.

LINK DNA packaging and its importance in allowing access to stored information are described in **Key Concept 11.4**.

In G1, S, and G2, chromosomes are relatively uncondensed and thus difficult to distinguish with a microscope. (From a distance, it is easier to see a wound-up spool of thread than a single strand.) However, during M phase, chromosomes are fully condensed and visible after staining with special dyes. Early in M phase, replicated chromosomes consist of two DNA molecules (and their associated packaging proteins), which are held together at the **centromere**,

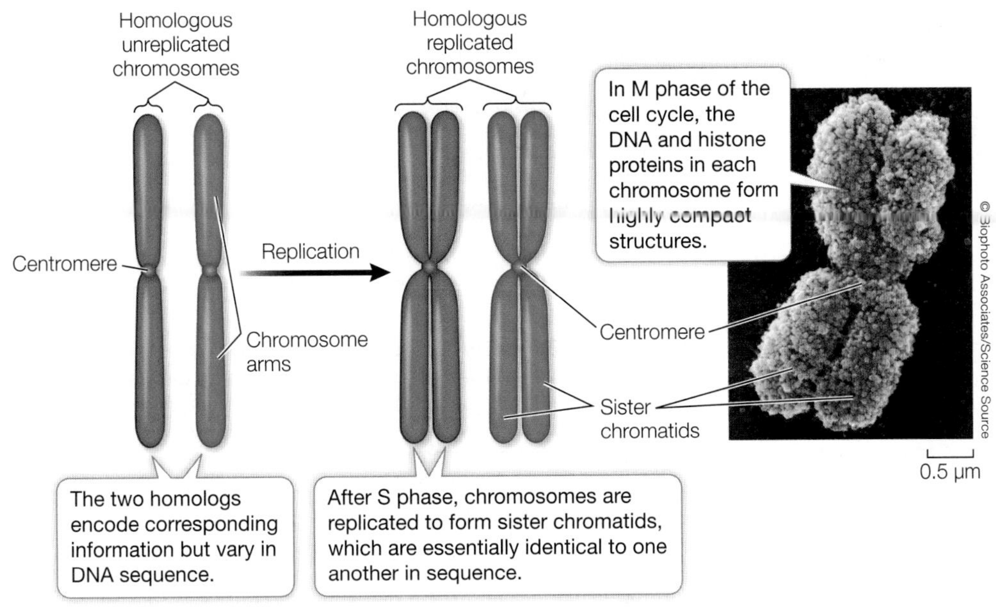

FIGURE 7.6 Eukaryotic Chromosomes Chromosomes are replicated during S phase of the cell cycle. In diploid cells, chromosomes are present as homologous pairs.

a structure used to orient and move chromosomes during cell division. The portion of the DNA molecule on each side of the centromere is called a **chromosome arm**. An unreplicated chromosome has two arms, whereas a replicated chromosome has four. The two essentially identical daughter DNA molecules are at this point called **sister chromatids** (**FIGURE 7.6**). Later in M phase, toward the end of mitosis, the two sister chromatids are separated into two chromosomes consisting of a single DNA molecule and its associated packaging proteins. For a given organism, the number and sizes of the condensed chromosomes constitute the **karyotype**. Each chromosome has a particular length, and the centromere is located at a particular position along its length. For example, humans have 46 chromosomes (23 homologous pairs) that differ in their sizes and centromere positions (**FIGURE 7.7**). (*Hint*: When you are counting the number of chromosomes in a cell, count the number of separate centromeres and not chromosome arms. That way you won't get confused regarding how to count replicated versus unreplicated chromosomes.)

Mitosis results in the production of genetically identical daughter nuclei

Many changes occur as the cell enters the first stage of mitosis, **prophase**. Chromosomes begin to condense, and then, in fully condensed prophase chromosomes, specialized protein structures

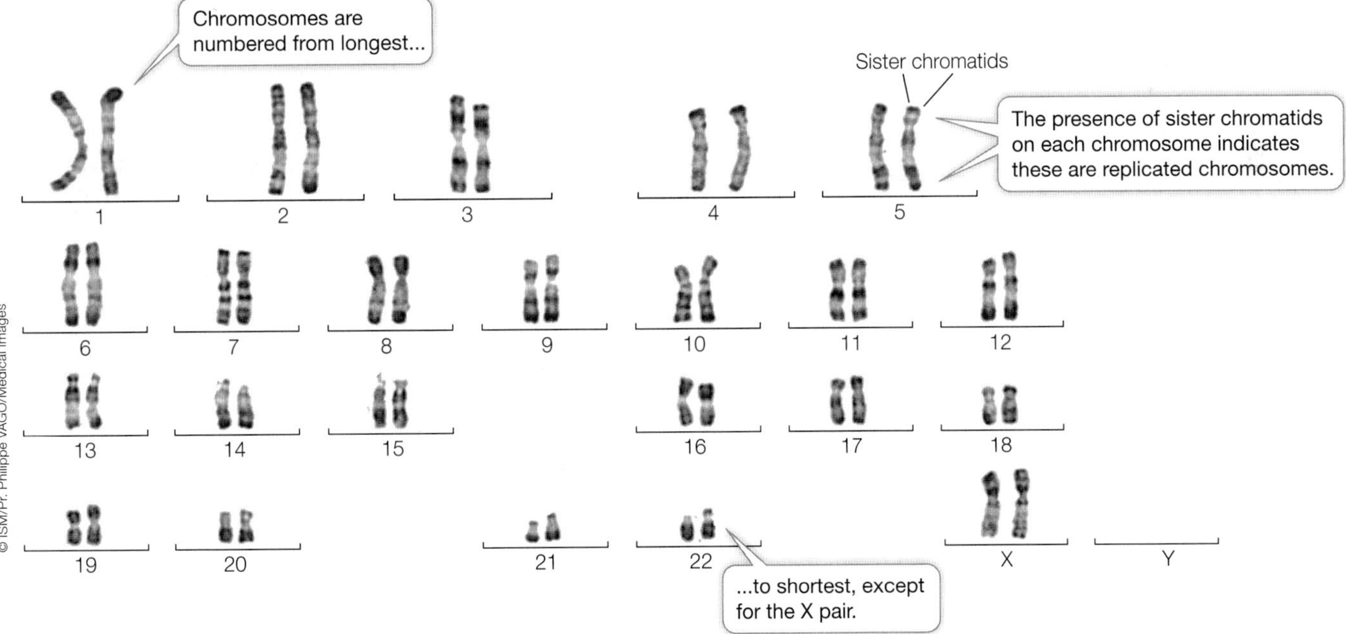

FIGURE 7.7 The Karyotype of a Human Female Humans have 23 pairs of homologous chromosomes, giving 46 total.

How do you think this diagram was made?

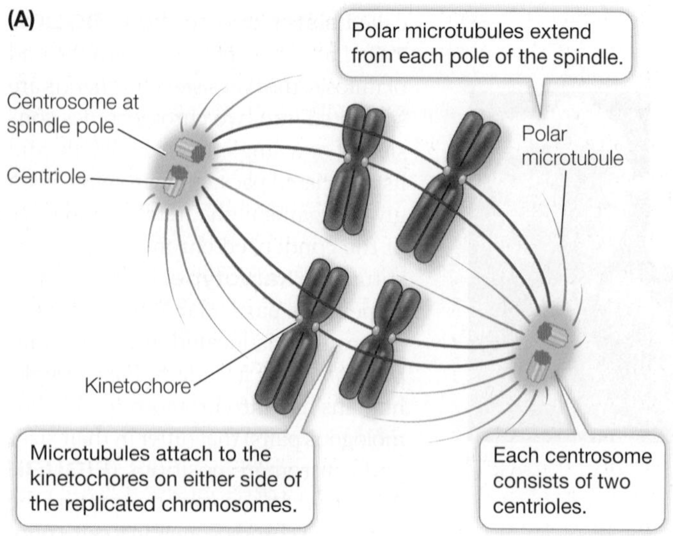

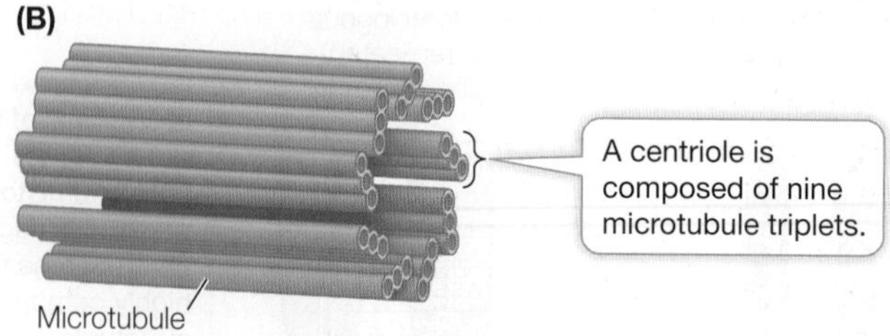

FIGURE 7.8 Formation of the Spindle The spindle begins to form in prophase. **(A)** Kinetochores that assemble on each side of the centromere are attachment sites for microtubules that originate at centrosomes. **(B)** A centrosome consists of two centrioles, each of which is composed of nine microtubule triplets that constitute the poles around which the spindle forms.

Activity 7.2 The Mitotic Spindle

called **kinetochores** appear on each side of the centromere, one per chromatid (**FIGURE 7.8**). Kinetochores are attachment sites for microtubules and are critical for chromosome movement. In addition to the condensing of chromosomes, two other events during prophase contribute to the orderly segregation of the replicated DNA: (1) centrosome orientation and (2) spindle formation.

In many eukaryotic organisms, structures called **centrosomes** are required for spindle formation. Centrosomes orient at both ends of the cell during prophase. A centrosome consists of a pair of **centrioles**, each one a hollow tube formed by nine triplets of microtubules. During S phase the centrosome becomes duplicated, and at the G2–M transition, the two centrosomes separate from one another, and eventually function as the "poles" toward which chromosomes move during segregation.

The positions of the centrosomes determine the plane at which the cell divides; therefore they determine the spatial

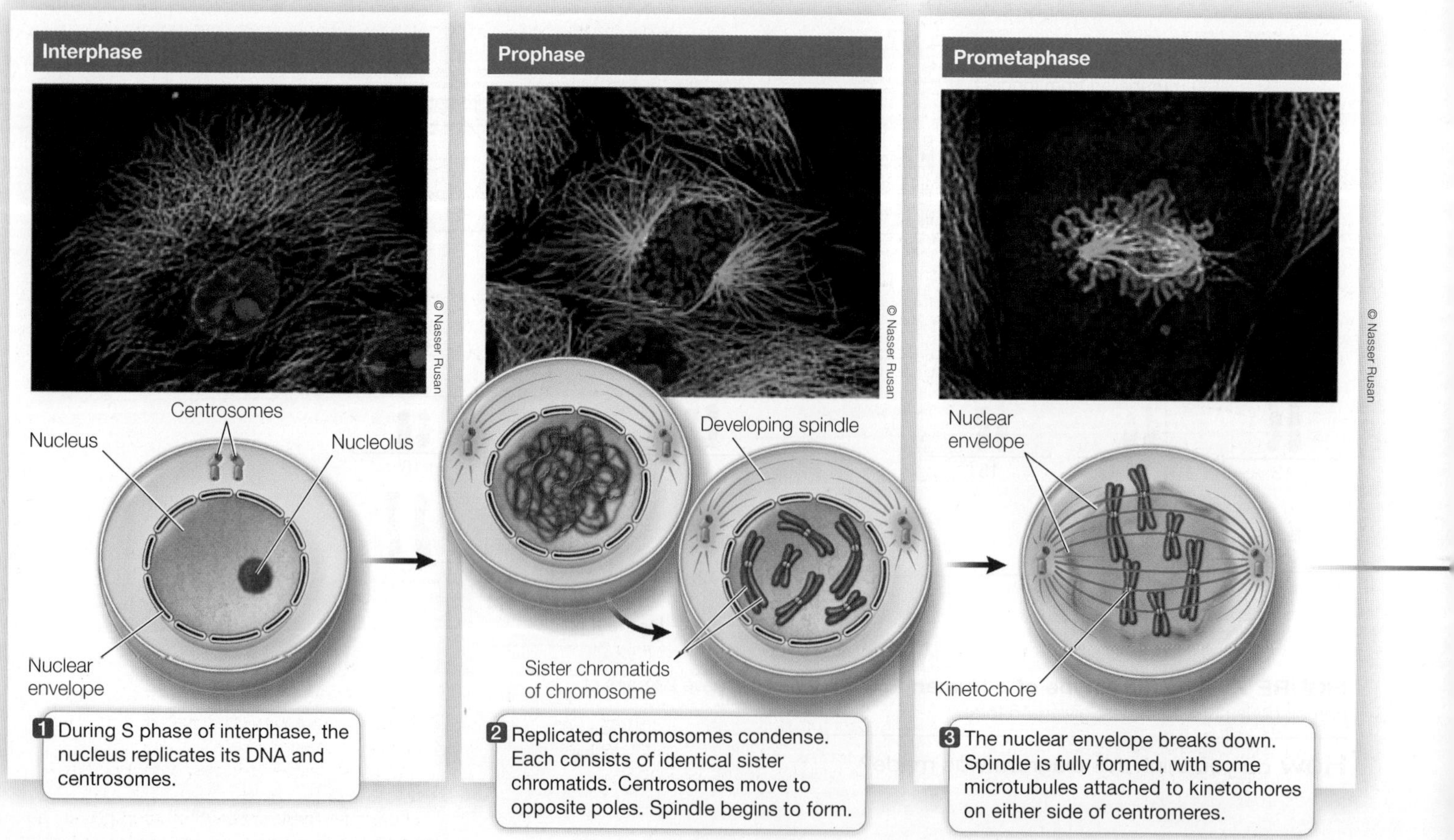

Interphase

Centrosomes
Nucleus
Nucleolus
Nuclear envelope

1 During S phase of interphase, the nucleus replicates its DNA and centrosomes.

Prophase

Developing spindle
Sister chromatids of chromosome

2 Replicated chromosomes condense. Each consists of identical sister chromatids. Centrosomes move to opposite poles. Spindle begins to form.

Prometaphase

Nuclear envelope
Kinetochore

3 The nuclear envelope breaks down. Spindle is fully formed, with some microtubules attached to kinetochores on either side of centromeres.

relationship between the two new cells that result from cell division. This relationship may be of little consequence to single free-living cells such as yeasts, but it is important for development in a multicellular organism. For example, during the development of an embryo, the daughter cells from some divisions must be positioned correctly to receive signals to form new tissues. The cells of plants and fungi lack centrosomes but have distinct microtubule organizing centers at each end of the cell, which play the same role.

Each of the two centrosomes serves as a pole from which tubulin dimers aggregate into microtubules that extend from the poles into the middle region of the cell, forming the spindle. The spindle forms during prophase and prometaphase, when the nuclear envelope breaks down. The microtubules are initially unstable, constantly forming and falling apart, until they contact kinetochores or microtubules from the other half spindle and become more stable.

FIGURE 7.9 shows the stages of mitosis. As the cell enters **prometaphase**, the spindle becomes fully formed, the nuclear envelope breaks down, and the replicated chromosomes become attached at their kinetochores to microtubules coming from the centrosome at the poles of the cell. Each replicated chromosome is attached on both sides of its centromere: the kinetochore on one side attaches to one pole, and the kinetochore on the other side attaches to the other pole.

By **metaphase** the pull of the microtubules has aligned the chromosomes on the metaphase plate, an equatorial position often at the midline of the cell. Microtubules attached to the two kinetochores of a replicated chromosome are engaged in a "tug-of-war," with a stalemate (no net movement) when the chromosome is exactly in the middle of the spindle (see metaphase panel of Figure 7.9). Once all of the chromosomes are aligned, the cell is ready for the next stage of mitosis. Note that each replicated chromosome aligns on the metaphase plate independently of its homolog. This is different than what we will see in meiosis in Key Concept 7.3.

The separation of sister chromatids into separate chromosomes occurs at the start of **anaphase**. The centromeres of each replicated chromosome divide, resulting in two centromeres, one per chromatid, each of which has a kinetochore on only one side. Since microtubules are only attached on one side, each of the now unreplicated chromosomes, formerly sister chromatids, rapidly moves poleward. The migration of the unreplicated chromosomes to the poles of the cell is a highly organized, active process. Two mechanisms may operate to move the chromosomes along. First, the kinetochores contain molecular motor proteins, including kinesin and dynein, which may use energy from ATP hydrolysis to move the chromosomes along the microtubules. Second, and probably

FIGURE 7.9 The Stages of Mitosis

Why do many of the chromosomes have a characteristic V shape in anaphase?

Activity 7.3 Images of Mitosis

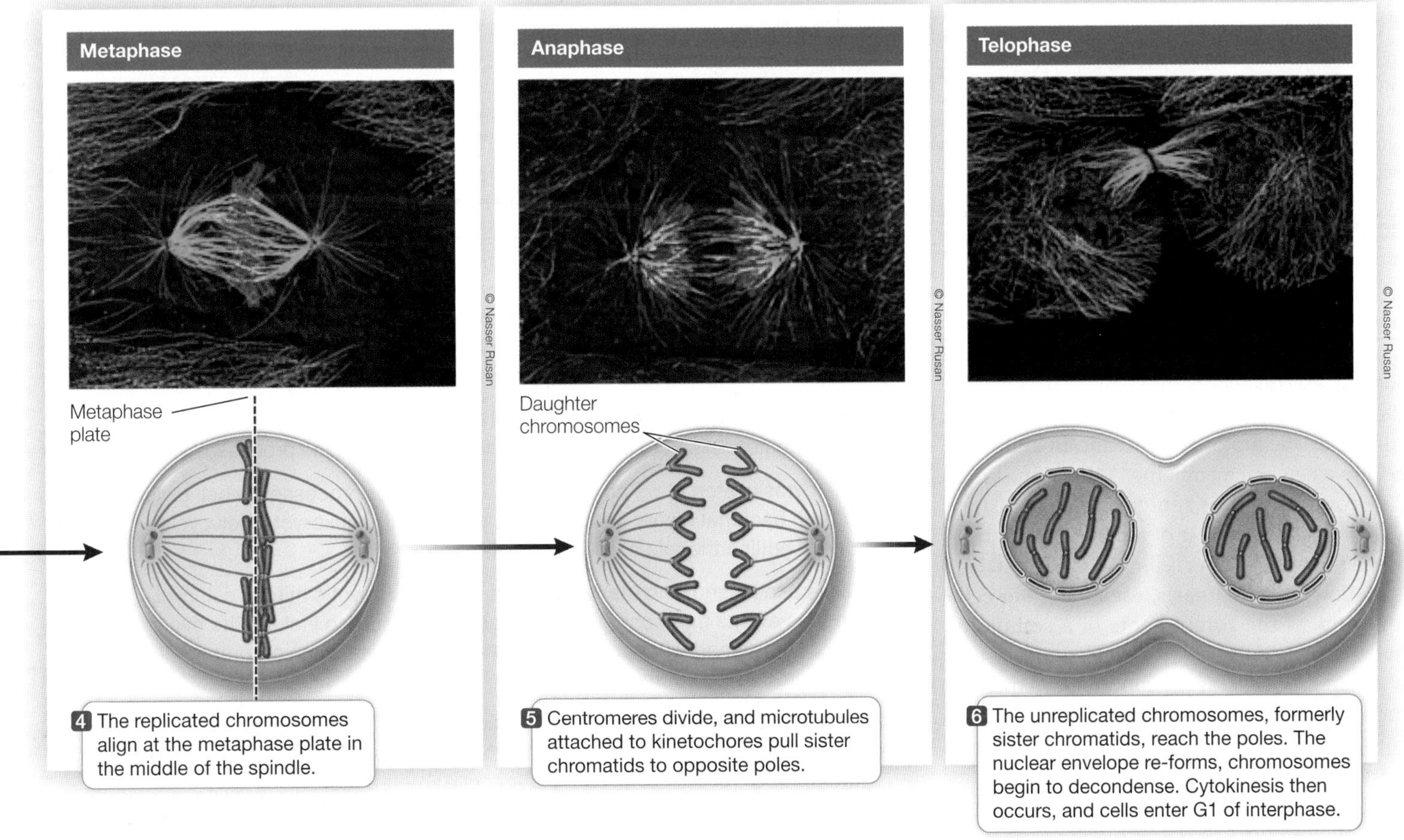

more important, the microtubules attached to the kinetochore and pole shorten from both ends, drawing the chromosomes toward the poles. At the end of anaphase, one of the sister chromatids from each of the replicated chromosomes that entered mitosis is present at each pole, indicating that the same genetic information is present at each pole.

LINK As discussed in **Key Concept 4.4**, the motor proteins dynein and kinesin, with energy derived from ATP hydrolysis, can change their three-dimensional shape. This action can be used to move attached organelles to new locations in the cell. See **Figure 4.14**.

Telophase occurs after the chromosomes have separated and is the last phase of mitosis. During telophase, a nuclear envelope forms around each set of new chromosomes, and the chromosomes become less compact. The spindle also disappears at this stage, as a result of depolymerization of microtubules. As a result, there are two genetically identical nuclei in a single cell.

Media Clip 7.1 Mitosis: Live and Up Close

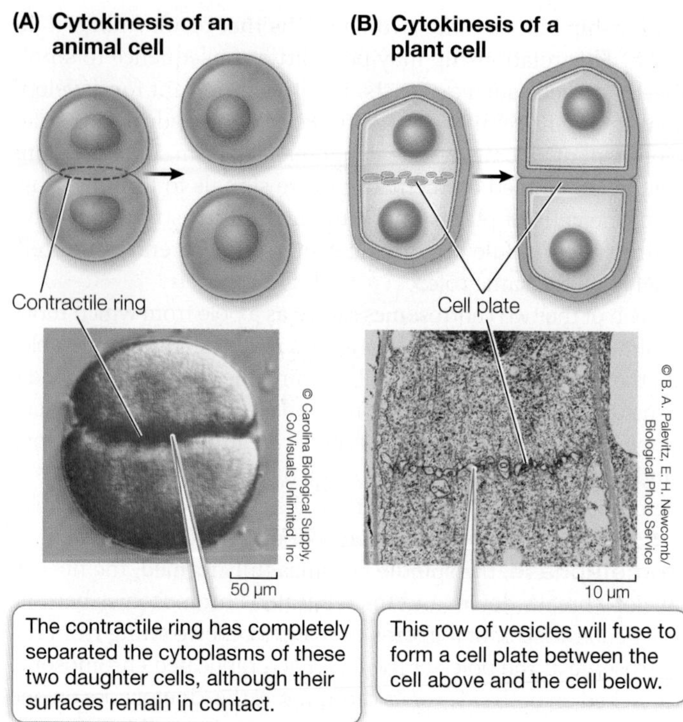

FIGURE 7.10 Cytokinesis: Division of the Cytoplasm Cytokinesis in animals **(A)** differs from cytokinesis in plants **(B)**, because of the plant cell wall.

Cytokinesis is the division of the cytoplasm

Mitosis refers only to the division of the nucleus. Cytokinesis, the division of the cell's cytoplasm, is the final stage of cell reproduction. This process occurs differently in plants and animals.

In animals, cytokinesis usually begins with a furrowing of the cell membrane, as if an invisible thread were cinching the cytoplasm between the two nuclei (**FIGURE 7.10A**), similar to binary fission in bacteria. This contractile ring is composed of microfilaments and other proteins, including a myosin motor protein, which form a ring on the cytoplasmic surface of the cell membrane. The ring produces a contraction, creating a cleavage furrow readily visible under the microscope. The cytoplasm divides differently in plant cells because they have rigid cell walls outside the cell membrane. As the spindle breaks down after mitosis, vesicles derived from the Golgi apparatus appear along the plane of cell division, roughly midway between the two daughter nuclei. The vesicles are propelled along microtubules by the motor protein kinesin and fuse to form a new cell membrane. At the same time they contribute their contents to a cell plate, which is the beginning of a new cell wall between the two daughter cells (**FIGURE 7.10B**).

Following cytokinesis, each daughter cell contains all the components of a complete cell. A precise distribution of chromosomes is ensured by mitosis. In contrast, organelles such as mitochondria and chloroplasts are not necessarily distributed equally, although at least one of each must be present in each daughter cell.

REVIEW & APPLY | 7.2

R Broadly speaking, cell division involves (1) signals that initiate cell division, (2) replication of DNA, (3) segregation of the replicated DNA, and (4) cytokinesis (division of the cytoplasm to produce two genetically identical daughter cells). In eukaryotes, cells in G1 progress to the S phase, where DNA synthesis occurs, then to the G2 phase and finally to the M phase, where mitosis (segregation) and cytokinesis occur. Replication and segregation are thus separate events. In contrast, DNA replication and segregation co-occur in binary fission. Mitosis is divided into stages involving the condensation of chromosomes into chromatin (prophase), the formation of a spindle (prometaphase), independent alignment of individual chromosomes (metaphase), the separation of sister chromatids to each pole (anaphase) and the formation of new nuclear envelopes (telophase).

A

1. Both binary fission and the mitotic cell cycle result in the production of identical daughter cells from a single maternal cell. How are the processes similar?
2. The cytoskeleton plays an active role in two major processes in cell division in both bacterial binary fission and M phase of the eukaryotic cell cycle. Briefly outline the role of the cytoskeleton.
3. Make a table showing the four phases of the eukaryotic mitotic cell cycle and the events that occur in each phase. Make sure to include the five stages of mitosis and what is occurring in each. Indicate when chromosomes are replicated versus unreplicated.
4. Haploid cells are those that have a single copy of each chromosome—there are no homologs present. Can a haploid cell undergo mitosis? Explain your answer with respect to events that occur at metaphase.

Binary fission and mitosis result in daughter cells with the same number of chromosomes as their parent cells. Sexual reproduction, however, requires a process of cell division in which the number of chromosomes is halved. We'll look at this process next.

7.3 Sexual Reproduction by Meiosis Halves the Number of Chromosomes and Generates Genetic Diversity

LEARNING OBJECTIVES

By the end of this key concept you should be able to:

7.3.1 **Describe and identify the stages of each of the two meiotic divisions.**

7.3.2 **Explain how the first meiotic division results in halving the number of chromosomes.**

7.3.3 **Describe how independent assortment and crossing over generate novel chromosomal combinations.**

In Key Concept 7.1 we described the role and importance of meiosis in sexual reproduction. Now we will see how this type of cell division gives rise to haploid cells. Production of a haploid cell is required for sexual reproduction, in which two haploid cells (gametes) fuse during fertilization. Meiosis consists of *two* nuclear divisions, **meiosis I** and **meiosis II**, that together produce four nuclei, each with the haploid number of unreplicated chromosomes. Although the nucleus divides twice during meiosis, the DNA is replicated only once, during S phase of the cell cycle. Unlike the products of mitosis, the haploid cells produced by meiosis are genetically different from one another and from the parent cell.

FIGURE 7.11 illustrates the events of meiosis. In the rest of this key concept we discuss some of the key features of meiosis.

Meiosis halves the number of chromosomes

Meiosis replaces mitosis in the eukaryotic cell cycle in diploid cells that are destined to produce haploid daughter cells. As noted above, meiosis consists of two nuclear divisions, meiosis I and meiosis II. Two features characterize meiosis I:

1. *Homologous chromosomes pair.* During prophase I, homologous, replicated chromosomes pair with each other along their entire lengths. Pairing is facilitated by special proteins that glue the homologs together. No such pairing of homologs occurs in mitosis (although sister chromatids do remain together until anaphase). This homolog pairing remains through metaphase I.

2. *There is no centromere division.* Instead of centromeres dividing as in mitosis, in anaphase I *the homologous chromosome pairs separate.* Anaphase I begins when the glue holding homologs together loosens. Homologs separate and then move to opposite poles.

Anaphase I is when the halving of chromosome number occurs; meiosis I is thus sometimes called the reduction division. A cell that begins with a diploid number of replicated chromosomes produces two cells after meiosis I, each of which has a haploid number of replicated chromosomes. To convince yourself of this, examine Figure 7.11 and count the number of chromosomes moving to each pole (remember to count centromeres and not chromosome arms).

After meiosis I ends, meiosis II can begin. There may be considerable time—even months to years!—between the two meiotic divisions. During the time between meiosis I and II, chromosomes may become less condensed than they were during meiosis I. But there is no replication of chromosomal DNA. However, centrosome replication does occur, allowing new spindles to be formed in each cell. Meiosis II is indistinguishable from mitosis: replicated chromosomes condense and attach to the spindle by their two kinetochores (prophase II), unpaired replicated chromosomes align on the metaphase II plate, centromeres divide and sister chromatids move to opposite poles (anaphase II), and the nuclear envelope re-forms and chromosomes uncondense (telophase II). Following cytokinesis, the products of meiosis are four cells, each containing the haploid number of unreplicated chromosomes. Importantly, *these four cells are not genetically identical.*

Animation 7.2 Meiosis

Independent assortment and crossing over generate diversity

In a diploid sexual organism, every somatic cell has two sets of chromosomes ($2n$): one set derived from its male parent (paternal chromosomes), the other from its female parent (maternal chromosomes). If the organism is multicellular, all of its cells are derived from many mitotic divisions of the initial zygote. Since mitosis results in identical daughter cells, every cell will be genetically identical (except for a few mutations). As mentioned above, however, the four daughter cells produced by meiosis differ from one another genetically. The shuffling of genetic material occurs by two processes: independent assortment and crossing over.

INDEPENDENT ASSORTMENT The DNA of a eukaryote is packaged into chromosomes, and the number of chromosomes varies with species. For example, the somatic cell chromosome number ($2n$) in a mosquito is 6, while a somatic cell of the fern *Ophioglossum reticulatum* has 1,260 chromosomes! The fact that the genetic material is packaged into multiple pairs of chromosomes has an important consequence: the cells formed after meiosis (which you know from earlier in this chapter are or will be gametes) differ from one another because of random alignment of homologous chromosomes on the metaphase I plate.

To understand this concept, consider the parental gametes that fused to make an individual. Suppose all the chromosomes received from one parent are colored red and the ones received from the other parent blue. At metaphase I, all the homologous pairs of chromosomes have one red and one blue replicated chromosome.

MEIOSIS I

Early prophase I

© C. A. Hasenkamp/Biological Photo Service

Centrosomes

1 The chromatin begins to condense following interphase.

Mid-prophase I

© C. A. Hasenkamp/Biological Photo Service

Pairs of homologs

2 Homologous chromosomes pair and condense further.

Late prophase I–prometaphase I

© C. A. Hasenkamp/Biological Photo Service

Chiasma

3 The chromosomes continue to coil and shorten. The chiasmata reflect crossing over, the exchange of genetic material between nonsister chromatids in a homologous pair. In prometaphase the nuclear envelope breaks down.

MEIOSIS II

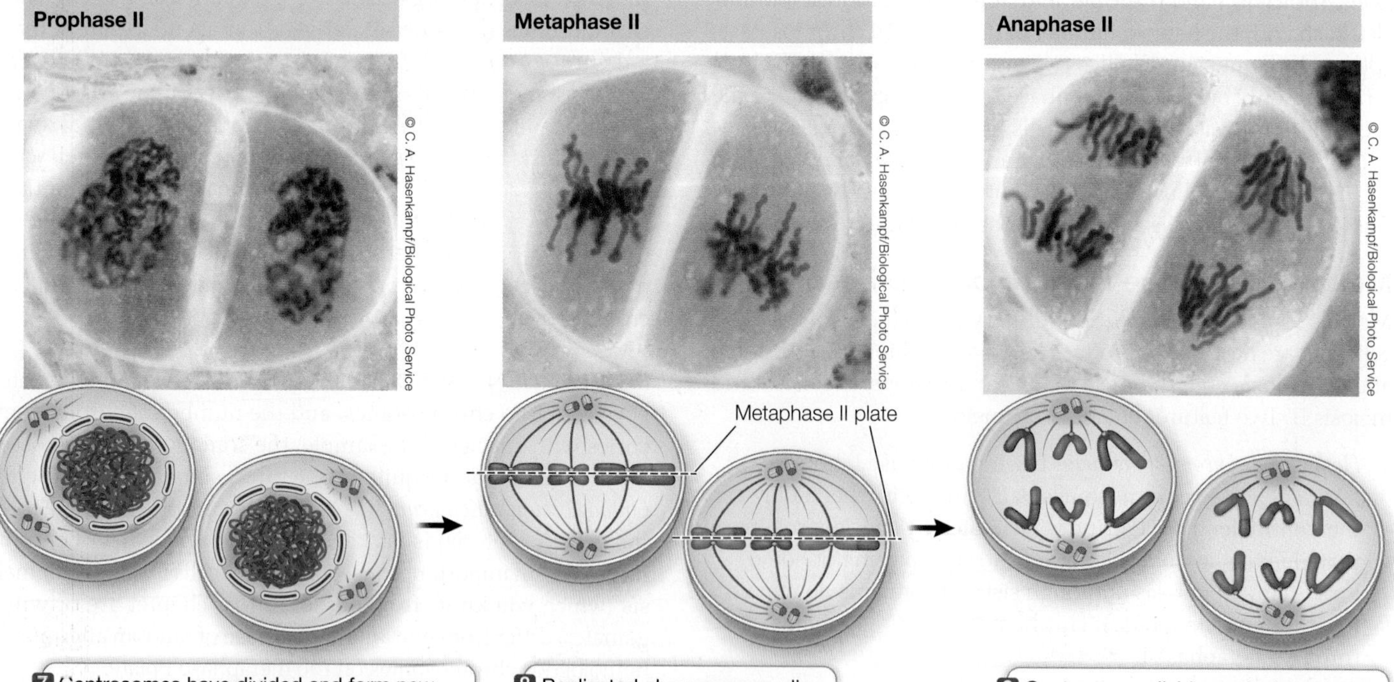

7 Centrosomes have divided and form new spindles (one in each cell). The nuclear envelope breaks down. Chromosomes condense again and microtubules attach at kinetochores on either side of centromere of each replicated chromosome.

8 Replicated chromosomes align on the metaphase II plate in the middle of the spindle.

9 Centromeres divide, and microtubules attached to kinetochores pull sister chromatids to opposite poles.

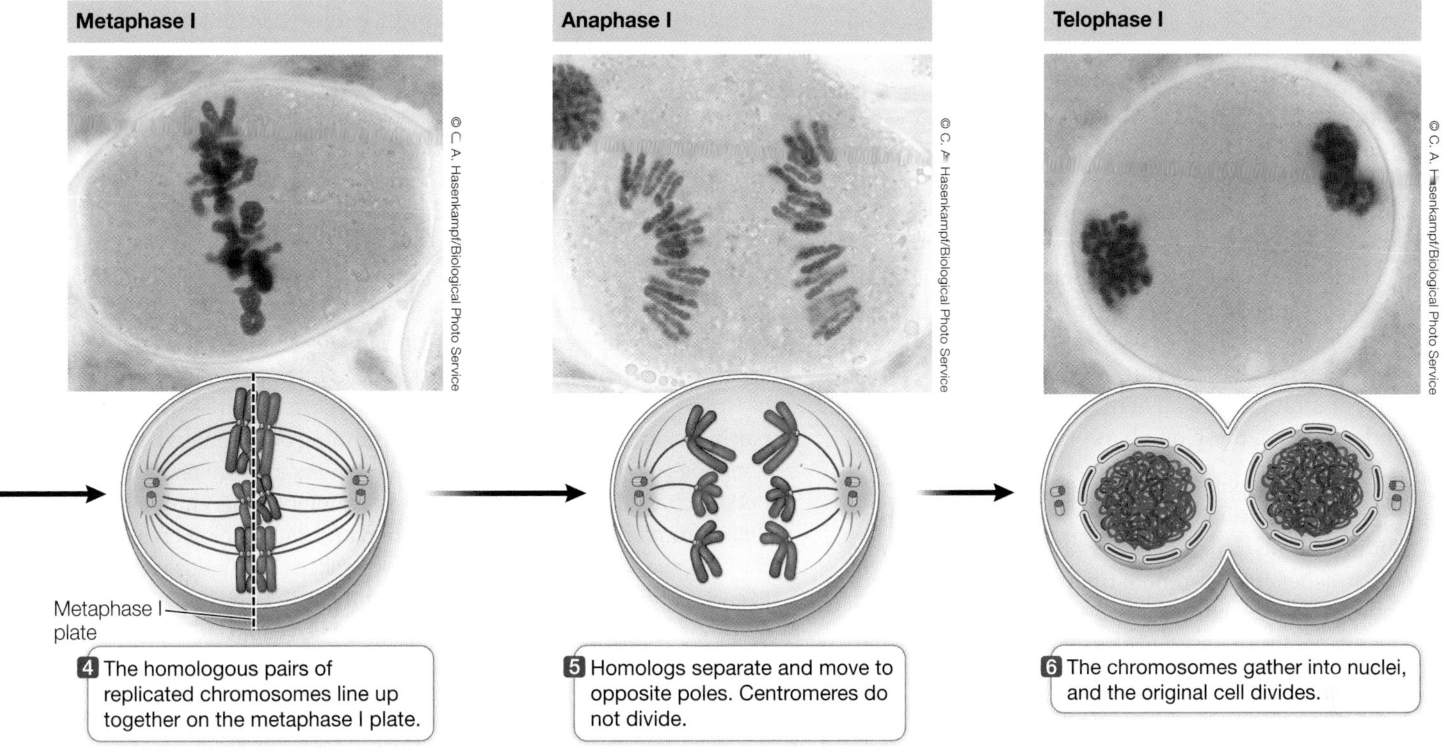

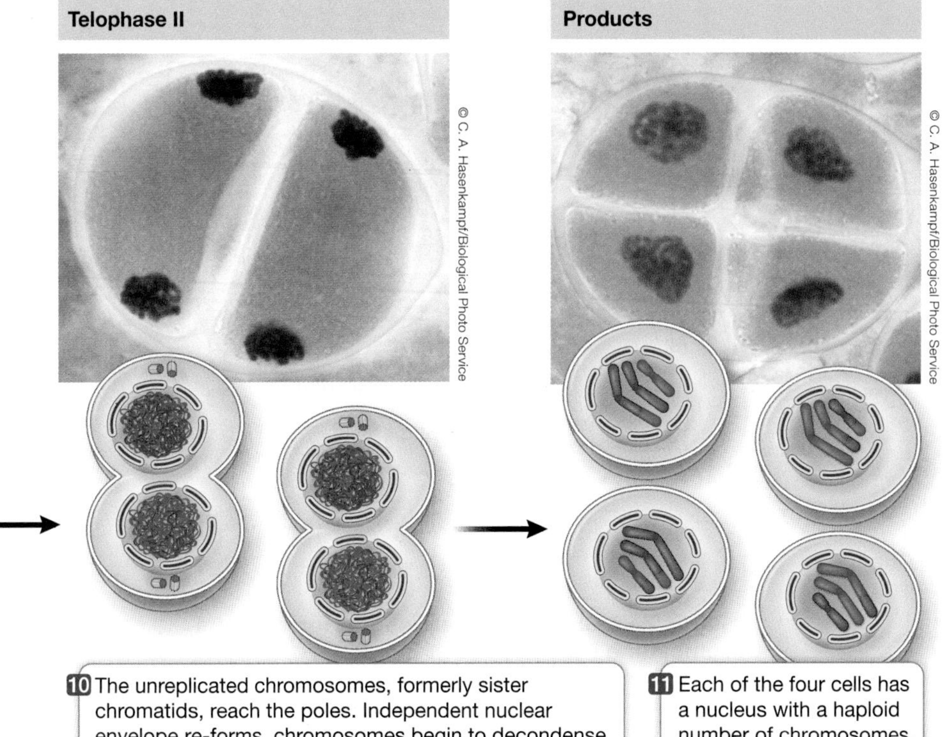

FIGURE 7.11 Meiosis: Generating Haploid Cells In meiosis I, the chromosome number is reduced from the diploid number to the haploid number. In meiosis II, chromosomes go from replicated to unreplicated, similar to what happens in mitosis. The end result of meiosis is the production of four daughter nuclei, each containing the haploid number of unreplicated chromosomes. The micrographs show meiosis during the production of male gametes in a lily; the diagrams illustrate corresponding phases to produce male gametes in an animal cell. (To help follow chromatids, chromosomes from one parent are colored blue and those from the other parent are red.)

Activity 7.4 Images of Meiosis

There is a 50 percent chance that the red member of a pair will be facing a particular pole. What proportion of the gametes produced will have only red or only blue chromosomes—that is, the same chromosomal makeup as the parental gametes?

The answer comes from the probability that all chromosomes align in the same orientation at metaphase I, which is $(1/2)^{(n-1)}$, where n is the number of chromosomes in a haploid set. Now, do the calculation for the two organisms mentioned above. For the mosquito, in which $n = 3$, the probability of getting gametes that contain the same chromosomes as a parent is $(1/2)^2 = 1/4$, or about 25 percent. For the fern, the number is much smaller: $(1/2)^{629} = 1/(2 \times 10^{189})$, a very small number indeed! Clearly, the more the chromosome pairs there are, the lower the probability that a product of meiosis will have all of the chromosomes that were present in one of the parental gametes. There are two key concepts essential to understanding these numbers:

1. When a homologous pair of chromosomes lines up on the metaphase plate during meiosis I and attaches to the spindle, it is random which member of the pair (paternal or maternal) attaches to the spindle fibers directed to a particular pole. When the chromosomes migrate to the poles during anaphase I, the paternally and maternally derived chromosomes move independently of one another (you can see this in Animation 7.2). This phenomenon is called **independent assortment** (**FIGURE 7.12**). (The consequences of independent assortment are described in more detail in Chapter 8.)
2. Because of genetic mutations, the DNA in the two homologous chromosomes is somewhat different for some genetic determinants. For example, earlier we mentioned two genetic types of pea seeds: round and wrinkled. In meiosis in a pea plant, one parent may have the genetic determinant for round seeds, and the other for wrinkled seeds. If there are many genes that have such diversity, the possibilities of the chromosomes carrying them occurring in new combinations in gametes due to independent assortment is very high.

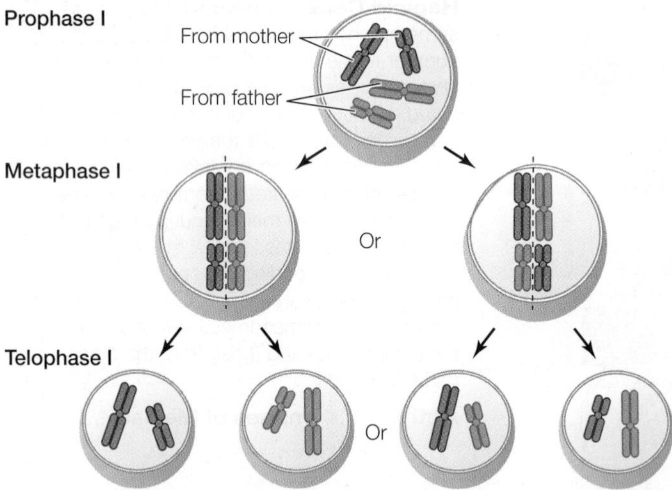

FIGURE 7.12 Independent Assortment: The Role of Chance in Genetic Diversity Chromosome pairs align independently at metaphase I with respect to the parent of origin of each chromosome, which is illustrated by color differences among homologs. To obtain gametes containing the same homologs as the parental gametes, the same-parent homologs must align on the same side of the metaphase I plate. The probability of this is $(1/2)^{(n-1)}$, indicating that the chance of getting parental-type gametes is very low, unless the chromosome number is small.

What is the probability of obtaining a parental-type gamete from independent assortment in a human (23 pairs of chromosomes)?

CROSSING OVER In addition to independent assortment, which generates diversity *among chromosomes* in haploid products, another process generates genetic diversity *within chromosomes* during meiosis. Meiosis I begins with a long prophase I (the first three panels of Figure 7.11), during which homologous chromosomes pair. This pairing lasts from prophase I to the beginning of anaphase I, when homologs move to opposite poles. While homologs are paired, genetic exchange occurs between nonsister homologous chromatids. One chromatid breaks at some location and its homolog breaks at the corresponding location, and the two broken chromatids reattach to each other. At a certain point during prophase I, the homologous chromosome pairs appear to repel each other, especially near the centromeres, but they remain attached. A major reason is the chromatid exchanges, which now appear as X-shaped attachment points called **chiasmata** (singular *chiasma*, "cross"). **FIGURE 7.13A** shows an electron micrograph of a homologous chromosome pair at prophase I that has two chiasmata.

A chiasma is a point where a chromatid breaks and rejoins to a nonsister homologous chromatid in a process called

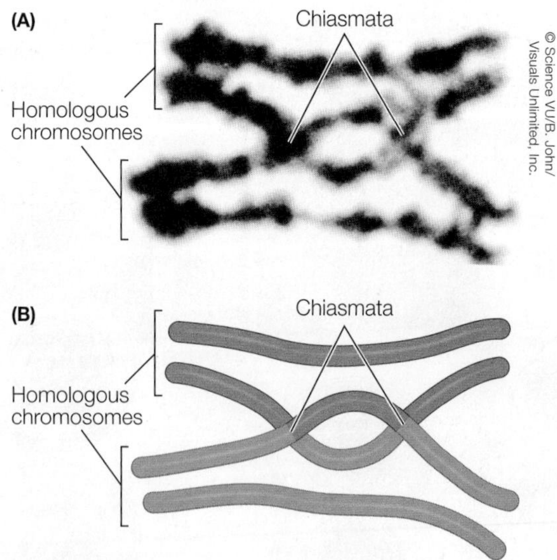

FIGURE 7.13 Chiasmata between a Homologous Chromosome Pair (A) The chiasmata or crossovers between these two chromosomes represent points where chromatids have broken and rejoined, resulting in exchange of genetic information. **(B)** Color changes show positions at which nonsister chromatids have broken and rejoined. In this example, two chromatids are involved in the two crossovers. The top red and bottom blue chromatids did not participate and are the same genetically as the homologs received from each parent of this individual, while the other two chromatids, with both red and blue sections, are not the same and are thus recombinants.

crossing over (**FIGURE 7.13B**). Crossing over involves an exchange of genetic material between the chromatids involved. For a particular chiasma, or crossover, either chromatid from one chromosome can participate with either chromatid from its homolog. In addition, a single chromatid can exchange material at more than one point along its length. Crossing over occurs shortly after pairing begins, but chiasmata do not become visible until later, when the homologs are repelling each other. Crossing over results in **recombinant chromatids**, which are genetically different from the chromosomes present in the haploid gametes received from each of the parents. Crossing over increases genetic variation among the products of meiosis by reshuffling genetic information between homologous chromosome pairs. Because crossovers can occur anywhere along the chromosome, each meiosis will have different numbers and locations of crossovers, indicating that *every haploid gamete produced by meiosis in a diploid organism may differ from every other.*

The combination of independent assortment and crossing over guarantees an enormous amount of diversity among gametes produced by a single individual. This generation of genetic diversity is one of the primary explanations for the prevalence of sexual life cycles across eukaryotic species.

Mitosis seldom takes more than an hour or two, but meiosis can take *much* longer. In human males, the cells in the testis that undergo meiosis take about a week for prophase I and about a month for the entire meiotic cycle. In females, prophase I begins in the developing egg cells (oocytes) in the ovary long before a woman's birth, during her early fetal development, but then pauses. Years later, meiosis resumes during the monthly ovarian cycle and is completed only after fertilization. Thus a single meiotic division may last over 40 years!

In summary, meiosis results in the production of four haploid daughter cells (gametes) that are genetically variable. Mitosis results in the production of two diploid daughter cells that are genetically identical. **FIGURE 7.14** compares the two processes.

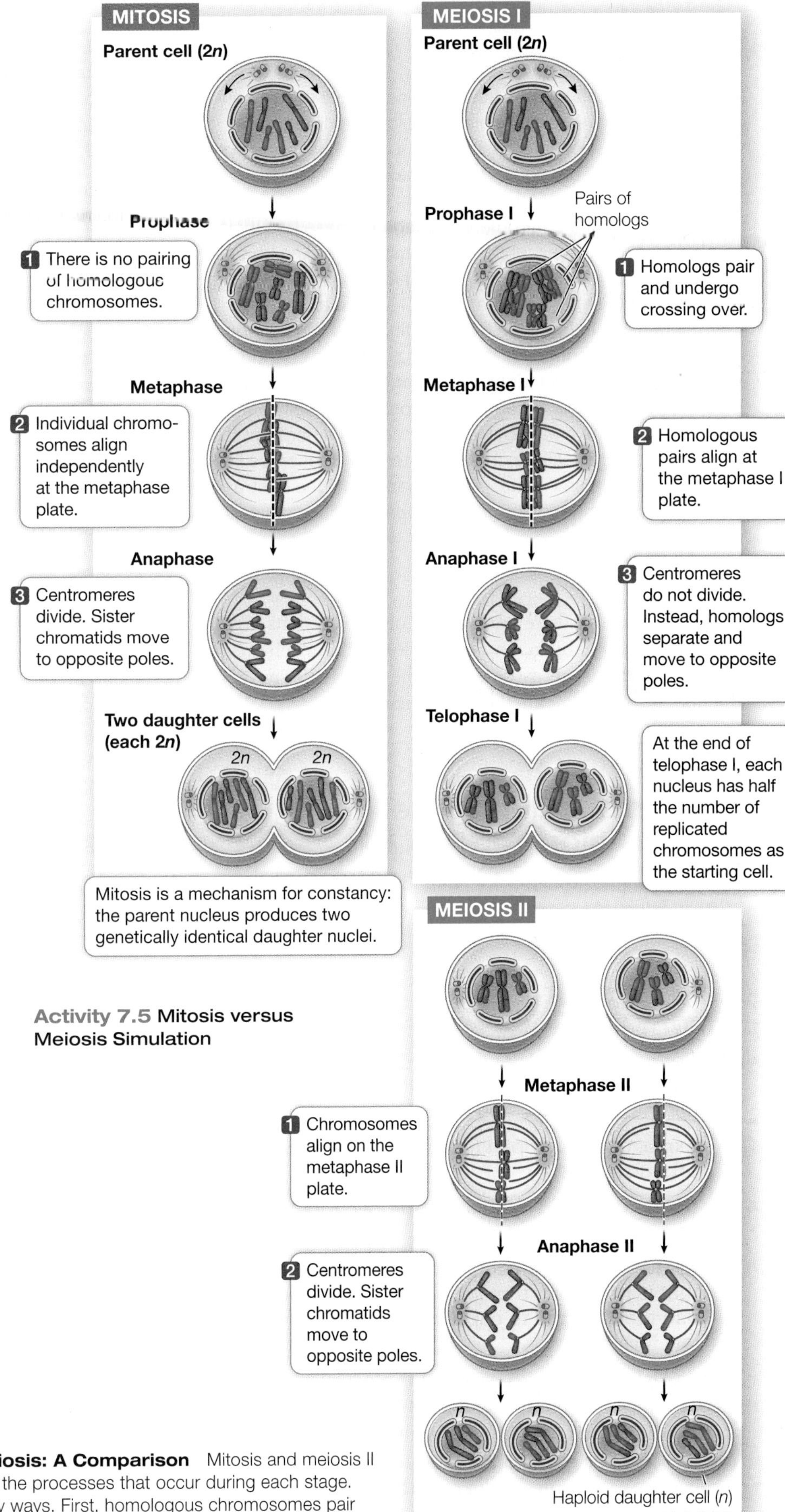

Activity 7.5 Mitosis versus Meiosis Simulation

FIGURE 7.14 Mitosis and Meiosis: A Comparison Mitosis and meiosis II are essentially the same in terms of the processes that occur during each stage. However, meiosis I differs in two key ways. First, homologous chromosomes pair during prophase I and align together on the metaphase I plate. Second, there is no centromere division during anaphase I; instead homologs separate and move to opposite poles, which results in halving the number of chromosomes per nucleus.

REVIEW & APPLY | 7.3

R One of the major characteristics of the sexual life cycle in eukaryotes is the presence of meiosis, which is a division that halves the number of chromosomes in a cell. Meiosis involves two sequential cell divisions. Prior to meiosis I, DNA is replicated, so each chromosome of a diploid pair has two chromatids attached at a centromere. During prophase I, homologous chromosomes pair, and in metaphase I they line up on the metaphase I plate of the spindle. Chromosomes, still with two chromatids, separate at anaphase I and go to the poles. After telophase I, each of the two resulting cells has half the number of chromosomes as the parental cell. The chromosome pairs go to poles independently of one another, and since the paternal and maternal chromosomes usually differ genetically, there is great possibility for cells to differ from the parental cells genetically. In meiosis II, the chromosomes line up on the metaphase II plate, and in anaphase II the centromeres duplicate and the chromatids separate. Thus meiosis produces four cells. In addition to independent assortment in anaphase I, parts of the paired homologous chromatids may exchange genetic material via crossing over during prophase I, creating even more genetic diversity.

A 1. State what is happening at each of the following stages of meiosis and how the meiotic stage is similar to, or different from, the mitotic stage listed: (a) prophase I versus prophase, (b) metaphase II versus metaphase, (c) anaphase I versus anaphase.

2. In Figure 7.14, four haploid gametes are produced by meiosis, each containing three non-homologous, unreplicated chromosomes. For each gamete, identify recombinant chromosomes versus parental chromosomes. To produce a gamete that is genetically identical to a parental gamete, what would have to occur during meiosis?

3. Why is meiosis I, rather than meiosis II, often called the reduction division of meiosis?

At any given moment, there are millions of cell divisions occurring in a complex organism. Our discussion of the mechanisms of mitosis and meiosis has assumed that these processes never go wrong. But they do. What are the consequences?

7.4 | **Errors during Cell Division** Can Result in Changes in Chromosome Number

LEARNING OBJECTIVES

By the end of this key concept you should be able to:

7.4.1 Define aneuploidy and draw the effects on daughter cells of nondisjunction occurring in mitosis, meiosis I, and meiosis II.

7.4.2 Explain why aneuploidy is rare at the whole-organism level.

7.4.3 Define polyploidy and state the two most common causes of it.

7.4.4 Draw what might occur at meiosis I in a species that is polyploid, triploid, or tetraploid, and use the diagram to explain why triploids have low fertility.

Mitosis in a diploid or haploid results in the production of daughter cells with the same number and identity of chromosomes in the parent cell. Meiosis in a diploid results in haploid cells with half the number but a genetically complete set of chromosomes, and fertilization regenerates diploid cells from haploid gametes. Errors in cell division or fertilization result in the production of cells carrying an incorrect number of chromosomes.

Errors in mitosis and meiosis can result in abnormal numbers of some chromosomes

One common way in which errors occur is by **nondisjunction**. This occurs in mitosis or meiosis when a pair of homologous chromosomes (in anaphase I of meiosis) or sister chromatids (in anaphase II of meiosis or anaphase of mitosis) fail to separate ("disjoin"). When nondisjunction occurs, the two resulting nuclei have incorrect numbers of chromosomes: one has an extra chromosome and one is missing a chromosome (**FIGURE 7.15**). For example, if two sister chromatids move to the same pole during mitosis in a diploid cell, the result is that one nucleus has three copies of that chromosome and the other nucleus has only one copy. Similarly, nondisjunction in meiosis results in gametes that contain the wrong number of copies of a chromosome. Note that only the chromosomes that undergo nondisjunction have incorrect copy numbers in the daughter cells; all other chromosomes have their usual numbers of copies. The condition of a cell, or organism, having an unequal number of copies of each chromosome is called **aneuploidy**. Aneuploid cells often have an incorrect copy number of only one chromosome. In a normal diploid cell, chromosomes occur in two copies and are said to be *disomic*. An aneuploid cell that should be diploid can be *nullisomic* (having no copies), *monosomic* (one copy), *trisomic* (three copies), *tetrasomic* (four copies), and so on, for one chromosome.

The effects of aneuploidy vary depending on how many and which cells of the organism are affected. Since every chromosome has DNA that carries genetic determinants for the functioning of the organism, a nullisomic cell, lacking a chromosome entirely, usually won't exist—it dies when it is formed. Other aneuploid cells are chromosomally imbalanced: they do not have the same number of copies of each chromosome as in a normal cell. This imbalance can cause significant problems. For example, aneuploidy is common in cancer cells and has been implicated in their abnormal functioning, but aneuploidy does not necessarily lead to cancer. Single-cell analyses of chromosome copy number indicate that a few percent of cells are aneuploid in otherwise normal mice and humans.

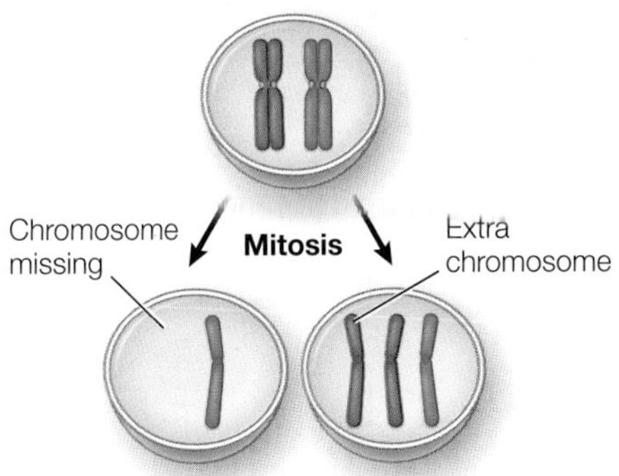

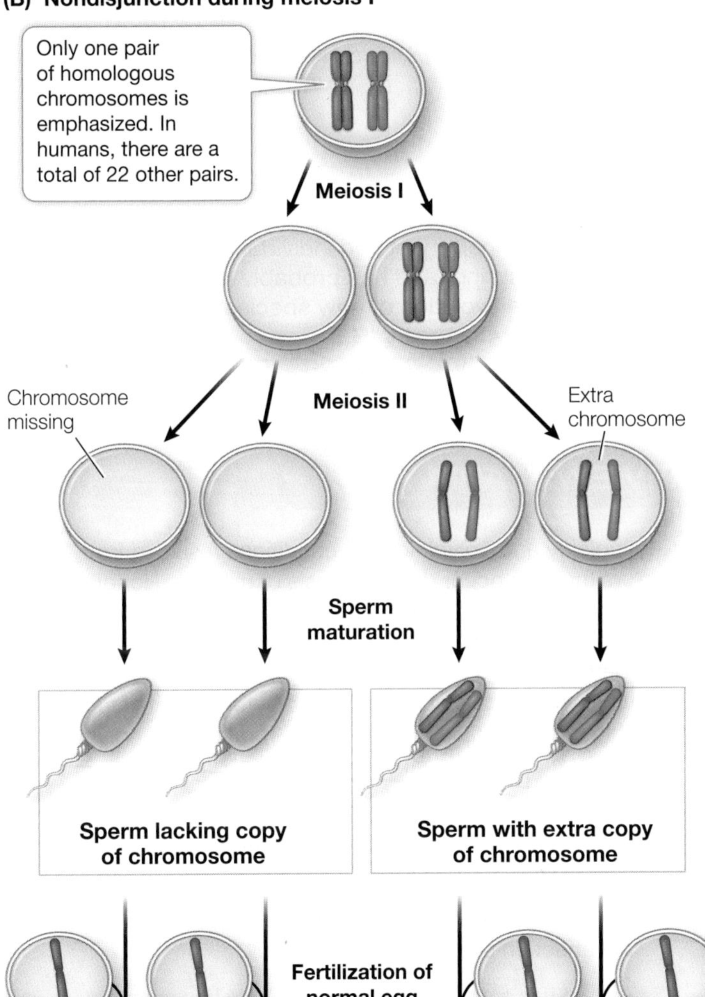

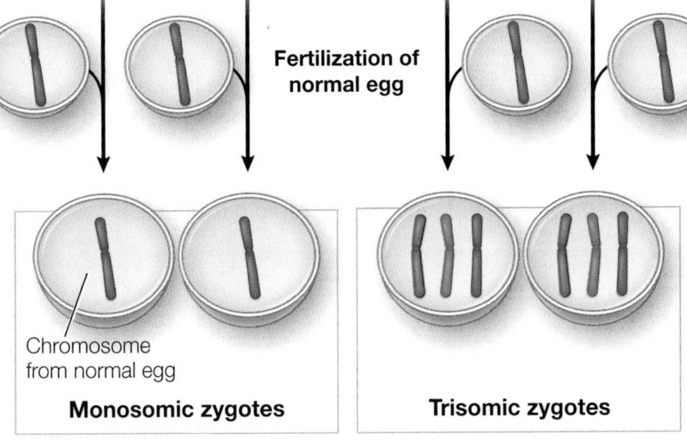

FIGURE 7.15 Aneuploidy Is Caused by Nondisjunction **(A)** In mitosis, nondisjunction results in both daughters being aneuploid: one cell has an extra copy of a chromosome and one cell lacks a copy. **(B)** Nondisjunction in meiosis I during sperm production results in four aneuploid gametes: two sperm have an extra copy of the chromosome and two sperm lack a copy. Fertilization of normal eggs produces aneuploid zygotes.

What is the effect of meiosis II nondisjunction on gametes?

Nondisjunction in meiosis results in aneuploid gametes (see Figure 7.15). If an aneuploid gamete is involved in fertilization, the resulting zygote (fertilized egg) will be aneuploid. Since the zygote gives rise to all the cells in the organism, the developing offspring will be aneuploid for every cell. Aneuploidy in all cells is usually harmful and thus is rare in newborn humans and other species. However, meiotic nondisjunction and the resulting aneuploidy are common in human gametes. About 10–30 percent of all human fertilized eggs show aneuploidy, but most of the embryos that develop from such zygotes do not survive. At least one-fifth of all recognized human pregnancies are spontaneously terminated (miscarried) during the first 2 months, largely because of trisomies and monosomies. Only a few aneuploidies are likely to survive to birth (in humans, trisomies for chromosomes 13, 16, 18, and 21), and only trisomy 21 individuals, who have Down syndrome (named for the British doctor who first described it in the nineteenth century), have high long-term survival. Trisomy 21 (**FIGURE 7.16A**) occurs in 1 in 1,000 births. In case you think this is a small number, consider that over 5 million people worldwide have Down syndrome. Individuals with Down syndrome have distinctive facial and growth characteristics and variable intellectual disabilities but can live long and productive lives. An exception to the harmfulness of aneuploidy is often observed for X and Y chromosomes, which occur in many species. As you will see in Chapter 8, the X and Y chromosomes are a homologous pair involved in sex determination, with males carrying an X and Y, and females carrying two Xs. Aneuploids that have only one X, or two Xs and a Y, or three Xs are all essentially normal—only by examining their chromosomes can their aneuploidy be detected.

Errors in cytokinesis or fertilization can affect all chromosomes

As described in Key Concept 7.1, most organisms are either diploid (for example, most animals) or haploid (for example, many fungi). However, organisms can have additional sets of chromosomes and be triploid ($3n$) (**FIGURE 7.16B**), tetraploid ($4n$), or show even higher-order **polyploidy**—the possessession of more than two entire sets of chromosomes. Polyploidy has various causes, including errors that occur during M phase. The two most common errors that occur during M phase that can lead to polyploid cells or organisms are:

1. Failure of the spindle to form, so that all the chromosomes remain in the same part of the cell after mitosis and thus are not separated by cytokinesis. If this occurs during mitosis, the result will be a cell that is tetraploid ($4n$).
2. Failure of cytokinesis to occur, so that the two daughter nuclei remain in the same cell. If this occurs during mitosis and the daughter nuclei fuse, the cell will again be tetraploid.

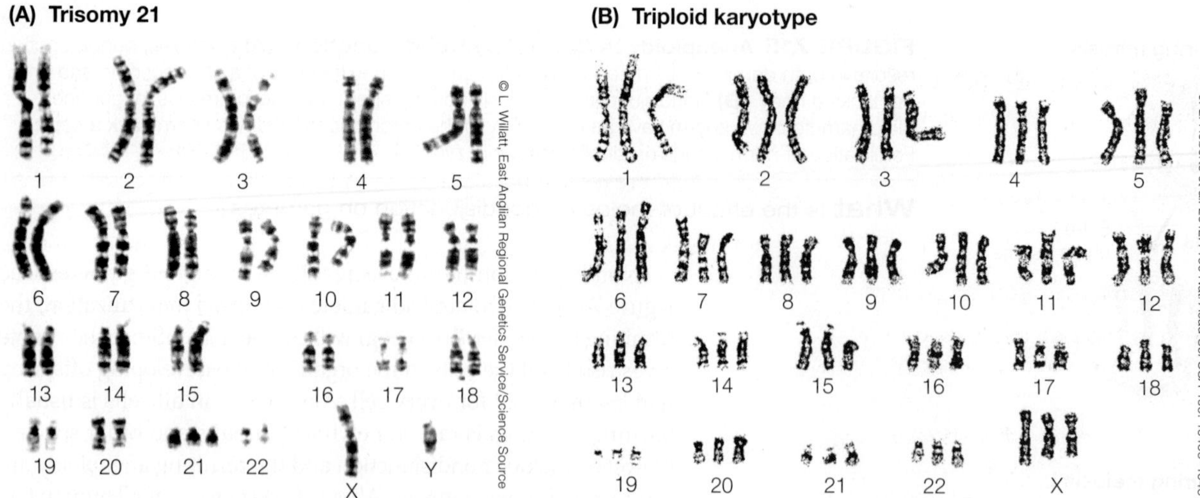

FIGURE 7.16 Abnormal Karyotypes in Humans **(A)** Trisomy 21 (here in a male) is a relatively common type of aneuploidy in humans and causes Down syndrome. **(B)** A triploid karyotype, caused by double fertilization, in a 7-week-old fetus.

These two mechanisms can also occur in meiosis, resulting in gametes that are diploid (instead of haploid). When a diploid gamete fuses with a haploid gamete, a triploid organism results. Another common mechanism that causes polyploidy, especially triploidy, is **polyspermy**, which occurs when more than one male gamete fertilizes a single female gamete.

Unlike aneuploid cells, polyploid cells are not imbalanced. All chromosomes have the same number of copies. In many organisms, especially plants and fungi, polyploidy has few, if any, harmful effects on the viability of a cell or organism. In fact, polyploidy is common in many plants and fungi. Most flowering plants and almost all ferns are polyploid. However, polyploidy is rarely beneficial in animals. Humans, for example, can produce triploid zygotes, but few survive to birth—about 5 percent of all miscarriages are triploids or higher-order polyploids. While polyploid species may have high viability, they may experience large reductions in fertility, which suggests problems in meiosis. In a normal diploid cell, with two sets of chromosomes, homologs pair at prophase I and separate during anaphase I. In a tetraploid or hexaploid ($6n$) cell, there is an even number of homologs for each chromosome, so every chromosome can pair with a homolog at prophase I. However, a triploid nucleus cannot undergo normal meiosis because one-third of the chromosomes lack partners. These chromosomes may go to either pole at anaphase I, or even be left on the metaphase I plate, giving rise to significantly unbalanced gametes (aneuploid for many chromosomes). These gametes may not be functional, and even if they are, they will produce unbalanced offspring, which are likely to have severely reduced viability.

Polyploidy has implications for agriculture, particularly in the production of hybrid plants. For example, the number of chromosomes must be taken into account in wheat breeding because there are diploid, tetraploid, and hexaploid wheat varieties. Polyploidy can be a desirable trait in crops and ornamental plants because it often leads to more robust plants with larger flowers, fruits, and seeds. In addition, triploid fruit varieties are desirable because they are infertile and therefore seedless.

LINK Polyploidy can lead to reproductive isolation (the inability of two individuals to produce fertile offspring) and has probably led to speciation (the evolution of new species), as described in **Key Concept 16.3**.

REVIEW & APPLY | 7.4

R Aneuploidy and polyploidy are conditions resulting from errors that can occur during mitosis or meiosis. Nondisjunction is a type of error in which a pair of homologous chromosomes or a pair of sister chromatids fail to separate properly, resulting in aneuploid nuclei and cells that have too many or too few chromosomes. Cytokinesis failure and double fertilization can result in the production of polyploid cells that have additional sets of chromosomes.

A
1. Compare the effects of nondisjunction in meiosis I versus meiosis II on the gametes produced.
2. A drug that prevents spindle formation is added to a zygote for a single round of the cell cycle. What will be the effect on chromosome number?
3. A plant has a diploid number of 6. State how many chromosomes are present in the following cells and which cells are imbalanced with respect to chromosome numbers: (a) a gamete, (b) a gamete with chromosome 2 disomy, (c) a leaf cell, (d) a tetraploid root cell.

Now that we have learned some of the details of cell division, we'll turn our attention to the control of the cell cycle.

7.5 The Cell Cycle and Cell Death Are Highly Regulated in Eukaryotes

LEARNING OBJECTIVES

By the end of this key concept you should be able to:

7.5.1 **Identify where in the cell cycle primary control of cell division occurs in prokaryotes and eukaryotes.**

7.5.2 **Explain how cyclins and cyclin-dependent kinases control the cell cycle.**

7.5.3 **Describe the function of checkpoints in eukaryotes and state where they occur in the cell cycle.**

7.5.4 **Differentiate between necrosis and apoptosis.**

7.5.5 **Describe how misregulation of cell division and cell death are defining features of cancer.**

In Key Concept 7.2 we noted that a cell division signal is the first requirement for cell division. In prokaryotes and single-celled eukaryotes, this signal is often tied to the environmental conditions: cell division will occur if there are sufficient resources for division. We also noted that since internal conditions in a multicellular organism are often suitable for cell division, signals for cell division in these organisms are not tied to the environmental conditions. Here we describe the signals that prompt cells of prokaryotic and eukaryotic organisms to initiate their cell cycles, and how progression through those cycles is tightly regulated.

Progression through binary fission is regulated in prokaryotes

In prokaryotes, when sufficient resources have been accumulated, cells begin their cell cycle by initiating DNA replication at the *ori* site on their chromosome. Initiation of DNA replication is the main control point determining whether the cell cycle will begin. A DNA-binding protein termed DnaA, when bound to ATP, binds to *ori* and initiates DNA replication. Initiation does not begin until sufficient levels of DnaA are in the cell and bound to *ori*. Immediately following initiation, DnaA levels are kept low for a large portion of the cell cycle to prevent reinitiation of DNA replication. Completion of cell division requires formation of the Z-ring of protein fibers that constrict the cell during cytokinesis, which is stimulated by the termination of DNA replication. Thus in a dividing bacterial cell, the initiation of cell division is caused by the initiation of DNA replication; termination of DNA replication occurs upon the replication of the *ter* site, and this triggers cytokinesis. In addition to these two steps, there are points where the cell cycle can arrest or pause based on internal and external signals. Specifically, sudden starvation will arrest DNA replication, even if it has already started, and DNA damage, which is an abnormal chemical change to a DNA molecule, such as loss of the base from a nucleotide, will prevent the formation of the Z-ring and cytokinesis (**FIGURE 7.17**).

LINK Changes to DNA that occur as part of normal DNA function, such as DNA methylation, are not classified as damage, and their function will be covered in **Chapter 11**.

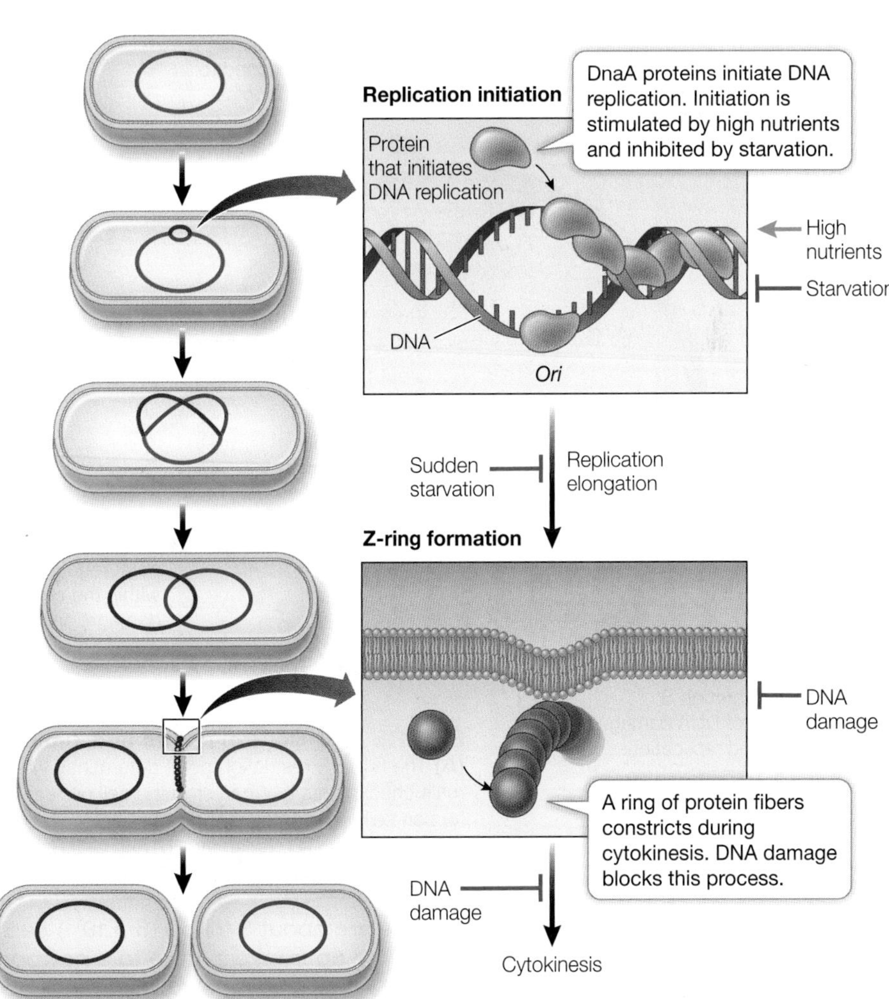

FIGURE 7.17 Bacterial Cell Cycle Regulation Binary fission is regulated at two primary points in the bacterial cell cycle: one regulating the initiation of DNA replication and one regulating Z-ring formation and cytokinesis after replication of the *ter* site, though DNA replication can stop if a cell is suddenly starved. Green arrows indicate stimulation, and red blunt end lines indicate inhibition of the particular protein. (After K. Jonas. 2014. *Curr Opin Microbiol* 18: 54–60.)

Cell cycle regulation differs in eukaryotes and prokaryotes

In eukaryotes, the cell cycle will not commence unless the entire cycle can be completed (see Figure 7.5). G1 can thus be thought of as the phase in which a cell is preparing enough resources to make it through the entire cell cycle. Whereas cell division in unicellular eukaryotes is often initiated by environmental signals (such as nutrient levels, as in prokaryotes), different types of cells in an adult multicellular organism divide at different times in the organism's life cycle and in different conditions. Careful control of cell division, even in high-nutrient environments, makes sense because continuous cell reproduction would make it difficult to maintain the forms and functions of different parts of the body.

The signals that stimulate cell division in multicellular eukaryotes are related to the needs of the entire organism. Mammals produce a variety of substances called **growth factors** that stimulate cell division and differentiation. For example, if you cut yourself and bleed, a blood clot eventually forms. Cell fragments called platelets in the blood vessels surrounding the clot secrete various growth factors that stimulate nearby cells to divide and heal the wound. Once a cell has received one or more signals to divide, progression through the cell cycle depends on a variety of internal signals.

LINK **Figure 6.10** shows a mitogen-activated protein kinase cascade that stimulates cell division.

The eukaryotic cell cycle is regulated internally

Key Concept 7.2 described the four phases of the eukaryotic cell cycle : G1, S, G2, and M. Progression through these phases is tightly regulated. For example, the G1 to S transition marks a key decision point for the cell: passing this point (called **R**, the **restriction point**) usually means the cell will proceed with the rest of the cell cycle and divide (**FIGURE 7.18**). The cell cycle responds to external signals to progress through the restriction point. Once past that point, internal signals take over for progression through the rest of the cell cycle.

What events cause a cell to enter S phase, passing the restriction point, or transition from one phase to the next? An indication that certain substances control these transitions came from cell fusion experiments. For example, an experiment involving the fusion of mammalian cells at G1 and S phase showed that a cell in S phase produces a substance that activates DNA replication (**INVESTIGATION FIGURE 7.19**). Similar experiments pointed to a cytoplasmic molecular activator for entry into M phase.

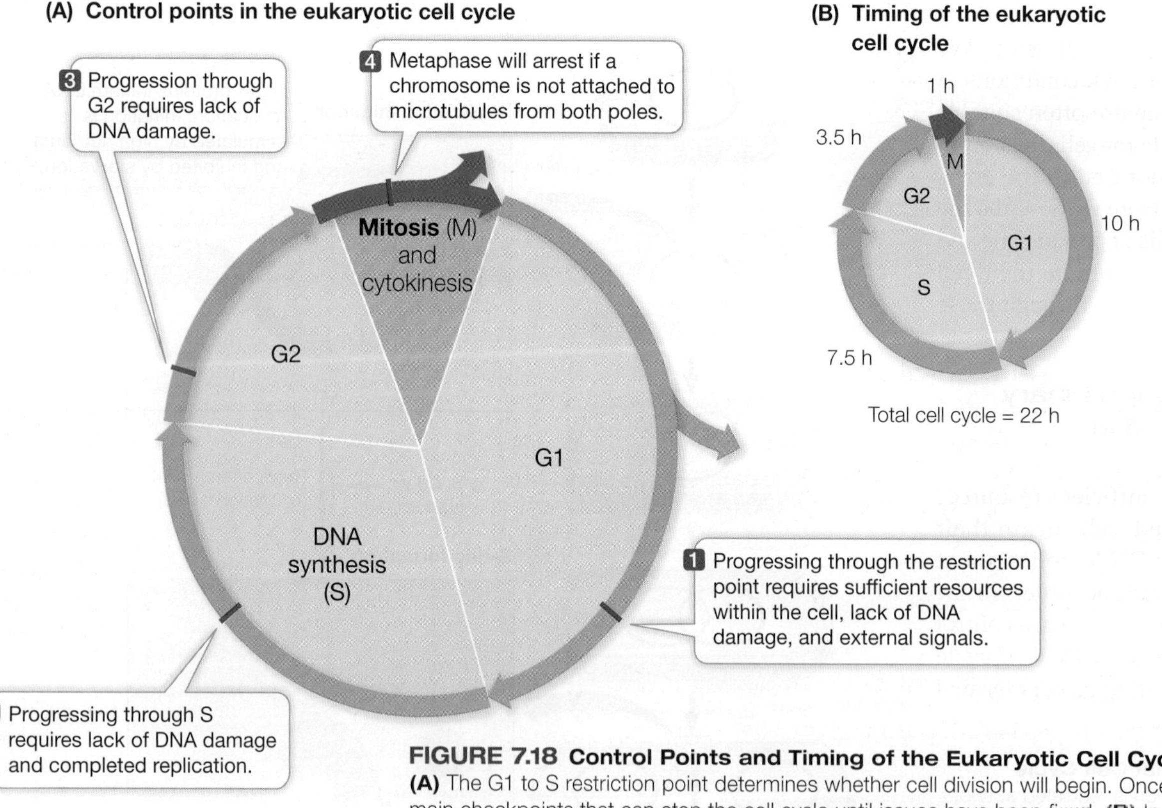

FIGURE 7.18 Control Points and Timing of the Eukaryotic Cell Cycle **(A)** The G1 to S restriction point determines whether cell division will begin. Once begun, there are three main checkpoints that can stop the cell cycle until issues have been fixed. **(B)** In the laboratory, typical human cells will complete the cell cycle in about 22 hours, with G1 and S being the longest phases. (After J. L. Corden and D. Tollervey. 2017. In *Cell Biology* 3rd ed., T. D. Pollard et al. [eds.], pp. 697–711. Elsevier: Philadelphia. https://doi.org/10.1016/C2014-0-00272-9)

If there are about 5 trillion (5×10^{12}) cells in a newborn baby, what is the fastest that one could go from a zygote to a newborn baby if cell divisions occurred back-to-back every 22 hours and there was no cell death?

INVESTIGATION

FIGURE 7.19 Regulation of the Cell Cycle Nuclei in G1 do not undergo DNA replication, but nuclei in S phase do. To determine if there is some signal in the S cells that stimulates G1 cells to replicate their DNA, cells in G1 and S phases were induced to fuse, creating cells with both G1 and S properties.[a]

HYPOTHESIS

A cell in S phase contains an activator of DNA replication.

METHOD

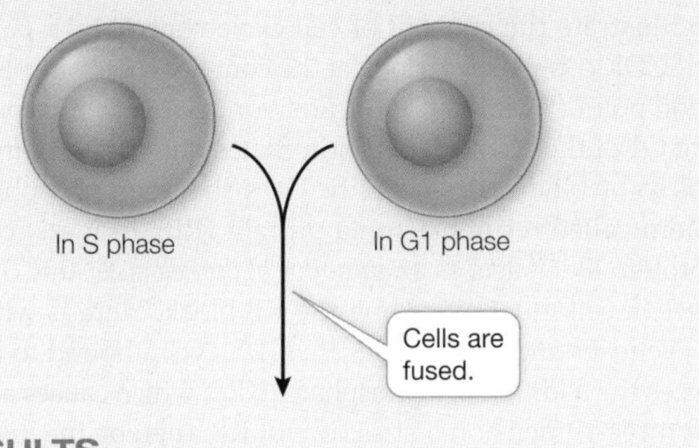

RESULTS

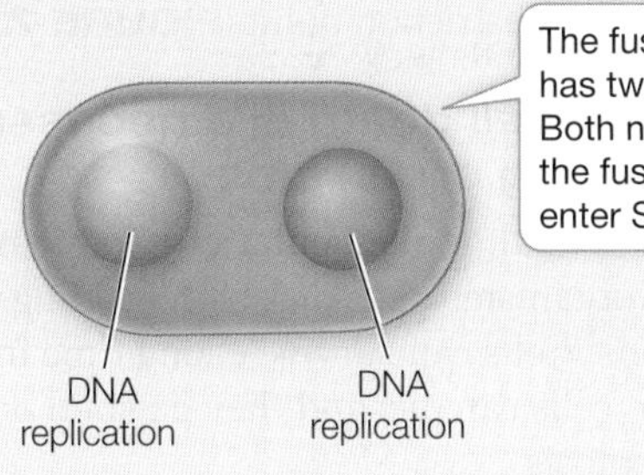

CONCLUSION

The S phase cell contains a substance that diffuses to the G1 nucleus and activates DNA replication.

ANALYZE THE DATA

The researchers used mammalian cells undergoing the cell cycle synchronously. Radioactive labeling and microscopy were used to determine which nuclei were synthesizing DNA; only nuclei that were synthesizing DNA became labeled:

Type of cells	Cells with labeled nuclei/total cells
Unfused G1	6/300
Unfused S	435/500
Fused G1 and S cells	17*/19

*Both nuclei labeled

1. What were the percentages of cells in S phase in each of the three experiments?
2. What does this mean in terms of control of the cell cycle?

[a]P. N. Rao and R. T. Johnson. 1970. *Nature* 225: 159–164.

A similar **Analyze the Data** exercise may be assigned in **Achieve**.

The eukaryotic cell cycle is controlled by cyclin-dependent kinases

The molecular activators revealed by the cell fusion experiments are protein kinases, which is perhaps not surprising given that kinases are common in cell signaling pathways (see Key Concept 6.3). The kinases involved in cell cycle regulation are called **cyclin-dependent kinases** (**CDKs**). They catalyze the phosphorylation of target proteins that regulate the cell cycle:

$$\text{Cell cycle regulator} + \text{ATP} \xrightarrow{\text{CDK}} \text{cell cycle regulator-P} + \text{ADP}$$

As their name implies, CDKs are activated by binding to **cyclin**, a protein subunit of the active kinase. This binding changes the shape of a CDK such that its active site is exposed, and is an example of allosteric regulation (**FIGURE 7.20**).

As **Key Concept 3.5** explains, allosteric regulation occurs when regulatory molecules bind to an enzyme at somewhere other than the active site to cause a change in conformation of the enzyme.

Several different CDKs function at specific phases of the cell cycle, called **cell cycle checkpoints**. At these points, progress of the cell cycle is regulated by the activity of specific CDKs. CDKs phosphorylate several target proteins, activating some and inactivating others, leading to appropriate events in each phase of the cell cycle. As in prokaryotic cell cycle control, various factors affect whether the eukaryotic cell cycle will proceed through a checkpoint. There are three checkpoints during interphase and one during mitosis that, if triggered, will arrest the cell cycle:

1. The G1, or G1 to S, checkpoint (restriction point), which is triggered by DNA damage
2. The S checkpoint, which is triggered by incomplete replication or DNA damage
3. The G2 checkpoint, which is triggered by DNA damage
4. The metaphase checkpoint, which is triggered by a chromosome that fails to attach to the spindle (see Figure 7.18)

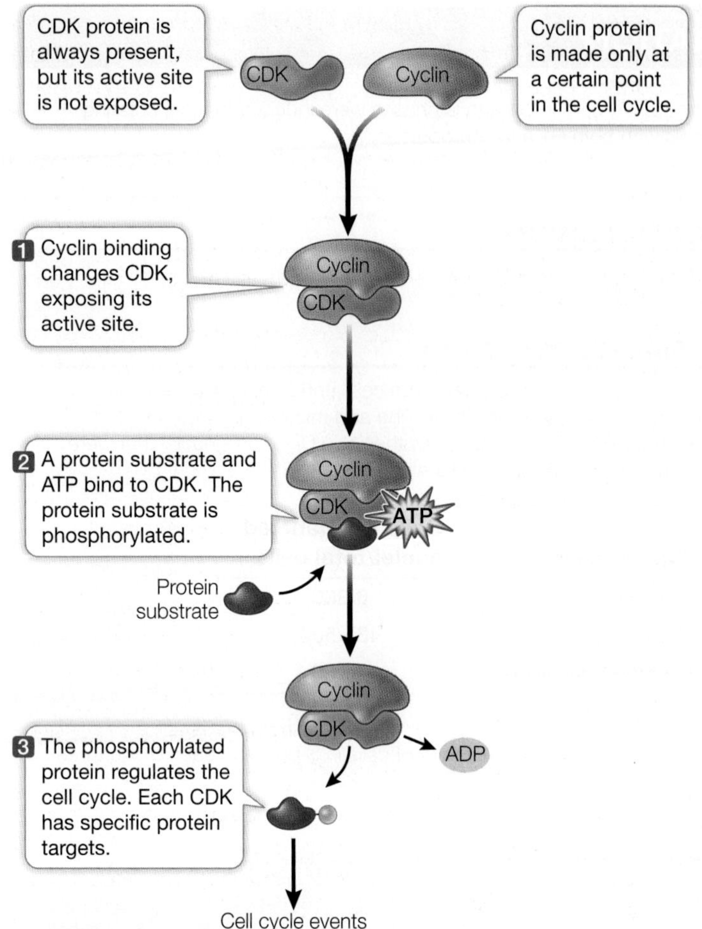

FIGURE 7.20 Binding of Cyclin to CDK Binding of a cyclin to its CDK activates the CDK enzyme activity, allowing it to phosphorylate target proteins.

Each CDK has a cyclin to activate it, and the cyclin is made only at the right time. After the CDK acts, the cyclin is broken down by a protease, an enzyme that hydrolyzes proteins (**FIGURE 7.21**). So a key event controlling the transition from one cell cycle phase to the next is the synthesis and subsequent breakdown of a particular cyclin. Cyclins are synthesized in response to various molecular signals, including growth factors. This starts a chain reaction:

Growth factor ⟶ cyclin synthesis ⟶ CDK activation ⟶ cell cycle phase-specific events

To illustrate the concept of cell cycle control by a particular cyclin–CDK complex, let's look at the complex that controls the restriction point (R) at the G1–S transition. G1/S cyclin is produced during G1 of the cell cycle and stimulates progression through R. It does so by binding to a specific CDK that then phosphorylates a protein called retinoblastoma protein (RB). In many cells, RB or a protein like it acts as an *inhibitor of the cell cycle* at R. To begin S phase, a cell must overcome the RB block. Here is where the G1/S cyclin-bound CDK comes in: G1/S cyclin bound to its CDK catalyzes the addition of a phosphate to RB, which causes a change in the three-dimensional structure of RB, thereby inactivating it (allosteric inactivation). With RB out of the way, the cell cycle can proceed though the restriction point (**FIGURE 7.22**).

Programmed cell death is a necessary process in living organisms

Many cells in an adult multicellular eukaryote never divide. In addition, in all organisms, cells may stop dividing under inappropriate conditions. Cells can survive a long time in a nondividing state. However, cells can die, and they do so in one of two ways, necrosis or apoptosis.

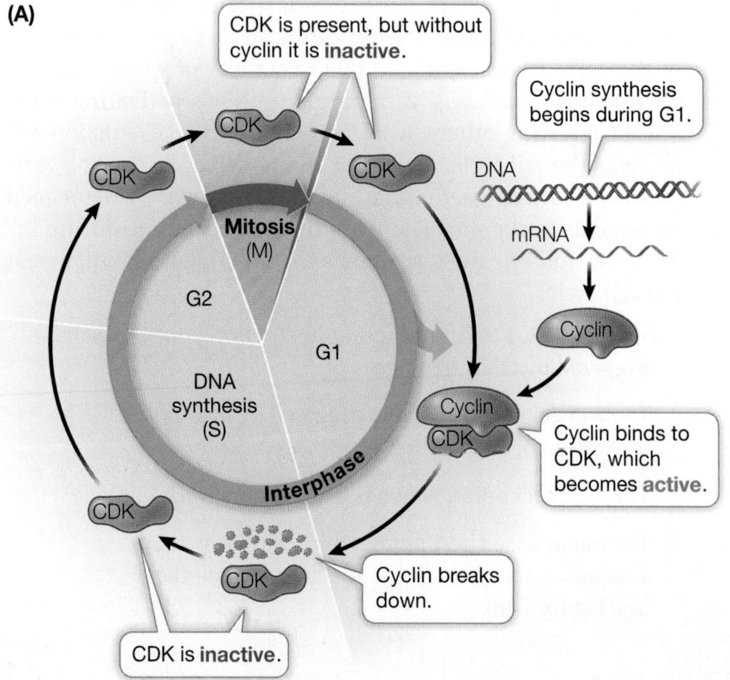

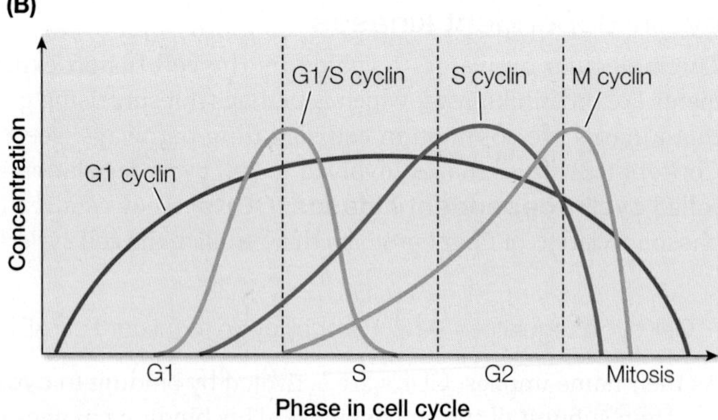

FIGURE 7.21 Cyclin Concentrations Change during the Cell Cycle **(A)** A particular cyclin is synthesized at a particular point during the cell cycle and, after activating a specific CDK, is degraded. **(B)** The concentrations of different cyclin signaling proteins determine CDK activity and cell cycle progression. (B after WikiMiMa/Michael Matthes.)

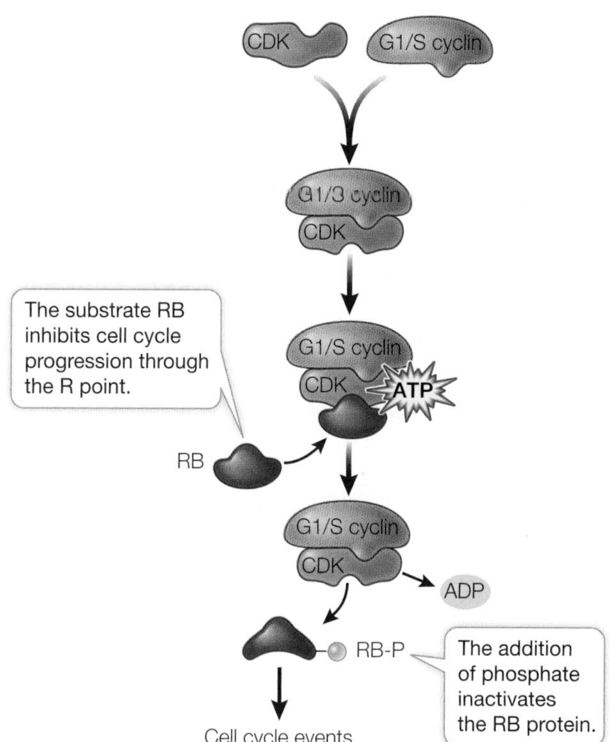

FIGURE 7.22 G1/S Cyclin Control of Progression through the Restriction Point G1/S cyclin is synthesized during G1 and binds to its target CDK, which then phosphorylates RB, inactivating it. RB is a suppressor of the cell cycle, preventing progression through the restriction point, so when RB is inactivated, progression through the restriction point is more likely to occur.

Necrosis occurs in multicellular organisms when a cell is damaged or is starved of nutrients, oxygen, or water. Necrotic cells often swell up and burst, releasing their contents into the extracellular environment. This process often results in inflammation, which also affects surrounding tissues. When a multicellular individual dies, all of its cells will soon succumb to this kind of death.

Apoptosis (Greek, "falling apart") is a *genetically programmed* series of events that result in cell death. Why would a cell initiate apoptosis, which is essentially cell suicide? In multicellular organisms there are at least four possible reasons:

1. The cell is no longer functioning appropriately. In many cases, cells are replaced as they age or if they are damaged. The longer a cell has been around, the more damage it has likely sustained. Cells are able to monitor their own health and initiate apoptosis when they show symptoms of reduced functionality.
2. The cell is no longer needed by the organism. In some cases, cells are present at certain times during development and then later are no longer needed. One example occurs before birth in humans. A human fetus has weblike hands, with connective tissue between the fingers. As development proceeds, this unneeded tissue disappears as the cells between the digits undergo apoptosis in response to specific signals, producing our familiar unwebbed hand.
3. Cells have reached the limit of how many times they can divide in laboratory cell cultures, a phenomenon known as the **Hayflick limit**. A major reason involves chromosome shortening at the ends during DNA replication (see Key Concept 9.2). In many organisms, populations of cells called stem cells divide without limit, as do cancer cells.
4. The cell is responding to an infection. Death of the cell prevents the infection from spreading throughout the rest of the organism. This is an especially important mechanism in plants as you will see in Chapter 27.

The events of apoptosis are similar in the cells of many organisms. The cell becomes detached from its neighbors, hydrolyzes its DNA into small fragments, and forms membranous lobes, or "blebs," that break up into cell fragments (**FIGURE 7.23**). In a remarkable example of the economy of nature, some surrounding living cells usually ingest the remains of the dead cell by phagocytosis. The remains are digested in the lysosomes, and the digestion products are recycled.

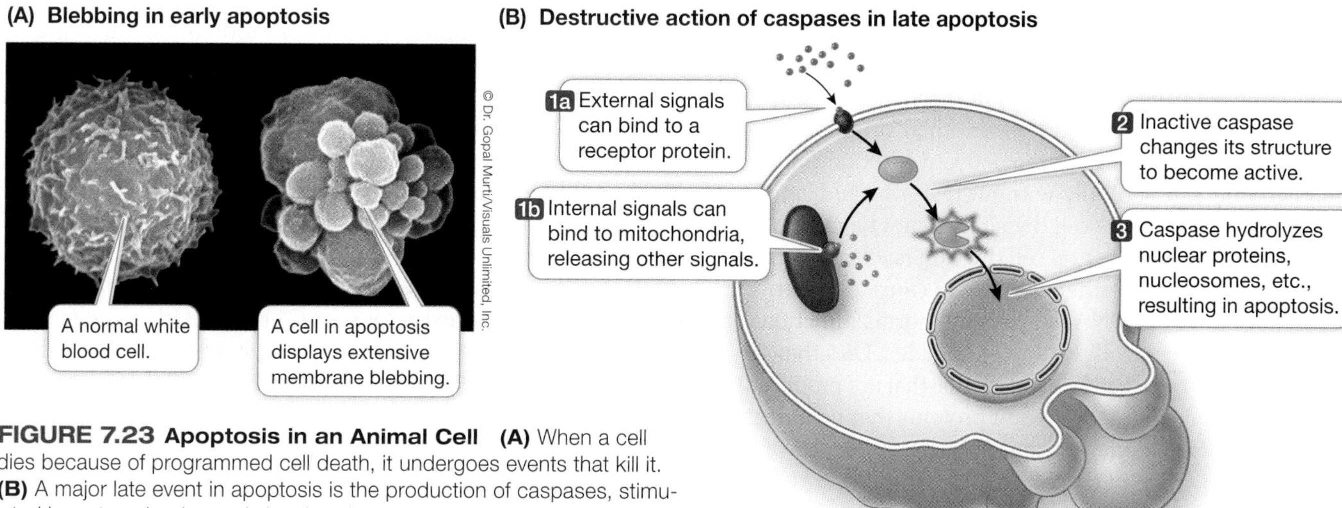

FIGURE 7.23 Apoptosis in an Animal Cell **(A)** When a cell dies because of programmed cell death, it undergoes events that kill it. **(B)** A major late event in apoptosis is the production of caspases, stimulated by external or internal signals, which hydrolyze various proteins.

THINK LIKE A SCIENTIST

Treating cancer

Process of Science

Science & Society

The goal of cancer treatment with drugs is remission—the elimination of nonstop cell division and an increase in apoptosis to shrink the tumor. One way to do this is to utilize processes of science to develop drugs that target working parts of cancer cells. Say, for example, you were able to isolate or make a drug that targets the microtubules of the spindle. Here you will explore the link between science and society as you use your knowledge of the cell cycle to answer questions about the effects of such a drug on cancer cells.

1. Would a dividing cell be able to continue through M phase if treated with a drug that interferes with the ability of the microtubules of the spindle to separate the chromosomes in anaphase when a cell divides? If not, what would the likely outcome be? What effect would such a drug treatment have on tumor cells in patients?

Paclitaxel (one version is called Taxol®) is an anticancer drug that works by binding to tubulin in microtubules, where it prevents the shortening necessary for chromosome attachment and movement (see Figure 7.24). This can result in a block at the mitotic cell cycle checkpoint and failure of cell division, including in cancer cells. Recent research has revealed that paclitaxel concentrations inside tumors of treated patients are too low to actually cause cell cycle arrest at M phase. Instead paclitaxel causes cells to produce tripolar or tetrapolar spindles (rather than the normal two poles).

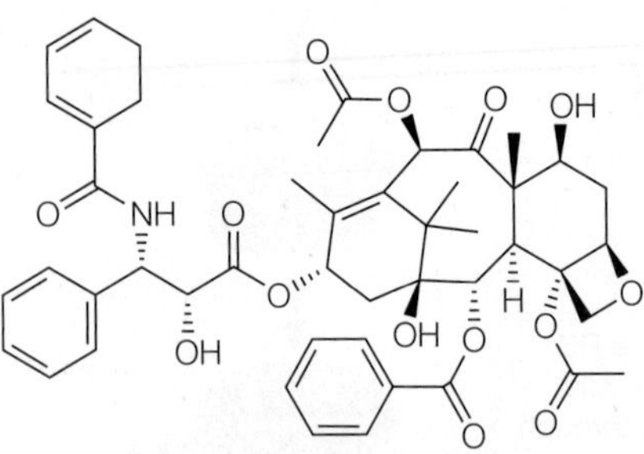

FIGURE 7.24 Structure of Paclitaxel

Microtubules from each pole are able to grab kinetochores and drag chromosomes during anaphase. At telophase, three or four nuclei form, and then cytokinesis may produce more than two cells. Even when two cells are produced, which chromosomes end up in each cell is random.

2. What effects might tripolar or tetrapolar spindles have on the resulting nuclei? Will the cells resulting from multipolar mitosis have any chromosomal imbalances? If so, do you expect cell survival to be reduced?
3. Regardless of the mechanisms of its action, paclitaxel is clearly bad news for cancer cells. But it also can be harmful to some normal cells. How is it possible that it can be so toxic to cancer cells, yet a patient survives?

In plant cells, because of their rigid cell walls, blebs do not form in the way they do in animal cells. Instead, plant cells digest their own cell contents, usually in the vacuole, and then release the digested components into the vascular (circulatory) system.

Cell division regulation is abnormal in cancer cells

Cancer cells have many characteristics not seen in most normal cells, including the ability to migrate from their original location and grow in other locations of the body. In addition, cancer cells often can divide essentially without restraint, indicating that external signals to initiate cell division are not needed. Nor do cancer cells heed signals to stop cell division (such as contact with other cells). These and other changes in cancer cells arise from mutations in the cells' DNA that encodes normal functions, including cell division controls. Examples include CDKs that do not need cyclins to be in an active state, cyclins that are produced without cell division signals, checkpoint proteins that no longer function to arrest the cell cycle, and apoptosis machinery that no longer functions. Any one of these changes will cause a cell to be more likely to undergo cell division. Put several such changes together and unrestricted cell division is guaranteed.

To understand some of the types of mutations seen in cancers, let's take a closer look at the regulation of RB activity by G1 cyclin and its CDK (**FIGURE 7.25**). G1 cyclin is the first cyclin to be produced in response to cell division signals (see Figure 7.21). G1 cyclin rises in concentration in G1 and remains at high concentration until the end of M phase. As with G1/S cyclin's CDK, one of the phosphorylation targets of G1 cyclin's active CDK is the RB protein, which it inactivates. Thus G1 cyclin helps a cell enter S phase by removing the restriction point block caused by the active RB protein. The expression of G1 cyclin is increased by the RAS protein. When growth factor ligands bind to their cell receptors, activation of RAS is a common effect (see Figure 6.10). Activated RAS then binds to effectors, some of which cause expression of G1 cyclin. Other cell factors, such as a protein called p16, inhibit the action of G1 cyclin's CDK, so that it is unable to phosphorylate RB. All of these proteins—RAS, p16, G1 cyclin and its CDK, and RB—have been found to be mutated in various cancers. However, the nature of the effects of mutations in these genes in cancer cells varies: p16

FIGURE 7.25 Regulation of RB Activity The RB protein inhibits progression through the restriction point. The RB protein is regulated by the p16 protein, which keeps it active, and the G1 cyclin CDK, which inactivates it by phosphorylation. The synthesis of G1 cyclin is, in turn, stimulated by growth factors acting through the RAS pathway. Green arrows indicate stimulation of a step and red blunt-end lines indicate inhibition of a step. All of the proteins shown have been found to be mutated in some cancers.

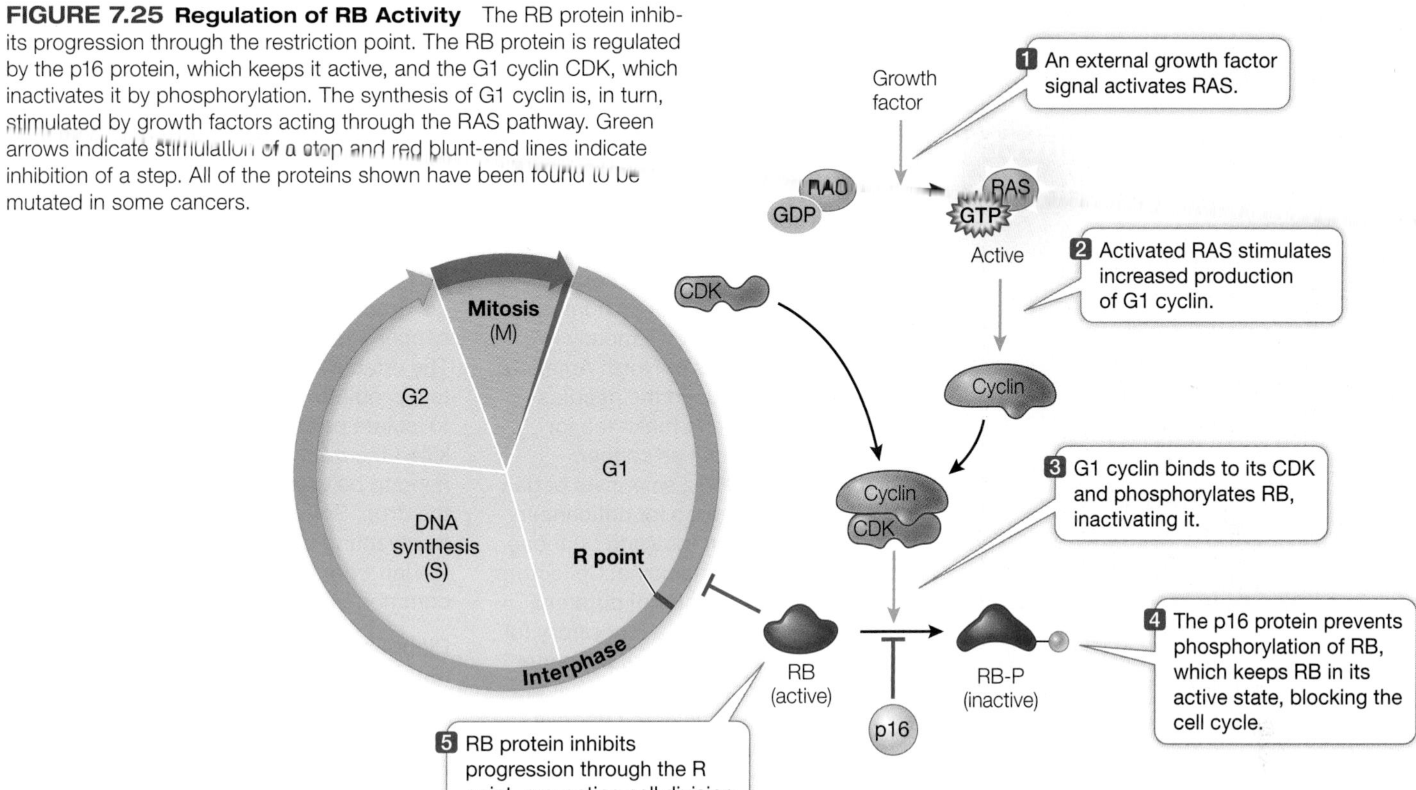

and RB mutations cause those proteins to be nonfunctional; mutations of RAS and CDK cause them to always be in their active state; and mutations of G1 cyclin cause increased expression, or greater stability, of that protein. The overall effect of these mutations is the uncontrolled proliferation of cells, since there is no restriction at the G1 checkpoint.

Since many different regulators affect cell cycle progression, it is not surprising that cancer cells have changes in many of these proteins, caused by mutations in the DNA. Since more mutations will be found in older cells that have undergone more cell divisions, cancer is much more prevalent in older organisms than in younger ones.

REVIEW & APPLY | 7.5

R Both prokaryotic and eukaryotic cell cycles are regulated by external signals that initiate the cycle, and by internal signals that keep the cycle going. In addition, there are checkpoints in the eukaryotic cell cycle that ensure that events in the cycle progress without errors. Cyclin-dependent kinases (CDKs) play critical roles in the regulation of cell cycle progression in eukaryotes. Cells can die from overall damage (necrosis) or from programmed cell death (apoptosis). The uncontrolled cell division that is a hallmark of cancer illustrates the importance of cell cycle controls, since the DNA that determines these controls is mutated in cancer cells and the controls are no longer functional.

A

1. Cultures of eukaryotic cells can be synchronized so that they are all in the same phase of the cell cycle at the same time. If you examined a culture at the beginning of G1, would the CDK that acts at the restriction point (R) be present? Would it be active? Would its cyclin be present? What would your answers be if the culture were at the restriction point, at the end of G1?
2. An essential role of cell division in complex eukaryotes is to replace cells that die. What causes cells to die?
3. In cancer cells, DNA sequences encoding proteins that control the cell cycle have been mutated. Examination of cells from multiple different tumors indicates that for some proteins, both copies of the genes encoding the proteins (the copy from the paternal gamete and the copy from the maternal gamete) are mutated such that they are nonfunctional, never just one. What does this tell us about the how the protein acts to regulate the cell cycle?

© Don Paulson Photography/Getty Images

Answer: Paclitaxel affects cancer cells by binding to microtubules in the spindle, which prevents the chromosomes from separating normally in a dividing cell and thereby kills the cell. The discovery of paclitaxel as an anticancer drug was based on a combination of traditional peoples' knowledge, systematic evaluation and testing using model systems, and clinical trials on humans. The Pacific yew tree grows mostly in Pacific coastal forests of North America. Native people have used the needles and twigs of the tree to make tea for treating diseases such as cancer. Beginning in the 1960s, scientists began to systematically search for anticancer drugs in plants, often focusing on those known in traditional medicine. Collectors would take different parts of different plants and send them to a laboratory for testing, first on various types of cancer cells to see if the cells stopped dividing. Any plant extracts that were effective were separated in the chemistry lab to isolate specific molecules, which then were also tested on cells. In this way, paclitaxel was isolated from yew tree bark and was shown to affect microtubules in the spindle. A next step was to see if the drug worked on cancer in laboratory animals. It did, and later was given to women with ovarian cancer in clinical trials. The results were spectacular, and paclitaxel was soon approved by government agencies to treat ovarian cancer and, later, other types of cancer. The effectiveness of this new drug led to a problem: A lot of bark was needed to isolate paclitaxel, and removing bark killed the trees. Whole forests would need to be destroyed to get enough of the drug. Scientists solved this problem by inventing ways to make paclitaxel in the lab from fungi growing on a more common plant and from cell cultures.

Plants are excellent places to look for novel, biologically active compounds because they have evolved numerous diverse compounds to defend against animals that eat them, as you will learn in Chapter 27.

7 VISUAL SUMMARY

7.1 Reproduction May Be Asexual or Sexual

Go to ACTIVITY 7.1

- **ASEXUAL REPRODUCTION** produces **CLONES**—genetically identical offspring (**FIGURE 7.1**).
- Sexual life cycles involve **MEIOSIS** and **FERTILIZATION** (**FIGURE 7.3**).

FIGURE 7.1

Asymmetrical cell divison

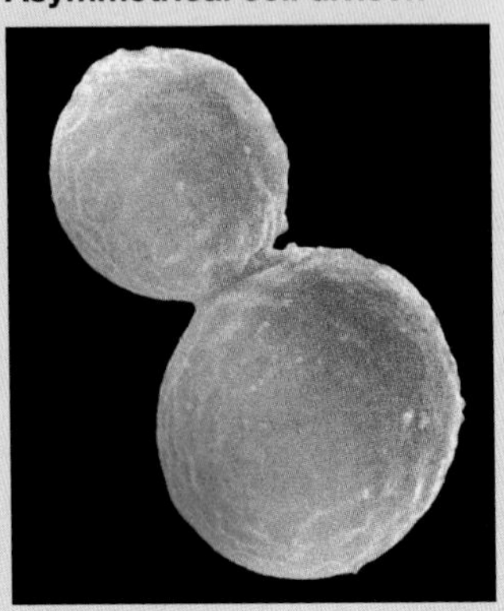
© Dr. Stanley Flegler/Visuals Unlimited, Inc.

FIGURE 7.3

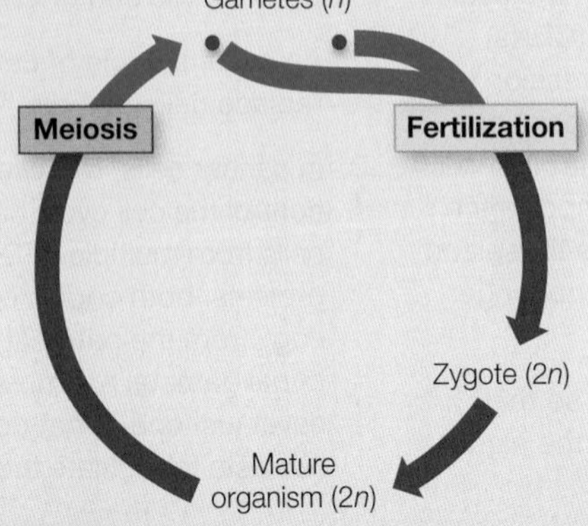

QUESTIONS

1. In what organisms are cell division and reproduction the same thing?
2. What is the difference between an asexual and a sexual life cycle?
3. What is the difference between a diplontic and a haplontic life cycle?

7.2 Asexual Reproduction Results in Genetically Identical Daughter Cells

FIGURE 7.4 **Binary fission in a bacterium**

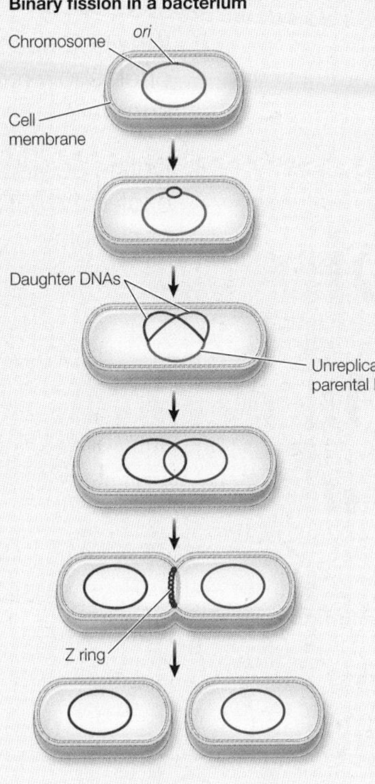

Go to ANIMATION 7.1 and ACTIVITIES 7.2 and 7.3

- **CELL DIVISION SIGNALS** lead to **DNA REPLICATION**, **DNA SEGREGATION**, and **CYTOKINESIS**.
- DNA replication and segregation co-occur in **BINARY FISSION** (**FIGURE 7.4**).
- **MITOSIS** has characteristic stages and involves a microtubule **SPINDLE** that moves condensed chromosomes.

QUESTIONS

1. What events occur during G1, S, G2, and M phases of the cell cycle?
2. What events occur at prophase, prometaphase, metaphase, anaphase, and telophase of mitosis?
3. How does binary fission differ from the mitotic cell cycle?

7.3 Sexual Reproduction by Meiosis Halves the Number of Chromosomes and Generates Diversity

FIGURE 7.14

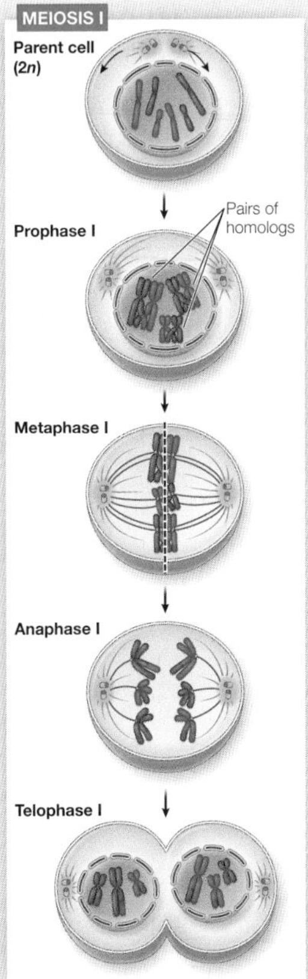

Go to ANIMATION 7.2 and ACTIVITIES 7.4 and 7.5

- The products of meiosis are four **HAPLOID** cells with unreplicated chromosomes.
- Meiosis involves two sequential cell divisions: **MEIOSIS I** (separation of **HOMOLOGOUS** chromosomes) and **MEIOSIS II** (separation of **SISTER CHROMATIDS**) (**FIGURE 7.14**).
- **INDEPENDENT ASSORTMENT** of chromosomes and **CROSSING OVER** produces genetic variation among the gametes produced by meiosis.

QUESTIONS

1. What events occur at prophase I, metaphase I, anaphase I, and telophase I of meiosis I?
2. How many alignments at metaphase I are possible when $2n = 6$?
3. What chromatids can be involved in crossing over?

7.4 Errors during Cell Division Can Result in Changes in Chromosome Number

- **NONDISJUNCTION** leads to **ANEUPLOIDY** (FIGURE 7.15).
- Cytokinesis failure and **POLYSPERMY** lead to **POLYPLOIDY** (FIGURE 7.16).

FIGURE 7.15

Nondisjunction during meiosis I

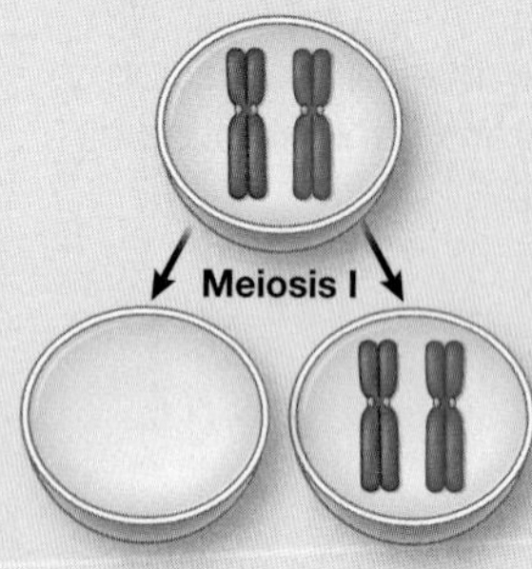

FIGURE 7.16

Triploid karyotype

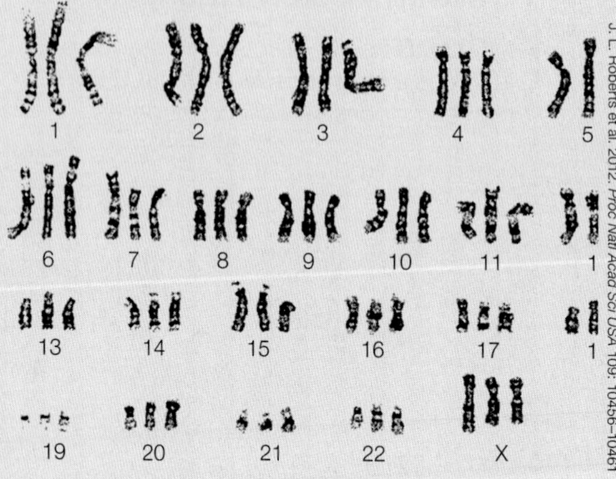

J. L. Roberts et al. 2012. *Proc Natl Acad Sci USA* 109: 10456–10461

QUESTIONS

1. What happens during nondisjunction?
2. Why do aneuploid cells often show reduced survival?
3. Define polyploidy and explain how it differs from aneuploidy.

 7.5 The Cell Cycle and Cell Death Are Highly Regulated in Eukaryotes

FIGURE 7.18
Control points in the eukaryotic cell cycle

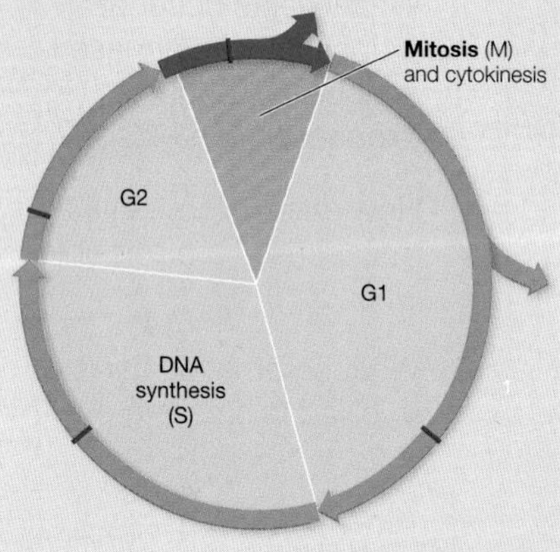

- The eukaryotic cell cycle has a **RESTRICTION POINT** (**R**) and three additional **CELL CYCLE CHECKPOINTS** (FIGURE 7.18).
- **CYCLINS** and **CYCLIN-DEPENDENT KINASES** (**CDKs**) play major roles in cell cycle regulation (**FIGURE 7.21**).
- Cells die due to damage, lack of nutrients, or programmed cell death.
- **MUTATIONS** in DNA coding for proteins involved in the cell cycle may lead to cancer.

FIGURE 7.21

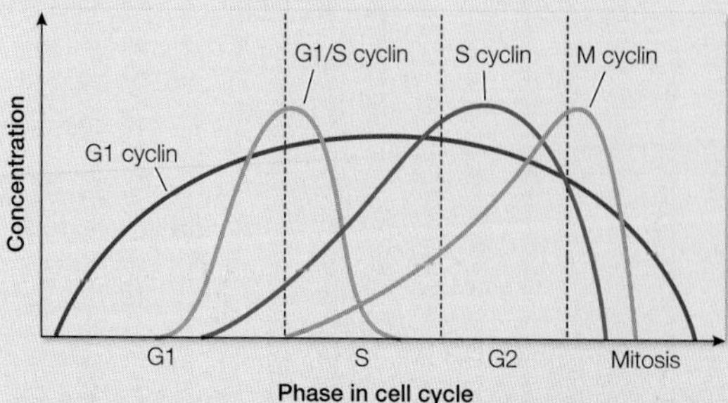

1. In eukaryotes, what factors can cause cell cycle arrest at the four checkpoints?
2. Why aren't all CDKs active throughout the cell cycle?
3. Why might a cell undergo apoptosis?
4. How is cell cycle control involved with cancer?

Go to **Achieve** for the e-book, animations, activities, and additional resources and assignments.

Inheritance, Genes, and Chromosomes

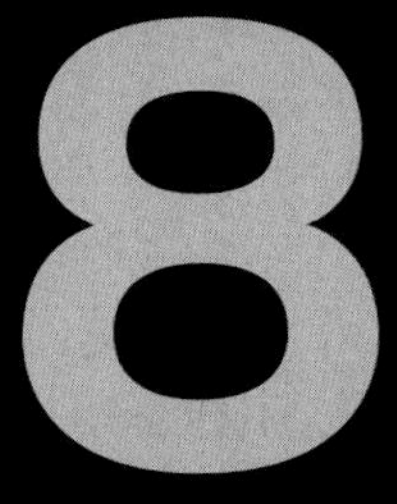

The 150 or so purebred varieties of dogs are incredibly diverse in behavior, coloration, and size.

How did the different characteristics of dogs arise, and how did selective reproduction lead to the emergence of new breeds?

You will find the answer to this question on page 200.

KEY CONCEPTS

8.1 Mendel Discovered Two Laws of Inheritance

8.2 Genes Are Inherited on Chromosomes

8.3 Alleles, Genes, and the Environment Interact to Produce Phenotype

8.4 Conjugation and Transformation Allow Exchange of Genetic Material between Prokaryotes

8.1 | Mendel Discovered Two Laws of Inheritance

LEARNING OBJECTIVES

By the end of this key concept you should be able to:

8.1.1 **Distinguish between phenotype and genotype.**

8.1.2 **State the law of segregation and explain how it demonstrates particulate inheritance.**

8.1.3 **Distinguish between dominant and recessive traits.**

8.1.4 **State and give an example of the law of independent assortment.**

8.1.5 **Use pedigree analysis to predict inheritance patterns in humans for a dominant or recessive trait.**

In all species, adult offspring tend to look like their parents—you are more similar to your biological parents than you are to other parents. This observation implies that information encoding various characteristics must be passed from parents to offspring. Genetics is the field of biology concerned with inheritance, and in one sense it has been around for a long time. There is good evidence that people were deliberately breeding animals (horses) and plants (the date palm tree) for desirable characteristics as long as 7,000 years ago. The general approach was to examine the variation among individuals of a species and breed those with the most desirable characteristics to one another and hope offspring showed these characteristics, indicating that they had inherited them. This method worked remarkably well. The many domesticated species with characteristics quite different from those found in their non-domesticated relatives show just how effective such breeding programs have been.

By the mid-nineteenth century, biologists had become keenly interested in how inheritance works. The prevailing hypothesis at the time was *blending inheritance*, which proposed that offspring are a blend of the characteristics of their parents. What was being blended was not specified, nor were the rules of blending: Do parents contribute equally? Does the mother contribute more since her gamete is larger? Do fathers contribute more to sons and less to daughters? If the blending hypothesis was correct, variation should decline over time as characteristics blend together over generations. But that is not what is observed. There is enormous variation across individuals within species. So it became clear that blending was not a viable explanation for inheritance, at least for some characteristics.

If blending was not correct, how might one investigate how inheritance works? In this key concept we look in detail at experiments performed in the 1860s by Gregor Mendel, an Austrian monk and scientist whose work clearly showed that inheritance, at least for some traits, was *particulate*—the factors underlying these traits were passed without alteration from one generation to the next, without blending. It is hard to overemphasize the importance of Mendel's work. Although it took 35 years for its importance to be fully realized, Mendel's theory of particulate inheritance revolutionized biology and is the foundation of the modern field of genetics.

Mendel used the scientific method to test his hypotheses

After entering the priesthood at a monastery in Brno, in what is now the Czech Republic, Gregor Mendel was sent to the University of Vienna, where he studied biology, physics, and mathematics. He returned to the monastery in 1853 to teach. Over the next 7 years, Mendel performed crosses (plant matings) using many thousands of pea plants to dissect the principles of inheritance. He presented his work, and conclusions, in two public lectures in 1865 and a detailed written publication in 1866, but it was ignored by mainstream scientists until 1900. By that time, the discovery of chromosomes had suggested to biologists that genes might be carried on chromosomes. When they read Mendel's work on particulate inheritance, the biologists connected the dots between genes and chromosomes.

In addition to figuring out some of the major principles of inheritance, Mendel's work illustrated how experiments into inheritance should be performed. In particular, he demonstrated the importance of working with organisms with the following features: (1) variability in one or more easily scorable traits, (2) large family sizes, (3) short generation times, (4) suitability for controlled matings, and (5) the ability to be grown easily and inexpensively in the field or lab. Mendel chose to study the common garden pea, a plant that can be grown inexpensively in large numbers and that can go through two generations in a single year. In addition, peas have flowers with both male and female sex organs (stamens and pistils, respectively), meaning they produce both pollen and ovules bearing male and female gametes (**FIGURE 8.1**). To perform a cross, pollen can be removed from stamens and manually placed onto pistils of another plant, resulting in **cross-fertilization**, or onto the pistils of the *same* plant, resulting in **self-fertilization**, or **selfing**. In cross-fertilization the stamens of the receiving flower are removed to prevent selfing.

More details on flower morphology and reproduction can be found in **Chapter 26**.

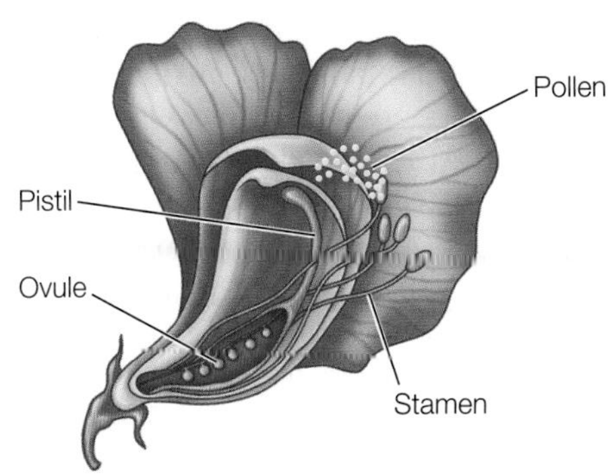

FIGURE 8.1 A Pea Flower Pea flowers have pollen-producing organs called stamens and an ovule-producing organ called the pistil.

Mendel's choice of the pea plant was ideal because there are many varieties available with variable characters that can be easily scored. A **character** is an observable physical feature, such as seed shape or flower color. Mendel examined seven characters, each of which had two different traits. A **trait** is a particular form of a character, such as round versus wrinkled seeds, or white versus purple flowers. Mendel first spent a few years setting up true-breeding varieties—plants that, when crossed among one another, or selfed, yield offspring with the same traits as the parents. True-breeding purple-flowered parental plants produced offspring that were purple-flowered, white-flowered parental plants produced white-flowered offspring, parents that grew from wrinkled seeds produced wrinkled seeds, and so on. Mendel reasoned that heritable characters in true-breeding varieties would be uniform in their underlying heritable makeup. He raised many generations of true-breeding varieties to ensure that the plants "bred true," and then set out to perform the crosses that changed the study of inheritance.

First, Mendel crossed plants that were true-breeding for different traits (purple versus white flower color, for example). When he performed a cross between two plants, he used each plant both as a male (pollen donor) and as a female (pollen recipient) in a so-called **reciprocal cross** (e.g., pollen from a true-breeding purple-flowered plant placed onto the pistil of a true-breeding white-flowered plant, and vice-versa). These plants were the **parental generation**, designated **P**. The offspring of these parental crosses—that is, the resulting seeds and the new plants that grew from them—constituted the **first filial generation**, or **F_1**. (The word "filial," from the Latin *filius*, "son," refers to the relationship between offspring and parents.) Mendel examined each F_1 plant to see which traits it bore and then recorded the number of F_1 plants expressing each trait. After he scored F_1 plants for their traits, he selfed them to produce a **second filial generation, F_2**, and the traits of the F_2 plants were scored. In all of Mendel's experiments, many crosses were performed, and hundreds of offspring scored, allowing Mendel to look for patterns of inheritance.

Mendel's first experiments involved monohybrid crosses

The term "hybrid" refers to the offspring of crosses between organisms that differ in one or more characters. Mendel's first set of experiments involved crosses of true-breeding parents that differed in a single character such as flower color. He planted the F_1 seeds from these crosses and allowed them to self-pollinate to produce the F_2 generation. These single-factor crosses are referred to as Mendel's **monohybrid crosses**.

Mendel performed monohybrid crosses for all seven pea characters he examined (**TABLE 8.1**). His method is illustrated in **FIGURE 8.2**, using flower color, which can be purple or white. When he crossed a true-breeding purple-flowered strain with a true-breeding white-flowered strain, all of the F_1 offspring had purple flowers; the white flower color trait seemed to have disappeared. However, when F_1 plants were allowed to self-pollinate

TABLE **8.1** | Mendel's Results from Monohybrid Crosses

Parental generation phenotypes			F_2 phenotypes			
Dominant		Recessive	Dominant	Recessive	Total	Ratio
Round seeds	×	wrinkled seeds	5,474	1,850	7,324	2.96:1
Yellow seeds	×	green seeds	6,022	2,001	8,023	3.01:1
Purple flowers	×	white flowers	705	224	929	3.15:1
Inflated pods	×	constricted pods	882	299	1,181	2.95:1
Green pods	×	yellow pods	428	152	580	2.82:1
Axial flowers	×	terminal flowers	651	207	858	3.14:1
Tall stems (1 m)	×	dwarf stems (0.3 m)	787	277	1,064	2.84:1

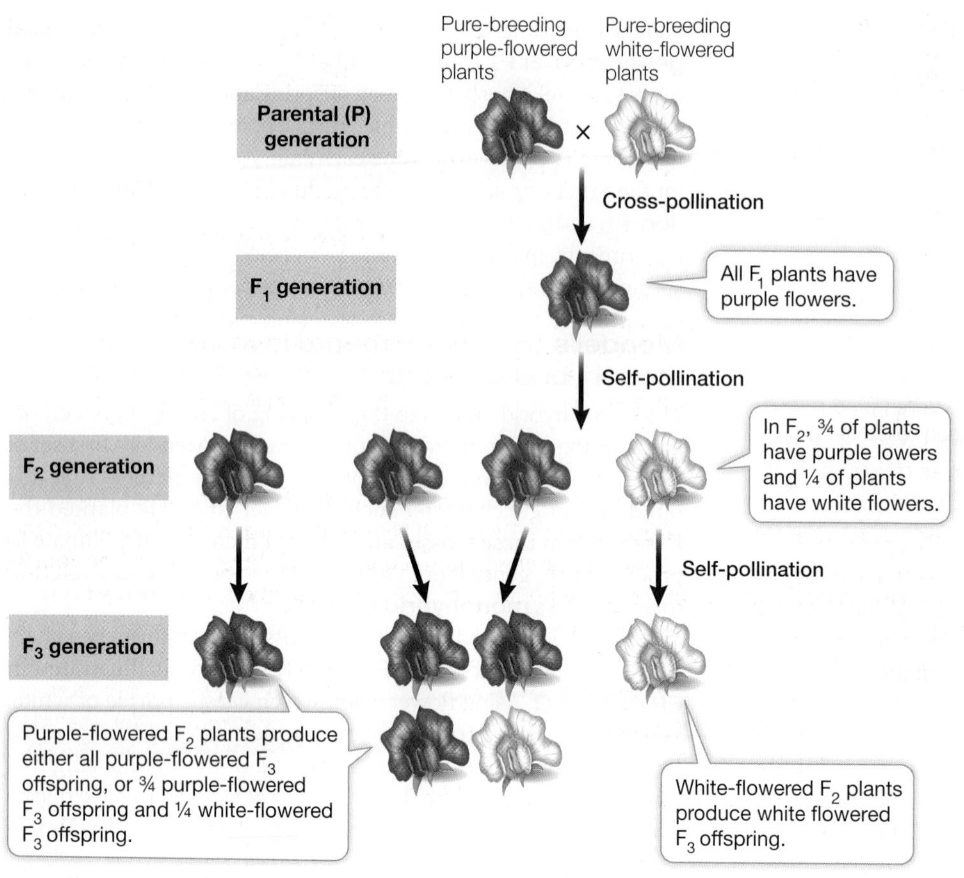

FIGURE 8.2 Mendel's Monohybrid Cross for Flower Color

to produce F_2 offspring, about three-fourths of the F_2 offspring had purple flowers and about one-fourth had white flowers, a 3:1 ratio. Mendel then allowed the F_2 offspring to self, producing an F_3. The white-flowered F_2 plants produced only white-flowered F_3 offspring, indicating that they were true-breeding. In contrast, the purple-flowered F_2 plants produced either all purple-flowered F_3 offspring, implying that they were true-breeding, or three-fourths purple-flowered F_3 offspring and one-fourth white-flowered F_3 offspring, just as in F_2 plants. The production of the F_3 allowed Mendel to realize that the F_2 consisted of the following ratios of different plant types: one-fourth were true-breeding white, half were just like the purple F_1, and one-fourth were true-breeding purple, a 1:2:1 ratio. Importantly, the frequencies of true-breeding purple and true-breeding white were equal in the F_2, implying that both traits were inherited similarly. In addition, Mendel found that the results were the same for reciprocal crosses; it did not matter whether a trait was present in the pollen donor or pollen recipient in the parental generation.

Identical patterns were observed for the other six characters: one trait disappeared in the F_1, only to reappear in one-fourth of the F_2. The F_2 always consisted of one-fourth true-breeding for the trait that was absent in the F_1, one-fourth true-breeding for the trait that was present in the F_1, and half like the F_1. These observations implied that (1) the missing trait (white flowers in the example shown in Figure 8.2) was present in the F_1, it was just not visible; (2) all characters obey the same transmission rule(s) since all seven gave a 3:1 F_2 ratio; and (3) parents contribute equally to offspring, regardless of whether they are the pollen donor or recipient, since reciprocal crosses give the same results.

What did these results imply about inheritance? Mendel reasoned that characters were determined by *pairs* of discrete, particulate determinants—what we now call genes, which are sequences of DNA carried on chromosomes, as we will discuss in Key Concept 8.2. Mendel did not use the term "gene," but we will do so for clarity. We know now that different traits arise because the DNA sequence of a gene for a particular character can vary, with a particular sequence resulting in one trait (e.g., white flowers in peas) and a slightly different sequence in the same gene resulting in another trait (e.g., purple flowers). We use the term **allele** to distinguish different DNA sequences of the same gene. There are two alleles for pea flower color: one that determines purple flowers (the purple allele) and another that determines white flowers (the white allele). The allelic makeup of an individual, either with respect to a single trait or to a larger set of traits, is termed its **genotype**. The outward expression of the genotype is the **phenotype**.

Let's review the genetics terms defined so far:

- A *character* is an observable physical feature in an organism (e.g., flower color).
- A *trait* is a particular form of a character (e.g., purple flower color).
- A *gene* is an inherited, particulate determinant of a character (e.g., the gene that determines flower color).
- An *allele* is a form of the gene that determines a particular form of character (e.g., the allele for purple flowers).
- The *genotype* is the allelic makeup of an individual (e.g., at the flower-color gene, a pea plant may have two white alleles, two purple alleles, or a white allele and a purple allele).
- The *phenotype* is the outward expression of the genotype (e.g., white or purple flowers).

Mendel reasoned that one allele of each gene pair is inherited from the male gamete and one from the female gamete; both parents contribute equally. This makes sense given that we now know that in diploid cells chromosomes exist in homologous pairs, whereas gametes have unpaired chromosomes (they are haploid) (see Key Concept 7.1). Genes occur at particular locations on chromosomes. In general, a specific location on a chromosome is called a **locus** (plural *loci*), which may refer to an entire gene, part of a gene, some other DNA sequence, or even to a single base pair. Different DNA sequences at a locus are termed alleles, even when the locus is not an entire gene. If the two alleles at a locus in an individual are the same—for example, both alleles consist of the DNA sequence that determines white flowers—the plant has a **homozygous** genotype with respect to that locus/gene. If the two alleles at that locus differ—one determines purple flowers and the other white flowers, for example—then the individual has a **heterozygous** genotype.

Returning now to Mendel's monohybrid experiments, let's address the question of why some traits seemed to disappear in the F_1 yet reappeared in subsequent generations. Mendel realized that heterozygous individuals that carry alleles for both traits of a character *exhibit* only one of them. He coined the term **dominant** to indicate the trait that is expressed (observed) in a heterozygous individual. In contrast, a **recessive** trait is the one that is not observed in a heterozygote. A heterozygote for the purple and white alleles has purple flowers, indicating that the purple allele is dominant to the white allele, which is recessive. *Only when an individual is homozygous for a recessive allele will the recessive trait be expressed,* explaining why Mendel consistently observed unequal ratios of the two trait values for each of the seven characters in the F_1 and F_2 generations in his monohybrid experiments (see Figure 8.2).

Alleles are designated by letters; dominant alleles are represented by an uppercase letter and recessive alleles by the corresponding lowercase letter. The assignment of these letters is somewhat flexible. For example, at the flower-color gene, we could use *w* for the recessive white allele and *W* for the dominant purple allele (as shown in Figure 8.3), or we could designate the purple allele as *P* and the white allele as *p*. If we decide to use *w* and *W*, then the parental cross can be represented as *ww* (white homozygote) crossed with *WW* (purple homozygote). These plants produce *w* and *W* gametes respectively that, upon fertilization, produce F_1 offspring that are *Ww* heterozygotes, and are purple because the purple allele (*W*) is dominant to the white allele (*w*).

Mendel's first law states that allele pairs segregate equally into gametes

How to explain the F_2 ratios? Mendel reasoned that during gamete formation in the F_1, each allele *segregates (separates) equally* from one another, such that half the gametes have one allele and half have the other. If those gametes *come together at random,* then the F_2 ratios could be easily obtained. One way to predict the genotypes from a cross is by writing down the allele combinations using a **Punnett square**, a method devised in 1905 by the British geneticist Reginald Punnett. Alleles from male gametes are arrayed along one edge of a square grid, and the corresponding alleles from female gametes are arrayed along the other edge. This method ensures that all possible random combinations of gametes are considered when calculating expected genotype frequencies in offspring. **FIGURE 8.3** shows the Punnett square for the F_1 cross for flower color.

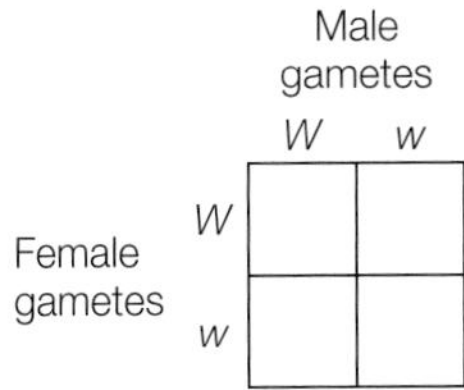

FIGURE 8.3 The Punnett Square A Punnett square for the F_1 monohybrid cross for flower color (*Ww* × *Ww*). Each parent can produce two haploid gamete genotypes (*W* or *w*), which can combine to produce four diploid F_2 genotypes, two of which are heterozygotes (*Ww* and *wW*) and two of which are homozygotes (*ww* and *WW*). If segregation is equal, such that *W* and *w* are each present in half of the gametes, and gametes fuse at random, then each square will have equal probability (= ¼).

In the F_1 *Ww* × *Ww* cross, four gamete combinations are possible: *W* or *w* in male gametes and *W* or *w* in female gametes. When these gametes come together, two (*W* male and *w* female, or vice versa) produce heterozygotes, and two (*w* male and *w* female, or *W* male and *W* female) produce homozygotes. If segregation is equal, such that *W* and *w* are present in half of the gametes, and gametes fuse at random, then each combination (each small square in the Punnett square) will have equal probability (= ¼). So one-fourth of the F_2 are homozygous for the white allele, have white flowers, and are true-breeding when selfed, and one-fourth of the F_2 are homozygous for the purple allele, have purple flowers, and are true-breeding when selfed. The other half of the F_2 are heterozygous, have purple flowers (because of dominance), and give rise to three-fourths purple offspring and one-fourth white offspring when selfed. Thus Mendel was able to explain the 3:1 phenotypic ratios and the 1:2:1 genotypic ratios (revealed by examining the F_3) in his monohybrid cross.

Mendel's data led him to his first conclusion about inheritance in pea plants, known as the **law of segregation**: *when an individual produces gametes, the two alleles of each pair separate equally, so half the gametes receive one allele of every gene and half receive the other allele.* An enormous number of monohybrid crosses in other species obtain the same ratios as those observed by Mendel, revealing that Mendel's conclusion holds in essentially all meiotic organisms.

Mendel tested his hypotheses by performing test crosses

Mendel verified equal segregation by performing **test crosses**, which involve crossing a plant expressing the dominant phenotype but with an unknown genotype to a recessive homozygous individual. For flower color, a purple-flowered plant has unknown genotype, as it can be *WW* or *Ww*. When crossed with a white-flowered plant, which must be a *ww* homozygote, two possible results are expected: (1) if the purple-flowered parent is a *WW* homozygote, then all of the offspring will be *Ww*, which are purple-flowered (**FIGURE 8.4, LEFT**); or (2) if the purple-flowered parent is a *Ww* heterozygote, then half of the offspring should be purple-flowered (*Ww*) and half should be white-flowered (*ww*) (a ratio of 1:1), because of equal segregation (**FIGURE 8.4, RIGHT**). The test cross thus allows determination of the genotype of an individual with a dominant phenotype, and it allowed Mendel to test for equal segregation in heterozygotes.

Probability is used to predict inheritance

Mendel's law of segregation allows predictions about the frequency of traits in the offspring of crosses. In a heterozygote ×

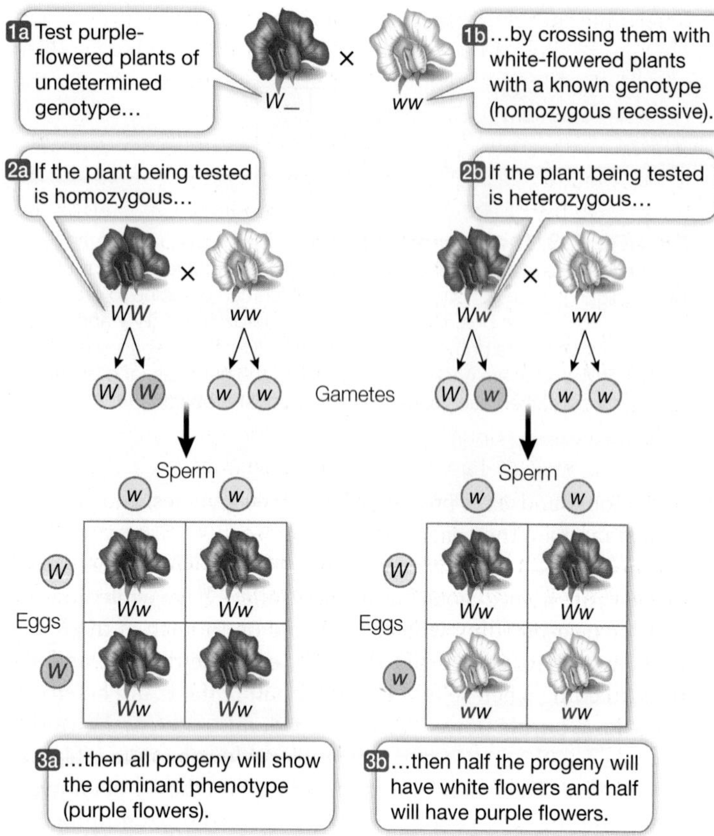

FIGURE 8.4 The Test Cross A test cross involves crossing an individual that has a (usually) unknown genotype with an individual that is homozygous for the recessive allele at the gene(s) affecting the character(s) being examined. In this example, a purple-flowered individual of unknown genotype (*WW* or *Ww*) is test crossed with a white-flowered individual (genotype *ww*). If the purple-flowered individual is *WW*, then all offspring will be purple-flowered (left side). If the purple-flowered individual is *Ww*, half the offspring will be purple-flowered and half will be white-flowered (right side).

Activity 8.1 Homozygous or Heterozygous?

heterozygote cross, three-fourths of the offspring are predicted to have the dominant phenotype and one-fourth the recessive phenotype. In a heterozygote × recessive homozygote cross, half of the offspring are predicted to have the dominant phenotype and half should have the recessive phenotype. Detecting patterns in frequencies requires the use of large sample sizes, and this was one of the keys to Mendel's success. By counting many offspring from each cross and accurately quantifying the frequency of each phenotype, he observed clear patterns of inheritance that allowed him to predict the underlying probability of an event occurring.

Probability plays a central role in predicting what occurs, in terms of ratios of genotypes and phenotypes, in the progeny of a given genetic cross. Here we briefly review some of the conventions and rules of probability that are often used to make predictions in genetic crosses. To give a concrete example, we'll use something with which we are all familiar: coin tosses.

Consider an event such as a single coin toss. If an outcome is absolutely certain, its probability is 1, and if it cannot possibly happen, its probability is 0. For example, if we toss a two-headed coin, which is a coin that is identical on both faces, the probability of getting a head is 1 and the probability of getting a tail is 0. Event outcomes that are not certain have probabilities between 0 (never) and 1 (always). Flipping a normal coin, with a head and a tail side, gives a head with probability ½, or a tail with probability ½. For any event, like a coin flip, the sum of the probabilities of every possible outcome (head or tail) is 1.

Two rules of probability—the multiplication rule and the addition rule—are very useful in predicting outcomes of coin tosses (and phenotype frequencies). If two coins are tossed, each acts independently of the other: the probability of getting a head with the second coin is unaffected by whether a head or tail was observed with the first coin. What is the probability of both coins coming up heads? The probability that the first coin will come up a head is ½. Likewise for the second coin. The multiplication rule of probability tells us that *the probability of two independent events occurring at the same time can be obtained by multiplying their individual probabilities*. The probability of getting two heads is thus ½ × ½ = ¼. The multiplication rule can be applied to the F_2 generation of a monohybrid cross. After the self-pollination of a *Ww* F_1 plant, the probability that an F_2 plant will have the genotype *WW* is ½ × ½ = ¼, because the chance that the sperm will have the genotype *W* is ½, and the chance that the egg will have the genotype *W* is also ½. Likewise, the probability of *ww* offspring is also ¼ (**FIGURE 8.5**).

What about the probability of getting a heterozygote? As you can see in Figures 8.3 and 8.5, there are *two* ways to get a *Ww* plant, or a head and a tail in a coin toss. In the case of the flower-color gene, the *W* allele can come from a sperm and the *w* from an egg (probability ¼), or the *W* allele could come from the egg and the *w* from the sperm (probability ¼). The addition rule of probability states that *when two events are mutually exclusive, the probability that either will occur is the sum of the probability of each event*. Thus the probability that an F_2 plant will be a heterozygote is equal to the sum of the probabilities of each of the two ways of forming a heterozygote: ¼ + ¼ = ½ (see Figure 8.5).

Probabilities predict expected outcomes: a normal coin is expected to come up a head half the time, and test-crossing a heterozygote is expected to give half recessive homozygous offspring. However, an expected outcome is not a guaranteed outcome. For example, if you toss a coin ten times, you expect five heads and five tails, but you might actually get four heads and six tails. Similarly, in a *Ww* × *ww* cross you might find five of eight offspring are *Ww*, though you expected only four. Do these outcomes imply that the probability of getting a head, or the probability of obtaining a *w* gamete from a *Ww* individual, is not equal to ½? No, of course not. It simply means that, by chance, you happened to flip four heads and six tails, or obtained five *W* and three *w* gametes from a heterozygote. As we make more and more tosses, the proportion of heads and tails will get closer and closer to ½. Determining whether a particular outcome (four heads and six tails, or five *Ww* and three *ww* offspring) is consistent with an expected outcome (five heads and five tails, or four *Ww* and four *ww* offspring) is a major objective of statistical tests (see Appendix B). Later in the chapter we will encounter one such test, the chi-square test for goodness of fit.

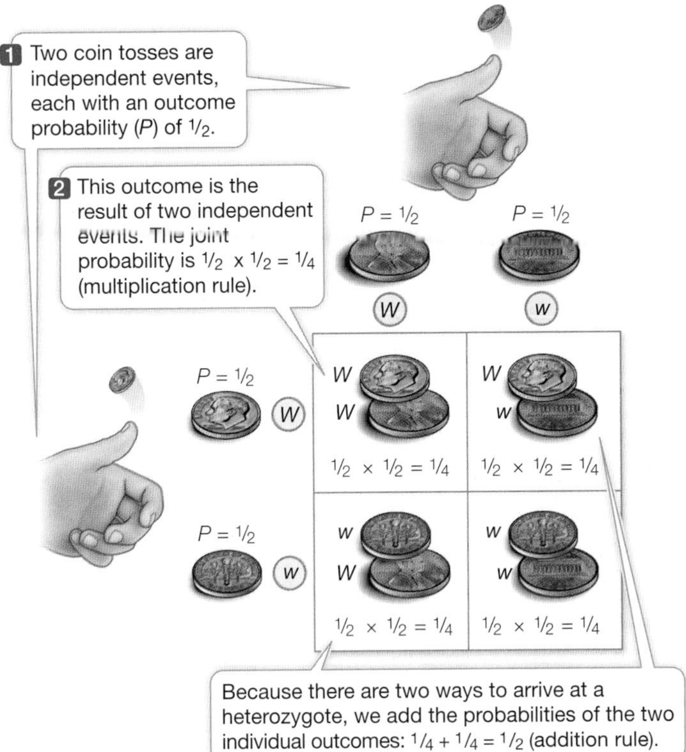

FIGURE 8.5 Multiplication and Addition Rules of Probability Flipping a coin and segregating alleles into gametes in a heterozygote are equivalent in terms of outcome probabilities. When a coin is flipped, the probability of getting a head (or tail) is ½. Likewise, the probability of a *Ww* plant producing a gamete containing the *W* (or *w*) allele is ½.

If you randomly choose two offspring from a *Ww* × *Ww* cross, what is the probability that both will have white flowers?

Mendel's second law states that allele pairs of different genes assort independently

Once Mendel understood inheritance for single characters, he wished to determine the rules for the inheritance of two, or more, characters at the same time. If characters behave independently, then he could predict the results of crosses involving two characters, a so-called **dihybrid cross**, using the probability rules we have just described.

Mendel's dihybrid crosses began with true-breeding peas that differed in two characters. For example, in one set of crosses he used seed shape (round, *R*, dominant to wrinkled, *r*) and seed color (yellow, *Y*, dominant to green, *y*). Note that both of these phenotypes are characteristics of the offspring: they are determined by the genotype of the seed, not the parent plant. One true-breeding parental variety with round yellow seeds (*RRYY*) was crossed with a true-breeding variety with wrinkled green seeds (*rryy*). A cross between these two varieties results in fusion of *RY* and *ry* gametes and produced *RrYy* F_1 offspring, which had round yellow seeds—as expected, because the round and yellow alleles are dominant.

Mendel selfed the F_1 plants to obtain F_2 seeds. All possible seed phenotypes were obtained in the F_2: round yellow, round green, wrinkled yellow, and wrinkled green (**FIGURE 8.6A**). As before, Mendel counted enough seeds that he was able to determine that the proportions of these four phenotypes were 9⁄16 round yellow, 3⁄16 round green, 3⁄16 wrinkled yellow, and 1⁄16 wrinkled green, a 9:3:3:1 dihybrid cross F_2 ratio. Given that the two characters are inherited independently of one another, this ratio is exactly what one would predict. For example, the probability of obtaining a round seed in a monohybrid F_2 is ¾. Similarly, the probability of obtaining a yellow seed is ¾. If these characters are inherited independently, such that transmission of seed shape is independent of the transmission of seed color, then the probability of getting a round yellow seed can be obtained using the multiplication rule and is ¾ × ¾ = 9⁄16, as observed. Similar calculations can be made to obtain the expected frequencies of the other three phenotype combinations (**FIGURE 8.6B**). These ratios can also be obtained from a Punnett square in which each parent can produce four equally common gametes: *RY*, *Ry*, *rY*, and *ry*, each with probability ¼ (**FIGURE 8.6C**).

Mendel obtained similar results using other combinations of pairs of characters. On the basis of these recurring patterns, he proposed that *alleles of different genes assort independently of one another during gamete formation*. This conclusion holds for many pairs of characters in many other organisms and is known as the **law of independent assortment**. As you will see in Key Concept 8.2, Mendel's law of independent assortment is not as universal as the law of segregation and can be broken, often spectacularly. If two genes are located near one another on the same chromosome, their alleles do not assort independently, a phenomenon called **linkage**. Thus Mendel's second law applies to pairs of genes on different chromosomes and to pairs of genes that are far apart on the same chromosome.

Mendel's laws can be observed in human pedigrees

Mendel developed his theories by performing specific crosses and counting many offspring. Neither of these approaches is possible with humans, so human geneticists rely on **pedigrees**, family trees that show the occurrence of inherited phenotypes in several generations of related individuals (**FIGURE 8.7**). Pedigree analysis is often done to trace the inheritance pattern of a particular trait, especially in cases where the rare allele causes a serious disease or abnormality.

Because humans have relatively few offspring, human pedigrees do not show the clear proportions of phenotypes that Mendel saw in his pea plants. For example, when a man and a woman who are both heterozygous for a recessive allele (say, *Aa*) have children together, each child has a ¼ probability of being a recessive homozygote (*aa*). But seldom does one couple have enough offspring to reliably show the one-fourth proportion. In a family with only two children, for example, both could be *aa* (or *AA*, or *Aa*) by chance. Instead of relying on one family, pedigree analyses examine multiple families and look for particular patterns. Figure 8.7A is a pedigree showing the pattern of inheritance of a rare dominant allele that causes an unusual phenotype. Two key features can be seen: (1) Every affected person (i.e., with the unusual phenotype) has an affected parent—the disorder does not skip generations looking backward. (2) About half of the offspring of an affected heterozygous individual will be affected. In Figure 8.7A, all affected individuals are heterozygous for the rare dominant allele.

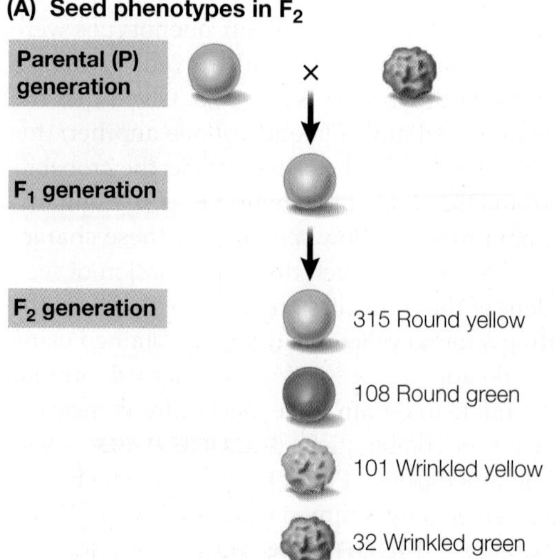

(B) Expected F_2 phenotypes obtained by combining phenotype ratios from single character crosses

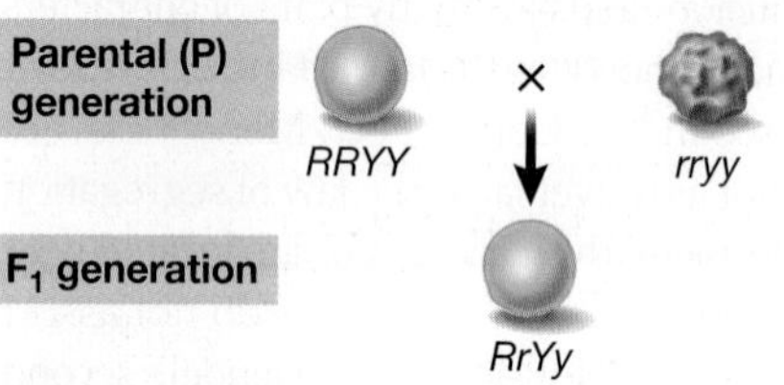

F_2 generation

Seed shape	Seed color	Product of monohybrid cross probabilities
Round ¾	Yellow ¾	¾ × ¾ = 9/16
Round ¾	Green ¼	¾ × ¼ = 3/16
Wrinkled ¼	Yellow ¾	¼ × ¾ = 3/16
Wrinkled ¼	Green ¼	¼ × ¼ = 1/16

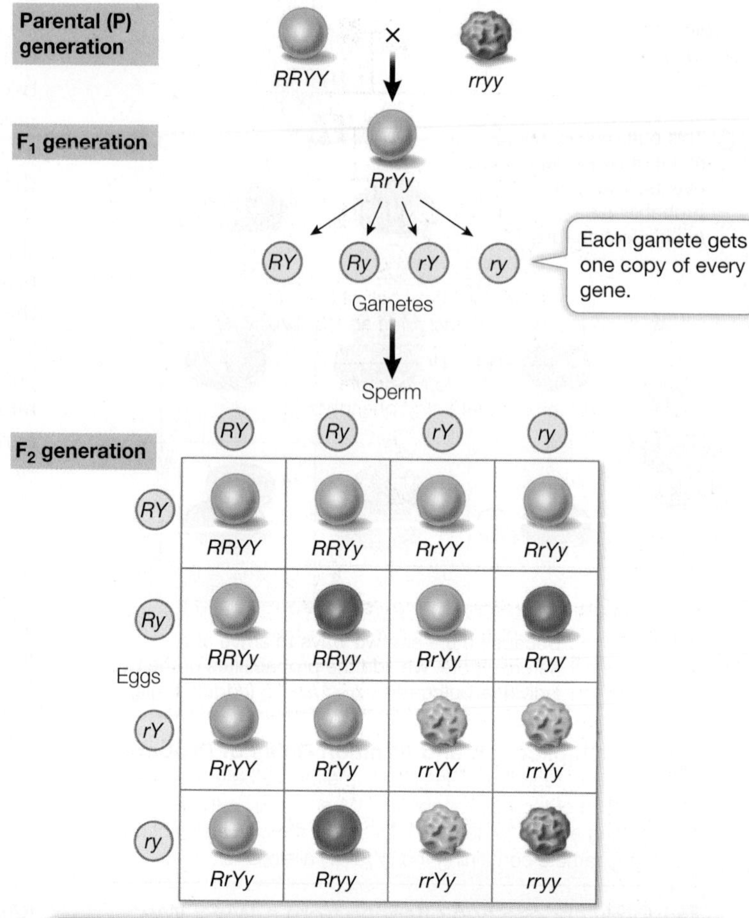

Results: When F_1 plants self-pollinate, the gametes combine randomly to produce an F_2 generation with four phenotypes in a 9:3:3:1 ratio.

FIGURE 8.6 Mendel's Dihybrid Cross (A) A true-breeding variety of peas with round yellow seeds was crossed with a true-breeding variety with wrinkled green seeds, producing an F_1 with round yellow seeds. F_1 plants were selfed and produced all four phenotypes in the F_2. The numbers obtained were approximately 9/16 round yellow seeds, 3/16 round green seeds, 3/16 wrinkled yellow seeds, and 1/16 round green seeds. Mendel interpreted these results in terms of independence of the inheritance of the two characters. **(B)** Independent combinations of phenotypes from monohybrid F_1 crosses predict the 9:3:3:1 ratio. **(C)** Independent assortment to produce four equally likely gametes (each with probability ¼) also predicts the 9:3:3:1 ratio, as seen in the Punnett square.

In this cross, 556 F_2 offspring were scored. Assuming independent assortment, how many of each phenotype were expected?

Media Clip 8.1 Mendel's Discoveries

Compare this pattern with the one shown in Figure 8.7B, which is typical for the inheritance of a rare recessive allele: (1) Affected individuals most often have two unaffected parents. (2) Only a small proportion of people are affected: about one-fourth of children whose parents are both heterozygous carriers of the recessive allele. For recessive traits, it is not possible just by phenotype to determine the genotype of individuals who are unaffected—they could be heterozygous carriers or homozygous for the dominant allele. To determine the genotype of individuals with the dominant phenotype in peas, Mendel did test crosses. You will see in Chapter 12 that in humans it is now possible to determine the alleles in an individual directly by DNA sequencing.

Several interesting patterns of inheritance can be predicted and studied. For example, some so-called sex-linked traits are inherited exclusively from the father or the mother. And some phenotypes reflect the influence of *both* alleles in a phenomenon called codominance. We'll look at such patterns in Key Concepts 8.2 and 8.3, after we discuss how the behavior of chromosomes in meiosis I explains Mendel's first and second laws.

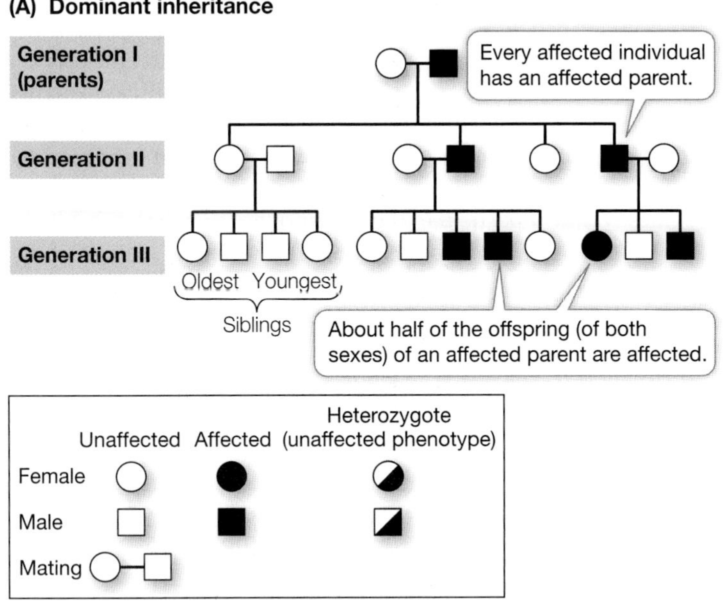

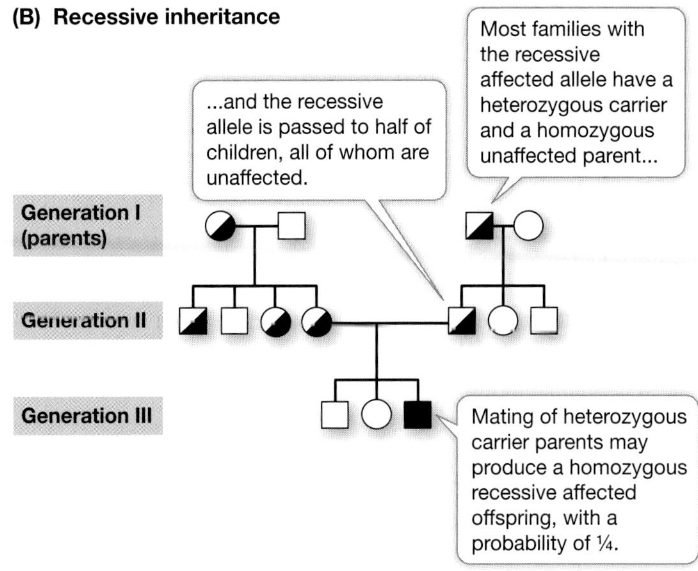

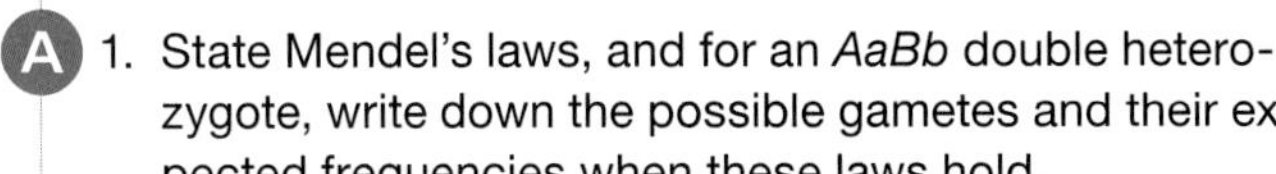

FIGURE 8.7 Human Pedigree Analysis **(A)** A pedigree for a rare dominant trait. Dominant traits do not skip generations looking backward: individuals who have the dominant trait must have a parent with the trait. Affected individuals are usually offspring of a heterozygous affected parent and a homozygous unaffected parent, a cross which produces half affected offspring. An example is Huntington's disease. **(B)** A pedigree for a rare recessive trait, indicating carriers (confirmed using direct genotyping, not by phenotype). Recessive traits often skip generations: individuals who have the recessive trait have parents who lack the trait. Affected individuals are usually offspring of two heterozygous unaffected parents, a cross which produces one quarter affected offspring. An example is cystic fibrosis.

In the pedigree in (A), are any of the individuals homozygous for either the dominant or the recessive allele?

Activity 8.2 Pedigree Analysis Simulation

REVIEW & APPLY | 8.1

R Mendel discovered that two parents contribute equally to offspring in terms of inheritance and that inheritance is determined by particulate entities that we now know as genes. Individuals carry two alleles—which may be identical or different—for each gene, and alleles segregate equally into gametes. Except in the case of genes that are located near one another on the same chromosome, the alleles of different genes assort independently of one another during gamete formation. Genotype frequencies can be predicted using Punnett squares and probabilities. In human genetics, pedigree analyses are used to determine inheritance patterns.

A

1. State Mendel's laws, and for an *AaBb* double heterozygote, write down the possible gametes and their expected frequencies when these laws hold.
2. Consider a character with two traits that are known to be caused by genotypic variation at a single gene. An individual with the dominant trait is test-crossed. What is the expected ratio of offspring phenotypes? An individual with the recessive trait is test-crossed. What is the expected ratio of offspring phenotypes?
3. A friend tells you that her biological father has an inherited disorder determined by a dominant allele but neither your friend nor her three siblings are affected. Is this possible? If so, what is the chance that it could happen?

Mendel's work had an impact on the science of genetics because the behavior of chromosomes during meiosis had been observed. It dawned on the scientific community that Mendel's laws of inheritance meant that genes reside on chromosomes. In the next key concept we will examine the chromosome theory of inheritance.

8.2 | Genes Are Inherited on Chromosomes

LEARNING OBJECTIVES

By the end of this key concept you should be able to:

8.2.1 **Describe how the behavior of chromosomes in meiosis I explains Mendel's first and second laws.**

8.2.2 **Explain how sex-linked inheritance was used to confirm the chromosome theory of inheritance.**

8.2.3 **Identify the expected pedigree pattern for a sex-linked dominant or sex-linked recessive trait.**

8.2.4 **For genes that are located on the same chromosome, explain how crossing over produces recombinant gametes during meiosis.**

Mendel's studies of pea plants supported the idea of particulate inheritance, which ultimately lead to the concept of a gene. But during Mendel's time the physical nature of genes and their location on chromosomes was unknown. By the time Mendel's work was rediscovered in 1900, dividing cells had been observed under the microscope, and the chromosome theory of inheritance—the notion that chromosomes are the carriers of genetic material—was proposed a few years later.

The behavior of chromosomes during meiosis is consistent with Mendel's laws

The separation of homologs during anaphase I of meiosis results in equal segregation of alleles: two of the four meiotic products will have one allele, and two will have the other allele. This happens *at every single meiosis*, and so Mendel's first law, the law of segregation, is true at every single meiotic cell division. For this reason, almost all genetic crosses in sexual species produce results that are consistent with Mendel's first law. Very rarely a genetic cross will produce offspring in proportions that deviate from equal segregation; such deviations are most often due to effects occurring after meiosis.

The behavior of chromosomes during meiosis also explains the independent assortment of alleles for two genes residing on different chromosomes. Independent assortment is caused by independent alignment of different pairs of homologous chromosomes. While a particular alignment of chromosomes results in the production of two gamete types from a single meiosis, if the two alignments are equally likely, then four possible gametes will occur with equal probability *when averaged over multiple meioses,* which explains Mendel's second law, the law of independent assortment (**FIGURE 8.8**). Mendel's second law does not apply to genes that are physically close together on the same chromosome because the two possible alignments are no longer equally likely.

LINK You can review the process of meiosis and the separation of chromosomes in **Key Concept 7.3.**

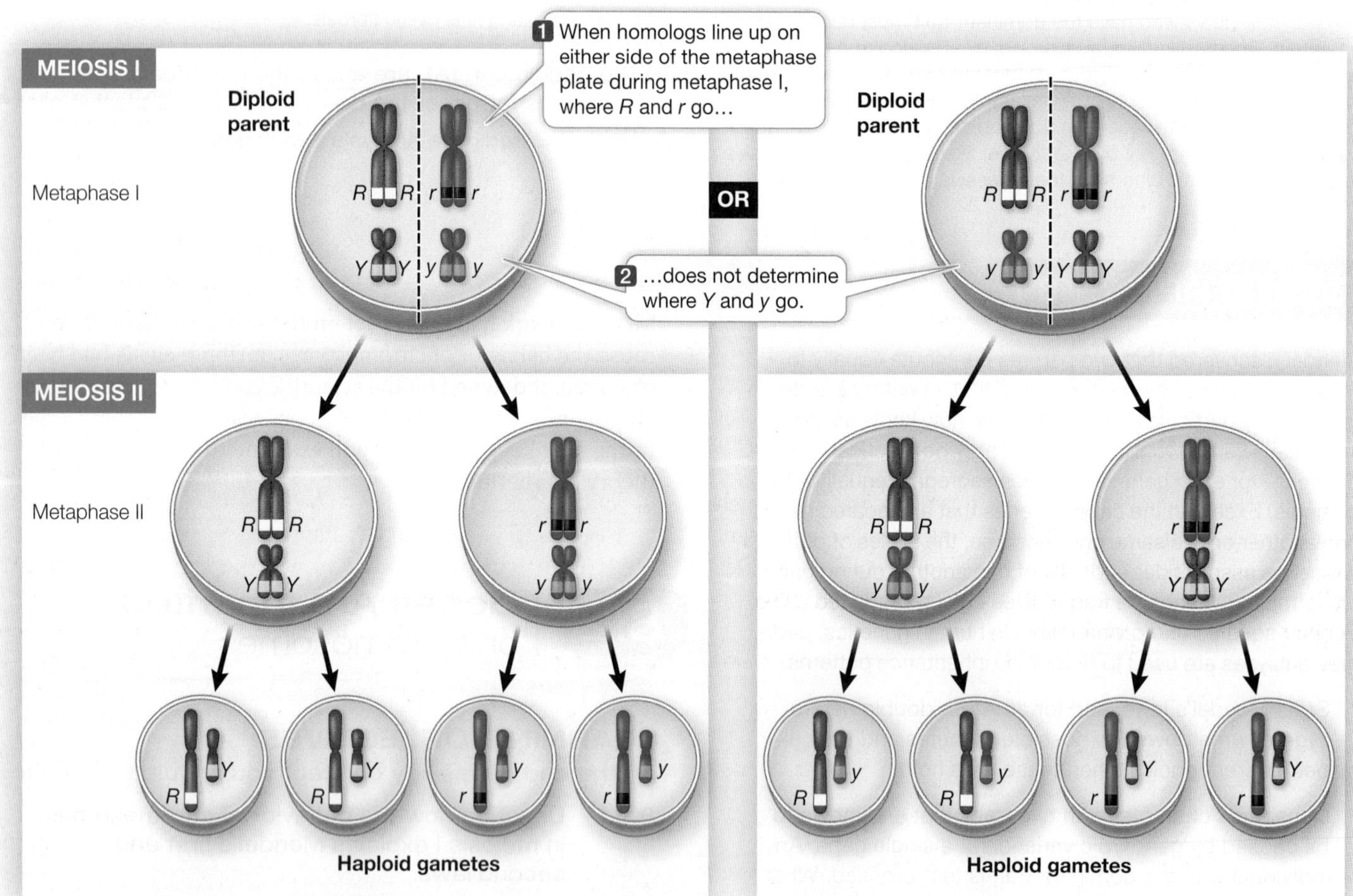

FIGURE 8.8 Meiosis and Mendel's Laws Mendel's law of segregation is explained by segregation of homologs during anaphase I of meiosis I. Notice how both the *R* and *r* alleles and the *Y* and *y* alleles have equally segregated into cells after meiosis I. Mendel's second law is explained by the two possible, and equally likely, alignments of homologs on the metaphase I plate. On the left side of the figure, *R* and *Y* chromosomes align on the same side of the metaphase I plate, resulting in half *RY* and half *ry* gametes. On the right side of the figure, *R* and *y* chromosomes align on the same side of the metaphase I plate, resulting in half *Ry* and half *rY* gametes. If both alignments are equally likely, with probability ½, the frequency of each of the four possible gametes will be ½ × ½ = ¼.

Animation 8.1 Independent Assortment of Alleles

While the behavior of chromosomes during meiosis was consistent with Mendel's laws, directly demonstrating that genes reside on chromosomes was not yet possible. However, two additional findings, sex linkage and crossing over, bolstered the chromosome theory. We will turn to these processes after we briefly describe the molecular nature of genes and alleles.

Alleles encode different versions of the same protein

When first advanced, the chromosome theory of inheritance stated that genes are on chromosomes, though it was not known what genes actually were. Genes are now known to be relatively short sequences of DNA (usually a few thousand base pairs in length) found on the much longer DNA molecules that make up chromosomes (which are often millions of base pairs long).

Genes determine phenotypes mostly by encoding proteins with particular functions. In many cases a dominant allele of a gene is expressed (transcribed and translated; see Key Concept 3.3) to produce a functional protein, whereas a recessive allele has been changed (mutated; see below) so that it either is no longer expressed or it encodes a protein that is nonfunctional. For example, the molecular nature of the wrinkled pea seed phenotype is the absence of an active enzyme called starch branching enzyme 1 (SBE1), which is essential for starch synthesis. With less starch, the developing seed has more sucrose, which causes an inflow of water by osmosis. When the seed matures, this water is lost, leaving a shrunken, wrinkled seed. This sequence of events explains the phenotype, but it does not address how the wrinkled allele arises.

Genes are subject to **mutations**, which are rare, stable, and inherited changes in the genetic material, DNA. In other words, an allele can mutate (change) to become a different allele (this can happen in several different ways, as we will detail in Key Concept 9.3). For example, we can envision that at one time all pea plants made round seeds and had the seed-shape allele *R*, which encodes an expressed, functional copy of the enzyme SBE1. At some point, a mutation in *R* resulted in a new allele, *r* (wrinkled seeds), which does not express a functional SBE1. If this mutation were present in a cell that underwent meiosis, some of the resulting gametes would carry the *r* allele, and some offspring of this pea plant would carry the *r* allele in all of their cells.

Geneticists usually define one allele of a gene as the **wild type**; this is the allele that is present in most individuals in nature ("the wild"). Other alleles of that gene are usually called mutant alleles, and they may produce different phenotypes. The wild-type and mutant alleles are inherited according to Mendelian laws. A gene with a wild-type allele that is present less than 99 percent of the time (the rest of the alleles being mutant) is said to be **polymorphic** (Greek *poly*, "many," + *morph*, "form"). It is important to emphasize that "wild type" and "dominant" are not the same thing. An allele that is uncommon in a population can be dominant or recessive in terms of its effect on a trait (see Figure 8.7). For example, in humans a mutant allele of the gene *MC1R* results in the phenotype of red hair and is inherited as a recessive, while a mutant allele of the gene *HTT* results in the phenotype of the neurological disorder Huntington's disease and is inherited as a dominant.

By producing phenotypic variety, mutations provide raw material for evolution. An allele may become more or less prevalent in a population, depending on its effect on the reproductive fitness of the individuals carrying it; see **Key Concept 13.2**.

Sex-linked inheritance is consistent with sex chromosome transmission

Much of our understanding of sex linkage and crossing over (see Key Concept 7.3) came from studies in the fruit fly *Drosophila melanogaster*. You may be familiar with this organism from your pantry, farmers' market, or school lab. The fruit fly was used by Thomas Hunt Morgan as an experimental organism for studying genetics soon after the rediscovery of Mendel. Fruit flies met many of the requirements of a good organism to study genetics (see Key Concept 8.1): they are inexpensive to rear in large numbers, have short generation times (an entire generation in a few weeks) and scorable phenotypes, and are easy to use for controlled matings. The fruit fly became a model organism for genetic research and remains so to this day.

The chromosomal makeup of the fruit fly genome is different than that of Mendel's peas in an important way—flies have a pair of **sex chromosomes**, which are the chromosomes involved in sex determination that differ in numbers between males and females. Chromosomes that are not sex chromosomes are termed **autosomes** and always appear in two copies regardless of the sex of the organism. Sex chromosomes, by contrast, do not always occur in homologous pairs. Female flies have a pair of large sex chromosomes that are homologs termed X chromosomes: females are XX. In contrast, male flies have a pair of sex chromosomes that differ from one another, consisting of a large X and a very small Y chromosome: males are XY. Like flies, human females are XX and males are XY. Sex chromosomes determine the sex in both flies and humans: XX individuals develop as females and XY individuals as males. Flies have three pairs of autosomes, each consisting of a pair of homologous chromosomes, and one pair of sex chromosomes.

In female fruit flies (and humans), the X chromosomes behave just like a pair of autosomes in meiosis, such that every egg ends up with one copy of the X chromosome. In males, the X and Y chromosomes, though different, do pair with one another and align together at the metaphase I plate. They then segregate at anaphase I, eventually resulting in half the sperm containing the X chromosome and half containing the Y chromosome. Fertilization of an X-bearing egg by an X-bearing sperm produces a daughter, whereas fertilization by a Y-bearing sperm produces a son.

The fact that the X is a large chromosome and the Y is small implies that many genes on the X chromosome are not expected to be present on the Y, which means that males have only one copy of these genes. Any gene that is present as a single copy in a diploid organism is called **hemizygous**. A male will express only the allele of his one X chromosome, whether that allele is dominant or not. The difference in copy number of genes on the X chromosome in males and females should result in unusual inheritance patterns for genes on the X chromosome. Finding such a pattern is evidence that genes are on chromosomes. Just such a pattern was seen for an eye-color mutant in *Drosophila* (**FIGURE 8.9**).

Morgan identified an eye-color gene in *Drosophila* in which the wild-type allele confers red eyes and a recessive mutant allele confers white eyes. Crosses between red-eyed flies and white-eyed flies of different sexes showed unusual patterns of inheritance that were consistent with the eye-color gene being on the X chromosome. The convention for writing genotypes for genes on sex chromosome is as follows: if the eye-color alleles are designated as *W* (red eyes) and *w* (white eyes), the presence of the alleles on the X chromosome can be designated by X^W and X^w.

Morgan crossed a homozygous red-eyed female (X^WX^W) with a hemizygous white-eyed male (X^wY; note that the Y does not carry the eye-color gene). All the male and female offspring from this cross had red eyes, because red (*W*) is dominant over white (*w*) and all the progeny inherited a wild-type X chromosome (X^W) from their mother (see Figure 8.9A). Note that this phenotypic outcome would have occurred even if the eye-color gene had been present on an autosome rather than a sex chromosome—it is the same pattern seen by Mendel in the F_1 monohybrid cross.

However, when Morgan performed the reciprocal cross, in which a white-eyed female (X^wX^w) was mated with a red-eyed male (X^WY), the results were unexpected: *all the male offspring were white-eyed and all the female offspring were red-eyed* (see Figure 8.9B). These results differed from those of reciprocal crosses performed on peas by Mendel, where the results were the same in both types of parental crosses. A key difference between Mendel's and Morgan's results was the fact that peas don't have sex chromosomes, whereas fruit flies have sex chromosomes containing genes. In the fruit flies, all sons from the reciprocal cross inherited their only X chromosome from their white-eyed mother and were therefore hemizygous for the white allele. The daughters, however, got both an X chromosome bearing the white allele (*w*) from their mother and an X chromosome bearing the red allele (*W*) from their father; therefore they were red-eyed heterozygotes. So Morgan's offspring ratios in reciprocal crosses could be explained by the eye-color gene being carried on the X chromosome. This further bolstered the chromosome theory that genes are carried on chromosomes.

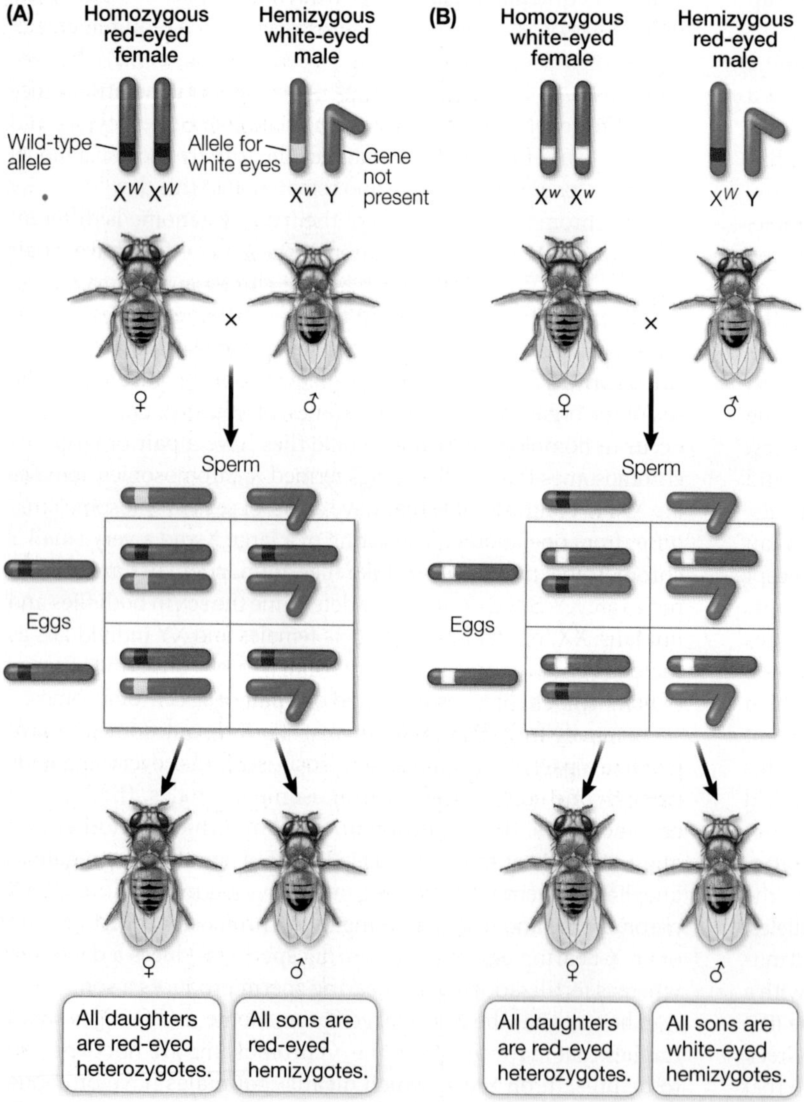

The term **sex-linked inheritance** defines the inheritance of a gene that is carried on a sex chromosome. (The term "sex-linked" can be misleading because this pattern of inheritance is not really linked to the sex of an organism—after all, both males and females carry X chromosomes. Instead it should be interpreted as "sex-chromosome linked." But the term "sex-linked inheritance" remains in the scientific literature.) Pedigree analyses in humans reveal characters that are inherited in a manner consistent with sex linkage. Most sex-linked characters involve rare X-linked recessive traits, such as the one in **FIGURE 8.10**. Inheritance of X-linked recessive traits reveals the following patterns (compare with the pedigrees for autosomal characters in Figure 8.7): (1) The recessive phenotype appears much more often in males than in females, because only one copy of the allele is needed for its expression in males, whereas two copies must be present in females. (2) An affected male passes the mutation to all of his daughters but to none of his sons, who instead get his Y chromosome. (3) Daughters who receive one X-linked mutation are heterozygous carriers. They are phenotypically normal, but they pass the mutant allele to (on average) half their sons and half their daughters (equal segregation). (4) Affected males often have affected maternal grandfathers.

FIGURE 8.9 The Inheritance of White Eye Color in Fruit Flies The gene controlling white versus red eye color is carried on the X chromosome in fruit flies. Reciprocal crosses give different phenotypic ratios in offspring.

What would be the result of crossing a heterozygous red-eyed female with a red-eyed male?

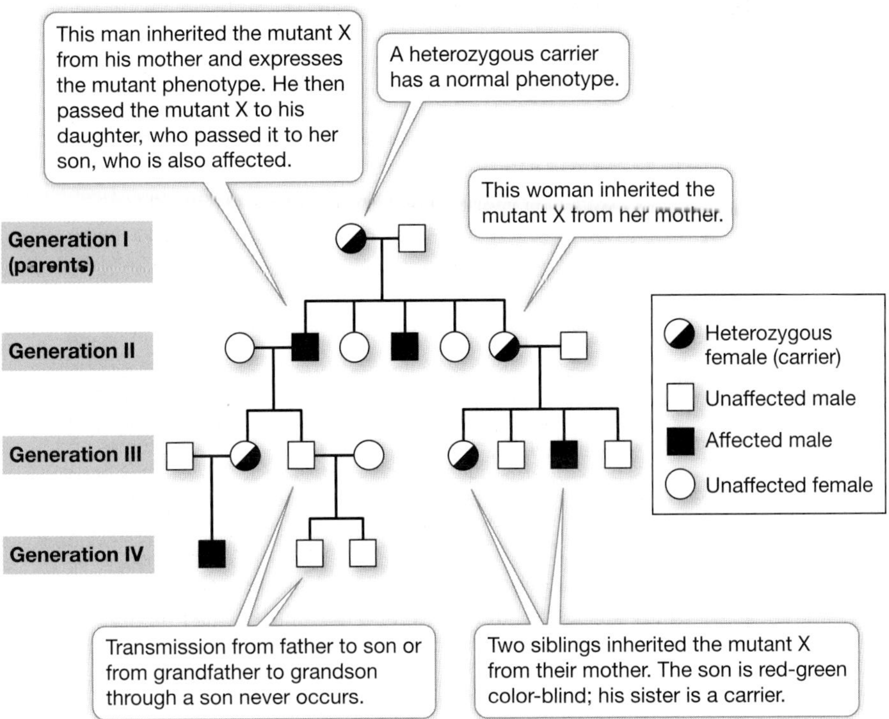

FIGURE 8.10 Red-Green Color Blindness Is an X-Linked Recessive Trait in Humans Color blindness is substantially more common in males than females. Male-to-male transmission never occurs, but males can transmit to grandsons (or later generations) through daughters.

X-linked dominant traits also show unusual inheritance patterns. Like all dominant characters, they never skip a generation (looking backward), but unlike with autosomal genes, affected males always have affected mothers, since there is no male-to-male transmission.

Animation 8.2 **Alleles That Do Not Assort Independently**

Genes on the same chromosome can be separated by crossing over in meiosis

Morgan performed many Mendelian crosses with fruit flies and found that Mendel's law of segregation always held true. In contrast, dihybrid crosses sometimes yielded phenotypic ratios that were not predicted by Mendel's law of independent assortment. For example, Morgan performed a cross between flies that varied in body-color and wing-size genes. The body-color allele *B* (wild-type gray body) is dominant over *b* (black body), and the wing-size allele *Vg* (wild-type wing) is dominant over *vg* (vestigial, or very small, wing). Morgan first crossed *BBVgVg* and *bbvgvg* flies and obtained F_1 flies that were heterozygous at both genes, *BbVgvg*. He then test-crossed these F_1 flies with *bbvgvg* flies to generate an F_2. Based on Mendel's second law, Morgan expected to see four phenotypes in a ratio of 1:1:1:1, but that is not what he observed. The body-color gene and the wing-size gene did not assort independently; instead, two phenotypes were common and two were rare (**INVESTIGATION FIGURE 8.11**).

These results became understandable when Morgan considered the possibility that the two genes are on the same chromosome and perhaps unable to assort independently as predicted by Mendel's law of independent assortment because they are physically linked together. What might we expect to see in this test cross if the body-color and wing size genes are on the same chromosome? Consider first the parents that produce the F_1 double heterozygote. One parent can produce only *BVg* gametes and the other can produce only *bvg* gametes. These come together to produce the double heterozygote, but importantly, the *B* allele is on the same homolog as the *Vg* allele, and the *b* allele is on the same homolog as the *vg* allele. We can illustrate this relationship by showing the F_1 genotype as *BVg/bvg*. When this double heterozygote undergoes meiosis, we might expect to see only *BVg* and *bvg* gametes. Such gametes are called **parental** because they are the same genotype as the gametes from the individual's parents that fused to make the double heterozygote. These parental gametes would produce F_2 offspring that were either gray flies with normal wings or black flies with vestigial wings in a 1:1 ratio in a test cross.

However, we also see the other two possible F_2 phenotypes, gray body with vestigial wings and black body with normal wings, indicating that the double heterozygote also produces *Bvg* and *bVg* gametes, which are termed **recombinant** because they differ from the two gametes that fused to make the double heterozygote. How are recombinant gametes produced if the two genes are on the same chromosome, such that alleles are physically located on the same DNA molecule?

Some of Morgan's F_2 flies displayed recombinant phenotypes, from recombinant gametes, because homologous chromosomes can physically exchange corresponding segments during prophase I of meiosis by crossing over (**FIGURE 8.12**; see also Figure 7.13). Each exchange event involves two of the four chromatids present in the paired homologs. In a crossover, a chromatid from one homolog and a nonsister chromatid from the other homolog both break at corresponding loci and then are rejoined to one another. Crossovers can occur at any point along the length of the chromosome. Both chromosomes involved in crossing over become recombinant (each chromatid ends up with alleles from both parental gametes) since chromosome segments are exchanged reciprocally.

A crossover event between the body-color and wing-size genes will produce recombinant gametes that do not have the parental genotypes. The probability of obtaining recombinant gametes is lower for genes that are physically closer together than for genes further apart on the chromosome. There is thus a relationship

INVESTIGATION

FIGURE 8.11 Alleles for Some Pairs of Genes Do Not Assort Independently Morgan showed that the genes for body color and for wing size are linked on the same chromosome, so that their alleles do not assort independently.[a]

HYPOTHESIS

Alleles at different genes always assort independently into gametes. (This hypothesis is based on Mendel's second law.)

METHOD

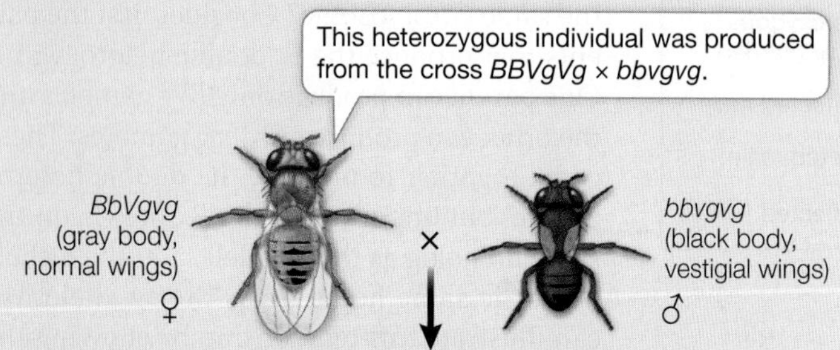

Cross true-breeding gray-bodied, normal-winged flies with true-breeding black-bodied, vestigial-winged flies to produce F_1 flies that are heterozygous at both genes (*BbVgvg*), then test cross the F_1 and count offspring phenotypes.

RESULTS

Genotypes	*BbVgvg* Gray, normal	*bbvgvg* Black, vestigial	*Bbvgvg* Gray, vestigial	*bbVgvg* Black, normal
Expected phenotypes if independent assortment occurs	575	575	575	575
Observed phenotypes (number of individuals)	965	944	206	185

These are the results expected from Mendel's second law (independent assortment)...

...but the actual results were inconsistent with the law.

Parental (first two columns); Recombinant phenotypes (last two columns)

CONCLUSION

The expected and observed numbers of offspring are substantially different. The hypothesis is rejected. The alleles for these two genes do not assort independently. The reason is that these genes are linked on the same chromosome and thus their alleles cannot align independently at metaphase I.

ANALYZE THE DATA

Mendel's work was "rediscovered" 35 years after its publication. Soon after the rediscovery, biologists began to find some exceptions to the rules of inheritance that Mendel had proposed. Thomas Hunt Morgan and his colleagues made dihybrid test crosses in fruit flies. In addition to looking for deviation from a 9:3:3:1 F_2 ratio of phenotypes in a Mendelian dihybrid cross, they proposed that the clearest way to test for linkage was to look for deviations from the 1:1:1:1 ratio expected from a double heterozygote × homozygous recessive (at both genes) test cross (see Figure 8.4). Morgan's group then hypothesized that linkage had a physical basis, namely that genes are linked together on chromosomes and that rare crossing over during meiosis gives rise to the less frequent phenotypes. Examination of actual chromosomal events confirmed this.

1. Morgan performed a dihybrid cross between black, normal-winged flies (*bbVgVg*) and gray, vestigial-winged flies (*BBvgvg*). The F_1 flies (*BbVgvg*) were interbred as in a classic Mendelian cross, yielding the F_2 phenotypes shown in the table below (Experiment 1). Compare these data with the expected data in a 9:3:3:1 ratio by using the chi-square test (see Appendix B for information about the chi-square test). Are there differences, and are they significant?
2. To quantify linkage, Morgan test crossed the *BbVgvg* F_1 females with black, vestigial-winged males (*bbvgvg*). (You should note that this is not the same test cross as the one shown in the experiment above. In that case, the original parents of the *BbVgvg* double heterozygote were *BBVgVg* and *bbvgvg*.) The results of this test cross are shown in the table below (Experiment 2). Compare the expected and observed numbers of each type of offspring using the chi-square test to determine if these genes are linked.
3. In a third experiment, Morgan crossed two strains of flies that were homozygous for the body color and wing genes. The resulting F_1 flies were all gray and normal-winged, and these were interbred. The results are shown in the table below (Experiment 3). What were the genotypes and phenotypes of the original parents that produced the F_1?

	Number of progeny showing each phenotype			
Experiment	Gray, normal	Black, normal	Gray, vestigial	Black, vestigial
1	2,316	1,146	737	0
2	578	1,413	1,117	307
3	246	9	65	18

[a]T. H. Morgan. 1912. *Science* 36: 719–720.

A similar **Analyze the Data** exercise may be assigned in **Achieve**.

between the physical distance between genes and the probability of obtaining a recombinant gamete. Geneticists, starting with Morgan, often calculate **recombination frequency** between two genes, which is calculated by dividing the number of recombinant progeny by the total number of progeny, to estimate how far apart the genes are on the chromosome (**FIGURE 8.13**). Recombination frequencies can be converted to a **genetic map distance**, in **map units** (also called centimorgans, cM), where 1 map unit equals 1 percent recombination frequency. Once distances among many pairs of genes have been calculated, they can

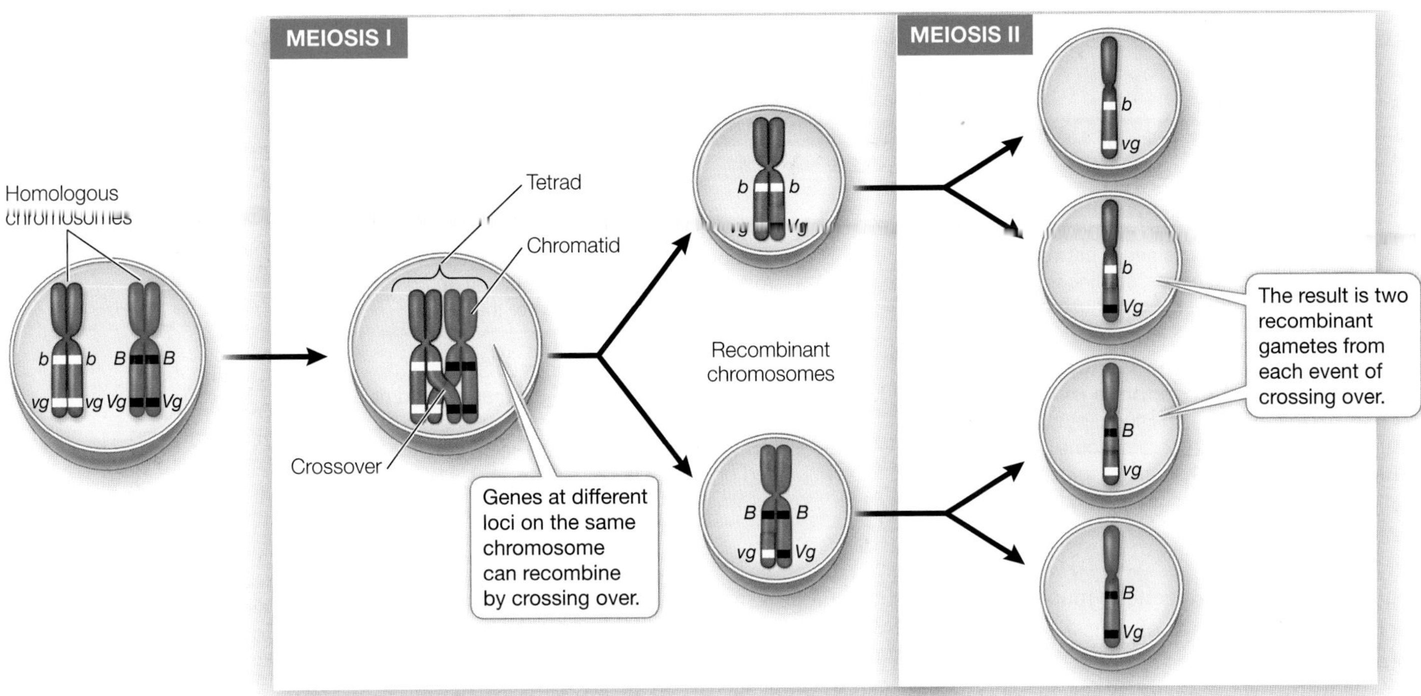

FIGURE 8.12 Crossing Over Produces Recombinant Gametes When genes are on the same chromosome, they can still produce recombinant gametes because of crossovers, which are physical breaks and reattachments that occur between nonsister chromatids.

Is it possible for a different *BbVgvg* double heterozygote to produce recombinant gametes that are *BVg* and *bvg*?

FIGURE 8.13 Calculating Genetic Map Distances Genetic maps indicate how far apart genes are on a chromosome. To calculate the map distance, the number of recombinant offspring (those with a phenotype different from either parental gamete) in a test cross is divided by the total number of offspring to obtain the frequency of recombinant offspring. The result is multiplied by 100 to give the percent of recombinant offspring. One percent recombination frequency equals 1 map unit. Recombinant versus parental gametes (and thus offspring) are scored with respect to the double heterozygote parent. The tester parent can only produce one type of gamete (*bvg*).

If the double heterozygote had been *Bvg*/*bVg* instead of *BVg*/*bvg* as shown, what would the expected frequencies of offspring have been in a test cross?

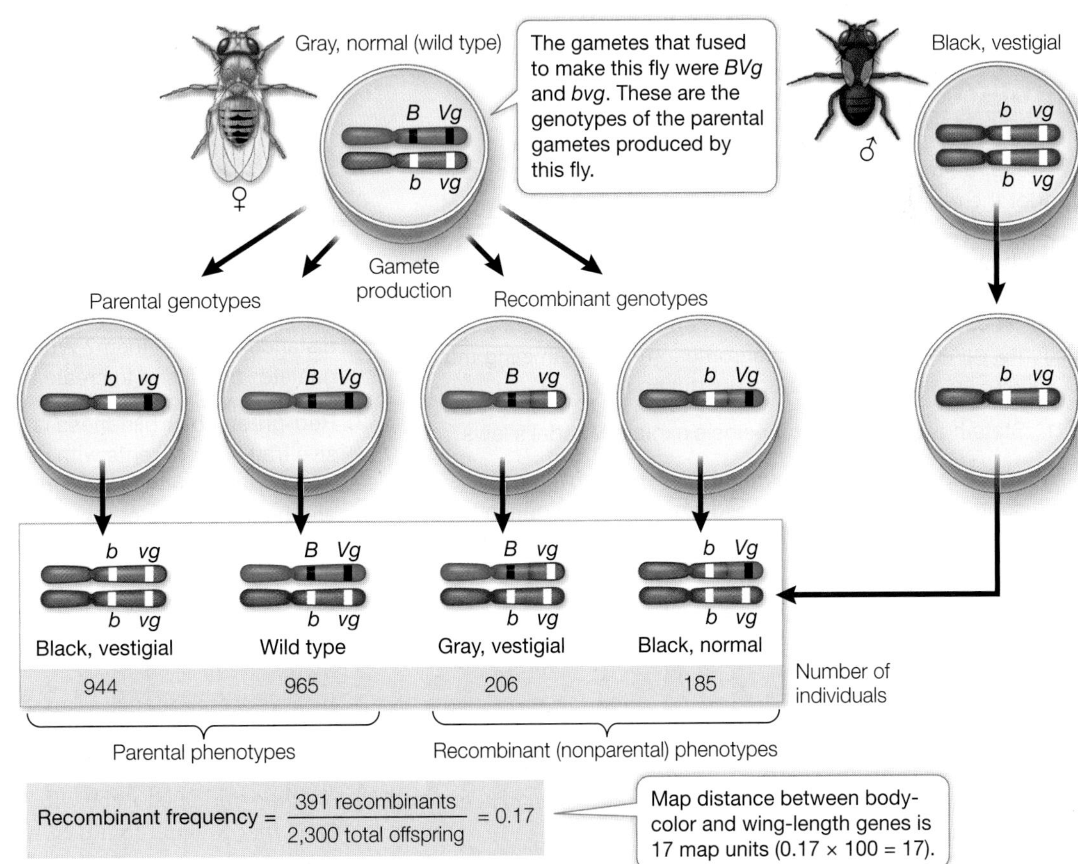

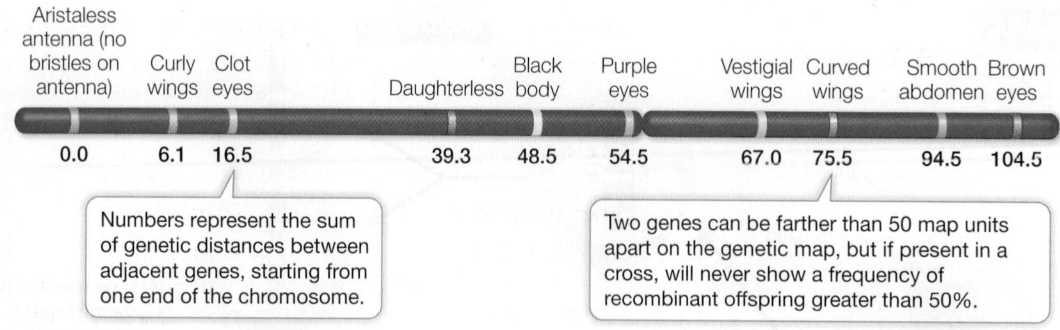

FIGURE 8.14 Part of the Genetic Map of Chromosome 2 of *Drosophila melanogaster* This map shows the positions, in map units (centimorgans), of several genes on chromosome 2. Hundreds of genes affecting various characters have been mapped to each of the four chromosomes found in this species.

(Data from *yourgenome.org*, website of Genome Research Limited. 2016. How do you map a genome? [online] Available at: https://www.yourgenome.org/facts/how-do-you-map-a-genome [Accessed 31 Oct. 2018. Last updated on 2016-06-02.])

be used to determine a genetic map for a chromosome. Genetic maps show the positions of genes/loci on a chromosome, where distances represent the sum of the distances between adjacent genes, starting from one end of the chromosome (**FIGURE 8.14**).

When genes are far apart on the same chromosome, there can be multiple crossovers between them. Multiple crossovers will randomize which alleles of the two genes end up on the same chromatid, which means that the frequency of parental and recombinant gametes will be equal. This indicates that, for two genes, the frequency of recombinant gametes is always less than or equal to the frequency of parental gametes. Mendel's law of independent assortment thus holds both for pairs of genes on different chromosomes and for pairs of genes on the same chromosome that are physically far apart.

Determining the genetic map distances among genes using crosses has been critically important for identifying the genes that cause a variety of phenotypes, from those important for plant and animal yield to those that cause disease in humans. While the modern era of genome sequencing (see Chapter 12) has made mapping somewhat less important in some areas of genetics research, it remains critically important for identifying genes underlying important traits in numerous species.

REVIEW & APPLY | 8.2

R The chromosome theory of inheritance was advanced soon after the rediscovery of Mendel's work, because his laws of segregation and independent assortment were consistent with the behavior of chromosomes during meiosis. The inheritance of sex-linked traits, which is consistent with the behavior of sex chromosomes in meiosis, further supported the theory. Linked genes can be mapped on a chromosome by examining the results of crossing over that occurs during meiosis I.

A

1. State how the events of meiosis explain Mendel's laws.
2. In *Drosophila*, a cross is made involving two genes: one controls eye color, with *S* (wild-type red eyes) dominant to *s* (mutant scarlet eyes), and another controls body color, with *E* (wild-type gray body) dominant to *e* (mutant ebony body). A double heterozygote, *SsEe*, is test-crossed and 1,000 offspring are obtained. These offspring have the following phenotypes: 140 red eyes and gray body, 361 red eyes and ebony body, 369 scarlet eyes and gray body, and 130 scarlet eyes and ebony body. What is the genotype of the test-cross parent? Are these two genes linked? If so, what is the map distance between them? What were the genotypes of the two gametes that fused to create the double heterozygote parent?
3. Red-green color blindness is inherited as an X-linked recessive trait. Two parents with normal color vision have a child who is red-green color-blind. Is the child a boy or girl? What are the genotypes of the child's parents? If the parents have another child, what is the probability that the child will be red-green color-blind?

The laws of inheritance as articulated by Mendel laid the groundwork for all future studies of genetics. As we will see next, the number of alleles at a single gene, and the relationship between genotype and phenotype with respect to dominance, can deviate from what Mendel found in his studies. In addition, genes can interact with one another to affect some phenotypes.

Alleles, Genes, and the Environment Interact to Produce Phenotype

LEARNING OBJECTIVES

By the end of this key concept you should be able to:

8.3.1 **Distinguish among dominance, incomplete dominance, and codominance.**

8.3.2 **Describe how epistasis is detected in a cross.**

8.3.3 **Give an example of how the environment can alter expression of a phenotype.**

Phenotypes do not always follow the simple patterns of inheritance shown by the pairs of alleles for Mendel's pea characters described in Key Concept 8.1. Existing alleles are subject to change by mutation and can give rise to new alleles—in fact, a single gene can have many alleles. In addition, alleles do not always show the dominant–recessive relationships observed by Mendel. A single allele may have multiple phenotypic effects, and a single character may be controlled by multiple interacting genes. The expression of a gene is generally affected by interactions with other genes and with the environment.

A single gene can have multiple alleles

Mendel developed his theories by studying just two alleles of each gene. But often a gene has multiple alleles (although any diploid individual can carry only two of them, one from each of the gametes that fused to make them). The alleles may show a hierarchy of dominance. An example is one of the coat-color genes in rabbits, which has four different alleles: *C* (dark gray), c^{chd} (chinchilla, a lighter gray), c^h (Himalayan, or point restricted), and *c* (albino) (**FIGURE 8.15**). These alleles show a dominance hierarchy with $C > c^{chd}$, $c^h > c$. Any rabbit with the *C* allele (paired with itself or another allele) is dark gray, and a *cc* rabbit is albino. The other colors result from different allele combinations, as shown in Figure 8.15. As this example illustrates, multiple alleles can increase the number of possible phenotypes determined by variation at a single gene.

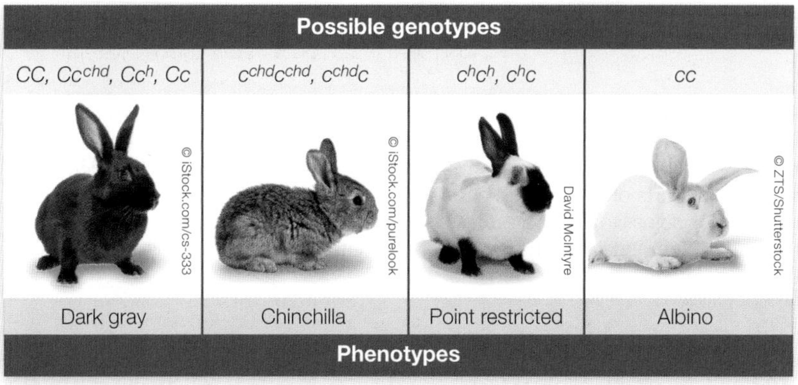

Possible genotypes			
CC, Cc^{chd}, Cc^h, Cc	$c^{chd}c^{chd}$, $c^{chd}c$	c^hc^h, c^hc	cc
© iStock.com/cs-333	© iStock.com/purelook	David McIntyre	© ZTS/Shutterstock
Dark gray	Chinchilla	Point restricted	Albino
Phenotypes			

FIGURE 8.15 Four Alleles of a Coat-Color Gene in Rabbits
These photographs show the phenotypes associated with nine of the possible genotypes. Note that an individual can have at most two alleles. Different combinations of two alleles result in different coat colors and pigment distributions. Multiple alleles at a given gene often show a hierarchy of dominance. In this case, $C > c^{chd}$, $c^h > c$, implying *C* is dominant to all other alleles and *c* is recessive to all other alleles.

Dominance is not always complete

Many genes have alleles that are neither fully dominant nor fully recessive to one another. Instead, the heterozygotes show an intermediate phenotype, a situation called **incomplete dominance**. For example, in eggplants a cross between a true-breeding purple-fruited plant and a true-breeding white-fruited plant produces an F_1 with violet (light purple) fruits. Recall that in Mendel's monohybrid crosses, he did not see F_1 offspring that had new phenotypes, different from those of the parents. In the F_2, eggplants produce a 1:2:1 phenotypic ratio (purple:violet:white), reflecting the frequency of the three genotypes (**FIGURE 8.16**). Thus the violet phenotype is specific to the heterozygote and is intermediate between the two homozygotes. Incomplete dominance simplifies genotyping since heterozygotes can be distinguished from homozygotes by their phenotype.

Sometimes two alleles of a gene both produce their phenotypes when present in a heterozygote—a phenomenon called **codominance**. An example is the ABO blood group in humans (this is also an example of multiple alleles). Blood groups are identified based on whether or not two particular surface glycoproteins—A and B—can be detected on red blood cells. The blood group can be A, B, AB, or O depending on whether A, B, both A and B, or neither A nor B glycoproteins are detected. A single gene, *I*, determines blood group by encoding an enzyme involved in the attachment of sugars to the glycoproteins. There are three alleles of the gene: I^A, which encodes an enzyme that attaches a particular sugar to make glycoprotein A; I^B, which encodes an enzyme that attaches a particular sugar to make glycoprotein B; and *i*, which does not make a functional enzyme. Both I^A and I^B are dominant to *i* such that I^AI^A and I^Ai are blood group A, and I^BI^B and I^Bi are blood group B. However, when both the I^A and I^B alleles are present in a heterozygote, both phenotypes (glycoproteins A and B) are expressed, resulting in the AB blood group. Thus I^A and I^B are codominant, and both are dominant to the *i* allele. Individuals who are *ii* homozygotes have blood group O since they lack glycoproteins A and B. This locus is critically important for blood transfusions because the A and B glycoproteins are antigenic: if a red blood cell with the A glycoprotein on its surface gets into the bloodstream of a person who lacks the I^A allele, the recipient produces antibodies against the "nonself" cells, which causes a variety of problems. It is thus important to match donors and recipients correctly. Individuals who have blood group AB do not produce antibodies against either glycoprotein and can receive blood from anyone. In contrast, individuals who have blood group O have neither glycoprotein and recognize both as nonself. Such individuals can receive only type O blood.

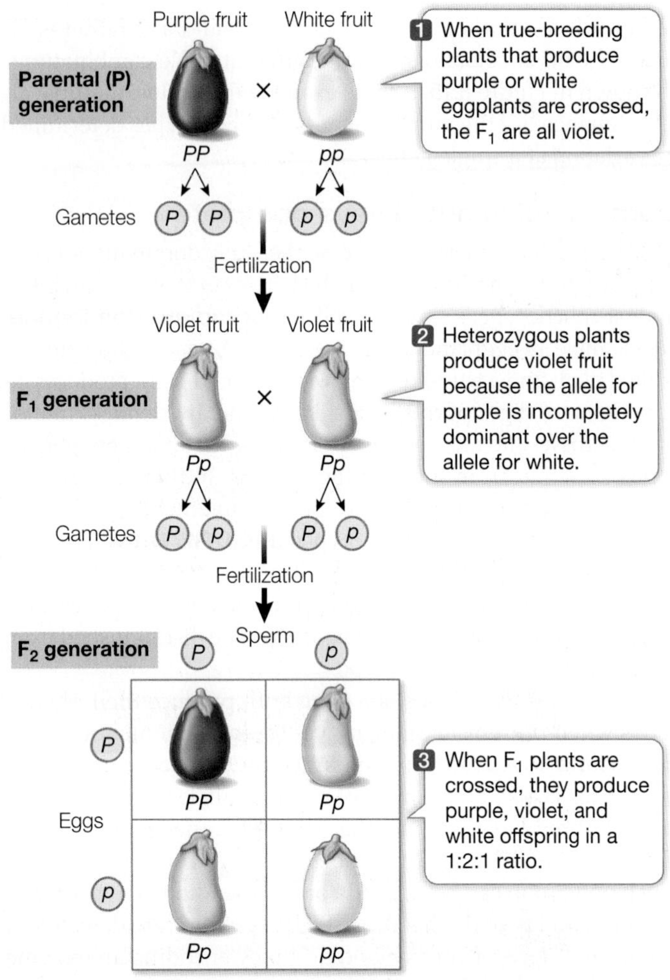

FIGURE 8.16 Incomplete Dominance in Eggplant Fruit Color A new phenotype, not present in parents, appears in the F_1. However, forming the F_2 makes it clear that the fruit-color gene still obeys Mendel's first law. The heterozygote has a phenotype that is intermediate between those of the two homozygotes, which is an example of incomplete dominance.

Reactions against A or B glycoproteins are an adaptive immune response, as described in **Key Concept 36.3**.

One gene can affect multiple characters, and one character can be affected by multiple genes

A single gene can produce a protein that functions in many different parts of an organism. Such genes are termed **pleiotropic**, and mutant alleles of them can have phenotypic effects on multiple characters. For example, cystic fibrosis is an autosomal recessive disorder (recall Figure 8.7B) that causes breathing problems, digestive problems, infertility, bone thinning, and imbalances in electrolytes (ions in blood) (**FIGURE 8.17**). Many proteins function in multiple parts of an organism, or at different times during the life of an organism, and for this reason pleiotropy is common.

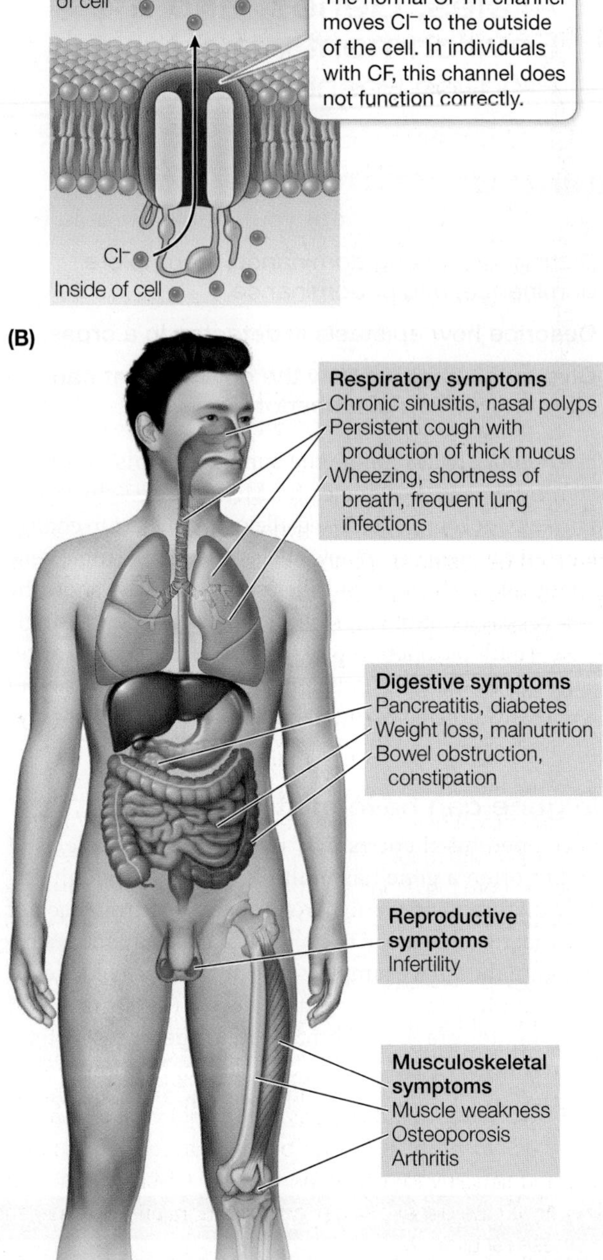

FIGURE 8.17 The Gene That Causes Cystic Fibrosis Is Pleiotropic **(A)** The cystic fibrosis gene (*CFTR*) encodes a transmembrane chloride ion (Cl^-) channel that is found in cells in various parts of the body. The recessive cystic fibrosis-causing allele encodes a defective channel protein that does not function correctly. **(B)** When an individual is homozygous for the mutant cystic fibrosis allele, many different characters are affected.

Proteins interact with one another in a variety of ways. For example, they might catalyze different steps of the same pathway, or they might be subunits of a larger protein. These interactions cause the phenotypic expression of particular genotypes at one gene to be affected by the genotypes present at a second gene, a phenomenon known as **epistasis**. One example of

THINK LIKE A SCIENTIST

Quantitative Reasoning

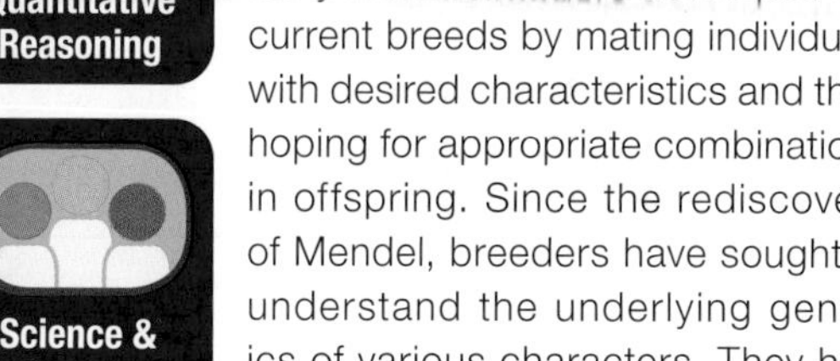

Science & Society

Coat color inheritance in Labrador retrievers

Early animal breeders developed our current breeds by mating individuals with desired characteristics and then hoping for appropriate combinations in offspring. Since the rediscovery of Mendel, breeders have sought to understand the underlying genetics of various characters. They have used the laws of inheritance discovered by Mendel to guide their studies—an example of the link between science and society. How have they done this?

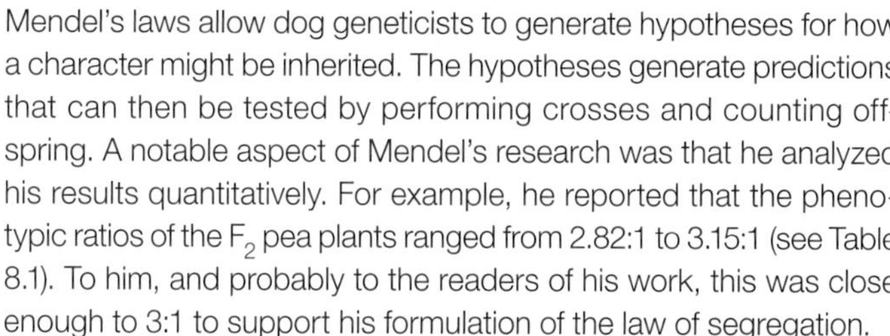

Mendel's laws allow dog geneticists to generate hypotheses for how a character might be inherited. The hypotheses generate predictions that can then be tested by performing crosses and counting offspring. A notable aspect of Mendel's research was that he analyzed his results quantitatively. For example, he reported that the phenotypic ratios of the F_2 pea plants ranged from 2.82:1 to 3.15:1 (see Table 8.1). To him, and probably to the readers of his work, this was close enough to 3:1 to support his formulation of the law of segregation.

Modern scientists have taken quantitative reasoning further by using a branch of mathematics called statistics. Tests have been developed to determine whether, in Mendel's data for example, 2.82:1 is really a 3:1 ratio with random fluctuations, or whether it is a significant deviation from the expected 3:1 ratio that therefore invalidates the biological explanation based on that ratio. (See Appendix B for various statistical tests that are used by biological scientists to test their hypotheses.)

Suppose a dog geneticist is interested in understanding the pattern of inheritance of three coat colors in Labrador retrievers: black, brown, and yellow (**FIGURE 8.18**). True-breeding strains can be developed for all three colors, so that crossing a true-breeding black to a true-breeding black always produces black puppies, and similarly for brown and yellow. Given enough true-breeding dogs, the geneticist performs monohybrid crosses and obtains these results:

Cross	F_1	F_2
Black × brown	18 black	33 black, 9 brown
Black × yellow	12 black	27 black, 10 yellow
Brown × yellow	14 brown	31 brown, 10 yellow

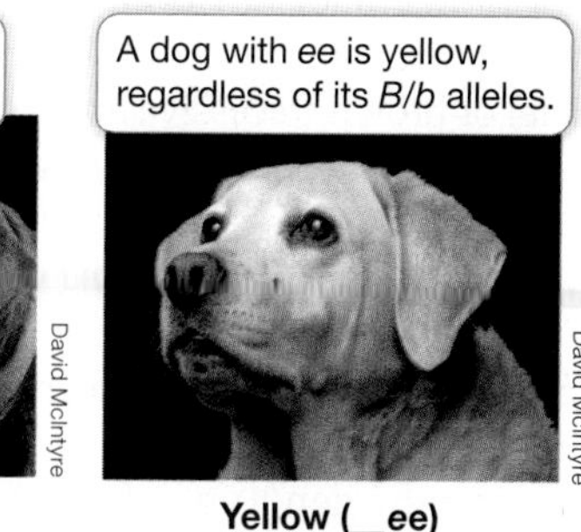

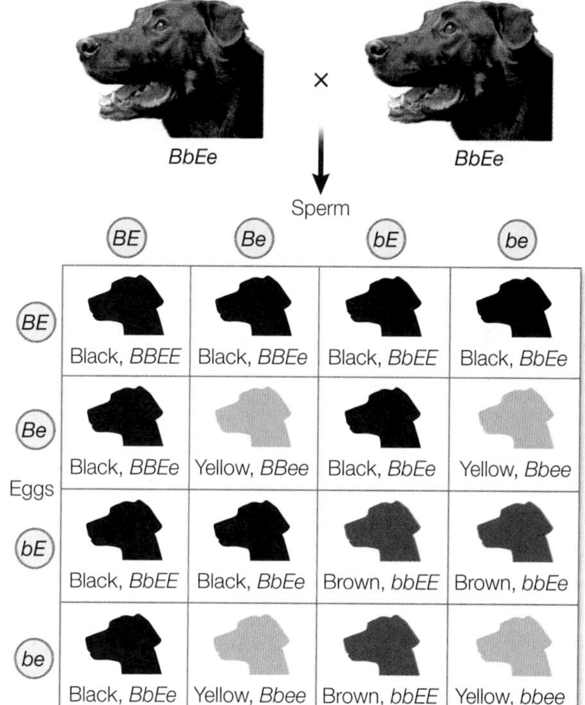

FIGURE 8.18 Coat Color in Labrador Retrievers
The unlinked coat-color and deposition genes interact epistatically to produce the coat-color phenotype. Using Mendel's laws and a Punnett square, the expected offspring phenotypes from a cross of two black dogs that are heterozygous at both genes (*BbEe*) can be determined. The result of the cross is a "modified 9:3:3:1," with 9/16 black, 3/16 brown and 4/16 (3/16 + 1/16) yellow.

In a test cross between *BbEe* and *bbee* Labrador retrievers, what are the phenotypes of the parents and what colors are the puppies expected to be?

1. Are these data approximately consistent with a single-gene, two-alleles inheritance pattern?
2. Using the chi-square test (see Appendix B and www.graphpad.com/quickcalcs/chisquared1.cfm), determine whether these data statistically support the hypothesis of coat color being determined by a single gene obeying Mendel's law of segregation.
3. The geneticist then made a cross between another group of true-breeding strains and obtained the data shown below. Is the F_1 result consistent with coat color being due to a single gene?

Cross	F_1	F_2
Brown × yellow	14 black	25 black, 8 yellow, 7 brown

epistasis is seen in the way in which two genes determine coat color in Labrador retrievers. One gene, *B*, is the pigment gene and determines whether a dog can produce black (dominant *B* allele) or brown (recessive *b* allele) pigment. The second gene, *E*, is the pigment deposition gene and determines whether pigment (black or brown) is incorporated into the hair of the dog. A dog that has the dominant *E* allele will incorporate the pigment and thus be black or brown, while a dog that is homozygous for the recessive *e* allele will not incorporate pigment into its coat and will be yellow (the color of the keratin protein in the hair). Thus knowing the genotype of the dog at the pigment gene is not enough information to determine the phenotype of the dog—its genotype at the deposition gene is also needed.

Some phenotypes are continuous

So far we have been considering characters that have **discrete traits**, meaning they are easily distinguished from one another and occur in two, or a few, different categories. Examples include purple versus white flower color, black versus gray body color, red versus white eyes, red-green color blindness versus normal vision, and A, B, or O blood groups. Many discrete characters have one or a few Mendelian genes that determine the observed phenotypic variation. In contrast, **quantitative traits** vary within a range. Examples include height, weight, number of bristles on a fly's body, and gallons of milk produced by a cow per day (**FIGURE 8.19**).

Quantitative traits have received a great deal of attention because they include many traits of practical significance (e.g., traits that influence seed production in crop plants and meat production in animals), and traits known to alter human health (e.g., blood sugar level). How do quantitative traits differ from discrete traits in terms of their underlying genetics? One way is that quantitative traits are often influenced by many genes—sometimes hundreds or more. And not all the genes associated with a quantitative trait influence it to the same degree. All the genes involved do, however, act in a Mendelian manner, with two alleles per gene per individual transmitted in the same manner as alleles at any other gene. Considerable research has been devoted to identifying genes affecting important quantitative traits, with the goal of altering the trait in a desired direction. For example, much genetic analysis has gone into finding ways to lower blood sugar level and increase crop yields. Once a gene that affects a quantitative trait has been identified, its specific function must be determined before interventions can be proposed and tested.

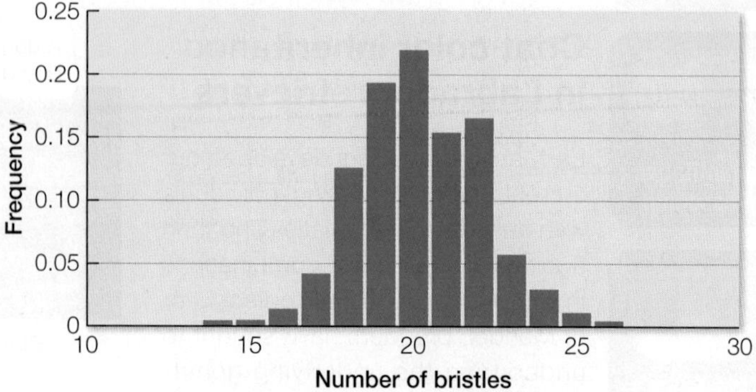

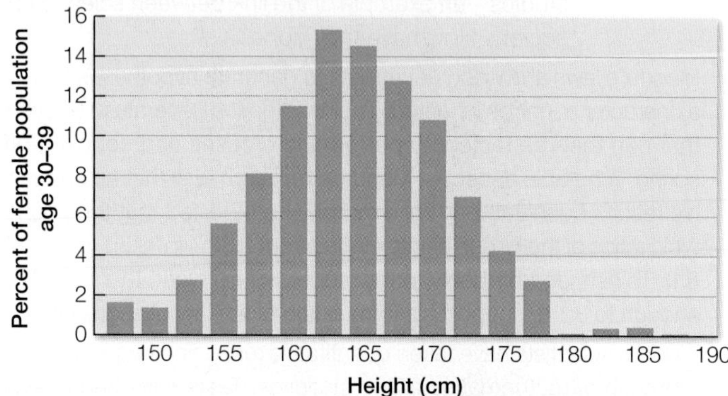

FIGURE 8.19 Quantitative Traits Can Have Many Different Values **(A)** The number of sensory bristles on a body segment in a set of more than 300 strains of *Drosophila melanogaster*. **(B)** Height of women in the United States aged 30–39. (A after T. F. Mackay and R. F. Lyman. 2005. *Philos Trans R Soc Lond B Biol Sci* 360: 1513–1527; B data from National Center for Health Statistics, based on National Health and Nutrition Examination Survey [NHANES].)

The environment affects gene action

The phenotype observed for a genotype of a particular gene may depend not only on the genotypes of other genes but also on the environment in which an organism lives. Environmental conditions include both abiotic factors, such as day length, temperature, humidity, and oxygen level, and biotic factors, such as the densities of competitors, predators, and parasites. When the phenotype expressed by a particular genotype is different in different environments, the genotype is said to show **phenotypic plasticity**. For example, the color of flowers in most hydrangeas is environmentally determined: they are blue in acidic soils and red in neutral soils (**FIGURE 8.20**).

Quantitative traits, such as height, are often affected by changes in the environment. For example, improved diet during childhood leads to a taller adult—one of the reasons that average height has increased in the last 100 years, especially in some parts of the world (up to 20-cm increases in some countries). Even for phenotypes caused by alleles at a single gene, changes in the environment can have a large effect. The rare, recessive, autosomal genetic disorder phenylketonuria (PKU) causes severe brain damage in homozygotes who eat a normal diet from birth. In contrast, lowering the level of phenylalanine (an amino acid) in the diet results in normal brain development. A change as simple as the presence of one amino acid in the diet completely changes the phenotype.

Understanding that environmental differences can affect phenotype leads us to two important realizations. First, because some of the phenotypic variation present within a population

FIGURE 8.20 Phenotypic Plasticity Flower color in most hydrangeas is determined by the pH of the soil.

will be nongenetic, it will respond less well to natural selection (see Chapter 13). Second, an individual's phenotype is determined not by genotype alone, but also by environmental conditions experienced over a lifetime. Many human traits, both physical and cognitive, are being investigated in the context of the "nature versus nurture" debate. A great deal of research has been conducted, for example, trying to understand the relative influences of genetics and the environment on intelligence and personality. The relatively new field of behavioral epigenetics investigates how the environment experienced during the lifetime of an organism can alter its behavior. Ongoing cancer research is revealing how the environment can affect the regulation of genes critical to the cell cycle. These and many other examples illustrate that many traits are influenced both by many genes and the environment.

REVIEW & APPLY | 8.3

R In Mendel's crosses, only two alleles occurred at a gene and the heterozygote always had the phenotype of one of the homozygotes. However, genes can sometimes have three or more different alleles, though an individual can only have two and a heterozygote can exhibit both homozygous phenotypes (codominance) or be intermediate between them (incomplete dominance). In addition, the phenotype exhibited by a particular genotype at a gene can be altered by both the genotype of other genes (epistasis) and the environment (phenotypic plasticity). Many complex phenotypes are determined by numerous genes.

A

1. A cross in dogs is made between a normal-coated individual and one with a double merle coat (patchy). The resulting F_1 have a merle coat (left), which is different than a double merle (right). If F_1 merle dogs are crossed with one another, the resulting puppies are one-fourth normal, half merle, and one-fourth double merle. What do these results indicate about the genotypes of normal, merle, and double merle individuals? Explain the dominance relationships for the alleles underlying these traits.

2. A daughter with blood type A has a mother with blood type O. What are their possible genotypes? What are the possible genotypes and blood types of the daughter's father?

3. If a dominant allele of one gene, *A*, is necessary for hearing in humans, and the dominant allele of another unlinked gene, *B*, results in deafness regardless of the genotype of other genes, what is the probability that a pairing of *AaBb* × *Aabb* individuals will result in a deaf child?

Sexual reproduction in eukaryotes produces haploid gametes that have novel allele combinations due to independent assortment and crossing over. Sexual reproduction is present in the vast majority of eukaryotes, suggesting that there is an evolutionary advantage to generating genetic diversity (some possible explanations will be covered in Key Concept 15.2). Prokaryotes do not have meiosis and so we might expect other mechanisms for them to generate genetic diversity, and that is exactly what we find.

8.4 Conjugation and Transformation Allow Exchange of Genetic Material between Prokaryotes

LEARNING OBJECTIVES

By the end of this key concept you should be able to:

8.4.1 **Describe how plasmids are transmitted between bacterial cells during conjugation.**

8.4.2 **Describe how the transfer of main chromosomal material differs from the transfer of plasmids during conjugation.**

As described in Key Concept 4.5, prokaryotic cells lack nuclei; they contain their genetic material mostly as a single, circular chromosome in a central region of their cells. Prokaryotes reproduce asexually by binary fission, a process that gives rise to progeny that are virtually identical genetically (see Key Concept 7.1)—that is, the offspring of cell reproduction in prokaryotes constitute a clone. How, then, do prokaryotes generate genetic diversity?

Bacteria transfer plasmid genes by conjugation

In addition to their main chromosome, many bacteria harbor additional smaller, circular DNA molecules called **plasmids** that replicate independently of the main chromosome. Plasmids typically contain at most a few dozen genes, often including genes for antibiotic resistance and genes that allow cells to transfer plasmids from one cell to another. This transfer of genes from one individual organism to another without sexual (meiotic) reproduction is called **lateral** (horizontal) **gene transfer**, to distinguish it from vertical gene transfer (gene transfer from parent to offspring). Along with mutation, lateral gene transfer generates genetic diversity among prokaryotes.

LINK The evolutionary consequences of lateral gene transfer and its role in identifying and classifying bacterial species are discussed in **Key Concepts 15.2** and **18.1**.

Electron microscopy shows that gene transfers between bacteria happen via physical contact between the cells (**FIGURE 8.21A**). Contact is initiated by a **sex pilus** (plural *pili*), a thin projection composed of proteins and phospholipids, many of which can extend from a cell (the donor). Once a sex pilus attaches to another cell (the recipient), it draws the two cells together and triggers the formation of a thin cytoplasmic bridge, the conjugation tube. Genetic material can then pass from the donor to the recipient through the conjugation tube. There is no reciprocal transfer of DNA from the recipient to the donor. This process is referred to as **bacterial conjugation**.

Plasmids move from one cell to another during conjugation, thereby transferring new genes to the recipient bacterium

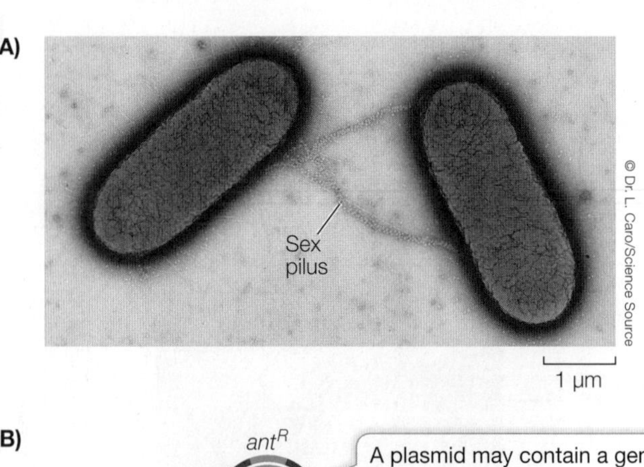

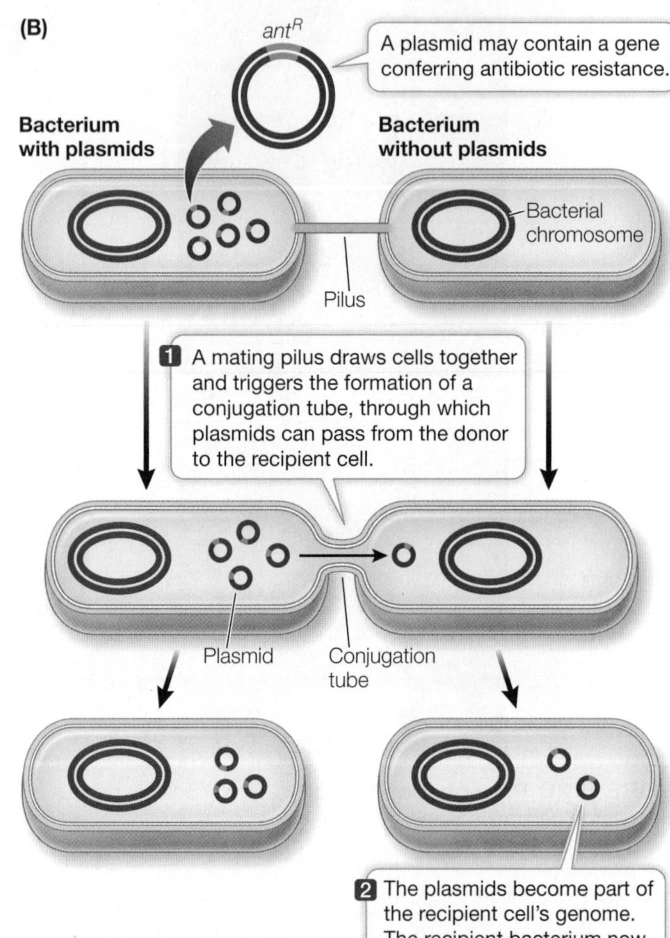

FIGURE 8.21 Bacterial Conjugation **(A)** Electron micrograph showing sex pili between two bacterial cells. **(B)** Pili bring cells together and cause the formation of a conjugation bridge. When plasmids enter a cell via conjugation, their genes can be incorporated into the recipient cell's genome. In the example shown here, one of the genes on the plasmid being transferred is a gene conferring antibiotic resistance (*antR*).

(**FIGURE 8.21B**). A single linear strand of the donor plasmid is transferred to the recipient, which is then followed by DNA synthesis of the other strand and circularization to produce an intact plasmid. The donor plasmid also has a second strand of DNA synthesized. Thus plasmids are able to transfer copies of themselves, and the genes they carry, rapidly among cells.

The evolution of drug-resistant bacteria is a major public health problem

Until the twentieth century, bacterial infections were a major scourge of humanity. With the discovery of antibiotics (particularly penicillin, which prevents the assembly of the bacterial call wall), many lethal infections were kept at bay. But over time mutations occurred in some bacteria that rendered them resistant to penicillin. When a person harboring these penicillin-resistant bacteria was treated with penicillin, the resistant bacteria were the ones that survived the drug and so they became widespread, particularly in hospitals. New antibiotics, such as methicillin, were developed that were effective against the resistant bacteria. Methicillin was successful for a while, but then the bacteria developed new mutations with a phenotype that was insensitive to methicillin. Humans are locked in an "arms race" with bacteria, and as is often mentioned in the popular press, bacteria appear to be winning. So-called superbugs that are resistant to many of our current antibiotics are showing up with increasing frequency.

How are bacteria able to evolve drug resistance so quickly? The answer is that resistance genes reside on plasmids. When a mutation occurs on this DNA, it can spread rapidly by conjugation in a population of bacteria, both within and between bacterial species (see Figure 8.21B). So conjugation, an efficient way for bacteria to gain new genes, also has public health consequences for humans.

Bacteria exchange chromosomal genes by conjugation

In addition to being able to transfer genes on plasmids by conjugation, bacteria are sometimes able to exchange genetic material present on their main chromosome during conjugation. This happens during conjugation, when the DNA strand that is transferred is part of the bacterial chromosome. While the entire main chromosome is seldom transferred, large pieces of chromosomal DNA may travel through the conjugation tube to the recipient cell and may then splice into the recipient cell's chromosome.

To illustrate what happens during this process, let's consider a bacterial donor strain of genotype *ABc* and a recipient strain of genotype *abC* (remember that bacteria have only one copy of each gene; they are effectively haploid), where these three genes occur next to one another on the chromosome. When the two strains are grown together in the laboratory, most of the cells produce clones and remain either *ABc* or *abC* in genotype. However, occasionally a few cells are found that have the genotype *ABC*. These cells arise from conjugation in which a piece of donor chromosomal DNA carrying the *A* and *B* alleles passes through the conjugation tube, lines up with the *a* and *b* region of the recipient chromosome, and then undergoes two crossover events (recombination) to swap out the alleles (**FIGURE 8.22**). A single crossover event, if it occurred, would result in the production of a large linear molecule, which would eventually be degraded, leading to the death of the cell.

Some bacteria recombine with DNA taken up from their environment

Some bacteria are able to take up DNA molecules from the environment. They do so by transporting a double-strand DNA molecule through a large protein complex in a process that is energy-requiring (recall active transport in Key Concept 4.2). Often, the DNA brought into the cell is hydrolized to nucleotides for use by the cell during DNA synthesis. However, in some cases the incoming DNA undergoes recombination (crossing over) and exchanges sequences with the recipient cell's chromosome if there are regions of homology (similar DNA sequences). If there are two crossovers between the linear DNA molecule that was taken up by the cell and the circular bacterial chromosome, new gene combinations can be produced, similar to conjugation involving the bacterial chromosome (see Figure 8.22). Genetic modification of a bacterial cell that occurs via the uptake of DNA from the environment is called **transformation**. Bacterial cells that are able to take up DNA from the environment (because they possess the DNA transporter) and be transformed are termed naturally competent. Not all bacterial cells have the DNA transporter and so are not naturally competent.

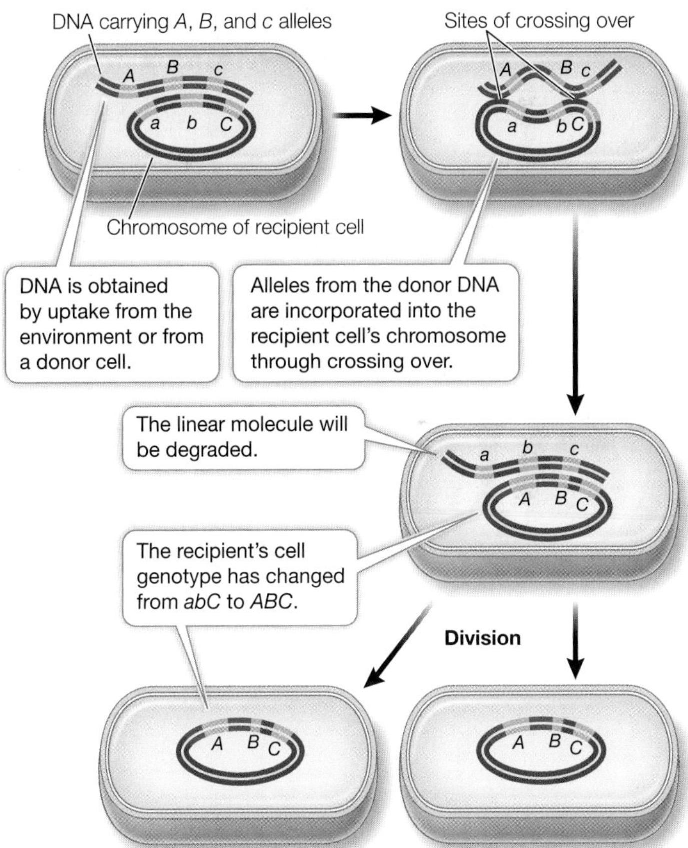

FIGURE 8.22 Alteration of Bacterial Chromosome Genotype by Conjugation and Crossing Over When a piece of linear bacterial chromosome is picked up from the environment or transferred from a donor to a recipient cell during conjugation, two crossovers at regions of homology can exchange alleles, generating new allelic combinations.

What would happen if the two crossovers occurred in a different location, such that one crossover was between the *A* and *B* gene and one was to the right of the *C* gene?

REVIEW & APPLY | 8.4

R Although they lack a sexual cycle involving meiosis and fertilization, bacteria can exchange genetic material through conjugation, a process by which plasmid genes are transferred from a donor cell to a recipient cell. Sometimes part of the bacterial main chromosome from the donor cell is transferred through the conjugation tube and may undergo crossing over with the recipient cell's chromosome. Similarly, in transformation, DNA taken up from the environment is integrated into the bacterial chromosome by crossing over.

A

1. Describe three ways in which the genes in a bacterial cell can be altered naturally. Explain whether each of these mechanisms can occur for any bacterial cell.
2. When resistance to a new antibiotic appears, it often spreads very rapidly. Why is this the case?
3. How does recombination after conjugation in prokaryotes differ from recombination during meiosis in eukaryotes?

How did the different characteristics of dogs arise, and how did selective reproduction lead to the emergence of new breeds?

Answer: Dogs were domesticated from wolves at least 15,000 years ago and, as such, represent the first domesticated species. Early dog breeders noticed the variations in body size, physical characteristics, and behaviors and then selectively bred dogs with desirable traits. Breeders interbred dogs with similar phenotypes to get animals that, like Mendel's peas, "bred true" for certain characteristics. For example, some dogs were selected and bred to be herders, resulting in shepherds and collies. Others were selected and bred to be hunters, such as retrievers and pointers. Since all breeds of dogs are the same species, they could be interbred and traits from different breeds were combined to make new breeds. Much of the history of dog breeds is being revealed through the Dog Genome Project, in which the DNA sequences of modern breeds are being analyzed to reconstruct the canine "family tree." A great deal of progress has been made determining the underlying genetics of numerous traits in dogs. In many cases, classic Mendelian crosses have been used to do so. However, as we will see in Chapter 12, DNA analysis can be used to determine the genetic basis of dog traits.

8.1 Mendel Discovered Two Laws of Inheritance

Go to ACTIVITIES 8.1 and 8.2

- **ALLELES** of a gene segregate equally into gametes, producing a 3:1 phenotypic ratio in the F_2 generation in a Mendelian **MONOHYBRID CROSS**.
- Alleles of different genes assort independently into gametes, producing a 9:3:3:1 phenotypic ratio in the F_2 generation of a Mendelian **DIHYBRID CROSS** (**FIGURE 8.6**).

FIGURE 8.6

Seed phenotypes in F_2

Parental (P) generation ×

F_1 generation

F_2 generation

315 Round yellow

108 Round green

101 Wrinkled yellow

32 Wrinkled green

Expected F_2 phenotypes obtained by combining phenotype ratios from single character crosses

Parental (P) generation *RRYY* × *rryy*

F_1 generation *RrYy*

F_2 generation

Seed shape	Seed color	Product of monohybrid cross probabilities
Round ¾	Yellow ¾	¾ × ¾ = 9/16
Round ¾	Green ¼	¾ × ¼ = 3/16
Wrinkled ¼	Yellow ¾	¼ × ¾ = 3/16
Wrinkled ¼	Green ¼	¼ × ¼ = 1/16

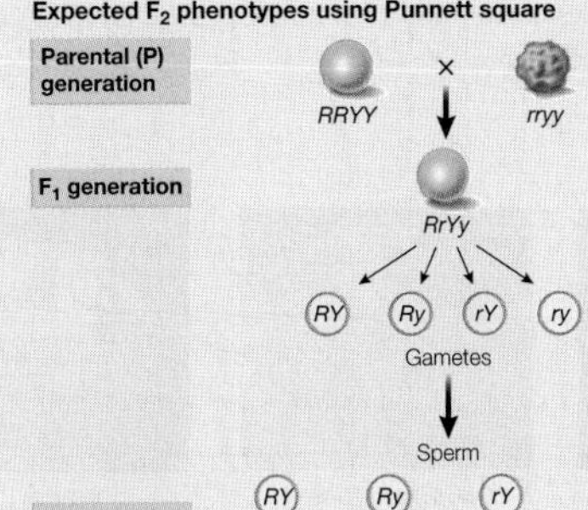

F_2 generation

Eggs \ Sperm	*RY*	*Ry*	*rY*	*ry*
RY	*RRYY*	*RRYy*	*RrYY*	*RrYy*
Ry	*RRYy*	*RRyy*	*RrYy*	*Rryy*
rY	*RrYY*	*RrYy*	*rrYY*	*rrYy*
ry	*RrYy*	*Rryy*	*rrYy*	*rryy*

Results: When F_1 plants self-pollinate, the gametes combine randomly to produce an F_2 generation with four phenotypes in a 9:3:3:1 ratio.

QUESTIONS

1. A test cross is made between a *BbCc* individual and a *bbcc* individual. Assuming Mendel's laws hold, what are the expected genotypes in the offspring?
2. A man with a rare Mendelian disorder has a father and grandmother who are also affected. No one else in his family, including his two children, is affected. What does this indicate about the inheritance of the disorder?

8.2 Genes Are Inherited on Chromosomes

Go to ANIMATIONS 8.1 and 8.2

FIGURE 8.9

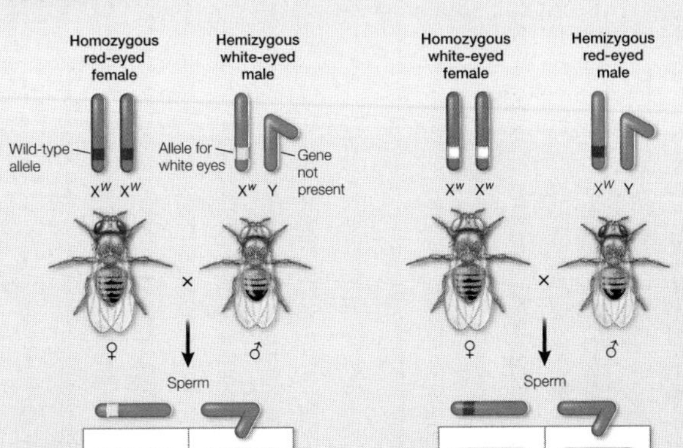

- Separation of homologs during anaphase I explains the **LAW OF SEGREGATION**.
- Independent alignment of chromosome pairs at metaphase I and crossing over at prophase I support the **LAW OF INDEPENDENT ASSORTMENT**.
- Sex-linked traits show unusual inheritance patterns consistent with being inherited on a **SEX CHROMOSOME** (**FIGURE 8.9**).

1. For genes on separate chromosomes, draw the chromosomal alignment at metaphase I for a cell in an *AaBb* individual that will give rise to two *Ab* and two *aB* gametes. How often do you expect this alignment to occur?
2. Explain why men are more commonly affected by X-linked recessive traits than women are.
3. Explain why men affected with a rare X-linked recessive trait seldom have affected sons.

8.3 Alleles, Genes, and the Environment Interact to Produce Phenotype

- Genes may have multiple alleles that show complete dominance, **INCOMPLETE DOMINANCE** (**FIGURE 8.16**), or **CODOMINANCE**.
- With **EPISTASIS**, the **PHENOTYPE** caused by one gene is affected by the **GENOTYPE** at one or more other genes (**FIGURE 8.18**).
- The observed phenotype can be altered by environmental conditions, a phenomenon called **PHENOTYPIC PLASTICITY** (**FIGURE 8.20**).

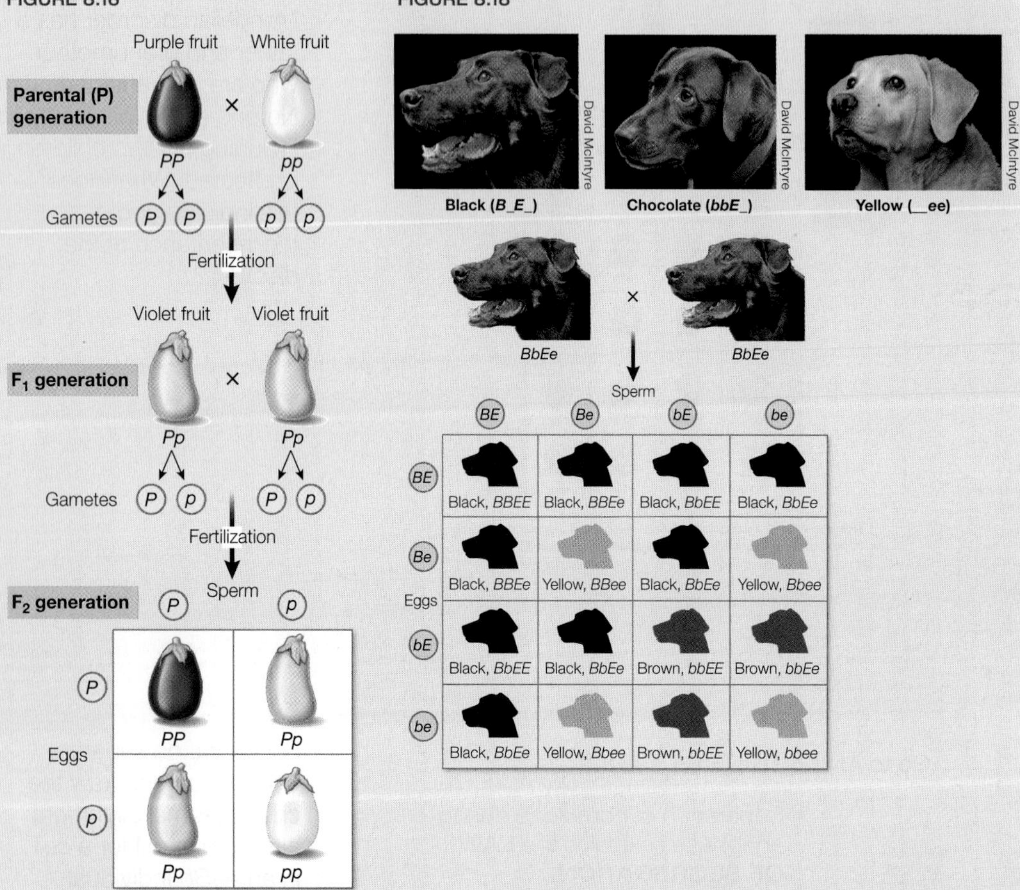

QUESTIONS

1. In a Mendelian monohybrid cross, the F_2 generation consists of one-fourth red-flowered, half pink-flowered, and one-fourth white-flowered individuals. What does this indicate?
2. A black lab is crossed with a yellow lab, and half the offspring are black and half are brown. What are the genotypes of these two dogs?
3. In winter, Siamese cats have darker faces than they do in the summer. What is one explanation for this observation?

8.4 Conjugation and Transformation Allow Exchange of Genetic Material between Prokaryotes

- **PLASMIDS** are passed to other cells via **BACTERIAL CONJUGATION** (**FIGURE 8.21**).
- Prokaryotes can exchange chromosomal material by **TRANSFORMATION** and conjugation (**FIGURE 8.22**).

FIGURE 8.21

FIGURE 8.22

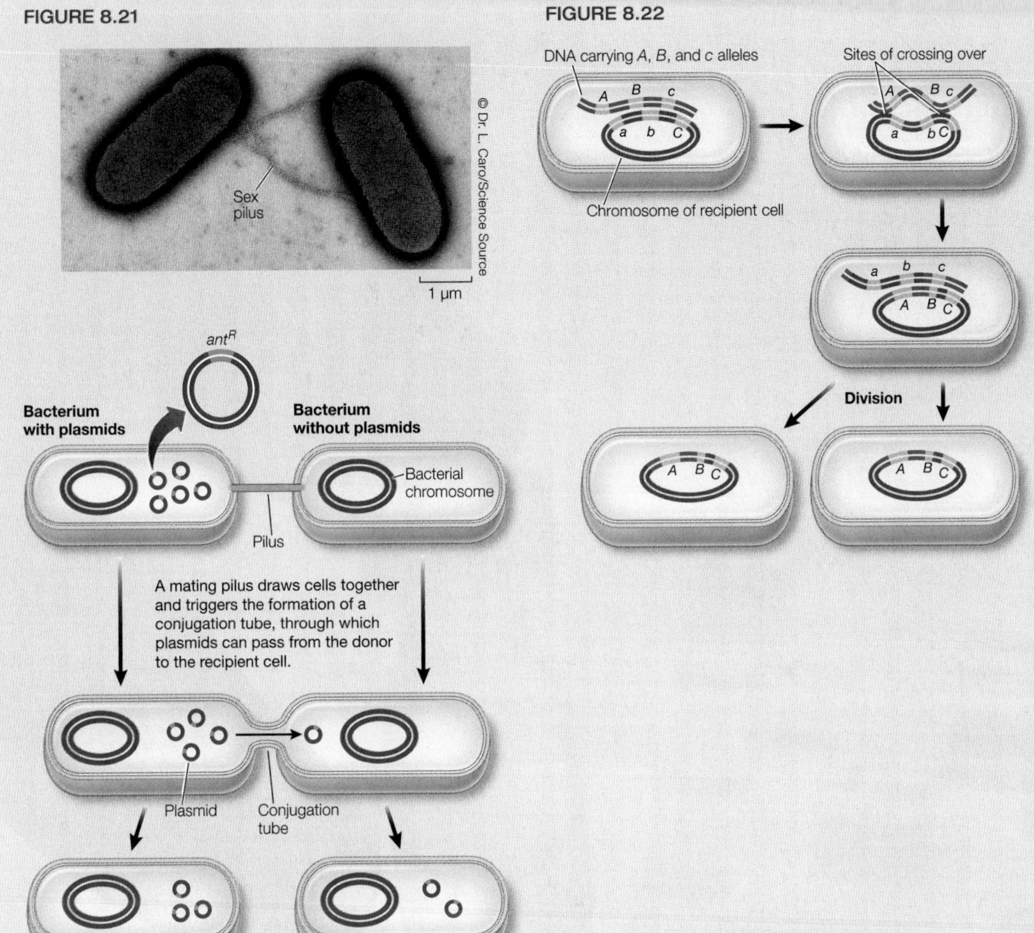

Activity 8.3 Concept Matching: Inheritance, Genes, and Chromosomes

QUESTIONS

1. What is a sex pilus?
2. What is the difference between the donor cell and the recipient cell in bacterial conjugation?
3. Are crossovers needed to transfer genetic material during conjugation and transformation?

Go to **Achieve** for the e-book, animations, activities, and additional resources and assignments.

9 DNA and Its Role in Heredity

The Combined DNA Index System (CODIS) is a national database in the United States that was established more than 20 years ago. It allows an individual to be uniquely identified from the DNA in a single drop of blood.

How can so much information be obtained from so little DNA?

You will find the answer to this question on page 224.

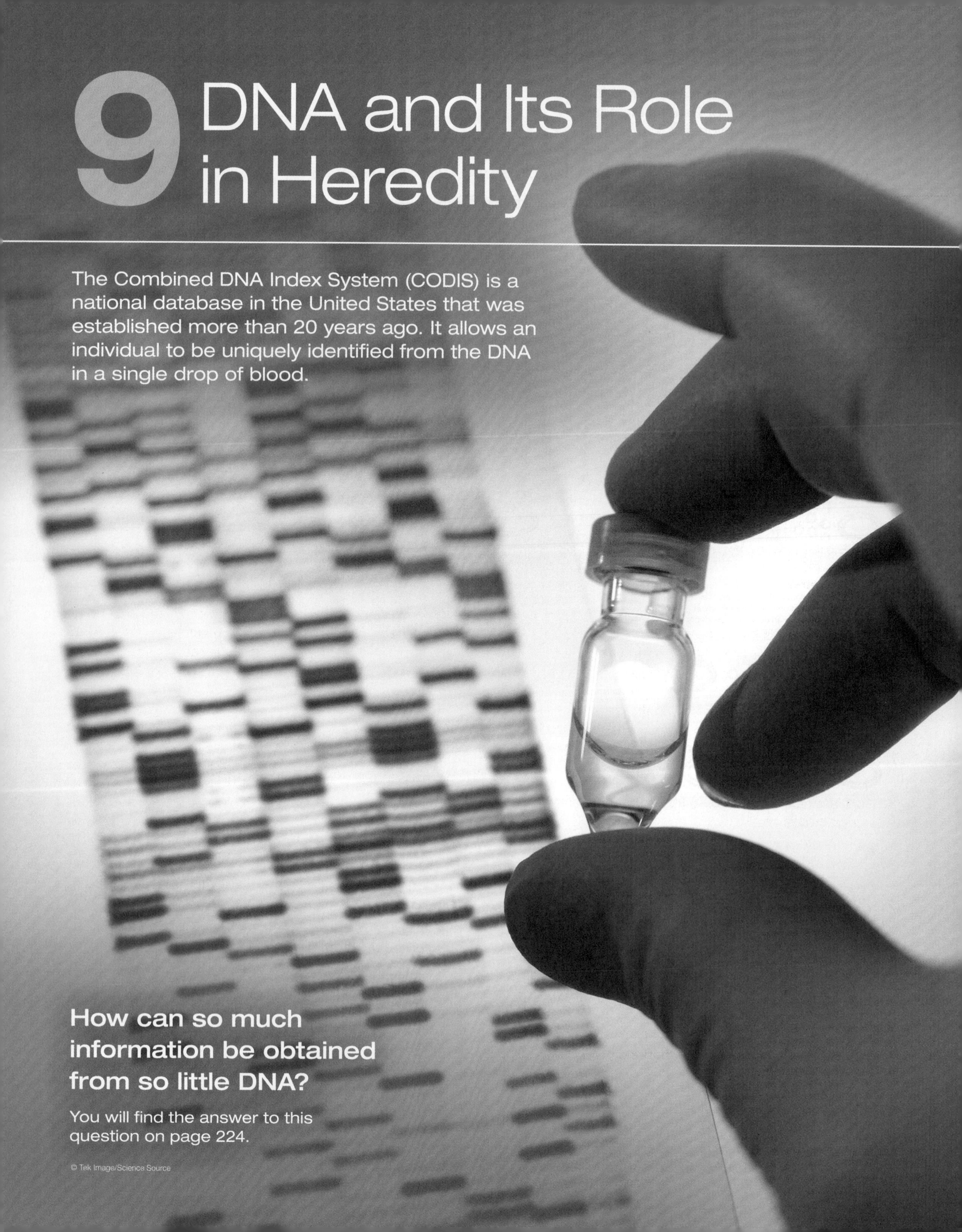

© Tek Image/Science Source

KEY CONCEPTS

9.1 DNA Is the Molecule of Inheritance

9.2 DNA Replication Is Semiconservative

9.3 DNA Mutations Alter DNA Sequence

9.1 DNA Is the Molecule of Inheritance

LEARNING OBJECTIVES

By the end of this key concept you should be able to:

9.1.1 **Describe how the Hershey–Chase experiment demonstrated that DNA is the molecule of inheritance.**

9.1.2 **State Chargaff's rules and what they suggest about the structure of DNA.**

9.1.3 **Describe the chemical structure of DNA.**

9.1.4 **Explain how the structure of DNA implies that the information in one strand is also present in the other strand.**

In Chapter 8 we outlined some of the evidence supporting the chromosome theory of inheritance. While it would take only a few years after the rediscovery of Mendel for the chromosome theory to be accepted, it would take decades to convincingly determine what molecule carries heritable information. Initially, proteins were the most likely candidates. After all, genes varied in their phenotypic effects (e.g., Mendel's peas), so perhaps proteins, with their variability, were genes. Ultimately, both circumstantial and experimental evidence pointed to DNA as the genetic material.

Circumstantial evidence suggested that DNA is the genetic material

Early observations pointed to the possibility that DNA is the genetic material. Scientists found that DNA: (1) is present in the cell nucleus and in condensed chromosomes, (2) doubles in abundance in the cell during S phase of the cell cycle, and (3) is twice as abundant in the diploid cells as in the haploid cells of a given organism. Let's look in more detail at these findings.

DNA was first isolated in 1868 by the young Swiss researcher–physician Friedrich Miescher, who isolated cell nuclei from white blood cells in pus from the bandages of wounded soldiers. When he treated these nuclei chemically, a fibrous substance came out of solution. He called it "nuclein" and found that it contained the elements C, H, O, N, and P. With no evidence except for finding it in the nucleus, Miescher boldly proposed that nuclein was the genetic material. His supervising professor was so astounded by Miescher's work that he repeated it himself in the laboratory, and finally allowed his student to publish it in a scientific journal.

In the early twentieth century dyes were developed that react specifically with DNA, showing color only when they bind to

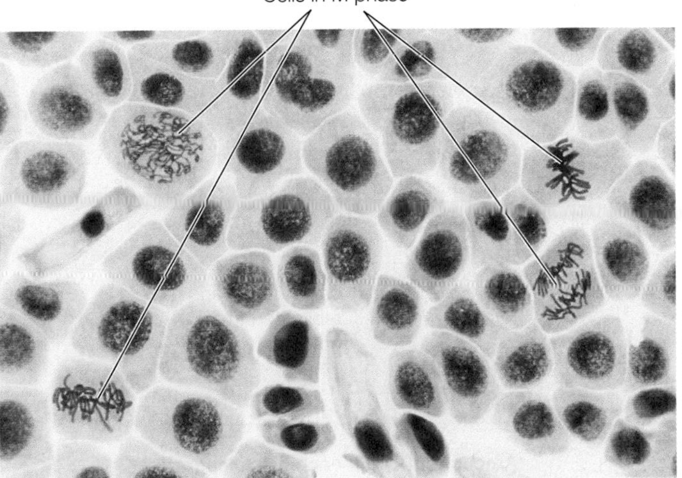

FIGURE 9.1 DNA Is in Chromosomes When dividing cells are treated with a chemical that specifically binds to DNA, staining is limited to the nucleus and, when visible, it is the chromosomes that are stained. Shown here is a group of dividing cells from onion root. Note that a few cells are in M phase, with distinct individual chromosomes visible.

Why are individual chromosomes visible in some of these cells but not others?

it. Individual cells could now be examined for the location and amount of DNA they contained. When dividing cells were treated with such a dye, staining was limited to the nucleus and, when they were visible, it was apparent that it was the chromosomes that were actually stained (**FIGURE 9.1**).

The amount of dye binding to DNA, and hence the intensity of color observed, was directly related to the amount of DNA present: the greater the color intensity, the more DNA. The intensity of color could be measured in an instrument called a flow cytometer. When the amount of DNA in each cell of a population of actively dividing cells was quantified, two populations of cells were generally seen: some cells, which were in S, G2, or M phases of the cell cycle, had twice as much DNA as most of the other cells, which were in G1 (**FIGURE 9.2**). Such staining experiments supported two other predictions for DNA as the genetic material:

1. Virtually all nondividing somatic cells of a particular organism have the same amount of nuclear DNA. This amount varies from species to species.

2. Cells resulting from meiosis have half the amount of nuclear DNA as somatic cells.

Experimental evidence confirmed that DNA is the genetic material

While circumstantial evidence suggested DNA might be the genetic material, experimental evidence was needed to confirm it. Chromosomes in eukaryotic cells contain DNA, but they also contain proteins that are bound to DNA. Therefore it was difficult to rule out the possibility that genetic information might be carried in proteins. Several experiments were performed

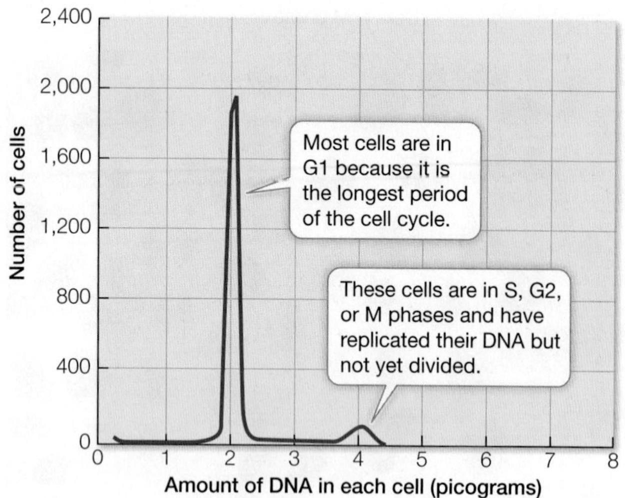

FIGURE 9.2 Dividing Cells Double Their DNA Content during the Cell Cycle When the DNA in dividing cells is stained and the staining intensity is used to estimate DNA content, two distinct populations of cells are seen: those in G1 (the bigger peak) and those in S, G2, or M phases (the smaller peak, with double the DNA content).

How could this technique be used to estimate genome size of a species, which is the length of the complete DNA sequence present in each of its cells?

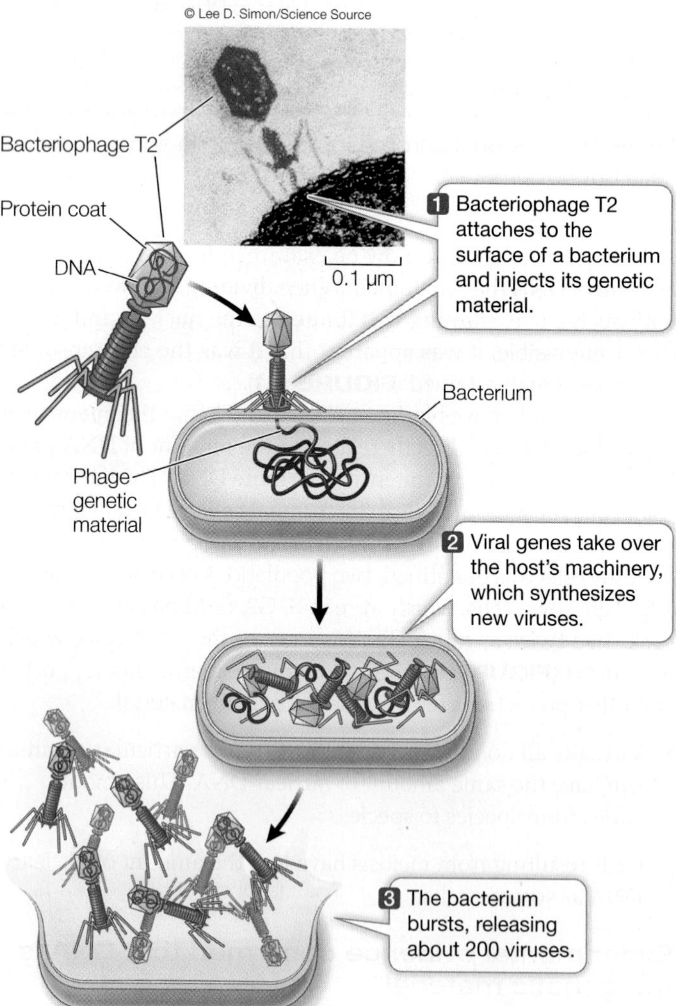

FIGURE 9.3 The Life Cycle of Bacteriophage Bacteriophage T2 infects *Escherichia coli* and depends on the bacterium to produce new viruses. The bacteriophage consists of DNA contained within a protein coat. When the virus infects an *E. coli* cell, it injects its genetic material into the host bacterium, turning the host cell into a viral replication machine.

from the 1920s to the 1950s to confirm that DNA was the genetic material.

In bacteria it was shown that transformation required DNA. Recall from Key Concept 8.4 that transformation occurs when a bacterium alters its genetic makeup by taking up DNA from its environment. Transformation was discovered by biologists working with two strains of a bacterium, termed R and S. The researchers showed that dead S cells could transform living R cells into the S strain. However, if the DNA of dead S cells was enzymatically degraded into individual nucleotides, transformation was no longer possible. Degrading other macromolecules, including proteins, lipids, polysaccharides, and RNA, did not alter the ability of the dead S strain to transform the R strain. Thus it was concluded that the ability to cause a heritable change in this bacterium, transforming the R strain into the S strain, was due to information carried on DNA.

Viruses provided another system to explore this question. Many viruses, including **bacteriophages** (viruses that infect bacteria), are composed of DNA and only one or a few kinds of protein. When a bacteriophage infects a bacterium, it takes about 20 minutes for the virus to hijack the bacterium's metabolic capabilities and turn the bacterium into a virus factory. Minutes later, the bacterium is dead and hundreds of viruses are released (**FIGURE 9.3**). Scientists thus realized that a virus contains information that is used to make more viruses. But did the viral DNA have the information for the change in the bacterium's functioning, or did the protein component of the virus play that role? Experiments performed by Alfred Hershey and Martha Chase showed that when a bacteriophage infects a bacterial cell, it injects only viral DNA. The injected DNA is then used to make progeny bacteriophage, implying that the information to do so is encoded on DNA (**INVESTIGATION FIGURE 9.4**).

Animation 9.1 The Hershey–Chase Experiment

Several types of information led to the discovery of the structure of DNA

Mendel suggested that genes are physically distinct entities, and research such as that conducted by Hershey and Chase identified DNA as the genetic material. The history of how the actual structure of DNA was deciphered is worth considering, as it illustrates how different kinds of data were critical for deciphering the structure, and was a landmark in our understanding of biology.

By the mid-twentieth century, the chemical makeup of DNA, as a polymer composed of nucleotide monomers, had been known for several decades. However, its three-dimensional structure had not been deciphered. Elucidating DNA's structure was a major

INVESTIGATION

FIGURE 9.4 The Hershey-Chase Experiment Hershey and Chase infected bacterial cells with radioactively labeled bacteriophage to investigate whether it was the protein component or the DNA component of the bacteriophage that hijacks the cell's genetic machinery to produce progeny phage.[a]

HYPOTHESIS

DNA carries the information to produce progeny phage.

METHOD

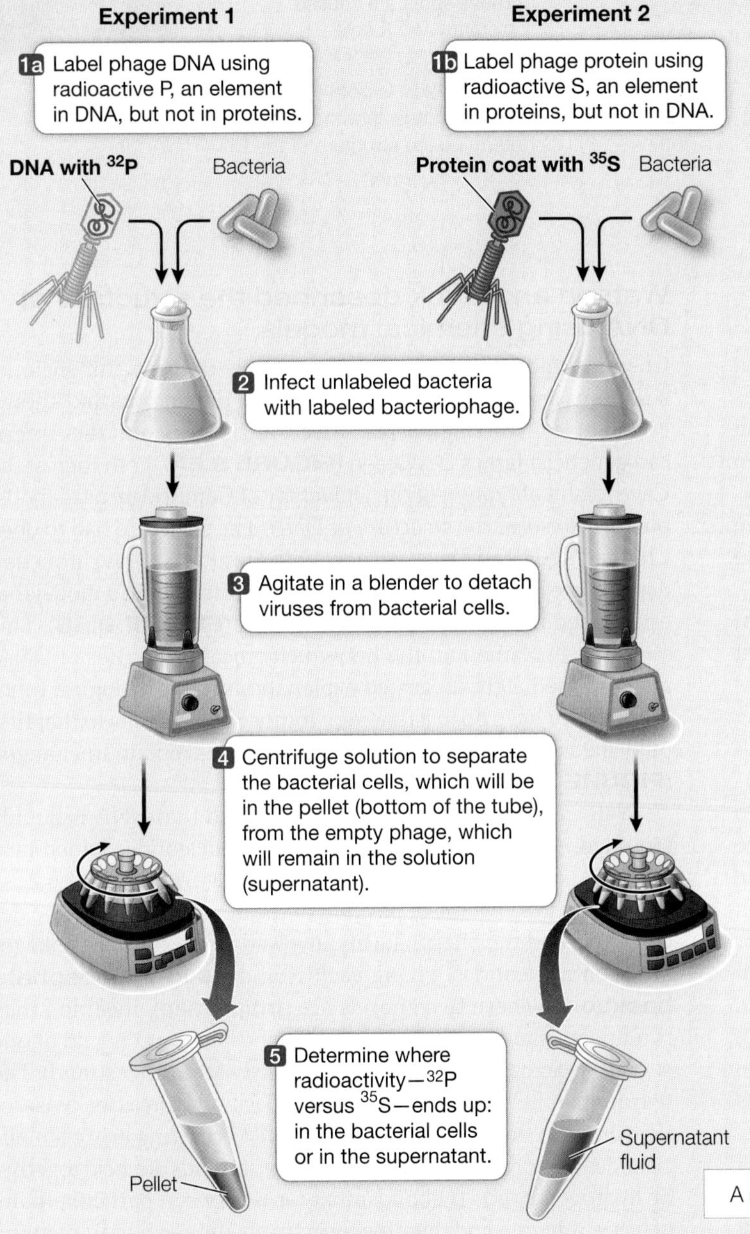

RESULTS

Time in blender (min)	Radioactivity in supernatant (%)	
	Radioactive phosphorus (^{32}P)	Radioactive sulfur (^{35}S)
0 (no agitation)	10	16
2.5	21	81

The data show that after centrifugation, most (81%) of the ^{35}S ends up in the supernatant. In contrast, a minor fraction (21%) of the ^{32}P ends up in the supernatant, indicating that most of it was in the pellet with the bacteria.

CONCLUSION

DNA is injected into the bacterial host cells, and proteins remain outside the cell. This supports the hypothesis that DNA carries the information to produce progeny phage.

ANALYZE THE DATA

1. Why were radioactive isotopes of phosphorus and sulfur chosen to label DNA and proteins, respectively? Why not radioactive nitrogen, oxygen, or carbon?
2. The first line of the table gives results for bacterial cells that were infected with bacteriophage but were not placed in the blender before centrifugation. For both the S and P labeling, some radioactivity ended up in the supernatant. What does this indicate?
3. Why did some phosphorus end up in the supernatant after 2.5 minutes of blending?
4. Why didn't all of the sulfur end up in the supernatant after 2.5 minutes of blending?

[a]A. D. Hershey and M. Chase. 1952. *J Gen Physiol* 36: 39–56.

A similar **Analyze the Data** exercise may be assigned in **Achieve**.

research question because it would provide insight into (1) how DNA is replicated between cell divisions, and (2) how protein sequence information is encoded.

The chemical composition of DNA provided important clues to its structure. Biochemists knew that DNA is a very long polymer of four different nucleotides. Each of these nucleotides consists of a molecule of the sugar deoxyribose, a phosphate group, and a nitrogen-containing base (see Figure 3.8). The only differences among the four nucleotides of DNA are their nitrogenous bases: the double-ring purines adenine (A)

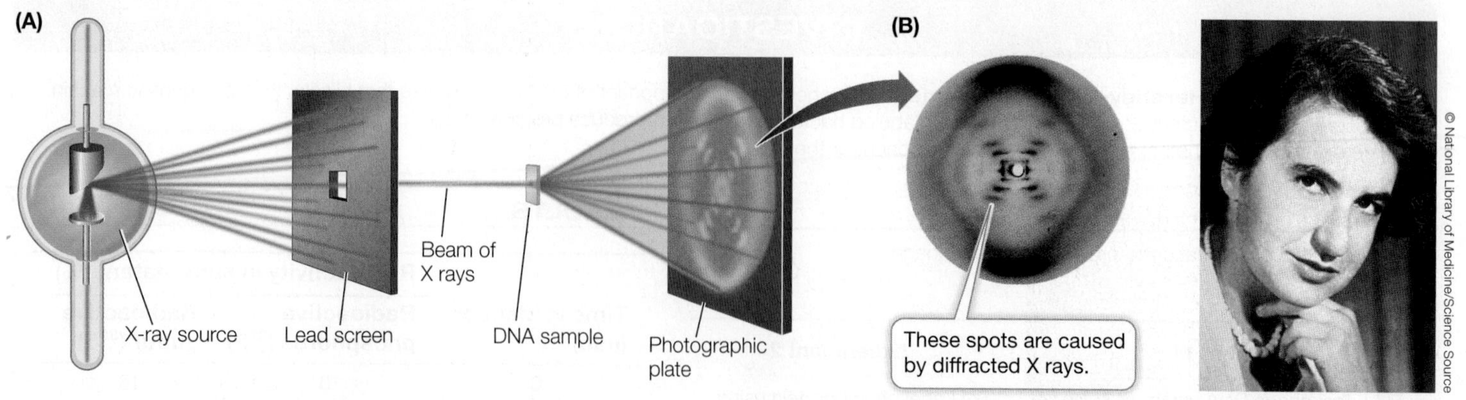

FIGURE 9.5 Rosalind Franklin and Her X-Ray Diffraction Image of DNA (A) The positions of atoms in a crystallized chemical substance can be inferred by the pattern of diffraction of X rays passed through it. The pattern in DNA is both highly regular and repetitive. **(B)** Rosalind Franklin's famous "photograph 51," shown here, helped other scientists determine the structure of the DNA molecule.

and guanine (G), and the single-ring pyrimidines cytosine (C) and thymine (T).

In 1950, biochemist Erwin Chargaff reported an important observation. He and his colleagues had found that DNA samples from many different species—and from different sources within a single organism—exhibited certain regularities with respect to the abundance of the different nucleotides. The following relationships, now known as Chargaff's rules, held for each sample: The amount of adenine equaled the amount of thymine (A = T), and the amount of guanine equaled the amount of cytosine (G = C), but the amounts of the other nucleotides were not all equal (A ≠ G, A ≠ C, T ≠ G, and T ≠ C). These relationships implied that the total abundance of purines (A + G) equaled the total abundance of pyrimidines (T + C).

Crucial evidence for the three-dimensional structure of DNA was obtained using X-ray crystallography. Some chemical substances, when they are isolated and purified, can be made to form crystals. The positions of atoms in a crystallized substance can be inferred from the diffraction pattern of X rays passing through the substance. A diffraction pattern can be visualized on a photographic plate as closely spaced light and dark spots or bands, formed when X rays are bent by objects (such as atoms) in their path. A diffraction pattern obtained by Rosalind Franklin in the early 1950s provided enormous insight into the structure of DNA (**FIGURE 9.5**). Franklin was an English chemist who was an exceptionally good X-ray crystallographer; her images from the 1950s, taken while working at King's College London, were the highest quality X-ray diffraction images ever taken of DNA. Franklin's work used DNA samples that contained very uniformly oriented DNA fibers from the English biophysicist Maurice Wilkins, also at King's College. Franklin's X-ray diffraction images suggested that the DNA molecule was long, helical, thin, of constant thickness, and consisting of two or perhaps three polymers, with nitrogenous bases on the interior.

With the chemical composition of DNA, Chargaff's rules, and Franklin's X-ray diffraction data, the stage was set for the three-dimensional structure of DNA to be fully described.

Watson and Crick described the structure of DNA using chemical models

Chemical model building is the assembly of three-dimensional structures using known relative molecular dimensions and known bond angles. The English physicist Francis Crick and the American geneticist James D. Watson (**FIGURE 9.6A**), both then at the Cavendish Laboratory of the University of Cambridge, used model building to solve the structure of DNA. The structure had to obey Chargaff's rules and be consistent with Franklin's X-ray diffraction data. In late February of 1953, Crick and Watson built a model that established the general structure of DNA (**FIGURE 9.6B**). This structure explained all the known chemical properties of DNA, and it immediately suggested explanations for its biological functions. There have been a few, very minor amendments to that first published structure, but its principal features remain unchanged (**FIGURE 9.6C**).

Watson and Crick's model hypothesized that DNA is double stranded, with each strand consisting of nucleotides joined by a phosphodiester bond between the 5′ phosphate on one deoxyribose and the hydroxyl at the 3′ position of the next ribose. Only the deoxyribose and the phosphate are involved in the covalent linkages between nucleotides, giving each strand a **sugar–phosphate backbone**. These two strands are **antiparallel**, meaning they run in opposite directions (the free 5′ phosphate at the end of one strand sits across from the free 3′ hydroxyl of the other strand). The two antiparallel strands twist around one another in a clockwise, or right-handed, helix, with a complete twist occurring every ten nucleotides (every 3.4 nanometers). The two strands are held together by hydrogen bonds (H bonds) that form between particular pairs of bases: adenine and thymine form two hydrogen bonds, guanine and cytosine form three hydrogen bonds. In DNA, all base pairs are either A-T or G-C, and these are termed Watson–Crick, or **complementary base pairs**. This base pairing explains Chargaff's rules. In addition, since both of these complementary base pairs involve a single-ring pyrimidine hydrogen bonding with a double-ring purine, the DNA molecule has constant thickness (equal to 2 nanometers), in agreement with Franklin's X-ray diffraction data.

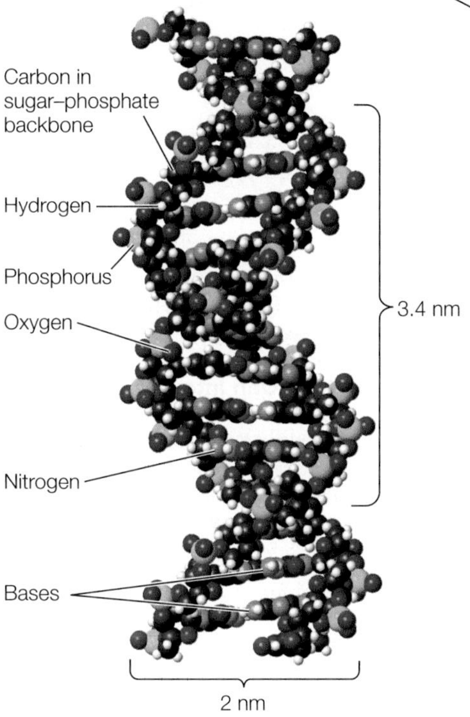

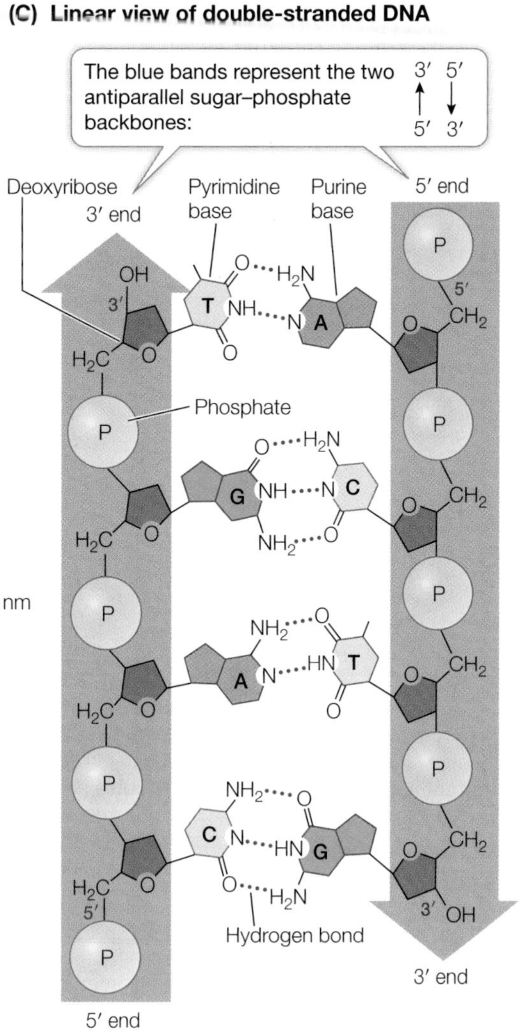

(A) Watson and Crick

(B) Space-filling model of DNA

FIGURE 9.6 Watson and Crick Described the Structure of DNA (A) James Watson (left) and Francis Crick and their DNA model in 1953. **(B)** A space-filling model showing the atoms in DNA (nm, nanometers). **(C)** The detailed chemical structure of DNA. Only N atoms involved in H-bonding are shown.

Media Clip 9.1 Discovery of the Double Helix

(C) Linear view of double-stranded DNA

base pair differs from a T-A with respect to available atoms in the minor versus major groove (see Figure 9.7). Thus the *surfaces of the base pairs are chemically distinct from one another both in identity and spatial orientation,* allowing other molecules, especially proteins, to recognize specific base-pair sequences and bind to them. The atoms and groups in the major groove are more accessible and tend to bind other molecules more frequently than those in the minor groove. *This binding of proteins to specific base-pair sequences is the key to protein–DNA interactions,* which are necessary for the replication and expression of the genetic information in DNA (see Key Concept 9.2 and Chapter 10).

The double-helical structure of DNA is essential to its function

The genetic material stores heritable information—the sequence of bases in DNA contains information that is passed from parent to offspring. The structure of DNA provided great insight into how this occurs. Every DNA molecule is a long sequence of base pairs—numbering in the thousands to millions (up to about 1 billion). Since the primary structures of proteins are long sequences of amino acids (see Key Concept 3.4), it seemed reasonable to expect that certain sequences of base pairs encode certain amino acids. How this coding might work immediately became one of the most important questions in biology following Watson and Crick's publication of the structure of DNA.

The double-helical structure of DNA suggested a way in which the information in DNA might be copied (*replicated*) so that it could be passed down to cells produced in mitosis and meiosis. Because of complementary base pairing, the information contained in a DNA molecule is fully contained in each of the two strands. Thus knowing the identity of a particular base on one strand (e.g., adenine), means that the base on the complementary strand is also known (e.g., thymine). DNA replication could be as simple as "unzipping" a DNA molecule into two single strands that could then be used as templates for synthesizing new strands. As Watson and Crick coyly pointed out in the original publication of their findings in 1953, "It has not escaped our notice that the specific pairing we have postulated immediately suggests a possible copying mechanism for the genetic material."

Watson and Crick's modeling showed that the bases are essentially perpendicular to the antiparallel strands and that adjacent bases (those next to each other on the same strand) stack like poker chips because of weak van der Waals interactions (see Chapter 2). In addition, when looked at from above, the hydrogen bonding in a base pair is such that the attachments of the bases to their sugars are not directly opposite one another—instead the two sugar–phosphate backbones are closer together on one side of the helix than on the other. The narrow side of the helix is termed the minor groove, and the wider side is the major groove. **FIGURE 9.7** shows the four possible configurations of the flat, hydrogen-bonded base pairs and the positions of the major and minor grooves.

Some of the O, N, and H atoms in the bases are involved in hydrogen bonding between bases. Other atoms on the outer edges of the bases are accessible for additional hydrogen bonding. Notice that the arrangements of unpaired atoms and groups differ in the A-T base pairs compared with the G-C base pairs, and that an A-T

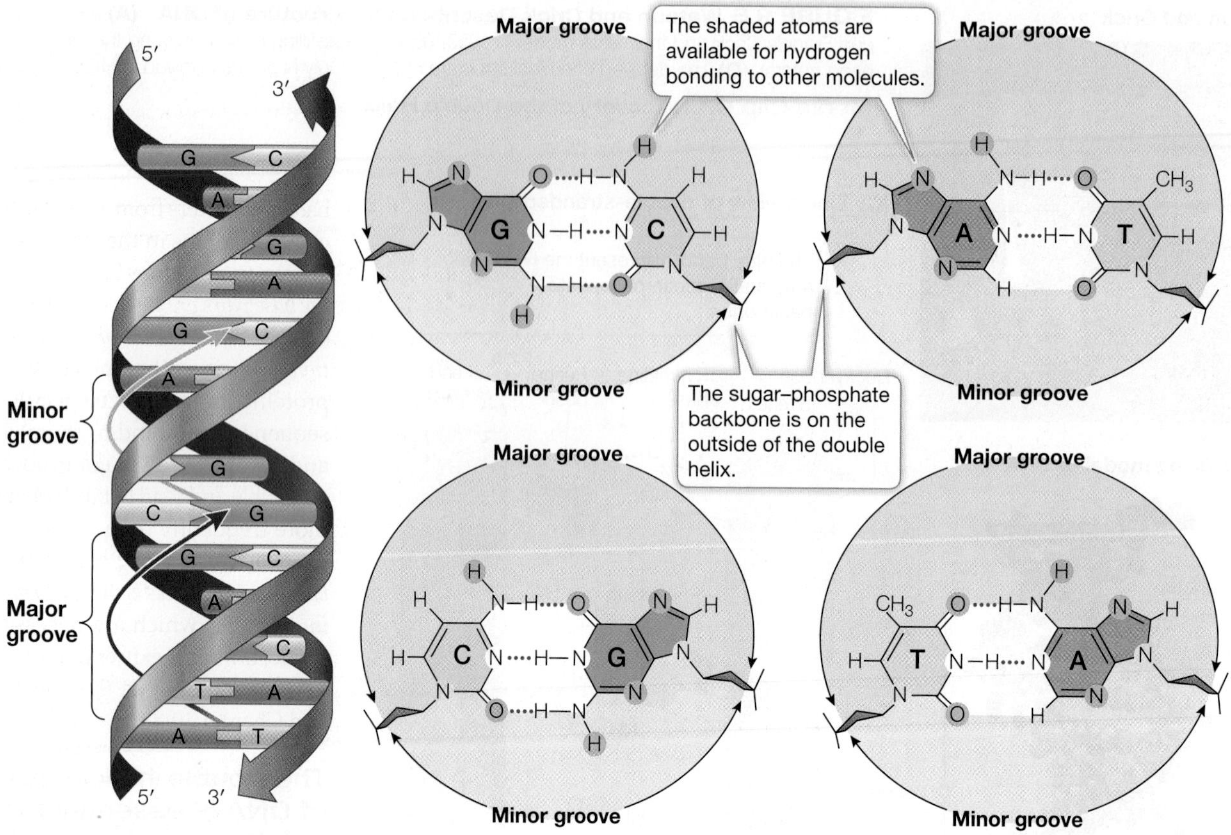

FIGURE 9.7 Base Pairs in DNA Can Interact with Other Molecules These diagrams show the four possible configurations of base pairs within the double helix viewed from above or below. The sugar–phosphate backbones are closer together on one side of the double helix than on the other side, resulting in the major and minor grooves. Atoms shaded in green are available for hydrogen bonding with other molecules, such as proteins, and are more accessible in the major groove.

REVIEW & APPLY | 9.1

R After decades of work that provided only circumstantial evidence, experimental evidence eventually demonstrated that DNA is the molecule of inheritance. An intense effort to determine the three-dimensional structure of DNA followed, with the goal of understanding how DNA functions. Chargaff's rules of base composition and Franklin's X-ray crystallography data helped Watson and Crick propose that DNA consists of two polynucleotide strands held together in an antiparallel, double-stranded helix by hydrogen bonds between complementary bases.

A

1. What were the unique chemical compositions of DNA and proteins that allowed Hershey and Chase to determine that DNA, and not proteins, is responsible for bacteriophage production after the virus infects a bacterial host cell?
2. A living R strain of bacteria can be transformed into an S strain when mixed with a dead S strain (see Key Concept 8.4 for a discussion of transformation). In this experiment, killing involves heating the S strain to 80°C–90°C, warm enough to break hydrogen bonds. What might happen to DNA molecules heated to this temperature? Why doesn't heating to this temperature prevent transformation by S-strain DNA?
3. When writing down a sequence of DNA, why is only the sequence of one strand usually given, for example 5′-ATCGG-3′?
4. If a DNA molecule is 20 percent G, what are the percentages of the other three bases?

We have seen that an individual DNA strand contains thousands or millions of nucleotides in a precise sequence. How is this huge amount of genetic information replicated before cell division? The elegant mechanism of DNA replication is the subject of the next key concept.

9.2 | DNA Replication Is Semiconservative

LEARNING OBJECTIVES

By the end of this key concept you should be able to:

9.2.1 **Describe how the Meselson-Stahl experiment demonstrated that DNA replication is semiconservative.**

9.2.2 **Describe the roles of various enzymes and other proteins in DNA replication.**

9.2.3 List the similarities and differences between DNA and RNA polymerases.

9.2.4 Explain the end-replication problem and how telomerase "solves" it.

DNA carries the hereditary information in every cell. As such, it is crucial that DNA is replicated *both completely and accurately* during the cell cycle. As mentioned above, Watson and Crick noted that the double-helical model of DNA suggested an obvious means of replication.

DNA replicates semiconservatively

Watson and Crick suspected each strand of a DNA double helix could be used as a **template** to synthesize a new strand of DNA using complementary base pairing. Biologists proposed three different models of replication based on how the original and newly synthesized strands would be arranged following replication of a DNA molecule (**FIGURE 9.8**):

1. **Semiconservative replication**: Each strand of the parental molecule could be used as a template for the synthesis of a new strand in each daughter molecule.
2. **Conservative replication**: The two parental strands could remain together (that is, could be "conserved") in one daughter molecule, while serving as a template for another daughter molecule consisting of two newly synthesized strands.
3. **Dispersive replication**: The parental molecule could end up dispersed among both strands in the two daughter molecules.

Original DNA | **Daughter molecules following replication**

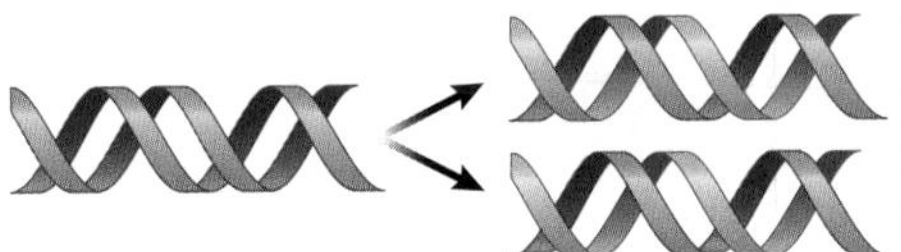

Semiconservative replication produces daughter molecules with both an original and a newly synthesized DNA strand.

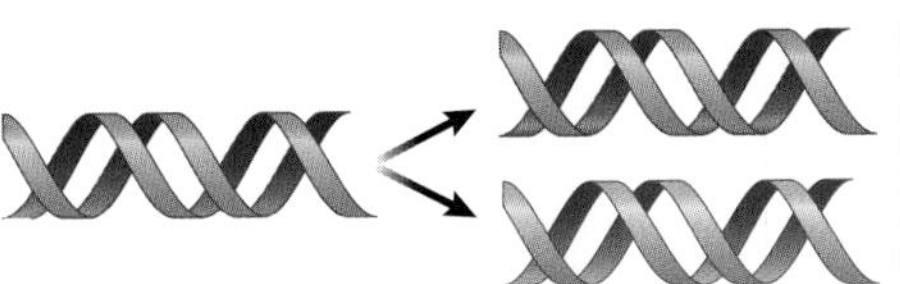

Conservative replication produces daughter molecules with either two original or two newly synthesized DNA strands.

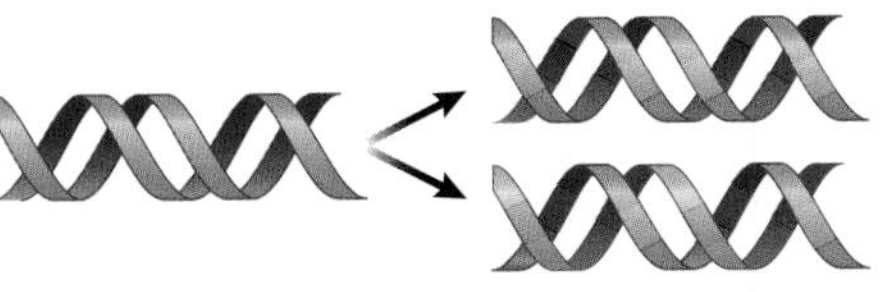

Dispersive replication produces daughter molecules with a mix of both original and newly synthesized DNA in each strand.

FIGURE 9.8 Three Models for DNA Replication Biologists hypothesized three different ways in which a parental DNA molecule might be replicated to produce two daughter molecules. In each model, the original DNA is shown in blue and the newly synthesized DNA is in red.

While semiconservative replication seemed to be the most likely way in which DNA would be replicated, as Watson and Crick had suggested, experiments were performed to distinguish among and rule out the other possibilities. One of these experiments, performed by Matthew Meselson and Franklin Stahl, followed DNA replication in bacteria that were grown on medium (food) containing different isotopes of nitrogen. This experiment elegantly showed that DNA replication is semiconservative.

Meselson and Stahl first grew bacteria for several generations on medium containing the ^{15}N isotope of nitrogen, which has an extra neutron in its nucleus and is a denser form of nitrogen than the more common ^{14}N isotope. After several generations the nitrogen atoms in the bacterial DNA were the ^{15}N isotope, giving the molecule a high density. They then transferred some of the bacteria to a medium containing only the ^{14}N isotope and allowed the bacteria to undergo two rounds of DNA replication. If DNA replication was semiconservative, one round of replication on the ^{14}N medium should result in one strand of the chromosomal DNA containing ^{14}N atoms and one strand containing ^{15}N atoms, resulting in molecules with intermediate density. After two rounds of DNA replication, half of the chromosomes should have one strand with ^{14}N and one strand with ^{15}N atoms and thus have intermediate density, while the other half of the chromosomes should have two strands with only ^{14}N atoms and thus have light density. After each round of replication, Meselson and Stahl determined the density of chromosomal DNA by measuring how far it traveled in a salt gradient, with farther movement indicating higher density. The results of their experiment are shown in **FIGURE 9.9**. The densities of chromosomal DNA predicted by semiconservative replication were observed, supporting the hypothesis of semiconservative replication and rejecting the other two possibilities.

Animation 9.2 The Meselson-Stahl Experiment

As we will see, DNA replication involves several different proteins. The overall process occurs in three steps:

1. **Initiation**, which involves unwinding (denaturing) the DNA double helix to separate the two strands and synthesizing of RNA primers
2. **Elongation**, which involves synthesizing new strands of DNA starting from the RNA primers and using each of the parental strands as templates
3. **Termination**, in which synthesis ends after each region of the DNA molecule has been replicated

During DNA replication, each incorporated nucleotide is covalently attached to the growing nucleic acid chain by a phosphodiester bond (see Figure 3.9). The base sequence of the new strand is complementary to the bases in the template strand. In addition, the newly synthesized strand is in the opposite orientation to the template strand. DNA synthesis always proceeds in the 5′-to-3′ direction, by adding nucleotides to the 3′ end of the growing strand, and

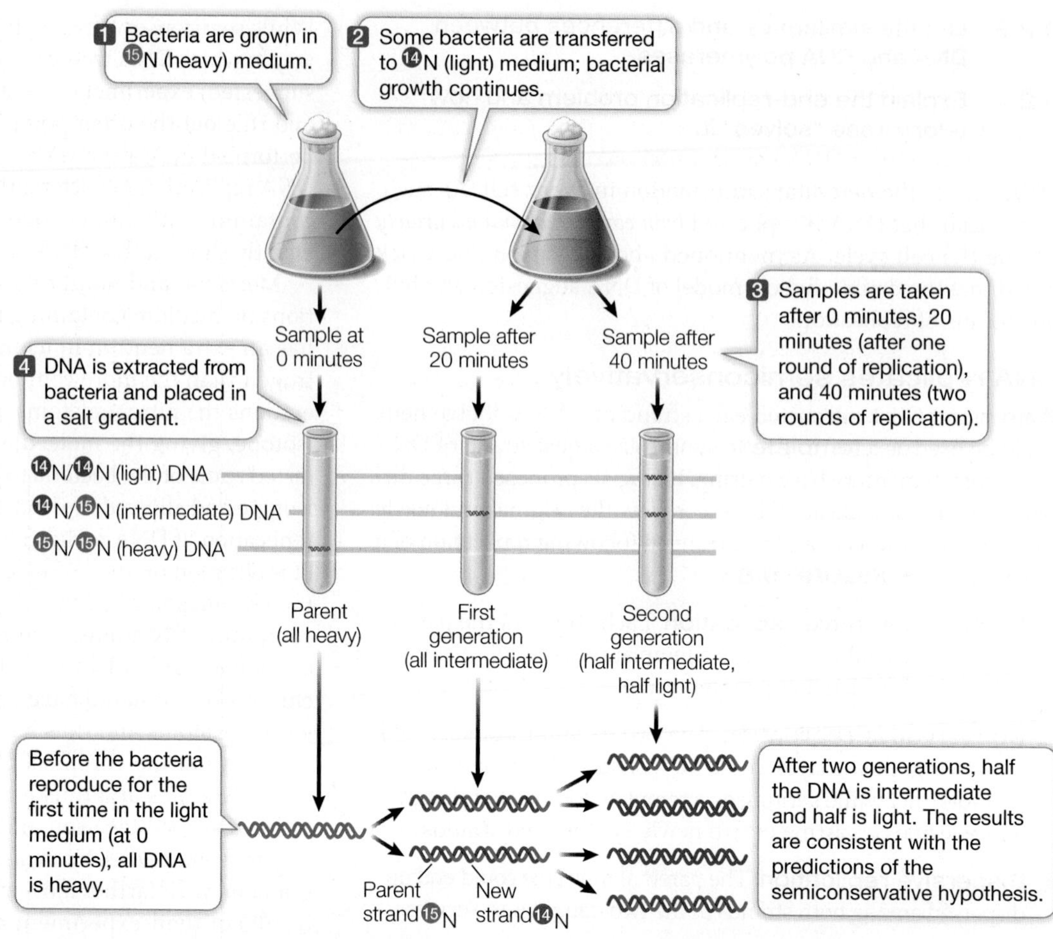

FIGURE 9.9 The Meselson–Stahl Experiment When centrifuged, DNA forms bands in the test tube according to density. If the semiconservative model of replication holds, first-generation daughter molecules should show intermediate density, reflecting one strand with ^{14}N and one strand with ^{15}N atoms. Second-generation daughter molecules should show both intermediate- and light-density molecules, reflecting half having one strand with ^{14}N and one strand with ^{15}N atoms, and half having both strands with only ^{14}N atoms. The results of the experiment were consistent with semiconservative replication only.

What are the predictions for the density of DNA after one and two rounds of replication under the conservative and dispersive models?

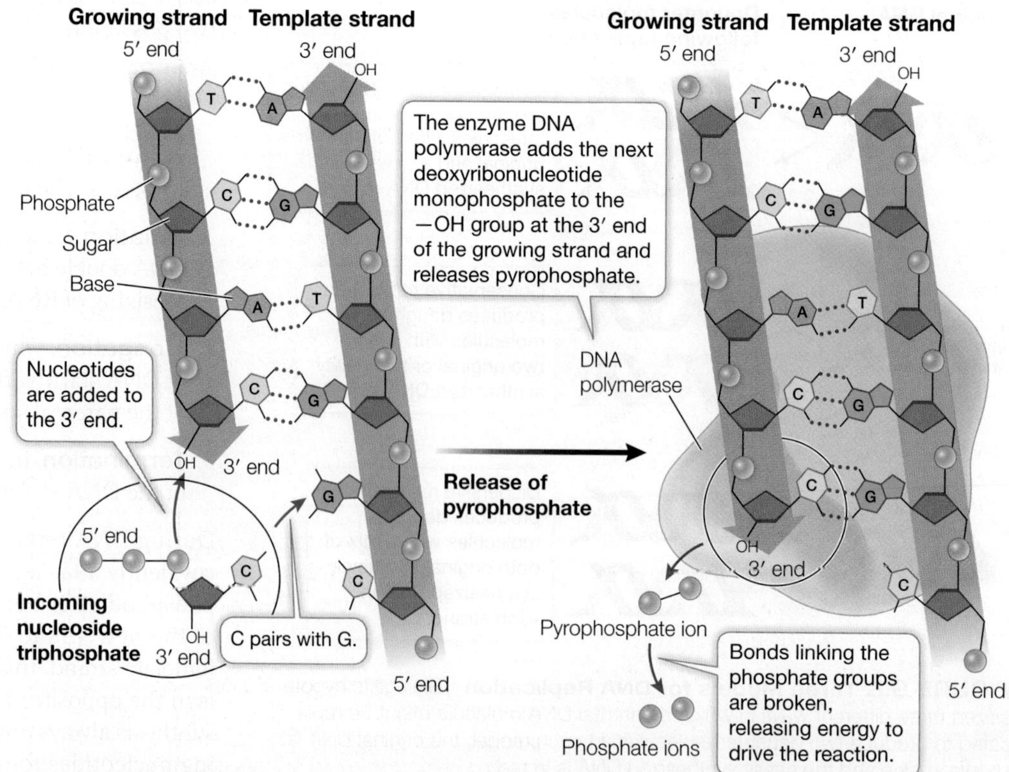

FIGURE 9.10 Nucleotides Are Added to the 3′ End during DNA Synthesis A nucleotide that complements the template base is added to a growing DNA strand at the 3′ end. The new strand thus extends in the 5′-to-3′ direction. Energy for the formation of the phosphodiester bond between nucleotides comes from the exergonic cleavage of the pyrophosphate from the triphosphate of the incoming nucleotide. The template strand is read in the 3′-to-5′ direction.

thus the template strand is always read in the 3′-to-5′ direction (**FIGURE 9.10**).

As we noted in Key Concept 3.3, a free nucleotide can have one, two, or three phosphate groups attached to its pentose sugar. The nucleotides used for DNA synthesis each have three phosphates attached at the 5′ carbon of the deoxyribose, and are thus called deoxyribonucleoside triphosphates, or dNTPs, where N indicates the specific nucleoside and is either adenosine (A), thymidine (T), cytidine (C), or guanosine (G). When added to the growing strand during DNA synthesis, the two outer phosphate groups are released together as pyrophosphate, and the resulting deoxyribonucleoside monophosphate (dAMP, dTMP, dCMP, or dGMP) is added to the growing nucleic acid chain (see Figure 9.10). The release of the two outer phosphate groups provides the energy for the formation of a phosphodiester bond between the inner phosphate group of the incoming nucleotide and the 3′ carbon on the deoxyribose at the end of the DNA chain (see Figure 3.9). The pyrophosphate is then hydrolyzed into two inorganic phosphates, releasing additional energy, which makes the overall free energy change of DNA polymerization more favorable (more negative) (see Figure 9.10).

Animation 9.3 DNA Replication and Polymerization

DNA replication involves several proteins

The first step of DNA replication is initiation. Initiation begins when the DNA helix unwinds, or denatures, so that the replication machinery can access it. DNA unwinding occurs when a large protein complex (the pre-replication complex) binds to specific DNA sequences on chromosomes called **origins of replication** (***ori***). The single circular chromosome of the bacterium *Escherichia coli* has a single *ori*. Unwinding of the DNA at the *ori* results in the formation of a replication bubble. The replication bubble consists of two **replication forks** (**FIGURE 9.11A**), each of which moves away from the *ori* during synthesis of the new DNA strands. The replication forks are where two strands of DNA are exposed once the double helix is opened, providing access to the proteins that perform DNA replication. The separation of the two stands of DNA at each fork is catalyzed by an enzyme called **DNA helicase**, which uses free energy from ATP hydrolysis to break hydrogen bonds between bases on the two strands. DNA helicase can be thought of as a wedge that is driven between the two strands of the DNA helix, forcing them apart. The two strands are prevented from re-forming hydrogen bonds by single-strand binding proteins that bind to each strand. Since DNA consists of a double helix, separating the two strands results in additional twisting of the helix in front of each replication fork. This additional twisting is relieved by an enzyme called topoisomerase. Topoisomerase works by breaking the

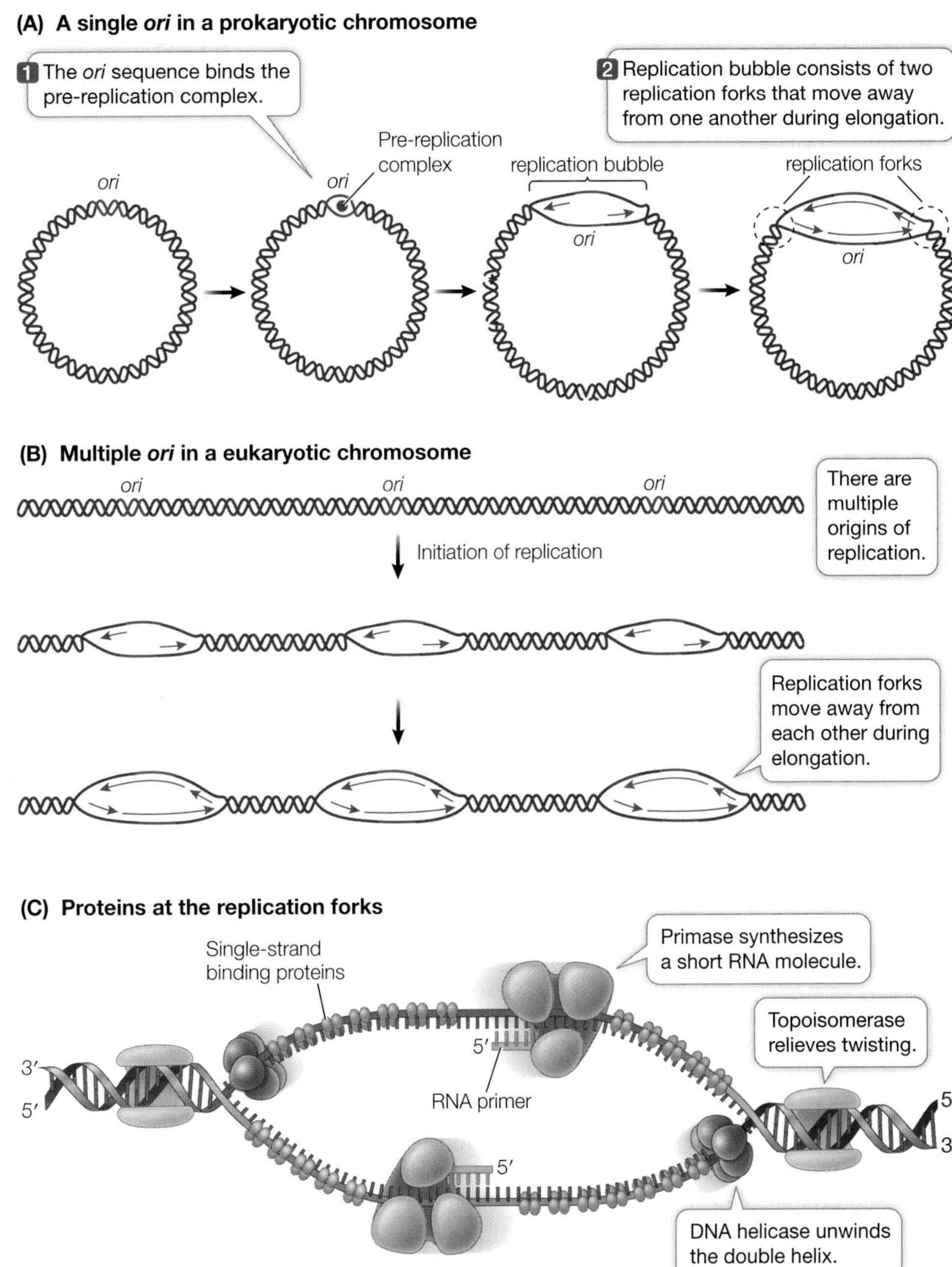

FIGURE 9.11 Initiation of DNA Replication in Prokaryotes and Eukaryotes **(A)** Initiation of DNA replication in most prokaryotes occurs at the single *ori* site. **(B)** Initiation in eukaryotes involves many—up to tens of thousands—of *ori* sites on each chromosome. **(C)** Initiation of replication in both prokaryotes and eukaryotes involves several proteins.

sugar–phosphate backbone of one strand, allowing the strand to untwist and then rejoining the backbone.

The replication rate in *E. coli* is approximately 1,100 base pairs per second, so it takes about 40 minutes to fully replicate the 5.4 million base pairs in the chromosome (with two replication forks). Eukaryotic chromosomes are much longer than those of prokaryotes—up to a billion base pairs—and are linear, not circular. In addition, their replication forks move much more slowly, on the order of tens of bases per second. If replication occurred from a single *ori*, it would take several weeks to fully replicate a single chromosome. So eukaryotic chromosomes have multiple *oris*, scattered at intervals of 10,000–40,000 base pairs, allowing them to replicate their chromosomes in a few hours or less (**FIGURE 9.11B**).

The goal of DNA replication is to generate two identical daughter molecules from each parent DNA molecule. It is thus critical that initiation from an *ori* occurs only once per round of replication. If an *ori* were used more than once, the DNA in that region (or the entire chromosome in the case of prokaryotes) would be replicated more than once. Several proteins are involved in controlling the initiation of replication at *ori* in both prokaryotes and eukaryotes to ensure they are used only once per round of replication.

Following denaturation at the *ori*, initiation proceeds with the synthesis of a short RNA molecule called a **primer**. The primer is complementary to the DNA template and is synthesized one nucleotide at a time by an enzyme called **primase** (**FIGURE 9.11C**). A polymerase is an enzyme that synthesizes nucleic acid polymers. RNA polymerase and DNA polymerase are used to assemble RNA and DNA molecules, respectively. In the case of primase, it is an RNA polymerase that synthesizes RNA, using a DNA template.

After initiation, the next step of DNA synthesis is elongation. The RNA primer is used by a DNA-dependent DNA polymerase that catalyzes the addition of deoxyribonucleotides to synthesize a new strand of DNA. The DNA polymerase continues to act until the replication of that section of DNA has been completed. Priming with a short RNA strand is required because DNA polymerases are unable to start polymerization using single-stranded DNA—they require a short region of double-stranded molecule to which they can add nucleotides. The ability to start polymerization without a primer is a major difference between RNA polymerases (no primer required) and DNA polymerases (primer required). The requirement of a primer for DNA polymerases allowed the development of two critical molecular techniques: the polymerase chain reaction (see Figure 9.13) and DNA sequencing (see Key Concept 12.1).

During elongation, a single replication fork moves in one direction, opening up the DNA double helix (**FIGURE 9.12A**). The resulting two strands will each be used as a template to synthesize a new DNA strand. One of these template strands—the **leading strand**—is synthesized continuously, while the other—the **lagging strand**—is synthesized discontinuously. To understand the difference between leading and lagging strand synthesis, it is important to remember two things:

1. The two DNA strands are *antiparallel*—that is, the 3′ end of one strand is paired with the 5′ end of the other.
2. The DNA polymerase synthesizes new DNA by adding nucleotides *only to the 3′ end of a new strand*.

The leading template strand is oriented so that DNA polymerase can add nucleotides to the 3′ end of the new strand in the same direction as fork movement and so polymerization occurs continuously. The lagging template strand is oriented so that DNA polymerase adds nucleotides to the 3′ end of the new strand in the direction away from the movement of the replication fork. The opposite direction of synthesis results in the lagging strand being synthesized discontinuously, in a series of fragments, with each fragment primed separately.

The lagging strand is initially a mess! It is synthesized in short stretches of a few hundred nucleotides (in eukaryotes) to a few thousand nucleotides (in prokaryotes) called **Okazaki fragments** (named after their discoverer, Reiji Okazaki). The lagging strand initially contains multiple RNA primers and has single-strand breaks between the 3′ end of the DNA of one fragment and the 5′ RNA primer on the next fragment (see Figure 9.12A). These problems arise because DNA polymerases require primers and can synthesize only in the 5′-to-3′ direction. Before the end of replication, the RNA primers must be removed and replaced with DNA, and the resulting DNA fragments must be joined together.

Animation 9.4 Leading and Lagging Strand Synthesis

Primer replacement involves a different DNA polymerase than is used for elongation. This DNA polymerase has two catalytic activities: removal of the RNA nucleotides one by one, and replacement with DNA nucleotides. For every nucleotide removed from the primer on the next fragment, the DNA polymerase adds a deoxynucleotide to the 3′ end of its fragment (**FIGURE 9.12B**). Eventually the RNA primer is completely removed and replaced, leaving a single unbonded break between the DNA that was at the primer location and the rest of the DNA strand synthesized from that primer. This single-strand break, or nick, is fixed by the enzyme **DNA ligase**, which catalyzes the formation of a phosphodiester bond, linking the fragments and making the lagging strand whole (**FIGURE 9.12C**).

The termination of DNA replication occurs when replication forks that are moving toward one another meet. In the circular DNA of prokaryotes, this occurs at a point that is opposite the *ori* where several termination sequences occur. Proteins bind to these sequences to prevent replication forks from progressing into regions that are already replicated. In eukaryotes, replication forks moving toward one another, having originated from different *ori*, terminate when they run into one another, or when one reaches the end of a linear chromosome, running out of template. When replication forks meet, they are prevented from progressing into already replicated regions.

Telomeres are not fully replicated in eukaryotic cells

When the replication fork reaches the end of a chromosome, the leading strand is copied all the way to the end, resulting in a complete double-stranded DNA helix. However, because replication of the lagging strand is in the opposite direction as replication fork movement, there is an RNA primer on the 5′ end of the new strand. While there are enzymes that can remove the RNA primer, no

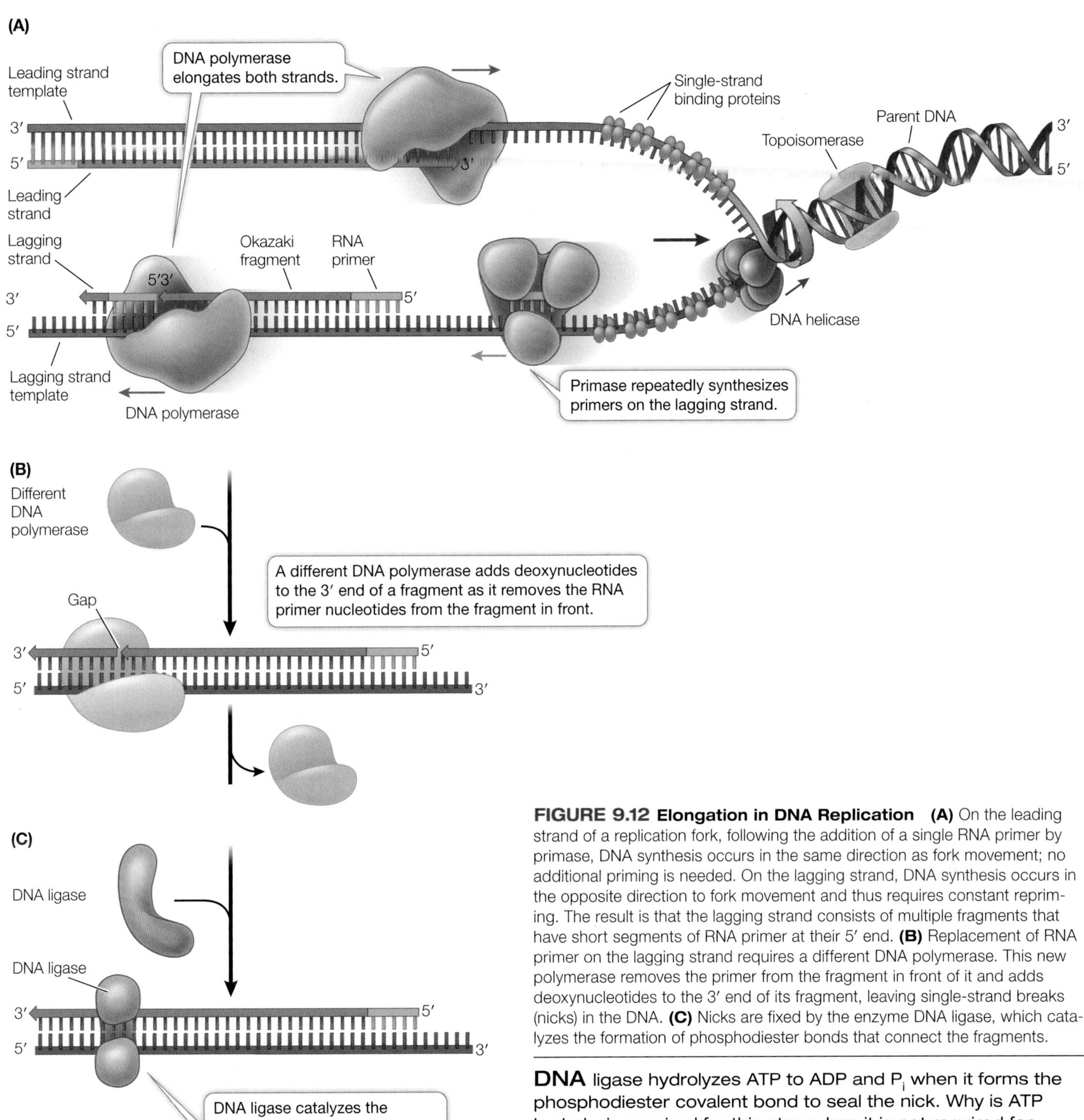

FIGURE 9.12 Elongation in DNA Replication **(A)** On the leading strand of a replication fork, following the addition of a single RNA primer by primase, DNA synthesis occurs in the same direction as fork movement; no additional priming is needed. On the lagging strand, DNA synthesis occurs in the opposite direction to fork movement and thus requires constant repriming. The result is that the lagging strand consists of multiple fragments that have short segments of RNA primer at their 5′ end. **(B)** Replacement of RNA primer on the lagging strand requires a different DNA polymerase. This new polymerase removes the primer from the fragment in front of it and adds deoxynucleotides to the 3′ end of its fragment, leaving single-strand breaks (nicks) in the DNA. **(C)** Nicks are fixed by the enzyme DNA ligase, which catalyzes the formation of phosphodiester bonds that connect the fragments.

DNA ligase hydrolyzes ATP to ADP and P_i when it forms the phosphodiester covalent bond to seal the nick. Why is ATP hydrolysis required for this step when it is not required for polymerization?

Activity 9.1 DNA Polymerase

DNA can be synthesized to replace the resulting gap because there is no 3′ end to extend. So the new chromosome has a short region of single-stranded DNA overhang at one end. This situation activates a mechanism for cutting off the single-stranded region, along with some of the intact double-stranded DNA. Thus the chromosome becomes slightly shorter with each cell division. This shortening is termed the "end-replication problem."

The end replication problem is serious: linear chromosomes get shorter every time the cell undergoes cell division. Since shortening is impossible to avoid, why haven't linear

THINK LIKE A SCIENTIST

Process of Science

Quantitative Reasoning

Science & Society

How can CODIS be used to identify suspects from a drop of blood?

Once the structure of DNA was discovered, various technologies were developed to score genetic variation. It was soon realized that some regions of genomes, including the human genome, were especially variable, possessing many alleles in a single population. Here you will use quantitative reasoning to determine how an individual can be uniquely identified from variable loci, see how the process of science was used to develop a technique for amplifying DNA from minute biological samples, and appreciate how science and society intersect with the now common use of DNA in forensic applications.

Loci with especially high levels of variation often have differences in the number of copies of a short repeated sequence. For example, a genetic locus may have a repeated AG sequence that occurs between 2 and 20 times. Each repeat number is an allele (e.g., 2 repeats, 3 repeats, 4 repeats, up to 20 repeats). An individual is usually heterozygous for the sequence (e.g., 6 repeats inherited from the mother and 10 from the father). These loci are seldom found in protein-coding regions of the genome.

At a locus with five equally frequent alleles in a population, the probability that any two individuals will have the same genotype by chance is about 7 percent. With more alleles, the probability of matching gets even smaller. Genotyping multiple loci makes the probability that two individuals will share the same genotype very low.

1. If the probability that two individuals will have the same genotype at a single locus is 7 percent, what is the probability that they will have the same genotype at 20 independent loci, assuming that the probability of matching is the same at each locus?

In the 1990s, the Combined DNA Index System (CODIS) was set up by the FBI as a registry of the DNA profiles of various individuals. CODIS collates information on the DNA of individuals at the local, state, and federal levels. For example, everyone who enters the armed forces submits a DNA sample, as do people arrested for a serious crime. This has led to a database that can be used in forensic investigations. A profile consists of an individual's genotype at 20 highly variable loci. Unless an individual has an identical twin, it is essentially impossible for two individuals to match at all 20 highly variable loci.

While CODIS allows comparisons of DNA samples, it requires getting information on the alleles at all 20 loci. How can a single drop of blood be used to genotype 20 loci? The answer is that the principles underlying DNA replication in cells were used to develop a laboratory technique that allows unlimited copying of small regions of DNA (up to a few thousand base pairs). This DNA amplification technique is called **polymerase chain reaction**, or **PCR**. PCR uses (1) a sample of double-stranded DNA to act as the template, (2) two short, artificially synthesized single-stranded DNA primers that are complementary to the ends of the sequence to be amplified, (3) the four dNTPs (dATP, dTTP, dCTP, and dGTP), (4) a DNA polymerase that can tolerate high temperatures without becoming denatured, and (5) salts and a buffer to maintain a near-neutral pH.

PCR is a cyclic process in which three steps are repeated over and over again to amplify a target sequence (**FIGURE 9.13**):

- Denaturation of DNA by heat to break hydrogen bonds and separate the two strands for primer binding
- Primer binding to the template
- Synthesis of new strand from primer by DNA polymerase

The initial template is often genomic DNA and can be at very low concentration, but both primers and nucleotides are in abundance. The first cycle produces new strands that are longer than the target sequence, but starting with the second cycle, copies that are exactly the length of the target are produced. With each cycle, the number of copies that are the correct length increases exponentially. In only 30 cycles, a single copy of the initial template can produce hundreds of millions of copies of the target sequence. Since DNA replication is fast even in a test tube, it takes only a few hours to go through this number of cycles. The PCR technique requires that the base sequences at each end of the target fragment be known ahead of

chromosomes disappeared? The answer is that various mechanisms to extend chromosome ends have evolved. The ends of chromosomes in eukaryotes have special structures called **telomeres** (**FIGURE 9.14**). Telomeres are DNA sequences that do not encode proteins. Their role is to protect the important protein-coding DNA in the chromosome from being lost. In many species, telomeres are long strings of short, repeated DNA sequences. In humans and other vertebrates, the repeated sequence is TTAGGG, and in humans it is repeated about 2,500 times in each telomere. At every cell division, the number of telomere repeats is reduced at one chromosome end or the other. However, there is an enzyme called **telomerase** that can add DNA repeats. Telomerase is yet another DNA polymerase: in this case, it uses an RNA template to synthesize new DNA. The RNA template is a component of the telomerase enzyme.

In addition to allowing chromosomes to be extended at their ends, which effectively solves the problem of shortening during replication, telomeres have another important function. When a chromosome undergoes a double-strand break, ends are created that trigger repair pathways in the cell. There is a repair pathway that rejoins the broken ends. It is critical that this pathway does not mistakenly act on chromosome ends and join them together. (Think about what might happen during mitosis if two chromosomes were joined.) To prevent this, special proteins bind to the telomere repeats so that they are not recognized as chromosome breaks. Telomeres thus protect chromosome ends from being treated as chromosome breaks.

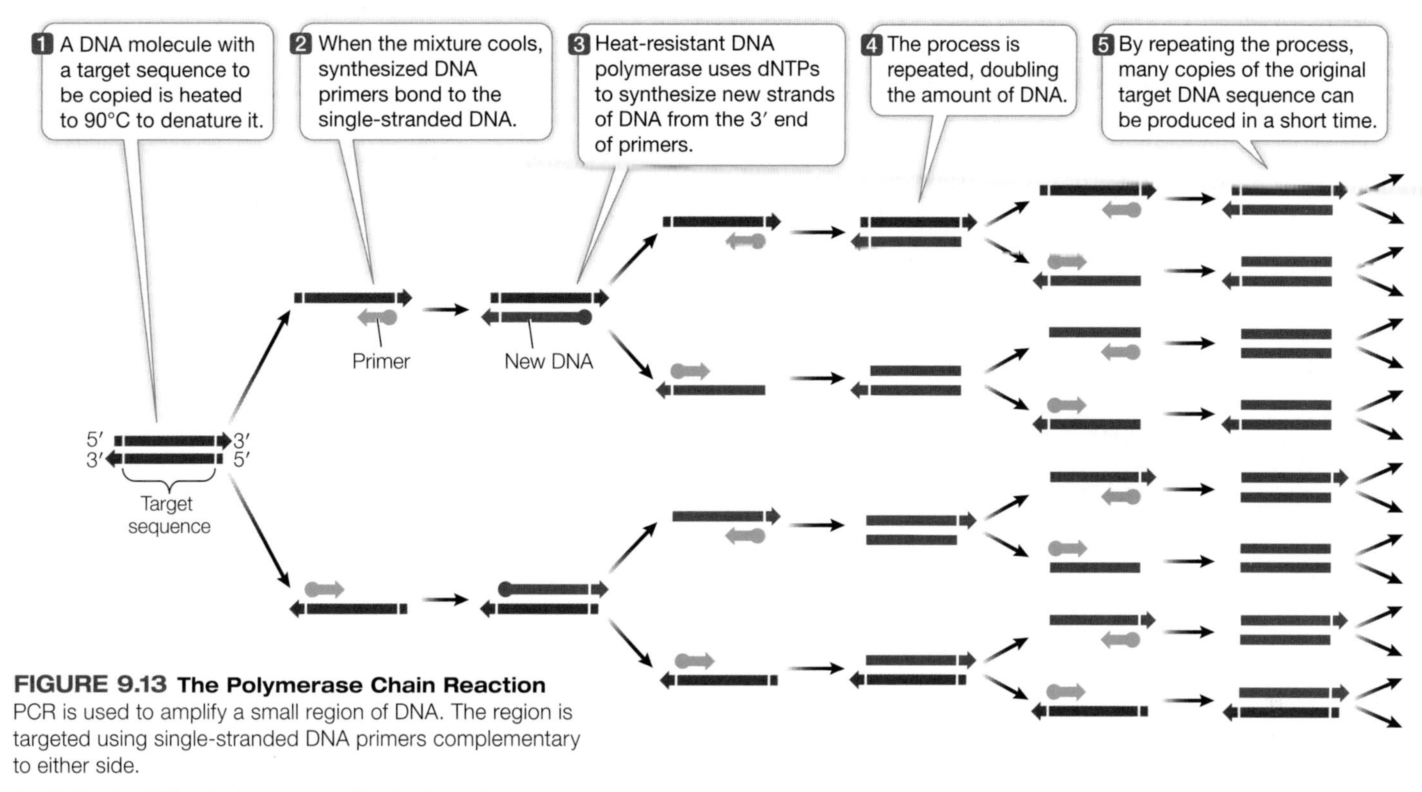

FIGURE 9.13 The Polymerase Chain Reaction
PCR is used to amplify a small region of DNA. The region is targeted using single-stranded DNA primers complementary to either side.

Activity 9.2 The Polymerase Chain Reaction Simulation

time, so that complementary single-stranded DNA primers, usually 15–30 bases long, can be synthesized in the laboratory.

2. How can a pair of primers of length 20 bases allow PCR to amplify a specific target sequence?
3. Why do you think the DNA polymerase used in PCR needs to be able to tolerate high temperature?

PCR can start with very few template molecules; as few as ten cells are needed to amplify all 20 CODIS loci. This means that a very small DNA sample left at crime scenes can be profiled, allowing identification of the person that left it. In addition to allowing DNA profiling from small samples, which revolutionized the use of DNA evidence in criminal proceedings, PCR also had an immediate and enormous impact on biology. The DNA of small pieces of tissue or cells can be readily amplified and analyzed. The importance of PCR led to a Nobel Prize in Chemistry for its inventor, Kary Mullis.

The importance of telomeres cannot be overstated. Each human chromosome may lose 50–200 base pairs of telomeric DNA after each round of DNA replication and cell division. After 20–30 cell divisions, the chromosome ends become short enough that telomeres can no longer fulfill their protective role. Apoptosis (programmed cell death) ensues, and the cell dies. Cell lineages that do not express telomerase do not last the entire lifetime of the organism because they lose telomere DNA after many cell divisions. By contrast, continuously dividing cells, such as bone marrow stem cells and gamete-producing cells, express telomerase and maintain their telomeric DNA.

There is a relationship between telomere length and aging: the older the cell lineage—that is, the more cell divisions it has undergone—the shorter its telomeres. Your parents' somatic cells have shorter telomeres that yours. In a laboratory dish, human cells will typically divide for about 30 generations, and then their telomeres are so short that the cells die. Adding telomerase to these cells or inducing them to express their own telomerase "fixes" this situation and the cells keep dividing indefinitely. Could getting humans to express telomerase in their somatic cells be important in reducing aging? Possibly, but such an intervention would possibly lead to an increase in cancer rate. Approximately 85 percent of cancer cell types express telomerase, allowing them to undergo many more cell divisions than normal cells. Drugs that affect telomerase activity are currently being developed both to help with aging and to fight cancers.

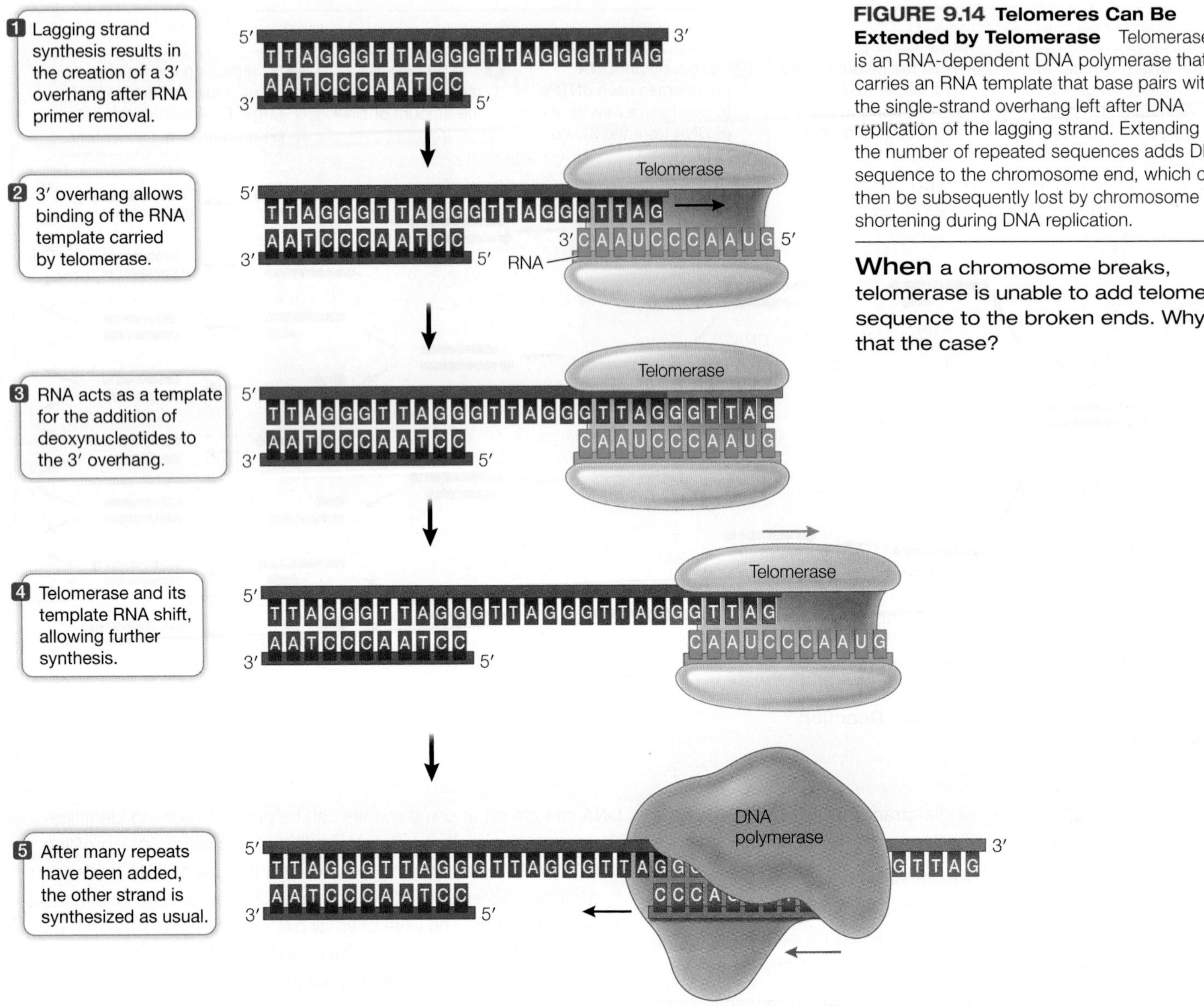

FIGURE 9.14 Telomeres Can Be Extended by Telomerase Telomerase is an RNA-dependent DNA polymerase that carries an RNA template that base pairs with the single-strand overhang left after DNA replication of the lagging strand. Extending the number of repeated sequences adds DNA sequence to the chromosome end, which can then be subsequently lost by chromosome shortening during DNA replication.

When a chromosome breaks, telomerase is unable to add telomere sequence to the broken ends. Why is that the case?

REVIEW & APPLY | 9.2

R The structure of DNA suggested, and experiments confirmed, that DNA replicates semiconservatively, using existing strands as templates for new strands. Replication starts at origins of replication (*ori*) and proceeds bidirectionally. At each replication fork where DNA opens up for the synthesis of new, complementary DNA strands, there is a continuously replicated leading strand and a discontinuously replicated lagging strand. The need for RNA priming implies that linear chromosomes will shorten at one end during each round of cell division. The enzyme telomerase compensates for chromosome shortening by replacing the repeated sequences that are lost when the RNA primer is removed.

A
1. How does semiconservative DNA replication differ from conservative replication?
2. Why is an RNA polymerase used during DNA replication? Use your answer to explain why the leading strand replicates continuously and the lagging strand discontinuously.
3. Suppose that DNA ligase were rendered nonfunctional for a single round of DNA replication in a prokaryote. How would the resulting daughter chromosomes be affected?
4. Telomeres on chromosomes in liver cells from older individuals are generally quite a bit shorter than those from younger individuals. What does this observation suggest about telomerase expression in liver cells? How might this observation relate to aging (decline in function with increased age)?

The complex process of DNA replication is amazingly accurate, but it is not perfect. In the next key concept we'll discuss what happens when things go wrong.

9.3 DNA Mutations Alter DNA Sequence

LEARNING OBJECTIVES

By the end of this key concept you should be able to:

9.3.1 **Explain why DNA mutations are ultimately the source of phenotypic variation.**

9.3.2 **Explain why many DNA mutations do not have a phenotypic effect.**

9.3.3 **List three ways in which DNA mutations can occur.**

9.3.4 **Identify how mutations are detected and repaired. Recognize that mutations often go through an intermediate stage that can be repaired and a mutation avoided.**

9.3.5 **Describe the types of chromosomal rearrangements.**

In Key Concept 9.2 we described how polymerases synthesize a new DNA strand that is complementary to the template strand. DNA replication is not perfect, however: errors do occur during the process and may result in mutations—permanent, inherited changes in DNA sequence. DNA is also subject to damage by chemicals and other environmental agents, which can also lead to mutations. Because mutations and the new phenotypes they encode provide the diversity that is the raw material for evolution, understanding how mutations occur is a key to understanding biology.

A common source of mutations is replication errors

The accurate transmission of genetic information is essential for the proper functioning and even the life of a single cell or multicellular organism. However, polymerases occasionally make mistakes in assembling polynucleotide strands. In eukaryotes, the **incorporation error rate**, which measures the probability that an incorrect base will be inserted into the new strand, is about 10^{-5} for DNA polymerases. In other words, during replication, DNA polymerase will insert 1 incorrect base in 100,000, resulting in a **mismatch** between complementary strands (for example, placing an A across from a C). While this rate may seem small, it really isn't. For example, in humans, with a diploid genome size of 6×10^9 base pairs, this rate would introduce 120,000 mismatches every cell division. This seems like a lot—and it is! However, when the daughter DNA molecules are examined after replication, there are typically only about six mismatches. What is going on here?

The errors by DNA polymerase do indeed occur. But *most of them are repaired*, either as they happen or shortly afterward. There are two ways in which this occurs:

1. **Proofreading** is performed by the DNA polymerases during replication. If a DNA polymerase recognizes that it has created a mismatch, it stops, backs up, removes the mismatched nucleotide, and then recommences polymerization (**FIGURE 9.15A**). DNA polymerases are very good at recognizing

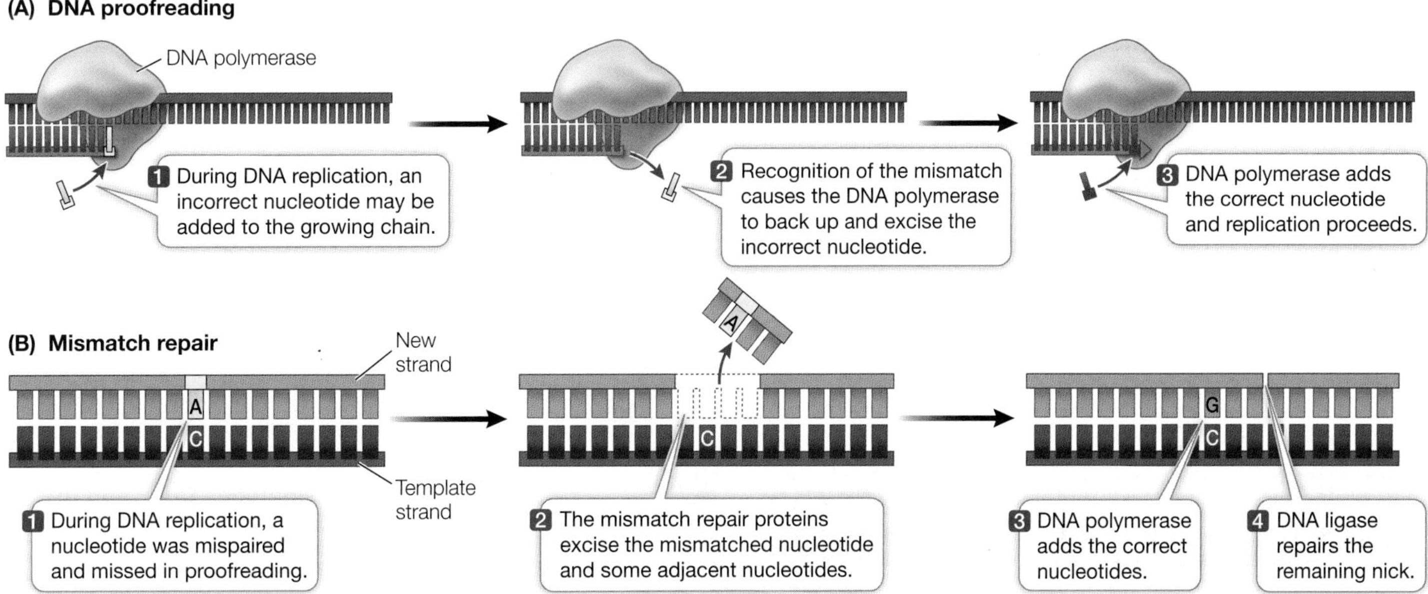

FIGURE 9.15 Proofreading and Mismatch Repair
(A) Incorporation of an incorrect nucleotide results in a mismatch that is usually recognized by proofreading and removed. **(B)** When a mismatch escapes proofreading, it is usually rapidly detected and correctly repaired by a mismatch repair complex. Repair involves deletion of a small segment of the new strand, after which a DNA polymerase and DNA ligase resynthesize and ligate the missing piece.

mismatches: 99 percent are recognized and removed by the enzyme. You can think of this process as akin to spell-check on a word-processing program.

2. **Mismatch repair** occurs after DNA has been replicated. The mismatch repair complex scans the newly replicated molecule looking for mismatched base pairs. It recognizes these because they have abnormal hydrogen bonding and, in many cases, alter the width of the DNA helix (when the mismatch involves two purines or two pyrimidines). If a mismatch is found, the mismatch repair proteins make two single-strand cuts on the new strand and remove the intervening fragment, and then a DNA polymerase resynthesizes the missing piece and DNA ligase seals the remaining single-strand break (**FIGURE 9.15B**). The mismatch repair complex finds and correctly repairs 99 percent of mismatches that escape DNA polymerase proofreading. You can think of this process as akin to a good editor or professor at the writing center.

While proofreading and mismatch repair together reduce the number of mismatches that are not removed and correctly repaired, a few mismatches will lead to **base-pair substitutions**. A base-pair substitution is a type **point mutation**—a mutation that substitutes, deletes or inserts a single base pair in a DNA molecule. Mismatches lead to base-pair substitutions in one of two ways: when the mismatch repair complex removes the original instead of the new strand, or when the mismatch is not repaired before the next round of DNA replication (**FIGURE 9.16**). In both cases the result is that the incorrect base in the new strand gets used as a template for DNA synthesis, resulting in substitution of the original base pair with another base pair. For example, if an A-C mismatch is formed by the DNA polymerase incorrectly adding an A opposite a C, then repair using the new strand (excising the original strand), or another round of DNA replication, will result in a base-pair substitution, with A-T instead of the original G-C base pair at that position.

Some mutations alter phenotype and the phenotypic change may be harmful or occasionally beneficial. In **Chapter 13**, you will learn that evolution would be impossible without mutations.

Chemical changes in bases can lead to DNA mutations

Replication errors by DNA polymerases are just one way in which mutations can occur in DNA. Other mechanisms involve spontaneous or induced chemical changes in the DNA molecule. Bases in DNA constantly undergo spontaneous chemical changes at a low rate that can alter base-pairing properties and result in point mutations. For example, each base can exist as two different structural isomers (see Key Concept 2.6) called tautomers, one of which is common and one rare. When a base temporarily forms its rare tautomer (a tautomeric shift) (**FIGURE 9.17A**) it can pair with a different base, leading to mismatches during DNA replication. The resulting mismatched base pair (e.g., A-C instead of A-T) produces a mutation after a second round of replication (see Figure 9.16). Another spontaneous change is the occasional removal of an amino group (NH_2) in cytosine, producing uracil (**FIGURE 9.17B**). If this deamination process occurs during, or just prior to DNA replication, an A will be inserted into the new DNA strand (because A pairs with U) instead of G (which pairs with the original C). Again, mutation will result after a second round of replication. **Spontaneous mutations** caused by polymerase errors or spontaneous chemical changes in bases cannot be avoided, though their rate is quite low. How important are spontaneous

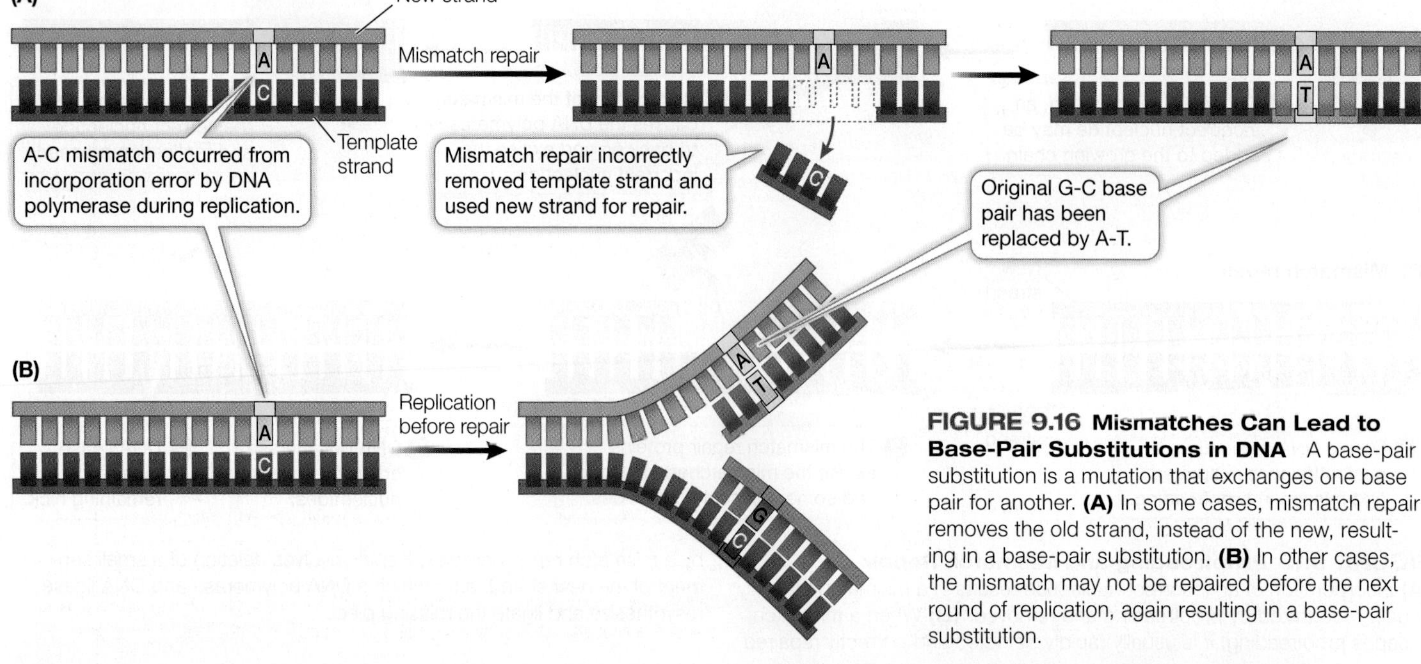

FIGURE 9.16 Mismatches Can Lead to Base-Pair Substitutions in DNA A base-pair substitution is a mutation that exchanges one base pair for another. **(A)** In some cases, mismatch repair removes the old strand, instead of the new, resulting in a base-pair substitution. **(B)** In other cases, the mismatch may not be repaired before the next round of replication, again resulting in a base-pair substitution.

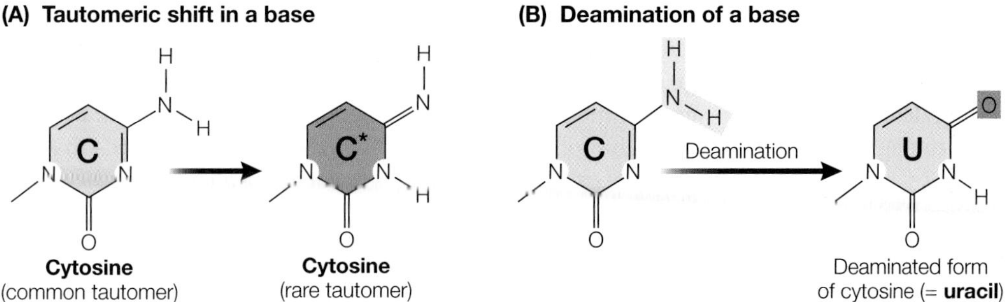

FIGURE 9.17 Spontaneous Chemical Changes Alter Base-Pairing Properties Two types of spontaneous chemical changes that alter base-pairing properties: a tautomeric shift, which creates a structural isomer of a base **(A)** and deamination (the loss of an -NH_2 group) **(B)**. In both of these examples, the altered C base can pair with A, which results in a C-G to T-A base-pair substitution if unrepaired.

mutations as a source of mutations? In humans, cancer is in large part a disease caused by genetic mutations. It is estimated that about 15–20 percent of cancers in humans are caused by spontaneous mutations. What causes the mutations that give rise to the remaining 80 percent of cancers?

Most cancer-causing mutations are caused by **mutagens**, substances that chemically alter DNA and lead to **induced mutations**. Cigarette smoke contains mutagens that chemically alter the bases in DNA and change their base-pairing properties, leading to mutations. Many of the actual molecules in smoke that are mutagenic have been identified. Smoking is responsible for about a third of all cancers. Another one-third of cancers appear to be related to diet. One dietary component in salted meat and fish forms chemicals upon cooking that are able to chemically alter bases such that they show mispairing, which leads to base-pair substitutions. Mutagens often cause DNA damage, which is an alteration of the molecule so that it has a chemical structure not normally found in DNA. Mutagens are not just chemicals. Radiation can also damage DNA. You are familiar with warnings about sunlight and skin cancer, as well as the risks of exposure to ionizing radiation from nuclear power plants and bombs. Both of these types of radiation damage DNA and can cause mutations.

What happens to damaged DNA? As noted above, DNA repair is an important function in cells for recognizing and repairing mismatches. In addition to proofreading and mismatch repair, there is a third DNA repair mechanism, called excision repair, which removes damaged nucleotides and replaces them with normal ones (**FIGURE 9.18A**). The excision repair proteins recognize damaged DNA and remove a fragment of the strand that includes the damaged nucleotide(s), and then DNA polymerase and DNA ligase fill the gap. For some types of DNA damage, the repair pathways recognize the specific damage and repair it directly. An example is the repair of thymine dimers, which are covalent linkages formed between adjacent thymines upon exposure to ultraviolet radiation. Dimers are converted back to two normal thymines by the action of photolyase, which uses light energy to make the repair (**FIGURE 9.18B**).

DNA damage cannot be avoided. In every cell, DNA damage is constantly occurring and being repaired, either by excision repair or using other repair pathways. Repair pathways thus play an enormously important role in reducing the rate at which mutations occur in DNA. However, sometimes repair does not occur, and damaged bases give rise to mutations that can affect various cellular processes in positive or negative ways, most often the latter.

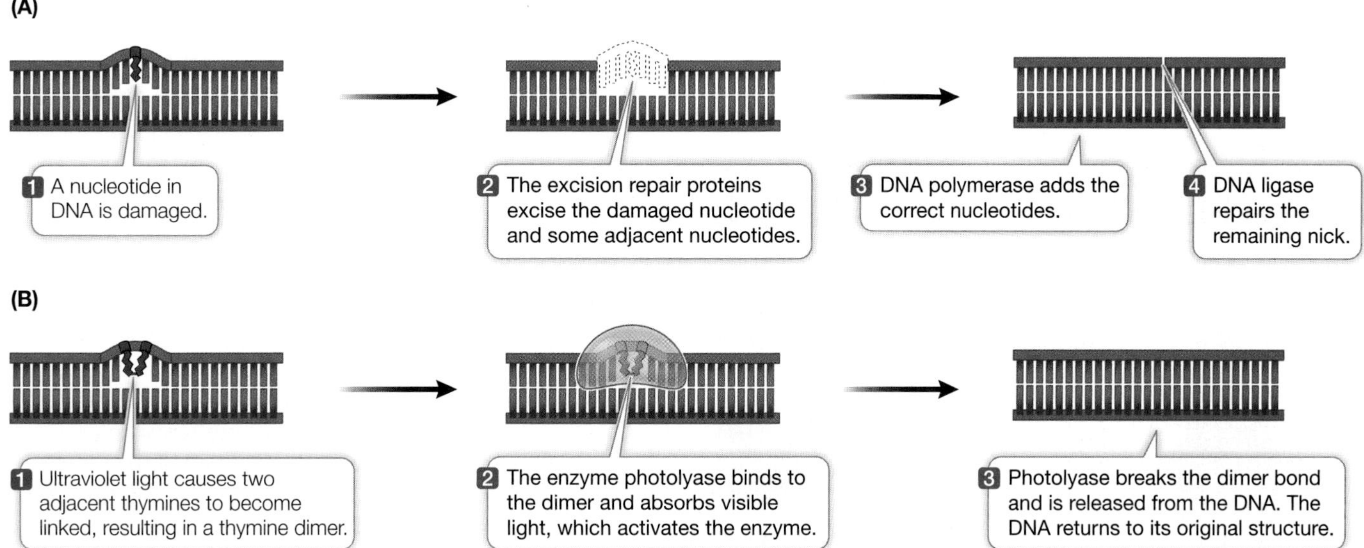

FIGURE 9.18 Repairing Damaged DNA **(A)** Excision repair can recognize and repair a variety of damaged regions of DNA. **(B)** Some repair pathways recognize only a single kind of DNA damage. One such pathway uses the enzyme photolyase, which reverses thymine dimer formation.

Many people associate mutagens with materials made by humans, but there are many naturally occurring mutagens. Plants (and to a lesser extent animals) make thousands of small molecules that serve a range of purposes, including defense against pathogens. Some of these molecules are highly mutagenic.

LINK In **Key Concept 27.4** we describe the diverse roles of plant molecules that fight pathogens.

As we have stated, mutations are important to the generation of diverse populations of organisms and thus to evolution. For humans who try to avoid mutagenesis that leads to changes to somatic cells (e.g., causing cancer) or to germ line cells (e.g., harming offspring), avoidance of mutagens is an ongoing concern. Spontaneous mutagenesis is not in our control: we cannot change our cells to prevent replication errors or bases that change in DNA due to their chemistry. However, we can certainly try to avoid mutagenic substances and radiation to reduce the rate of induced mutation.

A major public policy goal is to minimize the effects of both human-made (anthropogenic) and natural mutagens on human health. For example, the Montreal Protocol (the only international environmental agreement signed and adhered to by all nations) bans chlorofluorocarbons (CFCs) and other substances that deplete the ozone layer in Earth's upper atmosphere. The ozone layer screens out ultraviolet radiation from the Sun—radiation that can cause somatic mutations that lead to skin cancer. Similarly, bans on cigarette smoking have rapidly spread throughout the world. Cigarette smoking causes cancer because of the increased exposure of lung and other cells to the mutagens in tobacco smoke.

Mutations can have various phenotypic effects

In multicellular organisms, mutations can occur in the somatic cells, in which case they are not passed to offspring, or in the germ line cells (gametes), in which case they can be passed to offspring. Somatic mutations may have consequences for the phenotype of an individual. However, germ line mutations can have consequences for future generations, as they can be passed to offspring through gametes.

How likely is it that a mutation will affect a phenotype? This can happen if the mutation causes a functional change, likely caused by the new allele at a gene encoding a different amino acid sequence. Consider that an organism's genome is the total DNA sequence present in all of its chromosomes (or in its single chromosome, in the case of prokaryotes). Depending on the organism, its genome can consist of millions or billions of base pairs of DNA. Most genomes include both protein-coding genes (the main drivers of phenotype) and regions of DNA that are not expressed as proteins. Some sequences that do not encode proteins regulate which protein-coding genes are expressed at a certain time, so that mutations in these regions may have phenotypic effects. But much of the DNA in non-expressed regions has no real function, and so mutations in these regions have little or no phenotypic consequences. Mutations that occur in functional regions may or may not have phenotypic effects. Such mutations are often discussed in terms of their effects on protein-coding genes and their functions (**FIGURE 9.19**). Mutations in the region encoding the amino acid sequence in protein-coding genes may:

- not alter the protein's function (so-called **silent mutations**) either because they do not change the amino acid sequence, or because they alter the amino acid sequence without altering the function of the protein (if an amino acid is replaced by another with similar size and chemical properties)
- alter the protein's function by altering its amino acid sequence.

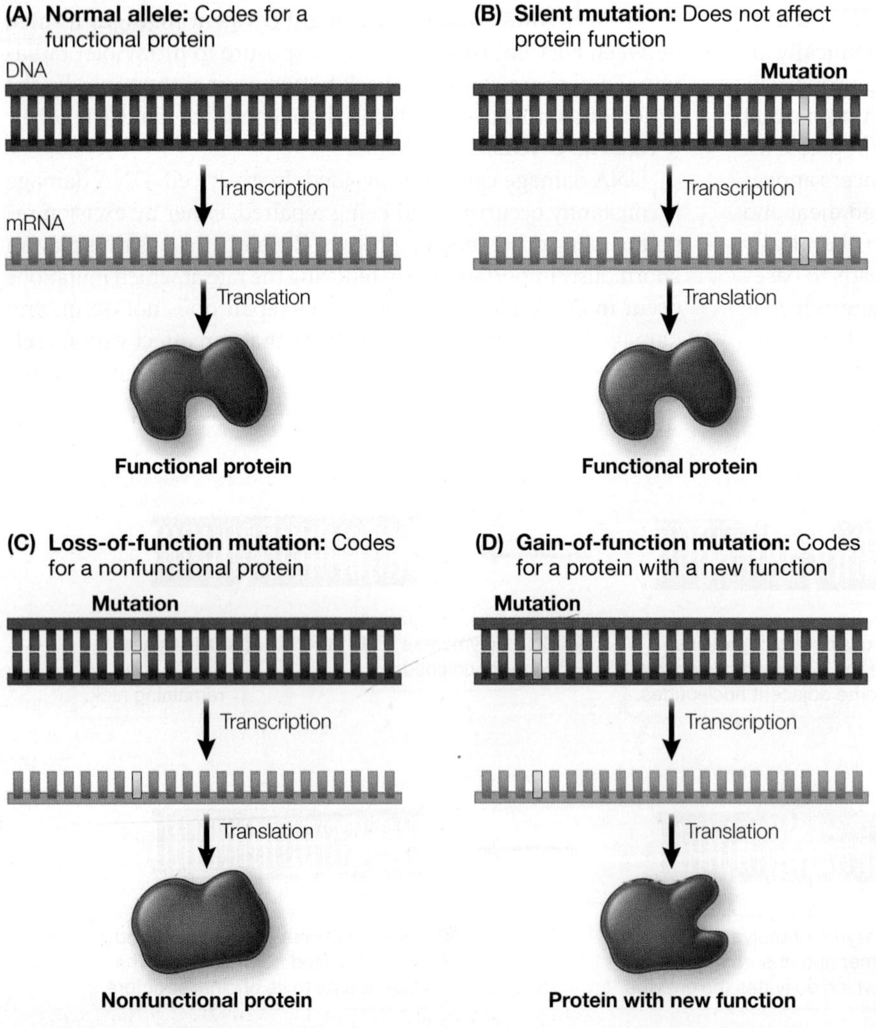

FIGURE 9.19 Effects of Mutations in Protein-Coding Genes Mutations may or may not affect the function of protein-coding genes.

LINK In **Chapter 10** we discuss how a base-pair substitution in a protein-coding gene can result in no change in amino acid sequence.

When a mutation causes a change in phenotype, the change can be subtle or large. **Loss-of-function** mutations occur either because expression of the gene is too low, or because a nonfunctional protein is produced (see Figure 9.19C). Loss-of-function mutations almost always show recessive inheritance in a diploid organism, because the presence of one wild-type allele usually results in sufficient functional protein. For example, the wrinkled seed phenotype studied by Mendel is due to a recessive loss-of-function mutation in the gene that encodes the protein starch branching enzyme 1 (SBE1). Even in plants with only one copy of the wild-type allele, there is enough SBE1 to produce the wild-type round phenotype.

Gain-of-function mutations lead to proteins with altered function (see Figure 9.19D). These rare mutations usually show dominant inheritance, because the presence of the wild-type allele does not prevent the mutant allele from functioning. This type of mutation is common in cancer. An example is a class of mutations in the human gene that encodes the tumor-suppressor protein p53. This protein normally functions to inhibit the cell cycle, but certain mutations cause p53 to promote the cell cycle and prevent programmed cell death. A p53 protein mutated in this way has a gain of oncogenic (cancer-causing) function.

Conditional mutations affect the phenotype only under certain environmental conditions. The wild-type phenotype is expressed under other conditions. Many conditional mutants are temperature-sensitive; that is, they show the altered phenotype only at a certain temperature. For example, there is a temperature-sensitive mutation that affects coat color in rabbits. In warmer parts of the body, the mutant protein is inactive, resulting in pale fur. In the cooler extremities of the body, such as the ears, nose, and feet, the protein is active, producing dark fur.

All mutations are alterations in the nucleotide sequence of DNA. So far we have considered small-scale mutations that alter only one nucleotide. We now turn to large-scale mutations in which entire segments of DNA are rearranged, duplicated, or irretrievably lost.

Chromosomal mutations are extensive changes in the genetic material

While point mutations such as base-pair substitutions affect only one nucleotide in a gene sequence, other kinds of mutations affect longer sequences of DNA. The most dramatic changes that can occur in the genetic material are chromosomal mutations. Although evolution is served by chromosomal mutations that produce genetic diversity upon which natural selection may act, chromosomal mutations are often deleterious. We have already discussed mutations that affect the number of chromosomes found in a cell, resulting in aneuploidy or polyploidy (see Key Concept 7.4). Another type of chromosomal mutation involves chromosomal rearrangements: deletions, duplications, inversions, and translocations.

Deletions are defined as the loss of a chromosome segment (**FIGURE 9.20A**). Their consequences can be severe or even fatal, especially when homozygous (because many genes will be absent). It is easy to imagine one mechanism that could produce a deletion: a DNA molecule might break at two points and the two end pieces might rejoin, leaving out the DNA between the breaks.

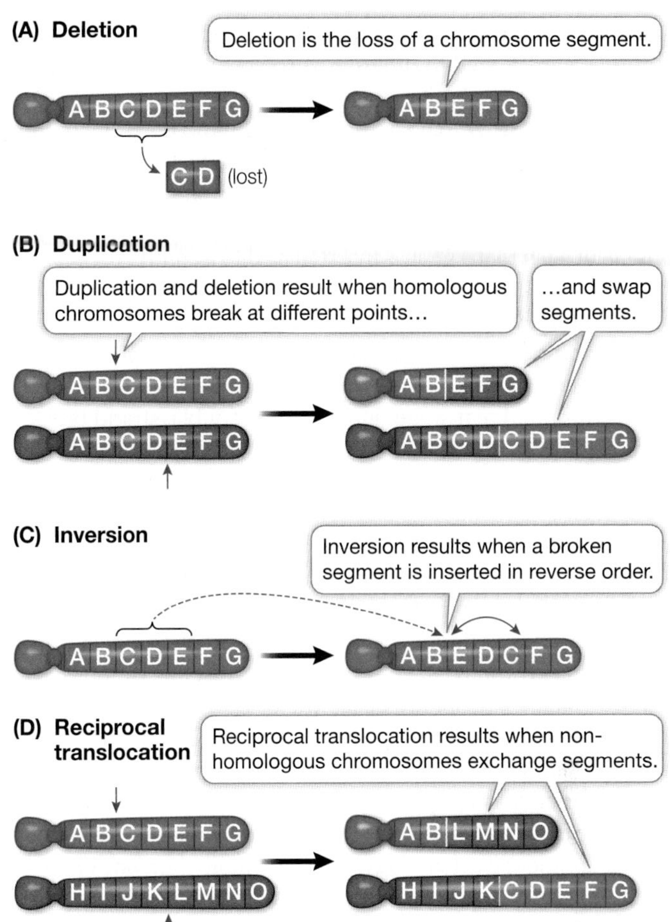

FIGURE 9.20 Types of Chromosomal Rearrangements

Duplications are mutations in which a segment of a chromosome is repeated, resulting in multiple copies of that segment (**FIGURE 9.20B**). Duplications occur if homologous chromosomes break at different positions and then reconnect to the wrong partner—a crossing-over error. One of the two chromosomes produced by this mechanism will lack a segment of DNA (it will have a deletion), and the other will have two tandem copies (a duplication) of the segment that was deleted from the first chromosome. Duplications often have relatively mild effects on phenotype. The main issue is that having extra copies of certain regions often leads to overexpression of genes in those regions.

Inversions are chromosomal mutations in which a segment of a chromosome is "flipped" so that its base sequence reads in reverse. Inversions happen when a segment of DNA is cut out, rotated 180 degrees, and then reinserted into the same location in the chromosome, so that the order of the chromosomal regions between the initial breaks is reversed (**FIGURE 9.20C**). If either break site occurs within a gene, it is likely to cause a loss-of-function mutation in that gene.

Translocations occur when segments of chromosomes break off and become joined to different chromosomes. Many translocations involve reciprocal exchanges of chromosome segments, as shown in **FIGURE 9.20D**. If the break points do not occur within a gene, then translocations are not likely to affect the phenotype because all the genes are present in the appropriate number of copies—it's just their locations that have changed. But some translocations occur such that a gene is inserted next to a control region that alters its expression. For example, in some blood cancers a gene that normally stimulates cell division and is rarely expressed is translocated to a control region that causes it to be highly expressed. The phenotype of a rapidly dividing cancer cell ensues.

Chromosomal rearrangements involve double-strand breaks. In some cases, these breaks occur during crossing over. As we discussed in Chapter 8, crossovers involve the breakage and religation of two homologous DNA molecules. Occasionally an aberrant crossover between homologous or nonhomologous chromosomes can lead to chromosomal rearrangements. In other cases, chromosome breaks represent DNA damage and may be caused by radiation or various chemicals. Ionizing radiation (including X rays, gamma rays, and particles emitted by unstable isotopes) can break the sugar–phosphate backbone of DNA, causing double-strand breaks. A double-strand break often triggers a repair pathway that attempts to rejoin the broken chromosome ends together. This repair pathway is termed **nonhomologous end joining** to emphasize the fact that any two broken ends can be joined—there is no requirement for homology. If there is more than one break in the nucleus, then this repair pathway might join pieces that were once separate, causing any of the possible rearrangements shown in Figure 9.20. This pathway does not join normal chromosomes together at their ends because telomeres prevent chromosome ends from being seen as chromosome breaks (see Key Concept 9.2). If there are enough chromosome breaks, a cell will likely die, either because many pieces will be lost or because apoptosis will be triggered. The ability to kill cells with ionizing radiation is one method that is used to reduce tumors.

REVIEW & APPLY | 9.3

R Base-pair substitutions are point mutations that go through a mismatch intermediate. Mismatches, and thus mutations, can occur spontaneously, such as when a DNA polymerase makes an error during replication or when bases change their chemical structure. Mutagens are substances that change DNA bases, ultimately causing mismatches and induced mutations. There are naturally occurring mutagens such as ultraviolet light from the Sun and plant defense chemicals, as well as human-made mutagens. DNA damage, while quite common, is repaired correctly almost all of the time, so that no mutation results. Some mutations have effects on the phenotype, while others do not. Chromosomal rearrangements, in which large pieces of DNA are lost, duplicated or moved about the genome, are caused by double-strand breaks and can have large or relatively subtle phenotypic effects.

1. If DNA polymerases were perfect, such that they never introduced an incorrect base, would mutations still occur?
2. Many mutations have no effect on the phenotype. Based on the information in this chapter, why might that be the case?
3. How do mutagens cause mutations?

Answer: A single drop of blood contains thousands of white blood cells, and thus tens of thousands of DNA molecules. While this might sound like a lot of DNA, it is actually not enough to determine the genotype of many loci; substantially more copies of DNA are needed. As described in Figure 9.13, the polymerase chain reaction (PCR) is a technique that is able to amplify (make additional copies of) a particular segment of DNA. DNA extracted from a drop of blood left at a crime scene can be split into multiple samples for PCR. Particular pairs of single-stranded DNA primers are then used in each sample to amplify each of the 20 CODIS loci. Successful amplification allows the genotype at each locus to be determined, resulting in a multi-locus genotype that is unique to the individual who left the sample. (Only in the case of identical twins would two individuals have the same genotype at 20 loci.) The development of the PCR technique, coupled with the discovery of highly variable loci, allows scientists to uniquely identify a person who leaves a small DNA sample at a crime scene.

9.1 DNA Is the Molecule of Inheritance

Go to ANIMATION 9.1

- DNA was shown to be the molecule of inheritance by the Hershey–Chase experiment in which **BACTERIOPHAGE** injected their DNA, and not proteins, into bacteria that then manufactured more phage (**FIGURE 9.4**).
- Watson and Crick came up with a three-dimensional model of DNA that obeyed Chargaff's rules and was consistent with Franklin's X-ray diffraction data (**FIGURE 9.6**).

FIGURE 9.4

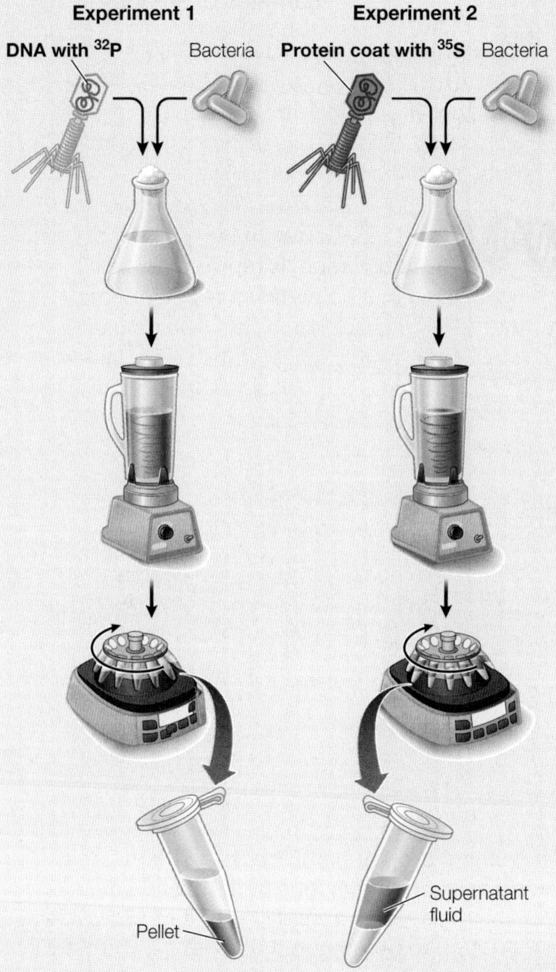

FIGURE 9.6

Watson and Crick

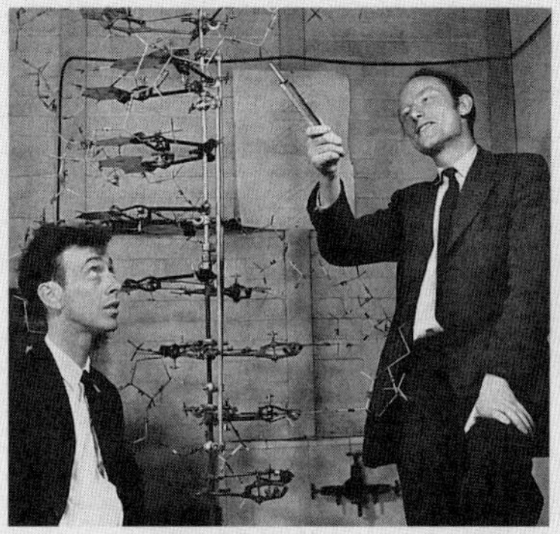

© A. Barrington Brown/Science Source

Space-filling model of DNA

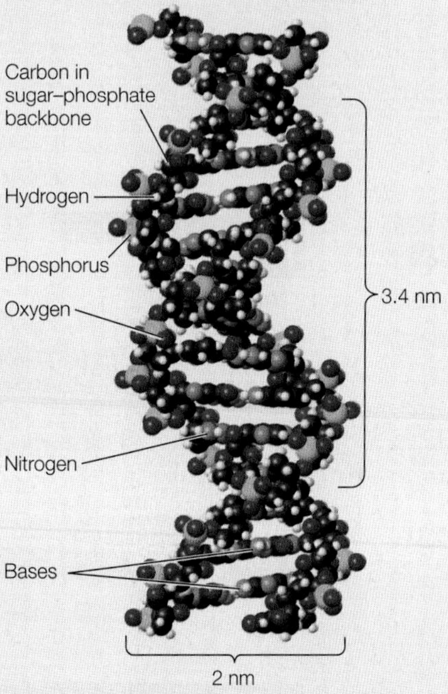

QUESTIONS

1. What did Chargaff discover about DNA that helped determine its structure?
2. If a strand of DNA has sequence 5′-ATCCGTTA-3′, what is the sequence of the complementary strand?
3. What does it mean to say that each strand in a DNA helix has directionality?

9.2 DNA Replication Is Semiconservative

Go to ACTIVITIES 9.1 and 9.2 and ANIMATIONS 9.2, 9.3, and 9.4

- Meselson and Stahl grew bacteria on media with heavy and light nitrogen to show that DNA replicates by semiconservative replication.
- DNA replication begins with **INITIATION** at *ori* sites and then RNA priming.
- Synthesis of the **LEADING STRAND** of the **REPLICATION FORK** is continuous, whereas synthesis of the **LAGGING STRAND** produces **OKAZAKI FRAGMENTS** (**FIGURE 9.12**).
- **TELOMERASE** functions to lengthen chromosome ends to compensate for the shortening that occurs with every cell division (**FIGURE 9.14**).

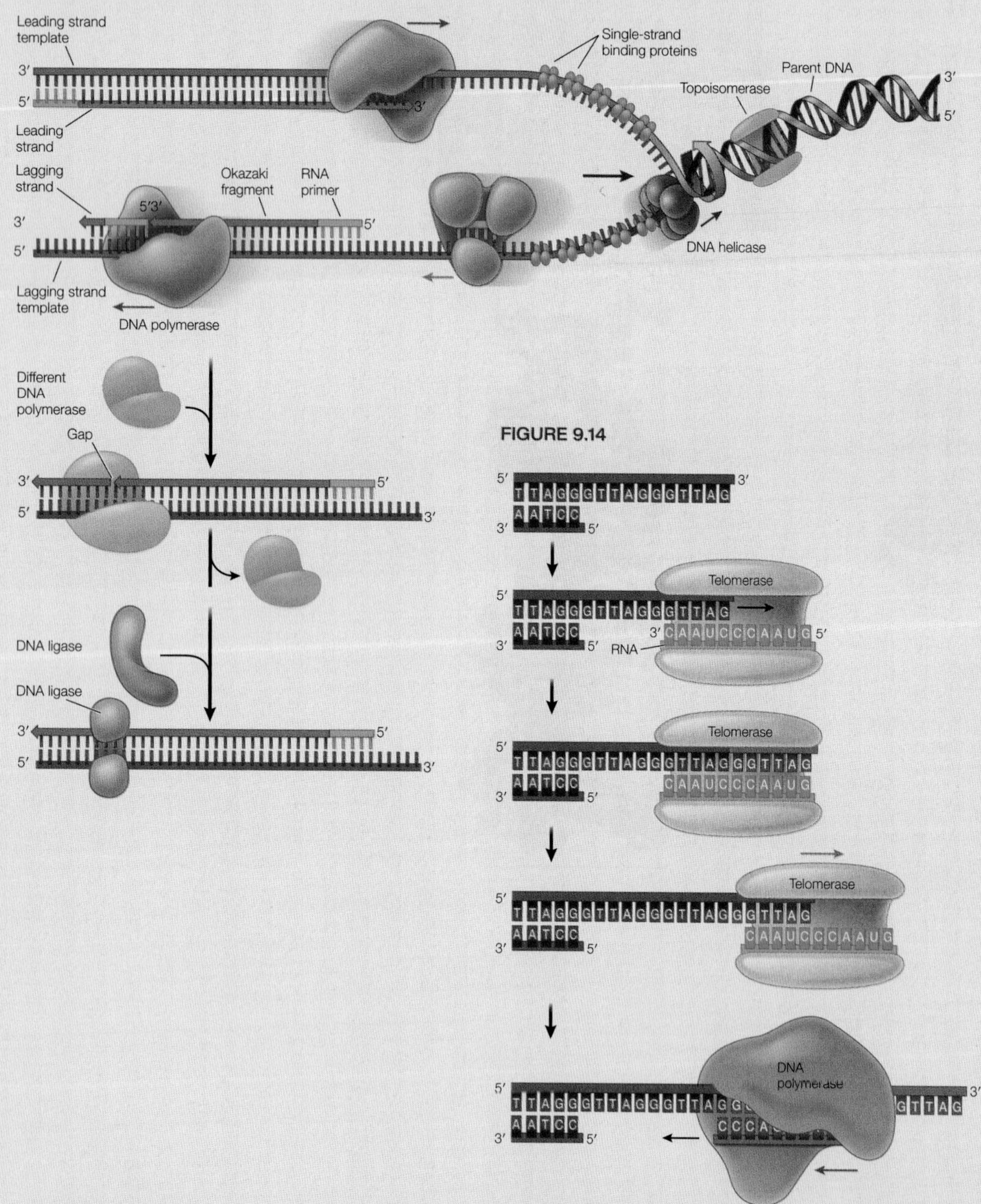

QUESTIONS

1. Why is it important that an *ori* be used only once during DNA replication?
2. What is the function of DNA ligase in replication?
3. Why are telomeres short in the liver cells but not in the gametes of an adult?

FIGURE 9.15

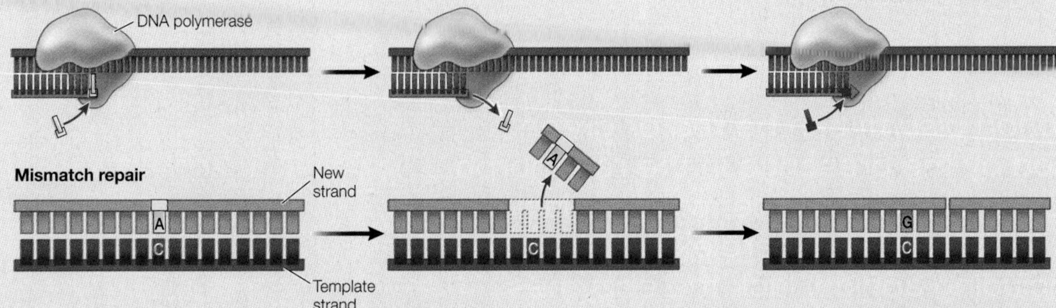

FIGURE 9.20

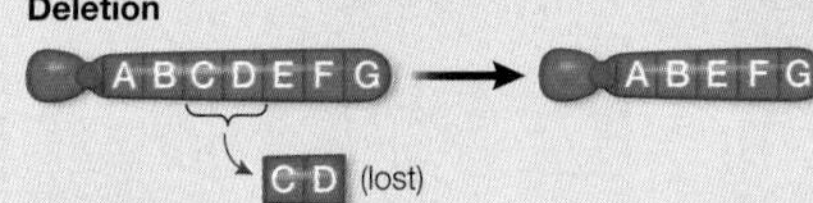

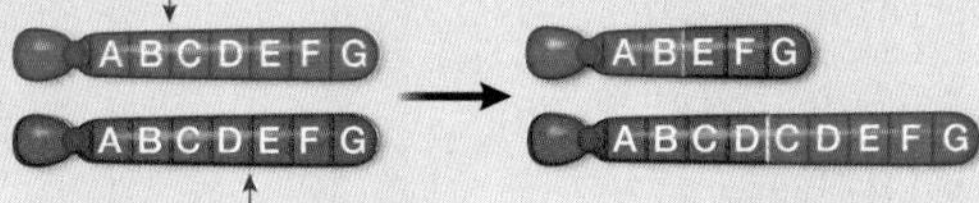

Inversion

A B C D E F G → A B E D C F G

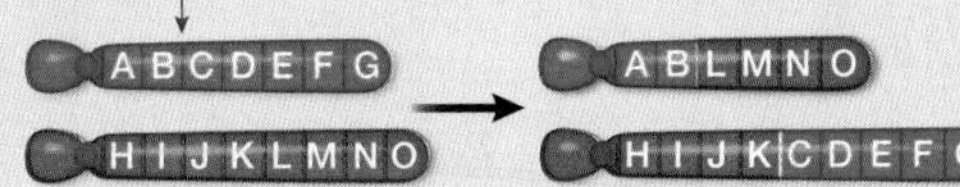

- DNA polymerases can make errors and introduce mismatches, which may be corrected by **PROOFREADING** or by **MISMATCH REPAIR** (**FIGURE 9.15**).
- Chromosomal rearrangements include **DELETIONS**, **DUPLICATIONS**, **INVERSIONS**, and **TRANSLOCATIONS** (**FIGURE 9.20**).

QUESTIONS

1. Under what conditions will a DNA polymerase error lead to a mutation?
2. What would be the effect on the organism of a genetic mutation that prevented any further DNA repair from occurring?
3. What are the two mechanisms that cause chromosomal rearrangements?

Go to **Achieve** for the e-book, animations, activities, and additional resources and assignments.

10 From DNA to Protein: Gene Expression

Pea aphids (*Acyrthosiphon pisum*) can be red or green. The difference is caused by the presence in the red morph of a carotenoid compound called torulene, which is lacking in the green morph. This difference is perplexing because normally all animals, including aphids, lack the enzymes necessary to synthesize carotenoids. A startling discovery regarding torulene synthesis in aphids supports the idea that proteins are encoded in the same manner in all organisms.

What discovery regarding torulene synthesis supported the hypothesis that protein sequences are encoded essentially identically in all organisms?

You will find the answer to this question on page 253.

KEY CONCEPTS

10.1 One Gene Encodes One Polypeptide

LEARNING OBJECTIVES

By the end of this key concept you should be able to:

10.1.1 **Describe how the study of biochemical pathways led to the identification of mutant genes encoding nonfunctional enzymes.**

10.1.2 **Describe what is meant by "one gene-one polypeptide."**

10.1.3 **Explain how a genetically determined single amino acid change can result in a functional change in a protein.**

10.1.4 **Describe the central dogma of molecular biology.**

10.1.5 **List the three most common types of RNAs in a cell.**

Mendel discovered the laws of inheritance (see Key Concept 8.1), and then the genetic material was identified as DNA (see Key Concept 9.1). In this chapter we will show that genes often code for proteins, and it is proteins that determine phenotypes. The connection between protein variation and phenotype variation was made before it was known that DNA is the genetic material.

Observations in humans led to the proposal that genes encode enzymes

The understanding that a protein is the expressed phenotypic product of a gene began with the study of a mutation. In the early twentieth century, the English physician Archibald Garrod saw several children with a rare disease. One symptom was that the patients' urine turned dark brown or black in air, and for this reason the disease was named alkaptonuria ("black urine") (**FIGURE 10.1A**). Examining the pedigrees of the children led Garrod to hypothesize that alkaptonuria was inherited as a recessive disorder (see Figure 8.7B). His insight came in 1902, just two years after the rediscovery of Mendel, and is another example of the importance of Mendel's work.

After many years of additional research, Garrod identified the biochemical abnormality in the affected children. He isolated an unusual substance, homogentisic acid, which accumulated in their blood, joints (where it crystallized and caused severe pain), and urine (which turned black when exposed to air). Enzymes as biological catalysts had just been discovered, and Garrod proposed that in healthy individuals homogentisic acid was converted to a

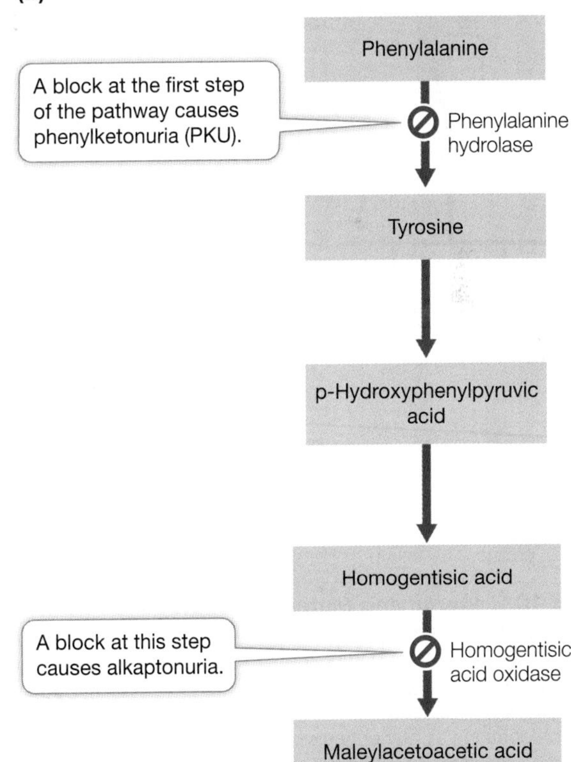

FIGURE 10.1 Alkaptonuria and Phenylketonuria
(A) The accumulation of homogentisic acid in the urine of patients with alkaptonuria causes their urine to turn black when exposed to air. **(B)** Both alkaptonuria and phenylketonuria are caused by abnormalities in specific enzymes in the pathway that breaks down the amino acid phenylalanine.

harmless product by an enzyme. He speculated that this enzyme might be missing or nonfunctional in affected individuals, allowing homogentisic acid to accumulate. Garrod hypothesized that the synthesis of the active enzyme is determined by a dominant wild-type allele of the gene that was mutated in his alkaptonuria patients. These and other studies led him to coin the term "inborn error of metabolism" to describe this kind of genetically determined biochemical disease.

It took several decades before the specific enzyme and the specific gene mutation that cause alkaptonuria were identified. In 1958, the enzyme was identified as homogentisic acid oxidase, and in 1996 the DNA mutation leading to alkaptonuria was described. Homogentisic acid is part of a catabolic pathway (see Chapter 5) that breaks down the amino acids phenylalanine and tyrosine (**FIGURE 10.1B**). Phenylketonuria (PKU), another genetic disease involving the same pathway, was discovered several decades after Garrod did his work. In phenylketonuria, the enzyme that converts phenylalanine to tyrosine is nonfunctional. PKU leads to elevated levels of phenylalanine, which causes significant intellectual disability if untreated. Fortunately it is straightforward, and now routine, to test for high phenylalanine levels in the blood of newborn infants, and if an affected child consumes a diet low in proteins containing phenylalanine, intellectual disability is avoided.

The realization that the presence of a nonfunctional enzyme in a linear metabolic pathway leads to the buildup of the enzyme's substrate was a critical insight for understanding the relationship between genes and proteins. Further investigation of this idea was done on model organisms, particularly the bread mold *Neurospora crassa.* George Beadle and Edward Tatum developed a method for studying mutations that knocked out metabolic pathways in *Neurospora* (**FIGURE 10.2**). ("Knocking out" is a widely used term that refers to blocking the expression of a gene.) They showed that the knockout mutation encoding a single enzyme was inherited in a Mendelian manner, suggesting that enzyme knockouts are nonfunctional alleles at genes that encode them. Adrian Srb and Norman Horowitz used Beadle and Tatum's methodology to characterize the number, order, and identity of enzymes involved in the biosynthesis of nutritionally critical compounds. **INVESTIGATION FIGURE 10.3** describes how Srb and Horowitz identified and ordered the enzymes in the arginine synthesis pathway.

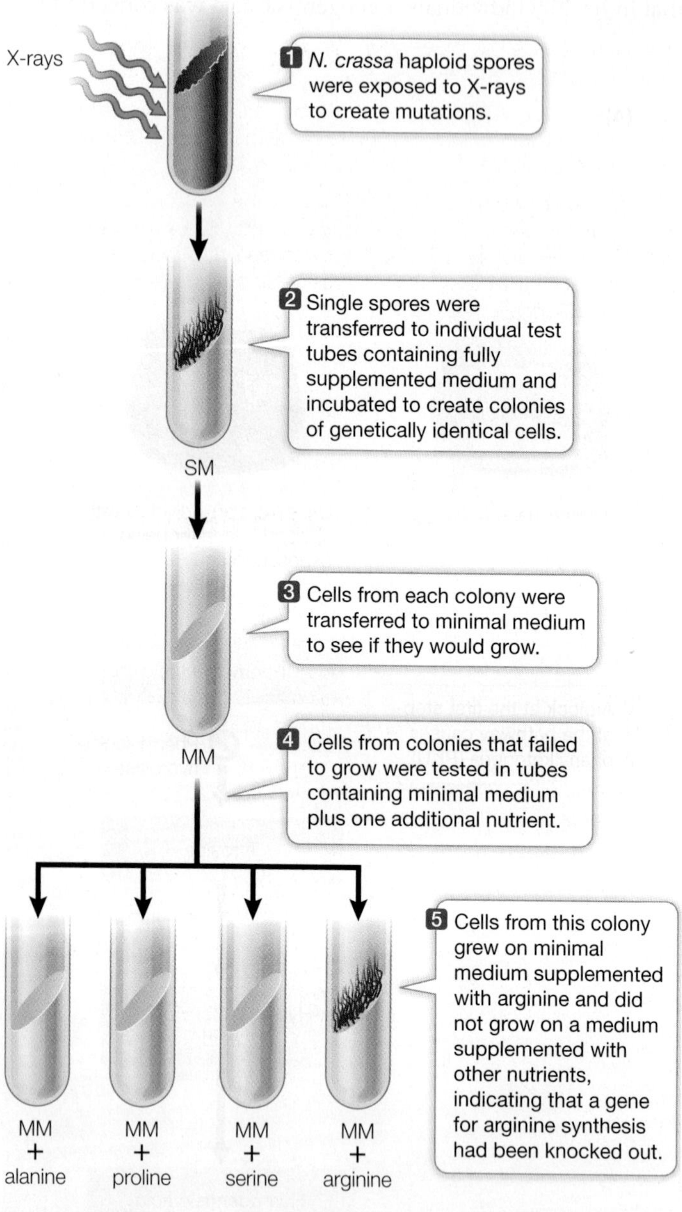

FIGURE 10.2 Identifying Mutants in Metabolic Pathways in *Neurospora crassa* George Beadle and Edward Tatum developed a method for identifying knockouts in metabolic pathways. They exposed haploid spores of *N. crassa* to radiation to create mutant spores and then germinated each spore on fully supplemented medium (SM), which contains all nutrients used by cells. They then identified single-spore strains that could grow on SM by not minimal medium (MM), which contains only the essential nutrients required for growth by a wild type strain. Further tests with MM supplemented with individual nutrients allowed them to identify the nutrient that could not be made (arginine is this example). They concluded that radiation had knocked out a gene for a critical enzyme in the metabolic pathway for the nutrient in *N. crassa* stains that could not grow without it. (After G. W. Beadle and E. L. Tatum. 1941. *Proc Natl Acad Sci USA* 27: 499–506.)

The concept of the gene has changed over time

The fact that enzyme knockouts are inherited in a Mendelian manner led to the "one gene–one enzyme" hypothesis. Given that the sequence of amino acids determines protein function, it was assumed that *a mutant phenotype arises from a change in the protein's amino acid sequence.* Later, when it was known that genes could also encode non-enzyme proteins, Beadle and Tatum's hypothesis was rephrased as **one gene–one polypeptide**. Recall from Key Concept 3.4 that some proteins have quaternary structure, meaning they are composed of more than one polypeptide. Such proteins are encoded at more than one gene.

An example in humans of a protein with quaternary structure is the oxygen-carrying protein hemoglobin. This protein consists of four polypeptide chains—two α-chains and two β-chains. Hemoglobin has been extensively studied because it is an abundant protein and there are numerous mutant versions in human

INVESTIGATION

FIGURE 10.3 One Gene-One Enzyme Adrian Srb and Norman Horowitz used Beadle and Tatum's technique to study knockouts in the arginine pathway. They showed that there are at least seven steps in the pathway, and they were able to order the intermediates ornithine and citrulline, and determine where in the pathway the knocked out enzymes occurred.[a]

HYPOTHESIS

The arginine biosynthetic pathway involves multiple metabolite intermediates, including ornithine and citrulline, with ornithine occurring first.

METHOD

1. Use Beadle and Tatum's technique to identify several mutant strains of *Neurospora crassa* that can grow only when supplemented with the amino acid arginine.
2. Fuse haploid strains to make diploids to determine how many different genes have been mutated.
3. Take strains that have different genes mutated and attempt to grow them on minimal medium supplemented with ornithine or citrulline.
4. Use growth data to determine the order in which ornithine and citrulline are employed in the arginine synthesis pathway and the steps in the pathway where different gene products (enzymes) are located.

RESULTS

	Neurospora growth after 5 days (mg dry wt)			
Strain	No supplement	Arginine added	Ornithine added	Citrulline added
1	0.9	37.2	29.2	37.6
2	0	20.9	10.5	18.7
3	0	16.7	7.7	15.2
4	1.1	33.2	25.5	30.0
5	1.0	37.6	0.8	34.1
6	2.3	35.0	2.5	42.7
7	0	20.4	0	0

Using strain fusions (not shown), they identified seven strains having mutations at different genes. The growth data indicate that strains 1–4 can grow when supplemented with arginine, ornithine, or citrulline; strains 5 and 6 can grow when supplemented with arginine or citrulline but not ornithine; and strain 7 can grow when supplemented with arginine but not with ornithine or citrulline.

CONCLUSIONS

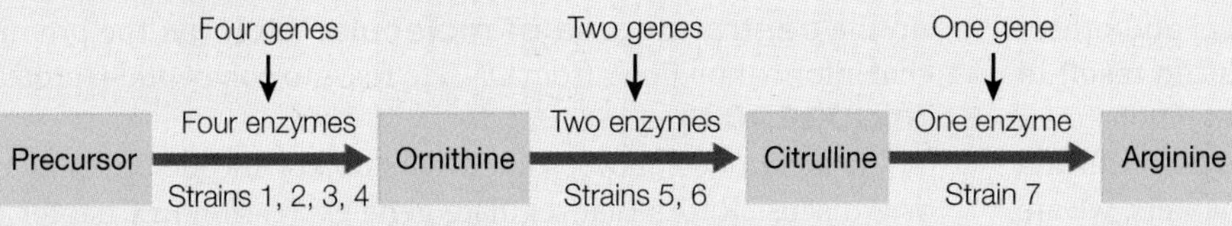

1. Since Srb and Horowitz isolated seven mutant strains that affected different genes, they proposed that the wild-type gene products of each mutant strain were enzymes that catalyzed seven separate steps in the arginine biosynthetic pathway.
2. Citrulline comes after ornithine in the arginine biosynthetic pathway, because it allows the growth of more of the mutant strains than ornithine does.
3. Strains 1–4 are mutants missing different active enzymes that lead to the formation of citrulline. Strains 5 and 6 have mutations affecting enzymes that catalyze steps between citrulline and ornithine, and strain 7 is a mutant missing an enzyme needed between citrulline and arginine.

ANALYZE THE DATA

1. To determine whether two strains that were unable to synthesize arginine had mutations at the same gene, Srb and Horowitz performed a fusion of two haploid, arginine-requiring strains to generate a diploid strain. What diploid genotype was generated, and what was its expected phenotype, assuming mutations were recessive, if the two strains (a) were mutants at the same gene or (b) were mutants at different genes?
2. If there is only a single step between citrulline and arginine, as indicated by the pathway diagram, what metabolite is expected to build up in strain 7?
3. If ornithine builds up in strain 6 but not in strain 5, can you order the two strains with respect to the two steps in the pathway between ornithine and citrulline?

[a]A. M. Srb and N. H. Horowitz. 1944. *J Biol Chem* 154: 129–139.

A similar **Analyze the Data** exercise may be assigned in **Achieve**.

populations (**FIGURE 10.4A**). For example, sickle-cell disease is caused by a point mutation in the gene for β-globin and is inherited as an autosomal recessive (carried on an autosome rather than a sex chromosome). In sickle-cell disease, one of the 146 amino acids in the β-globin chain is abnormal: at position 6, the normal glutamic acid has been replaced by valine. This replacement changes the charge of the protein (glutamic acid is negatively charged and valine is neutral), altering its structure in red blood cells. This, in turn, both impairs the ability of the blood to carry oxygen and causes the red blood cells to have defective, sickle shapes, which tend to block narrow capillaries, causing damage to tissues (**FIGURE 10.4B**). Thus a relatively

(A) Some mutations in the human β-globin gene

Variants of β-globin	Amino acid position (of 146)								
	2	6	7	16	24	26	56	63	95
A (wild type)	His	Glu	Glu	Gly	Gly	Glu	Gly	His	Lys
Tokuchi	Tyr								
S		Val							
C		Lys							
G			Gly						
J Baltimore				Asp					
Savannah					Val				
E						Lys			
Bangkok							Asp		
Zürich								Arg	
M Saskatoon								Tyr	
N Baltimore									Glu

Only three hemoglobin variants (S, C, and E) lead to clinical problems.

(B) Red blood cell phenotypes

FIGURE 10.4 Single Amino Acid Substitutions in β-globin Found in Human Populations **(A)** Each of these alleles codes for a protein with a single amino acid change in the 146-amino acid chain of β-globin. Only a few of the segregating alleles have clinical effects when homozygous. The sickle-cell disease allele ("S") is caused by a single amino acid change from glutamic acid to valine at position 6. **(B)** Individuals homozygous for the S allele have red blood cells that are prone to sickling, impeding blood flow.

Media Clip 10.1 Sickle Cells: Deformed by a Mutation

minor change, the substitution of a single amino acid at position 6, can have major effects on protein function and result in severe phenotypic effects.

Hundreds of other single amino acid alterations in β-globin have been reported, some with phenotypic effects and others without. Recall from Figure 9.19B that mutations can result in changes to the amino acid sequence of a protein but not its function. The one gene–one polypeptide relationship is a powerful and useful concept. However, as you will see in later chapters, we are learning of exceptions to the rule.

LINK Exceptions to the one gene–one polypeptide relationship include the alternative splicing of RNA, which can produce multiple functional polypeptides from a single gene; see **Key Concept 11.5**.

Mutations such as those that cause alkaptonuria and phenylketonuria result in alterations in amino acid sequences. But not all genes code for polypeptides. As we will see below and in Chapter 11, some DNA sequences are transcribed into RNA molecules that are *not* translated into polypeptides, but instead have other functions. Like all other DNA sequences, these RNA genes are subject to mutations, which may or may not affect the functions of the RNAs they produce.

Our understanding of genes and how they are expressed has increased dramatically since Watson and Crick first worked out the structure of DNA (see Key Concept 9.1).

Genes are expressed via transcription and translation

We discussed in Chapter 9 how genetic information in DNA is replicated so that it can be passed down to daughter cells. Expression of the genetic information stored in DNA is an entirely different process, although you will already be familiar with the differences between DNA and RNA, the sequence of amino acids in polypeptides, and the phenomenon of base pairing in nucleic acids. As we described briefly in Chapter 3 (see Figure 3.13), genes are expressed as RNAs, many of which are then translated into proteins. Gene expression involves **transcription**, whereby the information in a DNA sequence (a gene) is copied by base pairing into a complementary RNA sequence, and **translation**, whereby the RNA sequence is used to synthesize the amino acid sequence of a polypeptide. Replication, transcription, and translation describe the manner in which information is used and together make up the **central dogma of molecular biology**: the premise that information flows from DNA to RNA to polypeptide (protein) (**FIGURE 10.5**).

As mentioned above, some RNAs do not encode proteins. We will discuss RNAs that regulate gene expression in Chapter 11. Here we discuss three types of RNAs with regard to their roles in protein synthesis: messenger RNAs, ribosomal RNAs, and transfer RNAs. These three types of RNAs are the most abundant RNAs in cells.

Messenger RNAs (mRNAs) encode protein information. When a particular gene is expressed, the two strands of DNA unwind and separate into a **coding strand** and a **template strand**. The template strand is transcribed to produce a single-stranded mRNA molecule that is complementary to the template strand. In eukaryotic cells, the mRNA is altered during

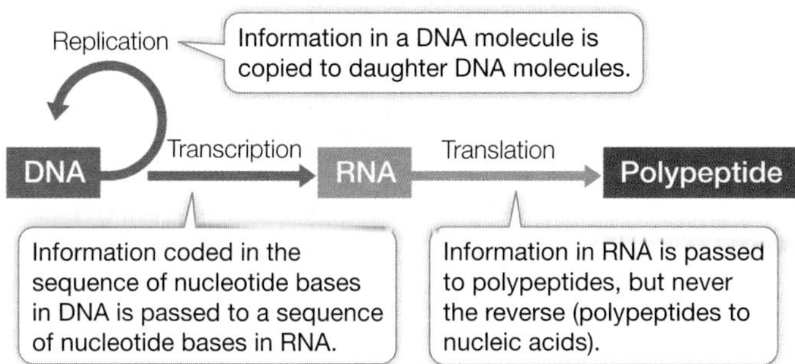

FIGURE 10.5 The Central Dogma of Molecular Biology Information is stored in DNA and copied during replication. Certain DNA sequences are then transcribed into RNA, and some RNA sequences are translated into amino acid sequences.

Have you seen any examples in previous chapters where information flow is not as shown in the central dogma?

processing in the nucleus and then moves to the cytosol, where it is translated into a polypeptide (**FIGURE 10.6**). No mRNA processing occurs in prokaryotes—as soon as the mRNA is transcribed, it is ready to be translated. The nucleotide sequence of the mRNA determines the ordered sequence of amino acids in the polypeptide chain, which is built by a ribosome. mRNAs are important, encoding all the proteins in a cell, and most genes encode mRNAs. Even so, mRNAs represent only about 5 percent or less of cellular RNA.

Several **ribosomal RNAs (rRNAs)** and multiple proteins make up ribosomes, which are essentially protein synthesis factories. One of the rRNAs catalyzes peptide bond formation between amino acids, to form a polypeptide. Ribosomes are common in cells, and ribosomal RNAs are extremely abundant, making up about 80 percent of total RNA in a cell.

Transfer RNAs (tRNAs) can both bind a specific amino acid and recognize a specific sequence of nucleotides in mRNA. It is the tRNAs that recognize which amino acid should be added next to a growing polypeptide chain. tRNAs are abundant, representing about 15 percent of total cellular RNA.

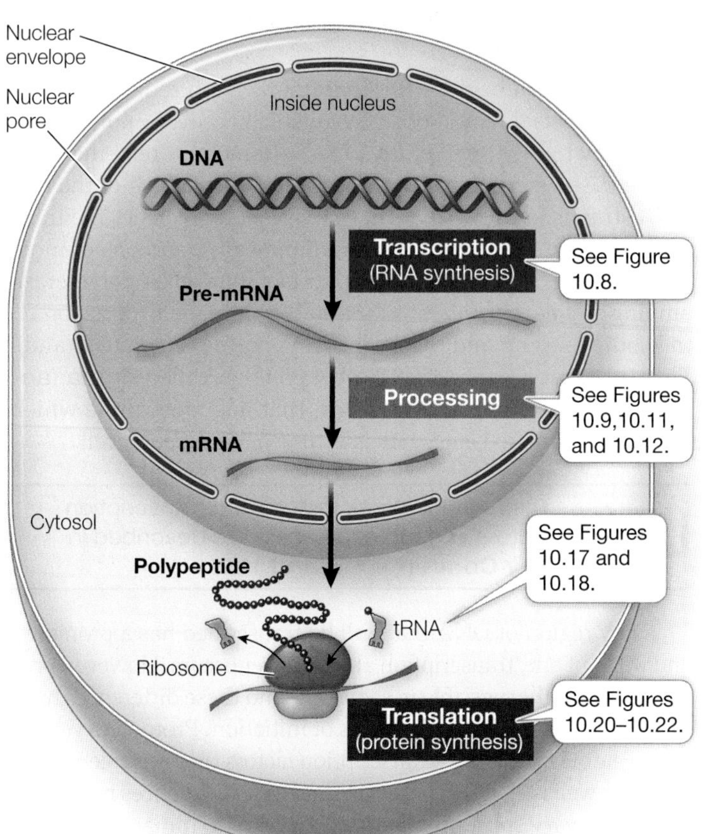

FIGURE 10.6 From Gene to Protein This diagram summarizes transcription and translation in a eukaryote. Note that the nucleus is typically smaller than shown here.

Activity 10.1 Eukaryotic Gene Expression

REVIEW & APPLY | 10.1

R Knockouts of metabolic pathways are inherited as single genes, suggesting that one gene encodes one polypeptide. An organism can survive the homozygous knockout of the gene for an enzyme in a critical metabolic pathway if it is supplied with the metabolites required in later stages of the pathway (after the block). Knocking out genes helped scientists learn the order in which enzymes act in metabolic pathways. The central dogma of molecular biology states that DNA replicates and is used as a template for transcription into RNA. Genes encoding proteins are transcribed into mRNAs, which are then translated into proteins with the help of rRNAs and tRNAs. Not all genes encode polypeptides. Some instead, encode RNAs as their final product, such as rRNA and tRNA genes.

A
1. Could a person inherit recessive alleles for both alkaptonuria and phenylketonuria and thus be a double homozygote for the mutant alleles? If so, what would the person's symptoms be?
2. What is the difference between the "one gene–one enzyme" and "one gene–one polypeptide" hypotheses?
3. A friend tells you that mutations that cause protein changes are deleterious. Is this statement correct?
4. With regard to the gene encoding β-globin, compare skin cells and developing red blood cells with respect to the presence or absence of DNA encoding the polypeptide and mRNA expressed from that sequence.

Having shown how the connection between genes and phenotypes can be understood in terms of DNA and proteins, we will now turn to some details of the process of gene expression, which is at the heart of what genes do.

10.2 Gene Expression Begins with Transcription of DNA into RNA

LEARNING OBJECTIVES

By the end of this key concept you should be able to:

10.2.1 Describe the initiation, elongation, and termination steps of transcription.

10.2.2 Distinguish between introns and exons in eukaryotic mRNAs and describe how introns are spliced out during mRNA processing.

10.2.3 Describe the functions of 5′ capping and 3′ poly A tailing of mRNA in eukaryotes.

Transcription—the formation of a specific RNA sequence from a specific DNA sequence—is performed by a DNA-dependent RNA polymerase. The process is essentially the same for the synthesis of mRNAs, tRNAs, and rRNAs. Like mRNAs, tRNAs and rRNAs are encoded by specific genes; their important roles in protein synthesis will be described in Key Concepts 10.3 and 10.4. There are also other kinds of RNA in the cell, with functions other than protein synthesis.

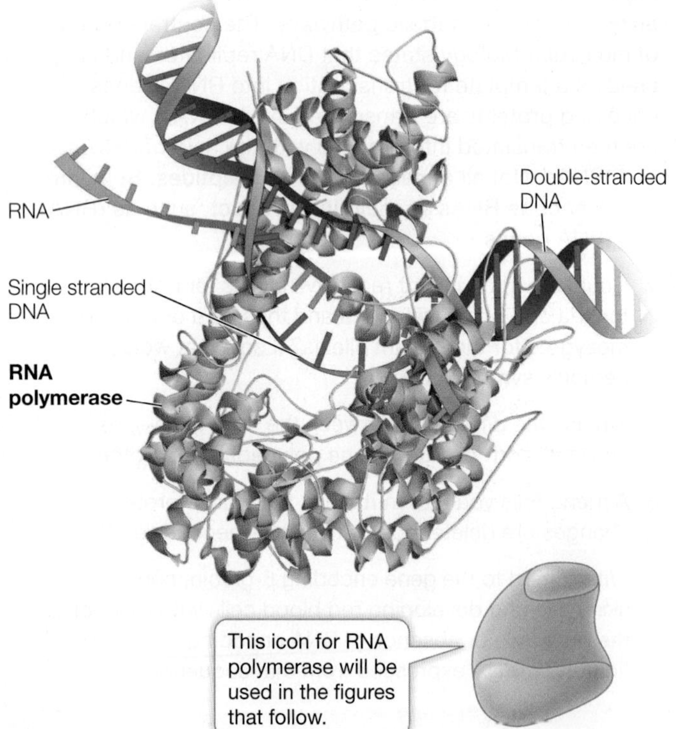

FIGURE 10.7 RNA Polymerase Interacting with DNA The RNA polymerase enzyme depicted (in yellow-brown) is from bacteriophage T7, but it is representative of most other RNA polymerases. The inset indicates how this enzyme will be represented in the figures that follow. (RNA polymerase by David McIntyre based on data from PDB ID: 1MSW. Y. W. Yin and T. A. Steitz. 2002. *Science* 298: 1387–1395.)

LINK Small nuclear RNAs are involved in processing mRNA after it is transcribed, and microRNAs play important roles in stimulating or inhibiting gene expression; see **Key Concept 11.5**.

RNA polymerases share common features

Like the DNA polymerases we have met so far, DNA-dependent RNA polymerases require a single-stranded DNA template. They read the template in the 3′-to-5′ direction and synthesize an RNA strand in the 5′-to-3′ direction. Instead of using deoxyribonucleoside triphosphates (dNTPs), RNA polymerases use one of the four ribonucleoside triphosphates (ATP, GTP, CTP, and UTP) to add the next nucleotide to the growing RNA strand. An important difference between DNA polymerases and RNA polymerases is that the latter do not need a primer to begin polymerization. This implies that they can start RNA synthesis anywhere, but as you will see, DNA sequences adjacent to genes and certain DNA binding proteins guide RNA polymerases to specific locations to initiate transcription. There is only one kind of RNA polymerase in bacteria and archaea, whereas there are several kinds in eukaryotes. However, they all share a common structure (**FIGURE 10.7**).

Transcription occurs in three steps

Like DNA replication, transcription can be divided into three distinct processes: initiation, elongation, and termination (**FIGURE 10.8**).

Transcription begins with initiation, which requires a **promoter**, a special region of DNA to which the RNA polymerase binds very tightly (see Figure 10.8A). Promoters "tell" the RNA polymerase where to begin transcription and which of the two DNA strands to transcribe. The core promoter includes the transcription start site, where transcription begins, and one or more short sequences of nucleotides that facilitate RNA polymerase binding and lie "upstream" of the start site (in the 5′ direction on the coding strand and the 3′ direction on the template strand). These sequences are bound by other proteins, called **sigma factors** and **transcription factors**, that help determine which genes are expressed at a particular time in a particular cell.

LINK The roles of sigma factors and transcription factors in gene expression are described in **Key Concepts 11.1 and 11.2.**

Every region of DNA that will be transcribed has a promoter, and without one, transcription almost never occurs. However, promoters are highly variable in sequence, and these differences have functional consequences in terms of initiation. Promoters vary in which sigma factors and transcription factors they are able to bind and how well they bind these factors. These differences mean that promoters vary substantially in their ability to initiate transcription. Furthermore, bacteria, archaea, and eukaryotes differ in the details of transcription initiation. Despite these variations, the basic mechanisms of initiation are the same throughout the living world and provide further evidence of the biochemical unity of life on Earth. At the end of initiation, an RNA polymerase is tightly bound to the promoter region, and the DNA is locally unwound

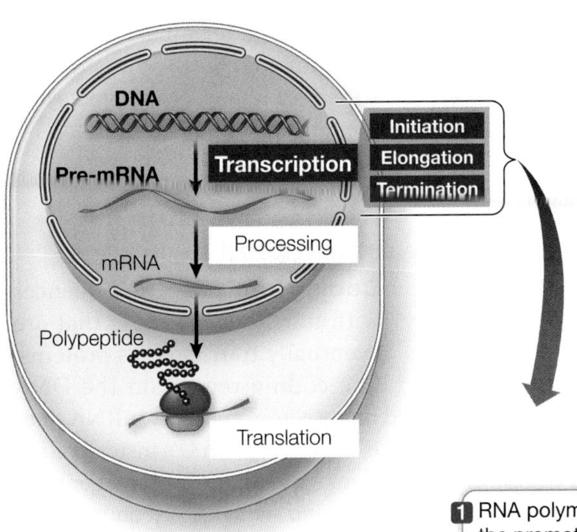

FIGURE 10.8 RNA Is Transcribed from DNA DNA is partially unwound by RNA polymerase so that one strand can serve as a template for RNA synthesis. The RNA transcript is formed and then peels away from the template strand, allowing the DNA that has already been transcribed to rewind into a double helix. Three distinct processes—initiation, elongation, and termination—constitute transcription. RNA polymerase is much larger in reality than indicated here, covering about 50 base pairs.

Animation 10.1 Transcription

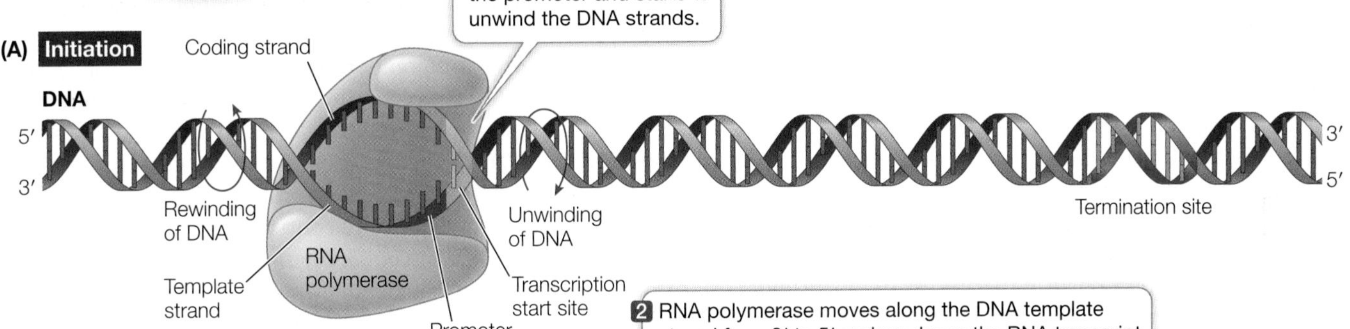

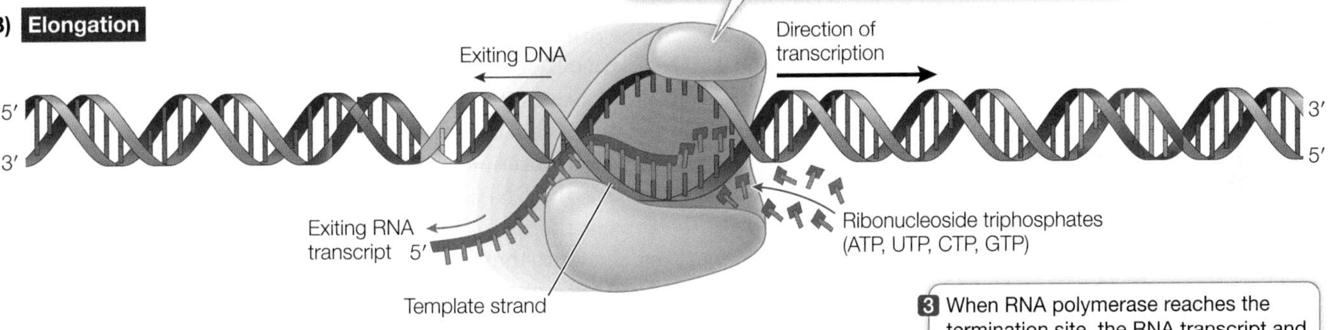

TABLE **10.1** | Differences between Prokaryotic and Eukaryotic Gene Expression

Characteristic	Prokaryotes	Eukaryotes
Transcription and translation	At the same time in the cytosol	Transcription in the nucleus, then translation in the cytosol
Gene structure	DNA sequence usually not interrupted by introns	Transcribed regions (exons) often interrupted by noncoding introns
Modification of mRNA after initial transcription but before translation	Usually none	Introns spliced out of pre-mRNA; 5′ cap and 3′ poly A tail added to mRNA

with the two strands separated, with the transcriptional start site of the template strand in the active site of the polymerase enzyme.

Transcription then enters the elongation phase (see Figure 10.8B). RNA polymerase synthesizes an RNA strand, reading the DNA template in the 3′-to-5′ direction and synthesizing RNA in the 5′-to-3′ direction, similar to DNA polymerases. The first nucleotide in the new RNA forms its 5′ end, and the RNA transcript is antiparallel to the DNA template strand. RNA polymerase synthesizes an RNA that is complementary to the template strand of the DNA, except that uracil is incorporated into RNA instead of thymine when an adenine is in the template strand. Transcription occurs when RNA polymerase catalyzes the formation of phosphodiester bonds between added nucleotides and the growing RNA chain, releasing pyrophosphate in the process (compare with Figure 9.10). As transcription progresses, the two DNA strands rewind behind the polymerase and the RNA grows as a single-stranded molecule.

Like DNA polymerases, RNA polymerases and associated proteins have mechanisms for proofreading during transcription, but the transcription mechanisms are not nearly as efficient as those for DNA replication. Examination of newly synthesized RNA molecules shows that an error occurs every 10^4–10^5 bases, similar to the error rate of DNA polymerases (see Key Concept 9.3). During DNA replication, these errors are almost all detected and repaired, but this is not the case for transcription. While one error per 100,000 bases is unacceptably high for DNA replication, it is less problematic for transcription, for several reasons:

- Many RNAs are only a few thousand base pairs in length, meaning they usually do not have any errors, or perhaps only one error.
- RNA errors are not passed on to the next generation of cells after division because RNA is constantly made anew, often at a rate of thousands of new molecules per minute, whereas DNA errors that become mutations are permanent and the mutated molecule is passed to daughter cells.
- Because many copies of RNA are synthesized from a gene, with each often having a relatively short life span, an RNA containing an error is rare and thus not as potentially harmful as a mutation in DNA, which causes all RNAs to be altered.

Elongation continues until a region encoded by particular base sequences specifies termination (see Figure 10.8C). The mechanisms of termination vary among different genes and organisms. In eukaryotes, multiple proteins are involved in recognizing the transcription termination site and separating the newly formed RNA strand from the DNA template and the RNA polymerase.

Eukaryotic coding regions are often interrupted by introns

Coding regions are sequences within a DNA molecule that are eventually translated as proteins. The coding region on the DNA template strand is transcribed into a complementary mRNA molecule, which has the same base sequence (with U's instead of T's) as the DNA coding strand. In prokaryotes, most of the genomic DNA is made up of coding regions, and the mRNA sequence (e.g., 5′-AUGAUAGCCCC…) can be found *without interruption* in the DNA coding strand (e.g., 5′-ATGATAGCCCC…). In eukaryotes the situation is often different (**TABLE 10.1**).

A diagram of the structure and transcription of a typical eukaryotic gene is shown in **FIGURE 10.9**. In prokaryotes and viruses, several adjacent genes sometimes share one promoter, but in eukaryotes each gene has its own promoter. The coding region of a prokaryotic gene is usually continuous (with no interruptions), but this is not true in eukaryotes. One of the most stunning discoveries in biology was the presence of noncoding sequences called **introns** (*int*ervening *regions*) in eukaryotic genes. That is, many eukaryotic gene sequences are interrupted by introns. The transcribed regions that are interspersed with the introns are called **exons** (*ex*pressed regi*ons*). Both introns and exons appear in the primary mRNA transcript, called the **precursor RNA**, or **pre-mRNA**, but the introns are removed by the time the mature mRNA leaves the nucleus. Pre-mRNA processing involves cutting introns out of the pre-mRNA transcript and splicing together the exon transcripts (see Figure 10.9). Introns *interrupt, but do not scramble*, the DNA sequence of a gene. The base sequences of the exons in the template strand, if joined in order, form a continuous sequence that is complementary to that of the mature mRNA. Most, but not all, eukaryotic genes and a very few prokaryotic genes contain introns. In eukaryotes, introns can be numerous (~8 per gene in humans) and enormous (>10,000 base pairs). If the occurrence of introns in genes is surprising to you, you are in good company. For scientists who were familiar with prokaryotic genes and gene expression, the discovery of introns in eukaryotic genes was entirely unexpected. **FIGURE 10.10** describes how introns were discovered.

One useful consequence of having introns in protein-coding genes, alternative splicing, is discussed in **Key Concept 11.5**.

Eukaryotic gene transcripts are processed before translation

The primary transcript of a eukaryotic gene is modified in several ways before it leaves the nucleus: introns are removed, and both ends of the pre-mRNA are chemically modified.

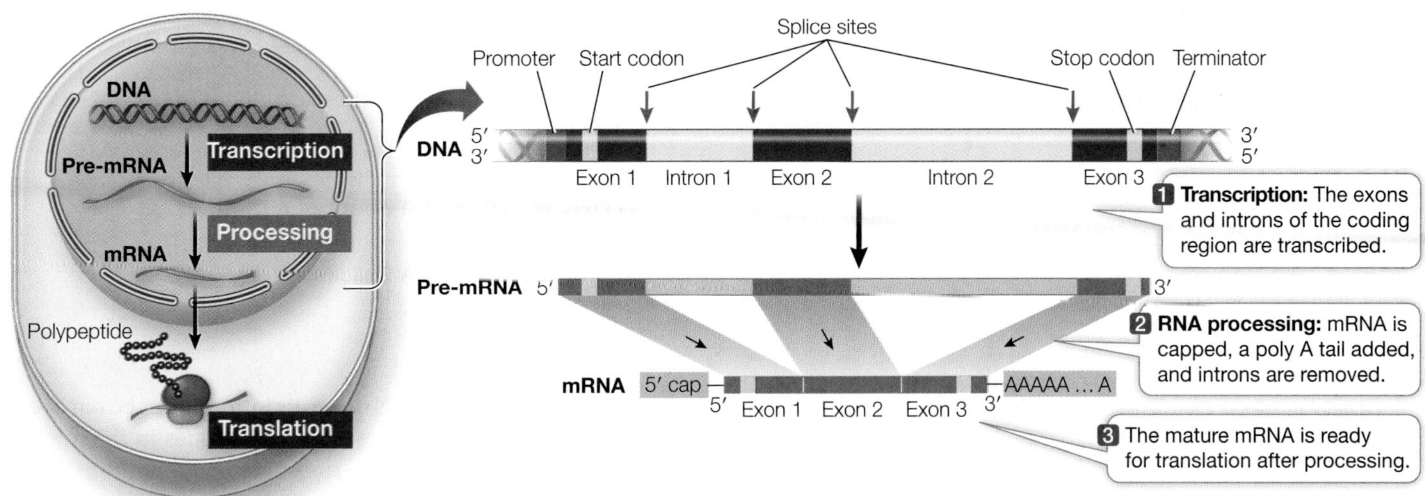

FIGURE 10.9 Transcription of a Eukaryotic Gene The β-globin gene diagrammed here is 1,733 base pairs long. The three exons contain information encoding 147 amino acids plus a signal (stop codon) indicating the end of translation. The two introns (noncoding sequences) totaling almost 1,000 base pairs of DNA are initially transcribed but then are spliced out of the pre-mRNA transcript during processing.

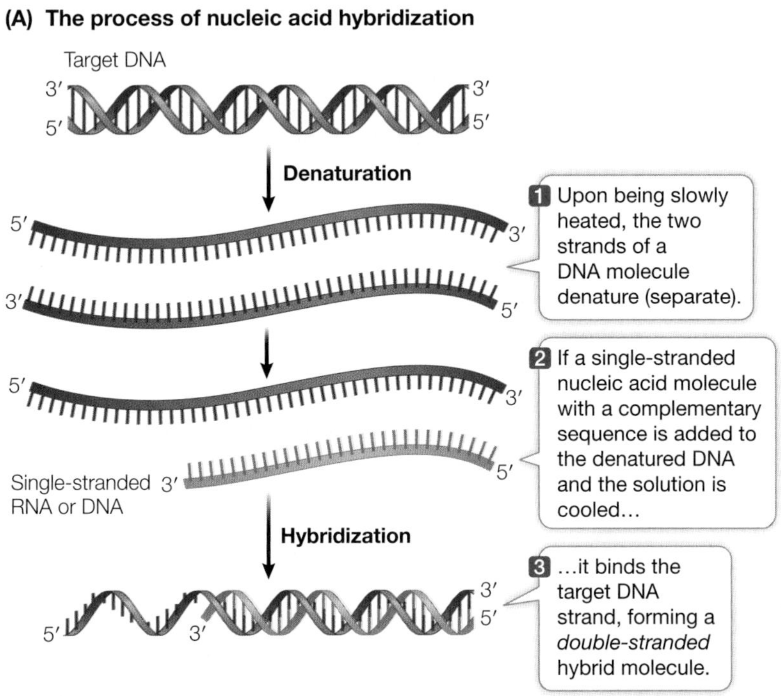

(A) The process of nucleic acid hybridization

Target DNA

3′ 3′
5′ 5′

Denaturation

1 Upon being slowly heated, the two strands of a DNA molecule denature (separate).

2 If a single-stranded nucleic acid molecule with a complementary sequence is added to the denatured DNA and the solution is cooled…

Single-stranded RNA or DNA

Hybridization

3 …it binds the target DNA strand, forming a *double-stranded* hybrid molecule.

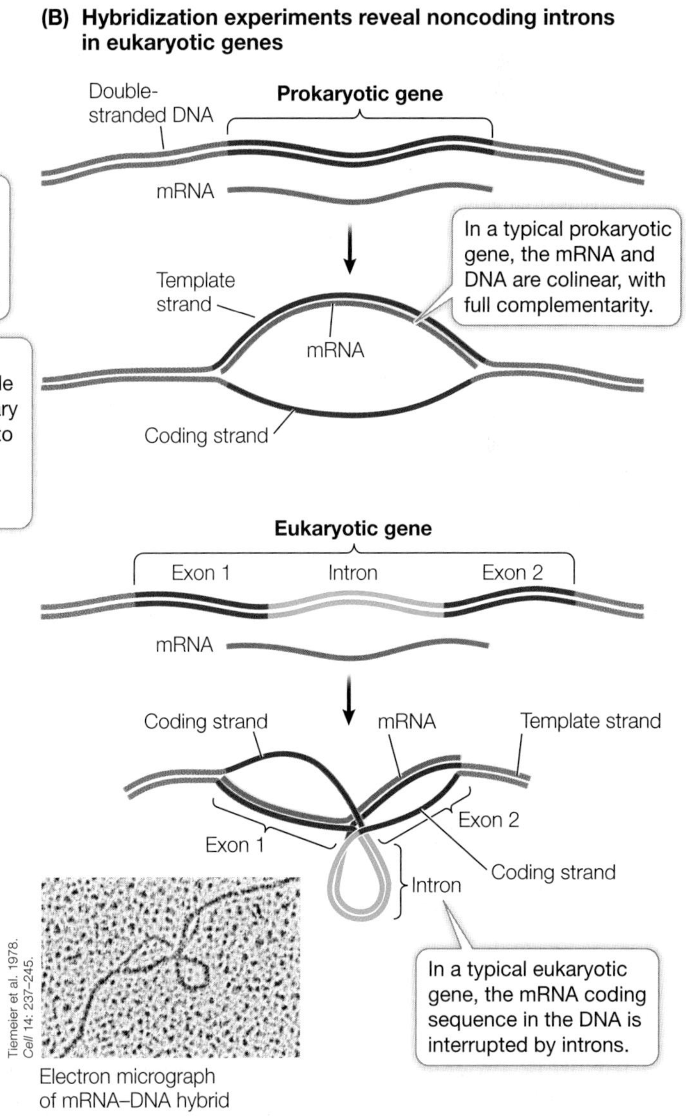

(B) Hybridization experiments reveal noncoding introns in eukaryotic genes

Double-stranded DNA

Prokaryotic gene

mRNA

Template strand

mRNA

Coding strand

In a typical prokaryotic gene, the mRNA and DNA are colinear, with full complementarity.

Eukaryotic gene

Exon 1 Intron Exon 2

mRNA

Coding strand

mRNA

Template strand

Exon 1

Exon 2

Intron

Coding strand

In a typical eukaryotic gene, the mRNA coding sequence in the DNA is interrupted by introns.

Tiemeier et al. 1978. *Cell* 14: 237–245.

Electron micrograph of mRNA–DNA hybrid

FIGURE 10.10 The Discovery of Introns **(A)** Single-stranded nucleic acid molecules from different sources will hydrogen bond with one another if they are complementary, in a process termed hybridization. **(B)** Mature mRNA will hybridize to the template strand of the gene from which it was transcribed. In prokaryotes, mRNA hybridizes fully to its template strand, implying that each of the base pairs in the DNA encode nucleotide sequence in the mRNA. However, when mature mRNA from a eukaryote is used, some unexpected looped structures are present in the hybrid molecules. These loops are stretches of DNA that do not have complementary nucleotide sequences on the mature mRNA. The observation of such loops was the first evidence for introns—regions of DNA that have no complementary sequence in mature mRNA. When pre-mRNA is used instead of mature mRNA, there is complete hybridization with no loops, revealing that the introns are part of the pre-mRNA transcript that are then removed during processing.

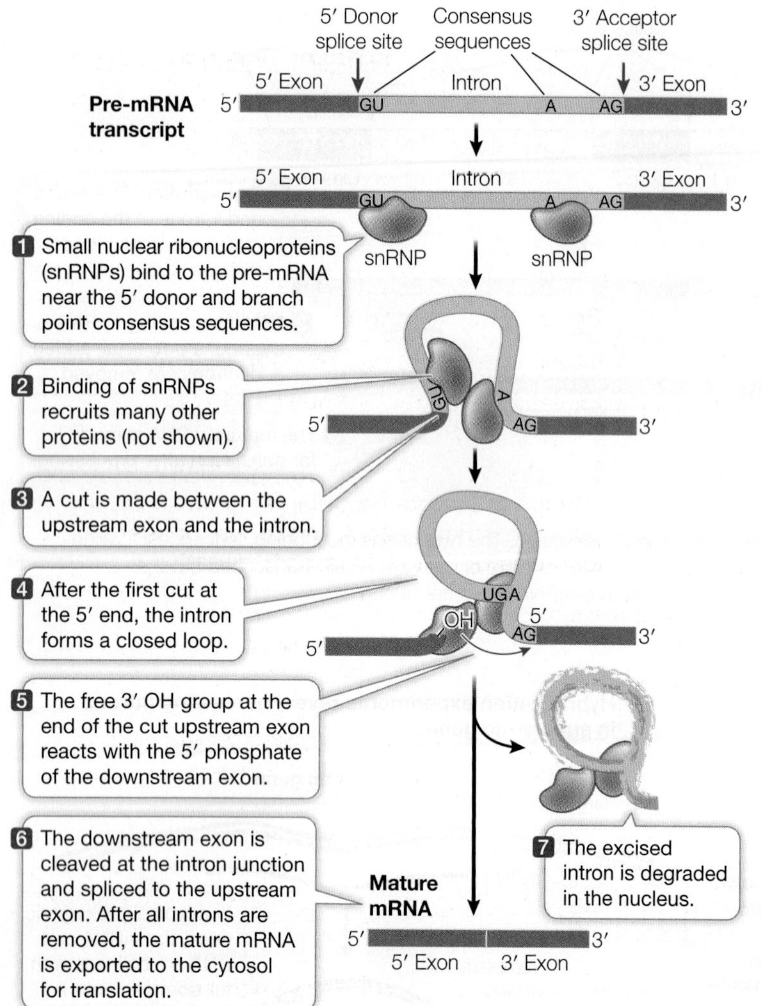

FIGURE 10.11 RNA Splicing Small nuclear ribonucleoprotein particles (snRNPs) bind to consensus sequences at the 5′ end of the intron and the branch point on pre-mRNA transcripts and recruit additional snRNPs to form a spliceosome. The intron is removed in two steps. First the 5′ end is cut and joined to the branch point. Then the 3′ end is cut and the two flanking exons are joined.

What would be the effect of a mutation at a splice site?

Animation 10.2 RNA Splicing

After the pre-mRNA is made, its introns must be removed. If this did not happen, the extra nucleotides in the mRNA would be translated by the ribosome and a nonfunctional protein would result. A process called **RNA splicing** removes the introns and splices the exons together. This process requires that the introns have certain characteristic **consensus sequences**—short stretches of DNA that appear with little variation ("consensus") across different introns. In the most common type of intron, the consensus sequence is GU at the 5′ end of the intron, AG at the 3′ end, and A followed by a sequence of pyrimidines (C's and U's) at the **branch point**—a sequence in the interior of the intron. As soon as the pre-mRNA is transcribed, these consensus sequences are bound by several **small nuclear ribonucleoprotein particles (snRNPs)**, which together make up a protein complex called the **spliceosome**. snRNPs are aggregates of RNA and proteins, and some of the RNAs have bases complementary to the transcribed consensus sequences in RNA in the intron, which facilitates correct identification of the intron. The spliceosome removes the intron in two steps. First the 5′ end of the intron is cut and joined to the branch site, and then the 3′ end is cut and the two exons are joined. Each intron is removed in this manner (**FIGURE 10.11**).

In some cases, the genetic mutation that causes a disease occurs at an intron consensus sequence. For example, some people with the autosomal recessive genetic disease beta thalassemia, which causes severe anemia, have a defect at the splice site in the β-globin gene. Consequently the β-globin pre-mRNA cannot be spliced correctly, which results in a defective β-globin mRNA and β-globin protein.

In addition to intron splicing, the pre-mRNA undergoes two additional processing steps, one at each end of the molecule (**FIGURE 10.12**). At the 5′ end of the pre-mRNA, a **5′ cap** (or G cap) is added as the pre-mRNA is being transcribed. The 5′ cap consists of guanosine triphosphate (GTP) that is methylated and added in a backward

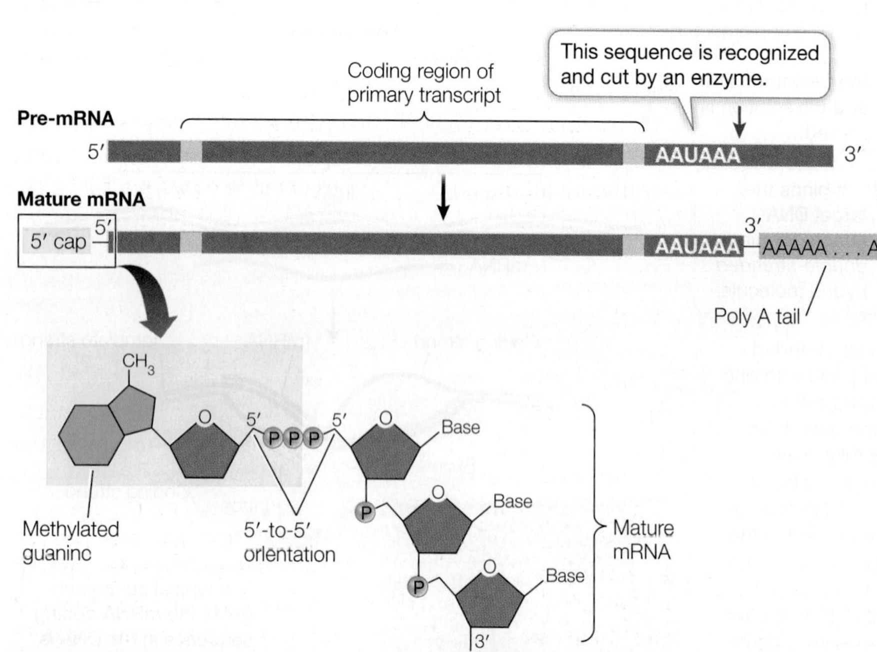

FIGURE 10.12 Processing the Ends of Eukaryotic Pre-mRNA Processing of pre-mRNA results in the addition of a 5′ cap and a 3′ poly A tail.

orientation (5′ phosphate to 5′ phosphate). The 5′-to-5′ phosphate bridge makes the 5′ end of the mRNA look very different from other RNA molecules in the cell and facilitates the binding of mRNA to the ribosome for translation. In addition, the 5′ cap protects the mRNA from being digested by ribonucleases (enzymes that break down RNAs), making the mRNA last longer in the cytosol.

The 3′ end of the pre-mRNA molecule is also modified, by having a sequence of 100–300 adenine nucleotides added by a *template-independent* RNA polymerase, to give a **poly A tail**. The poly A tail assists in the export of the mRNA from the nucleus, binding of proteins to initiate translation, and is also important for mRNA stability.

REVIEW & APPLY | 10.2

R Transcription of a gene begins with the RNA polymerase binding to the promoter region and locally denaturing the hydrogen bonds that hold the two strands of the helix together. Then RNA synthesis begins at the transcription start site and continues until the polymerase reaches a termination signal. In prokaryotes, mRNAs are ready to translate as soon as they are synthesized, but in eukaryotes they must first be processed in the nucleus before export to the cytosol for translation. Processing of the pre-mRNA involves intron removal—leaving only exons in the mature mRNA—and the addition of a 5′ cap and a 3′ poly A tail.

A

1. Part of a DNA coding strand has the sequence 5′-ATGGTGTACG-3′. What will be the sequence of the template strand and the RNA transcribed from this DNA? (Be careful to specify the 5′ and 3′ ends.)
2. Describe the consequences of the following:
   a. A mutation of a promoter sequence of a gene such that it no longer binds RNA polymerase.
   b. A mutation of the gene that encodes RNA polymerase, such that the polymerase is not made.
   c. A mutation that alters an intron splice site.
3. During processing, a pre-mRNA molecule is not capped. What will be the consequences of capping failure for this RNA molecule and for the phenotype of the cell?
4. The largest human gene encodes a muscle protein called titin; it is 305 kilobases (kb; 1 kb = 1,000 base pairs) in length, with 190 kb of introns and 115 kb of coding sequence in 363 exons, which together code for 38,138 amino acids. Can you deduce how many introns the titin gene has?

The transcription of a gene to produce mRNA is only the first step in gene expression. The next step is translation. We will look in more detail at the process of translation in Key Concepts 10.4 and 10.5, but first we will discuss the genetic code—the set of instructions that enable RNA to be translated into amino acids.

10.3 The Rules for Translation of RNA into Amino Acids Are Contained in the Genetic Code

LEARNING OBJECTIVES

By the end of this key concept you should be able to:

10.3.1 **Explain how the genetic code was cracked.**

10.3.2 **Explain what is meant by code degeneracy.**

10.3.3 **Explain the function of start and stop (nonsense) codons.**

10.3.4 **Describe the effects of a single DNA nucleotide change in a protein-coding sequence, distinguishing among silent, missense, nonsense, loss-of-stop, and frame-shift mutations.**

The translation of the nucleotide sequence of an mRNA into the amino acid sequence of a polypeptide occurs within the ribosome. In prokaryotes, transcription and translation are coupled: there is no nucleus, and ribosomes often bind to an mRNA as it is being transcribed in the cytosol. In eukaryotes, the nuclear envelope separates the locations of mRNA production and translation, with the latter occurring within ribosomes in the cytosol. In both cases, the key event is the decoding of one chemical "language" (the nucleotide sequence) into another (the amino acid sequence). For this reason, particular sequences in mRNA can be thought of as "code words" for particular amino acids.

The information for protein synthesis lies in the genetic code

The genetic information in an mRNA molecule is a series of sequential, non-overlapping three-letter "words" called **codons**. The "letters" are ribonucleotide bases in the mRNA. The **genetic code** relates codons to specific amino acids.

Molecular biologists cracked the genetic code in the early 1960s. The problem they addressed was perplexing: how could 20 different amino acids be specified using only four nucleotide bases (A, U, G, and C)? A triplet code with three-letter codons was considered likely because it was the shortest sequence with enough possible variations to encode all 20 amino acids. With four available bases, a triplet codon has $4 \times 4 \times 4 = 64$ variations (AAA, AAU, AAC, AAG, CAA, CAU, and so on). Since there are only 20 amino acids, scientists pondered whether some codons of a triplet code might not encode amino acids (20 codons would be needed for amino acids and the remaining 44 might not be used) or whether the genetic code might be degenerate (redundant), with multiple codons encoding the same amino acid.

Marshall Nirenberg and Heinrich Matthaei, at the U.S. National Institutes of Health, made the first decoding breakthrough in 1961 when they discovered they could synthesize protein in a test tube using a simple artificial RNA as a template for polypeptide synthesis. They could then identify the amino acids that were incorporated into the polypeptide. Their experiment, described in **INVESTIGATION FIGURE 10.13**, led to the identification of the first codons.

INVESTIGATION

FIGURE 10.13 Deciphering the Genetic Code
Nirenberg and Matthaei used a test tube protein synthesis system to determine the amino acids specified by synthetic mRNAs of known compositions.[a]

HYPOTHESIS

An artificial mRNA containing only one repeating base will direct the synthesis of a protein containing only one repeating amino acid.

METHOD

Nirenberg and Matthaei set up 20 translation systems, each of which included all the components needed to synthesize proteins, except mRNA, one radioactively labeled amino acid and 19 unlabeled amino acids. An mRNA consisting of a repeating nucleotide was added to each system. The synthesized protein was purified and its level of radioactivity measured. In the example shown here, the radiolabeled amino acid is phenylalanine.

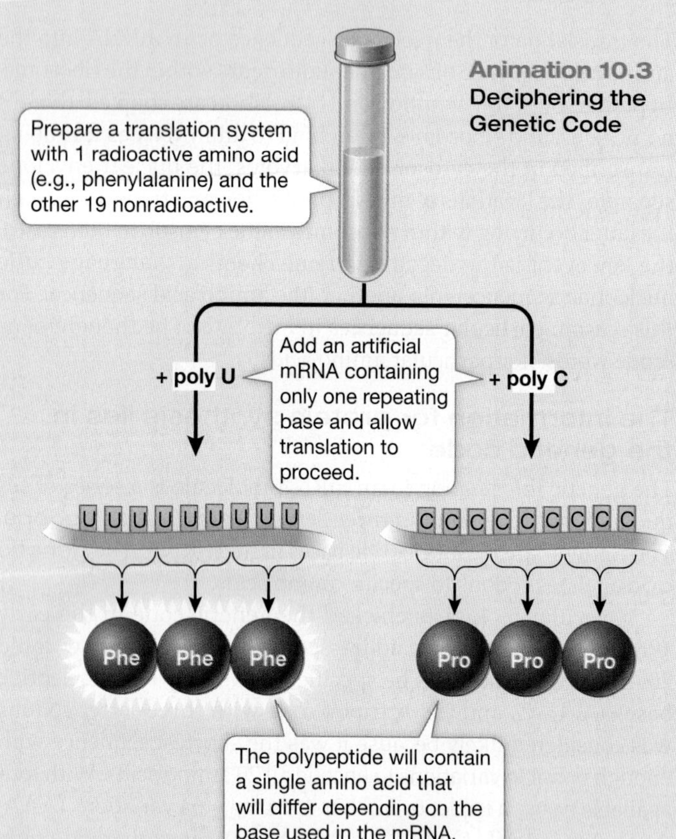

RESULTS

When the radiolabeled amino acid is phenylalanine a poly U mRNA results in a radioactive protein, but a poly C mRNA does not. Poly C mRNA only results in a radioactive protein when the labelled amino acid is proline.

CONCLUSIONS

Since the length of the codon was not yet determined, the conclusions were that:

1. One or more U nucleotides is the codon for phenylalanine only.
2. One or more C nucleotides is the codon for proline only.

ANALYZE THE DATA

1. Poly U, an artificial mRNA, was added to a test tube with all other components for protein synthesis ("Complete system"). Other test tubes differed from the complete system as indicated in the table. Samples were tested for radioactive phenylalanine incorporation, with the results shown in the table in counts/minute (units of radioactivity). Explain the results for each condition.

Condition	Radioactive phenylalanine in polypeptide (counts/minute)
Complete system	29,500
Minus poly U mRNA	70
Minus ribosomes	52
Minus ATP	83
Plus RNase (hydrolyzes RNA)	120
Plus DNase (hydrolyzes DNA)	27,600
Radioactive glycine instead of phenylalanine	33
Mixture of 19 radioactive amino acids minus phenylalanine	276

[a]M. W. Nirenberg and J. H. Matthaei. 1961. *Proc Natl Acad Sci USA* 47: 1588–1602.

A similar **Analyze the Data** exercise may be assigned in **Achieve**.

The main obstacle in Nirenberg and Matthaei's method was synthesizing artificial RNA templates of known sequence using more than one kind of nucleotide. While progress was made in the chemistry laboratory to overcome this obstacle, another breakthrough made solving the problem simpler. Scientists discovered that an artificial mRNA only three nucleotides long—amounting to one codon—could bind to a ribosome, and that the resulting complex could bind to a corresponding tRNA carrying a specific amino acid. (Key Concept 10.4 will discuss the specificity and role of tRNA; for now, just know that there is a tRNA for each amino acid and that it acts as a "bridge" between the mRNA and amino acid.) For example, an mRNA with the triplet AUC caused the tRNA carrying the amino acid isoleucine to bind to the ribosome. Using this test tube method, scientists could make any triplet codon and see which amino acid on its tRNA would bind to the ribosome. With this discovery, the complete deciphering of the genetic code happened quickly.

The genetic code is shown in **FIGURE 10.14**. Notice that 61 of the 64 possible codons encode amino acids—they are **sense codons**. So there is a great deal of degeneracy in the genetic code: almost all amino acids are encoded by at least two codons. A degenerate code should not be confused with an ambiguous code in which a single codon would encode two or more amino acids; instead each codon always encodes the same amino acid during translation. The three codons that do not encode amino acid information—UAA, UAG, and UGA—are **stop**, or **nonsense**, **codons** and cause termination of translation. When the translation machinery reaches one of these codons, translation stops and the polypeptide is released. AUG, which codes for methionine, is also the **start codon**, the initiation signal for translation.

Take a moment to examine the genetic code and you will quickly see patterns. Most amino acids are encoded by two or four codons, though there are a few exceptions having one codon (methionine and tryptophan), three codons (isoleucine), and six codons (leucine, arginine, and serine). When an amino acid is encoded by four codons, the first two letters are always the same. For example, proline is encoded by CCN, where N can be U, C, A, or G. When an amino acid is encoded by two codons, both begin with the same two letters and end either in U or C, or in A or G. Examples include cysteine, which is encoded by UGU and UGC, and lysine, which is encoded by AAA or AAG.

Why do these patterns exist? One favored evolutionary explanation is that the patterns in the genetic code arise as a consequence of natural selection to minimize the effects of unavoidable DNA mutations (see Key Concept 9.3). The patterns indicate that when more than one codon encodes a single amino acid, the codons are almost always a single substitution away from one another (AAA versus AAG, for example). This means some DNA mutations that cause a change in a codon do not alter the encoded amino acid—such mutations are termed **synonymous** and do not alter phenotype (see Figure 9.19). If the codons for a particular amino acid were randomly scattered around the genetic code table, synonymous mutations would be rare, so mutations would be more likely to alter phenotype. Another explanation is that the pattern results from the mechanics of translation. The manner in which the tRNA interacts with the mRNA allows for certain codon variants at the third position to interact with the same tRNA and thus encode the same amino acid (see Key Concept 10.4).

After the genetic code was described in bacteria, the immediate question was how the code might differ among different species. Amazingly, it was quickly discovered that the genetic code is (nearly) universal. *Essentially the same genetic code is used by all species.* When one of your ribosomes is translating an mRNA in the cytosol of a cell in your liver, it is using the same genetic code as a ribosome translating an mRNA in an archaeal cell, which is the same code used by a ribosome in a bacterial cell. Thus the code must be an ancient one that has been maintained intact throughout billions of years of evolution. This constancy implies either that there is a strong benefit for particular codons to encode particular amino acids, or that altering the genetic code must be extremely deleterious in most situations. The first possibility seems unlikely given that there is an enormous number of possible genetic codes that could be imagined that would work fine and possess the patterns observed in the actual genetic code. The second possibility seems reasonable when one considers what would happen if a codon suddenly encoded a different amino acid in all proteins in a cell.

While the genetic code is called universal, exceptions are known: within mitochondria and chloroplasts, and in one group of protists, a few codons differ, though most remain the same. The significance of the differences is not yet understood. What is clear is that the exceptions are few.

The universal genetic code indicates that all life came ultimately from a single common ancestor. The universal code also has profound implications for the production of **transgenic organisms**: organisms engineered to express a gene from another organism (**FIGURE 10.15**). The universal genetic code makes it substantially easier to produce transgenic organisms such as *Escherichia coli* cells that express the human protein insulin for the treatment of type I diabetes. The code for a protein that makes jellyfishes glow can be placed into the cells of other organisms to make them easier to study (see Appendix D). Indeed, the use of genetic engineering to answer "what if" questions is so routine now in biological research that it

First letter	Second letter: U	Second letter: C	Second letter: A	Second letter: G	Third letter
U	UUU Phenylalanine	UCU Serine	UAU Tyrosine	UGU Cysteine	U
	UUC Phenylalanine	UCC Serine	UAC Tyrosine	UGC Cysteine	C
	UUA Leucine	UCA Serine	**UAA** Stop codon	**UGA** Stop codon	A
	UUG Leucine	UCG Serine	**UAG** Stop codon	UGG Tryptophan	G
C	CUU Leucine	CCU Proline	CAU Histidine	CGU Arginine	U
	CUC Leucine	CCC Proline	CAC Histidine	CGC Arginine	C
	CUA Leucine	CCA Proline	CAA Glutamine	CGA Arginine	A
	CUG Leucine	CCG Proline	CAG Glutamine	CGG Arginine	G
A	AUU Isoleucine	ACU Threonine	AAU Asparagine	AGU Serine	U
	AUC Isoleucine	ACC Threonine	AAC Asparagine	AGC Serine	C
	AUA Isoleucine	ACA Threonine	AAA Lysine	AGA Arginine	A
	AUG Methionine; start codon	ACG Threonine	AAG Lysine	AGG Arginine	G
G	GUU Valine	GCU Alanine	GAU Aspartic acid	GGU Glycine	U
	GUC Valine	GCC Alanine	GAC Aspartic acid	GGC Glycine	C
	GUA Valine	GCA Alanine	GAA Glutamic acid	GGA Glycine	A
	GUG Valine	GCG Alanine	GAG Glutamic acid	GGG Glycine	G

FIGURE 10.14 The Genetic Code Genetic information is encoded in three-letter units—codons—that are read in the 5′-to-3′ direction on the mRNA. To decode a codon, find its first letter in the left column, then read across the top to its second letter, then read down the right column to its third letter. The amino acid the codon specifies is given in the corresponding row. For example, AUG codes for methionine, and GUA codes for valine.

Activity 10.2 The Genetic Code

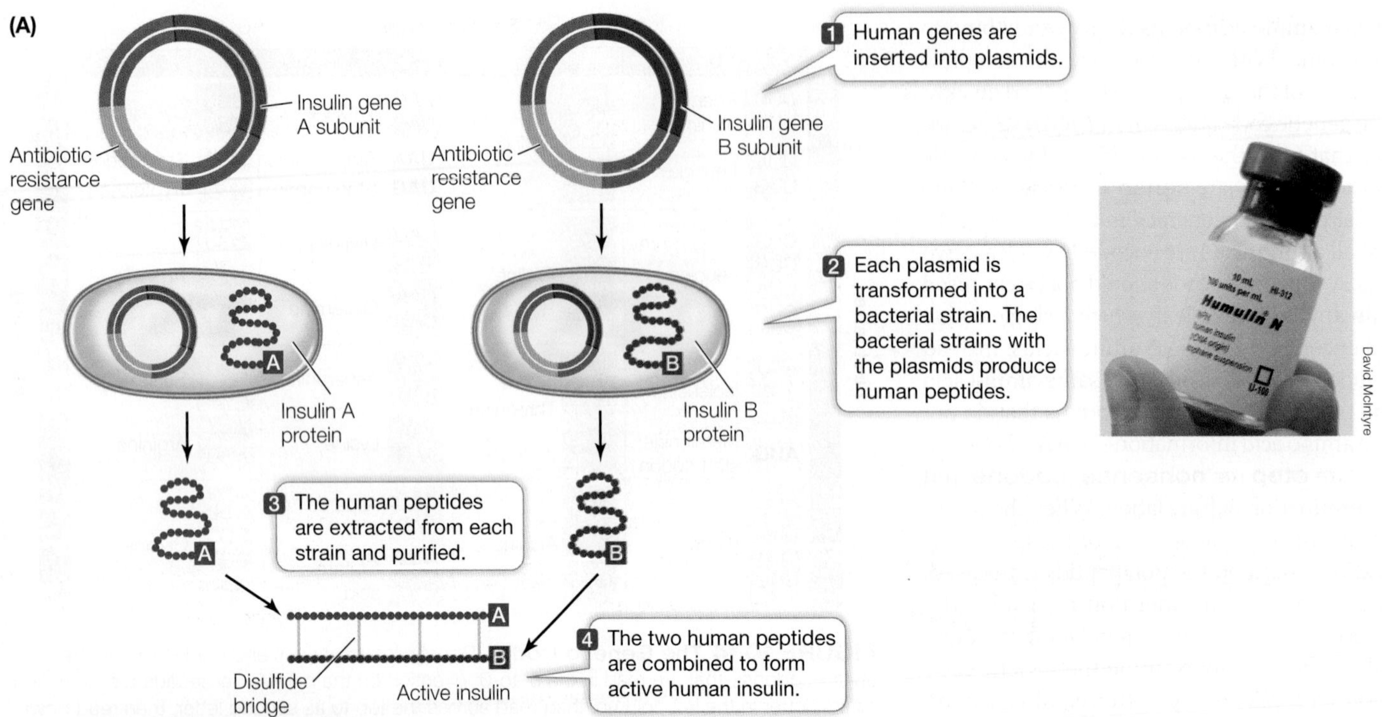

FIGURE 10.15 Transgenic Organisms Organisms have been engineered so that they express proteins from other species. The ability to make such genetically modified organisms is facilitated by the universality of the genetic code. **(A)** The DNA sequence encoding human insulin subunit A is put into a plasmid, which is transformed into a bacterial strain. Another bacterial strain is transformed with a plasmid containing the DNA sequence encoding human insulin subunit B (Appendix D describes how genes are placed into plasmids and transformed into bacterial cells). Each strain produces its respective human insulin subunit, which is extracted and purified, and then the two are combined to make active human insulin. **(B)** Fish containing a gene that has been inserted into their genome from jellyfish or corals that produce a fluorescent protein. These and other species of genetically modified fluorescent fish are now common in pet stores.

In (A), why is there an antibiotic resistance gene on the plasmid that is transformed into each bacterial strain?

is taken for granted. In all of these examples, the universality of the genetic code ensures that the newly introduced gene will be treated genetically in the same way as the DNA of the host organism.

The codons shown in Figure 10.14 are for mRNA. The base sequence of the template DNA strand is complementary and antiparallel to these codons. Thus, for example, 3′-ACC-5′ in the template DNA strand corresponds to tryptophan (which is encoded by the mRNA codon 5′-UGG-3′). However, the coding DNA strand has the same sequence as the mRNA, but with T's instead of U's (5′-TGG-3′ in the case of tryptophan). By convention, gene sequences are usually shown as the sequence of the coding strand, beginning at the 5′ end.

The genetic code predicts the effects of point mutations

Examination of the genetic code reveals that base-pair substitutions (point mutations) can have a variety of effects on protein sequence (**FIGURE 10.16**):

- **Synonymous mutations** occur when a DNA substitution alters the codon but does not alter the encoded amino acid. Synonymous mutations occur because of the degeneracy of the genetic code. For example, the codons CCC and CCU are both translated from mRNA as proline (Pro). So a change in the template strand of the DNA from 3′-GGG to 3′-GGA (a substitution from a G to an A) will not cause any change in amino acid sequence (see Figure 10.16B).
- **Missense mutations** (or replacement or nonsynonymous mutations) occur when a DNA substitution alters the codon so that it encodes a different amino acid. In most cases a missense mutation causes a single amino acid change in the protein, which may or may not cause a change in function. For example, GAU in mRNA is translated as aspartic acid (Asp), whereas a mutation that results in GUU is translated as valine (Val) (see Figure 10.16C). However, if the missense occurs at the start codon, such that the start codon is lost, the effect is more severe. In this case either the mRNA is not translated, or the ribosome

(A) Wild type (normal)

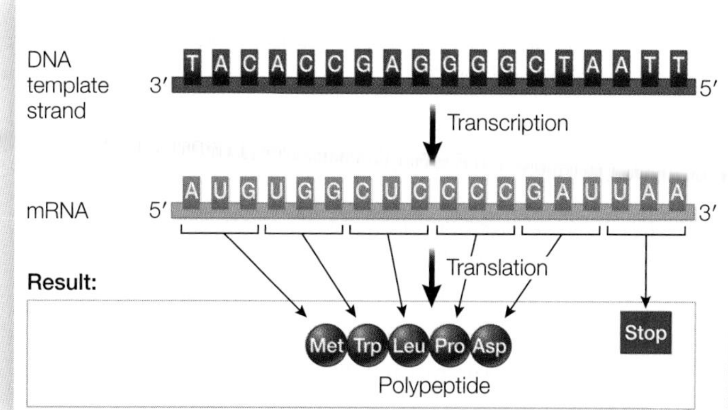

(B) Synonymous mutation

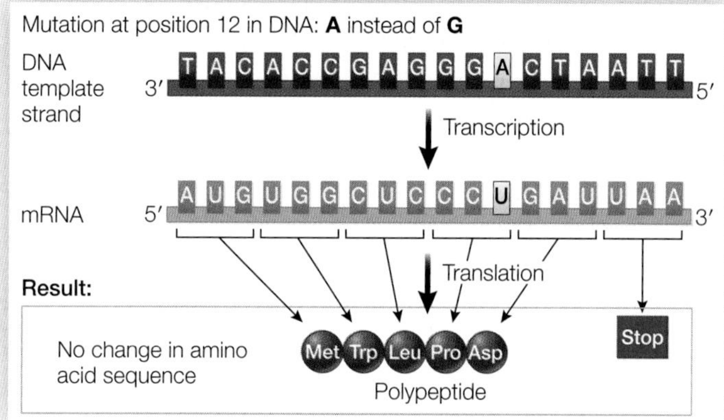

(C) Missense mutation

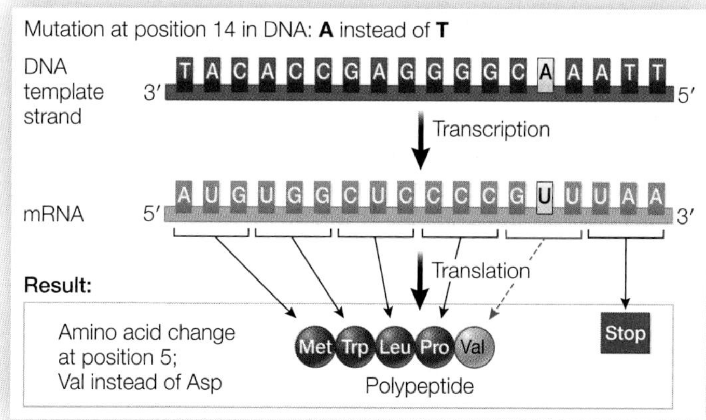

(D) Nonsense mutation

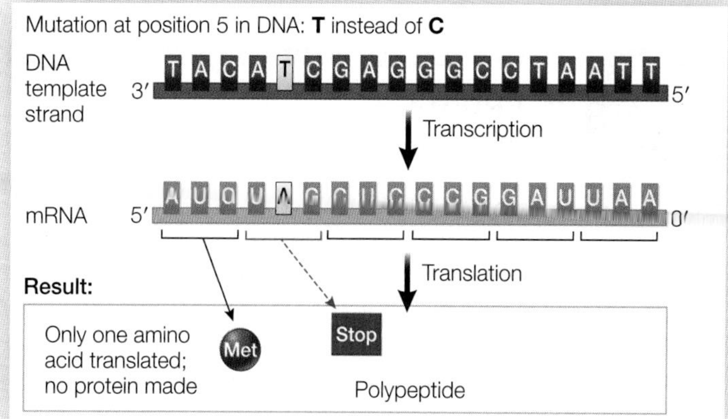

(E) Loss-of-stop mutation

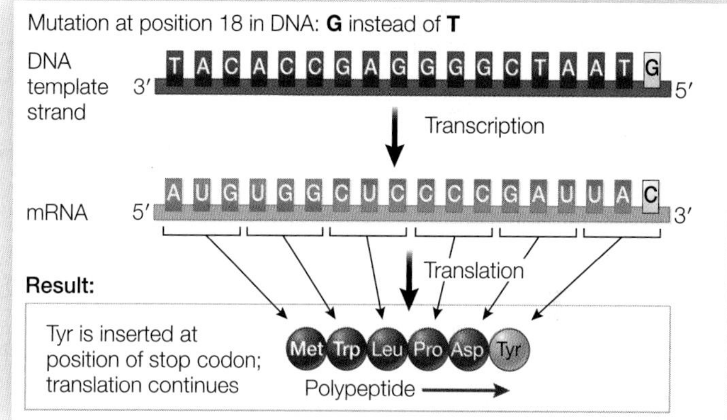

(F) Frame-shift mutation

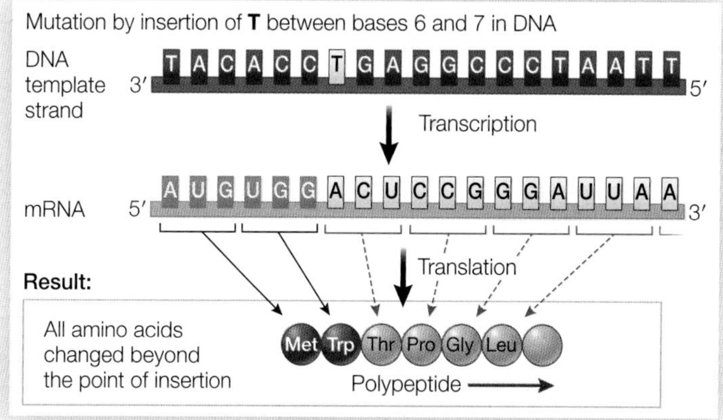

FIGURE 10.16 Point Mutations Single base pair changes in a coding region of DNA can have different effects on the protein the DNA encodes.

Activity 10.3 Point Mutations Simulation

What other kinds of point mutations might cause loss of function at a protein-coding locus?

uses another methionine codon to start translation, giving a different protein sequence.

- **Nonsense** (or **stop**) **mutations** result in a change from a sense codon to a stop (nonsense) codon, causing a premature termination of translation and a shortened protein. For example, the codon UGG is translated as the amino acid tryptophan (Trp). A point mutation could convert this to the stop codon UAG, which acts as a translation termination signal (see Figure 10.16D). Nonsense mutations are likely to be loss-of-function mutations.
- **Loss-of-stop mutations** result in a change from a stop codon to a sense codon, causing additional amino acids to be added to the end of the protein. For example, a substitution that converts UAA (stop) to UAC (tyrosine) allows the ribosome to continue translating (see Figure 10.16E). The effect of a loss-of-stop mutation depends on how many amino acids are added to

THINK LIKE A SCIENTIST

Evidence for lateral gene transfer in aphids

Red versus green coloration affects the survival of pea aphids differently in different environments, with the red morph being favored in some environments and the green morph in other environments. The color difference is due to the carotenoid tolurene, which is present in the red morph and absent in the green morph. A single gene inherited in a Mendelian manner, with a recessive green and a dominant red allele, determines whether torulene is synthesized and thus what color an aphid is.

1. What kind of information would allow researchers to show that red versus green body color is caused by different alleles at a single Mendelian gene?

Animals typically do not synthesize carotenoids but get them from their diet. It was thus possible that the gene affecting color determines how well aphids get torulene from their diet. However, analysis of carotenoid content indicated that the carotenoids in the aphids differed from those in the plants on which they feed. So where did the aphid carotenoids come from?

Once DNA sequencing technology was available (see Chapter 12), the entire genome of the pea aphid was sequenced and the genes identified. Surprisingly, several genes were found that encode enzymes that are extremely similar in sequence to fungal genes that encode the enzymes in the carotenoid synthesis pathway. Pea aphids are able to synthesize carotenoids using genes most likely obtained from fungi!

2. If the genetic code were not universal (or nearly so), would it be possible for genes that were transferred from fungi to encode functional enzymes in an insect?
3. The transferred genes (those acquired from fungi) in the aphid have introns that are also present (though smaller) in fungi. What does this imply about intron splicing?

DNA sequencing revealed that the green and red aphid DNAs differed by a missense mutation in a gene encoding an enzyme involved in the synthesis of torulene.

4. What is a missense mutation? Given that the green allele is a loss-of-function allele, what might you predict about the nature of the missense mutation?

This example of lateral gene transfer between widely different groups is only one of many that have been identified. Lateral gene transfer is rare, but when it occurs it can have enormous benefit for the recipient.

the end of the protein and how important that part of the protein is to function.

In addition to base-pair substitutions, insertions and deletions of one or a few bases can occur. If the number of bases being inserted or deleted is a multiple of three, the result is an addition or loss, respectively, of one or more amino acids from the protein. If the number of bases being inserted is not a multiple of three, then the result is a frame-shift mutation:

- **Frame-shift mutations** result from the insertion or deletion of one or more (but not a multiple of three) base pairs in the coding sequence. Since the genetic code is read as sequential, non-overlapping triplets, an insertion or deletion of a nucleotide in the coding region will cause the reading frame to be out of register, leading to different triplets being read, and an altered sequence of amino acids downstream of the frame shift in the resulting (often shortened) polypeptide (see Figure 10.16F). Frame-shift mutations are usually loss-of-function mutations because they affect so many amino acids in the protein.

Activity 10.4 Mutations and Function Simulation

REVIEW & APPLY | 10.3

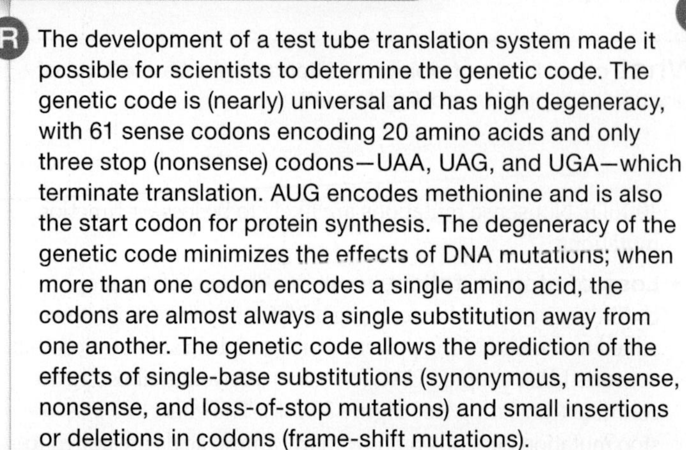

R The development of a test tube translation system made it possible for scientists to determine the genetic code. The genetic code is (nearly) universal and has high degeneracy, with 61 sense codons encoding 20 amino acids and only three stop (nonsense) codons—UAA, UAG, and UGA—which terminate translation. AUG encodes methionine and is also the start codon for protein synthesis. The degeneracy of the genetic code minimizes the effects of DNA mutations; when more than one codon encodes a single amino acid, the codons are almost always a single substitution away from one another. The genetic code allows the prediction of the effects of single-base substitutions (synonymous, missense, nonsense, and loss-of-stop mutations) and small insertions or deletions in codons (frame-shift mutations).

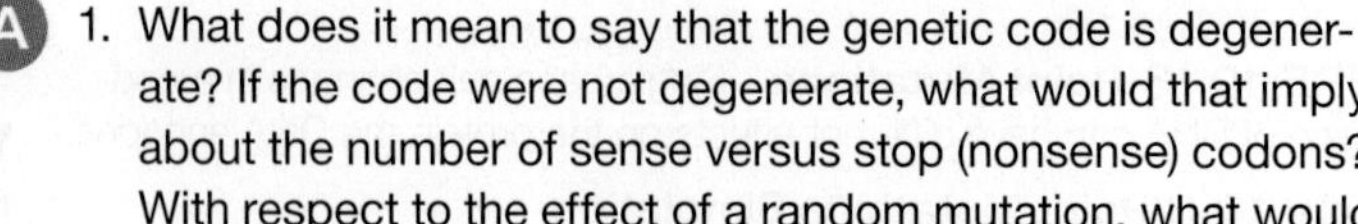

A

1. What does it mean to say that the genetic code is degenerate? If the code were not degenerate, what would that imply about the number of sense versus stop (nonsense) codons? With respect to the effect of a random mutation, what would a nondegenerate code imply?
2. An artificial mRNA is synthesized that is a repeat of the UA dinucleotide (UAUAUAUAUA...) and then is used in a test tube protein synthesis system. Assuming translation can begin anywhere on the mRNA, what would be the amino acid sequence of the resulting polypeptide chain?
3. A deletion of two consecutive base pairs in the coding region of DNA causes a frame-shift mutation. But a deletion of three consecutive base pairs causes the deletion of only one amino acid, with the rest of the polypeptide chain intact. Explain.

The mRNA with its coding information is translated into an amino acid sequence at the ribosome. We will now consider this process.

10.4 RNA Is Translated into Amino Acids by Ribosomes

LEARNING OBJECTIVES

By the end of this key concept you should be able to:

10.4.1 Describe the initiation, elongation, and termination steps of translation.

10.4.2 Describe the roles of the A, P, and E ribosome sites in polypeptide elongation.

10.4.3 Describe how the structure of tRNAs supports their function in translation.

The translation of mRNA into proteins requires molecules that can link the information contained in each mRNA codon with a specific amino acid. That function is performed by a set of transfer RNAs (tRNAs). Two key events must take place to ensure that the protein made is the one specified by the mRNA: (1) a tRNA must chemically read each mRNA codon correctly, and (2) the tRNA must deliver the amino acid that corresponds to the mRNA codon. Once the tRNAs "decode" the mRNA and deliver the appropriate amino acids, components of the ribosome catalyze the formation of peptide bonds between the amino acids.

Animation 10.4 Translation

Transfer RNAs carry specific amino acids and bind to specific codons

There is at least one specific tRNA molecule for each of the 20 amino acids. Each tRNA has three functions that are fulfilled by its structure and base sequence (**FIGURE 10.17**):

1. *tRNAs bind to particular amino acids.* Each tRNA binds to a specific enzyme that attaches it to only 1 of the 20 amino acids. This covalent attachment is at the 3′ end of the tRNA. We will describe the details of this vital process below. When it is carrying an amino acid, the tRNA is said to be "charged."
2. *tRNAs bind to mRNA.* At about the midpoint on the tRNA polynucleotide chain there is a triplet of bases called the **anticodon**, which is complementary to the mRNA codon for the particular amino acid that the tRNA carries. Like the two strands of DNA, the codon and anticodon bind together via hydrogen bonds, with the codon and anticodon being antiparallel. For example, the mRNA codon for methionine is 5′-AUG-3′, and the tRNA anticodon is 3′-UAC-5′.
3. *tRNAs interact with ribosomes.* The ribosome has several sites on its surface that just fit the three-dimensional structure of a tRNA molecule. Interaction between the ribosome and the tRNA is noncovalent.

Recall that 61 different sense codons encode the 20 amino acids in proteins (see Figure 10.14). If the codon and anticodon interacted using complementary base pairing, that would mean the cell would have to produce 61 different tRNA molecules, each with a different anticodon. In fact, cells get by with about two-thirds that many types of tRNA molecules. Thus there are fewer anticodon sequences than codon sequences. The reason this is possible is

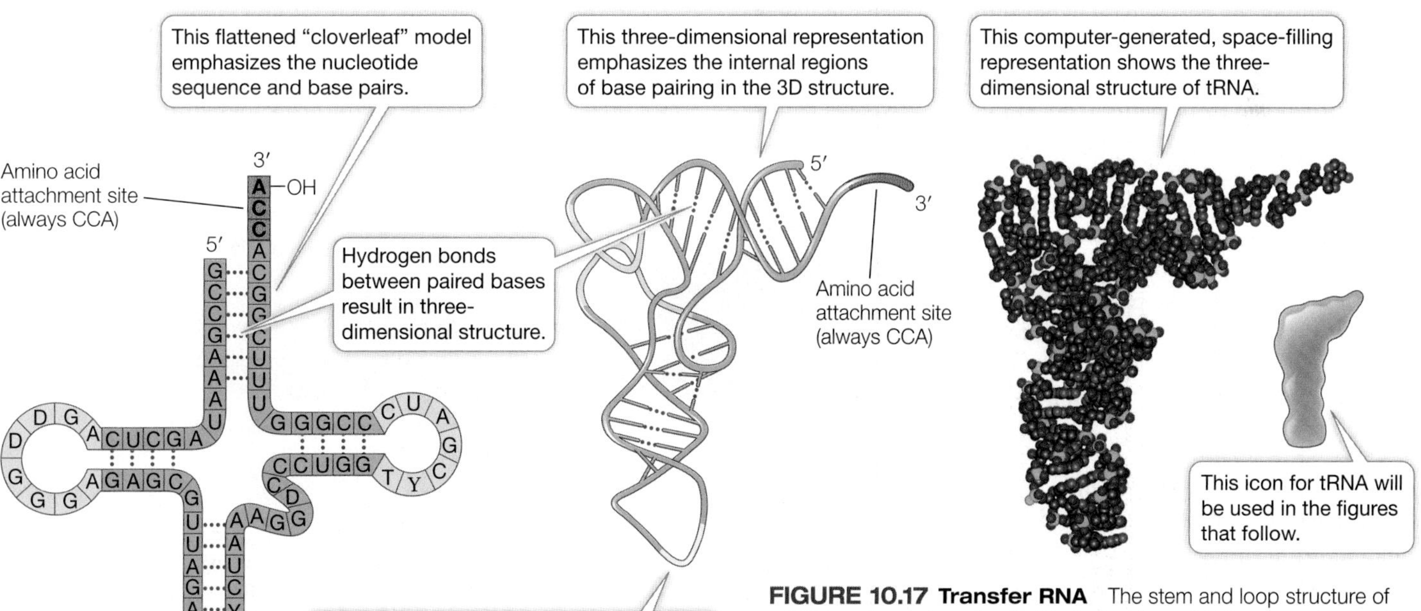

FIGURE 10.17 Transfer RNA The stem and loop structure of a tRNA molecule is well suited to its functions: binding to amino acids, base-pairing with codons on mRNA molecules, and interacting with ribosomes. Note that tRNAs possess several unusual nucleotides (i.e., D and Y). ("Cloverleaf" model and three-dimensional representation after H. Shi and P. B. Moore. 2000. *RNA* 6: 1091–1105; tRNA space-filling representation by David McIntyre based on data from PDB ID: 1EHZ. 2000. H. Shi and P. B. Moore. 2000. *RNA* 6: 1091–1105.)

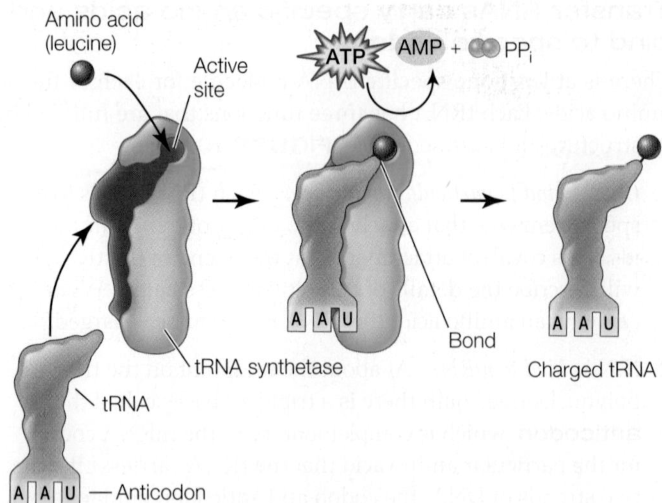

FIGURE 10.18 Charging a tRNA Molecule A particular tRNA synthetase binds 1 of the 20 amino acids, hydrolyzes ATP, then binds the corresponding tRNA and forms the covalent bond between the tRNA and the amino acid.

because the base pairing at the third position, which is the 3′ end of the codon and the 5′ end of the anticodon, is not strictly complementary. Instead the base pairing can exhibit "wobble," such that certain bases in the third position of the anticodon are able to pair with more than just their normal partner. Wobble allows a single tRNA to pair with two codons that start with the same two letters but end in A or G, or that end in C or T. Thus wobble explains, in part, the pattern of degeneracy seen in the genetic code.

In addition, the third position of the anticodon may be a nucleotide containing the purine base hypoxanthine, which can wobble pair with A, C, and U in the codon–anticodon context. (When hypoxanthine is attached to ribose it forms the ribonucleoside inosine and so is denoted by the letter I, not H.) Inosine's wobble pairing make it possible for the three isoleucine codons—AUU, AUC, and AUA—to be recognized by the same tRNA (with the anticodon 3′-UAI-5′). Wobble occurs in some tRNAs but not in others; of most importance, wobble base pairing does not allow the genetic code to be ambiguous. That is, *each mRNA codon is able to bind only to tRNA molecules carrying its specific amino acid.*

Each tRNA is specifically attached to an amino acid

The charging of each tRNA with its correct amino acid is achieved by a family of enzymes known as tRNA synthetases. Each enzyme is highly specific, binding only one amino acid and one corresponding tRNA. The reaction uses the energy in ATP to form a high-energy bond between the amino acid and the tRNA (**FIGURE 10.18**). The energy in this bond is later used in the formation of peptide bonds between amino acids in a growing polypeptide chain. Clearly, the specificity between the tRNA and its corresponding amino acid is extremely important.

A clever experiment by Seymour Benzer and his colleagues at Purdue University demonstrated the importance of the specificity of tRNA synthetases. They took the charged cysteine tRNA and chemically converted the cysteine into alanine, resulting in a cysteine tRNA charged with alanine. Which component—the amino acid or the tRNA—would be recognized when this hybrid charged tRNA was put into a protein synthesizing system? The answer was the tRNA. Everywhere in the synthesized protein where cysteine was supposed to be, alanine appeared instead. The cysteine-specific tRNA had delivered its cargo (alanine) to every mRNA codon for cysteine. This experiment showed that the protein synthesis machinery recognizes the anticodon of the charged tRNA, not the amino acid attached to it.

Translation occurs at the ribosome

The ribosome is the molecular machine where the translation of mRNA is accomplished with the help of tRNAs. All prokaryotic and eukaryotic ribosomes consist of two subunits (**FIGURE 10.19**). In eukaryotes, the large subunit consists of 3 different ribosomal RNA (rRNA) molecules and about 49 protein molecules arranged in a precise pattern. The small subunit consists of 1 rRNA molecule and about 33 proteins. These two subunits and several dozen other molecules interact noncovalently, fitting together like a jigsaw puzzle. If the hydrophobic interactions between the proteins and RNAs are disrupted, the ribosome falls apart, but it will reassemble perfectly when the disrupting agent is removed. When not active in the translation of mRNA, the ribosome exists as two separate subunits.

On the large subunit of the ribosome there are three sites to which a tRNA can bind, designated the A, P, and E sites (see Figure 10.19). The ribosome moves along the mRNA in the 3′ direction, and as it does so, a charged tRNA traverses these three sites, starting

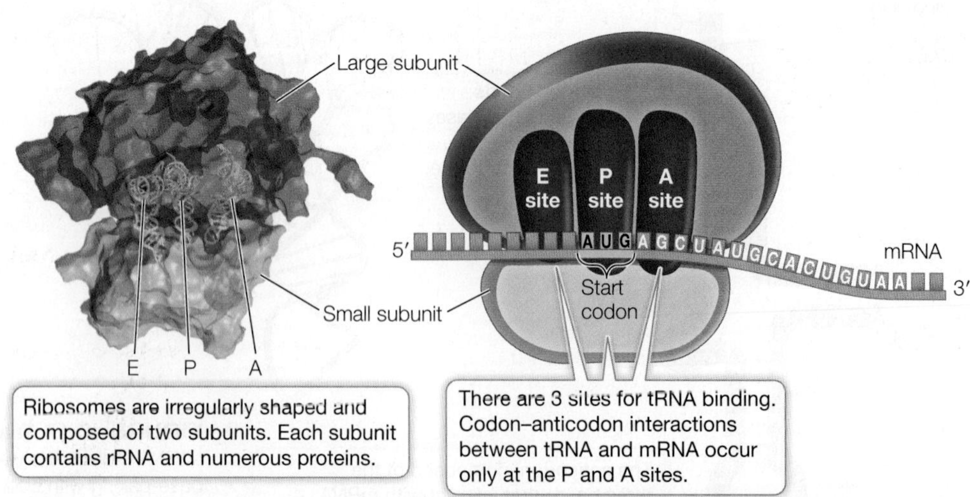

FIGURE 10.19 Ribosome Structure Each ribosome consists of a large and a small subunit. The subunits remain separate when they are not in use for protein synthesis. (Left by David McIntyre based on data from PDB 1GIX and 1G1Y. M. M. Yusupov et al. 2001. *Science* 292: 883–896.)

in the A site, progressing through the P site, and ending in the E site. The A (amino acid) site is where the charged tRNA anticodon binds to the mRNA codon, thus lining up the correct amino acid to be added to the growing polypeptide chain. The P (polypeptide) site is where the tRNA adds its amino acid to the polypeptide chain, or more accurately, the polypeptide chain is added to its amino acid. The E (exit) site is where the tRNA, having given up its amino acid and the rest of the peptide chain received in the P site, resides before being released from the ribosome. After being released, the uncharged tRNA is soon recharged by its tRNA synthetase and is ready to reenter the ribosome (see Figure 10.18).

The ribosome has a fidelity function, which ensures that a charged tRNA with the correct anticodon binds to the appropriate codon in the mRNA. When proper binding occurs, hydrogen bonds form between the three base pairs. The rRNA of the small ribosomal subunit plays a role in validating the three-base-pair match. Any tRNA that does *not* form hydrogen bonds with all three bases of the codon is ejected from the ribosome.

Translation takes place in three steps

Like replication and transcription, translation occurs in three steps: initiation, elongation, and termination.

The **initiation complex** consists of a charged tRNA and a small ribosomal subunit, both bound to the mRNA (**FIGURE 10.20**). Different groups of organisms have different ways to effect this binding. In prokaryotes, a short sequence of RNA, between the 5′ end of the mRNA and the start codon, binds to an rRNA molecule in the small subunit of the ribosome. In eukaryotes, a complex of proteins that includes the small subunit binds to the 5′ cap and then moves down the mRNA until it reaches a start codon. In both cases, once the small subunit is in the appropriate location, the anticodon of a methionine-charged tRNA binds to the start codon by complementary base pairing to complete the initiation complex. Thus the first amino acid in a new polypeptide chain is always methionine. (In bacteria, but not archaea, the first amino acid is a slightly modified form of methionine called formylmethionine.) However, not all mature peptides have methionine as their first amino acid. In many cases, the initial methionine is removed by an enzyme after translation.

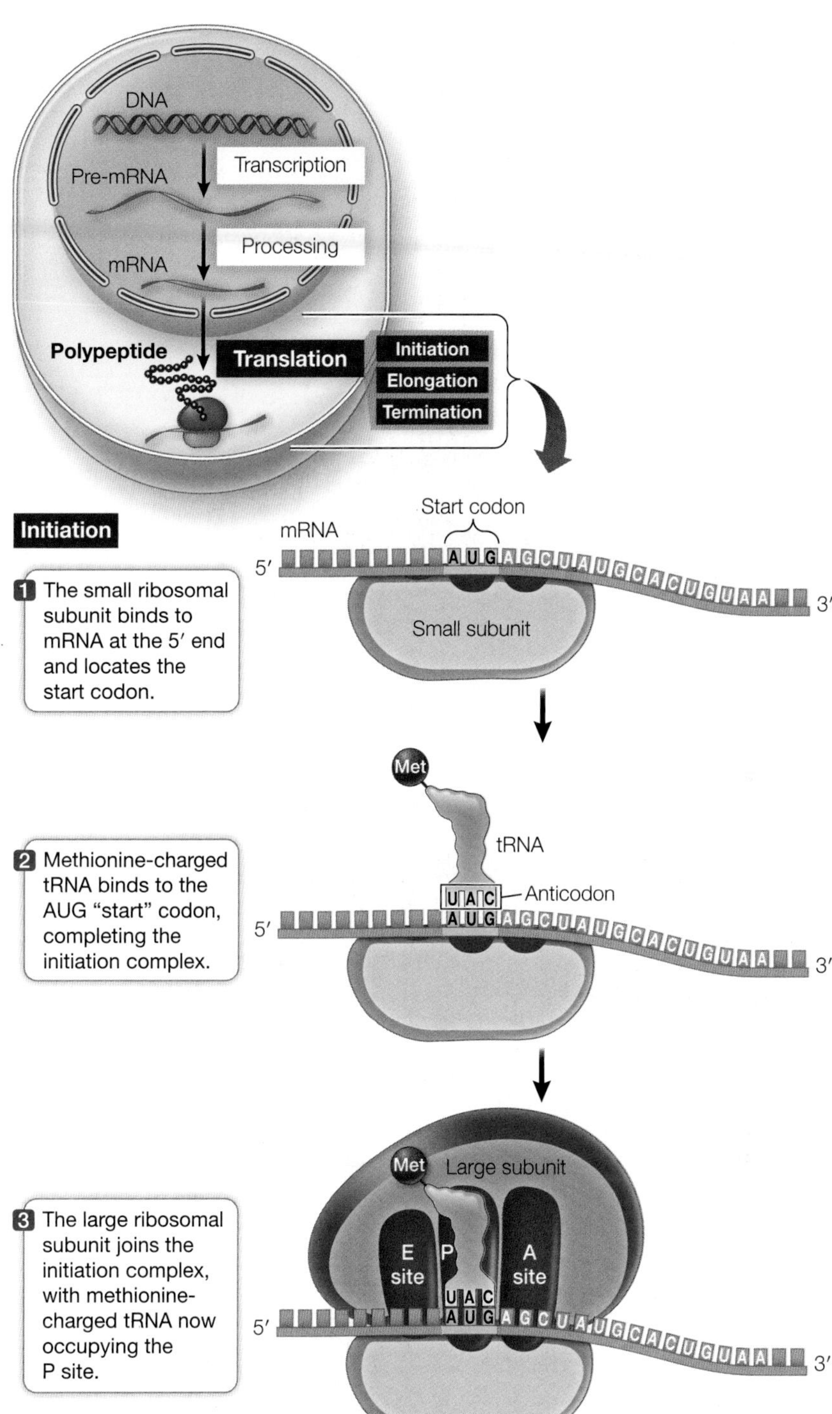

FIGURE 10.20 The Initiation of Translation Translation begins with the formation of an initiation complex.

After the methionine-charged tRNA has bound to the mRNA, the large subunit of the ribosome joins the complex. At this point the ribosome is fully assembled, the methionine-charged tRNA lies in the P site of the large subunit, and the A site is aligned with the second mRNA codon. The assembly of these components—mRNA, both ribosomal subunits, and methionine-charged tRNA—is facilitated by a group of proteins called **initiation factors**.

Once initiation is complete, synthesis of the peptide can occur during the elongation phase of translation. A charged tRNA whose anticodon is complementary to the second codon of the mRNA now enters the open A site of the large ribosomal subunit (**FIGURE 10.21**). The large subunit then catalyzes two reactions: (1) it breaks the bond between the methionine and its tRNA in the P site, and (2) it catalyzes the formation of a peptide bond between the methionine

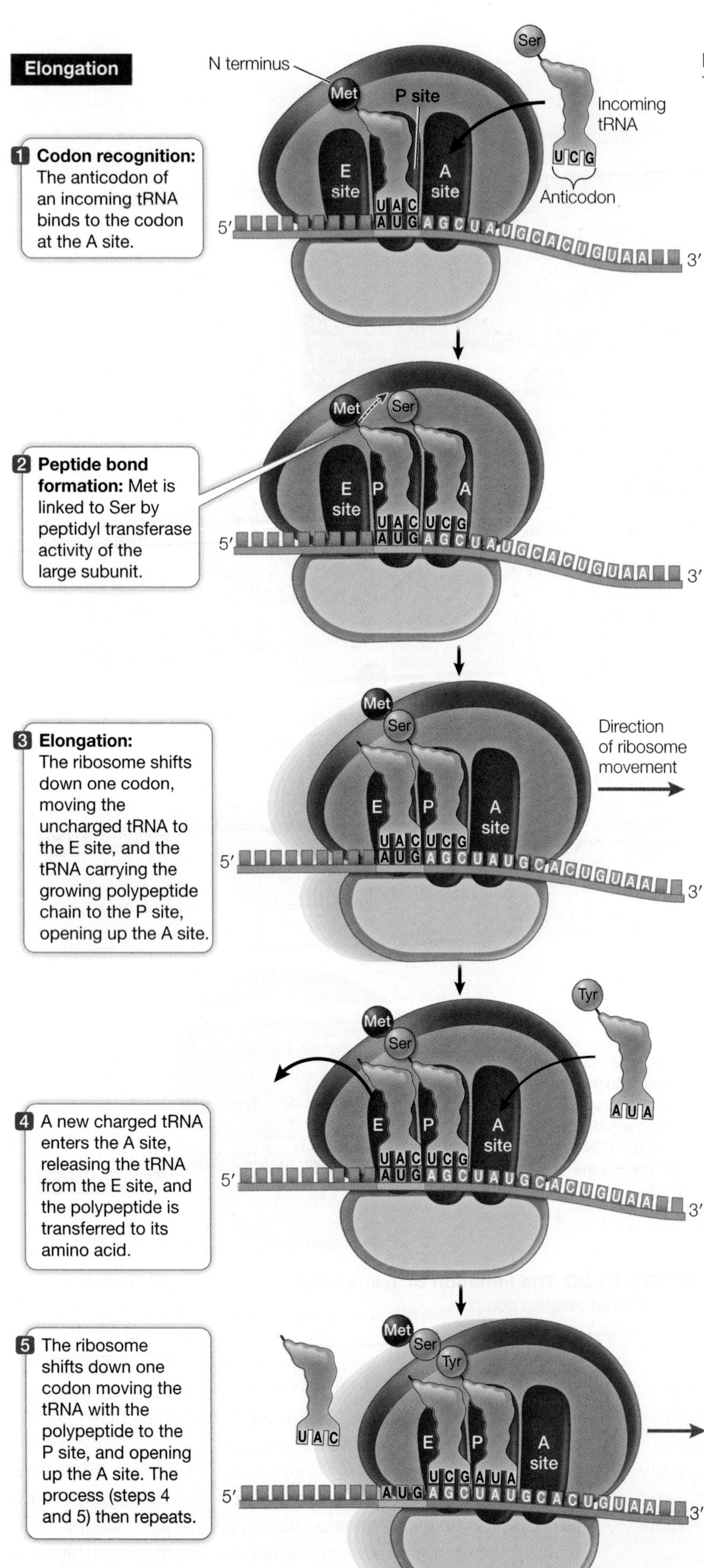

FIGURE 10.21 The Elongation Phase of Translation The polypeptide chain elongates as the mRNA is translated.

and the amino acid attached to the tRNA in the A site. Because the large ribosomal subunit performs these two actions, it is said to have peptidyl transferase activity. The component with this activity is actually one of the rRNAs in the ribosome, so the catalyst is an example of a **ribozyme** (from *ribo*nucleic acid and en*zyme*).

Methionine becomes the amino (N) terminus of the new protein (recall that polypeptides grow in the amino to carboxyl direction; see Key Concept 3.4). The second amino acid is now bound to methionine but remains attached to its tRNA in the A site. After the first tRNA releases its methionine, the ribosome moves exactly three bases down the mRNA (in the 3′ direction) so that the first tRNA is in the E site; the second tRNA, now carrying two amino acids (a dipeptide), is in the P site; and the A site is empty (see Figure 10.21, step 3). The next charged tRNA then enters the A site, and remains if it can pair appropriately with the codon (see Figure 10.21, step 4). This causes the uncharged tRNA that was in the E site to dissociate from the ribosome and return to the cytosol to become charged with another methionine. The growing peptide is transferred to the amino acid of the A site tRNA and then the ribosome again shifts down one codon and the uncharged tRNA that was in the P site moves to the E site. These steps are repeated, and the polypeptide chain grows as each new amino acid is added.

LINK You can review the structure and formation of peptide bonds in **Key Concept 3.4**, especially **Figure 3.16**.

Elongation continues until the ribosome shifts and a stop codon (UAA, UAG, or UGA) enters the A site. There are no tRNAs that can base pair with a stop codon, so the ribosome stalls and translation terminates (**FIGURE 10.22**). A stop codon can bind a protein release factor, which causes hydrolysis of the bond between the polypeptide chain and the tRNA in the P site. The newly completed polypeptide then separates from the ribosome, and the ribosome falls off, splitting into a large and a small subunit.

Media Clip 10.2 Protein Synthesis: An Epic on a Cellular Level

Polysome formation increases the rate of protein synthesis

Several ribosomes can simultaneously translate a single mRNA molecule, producing multiple polypeptides at the same time. As soon as the first

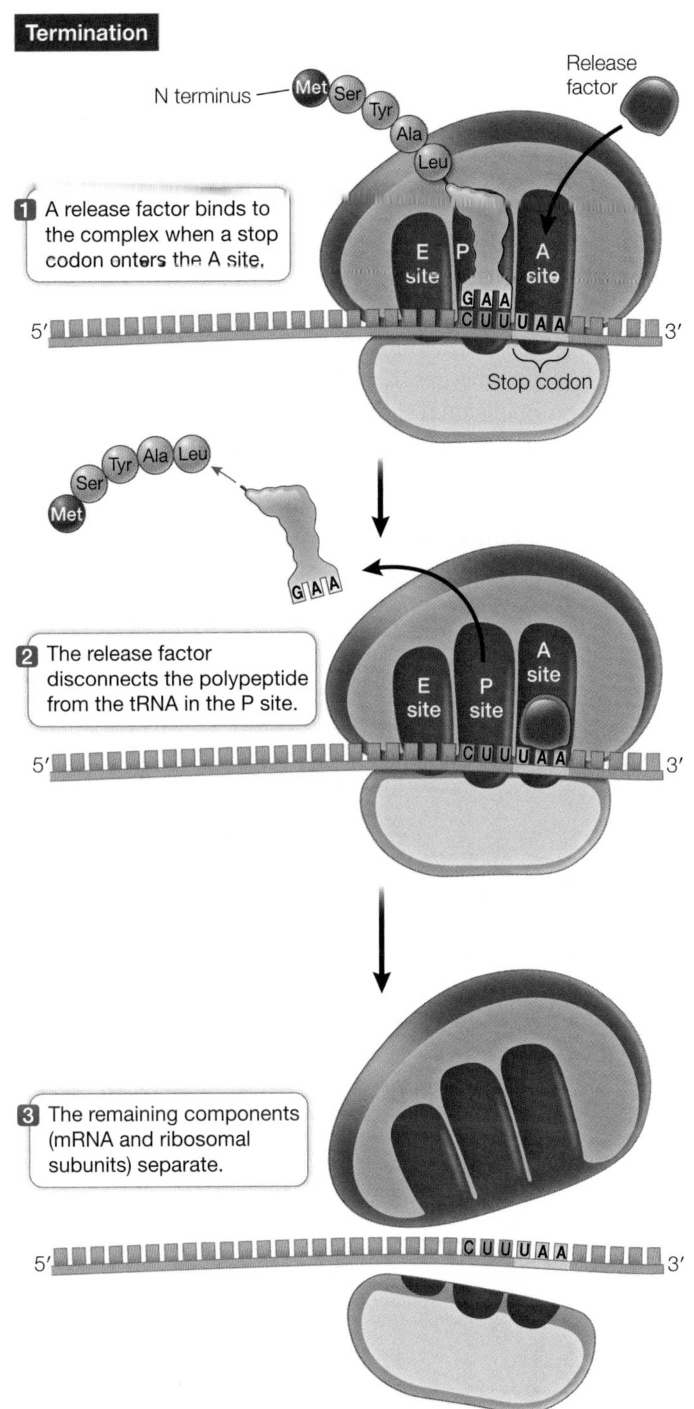

FIGURE 10.22 The Termination of Translation Translation terminates when the A site of the ribosome encounters a stop codon on the mRNA.

ribosome has moved far enough from the translation initiation site, a second initiation complex can form, then a third, and so on. An assemblage consisting of a strand of mRNA with its beadlike ribosomes and their growing polypeptide chains is called a polyribosome, or polysome (**FIGURE 10.23**). Cells that are actively synthesizing proteins contain large numbers of polysomes and few free ribosomes or ribosomal subunits.

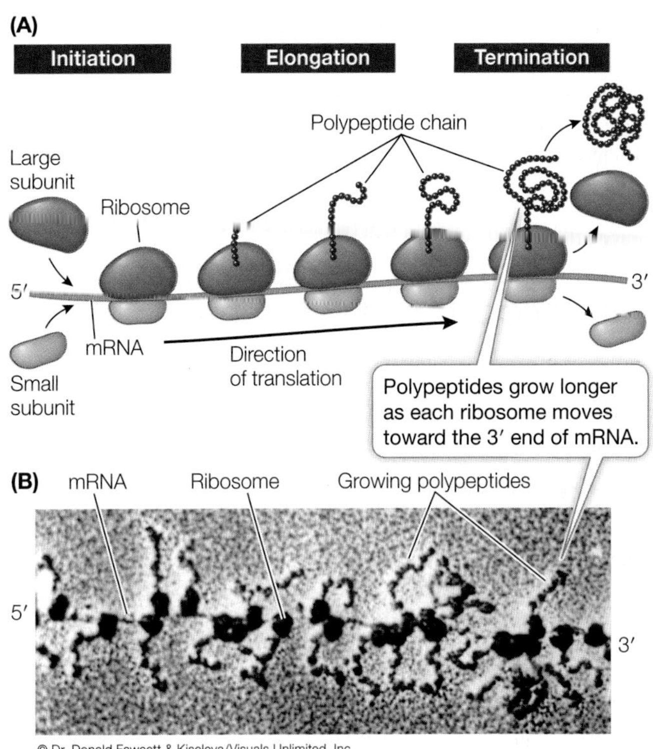

FIGURE 10.23 A Polysome **(A)** A polysome consists of multiple ribosomes and their growing polypeptide chains moving along an mRNA molecule. **(B)** An electron micrograph of a polysome.

REVIEW & APPLY | 10.4

R Each tRNA binds an amino acid at its 3′ end and interacts by base pairing (with wobble) of its anticodon with an mRNA codon. Translation proceeds in three steps. Initiation occurs when the small subunit binds to the 5′ end of the mRNA and the initiator met-tRNA binds to its AUG codon. Then the large subunit binds, leaving the met-tRNA in the P site. Elongation occurs as a charged tRNA enters the A site, peptidyl transferase transfers the peptide to the amino acid on A site tRNA, and then the ribosome moves one codon down the mRNA. Termination of translation occurs when the ribosome encounters a stop codon.

A

1. If you were to follow a tRNA in a cell, starting with it in an uncharged state in the cytosol, what events would it participate in?
2. Imagine a polypeptide whose second amino acid is tryptophan. Sketch a ribosome with the mRNA and the first two tRNAs for this polypeptide, noting their positions in the A, P, and E sites, assuming that the peptidyl transferase enzyme has not yet joined the two amino acids.
3. Every so often a tRNA synthetase makes a mistake and adds the incorrect amino acid to the tRNA. If this tRNA is then used during protein synthesis, what is the effect on protein sequence? What might be the effect on protein function?

The process of protein synthesis usually does not end with translation. Proteins can undergo covalent modifications both during and after translation, with chemical groups being added or parts of the polypeptide chains removed. We will now turn to these modifications.

10.5 | Proteins Are Sometimes Modified after Translation

LEARNING OBJECTIVES

By the end of this key concept you should be able to:

10.5.1 Explain how signal sequences function to target proteins to particular cellular locations.

10.5.2 Recognize that many proteins are post-translationally modified and describe three of these modifications.

Polypeptides are often modified after translation by the removal of amino acids or the addition of new chemical groups that contribute to the function of the mature protein. Once the final protein product has been made, it needs to be sent to its cellular destination. The site of a polypeptide's function in the cell may be far away from its point of synthesis at the ribosome, particularly in eukaryotes.

Signal sequences in proteins direct them to their cellular destinations

Protein synthesis always begins on free ribosomes floating in the cytosol, which is the "default" location for a protein. As the polypeptide chain emerges from the ribosome it may simply fold into its three-dimensional shape and perform its cellular role in the cytosol. However, some polypeptides contain a **signal sequence** (or signal peptide) consisting of a short stretch of amino acids that indicates where in the cell the polypeptide belongs. Different proteins are destined for different locations in the cell, such as the nucleus, mitochondria, plastids, peroxisomes, lysosomes, secretory vesicles, and so on. Proteins with different final

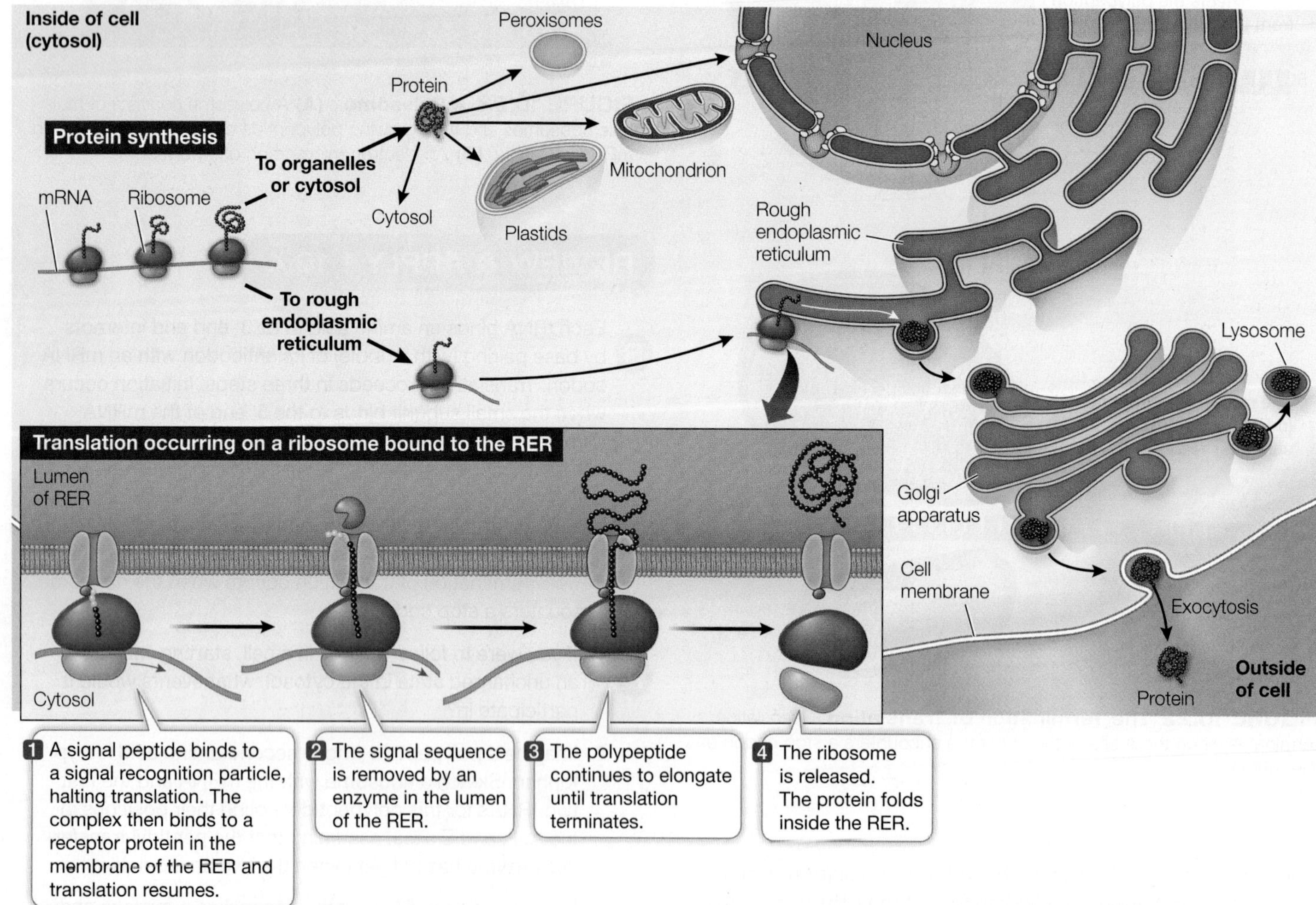

FIGURE 10.24 Destinations for Newly Translated Polypeptides in a Eukaryotic Cell Signal sequences on newly synthesized polypeptides bind to specific receptor proteins on the outer membranes of the organelles to which they are targeted. Once the protein has bound to it, the protein is directed to the appropriate organelle. Proteins destined for the rough endoplasmic reticulum (RER) are generally targeted during translation, whereas proteins destined for the mitochondria, plastids, and nucleus are targeted after translation.

destinations have different signal sequences that target them to the appropriate location.

Targeting to a specific location may occur during or after translation. Some signal sequences at the amino end of the targeted protein are bound by a signal-recognition particle (a protein-RNA complex) as soon as they exit the ribosome. Binding halts translation, which then resumes once the signal recognition particle binds to its specific receptor protein at the surface of the rough endoplasmic reticulum (RER; see Chapter 4). As translation proceeds, the polypeptide enters the RER and an enzyme cleaves off the signal sequence. Targeted proteins with a signal peptide that binds to the signal-recognition particle are thus put into the membrane or lumen (the inside) of the RER as they are being translated and may be retained there, or they may move elsewhere within the endomembrane system (Golgi apparatus, lysosomes, and cell membrane). Proteins in the RER that lack signal sequences for destinations within the endomembrane system are usually secreted from the cell via secretory vesicles that fuse with the cell membrane. This mechanism of entering the RER during translation is the usual route by which proteins enter the endomembrane system (see Key Concept 4.5).

For other proteins, translation is completed in the cytosol, and then a signal sequence—which may be at the amino or carboxyl end of the protein—is bound by other proteins that direct the protein to the correct organelle. This is the common route by which proteins enter the mitochondria, plastids, and nucleus, though it is also used by some endomembrane-bound proteins (**FIGURE 10.24**).

Determining the identity of a signal sequence involves adding or removing the proposed signal sequence from proteins and determining whether targeting is affected. For example, in the frog *Xenopus laevis* there is a nuclear localization signal (NLS) at the carboxyl end of a protein called nucleoplasmin. This protein is localized to the nucleus in *Xenopus* cells and will also localize to the nucleus if it is injected into primate cells in culture. Removal of the NLS prevents the correct localization of nucleoplasmin. Furthermore, addition of this signal to a protein that is usually cytosolic causes it to be transported into the nucleus after injection into primate cells (**INVESTIGATION FIGURE 10.25**).

INVESTIGATION

FIGURE 10.25 Nuclear Targeting by a Signal Sequence A series of experiments demonstrated that a nuclear localization signal (NLS) is sufficient to target a protein to the nucleus.[a]

HYPOTHESIS

A small sequence from the carboxyl end of the nucleoplasmin protein is sufficient to target it, or another protein, to the nucleus.

METHOD

Inject primate cells with nucleoplasmin protein from *Xenopus* or with chicken pyruvate kinase, with and without the putative NLS sequences. Nucleoplasmin is localized to the nucleus in *Xenopus* cells, and pyruvate kinase is localized to the cytosol in chicken cells.

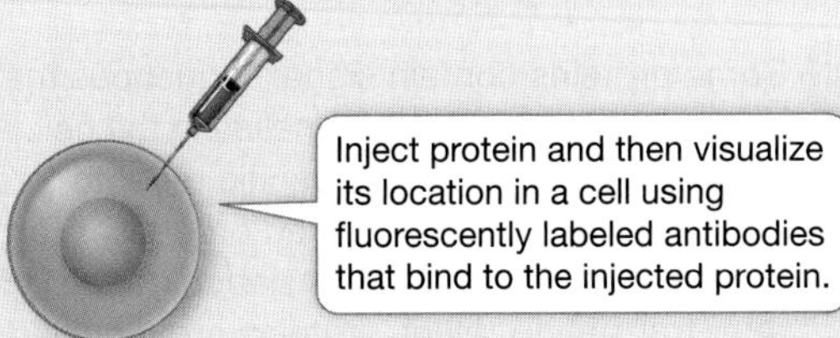

RESULTS

Injected protein:

Nucleoplasmin, a nuclear protein, with the NLS

NLS

Nucleoplasmin with the NLS removed

Pyruvate kinase, a cytosolic protein without the NLS

Pyruvate kinase, with attached NLS

The distribution of the protein in the cell is observed with a fluorescence microscope.

CONCLUSION

The putative NLS is sufficient and required for localization of a protein to the nucleus.

ANALYZE THE DATA

1. Why not use primate nucleoplasmin and primate pyruvate kinase to visualize with fluorescently labeled antibodies in this experiment?

A similar **Analyze the Data** exercise may be assigned in **Achieve**.

[a]C. Dingwall et al. 1988. *J Cell Bio* 107: 841–849.

Many proteins are modified during or after translation

Mature proteins are often not identical to the polypeptide chains that are translated from mRNA on the ribosomes. Instead, they are modified in any of several ways. Most modifications occur after translation (**FIGURE 10.26**) but some occur during translation. We have already met an example of the latter—the removal of the signal sequence in the lumen of the RER (see Figure 10.24). Another extremely common modification that occurs during translation is the removal of the initiator methionine. More than 80 percent of proteins will have their first methionine enzymatically removed during translation. There are a very large number of posttranslational modifications that proteins can undergo. Examples include:

- *Proteolysis*, which is the cutting of a polypeptide chain. Some mature proteins are actually made from polyproteins—long polypeptides containing the primary sequences of multiple distinct proteins—that are cut into final products by enzymes called proteases. Proteases are essential to some viruses, including human immunodeficiency virus (HIV), because the large viral polyprotein cannot fold properly unless it is cut. Certain drugs used to treat acquired immune deficiency syndrome (AIDS) work by inhibiting the HIV protease, thereby preventing the formation of proteins needed for viral reproduction.
- *Glycosylation* is the addition of carbohydrates to proteins to form glycoproteins. In both the ER and the Golgi apparatus, resident enzymes catalyze the addition of various oligosaccharides (short chains of monosaccharides; see Key Concept 3.2) to certain amino acid R groups on proteins. One such type of "sugar coating" is essential for directing proteins to lysosomes. Other types are important for protein conformation and for recognition functions at the cell surface. As we noted in Key Concept 8.3, different chains of sugars added to red blood cell proteins determine an individual's blood type. In many cases the attached oligosaccharides help stabilize proteins, such as those in the extracellular matrix, and those in the storage vacuoles of plants.
- *Phosphorylation* is the addition of phosphate groups to proteins and is catalyzed by protein kinases. The charged phosphate groups change the conformation of the protein, often exposing the active site of an enzyme or the binding site for another protein. Phosphorylation is especially important in cell signaling (see Key Concepts 6.2 and 6.3) and is often reversible, which gives rapid control over protein function—that is, an enzyme can be turned off or on rapidly.

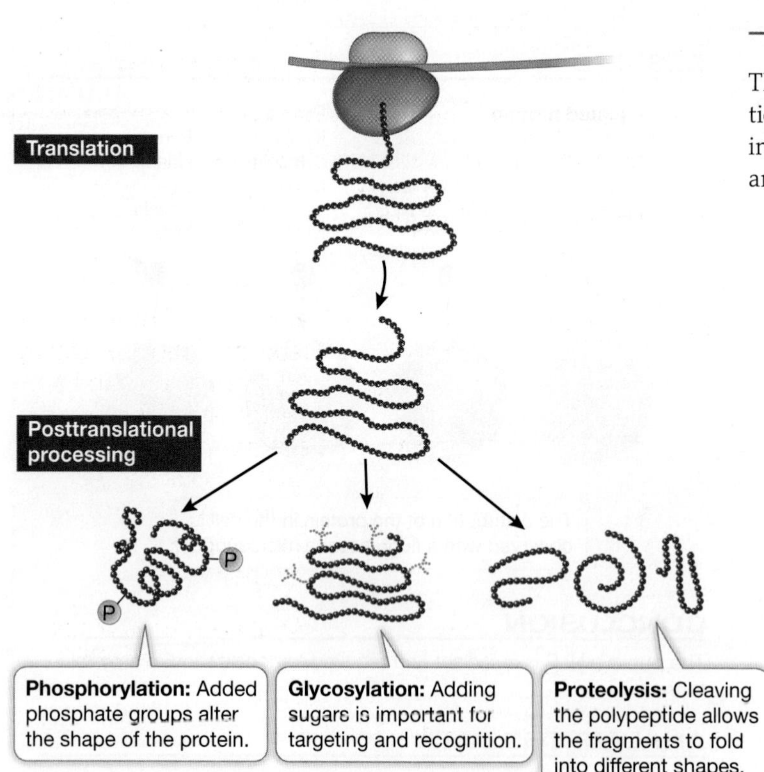

FIGURE 10.26 Posttranslational Modifications of Proteins
Most polypeptides must be modified after translation in order to become functional proteins.

LINK As emphasized in **Key Concept 3.4**, a protein's primary structure is defined by its amino acid sequence, and a protein will fold into its functionally correct three-dimensional configuration as long as its primary structure is not disrupted. Modifications to polypeptides made during and after translation significantly affect a protein's secondary, tertiary, and quaternary structure.

These examples should give you a sense of the kinds of modifications that can be made to proteins after they are synthesized. Keep in mind, however, that there are many more that are not listed but are essential to the final proper functioning of different proteins.

REVIEW & APPLY | 10.5

R Some proteins contain signal sequences that target them to particular organelles in the cell. A signal sequence binds to a specific receptor protein on the surface of the targeted organelle. Targeting can occur either during translation (to the RER) or after translation. Many proteins are modified by the removal of the initiator methionine, removal of signal sequences, proteolysis, phosphorylation, and glycosylation.

A
1. What are some ways in which posttranslational modifications alter protein structure and function?
2. Describe an experiment you would perform to test a proposed chloroplast-targeting signal sequence. Be specific about the type of cell and the proteins you would use.

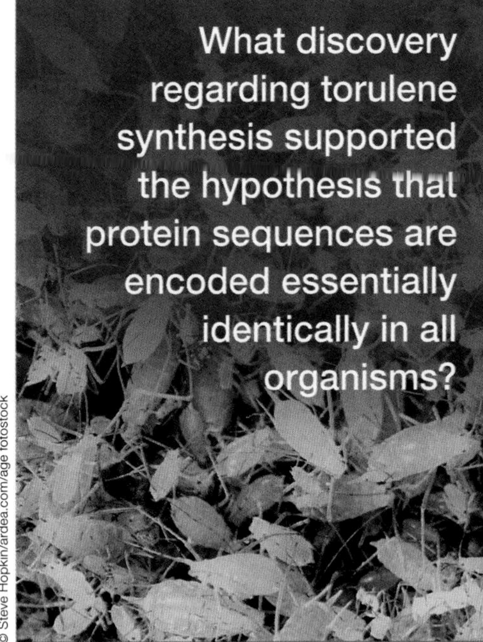

Answer: Pea aphids are unique among animals in having genes that encode enzymes for the biochemical pathway for the synthesis of carotenoids. Some of the aphids have all genes intact, and they make the red pigment, torulene. So these aphids are red. Other aphids have a mutation in one of the genes for carotenoid synthesis and cannot make torulene. These aphids are green. Remarkably, the DNA sequences of the genes encoding the carotenoid pathway in aphids are the same as the sequences of genes in a fungus that lives nearby. Fungi are well known to make carotenoids, as do plants. The best explanation for the arrival of fungus genes in an aphid is that some DNA was released from the fungus and was taken up by aphid tissue. This is called lateral gene transfer (as opposed to vertical gene transfer from parent to offspring). The fungal DNA was incorporated into the aphid genome and transmitted to aphid offspring. DNA uptake is not totally unusual in nature: recall from Key Concept 8.4 that bacteria can be transformed by DNA from their environment. What is remarkable is the uptake of DNA from such a vastly different organism. The facts that the fungus DNA is recognized and incorporated into the aphid genome, that it is expressed as proteins by the protein synthesis machinery of aphid cells, and even that its pre-mRNAs are spliced correctly illustrate the universality of the genetic code and the overall mechanisms of protein synthesis.

VISUAL SUMMARY 10

10.1 One Gene Encodes One Polypeptide

Go to ACTIVITY 10.1

- Experiments with knockouts in metabolic pathways led to the **ONE GENE–ONE POLYPEPTIDE** hypothesis.
- The **CENTRAL DOGMA OF MOLECULAR BIOLOGY** states that DNA is transcribed into **MESSENGER RNA**s, which are then translated into polypeptides (**FIGURE 10.5**).

FIGURE 10.5

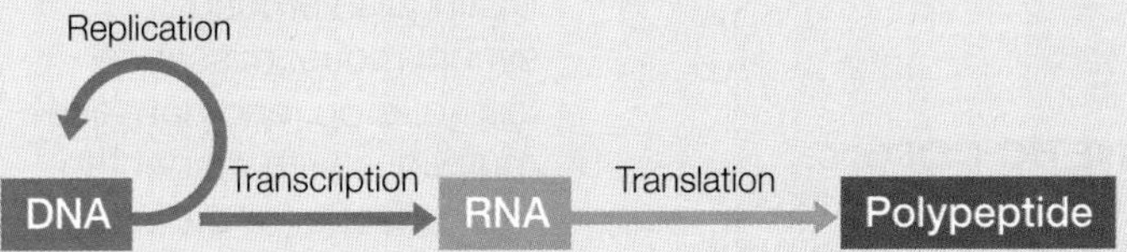

QUESTIONS

1. A different enzyme catalyzes each of the four steps of an amino acid biosynthetic pathway: A → B → C → D → amino acid. The four reactions are catalyzed by enzymes 1, 2, 3, and 4 respectively. What are the effects on growth of a homozygous knockout of enzyme 3? How might these effects be compensated?
2. What are the three most common kinds of RNAs in a cell?

10 VISUAL SUMMARY

10.2 Gene Expression Begins with Transcription of DNA into RNA

Go to ANIMATIONS 10.1 and 10.2

FIGURE 10.8

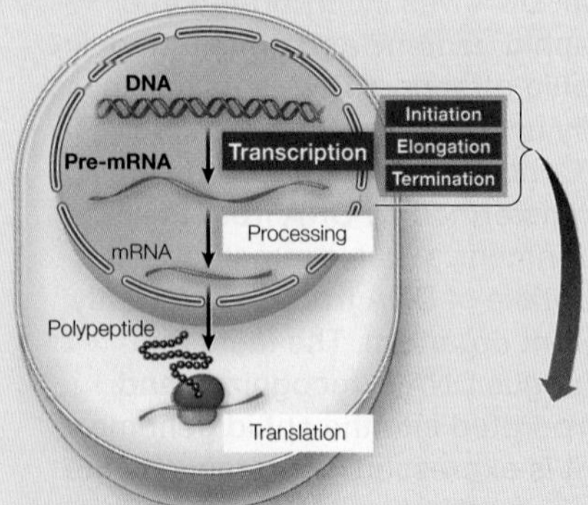

- **PROMOTERS** direct RNA polymerases to genes, initiating RNA synthesis (**FIGURE 10.8**).
- In the elongation phase of RNA synthesis, ribonucleotides that are complementary to the DNA template are added to the 3′ end of the RNA.
- Polymerases terminate when they reach a termination site.
- In eukaryotes, mRNA processing involves removing **INTRONS** and adding a **5′ CAP** and a 3′ **POLY A TAIL** to the ends of the **PRE-MRNA** transcript (**FIGURE 10.9**).

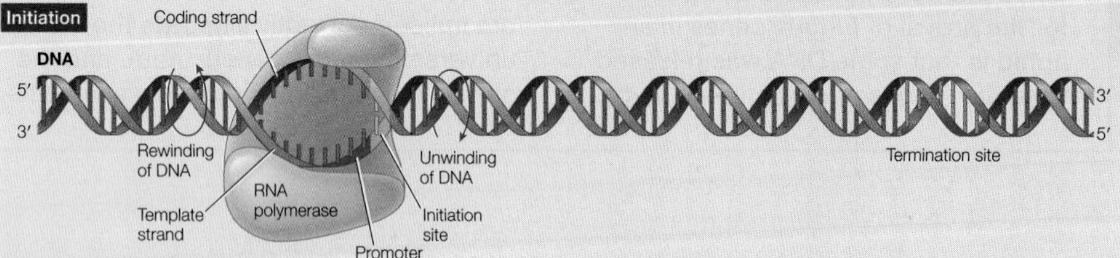

FIGURE 10.9

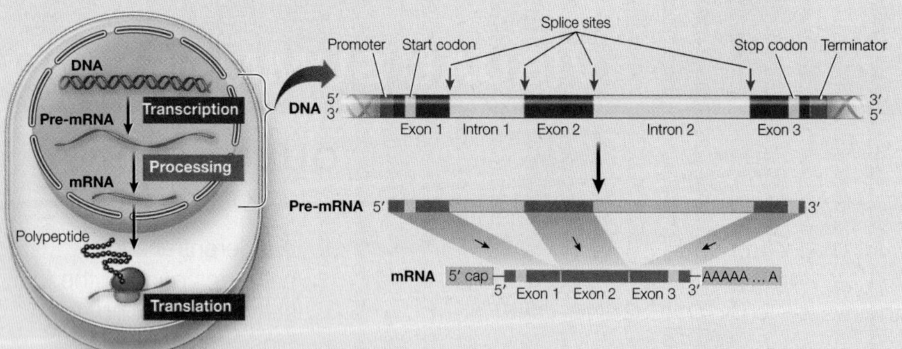

QUESTIONS

1. What three mRNA processing events occur in eukaryotes?
2. Will a mutation at a promoter always cause reduced gene expression?
3. How does the structure, transcription, and processing of a prokaryotic and eukaryotic protein-coding gene differ?

10.3 The Rules for Translation of RNA into Amino Acids Are Contained in the Genetic Code

Go to ACTIVITIES 10.2–10.4 and ANIMATION 10.3

FIGURE 10.14

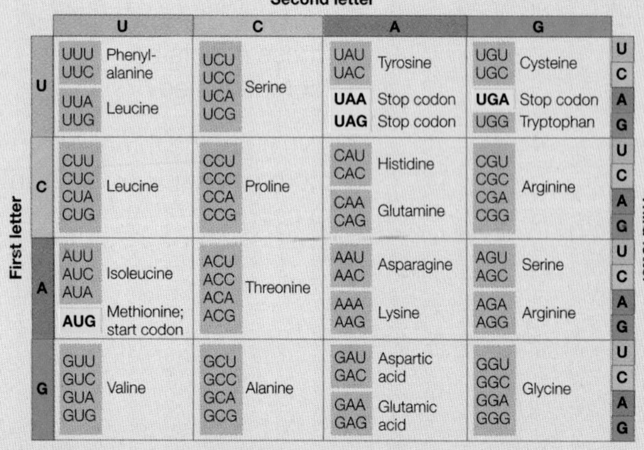

Second letter

First letter	U	C	A	G	Third letter
U	UUU Phenyl-alanine	UCU Serine	UAU Tyrosine	UGU Cysteine	U
U	UUC Phenyl-alanine	UCC Serine	UAC Tyrosine	UGC Cysteine	C
U	UUA Leucine	UCA Serine	**UAA** Stop codon	**UGA** Stop codon	A
U	UUG Leucine	UCG Serine	**UAG** Stop codon	UGG Tryptophan	G
C	CUU Leucine	CCU Proline	CAU Histidine	CGU Arginine	U
C	CUC Leucine	CCC Proline	CAC Histidine	CGC Arginine	C
C	CUA Leucine	CCA Proline	CAA Glutamine	CGA Arginine	A
C	CUG Leucine	CCG Proline	CAG Glutamine	CGG Arginine	G
A	AUU Isoleucine	ACU Threonine	AAU Asparagine	AGU Serine	U
A	AUC Isoleucine	ACC Threonine	AAC Asparagine	AGC Serine	C
A	AUA Isoleucine	ACA Threonine	AAA Lysine	AGA Arginine	A
A	**AUG** Methionine; start codon	ACG Threonine	AAG Lysine	AGG Arginine	G
G	GUU Valine	GCU Alanine	GAU Aspartic acid	GGU Glycine	U
G	GUC Valine	GCC Alanine	GAC Aspartic acid	GGC Glycine	C
G	GUA Valine	GCA Alanine	GAA Glutamic acid	GGA Glycine	A
G	GUG Valine	GCG Alanine	GAG Glutamic acid	GGG Glycine	G

- The **GENETIC CODE** is degenerate, with 61 **SENSE CODONS** and 3 **NONSENSE** (or **STOP) CODONS** (**FIGURE 10.14**).
- The effect of a base substitution at a **CODON** can be predicted using the genetic code.

1. Distinguish among synonymous, missense, loss-of-stop, and nonsense mutations with respect to their effect on a protein.
2. What is a frame-shift mutation, and what is its effect on a protein?

10.4 RNA Is Translated into Amino Acids by Ribosomes

- A tRNA is charged by adding the appropriate amino acid to its 3′ end and has an **ANTICODON** for pairing with a codon (**FIGURE 10.17**).
- Translation takes place in three stages: initiation (with methionine at AUG), elongation, and termination (at UAA, UGA, or UAG) (**FIGURE 10.21**).

FIGURE 10.17

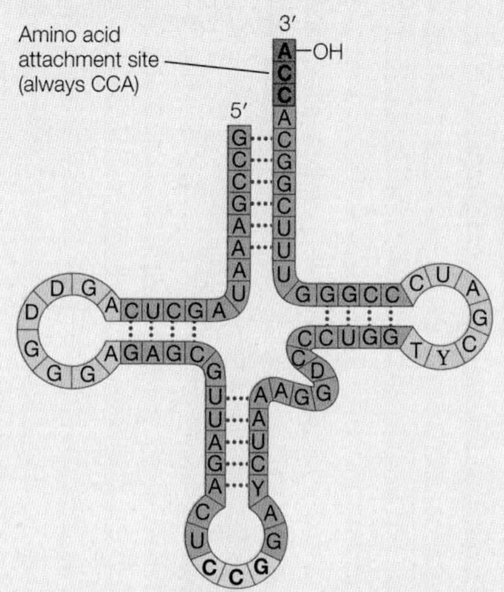

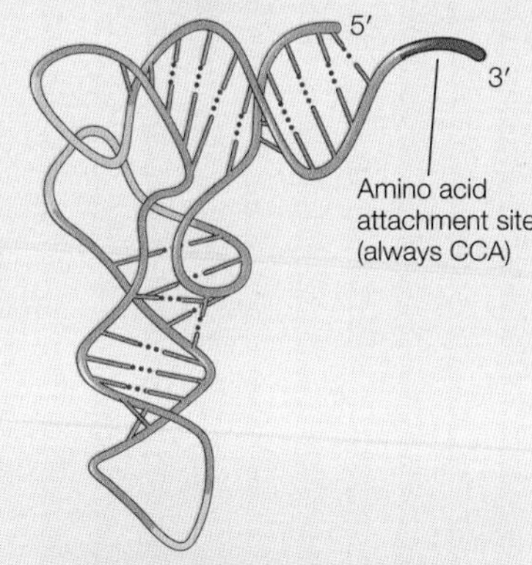

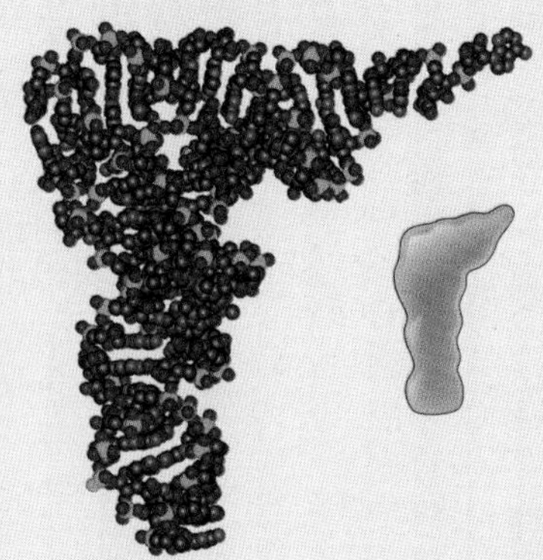

Go to ANIMATION 10.4

FIGURE 10.21

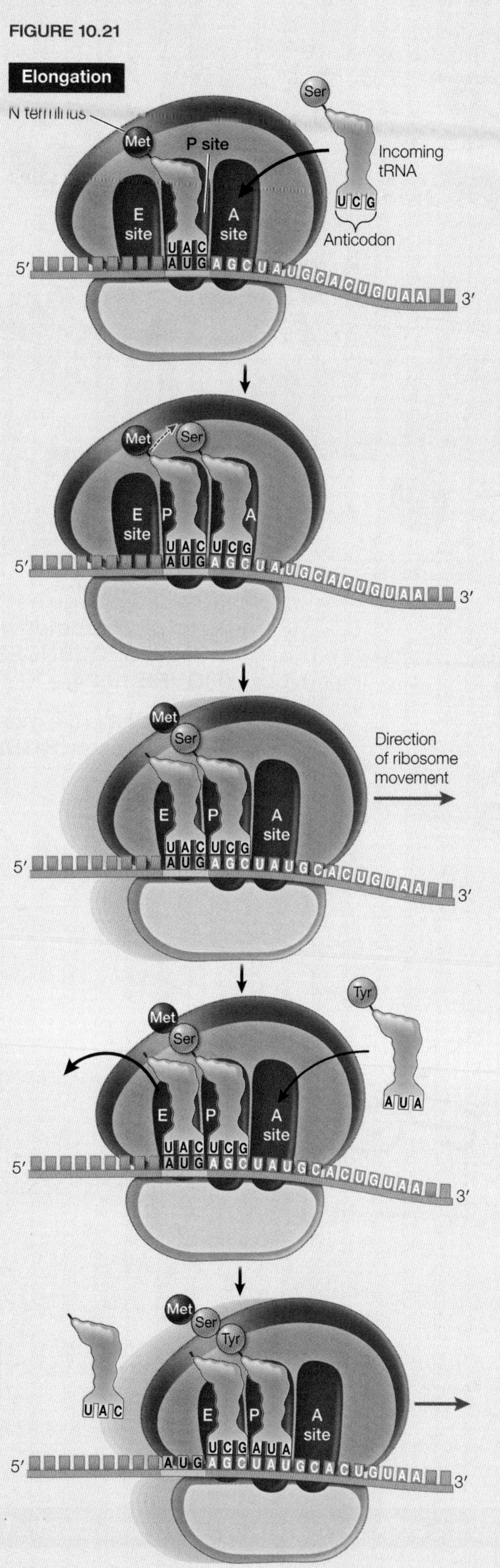

QUESTIONS

1. How is an mRNA recognized by a ribosome?
2. What are the functions of the three most common kinds of RNAs in a cell?

10.5 Proteins Are Sometimes Modified after Translation

FIGURE 10.24

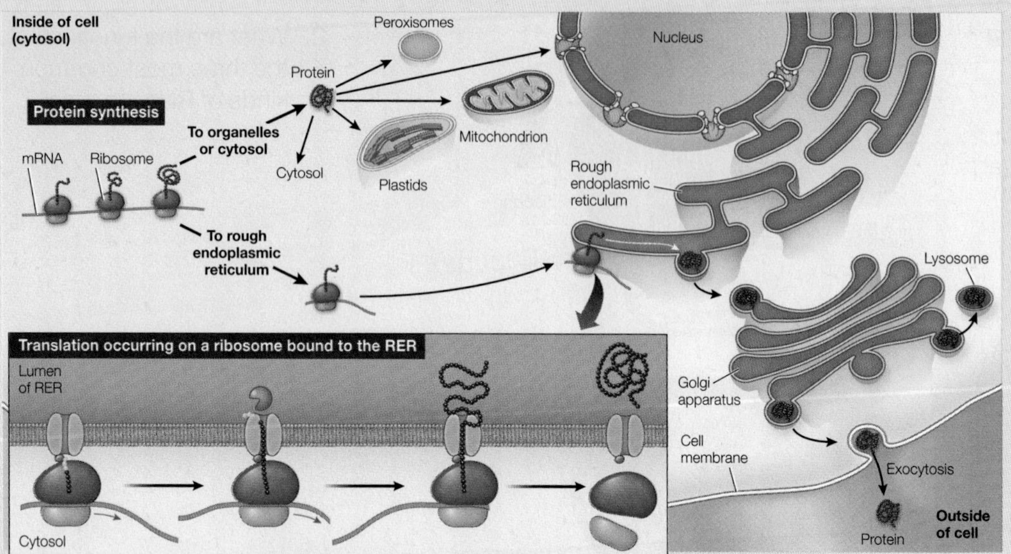

FIGURE 10.26

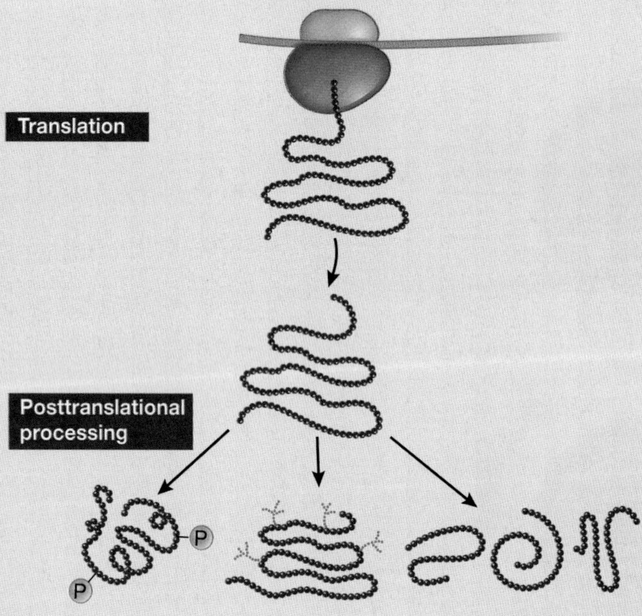

- Proteins are targeted to particular organelles with **SIGNAL SEQUENCES** (**FIGURE 10.24**).
- Many proteins are modified after translation (**FIGURE 10.26**).

QUESTIONS

1. What is the function of a signal sequence? Do all proteins have them?
2. In what ways does the functional form of a protein differ from the amino acid sequence encoded in the exons of its gene?

Go to **Achieve** for the e-book, animations, activities, and additional resources and assignments.

11 Regulation of Gene Expression

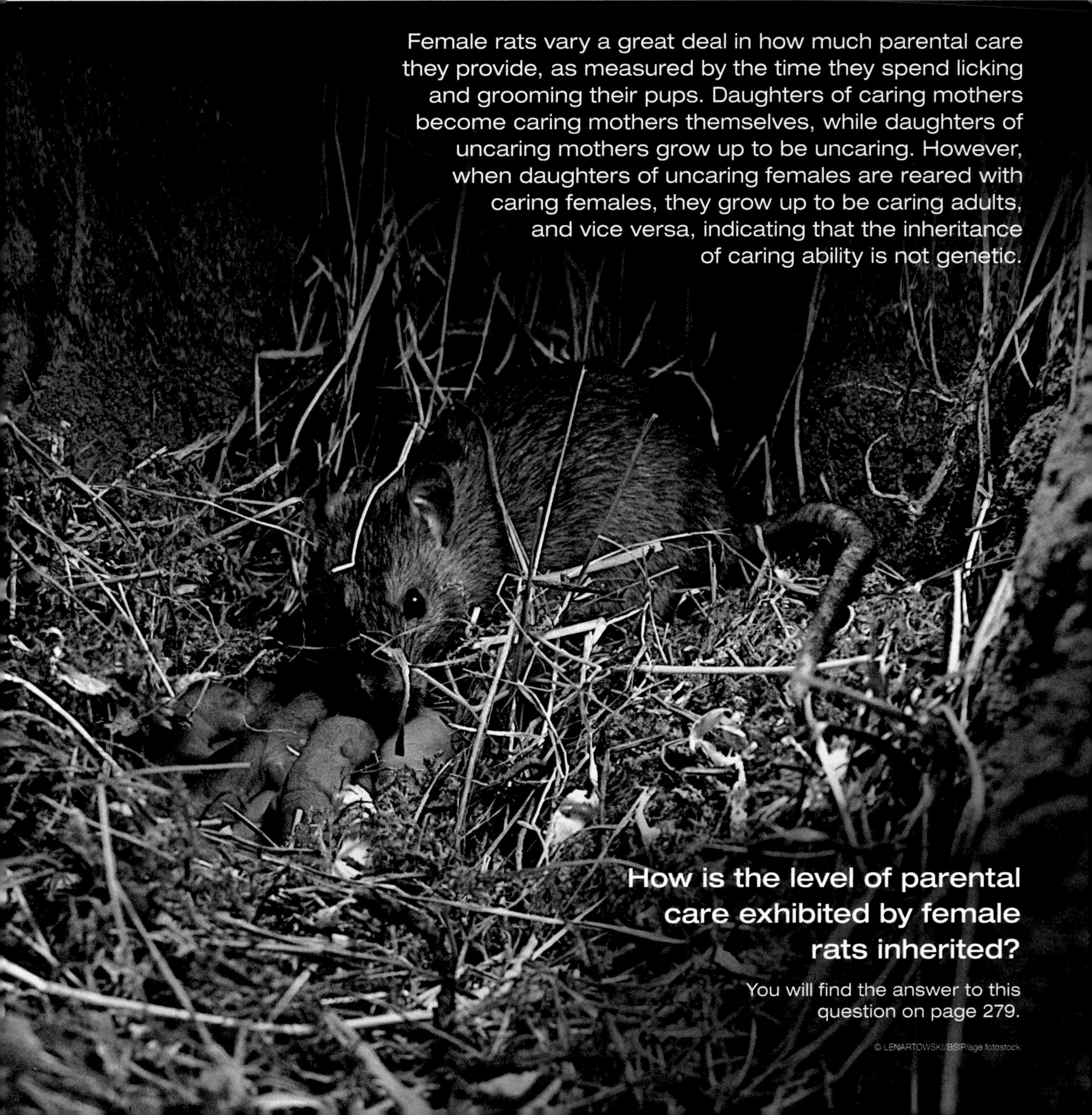

Female rats vary a great deal in how much parental care they provide, as measured by the time they spend licking and grooming their pups. Daughters of caring mothers become caring mothers themselves, while daughters of uncaring mothers grow up to be uncaring. However, when daughters of uncaring females are reared with caring females, they grow up to be caring adults, and vice versa, indicating that the inheritance of caring ability is not genetic.

How is the level of parental care exhibited by female rats inherited?

You will find the answer to this question on page 279.

© LENARTOWSKI/BSIP/age fotostock

KEY CONCEPTS

11.1 The Regulation of Gene Expression Occurs at Multiple Levels

11.2 Prokaryotic Gene Regulation Occurs Primarily at the Level of Transcription

11.3 Eukaryotic Transcription Is Regulated by General and Specific Transcription Factors

11.4 Transcription Can Be Regulated by Epigenetic Changes to DNA and Histones

11.5 Eukaryotic Gene Expression Can Be Regulated after Transcription

11.1 The Regulation of Gene Expression Occurs at Multiple Levels

LEARNING OBJECTIVES

By the end of this key concept you should be able to:

11.1.1 Describe situations in which gene expression might be regulated.

11.1.2 Identify the points at which gene expression can be regulated.

11.1.3 Distinguish among inducible, repressible, and constitutive genes.

11.1.4 Distinguish between positive and negative regulation of gene transcription.

The cells in your body vary tremendously even though they are essentially identical genetically. Such differences imply that the genes that are or are not expressed in a cell can vary. In this chapter we discuss some of the ways in which such variation in expression occurs.

Regulating gene expression allows genetically identical cells to have different phenotypes

In Chapter 10 we discussed how genes are expressed. For protein-coding genes, DNA is initially expressed as RNA, which is processed (in eukaryotes) and then translated into protein by ribosomes. Throughout this book we describe instances where gene expression is regulated so that the level of protein produced from a particular gene is changed. The expression of a gene is influenced by both internal factors, such as the stage of the cell cycle, and various biotic and abiotic external factors. Here are three examples we have met previously:

1. In Chapter 5: On a particular diet, certain enzymes involved in metabolic pathways may be upregulated or downregulated. For example, a low-glucose diet will lead to increased transcription (and translation) of genes involved in gluconeogenesis.
2. In Chapter 6: When an extracellular signal binds to its receptor on a eukaryotic cell, it sets in motion a signal transduction pathway that may end with some genes being activated (their expression switched on) and others being repressed (their expression switched off) in order to generate an appropriate cellular (and organismal) response.
3. In Chapter 7: During the cell cycle, particular cyclin proteins are synthesized only at specific stages. The genes for cyclins are not transcribed at other stages in the cycle.

These and many other examples indicate that gene expression is precisely regulated. Even though cells contain many thousands of protein-coding genes, not all of them are expressed at all times. When learning about gene regulation, keep in mind that the genes expressed by a cell determine the cell's phenotype. In certain conditions, certain cell phenotypes are appropriate, while others are inappropriate. For example, in a brain cell it makes sense to express genes that encode neurotransmitters, but those genes should not be expressed in liver cells. However, in both liver and brain cells, genes encoding the enzymes involved in "housekeeping" functions such as glycolysis should be expressed.

While this chapter focuses on the regulation of gene products, don't forget that enzymes are also regulated in their activity. For example, covalent modifications, such as phosphate addition or removal, or noncovalent binding at the active or an allosteric site can cause an enzyme to go from an active to an inactive state or vice versa.

As discussed in **Key Concept 3.5**, regulation of enzyme activity enables exquisite control of the concentration of products of metabolic pathways, and is usually accomplished by the binding of a regulatory molecule (e.g., an inhibitor) to either the active site or to an allosteric site on the enzyme.

Regulation occurs at all possible steps of gene expression

At every step of the way from DNA to protein, gene expression can be regulated (**FIGURE 11.1**). As you proceed through this chapter, you will see examples of gene regulation at pre-transcriptional, transcriptional, posttranscriptional, translational, and posttranslational levels. The rate of RNA synthesis is influenced by the accessibility of the gene's DNA (pre-transcriptional regulation) and by the rate of transcription. Posttranscriptional regulation occurs after transcription but before translation, and affects how RNA processing proceeds, whether or how rapidly mRNA is exported from the nucleus, and the stability of mRNA. Regulation at the level of translation alters whether or how rapidly an mRNA is translated. Posttranslationally, the way in which a protein folds and its stability can be regulated. In short, regulation can occur at every step in gene expression, but not all mechanisms are used for regulating a particular gene.

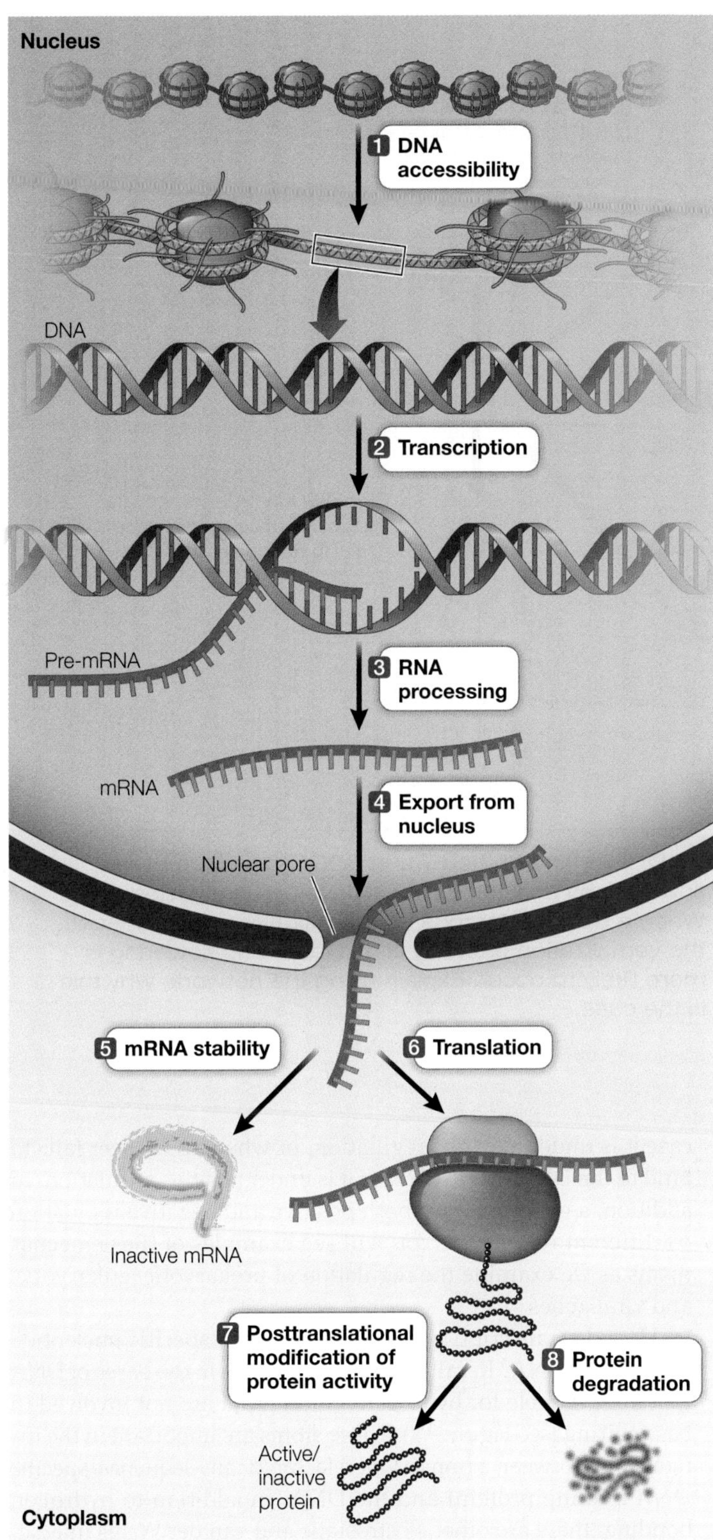

FIGURE 11.1 Potential Points for the Regulation of Gene Expression Gene expression can be regulated before transcription (1), during transcription (2), after transcription but before translation (3, 4, 5), during translation (6), or after translation (7, 8).

Activity 11.1 Eukaryotic Gene Expression Control Points

In addition to occurring at multiple possible points, the regulation of gene expression often involves multiple regulators, and the regulators themselves are subject to regulation, giving rise to **regulatory networks**—collections of interacting regulators that determine the expression level of a gene. Regulatory networks give cells the ability to use multiple factors to control gene expression, but also make understanding how a gene is regulated quite challenging. **FIGURE 11.2** shows a simplified transcriptional regulatory network for the control of flowering time in plants.

LINK **Key Concept 26.3** discusses several factors that initiate flowering, including photoperiodic cues (day length and night length).

Transcriptional regulation conserves energy

While the regulation of gene expression can occur at any point, it often occurs at the level of transcription. Regulation at transcription allows cells to conserve resources—every step from transcription to protein processing requires energy input. If a protein is not needed, it makes sense to shut off expression at the transcriptional level. When this happens, no RNA is synthesized, so no energy is wasted making RNA, processing it, transporting it out of the nucleus (for eukaryotes), translating it to form a protein, and processing the protein.

In general, compounds that stimulate the transcription of specific genes are called **inducers**, and genes that can be activated by inducers are called **inducible genes**. In contrast, compounds that reduce the transcription of specific genes are called **repressors**, and genes that can be turned off by repressors are called **repressible genes**. Some genes, called **constitutive genes**, are expressed most of the time at a constant rate in all cells. From what we have discussed so far in this book, you can probably come up with examples for all of these classes of genes. Some gene products are needed in all cells, such as tRNAs and ribosomal proteins that are required for protein synthesis, and are thus constitutive. Some enzymes, such as those required for triglyceride synthesis, are expressed only when glucose levels are high and are thus inducible in some tissues. Some enzymes, such as those involved in the synthesis of nonessential amino acids, are not needed if the diet has sufficient levels of the relevant amino acids and are likely to be repressible. Whether a gene is inducible or repressible gives us information on when a gene is expressed but does not indicate the mechanism by which expression is controlled. In general, inducible systems control catabolic pathways (see Key Concept 5.3), which are turned on only when the substrate is available, whereas repressible systems control anabolic pathways (see Key Concept 5.4), which are turned on until the concentration of the product becomes sufficient.

Genes are subject to positive and negative regulation

As we mentioned above, not all genes are active (being transcribed) at a given time. Two types of regulatory proteins—called **transcription factors**—control whether or not a gene is active: repressors and activators. Note that, confusingly, the term

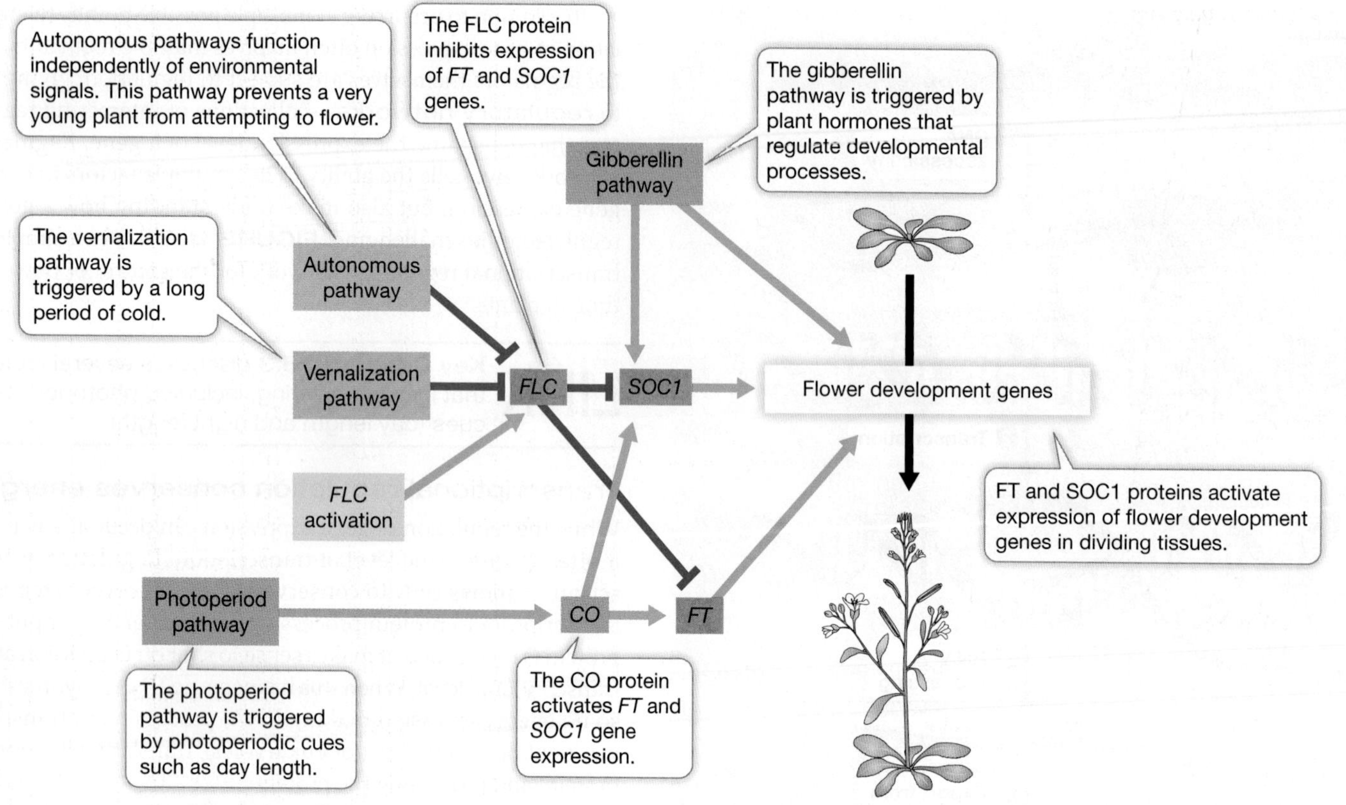

FIGURE 11.2 A Transcriptional Regulatory Network
The timing of flowering is controlled by multiple regulators that are produced in response to various internal and external signals. In the partial regulatory pathway shown here, which was worked out in *Arabidopsis thaliana*, some regulators reduce and others increase expression of target genes. Gene names are shown in italicized capital letters and numbers. (Based on I. Ausín et al. 2005. Environmental regulation of flowering. *Int J Dev Biol* 49: 689–705. doi: 10.1387/ijdb.052022ia)

Expression of the *FT* and *SOC1* genes stimulates flowering by activating flower development genes. When a plant undergoes vernalization (a period of cold), the vernalization pathway is activated and flowering is more likely to occur. Explain using the network, why this is the case.

repressor can correspond either to a compound that reduces gene expression or a transcription factor that reduces gene expression. In contrast, the term *inducer* refers to a compound that increases gene expression and the term *activator* refers to a transcription factor that increases gene expression. Transcription factor repressor and activator proteins bind to specific DNA sequences, which can be at or near the promoter or, in the case of eukaryotes, thousands of base pairs from the promoter. Binding can either increase or decrease the rate of transcription (**FIGURE 11.3**):

- In **negative regulation**, a transcription factor binds a specific site to reduce or prevent transcription. In this case, the transcription factor is a repressor.
- In **positive regulation**, a transcription factor binds a specific site to increase transcription. In this case, the transcription factor is an activator.

It is important to not confuse negative regulation with repressible genes and positive regulation with inducible genes. The method of regulation, positive or negative, does not indicate whether the gene is inducible or repressible. For example, a repressible gene may be turned off when a repressor binds to the DNA, in which case it is under negative regulation, or when an inducer fails to bind to the DNA, in which case it is under positive regulation. In addition, a gene can have both positive and negative regulation by different regulators. You will see examples of these mechanisms as we examine the regulation of prokaryotic, eukaryotic, and viral genes.

How do transcription factors recognize a specific nucleotide sequence in DNA? Recall that there are atoms in the bases of DNA that are available for hydrogen bonding but are not involved in base pairing (see Figure 9.7). These atoms are important in the interactions between a transcription factor (or any sequence-specific DNA binding protein) and the DNA. In addition to hydrogen bonding, there are other electrostatic and van der Waals interactions (see Key Concept 2.2) that can occur between DNA bases and amino acid side groups in the protein that stabilize binding. As for an enzyme and its substrate (see Key Concept 3.5), there is often an induced fit between the transcription factor and the DNA, such that the protein undergoes a conformational change after binding begins. Thus the spatial and chemical structures of the **DNA binding domain** of a transcription factor allow it to bind to specific DNA sequences but not to other sequences.

(A) Negative regulation

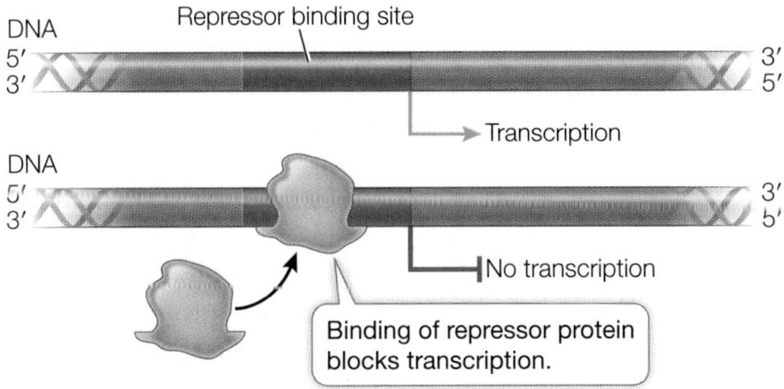

(B) Positive regulation

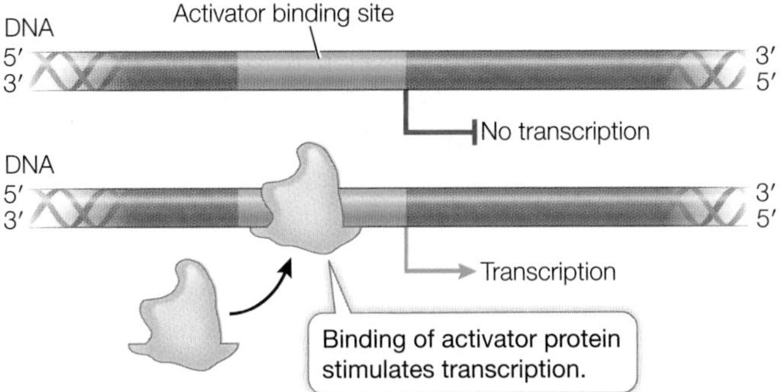

FIGURE 11.3 Positive and Negative Regulation of Transcription Transcription factors bind to DNA, causing **(A)** a decrease (negative regulation) or **(B)** an increase (positive regulation) in the rate of transcription of the gene.

REVIEW & APPLY | 11.1

R The regulation of gene expression allows cells to produce different phenotypes in response to internal and external factors. Genes may be constitutively expressed or may be inducible (capable of being activated by inducers) or repressible (capable of being repressed by repressors). While gene expression can be regulated at any point from before transcription to after translation, regulation at transcription is the most energy efficient. Transcriptional regulation involves binding of transcription factors that may increase (positive regulation by activators) or decrease (negative regulation by repressors) transcription rate.

A

1. For the following proteins, indicate whether the genes encoding them are likely to be inducible, repressible, or constitutive: (a) an enzyme in glycolysis, (b) DNA polymerase used during S phase of the cell cycle, and (c) an enzyme used in the biosynthesis pathway for a nonessential amino acid.
2. Why are some genes constitutive? Why doesn't a cell turn on these genes to produce the required amount of protein for the cell and then turn them off?
3. What is the difference between positive and negative regulation of transcription?

We'll now turn our attention to regulation of the genes encoding enzymes in metabolic pathways in prokaryotes. The regulation of such genes was the main focus of early work on gene expression.

11.2 Prokaryotic Gene Regulation Occurs Primarlly at the Level of Transcription

LEARNING OBJECTIVES

By the end of this key concept you should be able to:

11.2.1 **Define an operon.**

11.2.2 **Describe the conditions under which the *lac* operon will be transcribed at high levels.**

11.2.3 **Define an operator and describe how it interacts with a repressor to prevent or permit transcription.**

11.2.4 **Predict the effects on phenotype of mutations in various regions of the *lac* operon or in its regulatory genes.**

Prokaryotes often alter expression of multiple genes in response to their environment. They do this in a few ways.

Transcription regulation in prokaryotes can involve different sigma factors

As we noted in Chapter 10, RNA polymerase binds to specific DNA sequences at the promoter to initiate transcription. In prokaryotes, this binding is facilitated by transcription factors termed sigma factors. Different sigma factors may be present under different environmental conditions, allowing entire sets of genes to be regulated together. Genes that encode proteins with related functions may be at different locations in the genome but have the same promoter sequence. This allows them to be expressed at the same time and under the same physiological conditions.

In bacterial cells that are rapidly dividing under optimal conditions there is a relatively abundant sigma factor, $\sigma70$, which directs the RNA polymerase to bind to a promoter sequence that is shared by a large number of housekeeping genes involved in bacterial growth and cell division. The promoter recognition sequence for this sigma factor is TTGACA. If bacterial cells are exposed to an elevated temperature, they synthesize an additional sigma factor, $\sigma32$, which directs the RNA polymerase to several genes that share a common promoter sequence, CCTTGAA, and that encode proteins involved in responding to heat stress. Several other sigma factors are synthesized in response to other signals, such as starvation and nitrogen limitation. As we will see in Key Concept 11.3, this form of global gene regulation by proteins binding to RNA polymerase is also common in eukaryotes.

Among genes that are transcribed by an RNA polymerase bound to the same sigma factor, some are inducible and some

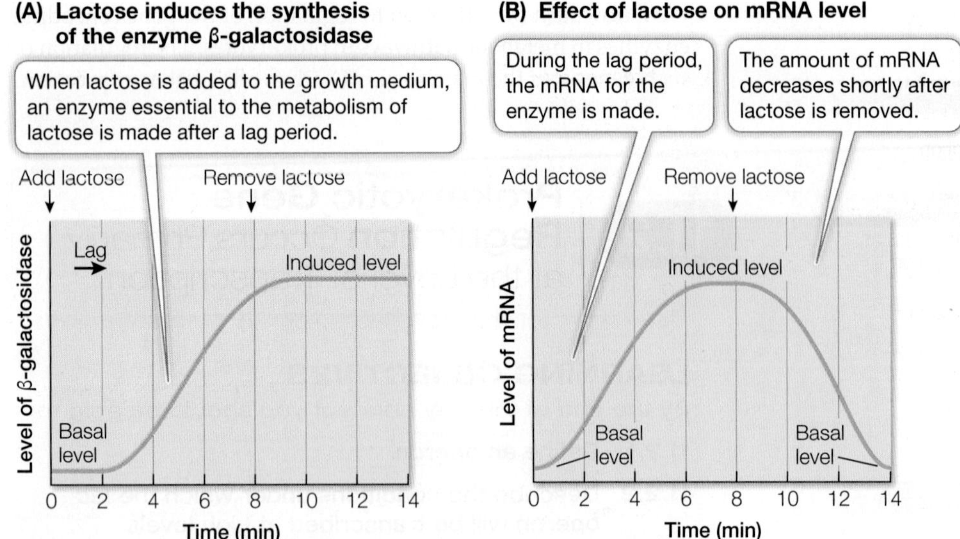

FIGURE 11.4 Lactose Induces Expression of β-Galactosidase Levels of the β-galactosidase protein **(A)** and mRNA **(B)** in *E. coli* cells before (basal), during, and after the addition of lactose to the medium. Note that protein and mRNA levels of permease and transacetylase show similar patterns of expression.

Why is the lag time for the protein longer than for the mRNA?

When *E. coli* is grown on a medium that contains glucose but no lactose, the basal (uninduced) levels of these three proteins are extremely low—only a few molecules per cell. But if the cells are transferred to a medium with lactose as the predominant sugar, they promptly begin making all three proteins after a short lag period, and within 10 minutes there are about 3,000 of each of these proteins per cell (the induced level) (**FIGURE 11.4A**). What causes this dramatic increase? A clue comes from measuring the concentration of mRNA for β-galactosidase. After lactose is added to the medium, the mRNA level increases *before* the level of β-galactosidase protein begins to rise (**FIGURE 11.4B**). The mRNA produced during the lag phase is then translated into protein. The high mRNA level depends on the presence of lactose, because if lactose is removed, the mRNA level goes down. Similar results are seen for permease and transacetylase. The response of the bacteria to lactose is clearly at the level of transcription.

repressible. To illustrate how regulation occurs for such genes, we will consider the regulation of proteins involved in lactose utilization and tryptophan synthesis in *E. coli*. We begin with the proteins involved in lactose utilization, which are encoded by inducible genes. Lactose is a β-galactoside—a disaccharide containing galactose linked to glucose. Three proteins are involved in the initial uptake and metabolism of lactose by *E. coli*:

1. Permease is a carrier protein in the bacterial cell membrane that moves lactose into the cell.
2. β-galactosidase is an enzyme that hydrolyzes lactose to glucose and galactose.
3. Transacetylase transfers acetyl groups from acetyl CoA to certain β-galactosides (but not lactose). Its role in the metabolism of lactose is not clear.

Operons are units of transcriptional regulation in prokaryotes

The genes that encode the three proteins for processing lactose in *E. coli* lie adjacent to one another on the *E. coli* chromosome. Only the first of the three genes contains a promoter. How, then, are the other two genes transcribed? This set of genes, like several others in prokaryotes but not in eukaryotes, shares a single promoter, and their DNA is transcribed into a single, continuous molecule of mRNA that contains the coding regions for the three proteins. Because this particular mRNA governs the synthesis of all three lactose-metabolizing enzymes, either all or none of these enzymes are made at any particular time.

A cluster of genes with a single promoter is called an **operon**, and the operon that encodes the three lactose-metabolizing proteins in *E. coli* is called the *lac* operon. The *lac* operon promoter

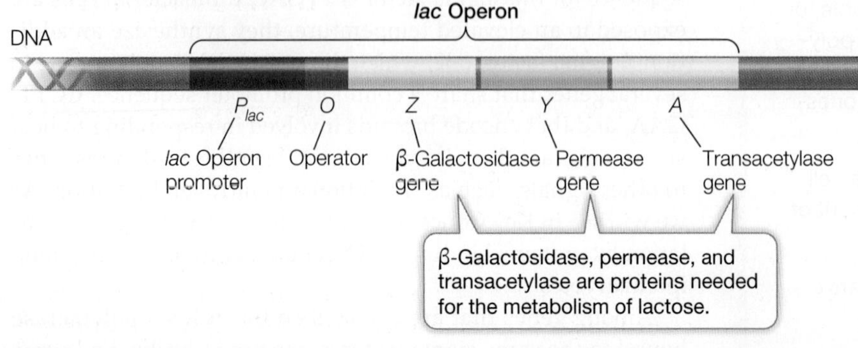

FIGURE 11.5 The *lac* Operon of *E. coli* The *lac* operon of *E. coli* has a promoter, operator, and three protein-coding genes. Note that in reality the promoter and operator are very small, just a few tens of bases, compared with the protein-coding genes, which are thousands of bases in length. The three genes controlled by the *lac* operon encode proteins that function in the transport of lactose across the cell membrane and its hydrolysis.

can be very efficient (the maximum rate of mRNA synthesis can be high), or its activity can be very low when the proteins are not needed. This example of transcriptional regulation, which we explore in more detail below, was worked out in the 1960s by Nobel Prize winners François Jacob and Jacques Monod.

Operator-repressor interactions regulate transcription in the *lac* and *trp* operons

The *lac* operon contains a short sequence of DNA sequence called an operator, which is near the promoter and controls transcription of the *lac* genes (**FIGURE 11.5**). An **operator** is a transcription factor binding site that when bound to a repressor reduces the rate of transcription. As such, regulation at operators is an example of negative regulation (see Figure 11.3). In an inducible operon, such as the *lac* operon, which is usually off, the repressor is usually bound to the operator. In contrast, in a repressible operon, such as the *trp* operon (described below), which is usually on, the repressor is usually not bound to the operator.

lac OPERON As we described above, the *lac* operon is transcribed at high levels only when the predominant sugar in the cell's environment is a β-galactoside such as lactose. When lactose first enters the cell, some of it is converted to a similar molecule, a structural isomer called allolactose. Allolactose is the inducer that switches on the expression of the *lac* operon. In the absence of allolactose, a repressor protein binds to the operator, preventing RNA polymerase from binding and thereby blocking transcription. When allolactose is present, the repressor detaches from the operator, allowing RNA polymerase to bind to the promoter and start transcribing the *lac* genes (**FIGURE 11.6**). The key to this regulatory system is the repressor protein, which is expressed from a constitutive gene and is thus always present in the cell in enough copies to keep the operator occupied. The repressor protein has two binding sites. One is a DNA-binding site that recognizes the DNA sequence in the operator and binds to it tightly. The other is an allosteric binding site for the inducer. When the inducer (allolactose) binds to the repressor, the repressor changes shape so that it can no longer bind DNA—an example of allosteric regulation (see Key Concept 3.5).

The gene for the *lac* repressor (gene *I* in Figure 11.6) is located just upstream of the *lac* operon on the *E. coli* chromosome. The *lacI* gene is referred to as a **regulatory gene** because it encodes a regulatory protein (a transcription factor). In contrast, a **structural gene** is any gene that encodes a protein that is not directly involved in gene regulation. The three genes that encode the lactose-metabolizing enzymes are structural genes.

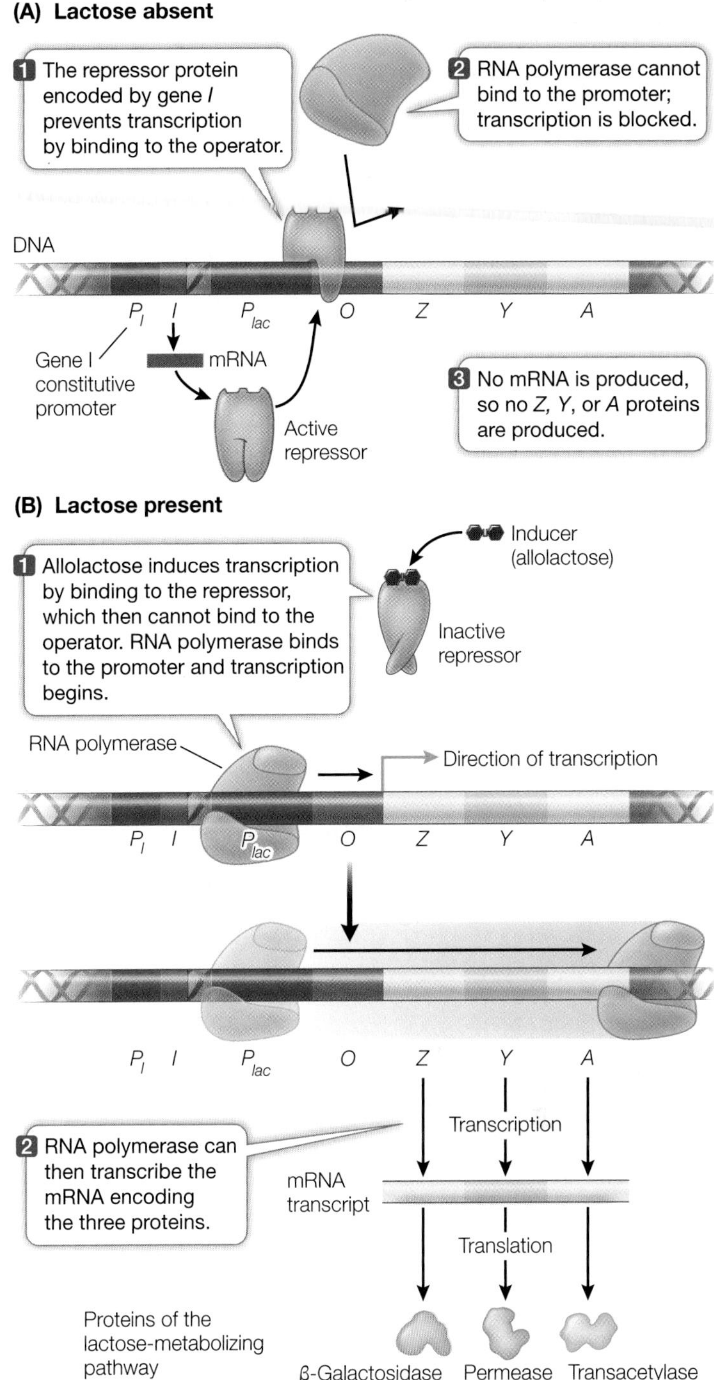

FIGURE 11.6 Induction of the *lac* Operon The repressor, which is encoded at the nearby *lacI* gene, controls transcription of the *lac* operon. **(A)** In the absence of allolactose, the repressor is bound to the operator. **(B)** When allolactose (the inducer) is present, it binds to the repressor protein and prevents the repressor from binding to the operator.

Animation 11.1 The *lac* Operon

trp OPERON The *trp* operon has a promoter and an operator and encodes five enzymes that catalyze the synthesis of tryptophan from a precursor molecule. The operon is repressible, meaning it is usually transcribed except when tryptophan levels are high. Like the inducible *lac* operon, the repressible *trp* operon is switched off when its repressor is bound to its operator. However, in this case the repressor binds to the DNA only in the presence of a **co-repressor**. The co-repressor is a molecule that binds to the repressor, causing it to change shape and allowing it to bind to the operator, thereby inhibiting transcription—another example of allosteric regulation. For the *trp* operon, the co-repressor is tryptophan. Being repressible makes sense for this operon. Tryptophan is always required for the synthesis of proteins, and so the default

THINK LIKE A SCIENTIST

Determining the regulation of the *lac* operon

In the late 1950s, François Jacob and Jacques Monod were interested in determining how the genes involved in lactose utilization were regulated in *E. coli*. *E. coli* can be grown on agar plates supplemented with nutrients and an energy source, such as glucose or lactose. Within a day, a single cell placed on a plate grows into a colony containing hundreds of millions of identical cells. β-galactosidase is expressed when lactose is being used, which occurs when glucose is absent and lactose is present (see Figure 11.4A).

Although Jacob and Monod used a different and more laborious method to identify the colonies' phenotypes, for illustrative purposes we show an experiment using a much easier method: β-galactosidase can hydrolyze a compound called X-Gal and produce a blue product. By adding X-Gal to the agar medium, it is thus straightforward to determine expression of β-galactosidase (**FIGURE 11.7**).

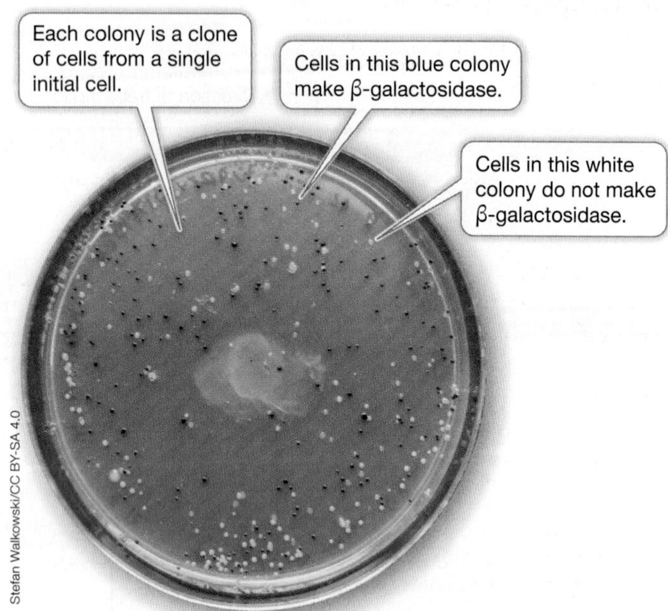

FIGURE 11.7 The Blue-White Screen Using X-Gal in the Medium

1. Under what conditions will bacterial colonies be blue in the presence of X-Gal?

Since lactose utilization genes respond to both glucose and lactose levels, we infer that there are two levels of regulation. In an attempt to find a regulatory mutant for the lactose genes, Jacob and Monod looked for mutant bacterial strains that would produce the enzymes involved in lactose utilization regardless of whether lactose was present or absent.

2. How might they have looked for these mutants?

Once they found mutant strains that always produced β-galactosidase regardless of lactose levels, they showed that the regulatory mutations were located near, but not in, the *lac* operon. Jacob and Monod reasoned that these mutants represented defects that knock out a repressor protein's function such that it could no longer bind to a region of the DNA (termed the operator) to prevent transcription.

To show that an operator exists, Jacob and Monod looked for regulatory mutants within the *lac* operon. In order to increase their chances of finding a mutant that was not in the repressor, they created an *E. coli* strain that had two copies (instead of one) of the repressor gene—all other genes were single copies.

3. How would having two copies of the repressor gene reduce the chance of finding a regulatory mutant in that gene, and thus increase their chance of finding an operator mutant?

This strategy allowed the scientists to find strains that were mutants for the operator and thus unable to bind the repressor, which supported their model of gene regulation. Jacob and Monod also discovered mutants that never produced functional β-galactosidase.

4. Some of these latter mutants were located in the *lac* operon. What kinds of mutations might they be?

5. One of the mutants that led to no β-galactosidase mapped to the repressor gene. How might that mutation have caused the observed phenotype (β-galactosidase always off)?

state should be to produce the enzymes to synthesize the amino acid. However, it is energy efficient for the cell to stop making the enzymes for tryptophan synthesis when tryptophan is present in adequate concentrations (**FIGURE 11.8**).

Both the *trp* and *lac* operons involve negative regulation—binding of the repressor protein turns off transcription. For the inducible *lac* operon, the substrate of a metabolic pathway (the inducer allolactose) interacts with a transcription factor (the repressor), rendering the repressor incapable of binding to the operator and thus allowing transcription. For the repressible *trp* operon, the final end-product of a metabolic pathway—in this case the co-repressor tryptophan—binds to the repressor protein, which is then able to bind to the operator and block transcription.

As we pointed out in Key Concept 11.1, positive regulation occurs when the regulatory protein is an activator that increases transcription when it binds. Positive gene regulation is common

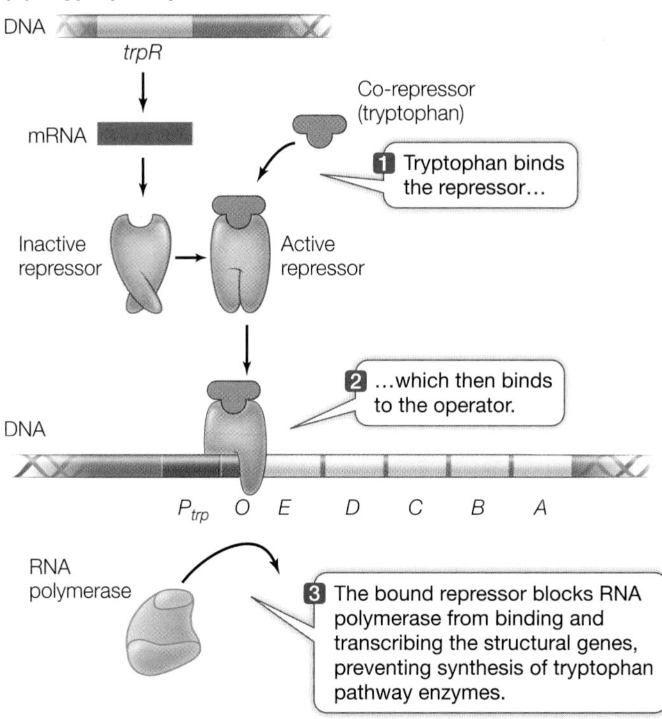

FIGURE 11.8 The *trp* Operon The tryptophan (*trp*) operon has a promoter, an operator, and five structural genes that encode the enzymes of the tryptophan biosynthetic pathway. A repressor protein, encoded at the *trpR* gene, produces an inactive repressor that becomes active when it binds to tryptophan and then is able to bind to the operator. The operon is thus transcribed when tryptophan levels are low **(A)** but not when tryptophan levels are high **(B)**.

Animation 11.2 The *trp* Operon

in eukaryotes and is frequently seen in bacteria. For example, when glucose is in a bacterial cell's environment, the *lac* operon is not transcribed at high levels, even when lactose, and hence the inducer allolactose, is present. Transcription of the *lac* operon is reduced because the transcription factor that binds the *lac* operon promoter can do so only in the absence of glucose. When glucose is absent, the transcription factor binds to the promoter, facilitating RNA polymerase binding and increasing transcription. In the presence of high glucose levels, the inactive transcription factor does not bind, and transcription remains at basal levels. Thus the *lac* operon is under both positive and negative regulation.

You have now seen two systems for regulating metabolic pathways, one that regulates gene expression and the other, described in Key Concept 3.5, in which an enzyme's activity can be altered by the binding to an allosteric site of a product of a pathway. **FIGURE 11.9** compares the two systems. Both involve allosteric changes in proteins. Of the two systems, the regulation of gene transcription is slower but results in greater savings of energy and resources.

Viruses use gene regulation strategies to hijack host cells

The immunologist Sir Peter Medawar once described a virus as "a piece of bad news wrapped in protein." As we described in Key Concept 9.1 for bacteriophage, once a virus's genetic material is in a host cell, it often turns that cell into a virus factory. This hostile takeover involves a radical change in the expression of numerous host cell genes, and often results in the death of the cell when new viral particles are released.

Viruses are not cells and do not carry out many of the processes characteristic of life. Instead, they are dependent on living host cells within which they reproduce. Unlike living cells, not all viruses use double-stranded DNA as their genetic material. A viral genome may consist of double- or single-stranded DNA, or double- or single-stranded RNA. Whether the genetic material is DNA or RNA, the viral genome often takes over the host's protein synthesis machinery within minutes of entering the cell. In what is called a **lytic cycle**, the host cell immediately begins to produce many new viral particles (virions), which are released as the cell breaks open, or lyses.

Some viruses that infect bacteria (that is, bacteriophages) reproduce by **lysogeny**, a process in which the viral genome becomes incorporated into the host cell genome and is replicated along with the host genome. The virus may survive in this way for many host cell generations, and the cell is unaffected and performs its normal functions. At some point, however, an environmental signal can cause the virus to enter the lytic cycle and produce viruses. Viruses that infect eukaryotic cells have life cycles that are essentially equivalent to the lytic or lysogenic life cycles of bacteriophages.

Viruses infect all cellular forms of life and are found everywhere on Earth. Several different types of viruses are described in **Key Concept 18.4**.

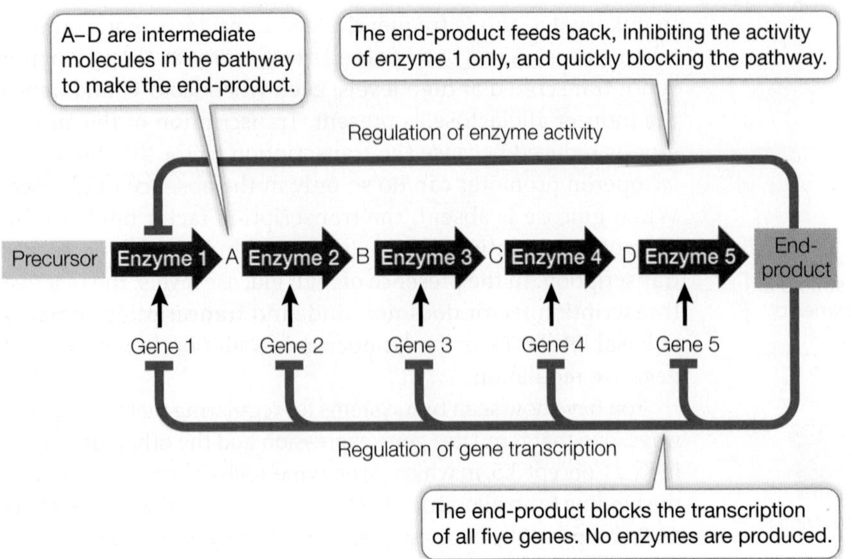

FIGURE 11.9 Systems to Regulate a Metabolic Pathway Feedback from the end-product of a metabolic pathway can block enzyme activity, or it can stop the transcription of genes that code for the enzymes in the pathway.

FIGURE 11.10 illustrates molecular events in the lytic life cycle of phage T4, a typical double-stranded DNA bacteriophage. At the molecular level, the lytic cycle has early, middle, and late stages at which particular phage genes are expressed. Upon infection, a set of early viral promoters are bound by the host RNA polymerase and then transcribed. These resulting RNAs encode viral proteins that have several effects. These include shutting down expression of host genes, stimulating viral genome replication, and activating the transcription of viral middle and late genes.

Viruses shut down host genes using three mechanisms:

1. A transcriptional mechanism involves the production of new sigma factors, which causes the RNA polymerase to no longer recognize bacterial promoters, instead recognizing phage middle and late promoters.
2. A posttranscriptional mechanism involves a virus-encoded RNA nuclease that degrades the host RNA before it can be translated.
3. Finally, a viral-encoded DNA nuclease (DNAse) digests the host's chromosome. Nucleoside monophosphates (e.g., GMP and TMP) produced from host RNA and DNA degradation are phosphorylated to give nucleoside triphosphates (e.g., GTP and TTP), which are used during transcription of viral genes and replication of the viral DNA genome.

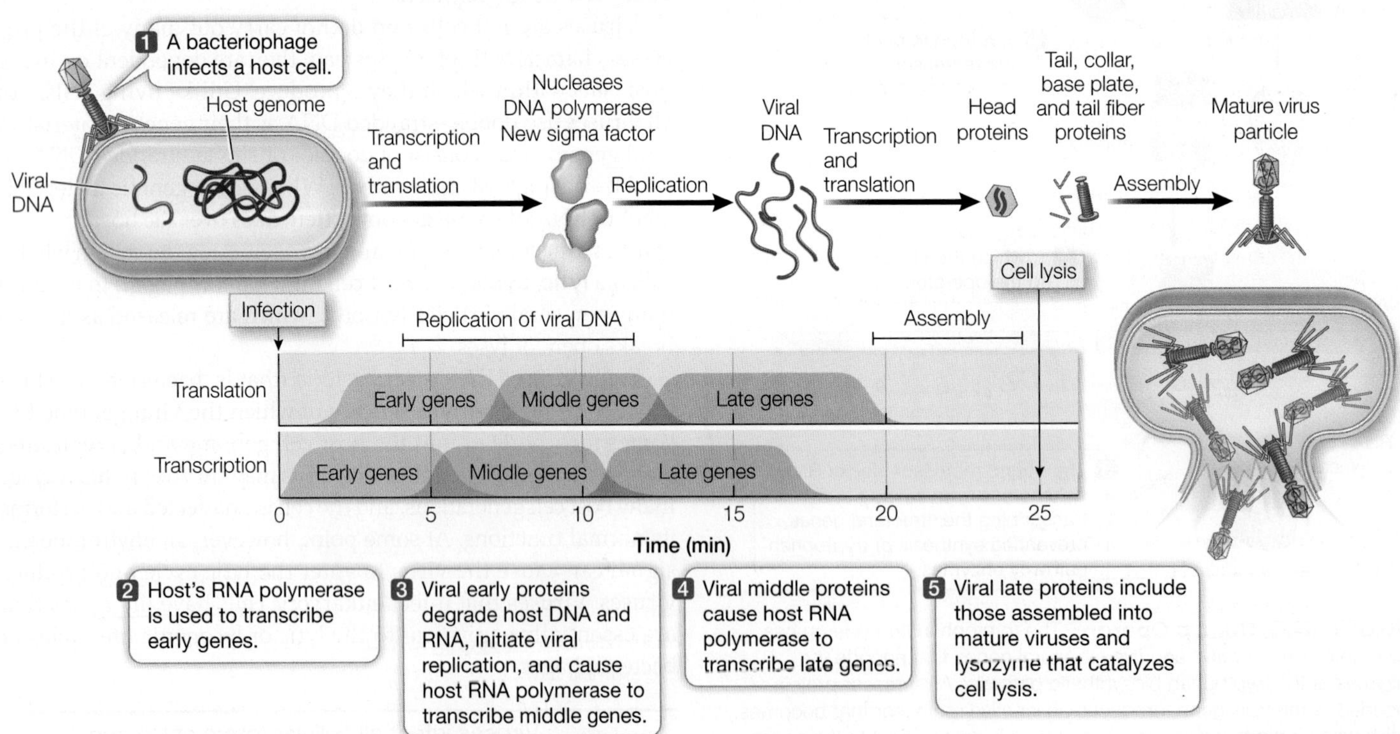

FIGURE 11.10 A Gene Regulation Strategy for Viral Reproduction In a host bacterial cell infected with a virus, the viral genome uses its early genes, transcribed by the host RNA polymerase, to shut down host transcription while it replicates itself. Once the viral genome is replicated, its middle genes are expressed, which in turn cause the activation of late genes to produce proteins that package the new viral genomes, and other proteins that lyse the host cell.

In the lytic life cycle, these processes begin within a few minutes after the virus first infects the cell. As the infection proceeds, viral late genes are transcribed, including ones that encode the proteins that package the new viral genomes and enzymes that lyse the host cell to release the new virions. Under ideal conditions, this entire process—from binding and infection to release of new phage—can be completed in only half an hour.

REVIEW & APPLY | 11.2

R Genes or operons that share the same promoter sequence can be co-regulated by synthesizing the appropriate sigma factor. For genes or operons using the same sigma factors, binding of positive and/or negative regulators allows control of levels of transcription. The inducible *lac* operon is under both positive regulation (for glucose control) and negative regulation (for lactose control). The repressible *trp* operon is under negative regulation. Having multiple genes in the same operon implies that regulation of their transcription will be identical, since they are all on the same mRNA. Bacteriophages shut down host genes and express early, middle, and late genes in sequence during the lytic cycle.

A

1. The bacterium *Klebsiella aerogenes* possesses a histidine utilization (*hut*) operon that encodes enzymes capable of using the amino acid histidine as an energy source. It also possesses a histidine biosynthetic (*his*) operon encoding the enzymes to synthesize histidine from precursors. Under what environmental conditions do you expect each these two operons to be transcribed?
2. What would be the effect on gene expression of a mutation in a sigma factor so that it could no longer bind to the RNA polymerase?
3. When a bacteriophage infects a bacterial host cell, why doesn't it degrade the host's RNA polymerase in order to shut off host transcription?
4. What would be the effects of the following mutations on regulation of the *lac* operon by lactose in the absence of glucose?
   a. Repressor mutation such that it no longer binds the operator
   b. Repressor mutation such that it no longer binds allolactose.
   c. Mutation in the lac operon promoter such that it no longer binds RNA polymerase.
   d. Mutation in the operator such that it no longer binds the repressor.

Studies of bacteria and bacteriophage provide a basic understanding of the mechanisms that regulate gene expression and of the roles of regulatory proteins in both positive and negative regulation. We will now turn to the control of gene expression in eukaryotes.

11.3 | Eukaryotic Transcription Is Regulated by General and Specific Transcription Factors

LEARNING OBJECTIVES

By the end of this key concept you should be able to:

- **11.3.1** **List the major differences between the control of transcription initiation in prokaryotes and eukaryotes.**
- **11.3.2** **Describe the basal transcription apparatus.**
- **11.3.3** **Identify similarities and differences between enhancers and silencers.**
- **11.3.4** **Explain how transcription factor expression differences allow cells to differ in their phenotype.**

As we mentioned in Key Concept 11.1, gene expression can be regulated at several different points in the process of transcribing a gene and translating the mRNA into a protein (see Figure 11.1). In this key concept we describe the mechanisms that result in the selective transcription of specific eukaryotic genes. Like prokaryotic cells, eukaryotic cells must precisely regulate the expression of their genes. Many eukaryotic genes are inducible and expressed only when needed. Inducibility is especially important in multicellular organisms with specialized cells and tissues—many genes are permanently off in some tissues. For example, the protein keratin is abundant in epithelial cells that give rise to our hair and nails but is absent in most other tissues, and hemoglobin is expressed only in developing red blood cells. In contrast, some genes are expressed in all human cells, such as those that encode enzymes needed for basic metabolic activities (such as glycolysis) and certain structural proteins such as actin (a component of the cytoskeleton).

The mechanisms for regulating transcription in eukaryotes are similar conceptually to those in prokaryotes. In both types of cells, DNA–protein interactions bring about negative and positive control of gene expression. However, there are significant differences in the regulation of transcription (**TABLE 11.1**).

Transcription factors act at eukaryotic promoters

Like bacterial promoters, a eukaryotic promoter is a region of DNA near the 5′ end of a gene where an RNA polymerase binds and initiates transcription. Eukaryotic promoters are extremely diverse in their sequences and difficult to characterize, but they each contain a core promoter sequence to which the RNA polymerase binds. The core promoters of many protein-coding genes contain a DNA sequence known as the **TATA box**—so called because it is rich in A-T base pairs.

RNA polymerase II is the polymerase that transcribes the protein-coding genes in eukaryotes. It cannot bind to the promoter and initiate transcription by itself. Rather, it does so only after various **general transcription factors**—regulatory proteins that affect transcription—have bound to the core promoter. General transcription factors bind to most promoters and are distinct from

TABLE **11.1** | Transcription in Bacteria and Eukaryotes

Characteristic	Prokaryotes	Eukaryotes
Locations of functionally related genes	Often clustered in operons	Often distant from one another with separate promoters
RNA polymerases	One	Three: RNA polymerase I transcribes rRNA RNA polymerase II transcribes mRNA RNA polymerase III transcribes tRNA and small RNAs
Promoters and other regulatory sequences	Few, simple	Many, diverse
Initiation of transcription	Binding of RNA polymerase with associated sigma factor	Binding of basal transcription apparatus, mediator, and specific transcription factors

transcription factors that have specific regulatory effects only at certain promoters or classes of promoters. The general transcription factors can be considered as acting similarly to the sigma factors in prokaryotic transcription initiation.

FIGURE 11.11 illustrates the assembly of the resulting transcription complex at a promoter containing a TATA box. First, the protein TFIID ("TF" stands for transcription factor) binds to the TATA box. Binding of TFIID changes both its own shape and that of the DNA, presenting a new surface that attracts the binding of other transcription factors. As you can see in the figure, RNA polymerase II binds only after several other proteins have bound to the complex. Then additional general transcription factors bind, setting the stage for transcription to occur. Together the RNA polymerase and general transcription factors make up the **basal transcription apparatus**.

If only the basal transcription apparatus binds to the core promoter sequence, the rate of transcription is quite low (basal levels). Transcription rates are modified by binding of specific transcription factors to other sequences that are found adjacent to the core promoter, in a region termed the regulatory promoter, or in nearby regions. Specific transcription factors interact with a protein complex called **mediator**, which facilitates the interaction between specific transcription factors and the basal transcription apparatus. Several different transcription factors can interact with mediator at the same time. The DNA binding sites for transcription factors that are not in the core or regulatory promoter are termed **enhancers** if binding increases transcription, and **silencers** if binding decreases transcription. While enhancers and silencers can be very distant from the promoter (and basal transcription apparatus)—sometimes tens of thousands of base pairs away—bending of the DNA helix allows the proteins that bind to them to come into close proximity for regulation (**FIGURE 11.12**).

Many DNA binding proteins are involved in the initiation of transcription, and the *combination* of factors present determines whether transcription is initiated. With more than 1,500 different transcription factors in humans, an enormous number of possible combinations of transcription factors could be used to

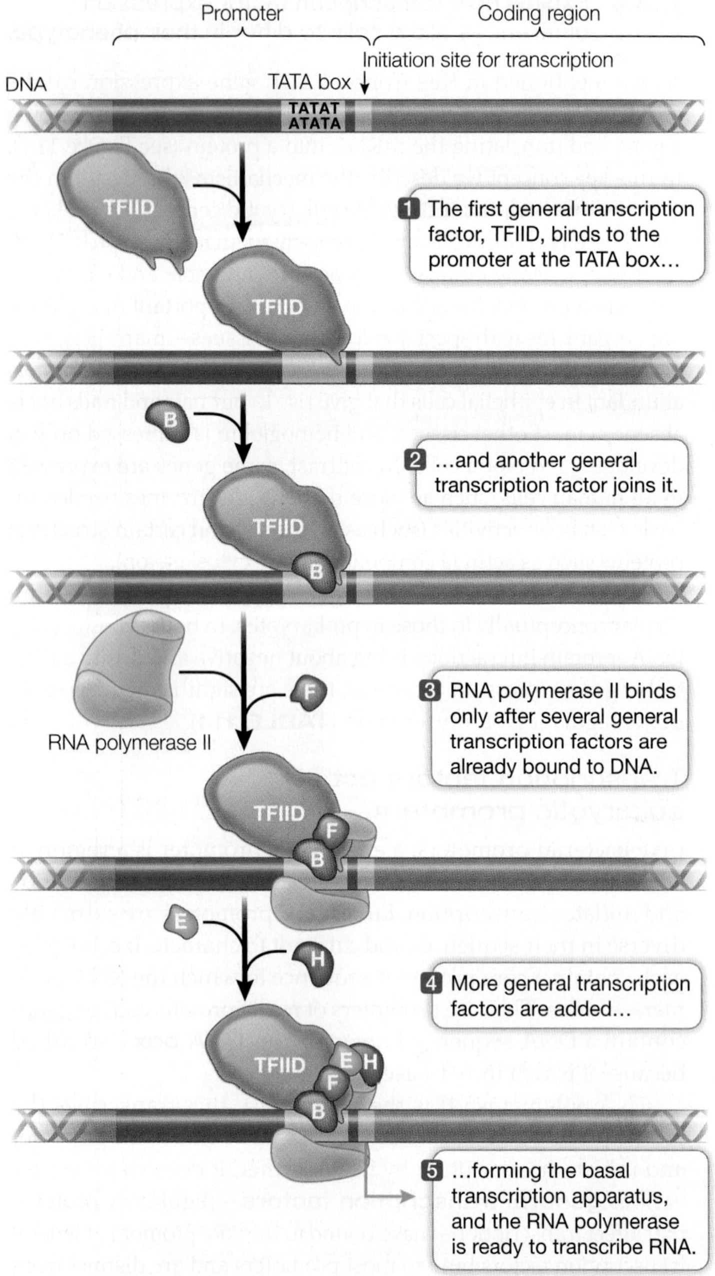

FIGURE 11.11 Early Steps in Eukaryotic Transcription Initiation Initiation of transcription requires the assembly of the basal transcription apparatus. Apart from TFIID, which binds to the TATA box, each transcription factor in this transcription complex has binding sites only for the other proteins in the complex and does not bind directly to DNA. B, E, F, and H are general transcription factors.

What would be the effect of knocking out a general transcription factor in a cell?

Animation 11.3 Initiation of Transcription

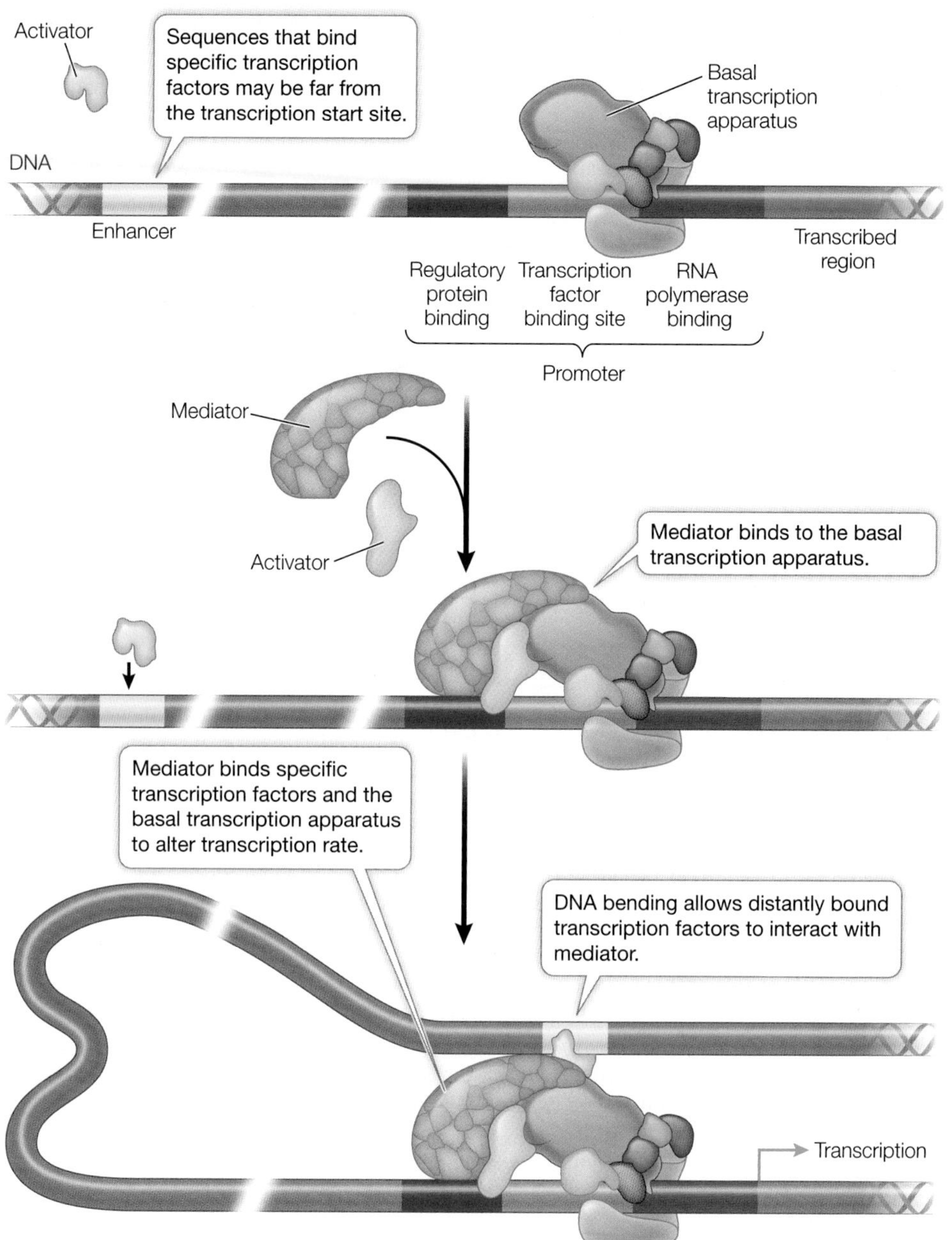

FIGURE 11.12 Later Steps in Eukaryotic Transcription Initiation Once the basal transcription apparatus is assembled, other transcription factors bind to the regulatory promoter and to enhancers and silencers and interact with mediator to alter the rate of transcription. This example shows binding of two activator transcription factors, one at the regulatory promoter and one at an enhancer element.

regulate a specific gene. So a major way in which cells can differ in phenotype is due to the combination of transcription factors that they express, which leads to induction of some genes and repression of others. Unlike prokaryotic and viral genes, eukaryotic genes are not organized into operons. Instead the expression of genes is coordinated via shared sequences in their promoters and regulatory elements, and the levels of the various transcription factors that bind those sequences. Coordinated gene expression allows cells to turn on, or off, one or a few transcription factors to achieve a response that may involve tens or hundreds of genes.

The expression of transcription factors underlies cell differentiation

During the development of a complex organism from fertilized egg to adult, cells become more and more differentiated (specialized). Differentiation is brought about in many cases by changes in gene expression, resulting from the activation (and inactivation) of transcription factors. Since virtually all differentiated cells contain the entire genome, their specific characteristics must arise from differential gene expression.

Currently there is great interest in cellular therapy: providing new, functional cells to patients who have diseases that involve the degeneration of certain cell types. An example is Alzheimer's disease, which involves the degeneration of neurons in the brain. Because of the possibility of immune system rejection (see Chapter 36), it would be optimal if patients could receive their own cells, modified in some way to be functional. Since specialized functions are under the control of transcription factors, turning readily available cells into a particular desired cell type might be achieved by altering transcription factor expression. Marius Wernig and his colleagues at Stanford University have made important progress toward this goal (**INVESTIGATION FIGURE 11.13**). They took skin fibroblasts from mice and manipulated the expression of transcription factors in the cells to change them into neurons. By repeating their experiments on human fibroblasts, they have brought cellular therapy closer to reality. Since their work, the number of studies demonstrating and refining methods for transforming one cell type into another has grown enormously.

Eukaryotic viruses can have complex life cycles

Eukaryotes are susceptible to infections by various kinds of viruses that have a variety of life cycle strategies. The viruses may have genomes of single- or double-stranded DNA or RNA. Sometimes these viral life cycles are similar to what we saw for the bacteriophage lytic cycle (see Figure 11.10). For example, the poliovirus completes its life cycle (from infection to release of new particles) in 4–6 hours, and each dying host cell can release up to 10,000 new particles. Compare this with the 24-hour cell cycle typical of dividing human cells. In other cases, viruses enter the genome of the eukaryotic host and remain in a quiescent state for long periods of time, similar to the lysogenic cycle of some bacteriophages. However, some eukaryotic viral life cycles can be quite complex.

INVESTIGATION

FIGURE 11.13 Expression of Specific Transcription Factors Turns Fibroblasts into Neurons Fibroblasts are cells that secrete abundant extracellular matrix and contribute to the structural integrity of organs. Neurons are highly specialized cells in the nervous system. Marius Wernig and his colleagues performed a series of experiments to find out whether expressing neuronal transcription factors in fibroblasts would be sufficient to cause the fibroblasts to become neurons.[a]

HYPOTHESIS

Expression of neuron-specific transcription factors in fibroblasts will turn fibroblasts into neurons.

METHOD

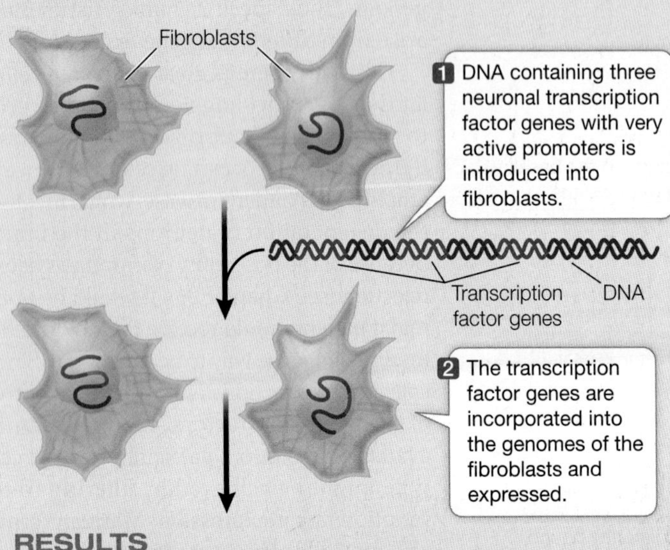

RESULTS

After 6 days, the fibroblasts develop into functional neurons, which form characteristic synapses with one another.

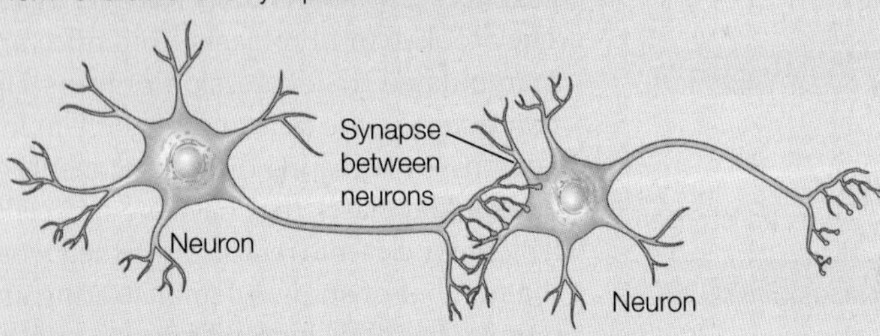

CONCLUSION

The expression of just three transcription factors is sufficient to transform a fibroblast into a neuron.

ANALYZE THE DATA

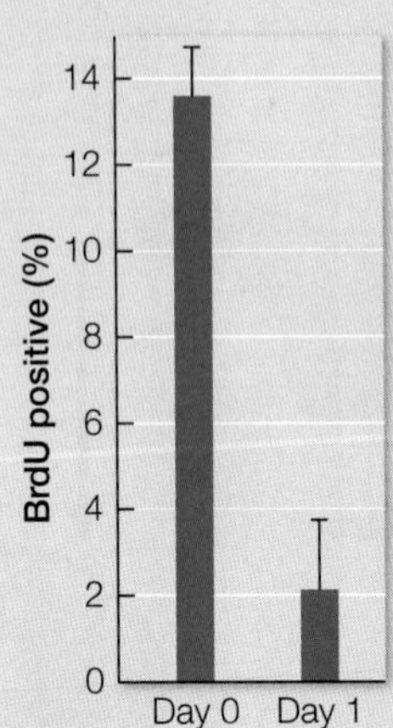

Fibroblasts are active in cell division; neurons are not. In addition to morphology, the lack of cell division was used to measure whether fibroblast cells expressing the three transcription factors were more similar to neurons. The rate of cell division can be estimated by growing cells on the labeled nucleotide bromodeoxyuridine (BrdU) and measuring its incorporation into DNA. The scientists compared BrdU incorporation from days 0 to 13 against incorporation from days 1 to 13 by adding BrdU at day 0 or 1. They reasoned that incorporation would be substantially lower starting at day 1 if expression of the three transcription factors reduced cell division. The percentage of labeled (hence dividing) cells is shown in the graph for cells with BrdU from days 0 to 13 ("Day 0") and from days 1 to 13 ("Day 1"). Error bars indicate ± one standard deviation.

1. Was cell division affected in the cells expressing the three transcription factors? Explain your answer, making sure you state what would have been expected if cell division was unaffected.
2. What statistical test would you use to show whether the difference between the two cell populations was significant? See Appendix B for a statistics primer.

[a]T. Vierbuchen et al. 2010. *Nature* 463: 1035–1041.

A similar **Analyze the Data** exercise may be assigned in **Achieve**.

As an example, we focus here on **human immunodeficiency virus** (**HIV**), the infective agent that causes acquired immunodeficiency syndrome (AIDS) in humans. HIV typically infects only cells of the immune system that express a surface receptor called CD4. The virion is enclosed within a phospholipid membrane derived from its previous host cell. Proteins in the membrane, encoded by the viral genome, are involved in the infection of new host cells, which HIV enters by direct fusion of the viral envelope with the host's cell membrane. The interaction between the proteins in the viral membrane and the receptors on the host cell that must occur for successful infection should remind you of the process of receptor-mediated endocytosis (see Figure 4.10). Thus HIV infection is limited to those cells in the body that have receptors to which HIV can bind.

HIV is a retrovirus: its genome is single-stranded RNA, and it carries within the virion an enzyme called reverse transcriptase. Shortly after infection, the **reverse transcriptase** makes a DNA strand that is complementary to the RNA, while at the same time degrading the RNA and making a second DNA strand that is complementary to the first (**FIGURE 11.14**). The resulting

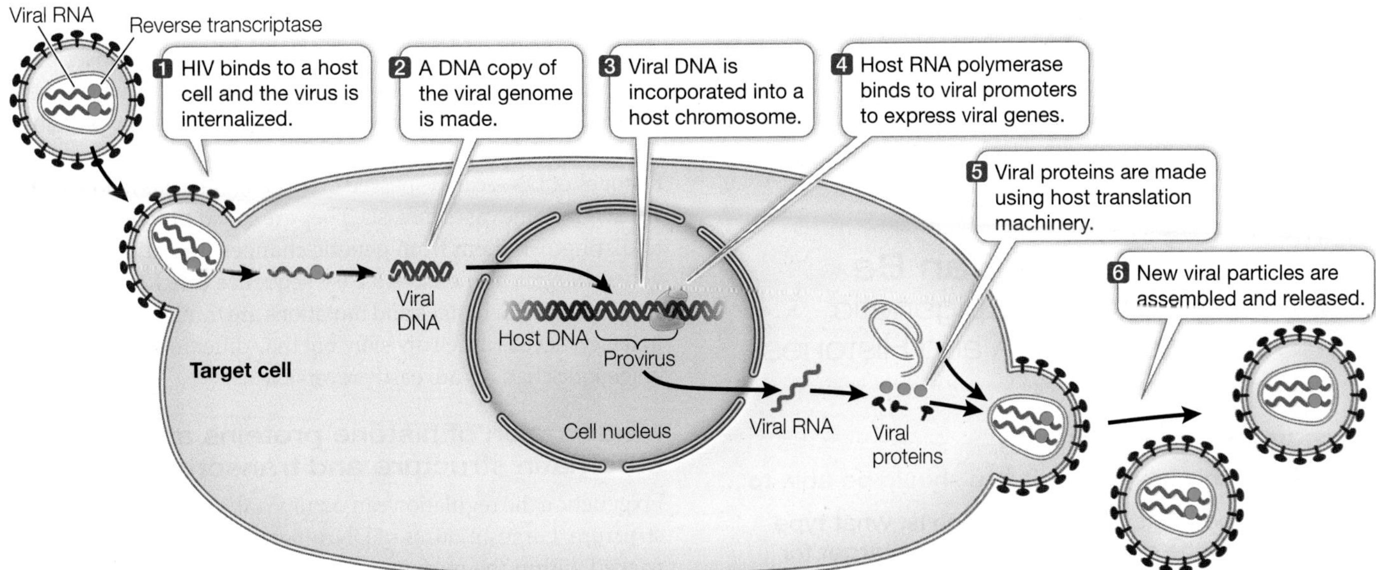

FIGURE 11.14 The Reproductive Cycle of HIV This retrovirus enters a host cell via fusion of its envelope with the host's cell membrane. Reverse transcription of retroviral RNA then produces a DNA provirus—a molecule of complementary DNA that becomes inserted into the host's genome. The inserted viral DNA directs the synthesis of new virus particles.

double-stranded DNA becomes integrated into the host's chromosome. The integrated viral DNA is called a provirus.

The provirus resides permanently in the host chromosome and can remain in an inactive state for years. During this time transcription of the viral DNA is initiated, but host cell proteins called termination factors prevent the RNA from elongating, and transcription is terminated prematurely. Under some circumstances, the rate of transcription initiation increases and some viral RNA is made. One of the viral genes encodes a protein called Tat (*Trans-activator of transcription*), which binds to the 5′ end of the viral RNA, preventing premature termination. As a result of Tat binding, the production of full-length viral RNA is dramatically increased, and the rest of the viral reproductive cycle is able to proceed. It was only after the discovery of this mechanism in HIV and similar viruses that researchers found that many eukaryotic genes are regulated at the level of transcription elongation.

REVIEW & APPLY | 11.3

R The initiation of transcription in eukaryotes involves the binding of general transcription factors to the core promoter and recruitment of the RNA polymerase. Once the basal transcription apparatus has formed, specific transcription factors bound to the regulatory promoter and to enhancers and silencers interact with the basal transcription apparatus through mediator to alter transcription rates. The large number of transcription factors and their associated binding sites give enormous flexibility to the control of gene expression. Different cells have different phenotypes because they express different sets of specific transcription factors. Genes within a cell are co-regulated by sharing common transcription factor binding sites. Viruses that infect eukaryotic cells take advantage of host cell proteins to gain entry into host cells and replicate within them. For example, HIV uses receptor proteins on the surface of host cells to fuse with them and, once it inserts into the host genome, host RNA polymerase II is responsible for transcribing its genes.

A

1. You engineer a gene to study where a newly discovered protein is expressed in mice. Your engineered gene has 500 base pairs of DNA from the upstream region of the protein-coding gene (and thus includes the promoter) attached to the coding region for green fluorescent protein (GFP). GFP is a jellyfish protein that is bright green when visualized using a fluorescent microscope (see Appendix D). You insert this gene into a mouse strain (which is easier said than done) and then examine various tissues for the presence of GFP using fluorescent microscopy. To your surprise, you find that all tissues show a very low level of fluorescence, but no tissues show high levels of expression. Explain these results.
2. A researcher knocks out a transcription factor in a mammalian cell line by deleting both copies of the gene. She then measures expression and finds no effect for most protein-coding genes. However, for a few genes there is a reduction, but not elimination, of gene expression. Explain her results, being sure to state what kind of transcription factor she has knocked out.
3. Tenofovir is used to treat patients with HIV. The drug acts by inhibiting reverse transcriptase. A patient is treated with tenofovir. What would be the effect in terms of progression of infection of the inhibition of reverse transcriptase on cells that were already infected with HIV versus newly infected cells?

We have discussed mechanisms used by cells to control gene transcription. These mechanisms involve the interaction of regulatory proteins with specific DNA sequences. As we'll see next, however, eukaryotes have other mechanisms for controlling gene expression that do not depend on specific DNA sequences.

11.4 | Transcription Can Be Regulated by Epigenetic Changes to DNA and Histones

LEARNING OBJECTIVES

By the end of this key concept you should be able to:

11.4.1 Describe what DNA methylation is, what type of site in the genome is a common target for methylation, and what the effects of methylation are on transcription.

11.4.2 Give an example of histone modification and how it can increase or decrease transcription.

11.4.3 Explain the main advantage of epigenetic regulation of gene expression.

11.4.4 Explain what is meant by "genomic imprinting."

11.4.5 Describe why epigenetic changes can be inherited in plants but not in mammals.

So far we have focused on regulatory events that involve binding of transcription factors to specific DNA sequences at or near a gene's promoter. However, in order for binding to occur, the DNA must be accessible. In eukaryotes, DNA is packaged with various proteins, especially histones, to form chromatin (see Key Concept 7.2). If the packaging is tight, it is difficult for transcription factors to access the DNA and assemble the basal transcription apparatus. In such cases, the chromatin is said to be closed. Transcription requires open chromatin, which is less tightly packaged. Thus packaging of DNA gives eukaryotic cells the ability to regulate transcription at the level of accessibility. Chromatin packaging is heavily influenced by two alterations: (1) chemical modification of histone proteins and (2) methylation of bases in the DNA molecule. These alterations of histones or DNA can be passed on to daughter cells following mitosis, and in some cases meiosis. They are called **epigenetic** changes to distinguish them from genetic changes (mutations), which involve changes to the DNA's base sequence (see Key Concept 9.3). Both epigenetic changes and mutations are transmitted from parent to daughter cells in cell division, but they differ in an important way: epigenetic changes are easily reversible.

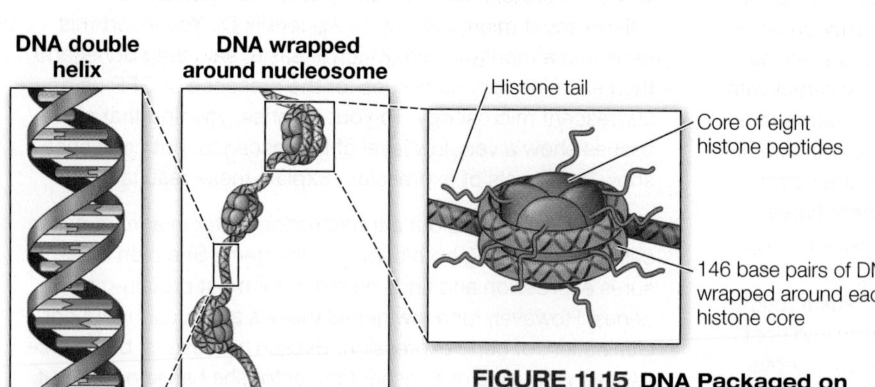

FIGURE 11.15 DNA Packaged on Nucleosomes Each nucleosome consists of a core of eight histone proteins with 146 base pairs of DNA helix wrapping around the core. Each nucleosome is separated from the next by 10–80 base pairs of DNA.

Modification of histone proteins affects chromatin structure and transcription

Epigenetic gene regulation can occur via the alteration of chromatin structure. Large amounts of DNA (nearly 2 meters in humans!) are packed within the nucleus (which has a diameter of about 5 μm). The basic unit of DNA packaging in eukaryotes is the **nucleosome**, a core of eight positively charged histone proteins around which DNA is wound (**FIGURE 11.15**). Approximately 146 base pairs of DNA are wrapped around each nucleosome, and each nucleosome is separated from the next by 10–80 base pairs of DNA.

Each histone protein has a "tail" of approximately 20 amino acids at its N terminus that sticks out of the compact structure and contains certain positively charged amino acids, notably lysine. Ordinarily there is strong ionic attraction between the positively charged histone tail proteins and DNA, which is negatively charged because of its phosphate groups. Because of this attraction, DNA wound on nucleosomes forms inaccessible closed chromatin, which is not transcribed. However, enzymes called **histone acetyltransferases** can add acetyl groups to these positively charged amino acids (acetylation), which neutralizes their charges (**FIGURE 11.16A**). This reduces the affinity of the histone tails for the DNA, loosening the compact nucleosome (**FIGURE 11.16B**), and facilitating binding of regulatory proteins. The majority of histone acetylation is found near gene promoters, but acetylated histones are also found throughout the transcribed regions of genes. Thus histone acetylation promotes both transcription initiation and elongation. Histone acetyltransferases can be recruited to promoters by transcription factors.

Another class of histone modifying enzymes, **histone deacetylases**, can remove the acetyl groups from histones and thereby repress transcription. Histones can also be modified in several other ways, including methylation (the addition of a methyl group) and phosphorylation (the addition of a phosphate group). Histone methylation can contribute to either the activation or repression of gene expression, depending on which lysine residue is methylated. Histone phosphorylation is involved in chromosome condensation during mitosis and meiosis, as

(A) Mechanism of acetylation

$$\cdots N{-}\underset{\underset{NH_3}{|}}{\underset{(CH_2)_4}{|}}{\overset{H}{\underset{}{C}}}{-}\overset{O}{\overset{\|}{C}}\cdots + CoA{-}S{-}\overset{O}{\overset{\|}{C}}{-}CH_3 \xrightarrow{HAT} \cdots N{-}C{-}\overset{O}{\overset{\|}{C}}\cdots + CoA{-}SH + H^+$$

Lysine in histone; Acetyl CoA; $(CH_2)_4$, $HN{-}\overset{}{C}({=}O){-}CH_3$: Acetyl-lysine; CoA

(B) Acetylation of histone tails

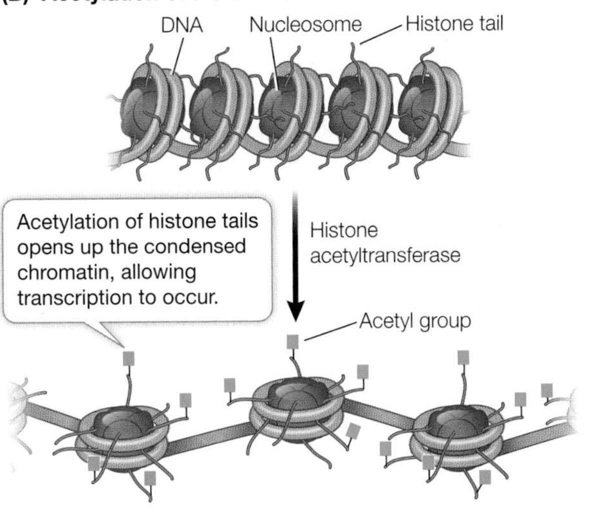

FIGURE 11.16 Epigenetic Modification of Histones Opens Chromatin for Transcription **(A)** Histone acetyl transferase (HAT) adds an acetyl group to lysine residues. **(B)** Histone acetylation neutralizes the charge of positively charged lysines on histone proteins, weakening their binding to DNA, and making the DNA more accessible to bind regulatory proteins.

What would be the effect of knocking out the histone acetyltransferase enzyme on expression of a gene that usually has acetylated histones in the promoter region?

well as affecting gene regulation. All of these modifications, and their resulting effects, are reversible, and so the transcriptional activity of a eukaryotic gene may be determined by the pattern of histone modification.

DNA methylation affects transcription

Many species methylate their DNA. Often it is cytosine bases that are methylated. Depending on the organism, from 0 to 5 percent of cytosines in the DNA are covalently modified by the addition of a methyl group (—CH_3), to form 5-methylcytosine (**FIGURE 11.17A**), in a reaction catalyzed by **DNA methyltransferase**. In mammals, methylation usually occurs on cytosines (C) that are adjacent to guanines (G). These are written as "CpG" sites, where "p" is for the phosphate in the DNA backbone, to emphasize that the two nucleotides are next to one another on the same strand. Each CpG dinucleotide is paired with a CpG dinucleotide on the opposite strand (reading 5′-to-3′ in both cases). A CpG site is said to be methylated when the C on both strands is methylated. When only one of the two Cs is methylated, a CpG site is said to be hemimethylated. In organisms that have DNA methylation, CpG sites are often methylated except when they are found in or near transcriptionally active promoters. Promoters usually contain regions of DNA that are rich in CpG sites, and such regions are called **CpG islands**. Thus unmethylated CpG islands are characteristic of a gene that is currently being expressed.

Methylated DNA binds specific proteins that are involved in the repression of transcription; thus heavily methylated genes tend to be inactive (silenced). Some of the proteins that bind to methylated DNA cause an increase in the compaction of chromatin. Whereas histone acetylation and deacetylation (see above) are dynamic processes resulting in short-term changes in gene expression, DNA methylation is usually a stable, long-term silencing mechanism. When DNA is replicated, a **maintenance methyltransferase** catalyzes the methylation of the Cs in the new DNA strands. This methyltransferase is able to identify the hemimethylated CpG sites that occur after a fully methylated CpG site is replicated, and methylate the nonmethylated C on the new strand (**FIGURE 11.17B**). Thus once a pattern of methylation is present in a cell's genome, it can be passed through cell division to daughter cells. However, methylation is reversible: a third enzyme, appropriately called **demethylase**, catalyzes the removal of the methyl group from 5-methylcytosine. Thus the pattern of cytosine modification can be altered. The enzymes involved in histone modification and in DNA methylation and demethylation interact to ensure that genes whose products are needed in the cell are kept unmethylated, and their associated histones acetylated.

Sometimes, large stretches of DNA or almost entire chromosomes are methylated. Under a microscope, two kinds of chromatin can be distinguished in the stained interphase nucleus: euchromatin and heterochromatin. The **euchromatin** appears diffuse and stains lightly; it contains the DNA that is transcribed into mRNA and has relatively lower DNA methylation levels. **Heterochromatin**, in contrast, is condensed and stains darkly; any genes it contains are generally not transcribed and are more heavily methylated.

A dramatic example of heterochromatin is the X chromosome in female mammals. A normal female mammal has two X chromosomes, whereas a normal male has an X and a Y (see Key Concept 8.2). Because each female cell has two copies of each X chromosome gene, the female has the potential to produce twice as much of each protein encoded on the X chromosome as the male. Nevertheless, for more than three-fourths of the genes on the X chromosome, the total amount of mRNA produced is essentially the same in females and in males. How does this happen?

In the early female embryo, one copy of X becomes heterochromatic and transcriptionally inactive in each cell, and the same X remains inactive in all of that cell's descendants. In a given female embryo cell, the "choice" of which X to inactivate is random. Recall that one X in a female comes from her father and one from her mother. Thus in one embryonic cell the paternal X might be inactivated, but in a neighboring cell the maternal X might be inactivated.

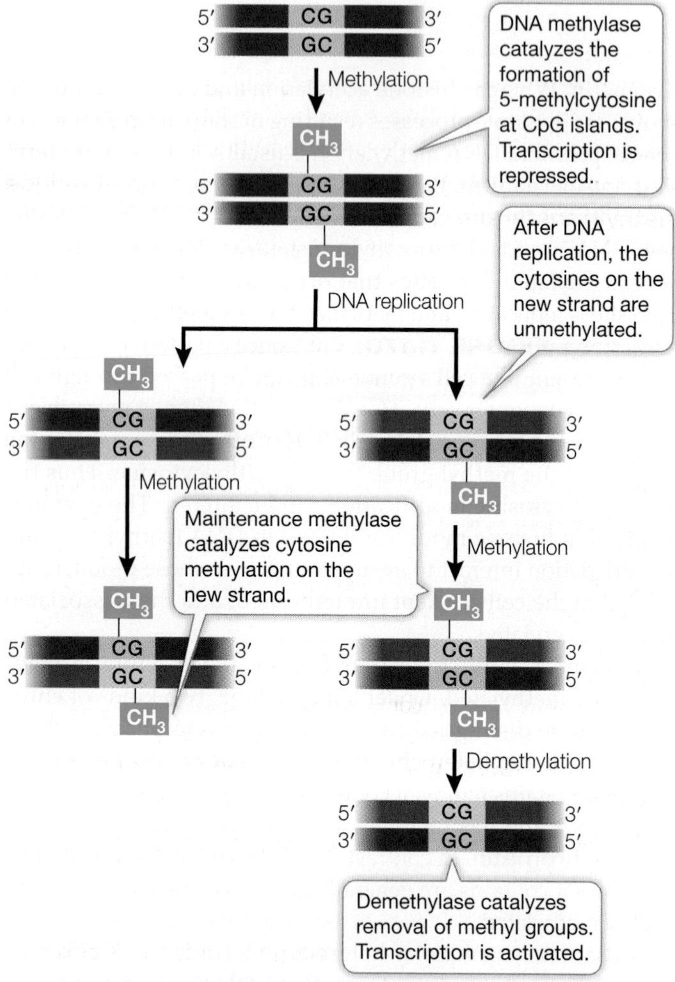

FIGURE 11.17 DNA Methylation: An Epigenetic Change The reversible formation of 5-methylcytosine in DNA **(A)** can alter the rate of transcription. **(B)** Inheritance of methylation involves methylation of the newly synthesized strand by maintenance methylase in hemimethylated DNA.

Because of X inactivation, only one X is expressed in every somatic cell of an adult female mammal, and the other X is silenced. Because all of a cell's descendants will have the same X inactivated, patches of adjacent cells will tend to have the same X inactivated. This can cause patchiness in phenotype. A familiar example is the calico, or tortoiseshell, cat (**FIGURE 11.18**). In these female cats, two alleles of an X-linked gene that contribute to coat color are orange (X^B) and black (X^b). In the embryo of a heterozygous female (X^BX^b), there is random X inactivation such that in some cells the orange allele is on the active X, and will thus be expressed, while in other cells the black allele is on the active X. These cells then form the progenitors of patches of skin in the animal.

The inactive heterochromatic X in the nucleus of a female mammal cell is so condensed that it is visible through a light microscope (similar to all chromosomes during M phase of the cell cycle; see Key Concept 7.2). The condensed X chromosome is called a Barr body after its discoverer, Murray Barr. Because one X is inactive in somatic cells of females, male and female cells have the same number of active Xs in each of their cells (one). Having the right "dosage" of transcriptionally active genes is important. The importance of the number of active chromosomes is seen in **aneuploidy**—a harmful condition in which an embryo has an unusual number of a particular chromosome (see Key Concept 7.4).

Epigenetic changes can be induced by the environment

In mammals, DNA methylation and histone modification patterns are reset both in the germ line and again very early during development. These resetting events imply that epigenetic modifications of DNA and histones early in development should be very similar, or identical, among individuals. However, epigenetic modifications in different individuals often differ. How much of the difference is due to genetic differences among individuals? A study of human monozygotic (identical) twins shows that DNA methylation can be altered during the lifetime of an individual, suggesting that the environment plays a large role.

Identical, or monozygotic, twins come from a single fertilized egg that divides to produce two separate individuals with identical

FIGURE 11.18 X Inactivation in Mammals A female calico cat is heterozygous for the orange and black alleles at the X-linked coat-color gene. In areas of the coat that are not white, calico cats show patches of orange and black fur, depending on which X was inactivated in the ancestral cell of a patch of cells.

Very rarely—about one time in several thousands—a calico cat is discovered that is a male. How is that possible? Does your explanation make a prediction regarding whether you would observe a Barr body in the nuclei of the cells of a male calico cat?

genomes. A comparison of hundreds of such twin pairs shows that in tissues of 3-year-olds, DNA methylation patterns are virtually the same between twins. But by age 50—when twins have usually been living apart and experiencing different environments for decades—the patterns are quite different, and different genes are expressed. This indicates that the *environment plays an important role in epigenetic modifications*, and therefore in the regulation of genes that these modifications affect.

What factors in the environment lead to epigenetic changes? Chemicals such as tobacco smoke and dietary components such as folic acid can affect DNA methylation patterns. Another factor might be stress: when rats are put in a stressful situation, such as removal from a caring mother (see the chapter opening), genes that are involved in important brain pathways become heavily methylated (and transcriptionally inactive).

Media Clip 11.1 **The Epigenetics of Identical Twins**

DNA methylation can result in genomic imprinting

As just mentioned, in mammals epigenetic changes are reset both in the germ line and during early development. For DNA methylation, this happens in two stages: first, the existing methyl groups are removed from the 5-methylcytosines by a demethylase, a process termed erasure, and then a DNA methyltransferase adds new methyl groups to particular sets of cytosines. This resetting of epigenetic patterns ensures that the appropriate pattern of gene expression for normal development will occur.

However, for a few genes, specific patterns of methylation develop in each sex during gamete formation, and these patterns, different in sperm and egg, are retained in the chromosomes of offspring after fertilization. The DNA methylation pattern in male gametes (sperm) differs from that in female gametes (eggs) at about 200 genes in the mammalian genome. That is, some genes in this group are methylated in eggs but unmethylated in sperm, and vice versa. For these genes, either the maternal or the paternal copy is transcriptionally active (not methylated), and the other is transcriptionally inactive. This is called **genomic imprinting**.

Imprinting of specific genes occurs primarily in mammals and flowering plants. Most imprinted genes are involved with embryonic development. An embryo must have both the paternally and maternally imprinted gene patterns to develop properly. In fact, attempts to make an embryo that has chromosomes from only one sex (for example, by chemically treating an egg cell to double its chromosomes) usually fail. So imprinting has an important lesson for genetics: *males and females may produce genetically similar gametes (except for the Y in half the sperm), but their gametes differ epigenetically.* Thus for a few tens of genes in your entire genome, you express only the copy you received from your mother, and for another few tens of genes you express only the copy you received from your father.

Transgenerational epigenetic inheritance is possible in some but not all organisms

As noted above, epigenetic changes can be passed from mother to daughter cells during mitosis. The maintenance methyltransferase enzyme makes inheritance of DNA methylation possible (see Figure 11.17). Can epigenetic changes be passed on from one organismic generation to the next like DNA mutations? If this occurred, the phenotype of an offspring of sexual reproduction would be due not just to the genes obtained from its parents but also to the epigenetic modifications passed through gametes that were determined in part by the environments the parents experienced. This possibility is unlikely in mammals for two reasons. First, epigenetic responses would have to occur in the germ line that produces gametes—changes that affect expression of genes in the brain, for example, would not be passed on to offspring. Second, when gametes are formed, histone modifications are removed and methylation of genes is largely removed as well. In short, the epigenetic changes are undone. However, in plants, germ cells are derived from tissues late in plant growth, during flower formation, not set aside early in development as they are in animals. In addition, epigenetic modifications are not erased in the germ line during gamete formation as they are in animals. It is thus not surprising that there is evidence for the transgenerational inheritance of epigenetic changes in plants.

REVIEW & APPLY | 11.4

R In order for proteins involved in the regulation of gene expression to act, they must be able to reach their DNA binding sites. How tightly nucleosomes are woven into chromatin controls the access of regulatory proteins to their binding sites. Chromatin condensation is altered by acetylation of histones and by methylation of DNA. Changes in chromatin accessibility can be passed from parent to daughter cells, resulting in so-called epigenetic inheritance of expression. Histone modifications and DNA methylation are reset in mammals, making transgenerational inheritance of environmentally induced epigenetic changes unlikely. In plants such resetting does not occur, and this allows epigenetic changes to be inherited across generations.

A

1. Why does histone acetylation tend to be associated with active transcription?
2. How is it possible for genetically identical cells to express different genes even though the transcription factors they currently express are the same?
3. Analysis of the cells in the germ line of heavy smokers indicate that some genes are methylated that are not methylated in nonsmokers. What sites (sequences of DNA) are likely methylated in these genes? What will be the effects of methylation on the expression of these genes? Do you think this unusual methylation pattern might be inherited by a smoker's offspring?

Thus far we have examined transcriptional gene regulation in viruses, prokaryotes, and eukaryotes. In the final key concept we will focus on the mechanisms occurring after transcription for regulating gene expression in eukaryotes.

11.5 Eukaryotic Gene Expression Can Be Regulated after Transcription

LEARNING OBJECTIVES

By the end of this key concept you should be able to:

11.5.1 Define alternative splicing.

11.5.2 Explain why the levels of mRNA in a cell do not always predict the levels of protein in a cell.

11.5.3 Describe the characteristics of microRNA and explain how miRNAs reduce gene expression.

11.5.4 List three ways in which translation can be regulated.

11.5.5 Explain the role of ubiquitin in a cell.

Gene expression involves transcription and then translation. So far we have described how eukaryotic gene expression is regulated at the transcriptional level. But as Figure 11.1 shows, there are many points at which regulation can occur after the initial gene transcript is made.

Different mRNAs can be made from the same gene by alternative splicing

Most primary mRNA transcripts in multicellular eukaryotes contain several introns (see Figure 10.9). We have seen how the splicing mechanism recognizes the boundaries between exons and introns. Consider what would happen if the β-globin pre-mRNA, which has two introns, were spliced from the start of the first intron to the end of the second: the middle exon would be spliced out along with the two introns. An entirely new protein (certainly not a β-globin) would be made, which would not have the same functions as normal β-globin. Through **alternative splicing**, particular exons of a gene may be included or excluded from the final mRNA transcript, allowing different proteins to be produced from that gene. Alternative splicing can occur by mistake and result in a defective protein, or it can be a deliberate mechanism for generating a family of different proteins with different activities and functions from a single gene (**FIGURE 11.19**).

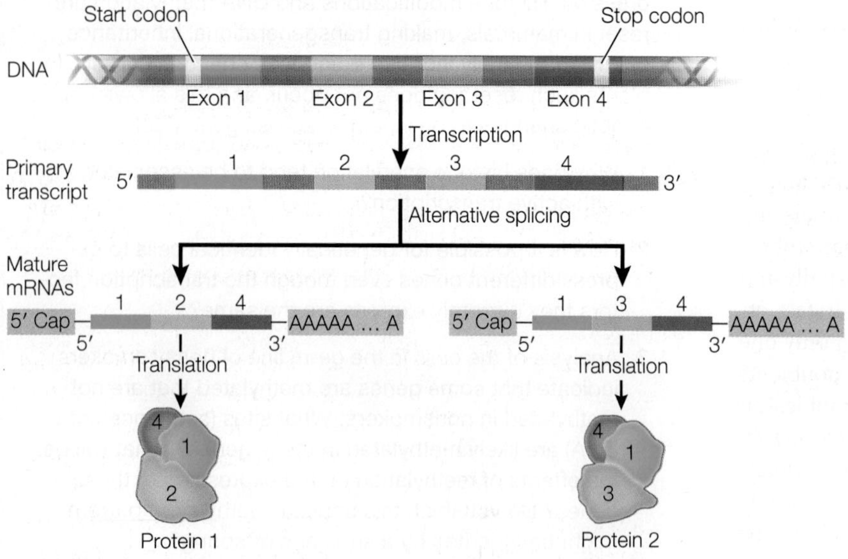

For some pre-mRNA transcripts, alternative splicing is common and can result in the same gene producing different proteins within a cell, or different proteins in different cells. For example, in *Drosophila* the *Sxl* gene plays a role in sexual development and undergoes different splicing in males and females. This gene has four exons, which we will designate 1, 2, 3, and 4. In the female embryo, splicing generates two active forms of the Sxl protein, one containing exons 1 and 2 and the other 1, 2, and 4. However, in the male embryo the protein contains all four exons (1, 2, 3, and 4) and is inactive. Regulation of the splicing of the *Sxl* pre-mRNA thus controls male versus female development.

Before the human genome was sequenced, most scientists estimated that they would find between 80,000 and 150,000 protein-coding genes. You can imagine their surprise when the actual sequence revealed only about 20,000 genes! In fact, there are many more mature mRNAs than there are genes in humans, and most of this excess comes from alternative splicing. Indeed, recent surveys show that more than 80 percent of all human genes are alternatively spliced.

FIGURE 11.19 Alternative Splicing Produces Different Mature mRNAs and Proteins Pre-mRNA can be spliced differently in different tissues, resulting in different proteins.

This gene has three introns. Assuming the first and last exons are always present in the mature mRNA, how many different proteins can be produced with alternative splicing?

Stability of mRNA can be regulated

Once an mRNA is transcribed and processed (in eukaryotes) it is ready for translation. How much translation occurs from a mature mRNA depends on (1) the mRNA's stability—how long it remains in the cell before it is degraded, and (2) the mRNA's rate of translation—how rapidly proteins are synthesized from the mRNA. Both mRNA stability and the rate of translation can be regulated.

Messenger RNAs are degraded by RNA nucleases that are present in the cytosol. Some sites in mRNAs are particularly susceptible to certain RNA nucleases. One method for altering stability of an mRNA is to produce a protein that binds to the mRNA in such a way that nuclease recognition sites are either more or less available for nuclease digestion.

Another way to alter an mRNA's stability is via regulatory RNAs. We have already met genes that do not encode proteins, namely those that encode ribosomal RNAs and transfer RNAs. Genomes also carry genes that encode other functional RNAs, some of which regulate gene expression. An important example is **microRNAs (miRNAs)**, very small RNAs that are involved in reducing gene expression in eukaryotes.

The first miRNA sequences were found in the worm *Caenorhabditis elegans*. This model organism goes through a set of well-defined cell divisions. As such, it has been an important organism for the study of development. Victor Ambros at the University of Massachusetts discovered mutations in a regulatory gene that caused abnormalities during larval development. The gene's role was to turn off expression of a developmental gene that allowed cells to progress to the next stage.

It was initially thought that the regulatory gene would encode a transcription factor that acted as a repressor. Further investigation showed that the gene encodes a 22-nucleotide miRNA that inhibits expression of the developmental gene posttranscriptionally by binding to its mRNA.

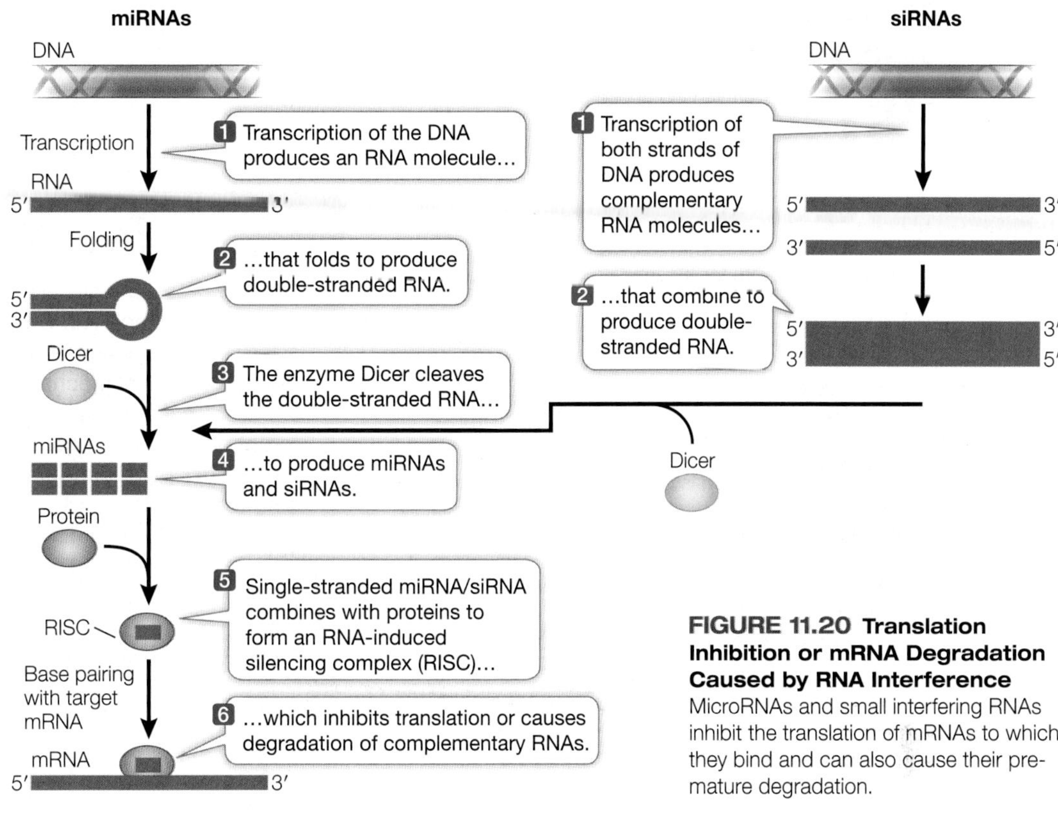

FIGURE 11.20 Translation Inhibition or mRNA Degradation Caused by RNA Interference MicroRNAs and small interfering RNAs inhibit the translation of mRNAs to which they bind and can also cause their premature degradation.

Thousands of miRNAs, and related small interfering RNAs (siRNAs), have been described in a variety of eukaryotes. Each is about 22 nucleotides long and can have dozens of mRNA targets. Any mRNAs with sequences that are complementary to all 22 bases of a particular miRNA are degraded. Messenger RNAs that are not perfectly complementary, containing one or a few mismatches when bound, are not degraded. Instead, the bound miRNA interferes with translation. Thus miRNAs affect gene expression by reducing mRNA stability or the rate of translation.

A microRNA is transcribed as a single RNA molecule that folds into a double-stranded RNA molecule. An siRNA is transcribed as two perfectly complementary RNA molecules that form a double-stranded molecule. In both cases, the double-stranded RNA is then processed into single-stranded molecules, which combine with a protein complex and bind to their target mRNA (**FIGURE 11.20**). This gene-silencing mechanism, called RNA interference, is found in most eukaryotes and is remarkably conserved, indicating that it is evolutionarily ancient and biologically important.

Translation of mRNA can be regulated

In addition to mRNA stability, the rate of translation of mRNAs will alter the levels of protein found in a cell. There are three known ways in which the translation of mRNA can be regulated:

1. *Binding of complementary RNA to the mRNA.* We saw an example of complementary RNA binding in our discussion of miRNAs above. There are several other examples of how complementary RNA binding to mRNA may interfere with initiation and elongation of translation, including some from prokaryotes. For example, an antisense RNA binds to the 5′ end of the plasmid gene *repB* in *Streptococcus pneumoniae*, preventing ribosome binding and initiation of translation. The RepB protein is involved in plasmid replication and the antisense regulation of translation is one mechanisms controlling plasmid copy number in bacterial cells.

2. *Modification of the 5′ cap.* In eukaryotes an mRNA usually has a chemically modified molecule of guanosine triphosphate (GTP) at its 5′ end (the cap). An mRNA that is capped with an unmodified GTP molecule is not translated. For example, the tobacco hornworm stores such inactive mRNAs in its egg cells. After an egg is fertilized, however, the caps are modified, allowing the mRNA to be translated to produce the proteins needed for early embryonic development.

3. *Inhibition of translation by protein binding.* Translational repressor proteins block translation by binding to mRNAs and preventing attachment to the ribosome or interfering with ribosome movement along the mRNA. In mammalian cells the rate of translation of the protein ferritin increases rapidly when the level of free iron ions (Fe^{2+}) increases in the cell. Iron is an essential nutrient, but the free ions can be toxic to the cell; ferritin binds the ions and stores them in a safe but accessible form. The amount of ferritin mRNA in the cell remains constant, but when the iron level is low, a translational repressor binds to the ferritin mRNA and prevents its translation. When the iron level rises, some of the excess Fe^{2+} ions bind to the repressor and alter its structure, causing the repressor to detach from the mRNA and allowing translation to proceed (**FIGURE 11.21**).

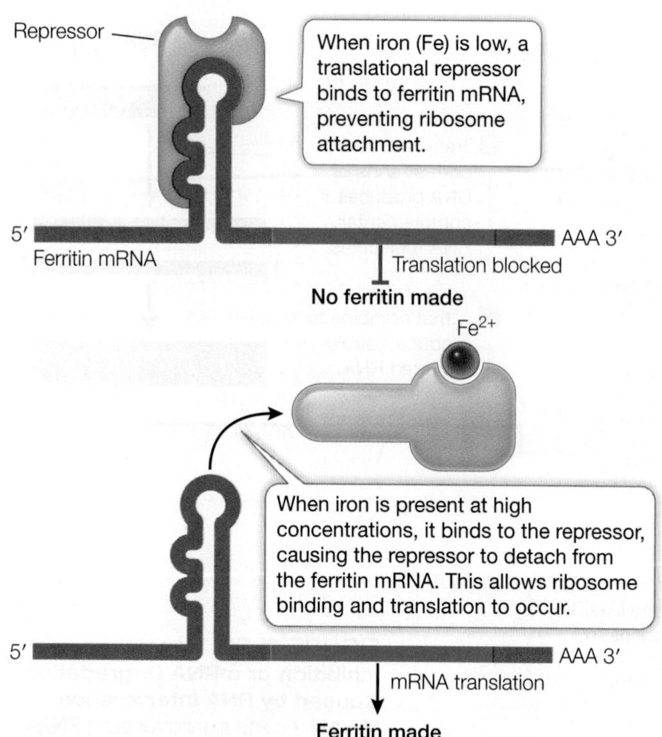

FIGURE 11.21 A Protein Repressor of Translation Binding of a translational repressor to mRNA blocks the mRNA from associating with the ribosome, so no protein is synthesized. Binding of Fe^{2+} by the repressor causes it to release from the mRNA, allowing translation and protein synthesis.

Protein stability can be regulated

The protein content of any cell at a given time is a function of both protein synthesis and protein degradation. Certain proteins can be targeted for destruction in a chain of events that begins when an enzyme attaches a 76-amino acid protein called **ubiquitin** (so named because it is ubiquitous, or widespread) to a lysine residue of the protein to be destroyed. Other ubiquitins then attach to the primary one, forming a polyubiquitin chain. The protein–polyubiquitin complex then binds to a huge protein complex called a **proteasome** (from *protease* + *soma*, "body") (**FIGURE 11.22**). Upon entering the proteasome, the polyubiquitin is removed and ATP energy is used to unfold the target protein. Three different proteases then digest the protein into small peptides and amino acids. You may recall from Chapter 7 that cyclins are proteins that regulate the activities of key enzymes at specific points in the cell cycle. Cyclins must be broken down at just the right time, and this is done by proteasomes. Degradation of transcription factors is very important for the regulation of gene expression in plants (see Key Concept 25.2).

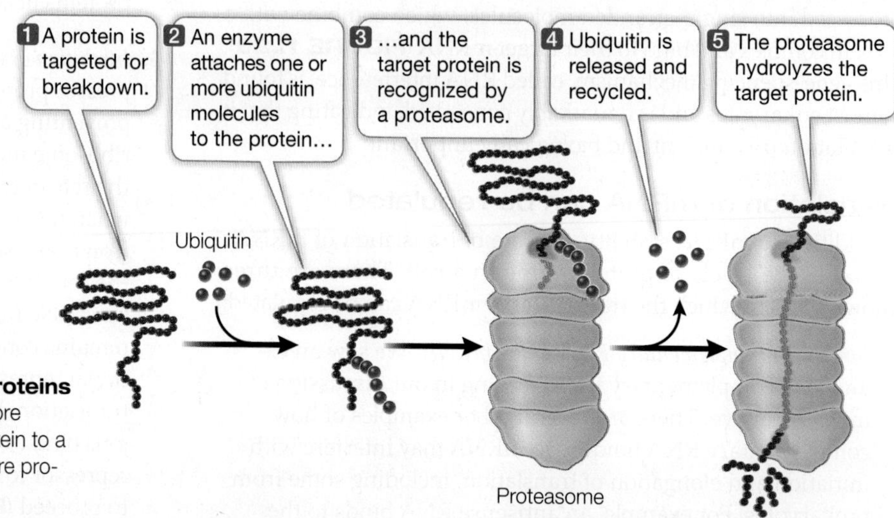

FIGURE 11.22 A Proteasome Breaks Down Proteins Proteins targeted for degradation are bound by one or more molecules of ubiquitin, which then direct the targeted protein to a proteasome. The proteasome is a complex structure where proteins are digested by several powerful proteases.

REVIEW & APPLY | 11.5

R Gene expression can be regulated after transcription, particularly in eukaryotes. Once an mRNA has been synthesized, the production of the protein it encodes can be moderated by several mechanisms. Alternative splicing, which is limited to eukaryotes, allows mature mRNAs encoding different proteins to be generated from the same pre-mRNA. Messenger RNA stability can be influenced by proteins that bind to the mRNA and, in eukaryotes, by complementary binding by 22-nucleotide miRNAs that target regions of the mRNA for digestion. The rate of translation is altered by proteins that bind to the mRNA, often preventing ribosome binding, and by miRNAs that imperfectly complement regions of the mRNA. Protein stability can also be regulated.

A

1. Two mutations are isolated in a gene that is expressed in the brain and liver of mice. One mutation is in the basal promoter, and the other is a nonsense mutation in the third exon. Mice that are homozygous for the nonsense mutation have normal brains but abnormal livers. Mice that are homozygous for the promoter mutation have both abnormal brains and livers. Why might the results differ for the nonsense versus promoter mutation?
2. Some bacteria that infect multicellular eukaryotes produce 22-nucleotide RNAs that they secrete from their cells. What might be the function of such RNAs?
3. A missense mutation in a gene found in yeast causes a lysine-to-serine substitution. One effect of this mutation is to cause levels of the protein to increase twofold. Why might a lysine-to-serine substitution cause such an effect?

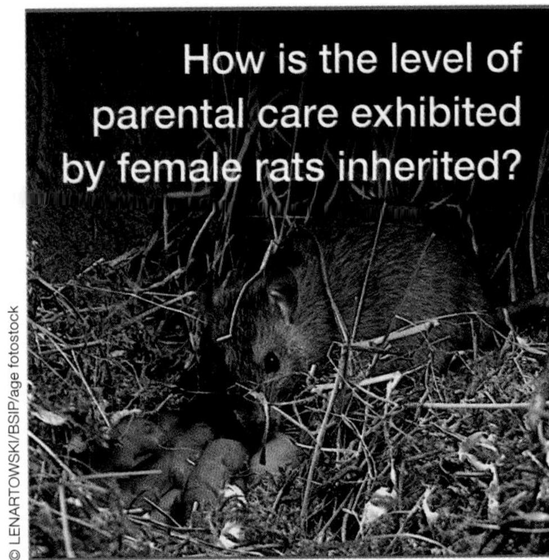

Answer: Since cross-reared rats exhibit the caring phenotype of their adoptive mothers, the inheritance cannot be genetic. One possibility is that the behavior is learned. However, pups raised by caring adoptive mothers also show a reduced response to stressful situations—a behavior they would not have observed in their adoptive mothers. Further work has revealed that the promoter of a gene involved in the stress response is unmethylated in pups reared with caring mothers but is methylated in pups reared with uncaring mothers. Pups with methylated promoters had low levels of the protein produced by this gene and grew up to be uncaring. Supporting the hypothesis that methylation causes the difference in caring propensity, treatment of rat pups to increase methylation caused them to be uncaring mothers, regardless of who raised them. Thus while epigenetic inheritance through gametes in mammals is unlikely because of resetting of epigenetic patterns during gamete formation and in early development, epigenetic inheritance can occur through a behavioral interaction between the mother and her offspring that modifies an offspring's epigenetic pattern of methylation.

VISUAL SUMMARY 11

11.1 The Regulation of Gene Expression Occurs at Multiple Levels

FIGURE 11.1

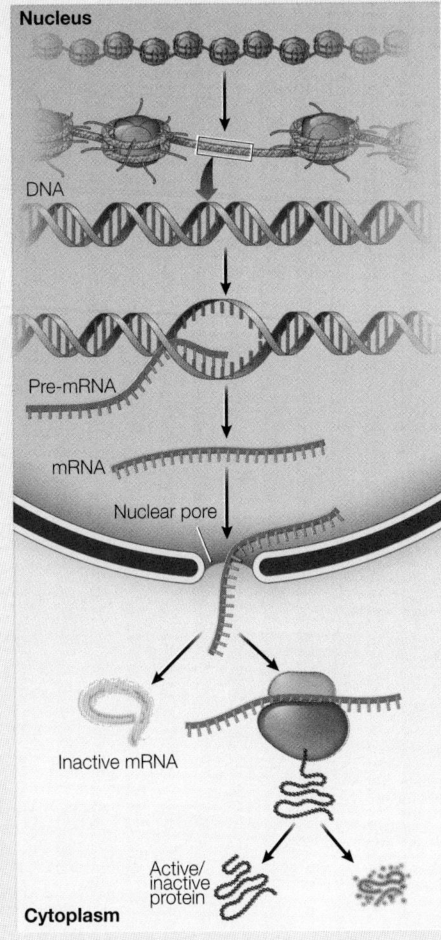

Go to ACTIVITY 11.1

- Gene expression can be regulated before transcription, during transcription, after transcription but before translation, during translation, and after translation (**FIGURE 11.1**).
- **CONSTITUTIVE GENES** are expressed at all times in all tissues. The expression of **INDUCIBLE** and **REPRESSIBLE GENES** can be turned on and off.
- Transcription can be controlled by **POSITIVE** or **NEGATIVE REGULATION**, or by both.

QUESTIONS

1. In bacteria, some genes are used to synthesize amino acids and some are used to break down amino acids as an energy source. Do you expect these genes to be inducible, repressible, or constitutive?
2. How can negative regulators be involved in the control of transcription both for inducible and for repressible genes?

11.2 Prokaryotic Gene Regulation Occurs Primarily at the Level of Transcription

Go to ANIMATIONS 11.1 and 11.2

- Many genes in prokaryotes are co-regulated in **OPERONS** (**FIGURE 11.5**).
- The *lac* operon is expressed when glucose levels are low and lactose levels are high.
- The *lac* operon has both negative regulation (lactose control) and positive regulation (glucose control) of transcription.

FIGURE 11.5

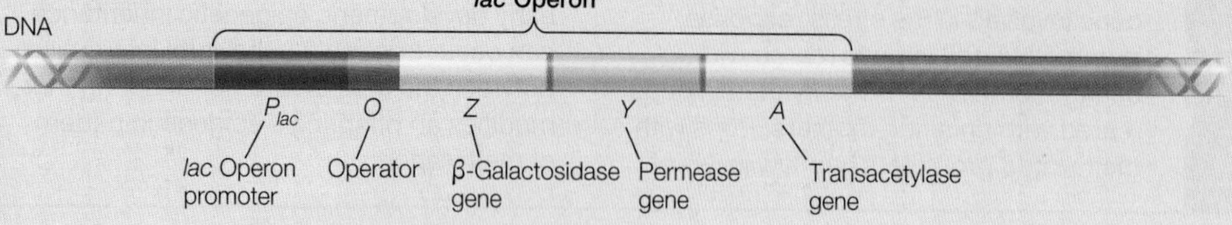

QUESTIONS

1. What is the defining characteristic of a prokaryotic operon?
2. Three strains of *E. coli* each contain a single mutation. Strain 1 contains a nonsense mutation in the *lacZ* gene, strain 2 has a nonsense mutation in the *lacI* gene, and strain 3 has a mutation that deletes the operator. State whether each strain would produce functional β-galactosidase when:
   a. Condition 1: Glucose is present and lactose is absent from the environment
   b. Condition 2: Glucose is absent and lactose is present
   c. Condition 3: Glucose is absent and lactose is absent

11.3 Eukaryotic Transcription Is Regulated by General and Specific Transcription Factors

Go to ANIMATION 11.3

- Initiation of transcription in eukaryotes involves assembly of the **BASAL TRANSCRIPTION APPARATUS** followed by interactions with specific transcription factors through **MEDIATOR** (**FIGURE 11.12**).
- **ENHANCERS** and **SILENCERS** bind specific transcription factors and may be far from the promoter.
- Differences in transcription factors underlie cell differentiation.

FIGURE 11.12

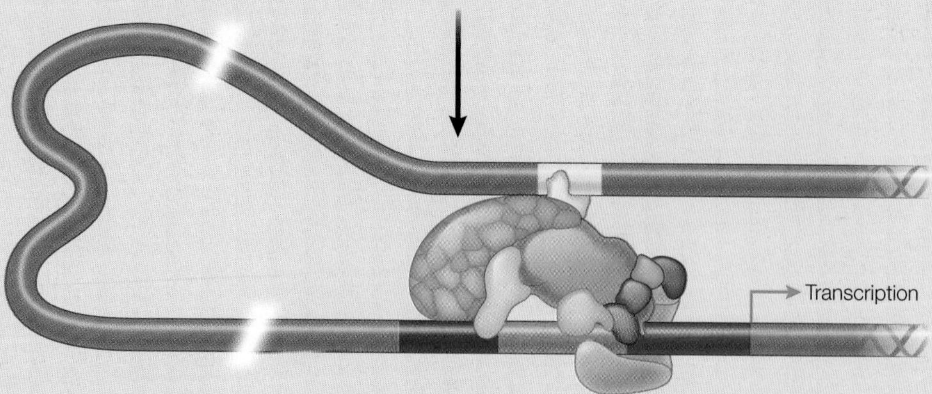

1. What is the difference between the promoter and enhancer and silencer sequences?
2. Explain how two cells with the same genome in the same organism can differ in the genes they express.

11.4 Transcription Can Be Regulated by Epigenetic Changes to DNA and Histones

- Proteins involved in the regulation of genes must be able to access their DNA binding sites.
- **EPIGENETIC** gene regulation can occur by loosening the compact **NUCLEOSOME** (**FIGURE 11.16**).
- DNA methylation tends to silence genes (**FIGURE 11.17**).

FIGURE 11.16

Acetylation of histone tails

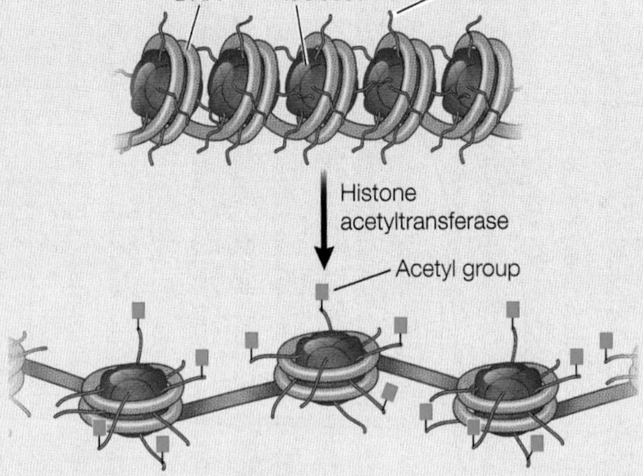

DNA
Nucleosome
Histone tail
Histone acetyltransferase
Acetyl group

FIGURE 11.17

Reversible methylation of cytosine

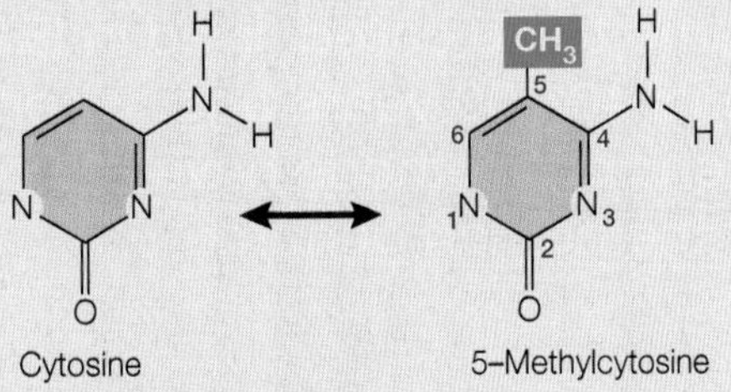

Cytosine
5–Methylcytosine

11.5 Eukaryotic Gene Expression Can Be Regulated after Transcription

- In eukaryotes, multiple proteins can be encoded by the same gene due to **ALTERNATIVE SPLICING** (**FIGURE 11.19**).
- mRNA stability and rate of translation can both be reduced by **microRNAS** (**FIGURE 11.20**).
- Protein stability alters the amount of functional protein in a cell.

FIGURE 11.19

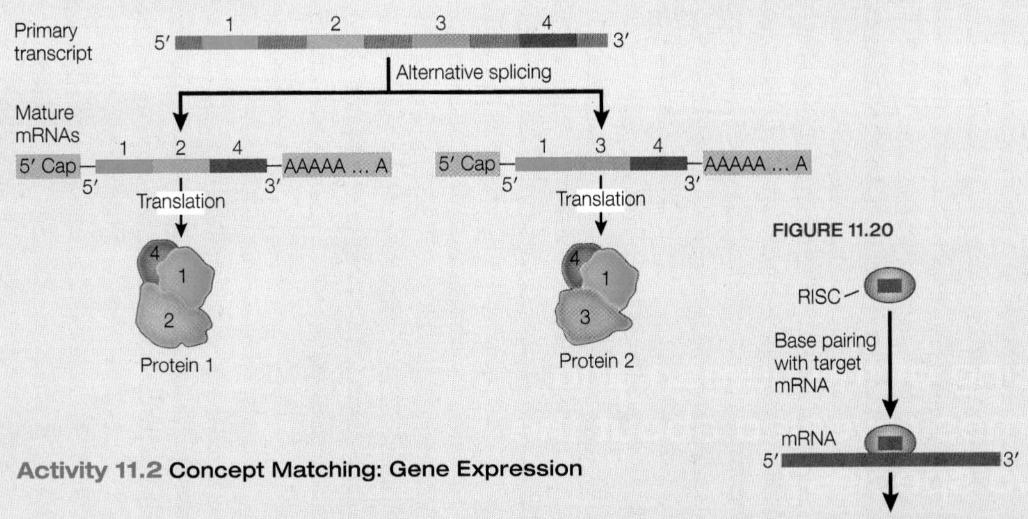

Activity 11.2 Concept Matching: Gene Expression

QUESTIONS

11.4

1. How is a pattern of DNA methylation passed from a parent cell to daughter cells?
2. What is genomic imprinting?
3. Why is epigenetic inheritance across generations unlikely in mammals?

11.5

1. In response to stress, a cell produces an miRNA that perfectly complements the mRNA from gene A and imperfectly complements the mRNA from gene B. What will be the effect on the mRNA and protein concentrations of these two genes?
2. A gene that produces high levels of mRNA in a cell has low levels of its encoded protein. Give two explanations for why this might be the case.

12 Genomes

More than 150 years ago, the fossil remains of an extinct relative of humans were found in the Neander Valley of Germany. Neanderthals, as they came to be called, lived in Europe and Asia from 450,000 until 40,000 years ago, when they died out. DNA sequencing methods have allowed a comparison of the human genome with that of Neanderthals.

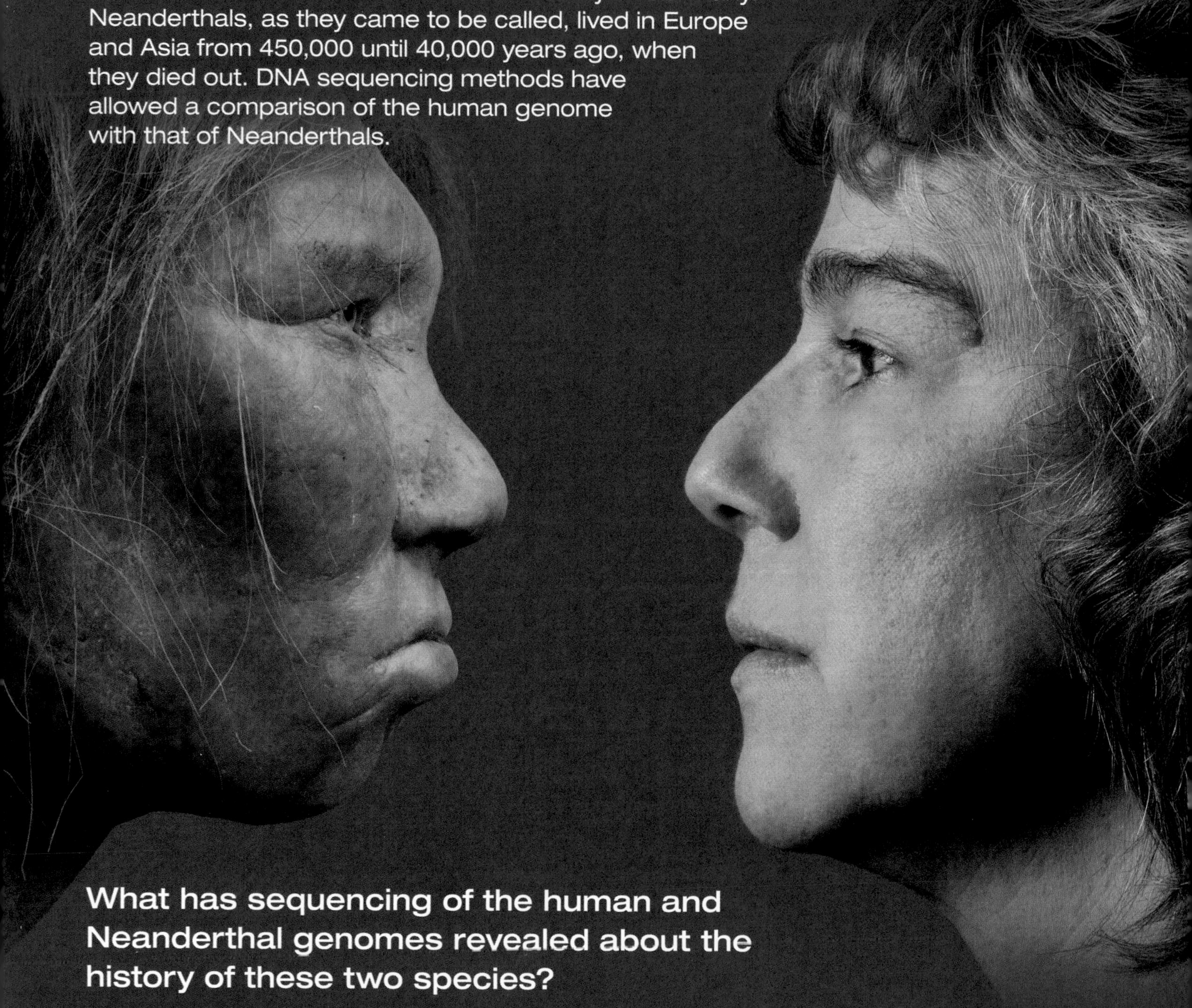

What has sequencing of the human and Neanderthal genomes revealed about the history of these two species?

You will find the answer to this question on page 304.

Neanderthal face: © Mark Thiessen/National Geographic Creative; Human face: M.H. Siddall

KEY CONCEPTS

12.1 The -omics Era Has Revolutionized Biology

LEARNING OBJECTIVES

By the end of this key concept you should be able to:

12.1.1 **Explain the role of dideoxynucleotides in Sanger sequencing.**

12.1.2 **Describe how high-throughput sequencing differs from Sanger sequencing.**

12.1.3 **Explain genome assembly and genome annotation.**

12.1.4 **Define genomics, transcriptomics, proteomics, and metabolomics.**

Following the rediscovery of Mendel's laws of inheritance in the early 1900s, the science of genetics involved studying one or a few genes at a time, typically in model organisms such as fruit flies and corn. Taking an approach now called **forward genetics**, scientists would identify a phenotype of interest that exhibited Mendelian inheritance, and then the gene underlying the phenotype would be identified by crosses and as much information about the gene as possible would be determined. For phenotypes known to be affected by multiple genes, such as height, forward genetics can identify the few genes that have the largest effect on such phenotypes. Once the genes are identified, molecular genetic approaches allow the DNA sequence of the genes to be determined, as described below.

While forward genetics starts with a phenotype and then identifies the underlying gene(s), **reverse genetics** does it the other way: it starts with the DNA sequence of a gene and then identifies the encoded phenotype. Reverse genetics uses the genetic code to predict the amino acid sequence of the protein, computational methods to predict protein structure and function, and experimental methods to mutate the gene and see the effect on the organism. Identifying genes for reverse genetics has been made easier through the enormous improvements in DNA sequencing techniques that have occurred over the last 20 years. It is now routine to sequence the entire genome of a species, not just one or a few genes. Sequencing an entire genome allows researchers to study the entirety of the genetic information in a species. In particular, it is possible to identify all, or at least most, of the genes and then use that information to study their functions and compare the genes with those in other, related species.

As you will see, the rise of genomics as a field of biology based on sequencing has led to a wide range of research and practical applications.

Sanger sequencing uses chain termination to sequence DNA

The British chemist and two-time Nobel Prize winner Frederick Sanger (**FIGURE 12.1A**) developed what has become known as the Sanger dideoxy sequencing method. This method leverages two important facts about DNA polymerases (see Key Concept 9.2):

1. DNA polymerases require a primer to begin polymerization.
2. A 3′ —OH group on the terminal nucleotide must be available on the growing strand for the addition of the next nucleotide.

In Sanger sequencing, a selected segment of DNA serves as a template for DNA synthesis. A single-stranded DNA primer is designed to be the starting point for DNA synthesis on the strand to be sequenced. Normal A, G, C, and T nucleotides, specifically deoxyribonucleoside triphosphates (dNTPs), are supplied for DNA synthesis. In addition, four dideoxynucleotides, specifically dideoxyribonucleoside triphosphates (ddNTPs), are included in the reaction mix, each modified to fluoresce in a specific color (**FIGURE 12.1B**). Each ddNTP lacks a 3′ —OH, and so DNA synthesis terminates when a ddNTP is incorporated into the growing chain of nucleotides. Sanger sequencing of DNA results in the production of new DNA strands that vary in length, in one-nucleotide increments, and are labeled in such a way that the end nucleotide can be identified (**FIGURE 12.1C**).

Until the invention of Sanger sequencing, the idea of sequencing an entire genome was far-fetched. Indeed, the first nucleic acid sequenced, an 80-base tRNA, took Robert Holley and his team of chemists 5 years to determine. In 1986 the Nobel laureate Renato Dulbecco and others proposed that the world scientific community be mobilized to undertake the enormous, expensive effort of sequencing the entire human genome. It was believed that a great deal would be learned about human disease, as well as numerous other interesting areas of human biology. Money for the publicly and privately funded **Human Genome Project** was obtained in 1990, and the plan was to be finished in 15 years. Determining the first human genome took almost that long (2003). Along the way, much was learned from sequencing smaller genomes from model organisms with which you are familiar.

Sanger sequencing allows DNA sequences of 300 to 1,000 bases to be obtained in a single reaction. It is especially useful for sequencing one or a few genes. However, for sequencing the millions to billions of nucleotides in whole genomes, Sanger sequencing is laborious. The key to genome, rather than single gene, sequencing is to generate many short fragments, each of which is sequenced. If a random method is used to generate the fragments, the fragments will be overlapping. For example, consider breaking the last phrase of the previous sentence:

(A) Frederick Sanger

NLM/NIH

(B) Normal nucleotides and dideoxynucleotides

Base (A, T, G, or C)

Deoxyribonucleoside triphosphate (dNTP) (normal nucleotide)

Base (A, T, G, or C)

label

Different flourescent label for each base

Dideoxyribonucleoside triphosphate (ddNTP) (chemically modified)

Absence of OH at the 3′ position means that additional nucleotides cannot be added.

(C) Synthesis stops at dideoxynucleotides

1 The single-stranded DNA fragment for which the base sequence is to be determined is isolated and serves as the template.

3′ ???????????????????? 5′

ddCTP ddGTP ddTTP ddATP

C G T A

2 Each of the ddNTPs is bound to a fluorescent dye.

3 A sample of DNA of unknown sequence is mixed with primer, DNA polymerase, dNTPs, and the fluorescent ddNTPs. Synthesis begins.

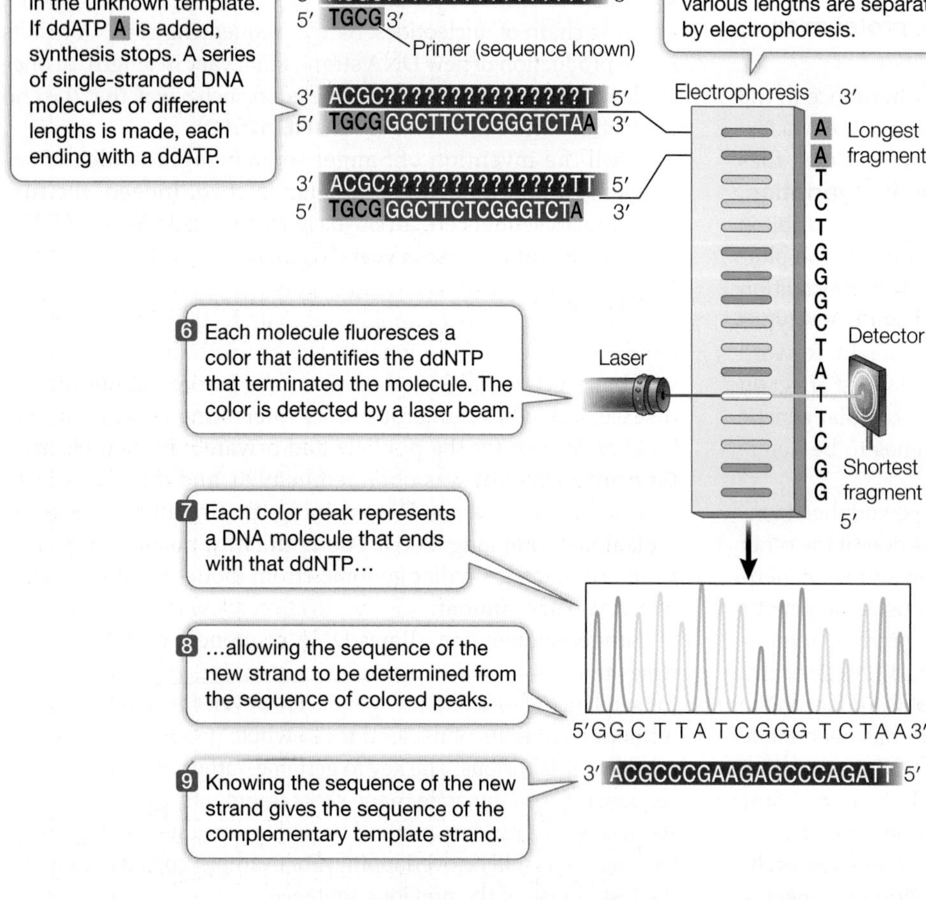

FIGURE 12.1 Sanger Sequencing Dideoxy sequencing, invented by Frederick Sanger **(A)**, causes synthesis to halt when a dideoxy nucleotide **(B)** is incorporated into the new strand. All terminated molecules of a particular length will end with the same fluorescently-labeled dideoxynucleotide that can then be scored and used to determine the sequence of the newly synthesized strand, and thus the template strand **(C)**.

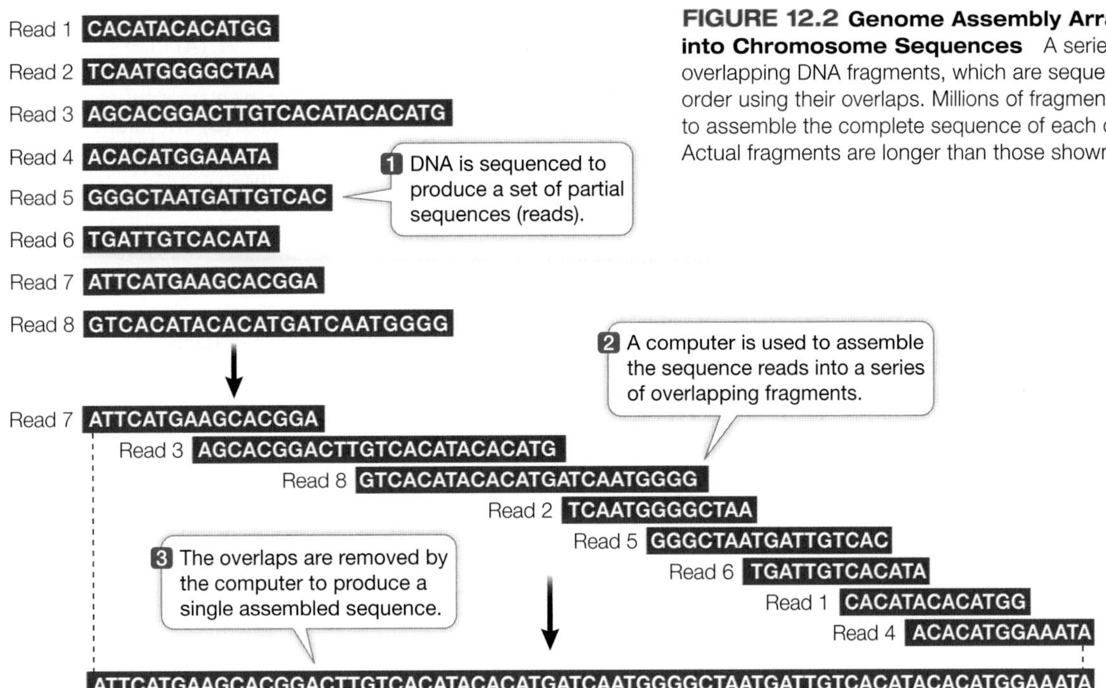

FIGURE 12.2 Genome Assembly Arranges DNA Fragments into Chromosome Sequences A series of random cuts generates overlapping DNA fragments, which are sequenced and then arranged in order using their overlaps. Millions of fragments are arranged in this way to assemble the complete sequence of each chromosome in a genome. Actual fragments are longer than those shown here.

First cut: thefrag ments willb eoverlap ping

Second cut: th efragm ent swill beo verla pping

Third cut: thefra gment swil lbeoverlap ping.....and so on

Given the overlaps, you can arrange these individual letters into the proper order. It is the same with DNA fragments, except that DNA fragments (which for genome sequencing are called "reads") are arranged by computer (**FIGURE 12.2**). Indeed, the development of DNA sequencing methods had to be paralleled by the development of computational methods to analyze them. It would be impossible to manually arrange millions of fragments from a genome. Putting the reads together to generate longer contiguous fragments and eventually entire chromosomes is called genome assembly.

DNA sequencing is now inexpensive at the genome scale

For whole genome DNA sequencing, the Sanger method has been supplanted by newer methods, collectively called **high-throughput** (or next-generation) **sequencing**. These methods use miniaturization techniques and robotics first developed for the electronics industry, as well as the principles of DNA replication and the polymerase chain reaction (PCR).

LINK You can review the processes of DNA replication and PCR in **Key Concept 9.2**.

FIGURE 12.3 outlines an example of high-throughput sequencing. In this method, more than 500 million reactions can occur on a surface about the size of a microscope slide. First, the DNA is prepared for sequencing (see Figure 12.3A):

1. Many copies of a large molecule of DNA is cut into small fragments of about 300 base pairs (bp) each. This can be done physically, using mechanical forces such as ultrasound or a syringe and needle to shear (break up) the DNA, or by using enzymes that hydrolyze the phosphodiester bonds between nucleotides at intervals in the DNA backbone (See Appendix D for a discussion of restriction enzymes).
2. The DNA fragments are denatured by heat, breaking the hydrogen bonds that hold the two strands together. Short, synthetic adapter sequences (oligonucleotides) are then attached to each end of each fragment.
3. The fragments are attached to the surface of a solid support, leaving a small amount of space between each molecule.
4. Single-stranded DNA primers complementary to the adapter sequences are used in PCR to produce many (~1,000) copies of each single-stranded DNA fragment, and its complementary single-stranded DNA molecule, each of which will be clustered around the original molecule. The copies of the original fragment in each cluster will be used as a template for the synthesis of new, complementary DNA. The multiple copies of template at a single location allow for easier detection of added nucleotides during the sequencing steps.

Once the clusters of template molecule have been generated, each is ready to be used as a template for sequencing (see Figure 12.3B):

1. The DNA fragments are heated to denature them. A universal primer (complementary to the adapter sequence at one end of the original single-stranded fragment), DNA polymerase, and the four deoxyribonucleoside triphosphates

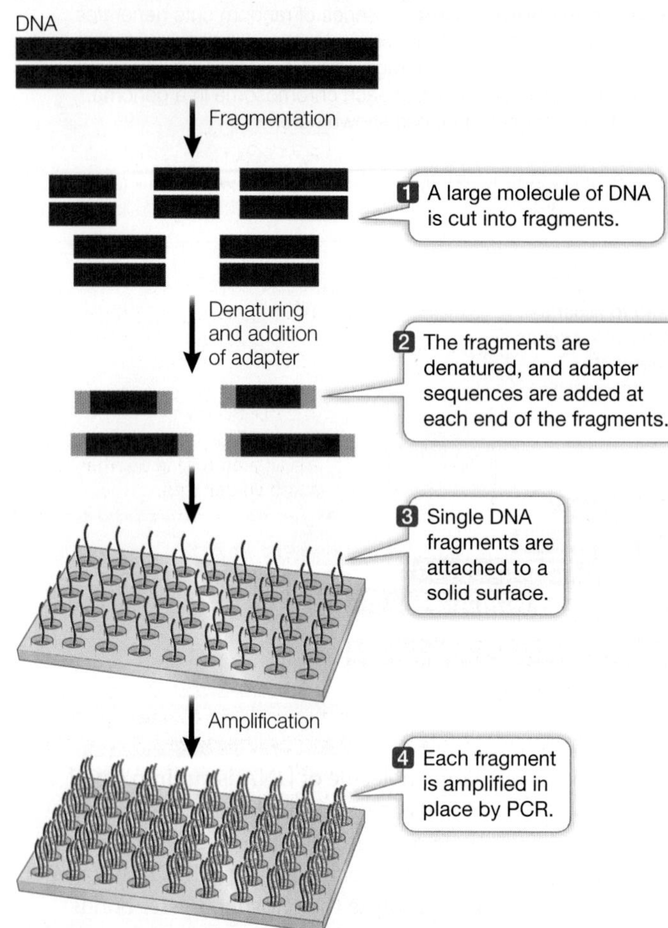

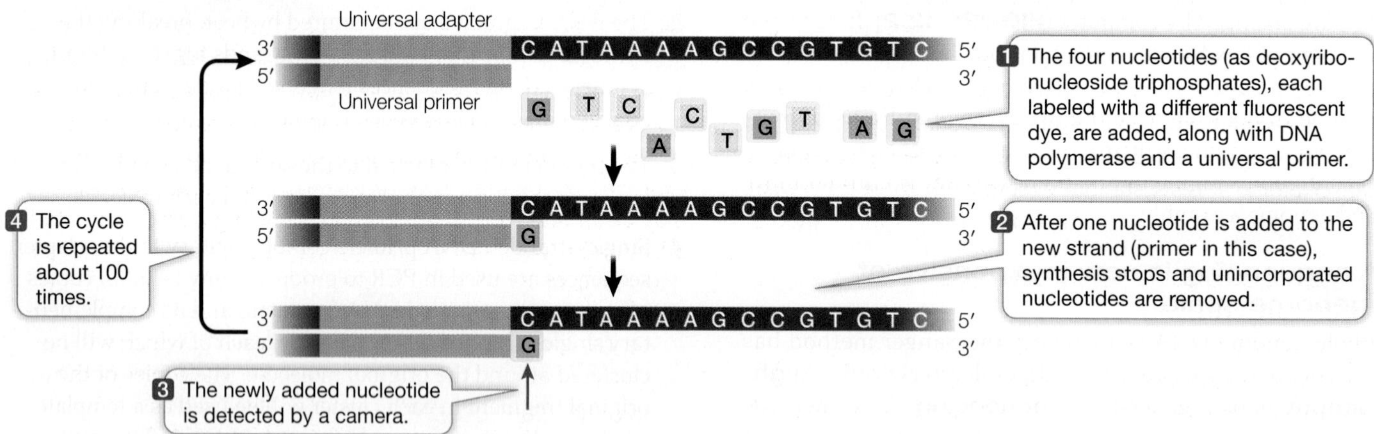

FIGURE 12.3 High-Throughput DNA Sequencing High-throughput sequencing is fast and inexpensive. **(A)** PCR amplification of individual single-stranded DNA fragments produces clusters of template molecules, half of which are identical and half complementary to the original molecule. **(B)** The strand complementary to only one of the two templates is sequenced one color-labeled nucleotide at a time, and a microscopic camera detects the color (and hence the identity) of the added nucleotide. Each cluster will produce a sequence that is 100–300 bp in length.

Why are half the molecules in a cluster complementary and half identical to the original single-stranded molecule?

Animation 12.1 Sequencing the Genome

Animation 12.2 High-Throughput Sequencing

(dNTPs; dATP, dGTP, dCTP, and dTTP) are added. (Note that unlike in Sanger sequencing, these are *not* dideoxynucleotides.) Each of the four dNTPs is labeled with a different colored fluorescent dye.

2. The DNA sequencing reaction is set up so that only one nucleotide at a time is added to the new DNA strand, which is complementary to the template strand.
3. The fluorescence of the newly added nucleotide at each location is detected with a camera. The color of the fluorescence indicates which of the four nucleotides was added.
4. The synthesis cycle is then repeated, with one more fluorescently-labeled nucleotide added during each repetition. Images are captured after each nucleotide is added, which gives the method its name: sequencing by synthesis. The series of

colors at each location indicates the sequence of nucleotides in the growing DNA strand at that location.

The power of this method derives from the fact that it is fully automated. More than 500 million reactions, each producing 100 or more bases of DNA sequence, can be run in a single machine in a few days. To put the sequencing advances in perspective, consider that the cost of sequencing the first human genome was $3 billion, or about $1 per base. The cost to sequence the entire human genome is now about $1,000, or $1 per 3 million bases. Sequencing an individual's genome is now inexpensive enough that it will likely become routine, and the potential health benefits are substantial (see Key Concept 12.4).

Transcriptomics identifies expressed genes

A genome is much larger than the **transcriptome**, which is the subset of the genome that is expressed as RNA in a particular cell or tissue at a particular time. In many cases it is the transcriptome that is of most interest, because expressed genes are likely to affect phenotype. As discussed in Chapter 11, the genomes of each somatic cell in the body are identical, but the sequences that are expressed are quite different from cell to cell and from tissue to tissue. The determination of which genes are expressed in what amounts can be made by **RNA sequencing (RNA-seq)**.

RNA sequencing is essentially the same as DNA sequencing except for the first steps. With RNA, it is necessary first to synthesize DNA copies of the RNAs. As you might remember from Chapter 11, RNA viruses possess a reverse transcriptase that they use to synthesize a double-stranded DNA molecule from an RNA molecule (see Figure 11.14). Biologists use this enzyme to generate DNA from fragmented cellular RNA, and then sequence the so-called complementary DNA (cDNA) using high-throughput sequencing. When these sequences are aligned with one another by computer, each of the resulting contiguous sequences represents a single RNA molecule. In addition, the cDNA sequences can be aligned with the genome to determine which genomic regions are expressed (**FIGURE 12.4**). Comparing transcriptomes across cell types allows researchers to determine similarities and differences in the RNAs being expressed. If a genomic DNA sequence appears in a transcriptome, it is substantially more likely to be functional than if it is only seen in the genome and not in the transcriptome.

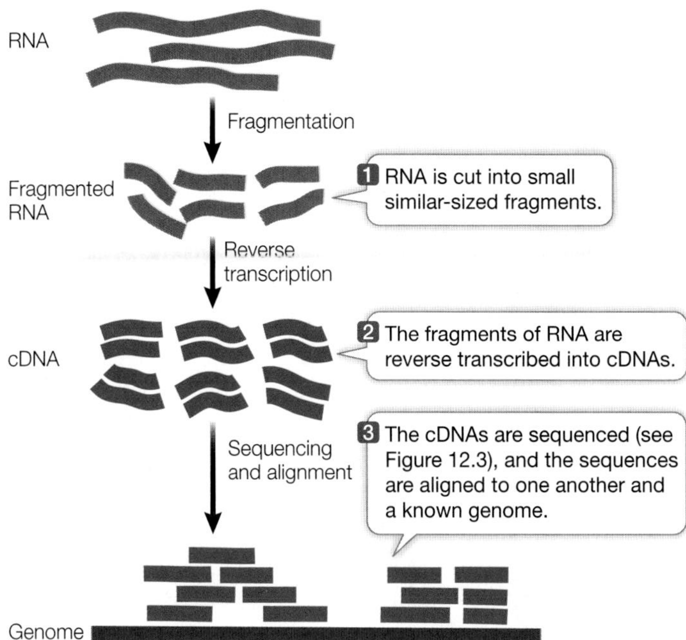

FIGURE 12.4 RNA Sequencing RNA is extracted from a sample, fragmented, and reverse transcribed into DNA, which is then sequenced. The sequences are aligned with one another and with a known genome if it is available. A gene with more fragments (per base) is expressed at a higher level than a gene with fewer fragments.

Suppose the figure represents the RNA-seq fragments aligning with a single gene. Why do some gene regions have no sequenced cDNA fragments, while others have many?

Genome sequences yield several kinds of information

New genome sequences are being published at an accelerating pace, creating a torrent of biological information. This information is used in two related fields of research, both focused on studying genomes. In **functional genomics**, biologists use sequence information to identify various parts of genomes and their functions (**FIGURE 12.5**). In particular, biologists seek to identify all of the protein-coding genes, RNA genes, regulatory sequences, and other functional sequences such as telomeres and centromeres. Identifying the various functional regions of the genome is termed genome annotation. Finding these sequences can be challenging, since most genes in most genomes have never been studied. The search for protein-coding genes leverages what is known about gene structure in prokaryotes and eukaryotes, the genetic code, and results from RNA-seq.

To find a protein-coding gene, biologists look for **open reading frames** (see Figure 12.5), which are sequences of DNA that, if transcribed and translated, would encode a reasonably long sequence of amino acids before reaching a stop codon. Such sequences are likely to encode a protein. An open reading frame that begins with an intron consensus sequence (boundary between exon and intron) and ends with an intron consensus sequence may be an internal exon. Some open reading frames (essentially all in prokaryotes) should begin with a start codon and/or end with a stop codon.

LINK As discussed in **Key Concept 10.2**, RNA splicing removes introns and splices the exons together. The process requires that the intron have certain characteristic consensus sequences—short stretches of DNA that appear with little variation ("consensus") across different introns.

Open reading frames allow identification of protein-coding genes. Once identified, the next step is to determine their function. Predicting the function of an encoded protein can be done using computational methods. Computer algorithms can scan the protein sequence and find motifs—regions of amino acids—that can be predicted to fold in certain ways or have enzyme activity. If a similar nucleotide or protein sequence has been discovered

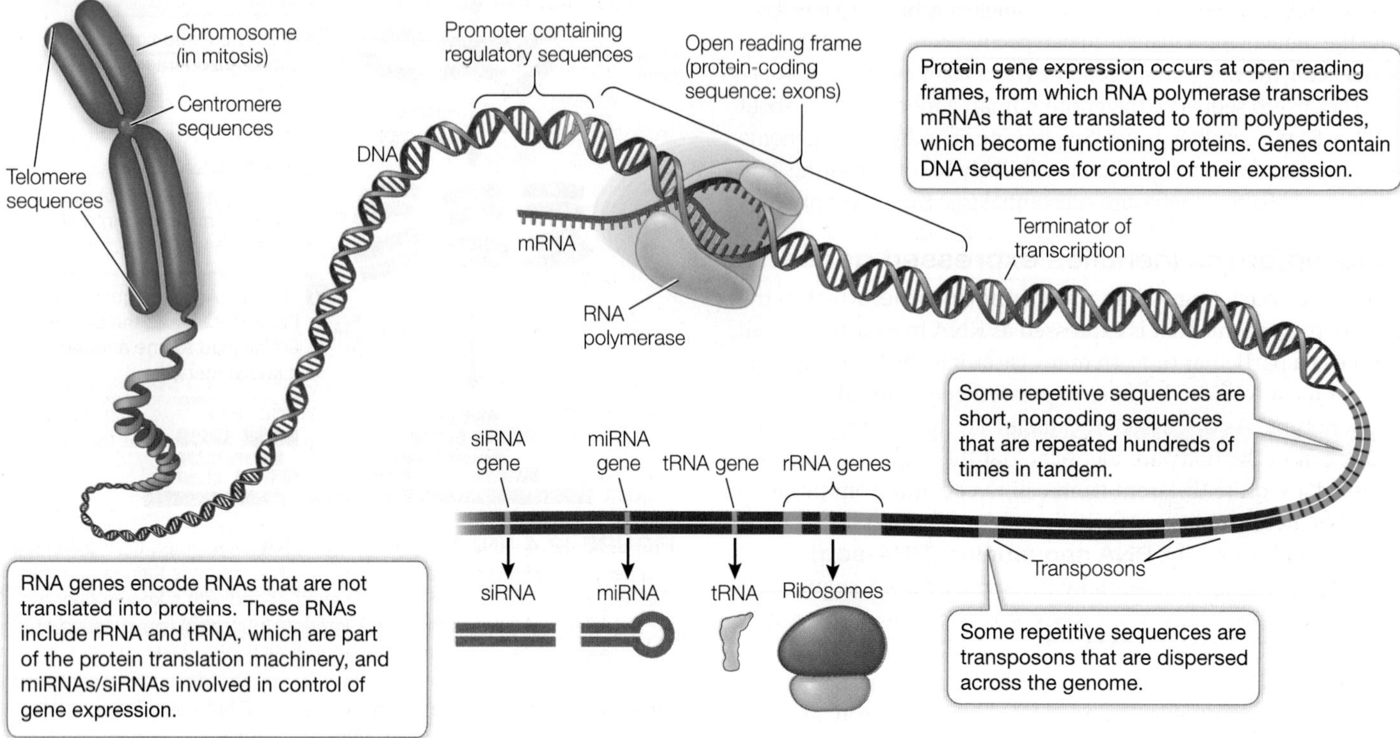

FIGURE 12.5 Functional Genomics Genome sequences contain many features, some of which are summarized in this overview. Elucidating all of the functional components contained in a genome sequence can help us understand how an organism functions.

previously in another organism, it is probable that the protein has a similar or identical function in the newly sequenced organism. For example, one might correctly guess that the RNA polymerase genes in humans and mice encode proteins with similar sequences and functions. An experimental approach to determine a gene's function is to insert it along with a promoter into a recipient cell or organism and look for a new phenotype. Conversely, knocking out the unidentified gene by mutation or deletion and looking for phenotypic changes may indicate the gene's normal function (see Appendix D for methods to make knock outs).

Regulatory sequences for genes, such as promoters and terminators for transcription, can be identified by characteristic consensus sequences in the vicinity of protein-coding regions. These regions are easier to find in prokaryotes because they occur in a more predictable location than in eukaryotes. Finding enhancers and silencers (see Key Concept 11.3) is especially difficult in eukaryotes because they may not be adjacent to the coding region of the gene.

In addition to functional genes, several other types of DNA sequences can be identified. These include centromere sequences (see Key Concept 7.2), telomere sequences (see Key Concept 9.2), transposons (see Key Concept 12.2), and other repetitive sequences.

Once genomes from different organisms were available, it allowed the development of the field of **comparative genomics**, which compares genomes both within and across species. Comparative genomics addresses several kinds of questions including what genes are present in all species, how rapidly nucleotide and amino acid sequences change for different regions of the genome, how and how often new genes arise or old genes are lost from genomes, how gene order changes due to chromosomal rearrangements or other processes, and how species are related to one another.

LINK The application of genome sequencing to reconstructing phylogenies (evolutionary trees) is described in **Chapters 14** and **15**.

Phenotypes can be analyzed using proteomics and metabolomics

Two new fields have emerged to complement genomics and provide a more complete snapshot of a cell or organism: proteomics and metabolomics.

PROTEOMICS **Proteomics** is the study of the **proteome**—the complete complement of proteins produced by an organism. Even for genes that encode only a single protein, the concentration of mRNA often does not predict protein abundance. In addition, many genes encode more than a single protein (**FIGURE 12.6A**). As described in Key Concept 11.5, alternative splicing leads to different combinations of exons in the mature mRNAs transcribed from a single gene. Posttranslational modifications also increase the number and the structural and functional diversity of proteins derived from one gene (see Figures 12.6A and 10.26). For these reasons, the only way to confidently determine the protein composition of a cell, rather than estimating it from mRNA sequences, is to measure it directly.

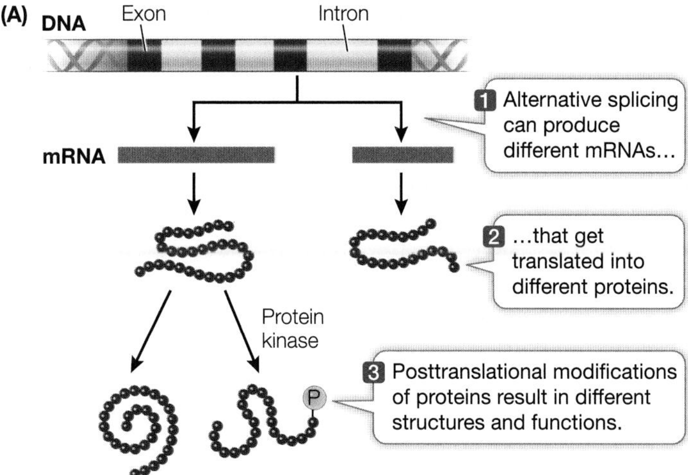

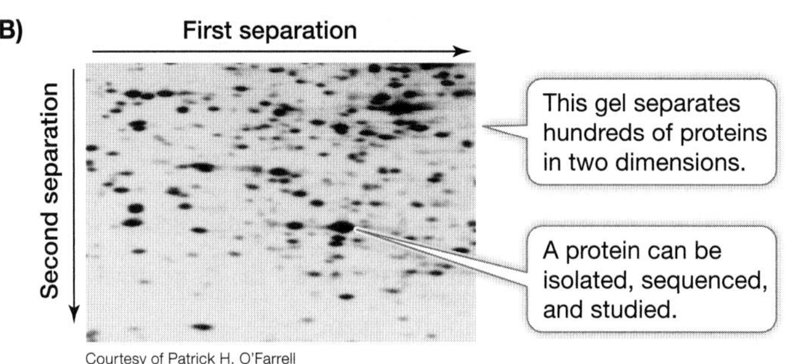

FIGURE 12.6 Proteomics **(A)** A single gene can code for multiple proteins due to alternative splicing and posttranslational modification, which is useful for examining the proteome. **(B)** Two-dimensional gel electrophoresis separates cell proteins on the basis of the pH at which a protein's overall charge is zero (isoelectric point, first separation), and size (second separation). Further analysis of each spot by mass spectrometry identifies different proteins.

Several approaches are commonly used to measure the proteome:

- Because of their unique amino acid compositions, most proteins can be separated using size and pH at which their overall charge is zero. On the basis of these two properties, they can be separated by two-dimensional gel electrophoresis (**FIGURE 12.6B**). Comparisons of gels can be used to identify differences in the proteome of different samples.
- Once they have been isolated from a two-dimensional gel, or using another method, individual proteins can be analyzed by mass spectrometry. This technique identifies molecules by the masses of their atoms.
- Antibodies can be used to isolate specific proteins or to detect them in cells or tissues.

None of the above methods are straightforward, especially on the genome scale. For this reason, the proteins produced by a cell, tissue, or organism are often inferred from RNA-seq data, using the universal genetic code to translate mature mRNAs into proteins computationally. This allows biologists to compare the primary amino acid sequence of proteins across organisms. However, it does not allow protein levels or posttranslational modifications to be determined.

METABOLOMICS **Metabolomics** is the comprehensive analysis of metabolites in a biological specimen. Studying genes and proteins gives a limited picture of what is going on in a cell. Many proteins are enzymes in metabolic pathways, and regulation of such pathways is often complicated, affected by both a cell's internal and external environments. It can thus be difficult to predict the identity and concentration of metabolites in a cell from the RNA and proteins present. For this reason, researchers often directly measure metabolites. The **metabolome** is the complete set of small molecules present in a cell, tissue, or organism. These include:

- *Primary metabolites* that are involved in normal cellular processes, such as intermediates in pathways such as glycolysis. This category also includes hormones and other signal molecules.
- *Secondary metabolites*, which are often unique to particular organisms or groups of organisms. They are often involved in special responses to the environment. Examples are antibiotics made by microbes, and the many chemicals made by plants that are used in defense against pathogens (disease-causing organisms) and herbivores.

Metabolomics aims to describe the metabolic profile of a cell, tissue, or organism under particular environmental conditions. Measuring metabolites involves sophisticated analytical instruments. If you have studied organic or analytical chemistry, you may be familiar with gas chromatography and high-performance liquid chromatography, which are used to separate molecules with different chemical properties. Mass spectrometry and nuclear magnetic resonance spectroscopy are then used to try to identify molecules. These measurements result in "chemical snapshots" of cells, tissues, or organisms, which can be related to physiological states.

Plant biologists have identified tens of thousands of secondary metabolites, many of them made in response to environmental challenges. The metabolomes of agriculturally important plants are being described, providing information that may be important in optimizing plant growth for food production. Most secondary metabolites in plants and in microbes evolved in defense against pathogens and predators and so are biologically active. For this reason, these metabolites are a rich resource for the development of pharmaceuticals, including antibiotics, pesticides, anticancer agents, seasonings and other flavorings, and everyday drugs such as aspirin (**TABLE 12.1**).

TABLE **12.1** | Some Examples of Secondary Metabolites

Secondary Metabolite	Examples (derived from)
Alkaloids	Morphine (opium poppy plant)
	Cocaine (coca plant)
	Caffeine (coffee plant)
Terpenoids	Paclitaxel/Taxol® (Pacific yew tree; see Chapter 7 opener)
Phenols	Salicylic acid in aspirin (willow tree)
	Capsaicin (chili pepper plants)
Antibiotics	Penicillin (*Penicillium glaucum* fungus)
	Streptomycin (*Streptomyces griseus* bacterium)

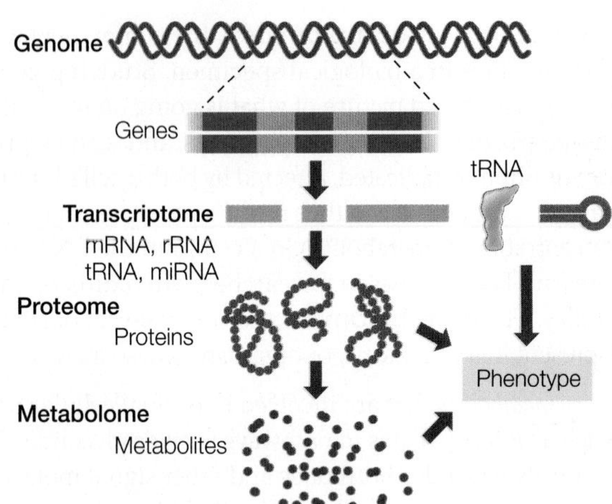

FIGURE 12.7 Genomics, Transcriptomics, Proteomics, and Metabolomics A combination of these approaches can give more comprehensive information about genotypes and phenotypes.

In addition to those listed here, other -omics fields exist, including epigenomics. What do you think is the area of interest for epigenomics? Would you expect to see variation in a multicellular organism's epigenome across different tissues?

Taken together, the genome, transcriptome, proteome, and metabolome can move biologists toward a more comprehensive picture of an organism's genotype and phenotype (**FIGURE 12.7**).

REVIEW & APPLY | 12.1

R Short stretches of DNA can be sequenced by Sanger dideoxy sequencing, involving chain termination using nucleotides missing the 3′ —OH group on deoxyribose. The development of high-throughput sequencing, using fluorescently labeled nucleotides and adding and identifying them one by one to a growing nucleotide chain, has enabled a great expansion of genome sequencing. In both methods, DNA is first sheared into short fragments (300–1,000 bp) and the overlapping fragments are individually sequenced and assembled by computer. Following assembly, genomes are annotated to describe their functional regions. Annotation identifies protein-coding regions and their associated sequences that control transcription and translation, RNA genes, and chromosome structures such as centromeres and telomeres. Transcriptomics uses RNA-seq, in which RNA is reverse transcribed into cDNA that is then sequenced, to identify genes being expressed in cells in a sample. Proteomics describes the totality of the proteins in an organism. Metabolomics seeks to identify and determine the abundance of the small molecules that participate in overall metabolism or are used in defense. Genomics, transcriptomics, proteomics, and metabolomics together give a detailed picture of the functions of an organism.

A

1. Using a table, compare genomics, transcriptomics, proteomics, and metabolomics with regard to the methods used and information obtained.
2. A DNA molecule is cut into the following fragments: AGTTT, TAGG, CGAT, and CCT. The same molecule is cut in a different way to produce TTCGA, TCCT, AGT, GG, and TA. A third cut produces TTCGAT, CCTT, AGG, and AGT. What is the sequence of this DNA molecule?
3. Many computer programs have been written that search for open reading frames (ORFs). These programs are very good at finding protein-coding genes in prokaryotes but perform substantially less well in eukaryotes. Why is that the case?

The first cellular genomes to be fully sequenced were those of prokaryotes. In the next key concept we will discuss these relatively small, compact genomes.

12.2 | Prokaryotic Genomes Are Small, Compact, and Diverse

LEARNING OBJECTIVES

By the end of this key concept you should be able to:

12.2.1 Describe what is meant by an essential gene.

12.2.2 Define the minimal genome.

12.2.3 Distinguish the core genome from the pan genome for a prokaryotic species.

12.2.4 Define metagenomics and give examples of its use.

12.2.5 Describe and compare transposon-mediated mutagenesis and CRISPR-Cas9 gene editing.

When DNA sequencing became possible in the late 1970s, the first genomes to be sequenced were those of simple viruses having only a few thousand nucleotides. The sequences quickly provided new information about how viruses infect their hosts and reproduce. The jump from sequencing viral genomes to bacterial genomes, with sizes in the millions of nucleotides, was substantial. It was not until 1995 that the first complete genome sequence of a free-living cellular organism, the bacterium *Haemophilus influenzae,* was published by a team led by Craig Venter and Hamilton Smith. We now know the genome sequences of thousands of different prokaryotic species, from which we have learned a great deal.

Prokaryotic genomes are streamlined and diverse

There are several notable features of bacterial and archaeal genomes:

- They are relatively small. Prokaryotic genomes range in size from about 160,000 to 12 million bp and are usually organized into a single circular chromosome.
- They are compact. Typically, more than 85 percent of the DNA consists of protein-coding regions or RNA genes, with only short sequences between genes.

TABLE **12.2** | Initial Estimates of Protein-Coding Genes in a Strain of Three Species of Bacteria from Genome Sequencing

	Number of genes in:		
Category	***E. coli***	***H. influenza***	***M. genitalium***
Total protein-coding genes	4,288	1,743	483
Biosynthesis of amino acids	131	68	1
Biosynthesis of cofactors	103	54	5
Biosynthesis of nucleotides	58	53	19
Cell envelope proteins	237	84	17
Energy metabolism	243	112	31
Intermediary metabolism	188	30	6
Lipid metabolism	48	25	6
DNA replication, recombination, and repair	115	87	32
Protein folding	9	6	7
Regulatory proteins	178	64	7
Transcription	55	27	12
Translation	182	141	101
Uptake of molecules from the environment	427	123	34
Other functions (including unknown)	2,314	869	205

Data from F. R. Blattner et al. 1997. *Science* 227: 1453–1462; R. D. Fleischmann et al. 1995. *Science* 269: 496–512; C. M. Fraser et al. 1995. *Science* 270: 397–404.

- Their genes usually do not contain introns. Exceptions include the rRNA and tRNA genes of archaea, which are frequently interrupted by introns.
- In addition to the main chromosome, they often carry smaller, circular DNA molecules called plasmids, which may be transferred between cells (see Key Concept 8.4).

Beyond these broad similarities, there is great diversity among prokaryotes in their genomes, reflecting the huge variety of environments where they are found. **TABLE 12.2** shows the functions of genes found in three species of bacteria. Functions are determined either from direct experimentation (especially in *E. coli*) or from similarities of gene sequences to genes from other species with known function. Note that even in prokaryotes, with their relatively small genomes, there are many genes with unknown function. This is especially true for species that represent groups that have not been studied extensively. When the first strain of *E. coli* was sequenced, its chromosome was estimated to contain 4,288 protein-coding genes (the current estimate is 4,140). At the time, only 2,656 (61%) of the open reading frames encoded proteins with known functions. Even now, the functions of over 10 percent of the protein-coding genes in the strain that was first sequenced remain unknown.

Insights into the biology of a species or strain can often be obtained from the list of genes that are present or absent. For example, the bacterium *Mycoplasma genitalium* has a substantially smaller genome than the other two bacterial species in Table 12.2, in part because it relies on its host's cells for many functions. It lacks genes for several of the biochemical pathways that other bacteria have. For example, *M. genitalium* lacks most of the enzymes needed to synthesize amino acids (see Table 12.2). This indicates that *M. genitalium* obtains its amino acids from its environment (usually the human urogenital tract).

In addition to learning about individual species or strains by examining the genes they possess or lack, comparative genomics across tens of thousands of prokaryotic strains has revealed several unexpected patterns:

- The number of genes shared across all prokaryotes is quite small, implying that only a few hundred genes are required for life.
- Comparisons across genetically differing strains of the same species reveal variation in gene content. This observation has led to the distinction between the **core genome**, containing genes common to all strains of a species, and the **pan genome**, the collection of all genes found in a species across all strains. In some species the core genome and the pan genome are similar in size, indicating that most strains contain the same number and identity of genes. However, in other species the difference can be substantial. For example, in *E. coli* the core genome has about 3,000 genes whereas the pan genome has more than 15,000 genes. Thus any two strains of *E. coli* will share about 3,000 genes but can differ for some or all of the remaining 1,000 or so genes. Variation across strains makes referring to "the" *E. coli* genome problematic. The solution is that "the" genome of *E.coli* refers to a particular strain, the so-called "reference" strain, which was the first strain sequenced. In general, all species have one genome that is categorized as the reference. Species with a large pan genome relative to their core genome, such as *E. coli*, often have extensive metabolic diversity, reflecting the variation in the environments in which different strains are found (**FIGURE 12.8**).
- Many examples have been found in which a gene in a particular species was clearly acquired from a different species—an event termed lateral gene transfer. We have already described lateral transfer of plasmids among bacterial cells (see Key Concept 8.4). In some cases plasmids can be transferred among species, which is one mechanism by which lateral gene transfer occurs. Another mechanism is transformation, in which a bacterial cell takes up DNA from the environment and then inserts it into its genome.

In summary, while some prokaryotic species have genomes that are astoundingly diverse in terms of their gene content, they generally remain small in size, with few or no introns, and have short regions between open reading frames. This pattern suggests that small genome size is strongly favored in these microbes.

Metagenomics reveals the diversity of viruses and prokaryotic organisms

If you take a microbiology laboratory course, you will learn how to identify various microorganisms on the basis of the conditions (such as aerobic versus anaerobic) under which they will

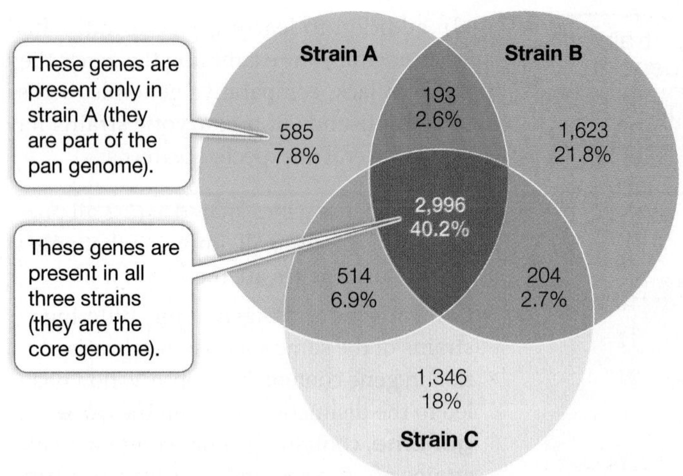

FIGURE 12.8 The Pan and Core Genomes in *Escherichia coli* Comparison of gene content across three strains of *E. coli*. Numbers indicate numbers of protein-coding genes; percentages indicate how much of the pan genome the numbers represent. Note that of the 7,461 protein-coding genes identified (the pan genome for these three strains), only 2,996 (40.2%) are found in all three strains (the core genome for these three strains). (After R. A. Welch et al. 2002. *Proc Natl Acad Sci USA* 99: 17020–17024. © 2002 National Academy of Sciences, U.S.A.)

grow in lab cultures. For example, staphylococci are a group of bacteria that inhabit skin and nasal passages. Unlike many bacteria, staphylococci can use the sugar alcohol mannitol as an energy source and thus can grow on a special medium containing mannitol and a high salt concentration. Often a dye is included in the medium, which changes color if the bacteria are pathogenic. Such culture methods have been the mainstay of microbial identification for more than a century and are still useful and important.

However, until recently it was not clear how many microbes cannot be cultured in the laboratory and were thus essentially unknown to science. With the advent of PCR and modern DNA analysis techniques, it is now possible to study microbes without culturing them in the laboratory. In 1985 Norman Pace, then at Indiana University, came up with the idea of isolating DNA directly from environmental samples. He used PCR to amplify specific sequences from the samples to determine which microbes were present. The PCR products were sequenced to explore their diversity. The term **metagenomics** was coined to describe this approach of analyzing genes without isolating the intact organism. It is now possible to do DNA sequencing with samples from almost any environment. The sequences can be used to detect the presence of previously unidentified organisms as well as known microbes, determine what metabolic diversity exists in a microbial community, and assemble genomes of previously unknown microbes (**FIGURE 12.9**). For example:

- Sequencing of DNA from 200 liters of seawater indicated that it contained 5,000 different viruses and 2,000 different bacteria, many of which had not been described previously.
- Sequencing RNA from nasal samples of COVID-19 patients revealed enormous variation in SARS-CoV-2 prevalence and the variable presence of other viruses and bacteria in their "infectomes."

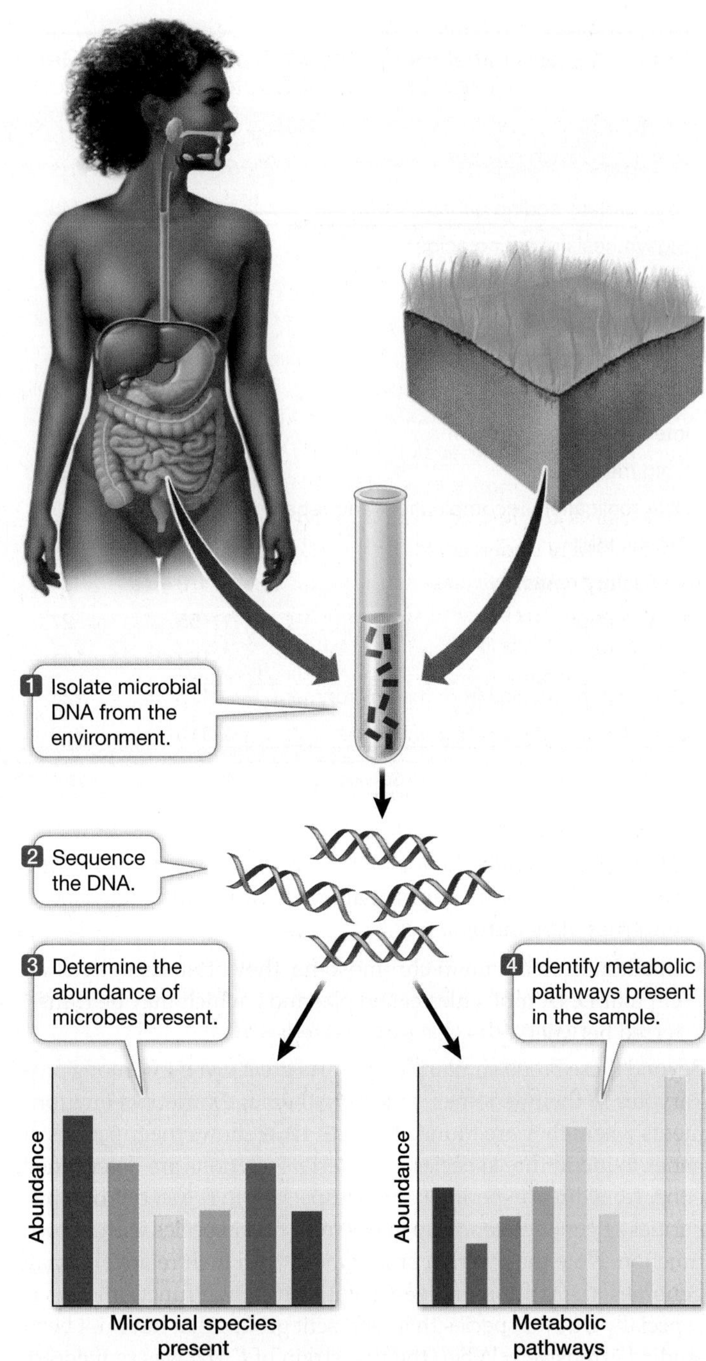

FIGURE 12.9 Metagenomics Microbial DNA is extracted from a sample and used to determine the identity and abundance of microbial species and the identity of metabolic pathways in the community. (After X. C. Morgan and C. Huttenhower. 2012. *PLoS Comput Biol* 8: e1002808; X. C. Morgan et al. 2013. *Trends Genet* 29: 51–58.)

- Water runoff from a mine contaminated with toxic chemicals contained many new species of prokaryotes thriving in this apparently inhospitable environment. Some of these organisms exhibited metabolic pathways that were previously unknown to biologists. These organisms and their capabilities may be useful in cleaning up pollutants from the water.
- Gut samples from 124 Europeans revealed that each person harbored at least 160 species of bacteria (constituting their gut

microflora or microbiome). Many of these species were found in all of the individuals, but the presence of other bacteria varied from person to person. Such variations in gut microflora may be associated with obesity or bowel diseases.

These and other discoveries are truly extraordinary and potentially very important. From various metagenomic studies, it is estimated that 90 percent of the microbial world has been invisible to biologists. Entirely new ecosystems of bacteria and viruses are being discovered and functionally characterized by examining the occurrence and abundance of genes encoding enzymes involved in different metabolic pathways. It is hard to overemphasize the importance of such an increase in our knowledge of the hidden world of microbes. This new knowledge underscores the remarkable diversity among prokaryotic organisms and will further our understanding of natural ecological processes. Furthermore, it has the potential to help us find better ways to manage environmental catastrophes such as oil spills, to remove toxic heavy metals from soil and water, and to discover new chemical compounds that have medical uses.

Some sequences of DNA can move about the genome

Genome sequencing of prokaryotes allowed scientists to study more broadly a class of DNA sequences that had been discovered by eukaryotic geneticists decades earlier. Segments of DNA called **transposons** (or transposable elements) can move from place to place in the genome and can even move from one piece of DNA (such as a chromosome) to another (such as a plasmid) in the same cell. A transposon might be at one location in the genome of one *E. coli* cell, and at a different location in another cell. The insertion of transposon DNA sequence from elsewhere in the genome into the middle of a protein-coding gene disrupts that gene. Any mRNA expressed from the disrupted gene will have the extra sequence, and the resulting protein will be abnormal. Transposons can also insert into regulatory regions, usually altering or knocking out expression. Consequently, movement of transposons can produce significant phenotypic effects by inactivating genes.

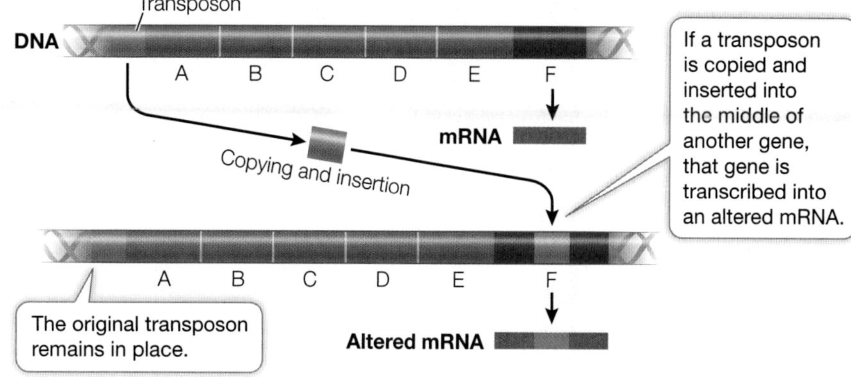

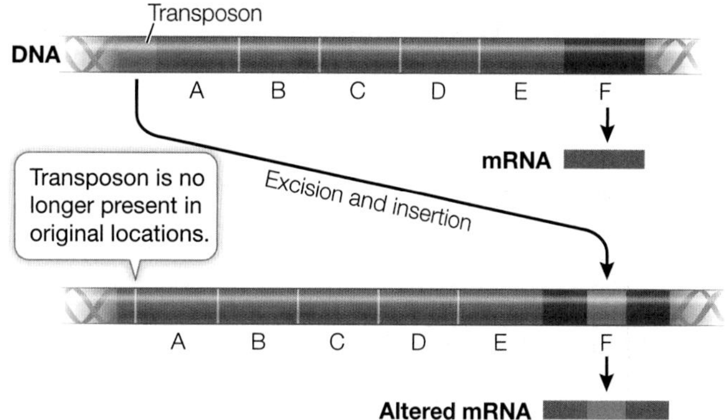

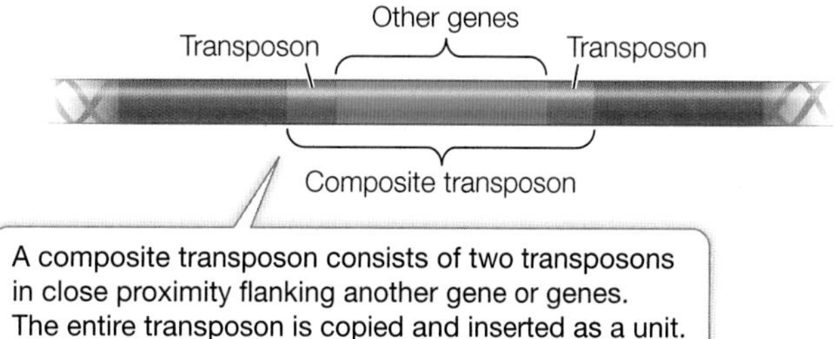

FIGURE 12.10 Transposons Can Move Transposons are DNA sequences that move from one location to another. **(A)** In replicative transposition, the DNA sequence is replicated and the copy inserts elsewhere in the genome. **(B)** In nonreplicative transposition, the DNA sequence is excised from one location and inserted elsewhere. **(C)** If two transposons are close enough, they may move together as a composite transposon, taking the intervening DNA sequences with them.

Transposons are often short sequences of 1,000 to 2,000 bp and are found at many sites in prokaryotic genomes. The mechanisms that allow them to move vary. For example, the transposon may be replicated, and then the copy inserted into another site in the genome (replicative transposition; **FIGURE 12.10A**), or the transposon might be excised from one location and move to another location (nonreplicative transposition; **FIGURE 12.10B**). If a transposon becomes duplicated, with two copies separated by one or a few genes, the result may be a single larger composite transposon (up to about 5,000 bp). In this case the additional genes can be carried to different locations in the genome (**FIGURE 12.10C**). Some composite transposons carry genes for antibiotic resistance, allowing these genes to move from the genome to plasmids, or vice versa. We will discuss transposons again in Key Concept 12.3.

The genes that are essential for cellular life can be identified experimentally

Comparative genomics in prokaryotes indicates that only a few genes are present in all of them (core genes common to prokaryotic species). Perhaps not surprisingly, many prokaryotic core genes are also found in the core genomes of eukaryotes—some core genes are universal. There are also some (nearly) universal gene segments that are present in many genes in many organisms. One example is a sequence encoding an ATP binding site, which

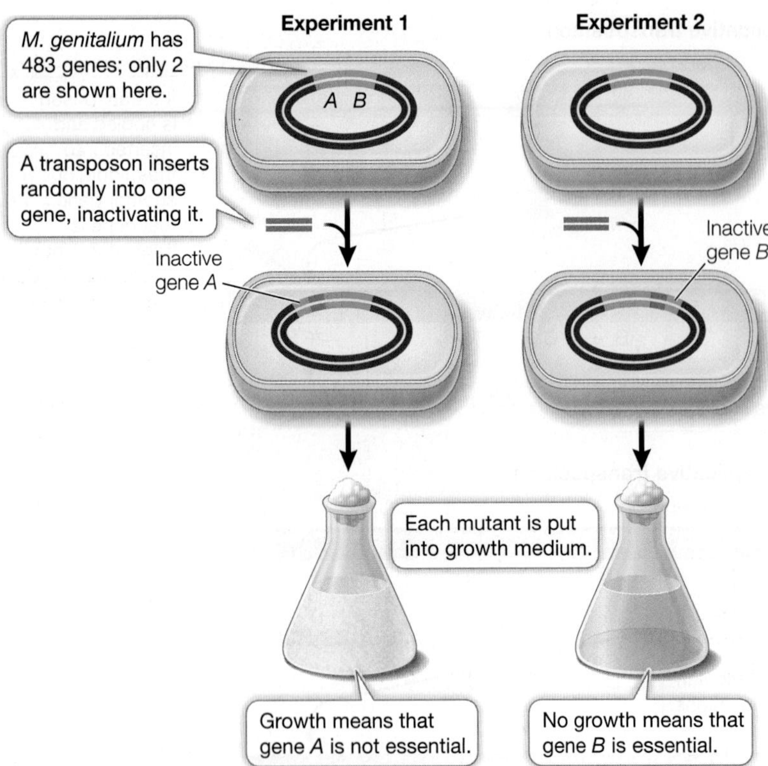

FIGURE 12.11 Transposon Mutagenesis to Determine the Minimal Genome *Mycoplasma genitalium* has one of the smallest known genomes of any prokaryote. But are all of its genes essential to life? By inactivating each of its genes, one by one, scientists determined which of them are essential for the cell's survival.

is a domain found in many proteins. These findings suggest that there is an ancient, minimal set of DNA sequences common to all cells that are absolutely required for life: the minimal genome. One way to identify these sequences is to look for them in computer analyses of sequenced genomes.

Another way to define the minimal genome is to take an organism with a simple genome and deliberately mutate one gene at a time to see what happens. *M. genitalium* has one of the smallest known genomes, with only 483 protein-coding genes. Even so, some of its genes are dispensable under some circumstances. For example, it has genes for metabolizing both glucose and fructose, but it can survive in the laboratory on a medium containing only one of these sugars. Under such circumstances, the bacterium doesn't need the genes for metabolizing the other sugar.

How do scientist determine which genes are essential? Researchers can address this question using transposons as mutagens. When transposons in a bacterium are activated, they insert themselves into the genome at random, so that particular strains contain mutations that have a random gene inactivated (**FIGURE 12.11**). The mutated bacteria can be tested for growth and survival, and DNA from interesting mutants sequenced to find out which genes are mutated. Genes for which mutations are never obtained are essential genes —they are required for life under laboratory conditions. When transposon mutagenesis was done with *M. genitalium*, the astonishing result was that only 381 of the protein-coding genes were needed for survival in the laboratory! Similar studies in different prokaryotic species reveal similar results—only a few hundred genes are essential for survival under laboratory conditions. Similar experiments are being carried out in eukaryotes as well to specify the minimal genome in those more complex species.

The ability to knock out genes experimentally is a powerful tool for determining their function. A new technique for doing so is called **CRISPR-Cas9 gene editing** (see Figure 12.12). The CRISPR acronym stands for Clustered Regularly Interspaced Short Palindromic Repeats and refers to short repeating sequences of DNA (about 30 base pairs in length) that were observed in certain locations in prokaryotic genomes. This technique is much more straightforward and specific than transposon mutagenesis.

Activity 12.1 CRISPR Simulation

REVIEW & APPLY | 12.2

R Prokaryotic genomes vary in size, though many contain about a few thousand genes. Among these genes is a set of genes found in all strains of a species —the core genome—and additional genes found in subsets of strains that together make up the pan genome. The pan genome can be about the same size as the core genome or substantially larger, depending on the diversity of environments in which the species is found. Comparing core genomes across species can help define an essential set of genes required for life. Another approach to identifying essential genes is to inactivate each gene one by one to determine whether the organism can survive without it. Transposon mutagenesis and CRISPR-Cas9 are two ways to inactivate genes. Prokaryotic genomes are usually streamlined, such that most base pairs have function: introns and sequences between genes are short or absent. Metagenomics has shown that microbial diversity is at least an order of magnitude (tenfold) greater than originally thought.

A

1. You have sequenced three strains of a newly discovered bacterium and annotated all of the genes in each one. From these genomes, how would you identify the core genome, the pan genome, and for one of the strains, the essential genes?
2. If the genetic code was not universal, how might the patterns observed in comparative genomics of prokaryotes differ?
3. You want to isolate a prokaryote that can live on discarded Styrofoam cups. Such an organism might live in a landfill where ground-up cups are discarded. How would you use metagenomics to identify such a bacterium?
4. What is the advantage of CRISPR-Cas9 gene editing over transposon mutagenesis for introducing knockout or other mutations into a genome in a cell?

THINK LIKE A SCIENTIST

Process of Science

Inactivation of specific genes using CRISPR-Cas9 gene editing

A major problem with transposon mutagenesis is that it creates mutations in the genome randomly—that is, scientists cannot predict which gene(s) will be inactivated by transposon insertion, and indeed, some genes may never get mutated in this way. Methods have been developed to target specific genes for mutation, the most powerful being CRISPR-Cas9 gene editing (**FIGURE 12.12**). Many prokaryotes (both bacteria and archaebacteria) have a CRISPR-based adaptive immune system that is used to protect cells against repeated infections by the same virus. A short sequence of the viral genome is incorporated into a special gene (called CRISPR) as a "spacer"—one of many spacers separated from one another by a repeated sequence. The gene is transcribed into an RNA that is processed to make "guide RNAs," which include RNA transcribed from both the spacer and adjacent repeated sequence. Guide RNAs can pair with DNA that is complementary to the spacer region and cause the DNA to be cleaved by an associated Cas9 protein, which is an endonuclease that causes double strand breaks.

Scientists use CRISPR-Cas9 to perform gene editing. In brief, custom-made guide RNAs can target the Cas9 endonuclease to any DNA sequence to cause DNA breaks. These breaks will then be repaired using either nonhomologous end joining (see Key Concept 9.3) or another DNA molecule that is supplied by the researcher. Nonhomologous end joining simply ligates the two cut DNA molecules. However, this repair system is not perfect and is error-prone—after broken ends are rejoined, the sequence often has a few insertions, deletions, or substitutions. Thus if a double-stranded chromosome break is made at a particular location and then nonhomologous end joining repairs the break, several mutations will likely be introduced. Alternatively, repair uses another DNA molecule as a template for repair, which allows the DNA sequence at the target location to be changed to another sequence. CRISPR-Cas9 gene editing can thus be used either to knock out or to alter the sequence of a gene. This opens up tremendous possibilities for altering life in directed ways. So it is not surprising that efforts are under way to alter crops for better growth and to repair genetic diseases in humans.

1. If a genome is 1 billion (10^9) bp in length, how long would a target sequence have to be to be unique? (Assume all four bases are equally frequent.)
2. If you wanted to knock out a eukaryotic gene by introducing one or two base-pair changes, which parts of the gene might you target?
3. Using CRISPR-Cas9 gene editing to knock out a gene in a multicellular diploid organism presents challenges not present in unicellular haploid organisms. What are these challenges, and how might they be overcome?

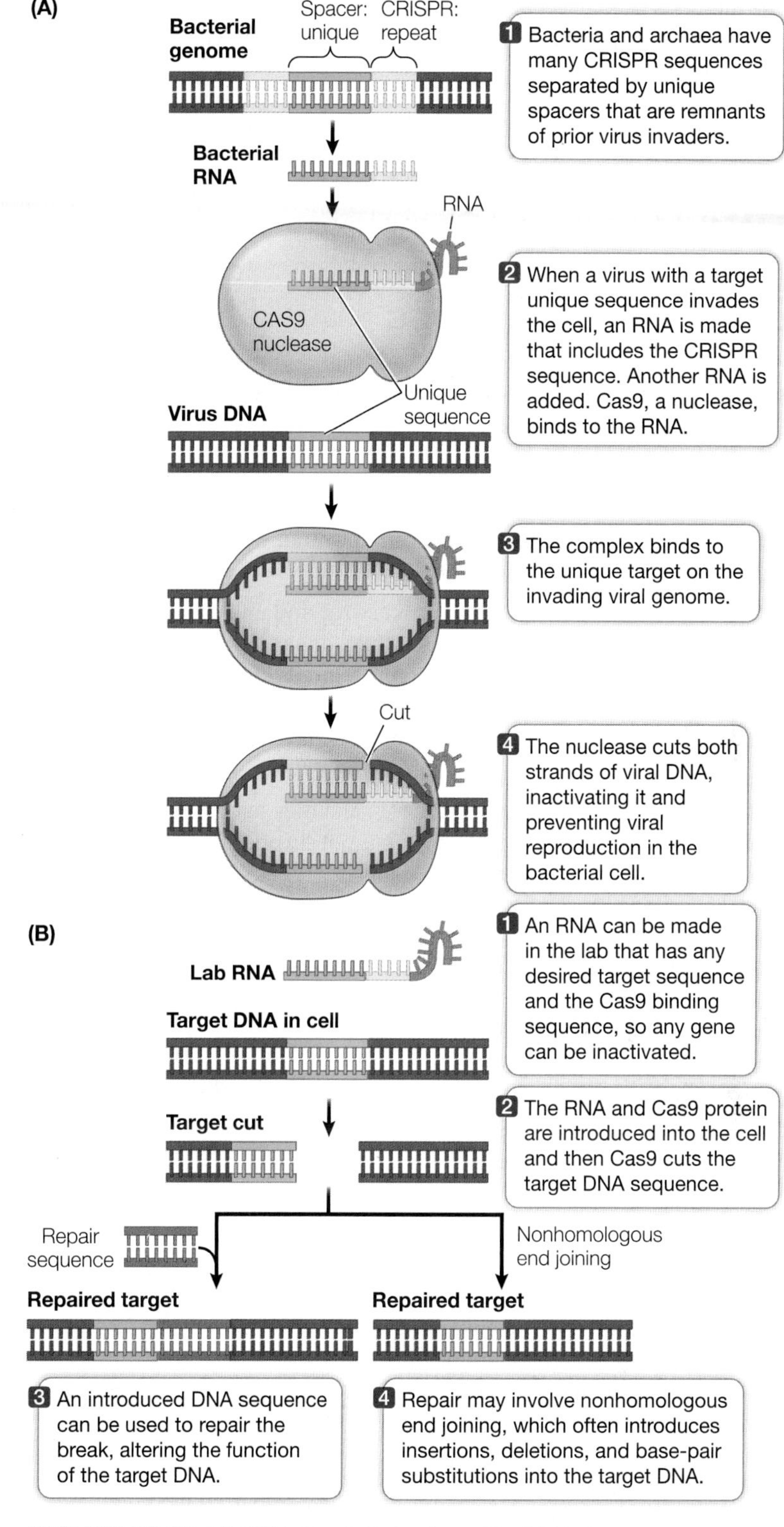

FIGURE 12.12 CRISPR-Cas9 Cuts Specific DNA Sequences **(A)** A simplified overview of CRISPR-mediated viral defense in prokaryotes. In brief, spacers allow the Cas9 RNA-directed DNA nuclease to recognize and degrade viral DNA that has previously infected it. **(B)** CRISPR-Cas9 gene editing. The spacer sequence is complementary to the sequence that is being targeted for a double-stranded break in the genome. Repair of the break can use nonhomologous end joining, which introduces mutations and likely renders the gene nonfunctional. Alternatively, an additional DNA molecule can be introduced so that repair results in a new gene sequence.

The methods used to sequence and analyze prokaryotic genomes have also been applied to eukaryotic genomes, which we will examine next.

12.3 | Eukaryotic Genomes Are Large and Complex

LEARNING OBJECTIVES

By the end of this key concept you should be able to:

12.3.1 List four differences between prokaryotic and eukaryotic genomes.

12.3.2 Explain gene duplication and the possible fates of a duplicate gene.

12.3.3 Describe the two kinds of repetitive sequences in eukaryotic genomes.

One obvious difference between prokaryotic and eukaryotic genomes is that eukaryotes possess multiple, linear chromosomes, while prokaryotes generally have a single circular chromosome (see Key Concept 4.5). However, as genomes have been sequenced and described, several additional differences have emerged:

- *Eukaryotic genomes, especially in multicellular species, are substantially larger than those of prokaryotes,* and they have more protein-coding genes. This difference is not surprising given that multicellular organisms have many cell types with specialized functions. Prokaryotes such as *E. coli* contain about 4,000 genes in a genome that is about 5 million bp. A rice plant, in contrast, has about 36,000 protein-coding genes in a genome of about 375 million bp. Other eukaryotic genomes are much larger, including human's at 3.1 billion bp, though they do not necessarily contain more protein-coding genes (about 20,000 in the human genome).
- *Eukaryotic genomes have more regulatory sequences*—and many more regulatory proteins—than prokaryotic genomes. The greater complexity of eukaryotes requires much more regulation, which is evident in the many points of control associated with the expression of eukaryotic genes (see Key Concepts 11.3–11.5 and Figure 11.1).
- *Much of eukaryotic DNA does not encode functional proteins.* Distributed throughout many eukaryotic genomes are various kinds of DNA sequences that are not transcribed into mature mRNA. Some of these sequences are genes for functional RNAs, such as rRNA, tRNA, and miRNA. Others are introns or regulatory sequences such as promoters, enhancers, and silencers. In addition, eukaryotic genomes contain large amounts of various kinds of repeated sequences that do not encode functional proteins.
- Unlike in prokaryotes, in a eukaryotic species different individuals usually have the same complement of genes; the core and pan genomes are essentially the same.

Model organisms reveal many characteristics of eukaryotic genomes

Among the first eukaryotic genomes to be sequenced were those from some of the model organisms that had been studied extensively. Model organisms are relatively easy to grow and study in a laboratory, and their short life cycles make them suitable for genetics studies. In addition, they exhibit characteristics representative of a larger group of organisms. Some model organisms, particularly the yeast *Saccharomyces cerevisiae*, the nematode (roundworm) *Caenorhabditis elegans*, the fruit fly *Drosophila melanogaster*, and the plant *Arabidopsis thaliana*, have relatively small genomes, less than 150 million bp, which makes them especially suitable for genome sequencing. **TABLE 12.3** shows some characteristics of the genomes of these four species and several others.

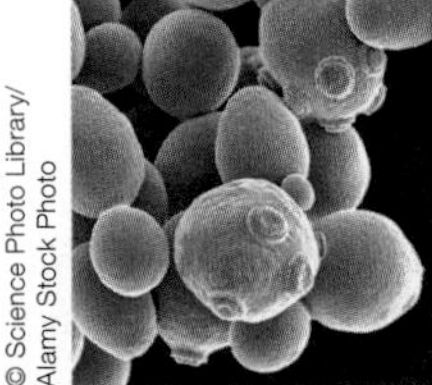

YEAST Baker's yeast, *Saccharomyces cerevisiae*, is a single-celled eukaryotic microbe and is the least complex of the eukaryotic model organisms. It has a microbial lifestyle like a prokaryote but possesses membrane-enclosed organelles like other eukaryotes. Yeast has 16

TABLE **12.3** | Sequenced Genomes of *E. coli* and Eleven Eukaryotic Species

Organism	Haploid genome size (Mbp)[a]	Number of protein-coding genes	Percentage of genome that encodes proteins
BACTERIA			
Escherichia coli	4.6	4,140	88
YEASTS			
Saccharomyces cerevisiae	12.2	6,600	70
PLANTS			
Arabidopsis thaliana	120	27,655	25
Oryza sativa (rice)	375	37,960	12
Populus trichocarpa (poplar)	434	41,335	9
Sorghum bicolor (great millet)	709	34,118	6
Glycine max (soybean)	978.5	56,897	7
Zea mays (corn)	2,135	39,591	2.5
ANIMALS			
Caenorhabditis elegans (nematode)	100	20,191	25
Drosophila melanogaster (fruit fly)	144	13,968	13
Mus musculus (house mouse)	2,728	22,481	1.3
Homo sapiens (human)	3,014	20,437	1.2

Data from Ensembl release 103–February 2021. © EMBL-EBI and https://www.ncbi.nlm.nih.gov.

[a]1 Mbp = 1,000,000 base pairs

linear chromosomes and 6,600 protein-coding genes. As in *E. coli*, most of the genes in yeast are not essential—only 20 percent or so are required for growth on rich medium. A striking difference between the yeast genome and that of *E. coli* is in the number of genes involved in secreting or targeting proteins to specific locations in the cell: yeast has 430 such genes, while *E. coli* has only 35. Both of these single-celled organisms appear to use about the same number of genes to perform the basic functions of cell survival. It is the compartmentalization of the eukaryotic yeast cell into organelles that requires it to have more genes. This finding is direct, quantitative confirmation of something we have known for a century: the eukaryotic cell is structurally and functionally more complex than the prokaryotic cell.

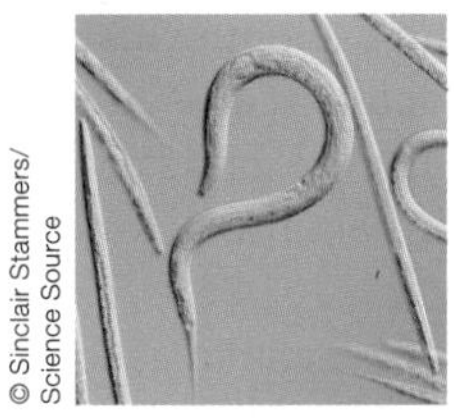
© Sinclair Stammers/ Science Source

NEMATODE The 1-millimeter-long nematode *Caenorhabditis elegans* normally lives in the soil but is easily reared in the laboratory. It is an important model organism for understanding tissue differentiation because it is a relatively simple multicellular eukaryote, with only 1,000 or so somatic cells in an adult and a transparent body. It develops over 3 days from a fertilized egg to an adult worm that has a nervous system, digests food, and reproduces sexually. Thousands can be kept on a standard petri dish and fed *E. coli*. The genome of *C. elegans* is eight times larger than that of yeast and has about three times as many protein-coding genes. Many of these extra genes encode proteins needed for cell differentiation, for intercellular communication, and for holding cells together to form tissues. As is similar in other organisms, most of the genes in the nematode are not essential—only 30 percent or so are required in a laboratory environment.

David McIntyre

FRUIT FLY The fruit fly *Drosophila melanogaster* is a famous model organism that has been used in genetics studies since shortly after the rediscovery of Mendel's work. Studies of fruit fly genetics resulted in the formulation of many basic principles of genetics (see Key Concept 8.2). The fruit fly is a much larger organism than *C. elegans* (it has ten times as many cells), and it is much more complex: it undergoes complicated developmental transformations from egg to larva to pupa to adult. These differences are reflected in the fruit fly genome, which has many genes encoding transcription factors needed for complex embryonic development. In general, the fruit fly genome has a distribution of coding sequence functions quite similar to those of many other complex eukaryotes. Most of the genes in fruit flies are not essential—only 25 percent or so are required in a laboratory environment.

David McIntyre (from seed provided by Tobias Baskin)

ARABIDOPSIS THALIANA About 250,000 species of flowering plants dominate the land and fresh water. Although there is generally more interest in the plants we use for food and fiber, scientists first sequenced the genome of a flowering plant with a relatively small genome. *Arabidopsis thaliana* is a member of the

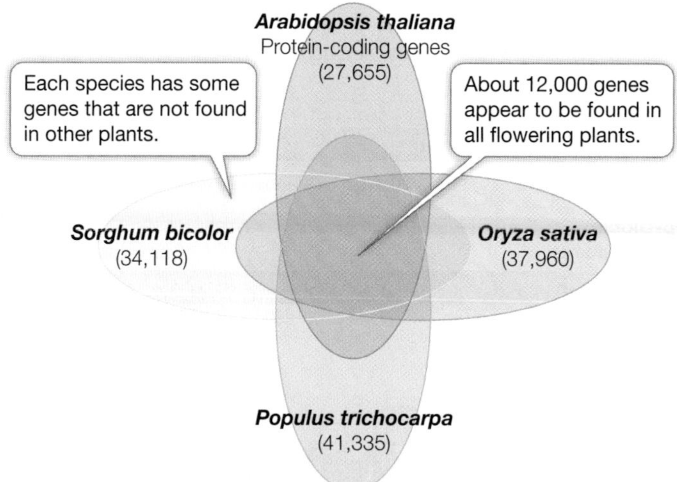

FIGURE 12.13 Plant Genomes Four plant genomes share a common set of approximately 12,000 genes that may comprise the plant core genome. The total numbers of protein-coding genes are shown in parentheses. Data from Ensembl release 103–February 2021. © EMBL-EBI.

mustard family and has long been a favorite model organism of plant biologists. It is small (a hundred could grow and reproduce in the space occupied by this page), easy to manipulate, and has a very short generation time (about 2 months). Its genome has 27,655 protein-coding genes, but many of these are duplicates, arising from chromosomal rearrangements. When these duplicate genes are subtracted from the total, about 15,000 unique genes are left—similar to the gene numbers found in fruit flies and nematodes. Indeed, many of the genes found in these animals have **orthologs**—genes in different species that are derived from a common ancestral gene—in *A. thaliana* and other plants, supporting the idea that plants and animals have a common ancestor.

A. thaliana has some genes, however, that are unique to plants, including genes involved in photosynthesis, in the transport of water throughout the plant, in the assembly of the cell wall, in the uptake and metabolism of inorganic substances from the environment, and in the synthesis of specific molecules used for defense against microbes and herbivores (secondary metabolites; see Key Concept 12.1). These plant defense molecules may be a major reason why the numbers of protein-coding genes in some plants are higher than in many animals.

Comparisons among diverse flowering plant species (including *A. thaliana*, rice, poplar, and sorghum) suggest that about 12,000 protein-coding genes are shared among all flowering plants (**FIGURE 12.13**). These may comprise the plant core genome.

Gene families are common in many eukaryotic species

In Chapter 9 we described gene duplication, which results in more than one copy of a given gene in the genome. Recall, for example, that chromosomal rearrangements can result in duplications (see Figure 9.20B). Sometimes only one or a few genes are duplicated at a time; sometimes an entire genome becomes duplicated (recall polyploidy from Chapter 7). In prokaryotes 90 percent of genes are

(A)

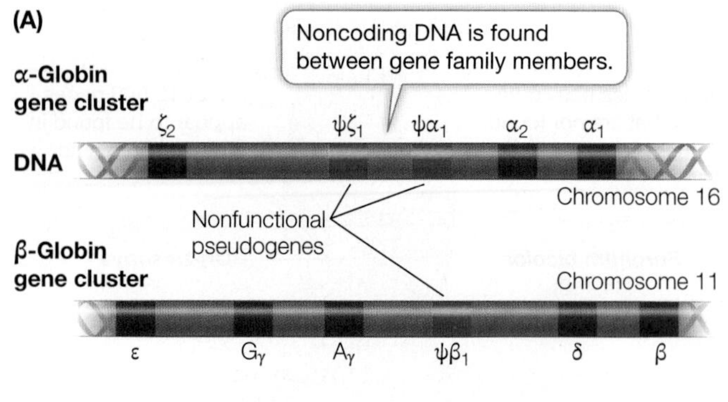

(B)

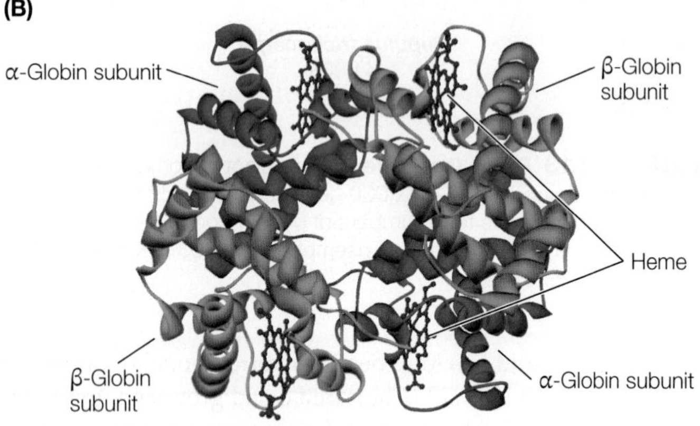

FIGURE 12.14 The Globin Gene Family **(A)** The α-globin and β-globin clusters of the human globin gene family are located on different chromosomes. The paralogous genes of each cluster are separated by noncoding DNA. The nonfunctional pseudogenes are indicated by the Greek letter psi (ψ). **(B)** Adult hemoglobin consists of two identical α-globin subunits, two identical β-globin subunits, and four heme cofactors. (Data from https://www.rcsb.org/structure/2HHB. G. Fermi et al. 1984. *J Mol Biol* 175: 159–174.)

likely to be unique, meaning they share little or no DNA or amino acid sequence identity with one another. In eukaryotes, however, the proportion of unique genes is lower, substantially so in some species. For example, in *A. thaliana* only 35 percent of genes are unique. The other 65 percent of genes are present in multiple copies. Multiple copies of a gene arise by gene duplication.

Following a gene duplication event, both the original gene and the new copy can be mutated. If the new copy has a mutation that renders its gene product (e.g., a protein) nonfunctional, it is called a **pseudogene**. The DNA sequence of a pseudogene may not differ greatly from that of the original gene. It may simply lack a promoter, for example, and thus fail to be transcribed. Or it may lack a recognition site needed for the removal of an intron, so that the transcript it makes is not correctly processed into a useful mature mRNA.

Over evolutionary time, a duplicate gene may become functionally important. For example, if increased expression of the original gene is beneficial, natural selection favors increasing its copy number by gene duplication. Alternatively, if a duplicate gene accumulates mutations that give it a novel, beneficial function, its maintenance in the genome will be favored by natural selection, as long as the original function is still encoded by the original copy.

Genes that arise via gene duplication within a species are termed **paralogs**, and two or more paralogs in a genome make up a group of closely related genes called a **gene family**. Some gene families, such as those encoding the globin proteins that make up hemoglobin, discussed below, contain only a few members. Other families, such as the genes encoding ribosomal RNA, have hundreds of members. Ribosomal RNA genes are identical in sequence, and the multiple copies allow substantially increased levels of gene expression (recall from Key Concept 10.1 that rRNA is typically the most abundant RNA in a cell).

In contrast to rRNA genes, the globin paralogs in vertebrates are not identical in sequence. These genes encode proteins that are found in hemoglobin and myoglobin (an oxygen-binding protein present in muscle). The globin genes arose from a single common ancestral gene. In humans there are three functional members and two pseudogenes in the α-globin cluster and five functional members and one pseudogene in the β-globin cluster (**FIGURE 12.14A**). Each hemoglobin molecule in an adult human is a tetramer containing two identical α-globin subunits, two identical β-globin subunits, and four heme cofactors (**FIGURE 12.14B**).

During human development, different genes of the globin gene cluster are expressed at different times and in different tissues. This differential gene expression has physiological significance. For example, the hemoglobin in the human fetus contains γ-globin, which binds O_2 more tightly than adult hemoglobin does. This specialized form of hemoglobin ensures that in the placenta, O_2 is transferred from the mother's blood to the developing fetus's blood. Just before birth, red blood cell precursors in the bone marrow stop synthesizing fetal hemoglobin and start making the adult forms (two α and two β). Thus hemoglobins with different binding affinities for O_2 are provided at different stages of human development.

Duplication events resulting in the evolution of various globin proteins are diagrammed in a gene family tree in **Figure 15.6**.

Eukaryotic genomes contain many repetitive sequences

Table 12.3 shows the genome size, number of protein-coding genes, and percentage of the genome that codes for proteins for 11 species of eukaryotes and *E. coli*. Examination of the data indicates that in multicellular eukaryotes less than half of the genome encodes proteins—only 1.2 percent in humans, for example. What about the rest of the genome? **FIGURE 12.15** provides an overview of the sequences that make up the human genome.

In addition to the 1–2 percent of the genome encoding protein-coding sequences, about 39 percent encodes genes or gene-related sequences. These include introns (30 percent of the genome) and regulatory regions, including promoters, RNA genes, and pseudogenes. That leaves about 60 percent of the human genome that is intergenic (not gene-related). Other multicellular eukaryotes have similarly high values of intergenic regions. Much of these intergenic regions is composed of repetitive DNA sequences.

Repetitive DNA in intergenic regions can be partitioned into two general types: tandem repeats, which are repeated sequences that are adjacent to one another at one or more locations in the genome, and dispersed repeats, consisting of transposons and related sequences.

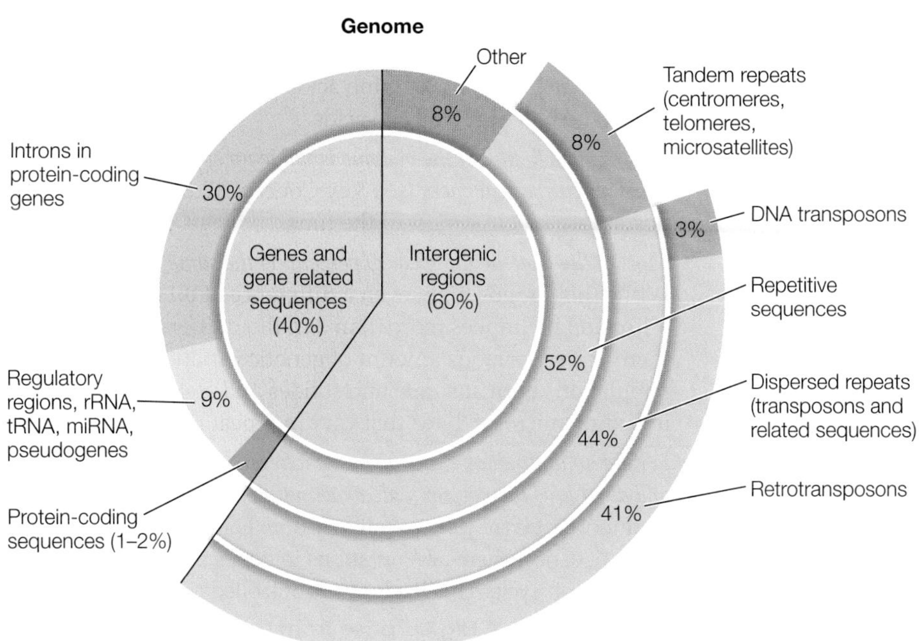

FIGURE 12.15 The Sequences That Make Up the Human Genome A tiny minority of the human genome codes for proteins: only 1.2 percent. As in other species that have large genomes, transposons, especially retrotransposons, are common in the human genome. (After T. A. Brown. 2002. *Genomes* 2e. Wiley-Liss: Oxford; Data from IHGSC. 2001. *Nature* 409: 860–921 and J. C. Venter et al. 2001. *Science* 291: 1304–1351.)

Tandem repeats include telomeres, centromeres, and microsatellites. You may recall that telomeres, which function to solve the end-replication problem in DNA replication, have a sequence of 6 bp that repeats several hundred times at the end of each chromosome (see Figure 9.14). Centromeres, the site of microtubule attachment during M phase of the cell cycle (see Key Concept 7.2), comprise thousands of repeats of a 170-bp sequence. **Microsatellites** (or short tandem repeats, STRs, or short sequence repeats, SSRs) are simple repeats whose length is usually 1–5 bp. The number of microsatellite repeats at a particular location varies among individuals and is inherited. Microsatellites are the basis of the FBI's Combined DNA Index System that is used to identify individuals from their DNA (see Chapter 9, Think Like a Scientist).

Dispersed repeats make up an astounding 44 percent of the human genome. There are many different kinds (or families) of transposons, but they are broadly grouped into two main classes: retrotransposons and DNA transposons.

Retrotransposons make RNA copies of themselves, which are then reverse transcribed into DNA before being inserted at new locations in the genome. They include two categories:

1. *LTR retrotransposons* have long terminal repeats (LTRs) of DNA sequence (100–5,000 bp) at each end. LTR retrotransposons constitute about 8 percent of the human genome.

2. *Non-LTR retrotransposons* do not have LTR sequences at their ends. They are further divided into two subcategories: SINEs and LINEs. SINEs (short *in*terspersed *e*lements) are up to 500 bp long and are transcribed but not translated. There are about 1.5 million of them scattered over the human genome, making up about 13 percent of the total DNA content. The most common SINE is the 300-bp *Alu* element, which is present in a million copies. LINEs (*l*ong *in*terspersed *e*lements) are up to 7,000 bp long, and some are transcribed and translated into proteins. They constitute about 20 percent of the human genome.

DNA transposons do not use RNA intermediates. Like some prokaryotic transposons, they are excised from the original location and become inserted at a new location without being replicated. They constitute about 3 percent of the human genome.

With so much of the human genome made up of transposons, it is tempting to think that transposons must have a beneficial role for the organism. Alternatively, these regions may have no benefit for the organism and could perhaps be better thought of as "junk" DNA. Many transposons increase in copy number when they move and insert into new locations in the genome. But most insert within introns or other regions that do not affect genes and so do not have a phenotypic effect.

REVIEW & APPLY | 12.3

R Sequencing of several model organisms revealed that eukaryotic genomes have more protein-coding genes and regulatory sequences than do prokaryotic genomes. Gene duplications and gene families are also relatively common in eukaryotic genomes. Eukaryotes also have surprisingly extensive amounts of non-protein-coding DNA. Much of the noncoding DNA is repetitive and includes both tandem repeats and dispersed repeats. Tandem repeats include simple microsatellites, centromeres, and telomeres. Retrotransposons and DNA transposons are extremely common dispersed repeats in the genomes of multicellular eukaryotes because they are able to make copies of themselves.

A

1. What are the differences between prokaryotic and eukaryotic genomes in terms of size, number of genes, amount of coding versus noncoding DNA, and chromosome structure?

2. Does the size of a genome determine how much information it contains? Explain your answer with reference to repetitive sequences and protein-coding genes.

3. In humans there is a particular SINE termed *Alu*, which is 300 bp in length. This element occurs in approximately 1 million copies per haploid human genome. What percentage of the genome is made up of the *Alu* element?

The analysis of eukaryotic genomes has resulted in an enormous amount of useful information, as we have seen. In the next key concept we will look more closely at the human genome.

12.4 Human Genomics Has Facilitated Advances in Many Areas

LEARNING OBJECTIVES

By the end of this key concept you should be able to:

12.4.1 Explain what a single nucleotide polymorphism (SNP) is.

12.4.2 Describe what is meant by "personal genomics."

12.4.3 Explain how an individual's genome can be used to discern ancestry.

12.4.4 Define pharmacogenomics.

12.4.5 Explain how cancer genomics is used in cancer treatment.

Sequencing of the first human genome was completed in 2003, at a cost of $3 billion. As costs to sequence genomes have declined, it has been possible to sequence many more human genomes. As of the time of this writing, the genomes of more than a million people have been sequenced, and the cost is less than $1,000 per person. An avalanche of information on human genomes has accumulated and continues to accumulate. While each person is different, some generalities about the human genome have emerged.

The human genome sequence held some surprises

The following are just some of the interesting facts we have learned about the human genome from sequencing multiple individuals and multiple tissues in the same individual:

- *Among the 3.1 billion bp in the haploid human genome, there are about 20,500 protein-coding genes.* This number is substantially smaller than expected. Before genome sequencing began, the diversity of human proteins suggested there would be 80,000–150,000 protein-coding genes. The actual number—not many more than in a nematode—means that posttranscriptional mechanisms (such as alternative splicing) must account for the observed number of proteins in humans. It turns out that most human genes encode multiple proteins via alternative splicing.
- In addition to the protein-coding genes, *there are another 24,000 non-protein-coding genes that are expressed in RNA,* a number that suggests their importance in processes such as posttranscriptional gene regulation (see miRNA/siRNA regulation in Key Concept 11.5).
- *The median size of a protein-coding gene is about 26,000 bp, and virtually all genes have many introns* (the median number is 7). (The median value is that at which there are equal numbers of smaller and larger values; see Appendix B.) Gene sizes vary greatly, from about 1,000 to 2.3 million bp. Variation in gene size is to be expected given that human proteins vary in size, from about 100 to 5,000 amino acids per polypeptide chain, with a median of 425 amino acids.
- *Approximately half of the genome is made up of transposons and other repetitive sequences* (see Key Concept 12.3). Most transposons are inactive most of the time.
- *Most of the genome (at least 99 percent) is the same in all people.* While this sounds high, even a difference of 0.1 percent over 3.1 billion bp implies more than 3 million differences! Thus there is an enormous amount of genetic variation in human populations both at single nucleotides and at other sequences, including microsatellites that vary in repeat number.
- *Within an individual, the genomes of different somatic cell lineages will diverge as they accumulate new mutations.* These changes can be important if they affect cell function. For example, in many types of cancer there are mutations in genes that regulate the cell division cycle. This leads to uncontrolled cell division.

Media Clip 12.1 A Big Surprise from Genomics

Comparisons among sequenced genomes from prokaryotes and eukaryotes have revealed some of the evolutionary relationships among genes. Some genes are present in both prokaryotes and eukaryotes; others are only in eukaryotes; still others are only in animals, or only in vertebrates (**FIGURE 12.16**).

Comparative genomics with the human genome is possible now that the genomes of other primates, including all the great apes and even Neanderthals (see the chapter opening), have been sequenced. Chimpanzees and bonobos are our closest living relatives, sharing nearly 99 percent of our DNA sequence. Gorillas and orangutans are next closest, with genomes that are about 98 percent and 97 percent similar to ours. Researchers have identified about 500 protein-coding genes that have undergone accelerated evolution in humans, chimpanzees, and gorillas, including some genes involved

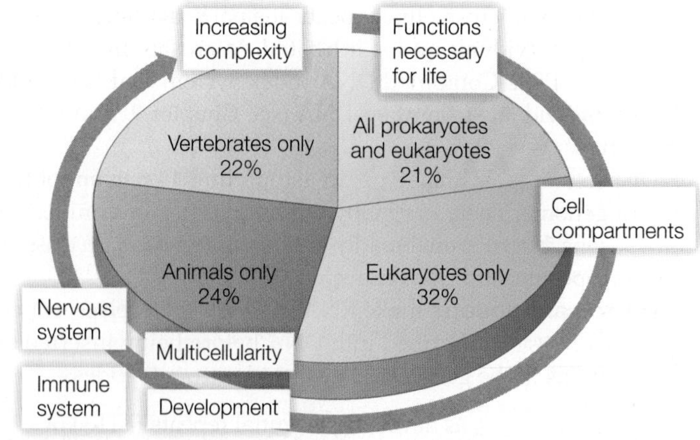

FIGURE 12.16 Evolution of the Genome A comparison of the human genome with other genomes has revealed how genes with new functions have been added over the course of evolution. Each percentage number refers to genes in the human genome shared by the listed group. Thus 21 percent of human genes have homologs in prokaryotes and other eukaryotes, 32 percent of human genes occur in all other eukaryotes, and so on. (After IHGSC. 2001. *Nature* 409: 860–921.)

in hearing and brain development. One goal of these analyses is to identify genes that have evolved rapidly in the human lineage and are possibly the cause of the traits that distinguish humans from other apes—the ones that "make humans human."

Human variation can be used to determine ancestry and disease risk

For a fee, it is possible to send your DNA to various companies that will then tell you about your genetic ancestry. Determining ancestry requires genotyping people at individual base pairs in the genome that are known to vary in human populations. These variable base pairs are termed **single nucleotide polymorphisms (SNPs)**. SNPs (pronounced "snips") initially arise as point mutations (see Key Concept 9.3) that then become common enough to occur at reasonable frequencies in human populations.

Most base-pair positions in the human genome do not have SNPs. For example, suppose a region of chromosomal DNA from 50 people is sequenced and the base pair at a certain position in every copy of a chromosome is noted. If there is an A-T pair at the position in every copy, there is no SNP. However, if at another position 74 of the 100 chromosomes have a G-C pair and 26 have a T-A pair, then that site has a SNP. Simply put, SNPs are alleles at a single base-pair locus and, for most SNPs, only two alleles will be present in a population (G-C versus T-A in this example). Usually only one of the two nucleotide identities is written, so this SNP would be called a G/T SNP, meaning the nucleotide on one strand is either a G or a T (equivalent to a C or an A on the complementary strand) (**FIGURE 12.17A**). At a G/T SNP, an individual could be a GG homozygote, a GT heterozygote, or a TT homozygote. To be clear, this means that both chromosomes of the diploid pair have a G-C pair at the particular position (GG homozygote), or a T-A pair (TT homozygote), or one chromosome with G-C and the other with T-A (GT heterozygote). The human genome contains more than 1.5 million SNPs.

SNPs can be used for a variety of studies, including ancestry. As you will see in Key Concept 15.1, SNPs accumulate in populations passively, and are typically not subject to natural selection. For example, a SNP within an intron would probably not have an effect on the phenotype. Some SNP alleles are common in certain human populations but rare in others. For example, if an individual has alleles at several SNPs that are only common in East Asian populations, then that indicates the individual's ancestors are from that region. In addition, given the large number of SNP loci, they can be used to determine relatedness among people. Identical twins obviously will have identical SNP genotypes. However, other relatives, such as siblings or first cousins, will also have more similar SNP genotypes than expected for unrelated individuals. This allows long-lost relatives to be found simply by comparing SNP genotypes at many loci. It also means that if enough DNA is left at a crime scene, it can be genotyped at many SNP loci and used to identify the suspect who left the sample (an exact match) or a relative, who could then be used to find the suspect.

SNPs can also be used to determine genetic causes of quantitative traits. Quantitative traits are determined not by single genes but by multiple genes interacting with the environment (see Key Concept 8.3). Several common diseases such as diabetes, heart

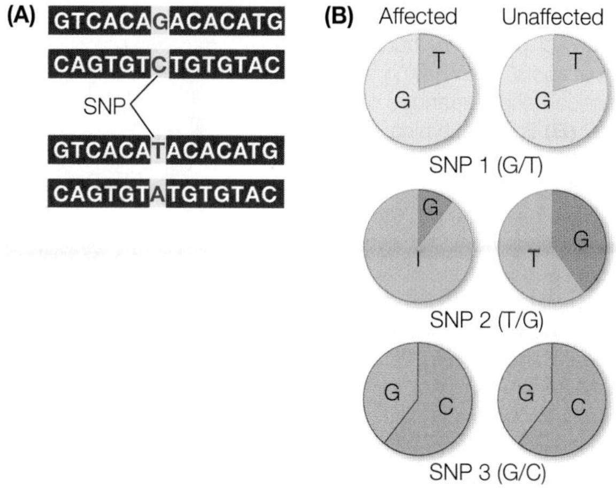

FIGURE 12.17 Single Nucleotide Polymorphisms (SNPs) **(A)** A single base-pair position in the genome that varies at the population level is a SNP. In this example we see a site that can be G-C or T-A, meaning it is a G/T, or equivalently a C/A, SNP. **(B)** The frequencies of two alleles, shown using pie charts, at each of three SNPs in affected and unaffected individuals reveal that the G allele at SNP 2 is associated with the affected phenotype, while alleles at SNPs 1 and 3 are not.

disease, and Alzheimer's disease are quantitative traits. A considerable amount of research involves attempting to identify the genes affecting such diseases. To find the causal genes, biologists genotype many SNPs across the genome and look for associations between the presence of a particular allele at a SNP and the disease. Examining multiple SNPs in multiple individuals, both with and without a disease of interest, to find an association between a particular SNP allele and the phenotype is termed a genome-wide association study (**FIGURE 12.17B**).

Given the usefulness of SNPs, being able to genotype them inexpensively is important. One obvious method for genotyping SNPs would be to sequence an individual's genome using high-throughput sequencing (see Figure 12.3). However, that approach involves sequencing many base pairs that are of little interest because they do not vary or are in regions that are unlikely to have genes that affect the phenotype, such as centromeres. A substantially less expensive approach that can target a few hundred thousand SNPs uses DNA microarrays.

A **DNA microarray** is a grid of microscopic spots of different oligonucleotides (short single-stranded DNA sequences) arrayed on a solid surface. It can be "probed" with an individual's DNA (after denaturation) or RNA; if the individual's DNA contains a sequence that is complementary to one of the oligonucleotides, the sequence will hybridize to (base-pair with) the DNA at that spot, a process called **nucleic acid hybridization**. Colored fluorescent dyes are used to detect spots that hybridize with a sequence in an individual's DNA. Often, adjacent spots have oligonucleotides that correspond to the two alleles at a particular SNP. For example, for an A/G SNP, one spot has the A allele oligonucleotide and the other has the G allele oligonucleotide. Extracting DNA from an individual, cutting it up, labeling it with a dye, and then hybridizing it to the microarray will give different patterns depending on the genotype (**FIGURE 12.18**).

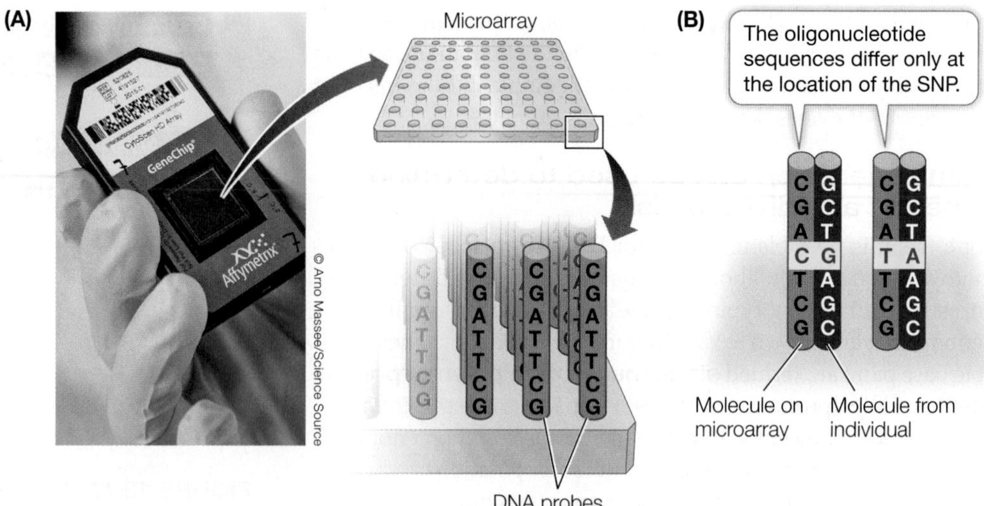

FIGURE 12.18 A DNA Microarray **(A)** A microarray chip contains hundreds of thousands of spots, each having a cluster of identical, single-stranded oligonucleotides. **(B)** For genotyping, two spots contain oligonucleotides that are identical, and complementary to the region of the genome containing the SNP, except at the location of the SNP, where each spot corresponds to one of the alleles (in reality, oligonucleotides are 25 bases in length). In this example, if DNA from an individual binds only to the left spot, that identifies the individual as a CC homozygote. If DNA binds only to the right spot, the individual is a TT homozygote. A heterozygous individual will have DNA binding equally to both spots.

INVESTIGATION

FIGURE 12.19 Genome-Wide Association Study of Common, Complex Diseases Genotyping SNPs across the genome, coupled with phenotype data, allows detection of regions that cause variation in complex phenotypes.[a]

HYPOTHESIS

There is a genetic component to many common, complex disorders, including coronary artery disease (CAD) and Crohn's disease (CD).

METHOD

1. Genotype 2,000 individuals with CAD, 2,000 with CD, and 3,000 who are unaffected by either disorder at 500,568 SNPs using a DNA microarray.
2. For each SNP locus, determine the allele frequency in individuals affected by each disorder versus the unaffected control group.
3. Determine the probability (*P*) of obtaining the observed difference in allele frequency between affected and unaffected groups assuming there is no relationship between the SNP genotype and the disorder. If this probability is very low, it suggests that the relationship observed is not caused by chance but represents a real relationship (implying it is likely biologically significant).

RESULTS

The graphs show the probability of obtaining the observed frequency difference in affected versus unaffected groups by chance for every SNP, versus the position of the SNP in the genome (by chromosome). The *y* axis is $-\log_{10}$ of the probability, so a larger value is a smaller probability. A *y*-axis value of 5 means there is a 1 in 100,000 chance of obtaining the result by chance. Points in green are considered to be statistically significant. With so many SNPs examined, rare events are expected, so probabilities have to be very low to be statistically significant.

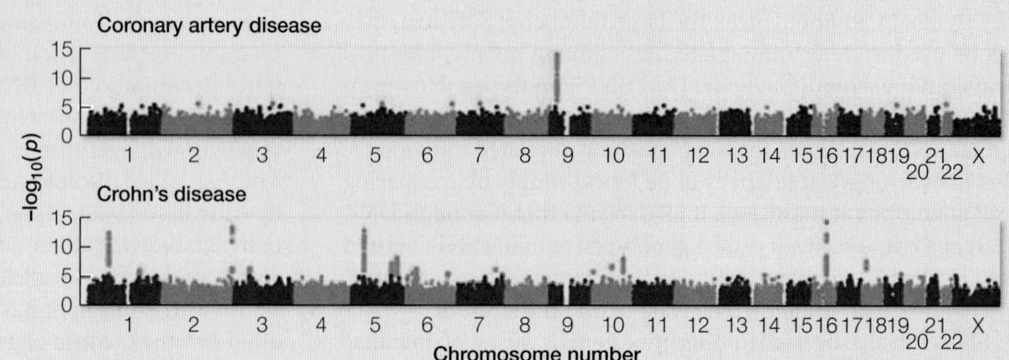

A similar **Analyze the Data** exercise may be assigned in **Achieve**.

CONCLUSION

Some regions of the genome have SNPs that are significantly associated with each of the two disorders.

ANALYZE THE DATA

1. The researchers found a single site in the genome that was strongly associated with CAD and a few others that were weakly associated. How many regions did they find that were strongly associated with CD?
2. For each SNP, the researchers calculated which allele was at higher frequency in affected individuals. They also determined the proportion of each genotype that was affected for each disorder. Suppose the C allele is the high-risk allele at a C/T SNP. What are the possible genotypes at this SNP locus, and which will have the highest and lowest proportion affected?
3. The most statistically significant SNP for CAD has a *y*-axis value of 15. What does this value mean?

[a]The Wellcome Trust Case Control Consortium. 2007. *Nature* 447: 661–678.

DNA microarrays allow rapid, inexpensive genotyping of hundreds of thousands of SNPs in thousands of individuals. Combining genotype data and medical records has been used to try to find the genetic basis of multiple disorders (**INVESTIGATION FIGURE 12.19**).

Animation 12.3 **DNA Microarray Technology**

Animation 12.4 **DNA Testing**

Personal genomics will soon be commonplace in medicine

Genome-wide association studies have revealed alleles that are associated with modestly increased risks for such diseases as breast cancer, diabetes, arthritis, obesity, and coronary heart disease. Private companies now offer to scan a human genome for SNP alleles. However, at this point it is unclear what a person without symptoms should do with the information, since multiple genes, environmental influences, and epigenetic effects all contribute to the development of many complex diseases. The idea that your disease risk can be determined by genotyping several thousand SNPs is one aspect of what is known as **personal genomics**. Soon physicians will be able to use your genotype at multiple SNPs to personalize health care—to suggest preemptive changes to diet or exercise and to screen for particular diseases for which your genotype makes you especially prone.

Pharmacogenomics, the study of how an individual's genotype affects his or her response to drugs or other agents, is another application of personalized genomics. Such analyses are done in the same manner as genome-wide association studies, with the objective of predicting whether a drug will be effective (**FIGURE 12.20**). Pharmacogenomics should also help reduce the incidence of adverse drug reactions.

Pharmacogenomics is also being used in cancer treatment. Analysis of gene expression in tumors from different patients has revealed a great deal of heterogeneity: not all breast cancers or prostate cancers express the same genes. Cancer diagnostics is moving away from classifying cancer on the basis of location (e.g., lung cancer) or cell type (e.g., non-small-cell lung cancer) to a molecular classification (e.g., a non-small-cell lung cancer expressing a certain growth factor receptor). This can lead to therapy in the form of a drug targeted to the particular protein expressed in the tumor. The key is identifying which patients will respond to the drug, and this is often done by gene expression or protein analysis.

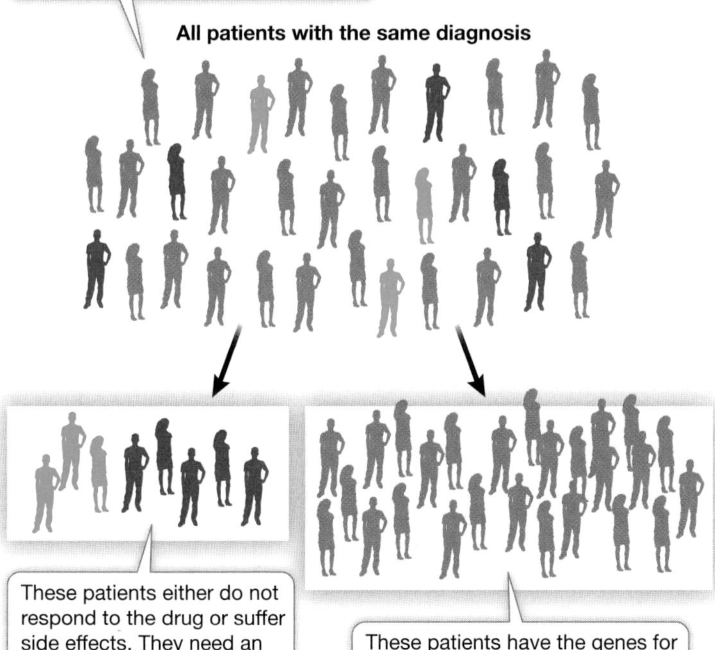

FIGURE 12.20 Pharmacogenomics Correlations between genotypes and responses to drugs will help physicians develop more personalized medical care. SNP analysis is used to identify people who will respond to a drug and those who will not. The different colors indicate individuals with different genotypes at SNPs affecting drug response. (After J. A. Johnson. 2003. *Trends Genet* 19: 660–666.)

REVIEW & APPLY | 12.4

R Sequencing of the human genome is now commonplace. The human genome contains substantially fewer protein-coding genes than expected from phenotypes, only 20,000 or so. The number of different proteins expressed by the genome is substantially larger, due to extensive alternative splicing. Human genomes vary among individuals at single nucleotide polymorphisms (SNPs), which can be genotyped inexpensively using DNA microarrays. SNP genotyping allows the determination of ancestry and is used in genome-wide association studies to identify genes underlying complex traits. Genomics allows personalized assessment of risks for various disorders and likelihood of drug responses.

A

1. The median length of a human gene is 26,000 bp. The median human polypeptide has 367 amino acids. Explain why these numbers are so different.
2. About 1,600 proteins found in humans are also found in yeast. What kinds of proteins might these be?
3. One often reads that humans and chimpanzees are 99 percent similar in their DNA sequence, and that is usually followed by a statement of how similar the two species are at the genetic level. Assuming this estimate applies to single base-pair differences between the two species, and that the genome size is the same size in both species, calculate the number of single base-pair differences between chimpanzees and humans.

What has sequencing of the human and Neanderthal genomes revealed about the history of these two species?

Answer: Sequencing of the Neanderthal genome and comparing it with modern human genomes reveals the striking fact that the vast majority—99.7 percent—of the two genomes is identical. This difference is about fourfold less than between humans and chimpanzees. However, recalling that the human genome has more than 3 billion bp, this still represents many differences. Another surprising finding is that the genomes of individuals of European or Asian ancestry have 1–7 percent Neanderthal alleles, especially SNPs. This suggests that the two species interbred. After all, they lived at the same time. But there is scant fossil evidence showing that the two species occurred together in the same place. A fascinating finding on the genomic comparison between modern humans and Neanderthals is in the gene *FOXP2* (forkhead box protein P2). This gene encodes a transcription factor expressed in the developing brain that is thought to be necessary for language —individuals with a mutated version of this gene suffer a severe speech disorder. The *FOXP2* gene in humans has two amino acid differences compared with the gene in other apes. It turns out that the Neanderthal genome has the human version of *FOXP2*, indicating that Neanderthals may have had language abilities similar to those of modern humans.

12 VISUAL SUMMARY

12.1 The -omics Era Has Revolutionized Biology

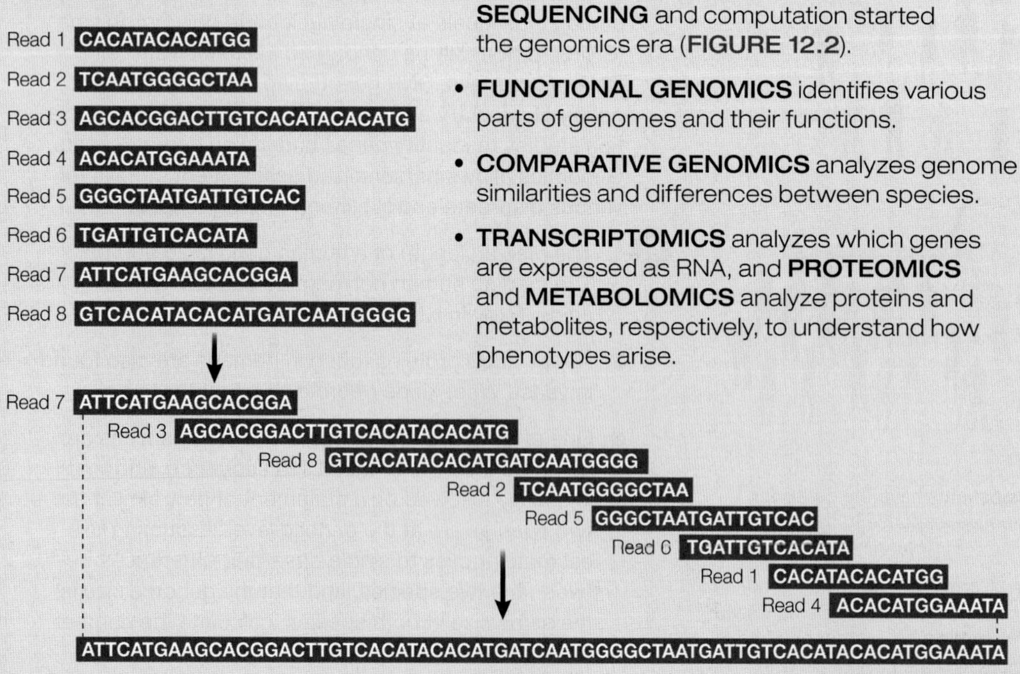

Go to ANIMATIONS 12.1 and 12.2

- Advances in **HIGH-THROUGHPUT SEQUENCING** and computation started the genomics era (**FIGURE 12.2**).
- **FUNCTIONAL GENOMICS** identifies various parts of genomes and their functions.
- **COMPARATIVE GENOMICS** analyzes genome similarities and differences between species.
- **TRANSCRIPTOMICS** analyzes which genes are expressed as RNA, and **PROTEOMICS** and **METABOLOMICS** analyze proteins and metabolites, respectively, to understand how phenotypes arise.

FIGURE 12.2

Read 1 CACATACACATGG
Read 2 TCAATGGGGCTAA
Read 3 AGCACGGACTTGTCACATACACATG
Read 4 ACACATGGAAATA
Read 5 GGGCTAATGATTGTCAC
Read 6 TGATTGTCACATA
Read 7 ATTCATGAAGCACGGA
Read 8 GTCACATACACATGATCAATGGGG

Read 7 ATTCATGAAGCACGGA
Read 3 AGCACGGACTTGTCACATACACATG
Read 8 GTCACATACACATGATCAATGGGG
Read 2 TCAATGGGGCTAA
Read 5 GGGCTAATGATTGTCAC
Read 6 TGATTGTCACATA
Read 1 CACATACACATGG
Read 4 ACACATGGAAATA

ATTCATGAAGCACGGACTTGTCACATACACATGATCAATGGGGCTAATGATTGTCACATACACATGGAAATA

QUESTIONS

1. What is the main way in which Sanger dideoxy sequencing differs from high-throughput sequencing by synthesis?
2. In genomics, what is the difference between genome assembly and annotation?
3. If we look computationally for open reading frames (ORFs) in random noncoding sequences of DNA, about how long will they be on average?

12.2 Prokaryotic Genomes Are Small, Compact, and Diverse

FIGURE 12.9

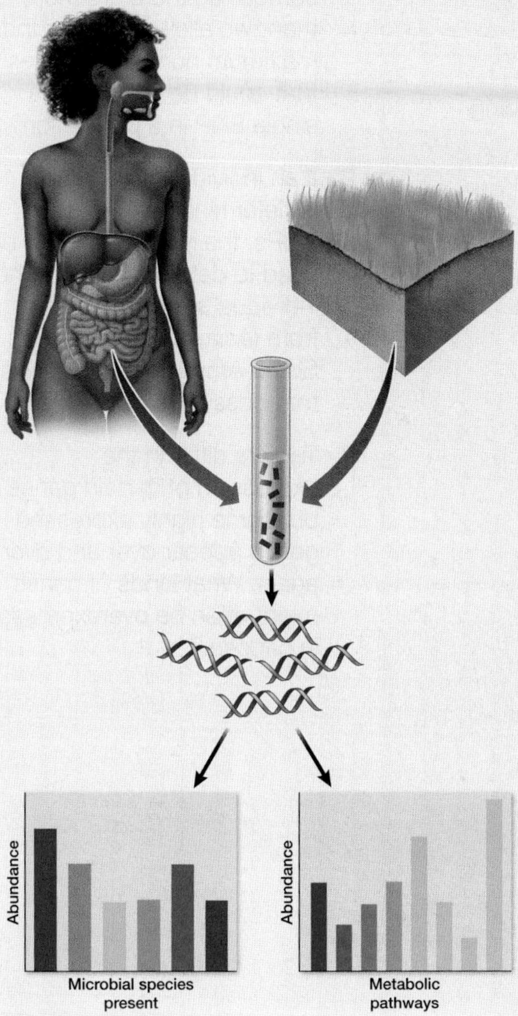

Abundance

Microbial species present

Abundance

Metabolic pathways

Go to ACTIVITY 12.1

- **METAGENOMICS**—sequencing DNA from environmental samples without isolating intact organisms—reveals enormous diversity in prokaryotic genomes (**FIGURE 12.9**).
- Comparative genomics has identified prokaryotic strains with only a few hundred genes in the **CORE GENOME** yet thousands of genes in the **PAN GENOME**.
- **TRANSPOSONS** and **CRISPR-CAS9 GENE EDITING** can be used to introduce mutations in studies of gene function.

QUESTIONS

1. If a gene is found to be essential in one species in nature, does that mean it will be essential in another species?
2. You sequence five strains of a newly discovered bacterium, each of which has 4,000 protein-coding genes. Comparisons across the five strains reveal that 500 genes are shared, though the total number of different genes is 10,000. In terms of protein-coding genes, how large is the genome of this bacterium? The core genome? The pan genome?
3. Metagenomics has discovered substantial prokaryotic diversity. Why was that diversity unknown previously?

12.3 Eukaryotic Genomes Are Large and Complex

FIGURE 12.15

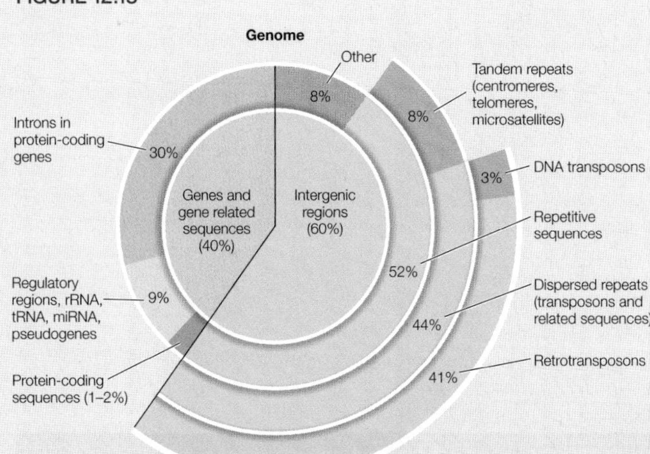

- Eukaryotic genomes encode more protein-coding genes than prokaryotic genomes and contain substantial quantities of noncoding and repetitive DNA (**FIGURE 12.15**).
- Gene duplication within a species results in **GENE FAMILIES**.
- Repetitive DNA may consist of dispersed repeats or tandem repeats.
- **TRANSPOSONS** are common in eukaryotic genomes.

QUESTIONS

1. Describe some attributes of model organisms for genomics and genetics studies.
2. What kinds of repetitive sequences are present in eukaryotic genomes?
3. What are the possible fates of a paralogous gene that arises by duplication?

12.4 Human Genomics Has Facilitated Advances in Many Areas

Go to ANIMATIONS 12.3 and 12.4

- Determining ancestry requires genotyping individual base pairs in the genome called **SINGLE NUCLEOTIDE POLYMORPHISMS** (**SNPS**) (**FIGURE 12.17**).
- SNPs can be identified using **DNA MICROARRAYS** (**FIGURE 12.18**).
- SNP genotyping allows identification of genes affecting quantitative traits, including disease risk and drug sensitivity, and personalization of healthcare.

FIGURE 12.17

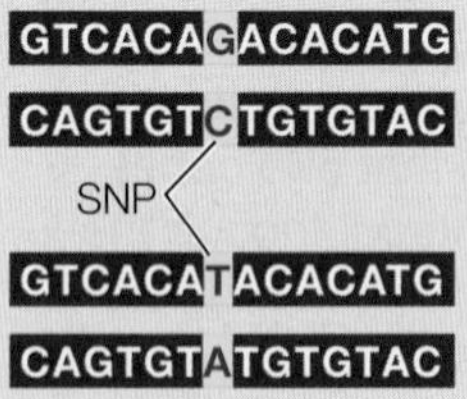

GTCACAGACACATG
CAGTGTCTGTGTAC
SNP
GTCACATACACATG
CAGTGTATGTGTAC

FIGURE 12.18

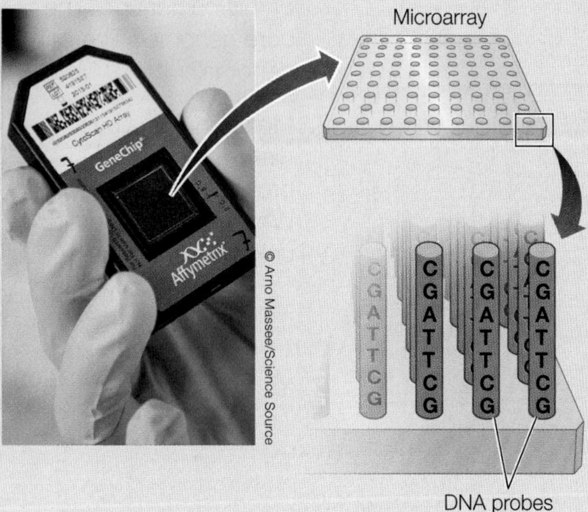

Activity 12.2 Concept Matching: Genomes

QUESTIONS

1. SNPs usually have only two alleles in a population, but sometimes there are more than two alleles. What is the maximum number of alleles that could be present at a single SNP in a population?
2. If an individual's genotype is determined at 500,000 SNPs, the information can be used to determine where the individual's ancestors were from (Asia, northern Europe, South America, etc.). How is that possible?
3. Tumors differ in the expression of certain genes, but some highly expressed genes appear over and over again. What kinds of genes might often be overexpressed in cancers?

Part

3 Evolution

13 Processes of Evolution

About 50 million years ago, bats evolved the ability to echolocate, which allows them to detect their prey in darkness. The bats produce ultrasonic sounds, and compare these outgoing sound pulses with the returning echoes of the pulses that bounce off their prey.

How do complex traits such as echolocation evolve in the first place?

You will find the answer to this question on page 327.

KEY CONCEPTS

13.1 Evolution Is Both Factual and the Basis of Broader Theory

13.2 Mutation, Selection, Gene Flow, Genetic Drift, and Nonrandom Mating Result in Evolution

13.3 Evolution Can Be Measured by Changes in Allele Frequencies

13.4 Selection Can Be Stabilizing, Directional, or Disruptive

13.5 Selection Can Maintain Polymorphisms in Populations

13.1 Evolution Is Both Factual and the Basis of Broader Theory

LEARNING OBJECTIVES

By the end of this key concept you should be able to:

13.1.1 **Differentiate between the everyday use of the word "theory" and a scientific theory.**

13.1.2 **Compare and contrast natural selection with the selection by human breeders for specific traits in domestic plants and animals.**

13.1.3 **Apply the principles of natural selection to predict and explain evolutionary outcomes.**

All biological populations change in their genetic makeup over time. Change in the genetic composition of populations over time is called **evolution**. We can, and do, observe evolutionary change on a regular basis, both in laboratory experiments and in natural populations. We measure the rate at which new mutations arise, observe the spread of new genetic variants through a population, and see the effects of genetic change on the form and function of organisms. In the fossil record, we observe the long-term morphological changes (which are the result of underlying genetic changes) that have occurred among living organisms. These underlying changes in the genetic makeup of populations drive the origin and extinction of species and fuel the diversification of life.

Since we can directly observe evolution, why do we speak of "evolutionary theory"?

In addition to observing and recording physical changes over evolutionary time, biologists have accumulated a large body of evidence about *how* these changes occur, and about *what* evolutionary changes have occurred in the past. The resulting understanding of the processes of evolutionary change is known as **evolutionary theory**.

Evolutionary theory has many useful applications. We constantly apply it, for example, to the study and treatment of diseases. Evolutionary theory is critical to the improvement of agricultural crops and practices, and to the development of industrial processes that produce new molecules with useful properties. At a more basic level, knowledge of evolutionary theory allows biologists to understand how life diversified. It also helps us make predictions about the biological world.

In everyday speech, people tend to use the word "theory" to mean an untested hypothesis, or even a guess. But evolutionary theory does not refer to any single hypothesis, and it certainly is not guesswork. A vast and rich array of geological, morphological, behavioral, and molecular data supports the factual basis of evolution. Observations of fossils and natural populations are supported by experiments that demonstrate the basic operation of evolutionary processes.

When we refer to "evolutionary theory," we are referring to our understanding of the processes that result in genetic changes in populations over time. We then apply that understanding to interpret the changes we observe in natural populations. We can directly observe the evolution of many living organisms. We can observe the yearly changes in the viruses that give us the flu or COVID-19, but it is evolutionary theory that allows us to apply our observations to the task of developing more effective vaccines against these viruses. Several processes of evolutionary change are recognized, and the scientific community is continually using evolutionary theory to expand its understanding of how and when these processes apply to particular biological problems.

Media Clip 13.1 **Watching Evolution in Real Time**

Darwin and Wallace introduced the idea of evolution by natural selection

In the early 1800s, it was not yet evident to many people that populations of living organisms evolve. But several biologists had suggested that the species living on Earth had changed over time—that is, that evolution had taken place. Jean-Baptiste Lamarck, for one, presented strong evidence for the fact of evolution in 1809, but his ideas about *how* it occurred were not convincing. At that time, no one had yet envisioned a viable process for evolution.

In the 1820s, a young Charles Darwin became passionately interested in the subjects of geology (with its new sense of Earth's great age) and natural history (the scientific study of how different organisms function and carry out their lives in nature). Despite these interests, he planned, at his father's behest, to become a doctor. But surgery conducted without anesthesia nauseated Darwin, and he gave up medicine to study at Cambridge University for a career as a clergyman in the Church of England. Always more interested in science than in theology, he gravitated toward scientists on the faculty, especially the botanist John Henslow. In 1831, Henslow recommended Darwin for a position on HMS *Beagle*, a Royal Navy vessel that was preparing for a survey voyage around the world (**FIGURE 13.1**).

Whenever possible during the 5-year voyage, Darwin went ashore to study rocks and to observe and collect plants and animals. He noticed striking differences between the species he saw in South America and those of Europe. He observed that the species of the temperate regions of South America (Argentina and Chile) were more similar to those of tropical South America (Brazil) than

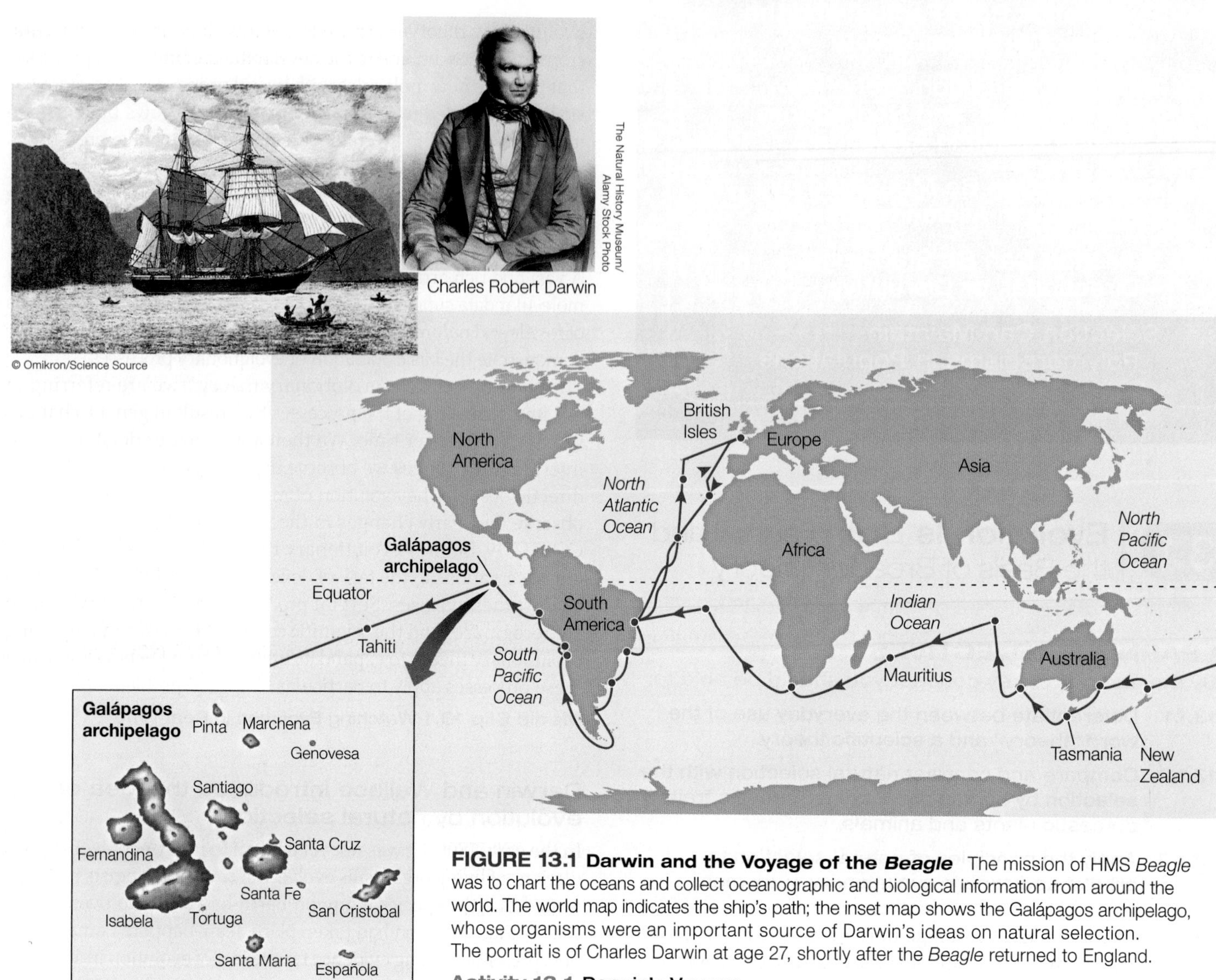

FIGURE 13.1 Darwin and the Voyage of the *Beagle* The mission of HMS *Beagle* was to chart the oceans and collect oceanographic and biological information from around the world. The world map indicates the ship's path; the inset map shows the Galápagos archipelago, whose organisms were an important source of Darwin's ideas on natural selection. The portrait is of Charles Darwin at age 27, shortly after the *Beagle* returned to England.

Activity 13.1 Darwin's Voyage

they were to temperate European species. When he explored the islands of the Galápagos archipelago, west of Ecuador, he noted that most of the animals were endemic to the islands (meaning they were unique and found nowhere else), although they were similar to animals found on the mainland of South America. Darwin also observed that the fauna of the Galápagos differed from island to island. He postulated that some animals had come to the archipelago from mainland South America and had subsequently undergone different changes on each of the islands. He wondered what might account for these changes.

When he returned to England in 1836, Darwin continued to ponder his observations. His thoughts were strongly influenced by the geologist Charles Lyell, who had recently popularized the idea that Earth had been shaped by slow-acting forces that are still at work today. Darwin reasoned that similar thinking could be applied to the living world. Within a decade, he had developed the framework of an explanatory theory for evolutionary change based on three major propositions:

1. Species are not immutable; they change over time.
2. Divergent species share a common ancestor and have diverged from one another gradually over time (a concept Darwin termed **descent with modification**).
3. Changes in species over time can be explained by **natural selection**: the increased survival and reproduction of some individuals compared with others, based on differences in their traits.

The first of these propositions was not unique to Darwin; several earlier authors had argued for the fact of evolution. A more revolutionary idea was his second proposition, that divergent species are related to one another through common descent. But Darwin is probably best known for his third proposition, that of natural selection.

Darwin realized that many more individuals of most species are born than survive to reproduce. He also knew that, although offspring usually resemble their parents, offspring are not identical to one another or to either parent. Finally, he was well aware that human breeders of plants and animals often selected their breeding stock based on the occurrence of particular traits that they found desirable. Over time, this selection by breeders resulted in dramatic changes in the appearance of the descendants of those plants or animals. In natural populations, wouldn't the individuals with traits that provided the best chances of survival and reproduction be the ones most likely to pass their traits on to the next generation? Darwin called this process natural selection, to emphasize that the "selection" in nature was based on traits that increased survival and reproduction of organisms. In essence, nature, rather than a human, was doing the "selection" for which organisms would contribute the most to the next generation. To emphasize this distinction, Darwin called the selective breeding process by humans **artificial selection**, and emphasized that humans often selected traits in breeding stock that would confer disadvantages to reproduction or longevity in nature.

In 1844, Darwin wrote a long essay describing the role of natural selection as a process of evolution. But he was reluctant to publish it, preferring to assemble more evidence first. Darwin's hand was forced in 1858, when he received a letter and manuscript from another traveling English naturalist, Alfred Russel Wallace, who was studying the plants and animals of the Malay Archipelago. Wallace asked Darwin to evaluate his manuscript, which included an explanation of natural selection almost identical to Darwin's. Darwin was at first dismayed, believing Wallace to have preempted his idea. Parts of Darwin's 1844 essay, together with Wallace's manuscript, were presented to the Linnean Society of London on July 1, 1858, thereby crediting both men for the idea of natural selection. Darwin then worked quickly to finish his full-length book, *On the Origin of Species*, which was published the following year.

Although Darwin and Wallace independently articulated the concept of natural selection, Darwin developed his ideas first. Furthermore, *On the Origin of Species* proved to be a stunning work of scholarship that provided exhaustive evidence from many fields supporting both the premise of evolution itself and the understanding of natural selection as a process of evolution. Thus both concepts are more closely associated with Darwin than with Wallace.

The publication of *On the Origin of Species* in 1859 stirred considerable interest (and controversy) among scientists and the public alike. Scientists spent much of the rest of the nineteenth century amassing biological and paleontological data to test evolutionary ideas and document the history of life on Earth. By 1900, the fact of biological evolution (defined at that time as change in the physical characteristics of populations over time) was established beyond any reasonable doubt. As biologists discovered the details of genetic inheritance in the twentieth century, the genetic mechanisms of evolution became clear. The development of methods for sequencing DNA in the late 1970s allowed biologists to document evolutionary changes within and between species with great precision. This technology led to explosive growth in the field of evolutionary biology. In the past three decades, well over a quarter of a million scientific papers on evolutionary observations, experiments, and theory have been published.

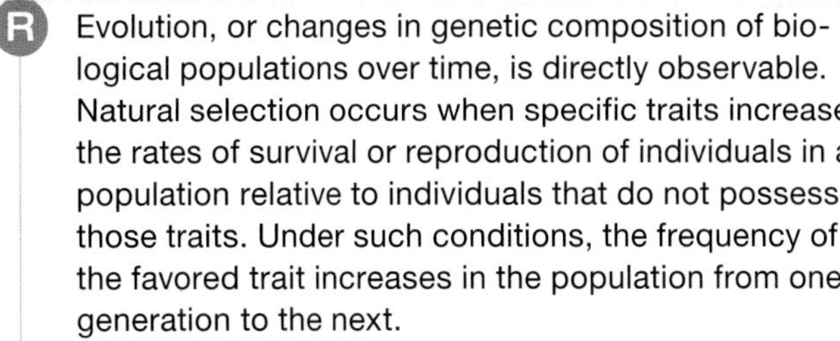

REVIEW & APPLY | 13.1

R Evolution, or changes in genetic composition of biological populations over time, is directly observable. Natural selection occurs when specific traits increase the rates of survival or reproduction of individuals in a population relative to individuals that do not possess those traits. Under such conditions, the frequency of the favored trait increases in the population from one generation to the next.

A

1. Why do biologists speak of "evolutionary theory" if the facts of evolution are not in doubt?
2. Antibiotics are drugs that kill most bacteria, but genetic mutations can allow some individual bacteria to survive short-term exposure to these drugs. How would you expect the frequency of resistant bacteria to change over time in populations of bacteria that were exposed to an antibiotic drug on a regular basis? Why do you think antibiotics come with a warning to take the full course of the treatment, rather than stopping after you begin to feel better?
3. In what ways does selection by humans in breeding agricultural crops differ from natural selection? Give an example of a trait that might be favored by artificial selection in agriculture, but selected against by natural selection in a wild population.
4. Natural selection cannot adapt populations to conditions they have not experienced. Yet many organisms appear to respond to natural events before they happen. For example, many mammals go into hibernation while it is still quite warm. Similarly, many birds leave the temperate zone for their southern wintering grounds long before winter has arrived. How do you think such "anticipatory" behaviors evolve?
5. As more humans live longer, many people face degenerative conditions such as Alzheimer's disease that (in most cases) are linked to advancing age. Assuming that some individuals may be genetically predisposed to successfully combat these conditions, is it likely that natural selection alone would act to favor such a predisposition in human populations? Why or why not?

Although the importance of natural selection to evolution has been confirmed in many thousands of scientific studies, it is not the only process that drives evolution. In the next section we'll consider a more complete view of evolutionary processes and how they operate.

13.2 Mutation, Selection, Gene Flow, Genetic Drift, and Nonrandom Mating Result in Evolution

LEARNING OBJECTIVES

By the end of this key concept you should be able to:

13.2.1 Distinguish between the everyday and scientific uses of the terms "adapt" and "evolve."

13.2.2 Describe how each of the five processes of evolution can result in changes in gene frequencies across generations of a population.

The word "evolution" is often used in a general sense to mean simply "change," but in a biological context "evolution" refers specifically to change in the genetic makeup of populations over time. Developmental changes that occur in a single individual over the course of the life cycle are not the result of evolutionary change. Evolution is genetic change across generations of a **population**—a group of individuals of a single species that live and interbreed in a particular geographic area at the same time. It is important to remember that *individuals do not evolve; populations do.*

The premise of natural selection was one of Darwin's principal insights and has been demonstrated to be an important process of evolution, but natural selection does not act alone. Four additional processes—mutation, gene flow, genetic drift, and nonrandom mating—affect the genetic makeup of populations over time. Before we consider how the other processes can change the frequencies of gene variants in a population, we need to understand how mutation brings such variants into existence.

Mutation generates genetic variation

The origin of genetic variation is mutation. As described in Key Concept 9.3 a mutation is any change in the nucleotide sequences of an organism's DNA. The process of DNA replication is not perfect, and some changes appear almost every time a genome is replicated. Mutations occur randomly with respect to an organism's needs; it is natural selection acting on this random variation that results in adaptation. Most mutations are either harmful to their bearers (deleterious mutations) or have no effect (neutral mutations). But a few mutations are beneficial, and even previously deleterious or neutral alleles may become advantageous if environmental conditions change. In addition, mutation can restore genetic variation that other evolutionary processes have removed. Thus mutation both creates and helps maintain genetic variation in populations.

Mutation rates can be high, especially in viruses and bacteria, which explains the rapid evolution of many pathogens. But in many protein coding genes of multicellular organisms, the mutation rate is very low (on the order of 10^{-8} to 10^{-9} changes per base pair of DNA per generation). Even low overall mutation rates, however, create considerable genetic variation, because each of a large number of genes may change, and populations often contain large numbers of individuals. For example, if the probability of a point mutation (an addition, deletion, or substitution of a single base) were 10^{-9} per base pair per generation, then each human gamete—the DNA of which contains 3×10^9 base pairs—would average three new point mutations ($3 \times 10^9 \times 10^{-9} = 3$), and each zygote would carry an average of six new mutations. The current human population of about 7 billion people would thus be expected to carry about 42 billion new mutations (i.e., changes in the nucleotide sequences of their DNA that were not present one generation earlier). So even though the mutation rate in humans is low, human populations still contain enormous genetic variation on which other evolutionary processes can act.

As a result of mutation, different forms of a gene, known as **alleles**, may exist at a particular chromosomal locus (see Key Concept 8.1). At any particular locus, a single diploid individual has no more than two of the alleles found in the population to which it belongs. The sum of all copies of all alleles at all loci found in a population constitutes that population's **gene pool** (**FIGURE 13.2**). (We can also refer to the gene pool for a particular chromosomal locus or loci.) The gene pool is the sum of the genetic variation in the population. The proportion of each allele in the gene pool is the **allele frequency**. Likewise, the proportion of each genotype among individuals in the population is the **genotype frequency**.

A simple experiment demonstrates how mutations accumulate in populations in a continual, almost constant fashion over time (**FIGURE 13.3**). Lines of the bacterium *E. coli* were grown in the laboratory for 20,000 generations, and the genomes were sequenced from individuals in the experimental lines at least once

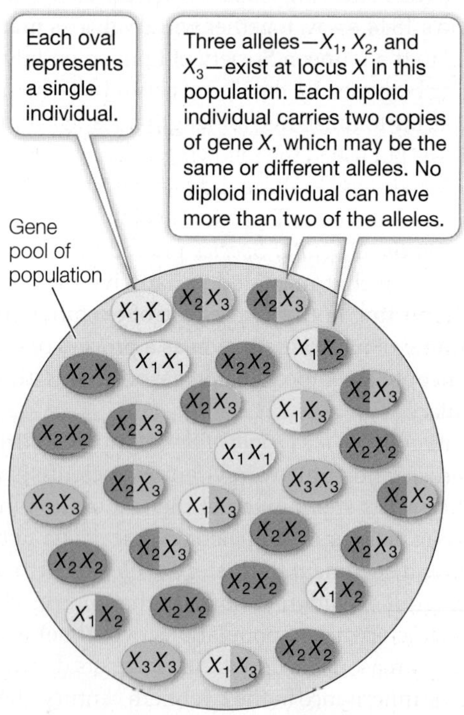

FIGURE 13.2 A Gene Pool A gene pool is the sum of all the alleles found in a population or at a particular locus in that population. This figure shows the gene pool for one locus, X, in a population of diploid organisms. The allele frequencies in this case are 0.20 for X_1, 0.50 for X_2, and 0.30 for X_3 (see Figure 13.11).

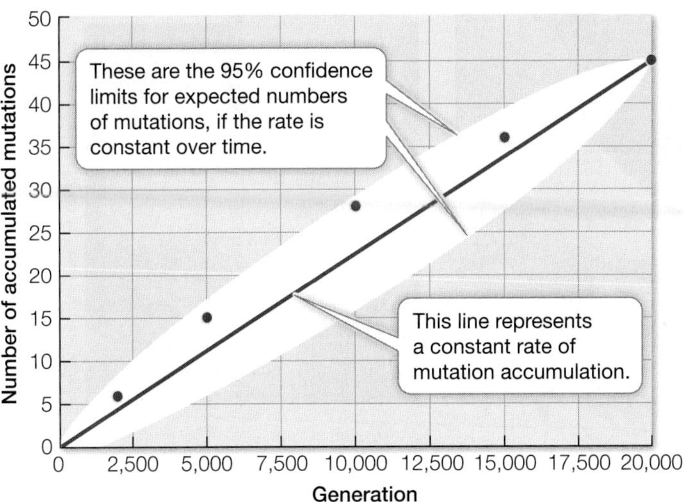

FIGURE 13.3 Mutations Accumulate Continually An experimental lineage of the bacterium *Escherichia coli* was propagated in the laboratory for 20,000 generations. Genomes were sequenced from individuals sampled at various points during the experiment and were compared with the genome of the ancestral clone. Note that mutations accumulated at a relatively constant rate throughout the experiment. (After J. E. Barrick et al. 2009. *Nature* 461: 1245–1247.)

every 5,000 generations. Over the experiment, the lines accumulated about 45 changes to their genomes, and these changes appeared at a fairly constant rate over time. Populations of all living organisms experience a similar accumulation of mutations over time (although the rate of change differs among species), and these changes provide the raw material for evolution.

As discussed in **Key Concept 8.2**, alleles of different genes assort independently in meiosis. The frequencies with which different allele combinations will be expressed in offspring can be calculated using a Punnett square or probability theory.

Selection acting on genetic variation leads to new phenotypes

As a result of mutation, the gene pools of nearly all populations contain variation for many traits. Selection that favors different traits can lead to distinctively different lineages that descend from the same ancestor. For example, artificial selection for different traits in a single European species of wild mustard produced many important crop plants (**FIGURE 13.4**). Agriculturalists were able to achieve these results because the original mustard population had genetic variation for the characters of interest (such as stem thickness or number of leaves).

Darwin compared artificial selection, which was commonly practiced by animal and plant breeders, with natural selection that occurred in natural populations. Many of Darwin's observations on the nature of variation and selection came from domesticated plants and animals. Darwin bred pigeons and thus knew firsthand the astonishing diversity in color, size, form, and behavior that breeders could achieve (**FIGURE 13.5**). He recognized close

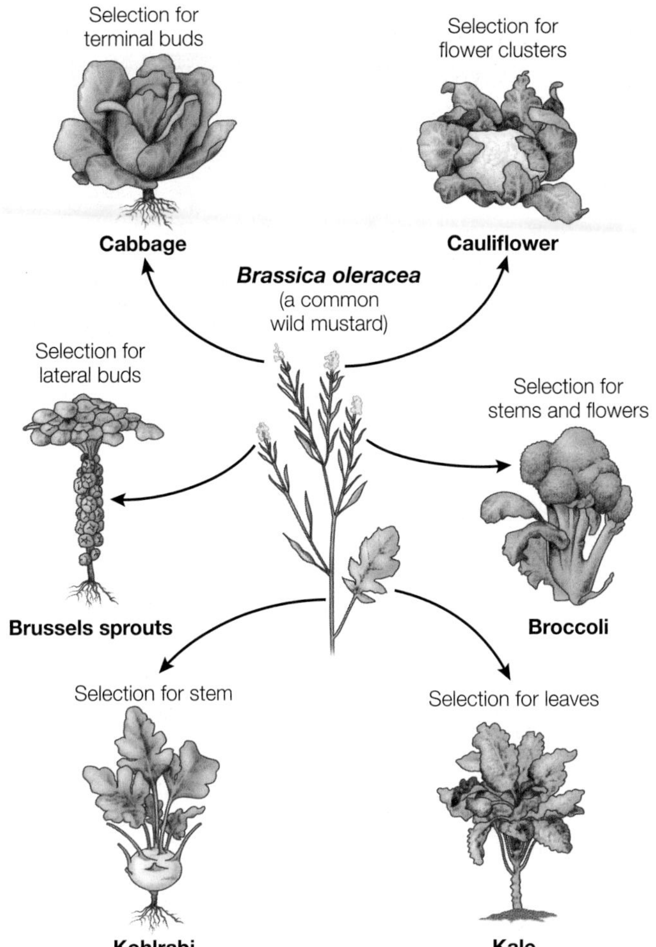

FIGURE 13.4 Many Vegetables from One Species All the crop plants shown here derive from a single wild mustard species. European agriculturalists produced these crop plants by selecting and breeding plants with unusually large buds, stems, leaves, or flowers. The results substantiate the vast amount of variation present in the gene pool of the ancestral species.

parallels between selection by breeders and selection in nature. Whereas artificial selection resulted in traits that were preferred by the human breeders, natural selection resulted in traits that helped organisms survive and reproduce more effectively. In both cases, selection simply increased the frequency of the favored trait from one generation to the next.

Laboratory experiments also demonstrate the existence of considerable genetic variation in populations, and show how this variation can lead to evolution through selection. In one such experiment, investigators bred populations of the fruit fly *Drosophila melanogaster* with high or low numbers of bristles on their abdomens from an initial population with intermediate numbers of bristles. After 35 generations, all flies in both the high- and low-bristle lineages had bristle numbers that fell well outside the range found in the original population (**FIGURE 13.6**). Selection for high and low bristle numbers resulted in new combinations of the many different genes that were present in the original population, so that the phenotypic variation seen in subsequent generations fell outside the phenotypic variation seen in the original population.

FIGURE 13.5 Artificial Selection Charles Darwin, who raised pigeons as a hobby, noted similar forces at work in artificial and natural selection. The "fancy" pigeons shown here represent 3 of the more than 300 varieties derived from the wild Rock Pigeon (*Columba livia*; far left) by artificial selection on characters such as color and feather distribution.

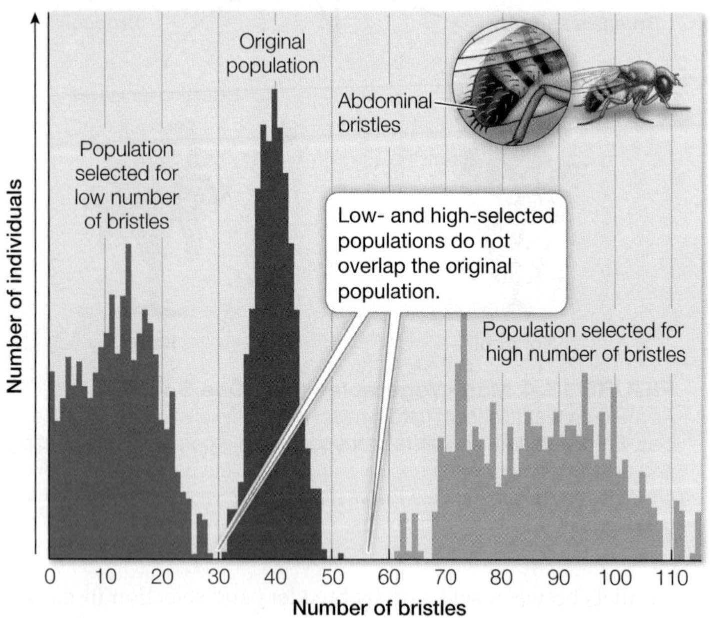

FIGURE 13.6 Artificial Selection Reveals Genetic Variation When investigators subjected *Drosophila melanogaster* to artificial selection for abdominal bristle number, that character evolved rapidly. The graph shows the number of flies with different numbers of bristles in the original population and after 35 generations of artificial selection for low and for high bristle numbers. (After G. A. Clayton et al. 1956. *J Genet* 55: 131–151.)

How did flies in the selected populations end up with fewer or more bristles than were observed in any of the flies in the original population?

Natural selection increases the frequency of beneficial mutations in populations

Darwin knew that far more individuals of most species are born than survive to reproduce. He also knew that, although offspring tend to resemble their parents, the offspring of most organisms are not identical either to their parents or to one another. He suggested that slight differences among individuals affect the chance that a given individual will survive and reproduce, which increases the frequency of the favored trait in the next generation. A favored trait that spreads through a population by natural selection is known as an **adaptation**; this word is used to describe both the trait itself and the process that produces the trait.

Biologists regard an organism as being adapted to a particular environment when they can demonstrate that a slightly different organism reproduces and survives less well in that environment. To understand adaptation, biologists compare the performances of individuals that differ in their traits.

In the opening of this chapter, we noted that bats use echolocation to detect their prey, such as flying moths. Some groups of moths have evolved the ability to detect ultrasounds of bats, which allows the moths to fly away from the bats or use evasive flight behavior to avoid predation. But many species of moths are unable to hear bats, so these moths have to rely on other strategies to avoid bat predation. As described in **INVESTIGATION FIGURE 13.7**, Jesse Barber and his colleagues noticed that several groups of non-hearing moths have long extensions of their wings ("wing tails"), and wondered if these structures might serve to distract echolocating bats and allow the moths to escape. To find out, the researchers cut the wing tails off some individual moths and compared the rates of bat predation and flight performance in the moths with and without wing tails. They found that the wing tails had very little effect on a moth's flying efficiency, but conveyed a huge advantage in reducing bat predation. The wing tails flutter as the moths fly, and the echolocation system of the bats is more likely to locate the fluttering, nonessential wing tail than the body of the moth, thereby allowing a moth to survive a bat attack. Thus the wing tails appear to be an adaptation that helps moths avoid bat predation.

Media Clip 13.2 Big Brown Bats (*Eptesicus fuscus*) versus *Actias luna*

Activity 13.2 Natural Selection Simulation

INVESTIGATION

FIGURE 13.7 Do Long Tails on Wings Help Moths Escape Bat Predation? Many groups of moths have evolved the ability to hear the ultrasounds that bats produce for echolocation. These moths can then use evasive behavior to avoid predation by the bats. But almost half of all moths lack ears to hear the sounds produced by bats. Jesse Barber and his colleagues noticed that many non-hearing moths have long, posterior wing extensions ("wing tails"), and wondered if these wing tails might serve to deflect attacks by echolocating bats.[a]

HYPOTHESIS

The fluttering of long tails on moth wings in flight serves to deflect the attacks of echolocating bats.

METHOD

1. Remove the wing tails from an experimental group of luna moths (*Actias luna*).
2. In an experimental setting, compare the capture rate by big brown bats (*Eptesicus fuscus*) of moths with and without wing tails.
3. Control for the effects of wing tail size by conducting comparable experiments on moths with various wing tail sizes, in species that naturally lack wing tails.
4. Compare the flying efficiency of luna moths with and without wing tails, to test the possibility that wing tails improve flying performance.

RESULTS

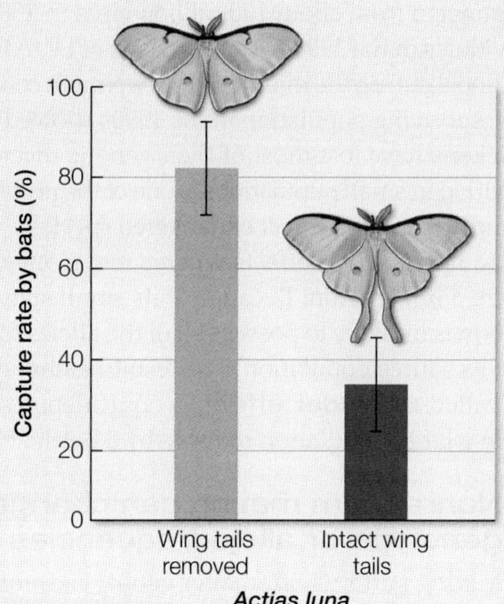

The successful capture rate by bats of luna moths was much higher in moths whose wing tails had been removed. Large wing tail size seemed to have an effect on reducing successful predation as well, but the long wing tails of luna moths provide a much greater benefit to predation avoidance than predicted by the increase in wing tail size alone. The removal of wing tails from luna moths had no significant effect on the flight performance of the moths. The graph shows successful capture rate by bats on luna moths, with and without wing tails. Bars show 95% confidence intervals.

CONCLUSION

The posterior extensions on the wings of luna moths reduce the rate of successful bat attacks on the moths. Bats attack the fluttering wing tails of luna moths rather than their bodies, allowing the moths to escape. The wing tails do not improve or impede flight performance but do help deter predation.

ANALYZE THE DATA

The experiment shows that the wing tails of luna moths reduce the rate of successful predation by bats. Is this innovation unique to luna moths? Jesse Barber and his colleagues conducted a phylogenetic analysis of moths to understand the evolutionary history of this innovation and how many times it has evolved in moths. We introduced phylogenetic trees in Chapter 1, and we will discuss them in greater detail in Chapter 14. These diagrams trace the evolutionary history of a group (such as moths) through time, as lineages diverge from a common ancestor. We show here a diagram of the phylogeny of some of the moth species as reconstructed by the investigators, with seven of the species illustrated. The colors of the branches indicate the reconstructed wing tail:body size ratio. Moths along blue branches lack obvious wing tails, whereas moths along green, yellow, or red branches have progressively longer wing tails.

1. Based on the phylogenetic tree shown, how many times do you think wing tail extensions have evolved in moths?
2. Within a group of moths that have evolved wing tails (such as the close relatives of luna moths), is there any evidence for increased wing tail length over time?

[a]J. R. Barber et al. 2015. *Proc Natl Acad Sci USA* 112: 2812–2816.

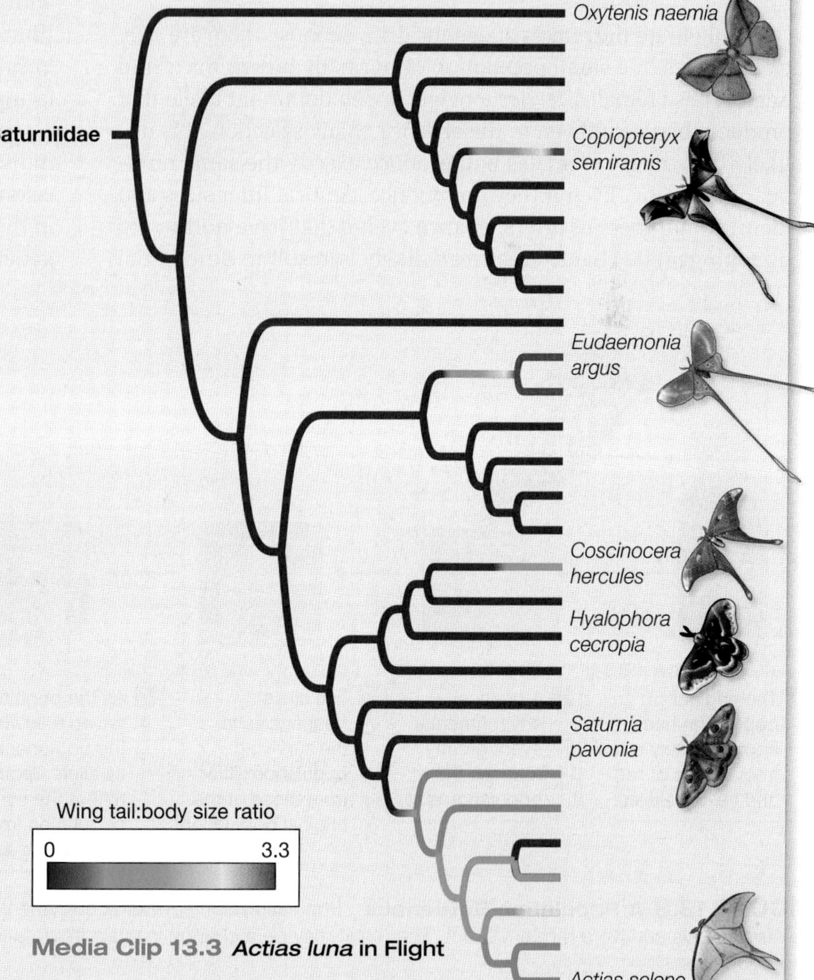

Media Clip 13.3 *Actias luna* in Flight

Gene flow may change allele frequencies

Few populations are completely isolated from other populations of the same species. Migration of individuals and movements of gametes (in pollen, for example) between populations—a phenomenon called **gene flow**—can change allele frequencies in a population. If the arriving individuals survive and reproduce in their new location, they may add new alleles to the population's gene pool, or they may change the frequencies of alleles present in the original population.

If gene flow between two populations stops, those populations may diverge and become different species; see **Key Concept 16.2**.

Genetic drift may cause large changes in small populations

In small populations, **genetic drift**—random changes in allele frequencies from one generation to the next—may produce large changes in allele frequencies over time. Harmful alleles may increase in frequency, and rare advantageous alleles may be lost. Even in large populations, genetic drift can influence the frequencies of neutral alleles (which do not affect the survival and reproductive rates of their bearers).

Activity 13.3 Genetic Drift Simulation

To illustrate the effects of genetic drift, suppose there are only two females in a small population of normally brown mice, and one of these females carries a newly arisen dominant allele that produces black fur. Even in the absence of any selection, it is unlikely that the two females will produce exactly the same number of offspring. Even if they do produce identical litter sizes and identical numbers of litters, chance events that have nothing to do with genetic characteristics are likely to result in differential mortality among their offspring. If each female produces one litter, but a flood envelops the black female's nest and kills all of her offspring, the novel allele could be lost from the population in just one generation. In contrast, if the brown female's litter is lost, then the frequency of the newly arisen allele (and phenotype) for black fur will rise dramatically in just one generation.

Genetic drift is especially potent when a population is reduced dramatically in size. Even populations that are normally large may occasionally pass through environmental events that only a small number of individuals survive, a situation known as a **population bottleneck**. The effect of genetic drift in such a situation is illustrated in **FIGURE 13.8**, in which red and yellow beans represent two alleles of a gene. Most of the beans in the small sample of the "population" that "survives" the bottleneck event are, just by chance, red, so the new population has a much higher frequency of red beans than the previous generation had. In a real population, the red and yellow allele frequencies would be described as having "drifted."

A population forced through a bottleneck is likely to lose much of its genetic variation. For example, when Europeans first arrived in North America, millions of Greater Prairie-Chickens (*Tympanuchus cupido*) inhabited the midwestern prairies. As a result of hunting and habitat destruction by the new settlers, the Illinois population of this species plummeted from about 100 million birds in 1900 to fewer than 50 individuals in the 1990s. A comparison of DNA from birds collected in Illinois during the middle of the twentieth century with DNA from the surviving population in the 1990s showed that Illinois prairie-chickens have lost most of their genetic diversity. Loss of genetic variation in small populations is one of the problems facing biologists who attempt to protect endangered species.

Genetic drift can have similar effects when a few pioneering individuals colonize a new region. Because of its small size, the colonizing population is unlikely to possess all of the alleles found in the gene pool of its source population. The resulting change in genetic variation, called a **founder effect**, is equivalent to that in a large population reduced by a bottleneck.

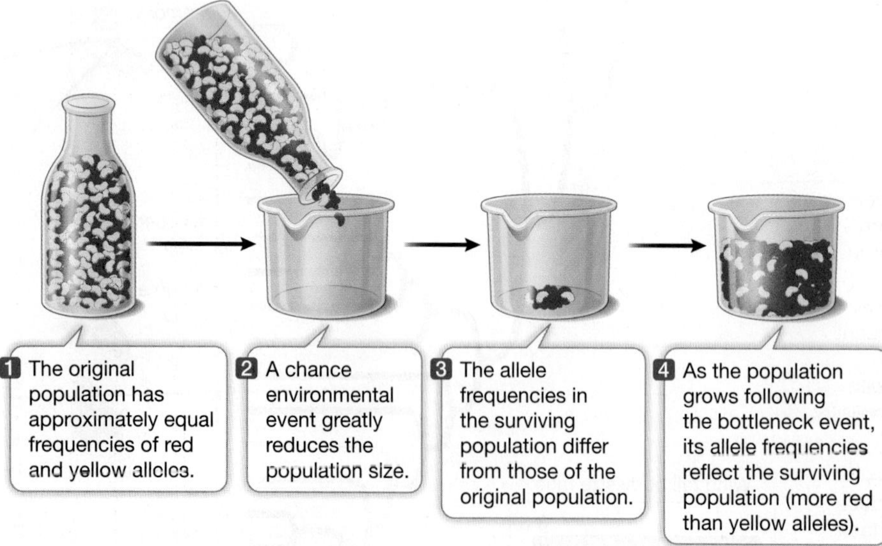

FIGURE 13.8 A Population Bottleneck Population bottlenecks occur when only a few individuals survive a random event. The result may be a change in allele frequencies within the population.

Nonrandom mating can change genotype or allele frequencies

Mating patterns often alter genotype frequencies because the individuals in a population do not choose mates at random. For example, self-fertilization is common in many groups of organisms, especially plants. Any time that individuals mate preferentially with other individuals of the same genotype (including themselves), homozygous genotypes will increase in frequency and heterozygous genotypes will decrease in frequency over time. The opposite effect (more heterozygotes, fewer homozygotes) is expected when individuals mate primarily or exclusively with individuals of different genotypes.

Many nonrandom mating systems produce changes in genotype frequencies but not in allele frequencies, and thus do not, by themselves, result in evolutionary change in a population.

Euplectes progne

FIGURE 13.9 What Is the Advantage? The extensive tail of the male African Long-tailed Widowbird inhibits its ability to fly. Darwin attributed the evolution of this seemingly nonadaptive trait to sexual selection.

However, nonrandom mating systems that result in different reproductive success among individuals do produce allele frequency changes from one generation to the next. One such process, **sexual selection**, occurs when individuals of one sex mate preferentially with particular individuals of the opposite sex rather than at random.

Sexual selection was first suggested by Charles Darwin, who developed the idea to explain the evolution of conspicuous traits that would appear to inhibit survival, such as bright colors and elaborate courtship displays in males of many species. He hypothesized that these features either improved the ability of their bearers to compete for access to mates (intrasexual selection) or made their bearers more attractive to members of the opposite sex (intersexual selection).

Darwin argued that while natural selection typically favors traits that enhance the survival of their bearers or their bearers' descendants, sexual selection is primarily about successful reproduction. An animal that survives but fails to reproduce makes no contribution to the next generation. Thus sexual selection may favor traits that enhance an individual's chances of reproduction even when these traits reduce its chances of survival. For example, females may be more likely to see or hear males with a given trait (and thus be more likely to mate with those males), even though the favored trait also increases the chances that the male will be seen or heard by a predator.

One example of a trait that Darwin attributed to sexual selection is the remarkable tail of the male African Long-tailed Widowbird (*Euplectes progne*), which is longer than the bird's head and body combined (**FIGURE 13.9**). Male widowbirds normally select, and defend from other males, a territory where they perform courtship displays to attract females. To investigate whether sexual selection drove the evolution of widowbird tails, a biologist clipped the tails of some captured male widowbirds and lengthened the tails of others by gluing on additional feathers. He then cut and reglued the tail feathers of still other males, which served as controls. Both short- and long-tailed males successfully defended their display territories, indicating that a long tail does not confer an advantage in male–male competition. However, males with artificially elongated tails attracted about four times more females than did males with shortened tails (**INVESTIGATION FIGURE 13.10**). Thus males with long tails pass on their genes to more offspring than do males with short tails, which leads to the evolution of this unusual trait.

INVESTIGATION

FIGURE 13.10 Sexual Selection in Action Behavioral ecologist Malte Andersson tested Darwin's hypothesis that excessively long tails evolved in male widowbirds because female preference for longer-tailed males increased their mating and reproductive success.[a]

HYPOTHESIS

Female widowbirds prefer to mate with the male that displays the longest tail; longer-tailed males thus are favored by sexual selection because they will father more offspring.

METHOD

1. Capture males and artificially lengthen or shorten tails by cutting or gluing on feathers. In a control group, cut and replace tails to their normal length (to control for the effects of tail cutting).
2. Release the males to establish their territories and mate.
3. Count the nests with eggs or young on each male's territory.

RESULTS

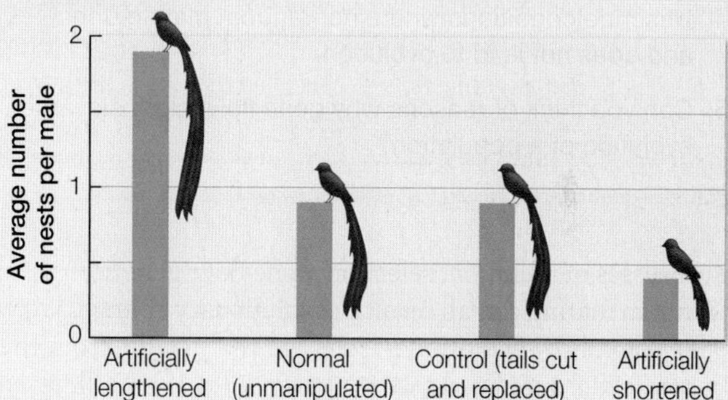

Male widowbirds with artificially shortened tails established and defended display sites successfully but fathered fewer offspring than did control or unmanipulated males. Males with artificially lengthened tails fathered the most offspring.

CONCLUSION

Sexual selection in widowbirds has favored the evolution of long tails in males.

ANALYZE THE DATA

Are the differences plotted above significantly different? To answer this question, see Analyze the Data 13.1 in LaunchPad for a simple method to test the statistical significance of the differences using the following data.

	Number of nests per male		
Group	Shortened tail	Control	Elongated tail
1	0	0	2
2	0	0	2
3	2	3	5
4	1	2	4
5	0	1	2
6	0	1	2
7	0	1	0
8	0	0	0
9	1	0	0

[a]M. Andersson. 1982. *Nature* 299: 818–820.

A similar **Analyze the Data** exercise may be assigned in **Achieve**.

REVIEW & APPLY | 13.2

R Evolutionary processes change the genetic structure of a population from one generation to the next. These processes include mutation, selection, gene flow, genetic drift, and nonrandom mating.

A
1. In everyday language, we speak of individuals "evolving" over their lifetime (if, for example, they change their opinion about an issue). This sense of "evolving" is clearly very different than what biologists mean when they refer to biological evolution. Explain why changes within an individual's lifetime are not examples of biological evolution.
2. How is mutation a necessary component of the evolutionary processes?
3. What kinds of traits would be more likely to evolve by genetic drift than by natural selection?
4. Describe conditions where nonrandom mating does and does not lead to evolution.
5. Can you think of reasons why gene flow might slow the evolution of a population?

The processes of mutation, selection, gene flow, genetic drift, and nonrandom mating can all result in evolutionary change. We will consider next how evolutionary change that results from these processes is measured.

13.3 | Evolution Can Be Measured by Changes in Allele Frequencies

LEARNING OBJECTIVES

By the end of this key concept you should be able to:

13.3.1 Calculate the allele frequencies for two alleles at a given locus in a diploid population.

13.3.2 Calculate observed and expected genotype frequencies from allele frequencies based on the principles of Hardy-Weinberg equilibrium.

13.3.3 Interpret data and formulate hypotheses to explain deviations from Hardy-Weinberg expectations in observed genotype frequencies.

Much of evolution occurs through gradual changes in the relative frequencies of different alleles in a population from one generation to the next. Major genetic changes can also be sudden, as happens when two formerly separated populations merge and hybridize, or when genes within a population are duplicated within the genome. But in most cases, we measure evolution by looking at changes in allele and genotype frequencies in populations over time.

A population's genetic structure can be described by frequencies of alleles and genotypes

To measure allele frequencies in a population precisely, we would need to count every allele at every locus in every individual in the population. Fortunately, we do not need to make such complete measurements because we can reliably estimate allele frequencies for a given locus by counting alleles in a reasonably sized sample of individuals from the population. The sum of all allele frequencies at a locus is equal to 1, so measures of allele frequency range from 0 to 1.

An allele's frequency is calculated using the following formula:

$$p = \frac{\text{number of copies of the allele in the population}}{\text{total number of copies of all alleles in the population}}$$

If only two alleles (we'll call them *A* and *a*) for a given locus are found among the members of a diploid population, those alleles can combine to form three different genotypes: *AA*, *Aa*, and *aa*. A population with more than one allele at a locus is said to be polymorphic ("many forms") at that locus. Applying the formula above, as shown in **FIGURE 13.11**, we can calculate the relative frequencies of alleles *A* and *a* in a population of *N* individuals as follows:

- Let N_{AA} be the number that are homozygous for the *A* allele (*AA*).
- Let N_{Aa} be the number that are heterozygous (*Aa*).
- Let N_{aa} be the number that are homozygous for the *a* allele (*aa*).

Note that $N_{AA} + N_{Aa} + N_{aa} = N$, the total number of individuals in the population, and that the total number of copies of both alleles present in the population is $2N$, because each individual is diploid. Each *AA* individual has two copies of the *A* allele, and each *Aa* individual has one copy of the *A* allele. Therefore the total number of *A* alleles in the population is $2N_{AA} + N_{Aa}$. Similarly, the total number of *a* alleles in the population is $2N_{aa} + N_{Aa}$. If *p* represents the frequency of *A*, and *q* represents the frequency of *a*, then

$$p = \frac{2N_{AA} + N_{Aa}}{2N}$$

and

$$q = \frac{2N_{aa} + N_{Aa}}{2N}$$

The calculations in Figure 13.11 demonstrate two important points. First, notice that for each population, $p + q = 1$, which means that $q = 1 - p$. So when there are only two alleles at a given locus in a population, we can calculate the frequency of one allele and obtain the second allele's frequency by subtraction. If there is only one allele at a given locus in a population, its frequency is 1: the population is then monomorphic at that locus, and the allele is said to be **fixed**.

Second, notice that population 1 (consisting mostly of homozygotes) and population 2 (consisting mostly of heterozygotes) have the same allele frequencies for *A* and *a*. Thus they have the same gene pool for this locus. Because the alleles in the gene pool are distributed differently among individuals, however, the genotype frequencies of the two populations differ.

RESEARCH TOOLS

FIGURE 13.11 Calculating Allele and Genotype Frequencies Allele and genotype frequencies for a gene locus with two alleles in the population can be calculated using the equations in panel 1. When the equations are applied to two populations (panel 2), we find that the frequencies of alleles *A* and *a* in the two populations are the same, but the alleles are distributed differently between heterozygous and homozygous genotypes.

1 In any population, where N is the total number of individuals in the population:

$$\text{Frequency of allele } A = p = \frac{2N_{AA} + N_{Aa}}{2N} \qquad \text{Frequency of allele } a = q = \frac{2N_{aa} + N_{Aa}}{2N}$$

Frequency of genotype $AA = N_{AA}/N$
Frequency of genotype $Aa = N_{Aa}/N$
Frequency of genotype $aa = N_{aa}/N$

2 Compute the allele and genotype frequencies for two separate populations of $N = 200$:

Population 1 (mostly homozygotes)	Population 2 (mostly heterozygotes)
$N_{AA} = 90$, $N_{Aa} = 40$, and $N_{aa} = 70$	$N_{AA} = 45$, $N_{Aa} = 130$, and $N_{aa} = 25$
$p = \frac{180 + 40}{400} = 0.55$	$p = \frac{90 + 130}{400} = 0.55$
$q = \frac{140 + 40}{400} = 0.45$	$q = \frac{50 + 130}{400} = 0.45$
Freq. AA = 90/200 = 0.45	Freq. AA = 45/200 = 0.225
Freq. Aa = 40/200 = 0.20	Freq. Aa = 130/200 = 0.65
Freq. aa = 70/200 = 0.35	Freq. aa = 25/200 = 0.125

The frequencies of the different alleles at each locus and the frequencies of the different genotypes in a population describe that population's **genetic structure**. Allele frequencies measure the amount of genetic variation in a population, whereas genotype frequencies show how a population's genetic variation is distributed among its members. Other measures, such as the proportion of loci that are polymorphic, are also used to measure variation in populations. With these measurements, it becomes possible to consider how the genetic structure of a population changes or remains the same over generations—that is, to measure evolutionary change.

Evolution will occur unless certain restrictive conditions exist

In 1908, the British mathematician Godfrey Hardy and the German physician Wilhelm Weinberg independently deduced the conditions that must prevail if the genetic structure of a population is to remain the same over time. If the conditions they identified do not exist, then evolution will occur. The resulting principle is known as **Hardy–Weinberg equilibrium**. Hardy–Weinberg equilibrium describes a model in which allele frequencies do not change across generations and genotype frequencies can be predicted from allele frequencies (**FIGURE 13.12**). The principles of Hardy–Weinberg equilibrium apply only to sexually reproducing organisms. Several conditions must be met for a population to be at Hardy–Weinberg equilibrium. Note that the following conditions correspond inversely to the five principal processes of evolution (discussed in Key Concept 13.2):

- *There is no mutation.* The alleles present in the population do not change, and no new alleles are added to the gene pool.
- *There is no selection among genotypes.* Individuals with different genotypes have equal probabilities of survival and equal rates of reproduction.
- *There is no gene flow.* There is no movement of individuals into or out of the population or reproductive contact with other populations.

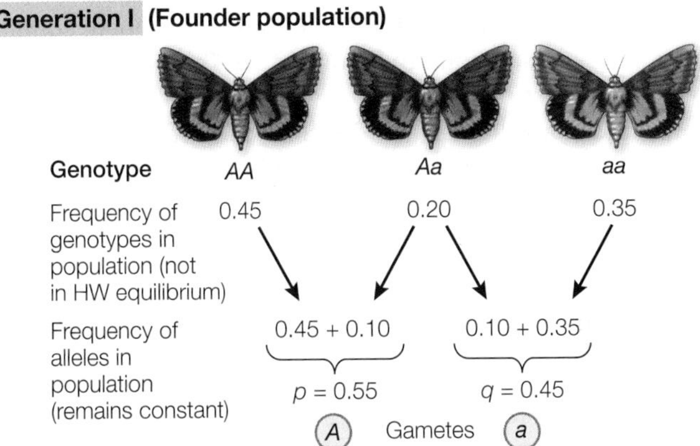

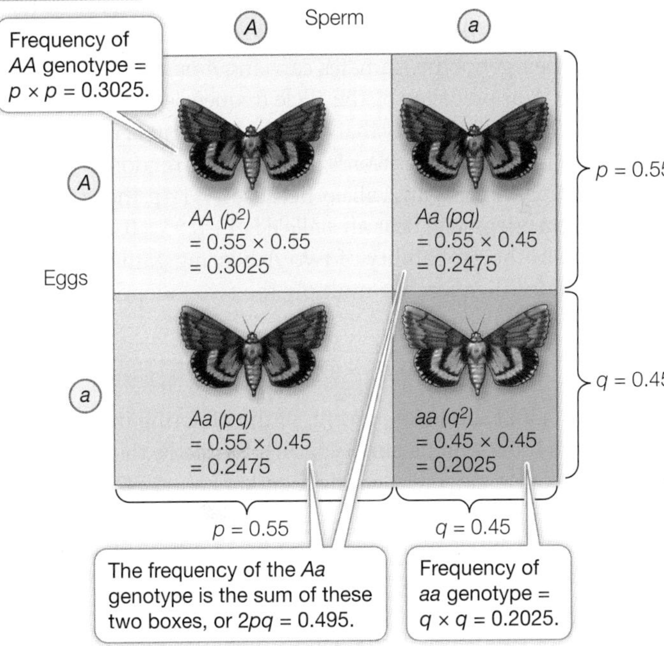

FIGURE 13.12 One Generation of Random Mating Restores Hardy-Weinberg Equilibrium Generation I of this population is made up of migrants from several source populations and so is not initially in Hardy–Weinberg equilibrium. After one generation of random mating, the allele frequencies are unchanged, and the genotype frequencies return to Hardy–Weinberg expectations. The lengths of the sides of each rectangle are proportional to the allele frequencies in the population; the areas of the rectangles are proportional to the genotype frequencies.

- *Population size is infinite.* The larger a population, the smaller will be the effect of genetic drift.
- *Mating is random.* Individuals do not preferentially choose mates with certain genotypes.

If these idealized conditions hold, two major consequences follow. First, the frequencies of alleles at a locus remain constant from generation to generation. Second, following one generation of random mating, the genotype frequencies occur in the following proportions:

Genotype	*AA*	*Aa*	*aa*
Frequency	p^2	$2pq$	q^2

To understand why these consequences are important, start by considering a population that is *not* in Hardy–Weinberg equilibrium, such as generation I in Figure 13.12. This could occur, for example, if the initial population is founded by migrants from several other populations, thus violating the Hardy–Weinberg assumption of no gene flow. In this example, generation I has more homozygous individuals and fewer heterozygous individuals than would be expected under Hardy–Weinberg equilibrium (a condition known as heterozygote deficiency).

Even with a starting population that is not in Hardy–Weinberg equilibrium, we can predict that after a single generation of random mating, and if the other Hardy–Weinberg assumptions are not violated, the allele frequencies will remain unchanged, but the genotype frequencies will return to Hardy–Weinberg expectations. Let's explore why this is true.

In generation I of Figure 13.12, the frequency of the *A* allele (p) is 0.55. If we assume that individuals select mates at random, without regard to their genotype, gametes carrying *A* or *a* combine at random—that is, as predicted by the allele frequencies p and q. Thus in this example, the probability that a particular sperm or egg will bear an *A* allele is 0.55. In other words, 55 out of 100 randomly sampled sperm or eggs will bear an *A* allele. Because $q = 1 - p$, the probability that a sperm or egg will bear an *a* allele is $1 - 0.55 = 0.45$.

To obtain the probability of two *A*-bearing gametes coming together at fertilization, we multiply the two independent probabilities of their occurrence:

$$p \times p = p^2 = (0.55)^2 = 0.3025$$

Therefore 0.3025, or 30.25 percent, of the offspring in generation II will have homozygous genotype *AA*. Similarly, the probability of two *a*-bearing gametes coming together is

$$q \times q = q^2 = (0.45)^2 = 0.2025$$

which means that 20.25 percent of generation II will have the *aa* genotype.

There are two ways of producing a heterozygote: an *A* sperm may combine with an *a* egg, the probability of which is $p \times q$; or an *a* sperm may combine with an *A* egg, the probability of which is $q \times p$. Consequently, the overall probability of obtaining a heterozygote is $2pq$, or 0.495. The frequencies of the *AA*, *Aa*, and *aa* genotypes in generation II of Figure 13.12 now meet Hardy–Weinberg expectations, and the frequencies of the two alleles (p and q) have not changed from generation I.

Under the assumptions of Hardy–Weinberg equilibrium, allele frequencies p and q remain constant from generation to generation. If Hardy–Weinberg assumptions are violated and the genotype frequencies in the parental generation are altered (say, by the loss of a large number of *AA* individuals from the population), then the allele frequencies in the next generation will be altered. However, based on the new allele frequencies, another generation of random mating will be sufficient to restore the genotype frequencies to Hardy–Weinberg equilibrium.

Activity 13.4 Hardy–Weinberg Equilibrium

Deviations from Hardy-Weinberg equilibrium show that evolution is occurring

You probably have realized that populations in nature never meet the stringent conditions necessary to be at Hardy–Weinberg equilibrium—which explains why all biological populations evolve. Why, then, is this model considered so important for the study of evolution? There are two reasons. First, the equation is useful for predicting the approximate genotype frequencies of a population from its allele frequencies. Second—and crucially—the model allows biologists to evaluate which processes are acting on the evolution of a particular population. The specific patterns of deviation from Hardy–Weinberg equilibrium can help us identify the various processes of evolutionary change.

REVIEW & APPLY | 13.3

R Hardy–Weinberg equilibrium describes the theoretical conditions required for evolution *not* to occur. Deviations from Hardy–Weinberg expectations provide information about how evolution *is* occurring in a given population.

A

1. The following sample lists the genotype at locus *A* for 10 individuals in a diploid population: *AA*, *AA*, *Aa*, *Aa*, *Aa*, *Aa*, *aa*, *aa*, *aa*, *aa*.
   a. Based on this sample, what is the observed frequency of allele *a*? The observed frequency of allele *A*?
   b. What are the observed frequencies of genotypes *aa*, *Aa*, and *AA*?
   c. After one generation of random mating, what would be the Hardy–Weinberg expectations for the frequencies of genotypes *aa*, *Aa*, and *AA*?
2. In a population of toads, allele *A* is present at a frequency of 0.2, and allele *a* is present at a frequency of 0.8. Based on Hardy–Weinberg expectations, what is the expected frequency of the genotype *Aa* in the population?
3. Assume that the observed frequency of the genotype *Aa* in the toad population in Question 2 is 0.15, rather than the expected frequency you calculated. What are some possible explanations?

Our discussion so far has focused on changes in allele frequencies at a single gene locus. Genes do not exist in isolation, however, but interact with one another (and with the environment) to produce an organism's phenotype. What effects can these interactions have on selection?

13.4 | Selection Can Be Stabilizing, Directional, or Disruptive

LEARNING OBJECTIVES

By the end of this key concept you should be able to:

13.4.1 Correlate the effect of each type of selection (stabilizing, directional, and disruptive) on phenotype evolution.

13.4.2 Distinguish conditions in which selection is likely to be stabilizing, directional, or disruptive.

Although evolution is defined as changes in the genetic makeup of a population from one generation to the next, natural selection acts directly on the phenotype—that is, on the physical features expressed by an organism with a given genotype—and therefore acts only indirectly on the genotype. The reproductive contribution of a phenotype to subsequent generations relative to the contributions of other phenotypes is called its **fitness**.

Changes in reproductive rate do not necessarily change the genetic structure of a population. For example, if all individuals in a population experience the same increase in reproductive rate (during an environmentally favorable year, for instance), the genetic structure of the population will not change. Changes in numbers of offspring are responsible for increases and decreases in the *size* of a population, but only changes in the *relative* success of different phenotypes in a population will lead to changes in allele frequencies from one generation to the next. The fitness of individuals of a particular phenotype is a function of the probability of those individuals surviving multiplied by the average number of offspring they produce over their lifetimes. In other words, the *fitness of a phenotype is determined by the relative rates of survival and reproduction of individuals with that phenotype.*

Until now, we have only discussed traits influenced by alleles at a single locus. Such traits are often distinguished by discrete qualities (black versus white, or smooth versus wrinkled) and so are called **qualitative traits**. Many traits, however, are influenced by alleles at more than one locus. Such traits are likely to show *continuous* quantitative variation rather than *discrete* qualitative variation, and so are known as **quantitative traits**. For example, body size is a continuous variable; the distribution of body sizes of individuals in a population, a trait that is influenced by genes at many loci as well as by the environment, is likely to resemble a continuous bell-shaped curve.

Natural selection can act on characters with quantitative variation in any one of several different ways, producing quite different results (**FIGURE 13.13**):

- **Stabilizing selection** preserves the average characteristics of a population by favoring average individuals.
- **Directional selection** changes the characteristics of a population by favoring individuals that vary in one direction from the mean of the population.
- **Disruptive selection** changes the characteristics of a population by favoring individuals that vary in both directions from the mean of the population.

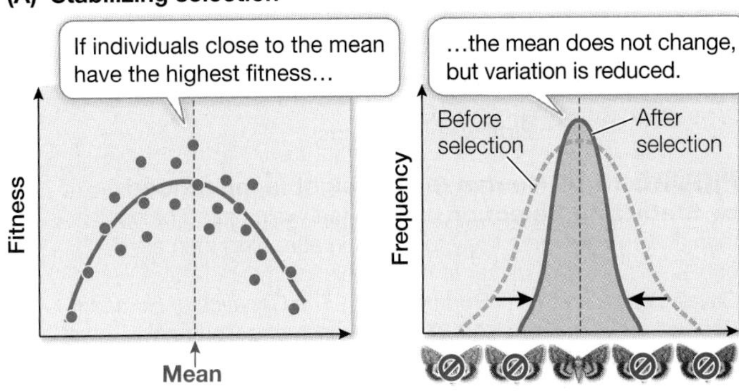

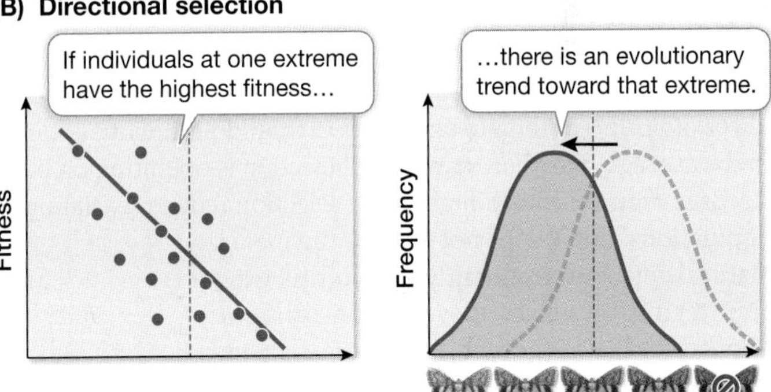

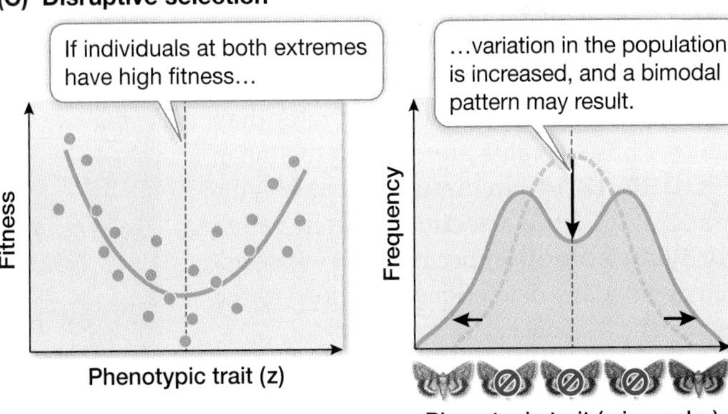

FIGURE 13.13 Natural Selection Can Operate in Several Ways The graphs in the left-hand column show the fitness of individuals with different phenotypes for the same trait. The right-hand graphs show the distribution of the phenotypes in the population before (dashed line) and after (solid line) the influence of selection.

Why might selection for a trait in a population shift from directional to stabilizing selection over time, even if the optimal value for the trait stays constant?

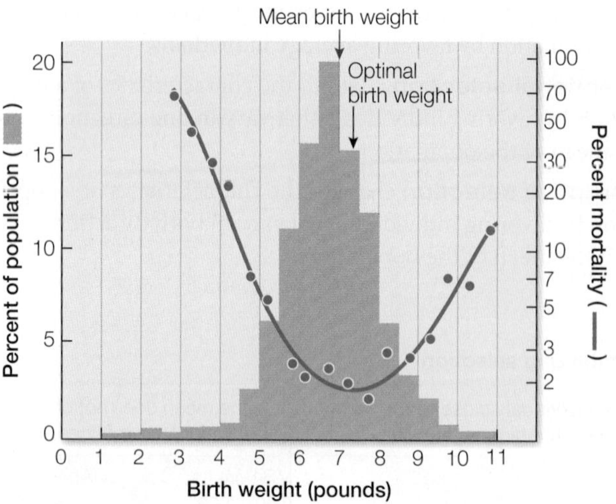

FIGURE 13.14 Human Birth Weight Is Influenced by Stabilizing Selection Babies that weigh more or less than average are more likely to die soon after birth than are babies with weights closer to the population mean. (After L. L. Cavalli-Sforza and W. F. Bodmer. 1971. *The Genetics of Human Populations*. San Francisco: W. H. Freeman and Company.)

Stabilizing selection reduces variation in populations

If the smallest and largest individuals in a population contribute fewer offspring to the next generation than do individuals closer to the average size, then stabilizing selection is operating on body size (see Figure 13.13A). Stabilizing selection reduces variation in populations, but it does not change the mean. Natural selection frequently acts in this way, countering increases in variation brought about by sexual recombination, mutation, or gene flow. Rates of phenotypic change in many species are slow because natural selection is often stabilizing. Stabilizing selection operates, for example, on human birth weight. Babies who are lighter or heavier at birth than the population mean die at higher rates than babies whose weights are close to the mean (**FIGURE 13.14**). In discussions of specific genes, stabilizing selection is often called **purifying selection** because there is selection against any deleterious mutations to the usual gene sequence.

Directional selection favors one extreme

Directional selection is operating when individuals at one extreme of a character distribution contribute more offspring to the next generation than other individuals do, shifting the average value of that character in the population toward that extreme.

In the case of a single gene locus, directional selection may result in favoring a particular genetic variant—referred to as **positive selection** for that variant. By favoring one phenotype over another, directional selection results in an increase in the frequencies of alleles that produce the favored phenotype.

Animation 13.1 Natural Selection

If directional selection operates over many generations, an evolutionary trend is seen in the population (see Figure 13.13B). Evolutionary trends often continue for many generations, but they can be reversed if the environment changes and different phenotypes are favored, or halted when an optimal phenotype is reached or trade-offs between different adaptational advantages oppose further change. The character then undergoes stabilizing selection.

The long horns of Texas Longhorn cattle (**FIGURE 13.15**) are an example of a trait that has evolved through directional selection. Texas Longhorns are descendants of cattle brought to the New World by Christopher Columbus, who picked up a few cattle in the Canary Islands and brought them to the island of Hispaniola in 1493. The cattle multiplied, and their descendants were taken to the mainland of Mexico. Spaniards exploring what would become Texas and the southwestern United States brought these cattle with them, and some of them escaped and formed feral herds. Populations of feral cattle increased greatly over the next few hundred years, but there was heavy predation from bears, mountain lions, and wolves, especially on the young calves. Cows with longer horns were more successful in protecting their calves against attacks, and over a few hundred years the average horn length in the feral herds increased considerably. In addition, the

FIGURE 13.15 A Result of Directional Selection In the American Southwest, long horns were advantageous for defending calves from attacks by predators, so cows with longer horns were more likely to raise calves successfully. As a result, horn length in feral herds of cattle increased between the early 1500s and the 1860s, leading to the Texas Longhorn breed. This evolutionary trend has been maintained by modern ranchers practicing artificial selection.

THINK LIKE A SCIENTIST

Process of Science

Observing and measuring phenotypic evolution

Anoles, an abundant and diverse group of lizards in the American tropics and subtropics, are common subjects of ecological and evolutionary studies. The Carolina (or green) anole, *Anolis carolinensis* (**FIGURE 13.16A**), is the only anole native to the southeastern United States. The Cuban brown anole, *Anolis sagrei* (**FIGURE 13.16B**), was introduced into south Florida in the late 1800s and has been expanding its range northward since then. When living alone, *A. carolinensis* occupies habitats from ground level to treetops, whereas *A. sagrei* lives on the ground and lower perches. The two species compete, and the more aggressive *A. sagrei* displaces *A. carolinensis* from its preferred perches. Where the two species live together, *A. carolinensis* is restricted to high treetops.

Across species of *Anolis*, there is a strong association between perch height and the structure of the toe pads that the lizards use to cling to branches (**FIGURE 13.16C**). Researchers wondered if the shift in perching habitat would lead to rapid evolutionary change in the toe pads of displaced *A. carolinensis* populations. In 1995, researchers deliberately introduced populations of *A. sagrei* onto five islands that previously supported only populations of *A. carolinensis*, and then followed the evolutionary changes in the lizard populations over the next 15 years. Researchers examined the size of the toe pads and the number of expanded scales with adhesive bristles (lamellae) on each pad. Other islands continued to support only *A. carolinensis*.

In 2010, the researchers measured toe-pad size and the number of lamellae on *A. carolinensis* collected from both sets of islands (with and without *A. sagrei*) (**FIGURE 13.16D, E**). The researchers also raised offspring of lizards from these islands in a "common garden experiment," which uses a common, controlled environment to confirm that any differences between the populations have a genetic basis. Here you will apply processes of science to investigate phenotypic evolution in these lizards.

1. What is the evidence for toe-pad evolution in the populations of *A. carolinensis* on the invaded islands? Why have toe pads in these populations of *A. carolinensis* changed so rapidly?
2. Why was the common garden experiment needed? How would the conclusions of the study have been different if the observed differences in the wild populations were not also evident in the lizards raised in a common, controlled environment?
3. Which evolutionary process or processes are most likely to explain the observed evolutionary changes? Which processes do you think are less important, and why?
4. Describe the ways that these populations of *A. carolinensis* deviate from assumptions of Hardy–Weinberg equilibrium (the conditions that would be necessary for the populations *not* to evolve).
5. Instead of *A. sagrei*, suppose another species of *Anolis* had been introduced to the Florida islands. Suppose this lizard were an aggressive treetop specialist that excluded *A. carolinensis* from the highest perches and restricted it to the ground and low perches. In this case, how would you expect the displaced populations of *A. carolinensis* to have evolved?

(A) ***Anolis carolinensis***

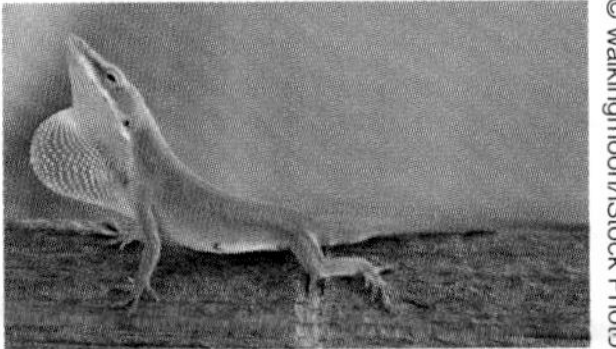

© walkingmoon/iStock Photo

(B) ***Anolis sagrei***

© NNehring/iStock Photo

(C)

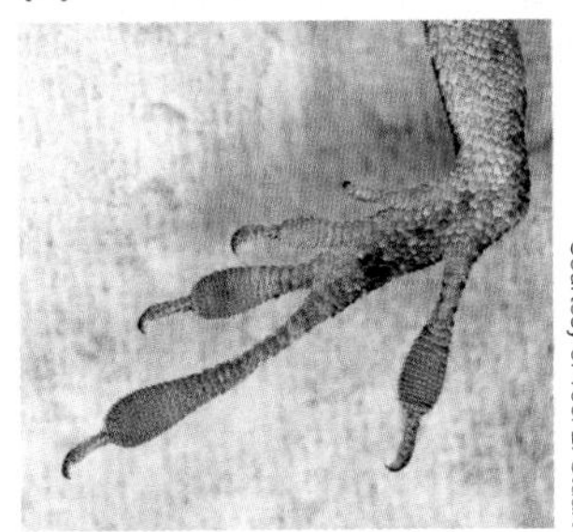

Courtesy of Yoel E. Stuart

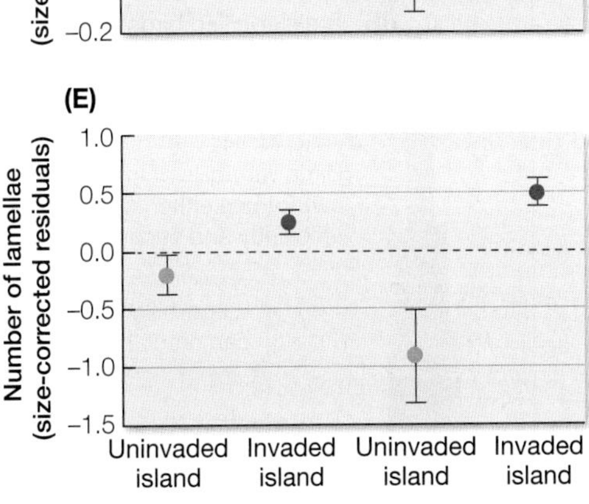

Both toe-pad area and number of lamellae evolved in the populations on invaded islands compared with populations on uninvaded islands.

Differences between wild-caught populations were retained in the common garden experiment. This shows that the observed changes were genetic rather than environmental, confirming that evolution (genetic changes in the population) occurred.

FIGURE 13.16 Lizard Competition and Evolution *A. carolinensis* **(A)** and *A. sagrei* **(B)** compete for habitat in areas where they occupy the same territory. **(C)** Foot and toe pads of *A. carolinensis*. **(D)** Toe-pad area of *A. carolinensis* from uninvaded and invaded islands. **(E)** Number of toe-pad lamellae of *A. carolinensis* from uninvaded and invaded islands. Bars in D and E indicate 95% confidence limits. (D, E after Y. E. Stuart et al. 2014. *Science* 346: 463–466.)

cattle evolved resistance to endemic diseases of the Southwest, as well as higher fecundity and longevity. Texas Longhorns often live and produce calves well into their twenties—about twice as long as many breeds of cattle that have been artificially selected by humans for traits such as high fat content or high milk production (which are examples of artificial directional selection).

Disruptive selection favors extremes over the mean

When disruptive selection operates, individuals at opposite extremes of a character distribution contribute more offspring to the next generation than do individuals close to the mean, which increases variation in the population (see Figure 13.13C).

The strikingly bimodal (two-peaked) distribution of bill sizes in the Black-bellied Seedcracker (*Pyrenestes ostrinus*), a West African finch (**FIGURE 13.17**), illustrates how disruptive selection can influence populations in nature. The seeds of two types of sedges (marsh plants) are the most abundant food source for these finches during part of the year. Birds with large bills can readily crack the hard seeds of the sedge *Scleria verrucosa*. Birds with small bills can crack *S. verrucosa* seeds only with difficulty; however, they feed more efficiently on the soft seeds of *S. goossensii* than do birds with larger bills. Young finches whose bills deviate markedly from the two predominant bill sizes do not survive as well as finches whose bills are close to one of the two sizes represented by the distribution peaks. Because there are few abundant food sources in the finches' environment, and because the seeds of the two sedges do not overlap in hardness, birds with intermediate-sized bills are less efficient in using either one of the species' principal food sources. Disruptive selection therefore maintains a bimodal bill size distribution.

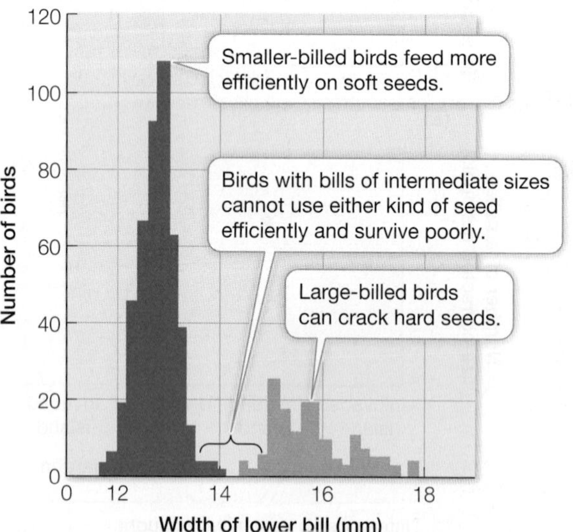

FIGURE 13.17 Disruptive Selection Results in a Bimodal Character Distribution The bimodal distribution of bill sizes in the Black-bellied Seedcracker (*Pyrenestes ostrinus*) of West Africa is a result of disruptive selection, which favors individuals with larger and smaller bill sizes over individuals with intermediate-sized bills. (After T. B. Smith 1993. *Nature* 363: 618–620.)

REVIEW & APPLY | 13.4

R Natural selection can both change and stabilize phenotypes within populations. Directional selection can result in continuous change to phenotypes; stabilizing selection slows change from a favored phenotype; and disruptive selection can result in one population that exhibits multiple distinct phenotypes.

A
1. Describe the different expected outcomes of stabilizing, directional, and disruptive selection.
2. Why would you expect selection on human birth weight to be stabilizing rather than directional?
3. Offer some examples of extreme phenotypes in animal or plant populations that could be explained by directional selection.

Genetic drift, stabilizing selection, and directional selection all tend to reduce genetic variation within populations. Nevertheless, as we have seen, most populations harbor considerable genetic variation. What processes produce and maintain genetic variation within populations?

13.5 | Selection Can Maintain Polymorphisms in Populations

LEARNING OBJECTIVES

By the end of this key concept you should be able to:

13.5.1 **Explain how genetic variation is useful for a species in the face of environmental variability.**

13.5.2 **Propose a process for the maintenance of deleterious alleles in a population and a method for testing the hypothesis.**

Genetic variation is the raw material on which mechanisms of evolution act. But if favored alleles increase in frequency in populations, then why aren't other alleles always lost from the population? In this section we discuss some factors that affect how genetic variation is established, how it is distributed among individuals, and how it is maintained within populations.

Frequency-dependent selection maintains genetic variation within populations

Natural selection often preserves variation as a polymorphism (the presence of two or more variants of a character in the same population). When the fitness of a given phenotype depends on its frequency in a population, a polymorphism may be maintained by a process known as **frequency-dependent selection**. *Perissodus microlepis*, a small fish that lives in Lake Tanganyika in East Africa, provides an example of frequency-dependent selection.

P. microlepis feeds on the scales of other fish, approaching its prey from behind and dashing in to bite off several scales from the prey's flank. Because of an asymmetrical jaw joint, the mouth of this scale-eating species opens either to the right or to the left; the direction is

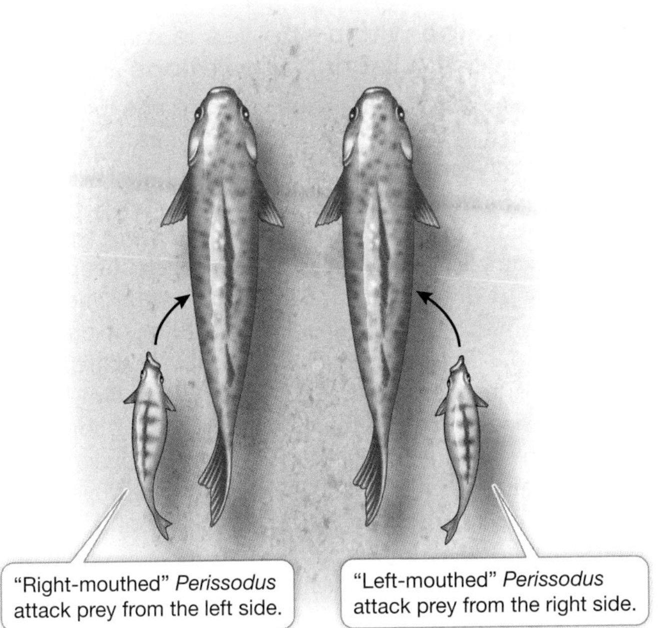

FIGURE 13.18 A Stable Polymorphism Frequency-dependent selection maintains equal proportions of left- and right-mouthed individuals of the scale-eating fish *Perissodus microlepis*.

genetically determined (**FIGURE 13.18**). "Right-mouthed" individuals always attack from the victim's left, and "left-mouthed" individuals always attack from the victim's right. The distorted mouth enlarges the area of teeth in contact with the prey's flank, but only if the scale-eater attacks from the appropriate side.

Prey fish are alert to approaching scale-eaters, so attacks are more likely to be successful if the prey must watch both flanks. Vigilance by prey thus favors equal numbers of right- and left-mouthed scale-eaters in a population, because if attacks from one side were more common than the other, prey fish would pay more attention to potential attacks from that side. Over an 11-year study of *P. microlepis* in Lake Tanganyika, the genetic polymorphism was found to be stable, and the two phenotypes of the scale-eaters remained at about equal frequencies.

Heterozygote advantage maintains polymorphic loci

In many cases, different alleles of a particular gene are advantageous under different environmental conditions. Most organisms experience a wide variety of environmental conditions over time. A night is dramatically different from the preceding day. A cold, cloudy day differs from a clear, hot one. Day length and temperature change seasonally. For many genes, a single allele is unlikely to perform well under all these conditions. In such cases, heterozygous individuals (with two different alleles) are likely to outperform homozygous individuals (with only one of those two alleles).

Colias butterflies of the Rocky Mountains live in environments where dawn temperatures are often too cold, and afternoon temperatures too hot, for the butterflies to fly. Populations of these butterflies are polymorphic for a gene that encodes the enzyme phosphoglucose isomerase (PGI), which influences how well a butterfly flies at different temperatures. Butterflies with certain PGI genotypes can fly better during the cold hours of early morning; others perform better during midday heat. The optimal body temperature for flight is 35°C–39°C, but some butterflies can fly with body temperatures as low as 29°C or as high as 40°C. During spells of unusually hot weather, heat-tolerant genotypes are favored; during spells of unusually cool weather, cold-tolerant genotypes are favored.

Heterozygous *Colias* butterflies can fly over a greater range of temperatures than homozygous individuals, which should give them an advantage in foraging and finding mates. A test of this prediction found that heterozygous males did indeed have a mating advantage, and further, that this advantage maintains the polymorphism in the population (**INVESTIGATION FIGURE 13.19**). Of course, the heterozygous genotype can never become fixed in the population, because the offspring of two heterozygotes will always include both classes of homozygotes in addition to heterozygotes.

INVESTIGATION

FIGURE 13.19 A Heterozygote Mating Advantage Among butterflies of the genus *Colias*, males that are heterozygous for two alleles of the PGI enzyme can fly farther under a broader range of temperatures than males that are homozygous for either allele. Does this ability give heterozygous males a mating advantage?[a]

HYPOTHESIS

Heterozygous male *Colias* will have proportionally greater mating success than homozygous males.

METHOD

1. For each of two *Colias* species, capture mated female butterflies in the field. In the laboratory, allow them to lay eggs.
2. Determine the genotypes of the females and their offspring, and thus the genotypes of the fathers.
3. Compare the frequency of heterozygotes among successfully mating males with the frequency of heterozygotes among all viable males (i.e., males captured flying with females).

RESULTS

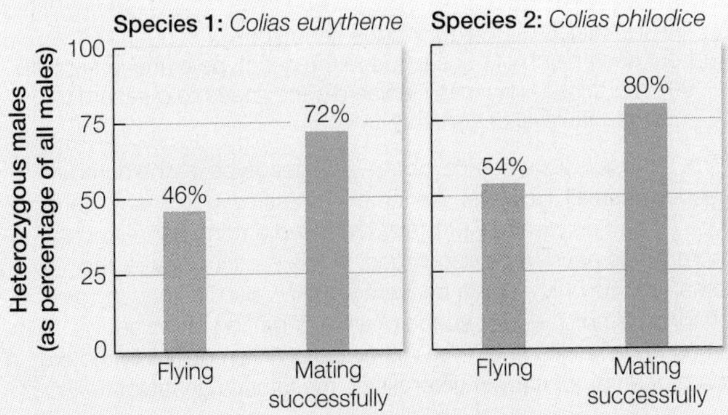

Male widowbirds with artificially shortened tails established and defended display sites successfully but fathered fewer offspring than did control or unmanipulated males. Males with artificially lengthened tails fathered the most offspring.

(Continued)

INVESTIGATION (continued)

CONCLUSION

Heterozygous *Colias* males have a mating advantage over homozygous males.

ANALYZE THE DATA

Ward Watt and his colleagues tested the hypothesis that males with two different alleles for the PGI enzyme (heterozygotes) were more likely to mate successfully with females than were homozygous males. They reasoned that the heterozygous males could fly farther under a broader range of temperatures than could homozygous males, and that this ability would give heterozygous males greater access to receptive females. The experiment above describes how they estimated the frequency of heterozygotes among the successful fathers with the frequency of heterozygotes among all viable males in the population. Samples of their data are given in the table.

Species	All viable males*	
	Heterozygous/ total	Percent heterozygous
C. philodice	32/74	43.2
C. eurytheme	44/92	47.8
	Mating males	
Species	**Heterozygous/ total**	**Percent heterozygous**
C. philodice	31/50	62.0
C. eurytheme	45/59	76.3

*"Viable males" are all males captured flying with females (hence with the potential to mate).

1. If we assume that the proportions of each genotype among mating males should be the same as the proportions seen among all viable males, what is the number of *mating males* expected to be heterozygous and homozygous in each sample?
2. Use a chi-square test (see Appendix B) to evaluate the significance of the difference in the observed and expected numbers of heterozygous and homozygous individuals among the mating males. The critical value ($P = 0.05$) of the chi-square distribution with one degree of freedom is 3.841. Are the observed numbers of genotypes among mating males significantly different ($P < 0.05$) from the expected numbers in these samples?
3. The investigators determined the genotypes of enough larvae from each batch of eggs to judge the genotype of the father with 99% certainty. How many larvae did they need to measure to achieve that level of certainty?

Hint: If the female is homozygous—say, genotype *ii*—the number needed is small. However, if a female is heterozygous—say, genotype *ij*—and only *ii* and *ij* progeny are found among her offspring, more larvae need to be genotyped. In this particular case, the father can be only *ii* or *ij*. If he were *ij*, the probability that any one offspring is *not jj* = 0.75, so the chance of getting only *ii* and *ij* among *N* offspring is 0.75^N. What value of *N* is required to reduce the probability of error in determining the father's genotype to 0.01?

[a] W. B. Watt et al. 1986. *Science* 12: 1187–1190; data from W. B. Watt et al. 1985. *Genetics* 109: 157–175.

A similar **Analyze the Data** exercise may be assigned in **Achieve**.

Genetic variation within species is maintained in geographically distinct populations

Much of the genetic variation within species is preserved as differences among members living in different places (populations). Populations often vary genetically because they are subjected to different selective pressures in different environments. Environmental conditions may vary significantly even over short distances. For example, in the Northern Hemisphere, temperature and soil moisture differ dramatically between north- and south-facing mountain slopes. In the Rocky Mountains of Colorado, the proportion of ponderosa pines (*Pinus ponderosa*) that are heterozygous for a particular peroxidase enzyme is particularly high on south-facing slopes, where temperatures fluctuate dramatically, often on a daily basis. This heterozygous genotype performs well over a broad range of temperatures. On north-facing slopes and at higher elevations, where temperatures are cooler and fluctuate less strikingly, a peroxidase homozygote, which has a lower optimal temperature, is much more frequent.

Plant species may also vary geographically in the chemicals they synthesize to defend themselves against herbivores. Some individuals of the white clover (*Trifolium repens*) produce the poisonous chemical cyanide. Poisonous individuals are less appealing to herbivores—particularly mice and slugs—than are nonpoisonous individuals. However, clover plants that produce cyanide are more likely to be killed by frost, because freezing damages cell membranes and releases cyanide into the plant's own tissues.

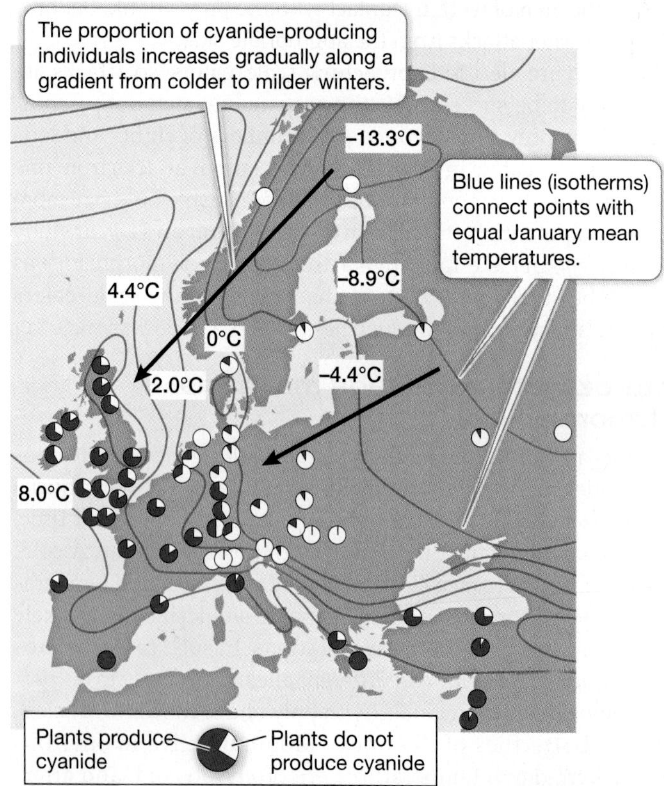

FIGURE 13.20 Geographic Variation in a Defensive Chemical The proportion of cyanide-producing individuals in European populations of white clover (*Trifolium repens*) depends on winter temperatures. (After H. Daday. 1954. *Heredity* 8: 61–78.)

In European populations of *Trifolium repens*, the frequency of cyanide-producing individuals increases gradually from north to south and from east to west (**FIGURE 13.20**). A pattern of gradual change in phenotype across a geographic gradient is known as **clinal variation**. In the white clover cline, poisonous plants make up a large proportion of populations only in areas where winters are mild. Cyanide-producing individuals are rare where winters are cold, even though herbivores graze clovers heavily in those areas.

REVIEW & APPLY | 13.5

R Frequency-dependent selection and heterozygote advantage act to maintain considerable genetic variation in most populations. Variation within species is also maintained among geographically distinct populations.

Historic range of malaria in human populations

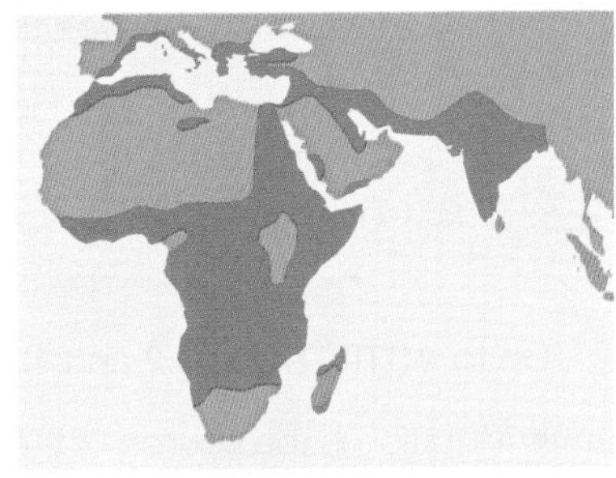

Distribution of sickle allele

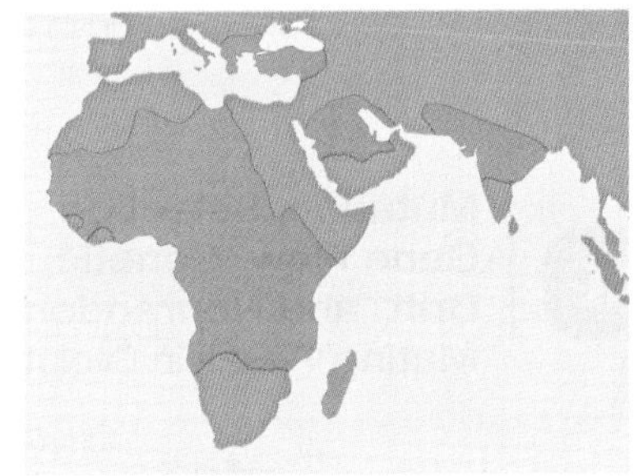

A

1. Human populations in central Africa have a high frequency of sickle-cell anemia, a genetic disorder that occurs in individuals who are homozygous for a particular allele of the hemoglobin gene. Despite strong selection against the homozygous sickle-cell condition, the sickle allele is maintained in relatively high frequencies in human populations that live in areas where malaria was historically common (compare the maps; *figure source*: Piel et al. 2010. *Nat Commun* 1: 104). Formulate a hypothesis that explains how this deleterious allele could be maintained in the population, even though individuals who are homozygous for the sickle allele are strongly disadvantaged.
2. How could you test the hypothesis that you formulated in Question 1?
3. How could the presence of a great deal of genetic variation within a population increase the chances that some members of the population would survive an unprecedented environmental change? Why is there no guarantee that this would be the case?

Activity 13.5 Gene Flow and Geographic Isolation Simulation

The processes of evolution have produced a remarkable variety of organisms. In Chapter 14 we will consider how biologists study and compare this great diversity of life using evolutionary history, or phylogenetic relationships.

Answer: For a trait such as echolocation to begin to evolve in a species, some rudimentary form of the trait has to be present before selection can act to refine this sensory mode. Often, species co-opt a trait that evolved for other purposes, and then selection refines it over time for a new purpose.

Most humans do not use echolocation. We rely heavily on sight, and most of us feel awkward moving around unfamiliar spaces in the dark. But what would happen if humans were forced to live in an environment where we could not see? We already hear, of course, so we have a system that could be co-opted for echolocation. We are also capable of making sounds. Echolocation simply requires the production of sounds, the ability to detect the reflected sounds, and the ability to process the information appropriately. Indeed, some blind people have trained themselves to use echolocation to sense their environment (see Media Clip 13.4). The learning of a behavior by an individual is not, however, the same as the evolution of a behavior in a species. Before natural selection can act, there has to be genetic variation in the population for the trait in question. The fact that a few humans are able to use simple forms of echolocation shows that variation for this trait is already present in our species. The same is true for many species, which is why it is not surprising that echolocation has evolved multiple times in species that live or feed in the dark.

Media Clip 13.4 Human Echolocation

13 VISUAL SUMMARY

QUESTIONS

13.1 Evolution Is Both Factual and the Basis of Broader Theory

Go to ACTIVITY 13.1

- **EVOLUTION** is genetic change in populations over time. Evolution can be observed directly in living populations as well as in the fossil record of life.
- **EVOLUTIONARY THEORY** refers to our understanding of the mechanisms of evolutionary change.
- Charles Darwin is best known for his ideas on the common ancestry of divergent species and on **NATURAL SELECTION** (the differential survival and reproduction of individuals based on variation in their traits) as a mechanism of evolution.

1. Explain the difference between "the fact of evolution" and what scientists mean by "evolutionary theory."
2. What did Darwin mean by "descent with modification"?
3. How did Darwin suggest that biological evolution could occur?

13.2 Mutation, Selection, Gene Flow, Genetic Drift, and Nonrandom Mating Result in Evolution

Go to ACTIVITIES 13.2 and 13.3

FIGURE 13.6

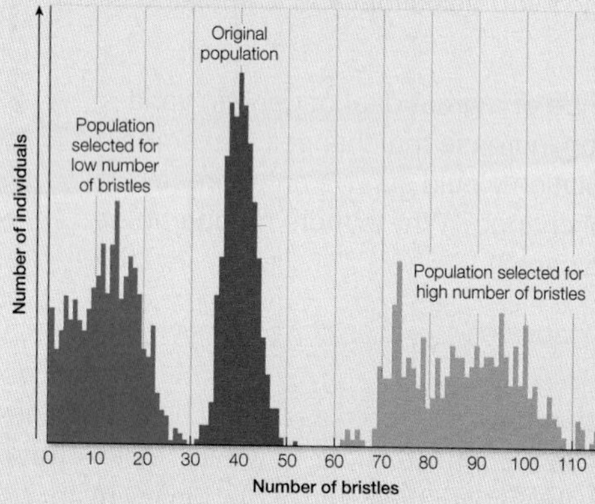

- Mutation is the source of the genetic variation on which mechanisms of evolution act.
- Within **POPULATIONS**, selection acts to increase the frequency of beneficial **ALLELES** and to decrease the frequency of deleterious alleles (**FIGURE 13.6**).
- **GENE FLOW**, **GENETIC DRIFT**, and nonrandom mating (as arises from **SEXUAL SELECTION**) can also result in the evolution of populations (**FIGURE 13.8**).

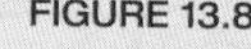

FIGURE 13.8

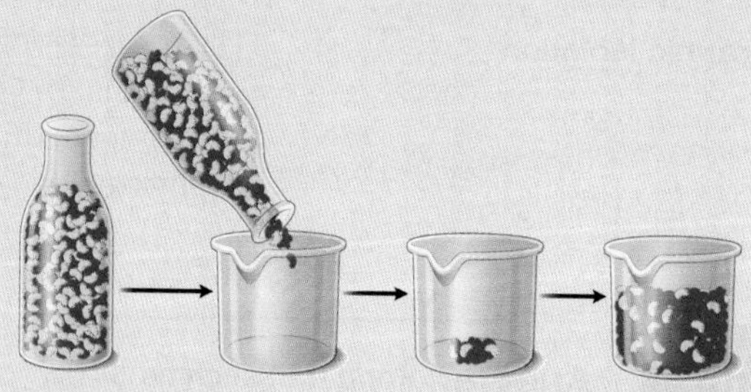

1. How does artificial selection differ from natural selection?
2. How can genetic drift oppose the forces of natural selection?
3. Contrast natural selection and sexual selection.

13.3 Evolution Can Be Measured by Changes in Allele Frequencies

Go to ACTIVITY 13.4

FIGURE 13.12

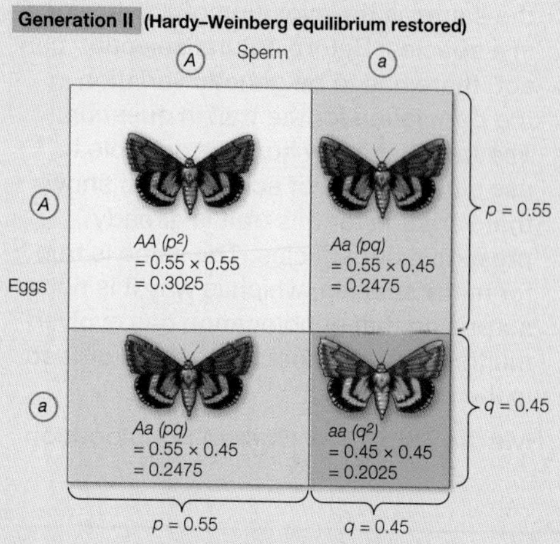

Generation II (Hardy–Weinberg equilibrium restored)

Sperm: A, a — Eggs: A, a

AA (p^2) = 0.55 × 0.55 = 0.3025

Aa (pq) = 0.55 × 0.45 = 0.2475

Aa (pq) = 0.55 × 0.45 = 0.2475

aa (q^2) = 0.45 × 0.45 = 0.2025

p = 0.55 q = 0.45

- Allele frequencies measure the amount of genetic variation in a population. Genotype frequencies show how a population's genetic variation is distributed among its members. Together, allele and genotype frequencies describe a population's **GENETIC STRUCTURE**.
- **HARDY-WEINBERG EQUILIBRIUM** predicts genotype frequencies from allele frequencies in the absence of evolution. Deviation from these frequencies indicates that evolutionary mechanisms are at work (**FIGURE 13.12**).

1. How many generations of random mating are required to achieve Hardy–Weinberg equilibrium expectations?
2. How do the assumptions of Hardy–Weinberg equilibrium correspond to the five principal processes of evolution?

QUESTIONS

13.4 Selection Can Be Stabilizing, Directional, or Disruptive

Go to ANIMATION 13.1

- **STABILIZING SELECTION** acts to reduce variation without changing the mean value of a trait (**FIGURE 13.13**).
- **DIRECTIONAL SELECTION** acts to shift the mean value of a trait toward one extreme (**FIGURE 13.13**).
- **DISRUPTIVE SELECTION** favors both extremes of a trait value, resulting in a bimodal character distribution (**FIGURE 13.13**).

FIGURE 13.13

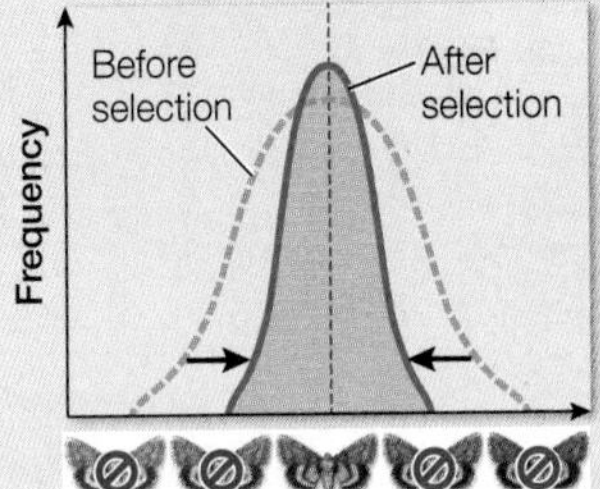

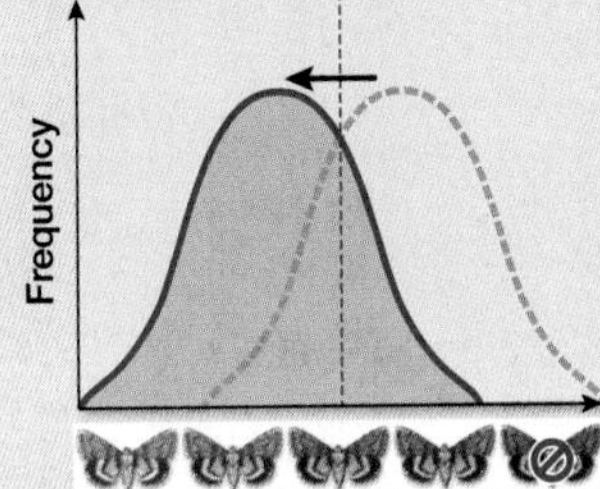

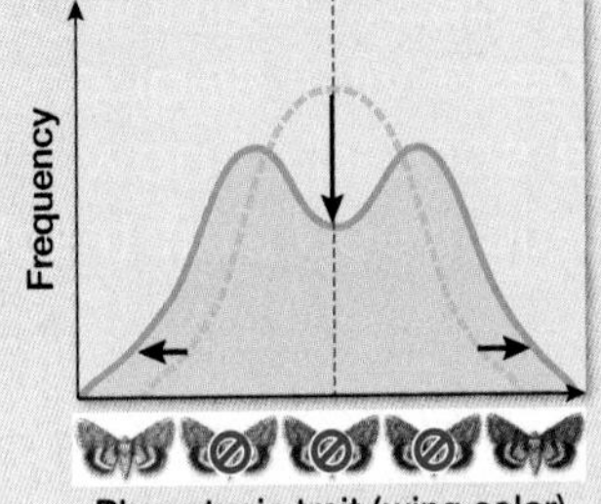

1. Which of the three types of selection (stabilizing, directional, or disruptive) does each of the following cases represent?
   a. Higher survival and reproduction of giraffes with longer necks compared with those with shorter necks.
   b. Higher survival and reproduction of cattle with intermediate calf weights compared with those with high or low calf weights.
   c. Higher survival and reproduction of grasshoppers with either small or large mouthparts, compared with those with intermediate-sized mouthparts.

13.5 Selection Can Maintain Polymorphisms in Populations

Go to ACTIVITY 13.5

- A polymorphism may be maintained by **FREQUENCY-DEPENDENT SELECTION** when the fitness of a genotype depends on its frequency in a population (**FIGURE 13.18**).
- A polymorphism may also be maintained by heterozygote advantage when the fitness of the heterozygote exceeds the fitness of either homozygote.
- Genetic variation within species may be maintained by the existence of genetically distinct populations over geographic space. A gradual change in phenotype across a geographic gradient is known as **CLINAL VARIATION** (**FIGURE 13.20**).

FIGURE 13.18

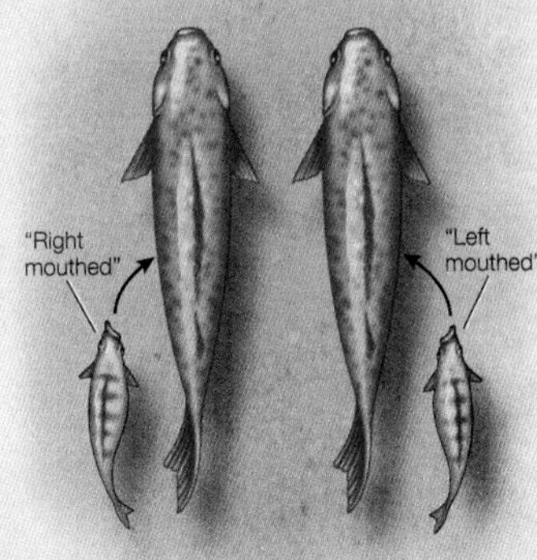

FIGURE 13.20

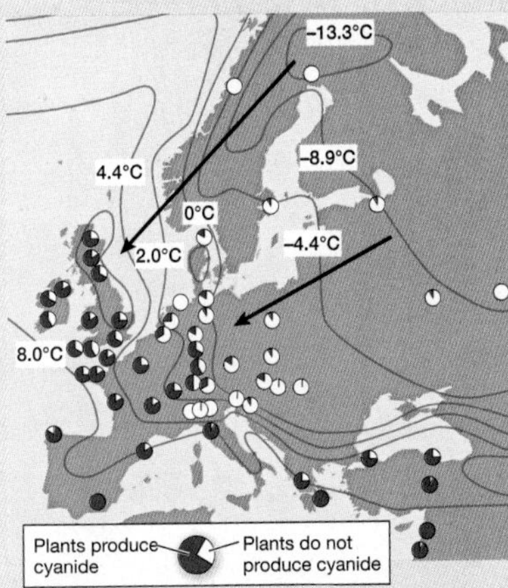

1. Which of the following are examples of frequency-dependent selection, heterozygote advantage, or clinal variation?
   a. Increased sprinting performance of individuals with two different alleles of an actin gene, compared with individuals with two copies of just one of the alleles.
   b. Increased survival of individuals with rare color patterns, compared with those with common color patterns.
   c. Geographic variation in the size of deer from northern to southern latitudes.

14 Reconstructing and Using Phylogenies

The COVID-19 pandemic resulted in hundreds of millions of cases of disease and millions of deaths worldwide. The virus responsible for the disease (SARS-CoV-2) was widely sampled and sequenced from infected individuals, beginning in the early days of the pandemic. These viral sequences allowed biologists to follow the spread of new variants by creating phylogenetic trees that track virus evolution.

Jan 2020 Apr 2020 Jul 2020 Oct 2020 Jan 2021 Apr 2021

How are phylogenetic methods used to understand the origin, evolution, and spread of new diseases?

You will find the answer to this question on page 348.

Data from GISAID using Augur/Auspice via Nextstrain

KEY CONCEPTS

14.1 All of Life Is Connected through Its Evolutionary History

LEARNING OBJECTIVES

By the end of this key concept you should be able to:

14.1.1 Draw and label the parts of a phylogenetic tree and explain the biological interpretation of each part.

14.1.2 Use a phylogenetic tree to determine relationships and infer character changes among species.

14.1.3 Explain how homoplasies (convergences and reversals of characters) are accounted for when constructing phylogenetic relationships.

The sequencing of complete genomes from many diverse species has confirmed what biologists have long suspected: all of life is related through a common ancestor. The common ancestry of life explains why the general principles of biology apply to all organisms. Thus we can learn much about how the human genome works by studying the biology of model organisms because we share a common evolutionary history with those organisms. The evolutionary history of these relationships is known as **phylogeny**, and a **phylogenetic tree** is a diagrammatic reconstruction of that history.

Phylogenetic trees depict evolutionary relationships among lineages

Phylogenetic trees are commonly used to depict the evolutionary history of species, populations, and genes. For many years such trees have been constructed based on physical structures, behaviors, and biochemical attributes of related organisms. Now, as genomes are sequenced for more and more organisms, biologists are able to reconstruct the history of life in ever-greater detail.

In Chapter 13 we discussed why we expect populations of organisms to evolve over time. We call a series of ancestor and descendant populations a **lineage**, which we can depict as a line drawn on a time axis, as shown in **FIGURE 14.1**. What happens when a single lineage divides into two? For example, a geographic barrier may divide an ancestral population into two descendant populations that no longer interbreed with one another. We depict such an event as a split, or **node**, in a phylogenetic tree (see Figure 14.1). Each of the descendant populations gives rise to a new lineage, and as these independent lineages evolve, new traits arise in each one. As the lineages continue to split over time, this history can be represented in the form of a branching tree that can be used to trace the evolutionary relationships from the ancient common ancestor of a group of species, through the various lineage splits, up to the present populations of the organisms.

A phylogenetic tree may portray the evolutionary history of all life forms. Phylogenetic trees can also depict the history of a major evolutionary group (such as the insects) or of a much smaller group of closely related species. In some cases, phylogenetic trees are used to show the history of individuals, populations, or genes within a species. The common ancestor of all the organisms in the tree forms the **root** of the tree.

The depictions of phylogenetic trees in this book are rooted at the left, with time flowing from left (earliest) to right (most recent) (**FIGURE 14.2A**). The timing of splitting events in lineages is shown by the position of nodes on a time axis. These splits represent events where one lineage diverged into two, such as a speciation event (for a tree of species), a gene duplication event (for a tree of

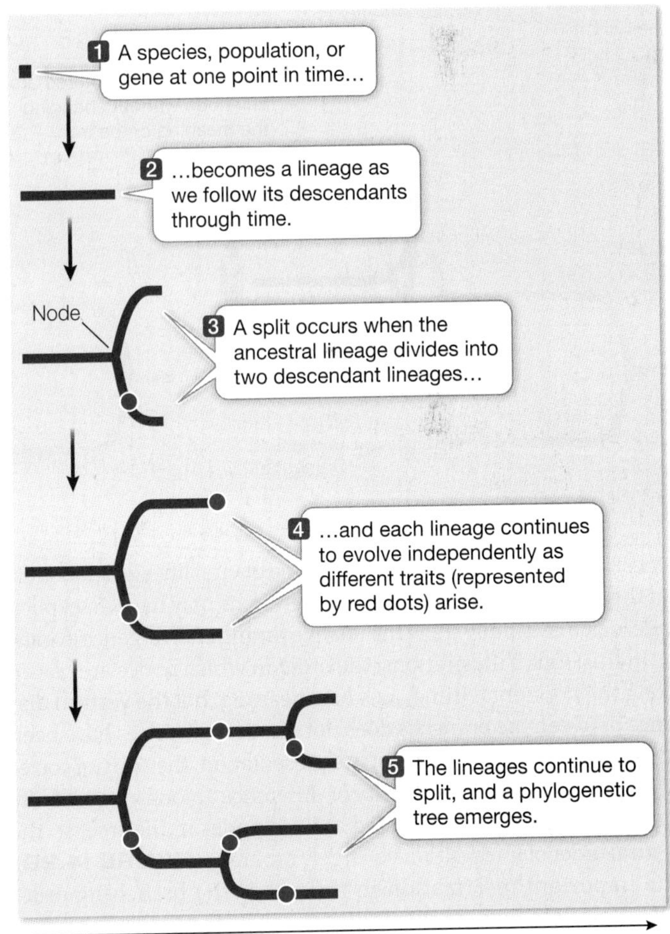

FIGURE 14.1 The Components of a Phylogenetic Tree Evolutionary relationships among organisms can be represented in a treelike diagram.

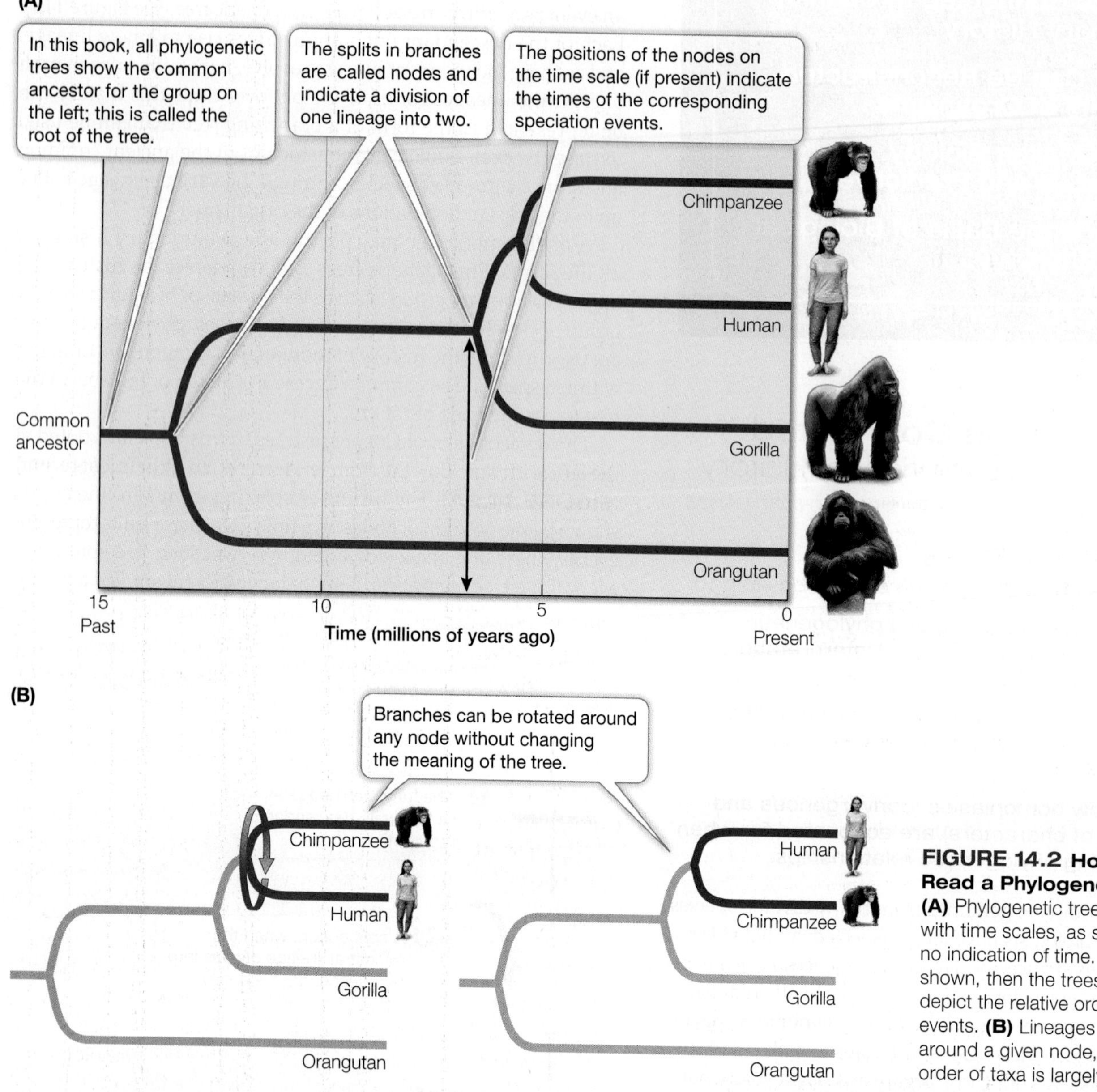

FIGURE 14.2 How to Read a Phylogenetic Tree **(A)** Phylogenetic trees can be produced with time scales, as shown here, or with no indication of time. If no time scale is shown, then the trees are only meant to depict the relative order of divergence events. **(B)** Lineages can be rotated around a given node, so the vertical order of taxa is largely arbitrary.

genes), or a transmission event (for a tree of viral lineages transmitted through a host population). The time axis may have an explicit scale, or it may simply show the relative timing of divergence events.

In this book's illustrations, the order in which nodes are placed along the horizontal (time) axis has meaning, but the vertical distance between the branches does not. Vertical distances have been adjusted for legibility and clarity of presentation; they do not correlate with the degree of similarity or difference among groups. Note too that lineages can be rotated around nodes in the tree, so the vertical order of lineages is also largely arbitrary (**FIGURE 14.2B**). The important information in the tree is the branching order along the time axis, as this indicates when the various lineages last shared a common ancestor.

Any group of species that we designate with a name is a **taxon** (plural *taxa*). Examples of familiar taxa include humans, primates, mammals, and vertebrates; in this series, each taxon is also a member of the next, more inclusive taxon. Any taxon that consists of all the evolutionary descendants of a common ancestor is called a **clade**. Clades can be identified by picking any point on a phylogenetic tree and from that point tracing all the descendant lineages to the tips of the terminal branches (**FIGURE 14.3**). Two species that are each other's closest relatives are called **sister species**. Similarly, any two clades that are each other's closest relatives are **sister clades**.

Before the 1980s, phylogenetic trees tended to be seen only in the literature on evolutionary biology, especially in the area of **systematics**—the study and classification of biodiversity. But almost every journal in the life sciences published during the last few years contains phylogenetic trees. Trees are widely used in molecular biology, biomedicine, physiology, behavior, ecology, and virtually all other fields of biology. Why have phylogenetic studies become so widespread?

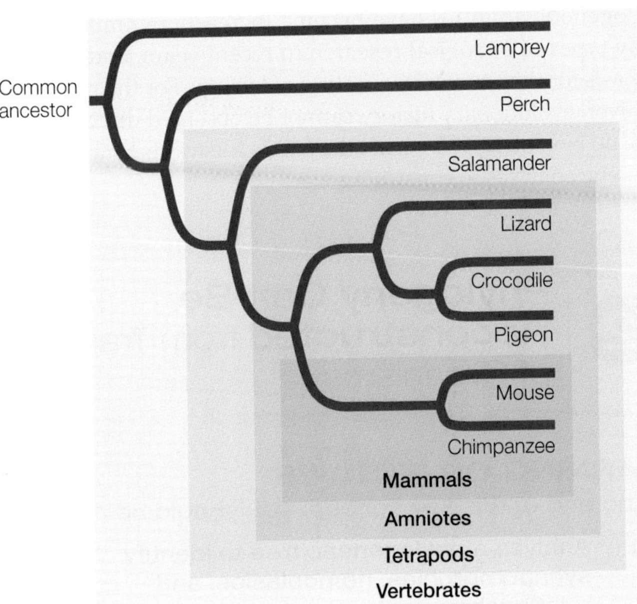

FIGURE 14.3 Clades Represent an Ancestor and All of Its Evolutionary Descendants All clades are subsets of larger clades, with all of life as the most inclusive taxon. In this example, the groups called mammals, amniotes, tetrapods, and vertebrates represent successively larger clades. Only a few species in each clade are represented on the tree.

Phylogenetic trees are the basis of comparative biology

In biology, we study life at all levels of organization—from genes, cells, organisms, populations, and species to the major divisions of life. In most cases, however, no individual gene or organism (or other unit of study) is exactly like any other gene or organism that we investigate.

Consider the individuals in your biology class. We recognize each person as an individual human, but we know that no two are exactly alike. If we knew everyone's family tree in detail, the genetic similarity of any pair of students would be more predictable. We would find that more closely related students have many more traits in common (from the color of their hair to their susceptibility or resistance to diseases). Likewise, biologists use phylogenies to make comparisons and predictions about shared traits across genes, populations, and species.

The evolutionary relationships among species, as represented in the **tree of life**, form the basis for biological classification. Biologists estimate that there are tens of millions of species on Earth. So far, however, only about 1.8 million species have been classified—that is, formally described and named. New species are being discovered all the time and phylogenetic analyses are constantly reviewed and revised, so our knowledge of the tree of life is far from complete. Yet knowledge of evolutionary relationships is essential for making comparisons in biology, so biologists build phylogenies for groups of interest as more data become available. The tree of life's evolutionary framework allows us to make many predictions about the behavior, ecology, physiology, genetics, and morphology of species that have not yet been studied in detail.

When biologists compare species, they observe traits that differ within the group of interest and try to understand when these traits evolved. In many cases, investigators are interested in how the evolution of a trait relates to environmental conditions or selective pressures. For instance, scientists have used phylogenetic analyses to discover changes in the genome of human immunodeficiency viruses (HIVs) that result in resistance to particular drug treatments. The association of a particular genetic change in HIV with a particular treatment provides a hypothesis about the evolution of resistance that can be tested experimentally.

Any features shared by two or more species that have been inherited from a common ancestor are said to be **homologous**. Homologous features may be any heritable traits, including DNA sequences, protein structures, anatomical structures, and even some behavior patterns. For example, all living vertebrates have a vertebral column, as did the ancestral vertebrate. Therefore the vertebral column is judged to be homologous in all vertebrates.

Derived traits provide evidence of evolutionary relationships

In tracing the evolution of a character, biologists distinguish between ancestral and derived traits. Each character of an organism evolves from one condition (the **ancestral trait**) to another condition (the **derived trait**). Derived traits that are shared among a group of organisms and are also viewed as evidence of the common ancestry of the group are called **synapomorphies** (*syn*, "shared"; *apo*, "derived"; *morph*, "form," referring to the "form" of a trait). Thus the vertebral column is considered a synapomorphy—a shared, derived trait—of the vertebrates. (The ancestral trait was an undivided supporting rod, as still occurs in some close relatives of vertebrates.)

Not all similar traits are evidence of relatedness. Similar traits in unrelated groups of organisms can develop for either of the following reasons:

- Superficially similar traits may evolve independently in different lineages, a phenomenon called **convergent evolution**. For example, although the *forelimb bones* of bats and birds are homologous, having been inherited from a common tetrapod ancestor, the *wings* of bats and birds are not homologous because they evolved independently from the forelimbs of different nonflying ancestors (**FIGURE 14.4**). Functionally similar structures that have independent evolutionary origins are called **analogous** characters.
- A character may revert from a derived state back to an ancestral state in an event called an **evolutionary reversal**. For example, the derived limbs of terrestrial tetrapods evolved from the ancestral fins of their aquatic ancestors. Then, within the mammals, the ancestors of modern cetaceans (whales and dolphins) returned to the ocean, and cetacean limbs evolved to once again resemble their ancestral state—fins. The superficial similarity of cetacean and fish fins does not suggest a close relationship between these groups. Instead, the similarity arises from evolutionary reversal.

Similar traits generated by convergent evolution and evolutionary reversals are called homoplastic traits or **homoplasies**.

Media Clip 14.1 Morphing Arachnids

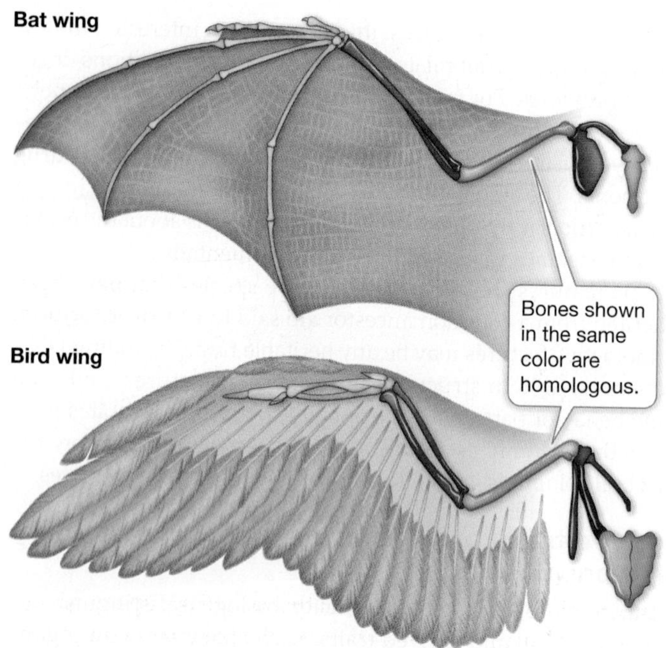

FIGURE 14.4 The Bones Are Homologous, the Wings Are Not The supporting bone structures of both bat wings and bird wings are derived from a common four-limbed ancestor and are thus homologous. However, the wings themselves—an adaptation for flight—evolved independently in the two groups.

A particular trait may be ancestral or derived, depending on our point of reference. For example, all birds have feathers. We infer from this that feathers (which are highly modified scales) were present in the common ancestor of modern birds. Therefore we consider the presence of feathers to be an ancestral trait for any particular group of modern birds, such as the songbirds. However, feathers are not present in any other living animals. In reconstructing a phylogeny of all living vertebrates, the presence of feathers is a derived trait found only among birds, and thus is a synapomorphy of the birds.

REVIEW & APPLY | 14.1

R A phylogenetic tree is a description of evolutionary relationships among organisms or their genes. All living organisms share a common ancestor and are related through the phylogenetic tree of life. Groups of related species that are descended from the same common ancestor are known as clades. Shared derived traits (known as synapomorphies) provide evidence for evolutionary relationships among species.

A
1. What biological processes are represented in phylogenetic trees?
2. Why is it important to consider only homologous characters in constructing phylogenetic trees?
3. Why might similar traits arise independently in species that are only distantly related? Can you think of examples among familiar organisms? How do biologists account for these homoplasies in reconstructing phylogenies?

Phylogenetic analyses have become increasingly important to many types of biological research in recent years, and they are the basis for the comparative nature of biology. For the most part, however, evolutionary history cannot be observed directly. How, then, do biologists reconstruct the past? One way is by using phylogenetic analyses to construct a tree.

14.2 | Phylogeny Can Be Reconstructed from Traits of Organisms

LEARNING OBJECTIVES

By the end of this key concept you should be able to:

14.2.1 **Analyze a phylogenetic tree to identify synapomorphies, homoplasies, and relationships among taxa.**

14.2.2 **Construct a phylogenetic tree from a data matrix of characters.**

To illustrate how a phylogenetic tree is constructed, consider the eight vertebrate animals listed in **TABLE 14.1**: lamprey, perch, salamander, lizard, crocodile, pigeon, mouse, and chimpanzee. We will initially assume that any given derived trait arose only once during the evolution of these animals (that is, there has been no convergent evolution), and that no derived traits were lost from any of the descendant groups (there has been no evolutionary reversal). For simplicity, we have selected traits that are either present (+) or absent (–).

Shared traits reflect common ancestry

In a phylogenetic study, the group of organisms of primary interest is called the **ingroup**. As a point of reference, an ingroup is compared with an **outgroup**: a species or group that is closely related to the ingroup but is known to be phylogenetically outside it. In other words, the root of the tree is located between the ingroup and the outgroup. Any trait that is present in both the ingroup and the outgroup must have evolved before the origin of the ingroup and thus must be ancestral for the ingroup. In contrast, traits that are present in only some members of the ingroup must be derived traits within that ingroup. As we will see in Chapter 22, a group of jawless fishes called the lampreys is thought to have separated from the lineage leading to the other vertebrates before the jaw arose. Therefore we have included the lamprey as the outgroup for our analysis. Because derived traits are traits acquired by other members of the vertebrate lineage *after* they diverged from the outgroup, any trait that is present in both the lamprey and the other vertebrates is judged to be ancestral.

We begin by noting that the chimpanzee and mouse share two traits—mammary glands and fur—that are absent in both the outgroup and in the other species of the ingroup. Therefore we infer that mammary glands and fur are derived traits that evolved in a common ancestor of chimpanzees and mice after that lineage

TABLE 14.1 | Eight Vertebrates and the Presence or Absence of Some Shared Derived Traits

Taxon	Jaws	Lungs	Claws or nails	Gizzard	Feathers	Fur	Mammary glands	Keratinous scales
Lamprey (outgroup)	–	–	–	–	–	–	–	–
Perch	+	–	–	–	–	–	–	–
Salamander	+	+	–	–	–	–	–	–
Lizard	+	+	+	–	–	–	–	+
Crocodile	+	+	+	+	–	–	–	+
Pigeon	+	+	+	+	+	–	–	+
Mouse	+	+	+	–	–	+	+	–
Chimpanzee	+	+	+	–	–	+	+	–

separated from the lineages leading to the other vertebrates. These characters are synapomorphies that unite chimpanzees and mice (as well as all other mammals, although we have not included other mammalian species in this example). By the same reasoning, we can infer that the other shared derived traits are synapomorphies for the various groups in which they are expressed. For instance, keratinous scales are a synapomorphy of the lizard, crocodile, and pigeon.

Table 14.1 also tells us that, among the animals in our ingroup, the pigeon has a unique trait: feathers. Feathers are a synapomorphy of birds and their extinct relatives. However, because we have only one bird in this example, the presence of feathers provides no clues concerning relationships among these eight species of vertebrates. However, gizzards are found in both birds and crocodiles, so this trait is evidence of a close relationship between birds and crocodilians.

By combining information about the various synapomorphies, we can construct a phylogenetic tree. We infer from our information that mice and chimpanzees—the only two animals that share fur and mammary glands—share a more recent common ancestor with each other than they do with pigeons and crocodiles. Otherwise we would need to assume that the ancestors of pigeons and crocodiles also had fur and mammary glands but subsequently lost them. There is no basis for making these additional assumptions.

FIGURE 14.5 shows a phylogenetic tree for the vertebrates in Table 14.1, based on the shared derived traits we examined.

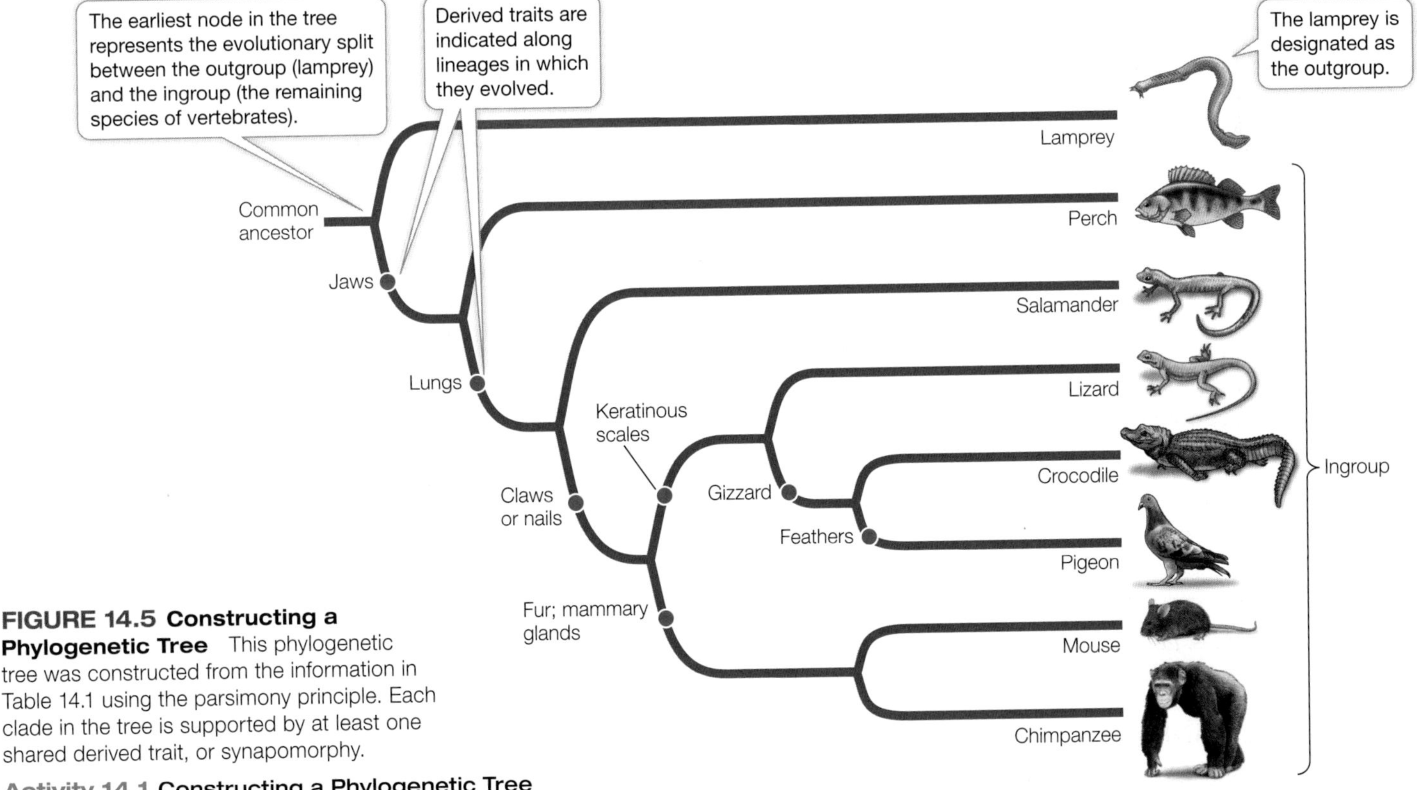

FIGURE 14.5 Constructing a Phylogenetic Tree This phylogenetic tree was constructed from the information in Table 14.1 using the parsimony principle. Each clade in the tree is supported by at least one shared derived trait, or synapomorphy.

Activity 14.1 Constructing a Phylogenetic Tree

This particular tree was easy to construct because it is based on a very small sample of traits, and the derived traits we examined evolved only once and were never lost after they appeared. Had we included a snake in the group, our analysis would not have been as straightforward. We would have needed to examine additional characters to determine that snakes evolved from a group of lizards that had limbs. In fact, the analysis of many characters shows that snakes evolved from burrowing lizards that became adapted to a subterranean existence.

Parsimony provides the simplest explanation for phylogenetic data

The phylogenetic tree shown in Figure 14.5 is based on only a very small sample of traits. Typically, biologists construct phylogenetic trees using hundreds or thousands of traits. With larger data sets, we would expect to observe traits that have changed more than once, and thus would expect to see convergence and evolutionary reversal. How do we determine which traits are synapomorphies and which are homoplasies? One way is to invoke the principle of parsimony.

In its most general form, the **parsimony principle** states that the preferred explanation of observed data is the simplest explanation. Applying the principle of parsimony to the reconstruction of phylogenies entails minimizing the number of evolutionary changes that need to be assumed over all characters in all groups in the tree. In other words, the best hypothesis under the parsimony principle is one that requires the fewest homoplasies. This application of parsimony is a specific case of a general principle of reasoning called *Occam's razor*: the best explanation is the one that best fits the data while making the fewest assumptions. More complicated explanations are accepted only when the evidence requires them. Phylogenetic trees represent our best estimates about evolutionary relationships, given our current knowledge. They are continually modified as additional evidence becomes available.

Phylogenies are reconstructed from many sources of data

Naturalists have constructed various forms of phylogenetic trees for more than 150 years. In fact, the only figure in the first edition of *On the Origin of Species* was a phylogenetic tree. Tree construction has been revolutionized, however, by the advent of computer software that allows us to consider far more data and analyze far more traits than could ever before be processed. Combining these advances in methodology with the massive comparative data sets being generated through studies of genomes, biologists are learning details about the tree of life at a remarkable pace (see Appendix A: The Tree of Life).

Any trait that is genetically determined, and therefore heritable, can be used in a phylogenetic analysis. Evolutionary relationships can be revealed through studies of morphology, development, the fossil record, behavioral traits, and molecular traits such as DNA and protein sequences. Let's take a closer look at the types of data used in modern phylogenetic analyses.

MORPHOLOGY An important source of phylogenetic information is **morphology**: the presence, size, shape, and other attributes of body parts. Since living organisms have been observed, depicted, collected, and studied for millennia, we have a wealth of recorded morphological data as well as extensive museum and herbarium collections of organisms whose traits can be measured. New technological tools, such as the electron microscope and computed tomography (CT) scans, enable systematists to examine and analyze the structures of organisms at much finer scales than was formerly possible.

Most species are described and known primarily by their morphology, and morphology still provides the most comprehensive data set available for many taxa. The morphological features that are important for phylogenetic analysis are often specific to a particular group. For example, the presence, development, shape, and size of various features of the skeletal system are important in vertebrate phylogeny, whereas floral structures are important for studying the relationships among flowering plants.

Morphological approaches to phylogenetic analysis have some limitations, however. Some taxa exhibit little morphological diversity, despite great species diversity. For example, the phylogeny of the leopard frogs of North and Central America would be difficult to infer from morphological differences alone, because the many species look very similar, despite important differences in their behavior and physiology. At the other extreme, few morphological traits can be compared across distantly related species (earthworms and mammals, for example). Furthermore, some morphological variation has an environmental (rather than a genetic) basis and so must be excluded from phylogenetic analyses. An accurate phylogenetic analysis often requires information beyond that supplied by morphology.

DEVELOPMENT Similarities in developmental patterns may reveal evolutionary relationships. Some organisms exhibit similarities only in early developmental stages. The larvae of marine creatures called sea squirts, for example, have a flexible gelatinous rod in the back—the notochord—that disappears as the larvae develop into adults. All vertebrate animals also have a notochord at some time during their development (**FIGURE 14.6**). This shared structure is one of the reasons for inferring that sea squirts are more closely related to vertebrates than would be suspected if only adult sea squirts were examined.

LINK Evolutionary developmental biology, discussed in **Chapter 15**, compares the developmental processes of different organisms to determine the ancestral relationship between them, and to discover how developmental processes evolved. Genetic toolkit genes are expressed in different ways in different species, resulting in major morphological differences among species. The existence of highly conserved genes involved in development makes it likely that similar traits will evolve repeatedly.

PALEONTOLOGY The fossil record is another important source of information on evolutionary history. Fossils show us where and when organisms lived in the past and give us an idea of what they looked like. Fossils provide important evidence that helps

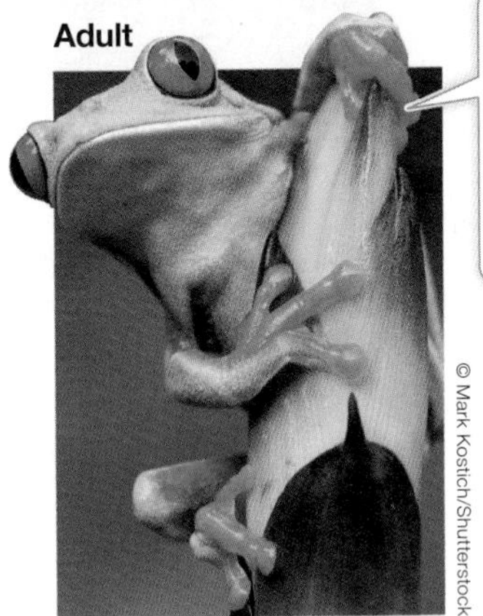

FIGURE 14.6 Development Reveals the Evolutionary Relationship between Sea Squirts and Vertebrates All chordates—a taxonomic group that includes sea squirts and frogs—have a notochord at some stage of their development. The larvae share similarities that are not apparent in the adults. Such similarities in development can provide useful evidence of evolutionary relationships. The notochord is lost in adult sea squirts. In adult frogs, as in all vertebrates, the vertebral column replaces the notochord as the support structure.

us distinguish ancestral from derived traits. The fossil record can also reveal when lineages diverged and began their independent evolutionary histories. Furthermore, in groups with few species that have survived to the present, information on extinct species is often critical to an understanding of the large divergences among the surviving species. The fossil record has limitations, however. Few or no fossils have been found for some groups, and the fossil record for many groups is fragmentary.

BEHAVIOR Some behavioral traits are culturally transmitted, and others are genetically inherited. If a particular behavior is culturally transmitted, it may not accurately reflect evolutionary relationships (but may nonetheless reflect cultural connections). Many bird songs, for instance, are learned and may be inappropriate traits for phylogenetic analysis. Frog calls, however, are genetically determined and appear to be acceptable sources of information for reconstructing phylogenies.

MOLECULAR DATA All heritable variation is encoded in DNA, and so the complete genome of an organism contains an enormous set of traits (the individual nucleotide bases of DNA) that can be used in phylogenetic analyses. In recent years, DNA sequences have become among the most widely used sources of data for constructing phylogenetic trees. Comparisons of nucleotide sequences are not limited to the DNA in the cell nucleus. Eukaryotes have genes in their mitochondria as well as in their nuclei. Plant cells also have genes in their chloroplasts. The chloroplast genome (cpDNA), which is used extensively in phylogenetic studies of plants, has changed slowly over evolutionary time, so it is often used to study relatively ancient phylogenetic relationships. Most animal mitochondrial DNA (mtDNA) has changed more rapidly, so mitochondrial genes are used to study evolutionary relationships among closely related animal species (the mitochondrial genes of plants evolve more slowly). Many nuclear gene sequences are also commonly analyzed, and now that entire genomes have been sequenced from many species, they too are used to construct phylogenetic trees. Information on gene products (such as the amino acid sequences of proteins) is also widely used for phylogenetic analyses.

Mathematical models expand the power of phylogenetic reconstruction

As biologists began to use DNA sequences to infer phylogenies in the 1970s and 1980s, they developed explicit mathematical models describing how DNA sequences change over time. These models account for multiple changes at a given position in a DNA sequence. They also take into account different rates of change at different positions in a gene, at different positions in a codon, and among different nucleotides. For example, nucleotide transitions (changes between two purines or between two pyrimidines) usually occur more often than transversions (changes between a purine and pyrimidine).

Mathematical models can be used to compute how a tree might evolve given the observed data. A **maximum likelihood** method will identify the tree that most likely produced the observed data, given the assumed model of evolutionary change. Maximum likelihood methods can be used for any kind of data, but they are most often used with molecular data, for which explicit mathematical models of evolutionary change are easier to develop. The principal advantages of maximum likelihood analyses are that they incorporate more information about evolutionary change than do parsimony methods, and they are easier to treat in a statistical framework. The principal disadvantages are that they are computationally intensive and require explicit models of evolutionary change (which may not be available for some kinds of character change).

The accuracy of phylogenetic methods can be tested

If phylogenetic trees represent reconstructions of past events, and if many of these events occurred before any humans were around to witness them, how can we test the accuracy of phylogenetic methods? Biologists have conducted experiments both in living organisms and with computer simulations that have demonstrated the effectiveness and accuracy of phylogenetic methods.

In one experiment designed to test the accuracy of phylogenetic analysis, a single viral culture of bacteriophage T7 was used as a starting point, and lineages were allowed to evolve from this ancestral virus in the laboratory (**INVESTIGATION FIGURE 14.7**). The initial culture was split into two separate lineages, one of which became the ingroup for analysis and the other of which became the outgroup for rooting the tree. The lineages in the ingroup were split in two after every 400 generations, and samples of the virus were saved for analysis at each branching point. The lineages were allowed to evolve until there were eight lineages in the ingroup. Mutagens were added to the viral cultures to increase the mutation rate so that the amount of change and the degree of homoplasy would be typical of the organisms analyzed in average phylogenetic analyses. The investigators then sequenced samples from the end points of the eight ingroups and one outgroup lineages, as well as from the ancestors at the branching points. They then gave the sequences from the end points of the lineages to other investigators to analyze, without revealing the known history of the lineages or the sequences of the ancestral viruses.

After the phylogenetic analysis was completed, the investigators asked two questions. Did phylogenetic methods reconstruct the known history correctly? And were the sequences of the ancestral viruses reconstructed accurately? The answer in both cases was yes. The branching order of the lineages was reconstructed exactly as it had occurred, more than 98 percent of the nucleotide positions of the ancestral viruses were reconstructed correctly, and 100 percent of the amino acid changes in the viral proteins were reconstructed correctly.

The experiment shown in Figure 14.7 demonstrated that phylogenetic analysis was accurate under the conditions tested, but it did not examine all possible conditions. Other experimental studies have taken other factors into account, such as the sensitivity of phylogenetic analysis to convergent environments and highly variable rates of evolutionary change. In addition, computer simulations based on evolutionary models have been used extensively to study the effectiveness of phylogenetic analysis. These studies have also confirmed the accuracy of phylogenetic methods and have been used to refine those methods and extend them to new applications.

REVIEW & APPLY | 14.2

R Using the parsimony principle, phylogenetic trees are constructed to find the simplest explanation for evolutionary changes in traits among a set of organisms. Maximum likelihood methods incorporate more explicit mathematical models of evolutionary change to reconstruct evolutionary history. Laboratory tests and simulations have confirmed the general accuracy of phylogenetic methods.

A The matrix below supplies data for seven land plants and an outgroup: an aquatic plant known as a stonewort. Each trait is scored as either present (+) or absent (–) in each of the plants.

	Trait						
Taxon	**Protected embryos**	**True roots**	**Persistently green sporophyte**	**Vascular cells**	**Stomata**	**True leaves**	**Seeds**
Stonewort (outgroup)	–	–	–	–	–	–	–
Liverwort	+	–	–	–	–	–	–
Pine tree	+	+	+	+	+	+	+
Bracken fern	+	+	+	+	+	+	–
Club moss	+	+	+	+	+	–	–
Sphagnum moss	+	–	–	–	+	–	–
Hornwort	+	–	+	–	+	–	–
Sunflower	+	+	+	+	+	+	+

1. Use this data matrix to reconstruct the phylogeny of land plants and to answer the questions below.
2. Which two of these taxa are most closely related?
3. Plants that produce seeds are known as seed plants. What is the sister group to the seed plants among these taxa?
4. Which two traits evolved along the same branch of your reconstructed phylogeny?
5. Are there any homoplasies in your reconstructed phylogeny?

Why do biologists expend the time and effort necessary to reconstruct phylogenies? Information about the evolutionary relationships among organisms is a useful source of data for scientists investigating a wide variety of biological questions. Next we will describe how phylogenetic trees are used to answer questions about the past, and to predict and compare traits of organisms in the present.

Activity 14.2 Phylogeny and Molecular Evolution Simulation

INVESTIGATION

FIGURE 14.7 Testing the Accuracy of Phylogenetic Analysis To test whether analysis of gene sequences can accurately reconstruct evolutionary phylogeny, we must have an unambiguously known phylogeny to compare against the reconstruction. Will the observed phylogeny match the reconstruction?

HYPOTHESIS

A phylogenetic tree reconstructed by analyzing the DNA sequences of living organisms can accurately match the known evolutionary history of the organisms.

METHOD

1. Researchers selected a single virus plaque to serve as the common ancestor for the experimental phylogeny.[a]
2. They then produced an unambiguous phylogeny of nine viral lineages, enhancing the mutation rate to increase variation among the lineages.
3. Viral sequences from the end points of each lineage (blue dots) were subjected to phylogenetic analysis by investigators who were unaware of the history of the lineages or the gene sequences of the ancestral viruses. These investigators reconstructed the phylogeny and ancestral DNA sequences based solely on their analyses of the descendants' genomes.

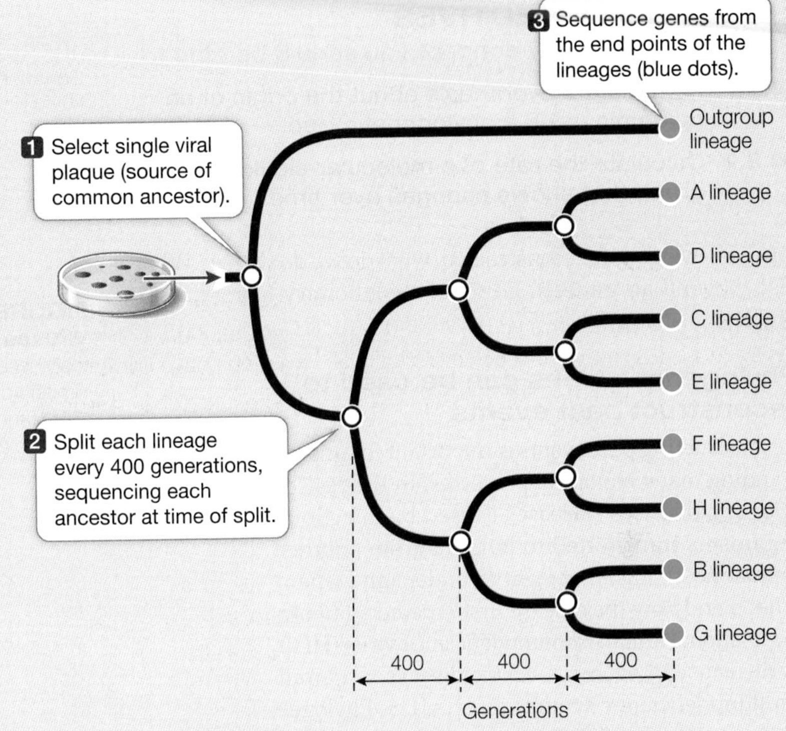

RESULTS

The true phylogeny and ancestral DNA sequences were accurately reconstructed solely from the DNA sequences of the viruses at the tips of the tree branches.

CONCLUSION

Phylogenetic analysis of DNA sequences can accurately reconstruct evolutionary history.

Animation 14.1 Using Phylogenetic Analysis to Reconstruct Evolutionary History

ANALYZE THE DATA

The full DNA sequences for the viral lineages produced in this experiment are thousands of nucleotides long. However, 23 of the nucleotide positions are shown in the table below, and you can use these data to repeat the researchers' analysis.[b] Each nucleotide position represents a separate character.

1. Construct a phylogenetic tree from the nucleotide positions using the parsimony principle (see Key Concept 14.2 and the examples in Table 14.1 and Figure 14.5). Use the outgroup to root your tree. Assume that all changes among nucleotides are equally likely.
2. Using your tree from Question 1, reconstruct the DNA sequences of the ancestral lineages.
3. Transitions are mutations that change one purine to the other (G ↔ A) or one pyrimidine to the other (C ↔ T), whereas transversions exchange a purine for a pyrimidine or vice versa (e.g., A ↔ C or T; C ↔ A or G). Which kind of mutation predominates in this phylogeny? Why might this be the case?

	Character at position																						
Lineage	**1**	**2**	**3**	**4**	**5**	**6**	**7**	**8**	**9**	**10**	**11**	**12**	**13**	**14**	**15**	**16**	**17**	**18**	**19**	**20**	**21**	**22**	**23**
A	T	C	G	G	G	C	C	C	C	C	C	C	A	A	C	C	G	A	T	A	C	A	A
B	C	C	G	G	G	T	C	C	C	T	C	C	G	A	T	T	A	G	C	G	T	G	G
C	C	C	G	G	G	C	C	C	T	C	C	T	A	A	C	C	G	G	T	A	C	A	A
D	T	C	A	G	G	C	C	C	C	C	C	C	A	A	C	C	G	A	T	A	C	A	A
E	C	T	G	G	G	C	C	C	C	C	C	T	A	A	C	C	G	G	T	A	C	A	A
F	C	T	G	A	A	C	C	C	C	C	C	C	G	A	C	T	G	G	C	G	C	G	G
G	C	C	G	G	G	T	T	C	C	T	C	C	G	A	T	T	A	G	C	G	C	G	G
H	C	C	G	G	A	C	C	C	C	C	C	C	G	C	C	T	G	G	C	G	C	G	G
Outgroup	C	C	G	G	G	C	C	T	C	C	T	C	G	A	C	C	G	G	C	A	C	G	G

[a] D. M. Hillis et al. 1992. *Science* 255: 589–592.

[b] J. J. Bull et al. 1993. *Evolution* 47: 993–1007.

A similar **Analyze the Data** exercise may be assigned in **Achieve**.

14.3 Phylogeny Makes Biology Comparative and Predictive

LEARNING OBJECTIVES

By the end of this key concept you should be able to:

14.3.1 Formulate a hypothesis about the origin of an epidemic using a phylogenetic tree.

14.3.2 Calculate the rate of a molecular clock from a graph that shows changes over time.

Once a phylogeny is reconstructed, what do we do with it? What else, beyond an understanding of evolutionary history, does phylogeny offer us?

Phylogenetic trees can be used to reconstruct past events

Reconstructing past events is important for understanding many biological processes. In the case of zoonotic diseases (diseases caused by infectious organisms transmitted to humans from another animal host), it is important to understand when, where, and how the disease first entered a human population. Human immunodeficiency virus (HIV) is the cause of such a zoonotic disease, acquired immunodeficiency syndrome, or AIDS. Phylogenetic analyses have become important for studying the transmission of viruses such as HIV. Phylogenies are also important for understanding the present global diversity of HIV and for determining the virus's origins in human populations. A broader phylogenetic analysis of immunodeficiency viruses shows that humans acquired these viruses from two different hosts: HIV-1 from chimpanzees, and HIV-2 from sooty mangabeys (**FIGURE 14.8**).

HIV-1 is the common form of the virus in human populations in central Africa, where chimpanzees are hunted for food, and HIV-2 is the common form in human populations in western Africa,

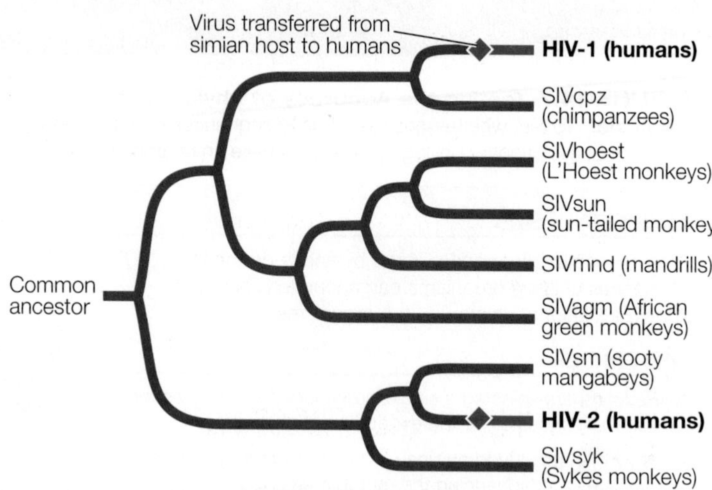

FIGURE 14.8 Phylogenetic Tree of Immunodeficiency Viruses The evolutionary relationships of immunodeficiency viruses show that these viruses have been transmitted to humans from two different simian hosts: HIV-1 from chimpanzees and HIV-2 from sooty mangabeys. (SIV stands for simian immunodeficiency virus.) (After B. Hahn et al. 2000. *Science* 287: 607–614.)

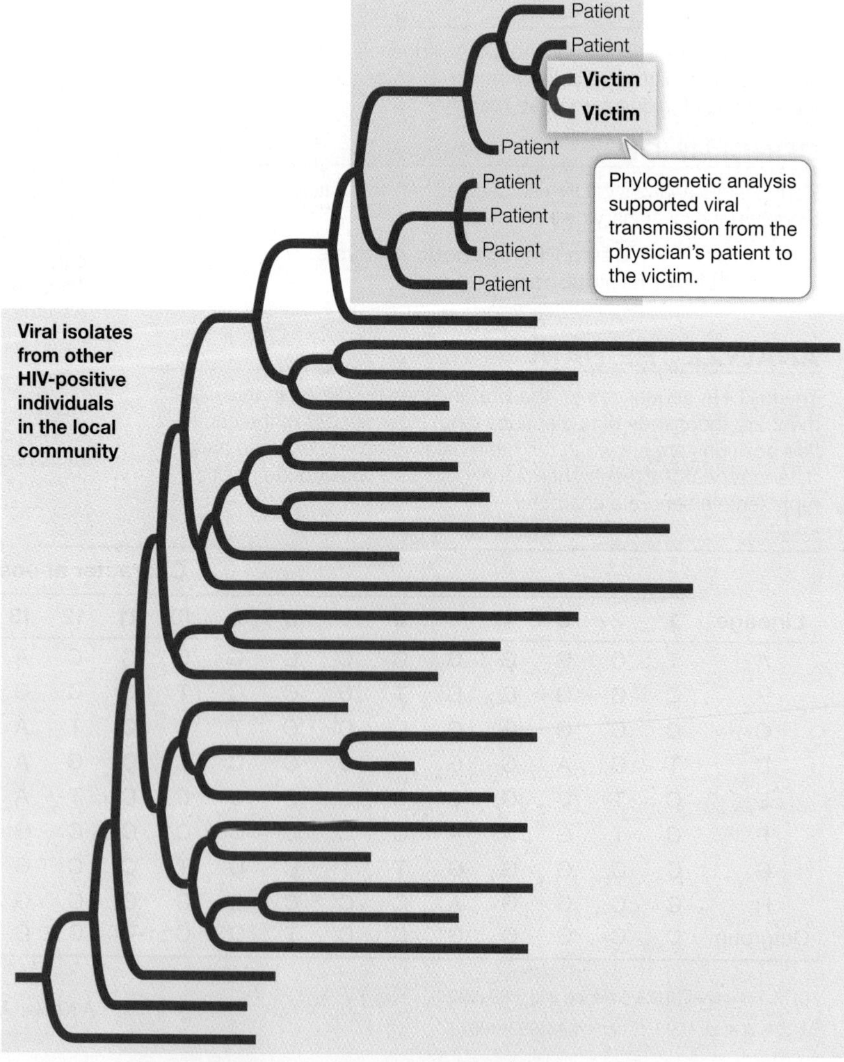

FIGURE 14.9 A Forensic Application of Phylogenetic Analysis This phylogenetic analysis demonstrated that strains of HIV present in a victim (shown in red) were a phylogenetic subset of viruses isolated from a physician's patient (shown in blue). This analysis was part of the evidence used to show that the physician drew blood from his HIV-positive patient and injected it into the victim in an attempt to kill her. A jury found the physician guilty of attempted murder. (After M. Metzker et al. 2002. *Proc Natl Acad Sci USA* 99: 14292–14297. © 2002 National Academy of Sciences, U.S.A.)

What hypothetical phylogenetic results could have exonerated the physician?

where sooty mangabeys are hunted for food. Thus it seems likely that these viruses entered human populations through hunters who cut themselves while skinning chimpanzees and sooty mangabeys. The global pandemic of AIDS occurred when these infections in local African populations rapidly spread through human populations around the world.

In recent years, phylogenetic analysis has become important in forensic investigations that involve viral transmission events. For example, phylogenetic analysis was critical for a criminal investigation of a physician who was accused of purposefully injecting blood from one of his HIV-positive patients into his former girlfriend in an attempt to kill her. The phylogenetic analysis revealed that the HIV strains present in the girlfriend were a subset of those present in the physician's patient (**FIGURE 14.9**). Other evidence was needed, of course, to connect the physician to this purposeful transmission event, but the phylogenetic analysis was important to support the viral transmission event from the patient to the victim.

THINK LIKE A SCIENTIST

Forensic phylogeny

Phylogenetic trees are used throughout biology, but only in recent years have they become important for forensic investigations. Here you will explore the relationship between science and society by applying your knowledge of phylogeny to a criminal court case.

A criminal case in Texas charged a defendant with knowingly and intentionally infecting a series of women with HIV. A phylogenetic analysis was used to demonstrate that the defendant transmitted HIV to his victims. (Other evidence was needed to prove knowledge and intent.) In this case, sequences of HIV isolated from the victims and the defendant, together with the closest sequences from an HIV database (the outgroup), were compared and used to construct a phylogenetic tree of the viruses (**FIGURE 14.10**). Viruses from each individual in the case are colored alike on the tree to the right. The labels are the codes for the individuals in the case. All of the individuals labeled CC01–CC08 are known to have engaged in sex; they represent an epidemiological cluster. (In forensic cases, samples are "blinded" to the investigators by assigning numbers to each sample, rather than using people's names. Only after the conclusions are finalized do other investigators decode the numbers to reveal the results.)

1. Which of the individuals labeled in the tree is consistent with being the source of this infection cluster? Why?
2. Why is the tree inconsistent with any of the other individuals being the source of infection within this cluster?
3. What was the purpose of including an outgroup made up of individuals who were outside the epidemiological cluster?

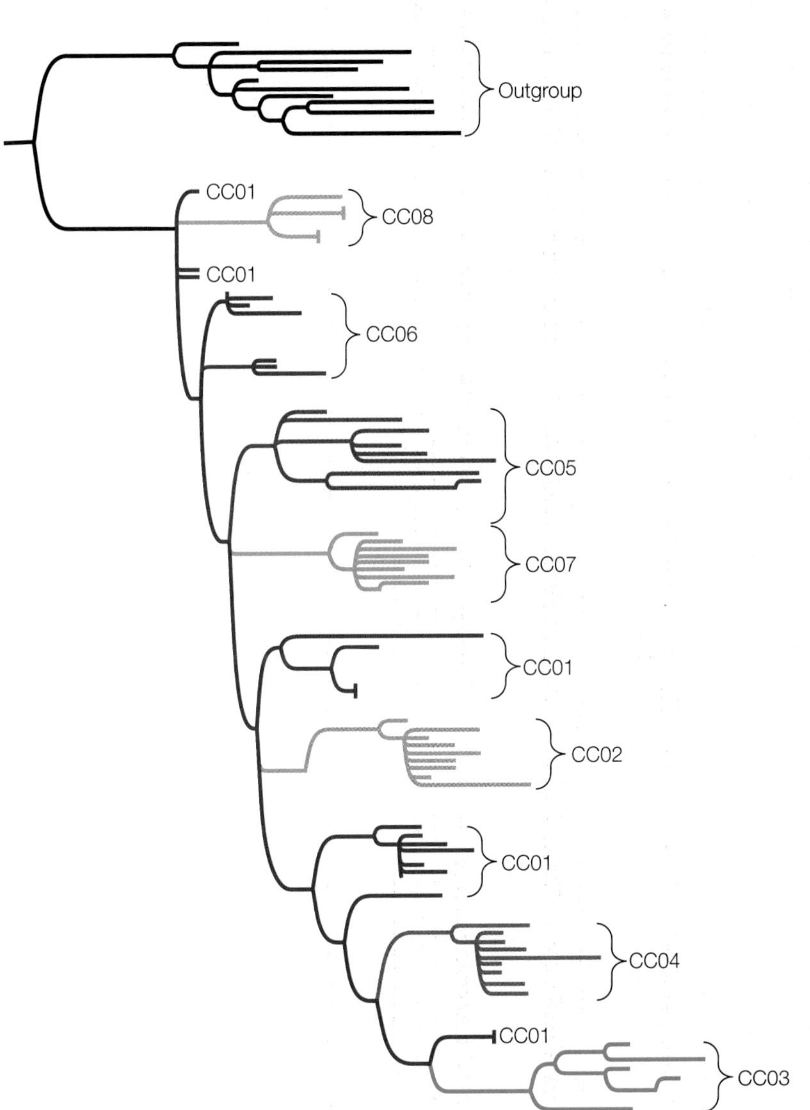

FIGURE 14.10 Forensic Phylogeny (After D. I. Scaduto et al. 2010. *Proc Natl Acad Sci USA* 107: 21242–21247.)

Phylogenies allow us to compare and contrast living organisms

Male swordtails—a group of fish in the genus *Xiphophorus*—have a long, colorful tail extension, and their reproductive success is closely associated with this appendage. Males with a long sword are more likely to mate successfully than are males with a short sword (an example of sexual selection; see Key Concept 13.2). Several explanations have been advanced for the evolution of this structure, including the hypothesis that the sword simply exploits a preexisting bias in the sensory system of the females. This sensory exploitation hypothesis suggests that female swordtails had a preference for males with long tails even before the tails evolved (perhaps because females assess the size of males by their total body length—including the tail—and prefer larger males).

To test the sensory exploitation hypothesis, phylogenetic analysis was used to identify the relatives of swordtails that had split most recently from their lineage before the evolution of swords. These closest relatives turned out to be fish in the genus *Priapella*. Even though male *Priapella* do not normally have swords, when researchers attached artificial swordlike structures to the tails of male *Priapella*, female *Priapella* preferred those males. This result provided support for the hypothesis that female *Xiphophorus* had a preexisting sensory bias favoring tail extensions even before the trait evolved (**FIGURE 14.11**). Thus a long tail became a sexually selected trait because of the preexisting preference of the females.

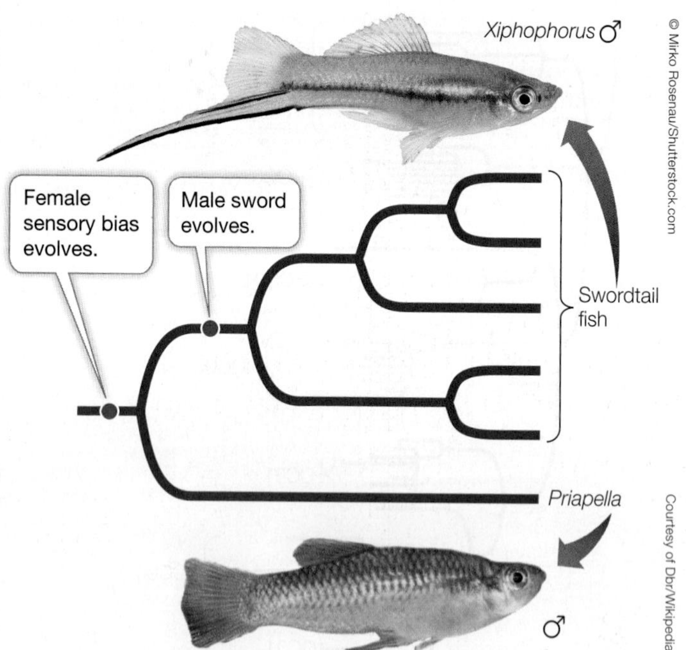

FIGURE 14.11 The Origin of a Sexually Selected Trait The tail extension of male swordtails (genus *Xiphophorus*) apparently evolved through sexual selection, as females mated preferentially with males that had long "swords." Phylogenetic analysis reveals that *Priapella* split from the swordtails before the evolution of the sword. The independent finding that female *Priapella* prefer male *Priapella* with an artificial sword further supports the idea that this appendage evolved as a result of a preexisting preference in females. (After A. L. Basolo. 1995. *Proc R Soc Lond B Biol Sci* 259: 307–311.)

The phylogeny allows us to understand when the trait evolved relative to the change in female preference.

Phylogenies can reveal convergent evolution

Like most animals, flowering plants (angiosperms) often reproduce by mating with another individual of the same species. But in many angiosperm species, the same individual produces both male and female gametes (contained within pollen and ovules, respectively). Self-incompatible species have mechanisms to prevent fertilization of the ovule by the individual's own pollen, and so must reproduce by outcrossing with another individual. Individuals of some species, however, regularly fertilize their ovules using their own pollen; they are self-fertilizing or selfing species, and their gametes are self-compatible.

Self-incompatibility in plants depends on the plant's ability to determine whether pollen is genetically different than its own pollen. How it does this is discussed in **Key Concept 26.2**.

The evolution of angiosperm fertilization mechanisms was examined in *Leptosiphon*, a genus in the phlox family that exhibits a diversity of mating systems and pollination mechanisms. The self-incompatible (outcrossing) species of *Leptosiphon* have long petals and are pollinated by long-tongued flies. In contrast, self-pollinating species have short petals and do not require insect pollinators to reproduce successfully. Using ribosomal DNA sequences, investigators reconstructed a phylogeny of this genus (**FIGURE 14.12**). They then determined whether each species was self-compatible by artificially pollinating flowers with the plant's own pollen or with pollen from other individuals and observing whether viable seeds formed.

The reconstructed phylogeny suggests that self-incompatibility is the ancestral state and that self-compatibility evolved three times within this group of *Leptosiphon*. The change to self-compatibility eliminated the plants' dependence on an outside pollinator and has been accompanied by the evolution of reduced petal size. Indeed, the striking morphological similarity of the flowers in the self-compatible groups once led to their being classified as members of a single species (*L. bicolor*). Phylogenetic analysis, however, shows them to be members of three distinct lineages. From this information we can infer that self-compatibility and its associated floral structure are convergent in the three independent lineages that had been called *L. bicolor*.

Ancestral states can be reconstructed

In addition to using phylogenetic methods to infer evolutionary relationships, biologists can use these techniques to reconstruct the morphology, behavior, or nucleotide and amino acid sequences of ancestral species (as was demonstrated for the ancestral sequence of bacteriophage T7 in Figure 14.7).

Reconstruction of ancient DNA sequences can also provide information about the biology of long-extinct organisms. For example, phylogenetic analysis was used to reconstruct an opsin protein in the ancestral archosaur (the most recent common ancestor of birds, dinosaurs, and crocodiles). Opsins are pigment proteins involved in vision; different opsins (with different amino acid sequences) are

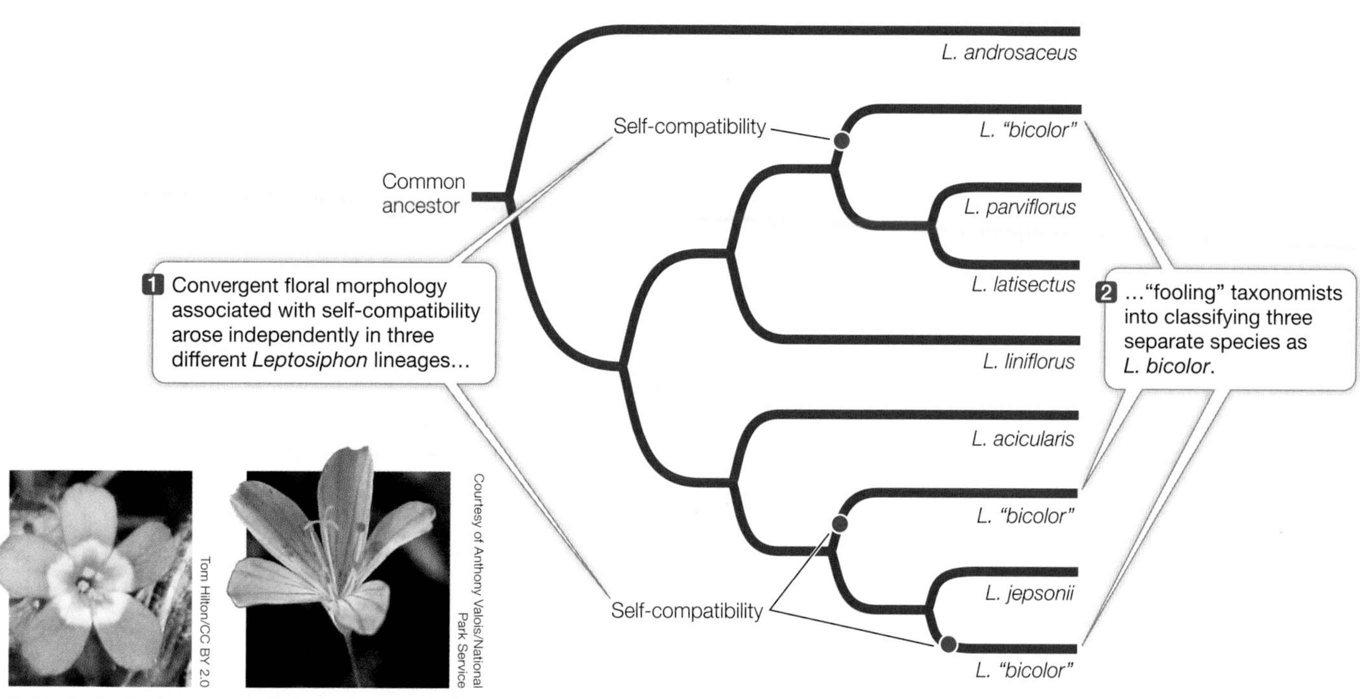

FIGURE 14.12 Phylogeny Reveals Convergent Evolution Self-compatibility apparently evolved independently three times among these species of the plant genus *Leptosiphon*. Because the appearance and structure of the flowers converged in the three selfing lineages, taxonomists mistakenly thought they were varieties of the same species. (After K. E. Holsinger. 2000. *Proc Natl Acad Sci USA* 97: 7037–7042.)

excited by different wavelengths of light. Knowledge of the opsin sequence in the ancestral archosaur would provide clues about the animal's visual capabilities and therefore about some of its probable behaviors. Investigators used phylogenetic analysis of opsin from living vertebrates to estimate the amino acid sequence of the pigment that existed in the ancestral archosaur. A protein with this same sequence was then constructed in the laboratory. The investigators tested the reconstructed opsin and found a significant shift toward the red end of the spectrum in the light sensitivity of this protein compared with that of most modern opsins. Modern species that exhibit similar sensitivity are adapted for nocturnal vision, so the investigators inferred that the ancestral archosaur might have been active at night. Thus reminiscent of the movies *Jurassic Park* and *Jurassic World*, phylogenetic analyses are being used to reconstruct extinct species, one protein at a time.

Molecular clocks help date evolutionary events

For many applications, biologists want to know not only the order in which evolutionary lineages split but also the timing of those splits. In 1965, Emile Zuckerkandl and Linus Pauling hypothesized that rates of molecular change were constant enough that they could be used to predict evolutionary divergence times—an idea that has become known as the **molecular clock** hypothesis.

Of course, different genes evolve at different rates, and there are also differences in evolutionary rates among species associated with differing generation times, environments, efficiencies of DNA repair systems, and other biological factors. Nonetheless, among closely related species, a given gene usually evolves at a reasonably constant rate. Therefore the protein encoded by the gene

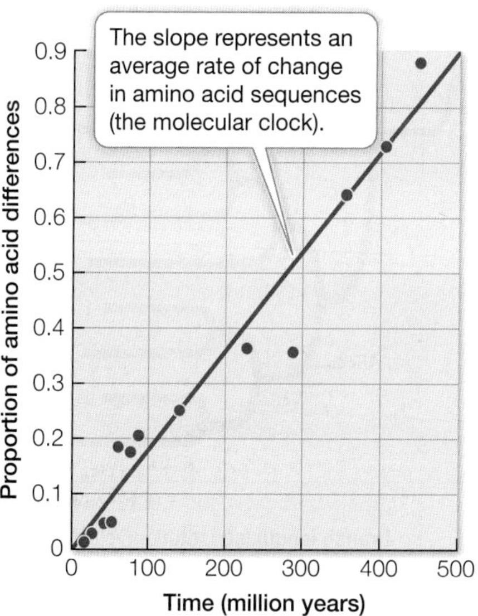

FIGURE 14.13 A Molecular Clock for the Protein Hemoglobin Amino acid replacements in hemoglobin have occurred at a relatively constant rate over nearly 500 million years of evolution. The graph shows the relationship between the time of divergence and the proportion of amino acids that have changed for 13 pairs of vertebrate hemoglobin proteins. The average rate of change represents the molecular clock for hemoglobin in vertebrates. (After R. E. Dickerson. 1971. *J Mol Evol* 1: 26–45.)

accumulates amino acid replacements at a relatively constant rate (**FIGURE 14.13**). A molecular clock uses the average rate at which a given gene or protein accumulates changes to gauge the time of divergence for a particular split in the phylogeny. Molecular clocks must be calibrated using independent data, such as the fossil record, known times of divergence, or biogeographic dates (e.g., the time of separations of continents). Using such calibrations, times of divergence have been estimated for many groups of species that have diverged over millions of years.

Molecular clocks are not only used to date ancient events; they are also used to study the timing of comparatively recent events. Most samples of HIV-1 have been collected from humans only since the early 1980s, although a few isolates from medical biopsies are available from as early as the 1950s. Biologists can use the observed changes in HIV-1 over the past several decades to project back to the common ancestor of all HIV-1 isolates, and estimate when HIV-1 first entered human populations from chimpanzees (**FIGURE 14.14**). This molecular clock was calibrated using the samples from the 1980s and 1990s, and then tested using the samples from the 1950s. As shown in Figure 14.14C, a sample from a 1959 biopsy is dated by molecular clock analysis at 1957 ± 10 years. Extrapolation back to the common ancestor of the samples suggested a date of origin for this group of viruses of about 1930. Although AIDS was unknown to Western medicine until the 1980s, this analysis shows that HIV-1 was present (probably at a very low frequency) in human populations in Africa for at least a half-century before its emergence as a global pandemic. Biologists have used similar analyses to conclude that immunodeficiency viruses have been transmitted repeatedly into human populations from multiple primates for more than a century (see also Figure 14.8). All of life is connected through evolutionary history, and the relationships among organisms provide a natural basis for making

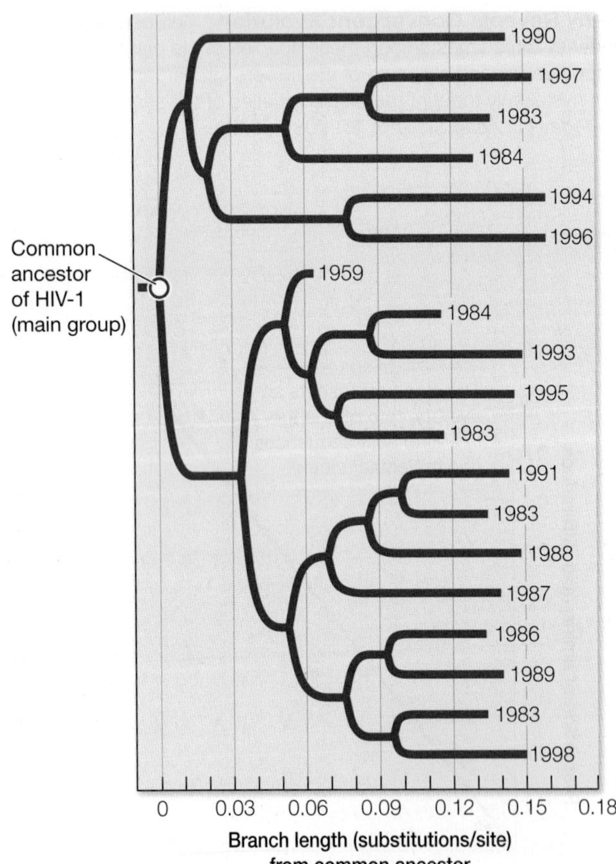

FIGURE 14.14 Dating the Origin of HIV-1 in Human Populations **(A)** A phylogenetic tree for samples of the main group of HIV-1 virus. The dates indicate the years in which the samples were taken. (For clarity, only a small fraction of the samples that were examined in the original study are shown.) **(B)** A plot of sample year versus genetic divergence from the common ancestor provided an average rate of divergence, or a molecular clock. **(C)** The molecular clock was used to date a sample taken in 1959 (as a test of the clock) and to estimate the date of origin of the HIV-1 main group (about 1930). Branch length from a common ancestor represents the average number of substitutions per nucleotide. (After B. Korber et al. 2000. *Science* 288: 1789–1796.)

Determine the expected branch length from the common ancestor of an HIV isolate from 1970.

REVIEW & APPLY | 14.3

R Phylogenetic trees are used to reconstruct the evolutionary history of lineages, to determine when and where traits arose, and to make biological comparisons among genes, populations, and species. They can also be used to reconstruct ancestral traits and to estimate the timing of evolutionary events.

A

1. West Nile virus kills birds of many species and can cause fatal encephalitis (inflammation of the brain) in humans and horses. In the 1930s the virus was first isolated in Africa, where it is thought to be endemic. By the 1990s it had been found throughout much of Eurasia. West Nile virus was not found in North America until 1999, when it was first detected in New York, but since that time it has spread rapidly across most of the United States. Use the phylogenetic tree of West Nile virus isolates shown at right to construct a hypothesis about the origin of the virus lineage that was introduced into the United States (*figure source*: R. S. Lanciotti et al. 1999. *Science* 1286: 2333–2337). The isolates are identified by their place and date of isolation.

2. Researchers examined the molecular divergence of a mitochondrial DNA gene (*cyt b*) among sister species of birds called honeycreepers on different Hawaiian Islands. They then plotted the measured molecular divergence against the estimated dates for separation of the islands (*figure source*: Fleischer et al. 1998. *Mol Ecol* 7: 533–545; bars represent the 95% confidence limits for the estimated molecular divergences). Calculate the average rate of change in *cyt b* (the molecular clock) from the graph.

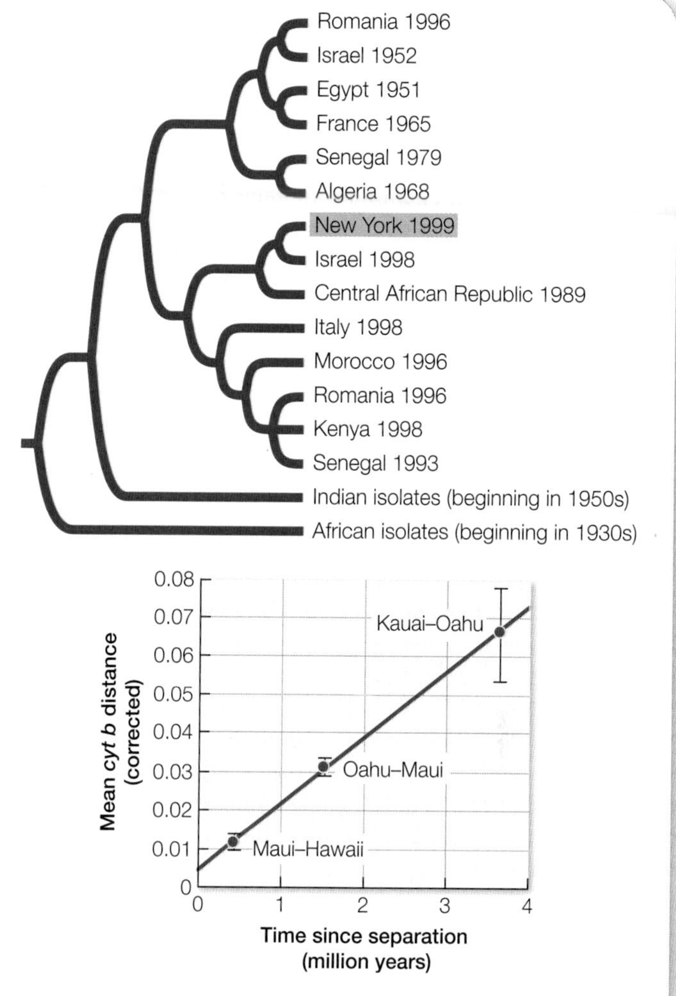

biological comparisons. For these reasons, biologists use phylogenetic relationships as the basis for organizing life into a coherent classification system.

14.4 Phylogeny Is the Basis of Biological Classification

LEARNING OBJECTIVES

By the end of this key concept you should be able to:

14.4.1 Use a phylogeny to classify a group of organisms.

14.4.2 Analyze a phylogenetic tree to identify monophyletic, polyphyletic, and paraphyletic groups.

The biological classification system in widespread use today is derived from a system developed by the Swedish biologist Carolus Linnaeus in the mid-1700s. Linnaeus developed a naming system called **binomial nomenclature** that has allowed scientists throughout the world to refer unambiguously to the same organisms by the same names.

Linnaean classification is based on standard taxon ranks

Linnaeus gave each species a two-part name, one part identifying the species itself and the other the genus to which it belongs. A **genus** (plural *genera*) is a group of closely related species. Optionally, the name of the taxonomist who first proposed the species name may be added at the end. Thus, *Homo sapiens* Linnaeus is the name of the modern human species. *Homo* is the genus to which the species belongs, and *sapiens* identifies the particular species in the genus *Homo*; Linnaeus proposed the species name *Homo sapiens*.

You can think of *Homo* as equivalent to your surname and *sapiens* as equivalent to your first name. The first letter of the genus name is capitalized, and the specific name is lowercase. Both of these formal designations are italicized. Rather than repeating the name of a genus when it is used several times in the same discussion, biologists often spell it out only once and abbreviate it to the initial letter thereafter (e.g., *D. melanogaster* rather than *Drosophila melanogaster*).

As we noted earlier, any group of organisms that is treated as a unit in a biological classification system, such as all species in the genus *Drosophila*, or all insects, or all arthropods, is called a taxon. In the Linnaean system, species and genera are further grouped into a hierarchical system of higher taxonomic categories. The taxon above the genus in the Linnaean system is the family. The names of animal families end in the suffix "-idae." Thus, Formicidae is the family that contains all ant species, and the family Hominidae contains humans and our recent fossil relatives, as well as our closest living relatives, the chimpanzees and gorillas. Family names are based on the name of a member genus; Formicidae is based on the genus *Formica*, and Hominidae is based on *Homo*. The same rules are used in classifying plants, except that the suffix "-aceae" is used for plant family names instead of "-idae." Thus, Rosaceae is the family that includes the genus *Rosa* (roses) and its relatives.

In the Linnaean system, families are grouped into orders, orders into classes, classes into phyla (singular *phylum*), and phyla into kingdoms. However, the ranking of taxa within Linnaean classification is subjective. Whether a particular taxon is considered, say, an order or a class is informative only with respect to the *relative* ranking of other related taxa. Although families are always grouped within orders, orders within classes, and so forth, there is nothing that makes a "family" in one group equivalent (in number of genera or in evolutionary age, for instance) to a "family" in another group.

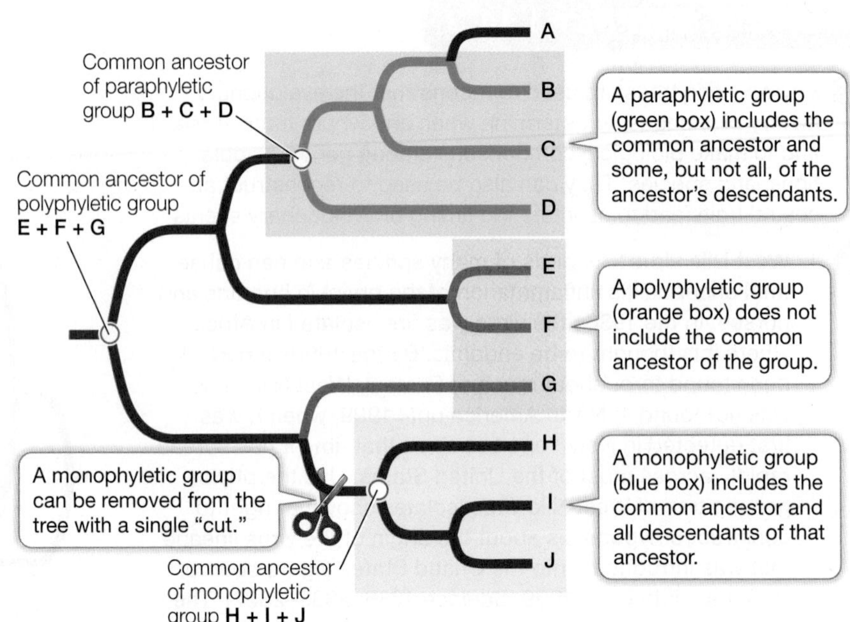

FIGURE 14.15 Monophyletic, Polyphyletic, and Paraphyletic Groups Monophyletic groups are the basis of taxa in modern biological classifications. Polyphyletic and paraphyletic groups are not appropriate for use in classifications because they do not accurately reflect evolutionary history.

Activity 14.3 Types of Taxa

Linnaeus recognized the overarching hierarchy of life, but he developed his system before evolutionary thought had become widespread. Biologists today recognize the tree of life as the basis for biological classification and often name taxa without placing them into the various Linnaean ranks.

Evolutionary history is the basis for modern biological classification

Today's biological classifications express the evolutionary relationships of organisms. Taxa are expected to be **monophyletic**, meaning that the taxon contains an ancestor and all descendants of that ancestor, and no other organisms. In other words, a monophyletic taxon is a historical group of related species, or a complete branch on the tree of life. As noted earlier, this is also the definition of a clade. A true monophyletic group can be removed from a phylogenetic tree by a single "cut" in the tree, as shown in **FIGURE 14.15**.

Note that there are many monophyletic groups on any phylogenetic tree, and that these groups are successively smaller subsets of larger monophyletic groups. This hierarchy of biological taxa, with all of life as the most inclusive taxon and many smaller taxa within larger taxa, down to the individual species, is the modern basis for biological classification.

Although biologists seek to describe and name only monophyletic taxa, the detailed phylogenetic information needed to do so is not always available. A group that does not include its common ancestor is **polyphyletic**. A group that does not include all the descendants of a common ancestor is referred to as **paraphyletic** (see Figure 14.15). Virtually all taxonomists now agree that polyphyletic and paraphyletic groups are inappropriate as taxonomic units because they do not correctly reflect evolutionary history. Some classifications still contain such groups because some organisms have not been evaluated phylogenetically. As mistakes in prior classifications are detected, taxonomic names are revised and polyphyletic and paraphyletic groups are eliminated from the classifications.

Several codes of biological nomenclature govern the use of scientific names

Several sets of explicit rules govern the use of scientific names. Biologists around the world follow these rules voluntarily to facilitate communication and dialogue. There may be dozens of common names for an organism in many different languages, and the same common name may refer to more than one species (**FIGURE 14.16**). The rules of biological nomenclature are designed so that there is only one correct scientific name for any single recognized taxon, and (ideally) a given scientific name applies only to a single taxon (that is, each scientific name is unique). Sometimes the same species is named more than once (when more than one taxonomist has taken up the task). In these cases, the rules specify that the valid name is the first name that was proposed. If the same name is inadvertently given to two different species, then the species that was named second must be given a new name.

Because of the historical separation of the fields of zoology, botany (which originally included mycology, the study of fungi), and microbiology, different sets of taxonomic rules were developed for each of these groups. Yet another set of rules emerged later for classifying viruses. This separation of fields resulted in duplicated taxon names in groups governed by the different sets of rules. *Drosophila*, for example, is both a genus of fruit flies and a genus of fungi, and some species in both groups have identical names. Until recently these duplicated names caused little confusion, since traditionally biologists who studied fruit flies were unlikely to read the literature on fungi (and vice versa). Today, given the prevalence of large, universal biological databases (such as GenBank, which includes DNA sequences from across all life), it is increasingly important that each taxon have a unique and unambiguous name. Biologists are working on a universal code of nomenclature that can be applied to all organisms, so that every species will have a unique identifying name or registration number. These efforts are helping to build an online Encyclopedia of Life (eol.org) that links all the information for all the world's species.

Phylogenetic methods are used by biologists to compare the evolutionary relationships across all of life and to study many evolutionary processes. In Chapter 15 we will use phylogenies as one of several tools to understand evolution at the level of genes and genomes.

(A) ***Asclepias tuberosa***

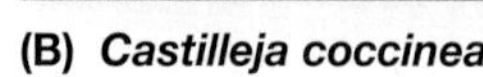

(B) ***Castilleja coccinea***

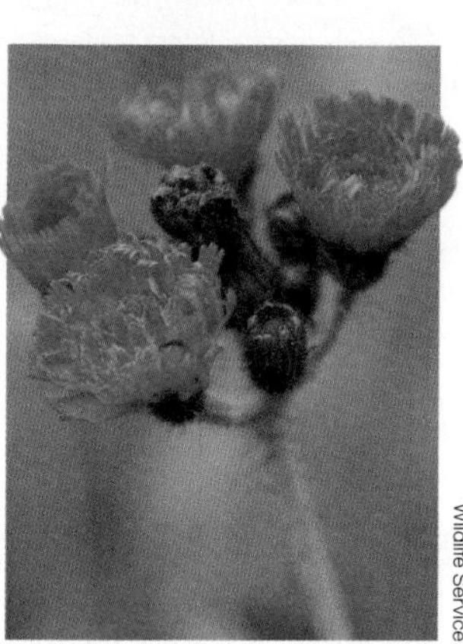

(C) ***Hieracium aurantiacum***

FIGURE 14.16 Same Common Name, Not the Same Species All three of these distinct plant species are called "Indian paintbrush." Binomial nomenclature allows us to avoid the ambiguity of such common names and communicate exactly what is being described. **(A)** *Asclepias tuberosa* is a perennial milkweed native to eastern North America. **(B)** *Castilleja coccinea* is also native to eastern North America, but is a member of a very different group of plants called scrophs. **(C)** *Hieracium aurantiacum* is a European species of aster that has been widely introduced into North America.

REVIEW & APPLY | 14.4

R Biologists organize and classify life by identifying and naming monophyletic groups. Several sets of rules govern the use of scientific names so that each species and higher taxon can be identified and named unambiguously.

A Consider the phylogeny and three possible classifications shown below.

Frogs
Salamanders
Caecilians
Mammals
Lizards
Turtles
Crocodiles
Birds

1. Which of these classifications contains a paraphyletic group?
2. Which of these classifications contains a polyphyletic group?
3. Which of these classifications is consistent with the goal of including only monophyletic groups in a biological classification?

Classification one:

Named group	Included taxa
Amphibia	Frogs, salamanders, and caecilians
Mammalia	Mammals
Reptilia	Lizards, turtles, and crocodiles
Aves	Birds

Classification two:

Named group	Included taxa
Amphibia	Frogs, salamanders, and caecilians
Mammalia	Mammals
Reptilia	Lizards, turtles, crocodiles, and birds

Classification three:

Named group	Included taxa
Amphibia	Frogs, salamanders, and caecilians
Homeothermia	Mammals and birds
Reptilia	Lizards, turtles, and crocodiles

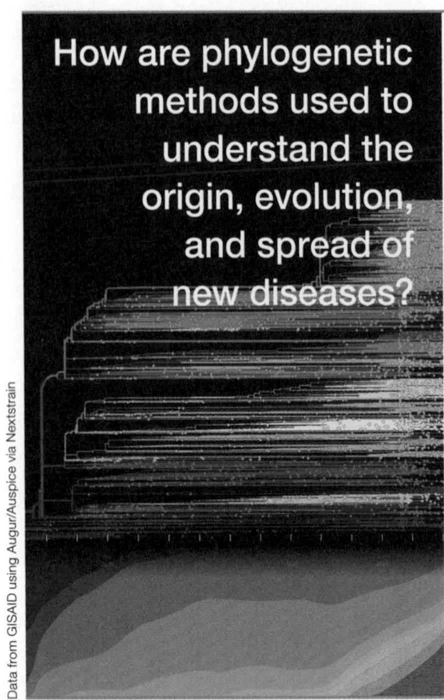

Answer: As we saw at the opening of this chapter, phylogenetic methods helped biologists track the spread and evolution of the virus that caused the COVID-19 pandemic. Comparative sequencing and phylogenetic analysis of related coronaviruses from other species showed that the virus that caused the COVID-19 pandemic originally came from a bat, possibly through an intermediate species. This transmission event likely took place in live animal markets in China, where many species are caged in close proximity. Knowing the source of new viruses helps us to understand how we can slow or prevent the movement of other viruses into human populations. Once an epidemic has started, phylogenetic analysis tells us how quickly the virus is spreading and changing, and helps us to identify new variants. In recent years, as more viruses have emerged in human populations, phylogenetic analysis has been essential for understanding where these viruses come from, how they move through human populations, and how quickly they evolve. This information is critical in the development of new and effective vaccines and strategies for slowing the spread of viruses that causes diseases such as AIDS, COVID-19, West Nile disease, and influenza.

14 VISUAL SUMMARY

14.1 All of Life Is Connected through Its Evolutionary History

FIGURE 14.1

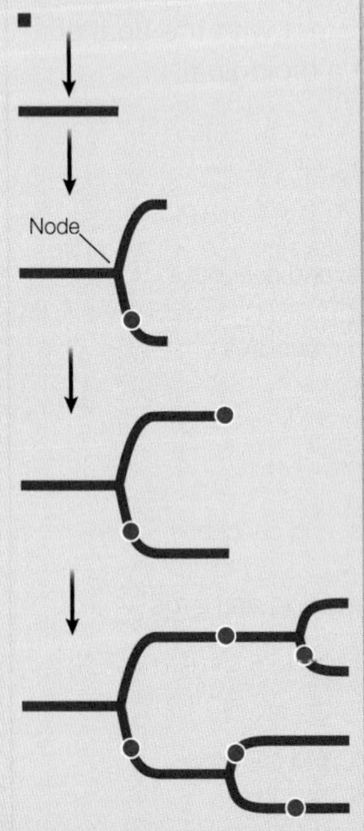

- **PHYLOGENY** is the history of evolutionary relationships among organisms or their genes. Groups of evolutionarily related species are represented as branches in a **PHYLOGENETIC TREE** (**FIGURE 14.1**).
- Named species and groups of species are called **TAXA**. A taxon that consists of an ancestor and all of its evolutionary descendants is called a **CLADE**. **HOMOLOGIES** are similar traits that have been inherited from a common ancestor.
- A derived trait that is shared by two or more taxa and is inherited from their common ancestor is called a **SYNAPOMORPHY**. Distantly related species may show similar traits that do not result from common ancestry. **CONVERGENT EVOLUTION** and **EVOLUTIONARY REVERSALS** can give rise to such traits, which are called **HOMOPLASIES** (**FIGURE 14.4**).

FIGURE 14.4

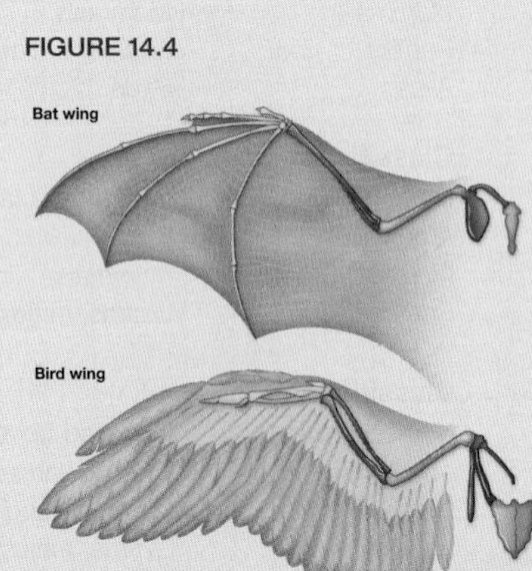

QUESTIONS

1. Draw a phylogenetic tree of humans, chimpanzees, gorillas, and orangutans. Label the root and nodes.
2. What does the root of the tree represent?
3. What do the nodes represent?

14.2 Phylogeny Can Be Reconstructed from Traits of Organisms

Go to ACTIVITIES 14.1 and 14.2 and ANIMATION 14.1

- Phylogenetic trees can be constructed from synapomorphies using the logic of **PARSIMONY** (**FIGURE 14.5**).

FIGURE 14.5

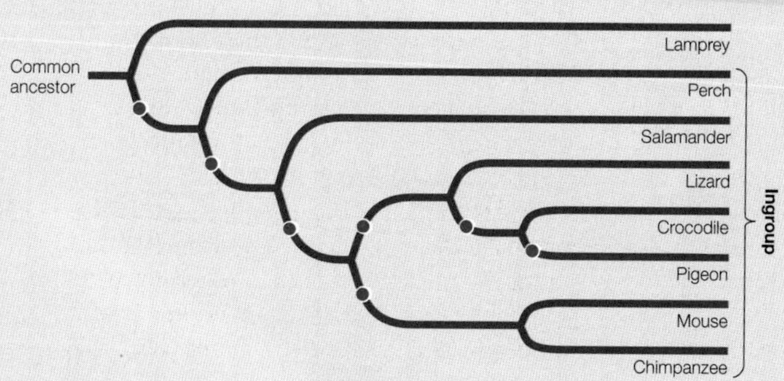

- Phylogenetic trees can also be constructed using **MAXIMUM LIKELIHOOD** methods, which find the tree most likely to have generated the observed data under a specific model of evolution.
- Phylogenetic methods have been tested in both experimental and simulation studies, and have been shown to be accurate under a wide variety of conditions.

QUESTIONS

1. What does the parsimony method minimize?
2. What is an example of a synapomorphy?
3. Why is it important that phylogenetic characters are genetically determined, rather than being influenced by the organism's environment?

14.3 Phylogeny Makes Biology Comparative and Predictive

FIGURE 14.11

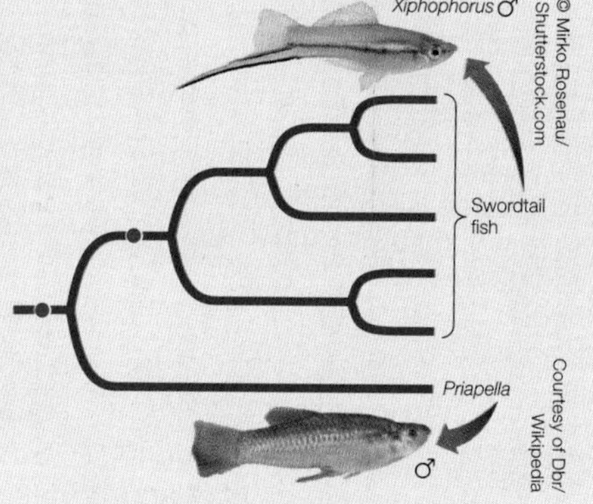

- Phylogenetic trees are used to make comparisons among living organisms.
- Phylogenetic trees are used to reconstruct the past and to understand the origin of traits (**FIGURE 14.11**).
- Phylogenetic trees may include estimates of divergence times of lineages determined by **MOLECULAR CLOCK** analysis (**FIGURE 14.13**).

FIGURE 14.13

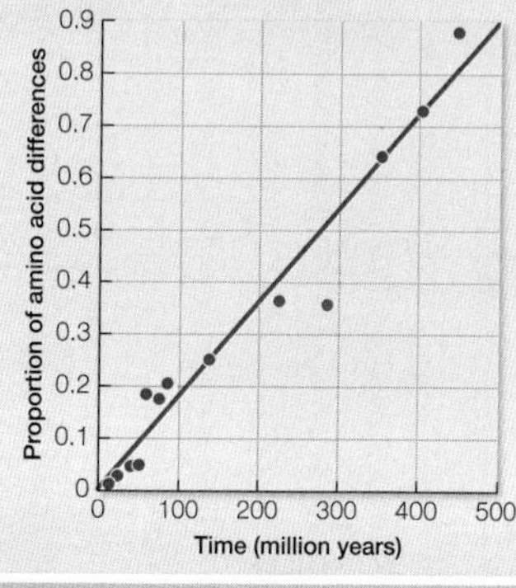

1. Draw a phylogenetic tree of HIV-1, HIV-2, and related SIVs from other primates.
2. Use the tree of HIVs and SIVs to explain the multiple origins of HIV in human populations.
3. How can phylogenetic trees be used to date at least one origin of HIV in humans?

14.4 Phylogeny Is the Basis of Biological Classification

Go to ACTIVITY 14.3

FIGURE 14.15

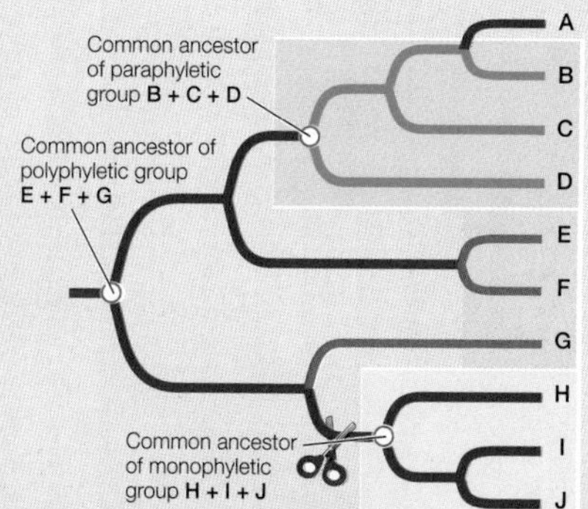

- Biologists use phylogenetic relationships to organize life into a coherent classification system.
- Taxa in modern classifications are expected to be **MONOPHYLETIC** groups. **PARAPHYLETIC** and **POLYPHYLETIC** groups are not considered appropriate taxonomic units (**FIGURE 14.15**).
- Several sets of rules govern the use of scientific names, with the goal of providing unique and universal names for taxa.

1. Draw a phylogenetic tree, with taxon labels A–F.
2. Label a monophyletic group and a paraphyletic group on the tree.
3. What are the advantages of scientific names for taxa, as compared with common names?

15 Evolution of Genes and Genomes

Several groups of both marine and freshwater fishes generate electric fields using an electric organ. A few generate a current strong enough to stun prey, but most, like this South American knifefish, use their electric field to sense their environment, to locate prey, and to communicate.

How have simple changes in gene duplication, gene sequences, and gene expression led to the repeated evolution of electric organs in fishes?

You will find the answer to this question on page 368.

KEY CONCEPTS

15.1 Genomes Reveal Both Neutral and Selective Processes of Evolution

LEARNING OBJECTIVES

By the end of this key concept you should be able to:

- **15.1.1** Explain how substitutions may be neutral or selective in a given organism.
- **15.1.2** Provide a mathematical explanation for why the rate of evolution of neutral mutations is independent of population size.
- **15.1.3** Evaluate gene sequences of protein-coding genes to identify positions that are evolving under purifying versus positive selection, and apply this knowledge to a biological problem.
- **15.1.4** Describe the role of noncoding DNA in the evolution of genome size.

An organism's **genome** is the full set of genes it contains, as well as any noncoding regions of the DNA (or in the case of some viruses, RNA). Most of the genes of eukaryotic organisms are found on chromosomes in the nucleus, but genes are also present in chloroplasts and mitochondria. In organisms that reproduce sexually, both males and females contribute nuclear genes, but mitochondrial and chloroplast genes are usually transmitted only via the cytoplasm of one of the two gametes (often from the female parent).

Evolution results from changes to gene sequences, location, and expression

Genomes must be replicated to be transmitted from parents to offspring. DNA replication does not occur without error, however. Mistakes in DNA replication—mutations—provide much of the raw material for evolutionary change. Even though the replication error rate per nucleotide is very small, most genomes are so large that a few errors are usually made every time a genome is replicated. These errors are actually essential for the long-term survival of life, because mutations are the initial source of the genetic variation that permits species to evolve in response to changes in their environment.

A particular allele of a gene will not be passed on to successive generations unless an individual carrying that allele survives and reproduces. The allele must function in combination with many other genes in the genome or it will quickly be selected against. Moreover, the degree and timing of a gene's expression are affected by its location in the genome. Evolutionary changes result from mutations to gene sequences, from gene duplications and rearrangements, and from changes in the timing, degree, and location of gene expression.

A genome, then, is not simply a random collection of genes in a random order along chromosomes. Rather, it is a complex set of integrated genes, regulatory sequences, and structural elements, interspersed with vast stretches of noncoding DNA that may have little direct function. The sequences of genes, as well as their location and expression, are subject to evolutionary change, as are the extent and location of noncoding DNA. All of these changes can affect the phenotype of an organism.

Biologists have now sequenced the complete genomes of a large number of organisms, including humans. The information in these sequences is helping us understand how and why organisms differ, how they function, and how they have evolved.

Most natural populations harbor far more genetic variation than we would expect to find if genetic variation were influenced by natural selection alone. This discovery, combined with the knowledge that many mutations do not change molecular function, stimulated the development of the neutral theory of molecular evolution.

Much of molecular evolution is neutral

To discuss the evolution of genes, we need to consider the specific types of mutations that are possible. A nucleotide substitution is a change in a single nucleotide in a DNA sequence (a type of point mutation; see Key Concept 9.3). Many nucleotide substitutions have no effect on phenotype, even if the change occurs in a gene that encodes a protein, because most amino acids are specified by more than one codon. A substitution that does not change the encoded amino acid is known as a **synonymous substitution** (also called a **silent substitution**) (**FIGURE 15.1A**). Synonymous substitutions do not affect the functioning of a protein (although they may have other effects, such as changes in mRNA stability or translation rates) and are therefore less likely than other types of substitutions to be influenced by natural selection.

A nucleotide substitution that *does* change the amino acid sequence encoded by a gene is known as a **nonsynonymous substitution** (a **missense substitution** changes an amino acid, and a **nonsense substitution** produces a new stop codon) (**FIGURE 15.1B**). In general, nonsynonymous substitutions are likely to be deleterious to the organism. But not every amino acid replacement alters a protein's shape and charge (and hence its functional properties). Therefore some nonsynonymous substitutions are selectively neutral, or nearly so. Another possibility is that a nonsynonymous substitution alters a protein in a way that confers an advantage to the organism and is therefore favored by natural selection.

LINK The genetic code determines the amino acid that is encoded by each codon; see **Figure 10.14**.

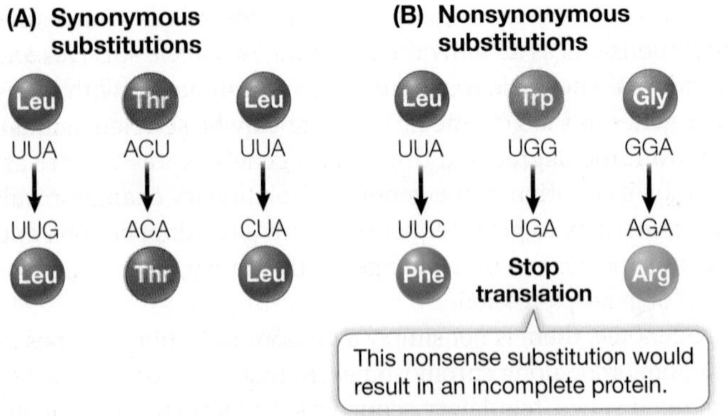

FIGURE 15.1 When One Nucleotide Changes
(A) Synonymous substitutions do not change the amino acid specified and do not affect protein function. Such substitutions are less likely than other substitutions to be subject to natural selection, although they contribute greatly to the buildup of neutral genetic variation in a population. **(B)** Nonsynonymous substitutions do change the amino acid sequence and are likely to have an effect (often deleterious, but sometimes beneficial) on protein function. Such nucleotide substitutions are targets for natural selection.

Synonymous substitutions accumulate over time in protein-coding genes at a much higher rate than nonsynonymous substitutions do. This is true even though the underlying mutation rates for both types of substitutions are equal (or at least similar). The reason is simple: as most nonsynonymous substitutions are harmful to the function of a gene, they are removed from the population over time through natural selection. Therefore, *substitution rates are higher at nucleotide positions that do not change the amino acid being expressed* (**FIGURE 15.2**). The rate of substitution is even higher in **pseudogenes**, which are copies of genes that are no longer functional (and so are not subject to strong selection).

Insertions, deletions, and rearrangements of DNA sequences are all mutations that may affect a larger portion of the gene or genome than do point mutations. Insertions and deletions of nucleotides in a protein-coding sequence interrupt its reading frame, unless they occur in multiples of three nucleotides (the length of one codon). Rearrangements may merely change the order of whole genes along chromosomes, or they may rearrange functional domains among individual genes.

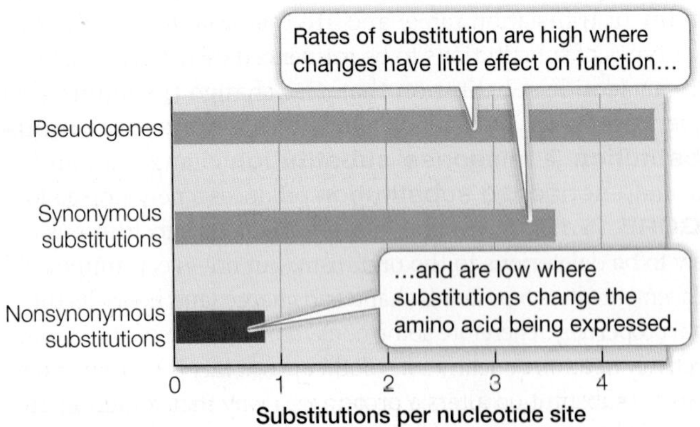

FIGURE 15.2 Rates of Substitution Differ Rates of nonsynonymous substitution are typically much lower than rates of synonymous substitution, and much lower than substitution rates in pseudogenes. This pattern reflects stronger stabilizing selection in functional genes than in pseudogenes.

When biologists began to examine the details of genetic variation of populations, they soon discovered many gene variants that had little or no effect on function. This gave rise to new ideas about how these neutral variants arise and spread in populations. Motoo Kimura proposed the **neutral theory** in 1968. He suggested that, at the molecular level, the majority of variants found in most populations are selectively neutral. That is, most gene variants confer neither an advantage nor a disadvantage on their bearers. Therefore these neutral variants must accumulate through genetic drift rather than through positive selection.

We saw in Key Concept 13.2 that genetic drift of *existing* gene variants tends to be greatest in small populations. However, the rate of fixation of *new* neutral mutations by genetic drift is independent of population size. To see why this is so, consider a population of size N and a neutral mutation rate of μ (mu) per gamete per generation at a particular locus. The number of new mutations would be, on average, $\mu \times 2N$, because $2N$ gene copies are available to mutate in a population of diploid organisms. The probability that a given mutation will be fixed by drift alone is its frequency, which equals $1/(2N)$ for a newly arisen mutation. We can multiply these two terms to get the rate of fixation of neutral mutations (m) in a given population of N individuals:

$$m = 2N\mu \frac{1}{2N}$$

But notice that the two effects of population size (N) cancel one another out:

$$m = \cancel{2N}\mu \frac{1}{\cancel{2N}}$$

Therefore the rate of fixation of neutral mutations depends only on the neutral mutation rate μ and is independent of population size. Any given mutation is more likely to appear in a large population than in a small one, but any mutation that does appear is more likely to become fixed in a small population. The rate of fixation of neutral mutations is thus equal to the mutation rate in populations of any size (i.e., $m = \mu$).

As long as the underlying mutation rate is constant, genes and proteins evolving in different populations should diverge from one another in neutral changes at a constant rate. The rate of evolution of particular genes and proteins is indeed often relatively constant over time, and therefore can be used as a molecular clock to calculate evolutionary divergence times between species (see Key Concept 14.3).

Neutral theory does not imply that most mutations have no effect on the individual organism, even though much of the genetic variation present in a population is the result of neutral evolution. Many mutations are rarely or never observed in populations because they are lethal or strongly detrimental, and the individuals that carry them are quickly removed from the population through natural selection. Similarly, because mutations that confer a selective advantage tend to be quickly fixed in populations, they also do not result in significant variation at the population level. Nonetheless, if we compare homologous proteins from different populations or species, some amino acid positions will remain constant under purifying selection, others will vary through neutral genetic drift, and still others will differ as a result of positive selection for change. How can these evolutionary processes be distinguished?

Positive and purifying selection can be detected in the genome

As we have just seen, substitutions in a protein-coding gene can be either synonymous or nonsynonymous, depending on whether they change the resulting amino acid sequence of the protein. The relative rates of synonymous and nonsynonymous substitutions are expected to differ in regions of genes that are evolving neutrally, or evolving under positive selection for change, or staying unchanged under purifying selection.

- If a given amino acid in a protein can be one of many alternatives (without changing the protein's function), then an amino acid replacement is neutral with respect to the fitness of an organism. In this case, the rates of synonymous and nonsynonymous substitutions in the corresponding DNA sequences are expected to be very similar, so the ratio of the two rates should be close to 1.
- If a given amino acid position is under **positive selection** for change, the observed rate of nonsynonymous substitutions is expected to exceed the rate of synonymous substitutions in the corresponding DNA sequences.
- If a given amino acid position is under **purifying selection**, then the observed rate of synonymous substitutions is expected to be much higher than the rate of nonsynonymous substitutions in the corresponding DNA sequences.

The evolution of lysozyme illustrates how and why particular codons in a gene sequence might be under different modes of selection. The enzyme lysozyme is found in almost all animals. It is produced in the tears, saliva, and milk of mammals and in the albumen (whites) of bird eggs. Lysozyme digests the cell walls of bacteria, rupturing and killing them. As a result, it plays an important role as a first line of defense against invading bacteria. Most animals defend themselves against bacteria by digesting them, which is probably why most animals have lysozyme. Some animals also use lysozyme in the digestion of food.

Among mammals, a mode of digestion called foregut fermentation has evolved twice. In mammals with this mode of digestion, the foregut—consisting of part of the esophagus and/or stomach—has been converted into a chamber in which bacteria break down ingested plant matter by fermentation. Foregut fermenters can obtain nutrients from the otherwise indigestible cellulose that makes up a large proportion of plant tissue. Foregut fermentation evolved independently in ruminants (a group of hoofed mammals that includes cattle) and in certain leaf-eating monkeys, such as langurs. We know that these evolutionary events were independent because both langurs and ruminants have close relatives that are not foregut fermenters.

In both mammalian foregut-fermenting lineages, lysozyme has been modified to play a new, nondefensive role. The modified lysozyme enzyme ruptures some of the bacteria that live in the foregut, releasing nutrients metabolized by the bacteria, which the mammal then absorbs. How many changes in the lysozyme molecule were needed to allow it to perform this function amid the digestive enzymes and acidic conditions of the mammalian foregut? To answer this question, Caro-Beth Stewart and her colleagues compared the lysozyme-coding sequences in foregut fermenters with those in several of their nonfermenting relatives. They determined which amino acids differed and which were shared among the species (**INVESTIGATION FIGURE 15.3**), as well as the rates of synonymous and nonsynonymous substitutions in lysozyme genes across the evolutionary history of the sampled species.

For many of the amino acid positions of lysozyme, the rate of synonymous substitution in the corresponding gene sequence was much higher than the rate of nonsynonymous substitution. This observation indicates that many of the amino acids that make up lysozyme are evolving under purifying selection. In other words, there is selection against change in the lysozyme protein at these positions, and the encoded amino acids must therefore be critical for lysozyme function. At other positions, several different amino acids function equally well, and the corresponding codons have similar rates of synonymous and nonsynonymous substitutions.

The most striking finding was that amino acid replacements in lysozyme happened at a much higher rate in the lineage leading to langurs than in any other primate lineage. The high rate of nonsynonymous substitution in the langur lysozyme gene shows that lysozyme went through a period of rapid change in adapting to the foreguts of langurs. Moreover, the lysozymes of langurs and cattle exhibit five instances of **convergent evolution**, a process whereby organisms that are not closely related independently evolve similar traits as a result of having to adapt to similar environments or ecological niches. More specifically, the lysozymes of langurs and cattle share five convergent amino acid replacements, all of which lie on the surface of the lysozyme molecule, well away from the enzyme's active site. Two of these shared replacements are changes from arginine to lysine, which make the protein more resistant to degradation by the stomach enzyme pepsin. By understanding the functional significance of amino acid replacements, biologists can explain the observed changes in amino acid sequences in terms of changes in the functioning of the protein.

LINK As explained in **Key Concept 14.1**, not all similar traits are evidence of relatedness. Similar traits may evolve independently in different lineages in the phenomenon called convergent evolution.

A large body of fossil, morphological, and molecular evidence shows that langurs and cattle do not share a recent common ancestor. However, langur and ruminant lysozymes share several amino acids that neither mammal shares with the lysozymes of its own closer relatives. The lysozymes of these two mammals have converged on some of the same amino acids despite their very different ancestry. The amino acids they share give these

INVESTIGATION

FIGURE 15.3 Convergent Molecular Evolution Langurs (a group of monkeys) and cattle are only distantly related, but both have evolved foregut fermentation. Both express the enzyme lysozyme in their foreguts to aid in breaking down bacteria that are involved in fermentation. Stewart and colleagues compared the gene sequences of lysozyme in mammals with and without foregut fermentation to see if there is convergence in the independently evolved amino acid sequences of lysozyme in langurs and cattle.[a]

HYPOTHESIS

Similar selective conditions in distantly related mammals have resulted in convergence of adaptations for foregut fermentation in the amino acid sequences of lysozyme.

METHOD

1. Isolate and sequence lysozyme from two distantly related mammal species with foregut fermentation (langur and cattle) as well as from other mammals that are more closely related to either langurs or to cattle but that lack foregut fermentation.

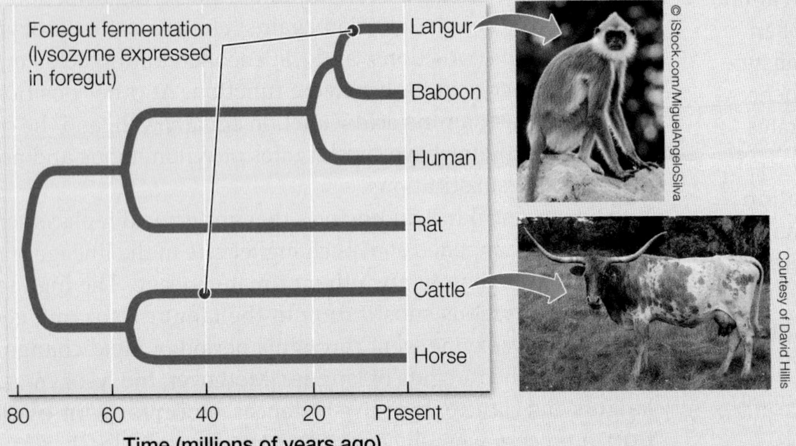

	Langur	Baboon	Human	Rat	Cattle	Horse
Langur		14	18	38	32	65
Baboon	0		14	33	39	65
Human	0	1		37	41	64
Rat	0	1	0		55	64
Cattle	5	0	0	0		71
Horse	0	0	0	0	1	

The lysozymes of langurs and cattle are convergent for 5 amino acids.

2. Tabulate the pairwise differences in the amino acid sequences. Plot the amino acid changes on the phylogenetic tree and count the number of convergent similarities between each pair of species. The results can then be plotted as a matrix.

RESULTS

The matrix shows the number of all pairwise amino acid differences above the diagonal and the number of convergent similarities below the diagonal.

CONCLUSION

The lysozyme sequences of the two species with foregut fermentation account for the majority of the convergent amino acid replacements observed among these species, demonstrating molecular convergence associated with the independent evolution of foregut fermentation.

ANALYZE THE DATA

Caro-Beth Stewart and her colleagues collected lysozyme sequences from six species of mammals. A small sample of the data they collected is shown in the table. The phylogeny of these six species is well supported from analysis of many genes and much morphological data.

1. Using the phylogenetic tree (at left), plot the amino acid changes across the phylogeny of the six mammals. Assume that the ancestral state is the amino acid present at the base of the tree.
2. Which amino acid positions show unique convergence between the langur and cattle lineages (i.e., the derived state is found only in cattle and langurs)?
3. Which additional position is convergent between cattle and the ancestor of langurs and baboons?
4. Did you detect any other convergent amino acid changes between any other pair of lineages? What does this suggest about the convergent changes you observed between cattle and langurs?

[a]C.-B. Stewart et al. 1987. *Nature* 330: 401–404.

	Amino acid position										
Species	**2**	**14**	**17**	**21**	**50**	**63**	**75**	**87**	**117**	**118**	**130**
Langur	I	K	L	K	E	Y	D	N	Q	N	V
Baboon	I	R	L	R	Q	Y	N	D	Q	N	V
Human	V	R	M	R	R	Y	N	D	Q	N	V
Rat	T	R	M	Y	Q	Y	N	D	K	N	V
Cattle	V	K	L	K	E	W	D	N	R	D	L
Horse	V	A	M	G	G	W	N	E	K	D	L
Ancestral state	V	R	M	R	Q	W	N	D	K	N	V

A similar **Analyze the Data** exercise may be assigned in **Achieve**.

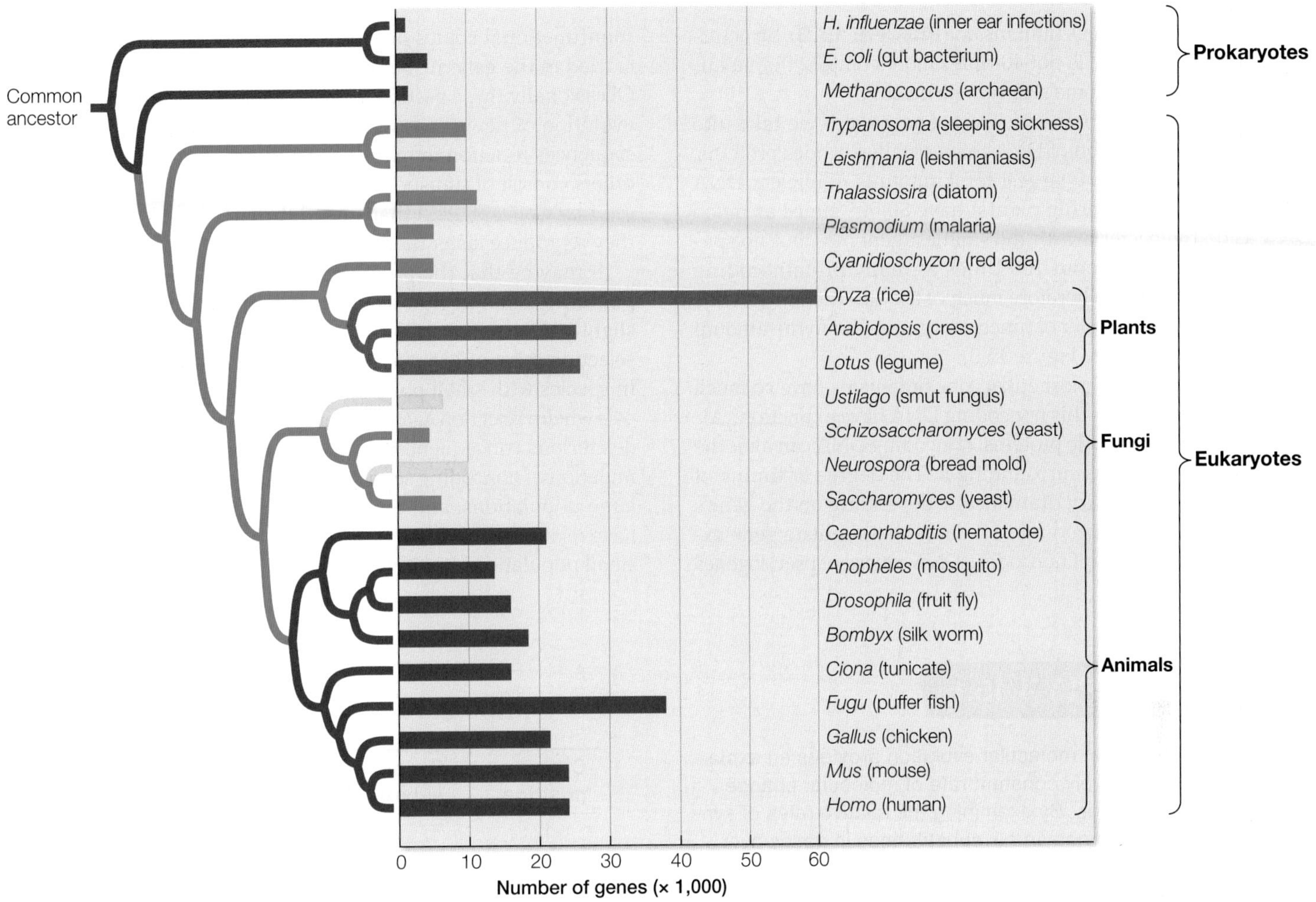

Common ancestor
H. influenzae (inner ear infections)
E. coli (gut bacterium)
Methanococcus (archaean)
Trypanosoma (sleeping sickness)
Leishmania (leishmaniasis)
Thalassiosira (diatom)
Plasmodium (malaria)
Cyanidioschyzon (red alga)
Oryza (rice)
Arabidopsis (cress)
Lotus (legume)
Ustilago (smut fungus)
Schizosaccharomyces (yeast)
Neurospora (bread mold)
Saccharomyces (yeast)
Caenorhabditis (nematode)
Anopheles (mosquito)
Drosophila (fruit fly)
Bombyx (silk worm)
Ciona (tunicate)
Fugu (puffer fish)
Gallus (chicken)
Mus (mouse)
Homo (human)
Prokaryotes
Plants
Fungi
Animals
Eukaryotes
0
10
20
30
40
50
60
Number of genes (× 1,000)

FIGURE 15.4 Evolution of Gene Number This figure shows the number of genes from a sample of organisms whose genomes have been fully sequenced, arranged by their evolutionary relationships. Bacteria and archaea (purple branches) typically have fewer genes than most eukaryotes. Among eukaryotes, multicellular organisms with tissue organization (plants and animals; dark green and blue branches, respectively) have more genes than single-celled organisms (aqua branches) or multicellular organisms that lack pronounced tissue organization (yellow branches). (Data from T. R. Gregory et al. 2007. *Nucleic Acids Res* 35 [Database issue]: D332–D338.)

lysozymes the ability to lyse the bacteria that ferment plant material in the foregut.

Genome size and organization also evolve

We know that genome size varies tremendously among organisms. Across broad taxonomic categories, there is some correlation between genome size and organismal complexity. The genome of the tiny bacterium *Mycoplasma genitalium* has only 470 genes. *Homo sapiens*, by contrast, has about 21,000 protein-coding genes. **FIGURE 15.4** shows the number of genes from a sample of organisms whose genomes have been fully sequenced, arranged by their evolutionary relationships. As this figure reveals, however, a larger genome does not always indicate greater complexity (compare rice with the other plants, for example). It is not surprising that more complex genetic instructions are needed for building and maintaining a large multicellular organism than a small single-celled bacterium. What is surprising is that some organisms, such as lungfishes, some salamanders, and lilies, have about

Percent of genome encoding functional genes
100
80
60
40
20
0
E. coli
Yeast
Drosophila
Arabidopsis
C. elegans
Human
Salamander
Lily
Lungfish
0.001
0.01
0.1
1
10
100
1,000
Genome size (× 10^9 base pairs)

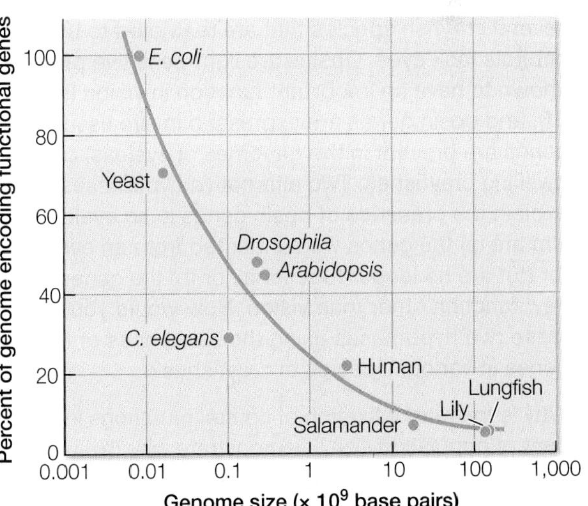

FIGURE 15.5 A Large Proportion of DNA Is Noncoding Most of the DNA of bacteria and yeasts encodes RNAs or proteins, whereas a large percentage of the DNA of multicellular species is noncoding. (Data from T. R. Gregory et al. 2007. *Nucleic Acids Res* 35 [Database issue]: D332–D338.)

40 times as much DNA as humans do (**FIGURE 15.5**). Structurally, a lungfish or a lily is not 40 times more complex than a human. So why does genome size vary so much?

Differences in genome size are not so great if we take into account only the portion of DNA that actually encodes proteins. The organisms with the largest total amounts of nuclear DNA (some ferns and flowering plants) have 80,000 times as much DNA as do the bacteria with the smallest genomes, but no species has more than about 100 times as many protein-coding genes as a bacterium. Therefore much of the variation in genome size lies not in the number of functional genes, but in the amount of noncoding DNA (see Figure 15.5).

Why do the cells of most eukaryotic organisms have so much noncoding DNA? Does this noncoding DNA have a function? Although it does not encode proteins, noncoding DNA can alter the expression of the genes surrounding it. The degree or timing of gene expression can vary dramatically depending on the gene's location relative to noncoding sequences that regulate gene expression. Other regions of noncoding DNA consist of pseudogenes (nonfunctional copies of former genes). Pseudogenes are often carried in the genome because the cost of doing so is very small. Occasionally these pseudogenes become the raw material for the evolution of new genes with novel functions. Other noncoding sequences function in maintaining chromosomal structure. Still others consist of transposable elements (also called transposons; see Key Concept 12.2) that spread through populations because they reproduce faster than the host genome.

It may be that the proportion of noncoding DNA is related primarily to population size. Noncoding sequences that are only slightly deleterious to the organism are likely to be purged by selection most efficiently in species with large population sizes. In species with small populations, the effects of genetic drift can overwhelm selection against noncoding sequences that have small deleterious consequences. Therefore selection against the accumulation of noncoding sequences is most effective in species with large populations, and thus species such as bacteria and yeasts have relatively little noncoding DNA compared with species with small populations (see Figure 15.5).

REVIEW & APPLY | 15.1

R The neutral theory of molecular evolution provides an explanation for the relatively constant rate of molecular change seen in many species. By examining the relative rates of synonymous and nonsynonymous substitutions in genes over time, biologists can distinguish the evolutionary mechanisms acting on individual genes. The noncoding portion of eukaryotic genomes is more variable in size across species than is the coding portion.

A

1. Over evolutionary history, many groups of organisms that inhabit caves have lost the organs of sight. For example, although surface-dwelling crayfishes have functional eyes, several crayfish species that are restricted to underground habitats lack eyes. Opsins are light-sensitive proteins known to have an important function in vision (see Chapter 31), and opsin genes are expressed in eye tissues. Opsin genes are present in the genomes of eyeless, cave-dwelling crayfishes. Two alternative hypotheses that might explain the presence of opsin genes in an eyeless organism are (a) the genes were inherited from an eyed ancestor but are no longer functional; or (b) the genes have a new function other than vision. How would you investigate these two hypotheses using the sequences of the opsin genes in various species of crayfishes?
2. Why is the rate of fixation of neutral mutations independent of population size? Demonstrate why this is true using mathematical expressions.
3. An investigator compared many gene sequences encoding surface proteins from influenza viruses sampled over time and collected the data shown in the table.

Codon position	Synonymous substitutions	Nonsynonymous substitutions
12	0	7
15	1	9
61	0	12
80	7	0
137	12	1
156	24	2
165	3	4
226	38	3

   a. Which positions encode amino acids that are likely to have changed as a result of positive selection? Explain your answer.

   b. Which positions encode amino acids that are likely to have changed as a result of purifying selection? Explain your answer.

   (*Hint*: To calculate rates of each substitution type, consider the number of synonymous and nonsynonymous substitutions present *relative to the number of possible substitutions of each type*. There are approximately three times as many possible nonsynonymous substitutions as there are synonymous substitutions.)

4. Suggest and contrast two hypotheses for the wide diversity of genome sizes among different organisms.

Most of our discussion so far has centered on changes in existing genes and phenotypes. Next we'll consider how new genes with novel functions arise in populations in the first place.

15.2 Rearrangements of Genomes Can Result in New Features

LEARNING OBJECTIVES

By the end of this key concept you should be able to:

15.2.1 **Compare and contrast the evolutionary advantages and disadvantages of sexual reproduction.**

15.2.2 **Describe how genes can move among different lineages via lateral gene transfer.**

15.2.3 **Using a phylogenetic tree of a gene family, identify gene duplications across the history of a group of species.**

15.2.4 **Describe how a duplicated gene provides opportunities for the evolution of new functions.**

Several evolutionary processes can result in the acquisition of major new characteristics in populations. Each of these processes results in larger and more rapid evolutionary changes than do single point mutations.

Sexual recombination amplifies the number of possible genotypes

In asexually reproducing organisms, each new individual is genetically identical to its parent unless there has been a mutation. When organisms reproduce sexually, however, offspring differ from their parents because of crossing over and independent assortment of chromosomes during meiosis, as well as the combination of genetic material from two different gametes, as described in Key Concept 8.2. Sexual recombination generates an endless variety of genotype combinations that increase the evolutionary potential of populations—a long-term advantage of sex. Although some species may reproduce asexually most of the time, most asexual species have some means of achieving genetic recombination.

The evolution of meiosis and sexual recombination was a crucial event in the history of life. Exactly how these processes arose is puzzling, however, because in the short term, sex has at least three striking disadvantages:

1. Recombination breaks up adaptive combinations of genes.
2. Sex reduces the rate at which females pass genes on to their offspring.
3. Dividing offspring into separate genders greatly reduces the overall reproductive rate.

To see why this last disadvantage exists, consider an asexual female that produces the same number of offspring as a sexual female. Let's assume that both females produce two offspring, but that 50 percent of the sexual female's offspring will be males (and thus contribute only sperm). In this next (F_1) generation, both asexual females will produce two more offspring each, but there is only one sexual F_1 female to produce offspring. Thus the effective reproductive rate of the asexual lineage is twice that of the sexual lineage.

The evolutionary problem is to identify the advantages of sex that can overcome short-term disadvantages. Several hypotheses have been proposed to explain the existence of sex, none of which is mutually exclusive. One hypothesis is that sexual recombination facilitates repair of damaged DNA, because breaks and other errors in DNA on one chromosome can be repaired by copying the intact sequence from the homologous chromosome.

Another advantage of sexual reproduction is that it permits the elimination of deleterious mutations. As explained in Key Concept 9.3, DNA replication is not perfect. Errors are introduced in every generation, and most of these errors result in lower fitness. Asexual organisms have no mechanism to eliminate deleterious mutations. Hermann J. Muller noted that the accumulation of deleterious mutations in a nonrecombining genome is like a genetic ratchet. The mutations accumulate—"ratchet up"—at each replication. A mutation occurs and is passed on when the genome replicates, then two new mutations occur in the next replication, so three mutations are passed on, and so on. Over time, the least-mutated class of individuals is lost from the population as new mutations occur. Deleterious mutations cannot be eliminated except by the death of the lineage or a rare back mutation. The accumulation of deleterious mutations in lineages that lack genetic recombination is known as **Muller's ratchet**. In sexual species, by contrast, genetic recombination produces some individuals with more of these deleterious mutations and some with fewer. The individuals with fewer deleterious mutations are more likely to survive. Thus sexual reproduction allows natural selection to eliminate particular deleterious mutations from the population over time.

Still another advantage of sex is the great variety of genetic combinations it creates in each generation. Sexual recombination does not directly influence the frequencies of alleles; rather, *it generates new combinations of alleles on which natural selection can act.* It expands variation in a character influenced by alleles at many loci by creating new genotypes. For example, genetic variation can be a defense against pathogens and parasites. Most pathogens and parasites have much shorter life cycles than their hosts and can rapidly evolve counteradaptations to host defenses. Sexual recombination can give the host's defenses a chance to keep up.

Lateral gene transfer can result in the gain of new functions

The tree of life is usually visualized as a branching diagram, with each lineage diverging into two (or more) lineages over time, from one common ancestor to the millions of species that are alive today. Ancestral lineages divide into descendant lineages, resulting in speciation events that are diagrammed on the tree of life. Species' genomes are altered as well by processes that result in **lateral gene transfer**—the horizontal movement of individual genes, organelles, or fragments of genomes from one lineage to another. Some species may pick up fragments of DNA directly from the environment. A virus may pick up some genes from one host and transfer them to a new host when the virus becomes integrated into the new host's genome. Hybridization between species also results in the lateral transfer of large numbers of genes.

Lateral gene transfer can be highly advantageous to the species that incorporates novel genes from a distant relative. Genes that confer antibiotic resistance, for example, are commonly transferred among different species of bacteria. Lateral gene transfer is another way, in addition to mutation and recombination, that species can increase their genetic variation.

The degree to which lateral gene transfer events occur in various parts of the tree of life is a matter of considerable current investigation and debate. Lateral gene transfer appears to be relatively uncommon among most eukaryotic lineages, although the two major endosymbioses that gave rise to mitochondria and chloroplasts involved lateral transfers of entire bacterial genomes to the eukaryotic lineage. Some groups of eukaryotes, most notably some plants, are subject to relatively high levels of hybridization among closely related species. Hybridization leads to the exchange of many genes among recently separated lineages of plants. The greatest degree of lateral transfer, however, occurs among bacteria. Many genes have been transferred repeatedly among bacteria, to the point that relationships and boundaries among species of bacteria are sometimes hard to decipher.

Many new functions arise following gene duplication

Gene duplication is yet another way that genomes can acquire new functions. When a gene is duplicated, one copy of that gene is potentially freed from having to perform its original function. The identical copies of a duplicated gene can have any one of four different fates:

1. Both copies of the gene may retain their original function (which can result in a change in the amount of gene product that is produced by the organism).
2. Both copies of the gene may retain the ability to produce the original gene product, but the expression of the genes may diverge in different tissues or at different times in development.
3. One copy of the gene may be incapacitated by the accumulation of deleterious mutations and become a functionless pseudogene.
4. One copy of the gene may retain its original function while the second copy changes and evolves a new function.

How often do gene duplications arise, and which of these four outcomes is most likely? Investigators have found that rates of gene duplication are fast enough for a yeast or *Drosophila* population to acquire several hundred duplicate genes over the course of a million years. They have also found that most of the duplicated genes that are still present in these organisms are very young. Many duplicated genes are lost from a genome within 10 million years—an eyeblink on an evolutionary time scale.

Many gene duplications affect only one or a few genes at a time, but in some cases entire genomes may be duplicated. Massive opportunities for new gene functions to evolve occur when all the genes in a genome are duplicated. Vertebrate evolution offers a case in point. The genomes of the jawed vertebrates have *four* diploid sets of many major genes, which leads biologists to conclude that two genome-wide duplication events occurred in the ancestor of these species. These duplications allowed considerable specialization of individual vertebrate genes, many of which are now highly tissue-specific in their expression.

Several successive rounds of duplication and sequence evolution may result in a **gene family**, a group of homologous genes with related functions, often arrayed in tandem along a chromosome. An example of a group of genes related by gene duplication is the globin gene family (**FIGURE 15.6**). Comparisons of the amino acid sequences among globins strongly suggest that this family of proteins arose via gene duplications.

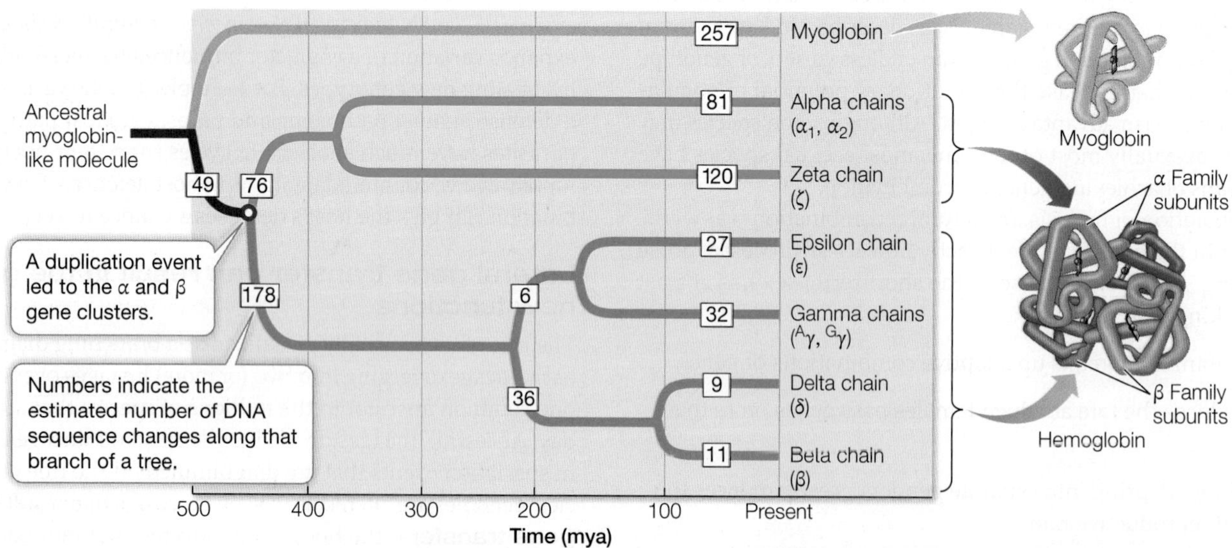

FIGURE 15.6 A Globin Family Gene Tree This gene tree suggests that the α-globin (blue) and β-globin (green) gene clusters diverged about 450 million years ago (mya) (open circle), soon after the origin of the vertebrates.

Activity 15.1 Gene Tree Construction

Hemoglobin is a tetramer (four-subunit molecule) consisting of two α-globin and two β-globin polypeptide chains. Hemoglobin carries oxygen in the blood. Myoglobin, a monomer, is the primary O_2 storage protein in muscle. Myoglobin's affinity for O_2 is much higher than that of hemoglobin, but hemoglobin has evolved to be more diversified in its role. Hemoglobin binds O_2 in the lungs or gills, where the O_2 concentration is relatively high, transports it to deep body tissues, where the O_2 concentration is low, and releases it in those tissues. With its more complex tetrameric structure, hemoglobin is able to carry four molecules of O_2 in the blood. Hemoglobin and myoglobin are estimated to have arisen through gene duplication about 450 million years ago.

REVIEW & APPLY | 15.2

R Sexual recombination produces new combinations of genes and facilitates the elimination of deleterious mutations from populations. Lateral gene transfer can result in the transfer of genetic functions between distantly related species. Gene duplication can lead to the evolution of new functions.

A

1. Contrast three disadvantages and three advantages of sexual reproduction compared with asexual reproduction. Why do you think that most species have some means of achieving genetic recombination among individuals?
2. What are some of the potential advantages of lateral gene transfer to the organisms that gain new genes by this mechanism?
3. Consider the following tree of a gene family sampled from the complete genomes of humans, chimpanzees, and gorillas. All members of the gene family are included in the tree. How many gene duplications and losses are likely to have occurred in the history of this gene family? Diagram their locations on the tree.

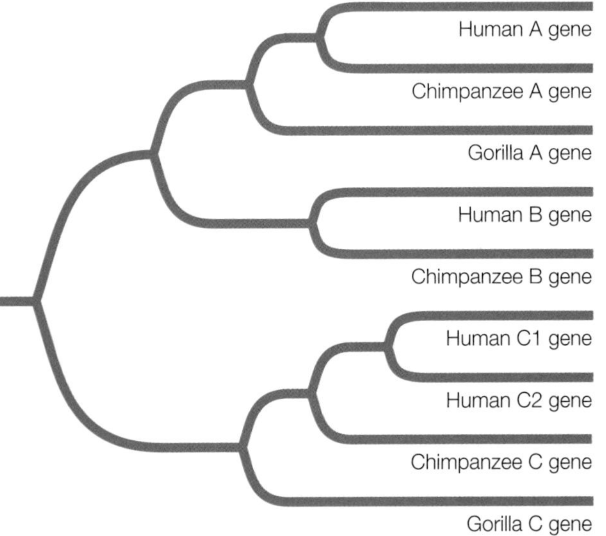

4. Why is gene duplication considered to be important for long-term evolutionary change?

We have discussed several ways that gene rearrangements and recombination can affect evolution. Next we'll consider how changes in gene expression are important to the evolution of body form in development.

15.3 | Changes in Gene Expression Often Shape Evolution

LEARNING OBJECTIVES

By the end of this key concept you should be able to:

15.3.1 **Describe factors that determine cell fate during development, using body segmentation in animals as an example.**

15.3.2 **Defend the claim "Genes controlling development are highly conserved across animals," using a specific example.**

15.3.3 **Provide examples of morphological differences among species that result from changes in the amount, timing, and location of gene expression.**

15.3.4 **Summarize how a mutation in a developmental gene led to a reduction in leg number in insects compared with other arthropods.**

Although a genome contains all of the genes needed to form a complete organism, these genes need to be expressed at specific times and in specific tissues. Given that every cell in the body contains the same genetic information, how are the right genes turned on to form a head at the anterior end of an animal, or a tail at the posterior end? Why don't heads or tails form elsewhere in the developing embryo? How do insects develop antennae on their head, but legs on their thorax? As biologists began to study and answer these questions, they realized that changes in gene expression might account for much of the evolution of diverse body forms we observe across living organisms. Thus was born the science of evo-devo, or evolutionary developmental biology.

A cascade of transcription factors establishes body segmentation in animals

Before we examine how changes in gene expression can account for evolution of body forms, we need to consider how body axis orientation and segmentation are controlled during development. This has been especially well studied in the fruit fly *Drosophila melanogaster*, but many of the genes that control developmental processes are conserved across a diversity of organisms, including ourselves.

The insect body is made up of segments that differ from one another. The adult fly has an anterior head (composed of several fused segments), three different thoracic segments, and eight abdominal segments at the posterior end. Each segment gives rise to different body parts: for example, antennae and eyes develop from head segments, wings from the thorax, and so on.

The life cycle of *Drosophila* from fertilized egg to adult takes about 2 weeks. The egg hatches into a larva, which then forms a pupa, which finally is transformed into the adult fly. By the time a larva appears—about 24 hours after fertilization—there are recognizable segments. Although the thoracic and abdominal segments all look similar, the fates of the cells to become different adult segments are already determined.

As in other organisms, fertilization in *Drosophila* leads to a rapid series of mitoses. However, the first 12 nuclear divisions do not result in the formation of new cells. A *multinucleate* embryo forms instead of a *multicellular* embryo. With no cell membranes to cross, gene inducers known as **morphogens** can diffuse easily within the embryo. Many of these morphogens affect transcription in the cell nuclei.

Genetic studies of *Drosophila* revealed a sequential pattern (cascade) of gene expression that results in the determination of each segment within 24 hours after fertilization. Three classes of genes are involved:

1. **Maternal effect genes** set up the major axes (anterior–posterior and dorsal–ventral) of the embryo.
2. **Segmentation genes** determine the boundaries and polarity of each of the segments.
3. **Hox genes** determine what organ will be made at a given location.

MATERNAL EFFECT GENES Like the eggs and early embryos of many other organisms, *Drosophila* eggs and larvae are characterized by unevenly distributed mRNAs and proteins, which are the products of specific maternal effect genes. These genes are transcribed in the cells of the mother's ovary, and the mRNAs are passed to the egg. These mRNAs and proteins establish the anterior–posterior axis and the dorsal–ventral axis of the developing embryo. Where some of these proteins are in highest concentrations in the embryo, transcription of additional genes is stimulated. For example, products of maternal effect genes stimulate the *hunchback* gene to be transcribed in the anterior end of the embryo, and the resulting gradient of Hunchback protein (itself a transcription factor) establishes the head region. A gradient of other proteins concentrated in the posterior region of the embryo establishes the anterior–posterior axis of the developing embryo.

The events that involve maternal effect genes begin before fertilization and continue after it, during the multinucleate stage, which lasts a few hours. At this stage the embryo looks like a bunch of indistinguishable nuclei under the light microscope. But the fates of the individual nuclei and the cells they will occupy have already begun to be determined. After the anterior and posterior ends have been established, the next step in pattern formation in fruit flies is the determination of segment number and locations.

SEGMENTATION GENES The number, boundaries, and polarity of the *Drosophila* larval segments are determined by proteins encoded by the segmentation genes. These genes are expressed when there are about 6,000 nuclei in the embryo (about 3 hours after fertilization). Three classes of segmentation genes act one after the other to regulate finer and finer details of the segmentation pattern:

1. **Gap genes** organize broad areas along the anterior–posterior axis. Mutations in gap genes result in gaps in the body plan—the omission of several consecutive larval segments.
2. **Pair rule genes** divide the embryo into units of two segments each. Mutations in pair rule genes result in embryos missing every other segment.
3. **Segment polarity genes** determine the boundaries and anterior–posterior organization of the individual segments. Mutations in segment polarity genes can result in segments in which posterior structures are replaced by reversed (mirror-image) anterior structures.

The expression of the genes is sequential (**FIGURE 15.7**). The products of the gap genes active pair rule genes, and the pair rule gene products activate segment polarity genes. By the end of this part of the cascade, nuclei throughout the embryo "know" which segment they will be part of in the adult fly. The next set of genes in the cascade determines the form and function of each segment.

Animation 15.1 Pattern Formation in the *Drosophila* Embryo

Media Clip 15.1 Spectacular Fly Development in 3D

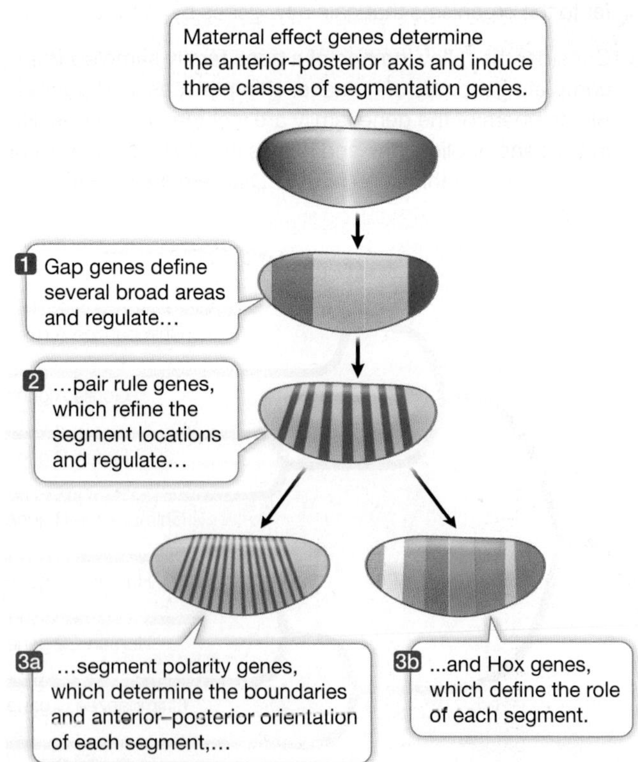

FIGURE 15.7 A Gene Cascade Controls Pattern Formation in the *Drosophila* Embryo Maternal effect genes induce gap, pair rule, and segment polarity genes. Hox genes then define the role of each segment. By the end of this cascade, a group of nuclei at the anterior of the embryo, for example, is determined to become the first head segment in the adult fly.

HOX GENES Hox genes encode a family of transcription factors that are expressed in different combinations along the length of the embryo, and help determine cell fates within each segment. Expression of certain Hox genes leads to the development of antennae in the head segment, whereas other Hox genes are expressed in the thorax to make wings, and so on. The *Drosophila* Hox genes occur in two clusters in the genome, in the same order as the segments whose function they determine (**FIGURE 15.8**). Hox genes are shared by all animals, and mutations in these genes have been linked to major changes in body structure among different animal groups.

How do we know that the Hox genes determine segment identity, and what structures are expressed in each segment? A clue comes from **homeotic mutations**—mutations that result in the replacement of one organ by another. For example, a mutation in the Hox gene *Antennapedia* causes legs to grow on the head of *Drosophila* in place of antennae (**FIGURE 15.9**). When another Hox gene, *bithorax,* is mutated, an extra pair of wings grows in a thoracic segment where wings do not normally occur. So the normal (wild-type) functions of the Hox genes must be to "tell" a segment what organ to form. Hox genes encode transcription factors and have a conserved 180-base-pair sequence called the **homeobox** (from which the genes get their name). The homeobox encodes a 60-amino acid sequence called the **homeodomain**. The homeodomain recognizes and binds to a specific DNA sequence in the promoters of its target genes, "turning on" the formation of specific structures.

LINK As discussed in **Key Concept 11.3** cell differentiation often is mediated by changes in gene expression resulting from the activation or deactivation of various transcription factors.

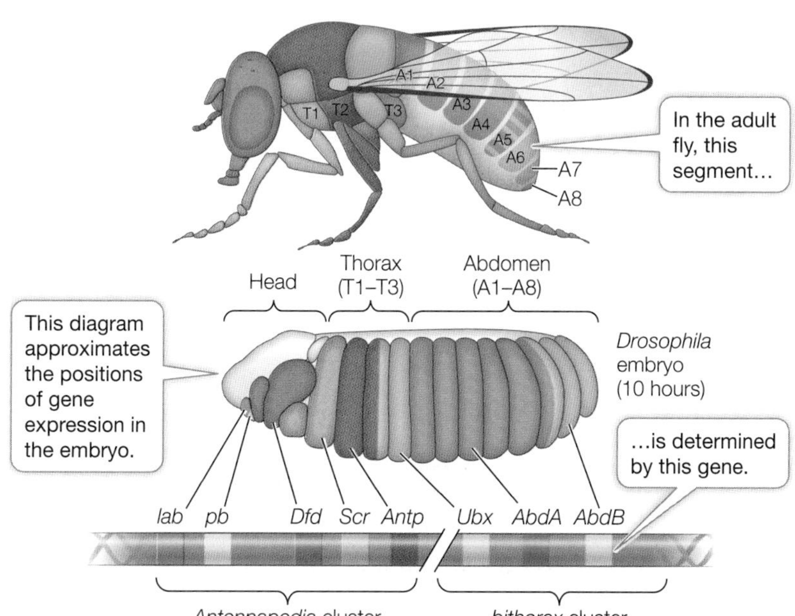

FIGURE 15.8 Hox Genes in *Drosophila* Determine Segment Identity Two clusters of Hox genes on chromosome 3 (bottom) determine segment function in the adult fly (top). These genes are expressed in the embryo (center) long before the structures of the segments actually appear. (Top after M. Peifer et al. 1987. *Genes Dev* 1: 891–898; bottom after S. D. Hueber et al. 2010. *PLoS ONE* 5: e10820. doi:10.1371/journal.pone.0010820.)

(A)

(B)

FIGURE 15.9 A Homeotic Mutation Replaces One Structure with Another **(A)** Normal development of antennae on the head of a fruit fly. **(B)** Shows the result of a homeotic mutation of the Hox gene *Antennapedia*, which controls formation of legs during development. The mutation causes legs, rather than antennae, to grow on the head.

A common "toolkit" of genetic mechanisms controls gene expression in development

About a dozen major kinds of eyes are found among animals, including the camera-like eyes of humans and the compound eyes of insects. Although the eyes of insects and vertebrates evolved independently, a remarkable discovery showed that common developmental pathways are involved in the embryonic formation of eyes in both groups.

More than a century ago, a mutation was described in *Drosophila* called *eyeless*, with a phenotype of—you guessed it—no eyes. This mutation remained a laboratory curiosity for many decades until its molecular nature was elucidated in the 1990s. Biologists isolated the protein product of the wild-type version of the *eyeless* gene, which they determined was a transcription factor that controls the expression of genes responsible for eye development. How did they know this? By making recombinant DNA constructs that allowed the wild-type *eyeless* gene to be expressed in different embryonic tissues of transgenic flies, the biologists were able to produce flies with extra eyes on various body parts such as the legs, the antennae, and under the wings.

The big surprise came when the biologists sequenced the *eyeless* gene and used computers to query a database of genes with known sequences. The *eyeless* gene sequence was similar to that of the *Pax6* gene in mice; *Pax6*, when mutated, leads to the development of abnormally small eyes. Could the extremely different eyes of flies and mice be variations on a common developmental theme? To test for functional similarity between the insect and mammalian genes, the biologists repeated their experiments on flies, but using the mouse *Pax6* gene instead of the fly *eyeless* gene. Once again, eyes developed at various sites on the transgenic flies. Thus a gene whose expression normally leads to the development of a mammalian eye now led to the development of the very different insect eye.

Fruit flies and mice last shared a common ancestor more than a half-billion years ago. Yet the *eyeless* and *Pax6* genes contain sequences that are highly conserved, not only in these two species but in others as well. Biologists call such genes **homologous**, meaning that they evolved from a gene present in a common ancestor.

In recent years a large number of homologous genes (the regulatory "toolkit") have been shown to control development in distantly related species. For example, the fruit fly homeotic genes such as *Antennapedia* and *bithorax* are similar to mouse (and human) genes that play similar developmental roles. This discovery indicates that the positional information controlled by these genes has been conserved, even as the structures formed at each position have changed. Remarkably, these genes are arranged along a chromosome in both fruit fly and mouse in the same order as they are expressed along the anterior–posterior axis of the embryos (**FIGURE 15.10**). These and other examples have led biologists to the idea that certain developmental mechanisms are controlled by specific sequences comprising a **genetic toolkit**, the contents of which have been changed and reshuffled over the course of evolutionary time to produce the amazing diversity of plants, animals, and other organisms that we see today.

Developmental modules based on a common set of genetic instructions can evolve separately within a species because **genetic switches** control how the toolkit is used. These switches turn genes on or off, or regulate the degree and timing of gene expression. Multiple switches control each gene, creating different expression patterns in different locations. In this way, elements of the genetic toolkit can be involved in multiple developmental processes and still allow individual modules to develop and evolve independently.

Animation 15.2 Modularity

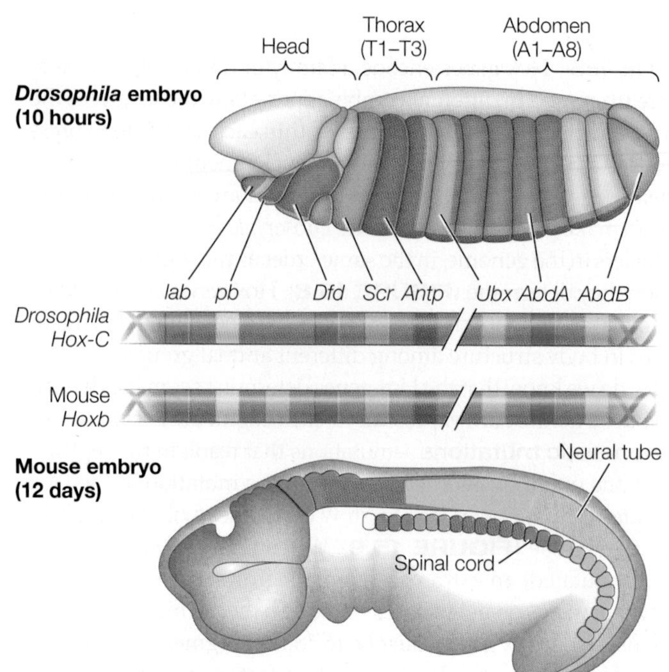

FIGURE 15.10 Regulatory Genes Show Similar Expression Patterns Homologous genes encoding similar transcription factors are expressed in similar patterns along the anterior–posterior axes of both insects and vertebrates. The mouse (and human) Hox genes are present in multiple copies; this prevents a single mutation from having drastic effects. (Top after S. D. Hueber et al. 2010 *PLoS ONE* 5: e10820. doi:10.1371/journal.pone.0010820; bottom after S. B. Carroll. 1995. *Nature* 376: 479–485.)

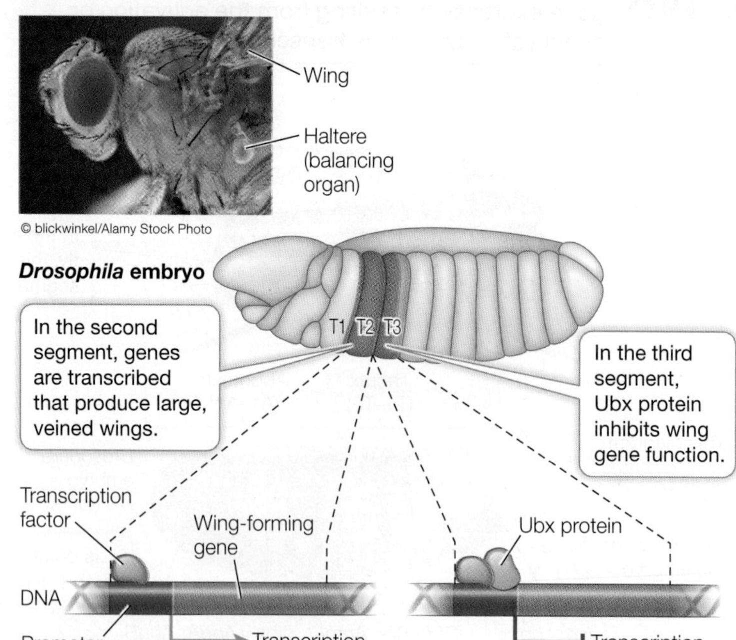

FIGURE 15.11 Segments Differentiate under Control of Genetic Switches The binding of a single protein, Ultrabithorax (Ubx), determines whether a thoracic segment produces full wings or halteres. (After S. D. Hueber et al. 2010. *PLoS ONE* 5: e10820. doi:10.1371/journal.pone.0010820.)

What would happen if a butterfly embyro were genetically engineered so that Ubx was expressed in all its thoracic segments?

During evolution, changes in the functions of genetic switches have led to changes in the forms and functions of organisms. To illustrate this, let's look at the development of wings in *Drosophila* and other insects. *Drosophila* species are members of the insect group Diptera, which means "two wings"—that is, they have a single pair of wings, whereas most insects have two pairs of wings (i.e., four wings). The single pair of wings of dipterans develops on the second thoracic segment, where the Hox gene *Antennapedia* (*Antp*) is expressed. *Antp* is also expressed in the third thoracic segment, but in that segment a pair of balancing organs called halteres develops in dipterans. A critical difference between thoracic segments 2 and 3 is that another Hox gene, *Ultrabithorax* (*Ubx*), is expressed along with *Antp* in segment 3 (**FIGURE 15.11**). *Ubx* represses *Antp* function in dipterans. If *Ubx* is inactivated by mutation, a second pair of wings forms in thoracic segment 3, as is typical of many other insect groups. Thus some major morphological differences among groups of animals can result from relatively small changes in gene expression.

The amount, timing, and location of gene expression control many morphological features

The modularity of development means that the molecular pathways for developmental processes such as organ formation operate independently of one another. For instance, the *Antennapedia* mutant fly shown in Figure 15.9 grew legs where antennae should be, but the rest of the segments formed structures normally. Modularity allows the timing and location of a developmental process to change without changing the whole organism. Major changes can result from differences in the amount, timing, and location of gene expression.

HETEROMETRY A striking example of **heterometry** ("different measure") is seen in beak development in Darwin's finches, in which the beaks of some species are large and robust for cracking seeds and in other species are thin and long for probing for food (see Figure 16.8). When he visited the Galápagos archipelago, Charles Darwin wrote that "seeing this gradation and diversity of structure in one small intimately related group of birds, one might really fancy that from an original paucity of birds in this archipelago, one species had been taken and modified for different ends."

Darwin had no idea of the genetic basis for such modification. Now we do. Beak shapes are evident when birds hatch from their eggs, so they must be determined during development. The beak develops from tissues at the anterior of the embryo that will form the facial bones. Cell divisions in this embryonic tissue are controlled by signaling proteins, one of which is called bone morphogenetic protein 4 (BMP4); another is the protein calmodulin. If BMP4 is present early and in large amounts, the beak becomes broad and deep. If calmodulin is present early and in large amounts, the beak grows longer and thinner (**FIGURE 15.12**). Thus beak structure is affected by changes in protein production.

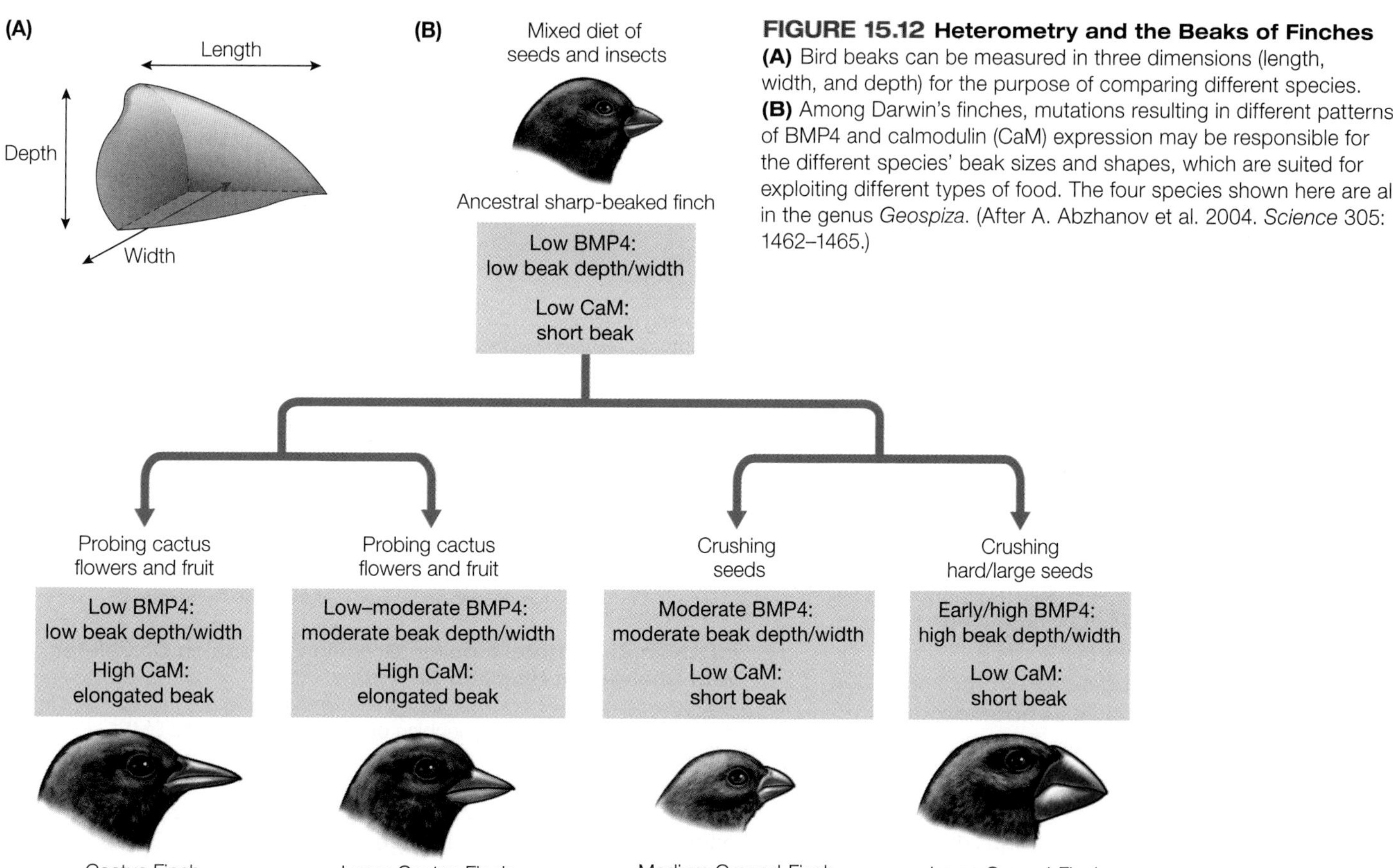

FIGURE 15.12 Heterometry and the Beaks of Finches **(A)** Bird beaks can be measured in three dimensions (length, width, and depth) for the purpose of comparing different species. **(B)** Among Darwin's finches, mutations resulting in different patterns of BMP4 and calmodulin (CaM) expression may be responsible for the different species' beak sizes and shapes, which are suited for exploiting different types of food. The four species shown here are all in the genus *Geospiza*. (After A. Abzhanov et al. 2004. *Science* 305: 1462–1465.)

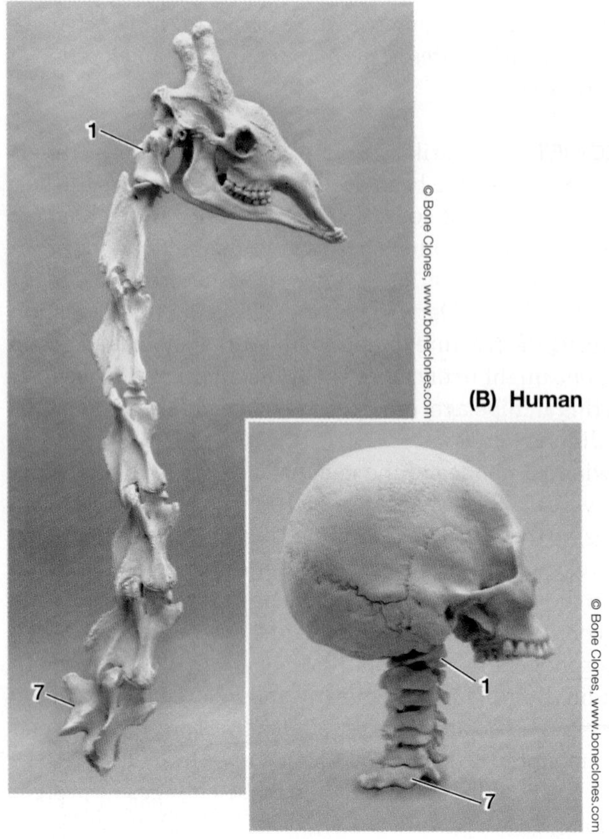

FIGURE 15.13 Heterochrony in the Development of a Longer Neck There are seven vertebrae in the neck of a giraffe **(A)** and human **(B)** (not to scale). But the vertebrae of the giraffe are much longer (25 cm versus 1.5 cm) because during development, growth continues for a longer period of time. This timing difference is called heterochrony.

HETEROCHRONY The evolution of the giraffe neck provides an example of **heterochrony** ("different time"). Giraffes, like most mammals, have seven cervical (neck) vertebrae. So giraffes did not get longer necks by adding more vertebrae. Instead, the cervical vertebrae of giraffes are much longer than those of other mammals (**FIGURE 15.13**).

Bone growth in mammals is the result of the proliferation of cartilage-producing cells called chondrocytes. Bone growth is stopped by a genetic signal that results in apoptosis, or cell death, of chondrocytes and calcification of the bone matrix. In giraffes this signaling process is delayed in the cervical vertebrae, so these vertebrae grow longer. Thus the evolution of longer necks resulted from changes in the timing of expression of the genes that control bone formation.

HETEROTOPY Spatial differences in the expression of a developmental gene are known as **heterotopy** ("different place"), exemplified by the different development of feet in ducks and chickens. The feet of all bird embryos have webs of skin that connect the toes. This webbing is retained in adult ducks (and other aquatic birds) but not in adult chickens (and other nonaquatic birds). The loss of webbing is controlled by the BMP4 signaling protein, which we noted is also involved in beak development (another example of use of a common genetic toolkit to produce different kinds of change).

BMP4 protein induces the cells that produce webbing to undergo apoptosis and thus eliminate the webbing between the toes. The hindlimbs of both duck and chicken embryos express the *BMP4* gene in the webbing between the toes. However, duck and chicken embryos differ in expression of the *Gremlin* gene, which encodes a protein that inhibits *BMP4* expression (**FIGURE 15.14**). In ducks, but not in chickens, *Gremlin* is expressed in the webbing cells and Gremlin protein inhibits *BMP4* expression. With no BMP4 protein to stimulate apoptosis, a webbed foot develops. If chick hindlimbs are experimentally exposed to Gremlin during development, the adult chicken will have ducklike webbed feet.

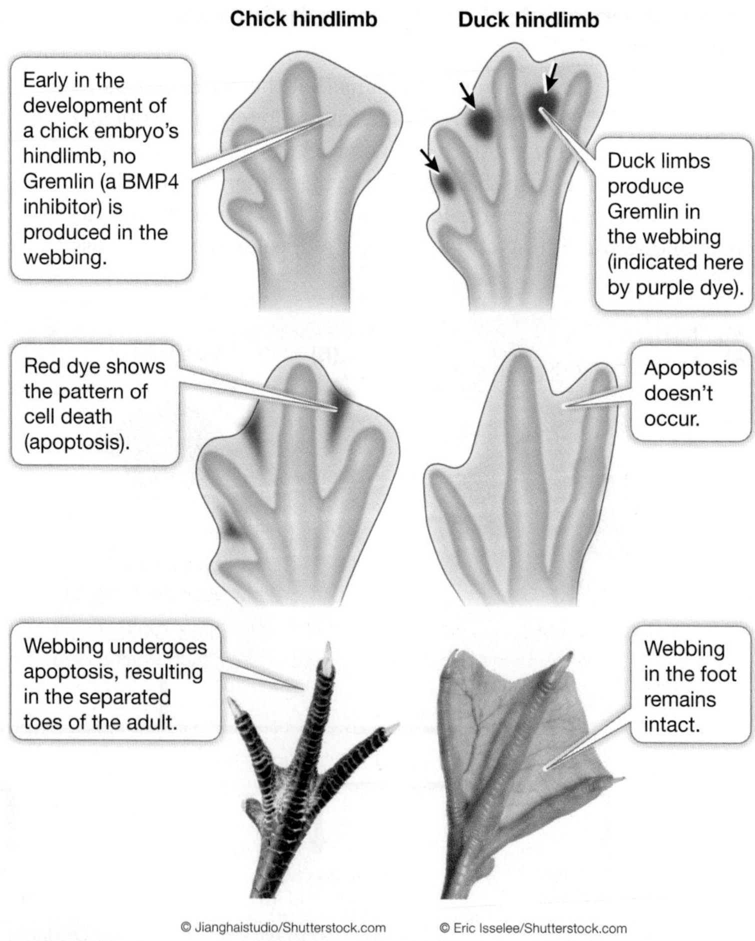

FIGURE 15.14 Heterotopy in Gremlin Expression Correlates with Changes in Hindlimb Structure The left column of photos shows foot development in a chicken; the right column shows foot development in a duck. Gremlin protein in the webbing of the duck foot inhibits BMP4 signaling, thus preventing the embryonic webbing from undergoing apoptosis.

What do you predict would happen if you blocked the expression of Gremlin in the feet of duck embryos?

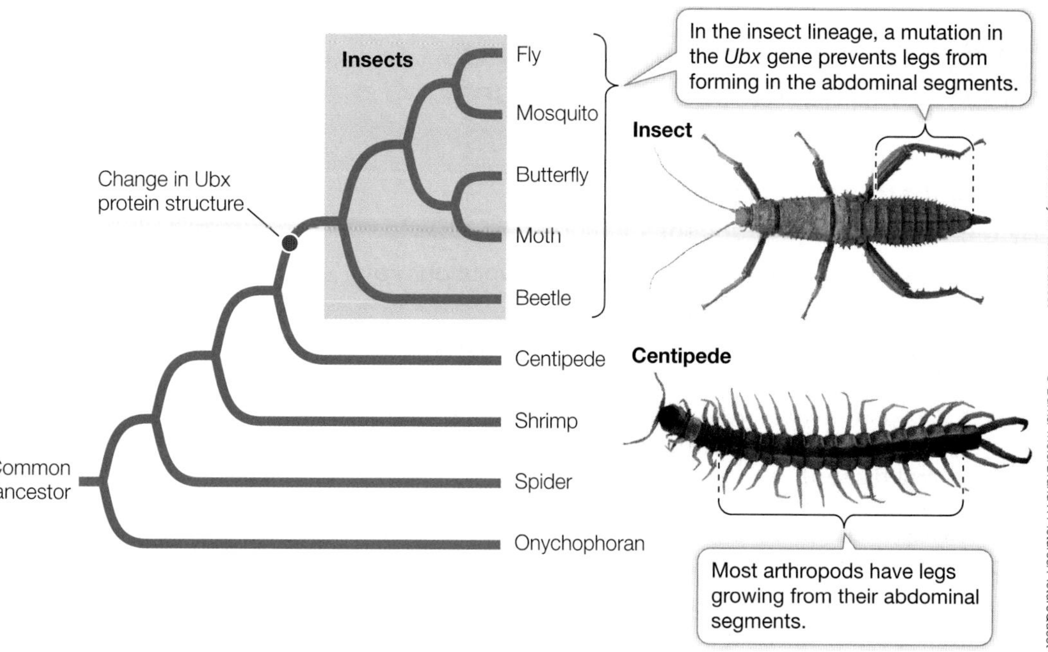

FIGURE 15.15 A Mutation in a Hox Gene Changed the Number of Legs in Insects In the insect lineage (blue box) of the arthropods, a change to the *Ubx* gene resulted in a protein that inhibits the *Dll* gene, which is required for legs to form. Because insects express this modified *Ubx* gene in their abdominal segments, no legs grow from these segments. Other arthropods, such as centipedes, produce an unmodified Ubx protein and do grow legs from their abdominal segments. (After R. Galant and S. Carroll. 2002. *Nature* 415: 910–913; M. Ronshaugen et al. 2002. *Nature* 415: 914–917.)

What do you predict the phenotype would be if you genetically engineered *Drosophila* to express the centipede version of *Ubx*?

Mutations in developmental genes can cause major morphological changes

Sometimes a major developmental change is caused by an alteration in the regulatory molecule itself rather than a change in where, when, or how much it is expressed. The gene that controls the number of legs in arthropods offers an excellent example. Arthropods all have head, thoracic, and abdominal regions with variable numbers of segments. Insects, such as *Drosophila*, have three pairs of legs on their three thoracic segments, whereas centipedes have many legs on both thoracic and abdominal segments. All arthropods express a gene called *Distal-less* (*Dll*) that controls segmental leg development. In insects, *Dll* expression is repressed in abdominal segments by the Hox gene *Ultrabithorax* (*Ubx*). *Ubx* is expressed in the abdominal segments of all arthropods, but it has different effects in different species. In centipedes, *Ubx* is co-expressed with *Dll* to promote the formation of legs. During the evolution of insects, a mutation in the *Ubx* gene sequence resulted in a modified Ubx protein that represses *Dll* expression in abdominal segments (**FIGURE 15.15**). A phylogenetic tree of arthropods shows that this change in *Ubx* occurred in the ancestor of insects, at the same time that abdominal legs were lost.

REVIEW & APPLY | 15.3

R A genetic toolkit consisting of highly conserved regulatory genes governs pattern formation in multicellular organisms. Genetic toolkit genes act independently among modules of a developing embryo. The expression of these developmental genes can differ among species in various ways, often resulting in major morphological differences. They can differ in amount of expression (heterometry), in the timing of expression (heterochrony), or in the location of expression (heterotopy). A mutation in a gene controlling a developmental pathway can result in major changes in morphology.

A

1. Using the story of the eye-determining genes *Pax6* and *eyeless*, support the claim that genes controlling development are highly conserved.
2. In a series of experiments on chick embryos, researchers applied different concentrations of BMP4 to the embryos' beak growth region. They measured the size of the beak cartilage at a later stage of development. Based on the data in the table below, what could you conclude about the role of BMP4 in beak growth?

Amount of BMP4	Cartilage diameter (mm)
None (control)	0.5
0.1 unit	0.7
0.3 unit	1.0
1.0 unit	1.8

3. How have diverse body forms in arthropods evolved by means of mutations in developmental genes?

Our understanding of molecular evolution has many practical applications across biology, and new ones are being developed every day. We'll now consider a few of these applications to medicine and industry.

15.4 | Molecular Evolution Has Many Practical Applications

LEARNING OBJECTIVES

By the end of this key concept you should be able to:

15.4.1 **Describe how studies of molecular evolution are used to understand protein function.**

15.4.2 **Explain how the methods of in vitro evolution correspond to natural evolutionary processes.**

15.4.3 **Evaluate evolutionary relationships among viral sequences to understand the transmission of a viral disease.**

Knowledge of gene evolution is used to study protein function

Earlier in this chapter we discussed the ways in which biologists can detect regions of genes that are under positive selection for change. What are the practical uses of this information? Consider the evolution of the family of genes encoding voltage-gated sodium channels. Sodium channels have many functions, including the control of muscle contractions as well as nerve impulses in the nervous system. Sodium channels can be blocked by various toxins, such as tetrodotoxin (TTX), a neurotoxin present in the tissues of some puffer fishes and several other animals. A human who eats those tissues of a puffer fish that contain TTX can become paralyzed and die because the toxin blocks sodium channels and prevents muscles and nerves from functioning.

LINK As described in detail in **Key Concept 31.2**, voltage-gated ion channels are stimulated to open or close by a change in the electric charge difference across a cell membrane, and thus let various ions into or out of cells.

But puffer fishes have sodium channels, too, so why doesn't the TTX cause paralysis in the puffer fishes themselves? The sodium channels of puffer fishes have evolved to become resistant to the toxin. Nucleotide substitutions in the puffer fish genome have resulted in changes in the proteins that make up the sodium channels, and those changes prevent TTX from binding to the sodium channel pore.

Several different substitutions that result in TTX resistance have evolved in the various duplicated sodium channel genes of the many species of puffer fishes. Many other changes that have nothing to do with the evolution of TTX resistance have occurred in these genes as well. Biologists who study the function of sodium channels can learn a great deal about how the channels work (and about neurological diseases that are caused by mutations in the sodium channel genes) by understanding which changes have been selected for TTX resistance. They do this by comparing the rates of synonymous and nonsynonymous substitutions across the genes in various lineages that have evolved TTX resistance. In a similar manner, molecular evolutionary principles are used to understand function and diversification of function in many other proteins.

In vitro evolution is used to produce new molecules

As biologists studied the relationships among selection, evolution, and function in macromolecules, they realized that molecular evolution could be employed in a controlled laboratory environment to produce new molecules with novel and useful functions. Thus were born the applications of **in vitro evolution**.

Living organisms produce thousands of compounds that humans have found useful. The search for such naturally occurring compounds, which can be used for pharmaceutical, agricultural, or industrial purposes, has been termed "bioprospecting." These compounds are the result of millions of years of molecular evolution across millions of species of living organisms. Biologists can imagine molecules that *could* have evolved but, lacking the right combination of selection pressures and opportunities, have not. For instance, we might want to have a molecule that binds a particular environmental contaminant so that the contaminant can be isolated and extracted from the environment. But if the contaminant is synthetic (i.e., not produced naturally), it is unlikely that any living organism will have evolved a molecule with the function we desire. This problem was the inspiration for the field of in vitro evolution.

The principles of in vitro evolution are based on the principles of molecular evolution that we have learned from the natural world. Consider the evolution of a new RNA molecule that was produced in the laboratory using the principles of mutation and selection. This molecule's intended function was to join two other RNA molecules. That is, the lab created a ribozyme—an RNA molecule capable of catalyzing a biochemical reaction. The process started with a large pool of random RNA sequences (10^{15} different sequences, each about 300 nucleotides long), which were then selected for any ligase activity (**FIGURE 15.16**). None were very effective ligases, but some were slightly better than others. The best of the ribozymes were selected and reverse-transcribed into cDNA (using the enzyme reverse transcriptase). The cDNA molecules were then amplified using the polymerase chain reaction (PCR; see Figure 9.13).

Key Concept 12.1 describes how complementary DNA (cDNA) is made by complementary base pairing to mRNA isolated from cells.

PCR amplification is not perfect, and it introduced many new mutations into the pool of sequences. These sequences were then transcribed back into RNA molecules using RNA polymerase, and the process was repeated. The ligase activity of the RNAs evolved quickly; after ten rounds of in vitro evolution, it had increased by about 7 million times. Similar techniques have since been used to create a wide variety of molecules with novel enzymatic and binding functions.

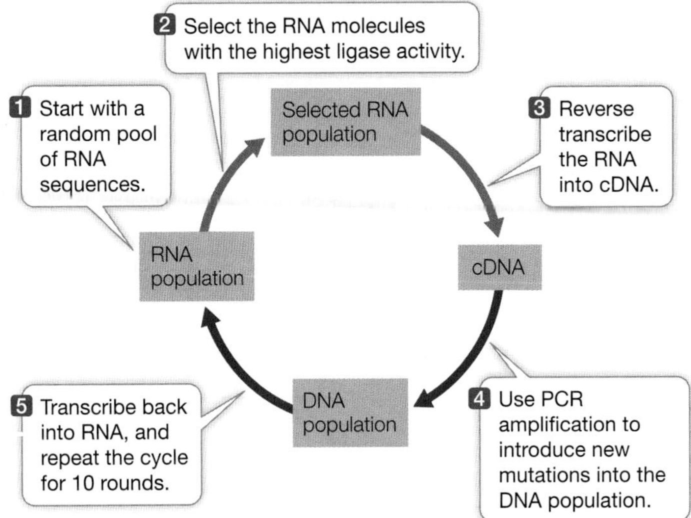

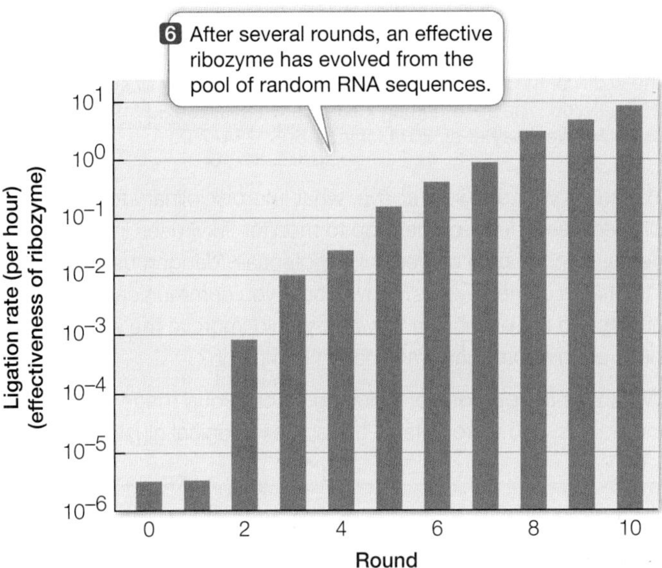

FIGURE 15.16 In Vitro Evolution Starting with a large pool of random RNA sequences, David Bartel and Jack Szostak of Massachusetts General Hospital produced a new ribozyme through rounds of mutation and selection for the ability to ligate (join) RNA sequences. (After D. P. Bartel and J. W. Szostak. 1993. *Science* 261: 1411–1418.)

Molecular evolution is used to study and combat diseases

Many of the most problematic human diseases are caused by living, evolving organisms that present a moving target for modern medicine. Historically, viral diseases such as influenza have been among the greatest factors in human mortality, and new viral diseases are constantly emerging in human populations, such as the coronavirus (SARS-CoV-2) that caused the COVID-19 pandemic. The control of these and many other human diseases depends on techniques that can track the evolution of pathogenic organisms over time.

Advances in transportation over the past century have allowed humans to move around the world with unprecedented speed and frequency. Unfortunately, this mobility has allowed pathogens to be transmitted among human populations, and between humans and other animals, at increasing rates. Cross-species transmission of viruses has led to the global emergence of many "new" diseases.

Studies of the genomes of many infectious agents have advanced our understanding and treatment of the diseases they cause. For example, rodent-borne hantaviruses were identified as the source of widespread respiratory illnesses, and the virus (and its host) that causes severe acute respiratory syndrome (SARS) was identified using evolutionary comparisons of genes. Studies of the origins, the timing of emergence, and the global diversity of many human pathogens depend on the principles of molecular evolution, as do the efforts to develop and use effective vaccines against these pathogens. For example, the techniques used to develop modern polio vaccines, as well as the methods used to track their effectiveness in human populations, rely on molecular evolutionary approaches.

In the future, understanding of molecular evolution will become even more critical to the identification of human (and other) diseases. Once biologists have collected data on the genomes of enough organisms, it will be possible to identify an infection by sequencing a portion of the infecting organism's genome and comparing this sequence with other sequences on an evolutionary tree. At present it is difficult to identify many common viral infections (those that cause "colds," for instance). As genomic databases and evolutionary trees increase, however, automated methods of sequencing and rapid phylogenetic comparison of the sequences will allow us to identify and treat a much wider array of human illnesses.

REVIEW & APPLY | 15.4

R Studies of molecular evolution have provided biologists with new tools for understanding the functions of macromolecules and how those functions can change over time. These tools can be used to develop synthetic molecules and to identify and combat human diseases.

A

1. Describe a specific example of how gene evolution can be used to study protein function.
2. List the key elements of in vitro evolution. How do these elements correspond to natural evolutionary processes?
3. How are principles of evolutionary biology used to identify emerging diseases?

Now that we have discussed how organisms and biological molecules evolve, we are ready to consider the process of speciation. Speciation is what leads to the branching events on the tree of life, and is the process that results in the millions of species that constitute biodiversity.

THINK LIKE A SCIENTIST

Science & Society

Interdisciplinary

Communicate & Collaborate

Why was the 1918–1919 influenza pandemic so severe?

The 1918–1919 influenza pandemic killed more than 50 million people worldwide—many more than the number of World War I-related combat deaths. Unlike flu epidemics before or since, the flu viruses that circulated in the 1918–1919 epidemic triggered an especially intense reaction in the human immune system.

Here you will address a critical issue facing society (pandemics) in addressing the question, What was different about the flu viruses circulating in the 1918–1919 epidemic that made them so lethal? You will then develop a plan to build interdisciplinary tools for monitoring flu epidemics, with the goal of identifying potentially dangerous strains. Finally, you will develop a strategy for communicating this information to the public, to improve the effectiveness of appropriate vaccination campaigns.

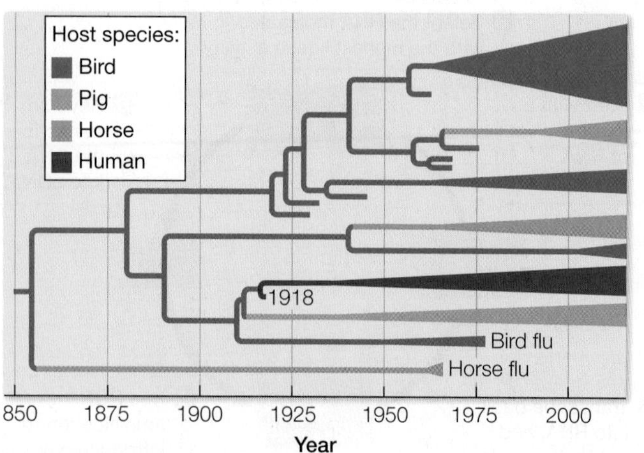

FIGURE 15.17 A Time-Calibrated Phylogeny Based on Genes Sequenced from Influenza Isolates Viruses represented by blue lineages were isolated from birds; green lineages from pigs; orange lineages from horses; and red lineages from humans. The triangles represent many closely related sequences sampled from the respective species in the time period indicated; the width of the triangle is proportional to the number of sequences analyzed. The 1918 human influenza virus is labeled "1918." (After M. Worobey et al. 2014. *Nature* 508: 254–257.)

1. All the influenza viruses shown in the phylogenetic tree (**FIGURE 15.17**) are thought to have arisen in one group of host animals. Which group of animals do you think is the original source of the virus, and why?
2. How many cross-species transmissions are shown on the tree? What hosts were involved (as source and recipient) of each transmission event?
3. What do you think is a likely explanation for the severity of the 1918–1919 influenza pandemic in humans?
4. Influenza was known in humans well before the 1918–1919 pandemic. Why do you think that no human influenza viruses are indicated on the tree before about 1918?
5. Based on your answers above, what interdisciplinary tools and procedures would you develop to monitor flu viruses in future years, with the goal of identifying potentially dangerous emerging strains of the flu virus? How would you communicate your findings to the public in a way that would improve the effectiveness of an appropriate vaccination campaign?
6. What was the purpose of including an outgroup made up of individuals who were outside the epidemiological cluster?

How have simple changes in gene duplication, gene sequences, and gene expression led to the repeated evolution of electric organs in fishes?

Answer: It might seem surprising that a complex structure such as an electric organ would evolve repeatedly in several different groups of fishes. Most electric organs are derived developmentally from muscle tissue. What changes are needed to convert muscle tissue into an electric organ?

As we discussed in Key Concept 15.4, voltage-gated sodium channels have a wide range of functions, including the control of muscle contractions and nerve impulses. Many different voltage-gated sodium channels arose through gene duplications. A duplicated gene is free to diverge from its ancestral function, since the original copy of the gene can continue to provide the original function. Early in the evolution of fishes, the entire genome was duplicated. Therefore, in most fishes, two different duplicated sodium channel genes are expressed in skeletal muscle. In electric fishes, one of these genes is turned off (not expressed) in skeletal muscle, but it *is* expressed in the tissue that becomes an electric organ. Several substitutions in the sodium channel gene allow a change in the function of the sodium channel it encodes: rather than producing a muscle contraction, a large number of these modified sodium channels in an electric organ produce an electric field. Similar changes have occurred repeatedly in different groups of electric fishes, but they are all based on parallel changes in the same duplicated gene. An ancestral gene duplication provided the opportunity for change, simple changes in the gene sequence dramatically altered the function, and a change in gene expression restricted the new function to the specialized electric organ. Through selection for these changes in function, a seemingly complex change (the electric organ) has evolved many times among fishes.

15.1 Genomes Reveal Both Neutral and Selective Processes of Evolution

FIGURE 15.1

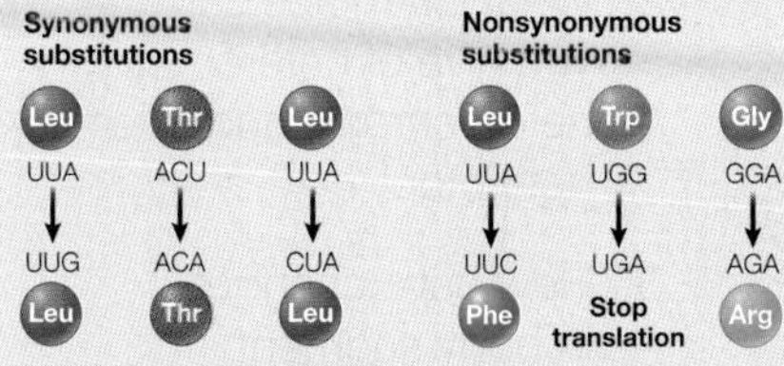

FIGURE 15.5

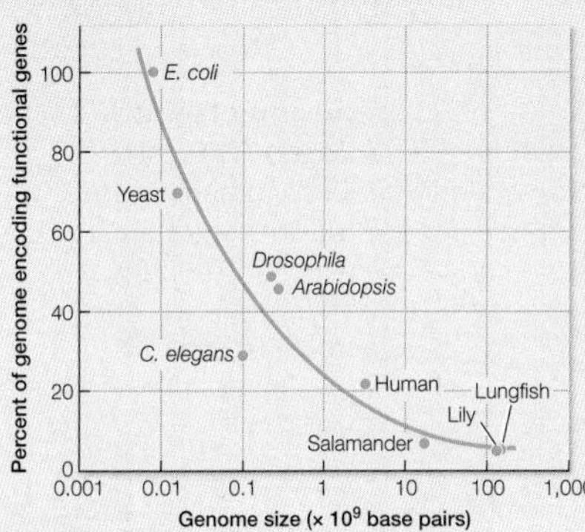

- **NONSYNONYMOUS SUBSTITUTIONS** of nucleotides result in amino acid replacements in proteins, but **SYNONYMOUS SUBSTITUTIONS** do not. Rates of synonymous substitution are typically higher than rates of nonsynonymous substitution in protein-coding genes (a result of stabilizing selection) (**FIGURE 15.1**).
- Much of the change in nucleotide sequences over time is a result of neutral evolution. The rate of fixation of neutral mutations is independent of population size and is equal to the mutation rate. Specific codons within a given gene sequence can be under different modes of selection.
- The total size of genomes varies much more widely across multicellular organisms than does the number of functional genes (**FIGURE 15.5**).
- Even though many noncoding regions of the genome may not have direct functions, these regions can affect the phenotype of an organism by influencing gene expression. Functionless **PSEUDOGENES** can serve as the raw material for the evolution of new genes.

QUESTIONS

1. How can biologists distinguish protein-coding genes that are evolving under positive selection from those evolving under stabilizing selection?
2. How would you support the statement that large genomes do not necessarily result in more complex organisms?
3. Which position in codons (first, second, or third) would you expect to evolve the fastest? Why?

15.2 Rearrangements of Genomes Can Result in New Features

Go to ACTIVITY 15.1

- Despite its short-term disadvantages, sexual reproduction generates countless genotype combinations that increase genetic variation in populations. In the absence of genetic recombination (as in some asexual organisms), deleterious mutations accumulate with each replication—a phenomenon known as **MULLER'S RATCHET**.
- **LATERAL GENE TRANSFER** can result in the rapid acquisition of new functions from distantly related species.
- Gene duplications can result in increased production of the gene's product, in divergence of the duplicated gene's expression, in pseudogenes, or in new gene functions. Several rounds of gene duplication can give rise to multiple genes with related functions, known as a **GENE FAMILY** (**FIGURE 15.6**).

FIGURE 15.6

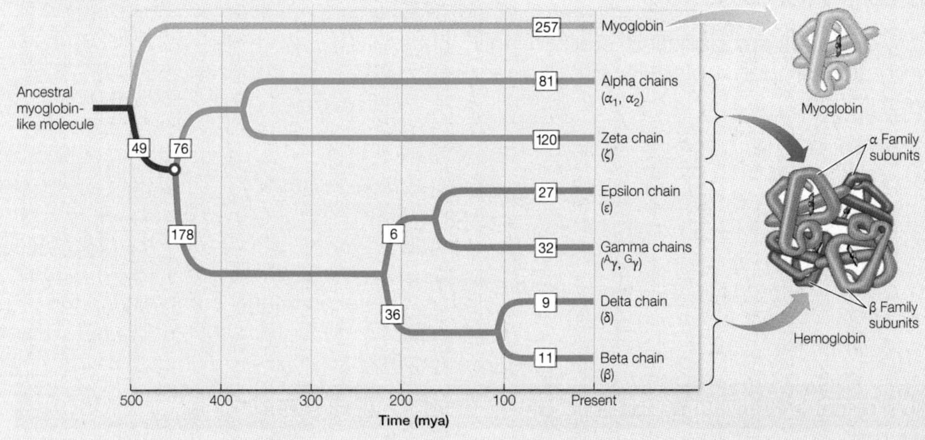

QUESTIONS

1. How does sexual recombination help keep deleterious mutations from accumulating in a population?
2. In what organisms is lateral gene transfer most common?
3. What is an example of a new function that arose following gene duplication?

15 VISUAL SUMMARY

15.3 Changes in Gene Expression Often Shape Evolution

FIGURE 15.10

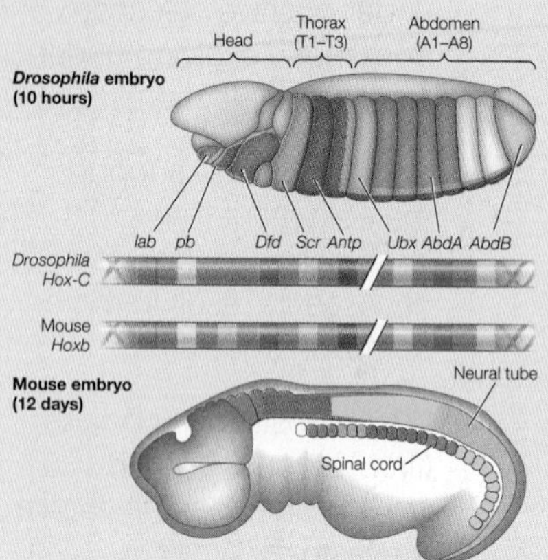

FIGURE 15.15

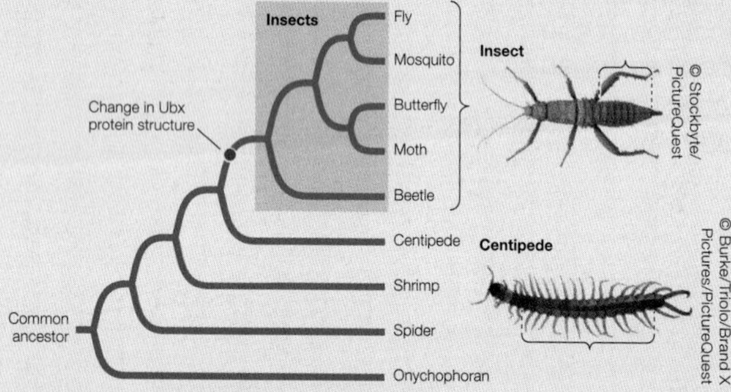

Go to ANIMATIONS 15.1 and 15.2

- A cascade of transcriptional activation sets up the axes of the embryo, the development of the segments, and the determination of cell fate in each segment. Hox genes help determine cell fate in the embryos of animals.
- Genes encoding transcription factors and other regulatory proteins that govern pattern formation in the developing bodies of multicellular organisms comprise what is called a **GENETIC TOOLKIT**. These regulatory genes have been highly conserved throughout evolution (**FIGURE 15.10**).
- Developmental genes can be expressed in a modular fashion in different amounts (**HETEROMETRY**), in different timing (**HETEROCHRONY**), or in different locations (**HETEROTOPY**). Mutations to genes that regulate development often result in major morphological changes (**FIGURE 15.15**).

QUESTIONS

1. What is an example of a homeotic mutation?
2. What examples can you give of heterometry, heterochrony, and heterotopy?
3. How have changes in body segment development been important in the evolution of arthropod appendages?

15.4 Molecular Evolution Has Many Practical Applications

FIGURE 15.16

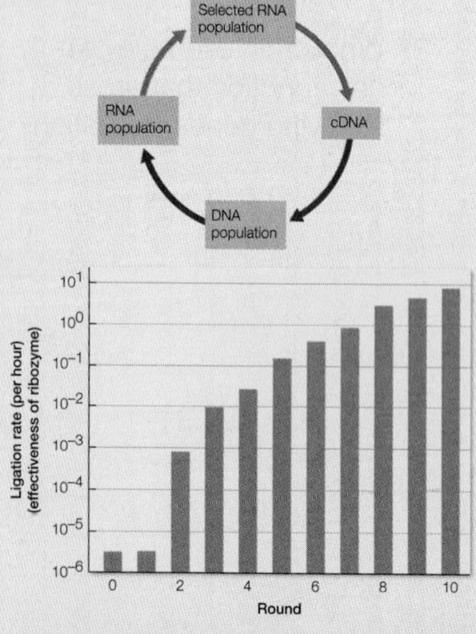

- Protein function can be studied by examining gene evolution. Detection of positive selection can be used to identify molecular changes that have resulted in functional changes.
- **IN VITRO EVOLUTION** is used to produce synthetic molecules with particular desired functions (**FIGURE 15.16**).
- Many diseases are identified, studied, and combated through molecular evolutionary investigations.

1. Why is positive selection associated with functional change?
2. What are the similarities and differences between in vitro evolution and natural selection?

Go to **Achieve** for the e-book, animations, activities, and additional resources and assignments.

Speciation 16

Lake Malawi, in the Great Rift Valley of East Africa, contains nearly 1,000 closely related species of haplochromine cichlid fishes that are found nowhere else in the world. This lake was formed less than 2 million years ago, so all of these species are thought to have arisen since then—a very rapid rate of speciation.

If speciation typically takes thousands to millions of years, how do biologists conduct experiments to study this process?

You will find the answer to this question on page 388.

KEY CONCEPTS

16.1 Species Are Reproductively Isolated Lineages on the Tree of Life

16.2 Speciation Is a Natural Consequence of Population Subdivision

16.3 Speciation May Occur through Geographic Isolation or in Sympatry

16.4 Reproductive Isolation Is Reinforced When Diverging Species Come into Contact

16.5 Speciation Rates Are Highly Variable across Life

16.1 Species Are Reproductively Isolated Lineages on the Tree of Life

LEARNING OBJECTIVES

By the end of this key concept you should be able to:

16.1.1 **Explain how biologists define species using morphology, reproduction, and evolutionary history.**

16.1.2 **Describe how reproductive isolation plays a role in each species concept.**

Biological diversity does not vary in a smooth, continuous way. People have long recognized groups of similar organisms that mate with one another, and they have noticed that there are usually distinct morphological breaks between these groups. Groups of organisms that share a suite of genetic and morphological attributes, and are reproductively isolated from other such groups, are known as **species** (note that this is both the plural and singular form of the word). Species are the result of the process of **speciation**: the divergence of biological lineages and the emergence of reproductive isolation between lineages.

Although "species" is a useful and common term, its usage varies among biologists who are interested in different aspects of speciation. Different biologists think about species differently because they ask different questions: How can we recognize and identify species? How do new species arise? How do different species remain separate? Why do rates of speciation differ among groups of organisms? In answering these questions, biologists focus on different attributes of species, leading to several different ways of thinking about what species are and how they form. Most of the various **species concepts** proposed by biologists are simply different ways of approaching the question "What are species?" Let's compare three major classes of species concepts to contrast the ways that biologists think about species.

We can recognize many species by their appearance

Someone who is knowledgeable about a group of organisms, such as birds or flowering plants, can usually distinguish the different species found in a particular area simply by looking at them. Standard field guides to birds, mammals, insects, and wildflowers are possible only because many species change little in appearance over large geographic distances (**FIGURE 16.1A**).

More than 250 years ago, Carolus Linnaeus developed the system of binomial nomenclature by which species are named today (see Key Concept 14.4). Linnaeus described and named thousands of species, but because he knew nothing about the genetics and little of the mating behavior of the organisms he was naming, he classified them on the basis of their appearance alone. In other words, Linnaeus used a **morphological species concept**, a construct that assumes that a species comprises individuals that "look alike" and that individuals that do not look alike belong to different species. Although Linnaeus could not have known it, the members of most of the groups he classified as species look alike because they share many alleles of the genes that code for their morphological features.

(A)

Dolichonyx oryzivorus
Male, Pennsylvania

Dolichonyx oryzivorus
Male, Illinois

(B)

Dolichonyx oryzivorus
Female

FIGURE 16.1 Not All Members of the Same Species Look Alike **(A)** It is easy to identify these two bobolinks (*Dolichonyx oryzivorus*) as members of the same species, even though they were photographed hundreds of miles apart, in Pennsylvania and Illinois. Despite their geographic separation, the two individuals are morphologically very similar. **(B)** Bobolinks are sexually dimorphic, which means the female looks quite different than the male.

(A) ***Hyla versicolor***

© iStock.com/SteveByland

(B) ***Hyla chrysoscelis***

Patrick Coin/CC BY-SA 2.5

FIGURE 16.2 Cryptic Species Look Alike but Do Not Interbreed These two species of gray tree frogs cannot be distinguished by their external morphology, but they do not interbreed even when they occupy the same geographic range. *Hyla versicolor* **(A)** is a tetraploid species (four sets of chromosomes), whereas *H. chrysoscelis* **(B)** is diploid (two sets of chromosomes). And although they look alike, the males have distinctive mating calls; female frogs recognize and mate with males of their own species based on these calls.

Using morphology to define species is practical and useful, but it also has limitations. Members of the same species do not always look alike. For example, males, females, and young individuals do not always resemble one another closely (**FIGURE 16.1B**). Furthermore, morphology is of little use in the case of **cryptic species**—instances in which two or more species are morphologically indistinguishable but do not interbreed (**FIGURE 16.2**). Biologists therefore cannot rely on appearance alone in determining whether individual organisms are members of the same or different species. Today, biologists use several additional types of information—especially behavioral and genetic data—to differentiate species.

Reproductive isolation is a key feature of most species concepts

The most important factor in the divergence of sexually reproducing lineages is the development of **reproductive isolation**, a state in which two groups of organisms can no longer exchange genes. If individuals of group A mate and reproduce only with one another, then group A constitutes a distinct species within which genes recombine. In other words, group A is an independent evolutionary lineage—a separate branch on the tree of life.

Recognizing the importance of reproductive isolation in maintaining species, evolutionary biologist Ernst Mayr proposed the **biological species concept**: *"Species are groups of actually or potentially interbreeding natural populations which are reproductively isolated from other such groups."* The phrase "actually or potentially" is an important element of this definition. "Actually" says that the individuals live in the same area and interbreed with one another. "Potentially" says that even though the individuals do not live in the same area, and therefore do not interbreed, other information suggests that they *would* do so if they were able to get together. This widely used species concept is not sufficient for all purposes, however. For example, it does not apply to organisms that reproduce asexually, and it is limited to a single point in evolutionary time.

The lineage approach to understanding species takes a long-term view

Evolutionary biologists often think of species as branches on the tree of life, an idea that articulates a **lineage species concept**. In this framework for thinking about species, one species splits into two descendant species, which thereafter evolve as distinct lineages. A lineage species concept allows biologists to consider species over evolutionary time.

A **lineage** is an ancestor–descendant series of populations followed over time. Each species has a history that starts with a speciation event (at which one lineage on the tree is split into two), and ends either at extinction or at another speciation event (at which the species produces two daughter species). The process of lineage splitting may be gradual, taking thousands of generations to complete. At the other extreme, an ancestral lineage may be split in two within a few generations (as happens with polyploidy, which we'll discuss in Key Concept 16.3). The gradual nature of some splitting events means that at a single point in time, the final outcome of the process may not be clear. In these cases, it may be difficult to predict whether the incipient species will continue to diverge and become fully isolated from their sibling species, or if they will merge again in the future.

The different species concepts are not mutually exclusive

Many named variants of these three major classes of species concepts exist. These various concepts are not incompatible; they simply emphasize different aspects of species or speciation. The morphological species concept emphasizes the practical aspects of recognizing species visually, although it sometimes results in underestimation or overestimation of the actual number of species. Mayr's biological species concept emphasizes that reproductive isolation is what allows sexual species to evolve independently of one another. The lineage species concept embraces the idea that sexual species are maintained by reproductive isolation, but extends the concept of a species as a lineage over evolutionary time. The species-as-lineage concept also accommodates species that reproduce asexually.

Virtually all species exhibit some degree of genetic recombination among individuals, even if recombination events are relatively rare and not associated with sexual reproduction. Significant reproductive isolation *between* species is necessary for lineages to remain distinct over evolutionary time. Furthermore, reproductive isolation is responsible for the morphological distinctiveness of most species, because mutations that result in morphological changes cannot spread between reproductively isolated species. Therefore no matter which species concept we emphasize, the evolution of reproductive isolation is important for understanding the origin of species.

REVIEW & APPLY | 16.1

R Species are distinct lineages on the tree of life. Speciation is usually a gradual process as one lineage divides into two. Over time, lineages of sexual species remain distinct from one another because they have become reproductively isolated.

A
1. Why do different biologists emphasize different attributes of species in formulating species concepts?
2. What makes reproductive isolation such an important component of each of the species concepts discussed here?
3. Why is the biological species concept not applicable to asexually reproducing organisms?

Although Charles Darwin titled his groundbreaking book *On the Origin of Species,* in fact it included very little discussion of speciation as we understand it today. Darwin devoted most of his attention to demonstrating that individual species are altered over time by natural selection. The remaining sections of this chapter will discuss the many aspects of speciation that biologists have learned about since Darwin's time.

16.2 Speciation Is a Natural Consequence of Population Subdivision

LEARNING OBJECTIVES

By the end of this key concept you should be able to:

16.2.1 Describe the Dobzhansky–Muller model of speciation and explain how genetic differences between species are correlated with reproductive isolation.

16.2.2 Evaluate data from two populations to determine if they are reproductively isolated.

Not all evolutionary changes result in new species. A single lineage may change over time without giving rise to a new species. Speciation requires the interruption of gene flow within a species whose members formerly exchanged genes. But if a genetic change prevents reproduction between individuals of a species, how can such a change spread through a species in the first place?

Incompatibilities between genes can result in reproductive isolation

If a new allele that causes reproductive incompatibility arises in a population, it cannot spread through the population because no other individuals will be reproductively compatible with the individual that carries the new allele. So how can one reproductively cohesive lineage ever split into two reproductively isolated species? Several early geneticists, including Theodosius Dobzhansky and Hermann Joseph Muller, developed a genetic model to explain this apparent conundrum (**FIGURE 16.3**).

The Dobzhansky–Muller model is quite simple. First, assume that a single ancestral population is subdivided into two daughter populations (by the formation of a new mountain range, for instance), which then evolve as independent lineages. In one of the descendant lineages, a new allele (*A*) arises and becomes fixed (see Figure 16.3). In the other lineage, another new allele (*B*) becomes fixed *at a different gene locus.* Neither new allele at either locus results in any loss of reproductive compatibility by itself. However, the two new forms of these two different genes have never occurred together in the same individual or population. Recall that the products of many genes must work together in an organism. It is possible that the new proteins encoded by the two new alleles will not be compatible with each other. If individuals from the two lineages come back together after these genetic changes, they may still be able to interbreed. However, the hybrid offspring will have a new combination of genes that may be functionally inferior, or even lethal (the combination of alleles *A* and *B* at the two loci in this example). This will not happen with all new combinations of genes, but over time, isolated lineages will accumulate many allele differences at many gene loci. Some combinations of these differentiated genes will not

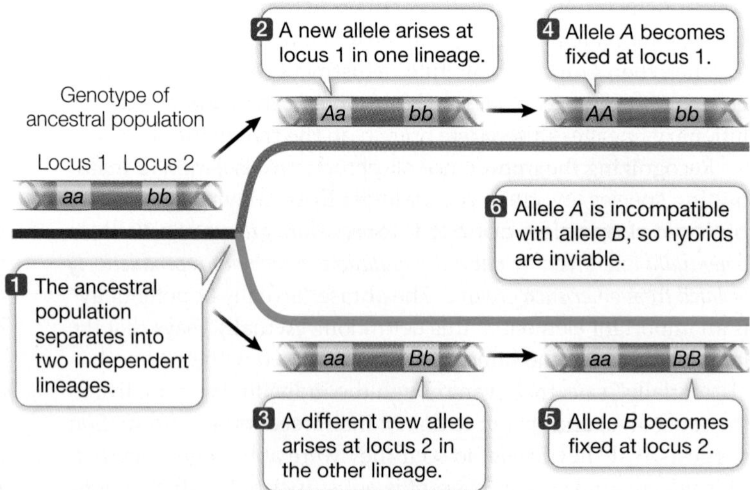

FIGURE 16.3 The Dobzhansky–Muller Model In this simple two-locus version of the model, two lineages from the same ancestral population become physically separated from each other and evolve independently. A new allele becomes fixed in each descendant lineage, but at two different loci. Neither of the new alleles is incompatible with the ancestral alleles, but the two new alleles in the two different genes are incompatible with each other. Thus the two descendant lineages are reproductively incompatible.

Activity 16.1 Speciation Simulation

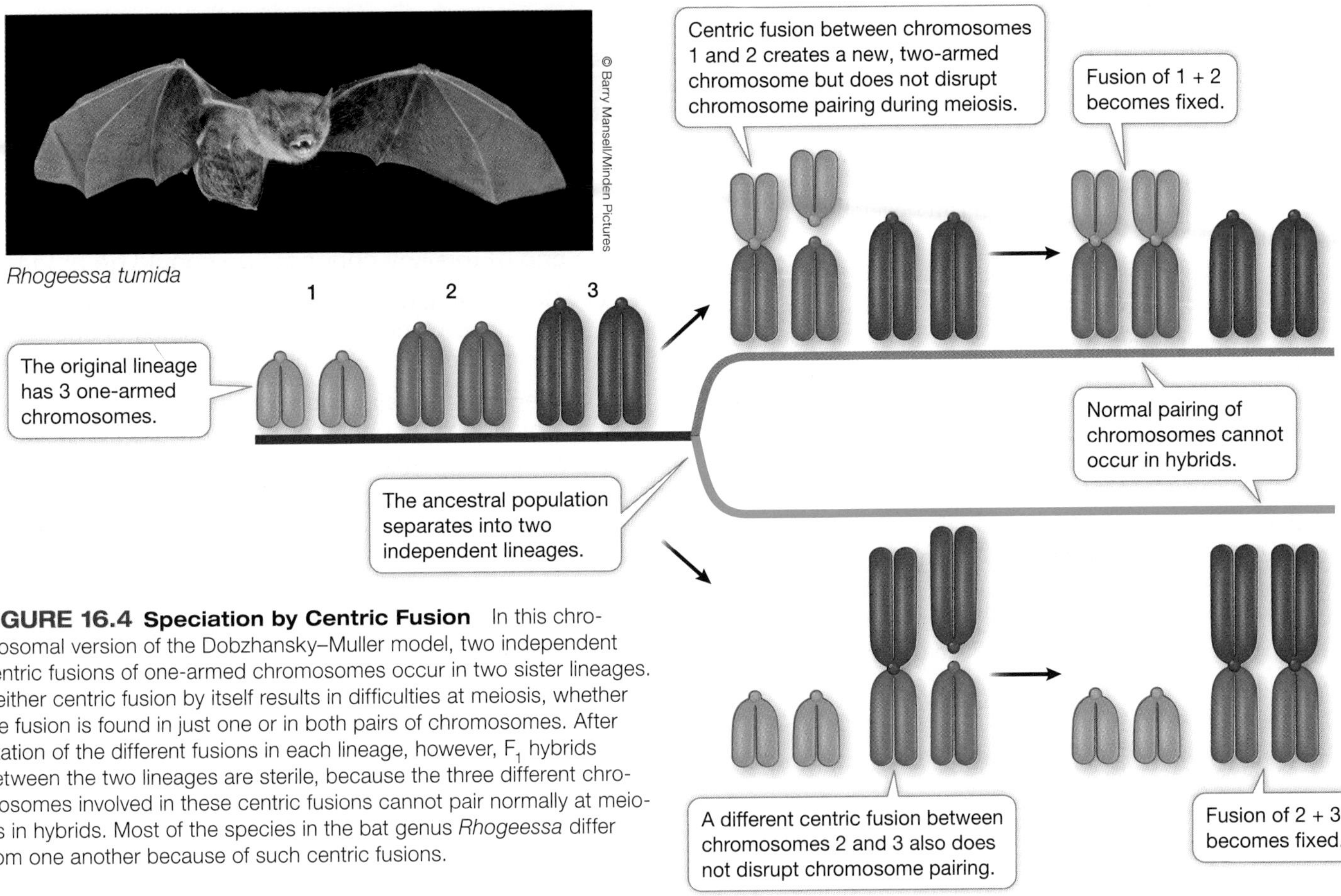

© Barry Mansell/Minden Pictures

Rhogeessa tumida

The original lineage has 3 one-armed chromosomes.

The ancestral population separates into two independent lineages.

Centric fusion between chromosomes 1 and 2 creates a new, two-armed chromosome but does not disrupt chromosome pairing during meiosis.

Fusion of 1 + 2 becomes fixed.

Normal pairing of chromosomes cannot occur in hybrids.

A different centric fusion between chromosomes 2 and 3 also does not disrupt chromosome pairing.

Fusion of 2 + 3 becomes fixed.

FIGURE 16.4 Speciation by Centric Fusion In this chromosomal version of the Dobzhansky–Muller model, two independent centric fusions of one-armed chromosomes occur in two sister lineages. Neither centric fusion by itself results in difficulties at meiosis, whether the fusion is found in just one or in both pairs of chromosomes. After fixation of the different fusions in each lineage, however, F_1 hybrids between the two lineages are sterile, because the three different chromosomes involved in these centric fusions cannot pair normally at meiosis in hybrids. Most of the species in the bat genus *Rhogeessa* differ from one another because of such centric fusions.

function well together in hybrids. Thus genetic incompatibility between the two isolated lineages will develop over time.

Many empirical and experimental examples support the Dobzhansky–Muller model. This model works not only for pairs of individual genes but also for some kinds of chromosomal rearrangements. Bats of the genus *Rhogeessa*, for example, exhibit considerable variation in centric fusions of their chromosomes. In other words, the chromosomes of the various species contain the same basic chromosomal arms, but in some species two acrocentric (one-armed) chromosomes have fused at the centromere to form larger, metacentric (two-armed) chromosomes. A polymorphism in centric fusion causes few, if any, problems in meiosis because the respective chromosomes can still align and assort normally. Therefore a given centric fusion can become fixed in a lineage. However, if a *different* centric fusion becomes fixed in a second lineage, then hybrids between individuals of each lineage will not be able to produce normal gametes in meiosis (**FIGURE 16.4**). Most of the closely related species of *Rhogeessa* display different combinations of these centric fusions and are thereby reproductively isolated from one another.

Reproductive isolation develops with increasing genetic divergence

As pairs of species diverge genetically, they become increasingly reproductively isolated (**FIGURE 16.5**). Both the rate at which reproductive isolation develops and the mechanisms that

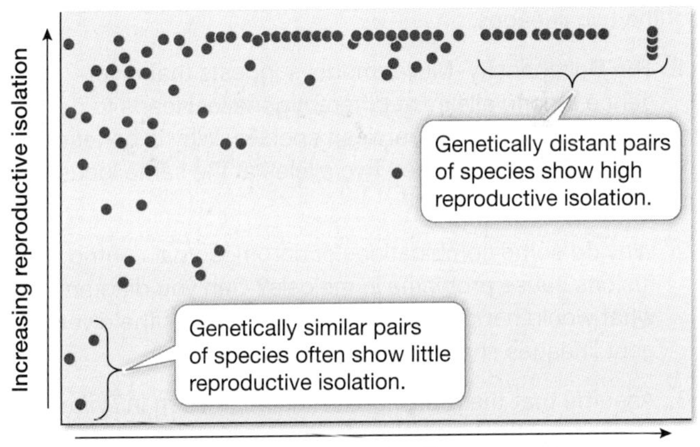

FIGURE 16.5 Reproductive Isolation Increases with Genetic Divergence Among pairs of *Drosophila* species, the more the species differ genetically (*x* axis), the greater their reproductive isolation from each other (*y* axis). Each dot represents a comparison of one species pair. Such positive relationships between genetic distance and reproductive isolation have been observed in many groups of plants, animals, and fungi. (After J. A. Coyne and H. A. Orr. 1989. *Evolution* 43: 362–381.)

Why might some genetically similar species pairs exhibit high levels of reproductive isolation?

produce it vary from group to group. Reproductive incompatibility has been shown to develop gradually in many groups of plants, animals, and fungi, reflecting the slow pace at which incompatible genes accumulate in each lineage. In some cases, complete reproductive isolation may take millions of years. In other cases (as with the chromosomal fusions of *Rhogeessa* described above), reproductive isolation can develop over just a few generations.

Partial reproductive isolation (reduced reproductive compatibility) has evolved in many strains of plants that have been artificially isolated by humans. In 1835, Thomas Drummond collected seeds of a newly discovered species of phlox in Texas and distributed them to nurseries in Europe. The European nurseries established more than 200 true-breeding strains of Drummond's phlox (*Phlox drummondii*) that differ in flower size, flower color, and plant growth form. The breeders did not select directly for reproductive incompatibility between strains, but in subsequent experiments in which strains were crossed and seed production was measured and compared, biologists found that reproductive compatibility between strains had been reduced by 14 to 50 percent, depending on the cross—even though the strains had been isolated from one another for less than two centuries.

REVIEW & APPLY | 16.2

R When two parts of a population become isolated from each other by some barrier to gene flow, they begin to diverge genetically. The Dobzhansky–Muller model describes how new alleles or chromosomal arrangements that arise in the two descendent lineages can lead to genetic incompatibility, and hence reproductive isolation, of the two lineages.

A

1. The Dobzhansky–Muller model suggests that divergence among alleles at different gene loci leads to genetic incompatibility between species. Why is genetic incompatibility between two alleles at the same locus considered less likely?
2. Why do some combinations of chromosomal centric fusions cause problems in meiosis? Can you diagram what would happen at meiosis in a hybrid of the divergent lineages shown in Figure 16.4?
3. Assume that the reproductive isolation seen in *Phlox* strains results from lethal combinations of incompatible alleles at several loci among the various strains. Given this assumption, why might the reproductive isolation seen among these strains be partial rather than complete?

We have now seen how the splitting of an ancestral population leads to genetic divergence and reproductive incompatibility in the two descendant lineages. Next we will consider ways in which the descendant lineages could have become separated in the first place.

16.3 | Speciation May Occur through Geographic Isolation or in Sympatry

LEARNING OBJECTIVES

By the end of this key concept you should be able to:

16.3.1 Compare and contrast allopatric and sympatric speciation, and recognize examples of speciation as being one or the other.

16.3.2 Diagram the relationship between geographic isolation and speciation, and interpret data to make and support predictions.

Many biologists who study speciation have concentrated on geographic processes that can result in the division of an ancestral species. Splitting of the geographic range of a species is one obvious way of achieving such a division, but it is not the only way.

Physical barriers give rise to allopatric speciation

Speciation that results when a population is divided by a physical barrier is known as **allopatric speciation** (Greek *allos*, "other," + *patria*, "homeland"). Allopatric speciation is thought to be the dominant mode of speciation in most groups of organisms. The physical barrier that divides the range of a species may be a body of water or a mountain range for terrestrial organisms, or dry land for aquatic organisms—in other words, any type of habitat that is inhospitable to the species. Such barriers can form when continents drift, sea levels rise or fall, glaciers advance or retreat, or climates change. The populations separated by such barriers are often, but not always, initially large. The lineages that descend from these founding populations evolve differences for a variety of reasons, including mutation, genetic drift, and adaptation to different environments in the two areas. As a result, many pairs of closely related **sister species**—species that are each other's closest relatives—may exist on either side of the geographic barrier. An example of a physical geographic barrier that produced many pairs of sister species was the Pleistocene glaciation that isolated freshwater streams in the eastern highlands of the Appalachian Mountains from streams in the Ozark and Ouachita Mountains (**FIGURE 16.6**). This splitting event resulted in many parallel speciation events among isolated lineages of stream-dwelling organisms.

Allopatric speciation may also result when some members of a population cross an existing barrier and establish a new, isolated population. Many of the more than 800 species of *Drosophila* found in the Hawaiian Islands are restricted to a single island. We know that these species are the descendants of new populations founded by individuals dispersing among the islands when we find that the closest relative of a species on one island is a species on a neighboring island rather than a species on the same island. Biologists who have studied the chromosomes of these fruit flies estimate that speciation in this group of *Drosophila* has resulted from at least 45 such founder events (**FIGURE 16.7**).

(A) Distribution of freshwater stream fishes in the Pliocene

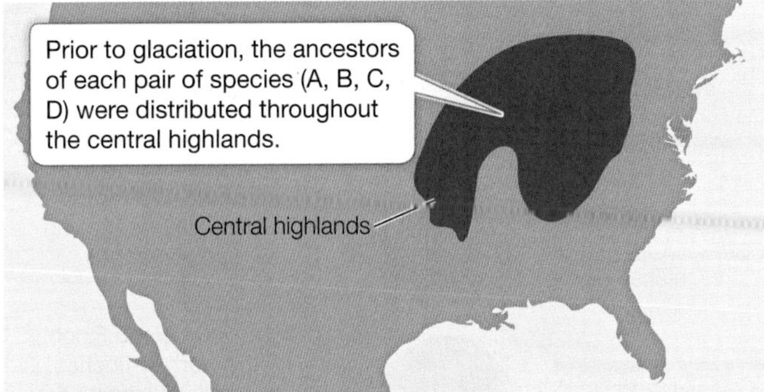

(B) Distribution of freshwater stream fishes in the Pleistocene

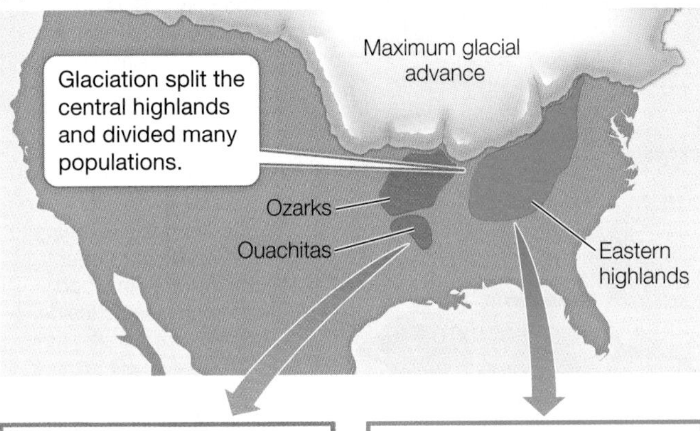

A_1 Missouri saddled darter *Etheostoma tetrazonum*

A_2 Variegated darter *E. variatum*

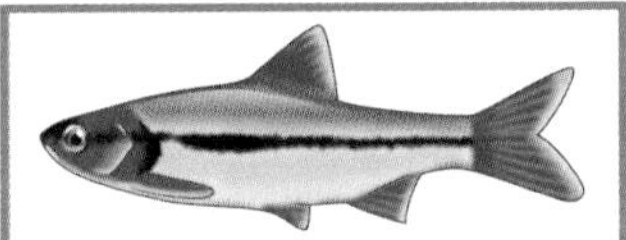

B_1 Bleeding shiner *Luxilus zonatus*

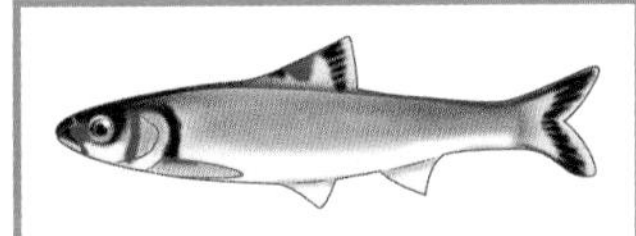

B_2 Warpaint shiner *L. coccogenis*

C_1 Ozark minnow *Notropis nubilus*

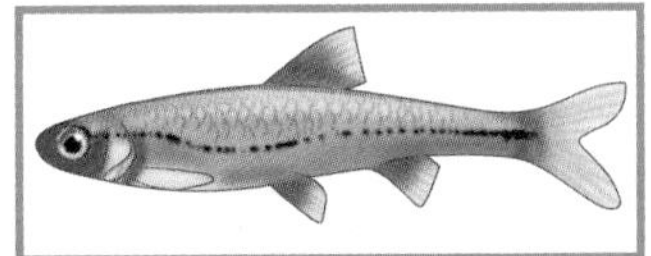

C_2 Tennessee shiner *N. leuciodus*

D_1 Ozark madtom *Noturus albater*

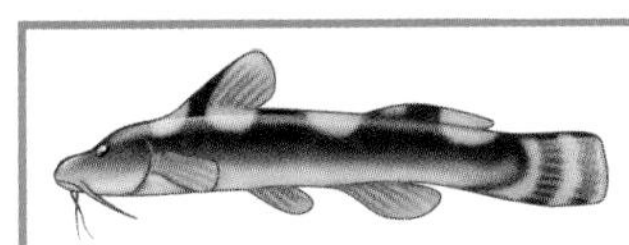

D_2 Elegant madtom *N. elegans*

FIGURE 16.6 Allopatric Speciation Allopatric speciation may result when an ancestral population is divided into two separate populations by a physical barrier and those populations then diverge. **(A)** Many species of freshwater stream fishes were distributed throughout the central highlands of North America in the Pliocene epoch (about 5.3–2.6 million years ago). **(B)** During the Pleistocene (about 2.6 million years ago–10,000 years ago), glaciers advanced and isolated fish populations in the Ozark and Ouachita Mountains to the west from fish populations in the highlands of the Appalachian Mountains to the east. Numerous species diverged as a result of this separation, including the ancestors of the four pairs of sister species shown here. (After R. L. Mayden. 1985. *Southwest Nat* 30: 195–211, R. L. Mayden. 1987. In *KS Geol Surv Guidebook Series 5, Quaternary Environments of KS*, W. C. Johnson [ed.] Lawrence, Kansas.)

After the retreat of the glaciers, why did the fish species in the Ozarks and Ouachitas remain reproductively isolated from those in the Appalachians to the east?

Animation 16.1 Speciation Mechanisms

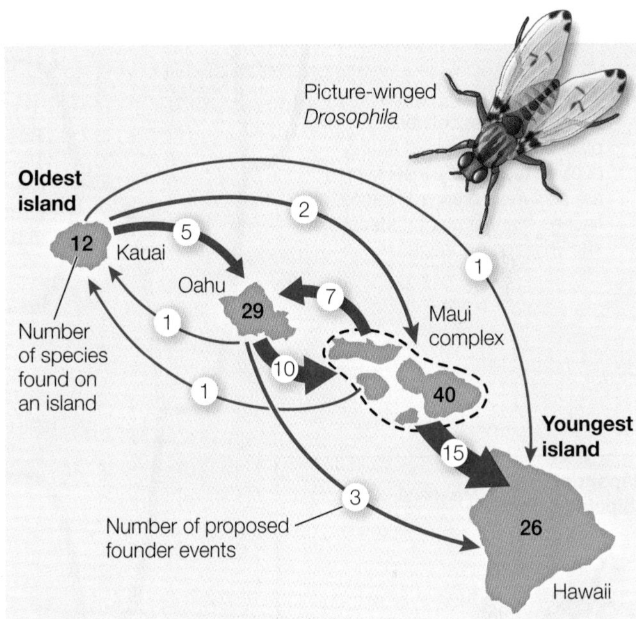

FIGURE 16.7 Founder Events Lead to Allopatric Speciation The large number of species of picture-winged *Drosophila* in the Hawaiian Islands is the result of founder events: the founding of new populations by individuals dispersing among the islands. The islands, which were formed in sequence as Earth's crust moved over a volcanic "hot spot," vary in age. (After H. L. Carson and A. R. Templeton. 1984. *Annu Rev Ecol Syst* 15: 97–131.)

Animation 16.2 Founder Events and Allopatric Speciation

The species of finches found on the islands of the Galápagos archipelago, some 1,000 kilometers off the coast of Ecuador, are among the most famous examples of allopatric speciation. Darwin's finches (as they are usually called, because Darwin was the first scientist to study them) arose in the Galápagos from a single South American finch species that colonized the islands. Today the Galápagos species differ strikingly not only from their closest mainland relative, but also from one another (**FIGURE 16.8**). The islands are sufficiently far apart that the birds move among them only infrequently. In addition, environmental conditions differ widely from island to island. Some islands are relatively flat and arid; others have forested mountain slopes. Sister lineages on different islands have diverged over hundreds of thousands of years, and several feeding specializations have arisen on different islands with different environments. Although finches occasionally fly between islands, an immigrant finch population is not likely to

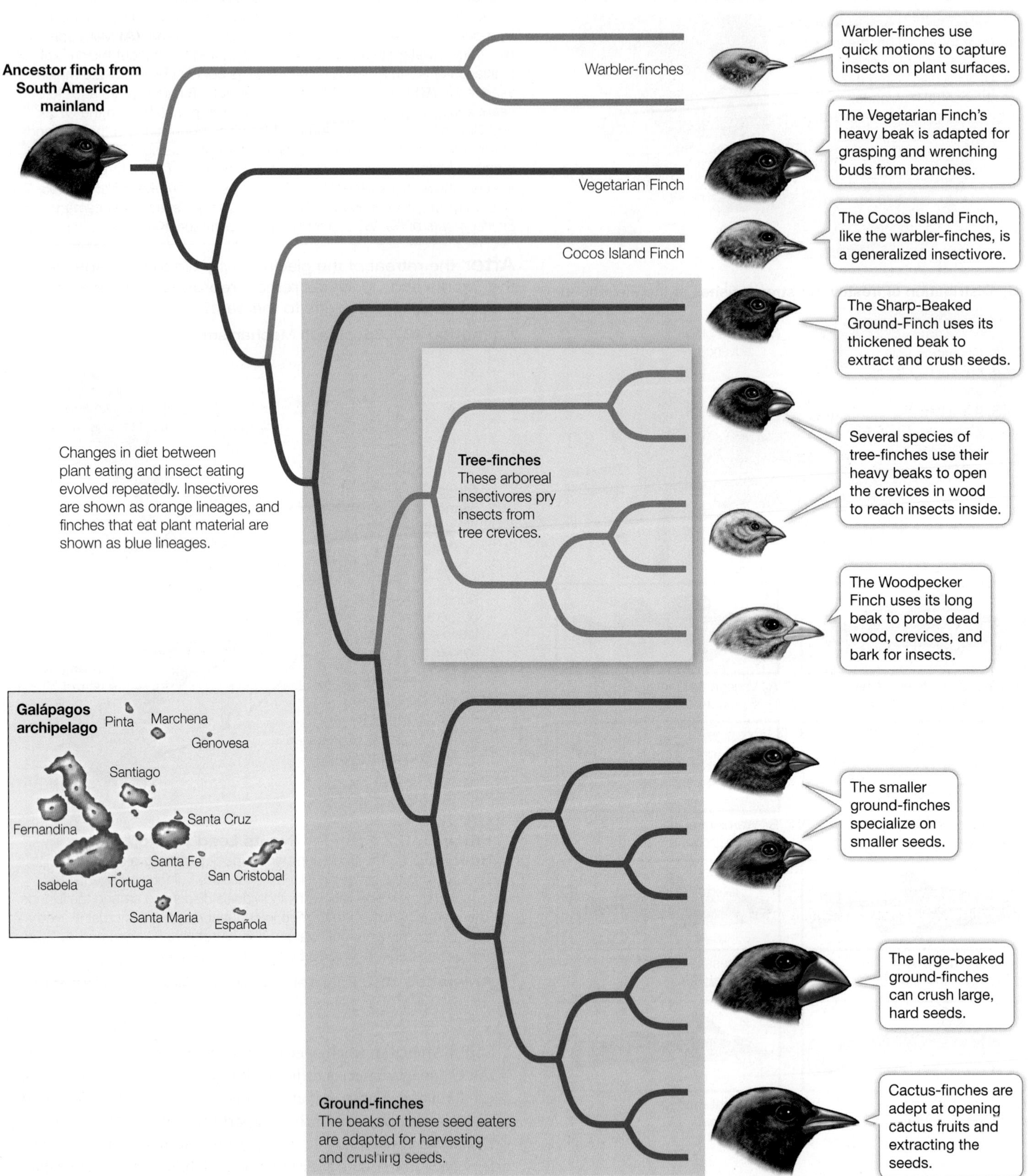

FIGURE 16.8 Allopatric Speciation among Darwin's Finches The descendants of the ancestral finch that colonized the Galápagos archipelago several million years ago evolved into many different species whose beaks are variously adapted to feed on buds, seeds, or insects. (After S. Lamichhaney et al. 2015. *Nature* 518: 371–375.)

Activity 16.2 Gene Flow and Geographic Isolation

become established unless the new environment is appropriate for its feeding specialization, and no other similar species are already present on the island. Each island now has from 1 to 4 species of finches, and biologists recognize between 14 and 18 species across the archipelago.

Sympatric speciation occurs without physical barriers

Geographic isolation is usually required for speciation, but under some circumstances speciation can occur in the absence of a physical barrier. Speciation without physical isolation is called **sympatric speciation** (Greek *sym*, "together with"). But how can such speciation happen? Given that speciation is usually a gradual process, how can reproductive isolation develop when individuals have frequent opportunities to mate with one another?

DISRUPTIVE SELECTION Sympatric speciation may occur with some forms of disruptive selection (see Key Concept 13.4) in which individuals with certain genotypes have a preference for distinct microhabitats where mating takes place. For example, sympatric speciation via disruptive selection appears to be taking place in the apple maggot fly (*Rhagoletis pomonella*) of eastern North America. Until the mid-1800s, *Rhagoletis* flies courted, mated, and deposited their eggs only on hawthorn fruits. About 150 years ago, some flies began to lay their eggs on apples, which European immigrants had introduced into eastern North America. Apple trees are closely related to hawthorns, but the smell of the fruits differs, and the apple fruits appear earlier than those of hawthorns. Some early-emerging female *Rhagoletis* laid their eggs on apples, and over time, a genetic preference for the smell of apples evolved among early-emerging insects. When the offspring of these flies sought out apple trees for mating and egg deposition, they mated with other flies reared on apples, which shared the same preferences.

Today the two groups of *Rhagoletis pomonella* in the eastern United States appear to be on the way to becoming distinct species. One group mates and lays eggs primarily on hawthorn fruits, the other on apples. The incipient species are partially reproductively isolated because they mate primarily with individuals raised on the same fruit and because they emerge from their pupae at different times of the year. In addition, the apple-feeding flies now grow more rapidly on apples than they originally did. Sympatric speciation that arises from such host-plant specificity may be widespread among insects, many of which feed on only a single plant species.

POLYPLOIDY The most common means of sympatric speciation is **polyploidy**, or the duplication of sets of chromosomes within individuals. Polyploidy can arise either from chromosome duplication in a single species (**autopolyploidy**) or from the combining of the chromosomes of two different species (**allopolyploidy**).

An autopolyploid individual originates when, for example, two accidentally unreduced **diploid** gametes (with two sets of chromosomes) combine to form a **tetraploid** individual (with four sets of chromosomes). Tetraploid and diploid individuals of the same species are reproductively isolated because their hybrid offspring are **triploid**. Even if these offspring survive, they are usually sterile; they cannot produce normal gametes because their chromosomes do not segregate evenly during meiosis (**FIGURE 16.9**). So a tetraploid individual usually cannot produce viable offspring by mating with a diploid individual—but it *can* do so if it self-fertilizes or mates with another tetraploid. Thus polyploidy can result in complete reproductive isolation in two generations—an important exception to the general rule that speciation is a gradual process.

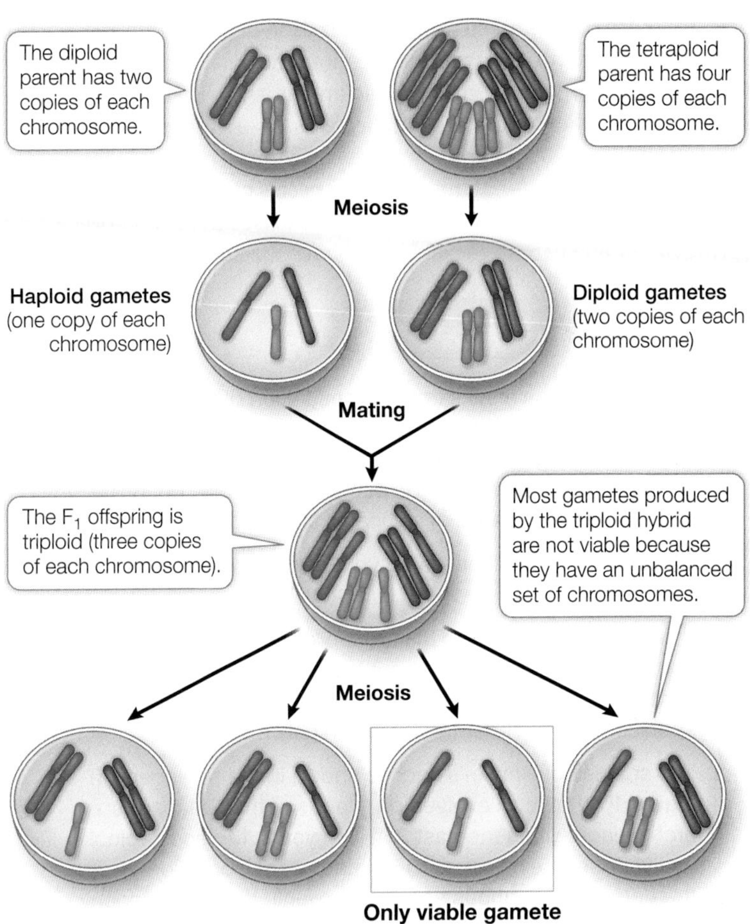

FIGURE 16.9 Tetraploids Are Reproductively Isolated from Their Diploid Ancestors Even if the triploid offspring of a diploid and a tetraploid parent survives and reaches sexual maturity, most of the gametes it produces will have aneuploid (unbalanced) numbers of chromosomes. Such triploid individuals are effectively sterile. (For simplicity, the figure shows only three homologous chromosomes. Most species have many more chromosomes, so viable gametes are extremely rare.)

Allopolyploids may be produced when individuals of two different (but closely related) species interbreed. Such hybridization often disrupts normal meiosis, which can result in chromosomal doubling. Allopolyploids are often fertile because each of the chromosomes has a nearly identical partner with which to pair during meiosis.

Speciation by polyploidy has been particularly important in the evolution of plants, although it has contributed to speciation in animals as well (such as the tree frogs in Figure 16.2). Botanists estimate that about 70 percent of flowering plant species and 95 percent of fern species are the result of recent polyploidization.

Some of these species arose from hybridization between two species followed by chromosomal duplication and self-fertilization. Other species diverged from polyploid ancestors, so that the new species shared their ancestors' duplicated sets of chromosomes. New species arise by polyploidy more easily among plants than among animals because plants of many species can reproduce by self-fertilization. In addition, if polyploidy arises in several offspring of a single parent, the siblings can fertilize one another.

REVIEW & APPLY | 16.3

R Allopatric speciation results from the separation of populations by geographic barriers; it is the dominant mode of speciation among most groups of organisms. Sympatric speciation may result from disruptive selection that results in ecological isolation, but polyploidy is the most common cause of sympatric speciation among plants.

A
1. Explain how speciation via polyploidy can happen in only two generations.
2. If allopatric speciation is the most prevalent mode of speciation, what do you predict about the geographic distributions of many closely related species? Does your answer differ for species that are sedentary versus highly mobile?
3. The species of Darwin's finches shown in the phylogeny in Figure 16.8 have all evolved on islands of the Galápagos archipelago within the past 3 million years. Molecular clock analysis (see Key Concept 14.3) has been used to determine the dates of the various speciation events in that phylogeny. Geological techniques for dating rock samples (see Key Concept 17.1) have been used to determine the ages of the various Galápagos islands. The table shows the number of species of Darwin's finches and the number of islands that have existed in the archipelago at several times during the past 4 million years (data from P. R. Grant. 2001. *Oikos* 92: 385–403).

Time (millions of years ago)	Number of islands	Number of finch species
0.25	18	14
0.50	18	9
0.75	9	7
1.00	6	5
2.00	4	3
3.00	4	1
4.00	3	0

   a. Plot the number of species of Darwin's finches and the number of islands in the Galápagos archipelago (dependent variables) against time (independent variable).
   b. Are the data consistent with the hypothesis that isolation of populations on newly formed islands is related to speciation in this group of birds? Why or why not?
4. If no more islands form in the Galápagos archipelago, do you think that speciation by geographic isolation will continue to occur among Darwin's finches? Why or why not? What additional data could you collect to test your hypothesis (without waiting to see if speciation occurs)?

Most populations separated by a physical barrier become reproductively isolated only slowly and gradually. If two incipient species once again come into contact with each other, what keeps them from merging back into a single species?

16.4 | Reproductive Isolation Is Reinforced When Diverging Species Come into Contact

LEARNING OBJECTIVES

By the end of this key concept you should be able to:

16.4.1 Create graphical representations of hybridization frequencies and analyze data to develop hypotheses about postzygotic isolation.

16.4.2 Design an experiment to investigate a given prezygotic or postzygotic isolating mechanism.

16.4.3 Relate the width of a hybrid zone to the fitness of hybrid offspring.

As we discussed in Key Concept 16.2, once a barrier to gene flow is established, reproductive isolation will begin to develop through genetic divergence. Over many generations, differences accumulate in the isolated lineages, reducing the probability that individuals from each lineage will mate successfully with individuals in the other lineage when they come back into contact. In this way, reproductive isolation can evolve as a by-product of the genetic changes in the two diverging lineages.

If reproductive isolation is incomplete when incipient species come back into contact, some hybridization is likely to occur. If hybrid individuals are less fit than non-hybrids, selection will favor parents that do not produce hybrid offspring. Under these conditions, selection will result in the strengthening, or **reinforcement**, of mechanisms that prevent hybridization.

Mechanisms that prevent hybridization from occurring are called **prezygotic isolating mechanisms**. Mechanisms that reduce the fitness of hybrid offspring are called **postzygotic isolating mechanisms**. Postzygotic isolating mechanisms result in selection against hybridization, which in turn leads to the reinforcement of prezygotic isolating mechanisms.

Prezygotic isolating mechanisms prevent hybridization

Prezygotic isolating mechanisms, which come into play before fertilization, can prevent hybridization in several ways.

MECHANICAL ISOLATION Differences in the sizes and shapes of reproductive organs may prevent the union of gametes from different species. With animals, the shapes of the reproductive organs of males and females of the same

FIGURE 16.10 Mechanical Isolation through Mimicry Many orchid species maintain reproductive isolation by means of flowers that look and smell like females of one—and only one—species of bee or wasp. A male insect of the correct species must land on the flower and attempt to mate with it; only males of this particular species are physically configured to collect and transfer the orchid's pollen. The constraints of this method of pollen transfer reproductively isolate the plant from related orchid species that attract different insect pollinators. The species shown here are the two players in one such interspecific relationship.

species often match, so that reproduction between individuals with mismatched reproductive structures is not physically possible. In plants, mechanical isolation may involve a pollinator. For example, some orchid species produce flowers that look and smell like the females of a particular species of bee or wasp (**FIGURE 16.10**). When a male bee or wasp visits and attempts to mate with the flower (thinking it is a female of his species), his mating behavior results in the transfer of pollen to and from his body by appropriately configured anthers and stigmas on the flower. Insects that visit the flower but do not attempt to mate with it do not trigger the transfer of pollen between the insect and the flower.

TEMPORAL ISOLATION Many organisms have distinct mating seasons. If two closely related species breed at different times of the year (or different times of day), they may never have an opportunity to hybridize. For example, in sympatric populations of three closely related leopard frog species, each species breeds at a different time of year (**FIGURE 16.11**). Although there is some overlap in the breeding seasons, the opportunities for hybridization are minimized.

BEHAVIORAL ISOLATION Individuals may reject, or fail to recognize, individuals of other species as potential mating partners. For example, the mating calls of male frogs of related species diverge quickly (**FIGURE 16.12**). Female frogs respond to mating calls from males of their own species but ignore the calls of other species, even closely related ones. The evolution of female preferences for certain male coloration patterns among the cichlids of Lake Malawi, some of which are illustrated at the beginning of this chapter, is another example of behavioral isolation.

Sometimes the mate choice of one species is mediated by the behavior of individuals of other species. For example, whether two plant species hybridize may depend on the food preferences of their pollinators. The floral traits of plants, including their color and shape, can enhance reproductive isolation either by influencing which pollinators are attracted to the flowers or by altering where pollen is deposited on the bodies of pollinators. A plant whose flowers are pendant will be pollinated by an animal with different physical characteristics than will a plant in which the flowers grow upright (**FIGURE 16.13A** and **B**). Because each pollinator prefers (and is adapted to) a different type of flower, pollinators rarely transfer pollen from one plant species to the other. Such isolation by pollinator behavior is seen in the mountains of California in two sympatric species of columbines (*Aquilegia*) that have diverged in flower color, structure, and orientation. *A. formosa* (**FIGURE 16.13C**) has pendant flowers with short spurs (spikelike, nectar-containing structures) and is pollinated by hummingbirds. *A. pubescens* (**FIGURE 16.13D**)

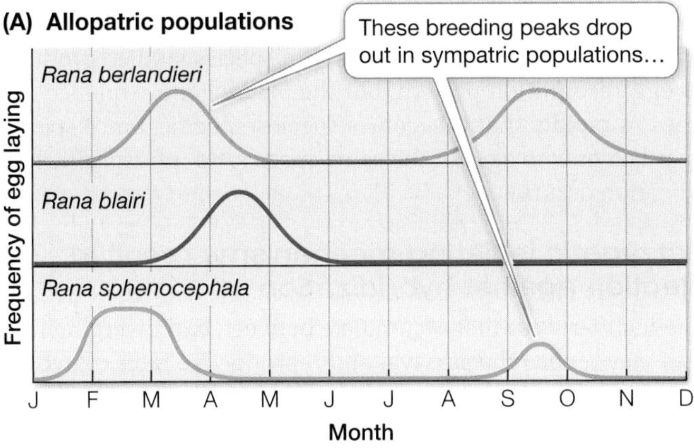

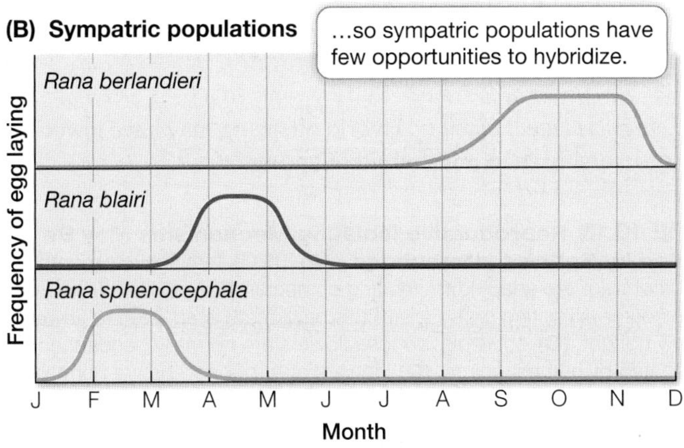

FIGURE 16.11 Temporal Isolation of Breeding Seasons (A) The peak breeding seasons of three species of leopard frogs (*Rana*) overlap when the species are physically separated (allopatry). **(B)** Where two or more species of *Rana* live together (sympatry), overlap between their peak breeding seasons is greatly reduced or eliminated. (After D. Hillis. 1981. *Copeia* 1981: 312–319.)

What means of selection helps reinforce this prezygotic isolating mechanism in areas of sympatry?

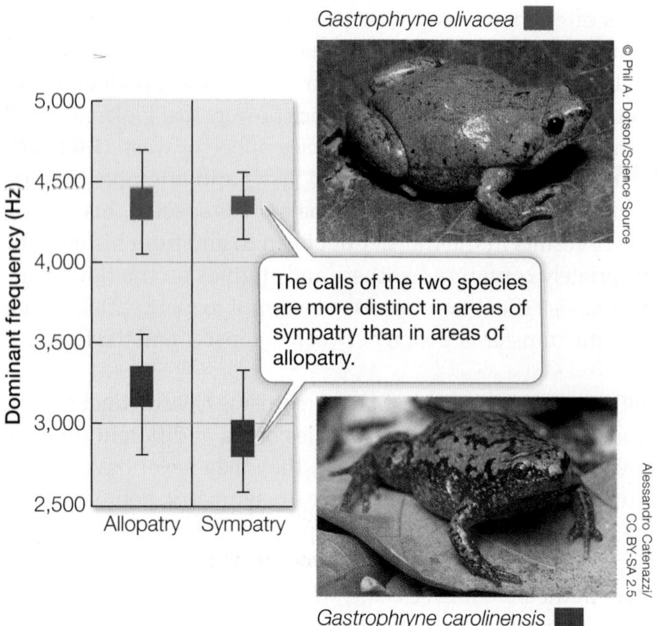

FIGURE 16.12 Behavioral Isolation in Mating Calls The males of most frog species produce species-specific calls. The calls of the two closely related frog species shown here differ in their dominant frequency (a high-frequency sound wave results in a high-pitched sound; a low frequency results in a low-pitched sound). Female frogs are attracted to the calls of males of their own species. Bars indicate 95% confidence limits for the mean frequencies of the calls. (After J. J. Loftus-Hillis and M. J. Littlejohn. 1992. *Evolution* 46: 896–906.)

Media Clip 16.1 Narrowmouth Toads Calling for Mates

FIGURE 16.13 Reproductive Isolating Mechanisms May Be Mediated by Species Interactions **(A)** This hummingbird's morphology and behavior are adapted for feeding on nectar from pendant flowers. **(B)** The nectar-extracting proboscis of this hawkmoth is adapted to flowers that grow upright. **(C)** *Aquilegia formosa* flowers are normally pendant and are pollinated by hummingbirds. **(D)** Flowers of *A. pubescens* are normally upright and are pollinated by hawkmoths. In addition, their long floral spurs appear to restrict access by some other potential pollinators.

has upright, lighter-colored flowers with long spurs and is pollinated by hawkmoths. The difference in pollinators means that these two species are effectively reproductively isolated even though they populate the same geographic range.

LINK Some plants and their pollinators become so tightly adapted to each other that they develop mutually dependent relationships, as described in **Key Concept 20.5**.

HABITAT ISOLATION When two closely related species evolve preferences for living or mating in different habitats, they may never come into contact during their respective mating periods. The *Rhagoletis* flies discussed in Key Concept 16.3 experienced such habitat isolation, as did the cichlid fishes that first adapted to different habitats in the newly formed Lake Malawi, which resulted in some of the speciation among the fishes shown at the beginning of this chapter.

GAMETIC ISOLATION The sperm of one species may not attach to the eggs of another species because the eggs do not release the appropriate attractive chemicals, or the sperm may be unable to penetrate the egg because the two gametes are chemically incompatible. Thus even though the gametes of two species may come into contact, the gametes never fuse into a zygote.

Gametic isolation is extremely important for many aquatic species that spawn (release their gametes directly into the environment). For example, gametic isolation has been extensively studied in spawning sea urchins. A protein known as bindin is found in sea urchin sperm and functions in attaching ("binding") the sperm to eggs. All sea urchin species studied produce this egg-recognition protein, but the *bindin* gene sequence diverges so rapidly that it becomes species-specific. Since sperm can attach only to eggs of the same species, no interspecific hybridization occurs.

Postzygotic isolating mechanisms result in selection against hybridization

Genetic differences that accumulate between two diverging lineages may reduce the survival and reproductive rates of hybrid offspring in any of three ways:

1. *Low hybrid zygote viability.* Hybrid zygotes may fail to mature normally, either dying during development or developing phenotypic abnormalities that prevent them from becoming reproductively capable adults.
2. *Low hybrid adult viability.* Hybrid offspring may have lower survivorship than non-hybrid offspring.

THINK LIKE A SCIENTIST

Reinforcement of reproductive isolation

Evolutionary biologists are particularly interested in reinforcement's role in speciation because this process directly connects natural selection to the origin of species. Here you will use processes of science to infer patterns and make predictions about reinforcement.

Different *Drosophila* species vary in the concentrations of different hydrocarbons on their cuticles. These hydrocarbons often serve as reproductive signals, with females preferring to mate with males that have the suite of hydrocarbons from the females' own species over those of other species. As they often are reproductive signals, the hydrocarbons may be the target of the selection involved in reinforcement.

Two species of *Drosophila*, *D. serrata* and *D. birchii*, are found on Australia's east coast and overlap in range. *D. serrata* populations from areas where the species encounters *D. birchii* (sympatric populations) differ from *D. serrata* populations that do not encounter the other species (allopatric populations), and these differences are likely due to reinforcement. The species do produce some hybrids, but the fitness of the hybrids is low. **FIGURE 16.14** shows concentrations of three different hypothetical hydrocarbons taken from allopatric and sympatric populations of *D. serrata* and *D. birchii*. (Numbers do not add to 100 because there are other hydrocarbons that are not shown.)

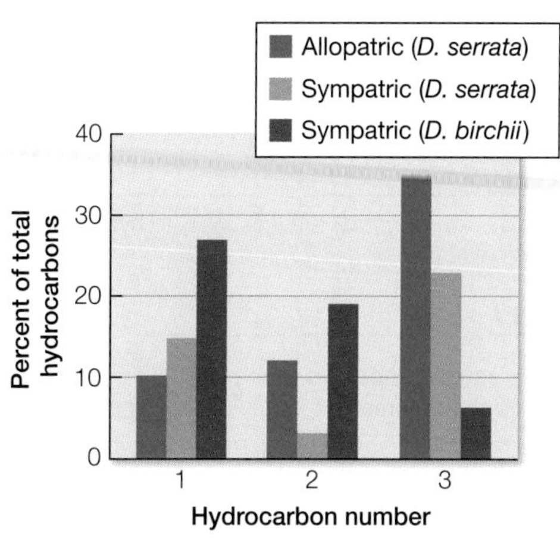

FIGURE 16.14 Reinforcement of Reproductive Isolation

1. Based on the data, which hydrocarbon(s) show a pattern that is consistent with reinforcement? Explain your answer.
2. Suppose the allopatric populations of *D. serrata* were taken into the lab and exposed to *D. birchii* for many generations. If the artificial selection in this experiment acted like the natural selection in reinforcement, what would you expect to happen to the concentrations of each of the hydrocarbons in the allopatric populations?
3. If the artificial selection did act like the natural selection involved in reinforcement, would you expect the frequency of hybridization between allopatric populations and *D. birchii* to increase or decrease? Why?
4. Would you expect the extent of postzygotic reproductive isolation to change as a result of the artificial selection in the experimental evolution study? Explain.

3. *Hybrid infertility*. Hybrids may mature into infertile adults. For example, the offspring of matings between horses and donkeys—mules—are typically sterile. Although otherwise healthy, sterile mules produce no descendants.

Natural selection does not directly favor the evolution of postzygotic isolating mechanisms. But if hybrids are less fit, individuals that breed only within their own species will leave more surviving offspring than will individuals that interbreed with another species. Therefore individuals that can avoid interbreeding with members of other species will have a selective advantage, and any trait that contributes to such avoidance will be favored.

Donald Levin of the University of Texas has studied reinforcement of prezygotic isolating mechanisms in flowers of the genus *Phlox*. Levin noticed that most individuals of *P. drummondii* in most of the range of the species in Texas have pink flowers. However, where *P. drummondii* is sympatric with its close relative, the pink-flowered *P. cuspidata*, most *P. drummondii* have red flowers. No other *Phlox* species has red flowers. The results of an experiment that Levin performed showed that reinforcement may explain why red flowers are favored where the two species are sympatric (**INVESTIGATION FIGURE 16.15**).

Likely cases of reinforcement are often detected by comparing sympatric and allopatric populations of potentially hybridizing species, as in the case of *Phlox*. If reinforcement is occurring, then sympatric populations of closely related species are expected to evolve more effective prezygotic reproductive barriers than do allopatric populations of the same species. As Figure 16.11 shows, the breeding seasons of sympatric populations of different leopard frog species overlap much less than do those of allopatric populations. Similarly, the frequencies of the frog mating calls illustrated in Figure 16.12 are more divergent in sympatric populations than in allopatric populations. In both cases, there appears to have been natural selection against hybridization in areas of sympatry.

INVESTIGATION

FIGURE 16.15 Flower Color Reinforces a Reproductive Barrier in Phlox Most *Phlox drummondii* flowers are pink, but in regions where the species is sympatric with *P. cuspidata*—which always has pink flowers—most *P. drummondii* individuals have red flowers. Most pollinators preferentially visit flowers of one color or the other. In this experiment, Donald Levin of the University of Texas explored whether flower color reinforces a prezygotic reproductive barrier, lessening the chances of interspecific hybridization.[a]

HYPOTHESIS

Red-flowered *P. drummondii* are less likely to hybridize with *P. cuspidata* than are pink-flowered *P. drummondii*.

METHOD

1. Introduce equal numbers of red- and pink-flowered *P. drummondii* individuals into an area with many pink-flowered *P. cuspidata*.
2. After the flowering season ends, measure hybridization by assessing the genetic composition of the seeds produced by *P. drummondii* plants of both colors.

RESULTS

Of the seeds produced by pink-flowered *P. drummondii*, 38% were hybrids with *P. cuspidata*. Only 13% of the seeds produced by red-flowered individuals were genetic hybrids. Bars indicate 95% confidence limits for the percent of seeds from hybrid matings.

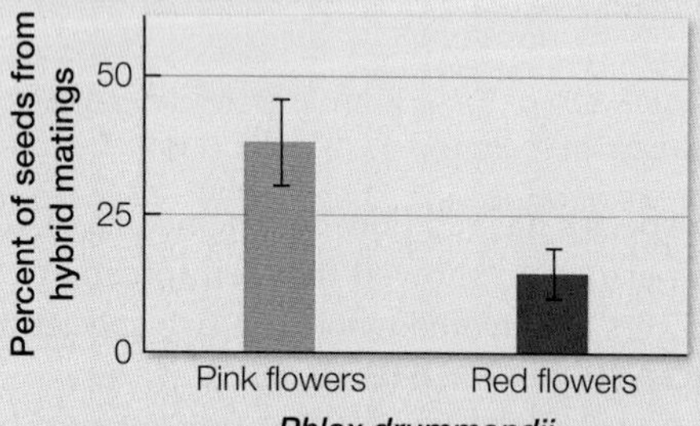

CONCLUSION

P. drummondii and *P. cuspidata* are less likely to hybridize if the flowers of the two species differ in color.

ANALYZE THE DATA

In the experiment described above, Donald Levin assessed the genetic composition of the seeds produced by *P. drummondii* at the end of the flowering season. The results are shown in the table below.

Morph (flower color)	Number of seeds (progeny)		
	P. drummondii	Hybrid	Total
Red	181 (87%)	27 (13%)	208
Pink	86 (62%)	53 (38%)	139

1. Check the 95% confidence intervals for the proportion of hybrid seeds in red- and pink-flowered *P. drummondii* in the graph in the results of the experiment. (See Appendix B's coverage of inferential statistics for a discussion of confidence intervals.) There are many websites available for calculating confidence intervals; a good one is the Vassar College statistical computation site, VassarStats.net. You can go to this site and select "Proportions" from the left-hand menu, then select "The Confidence Interval of a Proportion." What are the numerical values of the 95% confidence intervals?
2. Notice that the proportions of hybrids among the seeds of red- versus pink-flowered samples are significantly different, because the 95% confidence intervals do not overlap. To quantify the significance of this difference, use the website suggested in Question 1, but select "Significance of the Difference between Two Independent Proportions" from the "Proportions" menu. What null hypothesis are you testing in this case? (See Appendix B if you need help.) What is the *P*-value of getting results at least as different as these two samples if your null hypothesis is true?
3. How would you extend or improve the experimental design of this study? What kinds of additional test sites or conditions would you want to examine? How might replicate or control sites make the study more convincing?

[a]D. A. Levin. 1985. *Evolution* 39: 1275–1281.

A similar **Analyze the Data** exercise may be assigned in **Achieve**.

Hybrid zones may form if reproductive isolation is incomplete

Unless reproductive isolation is complete, closely related species may hybridize in areas where their ranges overlap, resulting in the formation of a **hybrid zone**. When a hybrid zone first forms, most hybrids are offspring of crosses between individuals of the two hybridizing species. However, subsequent generations include a variety of individuals with varying proportions of their genes derived from the original two species, so hybrid zones often contain recombinant individuals resulting from many generations of hybridization.

Detailed genetic studies can tell us why narrow hybrid zones may persist for long periods between the ranges of two species. The fire-bellied toad (*Bombina bombina*) lives in eastern Europe; the closely related yellow-bellied toad (*B. variegata*) lives in western and southern Europe. The ranges of the two species overlap in a long but narrow zone stretching 4,800 kilometers from eastern Germany to the Black Sea (**FIGURE 16.16**). Hybrids between the two species suffer from a range of defects, many of which are lethal. Those hybrids that survive often have skeletal abnormalities, such as a misshapen mouth, ribs that are fused to vertebrae, and a reduced number of vertebrae. By following the fates of thousands of toads from the hybrid zone, investigators found that a hybrid toad, on average, is only half as fit as an individual of either species. The hybrid zone remains narrow because there is strong selection against hybrids and because adult toads do not move over long distances. The zone has persisted for hundreds of years, however, because individuals of both species continue to move short distances into it, continually replenishing the hybrid population.

B. variegata (yellow-bellied toad)

© Rudmer Zwerver/Shutterstock.com

B. bombina (fire-bellied toad)

© Marek R. Swadzba/Shutterstock.com

Hybrid zone

FIGURE 16.16 A Hybrid Zone The narrow zone (shown in red) in which fire-bellied toads and yellow-bellied toads meet and hybridize has been stable for hundreds of years. (After J. Vörös et al. 2006. *Mol Phylogenet Evol* 38: 705–718.)

REVIEW & APPLY | 16.4

R Reproductive isolation may result from prezygotic or postzygotic isolating mechanisms. Lower fitness of hybrids can lead to the reinforcement of prezygotic isolating mechanisms.

A 1. As shown in Figure 16.11, the leopard frogs *Rana berlandieri* and *R. sphenocephala* usually have non-overlapping breeding seasons in areas of sympatry, but where they are allopatric, both species breed in both spring and fall. When new ponds are created where the ranges of the two species come close together, frogs from previously allopatric populations may colonize the new ponds and hybridize during their overlapping breeding seasons.

Imagine you have collected and tabulated data on hybridization between these two frog species. You have sampled various life stages of frogs and their tadpoles for 2 years after an initial spring breeding season at a newly established pond. Use the data below to answer the following questions.

Life stage	*R. berlandieri*	*R. sphenocephala*	F_1 hybrids
Recently hatched tadpoles (spring, year 1)	155	125	238
Late-stage tadpoles (summer, year 1)	45	55	64
Newly metamorphosed froglets (fall, year 1)	32	42	15
Adult frogs (year 2)	10	15	1

a. Create four pie charts (one for each life stage) showing the proportions of each species and of hybrids at each stage.

b. Offer reasons for the differences in the proportion of hybrids found at each life stage. Suggest some postzygotic isolating mechanisms that are consistent with your data.

c. Over time, what changes might you expect in the breeding seasons of the two species at this particular pond, and why? How would future pie charts differ from those you created above, if your predictions about breeding seasons are correct?

2. In each of the columbine (*Aquilegia*) species shown in Figure 16.13, the orientation of the flowers and the length of the flower spurs are associated with a particular type of pollinator (hummingbirds or hawkmoths). Columbine flowers vary in other ways as well—for example, in color and probably in fragrance. What experiments could you design to determine the traits that various pollinators use to distinguish among the flowers of different columbine species?

3. Why don't most narrow hybrid zones, such as the one between *Bombina bombina* and *B. variegata* shown in Figure 16.16, get wider over time?

Some groups of organisms have many species, others only a few. Hundreds of species of *Drosophila* evolved in the small area of the Hawaiian Islands over about 20 million years. In contrast, there are only a few species of horseshoe crabs in the world, and only one species of ginkgo tree, even though these latter groups have persisted for hundreds of millions of years. Why do different groups of organisms have such different rates of speciation?

16.5 | Speciation Rates Are Highly Variable across Life

LEARNING OBJECTIVES

By the end of this key concept you should be able to:

16.5.1 List four factors that commonly affect speciation rates.

16.5.2 Describe the conditions that favor adaptive radiation.

Many factors influence the likelihood that a lineage will split to form two or more species. Therefore rates of speciation (the proportion of existing species that split to form new species over a given period) vary greatly among groups of organisms. What are some of the factors that influence the probability of a given lineage splitting into two?

Several ecological and behavioral factors influence speciation rates

Many factors can influence rates of speciation across groups, including the diet, behavioral complexity, and dispersal abilities of the respective species.

DIET SPECIALIZATION Populations of species that have specialized diets may be more likely to diverge than those with more generalized diets. To investigate the effects of diet specialization on rates of speciation, Charles Mitter and colleagues compared species richness in some closely related groups of true bugs (hemipterans). The common ancestor of these groups was a predator that fed on other insects, but a dietary shift to herbivory (eating plants) evolved at least twice in the groups under study. Herbivorous bugs typically specialize on one or a few closely related species of plants, whereas predatory bugs tend to feed on many different species of insects. A high diversity of host-plant species can lead to a correspondingly high species diversity among herbivorous specialists. The study by Mitter and his colleagues showed that among these insects, the herbivorous groups do indeed contain many more species than do the related predatory groups (**FIGURE 16.17**).

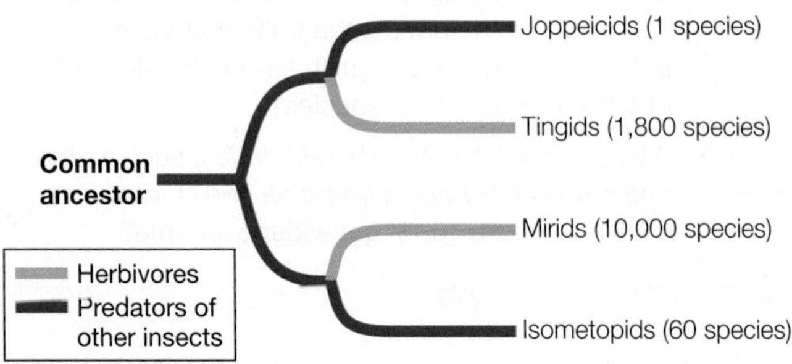

FIGURE 16.17 Dietary Shifts Can Promote Speciation Herbivorous groups of hemipteran insects have speciated several times faster than closely related predatory groups. (After C. Mitter et al. 1988. *Am Nat* 132: 107–128.)

POLLINATION Speciation rates are faster in animal-pollinated than in wind-pollinated plants. Animal-pollinated groups have, on average, 2.4 times as many species as related groups pollinated by wind. Among animal-pollinated plants, speciation rates are correlated with pollinator specialization. In columbines (*Aquilegia*), the rate of evolution of new species has been about three times faster in lineages that have long nectar spurs than in lineages that lack spurs. Why do nectar spurs increase the speciation rate? Apparently it is because spurs restrict the number of pollinator species that visit the flowers, thus increasing opportunities for reproductive isolation (see Figure 16.13).

SEXUAL SELECTION It appears that the mechanisms of sexual selection (see Key Concept 13.2) result in high rates of speciation. Some of the most striking examples of sexual selection are found in birds with polygynous mating systems (meaning one male may attract and mate with many different females). Birdwatchers travel thousands of miles to Papua New Guinea to witness the mating displays of male birds of paradise, which have brightly colored plumage (**FIGURE 16.18A**) and look distinctly different from females of their species—a phenomenon called sexual dimorphism. Males assemble at display grounds called leks, and females come there to choose a mate. After mating, the females leave the display grounds, build their nests, lay their eggs, and feed their offspring with no help from the males. The males remain at the lek to court more females.

The closest relatives of the birds of paradise are the manucodes (**FIGURE 16.18B**). Male and female manucodes differ only slightly in size and plumage (they are sexually monomorphic). They form monogamous pair bonds, and both sexes contribute to raising the young. There are only 5 species of manucodes, compared with 33 species of birds of paradise. By itself, this one comparison would not be convincing evidence that sexually dimorphic clades have higher rates of speciation than do monomorphic clades. However, when biologists examined all the examples of birds in which one clade is sexually dimorphic and the most closely related clade is sexually monomorphic, the sexually dimorphic clades were significantly more likely to contain more species. But why would sexual dimorphism be associated with a higher rate of speciation?

Animals with complex sexually selected behaviors are likely to form new species at a high rate because they make sophisticated discriminations among potential mating partners. They distinguish members of their own species from members of other species, and they make subtle discriminations among members of their own species on the basis of size, shape, appearance, and behavior. Such discriminations can greatly influence which individuals are most successful in mating and producing offspring, so they may lead to rapid evolution of behavioral isolating mechanisms among populations.

DISPERSAL ABILITY Speciation rates are usually higher in groups with poor dispersal abilities than in groups with good dispersal abilities because even narrow barriers can be effective in dividing a species whose members are highly sedentary. Until recently, the Hawaiian Islands had about 1,000 species of land snails, many of which were restricted to a single valley. Because snails move only short distances, the high ridges that separate the valleys were effective barriers to their dispersal. Unfortunately, introductions of other species and changes in habitat have resulted in the recent extinction of most of these unique Hawaiian land snails.

Invasion of new habitats and key innovations can lead to adaptive radiations

The rapid proliferation of a large number of descendant species from a single ancestor species is called an **evolutionary radiation**. Evolutionary radiations often occur when a species colonizes a new area, such as an island archipelago that contains no other closely related species, because of the large number of open ecological niches. If such a rapid proliferation of species results in an array of species that live in a variety of environments and differ in the characteristics they use to exploit those environments, it is referred to as an **adaptive radiation**.

Several remarkable adaptive radiations have occurred in the Hawaiian Islands. In addition to its 1,000 species of land snails, the native Hawaiian biota includes 1,000 species of flowering plants, 10,000 species of insects, and more than 100 bird species. However, there were no amphibians, no terrestrial reptiles, and only one native terrestrial mammal (a bat) on the islands until humans introduced additional species. The 10,000 known native species of insects on the islands are believed to have evolved from about 400 immigrant species; only 7 immigrant species are believed to account for all the native Hawaiian land birds. Similarly, as we saw earlier in this chapter, an adaptive radiation in the Galápagos archipelago resulted in the many species of Darwin's finches, which differ strikingly in the size and shape of their bills and, accordingly, in the food resources they use (see Figure 16.8).

The 28 species of Hawaiian sunflowers called silverswords are an impressive example of an adaptive radiation in plants (**FIGURE 16.19**). DNA sequences show that these species share a relatively recent common ancestor with a species of tarweed from the Pacific coast of North America. Whereas all mainland tarweeds are small, upright herbs (nonwoody plants such as *Madia sativa*; see Figure 16.19), the silverswords include shrubs, trees,

(A) ***Paradisaea minor***

© Tim Laman/National Geographic Creative/Getty Images

(B) ***Manucodia comrii***

Tim Laman/National Geographic Creative/Getty Images

FIGURE 16.18 Sexual Selection Can Lead to Higher Speciation Rates Birds of paradise (the brightly colored male is above the female on the branch) **(A)** and manucodes **(B)** are closely related bird groups of the South Pacific. Speciation rates are much higher among the sexually dimorphic, polygynous birds of paradise (33 species) than among the sexually monomorphic, monogamous manucodes (5 species).

Madia sativa (tarweed)

Dick Culbert/CC BY 2.0

Argyroxiphium sandwicense

© Pung/Shutterstock.com

Wilkesia gymnoxiphium

© Photo Resource Hawaii/Alamy Stock Photo

Dubautia menziesii

© Noble Proctor/The National Audubon Society Collection/Science Source

FIGURE 16.19 Rapid Evolution among Hawaiian Silverswords The Hawaiian silverswords, three closely related genera of the sunflower family, are believed to have descended from a single common ancestor (a plant similar to the tarweed *Madia sativa*) that colonized Hawaii from the Pacific coast of North America. The four plants shown here are more closely related than they appear to be based on their morphology.

and vines as well as both upright and ground-hugging herbs. Silversword species occupy nearly all the habitats of the Hawaiian Islands, from sea level to above the timberline in the mountains. Despite their extraordinary morphological diversification, all silverswords are genetically very similar.

The Hawaiian silverswords are more diverse in size and shape than the mainland tarweeds because their tarweed ancestors first arrived on islands that harbored very few plant species. In particular, there were few trees and shrubs because such large-seeded plants rarely disperse to oceanic islands. Trees and shrubs have evolved from nonwoody ancestors on many oceanic islands. On the mainland, however, tarweeds live in ecological communities that contain many tree and shrub species in lineages with long evolutionary histories. In those environments, opportunities to exploit the "tree" way of life have already been preempted.

REVIEW & APPLY | 16.5

R Dietary specialization, pollinator specialization, sexual selection, and poor dispersal abilities are correlated with high rates of speciation. Open ecological niches present opportunities for adaptive radiations.

A
1. How can pollinator specialization in plants and sexual selection in animals increase rates of speciation?
2. Why do adaptive radiations often occur when a founder species invades an isolated geographic area?

The result of 3.8 billion years of evolution has been many millions of species, each adapted to live in a particular environment and to use environmental resources in a particular way. In the next chapter we will look at the molecular processes that gave rise to that diversity.

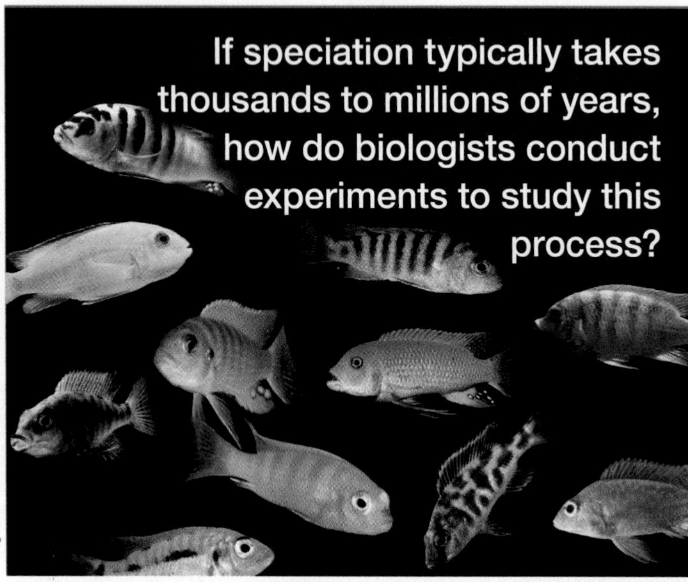

Photos by David McIntyre (courtesy of Exotic Fish and Pet World, Southampton, MA) and © Nantawat Chotsuwan, Andreas Gradin, Mikola Gyorgy, Grigorev Mikhail, Vladimir Wrangel/Shutterstock.com, © iStock.com/Vladimir Wrangel

Answer: Although speciation usually takes thousands to millions of years, and although it is typically studied in natural settings such as Lake Malawi or in field experiments such as the one described in Figure 16.15, some aspects of speciation can be studied and observed in controlled laboratory experiments. Most such experiments use organisms with short generation times, in which evolution is expected to be relatively rapid.

William Rice and George Salt conducted an experiment in which fruit flies (*Drosophila melanogaster*) were allowed to choose food sources in different habitats. The habitats—where mating also took place—were vials in different parts of an experimental cage (**FIGURE 16.20**). The vials differed in three environmental factors: (1) light; (2) the direction (up or down) in which the fruit flies had to move to reach food; and (3) the concentrations of two aromatic chemicals, ethanol and acetaldehyde. In just 35 generations, the two groups of flies that chose the most divergent habitats had become reproductively isolated from each other, having evolved distinct preferences for the different habitats.

The experiment by Rice and Salt (1988. *Am Nat.* 131: 911–917) demonstrated an example of habitat isolation as a prezygotic isolating mechanism. Even though the different habitats were in the same cage, and individual fruit flies were capable of flying from one habitat to the other, habitat preferences were inherited by offspring from their parents, and populations from the two divergent habitats did not interbreed. Similar habitat isolation is thought to have resulted in the early split between cichlids that preferred the rocky versus the sandy shores of Lake Malawi. In controlled experiments like this one, biologists can observe many aspects of the process of speciation directly.

Courtesy of William R. Rice

FIGURE 16.20 Evolution in the Laboratory For their experiments on the evolution of prezygotic isolating mechanisms in *Drosophila melanogaster*, Rice and Salt built an elaborate system of varying habitats contained within vials inside a large fly enclosure. Some groups of flies developed preferences for widely divergent habitats and became reproductively isolated within 35 generations.

16.1 Species Are Reproductively Isolated Lineages on the Tree of Life

Photos by David McIntyre (courtesy of Exotic Fish and Pet World, Southampton, MA) and © Nantawat Chotsuwan, Grigorev Mikhail/Shutterstock.com, © iStock.com/Vladimir Wrangel

- **SPECIATION** is the process by which one **SPECIES** splits into two or more daughter species, which thereafter evolve as distinct lineages.
- The **MORPHOLOGICAL SPECIES CONCEPT** distinguishes species on the basis of physical similarities; it often underestimates or overestimates the actual number of reproductively isolated species.
- The **BIOLOGICAL SPECIES CONCEPT** distinguishes species on the basis of **REPRODUCTIVE ISOLATION**.
- The **LINEAGE SPECIES CONCEPT** recognizes independent evolutionary lineages as species and allows biologists to consider species over evolutionary time.

QUESTIONS

1. What species concept or concepts would paleontologists likely use to group fossil remains into species?
2. How is reproductive isolation, the core of the biological species concept, also related to morphological and lineage species concepts?

16.2 Speciation Is a Natural Consequence of Population Subdivision

Go to ACTIVITY 16.1

- Genetic divergence results from the interruption of gene flow within a population.
- The **Dobzhansky-Muller model** describes how reproductive isolation between two physically isolated populations can develop through the accumulation of incompatible genes or chromosomal arrangements (**FIGURE 16.3**).
- Reproductive isolation increases with increasing genetic divergence between populations (**FIGURE 16.5**).

FIGURE 16.3

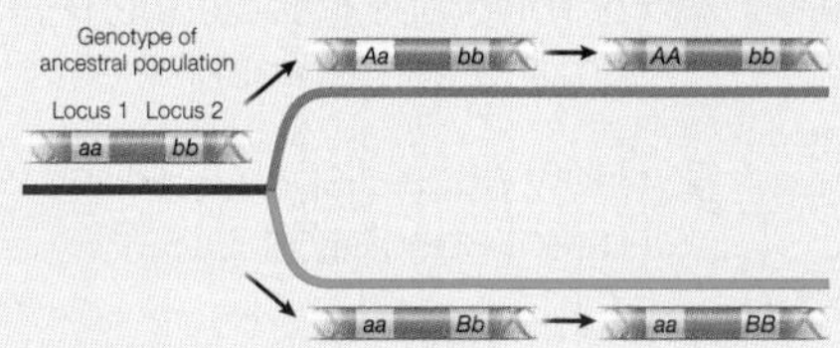

FIGURE 16.5

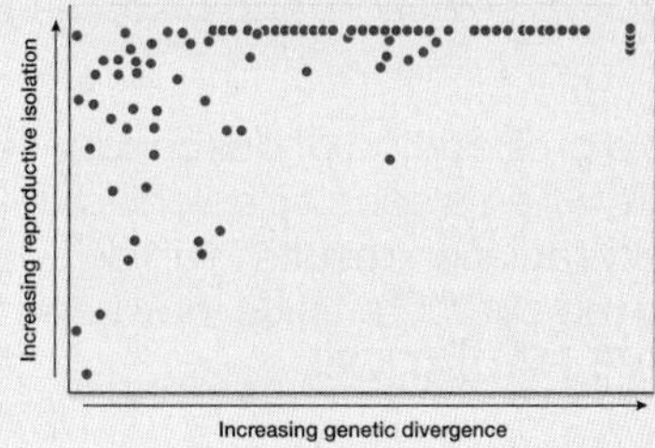

1. Why does the Dobzhansky–Muller model involve changes at multiple gene loci?
2. What kinds of chromosomal rearrangements work particularly well with the Dobzhansky–Muller model? Why?

16.3 Speciation May Occur through Geographic Isolation or in Sympatry

Go to ANIMATIONS 16.1 and 16.2

FIGURE 16.6

Distribution of freshwater stream fishes in the Pliocene

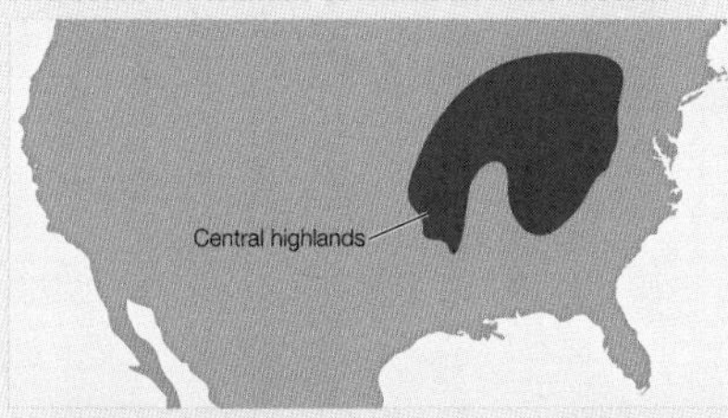

Distribution of freshwater stream fishes in the Pleistocene

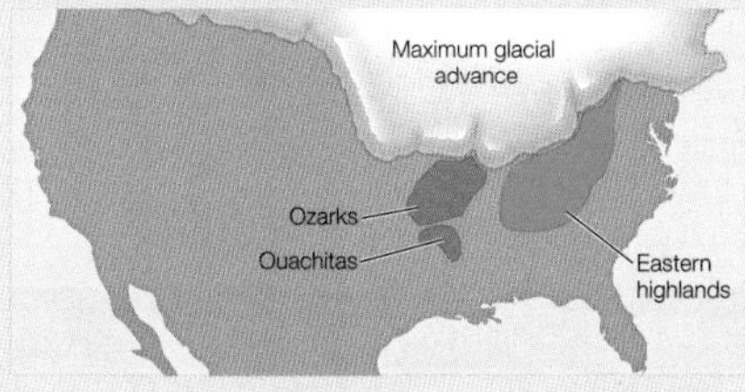

- **ALLOPATRIC SPECIATION**, which results when populations are separated by a physical barrier, is the dominant mode of speciation. This type of speciation may follow founder events, in which some members of a population cross a barrier and found a new, isolated population (**FIGURE 16.6**).
- **SYMPATRIC SPECIATION** results when the genomes of two groups diverge in the absence of physical isolation. It can result from disruptive selection in two or more distinct microhabitats.
- Sympatric speciation can occur within two generations via **POLYPLOIDY**, an increase in the number of chromosomes sets. Polyploidy may arise from chromosome duplications within a species (**AUTOPOLYPLOIDY**) or from hybridization that results in combining the chromosomes of two species (**ALLOPOLYPLOIDY**).

1. What examples of geographic events are likely to give rise to many species (in different groups) simultaneously via allopatric speciation?
2. How does speciation occur so quickly through polyploidy?

16 VISUAL SUMMARY

16.4 Reproductive Isolation Is Reinforced When Diverging Species Come into Contact

- **PREZYGOTIC ISOLATING MECHANISMS** prevent hybridization; **POSTZYGOTIC ISOLATING MECHANISMS** reduce the fitness of hybrids.
- Postzygotic isolating mechanisms lead to **REINFORCEMENT** of prezygotic isolating mechanisms by natural selection (**FIGURE 16.11**).
- **HYBRID ZONES** may form and persist if reproductive isolation between species is incomplete.

FIGURE 16.11

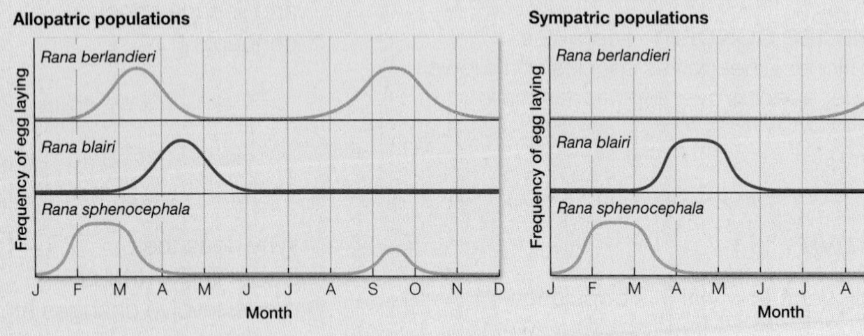

QUESTIONS

1. What are five major types of prezygotic isolating mechanisms?
2. What are three major types of postzygotic isolating mechanisms?

16.5 Speciation Rates Are Highly Variable across Life

- Dietary specialization, pollinator specialization, sexual selection, and dispersal ability all influence speciation rates (**FIGURE 16.17**).
- **EVOLUTIONARY RADIATION** refers to the rapid proliferation of descendant species from a single ancestor species. This often occurs when a species colonizes a new area with unoccupied ecological niches. If evolutionary radiation results in an array of species that live in a variety of environments and differ in the characteristics they use to exploit those environments, it is referred to as **ADAPTIVE RADIATION**.

FIGURE 16.17

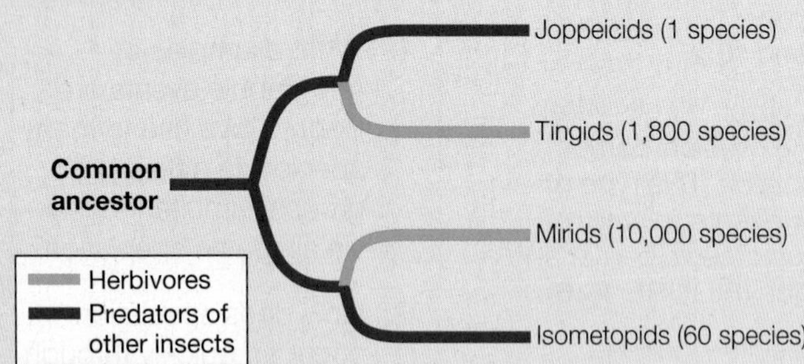

Activity 16.3 Concept Matching: Speciation

1. Why are speciation rates often higher among herbivorous insects than among carnivorous insects?
2. How does dispersal ability affect speciation rates?
3. What biogeographic event often precedes an evolutionary radiation?

Go to **Achieve** for the e-book, animations, activities, and additional resources and assignments.

The History of Life on Earth 17

Goliath beetles (*Goliathus* spp.) are among the largest species of flying insects alive today. But in earlier periods of Earth's history, there were much larger flying insect predators —the size of modern birds of prey. Such giant insects cannot live and fly in Earth's present environment.

What differences in Earth's past environment allowed flying insects to attain much larger sizes than they do today?

You will find the answer to this question on page 409.

KEY CONCEPTS

17.1 Events in Earth's History Can Be Dated

17.2 Changes in Earth's Physical Environment Have Affected the Evolution of Life

17.3 Major Events in the Evolution of Life Can Be Read in the Fossil Record

17.1 | Events in Earth's History Can Be Dated

LEARNING OBJECTIVES

By the end of this key concept you should be able to:

17.1.1 Construct a geological map indicating the ages of exposed rocks and use the map to search for fossils of a given age.

17.1.2 Select appropriate methods for dating fossils and rocks from different geological time periods.

17.1.3 Place important events in biological history onto a time line of Earth's history.

Some evolutionary changes happen rapidly enough to be studied directly and manipulated experimentally. Plant and animal breeding by agriculturalists provides examples of rapid, short-term evolution that we saw in Chapter 13. Other evolutionary changes, such as the appearance of new species and evolutionary lineages, usually take place over much longer time scales.

To understand long-term patterns of evolutionary change, we must not only think in time scales spanning many millions of years, but also consider events and conditions very different from those we observe today. Earth of the distant past was so unlike our present Earth that it would seem like a foreign planet inhabited by strange organisms. The continents were not where they are now, and climates were sometimes dramatically different from those of today. We know this because much of Earth's history is recorded in its rocks.

Geological strata tell us the relative ages of rocks

We cannot tell the ages of rocks just by looking at them, but we can visually determine the ages of rocks *relative to one another*. The first person to formally recognize this fact was the seventeenth-century Danish physician Nicolaus Steno. Steno realized that in undisturbed **sedimentary rocks** (rocks formed by the accumulation of sediments), the oldest layers of rock, or **strata** (singular *stratum*), lie at the bottom, and successively higher strata are progressively younger.

Geologists subsequently combined Steno's insight with their observations of fossils contained in sedimentary rocks. They developed the following principles of **stratigraphy** (the study of geological strata):

- Fossils of similar organisms are found in widely separated places on Earth.
- Certain fossils are always found in younger strata, and certain other fossils are always found in older strata.
- Fossil organisms found in younger strata are more similar to modern organisms than are those found in older strata.

These patterns revealed much about the relative ages of sedimentary rocks and the fossils they contain, as well as patterns in the evolution of life. But geologists still could not tell the actual, as opposed to the relative, ages of rocks. A method for assigning actual ages of rocks did not become available until after radioactivity was discovered at the beginning of the twentieth century.

Radioisotopes provide a way to date rocks

Radioactive isotopes of atoms—radioisotopes—decay in a predictable pattern over long periods. Over a specific time interval, known as a **half-life**, half of the atoms in a radioisotope decay to become a different, stable (nonradioactive) isotope (**FIGURE 17.1A**). The use of this knowledge to date fossils and rocks is known as **radiometric dating**.

To use a radioisotope to date a past event, we must know or estimate the concentration of that isotope at the time of that event, and we must know the radioisotope's half-life. In the case of carbon-14 (^{14}C), a radioisotope of carbon, the production of new ^{14}C in the upper atmosphere—by the reaction of neutrons with nitrogen-14 (^{14}N, a stable isotope of nitrogen)—just balances the natural radioactive decay of ^{14}C into ^{14}N. Therefore the ratio of ^{14}C to the more common stable isotope of carbon, carbon-12 (^{12}C), is relatively constant in living organisms and in their environment. As soon as an organism dies, however, it ceases to exchange carbon compounds with its environment. Its decaying ^{14}C is no longer replenished, and the ratio of ^{14}C to ^{12}C in its remains decreases over time. Paleontologists can use the ratio of ^{14}C to ^{12}C in fossil material to date fossils that are 60,000 years old or younger (and thus the sedimentary rocks that contain those fossils). If fossils are older than that, so little ^{14}C remains that the limits of detection using this particular isotope are reached.

Radiometric dating methods have been expanded and refined

Sedimentary rocks are formed from materials that existed for varying lengths of time before being weathered, fragmented, and transported, sometimes over long distances, to the site of their deposition. Therefore the radioisotopes in sedimentary rock do not contain reliable information about the date of its formation. Radiometric dating of rocks older than 60,000 years requires estimating radioisotope concentrations in **igneous rocks**, which are formed when molten material cools. To date sedimentary strata, geologists search for places where volcanic ash or lava flows have intruded into the sedimentary rock.

A preliminary estimate of the age of an igneous rock determines which radioisotopes can be used to date it (**FIGURE 17.1B**). The decay of potassium-40 (which has a half-life of 1.3 billion years) to argon-40, for example, has been used to date many of the ancient events in the evolution of life. Fossils in the adjacent

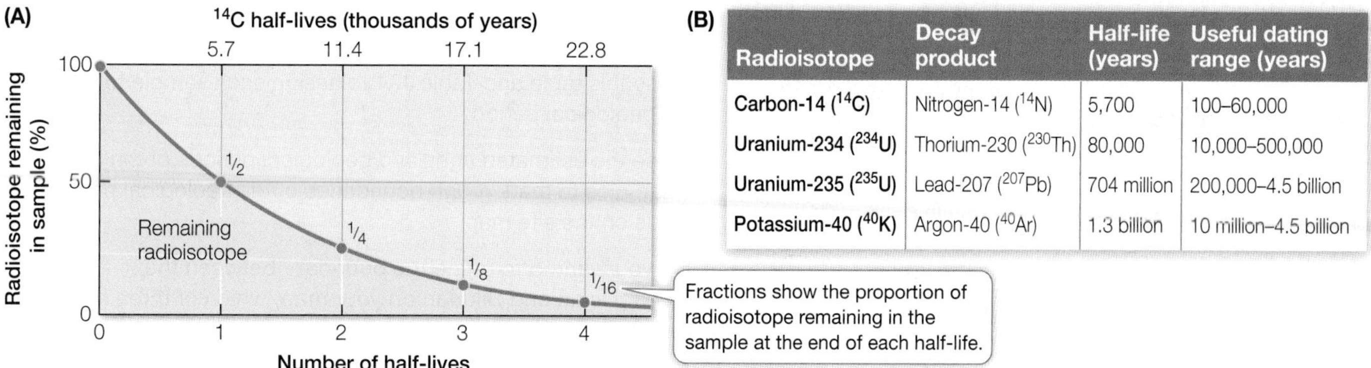

Radioisotope	Decay product	Half-life (years)	Useful dating range (years)
Carbon-14 (^{14}C)	Nitrogen-14 (^{14}N)	5,700	100–60,000
Uranium-234 (^{234}U)	Thorium-230 (^{230}Th)	80,000	10,000–500,000
Uranium-235 (^{235}U)	Lead-207 (^{207}Pb)	704 million	200,000–4.5 billion
Potassium-40 (^{40}K)	Argon-40 (^{40}Ar)	1.3 billion	10 million–4.5 billion

FIGURE 17.1 Radioactive Isotopes Allow Us to Date Ancient Rocks The decay of radioactive isotopes into stable isotopes happens at a steady rate. A half-life is the time it takes for half of the remaining atoms to decay in this way. **(A)** The graph demonstrates the principle of half-life using carbon-14 (^{14}C) as an example. The half-life of ^{14}C is 5,700 years. **(B)** Different radioisotopes have different characteristic half-lives that allow us to estimate the ages of many rocks.

sedimentary rock that are similar to those in other rocks of known ages provide additional clues to the rock's age.

Scientists have used several methods to construct a geological time scale

Radiometric dating of rocks, combined with fossil analysis, is the most powerful method of determining geological age. But in places where sedimentary rocks do not contain suitable igneous intrusions and few fossils are present, paleontologists turn to other dating methods.

One method, known as **paleomagnetic dating**, relates the ages of rocks to patterns in Earth's magnetism, which change over time. Earth's magnetic poles move and occasionally reverse themselves. Both sedimentary and igneous rocks preserve a record of Earth's magnetic field at the time they were formed, and that record can be used to determine the ages of those rocks. Other dating methods use information about continental drift, information about sea level changes, and molecular clocks.

LINK As explained in **Key Concept 14.3**, a molecular clock measures the average rate at which a given gene or protein product accumulates changes, and this rate of change can be used to deduce the geological time when two biological lineages diverged from one another.

Using all of these methods, geologists developed a **geological time scale** (**TABLE 17.1**). They divided the broad history of life into four **eons**. The Hadean refers to the time on Earth before life evolved. The early history of life occurred in the Archean, which ended about the time that photosynthetic organisms first appeared on Earth. Prokaryotic life diversified rapidly in the Proterozoic, and the first eukaryotes in the fossil record date from this time. These three eons are sometimes referred to collectively as Precambrian time, or simply the **Precambrian**. The Precambrian lasted approximately 3.8 billion years and thus accounts for the vast majority of geological time. It was in the Phanerozoic, however—a mere 542-million-year time span—that multicellular eukaryotes rapidly diversified. To emphasize the events of the Phanerozoic, Table 17.1 shows the subdivision of this eon into eras and periods. The boundaries between these divisions of time are based largely on the striking differences geologists observe in the assemblages of fossil organisms contained in successive strata. This geological record of life reveals a remarkable story of a world in which the continents and biological communities are constantly changing.

REVIEW & APPLY | 17.1

R The layering of sedimentary rock strata enables geologists to determine the relative ages of fossils. Assigning actual ages to these strata requires analysis of radioactive decay, paleomagnetic dating, and fossil comparisons across strata. Geologists divide the history of life into eons, eras, and periods based on assemblages of fossil organisms found in successive layers of rocks.

A Imagine you have been assigned the job of producing a geological map of rocks that were formed between 600 and 400 million years ago (mya). You collect a sample from each of ten sites (1–10 on the map below), determine the ratio of ^{206}Pb to ^{238}U for each sample, and use these ratios to estimate the ages of the rock samples, resulting in the table on the following page.

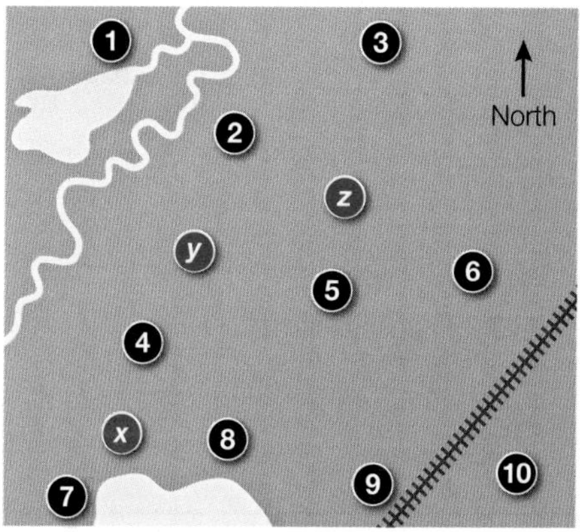

(Continued)

REVIEW & APPLY | 17.1 *(continued)*

Site	$^{206}Pb/^{238}U$ ratio	Estimated age (mya)
1	0.076	474
2	0.077	479
3	0.069	431
4	0.081	505
5	0.076	474
6	0.070	435
7	0.089	550
8	0.080	500
9	0.079	495
10	0.077	479

1. Use this table and Table 17.1 to assign each sample to a geological period.
2. Use the estimated ages and geological periods of the ten samples to mark rough boundaries of the geological periods on the site map.
3. If you wanted to refine the boundary between the Ordovician and Silurian on your map, which of three new sampling sites—*x*, *y*, or *z*—would you add to your analysis next?
4. If you were interested in finding fossils of terrestrial plants or animals, where on this map would you concentrate your searches?

TABLE **17.1** | Earth's Geological History

Eon	Era	Period	Onset[a]	Major physical changes on Earth
Phanerozoic (~0.5 billion years long)	Cenozoic	Quaternary (Q)	2.6 mya	Cold/dry climate; repeated glaciations
		Neogene (N)[b]	23.0 mya	Continents near current positions; North and South America became connected
		Paleogene (Pg)[b]	65.5 mya	Climate warms, then cools in response to changing carbon dioxide levels in Earth's atmosphere
	Mesozoic	Cretaceous (K)	145.5 mya	Laurasian continents attached to one another; Gondwana begins to drift apart; meteorite strikes near current Yucatán Peninsula at end of period
		Jurassic (J)	201.6 mya	Two large continents form: Laurasia (north) and Gondwana (south); warm climate
		Triassic (Tr)	251.0 mya	Pangaea begins to drift apart; hot/humid climate
	Paleozoic	Permian (P)	299 mya	Extensive lowland swamps; O_2 levels 50% higher than present; by end of period continents aggregate to form Pangaea, and O_2 levels drop rapidly
		Carboniferous (C)	359 mya	Climate cools; marked latitudinal climate gradients
		Devonian (D)	416 mya	Continents collide at end of period; giant meteorite probably strikes Earth
		Silurian (S)	444 mya	Sea levels rise; two large land masses emerge; hot/humid climate
		Ordovician (O)	488 mya	Massive glaciation; sea level drops 50 meters
		Cambrian (C)	542 mya	Atmospheric O_2 levels approach current levels
Proterozoic	Collectively called the Precambrian (~4 billion years long)		2.5 bya	Atmospheric O_2 levels increase from negligible to about 18%; "snowball Earth" from about 750 to 580 mya
Archean			3.8 bya	Earth accumulates more atmosphere (still almost no O_2); meteorite impacts greatly reduced
Hadean			4.5–4.6 bya	Formation of Earth; cooling of Earth's surface; atmosphere contains almost no free O_2; oceans form; Earth under almost continuous bombardment from meteorites

[a]mya, million years ago; bya, billion years ago.
[b]The Neogene and Paleogene are often referred to as the Tertiary when discussed together.

Media Clip 17.1 The Age of Earth

As geologists began to develop accurate ways to measure the age of Earth, they began to understand that Earth is far older than anyone had previously understood. During its 4.5-billion-year history, Earth has undergone massive physical changes. These changes have influenced the evolution of life, and life, in its turn, has influenced Earth's physical environment.

17.2 Changes in Earth's Physical Environment Have Affected the Evolution of Life

LEARNING OBJECTIVES

By the end of this key concept you should be able to:

17.2.1 **Characterize the relative impact of weather events and climate changes on Earth's physical conditions.**

17.2.2 **Explain how the production of atmospheric oxygen has affected Earth and its living organisms.**

17.2.3 **Analyze the connections between a specific geological event or environmental change (e.g., fluctuations in ocean currents or sea levels, volcanic eruptions, meteorite strikes) and an evolutionary trend among living organisms.**

Major events in the history of life
Humans evolve; many large mammals become extinct
Considerable movement of plants and animals between North and South America through the Panamanian Isthmus
Diversification of major vertebrate groups, flowering plants, and insects following the mass extinction at the end of the Cretaceous
Dinosaurs continue to diversify; mass extinction at end of period (~76% of species lost)
Diverse dinosaurs; radiation of ray-finned fishes; first fossils of flowering plants
Early dinosaurs; first mammals; marine invertebrates diversify; mass extinction at end of period (~65% of species lost)
Reptiles diversify; giant amphibians and flying insects present; mass extinction at end of period (~96% of species lost)
Extensive fern/horsetail/giant club moss forests; first reptiles; insects diversify
Jawed fishes diversify; first insects and amphibians; mass extinction at end of period (~75% of marine species lost)
Jawless fishes diversify; first ray-finned fishes; plants and animals colonize land
Mass extinction at end of period (~75% of species lost)
Rapid diversification of multicellular animals; diverse photosynthetic protists
Origin of photosynthesis, multicellular organisms, and eukaryotes
Origin of life; prokaryotes flourish
Life not yet present

As we saw in Key Concept 17.1, the Phanerozoic is notable for the rapid diversification of multicellular eukaryotes. The diversity of multicellular organisms has not, however, increased steadily through time. Throughout the history of life, new species have arisen while other species have gone extinct. Here we consider some of the physical changes on Earth that have resulted in dramatic changes in life's diversity.

The continents have not always been where they are today

The globes and maps that adorn our walls, shelves, and books give an impression of a static Earth. It would be easy for us to assume that the continents have always been where they are. But we would be wrong. The idea that Earth's land masses have changed their positions over the millennia, and that they continue to do so, was first put forth in 1912 by the German meteorologist and geophysicist Alfred Wegener. His idea, known as **continental drift**, was initially met with skepticism and resistance. By the 1960s, however, physical evidence and increased understanding of **plate tectonics**—the geophysics of the movement of major land masses—had convinced virtually all geologists of the reality of Wegener's vision. Plate tectonics provided the geological mechanism that explained Wegener's hypothesis of continental drift.

Earth's crust consists of several solid plates. Thick continental and thinner oceanic plates overlie a more malleable layer of Earth's mantle. Heat produced by radioactive decay deep in Earth's core sets up large-scale convection currents in the mantle (**FIGURE 17.2A**). New crust is formed as mantle material rises between diverging plates, pushing them apart.

Where oceanic plates and continental plates converge, the thinner oceanic plate is forced underneath the thicker continental plate, a process known as **subduction**. Subduction results in volcanism and mountain building on the continental boundary. For example, in the Pacific Northwest of North America, a series of volcanoes formed the Cascade mountain range as the Juan de Fuca oceanic plate has been subducted beneath a portion of the continental North American Plate (**FIGURE 17.2B**). When two oceanic plates collide, one is also subducted below the other, producing a deep oceanic trench and associated volcanic activity.

When two thick continental plates collide, neither plate is subducted. Instead, the plates push up against one another, forming high mountain chains. The highest mountain chain in the world, the Himalayas, was formed this way when the Indian Plate collided with the Eurasian Plate. When continental plates diverge, new crust forms in the intervening spaces, resulting in deep clefts called rift valleys in which large freshwater lakes typically form.

Many physical conditions on Earth have oscillated in response to plate tectonic processes. We now know that the movement of

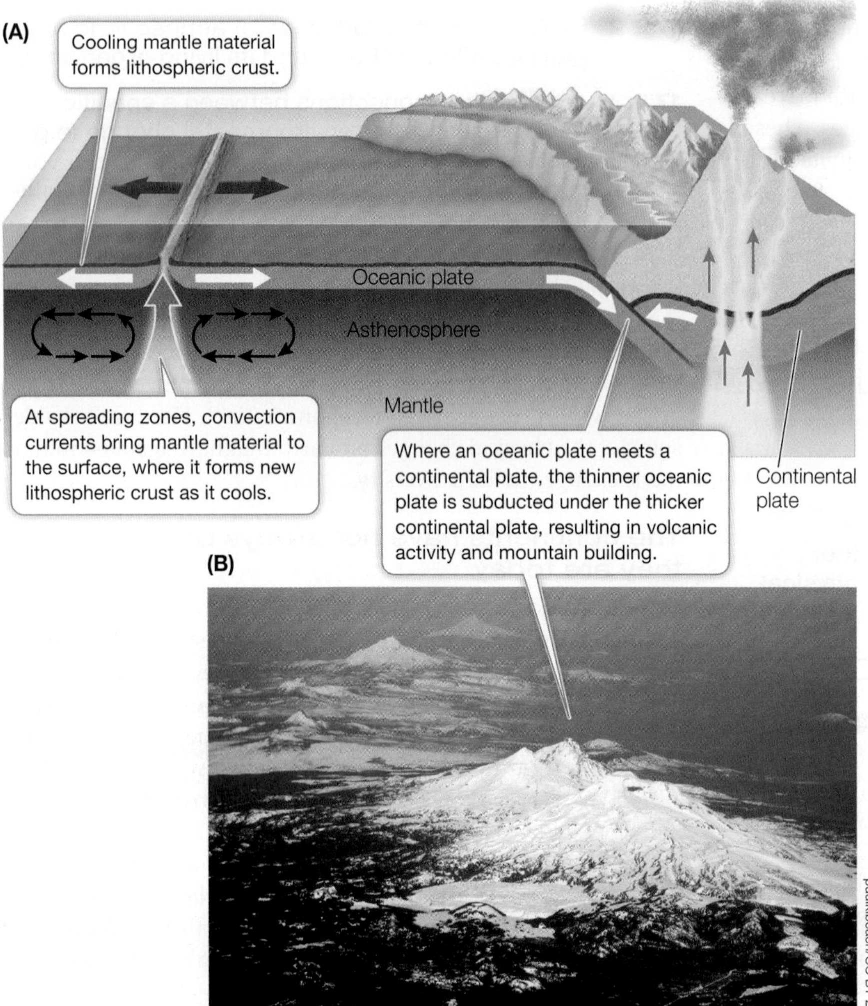

FIGURE 17.2 Plate Tectonics and Continental Drift (A) The heat of Earth's core generates convection currents in the viscous mantle material underlying the oceanic and continental plates. Those currents push the continental plates, along with the land masses they carry, together or apart. Where plates collide, one may slide under the other, creating mountain ranges and often volcanoes. **(B)** The Cascade Range of the Pacific Northwest of North America is an example of a mountain chain produced by subduction of an oceanic plate under a continental plate.

Why are both shores of the Pacific Ocean ringed with volcanic mountain ranges?

Media Clip 17.2 Lava Flows and Magma Explosions

the plates has sometimes brought continents together and at other times has pushed them apart, as seen in the maps across the top of Figure 17.12. The positions and sizes of the continents influence oceanic circulation patterns, global climates, and sea levels. Sea levels are influenced directly by plate tectonic processes (which can influence the depth of ocean basins) and indirectly by oceanic circulation patterns, which affect patterns of glaciation. As climates cool, glaciers form and tie up water over land masses; as climates warm, glaciers melt and release water.

Animation 17.1 Movement of Continents

Some of these dramatic changes in Earth's physical parameters resulted in **mass extinctions**, during which a large proportion of the species living at the time disappeared. These mass extinctions are the cause of the striking differences in fossil assemblages that geologists used to determine major geological time periods. After each mass extinction, the diversity of life rebounded, but recovery took millions of years.

Earth's climate has shifted between hot and cold conditions

Through much of Earth's history, the climate was considerably warmer than it is today, and temperatures decreased more gradually toward the poles. At other times, Earth was colder than it is today. Rapid drops in sea levels throughout the history of Earth have resulted mainly from increased global glaciation (**FIGURE 17.3**). Many of these drops in sea levels were accompanied by mass extinctions—particularly of marine organisms, which could not survive the disappearance of the shallow seas that covered vast areas of the continental shelves.

Earth's cold periods were separated by long periods of milder climates. Because we are living in one of the colder periods, it is difficult for us to imagine the mild climates that were found at high latitudes in the past. The Quaternary, which began 2.6 million years ago (mya) and extends into the present, has been marked by a series of glacial advances, interspersed with warmer interglacial intervals during which the glaciers retreated.

"Weather" refers to the daily events at a given location, such as individual storms and the high and low temperatures on a given day. "Climate" refers to long-term average expectations over the various seasons at a given location. Weather often changes rapidly, whereas climates typically change slowly. However, major climate shifts have taken place over periods as short as 5,000 to 10,000 years, primarily as a result of changes in Earth's orbit around the Sun. A few climate shifts have been even more rapid. For example, during one Quaternary interglacial period, the ice-locked Antarctic Ocean became nearly ice-free in less than 100 years. Some climate changes have been so rapid that the extinctions caused by them appear to be nearly instantaneous in the fossil record. Such rapid changes are usually caused by sudden shifts in ocean currents.

We are currently living in a time of rapid climate change caused largely by a buildup of atmospheric CO_2, primarily from the burning of fossil fuels by human populations. We are reversing the energy transformations accrued in the burial and decomposition of organic material that occurred (especially) in the Carboniferous, Permian, and Triassic, which gave rise to the fossil fuels we are using today. But we are burning these fuels over a few hundred years, rather than the many millions of years over which those deposits

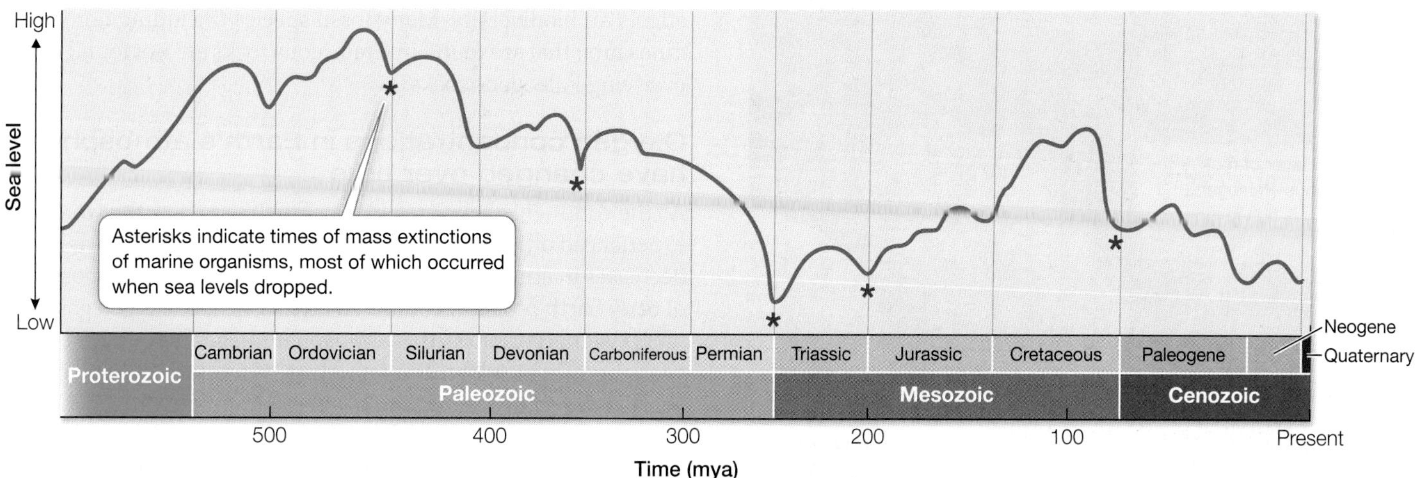

FIGURE 17.3 Sea Levels Have Changed Repeatedly Rapid drops in sea level are associated with periods of globally cooler temperatures and increased glaciation. Most mass extinctions of marine organisms have coincided with low sea levels. (After A. Hallum. 1984. *Annu Rev Earth Planet Sci* 12: 205–243.)

Why do cooler global temperatures result in lower sea levels?

accumulated. The current rate of increase of atmospheric CO_2 is unprecedented in Earth's history. A doubling of the atmospheric CO_2 concentration—which may happen during the current century—is expected to increase the average temperature of Earth, change rainfall patterns, melt glaciers and ice caps, and raise sea levels.

LINK The consequences of today's rapid climate changes are discussed in **Key Concept 42.5**.

Volcanoes have occasionally changed the history of life

Most volcanic eruptions produce only local or short-lived effects, but a few large volcanic eruptions have had major consequences for life. When Krakatau (a volcanic island in the Sunda Strait off Indonesia) erupted in 1883, it ejected more than 25 cubic kilometers of ash and rock, as well as large quantities of sulfur dioxide gas (SO_2). The SO_2 was ejected into the stratosphere and carried by high-altitude winds around the planet. Its presence led to high concentrations of sulfurous acid (H_2SO_3) in high-altitude clouds, creating a "parasol effect" so that less sunlight reached Earth's surface. Global temperatures dropped by 1.2°C in the year following the eruption, and global weather patterns showed strong effects for another 5 years. More recently, the eruption of Mount Pinatubo in the Philippines in 1991 (**FIGURE 17.4**) temporarily reduced global temperatures by about 0.5°C.

Although these individual volcanoes had only relatively short-term effects on global temperatures, they suggest that the simultaneous eruption of many volcanoes could have a much stronger effect on Earth's climate. What would cause many volcanoes to erupt at the same time? The collision of continents during the Permian, about 275 million years ago, formed a single, gigantic land mass and caused massive volcanic eruptions as the continental plates overrode one another (see Figure 17.2). Emissions from these eruptions blocked considerable sunlight, contributing to the advance of glaciers and a consequent drop in sea levels (see Figure 17.3). Thus volcanoes were probably responsible, at least in part, for the greatest mass extinction in Earth's history.

Courtesy of Dave Harlow, U.S. Geological Survey

FIGURE 17.4 Volcanic Eruptions Can Cool Global Temperatures The eruption of Mount Pinatubo in 1991 increased the concentrations of sulfurous acid in high-altitude clouds, which temporarily lowered global temperatures by about 0.5°C.

Extraterrestrial events have triggered changes on Earth

Asteroids are large rocky bodies in orbit around the Sun. If an asteroid, or even a rock too small to be called an asteroid, collides

FIGURE 17.5 Evidence of a Meteorite Impact The white layers of rock are Cretaceous in age, whereas the layers at the upper left were deposited in the Paleogene. Between the two is a thin, dark layer of clay that contains large amounts of iridium, a metal common in some meteorites but rare on Earth. Its high concentration in this sediment layer, deposited about 65.5 million years ago, suggests the impact of a large meteorite at that time.

with Earth and survives its passage through our atmosphere to hit the surface, then it is called a meteorite. At least 30 meteorites the of sizes between tennis and soccer balls strike Earth each year. Collisions of larger meteorites or comets are rare, but such collisions have probably been responsible for several mass extinctions. Several types of evidence tell us about these collisions. Their craters, and the dramatically disfigured rocks that result from their impact, are found in many places. Geologists have discovered compounds in these rocks that contain helium and argon with isotope ratios characteristic of meteorites, which are very different from the ratios found elsewhere on Earth.

A meteorite impact from the Chicxulub asteroid, which was 10 to 15 kilometers in diameter, caused or contributed to a mass extinction at the end of the Cretaceous (about 65.5 mya). The first clue that a meteorite was responsible came from the abnormally high concentrations of the element iridium found in a thin layer separating rocks deposited during the Cretaceous from rocks deposited during the Paleogene (**FIGURE 17.5**). Iridium is abundant in some meteorites, but it is exceedingly rare on Earth's surface. When scientists then discovered a circular crater 180 kilometers in diameter buried beneath the northern coast of the Yucatán Peninsula of Mexico, they constructed the following scenario. When it collided with Earth, the meteorite released energy equivalent to that of 100 million megatons of high explosives, creating great tsunamis. A massive plume of debris rose into the atmosphere, spread around Earth, and descended. The descending debris heated the atmosphere to several hundred degrees and ignited massive fires. It also blocked the Sun, preventing plants from photosynthesizing. The settling debris formed the iridium-rich layer. About a billion tons of soot with a composition matching that of smoke from forest fires was also deposited. These events had devastating effects on biodiversity. Many fossil species (including non-avian dinosaurs) that are found in Cretaceous rocks are not found in the overlying Paleogene rocks.

Oxygen concentrations in Earth's atmosphere have changed over time

As the continents have moved over Earth's surface, the world has experienced other physical changes, including large increases and decreases in atmospheric oxygen concentrations. The atmosphere of early Earth probably contained little or no free oxygen gas (O_2).

Oxygenation of Earth's atmosphere began about 2.5 billion years ago (bya), when the ancestors of modern cyanobacteria evolved the ability to use water as the source of hydrogen ions for photosynthesis. By chemically splitting H_2O, these bacteria generated O_2 as a waste product. They also made electrons available for reducing CO_2 to form the carbohydrate end-products of photosynthesis (see Key Concept 5.5). The O_2 they produced dissolved in water and reacted with dissolved iron. The reaction product then precipitated as iron oxide, which accumulated in alternating layers of red and dark rock known as banded iron formations (**FIGURE 17.6A**). These formations provide evidence for the earliest photosynthetic organisms. As photosynthetic organisms continued to release O_2, oxygen gas began to accumulate in the atmosphere.

One group of O_2-generating cyanobacteria formed rocklike structures called stromatolites, which are abundantly preserved in the fossil record (**FIGURE 17.6B**). To this day, cyanobacteria still form stromatolites in a few very salty places (**FIGURE 17.6C**). Cyanobacteria liberated enough O_2 to open the way for the evolution of oxidation reactions as the energy source for the synthesis of ATP.

As oxygen increased in the atmosphere, ultraviolet radiation from the Sun began to produce increased concentrations of ozone (O_3) in the stratosphere, 20 to 30 kilometers above Earth's surface. This **ozone layer** absorbs much of the Sun's damaging ultraviolet radiation and prevents it from reaching Earth's surface, which allowed expansion of life on land.

About 1.5 billion years ago, a cyanobacteria-like ancestor became symbiotic within eukaryotic cells, leading to the evolution of chloroplasts in photosynthetic plants and other eukaryotes. Eventually photosynthetic plants would cover much of the land, further increasing the production of atmospheric oxygen.

Thus the evolution of life irrevocably changed the physical nature of Earth. Those physical changes, in turn, influenced the evolution of life. When it first appeared in the atmosphere, O_2 was toxic to most of the anaerobic prokaryotes that inhabited Earth at the time. Over millennia, however, prokaryotes that evolved the ability to tolerate and use O_2 not only survived but gained the advantage. Aerobic metabolism proceeds more rapidly, and harvests energy more efficiently, than anaerobic metabolism. Organisms with aerobic metabolism replaced anaerobes in most of Earth's environments.

An atmosphere rich in O_2 also made possible larger and more complex organisms. Small single-celled aquatic organisms can obtain enough oxygen by simple diffusion even when dissolved oxygen concentrations in the water are very low. Larger single-celled organisms, however, have lower surface area-to-volume ratios. To obtain enough oxygen by simple diffusion, larger organisms must live in an environment with a relatively high oxygen

(A) Banded iron formations

(B) Vertical section through a fossil stromatolite

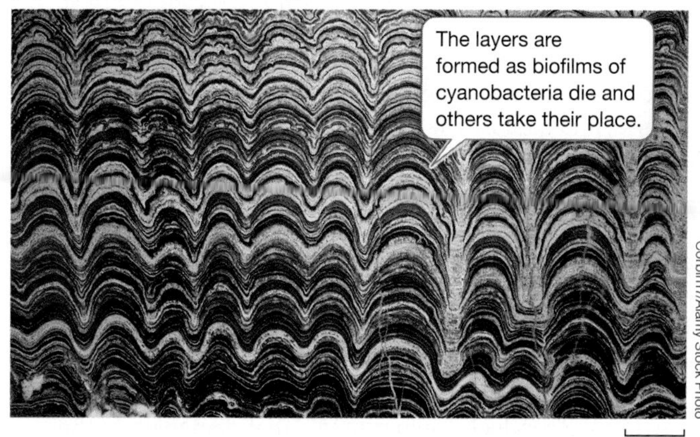

(C) Living stromatolites

FIGURE 17.6 Evidence of Early Photosynthesis **(A)** The alternating red and dark layers in this 2.25-billion-year-old sedimentary rock formation from Lake Superior resulted from a reaction between dissolved iron and the atmospheric oxygen produced by Earth's first photosynthetic organisms. The chemical reaction produced nearly pure iron oxide, or hematite, which forms the gray, metallic layers in this sample. The red bands are jasper tinged with much smaller amounts of iron oxide. **(B)** O_2-generating cyanobacteria formed multi-layered sedimentary deposits called stromatolites. This photo shows a vertical section through a fossil stromatolite. **(C)** These rocklike structures are living stromatolites that thrive in the very salty waters of Shark Bay in Western Australia.

concentration. Bacteria can thrive at 1 percent of the current oxygen concentration, but eukaryotic cells require levels that are at least 2–3 percent of the current concentration. For concentrations of dissolved oxygen in the oceans to have reached these levels, much higher atmospheric concentrations were needed.

Probably because it took many millions of years for Earth to develop an oxygenated atmosphere, only single-celled prokaryotes lived on Earth for more than 2 billion years. About 1.5 billion years ago, atmospheric O_2 concentrations became high enough for larger eukaryotic cells to flourish (**FIGURE 17.7**). Further increases in atmospheric O_2 concentrations in the late Precambrian enabled several groups of multicellular eukaryotes to evolve and diversify.

Oxygen concentrations increased again during the Carboniferous and Permian because of the evolution of large vascular plants. These plants lived in the expansive

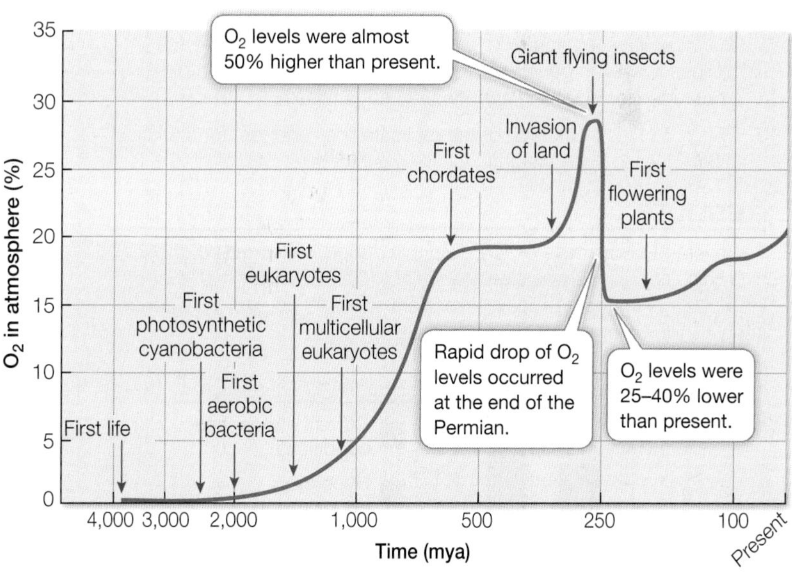

FIGURE 17.7 Atmospheric Oxygen Concentrations Have Changed over Time Changes in atmospheric oxygen concentrations have strongly influenced, and have been influenced by, the evolution of life. (Note that the horizontal axis of the graph is on a logarithmic scale.) (Data from R. A. Berner. 2006. *Geochimica et Cosmochimica Acta* 70: 5653–5664; R. Frei et al. 2009. *Nature* 461: 250–253.)

Why did oxygen levels decrease at the end of the Permian?

lowland swamps that existed at the time (see Table 17.1). Massive amounts of organic material were buried in these swamps as the plants died, leading to the formation of Earth's vast coal deposits. Because the buried organic material was not subject to oxidation as it decomposed, and because the living plants were producing large quantities of O_2, atmospheric O_2 increased to concentrations that have not been reached again in Earth's history (see Figure 17.7).

The drying of the lowland swamps at the end of the Permian reduced burial of organic matter as well as the production of O_2, so atmospheric O_2 concentrations dropped rapidly. Over the past 200 million years, with the diversification of flowering plants, O_2 concentrations have again increased, but not to the levels that characterized the Carboniferous and Permian.

Biologists have conducted experiments that demonstrate the changing selection pressures that can accompany changes in atmospheric O_2 concentrations. When fruit flies (*Drosophila melanogaster*) are raised in hyperoxic conditions (i.e., with artificially increased atmospheric concentrations of O_2), they evolve larger body sizes in just a few generations (**INVESTIGATION FIGURE 17.8**). The present atmospheric O_2 concentrations appear to constrain body size in these flying insects, whereas increases in O_2 appear to relax those constraints. This experiment demonstrates that the stabilizing selection on body size at present O_2 concentrations can quickly switch to directional selection for a change in body size in response to a change in O_2 concentrations.

Stabilizing selection resists change from a current state, whereas directional selection results in change away from a current state. See **Key Concept 13.4**.

INVESTIGATION

FIGURE 17.8 Atmospheric Oxygen Concentrations and Body Size in Insects C. Jaco Klok and his colleagues asked whether insects raised in hyperoxic conditions would evolve to be larger than their counterparts raised under today's atmospheric conditions. They raised strains of fruit flies (*Drosophila melanogaster*) under both conditions to test the effects of increased O_2 concentrations on the evolution of body size.[a]

HYPOTHESIS

In hyperoxic conditions, increased partial pressure of oxygen results in evolution of increased body size in flying insects.

METHOD

1. Separate a population of fruit flies into multiple lines.
2. Raise half the lines in current atmospheric (control) conditions; raise the other lines in hyperoxic (experimental) conditions. Continue all lines for seven generations. Raise the F_8 individuals of all lines under identical (current) atmospheric conditions.
3. Weigh 50 flies from each of the replicate lines and test for statistical differences in body weight.

RESULTS

The average body mass of F_8 individuals of both sexes raised under hyperoxic conditions was significantly ($P < 0.001$) greater than that of individuals in the control lines. Error bars show 95 percent confidence intervals for the mean (see Appendix B).

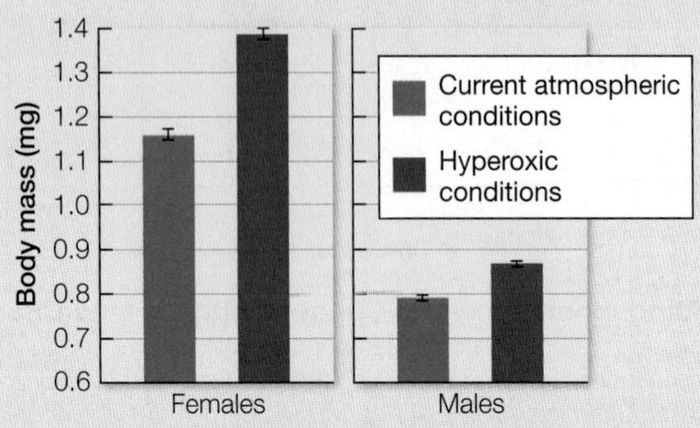

CONCLUSION

Increased O_2 concentrations led to evolution of larger body size in fruit flies, consistent with the trends seen among other flying insects in the fossil record.

ANALYZE THE DATA

The table shows the average body masses of the flies raised in hyperoxic conditions in the F_0 (i.e., before the first generation in hyperoxia), F_1, F_2, and F_8 generations.

	Average body mass (mg)	
Generation	**Males**	**Females**
F_0	0.732	1.179
F_1	0.847	1.189
F_2	0.848	1.254
F_8	0.878	1.392

1. Graph body mass versus generation for males and females.
2. Do the rates of evolution of larger body size appear to be constant throughout the experiment?
3. If you doubled the number of generations in the experiment, would you expect the increase in body mass seen under the hyperoxic conditions to double? Why or why not?

[a]C. J. Klok et al. 2009. *J Evol Biol* 22: 2496–2504.

A similar **Analyze the Data** exercise may be assigned in **Achieve**.

THINK LIKE A SCIENTIST

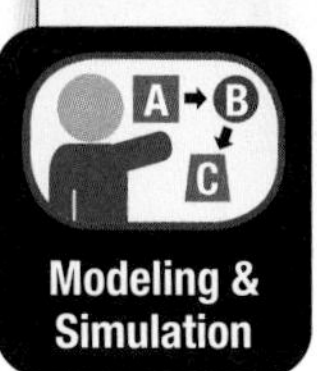

Changes in Earth's physical environment have affected the evolution of life

In the experiment shown in Investigation Figure 17.8, body mass of individuals in the experimental populations of *Drosophila* increased (on average) about 2 percent per generation in the high-oxygen environment (although the rate of increase was not constant over the experiment). In the Permian, giant flying insects, such as dragonflies the size of modern hawks, inhabited Earth. Is the rate of increase in body mass seen in Investigation Figure 17.8 sufficient to account for the giant insects of the Permian? How long would it take for giant insect body size to evolve?

Here you will use quantitative reasoning and a simple model of selection to estimate how quickly insect body size could have evolved in response to higher atmospheric oxygen concentrations.

1. Suppose that the average rate of increase in dragonfly size during the Permian was much slower than the rate observed in the experiment in Investigation Figure 17.8. We'll assume that the actual rate of increase for dragonflies was just 0.01 percent per generation, rather than the 2 percent observed over a few generations for *Drosophila*. We'll also assume that dragonflies complete only 1 generation per year (as opposed to 40 or more generations for *Drosophila*). Starting with an average body mass of 1 gram, calculate the projected increase in body mass over 50,000 years.*
2. What percentage of the Permian does 50,000 years represent? Use Table 17.1 for your calculation.
3. Do you think that increased oxygen concentrations during the Permian were sufficient to account for the evolution of giant dragonflies? Why or why not?
4. Critics of evolution sometimes argue that major groups appear very suddenly in the fossil record, without evidence of intermediate forms. Given your calculations above, would you expect to see sudden changes in the fossil record, even if the changes occurred gradually over 50,000 generations?

*This calculation is similar to computing compound interest for a savings account. Use the formula $W = S(1 + R)^N$, where W = the final mass, S = the starting mass (1 gram), R = the rate of increase per generation (0.0001 in this case), and N = the number of generations.

REVIEW & APPLY | 17.2

R Physical conditions on Earth have changed dramatically over time. Changes in Earth's climate and sea levels have had major effects on biological evolution. Continental drift, volcanic eruptions, and large meteorite strikes have contributed to major climate changes during Earth's history, and many of these climate shifts have resulted in mass extinction events. Changes in atmospheric concentrations of O_2 have also influenced the evolution of life, and the evolution of life has, in turn, affected the physical nature of the planet.

A
1. Explain why an occasional major winter blizzard is irrelevant to discussions of global climate warming.
2. Describe changes in the environment that may have favored the evolution of groups of multicellular organisms near the end of the Precambrian.

The many dramatic physical events of Earth's history have influenced the nature and timing of evolutionary changes among Earth's living organisms. We will now look more closely at some of the major events that characterize the history of life on Earth.

17.3 | Major Events in the Evolution of Life Can Be Read in the Fossil Record

LEARNING OBJECTIVES

By the end of this key concept you should be able to:

17.3.1 **List four major environmental changes in Earth's history and explain how they have affected biodiversity.**

17.3.2 **Explain what is meant by the "Cambrian explosion" and identify the factors that enabled rapid diversification of multicellular organisms.**

How do we know about the physical changes in Earth's environment and their effects on the evolution of life? To reconstruct life's history, scientists rely heavily on the fossil record. As we have seen, geologists divided Earth's history into eons, eras, and periods based on their distinct fossil assemblages (see Table 17.1). Biologists refer to the assemblage of all organisms of all kinds living at a particular time or place as a **biota**. All of the plants living at a particular time or place are its **flora**, and all of the animals are its **fauna**.

About 300,000 species of fossil organisms have been described, and the number steadily grows. The number of named species, however, is only a tiny fraction of the species that have ever lived. We do not know how many species lived in the past, but we have ways of making reasonable estimates. Of the present-day biota, about 1.8 million species have been named. The actual number of living species is probably well over 10 million, and possibly much higher, because many species have not yet been discovered and described by biologists. So the number of described fossil species is only about 3 percent of the estimated minimum number of living species. Life has existed on Earth for about 3.8 billion years. Many species last only a few million years before undergoing speciation or going extinct. From this we know that Earth's biota must have turned over many times during geological history. So the total number of species that have lived over evolutionary time must vastly exceed the number living today. Why have only about 300,000 of these tens of millions of species been described from fossils to date?

Several factors contribute to the paucity of fossils

Only a tiny fraction of organisms ever become fossils, and only a tiny fraction of fossils are ever discovered by paleontologists. Most organisms live and die in oxygen-rich environments in which they quickly decompose. Organisms are not likely to become fossils unless they are transported by wind or water to sites that lack oxygen, where decomposition proceeds slowly or not at all. Furthermore, geological processes transform many rocks, destroying the fossils they contain, and many fossil-bearing rocks are deeply buried and inaccessible. Paleontologists have studied only a tiny fraction of the sites that contain fossils, although they find and describe many new ones every year.

The fossil record is most complete for marine animals that had hard skeletons (which resist decomposition). Among the nine major animal groups with hard-shelled members, approximately 200,000 species have been described from fossils—roughly twice the number of living marine species in these same groups. Paleontologists lean heavily on these groups in their interpretations of the evolution of life. Insects and spiders are also relatively well represented in the fossil record because they are numerically abundant and have hard exoskeletons (**FIGURE 17.9**). The fossil record, though incomplete, is good enough to document clearly the factual history of the evolution of life.

FIGURE 17.9 Fossils in Amber Chunks of amber—fossilized tree resin—often contain detailed fossils such as this spider and three smaller insects, which were preserved when they became trapped in the sticky resin.

By combining information about physical changes during Earth's history with evidence from the fossil record, scientists have composed portraits of what Earth and its inhabitants may have looked like at different times. We know in general where the continents were and how life changed over time, but many of the details are poorly known, especially for events in the more remote past.

Precambrian life was small and aquatic

Life first appeared on Earth about 3.8 billion years ago (**FIGURE 17.10**). The fossil record of organisms that lived prior to the Phanerozoic is fragmentary, but it is good enough to establish that the total number of species and individuals increased dramatically in the late Precambrian.

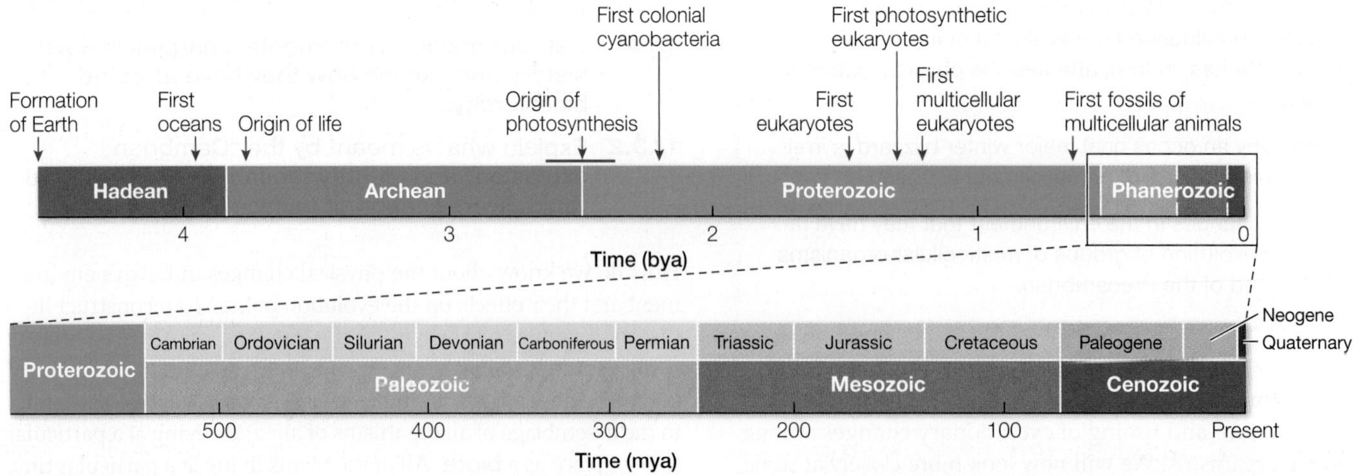

FIGURE 17.10 A Sense of Life's Time The top time line shows the 4.5-billion-year history of Earth. Most of this history is accounted for by the Precambrian, a 3.8-billion-year time span that saw the origin of life and the evolution of cells, photosynthesis, and multicellularity. The most recent 600 million years are expanded in the bottom time line and detailed in Figure 17.12.

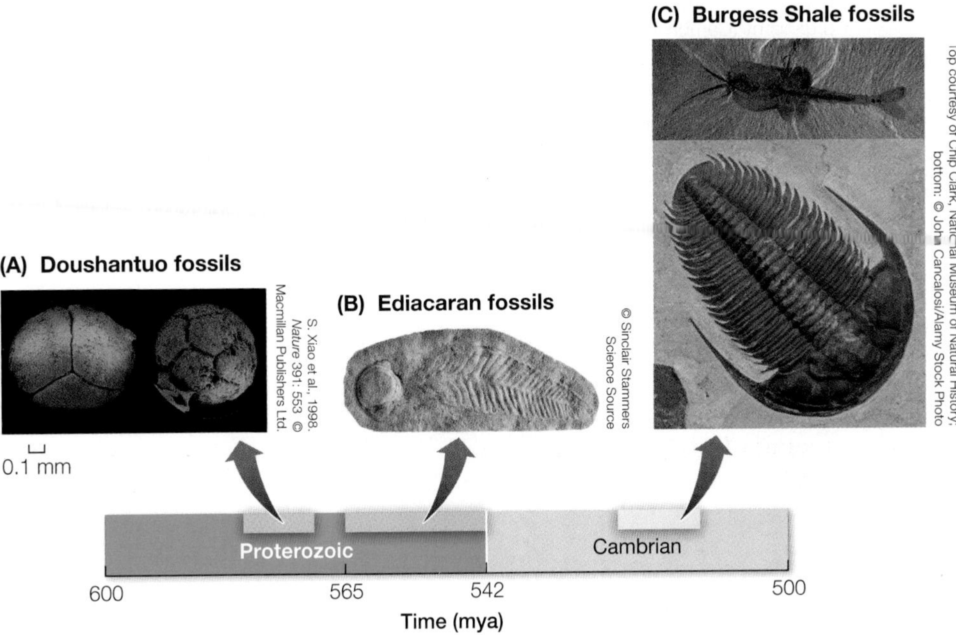

FIGURE 17.11 Diversification of Multicellular Organisms: The Cambrian Explosion Shortly after the end of Proterozoic glaciations (about 580 mya), several major radiations of multicellular organisms appear in the fossil record. **(A)** These microscopic fossils from the Doushantuo rock formation of China are the remains of tiny one-, two-, four-, and eight-celled stages of multicellular organisms. **(B)** Unusual soft-bodied marine invertebrates, unlike any animals alive at present, characterize the fossilized fauna preserved at Ediacara in southern Australia. **(C)** By the early Phanerozoic, fossilized faunas such as those preserved in Canada's Burgess Shale include extinct representatives of some of the major animal groups alive today.

For most of its history, life was confined to the oceans, and all organisms were small. For more than 3 billion years, all organisms lived in shallow seas. These seas slowly began to teem with microscopic prokaryotes. After the first eukaryotes appeared about 1.5 billion years ago, during the Proterozoic, unicellular eukaryotes and small multicellular animals fed on the microorganisms. Small floating organisms, known collectively as **plankton**, were strained from the water and eaten by slightly larger filter-feeding animals. Other animals ingested sediments on the seafloor and digested the remains of organisms within them. But it still took nearly a billion years before eukaryotes began to diversify rapidly into the many different morphological forms that we know today.

What limited the diversity of multicellular eukaryotes (in terms of their size and shape) for much of their early existence? It is likely that a combination of factors was responsible. We have already noted that oxygen levels increased throughout the Proterozoic, and it is likely that high atmospheric and dissolved oxygen concentrations were needed to support large multicellular organisms. In addition, geological evidence points to a series of intensely cold periods during the late Proterozoic, which would have resulted in seas that were largely covered by ice and continents that were covered by glaciers. The "snowball Earth" hypothesis suggests that cold conditions confined life to warm places such as hot springs, deep thermal vents, and perhaps a few equatorial oceans that avoided ice cover. The last of these Proterozoic glaciations ended about 580 million years ago, just before several major radiations of multicellular eukaryotes appear in the fossil record (**FIGURE 17.11**). Many of the multicellular organisms known from the late Proterozoic and early Phanerozoic were very different from any animals living today and may be members of groups that left no living descendants.

Life expanded rapidly during the Cambrian

The Cambrian Period (542–488 mya) marks the beginning of the Paleozoic, the first era of the Phanerozoic. The O_2 concentration in the Cambrian atmosphere was approaching its current level, and the glaciations of the late Proterozoic had ended nearly 40 million years earlier. A geologically rapid diversification of life took place that is often called the **Cambrian explosion**. This name is somewhat misleading, as the series of radiations it refers to actually began before the start of the Cambrian and continued for about 60 million years, into the early Cambrian (see Figure 17.11). Nonetheless, 60 million years represents a relatively short amount of time, especially considering that the first eukaryotes had appeared about a billion (= 1,000 million) years earlier. Many of the major animal groups represented by species alive today first appeared during these evolutionary radiations. **FIGURE 17.12** provides an overview of the numerous continental and biotic innovations that have characterized the Phanerozoic.

For the most part, fossils tell us only about the hard parts of organisms, but in some well-studied Cambrian fossil beds, the soft parts of many animals were preserved. Multicellular life was largely or completely aquatic during the Cambrian. If there was life on land at this time, it was probably restricted to microorganisms.

Many groups of organisms that arose during the Cambrian later diversified

Geologists divide the remainder of the Paleozoic into the Ordovician, Silurian, Devonian, Carboniferous, and Permian periods. Each period is characterized by the diversification of specific groups of organisms. Mass extinctions marked the ends of the Ordovician, Devonian, and Permian.

THE ORDOVICIAN (488–444 MYA) During the Ordovician, the continents, which were located primarily in the southern hemisphere, still lacked multicellular life. Evolutionary radiation of marine organisms, however, was spectacular during the early Ordovician, especially among animals, such as brachiopods and mollusks, that lived on the seafloor and filtered small prey from the water. At the end of the Ordovician, as massive glaciers formed over the southern continents, sea levels dropped about 50 meters,

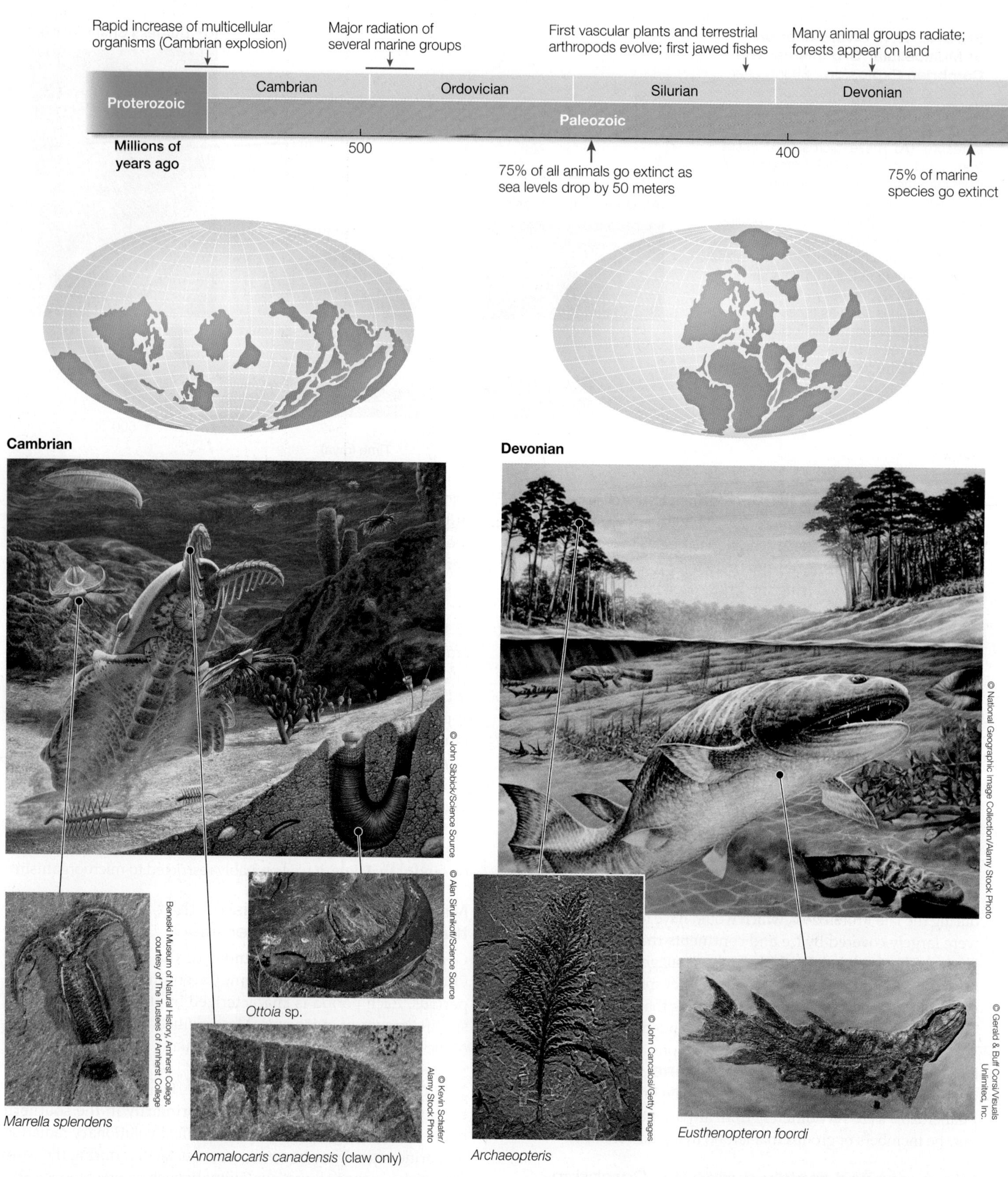

FIGURE 17.12 A Brief History of Multicellular Life on Earth The geologically rapid "explosion" of life shortly before and during the Cambrian saw the rise of many major animal groups that have representatives surviving today. The following three pages depict life's history through the Phanerozoic. The movements of the major continents during the past half-billion years are shown in the maps of Earth, and associated biotas for each time period are depicted. The artists' reconstructions are based on fossils such as those shown in the photographs. (After R. S. Dietz and J. C. Holden. 1970. *J Geophys Res* 75: 4939–4956; R. S. Dietz and J. C. Holden. 1970. *Sci Am* 223: 30–41.)

Extensive swamp forests produce coal; origin of amniotes; great increase in terrestrial animal diversity

Giant amphibians and flying insects; ray-finned fishes abundant in fresh water

On land, conifers become dominant plants; frogs and reptiles begin to diversify

First mammals appear

Dinosaurs, pterosaurs, ray-finned fishes diversify

First known flowering plant fossils

Carboniferous	Permian	Triassic	Jurassic
Paleozoic		Mesozoic	

300

200

Extinction of 96% of Earth's species; oxygen levels drop rapidly

Mass extinction event, including about 65% of all species

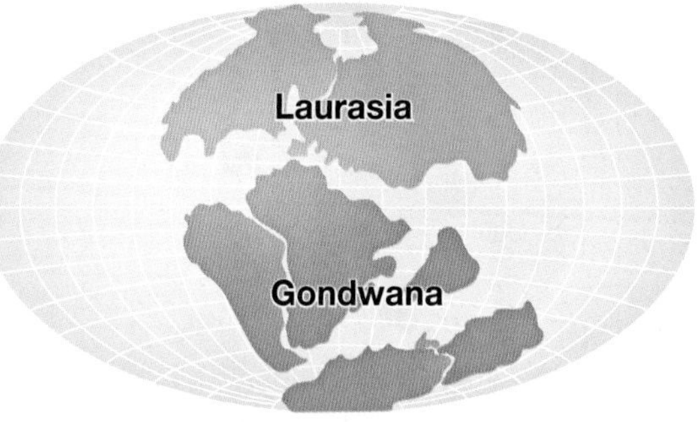

Permian

© John Sibbick/Science Source

Jurassic

© Sergey Krasovskiy/Stocktrek Images/Corbis

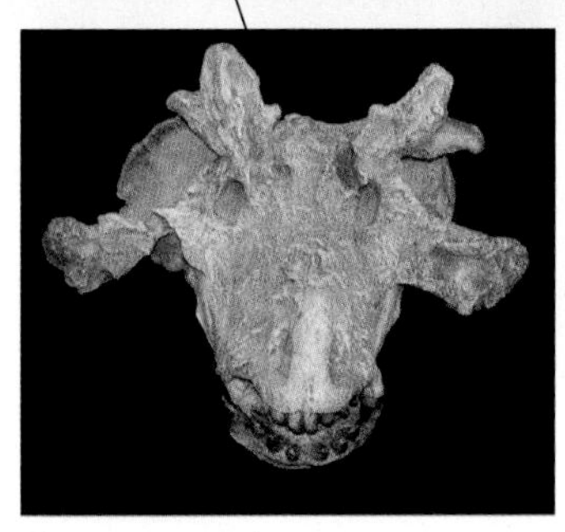
Courtesy of Gondwana Studios

Estemmenosuchus sp.

© John Cancalosi/Alamy Stock Photo

Equisetum sp.

© Biophoto Associates/Science Source

Ginkgo sp.

Europasaurus holgeri

dpa picture alliance archive/Alamy Stock Photo

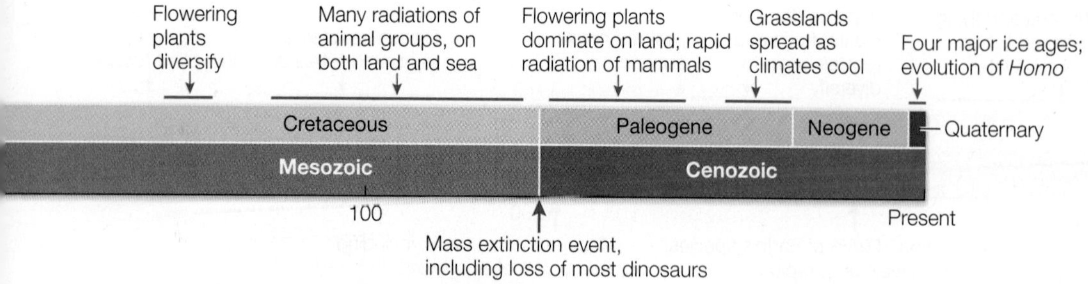

Flowering plants diversify
Many radiations of animal groups, on both land and sea
Flowering plants dominate on land; rapid radiation of mammals
Grasslands spread as climates cool
Four major ice ages; evolution of *Homo*
Cretaceous
Paleogene
Neogene
Quaternary
Mesozoic
Cenozoic
100
Present
Mass extinction event, including loss of most dinosaurs

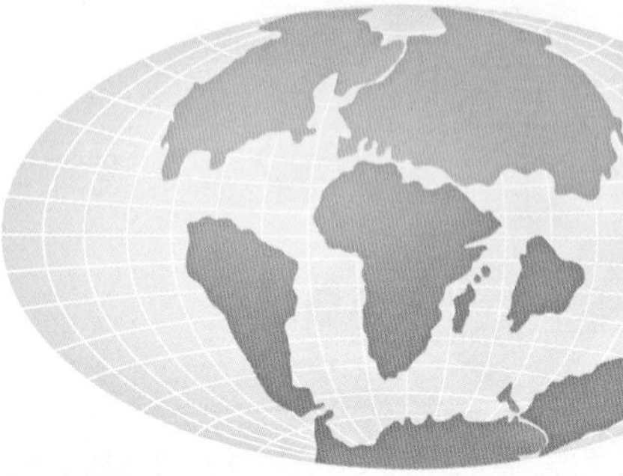

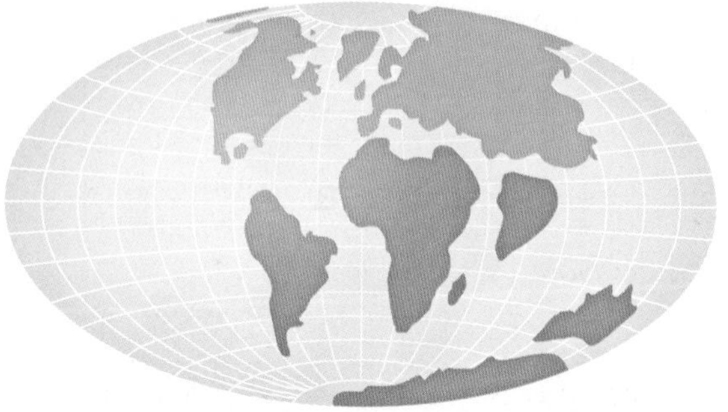

Cretaceous

© Anness Publishing/NHMPL

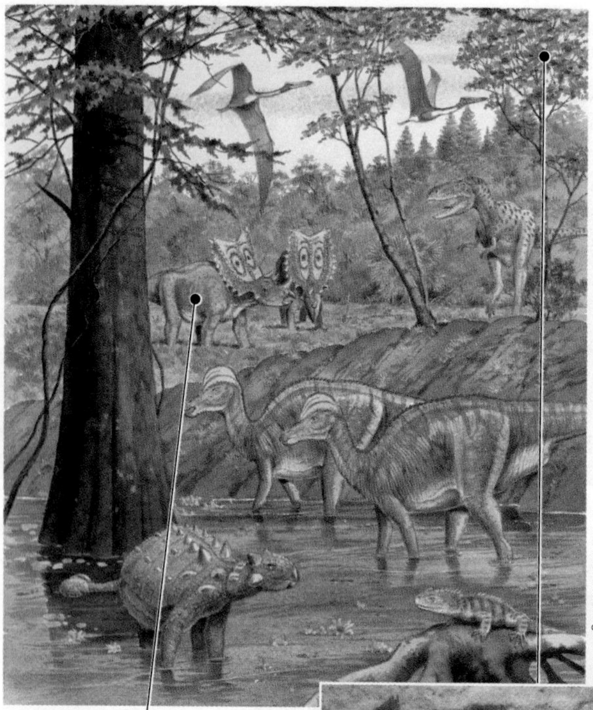

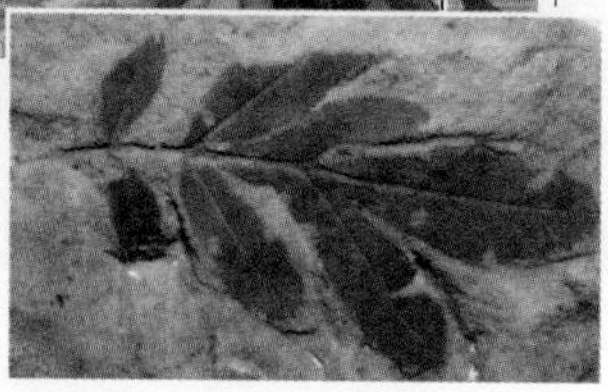

Sapindopsis belviderensis (leaves)

© Barbara J. Miller/Biological Photo Service

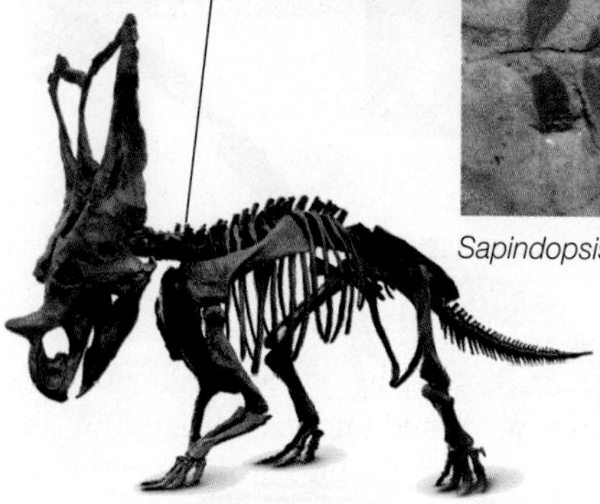

Chasmosaurus belli

© Oleksiy Maksymenko /Alamy Stock Photo

Paleogene

© Publiphoto/Science Source

Coryphodon

© Gilbert S. Grant/Science Source

Hyracotherium leporinum

Beneski Museum of Natural History, Amherst College, courtesy of The Trustees of Amherst College

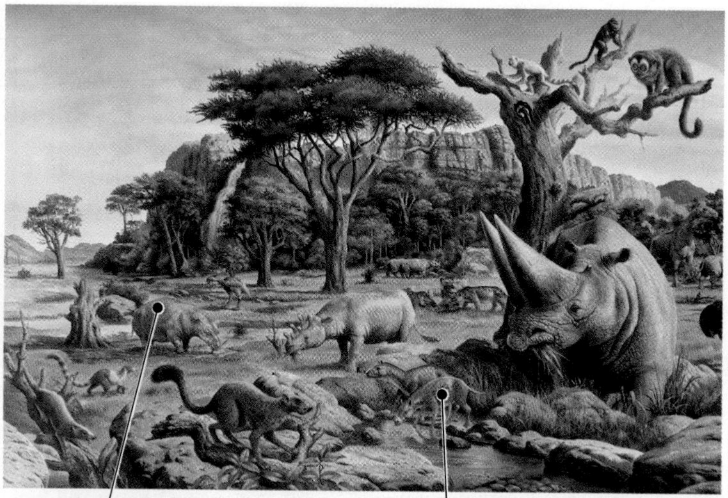

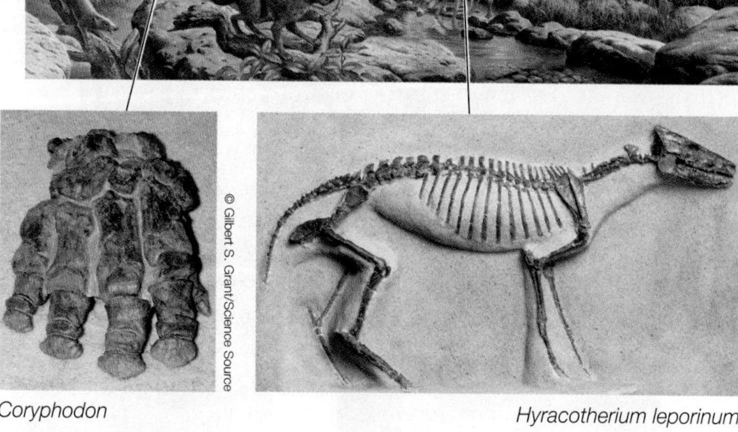

and ocean temperatures dropped. About 75 percent of all animal species became extinct, probably because of these major environmental changes.

THE SILURIAN (444–416 MYA) During the Silurian, the continents began to merge together. Marine life rebounded from the mass extinction at the end of the Ordovician. Animals able to swim in open water and feed above the ocean floor appeared for the first time. Fishes diversified as bony armor gave way to the less rigid scales of modern fishes, and the first jawed fishes and the first fishes with supporting rays in their fins appeared. The tropical sea was uninterrupted by land barriers, and most marine organisms were widely distributed. On land, the first vascular plants evolved late in the Silurian (about 420 mya). The first terrestrial arthropods —scorpions and millipedes—evolved at about the same time.

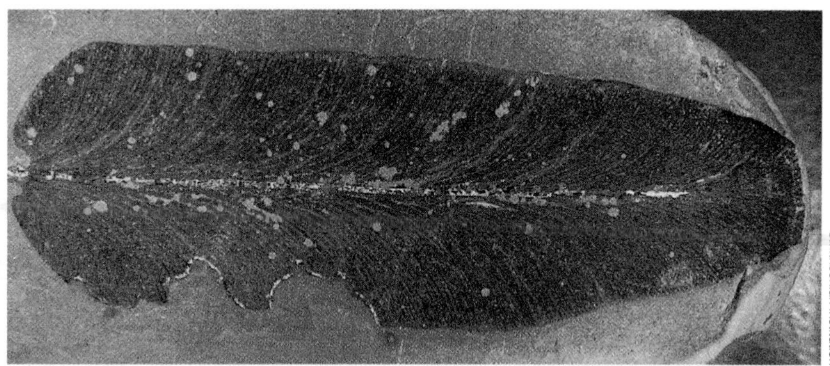

FIGURE 17.13 Evidence of Insect Diversification The margins of this fossil fern leaf from the Carboniferous have been chewed by insects.

THE DEVONIAN (416–359 MYA) Rates of evolutionary change accelerated in many groups of organisms during the Devonian. The major land masses continued to move slowly toward each other. In the oceans there were great evolutionary radiations of corals and of shelled, squid-like cephalopod mollusks.

Terrestrial communities changed dramatically during the Devonian. Club mosses, horsetails, and tree ferns became common, and some attained the size of large trees. Their roots accelerated the weathering of rocks, resulting in the development of the first forest soils. The first plants to produce seeds appeared in the Devonian. The earliest fossil centipedes, spiders, mites, and insects date to this period, as do the earliest terrestrial vertebrates.

A mass extinction of about 75 percent of all marine species marked the end of the Devonian. Paleontologists are uncertain about its cause, but two large meteorites collided with Earth at about that time and may have been responsible, or at least a contributing factor. The continued merging of the continents, with the corresponding reduction in the area of continental shelves, also may have contributed to this mass extinction.

THE CARBONIFEROUS (359–299 MYA) Large glaciers formed over high-latitude portions of the southern land masses during the Carboniferous, but extensive swamp forests grew on the tropical continents. These forests were dominated by giant tree ferns and horsetails with small leaves. Their fossilized remains formed the coal we now mine for energy. In the seas, crinoids (a group of echinoderms, related to sea stars and sea urchins) reached their greatest diversity, forming "meadows" on the seafloor.

The diversity of terrestrial animals increased greatly during the Carboniferous. Snails, scorpions, centipedes, and insects were abundant and diverse. Insects evolved wings, becoming the first animals to fly. Flight gave herbivorous insects easy access to tall plants, and plant fossils from this period show evidence of chewing by insects (**FIGURE 17.13**). The terrestrial vertebrates split into two lineages. The amphibians became larger and better adapted to terrestrial existence, while the sister lineage led to the amniotes—vertebrates with well-protected eggs that can be laid in dry places.

THE PERMIAN (299–251 MYA) During the Permian, the continents merged into a single supercontinent called **Pangaea**. Permian rocks contain representatives of many of the major groups of insects we know today. By the end of the period the amniotes had split into two lineages: the reptiles, and a second lineage that would lead to the mammals. Ray-finned fishes became common in the fresh waters of Pangaea.

Toward the end of the Permian, conditions for life deteriorated. Massive volcanic eruptions resulted in outpourings of lava that covered large areas of Earth. The ash and gases produced by the volcanoes blocked sunlight and cooled the climate. The death and decay of the massive Permian forests rapidly used up atmospheric oxygen, and the loss of photosynthetic organisms meant that relatively little new atmospheric oxygen was produced. In addition, much of Pangaea was located close to the South Pole by the end of the Permian. All of these factors combined to produce the most extensive continental glaciers since the snowball Earth times of the late Proterozoic. Atmospheric oxygen concentrations gradually dropped from about 30 percent to 15 percent. At such low concentrations, most animals would have been unable to survive at elevations above 500 meters, so about half of the land area would have been uninhabitable at the end of the Permian. The combination of these changes resulted in the most drastic mass extinction in Earth's history. Scientists estimate that about 96 percent of all multicellular species became extinct at the end of the Permian.

Geographic differentiation increased during the Mesozoic

The few organisms that survived the Permian mass extinction found themselves in a relatively empty world at the start of the Mesozoic Era (251 mya). As Pangaea slowly began to break apart in the Mesozoic, the biotas of the newly separated continents began to diverge. The oceans rose and once again flooded the continental shelves, forming huge, shallow inland seas. Atmospheric oxygen concentrations gradually rose as photosynthesis increased. Life once again proliferated and diversified, but different groups of organisms came to the fore. The three groups of phytoplankton (floating photosynthetic organisms) that dominate today's oceans—dinoflagellates, coccolithophores, and diatoms—became ecologically important at this time, and their remains are the primary origin of the world's oil deposits. Seed-bearing plants replaced the trees that had ruled the Permian forests.

The Mesozoic is divided into three periods: the Triassic, Jurassic, and Cretaceous. The Triassic and Cretaceous were terminated by mass extinctions, probably caused by meteorite impacts.

THE TRIASSIC (251–201.6 MYA) Pangaea remained largely intact through the Triassic. Many invertebrate groups diversified, and many burrowing animals evolved from groups living on the surfaces of seafloor sediments. On land, conifers and seed ferns were the dominant trees. The first frogs and turtles appeared. A great radiation of reptiles began, which eventually gave rise to crocodilians, dinosaurs, and birds. The first mammals appeared. The end of the Triassic was marked by a mass extinction that eliminated about 65 percent of the species on Earth.

THE JURASSIC (201.6–145.5 MYA) Late in the Jurassic, Pangaea became fully divided into two large continents: **Laurasia**, which drifted northward, and **Gondwana** in the south. Ray-finned fishes rapidly diversified in the oceans. The first lizards appeared, and flying reptiles (pterosaurs) evolved. Most of the large terrestrial predators and herbivores of the period were dinosaurs. Several groups of mammals made their first appearance, and the earliest known fossils of flowering plants are from late in this period.

THE CRETACEOUS (145.5–65.5 MYA) By the mid-Cretaceous, Laurasia and Gondwana had largely broken apart into the continents we know today (although the Indian subcontinent was still separated from Asia). A continuous sea encircled the tropics. Sea levels were high, and Earth was warm and humid. Life proliferated both on land and in the oceans. Marine invertebrates increased in diversity. On land, the reptile radiation continued as dinosaurs diversified further and the first snakes appeared. Early in the Cretaceous, flowering plants began the radiation that led to their current dominance of the land. By the end of the period, many groups of mammals had appeared.

As described in Key Concept 17.2, another meteorite-caused mass extinction took place at the end of the Cretaceous. In the seas, many planktonic organisms and bottom-dwelling invertebrates became extinct. On land, almost all animals larger than about 25 kilograms in body weight became extinct. Many species of insects died out, perhaps because the growth of the plants they fed on was greatly reduced following the impact. Some species in northern North America and Eurasia survived in areas that were not subjected to the devastating fires that engulfed most low-latitude regions.

Modern biotas evolved during the Cenozoic

By the early Cenozoic Era (65.5 mya), the continents were getting closer to their present positions, but the Indian subcontinent was still separated from Asia, and the Atlantic Ocean was much narrower. The Cenozoic was characterized by extensive radiations of amphibians, reptiles, and mammals, but other groups were also undergoing important changes.

Flowering plants diversified extensively and came to dominate world forests, except in the coolest regions, where the forests were composed primarily of gymnosperms. Novel changes in one group of plants (the legumes) allowed them to use atmospheric nitrogen directly by forming symbioses with a few species of nitrogen-fixing bacteria. The evolution of this symbiosis was the first "green revolution" and dramatically increased the amount of nitrogen available for terrestrial plant growth. This symbiosis remains fundamental to the ecological base of life as we know it today.

LINK The symbiosis between plants and nitrogen-fixing bacteria is covered in detail in **Key Concept 24.3**.

The Cenozoic is divided into the Paleogene, Neogene, and Quaternary periods, which are commonly subdivided into **epochs** (**TABLE 17.2**). The Paleogene and Neogene are often referred to as the Tertiary when they are discussed together, especially in the older literature.

THE PALEOGENE (65.5–23.0 MYA) During the Paleogene, the Indian subcontinent continued its northward drift. By about 55 million years ago it made initial contact with parts of southeastern Asia. By about 35 million years ago, the Indian Plate ran fully into the Eurasian Plate, and the Himalayas began to be pushed up as a result.

The early Paleogene was a hot and humid time, as carbon dioxide levels in Earth's atmosphere were high, which produced a greenhouse effect. The tropics were probably too hot to support rainforest vegetation and instead were clothed in low-lying vegetation. By late in the Paleogene, however, increased plankton growth in the oceans, combined with eventual burial of plankton in oceanic sediments, led to the removal of considerable carbon dioxide from Earth's atmosphere, and the climate became considerably cooler and drier. Many lineages of flowering plants evolved herbaceous (nonwoody) forms, and grasslands spread over much of Earth.

Amphibians, snakes, lizards, birds, and mammals all underwent extensive radiations during the Paleogene. The first of several waves of plants and terrestrial animals dispersed between Asia and North America across one of the several land bridges that have intermittently connected the two continents during the past 55 million years.

THE NEOGENE (23.0–2.6 MYA) During the Neogene, the continents began to take their current positions. North and South America became connected by the Panamanian Isthmus, which resulted in considerable interchange of plants and terrestrial animals between the two continents. For the same reason, the tropical areas

TABLE **17.2** | Subdivisions of the Cenozoic Era

Period	Epoch	Onset (mya)
Quaternary	Holocene (recent)	0.01 (~10,000 years ago)
	Pleistocene	2.6
Neogene	Pliocene	5.3
	Miocene	23
Paleogene	Oligocene	34
	Eocene	55.8
	Paleocene	65.5

of the Atlantic and Pacific oceans became isolated from one another, which resulted in divergence of the respective marine faunas of these oceans. Diversification of major plant and animal groups continued, as the biotas of the continents began to resemble those we see today.

THE QUATERNARY (2.6 MYA TO PRESENT) We are living in the Quaternary. It is subdivided into two epochs, the Pleistocene and the Holocene (the Holocene is also known as the Recent).

The Pleistocene was a time of drastic cooling and climate fluctuations. During 4 major and about 20 minor ice ages, massive glaciers spread across the continents, and the ranges of animal and plant populations shifted toward the equator. The last of these glaciers retreated from temperate latitudes less than 15,000 years ago. Organisms are still adjusting to this change. Many high-latitude ecological communities have occupied their current locations for no more than a few thousand years.

It was during the Pleistocene that divergence within one group of mammals, the primates, resulted in the evolution of the hominoid lineage. Subsequent hominoid radiation eventually led to the species *Homo sapiens*—modern humans. Many large bird and mammal species became extinct in Australia and in the Americas when *H. sapiens* arrived on those continents about 45,000 and 15,000 years ago, respectively. Many paleontologists believe these extinctions were the result of hunting and other influences of humans.

LINK The evolution of modern humans and their close relatives during the Pleistocene is discussed in **Key Concept 22.7**.

The tree of life is used to reconstruct evolutionary events

The fossil record reveals broad patterns in life's evolution. To reconstruct major events in the history of life, biologists also rely on the phylogenetic information in the tree of life. We can use phylogeny, in combination with the fossil record, to reconstruct the timing of such major events as the acquisition of mitochondria in the ancestral eukaryotic cell, the several independent origins of multicellular organisms, and the movement of life onto dry land. We can also follow major changes in the genomes of organisms, and we can even reconstruct many gene sequences of species that are long extinct.

Key Concept 14.3 describes how biologists reconstruct the gene sequences of extinct organisms.

Changes in Earth's physical environment have clearly influenced the diversity of organisms we see on the planet today. To study the evolution of that diversity, biologists examine the evolutionary relationships among species. Deciphering phylogenetic relationships is an important step in understanding how life has diversified on Earth. The next part of this book will explore the major groups of life and the different solutions these groups have evolved to major challenges such as reproduction, energy acquisition, dispersal, and escape from predation.

REVIEW & APPLY | 17.3

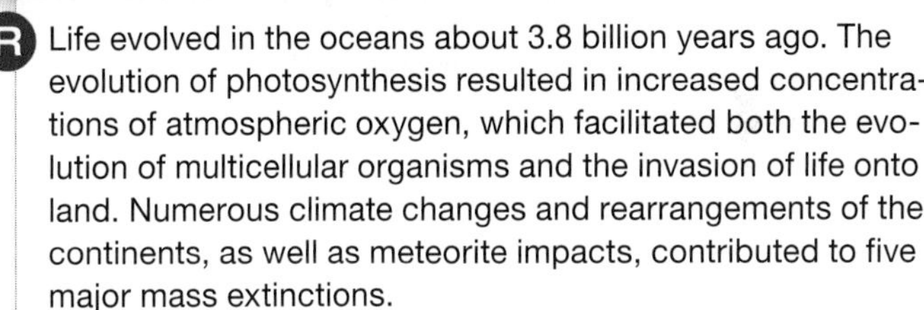
R Life evolved in the oceans about 3.8 billion years ago. The evolution of photosynthesis resulted in increased concentrations of atmospheric oxygen, which facilitated both the evolution of multicellular organisms and the invasion of life onto land. Numerous climate changes and rearrangements of the continents, as well as meteorite impacts, contributed to five major mass extinctions.

A

1. Give a likely reason for the onset of the "green revolution" of the Cenozoic, and explain why it enabled flowering plants to diversify.
2. Compare the tree of life in Appendix A with the chronology of major events in Table 17.1. To represent the diversity of life in one figure, the tree in the appendix is not drawn proportionally in relation to time. Which geological eon, era, or period seems most compressed on the tree, and why?

Answer: Almost anyone who has spent time around freshwater ponds is familiar with dragonflies. The largest dragonflies alive today have wingspans that can be covered by a human hand. Three hundred million years ago, however, there were dragonfly-like insects with wingspans of more than 70 centimeters —well over 2 feet, matching or exceeding the wingspans of many modern birds of prey. These giant insects were the largest flying predators of their time.

No flying insects alive today are anywhere near this size. But during the Carboniferous and Permian geological periods, 350 to 250 million years ago, many groups of flying insects contained gigantic members. None of these insects would be able to survive on Earth today. As we saw in Figure 17.7, oxygen concentrations in Earth's atmosphere were about 50 percent higher than than they are now, and those high oxygen concentrations are thought to have been necessary to support giant insects. Investigation Figure 17.8 shows how changes in atmospheric oxygen concentrations can affect the selection and evolutionary constraints on body size in flying insects, and result in rapid evolutionary changes.

17 VISUAL SUMMARY

17.1 Events in Earth's History Can Be Dated

- The relative ages of organisms can be determined by the dating of fossils and the **STRATA** of **SEDIMENTARY ROCKS** in which they are found.
- **RADIOMETRIC DATING** techniques use a variety of radioisotopes with different **HALF-LIVES** to date events in the remote past (**FIGURE 17.1**).
- Geologists divide the history of life into **EONS**, eras, and periods, based on major differences in the fossil assemblages found in successive strata (**TABLE 17.1**).

FIGURE 17.1

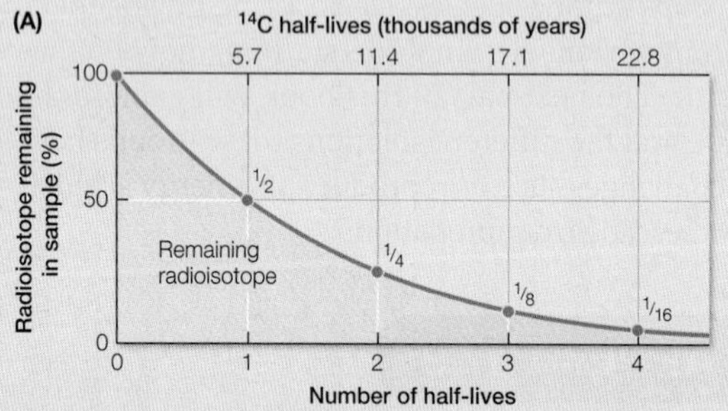

(B)

Radioisotope	Decay product	Half-life (years)	Useful dating range (years)
Carbon-14 (^{14}C)	Nitrogen-14 (^{14}N)	5,700	100–60,000
Uranium-234 (^{234}U)	Thorium-230 (^{230}Th)	80,000	10,000–500,000
Uranium-235 (^{235}U)	Lead-207 (^{207}Pb)	704 million	200,000–4.5 billion
Potassium-40 (^{40}K)	Argon-40 (^{40}Ar)	1.3 billion	10 million–4.5 billion

TABLE **17.1** I EARTH'S GEOLOGICAL HISTORY

Eon	Era	Period	Onset
Phanerozoic (~0.5 billion years long)	Cenozoic	Quaternary (Q)	2.6 mya
		Neogene (N)	23.0 mya
		Paleogene (Pg)	65.5 mya
	Mesozoic	Cretaceous (K)	145.5 mya
		Jurassic (J)	201.6 mya
		Triassic (Tr)	251.0 mya
	Paleozoic	Permian (P)	299 mya
		Carboniferous (C)	359 mya
		Devonian (D)	416 mya
		Silurian (S)	444 mya
		Ordovician (O)	488 mya
		Cambrian (C)	542 mya
Proterozoic	Collectively called the Precambrian (~4 billion years long)		2.5 bya
Archean			3.8 bya
Hadean			4.5–4.6 bya

QUESTIONS

1. How much of the evolutionary history of life was exclusively aquatic? When did plants and animals first colonize the continental land masses?
2. Why is carbon-14 dating limited to approximately the past 60,000 years?
3. What radioisotopes can be used to date much older rocks, up to the age of Earth?
4. How much of Earth's history does the Phanerozoic represent?

17.2 Changes in Earth's Physical Environment Have Affected the Evolution of Life

Go to ANIMATION 17.1

- Earth's crust consists of solid plates that move across the underlying mantle. **CONTINENTAL DRIFT** is caused by convection currents in the mantle, which move the plates and the continents that lie on top of them.
- Major physical events on Earth, such as continental collisions and volcanic eruptions, have affected Earth's climate, atmosphere, and sea levels. In addition, extraterrestrial events such as meteorite strikes have created sudden and dramatic environmental shifts and led to **MASS EXTINCTIONS** (**FIGURE 17.3**).
- Oxygen-generating cyanobacteria liberated enough O_2 to open the door to oxidation reactions in metabolic pathways. Aerobic prokaryotes were able to harvest more energy than anaerobic organisms and began to proliferate. Increases in atmospheric O_2 concentrations supported the evolution of large eukaryotic cells and, eventually, multicellular organisms (**FIGURE 17.7**).

FIGURE 17.3

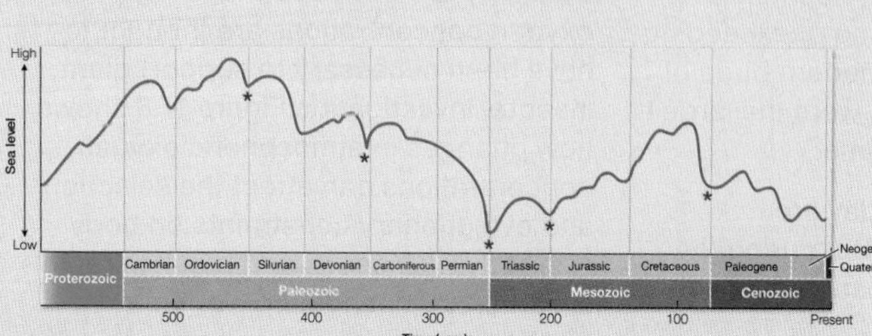

FIGURE 17.7

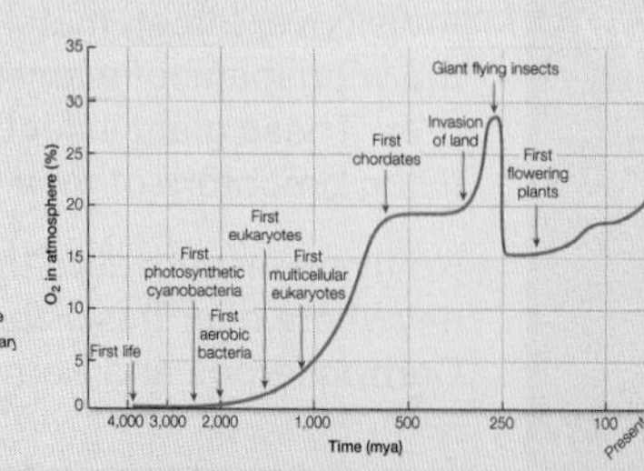

QUESTIONS

1. Explain how changes in atmospheric oxygen enabled the evolution of multicellular life.
2. What environmental changes resulted in the sudden drop in atmospheric oxygen at the end the Permian?
3. Describe the relationship between global temperature change and sea level fluctuations.
4. How can volcanic eruptions change the global climate?

17.3 Major Events in the Evolution of Life Can Be Read in the Fossil Record

- Paleontologists use fossils and evidence of geological changes to determine what Earth and its **BIOTA** may have looked like at different times (**FIGURE 17.10**).
- Before the Phanerozoic, life was almost completely confined to the oceans. Multicellular life diversified extensively during the **CAMBRIAN EXPLOSION**, a prime example of an evolutionary radiation (**FIGURE 17.11**).
- The periods of the Paleozoic Era were each characterized by the diversification of specific groups of organisms. During the Mesozoic Era, distinct terrestrial biotas evolved on each continent.
- The Cenozoic Era is divided into the Paleogene, Neogene, and Quaternary periods, which in turn are subdivided into **EPOCHS**. The Cenozoic saw the emergence of the modern biotas as vertebrates radiated extensively and flowering plants became dominant.
- The tree of life can be used to reconstruct the timing of evolutionary events.

FIGURE 17.10

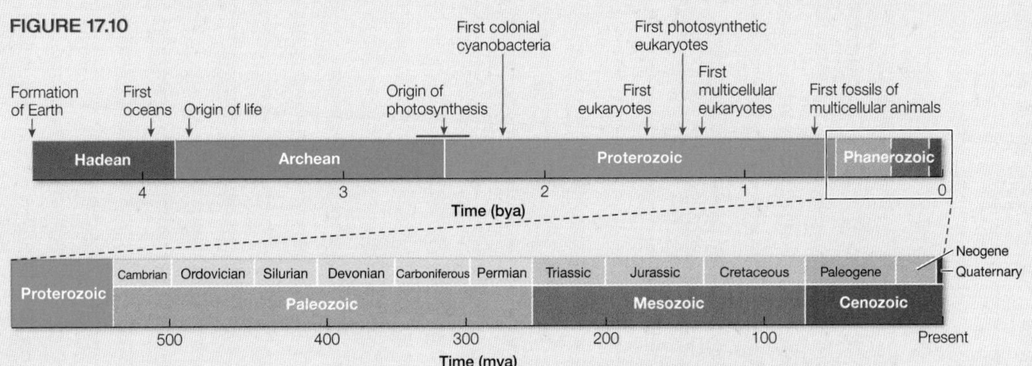

FIGURE 17.11

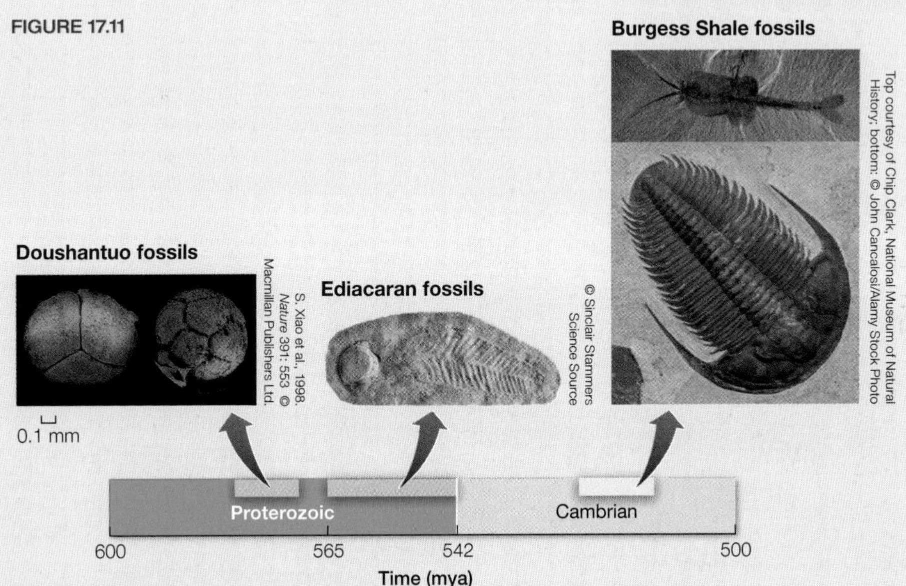

S. Xiao et al., 1998. *Nature* 391: 553 © Macmillan Publishers Ltd.

© Sinclair Stammers Science Source

Top courtesy of Chip Clark, National Museum of Natural History; bottom: © John Cancalosi/Alamy Stock Photo

Activity 17.1 Concept Matching: The History of Life on Earth

QUESTIONS

1. Why are the evolutionary changes in faunas near the beginning of the Cambrian sometimes described as an "explosion"?
2. What were two major causes of mass extinction events in Earth's history?
3. When did snowball Earth exist, and why might it be important in understanding the Cambrian explosion?
4. Describe the relationship of the modern continents to Gondwana, Laurasia, and Pangaea.

Go to **Achieve** for the e-book, animations, activities, and additional resources and assignments.

Part 4 Diversity

18 Bacteria, Archaea, and Viruses

The lights of cities are clear in this view of Earth. But what is that bright spot in the ocean off the Horn of Africa? It turns out that this is a giant bloom of bioluminescent bacteria, so large that the bacterial glow can be seen from space. Bacteria of the genus *Vibrio* produce this glow when their density grows to a critical level.

How and why do bacteria communicate with one another to assess their density?

You will find the answer to this question on page 437.

"Milky seas" data prepared by Steven Miller of the Naval Research Laboratory, Monterey, CA. Background data (Earth at Night image) courtesy Marc Imhoff (NASA/GSFC) and Christopher Elvidge (NOAA/NGDC); image by Craig Mayhew (NASA/GSFC) and Robert Simmon (NASA/GSFC)

KEY CONCEPTS

18.1 Bacteria and Archaea Are the Two Primary Divisions of Life

LEARNING OBJECTIVES

By the end of this key concept you should be able to:

18.1.1 **Compare and contrast the features of bacteria, prokaryotic archaea, and eukaryotes.**

18.1.2 **Describe how prokaryotic archaea and bacteria each contributed to the evolution of eukaryotes.**

18.1.3 **Draw a phylogenetic tree demonstrating the effects of lateral gene transfer.**

You may think that you have little in common with a bacterium. But all multicellular eukaryotes, including you, share many attributes with the **prokaryotes**—organisms that lack a nucleus (see Key Concept 4.5). For example, all organisms, whether eukaryotes or prokaryotes,

- have cell membranes and ribosomes (see Chapter 4)
- have a common set of metabolic pathways, such as glycolysis (see Chapter 5)
- replicate DNA semiconservatively (see Chapter 9)
- use DNA as the genetic material to encode proteins, and use similar genetic codes to produce those proteins by transcription and translation (see Chapter 10)

These shared features support the hypothesis that all living organisms share a common ancestor. If life had multiple origins, there would be little reason to expect all organisms to use overwhelmingly similar genetic codes or to share structures as distinctive as ribosomes. Furthermore, similarities in the DNA sequences of genes that are shared by all organisms confirm the monophyly of life.

Although all living things share many features, major differences have evolved across the diversity of life. Biologists have now sequenced the genomes of many living organisms, and these genomes allow us to reconstruct the details of evolutionary history. These studies clarify that there are two fundamental divisions of life: Bacteria and Archaea (**FIGURE 18.1**). Another major group evolved from within the Archaea: the Eukarya, which includes all plants, animals, and fungi.

When biologists speak of three major **domains** of life, they usually mean Bacteria, prokaryotic Archaea, and Eukarya. But if you examine Figure 18.1, you will see that eukaryotes evolved from within the prokaryotic Archaea. So in this sense, eukaryotes are a specialized group of Archaea that developed some important new features (including a cell nucleus) that allowed them to thrive. We call all the organisms that lack these specializations prokaryotes (Greek, "before the kernel," or before the evolution of a cellular nucleus). Prokaryotes are clearly not a monophyletic group, but the name "prokaryotes" is nonetheless a useful way of talking about all organisms that are not eukaryotes. In a similar manner, when most biologists talk about "archaea," typically they are referring to just the prokaryotic archaea.

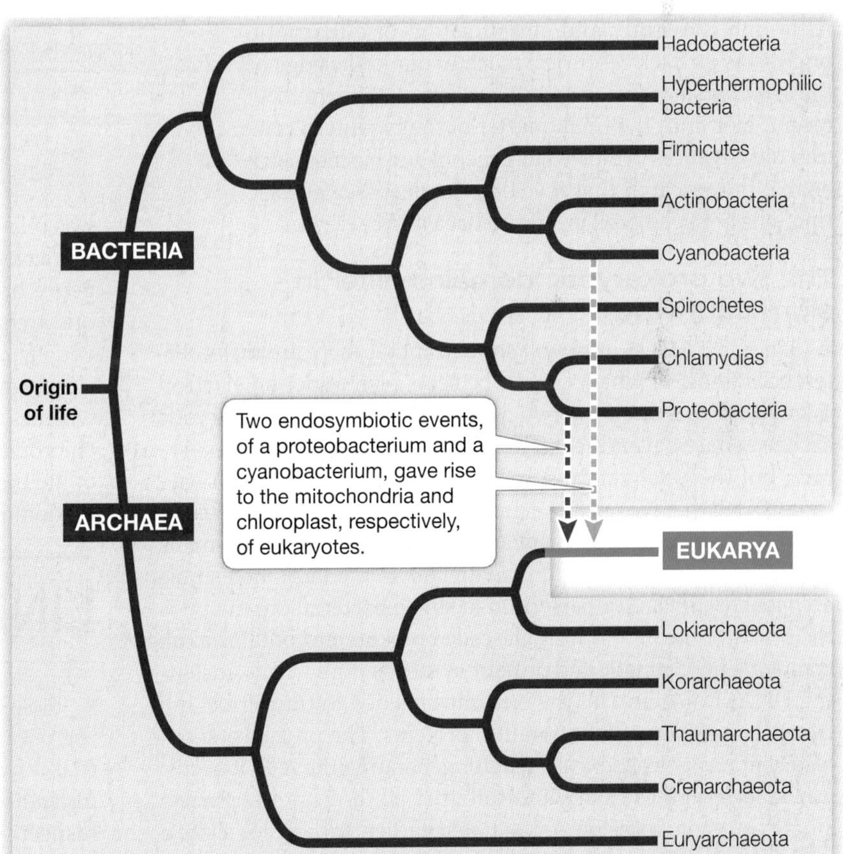

FIGURE 18.1 The Major Groups of the Living World This phylogenetic tree of Bacteria and Archaea shows their relationships to each other and to Eukarya. The relationships among the many clades of bacteria, most of which are not shown here, are incompletely resolved at this time.

Animation 18.1 The Primary Divisions of Life

All prokaryotic organisms are unicellular, although they may form large, coordinated colonies or communities consisting of many individuals. Eukaryotes, by contrast, include both unicellular as well as many multicellular life forms. As we saw in Chapter 4, prokaryotic cells differ from eukaryotic cells in some important ways:

- *Prokaryotic cells do not divide by mitosis.* Instead, after replicating their DNA, prokaryotic cells divide by their own method, binary fission (see Key Concept 7.2).
- *The organization of the genetic material differs.* The DNA of the prokaryotic cell is not organized within a membrane-enclosed nucleus. DNA molecules in prokaryotes are often circular. Many (but not all) prokaryotes have only one main chromosome and are effectively haploid, although many have additional smaller DNA molecules, called plasmids (see Key Concept 8.4).
- *Prokaryotes have none of the membrane-enclosed cytoplasmic organelles that are found in most eukaryotes.* However, the cytoplasm of a prokaryotic cell may contain a variety of infoldings of the cell membrane and photosynthetic membrane systems that are not found in eukaryotes.

Although the study and classification of eukaryotic organisms go back centuries, much of our knowledge of the evolutionary relationships of prokaryotes is extremely recent. Not until the final quarter of the twentieth century did advances in molecular genetics and biochemistry enable the research that revealed the deep-seated distinctions between prokaryotic archaea and bacteria.

TABLE **18.1** | Characteristics of Bacteria, Prokaryotic Archaea, and Eukarya

Characteristic	Bacteria	Prokaryotic Archaea	Eukarya
Membrane-enclosed nucleus	Absent	Absent	Present
Membrane-enclosed organelles	Few	Absent	Many
Peptidoglycan in cell wall	Present	Absent	Absent
Membrane lipids	Ester-linked Unbranched	Ether-linked Branched	Ester-linked Unbranched
Ribosomes[a]	70S	70S	80S
Initiator tRNA	Formylmethionine	Methionine	Methionine
Operons	Yes	Yes	Rare
Plasmids	Yes	Yes	Rare
Number of RNA polymerases[b]	One	One	Three
Ribosomes sensitive to chloramphenicol and streptomycin	Yes	No	No
Ribosomes sensitive to diphtheria toxin	No	Yes	Yes

[a] 70S ribosomes are smaller than 80S ribosomes.
[b] The structure of prokaryotic archaeal RNA polymerase is similar to that of eukaryotic polymerases.

The two prokaryotic domains differ in significant ways

A glance at **TABLE 18.1** will show you that there are major differences (most of which cannot be seen even under an electron microscope) between the two prokaryotic domains. Prokaryotic archaea share several features with their relatives the eukaryotes, but they also retain some ancestral similarities with bacteria. (Note that we use lowercase when referring to members of these domains and initial capitals when referring to the domains themselves.) The basic unit of an archaeon (the term for a single archaeal organism) or bacterium (a single bacterial organism) is the prokaryotic cell. Each single-celled prokaryote contains a full complement of genetic and protein-synthesizing systems, including DNA, RNA, and all the enzymes needed to transcribe and translate genetic information into proteins. The prokaryotic cell also contains at least one system for generating the ATP it needs.

Genetic studies clearly indicate that all forms of life share a single common ancestor. As we noted earlier, eukaryotes share a more recent common ancestor with certain groups of prokaryotic archaea than they do with bacteria (see Figure 18.1). However, the mitochondria of eukaryotes and the chloroplasts of photosynthetic eukaryotes (such as plants) originated through endosymbiosis with bacteria. Some biologists prefer to view the origin of eukaryotes as a fusion of two equal partners (one ancestor that was related to modern prokaryotic archaea and another that was more closely related to modern bacteria). Others view the divergence of the early eukaryotes from specific groups of prokaryotic archaea as an event separate from and earlier than the later endosymbioses. In either case, most eukaryotic genes are more closely related to those of specific groups of prokaryotic archaea, whereas other genes (especially genes related to mitochondria and chloroplasts) are most closely related to those of bacteria. The tree of life therefore contains some merging of lineages as well as the predominant divergence of lineages.

LINK The origin of mitochondria and chloroplasts by endosymbiosis is described in **Key Concept 19.1**.

Biologists estimate that the last common ancestor of all living organisms lived about 3 billion years ago. We can deduce that it had DNA as its genetic material, and that its machinery for transcription and translation produced RNAs and proteins, respectively. This ancestor likely had a circular chromosome. All living organisms are the products of billions of years of mutation, natural selection, and genetic drift, and they are all well adapted to present-day environments. The earliest prokaryotic fossils, which date back at least 3.5 billion years, indicate that there was considerable diversity among the prokaryotes even during those earliest days of life.

The small size of prokaryotes has hindered our study of their evolutionary relationships

Until about 300 years ago, nobody had even *seen* an individual prokaryote. Most prokaryotes remained invisible to humans until the invention of the first simple microscope. Prokaryotes are so small, however, that even the best light microscopes don't reveal much about them. It took advanced microscopic equipment and modern molecular techniques to open up the microbial world. (Microscopic organisms—both prokaryotes and eukaryotes—are often collectively referred to as "microbes.")

Before DNA sequencing became practical, taxonomists based prokaryote classification on observable phenotypic characters such as shape, color, motility, nutritional requirements, and sensitivity to antibiotics. One of the characters most widely used to classify prokaryotes is the structure of their cell walls.

The cell walls of almost all bacteria contain **peptidoglycan**, a cross-linked polymer of amino sugars that produces a firm, protective, meshlike structure around the cell. Peptidoglycan is a substance unique to bacteria; its absence from the cell walls of archaea is a key difference between the two prokaryotic domains. Peptidoglycan is also an excellent target for combating pathogenic (disease-causing) bacteria because it has no counterpart in eukaryotic cells. Antibiotics such as penicillin and ampicillin, as well as other agents that specifically interfere with the synthesis of peptidoglycan-containing cell walls, tend to have little, if any, effect on the cells of humans and other eukaryotes.

The **Gram stain** is a technique that can be used to separate most types of bacteria into two distinct groups. A smear of bacterial cells on a microscope slide is soaked in a violet dye and treated with iodine; it is then washed with alcohol and counterstained with a red dye called safranin. **Gram-positive bacteria** retain the violet dye and appear blue to purple (**FIGURE 18.2A**). The alcohol washes the violet stain out of **Gram-negative bacteria**, which then pick up the safranin counterstain and appear pink to red (**FIGURE 18.2B**). For most bacteria, the effect of the Gram stain is determined by the chemical structure of the cell wall:

- A *Gram-negative cell wall* usually has a thin peptidoglycan layer, which is surrounded by a second, outer membrane quite distinct in chemical makeup from the cell membrane (see Figure 18.2B). Together the cell wall and the outer membrane are called the cell envelope. The space between the cell membrane and the outer membrane (known as the periplasmic space) contains proteins that are important in digesting some materials, transporting others, and detecting chemical gradients in the environment.
- A *Gram-positive cell wall* usually has about five times as much peptidoglycan as a Gram-negative cell wall. Its thick peptidoglycan layer is a meshwork that may serve some of the same purposes as the periplasmic space of the Gram-negative cell envelope.

Shape is another phenotypic character that is useful for the basic identification of bacteria. The three most common shapes are spheres, rods, and spiral forms (**FIGURE 18.3**). Many bacterial names are based on these shapes. A spherical bacterium is called a **coccus** (plural *cocci*). Cocci may live singly or may associate in two- or three-dimensional arrays such as chains, plates, blocks, or clusters of cells. A rod-shaped bacterium is called a **bacillus** (plural *bacilli*). A spiral bacterium (shaped like a corkscrew) is called a **spirillum** (plural *spirilla*). Bacilli and spirilla may be single, form chains, or gather in regular clusters. Among the other bacterial shapes are long filaments and branched filaments.

Less is known about the shapes of prokaryotic archaea because many of these organisms have never been seen. Many prokaryotic archaea are known only from samples of DNA from

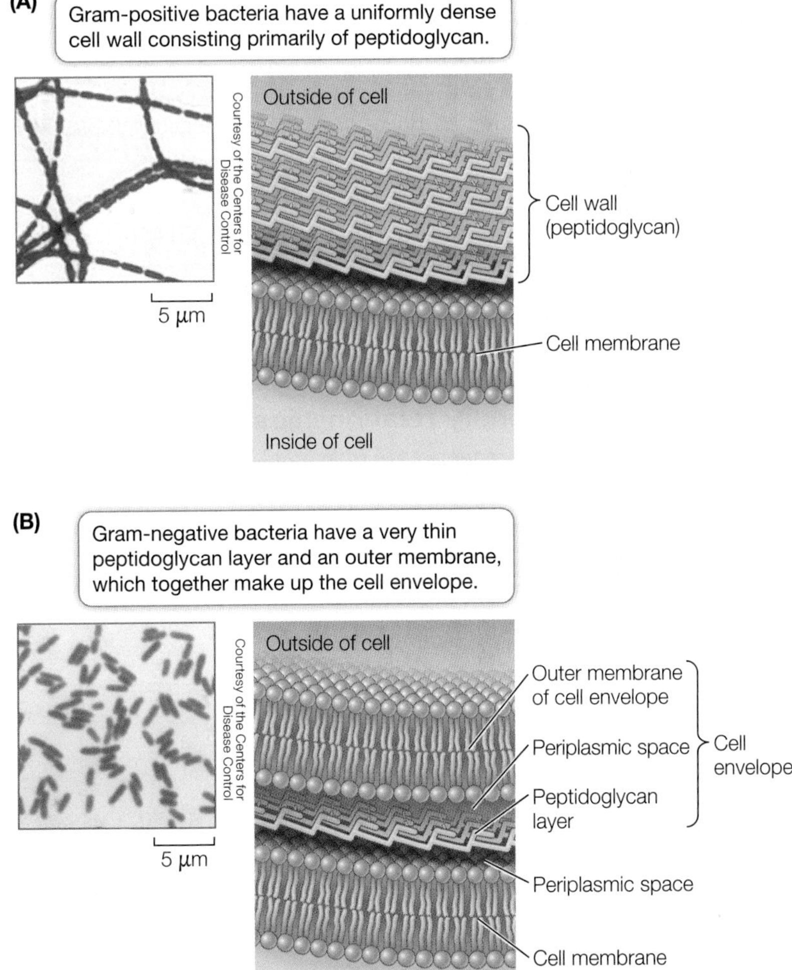

FIGURE 18.2 The Gram Stain and the Bacterial Cell Wall When treated with Gram-staining reagents, the cell walls of bacteria react in one of two ways. **(A)** Gram-positive bacteria have a thick peptidoglycan cell wall that retains the violet dye and appears deep blue or purple. **(B)** Gram-negative bacteria have a thin peptidoglycan layer that does not retain the violet dye, but picks up the counterstain and appears pink to red.

Activity 18.1 **Gram Stain and Bacteria**

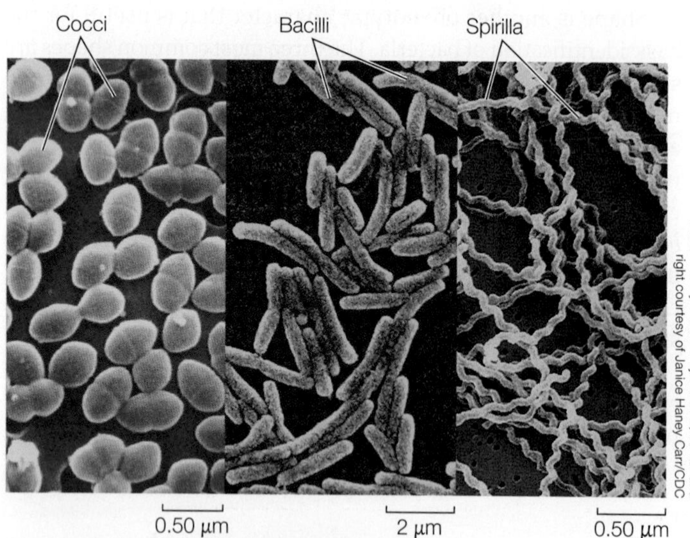

FIGURE 18.3 Bacterial Cell Shapes This composite, colorized micrograph shows the three most common bacterial shapes. Spherical cells are called cocci; those pictured are a species of *Enterococcus* from the mammalian gut. Rod-shaped cells are called bacilli; these *Escherichia coli* also reside in the gut. The helix-shaped spirilla are *Leptospira interrogans*, a human pathogen.

the environment. However, the species whose morphologies are known include cocci, bacilli, and even triangular and square species. Some flattened species grow on surfaces, arranged like sheets of postage stamps.

The nucleotide sequences of prokaryotes reveal their evolutionary relationships

Analyses of the nucleotide sequences of ribosomal RNA (rRNA) genes provided the first comprehensive evidence of evolutionary relationships among prokaryotes. Comparisons of rRNA genes are often used to identify microbes. For several reasons, rRNA is particularly useful for phylogenetic studies and identification purposes:

- rRNA was present in the common ancestor of all life and is therefore evolutionarily ancient.
- No free-living organism lacks rRNA, so rRNA genes can be compared across the tree of life.
- rRNA plays a critical role in translation in all organisms, so lateral transfer of rRNA genes among distantly related species is unlikely.
- rRNA has evolved slowly enough that gene sequences from even distantly related species can be aligned and analyzed.

Although studies of rRNA genes reveal much about the evolutionary relationships of prokaryotes, they don't always reveal the entire evolutionary history of these organisms. In some groups of prokaryotes, analyses of multiple gene sequences have suggested several different phylogenetic patterns. How could such differences among different gene sequences arise? Studies of whole prokaryotic genomes have revealed that even distantly related prokaryotes sometimes exchange genetic material.

Lateral gene transfer can lead to discordant gene trees

As noted earlier, prokaryotes reproduce by binary fission. If we could follow these divisions back through evolutionary time, we would be tracing the complete tree of life. At a much broader scale, these divisions of organisms lead to splits among the major evolutionary lineages, or species of life (represented in highly abbreviated form in Appendix A). Because binary fission is an asexual process that replicates whole genomes, we would expect phylogenetic trees of prokaryotes constructed from most gene sequences (see Chapter 14) to reflect these same relationships.

Even though binary fission is an asexual process, there are other processes—including transformation, conjugation, and transduction—that allow the transfer of genetic information between some prokaryotes without reproduction. Thus prokaryotes can transfer and recombine their DNA with that of other individuals (this is sex in the genetic sense of the word), but this genetic exchange is not directly linked to reproduction, as it is in most eukaryotes.

Prokaryotic exchange of genetic material by conjugation and transformation is described in **Key Concept 8.4** and **Appendix D**, respectively.

From early in evolution to the present day, some genes have been moving "sideways" from one prokaryotic species to another, a phenomenon known as **lateral gene transfer** (sometimes called horizontal gene transfer). Lateral gene transfers are well documented among closely related species, and some have been documented even across the domains of life.

Consider, for example, the genome of *Thermotoga maritima*, a bacterium that can survive extremely high temperatures. By comparing the 1,869 gene sequences of *T. maritima* with sequences encoding the same proteins in other species, investigators found that some of this bacterium's genes have their closest relationships not with the genes of other bacterial species, but with the genes of archaea that live in similar extreme environments.

When genes involved in lateral transfer events are sequenced and analyzed, the resulting **gene trees** will not match the organismal tree in every respect (**FIGURE 18.4**). The individual gene trees will vary because the history of lateral transfer events is different for each gene. Biologists can reconstruct the underlying organismal phylogeny by comparing multiple genes (to produce a consensus tree) or by concentrating on genes that are unlikely to be involved in lateral gene transfer events. For example, genes that are involved in fundamental cellular processes (such as the rRNA genes discussed above) are unlikely to be replaced by the same genes from other species because functional, locally adapted copies of these genes are already present.

What kinds of genes are most likely to be involved in lateral gene transfer? Genes that result in a new adaptation that confers higher fitness on a recipient species are most likely to be transferred repeatedly among species. For example, genes that produce antibiotic resistance are often transferred among bacterial species on plasmids, especially under the strong selection pressure such as that imposed by modern antibiotic medications. Improper or overly frequent use of antibiotics can select for resistant strains of

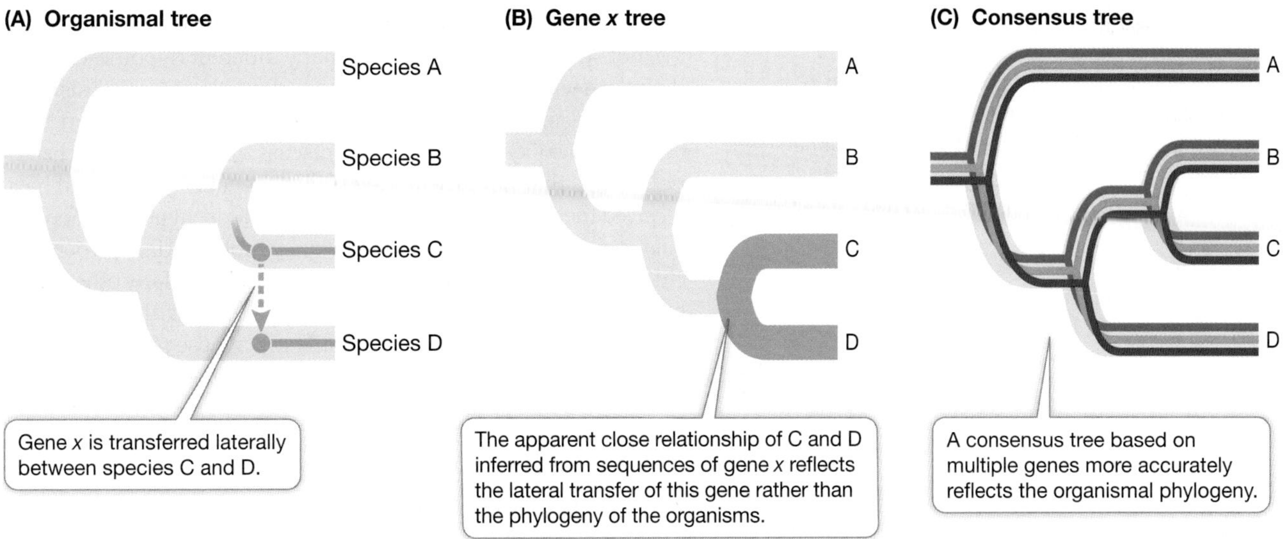

FIGURE 18.4 Lateral Gene Transfer Complicates Phylogenetic Relationships **(A)** The phylogeny of four hypothetical prokaryotic species, two of which have been involved in a lateral transfer of gene *x*. **(B)** A tree based only on gene *x* shows the phylogeny of the laterally transferred gene, rather than the organismal phylogeny. **(C)** A consensus tree based on multiple genes is more likely to reflect the true organismal phylogeny, especially if those genes come from a stable core of genes involved in fundamental processes.

Why are multiple lateral gene transfers between the same two branches on a phylogeny expected to be rare, at least compared with similarities inherited through the stable core?

bacteria that are much harder to treat. This selection for antibiotic resistance explains why informed physicians have become more careful in prescribing antibiotics.

It is debatable whether lateral gene transfer has seriously complicated our attempts to resolve the tree of prokaryotic life. Recent work suggests that it has not. Lateral gene transfer rarely creates problems at higher taxonomic levels, even though it may complicate our understanding of the relationships among individual species. Some species clearly obtain some of their genes from otherwise distantly related species, so evolutionary histories of individual genes may differ within a single organism. But it is now possible to make nucleotide sequence comparisons involving entire genomes, and these studies are revealing a stable core of crucial genes that are uncomplicated by lateral gene transfer. Gene trees based on this stable core more accurately reveal the organismal phylogeny (see Figure 18.4). The problem remains, however, that only a very small proportion of the prokaryotic world has been described and studied.

The great majority of prokaryotic species have never been studied

Most prokaryotes have defied all attempts to grow them in pure culture, causing biologists to wonder how many species, and possibly even major clades, we might be missing. A window onto this problem was opened with the introduction of a new way of examining nucleic acid sequences. When biologists are unable to work with the whole genome of a single prokaryotic species, they can instead examine genomes collected from an environmental sample (such as a scoop of sediment from the seafloor). This technique is known as **environmental genomics**.

Biologists now routinely isolate gene sequences, or even whole genomes, from environmental samples such as soil and seawater. Comparing such sequences with previously known ones has revealed that an extraordinary number of the sequences represent new, previously unrecognized species. Biologists have described only about 16,000 species of bacteria and fewer than 1,000 species of prokaryotic archaea (see Figure 1.4). The results of some environmental genomic studies suggest that there may be millions—perhaps hundreds of millions—of prokaryotic species. Other biologists put the estimate much lower, arguing that the high dispersal ability of many bacterial species greatly reduces endemism (i.e., the number of species restricted to a particular geographic area). Only the magnitude of these estimates differs, however; all sides agree that we have just begun to uncover Earth's prokaryotic diversity.

REVIEW & APPLY | 18.1

R Bacteria and Archaea are the two major divisions of the tree of life. Eukaryotes evolved from a group of prokaryotic archaea that captured and incorporated at least two different bacteria through endosymbiosis. Environmental genomic studies have revealed a much higher diversity of prokaryotes than was previously known.

A 1. What findings led to the establishment of Bacteria and Archaea as separate domains?

(Continued)

REVIEW & APPLY | 18.1 (continued)

2. The figure below shows an organismal tree in which gene *x* has undergone a lateral transfer event. Draw the phylogenetic tree you would expect based on gene *x*, as well as the phylogenetic tree you would expect based on a consensus of non-transferred genes.

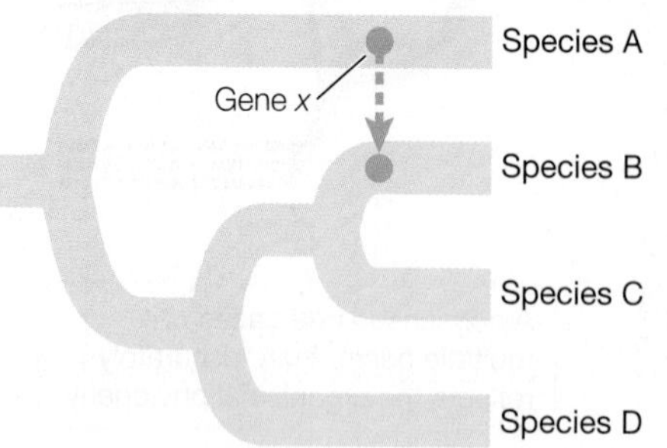

3. How did eukaryotes arise through evolutionary contributions from both prokaryotic archaea and bacteria?

Despite the challenges of reconstructing prokaryotic phylogeny, taxonomists are starting to establish evolutionary classification systems for these organisms. With a full understanding that new information requires periodic revisions in these classifications, we will next apply a current system of classification to organize our survey of prokaryotic diversity.

18.2 Prokaryotic Diversity Reflects the Ancient Origins of Life

LEARNING OBJECTIVES

By the end of this key concept you should be able to:

18.2.1 Interpret a phylogenetic tree of major lineages of Bacteria and Archaea and use it to explain the origins of eukaryotes.

18.2.2 Discuss two lines of evidence that support the origin of life in a high-temperature environment.

18.2.3 Explain the role of bacteria in changing the oxygen content of Earth's atmosphere.

The prokaryotes were alone on Earth for a very long time, adapting to new environments and to changes in existing environments. They have survived to this day, in massive numbers and incredible diversity, and they are found nearly everywhere. In numbers of individuals, prokaryotes are far more abundant than eukaryotes. Individual prokaryotes in the oceans number more than 3×10^{28}—more than the number of stars in the universe. Closer to home, the individual bacteria living in your intestinal tract outnumber all the humans who have ever lived.

Given our still-fragmentary knowledge of prokaryotic diversity, it is not surprising that there are many different hypotheses about the relationships of the major groups of prokaryotes. In this book we use a classification system that has considerable support from nucleotide sequence data. We will discuss a few of the major bacterial groups that have the broadest phylogenetic support and have received the most study, including hadobacteria, hyperthermophilic bacteria, firmicutes, actinobacteria, cyanobacteria, spirochetes, chlamydias, and proteobacteria (see Figure 18.1). Many other major groups of bacteria are known but are less thoroughly studied. We will then describe the prokaryotic archaea, whose great diversity is just beginning to be fully understood and appreciated.

Two early-branching lineages of bacteria live at very high temperatures

Several lineages of bacteria and archaea are **extremophiles**: they thrive under extreme conditions that would kill most other organisms. The **hadobacteria**, for example, are thermophiles (Greek, "heat lovers"). The group's name is derived from Hades, the ancient Greek name for the underworld. Hadobacteria of the genus *Deinococcus* are resistant to radiation and can degrade nuclear waste and other toxic materials. They can also survive extremes of cold as well as hot temperatures. Another hadobacterium, *Thermus aquaticus*, was the source of the thermally stable DNA polymerase that was critical for the development of the polymerase chain reaction. *Thermus aquaticus* was originally isolated from a hot spring, but it can be found wherever hot water occurs (including in many residential hot-water heaters).

The **hyperthermophilic bacteria** are another major group of extremophiles. Genera such as *Aquifex* live near volcanic vents and in hot springs, sometimes at temperatures near the boiling point of water. Some species of *Aquifex* need only hydrogen, oxygen, carbon dioxide, and mineral salts to live and grow. Species of the genus *Thermotoga* live deep underground in oil reservoirs as well as in other high-temperature environments.

Biologists have hypothesized that high temperatures characterized the ancestral conditions for life, given that most environments on early Earth were much hotter than those of today. Reconstructions of ancestral bacterial genes have supported this hypothesis by showing that the ancestral sequences functioned best at elevated temperatures. The presence of multiple lineages of extremophiles at the base of the bacterial tree (see Figure 18.1) also provides support for the origin of life in a high-temperature environment.

Firmicutes include some of the smallest cellular organisms

The **firmicutes** are also known as low-GC Gram-positive bacteria. The first part of this description (low-GC) derives from the relatively low ratio of G-C to A-T nucleotide base pairs in their DNA. The second part (Gram-positive) is less accurate: some of the firmicutes are in fact Gram-negative, and some have no cell wall at all. Despite these differences, phylogenetic analyses of DNA sequences support the monophyly of this bacterial group.

One group of firmicutes can produce resting structures called **endospores** (**FIGURE 18.5**). When a key nutrient such as

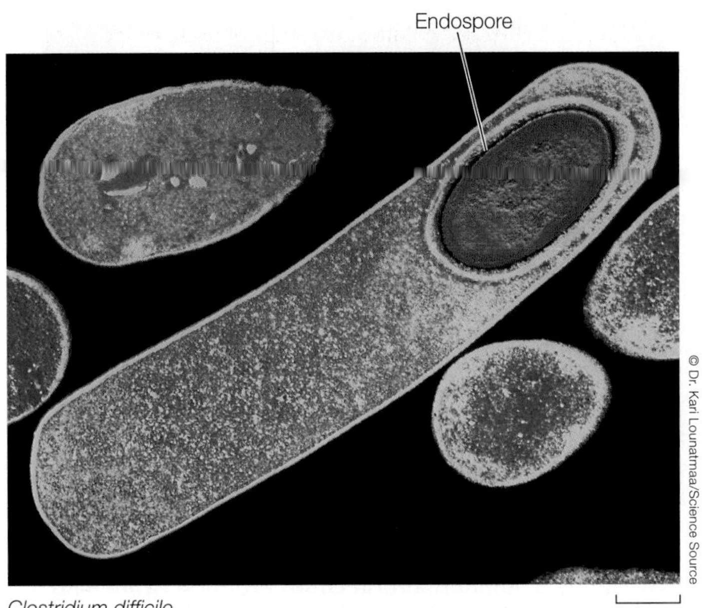

Clostridium difficile

FIGURE 18.5 A Structure for Waiting Out Bad Times Under harsh conditions, some firmicutes can replicate their DNA and encase it in an endospore. The parent cell then breaks down, and the endospore survives in a dormant state until conditions improve.

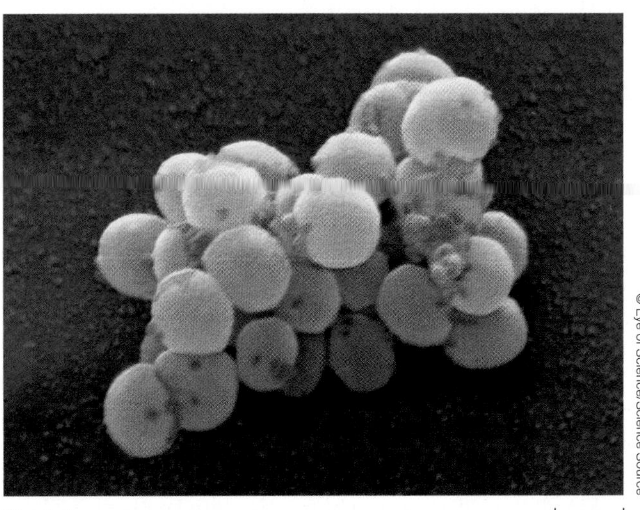

Staphylococcus aureus

FIGURE 18.6 Staphylococci "Grape clusters" are the usual arrangement of these firmicutes, which are often the cause of skin or wound infections.

nitrogen or carbon becomes scarce, the bacterium replicates its DNA and encapsulates one copy, along with some of its cytoplasm, in a tough endospore wall heavily thickened with peptidoglycan and surrounded by a spore coat. The parent cell then breaks down, releasing the endospore. Endospore production is not a reproductive process, as the endospore merely replaces the parent cell. The endospore, however, can survive harsh environmental conditions that would kill the parent cell, such as high or low temperatures or drought, because it is dormant—its normal metabolic activity is suspended. Later, if it encounters favorable conditions, the endospore becomes metabolically active and divides, forming new cells that are like the parent cell. Members of this endospore-forming group include the many species of *Clostridium* and *Bacillus*. Some of their endospores can be reactivated after more than 1,000 years of dormancy. There are even credible claims of reactivation of *Bacillus* endospores that are millions of years old.

Endospores of *Bacillus anthracis* are the cause of anthrax. Anthrax is primarily a disease of cattle and sheep, but it can be fatal in humans. When the endospores sense macrophage (a type of white blood cell that digests cellular debris and foreign substances in mammalian blood), they reactivate and release toxins into the bloodstream. *Bacillus anthracis* has been used as a bioterrorism agent because it is relatively easy to transport large quantities of its endospores and release them among human populations, where they may be inhaled or ingested.

Members of the genus *Staphylococcus*—the **staphylococci** (**FIGURE 18.6**)—are abundant on the human body surface; they are responsible for boils and many other skin problems. *Staphylococcus aureus* is the best-known human pathogen in this genus; it is present in 20 to 40 percent of normal adults (and in 50–70% of hospitalized adults). In addition to skin diseases, *S. aureus* can cause respiratory, intestinal, and wound infections.

Another interesting group of firmicutes, the **mycoplasmas**, lack cell walls, although some have a stiffening material outside the cell membrane. The mycoplasmas are among the smallest cellular organisms known (**FIGURE 18.7**). The smallest mycoplasmas have a diameter of about 0.2 μm. They are small in another crucial sense as well: they have less than half as much DNA as most other prokaryotes. It has been speculated that the DNA in a mycoplasma, which codes for fewer than 500 proteins, may be close to the minimum amount required to encode the essential properties of a living cell.

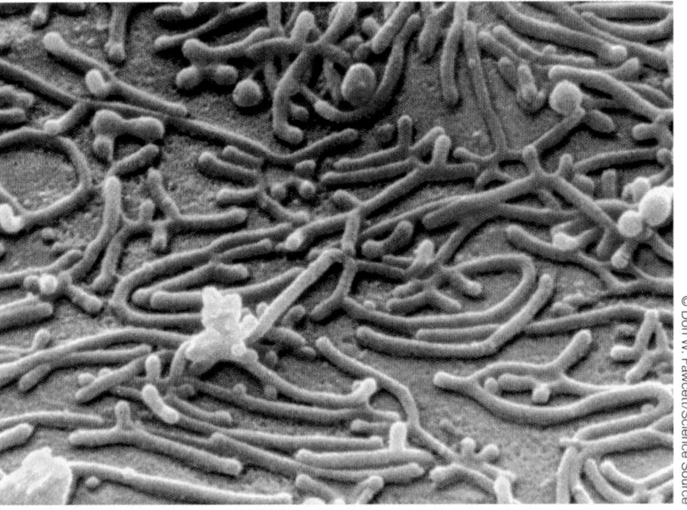

Mycoplasma sp.

FIGURE 18.7 Tiny Cells With about one-fifth as much DNA as *E. coli* has, mycoplasmas are among the smallest known bacteria.

Actinobacteria include major pathogens as well as valuable sources of antibiotics

Actinobacteria, also known as high-GC Gram-positive bacteria, have a higher ratio of G-C to A-T nucleotide base pairs than do the firmicutes (low-GC Gram-positives). These bacteria develop an elaborately branched system of filaments (**FIGURE 18.8**) that resembles the filamentous growth habit of fungi, albeit at a smaller scale. Some actinobacteria reproduce by forming chains of spores at the tips of the filaments. In species that do not form spores, the branched, filamentous growth ceases and the structure breaks up into typical cocci or bacilli, which then reproduce by binary fission.

The actinobacteria include several medically important bacteria. *Mycobacterium tuberculosis* causes tuberculosis, which kills 3 million people each year. Genetic data suggest that this bacterium may have been infecting our ancestors for almost 3 million years, making it the oldest known human bacterial pathogen. The genus *Streptomyces* produces streptomycin as well as hundreds of other antibiotics. We derive most of our antibiotics from actinobacteria.

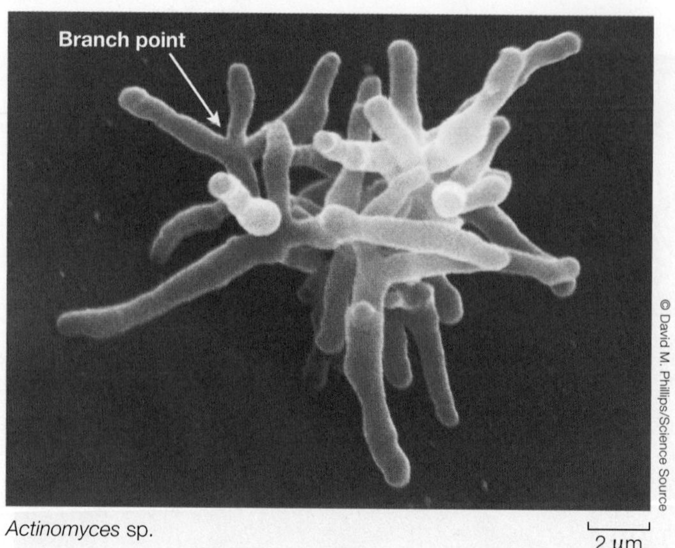

FIGURE 18.8 Actinobacteria Often Produce Branching Filaments The tangled, branching filaments seen in this scanning electron micrograph are typical of this medically important bacterial group.

Cyanobacteria were the first photosynthesizers

Cyanobacteria, sometimes called blue-green bacteria because of their pigmentation, are photosynthetic. They use chlorophyll *a* for photosynthesis and release oxygen gas (O_2); many species also fix nitrogen (which we will discuss in Key Concept 18.3). The production of oxygen by these bacteria transformed the atmosphere of early Earth, eventually leading to the oxygen-rich atmosphere we know today.

Cyanobacteria carry out the same type of photosynthesis that is characteristic of eukaryotic photosynthesizers. They contain elaborate and highly organized internal membrane systems called **photosynthetic lamellae**. As mentioned in Key Concept 18.1, the chloroplasts of photosynthetic eukaryotes are derived from an endosymbiotic cyanobacterium. So all photosynthesis is either directly from cyanobacteria, or from eukaryotes with chloroplasts that are derived from cyanobacteria. Cyanobacteria may live free as single cells or associate in multicellular colonies. Depending on the species and on growth conditions, these colonies may range from flat sheets one cell thick to filaments to spherical balls of cells. Some filamentous colonies of cyanobacteria differentiate into three specialized cell types: vegetative cells, spores, and heterocysts (**FIGURE 18.9**). **Vegetative cells** photosynthesize,

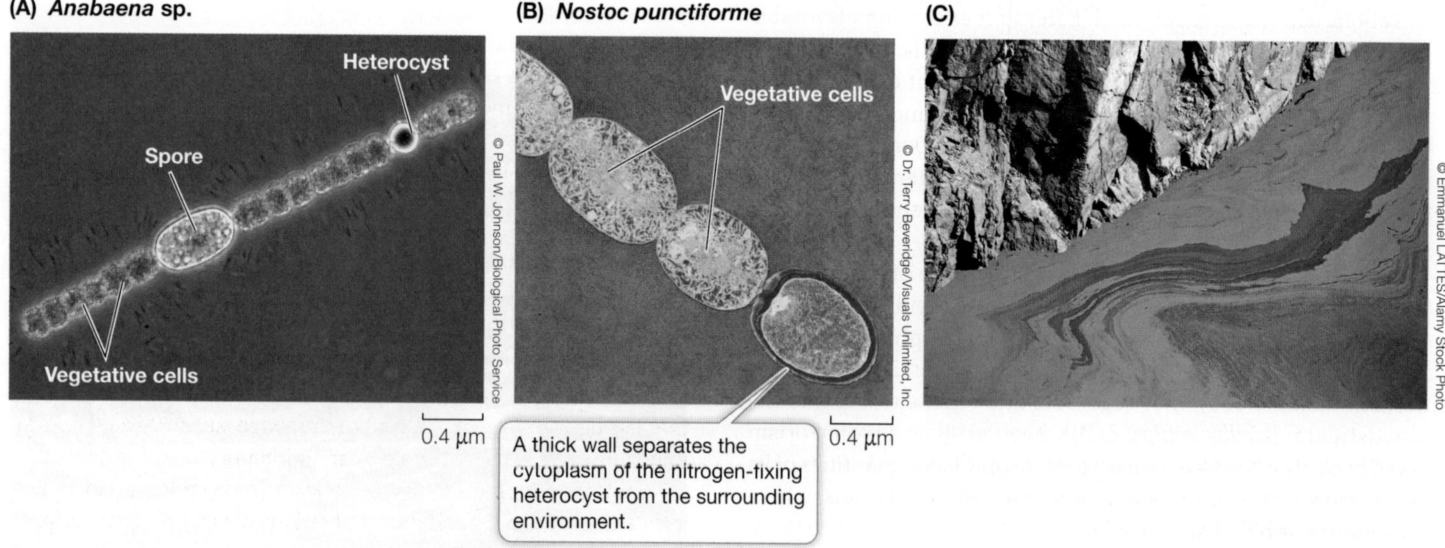

FIGURE 18.9 Cyanobacteria **(A)** Some cyanobacteria form filamentous colonies containing three cell types. **(B)** Heterocysts are specialized for nitrogen fixation and may serve as a breaking point when filaments reproduce. **(C)** This pond in Canada has experienced eutrophication: phosphorus and other nutrients generated by human activity have accumulated, feeding an immense green mat (commonly referred to as "pond scum") that is made up of several species of free-living cyanobacteria.

Media Clip 18.1 Cyanobacteria

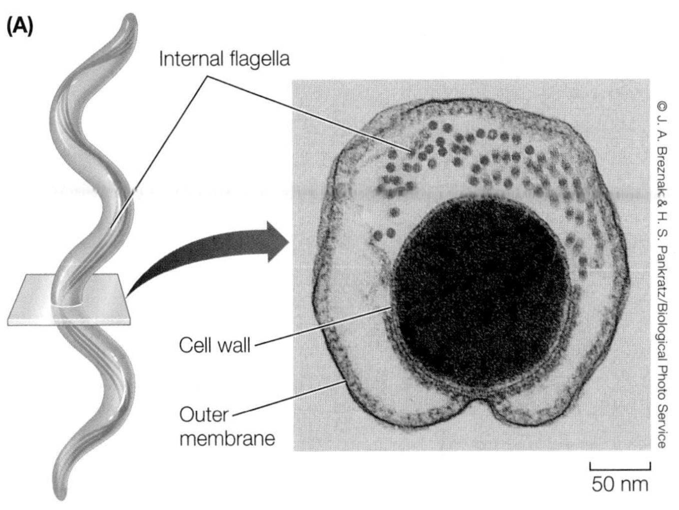

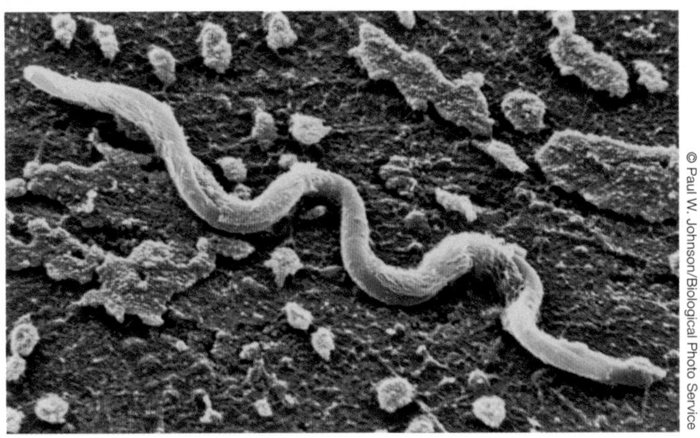

FIGURE 18.10 Spirochetes Get Their Shape from Axial Filaments (A) A spirochete from the gut of a termite, seen in cross section, shows the internal flagella that compose the axial filament, which these helical prokaryotes use to produce a corkscrew-like movement. **(B)** This spirochete species causes syphilis in humans.

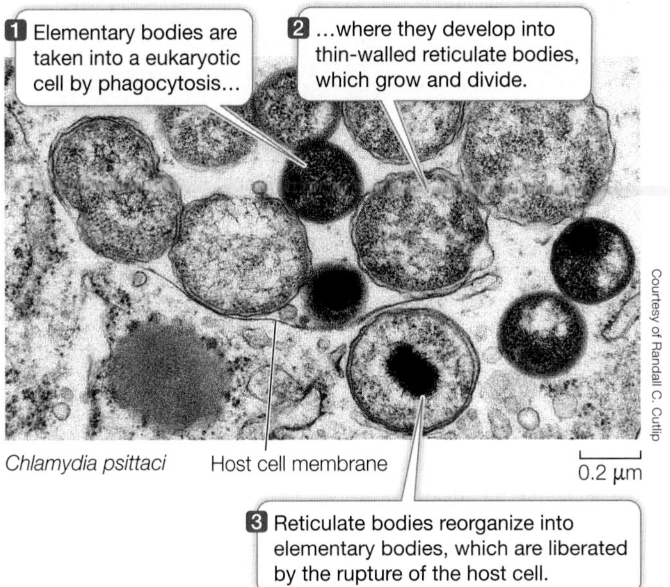

FIGURE 18.11 Chlamydias Change Form Elementary bodies and reticulate bodies are the two cell forms of the chlamydia life cycle.

spores are resting stages that can survive harsh environmental conditions and eventually develop into new filaments, and **heterocysts** are cells specialized for nitrogen fixation. All of the known cyanobacteria with heterocysts fix nitrogen. Heterocysts also have a role in reproduction: when filaments break apart to reproduce, the heterocyst may serve as a breaking point.

Spirochetes move by means of axial filaments

Spirochetes are Gram-negative, motile bacteria characterized by a unique structure known as an axial filament, which is composed of multiple internal flagella running through the periplasmic space (**FIGURE 18.10A**). The cell body is a long cylinder coiled into a helix. The internal flagella begin at either end of the cell and overlap in the middle. Motor proteins connect the axial filament to the cell wall, enabling the corkscrew-like movement of the bacterium. Many spirochetes are parasites of humans; a few are pathogens, including those that cause syphilis (**FIGURE 18.10B**) and Lyme disease. Others live free in mud or water.

Chlamydias are extremely small parasites

Chlamydias are among the smaller bacteria (0.2–1.5 μm in diameter). They are obligate parasites—that is, they can live only as parasites in the cells of other organisms. It was once believed that their obligate parasitism resulted from an inability to produce ATP—that chlamydias were "energy parasites." However, genome sequencing indicates that chlamydias have the genetic capacity to produce at least some ATP. They can augment this capacity by using an enzyme called a translocase, which allows them to take up ATP from the cytoplasm of their host in exchange for ADP from their own cells.

These tiny, Gram-negative cocci are unique among prokaryotes because of a complex life cycle that involves two different forms of cells, elementary bodies and reticulate bodies (**FIGURE 18.11**). Various strains of chlamydias cause eye infections (especially trachoma), sexually transmitted diseases, and some forms of pneumonia in humans.

The proteobacteria are a large and diverse group

By far the largest bacterial group, in terms of numbers of described species, is the **proteobacteria**. The proteobacteria include many species of Gram-negative photoautotrophs (see Key Concept 18.3) that use light-driven reactions to metabolize sulfur, as well as dramatically diverse bacteria that bear no phenotypic resemblance to the photoautotrophic species. Genetic and morphological evidence indicates that the mitochondria of eukaryotes were derived from a proteobacterium by endosymbiosis.

Among the proteobacteria are some nitrogen-fixing genera, such as *Rhizobium*, and other bacteria that contribute to the global nitrogen and sulfur cycles. *Escherichia coli*, one of the most studied organisms on Earth, is a proteobacterium. So, too, are many of

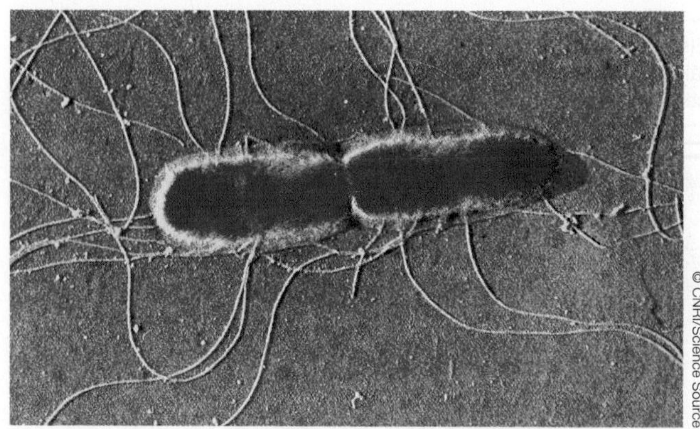

FIGURE 18.12 Proteobacteria Include Many Familiar Bacteria These conjugating (see Key Concept 8.4) cells of *Salmonella typhimurium* are exchanging genetic material. This pathogen causes a wide range of gastrointestinal illnesses in humans.

Media Clip 18.2 A Swarm of *Salmonella*

the most famous human pathogens, such as *Yersinia pestis* (which causes bubonic plague), *Vibrio cholerae* (cholera), and *Salmonella typhimurium* (gastrointestinal disease) (**FIGURE 18.12**).

The bioluminescent *Vibrio* shown at the opening of this chapter are also members of this group. There are many potential applications of the genes that encode bioluminescent proteins in bacteria. Already, these genes are being inserted into the genomes of other species in which the resulting bioluminescence is used as a marker of gene expression. Futuristic proposals for making use of bioluminescence in bioengineered organisms include crop plants that glow when they become water-stressed and need to be irrigated, and glowing trees that could light highways at night in place of electric lights.

FIGURE 18.13 Crown Gall Crown gall, a type of tumor shown here growing on the stem of a *Euonymus* shrub, is caused by the proteobacterium *Agrobacterium tumefaciens*.

Although fungi cause most plant diseases, and viruses cause others, about 200 known plant diseases are of bacterial origin. Crown gall, with its characteristic tumors (**FIGURE 18.13**), is one of the most striking. The causal agent of crown gall is *Agrobacterium tumefaciens*, a proteobacterium that harbors a plasmid used in recombinant DNA studies as a vehicle for inserting genes into new plant hosts.

Gene sequencing enabled biologists to differentiate Archaea from Bacteria

The original identification of Archaea as a group distinct from Bacteria was based on phylogenetic relationships determined from rRNA gene sequences. This separation was supported when biologists sequenced the first complete archaeal genome, which consisted of 1,738 genes—more than half of which were unlike any genes ever found in Bacteria.

Prokaryotic archaea are known for living in extreme habitats such as those with high salinity (salt content), low oxygen concentrations, high temperatures, and/or high or low pH (**INVESTIGATION FIGURE 18.14**). Many prokaryotic archaea are not extremophiles, however—they are common in soil, in many aquatic environments, and in the guts of animals, for example.

Recent studies are revealing many new lineages of Archaea, a few of which are shown in Figure 18.1: **Euryarchaeota**, **Crenarchaeota**, **Thaumarchaeota**, **Korarchaeota**, and **Lokiarchaeota**. The Lokiarchaeota were the first discovered members of the **Asgard archaea**—a group of particular interest because recent studies of their genomes show that they include the closest known prokaryotic relatives of eukaryotes.

Two characteristics shared by all prokaryotic archaea are the absence of peptidoglycan in their cell walls and the presence of lipids of distinctive composition in their cell membranes (see Table 18.1). The unusual lipids in the membranes of prokaryotic archaea are found in no bacteria or eukaryotes. Most lipids in bacterial and eukaryotic membranes contain unbranched long-chain fatty acids connected to glycerol molecules by **ester linkages**:

```
      O       H
      ‖       |
   —C—O—C—
              |
              H
```

In contrast, some lipids in membranes of prokaryotic archaea contain long-chain hydrocarbons connected to glycerol molecules by **ether linkages**:

```
    H       H
    |       |
  —C—O—C—
    |       |
    H       H
```

In addition, the long hydrocarbon chains in the lipids of prokaryotic archaea are branched. One class of archaeal lipids contains glycerol at *both* ends of the hydrocarbons (**FIGURE 18.15**). These lipids form a lipid monolayer structure that is unique to prokaryotic archaea. They still fit into a biological membrane because they are twice as long as

INVESTIGATION

FIGURE 18.14 What Is the Highest Temperature Compatible with Life? Can any organism thrive at temperatures above 120°C? This is the temperature used for sterilization, known to destroy all previously described organisms. Kazem Kashefi and Derek Lovley isolated an unidentified prokaryote from water samples taken near a hydrothermal vent and found it survived and even multiplied at 121°C. The organism was dubbed "Strain 121," and its gene sequencing results indicate that it is a prokaryotic archaeal species.[a]

HYPOTHESIS

Some prokaryotes can survive at temperatures above 120°C.

METHOD

1. Seal samples of unidentified, iron-reducing, thermal vent prokaryotes in tubes with a medium containing Fe^{3+} as an electron acceptor. Control tubes contain Fe^{3+} but no organisms.
2. Hold both tubes in a sterilizer at 121°C for 10 hours. If the iron-reducing organisms are metabolically active, they will reduce the Fe^{3+} to Fe^{2+} (as magnetite, which can be detected with a magnet).

RESULTS

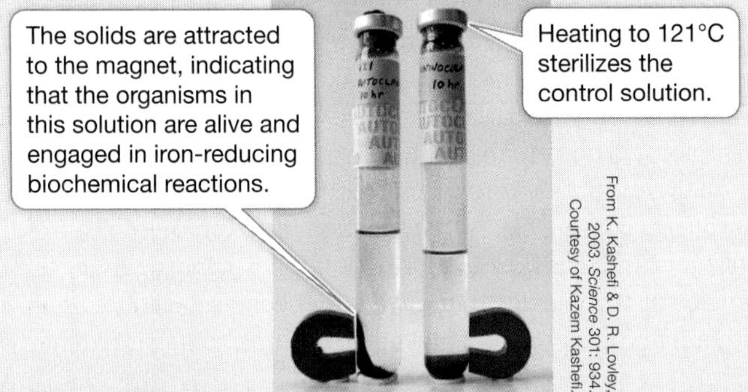

CONCLUSION

Prokaryotic archaea of Strain 121 can survive at temperatures above the previously defined sterilization limit.

ANALYZE THE DATA

After Kashefi and Lovley isolated Strain 121, they examined its growth at various temperatures. The table below shows generation time (time between cell divisions) at nine temperatures.

Temperature (°C)	Generation time (hr)
85	10
90	4
95	3
100	2.5
105	2
110	4
115	6
120	20
130	No growth, but cells not killed

1. Make a graph from these data showing generation time as a function of temperature.
2. Which temperature appears to be closest to the optimum for the growth of Strain 121?
3. Note that no growth occurred at 130°C, but that the cells were not killed. How would you demonstrate that these cells were still alive?

[a] K. Kashefi and D. R. Lovley. 2003. *Science* 301: 934.

A similar **Analyze the Data** exercise may be assigned in **Achieve**.

Some prokaryotic archaea have long-chain hydrocarbons that span the membrane (a lipid monolayer).

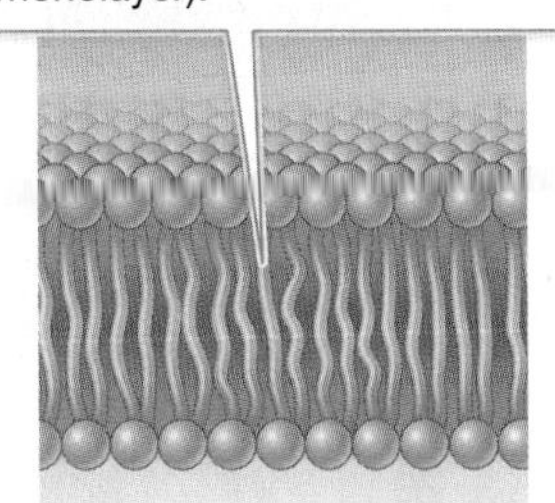

Other prokaryotic archaeal hydrocarbons fit the same template as those of bacteria and eukaryotes (a lipid bilayer).

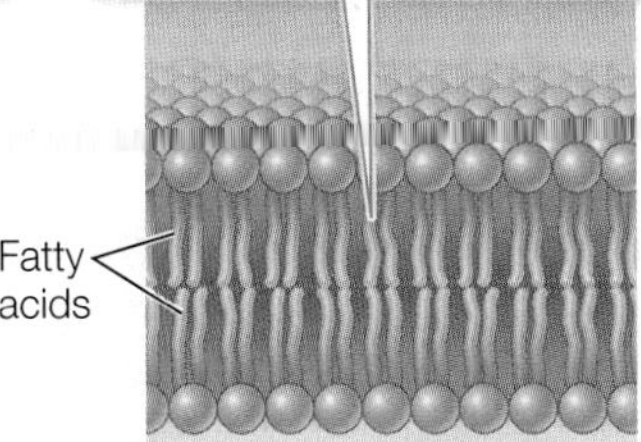

FIGURE 18.15 Membrane Architecture in Prokaryotic Archaea The long-chain hydrocarbons of many prokaryotic archaeal lipids have glycerol molecules at both ends, so that the membranes they form consist of a lipid monolayer. In contrast, the membranes of other archaea, bacteria, and eukaryotes consist of a lipid bilayer.

the typical lipids in the bilayers of other membranes. Lipid monolayers and bilayers are both found among the prokaryotic archaea.

Prokaryotic archaea live in extremely diverse environments

Most known crenarchaeotes are either thermophilic, acidophilic (acid loving), or both. Members of the genus *Sulfolobus* live in hot sulfur springs at temperatures of 70°C to 75°C. They become metabolically inactive at 55°C (131°F). Hot sulfur springs are also extremely acidic. *Sulfolobus* grows best in the range from pH 2 to pH 3, but some members of this genus readily tolerate pH values as low as 0.9. Most acidophilic thermophiles maintain an internal pH of 5.5 to 7 (close to neutral) in spite of their acidic environment. These and other crenarchaeotes thrive where very few other organisms can even survive (**FIGURE 18.16**).

Some species of euryarchaeotes are **methanogens**: they produce methane (CH_4) by reducing carbon dioxide as the key step in their energy metabolism. All of the methanogens are obligate anaerobes (see Key Concept 18.3). Comparison of their rRNA gene sequences has revealed a close evolutionary relationship among these methanogenic species, which were previously assigned to several different groups of bacteria.

Methanogenic euryarchaeotes release approximately 2 billion tons of methane gas into Earth's atmosphere each year, accounting for 80 to 90 percent of the methane that enters the atmosphere, including that produced in many animal digestive systems (including our own). Approximately one-third of this methane comes from methanogens living in the guts of ruminants such as cattle, sheep, and deer, and another large fraction comes from methanogens living in the guts of termites and cockroaches. Methane is increasing in Earth's atmosphere by about 1 percent per year and contributes to the greenhouse effect. Part of that increase is due to increases in cattle and rice production and the methanogens associated with both.

Another group of euryarchaeotes, the **extreme halophiles** (salt lovers), live exclusively in very salty environments. Because they contain pink carotenoid pigments, these archaea are sometimes easy to see (**FIGURE 18.17**). Extreme halophiles grow in the Dead Sea and in brines of all types. The reddish pink spots that

FIGURE 18.16 Some Crenarchaeotes Like It Hot Thermophilic crenarchaeotes can thrive in the intense heat of volcanic hot sulfur springs such as these in Yellowstone National Park.

can occur on pickled fishes are colonies of halophilic archaea. Few other organisms can live in the saltiest homes that the extreme halophiles occupy—most would "dry" to death, losing too much water to the hypertonic environment. Extreme halophiles have been found in lakes with pH values as high as 11.5. These are the most alkaline environments inhabited by living organisms, and almost as alkaline as household ammonia.

Some of the extreme halophiles have a unique system for trapping light energy and using it to form ATP—without using any form of chlorophyll—when oxygen is in short supply. They use the pigment retinal (also found in the vertebrate eye) combined with a protein to form a light-absorbing molecule called microbial rhodopsin.

Another member of the Euryarchaeota, *Thermoplasma*, has no cell wall. It is thermophilic and acidophilic, its metabolism is aerobic, and it lives in coal deposits. Its genome of 1,100,000 base pairs is among the smallest (along with that of the mycoplasmas) found in any free-living organism, although some parasitic organisms have even smaller genomes.

FIGURE 18.17 Extreme Halophiles Highly saline environments such as these commercial seawater evaporating ponds in San Francisco Bay are home to extreme halophiles. The prokaryotic archaea are easily visible here because of the rich red coloration of their carotenoid pigments.

Many of the known prokaryotic archaea are crenarchaeotes or euryarchaeotes, but studies of extreme environments have identified several lineages that are not closely related to either of these major groups. For example, the korarchaeotes are known only from DNA isolated directly from hot environments. The thaumarchaeotes were originally found in hot environments as well, but have since been found to be common in marine surface waters, where they oxidize ammonia and appear to play an important role in the nitrogen cycle.

The lokiarchaeotes were discovered in 2015 by sequencing environmental samples from near a hydrothermal vent called Loki's Castle deep in the Arctic Ocean. An organism's genome was detected and sequenced from this deep-sea sample. Sequence analysis revealed that the organism was a distinct lineage of archaea, which was named *Lokiarchaeum*. Of particular interest was the finding that the *Lokiarchaeum* genome contains a large number of genes with cell membrane–related functions—genes and functions that had previously been known to occur only in eukaryotes. After the discovery of *Lokiarchaeum*, biologists began to look for related archaea in other aquatic environments (both marine and freshwater), and discovered many new lineages of archaea with affinities to eukaryotes. Together, these close relatives of eukaryotes are known as the Asgard archaea (named after Asgard, one of the nine worlds of Norse mythology). The genomes of the Asgard archaea include many additional genes that were formerly thought to be restricted to eukaryotes, including genes that function in eukaryotes in controlling cell shape and cytoskeleton formation. These findings suggest that some of the properties that have long been associated exclusively with eukaryotes probably first arose within the Asgard archaea.

REVIEW & APPLY | 18.2

R Bacteria and Archaea are highly diverse groups that survive in almost every imaginable habitat on Earth. Many prokaryotes can survive and even thrive in habitats where no eukaryotes can live, including extremely hot, acidic, or saline conditions. Eukaryotes are most closely related to the Asgard archaea, although endosymbioses of bacteria within eukaryotic cells contributed to the evolution of eukaryotic organelles.

A

1. Consider the differences between prokaryotes and eukaryotes shown in Table 18.1. Why might eukaryotes be more like prokaryotic archaea in some features and more like bacteria in others?
2. Given that all species of life have evolved for the same amount of time since their common origin, how would you respond to someone who characterizes prokaryotes as "primitive"? Include at least two examples of major groups of prokaryotes to support your answer.
3. How have bacteria changed Earth's atmosphere over the past three billion years?
4. What two lines of evidence can you use to support the idea that eukaryotes evolved from a lineage of the Asgard archaea?

Prokaryotes are found almost everywhere on Earth and live in a wide variety of ecosystems. In the next key concept we will examine the contributions of prokaryotes to the functioning of those ecosystems.

18.3 Ecological Communities Depend on Prokaryotes

LEARNING OBJECTIVES

By the end of this key concept you should be able to:

18.3.1 **Describe what biofilms are, how they are formed, and why they present problems for humans.**

18.3.2 **Explain how bacteria communicate through quorum sensing.**

18.3.3 **Explain the critical role of prokaryotes in the nitrogen cycle.**

Prokaryotic cells do not usually live in isolation. Rather, they live in communities of many different species, often including microscopic eukaryotes. Whereas some microbial communities are harmful to humans, others provide important services. They help us digest our food, break down municipal waste, and recycle organic matter and chemical elements in the environment.

Many prokaryotes form complex communities

Some microbial communities form layers in sediments, and others form clumps a meter or more in diameter. Many microbial communities tend to form dense **biofilms**. Upon contacting a solid surface, the cells bind to that surface and secrete a sticky, gel-like polysaccharide matrix that traps other cells (**FIGURE 18.18**). Once a biofilm forms, the cells become more difficult to kill.

Biofilms are found in many places, and in some of those places they cause problems for humans. The material on our teeth that we call dental plaque is a biofilm. Pathogenic bacteria are difficult for the immune system—and modern medicine—to combat once they form a biofilm, which may be impermeable to antibiotics. Biofilms may form on just about any available surface, including contact lenses and artificial joint replacements. They foul metal pipes and cause corrosion, a major problem in steam-driven electricity generation plants. Fossil stromatolites—large, rocky structures made up of alternating layers of fossilized biofilm and calcium carbonate—are among the oldest remnants of life on Earth (see Figure 17.6C).

Some biologists are studying the chemical signals that prokaryotes use to communicate with one another and that trigger density-linked activities such as biofilm formation. We saw one example of this type of communication—called **quorum sensing**— is found in the bioluminescent *Vibrio* that are shown in the opening photo of this chapter. How does quorum sensing work? As demonstrated in **INVESTIGATION FIGURE 18.19**, individual *Vibrio* bacteria can excrete a signal that is detected by other individuals, and this signal

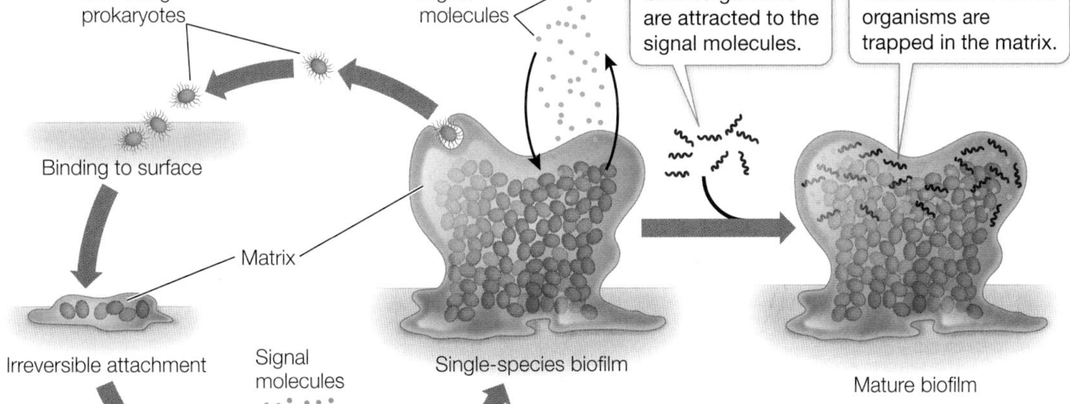

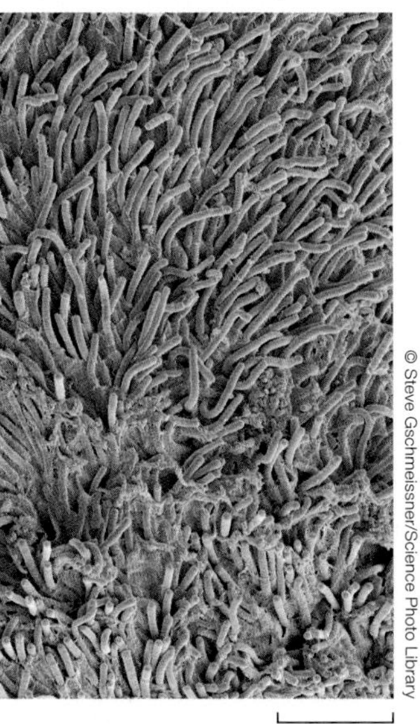

FIGURE 18.18 Forming a Biofilm **(A)** Free-living prokaryotes readily attach themselves to surfaces and form films that are stabilized and protected by a surrounding matrix. Once the population is large enough, the developing biofilm can send out chemical signals that attract other microorganisms. **(B)** Scanning electron micrography reveals a biofilm of dental plaque. The bacteria (red) are embedded in a matrix consisting of proteins from both bacterial secretions and saliva. (A after D. Monroe. 2007. *PLoS Biol* 5: e307.)

INVESTIGATION

FIGURE 18.19 How Do Bacteria Communicate with One Another? Bonnie Bassler and her colleagues at Princeton University investigated how *Vibrio fischeri* bacteria communicate with one another. These bacteria produce bioluminescence when they are present in sufficiently high densities. In a normal *V. fischeri* bacterium, the following pathway produces bioluminescence when a bacterial colony becomes dense enough to produce sufficient signal:

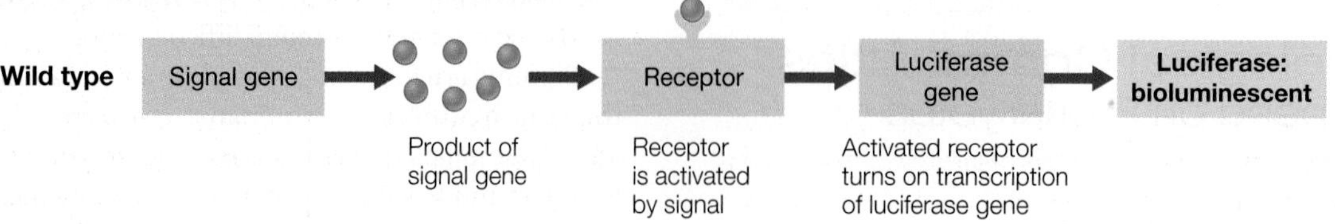

The fact that the bacteria emit light only when they are present in high densities suggests that the signal is used to communicate among nearby bacteria, alerting one another to their presence. But how can we tell that the signal produced by one bacterium is being received by another?[a]

HYPOTHESIS

V. fischeri can transmit the signal molecule from one individual to another, thus communicating about the presence of other nearby bacteria.

METHOD

1. Select two mutant strains of *V. fischeri* incapable of emitting light on their own. In strain A, a mutation for the signal receptor renders the bacteria unable to detect the signal molecule, so they do not bioluminesce:

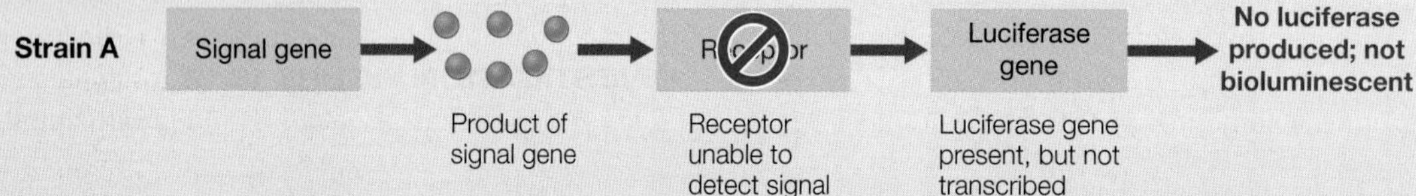

Strain B bacteria have a mutated signal gene, so they do not produce a signal molecule, although the receptor and the luciferase gene are normal:

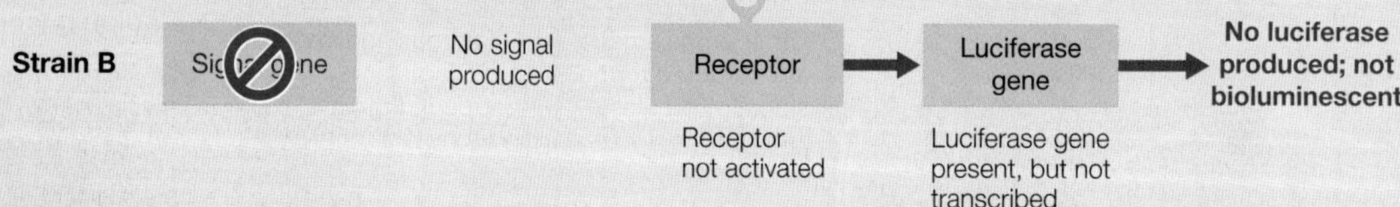

2. Incubate samples of *V. fischeri* on agar petri plates, as follows:
   Plate 1: Strain A only
   Plate 2: Strain B only
   Plate 3: Half of plate with strain A, half with strain B

RESULTS

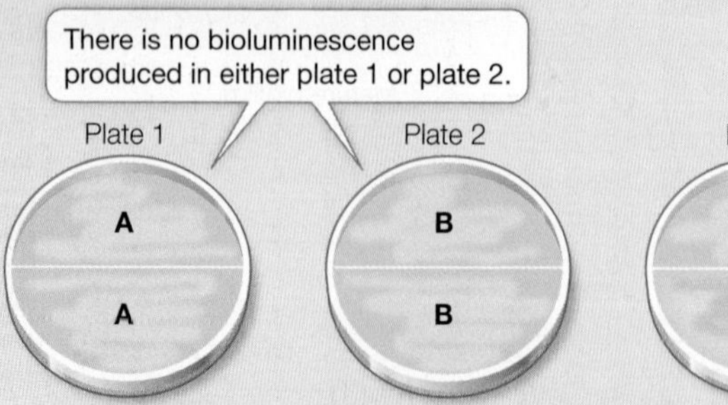

CONCLUSION

Vibrio bacteria can detect the signal molecule produced by other nearby bacteria and use this information to detect the presence of other bacteria.

[a]M. B. Miller and B. L. Bassler. 2001. *Annu Rev Microbiol* 55: 165–199.

then functions to turn on the genes that produce luciferase—an enzyme that produces bioluminescence when it is active.

Microbiomes are critical to the health of many eukaryotes

Although only a few bacterial species are pathogens, popular notions of bacteria as "germs" and fear of the consequences of infection cause many people to assume that most bacteria are harmful. Increasingly, however, biologists are discovering that the health of humans (as well as that of most other eukaryotes) depends in large part on the health of our **microbiomes**: the communities of bacteria and prokaryotic archaea that live in and on our bodies. Other communities of microbes live in close association with other multicellular organisms.

Every surface of your body is covered with diverse communities of bacteria (**FIGURE 18.20**). A recent study identified more than 1,000 species of bacteria that live on human skin. Inside your body, your digestive system teems with bacteria. When these communities are disrupted, they must be restored before the body can function normally.

Biologists are discovering that many complex health problems are linked to the disruption of our microbiomes. These diverse microbial communities affect the expression of our genes and play a critical role in the development and maintenance of a healthy immune system. When our microbiomes contain an appropriate community of beneficial species, our bodies function normally. But these communities are strongly affected by our life experiences, by the food we eat, by the medicines we take, and by our exposure to various environmental toxins. The recent rapid increase in the rate of autoimmune diseases in humans—diseases in which the immune system begins to attack the body—has been linked to the changing diversity and composition of our microbiomes.

The early acquisition of an appropriate microbiome is critical for lifelong health. Normally, a human infant acquires much of its microbiome at birth, from the microbiome in its mother's vagina. Other components of the microbiome are also acquired from the mother, especially through breast feeding. Recent studies have shown that babies born by cesarean section, as well as babies that are bottle-fed on artificial milk formula, typically acquire microbes from a wider variety of sources. Many of the bacteria acquired in this way are not well suited for human health. Biologists have discovered that the incidence of many autoimmune diseases is much higher in people who were born by cesarean section and in those who were fed on formula as infants, compared with individuals who were born vaginally and breast-fed as infants. The difference appears to be related to the composition of the individual's original microbiome.

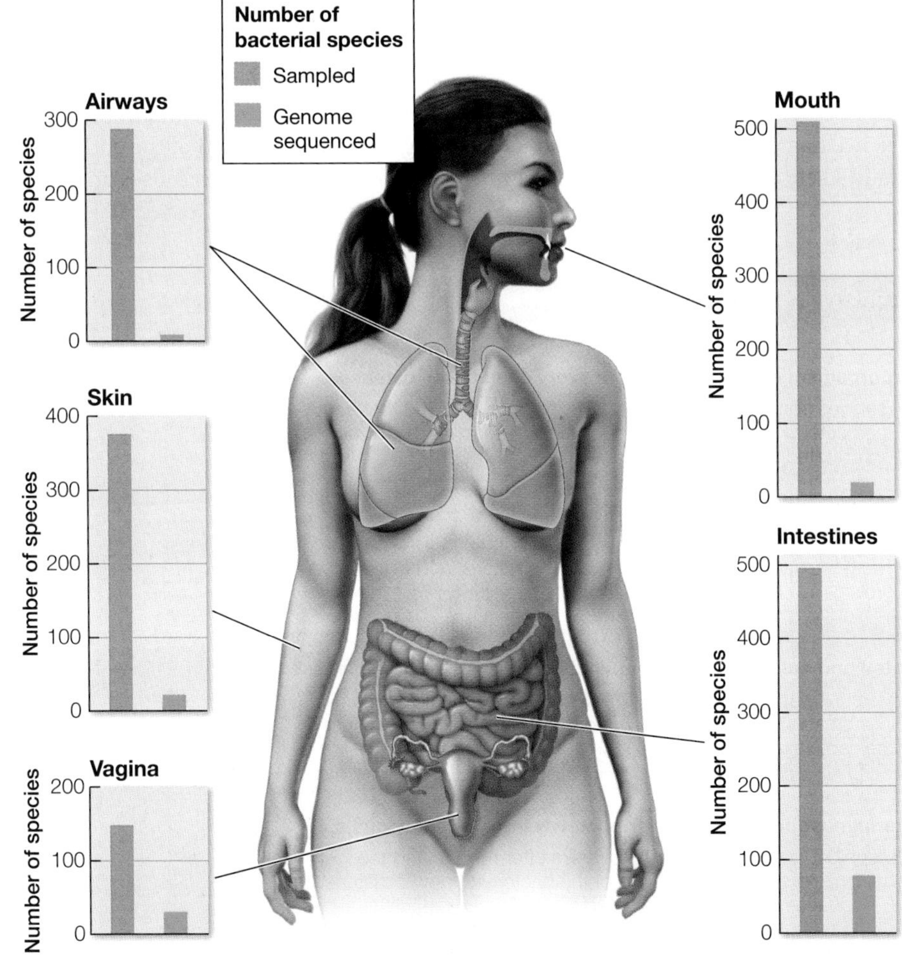

FIGURE 18.20 The Body's Microbiome Is Critical to the Maintenance of Health Surveys of the human microbiome have shown that this community includes thousands of diverse bacterial species that are adapted to grow in or on various parts of the body. Although we now know that the composition of this microbiome is closely associated with many aspects of human health, most of the component species are poorly characterized and remain largely unstudied by biologists. What has become clear is that, although the "subcommunities" in different parts of the body share similarities, each is a site-specific assemblage of many distinctive species. (After Y. K. Lee and S. K. Mazmanian. 2010. *Science* 330: 1768–1773.)

Humans use some of the metabolic products—especially vitamins B_{12} and K—produced by the microbiome living in the large intestine. Communities of bacteria line our intestines with a dense biofilm that is in intimate contact with the mucosal lining of the gut. This biofilm facilitates nutrient transfer from the intestine into the body, functioning like a specialized "tissue" that is essential to our health. This biofilm has a complex ecology that scientists have just begun to explore in detail—including the possibility that the species composition of an individual's gut microbiome may contribute to obesity.

Animals harbor a variety of microbes in their digestive tracts, many of which play important roles in digestion. Cattle depend on prokaryotes to break down plant material. Like most animals, cattle cannot produce cellulase, the enzyme needed to start the digestion of the cellulose that makes up the bulk of their plant food. However, bacteria living in a special section of the gut, called the rumen, produce enough cellulase to process the daily diet for the cattle.

A small minority of bacteria are pathogens

The late nineteenth century was a productive era in the history of medicine—a time when bacteriologists, chemists, and physicians proved that many diseases are caused by microbial agents. During this time, the German physician Robert Koch laid down a set of four rules for establishing that a particular microorganism causes a particular disease:

1. The microorganism is always found in individuals with the disease.
2. The microorganism can be taken from the host and grown in pure culture.
3. A sample of the culture produces the same disease when injected into a new, healthy host.
4. The newly infected host yields a new, pure culture of microorganisms identical to those obtained in the second step.

These rules, called **Koch's postulates**, were important tools in a time when it was not widely understood that microorganisms cause disease. Although modern medical science has more powerful diagnostic tools, Koch's postulates remain useful. For example, physicians were taken aback in the 1980s when stomach ulcers—long accepted and treated as the result of excess stomach acid—were shown by Koch's postulates to be caused by the bacterium *Helicobacter pylori* (**FIGURE 18.21**).

For an organism to be a successful pathogen, it must:

- arrive at the body surface of a potential host
- enter the host's body
- evade the host's defenses
- reproduce inside the host
- infect a new host

Failure to complete any of these steps ends the disease cycle of a pathogenic organism. Yet in spite of the many defenses available to potential hosts, some bacteria are very successful pathogens. Pathogenic bacteria are often surprisingly difficult to combat, even with today's arsenal of antibiotics. One source of this difficulty is their ability to form biofilms.

For the host, the consequences of a bacterial infection depend on several factors. One is the invasiveness of the pathogen: its ability to multiply in the host's body. Another is its toxigenicity: its ability to produce toxins (chemical substances that are harmful to the host's tissues). *Corynebacterium diphtheriae*, the agent that causes diphtheria, has low invasiveness and multiplies only in the throat, but its toxigenicity is so great that the entire body is affected. In contrast, *Bacillus anthracis*, which causes anthrax, has

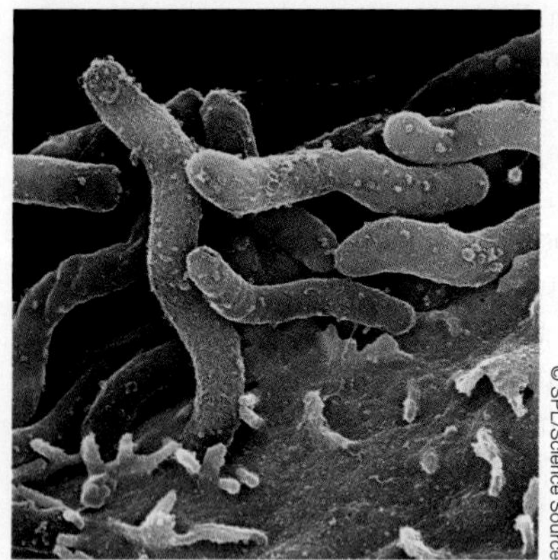

Marshall and Warren set out to satisfy Koch's postulates:

Test 1

The microorganism must be present in every case of the disease.

Results: Biopsies from the stomachs of many patients revealed that the bacterium was always present if the stomach was inflamed or ulcerated.

Test 2

The microorganism must be cultured from a sick host.

Results: The bacterium was isolated from biopsy material and eventually grown in culture media in the laboratory.

Test 3

The isolated and cultured bacteria must be able to induce the disease.

Results: Marshall was examined and found to be free of bacteria and inflammation in his stomach. After drinking a pure culture of the bacterium, he developed stomach inflammation (gastritis).

Test 4

The bacteria must be recoverable from newly infected individuals.

Results: Biopsy of Marshall's stomach 2 weeks after he ingested the bacteria revealed the presence of the bacterium, now christened *Helicobacter pylori*, in the inflamed tissue.

Conclusion

Antibiotic treatment eliminated the bacteria and the inflammation in Marshall's stomach. The experiment was repeated on healthy volunteers, and many patients with gastric ulcers were cured with antibiotics. Thus Marshall and Warren demonstrated that the stomach inflammation leading to ulcers is caused by *H. pylori* infections in the stomach.

FIGURE 18.21 Satisfying Koch's Postulates Robin Warren and Barry Marshall of the University of Western Australia won the 2005 Nobel Prize in Physiology or Medicine for showing that ulcers are caused not by the action of stomach acid but by infection with the bacterium *Helicobacter pylori*. (After B. J. Marshall and J. R. Warren. 1983. *The Lancet* 321: 1273–1275; B. J. Marshall et al. 1985. *Med J Aust* 142: 436–439.)

low toxigenicity but is so invasive that the entire bloodstream ultimately teems with the bacteria.

There are two general types of bacterial toxins: exotoxins and endotoxins. **Endotoxins** are released when certain bacteria grow or lyse (burst). Endotoxins are lipopolysaccharides (complexes consisting of a polysaccharide and a lipid component) that form part of the outer bacterial membrane. Endotoxins are rarely fatal to the host; they normally cause fever, vomiting, and diarrhea. Among the endotoxin producers are some strains of the proteobacteria *Salmonella* and *Escherichia*.

Exotoxins are soluble proteins released by living, multiplying bacteria. They are highly toxic—often fatal—to the host. Human diseases induced by bacterial exotoxins include tetanus (*Clostridium tetani*), cholera (*Vibrio cholerae*), and bubonic plague (*Yersinia pestis*). Anthrax is caused by three exotoxins produced by *Bacillus anthracis*. Botulism is caused by exotoxins produced by *Clostridium botulinum*; these exotoxins are among the most poisonous ever discovered. The lethal dose for humans of one exotoxin of *C. botulinum* is about one-millionth of a gram. Nonetheless, much smaller doses of this exotoxin, marketed under various trade names (e.g., Botox), are used to treat muscle spasms and for cosmetic purposes (temporary wrinkle reduction in the skin).

TABLE **18.2** | How Organisms Obtain Their Energy and Carbon

Nutritional category	Energy source	Carbon source
Photoautotrophs (some bacteria, some eukaryotes)	Light	Carbon dioxide
Photoheterotrophs (some bacteria)	Light	Organic compounds
Chemoautotrophs (some bacteria, many prokaryotic archaea)	Inorganic substances	Carbon dioxide
Chemoheterotrophs (some bacteria, some prokaryotic archaea, most eukaryotes)	Usually organic compounds; sometimes inorganic substances	Organic compounds

Prokaryotes have amazingly diverse metabolic pathways

Prokaryotic archaea and bacteria outdo the eukaryotes in terms of metabolic diversity. Although they are much more diverse in size and shape, eukaryotes draw on fewer metabolic mechanisms for their energy needs. In fact, much of the eukaryotes' energy metabolism is carried out in organelles—mitochondria and chloroplasts—that are endosymbiotic descendants of bacteria. The long evolutionary history of prokaryotes, during which they have had time to explore a wide variety of habitats, has led to the extraordinary diversity of their metabolic "lifestyles"—their use or nonuse of oxygen, their energy sources, their sources of carbon atoms, and the materials they release as waste products.

ANAEROBIC VERSUS AEROBIC METABOLISM Some prokaryotes can live only by anaerobic metabolism because oxygen is poisonous to them. These oxygen-sensitive organisms are called **obligate anaerobes**. By definition, an anaerobe does not use oxygen as an electron acceptor for its respiration. Other prokaryotes, called **facultative anaerobes**, can alternate between an anaerobic mode of metabolism (such as fermentation) and an aerobic mode (such as cellular respiration) as conditions dictate. **Aerotolerant anaerobes** cannot conduct cellular respiration, but they are not damaged by oxygen when it is present. At the other extreme from the obligate anaerobes, some prokaryotes are **obligate aerobes**, unable to survive for extended periods in the *absence* of oxygen. Obligate aerobes require oxygen for cellular respiration.

NUTRITIONAL CATEGORIES All living organisms face the same nutritional challenges: they must synthesize energy-rich compounds such as ATP to power their life-sustaining metabolic reactions, and they must obtain carbon atoms to build their own organic molecules. Biologists recognize four broad nutritional categories of organisms: photoautotrophs, photoheterotrophs, chemoautotrophs, and chemoheterotrophs. Prokaryotes are represented in all four groups (**TABLE 18.2**).

Photoautotrophs perform photosynthesis. They use light as their energy source and carbon dioxide (CO_2) as their carbon source. The cyanobacteria, like green plants and other photosynthetic eukaryotes, use chlorophyll *a* as their key photosynthetic pigment and produce oxygen gas (O_2) as a by-product of noncyclic electron transport.

There are other photoautotrophs among the bacteria, but these organisms use bacteriochlorophyll as their key photosynthetic pigment, and they do not produce O_2. Instead, some of these photosynthesizers produce particles of pure sulfur, because hydrogen sulfide (H_2S), rather than H_2O, is their electron donor for photophosphorylation. Many proteobacteria fit into this category. Bacteriochlorophyll molecules absorb light of longer wavelengths than the chlorophyll molecules used by other photosynthesizing organisms. As a result, bacteria using this pigment can grow in water under fairly dense layers of algae, using light of wavelengths that are not absorbed by the algae (**FIGURE 18.22**).

Photoheterotrophs use light as their energy source but must obtain their carbon atoms from organic compounds made by other organisms. Their "food" consists of organic compounds such as carbohydrates, fatty acids, and alcohols. For example, compounds released from plant roots (as in rice paddies) or from decomposing photosynthetic bacteria in hot springs are taken up by photoheterotrophs and metabolized to form building blocks for other compounds. Sunlight provides the ATP necessary for metabolism through photophosphorylation.

Chemoautotrophs obtain their energy by oxidizing inorganic substances, and they use some of that energy to fix carbon (convert inorganic carbon into organic carbon-based molecules). Some chemoautotrophs use reactions identical to those of the typical photosynthetic cycle, but others use alternative pathways for carbon fixation. Some bacteria oxidize ammonia or nitrite

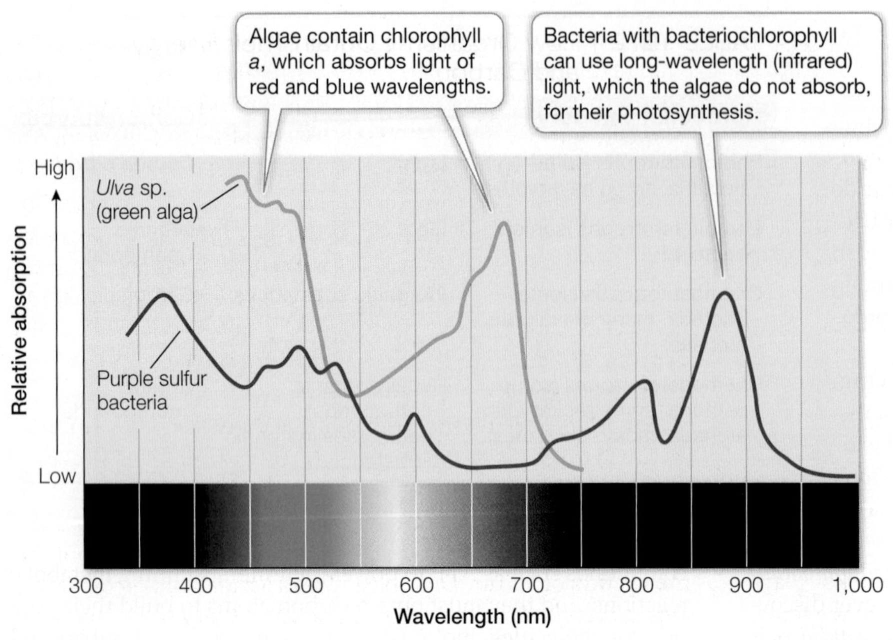

FIGURE 18.22 Bacteriochlorophyll Absorbs Long-Wavelength Light The green alga *Ulva* contains chlorophyll, which absorbs no light of wavelengths longer than 750 nanometers (nm). Purple sulfur bacteria, which contain bacteriochlorophyll, can conduct photosynthesis using longer infrared wavelengths. As a result, these bacteria can grow under layers of algae. (After F. Haxo and L. R. Blinks. 1950. *J Gen Physiol* 33: 389–422; S. Mehrabi et al. 2001. *Biomol Eng* 18: 49–56.)

THINK LIKE A SCIENTIST

Putting bacteria to work

Bacteria are increasingly used for a variety of purposes that range from treating medical conditions to mitigating toxic waste spills. Determining which bacterial species would be appropriate for a particular purpose requires being familiar with diverse phenotypic characters of these bacteria. Here you will explore the relationship between science and society by applying your knowledge of bacteria to environmental problems and issues of human health.

The table below lists the characteristics of four bacteria, *Clostridium novyi*, *Thermus aquaticus*, *Paracoccus denitrificans*, and *Trichodesmium thiebautii*. Use this information to address the questions.

1. The concentration of nitrates in a sample of municipal sewage water is well above acceptable levels. Which of these bacteria would you expect to be most appropriate for reducing nitrate concentrations? Explain your answer.
2. Because tumors often grow much more quickly than they can acquire blood supply, oxygen concentrations can be very low in cancerous tumors. These hypoxic conditions are generally not found elsewhere in the body. Biomedical researchers can take advantage of this unusual property to inject bacteria that would specifically target the tumor cells and have as little effect on the rest of the body as possible. Based on the information in the table, which bacterium would you expect to be most appropriate for this application? Explain your answer.
3. Adding small amounts of ammonium increased the productivity of an aquatic ecosystem, but the effect was short-lived. Based on that finding, which bacterium would be most likely to improve the productivity of the ecosystem?
4. The antibiotic vancomycin inhibits peptidoglycan synthesis in bacteria that have a thick cell wall with much peptidoglycan. Which bacterium would be most likely to be sensitive to this antibiotic? Explain your answer.

Characteristic	*C. novyi*	*T. aquaticus*	*P. denitrificans*	*T. thiebautii*
Metabolism	Obligate anaerobic	Obligate aerobic	Facultative anaerobic	Facultative anaerobic
Gram status	Positive	Negative	Negative	Negative
Optimal temperature (°C)	10°–40°	50°–80°	5°–30°	10°–30°
Typical habitat	Terrestrial	Aquatic	Aquatic	Aquatic
Additional information		Chemotroph	Denitrifier	Nitrogen fixer

ions to form nitrate ions. Others oxidize hydrogen gas, hydrogen sulfide, sulfur, and other materials. Many prokaryotic archaea are chemoautotrophs.

Finally, **chemoheterotrophs** obtain both energy and carbon atoms from one or more complex organic compounds that have been synthesized by other organisms. Most known bacteria and prokaryotic archaea are chemoheterotrophs—as are all animals and fungi and many protists.

Although most chemoheterotrophs rely on the breakdown of organic compounds for energy, some chemoheterotrophic prokaryotes obtain their energy by breaking down inorganic substances. Organisms that obtain energy from oxidizing inorganic substances (both chemoautotrophs as well as some chemoheterotrophs) are also known as lithotrophs (Greek, "rock consumers").

Prokaryotes play important roles in element cycling

The metabolic diversity of the prokaryotes makes them key players in the cycles that keep elements moving through ecosystems. Many prokaryotes are decomposers: organisms that metabolize organic compounds in dead organic material and return the products to the environment as inorganic substances. Prokaryotes, along with fungi, return tremendous quantities of carbon to the atmosphere as carbon dioxide, thus carrying out a key step in the carbon cycle.

The key metabolic reactions of many prokaryotes involve nitrogen or sulfur. For example, some bacteria carry out respiratory electron transport without using oxygen as an electron acceptor. These organisms use oxidized inorganic ions such as nitrate, nitrite, or sulfate as electron acceptors. Examples include the **denitrifiers**, which release nitrogen to the atmosphere as nitrogen gas (N_2). These normally aerobic bacteria, mostly species of the genera *Bacillus* and *Pseudomonas*, use nitrate (NO_3^-) as an electron acceptor in place of oxygen if they are kept under anaerobic conditions:

$$2\ NO_3^- + 10\ e^- + 12\ H^+ \rightarrow N_2 + 6\ H_2O$$

Denitrifiers play a key role in the cycling of nitrogen through ecosystems. Without denitrifiers, which convert nitrate ions back into nitrogen gas, all forms of nitrogen would leach from the soil and end up in lakes and oceans, making life on land much more difficult.

Nitrogen fixers convert atmospheric nitrogen gas into a chemical form (ammonia) that is usable by the nitrogen fixers themselves as well as by other organisms:

$$N_2 + 6\ H \rightarrow 2\ NH_3$$

All organisms require nitrogen in order to build proteins, nucleic acids, and other important compounds. Nitrogen fixation is thus vital to life as we know it. This all-important biochemical process is carried out by a wide variety of prokaryotic archaea and bacteria (including cyanobacteria) but by no eukaryotes, so we depend on these prokaryotes for our very existence.

For descriptions of the role of nitrogen in plant nutrition and in the global nitrogen cycle, see **Key Concepts 24.2, 24.3, and 42.3**.

Ammonia is oxidized to nitrate in soil and in seawater by chemoautotrophic bacteria called **nitrifiers**. Bacteria of two genera, *Nitrosomonas* and *Nitrosococcus*, convert ammonia (NH_3) to nitrite ions (NO_2^-), and *Nitrobacter* oxidizes nitrite to nitrate (NO_3^-), the form of nitrogen most easily used by many plants. What do the nitrifiers get out of these reactions? Their metabolism is powered by the energy released by the oxidation of ammonia or nitrite. For example, by passing the electrons from nitrite through an electron transport system, *Nitrobacter* can make ATP and, using some of this ATP, can also make NADH. With this ATP and NADH, the bacterium can convert CO_2 and H_2O into glucose.

We have already seen the importance of the cyanobacteria in the cycling of oxygen: in ancient times, the oxygen generated by their photosynthesis converted Earth's atmosphere from an anaerobic to an aerobic environment (see Key Concept 17.2). Other prokaryotes—both bacteria and archaea—contribute to the cycling of sulfur. Deep-sea hydrothermal vent ecosystems depend on chemoautotrophic prokaryotes that are incorporated into large communities of crabs, mollusks, and giant worms, all living at a depth of 2,500 m—below any hint of sunlight. These bacteria obtain energy by oxidizing hydrogen sulfide and other substances released in the near-boiling water flowing from volcanic vents in the ocean floor.

REVIEW & APPLY | 18.3

Many prokaryotes are beneficial and even necessary to other forms of life. Most animals, including humans, depend on a complex community of prokaryotes—a microbiome—to maintain health, especially of the immune and digestive systems. Prokaryotes play critical roles in cycling many elements through Earth's ecosystems, including carbon, nitrogen, and oxygen. Some bacteria are pathogenic—the direct causes of diseases. Finding cures for pathogenic diseases entails understanding how pathogens enter and reproduce in the body.

1. How do biofilms form, and why are they of special interest to researchers?
2. Why would elimination of all bacteria from a human gut be problematic from a health standpoint?
3. Why is nitrogen metabolism in prokaryotes vital to other organisms?

Before moving on to discuss the diversity of eukaryotic life, it is appropriate to consider another category of life that includes some pathogens: the viruses. Although they are not cellular, viruses are numerically among the most abundant forms of life on Earth. Their effects on other organisms are enormous. Where did viruses come from, and how do they fit into the tree of life? Biologists are still working to answer these questions.

18.4 Viruses Have Evolved Many Times

LEARNING OBJECTIVES

By the end of this key concept you should be able to:

18.4.1 **Explain why viruses are difficult to place within the tree of life.**

18.4.2 **Contrast the "escaped component" and "reduced parasite" hypotheses for the origin of viruses, and name major groups of viruses that best support each hypothesis.**

18.4.3 **Explain why phage therapy is an active area of research.**

Some biologists do not think of viruses as living organisms, primarily because they are not cellular and must depend on cellular organisms for basic life functions such as replication and metabolism. But viruses are derived from the cells of living organisms. They use the same essential forms of genetic information storage and transmission as do cellular organisms. Viruses infect all cellular forms of life—both prokaryotes and eukaryotes. They replicate, mutate, evolve, and interact with other organisms, often causing serious diseases in their hosts. Finally, viruses clearly evolve independently of other organisms, so it is almost impossible not to treat them as a part of life.

Viruses are abundant in many environments. In some freshwater and marine ecosystems, they can occur at densities of up to 10 million viruses per milliliter of water. Biologists estimate that there are approximately 10^{31} individual virus particles on Earth—about 1,000 times the number of cellular organisms on the planet. Viruses have an enormous effect on the ecology of the oceans. Every day, about one-half of the bacteria in the oceans are killed by viruses. Huge marine blooms of bacteria, such as the *Vibrio* bloom shown in the opening of this chapter, do not last for long because viral blooms soon follow the initial bacterial bloom. As the viruses increase, they begin to kill bacteria faster than the bacteria can reproduce.

Although viruses are found nearly everywhere on Earth and play an important role in many ecosystems, many aspects of their ecology and evolution are still poorly known. For example, several factors make virus phylogeny difficult to resolve. The tiny size of many virus genomes restricts the phylogenetic analyses that can be conducted to relate viruses to cellular organisms. Their rapid mutation rate, which results in rapid evolution of virus genomes, tends to cloud evolutionary relationships over long periods. There are no known fossil viruses (viruses are too small and delicate to fossilize), so the paleontological record offers no clues to virus origins. Finally, viruses are highly diverse (**FIGURE 18.23**). Several lines of evidence support the hypothesis that viruses have evolved repeatedly within each of the major groups of life. The difficulty in resolving deep evolutionary relationships of viruses makes a phylogeny-based classification difficult. Instead, viruses are placed in one of several functionally similar groups on the basis of the structure of their genomes (for example, whether the genomes are composed of RNA or DNA, and are double- or single-strand). Most of these defined groups are not thought to represent monophyletic taxa, however.

Many RNA viruses probably represent escaped genomic components of cellular life

Although viruses are now obligate parasites of cellular species, many viruses may once have been cellular components involved in basic cellular functions—that is, they may be "escaped" components of cellular life that now evolve independently of their hosts.

NEGATIVE-SENSE SINGLE-STRAND RNA VIRUSES An example of a likely "escaped" biochemical function comes from viruses with genomes composed of single-strand **negative-sense RNA**: RNA that is the complement of the mRNA needed for protein translation. Many of these negative-sense single-strand RNA viruses have only a few genes, including one for an RNA-dependent RNA polymerase that allows them to make mRNA from their negative-sense RNA genome. Modern cellular organisms cannot generate mRNA in this manner (at least in the absence of viral infections), but scientists speculate that single-strand RNA genomes may have been common in the distant past, before DNA became the primary molecule for genetic information storage.

A self-replicating RNA polymerase gene that began to replicate independently of a cellular genome could conceivably acquire a few additional protein-coding genes through recombination with its host's DNA. If one or more of these genes were to foster the development of a protein coat, the virus might then survive outside the host and infect new hosts. It is believed that this scenario has been repeated many times independently across the tree of life, given that many of the negative-sense single-strand RNA viruses that infect organisms from bacteria to humans are not closely related to one another. In other words, negative-sense single-strand RNA viruses do not represent a distinct taxonomic group, but rather exemplify a particular process of cellular escape that probably happened many different times.

Familiar examples of negative-sense single-strand RNA viruses include the viruses that cause measles, mumps, rabies, and influenza (see Figure 18.23A).

POSITIVE-SENSE SINGLE-STRAND RNA VIRUSES The genome of another type of single-strand RNA virus is composed of positive-sense RNA. Positive-sense genomes are already set for translation; no replication of the genome to form a complement strand is needed before protein translation can take place. Positive-sense single-strand RNA viruses (see Figure 18.23B) are the most abundant and diverse class of viruses. Most of the viruses that cause diseases in crop plants are members of this group. These viruses kill patches of cells in the leaves or stems of plants, leaving live cells amid a patchwork of discolored dead tissue (giving them the name of mosaic, or mottle, viruses) (**FIGURE 18.24**). Other viruses in this group infect bacteria, fungi, and animals. Human diseases caused by positive-sense single-strand RNA viruses include polio, hepatitis C, COVID-19, and the common cold. As is true of the other functionally defined groups of viruses, these viruses appear

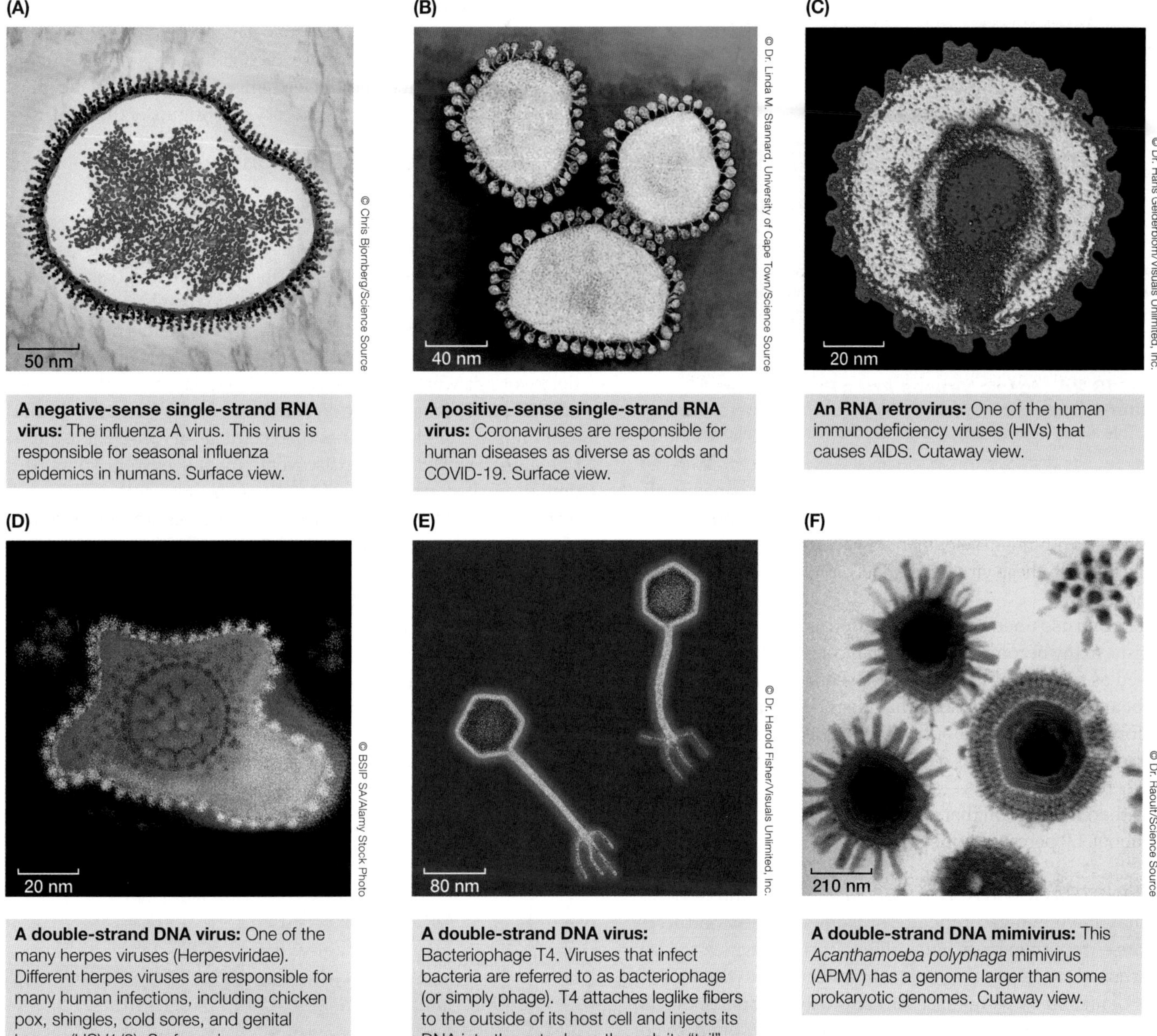

(A) **A negative-sense single-strand RNA virus:** The influenza A virus. This virus is responsible for seasonal influenza epidemics in humans. Surface view.

(B) **A positive-sense single-strand RNA virus:** Coronaviruses are responsible for human diseases as diverse as colds and COVID-19. Surface view.

(C) **An RNA retrovirus:** One of the human immunodeficiency viruses (HIVs) that causes AIDS. Cutaway view.

(D) **A double-strand DNA virus:** One of the many herpes viruses (Herpesviridae). Different herpes viruses are responsible for many human infections, including chicken pox, shingles, cold sores, and genital herpes (HSV1/2). Surface view.

(E) **A double-strand DNA virus:** Bacteriophage T4. Viruses that infect bacteria are referred to as bacteriophage (or simply phage). T4 attaches leglike fibers to the outside of its host cell and injects its DNA into the cytoplasm through its "tail" (pink structure in this rendition).

(F) **A double-strand DNA mimivirus:** This *Acanthamoeba polyphaga* mimivirus (APMV) has a genome larger than some prokaryotic genomes. Cutaway view.

FIGURE 18.23 Viruses Are Diverse Relatively small genomes and rapid evolutionary rates make it difficult to reconstruct phylogenetic relationships among viruses. Instead, viruses are classified largely by general characteristics of their genomes. The images here are computer artists' reconstructions based on cryo-electron micrographs.

to have evolved multiple times across the tree of life from different groups of cellular ancestors.

RNA RETROVIRUSES The RNA retroviruses are best known as the group that includes the human immunodeficiency viruses (HIVs; see Figure 18.23C). Like the previous two categories of viruses, RNA retroviruses have genomes composed of single-strand RNA and probably evolved as escaped cellular components.

Retroviruses are so named because reverse transcription (see Key Concept 11.3) is a necessary component of their reproduction. When the retrovirus enters the nucleus of its vertebrate host, viral reverse transcriptase produces complementary DNA (cDNA) from the viral RNA genome and then replicates the single-strand cDNA to produce double-strand DNA. Another virally encoded enzyme called integrase catalyzes the integration of the new piece of double-strand viral DNA into the host's genome. The viral genome is then replicated along with the host cell's DNA. The integrated retroviral DNA is known as a **provirus**.

Retroviruses are only known to infect vertebrates, although genomic elements that resemble portions of these viruses are

FIGURE 18.24 Mosaic Viruses Are a Problem for Agriculture Mosaic, or mottle, viruses are the most diverse class of viruses. This leaf is from an apple tree infected with a mosaic virus.

a component of the genomes of a wide variety of organisms, including bacteria, plants, and many animals. Several retroviruses are associated with the development of various forms of cancer, as cells infected with these viruses are likely to undergo uncontrolled replication.

As retroviruses become incorporated into the genomes of their hosts, many become nonfunctional copies that are no longer expressed as functional viruses. These sequences may provide a record of ancient viral infections that plagued our ancestors. Humans, for example, carry about 100,000 fragments of **endogenous retroviruses** in our genome. These fragments make up about 8 percent of our DNA—a considerably larger fraction of our genome than the fraction that comprises all of our protein-coding genes (about 1.2 percent of our genome).

DOUBLE-STRAND RNA VIRUSES Double-strand RNA viruses may have evolved repeatedly from single-strand RNA ancestors—or perhaps vice versa. These viruses, which are not closely related to one another, infect organisms from throughout the tree of life. Many plant diseases are caused by double-strand RNA viruses. Other viruses of this type cause many cases of infant diarrhea in humans.

Some DNA viruses may have evolved from reduced cellular organisms

Another class of viruses has a double-strand DNA genome (see Figure 18.23D–F). This group is also almost certainly polyphyletic (with many independent origins). Many of the common phage that infect bacteria are double-strand DNA viruses, as are the viruses that cause smallpox and herpes in humans.

Some biologists think that at least some of the DNA viruses may represent highly reduced parasitic organisms that have lost their cellular structure as well as their ability to survive as free-living species. For example, the mimiviruses, which are some of the largest DNA viruses (see Figure 18.23F), have a genome in excess of a million base pairs of DNA that encode more than 900 proteins. This genome is similar in size to the genomes of many parasitic bacteria and about twice as large as the genome of the smallest bacteria (**FIGURE 18.25**). Phylogenetic analyses of these DNA viruses suggest that they have evolved repeatedly from cellular organisms. Furthermore, recombination among different viruses may have allowed the exchange of various genetic modules, further complicating the history and origins of these viruses.

Viruses can be used to fight bacterial infections

Although some viruses cause devastating diseases, other viruses have been used to fight disease. Most bacterial diseases are treated today with antibiotics. Antibiotics were first discovered in the 1930s, but they were not widely used to treat bacterial diseases until the 1940s. So antibiotics were not yet available during World War I, when bacterial infections plagued the battlefields. Battlefield wounds were often infected by bacteria, and in the absence of antibiotics, these infections often led to the loss of limbs and lives. While trying to find a way to combat this problem, a physician named Felix d'Herelle discovered the first evidence of viruses that attack bacteria. He named these viruses bacteriophage, or "eaters of bacteria." Herelle extracted bacteriophage from the stool of infected patients. He then used these extracts to treat patients with deadly bacterial infections, including dysentery, cholera, and bubonic plague. This practice became known as **phage therapy**. After the war, phage therapy was widely used among the general public to treat bacterial infections of the skin and intestines.

Media Clip 18.3 Bacteriophage Attack *E. coli*

Phage therapy was mostly replaced by the use of antibiotics in the 1930s and 1940s as physicians grew concerned about treating

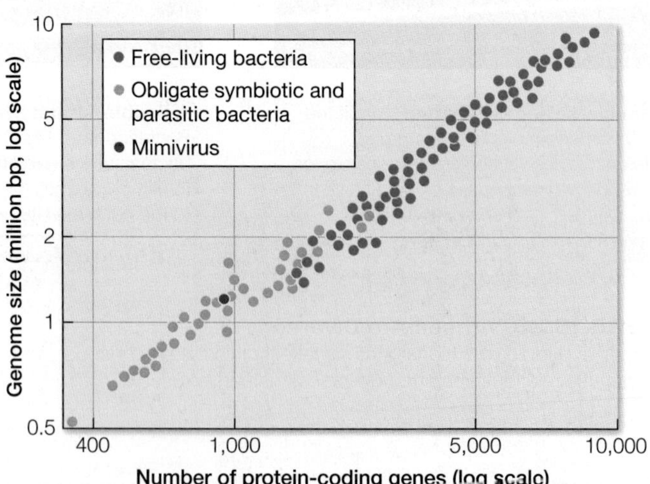

FIGURE 18.25 Mimiviruses Have Genomes Similar in Size to Those of Many Parasitic Bacteria The genome of *Acanthamoeba polyphaga* mimivirus contains 1,181,404 base pairs and encompasses 911 protein-coding genes. This observation is consistent with the hypothesis that this virus evolved from a parasitic bacterium. (After S. J. Giovannoni et al. 2005. *Science* 309: 1242–1245.)

Why do parasitic bacteria generally have smaller genomes and fewer protein-coding genes than do free-living bacteria?

patients with live viruses. Phage therapy continued to be used in the Soviet Union but largely disappeared from western medical practice. Today, however, many antibiotics are losing their effectiveness as bacterial pathogens evolve resistance to these drugs. Phage therapy is once again an active area of research, and it is likely that bacteriophage will become increasingly important as weapons against bacterial diseases. One advantage that bacteriophage may have over antibiotics is that, like bacteria, bacteriophage can evolve. As bacteria evolve resistance to a strain of bacteriophage, biologists can select for new strains of bacteriophage that retain their effectiveness against the pathogens. In this way, biologists are using their understanding of evolution to combat the problem of antibiotic-resistant bacteria.

It appears that the enormous diversity of viruses is, at least in part, a result of their multiple origins from many different cellular organisms. It may be best to view viruses as spin-offs from the various branches on the tree of life—sometimes evolving independently of cellular genomes, sometimes recombining with them. One way to think of viruses is as the "bark" on the tree of life: certainly an important component all across the tree, but not quite like the main branches.

REVIEW & APPLY | 18.4

R Viruses are highly diverse cellular forms of life that appear to have evolved independently from many different cellular organisms. Major groups of viruses are characterized by the nature of their genomes. Some viruses appear to have evolved from escaped components of cellular organisms, whereas other viruses may have evolved from parasitic cellular ancestors. Although some viruses are responsible for serious diseases, other viruses can be used to combat bacterial infections.

A

1. How does the size of viral genomes limit phylogenetic analyses of viruses?
2. What are the two main hypotheses of viral origins?
3. How can viruses be used to treat some human diseases?

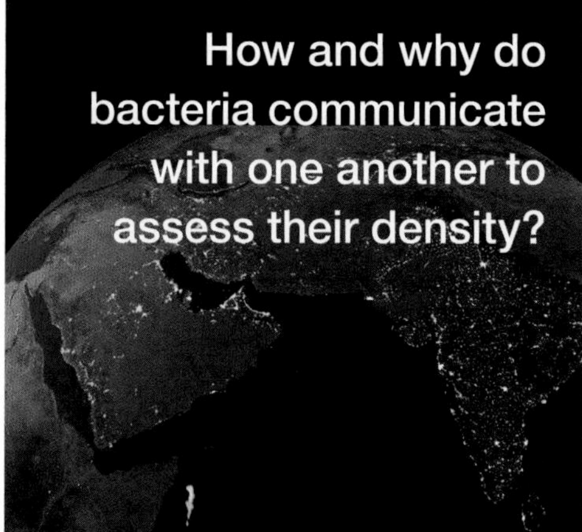

Answer: As we explored in Investigation Figure 18.19, *Vibrio* bacteria send a chemical signal that is received by other bacteria. This signal is needed to produce bioluminescence. As *Vibrio* populations increase in density, the bioluminescence becomes strong enough to be visible to humans and other animals. But how is bioluminescence an advantage to the bacteria? What is its function?

Although marine *Vibrio* are able to live independently, they truly thrive inside the guts of fishes and other marine animals. Inside a fish, *Vibrio* cells attach themselves to food particles, including phytoplankton, and are often expelled into the ocean as waste. How can they get back into their preferred environment? The bioluminescent glow produced by a dense colony of free-living *Vibrio* growing on phytoplankton attracts fishes, which consume the phytoplankton and thus ingest the bacteria—which gets the bacteria into a new host fish.

"Milky seas" data prepared by Steven Miller of the Naval Research Laboratory, Monterey, CA. Background data (Earth at Night image) courtesy Marc Imhoff (NASA/GSFC) and Christopher Elvidge (NOAA/NGDC); image by Craig Mayhew (NASA/GSFC) and Robert Simmon (NASA/GSFC)

18 VISUAL SUMMARY

18.1 Bacteria and Archaea Are the Two Primary Divisions of Life

Go to ANIMATION 18.1 and ACTIVITY 18.1

- The earliest split in the tree of life is between the Bacteria and Archaea. The Eukarya branched off from the Archaea, having evolved specializations (such as a nucleus and organelles) that differentiate the eukaryotes from the **PROKARYOTES**.
- Bacteria and archaea can be differentiated by their cell walls. The cell walls of bacteria contain **PEPTIDOGLYCAN**, which is a target for the action of many antibiotics.
- Although **LATERAL GENE TRANSFER** sometimes complicates our understanding of prokaryotic evolutionary history, prokaryotes are now principally classified by the nucleotide sequences of their genomes (**FIGURE 18.4**).

FIGURE 18.4

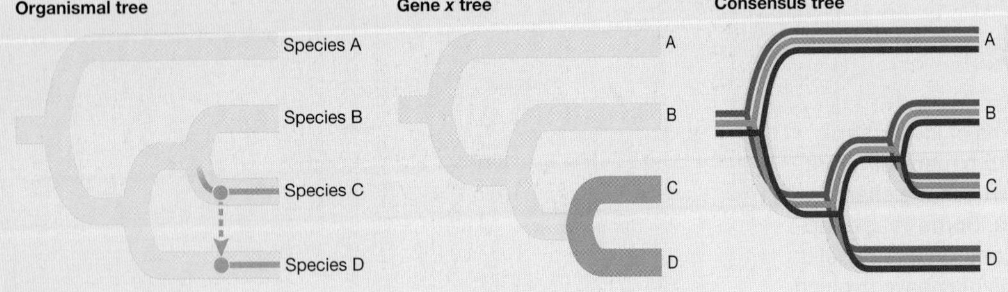

QUESTIONS

1. What evidence links the origin of eukaryotes with Archaea rather than Bacteria?
2. How is peptidoglycan relevant to the function of antibiotics?
3. How does lateral gene transfer complicate study of bacterial phylogeny?

18.2 Prokaryotic Diversity Reflects the Ancient Origins of Life

FIGURE 18.16

© Danita Delimont/Alamy Stock Photo

- Prokaryotes are the most numerous organisms on Earth, but only a small fraction of prokaryotic diversity has been characterized to date.
- Life on Earth likely originated in a high-temperature environment (**FIGURE 18.16**).
- Biotic communities depend on the action of bacteria, which provide such important ecological services such as nitrogen fixation and oxygen production through photosynthesis.
- Prokaryotic archaea live in highly diverse environments and include many **EXTREMOPHILES** and **METHANOGENS**.

QUESTIONS

1. What lines of evidence support the origin of life in a hot environment?
2. How have cyanobacteria been critical in changing Earth's environment over the past 3 billion years?
3. What genes present in the Asgard archaea were formerly thought to be found only in eukaryotic genomes?

18.3 Ecological Communities Depend on Prokaryotes

- **MICROBIOMES** are communities of prokaryotes that live in and on the bodies of multicellular organisms and are often important to the health of the hosts. Bacteria use processes such as **QUORUM SENSING** to recruit members to their communities (**FIGURE 18.18**).
- **KOCH'S POSTULATES** establish the criteria by which an organism may be classified as a pathogen. Relatively few bacteria—and no prokaryotic archaea—are known to be pathogens.
- Prokaryotic metabolism is diverse. Prokaryotes fall into four broad nutritional categories: **PHOTOAUTOTROPHS**, **PHOTOHETEROTROPHS**, **CHEMOAUTOTROPHS**, and **CHEMOHETEROTROPHS**.
- Prokaryotes play key roles in the cycling of elements such as nitrogen, oxygen, sulfur, and carbon.

FIGURE 18.18

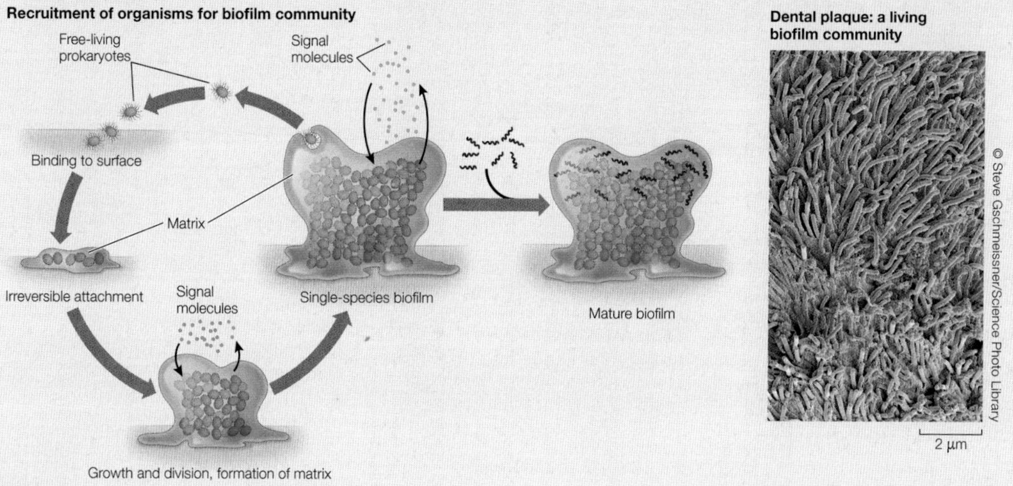

Recruitment of organisms for biofilm community

Free-living prokaryotes

Signal molecules

Binding to surface

Matrix

Irreversible attachment

Signal molecules

Single-species biofilm

Mature biofilm

Growth and division, formation of matrix

Dental plaque: a living biofilm community

2 μm

QUESTIONS

1. How are biofilms important to human health?
2. When and where do newborn humans typically acquire their microbiomes?
3. Which group—bacteria, prokaryotic archaea, or eukaryotes—contains no organisms that use light as an energy source?

18.4 Viruses Have Evolved Many Times

FIGURE 18.25

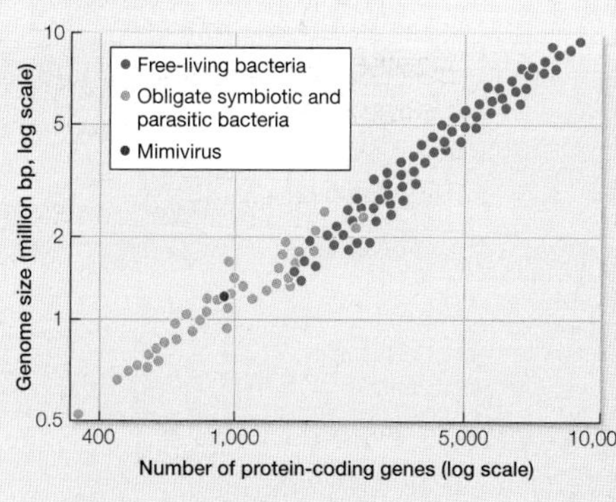

- Viruses have evolved many times from many different groups of cellular organisms. They are placed in groups according to the structure of their genomes, but these groups are not thought to represent monophyletic taxa (**FIGURE 18.25**).
- Viruses manipulate host genetic mechanisms in various ways to replicate themselves. **RETROVIRUSES**, for example, use reverse transcriptase to produce complementary DNA from their viral RNA genome and then insert their genetic material into the host's genome.
- Some viruses cause serious diseases in cellular organisms, but bacteriophage can be used to treat bacterial infections.

1. What molecules encode viral genomes?
2. How did endogenous retroviruses end up in vertebrate genomes?
3. Why did phage therapy fall out of use, and why is it once again of interest?

Go to **Achieve** for the e-book, animations, activities, and additional resources and assignments.

19 The Origin and Diversification of Eukaryotes

Blooms of several species of dinoflagellates are responsible for toxic red tides in many parts of the world. These protists produce powerful toxins that kill fishes and other marine life and produce aerosols that often cause asthma-like symptoms in humans onshore. The effects of a red tide can be devastating to local marine environments.

Red tides are harmful, but how are dinoflagellates beneficial to marine ecosystems?

You will find the answer to this question on page 460.

KEY CONCEPTS

19.1 Eukaryotes Acquired Features from Both Archaea and Bacteria

LEARNING OBJECTIVES

By the end of this key concept you should be able to:

19.1.1 **Explain the use of the term "protist" and describe the relationships of major protist groups to other eukaryotes.**

19.1.2 **List the five key events that led to the evolution of eukaryotes from a prokaryotic ancestor, and know how these events allowed for the establishment of unique cellular functions.**

19.1.3 **Explain how major changes in Earth's environment affected the evolution of the eukaryotic cell.**

19.1.4 **Classify the history of genes in a eukaryotic cell based on whether their origins are nuclear, mitochondrial, or chloroplastic.**

We easily recognize trees, mushrooms, and insects as plants, fungi, and animals, respectively. But there is a dazzling assortment of other eukaryotic organisms—many microscopic, but some, like kelp, that grow very large—that do not fit into these three groups. Eukaryotes that are not plants, animals, or fungi have traditionally been called **protists**. But phylogenetic analyses reveal that many of the groups we commonly refer to as protists are not, in fact, closely related. Thus the term "protist" does not describe a monophyletic group, but is a term of convenience for "all the eukaryotes that are not plants, animals, or fungi."

The unique characteristics of the eukaryotic cell lead scientists to conclude that the eukaryotes are monophyletic, and that a single eukaryotic ancestor diversified into the many different protist lineages as well as giving rise to the plants, fungi, and animals. As we saw in Key Concept 18.1, eukaryotes are a specialized group of archaea that acquired a cell nucleus. The mitochondria and chloroplasts of eukaryotes, however, are clearly derived from bacterial lineages (see Figure 18.1).

Biologists have hypothesized that the origin of eukaryotes from a specialized archaeal ancestor was followed by the endosymbioses with bacterial lineages that led to the origin of mitochondria and chloroplasts. Some biologists prefer to view the origin of eukaryotes as the fusion of lineages from the two prokaryotic groups. This difference is largely a semantic one that hinges on the subjective point at which we deem the eukaryotic lineage to have become definitively "eukaryotic." In either case, we can make some reasonable inferences about the events that led to the evolution of a new cell type, bearing in mind that the environment underwent an enormous change—from low to high availability of free atmospheric oxygen (O_2)—during the course of these events.

The modern eukaryotic cell arose in several steps

Several events were important in the origin of the modern eukaryotic cell (**FIGURE 19.1**):

- The origin of a flexible cell surface
- The origin of a cytoskeleton
- The origin of a nuclear envelope, which enclosed a genome organized into chromosomes
- The appearance of digestive vacuoles
- The acquisition of mitochondria and chloroplasts via endosymbiosis

FLEXIBLE CELL SURFACE We presume that ancient prokaryotic organisms, like most present-day prokaryotic cells, had firm cell walls. The first step toward the eukaryotic condition was the loss of the cell wall by a prokaryotic archaean. This wall-less condition occurs in some present-day prokaryotes.

Consider the possibilities open to a flexible cell without a firm wall, starting with cell size. As a cell grows larger, its surface area-to-volume ratio decreases (see Figure 4.11). Unless the surface area can be increased, the cell volume will reach an upper limit. If the cell's surface is flexible, however, it can fold inward and become more elaborate, creating more surface area for gas and nutrient exchange. With a surface flexible enough to allow infolding, the cell can exchange materials with its environment rapidly enough to sustain a larger volume and more rapid metabolism (see Figure 19.1, steps 1–2). Furthermore, a flexible surface can pinch off bits of the environment, bringing them into the cell by endocytosis. These infoldings of the cell surface, which also exist in some modern prokaryotes, were important for the evolution of large eukaryotic cells.

CHANGES IN CELL STRUCTURE AND FUNCTION Other early steps that were important for the evolution of the eukaryotic cell involved increased compartmentalization and complexity of the cell (see Figure 19.1, steps 3–7):

- The development of a more complex cytoskeleton
- The formation of ribosome-studded internal membranes, some of which surrounded the DNA
- The enclosure of the cell's DNA in a nucleus
- The formation of a flagellum from microtubules of the cytoskeleton
- The evolution of digestive vacuoles

Until a few years ago, biologists thought that cytoskeletons were restricted to eukaryotes. Improved imaging technology and molecular analyses have now revealed homologs of many cytoskeletal proteins in prokaryotes, so simple cytoskeletons evolved before the origin of eukaryotes. The cytoskeleton of a eukaryote, however, is much more developed and complex than that of a prokaryote. This greater development of microfilaments and microtubules supports the eukaryotic cell and allows it to manage changes in shape, to distribute daughter chromosomes, and to move materials from one part of its larger cell to other parts. In addition, the presence of microtubules in the cytoskeleton allowed some cells to develop the characteristic eukaryotic flagellum.

The DNA of a prokaryotic cell is attached to a site on its cell membrane. If that region of the cell membrane were to fold into the cell, the first step would be taken toward the evolution of a nucleus, a primary feature of the eukaryotic cell. The nuclear envelope appeared early in the eukaryotic lineage. The next step was probably phagocytosis—the ability to engulf and digest other cells.

ENDOSYMBIOSIS At the same time the processes outlined above were taking place, cyanobacteria were generating oxygen as a product of photosynthesis. The increasing concentrations of oxygen in the oceans, and eventually in the atmosphere, had disastrous consequences for most organisms of the time, which were unable to tolerate the newly oxidizing environment. But some prokaryotes evolved strategies to use the increasing oxygen, and—fortunately for us—so did some of the new phagocytic eukaryotes.

At about this time, endosymbiosis began to play a role in eukaryotic evolution (see Figure 19.1, steps 8–9). The theory of endosymbiosis proposes that certain organelles are the descendants of prokaryotes engulfed, but not digested, by ancient eukaryotic cells. One crucial event in the history of eukaryotes was the incorporation of a proteobacterium that evolved into the mitochondrion. Initially the new organelle's

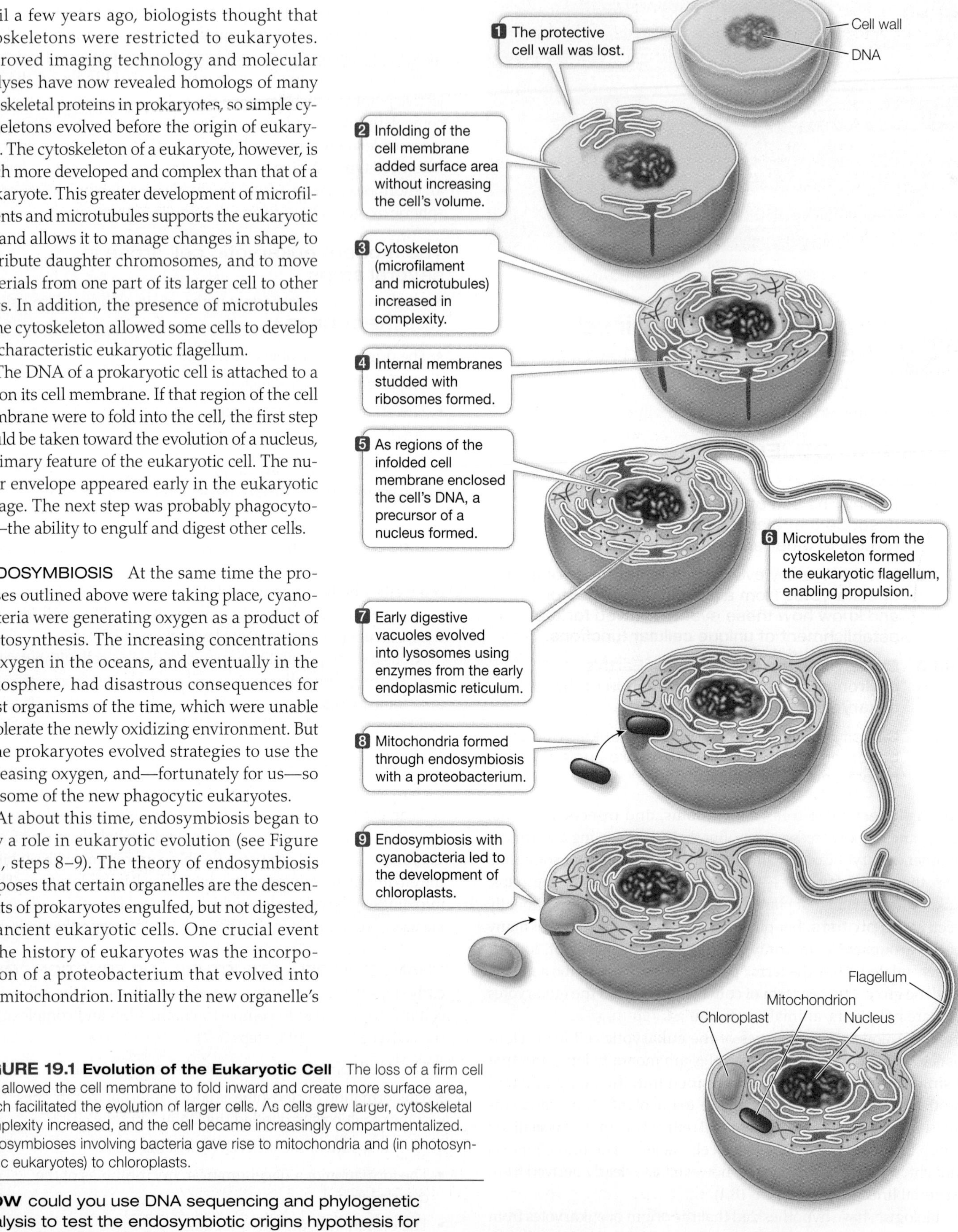

FIGURE 19.1 Evolution of the Eukaryotic Cell The loss of a firm cell wall allowed the cell membrane to fold inward and create more surface area, which facilitated the evolution of larger cells. As cells grew larger, cytoskeletal complexity increased, and the cell became increasingly compartmentalized. Endosymbioses involving bacteria gave rise to mitochondria and (in photosynthetic eukaryotes) to chloroplasts.

How could you use DNA sequencing and phylogenetic analysis to test the endosymbiotic origins hypothesis for mitochondria and chloroplasts?

primary function was probably to detoxify O_2 by reducing it to water. Later this reduction became coupled with the formation of ATP in cellular respiration. After this step, the essential eukaryotic cell was complete.

LINK You may wish to review the reactions of cellular respiration in **Key Concept 5.2**.

Photosynthetic eukaryotes are the result of yet another endosymbiotic step: the incorporation of a prokaryote related to today's cyanobacteria, which became the chloroplast.

Chloroplasts have been transferred among eukaryotes several times

Eukaryotes in several different groups possess chloroplasts, and groups with chloroplasts appear in several distantly related eukaryotic clades. Some of these groups differ in the photosynthetic pigments their chloroplasts contain. And not all chloroplasts are limited to a pair of surrounding membranes—in some microbial eukaryotes, chloroplasts are surrounded by three or more membranes. We now view these observations as evidence of a remarkable series of endosymbioses. This conclusion is supported by extensive evidence from electron microscopy and nucleic acid sequence comparisons.

All chloroplasts trace their ancestry back to the engulfment of one cyanobacterium by a larger eukaryotic cell. This event, the step that first gave rise to the photosynthetic eukaryotes, is known as **primary endosymbiosis** (**FIGURE 19.2A**). The cyanobacterium, a Gram-negative bacterium, had both an inner and an outer membrane (see Figure 18.2B). Thus the original chloroplasts had two surrounding membranes: the inner and outer membranes of the cyanobacterium. Remnants of the peptidoglycan-containing cell wall of the bacterium are present in the form of a bit of peptidoglycan between the chloroplast membranes of glaucophytes, the first eukaryotic group to branch off following primary endosymbiosis (as we will see in Chapter 20). Primary endosymbiosis also gave rise to the chloroplasts of the red algae, green algae, and land plants. The red algal chloroplast retains certain pigments of the original cyanobacterial endosymbiont that are absent in green algal chloroplasts.

Almost all remaining photosynthetic eukaryotes are the result of additional rounds of endosymbiosis. For example, the photosynthetic euglenids derived their chloroplasts from **secondary endosymbiosis** (**FIGURE 19.2B**). Their ancestor took up a unicellular green alga, retaining its chloroplast and eventually losing the rest of the constituents of the alga. This history explains why the photosynthetic euglenids have the same photosynthetic pigments as the green algae and land plants. It also accounts for the third membrane of the euglenoid chloroplast, which is derived from the euglenid's cell membrane (as a result of endocytosis). An additional round of endosymbiosis—**tertiary endosymbiosis**—occurred when a dinoflagellate apparently lost its chloroplast and took up another protist that had acquired its chloroplast through secondary endosymbiosis.

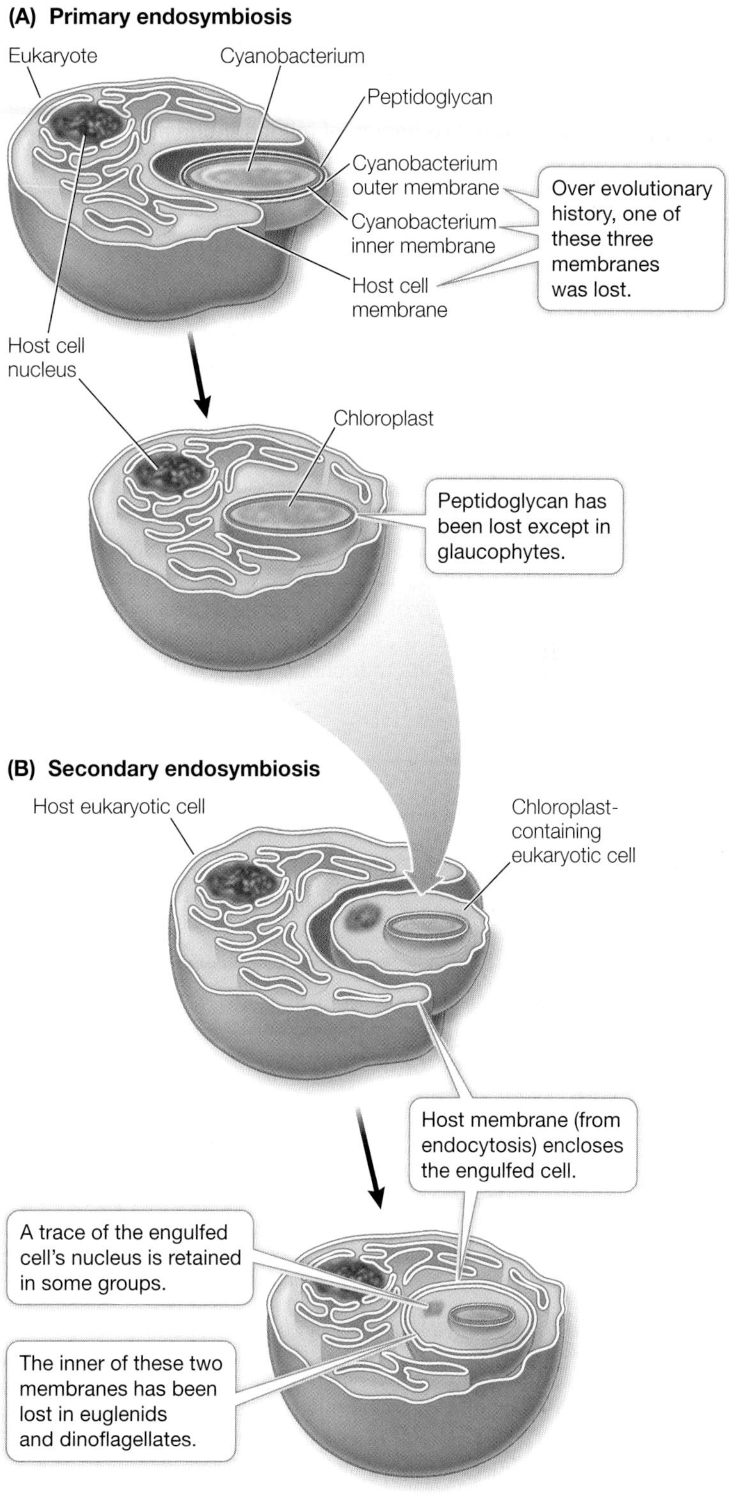

FIGURE 19.2 Endosymbiotic Events in the Evolution of Chloroplasts **(A)** A single instance of primary endosymbiosis ultimately gave rise to all of today's chloroplasts. **(B)** Secondary endosymbiosis—the uptake and retention of a chloroplast-containing cell by another eukaryotic cell—took place several times, independently.

Animation 19.1 **Family Tree of Chloroplasts**

REVIEW & APPLY | 19.1

R The modern eukaryotic cell probably arose from an ancestral prokaryotic archaean in several steps, including the origin of a flexible cell surface and the enclosure of the genetic material in a nucleus. Later, endosymbioses of proteobacteria and cyanobacteria led to the origins of mitochondria and chloroplasts, respectively.

A
1. Why was the development of a flexible cell surface a key event for eukaryotic evolution?
2. Explain how increased availability of atmospheric oxygen could have influenced the evolution of the eukaryotic cell.

Background for Questions 3–4: Ribosomal RNA (rRNA) genes are present in the nuclear genome of eukaryotes. There are also rRNA genes in the genomes of mitochondria and chloroplasts. Therefore photosynthetic eukaryotes have three different sets of rRNA genes, which encode the structural RNA of three separate sets of ribosomes. The gene tree below shows the evolutionary relationships among rRNA gene sequences isolated from the nuclear genomes of humans, yeast, and corn; from a prokaryotic archaean (*Halobacterium*), a proteobacterium (*E. coli*), and a cyanobacterium (*Chlorobium*); and from the mitochondrial and chloroplast genomes of corn.

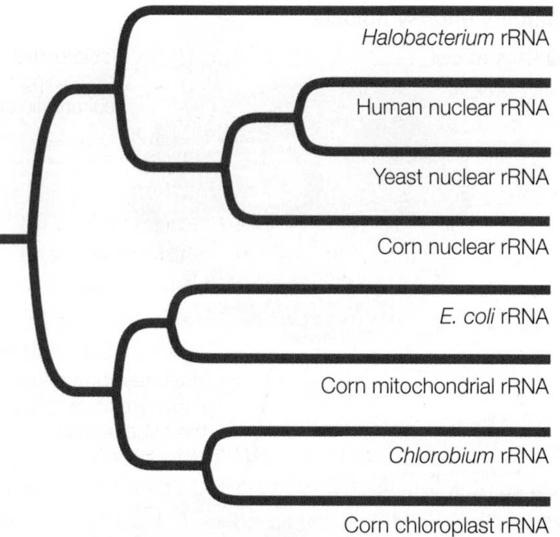

3. Why are the three rRNA genes of corn *not* one another's closest relatives?
4. Why is the relationship of the mitochondrial rRNA gene of corn to the rRNA gene of *E. coli* closer than it is to the nuclear rRNA genes of other eukaryotes? Can you explain the relationship of the rRNA gene from the chloroplast of corn to the rRNA gene of the cyanobacterium?
5. Explain why the term "protists" does not refer to a formal taxonomic group.

The features that eukaryotes gained from prokaryotic archaea and bacteria have allowed them to exploit many different environments. This led to the evolution of great diversity among eukaryotes, beginning with a radiation that started in the Precambrian.

19.2 | Major Lineages of Eukaryotes Diversified in the Precambrian

LEARNING OBJECTIVES

By the end of this key concept you should be able to:

19.2.1 **Distinguish among the major groups of microbial eukaryotes (alveolates, stramenopiles, rhizarians, excavates, and amoebozoans)**

19.2.2 **Justify the position that it is relatively easy for multicellularity to evolve.**

19.2.3 **Elaborate on why protists are important to our medical and economic concerns.**

Most eukaryotes can be placed in one of eight major clades that began to diversify about 1.5 billion years ago: alveolates, stramenopiles, rhizarians, excavates, plants, amoebozoans, fungi, and animals (**FIGURE 19.3**). Plants, fungi, and animals each have close protist relatives, which we will discuss along with those major multicellular eukaryotic groups in Chapters 20–22.

Each of the five major groups of protist eukaryotes covered in this chapter consists of organisms with enormously diverse body forms and nutritional lifestyles. Some protists are motile, whereas others do not move. Some protists are photosynthetic, whereas others are heterotrophic. Most protists are unicellular, but some are multicellular. Most protists are microscopic, but a few are huge (giant kelps, for example, can grow to half the length of a football field). We refer to the unicellular species of protists as **microbial eukaryotes**, but keep in mind that there are large, multicellular protists as well.

Multicellularity has arisen dozens of times across the evolutionary history of eukaryotes. Four of the origins of multicellularity resulted in large organisms that are familiar to most people: plants, animals, fungi, and brown algae (the last are a group of stramenopiles). In addition, there are dozens of smaller and less familiar groups among the eukaryotes that include multicellular species. Recent experimental studies have shown that artificial selection for multicellularity can produce repeated, convergent evolution of multicellular forms over just a few months in some normally unicellular eukaryotic species. In addition, many unicellular species retain individual identities but nonetheless associate in large multicellular colonies. There is a near continuum between fully integrated, multicellular organisms on the one hand and loosely integrated multicellular

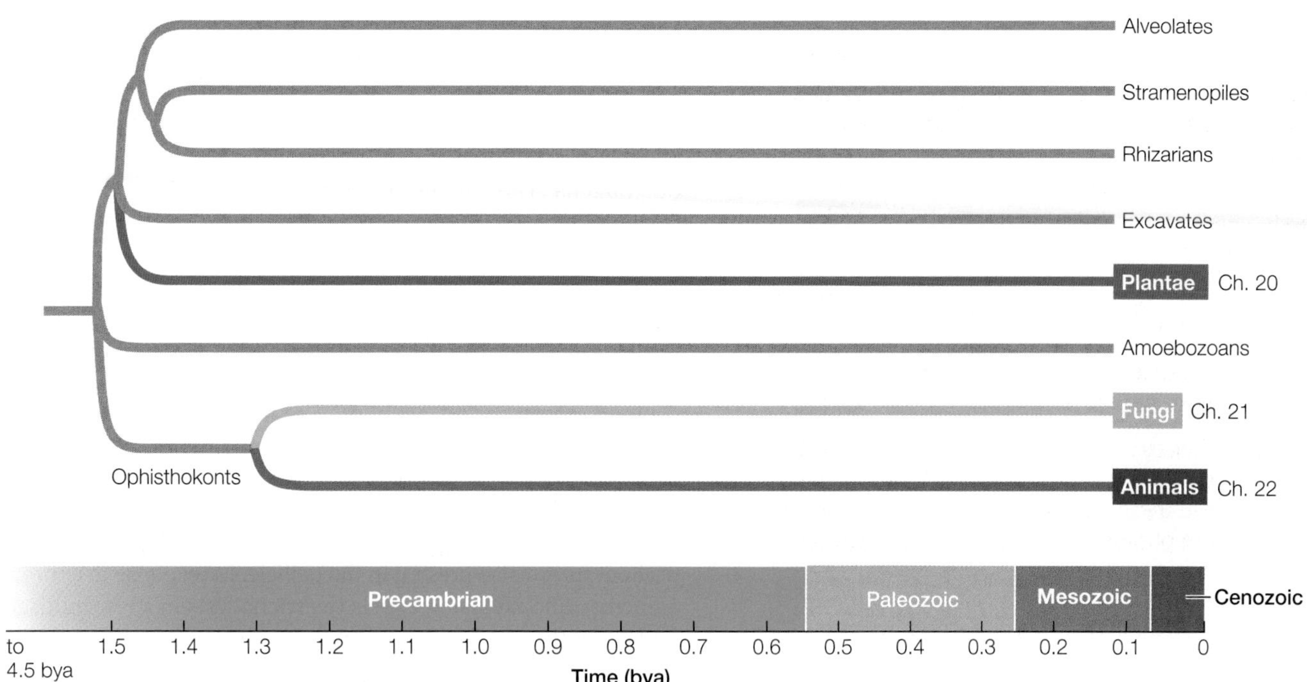

FIGURE 19.3 Precambrian Divergence of Major Eukaryotic Groups A phylogenetic tree shows a current hypothesis and estimated time line for the origin of the major groups of eukaryotes. The rapid divergence of major lineages between 1.5 and 1.4 billion years ago (bya) makes reconstruction of their precise relationships difficult. The major multicellular groups (tinted boxes) will be covered in subsequent chapters.

colonies of cells on the other. Biologists do not always agree on where to draw the line between the two.

Biologists used to classify protists largely on the basis of their life histories and reproductive features. In recent years, however, electron microscopy and gene sequencing have revealed many new patterns of evolutionary relatedness among these groups. Analyses of slowly evolving gene sequences are making it possible to explore evolutionary relationships among eukaryotes in ever-greater detail and with greater confidence. Nonetheless, some substantial areas of uncertainty remain, and lateral gene transfer may complicate efforts to reconstruct the evolutionary history of protists (as was also true for prokaryotes; see Key Concept 18.1). Today we recognize great diversity among the many distantly related protist clades.

Alveolates have sacs under their cell membranes

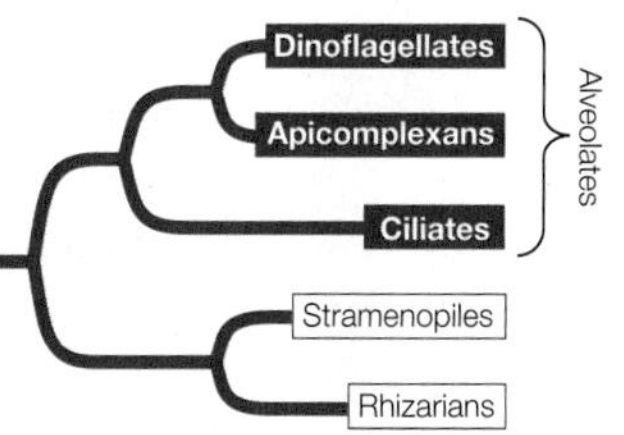

Alveolates are so named because they possess sacs, called alveoli, just beneath their cell membranes, which may play a role in supporting the cell surface. All alveolates are unicellular, and most are photosynthetic, but they are diverse in body form. The alveolate groups we will consider in detail here are the dinoflagellates, apicomplexans, and ciliates.

DINOFLAGELLATES Most **dinoflagellates** are marine and photosynthetic; they are important primary producers of organic matter in the oceans. Although fewer species of dinoflagellates live in fresh water, individuals can be abundant in freshwater environments. The dinoflagellates are of great ecological, evolutionary, and morphological interest. A distinctive mixture of photosynthetic and accessory pigments gives their chloroplasts a golden brown color. Some dinoflagellate species cause red tides, as shown in the opening photo of this chapter. Other species are photosynthetic endosymbionts that live within the cells of other organisms, including invertebrate animals (such as corals; see Investigation Figure 19.22) and other marine protists (see Figure 19.13A). Still others are nonphotosynthetic and live as parasites within other marine organisms.

Dinoflagellates have a distinctive appearance. They generally have two flagella, one in an equatorial groove around the cell, the other starting near the same point as the first and passing down a longitudinal groove before extending into the surrounding medium (**FIGURE 19.4**). Some dinoflagellates can take on different forms, including amoeboid ones, depending on environmental conditions. It has been claimed that the dinoflagellate *Pfiesteria piscicida* can occur in at least two dozen distinct forms, although this claim is highly controversial. In any case, this remarkable dinoflagellate, when present in large enough numbers, is harmful to fishes and can both stun and feed on them.

APICOMPLEXANS The exclusively parasitic **apicomplexans** derive their name from the apical complex, a mass of organelles contained in the apical end (the tip) of the cell. These organelles help the apicomplexan invade its host's tissues. For example, the apical complex enables *Plasmodium*, the causative agent of

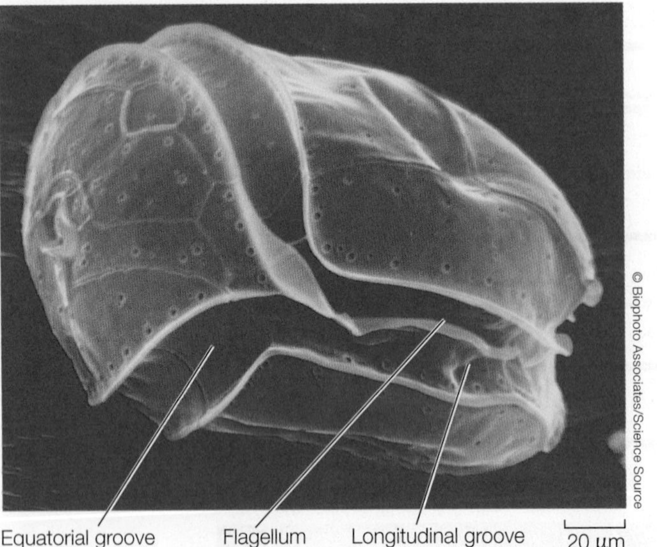

FIGURE 19.4 A Dinoflagellate The presence of two flagella is characteristic of many dinoflagellates, although these appendages are contained within deep grooves and thus are seldom visible. One flagellum lies within the equatorial groove and provides forward thrust and spin to the organism. The second flagellum originates in the longitudinal groove and acts like the rudder of a boat.

Media Clip 19.1 A Dinoflagellate Shows off Its Flagellum

malaria, to enter its target cells in the human body after transmission by a mosquito.

Like many obligate parasites, apicomplexans have elaborate life cycles featuring asexual and sexual reproduction through a series of very dissimilar life stages (see Figure 19.21). In many species, these life stages are associated with two different types of host organisms, as is the case with *Plasmodium*. Another apicomplexan, *Toxoplasma*, alternates between cats and rats to complete its life cycle. A rat infected with *Toxoplasma* loses its fear of cats, which makes it more likely to be eaten by, and thus transfer the parasite to, a cat.

CILIATES The **ciliates** are named for their numerous hairlike cilia, which are shorter than, but otherwise identical to, eukaryotic flagella. The ciliates are much more complex in body form than are most other unicellular eukaryotes (**FIGURE 19.5**). Their definitive characteristic is the possession of two types of nuclei (whose roles we will describe in Key Concept 19.3 when we discuss protist reproduction). Almost all ciliates are heterotrophic, although a few contain photosynthetic endosymbionts.

Paramecium, a frequently studied ciliate genus, exemplifies the complex structure and behavior of ciliates (**FIGURE 19.6**). The slipper-shaped cell is covered by an elaborate pellicle, a structure composed principally of an outer membrane and an inner layer of closely packed, membrane-enclosed sacs (the alveoli) that surround the bases of the cilia. Defensive organelles called trichocysts are also present in the pellicle. In response to a threat, a microscopic explosion expels the trichocysts in a few milliseconds, and they emerge as sharp darts, driven forward at the tip of a long, expanding filament.

The cilia provide *Paramecium* with a form of locomotion that is generally more precise than locomotion by flagella. A *Paramecium* can coordinate the beating of its cilia to propel itself either forward or backward in a spiraling manner. It can also back off swiftly when it encounters a barrier or a negative stimulus. The coordination of ciliary beating is probably the result of a differential distribution of ion channels in the cell membrane near the two ends of the cell.

Organisms living in fresh water are hypertonic to their environment. Many freshwater protists, including *Paramecium*, address this problem by means of specialized **contractile vacuoles** that excrete the excess water the organisms constantly take in by osmosis. The excess water collects in the contractile vacuoles, which then contract and expel the water from the cell.

Paramecium and many other protists engulf solid food by endocytosis, forming a **digestive vacuole** within which the food is

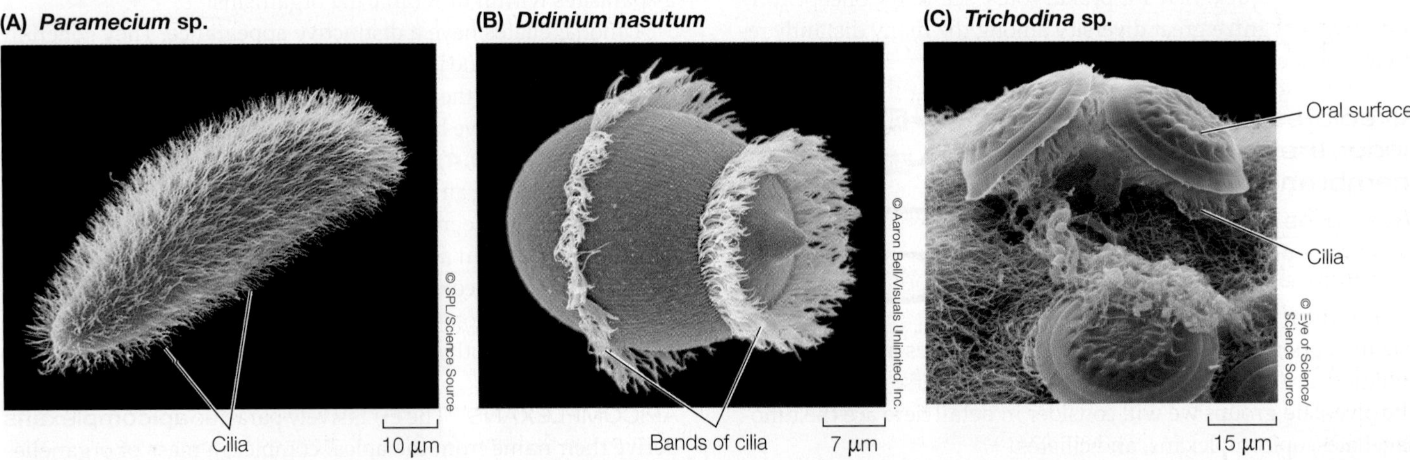

FIGURE 19.5 Diversity among the Ciliates **(A)** A free-swimming organism, this *Paramecium* belongs to a ciliate group whose members have many cilia of uniform length. **(B)** The barrel-shaped *Didinium nasutum* feeds on other ciliates, including *Paramecium*. Its cilia occur in two separate bands. **(C)** The parasite *Trichodina* creates suction with its ring of cilia that help attach it to the surface of a fish host.

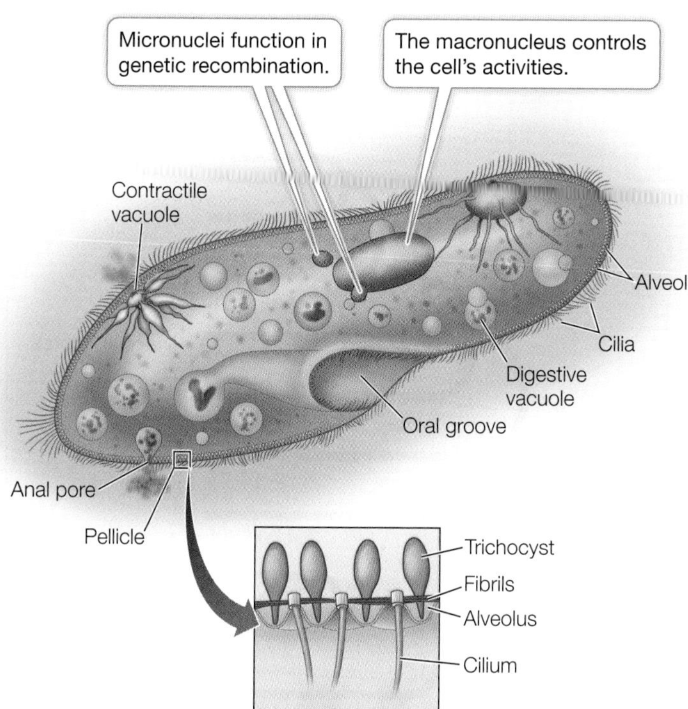

FIGURE 19.6 Anatomy of a *Paramecium* A *Paramecium*, with its many specialized organelles, exemplifies the complex body form of ciliates.

Activity 19.1 Anatomy of *Paramecium*

digested (**INVESTIGATION FIGURE 19.7**). Smaller vesicles containing digested food pinch away from the digestive vacuole and enter the cytoplasm. These tiny vesicles provide a large surface area across which the products of digestion can be absorbed by the rest of the cell.

Animation 19.2 Digestive Vacuoles

Stramenopiles typically have two unequal flagella, one with hairs

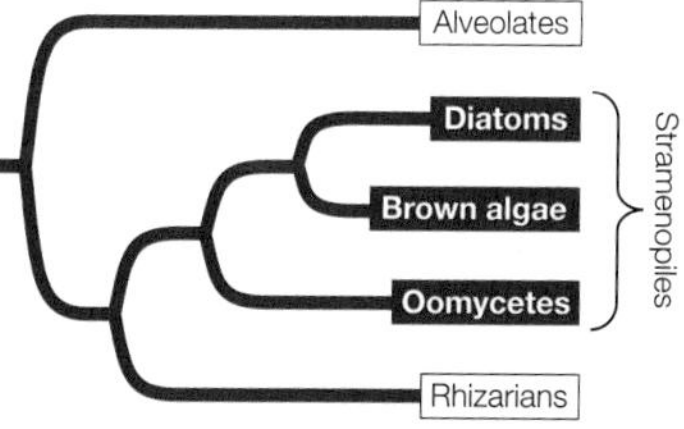

A morphological synapomorphy of most **stramenopiles** is the possession of rows of tubular hairs on the longer of their two flagella. Some stramenopiles lack flagella, but they are descended from ancestors that possessed flagella. The stramenopiles include the diatoms and the brown algae, which are photosynthetic, and the oomycetes, which are not.

DIATOMS All of the **diatoms** are unicellular, although some species associate in filaments. Many have sufficient carotenoids in their chloroplasts to give them a yellow or brownish color. All of them synthesize carbohydrates and oils as photosynthetic storage products. Diatoms lack flagella except in male gametes.

INVESTIGATION

FIGURE 19.7 The Role of Vacuoles in Ciliate Digestion An acidic environment is known to aid digestion in many multicellular organisms. Do ciliates also use acid to obtain nutrients?[a]

HYPOTHESIS

The digestive vacuoles of *Paramecium* produce an acidic environment that allows the organism to digest food particles.

METHOD

1. Feed *Paramecium* yeast cells stained with Congo red, a dye that is red at neutral or basic pH but turns green at acidic pH.
2. Under a light microscope, observe the formation and degradation of digestive vacuoles within the *Paramecium*. Note time and sequence of color (i.e., acid level) changes.

RESULTS

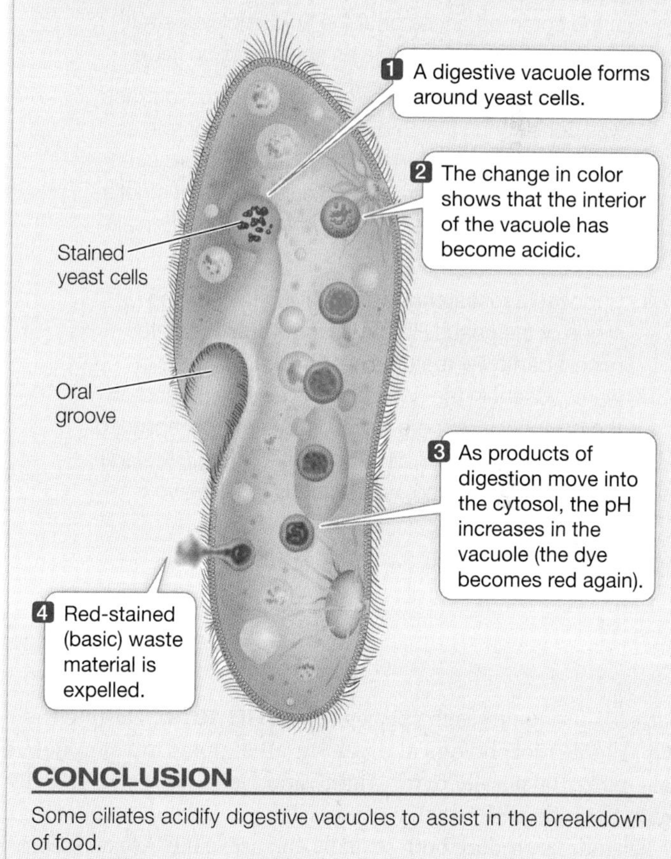

CONCLUSION

Some ciliates acidify digestive vacuoles to assist in the breakdown of food.

[a]S. O. Mast. 1947. *Biol Bull* 92: 31–72.

Architectural magnificence on a microscopic scale is the hallmark of the diatoms. Almost all diatoms deposit silica (hydrated silicon dioxide) in their cell walls. The cell wall of a diatom is constructed in two pieces, with the top overlapping the bottom like the top of a petri dish. The silica-impregnated walls have intricate

THINK LIKE A SCIENTIST

Using phylogenies to make predictions

Because closely related organisms are more likely to share characteristics than are more distantly related ones, biologists can use phylogenetic information to make predictions about organisms based on what is known about their relatives. Here you will develop interdisciplinary skills by combining phylogenetic reasoning (see Chapter 14) and information on the biology of alveolates (this chapter) to make predictions and apply information about alveolate evolution. **FIGURE 19.8** shows a phylogenetic tree for some of the alveolates, and below it is a table listing some of the characteristics for selected species.

1. Based on this information and what you learned in the text, what life history attribute was likely present in the common ancestor of the apicomplexans but not in other alveolates? Be as specific as possible.
2. Suppose a researcher examining the apicomplexans was considering which dinoflagellates to sample as an outgroup. Based on the phylogeny, is there a reason to favor choosing one particular genus of dinoflagellates over the others? If so, which genus? Explain your answer.
3. Suppose a researcher was studying the feeding mode of the genus *Pfiesteria* and wanted to determine if particular features evolved in this genus or were present in the ancestor of *Pfiesteria*. Based on the phylogeny, is there a reason to target sampling to a particular genus of dinoflagellate (outside *Pfiesteria*) over the others? If so, which genus? Explain your answer.

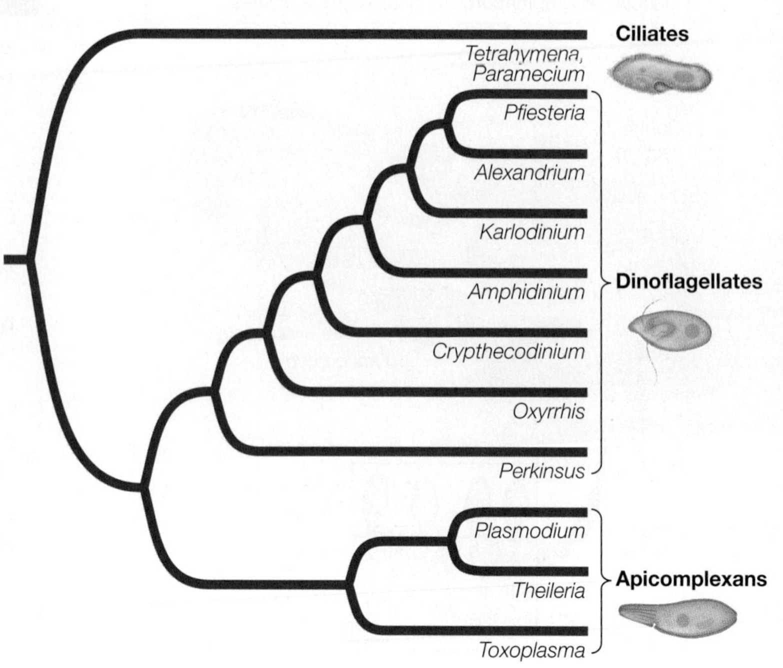

FIGURE 19.8 Using Phylogenies to Make Predictions

Species	Features
Theileria parva	Parasite of cattle, transmitted by ticks
Perkinsus marinus	Parasite of oysters; marine habitat
Crypthecodinium cohnii	Used to manufacture omega-3 fatty acids
Oxyrrhis marina	Omnivorous grazer; produces red tides
Pfiesteria shumwayae	Can be predatory on fishes; can use chloroplasts of ingested green algae for photosynthesis

patterns unique to each species (**FIGURE 19.9**). Despite their remarkable morphological diversity, all diatoms are symmetrical—either bilaterally (with "right" and "left" halves) or radially (with the type of symmetry possessed by a circle).

Diatoms reproduce both sexually and asexually. Asexual reproduction by binary fission is somewhat constrained by the stiff cell wall. Both the top and bottom of the "petri dish" become tops of new "dishes" without changing appreciably in size. As a result, the new cell made from the former bottom is smaller than the parent cell. If this process continued indefinitely, one cell line would simply vanish, but sexual reproduction largely solves this potential problem. Gametes are formed, shed their cell walls, and fuse. The resulting zygote then grows substantially in size before a new cell wall is laid down.

Diatoms are found in all the oceans and are frequently present in great numbers. They are major photosynthetic producers in coastal waters (see Key Concept 19.4) and are among the dominant organisms in the dense "blooms" of phytoplankton that occasionally appear in the open ocean. Diatoms are also common in fresh water and even occur on the wet surfaces of terrestrial mosses.

BROWN ALGAE The **brown algae** obtain their namesake color from the carotenoid fucoxanthin, which is abundant in their chloroplasts. The combination of this yellow-orange pigment with the green of chlorophylls *a* and *c* yields a brownish tinge. All brown algae are multicellular, and some are extremely large. Giant kelps such as those of the genus *Macrocystis* may be up to 60 meters long.

Media Clip 19.3 A Kelp Forest

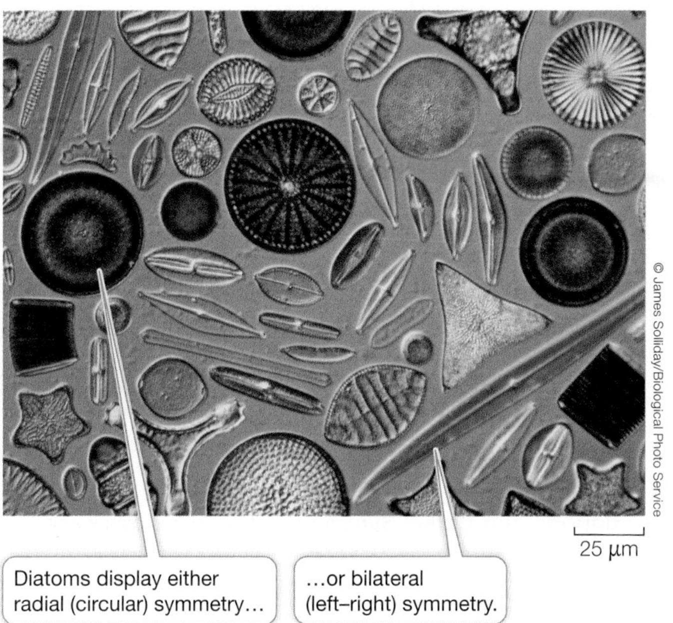

FIGURE 19.9 Diatom Diversity This bright-field micrograph illustrates the variety of species-specific forms found among the diatoms.

Media Clip 19.2 **Diatoms in Action**

FIGURE 19.10 Brown Algae (A) This seaweed illustrates the filamentous growth form of the brown algae. **(B)** Sea palms exemplify the leaflike growth form of brown algae. Sea palms and many other brown algal species are "glued" to the rocks by tough, branched structures called holdfasts that can withstand the pounding of the surf.

The brown algae are almost exclusively marine. They are composed either of branched filaments (**FIGURE 19.10A**) or of leaflike growths (**FIGURE 19.10B**). Some float in the open ocean; the most famous example is the genus *Sargassum*, which forms dense mats in the Sargasso Sea in the mid-Atlantic. Most brown algae, however, attach themselves to rocks near the shore. A few thrive only where they are regularly exposed to heavy surf. All of the attached forms develop a specialized structure, called a holdfast, that literally glues them to the rocks. The "glue" of the holdfast is alginic acid, a gummy polymer found in the walls of many brown algal cells. In addition to its function in holdfasts, alginic acid cements algal cells and filaments together. It is harvested and used by humans as an emulsifier in ice cream, cosmetics, and other products.

OOMYCETES The **oomycetes** are the water molds and their terrestrial relatives. Water molds are filamentous and stationary. They are **absorptive heterotrophs**—that is, they secrete enzymes that digest large food molecules into smaller molecules that they can absorb. They are all aquatic and **saprobic**—meaning they feed on dead organic matter. If you have seen a whitish, cottony mold growing on dead fishes or dead insects in water, it was probably a water mold of the common genus *Saprolegnia* (**FIGURE 19.11**).

Some other oomycetes, such as the downy mildews, are terrestrial. Although most of the terrestrial oomycetes are harmless or helpful decomposers of dead matter, a few are plant parasites that attack crops such as avocados, grapes, and potatoes.

Oomycetes were once classified as fungi. However, we now know that their similarity to fungi is only superficial, and that the oomycetes are more distantly related to the fungi than are many other eukaryotic groups, including humans (see Figure 19.3). For example, the cell walls of oomycetes are typically made of cellulose, whereas those of fungi are made of chitin.

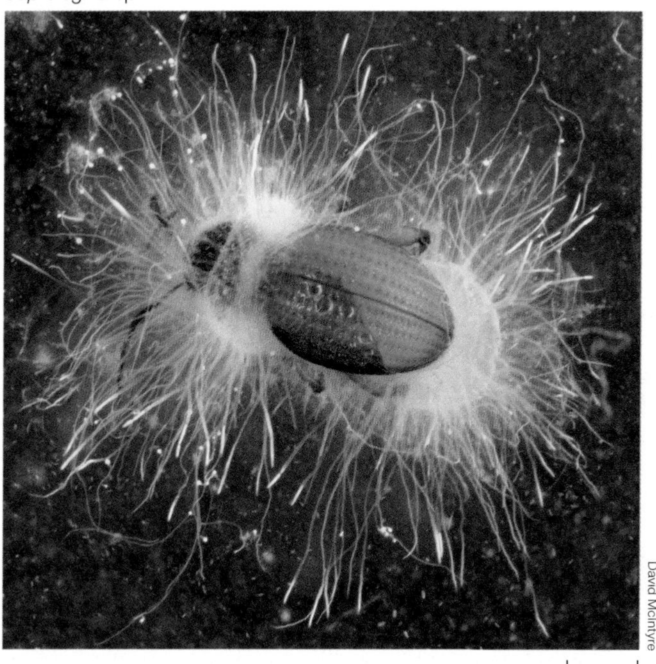

FIGURE 19.11 An Oomycete The filaments of a water mold radiate from the carcass of a beetle.

Rhizarians typically have long, thin pseudopods

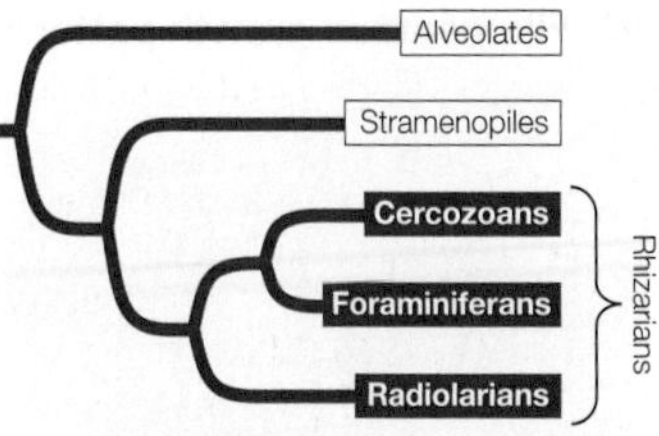

The three primary groups of **rhizarians**—cercozoans, foraminiferans, and radiolarians—are unicellular and mostly aquatic. The rhizarians have contributed their shells to ocean sediments, some of which have become terrestrial features over the course of geological history.

CERCOZOANS The **cercozoans** are a diverse group with many forms and habitats. Some are aquatic; others live in soil. One group of cercozoans possesses chloroplasts derived from a green alga by secondary endosymbiosis, and those chloroplasts contain a trace of the alga's nucleus.

FORAMINIFERANS Some **foraminiferans** secrete external shells of calcium carbonate (**FIGURE 19.12**). These shells have accumulated over time to produce much of the world's limestone. Some foraminiferans live as plankton; others live on the seafloor. Living foraminiferans have been found 10,896 meters down in the western Pacific's Challenger Deep—the deepest point in the world's oceans. At that depth, however, they cannot secrete normal shells because the surrounding water is too poor in calcium carbonate.

In living planktonic foraminiferans, long, threadlike, branched pseudopods extend through numerous microscopic apertures in the shell and interconnect to create a sticky, reticulated net, which the foraminiferans use to catch smaller plankton. In some foraminiferan species, the pseudopods provide locomotion.

FIGURE 19.12 Building Blocks of Limestone Some foraminiferans secrete calcium carbonate to form shells. The shells of different species have distinctive shapes. Over millions of years, the shells of foraminiferans have accumulated to form limestone deposits.

(A) *Amphilonche heteracantha*

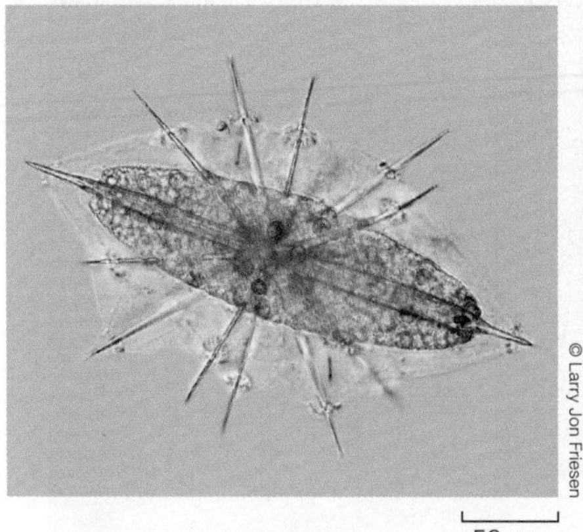

(B) *Thyrsocyrtis* sp.

FIGURE 19.13 Radiolarians Exhibit Distinctive Pseudopods and Radial Symmetry (A) The radiolarians are distinguished by their thin, stiff pseudopods and by their radial symmetry. The pigmentation seen at the center of this radiolarian's glassy endoskeleton is imparted by endosymbiotic dinoflagellates. **(B)** The endoskeleton secreted by a radiolarian.

RADIOLARIANS **Radiolarians** are recognizable by their thin, stiff pseudopods, which are reinforced by microtubules (**FIGURE 19.13A**). These pseudopods greatly increase the surface area of the cell, and they help the cell stay afloat in its marine environment.

Radiolarians also are immediately recognizable by their distinctive radial symmetry. Almost all radiolarian species secrete glassy endoskeletons (internal skeletons). The skeletons of the different species are as varied as snowflakes, and many have elaborate geometric designs (**FIGURE 19.13B**). A few radiolarians are among the largest of the unicellular eukaryotes, measuring several millimeters across.

Excavates began to diversify about 1.5 billion years ago

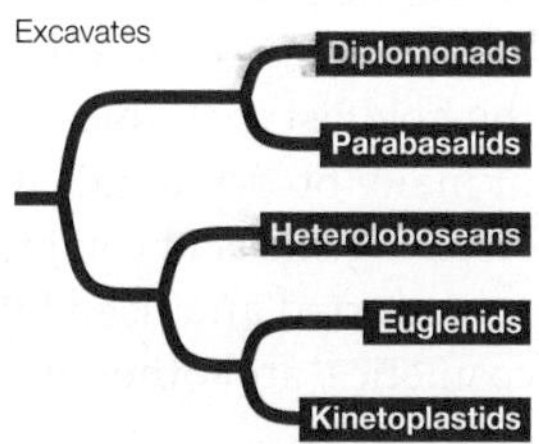

The **excavates** include several diverse groups that began to split from one another soon after the origin of eukaryotes. Several groups of excavates lack mitochondria, an absence that once led to the view that these groups might represent early-diverging eukaryotes that diversified before the evolution of mitochondria. However, the discovery of genes in the nucleus that are normally associated with mitochondria suggests that the absence of mitochondria is a derived condition in these organisms. In other words, ancestors of these excavate groups probably possessed mitochondria that were lost or reduced over the course of evolution. The existence of these organisms today shows that eukaryotic life is possible without mitochondria, at least among parasitic species.

DIPLOMONADS AND PARABASALIDS The **diplomonads** (**FIGURE 19.14A**) and the **parabasalids** (**FIGURE 19.14B**) are unicellular and lack mitochondria (although they have

(A) *Giardia muris*

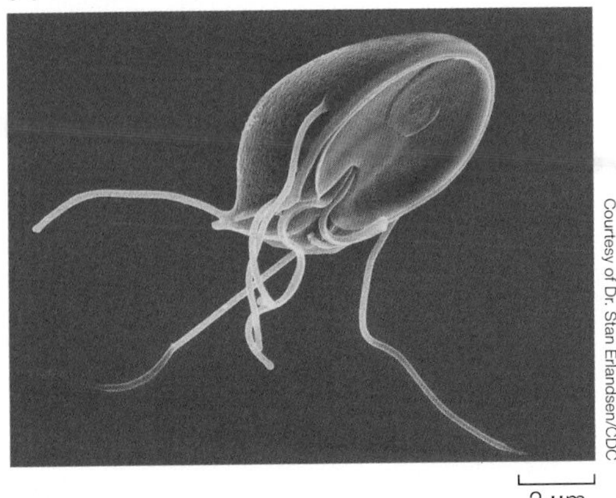

(B) *Trichomonas vaginalis*

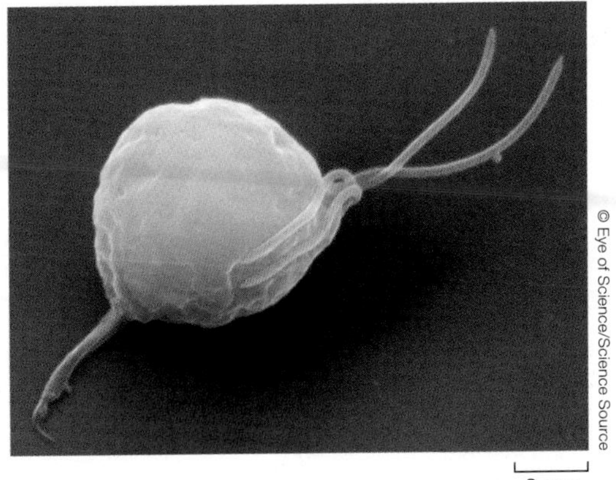

FIGURE 19.14 Some Excavate Groups Lack Mitochondria (A) *Giardia*, a diplomonad, has flagella and two nuclei. **(B)** *Trichomonas*, a parabasalid, has flagella and undulating membranes. Neither of these organisms possesses mitochondria.

reduced organelles that are derived from mitochondria). The parasitic *Giardia lamblia*, a diplomonad, causes the intestinal disease giardiasis. *Giardia* infections may result from contact with contaminated water. In the United States, such infections are common among hikers and campers using spring or stream water in recreational areas, as well as among children kept in close quarters (as in day-care centers). This tiny organism contains two nuclei bounded by nuclear envelopes, and it has a cytoskeleton and multiple flagella.

In addition to flagella and a cytoskeleton, the parabasalids have undulating membranes that also contribute to the cell's locomotion. *Trichomonas vaginalis* is a parabasalid responsible for a sexually transmitted disease in humans. Infection of the male urethra, where it may occur without symptoms, is less common than infection of the vagina.

HETEROLOBOSEANS The amoeboid body form appears in several protist groups that are only distantly related to one another. The body forms of **heteroloboseans**, for example, resemble those of loboseans, an amoebozoan group that is not at all closely related to heteroloboseans (see the next section). Amoebas of the free-living heterolobosean genus *Naegleria*, some of which can enter the human body and cause a fatal disease of the nervous system, usually have a two-stage life cycle, in which one stage has amoeboid cells and the other flagellated cells.

EUGLENIDS AND KINETOPLASTIDS The **euglenids** and **kinetoplastids** together constitute a clade of unicellular excavates with flagella. Their mitochondria contain distinctive disc-shaped cristae, and their flagella contain a crystalline rod not found in other organisms. They reproduce primarily asexually by binary fission.

The flagella of euglenids arise from a pocket at the anterior end of the cell. Spiraling strips of proteins under the cell membrane control the cell's shape. Some euglenids are photosynthetic. **FIGURE 19.15** depicts a typical cell of the genus *Euglena*. Like most other euglenids, this common freshwater organism has a complex cell structure. It propels itself through the water with the longer of its two flagella, which may also serve as an anchor to hold the organism in place. The second flagellum is often rudimentary.

Media Clip 19.4 **Euglenids**

The euglenids have diverse nutritional requirements. Many species are always heterotrophic. Other species, including species of *Euglena*, are fully autotrophic in sunlight, using chloroplasts to synthesize organic compounds through photosynthesis. When kept in the dark, these euglenids lose their photosynthetic pigment and begin to feed exclusively on dissolved organic material in the water around them. A "bleached" *Euglena* resynthesizes its photosynthetic pigment when it is returned to the light and becomes autotrophic again. But *Euglena* cells treated with certain antibiotics or mutagens lose their photosynthetic pigment completely; neither they nor their descendants are ever autotrophs again. However, those descendants function well as heterotrophs.

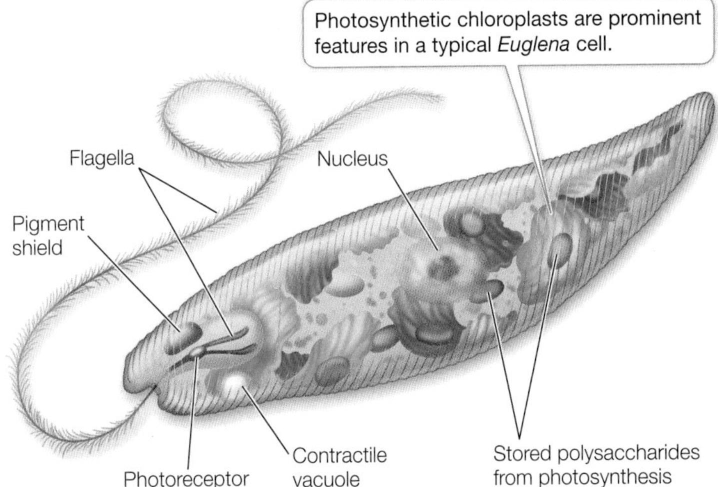

FIGURE 19.15 A Photosynthetic Euglenid In the *Euglena* species illustrated in this drawing, the second flagellum is rudimentary. Note that the primary flagellum originates at the anterior of the organism and trails toward its posterior.

TABLE **19.1** | Three Pathogenic Trypanosomes

	Trypanosoma brucei	*Trypanosoma cruzi*	*Leishmania major*
Human disease	Sleeping sickness	Chagas disease	Leishmaniasis
Insect vector	Tsetse fly	Assassin bugs (many species)	Sand fly
Vaccine or effective cure	None	None	None
Strategy for survival	Changes surface recognition molecules frequently	Causes changes in surface recognition molecules on host cell	Reduces effectiveness of macrophage hosts
Site in human body	Bloodstream; in final stages, attacks nerve tissue	Enters cells, especially muscle cells	Enters cells, primarily macrophages
Approximate number of deaths per year	7,000	11,000	63,000

The kinetoplastids are unicellular parasites with two flagella and a single, large mitochondrion. The mitochondrion contains a kinetoplast, a unique structure housing multiple circular DNA molecules and associated proteins. Some of these DNA molecules encode "guide proteins" that edit mRNA within the mitochondrion.

The kinetoplastids include several medically important species of pathogenic trypanosomes (**TABLE 19.1**). Some of these organisms are able to change their cell surface recognition molecules frequently, allowing them to evade our best attempts to kill them and thus eradicate the diseases they cause.

Amoebozoans use lobe-shaped pseudopods for locomotion

Amoebozoans

Loboseans

Plasmodial slime molds

Cellular slime molds

Amoebozoans appear to have diverged from other eukaryotes about 1.5 billion years ago (see Figure 19.3). It is not yet clear whether they are more closely related to opisthokonts (which include fungi and animals) or to other major groups of eukaryotes.

The lobe-shaped pseudopods of amoebozoans (**FIGURE 19.16**) are a hallmark of the amoeboid body form. Amoebozoan pseudopods differ in form and function from the slender pseudopods of rhizarians. We consider three amoebozoan groups here: the loboseans and two groups known as slime molds.

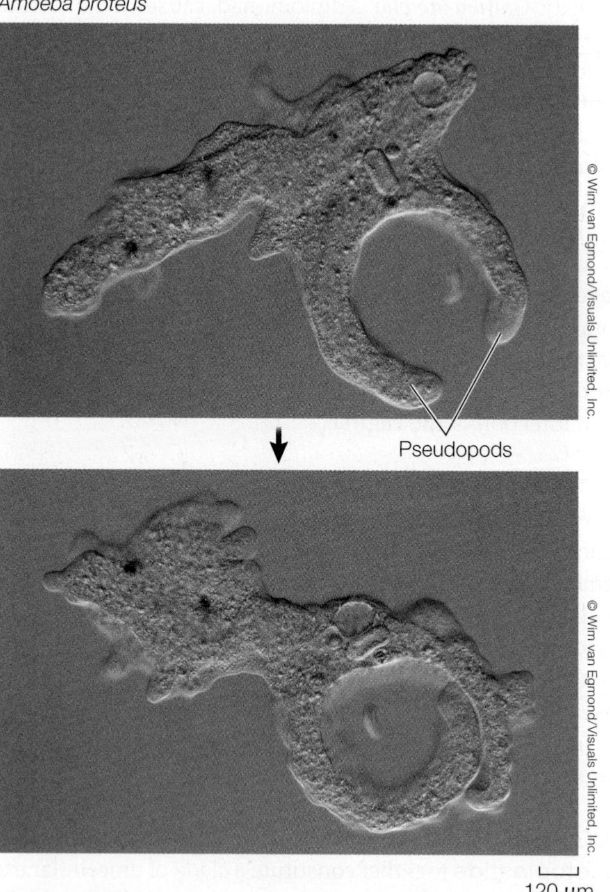

FIGURE 19.16 An Amoeba in Motion The flowing pseudopods of this "chaos amoeba" (a lobosean) are constantly changing shape as it moves and feeds.

Media Clip 19.5 Ameboid Movement

LOBOSEANS **Loboseans** are small amoebozoans that feed on other small organisms and particles of organic matter by phagocytosis, engulfing them with pseudopods. Many loboseans are adapted for life on the bottoms of lakes, ponds, and other bodies of water. Their creeping locomotion and their manner of engulfing food particles fit them for life close to a relatively rich supply of sedentary organisms or organic particles. Most loboseans exist as predators, parasites, or scavengers. Members of one group of loboseans, the testate amoebas, live inside shells. Some of these amoebas produce casings by gluing sand grains together (**FIGURE 19.17**). Other testate amoebas have shells secreted by the organism itself.

PLASMODIAL SLIME MOLDS If the nucleus of an amoeba began rapid mitotic division, accompanied by a tremendous increase in cytoplasm and organelles, but no cytokinesis, the

Nebela collaris

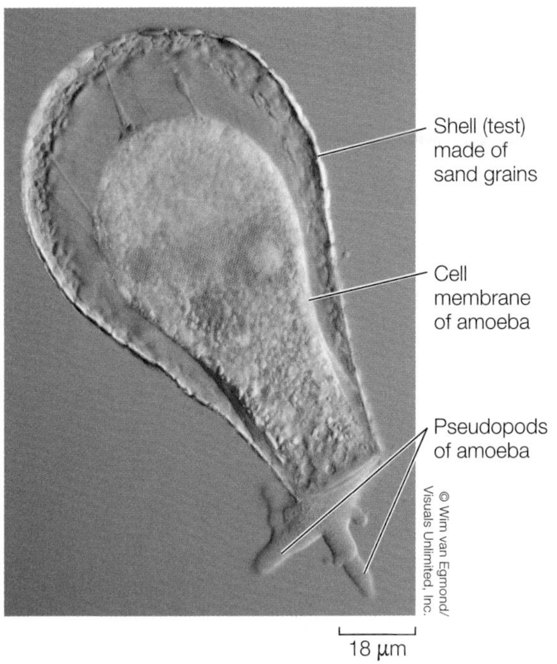

FIGURE 19.17 Life in a Glass House This testate amoeba has built a lightbulb-shaped shell, or test, by gluing sand grains together. Its pseudopods extend through the single aperture in the test.

resulting organism would resemble the multinucleate mass of a **plasmodial slime mold**. During its vegetative (feeding, nonreproductive) stage, a plasmodial slime mold is a wall-less mass of cytoplasm with numerous diploid nuclei. This mass streams very slowly over its substrate in a remarkable network of strands called a plasmodium (**FIGURE 19.18A**). The plasmodium is an example of a **coenocyte**: many nuclei enclosed in a single cell membrane. The outer cytoplasm of the plasmodium (closest to the environment) is normally less fluid than the interior cytoplasm and thus provides some structural rigidity.

Plasmodial slime molds provide a dramatic example of movement by **cytoplasmic streaming**. The outer cytoplasm of the plasmodium becomes more fluid in places, and cytoplasm rushes into those areas, stretching the plasmodium. This streaming reverses its direction every few minutes as cytoplasm rushes into a new area and drains away from a previous one, moving the plasmodium over its substrate. Sometimes an entire wave of plasmodium moves across a surface, leaving strands behind. Microfilaments and a contractile protein called myxomyosin interact to produce the streaming movement. As it moves, the plasmodium engulfs food particles by endocytosis—predominantly bacteria, yeasts, spores of fungi, and other small organisms as well as decaying animal and plant remains.

A plasmodial slime mold can grow almost indefinitely in its plasmodial stage as long as the food supply is adequate and other conditions, such as moisture and pH, are favorable. If conditions become unfavorable, however, one of two things can happen. First, the plasmodium can form an irregular mass of hardened cell-like components. This resting structure rapidly becomes a plasmodium again when favorable conditions are restored.

Alternatively, the plasmodium can transform itself into spore-bearing fruiting structures (**FIGURE 19.18B**). These stalked or branched structures rise from heaped masses of plasmodium. They derive their rigidity from walls that form and thicken between their nuclei. The diploid nuclei of the plasmodium divide by meiosis as the fruiting structure develops. One or more knobs, called sporangia, develop on the end of the stalk. Within a sporangium, haploid nuclei become surrounded by walls to form spores. Eventually, as the fruiting structure dries, it sheds its spores.

The spores germinate into wall-less, haploid cells called swarm cells, which can either divide mitotically to produce more haploid swarm cells or function as gametes. Swarm cells can live as separate individual cells that move by means of flagella or pseudopods, or they can become walled and resistant resting cysts when conditions are unfavorable; when conditions improve again, the cysts release swarm cells. Two swarm cells can also fuse to form a diploid zygote, which divides by mitosis (but without a wall forming between the nuclei) and thus forms a new coenocytic plasmodium.

CELLULAR SLIME MOLDS Whereas the plasmodium is the basic vegetative unit of the plasmodial slime molds, a single

(A) A plasmodium

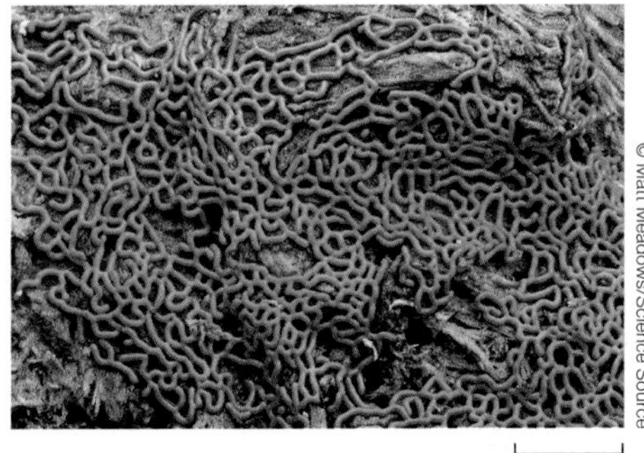

(B) Fruiting structures of a plasmodial slime mold

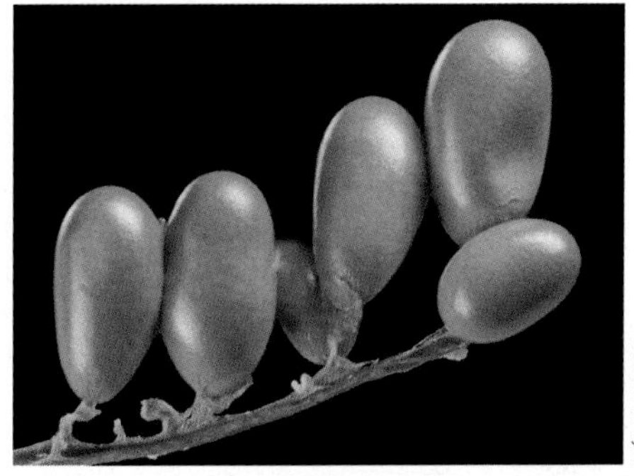

FIGURE 19.18 A Plasmodial Slime Mold **(A)** The plasmodial form of the slime mold *Hemitrichia serpula* covers rocks, decaying logs, and other objects as it engulfs bacteria and other food items; it is also responsible for the organism's common name of "pretzel mold." **(B)** Fruiting structures of *Leocarpus fragilis*.

Media Clip 19.6 Plasmodial Slime Mold Growth

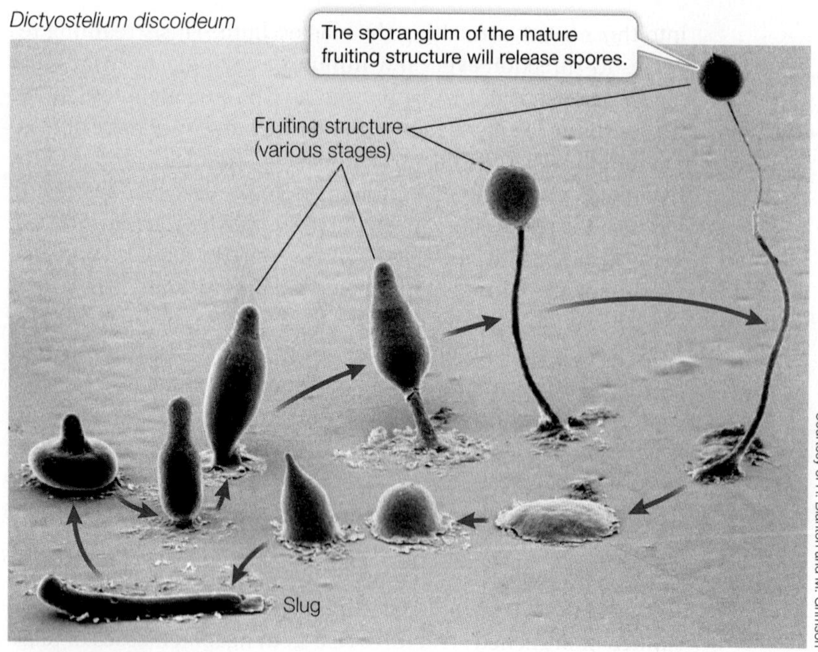

FIGURE 19.19 A Cellular Slime Mold
This composite micrograph shows the life cycle of a *Dictyostelium* slime mold.

amoeboid cell is the vegetative unit of the **cellular slime molds** (**FIGURE 19.19**). Cells called myxamoebas, which have single haploid nuclei, swarm together as they engulf bacteria and other food particles by endocytosis and reproduce by mitosis and fission. This simple life cycle stage, consisting of swarms of independent, isolated cells, can persist indefinitely as long as food and moisture are available.

When conditions become unfavorable, the cellular slime molds form fruiting structures, as do their plasmodial counterparts. The individual myxamoebas aggregate into a mass called a slug or **pseudoplasmodium**. Unlike the true plasmodium of the plasmodial slime molds, this structure is not simply a giant sheet of cytoplasm with many nuclei; the individual myxamoebas in the slug retain their cell membranes and therefore their identity.

A slug may migrate over a substrate for several hours before becoming motionless and reorganizing to construct a delicate, stalked fruiting structure. Cells at the top of the fruiting structure develop into thick-walled spores, which are eventually released. Later, under favorable conditions, the spores germinate, releasing myxamoebas.

The cycle from myxamoebas through slug and spores to new myxamoebas is asexual. Cellular slime molds also have a sexual cycle, in which two myxamoebas fuse. The product of this fusion develops into a spherical structure that ultimately germinates, releasing new haploid myxamoebas.

REVIEW & APPLY | 19.2

R The major lineages of eukaryotes began to diverge about 1.5 billion years ago. Major groups of eukaryotes are highly diverse in their habitat, nutrition, locomotion, and body form. Many protists are photosynthetic autotrophs, but heterotrophic lineages have evolved repeatedly. Although most protists are unicellular, multicellularity has arisen independently many times.

A

1. For each pair of groups below, describe how you could recognize members of the two groups and differentiate them from one another. Then describe features that the two groups in each pair share.
   a. Foraminiferans and radiolarians
   b. Ciliates and dinoflagellates
   c. Diatoms and brown algae
   d. Plasmodial slime molds and cellular slime molds
2. The fossil record of eukaryotes from the Precambrian is poor compared with that from the Cambrian and later geological periods, even though eukaryotes were diversifying for the last billion years of the Precambrian. Can you think of some reasons why the eukaryotic fossil record became more extensive in the Cambrian?
3. Give examples of alveolates, stramenopiles, and excavates that are important for medical or culinary reasons.
4. The red tides produced by *Oxyrrhis* and other dinoflagellates are not produced by apicomplexans. Similar red tides are produced by diatoms, which are distantly related to alveolates. Thus the production of red tides is convergent in these two groups. Describe another example from the text of convergent evolution in protists but outside alveolates.

The ancient origins of the major eukaryotic lineages and the adaptation of these lineages to a wide variety of lifestyles and environments resulted in enormous protist diversity. It is not surprising, then, that reproductive modes among protists are also highly diverse.

19.3 Protists Reproduce Sexually and Asexually

LEARNING OBJECTIVES

By the end of this key concept you should be able to:

19.3.1 Distinguish between sex (in the sense of exchange of genes leading to genetic recombination) and reproduction.

19.3.2 Predict the evolutionary consequences of reproduction without sex for many generations.

Although most protists engage in both asexual and sexual reproduction, sexual reproduction has yet to be observed in some groups. In some protists, as in all prokaryotes, the acts of sex and reproduction are not directly linked.

Several asexual reproductive processes have been observed among the protists:

- The equal splitting of one cell into two by mitosis followed by cytokinesis
- The splitting of one cell into multiple (i.e., more than two) cells
- The outgrowth of a new cell from the surface of an old one (known as **budding**)
- The formation of specialized cells (spores) that are capable of developing into new individuals (know as **sporulation**)

Asexual reproduction results in offspring that are genetically nearly identical to their parents (they differ only by new mutations that may arise during DNA replication). Such asexually reproduced groups of nearly identical organisms are known as **clonal lineages**.

Sexual reproduction among the protists takes various forms. In some protists, as in animals, the gametes are the only haploid cells. In others, the zygote is the only diploid cell. In still others, both diploid and haploid cells undergo mitosis, giving rise to alternating multicellular diploid and haploid life stages.

Some protists reproduce without sex and have sex without reproduction

As noted in Key Concept 19.2, members of the genus *Paramecium* are ciliates, which commonly have two types of nuclei in a single cell (one macronucleus and from one to several micronuclei; see Figure 19.6). The micronuclei are typical eukaryotic nuclei and are essential for genetic recombination. Each macronucleus contains many copies of the genetic information, packaged in units containing only a few genes each. The macronuclear DNA is transcribed and translated to regulate the life of the cell.

When paramecia reproduce asexually, all of the nuclei are copied before the cell divides. Paramecia (and many other protists) also have an elaborate sexual behavior called **conjugation**, in which two individuals line up tightly against each other and fuse in the oral groove region of the body. Nuclear material is extensively reorganized and exchanged over the next several hours (**FIGURE 19.20**). Each cell ends up with two haploid micronuclei, one of its own and one from the other cell, which fuse to form a new diploid micronucleus. A new macronucleus develops from that micronucleus through a series of dramatic chromosomal rearrangements. The exchange of nuclei is fully reciprocal: each of the two paramecia gives and receives an equal amount of DNA.

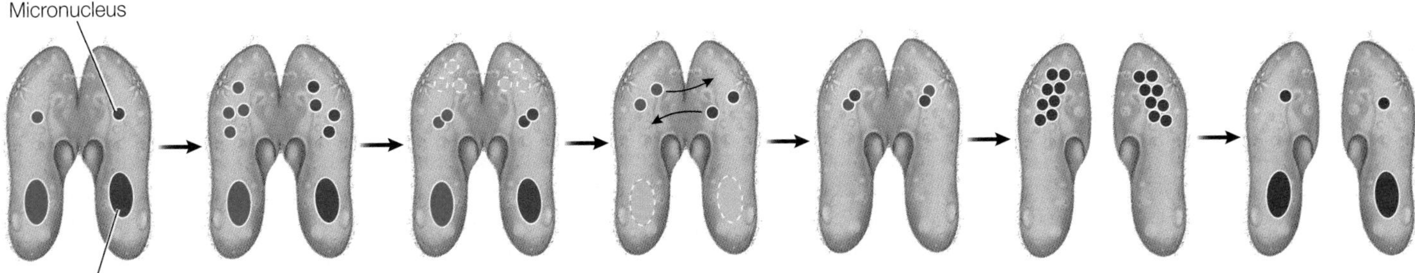

FIGURE 19.20 Conjugation in Paramecia The exchange of micronuclei by two conjugating *Paramecium* individuals results in genetic recombination. After conjugation, the cells separate and continue their lives as two individuals.

Why is conjugation considered sex without reproduction?

The two organisms then separate and go their own ways, each equipped with new combinations of alleles.

Conjugation in *Paramecium* is a sexual process of genetic recombination, but it is not a reproductive process. Two cells begin conjugation and two cells are there at the end, so no new cells are created. As a rule, each asexual clone of paramecia must conjugate periodically. Experiments have shown that if some species are not permitted to conjugate, the clones can live through only about 350 cell divisions before dying out.

Some protist life cycles feature alternation of generations

Alternation of generations is a type of life cycle found in many multicellular protists, all land plants, and some fungi (see Figure 20.4). A multicellular, diploid, spore-producing organism gives rise to a multicellular, haploid, gamete-producing organism. When two haploid gametes fuse, a diploid organism is produced. The haploid organism, the diploid organism, or both may also reproduce asexually. Note that alternation of generations is distinct from the familiar reproductive system of animals, in which the only haploid stages are unicellular gametes produced by multicellular, diploid adults.

The two alternating (spore-producing and gamete-producing) generations differ genetically (one has diploid cells, the other haploid cells), but they may or may not differ morphologically. In **heteromorphic** alternation of generations, the two generations differ morphologically; in **isomorphic** alternation of generations, they do not. Examples of both heteromorphic and isomorphic alternation of generations are found among the brown algae.

The gamete-producing generation does not produce gametes by meiosis because the gamete-producing organism is already haploid. Instead, specialized cells of the diploid spore-producing organism, called **sporocytes**, divide meiotically to produce four haploid spores. The spores may eventually germinate and divide mitotically to produce the multicellular haploid generation, which then produces gametes by mitosis and cytokinesis.

Gametes, unlike spores, can produce new organisms only by fusing with other gametes. The fusion of two gametes produces a diploid zygote, which then undergoes mitotic divisions to produce a diploid organism. The diploid organism's sporocytes then undergo meiosis and produce haploid spores, starting the cycle anew. Given the diversity of protists and of the environments in which they live, it is not surprising that they influence their environments in numerous ways.

REVIEW & APPLY | 19.3

R Protists reproduce both asexually and sexually, although sex occurs independently of reproduction in some species. Some multicellular protists exhibit alternation of generations, alternating between multicellular haploid and diploid life stages.

A

1. Why is conjugation between paramecia considered a sexual process but not a reproductive process?
2. Why do you think paramecia that are not allowed to conjugate begin to die out after about 350 rounds of asexual reproduction?
3. Although most diploid animals have haploid stages (e.g., eggs and sperm), their life cycles are not considered examples of alternation of generations. Why not?

19.4 | Protists Are Critical Components of Many Ecosystems

LEARNING OBJECTIVES

By the end of this key concept you should be able to:

- **19.4.1** Illustrate the complex life cycle of a parasite that involves multiple hosts.
- **19.4.2** Summarize the consequences of overproduction of some diatoms and dinoflagellates.
- **19.4.3** Describe some of the beneficial and harmful interactions that protists have on other species.
- **19.4.4** Provide two examples of how protists are important to humans.

Some protists are food for marine animals, while others poison those animals. Some are packaged as nutritional supplements for humans, and some are human pathogens. The remains of some form the sands of many modern beaches, and others are a major source of the oil that sometimes fouls those beaches.

Phytoplankton are primary producers

A single protist clade, the diatoms, performs about one-fifth of all photosynthetic carbon fixation on Earth—about the same amount as all of Earth's rainforests. These spectacular unicellular organisms (see Figure 19.9) are the predominant component of the oceanic phytoplankton, but the phytoplankton include many other protists that also contribute heavily to global photosynthesis. Like green plants on land, these "floating photosynthesizers" are the gateway for energy from the Sun into the rest of the living world; in other words, they are **primary producers**. These autotrophs are eaten by heterotrophs, including animals and many other protists. Those consumers are, in turn, eaten by other consumers. Most aquatic heterotrophs (with the exception of some species in the deep sea) depend on photosynthesis performed by phytoplankton.

Some microbial eukaryotes are deadly

Some microbial eukaryotes are pathogens that cause serious diseases in humans and other vertebrates. The best-known pathogenic protists are members of the genus *Plasmodium*, a highly specialized group of apicomplexans that spend part of their life cycle as parasites in human red blood cells, where they are the cause of malaria. In terms of the number of people affected, malaria is one of the world's three most serious infectious diseases: it infects more than 215 million people, and kills nearly half a million people, each year. On average, about two people die from malaria every minute of every day—most of them in sub-Saharan Africa, although malaria occurs in more than 90 countries.

Mosquitoes of the genus *Anopheles* transmit *Plasmodium* to humans. The parasites enter the human circulatory system when an infected female *Anopheles* mosquito penetrates the skin in search of blood. The parasites find their way to cells in the liver and the lymphatic system, change their form, multiply, and reenter the bloodstream, where they invade red blood cells.

The parasites multiply inside the red blood cells, which then lyse (burst), releasing new swarms of parasites. These episodes of bursting red blood cells coincide with the primary symptoms of malaria, which include fever, shivering, vomiting, joint pains, and convulsions. If another *Anopheles* bites the victim, the mosquito takes in *Plasmodium* cells along with blood. Some of the ingested cells develop into gametes that unite in the mosquito, forming zygotes. The zygotes lodge in the mosquito's gut, divide several times, and move into its salivary glands, from which they can be passed on to another human host. Thus *Plasmodium* is an extracellular parasite in the mosquito vector and an intracellular parasite in the human host (**FIGURE 19.21**). Such an organism—that

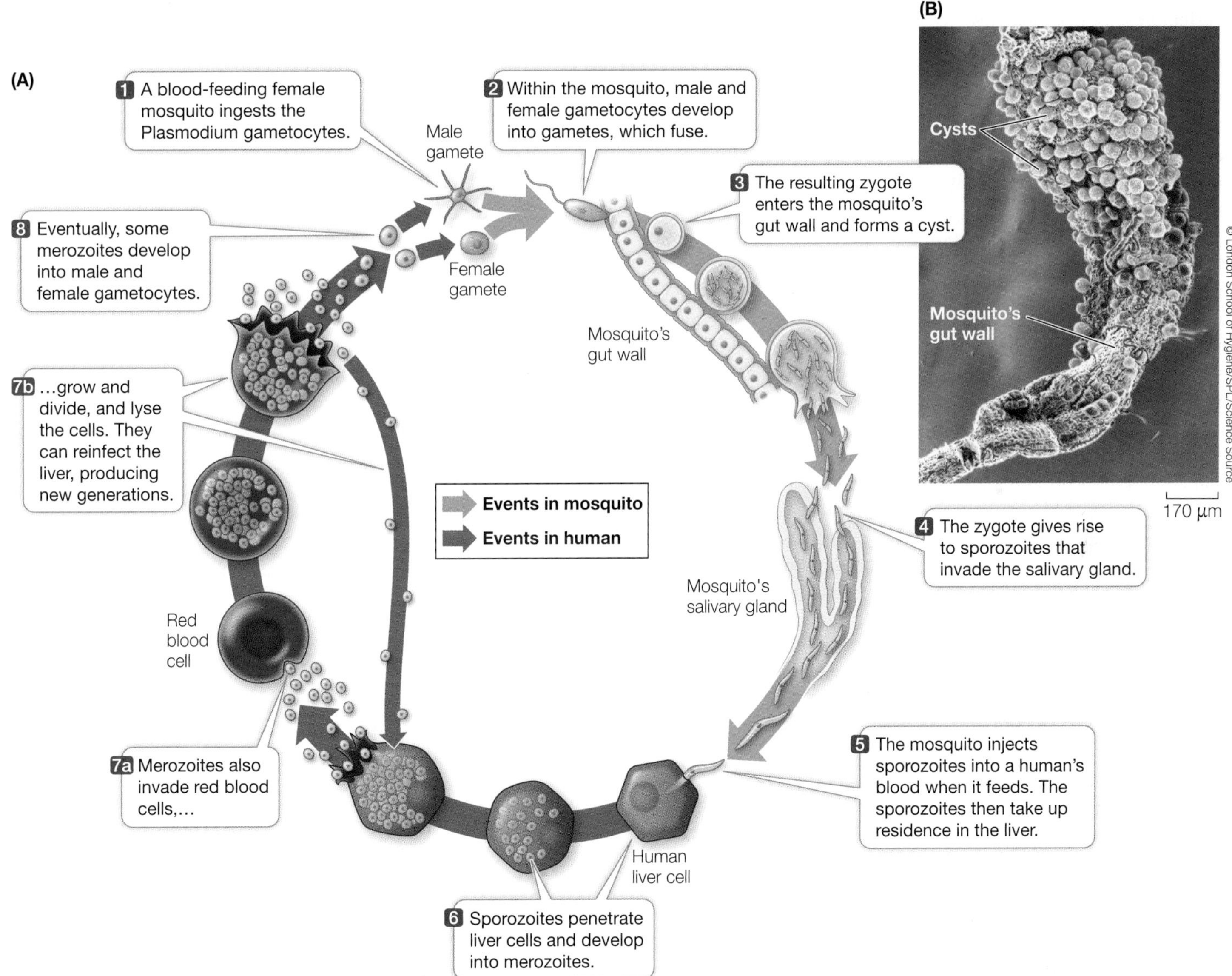

FIGURE 19.21 Life Cycle of the Malarial Parasite **(A)** Like many parasitic species, the apicomplexan *Plasmodium falciparum* has a complex life cycle, part of which is spent in mosquitoes of the genus *Anopheles* and part in humans. The sexual phase (gamete fusion) of this life cycle takes place in the insect, and the zygote is the only diploid stage. **(B)** Encysted *Plasmodium* zygotes (artificially colored blue) cover the stomach wall of a mosquito. Invasive sporozoites will hatch from the cysts and be transmitted to a human, in whom the parasite causes malaria.

Animation 19.3 Life Cycle of the Malarial Parasite

is, a parasite that requires more than one host—is said to have a **complex life cycle**.

Plasmodium has proved to be a singularly difficult pathogen to attack. The complex *Plasmodium* life cycle is best broken by the removal of stagnant water, in which mosquitoes breed. Using insecticides to reduce the *Anopheles* population can also be effective, but the benefits must be weighed against the ecological, economic, and health risks posed by the insecticides themselves.

Even some of the phytoplankton that are such important primary producers can be deadly, as noted in the caption of the photo that opens this chapter. Some diatoms and dinoflagellates reproduce in enormous numbers when environmental conditions are favorable for their growth. In the resulting red tides, the concentration of dinoflagellates may reach 60 million per liter of ocean water and produce potent nerve toxins that harm or kill many vertebrates, especially fishes.

Some microbial eukaryotes are endosymbionts

Endosymbiosis is common among the microbial eukaryotes, many of which live within the cells of animals. Many radiolarians harbor photosynthetic endosymbionts (see Figure 19.13A). As a result, these radiolarians, which are not photosynthetic themselves, appear greenish or golden, depending on the type of endosymbiont they contain. This arrangement is often mutually beneficial: the radiolarian can make use of the carbon compounds produced by its photosynthetic guest, and the guest may in turn make use of metabolites made by the host or receive physical protection. In some cases, the guest is exploited for its photosynthetic products while receiving little or no benefit itself.

Dinoflagellates are also common endosymbionts and can be found in both animals and other protists. Most, but not all, dinoflagellate endosymbionts are photosynthetic. Some dinoflagellates live endosymbiotically in the cells of corals, contributing the products of their photosynthesis to the partnership. Their importance to the corals is demonstrated when the dinoflagellates die or are expelled by the corals as a result of changing environmental conditions such as rising water temperatures or increased water turbidity. This phenomenon is known as **coral bleaching**. Unless the corals can acquire new endosymbionts, they are ultimately damaged or destroyed as a result of their reduced food supply (**INVESTIGATION FIGURE 19.22**).

High light and high temperature are two environmental stressors that disrupt the coral-alga symbiosis. See **Key Concept 1.5** for two experiments that examined the relationship between light and water temperature and coral bleaching.

We rely on the remains of ancient marine protists

Diatoms are lovely to look at, but their importance to us goes far beyond aesthetics, and even beyond their role as primary producers. Diatoms store oil as an energy reserve and to keep themselves afloat at the correct depth in the ocean. Over millions of years, diatoms have died and sunk to the ocean floor, where they have undergone chemical changes. In this way, they have become a major source of petroleum and natural gas, two of our most important energy supplies and political concerns.

Because the silica-containing cell walls of dead diatoms resist decomposition, some sedimentary rocks are composed almost entirely of diatom skeletons that sank to the seafloor over time. Diatomaceous earth, which is obtained from such rocks, has many industrial uses, such as insulation, filtration, and metal polishing. It has also been used as an "Earth-friendly" insecticide that clogs the tracheae (breathing structures) of insects.

Other ancient marine protists have also contributed to today's world. Some foraminiferans, as we have seen, secrete shells of calcium carbonate. After they reproduce (by mitosis and cytokinesis), the daughter cells abandon the parent shell and make new shells of their own. The discarded shells of ancient foraminiferans make up extensive limestone deposits in various parts of the world, forming a layer hundreds to thousands of meters deep over millions of square kilometers of ocean bottom. Foraminiferan shells also make up much of the sand of some beaches. A single gram of such sand may contain as many as 50,000 foraminiferan shells and shell fragments.

The shells of individual foraminiferans are easily preserved as fossils in marine sediments. Each geological period has a distinctive assemblage of foraminiferan species. Because the shells of foraminiferan species have distinctive shapes (see Figure 19.12) and because they are so abundant, the remains of foraminiferans are especially valuable in classifying and dating sedimentary rocks. In addition, analyses of the chemical makeup of foraminiferan shells can be used to estimate the global temperatures prevalent at the time when the shells were formed.

REVIEW & APPLY | 19.4

Protists have many effects, both positive and negative, on their environment. Some species are important primary producers, many are endosymbionts, and some are pathogens. Protists are among the most important producers of fossil fuels, and they are important components of sedimentary rocks.

1. What is the role of female *Anopheles* mosquitoes in the transmission of malaria?
2. Explain the roles of dinoflagellates in the two very different phenomena of coral bleaching and red tides.
3. What are two ways in which diatoms are important to human society?

The next three chapters will explore the major evolutionary radiations of multicellular eukaryotes, along with the protist ancestors from which they arose. Chapter 20 will describe the origin and diversification of plants, Chapter 21 will present the fungi, and Chapter 22 will provide a brief overview of the animals.

INVESTIGATION

FIGURE 19.22 Can Corals Reacquire Dinoflagellate Endosymbionts Lost to Bleaching? Some corals lose their chief nutritional source when their photosynthetic endosymbionts die, often as a result of changing environmental conditions. This experiment by Cynthia Lewis and Mary Alice Coffroth investigated the ability of corals to acquire new endosymbionts after bleaching.[a]

HYPOTHESIS

Bleached corals can acquire new photosynthetic endosymbionts from their environment.

METHOD

1. Count numbers of *Symbiodinium*, a photosynthetic dinoflagellate, living symbiotically in samples of a coral (*Briareum* sp.).
2. Stimulate bleaching by maintaining all *Briareum* colonies in darkness for 12 weeks.
3. After 12 weeks of darkness, count numbers of *Symbiodinium* in the coral samples; then return all colonies to light.
4. In some of the bleached colonies (the experimental group), introduce *Symbiodinium* strain B211—dinoflagellates that contain a unique molecular marker. Do not expose the others (the control group) to strain B211. Maintain both groups in the light for 6 weeks.

RESULTS

Error bars show 95% confidence limits for the means.

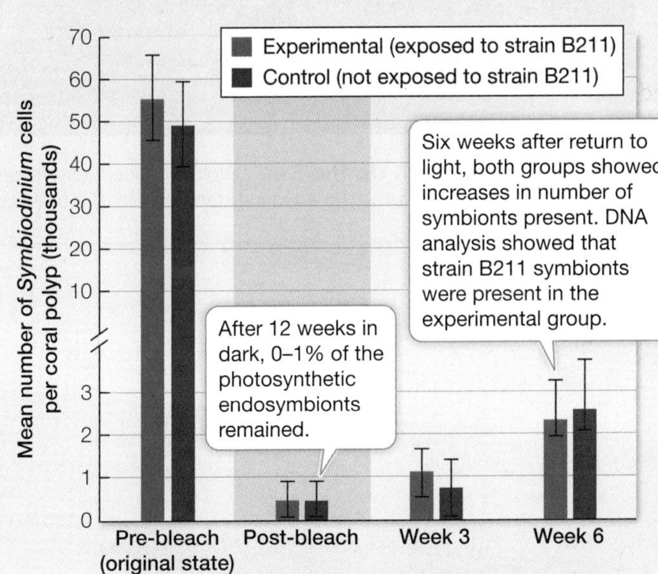

Pre-bleach

Courtesy of M. A. Coffroth and C. Lewis, University at Buffalo

Post-bleach

Courtesy of M. A. Coffroth and C. Lewis, University at Buffalo

CONCLUSION

Corals can acquire new endosymbionts from their environment following bleaching.

ANALYZE THE DATA

The data in the table below come from DNA analyses of *Symbiodinium* strains found in the experimental and control colonies of corals (*Briareum*) before and after bleaching. *Symbiodinium* strain B211 (which was not present before bleaching) was introduced to the experimental colonies after bleaching. Use these data to answer the questions below.

	***Symbiodinium* strain present (% of colonies)**			
	Non-B211	**B211**	**None[a]**	**Colony died**
Experimental colonies (strain B211 added)				
Pre-bleach	100	0	0	0
Post-bleach	58	0	42	0
Week 3	0	92	0	8
Week 6	8	58	8	25
Control colonies (no strain B211)				
Pre-bleach	100	0	0	0
Post-bleach	67	0	33	0
Week 3	67	0	33	0
Week 6	67	0	17	17

[a]Colonies remained alive, but no *Symbiodinium* were detected.

1. Are new strains of *Symbiodinium* taken up only by coral colonies that have lost all their original endosymbionts?
2. Does the acquisition of a new *Symbiodinium* strain always result in survival of a recovering *Briareum* colony?
3. In week 3, only strain B211 was detected in the experimental colonies, but in week 6, non-B211 *Symbiodinium* were detected in 8% of the experimental colonies. Can you suggest an explanation for this observation?

[a]C. L. Lewis and M. A. Coffroth. 2004. *Science* 304: 1490–1492.

© Don Paulson/Purestock/age fotostock

Answer: The photo at the beginning of this chapter illustrates a harmful red tide, or bloom of dinoflagellates. But not all dinoflagellate blooms produce problems for other species. Dinoflagellates are important components of many ecosystems, as we have seen throughout this chapter. Photosynthetic dinoflagellates also produce much of the atmospheric oxygen that most animals need to survive.

Corals and many other species depend on symbiotic dinoflagellates for food. In addition, as photosynthetic organisms, free-living planktonic dinoflagellates are among the most important primary producers in aquatic food webs. They are a major component of the phytoplankton and provide an important food source for many species.

Some dinoflagellates produce a beautiful bioluminescence (**FIGURE 19.23**). Unlike the bioluminescent bacteria shown at the start of Chapter 18, however, dinoflagellates cannot generate a steady bioluminescence, but produce bright flashes of light when disturbed, as people who swim in the ocean at night in certain regions often observe. What function do these flashes serve? Many light-emitting dinoflagellates are preyed on by other species, such as small crustaceans. When the dinoflagellates produce a bright flash, it functions like a "burglar alarm," and attracts secondary predators of the crustaceans. Experiments have shown that crustaceans reduce their feeding on dinoflagellates when they flash this signal. However, recent research shows another function of bioluminescence in toxic species of dinoflagellates. These toxic species produce a much lower intensity bioluminescence which may function as a warning of their toxicity to potential predators. So depending on the intensity and other details of the flash in different species, the bioluminescence can serve either to attract secondary predators or to warn off primary predators.

Media Clip 19.7 Flashing Dinoflagellates

© Per-Andre Hoffmann/Alamy Stock Photo

FIGURE 19.23 Light Up the Sea Bioluminescent dinoflagellates flash as an outrigger disturbs the ocean surface off the island of Bali.

QUESTIONS

19.1 Eukaryotes Acquired Features from Both Archaea and Bacteria

Go to ANIMATION 19.1

FIGURE 19.1 FIGURE 19.2

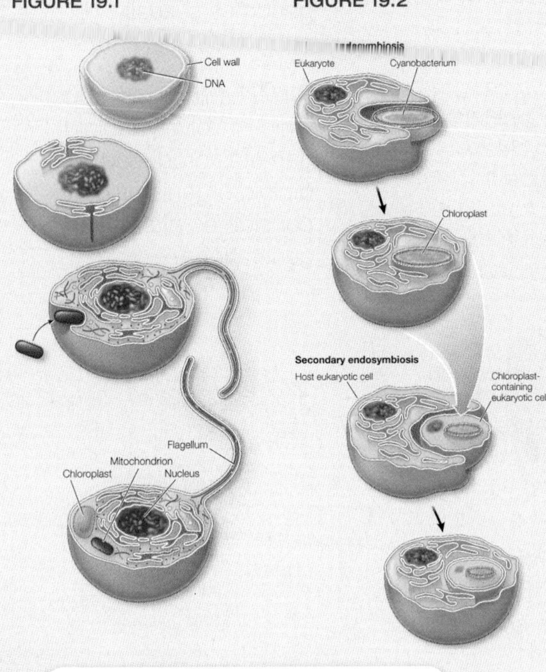

- The term **PROTIST** does not describe a formal taxonomic group. It is shorthand for "all eukaryotes that are not plants, animals, or fungi."
- Early events in the evolution of the eukaryotic cell included the loss of the firm cell wall and infolding of the cell membrane. Such infolding led to segregation of the genetic material in a membrane-enclosed nucleus (**FIGURE 19.1**).
- Mitochondria evolved by endosymbiosis with a proteobacterium.
- **PRIMARY ENDOSYMBIOSIS** of a eukaryote and a cyanobacterium gave rise to the first chloroplasts. **SECONDARY ENDOSYMBIOSIS** and **TERTIARY ENDOSYMBIOSIS** between chloroplast-containing eukaryotes and other eukaryotes gave rise to the distinctive chloroplasts of euglenids, dinoflagellates, and other groups (**FIGURE 19.2**).

1. Are all protists small in size (microbial)?
2. What is the phylogenetic evidence that mitochondria and chloroplasts arose through endosymbiosis?

19.2 Major Lineages of Eukaryotes Diversified in the Precambrian

Go to ACTIVITY 19.1 and ANIMATION 19.2

- Most eukaryotes can be placed in one of eight major clades that diverged about 1.5 billion years ago: alveolates, stramenopiles, rhizarians, excavates, plants, amoebozoans, fungi, and animals (**FIGURE 19.3**).
- **ALVEOLATES** are unicellular organisms with sacs (alveoli) beneath their cell membranes. Alveolate clades include the marine **DINOFLAGELLATES**, the parasitic **APICOMPLEXANS**, and the diverse, highly motile **CILIATES**.
- **STRAMENOPILES** typically have two flagella of unequal length, the longer one bearing rows of tubular hairs. Among the stramenopiles are the unicellular **DIATOMS**; the multicellular **BROWN ALGAE**; and the nonphotosynthetic **OOMYCETES**, many of which are **SAPROBIC**.
- **RHIZARIANS** are unicellular and aquatic. They include the **CERCOZOANS**; the **FORAMINIFERANS**, which secrete shells of calcium carbonate; and the **RADIOLARIANS**, which have thin, stiff pseudopods and glassy endoskeletons.
- The unicellular **EXCAVATES** include parasitic as well as free-living species, some of which have lost typical mitochondria.
- The **AMOEBOZOANS** move by means of lobe-shaped pseudopods and include both unicellular and multicellular members; examples are the **PLASMODIAL SLIME MOLDS** and **CELLULAR SLIME MOLDS**.

FIGURE 19.3

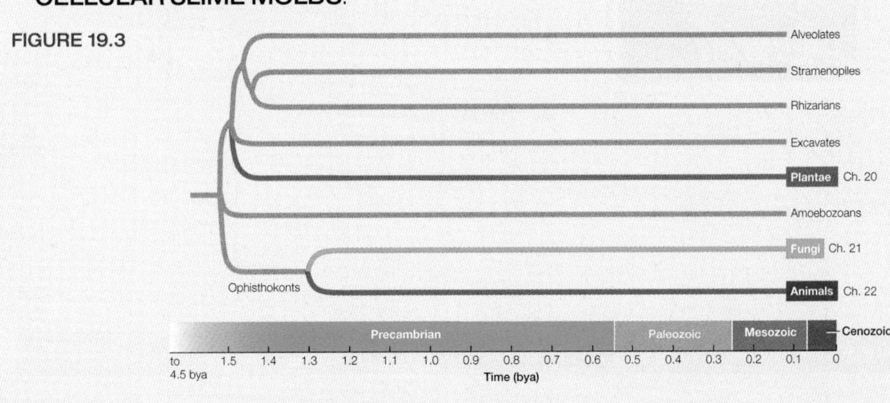

1. Name five of the major groups of eukaryotes in which multicellularity has arisen independently.
2. Which five of the major groups of eukaryotes contain numerous photosynthetic species?

19.3 Protists Reproduce Sexually and Asexually

- Asexual reproduction gives rise to **CLONAL LINEAGES** of organisms.
- **CONJUGATION** in *Paramecium* is a sexual process but not a reproductive one (**FIGURE 19.20**).
- **ALTERNATION OF GENERATIONS**, which involves cycles between a multicellular diploid stage and a multicellular haploid stage, is a feature of many multicellular protist life cycles (as well as those of some fungi and all land plants). The alternating generations may be **HETEROMORPHIC** or **ISOMORPHIC**.

FIGURE 19.20

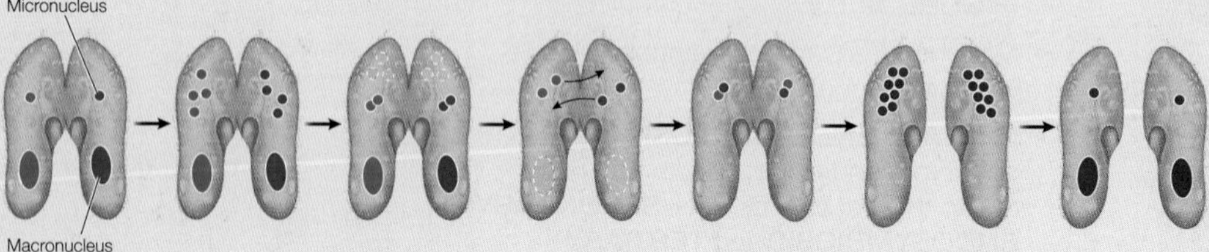

QUESTIONS

1. Explain how some species can recombine their genomes sexually, yet reproduce asexually.
2. Why is animal reproduction, which involves haploid gametes produced by diploid organisms, not considered alternation of generations?

19.4 Protists Are Critical Components of Many Ecosystems

Go to ANIMATION 19.3

- The diatoms are responsible for about one-fifth of the photosynthetic carbon fixation on Earth. They and other members of the phytoplankton are important **PRIMARY PRODUCERS** in the marine environment. Ancient diatoms are a major source of today's petroleum and natural gas deposits.
- Some protists are pathogens of humans and other vertebrates (**FIGURE 19.21**).
- Endosymbiotic relationships are common among microbial protists and often benefit both the endosymbionts and their protist or animal partners.

FIGURE 19.21

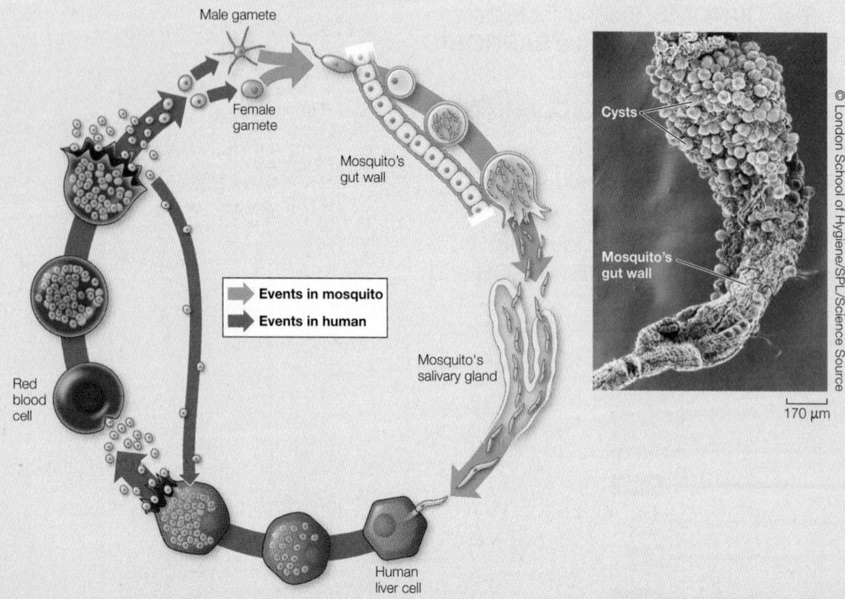

© London School of Hygiene/SPL/Science Source

1. Describe two distinct problems for human populations that are caused by protists.
2. What are two important benefits to humans from protists?

Go to **Achieve** for the e-book, animations, activities, and additional resources and assignments.

The Evolution of Plants 20

The living world depends largely on energy from the Sun, which drives photosynthesis. Using sunlight, plants convert carbon dioxide into carbohydrates and produce atmospheric oxygen as a by-product. Life on Earth as we know it today depends on the carbohydrates and oxygen produced by plants.

What percent of Earth's biomass do plants represent?

You will find the answer to this question on page 493.

KEY CONCEPTS

20.1 Primary Endosymbiosis Produced the First Photosynthetic Eukaryotes

20.2 Key Adaptations Permitted Plants to Colonize Land

20.3 Vascular Tissues Led to Rapid Diversification of Land Plants

20.4 Pollen, Seeds, and Wood Contributed to the Success of Seed Plants

20.5 Flowers and Fruits Led to Increased Diversification of Angiosperms

20.1 Primary Endosymbiosis Produced the First Photosynthetic Eukaryotes

LEARNING OBJECTIVES

By the end of this key concept you should be able to:

20.1.1 **Identify the common evolutionary origin of chloroplasts on the tree of life and explain the meaning of primary endosymbiosis.**

20.1.2 **Describe the ten major clades of Plantae and their evolutionary relationships to one another.**

Photosynthesis is critical to life as we know it. As detailed in Chapter 5, photosynthesis uses energy from sunlight to convert carbon dioxide and water into carbohydrates and oxygen. The reverse process, known as catabolism, provides organisms with the energy they need to function, grow, and reproduce. So almost all living organisms depend on the carbohydrates and oxygen produced by photosynthesis.

More than a billion years ago, when a cyanobacterium was first engulfed by an early eukaryote, the history of life was altered radically. The chloroplasts that resulted from primary endosymbiosis of this cyanobacterium (see Figure 19.2) were obviously important for the evolution of plants and other photosynthetic eukaryotes, but they were also critical to the evolution of all life on land. Until photosynthetic plants were able to move onto land, there was very little on land to support multicellular animals or fungi, and almost all life was restricted to the oceans and fresh waters.

Primary endosymbiosis is a shared derived trait—a synapomorphy—of the group known as **Plantae** (**FIGURE 20.1**). Although *Plantae* is Latin for "plants," in everyday language—and throughout this book—the unmodified common name "plants" is usually used to refer only to the land plants. However, the first several clades to branch off the tree of life after primary endosymbiosis are all aquatic. Most aquatic photosynthetic eukaryotes (other than those secondarily derived from land plants) are known by the common name **algae**. This name, however, is just a convenient way to refer to these groups, which are not all closely related. Many of the photosynthetic groups discussed in Chapter 19 (which acquired chloroplasts through secondary endosymbiosis) are also commonly called algae.

As noted in **Key Concept 14.1**, derived traits shared by a group of organisms that offer evidence of their common ancestry are called synapomorphies.

Several distinct clades of algae were among the first photosynthetic eukaryotes

The ancestor of Plantae was unicellular and may have been similar in general form to the modern **glaucophytes** (**FIGURE 20.2A**). These microscopic freshwater algae are thought to be the sister group to the rest of Plantae (see Figure 20.1A). The chloroplast of glaucophytes is unique in containing a small amount of peptidoglycan between its inner and outer membranes—the same arrangement found in cyanobacteria. Peptidoglycan has been lost from the remaining photosynthetic eukaryotes.

In contrast to the glaucophytes, almost all **red algae** are multicellular (**FIGURE 20.2B**). Their characteristic color is a result of the accessory photosynthetic pigment **phycoerythrin**, which is found in relatively large amounts in the chloroplasts of many red algae. In addition to phycoerythrin, red algal chloroplasts contain chlorophyll *a* as well as several other accessory pigments.

The red algae include species that grow in the shallowest tidal pools as well as the photosynthesizers found deepest in the ocean (as deep as 260 meters if nutrient conditions are right and the water is clear enough to permit light to penetrate). A few red algae inhabit fresh water. Most grow attached to a substrate by a structure known as a holdfast.

Despite their name, red algae don't always appear red in color. The ratio of two pigments—phycoerythrin (red) and chlorophyll *a* (green)—depends largely on the intensity of light that reaches the alga. In deep water, where light is dim, algae accumulate large amounts of phycoerythrin (which absorbs light at short wavelengths) and appear red. But many species growing near the surface contain a higher concentration of chlorophyll *a* and thus appear bright green.

The remaining algal groups in Plantae are the various "green algae." Like land plants, the green algae contain both chlorophylls *a* and *b* and store their reserve of photosynthetic products as starch in chloroplasts. All the groups that share these features are commonly called **green plants** because both of their photosynthetic pigments are green.

Three important clades of green algae are the **chlorophytes**, **coleochaetophytes**, and **stoneworts** (**FIGURE 20.3**). The chlorophytes are the sister group of all the other green plants, which are collectively called **streptophytes**. Among the streptophytes, the coleochaetophytes and stoneworts retain their eggs in the parental organism, as do land plants. They also share other cellular features with the land plants (see Figure 20.1A). Of these two groups, the stoneworts are thought to be the closest relatives of the land plants, based largely on similarities of their genes. The stoneworts also exhibit the branched apical growth that is typical of many land plants (see Figure 20.3C).

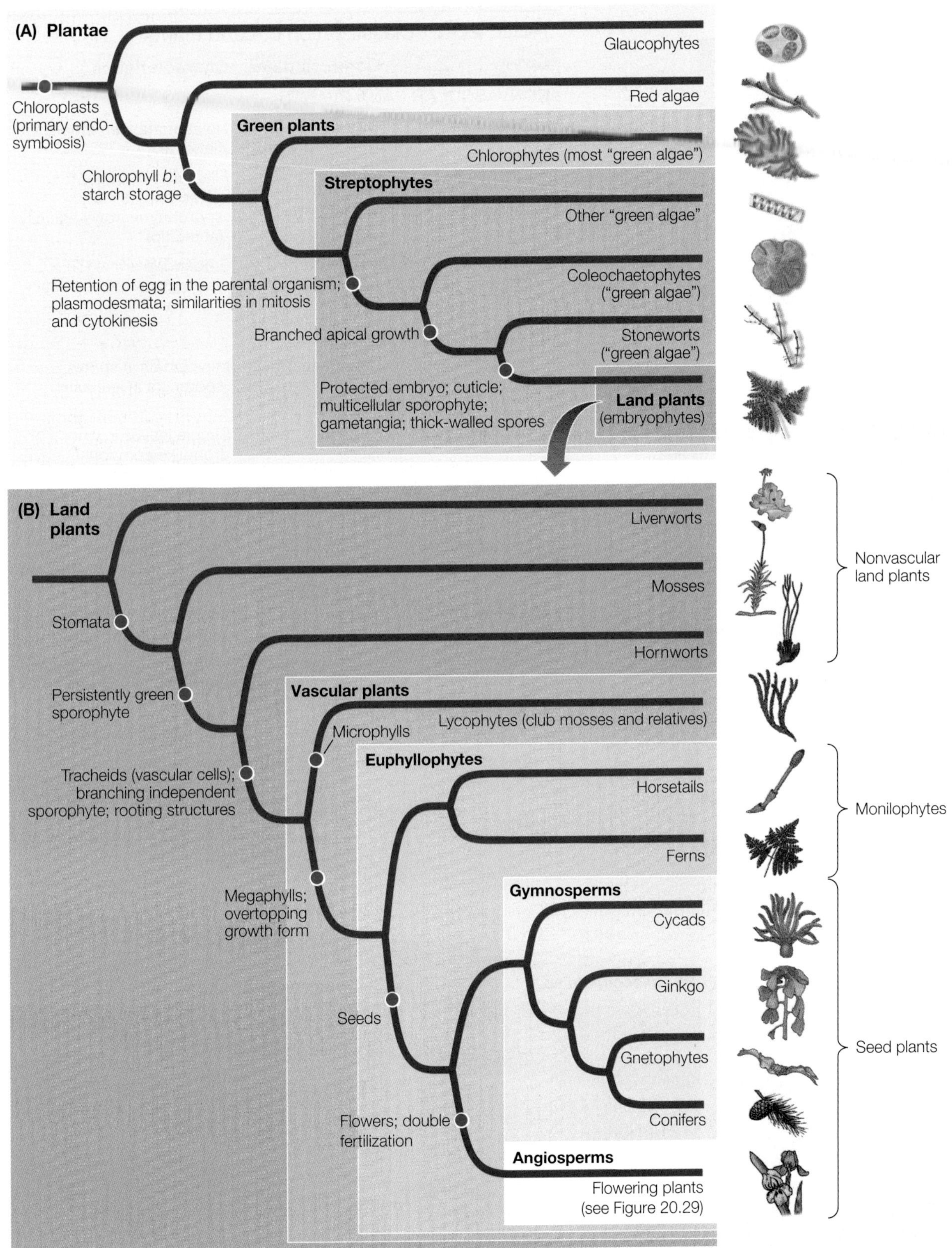

FIGURE 20.1 The Evolution of Plants **(A)** In its broadest definition, the term "plant" includes the glaucophytes, red algae, and green plants—all the groups descended from a common ancestor with chloroplasts. **(B)** Some biologists restrict the term "plant" to the green plants (those with chlorophyll *b*) or, even more narrowly, to the land plants. Three key characteristics that emerged during the evolution of land plants—protected embryos, vascular tissues, and seeds—led to their success in the terrestrial environment. (See Analyze the Data 20.1 in Achieve.)

Which two names of plant groups are used in this figure only as terms of convenience, and do not refer to clades?

(A) ***Glaucocystis* sp.**

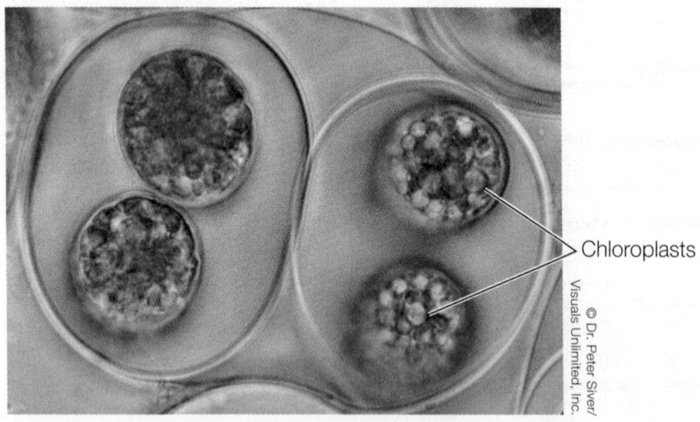

(B) ***Mastocarpus papillatus***

FIGURE 20.2 Early-Branching Groups of Plantae
(A) The chloroplasts of unicellular glaucophytes differ from chloroplasts of other Plantae in retaining a layer of peptidoglycan. This feature is thought to have been retained from the endosymbiotic cyanobacteria that gave rise to the chloroplasts of Plantae. This photograph shows a colony of four individuals, each with many chloroplasts. **(B)** The photosynthetic pigment phycoerythrin gives this red alga its rich hue, showing up vividly against a background of sand.

TABLE **20.1** | Classification of Land Plants

Group	Common name	Characteristics
NONVASCULAR LAND PLANTS		
Hepatophyta	Liverworts	No stomata; gametophyte flat or leafy
Bryophyta	Mosses	Filamentous stage; gametophyte leafy; sporophyte grows apically (at the tip)
Anthocerophyta	Hornworts	Embedded archegonia; sporophyte grows basally (i.e., from the ground)
VASCULAR PLANTS		
Lycopodiophyta	Lycophytes: club mosses and allies	Microphylls in spirals; sporangia in leaf axils
Monilophyta	Horsetails, ferns	Simple leaves in whorls or frondlike compound leaves
Seed plants		
Gymnosperms		
Cycadophyta	Cycads	Compound leaves; swimming sperm; seeds on modified leaves
Ginkgophyta	Ginkgo	Deciduous; fan-shaped leaves; swimming sperm
Gnetophyta	Gnetophytes	Vessels in vascular tissue; opposite, simple leaves
Coniferophyta	Conifers	Seeds in cones; needle- or scalelike leaves
Angiosperms	Flowering plants	Endosperm; carpels; gametophytes much reduced; seeds contained within fruits

(A) ***Ulva lactuca***

(B) ***Coleochaete* sp.**

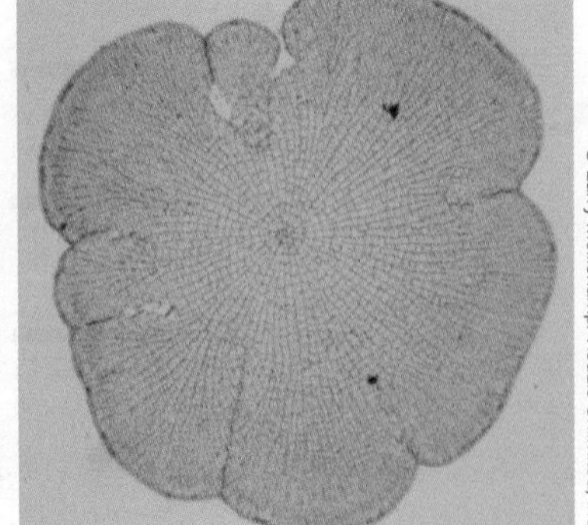

(C) ***Chara vulgaris***

FIGURE 20.3 "Green Algae" Consist of Several Distantly Related Groups **(A)** Sea lettuce, a chlorophyte, grows in ocean tidewaters. **(B)** The coleochaetophytes are thought to be the sister group of stoneworts plus land plants. **(C)** The land plants probably evolved from a common ancestor shared with stoneworts, which display the branching pattern we otherwise associate with land plants. A stonewort in the common freshwater genus *Chara* is shown here. The orange structures are male sex organs.

Media Clip 20.1 Reproductive Structure of Stoneworts

There are ten major groups of land plants

One of the key synapomorphies of the **land plants** is development from an embryo that is protected by tissues of the parent plant. For this reason, land plants are sometimes called **embryophytes** (*phyton*, "plant"). The green plants, the streptophytes, and the land plants each have been called "the plant kingdom" by different authorities, and others take an even broader view to include red algae and glaucophytes as "plants." To avoid confusion in this chapter, we will use modifying terms ("land plants" or "green plants," for example) to refer to the various clades of Plantae shown in Figure 20.1.

The land plants that exist today fall naturally into ten major clades (listed by their common names in the center column of **TABLE 20.1**). Members of seven of those clades possess well-developed vascular systems that transport materials throughout the plant body. We call these seven groups, collectively, the **vascular plants**, or **tracheophytes**, because they all possess fluid-conducting cells called **tracheids**. The remaining three clades (liverworts, mosses, and hornworts) lack tracheids and are referred to collectively as **nonvascular land plants**. Note, however, that the three groups of nonvascular land plants do *not* form a clade (unlike the vascular plants, which *are* a clade).

REVIEW & APPLY | 20.1

R Primary endosymbiosis is a synapomorphy of Plantae. The glaucophytes, the sister clade of the other Plantae, are unicellular algae that are similar to some of the earliest photosynthetic eukaryotes. The green plants contain chlorophyll *b* in addition to the chlorophyll *a* found in all Plantae.

A

1. Explain the different possible uses of the term "plant."
2. Why doesn't the term "algae" designate a formal taxonomic group?
3. What are some of the key differences among glaucophytes, red algae, and the various clades of green algae?
4. What evidence supports the phylogenetic relationship between land plants and the various groups of aquatic green algae?

The green algal ancestors of the land plants lived at the margins of ponds or marshes, ringing them with a mat of dense green. It was from such a marginal habitat, which was sometimes wet and sometimes dry, that early plants made the transition onto land.

20.2 | Key Adaptations Permitted Plants to Colonize Land

LEARNING OBJECTIVES

By the end of this key concept you should be able to:

20.2.1 Describe one or more key innovations in land plants that were most important to their transition to a terrestrial environment, and justify your choice(s).

20.2.2 Distinguish alternation of generations in plants from the life cycles of animals.

How did the land plants arise? To address this question, we can compare land plants with their closest relatives among the green algae. The features that differ between the two groups include the adaptations that allowed the first land plants to survive in the terrestrial environment.

Adaptations to life on land distinguish land plants from green algae

Land plants first appeared in the terrestrial environment between 500 and 450 million years ago (mya). How did they survive in an environment that differed so dramatically from the aquatic environment of their ancestors? While the water essential for life is everywhere in the aquatic environment, water is difficult to obtain and retain in the terrestrial environment.

No longer bathed in fluid, organisms on land faced potentially lethal desiccation (drying). Large terrestrial organisms had to develop ways to transport water to body parts distant from the source of the water. And whereas water provides aquatic organisms with support against gravity, a plant living on land must either have some other support system or sprawl unsupported on the ground. A land plant must also use different mechanisms for dispersing its gametes and progeny than its aquatic relatives, which can simply release them into the water.

Survival on land was facilitated by the evolution among plants of numerous adaptations, including:

- The *cuticle*, a coating of waxy lipids that retards water loss
- *Stomata*, small closable openings in leaves and stems that are used to regulate gas exchange and water loss
- *Gametangia*, multicellular organs that enclose plant gametes and prevent them from drying out
- *Embryos*, young plants contained within a protective structure
- Certain *pigments* that afford protection against the mutagenic ultraviolet radiation that bathes the terrestrial environment
- Thick *spore walls* containing a *polymer* that protects the spores from desiccation and resists decay
- A *mutually beneficial association with fungi* (mycorrhizae) that promotes nutrient uptake from the soil

The cuticle may be the most important—and the earliest—of these features. Composed of several unique waxy lipids that coat the leaves and stems of land plants, the cuticle has several functions, the most obvious and important of which is to keep water from evaporating from the plant body.

As ancient plants colonized the land, they not only adapted to the terrestrial environment, they also modified it by contributing to the formation of soil. Acids secreted by plants help break down rock, and the organic compounds produced by the breakdown of dead plants contribute nutrients to the soil. Such effects are repeated today as plants grow in new areas.

Life cycles of land plants feature alternation of generations

A universal feature of the life cycles of land plants is alternation of generations. Recall from Key Concept 19.3 the two hallmarks of alternation of generations:

1. The life cycle includes both a multicellular diploid stage and a multicellular haploid stage.
2. Gametes are produced by mitosis, not by meiosis. Meiosis produces **spores** that develop into multicellular haploid organisms.

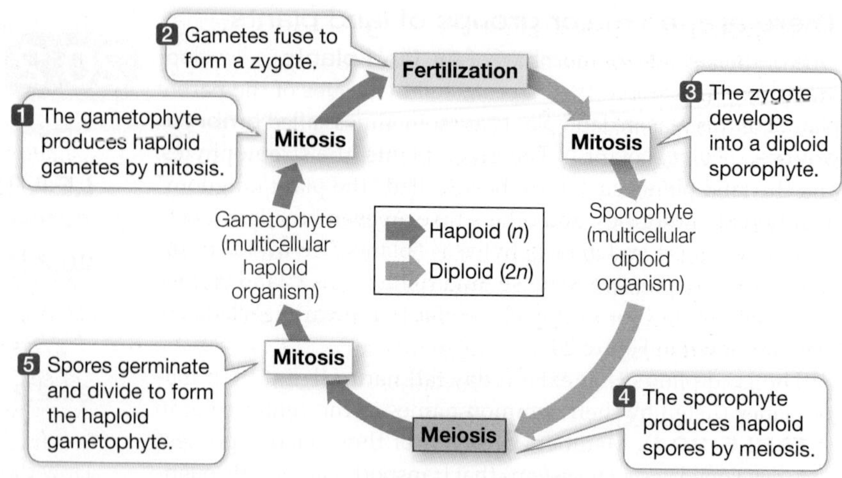

FIGURE 20.4 Alternation of Generations in Land Plants A multicellular diploid sporophyte generation that produces spores by meiosis alternates with a multicellular haploid gametophyte generation that produces gametes by mitosis.

What is a major difference in the haploid portion of the life cycle of plants and animals?

If we begin looking at the land plant life cycle at the single-cell stage—the diploid zygote—then the first phase of the cycle is the formation, by mitosis and cytokinesis, of a multicellular **embryo**, which eventually grows into a mature diploid plant. This multicellular diploid plant is called the **sporophyte** ("spore plant").

Cells contained within specialized reproductive organs of the sporophyte, called **sporangia** (singular *sporangium*), undergo meiosis to produce haploid, unicellular spores. By mitosis and cytokinesis, a spore develops into a haploid plant. This multicellular haploid plant, called the **gametophyte** ("gamete plant"), produces haploid gametes by mitosis. The fusion of two gametes (fertilization) forms a single diploid cell—the zygote—and the cycle is repeated (**FIGURE 20.4**).

The sporophyte generation extends from the zygote through the adult multicellular diploid plant and sporangium formation. In contrast, the gametophyte generation extends from the spore through the adult multicellular haploid plant to the gametes. The transitions between the generations are accomplished by fertilization and by meiosis. In all land plants, the sporophyte and the gametophyte differ genetically: the sporophyte has diploid cells, and the gametophyte has haploid cells.

There is a trend toward reduction of the gametophyte generation in plant evolution. In the nonvascular land plants, the gametophyte is larger, longer-lived, and more self-sufficient than the sporophyte. In those groups that appeared later in plant evolution, however, the sporophyte is the larger, more conspicuous, longer-lived, and more self-sufficient generation.

Nonvascular land plants live where water is readily available

The living species of nonvascular land plants are the liverworts, mosses, and hornworts. These three groups are thought to be similar in many ways to the earliest land plants. Most of these plants grow in dense mats, usually in moist habitats. Even the largest of these species are only about half a meter tall, and most are only a few centimeters tall or long. Why have they not evolved to be taller? The probable answer is that they lack an efficient vascular system for transporting water and minerals from the soil to distant parts of the plant body.

The nonvascular land plants lack the true leaves, stems, and roots that characterize the vascular plants, although they have structures analogous to each. Their growth form allows water to move through the mats of plants by capillary action. They have leaflike structures that readily catch and hold any water that splashes onto them. They are small enough that minerals can be distributed throughout their bodies by diffusion. As in all land plants, layers of maternal tissue protect their embryos from desiccation. Nonvascular land plants also have a cuticle, although it is often very thin (or even absent in some species) and thus is not highly effective in retarding water loss.

Most nonvascular land plants live on the soil or on vascular plants, but some grow on bare rock, on dead and fallen tree trunks, and even on buildings. Their ability to grow on such marginal surfaces results from a mutualistic association with fungi. The earliest association of land plants with fungi dates back at least 460 million years. This mutualism probably facilitated the absorption of water and minerals, especially phosphorus, from the first soils.

Land plants of many groups have mutualistic associations with fungi, as described in **Key Concept 21.2**.

Nonvascular land plants are widely distributed over six continents and even exist (albeit very locally) on the coast of the seventh, Antarctica. Most are terrestrial. Although a few species live in fresh water, these aquatic species are descended from terrestrial ones. None live in the oceans.

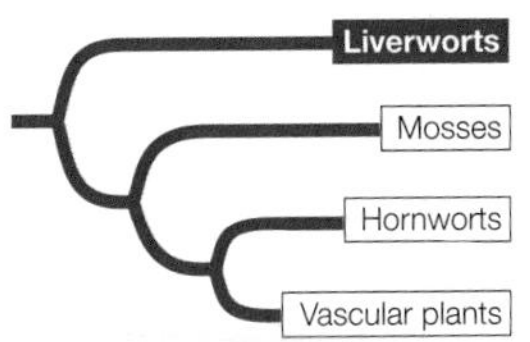

LIVERWORTS There are about 9,000 species of **liverworts**. Most liverworts have green, leaflike gametophytes that lie close to or flat on the ground (**FIGURE 20.5A**). The simplest liverworts are flat plates of cells a centimeter or so long with structures that produce sperm or eggs on their upper surfaces and rootlike filaments on their lower surfaces. The sporophyte remains attached to the larger gametophyte and rarely exceeds a few millimeters in length. Most liverworts can reproduce asexually (through simple division of the gametophyte) as well as sexually.

Media Clip 20.2 **Liverwort Life Cycle**

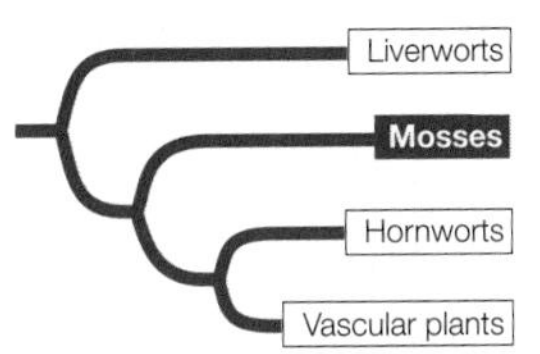

MOSSES The most familiar of the nonvascular land plants are the **mosses** (**FIGURE 20.5B**). These hardy little plants, of which there are about 15,000 species, are found in almost every terrestrial environment. They are often found on damp, cool ground, where they form thick mats.

The mosses are the sister lineage to the vascular plants plus the hornworts (see Figure 20.1B). They share with those lineages an advance over the liverworts in their adaptation to life on land: they have stomata, which are important for both water retention and gas exchange.

Some moss gametophytes are so large that they cannot transport enough water throughout their bodies solely by diffusion. Gametophytes and sporophytes of many mosses contain a type of cell called a hydroid, which dies and leaves a tiny channel through which water can travel. The hydroid is functionally similar to the tracheid, the characteristic water-conducting cell of vascular plants, but it lacks the lignin and the cell-wall structure that characterize tracheids. The possession of hydroids shows that the term "nonvascular land plant" is somewhat misleading when applied to mosses. Despite their simple systems of internal transport, however, the mosses are not considered vascular plants because they lack tracheids and other vascular tissues.

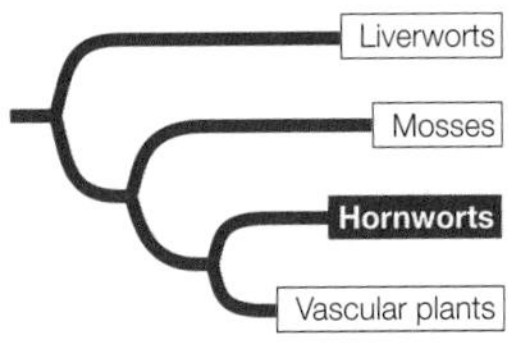

HORNWORTS The approximately 100 species of **hornworts** are so named because their sporophytes often look like little horns (**FIGURE 20.5C**). Hornworts have two characteristics that distinguish them from liverworts and mosses. First, the cells of hornworts each contain a single large, platelike chloroplast, whereas the cells of the other two groups contain numerous small, lens-shaped chloroplasts. Second, of the sporophytes in all three groups, those of the hornworts come closest to being capable of growth without a set limit.

(A) ***Marchantia polymorpha***

(C) ***Anthoceros*** **sp.**

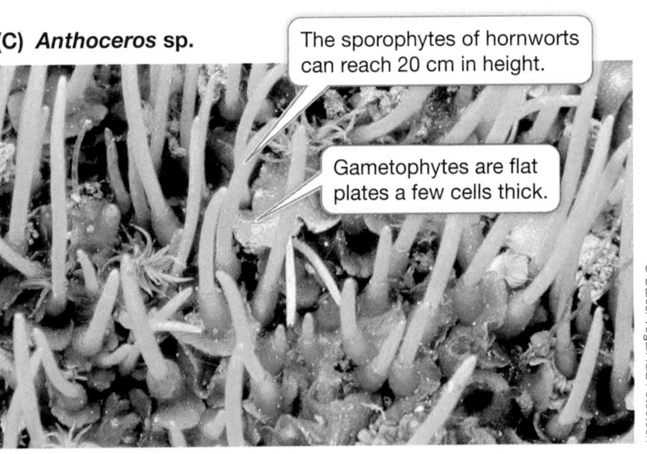

(B) ***Polytrichum commune***

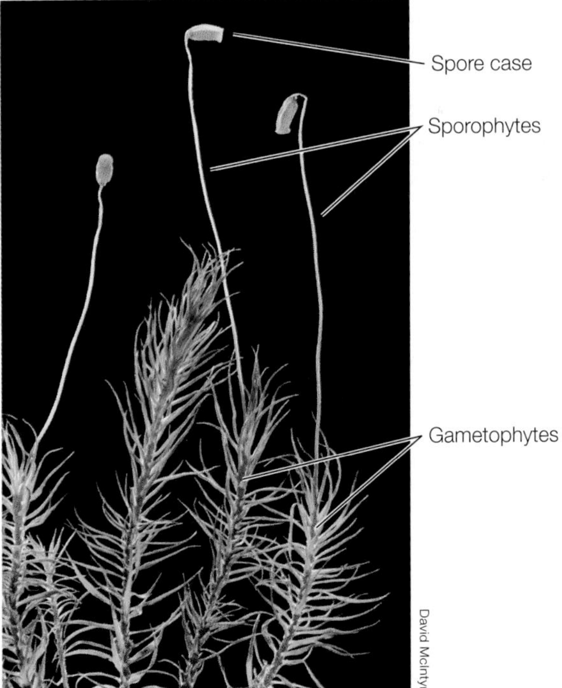

FIGURE 20.5 Diversity among Nonvascular Land Plants **(A)** Gametophytes of a liverwort. **(B)** The sporophytes and gametophytes are easily distinguished in this moss. **(C)** The sporophytes of many hornworts resemble little horns.

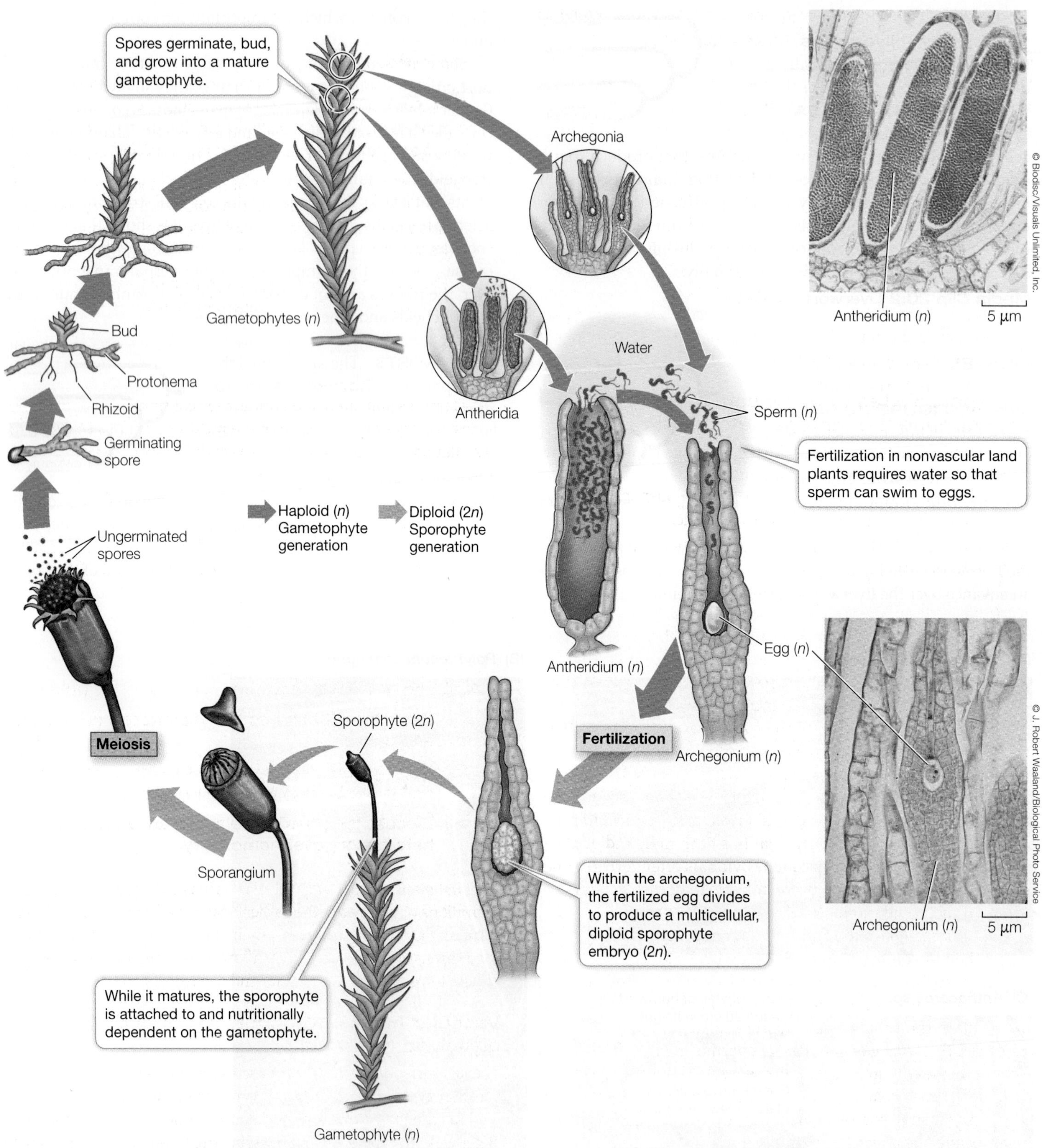

FIGURE 20.6 A Life Cycle Dependent on Water The life cycles of nonvascular land plants, exemplified here by that of a moss, are dependent on an external source of liquid water. The visible green structure of such plants is the gametophyte, which contains haploid archegonia and antheridia. Water carries sperm from the antheridia into an archegonium, inside which an egg is fertilized and grows into a multicellular, diploid sporangium.

Animation 20.1 Life Cycle of a Moss

Media Clip 20.3 Nonvascular Plant Reproduction

Liverwort and moss sporophytes have a stalk that stops growing as the spore-producing structure matures, so elongation of the sporophyte is strictly limited. The hornwort sporophyte, however, has no stalk, and it is persistently green (a trait shared with vascular plants). A basal region of the sporangium remains capable of indefinite cell division, continually producing new spore-bearing tissue above. The sporophytes of some hornworts growing in mild and continuously moist conditions can become as tall as 20 centimeters. Eventually, however, the sporophyte's growth is limited by the lack of a transport system.

Hornworts have evolved a symbiotic relationship that promotes their growth by providing them with greater access to nitrogen, which is often a limiting resource. Hornworts have internal cavities filled with mucilage, and the cavities are often populated by symbiotic cyanobacteria that convert atmospheric nitrogen gas into a form of nitrogen usable by their host plant.

The sporophytes of nonvascular land plants are dependent on the gametophytes

In the nonvascular land plants, the conspicuous green structure visible to the naked eye is the gametophyte. The gametophyte is photosynthetic and is therefore nutritionally independent. The sporophyte may or may not be photosynthetic, but it is always nutritionally dependent on the gametophyte and remains permanently attached to it.

FIGURE 20.6 illustrates a moss life cycle that is typical of the life cycles of nonvascular land plants. A sporophyte produces unicellular haploid spores as products of meiosis within a sporangium. When a spore germinates, it gives rise to a multicellular haploid gametophyte whose cells contain chloroplasts and are thus photosynthetic. Eventually gametes form within specialized sex organs, called the **gametangia**. The **archegonium** is a multicellular, flask-shaped female sex organ that produces a single egg. The **antheridium** is a male sex organ in which sperm, each bearing two flagella, are produced in large numbers. Both archegonia and antheridia are produced on the same individual, so each individual has both male and female reproductive structures. Adjacent individuals often fertilize one another's gametes, however, which helps maintain genetic diversity in the population.

Once released from the antheridium, the sperm must swim or be splashed by raindrops to a nearby archegonium on the same or a neighboring plant—a constraint that reflects the aquatic origins of the nonvascular land plants' ancestors. The sperm are aided on their journey by chemical attractants released by the egg or the archegonium. Before sperm can enter the archegonium, however, certain cells in the neck of the archegonium must break down, leaving a water-filled canal through which the sperm can swim to complete their journey. Notice that *all of these events require liquid water.*

Once sperm arrive at an egg, the nucleus of a sperm fuses with the egg nucleus to form a diploid zygote. Mitotic divisions of the zygote produce a multicellular, diploid sporophyte embryo. The sporophyte matures and produces a sporangium, within which meiotic divisions produce spores and thus the next gametophyte generation.

REVIEW & APPLY | 20.2

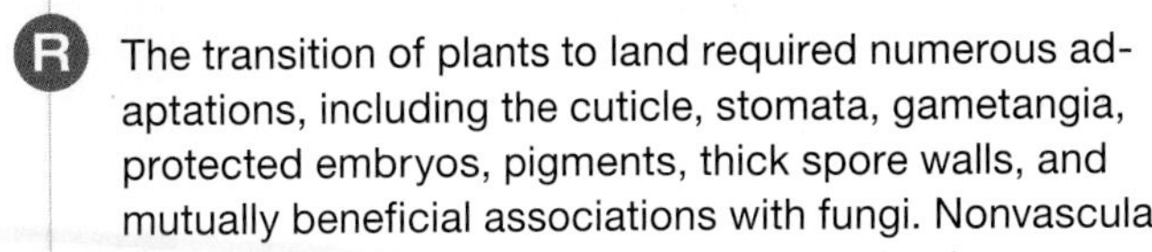

R The transition of plants to land required numerous adaptations, including the cuticle, stomata, gametangia, protected embryos, pigments, thick spore walls, and mutually beneficial associations with fungi. Nonvascular land plants rely on liquid water for reproduction.

A
1. Describe key adaptations of plants to the terrestrial environment, and describe the distribution of those adaptations among the liverworts, mosses, and hornworts.
2. Explain what is meant by "alternation of generations." Why do we not use this phrase for humans?

New features appeared in plants as they continued to adapt to the terrestrial environment. One of the most important of these was vascular tissues, the characteristic that defines the vascular plants.

20.3 Vascular Tissues Led to Rapid Diversification of Land Plants

LEARNING OBJECTIVES

By the end of this key concept you should be able to:

20.3.1 **Detail the differences between tracheids, xylem, and phloem.**

20.3.2 **List five important structures in vascular plants that probably first arose in lycophytes.**

20.3.3 **Discuss the possible selective advantages of heterospory over homospory.**

The first plants possessing vascular tissues did not arise until tens of millions of years after the earliest plants had colonized the land. But once vascular tissues arose, their ability to transport water and food throughout the plant body allowed vascular plants to spread to new terrestrial environments and to diversify rapidly.

Vascular tissues transport water and dissolved materials

The key synapomorphy of the vascular plants is a well-developed vascular system containing two types of tissues that are specialized for the transport of materials from one part of the plant to another. One type of vascular tissue, the **xylem**, conducts water and minerals from the soil to aerial parts of the plant. Because some of its cell walls contain a stiffening substance called lignin, xylem also provides support against gravity in the terrestrial environment. The other type of vascular tissue, the **phloem**, conducts the products of photosynthesis from sites where they are produced or released to sites where they are used or stored.

LINK The vascular tissues of plants are described in detail in **Chapter 23**.

Although the vascular plants are an extraordinarily large and diverse group, a particular event was critical to their evolution. Sometime during the Paleozoic, probably in the mid-Silurian (430 mya), a new cell type—the tracheid—evolved in sporophytes of the earliest vascular plants. The tracheid is the principal water-conducting element of the xylem in all vascular plants except the angiosperms (flowering plants) and one small group of gymnosperms—and tracheids persist even in these groups, along with a more specialized and efficient system derived from them.

The evolution of tracheids set the stage for the complete and permanent invasion of land by plants. First, these cells provided a pathway for the transport of water and mineral nutrients from a source of supply to regions of need in the plant body. And second, the cell walls of tracheids, stiffened by lignin, provided rigid structural support. This support is a crucial factor in a terrestrial environment because it allows plants to grow upward and thus compete for sunlight. A taller plant can intercept more direct sunlight (and thus conduct photosynthesis more readily) than a shorter plant, which may be shaded by the taller one. Increased height also improves the dispersal of spores.

The vascular plants featured another evolutionary novelty: a branching, independent sporophyte. A branching sporophyte body can produce more spores than an unbranched body, and it can develop in complex ways. The sporophyte of a vascular plant is nutritionally independent of the gametophyte at maturity. Among the vascular plants, the sporophyte is the large and obvious plant that one normally pays attention to in nature, in contrast to the relatively small, dependent sporophytes typical of most nonvascular land plants.

FIGURE 20.7 Reconstruction of an Ancient Forest Forests of the Carboniferous were characterized by abundant vascular plants such as club mosses, ferns, and horsetails, some of which reached heights of 40 meters. These forests were the source of modern coal deposits.

Diversification of vascular plants made land more suitable for animals

The initial absence of herbivores (plant-eating animals) on land helped make the first vascular plants successful. By the late Silurian (about 425 mya), the proliferation of land plants made the terrestrial environment more hospitable to animals. Arthropods, vertebrates, and other animals moved onto land only after vascular plants became established there.

Trees of various kinds appeared in the Devonian and dominated the landscape of the Carboniferous (359–299 mya). Forests of lycophytes (club mosses) up to 40 meters tall, along with horsetails and tree ferns, flourished in the tropical swamps of what would become North America and Europe (**FIGURE 20.7**). Plant parts from those forests sank into the swamps and were gradually covered by layers of sediment. Over millions of years, as the buried plant material was subjected to intense pressure and elevated temperatures, it was transformed into coal. Today that coal provides nearly 40 percent of the world's electricity. The world's coal deposits, although huge, are not infinite, and humans are burning coal deposits at a far faster rate than they were produced.

In the subsequent Permian, when the continents came together to form Pangaea, the continental interior became warmer and drier. The 200-million-year reign of the lycophyte–fern forests came to an end as they were replaced by forests of early gymnosperms.

The closest relatives of vascular plants lacked roots

The closest relatives of living vascular plants belonged to several extinct groups called **rhyniophytes**. The rhyniophytes were among a very few types of land plants in the Silurian. The landscape at that time probably consisted mostly of bare ground, with mats of nonvascular plants and stands of rhyniophytes in low-lying moist areas. Early versions of the structural features of the vascular plant groups appeared in the rhyniophytes of that time. These shared features strengthen the case for the origin of all vascular plants from a common nonvascular land plant ancestor.

Rhyniophytes did not have roots. They were apparently anchored in the soil by horizontal portions of stem called **rhizomes**, which bore water-absorbing unicellular filaments called **rhizoids**. These plants also bore aerial branches, and sporangia—homologous to the sporangia of mosses—were found at the tips of those branches. The branching pattern of rhyniophytes was **dichotomous**; that is, the apex (tip) of the shoot divided to produce two equivalent new branches, with each pair of branches diverging at approximately the same angle from the original stem.

Lycophytes are sister to the other vascular plants

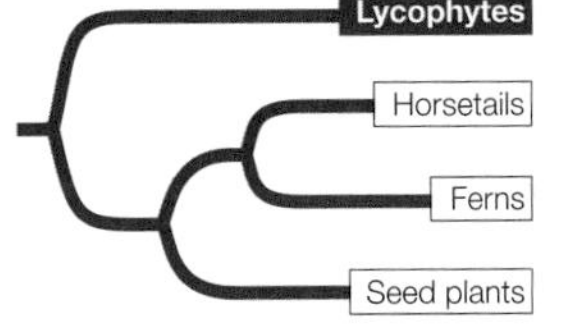

The club mosses and their relatives, the spike mosses and quillworts, are collectively called **lycophytes**. The lycophytes are the sister group to the remaining vascular plants (see Figure 20.1B). There are relatively few (just over 1,200) surviving species of lycophytes.

The lycophytes have true roots that branch dichotomously. The arrangement of vascular tissue in their stems is simpler than that in other vascular plants. They bear simple leaflike structures called **microphylls**, which are arranged spirally on the stem. Growth in lycophytes comes entirely from apical cell division. Branching in the stems, which is also dichotomous, occurs by division of an apical cluster of dividing cells.

The sporangia of many club mosses are aggregated in conelike structures called **strobili** (singular *strobilus*) (**FIGURE 20.8A**), which are clusters of spore-bearing microphylls attached to the end of the stem. Other club mosses lack strobili and bear their sporangia on (or adjacent to) the upper surfaces of specialized microphylls.

Horsetails and ferns constitute a clade

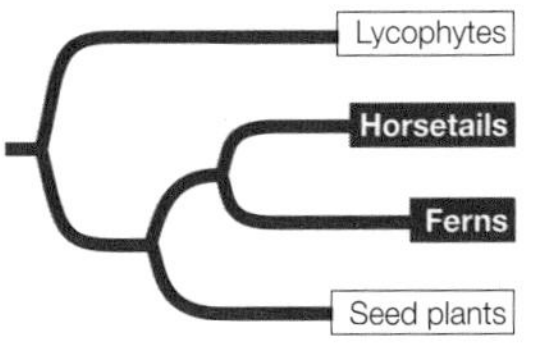

The horsetails and ferns were once thought to be only distantly related. As a result of genomic analyses, we now know that they form a clade: the **monilophytes**. In the monilophytes—as in the seed plants, to which they are the sister group (see Figure 20.1B)—there is differentiation between a main stem and side branches. This pattern contrasts with the dichotomous branching characteristic of the lycophytes and rhyniophytes, in which each split gives rise to two branches of similar size.

Today there are only about 15 species of **horsetails**, all in the genus *Equisetum*. The horsetails have reduced true leaves that form in distinct whorls (circles) around the stem (**FIGURE 20.8B**). Horsetails are sometimes called "scouring rushes" because rough silica deposits found in their cell walls once made them useful for cleaning. They have true roots that branch irregularly. Horsetails have a large sporophyte and a small gametophyte, each independent of the other.

The first **ferns** appeared during the Devonian. Today this group comprises more than 12,000 species. Analyses of gene sequences indicate that a few species traditionally allied with ferns may in fact be more closely related to horsetails than to ferns. Nonetheless, the majority of ferns form a monophyletic group.

Although most ferns are terrestrial, a few species live in shallow fresh water (**FIGURE 20.8C**). Terrestrial ferns are characterized by large leaves with branching vascular strands (**FIGURE 20.8D**). Some fern leaves become climbing organs and may grow to be as long as 30 meters.

(A) ***Lycopodium annotinum***

(B) ***Equisetum pratense***

(C) ***Marsilea mutica***

(D) ***Alsophilia spinulosa***

FIGURE 20.8 Lycophytes and Monilophytes **(A)** A strobilus is visible at the tip of this club moss. Club mosses have microphylls arranged spirally on their stems. **(B)** Horsetails have a distinctive growth pattern in which the stem grows in segments above each whorl of leaves. These are fertile shoots with sporangia-bearing structures at the apex. **(C)** The leaves of a species of water fern. **(D)** A tree fern on a mountain in India.

In the alternating generations of a fern, the gametophyte is small, delicate, and short-lived, but the sporophyte can be very large and can sometimes survive for hundreds of years (**FIGURE 20.9**). Ferns require liquid water for the transport of the male gametes to the female gametes, so most ferns inhabit shaded, moist woodlands and swamps. The sporangia of ferns typically are borne on a stalk in clusters called **sori** (singular *sorus*). The sori are found on the undersurfaces of the leaves, sometimes covering the entire undersurface and sometimes located at the edges.

Vascular plants branched out

Several features that were new to the vascular plants evolved in lycophytes and monilophytes. Roots probably had their evolutionary origins as a branch, either of a rhizome or of the aboveground portion of a stem. That branch presumably penetrated the soil and branched further. The underground portion could anchor the plant firmly, and even in this primitive condition it could absorb water and minerals.

The microphylls of lycophytes were probably the first leaflike structures to evolve among the vascular plants. Microphylls are

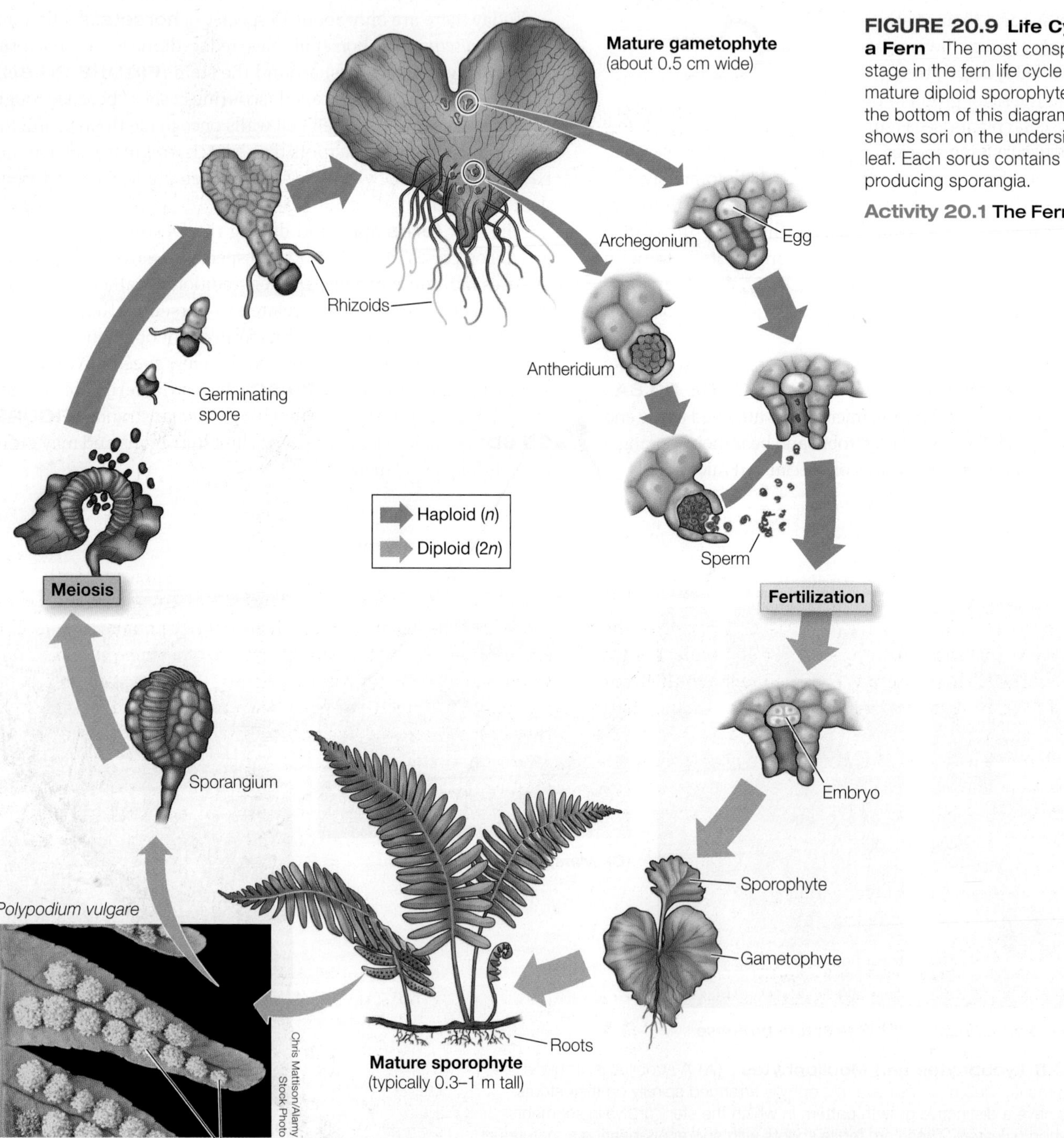

FIGURE 20.9 Life Cycle of a Fern The most conspicuous stage in the fern life cycle is the mature diploid sporophyte, shown at the bottom of this diagram. The inset shows sori on the underside of a fern leaf. Each sorus contains many spore-producing sporangia.

Activity 20.1 The Fern Life Cycle

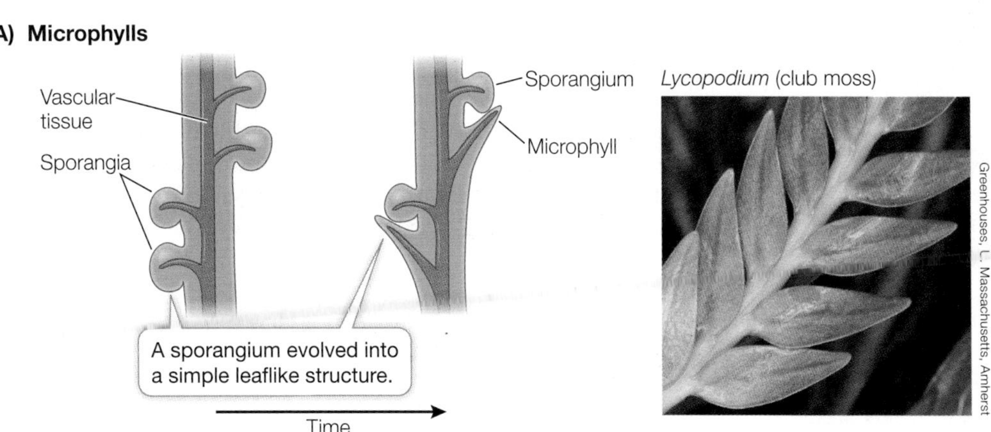

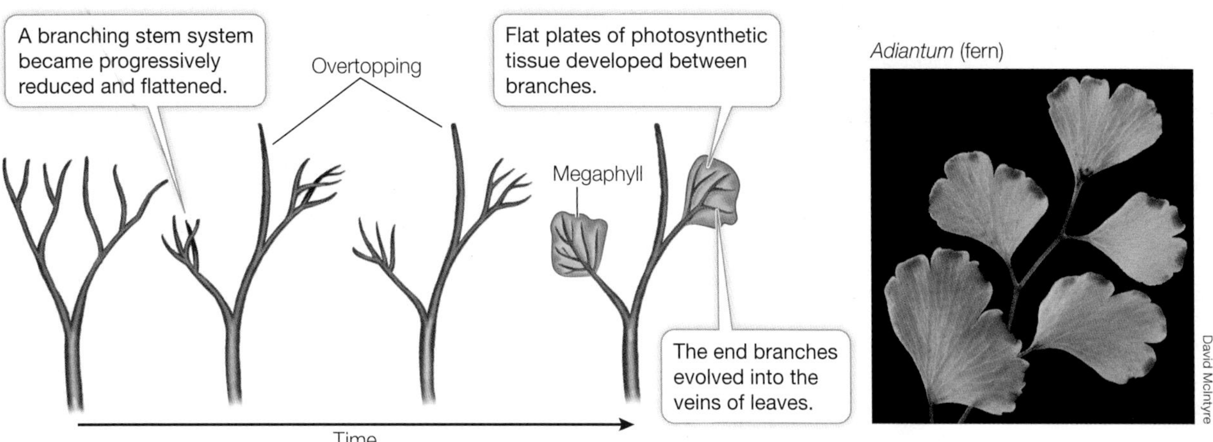

FIGURE 20.10 Evolution of Leaves **(A)** Microphylls are thought to have evolved from sterile sporangia. **(B)** The megaphylls of monilophytes and seed plants may have arisen as photosynthetic tissue developed between branch pairs that were "left behind" as dominant branches overtopped them.

usually small and only rarely have more than a single vascular strand, at least in existing species. Some biologists believe that microphylls had their evolutionary origins as sterile sporangia (**FIGURE 20.10A**). The principal characteristic of a microphyll is a vascular strand that departs from the vascular system of the stem in such a way that the structure of the stem's vascular system is scarcely disturbed. This pattern was evident even in the lycophyte trees of the Carboniferous, many of which had microphylls many centimeters long.

The monilophytes and seed plants constitute a clade called the **euphyllophytes** (*eu*, "true"; *phyllos*, "leaf"). An important synapomorphy of the euphyllophytes is **overtopping**, a growth pattern in which one branch differentiates from and grows beyond the others (**FIGURE 20.10B**). Overtopping would have given these plants an advantage in the competition for light, enabling them to shade their dichotomously branching competitors. The overtopping growth of the euphyllophytes also allowed a new type of leaflike structure to evolve. This larger, more complex leaf is called a **megaphyll**. The megaphyll is thought to have arisen from the flattening of a portion of a branching stem system that exhibited overtopping growth. This change was followed by the development of photosynthetic tissue between the members of overtopped groups of branches, which had the advantage of increasing the photosynthetic surface area of those branches.

Heterospory appeared among the vascular plants

In the lineages of present-day, seedless vascular plants that are most similar to their ancestors, the gametophyte and the sporophyte are independent, and both are usually photosynthetic. The spores produced by the sporophyte are of a single type and develop into a single type of gametophyte that bears both female and male reproductive organs. Such plants, which bear a single type of spore, are said to be **homosporous** (**FIGURE 20.11A**).

A system with two distinct types of spores evolved somewhat later. Plants of this type are said to be **heterosporous** (**FIGURE 20.11B**). In heterospory, one type of spore—the **megaspore**—develops into a specifically female gametophyte (a **megagametophyte**) that produces only eggs. The other type, the **microspore**, is smaller and develops into a male gametophyte (a **microgametophyte**) that produces only sperm. The sporophyte produces megaspores in small numbers in **megasporangia** and microspores in large numbers in **microsporangia**. Heterospory affects not only the spores and the gametophytes, but also the sporophyte plant itself, which must develop two types of sporangia.

The earliest vascular plants were all homosporous, but heterospory evidently evolved several times independently among later groups of vascular plants. The fact that heterospory evolved repeatedly suggests that it affords selective advantages. What advantages does heterospory provide? Heterospory allows the

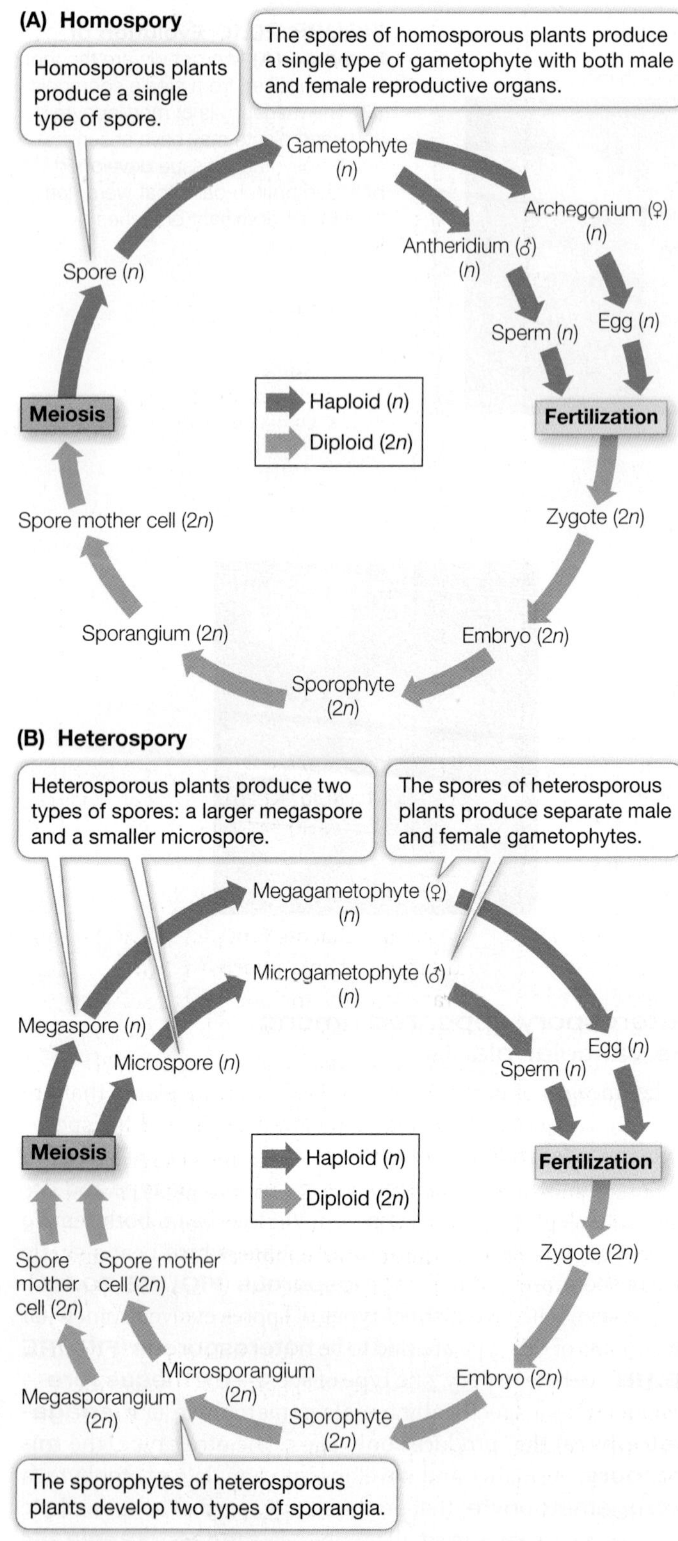

FIGURE 20.11 Homospory and Heterospory
(A) Homosporous plants bear a single type of spore. Each gametophyte has two types of sex organs, antheridia (male) and archegonia (female). **(B)** Heterosporous plants bear two types of spores that develop into distinctly male and female gametophytes.

Activity 20.2 Homospory
Activity 20.3 Heterospory

production of many small microspores, which are easily transported from plant to plant. Heterospory also results in the production of a few large megaspores in large megasporangia, which can provide nutrition and protection for the developing embryo. With this division, plants can increase the opportunities for long-distance cross-fertilization, and at the same time increase the chances for survival of the developing embryo.

REVIEW & APPLY | 20.3

R Vascular plants are characterized by a vascular system specialized for the transport of materials from one part of the plant to another. A new type of cell, the tracheid, marked the origin of this group. Later evolutionary events included the appearance of roots, leaves, and heterospory.

A
1. How do the vascular tissues xylem and phloem serve the vascular plants?
2. Describe the evolution and distribution of different kinds of leaves and roots among the vascular plants.
3. Explain the concept of heterospory. How does heterospory provide selective advantages over homospory?

All of the vascular plant groups we have discussed thus far disperse by means of spores. The embryos of these seedless vascular plants develop directly into sporophytes, which either survive or die, depending on environmental conditions. The spores of some seedless plants may remain dormant and viable for long periods, but the embryos of seedless plants are relatively unprotected (see Figure 20.9). Greater protection of the embryo evolved in the seed plants.

20.4 | Pollen, Seeds, and Wood Contributed to the Success of Seed Plants

LEARNING OBJECTIVES

By the end of this key concept you should be able to:

20.4.1 **Explain how the evolution of pollen enabled seed plants to thrive in terrestrial environments.**

20.4.2 **List three major advantages of having seeds.**

20.4.3 **Describe the earliest group of woody plants and list two advantages of woody plants.**

By the late Devonian, more than 360 million years ago, Earth was home to a great variety of land plants. The land plants shared the hot, humid terrestrial environment with insects, spiders, centipedes, and early tetrapods. These plants and animals evolved together, each acting as agents of natural selection on the other. In the Devonian, a new innovation appeared when some plants

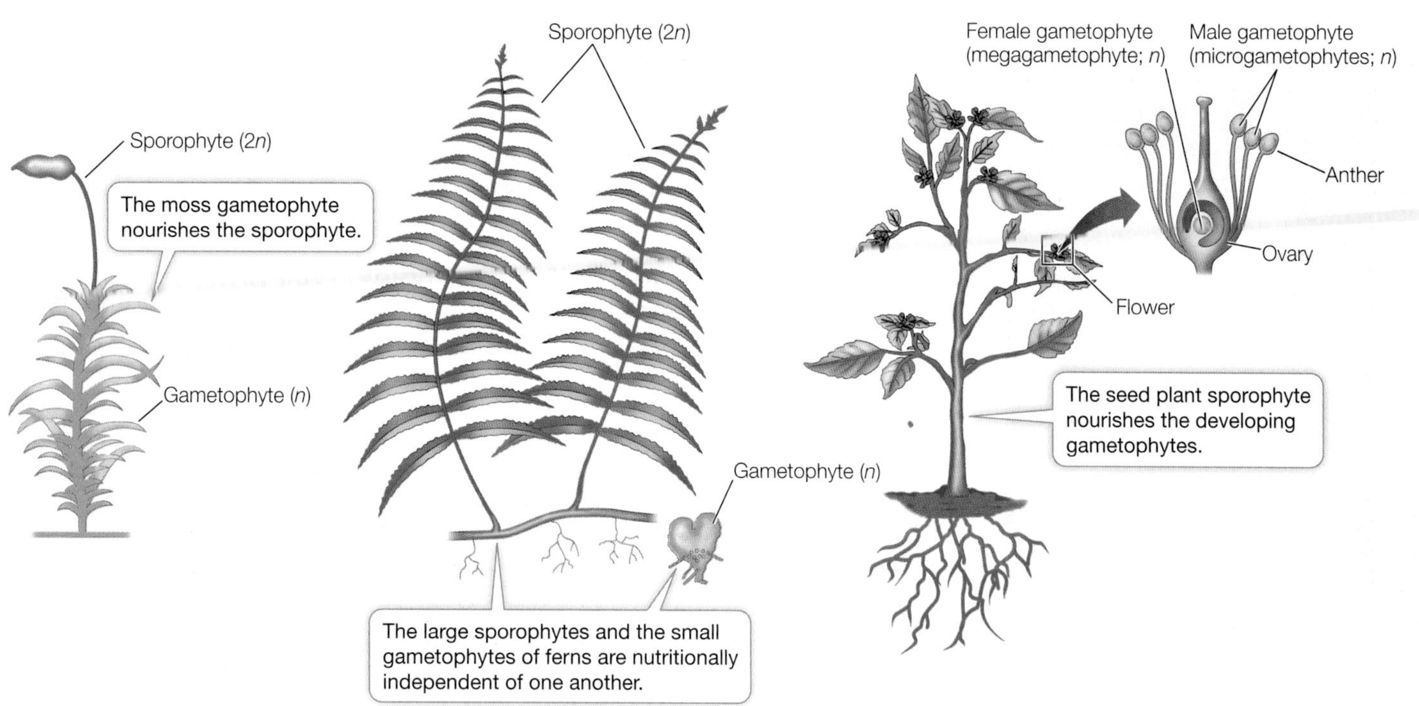

FIGURE 20.12 The Relationship between Sporophyte and Gametophyte In the course of plant evolution, the gametophyte (brown) has been reduced and the sporophyte (blue) has become more prominent.

When you see a growing moss, are you more likely to see the haploid or the diploid multicellular stage? Is your answer the same for a fern and for a flowering plant?

developed extensively thickened woody stems. Two other innovations, **pollen** and **seeds**, arose in the seed plants. Pollen increases reproductive opportunities in a terrestrial environment, and seeds provide a secure and lasting structure that protects the dormant stage of the embryo. A plant embryo may safely wait within its seed (in some cases for many years, or even centuries) until conditions are right for germination.

Features of the seed plant life cycle protect gametes and embryos

In Key Concept 20.2 we described a trend in plant evolution: the sporophyte became less dependent on the gametophyte, which became smaller in relation to the sporophyte. This trend continued with the seed plants, whose gametophyte generation is reduced even further than it is in the ferns (**FIGURE 20.12**). The haploid gametophyte develops partly or entirely while attached to and nutritionally dependent on the diploid sporophyte.

Among the seed plants, only the earliest-diverging groups of gymnosperms (including modern cycads and ginkgos) have swimming sperm. Even in these groups, sperm is transferred via pollen grains, so fertilization does not require liquid water outside the plant body. The evolution of pollen, along with the advent of seeds, gave seed plants the opportunity to colonize drier areas and spread over the terrestrial environment.

Seed plants are heterosporous (see Figure 20.11B)—that is, they produce two types of spores, one that becomes a microgametophyte (male gametophyte) and one that becomes a megagametophyte (female gametophyte). They form separate microsporangia and megasporangia on structures that are grouped on short stems, such as the stamens and carpels of an angiosperm flower.

Within the microsporangium, the meiotic products are microspores. Within its spore wall, a microspore divides mitotically one or a few times to form a multicellular male gametophyte called a **pollen grain**. Pollen grains are released from the microsporangium to be distributed by wind or by an animal pollinator (**FIGURE 20.13**). The spore wall that surrounds the pollen grain contains a substance called sporopollenin, the most chemically resistant biological compound known. Sporopollenin protects the pollen grain against dehydration and chemical damage—another advantage in terms of survival in the terrestrial environment.

In contrast to the microspores, the megaspores of seed plants are not shed. Instead they develop into female gametophytes within the megasporangia. These megagametophytes are dependent on the sporophyte for food and water.

In most seed plant species, only one of the meiotic products in a megasporangium survives. The surviving haploid nucleus divides mitotically, and the resulting cells divide again to produce a multicellular female gametophyte. The megasporangium is surrounded by sterile sporophytic structures, which form an **integument** that protects the megasporangium and its contents. Together, the megasporangium and integument constitute the **ovule**, which will develop into a seed after fertilization.

The arrival of a pollen grain at an appropriate landing point, close to a female gametophyte on a sporophyte of the same species, is called **pollination**. A pollen grain that reaches this point develops further. It produces a slender **pollen tube** that elongates and

FIGURE 20.13 Blown on the Wind Pollen grains are the male gametophytes of seed plants. The pollen of this hazel tree is dispersed by the wind. The grains may land near female gametophytes of the same or other hazel trees.

Media Clip 20.4 Pollen Transfer by Wind

digests its way toward the megagametophyte (**FIGURE 20.14**). When the tip of the pollen tube reaches the megagametophyte, sperm are released from the tube and fertilization occurs.

The resulting diploid zygote divides repeatedly, forming an embryonic sporophyte. After a period of embryonic development, growth is temporarily suspended (the embryo enters a dormant stage). The end product at this stage is the multicellular seed.

The seed is a complex, well-protected package

A seed contains tissues from three generations (**FIGURE 20.15**). A seed coat develops from the integument—the tissues of the diploid sporophyte parent that surround the megasporangium. Within the megasporangium is haploid tissue from the female gametophyte, which contains a supply of nutrients for the developing embryo. (This tissue is fairly extensive in most gymnosperm seeds. In angiosperm seeds it is greatly reduced, and nutrition for the embryo is supplied instead by a tissue called endosperm.) In the center of the seed is the third generation, the embryo of the new diploid sporophyte.

The seed is a well-protected resting stage. The seeds of some species may remain dormant but stay viable (capable of growth and development) for many years, germinating only when conditions are favorable for the growth of the sporophyte. During the

(A) A growing pollen tube

Pollen grains

The pollen tube elongates on its way to the megagametophyte.

© Dr. Jeremy Burgess/Science Source

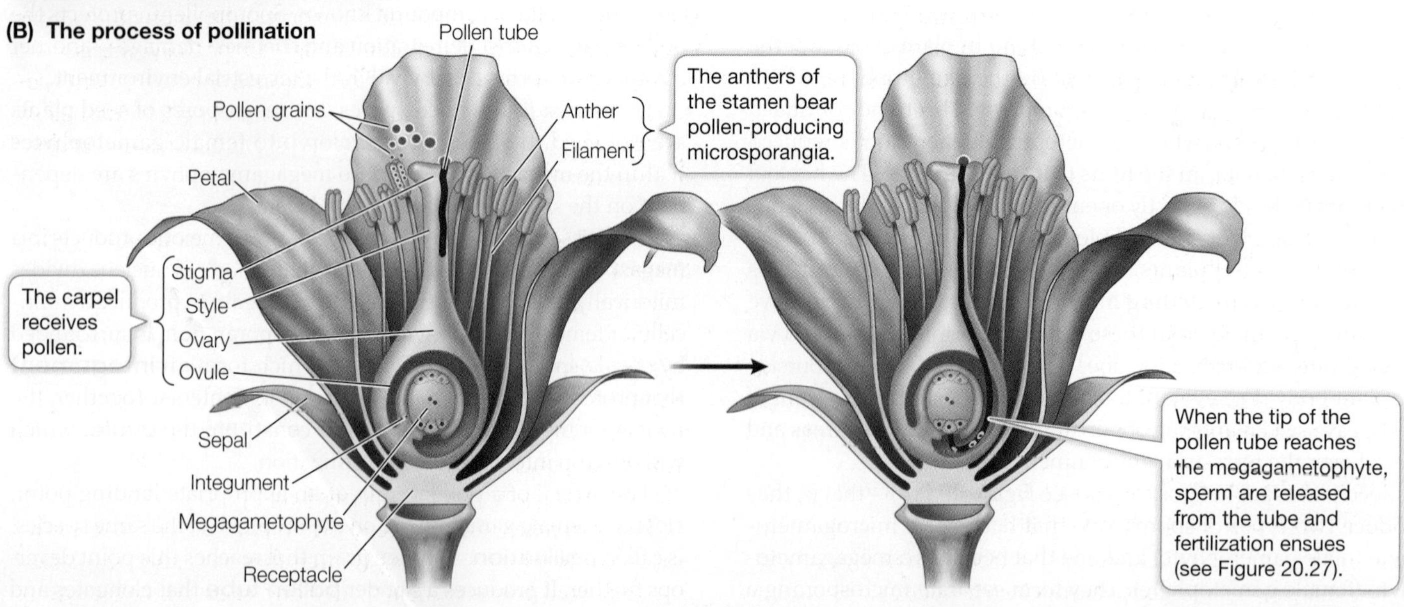

FIGURE 20.14 Pollination Is a Hallmark of the Seed Plants In most seed plants, a pollen tube grows from the pollen grain to the megagametophyte, where sperm are released. **(A)** Scanning electron micrograph of a pollen tube growing in a prairie gentian flower. **(B)** The process of pollination is diagrammed for a generalized angiosperm flower.

Activity 20.4 Flower Morphology

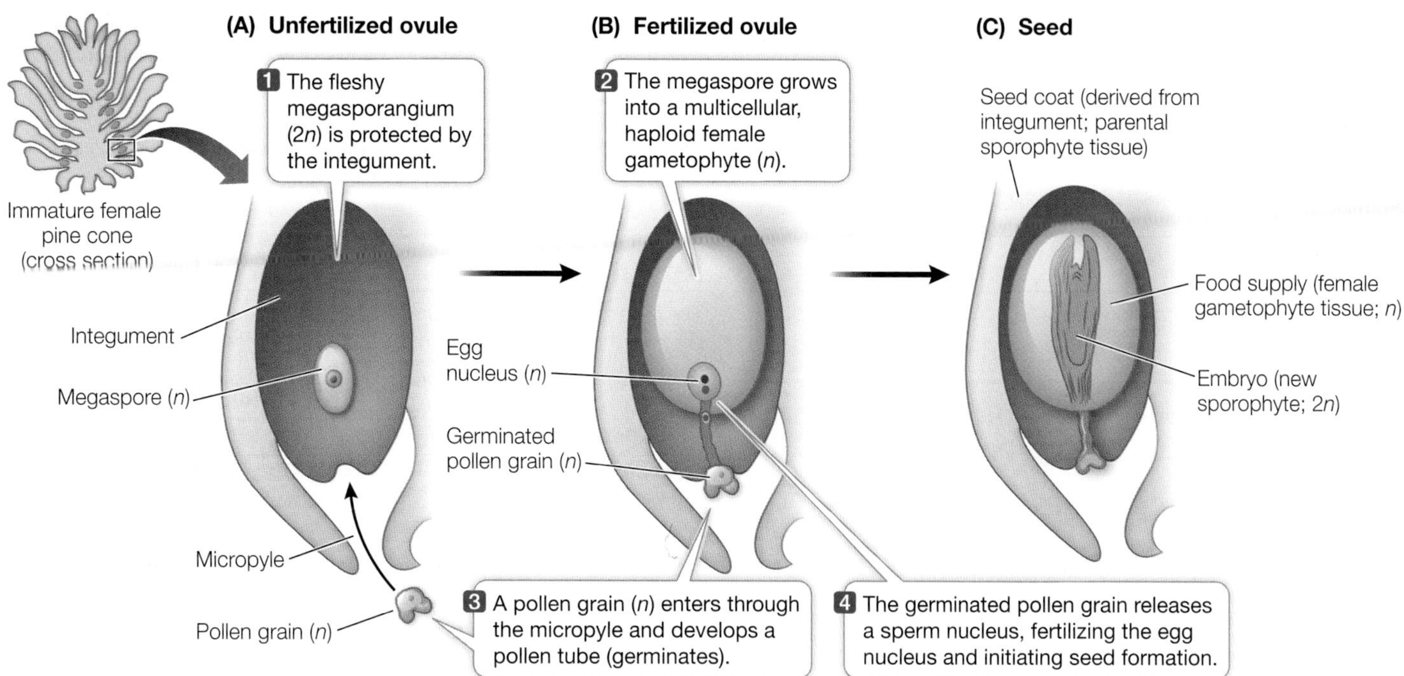

FIGURE 20.15 A Seed Develops These cross sections diagram the development of the ovule into a seed in a gymnosperm (*Pinus* sp.). Angiosperm seed development has differences (e.g., angiosperm integuments have two layers rather than one, and the angiosperm embryo is nourished by specialized tissue called endosperm) but follows the same general steps. **(A)** The haploid megaspore is nourished by tissues of the parental sporophyte (the diploid megasporangium). **(B)** The mature megaspore is fertilized by a pollen grain that penetrates the integument, germinates (grows a pollen tube), and releases a sperm nucleus. **(C)** Fertilization initiates production of a seed. A mature seed contains three generations: a diploid embryo (the new sporophyte), which is surrounded by haploid female gametophyte tissue that supplies nutrition, which is in turn surrounded by the seed coat (diploid parental sporophyte tissue).

dormant stage, the seed coat protects the embryo from excessive drying and may also protect it against potential predators that would otherwise consume the embryo and its nutrient reserves. Many seeds have structural adaptations or are contained in fruits that promote their dispersal by wind or, more often, by animals. When the young sporophyte resumes growth, it draws on the food reserves in the seed. The possession of seeds is a major reason for the enormous evolutionary success of the seed plants, which are the dominant life forms of most modern terrestrial floras.

How do we know how long most seeds remain viable? To find out, William J. Beal, a biologist at Michigan State University, decided to begin an experiment in 1879 that he could not hope to finish in his lifetime (**INVESTIGATION FIGURE 20.16**). He prepared 20 lots of seeds for long-term storage. Each lot consisted of 50 seeds from each of 23 species of plants. He mixed each lot of seeds with sand and placed the mixture in an uncapped bottle, then buried all the bottles upside down (so they would stay dry) on a sandy knoll. The seeds experienced normal temperature fluctuations for Michigan. At regular intervals ever since, other biologists have excavated a bottle and checked the viability of the seeds it contained. The seeds of most species remained viable for decades, whereas others have remained viable for more than a century.

A change in stem anatomy enabled seed plants to grow to great heights

The earliest fossil seed plants have been found in late Devonian rocks. These plants had extensively thickened woody stems, which developed through the proliferation of xylem. This type of growth, which increases the diameter of stems and roots in some modern seed plants, is called **secondary growth**, and its product is called secondary xylem, or wood.

The younger portion of the wood produced by secondary growth is well adapted for water transport, but older wood becomes clogged with resins or other materials. Although no longer functional in transport, the older wood continues to provide support for the plant. This support allows woody plants to grow taller than other plants around them and thus capture more light for photosynthesis.

Not all seed plants are woody. In the course of seed plant evolution, many groups lost the woody growth habit. Nonetheless, other advantageous attributes helped them become established in an astonishing variety of places.

Gymnosperms have naked seeds

Cycads
Ginkgos
Gnetophytes
Conifers
Angiosperms

The two major groups of living seed plants are the **gymnosperms** (such as pines and cycads) and the **angiosperms** (flowering plants; see Figure 20.1B). We'll discuss the flowering plants in Key Concept 20.5 and examine the gymnosperms here.

The gymnosperms are seed plants that do not form flowers or fruits. Gymnosperms (which means "naked-seeded") are so named because their ovules and seeds, unlike those of

INVESTIGATION

FIGURE 20.16 William Beal's Seed Viability Study William Beal began an experiment in 1879 to measure the long-term viability of seeds of several common plants. This ongoing experiment has been continued by biologists for well over a century. For the first 40 years of the experiment, Beal checked seed viability every 5 years. H. T. Darlington took over the experiment in 1915 and extended the sampling period to 10-year intervals beginning in 1920. R. S. Bandurski took over the experiment when Darlington retired, and extended the sampling period to 20 years in 1980, a century after the experiment began.[a]

QUESTION

How long do seeds of common plants remain viable?

METHOD

1. Collect 1,000 seeds of each of 23 species of plants.
2. Divide the seeds into 20 lots of 50 seeds of each of the 23 species.
3. Mix each of the lots of seeds with sand, and place the mixtures in uncapped bottles.
4. Bury the bottles upside down (to prevent the entry of water) on a marked, sandy knoll.
5. At regular intervals, excavate a bottle and check the viability of its contents.

© Kurt Stepnitz/Michigan State University

RESULTS

The table below shows the number of germinating seeds (of the original 50) from three of the species in years 50–100 of this ongoing experiment.

	Years after Burial					
Species	**50**	**60**	**70**	**80**	**90**	**100**
Oenothera biennis (evening primrose)	19	12	7	5	0	0
Rumex crispus (curly dock)	26	2	4	1	0	0
Verbascum blattaria (moth mullein)	31	34	37	35	10	21

Data from F. W. Telewski and J. Zeevaart. 2002. *Am J Bot* 89: 1285–1288.

CONCLUSION

Biologists have used this and similar experiments to estimate the maximum viability of seeds of many species of plants under near-natural environmental conditions.

ANALYZE THE DATA

The data presented in this experiment can be used to make predictions about seed viability. Use the data to answer the following questions.

1. Calculate the percent of viable seeds for these three species in years 50–100 and graph seed survivorship as a function of time buried.
2. No seeds of the first two species were viable after 90 years of the experiment. Assume 100% seed viability at the start of the experiment (year 0), and predict from your graph the approximate year when you think the last *Verbascum blattaria* seeds will germinate.
3. What factors do you think might influence the differences among the species in long-term seed viability?

[a]W. J. Beal. 1884. *Proc Soc Prom Agr Sci* 5: 44–46; H. T. Darlington. 1941. *Am J Bot* 28: 271–273; A. Kivilaan and R. S. Bandurski. 1981. *Am J Bot* 68: 1290–1292.

A similar **Analyze the Data** exercise may be assigned in **Achieve**.

angiosperms, are not protected by ovary or fruit tissue. Although there are probably fewer than 1,200 living species of gymnosperms, these plants are second only to the angiosperms in their dominance of the terrestrial environment. The gymnosperms can be divided into four major groups:

1. **Cycads** are palmlike plants of the tropics and subtropics (**FIGURE 20.17A**). Of the present-day gymnosperms, the cycads are probably the earliest-diverging clade. There are about 300 species, some of which grow as tall as 20 meters. The tissues of many species are highly toxic to humans if ingested.
2. **Ginkgos**, common during the Mesozoic, are represented today by a single genus and species: *Ginkgo biloba*, the maidenhair tree (**FIGURE 20.17B**). There are both male (microsporangiate) and female (megasporangiate) maidenhair trees. The difference is determined by X and Y sex chromosomes, as in humans. Few other plants have distinct sex chromosomes.

(A) *Encephalartos* sp.

(B) *Ginkgo biloba*

(C) *Welwitschia mirabilis*

(D) *Pinus longaeva*

FIGURE 20.17 Diversity among the Gymnosperms **(A)** Many cycads have growth forms that resemble both ferns and palms, although cycads are not closely related to either group. **(B)** The characteristic broad leaves of the maidenhair tree. **(C)** The straplike leaves of *Welwitschia*, a gnetophyte, grow throughout the life of the plant, breaking and splitting as they grow. **(D)** Conifers dominate many types of landscapes in the Northern Hemisphere. Bristlecone pines such as these are the longest-lived individual trees known.

3. **Gnetophytes** number about 90 species in three very different genera, which share certain characteristics analogous to ones found in the angiosperms. One of the gnetophytes is *Welwitschia* (**FIGURE 20.17C**), a long-lived desert plant with just two permanent straplike leaves, which split into many pieces that sprawl across the sand.
4. **Conifers** are by far the most abundant of the gymnosperms. There are about 700 species of these cone-bearing plants, including the pines, spruce, and redwoods (**FIGURE 20.17D**).

With the exception of the gnetophytes, the living gymnosperm groups have only tracheids as water-conducting and support cells in the xylem. Most gymnosperms lack the vessel elements and fibers (cells specialized for water conduction and support, respectively) that are found in angiosperms. While the gymnosperm water-transport and support system may thus seem somewhat less efficient than that of the angiosperms, it serves some of the largest trees known. The coastal redwoods of California are the tallest gymnosperms, with some individuals growing to well over 100 meters tall.

During the Permian, as environments became warmer and dryer, the conifers and cycads flourished. Gymnosperm forests changed over time as the gymnosperm groups evolved. Gymnosperms dominated the Mesozoic, during which the continents drifted apart and large dinosaurs lived. Gymnosperms were the principal trees in all forests until about 65 million years ago, and even today conifers are the dominant trees in many forests, especially at high latitudes and elevations. The oldest living single organism on Earth today is a gymnosperm

(A) Female cones (megastrobili) in *Pinus contorta*

(B) Male cones (microstrobili) in *Pinus contorta*

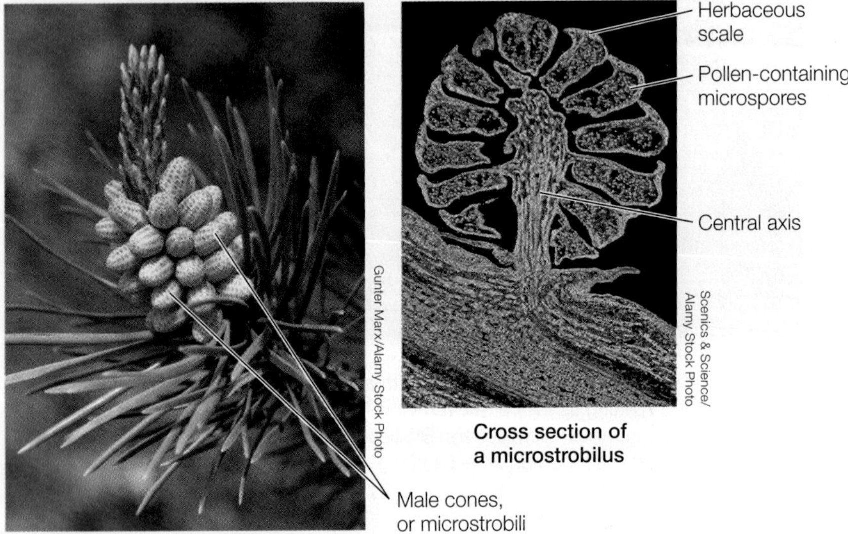

FIGURE 20.18 Female and Male Cones **(A)** The scales of female cones (megastrobili) are modified branches. **(B)** The scales of male cones (microstrobili) are modified leaves.

in California—a bristlecone pine that germinated about 5,000 years ago, at about the time the ancient Egyptians were starting to develop writing.

Conifers have cones and lack swimming sperm

The great Douglas fir and cedar forests found in the northwestern United States and the massive boreal forests of pine, fir, and spruce of the northern regions of Eurasia and North America, as well as on the upper slopes of mountain ranges elsewhere, rank among the great forests of the world. All these trees belong to one group of gymnosperms: the conifers, or cone-bearers.

Male and female **cones** contain the reproductive structures of conifers. The female (seed-bearing) cone is known as a **megastrobilus** (plural *megastrobili*). An example of a familiar megastrobilus is the woody cone of pine trees. The seeds in a megastrobilus are protected by a tight cluster of woody scales, which are modifications of branches extending from a central axis (**FIGURE 20.18A**). The typically much smaller male (pollen-bearing) cone is known as a **microstrobilus**. The microstrobilus is typically herbaceous rather than woody, as its scales are composed of modified leaves, beneath which are the pollen-bearing microsporangia (**FIGURE 20.18B**).

The life cycle of a pine illustrates reproduction in gymnosperms (**FIGURE 20.19**). The production of male gametophytes in the form of pollen grains frees the plant completely from its dependence on liquid water for fertilization. Wind, rather than water, assists conifer pollen grains in their first stage of travel from the microstrobilus to the female gametophyte inside a cone. A pollen tube provides the sperm with the means for the last stage of travel by elongating through maternal sporophytic tissue. When the pollen tube reaches the female gametophyte, it releases two sperm, one of which degenerates after the other unites with an egg. Union of sperm and egg results in a zygote. Mitotic divisions and further development of the zygote result in an embryo.

The megasporangium, in which the female gametophyte will form, is enclosed in a layer of sporophytic tissue—the integument—that will eventually develop into the seed coat that protects the embryo. The integument, the megasporangium inside it, and the tissue attaching it to the maternal sporophyte constitute the ovule. The pollen grain enters through a small opening in the integument at the tip of the ovule, the **micropyle**.

Most conifer ovules (which will develop into seeds after fertilization) are borne exposed on the upper surfaces of the scales of the cone (megastrobilus). The only protection of the ovules comes from the scales, which are tightly pressed against one another within the cone. Some pines, such as the lodgepole pine, have such tightly closed cones that only fire suffices to split them open and release the seeds. These species are said to be fire-adapted, and fire is essential to their reproduction.

About half of all conifer species have soft, fleshy modifications of cones that envelop their seeds. Some of these are fleshy, fruitlike cones, as in junipers. Others are fruitlike extensions of the seeds, as in yews. These tissues, although often mistaken for "berries," are not true fruits. As we will see when we discuss angiosperms, true fruits are a plant's ripened ovaries, which are absent in gymnosperms. Nonetheless, the fleshy tissues that surround many conifer seeds serve a similar purpose as that of the fruits of flowering plants, acting

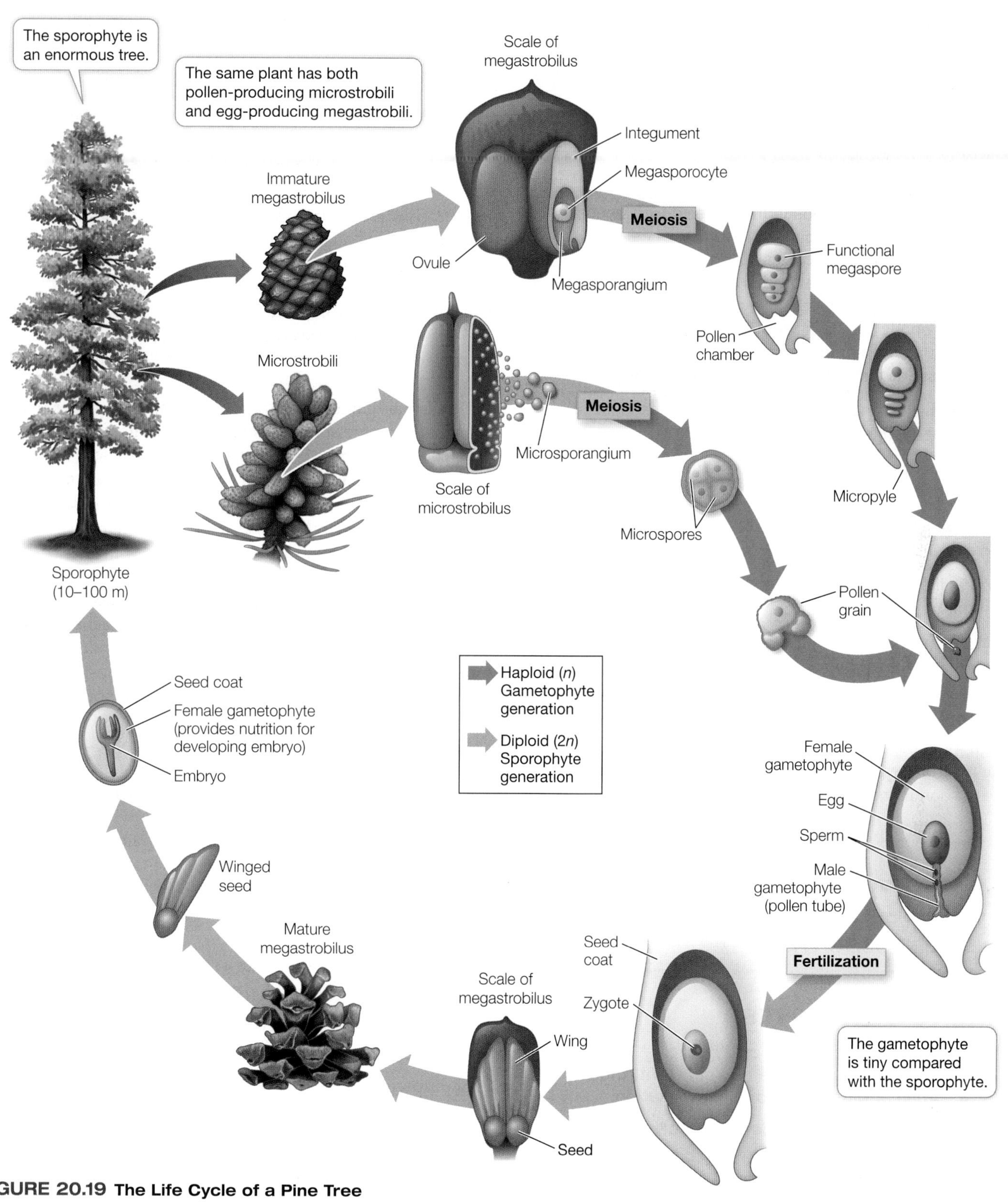

FIGURE 20.19 The Life Cycle of a Pine Tree
In conifers and other gymnosperms, the gametophytes are small and nutritionally dependent on the sporophyte generation.

Animation 20.2 Life Cycle of a Conifer

Activity 20.5 Life Cycle of a Conifer

as an enticement for seed-dispersing animals. Animals eat these fleshy tissues and disperse the seeds in their feces, often depositing the seeds considerable distances from the parent plant.

REVIEW & APPLY | 20.4

Today's living seed plants fall into two major groups, the gymnosperms and the angiosperms. Pollen grains, seeds, and wood are major evolutionary innovations of the seed plants. Protection of embryos is a hallmark of seed plants. Living gymnosperms can be divided into four major groups: cycads, ginkgos, gnetophytes, and conifers. All gymnosperms are woody and have seeds that are not protected by ovaries, although some have fleshy, fruitlike cones or extensions of seeds that entice animals to disperse the seeds.

A
1. Explain the importance of pollen in freeing seed plants from dependence on liquid water.
2. How do seeds and seed dormancy help plants survive in highly seasonal environments?
3. What part did the evolution of wood play in changing the terrestrial landscape?
4. How is fire necessary for the survival of some gymnosperms?
5. What is the function of the fleshy cones that surround the seeds of many gymnosperms?

Although gymnosperms still dominate the terrestrial landscape in some environments, angiosperms (the flowering plants) have become the dominant land plants across much of Earth.

20.5 | **Flowers and Fruits** Led to Increased Diversification of Angiosperms

LEARNING OBJECTIVES

By the end of this key concept you should be able to:

20.5.1 **Distinguish between the two different roles of sperm in angiosperm double fertilization.**

20.5.2 **Explain the difference between pollination and fertilization.**

20.5.3 **Predict the likely modes of seed dispersal based on the type of fruit a plant produces.**

20.5.4 **Outline the functions of flowers, fruits, and seeds.**

The most obvious feature defining the angiosperms is the **flower**, which is their sexual structure. Production of **fruits** is also a shared derived trait of angiosperms. After fertilization, the ovary of a flower (together with the seeds it contains) develops into a fruit that protects the seeds and can promote seed dispersal. As we will see, both flowers and fruits afford major reproductive advantages to angiosperms.

Angiosperms have many shared derived traits

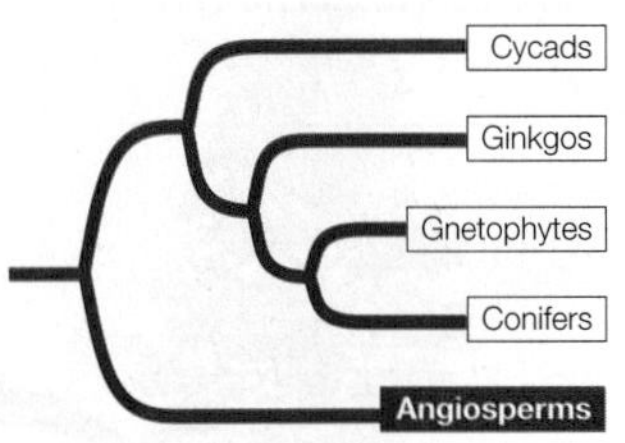

The name *angiosperm* ("enclosed seed") is derived from another distinctive trait of flowering plants that is related to the formation of fruits: the ovules and seeds are enclosed in a modified leaf called a **carpel**. Besides protecting the ovules and seeds, the carpel often interacts with incoming pollen to prevent self-fertilization, thus favoring cross-fertilization and increasing genetic diversity.

The female gametophyte of angiosperms is even more reduced than that of gymnosperms, usually consisting of only seven cells. Thus the angiosperms represent the current extreme of the trend we have traced throughout the evolution of vascular plants: the sporophyte generation becomes larger and more independent of the gametophyte, while the gametophyte becomes smaller and more dependent on the sporophyte.

The xylem of most angiosperms is distinguished by the presence of specialized water-transporting cells called **vessel elements**. These cells are larger in diameter than tracheids and connect with one another without obstruction, allowing easy water movement. A second distinctive cell type in angiosperm xylem is the **fiber**, which plays an important role in supporting the plant body. Angiosperm phloem possesses another unique cell type, called a companion cell. Like the gymnosperms, woody angiosperms show secondary growth, increasing in diameter by producing secondary xylem and secondary phloem.

A more comprehensive list of angiosperm synapomorphies, then, includes the following (some of these traits will be discussed later in this chapter):

- Flowers
- Fruits
- Ovules and seeds enclosed in a carpel
- Highly reduced gametophytes
- Germination of pollen on a stigma
- Double fertilization
- Endosperm (nutritive tissue for the embryo)
- Phloem with companion cells

The majority of these traits bear directly on angiosperm reproduction, which is a large factor in the success of this dominant plant group.

The sexual structures of angiosperms are flowers

Flowers come in an astonishing variety of forms—just think of some of the flowers you recognize. Flowers may be single, or they may be grouped together to form an **inflorescence**. Different

(A) ***Heracleum mantegazzianum***

(B) ***Zinnia elegans***

(C) ***Sorghum halepense***

FIGURE 20.20 Inflorescences **(A)** The inflorescence of Queen Anne's lace, a member of the carrot family, is a compound umbel. Each umbel bears flowers on stalks that arise from a common center. **(B)** Zinnias are members of the aster family; their inflorescence is a head. Within the head, each of the long, petal-like structures is a ray flower; the central portion of the head consists of dozens to hundreds of disc flowers. **(C)** Some grasses, such as quack grass, have inflorescences called spikes, which are composed of many individual flowers, or spikelets.

families of flowering plants have characteristic types of inflorescences, such as the compound umbels of the carrot family (**FIGURE 20.20A**), the heads of the aster family (**FIGURE 20.20B**), and the spikes of many grasses (**FIGURE 20.20C**).

If you examine any familiar flower, you will notice that the outer parts look somewhat like leaves. In fact, all the parts of a flower *are* modified leaves. The diagram in Figure 20.14B represents a generalized flower (for which there is no exact counterpart in nature). The structures bearing microsporangia are called **stamens**. Each stamen is composed of a **filament** bearing an **anther** that contains the pollen-producing microsporangia. The structures bearing megasporangia are called carpels. The swollen base of the carpel, containing one or more ovules (each containing a megasporangium surrounded by two protective integuments), is called the **ovary**. The stalk at the top of the carpel is the **style**, and the terminal surface that receives pollen grains is the **stigma**. Two or more fused carpels, or a single carpel if only one is present, are also called a **pistil**.

In addition, many flowers have specialized sterile (non-spore-bearing) leaves. The inner ones are called **petals** (collectively, the **corolla**) and the outer ones **sepals** (collectively, the calyx). The corolla and calyx can be quite showy and often play roles in attracting animal pollinators to the flower. The calyx more commonly protects the immature flower in bud. From base to apex, these floral organs—sepals, petals, stamens, and carpels—are usually positioned in circular arrangements or whorls and attached to a central stalk.

The generalized flower in Figure 20.14B has functional megasporangia and microsporangia. Such flowers are referred to as **perfect** (or hermaphroditic). Many angiosperms produce two types of flowers, one with only megasporangia and the other with only microsporangia. Consequently, either the stamens or the carpels are nonfunctional or absent in a given flower, and the flower is referred to as **imperfect**.

Species such as corn or birch, in which both megasporangiate (female) and microsporangiate (male) flowers occur on the same plant, are said to be **monoecious** ("one-housed"—but, it must be added, one house with separate rooms). Complete separation of imperfect flowers occurs in some other angiosperm species, such as willows and date palms; in these species, an individual plant produces either flowers with stamens or flowers with carpels, but never both. Such species are said to be **dioecious** ("two-housed").

Flower structure has evolved over time

The flowers of the earliest-diverging clades of angiosperms have a large and variable number of tepals (undifferentiated sepals and petals), carpels, and stamens (**FIGURE 20.21A**). Evolutionary change within the angiosperms has included some striking modifications of this early condition: reductions in the number of each type of floral organ to a fixed number, differentiation of petals from sepals, and changes in symmetry from radial (as in a lily or

(A) ***Nymphaea*** **sp.**

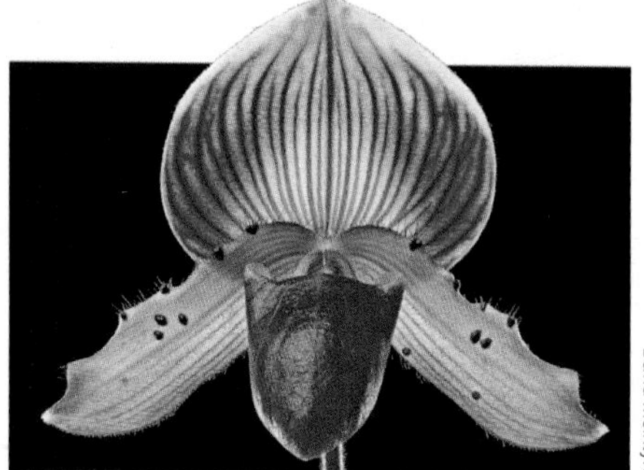

(B) ***Paphiopedilum*** **sp.**

FIGURE 20.21 Flower Form and Evolution **(A)** A water lily shows the major features of early flowers: it is radially symmetrical, and the individual tepals, stamens, and carpels are separate, numerous, and attached at their bases. **(B)** Orchids such as this Venus slipper have a bilaterally symmetrical structure that evolved much later than radial flower symmetry.

magnolia) to bilateral (as in a sweet pea or orchid), often accompanied by an extensive fusion of parts (**FIGURE 20.21B**).

According to one hypothesis, the first carpels to evolve were leaves with marginal sporangia, folded but incompletely closed. Early in angiosperm evolution, the carpels fused with one another, forming a single, multichambered ovary (**FIGURE 20.22A**). In some flowers, the other floral organs are attached at the top of the ovary, rather than at the bottom as in Figure 20.14B. The stamens of the most ancient flowers may have been leaflike (**FIGURE 20.22B**), with little resemblance to the stamens of the generalized flower in Figure 20.14B.

A perfect flower represents a compromise of sorts. On the one hand, by attracting a pollinating bird or insect, the plant is attending to both its female and male functions with a single flower type, whereas plants with imperfect flowers must create that attraction twice—once for each type of flower. On the other hand, the perfect flower can favor self-pollination, which is usually disadvantageous. Another potential problem is that the female and male functions might interfere with each other—for example, the stigma might be so placed as to make it difficult for pollinators to reach the anthers, thus reducing the export of pollen to other flowers.

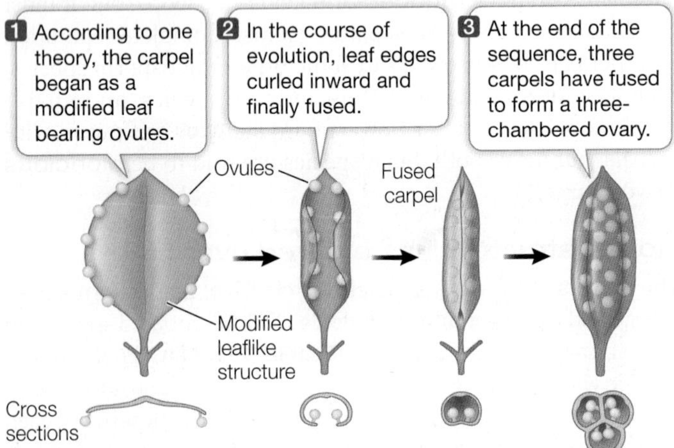

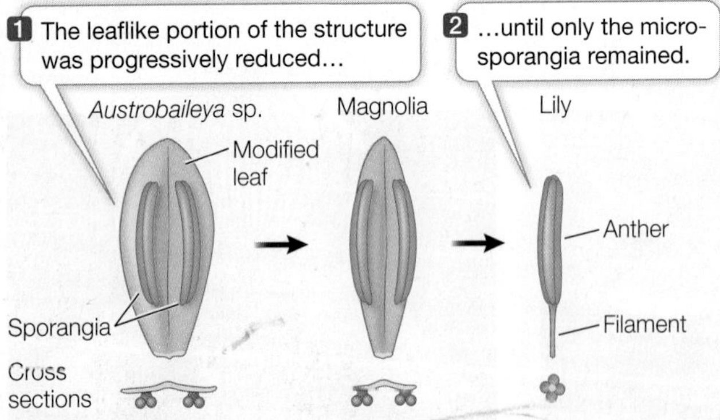

FIGURE 20.22 Carpels and Stamens Evolved from Leaflike Structures **(A)** Possible stages in the evolution of a carpel from a more leaflike structure. **(B)** The stamens of three modern plants show three possible stages in the evolution of that organ. (It is *not* implied that these species evolved from one another; their structures simply illustrate the possible stages.)

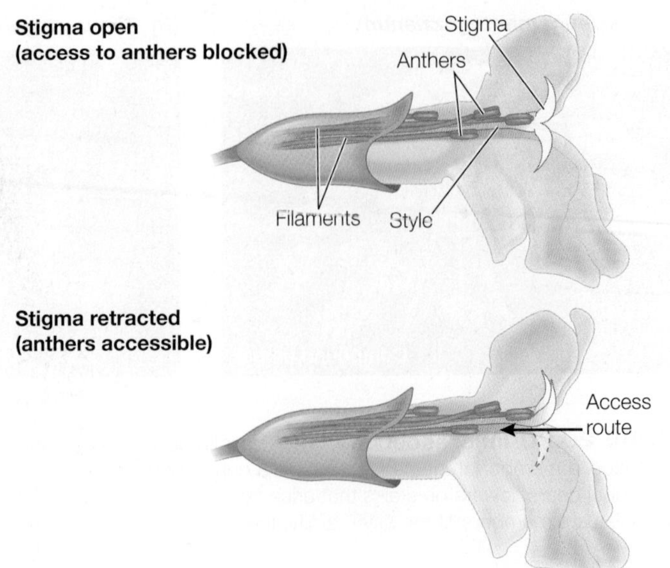

FIGURE 20.23 An Unusual Way to Prevent Self-Fertilization Both long stamens and long styles facilitate cross-pollination, but if these male and female structures are too close to each other, the likelihood of (disadvantageous) self-pollination increases. In *Mimulus aurantiacus*, the stigma is initially open, blocking access to the anthers. A hummingbird's touch as it deposits pollen on the stigma causes one lobe of the stigma to retract, creating a path to the anthers and allowing pollen dispersal. (After A. E. Fetscher. 2001. *Proc R Soc Lond B Biol Sci* 268: 525–529.)

Might there be a way around these problems? One solution is seen in the bush monkeyflower (*Mimulus aurantiacus*), which is pollinated by hummingbirds. Its flower has a stigma that initially serves as a screen, hiding the anthers (**FIGURE 20.23**). Once a hummingbird touches the stigma, one of the stigma's two lobes is retracted, so that subsequent hummingbird visitors pick up pollen from the previously screened anthers. Thus the first bird to visit the flower transfers pollen from another plant to the stigma, eventually leading to fertilization. Later visitors pick up pollen from the now-accessible anthers, fulfilling the flower's male function. **INVESTIGATION FIGURE 20.24** describes the experiment that revealed the function of this mechanism.

Angiosperms have coevolved with animals

Whereas most gymnosperms are pollinated by wind, most angiosperms are pollinated by animals. The many different mutualistic pollination relationships between plants and animals are vital to both parties.

Many flowers entice animals to visit them by providing food rewards. Pollen grains themselves sometimes serve as food for animals. In addition, some flowers produce a sugary fluid called nectar as a pollinator attractant, and some of these flowers have specialized structures to store and distribute it. In the process of visiting flowers to obtain nectar or pollen, animals often carry pollen from one flower to another or from one plant to another. Thus, in their quest for food, the animals contribute to the genetic

INVESTIGATION

FIGURE 20.24 The Effect of Stigma Retraction in Monkeyflowers Elizabeth Fetscher's experiments showed that the unusual stigma retraction response to pollination in monkeyflowers (illustrated in Figure 20.23) enhances the dispersal of pollen to other flowers.[a]

HYPOTHESIS

The stigma-retraction response in *M. aurantiacus* increases the likelihood than an individual flower's pollen will be exported to another flower once pollen from another flower has been deposited on its stigma.

METHOD

1. Set up three groups of monkeyflower arrays. Each array consists of one pollen-donor flower and multiple pollen-recipient flowers (with the anthers removed to prevent pollen donation).
2. In control arrays, the stigma of the pollen donor is allowed to function normally.
3. In one set of experimental arrays, the stigma of the pollen donor is permanently propped open (blocking access to the anthers).
4. In a second set of experimental arrays, the stigma of the pollen donor is artificially sealed closed (allowing access to the anthers).
5. Allow hummingbirds to visit the arrays, then count the pollen grains transferred from each donor flower to the recipient flowers in the same array.

RESULTS

Error bars show 1 standard error of the mean; the three groups are significantly different at $P < 0.05$ (see Appendix B).

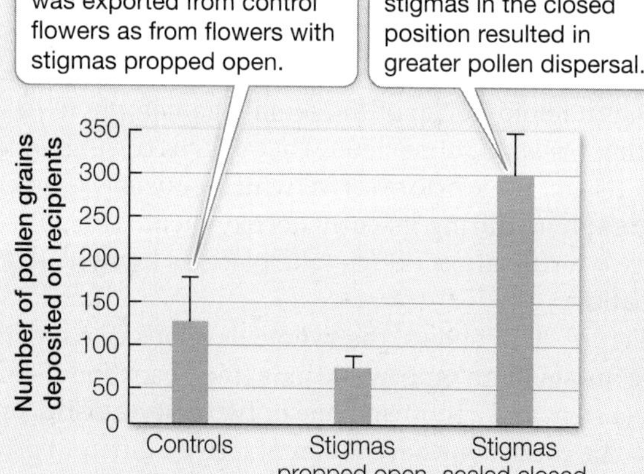

CONCLUSION

The stigma-retraction response enhances the male function of the flower (dispersal of pollen) once the female function (receipt of pollen) has been performed.

[a]A. E. Fetscher. 2001. *Proc R Soc Lond B Biol Sci* 268: 525–529.

diversity of the plant population. Insects, especially bees, are among the most important pollinators. Other major pollinators include some species of birds and bats.

Media Clip 20.5 Pollen Transfer by a Bat

For more than 150 million years, angiosperms and their animal pollinators have coevolved in the terrestrial environment. The animals have affected the evolution of the plants, and the plants have affected the evolution of the animals. Flower structure has become incredibly diverse under these selection pressures. Some of the products of coevolution are highly specific. For example, the flowers of some yucca species are pollinated by only one species of yucca moth, and that moth may exclusively pollinate just one species of yucca. Such specific relationships provide plants with a reliable mechanism for transferring pollen only to members of their own species.

Most plant–pollinator interactions are much less specific. In most cases, many different animal species pollinate the same plant species, and the same animal species pollinates many different plant species. However, even these less specific interactions have developed some specialization. Bird-pollinated flowers are often red and odorless. Many insect-pollinated flowers have characteristic odors, and bee-pollinated flowers may have conspicuous markings, called nectar guides, that are conspicuous only to animals, such as bees, that can see colors in the ultraviolet region of the spectrum (**FIGURE 20.25**).

Taraxacum officinale

FIGURE 20.25 See Like a Bee To normal human vision (top), the petals of a dandelion appear solid yellow. Ultraviolet photography reveals patterns that attract bees to the central region, where pollen and nectar are located (bottom).

THINK LIKE A SCIENTIST

Coevolution of plants and their pollinators

Recent studies show that switches in pollinators and large changes in flower morphology can evolve quickly. The two *Mimulus* species shown in **FIGURE 20.26A** are excellent examples of such changes. Despite being sufficiently closely related that they hybridize readily in the lab, these species differ greatly in morphology and color. They also differ in their pollinators, with *M. cardinalis* being pollinated exclusively by hummingbirds and *M. lewisii* being pollinated primarily by bees and other insects. (Note that other species may visit the flowers, but not pollinate them.) Here you will apply processes of science to determine how genetic variation that affects flower color can evolve in response to pollinators.

1. Given their pollinators, are the floral colors of the two species what you would expect? Explain your answer.
2. What differences are apparent between the flower corollas in the two species? Present a reasonable explanation for these differences, based on the pollinator differences.

Researchers found that variation at a single locus (called *yup*) was responsible for the color differences and that the *M. cardinalis* allele (*C*) is recessive, whereas the *M. lewisii* allele (*L*) is dominant. Thus *CC* genotypes are red, whereas both *LL* and *LC* genotypes are light pink. The researchers then examined the preferences of bees and hummingbirds for the different genotypes at the *yup* locus, keeping the rest of the genetic background of the plant the same. **FIGURE 20.26B** shows their findings; the error bars indicate 95% confidence limits. (Remember, these are visitation rates, not necessarily pollination rates.)

(A)

M. lewisii

M. cardinalis

(B)

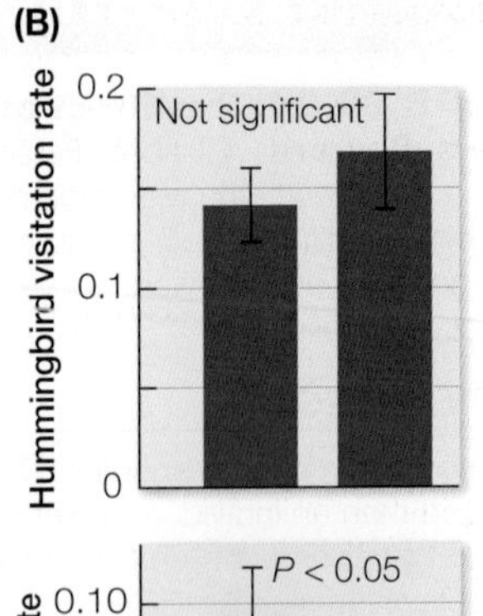

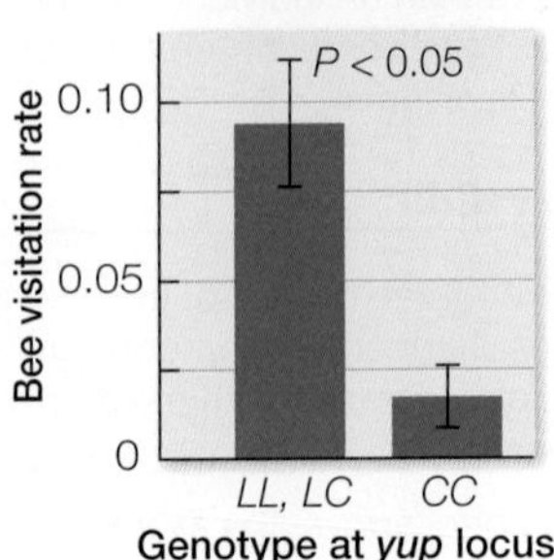

FIGURE 20.26 How Does Pollinator Preference Influence the Evolution of Flowers? (After D. W. Schemske and H. D. Bradshaw, Jr. 1999. *Proc Natl Acad Sci USA* 96: 11910–11915. © 1999 National Academy of Sciences, U.S.A.)

3. Based on these results, how does variation at the *yup* locus affect visitation by bees and hummingbirds?
4. Do the results presented in Question 3 show a significant preference for red or pink flowers by birds and bees, respectively? Given your answer, why do you think most bird-pollinated flowers are red?

The angiosperm life cycle produces diploid zygotes nourished by triploid endosperms

Like all seed plants, angiosperms are heterosporous. As we have seen, their ovules are contained within carpels rather than being exposed on the surfaces of scales, as in most gymnosperms. The male gametophytes, as in the gymnosperms, are pollen grains.

Pollination in the angiosperms consists of the arrival of a microgametophyte—a pollen grain—on a receptive surface in a flower (the stigma). As in the gymnosperms, pollination is the first in a series of events that results in the formation of a seed. The next event is the growth of a pollen tube extending to the megagametophyte. The third event is a fertilization process that, in detail, is unique to the angiosperms (**FIGURE 20.27**).

In nearly all angiosperms, *two* male gametes, contained in a single microgametophyte, participate in fertilization. The nucleus of one sperm combines with that of the egg to produce a diploid zygote, the first cell of the sporophyte generation. In most angiosperms, the other sperm nucleus combines with two other haploid nuclei of the female gametophyte to form a cell with a *triploid* ($3n$) nucleus. That cell, in turn, gives rise to triploid tissue, the **endosperm**, which nourishes the embryonic sporophyte during its early development. This process, in which two fertilization events take place, is known as **double fertilization**.

As Figure 20.27 shows, the zygote develops into an embryo, which consists of an embryonic axis (the "backbone" that will become a stem and a root) and one or two **cotyledons**, or "seed leaves." The cotyledons have different fates in different plants. In many, they serve as absorptive organs that take up and digest the endosperm. In others, they enlarge and become photosynthetic when the seed germinates. Often they play both roles.

The ovule develops into a seed containing the products of the double fertilization that characterizes angiosperms: a diploid zygote and a triploid endosperm (see Figure 20.27). The endosperm serves as storage tissue for starch or lipids, proteins, and other substances that will be needed by the developing embryo.

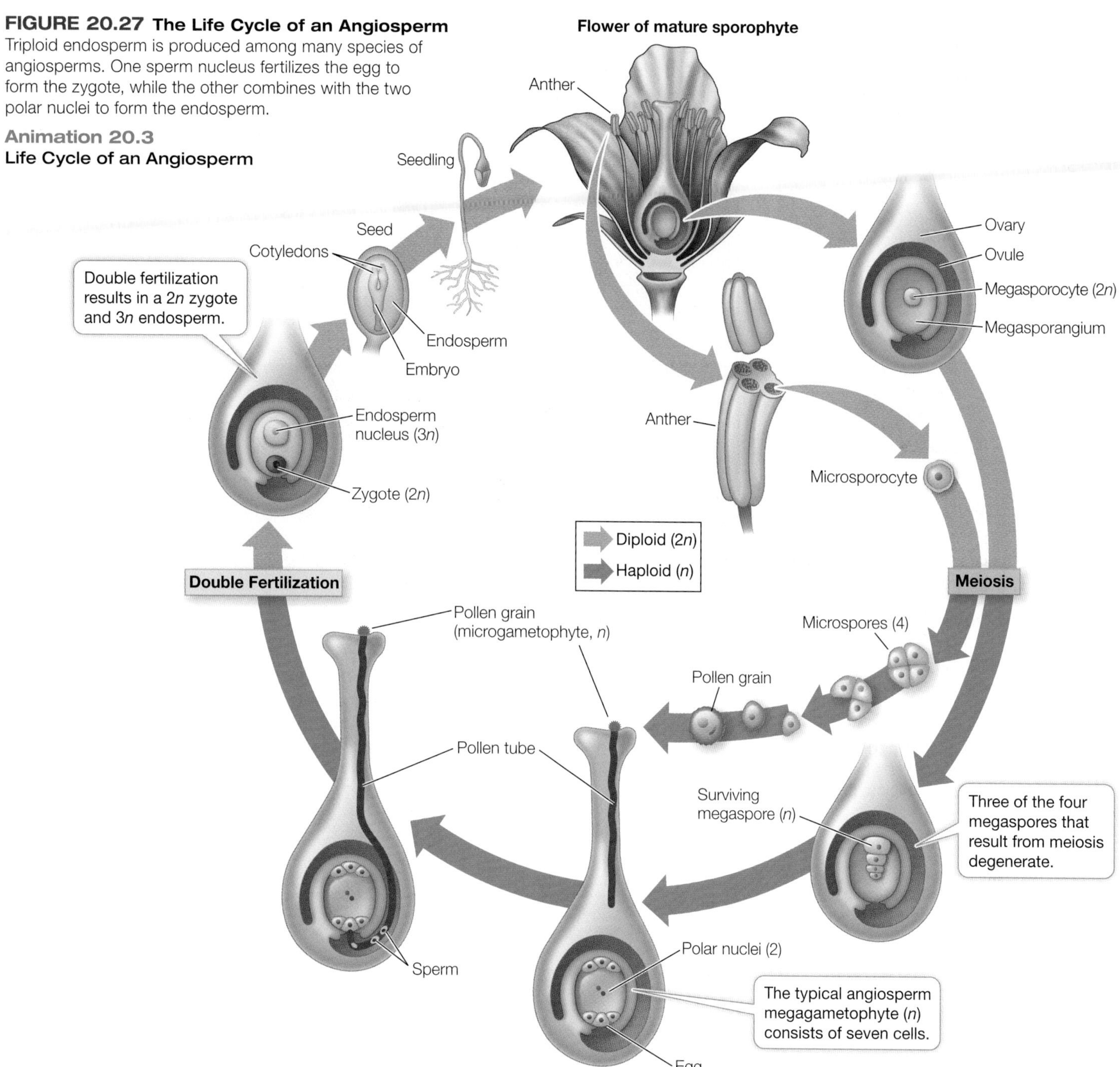

FIGURE 20.27 The Life Cycle of an Angiosperm Triploid endosperm is produced among many species of angiosperms. One sperm nucleus fertilizes the egg to form the zygote, while the other combines with the two polar nuclei to form the endosperm.

Animation 20.3 Life Cycle of an Angiosperm

Fruits aid angiosperm seed dispersal

Fruits typically aid in seed dispersal. Fruits may attach to or be eaten by an animal. The animal is then likely to move, after which the seeds may fall off or be defecated. Fruits are not necessarily fleshy. Fruits can also be hard and woody, or small and have modified structures that allow the seeds to be dispersed by wind or water.

A fruit may consist of only the mature ovary and its seeds, or it may include other parts of the flower or structures associated with it. A **simple fruit** is one that develops from a single carpel or several fused carpels, such as a plum or peach. A raspberry is an example of an **aggregate fruit**—one that develops from several separate carpels of a single flower. Pineapples and figs are examples of **multiple fruits**, formed from a cluster of flowers (an inflorescence). Fruits derived from parts in addition to the carpel and seeds are called **accessory fruits**—examples are apples, pears, and strawberries (**FIGURE 20.28**).

Media Clip 20.6 Flower and Fruit Formation

Recent analyses have revealed the phylogenetic relationships of angiosperms

FIGURE 20.29 shows the relationships among the major angiosperm clades. Recent molecular and morphological analyses have supported the hypothesis that the sister group of remaining

FIGURE 20.28 Fruits Come in Many Forms **(A)** The single seeds inside the simple fruits of cherries are dispersed by animals. **(B)** Each horse chestnut seed is covered by a hard, woody fruit (commonly called a nut) that allows it to survive drought. **(C)** The highly reduced simple fruits of dandelions are dispersed by wind. **(D)** A multiple fruit, the pineapple (*Ananas comosus*), has become one of the most economically significant fruit crops of the tropics. **(E)** An aggregate fruit (blackberry). **(F)** An accessory fruit (pear).

The large incisors of rodents help these animals penetrate the shells of woody nuts, which they consume as food. But unlike many animals that eat fruits without digesting the seeds, rodents destroy the seeds when they eat them. So how do rodents aid in dispersing the seeds of nut-bearing plants?

flowering plants is a single species of the genus *Amborella* (see Figure 20.29A). This woody shrub, with cream-colored flowers, lives only on New Caledonia, an island in the South Pacific. Other early-branching angiosperm groups include the water lilies, star anise and its relatives, and the **magnoliids** (**FIGURE 20.30**). The magnoliids include many familiar and useful plants, such as avocados, cinnamon, black pepper, and magnolias.

The two largest clades—the **monocots** and the **eudicots**—include the great majority of angiosperm species. The monocots are so called because they have a single embryonic cotyledon, whereas the eudicots have two.

Representatives of the two largest angiosperm clades are everywhere. The monocots (**FIGURE 20.31**) include grasses, cattails, lilies, orchids, and

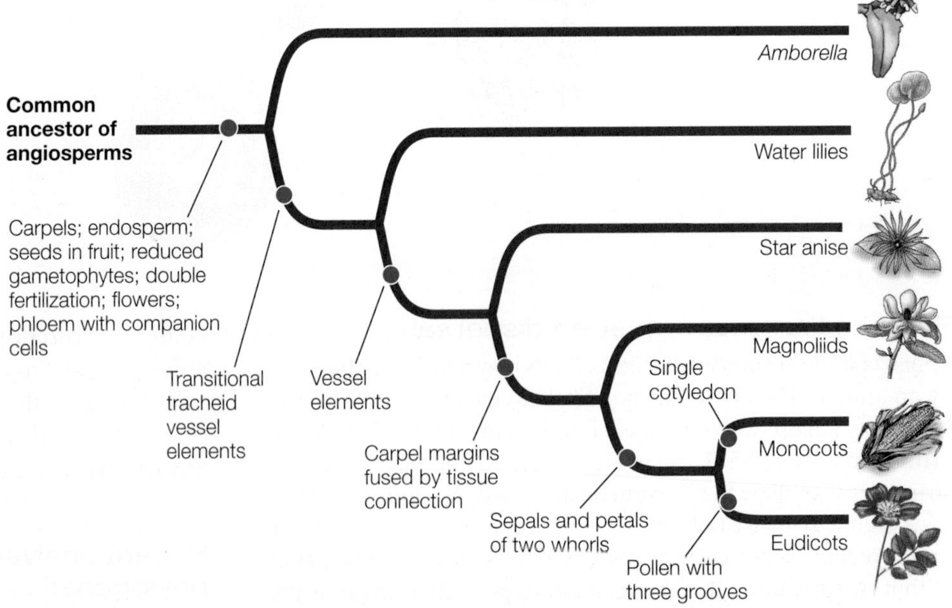

FIGURE 20.29 Evolutionary Relationships among the Angiosperms Recent analyses of many angiosperm genes have clarified the relationships among the major groups.

(A) ***Amborella trichopoda***

David McIntyre, courtesy of the U. Massachusetts Biology Department Greenhouses

(B) ***Nymphaea sp.***

© Yuriy Kulik/Shutterstock.com

© Rob & Ann Simpson/Visuals Unlimited, Inc.

(C) ***Illicium floridanum***

(D) ***Magnolia* sp.**

© iStock.com/Boogich

(E) ***Aristolochia ringens***

© foto76/Shutterstock

FIGURE 20.30 Monocots and Eudicots Are Not the Only Surviving Angiosperms (A) *Amborella*, a shrub, is sister to the remaining extant angiosperms. Notice the sterile stamens on this female flower, which may serve to lure insects that are searching for pollen. **(B)** The water lily clade was the next to diverge after *Amborella*. **(C)** Star anise and its relatives belong to another early-diverging angiosperm clade. **(D, E)** The largest clade other than the monocots and eudicots is the magnoliid complex, which includes magnolias and the group known as "Dutchman's pipe."

(A) ***Tulipa* sp. and *Narcissus* sp.**

Craig Roberts/Alamy Stock Photo

(B) ***Oryza sativa***

Prisma Bildagentur AG/Alamy Stock Photo

(C) ***Posidonia oceanica***

Images & Stories/Alamy Stock Photo

(D) ***Phoenix dactylifera***

blickwinkel/Alamy Stock Photo

FIGURE 20.31 Monocots (A) Monocots include many popular garden flowers such as these tulips (pink and white), and daffodils (yellow). **(B)** Monocot grasses such as rice feed the world; wheat, sugarcane, and corn (maize) are also grasses. **(C)** Seagrasses such as this Neptune's grass form "meadows" in the shallow, sunlit waters of the world's oceans. **(D)** Palms are among the few monocot trees. Date palms such as these are a major food source in some areas of the world.

palms. The eudicots (**FIGURE 20.32**) include the vast majority of familiar seed plants, including most herbs (i.e., nonwoody plants), vines, trees, and shrubs. Among the eudicots are such diverse plants as oaks, willows, beans, snapdragons, roses, and sunflowers.

(A) ***Malus*** **sp.**

© John N. A. Lott/Biological Photo Service

(B) ***Banksia coccinea***

Cygnis insignis

(C) ***Escobaria vivipara***

Courtesy of David Hillis

(D) ***Rafflesia arnoldii***

A & J Visage/Alamy Stock Photo

FIGURE 20.32 Eudicots **(A)** Eudicots include many trees, such as this crabapple tree. **(B)** Scarlet Banksia is a species of an Australian genus of eudicots that attracts a wide diversity of pollinators by producing large quantities of nectar. **(C)** Cacti comprise a large group of eudicots, with about 1,500 species in the Americas. Many, such as this spinystar cactus, bear large flowers for a brief period of each year. **(D)** *Rafflesia arnoldii*, found in the rainforests of Indonesia, bears the largest flower in the world. This plant is a root parasite of tropical vines and has lost its leaf, stem, and even root structures. It smells like decaying meat, which attracts its fly pollinators.

REVIEW & APPLY | 20.5

R Angiosperms are the dominant terrestrial plants of the Cenozoic. The synapomorphies of angiosperms include flowers, fruit, carpels, double fertilization, and endosperm. Most angiosperms also possess distinctive cells in the xylem and phloem. The largest angiosperm clades are the monocots and the eudicots.

A

1. What are the respective roles of the two sperm in double fertilization in angiosperms?
2. Explain the difference between pollination and fertilization.
3. What are some of the differences between fruits that are dispersed by attracting animals to a food resource and those that are dispersed by attaching themselves to animals? Give examples of each.
4. What are the different functions of flowers, fruits, and seeds?

Many fleshy fruits attract animals, which eat the fruit and then disperse the seeds in their feces. If all other factors are equal, large seeds have a better chance of producing a successful seedling than small seeds. So why isn't there selection for larger seeds in the fruits of all plants? In one study in Peru, the feces of the spider monkey *Ateles paniscus* were found to contain seeds from 71 species of plants. After eating fruit, the monkeys usually travel some distance before defecating, thus dispersing any undigested seeds.

If monkey feces are left undisturbed on the forest floor, rodents eat and destroy the vast majority of the seeds in the feces. To germinate successfully, the seeds in spider monkey feces need to be buried by dung beetles, which makes the discovery and destruction of seeds by rodents much less likely.

5. Ellen Andresen hypothesized that dung beetles were more likely to remove larger than smaller seeds from spider monkey dung before burying the dung. She added plastic beads of various diameters to spider monkey dung (to simulate seeds) and measured the percentage of beads buried with the dung by the beetles. Use her data to answer the questions below (E. Andresen. 1999. *Biotropica* 31: 145–158).

Bead diameter (mm)	2	4	6	8	10	12
Percent buried	100	76	52	39	20	4

   a. Plot bead size (the independent variable) versus percentage of beads buried by dung beetles (the dependent variable).
   b. Calculate a regression line for the relationship shown in your graph (see Appendix B). Approximately what percentage of beads with a diameter of 5 mm would you predict would be buried by the beetles? What about beads 14 mm in diameter?
   c. What other factors besides size might influence the probability of seed burial by dung beetles? Can you design an experiment to test your hypotheses?

6. Describe how changes in the population sizes of spider monkeys, rodents, and dung beetles would affect the reproductive success of various plant species.

Once life moved onto land, it was plants that shaped the terrestrial environment. Terrestrial ecosystems could not function without the foods and habitats provided by plants. Plants produce oxygen and remove carbon dioxide from the atmosphere. They play important roles in forming soils and renewing their fertility. Plant roots help hold soil in place, providing protection against erosion by wind and water. Plants also moderate local climates in various ways, such as by increasing humidity, providing shade, and blocking wind. All of these ecosystem services permit a great diversity of fungi and animals to exist on land.

Answer: In 2018, biologists estimated the biomass represented by all the major groups of life on Earth (**FIGURE 20.33**). They estimated gigatons of carbon (Gt C) that are incorporated into each major group. All of life represents about 550 Gt C. Of the total, about 82% of carbon (450 Gt C) is in plants, and most of the biomass of plants is present on land. The uncertainty in the estimate is relatively low for plants (the authors estimate 95% confidence limits are within 1.2-fold of their estimate for plants, or 375–540 Gt C), but the uncertainty of the estimates for the other groups is higher, especially for the prokaryotes. Despite the broad confidence limits in these estimates, it is clear that most life on Earth is green, and that it is this plant life that sustains (directly or indirectly) almost all other life on Earth.

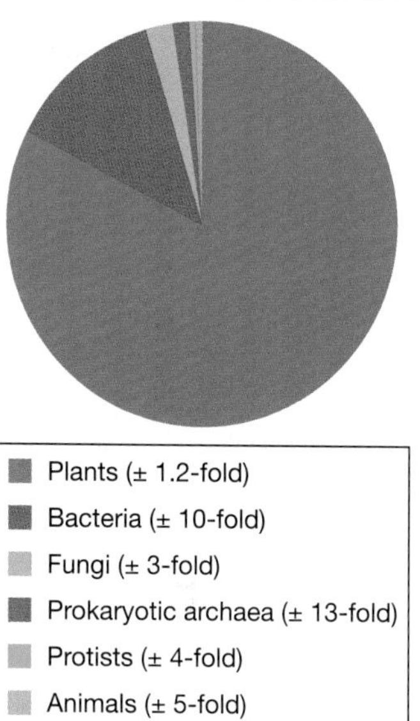

FIGURE 20.33 Most Life on Earth Is Green This pie chart shows the estimated percentages of carbon found in the major groups of living organisms on Earth; 95% confidence limits are shown as *x*-fold of the estimates. (After Y. M. Bar-On et al. 2018. *Proc Natl Acad Sci USA* May 21, 2018. 201711842; epub ahead of print.)

VISUAL SUMMARY 20

20.1 Primary Endosymbiosis Produced the First Photosynthetic Eukaryotes

- Primary endosymbiosis gave rise to chloroplasts and the subsequent diversification of **PLANTAE**. The descendants of the first photosynthetic eukaryote include **GLAUCOPHYTES**, **RED ALGAE**, several groups of green algae, and **LAND PLANTS**, all of which contain chlorophyll *a* (FIGURE 20.1).

FIGURE 20.1

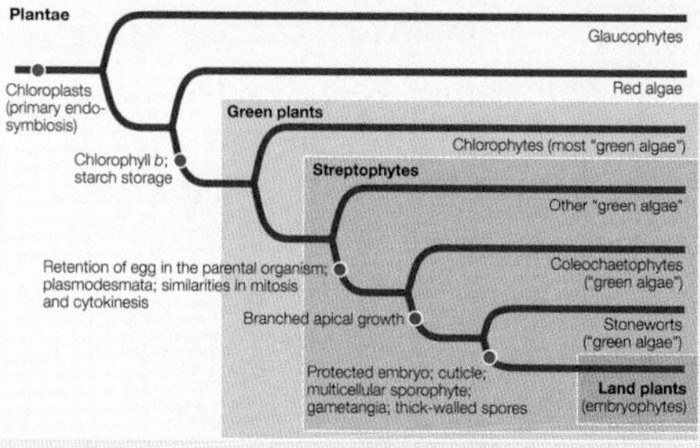

- **STREPTOPHYTES** include the land plants and two groups of green algae. **GREEN PLANTS**, which include the streptophytes and the remaining green algae, are characterized by the presence of chlorophyll *b* (in addition to chlorophyll *a*).
- Land plants, also known as **EMBRYOPHYTES**, arose from an aquatic green algal ancestor related to today's **STONEWORTS**. Land plants develop from embryos that are protected by parental tissue.

QUESTIONS

1. What kind of organism became endosymbiotic in plants and gave rise to chloroplasts?
2. Photosynthesis uses energy to convert carbon dioxide and water into carbohydrates and oxygen. What is the reverse of this process called?

20 VISUAL SUMMARY

20.2 Key Adaptations Permitted Plants to Colonize Land

Go to ANIMATION 20.1

- The acquisition of a cuticle, stomata, **GAMETANGIA**, a protected **EMBRYO**, protective pigments, thick spore walls with a protective polymer, and a mutualistic association with a fungus were all adaptations of land plants to terrestrial life.
- All land plant life cycles feature alternation of generations, in which a multicellular diploid **SPOROPHYTE** alternates with a multicellular haploid **GAMETOPHYTE** (**FIGURES 20.4** and **20.6**).
- The nonvascular land plants comprise the **LIVERWORTS**, **MOSSES**, and **HORNWORTS**. These groups lack specialized vascular tissues for the conduction of water or nutrients through the plant body, and their life cycles depend on liquid water.

FIGURE 20.4

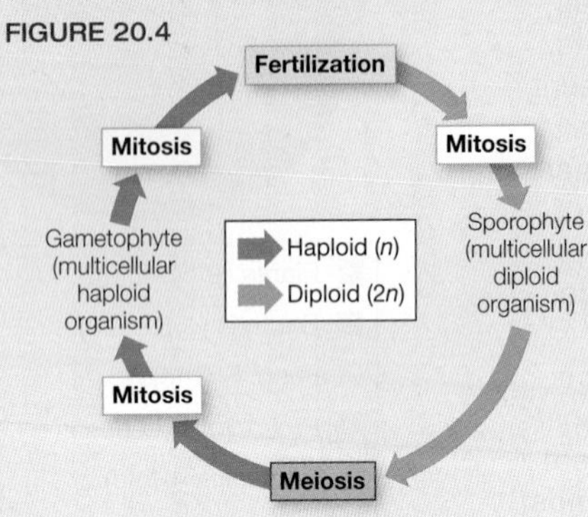

FIGURE 20.6

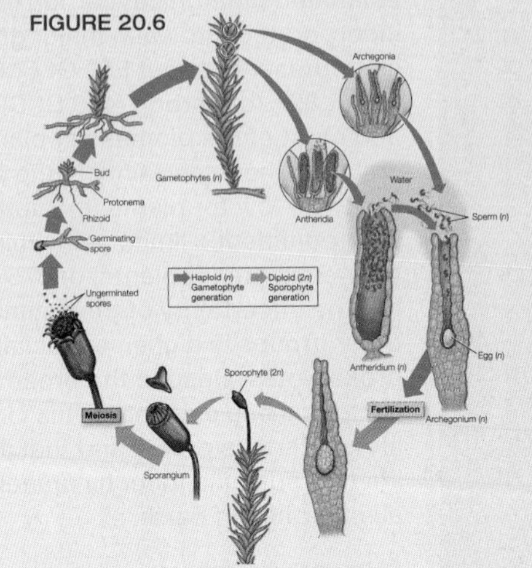

QUESTIONS

1. How does each of the adaptations to terrestrial life listed here convey an advantage to life on land?
2. When you look at a moss, which is most obvious—the haploid gametophyte generation, or the diploid sporophyte generation?
3. What part of the moss is the diploid sporophyte generation?

20.3 Vascular Tissues Led to Rapid Diversification of Land Plants

Go to ACTIVITIES 20.1, 20.2, and 20.3

- The vascular plants have a vascular system consisting of **XYLEM** and **PHLOEM** that conducts water, minerals, and products of photosynthesis through the plant body.
- Among living vascular plant groups, the **LYCOPHYTES** (club mosses and relatives) have only small, simple leaflike structures (**MICROPHYLLS**). True leaves (**MEGAPHYLLS**) are found in **MONILOPHYTES** (which include **HORSETAILS** and **FERNS**). The monilophytes and the seed plants are collectively called **EUPHYLLOPHYTES**.
- Roots may have evolved either from **RHIZOMES** or from stems. Microphylls probably evolved from sterile sporangia, and megaphylls may have resulted from the flattening and reduction of a portion of a stem system with **OVERTOPPING** growth (**FIGURE 20.10**).
- The earliest-diverging groups of vascular plants are **HOMOSPOROUS**, but **HETEROSPORY**—the production of distinct **MEGASPORES** and **MICROSPORES**—has evolved several times (**FIGURE 20.11**).

FIGURE 20.10 FIGURE 20.11

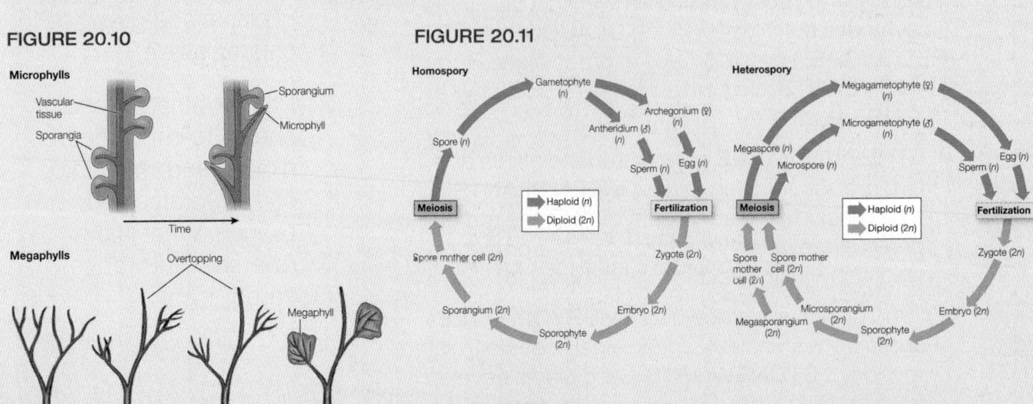

QUESTIONS

1. In the late Paleozoic, forests of giant club mosses, ferns, and horsetails covered large areas of land. What are these groups like today? Why do you think this is the case?
2. What is the advantage of heterospory over homospory that has facilitated heterospory's repeated evolution?
3. What is different about a fern gametophyte, compared with the gametophyte of a seed plant?

20.4 Pollen, Seeds, and Wood Contributed to the Success of Seed Plants

Go to ANIMATION 20.2 and ACTIVITIES 20.4 and 20.5

- Seed plants do not require liquid water for fertilization. **POLLEN**, the microgametophytes of seed plants, is carried to a megagametophyte by wind or by animals. Following **POLLINATION**, a **POLLEN TUBE** emerges from the pollen grain and elongates to deliver gametes to the megagametophyte (**FIGURE 20.14**).
- An **OVULE** consists of the seed plant megagametophyte and the **INTEGUMENT** of sporophytic tissue that protects it. The ovule develops into a **SEED** (**FIGURE 20.15**).
- The two major groups of living seed plants are the **GYMNOSPERMS** and **ANGIOSPERMS**, both of which include many woody species.
- The gymnosperms produce ovules and seeds that are not protected by ovary or fruit tissues. The major gymnosperm groups are the **CYCADS**, **GINKGOS**, **GNETOPHYTES**, and **CONIFERS**.

FIGURE 20.14 FIGURE 20.15

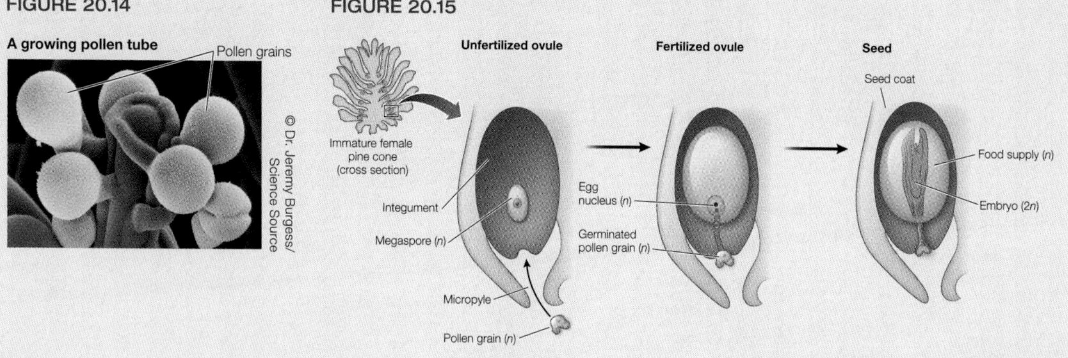

QUESTIONS

1. What is the difference between pollen and sperm?
2. When you look at a seed plant, which is most obvious—the haploid gametophyte generation, or the diploid sporophyte generation?

20.5 Flowers and Fruits Led to Increased Diversification of Angiosperms

Go to ANIMATION 20.3

FIGURE 20.29

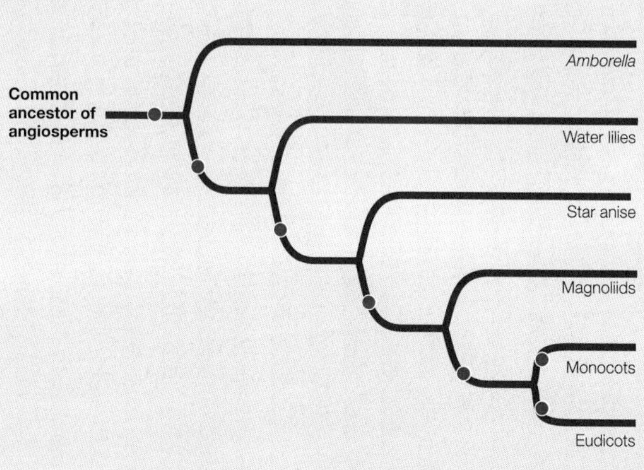

- **FLOWERS** and **FRUITS** are unique to the angiosperms, distinguishing them from the gymnosperms.
- The floral organs, from the base to the apex of the flower, are the **SEPALS**, **PETALS**, **STAMENS**, and one or more **CARPELS**. Stamens bear microsporangia in **ANTHERS**. The carpel includes an **OVARY** containing ovules and a receptive surface called the **STIGMA**.
- Flowers may be pollinated by wind or by animals. Many angiosperms have coevolved with their animal pollinators.
- The **MAGNOLIIDS** are the sister group to the **MONOCOTS** and **EUDICOTS** (the most species-rich angiosperm clades) (**FIGURE 20.29**).

QUESTIONS

1. What are three ways that different flowers attract insect pollinators?
2. Which of the following are true fruits: the red "berries" on a juniper tree; the parachute-like structures on a mature dandelion head; the acorns on an oak tree; a pineapple?

21 The Evolution and Diversity of Fungi

Alexander Fleming accidentally discovered antibiotics while he was studying the *Staphylococcus* bacteria and some of his plates of bacteria became infected with colonies of the fungus *Penicillium*. He noticed that bacterial growth was inhibited around the colonies of fungus, and from them he isolated a substance he would later name penicillin.

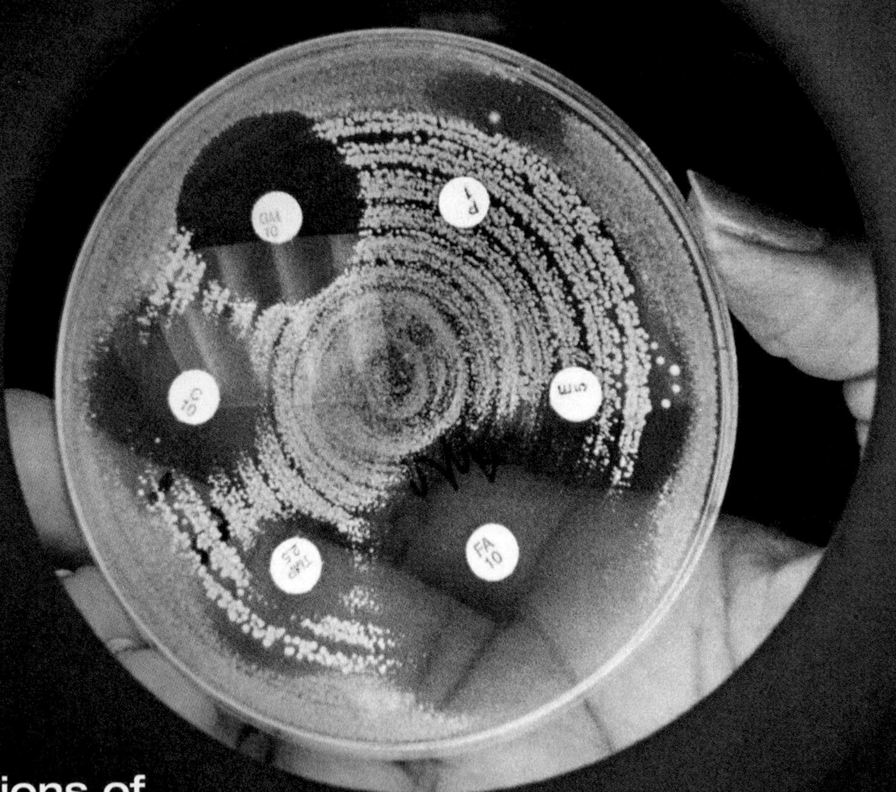

Why are the regions of inhibited bacterial growth around the various antibiotics so different in size, and how is that useful for human health care?

You will find the answer to this question on page 516.

KEY CONCEPTS

21.1 Fungi Digest Food Outside Their Bodies

LEARNING OBJECTIVES

By the end of this key concept you should be able to:

21.1.1 Explain how the structure of fungi facilitates their ability to decompose and consume other organisms.

21.1.2 Assess the ability of fungi to exist in different environments.

Fungi are distinctive in the way they digest their food. They secrete digestive enzymes to break down large food molecules in the environment, then absorb the breakdown products through their cell membranes in a process known as **absorptive heterotrophy**. This mode of nutrition allows them to be successful in a wide variety of environments. Many fungi are **saprobes**, meaning they absorb nutrients from dead organic matter. Others are parasites or predators, feeding on or consuming other living organisms. Still others are mutualists, living in intimate associations with other organisms that benefit both partners.

Modern fungi evolved from a unicellular protist ancestor that had a flagellum, although most fungi are not motile and have now lost this structure. The probable common ancestor of the animals was also a flagellated protist much like the living choanoflagellates (see Figure 22.2). Current evidence, including the sequences of many genes, suggests that the fungi, choanoflagellates, and animals share a common ancestor not shared by other eukaryotes. These three lineages form a group known as the **opisthokonts** (**FIGURE 21.1**). A synapomorphy (shared derived trait) of the opisthokonts is a flagellum that, if present, is posterior, as in animal sperm. The flagella of all other eukaryotes attach at the front or sides of the cell, rather than at the rear.

Synapomorphies that distinguish the fungi as a group among the opisthokonts include absorptive heterotrophy and the presence of **chitin**, a nitrogen-containing structural polysaccharide, in their cell walls. The fungi represent one of the four independent evolutionary origins of large multicellular organisms (plants, brown algae, and animals are the other three).

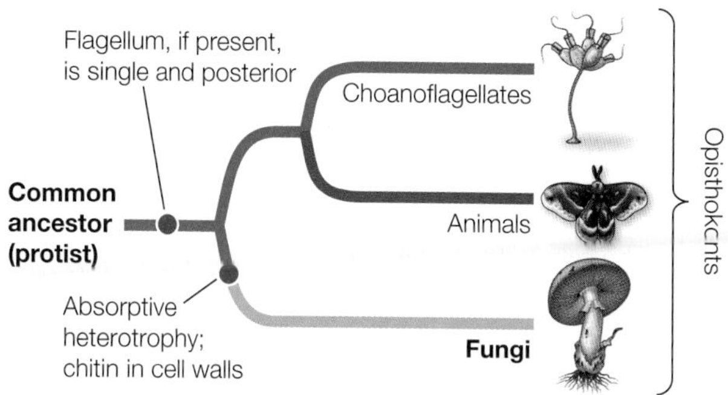

FIGURE 21.1 Fungi in Evolutionary Context Absorptive heterotrophy and the presence of chitin in their cell walls distinguish the fungi from other opisthokonts.

Yeasts are unicellular, free-living fungi

Most fungi are multicellular, but single-celled species are found in most fungal groups. Unicellular, free-living fungi are referred to as **yeasts** (**FIGURE 21.2**). Some fungi have both a yeast life stage and a multicellular life stage. Thus the term "yeast" does not refer to a single taxonomic group, but rather to a lifestyle that has evolved multiple times. Yeasts live in liquid or moist environments and absorb nutrients directly across their cell surfaces.

The ease with which many yeasts can be cultured, combined with their rapid growth rates, has made them ideal **model organisms** for study in the laboratory. They present many of the same advantages to laboratory investigators as do many bacteria, but because yeasts are eukaryotes, their genome structures and cells are much more like those of humans and other eukaryotes than are those of bacteria.

LINK Why we select and study a variety of model organisms is discussed in **Key Concept 12.1**. Each model organism has distinctive advantages and disadvantages. The features of yeasts make them ideal for some investigations but impractical for others.

FIGURE 21.2 Yeasts Unicellular, free-living fungi are known as yeasts. Many yeasts reproduce by budding—mitosis followed by asymmetrical cell division—as illustrated here.

Multicellular fungi use hyphae to absorb nutrients

The body of a multicellular fungus is called a **mycelium** (plural *mycelia*). A mycelium is composed of a branching mass of individual tubular filaments called **hyphae** (singular *hypha*) (**FIGURE 21.3A**), in which absorption of nutrients takes place primarily at each tip. The cell walls of the hyphae are greatly strengthened by microscopic fibrils of chitin. In some species of fungi, the hyphae are subdivided into cell-like compartments by incomplete cross-walls called **septa** (singular *septum*). These subdivided hyphae are referred to as **septate**. Septa do not completely close off compartments in the hyphae. Pores at the centers of the septa allow cytoplasm and organelles—sometimes even nuclei—to move in a controlled way between compartments (**FIGURE 21.3B**). In other species of fungi, the hyphae lack septa but may contain hundreds of nuclei. These multinucleate, undivided hyphae are referred to as **coenocytic**. The coenocytic condition results from repeated nuclear divisions without cytokinesis.

Certain modified hyphae, called **rhizoids**, anchor some fungi to their substrate (i.e., the dead organism or other matter on which they grow). These rhizoids are not homologous to the rhizoids of plants, and they are not specialized to absorb nutrients and water.

Fungi can grow very rapidly when conditions are favorable. In some species, the total hyphal growth of a fungal mycelium (not the growth of an individual hypha) may exceed 1 kilometer a day! The hyphae may be widely dispersed to forage for nutrients over a large area, or they may clump together in a cottony mass to exploit a rich nutrient source. The familiar mushrooms you may notice growing in moist areas are spore-producing fruiting structures (**FIGURE 21.3C**). In the fungal species that produce these structures, the mycelial mass is often far larger than the visible mushroom. The mycelium of one individual fungus discovered in Oregon covers almost 900 hectares underground and weighs considerably more than a blue whale (the largest animal). Aboveground, this individual is evident only as isolated clumps of mushrooms.

Fungi are in intimate contact with their environment

The filamentous hyphae of a fungus give it a unique relationship with its physical environment. The fungal mycelium has an enormous surface area-to-volume ratio compared with that of most large multicellular organisms. This large ratio allows the fungus to take in a great deal of water and mineral nutrients when soil moisture is high, which allows rapid growth in moist environments. The downside of the large surface area-to-volume ratio of the mycelium is its tendency to lose water rapidly in a dry environment. Thus fungi are most common in moist environments. You have probably observed the tendency of molds, toadstools, and other fungi to appear in damp places.

Another characteristic of some fungi is a tolerance for highly hypertonic environments (those with a solute concentration higher than their own; see Key Concept 4.2). Many fungi are more resilient than bacteria in hypertonic surroundings. Jelly, for example, does not support bacterial growth because its high sugar content makes it too hypertonic to those organisms, but it may

(A) Fungal hyphae in dry rot

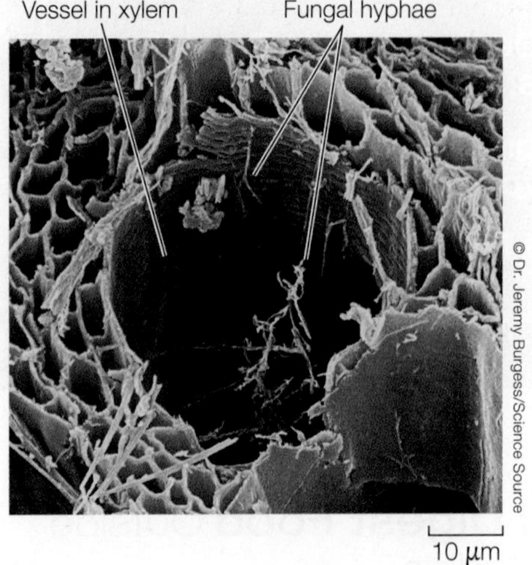

(B) Anatomy of hyphae

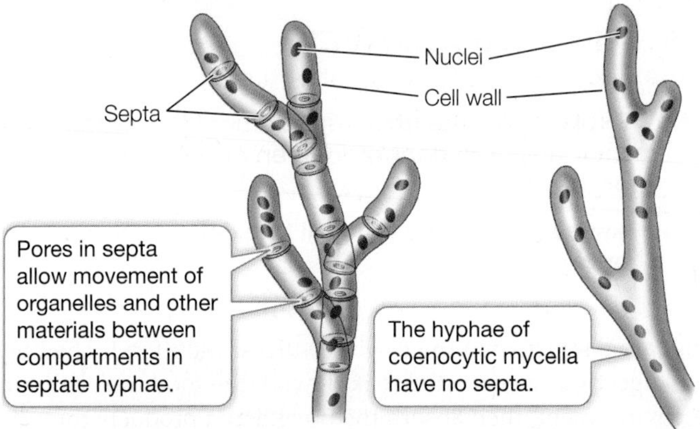

(C) The mycelium of a club fungus

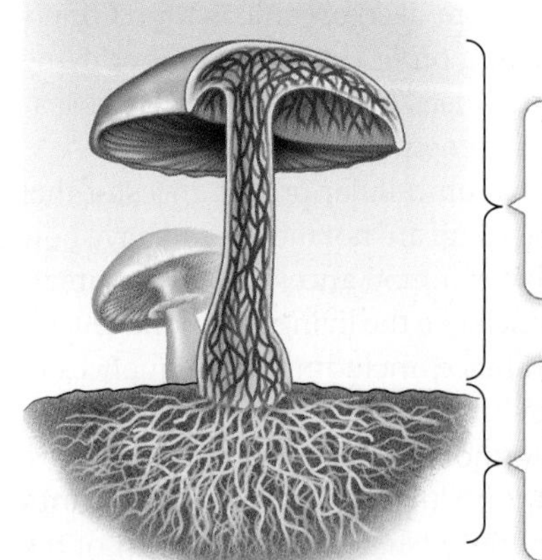

FIGURE 21.3 Mycelia Are Made Up of Hyphae (A) The minute individual hyphae of fungal mycelia can penetrate small spaces. In this artificially colored micrograph, hyphae (yellow structures) of a dry-rot fungus are penetrating the xylem tissues of a log. **(B)** The hyphae of septate fungal species are divided into organelle-containing compartments by porous septa. The hyphae of coenocytic fungal species have no septa. **(C)** The fruiting structure of a club fungus is short-lived, but the filamentous, nutrient-absorbing mycelium can be long-lived and cover large areas.

eventually harbor mold colonies. Mold in the refrigerator illustrates yet another trait of many fungi: tolerance of temperature extremes. Many fungi grow in temperatures as low as –6°C, and some can tolerate temperatures higher than 50°C.

REVIEW & APPLY | 21.1

R Fungi, like animals, are opisthokonts. Fungi are distinguished from other opisthokonts by absorptive heterotrophy and by the presence of chitin in their cell walls. Unicellular fungi called yeasts absorb nutrients directly across their cell surfaces. The body form of multicellular fungi—a mycelium made up of rapidly growing hyphae—allows them to practice absorptive heterotrophy efficiently in a variety of moist environments.

A
1. How does fungal structure facilitate absorptive heterotrophy?
2. What are the advantages and disadvantages to multicellular fungi of the large surface area-to-volume ratio of the mycelium? How does the mycelium restrict the types of environments in which fungi are generally found?

Fungi are important components of healthy ecosystems. They interact with other organisms in many ways, some of which are harmful and some of which are beneficial to those other organisms.

21.2 Fungi Are Decomposers, Parasites, Predators, or Mutualists

LEARNING OBJECTIVES

By the end of this key concept you should be able to:

21.2.1 **Explain how saprobic fungi are critical to Earth's carbon cycle.**

21.2.2 **Describe the relationship and summarize the benefits to each partner in a lichen association.**

21.2.3 **Describe the specific benefits that mycorrhizal interactions provide for fungi and their associated partner organisms.**

Without fungi, our planet would be very different. Picture Earth with only a few stunted plants and watery environments choked with the remains of dead organisms. Fungi do much of Earth's garbage disposal. Fungi not only help clean up the landscape and form soil, but also play a key role in recycling mineral nutrients. Furthermore, the colonization of the terrestrial environment was made possible in large part by associations that fungi formed with land plants and other organisms.

Saprobic fungi are critical to the planetary carbon cycle

Saprobic fungi, along with bacteria, are the major decomposers on Earth, contributing to the decay of nonliving organic matter and thus to recycling of the elements used by living things. In forests, for example, the mycelia of fungi secrete extracellular enzymes that degrade plant materials into simpler compounds that are then absorbed. Fungi are the principal decomposers of cellulose and lignin, the main components of plant cell walls (most bacteria cannot break down these materials). Other fungi produce enzymes that decompose keratin and thus break down animal structures such as hair and nails.

Were it not for the fungal decomposers, Earth's carbon cycle would fail. Great quantities of carbon atoms would remain trapped forever on forest floors and elsewhere. Instead, those carbon atoms are returned to the atmosphere in the form of CO_2 by fungal respiration, where they are again available for photosynthesis by plants.

LINK Earth's carbon cycle is discussed in detail in **Key Concept 42.3**. Fungi play a critical role in this cycle by decomposing other organisms and releasing carbon dioxide.

There was a time in Earth's history when populations of saprobic fungi declined dramatically. Vast tropical swamps existed during the Carboniferous and much of the Permian, as you saw in Chapter 17. When plants in these swamps died, they began to form peat. Peat formation led to acidification of the swamps. That acidity, in turn, drastically reduced the fungal population. The result? With the decomposers largely absent, large quantities of peat remained on the swamp floor and over time were converted into coal.

In contrast to their decline during the Carboniferous and much of the Permian, fungi did very well at the end of the Permian, a quarter of a billion years ago, when the aggregation of continents produced volcanic eruptions that triggered a global mass extinction. The fossil record shows that even as the majority of all multicellular species became extinct, fungi flourished—demonstrating both their hardiness and their role in recycling the elements in dead plants and animals.

Simple sugars and the breakdown products of complex polysaccharides are the favored sources of carbon for saprobic fungi. Most fungi obtain nitrogen from proteins or the products of protein breakdown. Many fungi can use nitrate (NO_3^-) or ammonium (NH_4^+) ions as their sole source of nitrogen. No known fungus can get its nitrogen directly from inorganic nitrogen gas, however, as can some free-living and plant-associated bacteria (that is, fungi cannot fix nitrogen; see Key Concept 18.3).

What happens when a fungus faces a dwindling food supply? A common strategy is to reproduce rapidly and abundantly. When conditions are good, fungi produce great quantities of spores, but the rate of spore production is commonly even higher when nutrient supplies go down. The spores may then remain dormant until conditions improve, or they may be dispersed to areas where nutrient supplies are higher.

Lycoperdon perlatum

FIGURE 21.4 Spores Galore Puffballs (a type of club fungus) disperse trillions of spores in great bursts. Few of the spores travel very far, however; some 99 percent of them fall within 100 meters of the parent puffball.

Not only are fungal spores abundant in number, but they are extremely tiny and easily spread by wind or water (**FIGURE 21.4**). These attributes virtually ensure that the spores will be scattered over great distances and that at least some of them will find conditions suitable for growth. The air we breathe contains as many as 10,000 fungal spores per cubic meter. No wonder we find fungi just about everywhere.

Some fungi engage in parasitic or predatory interactions

Whereas saprobic fungi obtain their energy, carbon, and nitrogen directly from dead organic matter, other species of fungi obtain their nutrition from parasitic—and even predatory—interactions.

PARASITIC FUNGI **Mycologists** (biologists who study fungi) distinguish between two classes of parasitic fungi based on their degree of dependence on their host. **Facultative parasites** can grow on living organisms but can also grow independently (including on artificial media). **Obligate parasites** can grow only on a specific, living host. The fact that their growth depends on a living host shows that obligate parasites have specialized nutritional requirements.

Plants and insects are the most common hosts of parasitic fungi. The filamentous structure of fungal hyphae is especially well suited to a life of absorbing nutrients from living plants. The slender hyphae of a parasitic fungus can invade a plant through stomata, through wounds, or in some cases, by direct penetration of epidermal cell walls (**FIGURE 21.5A**). Once inside the plant, the hyphae branch out to expand the mycelium. Some hyphae produce **haustoria**, branching projections that push through cell walls into living plant cells, absorbing the nutrients within those cells. The haustoria do not break through the cell membranes inside the cell walls, but instead invaginate into the membranes, so that the cell membrane fits them like a glove (**FIGURE 21.5B**). Fruiting structures may form, either within the plant body or on its surface.

Some parasitic fungi live in a close physical (symbiotic) relationship with a plant host that is usually not lethal to the plant. Others are *pathogenic*, weakening or even killing the host from which they derive nutrition.

Media Clip 21.1 Mind-Control Killer Fungi

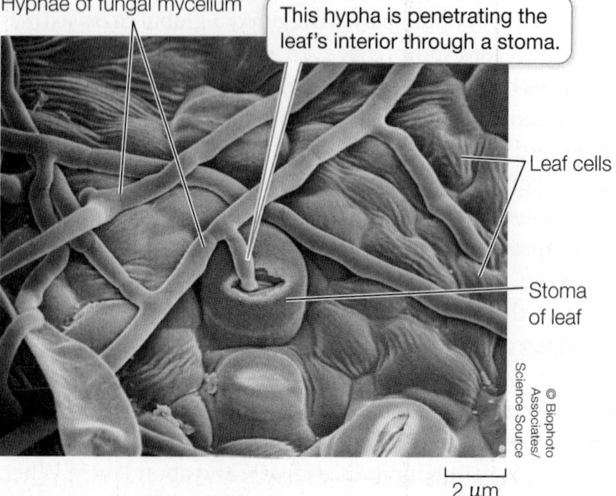

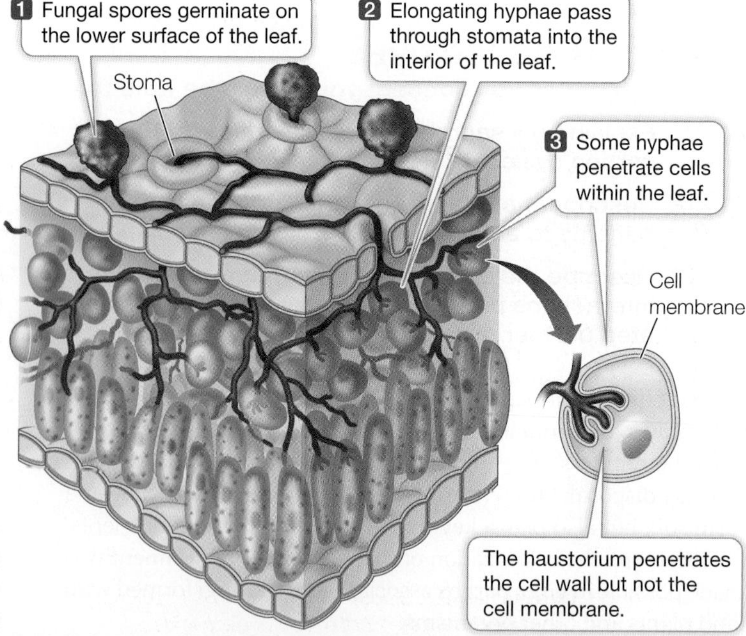

FIGURE 21.5 Invading a Leaf (A) Hyphae of the mildew *Phyllactinia guttata* growing on the underside of a hazel leaf. **(B)** Haustoria are fungal hyphae that push into the living cells of plants, from which they absorb nutrients.

PATHOGENIC FUNGI Although most human diseases are caused by bacteria or viruses, fungal pathogens are a major cause of death among people with compromised immune systems. Many people with AIDS die of fungal diseases, such as the pneumonia caused by *Pneumocystis jirovecii*. Even *Candida albicans* and certain other yeasts that are normally part of a healthy microbiome can cause severe diseases, such as esophagitis (which impairs swallowing), in individuals with AIDS and in individuals taking immunosuppressive drugs. Various fungi cause other, less threatening human diseases, such as ringworm and athlete's foot. The relatively close relationship between fungi and animals makes it difficult to find drugs that target unique features of fungal metabolism without harming human patients. As a result, fungal diseases are a growing international health problem.

The worldwide decline of amphibian species has been linked to the spread of a chytrid fungus, *Batrachochytrium dendrobatidis* (or *Bd* for simplicity). In some areas of the world where this fungus has been present for millennia, amphibian populations appear to be tolerant of *Bd* and it does not cause widespread die-offs. But a pathogenic strain known as the global pandemic lineage of *Bd* appears to have been introduced around the world in recent decades, likely through commercial trade of amphibians. These introductions of the global pandemic lineage of *Bd* are implicated in the loss or decline of many frog and salamander populations, and even the extinction of entire species.

Fungi are by far the most important plant pathogens, causing annual crop losses amounting to billions of dollars. Major fungal diseases of crop plants include black stem rust of wheat and other diseases of wheat, corn, and oats. The agent of black stem rust is *Puccinia graminis*, which has a complicated life cycle that involves two plant hosts (wheat and barberry). In an epidemic in 1935, *P. graminis* was responsible for the loss of about one-fourth of the wheat crop in Canada and the United States.

PREDATORY FUNGI Some fungi have adaptations that enable them to function as active predators, trapping nearby microscopic protists or animals. This feeding strategy is especially important in environments with lots of available carbon but little nitrogen, such as wood. The most common predatory strategy seen in fungi is to secrete sticky substances from the hyphae so that passing organisms stick to them. The hyphae then quickly invade the trapped prey, growing and branching within it, spreading through its body, absorbing nutrients, and eventually killing it.

A more dramatic adaptation for predation is the constricting ring formed by some species of soil fungi (**FIGURE 21.6**). When nematodes (tiny roundworms) are present in the soil, these fungi form three-celled rings with a diameter that just fits a nematode. A nematode crawling through one of these rings stimulates the fungus, causing the cells of the ring to swell and trap the worm. Fungal hyphae quickly invade and digest the unlucky victim.

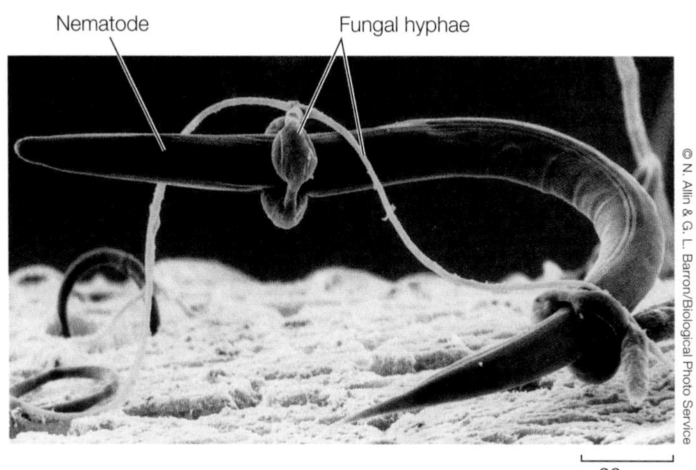

FIGURE 21.6 Fungus as Predator A nematode is trapped by hyphal rings of the soil-dwelling fungus *Arthrobotrys dactyloides*.

Mutualistic fungi engage in relationships that benefit both partners

Certain relationships between fungi and other organisms have nutritional consequences for both partners. Two relationships of this type are **symbiotic** (the partners live in close, permanent contact with each other) as well as **mutualistic** (the relationship benefits both partners).

LICHENS A **lichen** is not a single organism, but rather a meshwork of at least two radically different species: a fungus and a photosynthetic alga or cyanobacterium. Recent studies suggest that an additional fungal species (a unicellular yeast) is also often involved in this mutualism. Together the organisms that constitute a lichen can survive some of the harshest environments on Earth (although they are sensitive to poor air quality; see Key Concept 21.4). The biota of Antarctica, for example, features more than 100 times as many kinds of lichens as of plants. Relatively little experimental work has focused on lichens, perhaps because they grow so slowly—typically less than 1 centimeter in a year.

There are nearly 30,000 described "species" of lichens, each of which is assigned the name of its multicellular fungal component. These fungal components may constitute as many as 20 percent of all fungal species. Most of them are sac fungi (Ascomycota). Some of them are able to grow independently without a photosynthetic partner, but most have never been observed in nature other than in a lichen association. The photosynthetic component of a lichen is most often a unicellular green alga, but it can be a cyanobacterium, or may even include both.

Lichens are found in all sorts of exposed habitats: on tree bark, on open soil, and on bare rock. Reindeer moss (not a moss at all, but the lichen *Cladonia subtenuis*) covers vast areas in Arctic, sub-Arctic, and boreal regions, where it is an important part of the diets of reindeer and other large mammals.

The body forms of lichens fall into three principal categories. **Crustose** (crustlike) lichens adhere tightly to their substrate (**FIGURE 21.7A**). **Fruticose** lichens are highly branched and can grow upward like shrubs, or hang in long strands from tree branches or rocks (**FIGURE 21.7B**). **Foliose** (leafy) lichens are loosely attached and grow parallel to their substrate (**FIGURE 21.7C**).

A cross section of a typical foliose lichen reveals a tight upper region of fungal hyphae, a layer of photosynthetic cyanobacteria or

FIGURE 21.7 Lichen Body Forms The body forms of lichens fall into three principal categories. **(A)** Two crustose lichen species are growing together on this exposed rock surface. **(B)** The light green "shrubby" growth is a fruticose lichen. **(C)** Foliose lichens have a leafy appearance.

algae, a looser hyphal layer, and finally hyphal rhizoids that attach the entire structure to its substrate (**FIGURE 21.8**). The meshwork of fungal hyphae takes up mineral nutrients needed by the photosynthetic cells and also holds water tenaciously, providing a suitably moist environment. The fungus obtains fixed carbon from the photosynthetic products of the algal or cyanobacterial cells.

Within the lichen, fungal hyphae are tightly pressed against the photosynthetic cells and sometimes invade them without breaching the cell membrane (similar to the haustoria in parasitic fungi; see Figure 21.5). The photosynthetic cells not only survive these intrusions but continue to grow. Algal cells in a lichen "leak" photosynthetic products at a greater rate than do similar cells growing on their own, and photosynthetic cells taken from lichens grow more rapidly on their own than when associated with a fungus. On the basis of these observations, we could consider lichen fungi to be parasitic on their photosynthetic partners. In many places where lichens grow, however, the photosynthetic cells could not grow at all on their own.

Lichens can reproduce simply by fragmentation of the vegetative body (the **thallus**) or by means of specialized structures called **soredia** (singular *soredium*). Soredia consist of one or a few photosynthetic cells bound by fungal hyphae. They become detached from the lichen, are dispersed by air currents, and upon arriving at a favorable location, develop into a new lichen thallus (see Figure 21.8). Alternatively, the fungal partner may go through its sexual reproductive cycle, producing haploid spores. When these spores are discharged, however, they disperse alone, unaccompanied by the photosynthetic partner.

A soredium consists of one or a few photosynthetic cells surrounded by fungal hyphae.

Soredia detach from the parent lichen and travel in air currents, founding new lichens when they settle in a suitable environment.

Upper layer of hyphae

Photosynthetic cell layer

Loose layer of hyphae

Lower level of hyphal rhizoids

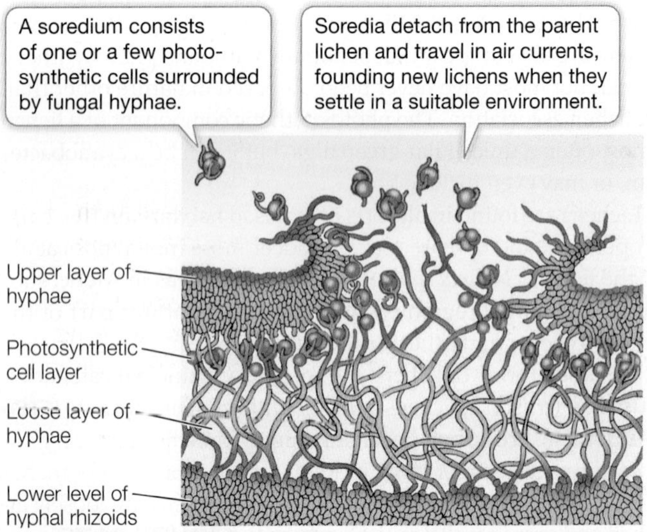

FIGURE 21.8 Lichen Anatomy Cross section showing the layers of a foliose lichen and the release of soredia.

Lichens are often the first colonists on new areas of bare rock. They get most of the mineral nutrients they need from the air and rainwater, augmented by minerals absorbed from dust. A lichen begins to grow shortly after a rain, as it begins to dry. As it grows, the lichen acidifies its environment slightly, and this acidity contributes to the slow breakdown of rocks, an early step in soil formation. With further drying, the lichen's photosynthesis ceases. The water content of the lichen may drop to less than 10 percent of its dry weight, at which point the lichen becomes highly insensitive to extremes of temperature.

MYCORRHIZAE Many vascular plants depend on a symbiotic association with fungi. This ancient association between plants and fungi was critical to the successful exploitation of the terrestrial environment by plants. Unassisted, the root hairs of many plants often do not take up enough water or minerals to sustain growth. However, the roots of such plants usually do become infected with fungi, forming an association called a **mycorrhiza** (plural *mycorrhizae*). There are two types of mycorrhizae, distinguished by whether or not the fungal hyphae penetrate the plant cell walls.

In **ectomycorrhizae**, the fungus wraps around the root tip, and its mass is often as great as that of the root itself (**FIGURE 21.9A**). The fungal hyphae penetrate the root and wrap around individual root cells but do not penetrate the cell walls. An extensive web of hyphae penetrates the soil in the area around the root, so that up to 25 percent of the volume near the root may be fungal hyphae. The hyphae attached to the root increase the surface area for the absorption of water and minerals, and the mass of hyphae in the soil acts like a sponge to hold water in the neighborhood of the root. Infected roots are short, swollen, and club-shaped, and they lack root hairs.

The fungal hyphae of **arbuscular mycorrhizae** enter the root and penetrate the cell walls of the root cells, forming

(A) An ectomycorrhizal fungus

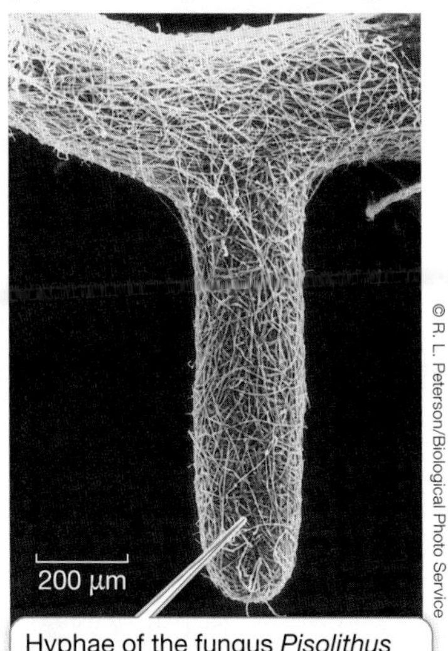

(B) An arbuscular mycorrhizal fungus

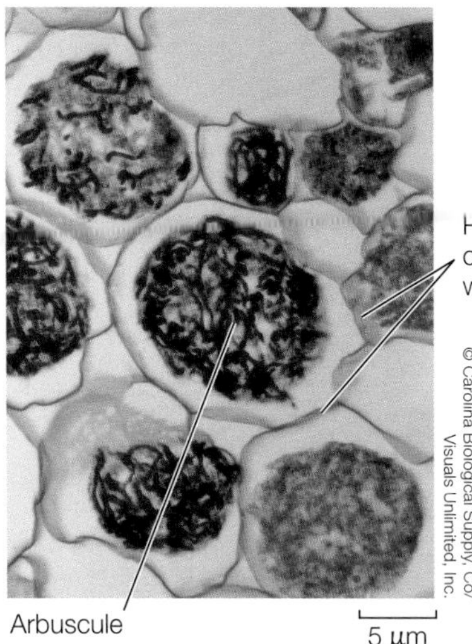

FIGURE 21.9 Mycorrhizal Associations **(A)** Ectomycorrhizal fungi wrap themselves around a plant root, increasing the area available for absorption of water and minerals. **(B)** Hyphae of arbuscular mycorrhizal fungi infect the root internally and penetrate the root cell walls, branching within the cells and forming a treelike structure, the arbuscule. (The cell cytoplasm has been removed to better visualize the arbuscule.)

arbuscular (treelike) structures inside the cell wall but outside the cell membrane. These structures, like the haustoria of parasitic fungi and the contact regions of fungal hyphae and photosynthetic cells in lichens, become the primary site of exchange between plant and fungus (**FIGURE 21.9B**). As in the ectomycorrhizae, the fungus forms a vast web of hyphae leading from the root surface into the surrounding soil.

LINK Mycorrhizae can vastly increase the root surface area in contact with essential nutrients in the soil. The importance of arbuscular mycorrhizae in plant nutrition is described in detail in **Key Concept 24.3**.

The mycorrhizal association is important to both partners. The fungus obtains needed organic compounds, such as sugars and amino acids, from the plant. In return, the fungus, because of its high surface area-to-volume ratio and its ability to penetrate the fine structure of the soil, greatly increases the plant's ability to absorb water and minerals (especially phosphorus). The fungus may also provide the plant with certain growth hormones and may protect it against attack by disease-causing microorganisms.

Plants that have active arbuscular mycorrhizae typically are a deeper green and may resist drought and temperature extremes better than plants of the same species that have little mycorrhizal development. Attempts to introduce some plant species to new areas have failed until some soil from the native area (presumably containing the fungus necessary to establish mycorrhizae) was provided. Trees without ectomycorrhizae do not grow well in the absence of abundant nutrients and water, so the health of our forests depends on the presence of ectomycorrhizal fungi. Many agricultural crops require inoculation of seeds with appropriate mycorrhizal fungi prior to planting. Without these fungi, the plants are unlikely to grow well, or in some cases at all. Certain plants that live in nitrogen-poor habitats, such as cranberry bushes and orchids, invariably have mycorrhizae. Orchid seeds will not germinate in nature unless they are already infected by the fungus that will form their mycorrhizae. Plants that lack chlorophyll always have mycorrhizae, which they often share with the roots of green, photosynthetic plants. In effect, these plants without chlorophyll are feeding on nearby green plants, using the fungus as a bridge.

Endophytic fungi protect some plants from pathogens, herbivores, and stress

In a tropical rainforest, 10,000 or more fungal spores may land on a single leaf each day. Some are plant pathogens, some do not affect the plant at all, and some invade the plant in a beneficial way. Fungi that live within aboveground parts of plants without causing obvious deleterious symptoms are called **endophytic fungi**. Recent research has shown that endophytic fungi are abundant in plants in all terrestrial environments.

Among the grasses, individual plants with endophytic fungi are more resistant to pathogens and to insect and mammalian herbivores than are plants lacking endophytes. The fungi produce alkaloids (nitrogen-containing compounds) that are toxic to animals. The alkaloids do not harm the host plant. In fact, some plants produce alkaloids (such as nicotine) themselves. The fungal alkaloids also increase the ability of grasses to resist stress of various types, including drought (water shortage) and salty soils. Such resistance is beneficial for some crops, although it can produce toxicity for grazing animals.

The role, if any, of endophytic fungi in most broad-leaved plants is unclear. They may convey protection against pathogens, or they may simply occupy space within leaves without conferring any benefit, but also without doing harm. The benefit, in fact, might be all for the fungus.

REVIEW & APPLY | 21.2

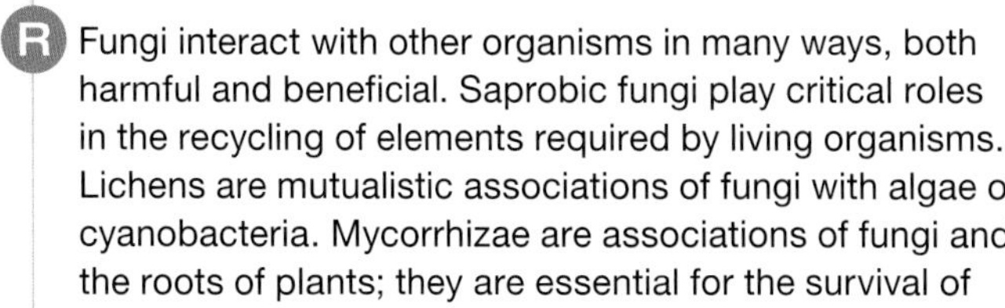

R Fungi interact with other organisms in many ways, both harmful and beneficial. Saprobic fungi play critical roles in the recycling of elements required by living organisms. Lichens are mutualistic associations of fungi with algae or cyanobacteria. Mycorrhizae are associations of fungi and the roots of plants; they are essential for the survival of most plant species.

A
1. What is the role of fungi in Earth's carbon cycle?
2. Describe the nature and benefits of the lichen association.
3. How does the mycorrhizal association benefit both the plant and the fungus?

Before molecular techniques widely used to study the phylogenetic relationships of fungi, one criterion used for assigning fungi to taxonomic groups was the nature of their life cycles—including the types of fruiting structures they produced. The next section will take a closer look at life cycles in the major groups of fungi.

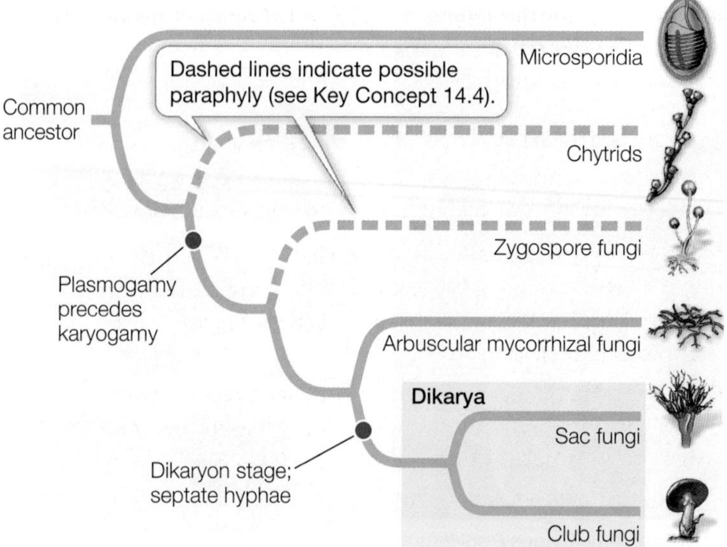

FIGURE 21.10 A Phylogeny of the Fungi Microsporidia are reduced parasites that are thought to be related to fungi, although their relationships are still debated. The dashed lines indicate that chytrids and zygospore fungi are possibly paraphyletic (see Table 21.1). The sac fungi and club fungi together form the clade Dikarya.

Activity 21.1 Fungal Phylogeny

21.3 Sex in Fungi Involves Multiple Mating Types

LEARNING OBJECTIVES

By the end of this key concept you should be able to:

21.3.1 Compare and contrast sexual reproduction in fungi with that in plants and animals.

21.3.2 Describe the dikaryon stage of the fungal life cycle, distinguish which major fungal groups include a dikaryon stage, and explain how the dikaryon stage differs from the haploid and diploid stages.

21.3.3 Compare and contrast the life cycles of zygospore fungi, sac fungi, and club fungi.

Major fungal groups were originally defined by their structures and processes for sexual reproduction and also, to a lesser extent, by other morphological differences. Although fungal life cycles are even more diverse than was once realized, specific types of life cycles generally distinguish five major groups of fungi: chytrids (Chytridiomycota and Blastocladiomycota), zygospore fungi (Zoopagomycota), arbuscular mycorrhizal fungi (Glomeromycotina), sac fungi (Ascomycota), and club fungi (Basidiomycota). In addition, a reduced parasitic group, microsporidia, is usually considered to be related to fungi, although the exact relationships of microsporidia within eukaryotes are debated. **FIGURE 21.10** diagrams the evolutionary relationships of these groups as they are understood today.

Chytrids and zygospore fungi may not represent monophyletic groups, as they each consist of several distantly related lineages that retain some ancestral features. The clades that are thought to be monophyletic within these two informal groupings are listed in **TABLE 21.1**. Most recent evidence from DNA analyses has

TABLE **21.1** | Classification of the Fungi

Group	Common name	Features
Microsporidia	Microsporidia	Intracellular parasites of animals; greatly reduced, among smallest eukaryotes known; polar tube used to infect hosts; placement among fungi is debated
Chytridiomycota and Blastocladiomycota	Chytrids[a]	Mostly aquatic and microscopic; zoospores and gametes have flagella in some species
Zoopagomycota	Zygospore fungi[b]	Reproductive structure is a unicellular zygospore with many diploid nuclei; hyphae coenocytic; no fleshy fruiting body
Glomeromycotina	Arbuscular mycorrhizal fungi	Form arbuscular mycorrhizae in plant roots; often reproduce asexually, but recent studies show sexual reproduction as well
Ascomycota	Sac fungi	Sexual reproductive saclike structure known as an ascus, which contains haploid ascospores; hyphae septate; dikaryon
Basidiomycota	Club fungi	Sexual reproductive structure is a basidium, a swollen cell at the tip of a specialized hypha that supports haploid basidiospores; hyphae septate; dikaryon

[a]The relationships between Chytridiomycota and Blastocladiomycota, both treated as chytrids here, are not yet well resolved.

[b]Two small groups traditionally considered zygospore fungi now appear to be more closely related to arbuscular mycorrhizal fungi.

supported the relationship of microsporidia to fungi, the possible paraphyly of chytrids and zygospore fungi, and the monophyly of sac fungi and club fungi.

Fungi reproduce both sexually and asexually

Both asexual and sexual reproduction occur among the fungi (**FIGURE 21.11**). Asexual reproduction takes several forms:

- The production of (usually) haploid spores within sporangia
- The production of haploid spores (not enclosed in sporangia) at the tips of hyphae; in sac fungi, such spores are called **conidia** (Greek *konis*, "dust")
- Cell division by unicellular fungi—either a relatively equal division of one cell into two (*fission*) or an asymmetrical division in which a smaller daughter cell is produced (*budding*)
- Mitotic growth and fragmentation of the mycelium

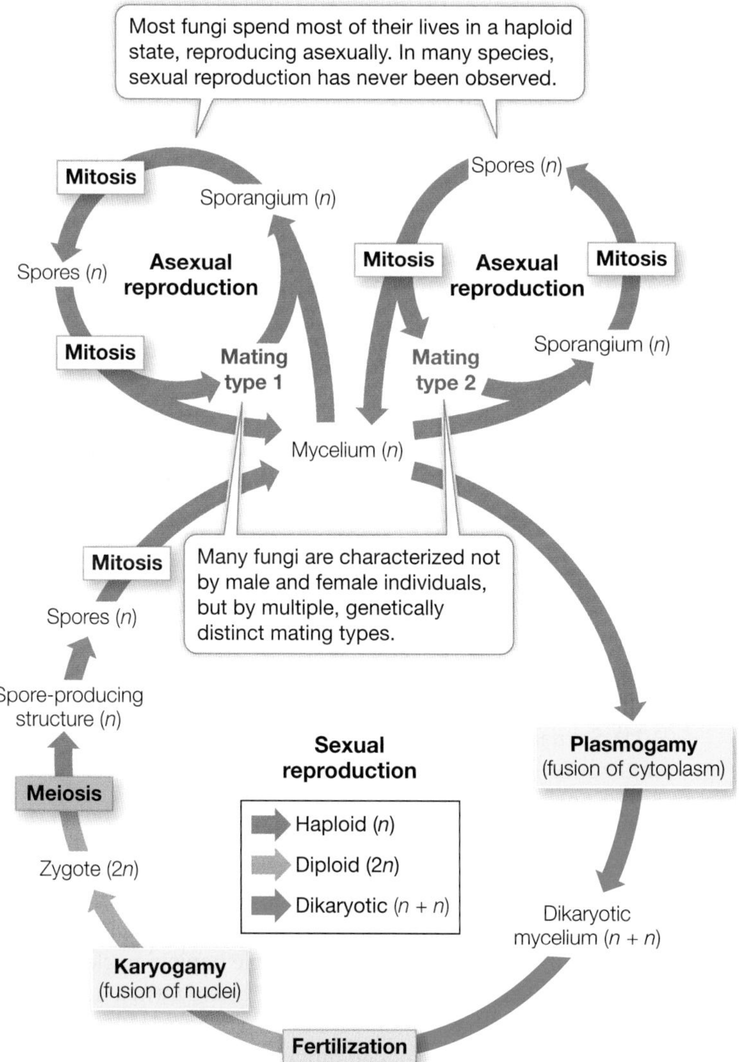

FIGURE 21.11 A Generalized Fungal Life Cycle Environmental conditions may determine which mode of reproduction—sexual or asexual—takes place at a given time.

Sexual reproduction is rare (or even unknown) in some groups of fungi but common in others. Sexual reproduction may occur so rarely in some species that biologists have simply never observed it. Species in which no sexual stage has been observed were once placed in a separate taxonomic group because knowledge of the sexual life cycle was considered necessary for classifying fungi. Now, however, these species can be related to other species of fungi through analysis of their DNA sequences.

Fungi mate in the most interesting of ways. They don't limit themselves to the confines of "males" and "females," but rather have many different **mating types**, any two of which can reproduce. And when they mate, they don't merely pass a few gamete cells between individuals like plants or animals do; instead, they fuse their bodies to become one and then exchange their nuclei. After fusing, many mating pairs remain fused for some time (a mushroom is basically two fused fungi getting it on), and the sexual act can thus last for weeks or months at a time. Eventually the nuclei fuse and recombine, and "genetic sex" finally takes place. After making millions of offspring, the exhausted fungi disintegrate their combined bodies and never text each other again.

Major groups of fungi differ in the details of when and how the nuclei from two individuals fuse and recombine, and in whether they produce a fleshy fruiting structure, as we will discuss below.

Microsporidia may be highly reduced, parasitic fungi

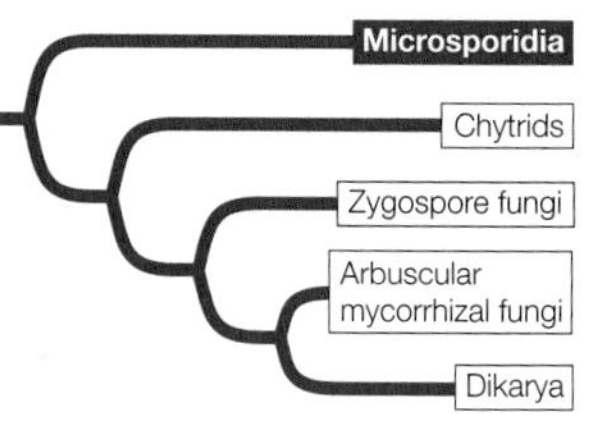

Microsporidia are highly reduced, unicellular parasites. They are among the smallest eukaryotes known, with infective spores that are only 1–40 micrometers (μm) in diameter. About 1,500 species have been described, but many more species are thought to exist. Their relationships among the eukaryotes have puzzled biologists for many decades, although the sequences of several microsporidian genes suggest a close relationship to fungi.

Microsporidia lack true mitochondria, although they have reduced structures, known as **mitosomes**, that are derived from mitochondria. Unlike mitochondria, however, mitosomes contain no DNA—the mitochondrial genome has been completely transferred to the nucleus. Because microsporidia lack mitochondria, biologists initially suspected that they represented an early lineage of eukaryotes that diverged before the endosymbiotic event from which mitochondria evolved. The presence of mitosomes, however, indicates that this hypothesis is incorrect. DNA sequence analysis, along with the fact that their cell walls contain chitin, has suggested that microsporidia are may be highly reduced, parasitic fungi, although their exact relationship to fungi is still being investigated.

Microsporidia are obligate intracellular parasites of animals, especially of insects, crustaceans, and fishes. Some species are known to infect mammals, including humans. Most

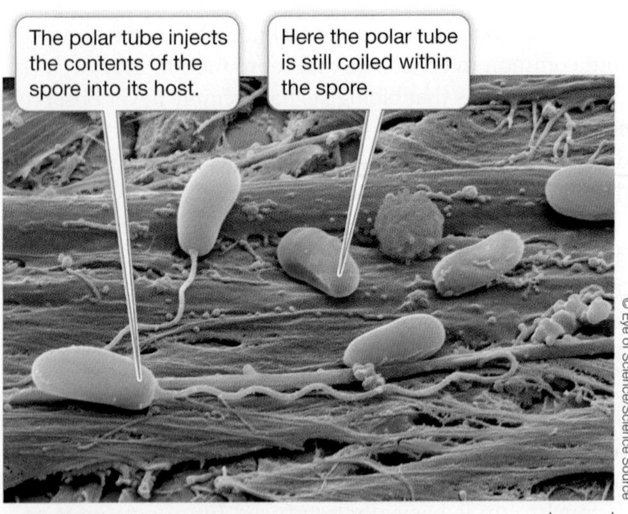

FIGURE 21.12 Invasion of the Microsporidia The spores of microsporidia grow polar tubes that transfer the contents of the spores into the host's cells. The species shown here infects many animals, including humans.

infections by microsporidia cause chronic disease in the host, with effects that include weight loss, reduced fertility, and a shortened life span. The host cell is penetrated by a polar tube that grows from the microsporidian spore. The function of the polar tube is to inject the contents of the spore, the sporoplasm, into the host (**FIGURE 21.12**). The sporoplasm then replicates within the host cell and produces new infective spores. The life cycle of some species is complex and involves multiple hosts, whereas other species infect a single host. In some insects, parasitic microsporidia are transmitted vertically (i.e., from parent to offspring). Only asexual reproduction is known for some microsporidia, but other species appear to have both asexual and sexual cycles.

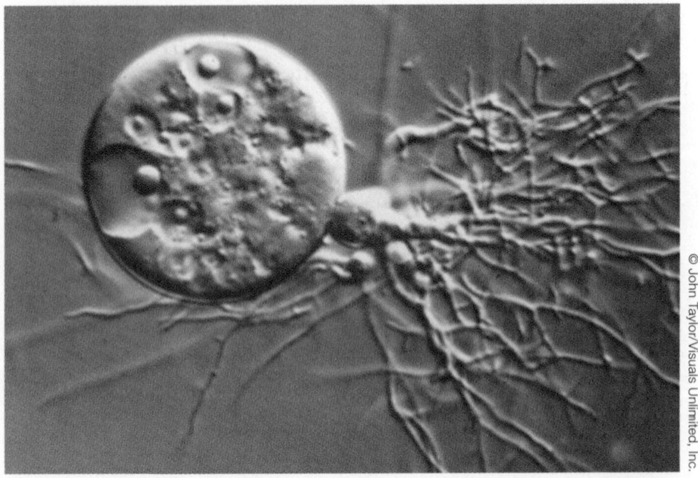

FIGURE 21.13 A Chytrid Branched rhizoids emerge from the sporangium of a mature chytrid.

Most chytrids are aquatic

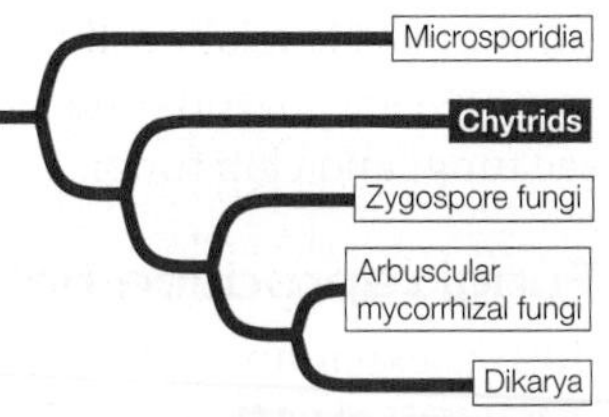

The **chytrids** (**FIGURE 21.13**) include several distinct lineages of mostly aquatic microorganisms once classified with the protists. However, morphological evidence (cell walls that consist primarily of chitin) and molecular evidence support their classification as early-diverging fungi. In this book we use the term "chytrid" to refer to both of the formally named clades listed as chytrids in Table 21.1, but some mycologists use this term to refer to only the Chytridiomycota. There are fewer than 1,000 described species in the two groups of chytrids.

Chytrids reproduce both sexually and asexually. Unlike other fungi, some chytrids that reproduce sexually possess flagellated gametes (similar to sperm in animals, except that both male and female gametes are flagellated in chytrids). Both the spores (called zoospores) and the gametes are flagellated (**FIGURE 21.14A**). Chytrids are the only fungi that include species with flagella at any life cycle stage. The retention of flagella reflects the aquatic environment in which fungi first evolved. As one type of gamete is larger than the other, the gametes can be distinguished as male (small gametes) and female (large gametes; see Figure 21.14A). Except for size, however, the two gametes are very similar.

The chytrids are diverse in form. Some are unicellular, others have rhizoids, and still others have coenocytic hyphae. They may be parasitic (on organisms such as algae, mosquito larvae, nematodes, and amphibians) or saprobic. Some have complex mutualistic relationships with foregut-fermenting animals such as cattle and deer. Many chytrids live in freshwater habitats or in moist soil, but some are marine.

Most fungal life cycles feature separate fusion of cytoplasms and nuclei

Zygospore fungi
Arbuscular mycorrhizal fungi
Ascomycota
Basidiomycota

Most members of the remaining four groups of fungi are terrestrial. Although the terrestrial fungi grow in moist places, they do not have motile gametes, so liquid water is not required for fertilization. Instead, the cytoplasms of two individuals of different mating types fuse (a process called **plasmogamy**) before their nuclei fuse (a process called **karyogamy**; see Figure 21.11). Sexual species of terrestrial fungi include some zygospore fungi, sac fungi, and club fungi.

Zygospore fungi reproduce sexually when adjacent hyphae of two different mating types release chemical signals that cause them to grow toward each other. These hyphae produce gametangia, which are specialized cells for reproduction that are retained as part of the hyphae. In the gametangia, nuclei replicate without cell division, resulting in multiple haploid nuclei in both gametangia. The two gametangia then fuse to form a unicellular zygosporangium that contains many haploid nuclei of each mating type. These haploid nuclei quickly pair up to form multiple diploid nuclei within the zygosporangium (**FIGURE 21.14B**). A thick, multilayered cell wall develops around the zygosporangium to form a well-protected resting stage that can remain dormant

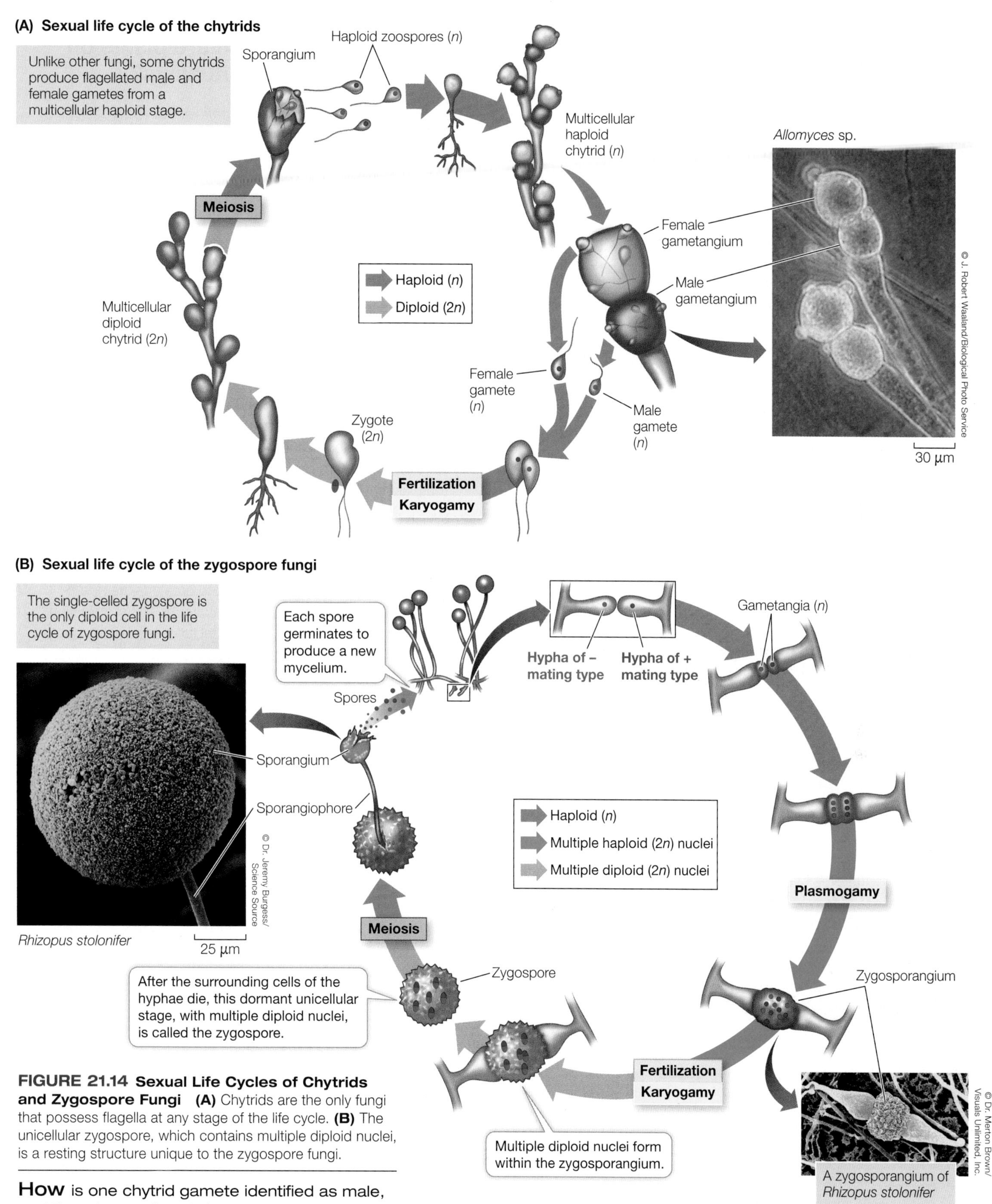

FIGURE 21.14 Sexual Life Cycles of Chytrids and Zygospore Fungi **(A)** Chytrids are the only fungi that possess flagella at any stage of the life cycle. **(B)** The unicellular zygospore, which contains multiple diploid nuclei, is a resting structure unique to the zygospore fungi.

How is one chytrid gamete identified as male, and the other as female?

for months. In harsh environmental conditions, this resting stage may be the only cell that survives as the surrounding cells of the hyphae die. At this stage the single surviving cell is known as a **zygospore**, which is the basis of the name of the zygospore fungi. When environmental conditions improve, the nuclei in the zygospore undergo meiosis and one or more stalked **sporangiophores** sprout, each bearing a sporangium. Each sporangium contains the products of meiosis: haploid nuclei that are incorporated into spores. These spores disperse and germinate to form a new generation of haploid hyphae.

The zygospore fungi include five major lineages of terrestrial fungi that live on soil as saprobes, as parasites of insects and spiders, or as mutualists of other fungi and invertebrate animals. Recent genome analyses suggest that two of these lineages may be more closely related to the arbuscular mycorrhizal fungi than they are to other groups treated here as zygospore fungi. Zygospore fungi produce no cells with flagella, and only one diploid cell—the zygospore—appears in the entire life cycle. Their hyphae are coenocytic. Most species do not form a fleshy fruiting structure. Instead, the hyphae spread in a radial pattern from the spore, with occasional stalked sporangiophores reaching up into the air (**FIGURE 21.15**).

More than 1,000 species of zygospore fungi have been described. One species you may have seen is *Rhizopus stolonifer*, the black bread mold. *Rhizopus* produces many stalked sporangiophores, each bearing a single sporangium containing hundreds of minute spores (see Figure 21.14B).

FIGURE 21.15 Zygospore Fungi Produce Sporangiophores These transparent structures are sporangiophores produced by a zygospore fungus growing on decomposing animal dung. The sporangiophores grow toward the light and end in tiny sporangia, which the stalked sporangiophores can eject as far as 2 meters. Animals ingest sporangia and then disseminate the spores in their feces.

Animation 21.1 Life Cycle of a Zygospore Fungus

THINK LIKE A SCIENTIST

How dependent are plants on their fungal mutualists?

Garlic mustard (*Alliaria petiolata*) is a highly invasive plant of Eurasian origin that causes much damage to native plants where it invades in North America. The manner by which this plant outcompetes native plants is unusual: it secretes chemicals that interfere with the mutualistic relationships the native plants have with arbuscular mycorrhizal fungi.

Some plants are more dependent than others on arbuscular mycorrhizal fungi. Researchers examined the effect of exposure to garlic mustard on several plant species and measured the effect that removal of arbuscular mycorrhizal fungi had on each plant. **FIGURE 21.16** shows the results of the study. Here you will use quantitative reasoning skills to analyze the effects of an introduced species on mycorrhizal fungi and their plant mutualists.

1. What is the relationship between the effect of exposure to garlic mustard on a plant species and its dependence on arbuscular mycorrhizal fungi? Is this relationship consistent with garlic mustard reducing plant growth by interfering with the plant's fungal mutualists?
2. Which types of plants are most dependent on their mycorrhizal mutualists? Explain your answer. What other observations can be made about habitat and reliance on fungal mutualists?
3. Climate change is expected to lead to greater extremes in temperature and precipitation. What are the likely implications for plants with arbuscular mycorrhizal fungi as climate change becomes more pronounced? Would these aspects of climate change likely increase or decrease the negative effects of exposure to garlic mustard?

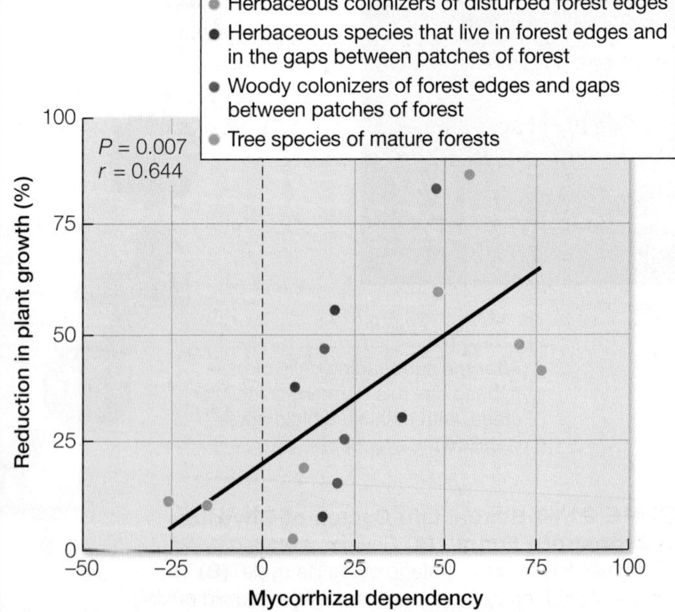

FIGURE 21.16 The Effect of an Introduced Species on Native Plants (After K. A. Stinson et al. 2006. *PLoS Biol* 4: e140; CC BY.)

Arbuscular mycorrhizal fungi form symbioses with plants

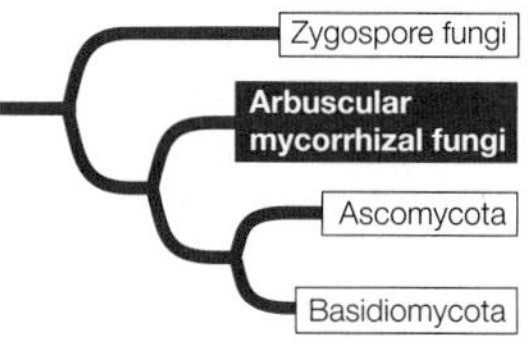

Arbuscular mycorrhizal fungi (Glomeromycotina) are terrestrial fungi that associate with plant roots in a symbiotic, mutualistic relationship (see Figure 21.9B). Only a few hundred species have been formally described, but 80 to 90 percent of all plants have associations with arbuscular mycorrhizal fungi, and most of the diversity of the group remains to be discovered.

The hyphae of arbuscular mycorrhizal fungi are coenocytic. These fungi use glucose from their plant partners as their primary energy source, converting it into other, fungus-specific sugars that cannot return to the plant. Until recently, arbuscular mycorrhizal fungi were only known to reproduce asexually, but recent studies demonstrate that they can undergo meiosis and genetic recombination.

The dikaryotic condition is a synapomorphy of sac fungi and club fungi

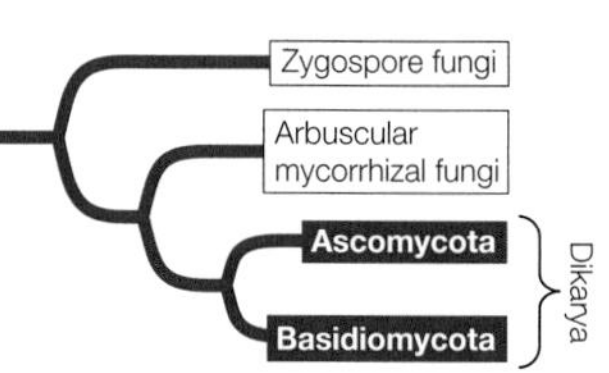

In the two remaining groups of fungi—the sac fungi and the club fungi—some stages have a nuclear configuration other than the familiar haploid or diploid states (**FIGURE 21.17**). In these fungi, karyogamy (fusion of nuclei) occurs long after plasmogamy (fusion of cytoplasm), so that *two genetically different haploid nuclei coexist and divide within each cell of the multicellular fruiting body*. This stage of the life cycle is called a **dikaryon** ("two nuclei"). The dikaryon is a synapomorphy of the sac fungi and club fungi, which are placed together in a clade called **Dikarya**.

Eventually, specialized cells form on the fruiting body, within which pairs of genetically dissimilar nuclei—one from each parent—fuse, giving rise to zygotes long after the original "mating." The diploid zygote nucleus then undergoes meiosis, producing four haploid nuclei. The mitotic descendants of those nuclei become spores, which germinate to give rise to the next haploid generation.

A life cycle with a dikaryotic stage has several unusual features. First, there are no gamete *cells*, only gamete *nuclei*. Second, the only true diploid structure is the zygote, although for a long period the genes of both parents are present in the dikaryon and can be expressed. In effect, the dikaryon is neither diploid ($2n$) nor haploid (n). Rather, it is dikaryotic ($n + n$). Therefore a harmful recessive mutation in one nucleus may be compensated for by a normal allele on the same chromosome in the other nucleus, and dikaryotic hyphae often have characteristics that are different from their n or $2n$ products. The dikaryotic condition is perhaps the most distinctive of the genetic peculiarities of the fungi.

The sexual reproductive structure of sac fungi is the ascus

The **sac fungi** (Ascomycota) are a large and diverse group of fungi found in marine, freshwater, and terrestrial habitats. There are approximately 64,000 known species, nearly half of which are the fungal partners in lichens. The hyphae of sac fungi are segmented by more or less regularly spaced septa. A pore in each septum permits extensive movement of cytoplasm and organelles (including nuclei) from one segment to the next.

Sac fungi are distinguished by the production of sacs called **asci** (singular *ascus*), which at maturity contain sexually produced haploid **ascospores** (see Figure 21.17A). The ascus is the characteristic sexual reproductive structure of the sac fungi. In the past, the sac fungi were classified on the basis of whether or not the asci are contained within a specialized fruiting structure known as an ascoma (plural *ascomata*) and on differences in the morphology of that fruiting structure. DNA sequence analyses have resulted in a revision of these traditional groupings, however.

SAC FUNGUS YEASTS Some species of sac fungi are unicellular yeasts. The 1,000 or so species in this group are among the most important domesticated fungi. Perhaps the best known is baker's, or brewer's, yeast (*Saccharomyces cerevisiae*; see Figure 21.2 and Key Concept 21.4), which metabolizes glucose obtained from its environment into ethanol and carbon dioxide by fermentation. Other sac fungus yeasts live on fruits such as figs and grapes and play an important role in the making of wine. Many others are associated with insects. In the guts of some insects, they provide enzymes that break down materials that are otherwise difficult for the insects to digest, especially cellulose.

Sac fungus yeasts reproduce asexually by budding. Sexual reproduction takes place when two adjacent haploid cells of dissimilar mating types fuse. In some species, the resulting zygote buds to form a diploid cell population. In others, the zygote nucleus undergoes meiosis immediately. When this happens, the entire cell becomes an ascus. Depending on whether the products of meiosis then undergo mitosis, a yeast ascus contains either eight or four ascospores, which germinate to become haploid cells. The sac fungus yeasts have lost the dikaryon stage.

FILAMENTOUS SAC FUNGI Most sac fungi are filamentous species, such as the cup fungi (**FIGURE 21.18A**), in which the ascomata are cup-shaped and can be as large as several centimeters across (although most are much smaller). The inner surfaces of the ascomata, which are covered with a mixture of specialized hyphae and asci, produce huge numbers of spores. The edible ascomata of some species, including morels (**FIGURE 21.18B**) and truffles, are regarded by humans as gourmet delicacies (and can sell at prices higher than gold). The underground ascomata of truffles have a strong odor that attracts mammals such as pigs, which then eat the fungi and disperse the spores. Humans sometimes take advantage of pigs' attraction to truffles and use pigs to locate truffles for human consumption.

The sexual reproductive cycle of filamentous sac fungi includes the formation of a dikaryon, although this stage is relatively brief compared with that in club fungi. Many filamentous sac fungi form multinucleate mating structures (see Figure 21.17A). Mating structures of two different mating types fuse and produce a dikaryotic mycelium, containing nuclei from both mating types. The dikaryotic mycelium often forms a cup-shaped

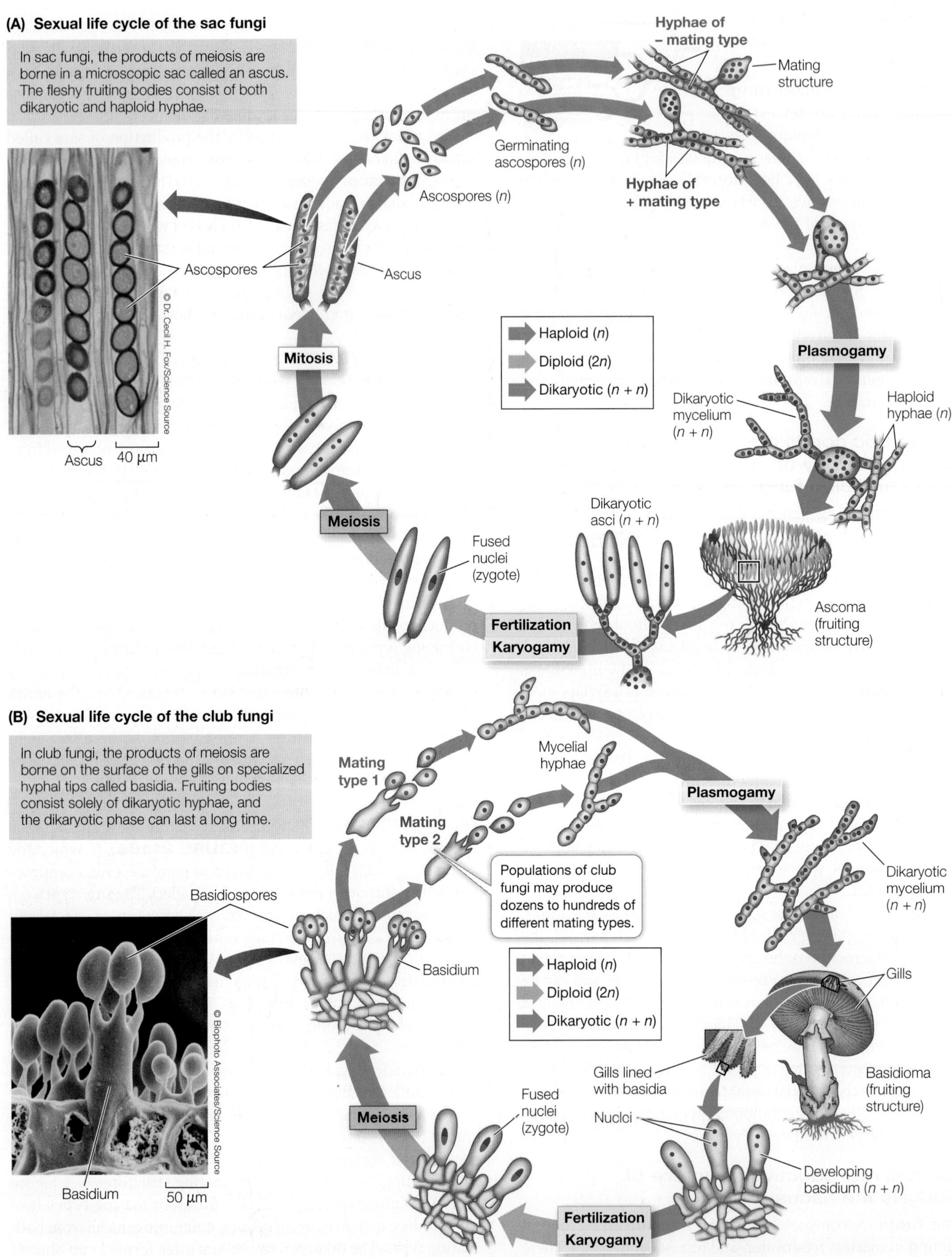
(A) Sexual life cycle of the sac fungi
In sac fungi, the products of meiosis are borne in a microscopic sac called an ascus. The fleshy fruiting bodies consist of both dikaryotic and haploid hyphae.
Ascospores
Ascus
40 μm
© Dr. Cecil H. Fox/Science Source
Hyphae of – mating type
Mating structure
Germinating ascospores (n)
Ascospores (n)
Hyphae of + mating type
Ascus
Haploid (n)
Diploid (2n)
Dikaryotic (n + n)
Mitosis
Plasmogamy
Dikaryotic mycelium (n + n)
Haploid hyphae (n)
Meiosis
Dikaryotic asci (n + n)
Fused nuclei (zygote)
Ascoma (fruiting structure)
Fertilization
Karyogamy
(B) Sexual life cycle of the club fungi
In club fungi, the products of meiosis are borne on the surface of the gills on specialized hyphal tips called basidia. Fruiting bodies consist solely of dikaryotic hyphae, and the dikaryotic phase can last a long time.
Mating type 1
Mycelial hyphae
Plasmogamy
Mating type 2
Populations of club fungi may produce dozens to hundreds of different mating types.
Dikaryotic mycelium (n + n)
Basidiospores
Basidium
Haploid (n)
Diploid (2n)
Dikaryotic (n + n)
Gills
© Biophoto Associates/Science Source
Gills lined with basidia
Basidioma (fruiting structure)
Fused nuclei (zygote)
Nuclei
Meiosis
Developing basidium (n + n)
Basidium
50 μm
Fertilization
Karyogamy

(A) *Cookeina tricholoma*

© Thomas Marent/Minden Pictures

(B) *Morchella esculenta*

© blickwinkel/Alamy Stock Photo

FIGURE 21.18 Sac Fungi **(A)** These brilliant red cups are the ascomata of a cup fungus. **(B)** Morels, which have a spongelike ascoma and a subtle flavor, are considered a culinary delicacy by humans.

ascoma, which bears the asci. Only after the formation of asci do the nuclei from the two mating types finally fuse. Both nuclear fusion and the subsequent meiosis that produces haploid ascospores take place within individual asci. The ascospores are ultimately released (sometimes shot off forcefully) by the ascus to begin the new haploid generation.

The sac fungi also include many of the filamentous fungi known as molds. **Molds** consist of filamentous hyphae that do not form large ascomata, although they can still produce asci and ascospores. Many molds are parasites of flowering plants. Chestnut blight and Dutch elm disease are both caused by molds. The chestnut blight fungus, which was introduced to the United States in the 1890s, had destroyed the American chestnut as a commercial species by 1940. Before the blight, this species accounted for more than half the trees in the eastern North American forests. Another familiar story is that of the American elm. Sometime before 1930, the Dutch elm disease fungus (first discovered in the Netherlands but native to Asia) was introduced into the United States on infected elm logs from Europe. Spreading rapidly—especially via bark beetles—the fungus destroyed great numbers of American elm trees.

Other plant pathogens among the sac fungi include the powdery mildews that infect cereal crops, lilacs, and roses, among many other plants. Mildews can be a serious problem for farmers and gardeners, and a great deal of research has focused on ways to control these agricultural pests.

The filamentous sac fungi can also reproduce asexually by means of conidia that form at the tips of specialized hyphae (**FIGURE 21.19**). Small chains of conidia are produced by the millions and can survive for weeks in nature. The conidia are often what give molds their characteristic colors.

The basidium is the sexual reproductive structure of club fungi

Club fungi (Basidiomycota) produce some of the most spectacular fruiting structures found among the fungi. These fruiting

◀ **FIGURE 21.17 Sexual Life Cycles among the Dikarya** **(A)** In sac fungi, the products of meiosis are borne in a microscopic sac called an ascus. The fleshy fruiting structure, the ascoma, consists of both dikaryotic and haploid hyphae. **(B)** The basidium is the characteristic sexual reproductive structure of the club fungi. The fruiting structures, called basidiomata, consist solely of dikaryotic hyphae, and the dikaryotic phase can last a long time.

How does the dikaryotic condition differ from diploidy?

Activity 21.2 Life Cycle of a Dikaryotic Fungus

© Nature's Geometry/Science Source

FIGURE 21.19 Conidia Chains of conidia (yellow) are developing at the tips of specialized hyphae arising from a *Penicillium chrysogenum* mold. This species is used to produce the antibiotic penicillin.

(A) ***Amanita muscaria***

© RF Company/Alamy Stock Photo

(B) ***Laetiporus sulphureus***

© blickwinkel/Alamy Stock Photo

FIGURE 21.20 Club Fungus Basidiomata **(A)** The fly agaric is a poisonous mushroom that contains hallucinogenic compounds. **(B)** A bracket fungus growing parasitically on a tree. Although this particular species is occasionally eaten when young, many similar-appearing bracket fungi are poisonous.

structures, called **basidiomata** (singular *basidioma*), include mushrooms of all kinds, puffballs (see Figure 21.4), and the bracket fungi often encountered on trees and fallen logs in a damp forest. About 30,000 species of club fungi have been described. They include about 4,000 species of mushrooms, including both poisonous (**FIGURE 21.20A**) and edible species. Bracket fungi (**FIGURE 21.20B**) play an important role in the carbon cycle by breaking down wood. They also do great damage to both cut lumber and timber stands. Some of the most economically damaging plant pathogens are club fungi, including the rust fungi and smut fungi that parasitize cereal grains. In contrast, other club fungi contribute to the survival of plants as fungal partners in ectomycorrhizae.

The hyphae of club fungi characteristically have septa with small, distinctive pores. As they grow, haploid hyphae of different mating types meet and fuse, forming dikaryotic hyphae, each cell of which contains two nuclei, one from each parent hypha. The dikaryotic mycelium grows and eventually, when triggered by rain or another environmental cue, produces a basidioma. The dikaryon stage may persist for years, or even centuries. This pattern contrasts with the life cycle of the sac fungi, in which the dikaryon is found only in the stages leading up to formation of the asci.

The **basidium** (plural *basidia*), a swollen cell at the tip of a specialized hypha, is the characteristic sexual reproductive structure of the club fungi (see Figure 21.17B). In mushroom-forming club fungi, the basidia typically form on specialized structures of the basidiomata known as gills. The basidium is the site of nuclear fusion and meiosis and thus plays the same role in the club fungi as the ascus does in the sac fungi and the zygosporangium does in the zygospore fungi.

After nuclei fuse in the basidium, the resulting diploid nucleus undergoes meiosis, and the four resulting haploid nuclei are incorporated into haploid **basidiospores**, which form on tiny stalks on the outside of the basidium. A single basidioma of the common bracket fungus *Ganoderma applanatum* can produce as many as 4.5 *trillion* basidiospores in one growing season. Basidiospores typically are forcibly discharged from their basidia and then germinate, and give rise to hyphae with haploid nuclei.

REVIEW & APPLY | 21.3

Sexual reproduction is common in some groups of fungi but has never been observed in others. Many fungal species have two or more genetically distinct but visually indistinguishable mating types. To mate, individuals of most sac fungi and club fungi fuse to produce dikaryotic fruiting bodies, in which two genetically different haploid nuclei coexist in each cell of the multicellular structure.

1. Explain the concept of mating types. How are they different from male and female sexes?
2. Describe how microsporidia infect the cells of their animal hosts.
3. What feature of the chytrids suggests that the fungi had an aquatic ancestor?
4. What is the role of the zygospore in the life cycle of zygospore fungi?
5. What distinguishes the fruiting bodies of sac fungi from those of club fungi?

Fungi are of special interest to biologists because of the roles they play in interactions with other organisms, including humans. But they are also useful as tools for studying many kinds of biological problems and for finding solutions to those problems.

21.4 Fungi Have Many Practical Uses

LEARNING OBJECTIVES

By the end of this key concept you should be able to:

21.4.1 **Describe four ways that fungi are used to produce food and drink for humans.**

21.4.2 **Provide an example of how fungi can be used for medical purposes.**

21.4.3 **Explain two ways that fungi can be used as bioindicators of environmental health.**

We've briefly noted the important part that fungi play in the production of human foods and beverages. We have also described the diverse roles that fungi play in natural ecosystems, from decomposers to pathogens to mutualistic partners. These diverse ecological roles have led to the use of fungi in studies of environmental change and in remediation of environmental pollution. Many fungi are also important model organisms for laboratory investigations of basic biological process. Others, as you saw at the opening of this chapter, have given us treatments for human diseases.

Fungi are important in producing food and drink

Grains from grasses provide most of the world's food supply for humans. But in most cases, we do not eat these grains directly as they are produced by the plants. Instead, we use them as a source of starch. To make the starch more pleasing and digestible for human consumption, we usually convert it to more complex and tasty forms of food and drink, often with the help of fungi.

Baker's (or brewer's) yeast (*Saccharomyces cerevisiae*) converts the starch from grain into ethanol. This process also forms carbon dioxide bubbles in bread dough, causing it to rise, which gives baked bread its light texture. The ethanol and carbon dioxide are baked away in bread making (which produces the pleasant aroma of baking bread). In contrast, the ethanol and carbon dioxide are retained when yeast is used to ferment grain into beer. The carbon dioxide gives beer its fizz, and the alcohol and yeast contribute to the taste and appeal of beer to those who enjoy it. Sugars, especially from fruits such as grapes, are also converted into alcohol and carbon dioxide by yeasts in the production of wine (although the carbon dioxide is not retained in most finished wine, as it is in beer). Many different strains of *S. cerevisiae* are used in wine production, which contributes to the distinctive nature of wine from different regions and wineries. Many other species of local, native yeasts are also used in producing distinctive local wines and beers. For example, fission yeast (*Schizosaccharomyces pombe*) was first isolated from African millet beer. Fission yeast takes its specific name (*pombe*) from the Swahili word for beer.

Media Clip 21.2 Time Lapse of Beer Fermentation

Brown molds of the genus *Aspergillus* are important in some human diets. *Aspergillus tamarii* acts on soybeans in the production of soy sauce, and *A. oryzae* is used in brewing the Japanese alcoholic beverage sake from rice. *Aspergillus niger* is the source of most commercial citric acid production. Citric acid gives food and soft drinks a tart taste and is also used as a food preservative. But some species of *Aspergillus* that grow on grains and on nuts such as peanuts and pecans produce extremely carcinogenic (cancer-inducing) compounds called aflatoxins. Aflatoxins can occur in high concentrations in foods such as peanut butter. In the United States and most other industrialized countries, moldy grain infected with *Aspergillus* is typically thrown out. In Africa, where food is scarcer, the grain is often eaten, moldy or not, and causes severe health problems, including high levels of certain cancers.

Penicillium is a genus of green molds, of which some species produce the antibiotic penicillin, as noted at the beginning of this chapter. But several species of *Penicillium* are important for food production as well. For example, *P. camemberti* and *P. roqueforti* are the organisms responsible for the characteristic strong flavors of Camembert and Roquefort cheeses, respectively.

Many fungi serve directly as a human food source. Mushroom enthusiasts seek out the delicious fruiting structures of a wide variety of edible sac and club fungi. In the United States, relatively few species of mushrooms are grown commercially, and wild mushrooms are collected mostly for personal consumption. But in many parts of the world, a wide variety of wild mushrooms are collected for sale and consumption. Fungi used for food are not limited to fruiting bodies such as mushrooms, however. Various species of lichens are eaten in Arctic regions as well as in parts of North America and Asia. In southwestern China, for example, several species of lichens are used as a primary ingredient in cooking (**FIGURE 21.21**).

Courtesy of David Hillis

FIGURE 21.21 Some Lichens Are Edible In southwestern China, several species of lichens that grow on tree bark serve as a primary ingredient in the local cuisine.

Fungi provide important weapons against diseases and pests

We started this chapter with a brief description of the discovery of penicillin. The discovery of antibiotics produced by fungi revolutionized medical treatment of bacterial diseases in humans and their domestic animals. Live strains of fungi are also used to combat various pest species of plants and animals.

In Africa, the parasitic plant witchweed (*Striga*) causes crop losses of about U.S. $7 billion every year. A group of Canadian biologists discovered that a strain of the mold *Fusarium oxysporum* could be applied to crops to control witchweed without harming the crop plants (**FIGURE 21.22**). Other strains of *Fusarium* that preferentially attack coca plants, the source of cocaine, have been proposed to combat illicit drug production. Still other fungi are used to attack various animal pests, such as termites and aphids, and even malaria-carrying mosquitoes.

Lichen diversity and abundance are indicators of air quality

Lichens can live in many harsh environments where few other species can survive, as you saw in Key Concept 21.2. In spite of their hardiness, however, lichens are highly sensitive to air pollution because they are unable to excrete any toxic substances they absorb. This sensitivity means that lichens are good biological indicators of air pollution levels. It also explains why they are not commonly found in heavily industrialized regions or in large cities.

FIGURE 21.22 A Pathogenic Fungus Attacks a Parasitic Plant The fungus *Fusarium oxysporum* is a potent pathogen of witchweed (*Striga*), a parasitic plant that attacks crops. The fungal spores are shown in blue; fungal filaments are in tan. Both colors were added to enhance this scanning electron micrograph.

Monitoring the diversity and abundance of lichens growing on trees is a practical and inexpensive system for gauging air quality around cities. Maps of lichen diversity provide environmental biologists with a tool for tracking the distribution of air pollutants and their effects. Sensitive biological indicators of pollution, such as lichen growth, allow biologists to monitor air quality without the use of specialized equipment. Lichens are naturally distributed across the environment, and they can also provide a long-term measure of the effects of air pollution across many seasons and years.

Fungi record and help remediate environmental pollution

Each year, biologists deposit samples of many groups of organisms in the collections of natural history museums. These museum collections serve many purposes, one of which is to document changes in the biota of our planet over time.

Collections of fungi made over many decades or centuries provide a record of the environmental pollutants that were present when the fungi were growing. Biologists can analyze these historical samples to see how different sources of pollutants were affecting our environment before anyone thought to take direct measurements (**INVESTIGATION FIGURE 21.23**). These long-term records are also useful for analyzing the effectiveness of cleanup efforts and regulatory programs for controlling environmental pollutants.

You have already seen that fungi are critical to the planetary carbon cycle because of their role in breaking down dead organic matter. Fungi are also used in remediation efforts to help clean up sites that have been polluted by oil spills or contaminated with toxic petroleum-derived hydrocarbons. Many herbicides, pesticides, and other synthetic hydrocarbons are broken down primarily through the action of fungi.

Reforestation may depend on mycorrhizal fungi

When a forest is cut down, it is not just the trees that are lost. A forest is an ecosystem that depends on the interaction of many species. As we have discussed, many plants depend on close relationships with mycorrhizal fungal partners. When trees are removed from a site, the populations of mycorrhizal fungi there decline rapidly. If we wish to restore the forest on the site, we cannot simply replant it with trees and other plants and expect them to survive. The mycorrhizal fungal community must be reestablished as well. For large forest restoration projects, a planned succession of plant growth and soil improvement is often necessary before forest trees can be replanted. As the community of soil fungi gradually recovers, trees that have been inoculated with appropriate mycorrhizal fungi in tree nurseries can be planted to reintroduce greater diversity to the soil fungal community.

Fungi are used as model organisms in laboratory studies

Much of what we know about many basic aspects of cell and molecular biology comes from the study of model organisms. Among

INVESTIGATION

FIGURE 21.23 Using Fungi to Study Environmental Contamination A. Russell Flegal and his colleagues analyzed museum samples of lace lichens (*Ramalina menziesii*) collected between 1892 and 2006 near San Francisco, California.[a] Their goals were to determine the sources of lead contamination and to evaluate the effectiveness of efforts to reduce this contamination.

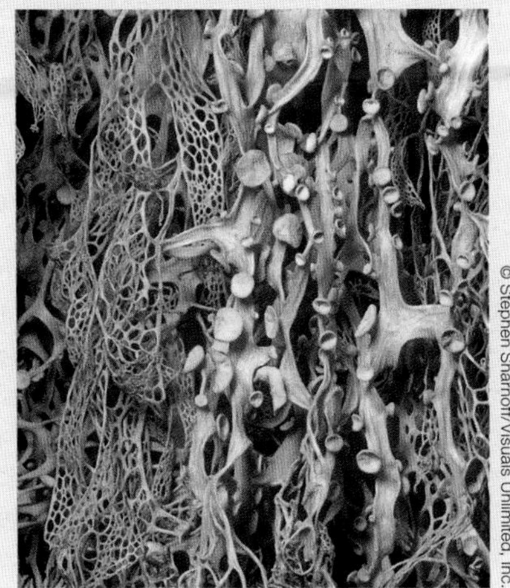

QUESTION

Is environmental lead contamination through time consistent with known sources of pollution, and have efforts to reduce lead pollution been effective?

METHOD

1. Obtain museum samples of lace lichen (*Ramalina menziesii*) collected near San Francisco.
2. Measure concentrations of lead (Pb) as well as the ratios of two lead isotopes, ^{206}Pb and ^{207}Pb.
3. Compare the ratio of the two isotopes in the lichen samples with that of the possible sources:
   a. Lead smelter, in operation from 1885 to 1971: $^{206}Pb/^{207}Pb$ ratio of 1.15–1.17
   b. Leaded gasoline, in use from 1930s to the early 1980s: $^{206}Pb/^{207}Pb$ ratio of 1.18–1.23
   c. Resuspension of historic lead contamination as atmospheric aerosols in recent decades: $^{206}Pb/^{207}Pb$ ratio of 1.16–1.19

RESULTS

Peak environmental lead contamination occurred between the late 1950s and 1980, with a rapid drop in the early 1980s after the closure of the smelter and the elimination of leaded gasoline. Average $^{206}Pb/^{207}Pb$ ratios of lead were below 1.17 before the 1930s, between 1.18 and 1.19 in the 1940s and 1950s, and >1.22 in the early 1980s. The ratios have declined since then.

CONCLUSIONS

1. The levels of environmental lead contamination and the ratios of lead isotopes are consistent with the identified sources.
2. The closure of the lead smelter and elimination of leaded gasoline have significantly reduced environmental lead contamination.

ANALYZE THE DATA

Sample	Year collected	Lead concentration (μg of Pb/g lichen)	$^{206}Pb/^{207}Pb$ ratio
1	1892	11.9	1.165
2	1894	4.0	1.155
3	1906	13.7	1.154
4	1907	22.9	1.157
5	1945	49.9	1.187
6	1957	34.2	1.185
7	1978	50.9	1.221
8	1982	10.0	1.215
9	1983	4.6	1.224
10	1987	1.0	1.198
11	1988	1.3	1.199
12	1995	1.9	1.202
13	2000	0.4	1.184
14	2006	1.8	1.184

Data from A. R. Flegal et al. 2010. *Environ Sci Technol* 44: 5613–5618.

1. Using the data in the table, plot lead concentration in the lichen samples against year of sample collection. Make a second plot, this one of $^{206}Pb/^{207}Pb$ ratio against year of sample collection. Calculate trend lines for your plots by plotting the average for each adjacent pair of observations, from 1892 to 2006. (In other words, plot the average for sample 1 and 2, then the average for sample 2 and 3, and so on, to produce a trend line for the data.)
2. Do your analyses support the conclusions of the study? How do you think you could improve the analysis?

[a]A. R. Flegal et al. 2010. *Environ Sci Technol* 44: 5613–5618.

A similar **Analyze the Data** exercise may be assigned in **Achieve**.

the eukaryotes, some fungi have numerous advantages over model plant and animal systems for laboratory investigations.

Of particular importance as model organisms are several species of sac fungi: *Aspergillus nidulans* (a brown mold), *Neurospora crassa* (a red bread mold), *Saccharomyces cerevisiae* (baker's, or brewer's, yeast), and *Schizosaccharomyces pombe* (fission yeast). These species can be cultured in large numbers in small spaces, and they have short generation times, so that genetic investigations can be conducted in days rather than years. Furthermore, their genomes are relatively small and encode relatively few genes compared with those of most plants and animals, so it is easier to elucidate the functions of the fungal genes responsible for basic biological functions.

REVIEW & APPLY | 21.4

R Fungi are important to humans in many ways. Some species are consumed directly as food, while others are important in food production. Fungi serve as important indicators of ecosystem health and are critical in reforestation and in pollution remediation efforts. Several species are important model organisms for studies of eukaryotic cell and molecular biology. Fungi are widely used to combat diseases and pests.

A

1. What are some of the ways in which fungi or fungal products contribute to the human food supply?
2. What are some advantages of using surveys of lichen diversity and museum collections of lichens to measure long-term changes in air quality, compared with direct measurements of atmospheric pollutants?
3. Can you develop a strategy for tree harvest that would ease the difficulty and expense of reforestation projects by retaining viable communities of mycorrhizal fungi?
4. Consider the data below for lichens at five survey sites. Predict the relative order of the sites with respect to their distance from the center of a large city. Other factors (besides distance to city center, such as prevailing wind direction) might affect your prediction. Can you think of two other major factors that might influence these results?

Site number	Number of lichen species	Tree branches covered in lichens (%)
1	5	38
2	1	2
3	3	15
4	8	75
5	13	100

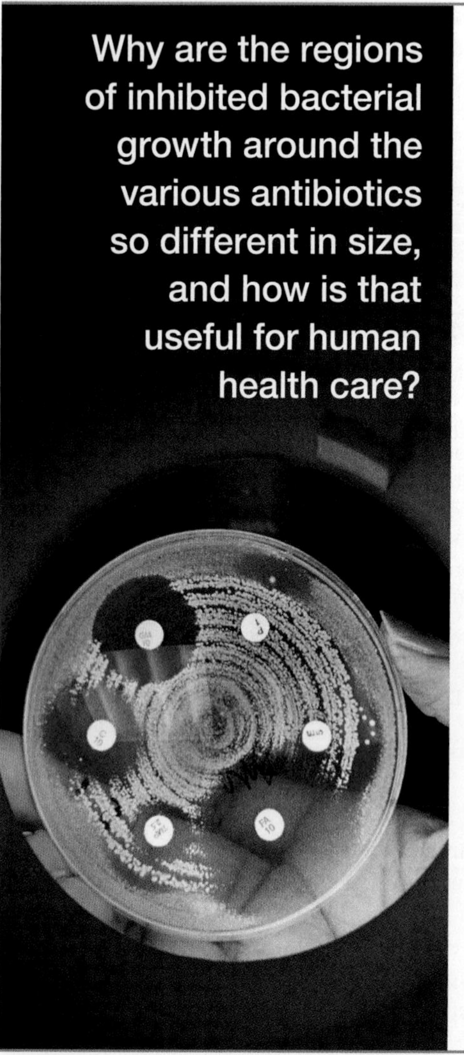

Why are the regions of inhibited bacterial growth around the various antibiotics so different in size, and how is that useful for human health care?

Answer: The variable regions of inhibited growth that surround the various antibiotic samples reflect the differential effectiveness of antibiotics to a particular bacterium. Physicians who treat a bacterial infection need to know which antibiotic will be most effective for clearing that infection. An antibiotic that might work well to fight one infection might work poorly to fight another. Experiments like the one shown identify which antibiotics are best at inhibiting growth of a tested bacterium. Bigger zones of inhibited growth indicate more effective antibiotics.

Beginning in the 1940s, antibiotics derived from fungi ushered in a "golden age" of freedom from bacterial infections. Today, however, many of these antibiotics are losing their effectiveness as pathogenic bacteria evolve resistance to them (**FIGURE 21.24**).

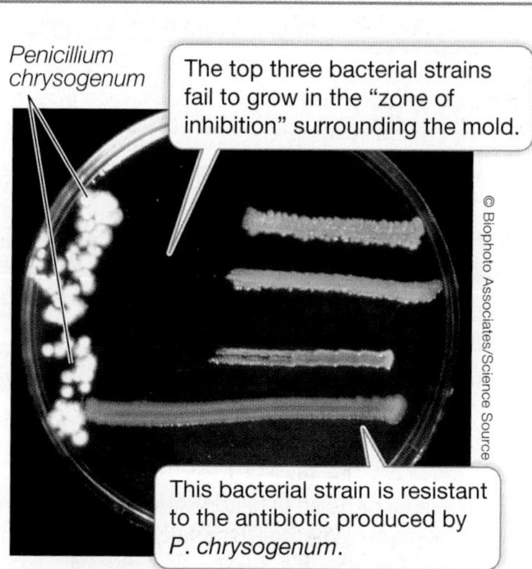

FIGURE 21.24 Penicillin Resistance In a petri dish similar to those in Alexander Fleming's lab, four strains of a pathogenic bacterium have been cultured along with *Penicillium* mold. One strain is resistant to the mold's antibiotic substance, as is evidenced by its growth up to the mold.

Most medical antibiotics are chemically modified forms of substances that are found naturally in fungi and other organisms. Fungi naturally produce antibiotic compounds to defend themselves against bacterial growth and to reduce competition from bacteria for nutritional resources. These naturally occurring compounds are usually chemically modified to increase their stability, improve their effectiveness, and facilitate synthetic production. From the late 1950s to the late 1990s, no new major classes of antibiotics were discovered. In recent years, however, three new classes of antibiotics have been synthesized based on information learned from naturally occurring, fungally derived antibiotics, leading to improved treatment of some formerly resistant strains of bacteria.

21.1 Fungi Digest Food Outside Their Bodies

- Fungi are distinguished from other **OPISTHOKONTS** by **ABSORPTIVE HETEROTROPHY** and by the presence of **CHITIN** in their cell walls.
- Some fungi are **SAPROBES**, others are parasites or predators, and some are mutualists.
- **YEASTS** are unicellular, free-living fungi.
- The body of a multicellular fungus is a **MYCELIUM**—a meshwork of filaments called **HYPHAE**. Hyphae may be **SEPTATE** (having **SEPTA**) or **COENOCYTIC** (multinucleate) (**FIGURE 21.3**).

FIGURE 21.3

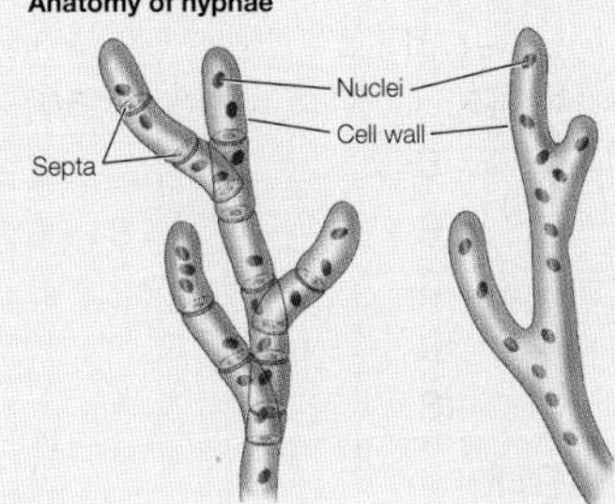

QUESTIONS

1. What is a fundamental difference in the way that fungi obtain their nutrition, as compared with plants and animals?
2. How do some fungi blur the distinctions between unicellularity and multicellularity?

21.2 Fungi Are Decomposers, Parasites, Predators, or Mutualists

- Saprobic fungi, which act as decomposers, make crucial contributions to the recycling of elements, especially carbon.
- Many fungi are parasites, harvesting nutrients from host cells by means of **HAUSTORIA** (**FIGURE 21.5**).
- Some fungi associate with unicellular green algae, cyanobacteria, or both to form **LICHENS**, which live on exposed surfaces of rocks, trees, and soil (**FIGURE 21.8**).
- **MYCORRHIZAE** are mutualistic associations of fungi with plant roots. They improve a plant's ability to take up nutrients and water. **ENDOPHYTIC FUNGI** live within plants and may provide their hosts with protection from herbivores and pathogens.

FIGURE 21.5

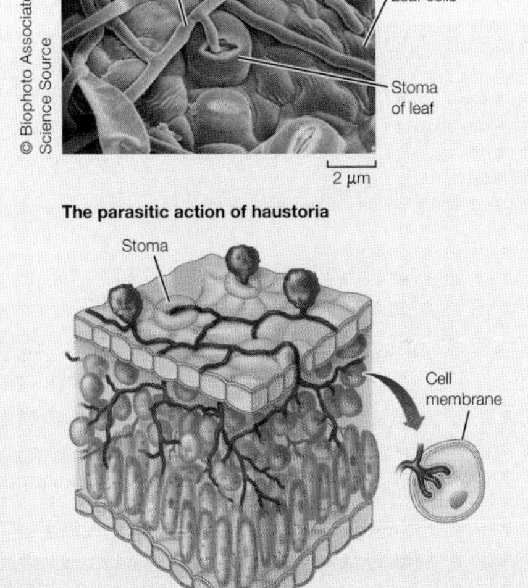

FIGURE 21.8

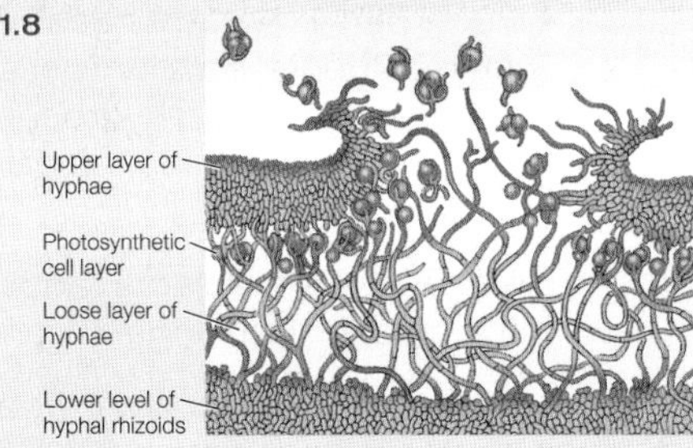

QUESTIONS

1. How would life on Earth be fundamentally different without fungi?
2. What are two distinct kinds of mycorrhizal mutualisms that form with plants?

QUESTIONS

21.3 Sex in Fungi Involves Multiple Mating Types

FIGURE 21.11

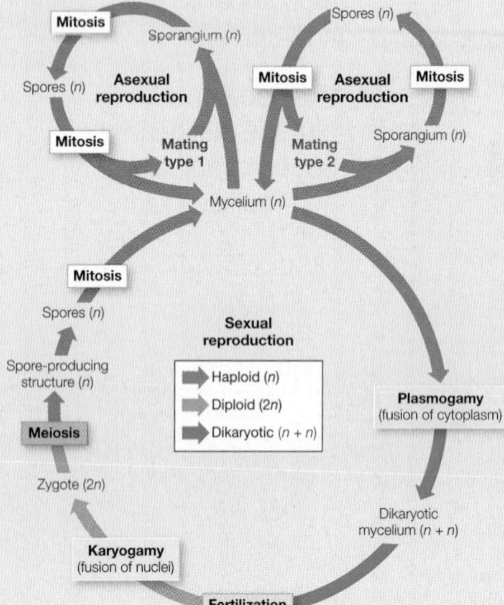

Go to ACTIVITIES 21.1 and 21.2 and ANIMATION 21.1

- The **MICROSPORIDIA** and **CHYTRIDS** diversified early in fungal evolution. The zygospore fungi, arbuscular mycorrhizal fungi, sac fungi, and club fungi form a monophyletic group, and the latter two groups form the clade **DIKARYA**.
- Many species of fungi reproduce both sexually and asexually. In many fungi, sexual reproduction occurs between individuals of different **MATING TYPES** (**FIGURE 21.11**).
- In the sexual reproduction of terrestrial fungi, **PLASMOGAMY** (fusion of cytoplasm) precedes **KARYOGAMY** (fusion of nuclei).
- **ZYGOSPORE FUNGI** have a resting stage known as a **ZYGOSPORE**, which contains many diploid nuclei.
- **ARBUSCULAR MYCORRHIZAL FUNGI** form mutualistic associations with plant roots and played an important role in the colonization of land by plants.
- In **SAC FUNGI** and **CLUB FUNGI**, a mycelium containing two genetically different haploid nuclei, called a **DIKARYON**, is formed. The dikaryotic ($n + n$) condition is unique to the fungi.

1. How is the dikaryotic condition different from haploid and diploid states?
2. Why are mating types in most fungi not identified as male and female?

21.4 Fungi Have Many Practical Uses

FIGURE 21.21

Courtesy of David Hillis

FIGURE 21.22

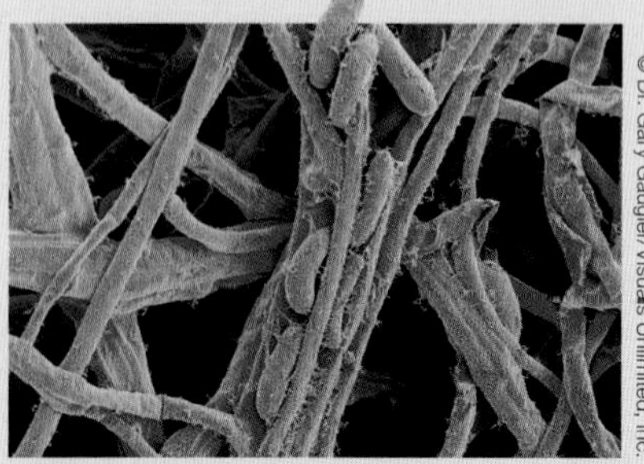

© Dr. Gary Gaugler/Visuals Unlimited, Inc.

- Some fungi are consumed as food by humans; other fungi are critical in baking, fermentation, and flavoring food (**FIGURE 21.21**).
- Fungi play important roles in cleaning up environmental pollutants such as synthetic petroleum-derived hydrocarbons. The diversity and abundance of lichen growth on trees are sensitive indicators of air quality.
- Reforestation projects require restoration of the mycorrhizal fungal community.
- Fungi provide important weapons against diseases and pests (**FIGURE 21.22**).

1. What critical function do fungi play in baking and brewing?
2. Why are lichens particularly useful as indicators of environmental pollution?

Go to **Achieve** for the e-book, animations, activities, and additional resources and assignments.

Animal Origins and Diversity 22

Of the 1.8 million species of life that have been discovered and named by biologists, about 1.3 million are animals. One group of animals, the insects, accounts for more than 1 million of these species, or more than half of all known species of living organisms. Despite these enormous numbers, biologists estimate that the number of undiscovered species of insects far exceeds the number of species that are known today.

Besides insects, which other groups of animals are thought to contain many more species than are known at present?

You will find the answer to this question on page 573.

KEY CONCEPTS

- **22.1** Distinct Body Plans Evolved among the Animals
- **22.2** Large Multicellular Animal Groups Arose in Parallel
- **22.3** Protostomes Have an Anterior Brain and a Ventral Nervous System
- **22.4** Arthropods Are Diverse and Abundant
- **22.5** Deuterostomes Include Echinoderms, Hemichordates, and Chordates
- **22.6** Life on Land Contributed to Vertebrate Diversification
- **22.7** Humans Evolved among the Primates

22.1 | Distinct Body Plans Evolved among the Animals

LEARNING OBJECTIVES

By the end of this key concept you should be able to:

22.1.1 **Explain how differences in body symmetry, body cavities, segmentation, appendages, and nervous systems are functionally related to movement and feeding.**

22.1.2 **Describe two important functions of nervous systems that allow many animals to have better control over mobility in the water, land, and air.**

How do we recognize an organism as an animal? The answer may seem obvious for many familiar animals, but less so for groups such as sponges, which were once thought to be plants.

Some of the general characteristics we associate with animals include:

- *Multicellularity*. In contrast to the bacteria, archaea, and most protists (see Chapters 18 and 19), all animals are multicellular. Animal life cycles feature complex patterns of development from a single-celled zygote into a multicellular adult.
- *Heterotrophic metabolism*. In contrast to most plants, all animals are heterotrophs. Animals are able to synthesize very few organic molecules from inorganic chemicals, so they must obtain the necessary organic molecules from their environment.
- *Internal digestion*. Although the fungi are also heterotrophs (see Chapter 21), animals and fungi digest their food differently. Whereas fungi digest food outside their bodies, most animals use internal processes to break down materials from their environment into the organic molecules they need. Most animals ingest food into an internal **gut** that is continuous with the outside environment and in which digestion takes place.
- *Movement and nervous systems*. In contrast to the majority of plants and fungi, most animals can move their bodies. This movement is often coordinated through a well-developed nervous system, which also typically functions as a sensory system. Animals must move to find food or bring food to them. Muscle tissue and nervous systems are unique to animals, and many animal body plans are specialized for movement and detection of prey.

Although these general features help us recognize animals, none is diagnostic for all animals. Some animals do not move, at least during certain life stages, and some plants and fungi do have limited movement. Some animals lack a nervous system, and the diffuse nerve nets (which lack a coordinating brain) of ctenophores and cnidarians are distinct and independently evolved from the **central nervous systems** (which have a coordinating brain) of bilaterians. Some animals lack a gut, and many multicellular organisms are not animals. So on what basis do we group all animals together in a single clade?

Animal monophyly is supported by gene sequences and cellular morphology

The most convincing evidence that all the organisms considered to be animals share a common ancestor comes from phylogenetic analyses of their gene sequences. Relatively few complete animal genomes are available, but more are being sequenced each year. Analyses of these genomes and of many individual gene sequences have shown that the animals are indeed monophyletic. The best-supported phylogenetic tree for the major animal groups is shown in **FIGURE 22.1**.

Surprisingly few morphological features are shared across all animal species. The morphological synapomorphies (shared, derived traits) that are present are evident primarily at the cellular level. They include a common set of extracellular matrix molecules, including collagen and proteoglycans, and several unique features of the junctions between animal cells. Although some animals in a few groups lack one of these cellular characteristics, it is believed that these traits were possessed by the ancestor of all animals and were subsequently lost in those groups.

Similarities in the organization and function of Hox and other developmental genes provide additional evidence of developmental mechanisms shared by a common animal ancestor. The Hox genes specify body pattern and axis formation, leading to developmental similarities across animals.

LINK Hox genes and their role in animal development and evolution are discussed in **Chapter 15**.

The common ancestor of animals may have been a colonial flagellated protist similar to existing colonial choanoflagellates (**FIGURE 22.2A**), which have similarities to the multicellular sponges (**FIGURE 22.2B**). Why did these early animal ancestors begin to form multicellular colonies? One hypothesis is that multicellular colonies are more efficient than are single cells

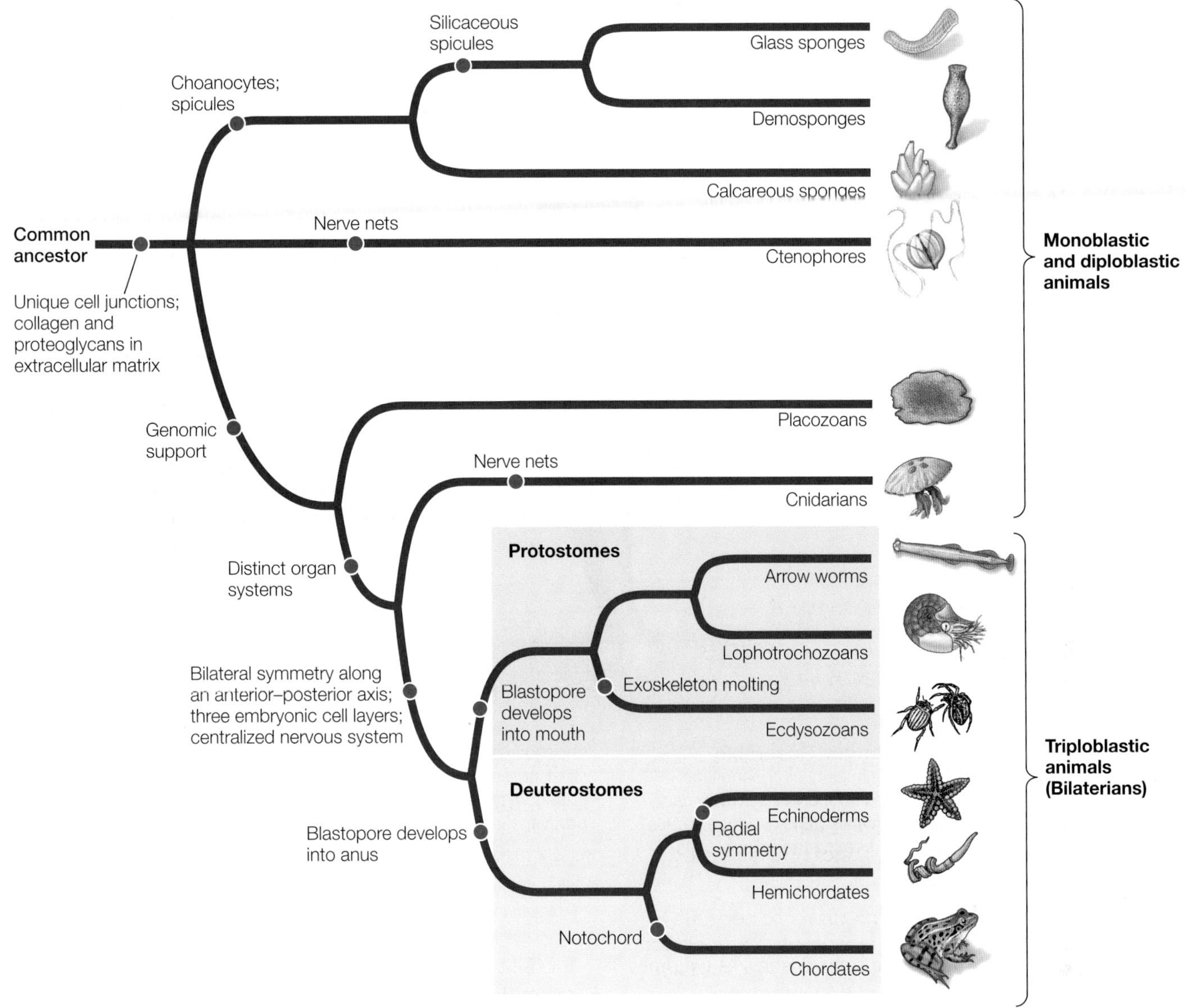

FIGURE 22.1 Animal Phylogeny This tree presents the best-supported current hypotheses of the evolutionary relationships among major groups of animals. Different analyses place either the sponges or the ctenophores as the sister group of the remaining animals. The traits highlighted by red circles will be explained as you read this chapter.

Based on this tree, which of the depicted traits evolved multiple times among animals, and in which lineages?

at capturing their prey. Experiments with living species of choanoflagellates show that they spontaneously form multicellular colonies in response to signaling compounds that are found on certain species of planktonic bacteria they eat (**INVESTIGATION FIGURE 22.3**).

Once multicellular colonies were formed in early animal lineages, it is likely that certain cells in the colony began to be specialized—some for movement, others for nutrition, others for reproduction, and so on. Once this functional specialization had begun, cells could continue to differentiate. Coordination among groups of cells could have improved by means of specific regulatory and signaling molecules that guided differentiation and migration of cells in developing embryos. Such coordinated groups of cells eventually evolved into the larger and more complex organisms that we call animals.

Nearly 80 percent of the 1.8 million named species of living organisms are animals, and millions of additional animal species await discovery. Evidence about the evolutionary relationships among animal groups can be found in fossils, in patterns of embryonic development, in the morphology and physiology of living animals, in the structure of animal proteins, and in gene sequences. Increasingly, studies of higher-level relationships have

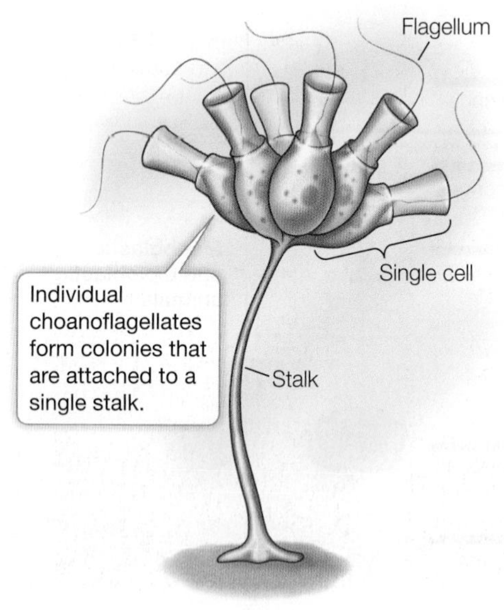

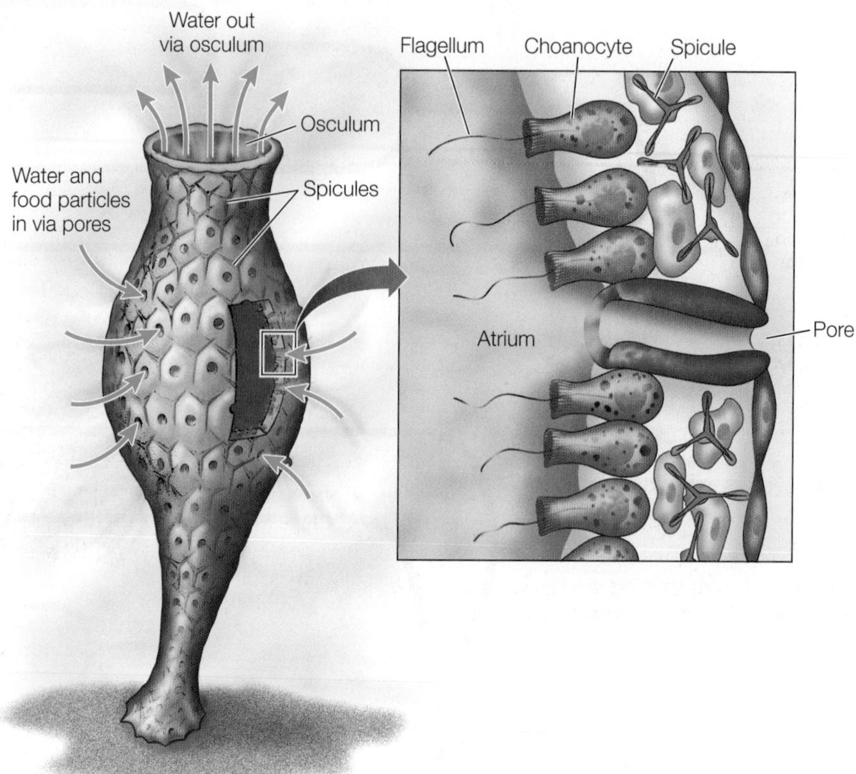

FIGURE 22.2 Choanocytes in Sponges Resemble Choanoflagellate Protists (A) The similarity of choanoflagellate protist colonies to sponge choanocytes supports an evolutionary link between this protist lineage and the animals. **(B)** A sponge moves food-containing water through its body by beating the flagella of its choanocytes—specialized feeding cells. Water enters the sponge through small pores and passes into water canals or an open atrium, where the choanocytes capture food particles from the water. The spicules are supportive, skeletal structures.

come to depend on genomic sequence comparisons, as genomes are ultimately the source of all inherited trait information.

A few basic developmental patterns differentiate major animal groups

The early cell divisions of an embryo are known as **cleavage**. Several different patterns of cleavage exist among animals. Although these patterns can be useful for characterizing major animal groups, genomic analyses have shown that many changes have occurred in cleavage patterns throughout animal evolution.

LINK The various patterns of embryonic cleavage and animal development are described in more detail in **Key Concept 35.2**.

Cleavage patterns are influenced by the configuration of the yolk, the acellular nutritive material that nourishes the growing embryo. The eggs of many animal groups contain a small amount of yolk that is evenly distributed throughout the egg cytoplasm. In some of these groups, the zygote and its descendant cells divide completely and evenly in a pattern known as **radial cleavage**. Radial cleavage is thought to be the ancestral condition for the bilaterian animals, as it is widely distributed among the major lineages. **Spiral cleavage**—a complicated permutation of radial cleavage—is found among many lophotrochozoans (a group that includes earthworms and clams). Lophotrochozoans with spiral cleavage are thus sometimes known as spiralians. The early branches of the ecdysozoans (molting animals, such as insects and nematodes) have radial cleavage, but most ecdysozoans have an idiosyncratic cleavage pattern that is neither radial nor spiral in organization.

Distinct layers of cells form during the early development of most animals. These cell layers differentiate into specific organs and organ systems as development continues. The embryos of **diploblastic** animals have two cell layers: an outer **ectoderm** and an inner **endoderm**. Embryos of **triploblastic** animals have, in addition to ectoderm and endoderm, a third distinct cell layer, **mesoderm**, which lies between the ectoderm and the endoderm. The existence of three cell layers in embryos is a synapomorphy of the triploblastic animals (which form a clade), whereas the diploblastic animals (ctenophores, placozoans, and cnidarians, which are not a clade) exhibit the ancestral condition (see Figure 22.1). Some biologists also consider sponges to be diploblastic, but since they do not have clearly differentiated

INVESTIGATION

FIGURE 22.3 What Induces Choanoflagellates to Form Multicellular Colonies? The choanoflagellate *Salpingoeca rosetta* forms multicellular, rose-shaped colonies (rosettes) at low frequencies in natural conditions. The rosette colonies are more efficient than are single cells at capturing some prey species of planktonic bacteria. Rosanna Alegado and her collaborators investigated the chemical signals that induce the choanoflagellates to form multicellular colonies when environmental conditions favor the rosette formations.[a]

HYPOTHESIS

Multicellular colonies of *S. rosetta* are induced by chemical compounds found on specific species of planktonic bacteria.

METHOD

1. Isolate individual species of bacteria from an environmental sample of *S. rosetta* and its associated bacterial community.
2. Use a combination of antibiotics to produce a bacteria-reduced culture line of *S. rosetta* that does not spontaneously form rosette colonies (the RCA line, for "rosette colonies absent").
3. Feed the RCA line with each of the isolated species of bacteria from step 1. Record any rosette colony formation.
4. Isolate compounds that induce multicellular colony formation from rosette-producing bacteria identified in step 3.

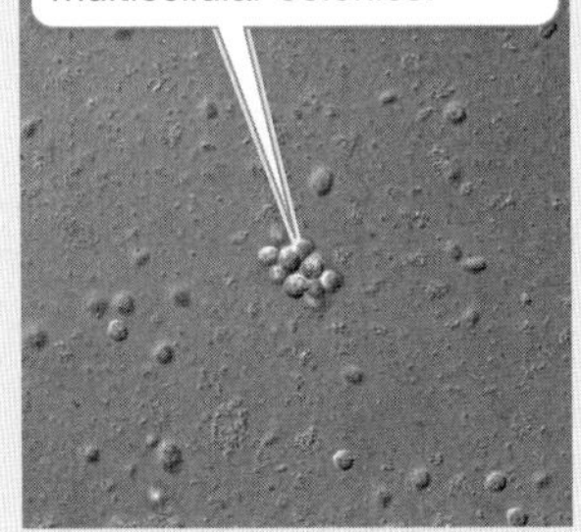

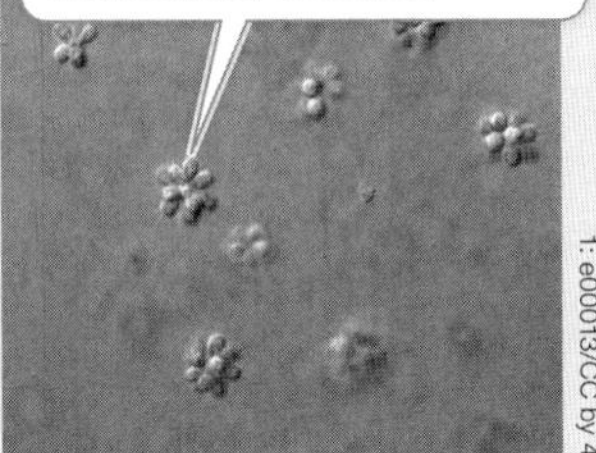

RESULTS

Alegado and colleagues tested various compounds from planktonic bacteria in the genus *Algoriphagus* and found that RIF-1, a sulfonolipid produced by the bacteria, is the signal that induces formation of multicellular colonies in *S. rosetta*.

CONCLUSION

A chemical signal (RIF-1) from a particular species of prey bacteria can induce the formation of multicellular colonies in some species of choanoflagellates.

[a]R. A. Alegado et al. 2012. *eLife:* 1: e00013.

A similar **Analyze the Data** exercise may be assigned in **Achieve**.

tissue types or embryonic cell layers, they are usually considered to be **monoblastic**.

During early development in many animals, in a process known as gastrulation, a hollow ball one cell thick indents to form a cup-shaped structure. The opening of the cavity formed by this indentation is called the blastopore (**FIGURE 22.4**). The overall pattern of gastrulation immediately after formation of the blastopore divides the triploblastic animals into two major groups:

1. In the **protostomes** (Greek, "mouth first"), the mouth arises from the blastopore, and the anus forms later.
2. In the **deuterostomes** ("mouth second"), the blastopore becomes the anus, and the mouth forms later.

Although the developmental patterns of animals are more varied than suggested by this simple dichotomy, sequencing data indicate that the protostomes and deuterostomes represent distinct animal clades. Together these two groups are known as the **bilaterians** (named for their usual bilateral symmetry), and they account for the vast majority of animal species.

FIGURE 22.4 Gastrulation Illuminates Evolutionary Relationships The blastopore is clear in this scanning electron micrograph of a sea urchin gastrula. Because sea urchins (echinoderms) are deuterostomes, this blastopore will eventually become the anal end of the animal's gut.

The general structure of an animal, the arrangement of its organ systems, and the integrated functioning of its parts are referred to as its **body plan**. As we discussed in Chapter 15, the regulatory and signaling genes that govern many major aspects of development (including body symmetry, body cavities, segmentation, and appendages) are widely shared among the different animal groups. Although the myriad animal body plans cover a wide range of morphologies, they can be seen as variations on five key features:

1. The *symmetry* of the body
2. The structure of the *body cavity*
3. The *segmentation* of the body
4. The existence and location of *external appendages* that are used for sensing, feeding, locomotion, mating, and other functions
5. The development of the *nervous system*

Each of these features affects how an animal interacts with its environment.

Most animals are symmetrical

The overall shape of an animal can be described by its **symmetry**. An animal is said to be symmetrical if it can be divided along at least one plane into similar halves. Animals that have no plane of symmetry are said to be asymmetrical. Placozoans and many sponges are asymmetrical, but most other animals have some kind of symmetry, which is governed by the expression of regulatory genes during development.

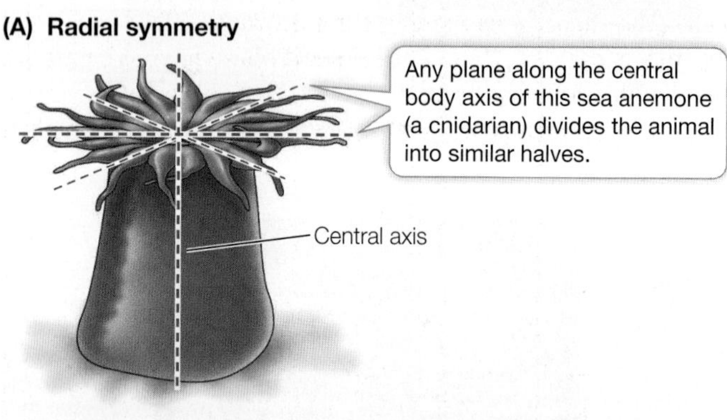

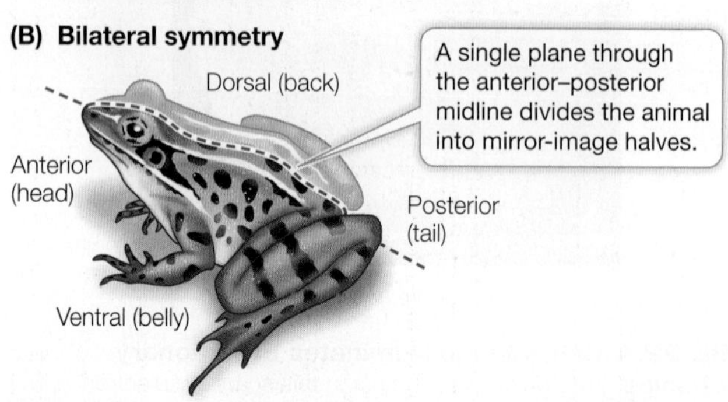

FIGURE 22.5 Body Symmetry Most animals are either **(A)** radially symmetrical or **(B)** bilaterally symmetrical.

In animals with **radial symmetry**, body parts are arranged around one main axis at the body's center (**FIGURE 22.5A**). Ctenophores (comb jellies) are radially symmetrical, as are many cnidarians (such as sea anemones and jellyfishes) and echinoderms. A perfectly radially symmetrical animal can be divided into similar halves by any plane that contains the main axis. However, most radially symmetrical animals—including the adults of echinoderms such as sea stars—are slightly modified, so that only some planes can divide them into identical halves. Some radially symmetrical animals are sessile (they remain fixed in one place) or drift with water currents. Others move about slowly but can move equally well in any direction.

Bilateral symmetry is characteristic of animals that have a distinct front end, which typically precedes the rest of the body as the animal moves. A bilaterally symmetrical animal can be divided into mirror-image (left and right) halves by a single plane that passes through the midline of its body. This plane runs from the front, or **anterior**, end of the body, to the rear, or **posterior**, end (**FIGURE 22.5B**). A plane at right angles to the midline divides the body into two dissimilar sides. The back of a bilaterally symmetrical animal is its **dorsal** surface; the underside is its **ventral** surface.

Bilateral symmetry is strongly correlated with **cephalization** (Greek *kephalos*, "head"), which is the concentration of sensory organs and nervous tissues at the anterior end of the animal. Cephalization has been favored by natural selection because the anterior end of a bilaterally symmetrical animal typically encounters new environments first.

The structure of the body cavity influences movement

The body plans of triploblastic animals can be divided into three types based on the presence and structure of an internal, fluid-filled **body cavity**:

1. **Acoelomate** animals such as flatworms lack an enclosed, fluid-filled body cavity. Instead, the space between the gut (derived from endoderm) and the muscular body wall (derived from mesoderm) is filled with masses of cells called mesenchyme (**FIGURE 22.6A**). These animals typically move by beating cilia.
2. **Pseudocoelomate** animals have a body cavity called a pseudocoel, a fluid-filled space in which many of the internal organs are suspended. A pseudocoel is enclosed by muscles (mesoderm) only on its outside; there is no inner layer of mesoderm surrounding the internal organs (**FIGURE 22.6B**).
3. **Coelomate** animals have a body cavity, the coelom, that develops within the mesoderm. It is lined with a layer of muscular tissue called the peritoneum, which also surrounds the internal organs. The coelom is thus enclosed on both the inside and the outside by mesoderm (**FIGURE 22.6C**).

The structure of an animal's body cavity strongly influences the ways in which it can move. The body cavities of many animals function as **hydrostatic skeletons**. Fluids are relatively incompressible, so when the muscles surrounding a fluid-filled body

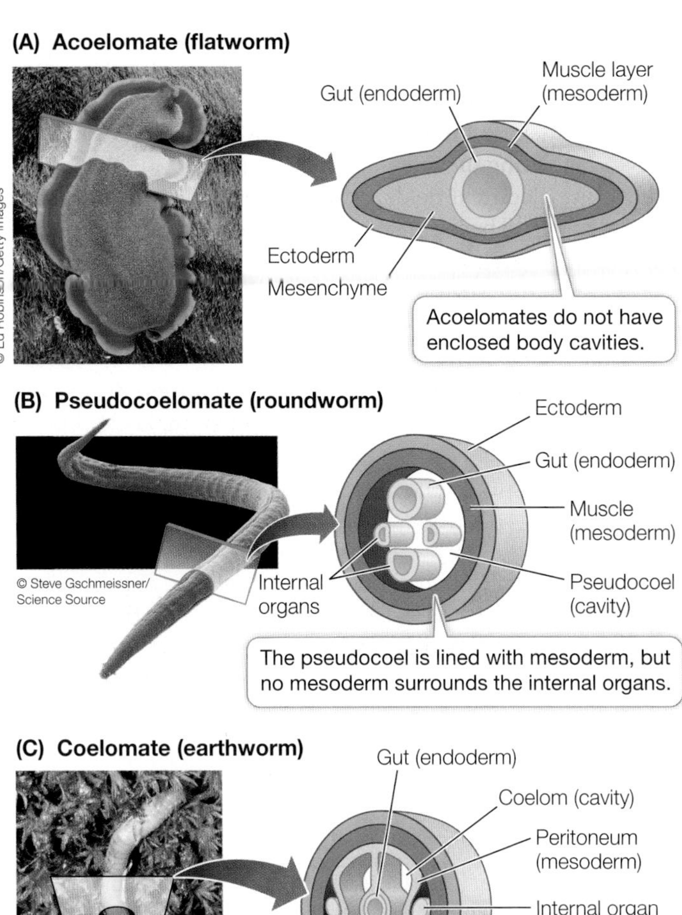

FIGURE 22.6 Animal Body Cavities **(A)** Acoelomates, like this marine flatworm, do not have an enclosed body cavity. **(B)** Pseudocoelomates have a body cavity bounded by endoderm and mesoderm. **(C)** Coelomates have a peritoneum surrounding the internal organs in a region bounded by mesoderm.

Activity 22.1 Animal Body Cavities

cavity contract, fluids shift to another part of the cavity. If the body tissues around the cavity are flexible, fluids squeezed out of one region can cause another region to expand. The moving fluids can thus move specific body parts. (You can see how a hydrostatic skeleton works by watching a snail emerge from its shell.) A coelomate animal has better control over the movement of the fluids in its body cavity than a pseudocoelomate animal does. An animal that has longitudinal muscles (running along the length of the body) as well as circular muscles (encircling the body cavity) has even greater control over its movement.

In terrestrial environments, the hydrostatic function of fluid-filled body cavities applies mostly to relatively small, soft-bodied organisms. Most larger animals (as well as many smaller ones) have hard skeletons that provide protection and facilitate movement. Muscles are attached to those firm structures, which may be inside the animal or on its outer surface (in the form of a shell or cuticle).

Segmentation facilitates specialization

Segmentation—the division of the body into segments—is seen in many animal groups. Segmentation facilitates specialization of different body regions (**FIGURE 22.7**). It also allows an animal to alter the shape of its body in complex ways and to control its movements precisely. If an animal's body is segmented, muscles in each individual segment can change the shape of that segment independently of the others. In only a few segmented animals is

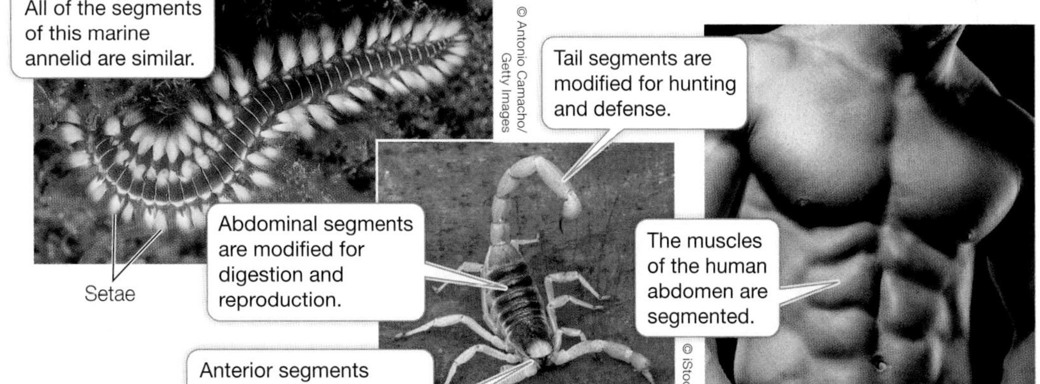

FIGURE 22.7 Body Segmentation **(A)** All of the segments of this marine clam worm, an annelid, are similar. Its appendages are tipped with bristles (setae) that are used for locomotion. **(B)** Segmentation allows the evolution of differentiation among the segments. The segments of this scorpion, an arthropod, differ in their form, function, and the appendages they bear. **(C)** Muscular segmentation is clearly visible in the abdomen of this body builder.

the body cavity separated into discrete compartments, but even partly separated compartments allow better control of movement. Segmentation occurs in several groups of protostomes and deuterostomes.

In some animals, such as annelids (earthworms and their relatives), similar body segments are repeated many times (**FIGURE 22.7A**). In other animals, including most arthropods, segments differ strikingly from one another (**FIGURE 22.7B**). The dramatic evolutionary radiation of the arthropods (including the insects, spiders, centipedes, and crustaceans) was based on changes in a segmented body plan that features muscles attached to the inner surface of an external skeleton, including a variety of external appendages that move these animals. In some animals, distinct body segments are not apparent externally—for example, the segmented vertebrae of vertebrates, including humans. Nonetheless, muscular segmentation is clearly visible in humans with well-defined, muscular bodies (**FIGURE 22.7C**).

Appendages have many uses

Getting around under their own power allows animals to obtain food, to avoid predators, and to find mates. Even some sedentary species, such as sea anemones, have larval stages that use cilia to swim, thus increasing the animal's chances of finding a suitable habitat.

Appendages that project externally from the body greatly enhance an animal's ability to move around. Many echinoderms, including sea urchins and sea stars, have large numbers of tube feet that allow them to move slowly across the substrate. Animals whose appendages have become modified into limbs are capable of highly controlled, more rapid movement. The presence of jointed limbs has been a prominent factor in the evolutionary success of the arthropods and the vertebrates. In four independent instances—among the arthropod insects and among the vertebrate pterosaurs, birds, and bats—body plans emerged in which limbs were modified into wings, allowing these animals to use powered flight.

Appendages also include many structures that are not used for locomotion. Many animals have antennae, which are specialized appendages used for sensing the environment. Other appendages (such as claws and the mouthparts of many arthropods) are adaptations for capturing prey or chewing food. In some species, appendages are used for reproductive purposes, such as sperm transfer or egg incubation.

Nervous systems coordinate movement and allow sensory processing

The bilaterian animals have a well-coordinated central nervous system. More diffuse nervous systems, called nerve nets, are present in some other animals, such as ctenophores and cnidarians. Nervous systems appear to be completely absent in a few animal groups, such as sponges and placozoans.

The central nervous system of bilaterians coordinates the actions of muscles, which allows coordinated movement of appendages and body parts. This coordination of muscles permits highly effective and efficient movement on land, in water, or through the air. The central nervous system is also essential for processing the sensory information gathered from a wide variety of sensory systems. Many animals have sensory systems for detecting light, for forming images of their environment (sight), for mechanical touch, for detecting movement, for detecting sounds (hearing), for detecting electrical fields, and for chemical detection (e.g., taste and smell). These sensory systems allow animals to find food, and the ability of animals to move allows them to capture or collect food from their environment. These same abilities also allow most animals to move to avoid potential predators or to search for suitable mates. Most animals can also assess the suitability of different environments and move appropriately in response to that information.

Media Clip 22.1 Nervous Systems Lead to Efficient Predators

REVIEW & APPLY | 22.1

R We generally recognize animals as multicellular heterotrophs with internal digestion, muscular systems that allow movement, and nervous systems. Animals are thought to be monophyletic because they share several derived traits, especially among their gene sequences. Major developmental differences also provide evidence of their evolutionary relationships. The body plans of animals are variations on patterns of symmetry, body cavity structure, segmentation, appendages, and nervous systems.

A

1. What general features distinguish animals from other major groups of living organisms?
2. Differentiate among the members of each of the following sets of related terms:
   a. radial symmetry/bilateral symmetry
   b. protostome/deuterostome
   c. diploblastic/triploblastic
3. Why is bilateral symmetry strongly associated with cephalization, the concentration of sensory organs in an anterior head?
4. Explain several ways in which body cavities, segmentation, and centralized nervous systems improve control over movement.

Variations in body symmetry, body cavity structure, life cycles, patterns of development, and survival strategies differentiate the many animal species. **TABLE 22.1** summarizes the living members of the major animal groups we will describe in the rest of this chapter.

TABLE 22.1 | Summary of Living Members of the Major Animal Groups

Group	Approximate number of living species described	Major subgroups, other names, and notes
Sponges	8,500	Demosponges, glass sponges, calcareous sponges
Ctenophores	250	Comb jellies
Placozoans	2	Additional species have been discovered but not yet formally named
Cnidarians	12,500	Anthozoans: Corals, sea anemones, and sea pens Hydrozoans: Hydras and hydroids Scyphozoans: Jellyfishes
PROTOSTOMES		
Arrow worms	180	Glass worms
Lophotrochozoans		
Bryozoans	5,500	Moss animals
Flatworms	30,000	Free-living flatworms; flukes and tapeworms (all parasitic); monogeneans (ectoparasites of fishes)
Rotifers and relatives	4,600	Rotifers, hairy-backs, spiny-headed worms, and jaw worms
Ribbon worms	1,200	Proboscis worms
Brachiopods	450	Lampshells
Phoronids	10	Sessile marine filter feeders
Annelids	19,000	Polychaetes (generally marine; may not be monophyletic) Clitellates: earthworms, freshwater worms, leeches
Mollusks	117,000	Chitons Bivalves: Clams, oysters, mussels Gastropods: Snails, slugs, limpets Cephalopods: Squids, octopuses, nautiloids
Ecdysozoans		
Nematodes	25,000	Roundworms
Horsehair worms	350	Gordian worms
Kinorhynchs	180	Mud dragons
Loriciferans	30	Brush heads
Priapulids	20	Penis worms
Onychophorans	180	Velvet worms
Tardigrades	1,200	Water bears
Arthropods		
Chelicerates	114,000	Horseshoe crabs, sea spiders, and arachnids (scorpions, harvestmen, spiders, mites, ticks)
Myriapods	12,000	Millipedes, centipedes
Crustaceans	67,000	Crabs, shrimps, lobsters and crayfishes, barnacles, copepods
Hexapods	1,056,000	Insects, springtails, two-pronged bristletails, and proturans
DEUTEROSTOMES		
Echinoderms	7,500	Crinoids (sea lilies and feather stars), brittle stars, sea stars, sea daisies, sea urchins, sea cucumbers
Hemichordates	120	Acorn worms and pterobranchs
Chordates		
Tunicates	2,800	Sea squirts (ascidians), salps, and larvaceans
Lancelets	35	Cephalochordates
Vertebrates	72,000	Hagfishes, lampreys, cartilaginous fishes, ray-finned fishes, coelacanths, lungfishes, amphibians, reptiles (including birds), and mammals

Large Multicellular Animal Groups Arose in Parallel

LEARNING OBJECTIVES

By the end of this key concept you should be able to:

22.2.1 Apply knowledge of phylogenetic relationships among major groups of animals to describe the evolution and loss of major structural features.

22.2.2 Compare the life histories of sponges, ctenophores, placozoans, and cnidarians and discuss their similarities and differences.

Looking at Figure 22.1, you can see that the protostome and deuterostome animals together comprise a monophyletic group known as the Bilateria. Some major traits that support the monophyly of bilaterians (in addition to genomic analyses) are strong bilateral symmetry, the presence of three distinct cell layers in embryos (triploblasty), and a centralized nervous system. There are, however, four animal groups—the sponges, ctenophores, placozoans, and cnidarians—that are not bilaterians.

The root of the animal tree is still debated

Biologists have proposed at least three different lineages as sister group of the remaining animals. Traditionally, sponges have been considered the sister group of other animals, based in large part on the similarity of some sponges to choanoflagellates (see Figure 22.2), and on the fact that sponges lack a nervous system. Placozoans also have very few tissue types, and also have no nervous system, so they have also been placed by most biologists as branching off near the base of the animal tree. However, some recent genomic studies suggest that the first split in the animal tree may have been between the lineage that leads to ctenophores and the rest of the animals (**INVESTIGATION FIGURE 22.8**).

Recent studies of the many genes that encode nervous systems indicate that these genes were duplicated and specialized separately in ctenophores, cnidarians, and bilaterians. This finding suggests that nervous systems evolved independently in these three groups in response to similar selection pressures associated with multicellularity, movement, and prey capture. Therefore the nervous systems of these groups are no longer considered homologous, or as evidence for a close evolutionary relationship. Most biologists now agree that the sponges and ctenophores were the first two lineages to split from the remaining animals, although there is still disagreement about which of these two lineages split first. The placozoans were then the next group to split from the remaining animals, followed by the split that gave rise to the cnidarians (see Figure 22.1).

Whichever of these groups was first to diverge, biologists think that the common ancestor of animals was very different from all animals that are alive today. An early split on the animal tree does not imply that sponges or ctenophores look anything like the ancestral animal. For example, all living ctenophores are quite closely related to one another, and they have been evolving for as long as all other animal lineages—ever since animals last shared a common ancestor more than 700 million years ago. Thus many traits of modern ctenophores (as well as modern sponges) are highly derived and specialized.

Molecular analyses of genomes place the earliest splits in animal phylogeny at more than 700 million years ago, well before the first large multicellular animals appear in the fossil record about 565 million years ago (see Figure 17.11). The first animals are thought to have had relatively few cells and to have been quite small. Then large multicellular animals, along with some of their characteristics (such as nervous systems), appear to have evolved several times in different animal lineages. As Earth warmed and oxygen concentrations increased in the oceans near the beginning of the Cambrian, several groups of animals evolved larger bodies and greater tissue complexity.

Sponges are loosely organized animals

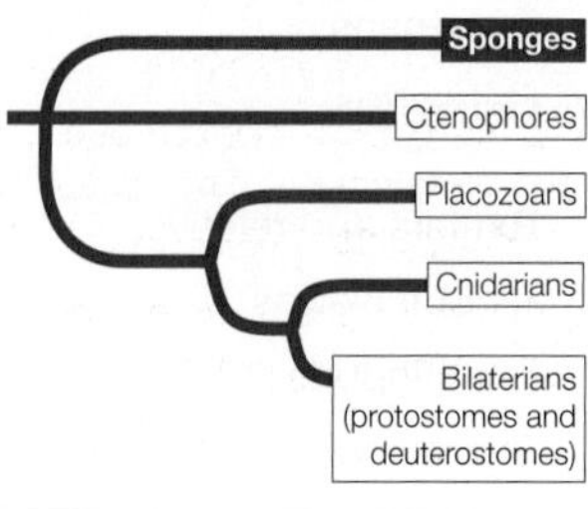

Although they have some specialized cells, **sponges** have no distinct embryonic cell layers and no true organs. Early naturalists classified sponges as plants because they were sessile and lacked body symmetry.

Sponges have hard skeletal elements called **spicules**, which may be small and simple or large and complex. Three major groups of sponges, which separated soon after the split between sponges and the rest of the animals, are distinguished by their spicules. Members of two groups (demosponges and glass sponges) have skeletons composed of silicaceous spicules made of hydrated silicon dioxide (**FIGURE 22.9A,B**). These spicules have greater flexibility and toughness than synthetic glass rods of similar length. Members of the third group, the calcareous sponges, take their name from their calcium carbonate spicules (**FIGURE 22.9C**).

The body plan of all sponges—even large ones, which may reach a meter or more in length—is an aggregation of cells built around a water canal system. Sponges bring water into their bodies by beating the flagella of their specialized feeding cells, the **choanocytes** (see Figure 22.2B). Water, along with any food particles it contains, enters the sponge by way of small pores and passes into the water canals or a central atrium, where choanocytes capture food particles.

A skeleton of simple or branching spicules, often combined with a complex network of elastic fibers, supports the body of most sponges. Sponges also have an extracellular matrix, composed of collagen, adhesive glycoproteins, and other molecules, that holds the cells together. Most sponges are filter feeders. A few species are carnivores that trap prey on hook-shaped spicules that protrude from the body surface.

Most of the 8,500 species of sponges are marine animals. Only about 50 species of sponges live in fresh water. Sponges come in a wide variety of sizes and shapes that are adapted to different movement patterns of water. Sponges living in intertidal or shallow subtidal environments are subject to strong wave action and attach firmly to the substrate. Most sponges that live in slowly flowing water are flattened and are oriented at right angles to the

INVESTIGATION

FIGURE 22.8 Reconstructing Animal Phylogeny from Protein-Coding Genes Several breakthroughs in our understanding of animal phylogeny have occurred in recent years as the sequences of genes and proteins have been compared across species. Casey Dunn and his colleagues compared sequences from many different proteins across a wide variety of animal groups to reconstruct their phylogeny.[a]

QUESTION

What was the earliest split in the animal tree? Are ctenophores, sponges, or placozoans the sister group of other animals?

METHODS

1. Collect mRNA from major groups of animals and choanoflagellate outgroups. (See Figure 22.2 to see why choanoflagellates were used as an outgroup.)
2. Reverse transcribe the mRNA into cDNA.
3. Sequence the cDNA, translate the genes into the amino acid sequences of the proteins, and align sequences of homologous proteins.
4. Reconstruct the phylogeny of the animal groups from the aligned protein sequences.

RESULTS

The tree below represents just a sample of the major lineages examined by Dunn and colleagues.

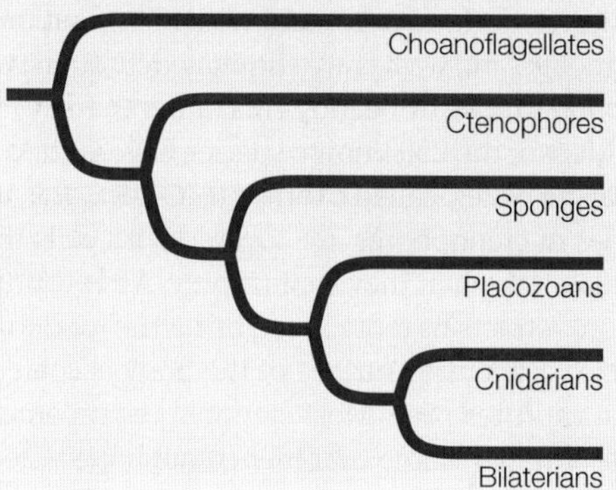

CONCLUSION

According to this data set, the earliest split in the animal tree was between ctenophores and the remaining animals.

ANALYZE THE DATA

The table below includes a small sample of the data collected and analyzed by Dunn and his colleagues. Use these representative data to reconstruct a tree of the represented species. In the original paper, Dunn and colleagues reported on 11,234 amino acid positions among 77 species of animals. Twenty-seven of these amino acid positions for 11 of those species are shown in the table. (See Figure 3.14 for the one-letter amino acid abbreviations.)

1. Construct a phylogenetic tree of these 11 species using the parsimony method. Use the outgroup (a choanoflagellate) to root your tree. Assume that all changes from one amino acid to another are equally likely. See Key Concept 14.2 for a review of the parsimony method of reconstructing phylogenetic trees.
2. How many character state changes (i.e., changes from one amino acid to another) occur along each branch on your tree?
3. Which characters support the ctenophores as the sister group of the remaining animals?
4. Which group on your tree represents the bilaterian animals? The protostomes? The deuterostomes?

[a] C. W. Dunn et al. 2008. *Nature* 452: 745–749.

A similar **Analyze the Data** exercise may be assigned in **Achieve**.

	Character state (amino acid at position)																										
Species	**1**	**2**	**3**	**4**	**5**	**6**	**7**	**8**	**9**	**10**	**11**	**12**	**13**	**14**	**15**	**16**	**17**	**18**	**19**	**20**	**21**	**22**	**23**	**24**	**25**	**26**	**27**
Clam	Y	S	T	G	L	H	E	N	Y	A	R	A	M	R	I	A	L	T	I	V	K	L	S	I	V	I	L
Earthworm	Y	A	T	G	L	H	E	N	Y	P	H	A	M	R	I	A	L	T	I	V	K	L	S	I	V	M	L
Tardigrade	Y	A	T	G	L	H	E	H	Y	K	R	A	M	R	V	A	T	S	I	V	R	L	N	L	V	L	L
Fruit fly	F	A	T	G	L	H	E	N	Y	K	R	A	M	R	I	A	L	S	I	V	S	L	D	L	V	L	L
Sea urchin	Y	A	T	G	L	L	E	N	Y	P	N	A	M	R	I	A	L	T	V	I	R	Q	N	L	T	V	K
Human	W	A	A	G	L	R	E	H	Y	P	K	A	I	R	I	S	V	T	V	I	R	Q	N	L	T	V	K
Chicken	W	A	A	G	L	R	E	H	Y	P	R	A	I	R	I	A	V	T	V	I	R	Q	N	L	T	V	K
Lancelet	Y	A	T	G	L	R	E	H	Y	P	K	A	M	R	I	A	V	T	V	I	R	L	N	L	T	V	K
Sponge	Y	G	L	S	L	R	P	N	F	P	K	S	M	S	V	A	L	T	V	I	R	Q	N	L	V	I	L
Ctenophore	Y	G	L	G	Q	D	P	N	F	P	K	S	M	S	V	A	L	T	V	I	R	Q	N	L	V	I	L
Choanoflagellate (outgroup)	Y	G	L	G	Q	D	P	N	F	P	K	S	F	S	V	A	L	T	V	I	R	Q	N	L	V	I	L

FIGURE 22.9 Sponge Diversity **(A)** The great majority of sponge species are demosponges, such as this barrel sponge. **(B)** As in the demosponges, the supporting structures of glass sponges are silicaceous spicules. The pores and water canals "typical" of the sponge body plan are apparent in both (A) and (B). **(C)** The spicules of calcareous sponges are made of calcium carbonate.

direction of current flow, allowing them to intercept water and the food it contains as it flows past them.

Sponges reproduce both sexually and asexually. In most species, a single individual produces both eggs and sperm, but individuals do not self-fertilize. Water currents carry sperm from one individual to another. Sponges also reproduce asexually by budding and fragmentation.

Activity 22.2 Sponge and Diploblast Classification

Ctenophores are radially symmetrical and diploblastic

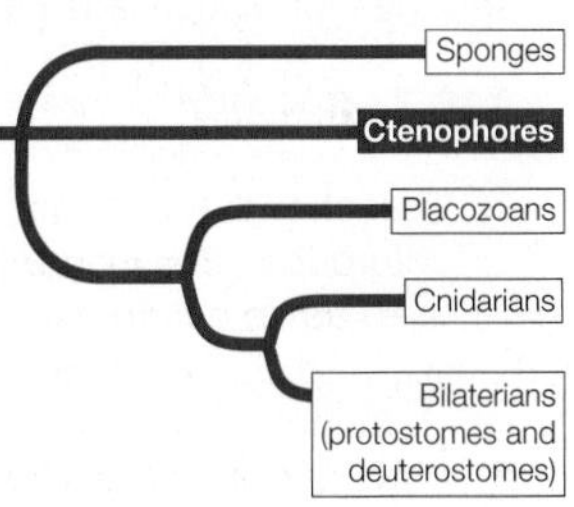

Ctenophores, also known as comb jellies, were until recently thought to be most closely related to the cnidarians (jellyfishes, corals, and their relatives). But ctenophores lack most of the Hox genes found in cnidarians and bilaterians, and recent studies of their genomes have indicated that ctenophores were among the earliest lineages to split from the remaining animals (see Investigation Figure 22.8).

Ctenophores have a radially symmetrical, diploblastic body plan. The two cell layers are separated by an inert, gelatinous extracellular matrix called **mesoglea**. Ctenophores, unlike sponges, have different openings for food entry and waste elimination: food enters through a mouth, and wastes are eliminated through two anal pores.

Ctenophores move by beating cilia rather than by muscular contractions. Most of the 250 known species have eight comblike rows of cilia-bearing plates, called **ctenes** (**FIGURE 22.10**). The feeding tentacles of ctenophores are covered with cells that discharge adhesive material when they contact prey. After capturing its prey, a ctenophore retracts its tentacles to bring the food to its mouth. In some species, the entire surface of the body is coated with sticky mucus that captures prey. Most ctenophores eat small planktonic organisms, although some eat other ctenophores. Ctenophores are common in open seas and can become abundant in protected bodies of water, where large populations may inhibit the growth of other organisms.

Ctenophore life cycles are uncomplicated. Gametes are released into the body cavity and then discharged through the mouth or the anal pores. Fertilization takes place in open seawater. In nearly all species, the fertilized egg develops directly into a miniature ctenophore, which gradually grows into an adult.

Media Clip 22.2 Ctenophores

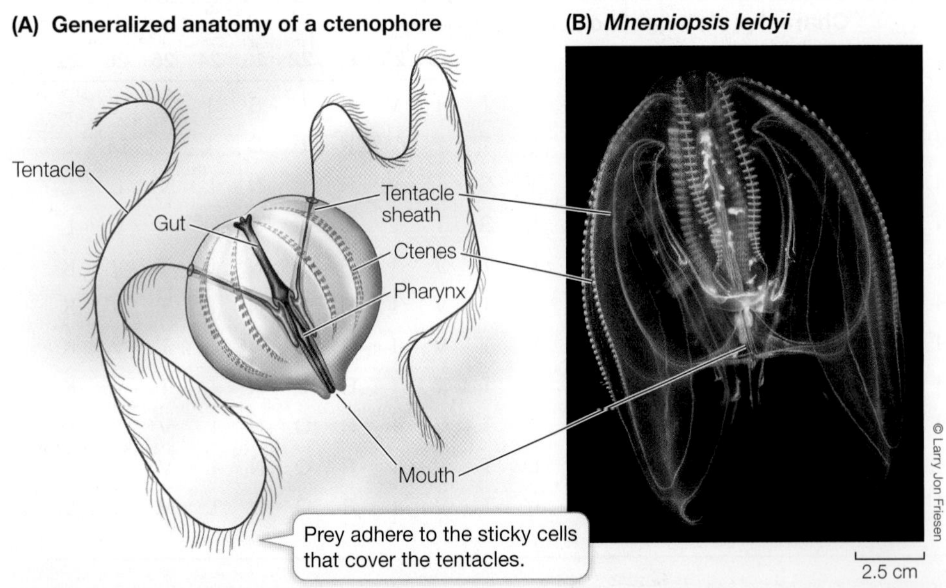

FIGURE 22.10 Comb Jellies Feed with Tentacles **(A)** The body plan of a typical ctenophore. In many species, long, sticky tentacles sweep through the water, efficiently harvesting small prey. **(B)** Other species, such as the one shown here, have short tentacles.

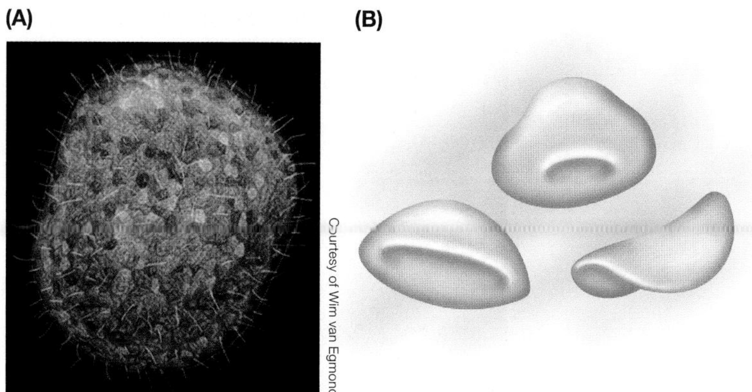

FIGURE 22.11 Placozoan Simplicity **(A)** As seen in this artist's rendition, adult placozoans are tiny (1–2 mm across), flattened, asymmetrical animals. **(B)** Recent studies have found a weakly swimming pelagic stage of placozoan to be abundant in many warm tropical and subtropical seas. (B after V. B. Pearse and O. Voigt. 2007. *Integr Comp Biol* 47: 677–692. Drawing by J. Keller and C. Patton, based on observations by V.B.P.)

Placozoans are abundant but rarely observed

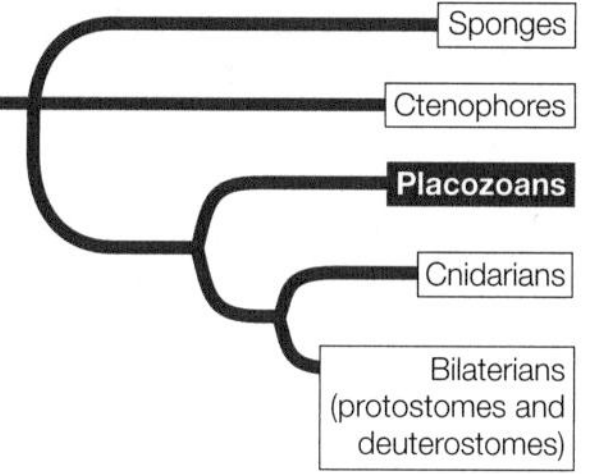

Placozoans are structurally very simple animals with only four distinct cell types (**FIGURE 22.11A**). Individuals in the mature life stage are usually observed adhering to surfaces (such as the glass of aquariums, where they were first discovered, or to rocks and other hard substrates in nature). Their structural simplicity—they have no mouth, gut, or nervous system—initially led biologists to suspect they might be the sister group of all other animals. Most phylogenetic analyses have not supported this hypothesis, however, and some aspects of the placozoans' structural simplicity may be secondarily derived. They are generally considered to have a diploblastic body plan, with upper and lower epithelial (surface) layers that sandwich a layer of contractile fiber cells.

Recent studies have found that placozoans have a pelagic (open-ocean) stage that is capable of swimming (**FIGURE 22.11B**), but the life history of placozoans is incompletely known. Most studies have focused on the larger adherent stages that can be found in aquariums, where they appear after being inadvertently collected with other marine organisms. The transparent nature and small size of placozoans make them very difficult to observe in nature. Nonetheless, it is known that placozoans can reproduce both asexually as well as sexually, although the details of their sexual reproduction are poorly understood. They are found in warm coastal seas around the world.

Cnidarians are specialized carnivores

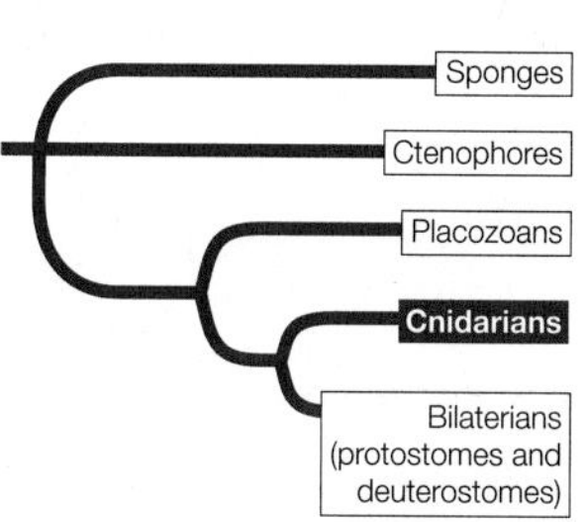

The **cnidarians** (jellyfishes, sea anemones, corals, and hydrozoans) make up the largest and most diverse group of non-bilaterian animals. The mouth of a cnidarian is connected to a blind sac called the **gastrovascular cavity**. The gastrovascular cavity functions in digestion, circulation, and gas exchange, and it also acts as a hydrostatic skeleton. The single opening serves as both mouth and anus.

The life cycle of many cnidarians has two distinct stages, one sessile and the other motile (**FIGURE 22.12**), although one or the other of these stages is absent in some groups. In the sessile **polyp** stage, a cylindrical stalk is attached to the substrate. The motile **medusa** (plural *medusae*) is a free-swimming stage shaped like a bell or an umbrella. It typically floats with its mouth and feeding tentacles facing downward.

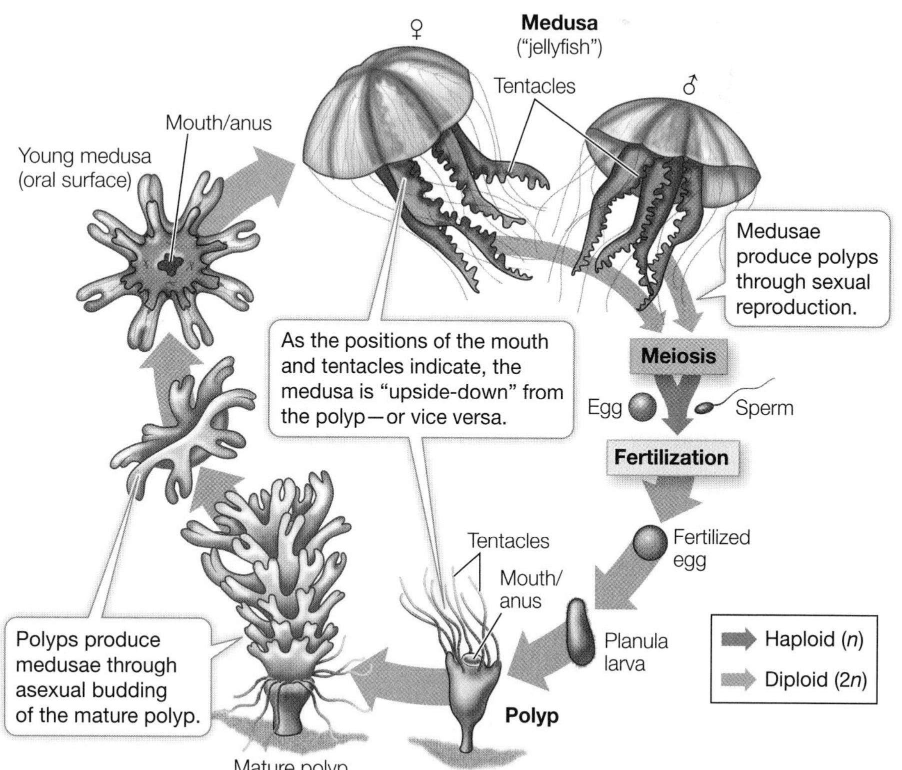

FIGURE 22.12 The Cnidarian Life Cycle Typically Has Two Stages The life cycle of a scyphozoan (jellyfish) exemplifies the typical cnidarian body forms: the sessile, asexual polyp; and the motile, sexual medusa.

Animation 22.1 Life Cycle of a Cnidarian

(A) *Diadumene leucolena*

© H. Wes Pratt/Biological Photo Service

(B) *Netrostoma setouchina*

© Reinhard Dirscherl/Alamy Stock Photo

(C)

© iStock.com/prasit chansarekorn

FIGURE 22.13 Diversity among Cnidarians **(A)** Sea anemones are sessile, living attached to a marine substrate. Water currents carry prey into the nematocyst-studded tentacles. **(B)** A jellyfish illustrates the complexity of a scyphozoan medusa. **(C)** Numerous corals species on a reef in Indonesia.

Mature polyps produce medusae by asexual budding. Medusae then reproduce sexually, producing eggs or sperm by meiosis and releasing the gametes into the water. A fertilized egg develops into a free-swimming, ciliated larva called a **planula**, which eventually settles to the bottom and develops into a polyp.

Cnidarians have epithelial cells with muscle fibers that contract and enable the animals to move, as well as simple nerve nets that integrate their body activities. They are specialized carnivores, using the toxin in harpoonlike structures called **nematocysts** to capture relatively large and complex prey. Some cnidarians, including many corals and anemones, gain additional nutrition from photosynthetic endosymbionts that live in their tissues.

LINK The importance of protist endosymbionts in coral colonies is discussed in **Key Concepts 1.5 and 19.4**.

Of the roughly 12,500 living cnidarian species, all but a few live in the oceans. The smallest cnidarians can hardly be seen without a microscope. The largest known jellyfish is 2.5 meters in diameter, and some colonial species can reach lengths in excess of 30 meters. Three clades of cnidarians have many species, some of which are shown in **FIGURE 22.13**. The anthozoans include sea anemones, sea pens, and corals. The scyphozoans are commonly known as jellyfishes or sea jellies. The less familiar hydrozoans include both freshwater and marine species.

REVIEW & APPLY | 22.2

R Early animal diversification led to five major lineages: sponges, ctenophores, placozoans, cnidarians, and bilaterians. Sponges and ctenophores split first from the remaining animals, followed closely by the placozoans. The earliest animals are thought to have been small and to have consisted of relatively few cells per individual. Large multicellular animals, with greater tissue differentiation, arose independently several times, giving rise to the five major groups of modern animals. Major components of nervous systems arose independently in three of these lineages (ctenophores, cnidarians, and bilaterians).

A

1. What is the basis for the claim that animal nervous systems evolved independently in ctenophores, cnidarians, and bilaterians?
2. Why are sponges and placozoans considered to be animals even though they lack the complex body structures found in most other animal groups?
3. The discoveries that placozoans in their pelagic stage are abundant in warm seas and that the mature stage settles on smooth surfaces suggest how these organisms might be collected and surveyed. What sampling procedures might you use to discover whether placozoans occur at a particular location along a coast?

In Key Concept 22.1 we noted that the name protostome means "mouth first," a developmental synapomorphy that links these animals. The vast majority of animals are protostomes.

22.3 Protostomes Have an Anterior Brain and a Ventral Nervous System

LEARNING OBJECTIVES

By the end of this key concept you should be able to:

22.3.1 Associate major morphological features of select protostomes (arrow worms, lophotrochozoans, ecdysozoans) with their life history and environment.

22.3.2 Compare cuticle thickness in different ecdysozoans and explain the consequences of such variation.

The protostomes are a highly diverse group of animals. Although their body plans are extremely varied, they are all bilaterally symmetrical animals whose bodies exhibit two major derived traits:

1. An anterior brain that surrounds the entrance to the digestive tract
2. A ventral nervous system consisting of paired or fused longitudinal nerve cords

Other aspects of protostome body organization differ widely from group to group. Although the common ancestor of the protostomes likely had a coelom, subsequent modifications of the coelom distinguish many protostome lineages. In at least one protostome lineage (the flatworms), the coelom has been lost (that is, the flatworms reverted to an acoelomate state). Some lineages are characterized by a pseudocoel (see Figure 22.6B). In two of the most prominent protostome groups, the coelom has been highly modified:

- Arthropods lost the ancestral condition of the coelom over the course of evolution. Their internal body cavity has become a hemocoel, or "blood chamber," in which fluid from an open circulatory system bathes the internal organs before returning to blood vessels.
- Most mollusks have an open circulatory system with some of the attributes of the hemocoel, but they retain vestiges of an enclosed coelom around their major organs.

The evolutionary relationships of one small group of protostomes, the **arrow worms** (see Figure 22.1), have been debated for many years. Although recent gene sequence studies clearly identify arrow worms as protostomes, there remains some question as to their exact placement within the protostomes. The 180 living species of arrow worms are small (3 mm–12 cm) marine predators of planktonic protists and small fishes.

The protostomes can be divided into two major clades—the lophotrochozoans and the ecdysozoans—largely on the basis of DNA sequence analysis. However, there are also some morphological characteristics that unite most members of these two groups.

Animation 22.2 An Overview of the Protostomes

Cilia-bearing lophophores and trochophore larvae evolved among the lophotrochozoans

Lophotrochozoans derive their name from two different features that involve cilia: a feeding structure known as a lophophore and a free-living larva known as a trochophore. Neither feature is universal for all lophotrochozoans, however.

Several distantly related groups of lophotrochozoans (including bryozoans, brachiopods, and phoronids) have a **lophophore**, a circular or U-shaped ring of ciliated, hollow tentacles around the mouth (**FIGURE 22.14A**). This complex structure is an organ for both food collection and gas exchange. The lophophore appears to have evolved independently several times, although it may be an ancestral feature that has been lost in many groups. Nearly all animals with a lophophore are sessile as adults, using the lophophore's tentacles and cilia to capture small floating organisms from the water.

Media Clip 22.3 Feeding with a Lophophore

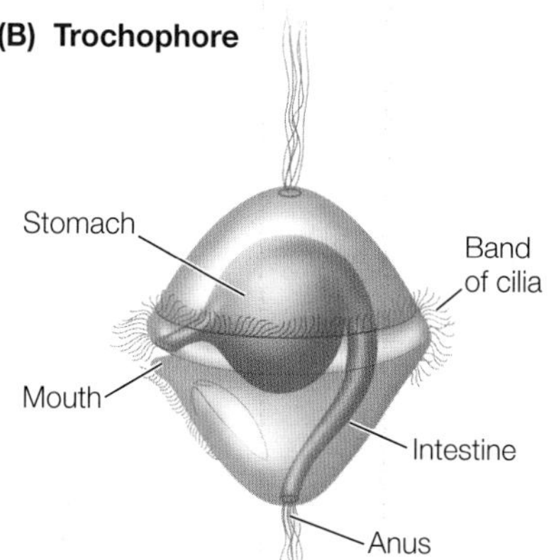

FIGURE 22.14 Lophophores and Trochophore Larvae Give Lophotrochozoans Their Name **(A)** The extended lophophore of a bryozoan is used to feed on suspended particles. **(B)** Many lophotrochozoans have a larval form known as a trochophore, which moves by beating a band of cilia.

Some lophotrochozoans, especially in their larval form, use cilia for locomotion. The larval form known as a **trochophore** moves by beating a band of cilia (**FIGURE 22.14B**). This movement of cilia also brings plankton closer to the larva, where it can capture and ingest them (its cilia are therefore similar in function to the cilia of the lophophore). Trochophore larvae are found among many of the major groups of lophotrochozoans, including the mollusks, annelids, ribbon worms, and bryozoans. This larval form was probably present in the common ancestor of lophotrochozoans, although it has been subsequently lost in several lineages.

Lophotrochozoans range from relatively simple animals with a blind gut (that is, a gut with a single opening that both takes in food and expels wastes) and no internal transport system to animals with a complete gut (having separate entrance and exit openings) and a complex internal transport system. Several of these groups exhibit wormlike bodies, but the lophotrochozoans encompass a wide diversity of morphologies, including a few groups with external shells. Included among the lophotrochozoans are species-rich groups such as flatworms, annelids, and mollusks, along with many less well known groups, some of which have only recently been discovered.

Reteporella couchii

FIGURE 22.15 Bryozoans Form Colonies The rigid orange tissue of this marine bryozoan colony connects and supplies nutrients to thousands of individual animals.

BRYOZOANS Most of the 5,500 species of **bryozoans** (moss animals) are colonial animals that live in a "house" made of material secreted by the external body wall. Almost all bryozoans are marine, although a few species occur in fresh or brackish water. A bryozoan colony consists of many small (1–2 mm) individuals connected by strands of tissue along which nutrients can be moved (**FIGURE 22.15**). The colony is created by the asexual reproduction of its founding member, and a single colony may contain as many as 2 million individuals. Rocks in coastal regions in many parts of the world are covered with luxuriant growths of bryozoans. Some bryozoans create miniature reefs in shallow waters. In some species, the individual colony members are differentially specialized for feeding, reproduction, defense, or support.

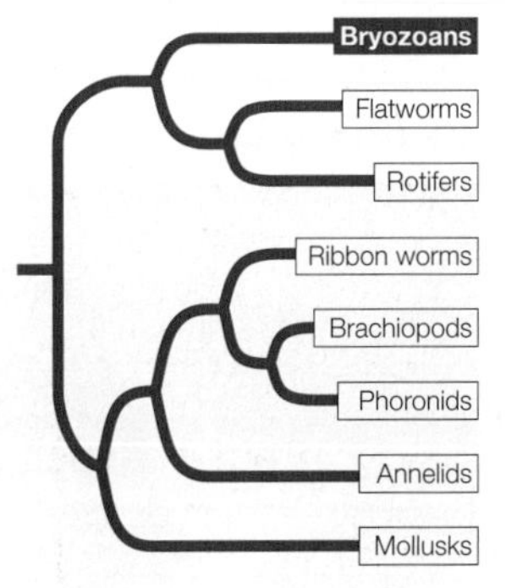

Bryozoans can also reproduce sexually. Sperm are released into the water, which carries the sperm to other individuals. Eggs are fertilized internally; developing embryos are brooded before they exit as larvae to seek suitable sites for attachment to the substrate.

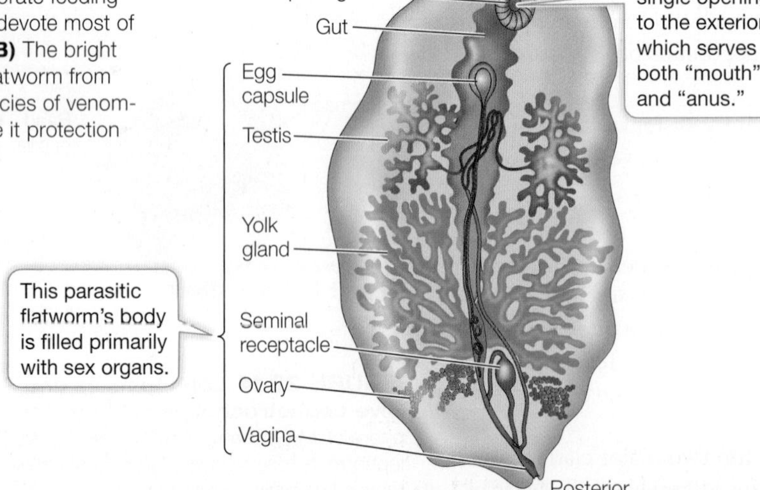

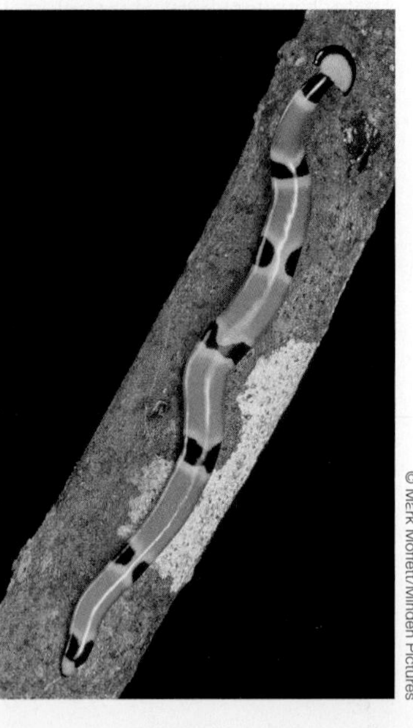

FIGURE 22.16 Flatworms (A) The fluke diagrammed here is representative of many parasitic flatworms. Absorbing nutrition from the host animal's gut, these internal parasites do not require elaborate feeding or digestive organs and can devote most of their body to reproduction. **(B)** The bright coloration of this terrestrial flatworm from Borneo resembles some species of venomous snakes, and may provide it protection from predation.

FLATWORMS Most of the 30,000 species of **flatworms** are tapeworms and flukes; members of these two groups are internal parasites, particularly of vertebrates (**FIGURE 22.16A**). Because they absorb digested food from the guts of their hosts, many parasitic flatworms lack digestive tracts of their own. Some cause serious human diseases, such as schistosomiasis, which is common in parts of Asia, Africa, and South America. The species that causes this devastating disease has a complex life cycle involving both freshwater snails and mammals as hosts. Other flatworms are external parasites of fishes and other aquatic vertebrates. The turbellarians include most of the free-living species (**FIGURE 22.16B**).

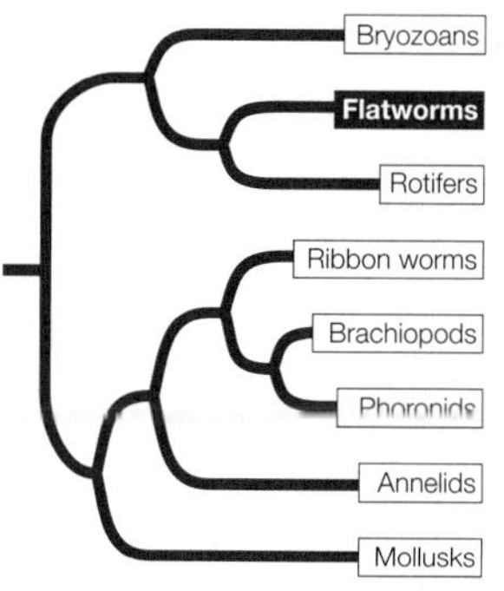

Flatworms lack specialized organs for transporting oxygen to their internal tissues. Lacking a gas transport system, each cell must be near a body surface, a requirement met by the dorsoventrally flattened body form. In flatworms that have a digestive tract, this consists of a mouth opening into a blind sac. The sac is often highly branched, forming intricate patterns that increase the surface area available for the absorption of nutrients. Most free-living flatworms are cephalized, with a head bearing chemoreceptor organs, two simple eyes, and a small brain composed of anterior thickenings of the longitudinal nerve cords. Free-living flatworms glide over surfaces, powered by broad bands of cilia.

ROTIFERS Most species of **rotifers** are tiny—50–500 μm long, smaller than some ciliate protists—but they have specialized internal organs (**FIGURE 22.17**). A complete gut passes from an anterior mouth to a posterior anus. The body cavity is a pseudocoel that functions as a hydrostatic skeleton. Rotifers typically propel themselves through the water by means of rapidly beating cilia rather than by muscular contraction.

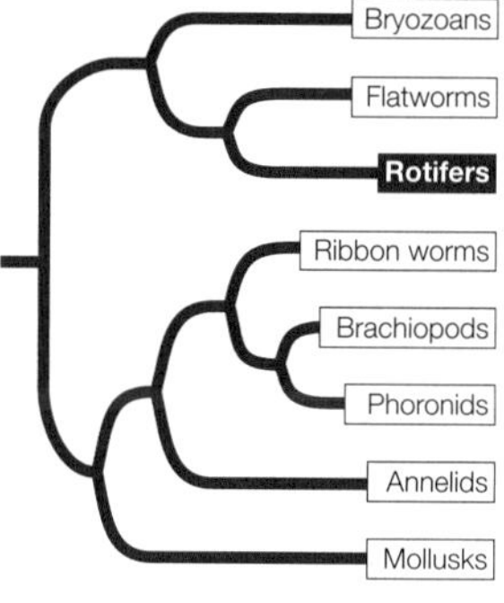

The most distinctive organ of rotifers is a conspicuous ciliated organ called the corona, which surmounts the head of many species. Coordinated beating of the cilia sweeps particles of organic matter from the water into the animal's mouth and down to a complicated structure called the mastax, in which food is ground into small pieces. By contracting muscles around the pseudocoel, a few rotifer species that prey on protists and small animals can protrude the mastax through their mouth and seize small objects with it.

Media Clip 22.4 Rotifer Feeding

Most species of rotifers live in fresh water. Some species rest on the surfaces of mosses or lichens in a desiccated, inactive state until it rains. When rain falls, they absorb water and become mobile, feeding in the films of water that temporarily cover the plants. Most rotifers live no longer than a few weeks.

Both males and females are found in some species of rotifers, but only females are known among the bdelloid rotifers (the *b* in "bdelloid" is silent). Biologists have concluded that the bdelloid rotifers may have existed for tens of millions of years without regular sexual reproduction. In general, lack of genetic recombination leads to the buildup of deleterious mutations, so long-term asexual reproduction typically leads to extinction (see Key Concept 15.2). However, recent studies indicate that bdelloid rotifers may avoid this problem by taking up fragments of genes directly from the environment during the desiccation–rehydration cycle. Such a mechanism allows genetic recombination among individuals in the absence of direct sexual exchange.

RIBBON WORMS **Ribbon worms** (proboscis worms) have simple nervous and excretory systems similar to those of flatworms. Unlike flatworms, however, they have a complete digestive tract with a mouth at one end and an anus at the other. Small ribbon worms move slowly by beating their cilia. Larger ones employ waves of muscle contraction to move over the surface of sediments or to burrow into them.

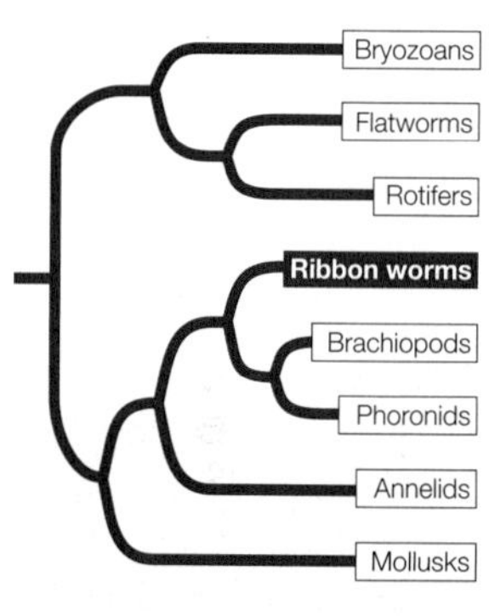

Within the body of nearly all of the 1,200 species of ribbon worms is a fluid-filled cavity called the rhynchocoel, within which lies a hollow, muscular **proboscis**. The proboscis, which is the worm's

(A) ***Philodina roseola***

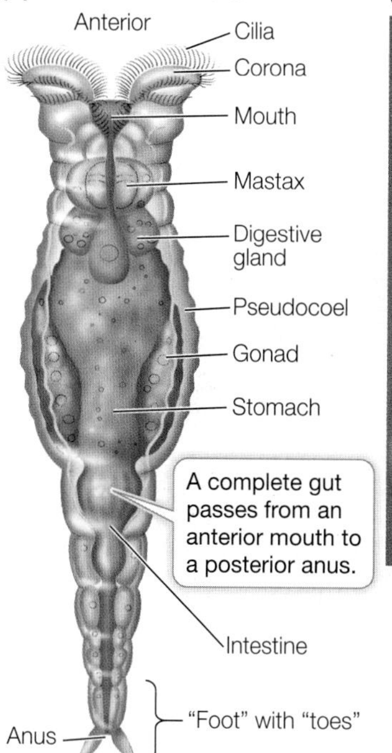

(B) ***Philodina*** **sp.**

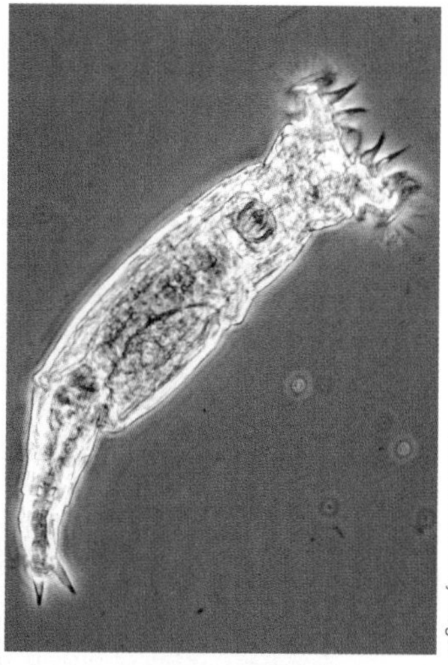

FIGURE 22.17 Rotifers **(A)** The individual diagrammed here reflects the general structure of many rotifers. **(B)** A light micrograph reveals the internal complexity of these tiny animals.

feeding organ, may extend much of the length of the body. Contraction of the muscles surrounding the rhynchocoel causes the proboscis to evert (turn inside out) explosively through an anterior pore (**FIGURE 22.18A**). The proboscis may be armed with sharp stylets that pierce prey and discharge paralytic toxins into the wound.

Media Clip 22.5 Explosive Extrusion of Ribbon Worm Proboscis

Ribbon worms are largely marine, although there are some freshwater and terrestrial species. Most ribbon worms are less than 20 centimeters long, but individuals of some species reach 20 meters or more. Some genera include species that are conspicuous and brightly colored (**FIGURE 22.18B**).

(A)

Floating in a cavity called the rhynchocoel, the proboscis can be everted rapidly.

Rhynchocoel

Proboscis

Proboscis retractor muscle

Proboscis pore

Mouth

Intestine

Anus

The tip of the everted proboscis bears a sharp, nail-shaped stylet.

(B) *Baseodiscus cingulatus*

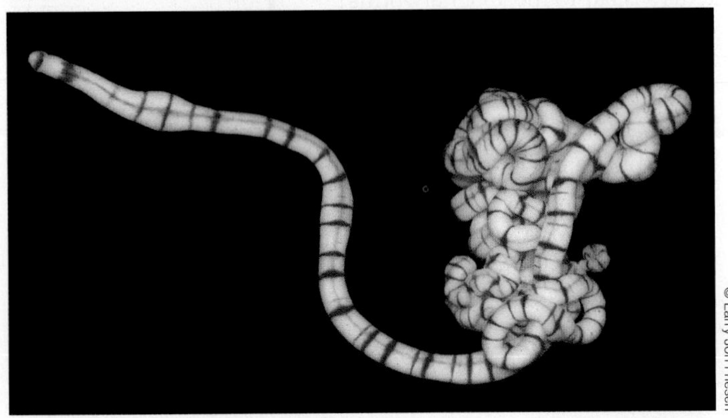

FIGURE 22.18 Ribbon Worms (A) The proboscis is the ribbon worm's feeding organ. **(B)** Although most ribbon worms are small, some marine species such as this one can reach several meters in length.

BRACHIOPODS AND PHORONIDS Like bryozoans (see Figure 22.14A), brachiopods and phoronids also feed using a lophophore, but this structure may have evolved independently in these groups. Although neither the brachiopods nor the phoronids are represented by many living species, the brachiopods—which have shells and thus leave an excellent fossil record—are known to have been much more abundant during the Paleozoic and Mesozoic.

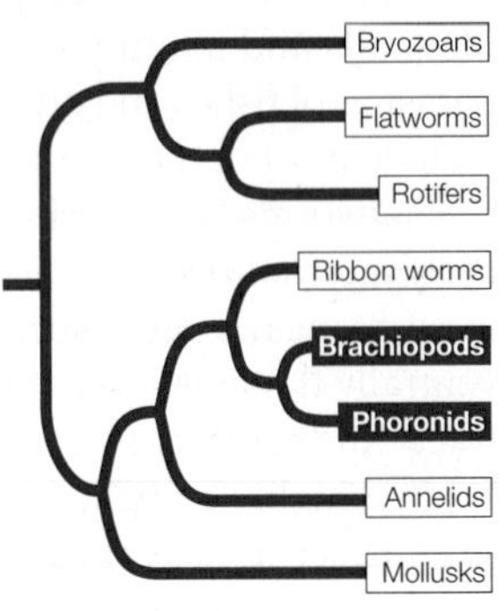

Brachiopods (lampshells) are solitary marine animals with a rigid shell that is divided into two parts connected by a ligament (**FIGURE 22.19**). Although brachiopods superficially resemble bivalve mollusks, shells evolved independently in the two groups. The two halves of the brachiopod shell are dorsal and ventral, rather than lateral as in bivalves. The lophophore is located within the shell. Most brachiopods are 4–6 centimeters long. More than 26,000 fossil brachiopod species have been described, but only about 450 species survive.

Laqueus sp.

FIGURE 22.19 Brachiopods The lophophore of this North Pacific brachiopod can be seen between the valves of its shell. Although its shell resembles those of the bivalve mollusks (see Figure 22.23D), it evolved independently.

The 10 known species of **phoronids** are small (5–25 cm long), sessile worms that live in muddy or sandy sediments or attached to rocky substrates. Phoronids are found in marine waters, from the intertidal zone to about 400 meters deep. They secrete tubes made of chitin, within which they live (**FIGURE 22.20**). Their cilia drive water into the top of the lophophore, and the water exits through the narrow spaces between the tentacles. Suspended food particles are caught and transported to the mouth by ciliary action. Eggs are fertilized internally, and the embryos are either released into the water or, in species with large embryos, retained in the parent's body, where they are brooded until they hatch.

ANNELIDS As discussed in Key Concept 22.1, segmentation allows an animal to move different parts of its body independently of one another, giving it much better control of its movement. A clear and obvious example of segmentation is seen in the body plan of the **annelids** (**FIGURE 22.21**).

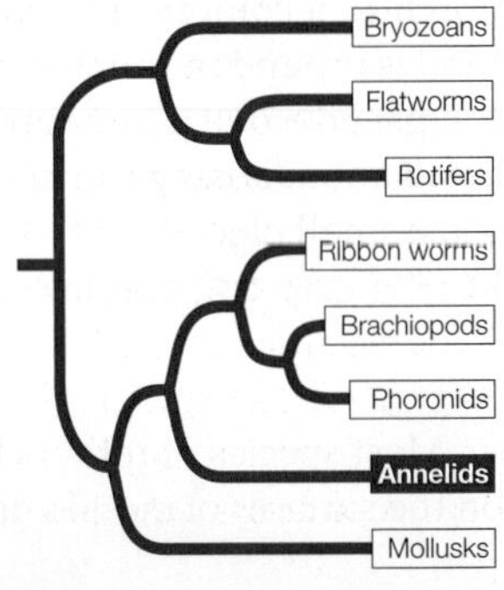

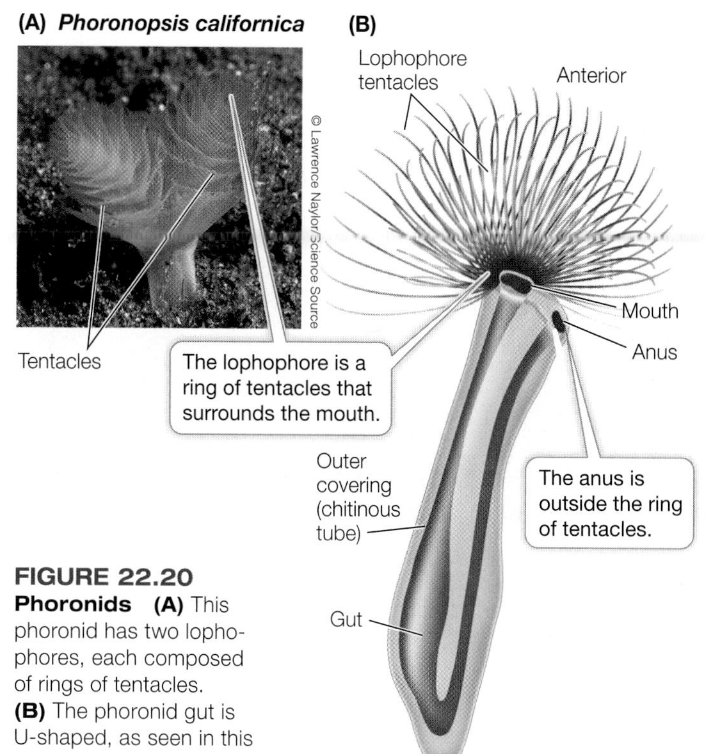

FIGURE 22.20 Phoronids (A) This phoronid has two lophophores, each composed of rings of tentacles. **(B)** The phoronid gut is U-shaped, as seen in this generalized diagram.

In most large annelids, the coelom in each segment is isolated from those in other segments. A separate nerve center called a ganglion (plural ganglia) controls each segment. Nerve cords that connect the ganglia coordinate their functioning. Most annelids lack a rigid external protective covering. Instead, they have a thin, permeable body wall that serves as a general surface for gas exchange. These animals are thus restricted to moist environments because they lose body water rapidly in dry air. The approximately 19,000 described annelid species live in marine, freshwater, and moist terrestrial environments.

Many annelids have one or more pairs of eyes and one or more pairs of tentacles (with which they capture prey or filter food from the surrounding water) at the anterior end of the body (**FIGURE 22.22A**). In some species, the body wall of most segments extends laterally as a series of thin outgrowths called parapodia. The parapodia function in gas exchange, and some species use them to move. Stiff bristles called setae protrude from each parapodium, forming temporary contact with the substrate and preventing the animal from slipping backward when its muscles contract.

Some annelids, such as the pogonophorans, secrete tubes made of chitin and other substances, in which they live (**FIGURE 22.22B**). Pogonophorans have lost their digestive tract (they have no mouth or gut). So how do they obtain nutrition? Part of the answer is that pogonophorans can take up dissolved organic matter directly from the sediments in which they live or from the surrounding water. Much of their nutrition, however, is provided by endosymbiotic bacteria that live in a specialized organ known as the trophosome. These bacteria oxidize hydrogen sulfide and other sulfur-containing compounds, fixing carbon from methane in the process. The uptake of the hydrogen sulfide, methane, and oxygen used by the bacteria is facilitated by hemoglobin in the pogonophorans' tentacles. It is this hemoglobin that gives the tentacles their red color (see Figure 22.22B).

Oligochaetes are annelids that have no parapodia or anterior tentacles, and they have only four pairs of setae bundles per segment. Earthworms—the most familiar oligochaetes—burrow in and ingest soil, from which they extract food particles. All oligochaetes are hermaphroditic; that is, each individual is both male and female. Sperm are exchanged simultaneously between two copulating individuals. Eggs and sperm are deposited outside the adult's body, in a cocoon secreted by the clitellum (see Figure 22.21). Fertilization occurs within the cocoon after it is shed, and when development is complete, miniature worms emerge and immediately begin independent life.

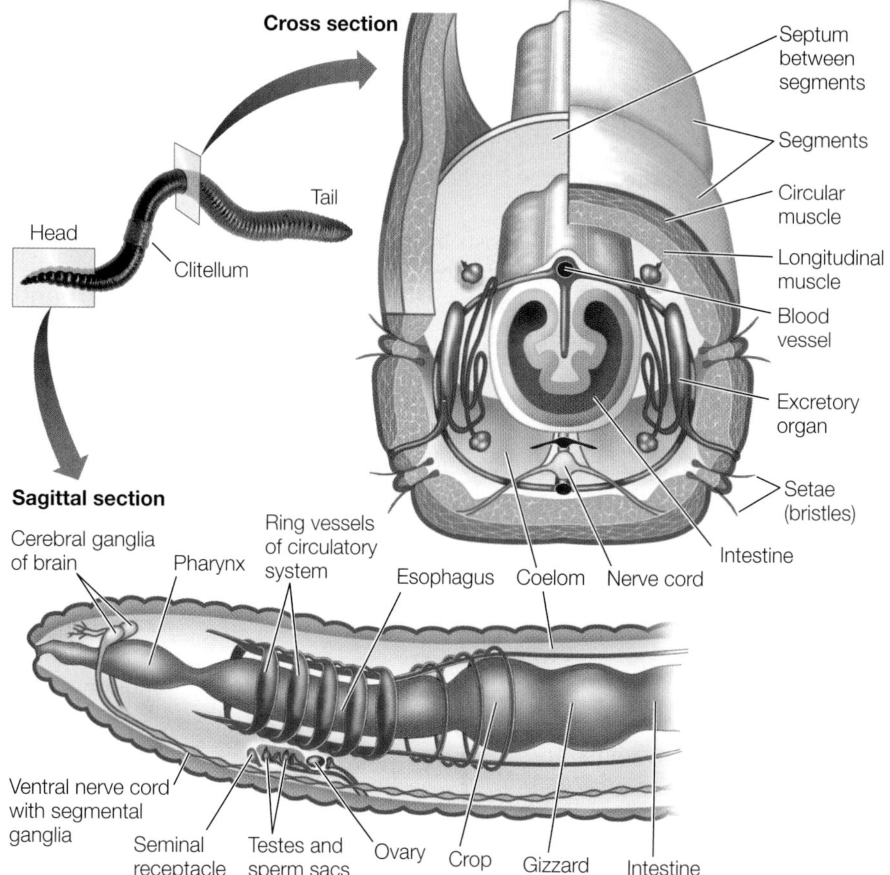

FIGURE 22.21 Annelids Have Many Body Segments The segmented structure of the annelids is apparent both externally and internally. Many organs of this earthworm, a common annelid, are repeated serially.

(A) ***Protula*** **sp.**

(B) ***Riftia*** **sp.**

(C) ***Placobdella parasitica***

FIGURE 22.22 Diversity among the Annelids **(A)** "Fan worms," or "feather duster worms," are sessile marine annelids that grow in chitinous tubes, from which their tentacles extend and filter food from the water. **(B)** Pogonophorans live around hydrothermal vents deep in the ocean. As in fan worms, their tentacles protrude from chitinous tubes. They do not possess a digestive tract, however, and obtain most of their nutrition from endosymbiotic bacteria. **(C)** This hermaphroditic freshwater leech is brooding a clutch of fertilized eggs.

Leeches, like oligochaetes, lack parapodia and tentacles (**FIGURE 22.22C**). They live in freshwater, marine, and terrestrial habitats. The leech coelom is not divided into compartments, and consists of a large, fluid-filled cavity. Groups of segments at each end of the body are modified to form suckers, which serve as temporary anchors that help the leech move. With its posterior sucker attached to a substrate, the leech extends its body by contracting its circular muscles. The anterior sucker is then attached, the posterior one detached, and the leech shortens itself by contracting its longitudinal muscles.

Many leeches feed on vertebrate hosts by making an incision from which blood flows. A feeding leech secretes an anticoagulant into the wound to keep the blood flowing. For centuries, medical practitioners used leeches to treat diseases they believed were caused by an excess of blood or by "bad blood." Although most leeching practices (such as inserting a leech in a person's throat to alleviate swollen tonsils) have been abandoned, *Hirudo medicinalis* (the medicinal leech) is used medically even today to reduce fluid pressure and prevent blood clotting in damaged tissues, to eliminate pools of coagulated blood, and to prevent scarring. The anticoagulants of certain other leech species contain anesthetics and blood vessel dilators and are being studied for possible medical uses.

Media Clip 22.6 Leeches Feeding on Blood

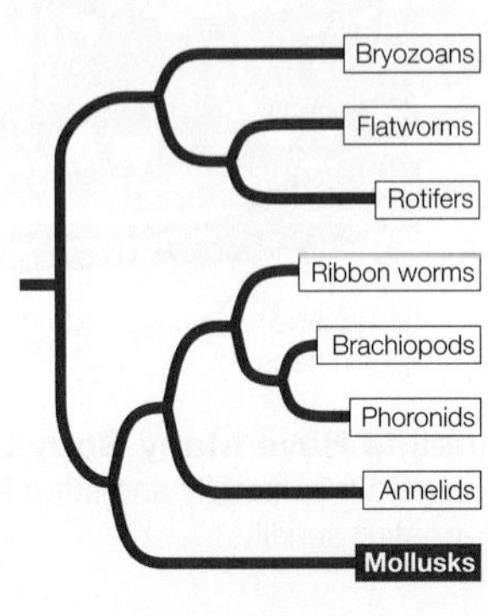

MOLLUSKS The most diverse group of lophotrochozoans are the **mollusks**, with about 117,000 species that inhabit a wide array of aquatic and terrestrial environments. There are four major clades of mollusks: chitons, gastropods, bivalves, and cephalopods. Although these groups differ dramatically in morphology, they all share the same three major body components: a foot, a visceral mass, and a mantle (**FIGURE 22.23A**).

- The molluscan **foot** is a muscular structure that originally was both an organ of locomotion and a support for internal organs. In cephalopods such as squids and octopuses, the foot has been modified to form arms and tentacles borne on a head with complex sensory organs. In other groups, such as clams, the foot is a burrowing organ. In some groups the foot is greatly reduced.
- The heart and the digestive, excretory, and reproductive organs are concentrated in a centralized, internal **visceral mass**.
- The **mantle** is a fold of tissue that covers the organs of the visceral mass. The mantle secretes the hard calcareous shell that is typical of many mollusks.

In most mollusks, the mantle extends beyond the visceral mass to form a mantle cavity. Within this cavity lie gills that are used for gas exchange. When cilia on the gills beat, they create a current of water. The gill tissue, which is highly vascularized (contains many blood vessels), takes up oxygen from the water and releases carbon dioxide. Many mollusks use their gills as filter-feeding devices. Other mollusks feed using a rasping structure known as the radula to scrape algae from rocks. In some mollusks, such as the marine cone snails, the radula has been modified into a drill or poison dart.

Except in cephalopods, molluscan blood vessels do not form a closed circulatory system. Blood and other fluids empty into a large, fluid-filled hemocoel, through which fluids move around the animal and deliver oxygen to the internal organs. Eventually the fluids reenter the blood vessels and are moved by a heart.

The approximately 1,000 living species of **chitons** (**FIGURE 22.23B**) are characterized by eight overlapping calcareous plates, surrounded by a structure known as the girdle. These plates and the girdle protect the chiton's internal organs and muscular foot.

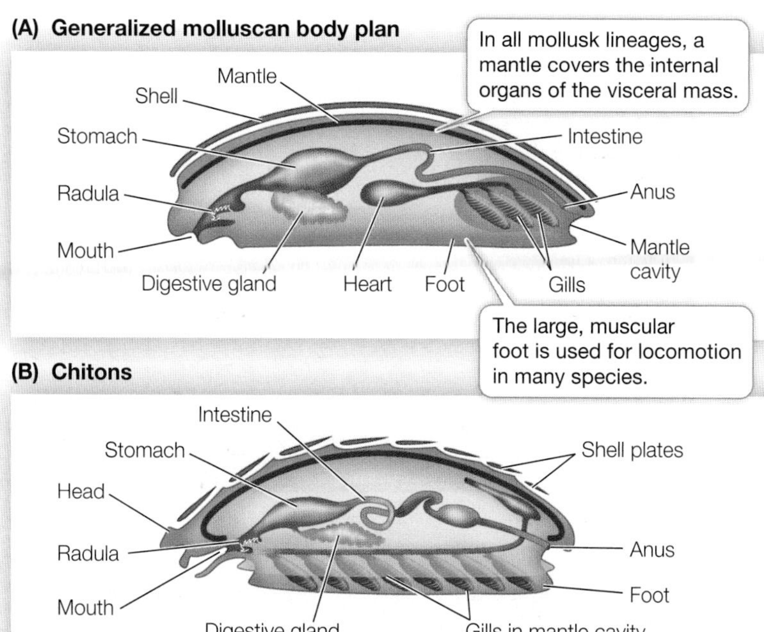

FIGURE 22.23 Organization and Diversity of Molluscan Bodies **(A)** The major molluscan groups display different variations on a general body plan that includes three major components: a foot, a visceral mass of internal organs, and a mantle. In many species, the mantle secretes a calcareous shell. **(B)** Chitons have eight overlapping calcareous plates surrounded by a girdle. **(C)** Most gastropods have a single dorsal shell, into which they can retreat for protection. **(D)** Bivalves get their name from their two hinged shells, which can be tightly closed. **(E)** Cephalopods are active predators; they use their arms and tentacles to capture prey.

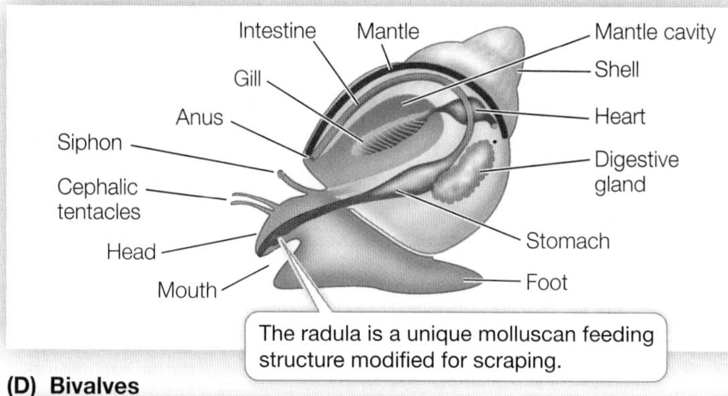

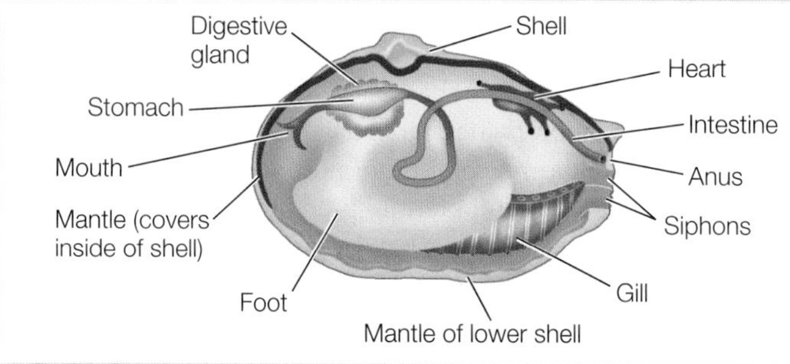

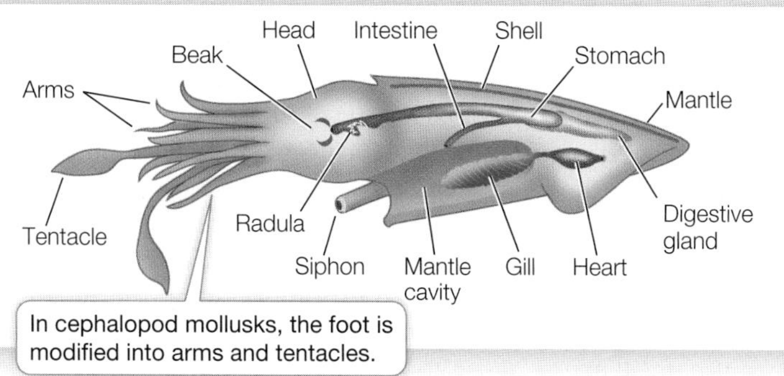

The chiton body is bilaterally symmetrical, and the internal organs, particularly the digestive and nervous systems, are relatively simple. Most chitons are marine omnivores that scrape algae, bryozoans, and other organisms from rocks with their sharp radula. An adult chiton spends most of its life clinging to rock surfaces with its large, muscular, mucus-covered foot. It moves slowly by means of rippling waves of muscular contraction in the foot. Fertilization in most chitons takes place in the water, but in a few species fertilization is internal and embryos are brooded within the body.

Gastropods are the most species-rich and widely distributed mollusks, with about 85,000 living species. Snails, whelks, limpets, slugs, nudibranchs (sea slugs), and abalones are all gastropods. Most species move by gliding on their muscular foot, but in a few species—the sea butterflies and heteropods—the foot is a swimming organ with which the animal moves through open ocean waters. The only mollusks that live in terrestrial environments—land snails and slugs—are gastropods (**FIGURE 22.23C**). In these terrestrial species, the mantle tissue is modified into a highly vascularized lung.

Clams, oysters, scallops, and mussels are all familiar **bivalves**. The 30,000 living species are found in both marine and freshwater environments. Bivalves have a very small head and a hinged, two-part shell that extends over the sides of the body as well as the top (**FIGURE 22.23D**). Many clams use their foot to burrow into mud and sand. Bivalves feed by taking in water through an opening called an incurrent siphon and filtering food from the water with their large gills, which are also the main sites of gas exchange. Water and gametes exit through the excurrent siphon. Fertilization takes place in open water in most species.

There are about 800 living species of **cephalopods**—squids, octopuses, and nautiluses. Their excurrent siphon is modified to allow the animal to control the water content of the mantle cavity. The modification of the mantle into a device for forcibly ejecting water from the cavity through the siphon enables these animals to move rapidly by "jet propulsion" through the water. With their greatly enhanced mobility, cephalopods (which first appeared early in the Cambrian) became the major predators in the open waters of the Devonian oceans. They remain important marine predators today.

Cephalopods capture and subdue prey with their tentacles (see Figure 22.24B). As is typical of active, rapidly moving predators, cephalopods have a head with complex sensory organs, most notably eyes that are comparable to those of vertebrates in their ability to resolve images. The head is closely associated with a large, branched foot that bears the tentacles and a siphon (**FIGURE 22.23E**). The large, muscular mantle provides an external supporting structure. The gills hang in the mantle cavity. Many cephalopods have elaborate courtship behavior, which can involve striking color changes.

Many early cephalopods had a chambered external shell divided by partitions penetrated by tubes through which gases and liquids could be moved to control the animal's buoyancy. Nautiluses are the only surviving cephalopods that have such external chambered shells, although squids and cuttlefishes retain internal shells.

Shells have been lost several times among the mollusks, as in several groups of gastropods, including the slugs and nudibranchs

(A) ***Chromodoris kuniei***

(B) ***Octopus bimaculoides***

FIGURE 22.24 Mollusks in Some Groups Have Lost Their Shells **(A)** Nudibranchs ("naked gills"), also called sea slugs, are shell-less gastropods. This species is brightly colored, alerting potential predators of its toxicity. **(B)** Octopuses have lost both the external and internal shell, which allows these cephalopods to squeeze through tight spaces.

(**FIGURE 22.24A**). These shell-less gastropods gain some protection from predation by being distasteful or toxic to many species. The often brilliant coloration of nudibranchs is aposematic, meaning it serves to warn potential predators of toxicity. Among the cephalopods, the octopuses have lost both external and internal shells (**FIGURE 22.24B**). Their lack of shells allows octopuses to escape predators by squeezing into small crevices.

Media Clip 22.7 **Octopuses Can Pass through Small Openings**

Ecdysozoans grow by shedding their cuticles

The distinguishing characteristic of **ecdysozoans** is their external covering, or **cuticle**, which is secreted by the underlying epidermis (the outermost cell layer). The cuticle provides these animals with both protection and support. Once formed, however, the cuticle cannot grow. How, then, can ecdysozoans increase in size? They do so by shedding, or **molting**, the cuticle and replacing it with a new, larger one. This molting process gives the clade its name (Greek *ecdysis*, "to get out of").

(A)

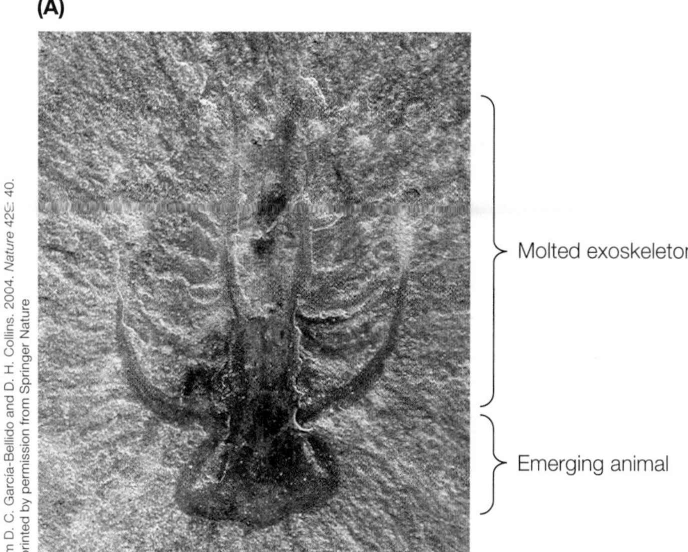

(B) ***Heterophrynus batesii***

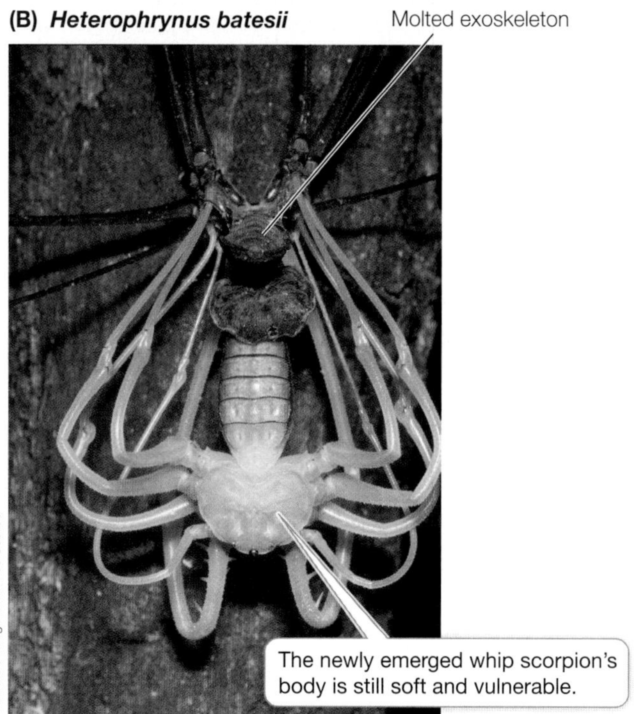

FIGURE 22.25 Molting, Past and Present **(A)** A 500-million-year-old fossil from the Cambrian captured an individual of a long-extinct arthropod species in the process of molting, demonstrating that molting is an evolutionarily ancient trait. **(B)** This tail-less whip scorpion has just emerged from its discarded exoskeleton and will be highly vulnerable until its new cuticle has hardened.

Media Clip 22.8 Molting a Cuticle

A soft-bodied arthropod from the Cambrian, fossilized in the process of molting, shows that molting evolved more than 500 million years ago (**FIGURE 22.25A**). An increasingly rich array of molecular and genetic evidence, including a set of Hox genes shared by all ecdysozoans, suggests they have a single common ancestor. Thus molting of a cuticle is a trait that may have evolved only once during animal evolution.

Before an ecdysozoan molts, a new cuticle is already forming underneath the old one. Once the old cuticle is shed, the new one expands and hardens. Until it has hardened, though, the animal is vulnerable to its enemies, both because its outer surface is easy to penetrate and because an animal with a soft cuticle moves slowly or not at all (**FIGURE 22.25B**).

In many ecdysozoans that have wormlike bodies, the cuticle is relatively thin and flexible. Such a cuticle offers the animal some protection but provides only modest body support. A thin cuticle allows the exchange of gases, minerals, and water across the body surface, but it restricts the animal to moist habitats. Many species of ecdysozoans with thin cuticles live in marine sediments from which they obtain prey. Some freshwater species absorb nutrients directly through their thin cuticles, as do parasitic species that live within their hosts.

The cuticles of other ecdysozoans, notably the arthropods, function as external skeletons, or **exoskeletons**. These exoskeletons are thickened by layers of protein and a strong, waterproof polysaccharide called **chitin**. An animal with a rigid, chitin-reinforced exoskeleton can neither move in a wormlike manner nor use cilia for locomotion. A hard exoskeleton also impedes the passage of oxygen and nutrients into the animal, presenting new challenges in other areas besides growth. New mechanisms of locomotion and gas exchange evolved in those ecdysozoans with hard exoskeletons.

To move rapidly, an animal with a rigid exoskeleton must have body extensions that can be manipulated by muscles. Such appendages evolved in the late Precambrian and led to the **arthropod** (Greek, "jointed foot") clade. Arthropod appendages exist in an amazing variety of forms. They serve many functions, including walking and swimming, gas exchange, food capture and manipulation, copulation, and sensory perception. Arthropods grasp food with their mouth and associated appendages and usually digest it internally. Their muscles are attached to the inside of the exoskeleton.

The arthropod exoskeleton has had a profound influence on the evolution of these animals. Encasement within a rigid body covering provides support for walking on dry land, and the waterproofing provided by the cuticle keeps the animal from dehydrating in dry air. These features have allowed arthropods to invade terrestrial environments several times.

The evolution of the diverse arthropods and their relatives will be described in greater detail in Key Concept 22.4. Here we describe the other major groups of ecdysozoans.

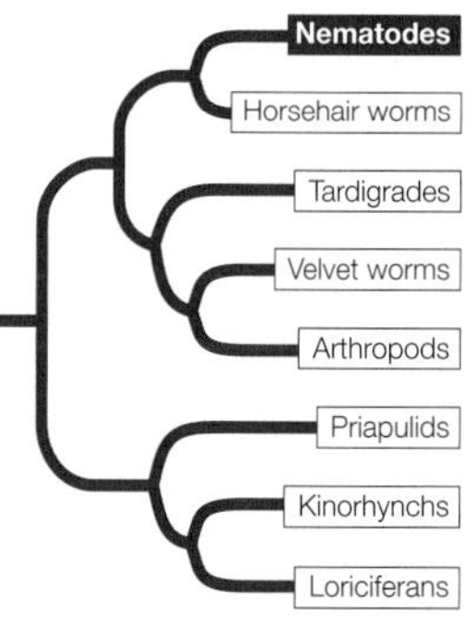

NEMATODES **Nematodes** (roundworms) have a thick, multilayered cuticle that gives their unsegmented body its shape. As a nematode grows, it sheds its cuticle four times. Nematodes exchange oxygen and nutrients with their environment through both the cuticle and the gut, which is only one cell layer thick.

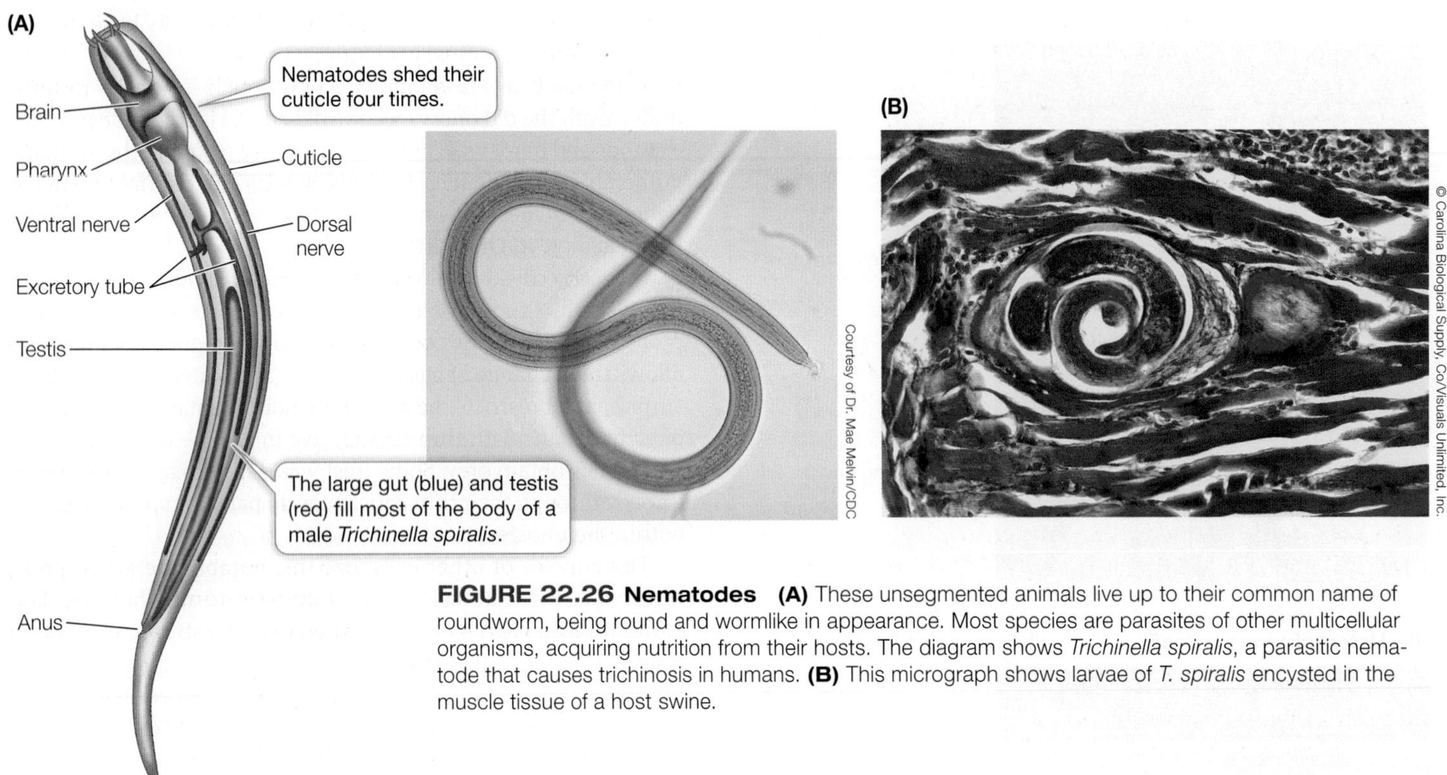

FIGURE 22.26 Nematodes (A) These unsegmented animals live up to their common name of roundworm, being round and wormlike in appearance. Most species are parasites of other multicellular organisms, acquiring nutrition from their hosts. The diagram shows *Trichinella spiralis*, a parasitic nematode that causes trichinosis in humans. **(B)** This micrograph shows larvae of *T. spiralis* encysted in the muscle tissue of a host swine.

Materials are moved through the gut by rhythmic contraction of a highly muscular organ, the pharynx, at the worm's anterior end. Nematodes move by contracting their longitudinal muscles.

Nematodes are probably the most abundant and universally distributed of all major animal groups. About 25,000 species have been described, but the actual number of living species may be in the millions. Many are microscopic, but the largest known nematode reaches a length of 9 meters (it is a parasite in the placentas of sperm whales). Countless nematodes live as scavengers in the upper layers of the soil, on the bottoms of lakes and streams, and in marine sediments. The topsoil of rich farmland may contain over 20 billion nematodes per hectare. A single rotting apple may contain as many as 90,000 individuals.

One soil-inhabiting nematode, *Caenorhabditis elegans*, serves as a model organism in the laboratories of many geneticists and developmental biologists. It is ideal for such research because it is easy to cultivate, matures in 3 days, and has a fixed number of body cells.

Many nematodes are predators, feeding on protists and small animals (including other roundworms). Most significant to humans, however, are the many species that parasitize plants and animals (**FIGURE 22.26A**). The nematodes that parasitize humans (causing serious diseases such as trichinosis, filariasis, and elephantiasis), domesticated animals, and economically important plants have been studied intensively in an effort to find ways of controlling them.

The structure of parasitic nematodes is similar to that of free-living species, but the life cycles of many parasitic species have special stages that facilitate the transfer of individuals among hosts. *Trichinella spiralis*, the species that causes the human disease trichinosis, has a relatively simple life cycle. A person may become infected by eating the flesh of an animal (usually a pig) that has *Trichinella* larvae encysted in its muscles (**FIGURE 22.26B**). The larvae are activated in the person's digestive tract, emerge from their cysts, and attach to the intestinal wall, where they feed. Later they bore through the intestinal wall and are carried in the bloodstream to muscles, where they form new cysts. If present in great numbers, these cysts can cause severe pain or death.

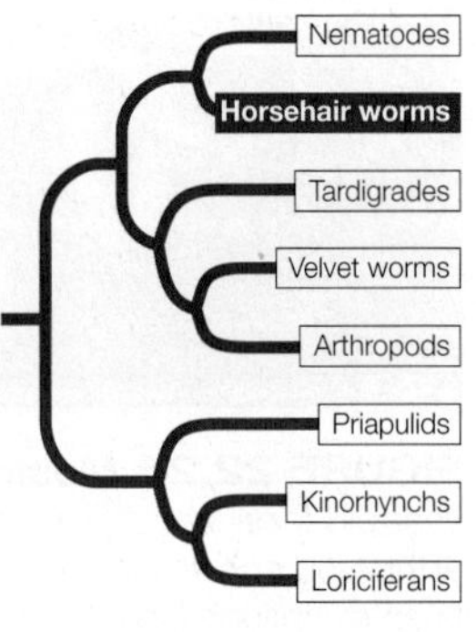

HORSEHAIR WORMS About 350 species of the unsegmented **horsehair worms** have been described. As their name implies, these animals are extremely thin in diameter; they range from a few millimeters up to a meter in length. Most adult horsehair worms live in fresh water among the leaf litter and algal mats that accumulate near the shores of streams and ponds. A few species live in damp soil.

Horsehair larvae are internal parasites of freshwater crayfishes and of terrestrial and aquatic insects (**FIGURE 22.27**). An adult horsehair worm has no mouth, and its gut is greatly reduced and probably nonfunctional. Some species feed only as larvae, absorbing nutrients from their hosts across the body wall. But other species continue to grow and shed their cuticles even after they have left their hosts, suggesting that some adult worms may be able to absorb nutrients from their environment.

FIGURE 22.27 Horsehair Worm Larvae Are Parasitic The larvae of this horsehair worm (*Paragordius tricuspidatus*) can manipulate its host's behavior. The hatching worm causes the cricket to jump into water, where the worm will continue its life cycle as a free-living adult. The insect, having delivered its parasitic burden, drowns.

SMALL MARINE CLADES There are several small, relatively poorly known groups of benthic marine ecdysozoans. The 20 species of **priapulids** are cylindrical, unsegmented, wormlike animals with a three-part body plan consisting of a proboscis, trunk, and caudal appendage ("tail"). It should be clear from their appearance why they were named after the Greek fertility god Priapus (**FIGURE 22.28A**). The 180 species of **kinorhynchs** live in marine sands and muds and are virtually microscopic; no kinorhynchs are longer than 1 millimeter. Their bodies are divided into 13 segments, each with a separate cuticular plate (**FIGURE 22.28B**). The minute (less than 1 mm long) **loriciferans** were not discovered until 1983. About 100 living species are known to exist, although only about 30 species have been described. The body is divided into a head, neck, thorax, and abdomen and is covered by six plates, from which the loriciferans get their name (Latin *lorica*, "corset") (**FIGURE 22.28C**).

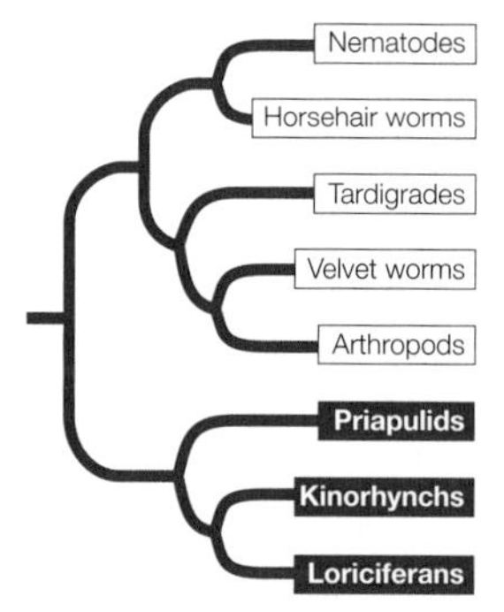

REVIEW & APPLY | 22.3

R The shared derived traits of protostomes include a blastopore that develops into a mouth, an anterior brain, and a ventral nervous system. Several lophotrochozoan groups are characterized by a filter-feeding ciliated structure known as a lophophore or by cilia-bearing larvae known as trochophores. Ecdysozoans, which have a body covering known as a cuticle, must molt periodically in order to grow.

A
1. How does an animal's body covering influence the way it exchanges gases, feeds, and moves?
2. What features make arthropods well adapted for colonizing terrestrial environments?
3. Why are most annelids restricted to moist environments?
4. Briefly describe how the basic body organization of mollusks has been modified to yield a wide diversity of animals.
5. What are three ways in which nematodes have a significant impact on humans?

FIGURE 22.28 Benthic Marine Ecdysozoans Members of these groups are marine bottom-dwellers. **(A)** Most priapulid species live in burrows on the ocean floor, extending the proboscis for feeding. **(B)** Kinorhynchs are virtually microscopic. The cuticular plates that cover their bodies are molted periodically. **(C)** Six cuticular plates form a "corset" around the minute loriciferan body.

We will turn now to the arthropods. Members of the four arthropod subgroups not only dominate the ecdysozoan clade but are also among the most diverse animals on Earth.

22.4 | Arthropods Are Diverse and Abundant

LEARNING OBJECTIVES

By the end of this key concept you should be able to:

22.4.1 Describe morphological innovations of arthropods, and their effects on arthropod diversity.

22.4.2 Distinguish among the two different forms of metamorphosis in arthropods.

22.4.3 Explain how the evolutionary innovations among arthropods are related to the habitats in which they live.

Nematodes
Horsehair worms
Tardigrades
Velvet worms
Arthropods
Priapulids
Kinorhynchs
Loriciferans

Arthropods and their relatives are ecdysozoans with paired appendages. Arthropods are an extremely diverse group of animals in numbers of species. Furthermore, the number of individual arthropods alive at any one time is estimated to be about 10^{18}, or a billion billion. Among the animals, only the nematodes are thought to exist in greater numbers.

Several key features have contributed to the success of the arthropods. Their bodies are segmented, and their muscles are attached to the inside of their rigid exoskeletons. Each segment has muscles that operate that segment and the jointed appendages attached to it. Jointed appendages permit complex movements, and different appendages are specialized for different functions. Encasement of the body within a rigid exoskeleton provides the animal with support for walking in the water or on dry land and provides some protection against predators. The waterproofing provided by chitin keeps the animal from dehydrating in dry air.

Representatives of the four major arthropod groups living today are all species-rich: the chelicerates (including the arachnids—spiders, scorpions, mites, and their relatives), myriapods (millipedes and centipedes), crustaceans (including shrimps, crabs, and barnacles), and hexapods (insects and their relatives). Phylogenetic relationships among arthropod groups are currently being reexamined in light of a wealth of new information, much of it based on gene sequences. These studies suggest that the chelicerates are the sister group to the remaining arthropods, and that the crustaceans may be paraphyletic (some crustaceans are more closely related to hexapods than to other crustaceans). There is strong support for the monophyly of arthropods as a whole.

The jointed appendages of arthropods gave the clade its name, from the Greek words *arthron*, "joint," and *podos*, "foot" or "limb." Arthropods evolved from ancestors with simple, unjointed appendages. The exact forms of those ancestors are unknown, but some arthropod relatives with segmented bodies and unjointed appendages survive today. Before we describe the modern arthropods, we will discuss those arthropod relatives.

Arthropod relatives have fleshy, unjointed appendages

Until fairly recently, biologists debated whether the **velvet worms** (onychophorans) were more closely related to annelids or arthropods, but molecular evidence clearly links them to the arthropods. Indeed, with their soft, fleshy, unjointed, claw-bearing legs, velvet worms may be similar in appearance to the arthropod ancestor (**FIGURE 22.29A**). The 180 species of velvet worms live in leaf litter in humid tropical environments. Their soft, segmented bodies are covered by a thin, flexible cuticle that contains chitin. They use their fluid-filled body cavities as hydrostatic skeletons. Fertilization is internal, and large, yolky eggs are brooded within the female's body.

Tardigrades (water bears) also have fleshy, unjointed legs and use their fluid-filled body cavities as hydrostatic skeletons (**FIGURE 22.29B**). Tardigrades are tiny (0.5–1.5 mm long) and lack both a circulatory system and gas exchange organs. The 1,200

(A) ***Macroperipatus torquatus***

© George Grall/National Geographic Creative/ Alamy Stock Photo

(B) ***Macrobiotus sapiens***

© Eye of Science/Science Source

FIGURE 22.29 Arthropod Relatives with Unjointed Appendages **(A)** Velvet worms have unjointed legs and use the body cavity as a hydrostatic skeleton. **(B)** Tardigrades can be abundant on the wet surfaces of mosses and plants and in temporary pools of water.

(A) *Pseudopallene* sp.

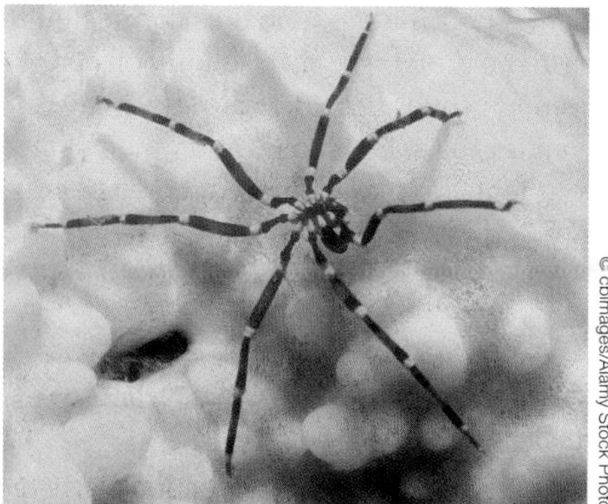
© cbimages/Alamy Stock Photo

(B) *Limulus polyphemus*

© Barbara J. Miller/Biological Photo Service

FIGURE 22.30 Two Small Chelicerate Groups **(A)** Although they are not spiders, it is easy to see why sea spiders were given their common name. **(B)** Horseshoe crabs are an ancient group that has changed very little in morphology over time.

known species live in marine sands and on temporary water films on plants, as well as in ephemeral pools. When these films and shallow pools dry out, the animals also lose water and shrink to small, barrel-shaped objects that can survive for at least a decade in a dormant state. Tardigrades have been found in densities as high as 2 million per square meter of moss.

Chelicerates are characterized by pointed, nonchewing mouthparts

In the **chelicerates**, the head bears two pairs of pointed appendages modified to form mouthparts, called chelicerae, which are used to grasp (rather than chew) prey. In addition, many chelicerates have four pairs of walking legs. The 114,000 described species are placed in three major clades: sea spiders, horseshoe crabs, and arachnids.

The sea spiders (pycnogonids) are a poorly known group of about 1,000 marine species (**FIGURE 22.30A**). Most are small, with leg spans less than 1 centimeter, but some deep-sea species have leg spans up to 60 centimeters. A few sea spiders eat algae, but most are carnivorous, eating a variety of small invertebrates.

There are four living species of horseshoe crabs, but many close relatives are known from fossils. Horseshoe crabs, which have changed very little morphologically during their long fossil history, have a large horseshoe-shaped covering over most of the body. They are common in shallow waters along the eastern coast of North America and the southern and eastern coasts of Asia, where they scavenge and prey on bottom-dwelling animals. Periodically they crawl into the intertidal zone in large numbers to mate and lay eggs (**FIGURE 22.30B**).

Arachnids are abundant in terrestrial environments. Most arachnids have a simple life cycle in which miniature adults hatch from internally fertilized eggs and begin independent lives almost immediately. Some arachnids retain their eggs during development and give birth to live young.

The most species-rich and abundant arachnids are the spiders, scorpions, harvestmen, mites, and ticks (**FIGURE 22.31**). More than 60,000 described species of mites and ticks live in soil, leaf litter, mosses, and lichens, under bark, and as parasites of plants and animals. Mites are vectors for wheat and rye mosaic viruses and cause mange in domestic animals as well as skin irritation in humans.

Spiders, of which about 50,000 species have been described, are important terrestrial predators with hollow chelicerae, which they use to inject venom into their prey. Some species have excellent vision that facilitates elaborate courtship displays and allows the spiders to hunt their prey. Others spin elaborate webs

(A) *Lycosa* sp.

© Peter J. Bryant/Biological Photo Service

(B) *Androctonus* sp.

© Oliver Gerhard/Alamy Stock Photo

(C) *Leiobunum rotundum*

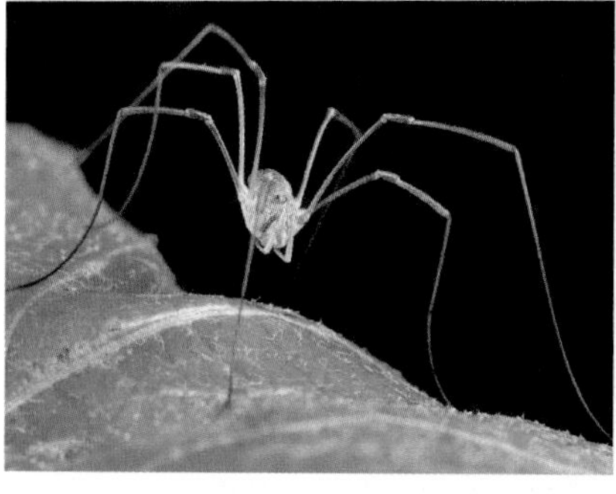
© Nigel Cattlin/Alamy Stock Photo

(D) *Brevipalpus phoenicis*

SEM by Eric Erbe; colorization by Chris Pooley/USDA ARS

FIGURE 22.31 Arachnid Diversity **(A)** Known for their web-spinning abilities, spiders are expert predators. **(B)** A desert scorpion assumes the animal's aggressive attack position. These arachnids are fearsome predators, and their sting can be dangerous to humans. **(C)** Harvestmen, also called daddy longlegs, are scavengers. **(D)** Mites include many free-living species as well as blood-sucking, external parasites.

made of protein threads in which they snare prey. The threads are produced by modified abdominal appendages connected to internal glands that secrete the proteins, which solidify on contact with air. The webs of different groups of spiders are strikingly varied, and this variation enables the spiders to position their snares in many different environments for many different types of prey.

Media Clip 22.9 Elaboration of Courtship Displays: Even Spiders Twerk

(A) ***Scolopendra hardwickei***

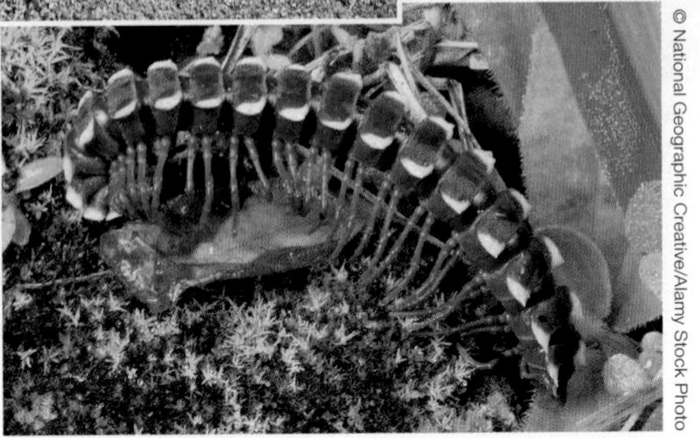

(B) ***Polydesmus*** **sp.**

FIGURE 22.32 Myriapods **(A)** Centipedes have modified appendages that function as poisonous fangs for capturing active prey. They have one pair of legs per segment. **(B)** Millipedes, which are scavengers and plant eaters, have smaller jaws and legs than centipedes do. They have two pairs of legs per segment.

Mandibles and antennae characterize the remaining arthropod groups

The remaining three arthropod groups—the myriapods, crustaceans, and hexapods—have mouthparts composed of mandibles rather than chelicerae, so they are together called **mandibulates**. Mandibles are often used for chewing as well as for biting and holding food. Another distinctive characteristic of the mandibulates is the presence of sensory antennae on the head.

The **myriapods** comprise the centipedes, millipedes, and their close relatives. Centipedes and millipedes have a well-formed head with the mandibles and antennae characteristic of mandibulates. Their distinguishing feature is a long, flexible, segmented trunk that bears many pairs of legs (**FIGURE 22.32**). Centipedes, which have one pair of legs per segment, prey on insects and other small animals. In millipedes, two adjacent segments are fused so that each fused segment has two pairs of legs. Millipedes are detritivores. More than 3,000 species of centipedes and 9,000 species of millipedes have been described, and many more

FIGURE 22.33 Crustacean Diversity **(A)** This decapod crustacean is a "Sally Lightfoot" crab from the Galápagos. **(B)** This pillbug, a terrestrial isopod, can roll into a tight ball when threatened. **(C)** Minute copepods such as this one are an important link in aquatic food chains. **(D)** Tadpole shrimp are branchiopods, not to be confused with the decapod crustaceans commonly called shrimp, or with the brachiopods (see Figure 22.19). **(E)** Barnacles are sessile as adults, attaching to a substrate by their muscular stalks and feeding with retractable feeding appendages (the delicate, tentacle-like structures seen here).

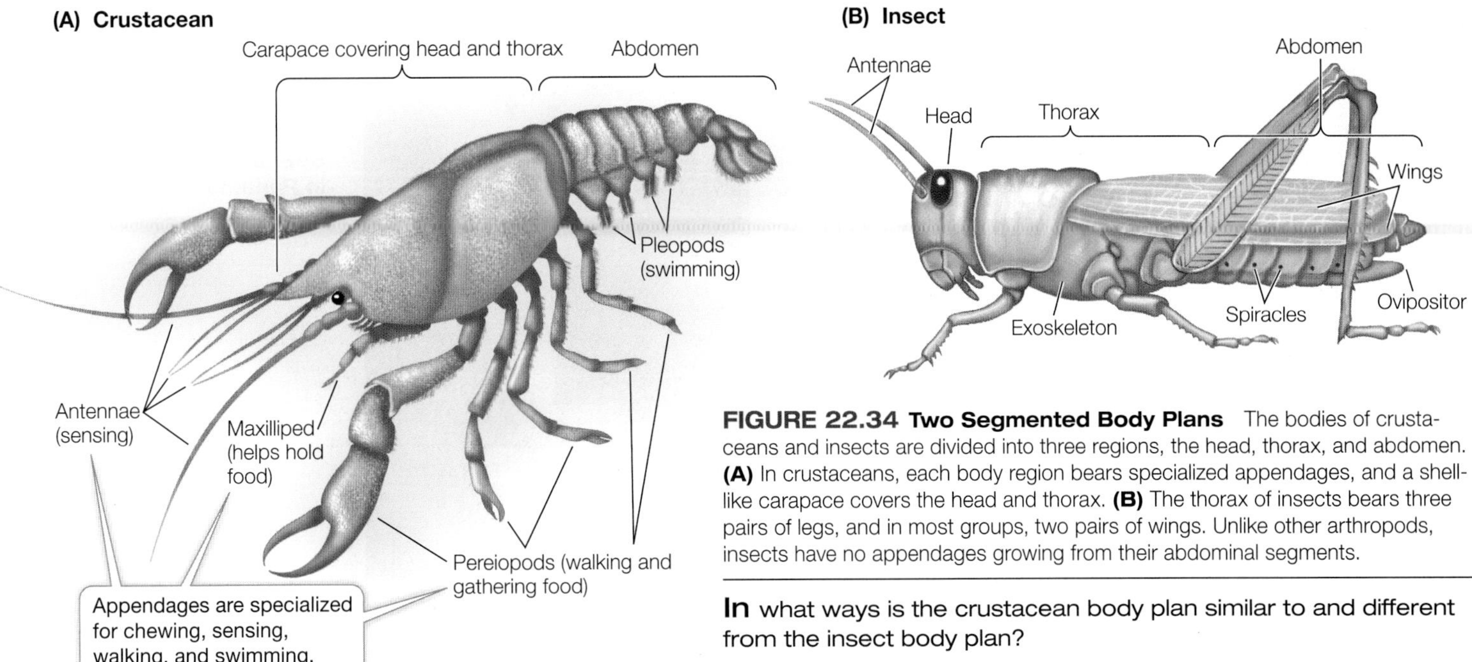

FIGURE 22.34 Two Segmented Body Plans The bodies of crustaceans and insects are divided into three regions, the head, thorax, and abdomen. **(A)** In crustaceans, each body region bears specialized appendages, and a shell-like carapace covers the head and thorax. **(B)** The thorax of insects bears three pairs of legs, and in most groups, two pairs of wings. Unlike other arthropods, insects have no appendages growing from their abdominal segments.

In what ways is the crustacean body plan similar to and different from the insect body plan?

species probably remain unknown. Although most myriapods are less than a few centimeters long, some tropical and subtropical species are ten times that size.

Crustaceans are the dominant marine arthropods today, and they are also common in freshwater and some terrestrial environments. The most familiar crustaceans are the shrimps, lobsters, crayfishes, and crabs (all decapods; **FIGURE 22.33A**) and the pillbugs (isopods; **FIGURE 22.33B**). Additional species-rich groups include the amphipods, ostracods, copepods (**FIGURE 22.33C**), and branchiopods (**FIGURE 22.33D**), all of which are found in freshwater and marine environments.

Barnacles are unusual crustaceans that are sessile as adults (**FIGURE 22.33E**). Adult barnacles look more like mollusks than like other crustaceans, but as the zoologist Louis Agassiz remarked more than a century ago, a barnacle is "nothing more than a little shrimp-like animal, standing on its head in a limestone house and kicking food into its mouth."

Media Clip 22.10 **Barnacles Feeding**

Most of the 67,000 described species of crustaceans have a body that is divided into three regions: head, thorax, and abdomen (**FIGURE 22.34A**). The segments of the head are fused together, and the head bears five pairs of appendages. Each of the multiple thoracic and abdominal segments usually bears one pair of appendages. The appendages on different parts of the body are specialized for different functions, such as gas exchange, chewing, capturing food, sensing, walking, and swimming. In many species, a fold of the exoskeleton, the carapace, extends dorsally and laterally back from the head to cover and protect some of the other segments.

The fertilized eggs of most crustacean species are attached to the outside of the female's body, where they remain during their early development. At hatching, the young of some species are released as larvae; those of other species are released as juveniles that are similar in form to the adults. Still other species release eggs into the water or attach them to an object in the environment.

More than half of all described species are insects

During the Devonian, more than 400 million years ago, some mandibulates colonized terrestrial environments. Of the several groups (including some crustacean isopods and decapods) that successfully colonized the land, none is more prominent today than the six-legged **hexapods**: the insects and their relatives. Insects are abundant and diverse in terrestrial and freshwater environments. Only a few insects live in salt water.

The wingless relatives of the insects—the springtails, two-pronged bristletails, and proturans—are probably the most similar of living forms to insect ancestors. These insect relatives have a simple life cycle; they hatch from eggs as miniature adults. They differ from insects in having internal mouthparts. Springtails can be extremely abundant (up to 200,000 per square meter) in soil, leaf litter, and on vegetation and are the most abundant hexapods in the world in terms of number of individuals (rather than number of species).

Like crustaceans, insects have a body with three regions—head, thorax, and abdomen. They have a single pair of antennae on the head and three pairs of legs attached to the thorax. In most groups of insects, the thorax also bears two pairs of wings. Unlike other arthropods, insects have no appendages associated with their abdominal segments (**FIGURE 22.34B**).

Insects can be distinguished from springtails and other hexapods by their external mouthparts and by antennae that contain a motion-sensitive receptor called Johnston's organ. In addition, insects have a derived mechanism for gas exchange in air: a system of air sacs and tubular channels called tracheae (singular trachea) that extend from external openings called spiracles inward to

THINK LIKE A SCIENTIST

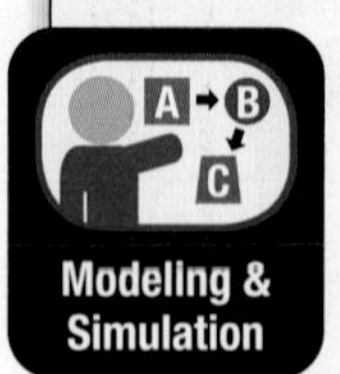

How do biologists estimate how many species are still undiscovered?

How can we estimate things that are not yet known? For example, in this book we've noted that there are thought to be far more species of insects that have not yet been discovered and described than the 1 million or so that are now known to science. How can we estimate the number of species that have not yet been discovered?

Entomologist Terry Erwin sought to answer this question and to make a rough estimate of the undescribed diversity of insects.[a] He based his estimate on data from the most diverse group of insects, the beetles. Many beetle species are restricted to a single species of host plant, so Erwin extrapolated results obtained by exhaustively collecting insects from a single species of host tree in Panama. Erwin sprayed insecticide into the canopy of *Luehea seemannii* trees, collected all the beetles from the trees, and sorted them into species based on their morphology (**FIGURE 22.35**). He then estimated how many of these beetle species were restricted to *L. seemannii* trees, how many beetle species were generalists that were found on other plants as well, and how many different species of host trees were available to beetles per hectare of Panamanian forest. Erwin also estimated the percentage of beetles that are tree-dwelling versus ground-dwelling, and the percentage of beetles among all known insect species. From the data in the following table, he was able to make a rough estimate of the total number of insect species on Earth. Here you will develop a simple model of species diversity, use Erwin's data to estimate parameters for this model, and then replicate Erwin's estimate for total insect diversity. You can also simulate different results from the model by changing your assumptions about these estimates.

Approximate number of beetle species collected from *Luehea seemannii* trees	1,200
Estimated number of host-specific beetles in this sample	163
Number of species of canopy trees per hectare of forest	70
Percentage of beetle species living in canopy (as opposed to ground-dwelling species)	75%
Percentage of beetles among all insect species	40%

FIGURE 22.35 Collecting Beetles from *Luehea Seemannii* Trees in Panama

1. From the data in the table, estimate the number of insect species in an average hectare of Panamanian forest. Assume that the data for beetles on *L. seemannii* are representative of the other tree species, and that all the species of beetles that are *not* host-specific were collected in the original sample. Remember to sum your estimates of the number of (a) host-specific beetle species in the forest canopy; (b) non-host-specific beetle species in the forest canopy; (c) beetle species on the forest floor; and (d) species of all insects other than beetles.
2. Use the following information to estimate the number of insect species on Earth. There are about 50,000 species of tropical forest trees. Assume that the data for beetles on *L. seemannii* are representative for other species of tropical trees, and calculate the number of host-specific beetles found on these trees. Add an estimated 1 million species of beetles that are expected across different species of trees (including in temperate regions). Estimate the number of ground-dwelling beetle species based on the percentage in the answer to Question 1. Now estimate the number of insect species found worldwide, based on the percentage of beetles among all insect species.
3. The estimates in the answers to Questions 1 and 2 are based on many assumptions. Do you think these assumptions are reasonable? Why or why not?

tissues throughout the body (see Figure 30.11). Insects use nearly all species of plants and many species of animals as food.

More than 1 million of the 1.8 million described species of living organisms are insects. Biologists have estimated that many millions of insect species remain to be discovered. **TABLE 22.2** lists the major insect groups. Two groups—the jumping bristletails and silverfishes—are wingless and have simple life cycles, like the springtails and other close insect relatives. The remaining groups are all pterygote insects. **Pterygotes** have two pairs of wings, except in some groups where one or both pairs of wings have been secondarily lost. Secondarily wingless groups include the parasitic lice and fleas, some beetles, and the worker individuals in many ants.

Hatchling pterygotes do not look like adults, and they undergo substantial changes at each molt. The immature stages of insects between molts are called **instars**. A substantial change that

TABLE 22.2 | The Major Insect Groups[a]

Group	Approximate number of described living species
Jumping bristletails (Archaeognatha)	560
Silverfishes (Thysanura)	580
PTERYGOTE (WINGED) INSECTS (PTERYGOTA)	
Mayflies (Ephemeroptera)	3,250
Dragonflies and damselflies (Odonata)	6,000
Neopterans (Neoptera)[b]	
Ice-crawlers (Grylloblattodea)	35
Gladiators (Mantophasmatodea)	15
Stoneflies (Plecoptera)	3,800
Webspinners (Embioptera)	500
Angel insects (Zoraptera)	40
Earwigs (Dermaptera)	2,000
Grasshoppers and crickets (Orthoptera)	24,000
Stick insects (Phasmida)	3,000
Cockroaches (Blattodea)	4,500
Termites (Isoptera)	3,000
Mantids (Mantodea)	2,500
Booklice and barklice (Psocoptera)	5,800
Thrips (Thysanoptera)	6,000
Lice (Phthiraptera)	5,000
True bugs, cicadas, aphids, leafhoppers (Hemiptera)	105,000
Holometabolous neopterans (Holometabola)[c]	
Ants, bees, wasps (Hymenoptera)	150,000
Beetles (Coleoptera)	400,000
Twisted-wing parasites (Strepsiptera)	650
Lacewings, ant lions, mantidflies (Neuropterida)	6,000
Dobsonflies, alderflies, fishflies (Megaloptera)	350
Snakeflies (Raphidoptera)	250
Scorpionflies (Mecoptera)	800
Fleas (Siphonaptera)	2,000
True flies (Diptera)	125,000
Caddisflies (Trichoptera)	12,000
Butterflies and moths (Lepidoptera)	180,000

[a]The hexapod relatives of insects include the springtails (Collembola; 3,000 spp.), two-pronged bristletails (Diplura; 600 spp.), and proturans (Protura; 10 spp.). All are wingless and have internal mouthparts.

[b]Neopteran insects can tuck their wings close to their bodies.

[c]Holometabolous insects are neopterans that undergo complete metamorphosis.

occurs between one developmental stage and another is called **metamorphosis**. If the changes between its instars are gradual, an insect is said to have **incomplete metamorphosis**. If the change between at least some instars is dramatic, an insect is said to have **complete metamorphosis**. In many insects with complete metamorphosis, the different life stages are specialized for different environments and use different food sources. In many species, the larvae are adapted for feeding and growing, whereas the adults are specialized for reproduction and dispersal.

Media Clip 22.11 Complete Metamorphosis

Pterygote insects were the first animals in evolutionary history to achieve the ability to fly. Flight opened up many new lifestyles and feeding opportunities that only the insects could exploit, and it is almost certainly one of the reasons for the remarkable numbers of insect species and individuals, and for their unparalleled evolutionary success.

Molecular data suggest that insects began to diversify about 450 million years ago, about the time of the appearance of the first land plants (although the earliest known fossil insects are only about 400 million years old). The earliest hexapods evolved in a terrestrial environment that lacked any similar organisms, which in part accounts for their remarkable success. But the success of the insects is also due to their wings. Homologous genes control the development of insect wings and crustacean appendages, suggesting that the insect wing evolved from a dorsal branch of a crustacean-like limb (**FIGURE 22.36**). The dorsal limb branch of crustaceans is used for gas exchange. Thus the insect wing probably evolved from a gill-like structure that had a gas exchange function.

The adults of most flying insects have two pairs of stiff, membranous wings attached to the thorax. True flies, however, have one pair of wings and a pair of stabilizers called haltares. In winged beetles, one pair of wings—the forewings—forms heavy, hardened wing covers. Flying insects are important pollinators of flowering plants.

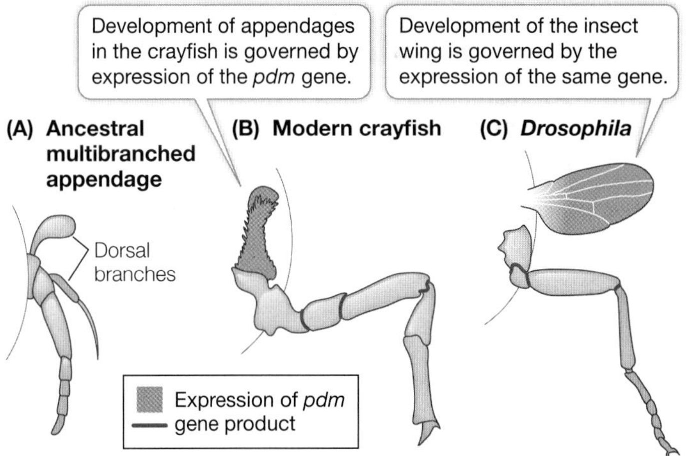

FIGURE 22.36 The Origin of Insect Wings? Insect wings may be derived from an ancestral appendage similar to that of modern crustaceans. **(A)** A diagram of the ancestral, multibranched arthropod limb. The uppermost dorsal branch may have been used for gas exchange. The *pdm* gene, a Hox gene, is expressed throughout the dorsal limb segment of the thoracic limb of a crayfish **(B)** and in the wings of *Drosophila* **(C)**.

If the hypothesized origin of insect wings is correct, then what was a possible function of the structure that would become a wing in insects?

(A) *Libellula depressa*, Odonata

(B) *Brachystola magna*, Orthoptera

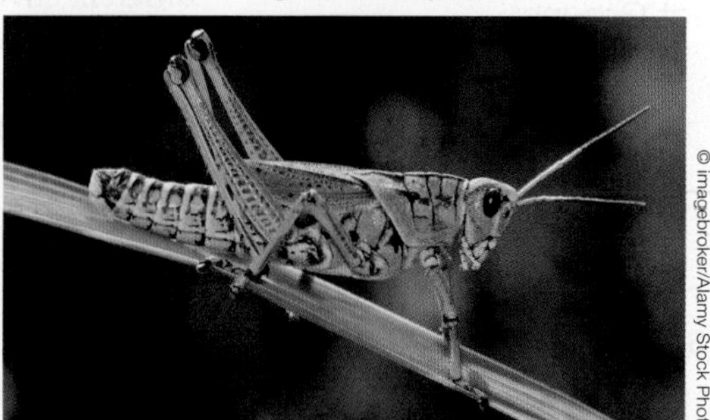

(C) *Graphosoma lineatum*, Hemiptera

(D) *Cetonia aurata*, Coleoptera

(E) *Gumaga* sp., Trichoptera

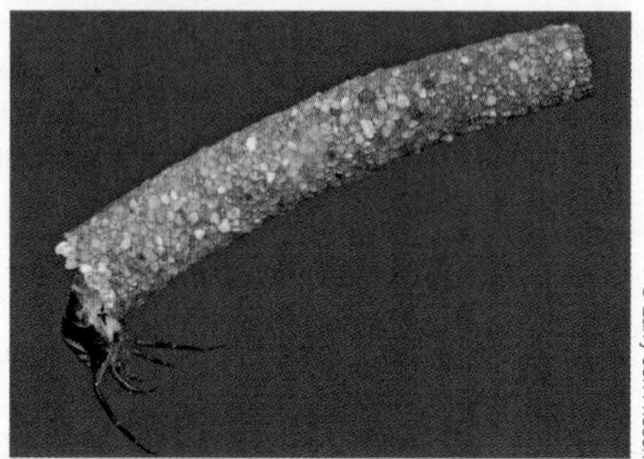

(F) *Papilio machaon*, Lepidoptera

(G) *Poecilanthrax willistoni*, Diptera

(H) *Mischocyttarus flavitarsis*, Hymenoptera

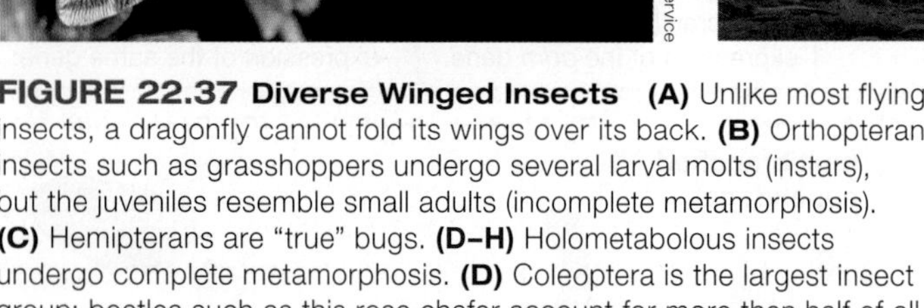

FIGURE 22.37 Diverse Winged Insects **(A)** Unlike most flying insects, a dragonfly cannot fold its wings over its back. **(B)** Orthopteran insects such as grasshoppers undergo several larval molts (instars), but the juveniles resemble small adults (incomplete metamorphosis). **(C)** Hemipterans are "true" bugs. **(D–H)** Holometabolous insects undergo complete metamorphosis. **(D)** Coleoptera is the largest insect group; beetles such as this rose chafer account for more than half of all holometabolous species. **(E)** A larval caddisfly emerges from its case constructed from silk and sand. **(F)** The swallowtail butterfly is found in most parts of North America. **(G)** This bee fly is a true fly. Although bee flies cannot sting, most potential predators avoid them because of their resemblance to bees. **(H)** This western paper wasp is a hymenopteran, a group in which most members display social behaviors.

Two groups of pterygotes, the mayflies and dragonflies (**FIGURE 22.37A**), cannot fold their wings against their bodies. This is the ancestral condition for pterygote insects, and the mayflies and dragonflies are not closely related to one another. Members of these groups have predatory or herbivorous aquatic larvae that transform into flying adults after they crawl out of the water. Dragonflies (and their relatives the damselflies) are active predators as adults. In contrast, adult mayflies lack functional digestive tracts. Mayflies live only about a day, just long enough to mate and lay eggs.

All other pterygote insects—the **neopterans**—can tuck their wings out of the way upon landing and crawl into crevices and other tight places. Some neopteran groups undergo incomplete metamorphosis, so hatchlings of these insects are sufficiently similar in form to adults to be recognizable. Examples include the grasshoppers (**FIGURE 22.37B**), roaches, mantids, stick insects, termites, stoneflies, earwigs, thrips, true bugs (**FIGURE 22.37C**), aphids, cicadas, and leafhoppers. They acquire adult organ systems, such as wings and compound eyes, gradually through several juvenile instars.

More than 80 percent of all insects belong to a subgroup of the neopterans called the **holometabolous** insects (see Table 22.2), which undergo complete metamorphosis (**FIGURE 22.37D–H**). The many species of beetles account for almost half of this group (Figure 22.37D). Also included are lacewings and their relatives; caddisflies (Figure 22.37E); butterflies and moths (Figure 22.37F); sawflies; true flies (Figure 22.37G); and bees, wasps, and ants, some species of which display unique and highly specialized social behaviors (Figure 22.37H).

Activity 22.3 **Protostome Classification**

REVIEW & APPLY | 22.4

R All arthropods have segmented bodies. Muscles in each segment operate that segment and the appendages attached to it. Jointed, specialized appendages permit complex patterns of movement, including, in insects, the ability to fly. With flight, insects took advantage of new feeding and lifestyle opportunities, which contributed to the unparalleled evolutionary success of this group.

A
1. What features have contributed to making arthropods among the most abundant animals on Earth, both in number of species and in number of individuals?
2. Describe the difference between incomplete and complete metamorphosis.
3. There are more described and named species of insects than of all other species on Earth combined. However, only a very few insect species live in marine environments, and those species are restricted to the intertidal zone or the ocean surface. What factors may have contributed to the insects' lack of success in the oceans?

The majority of Earth's animal species are protostomes, so it is not surprising that the protostome groups display a huge variety of different body plans and other characteristics. We will now consider the diversity of the deuterostomes, a group that contains far fewer species than the protostomes but which is even more intensively studied and, not incidentally, is the group to which humans belong.

22.5 | Deuterostomes Include Echinoderms, Hemichordates, and Chordates

LEARNING OBJECTIVES

By the end of this key concept you should be able to:

22.5.1 Describe how deuterostomes differ developmentally from protostomes.

22.5.2 Contrast the ways that echinoderms and hemichordates feed.

22.5.3 List the synapomorphies that characterize the three major deuterostome groups.

22.5.4 Describe how changes in appendages, jaws, teeth, and swim bladders led to new foraging opportunities for vertebrates.

It may surprise you to learn that you and a sea urchin are both deuterostomes. Adult sea stars, sea urchins, and sea cucumbers—the most familiar echinoderms—look so different from adult vertebrates (fishes, frogs, lizards, birds, and mammals) that it is difficult to believe all these animals are closely related. Two major pieces of evidence indicate that the deuterostomes share a common ancestor not shared with the protostomes:

1. Deuterostomes share a pattern of early development in which the mouth forms at the opposite end of the embryo from the blastopore, and the blastopore develops into the anus (this is opposed to the protostomes, in which the blastopore becomes the mouth).
2. Recent phylogenetic analyses of the DNA sequences of many different genes offer strong support for the shared evolutionary relationships of deuterostomes.

Note that neither of the above factors is apparent in the morphology of the adult animals.

Although there are far fewer species of deuterostomes than of protostomes (see Table 22.1), the deuterostomes are of special interest because they include many large animals—including humans—that strongly influence the characteristics of ecosystems. Many deuterostome species have been intensively studied in all fields of biology. Complex behaviors are especially well developed among some deuterostomes and are a vast and fascinating field of study in themselves (see Chapter 37).

There are three major clades of living deuterostomes:

1. **Echinoderms**: sea stars, sea urchins, and their relatives
2. **Hemichordates**: acorn worms and pterobranchs
3. **Chordates**: tunicates, lancelets, and vertebrates

All deuterostomes are triploblastic and coelomate (see Figure 22.6C). Skeletal support features, where present, are internal rather than external. Some species have segmented bodies, but the segments are less obvious than those of annelids and arthropods.

The earliest deuterostomes were bilaterally symmetrical, segmented animals with a pharynx that had slits through which water flowed. Echinoderms evolved their adult forms with unique symmetry (in which the body parts are arranged along five radial axes) much later, whereas other deuterostomes retained the ancestral bilateral symmetry.

The echinoderms and hemichordates (together known as **ambulacrarians**) have a bilaterally symmetrical, ciliated larva (**FIGURE 22.38A**). Adult hemichordates also are bilaterally symmetrical. Echinoderms, however, undergo a radical change in form as they develop into adults (**FIGURE 22.38B**), changing from a bilaterally symmetrical larva to an adult with **pentaradial symmetry** (symmetry in five or multiples of five). As is typical of animals with radial symmetry, echinoderms have no head, and they move equally well (but usually slowly) in many directions. Rather than having an anterior–posterior (head–tail) and dorsal–ventral (back–belly) body organization, echinoderms have an oral side (containing the mouth) and an opposite aboral side (containing the anus).

Animation 22.3 **An Overview of the Deuterostomes**

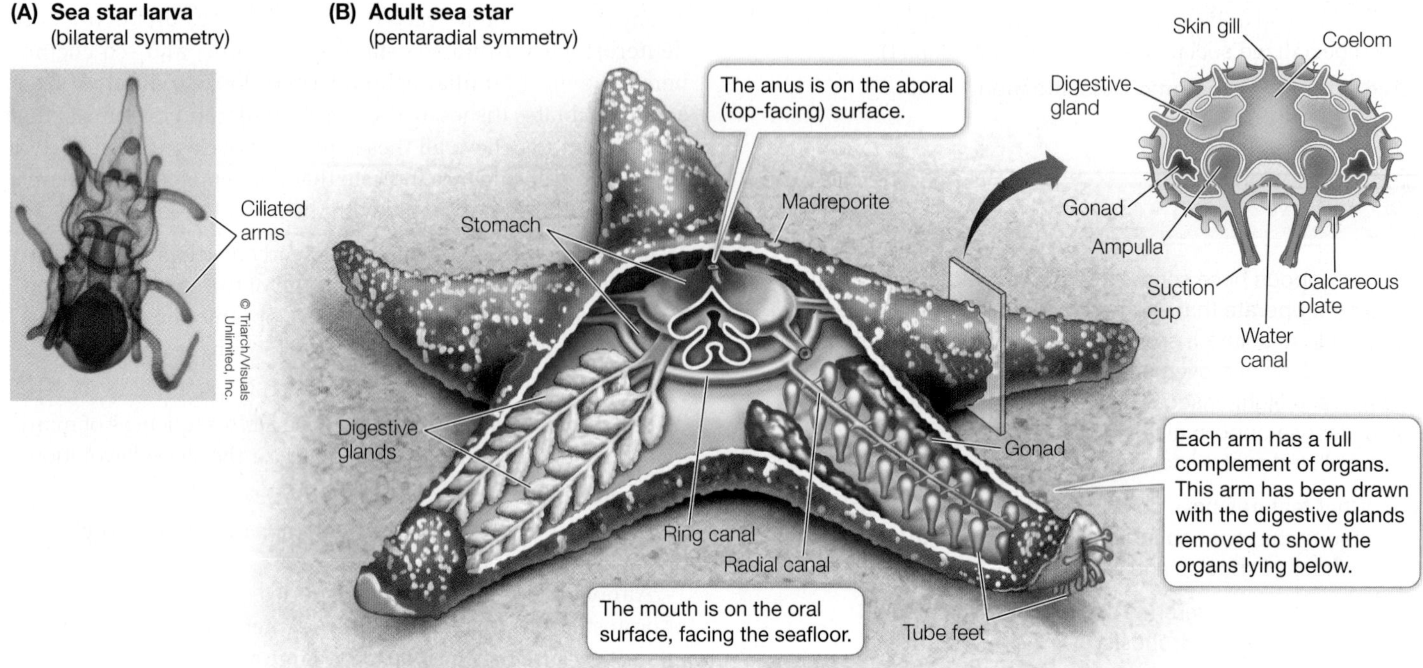

FIGURE 22.38 Bilaterally Symmetrical Echinoderm Larvae Become Radially Symmetrical Adults (A) The ciliated larva of a sea star has bilateral symmetry. Hemichordates have a similar larval form. **(B)** The radially symmetrical adult sea star displays the canals and tube feet of the echinoderm water vascular system, as well as the calcified internal skeleton. The body's orientation is oral–aboral rather than anterior–posterior.

Adult echinoderms have pentaradial symmetry

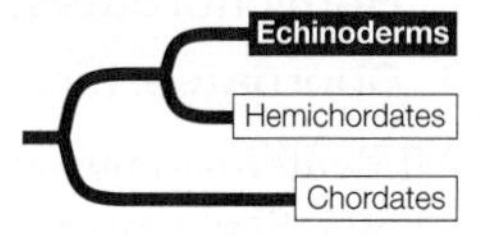

About 13,000 species of echinoderms in 23 major groups have been described from fossil remains. They are probably only a small fraction of those that actually lived. Only 6 of the 23 major groups known from fossils are represented by species that survive today. Many clades of echinoderms became extinct during the periodic mass extinctions that have occurred throughout Earth's history (see Table 17.1). Nearly all of the 7,500 extant species of echinoderms live only in marine environments.

In addition to having pentaradial symmetry, adult echinoderms have two unique structural features. One is a system of calcified internal plates covered by thin layers of skin and some muscles. The calcified plates of most echinoderms are thick, and they fuse inside the entire body, forming an internal skeleton. The other unique feature is a **water vascular system**, a network of water-filled canals leading to extensions called **tube feet**. This system functions in gas exchange, locomotion, and feeding (see Figure 22.38B).

Members of one major extant clade, the crinoids (sea lilies and feather stars), were more abundant and species-rich 500–300 million years ago than they are today. There are some 80 described living sea lily species, most of which are sessile organisms attached to a substrate by a stalk. Feather stars (**FIGURE 22.39A**) grasp the substrate with flexible appendages that allow for limited movement. About 600 living species of feather stars have been described.

Unlike the crinoids, most of the other surviving echinoderms are motile. The two main groups of motile echinoderms are the echinozoans (sea urchins and sea cucumbers; **FIGURE 22.39B,C**) and asterozoans (sea stars and brittle stars; **FIGURE 22.39D,E**).

The tube feet of the different echinoderm groups have been modified in a great variety of ways to capture prey. Sea lilies, for example, feed by orienting their arms in passing water currents. Food particles strike and stick to their tube feet, which are covered with mucus-secreting glands. The tube feet transfer food particles to grooves in the arms, where ciliary action carries the particles to the mouth. Sea cucumbers capture food with their anterior tube feet, which are modified into large, feathery, sticky tentacles that can be protruded from the mouth. Periodically, a sea cucumber withdraws the tentacles, wipes off the material that has adhered to them, and digests it.

Many sea stars use their tube feet to capture large prey such as annelids, gastropod and bivalve mollusks, small crustaceans such as crabs, and fishes. With hundreds of tube feet acting simultaneously, a sea star can grasp a bivalve in its arms, anchor the arms with its tube feet, and by steady contraction of the muscles in its arms, gradually exhaust the muscles the bivalve uses to keep its shell closed (see Figure 22.39D). To feed on a bivalve, a sea star can push its stomach out through its mouth and then through the narrow space between the two halves of the bivalve's shell. The sea star's stomach then secretes enzymes that digest the prey.

Media Clip 22.12 Sea Star Hunting Bivalves

(A) *Comanthina schlegeli*

© imagebroker/Alamy Stock Photo

(B) *Strongylocentrotus purpuratus*

(C) *Pseudocolochirus* sp.

(D) *Asterias rubens*

(E) *Ophiothrix spiculata*

FIGURE 22.39 Echinoderm Diversity **(A)** The flexible arms of this feather star are clearly visible. **(B)** Sea urchins are important grazers on algae in the intertidal zones of the world's oceans. **(C)** Sea cucumbers are unique among echinderms in having an anterior–posterior rather than an oral–aboral orientation of the mouth and anus. **(D)** Sea stars are important predators on bivalve mollusks such as mussels and clams. Suction tips on its tube feet allow a sea star to grasp both shells of the bivalve and pull them open. **(E)** The arms of the brittle star are composed of hard but jointed plates.

Most sea urchins eat algae, which they catch with their tube feet from the plankton or scrape from rocks with a complex rasping structure. Most of the 2,000 species of brittle stars ingest particles from the upper layers of sediments and assimilate the organic material from them, although some species filter suspended food particles from the water, and others capture small animals.

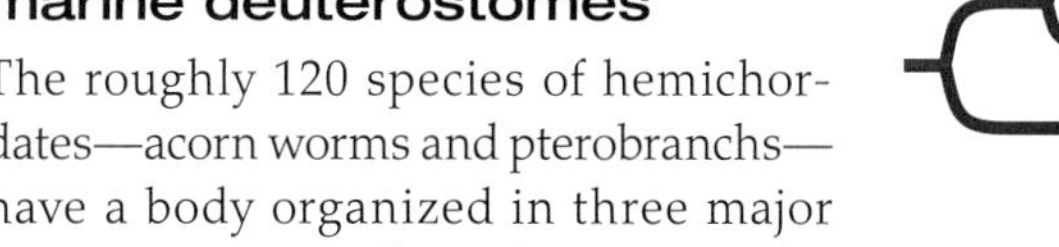

Hemichordates are wormlike marine deuterostomes

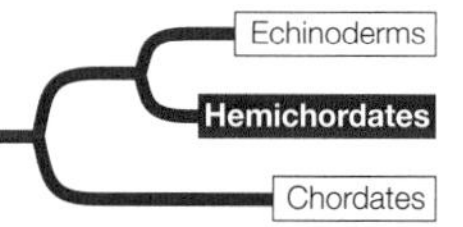

The roughly 120 species of hemichordates—acorn worms and pterobranchs—have a body organized in three major parts, consisting of a proboscis, a collar (which bears the mouth), and a trunk (which contains the other body parts). The 90 known species of acorn worms range up to 2 meters in length (**FIGURE 22.40A**). They live in burrows in muddy and sandy marine sediments. The digestive tract of an acorn worm consists of a mouth behind which are a muscular pharynx and an intestine. The pharynx opens to the outside through several pharyngeal slits through which water can exit. Highly vascularized tissue surrounding the pharyngeal slits serves as a gas exchange apparatus. Acorn worms breathe by pumping water into the mouth and out through the pharyngeal slits. They capture prey with the large proboscis, which is coated with sticky mucus to which small organisms in the sediment stick. The mucus and its attached prey are conveyed by cilia to the mouth. In the esophagus, the food-laden mucus is compacted into a ropelike mass that is moved through the digestive tract by ciliary action.

The 30 living species of pterobranchs are sedentary marine animals up to 12 millimeters long that live in a tube secreted by the proboscis. Some species are solitary; others form colonies of individuals joined together (**FIGURE 22.40B**). Behind the proboscis is a collar with anywhere from one to nine pairs of arms. The arms bear long tentacles that capture prey and function in gas exchange.

Shared chordate characteristics are most evident in larvae

As mentioned earlier, it is not obvious from examining the morphology of adult animals that echinoderms and chordates share a common ancestor. The evolutionary relationships among some chordate groups are not immediately apparent either. The features that reveal these evolutionary relationships are seen primarily in the larvae—in other words, it is during the early developmental stages that their evolutionary relationships are evident.

There are three principal chordate clades: the **lancelets** (also called cephalochordates), **tunicates** (also called urochordates), and **vertebrates**. Adult chordates vary greatly in form, but all

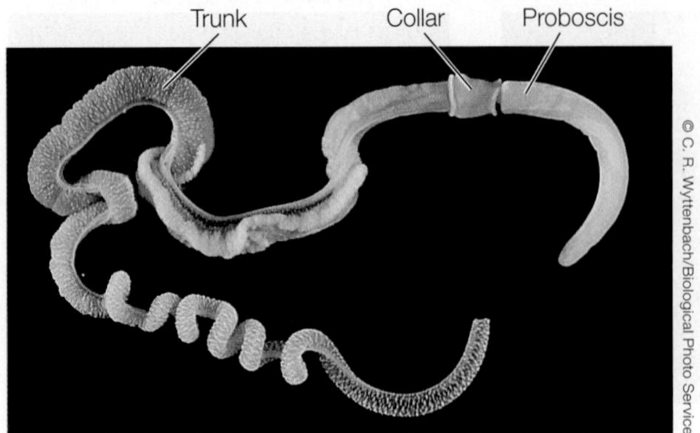

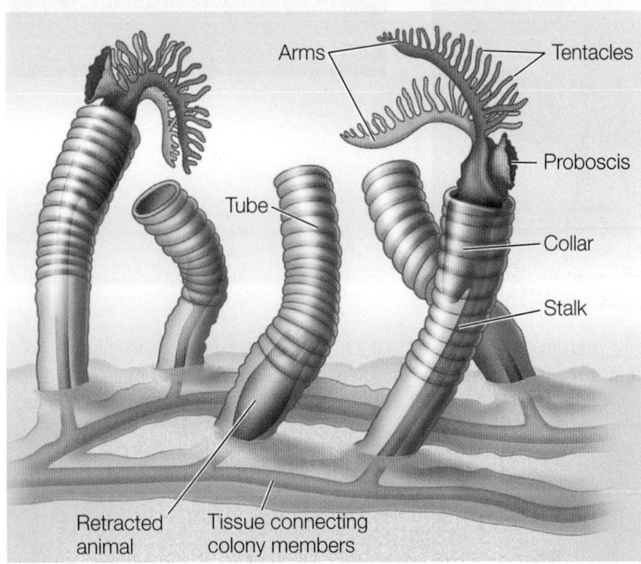

FIGURE 22.40 Hemichordates (A) The proboscis of an acorn worm is modified for burrowing. **(B)** The structure of a colonial pterobranch.

chordates display the following derived structures at some stage in their development (**FIGURE 22.41**):

- A dorsal hollow nerve cord
- A tail that extends beyond the anus
- A dorsal supporting rod called the **notochord**

The notochord is the most distinctive derived chordate trait. It is composed of a core of large cells with turgid fluid-filled vacuoles, which make it rigid but flexible. In most tunicates the notochord is lost during metamorphosis to the adult stage. In most vertebrate species, it is replaced during development by vertebrae that provide support for the body.

The ancestral pharyngeal slits (not a derived feature of chordates) are present at some developmental stage of chordates but are often lost in adults. The pharynx, which develops around the pharyngeal slits, functioned in chordate ancestors as the site for oxygen uptake and the elimination of carbon dioxide and water (as in acorn worms). The pharynx is much enlarged in some chordate species (as in the pharyngeal basket of the lancelet in Figure 22.41B).

Adults of most lancelets and tunicates are sessile

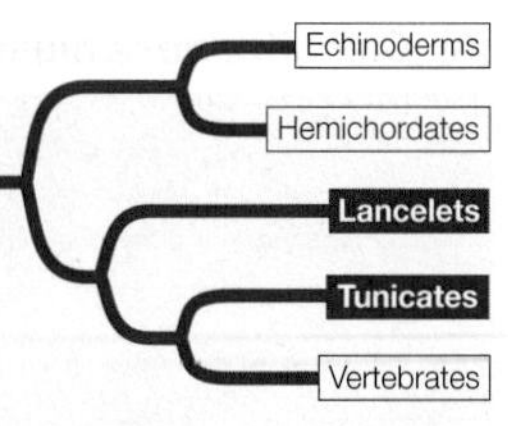

The 35 species of lancelets are small animals that rarely exceed 5 centimeters in length. The notochord, which provides body support, extends the entire length of the body throughout their lives (see Figure 22.41B). Lancelets are found in shallow marine and brackish waters worldwide. Most of the time they lie covered in sand with their head protruding above the sediment, but they can swim. They filter prey from the water with their pharyngeal basket.

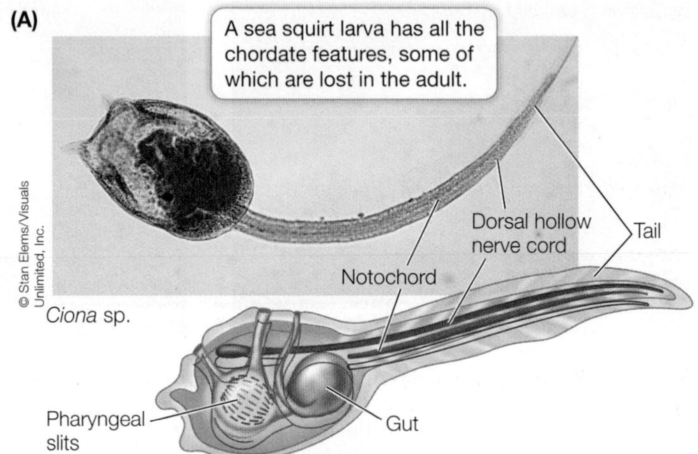

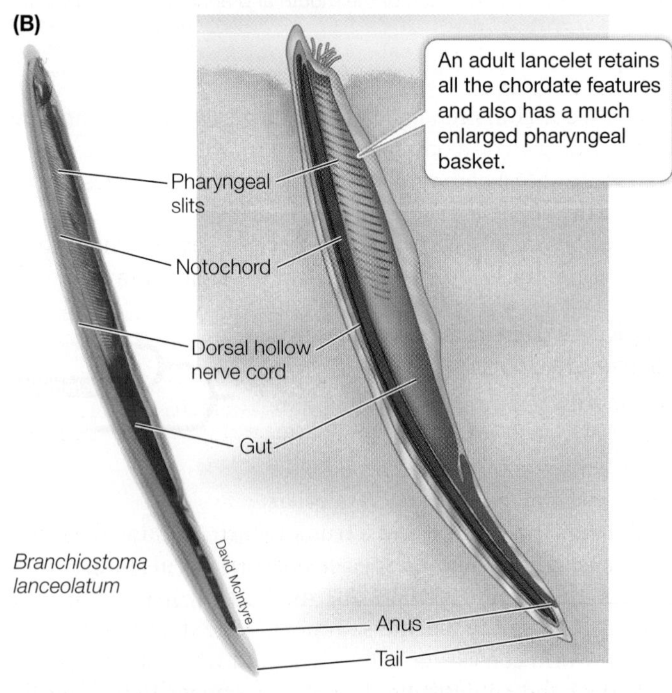

FIGURE 22.41 The Key Features of Chordates Are Most Apparent in Early Developmental Stages The pharyngeal slits of both tunicates and lancelets develop into a pharyngeal basket. **(A)** The sea squirt larva (but not the adult) has all three chordate features: a dorsal hollow nerve cord, post-anal tail, and notochord. **(B)** All three chordate synapomorphies are retained in the adult lancelet.

FIGURE 22.42 Adult Tunicates **(A)** The transparent tunic and the pharyngeal basket are clearly visible in this sea squirt. **(B)** A chainlike colony of salps floats in tropical waters.

All members of the three major tunicate groups—the sea squirts, salps, and larvaceans—live in marine environments. More than 90 percent of the 2,800 known species of tunicates are sea squirts (also called ascidians). Individual sea squirts range in length from less than 1 millimeter to 60 centimeters and attach themselves to hard substrates. Some sea squirts form colonies by asexual budding from a single founder. Colonies may measure several meters across. The baglike body of an adult sea squirt is enclosed in a tough tunic, which is the basis for the name "tunicate" (**FIGURE 22.42A**). The tunic is composed of proteins and a complex polysaccharide secreted by epidermal cells. The sea squirt pharynx is enlarged into a pharyngeal basket that filters prey from the water passing through it. Other tunicates, such as salps, are pelagic and live in chainlike colonies up to several meters long (**FIGURE 22.42B**). They float in tropical and subtropical oceans at depths down to 1,500 meters.

In addition to its pharyngeal slits, a sea squirt larva has a dorsal hollow nerve cord and a notochord that is restricted mostly to the tail region (see Figure 22.41A). Bands of muscle that surround the notochord provide support for the body. After a short time swimming in the plankton, the larvae of most species settle on the seafloor and transform into sessile adults. The swimming, tadpolelike larvae suggest a close evolutionary relationship between sea squirts and vertebrates (see Figure 14.6).

As we have discussed, the hemichordates, echinoderms, lancelets, and tunicates are marine animals. The lineage that led to the vertebrates is also thought to have evolved in the oceans, although probably in an estuarine environment (where fresh water meets salt water). Vertebrates have since radiated into marine, freshwater, terrestrial, and aerial environments worldwide.

The vertebrate body plan can support large, active animals

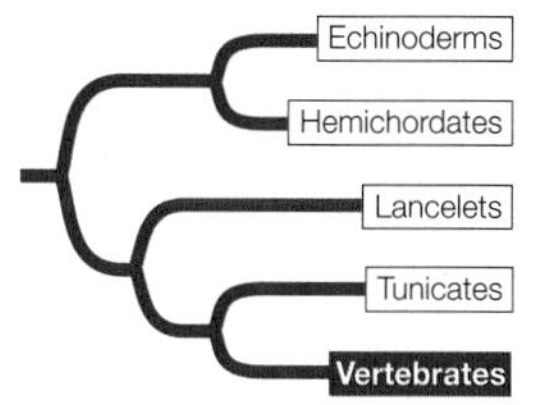

The vertebrates take their name from the unique, jointed, dorsal vertebral column that replaces the notochord during early development as their primary supporting structure. The individual elements in the vertebral column are called vertebrae. Four other key features characterize the vertebrates as well (**FIGURE 22.43**):

1. An anterior skull with a large brain
2. A rigid internal skeleton supported by the vertebral column, which also encloses the spinal cord
3. Internal organs suspended in a coelom
4. A well-developed circulatory system, driven by contractions of a ventral heart

The internal skeleton provides support for an extensive muscular system, which receives oxygen from the circulatory system and is controlled by the central nervous system.

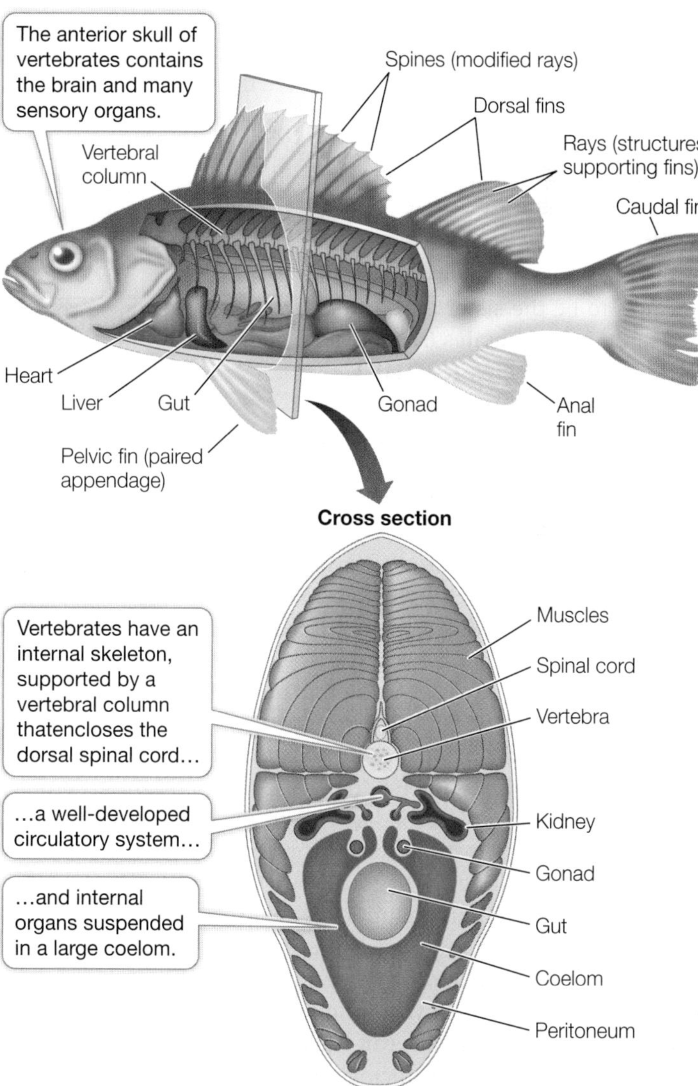

FIGURE 22.43 The Vertebrate Body Plan A ray-finned fish illustrates the structural elements common to all vertebrates. In addition to the paired pelvic fins, these fishes have paired pectoral fins on the sides of the body (not seen in this cutaway view).

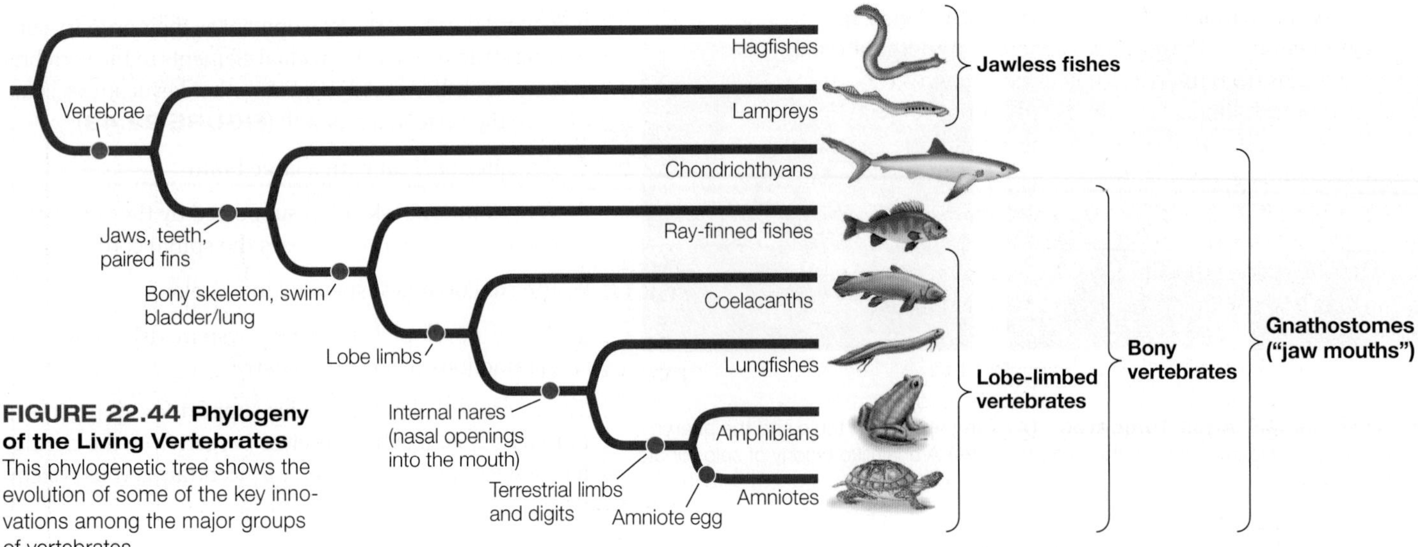

FIGURE 22.44 Phylogeny of the Living Vertebrates This phylogenetic tree shows the evolution of some of the key innovations among the major groups of vertebrates.

The evolution of these features allowed many vertebrates to become large, active predators, which in turn allowed the vertebrates to diversify widely (**FIGURE 22.44**).

Although the lampreys and hagfishes may look superficially similar (with elongate eel-like bodies and no paired fins), they differ greatly in their biology. Lampreys have a complete skull

The two groups of living jawless fishes are not closely related

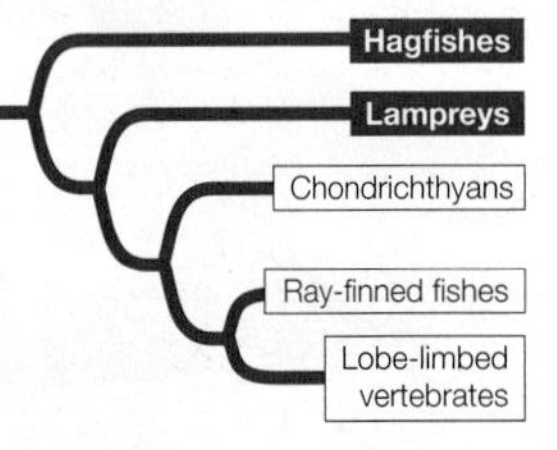

The **hagfishes** are thought by many biologists to be the sister group to the remaining vertebrates (see Figure 22.44). Hagfishes (**FIGURE 22.45A**) have a weak circulatory system with three small accessory hearts (rather than a single, large heart), a partial cranium or skull (containing no cerebrum or cerebellum, two main regions of the brain of other vertebrates), and no jaws or stomach. They also lack separate, jointed vertebrae and have a skeleton composed of cartilage. Thus some biologists do not consider hagfishes vertebrates, and use instead the term "craniates" to refer collectively to the hagfishes and the vertebrates. Some analyses of gene sequences suggest, however, that hagfishes may be more closely related to the vertebrate lampreys (**FIGURE 22.45B**); in this phylogenetic arrangement, the two groups are collectively called the cyclostomes ("circle mouths"). If in fact the hagfishes and lampreys do form a monophyletic group, then hagfishes must have secondarily lost many of the major vertebrate morphological features during their evolution.

The 80 known species of hagfishes are unusual marine animals that produce copious quantities of slime as a defense. They are virtually blind and rely largely on the four pairs of sensory tentacles around their mouth to detect food. Although they have no jaws, hagfishes have a tonguelike structure equipped with toothlike rasps that they can use to tear apart dead organisms and to capture their principal prey, annelid worms. Hagfishes have direct development (no larvae), and individuals may actually change sex from year to year (from male to female and vice versa).

Media Clip 22.13 Hagfish Slime

(A) ***Eptatretus stoutii***

(B) ***Lampetra fluviatilis***

FIGURE 22.45 Modern Jawless Fishes (A) Hagfishes burrow in the ocean mud, from which they extract small prey. They also scavenge on dead or dying fish. The degenerate eyes of hagfishes led to their being called (inaccurately) "blind eels." **(B)** Many species of lampreys are ectoparasites that attach to the bodies of living fishes and use their large, jawless mouths to suck blood and flesh.

and distinct and separate (although rudimentary) vertebrae, all cartilaginous rather than bony. Lampreys undergo a complete metamorphosis from filter-feeding larvae known as ammocoetes, which are morphologically quite similar in general structure to adult lancelets. The adults of many species of lampreys are parasitic, although several lineages of lampreys evolved to become nonfeeding as adults. These nonfeeding adult lampreys survive only a few weeks after metamorphosis—just long enough to breed. In the species that are parasitic as adults, the round mouth is a rasping and sucking organ that is used to attach to their prey and rasp at the flesh (see Figure 22.45B).

The nearly 50 species of lampreys either live in fresh water or live in coastal salt water and move into fresh water to breed. Some species of lampreys are critically endangered because of recent habitat changes and losses.

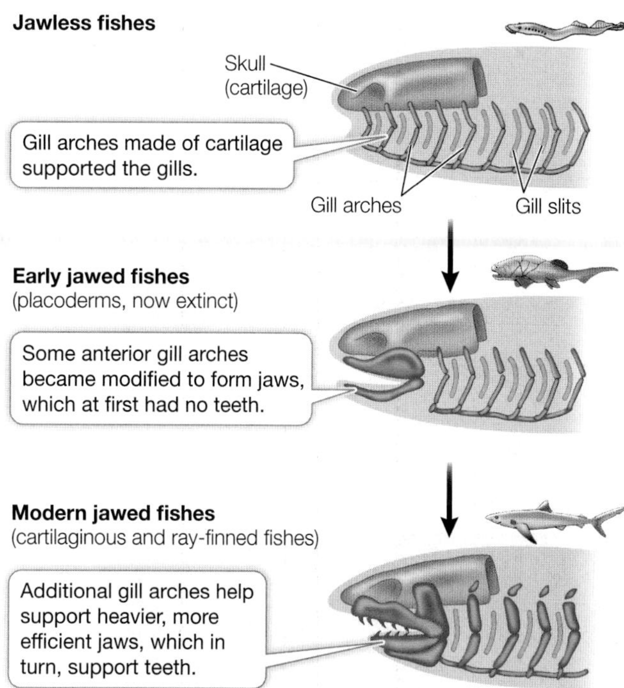

FIGURE 22.46 Verterbrate Jaw Evolution Jaws of vertebrates are derived from modifications of the anterior gill arches of jawless fishes.

Jaws and teeth improved feeding efficiency

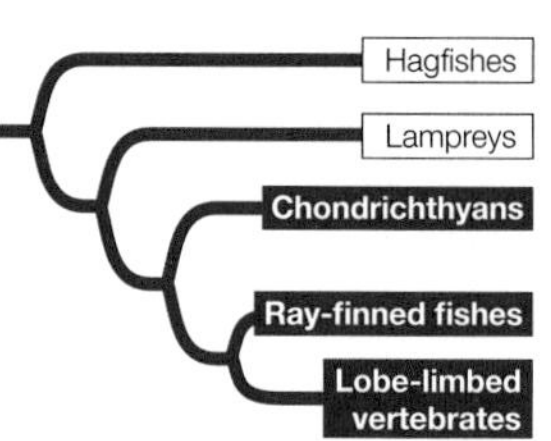

Many kinds of jawless fishes were found in the seas, estuaries, and fresh waters of the Devonian, but hagfishes and lampreys are the only jawless fishes that survived beyond the Devonian. Late in the Ordovician, some fishes evolved jaws via modifications of the skeletal arches that supported the gills (**FIGURE 22.46**). Those fishes and their descendants are referred to as **gnathostomes** (Greek, "jaw mouths"). Jaws greatly improved feeding efficiency, as an animal with jaws can grasp, subdue, and swallow large prey. Jawed fishes rapidly diversified during the Devonian, eventually replacing the jawless fishes in dominance of the seas.

The earliest jaws were simple, but the evolution of teeth made feeding even more efficient. In predators, teeth function crucially both in grasping and in breaking up prey. In both predators and herbivores, teeth enable an animal to chew both soft and hard body parts of their food. Chewing also aids chemical digestion and improves an animal's ability to extract nutrients from its food.

Fins and swim bladders improved stability and control over locomotion

Most jawed fishes have a pair of pectoral fins just behind the gill slits and a pair of pectoral fins anterior to the anus. Paired fins stabilize the position of jawed fishes in water (and in some cases, help propel them). Median dorsal and anal fins also stabilize the fish, or may be used for propulsion in some species. In many fishes, the caudal (tail) fin helps propel the animal and enables it to turn rapidly.

Several groups of gnathostomes became abundant during the Devonian. Among them were the **chondrichthyans**—sharks, skates, and rays (about 1,000 living species) and chimaeras (40 living species). Like hagfishes and lampreys, these fishes have a skeleton composed entirely of firm but pliable cartilage. Their skin is flexible and leathery, sometimes bearing scales that give it the consistency of sandpaper. Sharks move forward by means of lateral undulations of their body and caudal fin (**FIGURE 22.47A**). Skates and rays propel themselves by means of vertical undulating movements of their greatly enlarged pectoral fins (**FIGURE 22.47B**).

Most sharks are predators, but some feed by straining plankton from the water. Most skates and rays live on the ocean floor, where they feed on mollusks and other animals buried in the sediments. Nearly all cartilaginous fishes live in the oceans, but a few are estuarine or migrate into lakes and rivers. One group of rays is found in river systems of South America. The less familiar chimaeras (**FIGURE 22.47C**) live in deep-sea or cold waters.

One lineage of aquatic gnathostomes gave rise to the bony vertebrates, which soon split into two main lineages—the **ray-finned fishes** and the **lobe-limbed vertebrates**. Bony vertebrates have internal skeletons of calcified, rigid bone rather than flexible cartilage. In early bony vertebrates, gas-filled sacs that extended from the digestive tract supplemented the gas exchange function of the gills by giving the animals access to atmospheric oxygen. These features enabled those fishes to live where oxygen was periodically in short supply, as it often is in freshwater environments. These lunglike sacs evolved into **swim bladders** (organs of buoyancy), as well as into the lungs of tetrapods. By adjusting the amount of gas in its swim bladder, a fish can control the depth at which it remains suspended in the water while expending very little energy to maintain its position.

The outer body surface of most species of ray-finned fishes is covered with flat, thin, lightweight scales that provide some protection or enhance movement through the water. The gills of ray-finned fishes open into a single chamber covered by a hard flap, called an operculum. Movement of the operculum improves the flow of water over the gills, where gas exchange takes place.

(A) ***Carcharodon carcharias***

(B) ***Taeniura lymma***

(C) ***Hydrolagus colliei***

FIGURE 22.47 Chondrichthyans The fins of chondrichthyans lack supportive rays. **(A)** Most sharks, such as this great white shark, are active marine predators. **(B)** The fins of skates and rays, represented here by a ribbontail ray, are modified for feeding on the ocean bottom. **(C)** A chimaera, or ratfish.

Ray-finned fishes began to radiate during the Mesozoic and continued to radiate extensively throughout the Paleogene and Neogene (see Key Concept 17.3). Today there are about 36,000 known living species, encompassing a remarkable variety of sizes, shapes, and lifestyles (**FIGURE 22.48**). The smallest are less than 1 centimeter long as adults; the largest weigh as much as 900 kilograms. Ray-finned fishes exploit nearly all types of aquatic food sources. In the oceans they filter plankton from the water, rasp algae from rocks, eat corals and other soft-bodied colonial animals, dig animals from soft sediments, and prey on virtually all kinds of other fishes. In fresh water they eat plankton, devour insects, eat fruits that fall into the water in flooded forests, and prey on other aquatic vertebrates and, occasionally, terrestrial vertebrates.

(A)

(B) ***Phyllopteryx taeniolatus***

FIGURE 22.48 Ray-Finned Fishes **(A)** The two ray-finned species shown here are mutualists in their coral-reef ecosystem. The larger surgeonfish illustrates the body plan most commonly associated with this highly diverse clade. **(B)** This slow-moving weedy seadragon relies on its unusual morphology to provide camouflage in swaying seaweed.

REVIEW & APPLY | 22.5

R Adult echinoderms are characterized by pentaradial symmetry, an internal skeleton of calcified plates, and a unique water vascular system. Hemichordates have a bilaterally symmetrical body divided into three parts: proboscis, collar, and trunk. Both echinoderms and hemichordates are restricted to marine environments. Chordates are characterized by a dorsal hollow nerve chord, a post-anal tail, and a dorsal supporting rod called a notochord at some point during the life cycle. Specialized structures for support (a vertebral column), locomotion (such as fins), and feeding (jaws and teeth) evolved among aquatic vertebrates.

A
1. How does the body form of echinoderm larvae differ from that of echinoderm adults?
2. Describe some of the ways that echinoderms use their tube feet to obtain food, and contrast this with the feeding mechanism of hemichordates.
3. Describe the synapomorphies that characterize the chordates and the vertebrates, respectively.
4. How do the hagfishes differ from the lampreys in morphology? Why do some biologists contend that hagfishes are not vertebrates?
5. The body plan of most vertebrates is based on four appendages. What are the varied forms that these appendages take, and how are they used? In which lineages have two or more of these appendages been lost?

In some fishes, the lunglike sacs that gave rise to swim bladders became specialized for another purpose: breathing air. That adaptation set the stage for the vertebrates to move onto the land.

22.6 | **Life on Land** Contributed to Vertebrate Diversification

LEARNING OBJECTIVES

By the end of this key concept you should be able to:

- **22.6.1** **Explain how features of amphibians can make them especially vulnerable to environmental change.**
- **22.6.2** **Summarize the evidence that places the phylogenetic position of birds among the reptiles.**
- **22.6.3** **Describe how a mass extinction affected the diversification of mammals.**
- **22.6.4** **Outline the evolution of homeothermy, hair, and feathers across the phylogeny of amniotes.**

The evolution of lunglike sacs in fishes set the stage for the invasion of the land. Some early ray-finned fishes probably used those sacs to supplement their gills when oxygen levels in the water were low, as lungfishes and many groups of ray-finned fishes do today. But with their unjointed fins, those fishes could only flop around on land. Changes in the structure of the fins first allowed some fishes to support themselves better in shallow water and, later, to move better on land.

Jointed fins enhanced support for fishes

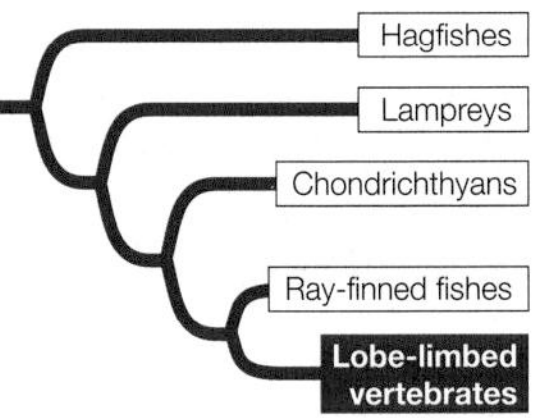

In the lobe-limbed vertebrates, the paired pelvic and pectoral fins developed into more muscular fins that were joined to the body by a single enlarged bone. The modern representatives of these lobe-limbed vertebrates include the coelacanths, lungfishes, and tetrapods.

The coelacanths flourished from the Devonian until about 65 million years ago, when they were thought to have become extinct. However, in 1938 a commercial fisherman caught a living coelacanth off South Africa. Since that time, hundreds of individuals of this extraordinary fish, *Latimeria chalumnae,* have been collected. A second species, *L. menadoensis,* was discovered in 1998 off the Indonesian island of Sulawesi. *Latimeria,* a predator of other fishes, reaches a length of about 1.8 meters and weighs up to 82 kilograms (**FIGURE 22.49A**). Its skeleton is composed mostly of cartilage, not bone. The cartilaginous skeleton is a derived feature in this clade because it had bony ancestors.

Media Clip 22.14 **Coelacanths in the Deep Sea**

Lungfishes were important predators in shallow-water habitats in the Devonian, but most lineages died out. The six surviving species live in stagnant swamps and muddy waters in South America, Africa, and Australia (**FIGURE 22.49B**). Lungfishes have lungs derived from the lunglike sacs of their ancestors as well as gills. When ponds dry up, individuals of most species can burrow deep into the mud and survive for many months in an inactive state while breathing air.

Evidence suggests that some early aquatic lobe-limbed vertebrates began to use terrestrial food sources, became more fully adapted to life on land, and eventually evolved to become ancestral **tetrapods** ("four legs"). How was this transition from an animal that swam in water to one that walked on land accomplished? Fossil evidence suggests that limbs able to prop up a large fish and make the front-to-rear movements necessary for walking evolved while these animals still lived in water. These limbs appear to have functioned in holding the animals upright in shallow water, perhaps even allowing them to hold the head above the water's surface. These same structures were then co-opted for movement on land, at first probably for foraging on brief trips out of water (**FIGURE 22.50**). The basic skeletal elements of those limbs can be traced through major changes in limb form and function among the terrestrial vertebrates.

(A) ***Latimeria chalumnae***

(B) ***Protopterus annectens***

FIGURE 22.49 The Closest Relatives of Tetrapods **(A)** The African coelacanth, discovered in deep waters of the Indian Ocean, represents one of two surviving species of a group that was once thought to be extinct. **(B)** All surviving lungfish species, such as this African lungfish, live in the Southern Hemisphere.

Eusthenopteron
380 mya
Fully aquatic; lobed fins

Tiktaalik
375 mya
Aquatic; lobed limbs intermediate between fins and legs

Acanthostega
365 mya
Semiterrestrial tetrapod

Humerus
Radius
Ulna
Distal elements of fin/wrist/hand

FIGURE 22.50 Tetrapod Limbs Are Modified Fins The major skeletal elements of the tetrapod limb were already present in aquatic lobe-limbed vertebrates some 380 million years ago (mya). The relative sizes and positions of these elements changed as lobe-limbed vertebrates moved to a terrestrial environment, where limbs were used to support and move the animal's body on land. (After I. Schneider and N. H. Shubin. 2013. *Trends Genet* 29: 419–426; M. C. Davis. 2013. *Integr Comp Biol* 53: 224–232.)

An early split in the tetrapod tree led to two main groups of terrestrial vertebrates: the **amphibians**, most of which remained tied to moist environments, and the **amniotes**, many of which adapted to much drier conditions.

Amphibians adapted to life on land

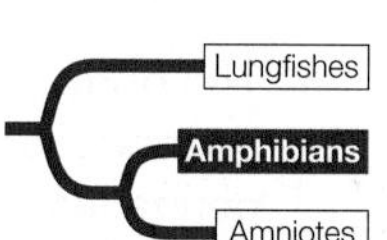

Most modern amphibians are confined to moist environments because they lose water rapidly through the skin when exposed to dry air. In addition, their eggs are enclosed within delicate membranous envelopes that cannot prevent water loss in dry conditions. In some amphibian species, adults live mostly on land but return to fresh water to lay and fertilize their eggs (**FIGURE 22.51**). The fertilized eggs give rise to larvae that live in water until they undergo metamorphosis to become terrestrial adults. However, many amphibians (especially those in tropical and subtropical areas) have evolved a wide diversity of additional reproductive modes and types of parental care. Internal fertilization, for example, evolved many times among the amphibians. Many species develop directly into adultlike forms from fertilized eggs laid on land or carried by the parents. Other species of amphibians are entirely aquatic, never leaving the water at any stage of their lives, and many of these species retain a larval-like morphology.

The more than 8,300 known species of amphibians living on Earth today belong to three major groups: the wormlike, limbless, tropical, burrowing or aquatic caecilians (**FIGURE 22.52A**), the tail-less frogs and toads (collectively called anurans; **FIGURE 22.52B**), and the tailed salamanders (**FIGURE 22.52C,D**).

Anurans are most diverse in wet tropical and warm temperate regions, although a few are found at very high latitudes. There are far more anurans than any other amphibians, with well over 7,000 described species and more being discovered every year. Some anurans have tough skins and other adaptations that enable them to live for long periods in very dry

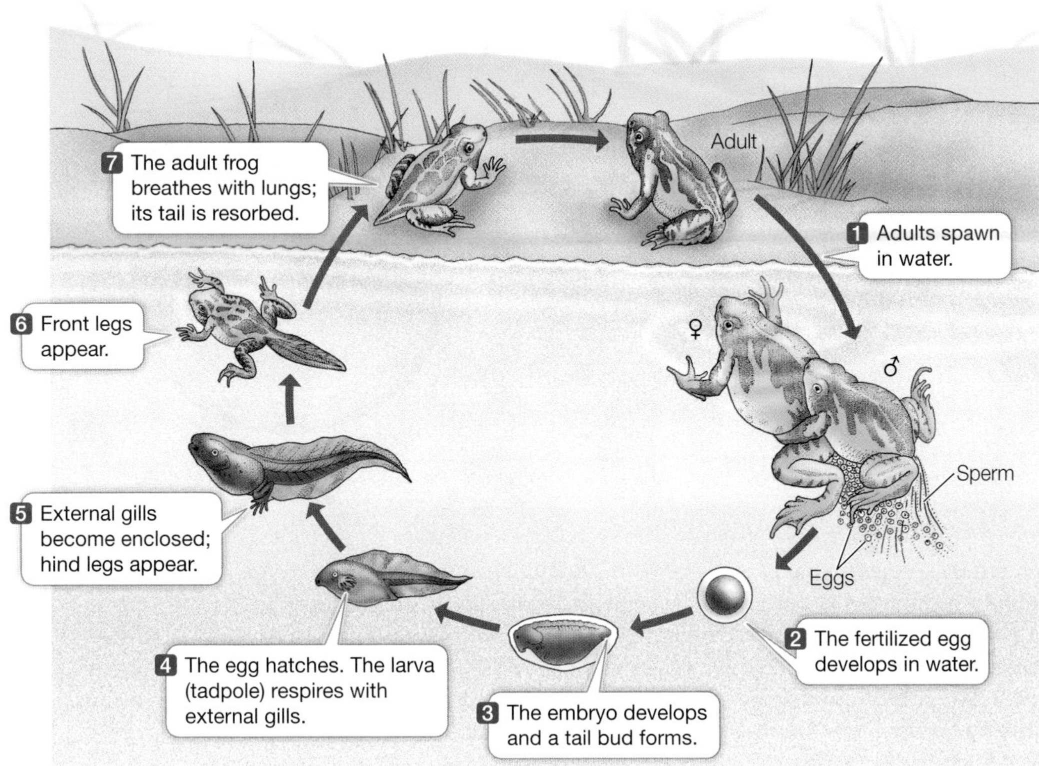

FIGURE 22.51 In and Out of the Water Most early stages in the life cycle of many amphibians take place in water. The aquatic tadpole transforms into a terrestrial adult through metamorphosis. Some species of amphibians, however, have direct development (with no aquatic larval stage), and others are aquatic throughout life.

What is the potential advantage of maintaining an aquatic tadpole stage, even though the adults are terrestrial?

Animation 22.4 Life Cycle of a Frog

deserts, whereas others live in moist terrestrial and arboreal environments. Some species are completely aquatic as adults. All anurans have a very short vertebral column, with a strongly modified pelvic region that is adapted for leaping, hopping, or propelling the body through water by kicking the hind legs.

The more than 600 described species of salamanders are most diverse in temperate regions of the Northern Hemisphere, but many species are also found in cool, moist environments in the mountains of Central America, and a few species penetrate into tropical regions. Many salamanders live in rotting logs or moist

(A) *Siphonops annulatus*

(B) *Incilius periglenes*

(C) *Salamandra salamandra*

(D) *Eurycea waterlooensis*

FIGURE 22.52 Diversity among the Amphibians **(A)** Burrowing caecilians superficially look more like worms than amphibians. **(B)** A golden toad pair in Costa Rica. This species has recently become extinct, one of many amphibian species to do so in the past few decades. **(C)** A European fire salamander. **(D)** This Austin blind salamander's life cycle remains aquatic; it has no adult terrestrial stage. The eyes of this cave dweller have become greatly reduced.

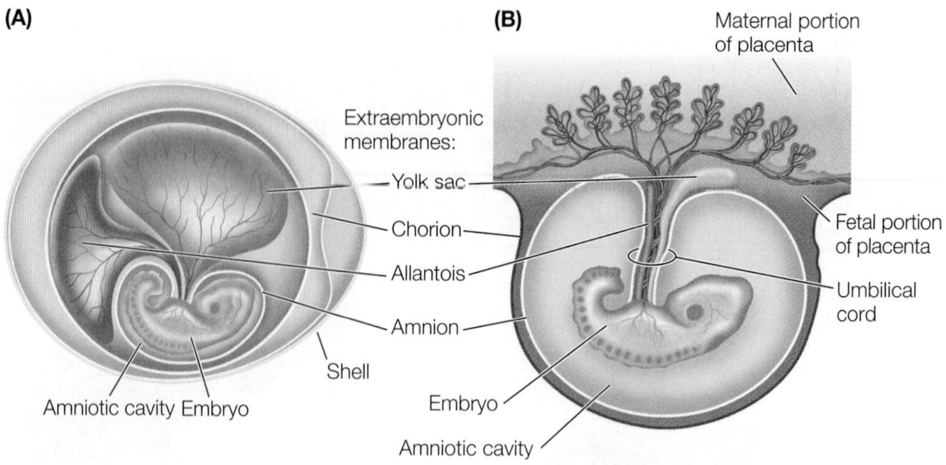

FIGURE 22.53 The Amniote Egg **(A)** The evolution of the amniote egg, with its water-retaining shell, four extraembryonic membranes, and embryo-nourishing yolk, was a major step in adaptation to the terrestrial environment. A chick egg is shown here. **(B)** In mammals, the developing embryo is retained inside the mother's body, with which the embryo exchanges nutrients and wastes via the placenta. Note the correspondence between the various membranes in (A) and (B).

How were the various membranes in the amniote egg modified in mammals to support viviparity?

Activity 22.4 The Amniote Egg

soil. One major group has lost lungs, and these species exchange gases entirely through the skin and mouth lining—body parts that all amphibians use in addition to their lungs. A completely aquatic lifestyle has evolved several times among the salamanders (see Figure 22.52D). These aquatic species have arisen through a developmental process known as **neoteny**, or the retention of juvenile traits (such as gills) by delayed somatic development. Most species of salamanders have internal fertilization, which is usually achieved through the transfer of a small jellylike, sperm-embedded capsule called a spermatophore.

Many amphibians have complex social behaviors. Most male anurans utter loud, species-specific calls to attract females of their own species (and sometimes to defend breeding territories), and they compete for access to females that arrive at the breeding sites. Many amphibians lay large numbers of eggs, which they abandon once they are deposited and fertilized. Some amphibians lay only a few eggs, which are fertilized and then guarded in a nest or carried on the backs, in the vocal pouches, or even in the stomachs of one of the parents. A few species of frogs, salamanders, and caecilians are viviparous, meaning they give birth to well-developed young that have received nutrition from the female during gestation.

Media Clip 22.15 Answering a Mating Call

Amphibians are the focus of much attention today because populations of many species are declining rapidly, especially in mountainous regions of western North America, Central and South America, and northeastern Australia. Worldwide, about one-third of amphibian species are now threatened with extinction or have disappeared completely in the last few decades because of habitat loss, disease, and climate change.

Amniotes colonized dry environments

Several key innovations contributed to the ability of amniotes to exploit a wide range of terrestrial habitats. The **amniote egg** (which gives the group its name) is relatively impermeable to water and allows the embryo to develop in a contained aqueous environment (**FIGURE 22.53A**). The leathery or brittle, calcium-impregnated shell of the amniote egg retards evaporation of the fluids inside but permits passage of oxygen and carbon dioxide. The egg also stores large quantities of food in the form of yolk, allowing the embryo to attain a relatively advanced state of development before it hatches. Within the shell are extraembryonic membranes that protect the embryo from desiccation and assist its gas exchange and excretion of waste nitrogen.

LINK The roles of the extraembryonic membranes of the amniote egg are discussed in **Key Concept 35.4**.

In several different groups of amniotes, the amniote egg became modified, allowing the embryo to grow inside (and receive nutrition from) the mother. For instance, the mammalian egg lost its shell whereas the functions of the extraembryonic membranes were retained and expanded (**FIGURE 22.53B**).

Other innovations evolved in the organs of terrestrial adults. A tough, impermeable skin, covered with scales or modifications of scales such as hair and feathers, greatly reduced water loss. Adaptations of the vertebrate excretory organs, the kidneys, allowed amniotes to excrete concentrated urine, ridding the body of waste nitrogen without losing a large amount of water in the process.

During the Carboniferous, amniotes split into two major groups: the **reptiles** and the lineage that eventually led to the **mammals** (**FIGURE 22.54**).

Reptiles adapted to life in many habitats

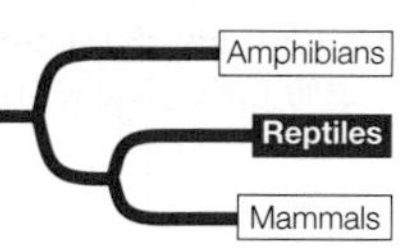

The lineage leading to modern reptiles began to diverge from other amniotes more than 300 million years ago. About 20,000 species of reptiles exist today, about half of which are birds. Birds are the only living members of the otherwise extinct dinosaurs, the dominant terrestrial predators of the Mesozoic.

The **lepidosaurs** constitute the second-most species-rich clade of living reptiles. This group is composed of the **squamates** (lizards, snakes, and amphisbaenians—the last a group of mostly legless, wormlike, burrowing reptiles with greatly reduced eyes)

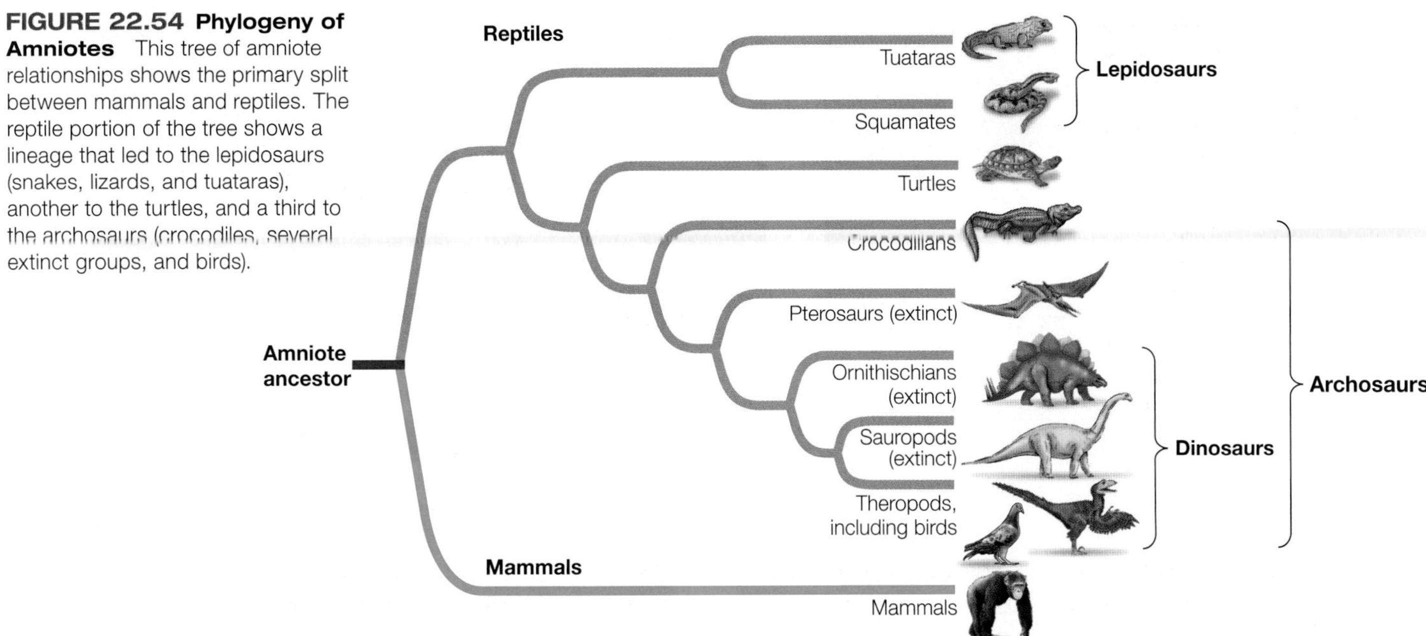

FIGURE 22.54 Phylogeny of Amniotes This tree of amniote relationships shows the primary split between mammals and reptiles. The reptile portion of the tree shows a lineage that led to the lepidosaurs (snakes, lizards, and tuataras), another to the turtles, and a third to the archosaurs (crocodiles, several extinct groups, and birds).

and the **tuataras**, which superficially resemble lizards but differ from them in tooth attachment and several internal anatomical features. Many species related to the tuataras lived during the Mesozoic, but today only two species, restricted to a few islands off New Zealand, survive (**FIGURE 22.55A**).

The skin of a lepidosaur is covered with horny scales that greatly reduce loss of water from the body surface. These scales, however, make the skin unavailable as an organ of gas exchange. Gases are exchanged almost entirely via the lungs, which are proportionally much larger in surface area than those of amphibians. A lepidosaur forces air into and out of its lungs by bellowslike movements of its ribs. The three-chambered lepidosaur heart partially separates oxygenated blood from the lungs from deoxygenated blood returning from the body. With this type of heart, lepidosaurs can generate high blood pressure and can sustain a relatively high metabolism.

Most lizards are insectivores, but some are herbivores; a few prey on other vertebrates. Many lizards walk on four limbs (**FIGURE 22.55B**), although limblessness has evolved repeatedly in the group, especially in burrowing and grassland species. The largest lizard, which grows as long as 3 meters and can weigh more than 150 kilograms, is the predaceous Komodo dragon of the East Indies.

Media Clip 22.16 **Komodo Dragons Bring Down Prey**

One major group of limbless squamates is the snakes (**FIGURE 22.55C**). All snakes are carnivores, and many can swallow objects much larger than themselves. Several snake groups evolved venom glands and the ability to inject venom rapidly into their prey.

The **turtles** comprise a reptilian group that has changed relatively little since the early Mesozoic. In these reptiles, dorsal and ventral bony plates form a shell into which the head and limbs can be withdrawn in many species (**FIGURE 22.55D**). The dorsal shell is a modification of the ribs. Most turtles live in aquatic environments, but several groups, such as tortoises and box turtles, are terrestrial. Sea turtles spend their entire lives at sea except when they come ashore to lay eggs. Human exploitation of sea turtles and their eggs has resulted in worldwide declines of these species, all of which are now endangered. A few species of turtles are strict herbivores or carnivores, but most species are omnivores that eat a variety of aquatic and terrestrial plants and animals.

Crocodilians and birds are living archosaurs

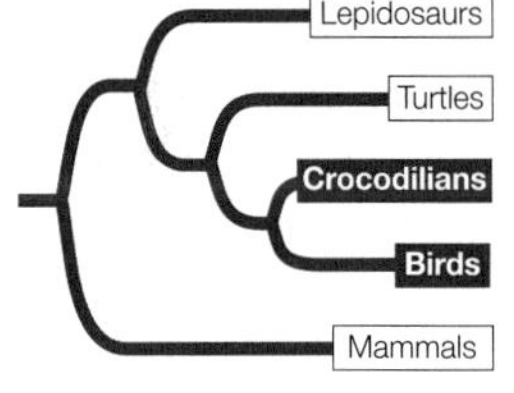

Another reptilian clade, the **archosaurs**, includes the crocodilians, the extinct pterosaurs and dinosaurs, and birds. Only the crocodilians and birds are represented by living species today. Archosaurs share such features as teeth (when present) set in sockets, two openings on each side of the skull (one in front of each eye and another on the jaw), and a prominent ridge on the femur. Genome sequences have confirmed the close relationship between crocodilians and birds among the living reptiles.

Modern **crocodilians**—crocodiles, caimans, gharials, and alligators—are confined to tropical and warm temperate environments (**FIGURE 22.56A**). All crocodilians are carnivorous; they eat vertebrates of all kinds, including large mammals. Crocodilians spend much of their time in water, but they lay their eggs in nests they build on land or on floating piles of vegetation. The eggs are warmed by heat generated by decaying organic matter that the female places in the nest. The female provides other forms of parental care as well: typically, she guards the eggs until they hatch, and in some species, she continues to guard and communicate with her offspring after they hatch.

(A) ***Sphenodon punctatus***

© Bruce Coleman, Inc./Alamy Stock Photo

(B) ***Trioceros jacksonii***

© Larry Jon Friesen

(C) ***Diadophis punctatus***

© Larry Jon Friesen

(D) ***Chelonoidis nigra abingdonii***

© Gordon Chambers/Alamy Stock Photo

FIGURE 22.55 Reptilian Diversity **(A)** This tuatara represents one of only two surviving species in a lineage that diverged long ago. **(B)** The Jackson's chameleon is native to eastern Africa but was introduced to Hawaii in 1972. **(C)** The ringneck snake of North America is harmless to humans. It coils its tail to reveal a bright orange underbelly, which distracts potential predators from the vital head region. **(D)** Galápagos tortoises (there are several closely related species) are the largest turtles and among the largest reptiles. They have been documented to live for more than 100 years in the wild.

Dinosaurs rose to prominence about 215 million years ago and dominated terrestrial environments for about 150 million years. However, only one group of dinosaurs, the **birds**, survived the mass extinction at the end of the Cretaceous. During the Mesozoic, most terrestrial animals more than a meter long were dinosaurs. Many were agile and could run rapidly; they had special muscles that enabled the lungs to be filled and emptied while the limbs moved. We can infer the existence of such muscles in dinosaurs

(A) ***Caiman crocodilus***

© blickwinkel/Alamy Stock Photo

(B) ***Struthio camelus***

Courtesy of Andrew D. Sinauer

FIGURE 22.56 Archosaurs **(A)** Caimans, alligators, and their relatives live in tropical and warm temperate climates. The spectacled caiman is the most common crocodilian throughout much of tropical Central and South America. **(B)** Birds are the other living archosaur group. This Common Ostrich is a palaeognath, a group that includes several flightless and weakly flying birds of the Southern Hemisphere. Flightless members of this early-diverging group lack the sternum keel structure that anchors the flight muscles of other bird groups.

from the structure of the vertebral column in fossils. Some of the largest dinosaurs weighed as much as 70,000 kilograms.

Biologists have long accepted the phylogenetic position of birds among the reptiles, although birds clearly have many unique, derived morphological features. In addition to the strong morphological evidence for this placement, fossil and molecular data emerging over the last few decades have provided definitive supporting evidence. Birds are a specialized group of **theropods**, a clade of predatory dinosaurs that shared such traits as a bipedal stance, hollow bones, a furcula ("wishbone"), elongated metatarsals with three-fingered feet, elongated forelimbs with three fingers, and a pelvis that points backward. Modern birds are homeothermic, meaning they regulate their body temperatures at a relatively constant temperature through physiological control. Although we cannot directly assess this physiological trait in extinct species, many fossil theropods share morphological traits that suggest they may have been homeothermic as well.

LINK Homeothermy and its physiological control are described in **Key Concept 28.3**.

The living bird species fall into two major groups that diverged about 90–80 million years ago from a flying ancestor. The few modern descendants of one lineage include a group of secondarily flightless and weakly flying birds, some of which are very large. This group, called the palaeognaths, includes the South and Central American tinamous and several large flightless birds of the southern continents—the rheas, Emu, kiwis, cassowaries, and the world's largest bird, the Common Ostrich (**FIGURE 22.56B**). The second lineage, the neognaths, has left a much larger number of descendants, most of which have retained the ability to fly.

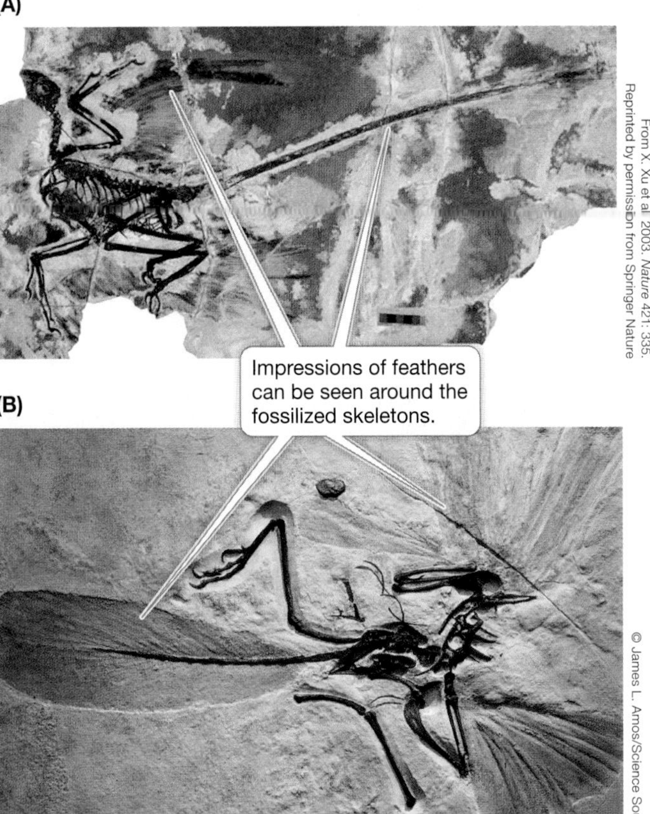

FIGURE 22.57 Mesozoic Bird Relatives The fossil record supports the evolution of birds from other dinosaurs. **(A)** *Microraptor gui* was a feathered dinosaur from the early Cretaceous (about 140 million years ago). **(B)** Dating from roughly the same time frame, *Archaeopteryx* is the oldest known birdlike fossil.

The evolution of feathers led to bird flight

During the Mesozoic, about 175 million years ago, a lineage of theropods gave rise to the birds. Recent fossil discoveries show that the scales of some theropod dinosaurs were modified to form feathers. The feathers of many of these dinosaurs were structurally similar to those of modern birds (**FIGURE 22.57**).

The evolution of feathers was a major force for diversification. Feathers are lightweight but are strong and structurally complex. The large quills of the flight feathers on the wings arise from the skin of the forelimbs to create the flying surfaces. Other strong feathers sprout like a fan from the shortened tail and serve as stabilizers during flight. The feathers that cover the body, along with an underlying layer of down feathers, provide birds with insulation that helps them survive in virtually all of Earth's climates.

The bones of theropod dinosaurs, including birds, are hollow with internal struts that increase their strength. Hollow bones would have made early theropods lighter and more mobile. Later, the hollow bones of theropods facilitated the evolution of flight. The sternum (breastbone) of flying birds forms a large, vertical keel to which the flight muscles are attached.

Media Clip 22.17 Falcons in Flight

Flight is metabolically expensive. A flying bird consumes energy at a rate 15–20 times faster than a running lizard of the same weight. Because birds have such high metabolic rates, they generate large amounts of heat. They control the rate of heat loss using their feathers, which may be held close to the body or elevated to alter the amount of insulation they provide. The lungs of birds allow air to flow through unidirectionally rather than by pumping air in and out. This flow-through structure of the lungs increases the efficiency of gas exchange and thereby supports an increased metabolic rate.

LINK You can read more about the effects of flying on metabolic rate in **Key Concept 28.2**. See **Key Concept 30.2** for a description of the breathing system of birds.

There are about 10,000 species of living birds, which range in size from the 150-kilogram Common Ostrich to a hummingbird weighing only 2 grams (**FIGURE 22.58**). The teeth so prominent among other dinosaurs were secondarily lost in the ancestral birds, but birds nonetheless eat almost all types of animal and plant material. Insects and fruits are the most important dietary items for terrestrial species. Birds also eat seeds, nectar and pollen, leaves and

(A) *Grus canadensis*

(B) *Aegolius acadicus*

(C) *Piranga olivacea*

FIGURE 22.58 Diverse Birds **(A)** Sandhill Cranes feed on plants, insects, small reptiles, and even small mammals. As in other crane species, their mating and courtship behavior includes an elaborate dance display. **(B)** Owls such as this Northern Saw-whet Owl are nighttime predators that can locate prey using their sensitive auditory systems. **(C)** Perching, or passeriform, birds such as this Scarlet Tanager comprise the most species-rich group of birds.

buds, carrion, and fishes and other vertebrates. By eating the fruits and seeds of plants, birds serve as major agents of seed dispersal.

Mammals radiated as non-avian dinosaurs declined

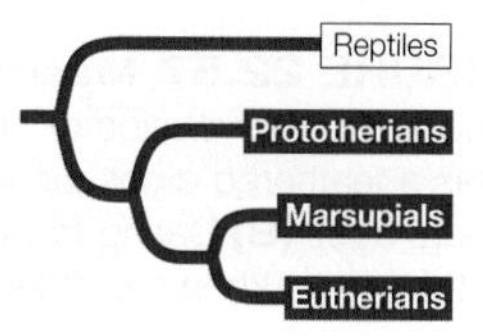

Small- and medium-sized mammals coexisted with the large dinosaurs throughout most of the Mesozoic, and most of the major groups that are alive today arose in the Cretaceous. After the non-avian dinosaurs disappeared during the mass extinction at the end of the Cretaceous, mammals increased dramatically in numbers, diversity, and size. Today mammals range in size from tiny shrews and bats weighing only about 2 grams to the blue whale, the largest animal on Earth, which measures up to 33 meters long and can weigh as much as 160,000 kilograms. Mammals have far fewer, but more highly differentiated, teeth than do fishes, amphibians, or reptiles. Differences among mammals in the number, type, and arrangement of teeth reflect their varied diets.

Four key features distinguish the mammals:

1. Sweat glands, which secrete sweat that evaporates and thereby cools an animal
2. Mammary glands, which in females secrete a nutritive fluid (milk) on which newborn individuals feed
3. Hair, which provides a protective and insulating covering
4. A four-chambered heart that completely separates the oxygenated blood coming from the lungs from the deoxygenated blood returning from the body (this last characteristic evolved independently in the archosaurs, including modern birds and crocodiles)

Mammalian eggs are fertilized within the female's body, and the embryos undergo a period of development in the female's body in an organ called the uterus. In the uterus, the embryo is contained in an amniotic sac that is homologous to one of four membranes found in the amniote egg (see Figure 22.53). The embryo is connected to the wall of the uterus by an organ called a placenta. The placenta allows for nutrient and gas exchange, as well as waste elimination from the developing embryo, via the female's circulatory system. Most mammals have a covering of hair (fur), which is luxuriant in some species but has been greatly reduced in others, including the cetaceans (whales and dolphins) and humans. Thick layers of insulating fat (blubber) replace hair as a heat-retention mechanism in the cetaceans. Humans learned to use clothing for heat retention when they dispersed from warm tropical areas.

The approximately 6,500 species of living mammals are divided into two primary groups: the **prototherians** and the **therians** (**TABLE 22.3**). Only five species of prototherians are known, and they are found only in Australia and New Guinea. These mammals, the duck-billed platypus and four species of echidnas, differ from other mammals in lacking a placenta, laying eggs, and having sprawling legs (**FIGURE 22.59A**). Prototherians supply milk for their young, but they have no nipples on their mammary glands. The milk simply oozes out and is lapped off the fur by the offspring.

Most mammals are viviparous

Members of the viviparous therian clade are further divided into the **marsupials** and the **eutherians**. Females of most marsupial species have a ventral pouch in which they carry and feed their offspring (**FIGURE 22.59B**). Gestation (pregnancy) in

FIGURE 22.59 Mammalian Diversity **(A)** The duck-billed platypus is one of the five surviving species of prototherians—mammals that lay eggs. **(B)** Female marsupials have a ventral pouch in which they nurture their offspring, which are extremely small at birth. The young of this opossum have grown large enough to leave her pouch. **(C–F)** Eutherian mammals. **(C)** Almost half of all eutherians are rodents, such as this North American porcupine. **(D)** Flight evolved in the common ancestor of bats. Virtually all bat species are nocturnal. **(E)** Many large mammals are important herbivores in terrestrial environments. Nowhere are these assemblages more spectacular than on the grass and brushlands of the African continent. **(F)** Bottlenose dolphins are cetaceans, a cetartiodactyl group that returned to the marine environment.

marsupials is brief; the young are born tiny but with well-developed forelimbs, with which they climb to the pouch. They attach to a nipple but cannot suck. The mother ejects milk into the tiny offspring until they grow large enough to suckle. Once her offspring have left the uterus, a female marsupial may become sexually receptive again. She can then carry fertilized eggs that are capable of initiating development and can replace the offspring in her pouch should something happen to them.

Eutherians include the majority of mammals. Eutherians are sometimes called placental mammals, but this name is inappropriate because some marsupials also have placentas. Eutherians are more developed at birth than are marsupials, and no external pouch houses them after they are born.

The approximately 6,100 living species of morphologically diverse eutherians are divided into 20 major groups (see Table 22.3). The relationships of the major groups of eutherians to one another have been difficult to determine because most of the major groups diverged in a short period of time during an explosive adaptive radiation. The two most diverse groups are the rodents (**FIGURE 22.59C**) and bats (**FIGURE 22.59D**), which together comprise

TABLE 22.3 | Major Groups of Living Mammals

Group	Number of described species	Examples
PROTOTHERIANS		
Monotremes (Monotremata)	5	Echidnas, duck-billed platypus
THERIANS		
Marsupials		
Diprotodonts (Diprotodontia)	158	Kangaroos, wallabies, possums, koala, wombats
New World opossums (Didelphimorphia)	127	Opossums
Carnivorous marsupials (Dasyuromorphia)	78	Quolls, dunnarts, numbat, Tasmanian devil
Omnivorous marsupials (Peramelemorphia)	30	Bandicoots and bilbies
Shrew opossums (Paucituberculata)	7	Andean rat opossums
Microbiothere (Microbiotherea)	3	Monito del monte
Marsupial moles (Notoryctemorphia)	2	Southern and northern marsupial moles
Eutherians		
Rodents (Rodentia)	2,643	Rats, mice, squirrels, woodchucks, ground squirrels, beavers, capybara
Bats (Chiroptera)	1,449	Fruit bats, echolocating bats
Shrews, moles, and relatives (Eulipotyphla)	554	Shrews, moles, solenodons
Primates (Primates)	519	Lemurs, monkeys, apes, humans
Even-toed hoofed mammals and cetaceans (Cetartiodactyla)	355	Deer, sheep, goats, cattle, antelopes, giraffes, camels, swine, hippopotamuses, whales, dolphins
Carnivores (Carnivora)	311	Wolves, dogs, bears, cats, weasels, pinnipeds (seals, sea lions, walruses)
Rabbits and relatives (Lagomorpha)	109	Rabbits, hares, pikas
African insectivores (Afrosoricida)	55	Tenrecs, golden moles
Spiny insectivores (Erinaceomorpha)	24	Hedgehogs
Tree shrews (Scandentia)	23	Tree shrews
Armadillos (Cingulata)	21	Armadillos
Long-nosed insectivores (Macroscelidea)	20	Elephant shrews
Odd-toed hoofed mammals (Perissodactyla)	18	Horses, zebras, tapirs, rhinoceroses
Anteaters and sloths (Pilosa)	16	Anteaters, tamanduas, sloths
Pangolins (Pholidota)	8	Asian and African pangolins
Hyraxes and relatives (Hyracoidea)	5	Hyraxes, dassies
Sirenians (Sirenia)	5	Manatees, dugongs
Elephants (Proboscidea)	3	African and Indian elephants
Colugos (Dermoptera)	2	Flying lemurs
Aardvark (Tubulidentata)	1	Aardvark

about two-thirds of the species. Rodents are traditionally defined by the unique morphology of their teeth, which are adapted for gnawing through substances such as wood. The bats probably owe much of their success to the evolution of flight, which allows them to exploit a variety of food sources and colonize remote locations with relative ease.

Media Clip 22.18 Bats Feeding in Flight

Grazing and browsing by members of several eutherian groups helped transform the terrestrial landscape. Herds of grazing herbivores feed on open grasslands, whereas browsers feed on shrubs and trees. The effects of herbivores on plant life favored the evolution of the spines, tough leaves, and difficult-to-eat growth forms found in many plants. In turn, adaptations to the teeth and digestive systems of many herbivore lineages allowed these species to consume many plants despite such defenses—a striking example of coevolution. A large animal can survive on food of lower quality than a small animal can, and large size evolved in several groups of grazing and browsing mammals (**FIGURE 22.59E**). The evolution of large herbivores, in turn, favored the evolution of large carnivores able to attack and overpower them.

Several lineages of terrestrial eutherians subsequently returned to the aquatic environments their ancestors had left behind (**FIGURE 22.59F**). The completely aquatic cetaceans—whales and dolphins—evolved from artiodactyl ancestors (cetaceans are closely related to the hippopotamuses). The seals, sea lions, and walruses also returned to the marine environment, and their limbs became modified into flippers. Weasel-like otters retain their limbs but have also returned to aquatic environments, colonizing both fresh and salt water. The manatees and dugongs colonized estuaries and shallow seas.

Activity 22.5 **Deuterostome Classification**

REVIEW & APPLY | 22.6

R The initial vertebrate colonization of dry land was facilitated by the evolution of lunglike sacs and jointed limbs. The amniotes also evolved impermeable body coverings, efficient kidneys, and the amniote egg, which resists desiccation.

A
1. Amphibians have survived and prospered for millions of years, but today many species are disappearing and populations of others are in serious decline. What features of their life histories might make amphibians especially vulnerable to the kinds of environmental changes now happening on Earth?
2. How has the diversification of mammals been influenced by mass extinction events and continental drift?
3. In the not-too-distant past, the idea that birds were reptiles was met with skepticism. Explain how fossils, morphology, and molecular evidence now support the position of birds among the reptiles.
4. Consider the phylogeny of the amniotes and the evolution of homeothermy, hair, and feathers. Which of these traits likely evolved convergently in more than one lineage? What is a likely functional relationship among these traits? Why do many paleontologists think that several extinct theropods were homeothermic?

The biology of one eutherian group—the primates—has been the subject of extensive research. The behavior, ecology, physiology, and molecular biology of the primates are of special interest to us because this lineage includes humans.

22.7 | Humans Evolved among the Primates

LEARNING OBJECTIVES

By the end of this key concept you should be able to:

22.7.1 **Summarize the diversity within the primates.**

22.7.2 **Articulate the relationship between bigger brains, smaller jaws, and neoteny.**

The **primates** evolved from a lineage of relatively small, arboreal, insectivorous eutherians. Grasping limbs with opposable digits are one of the major adaptations to arboreal life that distinguish primates from other mammals. Primates underwent extensive evolutionary radiation throughout the Paleogene and Neogene (**FIGURE 22.60**).

Two major lineages of primates split late in the Cretaceous

About 90 million years ago, late in the Cretaceous, the primates split into two clades: the wet-nosed primates (strepsirrhines) and the dry-nosed primates (haplorhines). **Wet-nosed primates**—lemurs, lorises, and galagos—once lived on all continents, but today they are restricted to Africa, Madagascar, and tropical Asia. All mainland wet-nosed primate species are arboreal and nocturnal. On the island of Madagascar, however, the site of a remarkable radiation of lemurs, there are also diurnal and terrestrial species. Tarsiers were once considered a part of this lineage as well, although today we know that they are more closely related to **simians**—monkeys and apes—than to lemurs, lorises, and galagos.

The second primate lineage, the **dry-nosed primates**—tarsiers, New World monkeys, Old World monkeys, and apes—began to diversify shortly after the mass extinction event at the end of the Cretaceous, in Africa or Asia. New World monkeys diverged from Old World monkeys and apes slightly later, but early enough that they may have originated in Africa and reached South America when those two continents were still close to each other. New World monkeys now live only in South and Central America, and all of them are arboreal. Many of them have a long, prehensile tail with which they can grasp branches. Many Old World monkeys are arboreal as well, but several species are terrestrial. No Old World monkey has a prehensile tail.

About 35 million years ago, a lineage that led to the modern apes separated from the Old World monkeys. Between 22 and 5.5 million years ago, dozens of species of apes lived in Europe, Asia, and Africa. The Asian apes—gibbons and orangutans—descended from two of these ape lineages. Orangutans are the

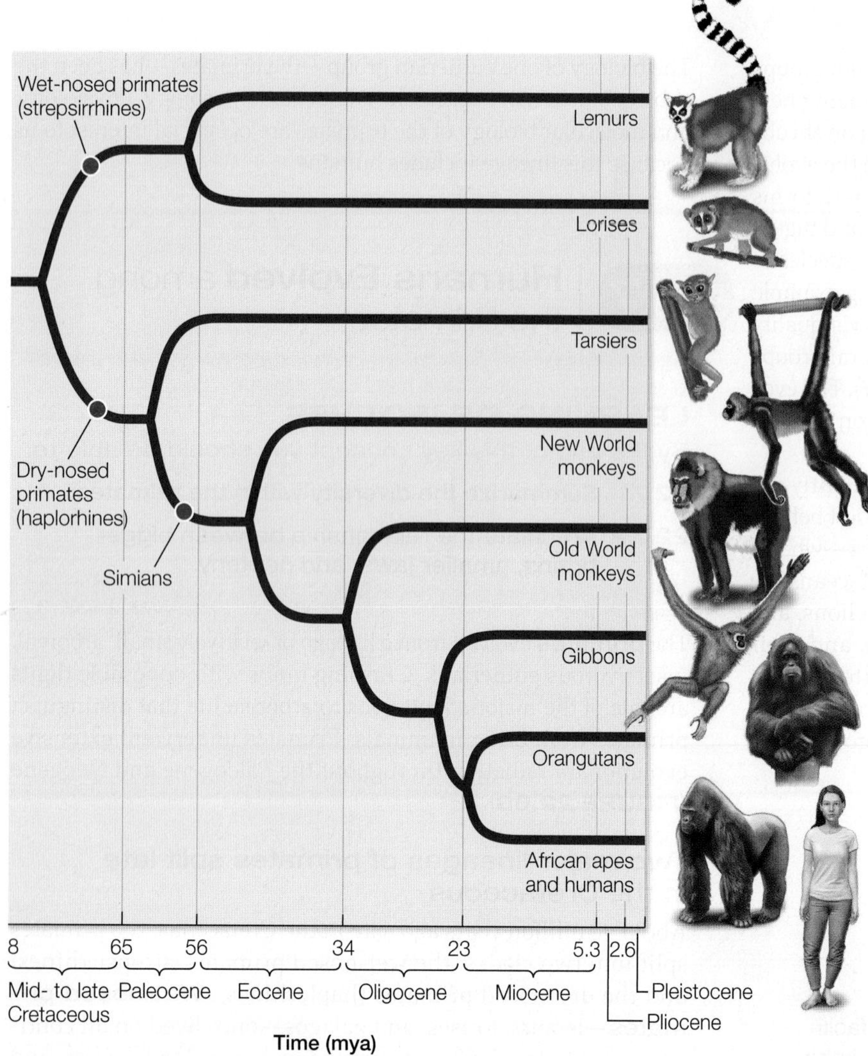

FIGURE 22.60 Phylogeny of the Primates The phylogeny of primates is among the best studied of any major group of mammals. This tree is based on evidence from many genes, morphology, and fossils. (mya, millions of years ago).

closest living sister group of the modern African apes: gorillas, chimpanzees, and humans.

Bipedal locomotion evolved in human ancestors

About 6 million years ago in Africa, a lineage split occurred that would lead to the chimpanzees on the one hand and to the **hominin** clade, which includes modern humans and their extinct close relatives, on the other.

The earliest protohominins, known as ardipithecines, had distinct morphological adaptations for **bipedal locomotion** (walking on two legs). Bipedal locomotion frees the forelimbs to manipulate objects and to carry them while walking. It also elevates the eyes, enabling the animal to see over tall vegetation to spot predators and prey. Bipedal locomotion is also energetically more economical than quadrupedal locomotion (walking on four legs). All three advantages were probably important for the ardipithecines and their descendants, the australopithecines (**FIGURE 22.61**).

The first australopithecine skull was found in South Africa in 1924. Since then australopithecine fossils have been found at many sites in Africa. The most complete fossil skeleton yet found was discovered in Ethiopia in 1974. The skeleton, approximately 3.5 million years old, was that of a young female who has since become known to the world as "Lucy." Lucy was assigned to the species *Australopithecus afarensis*. Fossil remains of more than 100 *A. afarensis* individuals have since been discovered, and there have been recent discoveries of fossils of other australopithecine species that lived in Africa 5–4 million years ago.

Experts disagree over how many species are represented by australopithecine fossils, but it is clear that multiple species of hominins lived together over much of eastern Africa several million years ago. A lineage of larger species (weighing about 40 kilograms) is represented by *Paranthropus robustus* and *P. boisei*, both of which died out between 1.5 and 1 million years ago. A lineage of smaller australopithecines gave rise to the genus *Homo*.

Early members of the genus *Homo* lived contemporaneously with *Paranthropus* in Africa for about a million years. Some 2-million-year-old fossils of an extinct species called *H. habilis* were discovered in the Olduvai Gorge, Tanzania. Other fossils of *H. habilis* have been found in Kenya and Ethiopia. Associated with these fossils were tools that these early hominins used to obtain food.

Another extinct hominin species, *Homo erectus*, arose in Africa about 1.6 million years ago. Soon thereafter it had spread as far as eastern Asia, becoming the first hominin to leave Africa. Members of *H. erectus* were nearly as large as modern people, but their brains were smaller and they had comparatively thick skulls. The cranium, which had thick, bony walls, may have been an adaptation to protect the brain, ears, and eyes from impacts caused by a fall or a blow from a blunt object. What would have been the source of such blows? Fighting with other *H. erectus* individuals is a possible answer.

Homo erectus used fire for cooking and for hunting large animals, and made characteristic stone tools that have been found in many parts of Africa and Asia. Populations of *H. erectus* survived until at least 250,000 years ago, although more recent fossils may also be attributable to this species. In 2004 some 18,000-year-old fossil remains of a small *Homo* were found on the island of Flores in Indonesia. Since then, numerous additional fossils of this diminutive hominin have been found on Flores, dating from about 500,000 to 12,000 years ago, when a major volcanic eruption on Flores likely led to its extinction. Many anthropologists think that this small species, named *H. floresiensis*, was most closely related to *H. erectus*.

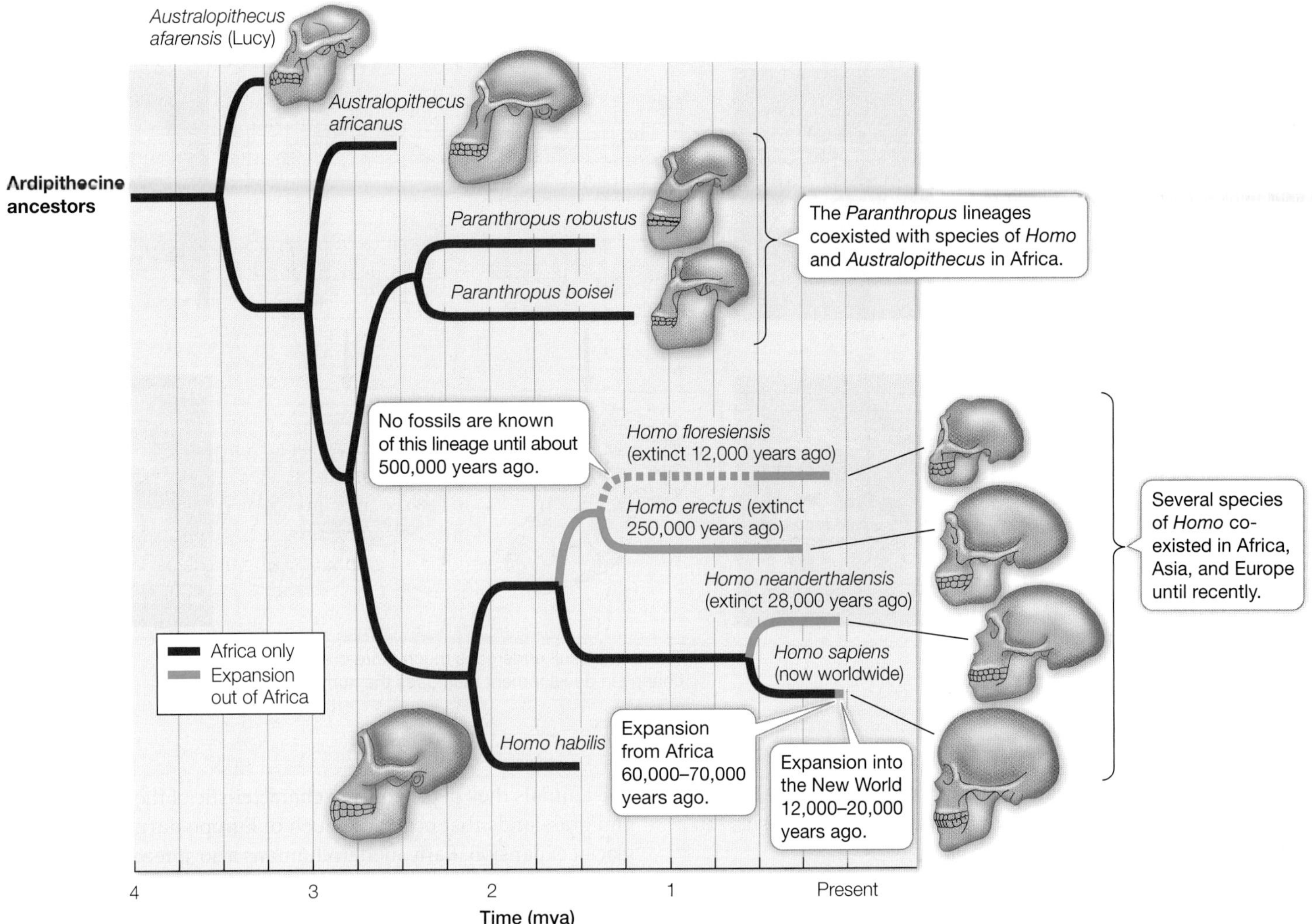

FIGURE 22.61 A Phylogenetic Tree of Hominins At times in the past, more than one species of hominin lived on Earth at the same time. Originating in Africa, hominins spread to Europe and Asia multiple times. All but one of these closely related species are now extinct; modern *Homo sapiens* have colonized nearly every corner of the planet.

Human brains became larger as jaws became smaller

In another hominin lineage that diverged from *H. erectus* and *H. floresiensis*, the brain increased rapidly in size, and the jaw muscles, which were large and powerful in earlier hominins, dramatically decreased in size. These two changes were simultaneous, which suggests that they might have been developmentally linked. These changes are another example of evolution by neoteny (which, you may recall from our discussion of amphibians, is the retention of juvenile traits through delayed somatic development). Human and chimpanzee skulls are similar in shape at birth, but chimpanzee skulls undergo a dramatic change in shape as the animals mature (**FIGURE 22.62**). In particular, the jaw grows considerably in relation to the brain case. As human skulls grow, relative proportions much closer to those of the juvenile skull are retained, which results in a large brain case and small jaw compared with those of chimpanzees. A mutation in a regulatory gene that is expressed only in the head may have removed a barrier that had previously prevented this remodeling of the human cranium.

The striking enlargement of the brain relative to body size in the hominin lineage was probably favored by an increasingly complex social life. Any features that allowed group members to communicate more effectively with one another would have been valuable in cooperative hunting and gathering as well as for improving one's status in the complex social interactions that must have characterized early human societies, just as they do ours today.

Several *Homo* species coexisted during the mid-Pleistocene, from about 1.5 million to about 250,000 years ago. All were skilled hunters of large mammals, but plants were important components of their diets as well. During this period another distinctly human trait emerged: rituals and a concept of life after death. Deceased individuals were buried with tools and clothing, supplies for their presumed existence in the next world.

One species, *Homo neanderthalensis*, was widespread in Europe and Asia between about 500,000 and 28,000 years ago. Neanderthals were short, stocky, and powerfully built. Their massive skull housed a brain somewhat larger than our own. They manufactured a variety of tools and hunted large mammals, which they

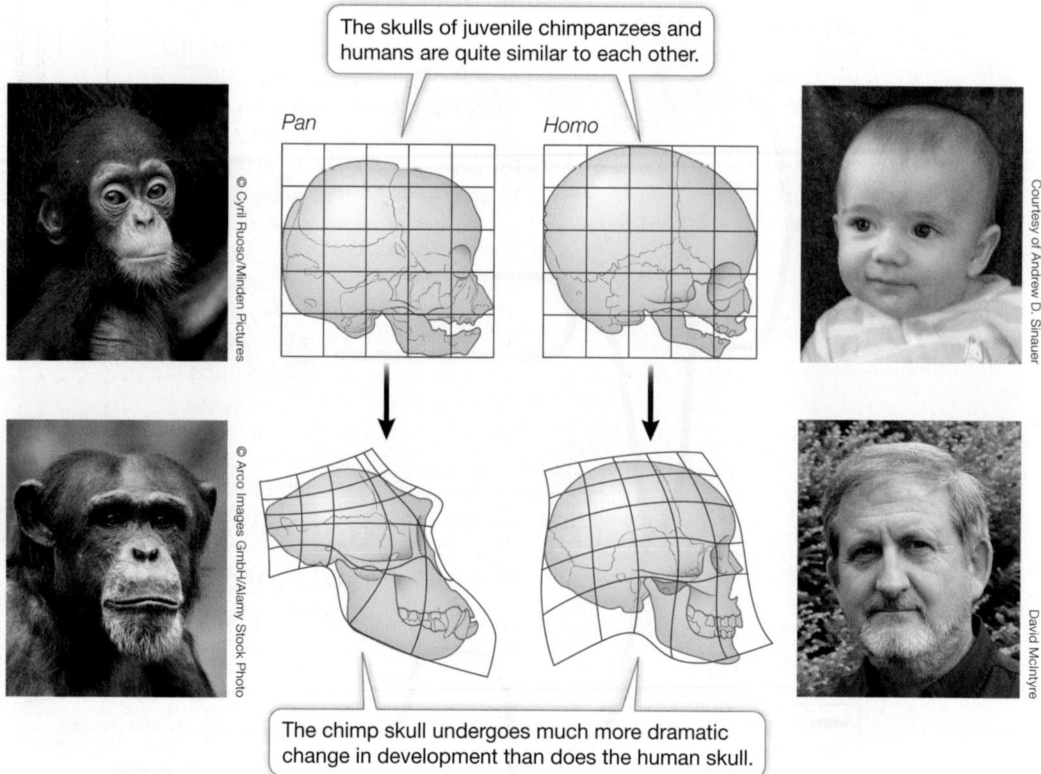

FIGURE 22.62 Neoteny in the Evolution of Humans The skulls and heads of juvenile chimpanzees and humans start out relatively similar in shape. The grid over the skulls shows how the various bony elements change in their relative proportions through maturation. The adult human skull retains a shape closer to its juvenile shape, resulting in a brain that is much larger relative to other parts of the skull (notably the jaw).

probably ambushed and subdued in close combat. A closely related lineage, the Denisovans, is known from less complete fossils, primarily from Asian caves.

Recently, biologists have been able to extract DNA from the bones of Neanderthals and Denisovans, and have compared the genomes of these species to those of modern humans. Early modern humans (*H. sapiens*) expanded out of Africa between 70,000 and 60,000 years ago. Then, about 35,000 years ago, *H. sapiens* moved into the range of *H. neanderthalensis* in Europe and western Asia, and into the range of the Denisovans in eastern Asia. Neanderthals abruptly disappeared about 28,000 years ago. Many anthropologists believe that Neanderthals and Denisovans were exterminated by those early modern humans. But comparisons of Neanderthal, Denisovan, and modern human genomes indicate that there was some limited interbreeding among these species while they occupied the same range. In humans with Eurasian ancestry, 1–4 percent of the genes in their genomes may be derived from Neanderthal ancestors. The genomes of some modern native people of Southeast Asia, Melanesia, and Australia contain even higher proportions of Denisovan genes. So all three of these lineages contributed in some degree to modern humans.

Early modern humans made and used a variety of sophisticated tools. They created the remarkable paintings of large mammals, many of them showing scenes of hunting, found in European caves. The animals they depicted were characteristic of the cold steppes and grasslands that occupied much of Europe during periods of glacial expansion. Early modern humans also spread across Asia, reaching North America perhaps as early as 20,000 years ago, although the date of their arrival in the Americas is still uncertain. Within a few thousand years, they had spread southward through North America to the southern tip of South America.

REVIEW & APPLY | 22.7

R Grasping limbs with opposable digits distinguish primates from other mammals. Bipedal locomotion and large brains evolved in the primate ancestors that led to humans, which facilitated the development of complex social lives and more effective communication.

A
1. Describe the differences between Old World and New World monkeys.
2. Explain how neoteny resulted in the development of humans with relatively large brains and small jaws.

Jochen Tack/Alamy Stock Photo

Besides insects, which other groups of animals are thought to contain many more species than are known at present?

Answer: It is perhaps easier to list the groups of animals for which a nearly complete inventory of living species has been completed than to list all the groups for which many new species remain to be described. Among the insects, the best studied group in terms of species is the butterflies, which are widely collected and studied. There are still many species of other lepidopterans (such as moths), however, remaining to be discovered. Most other major insect groups contain many undescribed species. Among the vertebrates, most species of birds and mammals have been described, although several new species of mammals are still discovered and named each year. Species discovery and description remain high for almost all other major groups of animals.

After insects, and perhaps even rivaling the insects in undiscovered diversity, are the nematodes. Although known nematode diversity is only about 1/40th that of the known insect diversity (in terms of number of described species), the taxonomy of nematodes is much more poorly studied than that of insects. Some biologists think there are likely to be species-specific parasitic nematodes of most other species of multicellular organisms. If so, then there may be as many species of nematodes as there are of plants, fungi, and other animals combined.

Most of the other diverse groups of animals contain many as yet undetected species, judging from the rate of new species descriptions. In particular, flatworms (especially the parasitic flukes and tapeworms), marine annelids, mollusks, crustaceans, myriapods, and chelicerates all contain large numbers of undescribed species. Among the vertebrates, the rate of discovery of new species remains particularly high for the fishes and amphibians.

VISUAL SUMMARY 22

22.1 Distinct Body Plans Evolved among the Animals

Go to ACTIVITY 22.1

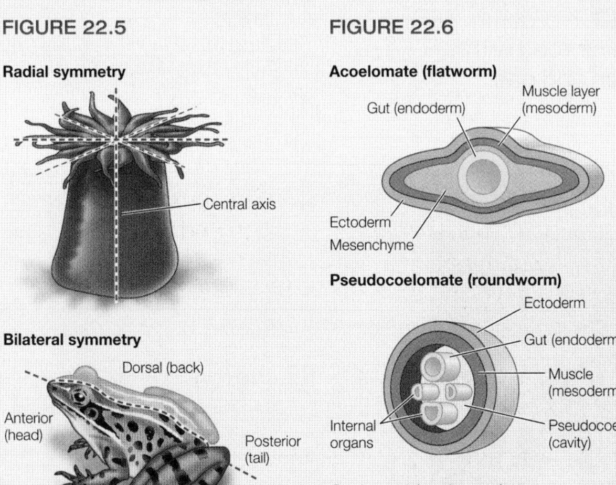

- Differences in their patterns of early development characterize two major clades of **TRIPLOBLASTIC** animals, the **PROTOSTOMES** and the **DEUTEROSTOMES**.
- Animal **BODY PLANS** can be described in terms of **SYMMETRY**, **BODY CAVITY** structure, **SEGMENTATION**, type of appendages, and type of nervous system (**FIGURE 22.5**).
- On the basis of their body cavity structure, animals can be described as **ACOELOMATES**, **PSEUDOCOELOMATES**, or **COELOMATES** (**FIGURE 22.6**).
- Segmentation takes many forms and improves control of movement, especially if the animal also has appendages.

QUESTIONS

1. What major difference in early development distinguishes protostomes from deuterostomes?
2. Which major groups of animals exhibit radial symmetry?
3. How do body cavities differ in acoelomates, pseudocoelomates, and coelomates?

22 VISUAL SUMMARY

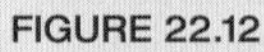

22.2 Large Multicellular Animal Groups Arose in Parallel

Go to ACTIVITY 22.2 and ANIMATION 22.1

FIGURE 22.12

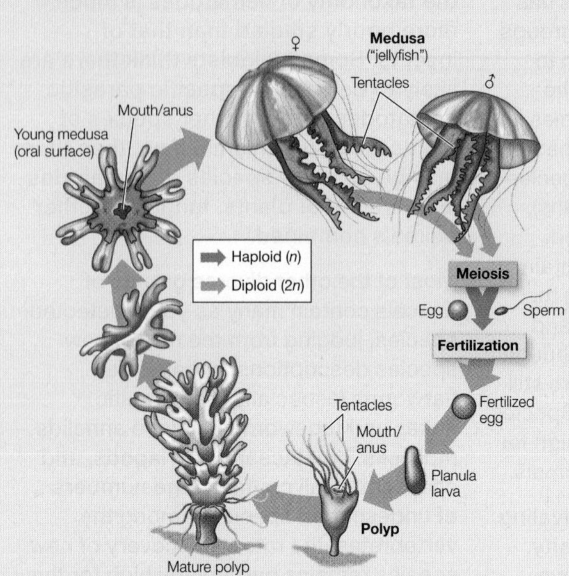

- Several large, multicellular groups diverged early in animal evolution, including **SPONGES**, **CTENOPHORES**, **PLACOZOANS**, **CNIDARIANS**, and bilaterians. The ancestral animal was unlike any of the modern groups and was likely small and composed of relatively few cells and no differentiated tissues.
- Complex nervous systems developed several times in animals, including in ctenophores, cnidarians, and bilaterians.
- The life cycle of most cnidarians has two distinct stages: a sessile **POLYP** stage and a motile **MEDUSA**. A fertilized egg develops into a free-swimming larval **PLANULA**, which settles to the bottom and develops into a polyp (**FIGURE 22.12**).

QUESTIONS

1. Which two groups are considered the most likely candidates to be the sister group of the remaining animals?
2. Which animal groups lack a nervous system?

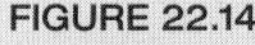

22.3 Protostomes Have an Anterior Brain and a Ventral Nervous System

Go to ANIMATION 22.2

- Protostomes ("mouth first") are bilaterally symmetrical animals that have an anterior brain surrounding the entrance to the digestive tract and a ventral nervous system. Protostomes comprise two major clades, the lophotrochozoans and the ecdysozoans.
- **LOPHOTROCHOZOANS** include a wide diversity of animals. Within this group evolved **LOPHOPHORES** (a complex organ for food collection and gas exchange) and free-living **TROCHOPHORE** larvae. The most species-rich groups of lophotrochozoans are the **FLATWORMS**, **ANNELIDS**, and **MOLLUSKS** (**FIGURE 22.14**).
- **ECDYSOZOANS** have a **CUTICLE** covering their body, which they must **MOLT** in order to grow. Some ecdysozoans, notably the **ARTHROPODS**, have a rigid cuticle reinforced with **CHITIN** that functions as an **EXOSKELETON**. New mechanisms of locomotion and gas exchange evolved among the arthropods (**FIGURE 22.25B**).

FIGURE 22.14

Lophophores Lophophore composed of tentacles

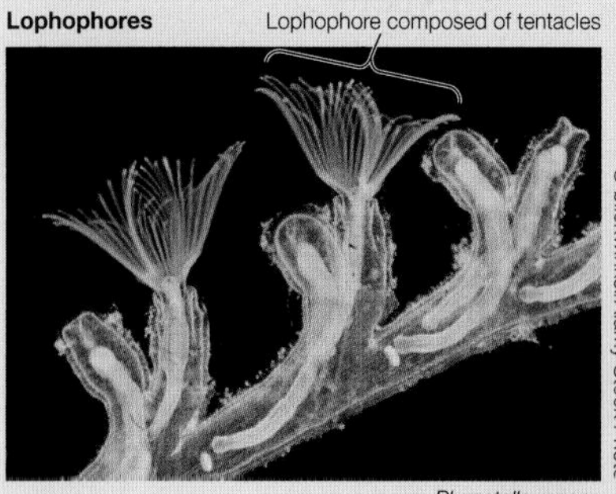

Plumatella repens

Trochophore

Stomach, Band of cilia, Mouth, Intestine, Anus

FIGURE 22.25B

Heterophrynus batesii Molted exoskeleton

QUESTIONS

1. Which three major groups of lophotrochozoans have lophophores?
2. What are the primary functions of the cuticle of ecdysozoans?

22.4 Arthropods Are Diverse and Abundant

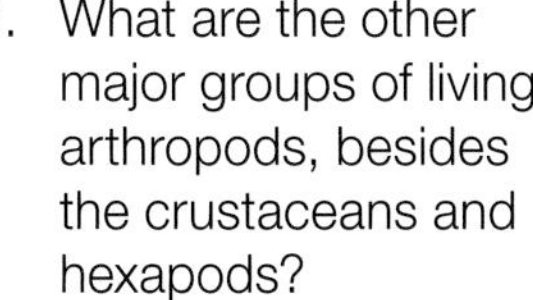

QUESTIONS

Go to ACTIVITY 22.3

FIGURE 22.34

Crustacean: Carapace covering head and thorax; Abdomen; Pleopods (swimming); Antennae (sensing); Maxilliped (helps hold food); Pereiopods (walking and gathering food)

Insect: Antennae; Head; Thorax; Abdomen; Wings; Exoskeleton; Spiracles; Ovipositor

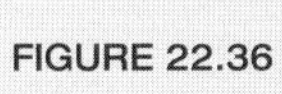

FIGURE 22.36

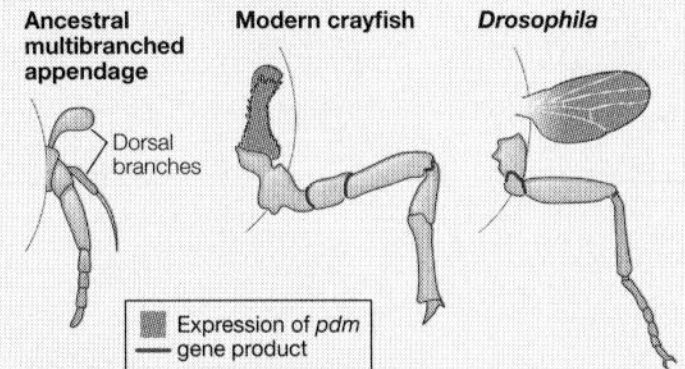

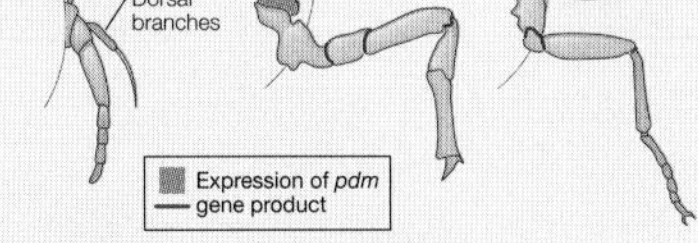

- Arthropods are the dominant animals on Earth in number of described species, and among the most abundant in number of individuals.
- Encasement within a rigid exoskeleton provides arthropods with support for walking as well as some protection from predators. The waterproofing provided by chitin keeps arthropods from dehydrating in dry air.
- Jointed appendages permit complex movements. Each arthropod segment has muscles attached to the inside of the exoskeleton that operate that segment and the appendages attached to it.
- **CRUSTACEANS** are the dominant marine arthropods and are also found in many freshwater and some terrestrial environments. Their segmented bodies are divided into three regions (head, thorax, and abdomen) with different, specialized appendages in each region (**FIGURE 22.34**).
- **HEXAPODS**—insects and their relatives—are the dominant terrestrial arthropods. They have the same three body regions as crustaceans, but no appendages form in their abdominal segments. Wings and the ability to fly first evolved among the insects, allowing them to exploit new lifestyles (**FIGURE 22.36**).

1. What are the other major groups of living arthropods, besides the crustaceans and hexapods?
2. What was the likely original function of the organ from which insect wings evolved?

22.5 Deuterostomes Include Echinoderms, Hemichordates, and Chordates

Go to ANIMATION 22.3

- **ECHINODERMS** and **HEMICHORDATES** both have bilaterally symmetrical, ciliated larvae.
- Most adult echinoderms have **PENTARADIAL SYMMETRY**. Echinoderms have an internal skeleton of calcified plates and a unique **WATER VASCULAR SYSTEM** connected to extensions called **TUBE FEET**.
- Chordates fall into three principal subgroups: **LANCELETS**, **TUNICATES**, and **VERTEBRATES**. At some stage in their development, all chordates have a dorsal hollow nerve cord, a post-anal tail, and a **NOTOCHORD** (**FIGURE 22.41**).
- The vertebrate body is characterized by a rigid internal skeleton, which is supported by a vertebral column that replaces the notochord, internal organs suspended in a coelom, a ventral heart, and an anterior skull with a large brain (**FIGURE 22.43**).
- The evolution of jaws from gill arches enabled individuals to grasp large prey and, together with teeth, cut them into small pieces (**FIGURE 22.46**).

FIGURE 22.41

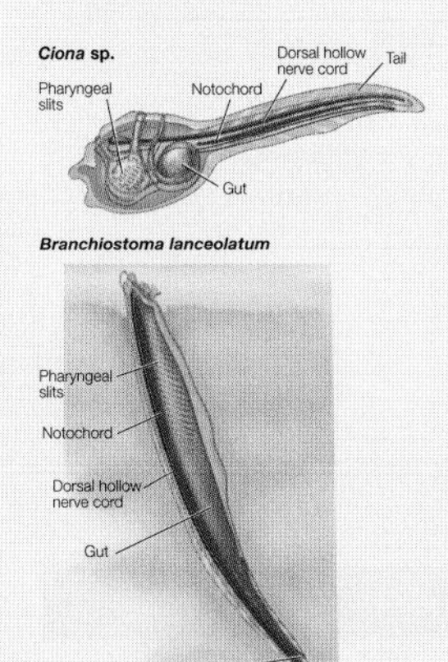

FIGURE 22.43

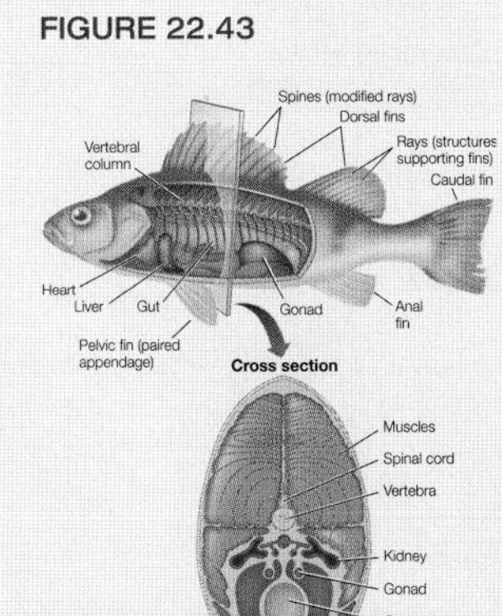

FIGURE 22.46

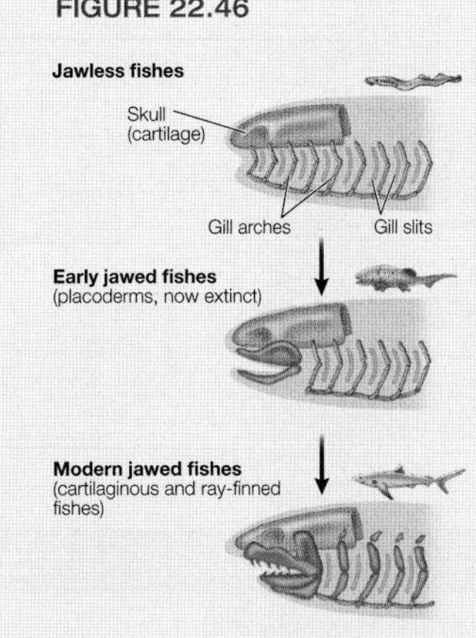

1. Draw the phylogenetic relationships among the following animals: hagfish, lamprey, shark, ray-finned fish, coelacanth, lungfish, frog, and turtle.
2. Name a shared morphological character that supports each of your groups in the tree from Question 1.

22 VISUAL SUMMARY

22.6 Life on Land Contributed to Vertebrate Diversification

FIGURE 22.50

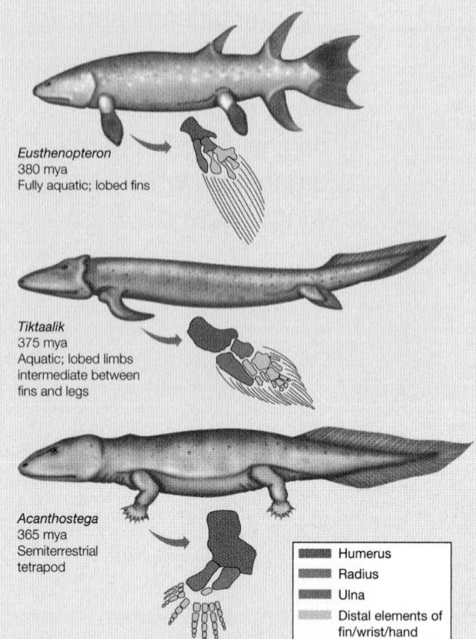

Go to ANIMATION 22.4 and ACTIVITIES 22.4 and 22.5

- Lungs and jointed appendages enabled one lineage of lobe-limbed vertebrates to colonize the land. This lineage gave rise to the **TETRAPODS** (**FIGURE 22.50**).
- The earliest split in the tetrapod tree is between the **AMPHIBIANS** and the **AMNIOTES** (**REPTILES** and **MAMMALS**).
- Most modern amphibians are confined to moist environments because they and their eggs lose water rapidly.
- An impermeable skin, efficient kidneys, and an egg that could resist desiccation evolved in the amniotes (**FIGURE 22.53**).

FIGURE 22.53

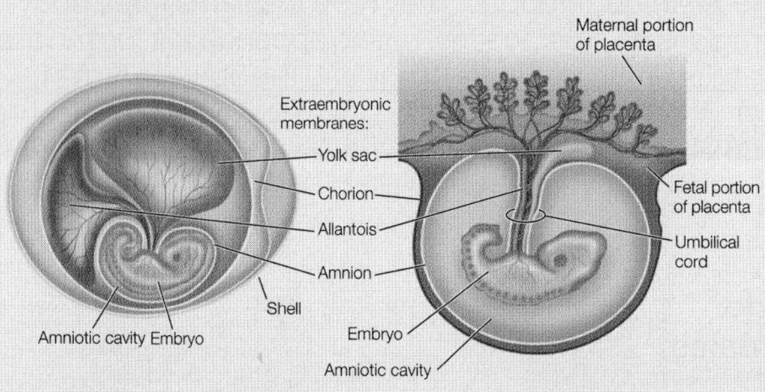

QUESTIONS

1. What are two examples of aquatic lobe-limbed vertebrates that are not tetrapods?
2. What are the two major groups of tetrapods that independently evolved the ability to maintain a relatively constant body temperature?
3. Which group of living tetrapods is most closely related to the extinct dinosaurs?

22.7 Humans Evolved among the Primates

FIGURE 22.62

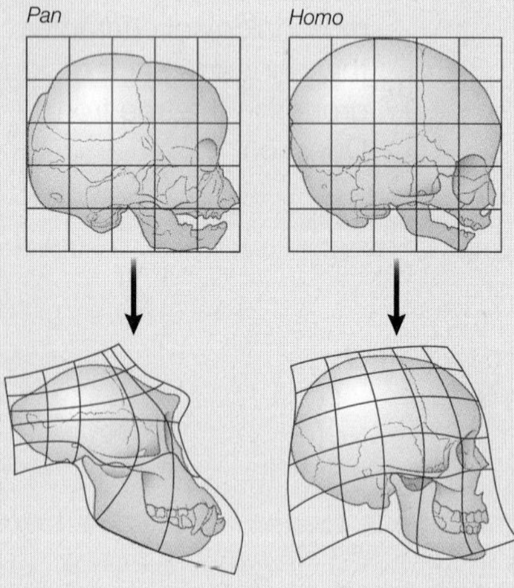

- Grasping limbs with opposable digits distinguish **PRIMATES** from other mammals. The **WET-NOSED PRIMATES** include the lemurs, lorises, and galagos; the **DRY-NOSED PRIMATES** include tarsiers, New World monkeys, Old World monkeys, and apes (including humans).
- **HOMININ** ancestors developed efficient **BIPEDAL LOCOMOTION**. Several species of *Homo* coexisted in parts of the world until recently.
- Within the *Homo* lineage, brains became larger as jaws became smaller. The two events appear to be developmentally linked and are an example of evolution via neoteny (**FIGURE 22.62**)

QUESTIONS

1. Several species of *Homo* are thought to have lived together on Earth until recently. Why is *Homo sapiens* thought to be the only species alive today?
2. What evidence suggests that *Homo sapiens* interbred to some extent with other *Homo* species in the past?

Go to **Achieve** for the e-book, animations, activities, and additional resources and assignments.

Part 5

Plant Form and Function

23 The Plant Body

The oldest living individual on Earth is a 5,000-year-old bristlecone pine tree. Such longevity is difficult to achieve because cells and organs become damaged over time. However, unlike mammals, in which the number of limbs is established before birth, these ancient trees continue to produce new body parts every growing season.

How do plants maintain the capacity to grow and develop throughout their lives?

You will find the answer to this question on page 596.

KEY CONCEPTS

23.1 The Plant Body Is Built from an Indeterminate Number of Simple Modules

LEARNING OBJECTIVES

By the end of this key concept you should be able to:

23.1.1 Sketch a eudicot and a monocot plant, indicating the locations of the apical meristems and organs.

23.1.2 Explain the extent to which different types of cells contribute to the structural support of a plant under different conditions of water availability.

23.1.3 Sketch a leaf, showing the major cell types and indicating their functions.

Flowering plants are organized into two systems: the root system and the shoot system (**FIGURE 23.1**). The parts of a plant that make up these systems, such as leaves and roots, are called **organs**, just as the parts of our bodies are. The organs that comprise the **shoot system** include leaves, stems, and, when the plant enters the reproductive phase, flowers. **Leaves** are the primary organ in which photosynthesis takes place. Leaves come in a variety of sizes and shapes, though most are broad and thin, a geometry that promotes light absorption. The leaves of many plants are attached to the stem by a stalk called a **petiole**. Common exceptions to this include bulbs in which leaves wrap around a short stem, and plants that grow in cold or windy environments in which heat retention is an advantage (such as alpine cushion plants). The **stem** holds and orients the leaves to the Sun and provides connections for the transport of materials between roots and leaves. The tip of the stem is made up of a small collection of cells called a **meristem** that has the capacity to produce new organs. Plants have several meristems, and each is named according to the position it occupies. The meristem at the top (apex) of the shoot is called the **shoot apical meristem**.

Because plants are stationary (sessile), they cannot move to find a more favorable environment or to avoid predators. Plants are assisted in meeting these challenges by their body plan, which consists of a series of repeating modules (units). Whereas animal bodies comprise a fixed number of parts (insects have six legs, for example), in plants the number of parts is not fixed. The ability to add new organs gives plants a striking ability to redirect their growth in response to their environments, as when new roots are formed in a nutrient-rich part of the soil, or when trees form new leaves in areas where light is abundant.

The organs that make up the shoot are arranged in repeating units called phytomers (see Figure 23.1). A **phytomer** consists of a leaf, the region of stem to which the leaf is attached (called the **node**) plus the portion of stem that extends to the next leaf (the **internode**), and one or more axillary buds. An **axillary bud** contains a shoot meristem tightly enclosed in leaves; its name indicates its position in the **axil**, the space that is formed where the leaf meets the stem. When the growing tip of a plant dies or is pruned away, the axillary bud that is closest to the cut site typically grows out and produces a new branch. This gives each phytomer the capacity to produce more units like itself. Removing

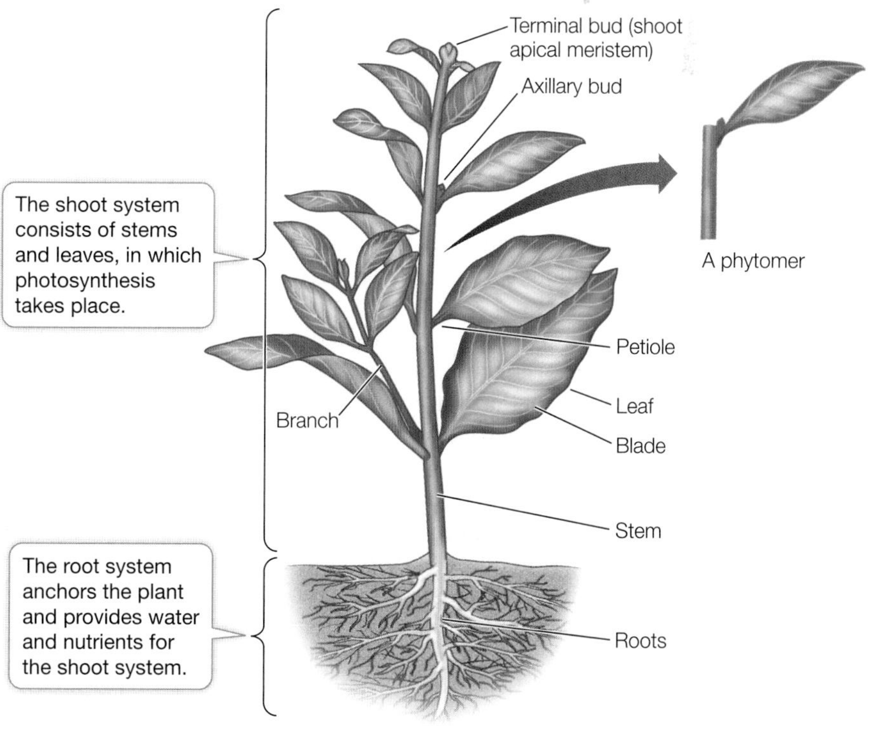

FIGURE 23.1 Vegetative Plant Organs and Systems of a Generalized Eudicot At the time of germination, a typical eudicot is organized around a single axis, with the root apical meristem on one end and the shoot apical meristem on the other. The shoot apical meristem gives rise to phytomers, the repeating units that form the vegetative (not yet flowering) shoot system.

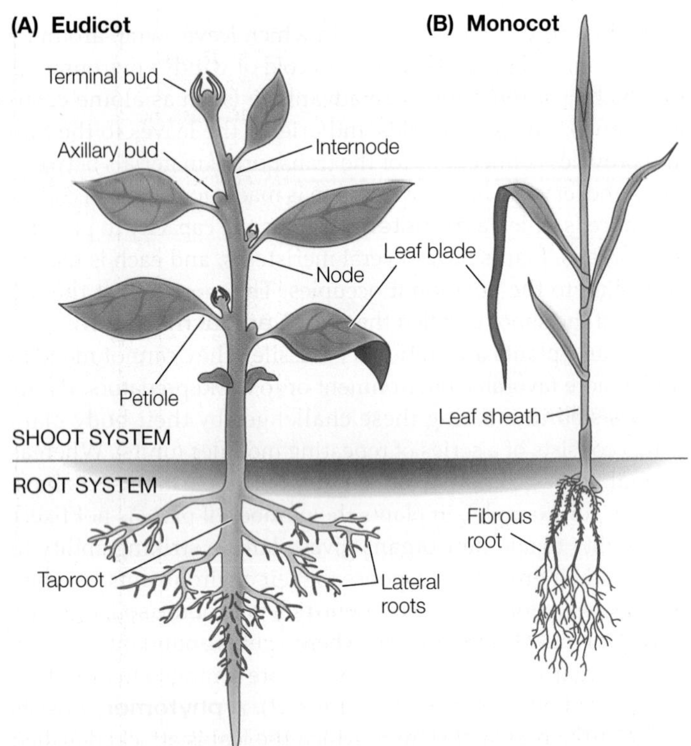

FIGURE 23.2 Eudicots and Monocots Typically Differ in Their Root Architecture **(A)** Eudicots typically have a taproot system with a single primary root that gives rise to lateral roots. **(B)** Monocots frequently have a fibrous root system.

the tip of a plant sometimes causes more than one axillary bud to grow out, with the result that some plants whose branches are nibbled by rabbits or pruned by gardeners become fuller or bushier over time.

The **root system** anchors the plant in place, absorbs water and dissolved minerals, and stores the products of photosynthesis. The extreme branching of roots and their high surface area-to-volume ratios allow them to absorb water and mineral nutrients efficiently. For example, the root system of a 4-month-old winter rye plant (*Secale cereale*) was found to be 130 times larger than the shoot system, with almost 13 million branches that had a cumulative length of more than 500 kilometers! The large size of root systems gives them an enormous surface area for absorbing water and nutrients from the soil.

The repeating units that make up a root system are called **lateral roots**, and a **root meristem** is located at the tip of each one. The meristem at the apex of the root that first emerged from the seed is also a root meristem, and is sometimes called the **root apical meristem**.

Most flowering plants belong to one of two major clades. **Monocots** are generally narrow-leaved plants such as grasses (including corn, wheat, and rice), lilies, and orchids. **Eudicots** are broad-leaved plants such as soybeans, roses, sunflowers, and maple trees. These two clades, which account for 97 percent of flowering plant species, differ in several basic characteristics. For example, their root systems are distinctly different in morphology (**FIGURE 23.2**). In eudicots, the root system is established by the primary root that emerged from the seed, called a taproot. This root can branch, forming lateral roots. In monocots, the primary root is short-lived and most of the roots arise from stem tissues above the initial root, resulting in what is called a fibrous root system. Monocot roots often form dense networks near the soil surface, binding the soil together. The fibrous root systems of

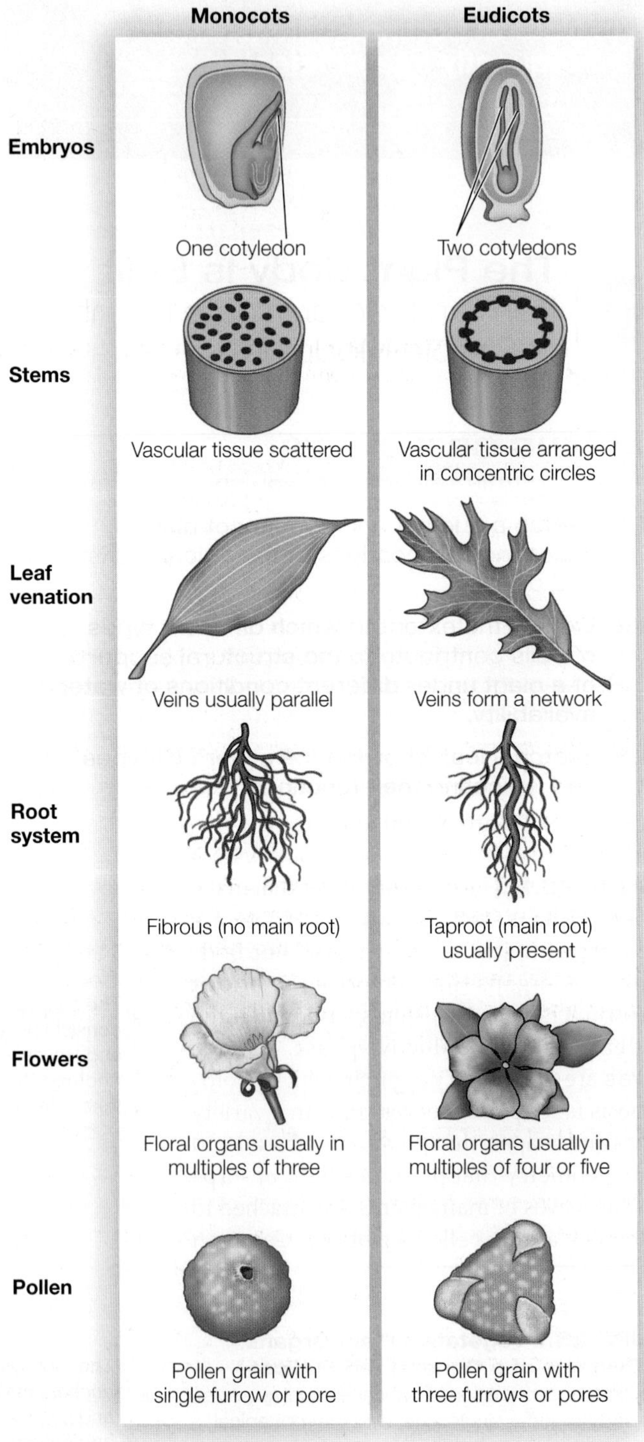

FIGURE 23.3 Comparing the Two Major Clades of Flowering Plants There are several structural differences between monocots and eudicots.

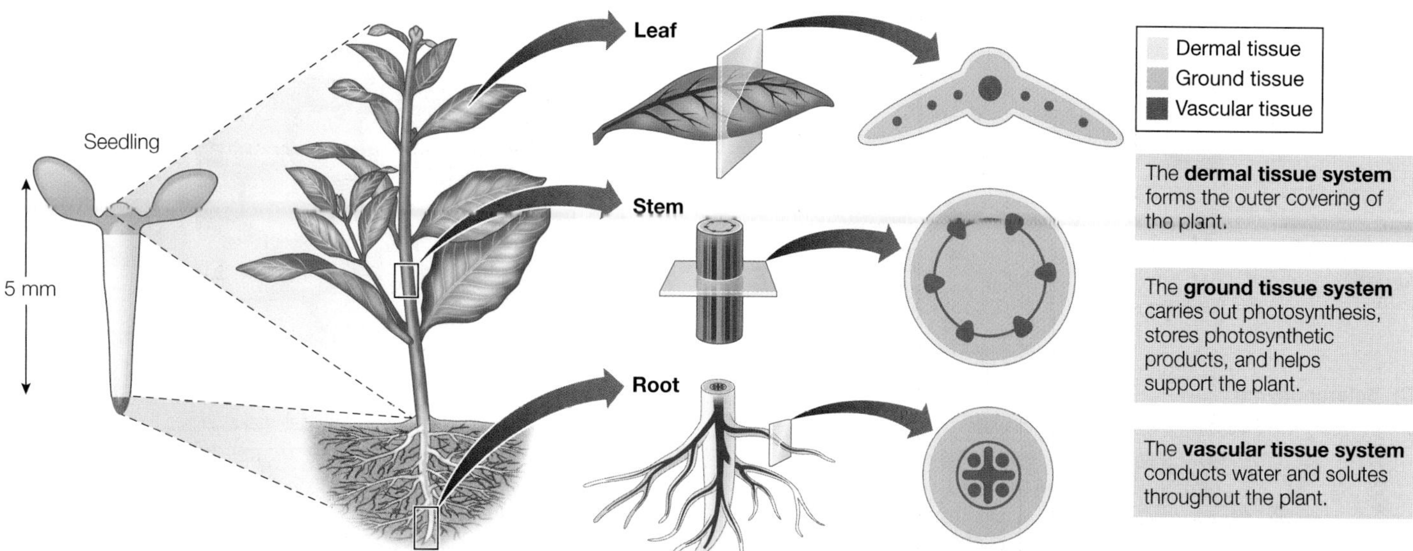

FIGURE 23.4 Three Tissue Systems Extend throughout the Plant Body The arrangement of dermal, ground, and vascular tissue shown here is typical of eudicots, but the three tissue systems are continuous in the bodies of all vascular plants.

grasses often protect steep hillsides where runoff from rain would otherwise cause erosion. **FIGURE 23.3** provides a more extensive list of differences between eudicots and monocots.

Pine trees, such as the one shown at the opening of this chapter, are gymnosperms, not flowering plants. Like flowering plants, gymnosperms have root and shoot systems that arise from meristems. Key Concept 20.4 describes how gymnosperms are classified separately based on differences in how they produce seeds.

Plants are composed of three tissue systems

Plant organs are organized into three tissues: dermal tissue, ground tissue, and vascular tissue (**FIGURE 23.4**). These tissues first arise in the embryo and persist throughout the life of the plant. The **dermal tissue system** develops in the plants' outermost layer of cells, called the epidermis. The **ground tissue system** comprises most of the tissue between the dermal tissue and the vascular tissue, so makes up most of the plant body. The **vascular tissue system** is the plant's plumbing, or transport, system. The presence of the vascular tissue system is the distinguishing feature of vascular plants.

Tissues are composed of cells

The cells that make up the tissue systems can be classified in many ways. One scheme is based on the types of cell walls that the cells produce. Unlike animals, plants have cell walls made of cellulose (a polysaccharide). As plants have neither an internal skeleton nor a hard exoskeleton, their cell walls serve the important structural function of holding the plant upright. Overly strong walls, however, can prevent cell elongation. Each species must strike a balance between thin walls, which permit growth, provide flexibility, and conserve resources, and thicker walls that provide more mechanical support. Different plants meet this challenge in different ways, involving three types of plant cells: parenchyma, collenchyma, and sclerenchyma (**FIGURE 23.5**).

Parenchyma cells are the most abundant cells in a plant and have large vacuoles and relatively thin cell walls. They perform a wide variety of functions, including photosynthesis (in the shoot), transport, synthesis and secretion of metabolites, and storage of protein in seeds and of starch in roots. Because their walls are thin, parenchyma cells depend on their water content for firmness: they are firm when they are **turgid** (full of water), somewhat like a water balloon. Parenchyma cells have **primary cell walls**, meaning that cell wall material is added only while the cell is expanding.

Collenchyma cells are elongated and have unevenly thickened cell walls. They are firm enough to provide support for growing tissues such as stems, yet flexible enough to permit growth. When a plant becomes dry, cells lose their turgor and wilting occurs in portions of the plant (generally the growing tips) that are composed of parenchyma and collenchyma. The familiar "strings" in celery stalks consist primarily of collenchyma cells. Like parenchyma cells, collenchyma cells have primary cell walls.

Sclerenchyma cells have thick walls reinforced with the polymer **lignin**. They form a primary cell wall when they are elongating, and then, after cell elongation stops, they add additional layers of material to the wall, forming a **secondary cell wall**. Most sclerenchyma cells then undergo programmed cell death (apoptosis; see Key Concept 7.5), but their strong cell walls remain and provide support for the plant even when the cells are not full of water. Sclerenchyma cells come in several types. **Fibers** are often organized into bundles and provide relatively rigid support to wood, bark, and other parts of the plant. **Sclereids** may pack together densely, as in a nut's shell or in some seed coats.

The three types of plant cells are distributed across the three tissue types. For example, the dermal tissue contains all three types of cells. To see how the different types of cells come together to form tissues and organs, we'll discuss the cells that make up the vascular tissue system, and then describe how cells and tissues give rise to the leaf.

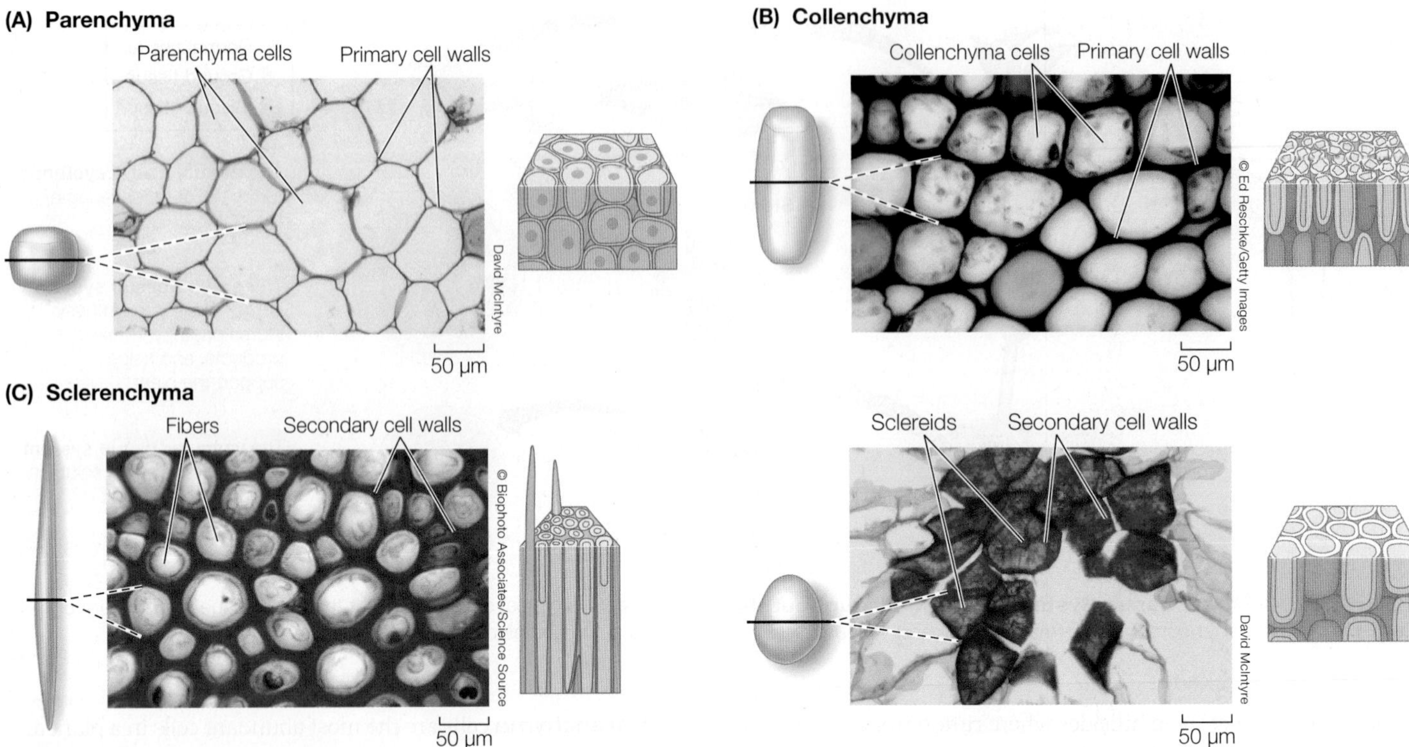

FIGURE 23.5 Ground Tissue Cell Types **(A)** Parenchyma cells in the petiole of coleus. Note the thin, uniform cell walls. **(B)** Collenchyma cells make up the outer cell layers of this spinach leaf vein. Their walls are thick at the corners of the cells and thin elsewhere. **(C)** Sclerenchyma cells can be classified as fibers or sclerids. Left: Fibers in a sunflower stem. The thick secondary walls are stained red. Right: Sclereids. The extremely thick secondary walls of sclereids are laid down in layers. They provide support and a hard texture in structures such as nuts and seeds.

Vascular tissue contains cells specialized for the transport of fluids

The vascular tissue system contains two types of tissues: **xylem**, which distributes water and mineral ions taken up by the roots to all the cells of the roots and shoots, and **phloem**, which consists of several types of living cells with various functions (**FIGURE 23.6**). Xylem is a complex tissue containing two types of water-conducting cells—tracheids and vessel elements—which, like most sclerenchyma cells, are dead when mature. These lignified (lignin-containing) cells also provide structural support for the

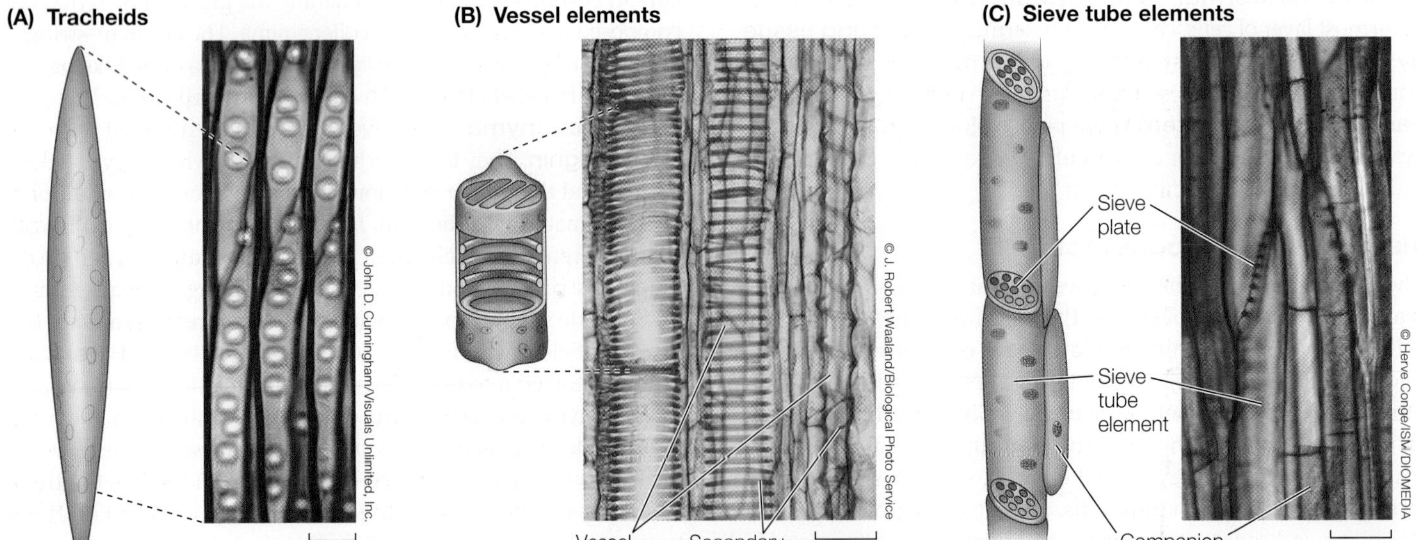

FIGURE 23.6 Vascular Tissue Cell Types **(A, B)** Tracheary elements of xylem. The thick cell walls are stained dark red. **(A)** Tracheids in pinewood. **(B)** Vessel elements in the stem of a squash. **(C)** Phloem sieve tube elements and companion cells in the stem of a cucumber.

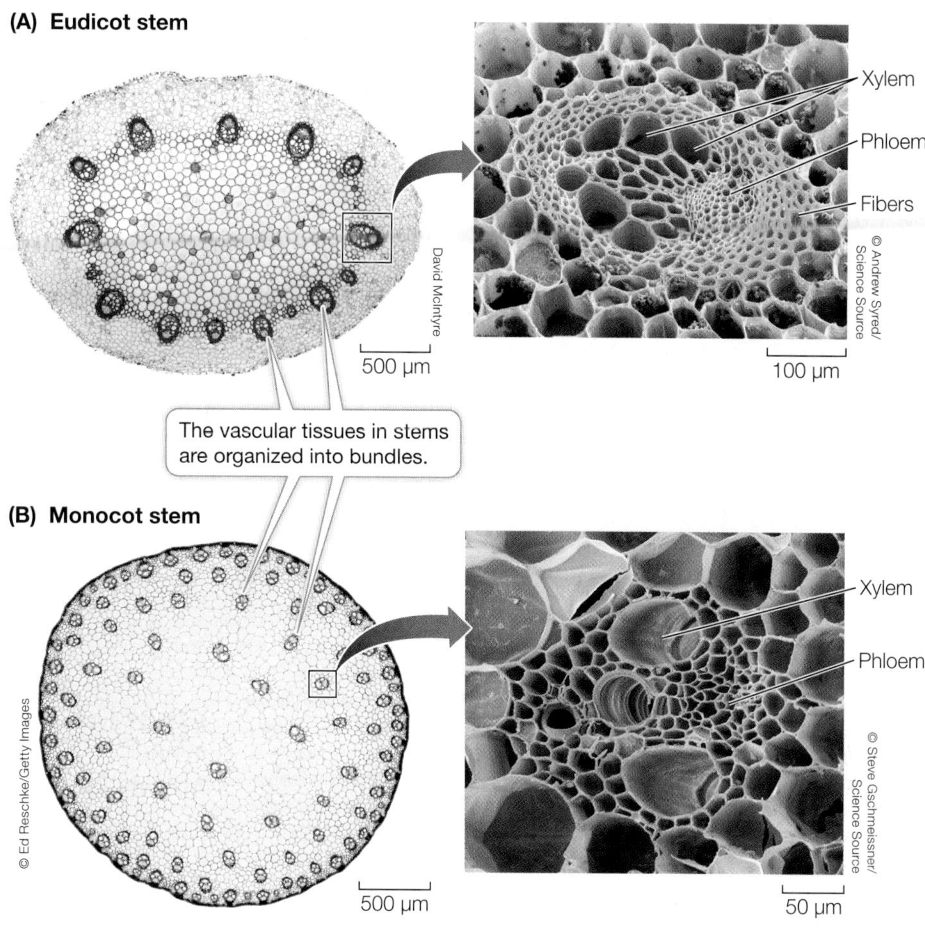

FIGURE 23.7 Stem Cross Sections Showing the Cells and Tissues They Comprise **(A)** In herbaceous and young woody eudicot stems, the bundles of vascular tissue (vascular bundles) are arranged cylindrically and set within the ground tissue. The outermost layer of the stem is the epidermis. **(B)** A scattered arrangement of vascular bundles is typical of monocot stems.

Activity 23.1 Eudicot Stem

Activity 23.2 Monocot Stem

plant. **Tracheids** are spindle-shaped cells, with thinner regions in the cell wall called pits through which water can move with little resistance from one tracheid to its neighbors. **Vessel elements** are larger in diameter than tracheids. They meet end-to-end and partially break down their end walls, forming an open pipeline (called a vessel) for water conduction. The xylem tissue also contains living parenchyma cells that regulate the movement of ions into the conducting cells, and fibers that provide structural support.

Phloem tissue generally consists of several types of living cells. Phloem transports carbohydrates (primarily sugars) from sites where they are produced (called **sources**) to sites where they are used or stored (called **sinks**). Sources are often photosynthetic organs (such as leaves) but also include nonphotosynthetic storage organs (such as tubers and seeds). Sinks include growing tissues, roots, and developing flowers and fruit. The carbohydrate-conducting cells of the phloem are called **sieve tube elements**, which, like vessel elements, meet end-to-end, forming sieve tubes that are connected by a set of pores called a **sieve plate**. Unlike vessel elements, the sieve tube elements are parenchyma cells. Although still alive, mature sieve tube elements have lost much of their cellular contents. Adjacent, fully functional **companion cells** are connected to the sieve tube elements by plasmodesmata and perform many of the phloem's metabolic functions. Phloem fibers, sclerenchyma cells that provide support for the plant, can also be present. **FIGURE 23.7** shows the arrangement of xylem and phloem tissues within the stem.

Leaves are photosynthetic organs containing all three tissue systems

An example of how cells and tissues come together to form an organ is found in the leaf. Leaf anatomy is beautifully adapted to carry out photosynthesis and to support that process by absorbing light, exchanging the gases O_2 and CO_2 with the environment, limiting evaporative water loss, and exporting the products of photosynthesis to the rest of the plant. **FIGURE 23.8** shows a section of a typical eudicot leaf in three dimensions. Virtually the entire leaf is covered on both its upper and lower surfaces by a layer of nonphotosynthetic epidermal cells. These cells secrete a waxy cuticle that is impermeable to water. Although the epidermis keeps water in the leaf, it also keeps out CO_2 that is needed for photosynthesis. Water vapor and gases are therefore exchanged through adjustable pores in the leaf called **stomata** (singular *stoma*). A pair of specialized cells called **stomatal guard cells** lies on either side of each stoma and controls the extent to which the pore is open (you'll learn more about how stomata open and close in Key Concept 24.4).

The leaf is elegantly structured for the reactions of photosynthesis. For more detail on photosynthesis and the role of leaf tissues, see **Key Concept 5.5**.

The ground tissue in most eudicot leaves includes two zones of photosynthetic parenchyma tissue called **mesophyll** (which means "middle of the leaf"). Cells in the upper layer(s) of the mesophyll are called the palisade mesophyll. **Palisade cells** typically contain many chloroplasts, which capture light. These cells are positioned just under the epidermis, where light is generally most abundant, and are long and narrow with slight spaces between the cells that gives them a large surface area for gas exchange. In contrast, the

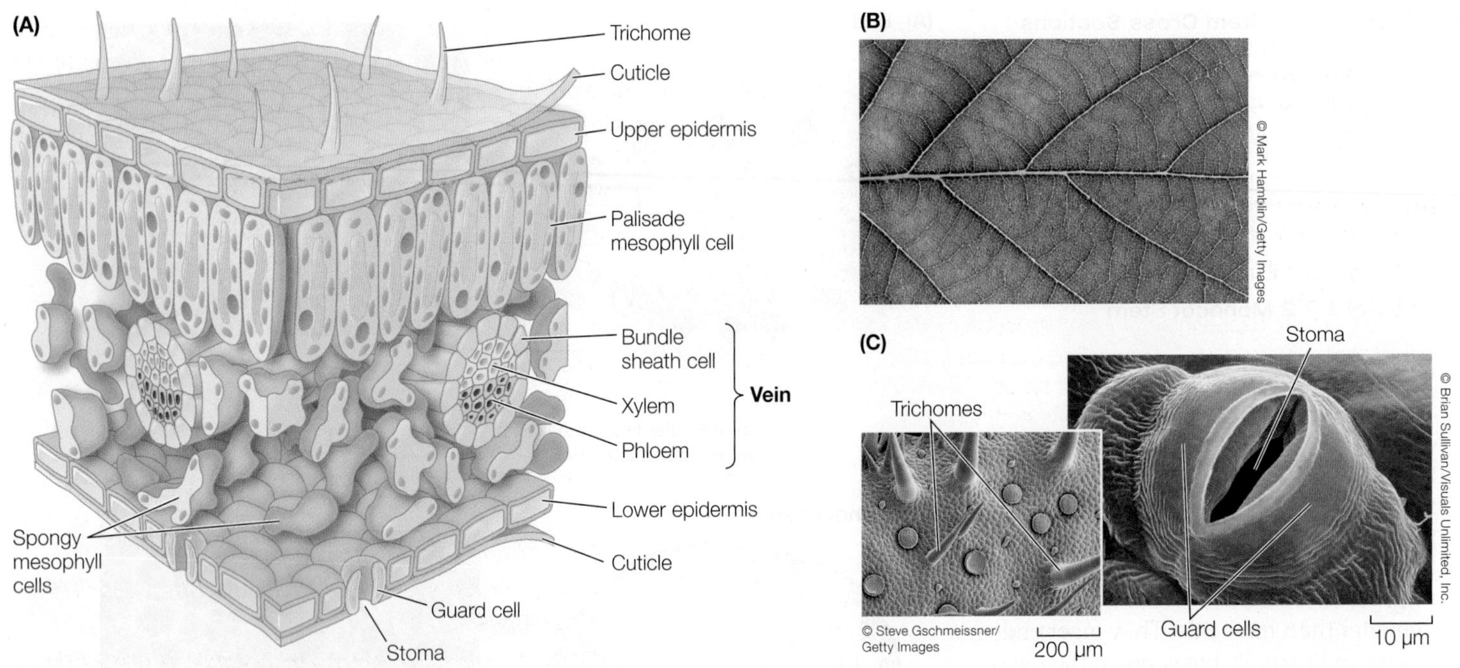

FIGURE 23.8 Eudicot Leaf Anatomy (A) This three-dimensional diagram shows a section of a eudicot leaf. **(B)** The network of fine veins in this maple leaf carries water to the mesophyll cells and carries photosynthetic products away from them. **(C)** Carbon dioxide enters the leaf through stomata, as on the epidermis of this eudicot leaf. Trichomes are epidermal hairs that can break up airflow, reducing loss of water vapor from the leaf when stomata are open and gas exchange is taking place.

What type of cell provides the leaf with the most structural support?

Activity 23.3 Eudicot Leaf

lower layer(s) of the leaf consist of irregularly shaped cells called **spongy mesophyll**. The irregular shape of the spongy mesophyll produces many points at which air and water are located next to one another. These junctions increase light scattering, and thus cause photons that were not absorbed in the upper layers to bounce around in the lower layers, giving the light more chances to be absorbed before exiting the leaf. Perhaps because of this, plants that grow in low light conditions tend to have a higher ratio of spongy to palisade mesophyll than do plants that grow in high light conditions. Vascular tissue is present throughout the leaf, forming a network of veins that extend to within a few cell diameters of all the cells of the leaf (see Figure 23.8B), ensuring that the mesophyll cells are well supplied with water and mineral nutrients by the xylem tissue, and adding mechanical support. The products of photosynthesis are loaded into the phloem tissue of the veins for export to the rest of the plant.

Pine needles differ from eudicot leaves in several ways, including the presence of features that adapt them to life in dry environments: a lower surface-to-volume ratio, a thick cuticle, and stomata that are recessed (sunken) below the epidermis (**FIGURE 23.9**). Sunken stomata decrease water loss because the moist air just above the stomatal opening is better retained (and thus less quickly blown away by the wind), decreasing the concentration gradient that drives water loss. Chloroplasts are present in the mesophyll tissue surrounding the stomatal openings. Vascular tissue (xylem and phloem) are located in a bundle in the center of the needle.

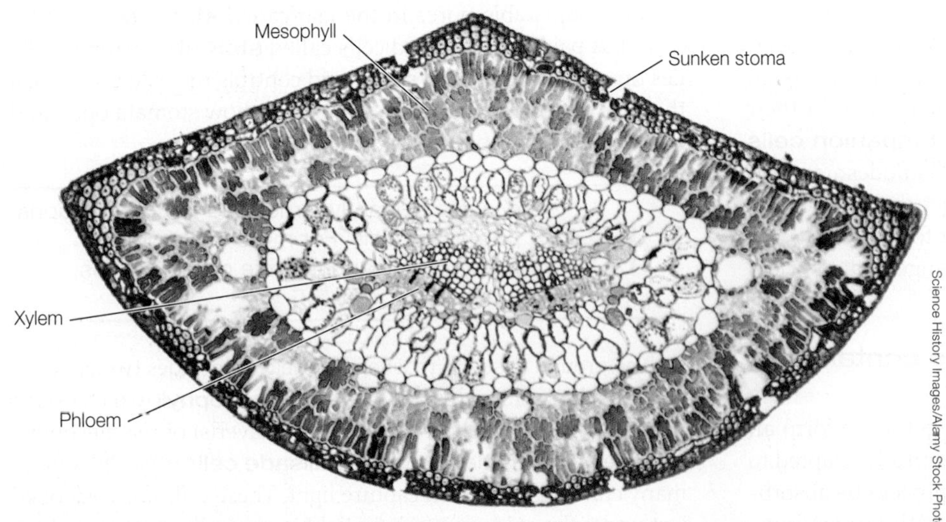

FIGURE 23.9 Anatomy of a Pine Needle Pine needles are adapted for water conservation by having a relatively small amount of surface area/volume, a thick epidermal layer, and sunken stomata.

REVIEW & APPLY | 23.1

R Flowering plants contain organs (e.g., roots, leaves, and, flowers) that are arranged into a root system and a shoot system. Plant meristems add to these systems by generating new organs throughout the plant's lifetime. Plant organs are composed of dermal, ground, and vascular tissues. These tissues are composed of specific cell types (parenchyma, collenchyma, and sclerenchyma) that are categorized primarily by the properties of their cell walls. The vascular tissue system is composed of xylem and phloem. The leaf contains dermal, ground, and vascular tissues built from specialized cell types, the combined properties of which promote photosynthesis.

A
1. Describe situations in which the mechanical support provided by collenchyma cells would be more advantageous than that provided by sclerenchyma cells, and vice versa.
2. Describe the types of cells in the vascular tissue system through which water and sugars are transported.
3. Give examples of tissues found in the leaf and explain how each is adapted to promote photosynthesis.

The modular structure of the plant body together with the ability to create additional subunits gives plants great flexibility in responding to the environment in which they grow. This ability to modulate the body plan in response to current conditions is facilitated by plants' meristem-based growth. How meristems function is described in the next Key Concept.

23.2 | Apical Meristems Build the Primary Plant Body

LEARNING OBJECTIVES

By the end of this key concept you should be able to:

23.2.1 **Explain how the shoot apical meristem gives rise to the shoot and its tissue systems.**

23.2.2 **Describe the structure and function of the root meristem.**

23.2.3 **Give examples of plants that achieve distinctly different forms by changing the relative shape, size, or number of their organs.**

Even before a seed germinates, the plant embryo inside the seed has **apical meristems** at the tips of the shoot and root (**FIGURE 23.10**). These meristems give rise to the rest of the plant directly

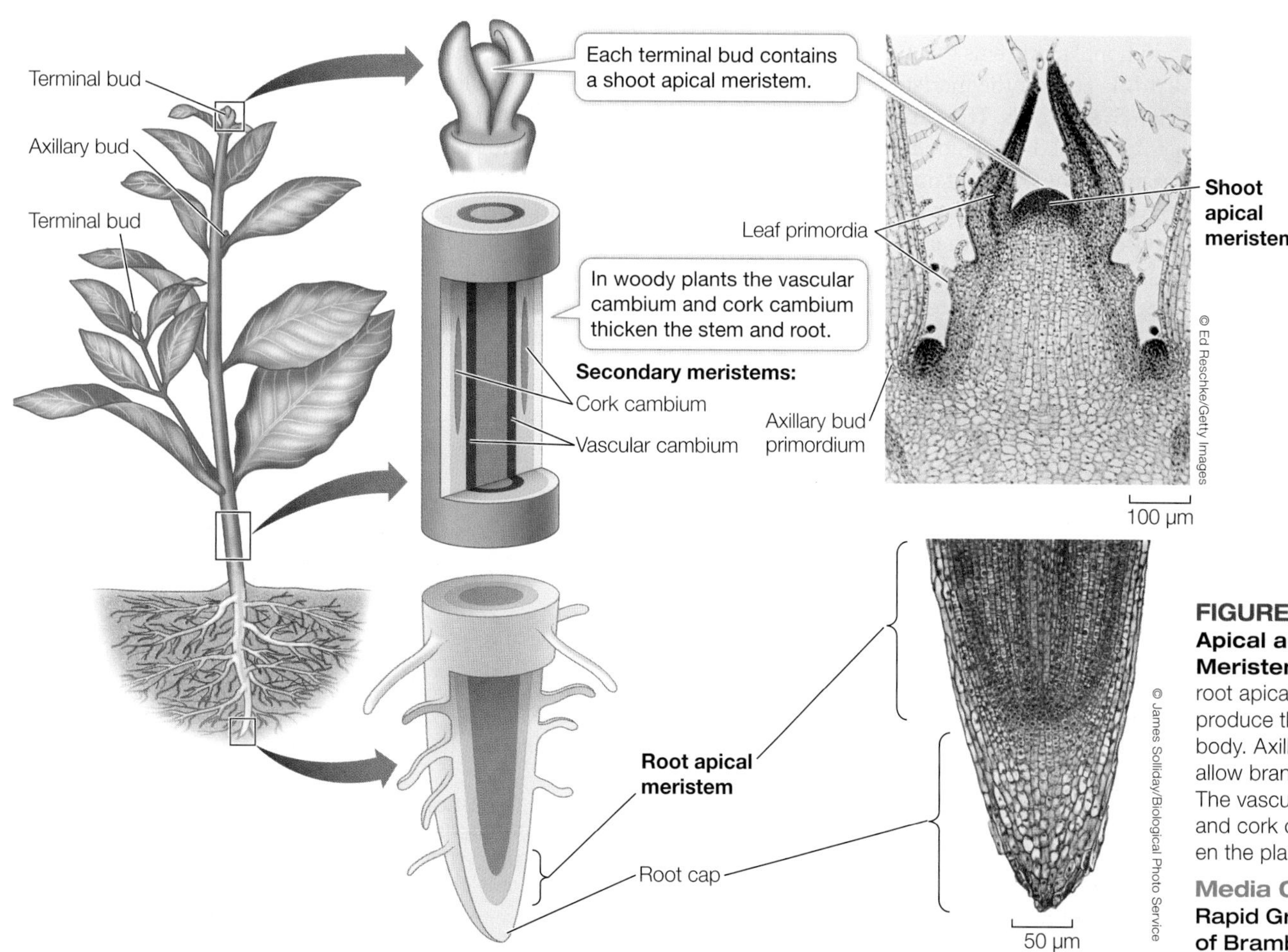

FIGURE 23.10 **Apical and Lateral Meristems** Shoot and root apical meristems produce the primary plant body. Axillary meristems allow branching to occur. The vascular cambium and cork cambium thicken the plant.

Media Clip 23.1 **Rapid Growth of Brambles**

FIGURE 23.11 Variegated Plants Have Been Used to Study Meristem Function This variegated plant carries a mutation that prevents synthesis of chlorophyll in some of its cells, leaving them white. Cells that lose their ability to make chlorophyll pass this trait on to all their descendants.

or through the creation of additional meristems that further elaborate the body plan. The dividing cells of the meristems serve as a reservoir for the production of new cells. They are comparable to animal stem cells in that upon cell division, one daughter cell is capable of differentiation, while the other retains its undifferentiated phenotype. Because plants can continue making organs throughout their lives, they are said to have an **indeterminate** body plan. This is in contrast to animals such as ourselves which have a fixed number of organs and hence a **determinate** body plan. Where animals might respond to an environmental challenge by changing their behavior, such as seeking shelter from the snow or sunlight, plants often respond to environmental challenges by changing the structure of their bodies.

The products of the shoot apical meristem become shoot tissues

The shoot apical meristem is located at the very top of a stem. This meristem can be called a **primary meristem** because the cells and tissues that arise from it form the primary plant body. This is in contrast to **secondary meristems**, discussed in Key Concept 23.3, which increase the width of the plant, as occurs when a tree trunk expands radially. The shoot apical meristem is typically dome-shaped and surrounded by leaves (see Figure 23.10). Cells in the center of the meristem divide infrequently, while cells around the edge divide more often. Division of cells along the sides of the meristem produces new cells that form bulges. Each bulge, called a leaf **primordium**, can form a leaf. (The word *primordium* comes from Latin and refers to the beginning, here, of an organ.) As the leaf forms, cells near the center of the meristem divide, replacing the cells that contributed to the new leaf and so maintaining the size of the meristem.

People learned more about how cells in the shoot apical meristem generate the adult body by examining plants in which some, but not all, of the cells have mutations that prevent them from making chlorophyll. If a cell in a wild-type leaf suffers a mutation that prevents it from making chlorophyll, the cell will be white, as will all the daughter cells derived from it (**FIGURE 23.11**). Because plant cells do not move past one another, the white patches serve as markers, allowing one to trace the path by which cells in the meristem gave rise to different parts of the adult plant. Using this type of analysis, called **clonal analysis**, it was determined that the shoot apical meristem has three cell layers. The outermost layer (layer 1, L1) produces the dermal layer. Layer two (L2) produces ground tissue. Vascular tissue arises from the third layer (L3). These layers are maintained because the cell divisions that take place in any one of the layers are typically oriented such that they add a new cell to that layer (**FIGURE 23.12**).

This type of analysis also allows one to determine the extent to which meristematic cells maintain themselves in their positions—an important factor in thinking about how a meristem persists over time. Cells occasionally divide in directions that interrupt the three-layer pattern. If a white L1 cell in an otherwise green plant (green L2 and L3) divides parallel to the surface of the meristem, it creates an extra layer. Counting from the outside, this section of the plant then has a white L1, a white L2 (the result of the recent atypically oriented cell division), and a green L3. This sort of change affects the appearance of the leaf because there is a tendency for the edges of leaves to contain cells derived from the L1 and L2 layers, while the center of a leaf is composed of cells from all three layers. Our example plant with its newly formed white L1, white L2, and green L3 would suddenly develop white edges on the leaves and perhaps a lighter shade of green in the center of the leaves. If a white L2 layer subsequently divided such that it gave rise to a white L3, subsequently formed leaves would be completely white. (The L1 is generally transparent because chlorophyll is produced only in the stomata). Through studies such as

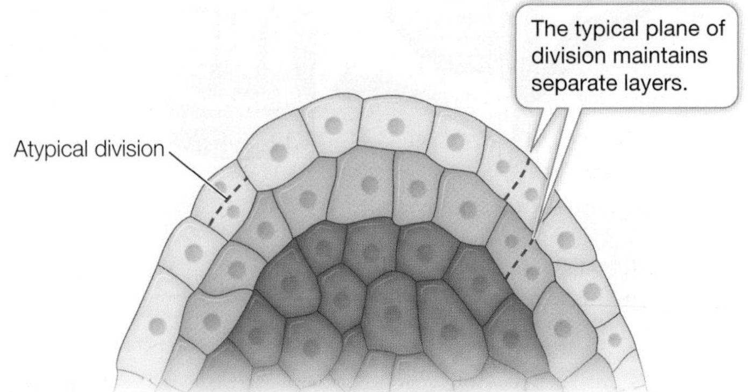

FIGURE 23.12 The Shoot Apical Meristem Is Composed of Three Layers of Cells Cells in each layer typically divide in a plane that maintains the separate layers. Occasional divisions in other planes can result in cells originally of one layer "invading" the next layer, as shown.

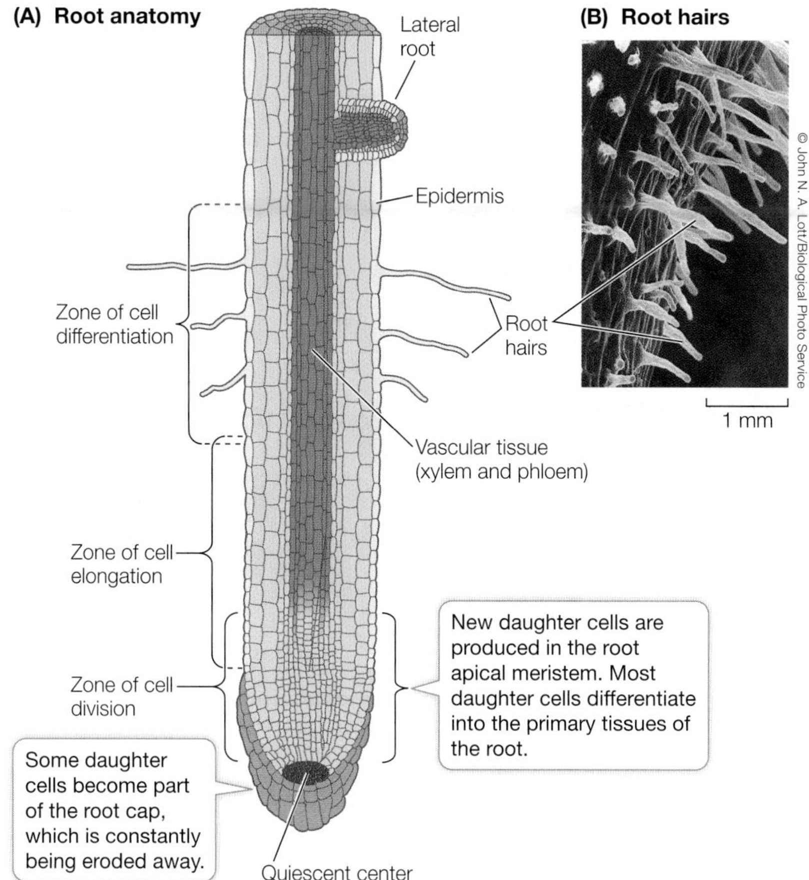

FIGURE 23.13 Tissues and Regions of the Root **(A)** Diagram of a longitudinal section of a root, showing how cell division creates the complex structure of the root. **(B)** Root hairs, seen with a scanning electron microscope (SEM).

The vasculature in lateral roots connects with the vasculature in the primary root. Why is this advantageous?

Media Clip 23.2 **Cell Expansion in Root Hairs**

these, researchers discovered that although cells in the meristem generally give rise to particular parts of the plant, there is some flux between cell lineages.

After cells of the shoot apical meristem divide, they can expand. Expansion of cells that lie underneath the meristem causes the shoot apical meristem to be pushed upward. This is how the stem elongates. Because the shoot apical meristem always remains on top, the ages of the cells in the stem progress from the youngest (newly formed) near the meristem to the oldest, most mature cells near the root.

The root meristem gives rise to the root cap and the primary root

The root meristem is organized around a small group of cells called the quiescent center (**FIGURE 23.13**). The cells in the quiescent center rarely divide. Surrounding these cells in a shape much like that of a donut are a set of initial cells that divide, giving rise to the rest of the cells in the root.

Initial cells on the rootward side of the meristem (i.e., those closest to the root tip) contribute to a **root cap**, which protects the delicate growing region of the root as it pushes through the soil. The anatomy of the root cap resembles a sock, pulled over the tip of the root. The root cap secretes a mucopolysaccharide (slime) that acts as a lubricant. Even so, the cells of the root cap are often damaged or scraped away and must therefore be replaced constantly. The root cap also detects the pull of gravity and thus dictates the directional growth of roots.

Initial cells on the shootward side of the meristem produce cells that form the main body of the root. When an initial cell divides, the daughter cell that is adjacent to the quiescent center remains an initial cell, while the daughter that is farther shootward enlarges, differentiates, and adds to the body of the root. Division of the epidermal initial gives rise to the dermal layer. The ground tissue arises from division of initial cells that lie interior to this, and the vascular tissue arises from division of initial cells located in the center of the root.

Because plant cells do not move past one another, the distance between a cell and the quiescent center is related to the length of time since the cell was created. This leads to several distinct zones in the root. The **meristematic zone** is the region close to the root tip in which cell division takes place. Just shootward of this zone is the **zone of cell elongation**, where the newly formed cells elongate and thus push the root tip farther into the soil. Shootward of that zone is the **zone of cell differentiation** (also called the maturation zone), where the cells take on their mature forms. **FIGURE 23.14** shows the different cell types contained in the mature portion of eudicot and monocot roots.

The formation of lateral roots (root branches) frequently begins in the **pericycle**, a ring of cells that surrounds the xylem and phloem. These cells retain some meristematic-like abilities and can contribute to thickening some roots. Division in the pericycle gives rise to a small group of cells called a lateral root primordium that pushes cells in the cortex and epidermis apart to reach into the soil. Some cells in the interior of the lateral root primordium differentiate into vascular cells that connect with the primary root, creating a continuous passageway for nutrients taken up from the soil to the rest of the plant. Mature lateral roots are highly similar to the primary root and have the capacity to form additional lateral roots from their own pericycle cell layers.

Some cells in the outermost layer (dermal tissue) of both primary roots and lateral roots form thin projections called **root hairs**. Root hairs greatly increase the surface area of the root system and so promote uptake of water and soil nutrients.

Diversity of form is generated by altering the relative shape, size, or number of basic modules

Differences in the relative shape, size, or number of the simple modules that make up a plant—roots, stems, leaves, meristems, and relatively few tissue and cell types—are responsible for the diversity in flowering plants that cover our planet. A single species, *Brassica oleracea* (wild mustard), is the ancestor of many morphologically

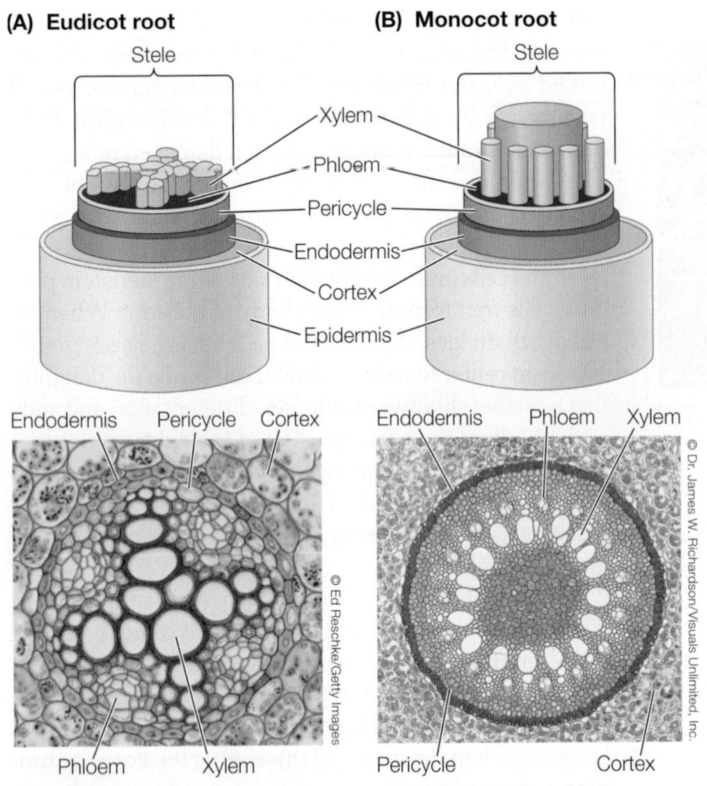

FIGURE 23.14 Products of the Root Meristem The arrangement of tissues in the stele (where the vascular tissue is) differs in eudicot roots **(A)** and monocot roots **(B)**. The photomicrographs show cross sections of the stele of a representative eudicot (buttercup, Ranunculus) and a representative monocot (corn, *Zea mays*), showing the arrangement of the primary root tissues.

Activity 23.4 Eudicot Root

Activity 23.5 Monocot Root

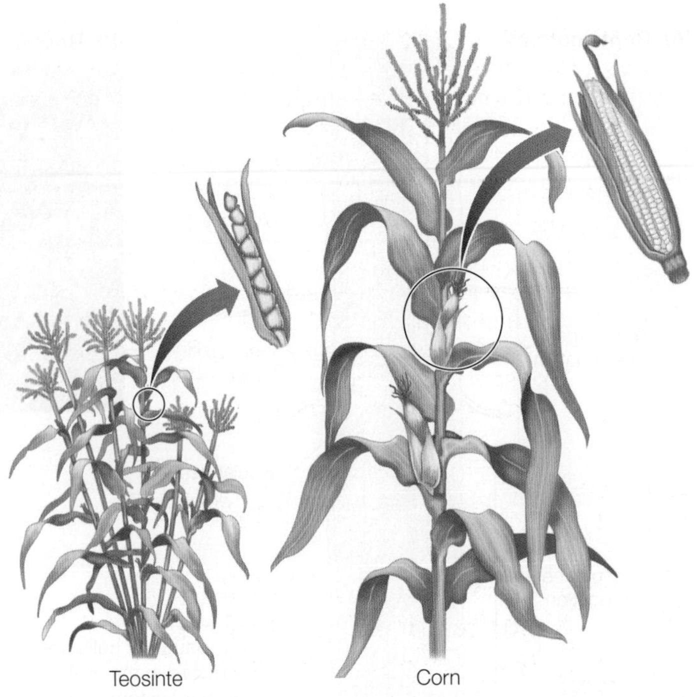

FIGURE 23.15 Corn Was Domesticated from the Wild Grass Teosinte Each teosinte plant has multiple branches. Beginning more than 8,000 years ago in Mexico, farmers favored plants with minimal branching. Reducing the number of branches results in fewer ears per plant but allows each ear to grow larger and produce more seeds.

diverse crops, including kale, broccoli, brussels sprouts, and cabbage (see Figure 13.4). From a genetic perspective, this observation suggests that minor differences in genomes or gene regulation can underlie dramatic differences in plant form. For example, corn was domesticated from the wild grass teosinte, which still grows in the hills of Mexico (**FIGURE 23.15**). One of the most conspicuous

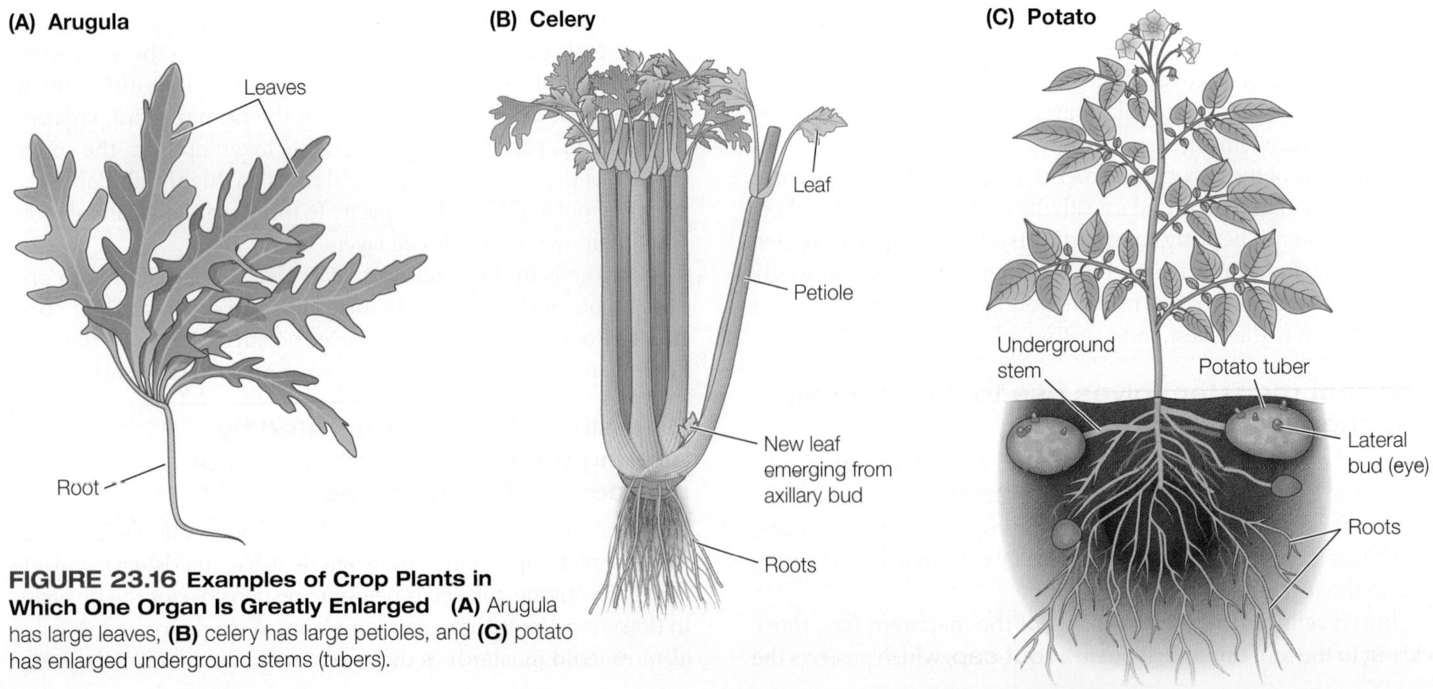

FIGURE 23.16 Examples of Crop Plants in Which One Organ Is Greatly Enlarged (A) Arugula has large leaves, **(B)** celery has large petioles, and **(C)** potato has enlarged underground stems (tubers).

differences between teosinte and domesticated corn is that teosinte, like other wild grasses, is highly branched, with many shoots, whereas domesticated corn has a single shoot. This morphological difference is due in large part to the activity of a single gene called *TEOSINTE BRANCHED 1* (*TB1*). The allele of *TB1* in domesticated corn represses branching by repressing the outgrowth of axillary meristems, whereas the allele in teosinte permits branching.

LINK The diversity of plants derived from wild mustard illustrated in **Figure 13.4** demonstrates how artificial selection—the purposeful selection for specific phenotypes by humans—can produce rapid evolutionary change.

Dramatic differences in plant morphology can sometimes make it difficult to know which part of a plant is which. For example, you may be surprised to learn that a potato is the stem of a potato plant. Extreme enlargement of plant organs is frequently seen in crop plants (**FIGURE 23.16**).

A good way to go about identifying a plant organ is to consider the relationship of organs to one another. For example, **simple leaves** (which have a single blade and petiole) are differentiated from **compound leaves** (in which several repeating leaf–petiole units are attached to a central stalk) on the basis of whether an axillary bud is present at the base of the organ. Axillary buds are found at the junction between leaf and stem in both simple and compound leaves, but not at the junction

THINK LIKE A SCIENTIST

How can one identify the anatomical parts of a plant if they appear unfamiliar?

How would a plant biologist who wants to identify plant organs that don't resemble those of plants with which she is familiar go about this task? The process of science involved in this investigation could include the following reasoning:

1. Plant organs can often be differentiated by determining the relationship of the organ in question to other organs. Carefully observe the plant and try to identify organs in terms of their spatial relationships to one another.
2. Because new knowledge is built on prior understanding, identification of a plant organ may be facilitated by finding species with characteristics that are intermediate between the known and unknown.
3. Defer judgment until you have a body of data that can be used as the basis for your hypothesis. The goal of science is to continually refine our understanding. We can be our own worst enemies in this process, because (as psychologists tell us) people find it difficult to change their minds. Instead of leaping to a conclusion that may lead you down the wrong path, collect data on the question at hand. Let's see how this works in the following example.

Pick up a large yellow onion. Can you say how it correlates with the image of a green plant shown in Figure 23.1? It may help to know that onions are monocots and that leeks are a close relative. So identifying the organs of an onion may be easier if we look also at leeks. The ideal way to do this exercise is to have the fresh objects in your hands. If that's not possible, look at the images in **FIGURE 23.17**.

Locate the leaves of a leek (see Figure 23.19A). What color are the tips of the leaves? What color are the bases of the leaves? While color can be a useful cue, does it seem to be central in this instance? What other observations may be pertinent? The position in which organs are placed relative to one another is more important than their color in terms of determining which part(s) of a plant

(A)

(B)

FIGURE 23.17 Anatomy of two plants in the onion family **(A)** A leek. **(B)** An onion (left) and a leek (right) cut open to reveal the stem.

you are looking at. Peel back the leaves to find the stem. Roughly speaking, how many times longer than the stem are the leaves? Because flowering plants have a single root–shoot axis, the roots will be found attached to the stem. Where are the roots of the leek?

Now pick up an onion (see Figure 23.19B). Where are the leaves? Remember that the onion in the store is no longer fresh and the green coloration in the top of the leaves may no longer be visible. If the leaves are not obvious, can you find the roots? You know that flowering plants have a single axis, with roots forming from the base of the stem, so once you locate the roots, you should be able to find the stem. Peeling back the leaves to find the stem is difficult here, so cut the onion in half lengthwise and then peel away the fleshy bases of the leaves (see Figure 23.19B). As you learned from investigating the leek, the leaves arise from the stem, so as you peel away the onion leaves, the stem should become apparent. Having found your way around the onion, you are ready to answer the question:

1. What part of the onion do you typically eat? (The answer is not what most people would first guess!)

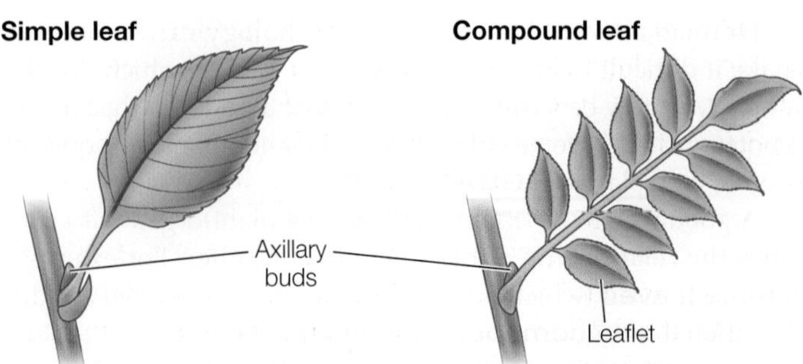

FIGURE 23.18 Simple versus Compound Leaves There is no axillary bud at the base of the leaflets of a compound leaf.

between leaflets (**FIGURE 23.18**). Identifying stems involves understanding that a plant stem serves as an axis around which new phytomeres can arise. Some types of turf (lawn) grass form dense mats of leaves that arise from the nodes of horizontal underground stems called **rhizomes**. A meristem located at the base of each leaf near the node allows these grass leaves to grow continuously—an exception to the general rule that leaf growth is determinate. Potato plants have rhizomes that become thickened at the ends, forming **tubers**. The eyes on potato tubers are axillary buds that, like any bud, can grow out to form a new shoot. Horizontal growth of a different kind is seen in the stems of strawberry plants, called runners or stolons, that extend aboveground for some distance before giving rise to a new set of leaves that appear as a vertically growing plant. Under some circumstances, plants may grow aboveground roots. These **adventitious roots** grow from tissues other than the root (**FIGURE 23.19**).

(A) Ivy

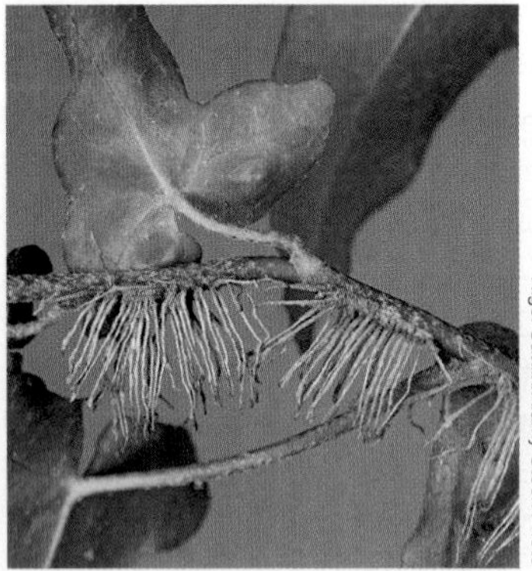

(B) Maize

FIGURE 23.19 Adventitious Roots Adventitious roots arise from parts of the plant other than the root, such as stems or branches. **(A)** Adventitious ivy roots help the vine attach to the surfaces on which it grows (climbs). **(B)** The prop roots of corn help stabilize the plant.

REVIEW & APPLY | 23.2

R The shoot and root apical meristems give rise to the entire post-embryonic plant. They are called primary meristems because their most direct impact is to increase the length (height) of the plant. The shoot apical meristem gives rise to the stem, leaves, and axillary meristems. The axillary meristems allow the shoot to branch. The root meristem gives rise to the tissues of the root. Some of the cells within the root (generally pericycle cells) can further divide, leading to formation of lateral root primordia. Lateral root primorida can mature into lateral roots, each of which contains its own meristem. Variation in the shape and organization of the plant body arises from modification of the relative size, shape, or number of a few basic units (stems, leaves, meristems).

1. (a) In which cell layer(s) would mutation generate the pattern of green and white seen on most of the plant in Figure 23.11? (b) What additional changes in the plane of cell division might have allowed the branch with all-white leaves to arise?
2. Why are turf grasses well suited for landscaping that involves frequent mowing?
3. Which part of the plant do we eat when we eat celery?
4. A research team investigated how long root cells continued to divide after being produced by the root apical meristem. They incubated germinating bean seeds for 2 hours with their roots suspended in water containing radioactive thymidine, which is taken up by dividing cells and incorporated into their DNA as it replicates (making the DNA radioactive). After this incubation period, the researchers removed some of the seeds (time 0 in the table) and took longitudinal slices from the roots (see Figure 23.13). They transferred the rest of the seeds to water containing nonradioactive thymidine and left them there for various times (12, 36, and 72 hours) before taking root slices. They tallied the number of radioactive cells found in each of the four regions of each root and compared those numbers with the total number of cells in each region. The table shows the results.

	Radioactive cells/total cells			
Time after transfer (hr)	**Root cap**	**Root apical meristem**	**Zone of cell elongation**	**Zone of cell differentiation**
0	0/63	90/115	0/32	0/22
12	2/58	95/110	16/29	0/27
36	36/60	45/119	14/30	12/31
72	12/24	0/116	2/31	30/32

a. At 0 hours, why did only root apical meristem cells contain radioactive thymidine?
b. For each region, calculate the percentage of radioactive cells and plot that percentage over time.
c. Explain the rise and fall in the percentage of radioactive cells in each region.

While the primary meristems give rise to the stem, leaves, and roots of a plant, they do little to increase its thickness. The manner in which trees and other woody plants increase in girth is discussed in the next Key Concept.

23.3 Many Eudicot Stems and Roots Undergo Secondary Growth

LEARNING OBJECTIVES

By the end of this key concept you should be able to:

23.3.1 Draw a cross section of a tree trunk showing primary and secondary xylem and phloem and the vascular cambium.

23.3.2 Illustrate how the vascular cambium contributes to growth of a tree trunk.

23.3.3 Compare the size of xylem cells produced during the spring versus the summer, and explain why the difference may be advantageous.

Most of the plants we have discussed are small, nonwoody ones that grow from their tips. But many plants—most trees, for instance—growth thicker as well as taller. The roots and stems of some eudicots develop a **secondary plant body**, the tissues of which we commonly refer to as wood and bark. These tissues are derived by secondary growth from two meristems called secondary (or lateral) meristems:

1. The **vascular cambium** produces cells that make up **secondary xylem** (wood) and **secondary phloem** (inner bark).
2. The **cork cambium** produces mainly waxy-walled protective cells near the exterior of the stem, supplying some of the cells that become outer bark.

Here we discuss how the vascular cambium leads to the increasing width of a tree trunk.

In the stem, the vascular cambium is initially a single layer of cells lying between the **primary xylem** and the **primary phloem** within the vascular bundles (**FIGURE 23.20**); the primary xylem and phloem are so named because they are produced by primary growth—before the formation of an active vascular cambium. Over time, the vascular cambium becomes a complete ring of cells, and typically divides such that it becomes several cell layers wide. A stem increases in diameter when the cells of the vascular cambium divide and expand. When the vascular cambium divides parallel to the outer edge of the stem, it produces two cells. If all of the cells that make up a single ring of vascular cambium divide in this way, a ring of cells is added to the plant—just as drawing a circle on a dart board adds an additional ring to the board. Frequently, the newly formed ring of cells is on the side of the vascular cambium closest to the center of the trunk. In this case, the new ring of cells closest to the center of the tree becomes part of the xylem, while the remaining ring of cells that is still in contact with the vascular cambium remains part of that meristem. Thus, secondary xylem has been produced while the vascular cambium has been maintained. On occasion, when the vascular cambium divides, the new ring of cells is located on the side of the vascular

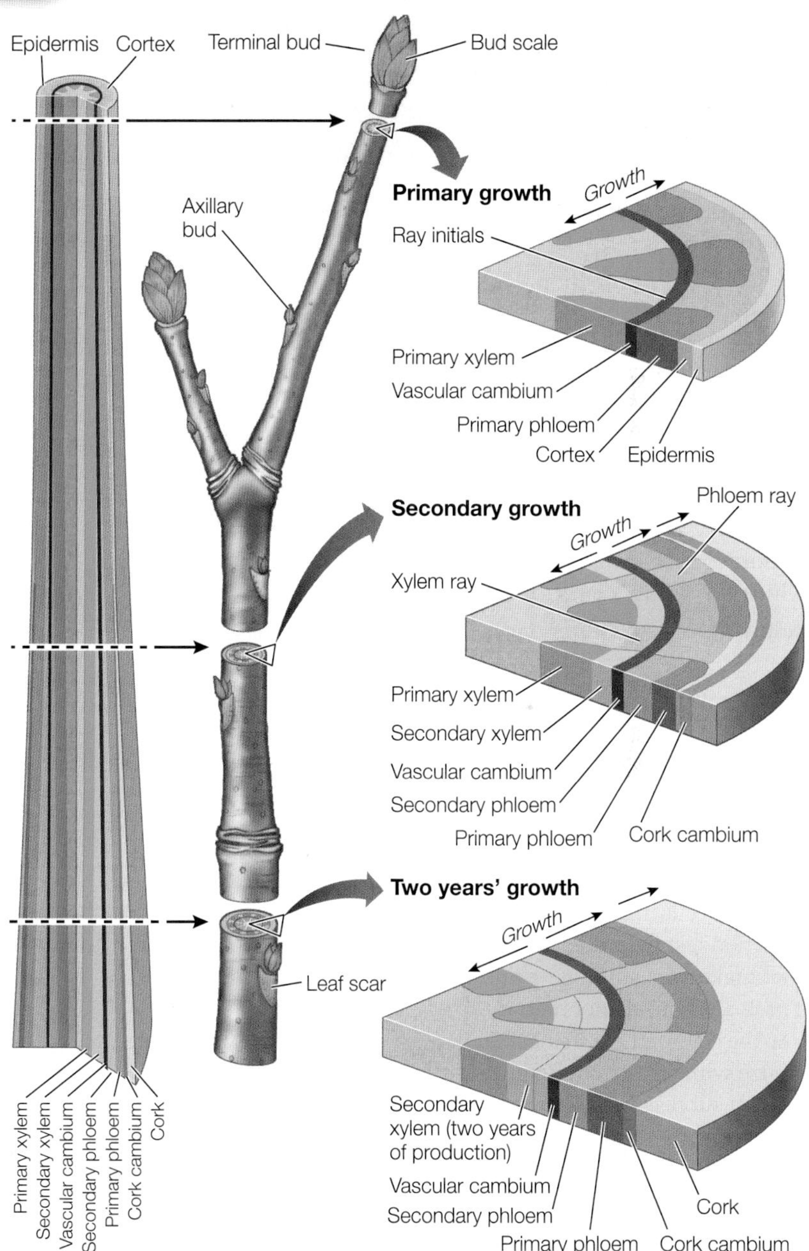

FIGURE 23.20 A Woody Twig Has Both Primary and Secondary Tissues The apical meristems in this dormant twig will produce primary growth in the spring. Lateral meristems are responsible for secondary growth. When the vascular cambium divides parallel to the outside of the stem, it forms a new ring of cells. New layers of cells on the interior side of the vascular cambium add to the xylem, while those exterior to the vascular cambium add to the phloem.

(A) Cross-section of tree trunk

(B) Spring and summer wood

Summer wood

Spring wood

Gary DeLong/Oxford Scientific/Getty Images

FIGURE 23.21 Annual Rings **(A)** Rings of secondary xylem are a noticeable feature of this cross section from a tree trunk. **(B)** Xylem cells that form in the early spring have a larger diameter than those that form later in the summer. This difference in diameter generates annual rings.

cambium closest to the bark. When this happens, the outermost ring of cells becomes phloem, while the remaining ring of cells remains part of the vascular cambium. Thus, the vascular cambium produces new layers of xylem and phloem, which then expand, further increasing the diameter of the stem. Secondary xylem cells are produced toward the inside of the stem, and secondary phloem cells toward the outside. Cells in the outer region of the secondary phloem may become cork cambium, which divides to form cells that mature into the outer bark.

Animation 23.1 Secondary Growth: The Vascular Cambium

The vascular cambium can also produce rays, groups of cells (often parenchyma) that extend from the xylem to the phloem (see Figure 23.20). Ray cells serve as sites where carbohydrates, frequently in the form of starch, can be stored.

The vascular cambium, which is a meristem, maintains itself and so persists for the life of the plant. On occasion, as the tree expands, cells in the vascular cambium divide at 90 degrees to the direction just described and thereby increase the number of cells that make up a single ring around the circumference of the tree. These divisions do not produce xylem or phloem—they simply allow the cells of the vascular cambium to remain roughly constant in size while the circumference of the tree trunk increases.

The successive layers of secondary xylem, which constitute the **wood**, contain many cell types. Some of these cells conduct water and minerals. It is important that new wood is produced continually, as the secondary xylem typically functions only for a few years. A given year's secondary phloem is active in transport of sugars only for that year; the sieve tube elements and companion cells, both of which are parenchyma cells, are crushed as the next year's layer of secondary phloem is formed to the inside and pushes outward.

Cross sections of most tree trunks (which are mature stems) from temperate-zone forests reveal annual rings in the wood (**FIGURE 23.21**), which result from seasonal variation in environmental conditions. Physics dictates that the amount of fluid that will flow through a given length of tubing depends on the difference in pressure at both ends of the tube and on the radius of the tube to the fourth power ($1/r^4$). Thus the amount of water that flows through the xylem conducting cells is expected to be profoundly influenced by the radius of the tube(s) through which it flows: large tubes, large flow; narrow tubes, low flow. (This same principle applies to the flow of blood through veins and capillaries in the human circulatory system; see Key Concept 30.6.) In spring, when water is relatively plentiful, the tracheids or vessel elements produced by the vascular cambium tend to be large in diameter and thin-walled with relatively thin secondary cell walls (although not as thin as the primary cell walls of parenchyma and collenchyma cells). This allows a mass of water to rapidly ascend the tree as leaves are formed during the spring. As water becomes less available in summer, narrower cells with thicker walls are produced, which makes this summer wood darker and sometimes denser than the wood formed in spring. Thus each growing season is recorded by a clearly visible annual ring. In wet years, more wide-diameter xylem conducting cells are produced than in drier years, so looking at tree rings allows us to infer something about the past history of precipitation. The rings of the bristlecone pines are particularly interesting because they give information on patterns of precipitation and drought over thousands of years.

Only eudicots and other non-monocot angiosperms, along with many gymnosperms, have a vascular cambium and a cork cambium and thus undergo secondary growth. The few monocots that form thickened stems—palms, for example—do so without secondary growth. Palms have a very wide apical meristem that produces a wide stem, and dead leaf bases add to the diameter of the stem. Flowering plants that lack secondary growth and many monocots, such as bamboo, accumulate large collections of fiber cells around each vascular bundle, which add significantly to the strength of the stem.

REVIEW & APPLY | 23.3

R Trees increase in width through the action of the vascular cambium, a secondary meristem that produces xylem to the inside and phloem to the outside. Xylem forms the wood of a tree. Because the xylem vessels formed in the spring are wider than those formed in the summer, trees that grow in temperate climates form annual rings. The secondary phloem, produced by the vascular cambium, forms the inner bark. The outer bark is formed from cells produced by the cork cambium.

A
1. Occasionally, trees can produce two annual rings in a single year. Can you think of a set of environmental conditions that might lead a tree to do this?
2. Compare and contrast the vascular cambium and the shoot apical meristem. What aspects of the vascular cambium define it as a meristem?

Apical and lateral meristems provide plants with the ability to grow in length, width, and number of branches, and thereby affect the form that the plant body acquires. The extent to which an individual is defined by the collection of these units is discussed in the next section.

23.4 In Natural Settings, Plants Are Frequently Interconnected

LEARNING OBJECTIVES

By the end of this key concept you should be able to:

23.4.1 Describe ways in which the stem of a plant can be connected with the stems of other plants of the same species.

23.4.2 Describe ways in which parts of different plant species can be interconnected.

It is intuitive to imagine that the many plants we see in a vast landscape exist as individuals—that the shoot system of a plant functions as a separate individual. What appear to be individual plants, however, are frequently interconnected. Interconnection occurs in many ways. Some plants, such as the turf grasses we mentioned earlier and bamboo, have rhizomes that connect the aboveground shoot systems. Other plants, such as aspen trees, give rise to new shoots via sprouts from the root system. Repeated root sprouting can result in individual plants that have a massive number of stems. One of the largest organisms on Earth is an aspen grove in Utah that includes about 40,000 stems and extends over 100 acres. The stems of a single individual tend to change color at the same time, a feature that contributes to the timing of beautiful fall foliage displays in the Rocky Mountains (**FIGURE 23.22A**).

Plants can also be involved in cross-species connections. Fully parasitic plants derive their carbon and water from other plants. Parasitic plants are often recognizable because they don't carry out photosynthesis and so can be entirely white or red, rather than green. Hemiparasitic plants carry out at least some of their own photosynthesis but derive water and other benefits from their hosts. Several flowers that flourish in semiarid portions of the Rocky Mountains fall into this category (**FIGURE 23.22B**). The roots of these hemiparasitic plants connect to the roots of their hosts via specialized structures called haustoria.

Plants can also be physically connected with one another via fungal partners. More than 90 percent of all flowering plants have associations with mycorrhizal (*myco*, "fungal," + *rhizoid*, "rootlike") fungi. A mycorrhizal interaction is a **mutualism** (an interaction benefitting both partners) in which the fine filamentous strands of the fungi extend the plant's root system and promote uptake of mineral nutrients, while the plant provides the fungi with carbon-based products derived from photosynthesis. Mycorrhizae often colonize more than one individual plant, and individual plants can be colonized by more than one type of mycorrhiza.

(A) Aspens

(B) Indian paintbrush

FIGURE 23.22 Plants Are Frequently Connected to other Plants Underground **(A)** An aspen grove is a collection of shoots, called ramets, that arise from a single root system. The extent of a clone can be seen as the leaves all turn color at the same time. **(B)** Paintbrush is a hemiparasitic plant whose roots are connected to those of other plants.

INVESTIGATION

FIGURE 23.23 Do Trees Share Carbon with One Another? Many plants are connected to fungal partners called mycorrhizae. In the forests of Canada, both paper birch and Douglas fir are associated with several of the same species of ectomycorrhizal fungi. These common fungal partners can create an underground network of fine fungal mycelia that connect individual birch and fir trees to one another. In contrast, cedar trees that grow in the area are associated with endomycorrhizal fungi and so are not part of the same fungal network. Suzanne Simard wondered whether birch and fir trees might use their shared fungal network to pass resources from plant to plant.[a]

HYPOTHESIS

Birch and fir trees that are part of the same fungal network use that shared connection to pass carbon from plant to plant.

METHOD

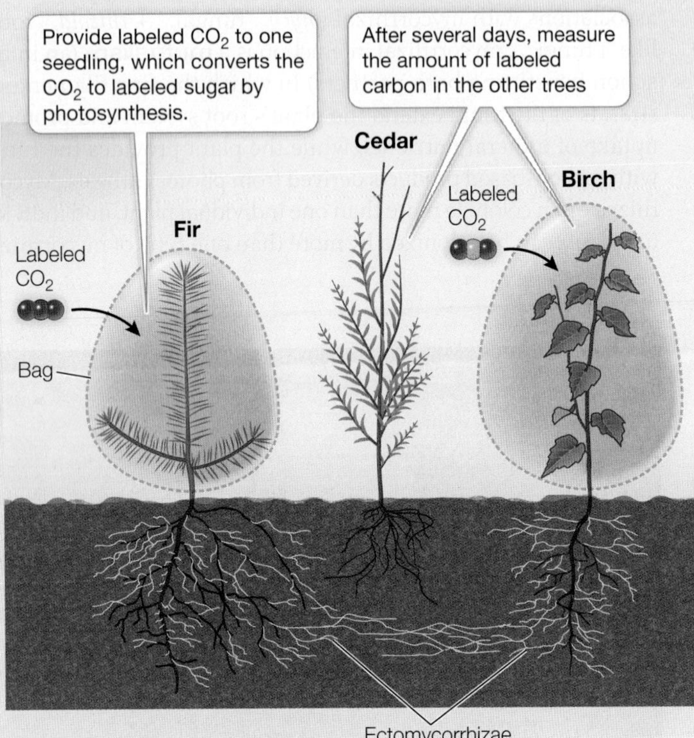

1. In a forest setting, grow paper birch, Douglas fir, and cedar trees in close proximity until they are 2 years old.
2. About a month before the start of the experiment, leave some of the fir trees alone and cover others with cloth, either partially or fully shading them.
3. To start the experiment, place a plastic bag over the top of each birch and fir tree. Inject labeled CO_2 into the air inside the bags. (Labeled CO_2 contains specific isotopes of carbon, e.g., ^{13}C and ^{14}C, that can be identified later on.) One tree gets one kind of label, and the other tree gets the other kind of label. Labeled CO_2 is not supplied to the cedar trees; they serve as controls for the experiment.
4. Two hours later, remove the bags. Several days later, harvest the seedlings and measure the amount and type of labeled carbon present in each one. Repeat the experiment 1 year later with seedlings that are 3 years old.
5. Determine the net amount of carbon transferred between trees by subtracting the total amount of carbon transferred from the firs to the birches from the amount transferred from the birches to the firs. Positive values indicate that the firs received more carbon; negative values indicate that the birches received more carbon.

RESULTS

Labeled carbon that was applied to birch trees was later found in Douglas fir trees, and vice versa. Much less of the labeled carbon ended up in the cedar trees.

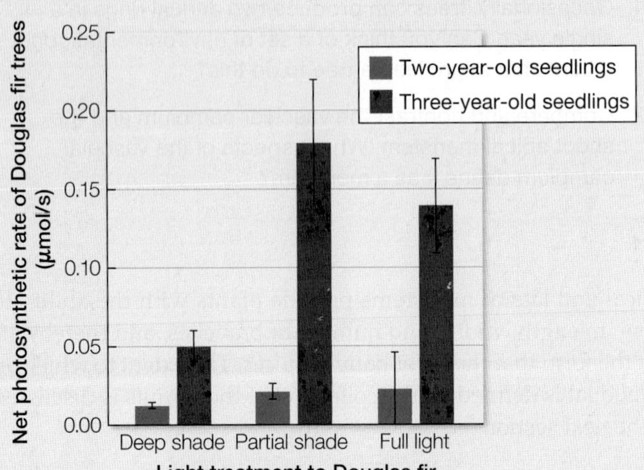

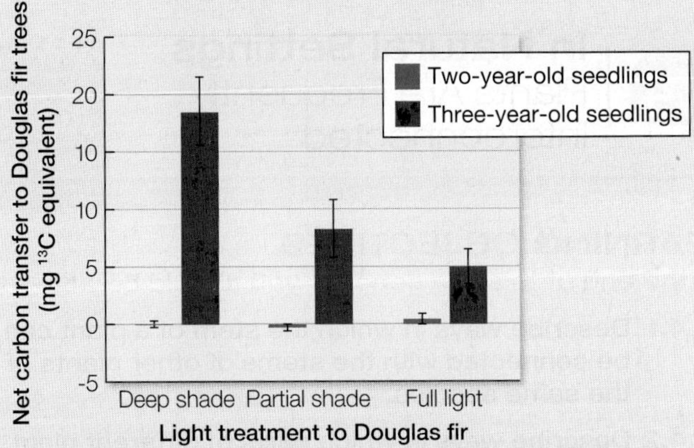

Values represent the mean ± 1 standard error of the mean.

CONCLUSION

These data strongly suggest that transfer of the carbon occurs via a common network of fungal mycelia. Thus individual trees are not as separate as they may appear. Rather, forest trees can be linked by an underground network of fungi that promote resource sharing.

ANALYZE THE DATA

1. Did shading the fir trees affect their rate of photosynthesis?
2. How would you explain the fact that the rate of photosynthesis was higher in the 3-year-old than in the 2-year-old fir trees?
3. Under what conditions did the fir trees receive the most carbon from the birches?

[a]S. W. Simard et al. 1997. *Nature* 388: 579–582.

A similar **Analyze the Data** exercise may be assigned in **Achieve**.

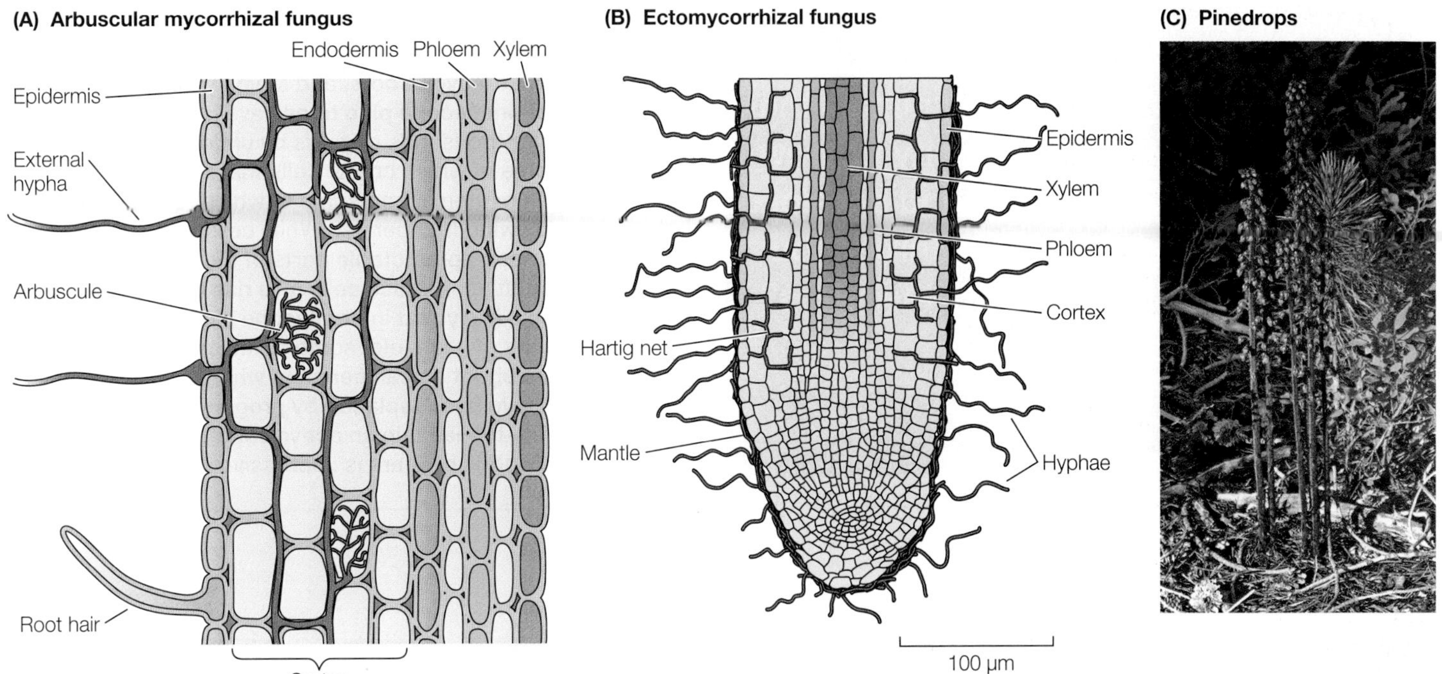

FIGURE 23.24 Most Plants Have Mycorrhizal Partners Plant root in association with **(A)** an arbuscular mycorrhizal fungus and **(B)** an ectomycorrhizal fungus. **(C)** While many mycorrhizal associations are mutualistic, some are not. The pinedrops shown here is a nonphotosynthetic plant that depends entirely on ectomycorrhizal fungi that connect to forest trees.

Thus mycorrhizae can form extensive, physically linked networks between different plant species.

Most mycorrhizae fall into two broad classes: **arbuscular mycorrhizae** (endomycorrhizae) and **ectomycorrhizae** (**FIGURE 23.24**). Both classes of fungi form extensive **mycelia**—thin strands of the fungus that extend through the soil. A major distinction between the two classes is found in their physical interaction with roots. Arbuscular fungi extend inside the walls of root cortical cells but not through their membranes. Inside the cortical cell walls, they form highly branched structures called arbuscules where phosphate is concentrated and ultimately transferred to the plant root cells. Arbuscular mycorrhizae are often found in association with herbaceous (nonwoody) plants. In contrast, woody plants such as forest trees are typically associated with ectomycorrhizae. The reproductive organs of ectomycorrhizae are mushrooms; the genera *Boletus* and *Russula* that are well known to many mushroom collectors are in this class. The association between ectomycorrhizae and the roots of certain plant species explains why particular mushrooms can be found near trees of a certain type. This author heard it said that morels, a particularly tasty mushroom, can often be found near the roots of dying apple trees—a correlation that proved true for a few years in her backyard. Ectomycorrhizae cover the plant root in a thick mass of mycelia. A few mycelia penetrate between the outer cell layers of the root, but they do not extend into the plant cells. The fungal filaments that extend from one plant often connect to other plants, including plants of different species. Thus the mycorrhizal networks in a forest literally form pathways through which plants share resources (**INVESTIGATION FIGURE 23.23**).

Mycorrhizal associations are not only common, they are ancient. Fossils of some of the earliest land plants (circa 400 million years ago) contain structures that look similar to modern arbuscules. Strikingly, the appearance of arbuscules in these plants predates their evolution of roots. The accumulated evidence strongly suggests that fungal associations coevolved with land plants.

The types of interactions described in this key concept complicate our view of what an individual plant is and are important to keep in mind because of their widespread nature.

REVIEW & APPLY | 23.4

R Shoots that appear to be one individual plant are often physically interconnected to other shoot systems. Connections can occur between individuals of the same species via organs such as rhizomes or root sprouts, or between species as when parasitic plants form haustoria that connect them to their hosts. Mutualistic associations between plant roots and fungal partners create mycorrhizal networks that connect different plants, even plants of different species, creating functionally connected communities.

A

1. Describe how you could sketch shoot systems of individual plants showing how they are physically interconnected.
2. If some trees grow better under one set of environmental conditions while other trees grow better under other environmental conditions, then having a network that shares resources could even out disparities in success. Which organism(s) in the network stand to gain by having a more reliable supply of photosynthate?

Answer: Meristems organize the formation of new organs, such as leaves, roots, and stems. The shoot apical meristems of bristlecone pine trees have maintained themselves for thousands of years by undergoing cell divisions, some of which create cells that become part of new leaves or stems, and some of which remain in the meristem, allowing it to persist. While cells in the meristem tend to give rise to predictable parts of the plant, there is some fluctuation in which cells give rise to which plant parts. It seems likely that in bristlecone pines that are now thousands of years old, some of the cells that were present in the shoot apical meristem when the plants first germinated have been replaced by progeny of nearby cells. Still, the ability to maintain an active set of meristematic cells for thousands of years is impressive indeed.

23 VISUAL SUMMARY

23.1 The Plant Body Is Built from an Indeterminate Number of Simple Modules

Go to ACTIVITIES 23.1, 23.2, and 23.3

FIGURE 23.1

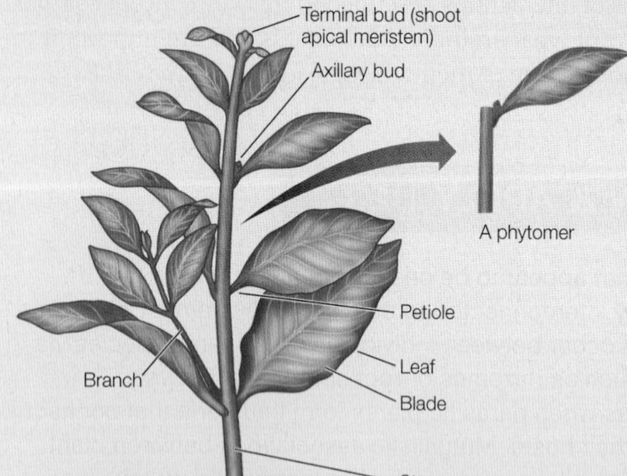

- Groups of cells called **MERISTEMS** divide and give rise to the rest of the plant body.
- Plant growth is **INDETERMINATE**: the number of organs depends on the environment.
- The **SHOOT SYSTEM** is built from repeating units called **PHYTOMERS** (**FIGURE 23.1**).
- Plant structures are built from three tissue systems: the **DERMAL**, **GROUND**, and **VASCULAR TISSUE SYSTEMS**.
- Three types of plant cells: **PARENCHYMA**, **COLLENCHYMA**, and **SCLERENCHYMA** are classified based primarily on the properties of their cell walls.

QUESTIONS

1. How is having an indeterminate body plan advantageous for (a) plants that live near grazing animals and (b) plants that grow in soils in which the distribution of water and nutrients changes over time?
2. If a houseplant is not watered, which part of the plant wilts first, the growing tips or the portion near the roots? Based on your answer, predict which part of the plant contains more collenchyma cells and which contains more sclerenchyma cells.

VISUAL SUMMARY 23

23.2 Apical Meristems Build the Primary Plant Body

Go to ACTIVITES 23.4 and 23.5

- The **SHOOT APICAL MERISTEM** gives rise to the **STEM** and typically an **INDETERMINATE** number of **LEAVES**. Branches form indirectly from axillary meristems.
- The planes of division within the shoot apical meristem give rise to separate layers of cells that form the tissues (**FIGURE 23.12**).
- Cells in the **ROOT MERISTEM** divide, giving rise to new cells that elongate and then differentiate, forming three distinct zones in the root. **LATERAL ROOTS** form through division of **PERICYCLE** cells (**FIGURE 23.13A**).
- The diversity of plant form is generated by changes in the relative shape, size, or number of a few repeating elements.

FIGURE 23.13

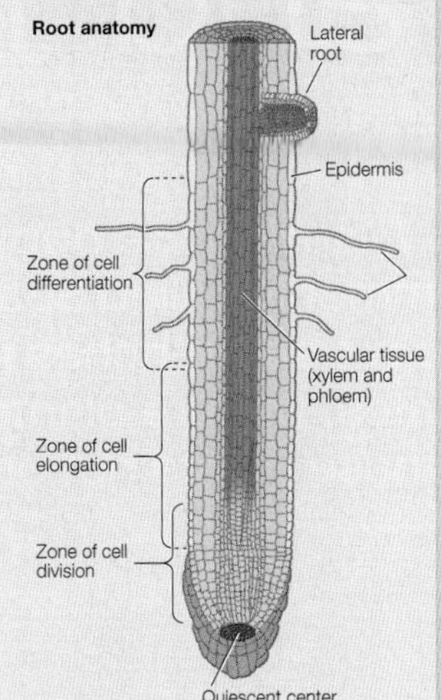

FIGURE 23.12

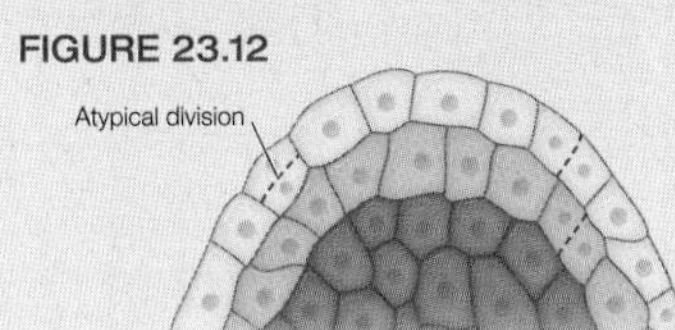

QUESTIONS

1. How does the pattern of cell division in the shoot apical meristem relate to the existence of a dermal tissue layer?
2. The initial cells that divide to form all the cells in the root surround a few cells called the quiescent center. Some initial cells divide to form the root cap, while other initial cells divide to form the primary root. Given that plant cells do not move past one another, where must the quiescent center be located?
3. What anatomical part of the plant are brussels sprouts (see Figure 13.4) and garlic cloves?

23.3 Many Eudicot Stems and Roots Undergo Secondary Growth

Go to ANIMATION 23.1

- Tree stems grow thicker through the action of the **VASCULAR CAMBIUM**.
- Division of the vascular cambium causes one ring of cells to be divided into two concentric rings, one of which remains vascular cambium. The other ring becomes **SECONDARY XYLEM** (if produced toward the interior) or **SECONDARY PHLOEM** (if produced toward the exterior) (**FIGURE 23.20**).
- When the xylem conducting cells produced at the end of the summer are narrower than those formed in the spring, an annual ring is produced.

FIGURE 23.20

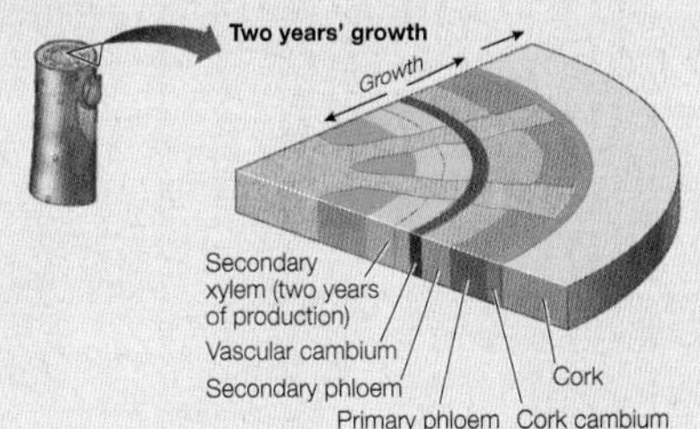

QUESTIONS

1. In the diagram on the left, which part of the xylem is the most recently formed, the light teal or the darker teal?
2. What is the advantage of producing wider xylem conducting cells? What is the disadvantage?
3. Why is production of annual rings less common in trees that grow near the equator?

23.4 In Natural Settings, Plants Are Frequently Interconnected

FIGURE 23.24

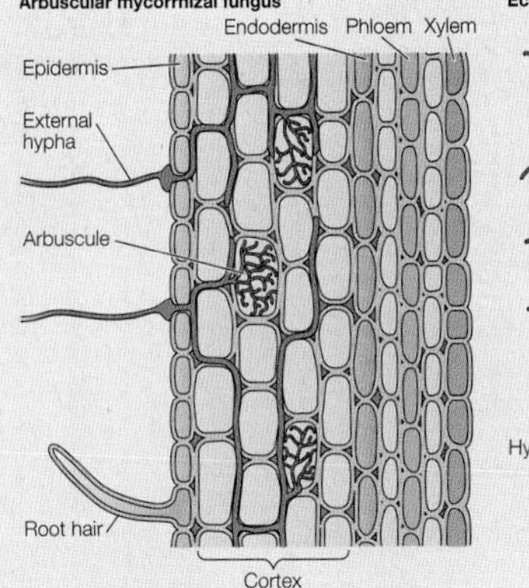

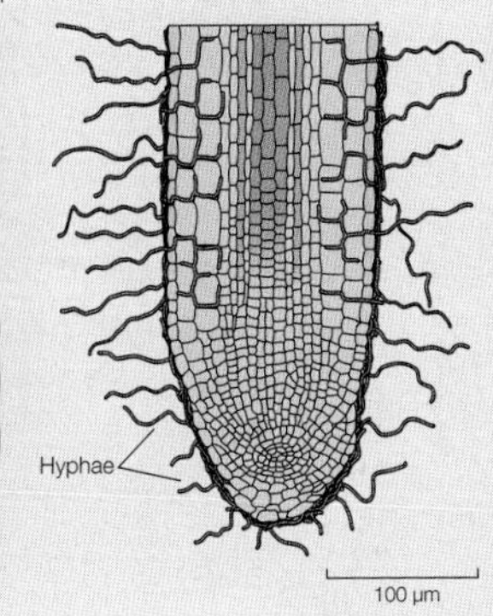

- Plants that appear to be separate individuals are frequently interconnected underground.
- Almost all flowering plants have fungal partners called **MYCORRHIZAE**. The fungi enhance the plants' uptake of nutrients from the soil, while plants provide the fungi with carbon-based products from photosynthesis (**FIGURE 23.24**).
- Because mycorrhizae frequently connect to multiple plants, the fungi provide conduits by which individual plants share their resources with one another.

QUESTIONS

1. How do arbuscular mycorrhizae and ectomycorrhizae differ?
2. Under what conditions might an individual plant benefit from being part of a multispecies network? Under what conditions might such a connection be harmful?

24 Plant Nutrition and Transport

"Air plants" in the genus *Tillandsia* are popular houseplants. *Tillandsia usneoides*, commonly called Spanish moss, can also be seen hanging from tree branches in parts of the south-eastern United States. Mature plants of this species lack roots—yet these plants not only survive, they grow.

How can a plant acquire the molecules it needs to permit growth if it doesn't take in water and nutrients through its roots?

You will find the answer to this question on page 617.

KEY CONCEPTS

24.1 Essential Plant Nutrients Include the Products of Photosynthesis

LEARNING OBJECTIVES

By the end of this key concept you should be able to:

24.1.1 Describe how plants obtain the elements that serve as building blocks for their bodies.

24.1.2 Describe common plant nutrient deficiencies, based on the biological molecules whose production is limited and the visible symptoms.

Plants derive energy that can be used to power the activities of their cells by breaking down carbohydrates that are produced during photosynthesis. They do this by using cellular respiration, the same process that we use when our bodies convert sugar into chemical-bond energy in the form of ATP. But plants and animals obtain their carbohydrates in fundamentally different ways. Animals such as ourselves eat them. Plants, by contrast, open their stomata, an act that allows carbon dioxide (CO_2) in the air to enter plant tissues (**FIGURE 24.1**). Together with energy from the Sun, these gaseous CO_2 molecules are converted to carbohydrates by the process of photosynthesis. Because these carbohydrates are the product of photosynthesis, they are often called fixed carbon or **photosynthate**. In addition to serving as a source of energy, these fixed carbon molecules form the physical building blocks from which the plant body is constructed. The elements that contribute most to the dry weight (after water is removed) of a plant are carbon and oxygen—meaning that most of the mass of a plant comes from the air.

Key Concept 5.5 discusses the carbon-fixation reactions by which the captured energy of sunlight is used to convert CO_2 and H_2O into carbohydrates.

The plant body contains several elements besides carbon. Nitrogen (N) is essential for the synthesis of proteins and nucleic acids, and plants obtain it, directly or indirectly, through interactions with bacteria and fungi, as we will see later in this chapter. In addition, nucleic acids contain phosphorus (P), many proteins contain sulfur (S), and chlorophyll contains magnesium (Mg). Metal ions also play important biochemical roles. For example, iron (Fe^{3+}) is used as a cofactor in some proteins, including those that facilitate electron transport in mitochondria and chloroplasts. Plants usually obtain these and other mineral nutrients from the soil.

Nutrients can be defined by their deficiency

A plant nutrient is called an **essential element** if its absence causes severe disruption of normal plant growth and reproduction. An essential element cannot be replaced by any other element. The essential elements for plants were largely identified by growing plants **hydroponically**—that is, with their roots suspended in nutrient solutions instead of soil. Growing plants in this manner allows for greater control of nutrient availability than is possible in a complex medium such as soil. In the first successful experiments of this type, performed a century and a half ago, plants grew seemingly normally in solutions containing only calcium nitrate [$Ca(NO_3)_2$], magnesium sulfate ($MgSO_4$), and potassium phosphate (KH_2PO_4). A solution missing any of these compounds could not support normal plant growth. Tests with other compounds that included various combinations of these same elements soon established the six macronutrients listed below.

- **Macronutrients** have concentrations that are typically at least 1 gram per kilogram of the plant's dry matter. There are six macronutrients:
  1. Nitrogen (taken up as NO_3^- or NH_4^+)
  2. Phosphorus (taken up as PO_4^{3-})
  3. Potassium (taken up as K^+)
  4. Sulfur (taken up as SO_4^{2-})
  5. Calcium (taken up as Ca^{2+})
  6. Magnesium (taken up as Mg^{2+})
- **Micronutrients** have concentrations of less than 0.1 gram per kilogram of the plant's dry matter. They include:
  - Iron (Fe^{2+})
  - Chlorine (Cl^-)
  - Manganese (Mn^{2+})
  - Zinc (Zn^{2+})
  - Copper (Cu^{2+})
  - Nickel (Ni^{2+})
  - Boron (BO_3^{3-})
  - Molybdenum [$Mo(O_4)^{2-}$]

Identifying micronutrients using hydroponic experiments is difficult because of the small amounts involved. Sufficient amounts of a micronutrient can be present as contaminants in the environment used to grow plants. Therefore nutrition experiments must be performed in tightly controlled laboratories with special air filters that exclude microscopic mineral particles, and they must use only the purest available chemicals. Furthermore, a seed may contain enough of a micronutrient to supply the embryo and the resultant second-generation plant throughout its lifetime. There might even be enough left over to pass on to the

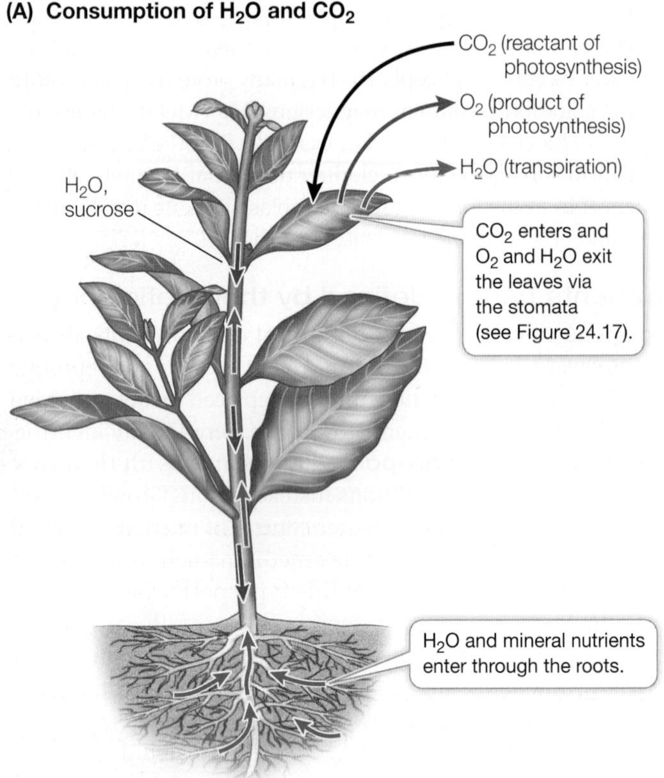

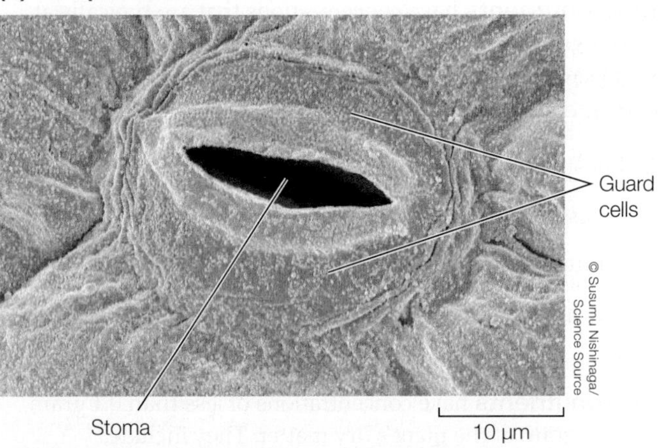

FIGURE 24.1 Plants Gain Mass When Carbon Dioxide from the Air Is Converted to Carbohydrates by Photosynthesis **(A)** Diagram of water and carbon flow in a typical land plant. Water travels through the plant to the atmosphere, exiting through stomata, with only a small fraction used within the plant. **(B)** Plants take in CO_2 through stomata, which share a functional similarity with animal mouths **(C)**.

The word "stoma" comes from the Greek word for "mouth." How are plant stomata functionally similar to the mouths of animals?

next generation of plants! Thus even if a plant is completely deprived of a micronutrient, it may take a long time for deficiency symptoms to appear.

Deficiencies in the essential elements frequently can be identified based on the appearance of the affected plant (**FIGURE 24.2**). Chlorophyll, the light-absorbing molecule that gives leaves their green color, contains magnesium. When plants lack sufficient magnesium, the older leaves may turn yellow. Plants that have low nitrogen or iron content also have yellow leaves. The location in which a deficiency shows up may differ depending on the location in which the element is required and the extent to which the element can move within the plant.

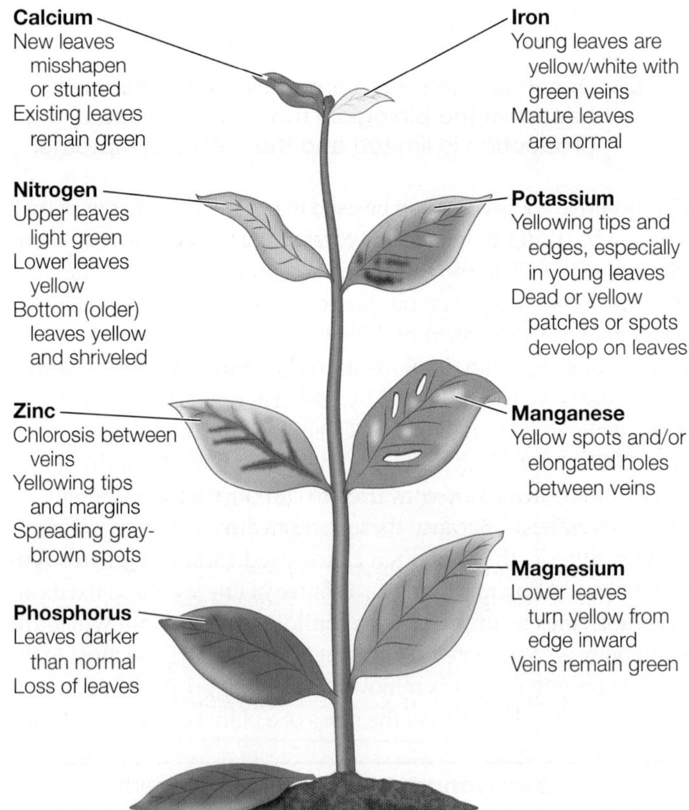

FIGURE 24.2 Mineral Nutrient Deficiency Symptoms Deficiencies of several nutrients are depicted in this diagram. Of course, it is very unlikely that a single plant would show all of these symptoms.

Animation 24.1 Nitrogen and Iron Deficiencies

REVIEW & APPLY | 24.1

R One of the most pervasive misconceptions about plants is that they obtain most of their dry mass from the soil. In fact, the vast majority of the dry mass of a plant comes from CO_2 in the air. Plants also contain nitrogen and other minerals that typically come from the soil, but these elements are present in much lower quantities than carbon and oxygen.

A
1. In the mid-1600s, Jan Baptista van Helmont ran an experiment with the goal of determining where the mass of a plant came from. He put 200 pounds of soil in a pot, watered it, and added a 5-pound willow tree. He continued to water the tree in its pot for 5 years. At the end of the experiment, he dried the soil and found the weight of the soil + pot to be about 2 ounces less than it had been at the start of the experiment. The tree, however, now weighed about 164 pounds more than it did at the start. He concluded that the additional 164 pounds came from water. Was he correct? Explain.
2. Nitrogen is essential for plant growth. What molecules that are essential to plant growth contain nitrogen?

While the dry weight of a plant comes largely from CO_2 in the air, many essential elements are taken up from the soil as well. In the next key concept we'll examine the characteristics of the soil that promote plant growth.

24.2 Mineral Nutrients Are Commonly Acquired from the Soil

LEARNING OBJECTIVES

By the end of this key concept you should be able to:

24.2.1 Describe how the proton-excreting activity of roots promotes nutrient uptake.

24.2.2 Explain the significance of membrane proteins in regulating the molecules that reach the vascular tissue.

24.2.3 Discuss issues around the use of nitrogen fertilizer from more than one perspective.

Most terrestrial plants grow in soil, which provides plants with several benefits:

- Anchorage for mechanical support of the shoot
- Mineral nutrients and water from the soil solution
- O_2 for root respiration, which is provided by air spaces between soil particles
- The services of other soil organisms

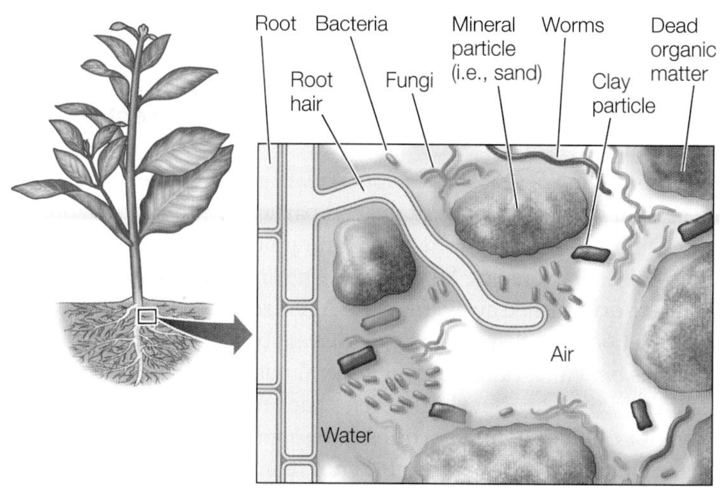

FIGURE 24.3 The Complexity of Soil Soils favorable for plant growth contain both clay and larger mineral particles, as well as water, air, and organic matter. Other organisms are also present.

Soils have both living and nonliving components (**FIGURE 24.3**). The living components include plant roots as well as populations of bacteria, fungi, protists, and animals such as earthworms and insects. The nonliving components include rock fragments ranging in size from large stones to clay particles 2 micrometers or less in diameter. Soil also contains water and dissolved mineral nutrients, air spaces, and dead organic matter.

Although soils vary greatly, almost all of them have a soil profile consisting of several horizontal layers, called **horizons**. Soil scientists recognize three major horizons—termed A, B, and C—in a typical soil (**FIGURE 24.4**). The A horizon is the **topsoil**,

FIGURE 24.4 A Soil Profile The A, B, and C horizons of a soil can sometimes be seen at construction sites such as this one in Massachusetts. The darker upper layer (the A horizon) is home to most of the living organisms in the soil.

which contains most of the soil's living and dead organic matter. It supports the plant's mineral nutrient needs. The B horizon is the **subsoil**, which accumulates materials from the topsoil above it and the parent rock below. The C horizon is the **parent rock** from which the soil arises.

The soil's fertility (ability to support plant growth) is determined by several factors. The size of the soil particles in the A horizon, for example, is important in determining nutrient retention and other soil characteristics. Mineral nutrients in the soil solution can be **leached** (washed away) from the A horizon downward by rain (or irrigation) and thereby become unavailable to plants. Tiny clay particles bind a lot of water, and mineral nutrients are not readily leached from them, but clay particles pack tightly and leave little room for air. Large sand particles have opposite properties; they provide plenty of air spaces but do not retain water or nutrients. A **loam** is a soil that is an optimal mixture of sand, clay, and silt (mineral particles intermediate in size between sand and clay) and thus has sufficient supplies of air, water, and nutrients for plants.

In addition to mineral particles, soils contain dead organic matter, largely from plants. Soil organisms break down dead leaves and other organic materials into a substance called **humus**. This material is used as a food source by microbes that break down complex organic molecules and release simpler molecules into the soil solution. Humus also improves the texture of soil and provides air spaces that increase oxygen availability to plant roots.

Cation exchange makes nutrients available to plants

Chemical weathering results in clay particles covered with negatively charged chemical groups (see Figure 24.5). These negatively charged clay particles bind the positively charged ions (cations) of minerals that are important for plant nutrition, such as potassium (K^+), magnesium (Mg^{2+}), and calcium (Ca^{2+}). This binding prevents the cations from being leached out of the soil, but to become available to plants, they must be released from the clay particles into the soil solution.

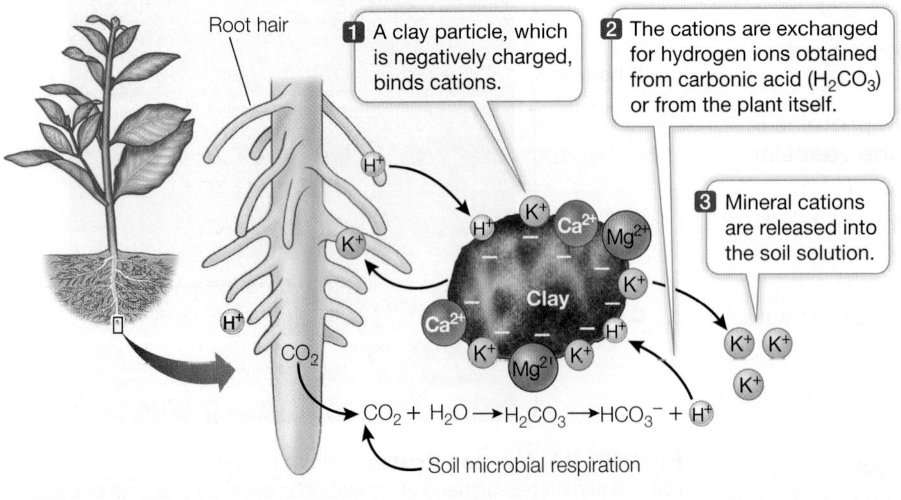

Recall that the root surface is covered with epidermal cells bearing root hairs (see Figure 23.13B). Proteins in the cell membrane of these cells actively pump protons (H^+) out of the cells. In addition, cellular respiration in the roots and by organisms in the soil releases CO_2, which dissolves in the soil water and reacts with it to form carbonic acid. This acid ionizes to form bicarbonate and free protons:

$$CO_2 + H_2O \leftrightharpoons H_2CO_3 \leftrightharpoons H^+ + HCO_3^-$$

Proton pumping by the root cells and ionization of carbonic acid both act to increase the proton concentration in the soil surrounding the root. The protons bind to clay particles, displacing cations that would otherwise bind to the surface of the clay; in essence, the protons trade places with the cations in a process called **cation exchange** (**FIGURE 24.5**). Cation exchange releases mineral nutrient cations into the soil solution, where they are available to be taken up by roots. Soil fertility is determined in part by the soil's ability to provide nutrients in this manner.

There is no comparable mechanism for binding and releasing negatively charged ions (anions). As a result, important anions such as nitrate (NO_3^-) and sulfate (SO_4^{2-}) may be leached from the A horizon.

Proteins regulate uptake of mineral nutrients

Regulation of the internal contents of a cell is essential to maintain the activity of life: toxins are excluded and nutrients are selectively taken in. Thus the chemical composition inside a plant cell is not identical to that in the soil water. Plants regulate what they take in from the soil by using transport mechanisms that selectively move beneficial molecules across the cell membrane.

Root hairs and mycorrhizae give roots a large surface area across which soil water and any solutes that it contains can enter the walls of the epidermal cells. The walls of the cells toward the outside of the root are not water resistant, so water and minerals from the soil can travel freely through these walls and the spaces between cells (called the apoplast) as far as the **endodermis** (a ring of cells that surrounds the vascular tissue, just interior to the cortex) (**FIGURE 24.6**). The uptake of water in roots is somewhat similar to what happens if you put a cardboard box next to a source of water: just as water seeps into cardboard, the walls of root cells passively soak up water until it reaches the endodermal cell layer. The walls of mature endodermal cells are modified in a way that seals off the spaces between neighboring endodermal cells and limits movement through them. Specifically, the endodermal cell walls contain a region called the **Casparian strip** that recent studies have shown to be composed of lignin. This lignin-rich strip of wall material is surrounded by a waxy layer of **suberin** (a hydrophobic substance largely composed of fatty acids). Water and ions that don't move through these walls can only enter the plant by passing through a cell membrane. Because many compounds cannot pass through a cell membrane on their own, this gives the

FIGURE 24.5 Cation Exchange Plants obtain mineral nutrients from the soil primarily in the form of positive ions; potassium (K^+) is the example shown here.

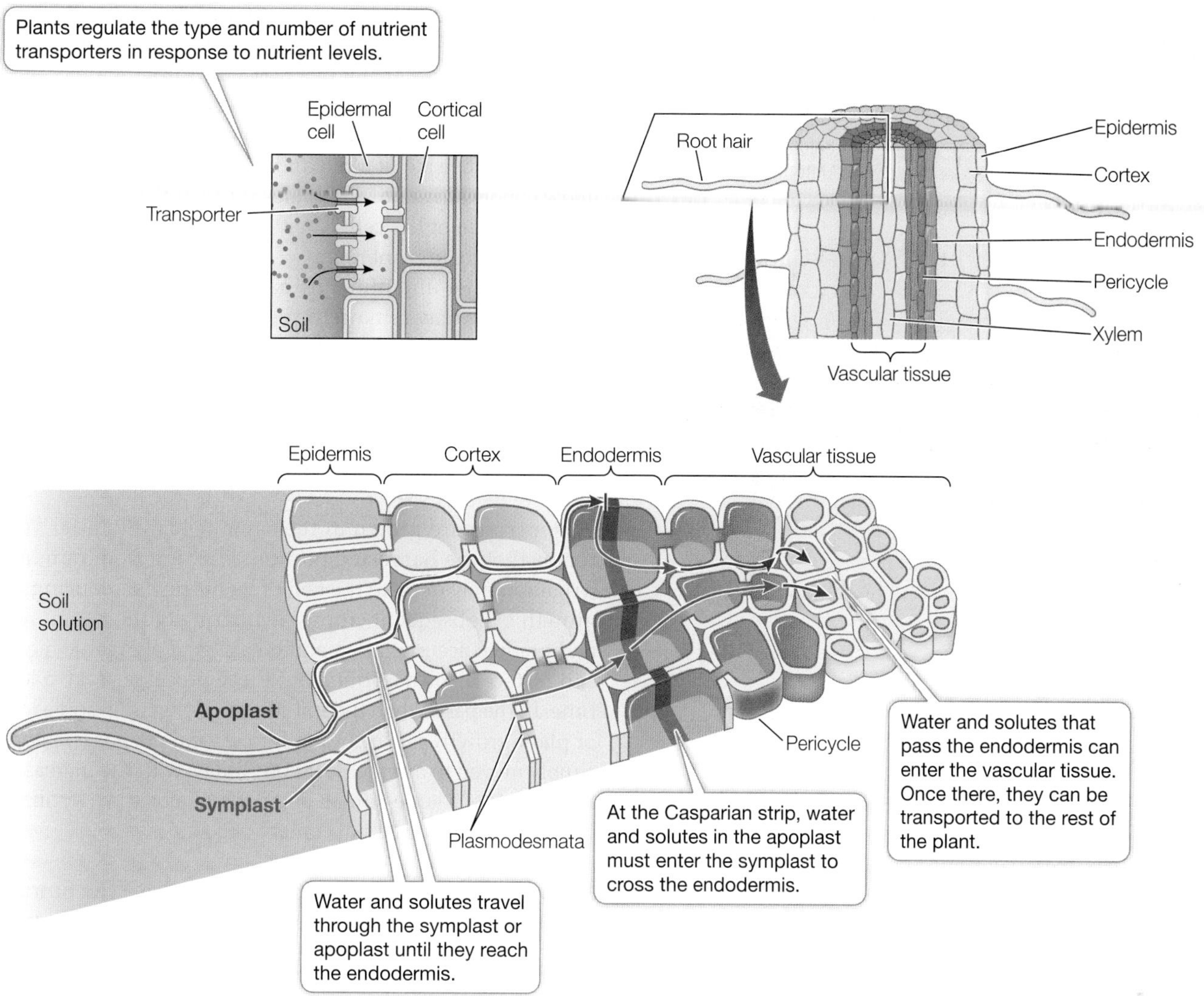

FIGURE 24.6 Pathways of Nutrient Uptake Water and solutes can move into the root through the apoplast or symplast until they reach the endodermis (shown in dark green). Water and solutes that do not pass the Casparian strip must enter the plant by crossing a cell membrane; transport proteins in cell membranes allow this process to be regulated. Inside the vascular tissue, the water and solutes enter the xylem (light blue).

Where might transport proteins that permit nutrients to enter the symplast, and thereby bypass the barrier presented by the endodermal cell walls, be located?

Activity 24.1 Apoplast and Symplast of the Root

plant an opportunity to regulate its internal contents (called the **symplast**). For example, the Casparian strip appears to limit uptake of sodium (Na^+), a compound that, in high concentrations, is toxic to plants. In summary:

- *The* **apoplast** *(Greek* apo, *"away from"*; plast, *"living material") consists of the cell walls, which lie outside the cell membranes, and intercellular spaces (spaces between cells).* Water and solutes can flow through the apoplast without ever having to cross a membrane.
- *The symplast (Greek* sym, *"together with") consists of the cytoplasm of the living cells, which are connected to one another by small pores called* **plasmodesmata** (see Figure 4.20C). The selectively permeable cell membranes of the root cells control access to the symplast, so movement of water and solutes into the symplast is tightly regulated.

LINK The Casparian strip is functionally equivalent to the tight junctions found in some animal tissues; see **Figure 4.19**.

Mineral nutrients that can't pass through a membrane on their own can enter the symplast via specialized transport proteins. The plant *Arabidopsis thaliana* has more than 50 genes that encode nitrate (NO_3^-) transporters and at least 4 genes for phosphate (PO_4^{3-}) transporters. Nutrient uptake is highly regulated

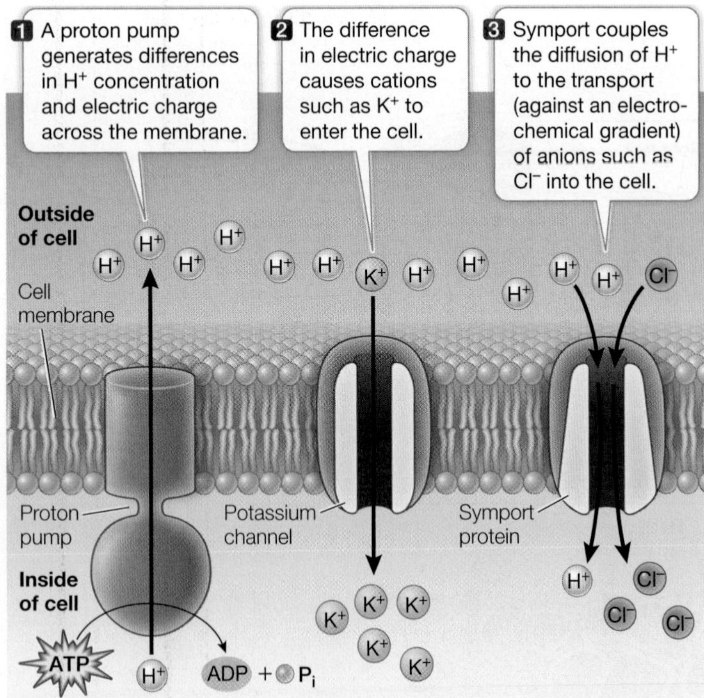

FIGURE 24.7 The Proton Pump in Transport of K^+ and Cl^-
The active transport of hydrogen ions (H^+) out of the cell by the proton pump (1) drives the movement of both cations (2) and anions (3) into the cell.

because the levels of ions inside cells must be maintained at constant levels. The genes for ion transporters are regulated at the transcriptional level in response to the amounts of each nutrient inside cells: low nutrient levels stimulate transcription, whereas high levels repress transcription. In addition, the activity of the transport proteins is also regulated (for example, by phosphorylation). Protein channels called aquaporins allow water, which can move through a membrane slowly on its own, to move much more rapidly. The abundance and activity of aquaporins are regulated so that the rate of water movement can be controlled, but the direction is not affected. Aquaporins are an example of facilitated diffusion, as they allow water to more rapidly move to its equilibrium state (see Key Concept 4.2).

In many cases, movement of a molecule into a cell requires energy, and thus depends on active transport. Root cell membranes contain **proton pumps**, which use energy from ATP to move protons out of the cell against a proton concentration gradient (**FIGURE 24.7, STEP 1**). Because protons (H^+) are positively charged, their accumulation outside the cell creates an electrical gradient. The region just outside the cell becomes more positively charged than the inside of the cell. With the inside of the cell more negative than the outside, cations such as potassium (K^+) move into the cell down the electrical gradient through specific membrane channels (**FIGURE 24.7, STEP 2**). Anions such as chloride (Cl^-) are moved into the cell against an electrochemical gradient by a membrane transport protein that couples their movement with that of H^+ (**FIGURE 24.7, STEP 3**). These processes are examples of secondary active transport.

Directional movement of a molecule through the body of the plant, such as movement of minerals toward the center of the root, can also be facilitated by having uptake proteins located on one side of a cell and distinct efflux proteins that allow export of various molecules through the membrane on the other side of the cell. A few examples of this have been found recently in plant roots.

Once water and minerals have passed the endodermal barrier, they may remain in the symplast until they reach parenchyma cells in the pericycle or xylem. These cells then actively transport mineral ions into the vascular tissue, from whence they can be carried to the rest of the plant.

Fertilizers can be used to add nutrients to soil

When crops are grown in the same fields year after year, removal of the harvested portion, leaching from rain, and irrigation may deplete the soil of its nutrients, reducing the growth of subsequent crops. Weathering of the parent rock and gradual accumulation and breakdown of organic matter can restore a soil's fertility, but these processes take a long time. In the past, when the soil could no longer support a level of plant growth sufficient for agricultural purposes, people simply moved to another location. Today, with the increasing food needs of a large human population, people use fertilizers to improve the nutritional content of soil.

ORGANIC FERTILIZERS Farmers can increase the nutrient content of soil by adding organic materials such as compost (partially decomposed plant material) or manure (the waste from farm animals), which is a particularly good source of nitrogen. These organic fertilizers add nutrients to the soil but still allow for a slow release of ions, with little leaching, as the materials decompose. Humus is used as a food source by soil organisms, which in turn release simpler molecules to the soil solution. For example:

$$\text{Leaf proteins} \xrightarrow{\text{Bacteria}} NH_4^+ \xrightarrow{\text{Bacteria}} NO_3^-$$

These simpler molecules can dissolve in soil water and be taken up by plant roots.

INORGANIC FERTILIZERS The discovery and use of inorganic fertilizers greatly increased crop yields around the world. Nitrogen is the most frequently limiting nutrient for plant growth, although additions of phosphate and potassium frequently boost plant growth as well. Inorganic fertilizers are characterized by their "N-P-K" percentages. A 5-10-10 fertilizer, for example, contains 5 percent nitrogen (typically NH_4^+ or NO_3^-), 10 percent phosphate (P_2O_5), and 10 percent potash (K_2O) by weight (of the nutrient-containing compound, not as weights of the elements N, P, and K). Sulfur, in the form of ammonium sulfate [$(NH_4)_2SO_4$], is also occasionally added to soils.

(A) ***Dionaea muscipula***

(B) A tank *Tillandsia*

(C) *Tillandsia* trichomes

FIGURE 24.8 Nutrients from Sources Other than Soil **(A)** The Venus flytrap obtains nitrogen from the bodies of arthropods trapped inside the plant when its specialized leaves snap shut. **(B)** A tank bromeliad collects water and debris, including insects, that fall into the region at the center of the leaves. **(C)** The specialized hairs (called trichomes) on the leaves of atmospheric *Tillandsia* plants wick water and dissolved nutrients toward specialized cells (here shown in light blue) where they are absorbed. The magnification of this image is very much greater than that of the other two images in this panel. 100 μm = 0.1 mm.

Media Clip 24.1 **A Venus Fly Trap "Snaps to It"**

To return to our example of nitrogen, inorganic fertilizers reach plants by this process:

$$\text{Chemical manufacture} \longrightarrow NH_4^+ \xrightarrow{\text{Bacteria}} NO_3^-$$

Manufacture of nitrogen fertilizer begins with an energy-intensive industrial process in which N_2 from the atmosphere is combined with H_2 at high temperatures and pressures in the presence of a catalyst to form ammonia (NH_3). This process accounts for about 2 percent of the world's energy use, mostly in the form of natural gas. Although organic and inorganic fertilizers have the same effects on crop nutrition, their costs are vastly different: manure and compost are free, whereas NH_4^+ made in a factory can be very expensive.

Use of nitrogen fertilizer is widely credited as a major reason for increased crop production over the past 50 years. It is not, however, without its costs. When the amount of nitrogen-rich fertilizer (organic or inorganic) applied exceeds that which the crops take in, leaching of nitrogen from farm fields to rivers, lakes, and oceans can affect the balance of those ecosystems, ultimately leading to "dead zones" in which levels of oxygen in the water are too low to support fish and other life forms. A large dead zone now forms each summer in the Gulf of Mexico. In addition, soil bacteria can convert a considerable amount of the nitrogen that is not taken up by plants to nitrous oxide (N_2O), a potent greenhouse gas that is associated with global warming (see Key Concept 42.4). Many strategies have been devised to help farmers optimize the amount and timing of nitrogen application. Based on the fact that nitrogen deficiency results in yellowing of leaves, a simple leaf color chart has been developed that smallholder farmers can hold up next to rice leaves to determine when an additional application of nitrogen fertilizer may be effective.

The cycling of nitrogen through biological systems is described in **Key Concept 42.3**. Runoff of excess fertilizer can stimulate growth of phytoplankton and algae. Respiration by these primary producers and by the decomposers that process their dead bodies can reduce oxygen levels below the tolerances of many aquatic organisms, resulting in "dead zones" devoid of oxygen-requiring aquatic life.

Some plants obtain nutrients through their leaves

Although the majority of plants obtain their mineral nutrients from the soil, some plants get their nutrients from other sources. For example, there are about 500 known carnivorous plant species that obtain some of their nutrients by digesting arthropods. These plants typically grow in boggy soils in which little nitrogen or phosphorus is available. Digestion (hydrolysis) of arthropod prey helps provide those missing nutrients. A well-known example of a carnivorous plant is the Venus flytrap (genus *Dionaea*; **FIGURE 24.8A**), which has a modified leaf with two halves that fold together. When an insect touches trigger hairs on the leaf, its two halves quickly come together, and their spiny margins interlock and trap the insect before it can escape. The leaf then secretes enzymes that digest the prey.

Bromeliads are another group of plants that obtain nutrients through their leaves. Some bromeliads have leaves that wrap tightly around one another, creating tanks that can be filled with rainwater (**FIGURE 24.8B**). Debris such as dead leaves and arthropods such as ants also collect in the tanks, where they decompose and supply mineral nutrients that are taken in by specialized leaf hairs.

INVESTIGATION

FIGURE 24.9 Plants Can Obtain Nutrients from Materials Present in the Air and Rain Some plants, including *Tillandsia capillaris*, live on other plants, typically trees. These epiphytic plants rely on their hosts for physical support and receive their nutrition from rainwater, particles that fall on the leaves, and nitrogen-fixing bacteria (nitrogenase activity was detected on the leaves). Some *T. capillaris* individuals have been observed living on power lines, a location in which they don't receive as much organic debris as they would if they lived on a host tree.[a]

© Luiz Claudio Marigo/Minden Pictures

HYPOTHESIS

Tillandsia capillaris plants do not require organic material from a host plant.

METHOD

Collect, wash, and dry plants growing in a forest and plants growing on power lines located along a road that bounds the forest. Measure the chemical composition of the leaves and the particles that have fallen on the leaves.

RESULTS

	Chemical composition of *T. capillaris* plant tissues (mg per gram of leaf)			Weight of material collected on the leaf surface (mg per gram of leaf)				
	Carbon (C)	Phosphorus (P)	Nitrogen (N)	Total weight of material	PO_4^{3-}	SO_4^{2-}	NH_4^+	NO_3^-
Plants growing on power lines	587.2	4.3	24.4	12.1	0.00012	0.0012	37.6	0.050
Plants growing on trees	624.4	3.1	14.0	34.0	0.00041	0.00011	37.2	0.046

CONCLUSION

The concentration of nitrogen in the plants that were growing on the power line was not less (in fact, it was more) than it was in the plants growing on trees. Thus organic material from a host tree is not required to support the growth of *T. capillaris*; sources other than the host trees must provide nitrogen to the plants. How then did the plants that grew on power lines acquire more nitrogen? It was further hypothesized that the higher nitrogen content of the plants growing on the power lines might result from vehicle emissions associated with the road along which they were growing.

ANALYZE THE DATA

1. What percent of the total material that fell on the leaves was PO_4^{3-} for plants on power lines versus trees? Did this vary much or was the difference less than 50%?
2. What percent of the total material that fell on the leaves was SO_4^{2-} for plants on power lines versus trees? Did this vary much or was the difference less than 50%?
3. The material that collected on the plants growing on power lines contained about 30 times more sulfur than did the material collected on the plants growing on trees. Emissions from gasoline and diesel-driven vehicles include sulfur. Does this fit with the data presented in the table below?

[a]A. B. Abril and E. H. Bucher. 2009. *J Arid Environ* 73: 393–395.

A similar **Analyze the Data** exercise may be assigned in **Achieve**.

A subset of the bromeliads (such as the one in the chapter opening photo) are called atmospherics or air plants. Air plants do not form large tanks but have spectacularly modified leaf hairs that aid in the absorption of water and nutrients. When dry, these hairs are shaped like a flat sheet that forms a bit of a funnel around a set of leaf cells that are adapted for taking in water (**FIGURE 24.8C**). When wet, these hairs fall back onto the surface of the leaf, and water wicks rapidly through the space between the leaf hair and the leaf toward the cells that will absorb it. Thus the water from mist, fog, and rain, along with the nutrients it contains, is rapidly delivered to the interior of the leaf. Modifications such as this help explain how **epiphytes** (*epi*, "upon"; *phyte*, "plant")—nonparasitic plants that grow on other plants and sometimes even on hard substrates such as rocks or power lines—can absorb nutrients even though they may not contact the soil. **INVESTIGATION FIGURE 24.9** examines this topic more thoroughly.

THINK LIKE A SCIENTIST

Testing new analytical methods: Might *Tillandsia* make useful air pollution monitors?

Advances in biology often follow the arrival of new techniques that allow us to measure things that were previously hidden. The faster, easier, and less expensive the techniques are, the more new information can be gained. With air pollution a rising concern around the world, scientists have begun to wonder if species of *Tillandsia* that are adapted for uptake of compounds from the air might make useful sensors for air pollution.

Effective sensors or indicators have a reproducible relationship between the amount of the thing being measured (here, the air pollutant) and the change in the indicator (here, the amount of pollutant in the plant). For example, the concentration of protein in a sample can be determined by adding a dye that binds to protein and turns a specific color.

The amount of color can be quantified in a spectrophotometer. In an ideal world, the relationship between the amount of protein and the amount of new color produced would be a straight line with no deviations at any point; in reality, however, there is usually a limited range over which an indicator is useful. Adding dye to a series of samples that have a known protein concentration makes it possible to calibrate the reaction—that is, to obtain a standard curve (**FIGURE 24.10**). After obtaining a standard curve, the concentration of protein in an unknown sample can be determined by adding the dye, measuring the resulting color of the sample, and then comparing that to the standard curve (see yellow star in Figure 24.10). This works even if the relationship is not linear, as long as the relationship between protein concentration and formation of the new color is correlated in a predictable and known way.

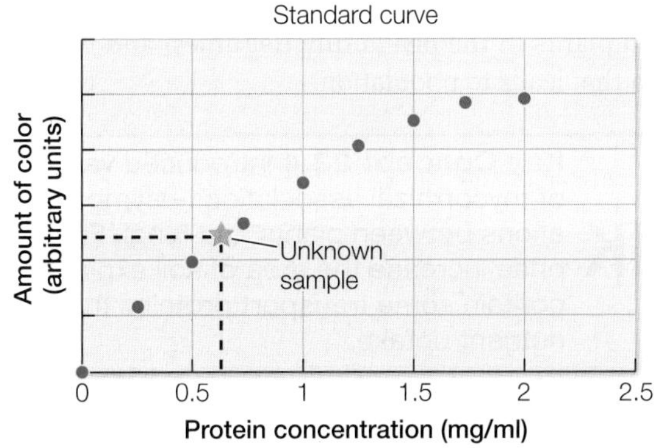

FIGURE 24.10 A Standard Curve A standard curve plots the amount of product (here, the amount of color produced) when a sample of known concentration is reacted with an indicator such as a dye.

A research team lead by Eva Shreck investigated the potential for *Tillandsia capillaris* plants to serve as monitors of metal contamination in the air by transplanting several of these plants to a heavily contaminated site near a smelter. Simple filters were placed next to the plants. Each month thereafter, the amount of metal that accumulated in the filters and in the (washed) plants was determined. Results for two of the metals in the study, zinc and cadmium, are shown in **FIGURE 24.11**.

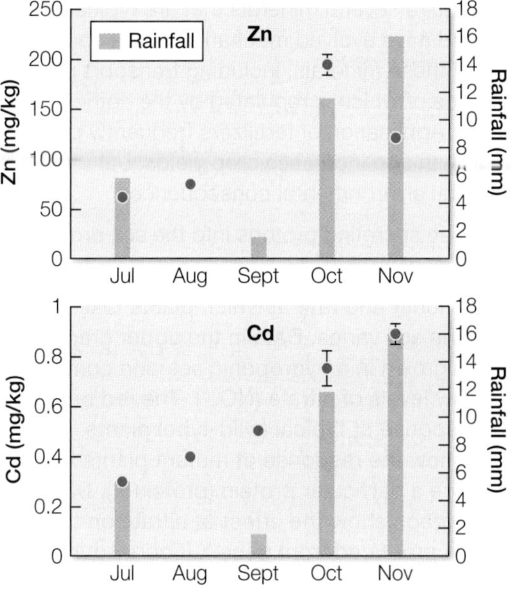

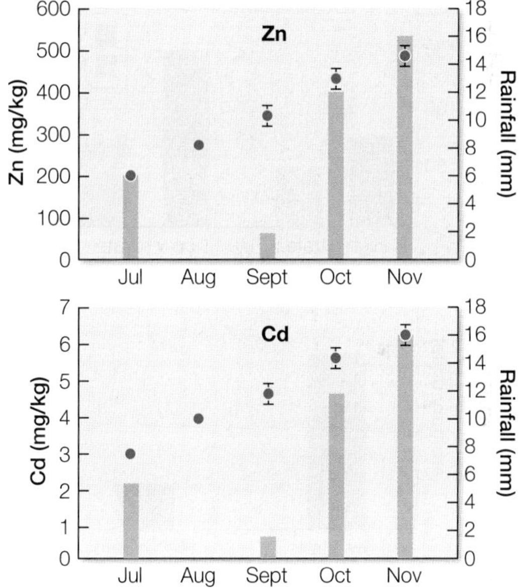

FIGURE 24.11 Test of the Suitability of *T. capillaris* as a Biosensor Pollutant accumulation in **(A)** *T. capillaris* versus **(B)** filters. (Zn, zinc; Cd, cadmium.) Error bars indicate standard deviation. (After E. Shreck et al. 2016. *Ecol Indic* 67: 227–237.)

1. For which of these metals would *Tillandsia capillaris* be a useful bioindicator? Explain your reasoning. (In September, the amount of zinc in the plants was not determined; we shouldn't make any assumptions about what that value was. It is simply a missing data point.)
2. If there is a difference in the pattern of metal accumulation in plants versus filters, construct a hypothesis as to what process(es) may have generated that difference.
3. Describe a process that could be used to establish a standard curve for levels of airborne cadmium, using a filter as an indicator.

REVIEW & APPLY | 24.2

R Plants require several minerals that are typically found in soil, and have evolved mechanisms that can promote uptake of these minerals, including transport proteins, the abundance of which is regulated by the nutrient status of the plant. Application of fertilizers frequently promotes plant growth and increases crop yields, but can have detrimental environmental consequences.

A 1. How can secreting protons into the soil promote uptake of magnesium (Mg^{2+})?

2. The amount and rate at which plants take up nitrogen from the soil varies. Data in the upper graph are from plants grown in a hydroponic solution containing high and low levels of nitrate (NO_3^-). The red bars show the response of typical (wild-type) plants. The orange bars show the response of mutant plants that do not produce a particular protein (protein X). Data in the lower graph show the effect of nitrate on the amount of protein produced from gene X (*figure source*: M. Orsel et al. 2004. *Planta* 219: 714–721).

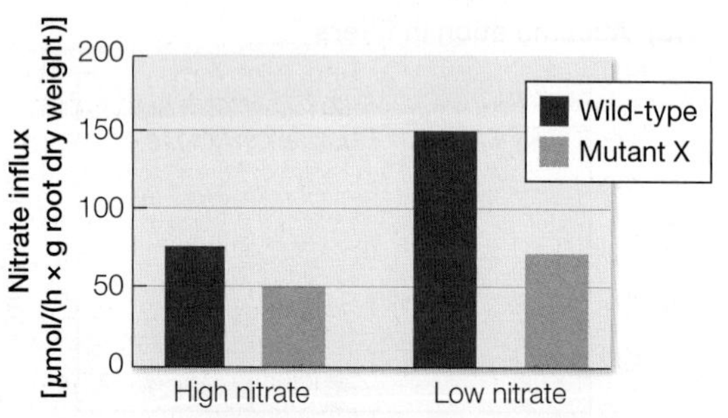

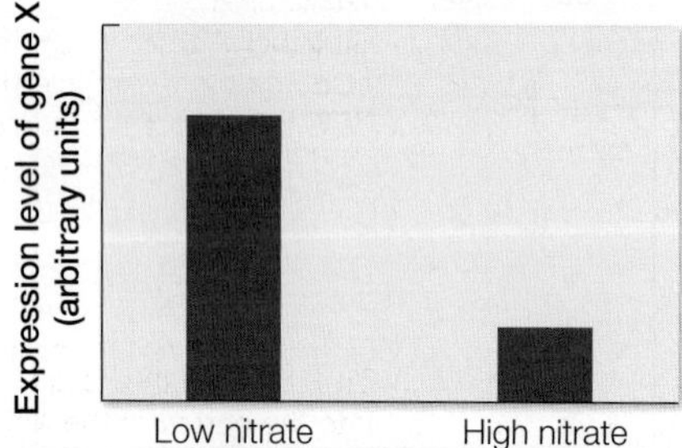

a. Based on the data, what type of activity does protein X have?

b. What advantage might a plant gain by regulating the amount of a nutrient transporter produced, rather than maintaining high levels of uptake proteins at all times?

c. Food for thought: How do you weigh the advantages and disadvantages of fertilizer use when the costs and benefits are dissimilar?

As described above, the chemical and physical properties of soil affect plants' ability to acquire nutrients from it, but the underground world is far more dynamic than has been so far explained. In the next key concept we'll describe how plant roots and other organisms that live in the soil affect the extent to which nutrients are available to plants.

24.3 | Root Exudates Affect Soil Organisms and Nutrient Uptake

LEARNING OBJECTIVES

By the end of this key concept you should be able to:

24.3.1 **Explain what benefits plants obtain from releasing some of their photosynthate into the rhizosphere.**

24.3.2 **Describe the process of nodulation and how it promotes nitrogen fixation.**

One gram of soil contains from 6,000 to 50,000 bacterial species and up to 200 meters of fungal hyphae, although both are largely invisible to the naked eye. Plants spend a significant amount of their photosynthate to produce and exude (secrete) a wide range of compounds from their roots into this complex environment, and as a result, plants influence the type and abundance of the organisms that live near their roots. These secretions are called **exudates**, and the region of soil close enough to a root to be affected by them is called the **rhizosphere**. The composition of the rhizosphere changes over time depending on the environment and the state of the plant. When plants are attacked, their roots may secrete molecules that increase the abundance of beneficial bacteria. Some of these bacteria prime the plant's defenses, helping protect the plant against future attacks. Other secretions can attract symbiotic partners or can poison competitors. The existence of a dynamic bacterial community in the rhizosphere has led to comparisons between the bacteria in the rhizosphere and the bacterial communities in the human gut.

Two relationships that are initiated by chemical signals sent by plants to other soil organisms are (1) the association between most land plants and arbuscular mycorrhizae, and (2) the association between plants in the pea family (legumes) and bacteria called rhizobia that leads to nodulation.

LINK **Key Concept 23.4** introduced various types of mycorrhizal associations—symbiotic associations between plants and fungi. Fungal hyphae increase the area of soil explored and contain some transport proteins that assist in nutrient uptake.

Chemical signals from plants stimulate the formation of arbuscular mycorrhizae

An association with arbuscular mycorrhizae can be initiated by the production and release of chemical signal molecules called **strigolactones** from plant roots. Strigolactones stimulate rapid growth of fungal hyphae toward the root (**FIGURE 24.12A**). In response, the fungus signals the plant to form a **prepenetration apparatus** (**PPA**), which guides the growth of the fungal hyphae into the root cortex. The sites of nutrient exchange between fungus and plant are the arbuscules, which form inside root cortical cells. Mycorrhizae expand the root surface area from 10-fold to 1,000-fold and are important in increasing the plant's uptake of

FIGURE 24.12 Roots Send Signals for Colonization Plant roots send chemical signals to arbuscular mycorrhizal fungi **(A)** and nitrogen-fixing bacteria **(B)** to stimulate colonization. Neither process occurs in all plants.

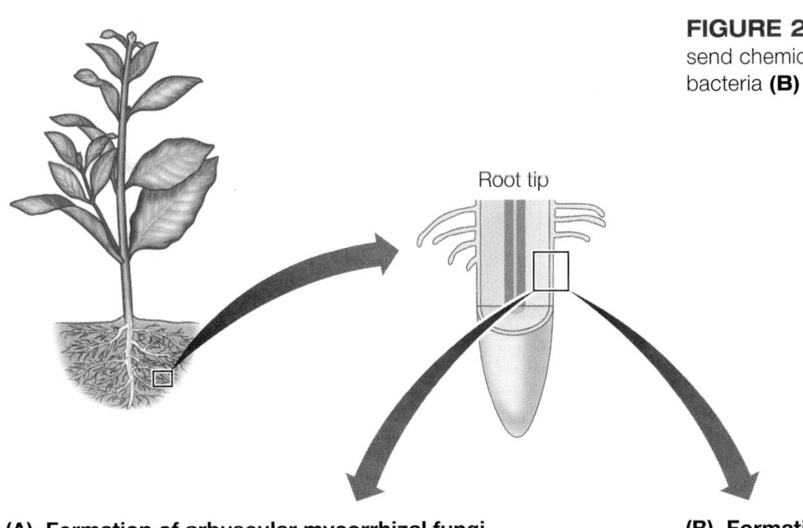

(A) Formation of arbuscular mycorrhizal fungi

Cortical cells
Epidermis
Strigolactones
Spore
Hypha

1 Plant roots produce strigolactones that stimulate rapid growth of fungal hyphae toward the root.

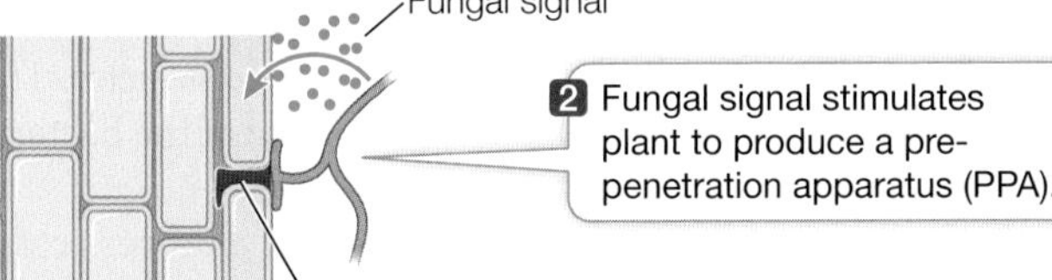

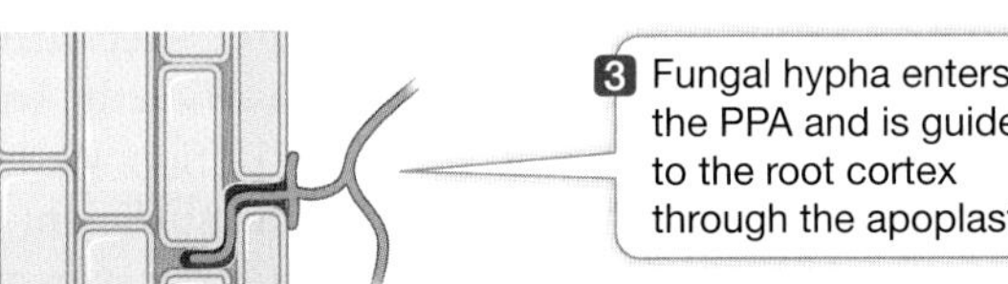

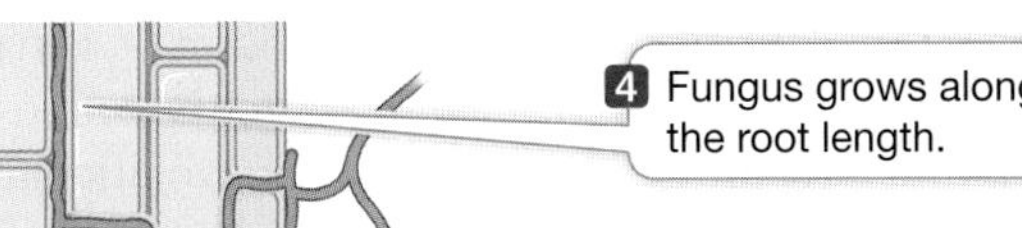

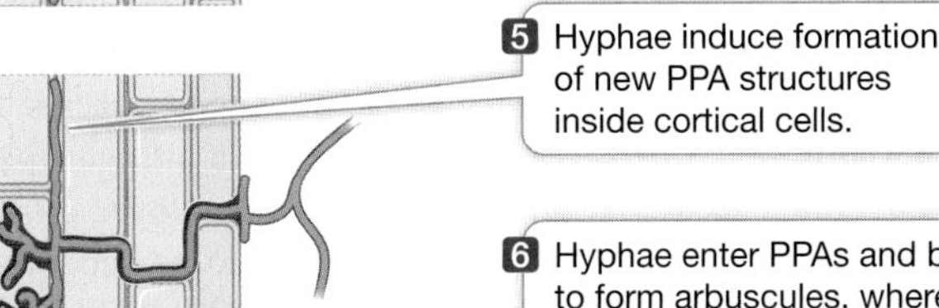

(B) Formation of a nitrogen-fixing root nodule

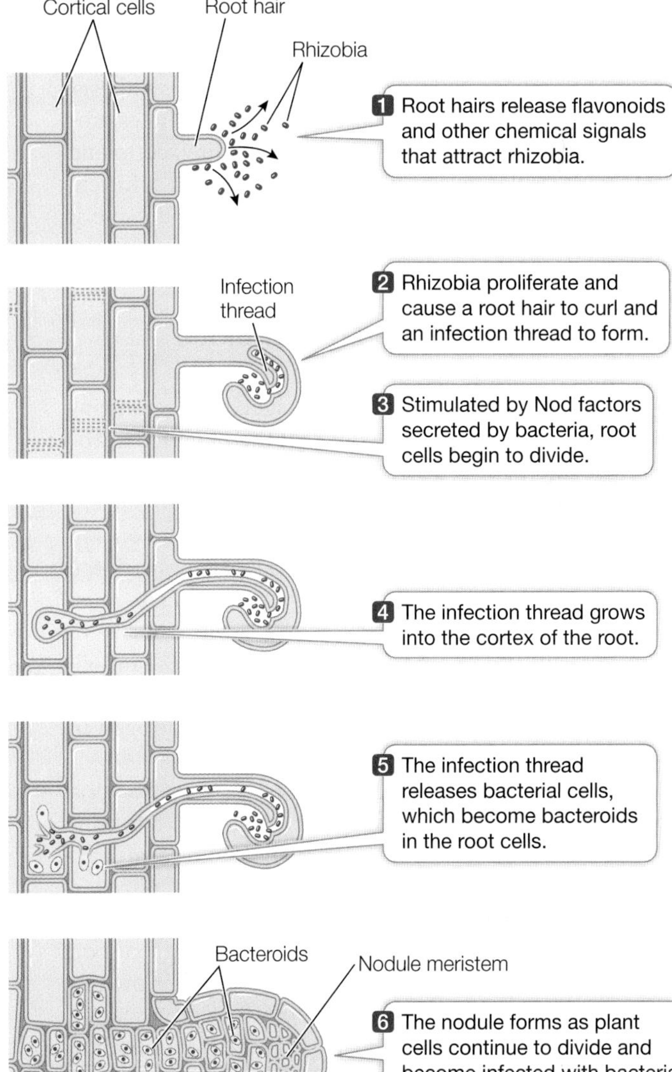

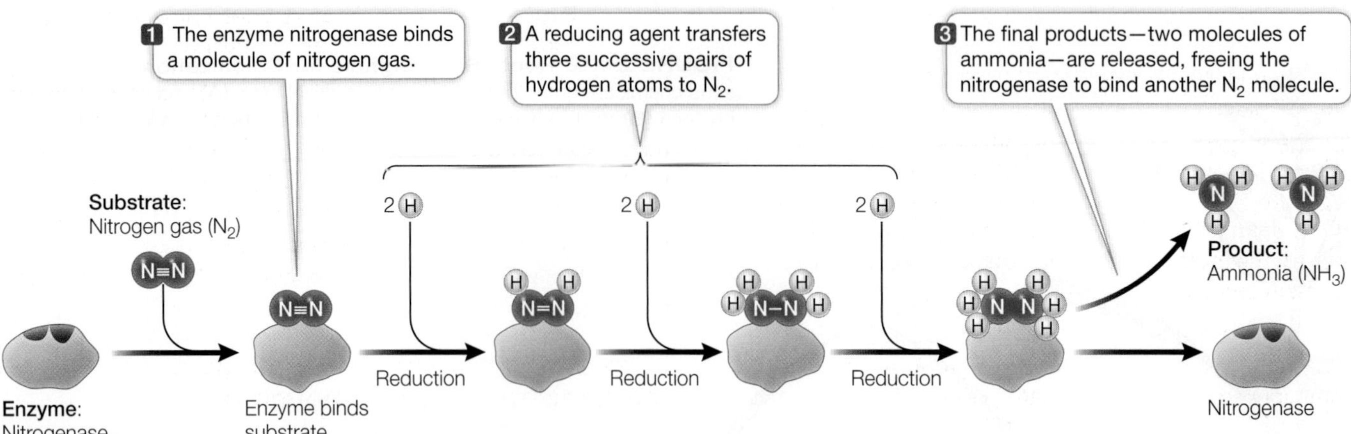

FIGURE 24.13 Nitrogenase Fixes Nitrogen Throughout the chemical reactions of nitrogen fixation, the reactants are bound to the enzyme nitrogenase. A reducing agent transfers hydrogen atoms to a molecule of nitrogen gas (N_2), and eventually the final product—ammonia (NH_3)—is released. This reaction requires a large input of energy: about 16 ATPs.

nutrients, particularly phosphate. In exchange for this service, the fungus obtains an energy source: the products of photosynthesis. In fact, up to 20 percent of the photosynthate (carbohydrates produced by photosynthesis) of terrestrial plants is directed to and consumed by arbuscular mycorrhizal fungi. Plants with sufficient levels of phosphate secrete less strigolactone, meaning the rate at which this energetically costly association occurs is tuned to the nutritional state of the plant.

Nitrogen-fixing bacteria have symbiotic relationships with plants

An association between plants and nitrogen-fixing bacteria involves an exchange of photosynthate from the plant for nitrogen from the bacteria. It is somewhat surprising that nitrogen is often limiting for plant growth, given that almost 80 percent of the air is composed of N_2 gas. However, the two N atoms in N_2 are linked by a very stable triple bond, and plants lack the machinery needed to break that bond. As a result, plants must take in nitrogen in the form of NH_4^+ or NO_3^-. These forms of nitrogen are initially present in the dead organic material in soil, but soil bacteria rapidly convert them back to gaseous forms that are released to the air, so the reserves of available nitrogen in the soil can be low.

While plants can't break the triple bond in N_2, some bacteria have an enzyme called **nitrogenase** that enables them to convert N_2 into a more reactive and biologically useful form, ammonia (NH_3)—that is, microorganisms are capable of nitrogen fixation. (Recall that the process of taking in gaseous carbon [CO_2] and incorporating it into solid molecules is called carbon fixation.) Some bacteria that fix nitrogen live on their own in soil or water (e.g., *Azotobacter* bacteria and *Nostoc* cyanobacteria). Other types of nitrogen-fixing bacteria (rhizobia) enter into symbiotic relationships with plants in the pea family (legumes). Nitrogen fixation requires a great deal of energy. Plants can provide this energy in exchange for useable nitrogen, but it costs the plants to do so. The association between plants and rhizobia is strictly controlled and occurs more frequently when the nitrogen levels in plants are low.

Roots of legumes release chemical signals that attract the rhizobia (**FIGURE 24.12B**). These signals stimulate the bacteria to secrete Nod (nodulation) factors, which in turn signal cells in the root cortex to divide and form the plant tissue that constitutes the root **nodule**. Bacteria enter the root via an **infection thread**, analogous to the prepenetration apparatus in mycorrhizal associations, and eventually reach cells inside the root nodule. There the bacteria are released into the cytoplasm of the nodule cells, enclosed in plant membrane–derived vesicles. The bacteria differentiate into **bacteroids**—the form of the bacteria that can fix nitrogen.

Inside the nodule, nitrogen gas is reduced in a process catalyzed by the bacterial enzyme nitrogenase (**FIGURE 24.13**). Nitrogenase is strongly inhibited by oxygen. The environment in the plant root tissues is typically aerobic, but the specialized nodules that form when rhizobia infect a plant provide the bacteria with the conditions necessary for nitrogen fixation. Within a nodule, O_2 is maintained at a low level that is sufficient to support aerobic respiration by the bacteria (which is necessary to supply energy for the fixation reaction) but not so high as to inactivate nitrogenase. The O_2 level is regulated by a plant-produced protein called **leghemoglobin**, which is an O_2 carrier. Leghemoglobin is a close relative of hemoglobin, the red, oxygen-carrying pigment of animals, and is thus an evolutionarily ancient molecule.

As in the association with arbuscular mycorrhizae, maintaining nitrogen-fixing bacteria is costly to the plant, consuming as much as 20 percent of all of the energy stored in photosynthate. Relatively few plant species are legumes, but those species are important to humans because they can be grown in nitrogen-poor soil. They include peas, soybeans, black-eyed peas, clover, alfalfa, and some tropical shrubs and trees. Most other plants depend on nitrogen fixed by free-living bacteria in the soil or released by the breakdown of proteins present in dead organic matter.

REVIEW & APPLY | 24.3

R Plant roots release chemicals into the soil that influence the abundance and activity of other organisms living there. For example, many plants secrete strigolactones that stimulate arbuscular mycorrhizae and promote mutualistic associations with them. Some plants, particularly legumes, send out signals that attract nitrogen-fixing bacteria which then infect the roots, establishing nodules in which nitrogen fixation can occur.

A

1. Compare the events that occur when arbuscular mycorrhizal fungi and rhizobia infect a plant root. What is the nutritional advantage to the plant in each case?
2. Researchers grew lemon seedlings in soils that contained phosphate fertilizer, arbuscular mycorrhizal fungi, or both. The table shows the mean dry weight of the seedlings after 6 months.

Phosphate fertilizer added (g)	Mycorrhizae present	Seedling dry weight (g)
0	No	1
0	Yes	10
12	No	28
12	Yes	166
24	No	20
24	Yes	210

   a. What do these results tell us about the role of phosphate fertilizer in the growth of lemon plants?
   b. Why didn't the mean weight of the seedlings increase when the amount of phosphate fertilizer, in the absence of mycorrhizae, was doubled?

Materials taken up from the soil by mycorrhizal hyphae, root hairs, and other parts of the root must move from the point of uptake to the rest of the plant. Next we'll explain the mechanisms by which water and the compounds dissolved in it—which include nutrients from the soil, amino acids, and some hormones—move from the roots to the leaves.

24.4 | Water and Solutes Are Transported in the Xylem

LEARNING OBJECTIVES

By the end of this key concept you should be able to:

24.4.1 Predict which way water will flow based on information about the components of water potential.

24.4.2 Describe how water is transported in the xylem by the transpiration-cohesion-tension mechanism.

24.4.3 Explain the function of stomata and how they regulate water and gas exchange.

Most plants take in water from their roots, and we know from observation that plants wilt if water is not present in their leaves, but how does water move from the roots to the leaves? In our bodies, the heart provides a regular force that powers the movement of blood. Plants don't have a comparable large muscle. Instead, they rely on physical principles that govern water movement, including the fact that water has a high tendency to evaporate (move from the wet plant to the dry air).

Differences in water potential govern the direction of water movement

The tendency of water to move between two wet plant cells that are at roughly the same height (i.e., where differences in gravity are not a major factor) depends on the balance between two factors: the concentration of substances that are dissolved in the water (solutes) and the pressure. Water moves toward high concentrations of solutes. You may have observed this effect if you ever put sugar on the outside of a strawberry or salt on the outside of a slug. Water moves from the cells toward the region that has a high concentration of sugar or salt. In contrast, water moves away from regions of high pressure. You can see this by moving your hand across the surface of a tub of water: the movement of water away from your hand creates a wave. The combined effect of these two tendencies is reflected in a concept called water potential.

In plant cells, the direction of water movement is determined by the **water potential** (represented by the Greek letter psi, Ψ), defined as the tendency of a solution (water plus solutes) to take up water from pure water across a membrane. Whenever water moves across a selectively permeable membrane by osmosis, it moves toward the region of lower (more negative) water potential (**FIGURE 24.14A, LEFT**). Water potential is expressed in megapascals (MPa), a unit of pressure. Atmospheric pressure—"one atmosphere"—is about 0.1 MPa, or 14.7 pounds per square inch (a typical pressure in an automobile tire is about 0.2 MPa).

You can review the structure and function of the cell membrane in **Chapter 4**; **Key Concept 4.2** describes osmosis.

The components of water potential are:

- **Solute potential (Ψ_s)**: As solutes are added to one side of a semi-permeable membrane, the tendency of water to move toward that side of the membrane increases and water potential decreases. (If the solute concentration outside a cell is greater than inside, the solution is hypertonic.)
- **Pressure potential (Ψ_p)**: When any closed compartment takes up water, it tends to swell. The walls of the compartment, however, resist that swelling (think of blowing up a balloon). The result is an increase in internal pressure in the compartment, which decreases the tendency to take up more water (increases the water potential; **FIGURE 24.14A, RIGHT**).

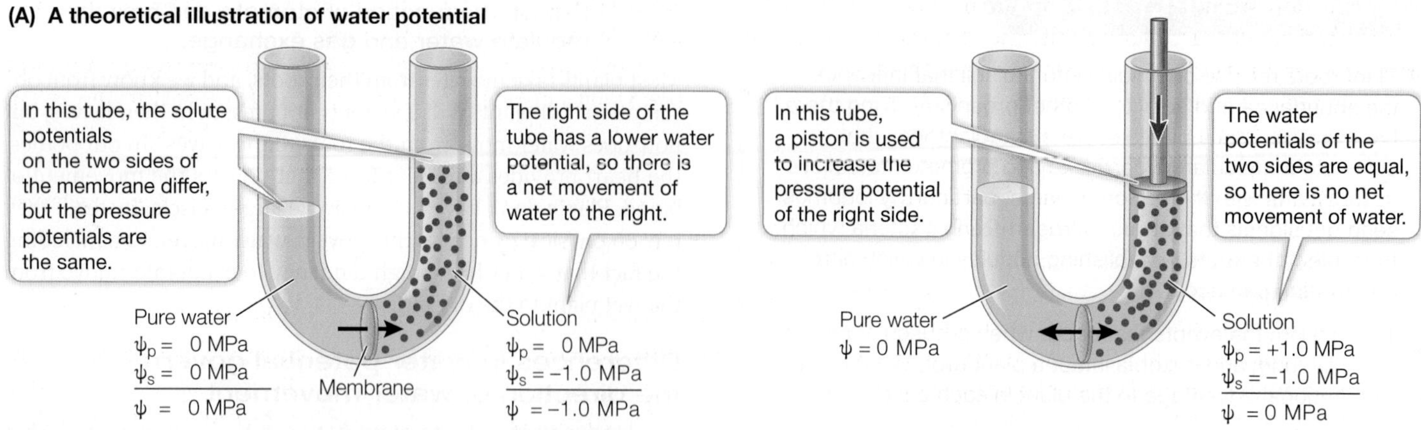

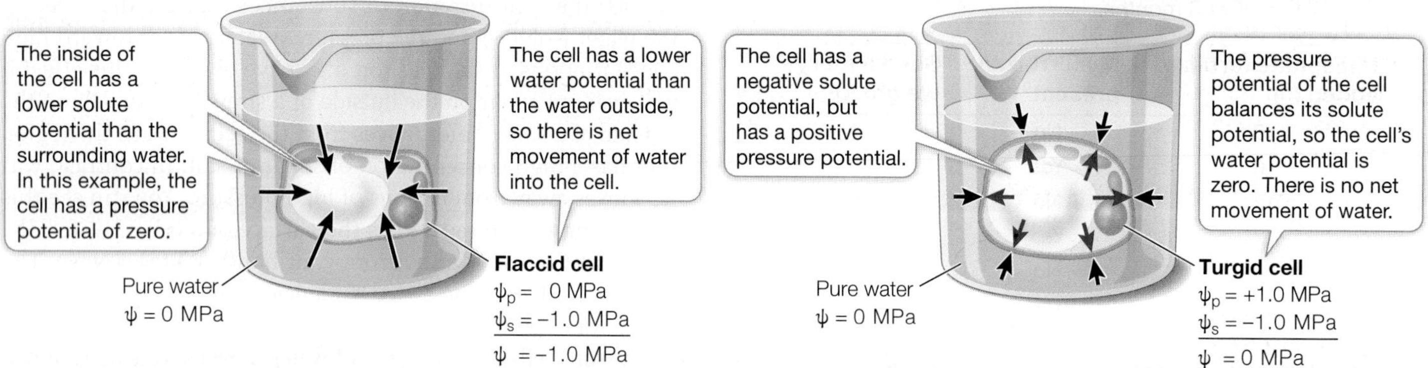

FIGURE 24.14 Water Potential, Solute Potential, and Pressure Potential
Activity 24.2 Water Uptake in Plants

We can express the effects of solute potential and pressure potential on water potential in the form of an equation:

$$\Psi = \Psi_s + \Psi_p$$

By definition the solute potential of pure water is zero; because added solutes decrease water potential, solute potential is usually negative. Pressure potential is defined as zero when it equals atmospheric pressure. Pressure potentials less than atmospheric pressure are negative and those greater than atmospheric pressure are positive.

In a plant cell (**FIGURE 24.14B**), the cell wall resists the swelling that would otherwise occur as the cell takes up water (see also Figure 4.3). The resulting **turgor pressure** is equivalent to the pressure potential exerted by the piston in Figure 24.14A. Water will enter plant cells by osmosis until the pressure potential exactly balances the solute potential. At this point the cell is said to be "turgid"—that is, it has a significant positive pressure potential. The physical structure of many plants is maintained by the (positive) pressure potential of their cells; if the pressure potential drops (for example, if the plant does not have enough water), the plant wilts (**FIGURE 24.15**).

Water moves through the xylem by the transpiration–cohesion–tension mechanism

Once water has arrived in the xylem, it is all uphill from there! Consider the magnitude of what xylem accomplishes. A single maple tree 15 meters tall has been estimated to have some 177,000 leaves,

FIGURE 24.15 A Wilted Plant A plant wilts when the turgor pressure in its cells is low.

with a total leaf surface area of 675 square meters—1.5 times the area of a basketball court. During a summer day, that tree loses 220 liters of water *per hour* to the atmosphere by evaporation from its leaves. To prevent wilting, the xylem must transport 220 liters of water 15 meters from the roots up to the leaves every hour.

How is the xylem able to move so much water to such great heights? Part of the answer lies in the structure of the xylem and its several cell types, which we introduced in Chapter 23 (see Figure 23.6). Xylem vessels consist of a long "straw" of cell walls, which provide both structural support and the rigidity needed to maintain a pressure gradient within the xylem sap. Adhesion of water to the xylem cell walls prevents water from falling back to the roots when the driving force for its upward movement stops. Early experiments ruled out two possible ways that had been hypothesized for moving water up through these vessels from roots to leaves: upward pressure and capillary action.

Upward pressure would move water by "pushing the bottom." A simple experiment in 1893 demonstrated that this mechanism does not suffice to get water to the top of a tree: A tree was cut at its base, and the sawed-off upper stem was placed in a vat of poison that killed living cells. (Recall that mature xylem is made up of dead cells.) The poison continued to rise through the trunk, even though cells were killed along the way. This observation showed that a living "pump" is not required to push the xylem sap up a tree. Furthermore, the roots were absent, yet upward flow in the trunk continued, demonstrating that the roots do not provide upward pressure. When the poison reached the leaves and killed them, however, all further movement up the trunk stopped, demonstrating that leaves must be alive for water to move in the xylem.

Capillary action was also ruled out as the mechanism that makes water rise in the xylem. Water molecules have strong cohesion, so a column of water can indeed rise a short distance in a thin tube (e.g., a straw or a xylem vessel) by capillary action, but calculations have shown that xylem vessels, at 100 micrometers (μm) in diameter, are simply too wide to get water to the top of a 15-meter tree in this fashion. In fact, the maximum height for a water column raised by capillary action alone in a 100-μm-diameter vessel is only 0.15 meters.

Almost a century ago, the **transpiration-cohesion-tension** theory was proposed to explain water transport in the xylem (**FIGURE 24.16**). According to this theory, the key elements of xylem transport are:

- *Transpiration*: evaporation of water from cells within the leaves
- *Cohesion* of water molecules in the xylem sap as a result of hydrogen bonding
- *Tension* on the xylem sap resulting from transpiration

The concentration of water vapor in the atmosphere is lower than that in the air spaces between the cells of a leaf. Because of this difference, water vapor diffuses out of the leaf in a process called **transpiration**. Within the leaf, water evaporates from the moist walls of the mesophyll cells and enters the intercellular air spaces. As water evaporates from the aqueous film coating each cell, the film shrinks back into tiny spaces in the cell walls, increasing the curvature of the water surface and thus increasing its surface tension. Because of hydrogen bonding, water molecules have cohesion, and therefore the increased tension (negative pressure

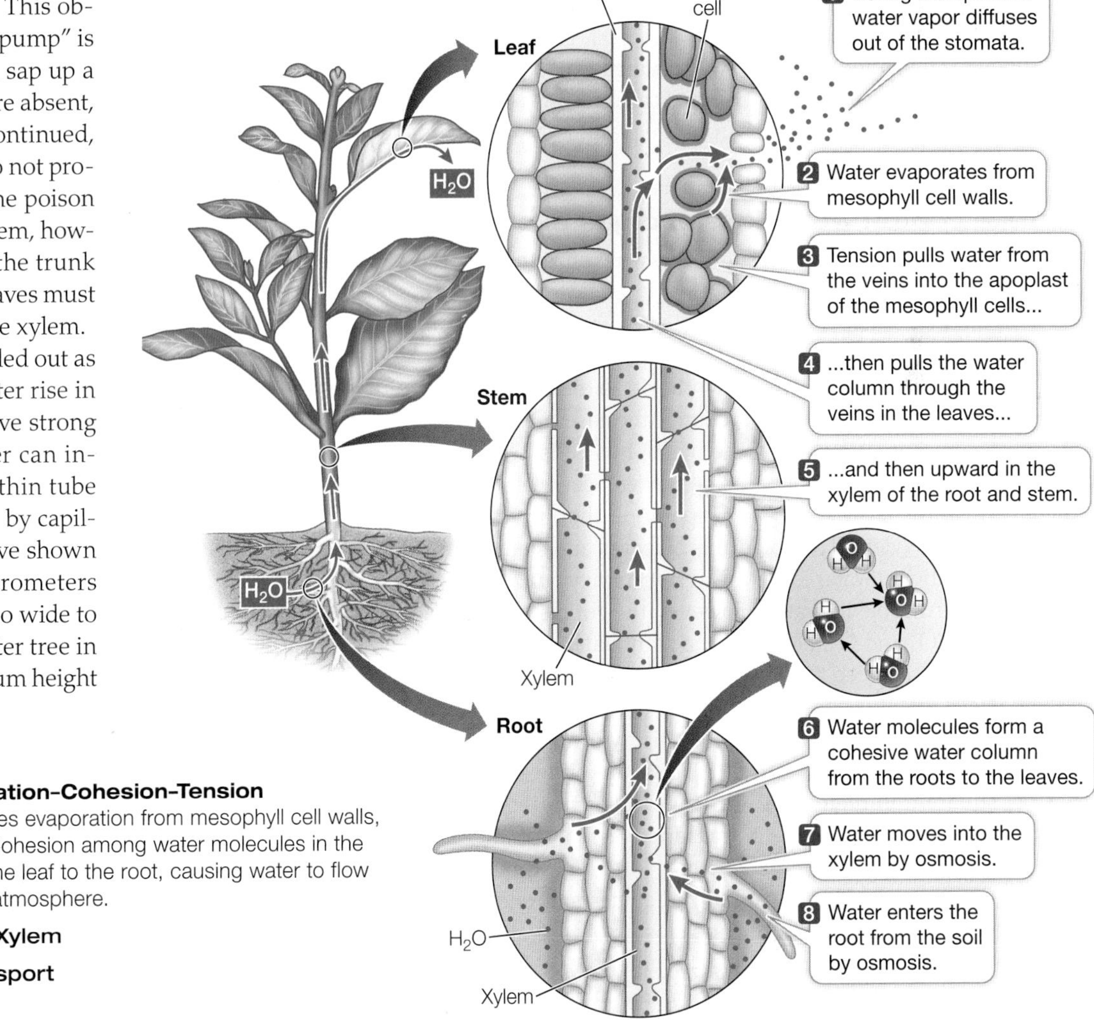

FIGURE 24.16 The Transpiration-Cohesion-Tension Mechanism Transpiration causes evaporation from mesophyll cell walls, generating tension on the xylem. Cohesion among water molecules in the xylem transmits the tension from the leaf to the root, causing water to flow in the xylem from the roots to the atmosphere.

Media Clip 24.2 **Inside the Xylem**

Animation 24.2 **Xylem Transport**

potential) in the surface film draws more water into the cell walls, replacing that which was lost by evaporation. The resulting tension in the mesophyll draws water from the xylem of the nearest vein into the apoplast surrounding the mesophyll cells. The removal of water from the veins, in turn, establishes tension on the entire column of water contained in the xylem, so that the column is drawn upward all the way from the roots.

Each part of this model is supported by evidence:

- The difference in water potential between the soil solution and air is huge, on the order of –100 MPa. This difference generates more than enough tension to pull a water column up the tallest tree.
- There is a continuous column of water in the xylem, which is caused by cohesion.
- Measurements of xylem pressures in cut stems show a negative pressure potential, which indicates considerable tension.

In addition to its role in xylem transport, transpiration has the added benefit of cooling a plant's leaves (much as humans sweat to cool off). The evaporation of water from mesophyll cells consumes heat, thereby decreasing the leaf temperature. A farmer can hold a leaf between thumb and forefinger to estimate its temperature; if the leaf doesn't feel cool, that means transpiration is not occurring and it must be time to water.

Although transpiration provides the driving force for the transport of water and minerals in the xylem, it also results in the loss of tremendous quantities of water from the plant. How do plants control this loss?

Stomata control water loss and gas exchange

In Chapter 5 we described photosynthesis, which has the general equation

$$CO_2 + H_2O \longrightarrow \text{carbohydrate} + O_2$$

Leaves are the chief organs of photosynthesis, so they must have a large surface area across which to exchange the gases O_2 and CO_2. This surface area is provided by the mesophyll cells being surrounded by abundant air spaces inside the leaf (see Figure 23.8), which are also the location of the evaporation that drives water movement in the xylem. The epidermis of leaves and stems minimizes transpirational water loss by secreting a waxy cuticle, which is largely impermeable to water. However, the cuticle is also impermeable to CO_2 and O_2. How can the plant balance its need to retain water with its need to obtain CO_2 for photosynthesis?

Plants have met this challenge with the evolution of **stomata** (singular *stoma*): pores in the leaf epidermis, typically on the underside of the leaf (**FIGURE 24.17**). There are up to 1,000 stomata per square millimeter in some leaves, constituting up to 3 percent of the total surface area of the shoot. Such an abundance of stomata could lead to excessive water loss. But stomata are not static structures. A pair of specialized epidermal cells, called **guard cells**, controls the opening and closing of each stoma.

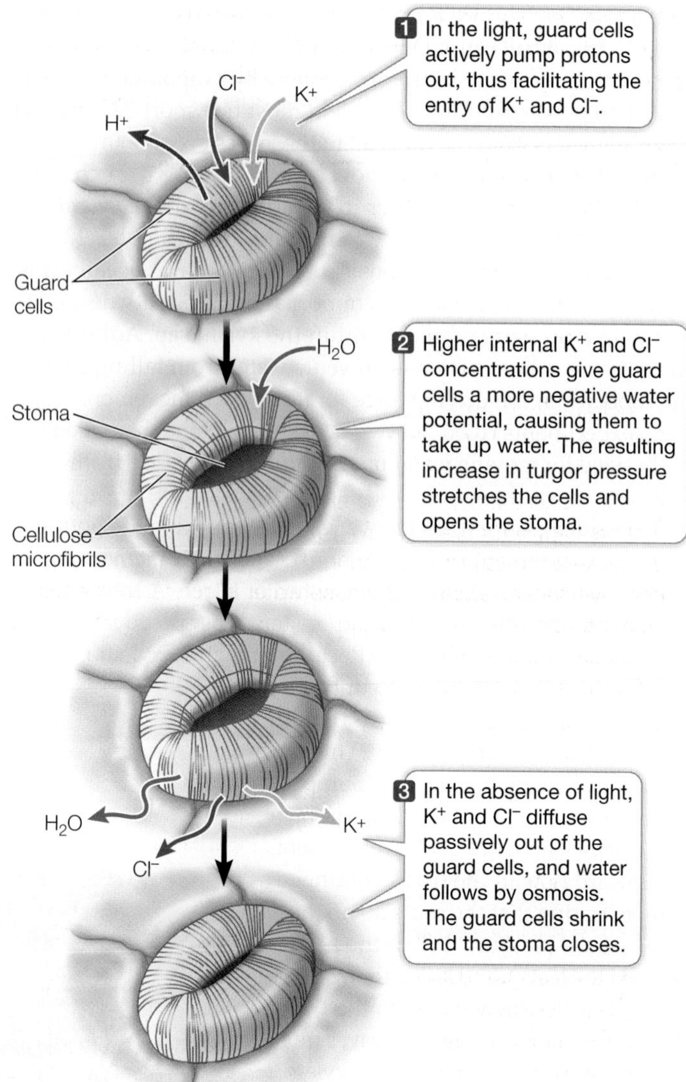

FIGURE 24.17 Stomatal Opening Is Controlled by the Water Potential of the Guard Cells Ion concentrations affect the water potential of the guard cells (shown in green), controlling the opening and closing of stomata. When potassium ions (K^+) enter guard cells, along with negatively charged ions (e.g., Cl^-) that maintain the electrical balance, the water potential decreases and stomata open.

When the stomata are open, CO_2 can enter the leaf by diffusion, but water vapor diffuses out of the leaf at the same time. Closed stomata prevent water loss but also exclude CO_2 from entering the leaf. Most plants open their stomata when the light intensity is sufficient to maintain a moderate rate of photosynthesis. At night, when darkness precludes photosynthesis, their stomata are closed; no CO_2 is needed at this time, and water is conserved. Even during the day, the stomata close if water is being lost at too great a rate.

Guard cells can respond to changes in light and CO_2 concentration in a matter of minutes by changing their solute potential. The absorption of light by a pigment in the guard cell's cell membrane activates a proton pump (see Figure 24.7), which actively transports H^+ out of the guard cells and into the apoplast of the surrounding epidermis. The resulting electrochemical gradient drives K^+ into the guard cells, where it accumulates (see Figure 24.17). Negatively charged chloride (Cl^-) ions and organic ions also move into and out of the guard cells along with the K^+ ions, maintaining electrical balance. The increased concentration of K^+ and other solutes inside the guard cells makes the solute potential of the guard cells more negative. Water then enters by osmosis, increasing the turgor pressure of the guard cells. The guard cells change their shape, becoming more turgid in response to the increase in pressure potential, so that a space—the stoma—appears between them. This resembles the change in shape of a bicycle inner tube that opens to a wide circle as air is pumped in. The stoma closes in the absence of light: the proton pump becomes less active, K^+ ions diffuse passively out of the guard cells, water follows by osmosis, the pressure potential decreases, and the guard cells lose turgor and sag together, sealing off the stoma. Guard cell membranes are particularly rich in aquaporins, making guard cells well adapted for the rapid water movements involved in stomatal responses.

REVIEW & APPLY | 24.4

R The direction of water movement in a plant is determined by water potential. For moist plant cells sitting next to one another, water potential is determined by the sum of the solute potential and the pressure potential. Water moves to regions that have lower (more negative) water potentials. Water transpires (evaporates from the leaves) because the air has a very negative water potential. Because water is cohesive, the tension created when water evaporates pulls a long chain of water molecules up the xylem. Thus water moves through the xylem via transpiration, cohesion, and tension.

A 1. Consider the two cells, A and B, below. Assuming the cells are sitting next to one another, which way will the water flow: toward cell A or toward cell B? Explain your answer.

Cell A	Cell B
Ψ_p = 0.32 Mpa	Ψ_p = 0.40 MPa
Ψ_s = −0.90 MPa	Ψ_s = −0.95 MPa

2 *Tillandsia fasciculata* plants are epiphytic, living on the trunks of other trees. Their roots are not in the soil. To determine the water relations in this species, leaves were collected just before dawn on the day(s) indicated in the graph. The water potential and solute potential of living plants are always negative; note that the values on the *y*-axis on the graph carry a negative sign. Blue circles represent the water potential, and red circles represent the osmotic potential (solute potential) (*figure source*: G. Zotz and J. L. Andrade. 1998. *J Plant Physiol* 152: 545–554).

a. Use the graph to determine the water potential and solute potential of these plants in mid-August (in the wet season) and in mid-February (in the dry season). How much did each change between August and February?

b. Estimate the pressure potential in mid-August and in mid-February.

c. Which component of water potential, solute potential or pressure potential, contributed the most to the decrease in water potential observed in the dry season? Based on

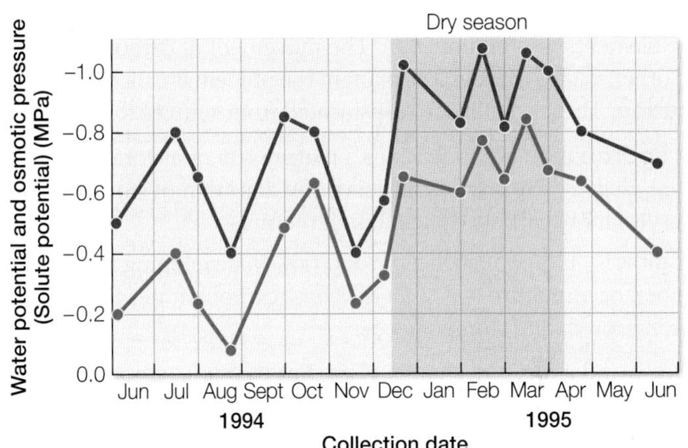

this observation, do you expect that these plants would have appeared more wilted in the dry season?

3. “Typical” plants open their stomata during the day when sunlight that provides the energy for carbon fixation is available. Some species have a special adaptation (called CAM; see Key Concept 5.5) that includes opening stomata only at night, when it is cooler. This adaptation causes less water to be lost. It also limits the amount of carbon obtained from the air and so results in slow growth—a functional compromise in dry environments. Some plants in the genus *Tillandsia* are “typical,” while other species have the CAM adaptation. The table below lists the negative pressure (otherwise known as tension) in the xylem in the day and night for representatives of two *Tillandsia* species. Which species, A or B, opens its stomata only at night? Explain.

Tillandsia species	Xylem pressure in the day (MPa)	Xylem pressure in the night (MPa)
A	−0.51	−0.19
B	−0.21	−0.36

Source: J. A. C. Smith et al. 1985. *Oecologia* 67: 475–485.

Now that we understand how leaves obtain the water and CO_2 required for photosynthesis, let's examine how the products of photosynthesis are transported to other parts of the plant where they are needed.

24.5 Solutes Are Transported in the Phloem

LEARNING OBJECTIVE

By the end of this key concept you should be able to:

24.5.1 Describe how solutes are transported in the phloem by pressure flow.

As photosynthesis occurs in the leaf, the carbohydrate products of photosynthesis (photosynthate; mainly sucrose) diffuse to the nearest small vein, where they are actively transported into sieve tube elements (see Figure 23.6). The movement of carbohydrates and other solutes through the plant in the phloem is called **translocation**. These solutes are translocated from sources to sinks:

- A **source** is an organ (such as a mature leaf or a storage root) that *produces* (by photosynthesis or by digestion of stored reserves) more carbohydrates than it requires.
- A **sink** is an organ (such as a root, flower, developing fruit or tuber, or immature leaf) that *consumes* carbohydrates for its own growth and storage needs.

Sources and sinks can change roles. For example, storage roots (such as sweet potatoes) are sinks when they accumulate carbohydrates, but they are sources when their stored reserves are mobilized to nourish other organs in the plant.

Sucrose and other solutes are carried in the phloem

Evidence that the phloem carries sucrose and other solutes initially came in the 1600s, when Marcello Malpighi removed a ring of bark (containing the phloem) from the trunk of a tree, while leaving the xylem intact—that is, he "girdled" the tree (**FIGURE 24.18**). Over time, the bark in the region above the girdle swelled. Malpighi correctly concluded that the solution coming from the leaves above the girdle was trapped in the bark. Later, the bark below the girdle died because it was no longer receiving sugars from the leaves. Eventually the roots, and then the entire tree, died.

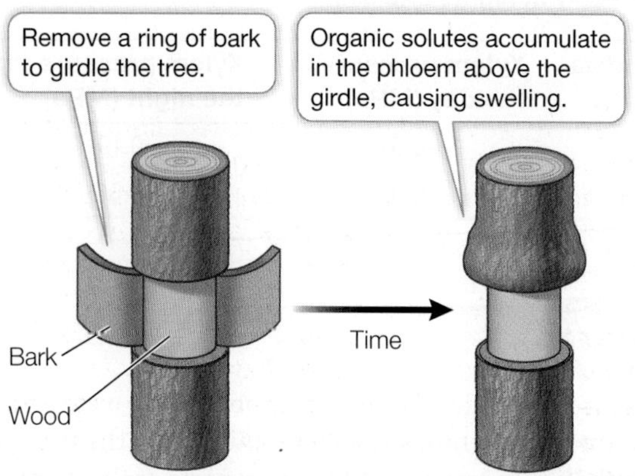

FIGURE 24.18 Tree Girdling Demonstrates Phloem Flow

More recently plant biologists have analyzed the contents of phloem with the help of aphids. These insects feed on plants by drilling into sieve tube elements with a specialized organ, the stylet. The pressure potential in the sieve tube is higher than that outside the plant, so the phloem sap is forced through the stylet and into the aphid's digestive tract. So great is the pressure that sugary liquid is forced through the insect's body and out its anus. If an aphid is frozen in the act of feeding, its body can be broken off the plant stem, leaving the stylet intact. Phloem sap continues to flow out of the stylet for hours.

Several observations have come from analyzing phloem sap in this way, as well as from other experiments:

- Ninety percent of the phloem solutes consist of sucrose, with the rest being hormones, other small molecules such as mineral nutrients and amino acids, and viruses.
- The flow rate can be very high, as much as 100 centimeters per hour.
- Different sieve tube elements conduct their contents in different directions (up or down the stem, for example). Movement in the phloem as a whole is bidirectional, but it is unidirectional within an individual sieve tube.
- Movement of sap in the phloem requires living cells, in contrast to movement in the xylem.

The pressure flow model describes the movement of fluid in the phloem

The observations described above led to the **pressure flow model** as an explanation for translocation in the phloem (**FIGURE 24.19**). According to this model, sucrose is actively transported at a source into companion cells, from which it flows through plasmodesmata into the sieve tube elements. This gives those cells a higher sucrose concentration than the surrounding cells (a more negative solute potential and therefore a lower water potential), and water therefore enters the sieve tube elements from the xylem by osmosis. The entry of this water increases turgor pressure (causes a more positive pressure potential) in the sieve tube at the source. This pressure pushes the entire fluid content of the sieve tube toward the sink end of the tube; molecules that are present in the phloem are pushed along with the water, much as a branch thrown on a stream will be carried by the water. In other words, the sap moves in response to a pressure gradient. In the sink, the sucrose is unloaded both passively and by active transport, and water moves back into the xylem. In this way, the gradient of solute potential and pressure potential needed for the movement of phloem sap (translocation) is maintained.

Two steps in phloem translocation require metabolic energy:

- *Loading*: Transport of sucrose and other solutes from sources into companion cells and then into the sieve tubes
- *Unloading*: Transport of solutes from the sieve tubes into some sinks

The need for metabolic energy is why phloem transport, unlike xylem transport, requires living cells.

Two general routes can be taken by sucrose and other solutes as they move from the mesophyll cells into the phloem. In many

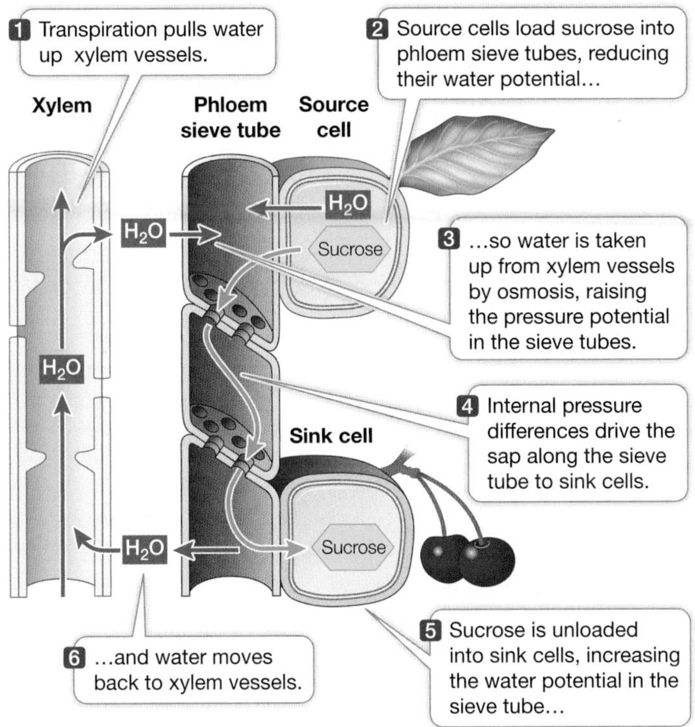

FIGURE 24.19 The Pressure Flow Model Water potential differences produce a pressure gradient and bulk flow of phloem sap from sources to sinks.

Animation 24.3 The Pressure Flow Model

plants, sucrose and other solutes follow an apoplastic pathway: they leave the mesophyll cells and enter the apoplast before they reach the sieve tube elements. Specific sugars and amino acids are then actively transported into cells of the phloem. Because the solutes cross at least one selectively permeable membrane in the apoplastic pathway, selective transport can be used to regulate which specific substances enter the phloem. In other plants, solutes follow a symplastic pathway: the solutes remain within the symplast all the way from the mesophyll cells to the sieve tube elements. Because no membranes are crossed in the symplastic pathway, a mechanism that does not involve membrane transport is used to load sucrose into the phloem.

In sink regions, the solutes are transported *out* of the sieve tube elements and into the surrounding tissues. This unloading serves several purposes: it helps maintain the gradient of solute potential, and hence of pressure potential, in the sieve tubes; it moves nutrients to parts of the plant that are growing rapidly, such as young leaves or flowers, and can help build up high concentrations of carbohydrates in storage organs, such as developing fruits and seeds.

REVIEW & APPLY | 24.5

R The pressure flow model proposes that the contents of the phloem move from sources (areas that produce excess carbohydrates) to sinks (areas that use more carbohydrate than they produce) by pressure-driven bulk flow. High concentrations of sucrose at the source cause water to move into the phloem, increasing the pressure in that area. When sucrose is unloaded at the sinks, the pressure in the sinks declines, which creates a pressure difference that drives the phloem from source to sink.

A

1. Deer sometimes graze on evergreen trees by eating bark during the winter when other food is scarce. In the spring, those trees may die. What problem arises from having the bark eaten?
2. Movement of blood through our bodies is driven by the heart, a single large muscle. In contrast, movement of liquid and solutes through the phloem has been described as being driven by a large number of tiny "pumps." Where in the phloem transport system are these small pumps located?
3. Make a table that compares the flow of fluids in xylem and in phloem with respect to the driving force for movement. Include columns to indicate the tissue types in which flow occurs, and rows indicating the driving force, the cell types, whether the cells are alive, and whether the driving pressure is positive or negative.

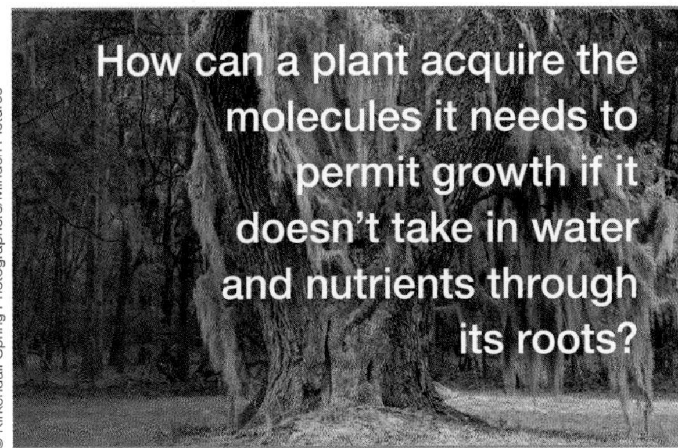

Answer: Air plants are indeed remarkable in surviving without using roots for absorption. They obtain CO_2, water, and mineral nutrients through their leaves. Specialized hairs, called trichomes, funnel water toward the surface of the leaf where it is absorbed. Water loss is reduced by a combination of low stomatal densities and by the trichomes, which can reflect light. For many plants that grow in high-light environments, reflecting light decreases leaf temperature and minimizes transpiration. Some studies suggest that nitrogen is supplied through rain or fog; nitrogen-fixing bacteria have also been observed on or around the leaves of *Tillandsia* plants.

24 VISUAL SUMMARY

QUESTIONS

24.1 Essential Plant Nutrients Include the Products of Photosynthesis

FIGURE 24.1

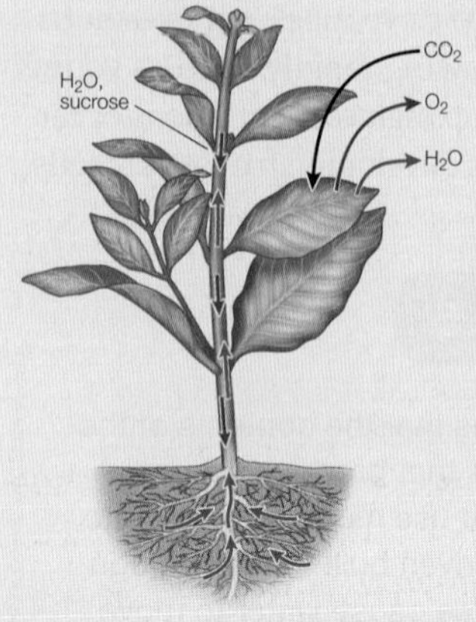

Go to ANIMATION 24.1

- The dry weight of a plant comes primarily from carbon in the air (**FIGURE 24.1**).
- The products of photosynthesis are broken down to provide energy for life.
- The **ESSENTIAL ELEMENTS** that plants require are classified as **MACRONUTRIENTS** or **MICRONUTRIENTS** depending on the amount needed. These are generally taken in from the soil.

1. What is the source of the carbon and nitrogen in a plant?
2. Does a plant contain more carbon or more nitrogen?
3. What types of molecules contain carbon? Nitrogen? Iron?
4. How might you identify a nutrient deficiency in a plant?

24.2 Mineral Nutrients Are Commonly Acquired from the Soil

FIGURE 24.6

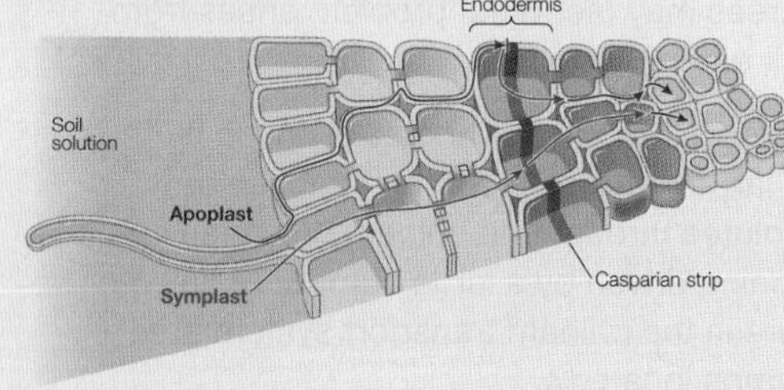

Go to ACTIVITY 24.1

- **LEACHING** of minerals from the **TOPSOIL** (the A **HORIZON**, where the organic material resides) is affected by soil particle size.
- **CATION EXCHANGE** occurs when plants pump H^+ into the soil, liberating cations that plants can then take up.
- Movement of compounds in the soil water through the plant's **APOPLAST** is restricted by specialized cell walls in the **ENDODERMIS** called the **CASPARIAN STRIP** (**FIGURE 24.6**).

1. How might a plant with a deficiency of a particular cation increase its uptake of that cation? What is the cost associated with this strategy?
2. How does the Casparian strip affect Na^+ uptake?

24.3 Root Exudates Affect Soil Organisms and Nutrient Uptake

FIGURE 24.12

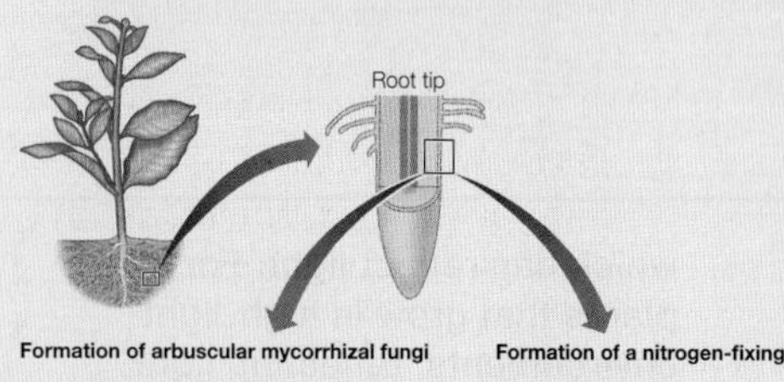

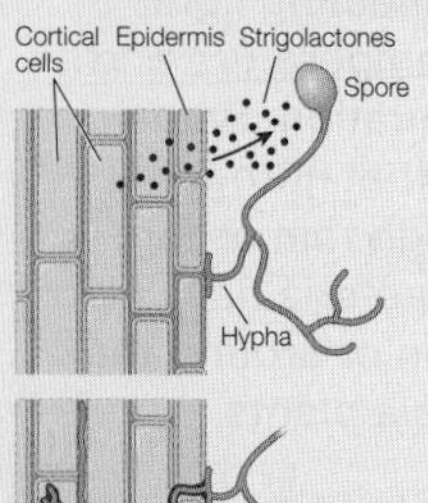

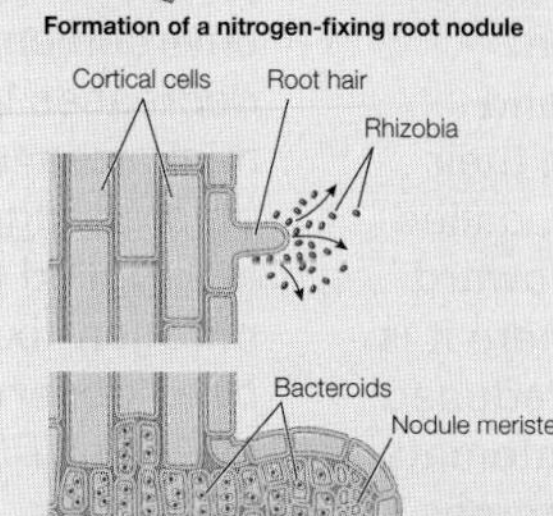

- Roots secrete **EXUDATES** into a region of the soil called the **RHIZOSPHERE**. Exudates alter the abundance and activity of beneficial microorganisms (**FIGURE 24.12**).
- Plants send signals that stimulate the formation of arbuscular mycorrhizae, which expand the root system.
- Nitrogen fixation occurs in root **NODULES** with the aid of bacteria that produce the enzyme **NITROGENASE**.

1. In what ways can the presence of microorganisms in the soil benefit plant growth and development?
2. What properties of nodules promote nitrogen fixation?
3. In what ways might the root–soil interface be similar to the interior of the human gut?

24.4 Water and Solutes Are Transported in the Xylem

Go to ACTIVITY 24.2 and ANIMATION 24.2

FIGURE 24.16

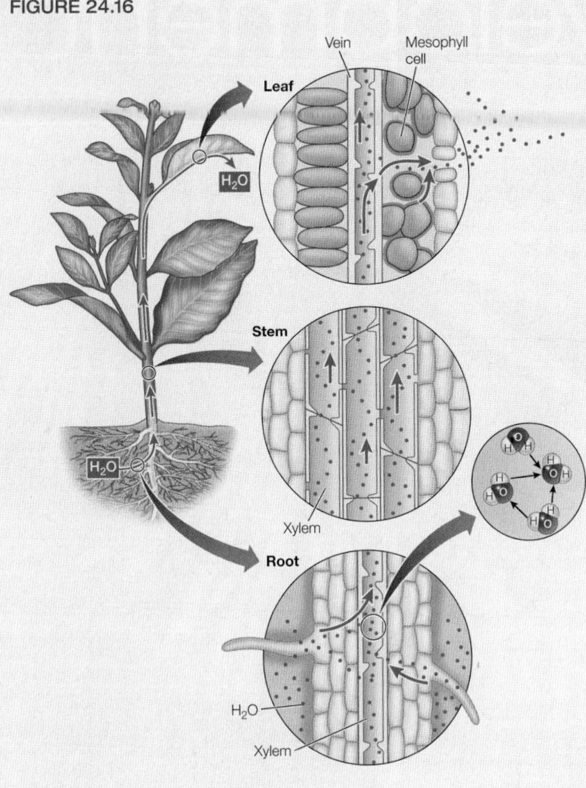

- **WATER POTENTIAL** is the sum of the **SOLUTE POTENTIAL** and the **PRESSURE POTENTIAL** ($\Psi = \Psi_s + \Psi_p$). Water tends to move toward regions of high solute concentration and away from regions of high pressure.
- Loss of water from leaves creates tension (negative pressure) that pulls a column of water through the xylem. This is called the **TRANSPIRATION–COHESION–TENSION** mechanism for movement of xylem sap (**FIGURE 24.16**).
- **STOMATA** open and close to allow a balance between intake of CO_2 and the loss of water.

QUESTIONS

1. In many agricultural systems that rely on irrigation, the solutes in the irrigation water remain in the field after the water has evaporated. As a result, the fields become increasingly salty over time. Explain why it is more difficult for a plant to grow in soil that contains a high concentration of salts.
2. Some mangroves live with their roots submerged in (salty!) seawater. What change in their root cells would make their water potential lower than that of the sea, allowing them to take up water from the sea?
3. Some artificial plants look quite real. What information do you gain from touching the leaf that can allow you to rapidly determine if the plant is real or not?

24.5 Solutes Are Transported in the Phloem

Go to ANIMATION 24.3

FIGURE 24.19

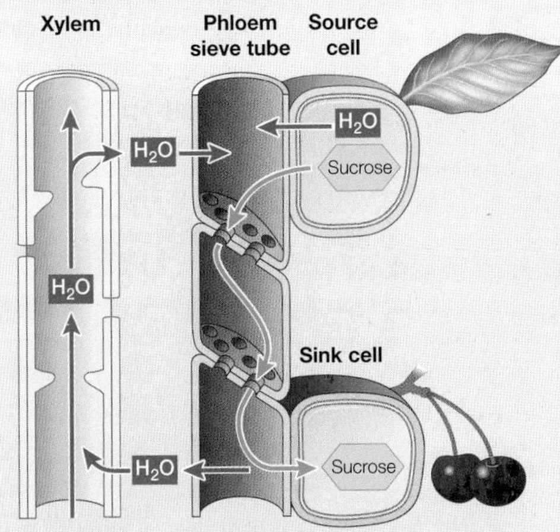

- The phloem moves photosynthate and other solutes from **SOURCE** to **SINK** (**FIGURE 24.19**).
- According to the **PRESSURE FLOW MODEL**:
  - High concentrations of solutes at the source draw water, increasing pressure that pushes the fluid and its contents toward the sink.
  - As solutes are unloaded at sinks, water moves out, and pressure drops.
  - Water leaving the phloem at a sink may reenter the xylem.

1. Some people say that the contents of the xylem move up, toward the leaves, while the contents of the phloem move down, toward the roots. What is wrong with this statement?
2. In the phloem, water flows away from, rather than toward, a high concentration of solutes. Explain how this is possible.
3. What is the driving force that causes the contents of the phloem to pass through an aphid stylet?

25 Plant Growth and Development

Plants survive in unfavorable environments in part because they have a remarkable ability to grow and develop new organs throughout their lives—something that most animals cannot do. Sometimes, however, the processes that allow plants to produce new body parts go a bit awry.

What happened to the plant shown here?

You will find the answer to this question on page 636.

KEY CONCEPTS

25.1 Cell Differentiation Occurs Progressively as Meristematic Cells Mature

LEARNING OBJECTIVES

By the end of this key concept you should be able to:

25.1.1 **Explain how expression of a transcription factor can establish the identity of a cell.**

25.1.2 **Hypothesize as to why retaining the ability to undergo redifferentiation is advantageous to plants.**

25.1.3 **Describe how plant cells increase in size by taking in water.**

25.1.4 **Describe mechanisms that affect the shape of plant cells and tissues.**

As organisms move through their life cycles, they change continually. A gardener who plants a seed may talk about the resulting plant as if it were a single thing, but the size and structure of the seed, seedling, leafy plant, and fruit-bearing plant are always different. **Development** is the term used to describe the process by which a plant or other organism changes, taking on the successive forms that characterize its life cycle. Plant development occurs through a combination of **growth**, an increase in size of the body and its organs, and **differentiation**, the process by which cells become different from one another. Because plants are rooted in place, their survival depends on their ability to cope with the environment around them. They can't run away or put on a coat as we can. Their remarkable ability to produce new organs and reshape their bodies throughout their lives plays a significant role in their ability to thrive in a variety of environmental circumstances.

Cell fates become progressively more restricted during development

A zygote is a single cell that gives rise to an embryo and, from that, an organism. The cells of an early embryo are **totipotent** (*toti*, "all"; *potent*, "capable"), meaning they have the potential to differentiate into any cell type, including more embryonic cells. As the embryos of many organisms mature, cells become progressively restricted in their potential, eventually being limited to producing just one cell type. When a cell has taken on the characteristics of a single specific cell type, we say that it has differentiated. That differentiated state reflects the **cell fate**.

While plant cells undergo progressive changes in their identity as the plant develops, plant cells are remarkable in the extent to which they remain capable of redifferentiation. It is possible to cut sections of mature leaves or roots and grow them in culture. Culturing plant cells causes them lose their prior differentiated state and form **callus**—a cluster of dividing cells. Given appropriate hormonal treatments, callus cells can redifferentiate to form a whole new plant. In 1958 Frederick Steward at Cornell University showed that if he isolated cells from a carrot root and maintained them in a suitable nutrient medium, he could induce them to divide and give rise to masses of cells called calli (singular callus), which no longer had the cell-type characteristics that they started with. These callus cells could be maintained in culture indefinitely. Furthermore, if the callus cells were provided with the right chemical cues, they could develop into embryos and eventually into complete new plants (**FIGURE 25.1**).

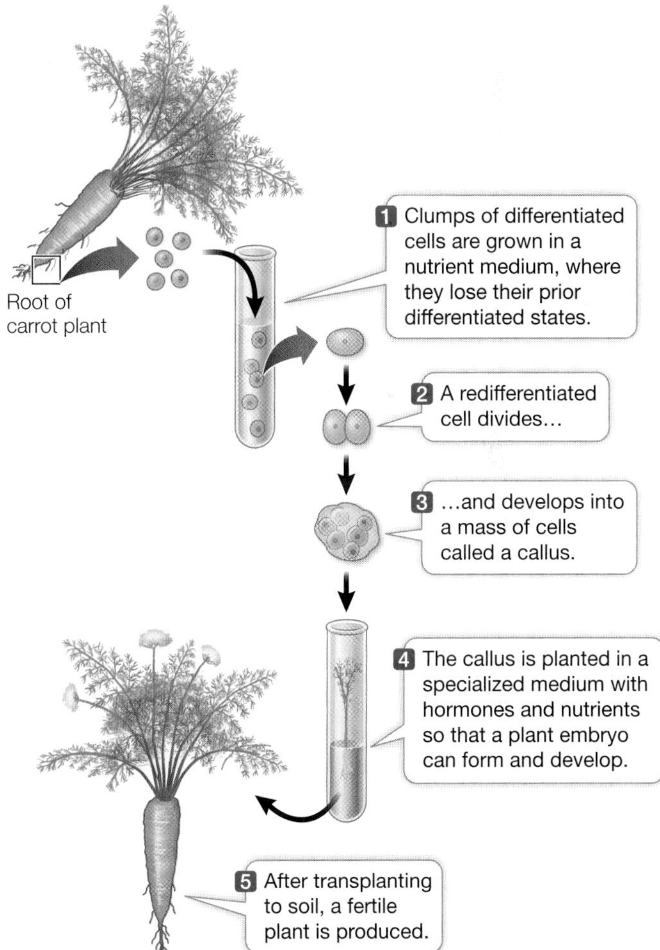

FIGURE 25.1 Differentiated Plant Cells Can Be Totipotent When differentiated plant cells are removed from the plant and placed in medium with nutrients and specific growth hormones, they can redifferentiate into callus and then be induced to generate an entire new plant. (After F. C. Steward. 1958. *Am J Bot* 45: 705–709.)

The new plants in Steward's experiments were genetically identical to the cells from which they came—that is, they were **clones** of the original carrot plant. The ability to produce clones is evidence that the somatic cells in a plant have all the genetic information needed to become any cell in the plant.

Many types of cells from other plant species show similar behavior in the laboratory. The ability to generate a whole plant from groups of cells or even a single cell has been invaluable in agriculture and forestry. For example, trees from planted forests are used in making paper, lumber, and other products. To replace the trees reliably, forestry companies regenerate new trees from the leaves of selected trees with desirable traits. The characteristics of these clones are more uniform and predictable than those of trees grown from seeds.

Both the meristem-based growth of plants (described in Chapter 23) and the ability of their cells to redifferentiate give plants an extraordinary ability to respond to environmental factors in a developmental way. Animals lack the extreme flexibility that plant cells display in terms of their ability to redifferentiate. It's still not possible to generate a whole new person from a muscle cell taken from one's arm. However, animals can respond to challenging environments by moving away from them, or using other aspects of behavior to change the situation. When an animal is thirsty, it can walk to a location where it can get a drink. Plants can't do that because they are rooted in place. However, a plant that is growing in a patch of soil lacking sufficient nutrients can form new roots that reach into more favorable regions. This observation has given rise to the saying "Where animals respond behaviorally, plants respond developmentally."

Changes in gene expression underlie cell fate determination and differentiation

A central component of development is the ability to form distinct cell types. The cells in human eyes contain proteins that allow us to see light, while intestinal cells produce digestive enzymes; if eyes produced digestive enzymes, they would not work very well! The same is true for plants. Epidermal cells in the root contain proteins that span the cell membrane and promote uptake of nutrients from the soil, while leaves contain proteins that promote photosynthesis. Given that each cell in a multicellular organism generally contains the same genes, how do the differences between cell types arise? The answer is that the distinction between cells largely arises from the proteins they contain.

LINK As described in **Key Concept 10.2**, the process by which proteins are produced from DNA is called gene expression. Gene expression involves transcription of a gene, resulting in production of mRNA, followed by translation (protein synthesis). Gene expression can be regulated at several different points in this process. **Key Concept 11.3** describes how transcription factors are involved in regulating production of mRNA from specific genes.

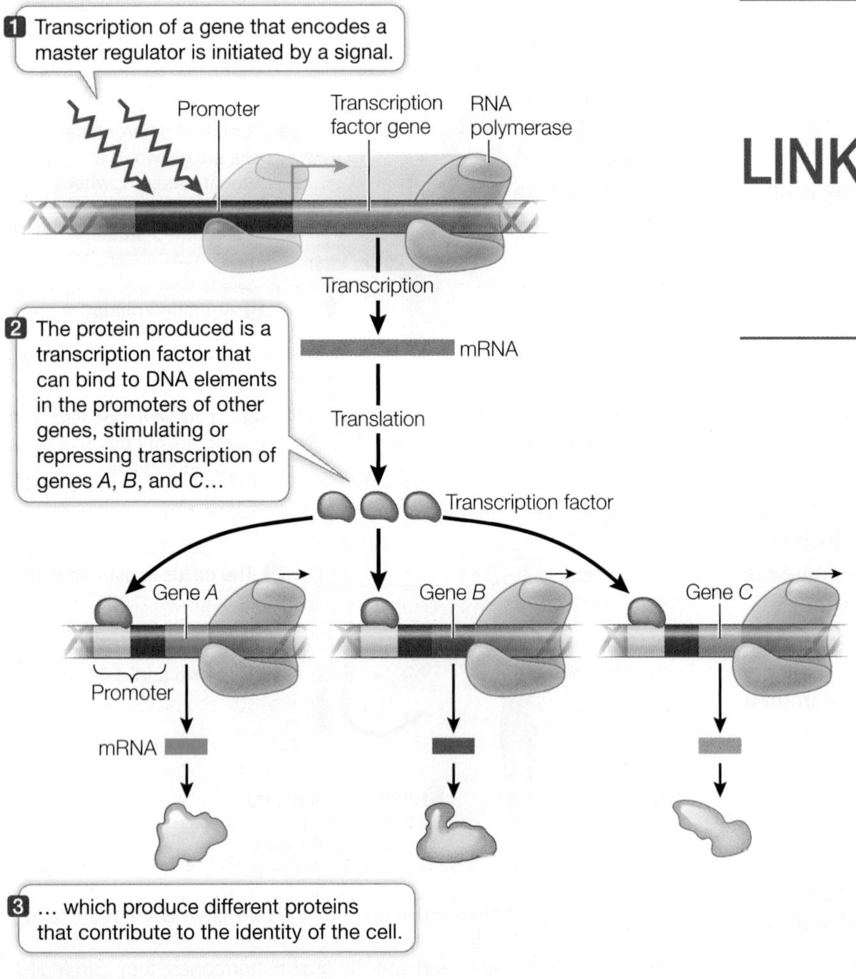

Transcription factors, a class of proteins that bind to regulatory sequences in DNA, play a central role in regulating gene expression. When a transcription factor binds to DNA, it forms part of a complex that determines the likelihood that RNA polymerase will be able to bind and carry out transcription of the associated gene, forming mRNA (**FIGURE 25.2A**). Whether a transcription factor acts as an activator or a repressor varies depending on the identity of the transcription factor and on which other proteins are present in the protein–DNA complex. A single type of transcription factor typically binds to one specific sequence of DNA, but this sequence can be present near many different genes. Thus a single type of transcription factor can affect the expression of many different genes (**FIGURE 25.2B**). If a transcription factor affects the expression of other transcription factors, a cascade is generated, amplifying the effect. Some transcription factors that sit at the head of such cascades are called **master regulators** because their

FIGURE 25.2 Coordinating Gene Expression A single signal causes the synthesis of a transcription factor that acts on many genes.

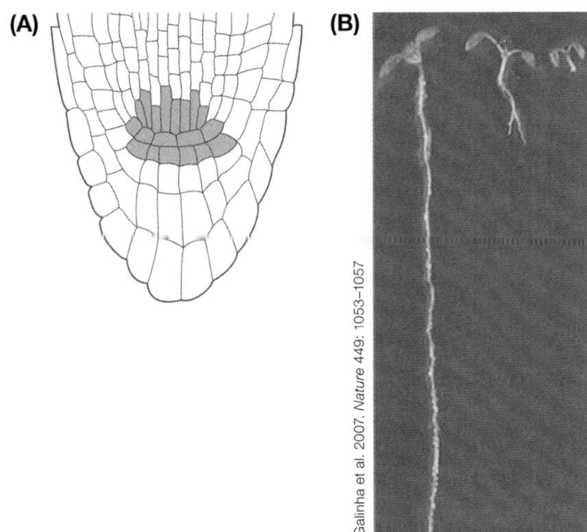

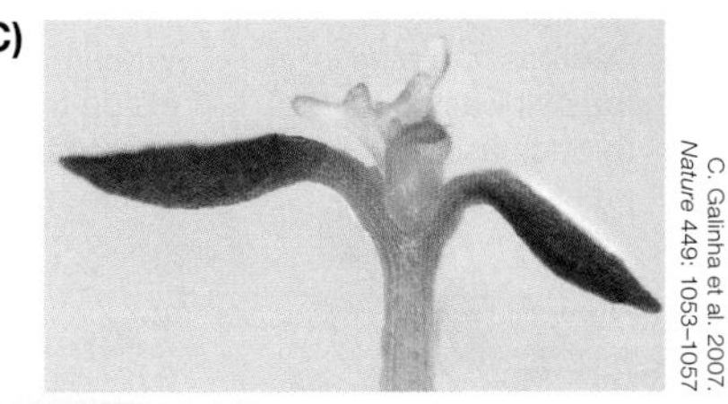

FIGURE 25.3 PLETHORA Expression Maintains Root Stem Cells Loss of PLETHORA (PLT) function results in failure to maintain the stem cells in the root. **(A)** Diagram showing the stem cells and the quiescent center in the root meristem. **(B)** A wild-type root is shown on the left; a *plt1,plt2* double mutant in the center; and a *plt1,plt2,plt3* triple mutant on the right. When stem cells are not maintained, the root in the mutant plants does not continue to elongate. (Wild-type gene names are written in uppercase, mutant gene names in lowercase.) **(C)** Inducing expression of PLETHORA 2 in the shoot apical meristem results in the formation of a rootlike structure there. (B and C from C. Galinha et al. 2007. *Nature* 449: 1053–1057.)

What does it mean to maintain stem cells?

expression can cause a cell to produce or repress so many proteins that the cell takes on a new cell fate. Development can thus be seen as a question of how cells come to express different transcription factors, particularly the master regulators that cause a cell to differentiate into its final cell fate.

An example of plant transcription factors that act to specify cell fate is found in the PLETHORA (PLT) family. **Stem cells** are cells that both maintain themselves and give rise to the cells that produce differentiated cell types. In plants, stem cells can be found in the meristems: think of the vascular cambium that maintains itself while producing cells that form the xylem and phloem, or of the divisions that take place in the center of the shoot apical meristem that give rise to the cells that form the stem and leaves. In the root meristem, stem cells surround the quiescent center, a small group of cells that rarely divide (**FIGURE 25.3A**). If the root meristem did not contain stem cells, the cell production that leads to regular growth of the root would stop. *PLETHORA* genes encode transcription factors that are required for the formation and maintenance of stem cells. Experiments examining mutant plants that have lost the function of multiple PLETHORA proteins have demonstrated that PLETHORA proteins confer the stem cell identity. Plants lacking functional copies of *PLT1* and *PLT2* have short roots, while embryos lacking *PLT1, PLT2,* and *PLT3* form no root at all (**FIGURE 25.3B**). In contrast, when high levels of PLT2 protein are induced in the shoot apical meristem, the meristem gives rise to a structure that resembles a root (**FIGURE 25.3C**)! A mutation in which one type of organ is replaced by another is called a **homeotic mutation**.

Key Concept 26.3 describes how various combinations of transcription factors direct flower development.

Cell expansion increases the size of a plant cell and gives it shape

While changes in gene expression affect the proteins that are present in specific cells, the overall form of a plant results from growth of the cells that make up the plant. A common misconception is that growth occurs because of cell division, but plant growth actually occurs by *enlargement* of existing cells. Cell division partitions the existing material into smaller units, without directly increasing the plant's size. The size of a plant increases when cells expand. Cell expansion often leads to cell division, so increases in size and cell number are highly correlated.

Growth of a plant cell is restricted by the cell wall that lies outside the membrane. Plant cells increase in size by taking in water in what is called the **stress relaxation cycle** (**FIGURE 25.4**). The cycle begins when pressure from the cell contents (turgor pressure) acts on the cell wall. This physical stress alters the structure of the material that makes up the cell wall. Some of the changes that occur in the cell wall are reversed when the pressure decreases, but some of the changes are not reversible —much as when someone pulls on a swimsuit that is too small and the material becomes permanently stretched out. Once the structure of the cell wall changes, the wall "relaxes." Relaxation of the wall causes the internal turgor pressure to fall—because the forces between the cell and its wall are equal and opposite. Having a lower internal pressure can result in more water entering the cell, continuing the cycle.

The rate of cell wall expansion can be regulated by changing the chemical composition of the wall. Some plant hormones increase the rate of cell expansion by making the wall easier to extend ("looser"). That is, they change the composition of the wall such that a given amount of pressure results in a larger expansion of the cell. These hormones promote growth.

Growing plant cells must synthesize new wall material that adds to the thickness of their walls; otherwise their walls would continually become thinner. Most plant cell walls are composed of long filaments called **cellulose microfibrils** that are embedded in a network of polysaccharides and proteins. Cellulose is a polymer composed of repeating glucose units joined in a straight chain (see Figure 3.7). Cellulose microfibrils are synthesized by protein complexes called **cellulose synthases** that sit in the cell membrane, receiving subunits from the cytoplasm and releasing

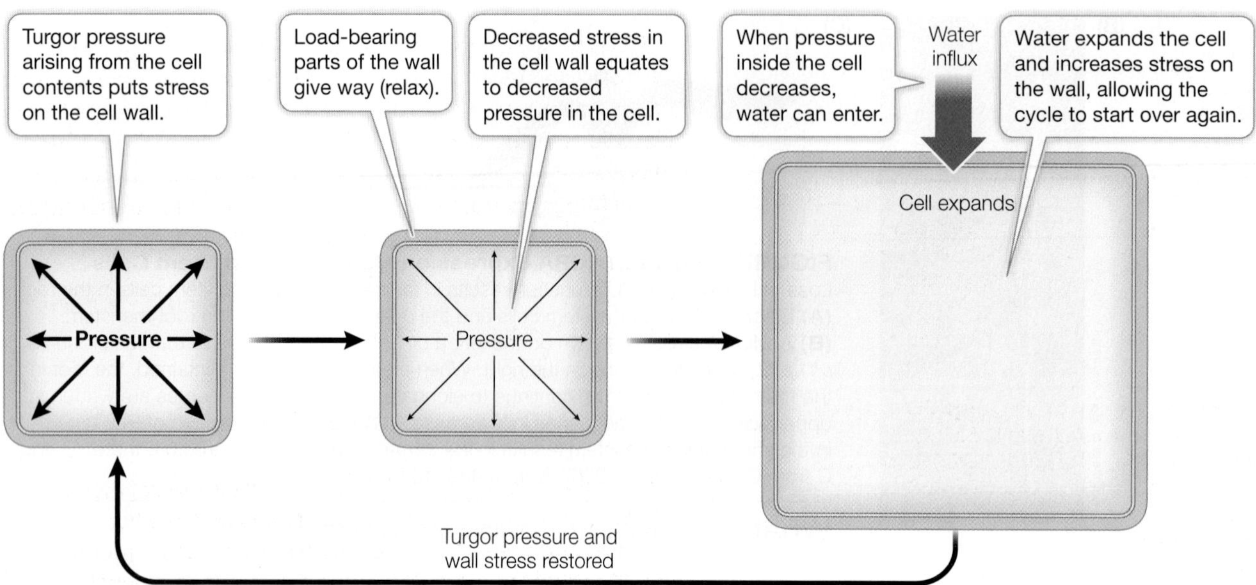

FIGURE 25.4 Plant Cell Expansion Occurs via Stress Relaxation
(After D. J. Cosgrove. 2015. *J Exp Bot* 67: 463–476.)

microfibril chains toward the cell wall (**FIGURE 25.5**). The cellulose synthase complexes synthesize multiple chains at one time, and these chains pack together to form a single microfibril. Cellulose microfibrils provide strength and rigidity to the wall.

If plant cells lacked a stiff restraint, they would grow as spheres and plants might resemble beach balls. Neither plants nor plant cells are entirely spherical, however. Epidermal cells on the leaf can be shaped like pieces of a jigsaw puzzle, while cells in an elongating stem are often cylindrical. Nonspherical shapes arise because the plant has the ability to regulate the extent of growth in one area versus another. Typically, the wall in one part of a cell is softer or easier to loosen than the wall in other parts of the cell.

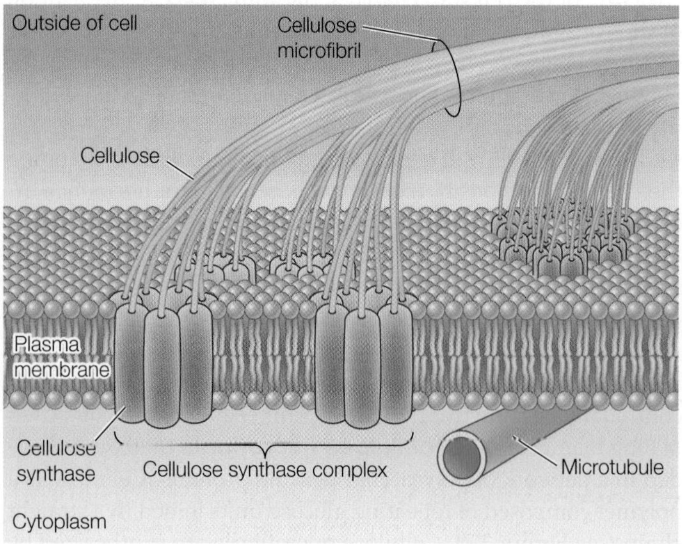

FIGURE 25.5 Cellulose Synthase Complexes Cellulose synthase complexes form chains of glucose (glucan) that come together in the cell wall space to form cellulose microfibrils.

These differences give an asymmetry to cell expansion that affects the overall shape of a plant.

Whether a cell continues to expand in the same manner or changes its direction of expansion is of central importance to the development of the organism. Changes in the direction of expansion are responsible for the formation of branches and complex shapes. The direction of maximal cell expansion usually can be predicted by the arrangement of **cortical microtubules**, cytoskeletal proteins that are anchored to the cell membrane. Cortical microtubules are linked to the protein complexes that synthesize cellulose, and the orientation of the cortical microtubules appears to direct the orientation in which cellulose microfibrils are laid down in the cell wall. For elongated cells in an growing stem, both the microtubules and cellulose microfibrils wrap around the narrow axis of the cell, like hoops in a wine barrel (**FIGURE 25.6**). Cellulose microfibrils resist expansion, resulting in there being more growth in length than in width. The orientation of microtubules is highly regulated by the proteins that bind to them and gives the plant a mechanism for controlling the shape of its body.

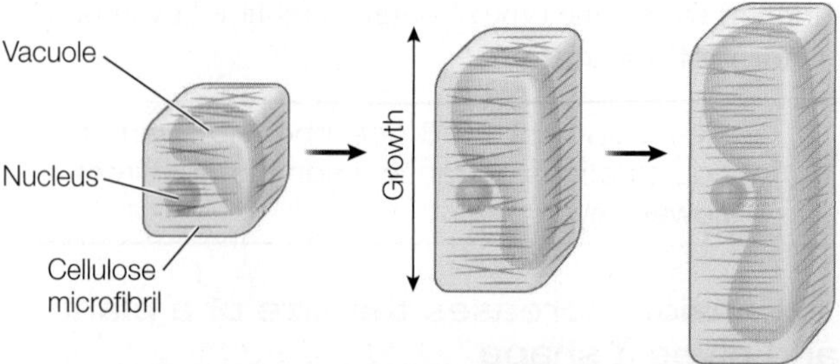

FIGURE 25.6 The Orientation of Cellulose Microfibrils Affects the Direction of Cell Expansion Transverse cellulose microfibrils restrict lateral expansion, resulting in an elongated shape.

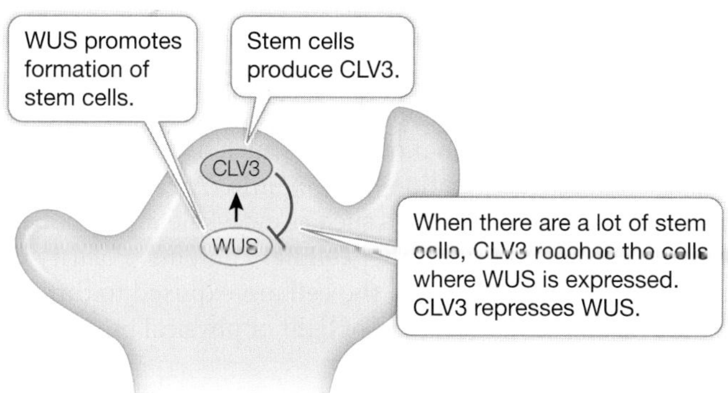

FIGURE 25.7 A Negative-Feedback Loop Controls the Number of Stem Cells and Meristem Size A negative-feedback loop between a transcription factor (WUSCHEL, WUS) that promotes stem cell formation and a mobile peptide (CLAVATA 3, CLV3) that is produced by stem cells keeps the number of stem cells and thereby the size of the shoot meristem in balance.

Control of stem cell number maintains plant proportions

Another factor that affects the shape of the plant body is the size of its meristems. If the meristem that produces a flower is larger than normal, the resulting flower may have extra petals. If the shoot meristem is larger and wider than normal, it can give rise to stems that are also wider than normal. The size of the shoot meristem is regulated by a negative-feedback loop. Stem cells in the top layers of the meristem produce a small peptide, called CLAVATA 3 (CLV3), that moves down into the stem where a group of target cells are located. The target cells produce a transcription factor, called WUSCHEL (WUS), the activity of which promotes formation of stem cells in the meristem. When CLV3 secreted from the meristem reaches the target cells, production of WUS is inhibited, and the production of stem cells declines (**FIGURE 25.7**). When the number of stem cells at the top of the meristem increases beyond its normal range, the concentration of CLV3 that reaches the target cell is large and WUS expression is turned down, decreasing the rate at which stem cells are added to the meristem. When the number of stem cells at the top of the meristem is smaller than normal, WUS is not repressed and more stem cells are added to the meristem. This system allows the meristem to maintain itself around a set size.

FIGURE 25.8 Fasciation The contorted shapes of these cockscomb (*Celosia cristata*) flowers are a result of fasciation.

Occasionally the mechanisms that regulate the size and shape of the meristem fail. When a shoot apical meristem becomes broader and flatter than normal, it can produce stems that are shaped more like a ribbon than a cylinder. If a floral meristem is similarly affected, it may result in flowers with extra petals or a broad, crested shape. Plants that grow from such meristems are said to be **fasciated**. Cockscomb plants regularly form fasciated flowers (**FIGURE 25.8**), and some astute observers may have seen the same pattern in an occasional plant of some other type, such as a dandelions. Mutation of *CLAVATA* genes can promote fasciation.

REVIEW & APPLY | 25.1

R Plants are unusually flexible in terms of their development. Plant cells retain the ability to redifferentiate into other cell types in response to specific environmental cues throughout the life of the plant. At any given moment, the identity of a cell depends on which master regulator genes are expressed and thus which transcription factors are present. One attribute of cells that varies with their identity is the size and shape to which they grow; this is regulated by the orientation of cellulose microfibrils in their cell walls. The extent and direction of the expansion of individual cells, together with the number of cells in the meristem(s), give rise to the overall shape of a plant.

A
1. Some leaf cells have small hairs called trichomes. What might happen to the structure and appearance of the leaf if there were a loss-of-function mutation in the *GL2* gene, a master regulator of trichome production?
2. Predict the direction(s) of minimal and maximal expansion for the epidermal cell shown here; the lines indicate the location of cortical microtubules. How would cell expansion be altered if the cell were treated with a chemical that caused microtubules to disassemble? (*Figure source*: A. Sampathkumar et al. 2014. ELife 3: e01967; CC BY 3.0.)

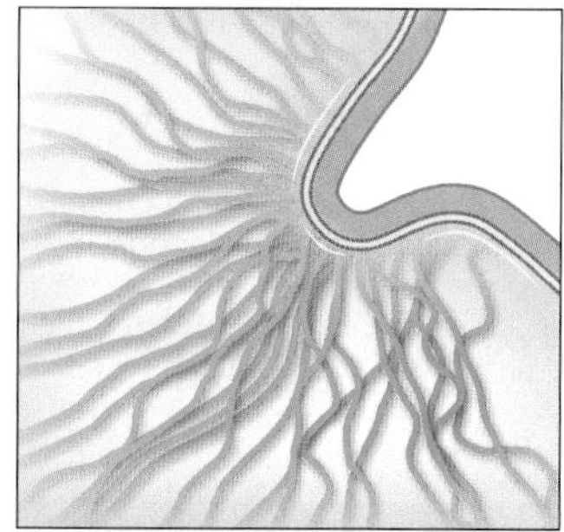

3. How would the size and appearance of a shoot apical meristem change if:
   a. the plant had a loss-of-function mutation for *CLV3*?
   b. the plant had a loss-of-function mutation for *WUS*?

Much of development is a sequential cascade, in which an initial difference in transcription factor abundance leads to a second event, which leads to a third event, and so on. However, for cells to become different from one another, something must generate an initial difference between them. In the next key concept we consider ways in which two initially identical cells can become distinct.

25.2 Multiple Mechanisms Allow Cells That Are Initially Identical to Become Distinct from One Another

LEARNING OBJECTIVES

By the end of this key concept you should be able to:

25.2.1 Describe three mechanisms by which cells can become different from one another.

25.2.2 Give an example of positional information that instructs plant cells as to their identity.

Flowering plant embryos have a distinct top and bottom corresponding to what will become the shoot and root systems. Such a difference between the ends (or poles) of a cell, organ, or organism is called **polarity**. Within a cell, polarity can arise when certain materials in the cell called **cytoplasmic determinates** are distributed unequally in the cytoplasm. Cytoplasmic determinants can be signal molecules such as hormones, proteins (including transcription factors), mRNA or small regulatory RNAs; or larger structures such as cytoskeletal components (microtubules or actin) or other cellular contents. Polarity can also arise if different sides of the cell are exposed to different conditions, such as the amount of light or physical pressure, including gravitational pull.

Cell polarity is important because it represents a difference that can be amplified by subsequent developmental events, eventually giving rise to distinct cell types. For example, in flowering plants the egg cell itself is nonsymmetrical: one end of the egg cell is attached to the parent plant while the other end of the egg cell is exposed to the cytoplasmic space surrounding the polar nuclei (**FIGURE 25.9A**). This difference is propagated by the first cell division of the fertilized egg, which results in a larger basal daughter cell (attached to the parent plant) and a smaller apical daughter cell on the other end. The larger cell forms a stalk called the **suspensor**, while progeny of the smaller cell form the embryo proper (**FIGURE 25.9B** and **C**). As the embryo starts to develop, the hormone auxin (discussed below) accumulates in the apical cell. Later on, auxin transport proteins focus the flow of auxin downward (in the direction of the suspensor), and accumulation of high concentrations of auxin at this end of the embryo contributes to it being specified as the root meristem (**FIGURE 25.9D**). Auxin is then transported through the outer cell layers such that auxin accumulates in two additional regions on the top (shoot side) of the embryo. These two small regions then become specified as the cotyledons (**FIGURE 25.9E**).Thus the formation of the root–shoot axis occurs by a series of chain-reaction events, set off by the original polarity of the egg.

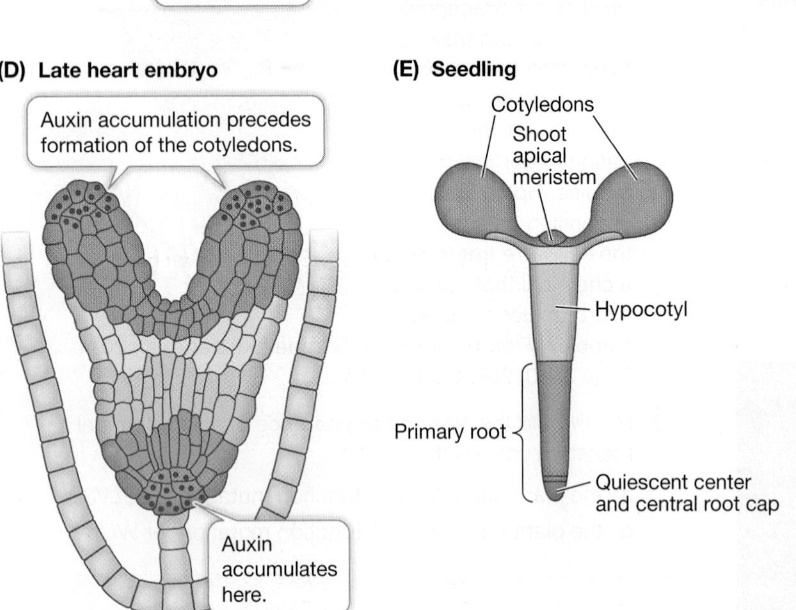

FIGURE 25.9 Initial Asymmetries can be Propagated, Producing Large Differences In flowering plants such as *Arabidopsis thaliana*, formation of the root-shoot axis occurs via a series of sequential steps that have their origin in the initial asymmetry of the egg cell. (After T. Laux et al. 2004. *Plant Cell* 16: S190–S202 and J. Friml et al. 2003. *Nature* 426: 147–153.)

Activity 25.1 Stages of Plant Development

While the embryo arises from an egg cell that has an inherent polarity, there are cases in which differences between cells arise de novo (newly), in the absence of a preexisting pattern. Consider, for example, cells in a leaf. Some of the epidermal cells will form stomatal guard cells while others will not. How do the epidermal cells that descend from the leaf primordium become different from one another? What factor(s) can trigger the difference?

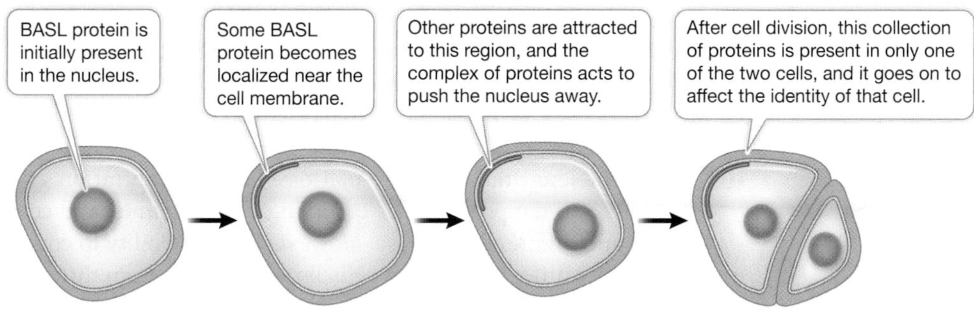

FIGURE 25.10 Asymmetrical Cell Division In the process leading to the formation of stomatal guard cells, a leaf cell first becomes polarized (asymmetrical). The two new cells that result from the subsequent asymmetrical division are different in size, and one of their proteins (called BASL) is localized in different ways. Because of these differences, the two cells differentiate into distinct types of cells. (After J. Dong et al. 2009. *Cell* 137: 1320–1330.)

Cells that undergo asymmetrical cell division can produce daughter cells with an identity different from that of their parents

Cells can become distinct from one another at the point of cell division. For example, a meristematic cell may produce one daughter cell that remains in the meristem and one daughter that differentiates into something else. The daughter cells can take on different identities if their internal contents (types or levels of mRNA, protein, etc.) are distinct. But what causes the original cell to divide in an asymmetrical manner? This question has been studied for a leaf cell that divides asymmetrically, producing cells of different sizes (**FIGURE 25.10**). The smaller cell is called a **meristemoid**, and subsequent divisions of this cell lead to formation of stomatal guard cells. Prior to asymmetric cell division of the precursor cell, a protein called BASL is located in the nucleus. When BASL interacts with a specific enzyme that has kinase activity, BASL becomes phosphorylated and can accumulate at the periphery of the cell. BASL then binds to additional proteins forming a protein complex that is asymmetrically localized near the cell membrane. A **positive feedback loop**, in which the presence of some BASL recruits more BASL to the region, may promote the formation of this complex in just one part of the cell. Microtubules help move the nucleus away from the BASL-containing region. Hence, when the nucleus divides and a new cell wall is laid down, only one of the two daughter cells inherits the BASL-containing protein complex. In these cells, the presence of the complex leads to destruction of a transcription factor that affects cell fate. Thus, BASL protein asymmetry results in the daughter cells having different identities.

Amplification of small stochastic differences allows cells to become distinct from one another

If two cells are initially identical, the rates at which specific transcription factors are synthesized and degraded in them will be the same. On average this will result in the concentration of transcription factors in the two cells being the same. On an individual basis, however, the timing of the synthesis and degradation of transcription factors varies: one cell might finish transcribing a particular sequence slightly before the other cell does, and so in a stochastic (random) manner, one cell might have a slightly higher concentration of some particular factor than the other cell. Over time that difference would be expected to average out. However, if the cells have a mechanism by which an initial difference can be amplified, then a tiny initial difference can lead to a large difference between the cells.

Positive-feedback loops, in which some factor leads to production of more of itself, amplify small initial differences. In the context of development, positive-feedback loops cause cells that are undergoing a change in fate do so in an all-or-none manner. An example occurs in fruit ripening. Once a fruit begins producing **ethylene**, a gaseous plant hormone that stimulates ripening, the ethylene induces more of its own production in an autocatalytic cycle (**FIGURE 25.11**). This leads to a burst of ethylene production and ensures that a fruit ripens all at once. Because ethylene production can be induced by wounding a plant, and diffuses through the air to affect other fruits nearby, it also explains the old saying "One bad apple spoils the bunch."

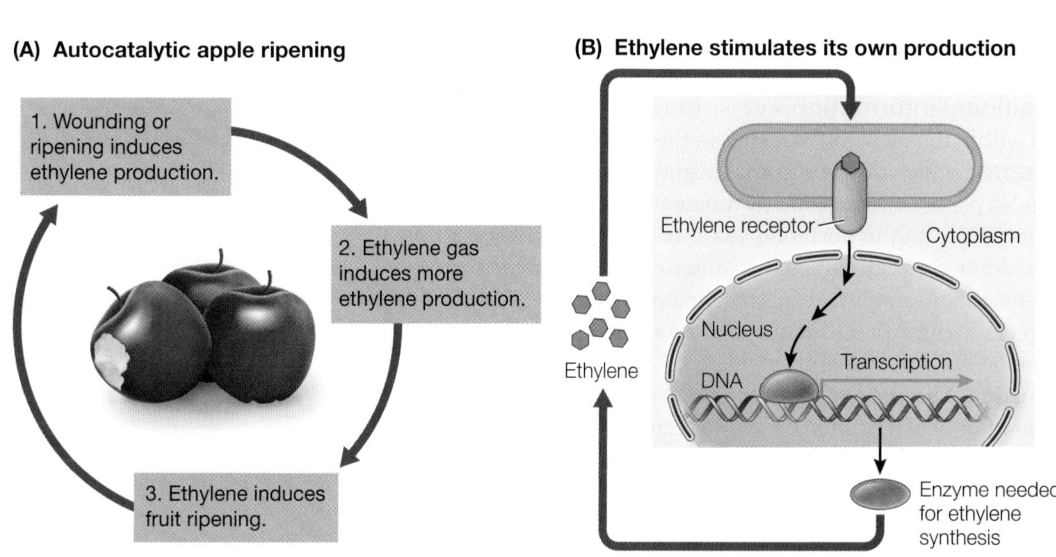

FIGURE 25.11 Positive-Feedback Loops Allow for All-or-Nothing Responses

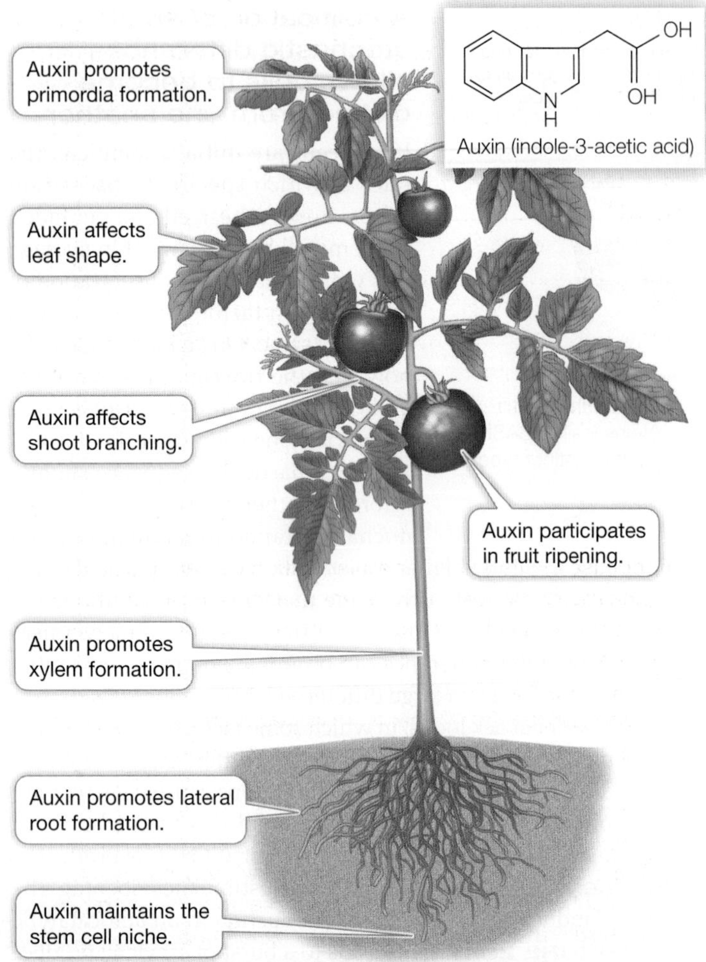

FIGURE 25.12 Auxin Induces Many Different Developmental Responses

Exposure to different environments can induce cells to become different from one another

Environmental influences can be external to the plant (such as light, temperature, or pressure) or they can come from positional differences within the plant. The key cellular question "What am I (or what will I be)?" is often answered in part by "Where am I?" **Positional information**, cues arising from the location of the cell within the plant, often comes in the form of a signal called an **effector**, which diffuses from one group of cells to surrounding cells. In plants, the fate of many different types of cells is conferred by exposure to the hormone auxin (**FIGURE 25.12**). **Auxin**, indole-3-acetic acid, is a small organic molecule about the size of an amino acid. As might be expected for a signal whose accumulation triggers changes in cell identity, the synthesis, transport, and degradation of auxin are highly regulated. Accumulation of high levels of auxin in specific regions of the shoot apical meristem promotes formation of leaf primordia, while accumulation of auxin in the lower part of the embryo specifies the root. Long-term exposure to auxin can induce production of the PLETHORA proteins, discussed earlier, that specify stem cell fate. The PLETHORA proteins also induce auxin synthesis and regulate auxin transport, suggesting that development of regions (called niches) in which stem cells reside may be self-reinforcing. Many other types of cells also respond to auxin by changing their cell fate, making this hormone one of the central regulators of plant development.

How does one signal molecule produce so many different responses? One possibility is that cells respond differently based on the concentration of signal they are exposed to. Another is that the length of exposure matters. Both concentration gradients and temporal pulses of auxin occur in plants, and cells respond differently to these. However, auxin's ability to confer so many different fates on cells largely derives from different sets of proteins contained within the target cells. The interaction of auxin with a cell usually leads to the degradation of specific transcription factors. Frequently, the proteins targeted for degradation are transcriptional repressors (**FIGURE 25.13**). Degradation of a transcriptional repressor means that transcription can no longer be repressed—auxin induces gene expression by removing the repressor. The effect of degrading a transcriptional repressor depends entirely on which repressor it was and on

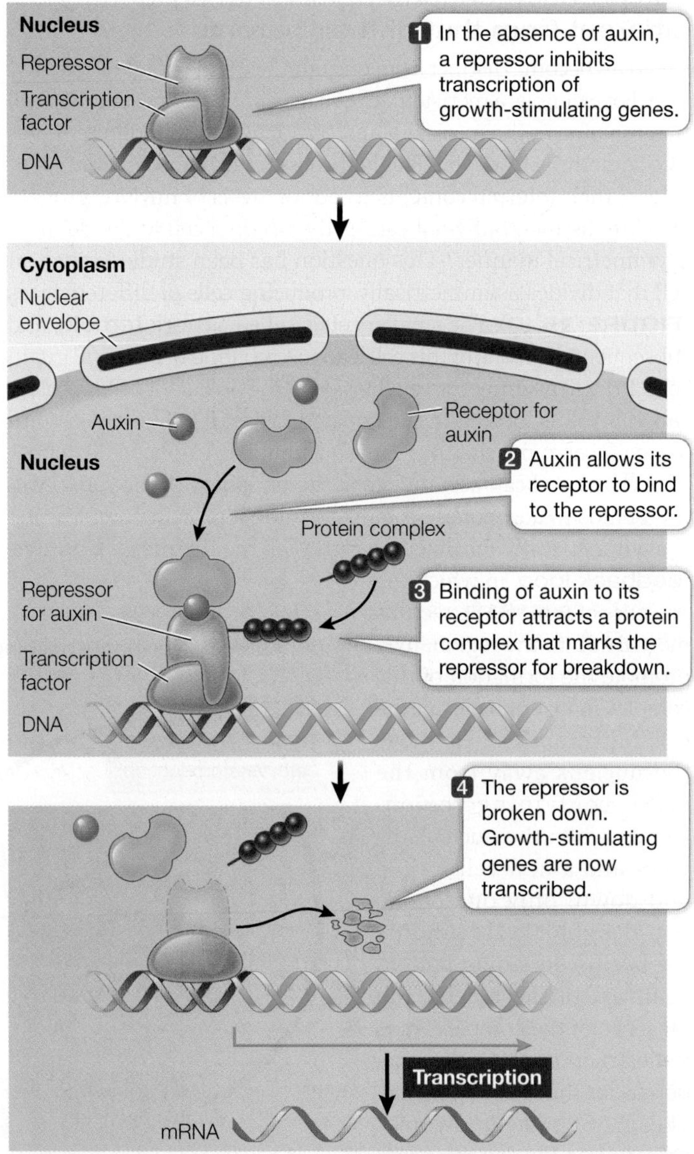

FIGURE 25.13 Auxin Signal Transduction Auxin can stimulate gene transcription by inducing degradation of a transcriptional repressor.

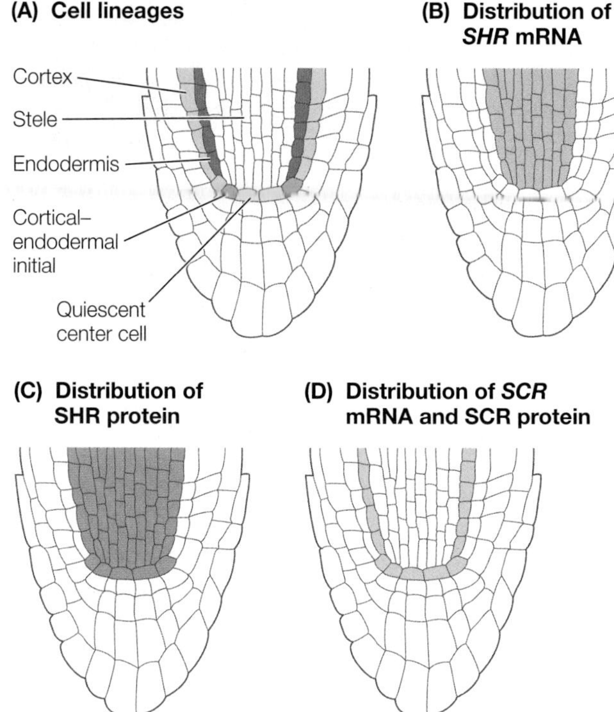

FIGURE 25.14 Cell Fate Can Be Determined by Positional Information Specification of the endodermal cell fate is jointly controlled by the SHORTROOT (SHR) and SCARECROW (SCR) transcription factors. **(A)** Longitudinal division of the cortical–endodermal initial cell (blue-green) gives rise to the two cell files that differentiate as the endodermis (dark green) and cortex (light green). **(B)** *SHR* mRNA is transcribed and translated in the stele (light purple), but **(C)** the SHR protein (dark purple) moves outward from the stele into the cell file that will form the endodermis. **(D)** When SHR enters the nucleus of this outer cell file, it is bound by a protein complex that includes SCR (orange), trapping it in that cell file and specifying it as the endodermis. (After Nakajima et al. 2001. *Nature* 413: 307–311.)

In which cells is the SHR-SCR complex present?

what other proteins are present in the target cell—factors that vary depending on the position and state of the target cell prior to auxin signaling. Similar mechanisms have been found to underlie the action of other hormones, such as the **gibberellins**, an important group of hormones that promote seed germination and stem growth and regulate several other aspects of plant development. **Media Clip 25.1 Gibberellin Binding to Its Receptor**

A different way of imparting positional information is found in the transcription factor network that establishes the identity of endodermal cells, the cells that contain the Casparian strip (see Figure 24.6). In the model plant *Arabidopsis thaliana*, the endodermis and the cortex are both one cell wide and arise from division of a stem cell called the cortical–endodermal initial (**FIGURE 25.14**). Loss-of-function mutations in either of two genes, *SHORTROOT* (*SHR*) or *SCARECROW* (*SCR*), results in failure of this division and loss of the two distinct layers. SHR is a transcription factor that is expressed in the central vasculature, interior to (not including) the cortical–endodermal initial. The SHR protein can move from cell to cell and tends to move outward into the file of cells that will give rise to the endodermis. Once there, the protein enters the nucleus and binds to a complex that includes SCR. Formation of the SHR–SCR complex leads to asymmetrical cell division of the cortical–endodermal initial and specifies the inner layer as the endodermis. The division is asymmetrical because the two new cells have different neighbors and thus are exposed to different environments. Because SCR is expressed only in the cell layer next to the vasculature, the one that will form the endodermis, this mechanism ensures that the endodermis will be located next to the vasculature. Because binding of SHR to SCR prevents SHR from moving outward to additional cell layers, the mechanism also ensures that the endodermis will be only one cell wide.

REVIEW & APPLY | 25.2

R Because plants are multicellular, their cells can become specialized and take on different identities. How do two cells that are initially identical become differentiated? One way is via asymmetrical cell division. If cell division partitions the cell contents (e.g., protein, mRNA, organelles) asymmetrically, or if it creates two cells that are in different local environments by placing them in contact with different neighbors, then the two new cells will be different from one another. Differences can also arise after cell division, under the influence of internal or external factors. Internal differences can arise stochastically (randomly) because of tiny fluctuations in the rate at which events such as protein synthesis take place. Usually these differences average out over time, but if mechanisms exist that amplify the differences, the differences may become large enough that the cells take on new fates. The external environment (e.g., factors such as light or temperature) and the internal environment (e.g., presence of hormones such as auxin) can also trigger changes in cell identity.

A

1. Can two cells that are initially the same and continue to be exposed to the exact same environment differentiate from one another? If so, explain how this difference can arise.
2. Explain how the interaction between auxin and *PLETHORA* genes might be self-reinforcing.
3. Explain how the ability of PLETHORAs to regulate auxin synthesis and transport could impart positional information on neighboring cells.
4. Ripe bananas and avocados produce large amounts of ethylene gas. Explain why putting a banana in a paper bag with other fruits, such as apples, causes the apples to ripen more rapidly than they otherwise would.
5. Explain how the mechanism that specifies the endodermis results in it being only one cell layer wide.

Translating environmental changes into developmental changes requires a coordinated response by the plant. The next key concept addresses the role that plant hormones play in this process.

Developmental Changes Can Be Mediated by Hormones

LEARNING OBJECTIVES

By the end of this key concept you should be able to:

25.3.1 **Explain what a photoreceptor is and give two examples of photoreceptors found in plants.**

25.3.2 **Explain how hormones can carry a signal from the point of perception to the point of response.**

25.3.3 **Describe how auxin moves through a plant.**

When plants respond to environmental factors by changing their development, it is important that the changes be coordinated across the plant body. **Hormones**, chemical signals that act at very low concentrations at sites often distant from where they are produced, provide a mechanism for achieving this. Plant hormones fall into many of the same classes as animal hormones. Some plant hormones are peptides, some are steroids, some are gases, and others are small molecules with a variety of chemical structures (**TABLE 25.1**). Plant hormones are notable, however, in that each hormone tends to have multiple different effects on different types of cells. When hormones are involved in response to some external environmental signal, they provide a mechanism for generating a coordinated response. In this key concept we consider two specific examples of how hormones operate downstream of light perception to affect plant growth and development.

Phototropism is mediated by a blue-light receptor and the hormone auxin

Plants are completely dependent on sunlight for energy, so it is not surprising that they have multiple ways to sense light and respond appropriately. Plants contain **photoreceptors**, proteins that absorb specific wavelengths of light and pass a signal on to the rest of the cell. After light has been perceived, plants can alter their development in response to it, a process called **photomorphogenesis**. The set of photoreceptors present in plants gives them the ability to respond to two aspects of light: its *quality*—the wavelengths of light to which the plant is exposed; and its *quantity*—the intensity and duration of light exposure.

FIGURE 25.15 Phototropism Seedlings grow toward light when cells on the side of the plant that faces away from the light (the shaded side) elongate more than cells on the lit side (see also Figure 25.19A).

Animation 25.1 Tropisms

> **LINK** As discussed in **Key Concept 5.5**, single photons of sunlight travel with different wavelengths—the shorter the wavelength, the higher the energy in the photon. The different wavelengths associated with visible light are responsible for the different colors we see. For example, blue light has a shorter wavelength than red light. **Figure 5.16** shows the wavelengths of the electromagnetic spectrum.

Many plants respond to blue light by growing toward it, a response called **phototropism** (**FIGURE 25.15**). You may have observed this response in houseplants that grow toward a window. To find out more about the photoreceptor that mediates this response, scientists determined the wavelengths of light that were most effective in generating this response. A graph that shows the degree of response produced by different wavelengths is called an **action spectrum**. The action spectrum for the phototropic response of a coleoptile (the organ that covers the first leaves of a seedling as it pushes up through the soil) demonstrates that blue light at a wavelength of 436 nanometers (nm) is the most effective in stimulating the growth of this

TABLE **25.1** | Examples of Plant Hormones

Structural class	Specific hormones and some activities they affect
Small organic molecules	**Auxin:** Cell elongation and tropic responses, cell division, stem cell maintenance, initiation of lateral roots and leaf primordia, xylem formation, fruit ripening
	Cytokinin: Cell division, inhibition of leaf senescence, shoot formation; often acts in opposition to auxin in a concentration dependent manner
	Gibberellin: Seed germination, flowering, stem elongation
	Abscisic acid: Seed dormancy, drought tolerance, stress response; often acts in opposition to gibberellin
	Strigolactone: Regulates shoot branching, attracts mycorrhizal fungi
	Salicylates: Defense responses to pathogen attack
	Jasmonic acid: Wound response and plant defense against herbivores
Gases	**Ethylene:** Fruit ripening, stem elongation, leaf and petal senescence
Steroids	**Brassinolide:** Increased plant growth
Peptides	**Systemin:** Plant defense response
	CLAVATA3: Meristem size regulation

THINK LIKE A SCIENTIST

Correlation and causation

Scientists identified proteins in plants that absorb blue light and carry out a reaction in response to it, but how could they determine if those proteins are the one(s) actually involved in phototropism? When testing new theories, scientists often begin by looking to see if events or reactions are correlated. As part of the process of science, this can be useful as a way to potentially disprove a hypothesis. If, for example, phototropism is most strongly activated by blue light, then a protein that effectively absorbs green light but does not absorb blue light would not be a candidate for the photoreceptor that mediates phototropism. But does this correlation prove that the photoreceptor for the phototropic response has been identified? The answer is no. The world is full of correlations that do not arise from causal connections. For example, many people fall asleep at night after lying down close to an alarm clock. This does not mean that the presence of an alarm clock is the driving factor that causes people to fall asleep.

How do scientists determine whether correlated events are causally related? If A causes B, then we will see a correlation between the level or amount of A and B. To show a causal relationship, we can add or remove the item in question (e.g., the alarm clock) and see what effect this has. This approach is taken when evaluating the effects of genes by examining loss-of-function mutants and overexpression of a gene of interest.

A powerful technique for searching for a gene that is causally involved in some aspect of plant growth or development is to conduct a genetic screen (**FIGURE 25.16**). This works even if one does not have a specific protein in mind (and was commonly used before the recent invention of CRISPR technology). To begin, seeds are exposed to a chemical that induces mutations.

A successful genetic screen identifies mutants that fail to exhibit a normal response, or that exhibit an unusual response. In the screen described in Figure 25.16, the genes identified will not all code for photoreceptors; the proteins that the mutated genes encode could act anywhere in the pathway between light perception and the final response. Some of these genes may be directly involved, and others indirectly involved. What is certain, however, is that all of the genes identified in a genetic screen will be causally related to the observed response.

1. From what you read about phototropism, what genes might you expect to uncover in the screen described in Figure 25.16? If you don't know the names of the genes, describe the pathway(s) in which the affected proteins would be involved.
2. Imagine that you want to identify genes that contribute to the regulation of the size of the flower meristem. How might you do this? What phenotypes would you look for? Explain your reasoning.

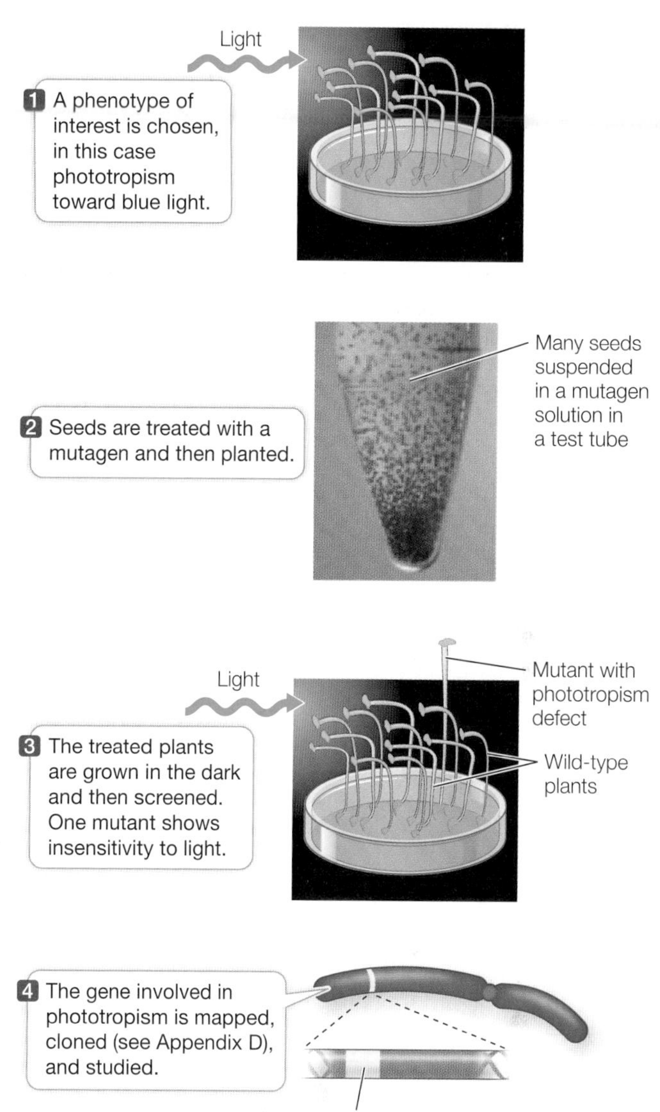

FIGURE 25.16 Genetic Screens Can Be Used to Determine whether Certain Proteins Cause Specific Effects *Arabidopsis* seeds were exposed to a chemical that induced mutations. Then the seeds were grown and exposed to blue light to determine which, if any, failed to respond. Using this genetic screen, researchers were able to identify the gene for a blue-light receptor protein called phototropin. (From J. M. Alonso and J. R. Ecker. 2006. *Nat Rev Genet* 7: 524–536.)

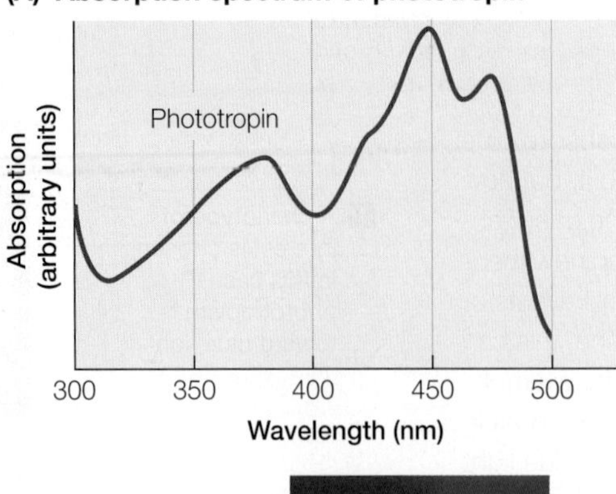

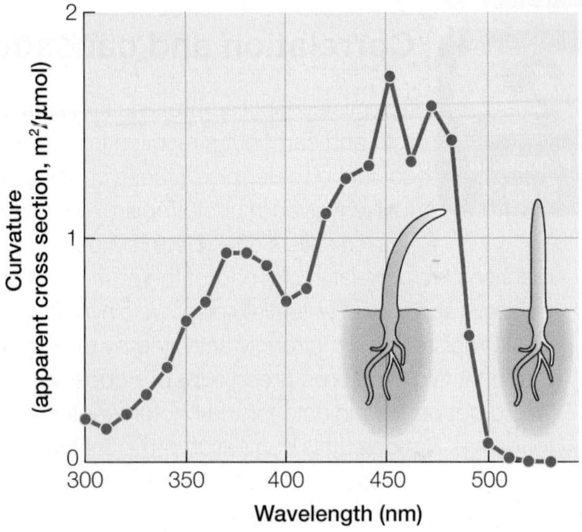

FIGURE 25.17 Action Spectrum for Phototropism The absorption spectrum for phototropin **(A)** is similar to the action spectrum for the bending of a coleoptile toward light **(B)**. Maximum bending occurs when coleoptiles are exposed to blue light. (A after W. R. Briggs and J. M. Christie. 2002. *Trends Plant Sci* 7: 204–210; B after T. Baskin and M. Iino. 1987. *Photochem Photobiol* 46: 127–136.)

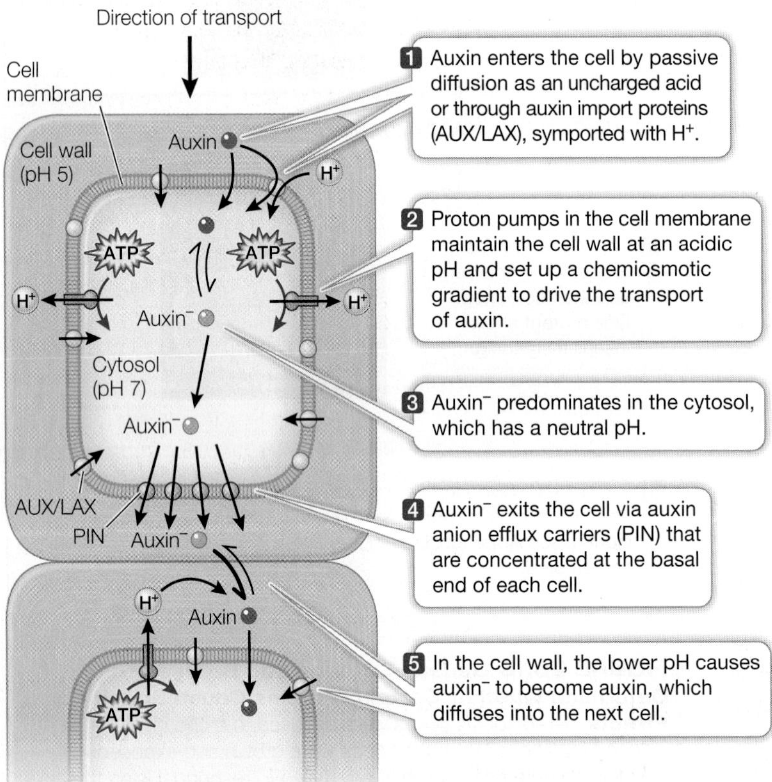

FIGURE 25.18 Polar Transport of Auxin The direction of auxin transport is controlled by the location of auxin transport proteins. Auxin enters cells on its own or via AUX/LAX H^+-auxin symporters, which are evenly distributed in the membrane. Once in a cell, auxin loses a proton (H^+) and becomes charged (auxin$^-$) and can exit only through auxin anion efflux carriers (PIN proteins), which are restricted to specific parts of cells. In the stem, PIN proteins are located on the basal side of cells, leading to a net movement of auxin toward the roots.

What would happen if you cut a section of stem tissue out of a plant and turned it upside down: would auxin move down, in the direction of gravity, or would it move up toward the new "top" side of the section?

organ toward light (**FIGURE 25.17**). Using a genetic approach with *Arabidopsis thaliana*, researchers were able to identify the gene for a blue-light receptor protein called **phototropin**, which is located in the cell membrane. Phototropin is composed of a polypeptide and a small organic molecule (a flavin mononucleotide); the flavin is the part that absorbs blue light.

How does the absorption of blue light cause plants to grow toward it? Charles Darwin and his son Francis investigated the phenomenon in canarygrass (*Phalaris canariensis*) seedlings grown in the dark. To find the part of the plant that perceives the light, they covered different parts of the plant and then exposed them to light from one side. Coleoptiles grew toward the light whenever their tips were exposed. If the tip was covered, however, there was no response. The Darwins concluded that the tip contains the photoreceptor that responds to light.

The actual bending toward the light occurs in a growing region a few millimeters below the tip. Therefore the Darwins reasoned that some type of signal must travel from the tip of the coleoptiles to the elongating region. This substance is now known to be the hormone auxin.

Animation 25.2 Went's Experiment

Auxin moves from cell to cell via specific transporters (**FIGURE 25.18**). These transporters are located on specific parts of the cells, and their combined effect imposes direction on the flow of auxin through the tissue, so it is called **polar auxin transport**. Specific transporters are necessary because at the pH found in the cytoplasm, the auxin molecule has a negative charge so it cannot exit a cell without assistance (see Figure 25.18, steps 1–3). Protein channels called PIN proteins allow auxin to exit a cell down auxin's concentration gradient (that is, by facilitated diffusion; see Figure 25.18, step 4).

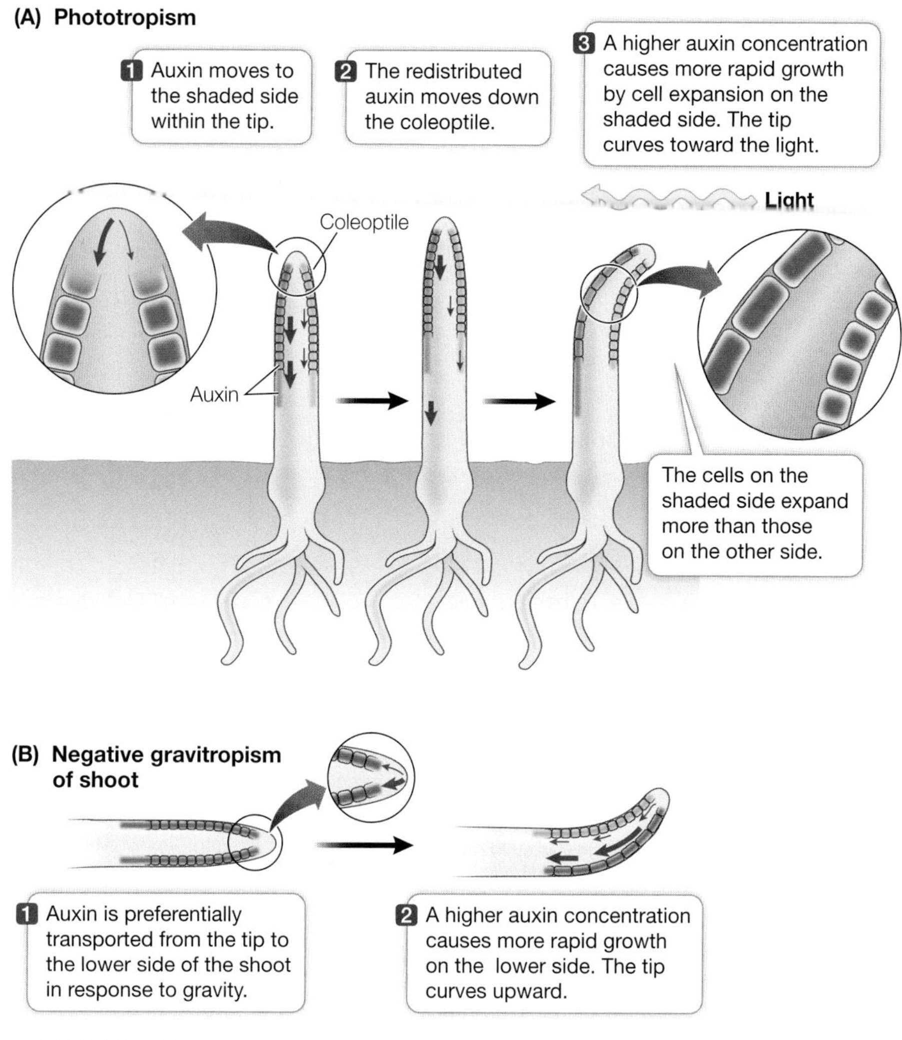

FIGURE 25.19 Plants Respond to Light and Gravity by Altering the Distribution of Auxin in Their Tissues Phototropism **(A)** and gravitropism **(B)** occur in shoot apices in response to a redistribution of auxin.

Animation 25.3 **Auxin Affects Cell Walls**

In the stems of *Arabidopsis thaliana* plants, these auxin efflux proteins are located on the basal side of cells, the side closer to the roots. A family of channel proteins called ABCB proteins that are present throughout the entire membrane also promotes auxin efflux. The pH outside the cell is more acidic (higher concentration of H^+), so when auxin, which is a weak acid, moves outside the cell, it picks up an H^+ and becomes neutral (see Figure 25.18, step 5). As a small neutral molecule, some auxin can cross the membrane and enter the next cell. More rapid uptake is driven by H^+-auxin symport with the assistance of auxin import proteins called AUX/LAX proteins.

Phototropism involves movement of auxin from the sunlit side of a stem or coleoptile to the shaded side. When light strikes a canarygrass coleoptile on one side, auxin at the tip moves laterally toward the shaded side This asymmetry in auxin is maintained as polar transport moves auxin down the coleoptile, so that in the growing region below, the auxin concentration is highest on the shaded side. Auxin stimulates cell elongation in the stem, so there is more growth on the shaded side, causing the coleoptile to bend toward the light (**FIGURE 25.19A**).

Light is not the only signal that can cause the redistribution of auxin. Auxin moves to the lower side of a shoot that has been tipped sideways, causing more rapid growth on the lower side and hence upward bending of the shoot. Such growth in a direction determined by gravity is called gravitropism (**FIGURE 25.19B**). Gravitropism in roots also involves differential growth caused by lateral movement of auxin, but the auxin that accumulates on the lower side of the root inhibits rather than promotes elongation, resulting in roots growing down toward gravity, rather than upward as shoots do.

Seed germination is promoted by phytochrome and the hormone gibberellin

Lettuce seeds spread on soil will germinate only in response to light. Even a brief flash of dim light will suffice. This response prevents the seeds from germinating when they are so deep in the soil that the seedlings would not be able to reach the light before their seed reserves ran out. The action spectrum of this process shows that it is induced by red light (650–680 nm). This indicates that plants must have a photoreceptor that absorbs red light.

What is especially remarkable about this response is that *it is reversible by far-red light* (710–740 nm). For example, if lettuce seeds are exposed to brief, alternating periods of red and far-red light in close succession, they respond only to the final exposure. If it is red, the seeds germinate; if it is far-red, they remain dormant (**INVESTIGATION FIGURE 25.20**). This reversibility of the effects of red and far-red light regulates many other aspects of plant development, including flowering and shoot development.

The basis for the effects of red and far-red light is **phytochrome**, a photoreceptor located in the cytosol of plants that exists in two interconvertible states (**FIGURE 25.21A**). The phytochrome protein consists of a polypeptide and a small organic molecule called a **chromophore**. When the chromophore absorbs light of specific wavelengths, the resulting change in its electronic configuration causes the phytochrome polypeptide to change its shape. The default, or "ground," state of phytochrome absorbs principally red light and is called $\mathbf{P_r}$, for the red-absorbing form (**FIGURE 25.21B**). When P_r absorbs a photon of red light, it is converted into the other isoform, $\mathbf{P_{fr}}$.

P_{fr} is the active isoform of phytochrome—the form that triggers important biological processes in various plants. These processes include chlorophyll synthesis, the unfolding of the apical hook that

INVESTIGATION

FIGURE 25.20 Sensitivity of Seeds to Red and Far-Red Light Lettuce seeds will germinate if exposed to a brief period of light. An action spectrum indicated that red light was most effective in promoting germination, but far-red light would reverse the stimulation. An experiment by Harry Borthwick and his colleagues asked what effect multiple, alternating flashes of red and far-red light would have if the flashes were given one right after the other.[a] In each trial, the final exposure determined the germination response.

HYPOTHESIS

The effects of red and far-red light on lettuce seed germination are mutually reversible.

METHOD

Expose lettuce seeds to alternate periods of red light R for 1 minute and far-red light FR for 4 minutes.

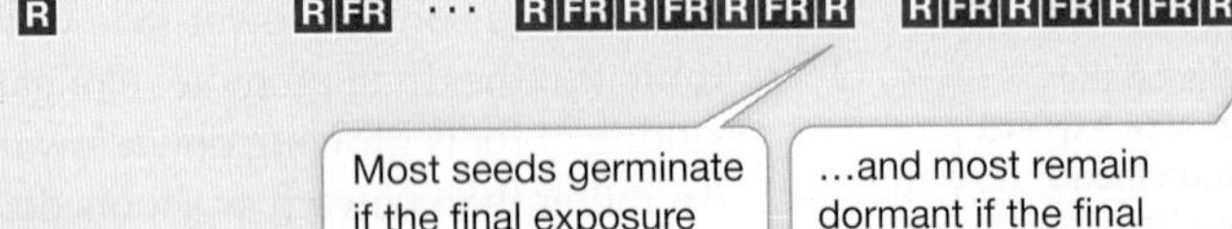

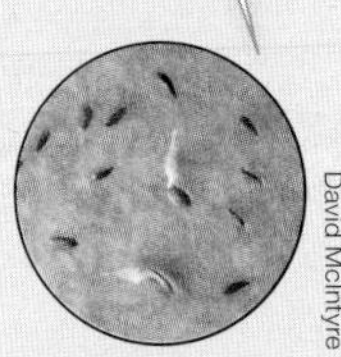

RESULTS

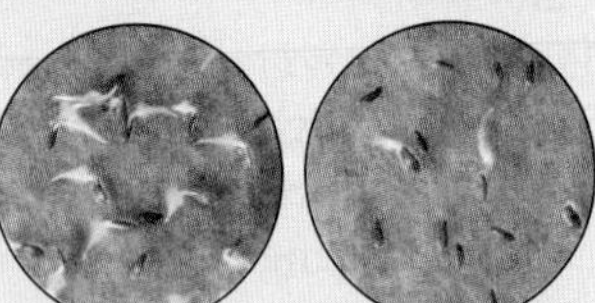
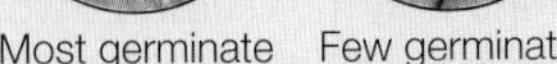

CONCLUSION

Red light and far-red light reverse each other's effects. We now know that phytochrome is responsible for these effects.

ANALYZE THE DATA

Seven groups of 200 lettuce seeds each were incubated in water for 16 hours in the dark. One group was then exposed to white light for 1 minute. A second group (controls) remained in the dark. Five other groups were exposed to red (R) and/or far-red (FR) light. All the seeds were then returned to darkness for 2 more days. The table shows the number of seeds that germinated in each group.

Condition	Seeds germinated
1. White light	199
2. Dark	17
3. R	196
4. R then FR	108
5. R then FR then R	200
6. R then FR then R then FR	86
7. R then FR then R then FR then R	198

1. Calculate the percentage of seeds that germinated in each case.
2. What can you conclude about the photoreceptors involved?

[a]H. Borthwick et al. 1952. *Proc Natl Acad Sci USA* 38: 662–666.

A similar **Analyze the Data** exercise may be assigned in **Achieve**.

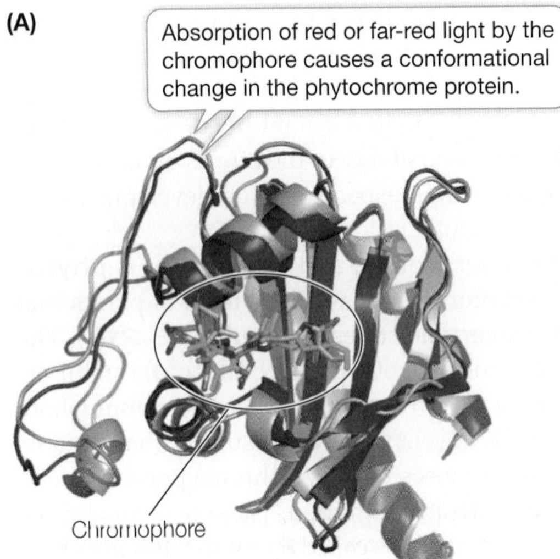

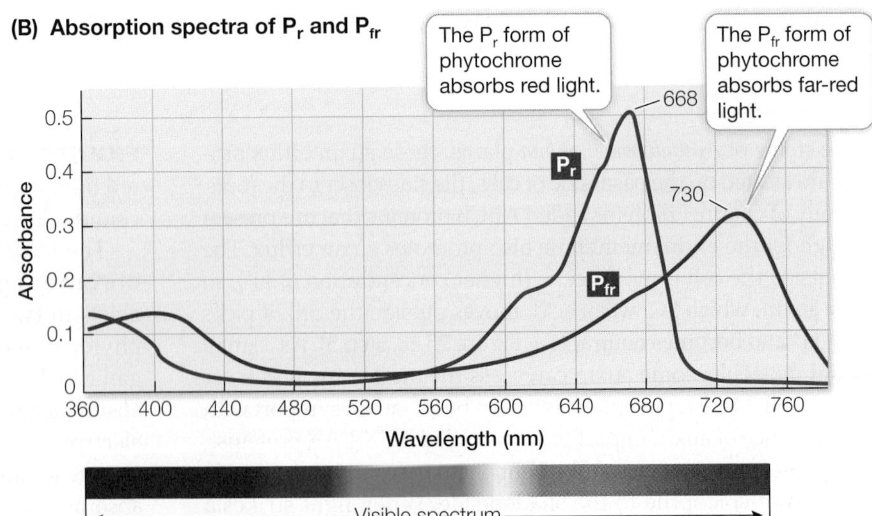

FIGURE 25.21 Phytochrome Changes Its Conformation upon Photoexcitation (A) Three-dimensional structure showing part of the phytochrome protein with its attached chromophore in the center. The chromophore can absorb red or far-red light. The chromophore of P_r preferentially absorbs red light, and when it does so its electronic configuration changes, as shown by the gray (P_r) and blue (P_{fr}) images of the chromophore. This change in the chromophore induces a change in the shape of the phytochrome protein from the P_r state (gray) to the P_{fr} state (magenta). Conversely, absorption of far-red light by the chromophore of P_{fr} returns the phytochrome to its P_r state. **(B)** Absorption spectra of the P_r and P_{fr} states of phytochrome. (A from A. T. Ulijasz et al. 2010. *Nature* 463: 250–254.)

protects the delicate shoot as the stem grows up out of the soil, and leaf expansion (**FIGURE 25.22**). The P_{fr} isoform can be converted back into the P_r form by absorbing far-red light. P_{fr} can also revert to the P_r form in darkness in a process called dark reversion, though this occurs much more slowly than the light-induced changes.

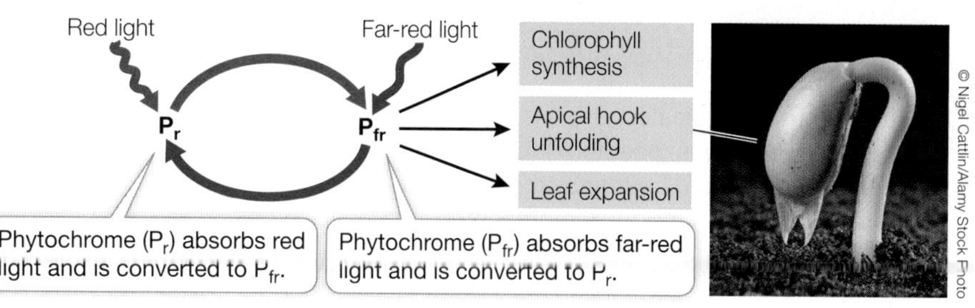

FIGURE 25.22 Phytochrome Is a Photoreceptor with Two Interconvertible States Phytochrome is synthesized in the P_r form. Upon exposure to red light, it takes on the active P_{fr} form. P_{fr} moves to the nucleus where it binds to transcription factors. It also acts as a protein kinase (an enzyme that adds phosphate groups to other proteins). In the dark (not shown), P_{fr} gradually reverts to the P_r form. The dark reversion is temperature-dependent.

The ratio of red to far-red light determines whether a phytochrome-mediated response will occur. For example, in full daylight there is more red than far-red light (the ratio is about 1.2:1) and the P_{fr} isoform predominates. But for a plant growing in the shade of other plants, the ratio may be as low as 0.13:1, in which case most phytochrome is in the P_r isoform. The low ratio of red to far-red light in the shade results from absorption of red light by chlorophyll in the leaves overhead, so less of the red light gets through to the plants below. Shade-intolerant species respond by stimulating cell expansion in the stem and thus growing taller to escape the shade (similar to the behavior of seedlings grown in the dark). Similarly, shade cast by other plants prevents germination of seeds that require red light to germinate.

How does the active form of phytochrome—P_{fr}—work? In the inactive P_r state, phytochrome is a cytoplasmic protein. When P_r absorbs red light, the chromophore changes shape, which leads to a change in the conformation of the phytochrome protein from the P_r isoform to the P_{fr} isoform (see Figure 25.21A). Conversion to the P_{fr} isoform results in movement of P_{fr} from the cytosol to the nucleus. Once in the nucleus, P_{fr} interacts with specific transcription factors, promoting their degradation and thereby modulating expression of genes involved in photomorphogenesis. In *Arabidopsis thaliana* phytochrome affects an amazing 2,500 genes (10 percent of the entire genome) by either increasing or decreasing their expression. Some of these genes are related to other hormones. For example, when P_{fr} is formed in seeds, genes for gibberellin synthesis are activated and genes for gibberellin breakdown are repressed. As a result, gibberellins accumulate.

The mechanism by which gibberellins promote seed germination was extensively studied in barley. Dormant barley seeds contain an embryo, nutritive tissue (endosperm), and a seed coat. Barley seeds contain very little water: only 5–15 percent of their weight is water, compared with 80–95 percent for most other plant parts, but their low water potential means that they will take up water if the seed coat is permeable. Once a seed takes in water, gibberellins diffuse through the endosperm to a surrounding tissue layer called the aleurone layer, which lies underneath the seed coat. The gibberellins cause cells in the aleurone layer to synthesize and secrete enzymes that digest proteins and starch. Degradation of these macromolecules releases amino acids and sugars that the embryo uses as it begins to grow and break through the seed coat (**FIGURE 25.23**). These observations have practical importance: in the beer brewing industry, gibberellins are used to enhance the "malting" (germination of barley and the breakdown of its endosperm), producing sugar that is fermented to alcohol.

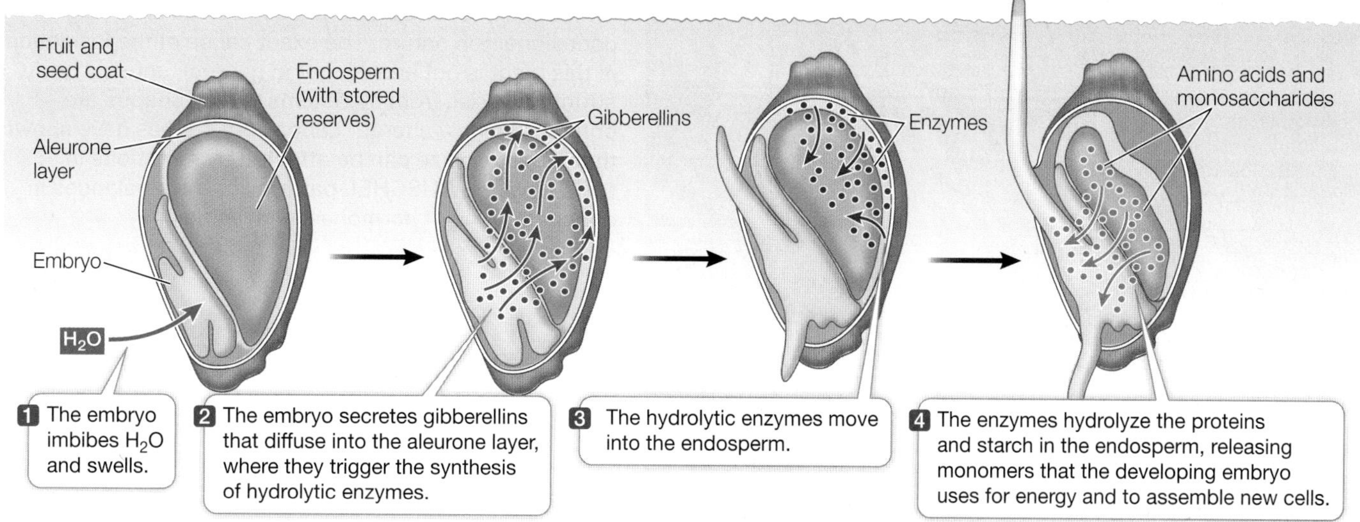

FIGURE 25.23 Gibberellins and Seed Germination During seed germination in cereal crops, gibberellins trigger a cascade of events that result in the conversion of starch and protein stores in the endosperm into monomers that can be used by the developing embryo.

Activity 25.2 Events of Seed Germination

REVIEW & APPLY | 25.3

R Plants mount coordinated responses to the environment by employing numerous receptors that enable them to perceive the environment, and hormones that carry the resulting signal throughout the plant body. Photoreceptors, pigment-containing proteins that change shape (conformation) when a photon of light is absorbed, allow plants to perceive light. Upon receipt of a light signal, the entire plant responds in a coordinated way. For example, if a plant is directing its resources into growing taller to get past the shade of a neighbor, it also waits until the shade has been avoided before expanding its leaves. Phototropism, the growth of plants toward light, occurs when light absorption by phototropin causes the hormone auxin to accumulate on the shaded side of the stem, resulting in that side growing longer. Seed germination is promoted by phytochrome, a photoreceptor that switches between two states depending on what color light it absorbs. Absorbing red light results in the formation of active phytochrome that affects gene expression and increases the amount of gibberellins in the seed. Gibberellins promote seed germination by inducing expression of enzymes that break down stored nutrients, releasing energy and amino acids that support growth of the embryo.

A 1. The graph below shows the results of applying auxin to the lit side of a corn coleoptile. Positive curvatures are toward the light, negative curvatures away from the light. The *y* axis on the left side (blue) shows how rapidly the lit side is growing (blue graph line). The *y* axis on the right side (purple) indicates how much the seedlings have curved (purple graph line). (a) Describe in words what the graph shows. (b) Explain why increasing concentrations of auxin affect curvature in the manner shown. (*Figure source*: T. I. Baskin et al. 1986. *Plant Phys* 81: 306–309.)

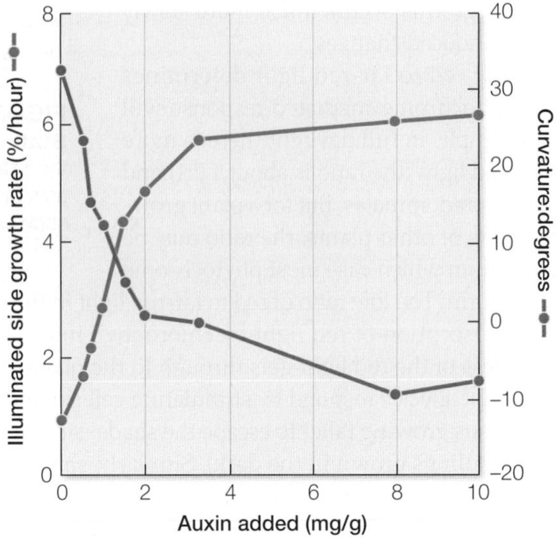

2. Compare the wavelengths of light absorbed by chlorophyll (see Figure 5.18) and by phytochrome (see Figure 25.21B). How does the difference influence a plant's responses to light when a plant grows beneath a forest canopy?

3. How would you show that a physiological process in a plant is brought about by phytochrome?

Answer: The plant has a stem that is wider and thinner than normal, a condition referred to as fasciation. Although uncommon, plants such as this are seen occasionally in nature. The exact cause of the fasciation in this plant is not known, but in general, unusual body structures arise from meristems whose shapes are correspondingly altered. Laboratory studies have shown that meristem size can be affected by mutations in the CLAVATA–WUSCHEL pathway or by imbalances in concentrations of hormones such as auxin.

25.1 Cell Differentiation Occurs Progressively as Meristematic Cells Mature

FIGURE 25.2

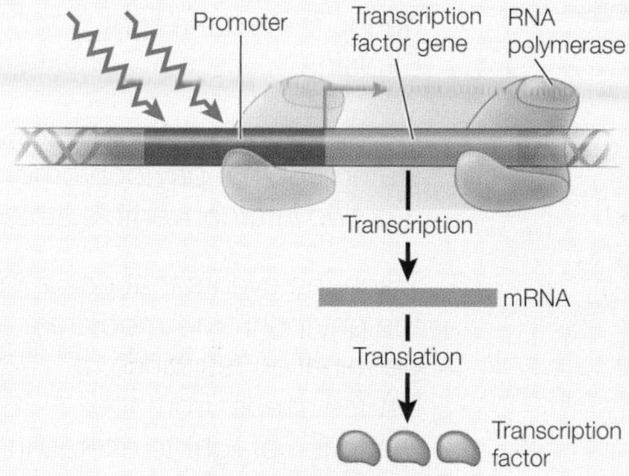

FIGURE 25.3

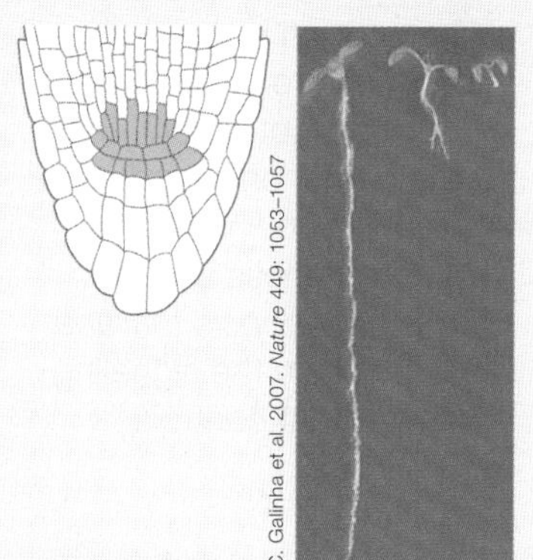

- Transcription factors called **MASTER REGULATORS** cause cells to **DIFFERENTIATE** into distinct types of cells (**FIGURE 25.2**).
- Cells grow by taking in water, which adds pressure to the cell wall, expanding it.
- The direction of cell growth is regulated by the orientation of **CORTICAL MICROTUBULES** and **CELLULOSE MICROFIBRILS**.
- Expression of PLETHORA transcription factors in the root meristem prevents loss of stem cells and so promotes root formation. Expression of PLETHORA 2 in the shoot meristem results in formation of root-like organs from the shoot (**FIGURE 25.3**).

QUESTIONS

1. Explain how the activity of a transcription factor can determine cell type. What is it that the transcription factor does?
2. Plants exhibit great developmental flexibility. Unlike most animals, plants remodel their bodies, creating new organs throughout their life cycle. They also retain the capacity to rebuild themselves from almost any cell. Provide an argument as to why this plasticity could be particularly advantageous.
3. Describe how cell expansion occurs in the context of the stress relaxation cycle.
4. Explain how the mobile peptide CLV3 helps regulate meristem size.

25.2 Multiple Mechanisms Allow Cells That Are Initially Identical to Become Distinct from One Another

Go to ACTIVITY 25.1

- Cells can become different from one another because of asymmetrical cell division, amplification of initially small differences by **POSITIVE-FEEDBACK** loops (**FIGURE 25.11**), or the influence of various environmental factors.
- **AUXIN** regulates many aspects of plant development by inducing changes in cell fate.
- **POSITIONAL INFORMATION** contributes to establishing cell fate.

FIGURE 25.11

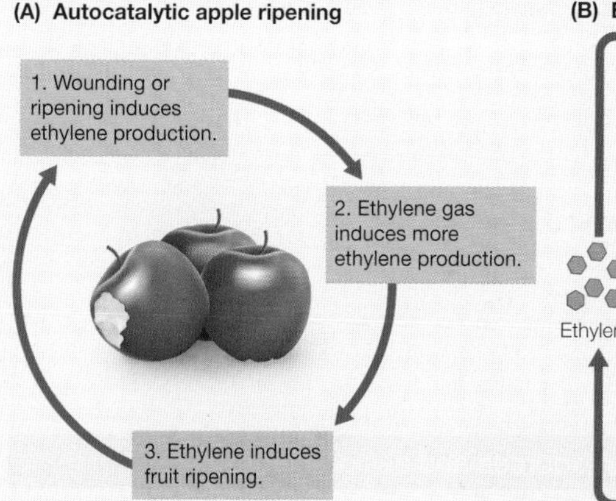

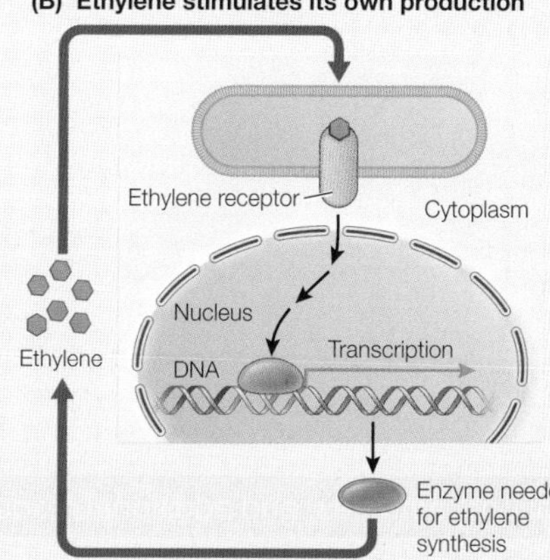

QUESTIONS

1. Describe what is meant by a "stochastically arising difference," and how one might be reinforced.
2. How can a cell's position in a plant affect its identity?
3. Homeostasis, the maintenance of something as it is, is frequently established by negative-feedback loops, whereas developmental switches are typically promoted by positive-feedback loops. What advantages do positive-feedback loops have in terms of inducing developmental changes?

25.3 Developmental Changes Can Be Mediated by Hormones

Go to ANIMATIONS 25.1–25.3 and ACTIVITY 25.2

FIGURE 25.18

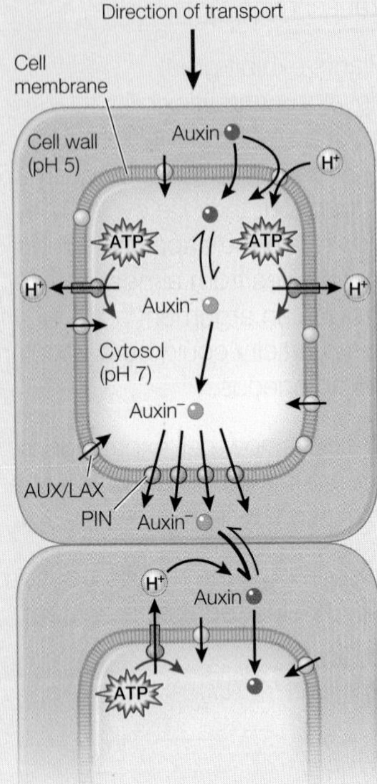

- The location of auxin transport proteins within cells results in directional movement of auxin, called **POLAR AUXIN TRANSPORT** (**FIGURE 25.18**).
- **PHOTOTROPISM** occurs when perception of blue light by **PHOTOTROPIN** leads to lateral redistribution of auxin. More auxin on the shaded side leads to more elongation on that side, causing the plant to grow toward the light (**FIGURE 25.19**).
- **PHYTOCHROME** is a **PHOTORECEPTOR** that has two interconvertible states, $\mathbf{P_r}$ and $\mathbf{P_{fr}}$. The fraction that is in the active (P_{fr}) state depends on the ratio of red to far-red light to which the plant is exposed. When the **CHROMOPHORE** of phytochrome absorbs red light, the conformation of phytochrome changes from the P_r state to the P_{fr} state, which influences transcriptional activity affecting many genes.

FIGURE 25.19

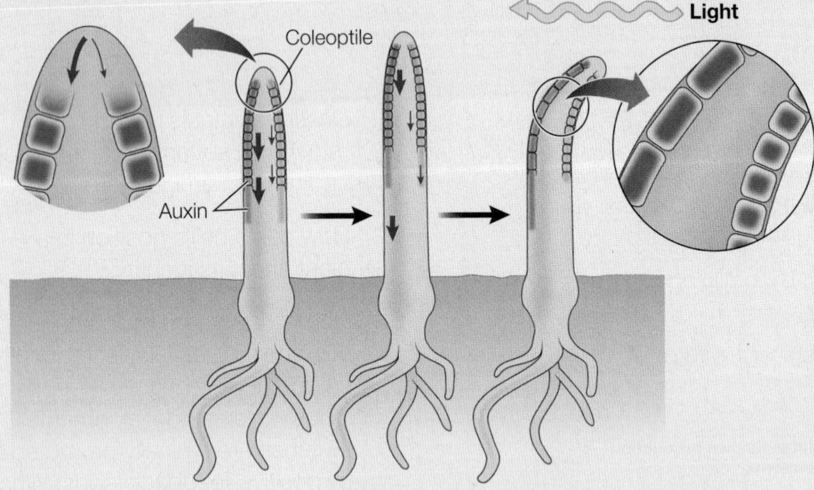

QUESTIONS

1. Even though auxin moves *away* from the lit side of a coleoptile tip, the coleoptile bends *toward* the light. Explain.
2. Explain what is meant by the statement that phytochrome has two interconvertible states.
3. Trace the events that occur in the phytochrome-mediated germination response of lettuce seeds, from the reception of light to alteration in gene expression.
4. When grown in the dark, seedlings are tall, thin, and white (not green), a phenomenon called etiolation. How would you set up a genetic screen using *Arabidopsis thaliana* to investigate the signaling pathway that controls etiolation?

Go to **Achieve** for the e-book, animations, activities, and additional resources and assignments.

Reproduction of Flowering Plants 26

The flowers of many plants, such as these sunflowers, appear all at once at certain times of the year.

What triggers plants of a given species to open their flowers all at the same time?

You will find the answer to this question on page 654.

KEY CONCEPTS

26.1 Flowering Plants Can Reproduce Asexually

LEARNING OBJECTIVES

By the end of this key concept you should be able to:

26.1.1 Explain what it means for a plant to reproduce asexually.

26.1.2 List the advantages of vegetative reproduction to the plant and to agriculture.

Flowering plants (angiosperms) use many forms of reproduction. One form they frequently use is **asexual reproduction**—that is, they reproduce in ways that produce genetically identical progeny (clones). Clones can be propagated from fragments of vegetative organs such as roots, stems, or leaves in what is called **vegetative reproduction**. Strawberries, for example, produce horizontal stems, called **stolons** or runners, that grow along the soil surface, form roots at intervals, and establish potentially independent plants (**FIGURE 26.1A**). Garlic can reproduce vegetatively because each clove in a garlic bulb is a modified stem that can give rise to a new genetically identical garlic plant (**FIGURE 26.1B**). Fleshy underground stems, such as potato **tubers**, can produce plants from buds called "eyes" (see Figure 23.16C). Cutting tubers into pieces allows farmers to grow several new plants from a single tuber. Leaves can be the source of new plants, as in the succulent plant, mother of thousands (**FIGURE 26.1C**).

Even reproduction that occurs via flowers may be asexual. For example, in the reproductive process called **apomixis**, diploid cells in the female part of the flower form embryos without undergoing meiosis and fertilization, processes that are typically part of sexual reproduction. Apomictic embryos are genetic clones of their parent. Relatively few species use apomixis as a regular means of reproduction, but dandelions, blackberries, and some citrus trees are examples of plants that reproduce in this manner.

While many plants are prolific in their creation of new plants by asexual means, most are also capable of flowering and going through **sexual reproduction** (reproduction involving the union of gametes from two individuals). Why would both mechanisms of reproduction have evolved in a plant? Asexual reproduction may be optimal for plants that are well adapted to a stable environment, as they can pass on a combination of alleles that functions well in those conditions. Furthermore, in using asexual reproduction, the plant typically avoids the costs of producing flowers and the potentially unreliable processes of cross-pollination and seed germination. However, asexual reproduction does not generate the diversity of offspring that sexual reproduction can produce. Sexual reproduction increases genetic diversity because it includes **meiosis**, a reductive cell division during which chromosomes reassort and may cross over. Diversity can be particularly advantageous when plants face changing environmental conditions. For example, the English elm tree (*Ulmus minor*) was widely reproduced by suckers and planted around cities. In 1967 a fungal pathogen called Dutch elm disease struck English elms. The population lacked

FIGURE 26.1 Vegetative Organs Modified for Reproduction **(A)** Stolons. **(B)** Bulbs. **(C)** Plantlets.

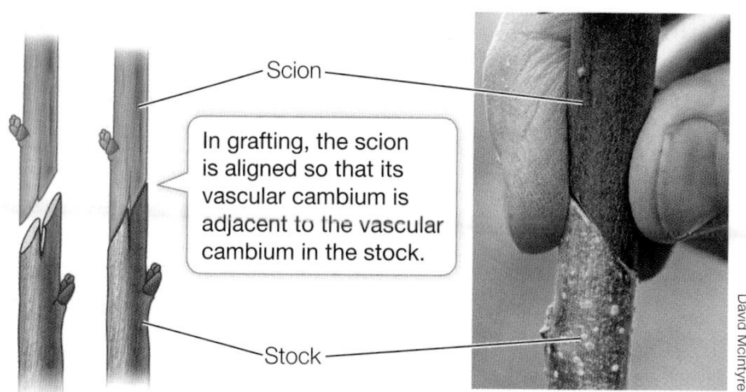

FIGURE 26.2 Grafting Grafting—attaching a piece of a plant to the root or root-bearing stem of another plant—is a common horticultural technique. In the photo, a scion of one apple variety is being grafted onto a stock of another variety.

genetic diversity, and no individuals had genes that would confer resistance to the pathogen. Today the English elm is all but gone from England.

The advantages and disadvantages of asexual reproduction often vary depending on the substrate on which the plants are growing. Plants that reproduce vegetatively are frequently found in environments where conditions for seed germination are unreliable. Plants with stolons or underground stems called **rhizomes**, such as beach grasses, rushes, and sand verbena, are common on coastal sand dunes. Rapid vegetative reproduction enables these plants, once established, to stay ahead of the shifting sand; in addition, the dunes are stabilized by the extensive network of rhizomes or stolons. Vegetative reproduction is also common in some deserts, where the dry environment is often unsuitable for seed germination and the establishment of seedlings.

Vegetative reproduction is important in agriculture

Vegetative reproduction is particularly advantageous in agriculture because it provides a means of multiplying plants that have desired attributes without altering their genetic composition. One of the oldest methods of vegetative reproduction is to make cuttings of stems, insert them in soil or water, and wait for them to form roots and become autonomous plants. Micropropagation, in which sections of a single plant are used to grow multiple new plants, provides a method for mass production of rare plants.

Woody plants can be propagated asexually by **grafting**: attaching a bud or a piece of stem from one plant to a root or root-bearing stem of another plant. The part of the resulting plant that comes from the root-bearing "host" is called the **stock**; the part added to it is the **scion** (**FIGURE 26.2**). The vascular cambium (see Figure 23.20) of the scion associates with that of the stock, forming a continuous cambium that produces xylem and phloem. The cambium allows the transport of water and minerals to the scion and of photosynthate to the stock. Much of the fruit grown for market in the United States is produced on grafted trees. Wine grapes, which grow on woody vines, are also commonly grafted onto hardy rootstock. One advantage of grafting is that a desirable scion (one that produces good fruit, for example) can be grafted onto a genetically different stock that has other desirable qualities, such as disease-resistant roots.

Another method widely used for asexual plant propagation is **meristem culture**, in which pieces of shoot apical meristem are cultured on growth media to generate plantlets, which can then be planted in the field. This strategy is vital when uniformity is desired, as in forestry, or when virus-free plants are the goal, as with strawberries and potatoes.

Today many plant breeders are interested in the possibility of introducing an apomictic reproductive system into crop plants. Were that possible, the chromosomes of an agricultural line with particularly favorable attributes could be passed on in the form of seed. Accomplishing this goal may be complicated by the fact that the process of sexual reproduction often "resets" the extent of DNA methylation in the genome, resulting in changes in gene expression.

LINK As discussed in **Key Concept 11.4**, epigenetic changes (such as DNA methylation and histone modifications) can affect the rate at which individual genes are transcribed into mRNA. Epigenetic modifications of DNA thus affect gene expression and alter the phenotype of organisms.

REVIEW & APPLY | 26.1

R Asexual reproduction produces clones of the parent plant, a feature that can be advantageous if a plant is well adapted to a relatively stable environment. Apomixis is a relatively uncommon form of asexual reproduction in which embryos form without going through a reductive meiotic division. Vegetative reproduction is particularly useful in agricultural settings as it allows cultivars or improved lines of plants to be faithfully reproduced.

A

1. Which form of reproduction, sexual or asexual, is more likely to be successful in the face of global climate change?
2. Many groceries sell seedless grapes. How are these most likely to be propagated?
3. Most of the navel orange trees that once grew all over southern California are derived by grafting and apomixis from a single tree that still exists in Riverside, California. How would you prove this?

Most flowering plants have evolved to reproduce sexually because this strategy results in the genetic diversity that is the raw material for evolution. In the next key concept we outline the process of sexual reproduction in flowering plants.

26.2 Most Flowering Plants Reproduce Sexually

LEARNING OBJECTIVES

By the end of this key concept you should be able to:

26.2.1 Outline the life cycle of a flowering plant, including the steps that give rise to the haploid and diploid stages.

26.2.2 Explain why each of the two fertilization events in plants is advantageous.

26.2.3 Describe three mechanisms that plants employ to avoid self-fertilization.

The plant life cycle differs from our own in having two separate multicellular stages: a haploid **gametophyte** that produces **gametes** (egg and sperm) and a diploid **sporophyte** that produces **spores**. Each stage of the life cycle gives rise to the other, so this two-stage system is called **alternation of generations** (**FIGURE 26.3**).

LINK **Chapter 20** offers an overview of several types of plant life cycles as well as an introduction to alternation of generations. **Figure 20.6** illustrates the life cycle of a moss, and **Figure 20.9** illustrates the life cycle of a fern.

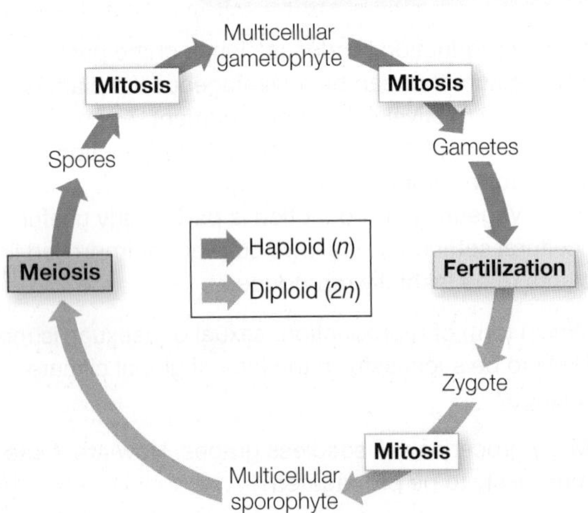

FIGURE 26.3 Alternation of Generations The plant life cycle includes two separate generations: a haploid gametophyte and a diploid sporophyte. Cells in the diploid sporophyte (spore-forming plant) undergo meiosis to produce haploid spores. These spores divide by mitosis to create a multicellular haploid organism called a gametophyte. Cells in the haploid gametophyte (gamete-forming plant) differentiate as gametes (eggs and sperm) that can undergo fertilization, forming a new sporophyte.

Human sperm and eggs are formed by meiosis directly from the diploid organism. Are plant gametes formed directly from meiosis of the diploid organism? If not, what structure do they arise from?

In animals, meiosis usually produces gametes directly, so there is no multicellular haploid stage. In plants, meiosis leads to the production of haploid spores. These haploid spores go on to do something that our gametes do not do: they divide by mitosis, producing new cells that are also haploid. These cell divisions lead to the formation of a new multicellular entity called the gametophyte. As a gametophyte matures, some of its cells take on different characteristics—that is, they differentiate. In male gametophytes, some of the cells form sperm. In female gametophytes, some of the cells form eggs. When haploid sperm from a male gametophyte fertilize a haploid egg from a female gametophyte, they form a diploid sporophyte. In flowering plants, the plants we see are sporophytes. The gametophytes are tiny and contain only a few cells, but the gametes that they produce remain essential to the life cycle. Thus the plant life cycle includes separate multicellular stages: haploid ones and diploid ones. This can be difficult to grasp because the human life cycle does not contain a multicellular haploid stage.

Plant and animal life cycles also differ in how the **germline** cells—the cells that give rise to the gametes—originate. In animals, the germline cells originate and are set aside before birth. In plants, the cells that will form gametes develop in the adult organism, usually in response to environmental conditions, and so are not set aside.

The flower is the reproductive organ of angiosperms

FIGURE 26.4 shows the life cycle of a flowering plant. While some plants have a single type of gametophyte that forms both egg and sperm, flowering plants have two separate types of gametophytes: the **female gametophyte** gives rise to the egg, and the **male gametophyte** gives rise to the sperm. Both male and female gametophytes form in the flower. In Figure 26.4, fertilization of the egg is accomplished by sperm that are created in the same flower, a process called **self-pollination**. Modifications that promote fertilization by pollen from another plant (**outcrossing**) will be discussed in Key Concept 26.3. Below we trace the events that occur in the female and male gametophytes before the pollen meets the egg, as well as the events following fertilization that lead to the development of seed and fruit. The overall process is called **double fertilization** because it involves two fertilization events: (1) the fertilization of the egg and (2) the formation of a nutrient substance called the endosperm.

FEMALE GAMETOPHYTE The female gametophyte forms when cells inside a part of the flower called the carpel undergo meiosis. Figure 26.4 begins with a megasporocyte that will undergo meiosis. Of the four haploid cells resulting from meiosis, three undergo programmed cell death (see Figure 26.4, steps 1 and 2). Typically, the surviving cell undergoes three mitotic divisions without cytokinesis, producing eight haploid nuclei, all initially contained in the **embryo sac** in the ovary. This embryo sac and its contents comprise the female gametophyte. Three haploid nuclei cluster at one end of the sac, another three at the opposite end of the embryo sac, and the remaining two, called **polar nuclei**, are located in the middle (see Figure 26.4, step 3). The

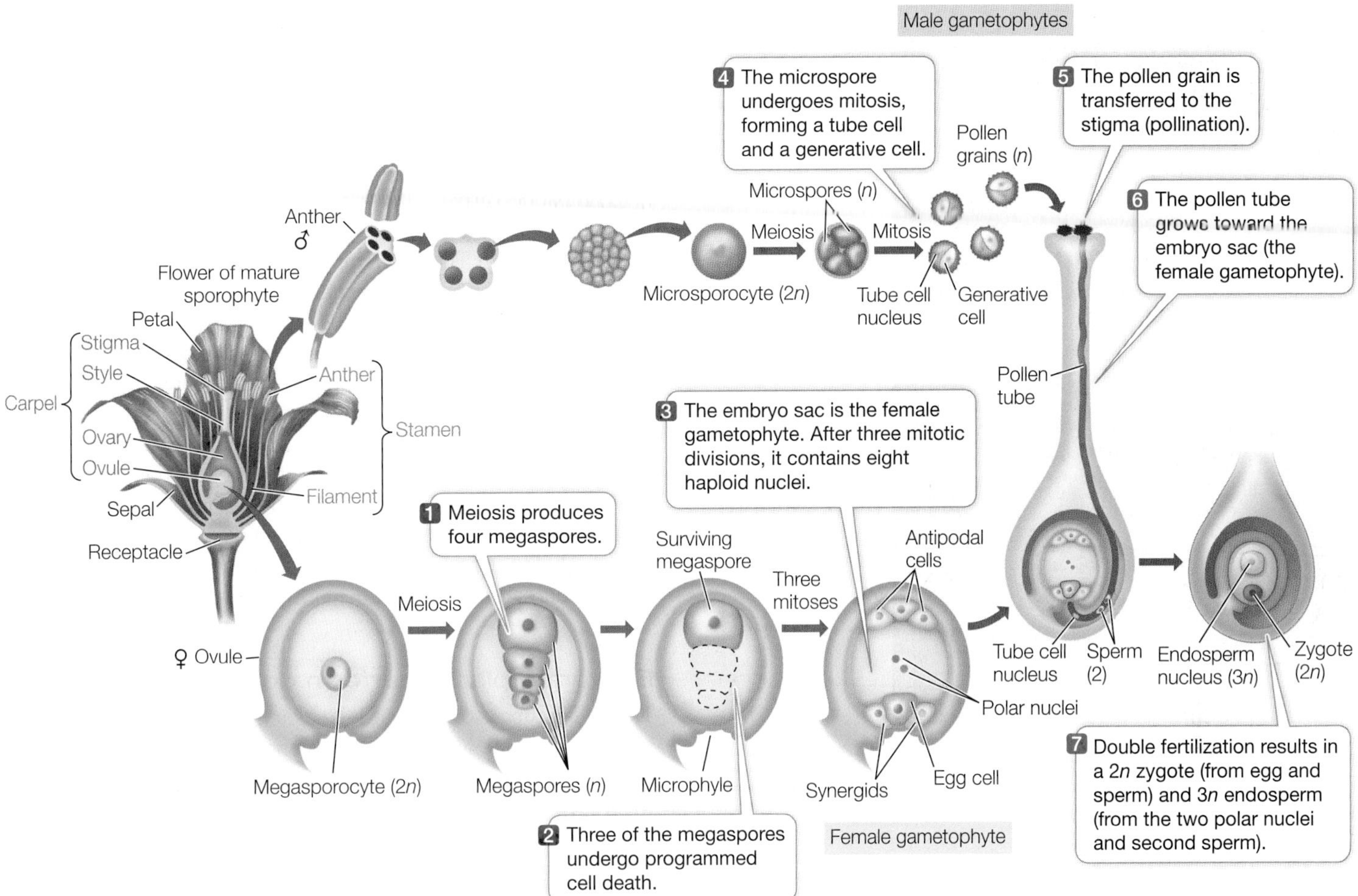

FIGURE 26.4 Sexual Reproduction in Flowering Plants The embryo sac is the female gametophyte; the pollen grain is the male gametophyte. The male and female gametes meet and fuse in the embryo sac. Flowering plants have double fertilization, in which a zygote and an endosperm nucleus form from separate fusion events. The zygote forms by the fusion of one sperm cell with the egg cell. The endosperm forms after the other sperm cell fuses with the central cell, which contains two nuclei. The three nuclei fuse, forming a triploid cell.

Activity 26.1 Sexual Reproduction in Flowering Plants

three haploid cells that are closest to the **micropyle** (a small opening at the tip of the ovule) play important roles in fertilization. One of these, the **egg cell**, is the female gamete; the two haploid cells on either side of the egg cell are called **synergids**, which promote fertilization of the egg cell by secreting chemicals that attract the **pollen tube** through which pollen will eventually reach the egg cell.

MALE GAMETOPHYTE The male gametophytes arise from meiosis that takes place in the anther, which is part of the stamen. The four haploid products of meiosis in the stamen, the **microspores**, each develop a cell wall and undergo a single mitotic division, producing four two-celled **pollen grains** (male gametophytes) that are released into the environment. The two cells in a pollen grain have different roles (see Figure 26.4, step 4). One, called the **tube cell**, forms the elongating pollen tube that delivers the sperm to the embryo sac after the pollen grain is transferred to a stigma (part of the female reproductive organ) in the process of **pollination** (see Figure 26.4, steps 5 and 6). The other, called the **generative cell**, divides by mitosis to form two sperm cells that participate in double fertilization (see Figure 26.4, step 7). (In some plants this division takes place before pollen is released from the anther, resulting in three-celled pollen grains.)

The evolution of pollen in the seed plants made it possible for male gametes to reach the female gametophyte in the absence of liquid water. In ferns, fertilization takes place when motile sperm move through water toward eggs. In seed plants, fertilization generally occurs after wind (or an animal) carries pollen from one seed plant to another through the air. Wind-pollinated flowers have sticky or featherlike stigmas, and they produce a great number of pollen grains because of the low probability of an individual grain encountering a suitable stigma. More recently in evolutionary time, most flowering plants have come to rely on animals (e.g., insects, birds, bats) for pollination. Pollen transport by animals greatly increases the probability that pollen will get to a female gametophyte of the same species.

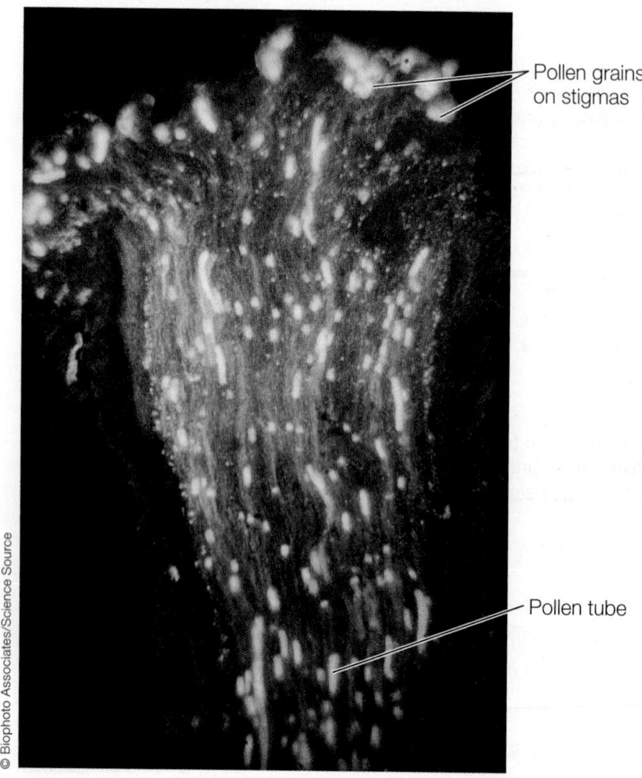

FIGURE 26.5 Pollen Tubes Begin to Grow Staining pollen with a dye allows it to be seen through a microscope. These pollen grains have landed on the stigmas of a crocus.

LINK Natural selection has favored floral features that increase the likelihood of successful pollination. **Key Concept 20.5** describes how angiosperms coevolved with animal pollinators.

When a genetically appropriate pollen grain lands on the stigma of a compatible carpel, it takes in water from the stigma and germinates. Germination involves the development of a pollen tube (see Figure 26.4, step 4). The pollen tube elongates and moves through the carpel toward the ovary, where the female gametophytes are located (**FIGURE 26.5**). The pollen tube typically grows at a rate of 1.5–3 millimeters per hour, taking just 1–2 hours to reach its destination.

The direction in which the pollen tube grows is guided by a chemical signal produced by the synergids in the female gametophyte. If one synergid is destroyed, the ovule still attracts pollen tubes, but destruction of both synergids renders the ovule unable to attract pollen tubes, and fertilization does not occur. The attractant appears to be species-specific; in some cases, isolated female gametophytes attract only pollen tubes of the same species.

Media Clip 26.1 Pollen Germination in Real Time

Flowering plants perform double fertilization

The pollen tube grows toward the ovary under the direction of the haploid tube cell. Meanwhile the generative cell divides once (if it hasn't already done so) to form two haploid sperm cells. Once the pollen tube reaches the embryo sac, one of the two synergids degenerates, and the pollen tube releases its two sperm cells into the embryo sac. Two fertilization events occur (**FIGURE 26.6**):

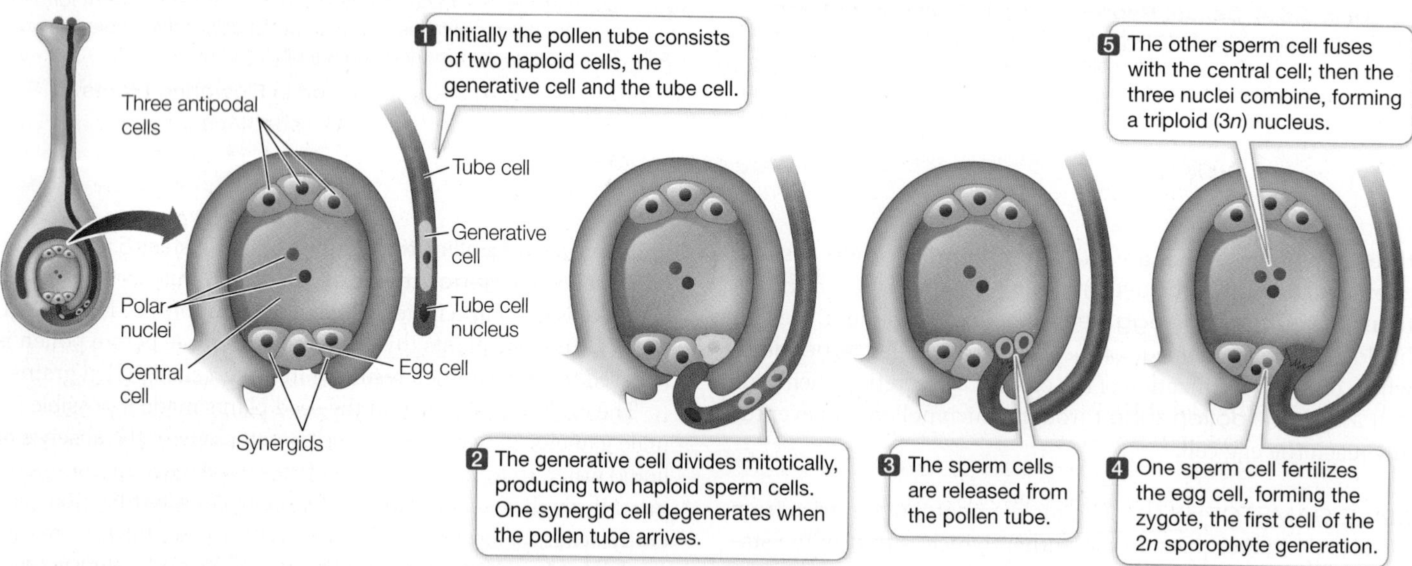

FIGURE 26.6 Double Fertilization Two sperm are involved in two cell fusion events, hence the term "double fertilization." One sperm is involved in the formation of the diploid zygote, and the other sperm results in the formation of the triploid endosperm cell that divides to form endosperm. Double fertilization is a characteristic feature of angiosperm reproduction.

Animation 26.1 Double Fertilization

How much genetic diversity is expected between the two polar nuclei, given that the entire female gametophyte forms by mitotic division of a single haploid cell? What would happen if the haploid polar nuclei contained a loss-of-function allele for a gene whose activity was required for endosperm formation? What advantage results from formation of a triploid endosperm?

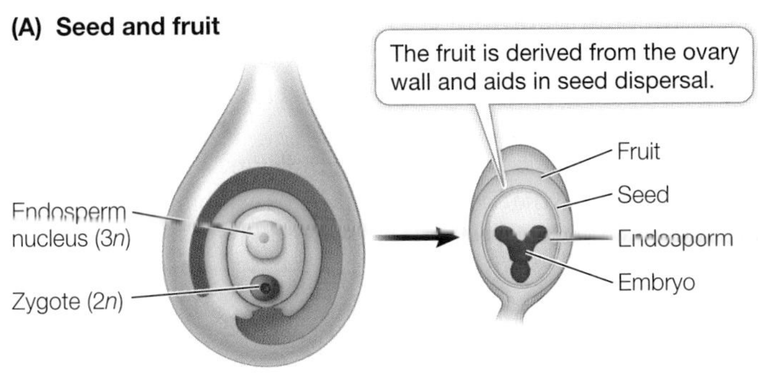

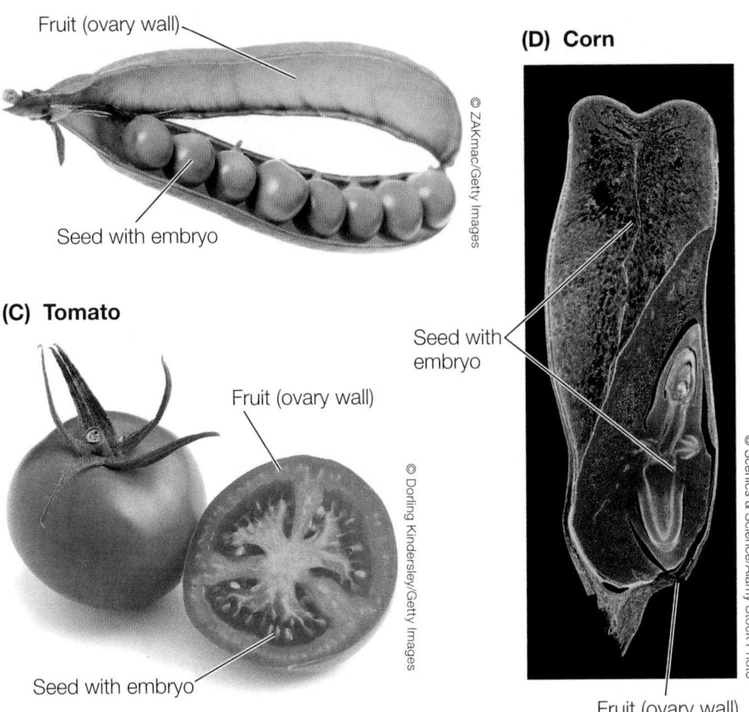

FIGURE 26.7 Angiosperm Fruits There are a variety of fruits, but all have a seed containing the embryo, surrounded by a fruit that usually comes from the wall of the ovary.

- One sperm cell fertilizes the egg cell, producing the diploid ($2n$) **zygote**, which forms the new sporophyte embryo.
- The other sperm cell fertilizes the central cell of the embryo sac, when the sperm fuses with the two polar nuclei. This fusion forms a triploid ($3n$) nucleus that undergoes rapid mitosis to form a specialized tissue, the **endosperm**, which provides nourishment for the developing embryo. In some plants, the endosperm persists in the mature seed and contains reserves that provide chemical building blocks for the germinating seedling while it is underground and cannot perform photosynthesis.

Double fertilization is a characteristic feature of flowering plant reproduction. The remaining cells of the female gametophyte—the antipodal cells and the remaining synergid—eventually degenerate, as does the pollen tube nucleus.

In plants that reproduce by apomixis, embryos form within the flower without their precursor cells having gone through meiosis or fertilization. If, for example, the cells that would typically undergo meiosis fail to do so, and the rest of the developmental process continues in the typical sequence, a diploid egg cell ends up being located inside an ovule. If this diploid cell then begins to divide mitotically, it can form a diploid embryo that is genetically identical to the cells from which it arose. Reproduction without meiosis can also occur if the diploid cells that surround the embryo sac are stimulated to form an embryo. In both cases, seed and fruit development proceeds normally. But the genetic consequences are profound: *apomixis produces clones.*

Embryos develop within seeds contained in fruits

Fertilization initiates the highly coordinated growth and development of the embryo, endosperm, and surrounding tissues. The tissue layers immediately surrounding the ovule develop into the **seed coat**. As seeds develop, they prepare for dispersal and dormancy by losing up to 95 percent of their water content. You can see this desiccation by comparing corn grains (e.g., popcorn) with ripe corn from the cob or a can. The embryo inside a dry seed contains very little water but is still alive; it has protective proteins that keep its cells in a viscous state.

Fertilization also triggers the ovary (and sometimes other parts of the flower) to develop into a **fruit** that encloses the seed (**FIGURE 26.7**). Fruits aid in seed dispersal and protect seeds from damage by animals and infection by microbial pathogens. If a plant has successfully grown and reproduced, its location is likely to be favorable for the next generation too. So some offspring stay near the parent, as is the case in many tree species whose seeds simply fall to the ground. However, if the species is a perennial, offspring that germinate near their parent will be competing with their parent for resources, which may be too limited to support a dense population. Furthermore, even though local conditions were good enough for the parent to produce at least some seeds, there is no guarantee that conditions will still be good the next year, or that they won't be better elsewhere. Thus in many cases, seed dispersal is vital to a species' survival.

Because plants are not mobile, many fruits are modified in ways that increase their chances of being carried to new locations. Some species, such as peaches and tomatoes, produce fleshy, edible fruits that entice animals to eat them. Being eaten can be to the plant's advantage if the seeds are spit out or, if swallowed, survive their passage through the digestive tract and are later excreted in new locations. The fruits of other species are dry or inedible. Some, such as burrs, attach themselves to animals or to your clothes and shoes. Others, like those of the thistle, are carried to new locations by wind. Water also disperses some fruits; coconuts have been known to float thousands of miles between islands. Dispersing seeds over substantial distances may increase the probability that at least a few of the many seeds produced by a plant will find suitable conditions for germination and growth to sexual maturity.

(A) Jack-in-the-Pulpit

(B) A male inflorescence

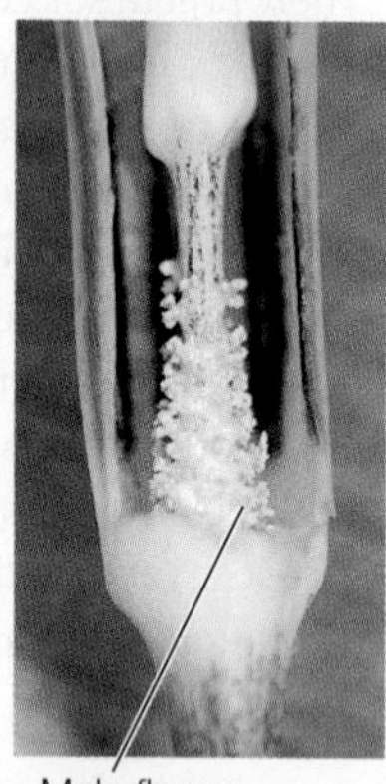

(C) A female inflorescence

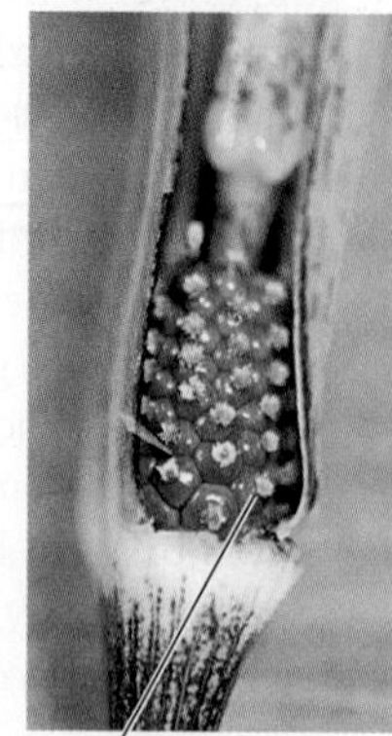

FIGURE 26.8 Sexual Fluidity Jack-in-the-pulpit plants **(A)** can switch between producing all-male **(B)** and all-female **(C)** flowers, depending on the environment in which they are grown.

Flowering plants have mechanisms to prevent inbreeding

As shown in Figure 26.4, some plants self-pollinate: that is, they can be fertilized by their own pollen. Pollen from a given plant may land on the stigma of one of its own flowers—even on the very flower that produced it. Mendel's garden peas (see Chapter 8), for example, reproduced in this way; he removed the pollen-forming parts of the flower to control the parentage of his genetic crosses. Self-pollination mitigates the difficulties of wind or animal pollination of a distant flower. However, "selfing" leads to homozygosity, which can reduce the reproductive fitness of offspring (known as inbreeding depression). Most plants have evolved mechanisms that prevent self-pollination and thus self-fertilization.

SEPARATION OF MALE AND FEMALE GAMETOPHYTES Some plant species form two separate types of flowers: male and female. These can be separate flowers on the same plant (called monoecious species), or on entirely separate plants (dioecious species). If an individual has flowers of only one sex, self-fertilization does not occur. Some interesting perennial species, such as jack-in-the-pulpit (*Arisaema triphyllum*), have somewhat fluid sex determination. Within a given year, a single plant produces only male or female flowers, but the sex of an individual can switch between years. In general, jack-in-the-pulpit plants grown in nutrient-rich conditions bear female flowers (**FIGURE 26.8**).

Separating pollen from a receptive stigma decreases the chance that self-fertilization will occur. But how is self-fertilization prevented in flowers that bear both the pollen-forming stamen and ovule-bearing carpel? In many such species, the stigma matures and becomes receptive to pollen at a different time than the period in which pollen ripens and is shed.

GENETIC SELF-INCOMPATIBILITY A pollen grain that lands on a stigma of the same plant will germinate and fertilize the female gamete only if the plant is self-compatible (capable of self-pollination). To prevent self-fertilization, many plants are genetically **self-incompatible**. This mechanism depends on the ability of a plant to determine whether pollen is genetically similar or genetically different from "self." Rejection of "same-as-self" pollen prevents self-fertilization.

Self-incompatibility in plants is controlled by a cluster of tightly linked genes called the *S* locus (for self-incompatibility). The *S* locus encodes proteins in the pollen and style that interact during the recognition process. A self-incompatible species typically has many alleles of the *S* locus. The pollen phenotype may be determined by its own haploid genotype or by the diploid genotype of its parent plant. In either case, if the pollen expresses an allele that matches one of the alleles of the recipient carpel, the pollen is rejected. Depending on the type of self-incompatibility, the rejected pollen either fails to germinate or the pollen tube is prevented from growing through the flower; either way, self-fertilization is prevented.

REVIEW & APPLY | 26.2

R The flowering plant life cycle includes both haploid (gametophyte) and diploid (sporophyte) stages. Meiosis that takes place in the anther gives rise to pollen. Meiosis that takes place in the carpel gives rise to a female gametophyte that divides by mitosis to form an embryo sac with eight nuclei that are distributed among seven cells, including the egg cell and the central cell that contains two polar nuclei. Pollen grains that land on receptive stigmas germinate. The male gametophyte forms a pollen tube that grows toward the embryo sac. The pollen tube contains a tube cell and a generative cell. The generative cell of the male gametophyte divides mitotically to produce two haploid sperm in the pollen tube, one of which fertilizes the egg, forming an embryo. The other sperm fertilizes the two polar nuclei, forming a nutritive 3*n* tissue called the endosperm. The embryo develops into a seed surrounded by a fruit. Many plants increase the genetic diversity achieved through sexual reproduction by employing mechanisms that reduce or prevent self-pollination.

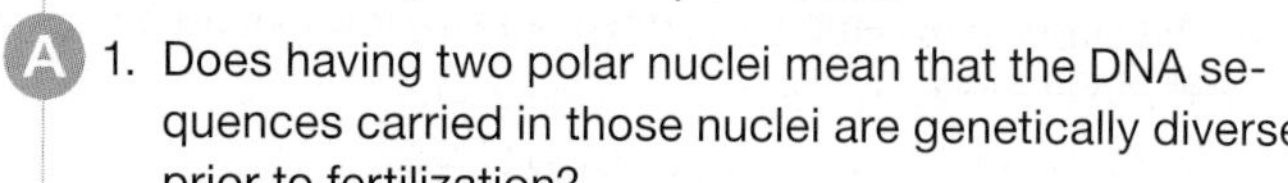

A

1. Does having two polar nuclei mean that the DNA sequences carried in those nuclei are genetically diverse prior to fertilization?
2. The two sperm that participate in double fertilization are genetically identical. Are the two products of fertilization genetically identical?
3. In the human life cycle, single-celled gametes (eggs and sperm) are produced directly by meiosis of the diploid organism. (a) Which stage of the plant life cycle do humans lack? (b) If you were constructing an imaginary world in which the human life cycle had the same two stages as a plant life cycle, what would need to be added? Your answer should include a description of how the egg and sperm would arise.
4. Producing two separate male and female flowers requires more energy than producing just one type of flower. What advantage does producing two separate flowers confer that might make this a reasonable trade-off?

Now that we have described the process of sexual reproduction in flowering plants, let's turn to the events that signal a transition from vegetative growth of the shoot to the formation of flowers.

26.3 | Hormones and Signaling Determine the Transition from the Vegetative to the Reproductive State

LEARNING OBJECTIVES

By the end of this key concept you should be able to:

26.3.1 **Compare and contrast vegetative and reproductive meristems in terms of their location and the organs formed.**

26.3.2 **Explain how floral organs are specified.**

26.3.3 **Explain how flowering is triggered in a day-length-sensitive plant.**

26.3.4 **Describe the experimental approach that led to the identification of flowering genes.**

The time at which a plant flowers (also called blooming) is an important determinant of its success. If a plant begins flowering too late in the year, the process of making seeds may not be completed before cold sets in, whereas too early a start might leave the plant without sufficient resources to complete the flowering process. Additionally, blooming at the same time as neighboring individuals increases the chance of outcrossing—having eggs fertilized by sperm from a different individual. In agricultural settings, the timing of flowering is a major determinant of yield. Thus understanding how plants transition from **vegetative growth** (stem and leaf production) to **reproductive growth** (flower production) has been a major focus of plant biologists.

Shoot apical meristems can become inflorescence meristems

The transition from vegetative growth to the flowering state involves a change in the identity of one or more of the meristems in the shoot. Prior to flowering, the shoot apical meristems produce leaves, stems, and axillary buds—which contain meristems called axillary meristems (**FIGURE 26.9A**). At the transition to flowering, a new set of genes is expressed in the shoot meristem, causing it to become respecified as an **inflorescence meristem**. An inflorescence meristem gives rise to **floral meristems** and to smaller leaflike structures called **bracts**, as well as to new meristems in the angles between the bracts and the stem (**FIGURE 26.9B**). Each floral meristem gives rise to a single flower, typically including four consecutive **whorls** (rings) of organs—the sepals, petals, stamens, and carpel—encircling a central axis (**FIGURES 26.9C** and **26.10A**). The whorls are separated by very short internodes, which keep the flower compact. Thus floral meristems are responsible for determinate growth—growth that is limited in extent, like that of many leaves.

The identity of floral organs is specified by the combination of floral identity genes

In the model plant *Arabidopsis thaliana*, each flower begins as a floral meristem of about 700 cells arranged in a dome, and the four whorls develop from this meristem. The organs that make up the flower are evolutionarily derived from leaves. Their identity is determined by

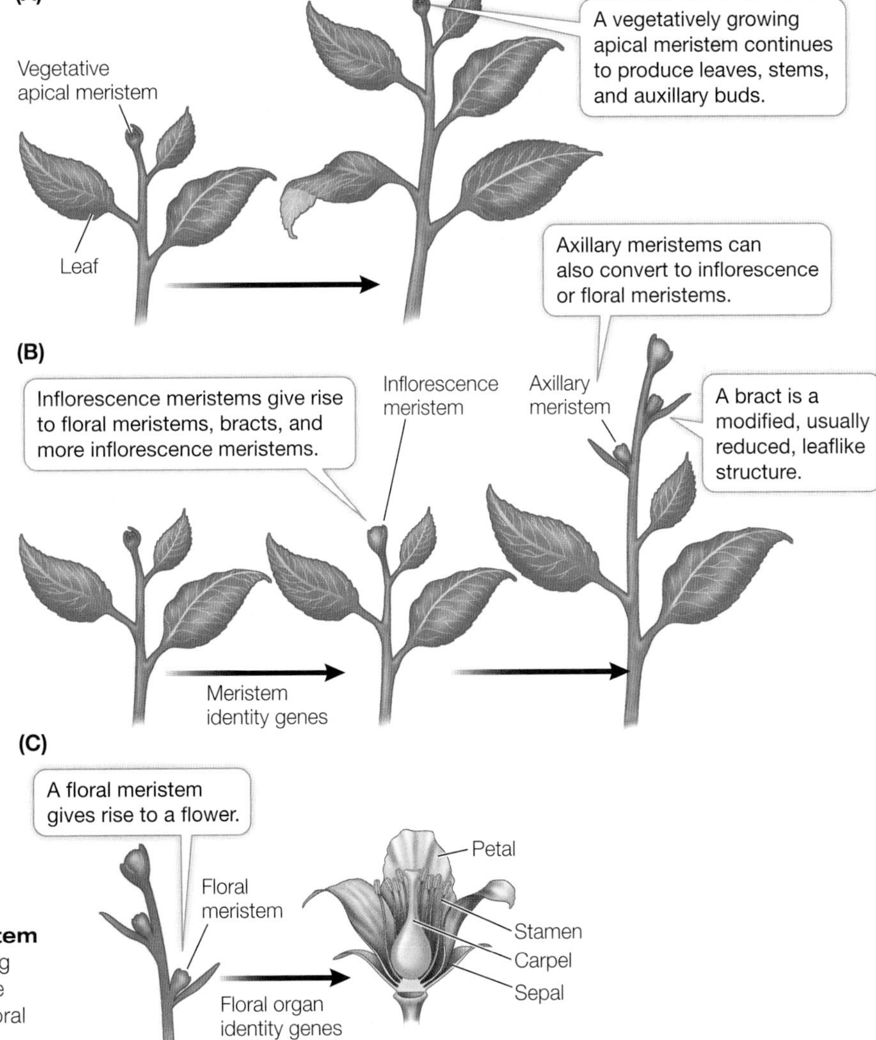

FIGURE 26.9 Flowering and the Apical Meristem A vegetative apical meristem **(A)** grows without producing flowers. Once the transition to the flowering state is made **(B)**, inflorescence meristems give rise to bracts and to floral meristems **(C)**, which give rise to the flowers.

(A) The four organs of a flower

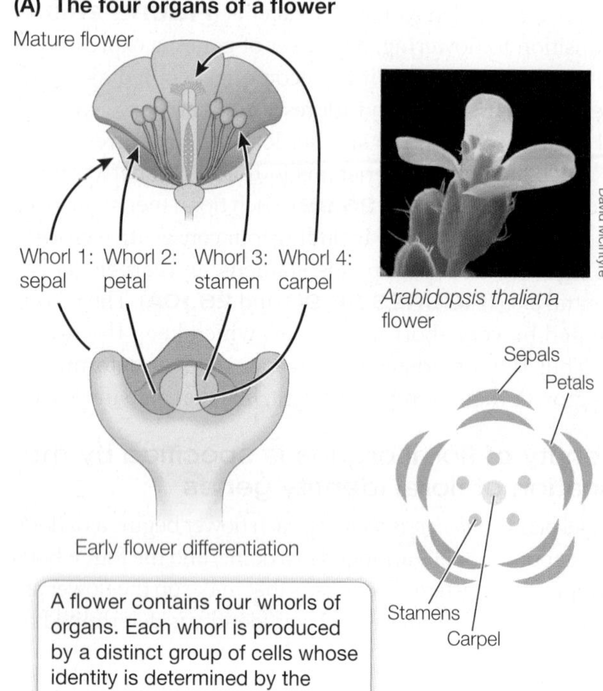

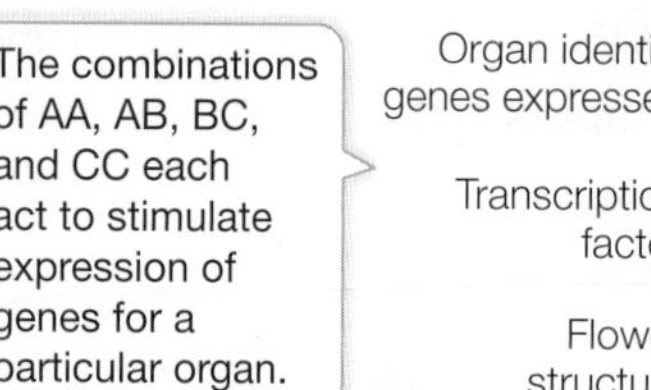

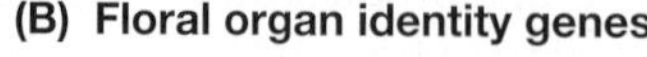

(B) Floral organ identity genes

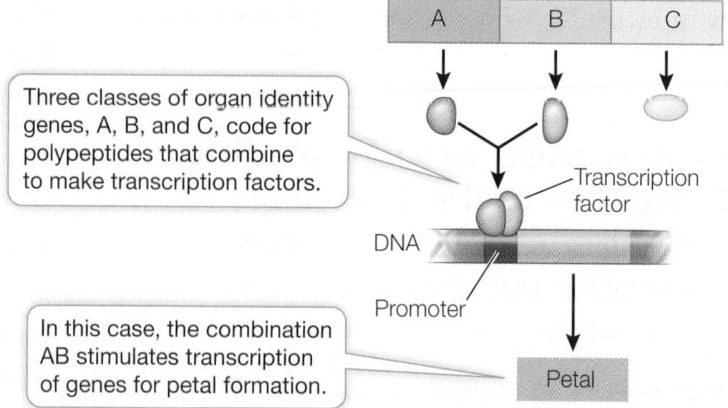

(C) Expression of identity genes

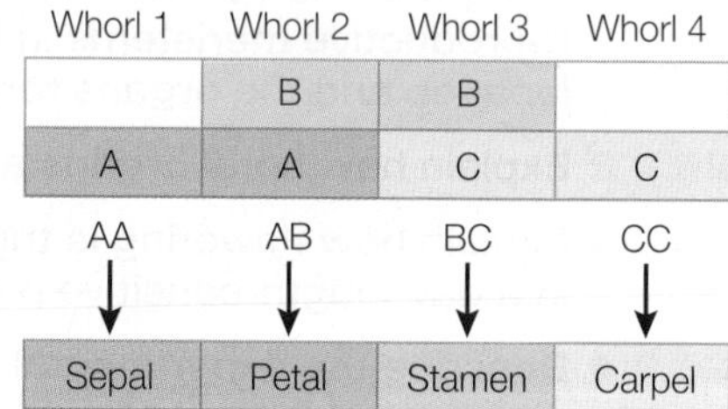

	Whorl 1	Whorl 2	Whorl 3	Whorl 4
Organ identity genes expressed		B	B	
	A	A	C	C
Transcription factor	AA	AB	BC	CC
Flower structure	Sepal	Petal	Stamen	Carpel

FIGURE 26.10 ABC Model for Flower Formation (A) The four organs of a flower—sepals (pink), petals (purple), stamens (green), and carpel (yellow)—grow in whorls that develop from the floral meristem. **(B)** Floral organs are determined by three classes of organ identify genes, the polypeptide products of which combine to form transcription factors. **(C)** Combinations of polypeptide subunits in transcription factors activate gene expression for specific genes.

Activity 26.2 Genes and Development Simulation

three classes of **organ identity genes**—classes A, B, and C—that encode transcription factors (**FIGURE 26.10B**). These proteins act in combination to specify the type of organ that will be produced. For example, a combination of transcription factors that gives rise to purple petals would do so in part by leading to the expression of enzymes required for synthesis of purple pigment. Expression of class A genes alone gives rise to sepals, a combination of class A and class B genes gives rise to petals, a combination of class B and class C genes gives rise to stamens, and class C genes alone give rise to carpels. The genes that are expressed in a whorl generate that whorl's specific organ identity (**FIGURE 26.10C**).

Modifications of this basic structure can result in great diversity. For example, expression of class B genes in the outermost whorl as well as in whorls 2 and 3 causes what would have been sepals to look like petals. This explains how tulip flowers arise (**FIGURE 26.11A**). The multiple petals in cultivated roses are derived from wild relatives with one whorl of five petals and many stamens. A simple conversion of stamens to petals can be achieved by changing the boundary of class A gene expression. Orchids can be particularly remarkable in having multiple types of unusually shaped petals (**FIGURE 26.11B**). These appear to have arisen through diversification of the organ identity genes, creating a situation in which individual orchid petals express different versions of the identity genes. Flowers that lack specific organs, such as flowers that are all male or all female, can arise through selective abortion of particular whorls).

Photoperiodic cues can initiate flowering

Early physiological experiments showed that for some plants the signal that sets the plants' shoot apical meristems on the path to flowering was the length of daylight, or **photoperiod**. This allows plants to flower at a particular time of the year, which can be advantageous in many ways. For example, plants that flower in short spring days are likely to have a long period during which they can grow before setting seed for the next year. They are also less likely to bloom early if there is an unseasonably warm period during the winter—an event that can lead to death of the flowers

(A)

(B)

FIGURE 26.11 Variation in Floral Form (A) Tulip. **(B)** Orchid.

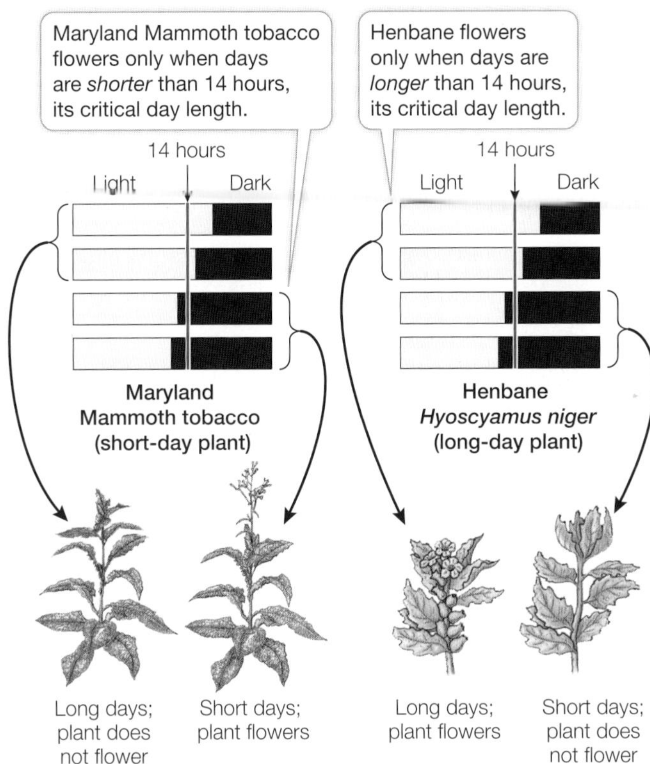

FIGURE 26.12 Day Length and Flowering Flowering of Maryland Mammoth tobacco is initiated when the days become shorter than a critical length. Maryland Mammoth tobacco is thus called a short-day plant. Henbane, a long-day plant, shows an inverse pattern of flowering. (After W. W. Garner and H. A. Allard. 1921. In *Yearbook 1920*, USDA, pp. 377–400. US Government Printing Office: Washington, DC.)

Activity 26.3 Flowering and Day Length Simulation

when the typical cold winter temperatures return. Plants whose flowering time is not affected by day length are called day-neutral plants. Plants that flower in response to photoperiodic stimuli fall into two main classes (**FIGURE 26.12**), although there are variations on these patterns:

1. **Short-day plants** (**SDPs**) flower only when the day is shorter than a critical maximum. They include poinsettias and chrysanthemums. Thus we see chrysanthemums in nurseries in fall and poinsettias in winter.
2. **Long-day plants** (**LDPs**) flower only when the day is longer than a critical minimum. Spinach and clover are examples of LDPs. Spinach tends to flower and become bitter in the summer and is therefore normally planted in early spring.

The terms "short-day plant" and "long-day plant" assume that the key events that induce flowering occur in the daytime. What about the night? After all, short-day plants could just as easily be called "long-night plants," and long-day plants could be called "short-night plants." As it turns out, *night length is indeed the critical factor that induces flowering*, as a series of greenhouse experiments confirmed. In a greenhouse, the overall length of a day or night can be varied irrespective of the 24-hour natural cycle. For example, if cocklebur, a SDP, is exposed to several long periods of light (16 hours each), it will still flower as long as the dark period between them is 9 hours or longer. This 9-hour inductive dark period also induces flowering even if the light period varies from 8 hours to 12 hours.

Biologists noticed that when the inductive dark period was interrupted by a brief period of light, the flowering signal generated by the long night disappeared. It took several days of long nights for the plant to recover and initiate flowering. Interrupting the day with a dark period had no effect on flowering. A clue as to what occurred in the plant when the flash of light was given came when biologists determined the action spectrum for the response—that is, they produced a graph showing which wavelengths of light produced what levels of response. As with lettuce seed germination (see Investigation Figure 25.20), red light was most effective at breaking the "night" stimulus, and its effect was reversible by far-red light. This indicated that the photoreceptor involved in regulating flowering is phytochrome.

Animation 26.2 The Effect of Interrupted Days and Nights

LINK The properties of phytochrome as a photoreversible photoreceptor of red and far-red light are described in **Key Concept 25.3**.

The flowering stimulus originates in the leaf

The photoperiodic receptor for flowering is located in the leaf, not in the shoot apical meristem. If a single leaf of a SDP growing under short-night conditions is covered to simulate a long-night exposure, the shoot apical meristems will transition to flowering as if the entire plant were exposed to long nights (**INVESTIGATION FIGURE 26.13**). Masking the buds (leaving the leaves exposed) does not induce flowering.

Because the receptor of the photoperiodic stimulus is physically separated from the tissue on which the stimulus acts (the shoot apical meristem), the inference can be drawn that a signal travels from the leaf through the plant's tissues to the shoot apical meristem. Biologists have found additional evidence that a diffusible signal travels from the leaf to the shoot apical meristem:

- If a single leaf is photoperiodically induced and then immediately removed after the inductive dark period, the plant does not flower. If the induced leaf remains attached to the plant for several hours, however, the plant flowers. This result suggests that something is synthesized in the leaf in response to the inductive dark period, and then moves out of the leaf to induce flowering.
- In several species, if an induced leaf from one species is grafted onto another, noninduced plant of a different species, the recipient plant flowers. This indicates that the same diffusible chemical signal is used by both species.

This diffusible signal was given a name, **florigen**, meaning "flower inducing," decades ago, but attempts to identify the signal by isolating compounds that were transported through the

INVESTIGATION

FIGURE 26.13 The Flowering Signal Moves from Leaf to Bud Phytochrome, the receptor for photoperiod, is in the leaf, but flowering occurs in the shoot apical meristem. To investigate whether there is a diffusible substance that travels from leaf to bud, James Knott exposed a single leaf of cocklebur plants to their inductive dark period.[a]

HYPOTHESIS

The leaves measure the photoperiod.

METHOD

Grow cocklebur plants under long days and short nights. Mask a leaf on some plants and see if flowering occurs.

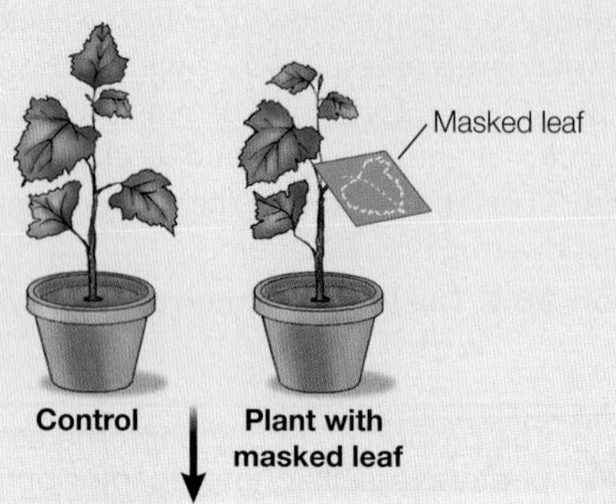

RESULTS

If even one leaf is masked for part of the day—thus shifting that leaf to short days and long nights—the plant will flower.

CONCLUSION

The leaves measure the photoperiod. Therefore some signal must move from the induced leaf to the flowering parts of the plant.

ANALYZE THE DATA

In related experiments, leaves were removed from plants before the plants were exposed to the inductive dark period. There were six plants in each condition.[b]

Condition	Number of plants that flowered
No inductive dark period, intact plant	0
Inductive dark period, intact plant	6
Inductive dark period, all leaves removed	0
Inductive dark period, all but one leaf removed	6

1. Do the data support the conclusion of the experiment by Knott described above? Explain your answer.
2. How would you modify the experiments shown in the table to find out how many days of the inductive dark period are required before the signal is produced? Would you expect the results to be different for the intact plants and plants with only one leaf? Why or why not?

[a]J. E. Knott. 1934. *Proc Soc Hort Sci* 31: 152–154.

[b]K. C. Hamner and J. Bonner. 1938. *Bot Gaz* 100: 388–431.

A similar **Analyze the Data** exercise may be assigned in **Achieve**.

plant were not successful. Thus plant biologists began using a genetic approach. Seeds of *Arabidopsis thaliana* were exposed to mutagens, compounds that cause changes in DNA, and mutant plants that flowered early or late were examined to determine which genes were affected.

Florigen is a small protein

A. thaliana is a facultative LDP: it can flower in short days, but it flowers at a much younger age when it is grown in long days. One of the genes that was shown to be important for its response to day length encodes a protein called **CONSTANS** (**CO**). The amount of CO protein in leaves builds up in the afternoon under long-day conditions, but not under short-day conditions. Plants have a circadian clock that causes the level of some proteins to oscillate in a 24-hour cycle. Part of the reason that CO protein accumulates under long days is that when one particular protein that is highly expressed in the afternoon is also exposed to light, it forms a complex that increases the stability of CO, thereby increasing the level of CO that accumulates. CO is a transcription factor, and when its level is sufficiently high it leads to production of another protein called **FLOWERING LOCUS T** (**FT**) (**FIGURE 26.14**).

FT protein is the long-sought "florigen." FT travels in the phloem from the leaf to the shoot apical meristem, where, together

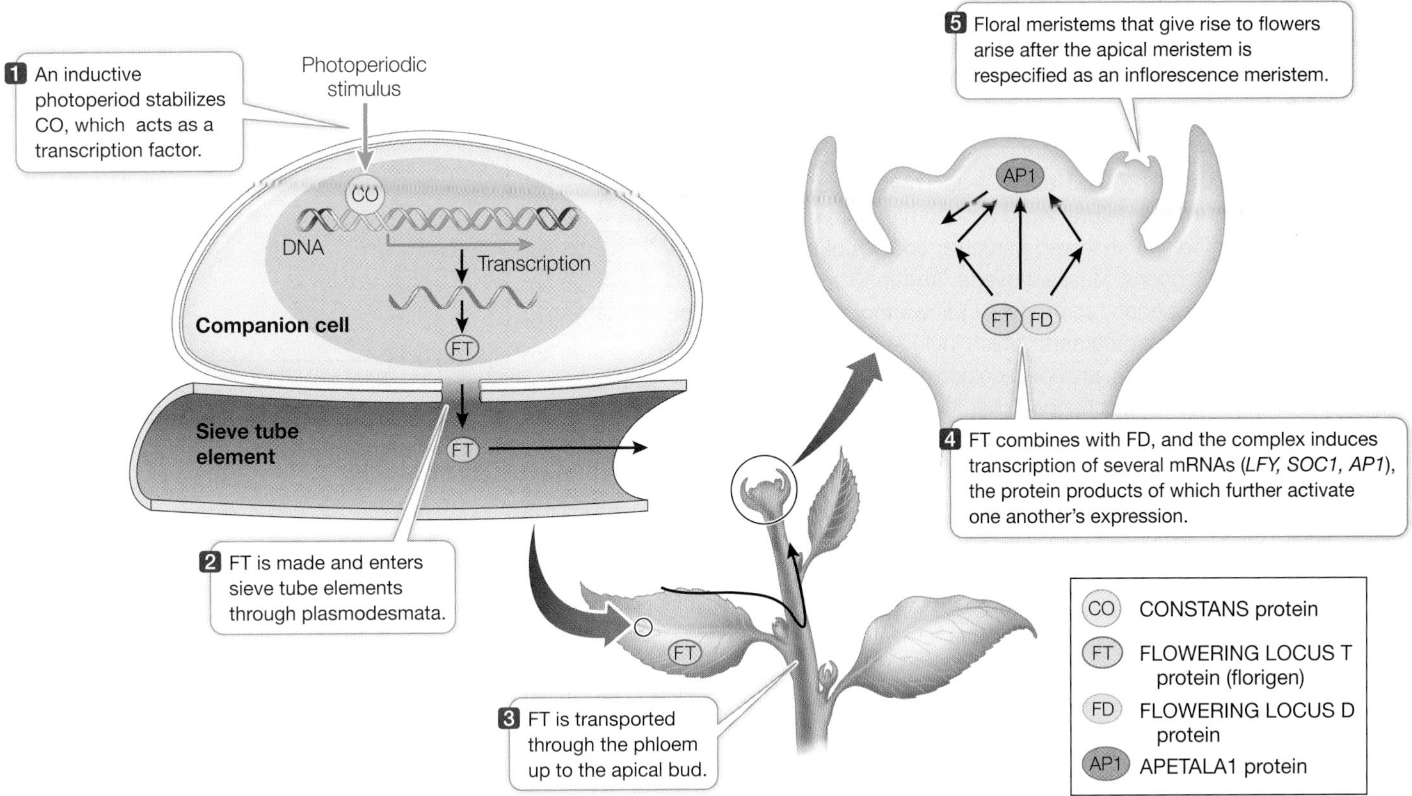

FIGURE 26.14 Florigen Induces Gene Expression That Leads to Flower Formation Florigen (FT) is a protein made in the phloem companion cells. It travels in the sieve tubes from the leaf to the bud meristem. There, FT combines with the protein flowering locus D (FD) to stimulate transcription of genes that initiate flower formation.

with a protein called flowering locus D (FD), it initiates transcription of a set of genes that code for transcription factors, including a class A floral organ identity gene called ***APETALA 1*** (***AP1***). The transcription factors that are upregulated by FT–FD activate the expression of one another, resulting in a positive feedback loop. This results in strongly increased levels of these proteins, which together cause the plant to enter the reproductive phase, producing floral meristems rather than leaves.

Having a positive feedback loop at the center of a developmental transition is advantageous because it can act as a switch between two distinct programs. The meristem is either vegetative, producing leaves, or the positive feedback loop has been activated and the meristem is reprogrammed such that it gives rise to flowers. There is little chance of an in-between state. The advantage of having an all-or-none system is seen in some mutant plants that fail to produce proteins involved in the positive feedback loop; one such mutant ends up producing "flowers" that are composed of small green leaves—an outcome that is unlikely to promote reproductive success.

Having multiple steps along the pathway leading to flowering also opens up opportunities for regulation. Any environmental factor that affects the expression of CO, FT, or the genes that regulate these proteins can affect the time of flowering. Likewise, altered levels of any of the genes involved in the positive feedback loop can affect the time at which plants begin to flower.

Given the importance of flowering time to reproductive success, it is perhaps not surprising that both environmental and internal (endogenous) cues affect the flowering pathway. We discuss a few examples of such cues below.

Flowering can be induced by temperature cues

Another cue used to trigger flowering in many plant species is temperature. Neighboring plants are highly likely to experience similar temperatures, so like day length, temperature provides a mechanism that can promote synchronous flowering and thus promote outcrossing. As Earth's climate gets warmer, the time of year when temperatures are warm enough to trigger flowering has gradually advanced. Recently, biologists found that the ability of arabidopsis plants to flower earlier in warmer temperatures (27°C vs. 22°C) depends on a transcription factor that increases the expression of the florigen gene *FT*. Surprisingly, this transcription factor (PIF4) is one that binds to phytochrome, a photoreceptor involved in day-length perception, a finding that caused scientists to wonder whether phytochrome might play a role in response to temperature (see Chapter 27 for more details).

In some plant species, the environmental signal for flowering is exposure to cold temperatures, or **vernalization** (Latin *vernus*, "spring"). For example, if tulips are grown in a region that does not have sufficiently cold winters, they will not flower unless the

THINK LIKE A SCIENTIST

Quantitative Reasoning

Science & Society

Communicate & Collaborate

Impact of temperature on the bloom time of plants near Walden Pond in Concord, Massachusetts

In the mid-1800s, Henry David Thoreau kept detailed records of the natural word around him, including the dates when he first observed particular species of plants in flower near Concord, Massachusetts. About 30 years later, Alfred Hosmer began recording first flowering dates in the same area. These data sets, which include 32 species of spring flowering plants, provide a golden opportunity to explore how bloom time has changed over the years.

A team led by Elizabeth Ellwood collected data for the same species in the same area in the early 2000s and generated a graph of all three data sets showing the mean date on which spring flowers first appeared (**FIGURE 26.15A**).

1. By approximately how many days has the date on which the first flowers appear in Concord, MA, advanced since 1852?

The Ellwood data indicate that the mean date of flowering has advanced (become earlier) over time. Similar observations have been made throughout the world.

Are all species equally responsive to temperature, or does temperature affect the bloom time of some species more than others? One approach to answering this question is to determine how strongly the bloom time of each species correlates with temperature, and then to plot that measure (called a correlation coefficient; see Appendix B) against the time of year when the species typically blooms. This allows us to see whether plants that bloom early or late in the season are equally affected by temperature. **FIGURE 26.15B** shows the results of such a study of almost 300 species in Massachusetts. In this graph, a correlation coefficient of zero means there is no correlation between temperature and bloom time for that species. A positive correlation coefficient indicates that warmer temperatures correlate with later bloom times; the closer the value is to 1, the stronger the correlation. A negative correlation coefficient indicates that warmer temperatures correlate with earlier bloom times. (A more detailed description of correlation coefficients can be found in Appendix B.)

2. For which group of plants (spring-blooming or summer-blooming) is bloom time most highly correlated with temperature?
3. The bloom time of some plants is not greatly affected by temperature. What other factor(s) might influence bloom time?

When the mean first flowering dates of the 32 spring-flowering species in Concord, MA, were plotted against the mean spring temperature (March–May), a strong correlation was observed: warmer spring temperatures correlate with earlier flowering (**FIGURE 26.15C**).

4. Based on the data in Figure 26.15C, by approximately how many days would a 2-degree-Celsius increase in temperature from a mean spring temperature of 9°C to 11°C advance the date on which the first flowers appear?

To learn more about how flowering time is affected by climate, or to participate in a citizen science project by documenting flowering times in your area, see https://budburst.org/.

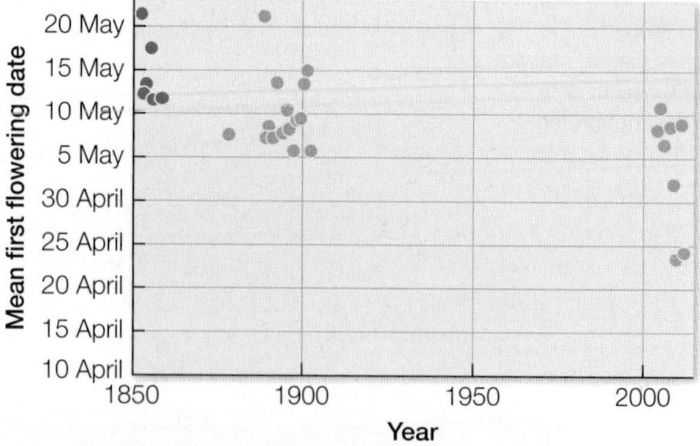

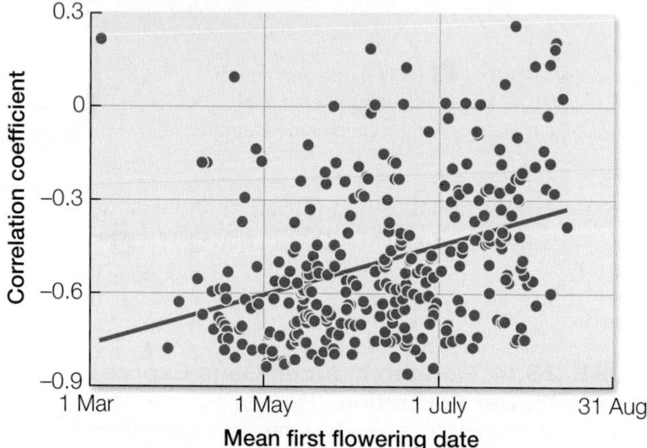

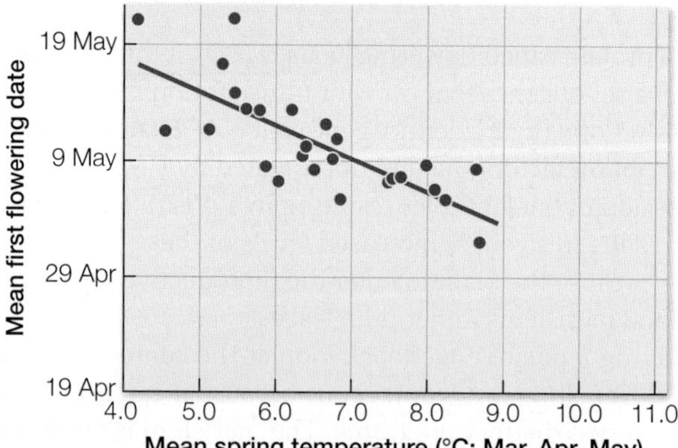

FIGURE 26.15 Temperature and the Onset of Flowering **(A)** Data plotted by Elizabeth Ellwood and her team. Colors reflect the period in which data were collected, and each data point represents mean first bloom time for all 32 species in that year. **(B)** The strength of the correlation between temperature and onset of flowering plotted against a species' mean first flowering date. The species are arranged along the *x* axis based on the month in which they bloom (e.g., spring-blooming plants are located near May, summer-blooming plants near July). **(C)** Symbols represent mean first flowering date plotted against mean spring temperature for 32 species in Concord, MA. (A and C after E. R. Ellwood et al. 2013. *PLoS One* 8: e53788; CC BY; B after A. J. Miller-Rushing and R. B. Primack. 2008. *Ecology* 89: 332–341.)

bulbs are dug up and put in cold storage (such as the vegetable crisper in a refrigerator) for several weeks. Likewise, winter wheat that is not exposed to cold will not flower normally in the next growing season. Thus winter wheat is sown in the fall, grows into a seedling, overwinters (often covered by snow), and flowers early in the following summer. This feature allows some farmers to grow two crops in a year.

Vernalization evolved in plants adapted to grow in climates with harsh winters and a short growing season. Some strains of *Arabidopsis thaliana* require vernalization (**FIGURE 26.16**), and those strains have been useful in figuring out the molecular pathway involved. The gene *FLOWERING LOCUS C* (*FLC*) encodes a transcription factor that inhibits expression of the flowering-inducer *FT*. Transcription of *FLC*, or of any other gene, requires that the region of DNA in which the gene resides opens up such that RNA polymerase can access the coding strand. Cold temperatures inhibit the synthesis of FLC protein epigenetically: the chromatin is modified so that the region of DNA in which the *FLC* gene resides is not accessible for transcription, so expression of the gene is silenced. The longer the cold lasts, the more cells there are in which the *FLC* gene is silenced. Thus long vernalization periods allow FT to be produced and flowering to proceed (**FIGURE 26.17**). Similar proteins control some steps in vernalization in cereals.

Media Clip 26.2 How Plants Remember

LINK Chromatin modification is a common way to regulate gene expression; see **Key Concept 11.3**.

Flowering can be induced by age and hormone levels

Plants, like animals, go through phases of development and do not enter the reproductive phase until they have completed earlier, juvenile phases. For example, many fruit trees do not flower until they are several years old. In *A. thaliana*, flowering is actively repressed during the juvenile phase by direct binding of a transcription factor to the promoter of *FT*, repressing its expression. As the plants age, other transcription factors that promote the expression of *FT* increase in abundance, giving the plants a greater tendency to flower when they are older. In addition, flowering of *A. thaliana* plants can be induced by the hormone gibberellin. Direct application of gibberellins to buds in *A. thaliana* results in activation of one of the meristem identity genes that forms part of the positive feedback loop leading to flowering.

FIGURE 26.16 Vernalization Flowering of a genetic strain of *Arabidopsis thaliana* is promoted by vernalization. Without vernalization, flowering is greatly delayed, and the plants grow many more leaves than normal before the first flower is produced (left). If plants of the same genotype are given a cold period, flowers appear when the plants are much younger and smaller (right).

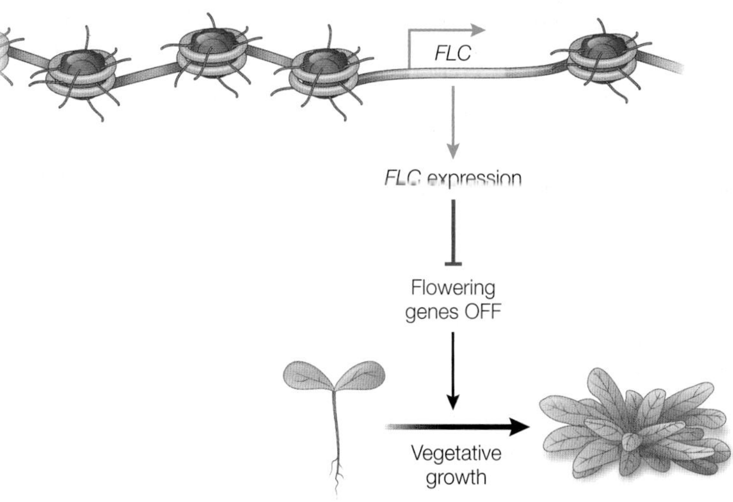

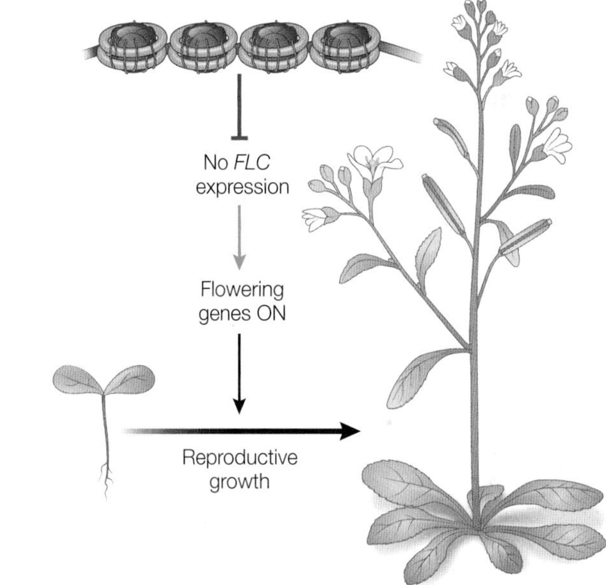

FIGURE 26.17 Chromatin Remodeling during Vernalization **(A)** In the absence of vernalization, a gene called *FLOWERING LOCUS C* (*FLC*) is expressed and flowering is repressed. **(B)** Prolonged cold weather leads to chromatin remodeling that represses expression of *FLC*. The absence of FLC protein allows flowering genes to be expressed.

REVIEW & APPLY | 26.3

R Plants switch from a vegetative phase in which the in shoot produces leaves to a reproductive one in which they produce flowers by expressing a new set of genes in the shoot apical meristem. Expression of genes that promote reproductive development is initiated by the arrival of florigen (FT protein) in the shoot meristem. In many plant species, FT is produced in the leaves in response to inductive day lengths (long days or short days, depending on the species). FT can also be induced in response to warm temperatures and increasing plant age. Some species require a long-term cold treatment (vernalization) to silence *FLC*, a repressor of *FT* expression, prior to flowering.

A
1. How do shoot apical meristems, inflorescence meristems, and floral meristems differ, and what are the genes that control the transitions between them?
2. Construct a diagram showing how expanding the location in which particular class A, B, and C genes are expressed could give rise to a tulip that has two outer whorls each of which looks like petals.
3. An experiment was done using a long-day plant species that normally flowers when days are 16 hours (or longer) and nights are 8 hours (or shorter). These plants were subjected to the light regimes listed in the table.

A. 14 hours light + 6 hours dark	G. 8 hours light + 8 hours dark
B. 14 hours light + 7 hours dark	H. 10 hours light + 8 hours dark
C. 14 hours light + 8 hours dark	I. 12 hours light + 8 hours dark
D. 14 hours light + 10 hours dark	J. 8 hours light + 10 hours dark
E. 14 hours light + 12 hours dark	K. 10 hours light + 10 hours dark
F. 14 hours light + 14 hours dark	L. 12 hours light + 10 hours dark

   a. List the light regimes that should have resulted in flowering.
   b. Predict the effect of adding a brief period of far-red light in the middle of the dark period of each light regime. Then predict the effect of adding a brief period of red light in the middle of the dark period of each light regime.
4. Describe the mutations that could be involved in each of the following observations:
   a. A mutant plant flowers even when not exposed to its inductive photoperiod. When a leaf from this mutant plant that has been exposed to the inductive photoperiod is grafted onto an unexposed wild-type plant, the recipient plant flowers.
   b. If a gene is coupled to an active promoter and expressed at high levels in the shoot apical meristem, flowering is induced even in the absence of an appropriate photoperiodic stimulus.

What triggers plants of a given species to open their flowers all at the same time?

Answer: Having many plants of the same species bloom all at the same time is evolutionarily advantageous because it promotes outcrossing, which increases genetic diversity. Mechanistically, bloom times are synchronized when plants of the same species respond to environmental cues that affect all of the plants in one area in the same way. For example, many plants measure day length and bloom when the days reach some predetermined length. Plants also bloom in response to temperature cues.

26.1 Flowering Plants Can Reproduce Asexually

- **VEGETATIVE REPRODUCTION** produces genetically identical plants (**FIGURE 26.1**). Cuttings, **GRAFTING**, micropropagation, and **MERISTEM CULTURE** are forms of **ASEXUAL REPRODUCTION** commonly used in agriculture.
- **APOMIXIS** is a form of asexual reproduction that produces seeds without fertilization.

FIGURE 26.1

Stolons

Plantlets

QUESTIONS

1. Johnny Appleseed is described in many legends as walking about the midwestern United States in the 1800s as it was being settled, distributing apple seeds. What might you predict about the quality of the apples from the trees he planted? (*Hint*: If you have an apple and plant its seeds, will the tree that grows from that seed be a faithful replica—that is, a clone—of the original tree?)
2. What if you want a better variety of apples? What is the best way to create a new variety?
3. If you find an apple tree that produces better apples than all of your other trees, what is the best way to reproduce it?

26.2 Most Flowering Plants Reproduce Sexually

Go to ACTIVITY 26.1 and ANIMATION 26.1

- The life cycle of flowering plants has two stages: a haploid **GAMETOPHYTE** and a diploid **SPOROPHYTE** (**FIGURE 26.4**).
- Gametophytes produce **GAMETES** (eggs and sperm); sporophytes produce **SPORES**.
- In **DOUBLE FERTILIZATION**, one sperm fertilizes an egg, forming the $2n$ **ZYGOTE**, and the other sperm fertilizes the **POLAR NUCLEI** to form the $3n$ **ENDOSPERM**.

FIGURE 26.4

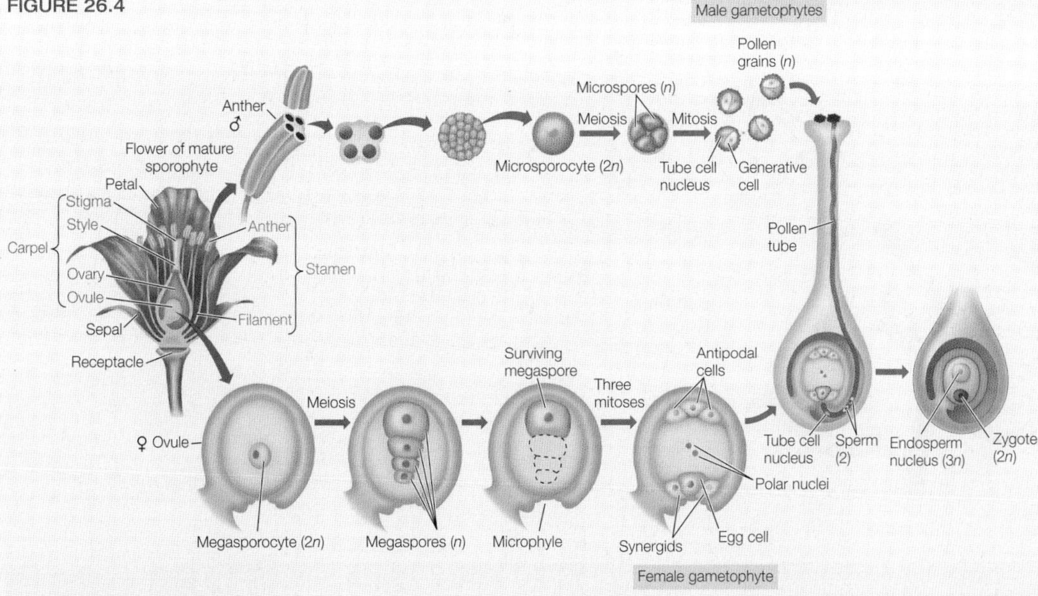

QUESTIONS

1. Are the grasses, bushes, and flowering trees that one sees sporophytes or gametophytes?
2. Does the human life cycle have a generation equivalent to the plant gametophyte?
3. In what ways are pollen tubes similar to fern gametophytes?

26.3 Hormones and Signaling Determine the Transition from the Vegetative to the Reproductive State

Go to ACTIVITIES 26.2 and 26.3 and ANIMATION 26.2

- During **VEGETATIVE GROWTH**, plants produce leaves from the shoot apical meristem.
- During **REPRODUCTIVE GROWTH**, the shoot apical meristem expresses a different set of transcription factors and becomes an **INFLORESCENCE MERISTEM**. The inflorescence meristem gives rise to **FLORAL MERISTEMS**, each of which produce a flower.
- Flowers are determinate structures that typically contain four **WHORLS** of organs: sepals, petals, stamens, and carpel. The identity of each whorl is determined by the transcription factors present (**FIGURE 26.10**).
- Day-length-sensitive plants flower when **FLORIGEN** (**FT** protein) is produced in the leaf and moves to the shoot apical meristem (**FIGURE 26.14**).

FIGURE 26.10

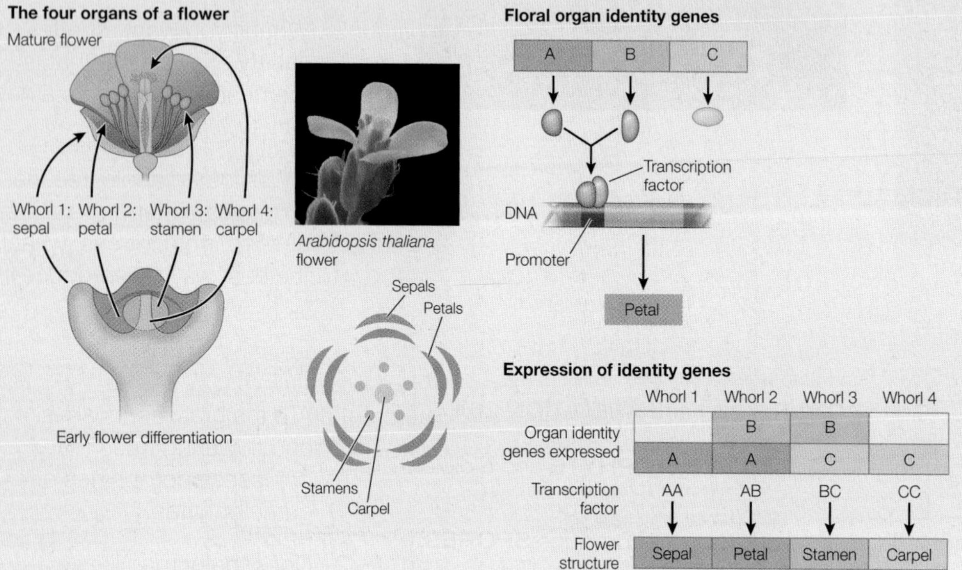

FIGURE 26.14

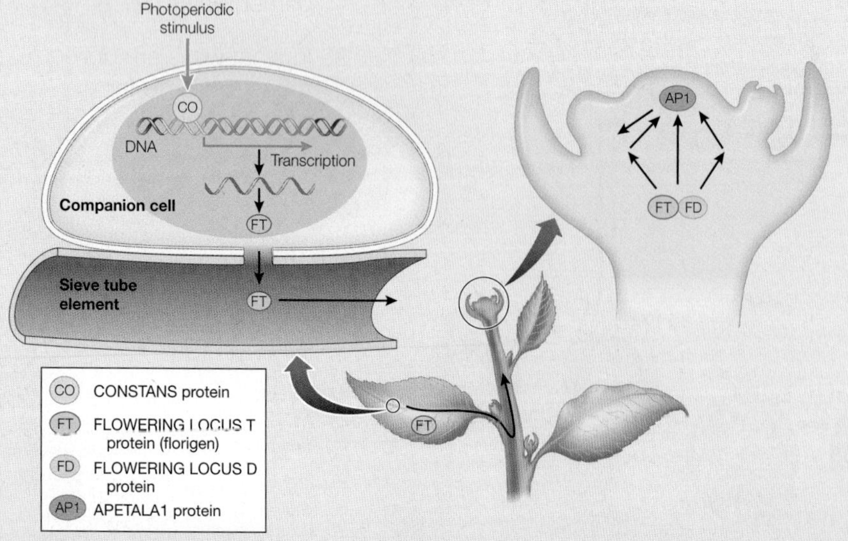

QUESTIONS

1. How can expression of a transcription factor affect the identity of an organ?
2. Explain how changing the location in which some of the class A, B, and C genes are expressed would cause the numerous stamens present in a wild rose to be converted into the numerous petals of a cultivated rose.
3. If a short-day plant (SDP) with a critical day length of 12 hours is given an 11-hour day and a 13-hour night, it will flower. Will the same plant flower if it is given a flash of red light in the middle of the night?
4. How did scientists determine what type of compound florigen is?

Plants in the Environment 27

Global climate change encompasses a wide range of environmental factors, including increases in atmospheric carbon dioxide, temperature, the severity of storms, and the incidence of drought. People sense changes in the environment and respond by altering their behavior. We seek shelter from storms, mount rescue parties after floods, fight wildfires, and make snow at ski resorts. Plants cannot move to avoid stressful conditions, but they can perceive and change in response to their environment.

What impact will global climate change have on crop production?

You will find the answer to this question on page 673.

KEY CONCEPTS

27.1 Plants Respond to the Availability of Water and Carbon Dioxide

27.2 Plants Respond to Differences in Soil Composition

27.3 Plants Respond to Temperature

27.4 Plants Have Constitutive and Induced Responses to Pathogens and Herbivores

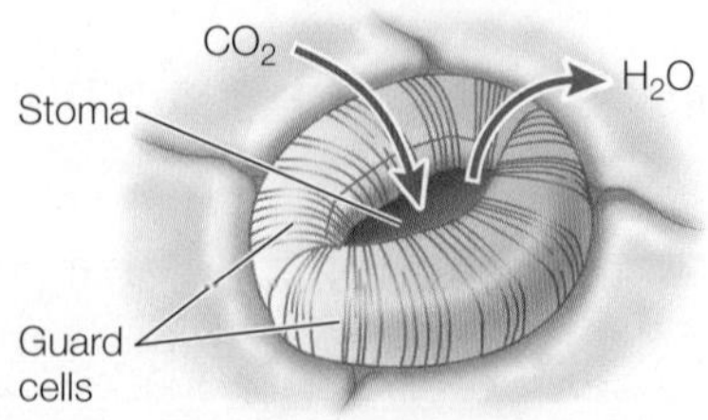

FIGURE 27.1 Stomatal Compromise Opening stomata increases gas exchange, promoting CO_2 uptake, but also increases water loss.

27.1 Plants Respond to the Availability of Water and Carbon Dioxide

LEARNING OBJECTIVES

By the end of this key concept you should be able to:

27.1.1 **Explain the trade-off between having stomata open and closed and describe advantages conferred by each condition.**

27.1.2 **Describe stomatal changes that occur in response to increased levels of atmospheric CO_2.**

27.1.3 **Outline the signaling pathway that allows plants to respond to low soil moisture.**

27.1.4 **Distinguish between the response an individual plant may have to an environmental condition and an adaptation to that environment.**

Plants face a continual challenge: every time stomata open to let in carbon dioxide (CO_2) that is needed for photosynthesis, water is lost through those same openings. CO_2 makes up a relatively small fraction of the molecules in air. Currently, the concentration of CO_2 is about 411 parts per million (0.04% of the air), so getting substantial amounts of carbon into a plant requires that the inside of the leaf exchange gases with a large volume of outside air. During this exchange, water from the moist leaf interior is lost by evaporation. On average, for every 1 gram of carbon fixed, 266 grams of water move from the roots through the plant and out to the atmosphere. The amount of water lost through this process, called transpiration (see Key Concept 24.4), is hundreds of times greater than the water required for other processes that take place in the plant. As a result, control of when, how much, and under what conditions stomata will open is highly regulated by the plant. **Stomatal compromise** refers to the balance between two opposing processes that promote plant survival: stomatal opening, which allows uptake of CO_2, and stomatal closing, which limits water loss (**FIGURE 27.1**). The compromise is arrived at through integration of signals that reflect the availability of light, CO_2, and water.

Plants respond to the concentration of CO_2 in the leaf

Because CO_2 is the substrate for photosynthesis, increases in its concentration should increase carbon fixation. For plants in well-watered, nutrient-sufficient conditions, the amount of carbon fixation can limit plant growth. Multiple studies in which CO_2 is pumped out over a small field or plot of land—a method called *free-air carbon dioxide enrichment* (FACE)—have shown that a fertilization effect resulting in more plant growth occurs if water and nutrients are available in sufficient quantities. However, in natural settings, water and nutrients often become limiting. Increased seed yield, an important metric for crop production, is also seen in FACE experiments, although the proportion of the seed that is made up of protein may decrease as more carbohydrate is produced. Some plant species, including corn, that grow in hot climates have evolved mechanisms to concentrate CO_2 (**FIGURE 27.2**). Called C_4 plants, these species may benefit less than others from increased levels of CO_2. While experimental data indicate that plants can effectively remove (sequester) carbon from the atmosphere, thereby decreasing the levels of atmospheric CO_2, the extent to which plants do this does not compensate for increases in atmospheric CO_2.

Media Clip 27.1 Leaves for Every Environment

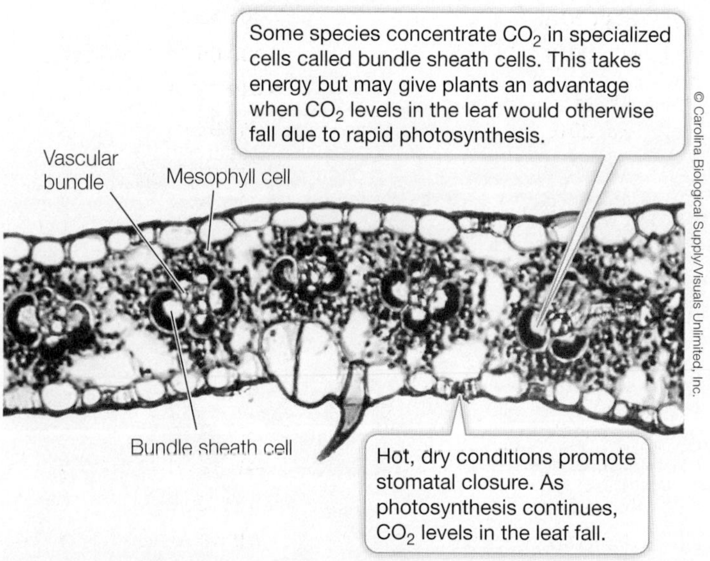

FIGURE 27.2 C_4 Photosynthesis Many plants adapted for C_4 photosynthesis have altered leaf anatomy, allowing CO_2 to be concentrated in specialized cells called bundle sheath cells.

THINK LIKE A SCIENTIST

Modeling Earth's carbon cycle

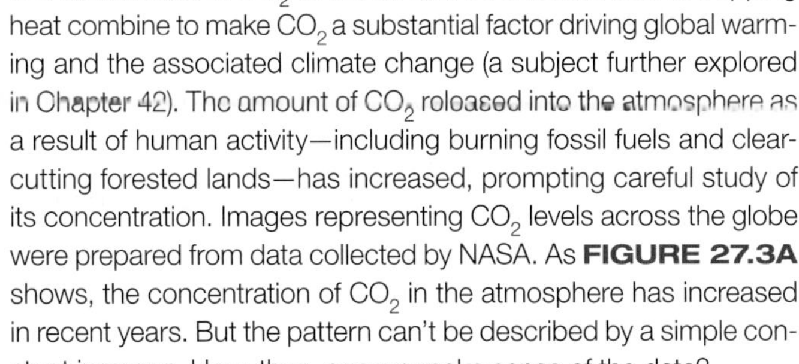

The abundance of CO_2 in the air and its effectiveness at trapping heat combine to make CO_2 a substantial factor driving global warming and the associated climate change (a subject further explored in Chapter 42). The amount of CO_2 released into the atmosphere as a result of human activity—including burning fossil fuels and clear-cutting forested lands—has increased, prompting careful study of its concentration. Images representing CO_2 levels across the globe were prepared from data collected by NASA. As **FIGURE 27.3A** shows, the concentration of CO_2 in the atmosphere has increased in recent years. But the pattern can't be described by a simple constant increase. How, then, can we make sense of the data?

Modeling can assist in understanding the behavior of complex systems by identifying key parts of the system. The behavior of these central components can then be understood individually before any interactions are considered. Useful models are simplified descriptions that, while not perfectly accurate, make it easier to understand the central aspects that drive a system.

A common approach to modeling is to find a simple function that best fits the observed behavior of the system being studied. To do this, one determines how closely the observed data fit the graphs of various functions. For example, if the amount of something increases over time, one can try to determine if that increase is primarily linear, periodic (as in a sine wave), or asymptotic (a curve whose change slows as it approaches some particular value) (**FIGURE 27.3B**). Some models take this a step further by showing the combined results of two or more simple functions.

Another way to simplify measurements of CO_2 is to restrict the system being modeled to one location. Mauna Loa Observatory has been monitoring the level of CO_2 in the atmosphere at its location in Hawaii since 1958. A subset of recent data is shown below. Measurements are in parts per million (ppm).

Date	Mean concentration of CO_2 at Mauna Loa for the indicated month (ppm)
Sept 2018	405.52
Nov 2018	408.05
Jan 2019	410.85
Mar 2019	411.93
May 2019	414.76
July 2019	411.78
Sept 2019	408.48
Nov 2019	410.22
Jan 2020	413.31
Mar 2020	414.44
May 2020	417.10
July 2020	414.42
Sept 2020	411.19
Nov 2020	412.88
Jan 2021	415.16

Source: Dr. Pieter Tans, NOAA/ESRL (www.esrl.noaa.gov/gmd/ccgg/trends/) and Dr. Ralph Keeling, Scripps Institution of Oceanography (scrippsco2.ucsd.edu/).

(A) Increasing CO_2 levels around the world

March 2014

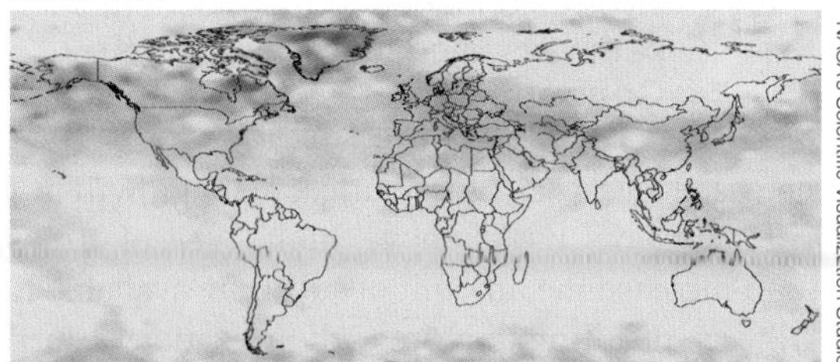

September 2014

March 2015

Atmospheric CO_2 (ppm)

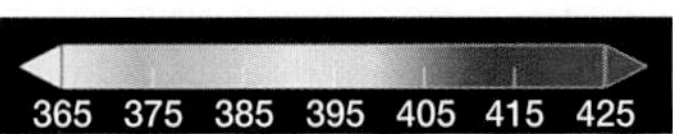

(B) Three functions associated with an increasing variable

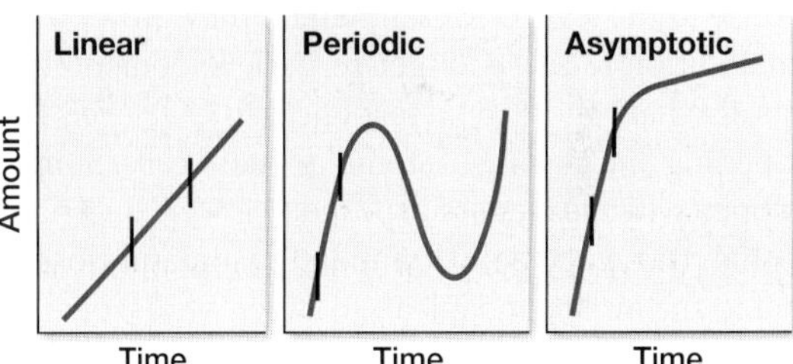

FIGURE 27.3 Modeling Earth's Carbon Cycle **(A)** These NASA images show the concentration of CO_2 in the atmosphere in the spring and fall, measured in parts per million (ppm). **(B)** Three functions associated with an increasing variable.

1. Graph the data shown at left.
2. Which model best fits your graph: (a) flat line, (b) linear, increasing, (c) sine wave, (d) linear increasing + sine wave, or (e) asymptotic?
3. Given what you have learned about CO_2 levels in the atmosphere, formulate a hypothesis as to what processes are responsible for each of the components in the model that you described above. Remember that the data you graphed are not a global average; they come from one specific location in the Northern Hemisphere.

LINK C_4 photosynthesis is an adaptation that involves changes to the biochemical pathways used to fix CO_2. See **Key Concept 5.5**.

Increased levels of CO_2 can affect plant growth in distinct ways. First, as discussed above, high levels of atmospheric CO_2 can increase the amount of substrate available for photosynthesis, resulting in more carbon fixation occurring via existing biochemical pathways. Second, plants can sense the concentration of CO_2 and mount specific responses to it. As discussed in Key Concept 6.1, a typical sensory system includes a signal, a receptor that interacts with that signal, a series of downstream steps called the signal transduction pathway, and the ultimate response.

Evidence that plants sense and respond to CO_2 levels is found in changes in the number of stomata that are produced on new and developing leaves. Adding CO_2 to the air around a growing plant typically leads to a decrease in the number of stomata per area on the leaf. To see if this effect, which was observed in the laboratory, might also be active in the natural world, scientists turned to herbaria. An herbarium is a facility that houses indexed collections of dried plant material. The density of stomata on leaves collected prior to the Industrial Revolution, a period that had lower levels of CO_2, was compared with that on leaves collected recently. For many species, the density of stomata on preserved leaves was observed to have declined during this period when atmospheric CO_2 was increasing. Examination of fossil plants further confirms this observation. These data from wild plants support the laboratory observation that plants alter their stomatal density in response to the level of CO_2 in which they grow.

In addition to these longer-term developmental changes, plants also have more rapid responses to CO_2 levels that affect the opening and closing of stomatal guard cells. Stomata typically open when leaves are exposed to lower concentrations of CO_2 than those normally found in the air, and close if CO_2 is added to the air around a leaf. In other words, opening and closing of stomata regulates conductance—the ease with which gases inside and outside the plant are exchanged.

Mounting a response to the level of CO_2 requires that the plant first sense the concentration of CO_2 and then initiate a signaling pathway. Identifying the receptor and the pathway is an area of active research. Some studies suggest that stomatal response to CO_2 levels may involve the enzyme carbonic anhydrase (**FIGURE 27.4**). Carbonic anhydrase catalyzes the conversion of gaseous CO_2 and water into bicarbonate and protons, both of which are present in the soluble portion of the cytoplasm. It is hypothesized that the products of the reaction carried out by carbonic anhydrase initiate a signal transduction cascade within plant cells.

(A) Conductance in normal plants (red) versus plants with reduced carbonic anhydrase activity (blue)

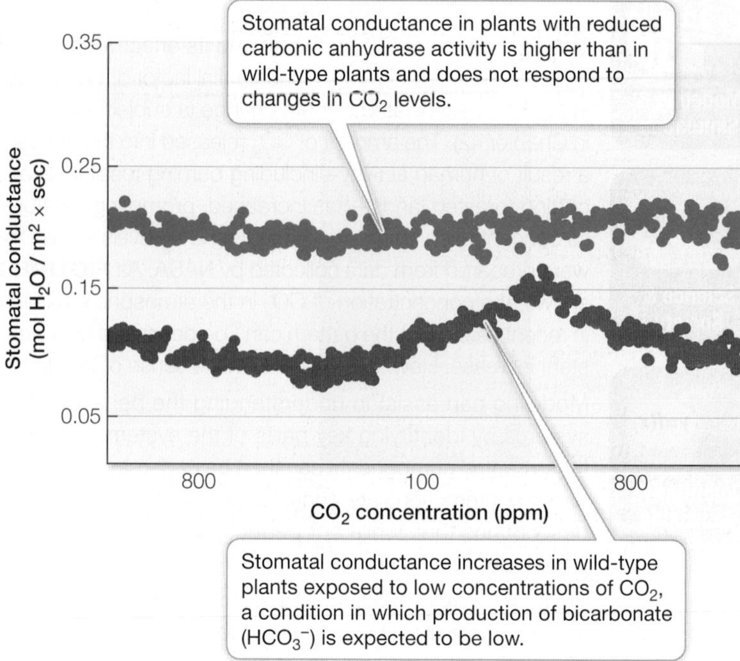

(B) The reaction catalyzed by carbonic anhydrase

$$CO_2 + H_2O \xrightarrow{\text{Carbonic anhydrase}} HCO_3^- + H^+$$

FIGURE 27.4 Plants May Use Bicarbonate Levels as an Indicator of CO_2 Levels **(A)** Stomatal opening is affected by the activity of carbonic anhydrase. **(B)** Carbonic anhydrase converts gaseous CO_2 and water to soluble bicarbonate and protons. (A after. C. B. Engineer et al. 2016. *Trends Plant Sci* 21: 16–30.)

Do high levels of bicarbonate and H^+ appear to promote or repress stomatal opening?

Plant growth is limited by water availability

Inadequate water supply is one of the primary factors that limits plant growth in natural habitats. Water limitation also affects production of our most important food crops. The impact of water deficit on the growth of flowering plants is expected to increase in the future as water resources become more scarce. Additionally, global climate change is expected to increase the extent of water limitation on plant growth through increased temperatures, which promote evaporation, and through an increase in drought, particularly in regions that already tend to be dry. (See Chapter 42 for a more complete discussion of global climate change and its effects.)

When the weather is abnormally dry, the water potential (see Key Concept 24.4) of the soil declines (becomes more negative). Water moves to the region with the lowest water potential, so a low water potential in the soil makes it more difficult for a plant to take in water. If a plant's roots cannot take up water from the soil, the plant cannot grow. Water is needed to keep cells hydrated, to drive the transport of nutrients and sugars through the plant, and to produce the turgor pressure that drives cell expansion. Extreme water deficits in plant cells also reduce membrane integrity as the polar–nonpolar forces that orient the lipid bilayer are reduced, and lead to changes in the three-dimensional structures of proteins. In some cases, the changes cause proteins to form nonfunctional aggregates (clumps). A plant can suffer irreversible damage when the structure of its cells is compromised in these ways.

Hormonal signaling promotes response to drought stress

When plants sense water deficit, a signaling pathway is set in motion that initiates several measures to conserve water and maintain cellular integrity. This pathway begins with the production of the hormone **abscisic acid (ABA)** in leaves and roots (**FIGURE 27.5**). ABA affects the cells that it reaches by binding to a soluble protein receptor. Binding of ABA to its receptor sets off a series of reactions that end up (a) affecting the opening and closing of specific ion channels and (b) regulating the expression of specific genes. The signal transduction cascade that leads to these changes is initiated when ABA binding to its receptor causes the conformation (shape) of the receptor to change. In this new conformation, the ABA receptor can bind to and inactivate a protein phosphatase (an enzyme that removes phosphate groups from other proteins). Inactivation of this protein phosphatase affects the structure and activity of many proteins, and so impacts the activities of the cell. In the end, high levels of ABA promote stomatal closure and, in some cases, increase the resistance to water flow through the rest of the plant. As a result, the amount of water lost via transpiration decreases. ABA also affects root growth such that mild drought often results in an increase in root biomass.

Many plant genes whose expression is altered by drought stress have been identified, largely through research using DNA microarrays (see Figure 12.18), proteomics (see Key Concept 12.1), and other molecular approaches. One group of proteins in which production is upregulated during drought stress is the *l*ate *e*mbryogenesis *a*bundant (LEA; pronounced "lee-yuh") proteins. These hydrophobic proteins also accumulate in maturing seeds as they dry out (hence their name). The LEA proteins bind to membrane proteins and other cellular proteins to stabilize them, preventing them from clumping together and losing function as seeds dry out.

Some plants have special adaptations for life in very dry conditions

In contrast to short-term **acclimation** to the environment that occurs via induced responses during the life of an individual plant, some plant species have evolved constitutive **adaptations**—defenses that are always present—that allow them to survive in low-water environments. Many of these adaptations prevent or reduce the inevitable water loss through transpiration that occurs during active photosynthesis. Other adaptations help plants tolerate the high levels of light and heat that are often found in deserts. Understanding the constitutive adaptions present in plants that grow in dry environments is of particular interest to scientists who want to increase the drought resistance of crop plants in hopes of better suiting them to present and future climatic conditions. This remains a challenging goal, however, as adaptations that improve drought tolerance often decrease plant yield. Understanding plant adaptations is also valuable in adding to a general appreciation of how plants survive. Below we describe several examples of constitutive adaptations.

WATER-STORING STRUCTURES Plants adapted to dry environments are called **xerophytes** (Greek *xeros*, "dry," + *phyte*, "plant"). **Succulence**—the possession of fleshy, water-storing leaves or stems—allows plants to take up large amounts of water when it is available (such as after a brief thunderstorm) and then draw on the stored water during subsequent dry periods. Cacti are an example of succulent plants. Succulents can also have a reduced number of stomata, which further reduces water loss.

LEAF STRUCTURES Three structural adaptations are found in the leaves of many xerophytes:

1. Specialized leaf anatomy, associated with different forms of photosynthesis, that reduces water loss
2. A thick cuticle and a profusion of trichomes (leaf hairs) over the leaf epidermis, which retard water loss by trapping moist air next to stomatal openings
3. Trichomes that diffract and diffuse sunlight, thereby decreasing the intensity of light impinging on the leaves and the risk of damage to the photosynthetic apparatus by excess light

In some xerophytes, the stomata are located in sunken cavities below the leaf surface, known as stomatal crypts, where they are

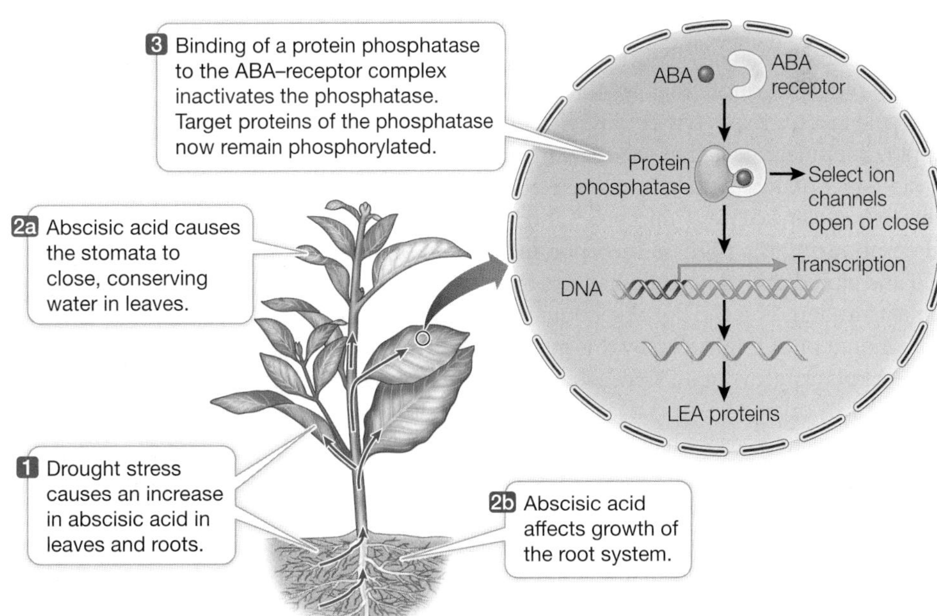

FIGURE 27.5 Plants Respond to Water Stress by Producing the Hormone Abscisic Acid (ABA) ABA protects plants against water stress by closing stomata and altering root architecture.

(A) Cross-section of a xerophyte leaf

Upper side of leaf

Stomatal crypt

Lower side of leaf

0.5 mm

© John N. A. Lott/Biological Photo Service

(B) Close-up of a stomal crypt

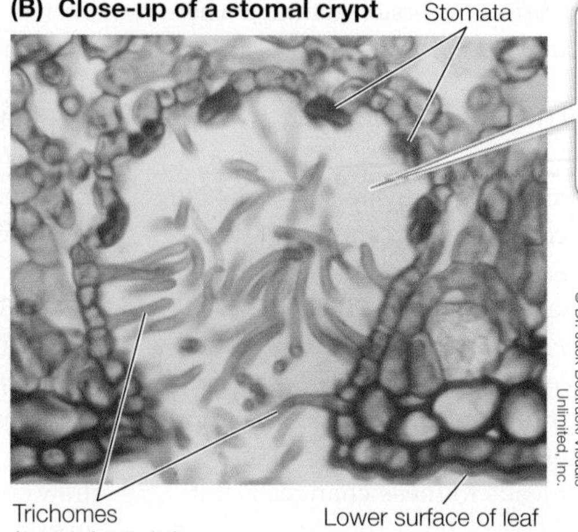

A section through a leaf's surface shows stomata sunken in crypts protected by trichomes.

FIGURE 27.6 Stomatal Crypts Hair-filled cavities that surround stomata in some xerophytes decrease the rate of water loss by retaining moist air around the stomata. **(A)** Cross-section of a leaf showing three stomatal crypts. **(B)** Close-up of one stomatal crypt, showing stomata that line the crypt.

sheltered from the drying effects of air currents (**FIGURE 27.6**). Trichomes surrounding the stomata slow air currents further. Cacti and similar plants have spines rather than typical leaves, and photosynthesis is confined to the fleshy stems. The spines may help the plants cope with desert conditions by reflecting solar radiation or by dissipating heat. Spines also deter herbivores.

PHOTOSYNTHETIC ADAPTATIONS The primary photosynthetic pathway described in Chapter 5 is called C_3 photosynthesis. Some plants, including some xerophytes that are adapted to life in hot climates, have an adaptation called C_4 photosynthesis that allows them to bind CO_2 in cells located near the stomata and then release it in cells surrounding the vasculature (the bundle sheath cells; see Figure 27.2 and also Figure 5.24A). Because this system concentrates CO_2 in the inner cells, it requires some energy, but it avoids some of the added costs incurred when stomatal closure causes CO_2 levels in the leaf to fall. C_4 photosynthesis reduces the amount of water lost for each molecule of CO_2 fixed. Other plants, generally succulents, employ crassulacean acid metabolism (CAM) for photosynthesis, which allows them to open their stomata only at night when it is cooler (see Key Concept 5.5 and Figure 5.24B). CO_2 taken in during the night is incorporated into an organic acid and stored until daylight when it is re-released in chloroplasts undergoing the light reactions (see Figure 5.24B). The amount of CO_2 taken up by this mechanism is reduced, so CAM plants grow more slowly, but because the amount of water lost during CAM is drastically reduced, it can be advantageous in areas such as deserts where growth would not otherwise occur.

ROOT SYSTEMS THAT MAXIMIZE WATER UPTAKE Roots may be adapted to dry environments. Xerophytes and other plants that must cope with inadequate water supplies may accumulate high concentrations of the amino acid proline or of secondary metabolites in their vacuoles. This solute accumulation lowers the water potential in a plant's cells below that in the soil, which allows the roots to take up water via osmosis. Plants living in saline environments share this and several other adaptations with xerophytes.

DROUGHT ESCAPE Some desert plants have no special structural adaptations for water conservation. Instead, these desert annuals, called drought escapers, simply evade periods of drought. Drought escapers carry out their entire life cycle—from seed to seed—rapidly to exploit the brief period in which rainfall makes the surrounding desert soil sufficiently moist for growth and reproduction (**FIGURE 27.7**). A different strategy is seen in deciduous perennial plants, particularly in Africa and South America, that reduce their surface area by shedding their leaves in response to drought, and so conserve water. These plants remain dormant until conditions are again favorable for growth.

FIGURE 27.7 Desert Annuals Escape Drought The seeds of many desert annuals lie dormant for long periods, awaiting conditions appropriate for germination. When they do receive enough moisture to germinate, they grow and reproduce rapidly before the short wet season ends. During long dry spells, only dormant seeds remain alive.

REVIEW & APPLY | 27.1

R Plants strike a balance between the positive effect of taking in carbon for photosynthesis (in the form of CO_2) and the negative effects of water loss. The amount of stomatal opening that achieves a functional balance (the stomatal compromise) is arrived at through integration of multiple signals, including the concentration of CO_2 in the leaf and water availability. In response to dry environments, plants synthesize the hormone abscisic acid (ABA). ABA induces stomatal closing and other responses that allow plants to better survive water stress, but generally limits growth compared with what could occur in well-watered conditions. Some plant species have also evolved specific adaptations that allow them to survive in dry environments.

A

1. List some constitutive and induced mechanisms that allow plants to cope with drought conditions.
2. Colleagues in plant physiologist José Dinneny's laboratory discovered that when Arabidopsis thaliana roots were exposed to different environments, one wet and one dry, lateral roots formed preferentially on the wet side. They termed this process hydropatterning. Is hydropatterning an example of an adaptation or an acclimation?
3. Data in the accompanying graph were obtained from three tomato plants as they dried out. (After Skelton et al. 2017. Casting light on xylem vulnerability in an herbaceous species reveals a lack of segmentation. *New Phytol* 214: 561–569.)
   a. If a larger number of stomata were open, or individual stoma were open for a larger fraction of the time, would that tend to increase or decrease stomatal conductance?
   b. Which end of the leaf water potential scale shown here represents a fully moist leaf?
   c. How does it benefit a plant to have a low stomatal conductance when the leaf water potential is low?
   d. What is the disadvantage to having a low stomatal conductance?
4. If an inhibitor of ABA synthesis was sprayed on a plant and the plant was exposed to drought conditions, what would happen?

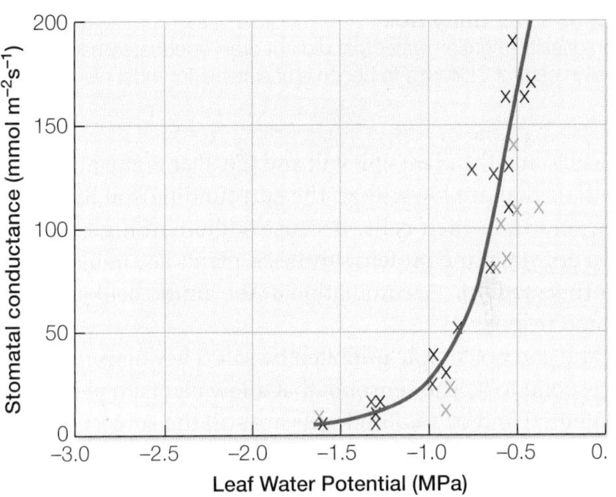

Because access to water and CO_2 are central requirements of plant productivity, plants have specialized systems for sensing and responding to them. How plants interact with the soil environment is the subject of the next key concept.

27.2 Plants Respond to Differences in Soil Composition

LEARNING OBJECTIVES

By the end of this key concept you should be able to:

27.2.1 **Give two reasons why salty soils pose a challenge for many plants.**

27.2.2 **Describe ways in which plants respond to nutrient depletion of the soil.**

27.2.3 **Explain the concept of phytoremediation.**

Soil fertility has been a major determinate of the success of civilizations because while plants, and therefore crops, gain considerable mass from the CO_2 in the air, they also require water and some nutrients from the soil. With the amount of CO_2 in the air increasing, nutrient supplies in the soil, including nitrogen, increasingly become limiting for plant growth.

Salty soils pose a challenge to plant growth

Salty, or saline, environments (high in Na^+, K^+, Ca^{2+}, and Cl^-) are found in nature in diverse locales, from hot, dry deserts to moist, cool coastal marshes. Rising sea levels and storm surges bring salty water to coastal plant communities. In addition, agricultural land can become increasingly saline as a result of irrigation as sodium that is dissolved in the irrigation water is left behind when the water evaporates. In ancient times, the region between the Tigris and Euphrates Rivers (in what is now Iraq) was once highly fertile, but extensive irrigation canals brought salts that accumulated and raised the water table, which was salty. Ancient records show a shift in production from salt-sensitive to salt-resistant crops and a decline in yield. Salinization, which can eventually make land unsuitable for farming, continues to be a problem worldwide today (**FIGURE 27.8**).

Because of its high salt concentration, a saline environment has a very negative water potential. To obtain water from such an environment, a plant must have an even more negative water potential (see Key Concept 24.4); otherwise water will move out

Courtesy of Scott Bauer/USDA ARS

FIGURE 27.8 Salty Soil Accumulation of salt from repeated evaporation of salty irrigation water that did not drain adequately has caused this soil in central California to become unsuitable for most plant growth.

© Jurgen Freund/Minden Pictures

FIGURE 27.9 Excreting Salt This saltwater mangrove plant has special glands that excrete salt, which appear here as crystals on the leaves.

of its cells, and the plant will wilt and die. Plants can attain water potentials that are lower than the surrounding soil by accumulating solutes in their cells. Because sodium at high levels can inhibit enzymes and protein synthesis, plants accumulate solutes other than sodium. Accumulation of the amino acid proline is a common response.

Plants are not simply inhibited by salt. They actively perceive and respond to it. The sensor(s) that allow plants to perceive water potential and/or sodium levels are still the subject of debate. Because exposure to low water potentials will decrease the turgor pressure inside cells, proteins in the cell membrane that are sensitive to how stretched the membrane is or that measure the distance between the membrane and the cell wall are possible options for sensors. Whatever the sensor, ABA levels often increase in response to salty soils and can have species-specific effects on root architecture. Plants can also respond to increasing concentrations of salt by growing away from them, a process called **halotropism** (*halo*, "salt"). Halotropism is mediated by auxin. Auxin becomes redistributed in the root such that roots grow away from the region of highest salt concentration. Other examples of auxin-mediated tropisms are shown in Figure 25.19.

Plants that are adapted for survival in saline soils are called **halophytes**. Most halophytes take up Na^+, and many take up Cl^-, into their roots and transport those ions to their leaves, where they accumulate in the central vacuoles of leaf cells, away from more sensitive parts of the cells. The accumulated salts in the tissues of halophytes make their water potential more negative than the soil solution and allow them to take up water from their saline environment. Some halophytes have **salt glands** in their leaves that excrete salt, which collects on the leaf surface until it is removed by rain or wind (**FIGURE 27.9**). This adaptation, which reduces the danger of poisoning by accumulated salt, is found in some desert plants and in some plants that grow in saline habitats. The negative water potential in the salt-laden leaves also promotes water flow from the roots up through the xylem by transpiration–cohesion–tension (see Key Concept 24.4).

Plants alter the architecture of their root systems in response to soil nutrient levels

Plant growth is limited when soils are deficient in nutrients such as phosphorus and nitrogen. These nutrients are distributed unevenly in the soil across space and time, and plants respond to these changes in ways that promote their success. Phosphate has low mobility in the soil and is primarily found in the topsoil (A horizon; see Figure 24.4), along with the bulk of the organic matter. When *Arabidopsis thaliana* plants are grown on a controlled medium that is deficient in phosphorus, they produce many lateral roots, as well as more and longer root hairs, but this growth is limited to the upper portions of the root system (**FIGURE 27.10A**, middle panel). These responses are mediated by increased expression of the auxin receptor (TIR1) in roots of phosphate-deficient plants. Collectively, these changes in the root system promote foraging for nutrients in the topsoil.

In contrast, the form of nitrogen most commonly taken up by plants, nitrate (NO_3^-), is relatively mobile in soil and is more

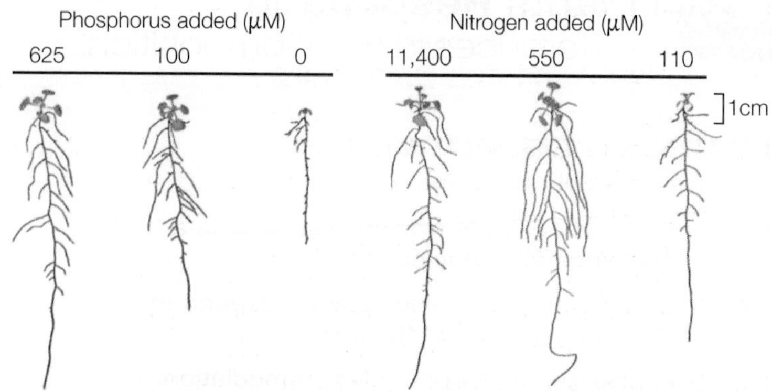

FIGURE 27.10 Plants Alter the Architecture of Their Root Systems in Response to Soil Nutrients Plants shown here were grown on media with uniformly low levels of either phosphorus **(A)** or nitrogen **(B)**. (After B. D. Gruber et al. 2013. *Plant Physiol* 163: 161–179.)

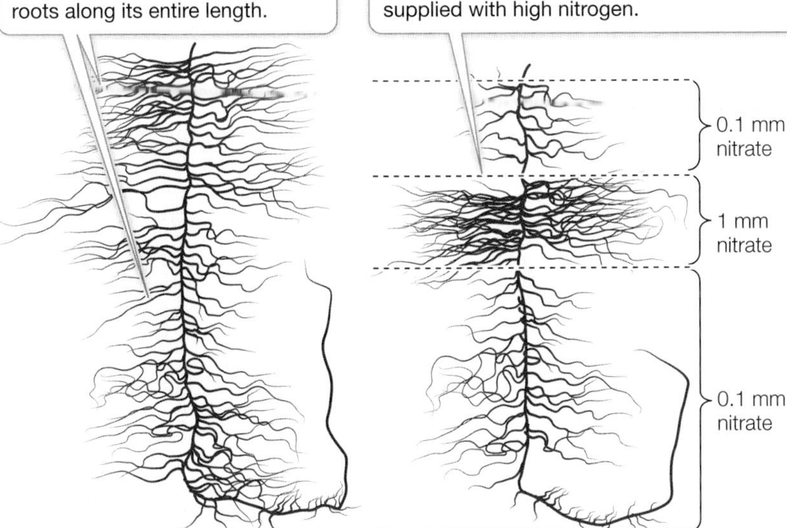

FIGURE 27.11 Low-Nitrate Plants Can Grow Branch Roots in a Local Region That Has High Nitrate Levels (After M. C. Drew et al. 1974. *J Exp Bot* 24: 1189–1202.)

prevalent in lower regions of the soil. When *A. thaliana* plants are grown on a uniform medium that contains moderately low levels of nitrogen, the primary root is longer than that of plants grown with higher concentrations of nitrogen, and the lateral roots are more vertically oriented (**FIGURE 27.10B**, middle panel). In other words, low nitrogen levels can cause the root system to grow deeper into the soil where the likelihood of reaching nitrate-rich patches of soil may increase. In addition, when nitrogen-starved plants are exposed to patches of soil that have high concentrations of nitrate, they can mount a different, local response that results in proliferation of roots just in the region where the nitrate concentration is high (**FIGURE 27.11**). Nitrate-responsive changes in root architecture are accompanied by changes in the expression of many genes, including genes that code for proteins involved in nitrogen transport and assimilation. Together, these systemic and local responses allow plants to respond to the changing soil environment.

Some plants can help clean soils

High concentrations of some metal ions are toxic to many plants. Some soils are naturally rich in toxic metals as a result of normal geological processes or acid rain. The mining of metallic ores also leaves localized areas with high concentrations of heavy metals and low concentrations of nutrients. Such sites are hostile to most plants, and seeds falling on them generally do not produce adult plants.

Some plants survive in the presence of toxic metals by accumulating them in concentrations that would kill most plants. More than 200 plant species have been identified as **hyperaccumulators** that store large quantities of metals such as arsenic (As), cadmium (Cd), nickel (Ni), aluminum (Al), and zinc (Zn).

Perhaps the best-studied hyperaccumulator is alpine pennycress (*Thlaspi caerulescens*). Before the advent of chemical analysis, miners used the presence of this plant as an indicator of mineral-rich deposits. A *T. caerulescens* plant may accumulate as much as 30 grams per kilogram dry weight Zn (most plants contain 0.1 g/kg dry weight) and 1.5 grams per kilogram dry weight Cd (most plants contain 0.001 g/kg dry weight). Studies of *T. caerulescens* and other hyperaccumulators have revealed the presence of several common adaptations.

- Increased ion transport into the roots
- Increased rates of translocation of ions to the leaves
- Accumulation of ions in vacuoles in the shoot
- Resistance to the ions' toxicity

Knowledge of these hyperaccumulation mechanisms and the genes underlying them has led to the emergence of **phytoremediation**, a form of bioremediation that uses plants to clean up environmental pollution in soils. Some phytoremediation projects use natural hyperaccumulators, whereas others use genes from hyperaccumulators to create transgenic plants that grow more rapidly in and are better adapted to a particular polluted environment. In either case, the plants are grown in the contaminated soil, where they act as natural "vacuum cleaners" that take up the contaminants (**FIGURE 27.12**). The plants are then harvested and safely disposed of to remove the contaminants. Perhaps the most dramatic use of phytoremediation occurred after an accident at the nuclear power plant at Chernobyl, Ukraine (then part of the Soviet Union), in 1986, when sunflower plants were used to remove toxic and radioactive uranium from the nearby soil. Sunflowers are now being used in Japan to take up radioactive substances from the Fukushima Daiichi reactor accident in 2011. Sunflowers are now the international symbol of nuclear disarmament.

FIGURE 27.12 Phytoremediation Plants that accumulate heavy metals can be used to clean up contaminated soils. This image of sunflowers was chosen for the visual link it provides between sunflowers and detoxification, although the soils in this photo are not known to be contaminated.

REVIEW & APPLY | 27.2

R Plants respond and adapt to the composition of the soil. Plants can respond to regions of increasing salt through halotropism, auxin-mediated growth away from a high concentration of salt. Many plants respond to low phosphorus and nitrogen by altering the shape of their root systems. Some species have constitutive adaptations that allow them to survive in challenging environments, such as the ability to pack toxins into vacuoles or to excrete salt through salt glands. The ability of certain plants to take up heavy metals and other toxic compounds from the soil is being used in phytoremediation efforts to clean up uranium-contaminated soils near nuclear reactors.

A 1. Quinoa (*Chenopodium quinoa*) is a crop plant that produces protein-rich seeds. Five different cultivars of quinoa were grown in control (non-salty) and salty soil, after which the density of stomata was measured. Describe what the graph below shows and hypothesize as to why a lower density of stomata would be more advantageous in some conditions than in others. Bars represent standard errors. (After S. Shabala et al. 2013. *J Plant Physiol* 170: 906–914.)

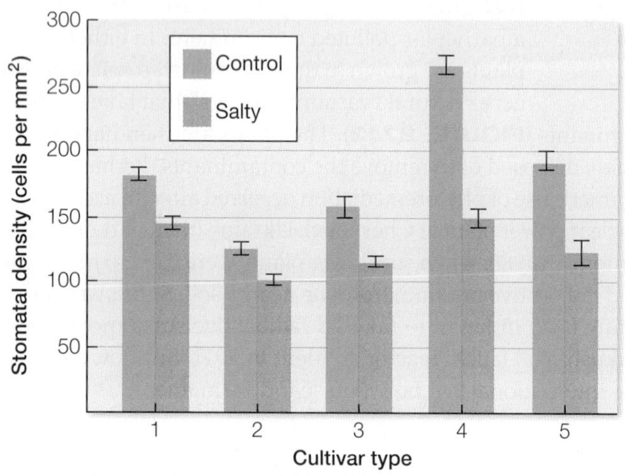

2. The effect of the *Arabidopsis thaliana* gene *NRT1.1* on nitrate response was investigated by growing plants on agar containing low levels of nitrate and then placing the tip of one lateral root onto agar that contained a high concentration of nitrate. The results are shown below. (From T. Remens et al. 2006. *Proc Natl Acad Sci USA* 103: 19206–19211. © 2006 National Academy of Sciences, U.S.A.)

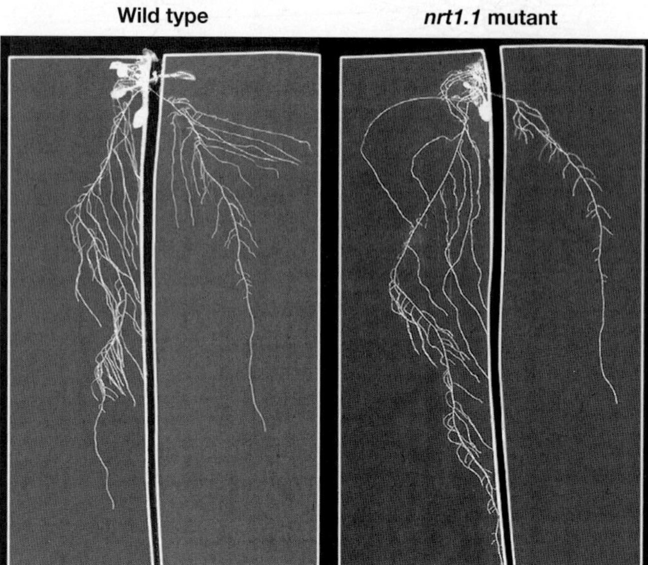

a. What aspect of root development is altered in the mutant versus the wild type when grown on high nitrate? In composing your answer, note that the root on the high-nitrate is a lateral root, not the main root. So it should be compared with lateral roots on the wild type.

b. Rephrase your answer to (a) to describe the effect that the functional NRT1.1 protein has on root development.

3. TIR1 is an auxin receptor whose expression is upregulated in response to low phosphate. How might the roots of a plant growing in a low-phosphate environment be affected if the phosphate receptor itself were upregulated?

4. Describe a mechanism shared by plants that tolerate high concentrations of salts and plants that tolerate heavy metals.

We've now described several ways in which plants sense and respond to their environment, including their responses to salt, water level, and CO_2. In the previous chapter, we described how photoreceptors (including phytochrome) are involved in sensing light. In the next key concept we will explore in more detail how plants sense another environmental variable: temperature.

27.3 | Plants Respond to Temperature

LEARNING OBJECTIVES

By the end of this key concept you should be able to:

27.3.1 Illustrate ways in which plants respond to extreme temperatures to minimize their harmful effects.

27.3.2 Define attributes of a successful plant temperature sensor.

27.3.3 Assess how well the rate of dark reversion of phytochrome fits the criteria of what a temperature sensor should be.

Plant growth and development are highly responsive to temperature. As discussed in Chapter 26, warm temperatures promote early flowering of many spring-blooming species. Warm temperatures also increase the rate of stem elongation in many species. More rapid flowering and stem elongation are not simply passive accelerations of processes that were occurring at lower temperatures—they are active cellular responses that involve changes in gene expression. A signal can only induce a response if it is perceived, so the existence of temperature responses means that plants must have a receptor that allows them to perceive the current temperature. When plants receive a signal in the form of a soluble molecule such as ABA, auxin, or nitrate, binding of the signal molecule to its receptor changes the conformation and chemistry of the receptor in ways that affect its interaction with other molecules. When the signal is light, photoreceptors, proteins that contain pigments that absorb photons, similarly change their conformation and set off signal transduction pathways. What happens when the signal is temperature? What sorts of receptors can detect a change in temperature?

Dark reversion of phytochrome may serve as a temperature sensor

Exciting new data suggest that phytochrome may act as a temperature sensor as well as a light sensor. The amount of active phytochrome (P_{fr}) in a plant depends on the intensity and wavelengths (colors) of light to which it is exposed (**FIGURE 27.13A**), and on the rate at which the active P_{fr} reverts to the inactive P_r state. This is like saying that the amount of water in a bathtub depends on the rate at which water flows into it *and* the rate at which the tub drains. Reversion of the active P_{fr} to the inactive P_r (draining of the tub) can occur if the plant is exposed to far-red light. In addition, P_{fr} can revert to P_r on its own without the addition of far-red light; this process is called dark reversion because it does not require light. However, dark reversion actually occurs both in the light and the dark. If the rate at which P_{fr} dark reverts to P_r speeds up as the temperature rises, then the amount of active phytochrome (P_{fr}) in a plant will decline, even if the light exposure stays the same. This idea was tested by exposing plants to continuous red light, which produces the active P_{fr} at a constant rate (e.g., the rate at which the tub filled was held constant). When this experiment was done at different temperatures, higher temperatures resulted in lower amounts of P_{fr} (e.g., higher temperatures must have increased the rate at which the tub drained) (**FIGURE 27.13B**). Might the rate at which phytochrome dark-reverts from the active to the inactive state convey information about the temperature to the plant and thus serve as a temperature sensor?

To determine whether phytochrome has a causal effect on plant response to temperature, researchers compared the ability of normal (wild-type) plants to respond to temperature with that of plants in which phytochrome was disrupted. Two responses that plants have to temperature were observed:

1. In wild-type plants, stem elongation occurs more rapidly in warm temperatures than in cold temperatures. In plants that lack functional phytochrome, this temperature response is disrupted. Stems of phytochrome mutants grown in cold temperatures are longer than normal, indicating that the presence of phytochrome represses stem elongation (**FIGURE 27.14**).
2. Wild-type plants also respond to warm temperatures by increasing the expression of some genes during the night. Might phytochrome, which can act as a transcription factor, be responsible for some of the temperature-dependent changes in gene expression that occur during the night?

(A)

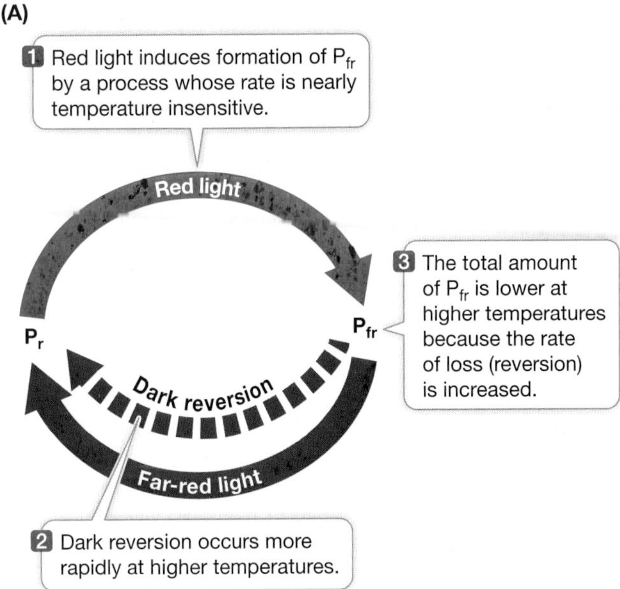

(B)

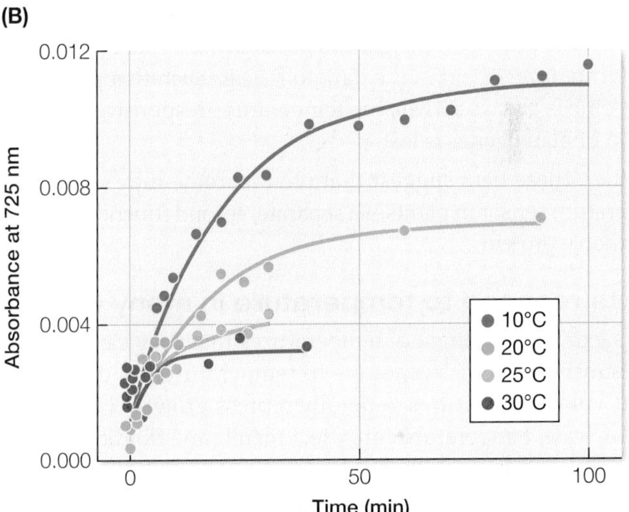

FIGURE 27.13 Dark Reversion of Phytochrome Is Temperature-Sensitive **(A)** The P_{fr} form of phytochrome can revert to the P_r form by exposure to far-red light or by a light-independent process called dark reversion. **(B)** The total amount of P_{fr} phytochrome present when continually illuminated with dim red light depends on the temperature, as would be expected if the rate of dark reversion is temperature-dependent. (B after M. Legris et al. 2016. *Science* 354: 897–900.)

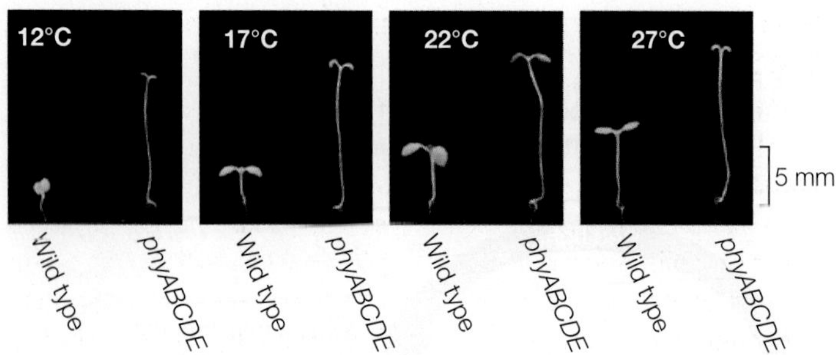

FIGURE 27.14 Temperature-Dependent Stem Elongation Requires Phytochrome The *phyABCDE* mutant does not make functional phytochrome and does not exhibit the temperature-dependent elongation response seen in wild-type plants. (From J. H. Jung et al. 2016. *Science* 354: 886–889.)

Are the mutants that lack phytochrome shorter or taller than wild-type plants when grown in the cold? Does phytochrome promote or inhibit stem elongation?

When the phytochrome gene is mutated such that P_{fr} forms but can never revert in the dark to P_r (e.g., such that phytochrome is always active), the temperature-responsive expression of many genes is lost.

Together, these data suggest that phytochrome may serve as a temperature sensor in plants—a separate, second function for this remarkable protein.

Plants respond to temperature in many ways

Plants expand the range of temperatures that they can survive by mounting specific responses to temperature. As discussed above, cold temperatures generally repress growth. On a subcellular scale, temperature can affect membrane fluidity and alter membrane permeability to solutes. High temperatures cause membranes to become more fluid, while low temperatures cause membranes to become less fluid. These changes have strong effects on cellular function: imagine the impact on cell activity if the ATP-generating enzyme ATP synthase could not move rapidly in the membrane! Many plant species can adjust to cold temperatures through a process called **cold-hardening**, which requires repeated exposure to moderately cool temperatures over many days. A key change during cold-hardening is an increase in the proportion of unsaturated fatty acids in cell membranes, which allows them to retain their fluidity and function normally at cooler temperatures (see Figure 3.1). Plants have a greater ability to modify the degree of saturation of their membrane lipids than animals do. In addition, low temperatures induce the formation of proteins that protect against chilling injury.

If ice crystals form within plant cells, they can kill the cells by puncturing organelles and cell membranes. Furthermore, the growth of ice crystals outside the cells can draw water from the cells and dehydrate them. Freeze-tolerant plants have a variety of adaptations to cope with these problems, including the production of antifreeze proteins that slow the growth of ice crystals.

INVESTIGATION

FIGURE 27.15 Effect of Temperature on Corn Yields
Earth's temperature is warming, and understanding how increased temperatures affect crop production will be important in designing strategies to mitigate negative impacts on our food supply. Correlative studies have revealed that corn yield declines at higher temperatures, with the drop-off becoming particularly steep above 30°C. These observations have prompted further investigation into the mechanisms that underlie the change in corn yield.[a]

HYPOTHESIS

The effect of temperature on corn yield depends on the time at which the temperature stress occurs.

METHOD

1. Grow corn plants in two separate greenhouses. During the day, maintain the temperature at about 25°C (cool) in one greenhouse and heat the other one to 35°C (hot). At night, open the greenhouses and allow them to come to the outside temperature. Keep plants well-watered.
2. Move plants from the cool greenhouse to the hot greenhouse for a period of 1 week, then move them back to the cool greenhouse. Transfer groups of plants at different points during their development. Leave control plants in the cool greenhouse throughout the experiment.
3. Allow plants to complete development, harvest their first ears, and measure kernel weight, kernel number, and total kernel weight per ear.

RESULTS

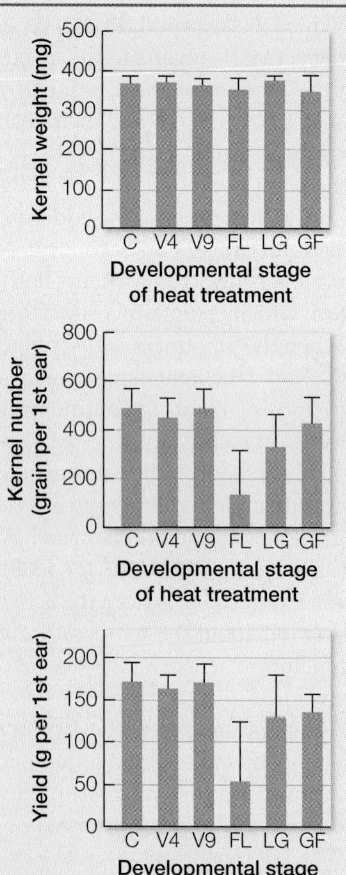

Bars represent different groups of plants that were heat stressed at progressively more mature stages of development. C, control plants; V4 and V9, stages of vegetative (leaf) development; FL, pollen-shedding phase; LG, lag phase; GF, grain-filling phase. Error bars represent half a standard deviation.

CONCLUSION

Yield was most strongly affected when plants were heat stressed around the time that pollen was shed.

ANALYZE THE DATA

To analyze the effect of heat on pollen, the topmost ear of plants grown in cool and hot greenhouses was hand-pollinated using pollen from either the hot or cool greenhouse or from corn plants that were grown just outside the greenhouse (control pollen). The resulting data are shown in the graphs below.
All data in this figure are for the topmost ear of corn. Error bars represent half a standard deviation.

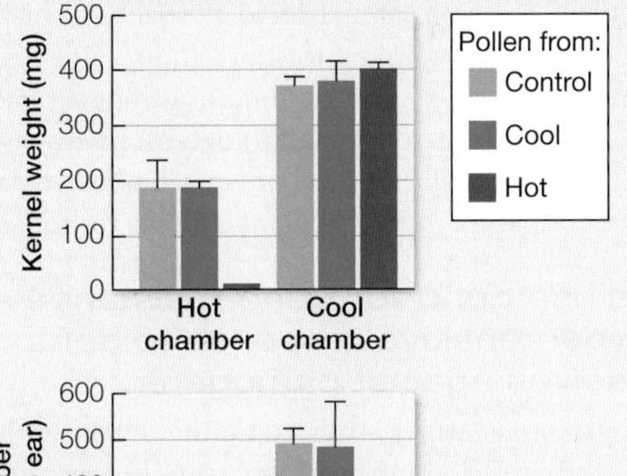

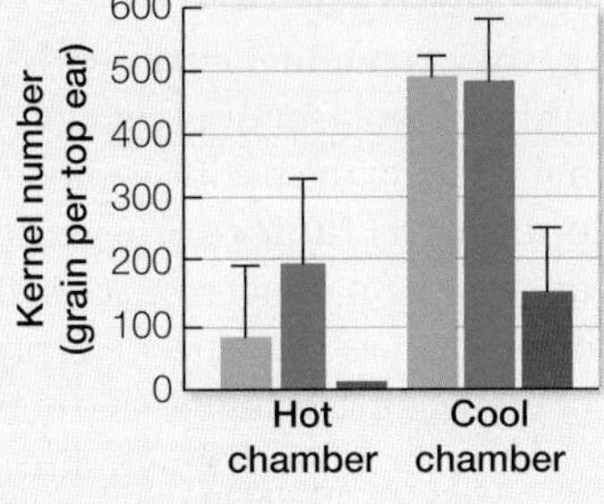

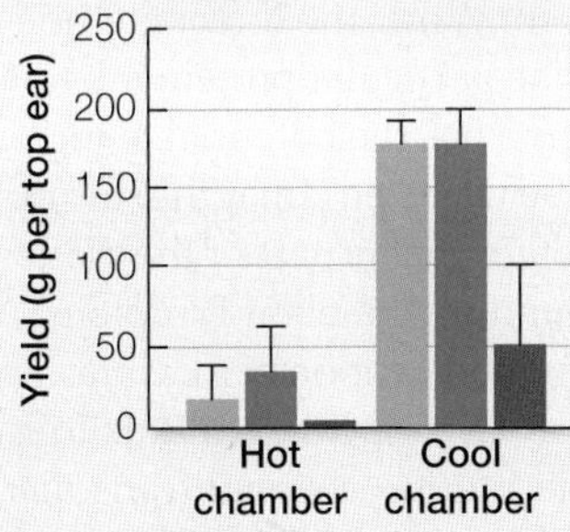

1. Formation of a kernel of corn requires successful pollination. Did the temperature in which the pollen developed affect its ability to produce new kernels? If so, which pollen was least effective?
2. Did the source of the pollen affect the average weight of the kernels that formed?

[a]J. I. Lizaso et al. 2018. *Field Crops Res* 216: 129–140.

A similar **Analyze the Data** exercise may be assigned in **Achieve**.

Plants have inducible responses to hot temperatures

Plant species differ in their sensitivity to temperature, but all plants have their limits. High temperatures destabilize membranes and denature many proteins, especially some of the enzymes of photosynthesis. Heat stress also has a strongly negative effect on pollination (**INVESTIGATION FIGURE 27.15**). The plant inducible response to heat stress is similar to the response to drought stress in that new proteins are made, often under the direction of an abscisic acid–mediated signaling pathway. Within minutes of experimental exposure to raised temperatures (typically an increase of 5–10 degrees Celsius), plants synthesize several kinds of **heat shock proteins**. Some heat shock proteins (called chaperonins) help maintain the activities of the cytoplasm by assisting other proteins in maintaining their folded structure. Threshold temperatures for the production of heat shock proteins vary, but 39°C is sufficient to induce them in most plants.

REVIEW & APPLY | 27.3

R Plants respond to a wide range of environmental factors. A common misconception is that because plants do not have the same sensory organs that people do, plants are not aware of their environments. However, the ability of plants to modulate their growth rates, gene expression, membrane fluidity, and other factors in response to temperature clearly indicates that plants sense temperature. Plant scientists are working to discover the various sensors that permit plants to perceive these stimuli. Recent evidence suggests that phytochrome may act as a temperature sensor, because the rate of dark reversion is temperature-sensitive. This would be a second function for phytochrome (light perception being the first).

A
1. What attributes does something need to have in order to qualify as a receptor? To what extent does the dark reversion of phytochrome meet those standards?
2. A phytochrome mutant that can form P_{fr} but does not undergo dark reversion has been identified. Draw a diagram indicating the extent of stem elongation expected for wild-type plants and for plants carrying this non-reverting phytochrome mutation.
3. Describe one aspect of cell structure that is particularly sensitive to heat, and another that is particularly sensitive to cold.

So far this chapter has discussed how plants sense and respond to abiotic (nonbiological) aspects of the environment, such as temperature and chemical constituents of the air and soil. Next we'll discuss how plants respond to the biotic (living) environment—which, like the abiotic environment, is also being affected by climate change.

27.4 Plants Have Constitutive and Induced Responses to Pathogens and Herbivores

LEARNING OBJECTIVES

By the end of this key concept you should be able to:

27.4.1 Compare and contrast plants' physical and chemical means of defense.

27.4.2 Describe one way in which plants recognize and respond to viral infection.

27.4.3 Describe the process by which jasmonate defends plants against herbivory.

It is remarkable that plants, which are immobile and filled with sugars and other carbohydrates, persist in the natural world despite the presence of other living beings that might gain energy by eating them. Successful infection by a **pathogen**—a disease-causing agent, which for plants includes fungi, bacteria, protists, and viruses—can significantly reduce photosynthesis and cause massive cell and tissue death. **Herbivores**, which eat plants, can remove large portions of a plant in a very short time. However, plants have evolved together with those that would attack them and have a range of defense mechanisms. Through a continuing "arms race" in which pathogens evolve mechanisms with which to attack plants, and plants evolve mechanisms for defending themselves against those attacks, both plants and their attackers persist. This arms race has resulted in a highly diverse array of plant receptors and defense responses that overlap and intersect in complex ways. Like the responses of the human immune system (see Chapter 36), the responses by which plants fight off infection can be either constitutive or induced. Unlike humans, however, plants do not make antibodies. Instead, they use a range of other tactics to recognize different types of attackers.

Physical barriers form constitutive defenses

As with humans and their skin, a plant's first line of defense is its outer surfaces, which can prevent the entry of pathogens. The parts of plants that are exposed to the outside environment are covered with cutin, suberin, and waxes. These substances not only prevent water loss by evaporation but can also prevent fungal spores and bacteria from entering the underlying tissues. The plant cell wall is also an important physical barrier to pathogens and can be strengthened further as an induced defense response (as we will see below). Thick cell walls and tree bark may also deter some herbivores. Small herbivores can be deterred by trichomes (leaf hairs), and larger ones by thorns and spines.

Chemical deterrents can be constitutive or induced

In addition to physical barriers, plants use "chemical warfare" to repel or inhibit other herbivores, using special chemicals called **secondary metabolites**. These substances are referred to as "secondary" because, unlike molecules such as sugars and amino acids, they are not used for basic cellular processes. The more than 10,000 known secondary metabolites range in atomic mass from about 70 to more than 390,000 daltons, but most are small. Examples of secondary metabolites that protect plants from being eaten include terpenoids (including volatile compounds that contribute to the smell and character of mint plants and spices such as basil and oregano) and alkaloids (nitrogen-containing compounds such as capsaicin, which makes some peppers taste hot, and nicotine in tobacco). Some secondary metabolites are produced by only a single plant species, whereas others are characteristic of entire plant genera or even families.

The effects of defensive secondary metabolites on other organisms are diverse. Some act on the nervous systems of herbivorous insects, mollusks, or mammals. Others mimic the natural hormones of insects, causing some larvae to fail to develop into adults. Still others, such as cyanogenic glycosides, break down when eaten and produce toxins that deter feeding. The presence of cyanogenic glycosides is why cassava roots must be carefully prepared before being eaten. Iridoid glycosides, which are glucose molecules covalently linked to short-chain fatty acids produced by plantains (*Plantago*), inhibit the growth of fungal pathogens. When constitutive defenses fail to deter a pathogen, plants can initiate induced resistance mechanisms. Below we describe a few specific examples of inducible plant defense responses.

Plants use gene-for-gene resistance and the hypersensitive response to defend themselves against pathogens

When pathogens attack, they come in contact with plant cells. Receptors that reside in the cytoplasm of plant cells called **resistance (R) proteins** interact with specific components of the pathogen called **avirulence (Avr) factors**. There are hundreds of Avr factors, which vary among pathogen species and strains. There are also hundreds of *R* genes that encode R proteins that recognize the presence of Avr factors. Recognition can be direct, as when a plant R protein binds to an Avr factor, or indirect, meaning a third protein is involved. While the direct method requires that the pathogen produce one particular protein in order to be recognized, the indirect method has the potential to increase the range of pathogens that can be detected. Imagine, for example, that a third (intermediate) protein that is part of the plant's constitutive defense response is bound by the plant R protein. Any infection that results in enzymatic modification of this intermediate protein such that the plant R protein no longer binds to it will liberate the R protein, setting in motion a signal transduction pathway that triggers the plant's specific immunity. This type of resistance, whether direct or indirect, is called **gene-for-gene resistance** because it involves a recognition event between two specific molecules: one determined by an *Avr* gene in the pathogen and the other determined by an *R* gene in the plant (**FIGURE 27.16**).

If the plant does not have a receptor that can recognize infection by a particular species or strain of pathogen, the plant does not turn on specific inducible defenses. The lack of a receptor makes the plant especially susceptible to that pathogen. A major goal of plant breeders for the past 50 years has been to identify *R* and *Avr* genes and to breed new *R* genes into crops to make them more resistant

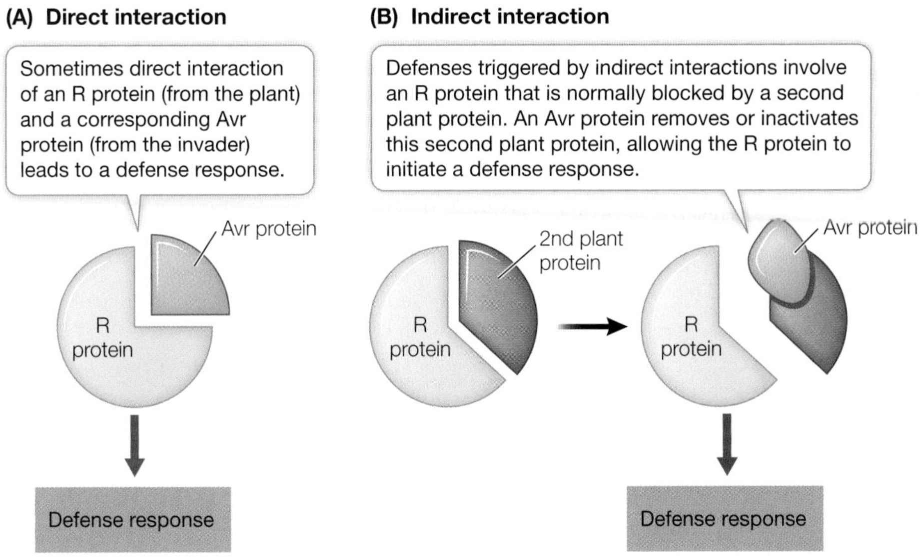

(A) Direct interaction

Sometimes direct interaction of an R protein (from the plant) and a corresponding Avr protein (from the invader) leads to a defense response.

Avr protein
R protein
Defense response

(B) Indirect interaction

Defenses triggered by indirect interactions involve an R protein that is normally blocked by a second plant protein. An Avr protein removes or inactivates this second plant protein, allowing the R protein to initiate a defense response.

2nd plant protein
R protein
Avr protein
R protein
Defense response

FIGURE 27.16 Gene-for-Gene Resistance Plants initiate defense responses when they recognize proteins produced by the pathogen. Recognition of pathogen-produced (Avr) proteins by plant R proteins can be by **(A)** direct interaction or **(B)** indirect interaction. (After J. L. M. Soosaar et al. 2005. *Nat Rev Microbiol* 3: 789–798.)

Animation 27.1 Signaling between Plants and Pathogens

to pathogens. This effort speeds up the naturally occurring coevolutionary arms race between pathogens attempting to evade detection by plants and plants evolving ways to detect them.

Once a pathogenic infection is recognized, the plant mounts a response. One common response is surprising: rather than minimizing damage to an infected cell, the plant cells around the site of infection undergo a **hypersensitive response** entailing programmed cell death. The hypersensitive response leads to the sacrifice of a few cells to protect the rest of the plant, resulting in necrotic lesions (**FIGURE 27.17**). This strategy is particularly effective against viruses because viral replication does not take place without the proteins and metabolic products of a living plant cell (see Key Concept 18.4). Viruses are typically carried to a plant by insects (often aphids), but extensive infection requires that the virus spread from cell to cell within the plant. Death of a small group of infected cells can help seal off the site of infection, preventing spread of the pathogen. Surrounding (still living) cells that produce cell wall polysaccharides assist by sealing off plasmodesmata. Some of the responding cells also produce toxic chemicals, including **phytoalexins**, small antibiotic-like molecules, before they die. The resulting necrotic lesion contains and isolates what is left of the infection. The rest of the plant remains free of the infecting pathogen.

RNA silencing is an induced defense mechanism

Another type of induced defense that acts against viruses is mediated by small RNAs. When RNA viruses invade plant cells and begin to replicate, double-stranded RNA (dsRNA) is formed. The plant has enzymes that chop dsRNA into small pieces called **small interfering RNAs**, or **siRNAs** (**FIGURE 27.18**). Viral mRNAs that contain sequences complementary to the siRNAs are

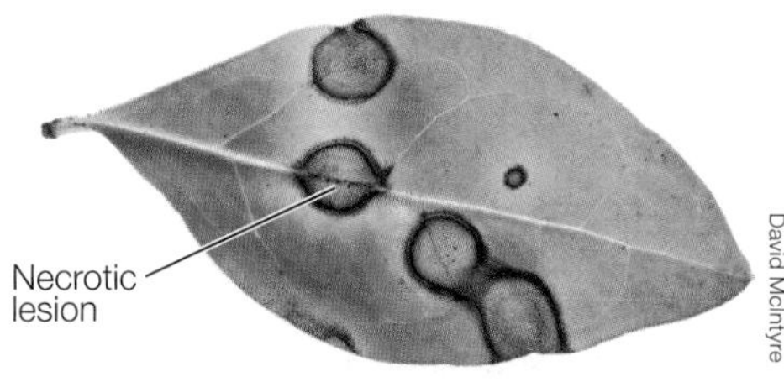

FIGURE 27.17 The Hypersensitive Response Defends against the Spread of Viral Infection

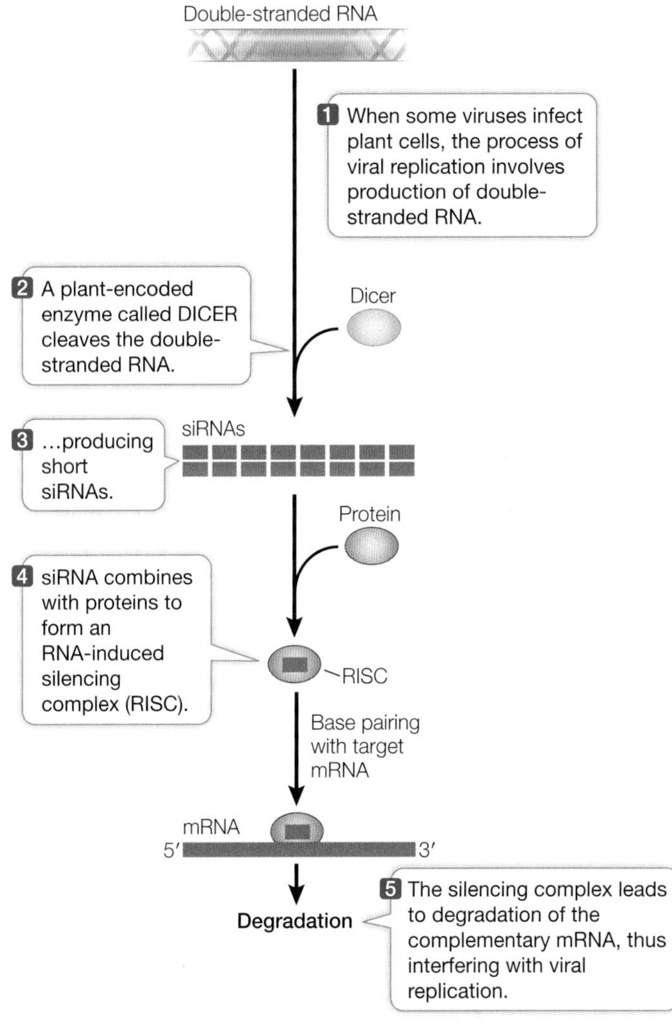

FIGURE 27.18 Production of Small Interfering RNAs Can Protect Plants by Blocking Translation of Virally Encoded mRNAs

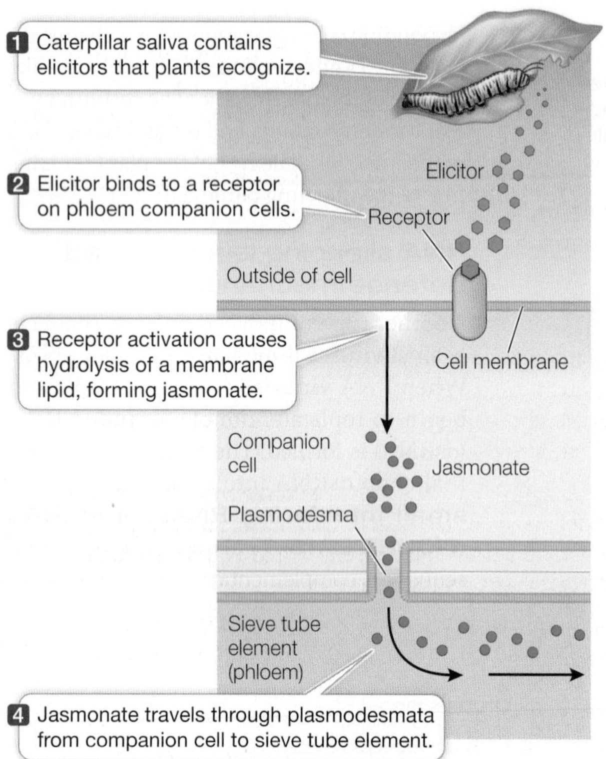

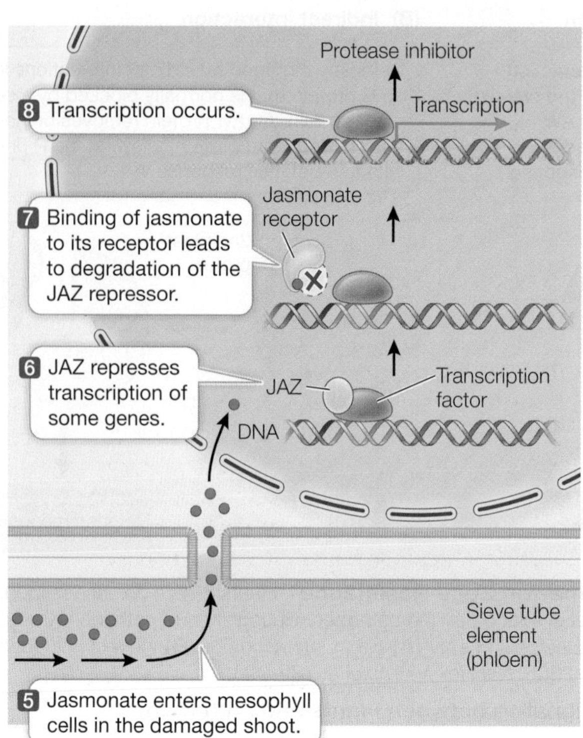

FIGURE 27.19 Response to Caterpillar Attack

Activity 27.1 Plant Defenses Simulation

then bound and degraded by machinery in the plant cell, which blocks viral replication. This phenomenon is an example of **RNA interference (RNAi)**. The siRNAs can spread rapidly through the plant by way of plasmodesmatal connections between cells, providing resistance throughout the plant.

Jasmonate is produced in response to herbivory

Attacks by herbivores require a different kind of response. When caterpillars (small herbivores) chew on a plant, substances in the insect's saliva and the physical wounding of the plant cells serve as signals to which the plant can respond. When a plant senses an herbivore-produced elicitor, it makes **jasmonate** (the ionized form of jasmonic acid) and a variety of jasmonate derivatives. These molecules trigger many plant defenses at the site of herbivore attack and throughout the plant (**FIGURE 27.19**). Jasmonate induces changes in gene expression by binding to a transcriptional inhibitor called a JAZ protein. After binding to jasmonate, the JAZ protein is targeted for degradation, and genes that were previously inhibited can be transcribed and expressed. An important group of plant proteins that are expressed in response to insect attack are protease inhibitors. Once inside an insect's gut, protease inhibitors interfere with proteases (enzymes that degrade proteins) that are part of the caterpillar's digestive system and thus stunt the insect's growth by limiting the caterpillar's ability to digest the plant material it eats.

Jasmonates can also be part of a "call for help" by triggering the formation of volatile compounds that attract insects that prey on herbivores attacking the plant. When corn is attacked by one particular moth larva, for example, the plant produces volatile compounds that attract wasps that parasitize herbivorous caterpillars.

Media Clip 27.2 Plants Can Hear Danger

Why don't the defensive chemicals that are so toxic to herbivores and pathogens kill the plants that produce them? In some cases, the defensive chemicals are directed at organs or systems that are not found in plants, such as the nervous, digestive, or endocrine systems of animals. In addition, plants that produce toxic defensive chemicals often sequester them in vacuoles or store them as harmless precursors until used.

Plant–pathogen interactions are affected by environmental conditions

Environmental conditions affect the health of both plants and pathogens, altering their ability to fend off or mount an attack. Warmer temperatures increase the metabolic rates of insects, so the amount of plant material that an individual insect eats increases as the temperature rises. Environmental conditions can also directly affect the likelihood that an interaction will take place. For example, wind, which can spread fungal spores long distances, promotes fungal infection. Hot, humid conditions can also favor infection by fungi that enter leaves through open stomatal pores. Plant–pathogen interactions require that both of the organisms be present in the same location at the appropriate point(s) in their life cycle. If changes in climate affect the timing of plant and pathogen life cycles in an asynchronous manner, their interactions could be strongly affected. For example, if warmer winter temperatures increase the percent of pathogens overwintering more than they affect plant populations, plant infection rates may increase, while other sorts of mismatches in the timing of life cycles may create temporal separations that decrease interactions. The amount of damage caused by a pathogen depends on the environment as well as on the pathogen and the plant.

REVIEW & APPLY | 27.4

R Plants persist in the environment only to the extent that they can ward off infection-causing pathogens and limit the damage by herbivores that eat them. They do this through production of physical and chemical deterrents. The precise defenses involved are constantly evolving as plants benefit when new defenses arise, and attackers benefit when they evolve ways to get around those defenses. This evolutionary arms race has resulted in a complex web of responses. Some defenses, such as the hypersensitive response, are activated only when a pathogen is recognized. In some cases, injured plants release volatiles that prime neighboring plants to mount defenses that reduce damage by pathogens. In other cases, plants release volatiles that function as a "call for help" by attracting predators that prey on the plants' own predators.

A

1. Because you are a biology student, your nonscientist neighbors are confident that you know all about plants. A neighbor brings you a potted plant that he bought 3 weeks ago from a nursery. The leaves have brown spots. He bought another plant of the same species a few weeks prior. That plant has no spots.
   a. You suspect that the spots are caused either by a nutrient deficiency or a disease. Given a large supply of the plants and soils involved, how would you distinguish between these possibilities?
   b. Assuming the spots are caused by a fungal disease, how might you try to create plants that are resistant to the disease?
   c. You go to the nursery where the neighbor bought the infected plant and find several plants with no brown spots on their leaves. What plant characteristics might account for the absence of spots?
2. Generalize how plants can mount a response after receiving a chemical signal.

© Mitchell Krog/Getty Images

Answer: Climate change has already had negative effects on crop production in some regions, particularly for corn and wheat. Crop production declines because the fertilizing effect of increased levels of CO_2 is outweighed by the negative effects of high temperature and changes in precipitation on yield and protein content. High temperature in combination with drought and elevated CO_2 can be particularly challenging because the resulting stomatal closure decreases evaporation and thereby increases heat stress in the leaf. Models further predict that an increase in average global temperature of about 2 degrees Celsius over preindustrial levels would increase the damage done by pests by close to 50 percent for wheat and about 30 percent for corn (C. A. Deutsch et al. 2018. *Science* DOI: 10.1126/science.aat3466). While the future impact of climate change will depend on the extent of warming and the degree to which people are able to adapt by growing new or different crops, climate change makes global food security an increasingly serious challenge, especially for a world in which population continues to increase.

27 VISUAL SUMMARY

Plants Respond to the Availability of Water and Carbon Dioxide

- **STOMATAL COMPROMISE** balances the uptake of CO_2 (used for photosynthesis) with water loss (**FIGURE 27.1**).
- Plants sense CO_2 levels and respond by regulating stomatal conductance.
- Plants respond to water stress by producing the hormone **ABSCISIC ACID** (**ABA**) (**FIGURE 27.5**).
- Constitutive **ADAPTATIONS** to drought stress include succulence and crassulacean acid metabolism (CAM).

FIGURE 27.1 FIGURE 27.5

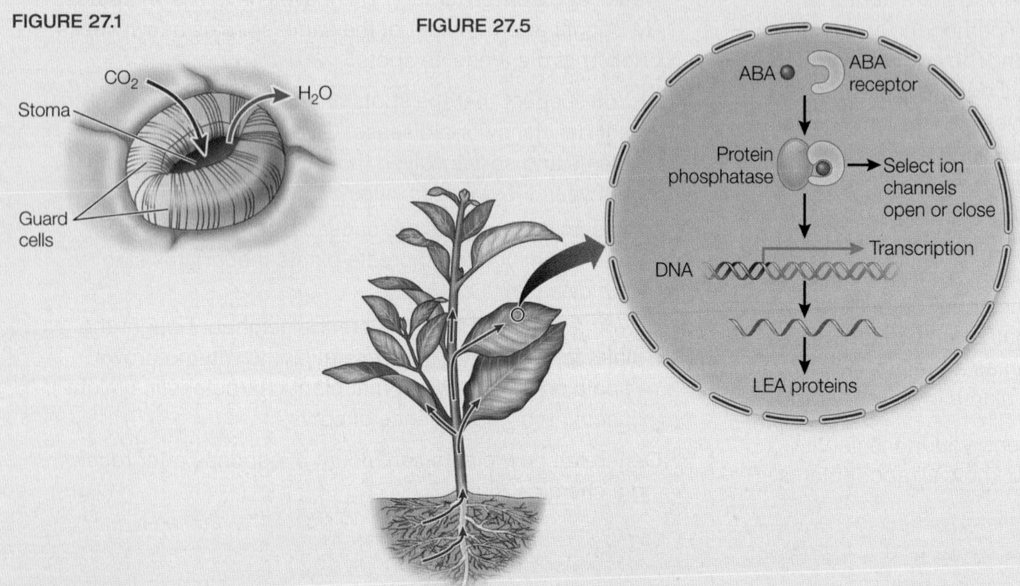

QUESTIONS

1. Explain the trade-off between having stomata open and closed.
2. Plants can perceive ABA because they have receptors for it. What type of molecule serves as an ABA receptor? How does binding to ABA change such a receptor?
3. What is the difference between a response that a plant may have to an environmental condition and an adaptation to that environment?
4. Plant cells obtain energy by breaking down the carbohydrates that are produced during photosynthesis. This process, called respiration, releases CO_2. Respiration takes place continually, in both the day and night, while in most species photosynthesis occurs only in the day. Predict how nighttime respiration will affect the tendency of stomata to open during the night.

Plants Respond to Differences in Soil Composition

- Root systems are shaped by local and systemic responses to soil nutrient levels (**FIGURE 27.10**).
- **HALOTROPISM**, growth of roots away from regions of high salt concentration, is mediated by differences in auxin transport in the root.
- **PHYTOREMEDIATION** is performed when plants that **HYPERACCUMULATE** heavy metals are used to clean up soils.

FIGURE 27.10

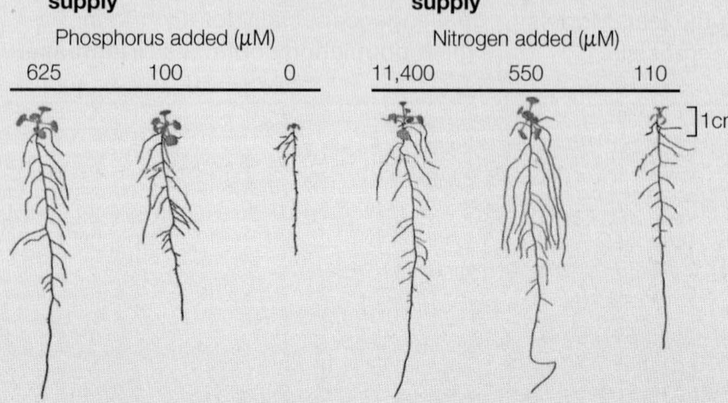

QUESTIONS

1. How do salty soils pose challenges for plants?
2. Give an example of a mechanism that allows plants to *respond* to salty soils and an example of an *adaptation* to those same soils.
3. Plants don't have the same sensory organs (eyes, ears, nose) that people do, but they still sense and respond to the environment. Give examples of plant responses that are similar to (a) taste, in that they allow plants to sense the presence of chemicals in the environment, and (b) smell, in that they allow plants to sense the presence of gaseous molecules.

27.3 Plants Respond to Temperature

- Plants can respond to temperature by changing their growth rate, flowering time, and membrane fluidity.
- A receptor or sensor should interact with the signal and be changed by that interaction in a way that affects downstream signaling events.
- Phytochrome appears to be a temperature sensor (as well as a light receptor) (**FIGURES 27.13** and **27.14**).

FIGURE 27.13

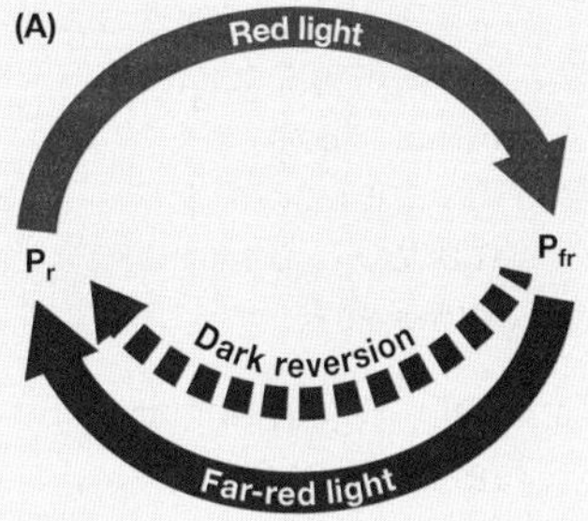

FIGURE 27.14

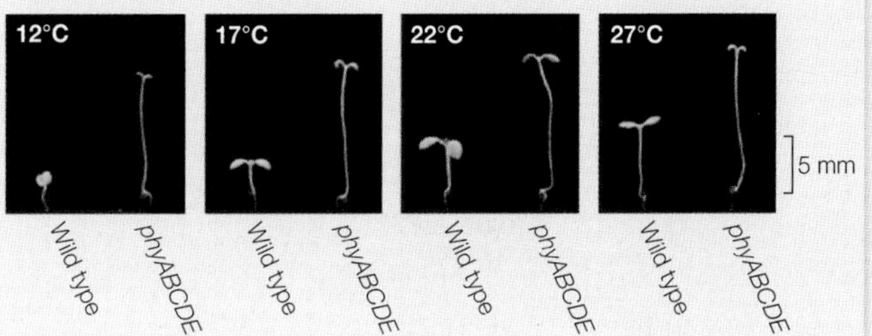

QUESTIONS

1. Provide evidence to support the claim that phytochrome is a temperature sensor. Indicate which part of the requirements for a receptor (or sensor) each piece of evidence fills.
2. Construct an argument as to why it might be adaptive for a spring-flowering plant to wait until it is quite warm before it begins growing rapidly.
3. What part of the plant life cycle is most sensitive to heat stress?

27.4 Plants Have Constitutive and Induced Responses to Pathogens and Herbivores

Go to ANIMATION 27.1 and ACTIVITY 27.1

- Plants have physical and chemical means of defense against **PATHOGENS** and **HERBIVORES**.
- Plants that recognize infection using *R* gene–mediated receptors can mount a local **HYPERSENSITIVE RESPONSE**.
- **JASMONATE** induces defenses throughout the plant, including production of proteinase inhibitors (**FIGURE 27.19**).

FIGURE 27.19

Elicitor
Receptor
Outside of cell
Cell membrane
Companion cell
Jasmonate
Plasmodesma
Sieve tube element (phloem)
Protease inhibitor
Transcription
Jasmonate receptor
JAZ
Transcription factor
DNA
Sieve tube element (phloem)

Activity 27.2 Concept Matching: Plant Adaptations

1. What is the evolutionary "arms race"?
2. Outline the pathway from signal to response by which a hypersensitive response is mounted.
3. Explain how production of proteinase inhibitors can protect plants from herbivory by caterpillars.

Go to **Achieve** for the e-book, animations, activities, and additional resources and assignments.

Part 6
Animal Form and Function

28 Transformations of Energy and Matter: Nutrition, Temperature, and Homeostasis

The Arctic fox (*Vulpes lagopus*, sometimes called *Alopex lagopus*) occurs in the Far North all around the world. It is active throughout the year and during the winter routinely encounters air temperatures lower than –30°C.

Imagine a cell deep in this fox's body, such as in its heart muscle. Think of the fox's outdoor environment in winter, and think of the immediate surroundings of the cell (the cell's environment). How does the cell's environment compare with the outdoor environment? If these two environments are different, what properties have evolved in Arctic foxes to maintain the difference?

You will find the answer to these questions on page 702.

KEY CONCEPTS

28.1 Animals Eat to Obtain Chemical Building Blocks and Energy

28.2 An Animal's Energy Needs Can Be Quantified

28.3 Responses to Temperature Help Clarify Homeostasis

28.4 Animal Bodies Exhibit Division of Labor, but Each Cell Must Make Its Own ATP

28.1 Animals Eat to Obtain Chemical Building Blocks and Energy

LEARNING OBJECTIVES

By the end of this key concept you should be able to:

28.1.1 Specify some of the types of organic molecules that animals must obtain in their foods.

28.1.2 Explain why animals must eat essential nutrients throughout their lives.

28.1.3 Explain the fundamental reasons that animals need energy.

The animals are a large and diverse group: about 80 percent of all the named species of living organisms are animals (see Chapter 22). In this part of the book, we focus mostly on the ways that animals *function*. Two properties of animals stand out when their function is considered. First, animals typically have muscles and can move —such as by walking, flying, or swimming. Second, animals are heterotrophs: organisms that require preformed organic molecules as sources of energy and chemical building blocks. (In contrast, the most common *autotrophs*—plants and algae—obtain energy from sunlight and synthesize organic compounds from inorganic precursor molecules.) Heterotrophs obtain energy by breaking the chemical bonds of organic compounds obtained from other organisms, and they build their tissues from the matter present in preexisting organic compounds obtained from other organisms.

The complex food webs that arise through interactions between heterotrophs and autotrophs fall into the domain of ecology, and will be discussed in Key Concept 40.3. Here we discuss the transformations of matter and energy that take place within the individual animal body.

Animals need chemical building blocks to grow and to replace chemical constituents throughout life

Why does an animal need the chemical building blocks it obtains by eating? One reason is growth. To grow, an animal requires chemical building blocks—such as fatty acids and amino acids—to build each new cell it adds to its body. There is also a second, less obvious reason.

Even after animals are fully grown, they typically lose chemical building blocks from their cells and tissues every day and thus need new chemical building blocks to replace what they've lost, throughout their lives. This important fact was discovered by using isotopic forms of carbon atoms and nitrogen atoms (see Key Concept 2.1 for a discussion of isotopes). Suppose you make some food in which the nitrogen atoms are nitrogen-15 instead of the usual isotope, nitrogen-14. Suppose you then feed that food to a full-grown, adult dog (or person). After a few weeks you will find that many of the nitrogen-14 atoms in the dog's body have been replaced with nitrogen-15 atoms—even though the dog looks no different. If you then start to feed the dog regular food (with nitrogen-14 atoms) again, you will find that the nitrogen-15 atoms in the dog's body are gradually replaced with nitrogen-14 atoms.

Experiments such as this demonstrate that the atoms inside an adult animal's body are constantly exchanged with atoms in the animal's surrounding environment. For example, our red blood cells live for only about 4 months. When a red blood cell is at the end of its life span, it is broken down and replaced. As it breaks down, some of its constituents are excreted. For example, some of the iron from the cell is excreted. To replace the cell, our bodies need to get replacement constituents—such as replacement iron—from our environment.

Nearly all the cells in our body are engaged in the process of breakdown and replacement (although not necessarily on a 4-month schedule). As the process takes place, some atoms originally inside our body are lost, and some atoms are taken in from the environment. In people, 2–3 percent of body protein is broken down and rebuilt every day! Some amino acids are lost when proteins are broken down, explaining why we need to eat amino acid–containing foods all our lives, even after we are fully grown.

These considerations lead us to one of the most important principles in the study of animal function: Over time, an animal is not defined by its atomic building blocks (because individual atoms constantly come and go). Instead an animal is defined by the organization of its body.

Animals need chemical-bond energy to maintain the organization of their body

Organization is the most essential attribute of life. All living things are organized in a great many ways. Consider, for example, the Arctic fox and its surroundings in our opening photo. The temperature inside the fox's body differs from the outside temperature, and this illustrates a type of organization. Thermal energy is not distributed at random. Instead, in the case of a warm-bodied animal, it is more concentrated inside than out. **FIGURE 28.1** illustrates other examples of organization in the body: the organization of electric charge across a cell membrane and the organization of cells into tissues and organs. The distribution of electric charges across a cell membrane is not random (see Figure 28.1A). As you saw in Key Concept 5.2, the movement of ions across membranes is essential to many biological processes, and an electrochemical gradient is maintained across the cell membrane, with positive charges being more concentrated on the outer surface than the inner surface. More obvious types of organization are

(A) An organized distribution of charge across the cell membrane

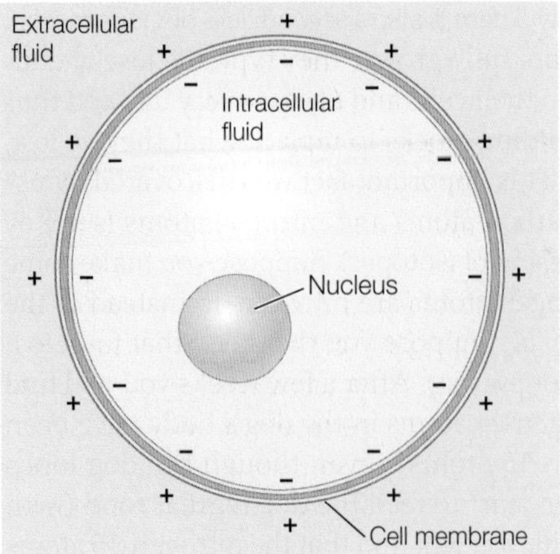

(B) Cell types organized into layers within the retina

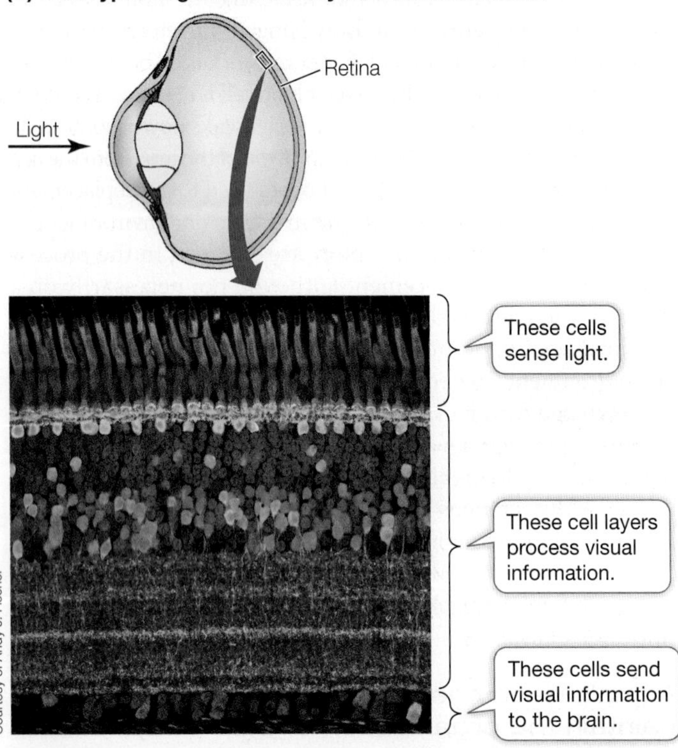

FIGURE 28.1 Two Types of Organization in an Animal's Body **(A)** At the cell membrane of a single cell, the balance of charges on the outer surface differs from that on the inner surface. There is a net positive charge on the outside but a net negative charge on the inside. The relevant charges are associated with ions; for example, a sodium ion (Na^+) contributes a positive charge. **(B)** A highly organized layering of cells is evident in this magnified and colorized image of cells in the retina of a chick's eye. The cells in the top layer are the rods and cones responsible for light sensing, and the layers of cells beneath are responsible for successive steps of image processing in the retina before visual information is sent to the brain.

structural. In individual cells, atoms are not distributed at random. Instead, they are organized into molecules such as phospholipids and proteins. In turn, these individual molecules are organized into cellular structures such as the cell membrane and into cellular biochemical processes such as sequences of enzyme-catalyzed reactions. Moreover, cells are organized relative to each other. As Figure 28.1B shows, for instance, individual nerve cells are highly organized in relation to each other in the retina of the vertebrate eye. This structural organization is essential for vision.

The second law of thermodynamics tells us that any organized system, left to itself, tends to lose organization and become more random as time passes. Animals must combat this tendency because any loss of organization in an animal's body threatens the animal's well-being or even life itself.

LINK The second law of thermodynamics has important implications for the use of energy by animals. You can review the laws of thermodynamics in **Key Concept 2.3**.

For biologists, the best definition of energy is that it is the capacity to create or maintain organization. In essence, the reason animals need to obtain energy is so they can do work (of several types) to maintain their organization by combating the effects of the second law of thermodynamics.

To fully understand the need for energy, we need to recognize that energy exists in several forms, and that only some forms can do work. In living organisms, *heat* is a form of energy that is unable to do work. To do work, animals need a different form of energy, notably *chemical-bond energy*, also known as *chemical energy*. Moreover, when animals use chemical-bond energy to do work, they convert that energy to heat, which they are unable to convert back to chemical-bond energy. These principles explain why animals need to eat to obtain new chemical-bond energy from their foods throughout their lives. As animals use chemical-bond energy to do work, they convert it to heat. Thus they continually need more chemical-bond energy, which they get by ingesting organic molecules that they can break down.

Food provides a great variety of nutrients, some of which are essential

Animals collect food in many ways. They may catch and devour other animals (**FIGURE 28.2A**), or they may consume plants, such as by grazing on grasses or eating the fruits of trees. Chapter 22 described some of the diverse feeding strategies that have evolved among animals. To us, predation and herbivory are the most familiar forms of obtaining food. However, many animals obtain their foods in ways that are less familiar. The two most important of these alternative feeding mechanisms are *suspension feeding* and *obtaining food molecules from microbial symbionts*.

Suspension feeding (sometimes called filter feeding) is used mostly by aquatic animals, and it refers to collecting small food particles suspended in the water. In both fresh water and the ocean, the water is often rich with tiny suspended objects that can be used as food, such as algal cells, tiny animal larvae, or minute bits of dead organic matter. Suspension feeders are able to collect these small food particles in great numbers, thereby meeting their food needs even though each individual food item is tiny. Clams, many aquatic insects, and many fishes (such as sardines and anchovies) are suspension feeders. A mussel (**FIGURE 28.2B**), for example, eats algal cells—suspended in the surrounding water—that are so small as to be invisible to our

(A) Predation

(B) Suspension feeding

(C) Symbiosis

FIGURE 28.2 Three Types of Feeding **(A)** Predation is illustrated by this osprey, which dives from the sky to catch fishes. **(B)** These mussels (mollusks similar to clams) engage in suspension feeding by pumping water from the environment through their bodies to collect invisible algal cells from the water in such numbers that the mussels' food needs are met. **(C)** Reef corals acquire food from microbial symbionts. Symbiotic algae in the corals' tissues carry out photosynthesis and make food molecules that are shared with the animal cells that harbor the algae.

eyes, but the feeding mechanism of a mussel can collect millions of these invisible cells per day.

Some animals meet much of their food needs by using food molecules from microbial symbionts. Symbiotic relationships have evolved between animals and microbes in which microbes live within the animals' bodies and synthesize food molecules that are directly transferred to the animals as food. All the reef-building corals, for example, house symbiotic algal cells within their tissues (**FIGURE 28.2C**; see also Figures 1.14, 1.15, and 19.22). The algal symbionts synthesize food molecules (such as glucose) by photosynthesis, and these food molecules then move out of the algal cells directly into the animal cells of the coral. This mode of feeding explains why reef-building corals need clear waters where they receive lots of sunlight.

Regardless of how an animal feeds, the animal must obtain foods that meet its needs for both chemical substances and energy. **Nutrition** is the study of how animals meet these challenges. Here we will first consider the chemical aspects of nutrition, and then we will focus on energy.

The molecules in an animal's body are made from the molecules in the animal's food. Thus, as animals gather food, they must obtain specific chemical building blocks that they are capable of using to build their body parts—such as amino acids to build proteins, fatty acids to build cell membranes, and calcium to build a skeleton. For carrying out these processes, animals have enormous capacities to interconvert molecules. For example, lipid molecules in their foods can be converted to carbohydrate molecules in their bodies, and certain amino acids can be converted to others (see Key Concept 5.3).

The interconversion processes possessed by animals are not all-powerful, however; and animals typically cannot synthesize all the molecules they need from other molecules. Thus there are molecules or categories of molecules that an animal must obtain (already fully formed) in its food. As we discuss this subject, you'll see that the word "essential" has a unique meaning in the study of nutrition.

An **essential nutrient** is defined as a type of molecule that an animal needs but cannot synthesize for itself. Animals do not all have the same essential nutrients, because animals of different kinds can differ in what nutrients they need, and they can also differ in the interconversion processes they have evolved. We will start our discussion of essential nutrients by focusing on amino acids.

Recall that the **standard amino acids** are the 20 kinds of amino acids that animals use to build proteins (see Key Concept 3.4). Biochemists have discovered that animals can synthesize some of the 20 standard amino acids but not others. Adult humans, for example, can synthesize 12 of the standard amino acids by rearranging the subparts (carbon skeletons and functional groups) of other molecules in their diet. Those 12 amino acids are *nonessential* (although we need them, they do not qualify as *essential* because we can make them). The remaining eight amino acids cannot be synthesized by adult humans. These **essential amino acids** —isoleucine, leucine, lysine, methionine, phenylalanine, threonine, tryptophan, and valine—must be obtained fully formed in the diet. The list of essential amino acids tends to be similar from one type of animal to another, but there are important differences. Infant humans, for example, require histidine in their diet as well as the eight amino acids that are essential for adults.

You might wonder: Where do the essential amino acids come from? Plants and algae can synthesize all 20 standard amino acids. For this reason, animals can obtain their essential amino acids by eating plants or algae. Different plants and algae vary in their proportions of amino acids. For an animal to get all of its essential amino acids in adequate quantities, it may therefore need to eat more than one kind of plant or alga. For adult humans, combinations of grain foods and legume foods—such as corn and beans—are often effective in providing all the essential amino acids (**FIGURE 28.3**).

Foods of animal origin—such as milk, eggs, and meat—tend to be good sources of all eight essential amino acids. Cows, chickens, and other farm animals get their essential amino acids by eating plants. They build those amino acids into proteins in their bodies or into the products they make, such as milk and eggs. We then get the amino acids when we eat these foods.

Animals also need certain **essential fatty acids** in their foods. There are a great many kinds of fatty acids, but in humans and many other animals only two commonly qualify as being essential nutrients: alpha-linolenic (α-linolenic) acid and linoleic acid. If α-linolenic acid is essential for an animal, the animal must either eat it in its diet or eat very similar fatty acids that can be converted to it by simple reactions. Alpha-linolenic acid and the other fatty acids that can be converted to it are called the *omega-3 fatty acids*. Linoleic acid and the other fatty acids that can be converted to it are the *omega-6 fatty acids*.

Vitamins are another group of essential nutrients. Vitamins are difficult to define chemically because needs for various vitamins have evolved independently or relatively independently, meaning that different vitamins often have unrelated chemical structures. Vitamin A (**FIGURE 28.4**) provides an excellent illustration of the evolutionary history of a vitamin. When animals evolved vision, they used a slightly modified version of this complex chemical structure for light collection in their eyes. But they did not evolve an ability to make the structure! Instead, they depended on getting the structure, already made, from plants and algae that synthesize and use it to help collect light for photosynthesis. To this day, humans and many other animals must obtain vitamin A as a fully formed ("prefabricated") chemical structure in their diets.

FIGURE 28.3 A Common Food Combination Provides All Eight Essential Amino Acids All around the world, cultures exist in which corn and beans are major parts of the daily diet. We mimic this diet when we snack on tortilla chips and bean dip. Eaten together, these two foods provide all the essential amino acids required by adult humans. (Corn by itself often does not provide enough lysine or isoleucine, and beans often do not provide enough methionine or tryptophan.) When two or more foods must be eaten to get all eight essential amino acids, it is important to eat the foods at about the same time.

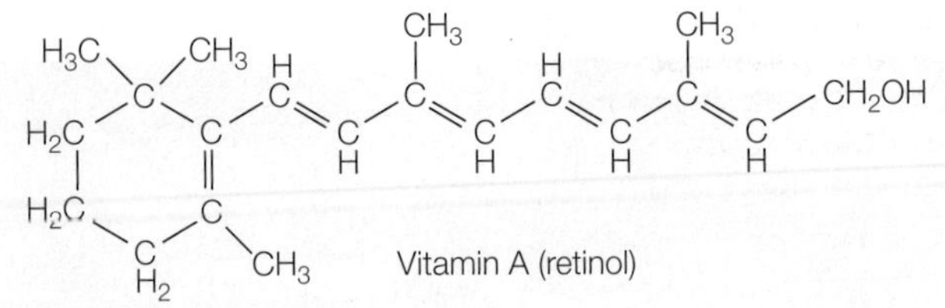

Vitamin A (retinol)

FIGURE 28.4 The Chemical Structure of Vitamin A

All the vitamins are carbon compounds, and they are needed in just tiny amounts. Thus we can quickly define a vitamin as an essential carbon compound required in tiny amounts. The list of vitamins varies a bit among animals. For example, most mammals can make their own ascorbic acid, so it is not a vitamin for them. However, primates—including humans—cannot synthesize ascorbic acid, and for them it is a vitamin: vitamin C. Vitamin D is a special case because humans can synthesize it if their skin is exposed to sunlight but otherwise cannot. Vitamin D is essential for people only if they have inadequate sun exposure, as is often true in winter when people stay indoors or wear heavy clothing.

The **essential minerals**, another important category of essential nutrients, are defined as chemical elements that animals require in addition to carbon, oxygen, hydrogen, and nitrogen (the most abundant elements in organic molecules). Familiar examples of essential minerals are calcium, phosphorus, iodine, and iron. Calcium must be obtained in the diet for animals to synthesize calcified skeletons. Phosphorus is also often required for making the skeleton, and in addition it is required to synthesize the nucleic acids (DNA and RNA) and ATP. Iodine is required by vertebrate animals because it is necessary for synthesizing the thyroid hormones. Iron is universally required because it is a key building block of proteins in the electron transport system, and it is required as well to synthesize hemoglobin in the animals (such as earthworms as well as vertebrates) that use hemoglobin as an oxygen-transport compound.

An important question to consider is: Why can't we take in enough of all the essential nutrients when we first reach our adult size and then forget about them for the rest of our lives? Why do we need to worry about getting enough of the essential amino acids, fatty acids, minerals, and vitamins during every week of our lives? The answer is the ceaseless turnover of body constituents we discussed earlier in this chapter. Throughout our lives, we lose essential molecules from our tissues and must replace them. Recall, for example, that in people, 2–3 percent of body protein is broken down and rebuilt every day! Essential amino acids are lost to our environment when proteins are broken down, explaining why we need to eat foods containing the essential amino acids all our lives, even after we are fully grown.

Food provides energy

Animals need chemical-bond energy to live. That is, animals must obtain in their food organic molecules that they can break down to release energy—which they then can use to move about, pump blood, synthesize body parts, and carry out all the other energy-demanding requirements of life.

How much energy does an animal need, and how is this measured? The energy content of a food can be quantified by measuring the amount of heat the food produces when it is completely

FIGURE 28.5 Food Energy Values Are Measured by Explosive Combustion in a Bomb Calorimeter

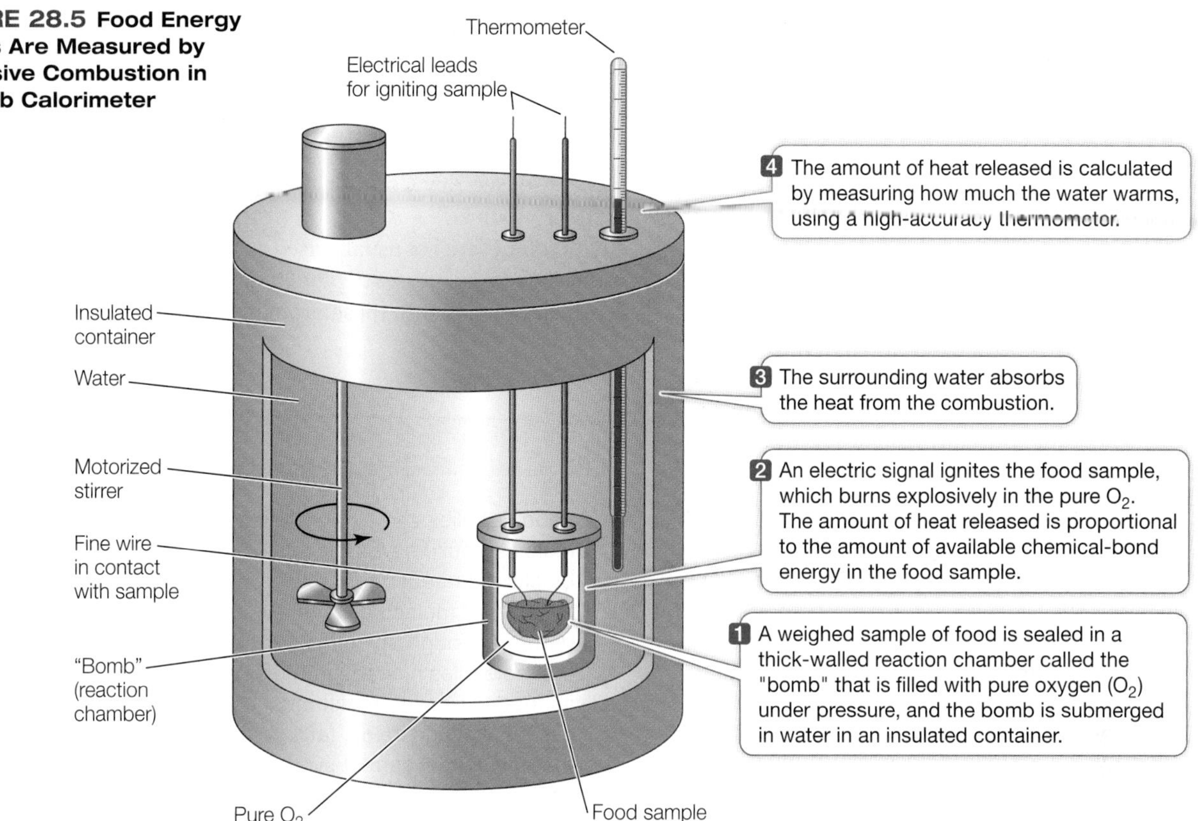

burned in the presence of oxygen (O_2), producing CO_2, H_2O, and other fully oxidized chemical products (**FIGURE 28.5**). We can determine the energy content of a food this way because both burning and aerobic animal respiration follow the same chemical rules. Burning and aerobic respiration involve different series of chemical steps, but the products of the two processes are almost identical. Heat is one product. The amount of heat released by burning a food is close to that produced when an animal metabolically breaks down the same food and uses the energy released (and we can correct for any differences by use of chemical rules). To fully understand this point, it is important to recognize that when an animal *initially* breaks down a food, it obtains energy in the form of chemical-bond energy, which it can use to do work, but then ultimately as it uses the chemical-bond energy, it converts the energy to heat. We are saying here that the amount of heat released by burning is the same as that *ultimately* released in the body.

Food energy values in the United States are typically expressed in calories and kilocalories. A **calorie** (**cal**) is the amount of heat required to raise the temperature of 1 gram of water by 1 degree Celsius, and a **kilocalorie** (**kcal**) is 1,000 times greater. We can use these units for any type of energy, not just heat, because all forms of energy are proportional to one another.

Confusion arises because people often refer to kilocalories as "calories" in everyday life. When you see the word "calories," you should check whether calories or kilocalories are being used. In the Nutrition Facts on food packages in the United States, for example, the energy values reported as "calories" are actually kilocalories (**FIGURE 28.6**). The **joule** is an alternative, metric unit

Nutrition Facts

Serving Size 1 cup (228 g)
Servings Per Container 2

Amount Per Serving	
Calories 250	Calories from Fat 110
	% Daily Values*
Total Fat 12g	**18%**
Saturated Fat 3g	**15%**
Trans Fat 0g	
Cholesterol 30mg	**10%**
Sodium 470mg	**20%**
Total Carbohydrate 31g	**10%**
Dietary Fiber 0g	**0%**
Sugars 5g	
Protein 5g	**10%**
Vitamin A 4% •	Vitamin C 2%
Calcium 20% •	Iron 4%

*Percent Daily Values are based on a 2,000 calorie diet. Your Daily Values may be higher or lower depending on your calorie needs.

FIGURE 28.6 A Nutrition Label of the Sort Used in the United States This is the label for a particular brand of macaroni and cheese. Although this label refers to "calories," the values on it are in fact kilocalories.

Why is it important to know that kilocalories are the unit being used?

for measuring energy (1 calorie = 4.2 joule). Most countries use the joule for expressing food energy values, and scientists almost always use it in their scientific work.

Let's consider a meal of macaroni and cheese for which the Nutrition Facts are presented in Figure 28.6. As you can see, each serving supplies 250 kcal. When the meal is sitting on your plate waiting to be eaten, there are 250 kcal of energy in the chemical bonds of the starch, fats, and other molecules in the meal. When the molecules are broken down by the citric acid cycle and the electron transport chain, some of the chemical-bond energy of the food will be used to make ATP. Then the energy of ATP will be used to do work, such as to power the heart. In these ways the chemical-bond energy in the meal will ultimately turn to heat, and—in total—250 kcal of heat will be produced.

LINK As explained in **Key Concept 5.2**, carbohydrate catabolism in the presence of oxygen releases a large amount of energy. The citric acid cycle initiates the oxidation of glucose, and the energy harvested by catabolism is processed by the electron transport chain to produce ATP (see Figure 5.9).

How much chemical-bond energy does an animal consume each day, converting it to heat? The answer depends on the animal's metabolic rate. An animal's **metabolic rate**, expressed per day, is the amount of energy the animal converts to heat per day. A typical person in Western society, for example, converts an average of 2,000–2,500 kcal per day in his or her daily mix of rest and activity. For general purposes, if you don't want to overeat and gain weight, add up the energy values of all the foods you eat each day to make sure the total is no more than 2,000–2,500 kcal.

The three major types of food molecules—lipids (fats and oils), carbohydrates, and proteins—differ in energy content per unit of weight, with lipids being about twice as dense with energy as carbohydrates or proteins. Lipids yield about 9.4 kcal per gram when they are broken down, compared with 4.1 kcal/g for carbohydrates and 4.5 kcal/g for proteins. As you might expect, therefore, foods containing lots of lipids tend to provide lots of kilocalories per serving.

When an animal eats foods containing more chemical-bond energy than the animal needs, the animal often stores some or all of the extra chemical-bond energy for future use. Animals store energy mostly in the form of lipids. This makes sense because lipids are particularly energy dense. Stored materials need to be carried around until they are used; for instance, if a bird stores energy for use while it migrates, it must carry the stored energy until it uses it. For each gram of lipid that an animal stores and carries, it gets more than 9 kcal of energy—a large amount—when it finally uses the lipid. Animals do not need to eat lipids to store lipids. They can convert carbohydrates and proteins to lipids to store chemical-bond energy.

Animals also store glycogen, a type of carbohydrate (specifically, polymerized glucose), but they do not store nearly as much energy as glycogen as they store as lipid. Glycogen is stored because some tissues require carbohydrates as their energy source. When glycogen is broken down, it yields the sugar glucose. The vertebrate brain gets its energy almost entirely from glucose. Thus glycogen stores can directly meet the brain's energy needs. When skeletal muscles employ anaerobic glycolysis to make ATP for exercise (see Key Concept 28.4), they also need glucose as their energy source.

Proteins are not usually used as storage compounds. Although there are large masses of protein in an animal's body (muscles are principally protein), these proteins are in use, rather than being stored for future use.

REVIEW & APPLY | 28.1

R Food supplies chemical building blocks from which animals can build new molecules and cells. Chemical building blocks are needed for growth and also for the ceaseless turnover of molecules and cells that adult animals undergo. Although an animal is able to synthesize many of the molecules it needs from the chemical constituents of its food, an animal's diet must contain certain fully formed essential amino acids, fatty acids, vitamins, and minerals that the animal cannot synthesize. An animal is defined by the organization of its body rather than by the matter that composes its body. An animal requires energy (the capacity to create or maintain organization) to maintain the organized state of its body. It acquires this energy from the chemical-bond energy in its food. The chemical-bond energy used by an animal in maintaining its body is ultimately converted to heat—necessitating that new chemical-bond energy be obtained.

A

1. To eat, some animals capture other animals or consume plants. Not all food sources are highly visible. Name two ways by which animals acquire foods that are not readily seen.
2. In what form does energy enter a heterotroph's body, and in what form does most of that energy inevitably leave?
3. When early explorers first traveled overland to the North and South Poles by using skis and sleds pulled by sled dogs, they packed mostly fats as food for their journey. Explain why, taking full account of the particulars of polar travel at the time.
4. When Alexander Hamilton was buried, many of the atoms that had been in his body while he was alive were left aboveground. Explain.

Animals vary tremendously in how large their energy needs are. Among animals of any particular body size, some need 10 or even 50 times as much energy as others. Next we will discuss how energy needs are measured and some of the reasons for the wide variation.

THINK LIKE A SCIENTIST

Using quantitative reasoning to communicate with nonscientists about "burning off" extra food calories

Research on science communication shows that people often have private theories about everyday questions. For example, people watch balls bounce from an early age, and they often devise private theories of the physics of bouncing. When people then take university physics and listen to a lecture on the physics of bouncing, they may stick with their private theories. Research shows that even if a lecturer is telling them something very different from their private theory, they may often think the lecturer is simply repeating their private theory.

This phenomenon is one of the obstacles to science communication. Everyone who aspires to educate people about scientific questions must be aware of this potential obstacle.

Consider, for example, private theories about overeating and exercise. Suppose Jack agonizes over whether he should have an extra serving (100 g = 3.5 ounces) of his all-time favorite cheesecake with caramel-fudge sauce: a high-calorie desert. As Jack cuts the cake and puts it on his plate, he rationalizes: "You only live once. I'm going to have it. I'll just take a run and burn off the calories in the morning." He figures that since exercise can be strenuous and involves sweating and hard breathing, it's bound to expel unneeded calories and fat from his body.

As scientists, we recognize that we can analyze Jack's plan using quantitative reasoning. We can measure the energy content of his serving of cheesecake (see Figure 28.5), and we can measure his energy use per minute as he runs (see Figure 28.7). Let's do a quantitative calculation.

You can look up online that a 3.5-ounce slice of cheesecake with caramel-fudge sauce might well contain 1,600 kcal of chemical-bond energy. As to the energy turned into heat by exercise, we can get quantitative values from Figure 28.8A (or we could put Jack on a treadmill). Assume Jack weights 80 kg (176 pounds). If we use this body weight and the quantitative values in Figure 28.8A, we can create the table below (you can create the table from the graph yourself; you just need to know that during aerobic metabolism 1 liter of O_2 consumed corresponds to about 4.8 kcal of heat produced):

Speed	Jack's metabolic rate
4.5 miles per hour (= 120 m/min)	11 kcal/min
5.6 mph	13 kcal/min
6.7 mph	15 kcal/min
7.8 mph	17 kcal/min
8.9 mph (= 240 m/min)	19 kcal/min

1. How long will Jack need to jog to burn off the calories in his serving of cheesecake if he jogs at a fairly comfortable pace of 5.6 mph?
2. Suppose Jack is busy and wants to invest less time and decides to run at a pace of 8.9 mph? How long will it take Jack to burn off 1,600 kcal?

If Jack doesn't like sweating, he might decide he can walk off the kilocalories. At a walking pace, he'd likely have a metabolic rate near 5 kcal/min and need 5 hours to turn the cheesecake kilocalories to heat. Because the cheesecake is composed mostly of lipids and carbohydrates, Jack's body will tend to store any kilocalories he fails to turn to heat through exercise. Thus, if Jack loses his resolve to exercise by the time the next day dawns, he will have added body fat.

3. Suppose you are asked to talk to a general audience about overeating and exercise. As you prepare your talk, you will need to recognize that some people in your audience will probably have strongly felt private theories on the question. Provide an outline of your speech.

28.2 An Animal's Energy Needs Can Be Quantified

LEARNING OBJECTIVES

By the end of this key concept you should be able to:

28.2.1 **Explain what an animal's "metabolic rate" is, and how it can be measured.**

28.2.2 **Explain how the metabolic rates of animals are affected by exercise and body size.**

28.2.3 **Specify the advantages and disadvantages of both regulation and conformity.**

How much energy does an animal need? This is a complicated question because animals use energy for many different purposes. Every cell and tissue in an animal's body uses chemical-bond energy every day to maintain its own internal organized state. Cells and tissues also use energy to do additional types of physiological work. When animals run, for example, their skeletal muscles use energy to generate mechanical forces for locomotion. An animal's heart muscle uses energy to pump blood.

An animal's metabolic rate is quantified by measuring its rate of heat production or O_2 consumption

To discuss metabolic rate, we need to begin by discussing the word "consumed" because it has a specific meaning in discussions of energy. The word can seem confusing sometimes because it is used in more than one way in the study of animals. For discussing energy, however, its meaning is easy to state. When we say energy is "consumed" by an animal, we mean the energy is converted to heat.

In what way is it accurate to say that energy converted to heat is "consumed"? Energy cannot be destroyed; thus "consumed" does not refer to destruction. However, whereas chemical-bond energy can do physiological work, heat cannot do physiological work. Thus the conversion of chemical-bond energy to heat has a huge consequence: the energy loses its ability to do work for the organism. This is why biologists say that energy converted to heat is "consumed" or used up.

An animal's *metabolic rate* is defined to be its rate of energy consumption. We mentioned this before, but now you can see more exactly what "energy consumption" means: an animal's metabolic rate is the rate at which it consumes chemical-bond energy and converts it to heat.

Why are metabolic rates important? The main reason is that when energy is consumed, it needs to be replaced. An animal with a high metabolic rate needs to find and eat a lot of food per unit of time to replace the chemical-bond energy it is consuming. Animals with lower metabolic rates have lower food needs. All living animals have measurable metabolic rates (that is, all are found to produce heat when measurements are made with appropriate instruments). An animal not producing heat is a dead animal! How can it be, then, that many living animals—such as frogs and earthworms—do not feel warm when we touch them? Such animals are producing heat. They do not warm up, however, because they are losing heat as fast as they produce it.

How can we quantify an animal's metabolic rate? Because consumed energy becomes heat, measuring an animal's rate of heat production would be one way to assess its metabolic rate. But measuring heat production directly is difficult. Thus biologists usually take advantage of a simple relationship between heat production and use of oxygen (O_2). Oxygen is used in the oxidation of organic matter during aerobic metabolism (see Key Concept 5.2):

$$\text{Organic compound} + O_2 \rightarrow CO_2 + H_2O + \text{heat}$$

There is an approximately one-to-one relationship between the rate at which O_2 is used and the rate at which heat is produced. To quantify an animal's metabolic rate, we can therefore measure its rate of O_2 use. Biologists nearly always use this method to quantify metabolic rates. The rate of *O_2 use* is usually called the rate of *O_2 consumption*. (Note that "consumption" here refers to the combination of O_2 with other atoms. Thus the word "consumption" is being used here in a different way than when we talk about energy consumption.)

Physical activity increases an animal's metabolic rate

Physical activity often exerts a strong effect on an animal's metabolic rate. For example, when people or other vertebrates engage in sustained exercise, their metabolic rate typically goes up by approximately a factor of 10 as they transition from rest to all-out exertion.

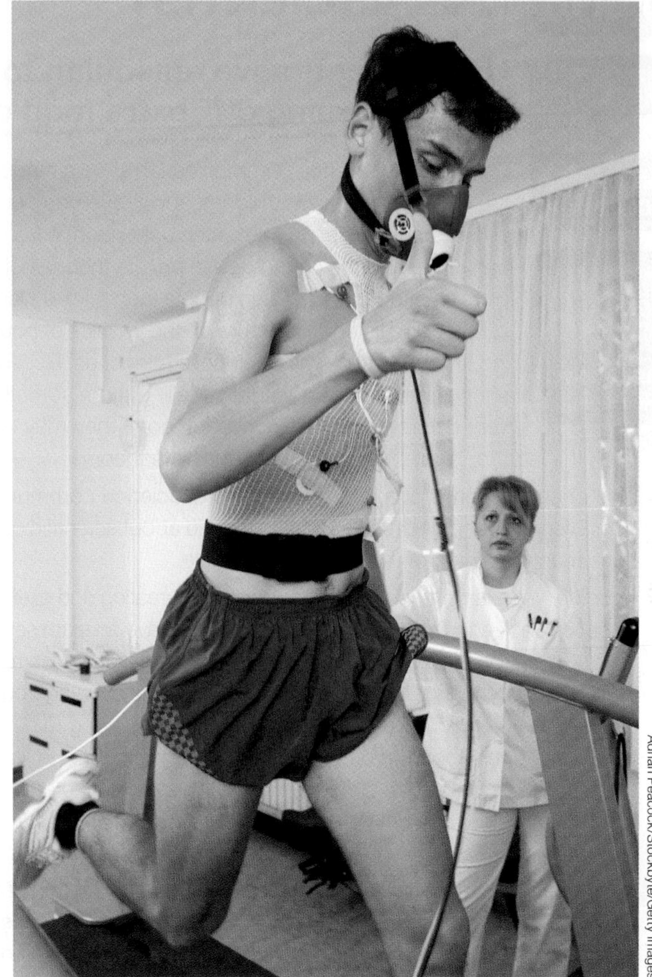

FIGURE 28.7 Measuring the Energy Cost of Exercise on a Treadmill This man is breathing from a device that measures his rate of O_2 consumption. His metabolic rate can be calculated by converting his rate of O_2 consumption to his rate of heat production. By varying the rate that the treadmill surface moves, investigators can vary the speed at which the man must run.

The effect of physical activity on metabolic rate is revealed in treadmill studies (**FIGURE 28.7**). Suppose, for example, that we carry out such a study on a college student. While running on the treadmill the student will breathe from a device that measures his or her rate of O_2 consumption, and the student's speed of running will be varied by varying the speed of the treadmill. In this way, we will be able to correlate variations in metabolic rate with running speed. When people run, their metabolic rate typically increases linearly with their speed (**FIGURE 28.8A**). Many types of animals—even lions and cockroaches—have been trained to run on treadmills to study the effect of physical activity on their metabolic rates. A linear relationship between metabolic rate and speed is typical for animals that run or walk.

The relationship between metabolic rate and speed is different for animals that fly or swim. Wind tunnels have been used to study the relationship in flying animals. A bird in a wind tunnel flies into a wind produced by a fan, and the bird's flight speed

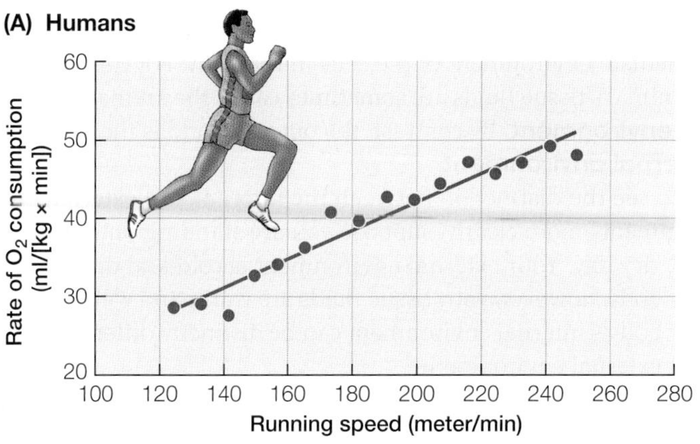

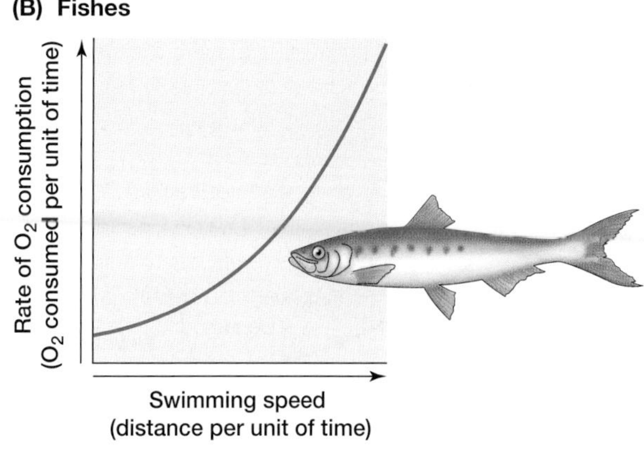

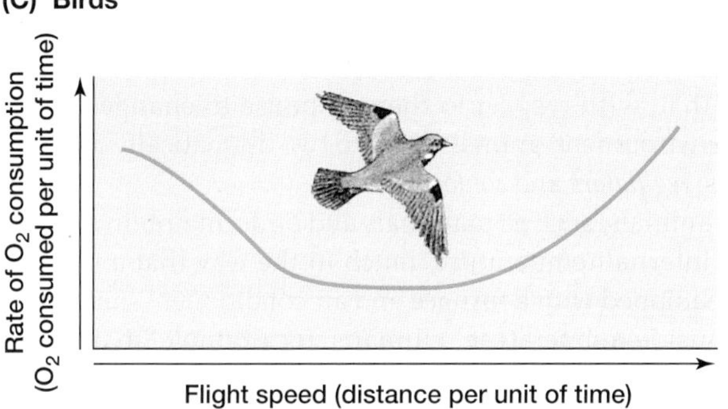

FIGURE 28.8 Cost of Exercise Rate of O_2 consumption as a function of forward speed for **(A)** people running, **(B)** fishes swimming, and **(C)** birds flying. (A after J. L. Mayhew. 1977. *Br J Sports Med* 11: 116–121; B after J. R. Brett. 1964. *J Fish Res Board Can* 21: 1183–1226; C after V. Tucker. 1968. *J Exp Biol* 48: 67–87.)

must match the wind speed. Devices that work on a similar principle have been designed to study swimming in fishes, which swim into a water current produced by a pump and must swim as fast as the current.

Experiments such as these tell us that when fishes swim, their metabolic rate typically increases exponentially as their speed increases, because the resistance posed by the water increases exponentially with speed (**FIGURE 28.8B**). Birds, however, often show a U-shaped relationship between metabolic rate and flight speed (**FIGURE 28.8C**). At low speeds, a bird must maintain a high metabolic rate just to stay airborne. As it speeds up, its metabolic rate decreases because flight is aided by the rapid flow of air over its wings. As the bird flies even faster, however, its metabolic rate rises again, because flying at a high speed requires more effort than flying at a medium speed. Airplanes exhibit a similar type of relationship between their rate of energy consumption and speed.

Believe it or not, birds flying and fishes swimming can cover distance much more cheaply—in terms of energy used per kilometer (or mile)—than mammals running. This may help explain why long-distance annual migrations have evolved principally in birds and fishes rather than in mammals. The metabolic cost to cover distance by swimming is much higher for humans than for fishes, in part because of the difference between swimming at the surface and swimming underwater. A person on a bicycle covers distance with roughly the same energy efficiency as a human-sized fish swimming. So to get a subjective sense of a fish's energy cost of covering distance, hop on a bicycle!

Among related animals, metabolic rate usually varies in a regular way with body size

Among terrestrial mammals, species vary widely in body size. Some species of adult mice weigh only about 20 grams (indeed, some shrews weigh only 2 g), whereas at the other end of the mammalian size spectrum, adult elephants can weigh more than 5,000 kilograms (5,000,000 g). Considering this wide range of mammalian body sizes, might the metabolic rates of mammals vary in some predictable way with respect to size?

To answer this question, we need to measure the metabolic rates of a wide range of species under standardized conditions. In studies of mammals and birds, a standardized measurement of metabolic rate is called the **basal metabolic rate** or **BMR**. By definition a BMR is measured when animals are resting in a comfortable thermal environment and have not recently eaten.

The simplest way to compare the BMRs of mammals of different body sizes is to express the rates per unit of body weight. To do this, we take the BMR for each animal and divide it by the animal's body weight to obtain BMR per gram, or BMR/g. You might expect that the BMR/g would be the same in all terrestrial mammals, from mice to elephants, but it is not even close to being the same.

In fact, BMR/g is almost 20 times higher in a mouse than in an elephant! **FIGURE 28.9**—often termed the "mouse-to-elephant curve"—shows how BMR/g varies over the entire range of body weights between mice and elephants. If you start with mice and look at bigger and bigger species of mammals, BMR/g gets lower and lower. Relationships of this sort—in which

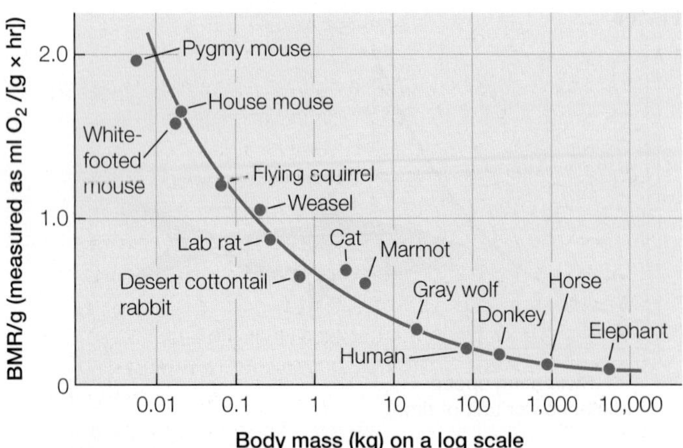

FIGURE 28.9 The Mouse-to-Elephant Curve in Basal Metabolic Rate Basal metabolic rate (measured as the rate of O_2 consumption) per gram of body weight in terrestrial mammals decreases dramatically as body size increases. The *x* axis is a logarithmic plot of body weight. (After K. Schmidt-Nielsen. 1997. *Animal Physiology: Adaptation and Environment*. Cambridge, UK: Cambridge University Press.)

animal characteristics are examined as functions of body size —are called **scaling relationships**.

The scaling relationship between BMR/g and body size in mammals has major implications. For example, it exerts a strong effect on how much food an individual must eat per day. Little mammals need much more food per gram of body weight than large mammals do, an important fact for both pet owners and zoo keepers to recognize. Food requirements in relation to body size also strongly influence the structure of ecological communities. A given environment must be able to supply enough food to support coexisting species within it (see Key Concept 40.3). Suppose there are 100 mice and 1 rabbit living in a field or forest. If each mouse weighs 20 g and the rabbit weighs 2,000 g, the mice collectively weigh as much as the rabbit—2,000 g. You might therefore expect that both the mice and the rabbit would eat similar amounts of food each day. However, the mice will in fact eat about three times as much food as the rabbit, because the BMR/g of a mouse is approximately triple the BMR/g of a medium-sized rabbit.

The same type of scaling relationship between metabolic rate per gram and body size (i.e., the same shape as seen in the mouse-to-elephant curve) is seen in almost all types of animals, including crabs, fishes, lizards, and birds. *Within each group of related species*, metabolic rate per gram tends to decrease in a major way as body size increases.

Among animals of a single body size, metabolic rate depends on the ways animals relate to their environments

To discuss this important topic, we need to start with a little background information. Suppose we ask, what is the immediate environment of the cells inside an animal's body? The cells are bathed with body fluids called the **tissue fluids** (or more formally, the **interstitial fluids**). Thus the tissue fluids are the immediate environment of the cells in the body. For this reason, an animal's tissue fluids are sometimes called the animal's **internal environment**. By contrast, the outside world is the animal's **external environment**.

To see the distinction between the internal and external environments more clearly, suppose you are standing outside on a cold, dry day. Your external environment is cold and dry. Inside your body, however, your tissue fluids are warm and wet. That is, your body's internal environment can be distinctly different from your external environment.

What happens inside an animal when its external environment changes? Let's consider the question in the context of temperature. What happens to the temperature of the *internal environment* (the temperature of the tissue fluids) when the temperature of an animal's *external environment* changes? Studies of this question have revealed that, with respect to their response to changes in the external environment, animals fall into two dramatically different categories: *regulators* and *conformers*.

Some animals, such as mammals and birds, maintain a nearly constant internal temperature, much in the way that a modern house—equipped with a furnace and air conditioner—maintains a steady inside temperature. Humans, for example, maintain a nearly constant internal temperature of about 37°C (98.6°F). In these types of animals, called **regulators**, the internal temperature does not change even if the temperature of the external environment changes. By definition, **regulation** occurs when the internal environment stays constant even as the external environment changes (**FIGURE 28.10A**). In **conformers** (such as fishes and frogs), however, if the temperature of the external environment changes, the internal temperature changes to match the new external temperature. By definition, **conformity** occurs when the internal environment varies to match the external environment (**FIGURE 28.10B**).

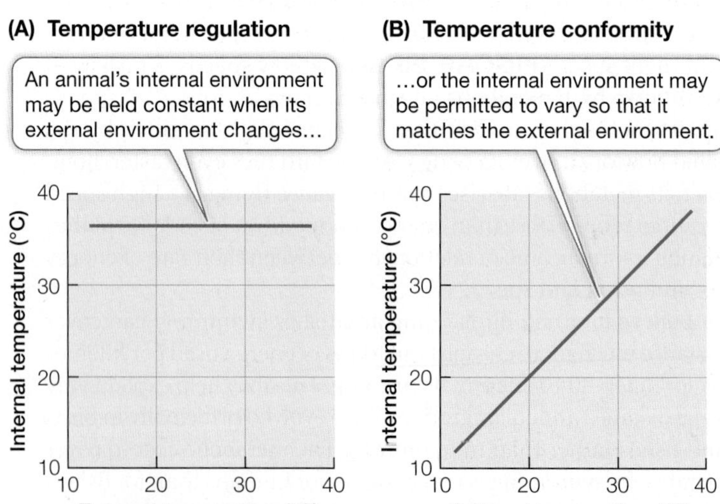

FIGURE 28.10 Regulation versus Conformity These graphs from the study of temperature illustrate the general principles of **(A)** regulation and **(B)** conformity.

With this background, we gain a better understanding of the Arctic fox in our opening photo. Regardless of how cold the fox's external environment becomes, the internal environment of a cell deep in the fox's body remains warm—and almost constant in temperature.

The concepts of regulation and conformity apply to all characteristics of an animal's internal and external environments, not just to temperature. Moreover, an animal that conforms in some characteristics may regulate in others. We've seen, for example, that fishes are conformers in regard to temperature. They are regulators, however, in terms of the concentrations of sodium (Na^+) and chloride (Cl^-) ions in their tissue fluids. When salmon migrate out of rivers into the ocean, for example, the saltiness of their tissue fluids remains almost constant. Thus their cells have a nearly constant internal salinity environment even as the water outside their body changes from fresh to the saltiness of the ocean. Whereas the fishes are conformers in temperature, they are regulators in ion concentrations.

In terms of energy cost, regulation is far more expensive than conformity. We can see this (as an analogy) if we compare a modern house to a simple log cabin that has no stove or furnace. The interior temperature of the modern house is regulated. The internal temperature of the cabin, however, simply conforms to the external temperature. Whereas the owner of the house pays steep utility bills to meet the energy costs of regulating the temperature in the house, the owner of the cabin has no fuel costs at all. That is, regulation is costly, compared with conformity. We also realize, however, that the cabin is not entirely superior. Whereas the temperature in the house is always comfortable, if we move into the cabin, we may boil in hot weather and freeze in cold weather. This contrast helps us understand the advantages and disadvantages of regulation and conformity in animals.

Regulation is energetically expensive but provides internal stability. Conformity is energetically cheap but does not provide internal stability. Regulators pay a high price in energy but get a payback: their body cells enjoy constancy in their surroundings. The cells of conformers, by contrast, must cope throughout life with considerable variation in their immediate surroundings. Where we speak here of energy costs, we are in fact also speaking of metabolic rates. Animal metabolic rates are greatly affected by regulation and conformity,

How large is the difference between the rates of energy use by regulators and conformers? As we will see in the next key concept, the average annual metabolic rates of mammals and birds living in their natural environments are often more than 10 times higher than the average annual metabolic rates of fishes and lizards in their natural environments (comparing animals of equal body size). The principal reason for this difference is that mammals and birds regulate their internal temperature, whereas fishes and lizards do not.

REVIEW & APPLY | 28.2

R An animal's metabolic rate is defined as its rate of energy consumption—the rate at which the animal uses chemical-bond energy, converting it to heat. Metabolic rate is quantified by measuring an animal's rate of O_2 consumption or (far less commonly) its rate of heat production. Physical activity increases an animal's metabolic rate; the increase is a linear function of running speed in mammals but an exponential function of swimming speed in fishes, and is often U-shaped in flying birds. Within sets of related animals, the resting metabolic rate depends on body size: small-bodied species typically have higher resting metabolic rates per gram of body weight than related large-bodied species do. Another profound determinant of metabolic rate is whether an animal exhibits regulation of body temperature or conformity of body temperature. Regulation is more expensive than conformity.

1. Why does an animal's need for food depend on its metabolic rate?
2. What is an animal's internal environment? How do the internal environments of regulators and conformers differ?
3. Explain the two ways that the word "consumed" is used in discussing energy and O_2.
4. Which is likely to require more food over a given time period, a group of 60 skates, each weighing 1 kilogram, or a pair of 30-kilogram sharks? Skates and sharks both belong to the same group of animals, the cartilaginous fishes.

Now the stage is set for us to discuss one of the most important concepts in all of biology: *homeostasis*. The responses of animals to temperature are important in their own right and also provide a useful context for developing a full understanding of homeostasis.

28.3 | Responses to Temperature Help Clarify Homeostasis

LEARNING OBJECTIVES

By the end of this key concept you should be able to:

28.3.1 **Define homeostasis and give an example of it.**

28.3.2 **Specify how homeotherms and poikilotherms (ectotherms) differ in their function and energy needs.**

28.3.3 **Describe how a negative-feedback control system works.**

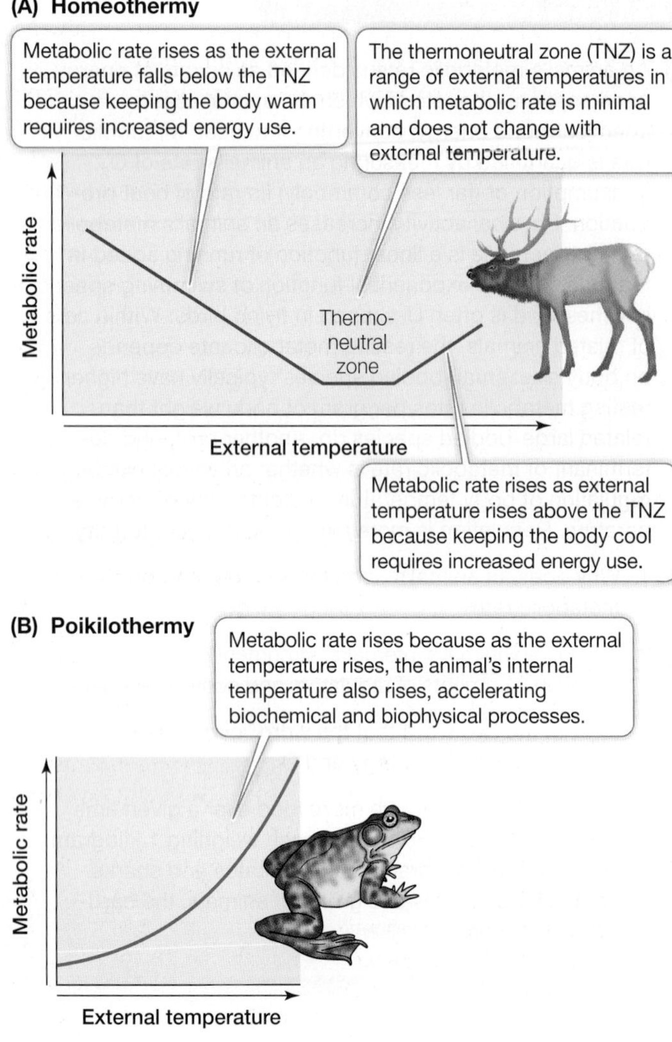

FIGURE 28.11 The Relationship between Metabolic Rate and External Temperature **(A)** The relationship in homeotherms such as mammals and birds. **(B)** The relationship in poikilotherms such as frogs, fishes, lizards, sea stars, and resting insects.

As the nineteenth century ended and the twentieth century began, physiologists (scientists who study body mechanisms that keep organisms alive and functioning) were focused on the functional biology of human beings and that of "model" nonhuman animals (such as dogs) used in biomedical research. These investigators gradually realized that when people (or dogs) are healthy, characteristics of their blood and tissue fluids remain remarkably consistent over time. For example, examinations of **blood plasma**, which is the blood solution in which the red blood cells and other blood cells are suspended, revealed that the concentration of glucose ("blood sugar") was nearly constant in the blood plasma and tissue fluids, regardless of when a person had last eaten a meal. The scientists likewise discovered that temperature, pH, and the concentrations of major ions are nearly constant in the blood plasma and tissue fluids of healthy people.

The famous physiologist Walter Cannon (1871–1945) highlighted the importance of the *stability of the internal environment*: the tendency of the blood plasma and tissue fluids to have stable characteristics. He coined the term **homeostasis** to refer not only to the stability of the internal environment but also the mechanisms that maintain the stability. To this day, homeostasis is an exceedingly important organizing concept in medicine, veterinary medicine, and many other fields of biology. The meaning of homeostasis has shifted a little over time; the term is sometimes used today to refer to any sort of stability or constancy in an animal.

Homeostasis is a parallel concept to regulation, and homeostasis has the same pros and cons as regulation: homeostasis ensures a stable environment for the cells in the body, but it is energetically expensive. Recognizing that homeostasis has both pros and cons, we are not surprised to find—when we consider the entire animal kingdom—that homeostasis is not universal. Humans, dogs, and many other animals exhibit high degrees of homeostasis. However, many types of animals exhibit less homeostasis, and there are even animals that exhibit little homeostasis.

Animals are classed as homeotherms or poikilotherms based on their thermal relationships with their external environments

Animals that maintain an approximately constant internal body temperature—such as mammals and birds—are said to exhibit **thermoregulation** and are called **homeotherms**. They are also said to exhibit **homeothermy**. Homeothermy, a topic that we will soon discuss, is a type of homeostasis. Sometimes homeotherms are called endotherms, although in this book we stick with the term homeotherms.

In most homeotherms, if a resting individual is exposed to a variety of external temperatures, its body temperature remains nearly constant, but its metabolic rate varies with external temperature as seen in **FIGURE 28.11A**. To describe the way that a homeotherm's metabolic rate varies, let's start by noting (see the figure) that over a range of relatively mild external temperatures, called the **thermoneutral zone** (**TNZ**), the animal's metabolic rate is constant. However, its metabolic rate rises outside the TNZ. *At external temperatures below the TNZ*, a homeotherm's metabolic

FIGURE 28.12 Some Poikilotherms Can Tolerate Extreme Tissue Temperatures This tolerance allows them to prosper in environments where most animals cannot live.

FIGURE 28.13 Poikilotherms Often Use Behavior to Regulate Body Temperature The body temperature of a lizard depends on the temperature of its external environment, but the lizard can regulate its body temperature by moving from place to place within its environment. This lizard has found a place bathed with sunlight within the deep shade of its forest environment. It is basking in the sunlight, a behavior that will gradually elevate its body temperature. When the lizard has warmed to its preferred body temperature, it will move into shade for a while to prevent its body temperature from rising too high.

Is the lizard likely to regulate its body temperature throughout the 24-hour day?

rate increases as the external temperature falls. This happens for the same reason that the furnace in a house burns fuel faster as the weather outdoors becomes colder. A homeotherm must progressively increase its rate of metabolic heat production as the external temperature becomes progressively colder, to offset its progressively increasing rate of heat loss. *At external temperatures above the TNZ*, a homeotherm's metabolic rate increases as the external temperature rises. This happens for the same reason that the air conditioner in a house works harder as the weather outdoors becomes hotter. As the animal's external temperature becomes progressively warmer, the animal must work progressively harder to keep its body temperature from rising. Its rising metabolic rate reflects this progressively harder work.

Most animals are not homeotherms. Most are **poikilotherms**, or **ectotherms**, defined as animals that permit their body temperature to match their external temperature. Frogs, sea stars, crayfishes, the great majority of fishes, and many other kinds of animals are examples. The words "poikilotherm" and "ectotherm" reflect two complementary characteristics of these animals: the animals have body temperatures that are *variable* and determined by the *external* temperature. The prefix *poikilo* means "variable," and *ecto* means "external." In this book we favor the term poikilotherm.

The relationship between metabolic rate and external temperature in a poikilotherm is different than in a homeotherm. Over a broad range of external temperatures, when the external temperature of a poikilotherm rises, the animal's metabolic rate also rises—in an approximately exponential manner (**FIGURE 28.11B**). This occurs because as the external temperature rises, the animal's internal temperature also rises. A poikilotherm's tissues are cold in a cold environment and warm in a warm environment. Low tissue temperatures slow both biochemical processes (such as enzyme-catalyzed reactions) and biophysical processes (such as diffusion), thereby slowing metabolism. High tissue temperatures do the opposite.

An important question about poikilotherms is how high their body temperature can rise—and how low it can fall. Poikilotherms vary widely in their body temperature limits (**FIGURE 28.12**).

Poikilotherms often exert some control over their tissue temperature by means of behavior, a process termed *behavioral thermoregulation*. Because a poikilotherm's body temperature is determined by the temperature of its environment, the animal can adjust its body temperature by positioning itself in a suitable environment. For example, if a poikilotherm's body temperature is too low, the animal can raise it by moving to a warmer place in its habitat (if a warmer place is available), such as a patch of sunlight (**FIGURE 28.13**). In some cases poikilotherms maintain a stable body temperature for many continuous hours by using behaviors to exploit warming and cooling opportunities in their environment.

As we have already suggested, homeothermy is more energetically expensive than poikilothermy. To explore this difference, **FIGURE 28.14** shows the metabolic-rate-versus-external-temperature relationship for a vertebrate homeotherm on the same plot as that of a vertebrate poikilotherm of the same body size. Notice that at each external temperature, the metabolic rate of the homeotherm is far higher than that of the poikilotherm.

Over the course of an entire year, the average metabolic rates of mammals and birds living in their natural environments are typically 15–30 times higher than the average metabolic rates of fishes and lizards in their natural environments (comparing animals of equal body size). This contrast dramatizes the advantages and disadvantages of homeothermy (i.e., thermal homeostasis). The cells inside the body of a homeotherm function at approximately one temperature regardless of whether the outside environment

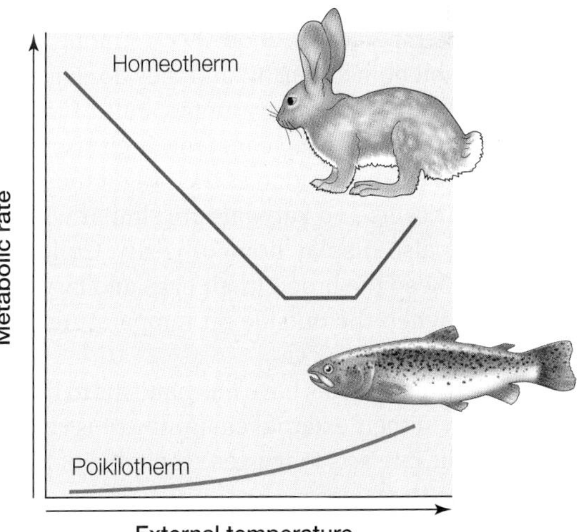

FIGURE 28.14 Metabolic Rate in a Vertebrate Homeotherm and a Vertebrate Poikilotherm of the Same Body Size Notice that at every temperature, the metabolic rate of the homeotherm is far higher than that of the poikilotherm.

ACTIVITY 28.1 Thermoregulation

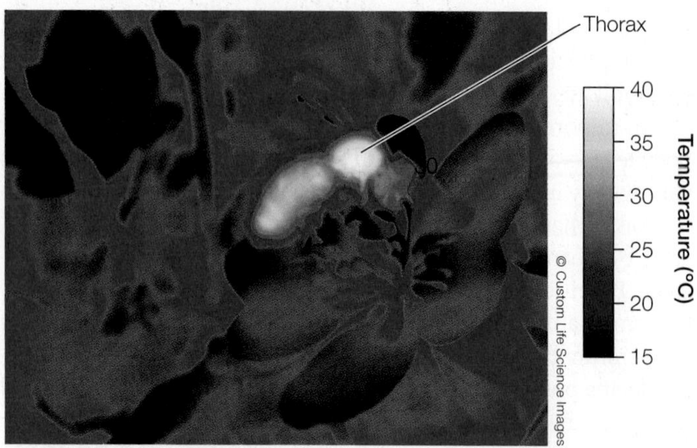

FIGURE 28.15 Thermoregulation in a Honey Bee This worker bee is foraging on a flower on a chilly day. During such behavior, the thorax is kept at a high, relatively stable temperature by heat produced by the flight muscles. The temperatures of the external surfaces of the bee, flower, and other plant parts are color-coded in this image, which was made by visualizing the surfaces at infrared wavelengths.

is cold or warm. The price of this regulation of body temperature is a 15- to 30-fold increase in an animal's food needs!

As explained in **Key Concept 39.3**, maintenance of homeostasis can affect life history trade-offs—the amount of resources an animal can allocate to reproduction, growth, and defense.

Homeotherms have evolved thermoregulatory mechanisms

Mammals, birds, and insects are the major groups of homeothermic animals. Placental mammals have core body temperatures of about 37°C whether they live in cold or warm climates. Birds have higher body temperatures, about 39°C. Among insects, only some species exhibit homeothermy, and they do so just at certain times, such as during flight. Many moths, butterflies, bees, and dragonflies maintain relatively constant, high tissue temperatures in their thorax—the location of their flight muscles—when they are flying. Crickets and katydids are similar when singing. Honey-bee hives also display homeothermy. Collectively, the workers in a hive keep the brood (their eggs and larvae) at about 32°C–36°C even when the outside air temperature is far lower than freezing or as hot as 50°C.

All homeothermic animals have mechanisms to increase their metabolic rate when their external environment is cold. Their increase in metabolic rate acts to increase their rate of heat production and thus keep them warm.

For increasing their metabolic rate, insects use their flight muscles. Even when insects are not flying, such as when they are foraging on flowers, the flight muscles in their thorax can increase their metabolic rate and provide the heat needed to stay warm (**FIGURE 28.15**); to do so, the muscles contract rapidly in a way that does not produce flight. Such contraction uses ATP, releasing ATP-bond energy as heat.

Mammals and birds **shiver** to increase heat production in a cold environment. During shivering, skeletal muscles contract in such a way that they produce only subtle, quivering motions. The energy for these contractions comes from the chemical bonds of ATP (see Key Concept 5.1). As molecules of ATP are used for contraction by the quivering muscles, energy from chemical bonds in ATP is converted to heat at a heightened rate.

Many small and medium-sized mammals, but not birds, are capable of **nonshivering thermogenesis**, in which thermogenesis (heat production) occurs without shivering (**INVESTIGATION FIGURE 28.16**). This process takes place in a specialized type of fatty tissue (related to muscle tissue) called **brown adipose tissue** or **BAT** (**FIGURE 28.17**). The mechanism used by BAT to make heat is uncoupling of oxidative phosphorylation. In this process, no ATP is made. Instead, chemical-bond energy of food molecules is released directly as heat in the mitochondria because the mitochondrial proton gradient is short-circuited.

LINK

As discussed in **Key Concept 5.2**, oxidative phosphorylation is a process by which a proton gradient is established across the inner mitochondrial membrane to drive the synthesis of ATP. In cases of uncoupling, proteins termed *uncoupling proteins* in the mitochondria disrupt the proton gradient. ATP cannot be made (or is made to a reduced extent), but the oxidation of NADH still occurs, releasing the energy available from NADH in the form of heat.

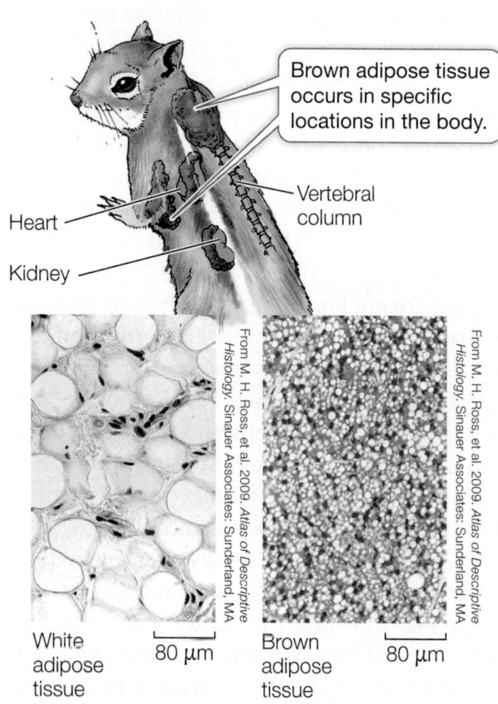

FIGURE 28.17 Brown Adipose Tissue In many mammals, specialized brown adipose tissue produces heat. Cells of white adipose tissue (left) contain large droplets of lipid and have a limited blood supply. Cells of brown adipose tissue (right) have many tiny lipid droplets, are packed with mitochondria, and are richly supplied with blood.

INVESTIGATION

FIGURE 28.16 Body Temperature and Rate of O_2 Consumption in Warm- and Cold-Acclimated Rats Injected with Curare Rats usually maintain an internal body temperature near 37°C and will shiver to maintain this body temperature when suddenly placed in a cold room. W. H. Cottle and L. D. Carlson noticed, however, that rats left in the cold for several weeks stopped shivering and yet were able to stay warm. Cottle and Carlson hypothesized that the rats had developed an alternative, *nonshivering* way to produce heat for thermoregulation, one not dependent on muscle contraction.

HYPOTHESIS

As rats acclimate to a cold environment, they develop a nonshivering mechanism to produce the heat they require to maintain a high body temperature.

METHOD

1. Prepare two groups of rats (8 individuals in each): (1) *warm-acclimated rats* that have been living continuously for many weeks in a warm room, and (2) *cold-acclimated rats* that have been living in a cold room for 60 days.
2. Anesthetize the rats and treat them with curare, a drug that prevents shivering.
3. Expose the rats to an environmental temperature of 28°C for about an hour, then lower the temperature to 5°C. Record the body temperature and rate of O_2 consumption of the rats for the duration of the experiment.

RESULTS

The graph shows continuous records of body temperature and the rate of O_2 consumption in the two groups during the experiment. The scale for temperature is on the right, whereas that for O_2 consumption is on the left. The environmental temperature was changed during the experiment, as shown. Symbols (solid circles) are averages. Vertical bars are standard deviations of the data for the eight individual rats in each group. The cold-acclimated rats were able to maintain their body temperature after the room was cooled, even though shivering was prevented. The warm-acclimated rats were not able to maintain their body temperature. The rate of O_2 consumption rose in cold-acclimated rats when the room was cooled, whereas the rate of O_2 consumption in the warm-acclimated rats fell.

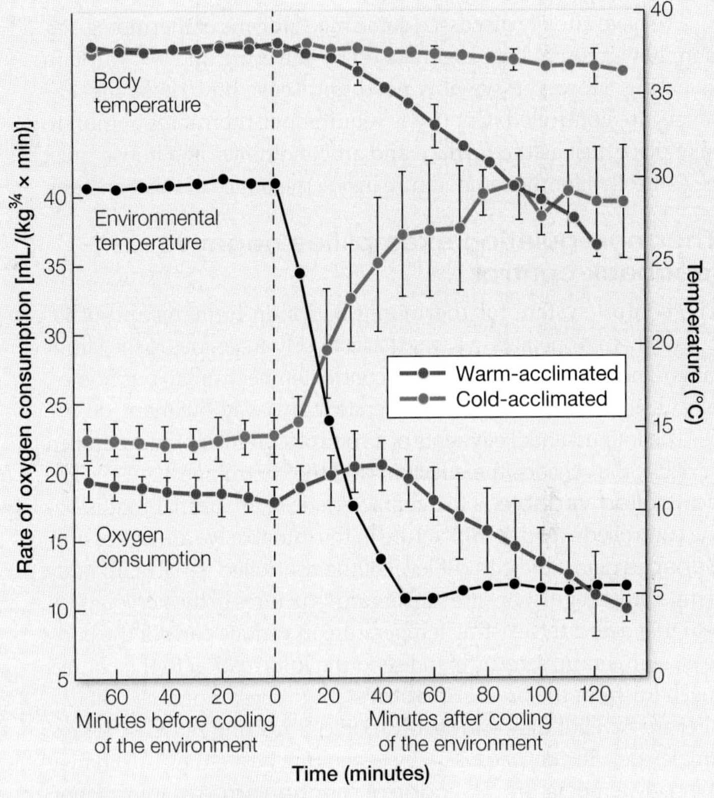

CONCLUSION

The results support the hypothesis that cold-acclimated rats are capable of nonshivering thermogenesis.

ANALYZE THE DATA

1. Why did the warm-acclimated rats respond as they did when the environmental temperature was lowered to 5°C?
2. Explain why the results suggest that the hypothesis is correct.
3. If the results indicate the presence of nonshivering thermogenesis, do they tell us *where* in the body this process takes place?

[a]W. H. Cottle and L. D. Carlson. 1956. *Proc Soc Exp Biol* Med 92: 845–849.

A similar **Analyze the Data** exercise may be assigned in **Achieve**.

Body insulation—for example, insulation provided by fat, fur, and feathers—is another important way homeotherms maintain high tissue temperatures in cold environments. Heat is also often retained by specialized blood flow patterns, which we will discuss in Key Concept 29.1. The Arctic fox featured at the start of this chapter has both these types of adaptation: it has highly effective fur insulation and employs specialized blood flow patterns in its limbs. These adaptations dramatically reduce heat loss, and the fox doesn't need to increase its metabolic rate (its rate of heat production) unless the air temperature is colder than –20°C, when it can turn to shivering to stay warm.

In hot environments, homeotherms often need to increase their rate of heat loss to keep from overheating. The most common way of doing this is to increase the water evaporation rate. When water turns from a liquid to a gas (water vapor), it absorbs a great deal of heat per gram (see Key Concept 2.5). This heat is removed from the animal when the water vapor travels away in the atmosphere. Many large mammals—such as humans, horses, and camels—sweat to increase their rate of evaporation. Dogs pant. Honey bees collect water and spread it on interior surfaces of their hive, then fan the moistened surfaces with their wings to speed evaporation.

Thermoregulation illustrates that homeostasis requires a control system

For understanding the fundamental principles of the *control* of homeostasis, thermoregulation is often considered the best example, because it illustrates the operation of a **control system** that can sense when conditions deviate from the norm and can initiate physiological mechanisms in response. Control systems are essential for homeostasis.

We have already discussed some mechanisms of thermoregulation: how internal heat is generated by shivering and lost through sweating. However, we now need to discuss how these mechanisms are controlled. Control is required for thermoregulation to take place, just as the furnace and air conditioner in a house must be controlled for the temperature inside the house to stay constant.

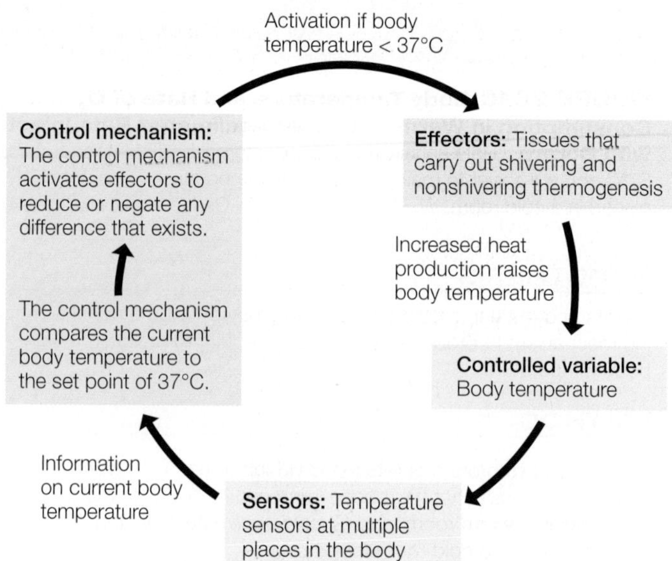

FIGURE 28.18 Negative Feedback: A Systems Diagram of the Temperature Control System of a Mammal The major components of this system are the controlled variable (body temperature), sensors, effectors, and the control mechanism. The system stabilizes body temperature by negative feedback.

Why is "negative feedback" called "negative"?

Thermoregulation exemplifies negative-feedback control

The control system for thermoregulation in humans and other homeotherms operates in a way that is closely analogous to the familiar engineered systems we use to control temperature in our homes. Not all animal control systems operate in this way, but many do.

The four essential elements of a control system are the controlled variable, the sensors, the effectors, and the control mechanism. The **controlled variable** is the characteristic of an animal that is being controlled—body temperature in the example we are discussing. **Sensors** detect the current level of the controlled variable. In body temperature regulation, the sensors are structures in the nervous system that detect the existing temperature in various parts of the body (e.g., skin, brain, scrotum) and send the information to the control mechanism in the brain. **Effectors** are tissues or organs that can alter the level of the controlled variable. Shivering muscles, for example, are effectors because—by generating heat—they can change the body temperature. The **control mechanism** uses information from the sensors to determine which effectors to activate—and how intensely—to modify the controlled variable. In vertebrate animals, the control mechanism for body temperature is composed of nerve cells located in and near the hypothalamus of the brain.

ANIMATION 28.1 The Hypothalamus

FIGURE 28.18 provides a systems diagram that can serve as a guide as we discuss the operation of the control system for body temperature in more detail. As diagrammed, the control mechanism receives information on the existing body temperature from the sensors, and it compares the existing body temperature with a *set-point temperature*—the temperature that is supposed to prevail, in this case 37°C (the usual body temperature of a placental mammal). The control mechanism then activates effectors *to reduce or remove the difference between the actual body temperature and the set-point temperature*. If the body temperature is lower than the set point, the control mechanism activates effectors that will raise the rate of heat production and thereby increase the body temperature until it reaches approximately 37°C. These effectors may include muscles that shiver or brown fat that undergoes nonshivering thermogenesis. If the body temperature is higher than the set point, the control mechanism activates responses (e.g., sweating or panting) that will raise the rate of heat loss, thereby reducing or removing the difference. In this case the effectors may include sweat glands that secrete sweat or breathing muscles that produce panting movements.

ACTIVITY 28.2 Temperature Regulation System Simulation

This control mechanism is an example of a **negative-feedback** system. The process is called "feedback" because after the control mechanism issues commands to effectors, it receives information from the sensors on how its commands have altered the body temperature (the controlled variable). Information is "fed back" to the control mechanism. The process is described as "negative" because the control mechanism activates effectors in ways that reduce or negate any difference that exists between the body temperature and the set point. Negative feedback is a stabilizing influence and is often involved in maintaining homeostasis. In this case, it keeps the body temperature near 37°C. Another important negative-feedback system is the system that regulates the concentration of glucose (blood sugar) in a mammal's blood plasma, as we'll discuss in Key Concept 28.4.

Negative feedback systems are just one type of system that operates within and among organisms and their environments. See **Key Concept 1.2**.

We will return to thermoregulation in Chapter 29, where we will look at the roles that thermoregulation plays in adaptation to the environment. Hibernation will be discussed there, for example.

You may recall from Chapter 1 that *positive feedback* sometimes plays a role in the control of biological systems. Negative and positive feedback operate in very different ways, meaning it is important to distinguish between them. **Positive feedback** occurs when deviations of a controlled variable from its typical level are increased or amplified by the action of the control mechanism. Runaway positive feedback can destabilize a system and thus is not a feature of homeostasis. However, destabilization can sometimes be advantageous, provided it is ultimately brought under control. A nerve impulse is an example of positive feedback that lasts for a very brief time and then is terminated (see Figure 31.15). We will discuss in Chapter 34 how the birth process of mammals is controlled in part by positive feedback.

REVIEW & APPLY | 28.3

R The concept of homeostasis refers to the stability of the internal environment and the mechanisms that maintain stability. Various characteristics of the internal environment are regulated in order to maintain homeostasis; internal temperature is an example. Homeotherms (which include mammals, birds, and some insects during flight) are capable of thermoregulation, which enables them to maintain a nearly constant internal temperature regardless of the temperature of their external environment. Most animals are poikilotherms (also called ectotherms), in which internal temperature varies with, and matches, the external temperature. Homeotherms have evolved a variety of mechanisms that allow them to thermoregulate. These include shivering, nonshivering thermogenesis, and sweating. They also have a control mechanism in the brain that controls the other mechanisms in a negative-feedback fashion to stabilize the internal temperature. Poikilotherms often exert some control over their internal temperature through behavior, by choosing favorable locations. In animals of similar body size, homeotherms have far higher energy needs than poikilotherms.

1. Describe how nonshivering thermogenesis contributes to homeostasis.
2. Being homeotherms, we humans often think that homeothermy is the best way to live. The great majority of animals, however, are poikilotherms. Why? Why don't homeotherms outcompete the poikilotherms and drive them to extinction? In answering, provide quantitative information.
3. Imagine a house with an air conditioner, furnace, and electric thermostat consisting of a thermometer and simple computer. For this control system, identify the controlled variable, sensor(s), effector(s), and control mechanism. Describe how you would draw a negative-feedback loop for cooling a house using these components.
4. Does negative feedback always have a diminishing effect on the controlled variable? That is, because it's "negative," does the feedback always make the level of the controlled variable very low?

Now that we've discussed the ways that animals use energy, let's return to the importance of body organization and consider some of the ways that structure and function are related in physiological systems. Then we'll consider the structure and function of the systems responsible for processing food and delivering food molecules to all the cells of the body.

28.4 Animal Bodies Exhibit Division of Labor, but Each Cell Must Make Its Own ATP

LEARNING OBJECTIVES

By the end of this key concept you should be able to:

28.4.1 Describe three fluid compartments in animals.

28.4.2 State what is meant by "division of labor," and specify how a circulatory system can be crucial for division of labor.

28.4.3 Identify a way in which form and function are closely related in the digestive system.

28.4.4 Explain how chemical-bond energy from foods is delivered to cells, and specify where this energy is used to make ATP.

Throughout the living world, structure and physiology—form and function—are intimately related. Often, form cannot be understood without understanding function, and vice versa. The goal of this key concept is to highlight some basic patterns in animal structure and discuss closely related physiological (functional) themes.

An important point about the structure of animals is that their bodies typically consist mostly of water. Adult mammals (including us), for example, are typically about 60 percent water. Water is distributed in various body *compartments,* which by definition are *categories of location* rather than specific organs or specific places. Typically, most of the fluid in an animal's body is located inside its cells as **intracellular fluid**, and the water in all the intracellular fluid throughout the body is said to be in the *intracellular compartment.* The rest of the fluid in the body is the **extracellular fluid**, said to be in the *extracellular compartment.* There are two subcategories of extracellular fluid, which we have already mentioned: the fluid portion of blood—termed the blood plasma—and the extracellular fluid found between cells in tissues throughout the body—termed the tissue fluid or, more formally, interstitial fluid.

Fluid compartments are separated from one another by epithelia and cell membranes

How are the various fluid compartments kept separate in the body? There are two different scales of space we need to consider to answer this question, and they correspond to separation by *epithelia* and by *cell membranes.*

EPITHELIA Consider the following three examples in which fluids of two types lie close together:

1. Blood plasma inside a blood capillary and tissue fluid surrounding the outside of the capillary
2. Watery fluid inside the lumen (central cavity) of a vertebrate's intestines and tissue fluid in the intestinal wall
3. Pond water on the outside of a submerged frog's skin and tissue fluid in the tissue of the frog's skin

In all three cases, the two fluids are separated by an **epithelium** (plural *epithelia*), which is a sheet of cells (epithelial cells) that covers a body surface or organ or lines a body cavity. The simplest and most common type of epithelium, called a **simple epithelium**, consists of just a single layer of cells that rests on a *nonliving* and highly permeable *basement membrane* (also called a basal lamina) that the cells help synthesize (**FIGURE 28.19**). Simple epithelia are exceedingly common. In our bodies, for example, a simple epithelium lines all our blood vessels and our intestines, as in the cases mentioned above. A simple epithelium also lines our kidney tubules, mammary glands, and sweat glands. In a frog, the skin consists entirely of living cells (unlike our skin, which has dead cells on the outside), and the outermost part of the skin is an epithelium. In an insect or crayfish, the outermost layer of living cells (inside the exoskeleton) is an epithelium.

An epithelium typically does much more than simply compartmentalize the body by separating fluids. An epithelium typically has numerous functional capabilities and plays major physiological roles. The cells of an epithelium, for example, commonly have mechanisms to pump ions between the fluids on either side of the epithelium. Some epithelia secrete substances such as hormones, mucus, digestive enzymes, milk, or sweat. Some absorb nutrients from the gut. Other epithelia serve sensory functions, including smell and taste.

CELL MEMBRANES Cell membranes function at a much smaller scale to separate fluid compartments. As described in Key Concept 4.1, the cytoplasm of each cell is separated from the extracellular fluid (tissue fluid) by the cell membrane. Compartmentalization of the organelles within a eukaryotic cell is also accomplished by membranes similar in structure to the cell membrane (see Key Concept 4.5). Like epithelia, cell membranes have numerous functional capabilities and play major physiological roles. Cell membranes pump ions between the intracellular fluid and extracellular fluid. They often also play key roles in producing and receiving physiological signals (signaling).

LINK Cell membranes are exquisitely constructed to control the transport of water and solutes into and out of a cell. Diffusion and osmosis tend to equalize the concentrations of solutes on either side of a membrane. Crucially, however, the concentrations of specific solutes within a cell are usually different from those in the extracellular fluid bathing the cell. To maintain these differences, a cell must sometimes use active transport to move solutes against their electrochemical gradients across the cell membrane. See **Key Concept 4.2**.

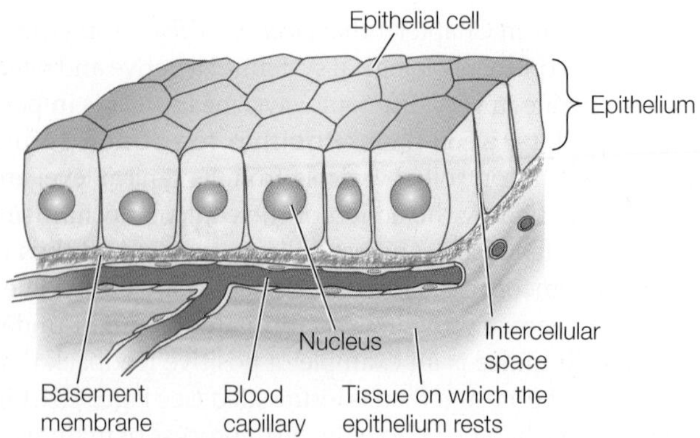

FIGURE 28.19 A Simple Epithelium A simple epithelium consists of a single layer of cells. The cells rest on a nonliving basement membrane that is highly permeable to gases, water, and ions. Blood capillaries do not penetrate into the epithelium itself but instead are under the basement membrane.

Animals exhibit a high degree of division of labor

In animals (and other multicellular organisms), individual cells need not perform for themselves all the functions required for life. Instead, cells can become specialized to perform just certain functions and depend on other cells in the body to perform others, a phenomenon termed *division of labor*.

The animal body is organized into a hierarchy of increasingly complex structures: cells, tissues, organs, and multi-organ systems. A **tissue** is an assemblage of cells of similar type. An **organ** consists of two or more types of tissue with a defined structural relationship to each other. A **multi-organ system** consists of multiple organs working together. Because animals' need for food has been a recurring theme in this chapter, let's look at the digestive system of mammals for an illustration of this hierarchy, emphasizing the *midgut* (also called the *small intestine*). The midgut is a particularly important part of a mammal's digestive system because it is where most digestion and absorption of food occur.

As **FIGURE 28.20** shows, the midgut—an organ—consists of several tissue types: muscle layers, nervous tissue, connective tissue, an epithelium lining the lumen (central cavity), and so forth. Broadly speaking, each of these tissue types consists principally of a single type of cell, such as smooth muscle cells in the muscle layers and nerve cells in the nervous tissue.

Structurally, one of the most important features of the midgut is the sheer size of the epithelium that lines the lumen. The midgut is long—about 6 meters (20 feet) long in adult humans—and this helps provide for a large epithelium. Even more important, the inner surface of the midgut bears huge numbers of tiny fingerlike extensions (called villi, each about 1 mm long) that project into the lumen. Each of these extensions is covered with epithelial cells, greatly enlarging the total surface area of the epithelium. Why is the large size of the epithelium important? The quick answer is the point we have already stressed: *form* and *function* are intimately related. The epithelial cells of the midgut lining carry out most of the digestion and absorption that occur in the organ. **Digestion** is the breakdown of food molecules by enzyme action into smaller chemical components, such as glucose. **Absorption** is the uptake into the blood of

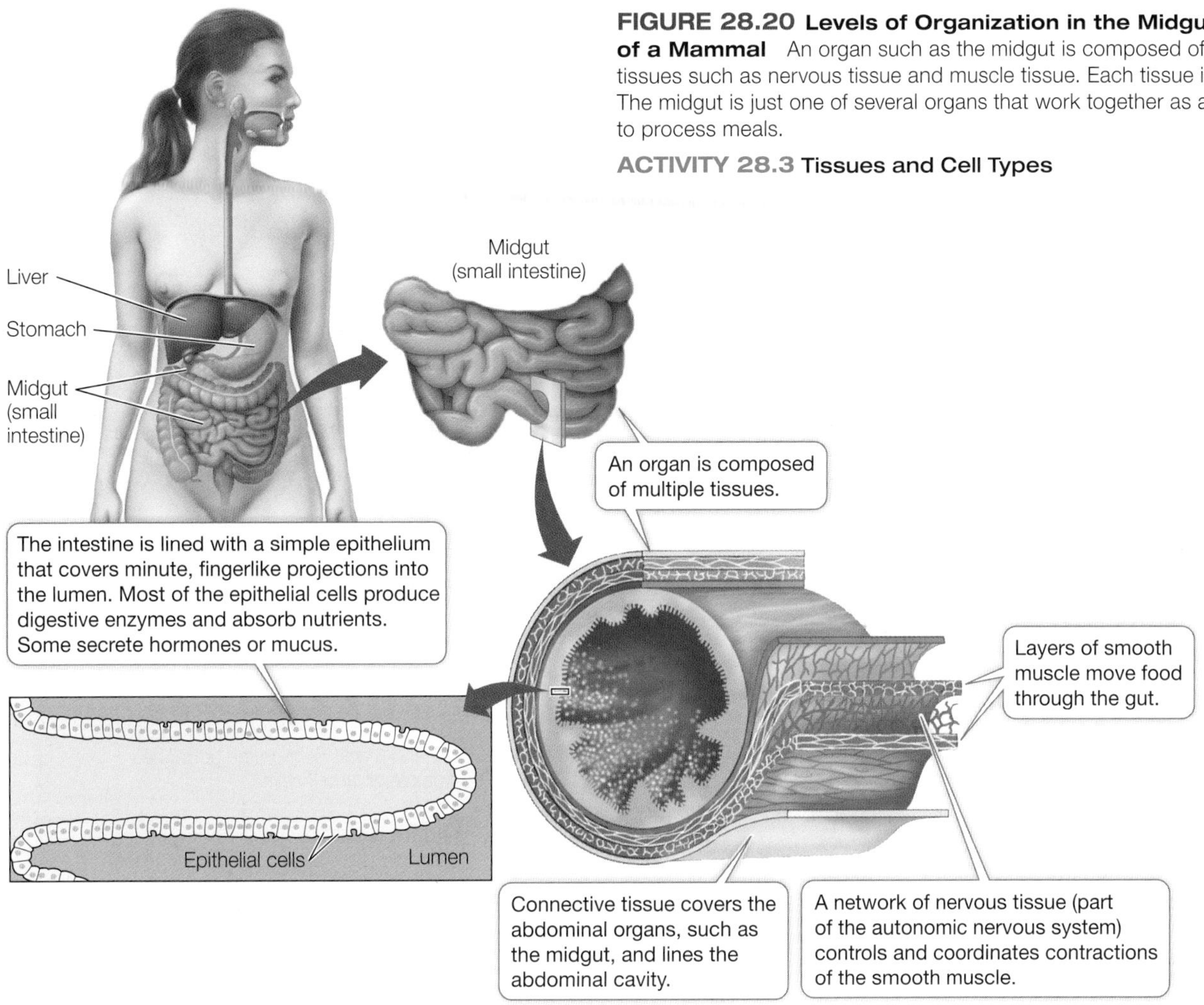

FIGURE 28.20 Levels of Organization in the Midgut (Small Intestine) of a Mammal An organ such as the midgut is composed of multiple types of tissues such as nervous tissue and muscle tissue. Each tissue is composed of cells. The midgut is just one of several organs that work together as a multi-organ system to process meals.

ACTIVITY 28.3 Tissues and Cell Types

the products of digestion. The cells of the midgut epithelium help with digestion because they bear molecules of digestive enzymes on their cell surfaces, and they help with absorption because they have transport mechanisms (such as membrane proteins) that enable them to take up the products of digestion and pass them to the blood (**FIGURE 28.21**), which carries them throughout the body. With a huge epithelium, the midgut is able to carry out its roles in digestion and absorption effectively.

During its normal function, the midgut receives partially digested food from the stomach. In addition to its inherent digestive and absorptive capabilities just discussed, the midgut also receives digestive enzymes from the pancreas, and from the liver it receives an emulsifying agent called bile, which is important in the digestion and absorption of lipids.

With these considerations in mind, you can see that the midgut does not work by itself. Instead it is part of a multi-organ system responsible for food processing (**FIGURE 28.22**). In addition to the organs we have mentioned, nervous and hormonal controls are also essential parts of this system, responsible for coordinating the functions of all the organs and tissues involved. For example, the nerve cells in the midgut (see Figure 28.20) control the midgut muscles to produce gut movements, such as waves of muscle contraction (peristalsis) that move the digesting food forward.

Here we have looked at just a single system and its organs, but we again underscore that the successful functioning of *all* body systems depends on a division of labor. Every organ in an animal's body is specialized to perform certain functions and is utterly

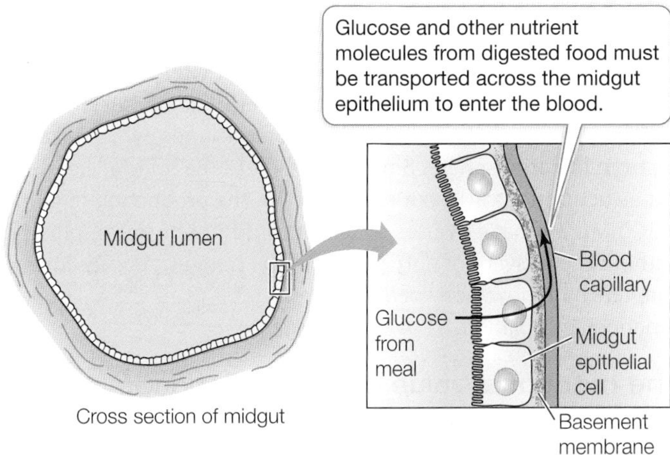

FIGURE 28.21 The Simple Epithelium Lining the Midgut Absorbs Nutrients Such as Glucose from Digested Food

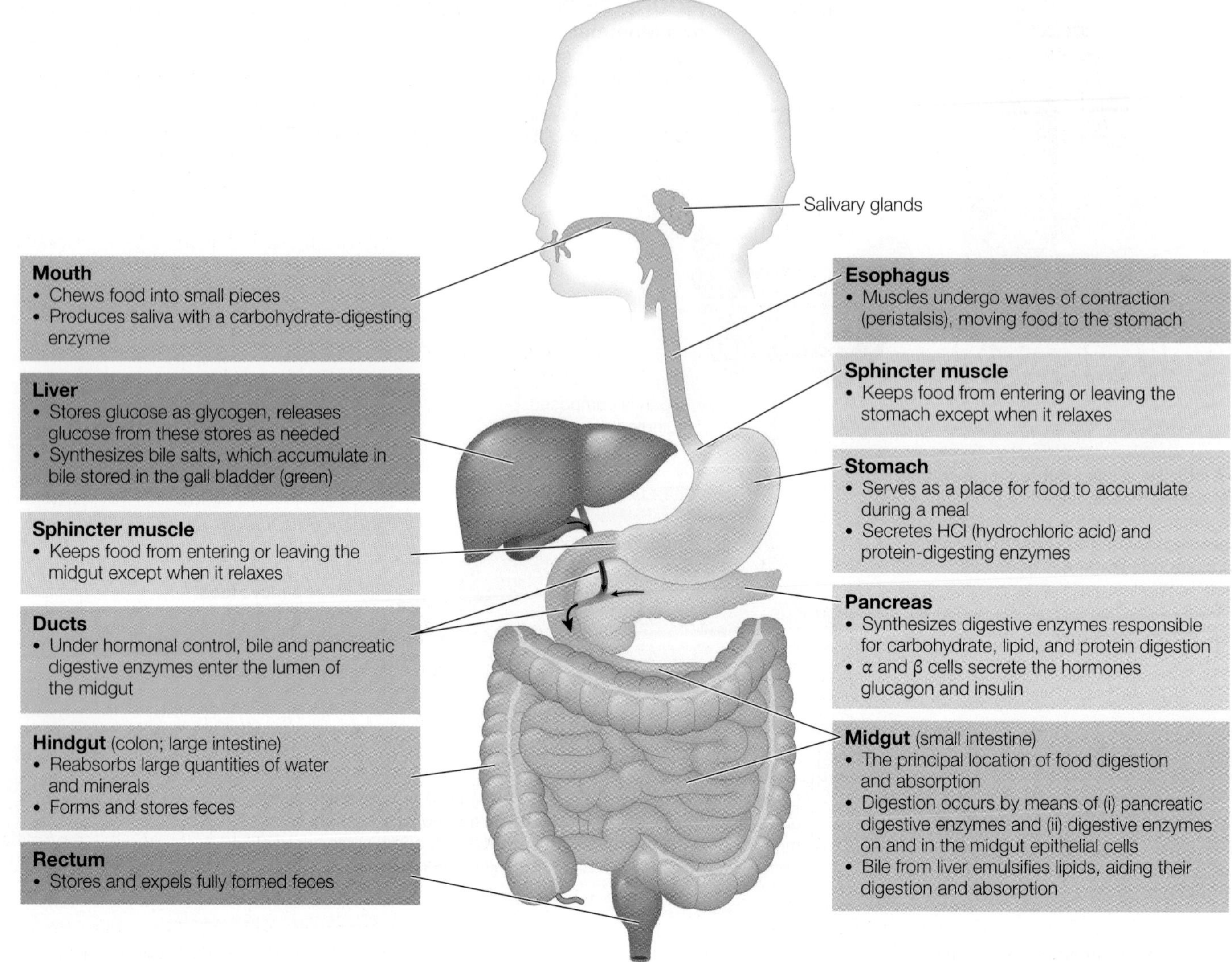

FIGURE 28.22 Overview of a Multi-Organ System: The Human Digestive System In addition to the organs shown, the brain and a variety of endocrine cells play key control roles in the digestive system. A sphincter muscle is a circular muscle that can contract tightly and steadily for long periods, thereby preventing exchange between two chambers.

ACTIVITY 28.4 The Human Digestive System

MEDIA CLIP 28.1 Following Food from Mouth to Gut

dependent on other organs to perform other vital functions. The midgut of a mammal, for example, requires O_2 for its survival and thus has a life-and-death dependency on the lungs, which take up O_2 from the atmosphere. Similarly, whereas the midgut clearly is not structured to carry out reproduction, the production of gametes in an animal's ovaries or testes could not be accomplished without the delivery of nutrients from the midgut. As subsequent chapters will show, none of the organ systems in an animal's body functions in isolation.

The nutritional value of a food depends on whether an animal can digest it

The molecules in the food an animal ingests are typically large molecules that must be broken into smaller pieces by digestion before absorption can occur. For example, protein molecules in the ingested food typically are far too large to be absorbed. They must be broken down into individual amino acids (or short chains of amino acids) before they can be absorbed into the blood from the lumen of the midgut. Digestive enzymes synthesized by the animal carry out this breakdown. Each of these enzymes is specific in the type of breakdown it can catalyze. Different animals have evolved different sets of digestive enzymes, and the set of enzymes synthesized by each species determines what foods the species can process.

With all these points in mind, you can see that the digestive system helps determine the food value of ingested foods. If an animal eats a large molecule that its digestive system cannot break down because it does not synthesize the specific enzymes required for breakdown, that molecule has no food value. The molecule simply passes through the length of the digestive

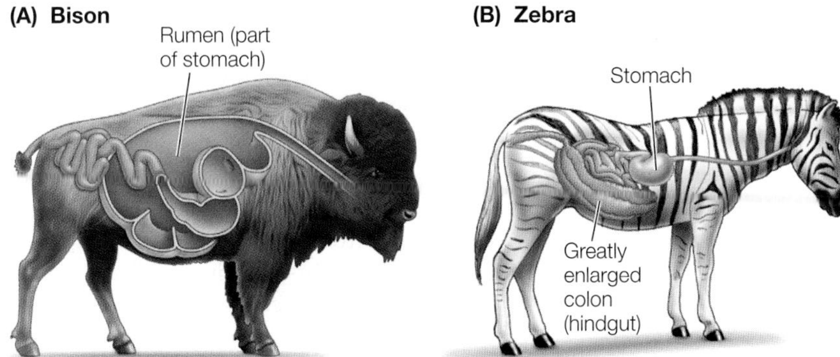

FIGURE 28.23 Mammals with Specialized Gut Compartments Housing Mixed Microbial Communities All animals, including humans, have functionally significant microbes, called the gut microbiome (see Chapter 38), living in the lumen of their digestive system. Some herbivorous animals have evolved specialized gut compartments to house mixed microbial communities that play dramatically important roles in the processing of food. **(A)** The ruminant mammals, such as the bison seen here, have a four-chambered stomach. Of these four chambers, the rumen is by far the largest. It does not secrete acid. Instead, the rumen provides a benign environment for massive populations of microbes that process the grass, twigs, and other plant material a ruminant mammal eats. **(B)** Horses and zebras have a grossly enlarged colon (hindgut). Their cecum (homologous to the human appendix) is also highly enlarged. The colon and cecum provide a benign environment for massive microbe populations.

system and comes out in the feces. By contrast, another species of animal might be able to eat the same large molecule and gain great food value from it if the animal can synthesize the appropriate digestive enzymes required to break down the large molecule into small, absorbable pieces.

In their evolution, animals have needed to evolve digestive enzymes suited to the foods they eat. Consider mammals that eat lots of insects, for example. The blood sugar of insects is an unusual sugar called trehalose that must be broken down to be absorbed. Most species of mammals do not eat great numbers of insects and have not evolved the digestive enzyme required for trehalose breakdown. Species of bats that feed principally on insects have, however, evolved the ability to synthesize this enzyme, and therefore they get food value from the blood sugar of the insects they eat.

One of the most abundant organic molecules on earth is cellulose, which is synthesized by land plants to provide structural support, so the plants can stand erect, exposing their leaves to sunlight. Terrestrial herbivores take in a lot of cellulose as they munch on grasses, bushes, and trees. Yet no vertebrate animal has evolved the necessary digestive enzymes to digest cellulose. Nor have most invertebrates. Thus this extremely abundant material has no food value if we simply consider the digestive enzymes of animals. In fact, when people eat cellulose, it passes through the digestive tract and comes out in the feces undigested.

The terrestrial mammals called *ruminants* have evolved a way to break down and absorb cellulose: they have evolved symbiotic relationships with microbes that can break down cellulose by anaerobic fermentation reactions (see Key Concept 5.2). Ruminants—such as cows, bison, and antelopes—have a greatly enlarged, four-chambered stomach. The largest of their stomach chambers, called the rumen, houses a thriving community of bacteria, protists, and yeasts (**FIGURE 28.23A**). This mixed microbial community ferments cellulose, producing small organic molecules (e.g., acetic acid) that enter the blood and are delivered throughout the body to cells, where the products of microbial fermentation can be used to make ATP. Horses and zebras have a highly enlarged colon (hindgut) where mixed microbial communities live (**FIGURE 28.23B**). Once again we see that the digestive system helps determine the food value of foods.

Division of labor requires a rapid transport system: the circulatory system

As division of labor evolved in animals, this created a need for rapid transport of materials from one organ to another. As we've said, for example, the midgut must get its O_2 from the lungs. Thus a transport system is required to carry O_2 from lungs to midgut. The molecules that the midgut absorbs from food also need to be transported to other organs where the molecules are used.

These considerations help us understand the evolution of the circulatory system (**FIGURE 28.24**), in which blood flows rapidly from one region of the body to another. A rapid transport system is required for division of labor to be successful, and a circulatory (cardiovascular) system has evolved in most types of animals to meet this need.

In many cases, the need for O_2 transport is the most pressing of all transport needs in an animal. For this reason, the sophistication of the circulatory system is usually correlated with metabolic rate, a correlation we'll discuss in more detail in Chapter 30. Animals with high metabolic rates require high rates of O_2 transport and have evolved circulatory systems that meet this need.

Chemical-bond energy is transported by the blood in the form of glucose, other small carbohydrates, and fatty acids

One of the chief functions of eating is to acquire energy: each cell in an animal's body must receive chemical-bond energy from the animal's food. What are the processes by which that energy is delivered to the cells? Although these processes are intricate, some of their most important features are easy to understand.

As a meal is processed, sugar molecules and fatty acids are produced by digestion and absorbed into the blood These absorbed molecules serve as the chief vehicles by which chemical-bond energy from food is transported by blood flow throughout the body. Each cell in an animal's body can transport sugar molecules and fatty acids from the blood into its cytoplasm and use them to make ATP. In insects, trehalose is the chief transport sugar; in mammals it is glucose. Here our principal focus is on mammals and glucose transport.

Mammals alternate between periods of eating and not eating. For example, in many human cultures, people eat three times a day but do not eat at other times. Accordingly, there are times each day when large quantities of glucose enter the blood from the midgut, but there are other times when the midgut delivers

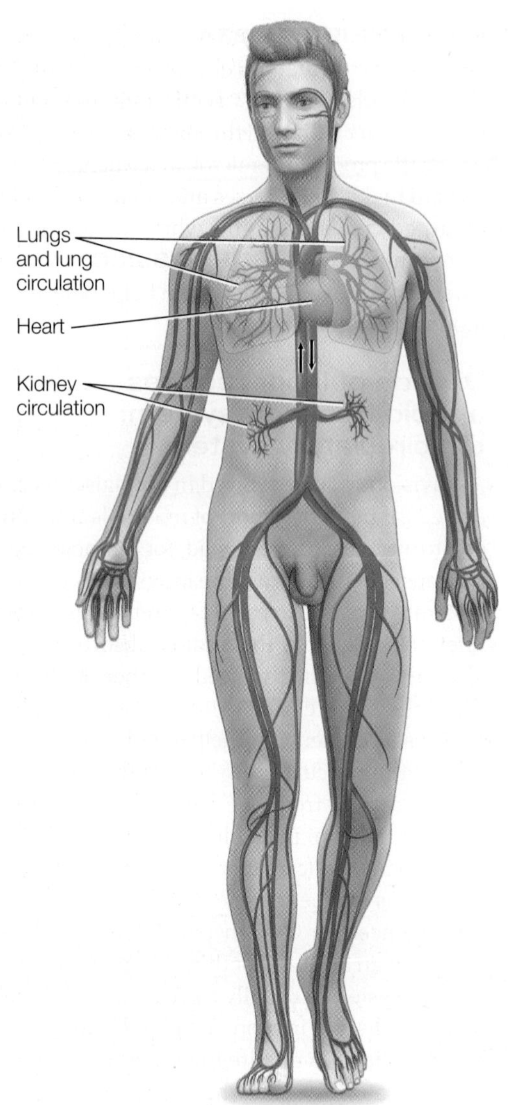

FIGURE 28.24 The Body's Transport System Driven by the pumping of the heart, blood moves rapidly throughout the body in the circulatory system, carrying materials from organ to organ. Blood leaves the heart in arteries (red) and returns to the heart in veins (blue).

little or no glucose to the blood. The blood glucose concentration, however, does not rise to a very high level after each meal and fall to a very low level between meals. Instead, mammals have evolved mechanisms that keep the blood glucose concentration nearly constant. That is, they have evolved homeostasis of blood glucose concentration.

Maintaining a relatively constant blood glucose concentration is important for several reasons, one being that abnormally high concentrations of glucose can damage the kidneys and the retinas of the eyes, and can lead to abnormal brain function, even coma. Homeostasis of blood glucose concentration ensures that when an animal has not eaten for some time, blood glucose concentration does not fall to zero but remains at a level that sustains cells.

Glucose homeostasis is maintained principally by two hormones secreted by the pancreas. Like most organs, the pancreas has more than one function. Some of its cells make digestive enzymes that enter the lumen of the midgut through small ducts. Other cells in the pancreas make hormones that are secreted into the blood.

There are two major types of hormone-secreting pancreatic cells. The **α cells** (alpha cells) secrete the hormone **glucagon**. The **β cells** (beta cells) secrete the hormone **insulin**. These hormones and the cells that secrete them constitute a control system for blood glucose. This control system provides our second example of negative-feedback control.

Let's now look in some detail at this negative-feedback control system (**FIGURE 28.25**). As we describe the system, keep in mind that cells throughout the body continuously remove glucose from the blood, although the rate at which they do so varies. During the digestion and absorption of a meal, glucose enters the blood rapidly, and the blood glucose concentration tends to increase. This increase is sensed by the β cells, and they respond by increasing their secretion of insulin. Insulin, in turn, acts to drive the blood glucose concentration down—a negative-feedback effect that tends to stabilize the blood glucose concentration. Insulin does this by increasing the rate at which many types of body cells take up glucose from the blood. It also increases the rate at which liver cells store glucose in the form of glycogen, a polymerized form of glucose (the liver is a major glucose storage location). Moreover, insulin increases the rate at which fat cells take up glucose and convert it to lipid, which they store.

After a meal is processed, blood glucose concentration tends to fall because glucose is no longer entering the blood from the meal. The control system again acts, this time to ensure that the blood glucose concentration does not fall too low. As the blood glucose concentration starts to fall, the β cells reduce insulin secretion, a response that lowers the rates at which cells throughout the body remove glucose from the blood, including the rates at which liver and fat cells store glucose. At the same time, the α cells secrete glucagon, the most important effect of which is to stimulate glucose addition to the blood by the liver. Glucagon stimulates the liver to convert stored glycogen to glucose and to synthesize glucose from other types of molecules such as amino acids. The addition of glucose to the blood by the liver—stimulated by the decline in blood glucose concentration between meals—prevents the glucose decline from becoming excessive.

Diabetes is perhaps the most famous of all hormonal diseases. In the simplest type of diabetes to understand, the β cells are unable to secrete insulin at rates required for health. Because insulin is the only hormone that is capable of lowering blood glucose concentration, other hormones cannot make up for insulin deficiency. With inadequate insulin, therefore, the blood glucose concentration tends to be too high, and the excessive glucose can damage many tissues, as we have already mentioned. In the 1920s, insulin was discovered, and physicians started to inject synthetic insulin as a way of treating people with insulin deficiency. Since then great progress has been made in treating diabetes.

Each cell must make its own ATP

One of the most important principles in the study of animals is this: each cell must make its own ATP. ATP serves as the immediate energy source for most cellular functions, and is thus one of the most vitally important molecules in the life of an animal.

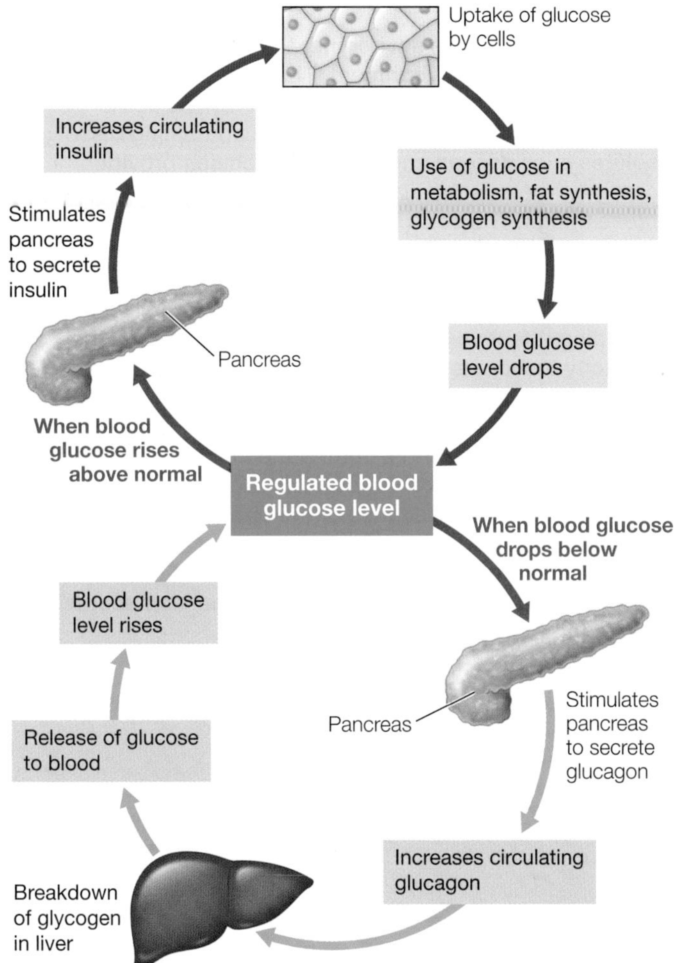

FIGURE 28.25 Regulation of Blood Glucose Levels The hormone insulin acts to promote glucose utilization and storage, thereby lowering blood glucose (see upper circle). Glucagon (see lower circle) stimulates the liver to break down glycogen and release glucose into the blood, which raises blood glucose.

ANIMATION 28.2 Insulin and Glucose Regulation

However, ATP is not transported from cell to cell. The rate of ATP supply in each cell depends on that particular cell's ability to synthesize ATP.

As we have seen, energy is transported in an animal's body—from organ to organ or from cell to cell—in the form of food compounds such as glucose and fatty acids. Each cell in the body takes up glucose and fatty acids from the blood and then breaks these compounds down internally to make ATP.

Most animal cells make most of their ATP by aerobic processes that require O_2. Aerobic ATP production takes place mostly in the mitochondria of a cell, by the processes of electron transport and oxidative phosphorylation (see Chapter 5). During aerobic ATP production, all types of food molecules—sugars, lipids, and proteins—can be used as fuels, that is, as sources of chemical-bond energy for ATP synthesis.

Some cells also have processes for making ATP without O_2. These mechanisms—termed *anaerobic*, meaning without oxygen—are not universally present. Brain cells in nearly all vertebrates, for example, have little ability to make ATP anaerobically. This explains why the vertebrate brain is quickly damaged by the absence of O_2 (as, for example, during a heart attack or stroke). However, many types of cells—including muscle, kidney, and gut cells—have well-developed mechanisms for making ATP anaerobically.

REVIEW & APPLY | 28.4

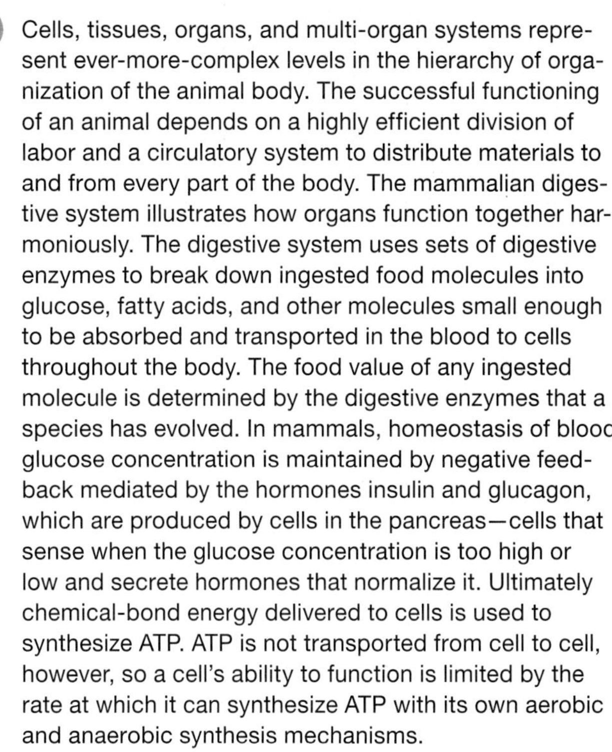

R Cells, tissues, organs, and multi-organ systems represent ever-more-complex levels in the hierarchy of organization of the animal body. The successful functioning of an animal depends on a highly efficient division of labor and a circulatory system to distribute materials to and from every part of the body. The mammalian digestive system illustrates how organs function together harmoniously. The digestive system uses sets of digestive enzymes to break down ingested food molecules into glucose, fatty acids, and other molecules small enough to be absorbed and transported in the blood to cells throughout the body. The food value of any ingested molecule is determined by the digestive enzymes that a species has evolved. In mammals, homeostasis of blood glucose concentration is maintained by negative feedback mediated by the hormones insulin and glucagon, which are produced by cells in the pancreas—cells that sense when the glucose concentration is too high or low and secrete hormones that normalize it. Ultimately chemical-bond energy delivered to cells is used to synthesize ATP. ATP is not transported from cell to cell, however, so a cell's ability to function is limited by the rate at which it can synthesize ATP with its own aerobic and anaerobic synthesis mechanisms.

A

1. What is a simple epithelium, and where might you find one in a person?
2. Could a single cell extracted from a multicellular organism survive on its own, similar to the way the lone cell of a unicellular organism does? Explain.
3. In what ways do the α and β cells of the mammalian pancreas exert negative-feedback control over the blood glucose concentration?
4. In mammals that have evolved a symbiotic relationship with a mixed microbial community in their digestive tract, the microbes in the community multiply and tend to increase in numbers. Bacteria multiply, for example, producing increased bacterial biomass. The question then arises of whether the mammal can digest this bacterial biomass to obtain food value from it. How would you expect horses and cows to differ in the answer? [*Hint*: Note that the location of the mixed microbial community is different in horses and cows.]

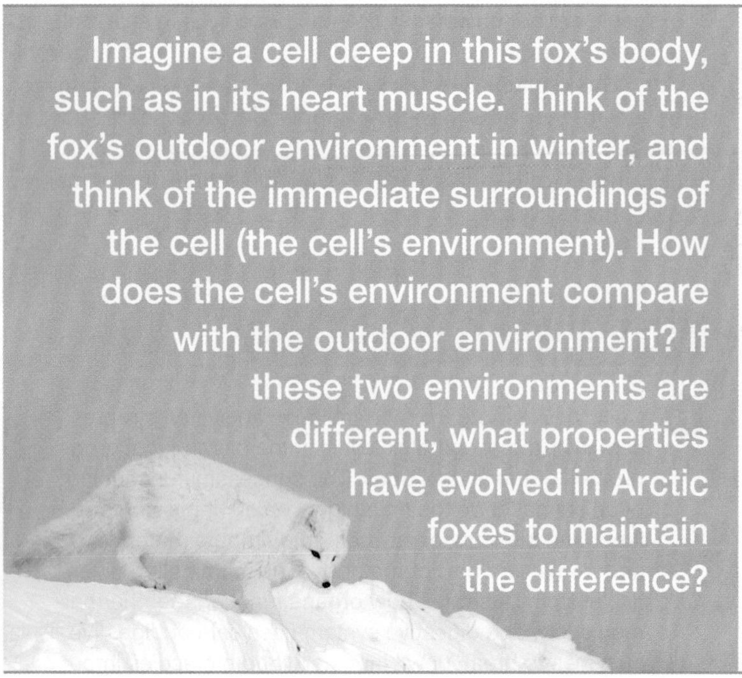

Imagine a cell deep in this fox's body, such as in its heart muscle. Think of the fox's outdoor environment in winter, and think of the immediate surroundings of the cell (the cell's environment). How does the cell's environment compare with the outdoor environment? If these two environments are different, what properties have evolved in Arctic foxes to maintain the difference?

Answer: The cell's environment is very different from the fox's outdoor environment. That is, the internal and external environments are very different. The cell is surrounded by a warm liquid solution containing dissolved solutes such as ions and sugars (which can provide energy), whereas the atmosphere does not contain any of these things. The cell's temperature is relatively constant: approximately 37°C—unlike the outdoor temperature, which is variable and can be very low. The cell's temperature is maintained by an elaborate thermoregulatory system, including effective insulation (provided by fur) and the capacity to increase the rate of internal heat production by shivering. The thermoregulatory system also includes a control system consisting of sensors, effectors (e.g., shivering muscles), and a control mechanism in the brain that stabilizes the fox's internal temperature through negative feedback. Because of the operation of the control system, the fox increases its metabolic rate as the external environment becomes colder at air temperatures below thermoneutrality, thereby ensuring thermal homeostasis inside the fox.

28 VISUAL SUMMARY

28.1 Animals Eat to Obtain Chemical Building Blocks and Energy

FIGURE 28.4

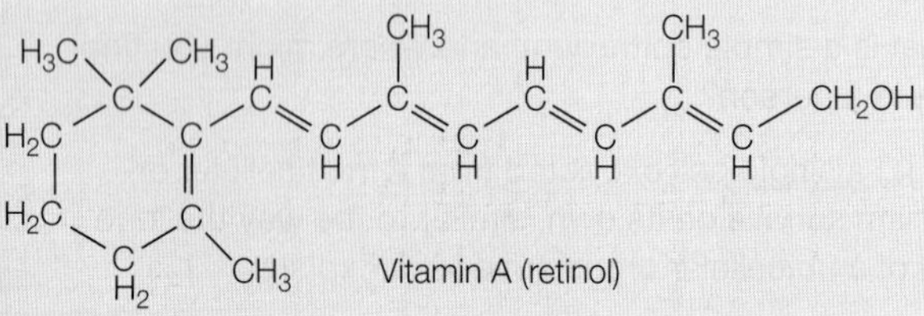

- Because of turnover of atoms, the defining characteristic of an animal is the organization of its body, not the particular atoms of which its body is composed.
- **ESSENTIAL NUTRIENTS**, such as **VITAMINS** and **ESSENTIAL AMINO ACIDS**, are nutrients an animal cannot synthesize and therefore must obtain in its diet (**FIGURE 28.4**).
- As an animal uses chemical-bond energy, it converts the energy to heat, explaining why animals produce heat and why they must obtain new chemical-bond energy by eating throughout their lives.

QUESTIONS

1. What is the advantage of eating corn and beans at the same time?
2. Why is the second law of thermodynamics important for understanding animals?
3. Why do migrating animals typically store lipids rather than carbohydrates as internal sources of energy for their migration?

28.2 An Animal's Energy Needs Can Be Quantified

FIGURE 28.8

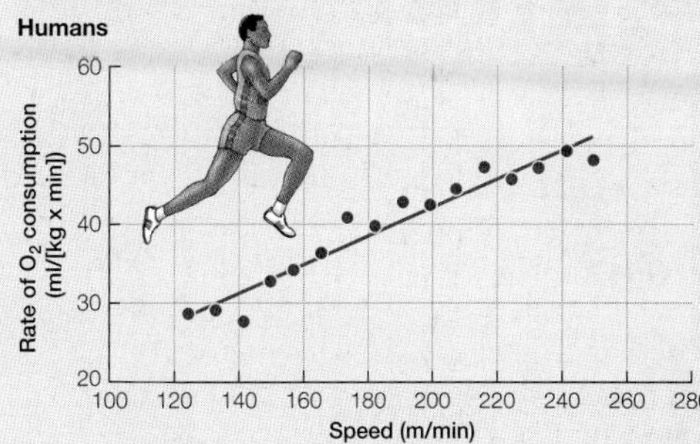

- When animals exercise, they increase their metabolic rate (**FIGURE 28.8**). Among vertebrates, it is typical for the metabolic rate during all-out aerobic exercise to be about 10 times higher than the metabolic rate at rest.
- Among sets of related species of animals, large-bodied species typically have lower weight-specific metabolic rates than small-bodied species when all are measured under similar conditions (such as when **BASAL METABOLIC RATES** of all are compared).
- **REGULATION** is more energetically expensive than **CONFORMITY**.

QUESTIONS

1. Regarding the basal metabolic rate per gram of body weight (BMR/g), does a mouse have a BMR/g that is about 1/10th that of an elephant, 3 times that of an elephant, or 20 times that of an elephant? Why are scaling relationships important in regard to how energy is used in an ecological community?
2. What are the relative advantages and disadvantages of regulation versus conformity?
3. Biologists commonly measure the rate of O_2 consumption by animals. Why is this a common measurement?

28.3 Responses to Temperature Help Clarify Homeostasis

Go to ACTIVITIES 28.1 and 28.2 and ANIMATION 28.1

FIGURE 28.11

Homeothermy

Metabolic rate

Thermo-neutral zone

External temperature

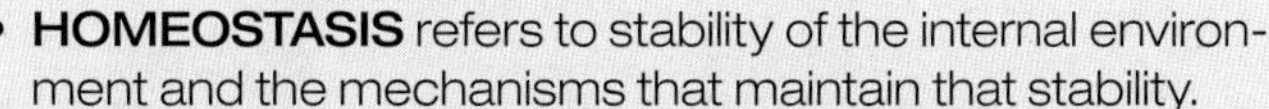

- **HOMEOSTASIS** refers to stability of the internal environment and the mechanisms that maintain that stability.

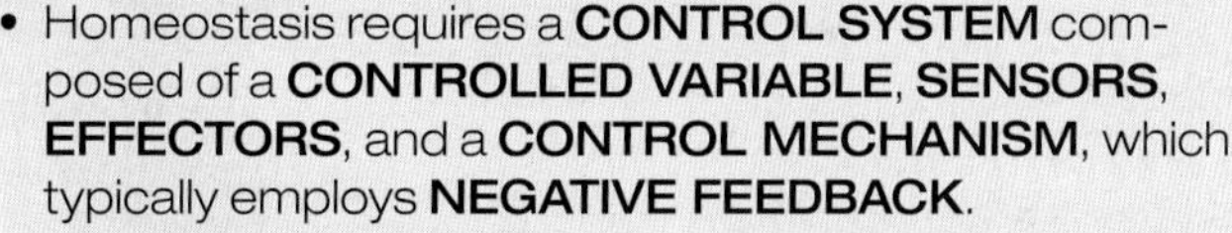

- Homeostasis requires a **CONTROL SYSTEM** composed of a **CONTROLLED VARIABLE**, **SENSORS**, **EFFECTORS**, and a **CONTROL MECHANISM**, which typically employs **NEGATIVE FEEDBACK**.
- **HOMEOTHERMS** maintain a steady internal temperature. Mammals do this by employing mechanisms such as **SHIVERING**, **NONSHIVERING THERMOGENESIS**, sweating, and panting (**FIGURE 28.11**).
- **POIKILOTHERMS** (**ECTOTHERMS**) permit their internal temperature to match the external temperature but may nonetheless use behavior to control their internal temperature.

1. In a control system, what is an effector? Give an example.
2. What is the mechanism by which evaporation of water cools an animal?
3. In what way do "ectotherm" and "poikilotherm" represent two complementary terms that emphasize different aspects of the same type of temperature relationship?

28.4 Animal Bodies Exhibit Division of Labor, but Each Cell Must Make Its Own ATP

Go to ACTIVITIES 28.3 and 28.4, and ANIMATION 28.2

FIGURE 28.19

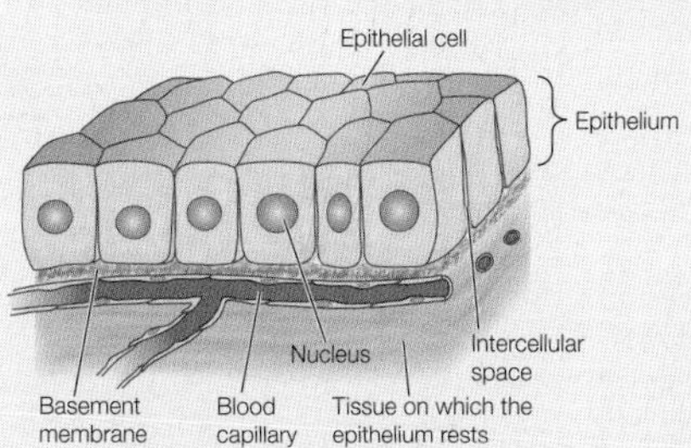

- The fluid compartments of the body are separated by **EPITHELIA** and cell membranes, both of which carry out functions such as solute transport (**FIGURE 28.19**).
- The circulatory system plays a key role in division of labor because it can carry materials rapidly between **TISSUES** and **ORGANS** that depend on each other.
- Chemical-bond energy is transported from organ to organ in the form of nutrient molecules such as glucose or fatty acids, not by transport of ATP. Each cell must make its own ATP.

1. What are the three principal fluid compartments of an animal's body?
2. From a *structural* viewpoint, how is the area of the midgut epithelium enhanced? Why is the large area of the epithelium of *functional* importance?
3. How do an animal's digestive enzymes determine the food value of molecules the animal eats?

Go to **Achieve** for the e-book, animations, activities, and additional resources and assignments.

29 Animals in Their Environments

These salmon have been living in the ocean for the last several years. They recently entered a freshwater river and are swimming upstream to calmer, shallower water where they will spawn. Their offspring will hatch and undergo their early growth in fresh water, then migrate to the sea.

Water (H_2O) and dissolved ions (such as Na^+ and Cl^-) move constantly by osmosis and diffusion across the permeable gill membranes of fishes. How do these movements of water and ions change as a salmon switches between seawater and freshwater habitats? What concentrations do salmon maintain in their tissue fluids, and how are these concentrations relevant? Do the fishes maintain homeostasis? How?

You will find the answer to these questions on page 726.

KEY CONCEPTS

29.1 Animals Prosper in Diverse Thermal Environments

LEARNING OBJECTIVES

By the end of this key concept you should be able to:

29.1.1 Explain the importance of microenvironments for the success of animals in diverse environments.

29.1.2 Give examples of behavioral and physiological adaptations to thermal environments.

29.1.3 Explain how molecular evolution can play a critical role in animal diversification.

Animals have diversified into virtually all environments on Earth. In this chapter we build on concepts introduced in Chapter 28 to explore the functional adaptations involved in this diversification. In their natural environments, animals have evolved environment-specific adaptations. Thus, for example, Arctic species of mammals have evolved different adaptations from tropical mammals. Always, animal adaptations can only be understood in relation to the specific environments where the animals live and have evolved. We start the chapter by considering adaptations to environmental temperature and then move on to consider adaptations of aquatic animals to environmental salinity, how individual animals respond to long-term environmental changes, and the internal timing mechanisms that help animals remain synchronized with regular variations in their environment such as the cycle of day and night.

The thermal environments on Earth are extremely diverse. At one extreme, in the Arctic, the air temperature falls as low as –50°C (about –60°F), sometimes even lower. At the other extreme, in some of the hot deserts, the air temperature reaches +40°C–50°C (about 100°F–120°F), and radiant heating by the Sun adds additional heat stress. The range of temperatures in aquatic habitats is less extreme. Nonetheless, the temperature of the upper waters of the ocean is –2°C (28°F) in broad regions of the polar seas and can average +30°C (86°F), or even higher, in the tropics. One of the most amazing attributes of life is that, as animal species have diversified, some have adapted to prosper in the coldest and hottest environments and all the environments in between. This remarkable fact has inspired great curiosity about the particular adaptations that make such interspecific diversity possible.

Small- and large-bodied animals differ in behavioral options

In a forest, lake, or any other large habitat, small places inevitably exist where physical conditions are significantly different from the average in the habitat. Small-bodied animals may enter such small places. Large-bodied animals, however, typically cannot enter small subparts of the habitat and are exposed to the statistically average conditions of the larger habitat. Places within a habitat that potentially differ from the habitat at large are called **microhabitats**, or **microenvironments**. A related concept is that of microclimates. A **microclimate** is the set of climate conditions (temperature, humidity, wind speed, etc.) prevailing in a microenvironment.

To exemplify the different circumstances faced by small- and large-bodied animals, let's consider an Arctic habitat in winter. The air temperature typically increases with depth in deep snow. As shown in **FIGURE 29.1**, a small mammal can therefore find a microenvironment of higher temperature by burrowing and occupying a small cavity deep in the snow. The graph in the figure shows data for a winter night in Sweden; on that night, when the

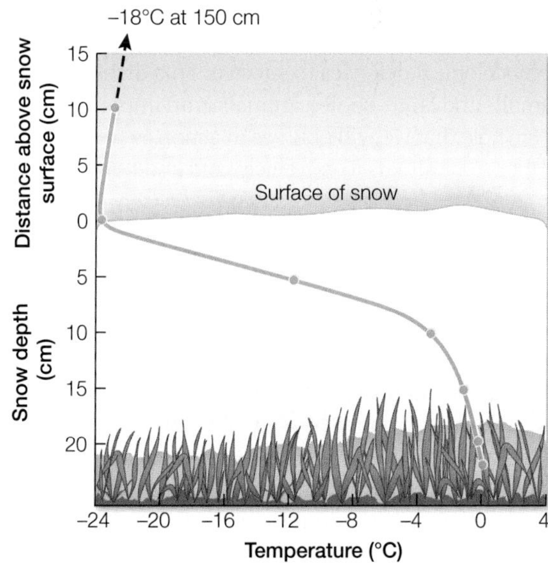

FIGURE 29.1 A Small Animal Can Enter a Protective Microenvironment Here a lemming is living in a chamber within deep snow, where the air temperature is higher than in the open air and the air is still. The graph shows how the air temperature varied with snow depth on a March night in Sweden. (Graph after C. C. Coulianos and A. G. Johnels. 1962. *Arkiv för zoologi* 15: 363–370.)

FIGURE 29.2 Reindeer Prosper in Severe Cold

air temperature in the general atmosphere above the snow was far below freezing (–18°C; 0°F), the temperature 25 centimeters (10 inches) deep was simply at the freezing point (0°C; 32°F). Thus a mouse, lemming, or ptarmigan (an Arctic bird) burrowing that deep into the microenvironment under the snow would have experienced 0°C air without wind. Simultaneously, a person, caribou, or other large mammal would have experienced –18°C air, possibly with a stiff wind.

In a stressful environment, we can think of animals as having *behavioral* and *physiological* defenses against the stresses they face. As a broad generalization, small- and large-bodied animals differ in their relative use of these two types of defenses. Small-bodied animals often deal with the stresses of an extreme environment by using behavior to evade the stresses rather than by depending on physiological defenses. Large-bodied animals, by contrast, often have fewer behavioral options; they therefore often must face the full brunt of the environment and depend more on physiological defenses to survive. This distinction between small- and large-bodied animals is an important organizing principle that we will now see in action as we focus on adaptations of animals to extremely cold and hot environments.

Large mammals in the Arctic have evolved specialized physiological defenses against cold

To gain insight into the adaptation of large mammals to the Arctic, let's focus on reindeer (*Rangifer tarandus*), which have a long and successful evolutionary history in the Arctic and sub-Arctic (**FIGURE 29.2**). To clarify their specialized physiological adaptations, we can compare them with humans—which are similar in body size but have a very different evolutionary history. Reindeer and humans have the same average core body temperature—about 37°C (98.6°F). Virtually all placental mammals are similar in this regard. For reasons that are not well understood, as the placental mammals diversified across the face of Earth, they retained the same fundamental core body temperature.

In Chapter 28 we discussed the typical relationship between metabolic rate and environmental temperature in homeotherms (see Figure 28.11A). **FIGURE 29.3** compares this relationship in reindeer and unclothed humans. The figure includes two plots for reindeer because they exhibit dramatic *phenotypic plasticity* between summer and winter (Key Concept 29.3 discusses phenotypic plasticity more fully). An unclothed human has a very high metabolic rate at temperatures colder than freezing and can survive for only a short time. Reindeer have far lower metabolic rates. For example, at –40°C, the metabolic rate of a winter reindeer is only about 150 watts, whereas a person (who could not survive at that temperature) would have a metabolic rate of about 800 watts!

How can we explain the difference in energy costs between species? There are two principal explanations. First, reindeer have evolved far better insulation. Their winter pelage is almost 4 centimeters (1.5 inches) thick, and the individual hairs are hollow. Nonmoving air is the best insulating material known in the natural world, and as **FIGURE 29.4** shows, a reindeer's hairs are filled with open spaces full of nonmoving air. Second, reindeer have evolved a sophisticated mechanism—called regional

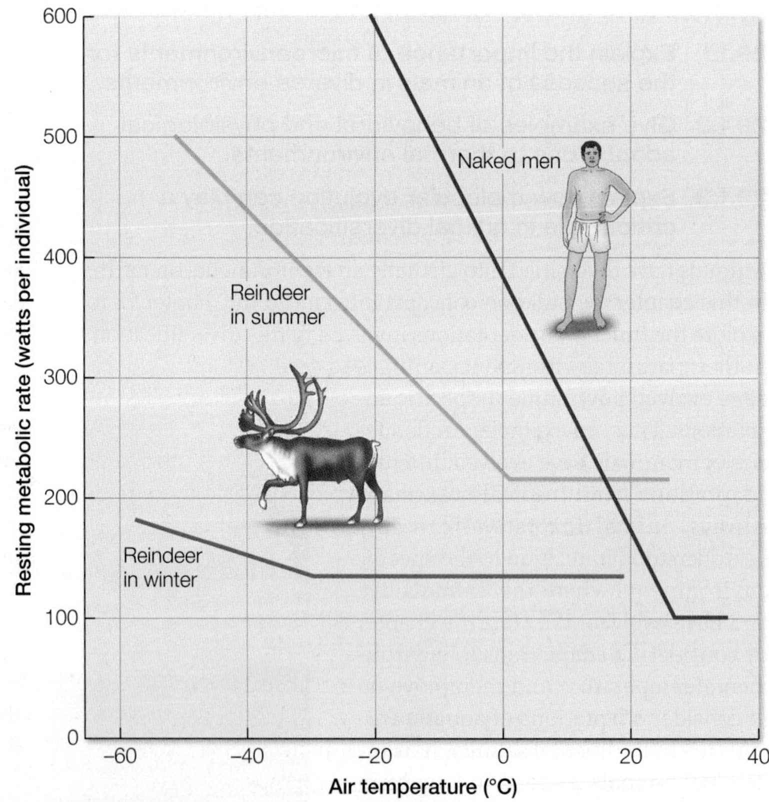

FIGURE 29.3 The Relationship between Metabolic Rate and Environmental Temperature in Humans, Summer-Acclimatized Reindeer, and Winter-Acclimatized Reindeer Only temperatures in the thermoneutral zone and lower than the thermoneutral zone are included. The humans were men who wore only boxer shorts. By being approximately in their natural state, they could be compared directly with the reindeer. The reindeer were from a Norwegian population in which the average adult body weight is about 70 kilograms—similar to the average adult weight of non-obese men. (After R. W. Hill et al. 2013. *PLoS ONE* 8: e76238; K. J. Nilssen et al. 1984. *Am J Physiol* 247: R837–R841.)

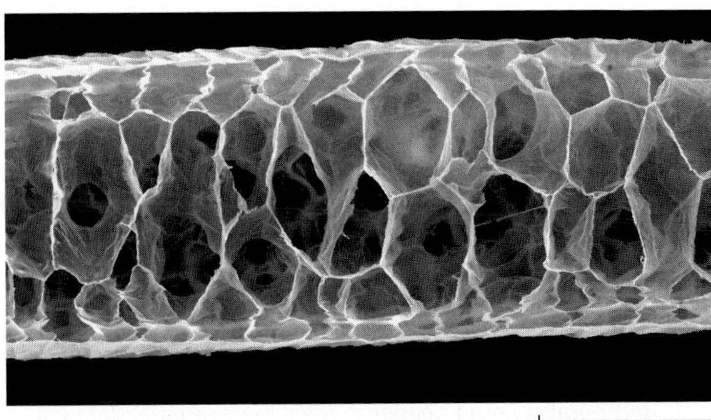

FIGURE 29.4 A Reindeer Hair in Longitudinal Section This image, obtained by scanning electron microscopy, shows that the inside of the hair is occupied by open, air-filled spaces. (Thanks to Nigel D. Meeks and Caroline R. Cartwright, Department of Scientific Research at The British Museum. © The British Museum 2005.)

Why are air-filled spaces in the hairs important?

hypothermia (see below)—that allows the tissues in their appendages to be cooler than the core tissues deep in their body (e.g., heart, brain, gut). For example, in one study, the tissues in a reindeer's lower legs were measured to be at 8°C while simultaneously the core tissues were at 37°C. Appendage tissues also tend to cool off in humans, but the mechanism in humans is not sophisticated enough to guarantee safety. Tissue freezing (frostbite) occurs in humans and can be devastating. In reindeer and many other Arctic mammals, the control mechanism for tissue cooling ensures that the appendage tissues do not freeze even when the outside environment is as cold as –50°C. With their highly insulating fur and their ability to allow appendage tissues to cool off without risk, reindeer can prosper in the Arctic with vastly lower metabolic rates than would be required of an unclothed person.

Meanwhile, the small-bodied mammals in the Arctic, such as mice and lemmings, behaviorally avoid exposure to extreme cold by occupying a protective microenvironment in the snow (see Figure 29.1).

Regional hypothermia is achieved by the circulatory system and its controls

Among large-bodied Arctic mammals, **regional hypothermia** in the appendages—a phenomenon in which appendage tissues are cooler than body core tissues—is common. This phenomenon is called "hypothermia" because the tissue temperatures are low, and "regional" because the low tissue temperatures prevail in just subparts of the body. It has been best studied in Arctic foxes (see the opening photo in Chapter 28), which we use as our example here. The mechanism in reindeer is believed to be closely similar.

When an Arctic fox walks over the snow and ice in winter, the tissues in its lower legs are much cooler than 37°C (the temperature in its body core). In fact, the temperature in the tissues of its foot pads is approximately 0°C. Body heat is conserved by not keeping the lower legs as warm as the body core. Because the temperature in its feet is near freezing, the fox loses little heat across its feet, and its feet are kept dry because they are too cold to melt the snow or ice. The control system for tissue temperature in Arctic foxes is amazingly exacting. Heat is metered into the foot pads so that the tissues are allowed to cool to near the freezing point but never allowed to get colder than freezing, regardless of how cold the environment gets. An Arctic fox therefore does not risk getting frostbite.

How is the cooling of the tissues achieved? The cells in the tissues of appendages need O_2 like other cells do. Thus blood flow to the appendages can't simply be stopped to keep body heat from entering them. Many types of mammals (and birds) have evolved an arrangement of arteries and veins in their appendages that conserves heat by a mechanism called countercurrent heat exchange. As we discuss this process, keep in mind that the blood vessels involved are arteries and veins (not capillaries). The walls of arteries and veins are far too thick for O_2 to pass through them. Heat, however, can pass through.

FIGURE 29.5A shows two arrangements of the arteries and veins in an appendage. The diagrams show only one artery and one vein, but that's just for simplicity. In the arrangement on the left, the arteries and veins are far away from each other and do not interact. On the right, the arteries and veins are touching. The right-hand arrangement is termed countercurrent—a term that refers to two closely juxtaposed fluid streams flowing in opposite directions. In the countercurrent arrangement, heat—traveling through the walls of the blood vessels—can move from one fluid stream to the other—that is, **countercurrent heat exchange** occurs.

What happens in an appendage with countercurrent heat exchange? As warm, arterial blood flows into the appendage, the warm blood comes close to venous blood that is leaving the appendage. This venous blood is cooler because it is returning from the far end of the appendage. Heat travels out of the warm arterial blood into the cooler venous blood (**FIGURE 29.5B**). This process continues as the arterial blood flows farther into the appendage—because even though the arterial blood gets cooler, the venous blood next to it also gets cooler. In this way, a lot of heat moves out of the arterial blood and into the venous blood in the upper parts of the appendage. Moreover—and this is very important—this heat is immediately transported back into the animal's body core as the venous blood flows back into the body. This heat is never carried by the blood to the outer end of the appendage, and therefore cannot be easily lost to the cold environment. The outer end of the appendage remains at a low temperature because the heat never gets to it. Meanwhile, the blood flow carries O_2 to the outer end because O_2 cannot pass through the vessel walls.

The effectiveness of countercurrent heat exchange depends on how extensively the arteries and veins make contact with each other. With extensive, highly controlled interaction between the two sets of vessels, the tissues at the outer end of an appendage can be kept just above the freezing point, as in Arctic foxes.

In hot deserts and other similar habitats, small- and large-bodied mammals again differ

A desert is, by definition, a relatively dry place where water shortage exerts a dominant, controlling influence on life. Hot deserts

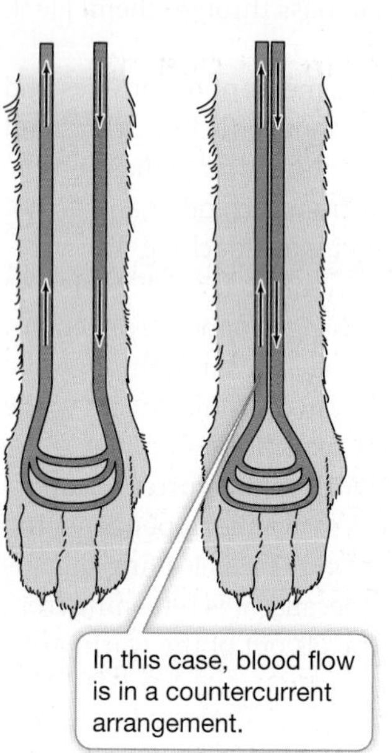

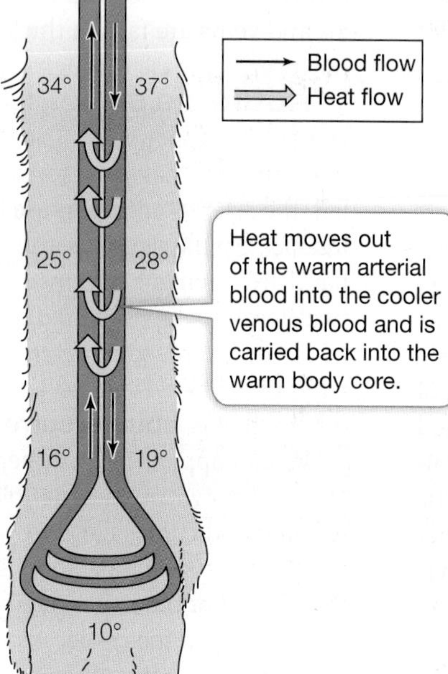

FIGURE 29.5 Countercurrent Heat Exchange Here we look at blood flow and heat flow in a leg of a mammal standing in a cold environment. **(A)** Two different arrangements of arteries (red) and veins (blue) are possible in the leg. In the countercurrent arrangement, the arteries and veins—carrying blood in opposite directions—are closely juxtaposed. **(B)** During counter-current heat exchange, blood flows all the way to the outer end of an appendage and back, but heat does not. Heat takes a shortcut, enabling it to move back into the body core before it can be readily lost to the environment at the exposed outer end of the appendage.

therefore present dual challenges. First, they are thermally stressful because of high air temperatures and the Sun beating down. Second, they are places where water can be difficult to obtain. As we consider the mammals that prosper in hot deserts and other similar habitats, it is important to recall that their fundamental core body temperature is 37°C: the invariant fundamental body temperature seen in virtually all placental mammals.

Small-bodied mammals—notably rodents—are often abundant in hot deserts. Some of the best-studied examples are the kangaroo rats of the American Southwest (**FIGURE 29.6**). Behavioral evasion of high environmental temperatures is a key to the success of the small-bodied mammals. They are mostly nocturnal (active at night) and live in burrows underground during their daytime resting period. They thus use two behavioral specializations—nocturnality and burrowing—to minimize exposure to the stresses that we humans associate with deserts. During the daytime in a hot desert, an animal burrowing down into the soil encounters cooler temperatures the deeper it goes. For example, data gathered in the Arizona desert showed that when the air temperature was 45°C–50°C (113°F–122°F) at the height of a person (and the Sun was beating down), a mouse could find a microenvironment where the air temperature was 26°C (79°F), free of solar heating, by burrowing 2 meters deep into the soil. At night the desert surface often cools off quickly. Thus when rodents venture out in the night, they again encounter moderate temperatures. Because the animals avoid high temperatures both day and night, they do not use water for evaporative cooling—an achievement that greatly reduces their water needs.

Unlike small-bodied mammals, large-bodied species must face the full brunt of the hot environment when living in hot deserts and similar habitats, and species adapted to these habitats have evolved physiological specializations that permit them to succeed. For example, consider Grant's gazelle (*Nanger granti*), a well-studied antelope that is similar in body size to reindeer and humans. Grant's gazelle is abundant in the grasslands of equatorial East Africa (**FIGURE 29.7**). Each year these grasslands experience a dry season lasting 6 months, during which no rain falls and all the streams dry up. Thus the gazelles not only live in a hot place, but also have no

FIGURE 29.6 A Desert Kangaroo Rat Kangaroos rats, such as *Dipodomys ordii* shown here, occur in the deserts of the American Southwest and northern Mexico. They hop about on their hind legs.

FIGURE 29.7 Grant's Gazelle Thrives in High Heat and Drought

access to drinking water for months at a time. Dromedary camels face a similar challenge, living in hot deserts.

To understand the adaptations of these animals, it is instructive first to study the reactions of humans to a hot environment. We humans exhibit a rigid type of homeostasis in regard to our body temperature. When we are at rest and healthy (i.e., free of fever), we rigidly maintain a deep body temperature close to 37°C. Thus, if a human is placed outdoors in the daytime in the hot habitat of Grant's gazelle, the human starts immediately to sweat. The evaporation of sweat cools the body sufficiently to get rid of excess environmental heat, thereby keeping the body temperature at 37°C. However, this solution to the environmental challenge requires a lot of water; a person might sweat as much as 2 liters of water per hour in the daytime heat. Thus drinking water is essential. The gazelles, lacking drinking water for long periods, cannot use the same strategy as humans.

To prosper in its hot, dry environment, Grant's gazelle has evolved a far less rigid type of temperature homeostasis than humans. When the gazelles are not under heat stress, they have a core body temperature near 37°C. However, during the heat of the day, they can tolerate a rise in core body temperature to about 46°C (115°F)—hot enough to kill a human quickly. Their heart and other tissues must have evolved adaptations to tolerate such temperatures. With this tolerance, however, the gazelles have greatly reduced water requirements. A gazelle's body temperature is about 37°C in the morning. As its environment heats up, the gazelle does not sweat but instead allows heat to accumulate in its body, driving up its body temperature. During the night, the environment cools off and the gazelle loses its accumulated heat without sweating. Thus the same mechanism can be used again the next day. Dromedary camels have evolved a similar mechanism for living in their hot, dry environment. The unusually high core body temperature of these animals during the day is termed *hyperthermia*. Compared with humans, these animals have far more variable body temperatures, but their hyperthermia each day greatly reduces their water need.

Of all the tissues in the body, the brain is particularly vulnerable to high temperatures. Grant's gazelle has evolved a way to keep its brain tissue 2–3 degrees Celsius cooler than the other deep tissues of its body. The mechanism of brain cooling depends on close juxtaposition of arteries and veins (and in fact is a countercurrent system). The arteries that carry blood to the brain are juxtaposed with veins carrying blood cooled by evaporation in the nasal passages. In this way the arterial blood is cooled before it reaches the brain.

Media Clip 29.1. Thermoregulation in Animals

Lizards and insects prosper in hot deserts

Certain species of lizards and insects are among the most successful animals in hot deserts. Being small-bodied, they can exploit favorable microenvironments. Doing so is a key to their success.

A striking feature of desert lizards is that, unlike most desert rodents, they are diurnal (active during the daytime) and venture out on the desert surface in the heat of the day. During the daytime they are effective behavioral thermoregulators (see Key Concept 28.3): each species uses behavior to maintain a species-specific, relatively constant body temperature. Thus a lizard's body temperature is under control and not free to skyrocket upward toward dangerous levels. The most important mechanism of this life-protecting thermoregulation is exploitation of microenvironments. The lizards use burrows. In addition, being small, they can fit into patches of shade cast by vegetation or rocks and thus can choose sunshine or shade. Moreover, there is a steep drop in air temperature from the desert surface up to a height of 30 centimeters; thus the lizards can find different temperatures by staying at the ground surface or climbing bushes. The lizards behaviorally exploit all of these variations in microenvironment to thermoregulate, as the spiny lizard *Sceloporus serrifer* illustrates (**FIGURE 29.8**).

Insects are also among the most successful animals in hot deserts. Some of the most thoroughly studied species are desert ants. Being social insects, they live in large colonies, which are located in benign microenvironments underground. In some species, the workers venture out on the desert surface in the heat of the day to find insects of other species that have recently died. To get needed water from the bodies of these prey, the workers must find them soon after death, before they dry out—accounting for the ants' daytime foraging. One adaptation of the ants to their hot environment is that they have evolved one of the most sophisticated animal navigation systems known, as we will discuss in Chapter 37 (see Figure 37.10). Because of their navigation abilities, workers foraging on the desert surface can always find their underground microenvironment immediately. They thus can save their lives if they are in danger of being killed by the heat of the Sun.

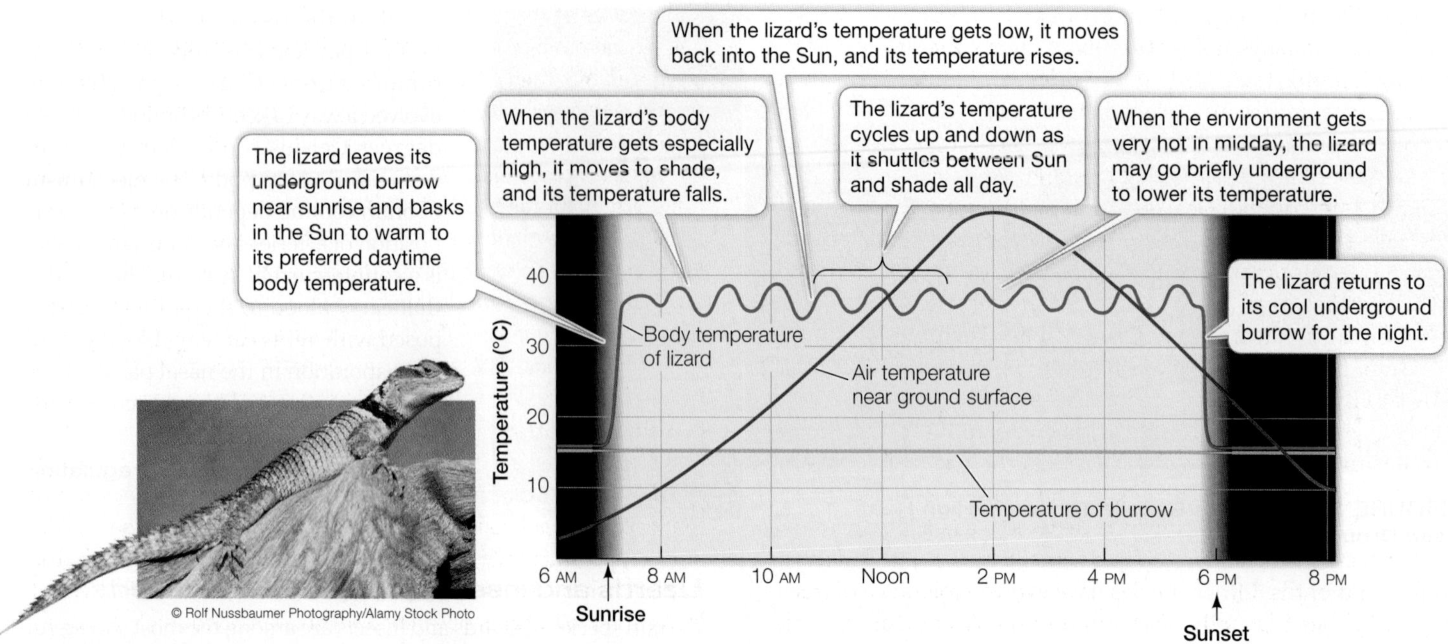

FIGURE 29.8 Behavioral Thermoregulation by a Desert Spiny Lizard (*Sceloporus serrifer*)

In cold climates, many small mammals hibernate in winter, and insects tolerate freezing or supercooling

Most of the species of mammals that undergo hibernation in winter are relatively small: 5 kilograms (the size of a groundhog) or smaller. People often think of hibernation as being such an extraordinary adaptation to winter that all mammal species "should" engage in it. However, in the cold places on Earth (e.g., Alaska), only about half of the mammal species in the size range specified hibernate. For example, ground squirrels and marmots hibernate, but lemmings and tree squirrels do not. This suggests that hibernation has both pros and cons. Here we emphasize its advantages.

Hibernation is a state of low body temperature and thermal conformity that persists for a long period of time in the winter. The usual pattern for a hibernating species is that the animal functions as a homeotherm—regulating its body temperature near 37°C—during the warm months of the year, but during the cold months the animal suspends homeothermy and exhibits temperature conformity (see Key Concept 28.2). During the stage of conformity, the animal allows its core body temperature to match the external temperature over a wide range of temperatures, meaning in many cases that the core body temperature falls as low as 0°C–10°C (**FIGURE 29.9**).

During hibernation an animal's metabolic rate is very low. This is true in part because its tissues are cold, but in addition the animal exhibits a biochemically induced reduction of metabolic rate called *metabolic depression* (which is poorly understood). In some cases, the metabolic rate during hibernation is only 1 percent as high as it would be if the animal were not hibernating!

By reducing energy needs, hibernation allows an animal to spend the winter continuously in a relatively protective and safe microenvironment, such as an underground burrow. The amount of metabolic fuel the animal requires is reduced to the point that the animal need not roam around its habitat in search of food (possibly not finding enough). Instead, it can live entirely on body fat or on stored food it collected the previous summer.

Hibernators, in a sense, have evolved a way to reap the benefits of both regulation (homeothermy) and conformity

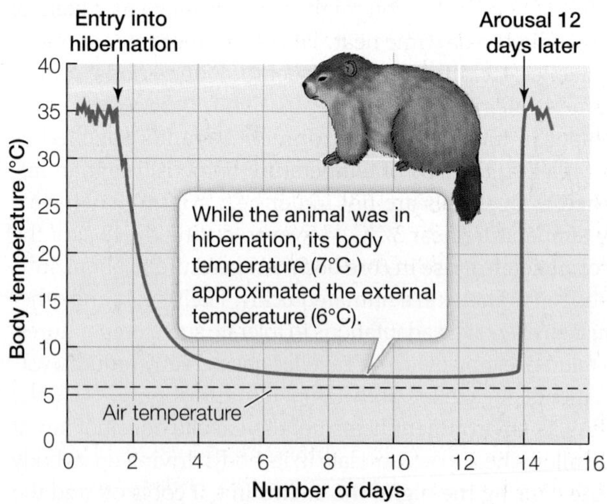

FIGURE 29.9 A Bout of Hibernation in a Groundhog Hibernators undergo hibernation in burrows or other protective microenvironments. This groundhog (*Marmota monax*), also known as a woodchuck, was living where the external temperature was well above freezing, 6°C. Its body temperature was approximately 35°C before and after the bout of hibernation, but during the 12 days of hibernation its body temperature was very close to the external temperature. (After K. B. Armitage, et al. 2000. In G. Heldmaier and M. Klingenspor [eds.], *Life in the Cold*, pp. 73–80. Springer-Verlag: Berlin Heidelberg.)

(poikilothermy) (see Key Concept 28.3). In summer, when a hibernator displays homeothermy, it enjoys the advantages of having a high, stable body temperature. In winter it reaps the advantage of conformity: a very low metabolic cost of life. Hibernators essentially switch between two metabolism–temperature curves resembling those in Figure 28.11.

Insects in winter differ from mammals in a critical way: when they are at rest and not flying, they have no ability at all to keep their body temperature higher than the external temperature. Many insect species inhabit places where the external temperature can fall below freezing. Their tissues then, quite simply, are at risk of freezing. Many insect species have evolved a remarkable ability to stabilize a supercooled state in which their body fluids do not freeze despite being cold enough to freeze. For example, certain of the insects that spend the winter inside plant stems in Alaska exhibit this trait. The stems stick up in the air and can be exposed to temperatures as low as –50°C, or colder. The insects inside supercool and remain in that state—unfrozen—for months. Other species have evolved tolerance of freezing. A species of woolly bear found in the Arctic, for example, simply curls up on the ice surface and freezes solid—yet it has evolved a way to survive and recover unharmed.

Fishes are models of molecular adaptation to temperature

As animals have confronted diverse environmental temperatures over evolutionary time, some of their most important adaptations have been at the molecular level. Molecular adaptation is understood particularly well in ocean fishes. There are more than 20,000 species of ocean fishes, and all the thermal habitats in the ocean—from polar seas to tropical lagoons—are occupied by many species. With this richness of animals to study, some stunning discoveries have been made.

Although the polar waters surrounding Antarctica are at a temperature of –2°C (unfrozen because of the saltiness of the water), many species of fishes prosper there. These fishes are not slowed to a standstill by the low temperature. Their populations are large enough—and they reproduce and grow rapidly enough—that the fishes account for much of the food of the fantastic Antarctic assemblages of penguins and seals. A thought-provoking clue about the physiology of the polar fishes is that many species die of overheating if they are warmed to about +5°C—signifying that the fishes are specialized to function at very low tissue temperatures. The same is true of the Antarctic krill (shrimp-like crustaceans) that also are sufficiently prosperous in the frigid polar seas that they provide massive quantities of food for seals and penguins. The krill cannot survive if the water is warmer than about +5°C.

At the other extreme, many species of fishes prosper in tropical waters, as in coral-reef ecosystems. These fishes die from being excessively chilled if they are cooled to +5°C. As astounding as it seems, a single water temperature—about +5°C—can be too warm for Antarctic fishes to survive and yet too cold for tropical fishes to survive. How can this be?

Decades of research have revealed that molecular evolution has played a major role in adapting fishes to the diverse thermal environments they occupy. All fish species have the enzymes for the citric acid cycle, for anaerobic glycolysis (fermentation), and for many other metabolic functions (see Chapter 5). However, many of these enzymes exist in *multiple molecular forms*. For example, there are many molecular forms of malate dehydrogenase, a key enzyme in the citric acid cycle, and there are many molecular forms of lactate dehydrogenase, the key enzyme for anaerobic ATP production. Fishes that have evolved in different thermal environments have evolved different molecular forms of at least some of the key enzymes. The polar species have molecular forms that function well at polar temperatures. But these forms do not always function well at warmer temperatures. Thus the polar-adapted molecular forms do two things. They make possible the life of polar fishes at polar temperatures. They also, however, require that polar fishes be at polar temperatures because those are the only temperatures where all the enzymes work well. Similarly, the tropical fishes have evolved molecular forms that function well at tropical temperatures but may not function adequately at colder temperatures. Thus the tropical-adapted molecular forms not only permit the tropical fishes to live at tropical temperatures but also require them to live at tropical temperatures.

Molecular evolution of metabolic enzymes has occurred throughout the range of thermal environments that fishes occupy, not just at the extremes of temperature. For example, there are many species of barracudas (**FIGURE 29.10**) in the warm ocean waters of the world. Four species that have been studied in detail live in thermal environments that differ from one another by about 3–4 degrees Celsius; that is, one species may be found in waters that are 3–4 degrees Celsius warmer on average than those of another species. All four of the species have evolved different molecular forms of lactate dehydrogenase. This fact indicates that only relatively small differences in habitat (body) temperature are sufficient for evolution to favor molecular adaptation.

WaterFrame/Alamy Stock Photo

FIGURE 29.10 A School of Barracudas Each barracuda species has enzyme forms that adapt it to its particular thermal environment.

THINK LIKE A SCIENTIST

Science & Society

Process of Science

Is global warming affecting animal life or not?

Is global warming having an impact on animals or not? Faced with this question, some people wonder how they might spin the information available, whereas scientists wonder how they can get *direct* and *dispassionate* evidence on the question.

Suppose you are going to offer formal advice to a governmental agency on the question and its implications for society. Suppose also that you have decided to focus on fishes and you want to base your advice on directly measured evidence. Specifically, you plan to use the scientific process: state a hypothesis and test it using data gathered from natural habitats.

FORMING A HYPOTHESIS: For each fish species, the *geographic range* is the region (e.g., longitudes and latitudes) where the species is found.

1. Considering fishes in the Northern Hemisphere, how would you hypothesize that their ranges would change during a period of global warming if in fact they are affected by the warming? In other words, what would you predict? In answering, make explicit use of what you've learned in this chapter about molecular evolution in fishes.

TESTING THE HYPOTHESIS: How would you test whether fishes are adhering to your prediction? Using mathematics, you can calculate the center of a species' range. **FIGURE 29.11** shows the range trend for codfishes (*Gadus morhua*) living in the North Sea between 1977 and 2001—a period during which thermometers measured a small, gradual increase in water temperature. There are 24 dots, each one corresponding to 1 year. The *y* coordinate of the dot is the latitude of the range center in that year, and the *x* coordinate is the average winter water temperature (near the ocean bottom) in that year.

2. What does Figure 29.11 demonstrate (assume statistical significance)?

3. Does Figure 29.11 demonstrate that global warming affects codfishes? Explain.

FIGURE 29.11 Latitude of the Range Center as a Function of Water Temperature in the North Sea Cod Population between 1977 and 2001 Each dot corresponds to 1 year. For measuring water temperature, thermometers were placed at fixed locations over the entire study period. The line is fitted statistically. (After A. L. Perry et al. 2005. *Science* 308: 1912–1915.)

The investigators who produced the graph in Figure 29.11 analyzed 35 additional North Sea fish species in the same way, over the same 24 years. The table below summarizes their results (including those for codfishes).

Category	Number of fish species in the category
Species that showed a statistically significant positive relationship between latitude and water temperature	12
Species that showed a statistically significant negative relationship between latitude and water temperature	2
Species that showed no statistically significant relationship between latitude and water temperature	22

4. Why did the investigators go to all the trouble of gathering data on 35 additional species rather than simply analyzing codfishes? Based on the table, what would be your advice to the government agency regarding the possible implications of global warming?

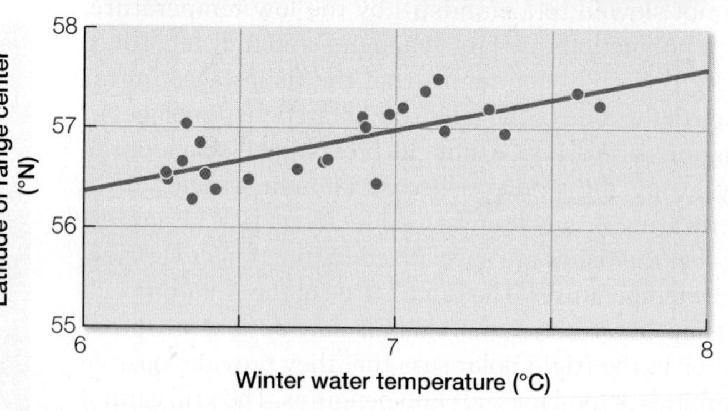

REVIEW & APPLY | 29.1

R Animals collectively occupy a wide range of thermal environments, although each species tends to be specialized for a limited range. Large-bodied species face the full extremes of the environments where they live and must cope physiologically. In contrast, small-bodied species—while having physiological defenses—often use behavioral strategies to occupy microenvironments where they can escape harsh conditions. Some mammals use hibernation, a dramatic mechanism in which body temperature conforms to the low environmental temperature, to survive winter. Ocean fishes demonstrate that adaptation to temperature may occur at the molecular level. Each species of fish has evolved molecular forms of key metabolic enzymes suited for function in its particular thermal environment. Enzyme molecular forms specialized for one type of environment may preclude success in alternative environments.

A

1. What are microenvironments and microclimates? Give an example of an organism that evades temperature extremes by exploiting a microenvironment or microclimate.
2. Consider this statement: A particular enzyme, such as lactate dehydrogenase, often exists in multiple molecular forms in the animal kingdom. What is meant by this statement, and why is the phenomenon of potential importance?
3. Does "homeostasis" always refer to maintaining exactly the same level of internal constancy, or when we think about internal constancy, are there different degrees of homeostasis? Approach your answer critically, and cite examples for your principal assertions.

Temperature is just one of the major ways in which animal environments differ. In aquatic environments, another major variable with which animals must contend is salinity.

29.2 Animals Live in the Ocean, Fresh Water, and Intermediate Salinities

LEARNING OBJECTIVES

By the end of this key concept you should be able to:

29.2.1 **Explain how a motionless animal can provide a perfect example of the highly dynamic nature of life.**

29.2.2 **Explain the principal mechanisms by which an aquatic animal maintains in its body fluids a consistent salinity that differs from the salinity of its environment.**

29.2.3 **Describe the typical blood osmotic pressures of ocean animals, freshwater animals, and terrestrial vertebrates, and outline the hypothesized evolutionary history of these patterns.**

Aquatic environments account for more than two-thirds of Earth's surface. By this standard, they are the major environments on Earth. There are two principal types of aquatic environment: bodies of saltwater (principally the oceans) and bodies of fresh water (lakes, rivers, and streams). These two principal types are dramatically unequal in size, however. The oceans cover about 70 percent of Earth's surface. Bodies of fresh water cover about 1 percent. (Land covers 29%.) Fresh water plays an outsize role in human affairs because of the close proximity of lakes, rivers, and streams to the places where we live—and because of our dependence on fresh water for drinking. Nonetheless, fresh water is the smallest of the major environments on Earth, dramatically smaller than the others. In part because of its smallness, fresh water is by far the most threatened major environment—often used as a place to dump pollutants, often redirected, often just plain used up.

Our discussion of aquatic environments focuses on both the water itself and the salts (e.g., Na^+ and Cl^-) dissolved in the water. In tandem, we consider the body fluids of the animals. As emphasized in Key Concept 28.2, an animal's tissue fluids are the immediate environment of its cells: the animal's internal environment.

All aqueous solutions are characterized by osmotic pressure, ionic composition, and volume

Three properties of aqueous solutions—osmotic pressure, ionic composition, and volume—strongly affect how aquatic animals interact with their environments. These properties are interrelated, but often it is helpful to think of them as three distinct characteristics.

OSMOTIC PRESSURE The **osmotic pressure** of an aqueous solution is a measure of the total concentration of solutes (dissolved matter) in the solution. Each individual dissolved entity makes an approximately equal contribution to the osmotic pressure. For example, if a body fluid has glucose, Na^+, and Cl^- dissolved in it, the individual dissolved entities are the individual glucose molecules, Na^+ ions, and Cl^- ions. Each individual dissolved entity—regardless of its type—makes a roughly equal contribution to the osmotic pressure. A solution having a mole of dissolved entities per liter (i.e., 6×10^{23} entities per liter) is said to be a 1-osmolar solution, symbolized 1 *Osm*. A 1-osmolar solution can also be described as a 1,000-milliosmolar solution, symbolized 1,000 m*Osm*. Osmotic pressures (which are sometimes described as "osmotic concentrations") are expressed in this osmolar system of units. For example, if a solution has one-quarter as many dissolved entities per liter as a 1-osmolar solution, it is 0.25 osmolar or 250 milliosmolar.

Why are osmotic pressures important? This is a key question that can lead to a lot of confusion. Thus it is important to recognize and remember that osmotic pressures typically have one and only one "claim to fame": *they determine the direction of osmotic water movement*. Water moves by osmosis from regions where the osmotic pressure (total solute concentration) is low to regions

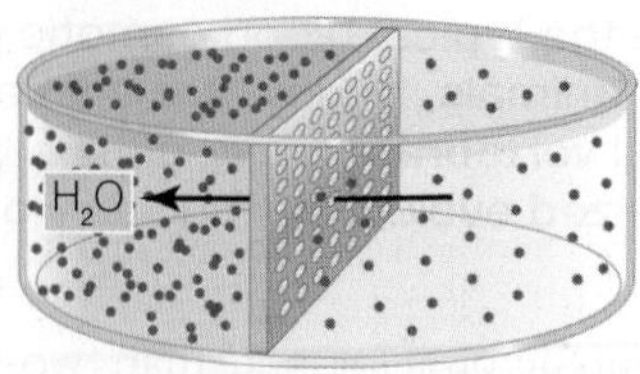

FIGURE 29.12 Water Moves by Osmosis from a Region Where the Osmotic Pressure Is Low to One Where It Is High Dots symbolize individual dissolved entities.

where the osmotic pressure is high (**FIGURE 29.12**; see also Key Concept 4.2). A good way to remember this is to recognize that a solution low in solutes is high in water, and vice versa. Water moves by osmosis from where it is high in concentration to where it is low in concentration: from low osmotic pressure to high osmotic pressure.

For discussing osmotic water movements, the terms *isosmotic*, *hyperosmotic*, and *hyposmotic* (see Key Concept 4.2) are helpful. These terms refer to the relative osmotic pressures of two solutions. Two solutions that have the same osmotic pressure are **isosmotic** (short for iso-osmotic; *iso*, "equal"). When two solutions have different osmotic pressures, the solution with the higher osmotic pressure is described as **hyperosmotic** (*hyper*, "higher"), and the solution with the lower osmotic pressure is described as **hyposmotic** (short for hypo-osmotic; *hypo*, "lower"). Always remember that these terms are relative. They mean nothing unless a comparison is being made between two specific solutions.

IONIC COMPOSITION Two solutions of the same osmotic pressure can have very different ion concentrations (e.g., one may be rich in Na^+, the other in K^+). This fact emphasizes that ionic composition is itself a property of significance. Ion concentrations are important for several reasons. One of the most important is that they help determine the direction of ion diffusion. Each particular type of ion (e.g., Na^+ or K^+) diffuses in response to its particular concentration differences from place to place.

LINK Diffusion is the process of random movement of dissolved entities toward a state of equilibrium. In the case of ions, two factors determine the equilibrium state: ion concentrations and the prevailing electrical gradient. Diffusion is a net movement toward electrochemical equilibrium. You can review the process of diffusion in **Key Concept 4.2**.

VOLUME When we talk about the body fluids of an animal, volume is a third property of great significance. We experience the importance of body fluid volume every time we get a bruise that causes tissue swelling. The body fluids in the bruised tissue may be perfectly normal in their osmotic pressure and ionic composition. Yet something is not normal: namely volume. Tissue swelling happens because of accumulation of an excess volume of body fluids.

The fluxes of water and ions between a freshwater animal and its environment are often highly dynamic

Let's now apply the principles just discussed by focusing on a freshwater animal in its environment. Specifically, let's consider a 100-gram goldfish in a goldfish bowl (**FIGURE 29.13**). In the process we will make a major discovery.

All freshwater animals, including our goldfish, have body fluids that are much more concentrated in solutes than fresh water is. That is, freshwater animals are markedly *hyperosmotic* to fresh water, and the concentrations of major ions—notably Na^+ and Cl^-—in their blood plasma and tissue fluids are far higher than the concentrations of the same ions in fresh water. Freshwater animals maintain their hyperosmotic state by use of regulatory mechanisms and thus are classified as **hyperosmotic regulators**.

The gill membranes of freshwater animals are permeable to water (H_2O) and ions, whereas their general body surfaces (e.g., the skin of a fish or the shell-covered integument of a crayfish) are not permeable. The gill membranes must be permeable to O_2 (so the animal can breathe and take up O_2), and this means they are permeable to water and ions as well. Often some of the other delicate membranes, such as the mouth membranes, are also at least somewhat permeable to water and ions.

Because the body fluids of a freshwater animal are hyperosmotic to fresh water, water (H_2O) passes steadily inward across the animal's gill membranes by osmosis, traveling from the environmental water into the animal's blood plasma. Simultaneously, because Na^+ and Cl^- are more concentrated in the blood plasma than in fresh water, these ions—the major ions in the blood plasma—pass steadily outward across the gill membranes by diffusion (traveling from the blood plasma into the environment). **FIGURE 29.14A** shows these processes and summarizes other important processes that we'll now discuss.

FIGURE 29.13 Invisible Dynamism The water and salt exchanges between an animal and its environment often occur at high rates despite being invisible. A goldfish in its bowl may look almost inert. In fact, water is moving rapidly by osmosis into the fish's body from its environment, and the fish is producing urine equally rapidly to keep the volume of its internal body fluids constant.

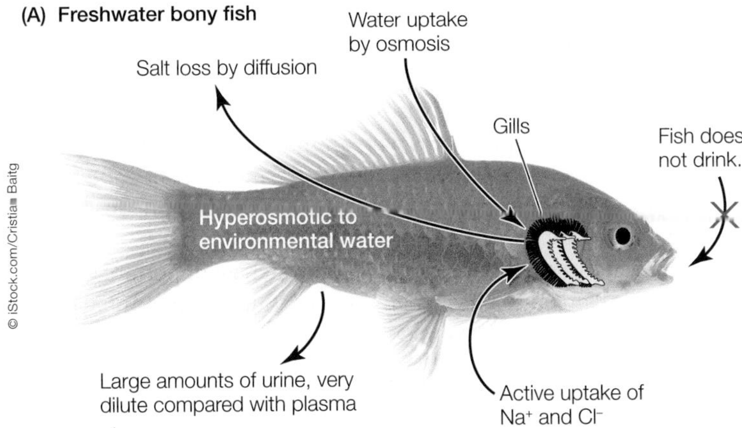

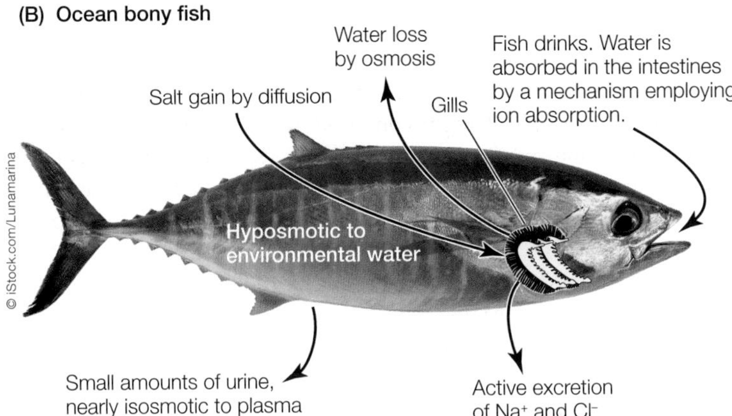

FIGURE 29.14 Principal Water and Salt Exchanges in Bony Fishes The water and salt exchanges in a freshwater fish **(A)** are in many ways the opposite of those in an ocean fish **(B)**. The "bony" fishes are the fishes with bony skeletons rather than cartilaginous skeletons. Virtually all freshwater fishes are bony. Most ocean fishes are bony, although there are also many cartilaginous fishes (e.g., sharks and skates) in the ocean.

Freshwater animals, including our goldfish, must invest energy to correct the problems created by osmosis and diffusion. To void the water they are gaining by osmosis, their kidneys produce a large volume of very dilute urine per day. Producing a large volume of dilute urine costs energy, but it means the kidneys keep the blood at its correct volume and osmotic pressure.

Ions (salts) lost by outward diffusion across the gills must also be replaced. Freshwater fishes have specialized active-transport cells in their gill membranes that use ATP to pump Na^+ and Cl^- directly from their freshwater environment into their blood. (Similar active-transport cells occur in the gills of crayfishes and the skin of adult frogs.) These ion pumps are so effective that they can take up a lot of Na^+ and Cl^- per day from the fresh water in the environment and in that way replace the ions that are diffusing outward.

LINK Animals sometimes maintain internal solute concentrations that are higher than those of the surrounding environment. To move substances against their concentration gradients, cells use active transport, which requires energy. See **Key Concept 4.2**.

Watching a goldfish in a goldfish bowl, we might guess that nothing is going on. We might watch the goldfish for hours and see no activity. Actually, however, measurements reveal that a 100-gram goldfish gains about 30 grams of water per day by osmosis. Because the fish is hyperosmotic to its environment, osmosis occurs steadily, without interruption, throughout every minute of the 24-hour day. The fish absorbs so much water from its environment that if the water were simply to accumulate in its body, the fish would be one-third bigger each morning than it was the morning before! This is prevented because the kidneys excrete an amount of urine equal to one-third of the fish's body weight in the same period of time. Despite appearances, a goldfish is a highly dynamic animal, with large volumes of water coming and going all the time. Similarly, large quantities of ions (salts) are lost each day because of the steady diffusion of ions outward, and equal amounts are taken up by the ion pumps.

Now we will turn to the aquatic animals in other habitats than fresh water. Then, after we've considered how different aquatic organisms cope with differences in water salinity, we'll ponder how these mechanisms evolved.

Most ocean invertebrates are isosmotic to seawater

The ocean is the cradle of life. Animal life first appeared there. We can hope, therefore, that by studying the water and salt relations of today's ocean invertebrates, we can gain insight into the original water and salt relations of animals.

Nearly all invertebrates in the ocean today are isosmotic to seawater. In **FIGURE 29.15** we can see several types of ocean invertebrates, including a sea star and soft corals. All are isosmotic to the seawater in which they live (put loosely, their blood plasma is as salty as the seawater). Seawater has an osmotic pressure of almost exactly 1 *Osm*. The blood and other body fluids of most ocean invertebrates also have an osmotic pressure of about 1 *Osm*.

FIGURE 29.15 Think about This Ocean Scene from the Viewpoint of the Water-Salt Relations of the Animals All the invertebrates are isosmotic to the surrounding seawater. The fish, however, is strongly hyposmotic.

This isosmotic relationship is a simple and cheap way to live. Little water moves into or out of these animals by osmosis because the osmotic pressure is the same inside and out. Thus the animals have little need to expend energy to correct water gains or losses. The fact that the isosmotic relationship is almost universal among today's ocean invertebrates—coupled with the fact that it is energetically cheap—strongly suggests that the earliest animals were isosmotic to seawater. Today's ocean invertebrates have spent their entire evolutionary histories in the oceans. We believe they retain the isosmotic condition of their distant ancestors.

Ocean bony fishes are strongly hyposmotic to seawater

There are more than 20,000 species of fishes with bony skeletons in the ocean, as we noted before. The body fluids of these bony fishes are far more dilute, in terms of both osmotic pressure and the concentrations of major ions, than the seawater in which they swim —that is, the fishes are hyposmotic to seawater. The osmotic pressure of their blood plasma and other body fluids is 0.3–0.5 *Osm*, whereas that of seawater is 1 *Osm*. The fishes maintain this difference by use of regulatory mechanisms and thus are classified as **hyposmotic regulators**. In Figure 29.15, the *fish* in the scene is the only animal that has body fluids more dilute than the seawater.

Just as in freshwater fishes, the gill membranes of ocean bony fishes (and sometimes other limited parts of their body, such as the mouth membranes) are permeable to water (H_2O) and salts. Because of this permeability and their dilute body fluids, ocean bony fishes face two relentless, never-ending problems:

1. *They tend to lose water* by osmosis from their dilute body fluids into the surrounding, more concentrated seawater (**FIGURE 29.14B**). Their steady loss of water tends to dehydrate them (living in the ocean is like living in a desert for these fishes).
2. *They tend to gain ions*, such as Na^+ and Cl^-, by diffusion from the seawater into their more dilute body fluids. Their steady losses of water and gains of ions tend to make their blood plasma and tissue fluids become too concentrated.

Ocean bony fishes expend about 8–17 percent of their metabolic energy each day to fix these problems. To replace the water they are losing, they drink the seawater, and when the ingested seawater gets into their intestines, they expend energy to pump ions out of the seawater so that H_2O from the seawater enters their blood plasma. They also use energy to excrete excess ions. Fishes cannot make highly concentrated urine. Thus the ocean bony fishes depend on *extrarenal salt excretion* (salt excretion by organs or tissues other than the kidneys) to get rid of the excess ions. Their most important extrarenal mechanism is that specialized cells in their gill membranes, known as mitochondria-rich cells or chloride cells, use ATP for active ion transport. These cells pump Na^+ and Cl^- out of the blood plasma directly into the more concentrated seawater.

Evolutionary history explains why body fluid concentrations vary among animals

Why are the body fluids of ocean fishes more dilute than seawater? Why do freshwater animals have the body fluid concentrations they do? Biologists do not know the answers to such questions for certain. However, there is a well-developed hypothesis.

As we have said, animal life began in the oceans, and early animals were almost certainly isosmotic to seawater (1 *Osm*). During the long history of life, however, animals of many phyla invaded freshwater rivers and streams. Today's freshwater fishes and freshwater invertebrates (such as crayfishes and snails) are descended from ocean ancestors that invaded fresh water.

When these animals entered fresh water, an important general principle came into play: when the body fluids of an animal differ from the animal's environment in some way, the cost of maintaining this difference tends to be greater when the difference is large than when it is small. (We are familiar with an analogy in our apartments and homes. The cost of keeping them warm in winter depends on how warm we keep them. Maintaining a big difference between the temperatures inside and outside requires more energy than keeping the inside just slightly warmer than the outside.) When invertebrates and early fishes first invaded fresh water from the ocean, they encountered a large difference between their internal osmotic pressure, 1 *Osm*, and the external osmotic pressure, which is about 0 *Osm* in fresh water. Maintaining this difference required a high rate of energy use. Biologists hypothesize that as time passed, these early invaders evolved lower and lower internal concentrations—making their internal concentrations more similar to that of fresh water and thereby lowering the animals' energy costs. Today's crayfishes and freshwater fishes have internal osmotic pressures of about 0.4–0.5 *Osm* (crayfishes) and 0.3–0.35 *Osm* (fishes).

After the freshwater fishes evolved a body fluid osmotic pressure of about one-third osmolar, two major evolutionary developments are hypothesized to have occurred:

1. First, *certain freshwater fishes invaded the land, giving rise to all the groups of terrestrial vertebrates*. Many types of evidence point to this conclusion. For one, the body fluids of terrestrial vertebrates today tend to have osmotic pressures of about one-third osmolar. For example, the osmotic pressure of human blood plasma is about 0.3 *Osm*.
2. Second, *certain freshwater fishes are hypothesized to have reinvaded the oceans*. Moreover, the fishes that re-invaded the oceans from fresh water are hypothesized to have given rise to virtually all the ocean bony fish species alive today. This is believed to explain why today's ocean bony fishes are not isosmotic to seawater like the ocean invertebrates, but instead have body fluids that are far more dilute than seawater.

LINK For more on the evolutionary origins of terrestrial animals, see **Key Concept 22.6**.

What about the air-breathing vertebrates that live in the ocean (or in association with the ocean) today: the whales, seals, seabirds, ocean lizards, and sea turtles? Strong evidence indicates that these ocean vertebrates are descended from terrestrial vertebrates. As we have seen, the terrestrial vertebrates have blood plasma osmotic pressures of about one-third osmolar. The ocean mammals, birds, lizards, and sea turtles have inherited this trait

FIGURE 29.16 This Gull Living by the Ocean Gets a Salt Load from Its Food

and have similar internal osmotic pressures. Accordingly, when these animals feed on ocean invertebrates—or on seaweeds, which are as salty as ocean invertebrates—they obtain a substantial salt load (excess salt). For example, consider a gull living by the ocean and feeding on a sea star (**FIGURE 29.16**). The gull's body fluids are about one-third osmolar. The body fluids of a sea star, however, are 1 *Osm*—basically three times saltier. The kidneys of birds, lizards, and turtles cannot excrete excess salts sufficiently to deal with this salt load, and all the seabirds, ocean lizards, and sea turtles have evolved extrarenal organs termed **salt glands** that use energy from ATP to get rid of the excess salts.

The salt glands are located in the head and secrete highly concentrated salt solutions. In birds, the secretions exit by way of the nostrils, whereas in sea turtles they exit as tears (**FIGURE 29.17**). Videos of mother sea turtles laying their eggs often zoom in to show the tears. They are not tears of emotion.

Some aquatic animals experience varying environmental salinities

In the course of their life cycle, some bony fishes—notably salmon, sturgeon, and many eels—migrate between rivers and the ocean. Salmon, for example, are conceived and undergo their early development in fresh water. Then they migrate to the ocean and stay there for a long period during which they eat and grow. After that, they return to fresh water to spawn, and—depending on the species—they either die or undertake a new ocean migration. As a salmon undergoes its migrations, it switches between being a typical hyperosmotic regulator when in fresh water and being a typical hyposmotic regulator when in seawater. Look back at Figure 29.14. Individual salmon function like a freshwater fish (see Figure 29.14A) when in fresh water and like an ocean fish (see Figure 29.14B) when in seawater. In this process many functions switch direction—or switch on or off:

- When a salmon is in fresh water, its gills use ATP to pump ions into its blood plasma from the surrounding water. In the ocean, its gills use ATP to pump ions out of its blood plasma into the seawater.
- A salmon in fresh water does not drink because it is already overloaded with water from osmosis. Upon entering the ocean, it starts to drink and activates intestinal ion transport mechanisms required to take water into its blood plasma from the ingested seawater in its gut.

(A) Gull skull with salt glands

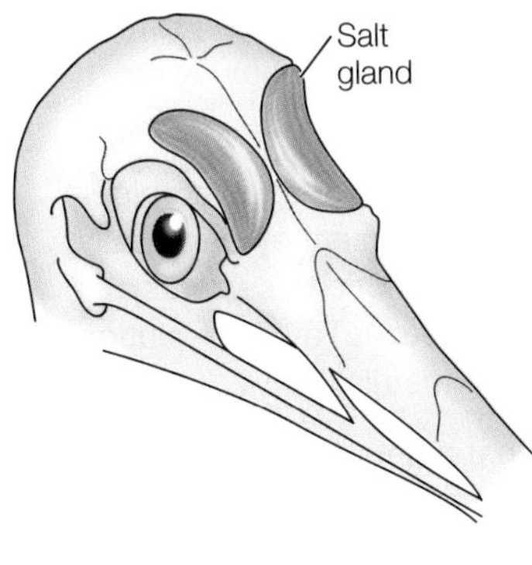

(B) A seabird voiding excess salt

(C) A sea turtle voiding excess salt

FIGURE 29.17 Salt Glands in Seabirds and Sea Turtles
(A) The position of the salt glands in a gull. In a bird, the fluid secreted by each gland enters a duct that empties into the nasal passages. The fluid then exits by way of the nostrils. **(B)** Fluid secreted by the salt glands exits the external tubular nostrils of a Northern Fulmar (*Fulmarus glacialis*) and then drips off the tip of the bird's bill. **(C)** Sea turtles, such as this leatherback sea turtle (*Dermochelys coriacea*), have salt glands, which secrete fluids that drip like tears from the area around the eyes.

Salt Glands have often evolved in grasses, bushes, and trees that live along coastlines in contact with seawater. Why do you suppose this dramatic evolutionary convergence between plants and animals has occurred?

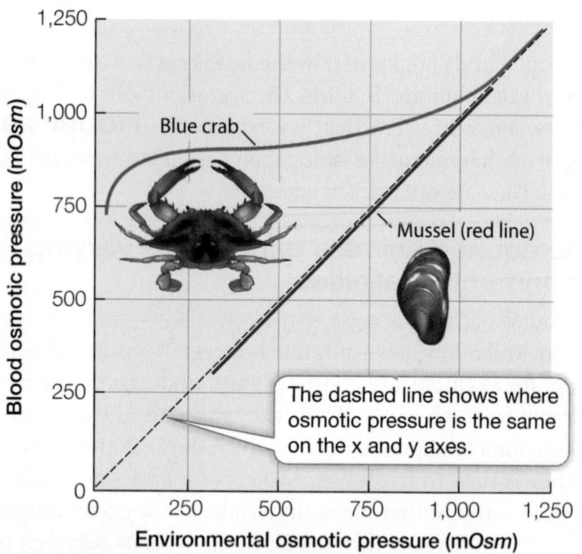

FIGURE 29.18 Osmotic Conformers and Osmotic Regulators When ocean invertebrates enter waters where the salinities are lower than in the ocean, osmotic conformers (such as mussels) allow their blood osmotic pressure to match the environmental osmotic pressure. Osmotic regulators (such as many crabs) keep the osmotic pressure of their blood from falling as low as the osmotic pressure in the environmental water, and their internal osmotic pressure is therefore more stable than the environmental osmotic pressure. The osmotic pressure of seawater is 1,000 m*Osm*. (Mussel: After W. T. W. Potts and G. Parry. 1964. *Osmotic and Ionic Regulation in Animals*. Pergamon, Oxford. Blue crab: After L. B. Kirschner. 1991. In C. L. Prosser [ed.] *Environmental and Metabolic Animal Physiology*, 4th ed., pp. 13–107. Wiley-Liss, New York, with data from B. S. Ballard and W. Abbott. 1969. *Comp Biochem Physiol* 29: 371–687.)

- In fresh water, a salmon's kidneys produce abundant urine that is very dilute compared with the blood plasma. In sea water, the kidneys produce just a small volume of urine that is as concentrated as a fish can make (isosmotic to blood plasma).

These changes are mostly controlled hormonally, with one of the principal hormones being prolactin.

In the ocean near the coastlines of continents, resident fishes, crabs, shrimps, and other animals often encounter variations in the salinity of their environment even though they do not migrate as salmon do. This happens because fresh water and seawater mix along coastlines, where rivers and streams flow into the ocean. If an individual moves close to the coastline, it may be immersed in very dilute water, even fresh water. If it moves far from the coastline, it may be in full-strength seawater. In between, the animal may encounter many different salinities. Even if an animal is attached to the bottom and thus lives at a fixed location, it may experience changing salinities because tides and currents move coastal waters around.

Most ocean invertebrates that live near coastlines are **osmotic conformers** (osmoconformers): as they experience environments that vary in salinity, the osmotic pressure of their body fluids varies in tandem with the osmotic pressure of the environmental water. The marine mussel in **FIGURE 29.18** illustrates this pattern. We know this type of response is stressful because most osmotic conformers do not survive in water that is much more dilute than seawater. Marine mussels are an exception.

LINK Key Concept 28.2 discusses the general principles of regulation versus conformity.

Some ocean invertebrates that live near coastlines are **osmotic regulators** (osmoregulators), meaning they maintain a relatively constant internal osmotic pressure despite experiencing a range of external salinities. Most of the highly successful invertebrates found in coastal waters are osmotic regulators. An example is provided by the blue crab (*Callinectes sapidus*; see Figure 29.18) which is of great commercial importance because of being one of the most successful coastal species.

REVIEW & APPLY | 29.2

R Freshwater animals are hyperosmotic to fresh water, and therefore steadily gain water from their freshwater environment by osmosis and lose ions to the environment by diffusion. To counteract these processes, the animals make dilute urine at a high rate and actively transport ions into their blood plasma. Ocean invertebrates are isosmotic to seawater. However, ocean bony fishes (because they are descended from freshwater ancestors) are hyposmotic to seawater, and therefore lose water by osmosis and gain ions by diffusion. To counteract these processes, ocean bony fishes drink seawater (gaining H_2O by transporting ions out of the ingested seawater) and actively transport ions out of their blood plasma. Terrestrial vertebrates are descended from freshwater ancestors, as are (indirectly) the air-breathing ocean vertebrates such as seals and seabirds. The osmotic pressure of the blood plasma of these animals is thus about one-third that of seawater. Salt glands provide a way for seabirds and ocean lizards and turtles to excrete excess salt.

A

1. Some marine mussels are of exceptional commercial importance, being considered gourmet foods. From the viewpoint of water and salt balance, in what way are mussels exceptional?
2. The famous marine iguanas of the Galápagos archipelago are unique among lizards because they principally eat ocean seaweeds that are isosmotic to seawater. Periodically a marine iguana "sneezes" a salty solution, made by salt glands, out of its nostrils. Why?
3. In Chapter 28 we stressed that animals use energy to maintain the organization of their bodies. Focusing on water and salts (rather than other types of organization), how is this point illustrated by (a) freshwater fishes and (b) ocean bony fishes?

Next we discuss the important phenomenon that the phenotype of an individual animal can undergo change during the animal's lifetime, as the animal lives in different environments.

29.3 The Phenotypes of Individual Animals Can Change in Response to Environmental Change

LEARNING OBJECTIVES

By the end of this key concept you should be able to:

29.3.1 Define the concept of phenotypic plasticity.

29.3.2 Give examples of phenotypic plasticity in both biochemistry and organ function.

The environment of an individual animal can undergo long-term changes (i.e., changes lasting at least several days). Seasonal changes are one type of long-term change: an individual terrestrial mammal may live at air temperatures below freezing for many months in winter and later live at air temperatures higher than 20°C for many months in summer.

Animals have evolved mechanisms whereby individuals can alter their phenotype in response to long-term environmental changes. The rats we discussed in Investigation Figure 28.16 provide an example. When the rats were living on a long-term basis in a warm environment, they showed little or no capacity for nonshivering thermogenesis and depended on shivering to keep warm. However, when they lived on a long-term basis in a cold environment, their phenotype changed: they developed a large amount of brown adipose tissue and a high capacity for nonshivering thermogenesis. When an individual's phenotype changes as a result of long-term exposure to an environment, the individual is said to **acclimate** or **acclimatize** to that environment.

Phenotypic plasticity refers to an individual's ability to display two or more different phenotypes at different times during its life. An individual's genotype—or genetic constitution—is constant for the individual's entire life. Thus, in cases of phenotypic plasticity, one genotype expresses two or more phenotypes.

We often get the impression that natural selection favors fixed traits. For example, natural selection has favored a four-chambered heart in mammals, and all mammals have a four-chambered heart: a fixed trait. Natural selection has favored red feathers in male cardinals, and virtually all male cardinals have red feathers: a fixed trait. However, this is a limited view of natural selection. Natural selection can also favor the evolution of patterns of response.

Phenotypic plasticity is a type of evolved genetic programming. During the evolution of phenotypic plasticity, natural selection—rather than favoring a fixed trait—instead favors a *genetic program of responses* (such as programmed changes in gene expression) so that some environments lead to one phenotype and other environments lead to other phenotypes. In this context, individuals with one genetic program of response may survive and reproduce more successfully than other individuals in the same population with a different program of response. Natural selection will then favor the former program of response, and that program of response will evolve.

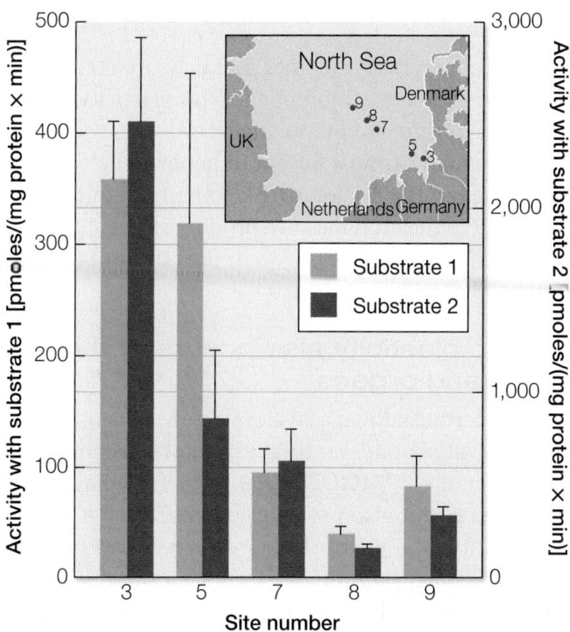

FIGURE 29.19 Fishes Exposed to Pollutants Have Higher Activities of Cytochrome P450 Enzymes North Sea dab (*Limanda limanda*) were sampled from several sites along a transect off the Baltic coast of Germany. Site 3 was close to the polluted effluent of the Weser and Elbe Rivers; site 9 was farthest from shore, and sites 5, 7, and 8 were in between. The activity of a liver cytochrome P450 enzyme was assayed at five of the sites using two different substrates. The *activity* of an enzyme is the rate at which it catalyzes the conversion of substrate to product under standardized conditions. Error bars show the standard error of the mean. (After K. W. Renton and R. F. Addison. 1992. *Mar Environ Res* 34: 151–155.)

Phenotypic plasticity is common at the biochemical level

Phenotypic plasticity occurs often at the biochemical level, and one of the most important ways it does so is by the action of **inducible enzymes**. An enzyme is called inducible if its concentration in an animal's cells, rather than being constant, depends on the animal's environment or experiences.

Many fishes and aquatic invertebrates, for example, have inducible enzymes that respond to long-term changes in body temperature. When winter comes and body temperature falls because the water temperature falls, these animals acclimatize over days or weeks by producing higher concentrations of enzymes that play key roles in energy metabolism. The fishes and invertebrates thus have high concentrations of the enzymes in winter and lower concentrations in summer. These enzyme changes make the metabolic rates of the animals less sensitive to long-term changes in temperature than they otherwise would be.

Detoxification enzymes also provide outstanding examples of plasticity in enzyme phenotype in response to environmental change. For example, enzymes known as *cytochrome P450 enzymes*, found throughout the animal kingdom, play important roles in detoxifying a wide range of environmental toxins such as halogenated aromatic hydrocarbons (HAHs). Fishes that live in polluted waters often have higher levels of P450 enzymes—and are more capable of detoxifying toxins—than fishes of the same species that live in pristine waters (**FIGURE 29.19**).

Similarly, human liver enzymes that detoxify ethanol (the alcohol in beer and wine) exhibit phenotypic plasticity, depending on a person's exposure to ethanol ingestion. These detoxification enzymes are usually almost absent in people who do not drink ethanol. However, if a person drinks ethanol regularly, higher concentrations of the enzymes are produced. This helps explain why people who regularly drink are more resistant to drunkenness than people who don't drink.

Phenotypic plasticity also occurs at the scale of tissues and organs

The tiny crustaceans known as water fleas (*Daphnia*), often abundant in freshwater ponds and lakes, display visibly dramatic phenotypic plasticity (**FIGURE 29.20**). This plasticity has a biochemical basis but also has a large impact at the scale of tissues and organs. The water fleas are pale-colored when their environmental water has a high O_2 concentration because they contain almost no O_2-transporting hemoglobin. However, the water fleas turn bright red when they experience a low environmental O_2 concentration because they then synthesize and accumulate hemoglobin.

An important question that arises is, why do the water fleas not maintain high levels of hemoglobin all the time? Although the question does not have a definitive answer, biologists in general hypothesize that the maintenance of a useless trait imposes costs that are best avoided. If a high concentration of hemoglobin were a fixed trait, water fleas would expend resources to keep the hemoglobin concentration high even in O_2-rich environments where the hemoglobin is unnecessary. Phenotypic plasticity solves this problem. With phenotypic plasticity, no resources need be invested in hemoglobin in O_2-rich environments, yet the animals can enjoy the benefits of hemoglobin in O_2-poor environments.

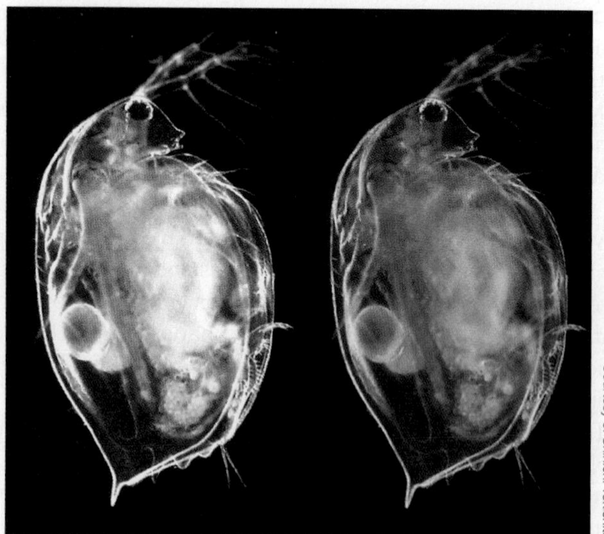

Courtesy of Shinichi Tokishita

FIGURE 29.20 Phenotypic Plasticity During their lifetimes, individual water fleas (*Daphnia*) can be pale (left) or bright red (right), depending on whether they have been living in O_2-rich or O_2-poor water. The color change reflects how much hemoglobin the animals are synthesizing.

Already we have seen other instances of phenotypic plasticity at the scale of tissues and organs. For example, reindeer display phenotypic plasticity—between summer and winter—in the thickness of their pelage and in the shape of their metabolism–temperature relationship (see Figure 29.3). Similarly, individual small mammals—such as rats, mice, and squirrels—that have been living in a warm environment have small amounts of brown adipose tissue (BAT; see Figure 28.17), but when they are exposed to a cold environment, their BAT grows dramatically. Because of this plasticity, the animals need not invest in maintaining BAT when their environment is warm, but they can have a high rate of BAT heat production when their environment is cold.

REVIEW & APPLY | 29.3

R Phenotypic plasticity refers to an individual's ability to display two or more different phenotypes at different times during its life. At a biochemical level, phenotypic plasticity often involves inducible enzymes. Phenotypic plasticity also occurs at the scale of tissues and organs, as illustrated by hemoglobin in water fleas and brown adipose tissue in small mammals. The capacity for phenotypic plasticity is genetically based and subject to natural selection.

A

1. What is an inducible enzyme, and what role do inducible enzymes play in phenotypic plasticity?
2. How might a scientist justify determining the relative purity of a stream by measuring the levels (e.g., concentrations) of P450 enzymes in the fishes that inhabit it?

To close the chapter, we now turn to animal responses to repeating cycles in their environment, particularly the cycle of day and night. The day-night cycle is of extreme importance for many animals, and animals have evolved internal timekeeping mechanisms that assist synchronization between their functional cycles and the environmental day-night cycle.

29.4 | Animals Have Biological Clocks Tuned to Cycles in Their Environment

LEARNING OBJECTIVES

By the end of this key concept you should be able to:

29.4.1 Distinguish between exogenous and endogenous timing.

29.4.2 Explain why it is true to say that circadian clocks do not keep exact time and need to be reset to work accurately.

29.4.3 Specify some benefits an animal may gain by having an endogenous circadian clock.

As Earth spins on its axis, each region of the planetary surface alternately faces toward the Sun and away from the Sun—giving rise to a rhythm of day and night that has existed since life first appeared. The length of the day–night cycle has been close to 24 hours for tens of millions of years and thus relatively constant during much of animal evolution. Although in everyday language people sometimes use the word "day" to refer to just the daylight hours of the 24-hour cycle, in our discussion here we will use "day" to mean the entire 24-hour day, and "daytime" to mean the sunlit hours of each day.

Many important characteristics of animals and plants are synchronized with the 24-hour daytime–nighttime cycle in the environment. For example, many animals sleep in the nighttime and are awake in the daytime. It is easy to see why life would have evolved rhythms synchronized with the ever-present daytime–nighttime rhythm. The daytime–nighttime rhythm is as much a feature of the environment as temperature, and thus we expect life to have responded to it and taken advantage of it. There are two radically different mechanisms by which synchronization could be achieved:

1. An animal could synchronize its daytime–nighttime rhythms with the environmental daytime–nighttime rhythm by use of *external* (also called *exogenous*) *cues*. Consider the sleep–wake cycle, for example. Each day, a person (or other animal that sleeps at night) could notice when the environment turns dark and use the onset of darkness as an external cue to go to sleep. Later, the person could notice the return of daylight and use that as an external cue to stop sleeping.

2. Alternatively, an animal could have a built-in clock: an *endogenous ability to keep track of time*. The animal would synchronize its endogenous clock with the daytime–nighttime cycle in the environment (equivalent to setting a watch). Then the animal would not need an external cue to know when to go to sleep or wake up. It could sleep and wake up at appropriate times in a purely self-timed way.

For much of history, biologists assumed that the daytime–nighttime rhythms of animals and plants were controlled by use of external cues. The first evidence to the contrary came in 1729 when a Frenchman named de Mairan placed a leaf-folding plant in a lightless cave. The plant, probably a type of mimosa, folded its leaves in the nighttime and opened them in the daytime when living in its natural environment. de Mairan observed that the plant continued to fold and open its leaves while in the cave, where the environment was continuously dark, suggesting that a living organism might have an endogenous ability to keep track of time.

More than two centuries passed before biologists reached a consensus on this question. To understand why two centuries could pass, consider the types of uncertainties that had to be resolved. Over the years following de Mairan's study, many experiments were done by placing animals and plants in places where they could not see daytime or nighttime, and it was often observed that the animals and plants continued to show daytime–nighttime rhythms. But did that mean their rhythms continued without any external cues? Might the temperature be rising in daytime and falling at night (perhaps slightly, but enough to be detected)? Might there be changes in the environmental noise level as daytime came to an end? Might a factor such as the intensity of bombardment by cosmic rays be different in the daytime than the nighttime? An animal or plant housed where it could not see daylight might use such alternative *external* cues to determine when daylight was occurring, and in that way continue its daytime–nighttime rhythms without an endogenous timekeeping mechanism.

Biologists did not become completely convinced of the existence of endogenous timekeeping until the last 20–30 years of the twentieth century when a *mechanism* of endogenous timekeeping was discovered (**INVESTIGATION FIGURE 29.21**). Finally, with that discovery, biologists not only had examples of timekeeping that seemed to be endogenous, but also they knew *how* living cells could keep track of time. Today we recognize that—whereas daytime–nighttime rhythms are sometimes timed using external exogenous cues—they are often timed by endogenous biological clocks.

Biological clocks are endogenous but must be synchronized with environmental time

What do we mean by a "biological clock"? A **biological clock** is a self-contained, metabolic mechanism of keeping track of time, not requiring outside information to function. The most common biological clocks are *daily* (also called *circadian*, as we will soon discuss)—that is, they provide timing information for processes that recur every 24 hours (i.e., every solar day). In addition, there are *lunar-day clocks*, which provide timing information for processes recurring every lunar day (24.8 hours), and *annual clocks*, which provide timing information for processes that recur every year. Although here we focus on clocks in animals, biological clocks also occur in plants and microbes. For example, the single-celled green alga *Chlamydomonas reinhardtii* shows endogenous daily rhythms in its movements toward light and nitrogen sources, demonstrating in a vivid way that the timing mechanism can exist within the confines of a single cell.

Biological clocks are control mechanisms. They provide information to control systems, permitting the control systems to function in different ways at different times. Consider, for example, the negative-feedback system that controls a mammal's body temperature. A biological clock can alter the set point of the system, so that the set point is higher in the daytime than the nighttime. The body temperature will then be controlled (as it often is in a diurnal species) to be a bit higher during daytime.

A type of experiment that can tell us a lot about how daily biological clocks function is one in which a mouse or squirrel is housed in a cage with a running wheel and a record is kept of the turning of the wheel, telling us when the animal is running. **FIGURE 29.22** shows data of this type for a species of nocturnal squirrel. The data are for a single, typical individual studied for 10 days.

First the squirrel was put into a day–night cycle environment: The lights in the squirrel's room were turned on for half of each day and off for the other half. Figure 29.22A shows the squirrel's running behavior. Each horizontal line in the graph corresponds to a day (the days are numbered at the left). Wherever there is a short vertical line on the graph, the squirrel was running. At night these vertical lines—colored gray—were so close together that long bars of solid gray were recorded; during these times, the squirrel was active continuously. Notice that on each of the 10

INVESTIGATION

FIGURE 29.21 The Discovery of the Genetic Basis of the Endogenous Circadian Timekeeper in *Drosophila melanogaster* Several decades passed in the twentieth century when strong circumstantial evidence accumulated for the existence of endogenous timekeeping in animals and plants. A fierce debate continued over the existence of biological clocks, however, because biologists knew of no mechanism by which a cell could endogenously keep track of time. Many laboratories had engaged in an almost heroic effort to find a mechanism, but without success. Researchers Ronald Konopka and Seymour Benzer had the idea of trying to find individuals with genetic mutations that affected their biological rhythms. Such mutations were unknown at the time in any organism. Konopka and Benzer reasoned that if such mutations could be found, then the genes controlling biological rhythms could be identified and in turn the mechanism underlying biological clocks could also be identified. Konopka and Benzer worked on the fruit fly *Drosophila melanogaster*, which was known already to exhibit several types of biological rhythms. The researchers focused on the *eclosion* rhythm and an *activity* rhythm. The term "eclosion" refers to the emergence of the adult fly from the pupa, an event that tends to occur in the morning in normal flies. The activity rhythm they studied was simply the pattern of activity of individual adult flies: they are more active at certain times of the day than at other times, and on successive days activity occurs at the same time of day.[a]

HYPOTHESIS

Possibly there are mutant flies that (compared with ordinary flies) exhibit changes in their endogenous eclosion or activity rhythms. If such mutant flies exist, they might be used to understand the mechanism by which biological rhythms are endogenously timed.

METHOD

1. Expose flies to a mutagen: a chemical that can modify DNA and therefore cause gene mutation.
2. Place pupae from 2,000 hereditary lines in complete darkness to isolate them from external cues.
3. Look for mutant lines showing evidence of a disrupted biological clock function, specifically, changes in the flies' free-running eclosion rhythm.

RESULTS

Of the 2,000 lines of flies analyzed, 3 lines showed evidence of a mutation affecting the biological rhythms of the flies. These were the first "rhythm mutants" ever known to science. Three aspects of the results were especially significant. First, as the graphs at right show, the free-running eclosion rhythms of three lines of flies (shown in B, C, and D) exhibit three distinct modifications from the normal free-running eclosion rhythm (shown in A). Second, in each of the three mutant lines of flies, the free-running activity rhythm was also modified (data not shown), and it was modified in a way that matched the change in the eclosion rhythm of that particular line. Third, when the investigators did experiments to determine the physical locations of the mutated genes on the chromosomes, they found that in all three of the mutant lines the location was almost exactly the same.

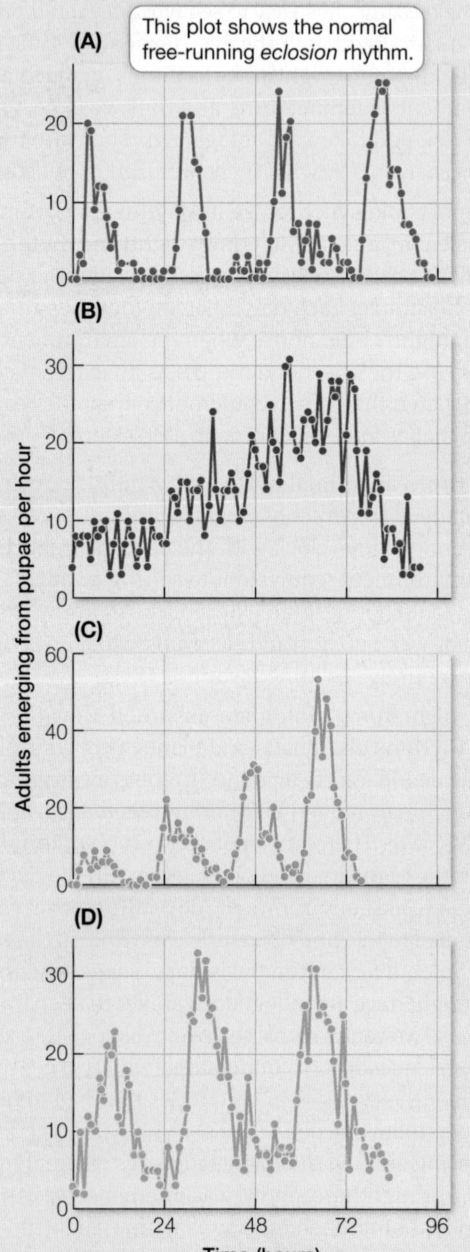

CONCLUSION

The function of the biological clock that controls the eclosion and activity rhythms is genetically determined.

ANALYZE THE DATA

1. In what way do these results strongly suggest that the flies were not using external cues of any kind to time their biological rhythms?
2. Making measurements with a ruler, describe the three mutant rhythms shown in parts B, C, and D of the figure. The *period* of rhythm is defined to be the time that passes between one peak and the next peak. Include measurements of the period in your answer.
3. Make and defend a summary statement regarding the gene or genes that are mutated, taking account of all the results.
4. In brainstorming mode, explain ways that these results could be used to discover the mechanism of endogenous timekeeping.

[a]R. J. Konopka and S. Benzer. 1971. *Proc Natl Acad Sci USA* 68: 2112–2116.

A similar **Analyze the Data** exercise may be assigned in **Achieve**.

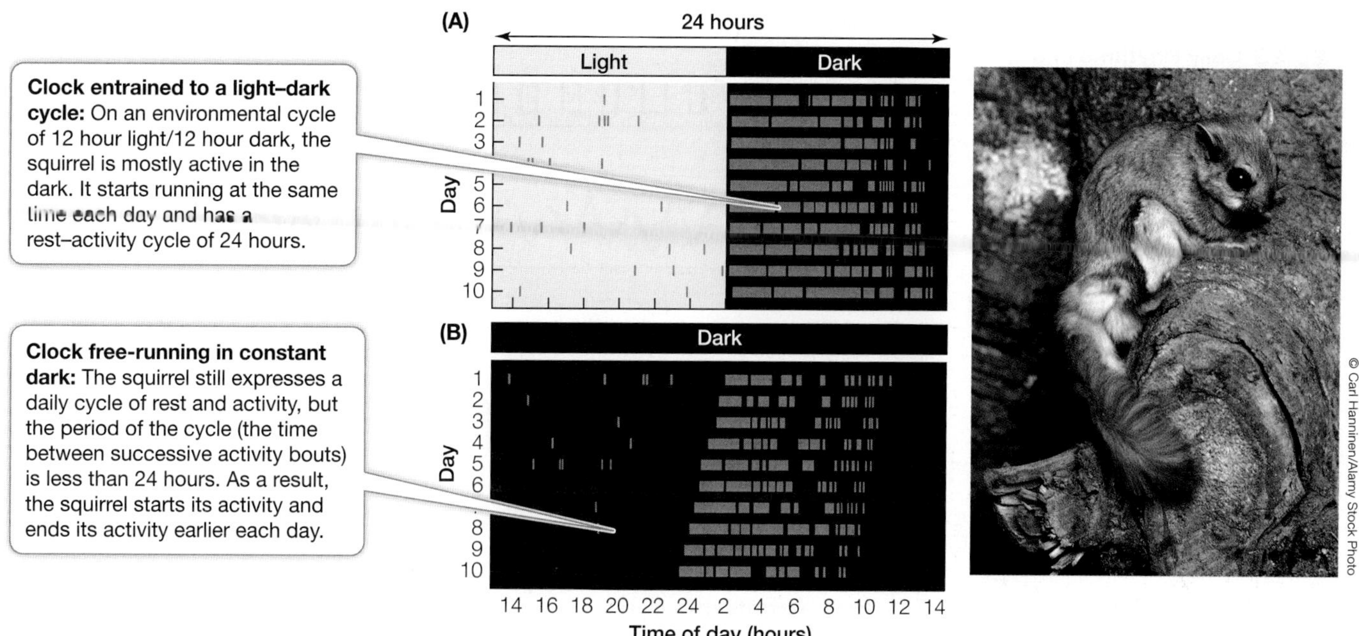

FIGURE 29.22 Control of Running Activity by a Circadian Biological Clock The rest–activity cycle of a flying squirrel (shown) was studied **(A)** in the presence of a light–dark cycle and **(B)** in continuous darkness. (A from P. J. Decoursey. 1972. *J Comp Physiol* 78: 221. https://doi.org/10.1007/BF00697656.)

Animation 29.1. **Circadian Rhythms**

studied days, the squirrel started to be active soon after the lights went off. Thus from one night to the next, the squirrel started its activity at the same time. To summarize: The squirrel was provided with *external timing cues* in part A, and it rested and became active in exactly a 24-hour cycle.

Next the squirrel was studied in continuous darkness (see Figure 29.22B). In this environment, the squirrel did not receive external cues about the time of day (the environment, in addition to being dark all the time, was isolated from outside temperature variations, outside sounds, and any other outside cues imaginable). The squirrel's activity, therefore, was timed by its endogenous biological clock. Under conditions like this—when there are no external timing cues—a biological clock is said to **free-run**. ("Free-run" is a general term that describes the clock and does not refer to running behavior.) The squirrel continued to alternate between periods of rest and activity on an approximately 24-hour cycle, but notice that the free-running cycle was not exactly a 24-hour cycle. The squirrel's period of continuous activity started earlier each day. Thus the biological clock—operating on its own (without external cues)—did not keep perfect time.

A universal property of biological clocks is that they do not keep perfect time when they free-run. Instead, they run fast or slow relative to actual time. Consider a free-running biological clock that goes through an entire timing cycle in 23.8 hours. After 1 day of free-running, the clock will be 0.2 hour off, compared with the actual 24-hour solar-day cycle. After 2 days it will be 0.4 hour off, and after 5 days it will be an entire hour off.

In natural environments, this long-term drift is prevented because animals receive external timing cues and their biological clocks do not free-run. Each day, an animal's biological clock is reset to be synchronous with actual time—a process termed **entrainment**. (This phenomenon is analogous to having a watch that runs fast or slow. If you own such a watch and you want it to show true time, you might reset it to actual time each day.) Sunrise and sunset are the most commonly used cues for entrainment. In a natural environment, animals observe sunrise and sunset each day, and they use these indicators to synchronize their biological clocks with the actual 24-hour solar-day rhythm.

In formal terminology, biological clocks that have free-running timing cycles of about 24 hours are called **circadian clocks**. The word "circadian" is not simply a fancy synonym of "daily" (which is a vaguer term). The prefix "circa" means "about." "Circadian" means "about a day" and emphasizes that the self-contained, free-running periodicities of these clocks are not exactly 24 hours. Just like circadian clocks, the endogenous clocks that animals use to time processes on annual and lunar-day cycles do not keep exact time. They are called *circannual* and *circatidal clocks* for this reason. They require entrainment to keep exact time.

Circadian clocks are believed to be virtually universal among animals. In an individual animal, most or all cells have their own circadian clocks. These cell clocks are coordinated by a "master clock" that sends out signals to cells in the individual's body, keeping the body as a whole synchronized. In mammals, the master clock consists of nerve cells in the parts of the brain called the suprachiasmatic nuclei.

As human beings, we of course wonder about our own circadian clocks. **FIGURE 29.23** shows data for a young man who lived by himself in an apartment without clocks, phones, windows, or other avenues by which he could know the time of day in the outside world. He could turn the lights on and off, prepare food, go to bed, urinate, and engage in other processes of daily living, but his timing of those activities was based entirely on his

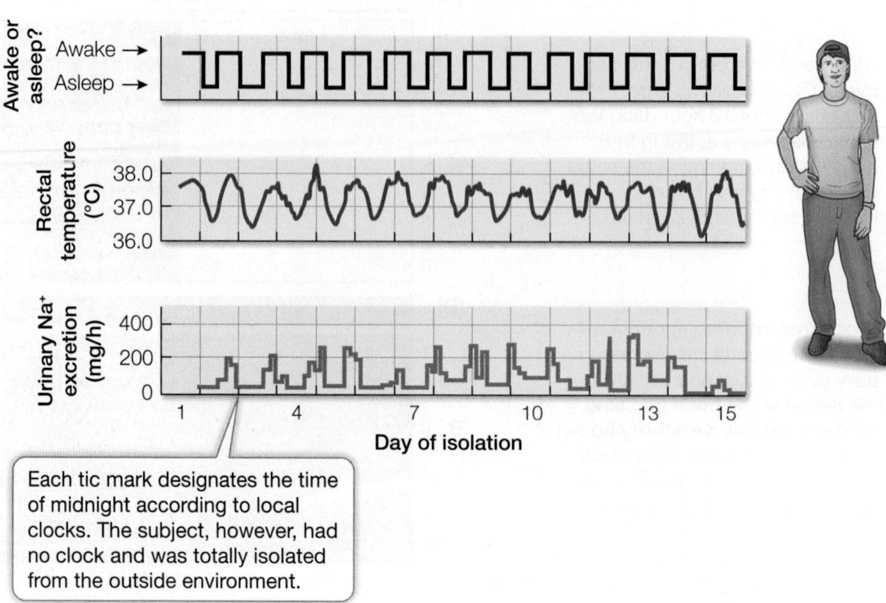

FIGURE 29.23 Daily Rhythms in a Young Man Living without Environmental Information on the Time of Day The graphs show the man's patterns of variation in three remotely monitored physiological functions during the first 15 days of the study. Over the period of study, the man continued to show rhythmicity in the properties investigated, but the exact timing of events gradually drifted away from the initial timing. (After R. A. Wever. 1979. In *The Circadian System of Man: Results of Experiments Under Temporal Isolation*, pp. 100–114. Springer-Verlag: New York.)

endogenous sense of time. The figure shows his actual sleep-wake cycle and his actual cycles of rectal temperature and urinary sodium excretion during his first 15 days. His rhythms continued, but his circadian clock ran slow, as you can demonstrate with a ruler. Relative to actual time, the man's activities became later and later as the days passed. For example, although he went to sleep at midnight on day 1, he did not go to sleep until about 16 hours later than midnight (i.e., mid-afternoon) on day 15.

Biological clocks permit anticipation

What are biological clocks good for? Why have animals evolved endogenous clocks instead of simply relying on external cues (like sunrise and sunset) to obtain information on time? This is a complicated question, and here we discuss it in just a limited way.

One of the most important advantages of endogenous clocks is that they allow animals to anticipate future events, just as our watches let us anticipate when future events will occur. For example, imagine a nocturnal bat that during the daytime lives in a deep, dark cave where it cannot see if the Sun is up or not. Being nocturnal, the bat would sleep in the daytime. During its sleeping time, if the bat lacked an endogenous clock, it would have no idea when to wake up. Without an endogenous clock, the bat would—now and then—need to take a break from sleep and fly to its cave entrance to see if night had arrived. With an endogenous clock, however, the bat has an endogenous sense of time: it knows the time of sunset without needing to see the Sun. Investigators have studied bats in their caves, and from these studies we know that bats start flying around—deep in their dark caves—in advance of sunset. They anticipate nightfall.

Circannual clocks are known to control annual hibernation rhythms in some species of mammals, and they control annual reproductive rhythms in some species, such as sheep. In cases like these, the animals can anticipate when they will be hibernating or reproducing. Thus they can prepare.

Biological clocks permit some animals to have a sun compass

There are several more-subtle ways that endogenous clocks provide animals with abilities that they could not have if they were completely dependent on external cues to know the time. Here we discuss one of these, the sun compass.

Many types of animals have a **sun compass**, meaning that they can determine north, south, east, and west from observations of the Sun. Homing pigeons provide an excellent example. These pigeons fly back to their home nests even after being transported tens of kilometers away. For instance, during both World Wars, homing pigeons were extensively used to carry messages. They were reared in home nests near command centers. They then were carried into battle and released with critical messages that they transported back to the command centers. A sun compass is one mechanism (of several) that homing pigeons use to find home.

To understand the challenge of using the Sun as a compass, suppose for a moment that you are totally ignorant of the time of day and that you are suddenly dropped at an unfamiliar location. Suppose also that you see the Sun just above the horizon. With the Sun in that position, you would reason that when you face the Sun, you are facing either east or west. Without knowing the time of day, however, you would not know which: east or west? From thought exercises like this, researchers hypothesized that when homing pigeons use the Sun as a compass, the birds not only must observe the position of the Sun but also must know the time of day. The birds need timing information to calculate compass directions from the Sun's position.

The researchers hypothesized that the pigeons use their circadian clock to determine the time of day. To test this hypothesis, the researchers (working in the Northern Hemisphere) released homing pigeons at a place from which the birds needed to fly south to return home. The researchers observed that if the

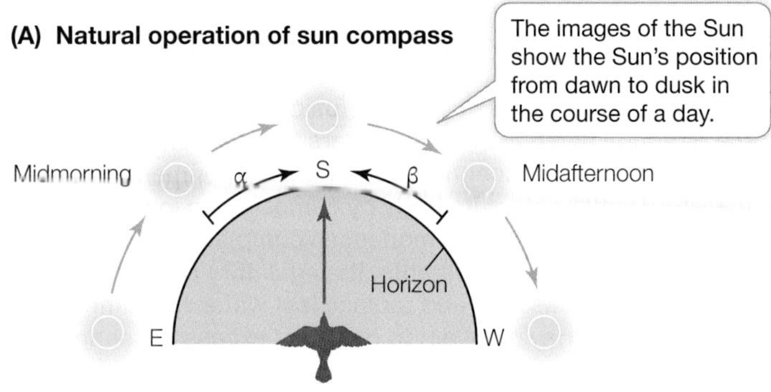

(B) Operation of sun compass in a pigeon in which the circadian clock is entrained to be 6 hours off

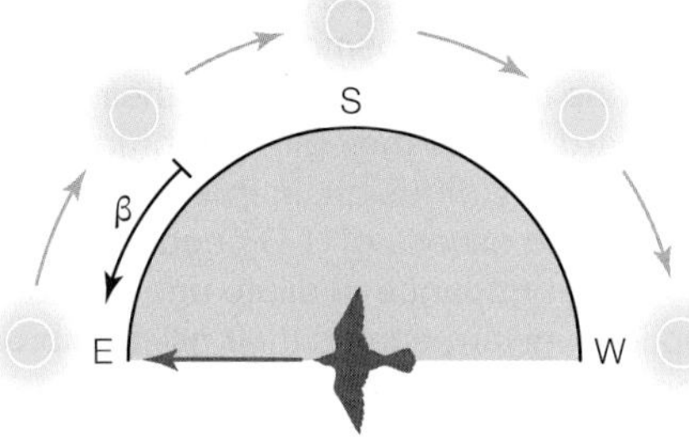

FIGURE 29.24 An Experiment Demonstrates that the Sun Compass of a Homing Pigeon Uses Information from the Circadian Clock A pigeon that had to fly south to return home was tested in the Northern Hemisphere. Here we see a bird's-eye view of the southern horizon. **(A)** The pigeon with its circadian clock functioning normally, showing appropriate flight angles at midmorning and midafternoon. **(B)** The pigeon with its circadian clock manipulated so the pigeon interpreted midmorning to be midafternoon.

Animation 29.2 **Time-Compensated Solar Compass**

pigeons were released *at any time of day,* they flew south. Using **FIGURE 29.24A**, let's analyze the challenge a pigeon faces. To get started, think about the compass position of the Sun as it crosses the sky each day. (The Sun's compass position can be determined by dropping an imaginary vertical line from the Sun to the horizon and noting where the line intersects the horizon between east and west.) The Sun is in the east at sunrise, southeast at midmorning, south at noon, southwest at midafternoon, and west at sunset. Thus a pigeon can't simply fly at a fixed angle relative to the Sun to fly south; it must adjust the angle according to the time of day. As Figure 29.24A shows, if a pigeon is released to fly home in the midmorning, it must fly at angle α to the right of the Sun's position. But if the pigeon is released to fly home in the midaftenoon, it must fly at angle β to the left of the Sun's position. To determine the correct angle, the pigeon must know the time of day. The researchers hypothesized that a pigeon uses its circadian clock to determine the time.

To test their hypothesis, the researchers manipulated pigeons so that their circadian clock was 6 hours "off,' meaning the time of day measured by the bird's clock was 6 hours different from the actual time of day. They did this by putting the pigeons in rooms where the lights were turned on 6 hours before sunrise and turned off 6 hours before sunset. The pigeons' circadian clocks became entrained to this artificial light–dark cycle after 2 weeks or so. The researchers reasoned that if their hypothesis was correct, these pigeons would not be able to find home using the Sun because —using their wrongly set circadian clock—they would calculate the flight angle incorrectly.

The pigeons in fact performed exactly as the hypothesis predicted when they were taken outdoors again and tested for their ability to return home. For example, when released in midmorning, they flew east! With their timekeeping mechanism off by 6 hours, they interpreted the time of day to be midafternoon (6 hours later than the actual time, midmorning). They thus flew at an angle β to the left of the Sun and headed east (**FIGURE 29.24B**). This result provided strong evidence that the birds' circadian clock was a key part of their sun compass.

REVIEW & APPLY | 29.4

R Animals have self-contained, metabolic mechanisms (i.e., endogenous mechanisms) of keeping track of time, known as biological clocks. Biological clocks that have free-running cycles of about 24 hours are called circadian clocks (*circa*, "about"). Some animals also have circannual or circatidal clocks. A circadian clock does not keep perfect time; it times events on only an approximate 24-hour cycle. Thus circadian clocks run fast or slow, and must be entrained by external cues (most commonly sunrise or sunset) to time events on an exact 24-hour cycle. A biological clock enables an animal to anticipate, and therefore prepare for, future events. A circadian clock also enables an animal to have a sun compass—an ability to determine compass directions by observing the position of the Sun.

1. How does exogenous timing differ from endogenous timing?
2. Suppose an animal wakes up each day at a time controlled by a circadian clock. Over a period of a month, how will the animal's wake-up time change if entrainment is prevented?
3. Ricky, who prides himself on skepticism, read online that daily biological clocks do not come even close to keeping correct time. They can be off by 10 minutes every day, his source said. Trying to impress his latest date, he was overheard to say, "So it's obvious, right? They're completely useless." With Ricky's wisecrack in mind, think about organisms before biological clocks evolved. If mutations that could give rise to an early form of clock occurred, why might natural selection have favored them?

Water (H_2O) and dissolved ions (such as Na^+ and Cl^-) move constantly by osmosis and diffusion across the permeable gill membranes of fishes. How do these movements of water and ions change as a salmon switches between seawater and freshwater habitats? What concentrations do salmon maintain in their tissue fluids, and how are these concentrations relevant? Do the fishes maintain homeostasis? How?

Answer: When a salmon is in seawater, its body fluids are more dilute that the environmental water. Thus water leaves its body by osmosis and ions enter its body by diffusion. These movements reverse direction when the salmon enters fresh water, where its body fluids are more concentrated than the environmental water and therefore water enters its body by osmosis and ions leave its body by diffusion. Thus the salmon faces completely opposite challenges in the two environments. Its regulatory processes must act to get rid of excess ions and replace lost water when the fish is in seawater, but they must act to get rid of excess water and replace lost ions when the fish is in fresh water. Salmon maintain almost the same osmotic pressure and ion concentrations in their tissue fluids whether they are living in seawater or fresh water. They thus maintain homeostasis, which is defined to be internal stability. In seawater, they drink and absorb H_2O from the ingested seawater to replace H_2O they lose by osmosis. They also produce only a scant amount of urine (helping keep H_2O in their body), and they pump ions outward across their gill membranes, helping excrete excess ions they gain by diffusion. In fresh water, they do not drink (they already have an excess of H_2O because of inward osmosis). They produce an abundance of dilute urine (voiding excess H_2O) and pump ions inward across their gill membranes (helping replace the ions they lose by diffusion).

29 VISUAL SUMMARY

29.1 Animals Prosper in Diverse Thermal Environments

- Mammals have evolved physiological specializations to survive in extreme environments, including enhanced body insulation, **HIBERNATION**, and finely controlled **REGIONAL HYPOTHERMIA** in cold environments (**FIGURES 29.2** and **29.5**), and the ability to tolerate hyperthermia during the daytime in hot, dry environments.
- Small-bodied animals rely strongly on behavioral specializations, including the use of **MICROENVIRONMENTS** and **MICROCLIMATES**, to evade extreme environments. In hot environments, for example, behavioral specializations include burrowing and nocturnality.
- Fishes that have evolved in different thermal environments have evolved different molecular forms of key metabolic enzymes.

FIGURE 29.2

FIGURE 29.5

Countercurrent heat exchange

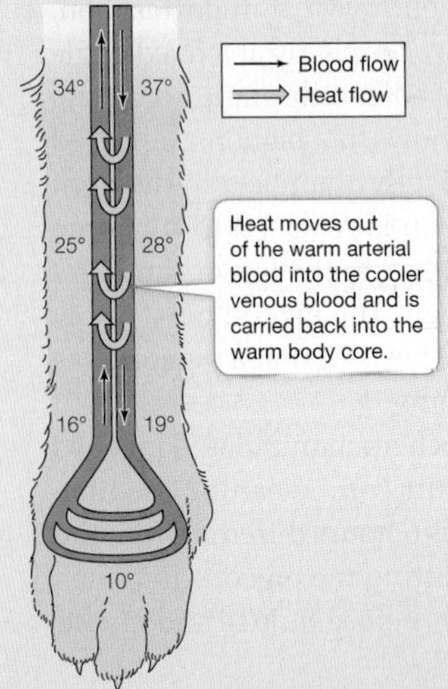

QUESTIONS

1. Mammalian hibernators are sometimes said to alternate between temperature regulation and conformity. Explain.
2. Why is the cooling mechanism of sweating inappropriate for Grant's gazelles? How do these animals adjust to their hot, dry environment?
3. How can a polar fish species die of heat stress at +5°C and a tropical species die of cold stress at +5°C?

29.2 Animals Live in the Ocean, Fresh Water, and Intermediate Salinities

- Animals living in aquatic environments often experience steady gains or losses of H_2O by osmosis, and steady gains or losses of ions (e.g., Na^+, Cl^-) by diffusion (**FIGURE 29.14**). They must counteract these processes, at an energy cost.
- Over evolutionary time, animals have gone through a series of invasions into new habitats that have affected the concentrations of body fluids in modern animals and the particular regulatory challenges animals face.

FIGURE 29.14

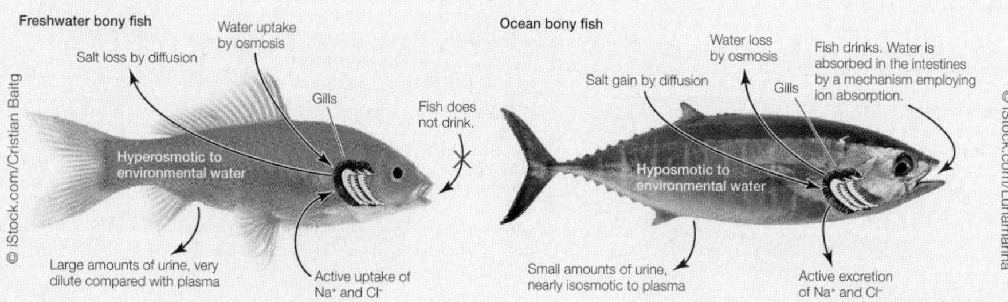

© iStock.com/Cristian Baitg

© iStock.com/Lunamarina

QUESTIONS

1. How does the evolutionary history of ocean bony fishes explain why they are not isosmotic to seawater?
2. Explain why the blood plasma of humans has an osmotic pressure of about 0.3 *Osm*.
3. Freshwater fishes in general do not drink. Why not?

29.3 The Phenotypes of Individual Animals Can Change in Response to Environmental Change

FIGURE 29.20

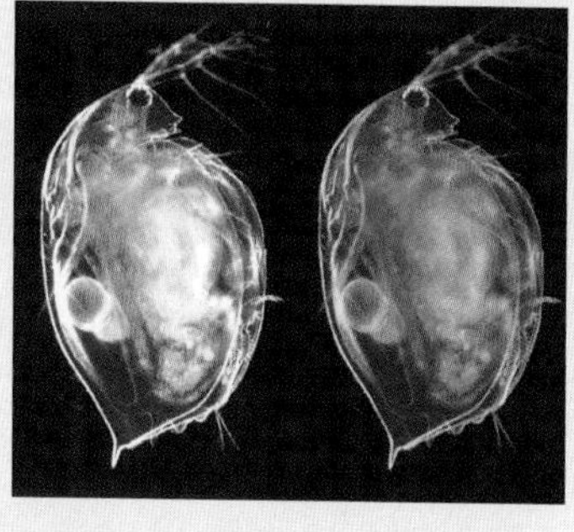

Courtesy of Shinichi Tokishita

- An individual animal can exhibit more than one phenotype during its life—a phenomenon termed **PHENOTYPIC PLASTICITY**—because one genotype can express two or more phenotypes (**FIGURE 29.20**).
- Phenotypic plasticity is often displayed by individuals that experience long-term changes of their environment, to which they **ACCLIMATE** or **ACCLIMATIZE**.

1. Give an example of phenotypic plasticity in which inducible enzymes are involved.
2. Explain why it might be advantageous for a small mammal to exhibit phenotypic plasticity in brown adipose tissue (BAT), rather than constantly having a certain amount of BAT as a fixed trait.

29.4 Animals Have Biological Clocks Tuned to Cycles in Their Environment

Go to ANIMATIONS 29.1 and 29.2

FIGURE 29.22

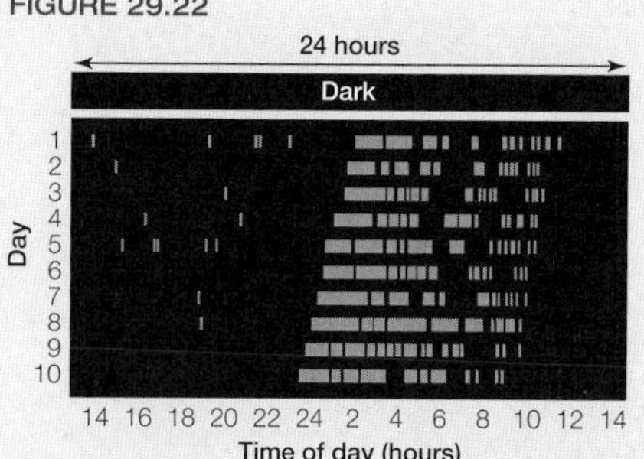

- Animals must coordinate their daytime–nighttime rhythms with the daytime–nighttime rhythm in the environment. They sometimes use external cues. They also have evolved endogenous **BIOLOGICAL CLOCKS** that help them remain synchronized with the regular variations in their environment (**FIGURE 29.22**).
- In individual animals, most or all cells have their own **CIRCADIAN CLOCK**. These cell clocks are coordinated by a master clock, which in mammals is located in the suprachiasmatic nuclei in the brain.
- The **FREE-RUNNING** periodicities of circadian clocks are different than 24 hours, meaning **ENTRAINMENT** is required for accuracy.

Activity 29.1. Concept Matching: Animals in Their Environment

1. An animal's biological clocks are part of its control system. Explain what this means, possibly by use of an example.
2. Why are circadian clocks termed "circadian"?
3. Name two advantages animals can gain by having endogenous timekeeping.

30 Breathing and Circulation

Tunas rank with the elite endurance athletes on Earth. Like wolves, they are carnivores that hunt in groups and depend on speed, strength, and awesome teeth to attack prey, which often themselves are fast-moving. Searching for prey, tunas swim continuously, day and night, at speeds of one to two body lengths per second. For bluefin tunas, this means traveling continuously at speeds of 2–6 meters per second.

The swimming muscles of a tuna require O_2 at a high, steady rate to synthesize ATP fast enough—and consistently enough—to support the tuna's strong, sustained swimming. A tuna's source of O_2 is the seawater of its external environment. What properties would you expect to find in the breathing and circulatory systems of a tuna, and why?

You will find the answer to this question on page 763.

KEY CONCEPTS

30.1 The Pathway for Oxygen: Oxygen Must Travel from Environment to Mitochondria

LEARNING OBJECTIVES

By the end of this key concept you should be able to:

30.1.1 **Explain the pros and cons of diffusion and bulk flow as mechanisms of respiratory gas transport.**

30.1.2 **Explain why diffusion and bulk flow can be said to alternate in the pathway for O_2 delivery to the mitochondria.**

30.1.3 **State Fick's law and verbally outline what it means for O_2 diffusion.**

Oxygen (O_2) is used in the mitochondria, and that's where it must be delivered to perform its crucial role in aerobic metabolism. Recall why cells require O_2. When a cell carries out the aerobic breakdown of glucose or other food molecules (see Key Concepts 5.2 and 28.4) to obtain energy for ATP synthesis, electrons are removed from the food molecules that are being oxidized. These electrons must have a final resting place after they pass through the electron transport chain in the cell's mitochondria. O_2 serves as the final resting place. The electrons combine with O_2 to reduce the O_2 and make water (see Figure 5.9), which then mingles with the rest of the water in an animal's body. If you are puzzling over how electrons and O_2 can make H_2O, the answer is that H^+ ions (protons) from the surroundings of the reacting molecules join in the reaction.

An animal obtains its O_2 from its environment. Thus O_2 must travel the *pathway for oxygen* from the environment to the mitochondria. Think, for example, about the pathway an O_2 molecule must travel from the atmosphere to a mitochondrion inside a muscle cell in your calf (**FIGURE 30.1**). First the O_2 molecule must travel deep into your lungs (step 1 in the figure). Then (step 2) it must cross two simple epithelia to enter your blood: the alveolar epithelium forming your lung wall and the capillary epithelium forming the wall of a blood capillary (recall from Chapter 28 that a *simple* epithelium consists of just one cell layer). The O_2 molecule must then travel in the circulating blood from your thorax to your leg (step 3) to reach a blood capillary next to the muscle cell. Finally (step 4), the molecule must cross the epithelial wall of the blood capillary, pass through the cell membrane to enter the muscle cell, and then travel through the cell cytoplasm to reach and enter the mitochondrion.

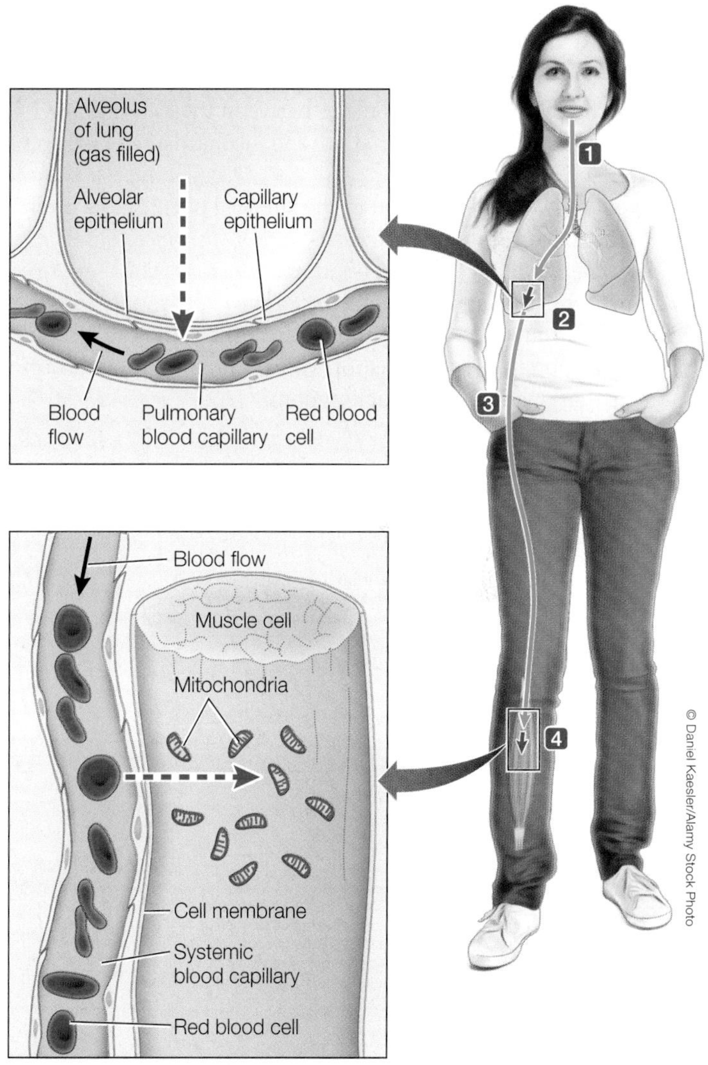

FIGURE 30.1 The Pathway for Oxygen The pathway that O_2 molecules travel from the atmosphere to an animal's mitochondria—in this example inside a human muscle cell—entails four steps that are detailed in the text. Mitochondria are actually much smaller relative to the muscle cell than shown here. The pathway in an aquatic animal such as a tuna is similar to that pictured here, except that the starting point is the water in the tuna's environment, and O_2 enters the blood in the gills instead of the lungs.

The aerobic breakdown of food molecules produces carbon dioxide (CO_2), because CO_2 is one of the products when molecules move through the citric acid cycle (see Figure 5.8). Unlike the H_2O produced when electrons combine with O_2, CO_2 cannot be permitted to accumulate in an animal's body water. CO_2 tends to lower the pH of body fluids if it accumulates in them (explaining why it is often called a "gaseous acid"), and it can have other disadvantageous effects. Thus as an animal takes in O_2 from its environment and transports it to cells, the animal must also void CO_2 by transporting it from its cells to the environment. Together, O_2 and CO_2 are called the **respiratory gases**.

LINK You can review aerobic cellular respiration and the production of ATP in **Key Concept 5.2**.

O_2 and CO_2 are transported by diffusion and bulk flow

How do the respiratory gases move between the external environment and cells? That is, what are the mechanisms of O_2 and CO_2 transport?

Given that oxygen is crucial to the survival of cells, it is often surprising to learn that *there are no active transport mechanisms for* O_2. This means that chemical-bond energy from ATP—energy from metabolism—cannot be used to drive O_2 through an epithelium or through a cell membrane. Much the same can be said for CO_2, although active transport of CO_2 is known in certain limited situations in the animal kingdom.

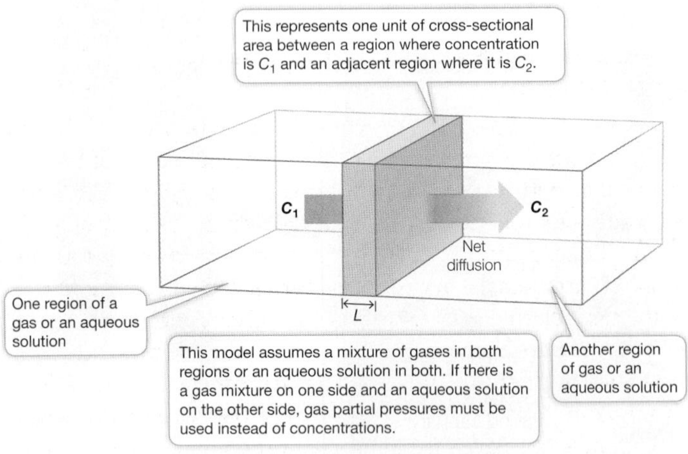

FIGURE 30.2 A Physical Model of Gas Diffusion C_1 and C_2 are concentrations, with C_1 assumed to be higher than C_2. L is the distance separating the region of concentration C_1 from the region of concentration C_2.

With no active transport for O_2, the transport of O_2 can be achieved by only two mechanisms: diffusion—which we discussed in Key Concept 4.2—and bulk flow. **Bulk flow** refers to a flow of matter from one place to another. Winds and water currents are forms of bulk flow in the physical world. When you breathe, you create a bulk flow of air in your lung passages, and O_2 can be carried from place to place in the flow. Similarly, the beating of your heart causes a bulk flow of blood through your blood vessels, and O_2 can be transported in that flow. CO_2 can also be transported by diffusion and bulk flow.

The diffusion of gases can be highly effective if distances are very short, but it has important limitations

Diffusion results from random molecular motions (see Key Concept 4.2). If a certain type of molecule is more concentrated in one region of a solution than in neighboring regions, it is a simple statistical fact that random motions will carry more molecules away from the concentrated region than into it. In this way, the random motions cause a net transport of molecules out of the concentrated region and into neighboring dilute regions. When we say that molecules "diffuse," we mean they move in a net fashion from place to place by this mechanism.

Within an aqueous solution or within a gas, the rate of diffusion follows a simple and intuitively reasonable equation known as **Fick's law**:

$$\text{Rate of diffusion per unit of cross-sectional area} \propto \frac{C_1 - C_2}{L}$$

where C_1 represents the concentration of a certain type of molecule in a defined three-dimensional region of an aqueous solution or gas, C_2 represents the concentration of that type of molecule in an adjacent region, and L (representing "length") is the distance between the two regions. To understand the equation, let's think of O_2 as the molecule that is undergoing diffusion (we could just as easily use any other specific molecule). Now, picture an imaginary plane that is perpendicular to the direction of net O_2 diffusion between the two regions. The equation allows us to calculate the rate of diffusion per unit of cross-sectional area in this plane (**FIGURE 30.2**). What exactly do we mean by *rate of diffusion*? It is the rate at which O_2 molecules move in a *net fashion* from the higher-concentration region to the lower-concentration region (e.g., the net number of O_2 molecules that move per minute).

The equation for Fick's law tells us that the rate of diffusion depends on the difference in concentration: molecules move faster from one region to the other if there is a large concentration difference between the regions rather than a small difference. The equation also tells us that the rate depends on how far apart the regions are: molecules move faster from one region to the other if there is a short distance separating the regions rather than a longer distance.

Let's now ask a key question: Can diffusion move lots of O_2 (or CO_2) from one place to another in a short period of time? The answer is both yes and no, and this distinction is one of the most important concepts to understand about diffusion. The answer depends on the distance involved.

Suppose O_2 on the outside of a cell is diffusing across the cell membrane into the cell. The distance, *L*, for this diffusion is very short: the thickness of a cell membrane—about 0.01 micrometer (μm). Thus this diffusion will be very fast. In fact, if you were to arrange to have a high O_2 concentration outside a cell and a low concentration inside, diffusion would move enough O_2 into the cell in just 0.1 microsecond to halve the concentration difference between outside and inside! This calculation explains why all cells depend simply on diffusion for O_2 to pass into them through their cell membrane (remember: there is no active transport for O_2).

By contrast, suppose O_2 present in body fluids near your lungs is diffusing through your body fluids to a muscle cell in your calf. For this diffusion, *L* is much longer—about 1 meter. Diffusion across this distance therefore will be slow—far slower than you might imagine. If you had a high O_2 concentration near your lungs and a low concentration in your calf muscle, diffusion would require more than *30 years* to halve the concentration difference! Diffusion would be totally useless for getting O_2 to your calf muscle.

We see, then, that diffusion can be highly effective for moving O_2 over very short distances, but it is not effective over long distances. The founder of modern respiratory physiology, August Krogh, calculated an important rule of thumb for understanding diffusion. He considered O_2 diffusing through a tissue to reach mitochondria that are carrying out cellular respiration at an average rate for animals. Then he asked, how thick can the tissue be and still have the mitochondria receive O_2 at the rate they need? He obtained his answer—about 0.5 millimeter—by doing a calculation with Fick's law. Today physiologists still use Krogh's rule of thumb and view 0.5 millimeter as the greatest distance over which O_2 can diffuse fast enough through tissue or tissue fluids to meet metabolic needs.

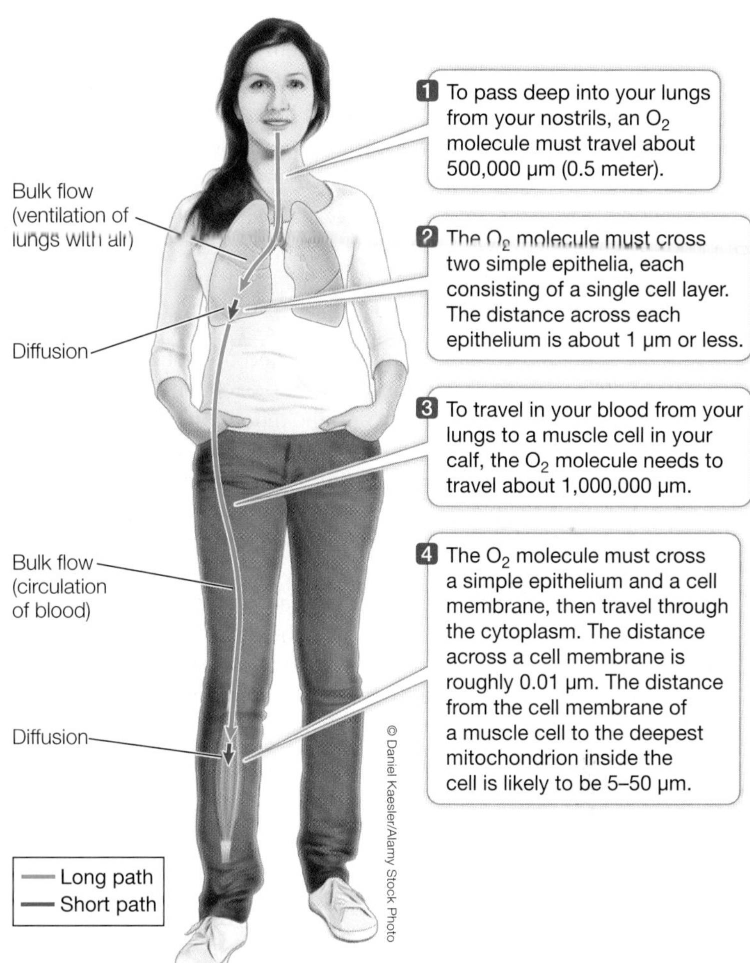

FIGURE 30.3 O_2 Transport into an Animal Often Occurs by Alternating Bulk Flow and Diffusion This alternation is clear if you read from top to bottom. Compare this figure with Figure 30.1.

The pathway for oxygen often consists of alternation of bulk flow and diffusion

Krogh's rule of thumb highlights a limitation of diffusion: diffusion is not suitable for long steps in the pathway for oxygen. This limitation explains why many animals have evolved mechanisms of bulk flow to transport O_2 over long distances.

Diffusion and bulk flow alternate along the pathway for oxygen in humans and other mammals (**FIGURE 30.3**). As O_2 first enters the body, it is carried by bulk flow over the long distance from the atmosphere to the depths of the lungs. This bulk-flow system is what we usually call *breathing*. Next, O_2 diffuses across the thin epithelia in the lungs to pass into the blood. The O_2 is then carried by a second bulk-flow system—which we usually call *circulation of the blood*—to reach the legs. Then O_2 exits the blood in the legs by diffusing across the epithelium of a blood capillary and across a cell membrane, and finally it diffuses through the cytoplasm to reach a mitochondrion.

Partial pressures are often used to study gas diffusion

The version of Fick's law we have discussed—which uses gas concentrations—applies when diffusion occurs between two regions filled with gas mixtures or between two regions filled with aqueous solutions (see Figure 30.2). However, it does not apply when diffusion occurs between a gas mixture and an aqueous solution. This is because diffusion is governed by the *chemical activity* of the gas molecules that are diffusing, and concentration does not directly measure chemical activity. Instead, the chemical activity is measured by **partial pressure**. Partial pressure is easily defined in a mixture of gases (not so easily in aqueous solutions): the partial pressure of a particular gas is the portion of the total pressure of a gas mixture that is exerted by that gas. For example, the partial pressure of O_2 in a mixture of gases is the portion of the total gas pressure that the O_2 in the mixture exerts.

As we go forward in this chapter, we will often use partial pressures. For our purposes, it will be sufficient to know that a gas always diffuses from regions where its partial pressure is high to ones where its partial pressure is low. This is an entirely general principal that applies whether diffusion is occurring between two gas mixtures, between two aqueous solutions, or between a gas mixture and an aqueous solution. Fick's law, in fact, can be rewritten by using partial pressures in place of concentrations, and then it becomes a general law that applies in all these circumstances.

REVIEW & APPLY | 30.1

R Most cells require a constant supply of O_2 and continuous removal of CO_2. These respiratory gases are exchanged between an animal's environment and the mitochondria in its cells by a combination of diffusion and bulk flow. Diffusion—described by Fick's law—can carry gases extremely rapidly over microscopic distances but is very slow over long distances. Bulk flow can carry gases rapidly over long distances. In humans and other mammals, bulk flow and diffusion alternate in the pathway for oxygen from environment to mitochondria. O_2 is carried from the environment to the gas exchange membrane in the lungs by bulk flow (breathing), crosses the gas exchange membrane by diffusion, is carried from the lungs to cells throughout the body by bulk flow (circulating blood), and diffuses from the blood to the mitochondria.

A

1. Where in the human body do inhaled O_2 molecules first cross a cell membrane?
2. Review the equation for Fick's law. What does *L* mean, and what happens to the rate of diffusion as *L* increases?
3. Many fishes live in waters where the O_2 concentration (and partial pressure) is sometimes high and sometimes low. When O_2 is low in the environmental water, why can't a fish simply expend energy and do metabolic work to take up the O_2 it needs across its gill membranes?

Now we will shift our focus to the structure and function of the diverse breathing organs animals have evolved.

30.2 | Animals Have Evolved Diverse Types of Breathing Organs

LEARNING OBJECTIVES

By the end of this key concept you should be able to:

30.2.1 **State the important ways that water and air differ as respiratory environments.**

30.2.2 **Contrast the role of the circulatory system in meeting the O_2 needs of a mammal and an insect.**

30.2.3 **Describe how gills differ from lungs.**

30.2.4 **Explain how the specific directions of water and blood flow in a fish's gills affect how well O_2 is taken up.**

What is breathing? To answer, we first need to define the gas exchange membranes. The **gas exchange membranes** of an animal consist of the thin layers of tissue—usually composed of just one or two simple epithelia—where the respiratory gases move between the animal's environmental medium—air or water—and its internal tissues. Our own gas exchange membranes, for example, are the membranes deep in our lungs where O_2 diffuses out of the air and enters our blood and where CO_2 leaves our blood and enters the air. All animals have gas exchange membranes of some type.

Breathing—also called **external respiration**—is the process by which O_2 from the outside environment gets to the gas exchange membranes and the process by which CO_2 moves from those membranes to the outside world. In a fish, for example, breathing transports O_2 and CO_2 between the watery environment in which the fish lives and the outer surfaces of the gills. In people, breathing transports O_2 and CO_2 between the general atmosphere and the outer surfaces of the lung membranes where exchange with the blood occurs.

Animals have evolved specialized breathing organs

In their breathing organs, animals typically have evolved a *large surface area* of thin gas exchange membrane. The large surface area is achieved by complex patterns of tissue folding and branching. Consider your own lungs. The area of the gas exchange membranes in them is about 130 square meters! That's the same as the floor area of a classroom for 80 students. All that surface area can fit into the small volume of your lungs because of branching and folding. The large surface area is important because it enhances the rate at which O_2 and CO_2 can diffuse across the gas exchange membranes. Recall that Fick's law tells us the rate of diffusion *per unit of cross-sectional surface area*. If a gas exchange membrane has a large surface area, diffusion will occur faster across the whole membrane than if it were smaller in surface area. Remember also that Fick's law tells us that diffusion occurs faster if the distance to be covered is short rather than long. As we have stressed, gas exchange membranes are thin.

Physiologists categorize breathing organs into two fundamental types: lungs and gills (**FIGURE 30.4**). In **lungs** the branchings and foldings of the gas exchange membrane are invaginated (folded inward) into the body, and the passages within the breathing organ *contain* the environmental medium (air or water). In **gills** the branchings and foldings of the gas exchange membrane are evaginated (folded outward) from the body and are *surrounded* by the environmental medium. In general, air-breathing animals have lungs and water-breathing animals have gills. Exceptions occur, however.

Ventilation refers to bulk flow of air or water between the gas exchange membrane and the outside world. Breathing often occurs by ventilation. Humans breathe in this way: we create a bulk

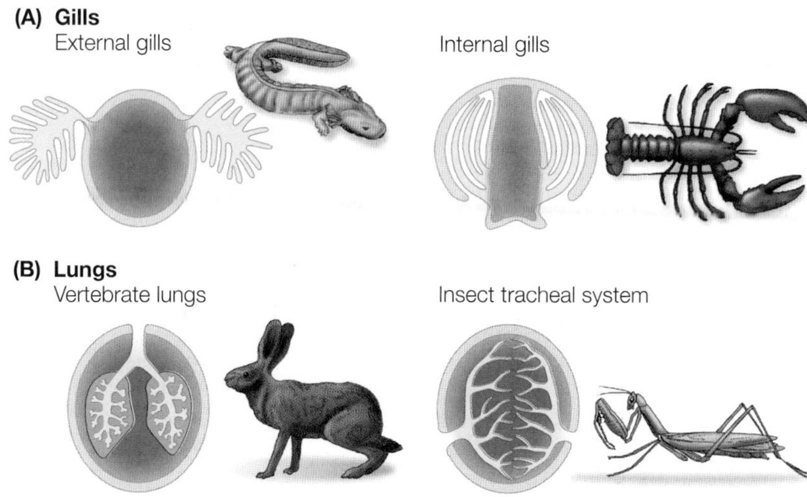

FIGURE 30.4 Gills and Lungs Physiologists classify all specialized breathing organs (blue in these diagrams) as either gills or lungs, depending on whether they are evaginated structures **(A)** or invaginated structures **(B)**. The use of the word "lungs" does not always match everyday usage. For example, the tracheal breathing system of an insect (discussed later in this chapter) is a lung because it is invaginated and contains air (rather than being surrounded by air), but it is rarely called a lung in everyday language. Gills are subdivided into external gills (which project directly into the animal's environment) and internal gills (which project into a protective gill chamber that is ventilated with water).

flow of gas into and out of our lungs when we inhale and exhale. However, breathing can also occur entirely by diffusion (termed *diffusion breathing*) if the distances to be covered are short. Many tiny insects, such as gnats, are believed to breathe entirely by diffusion rather than ventilation.

When lungs are ventilated, air usually moves both in and out by the same lung passages. This type of ventilation is described as **tidal** because air flow occurs first in one direction and then in the opposite direction in the same airways—in and out like the ocean tides. Humans exemplify tidal ventilation.

In animals having gills, ventilation usually involves pumping water in a one-way stream through the passages that lead to the gills, over the gill surfaces, and through the passages that lead away from the gills. This ventilation is termed **unidirectional**.

How else are lungs and gills specialized? Most types of breathing organs need circulatory specializations because cells throughout the body depend on blood flow to receive O_2 taken up in the lungs or gills. **Perfusion** refers to blood flow through the capillaries or other small blood vessels of a tissue. Perfusion of the gas exchange membrane in a breathing organ must be rapid enough for the blood to pick up O_2 as fast as it is delivered by breathing. Then blood must circulate throughout the body at a high enough rate to deliver O_2 by bulk flow from the lungs or gills to meet the O_2 requirements of all the cells.

All things considered, we can say that most types of specialized breathing organs have three elements:

1. A ventilation system that brings air or water rapidly by bulk flow to the gas exchange membrane
2. A thin gas exchange membrane of large surface area, with air or water on one side and blood on the other side
3. A high rate of blood perfusion provided by the circulatory system, which pumps blood rapidly between the lungs or gills and the rest of the body

Air and water are very different respiratory environments

Before we look in detail at some of the specialized breathing organs that animals have evolved, it is important to recognize that air and water are very different respiratory environments. Let's start with diffusion in discussing the differences. Diffusion takes place much more rapidly through air than water. The rate of diffusion of O_2, for example, is about 200,000 times faster through air than through water if all other relevant factors (e.g., distance and area) are kept constant.

Sea turtle nests that become flooded provide a striking illustration of the difference in diffusion rates through air and water (**FIGURE 30.5**). Loggerhead

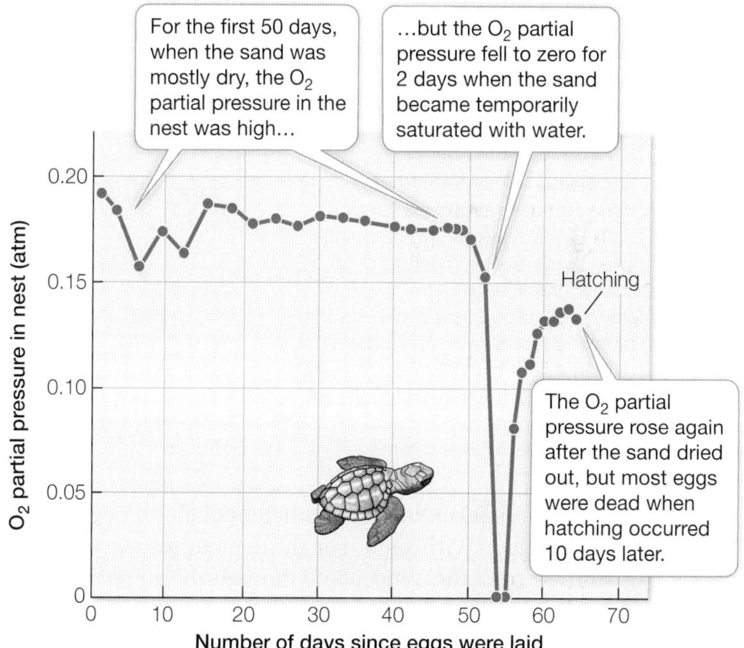

FIGURE 30.5 The O_2 Supply for Sea Turtle Eggs Depends on whether the Sand in which They Are Buried Is Dry or Wet The O_2 partial pressure in a nest of sea turtle eggs depends on how fast the developing turtles use O_2 and how fast O_2 is resupplied by diffusion from the atmosphere. During the first 50 days of development in the nest studied here, the sand was dry and O_2 diffused into the nest as fast as the developing turtles used O_2—keeping the O_2 partial pressure in the nest high. When the sand became soaked with water for 2 days, the O_2 the developing turtles used was not replaced, and the O_2 partial pressure in the nest fell to zero. (atm = atmosphere, a unit of pressure.) (Data kindly provided by Nathan A. Miller.)

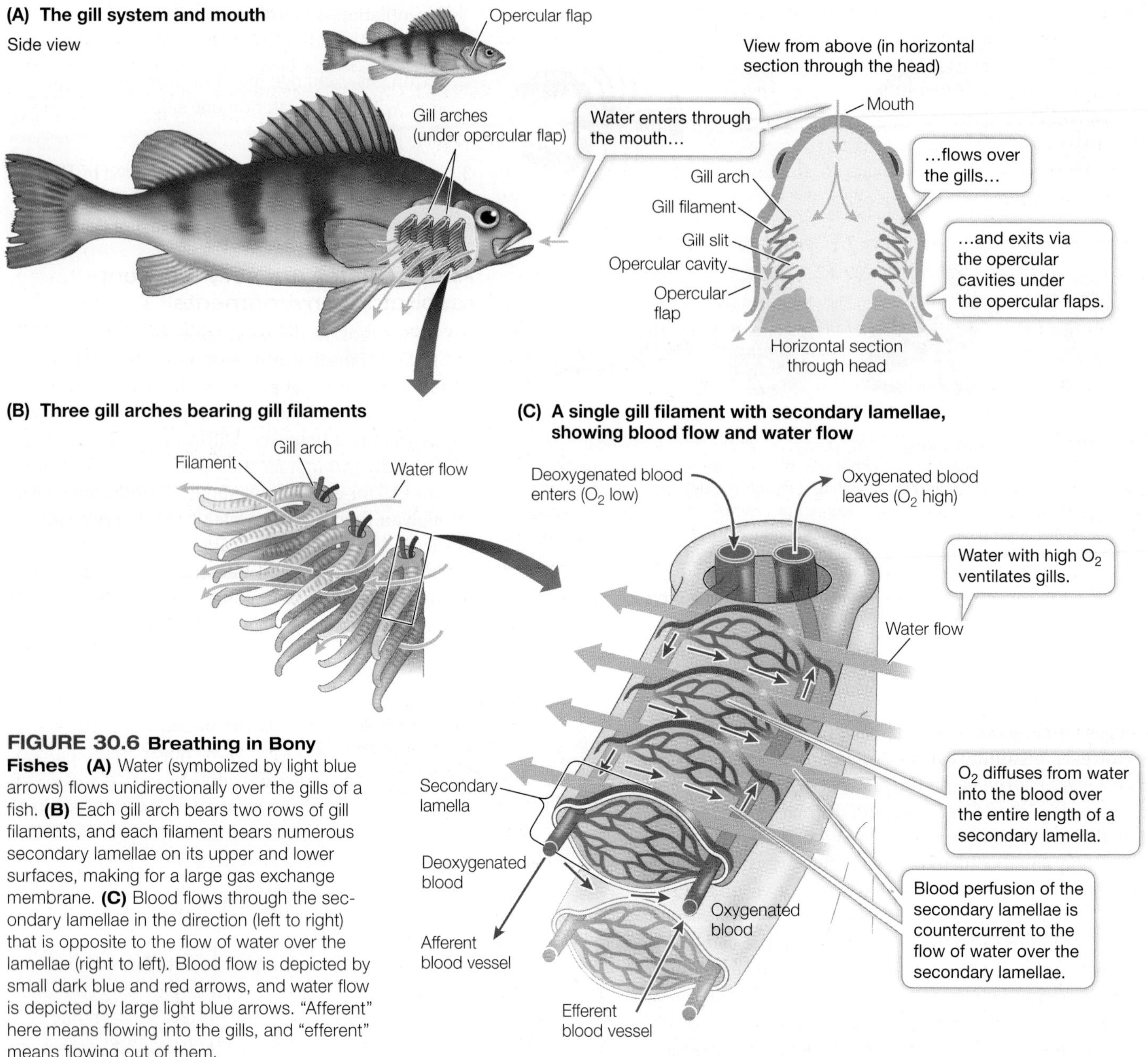

FIGURE 30.6 Breathing in Bony Fishes **(A)** Water (symbolized by light blue arrows) flows unidirectionally over the gills of a fish. **(B)** Each gill arch bears two rows of gill filaments, and each filament bears numerous secondary lamellae on its upper and lower surfaces, making for a large gas exchange membrane. **(C)** Blood flows through the secondary lamellae in the direction (left to right) that is opposite to the flow of water over the lamellae (right to left). Blood flow is depicted by small dark blue and red arrows, and water flow is depicted by large light blue arrows. "Afferent" here means flowing into the gills, and "efferent" means flowing out of them.

sea turtles lay their eggs about 40 centimeters deep on ocean beaches and cover them with sand. For the eggs to receive O_2, the O_2 must diffuse from the atmosphere through the air spaces in the sand. Nathan Miller outfitted nests with tiny probes that continually monitored the partial pressure of O_2 in the nests. The partial pressure remained high when the sand was dry (and the spaces between sand grains were therefore air-filled), indicating that diffusion of O_2 from the atmosphere kept up with O_2 use by the eggs (Krogh's rule does not apply to diffusion through air). However, if the sand became flooded with water—so the spaces between sand grains were water-filled—the O_2 partial pressure in a nest could fall rapidly to zero, causing the death of many eggs.

In addition to its slow rate of diffusion through water, O_2 is not very soluble in water. *That is, only a limited amount of O_2 can be dissolved in each unit volume of water.* Consider that when you put sugar in water, the sugar molecules become "invisible"—we say they are *dissolved*. O_2 does exactly the same thing. When it dissolves in water, the individual O_2 molecules are dispersed among water molecules and become invisible (bubbles, regardless of how tiny, are not in solution). How much O_2 can be dissolved in water? Suppose we have some fresh water in a jar and we bubble it continuously with air so that the water and air are at equilibrium (a state in which they have the same O_2 partial pressure). If we do this study at 12°C, the air will contain about 200 milliliters (ml) of O_2 in each liter. The water, however, will contain only about 8 ml of dissolved O_2 in each liter! The solubility of O_2 in water is so low that the amount present per liter is only about 4 percent as high as in air—even when the water is steadily bubbled with air.

The solubility of O_2 in water decreases as water temperature rises. Thus the concentration of O_2 in water tends to decrease as water warms. This is a substantial effect: whereas fresh water at 0°C dissolves 10.2 ml of O_2 per liter, the same water warmed to 24°C dissolves only 6.2 ml per liter.

All of the animal species alive today that have very high metabolic rates are air breathers. Biologists think this is true because water is a challenging place to live in terms of getting enough O_2.

Fishes have elaborate gills in which gas exchange occurs across secondary lamellae

The gills of a bony fish (a fish with a bony skeleton) are supported by two sets of bone-reinforced gill arches, located on each side of the head (**FIGURE 30.6A**). The gill arches resemble vertical pillars with spaces—termed gill slits—in between. The mouth cavity is on the inner side of each set of gill arches. A gill flap, called the *opercular flap*, covers each set of gill arches on the outer side. A fish pumps water unidirectionally: first into its mouth, then through its mouth cavity and through its gill slits, over its gills, and finally out from under the opercular flaps (see Figure 30.6A).

Fish gills have an enormous surface area for gas exchange because they are extensively branched and folded. Each gill of a bony fish consists of two rows of long, flat filaments called gill filaments (**FIGURE 30.6B**). The upper and lower surfaces of each gill filament are folded into many flat folds—called secondary lamellae (singular *lamella*)—that run perpendicular to the long axis of the filament. The secondary lamellae are the principal sites of respiratory gas exchange (**FIGURE 30.6C**). Blood flows through small blood passages within each lamella, and these blood passages are separated from the water flowing over the gills by just a thin epithelium. This epithelium is the gas exchange membrane. Its total surface area is enormous because of the great numbers of filaments and secondary lamellae.

Although all bony fishes share these same structural features, the gills of bony fishes also illustrate an important, common phenomenon: related species are often variations on a theme—not cookie-cutter replicas—in terms of their specific characteristics. When researchers have painstakingly measured the total area and thickness of the gas exchange membrane in many different species, they have found a great deal of variation among species. Yellowfin tunas have evolved more than five times as much gill surface area as trout of similar body size, and the average thickness of the gas exchange membrane is about one-tenth as great in the tunas as in other measured fishes. These specializations in tunas are believed to be adaptations for sustaining their very active lives.

Blood flows unidirectionally *through* each secondary lamella, as Figure 30.6C shows. The direction of water flow over the *outside* surface of the lamella is opposite. That is, *blood and water flow in opposite directions along the gas exchange membrane—a countercurrent* arrangement (see Key Concept 29.1) that is important for efficient gas exchange, as we discuss below.

What drives the flow of water? Most fishes ventilate their gills by means of a two-pump mechanism in which pumping is achieved by contractions of breathing muscles. One pump occurs in the mouth cavity. Muscles first expand the mouth cavity so it fills with water through the mouth. Muscles then contract the mouth cavity with the mouth closed, driving the water through the gill slits. The second pump occurs in the opercular cavities—the spaces under the opercular flaps. Muscles pull the opercular flaps away from the body wall, creating a sucking force that draws water through the gill slits into the opercular cavities. Muscles then pull the opercular flaps toward the body wall, discharging water.

Some highly active fishes—notably tunas and some sharks—ventilate their gills in a different way by driving water across the gills as they swim, a process called *ram ventilation*. A swimming tuna holds its mouth open so that water is rammed into its mouth and across its gills by the fish's forward motion. Fishes of this type can die if prevented from swimming! Ram ventilation is theorized to help reduce the cost of ventilating the gills because the swimming muscles accomplish the process as an incidental effect of their primary function: swimming.

The directions of ventilation and perfusion can greatly affect the efficiency of gas exchange

Now that we have looked at a specific example of a breathing system, let's discuss an important general principle, namely that the directions of ventilation and perfusion can affect the efficiency of gas exchange. We illustrate this principle here with fish breathing. Whereas in this case the environmental medium is water, the same principle can apply when the environmental medium is air.

Think of blood flow along the gas exchange membrane as a fluid stream, and of water flow along the membrane as a second fluid stream. If the two fluid streams flow in the same direction, the entire system of blood and water flow is a *cocurrent system* (sometimes called a *concurrent system*). If the two fluid streams flow in opposite directions, it is a *countercurrent system*. Gas exchange between blood and water is called **cocurrent gas exchange** in a cocurrent system and **countercurrent gas exchange** in a countercurrent system.

Keep in mind that as blood circulates in the body, it flows back and forth between the breathing organ (gills in a fish) and the systemic tissues. The **systemic tissues** in any animal are all the tissues other than those of the breathing organ. In the systemic tissues, the blood releases O_2 to the cells and picks up CO_2. Then the blood returns to the breathing organ to unload CO_2 and pick up more O_2.

Also keep in mind Fick's law, which in its general form states that a gas diffuses from where its partial pressure is high to where its partial pressure is low—and at a rate determined by the difference in partial pressure. The partial pressures of interest here are those in the blood plasma and water at the gas exchange membrane. The rate of diffusion of O_2 across the gas exchange membrane into the blood is proportional to the difference between the O_2 partial pressure of water and that of blood.

Now let's follow the blood from the moment it arrives at the gas exchange membrane. At that moment, the blood has a low O_2 partial pressure because it has just returned from the systemic tissues. What is the O_2 partial pressure of the water the blood first encounters? In a cocurrent system (**FIGURE 30.7A**), this water is fresh and has a high O_2 partial pressure. (In our discussion here, "fresh" does not mean salt-free, but rather "unused" or pristine with regard to O_2 level.) Thus there is a large difference in O_2 partial pressure between the water and blood, and O_2 diffuses

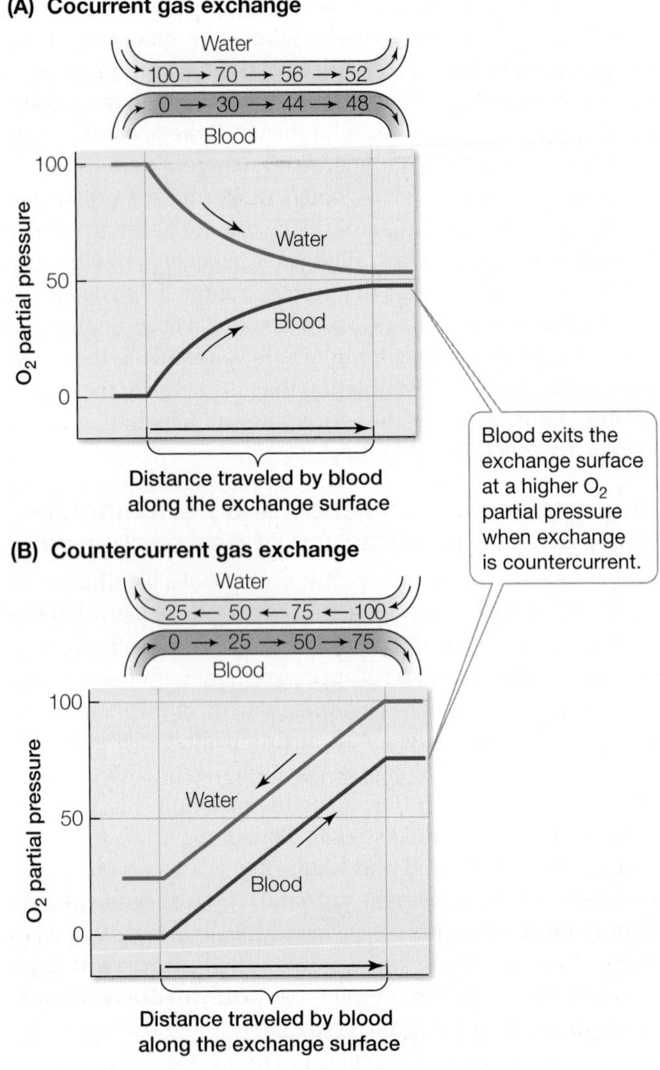

FIGURE 30.7 Countercurrent Gas Exchange Is More Efficient Than Cocurrent **(A)** Cocurrent gas exchange. **(B)** Countercurrent gas exchange. The diagram above each graph depicts the flow of blood and environmental water along the gas exchange membrane. Numbers are O_2 partial pressures in arbitrary units. The blood reaches a higher O_2 partial pressure when gas exchange is countercurrent.

rapidly into the blood. However, as the blood and water flow in the same direction along the gas exchange membrane in a cocurrent system, the O_2 partial pressure of the blood rises whereas that of the water falls, so the difference in partial pressure becomes smaller and smaller. Accordingly, the rate of O_2 diffusion into the blood becomes lower and lower.

A key question to ask is, when the blood finally leaves the gas exchange membrane, what is the O_2 partial pressure of the water with which it last undergoes gas exchange? As Figure 30.7A shows, that O_2 partial pressure is far lower than that in fresh environmental water. The O_2 partial pressure of the blood cannot be any higher. Thus the blood O_2 partial pressure will always be far lower than the O_2 partial pressure of fresh environmental water.

A countercurrent gas exchange system works in a very different way (**FIGURE 30.7B**). In a countercurrent system, the water that blood first encounters is stale (i.e., already O_2-depleted) and has a low O_2 partial pressure in comparison with fresh environmental water. Thus at the spot where blood first arrives at the gas exchange membrane, there is just a moderate difference in O_2 partial pressure between the water and blood in comparison with the initial difference in the cocurrent system, and O_2 diffuses into the blood at only a moderate rate by comparison. However, this moderate difference is *maintained* in the countercurrent system rather than becoming smaller as it does in the cocurrent system: as the blood flows along the gas exchange membrane and its O_2 partial pressure rises, the blood keeps encountering fresher and fresher water with a higher and higher O_2 partial pressure. For this reason, at every step of the way the O_2 partial pressure of the water is substantially higher than that of the blood. As a consequence, O_2 keeps diffusing into the blood at a substantial rate, even as the blood becomes richer and richer in O_2.

Now we come back to the same key question we asked earlier: When the blood finally leaves the gas exchange membrane, what is the O_2 partial pressure of the water with which it last undergoes gas exchange? In the countercurrent system, this water is entirely fresh environmental water and has a high O_2 partial pressure. The O_2 partial pressure of the blood can accordingly rise high enough to be relatively close to that of fresh environmental water.

In conclusion, blood leaves the gas exchange membrane with a higher O_2 partial pressure when exchange is countercurrent than when it is cocurrent. This difference occurs because, as blood flows through the breathing organ, the blood takes up more O_2 from the water in a countercurrent system, where uptake continues at a substantial rate every step of the way along the whole length of the gas exchange membrane. A countercurrent system is more efficient in transferring O_2 from the environmental medium to the blood.

LINK Countercurrent exchange is also important for thermoregulation in some animals; see **Figure 29.5**.

Gas exchange is countercurrent in the gills of fishes (see Figure 30.6C). If we focus on the spot where blood leaves a secondary lamella, we see that the water on the outside of the lamella at that spot is water that has just started to flow across the gills. That water is accordingly high in O_2, ensuring that the blood is brought to the highest possible O_2 partial pressure before exiting the lamella and traveling to the systemic tissues. The gills of many aquatic invertebrates, such as clams and aquatic snails, also employ countercurrent gas exchange.

Most terrestrial vertebrates have tidally ventilated lungs

The lungs of terrestrial vertebrates evolved as saclike outpocketings of the digestive tract. This ancient heritage is still reflected today in the fact that, except in birds and some crocodilians, the finest branches of the tubular parts of the lungs end in *closed, dead-end sacs of gas exchange membrane*. Because of this structure, ventilation of the gas exchange membranes cannot be unidirectional but must be tidal: fresh air flows in and exhaled gas flows out by the same route. Tidal ventilation has its own implications for gas

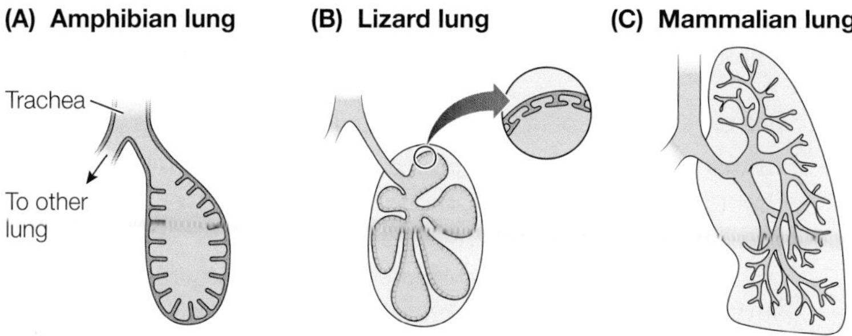

FIGURE 30.8 Three Levels of Complexity in Lungs of Terrestrial Vertebrates

exchange, distinct from cocurrent and countercurrent exchange (both of which require unidirectional ventilation).

The lungs of modern amphibians are relatively simple and help us understand what vertebrate lungs probably were like early in their evolution (**FIGURE 30.8A**). An amphibian lung has an open central cavity, which is surrounded by intricately folded, well-perfused membranes in the lung walls. Compared with amphibians, some lizards—notably the monitor lizards—have lungs that are considerably more complex (**FIGURE 30.8B**). Each lung is subdivided into several separate parts served by separate airway tubes, instead of having a single, open central cavity. This structure enlarges the surface area of gas exchange membrane. The intricate branching of airways in the mammalian lung—which we discuss in detail in Key Concept 30.3—represents a further, more elaborate step in the same direction (**FIGURE 30.8C**). The area of gas exchange membrane in a lung of a mouse or rat is about ten times greater than in a lung of a mouse- or rat-sized lizard.

The inside surfaces of vertebrate lungs are coated with a microscopically thin layer of lipids and proteins called **lung surfactant**. By reducing surface tension at the surfaces of the lung epithelium, this material helps stabilize the complex three-dimensional structure of the gas exchange membranes. Infants born without adequate lung surfactant are in danger of dying unless they receive medical care because the interior surfaces of their lung membranes tend to stick together, requiring lots of effort to pull them apart with each inhalation.

How is ventilation achieved? In amphibians, the lungs are inflated by *positive pressure*. During each inhalation, the mouth cavity is first filled with air from the atmosphere. The mouth and nostrils are then sealed, and muscles contract the mouth cavity, raising the pressure high enough to drive air into the lungs. This mechanism is the same as that used by air-breathing fishes (such as lungfishes) to fill their air-breathing organs, suggesting that amphibians carry over the mechanism from their fishlike ancestors.

With the evolution of reptiles, a major transition in ventilation took place. In many types of reptiles, such as lizards, the lungs are inflated by *suction*: the lung passages are expanded during inhalation, creating a lower air pressure inside the lungs than in the atmosphere, thereby sucking air in. This is also true in mammals (see Key Concept 30.3 for greater detail in humans).

Birds have rigid lungs ventilated unidirectionally by air sacs

Birds and mammals both have high metabolic rates and thus require sophisticated lungs that can supply O_2 at high rates to the rest of the body. The lungs of birds and mammals, moreover, are both believed to have evolved from lungs similar to those seen in today's lizards. The lungs of birds, however, are radically different from those of mammals.

The lungs of birds are often described as "rigid." This does not mean they are solid or hard, but rather that they do not change much in volume as a bird inhales and exhales.

In addition to the lungs, the breathing system of a bird includes air sacs (**FIGURE 30.9A**). These thin-walled sacs, which fill much of the internal space in a bird's body, connect to airways in the lungs. The air sacs do not participate in exchange of O_2 and CO_2 between the air and blood. Instead, their function is ventilation. They expand and

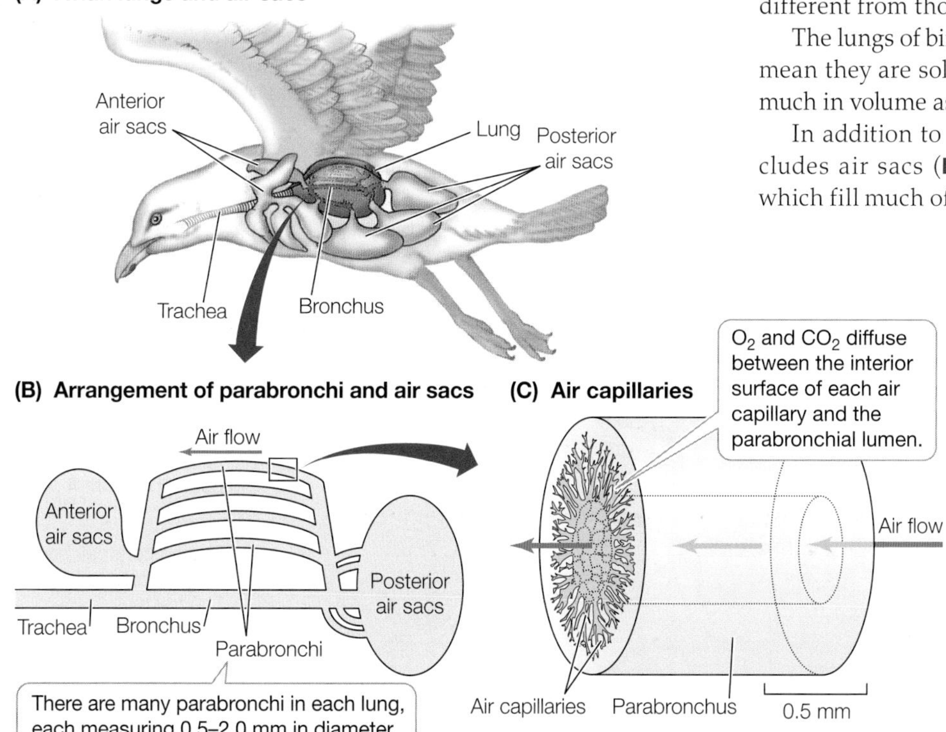

FIGURE 30.9 The Breathing System of a Bird **(A)** The lungs in the thorax are relatively rigid. The air sacs—usually nine in number—are located in the thorax and abdomen and serve as bellows to ventilate the lungs. **(B)** Arrays of numerous parabronchi in the lungs are the principal sites where gas exchange takes place. Air flows through the parabronchi unidirectionally. **(C)** Blind-ended air capillaries branch from the open central airway of each parabronchus. The interior surfaces of the air capillaries are the gas exchange membranes.

contract during inhalation and exhalation, acting like bellows and forcing air to flow past the gas exchange membranes in the rigid lungs.

Numerous long tubes of small diameter called *parabronchi* (singular *parabronchus*) are the most important airways for gas exchange in the lungs of a bird (**FIGURE 30.9B**). As astounding as it may sound, air flows through most of a bird's parabronchi unidirectionally, from posterior to anterior, during all phases of the breathing cycle, whether the bird is inhaling or exhaling! Air flows unidirectionally through the parabronchi because of coordinated actions of thoracic and abdominal muscles that expand and contract the anterior and posterior sets of air sacs (**FIGURE 30.10**). When a bird inhales, some of the incoming fresh air travels to the posterior of each lung and enters the parabronchi. That air then flows through the parabronchi from posterior to anterior and exits into anterior air sacs. Some of the incoming fresh air also inflates the posterior air sacs. Then, when the bird exhales, this fresh air in the posterior air sacs is driven into the parabronchi, again flowing from posterior to anterior. As this occurs, stale air exiting the parabronchi at the anterior end, plus stale air from the anterior air sacs, is exhaled into the atmosphere.

Animation 30.1 Airflow in Birds

Many microscopically tiny *air capillaries* extend into the walls of each parabronchus from the parabronchial lumen (**FIGURE 30.9C**). The walls of the air capillaries are richly perfused with blood and serve as the gas exchange membranes. As air moves by bulk flow through the parabronchi, O_2 and CO_2 are thought to move in and out of the tiny air capillaries by diffusion. Gas exchange with the blood is termed "cross-current"—a type of exchange intermediate in efficiency between cocurrent and countercurrent.

Insects have airways throughout their bodies

In insects (and some spiders), a breathing system has evolved that is unlike that of any other animals (**FIGURE 30.11**). This breathing system is a type of lung because it consists of gas-filled tubules that are invaginated into the body from the body surface. The tubules branch everywhere in an insect's body. In fact they branch in such fantastically intricate ways that a gas-filled tubule comes close to every cell. In some metabolically active tissues, tubules grow into cells, indenting the cell membrane inward. The tubules connect to the atmosphere at several openings, called *spiracles*, on the outside surface of an insect's body. O_2 makes its way directly from the atmosphere, through the gas-filled tubules, to each cell. CO_2 moves directly from all cells to the atmosphere. Because of this unique breathing system, blood flow is generally unimportant in delivering O_2 and getting rid of CO_2.

Most of the gas-filled tubules are called **tracheae**, and for this reason, insects are said to have a **tracheal breathing system**. The tracheae are lined with a thin, gas-permeable layer of chitin, which provides structural support and keeps them from collapsing shut. The finest tubules are called **tracheoles**. Air sacs are connected to the tracheae in many insects.

The air inside the tracheal breathing system is believed to be completely still in some insects. In these cases O_2 and CO_2 travel everywhere by diffusion. Keep in mind that Krogh's rule of thumb refers just to diffusion through water. Diffusion in air is much faster (200,000 times faster in the case of O_2). Because of the fast diffusion that can occur in air, insects can be moderately large (e.g., 5 mm long) and still have their gas exchange needs

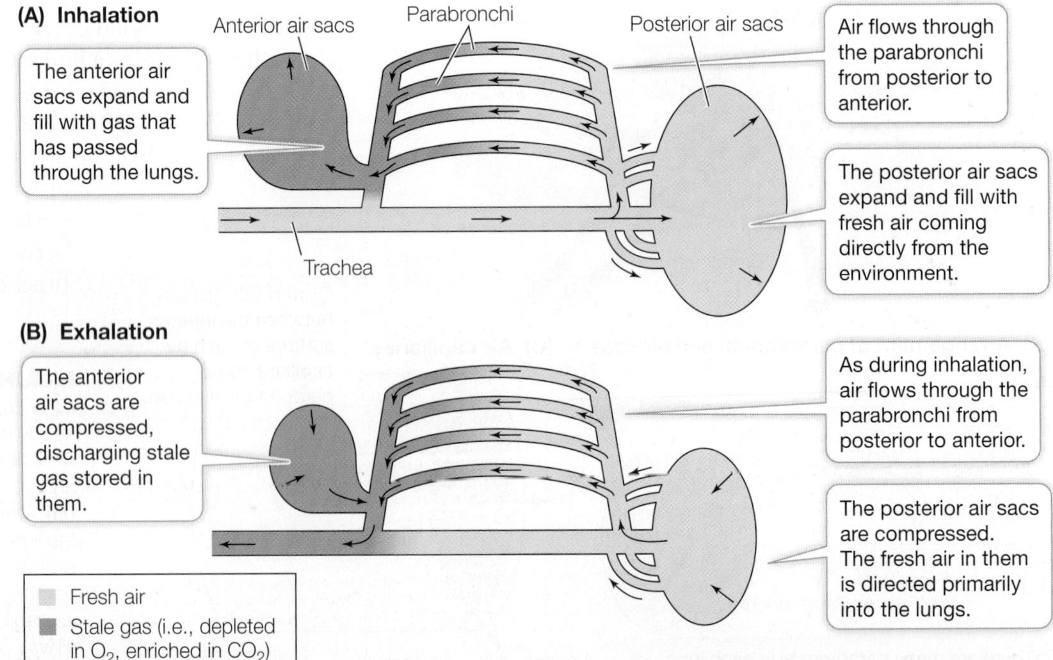

FIGURE 30.10 Airflow in the Lungs and Air Sacs of Birds Airflow during inhalation, when air sacs expand **(A)**, and during exhalation, when air sacs contract **(B)**. Air flows from posterior to anterior through the parabronchi during both inhalation and exhalation.

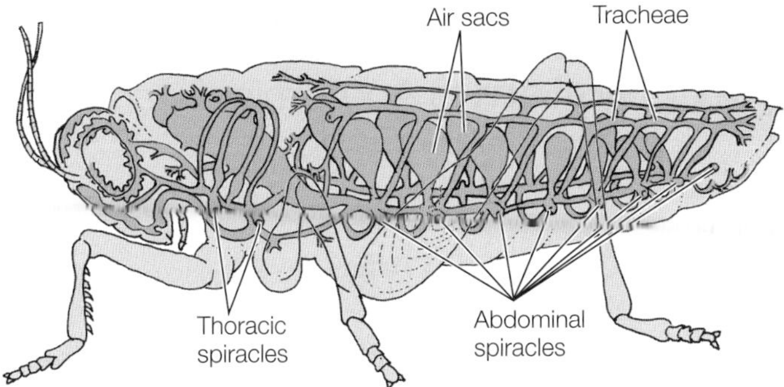

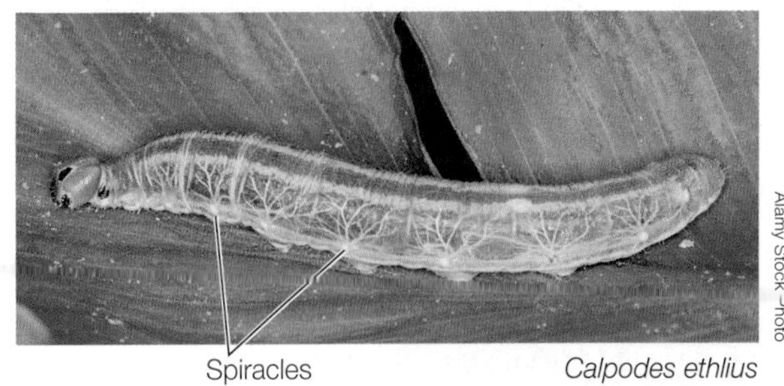

© Rick & Nora Bowers / Alamy Stock Photo

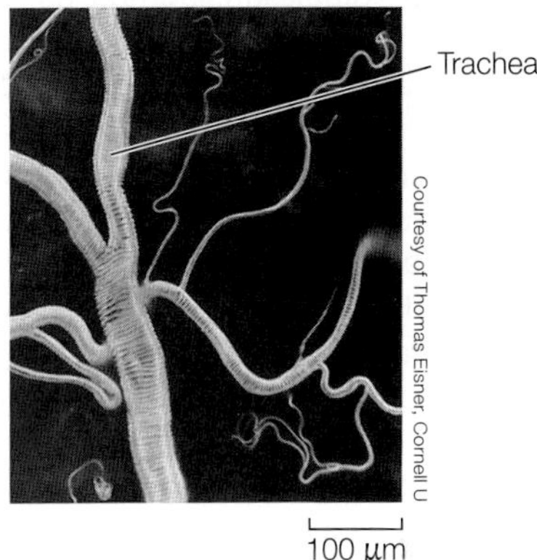

Courtesy of Thomas Eisner, Cornell U

FIGURE 30.11 The Tracheal Breathing System of Insects **(A)** The insect breathing system consists of elaborate arrays of branching, gas-filled tubes called tracheae, which open to the atmosphere through pores called spiracles. In many insects, air sacs connect to the tracheae and can act as bellows to create tracheal ventilation. This drawing shows only the largest structures and thus omits the huge numbers of very fine tracheae and tracheoles that branch throughout all tissues. **(B)** Large tracheae can be seen inside the body of this larva of the Brazilian Skipper (*Calpodes ethlius*) because the larva has a transparent body wall. Each array of tracheae branches inward from a spiracle on the body surface. As in A, fine structures cannot be seen. **(C)** This scanning electron micrograph shows an insect trachea dividing into smaller tracheae and tracheoles. All cells in an insect's body are close enough to one of these branches to receive O_2 directly from the tracheal system by diffusion.

met by diffusion in the tracheal system, provided they are not very active.

In insects that are large or active, mechanisms have evolved to create bulk flow of air inside the tracheal system. Grasshoppers, for example, pump air through their tracheae by alternately contracting and elongating their bodies, motions that exert forces on the air sacs, which then act as bellows (see Figure 30.11A). The tracheae throughout an insect's body sometimes steadily undergo rhythmic contractions.

Insect flight muscle, during flight, has one of the highest metabolic rates per gram known in any form of life. The tracheal breathing system—despite being so very different from our own—is able to supply O_2 to the flight muscle cells at the extraordinary rate required.

Some animals lack specialized breathing organs

Some types of animals—such as flatworms and sponges—prosper without specialized breathing organs. Instead, they take up O_2—and void CO_2—across their body surfaces. These animals have two properties that help us understand why they have not evolved specialized breathing organs: they have relatively low metabolic rates, and they have body plans in which most of their cells are close to a body surface.

Flatworms (**FIGURE 30.12A**) have a flat, thin body plan, which ensures that each cell is near a body surface—often within 0.5 millimeter. Thus Krogh's rule of thumb is not a limitation. O_2 can diffuse from the body surface through the tissues and tissue fluids directly

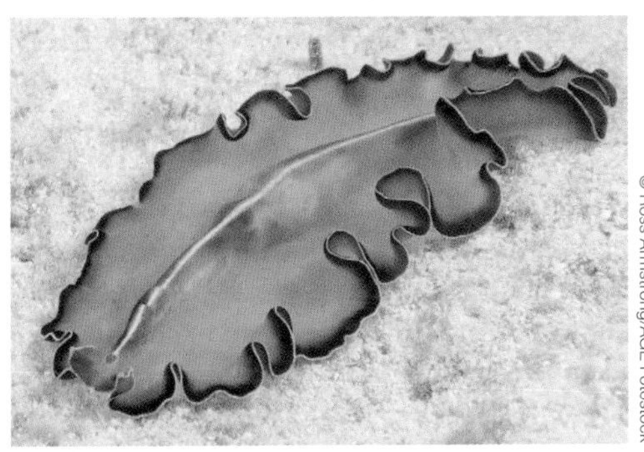

© Ross Armstrong/AGE Fotostock

© Pete Oxford/Minden Pictures

© Pete Oxford/Minden Pictures

FIGURE 30.12 Specialized Breathing Organs Are Not Required with Certain Body Plans **(A)** Flatworm. **(B)** Sponge.

to each cell at a rate high enough to meet the cell's metabolic needs. The bodies of sponges (**FIGURE 30.12B**) are organized around an extensive system of internal passageways through which water flows continually, driven by beating flagella. All of a sponge's cells are close enough to one of these passageways to receive O_2 by diffusion from the environmental water flowing through.

REVIEW & APPLY | 30.2

R Air and water differ considerably as respiratory environments. In aquatic environments, gas exchange is limited by the low diffusion rate and low solubility of O_2 in water relative to air. Breathing organs are classified based on their structure: lungs fold inward into the body; gills fold outward from the body. All breathing organs have a gas exchange membrane that separates the environmental medium from the tissues of the animal; a large, thin membrane facilitates gas exchange. Breathing is defined to be the process that carries O_2 and CO_2 between the outer environment and the gas exchange membrane and can occur by bulk flow (ventilation) or diffusion. Ventilation can be tidal or unidirectional. In the latter case, highly efficient countercurrent gas exchange occurs if blood perfuses the gas exchange membrane in the opposite direction to the direction of air or water flow. Fishes employ countercurrent gas exchange in their gills. Mammals breathe with tidally ventilated lungs. The breathing system of birds includes air sacs that act as bellows to ventilate the parabronchi unidirectionally. Insects employ a tracheal breathing system in which branching gas-filled tubules (tracheae and tracheoles) come close to all cells of the body.

A

1. Why is water a more difficult environment than air for an animal to obtain adequate O_2?
2. As blood flows along the gas exchange membrane, how does the difference in O_2 partial pressure between the environmental medium and blood vary in countercurrent and cocurrent systems?
3. Describe how air flows through the gas exchange airways of a bird lung in one direction, whether the bird is inhaling or exhaling.
4. Mammals, birds, and insects are the three groups of animals in which species with the highest known weight-specific metabolic rates occur. Of these three groups, two—mammals and birds—have high-performance circulatory systems that can circulate blood rapidly throughout the body and perfuse tissues at high rates. Insects do not have high-performance circulatory systems. Explain.

Now let's take a closer look at the anatomy and function of mammalian lungs, focusing on the human breathing system.

30.3 | The Mammalian Breathing System Is Anatomically and Functionally Elaborate

LEARNING OBJECTIVES

By the end of this key concept you should be able to:

30.3.1 **State why resting human inhalation is described as active, whereas resting exhalation is described as passive.**

30.3.2 **Explain why control of ventilation by blood CO_2 is an example of a negative-feedback system.**

30.3.3 **State the physiological reason why a broken neck can cause immediate suffocation.**

Mammals have high metabolic rates and thus require sophisticated lungs with a high surface area of gas exchange membrane. This high surface area is achieved structurally by an intricate branching pattern. One way to understand this branching pattern is to think of your lungs as an upside-down tree, where your trachea, or windpipe, corresponds to the tree trunk (**FIGURE 30.13A**). Your trachea branches to form two tubes, then each of those tubes branches, and those branches also branch—and so on until the final branches are 23 steps removed from the trunk—the trachea. Each final branch ends in a closed sac of gas exchange membrane that is itself highly folded to increase its surface area. The result is 130 square meters of thin membrane with air on one side and blood on the other side—all assembled within the volume of your two lungs. By virtue of both its large area and thinness, this gas exchange membrane is ideally suited for O_2 to diffuse rapidly into your blood and for CO_2 to diffuse rapidly out.

Let's focus briefly on the terminology of the branching airway system. The two airway tubes formed by branching of the trachea—one of which enters each lung—are termed *primary bronchi* (singular *bronchus*) (see Figure 30.13A). Within each lung, the primary bronchus branches to form *secondary bronchi*, which branch still further, ultimately giving rise to fine tubes called *bronchioles*, which also branch (**FIGURE 30.13B**). At the farthest reaches of the branching airways, each tiny branch ends blindly in an **alveolar sac** of epithelial tissue (**FIGURE 30.13C**). The wall of each alveolar sac consists of numerous semispherical outpocketings called **alveoli** (singular *alveolus*) (**FIGURE 30.13D**), which are the principal sites of gas exchange between air and blood (**FIGURE 30.13E**).

The trachea, primary bronchi, and most of the succeeding branches are not involved in the exchange of O_2 and CO_2 between the air and blood. They are called the **conducting airways** of the lung because their principal function is to conduct the bulk flow of air in and out of the lungs during inhalation and exhalation. The wall of the trachea and the walls of many further airway branches contain rings of cartilage, which help prevent airway collapse.

The **respiratory airways** are those in which O_2 and CO_2 are exchanged between the air and blood. They include the last few branches of the branching airway system—termed *respiratory*

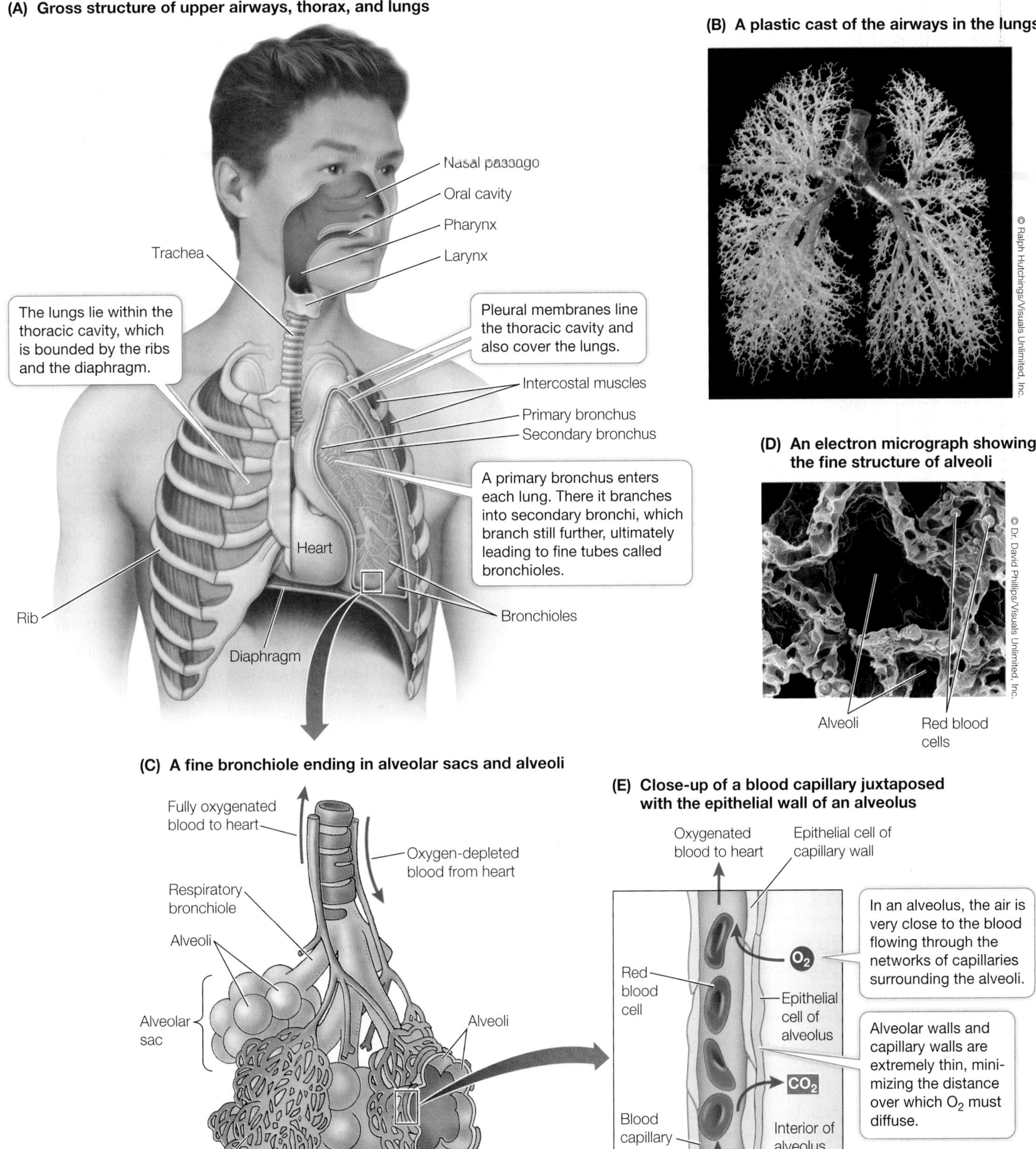

FIGURE 30.13 The Human Breathing System The hierarchy of human breathing structures extends from the nose and mouth to the minuscule alveoli. Because lung capillaries are a bit larger in diameter than systemic capillaries, red blood cells pass through the former in the manner shown in (E).

Activity 30.1 The Human Respiratory System

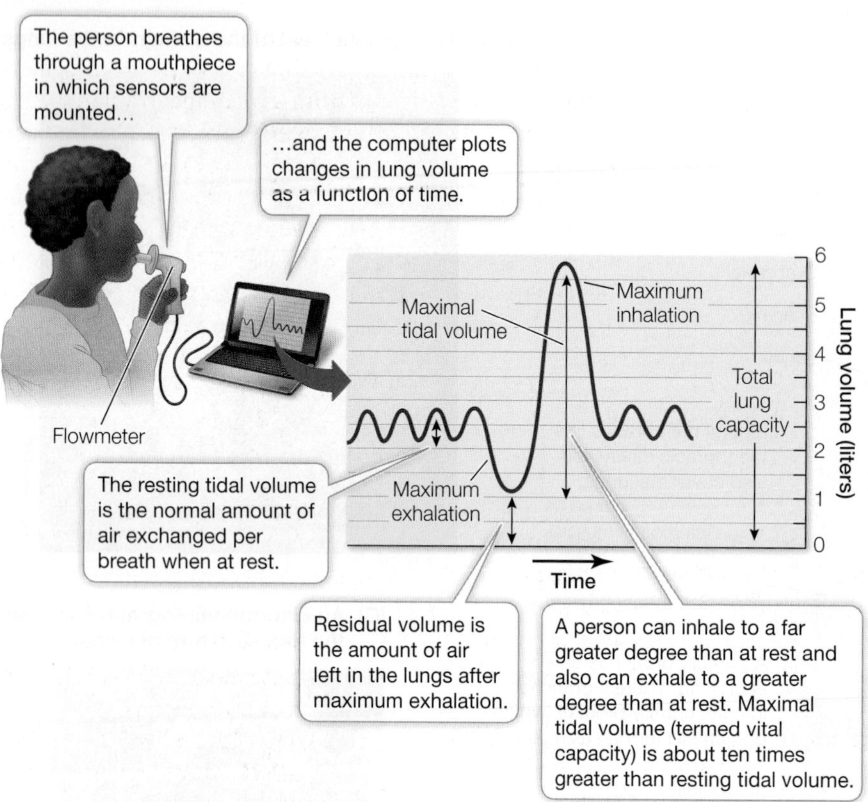

FIGURE 30.14 Lung Ventilation Can Vary Over a Wide Range Ventilation of the lungs can be studied by having an individual inhale and exhale through a mouthpiece in which sensors measure rate, amplitude, and other properties of interest.

bronchioles—and the alveolar sacs, including the alveoli. In healthy human lungs, blood and air are separated by only about 0.5 micrometer in these airways. The respiratory bronchioles are ventilated by bulk air flow as a mammal inhales and exhales. However, because the alveoli are dead ends, the air inside them does not move, and O_2 and CO_2 must diffuse across the central cavities of the alveoli. This diffusion occurs rapidly because the distance covered is minute.

Ventilation is adjusted over a wide range to meet needs of rest and exercise

The amount of air that moves in and out of a mammal's lungs per breath is called the **tidal volume** (**FIGURE 30.14**). When an individual is at rest, it is called the *resting tidal volume*—about 0.5 liter in an adult person. The *maximal tidal volume*—also called vital capacity—is the amount of air that moves in and out per breath when an individual inhales and exhales as vigorously as possible. It is about 5 liters in an adult person. We use only about 10 percent of our maximal tidal volume when we breathe at rest.

An individual's **respiratory minute volume** is the total volume of air that the individual inhales and exhales per minute. It is calculated by multiplying tidal volume (the volume exchanged per breath) by the number of breaths per minute. Mammals increase their minute volume when they exercise—a response that helps increase their rate of O_2 uptake to match the increased rate at which O_2 is used. At rest adult people breathe about 12 times per minute. With a resting tidal volume of about 0.5 liter, our resting minute volume is therefore about 6 liters per minute. Well-conditioned athletes can breathe at least 30 times per minute at a tidal volume of over 3 liters, meaning they can increase their minute volume by a factor of more than 15!

Mammals and other animals with tidally ventilated lungs do not empty their lungs completely when they exhale. Some "stale" air is always left behind in the lungs, even following maximal exhalation (see Figure 30.14). This means that when a mammal inhales, the fresh air always mixes with some stale air from the previous breath—a significant consequence of tidal ventilation.

The lungs are ventilated by expansion and contraction of the thoracic cavity

Mammalian lungs are suspended in the thoracic cavity (chest cavity), which is bounded by the rib cage and—on the bottom—by a sheet of muscle, unique to mammals, called the **diaphragm** (see Figure 30.13A). Any increase in the volume of the thoracic cavity pulls on the walls of the lungs, thereby expanding the lungs and causing inhalation.

Two muscles or sets of muscles are critical for breathing: the diaphragm and the intercostal (*costa*, "rib") muscles that run between adjacent ribs (see Figure 30.13A). The diaphragm is dome-shaped when it is not contracting, protruding upward from the abdomen into the thoracic cavity. It flattens when it contracts, so it protrudes less into the thoracic cavity, expanding the volume of the cavity. There are two sets of intercostal muscles. One set, when it contracts, changes the positions of the ribs in a way that enlarges the volume of the thoracic cavity. The other set makes the thoracic cavity smaller when it contracts.

When humans are at rest, inhalation is active but exhalation is passive. **FIGURE 30.15** shows the positions of the rib cage and diaphragm at the end of exhalation. As the next inhalation begins, the diaphragm and certain intercostal muscles actively contract (explaining why inhalation is termed "active"). These contractions (1) flatten the diaphragm, so it moves to the position shown by the bottom dashed line in Figure 30.15, and (2) expand the rib cage to the positions shown by the dashed lines on the two sides. Both actions expand the thoracic cavity and thus expand the lungs, which fill by suction (a lower pressure in the lungs than in the atmosphere). During exhalation at rest, all these muscles simply relax (explaining why resting exhalation is termed "passive"). The relaxation of the muscles allows the elastic recoil of the lungs and thoracic cavity to restore the relaxed lung volume.

During exercise, muscle contractions are employed not only during inhalation but also during exhalation, making exhalation faster and more forceful. Compared with rest, additional

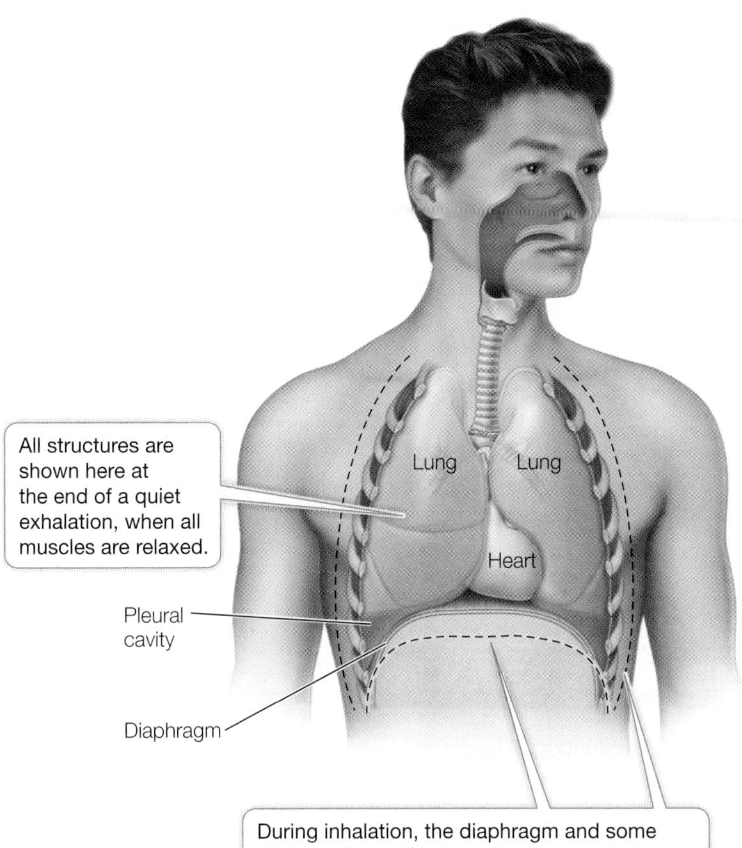

FIGURE 30.15 Resting Human Ventilation The drawing shows the end of an exhalation. Structures move to the positions of the dashed lines at the end of inhalation. In a person at rest, inhalation is an active process driven by contraction of the diaphragm and some of the intercostal muscles. However, exhalation is a passive process that occurs by elastic recoil; the thoracic cavity and lungs elastically return to their relaxed volume as soon as muscle contractions stop maintaining them in an expanded state. Considered collectively, the ribs are often termed the rib cage because they surround the lungs.

Animation 30.2 Airflow in Mammals

breathing muscles come into play. These changes increase the tidal volume and the number of breaths per minute, providing an increased respiratory minute volume.

The breathing rhythm depends on nervous stimulation of the breathing muscles

All the muscles involved in breathing require nervous stimulation to contract. The nerve impulses originate in the brain. These facts explain why a broken neck can kill a person within seconds. If the nerves carrying the breathing impulses from the brain to the breathing muscles are severed, the muscles stop contracting. Without artificial respiration, the person suffocates.

Physiologists have long sought the exact places in the brain where the breathing rhythm originates. Investigators recently identified these rhythm-generating (rhythmogenic) centers by studying thin slices of brain tissue taken from nestling rodents. Although these slices are cut off from all their normal connections with other parts of the nervous system, they remain functional for some time. Investigators have shown that two neural complexes in the medulla oblongata of the brain (a part of the brainstem; see Figure 31.38) generate a breathing rhythm under these circumstances (**FIGURE 30.16**). These rhythm-generating centers, called the pre-Bötzinger complexes, are believed to be the beginning points for the nerve impulses that activate the muscles we use in breathing.

Breathing is under negative-feedback control by CO_2

You might expect that a mammal's breathing would be regulated by the blood's O_2 content. In fishes, the blood O_2 partial pressure is indeed the primary feedback stimulus for gill ventilation. However, mammals are much more sensitive to changes in blood CO_2 than to changes in blood O_2.

To fully understand control of breathing by CO_2, it is important to recognize that CO_2 dissolved in the blood reacts with blood H_2O, as summarized by this reaction:

$$CO_2 + H_2O \rightleftharpoons H_2CO_3 \rightleftharpoons H^+ + HCO_3^-$$

An increase in CO_2 in the blood tends to drive the reaction to the right and raise the concentration of H^+, making the blood a bit more acidic.

What happens when the cells of a mammal's body increase their metabolic rate and therefore add CO_2 to the blood at an increased rate? There are two immediate consequences. First, the partial pressure of CO_2 in the blood plasma rises, and second, the concentration of H^+ in the blood plasma also rises. Chemosensory (chemical-sensing) cells located on the ventral surface of the medulla oblongata in the brain detect both of these changes. This information is then used to control breathing in a negative-feedback manner. Breathing is stimulated so that the respiratory minute volume increases. This increased ventilation increases the rate of exhalation of CO_2. In this way, CO_2 is voided more rapidly when the cells of the mammal's body have increased their rate of CO_2 production. The increased ventilation also tends to return both the blood CO_2 partial pressure and H^+ concentration back toward their values that existed before the increase in CO_2 production.

What if, under different circumstances, a mammal's cells decrease the rate at which they add CO_2 to the blood? In this case, the blood CO_2 partial pressure and H^+ concentration initially tend to fall. The chemosensory cells detect these changes, leading again to negative-feedback control of breathing. In this instance, breathing is inhibited, lowering the rate of CO_2 exhalation and causing the blood CO_2 partial pressure and H^+ concentration to rise.

These controls serve to adjust the respiratory minute volume so that the rate of CO_2 exhalation approximately matches the rate

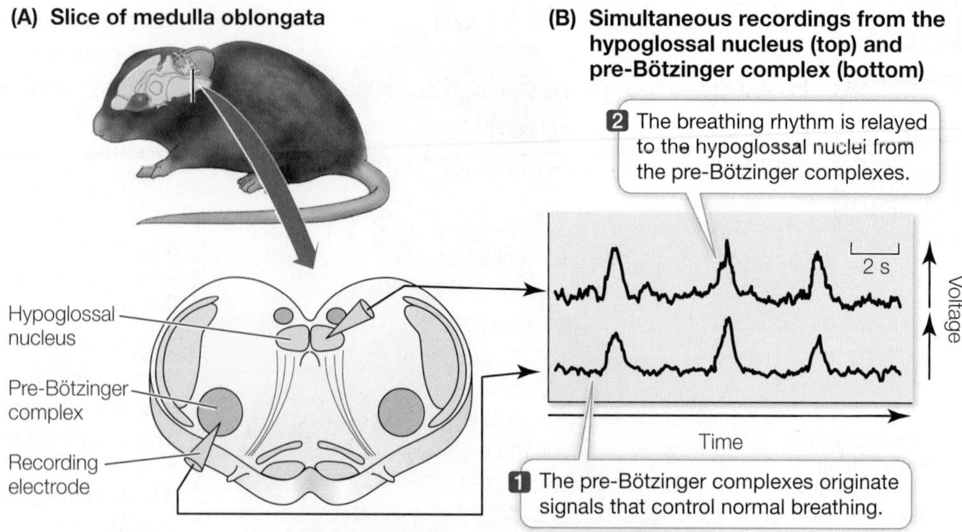

FIGURE 30.16 Where Does the Breathing Rhythm Originate? To find the neural tissue where the breathing rhythm starts, physiologists studied thin slices of living brain tissue from nestling mice. **(A)** This slice, cut from the medulla oblongata, has no connections to other parts of the nervous system. Its neural activity is therefore completely self-contained. **(B)** Electrical recordings were made from one of the two clusters of nerve cells (orange) called the pre-Bötzinger complexes. The investigators found that these complexes generate electric signals that control normal breathing. From simultaneous recordings from one of the two other nerve cell clusters (light purple) called the hypoglossal nuclei, the investigators found that the breathing rhythm that starts in the pre-Bötzinger complexes is relayed to the hypoglossal nuclei. In an intact mouse, the breathing rhythm would then be relayed to the breathing muscles via the hypoglossal nerves. (After S. P. Lieske et al. 2000. *Nat Neurosci* 3: 600–607.)

at which the cells of the body produce CO_2. The controls also stabilize the blood CO_2 partial pressure and H^+ concentration (an example of homeostasis), by tending to bring them down if they start to rise and tending to bring them up if they start to fall.

Media Clip 30.1 Deep Diving on a Single Breath

Breathing is also controlled in part by body motion and blood O_2

You may have noticed that when you suddenly start running, your breathing rate increases almost immediately—far sooner than your blood CO_2 partial pressure might be expected to change. This everyday observation suggests that body motion may serve to activate increased ventilation during exercise, a hypothesis that researchers have investigated. The general consensus today is that information from receptors that detect motion in muscles and joints helps control the intensity of ventilation, although controversy remains about the relative roles of motion and CO_2.

Blood O_2 partial pressure is also sometimes used in controlling the intensity of ventilation, particularly in circumstances—such as life at high elevation—when a large drop in blood O_2 partial pressure can occur. Two kinds of oxygen-sensing chemosensory bodies, called *carotid bodies* and *aortic bodies*, monitor the O_2 partial pressure of arterial blood. When these sensors detect a decline in the O_2 partial pressure, the rate of ventilation increases. Humans have only carotid bodies, which are found along the carotid arteries in the neck. Aortic bodies are found in some other mammals, positioned along large arteries leaving the heart.

REVIEW & APPLY | 30.3

R In human lungs, 23 levels of branching in the airways arise from the trachea. The finest airways end blindly in alveolar sacs, the walls of which consist of many alveoli. Branching creates an enormous area of thin gas exchange membrane. In a resting person, inhalation occurs when contractions of the diaphragm and some of the intercostal muscles expand the thoracic cavity, whereas exhalation occurs passively, by elastic recoil when the inhalation muscles relax. During exercise, both exhalation and inhalation result from active muscular contraction. The tidal volume is the amount of air that moves in and out of the lungs per breath, and the respiratory minute volume is the total volume of air that is inhaled and exhaled per minute. The most important feedback stimulus for breathing is the partial pressure of CO_2 in the blood, which is detected (along with pH changes it causes) by chemosensory cells on the surface of the medulla oblongata. Breathing rate can also be influenced by body motion and by large changes in arterial blood O_2 partial pressure.

A
1. The alveoli can become filled with tissue fluid in several types of diseases, such as chronic heart weakening that causes pulmonary edema. This condition is a health emergency. Taking account of the particulars of alveolar O_2 transport, explain why.
2. In a healthy individual, how does the respiratory minute volume change in response to a decrease in blood pH?
3. How could a neck injury that leaves the trachea and other air passageways intact result in an inability to breathe?

Activity 30.2 Concept Matching: Breathing

Now we will make a major transition and turn to the circulation of the blood. In addition to ventilation, the circulation is the second step in the pathway for oxygen that entails bulk flow.

THINK LIKE A SCIENTIST

Quantitative Reasoning

How does a person's maximal rate of O_2 consumption vary with elevation in the mountains?

When a person who has been living at sea level travels to a high altitude and is breathing the atmospheric air (as opposed to compressed air in a tank), their body autonomically responds by increasing the rate of ventilation of the lungs. We say we "breathe harder." Despite this response, people cannot exercise as intensely at high elevations as at sea level. For understanding a person's maximal capability to exercise, a useful measurement is $\dot{V}_{O_2max}$ (pronounced "vee-oh-two-max"), the maximal rate at which an individual can take up O_2 by breathing and use the O_2 in metabolism. A person can exercise only as intensely as their rate of ATP production permits. Moreover, the rate of ATP production can be only as high as $\dot{V}_{O_2max}$ permits. Thus if $\dot{V}_{O_2max}$ is reduced, a person's maximal intensity of exercise is reduced. Moreover, if $\dot{V}_{O_2max}$ is reduced, all types of exercise become harder to perform and more fatiguing.

A scientist interested in exercise at elevation wants to know how $\dot{V}_{O_2max}$ varies with elevation. You can imagine that, to get the answer, a focused study at many elevations on statistically meaningful numbers of people could be prohibitively expensive. Charles Fulco and his colleagues obtained the needed data by carrying out a meta-analysis—a statistical approach that assembles a large data set by seeking out and combining data from many smaller data sets.

Searching the scientific literature, Fulco was able to find 67 different studies in which researchers had measured $\dot{V}_{O_2max}$ at various elevations or simulated elevations. Some of these studies had subparts (e.g., data from two different elevations); in total there were 146 different subparts, each of which involved measures on multiple individual people. For each subpart, Fulco calculated the mean $\dot{V}_{O_2max}$ and the mean elevation of the study. Then he and his team did a meta-analysis by plotting all 146 results on one plot (**FIGURE 30.17**).

1. What does the meta-analysis reveal about $\dot{V}_{O_2max}$ and human exercise at high elevation? Explain.

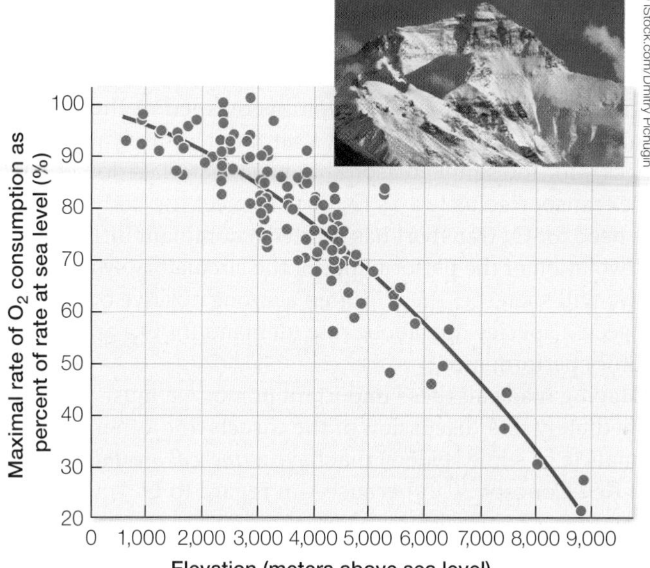

FIGURE 30.17 Maximal Rate of O_2 Consumption as a Function of Elevation in People Breathing Atmospheric Air: Results of a Meta-Analysis Each point represents the average result of a study on multiple individuals. The red line is statistically fitted to all the data. (After C. S. Fulco et al. 1998. *Aviat Space Environ Med* 69: 793–801.)

2. Focusing on breathing, what is the likely explanation for the observed results?
3. Reinhold Messner is one of the first two people known to reach the peak of Mount Everest (elevation 8,848 m) while breathing atmospheric air. Although in superb physical condition, he required an entire hour to climb the final 100 meters! That is, during his last hour of climbing, he covered only 1.7 meters per minute. Explain his slow progress, basing your answer on the meta-analysis.
4. Thinking as broadly as possible, what are the pros and cons of a meta-analysis? .

30.4 Animals Have Evolved Circulatory Systems for Rapid Internal O_2 Transport

LEARNING OBJECTIVES

By the end of this key concept you should be able to:

30.4.1 State the difference between closed and open circulatory systems.

30.4.2 Explain how hemoglobin increases its O_2 delivery when a person who has been resting starts to exercise.

30.4.3 Explain how blood flow in a vertebrate is redirected to various organs by the circulatory system to accommodate changes in organs' relative needs for blood flow.

After O_2 has entered the tissues of an animal in the breathing organ, the next step in the pathway for oxygen is to deliver the O_2 to cells throughout the body. In most types of animals, bulk flow via a circulatory system is the method by which O_2 is transported rapidly enough to meet metabolic needs for O_2. A **circulatory system** typically consists of a muscular pump (heart), a fluid (blood), and a series of conduits (blood vessels or channels) through which the blood is pumped. The flowing blood transports O_2 rapidly by bulk flow to the vicinity of all cells. The circulatory system also transports many other materials throughout the body by bulk flow, including nutrient molecules of many kinds (see Key Concept 28.1), hormones, heat, and metabolic wastes.

The transport of O_2 is the most pressing and urgent function of the circulatory system in most types of animals. In these animals, if the circulatory system were freed of the need to transport O_2, it could operate at a far more leisurely pace than it does—for the simple reason that most materials do not have to be transported as fast as O_2 does to meet the body's needs. The need for O_2 transport has exerted a dominant influence on the evolution of the performance of the circulatory system, and (as we will soon see) there is often a strong positive correlation between a species' metabolic rate (demand for O_2) and its circulatory performance.

Having made all these important points, we must now turn immediately to a discussion of the insects and other types of animals (e.g., some spiders) that have a *tracheal breathing system* (see Key Concept 30.2) because—in regard to O_2 transport—these types of animals are major exceptions to the usual rules. Besides being exceptions, they are also the most numerous animals on Earth!

As we discussed in Key Concept 30.2, in insects—with their tracheal breathing system—O_2 is bought to each cell in the body by a branch of the breathing system, not by blood circulation. In what way might this fact be of theoretical interest? We have just theorized that in many types of animals the circulatory system could operate at a far more leisurely pace were it not required to transport O_2. Insects provide a test of this proposition since the insect circulatory system in fact does not need to transport O_2 to any great extent. As we would predict from our proposition, the insect circulatory system is comparatively a low-performance system. The insect system serves mostly to transport materials other than O_2 (e.g., nutrients and hormones) and is relatively unsophisticated compared with the circulatory systems in mammals and many other types of animals.

Animals have evolved two basic types of circulatory systems: closed and open. A circulatory system is **closed** if the blood is always contained within blood vessels as it circulates. It is **open** if the blood exits blood vessels as it flows through the body.

A closed circulatory system moves blood through blood vessels

Vertebrates have a closed circulatory system. Other animals with a closed system include the annelid worms, such as earthworms, and the cephalopod mollusks, such as squid and octopuses. The vertebrates provide excellent examples and are our focus here.

Vertebrates nicely illustrate the relationship between metabolic rate and circulatory function. Mammals and birds typically have far higher rates of blood flow and higher blood pressures (which are responsible for driving the blood flow) than lizards, frogs, and fishes of the same body size. This difference in circulatory performance correlates with differences in metabolic rates. Mammals and birds have far higher metabolic rates than lizards, frogs, and fishes (see Key Concept 28.3) and therefore require circulatory systems that can deliver O_2 far faster. At a finer taxonomic scale, a similar relationship is seen among fish species. Fish species with high metabolic rates, such as tunas, have circulatory systems capable of higher performance than those of sluggish fish species.

The heart of a vertebrate pumps blood into large blood vessels called *arteries*. The arteries branch like trees into smaller and smaller blood vessels as they carry the blood to tissues and organs throughout the body. The branching gradually leads to great numbers of vessels so small in diameter that our eyes can hardly see them (**FIGURE 30.18**). And then, as the branching continues, it leads to microscopic vessels that we cannot see with our naked eye. After blood has passed through these vessels, it flows back toward the heart through vessels that merge into vessels of progressively larger diameter—the largest being called *veins*. **FIGURE 30.19A** shows the anatomy of arteries and veins and their relationship in the circulatory system.

This is a good point to say a few preliminary words about the structure and naming of blood vessels. The open passageway in the center of a blood vessel, through which blood flows, is termed the **lumen**. In all types of vertebrate blood vessels, the lumen is lined with a simple epithelium (see Figure 28.19), which for historical reasons is termed the **vascular endothelium**. This layer of cells is in direct contact with the blood and therefore of great significance. Some blood vessels—notably capillaries (soon to be discussed)—consist of only the vascular endothelium. However, the walls of

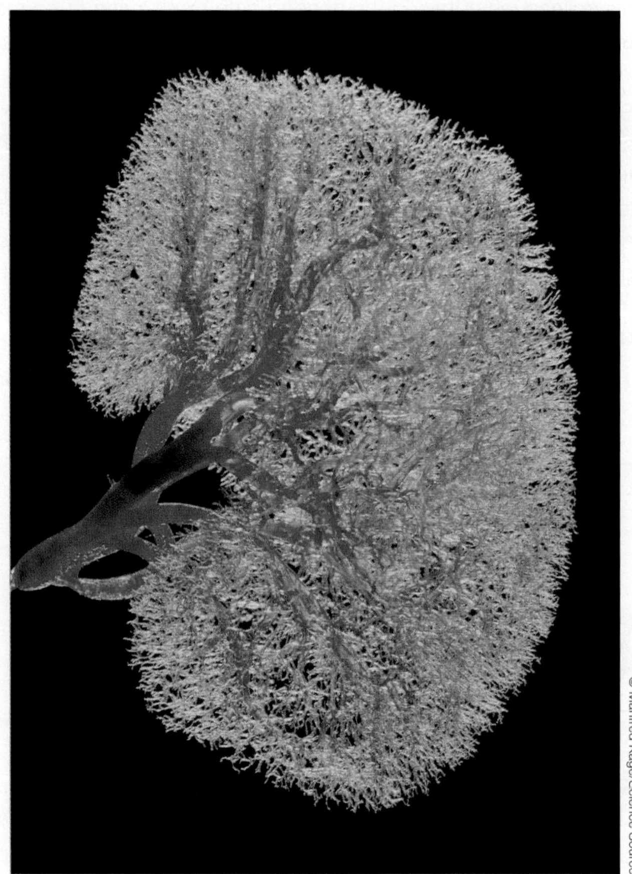

FIGURE 30.18 Arteries of Visible Size That Carry Blood to the Human Kidney The blood vessels visualized here are only the ones that can be seen with the naked eye. These vessels lead to microscopic vessels, which are even more profuse.

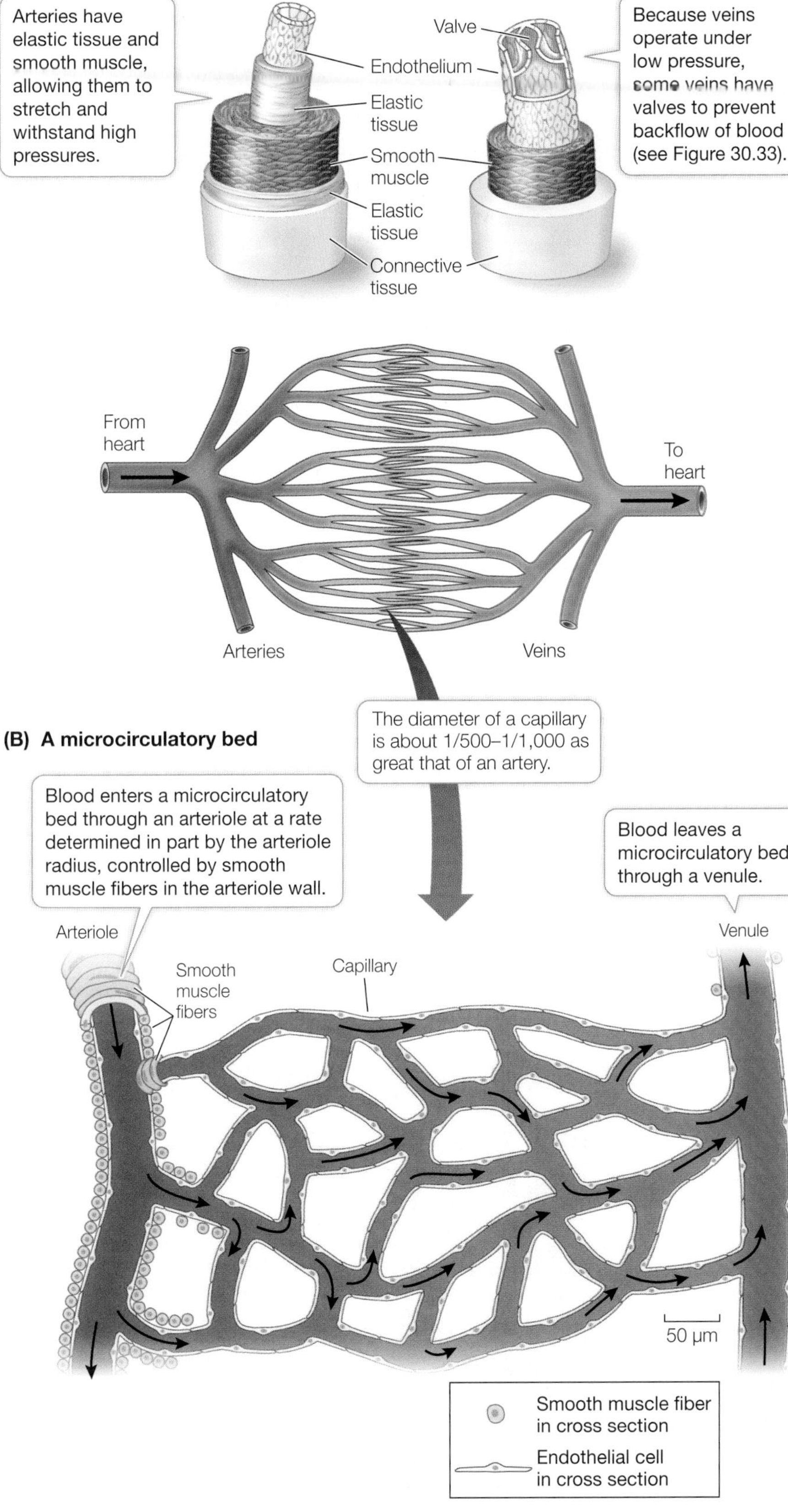

FIGURE 30.19 Blood Vessels in a Closed Circulatory System The large vessels—the arteries and veins—are far larger in diameter than the capillaries and other vessels in the microcirculation. **(A)** In this schematic diagram of the vertebrate circulatory system, red symbolizes oxygenated (O_2-rich) blood and blue symbolizes deoxygenated (O_2-poor) blood. A diagram such as this cannot accurately show the relative sizes of all vessels, because the smallest vessels are very tiny compared with the largest. **(B)** A microcirculatory bed composed of an arteriole, a network of capillaries, and a venule. Individual endothelial cells are shown in the walls of the vessels.

Activity 30.3 Structure of Blood Vessels

most types of blood vessels have additional layers of tissue—such as muscle—surrounding the endothelium (see Figure 30.19A). **Arteries** are vessels that carry blood away from the heart. **Veins** are vessels that carry blood toward the heart.

For understanding the properties of a closed circulatory system, it is particularly important to focus on the microscopically tiny blood vessels—named arterioles, capillaries, and venules—which are collectively termed the vessels of the **microcirculation** (**FIGURE 30.19B**). Open circulatory systems lack these vessels or have only incomplete sets of them. In a closed circulatory system, however, the blood remains within these minute vessels as it passes through the tissues and organs, and the properties of the vessels are of profound importance. As blood flows into a microcirculatory bed (a local set of microscopic vessels), it first passes through an **arteriole** which has smooth muscle fibers (cells) in its walls. The blood then flows through **capillaries**, the walls of which—as we have said—consist only of vascular endothelium. The blood then leaves the capillaries in a **venule**.

The capillaries are the principal vessels in which O_2, CO_2, and other materials are exchanged between the blood and the cells of the body. The capillaries play this role for two major reasons. First, in each tissue of the body, capillaries are found near every tissue cell. For instance, capillaries are positioned near every brain cell and every muscle cell. The number of capillaries is so great as to be almost unimaginable. Added together, for example, the capillaries in a single cubic centimeter of mammalian skeletal muscle or heart muscle are 10–30 meters long! Second, capillary walls, consisting of only flattened endothelial cells, are less than 1 micrometer thick, and in many tissues, there are gaps (pores) between

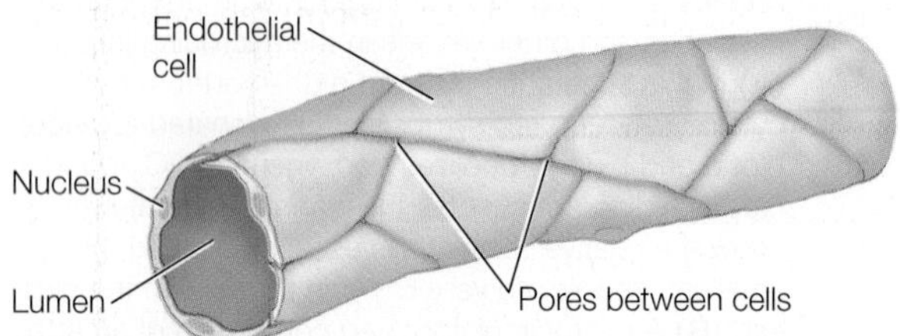

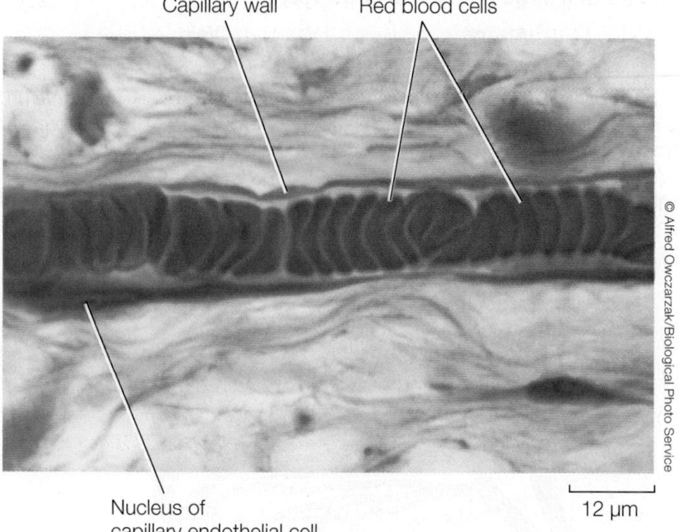

FIGURE 30.20 Vertebrate Blood Capillaries The capillaries may be small, but they are where blood flow accomplishes most of its missions. **(A)** The wall of a capillary consists of a single layer of highly flattened epithelial cells (called endothelial cells in this context). Pores (often termed fenestrations) are found between cells in the capillaries of some tissues but not others. **(B)** The diameter of the lumen of a capillary is often small enough that red blood cells must move through in single file.

Media Clip 30.2 Capillary Flow: A Tight Squeeze

adjacent endothelial cells (**FIGURE 30.20A**). These properties of capillaries mean that O_2, hormones, and nutrient molecules—but not blood cells—can readily move outward from the blood to surrounding cells and tissues by diffusion; CO_2 and other waste products can move inward; and water and dissolved ions can move in or out. The lumen of a capillary is often just slightly larger in diameter than individual red blood cells. For this reason, red blood cells (~7–8 μm in diameter in a human) must pass through capillaries in single file (**FIGURE 30.20B**).

Arterioles are as important for the function of the circulatory system as capillaries are. The walls of an arteriole consist of an inner layer of vascular endothelium surrounded by an outer layer of smooth muscle fibers (see Figure 30.19B) that are under the control of the autonomic nervous system (see Key Concept 31.5) and other control mechanisms. By contracting and relaxing, the muscle fibers in the wall adjust the diameter of the arteriole's lumen. Contraction of the muscle cells causes *vasoconstriction*, reducing the lumen diameter. Relaxation of the muscle cells allows the lumen diameter to increase, a process called *vasodilation*. Control of the arteriole lumen diameter in these ways is called **vasomotor control**.

Why do changes in the lumen diameter of arterioles matter? They are of enormous importance because they *control where blood flows*. The famous French physiologist Jean Poiseuille (pronounced *Pwa-sul*) demonstrated that the rate of fluid flow through a tube is extraordinarily sensitive to the diameter of the tube lumen because this rate varies as a function of the fourth power of the diameter. If the diameter of the tube lumen is doubled, the flow rate through the tube increases by a factor of 16! Conversely, if the diameter is halved, the flow rate drops to 1/16th of its previous value. In the circulatory system, an arteriole precedes each capillary bed (set of interconnected capillaries) and controls the flow of blood into it (**FIGURE 30.21**). By controlling the diameters of arterioles, the autonomic nervous system and other systems can enhance flow into some capillary beds while impeding flow into other capillary beds.

Several everyday examples illustrate arteriolar control of blood flow. The skin of people with fair complexions is often more pinkish in warm than in cold weather. This change reflects vasomotor control. Skin arterioles dilate in warm weather, permitting vigorous blood flow to the capillary beds near the skin surface—a response that makes the skin turn pink and helps metabolic heat leave the body. When people exercise, arterioles in their exercising skeletal muscles dilate, and arterioles in many other parts of their

FIGURE 30.21 An Arteriole Exerts Vasomotor Control over the Rate of Blood Flow through a Capillary Bed **(A)** When the lumen of the arteriole is large in diameter, blood flows rapidly through the capillary bed. **(B)** The rate of blood flow drops dramatically when the arteriole diameter decreases.

body constrict. These changes increase blood flow to the exercising muscles. Erection of the penis is one of the most familiar cases of arteriolar control. During sexual arousal, dilation occurs in arterioles that supply blood to the spongy tissue of the penis, permitting the spongy tissue to inflate with blood under high pressure (see Key Concept 34.2).

In an open circulatory system, blood leaves blood vessels

A circulatory system is open if the blood exits blood vessels as it flows through the body. Arthropods (e.g., crabs and lobsters) and most mollusks (e.g., snails and clams) have open circulatory systems. In an open circulatory system, the heart typically pumps blood into arteries that carry the blood for at least a short distance. However, the vessels end (often without any microcirculatory vessels), and the blood pours out into spaces that are surrounded not by vascular endothelium but by ordinary tissue cells. After this, the blood makes its way back to the heart, not in veins but by flowing through channels among ordinary tissue cells.

In an open circulatory system, there is no distinction between blood (the fluid pumped by the heart and flowing in the blood vessels) and interstitial fluid (the fluid found between cells in the ordinary tissues—also called tissue fluid) (see Chapter 28). Physiologists debate what the correct name for the fluid in an open circulatory system should be. Here we call it blood. Some biologists prefer to call the fluid *hemolymph* because it has attributes of both blood (*hemo*, "blood") and tissue fluid (lymph). Because tissue cells are directly bathed by the blood, they readily take up O_2 and other materials from the blood and release wastes into it.

Open circulatory systems are not very well understood, although we know they sometimes have very high rates of blood flow. These high flow rates do not require high pressures, however, because the blood does not need to be forced through tiny vessels in capillary beds. Open circulatory systems are thought to be limited in how well they can control where the blood flows.

Although all arthropods have an open circulatory system, the system plays different roles in some groups than in others. As we have already stressed, the circulatory system plays little role in O_2 transport in insects. However, it plays a major role in O_2 transport in the gill-breathing arthropods such as crabs, crayfishes, and lobsters because O_2 taken up by the gills enters the blood and then must be distributed throughout the body by blood flow (similar to O_2 transport in fishes) (**FIGURE 30.22**).

Blood often contains respiratory pigments that enhance its O_2 transport capability

Let's now focus again on blood flow as a key step in the pathway for oxygen (see Figure 30.1, step 3). As blood flows from the breathing organ to tissues throughout the body, the rate at which it transports O_2 depends on two factors: (1) the rate of blood flow and (2) how much O_2 is carried by each unit volume of blood. As we pointed out earlier, the amount of O_2 that can *dissolve* in each unit of volume is not very great because O_2 is not very soluble in water (or in aqueous solutions).

Animals have commonly evolved **respiratory pigments** that are present in the blood and that greatly increase the amount of O_2 that can be carried by each unit of blood volume. **Hemoglobin** is the respiratory pigment in humans. As blood flows out of our lungs to travel to the rest of our body, only about 1/50th of the O_2 it carries is dissolved. The rest is chemically combined with hemoglobin. Thus *the presence of the respiratory pigment results in a 50-fold increase in the amount of O_2 carried by each unit of blood volume.*

Before going further, let's say a few words about the composition of blood. Blood consists of two parts: the blood plasma and the blood cells. Blood plasma is obtained by removing the cells. Thus the **blood plasma** is the "blood solution"—free of cells—that consists of water and materials (solutes) dissolved in the water, such as glucose and other nutrient molecules, ions, waste products, hormones, clotting proteins, and O_2. In vertebrates, most of the cells suspended in the blood plasma are **red blood cells**, also called **erythrocytes**, which contain hemoglobin and carry O_2. The blood of vertebrates also contains a smaller number of **white blood cells** or **leukocytes**—cells of the immune system discussed in Chapter 36—as well as **platelets** (pinched-off fragments of cells), which are involved in blood clotting.

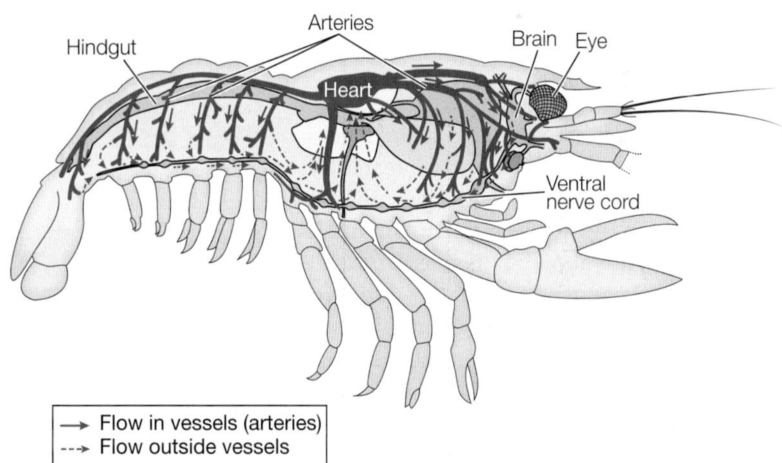

FIGURE 30.22 Blood Flow in the Open Circulatory System of a Crayfish or Lobster Crayfishes and lobsters have many arteries, which extend forward and backward from the heart. The heart pumps blood into these arteries, which transport the blood to most parts of the body. The arteries are open-ended, however; the blood is released, flowing into channels among ordinary tissue cells. Then blood from the anterior parts of the body flows posteriorly, and blood from the posterior parts of the body flows anteriorly, converging on a ventral sinus (cavity) at the bases of the principal legs. From there the blood returns to the heart. Although not shown here, the blood passes through the gills (which are attached at the bases of the principal legs) in this final step.

The respiratory pigments are proteins that undergo *reversible combination* with molecular oxygen, O_2. Thus they can pick up O_2 in the lungs and release O_2 in the systemic tissues (recall that the systemic tissues are all the tissues other than those in the breathing organ). These proteins are called "pigments" because they are colored and impart color to the blood. When a respiratory pigment combines with O_2, the process is called **oxygenation**, and the pigment is said to be oxygenated. This choice of words is important. A respiratory pigment is not chemically oxidized when it combines with O_2, so the process is termed oxygenation, not oxidation. When a pigment releases O_2, the process is **deoxygenation**, and the pigment becomes deoxygenated.

In the animal kingdom, there are two principal blood respiratory pigments: hemoglobin and hemocyanin. Hemoglobin is an iron-containing protein to which O_2 binds at the molecular sites (called heme sites) where iron atoms occur. Hemoglobin has evolved independently many times and is by far the most widespread blood respiratory pigment. It is found in almost all vertebrates, in annelid worms (such as earthworms), and in some members of at least seven other phyla. In mammals and most other vertebrates, each hemoglobin molecule is an assembly of four protein subunits, each of which contains one iron atom (in its heme subpart). In all vertebrates, hemoglobin is found inside red blood cells, where it is responsible for their red color. However, in earthworms and other annelids, the hemoglobin is dissolved in the blood plasma.

Hemoglobin changes color as it takes up and releases O_2. Fully oxygenated hemoglobin is bright crimson red. Fully deoxygenated hemoglobin is purple-red. The finger probes (pulse oximeters) that hospital patients wear to track blood oxygenation work by detecting how much hemoglobin is one color and how much is the other.

Red blood cells turn over rapidly, especially in mammals. In humans, each red cell circulates in the blood for only about 120 days before it is broken down and replaced by a new cell synthesized in the bone marrow. Under normal conditions, your blood contains about 5 billion red blood cells per milliliter, and your bone marrow produces about 2 million new cells every second.

Hemocyanin is the second most widely distributed respiratory pigment. Like hemoglobin, hemocyanin is a metal-containing protein to which O_2 binds at the molecular sites where metal atoms occur. The metal, however, is copper rather than iron. Hemocyanin is never found inside blood cells. Instead it is always simply dissolved in the blood plasma. It is found widely in mollusks and in arthropods such as crabs, lobsters, and horseshoe crabs (but not insects).

Like hemoglobin, hemocyanin changes color as it takes up and releases O_2. It is blue when oxygenated but colorless when deoxygenated. Hemocyanin is usually not concentrated enough in the blood to give the blood a strong color, but in some animals that have lots of hemocyanin, the blood leaving the gills is a bright cobalt blue!

Respiratory pigments combine with O_2 reversibly

What gives the molecules of a respiratory pigment their ability to transport O_2? The most important functional characteristic of these molecules is that they combine *reversibly* with O_2. Both their combination with O_2 and their release of O_2 occur automatically, and almost instantly, as a simple consequence of their chemical properties. The molecules take up O_2 when O_2 is abundant in their surroundings. They release O_2 when their surroundings are O_2-depleted.

The graph shown in **FIGURE 30.23**, called an **oxygen equilibrium curve**, describes the uploading and offloading of O_2 in adult human arterial blood. The *x* axis shows the partial pressure of O_2 in the blood, measured in millimeters of mercury (mm Hg) (a unit for measuring pressure). Partial pressure is used in plots of this type because it measures the chemical activity of O_2, which is the factor of key importance in driving the uptake or release of O_2. The *y* axis shows the amount of O_2 bound to

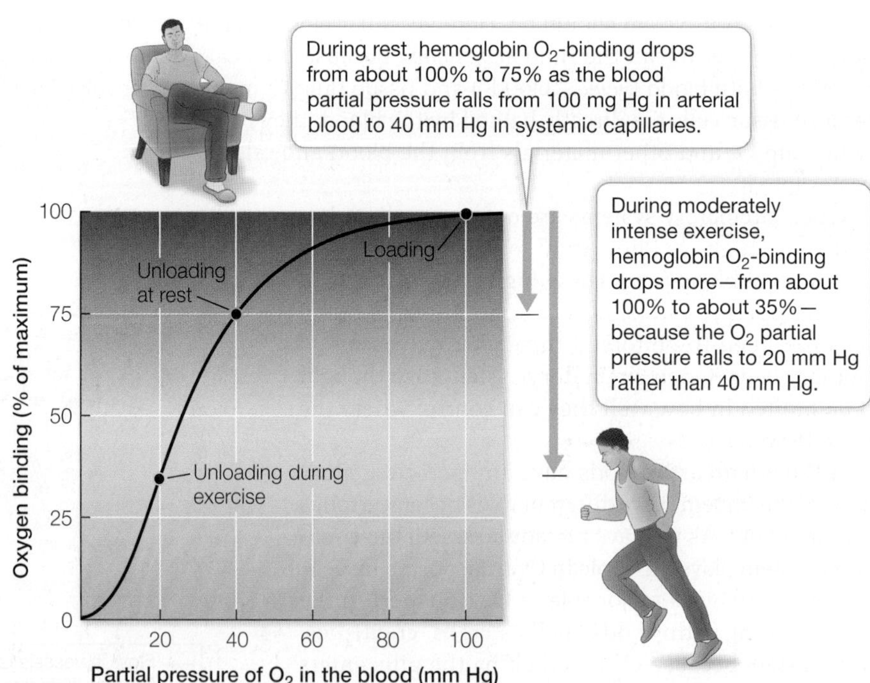

FIGURE 30.23 Loading and Unloading of O_2 by Human Hemoglobin This curve—known as an oxygen equilibrium curve—shows the chemical behavior of human hemoglobin in arterial blood. Hemoglobin almost instantly adjusts its O_2 binding to match changes in the prevailing O_2 partial pressure, taking up or releasing O_2 as needed. The vertical green arrows show how much O_2 is unloaded by hemoglobin as it flows from the lungs through the systemic tissues during rest and exercise. The blood O_2 partial pressure is expressed in millimeters of mercury (mm Hg) of pressure.

Activity 30.4 Oxygen-Binding Curves

hemoglobin at each partial pressure, as a percentage of the maximum amount that the hemoglobin can bind.

An oxygen equilibrium curve is of great interpretive value. To see this, let's use Figure 30.23 and focus on blood O_2 transport in a person. The O_2 partial pressure in the alveoli of a person's lungs is kept near 100 mm Hg, which means that the O_2 partial pressure in blood oxygenated in the alveoli is also kept near 100 mm Hg. As shown by the "Loading" point on the graph, human hemoglobin combines with a maximum amount of O_2—about 100 percent of the O_2 it can possibly carry—when the O_2 partial pressure is near 100 mm Hg. In short, human hemoglobin becomes highly oxygenated in the lungs.

In the systemic tissues of a person at rest, the O_2 partial pressure, on average, is about 40 mm Hg. The "Unloading at rest" point on the graph shows the consequences. O_2 binding is only about 75 percent of maximum when human hemoglobin is at 40 mm Hg. When highly oxygenated blood from the lungs first flows into the capillaries of the systemic tissues, its oxygen binding is about 100 percent, but as soon as it is exposed to the partial pressure of 40 mm Hg that prevails in the systemic tissues, it releases some O_2 so that its oxygen binding falls to 75 percent. This is the process by which hemoglobin releases O_2 to the tissues. Later, when the blood flows back into the lungs, the hemoglobin is exposed once more to a partial pressure of about 100 mm Hg and therefore takes up O_2, so that its oxygen binding is once again about 100 percent.

During exercise, hemoglobin can greatly increase the amount of O_2 it releases in the exercising muscles. The O_2 partial pressure in exercising muscles often falls to about 20 mm Hg during moderately intense exercise because the muscle mitochondria are using O_2 rapidly. When hemoglobin is exposed to this partial pressure, its oxygen binding falls to only 35 percent (see "Unloading during exercise" in Figure 30.23). The hemoglobin thus releases much more of its O_2 than it does in resting tissues.

When we exercise, there are two major ways that our circulatory system increases O_2 delivery to our muscles. First, our heart beats faster—and pumps more blood per beat—so our blood circulates faster. Second, our hemoglobin releases more than twice as much O_2 each time it flows from our lungs to our exercising muscles.

Activity 30.5 Hemoglobin Loading and Unloading Simulation

REVIEW & APPLY | 30.4

R The circulatory system transports respiratory gases and other materials by rapid bulk flow of blood. O_2 transport is the most urgent function of the circulatory system in most types of animals. Animals with relatively high rates of O_2 consumption tend to have relatively high-performance circulatory systems. Insects are an exception because their tracheal breathing system fully meets their needs for O_2 transport. In closed circulatory systems, the blood never leaves vessels. In open circulatory systems, the blood (sometimes called hemolymph) leaves vessels and bathes tissue cells directly. One advantage of closed circulatory systems is their ability to direct blood selectively to specific tissues; blood flows through microcirculatory beds where arterioles exert vasomotor control over which microcirculatory beds receive flow. Respiratory pigments that bind O_2 reversibly—such as hemoblobin and hemocyanin—greatly increase the amount of O_2 that blood can deliver to tissues. A respiratory pigment binds O_2 when the partial pressure of O_2 is high (as in the breathing organ) and releases it when the partial pressure is low (as in the systemic tissues).

A

1. What do we mean when we say that hemoglobin and hemocyanin combine reversibly with O_2, and what controls this reversible combination?

2. As stated in the text, "Open circulatory systems are thought to be limited in how well they can control where the blood flows." How does a closed circulatory system control where the blood flows, and why might an open circulatory system have only a limited ability to exert such control?

3. Atherosclerosis ("hardening of the arteries") is a potentially deadly chronic disease. The immediate problem is that fatty deposits—composed of lipids such as cholesterol and triacylglycerols—build up in arteries. Focusing on these drawings of small arteries, take out a ruler and, using the quantitative principle developed by Poiseuille, estimate how much more slowly blood will pass through the atherosclerotic artery on the right, compared with the healthy artery on the left, assuming everything else is the same except the fatty deposits. People start their lives with arteries like the one on the left.

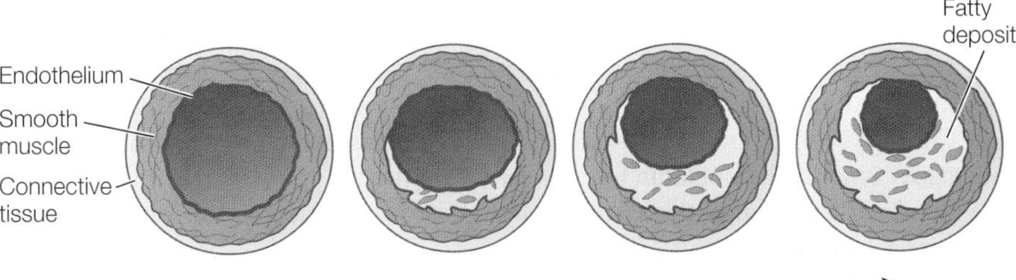

Having completed our initial overview of the circulatory system, we now take a closer look at the structure and function of hearts.

30.5 | A Beating Heart Propels the Blood

LEARNING OBJECTIVES

By the end of this key concept you should be able to:

30.5.1 **Explain how a wave of contraction spreads through a vertebrate heart during each heartbeat.**

30.5.2 **Explain why the coronary blood vessels are vitally important.**

30.5.3 **Describe how the heart increases the rate of blood flow through the circulatory system during exercise.**

A **heart** is a discrete, localized pumping structure. Some animals with a circulatory system lack a heart. In earthworms and other annelid worms, for example, the blood is propelled entirely by peristaltic waves of contraction in muscular blood vessels. Some animals, by contrast, have more than one heart. Crayfishes and lobsters, for instance, have a small accessory heart that helps perfuse their brain, in addition to their large, principal heart. Squid and octopuses have accessory hearts that perfuse their gills.

The muscle tissue of a heart is termed the **myocardium** (*myo*, "muscle"; *cardi*, "heart"). As discussed in Key Concept 33.3, in vertebrates the cardiac muscle cells in the myocardium have distinctive properties, compared with skeletal muscle cells or smooth muscle cells.

The function of a heart is often analyzed in terms of its stroke volume, heart rate, and cardiac output. A heart's **stroke volume** is the volume of blood it pumps per beat. A heart's **cardiac output** is the volume of blood it pumps per minute, calculated by multiplying its stroke volume by the number of beats per minute, the heart rate. For example, when you are at rest, your heart has a stroke volume of about 70 milliliters per beat (volume pumped into the aorta) and beats about 70 times per minute. Your resting cardiac output is thus about 5 liters per minute (70 ml/beat × 70 beats/min = 4,900 ml/min, or about 5 liters/min).

To increase the rate of O_2 transport during exercise, the human heart increases both stroke volume and heart rate. In an average young person, these changes, working together, can increase cardiac output about fourfold over the resting value. As we discussed in Key Concept 30.4, O_2 transport is also increased during exercise by deoxygenating hemoglobin to a greater extent than at rest as the blood passes through the systemic tissues, especially the working muscles (see Figure 30.23).

Vertebrate hearts are myogenic and multi-chambered

Hearts are classified based on the answers to two questions: how is each beat initiated, and how many chambers are present?

Regarding the first question, vertebrate hearts are classified as **myogenic**, meaning "beginning in muscle," because heartbeats originate in the muscle tissue and do not require nervous stimulation to be initiated. Modified muscle cells within the myocardium initiate each heartbeat, and a vertebrate heart continues to beat even after all connections to other tissues of the body are severed. Mollusks, too, have myogenic hearts. Crustaceans do not, as we will see later.

Vertebrate hearts are also classified as *multi-chambered* because they consist of two or more distinct chambers. In most fishes, the heart consists of two primary, muscular chambers: an **atrium** (plural *atria*)—which is a thin-walled chamber through which blood passes on its way to the principal pumping chamber—and a thick-walled, highly muscular **ventricle**. Amphibian hearts have three chambers (two atria that empty into one partially divided ventricle). The hearts of mammals and birds have four chambers. One-way valves ensure that blood flows through the chambers in the correct order.

As we now start to discuss patterns of blood flow throughout the body, driven by the heart, it will be helpful to introduce the concepts of the *systemic* circuit and the *breathing-organ* circuit. The **systemic circuit** consists of the blood vessels (or blood channels in the case of an open circulatory system) that carry blood to and from the systemic tissues, and through those tissues. The **breathing-organ circuit** consists of the blood vessels that carry blood to and from the breathing organ, and through the organ. The breathing-organ circuit is sometimes called the *pulmonary circuit* or *lung circuit* in animals that breathe with lungs, and the *gill circuit* in animals that breathe with gills.

The mammalian heart consists of four chambers: two highly muscular ventricular chambers, which contract together, and two less-muscular atria, which also contract together (**FIGURE 30.24**). O_2-depleted blood returning to the heart from the systemic tissues enters the right atrium and then flows through a one-way valve into the right ventricle. When the ventricles contract, the O_2-depleted blood is pumped into the pulmonary arteries leading to the lungs and then through the pulmonary circuit, where it becomes oxygenated. The O_2-rich blood delivered to the heart from the lungs enters the left atrium and then passes through a one-way valve into the left ventricle. When the ventricles contract, this O_2-rich blood is pumped into a massive artery, the **aorta**, from which it flows through branching arteries into all parts of the systemic circuit. The blood becomes partially depleted of O_2 as it passes through the systemic tissues, thereafter returning to the right atrium.

The cycle of contraction and relaxation of the heart is called the **cardiac cycle**. This cycle is divided into two phases: **systole** (pronounced sís-toll-ee), when the ventricles contract, and **diastole** (die-ás-toll-ee), when the ventricles relax (**FIGURE 30.25**).

The ventricles pressurize the blood inside them when they contract, and this positive pressure—developed by contraction of the ventricular myocardium—is the principal force that drives blood through the circulatory system. The positive pressure developed by the right ventricle forces blood to flow through the pulmonary circuit to the lungs and back to the heart. The positive pressure developed by the left ventricle forces blood to flow through the systemic circuit. The left ventricle is more muscular than the right (see Figure 30.24) and develops higher pressures, which help ensure flow through the larger systemic circuit.

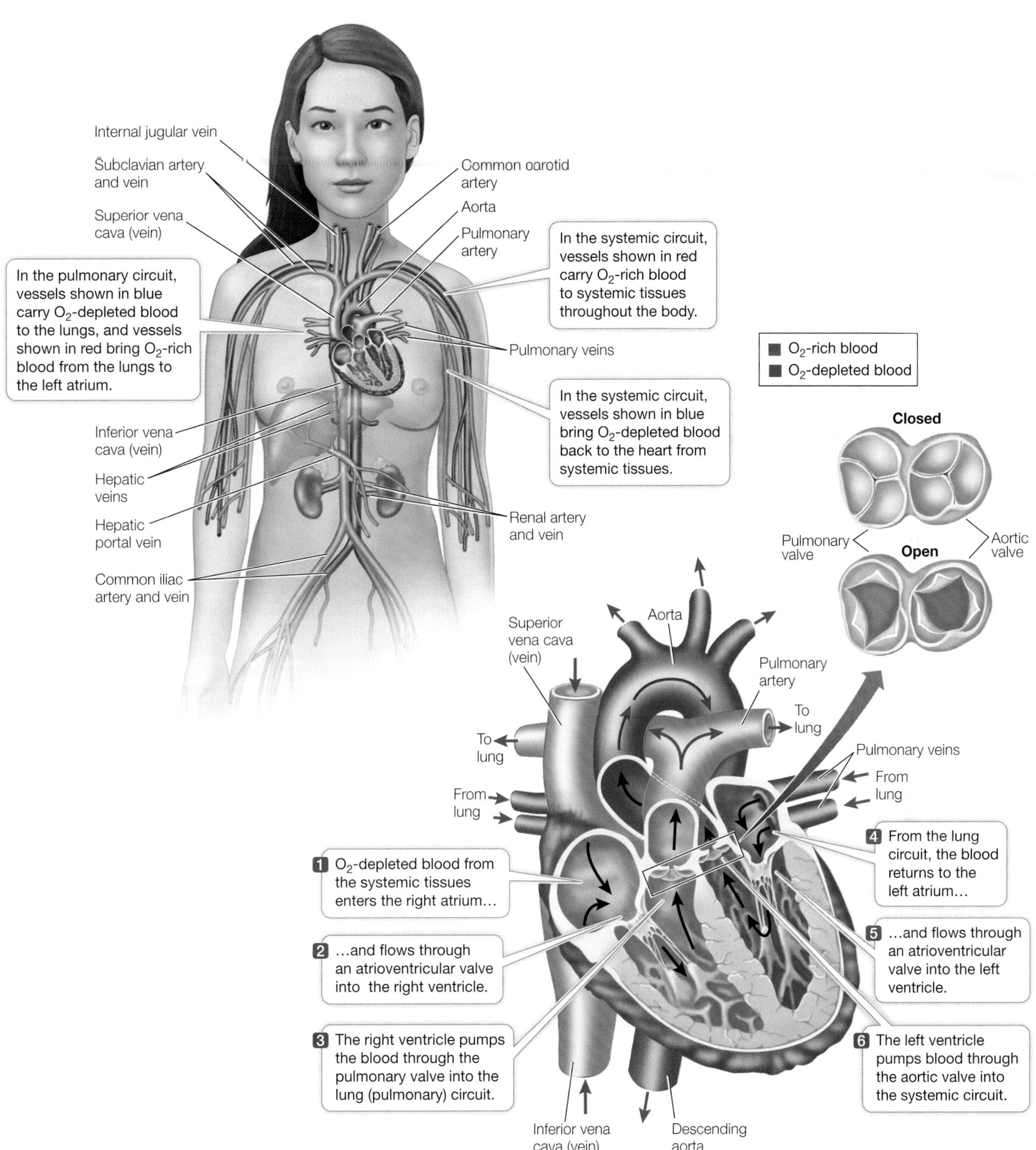

FIGURE 30.24 The Human Heart and Circulation In the human heart, blood flows from the right heart to the lungs, then to the left heart, and from there to the systemic tissues. Color indicates whether the blood is O_2-rich (red) or O_2-depleted (blue). Thus the pulmonary arteries (carrying blood to the lungs) are blue, and the pulmonary veins (carrying blood to the heart from the lungs) are red, although arteries in the systemic circuit are red and veins in that circuit are blue. The atrioventricular valves prevent blood from flowing back into the atria when the ventricles contract. The pulmonary and aortic valves prevent blood from flowing back into the ventricles from the pulmonary artery and the aorta, respectively, when the ventricles relax.

Activity 30.6 The Human Heart

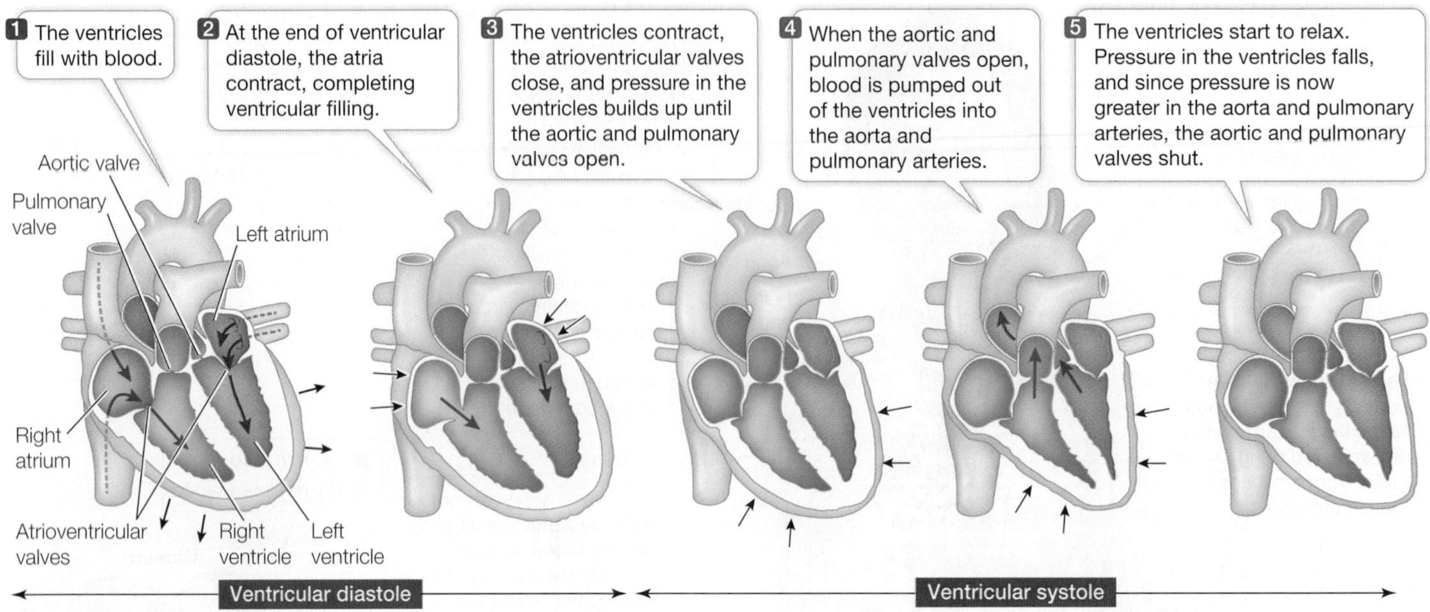

FIGURE 30.25 The Cardiac Cycle Within the heart, red (including red arrows) represents O_2-rich blood. Blue (including blue arrows) represents O_2-depleted blood. Black arrows symbolize changes in the size of the ventricles.

Animation 30.3 The Cardiac Cycle

The **blood pressure** at any point in the circulatory system is the extent to which the pressure in the blood exceeds the pressure in the environment of the animal. As you might imagine, the blood pressure in the aorta rises and falls during the cardiac cycle, even though the aortic walls are elastic and therefore tend to dampen pressure changes. In healthy young people, the peak aortic pressure during systole is about 120 mm Hg (millimeters of mercury, referring to the height of a vertical column of mercury that the pressure can support). The lowest aortic pressure during diastole is about 80 mm Hg. These pressures undergo little change as blood flows through major arteries and can therefore be measured in the arm, as is usually done. They are typically quoted as a ratio: "120 over 80." **Hypertension** ("high blood pressure") is a disease in which the pressures are significantly higher than this norm.

Each heartbeat is initiated by **pacemaker cells**, which are modified muscle cells found in a localized part of the right atrial wall termed the *sinoatrial (S-A) node* (**FIGURE 30.26A**). As we will discuss in detail in Key Concept 33.3, the cardiac muscle cells in the myocardium are electrically coupled, meaning that when any particular cell undergoes electrical depolarization (a reversal of the electric charge difference across its cell membrane) and consequently contracts, it directly initiates electrical depolarization and contraction in the cells next to it. Accordingly, when electrical depolarization and contraction are initiated at any one point in a mass of heart muscle, they spread rapidly throughout the mass of muscle by moving from cell to cell. This process ensures that all the cells within a large mass of muscle depolarize and contract almost simultaneously. The pacemaker cells in the S-A node rhythmically depolarize; they do this completely on their own, without any need for an external stimulus. Each time they depolarize, the depolarization rapidly spreads throughout the atria (**FIGURE 30.26B**, Steps 1 and 2), bringing about atrial contraction. The depolarization and contraction do not instantly and directly spread into the ventricles, however, because a sheet of fibrous connective tissue lies between the atrial and ventricular muscle masses (see Figure 30.26A), and the two muscle masses are therefore electrically isolated.

Electrical depolarization in muscle cells is described in **Key Concept 33.1. Key Concept 33.3** provides more details about the special characteristics of cardiac muscle.

How, then, does electrical depolarization move from the atria to the ventricles and cause the ventricles to contract? A cluster of modified muscle cells termed the *atrioventricular (A-V) node* is situated in the atrial muscle near the junction of the atria and ventricles (see Figure 30.26A). Electrical depolarization of the atria causes depolarization in this node after a slight delay. This delay is important because it ensures that the ventricles contract *after* the atria. Once depolarized, the cells of the A-V node relay the electrical depolarization into a system of modified muscle cells called the **conducting system** (see Figure 30.26A), in which the depolarization travels very rapidly from cell to cell, sweeping at high speed into the right and left ventricles (see Figure 30.26B, Step 3). As soon as muscle cells in the ventricles start to depolarize and contract, the wave of depolarization and contraction spreads rapidly from cell to cell within the ventricles (see Figure 30.26B, Step 4). The result is a strong, coordinated ventricular contraction.

Even though the heart beats rhythmically on its own, it is subject to outside controlling influences termed *extrinsic controls*. Blood-borne hormones can strongly modify and control heart action. For example, we've all felt our heart pound after our adrenal glands secrete a rush of epinephrine and norepinephrine (adrenaline and noradrenaline) in response to fright or stress. The autonomic

(A) The conducting system and sinoatrial node

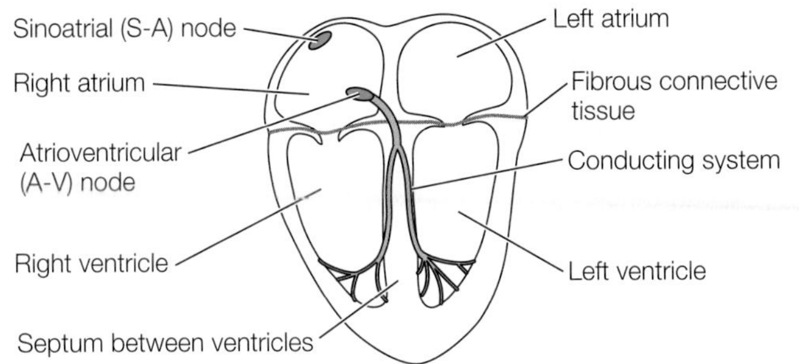

(B) The initiation and spread of depolarization during a heartbeat

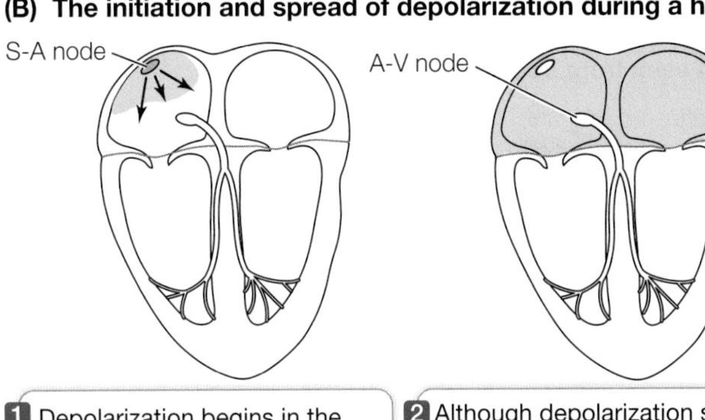

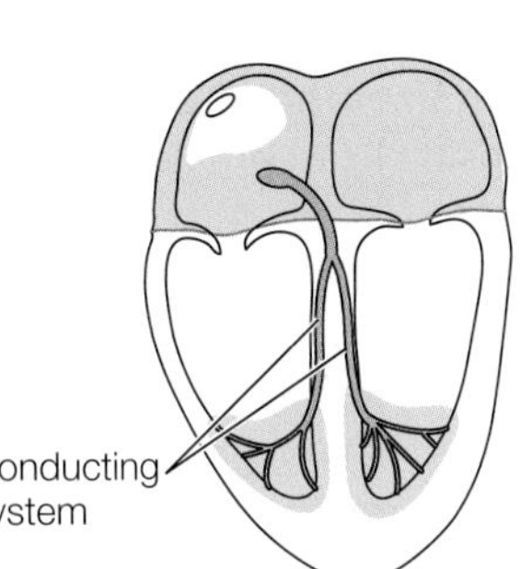

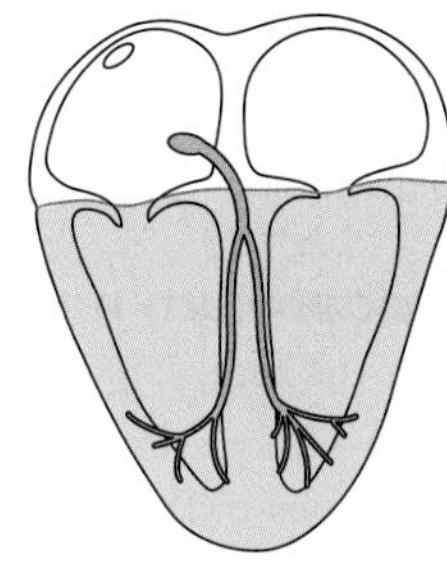

1 Depolarization begins in the S-A node and spreads outward through atrial muscle.

2 Although depolarization spreads rapidly throughout the atrial muscle, its spread into the A-V node is delayed. The depolarized atria start to contract.

3 Once the A-V node becomes depolarized, the depolarization spreads very rapidly into the ventricles along the conducting system. Atrial muscle starts to repolarize.

4 The nearly simultaneous depolarization of cells throughout the ventricular myocardium leads to forceful ventricular contraction.

Depolarized
Not depolarized

FIGURE 30.26 The Process of Activation of the Mammalian Heart during Each Heartbeat The muscle cells of the heart, like virtually all muscle cells, are activated to contract by electrical depolarization of their cell membrane. **(A)** Structures composed of modified muscle cells—the S-A node, A-V node, and conducting system—initiate and coordinate the spread of depolarization through the heart during each heartbeat. **(B)** In each of the successive stages shown, muscle cells in tan areas are depolarized, whereas cells in white areas are not depolarized. White areas may be awaiting depolarization or may have repolarized following depolarization. (After R. F. Rushmer. 1961. *Cardiovascular Dynamics,* 2nd ed. Saunders: Philadelphia.)

Why is the sheet of fibrous connective tissue important?

nervous system (ANS) also exerts strong controlling influences. Both the pacemaker cells in the S-A node and the ordinary muscle cells of the myocardium are highly innervated by the ANS. Stimulation by the sympathetic division of the ANS causes the S-A node to depolarize more frequently (increasing the heart rate) and causes ordinary myocardial cells to increase their force of contraction.

Activity 30.7 Blood Pressure and Heart Rate Simulation

LINK The autonomic nervous system controls many involuntary processes, as discussed in **Key Concept 31.5**. **Key Concept 32.2** describes the numerous effects of epinephrine and norepinephrine.

An electrocardiogram records the electrical activity of the heart

Great numbers of muscle cells depolarize simultaneously each time a heart contracts (see Figure 30.26B). Because the cells undergo electrical depolarization simultaneously, their individual voltage signals add together to produce a large voltage signal. This voltage signal is so large that it often can be readily detected on the surface of the body! An **electrocardiogram** (**ECG** or **EKG**) is simply a record of heart-related voltage differences on the surface of the body over time (**FIGURE 30.27**). EKGs can

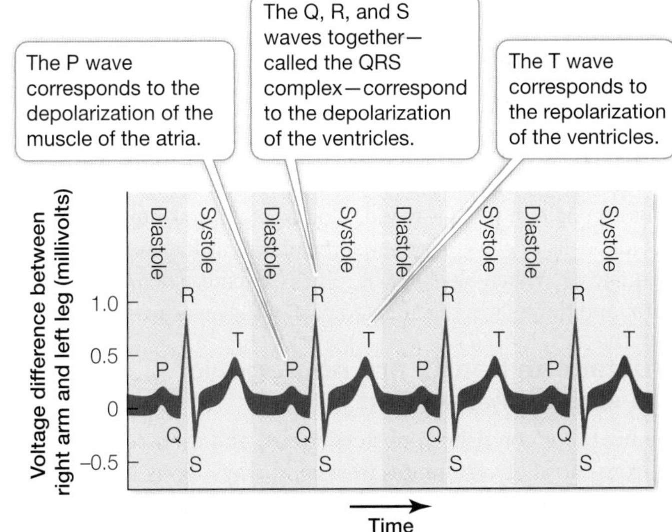

FIGURE 30.27 The Electrocardiogram (EKG) Is an Electrical Signature of Heart Action A normal EKG recorded from a person, showing four heartbeats. A series of electrical events—consisting of defined waveforms—occurs during each heartbeat. The waveforms are named P, Q, R, S, and T. To record this EKG, two electrodes were placed in contact with the skin on the right arm and left leg. The voltage difference between them was recorded as a function of time. Vertical color bands identify times of ventricular systole and diastole.

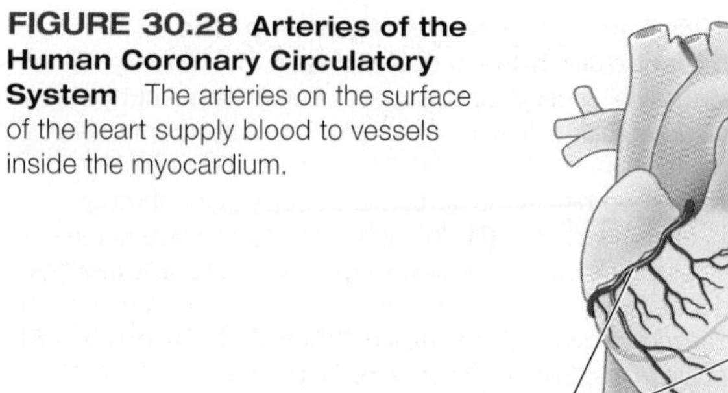

FIGURE 30.28 Arteries of the Human Coronary Circulatory System The arteries on the surface of the heart supply blood to vessels inside the myocardium.

be recorded from all sorts of animals, even lizards and octopuses. They are an important tool for diagnosing heart problems in people and pets.

The myocardium must receive O_2

The cells of the heart muscle in all animals must have a way to obtain O_2. We know the most about this process in vertebrates.

In most types of fishes, the myocardial cells get their O_2 directly from the blood flowing through the heart chambers. This is believed to be the primitive mechanism of oxygenating the myocardium in vertebrate animals. However, this mechanism cannot meet cellular O_2 needs in hard-working, high-performance hearts, explaining the evolution of a coronary circulatory system in some highly active fishes (such as tunas) and in mammals and birds.

A **coronary circulatory system** consists of arteries, capillaries, and veins inside (or closely associated with) the myocardium. In animals with a coronary circulation, the heart pumps O_2-rich blood into its own coronary arteries (**FIGURE 30.28**). This blood then flows to the coronary capillary beds in the myocardium, providing O_2 to all the cardiac muscle cells. Mammals and birds depend entirely on a coronary circulation for their cardiac muscle cells to receive O_2. Fishes with high-performance hearts, such as tunas and salmon, depend in part on a coronary circulatory system.

Coronary artery disease is a common, life-threatening problem for people in today's world. Blockage of a coronary artery, by fatty deposits or clots, deprives some of the cardiac muscle cells of adequate O_2. When blockage becomes extreme, a heart attack can occur, and the cells in the affected part of the myocardium can die.

Crustacean hearts are neurogenic and single-chambered

The hearts of crayfishes, lobsters, crabs, and their relatives differ from those of vertebrates in several ways. Thus the study of crustacean hearts helps us appreciate that evolution has produced diverse types of circulatory systems in animals.

The muscle cells of a crustacean heart require nervous stimulation to contract. Crustacean hearts are thus classified as **neurogenic**, meaning that beating "begins in neural tissue." Some scorpions and spiders also have neurogenic hearts. An additional property of the crustacean heart is that it consists of just a single chamber.

A ganglion—a cluster of many nerve cells—associated with the dorsal wall of the heart is responsible for initiating and controlling heartbeats (**FIGURE 30.29A**). One of the nerve cells in this cardiac ganglion acts as a pacemaker cell, rhythmically initiating nerve impulses. These impulses are then relayed by nerve cells to every muscle cell in the myocardium. The muscle cells in a crustacean heart contract simultaneously during a heartbeat because nerve cells stimulate all of them simultaneously—a very different mechanism than in vertebrate hearts. If the cardiac ganglion and the heart muscle are dissected apart, the ganglion continues to produce rhythmic nerve impulses, but the muscle cells stop contracting.

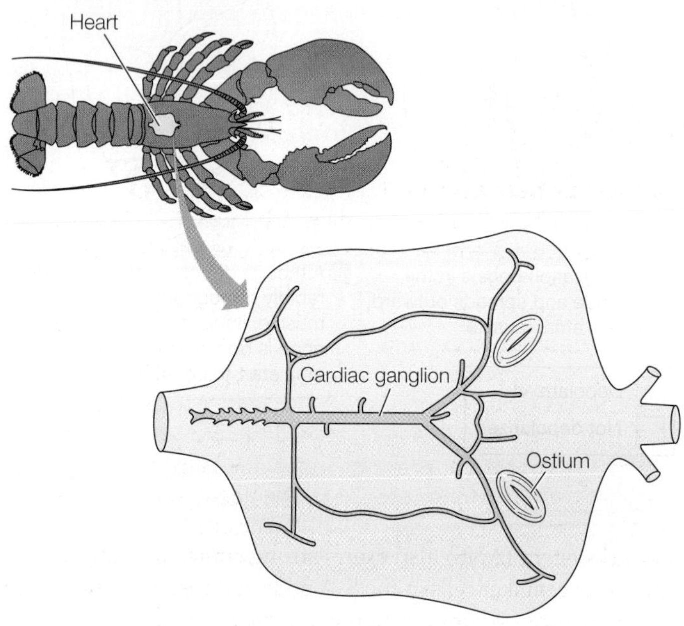

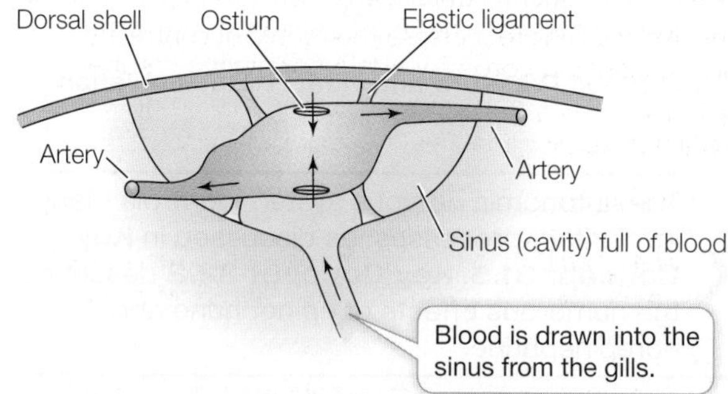

FIGURE 30.29 The Heart of a Lobster **(A)** The lobster heart is stimulated to beat by nerve impulses that originate in a cardiac ganglion (yellow) associated with the dorsal heart wall. **(B)** Blood enters the single heart chamber via slits in the heart wall, called ostia, that act as one-way valves, allowing blood to enter the heart chamber but not exit. When the heart contracts, blood inside is forced into the arteries. (After D. K. Hartline. 1967. *J Exp Biol* 47: 327–346.)

Why are only arteries shown connecting to the heart? Explain.

Both positive pressure and suction are crucial for heart function in crustaceans. The single-chambered heart is suspended by elastic ligaments in a blood-filled cavity (**FIGURE 30.29B**). All the vessels attached to the heart are arteries and carry blood away from the heart in many directions (see Figure 30.22). Three pairs of slits, called ostia (singular *ostium*), penetrate the heart wall and act as one-way valves that allow blood to flow into but not out of the heart chamber. When the heart contracts, the elastic ligaments are stretched like rubber bands, and the blood inside the heart is pressurized, forcing the blood to flow out into the arteries. Then, when the heart relaxes, the stretched ligaments shorten elastically, pulling the walls of the heart chamber outward and tending to cause suction inside the heart. This suction pulls blood into the heart chamber through the ostia, filling the heart in preparation for the next contraction.

REVIEW & APPLY | 30.5

R Vertebrate hearts are multi-chambered and myogenic. The simplest type of vertebrate heart, found in fishes, has two muscular chambers: an atrium that receives blood from the body and a ventricle that pumps blood out of the heart. Mammals and birds have four-chambered hearts (two atria and two ventricles) in which blood flows from the right atrium and ventricle to the lungs, then to the left atrium and ventricle, and then to the rest of the body in what is called the systemic circuit. The cardiac cycle has two phases: systole, when the ventricles contract, and diastole, when the ventricles relax. The heartbeat in mammals is initiated by specialized muscle cells in the right atrium—the pacemaker cells—that spontaneously depolarize, triggering a wave of depolarization that spreads directly from muscle cell to muscle cell in the atria and eventually, with a brief delay, into the ventricles. In most fishes, the myocardium (heart muscle) gets its O_2 from the blood flowing through the heart chambers. In some highly active fishes a coronary circulatory system supplies O_2 to part of the myocardium, and in mammals the myocardium is serviced entirely by a coronary system.

A

1. What is a coronary artery?
2. In the human heart, why do the ventricles contract after the atria, even though cardiac muscle cells directly activate neighboring cells?
3. What would you observe if you obtained a living crayfish heart and separated the cardiac ganglion from the heart muscle and then studied the two parts? What conclusions would you draw from the results?

Now that we have discussed the structure and function of hearts, we will complete our examination of the circulatory system by focusing on the blood vessels.

30.6 | The Vascular System Plays Many Roles

LEARNING OBJECTIVES

By the end of this key concept you should be able to:

30.6.1 **Explain why a series circulatory plan is efficient in transporting O_2 from the breathing organ to the systemic tissues.**

30.6.2 **Explain the challenges of perfusing the microcirculation in a vertebrate circulatory system.**

30.6.3 **Explain how the vascular system is specialized in some organs compared with other organs—and in some types of animals compared with other types.**

The circulatory system is often called the **cardiovascular system**, recognizing the profound importance of the blood vessels (collectively known as the *vascular system*) as well as the heart. Of all the "big picture" characteristics of a circulatory system, the most important is the *circulatory plan*, which refers to the organization and structure of the whole system. The circulatory plan depends on the relationship that has evolved among the heart and blood vessels.

A series circulatory plan is most common

The most common type of circulatory plan in animals is for the systemic and breathing-organ circuits to be connected in *series*, meaning that blood flows through the two circuits sequentially: first through one circuit, then through the other, then back through the first (**FIGURE 30.30A**). Fishes display such a series arrangement (**FIGURE 30.30B**), the only exceptions being air-breathing species. In a typical fish, the heart pumps blood into arteries that lead directly to the gills. After the blood perfuses the gills and exits them, it is collected into additional arteries, which branch to supply blood to all the systemic tissues, ranging from the brain in the head to the kidneys and swimming muscles in the abdomen. After the blood perfuses the systemic tissues, it enters veins that take it back to the heart.

A series arrangement such as that seen in fishes is considered to be the most efficient for delivering O_2. It is efficient because all the blood pumped to the systemic tissues is O_2-rich, having just passed through the gills. Moreover, all the blood pumped to the gills is O_2-depleted, having just exited the systemic tissues. Energy is not wasted in sending any O_2-rich blood to the gills or any O_2-depleted blood to the systemic tissues.

Natural selection has favored the evolution of a series arrangement in many types of animals. We have already seen, for example, that crustaceans have series circulation in their open circulatory system (see Figure 30.22). After the blood of a crayfish or crab passes through the gills, it flows directly to the heart, and the heart pumps it through the systemic circuit, after which the blood returns to the gills. Squid and octopuses, which have closed circulatory systems, also have the gill circuit and systemic circuit connected in series.

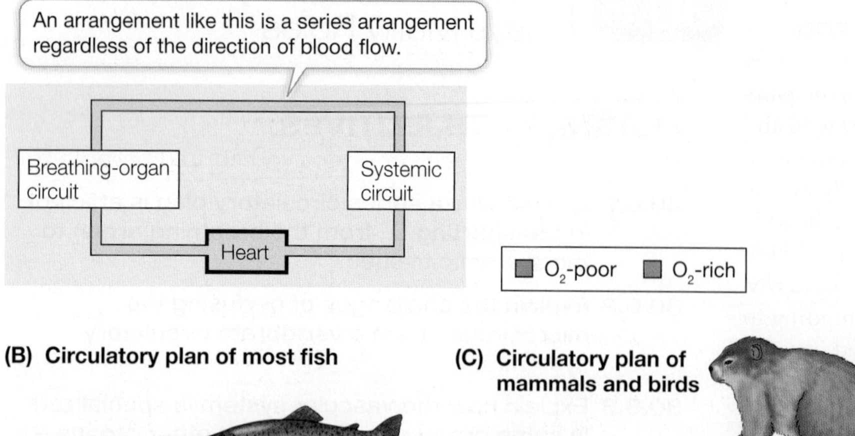

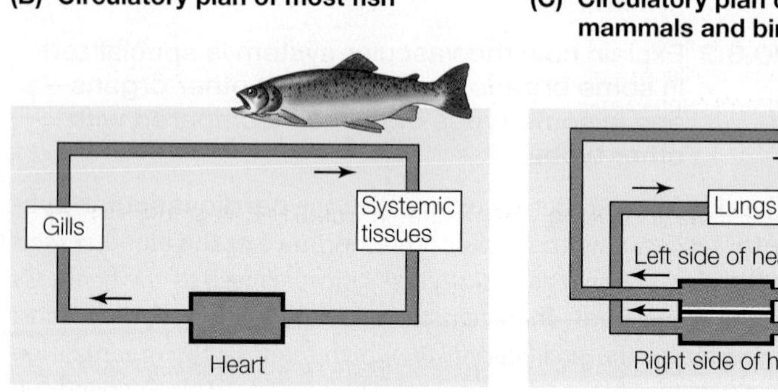

FIGURE 30.30 Series Circulatory Plans **(A)** A schematic diagram of a circulatory system in which the systemic and breathing-organ circuits are connected in series. The circulatory plan **(B)** in most fishes (only the air-breathing fishes are exceptions) and **(C)** in mammals and birds.

Activity 30.8 Vertebrate Circulatory Systems

The presence of a four-chambered heart in mammals and birds adds complexities to understanding their circulatory plans. However, like fishes, mammals and birds have the systemic circuit and breathing-organ circuit connected in a high-efficiency series arrangement.

In a mammal or bird, blood passes through the heart twice to make a complete trip through the circulatory system. As we have seen, the heart is completely divided into right and left sides (blood flowing through either side cannot mix with blood flowing through the opposite side). As blood travels from the systemic circuit to the lung circuit, it passes through the right side of the heart. As it travels from the lung circuit to the systemic circuit, it passes through the left side of the heart. **FIGURE 30.30C** shows the circulatory plan. If you trace the flow of blood from any particular point back to the same point, you will see that blood passes from the systemic circuit to the lung circuit, then back to the systemic circuit, and so forth: a series circulation.

A limitation faced by fishes is that, after their heart pumps their blood, the blood must pass through both the gill and systemic circuits without receiving additional propulsive energy (an added push) from the heart. Mammals and birds have evolved circulatory plans without this limitation. In a mammal or bird, the heart pumps the blood just before it travels to the lungs, and it pumps it again just before it travels to the systemic circuit.

The hearts of amphibians, lizards, snakes, and turtles have been extensively studied because they are structured in ways that do not anatomically guarantee a series circulatory plan. Although the hearts of these animals have two atria, they have just a single ventricle. Thus O_2-depleted blood flowing into the ventricle from the systemic circuit can in principle be pumped back into the systemic circuit, and O_2-rich blood arriving from the lungs can in principle be pumped back into the pulmonary circuit. The structure of the ventricle in these animals is very intricate, however, making its function difficult to predict from its structure. Recent functional studies have shown conclusively that in some lizards and snakes a series circulatory plan is in fact maintained. For example, it some cases, when the ventricle contracts, its shape changes in such a way that an internal physical barrier is temporarily formed, keeping the O_2-depleted blood separate from the O_2-rich blood.

As we now discuss some additional aspects of the vascular system, we will narrow our focus to vertebrates, particularly mammals, and thus will be concerned only with closed circulatory systems.

The nature of blood vessels varies with their position

Functional demands on the vessels of the vascular system vary with the vessels' positions in the system, and the structures of the vessels vary in parallel. To explore this subject, let's briefly review the vessels in the systemic circuit in the order in which blood flows through them.

Because blood emerges from the heart into the arteries, the arteries must withstand high and variable pressures. Structurally, the arteries have thick walls composed of elastic tissue and smooth muscle (see Figure 30.19A). This structure allows them to withstand the high pressures. Moreover, wall elasticity allows the arteries to play an important pressure-smoothing role. With each heartbeat, the walls of the major arteries stretch a bit, dampening the pressure surge. Then, during the next diastole, the stretched walls return to their unstretched state, in the process squeezing the blood and helping maintain blood pressure while the heart is between beats. In this way the elasticity of the arterial walls helps smooth out the ups and downs of arterial blood pressure. After blood flows through arteries of ever-smaller diameter, it flows into arterioles and capillaries. The tiny arteries and arterioles are equipped with smooth muscle, and as we discussed earlier, they exercise vasomotor control over the flow of blood to capillary beds, so that—in an ever-changing way—some capillary beds receive more flow than others. After blood passes through the capillaries—the walls of which are only one cell thick—the blood enters venules, then tiny veins, then ever-larger veins. Blood pressure is far lower in the veins than in the arteries, and the walls of the veins are thinner and expand more easily (see Figure 30.19A). A major role of the veins is to serve as a blood volume reservoir. Any blood that is not held elsewhere in the circulatory system is held in the veins.

The vasculature is not the same in all organs and tissues. Instead the vasculature in particular organs can be specialized in its structure and functional characteristics. In the systemic circuit, two cases are important enough to deserve mention.

In the brain and spinal cord, the capillaries—although still composed of just a single layer of vascular endothelial cells—are structurally different from capillaries elsewhere in the body and have different relationships with juxtaposed cells. As a consequence, many types of solutes cannot readily cross the capillary walls to move between the blood plasma and the tissue fluids that bathe the neurons. This obstacle to free exchange is called the **blood–brain barrier**. One reason for the barrier is that the walls of the capillaries in the brain and spinal cord completely lack pores. The blood–brain barrier helps protect the neurons in the brain from exposure to toxins in the circulating blood. It also permits particularly tight control of solute exchange, because many types of solutes require specific transport mechanisms to cross the capillary walls rather than being able to leak across.

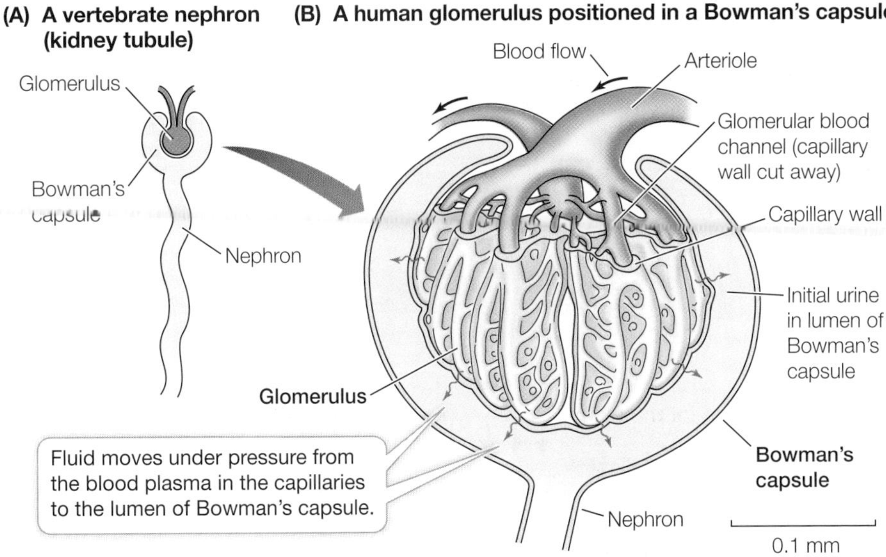

FIGURE 30.31 Urine Formation Depends on the Specialized Vasculature of the Kidneys **(A)** At its closed end, each vertebrate nephron (kidney tubule) begins with a Bowman's capsule containing a cluster of specialized capillaries called a glomerulus. **(B)** Urine is initially formed by pressure-driven flow through the walls of the capillaries and the wall of the Bowman's capsule. Later, as the initial urine flows the length of the nephron, it is modified in composition and volume to form the final urine, which is excreted. (After H. Elias et al. 1960. *J Urol* 83: 790–798.)

Activity 30.9 The Vertebrate Nephron

In the kidneys, blood pressure plays a role of life-and-death importance. A kidney consists mostly of nephrons, fine tubular structures that produce the urine. Each nephron starts with an invaginated cuplike structure called a Bowman's capsule (**FIGURE 30.31A**). Sitting in the Bowman's capsule is a cluster of interconnecting blood capillaries termed a *glomerulus* (**FIGURE 30.31B**). The properties of the glomerular capillaries are almost the diametric opposite of those of brain capillaries, as the glomerular capillaries have many pores and readily permit many types of solutes to pass through their walls. Blood pressure in these capillaries drives fluid—carrying solutes of many kinds—through the capillary walls and into the lumen of each kidney tubule (see Figure 30.31B). This is the mechanism by which urine formation begins. The mechanism depends on a relatively high blood pressure in the capillaries, and the blood pressure is higher in these capillaries than in most because of the way the vasculature is constructed.

Pressure and linear velocity vary greatly as blood flows through the vascular system

FIGURE 30.32 offers two major insights into blood flow through the systemic vascular system. First, most of the work done by the heart is expended in forcing the blood through the microcirculation! The blood pressure created by the heart in the aorta represents potential energy. This potential energy is converted to energy of motion as the blood flows, and then it is lost as heat through a type of friction that occurs during blood flow. As a result, the blood pressure falls as the blood flows. Moreover, the pressure falls in direct proportion to the amount of work that must be done to make the blood flow. With this background, you can see by studying the blood pressure data in Figure 30.32 that little work is done to make the blood flow through the big arteries (pressure does not drop much there). In dramatic contrast, a great deal of work must be done to make blood flow through the arterioles, capillaries, and venules (pressure drops from a high level to almost to zero in those vessels).

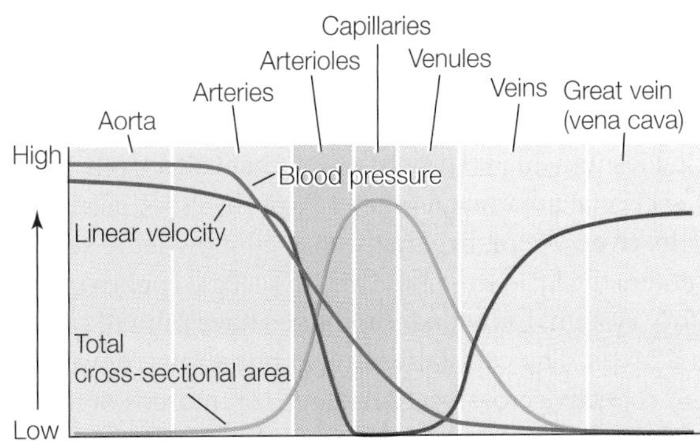

FIGURE 30.32 Blood Pressure and Linear Velocity Change Dramatically as Blood Flows through the Systemic Vascular System This graph shows how three critical characteristics of the circulatory system change from one end of the human systemic vascular system to the other: the mean blood pressure (pressure averaged over the full cardiac cycle), the total cross-sectional area of the blood vessels, and the linear velocity of blood flow. (After E. Feigl. 1974. In *Physiology and Biophysics*, 20th ed., vol. 2. T. C. Ruch and H. D. Patton [eds.], pp. 10–22. Saunders: Philadelphia.)

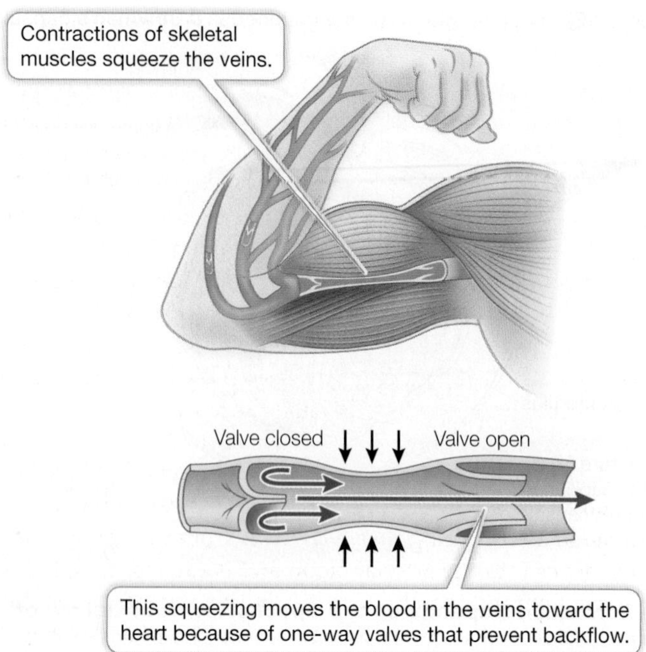

FIGURE 30.33 Veins Have Valves That Allow Skeletal Muscle Contractions to Aid Circulation

You can see in Figure 30.32 how low the blood pressure becomes by the time the blood flows into the veins. This low blood pressure helps explain why one-way valves have evolved in the veins (valves do not occur in arteries). The pressure from the beating of the heart is reduced to such a low level in the veins that it may not always be sufficient to make blood flow back to the heart. With one-way valves, contractions of skeletal muscles can help propel blood to the heart (**FIGURE 30.33**).

The second major insight that we obtain by studying Figure 30.32 comes from study of the linear velocity data: although the blood flows very rapidly through large arteries, it slows to a crawl as it moves through the capillaries. This slow movement through the capillaries is absolutely essential. The linear velocity of the blood is defined as the linear distance the blood travels per second. Suppose that a river suddenly divides into 10,000 little streams that collectively have a cross-sectional area much greater than the cross-sectional area of the river. We realize that water will flow more slowly after it enters the little streams. This same logic applies in the circulatory system. Individual capillaries have minute cross-sectional areas. The capillaries are so numerous, however, that their collective cross-sectional area is hundreds of times greater than the cross-sectional area of the aorta (see the graph of cross-sectional area in Figure 30.32). Accordingly, as blood flows from the aorta to the capillaries, it slows from an astounding linear velocity of more than 40 centimeters per second to almost zero.

Why does this slowing matter? Capillaries are less than 1 millimeter long. For the blood to have time to exchange materials with the tissues, it must be moving very slowly as it passes through the capillaries.

In a typical systemic capillary bed, blood flow leaves behind fluid that the lymphatic system picks up

The capillaries are sufficiently permeable in a typical capillary bed in most parts of the systemic circuit that, as blood first flows into a capillary, blood pressure drives some fluid (far less than in the kidneys) out of the blood plasma, across the capillary walls, and into the surrounding interstitial fluids. Then, before the blood leaves the capillary, some of this fluid returns to the blood plasma, but not all of it. These processes were first clarified by the physiologists E. H. Starling and E. M. Landis.

Starling and Landis realized that blood plasma in the capillaries has a higher osmotic pressure than the interstitial fluids, for the simple reason that the blood plasma contains more dissolved proteins than the interstitial fluids. Water thus tends to be drawn *into* the blood plasma in a capillary by osmosis across the capillary walls (see Key Concept 4.2). Water also tends to be driven *out of* the blood plasma in a capillary by the blood pressure inside the capillary. At the beginning of a capillary, where blood has just entered it, the effect of the blood pressure exceeds the effect of the osmotic pressure, and fluid moves out of the blood plasma (**FIGURE 30.34**). Blood pressure falls, however, as blood flows through a capillary. Consequently, at the end of the capillary, the effect of the osmotic pressure exceeds the effect of the blood pressure, and fluid moves from the interstitial fluids into the blood plasma.

The net effect of these dynamic processes is a small loss of fluid from the blood plasma (see Figure 30.34). How great is the fluid loss per day in an adult person, considering all the capillaries in

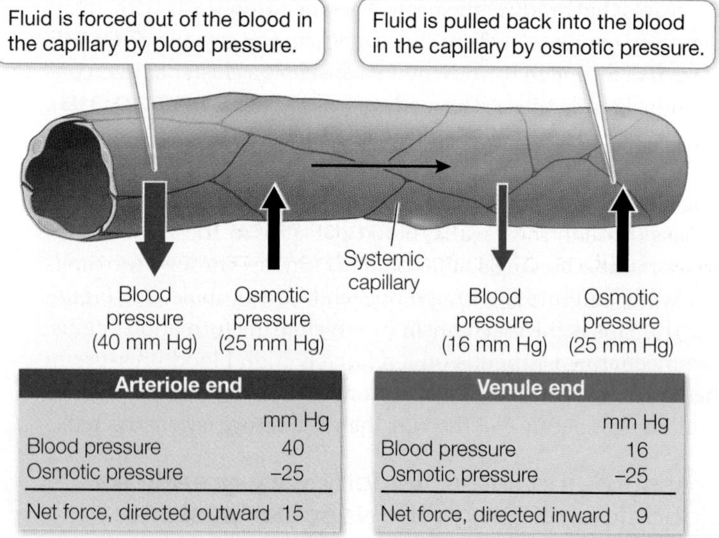

FIGURE 30.34 Blood Plasma Loses and Regains Fluid in Typical Systemic Capillaries Quantitative values are for humans. The widths of the thick red and black arrows symbolize relative magnitudes of the forces at work in typical capillaries. When blood pressure is greater than osmotic pressure, fluid leaves the blood in the capillaries, and when blood pressure falls below osmotic pressure, fluid returns to the blood in the capillaries. The balance of these two forces changes over the length of a capillary as blood pressure falls. The phenomena shown here occur in the systemic tissues. They do not normally occur in the lungs (where the alveoli could be flooded) because blood pressure is far lower in the lung circuit than in the systemic circuit.

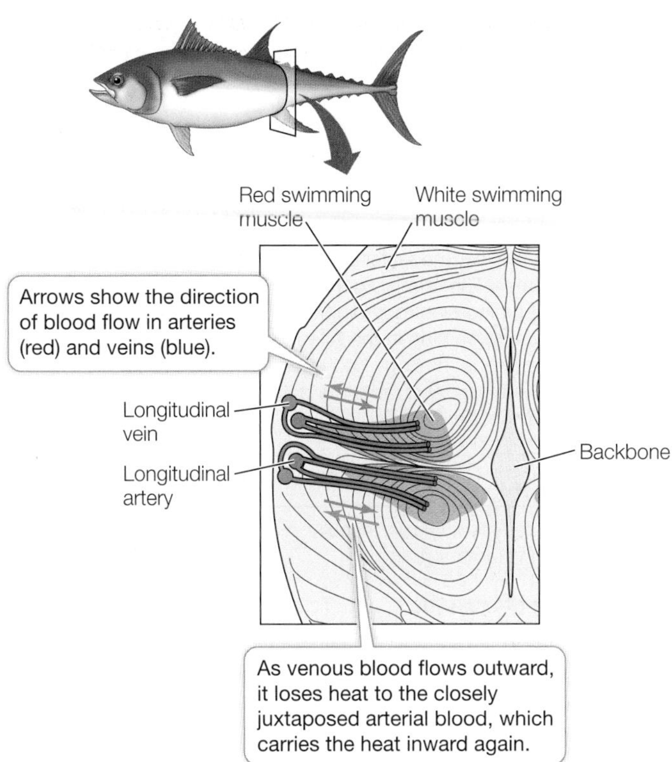

FIGURE 30.35 Countercurrent Heat Exchange Warms the Red Swimming Muscles of Tunas

the body and all the times that blood passes through them? Blood flow through the systemic capillaries leaves behind about 2–4 liters of fluid in a person's systemic interstitial fluids each day. A widespread system of vessels called the **lymphatic system** (see Figure 36.4) collects this excess fluid—called *lymph*—and returns it to the blood. Some diseases cause imbalances in the production or removal of lymph and can lead to chronic tissue swelling because of lymph accumulation.

The vasculature plays major roles in adaptations of animals to their environments

As animals have adapted to their environments over evolutionary time, changes in the structure or function of the vascular system have sometimes played major roles. Some animals, for example, have evolved specialized, parallel arrays of blood vessels that help create large differences of temperature from place to place within their body. A second example is that diving mammals have evolved specialized vasomotor controls that enable them to stay submerged for long periods of time. Let's take a closer look at these cases.

We discussed *countercurrent heat exchange* in the appendages of mammals and birds in Key Concept 29.1. This type of exchange, which helps the animals conserve heat, depends on parallel arrays of arteries and veins in the appendages (see Figure 29.5).

Tunas use countercurrent heat exchange—again based on parallel arrays of arteries and veins—to keep their red swimming muscles warm (**FIGURE 30.35**). In the vast majority of fish species, the temperature of all tissues is the same as the water temperature. The warming of the red swimming muscles in tunas is thus a specialized characteristic. The red muscles, which are the principal muscles used in steady swimming (see Key Concept 33.2), are positioned in an unusual place in tunas, deep in their body, nearer their backbone than their skin (see Figure 30.35). Tunas also have evolved an unusual arrangement of arteries and veins. The longitudinal arteries and veins that carry blood along the length of their body are just under their skin. Branches from each longitudinal artery travel *inward* to the red swimming muscles (see Figure 30.35). Veins from the red muscles travel *outward* to connect with a longitudinal vein. These arteries and veins—which are carrying blood to and from the red muscles—are exceedingly numerous and closely juxtaposed with each other in parallel arrays. Because the red swimming muscles work hard, they produce heat at a relatively high rate. The venous blood, flowing outward from the muscles, tends to carry this heat away. However, the arteries carrying inward-flowing arterial blood are touching the veins. Heat from the venous blood thus moves into the arterial blood and is carried directly back to the swimming muscles. Because of this countercurrent heat exchange, the muscles tend to retain the heat they make and are warmed by it. Their elevated temperature probably helps the muscles maintain a high power output for swimming.

One of the marvels of life on Earth is the ability of air-breathing mammals such as seals and whales to stay underwater for long periods (up to 2 hours in some cases) (**FIGURE 30.36**). Decades ago, Laurence Irving had a good idea about the mechanism involved, and he tested his idea with experiments on harbor seals that he studied in the swimming pool at Swarthmore College. He reasoned that the brain has an absolute need for O_2. Thus for a seal to stay underwater, the seal must meet its brain's O_2 need. He also reasoned that most other tissues can turn to anaerobic metabolism for many minutes or tens of minutes. He thus hypothesized that a vasomotor mechanism could permit

FIGURE 30.36 A Gray Seal Diving Gray seals (*Halichoerus grypus*) can remain submerged for more than 20 minutes even though they are air breathers.

INVESTIGATION

FIGURE 30.37 Are the Cardiovascular Responses of Seals during Natural Diving All-or-None or Graded? When the cardiovascular responses of seals during diving were first discovered, the studied seals were suddenly forced underwater, and their responses were all-or-none: although not at all evident when the seals were breathing air, the cardiovascular responses occurred immediately and maximally as soon as the seals were submerged. Investigators wondered whether these responses are all-or-none in wild seals that dive voluntarily in their natural habitats. But how could free-living, wild seals be studied to find out? Even to this day, only a few studies of this subject have been carried out on free-living seals because of the challenges of working on wild animals free in the open ocean. Here we discuss two experiments. The first is by Randall W. Davis and colleagues, who studied wild Weddell seals (*Leptonychotes weddellii*) diving under the ice in Antarctica.[a] When the seals came to holes in the ice to breathe, Davis et al. made injections and took blood samples. They injected seals with inulin (not insulin), a compound used in human medicine to measure how fast the kidneys make urine. They then measured how fast the inulin was removed from the blood: the faster inulin was removed, the faster urine was being made (only the kidneys remove inulin). The other experiment is by Dave Thompson and Mike Fedak, who studied electrocardiograms (EKGs) in wild gray seals (*Halichoerus grypus*) swimming in the Atlantic Ocean near Scotland.[b] By the early 1990s, the technology had been developed to put an EKG monitor in a free-living seal and radio the data to investigators. Figure 30.36 shows a gray seal diving.

HYPOTHESIS

Cardiovascular responses are all-or-none during voluntary diving in wild seals.

METHOD

Kidney function in Weddell seals

1. Inject wild Weddell seals with inulin before their dives.
2. When a seal returns from its dive, note the length of the dive, take a blood sample, and measure the amount of inulin in the blood.
3. Calculate the rate of inulin removal from the blood.

Heart rate in gray seals

1. Attach EKG monitors to wild gray seals.
2. Collect data on heart rate and diving duration using sonic and radio telemetry.

RESULTS

(A) Kidney function in Weddell seals (*Leptonychotes weddellii*)

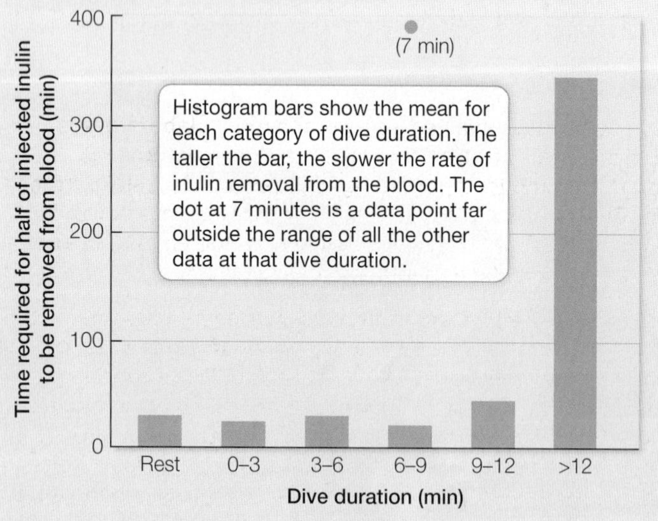

(B) Heart rate in gray seals (*Halichoerus grypus*)

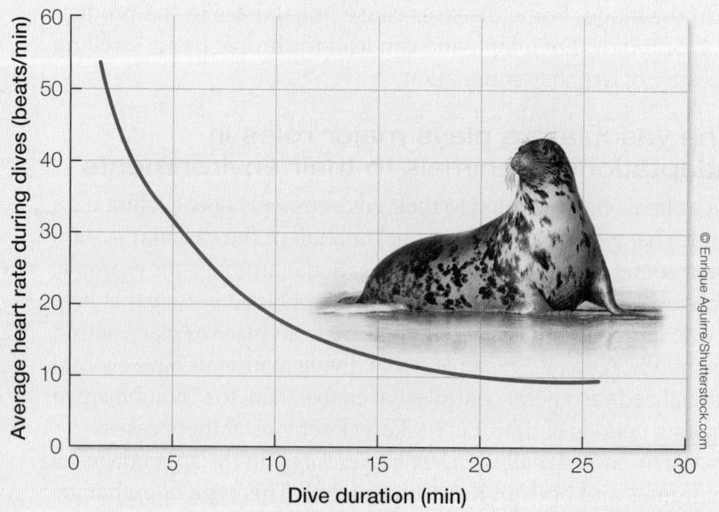

CONCLUSION

The hypothesis is rejected. Cardiovascular responses are graded, not all-or-none, during voluntary diving.

ANALYZE THE DATA

1. Explain how the inulin data point to graded activation of the cardiovascular responses.
2. Explain how the heart rate data point to graded activation of the cardiovascular responses.
3. Although seals have strong cardiovascular responses that can help them stay underwater for long periods of time, why might it be to their advantage not to use their maximal responses during short dives?

[a]R. W. Davis et al. 1983. *Am J Physiol* 245: R743–R748; [b]D. Thompson and M. A. Fedak. 1993. *J Exp Biol* 174: 139–164.

A similar **Analyze the Data** exercise may be assigned in **Achieve**.

seals to stay underwater for long periods: during a dive, he reasoned, the small arteries and arterioles carrying blood to the skin, kidneys, intestines, and most muscles could strongly contract, preventing blood flow to all these tissues that can get along for a while without O_2. In this way, the O_2 available in a seal's body would not be used by those tissues, and thus the available O_2 would be conserved and could meet the brain's O_2 requirement for a long time.

Irving tested his hypothesis by pushing seals underwater in the pool and studying their metabolism and blood flow, and he collected abundant data showing that his hypothesis was correct. One of his most unexpected and stunning observations, however, was that the changes in the circulatory system were massive and sudden. When he shoved a seal underwater, the seal immediately produced a maximal cardiovascular response. One of the easiest properties for Irving to study was heart rate. A seal pushed underwater would immediately reduce its heart rate to a small fraction of the rate in air. Irving reasoned (and others later demonstrated) that the key process that was taking place was that vasomotor controls were immediately cutting off blood flow to much of the body. The heart rate was merely being adjusted to be compatible with this change. In brief, when blood flow was cut off to much of the body, a reduced heart rate could meet the remaining need for blood flow, and that was the reason Irving observed the sharp drop in heart rate.

In recent decades, investigators have wondered whether seals that are diving voluntarily in their natural habitats display this type of immediate and maximal response. **INVESTIGATION FIGURE 30.37** discusses research on this question.

REVIEW & APPLY | 30.6

R In most animals the systemic and breathing-organ circuits are connected in series—an arrangement that is typically the most efficient for delivering O_2 to the systemic tissues. The functional demands and structure of the blood vessels vary with the vessels' positions. Arteries have thick, elastic walls that enable them to resist high pressures and play a pressure-smoothing role. Veins have thinner walls and one-way valves and have a high capacity for storing blood. The structure and properties of blood vessels can differ from one organ to another (e.g., the blood–brain barrier in the brain vs. the permeable capillaries in the kidney).

In the closed vertebrate circulatory system, most of the heart's energy is used to drive blood through the microcirculation. In ordinary systemic capillary beds, blood pressure forces fluid out of the blood as it travels through the initial part of a capillary, but osmotic pressure draws some of the fluid back into the blood as blood pressure falls. The net result is a small loss of fluid, which is picked up by the lymphatic system and returned to the blood.

Evolutionary adaptation to the environment has led to vascular specializations, such as parallel arrays of blood vessels that provide for countercurrent heat exchange in mammalian limbs and in red swimming muscle in tunas. Diving mammals have evolved specialized vasomotor control that permits blood flow to be curtailed to large parts of the body, conserving O_2.

A

1. Why is a series arrangement of the systemic and breathing-organ circuits efficient for O_2 transport?
2. In which vessels of the mammalian circulatory system does the blood move most slowly? Why, and why is this important?
3. Certain parasitic worms can block lymphatic vessels. Suppose the lymphatic vessels draining a person's leg were completely blocked by a parasite infection. What do you think would happen to the leg? Explain.
4. In what ways is the circulatory system of the mammalian kidney specialized for urine formation?

The swimming muscles of a tuna require O_2 at a high, steady rate to synthesize ATP fast enough—and consistently enough—to support the tuna's strong, sustained swimming. A tuna's source of O_2 is the seawater of its external environment. What properties would you expect to find in the breathing and circulatory systems of a tuna, and why?

Answer: Tunas have evolved some of the most elaborate gills known in fishes: gills in which the gas exchange membrane is particularly large and thin, permitting high rates of O_2 uptake from seawater. Tunas also employ ram ventilation to ventilate their gills, a type of ventilation that is thought to minimize the amount of energy that is used for gill ventilation rather than swimming.

Tunas have a high-performance circulatory system—characterized by high blood flow rates—that can transport O_2 at high rates from the gills to the swimming muscles. The myocardium in the heart is partly served by a coronary circulatory system, rather than having to get all of its O_2 from the blood flowing through the heart chambers as in most fishes. Strong comparative evidence indicates that a coronary circulatory system is a superior way to provide O_2 to the myocardial cells, which provide the power for circulation of the blood.

Tunas belong to the small subset of fishes that metabolically warm some of their tissues. Because of a specialized vascular anatomy and countercurrent heat exchange, the red swimming muscles of tunas are kept at relatively high temperatures. This warming is believed to help the muscles maintain high power outputs.

30 VISUAL SUMMARY

30.1 The Pathway for Oxygen: Oxygen Must Travel from Environment to Mitochondria

FIGURE 30.1

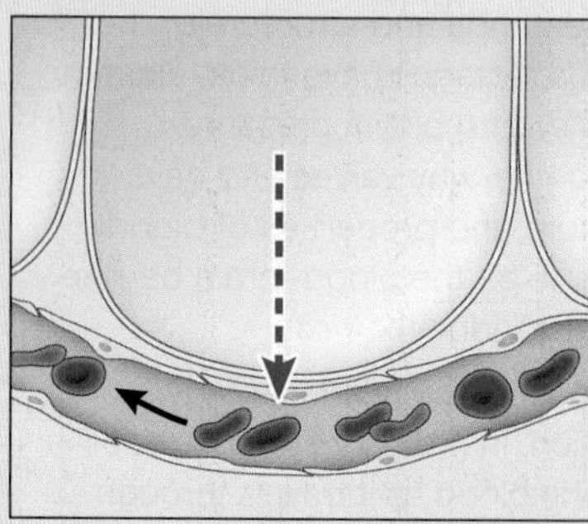

- Transport of O_2 occurs by diffusion and **BULK FLOW**, never by active transport.
- Diffusion is effective only over very short distances. **FICK'S LAW** governs the rate of diffusion per unit of cross-sectional area.
- The pathway for oxygen consists of alternation of bulk flow and diffusion: bulk flow by **VENTILATION** of the **BREATHING ORGAN**, followed by diffusion across the **GAS EXCHANGE MEMBRANE**, followed by bulk flow by blood circulation, then diffusion to the mitochondria.

QUESTIONS

1. What is the physical cause of diffusion?
2. Why do cells need O_2?
3. What is Krogh's rule of thumb regarding diffusion distance?

30.2 Animals Have Evolved Diverse Types of Breathing Organs

Go to ANIMATION 30.1

FIGURE 30.9

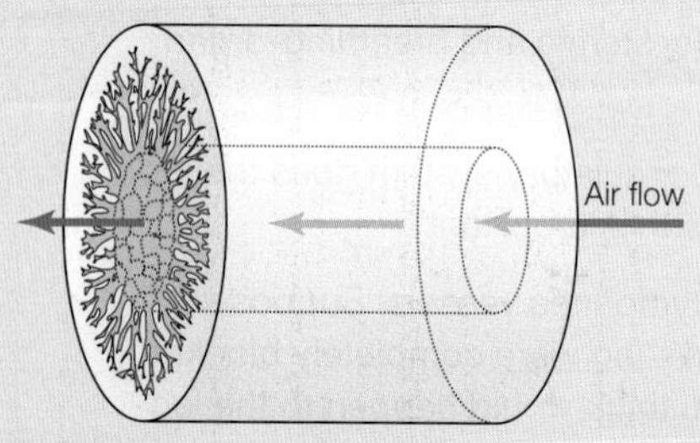

- Breathing organs have two fundamental characteristics: a thin **GAS EXCHANGE MEMBRANE** and a large membrane area, typically achieved by branching and folding.
- When **VENTILATION** is **UNIDIRECTIONAL**, **COUNTERCURRENT GAS EXCHANGE** is more effective than **COCURRENT GAS EXCHANGE**.
- Insects have a **TRACHEAL BREATHING SYSTEM** in which the circulatory system plays little role in O_2 and CO_2 transport.

1. How can flatworms and sponges acquire O_2 without specialized breathing organs?
2. How do gills differ from lungs?
3. In terms of Fick's law, why are thin gas exchange membranes with a large surface area important?

30.3 The Mammalian Breathing System Is Anatomically and Functionally Elaborate

Go to ACTIVITIES 30.1 and 30.2 and ANIMATION 30.2

FIGURE 30.13

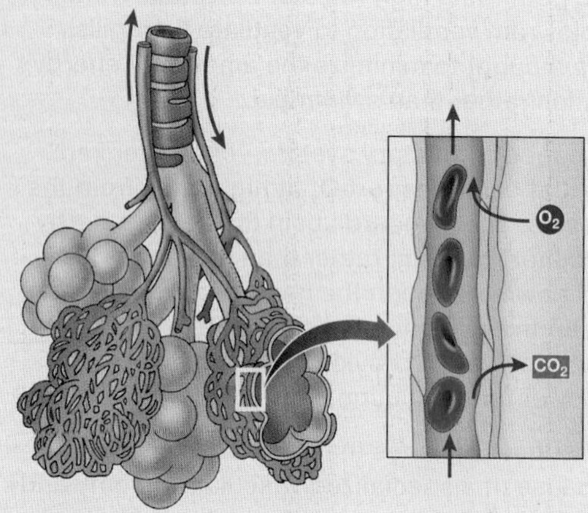

- In the highly branched mammalian lung, most airways are **CONDUCTING AIRWAYS**. The **RESPIRATORY AIRWAYS** end blindly in **ALVEOLAR SACS** with **ALVEOLI**, across which gases move by diffusion (**FIGURE 30.13**).
- In people at rest, inhalation is active (caused by contraction of the **DIAPHRAGM** and certain intercostal muscles), whereas exhalation is passive (caused by elastic recoil).
- The breathing muscles are stimulated to contract by nerve impulses that originate in the pre-Bötzinger complexes in the medulla oblongata of the brain.

1. During mammalian exercise, what properties of breathing change in comparison with being at rest?
2. Why is mammalian ventilation referred to as "tidal"?
3. How are blood CO_2 concentration and pH related, and what roles do they play in controlling ventilation?

30.4 Animals Have Evolved Circulatory Systems for Internal Oxygen Transport

Go to ACTIVITIES 30.3, 30.4, and 30.5

FIGURE 30.18

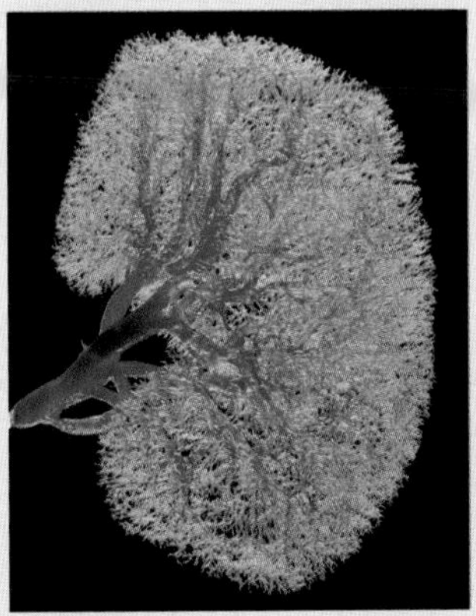

- In a **CLOSED CIRCULATORY SYSTEM**, blood stays within blood vessels, which profusely branch (**FIGURE 30.18**). Blood leaves the vessels in an **OPEN CIRCULATORY SYSTEM**.
- In the **MICROCIRCULATION** of a closed circulatory system, **CAPILLARIES** are the principal sites of exchange between blood and surrounding cells.
- **HEMOGLOBIN** combines reversibly with O_2, increasing O_2 transport. It is **OXYGENATED** in the breathing organ and **DEOXYGENATED** in systemic tissues.

QUESTIONS

1. Where are hemoglobin and hemocyanin found in the blood?
2. Outline the flow of blood through the body of a crustacean such as a crayfish or lobster.
3. Why is the relationship discovered by Poiseuille—concerning rate of fluid flow and diameter of a tube lumen—important for understanding blood flow through the circulatory system?

30.5 A Beating Heart Propels the Blood

Go to ACTIVITIES 30.6 and 30.7 and ANIMATION 30.3

- Vertebrate **HEARTS** are multi-chambered (four-chambered in mammals and birds) and **MYOGENIC**, whereas crustacean hearts are single-chambered and **NEUROGENIC**.
- In the mammalian heart, modified muscle cells in the sinoatrial (S-A) node initiate a wave of depolarization at the start of each heartbeat. After a delay at the atrioventricular (A-V) node, the depolarization spreads via the **CONDUCTING SYSTEM** into the ventricles (**FIGURE 30.26**).
- The **MYOCARDIUM** in a mammalian heart receives O_2 by circulation of blood through the **CORONARY CIRCULATORY SYSTEM**.

FIGURE 30.26

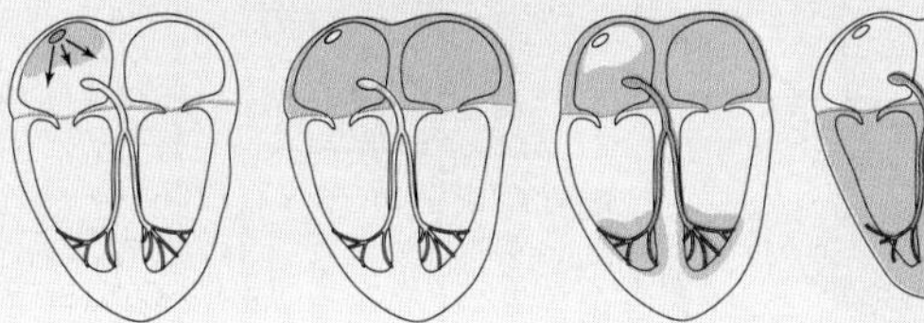

1. What is hypertension?
2. Why does the action of the heart from one beat to the next give rise to voltage differences that we can measure at the body surface in the form of an electrocardiogram?
3. Explain why, in a vertebrate, depolarization in one cell in a mass of cardiac muscle leads to contraction throughout.

30.6 The Vascular System Plays Many Roles

Go to ACTIVITIES 30.8 and 30.9

- The systemic circuit is usually connected in series with the breathing-organ circuit.
- In a closed circulatory system, a great deal of energy is required to force blood to flow through the microcirculation (**FIGURE 30.34**).
- Specializations of the circulatory system play vital roles in urine formation, heat conservation (e.g., in the swimming muscles of tunas), and O_2 conservation during diving by seals and whales.

FIGURE 30.34

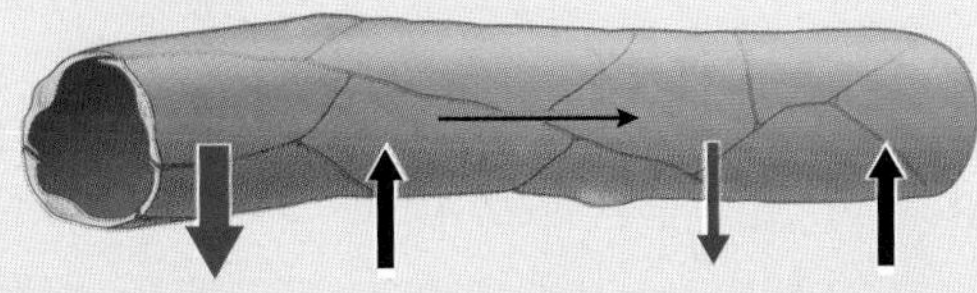

1. What is the blood–brain barrier?
2. How does countercurrent heat exchange warm the red swimming muscles of tunas?
3. How is vasomotor control hypothesized to permit long dives by seals?

Go to **Achieve** for the e-book, animations, activities, and additional resources and assignments.

31 Neurons, Sense Organs, and Nervous Systems

Ospreys (*Pandion haliaetus*) eat live fish, which they capture by diving from above. They are patient, visual hunters, often spending many minutes looking down—as they hover or fly slowly, perhaps 10 meters above the water—before choosing a target and a moment to dive. At that moment, the Osprey stops flapping, pulls in its wings, and goes into a headfirst free fall, making mid-dive course corrections as it rapidly descends. The talons are brought to the fore—outstretched. If all goes well, the talons strike and grasp a fish.

As an Osprey searches and dives, what processes must take place for the Osprey to choose a fish as a target and to adjust its trajectory during its high-speed dive so that it strikes its target?

You will find the answer to this question on page 802.

KEY CONCEPTS

31.1 Nervous Systems Are Composed of Neurons and Glial Cells

31.2 Neurons Generate Electric Signals by Controlling Ion Distributions

31.3 Neurons Communicate with Other Cells at Synapses

31.4 Sensory Processes Provide Information on an Animal's External Environment and Internal Status

31.5 Neurons Are Organized into Nervous Systems

31.1 Nervous Systems Are Composed of Neurons and Glial Cells

LEARNING OBJECTIVES

By the end of this key concept you should be able to:

31.1.1 **Describe the anatomical and functional parts of a neuron (nerve cell).**

31.1.2 **Explain why signal communication by neurons is described as "fast" and "addressed."**

31.1.3 **Explain why neurons are said to "integrate" signals.**

31.1.4 **List three functions that glial cells perform.**

Animals need a way to transmit signals at high speeds from place to place within their bodies. Suppose, for example, that you are crowded with other people in the kitchen during a party, and being distracted, you put your hand behind you and lean on a hot stove. You detect the threat and respond immediately—pulling your hand away—because of the function of your nervous system.

The major parts of the human nervous system are shown in **FIGURE 31.1**. The brain and spinal cord are termed the **central nervous system** (**CNS**), whereas all the other parts are the **peripheral nervous system** (**PNS**). All parts are composed of specialized **nerve cells**, or **neurons**, which often (but not always) have a highly elongated structure. Like most other individual cells, neurons are invisibly tiny—so small that the brain and spinal cord contain billions of them. In the PNS, neurons are organized into **nerves**, each of which is a bundle of hundreds, thousands, or tens of thousands of neurons (**FIGURE 31.2**). In all vertebrates, **spinal nerves** (see Figure 31.1) arise in pairs—one going toward each side of the body—along the length of the spinal cord. Among the neurons within a single nerve, it is common for some neurons to carry signals toward the CNS, whereas others carry signals out of and away from the CNS (although each neuron carries signals in just one direction). This is true for, example, in the spinal nerves. Some neurons are termed **sensory neurons** because they carry signals from sense organs to the CNS. Other neurons are **motor neurons** because they carry signals to muscle cells, stimulating contraction.

A fundamental property of neurons is that they transmit signals at high speeds. Mammalian neurons, for example, routinely transmit signals at 20–100 meters per second! If you accidently touch a hot surface, sensory neurons in your fingers are activated, initiating danger signals that travel to the CNS. Each of these neurons (which are single cells) is so highly elongated that it extends from a finger to the spinal cord (**FIGURE 31.3**). Because of the high speed of signal transmission in neurons, the danger signal travels from your finger to your spinal cord in less than 0.02 second, the first step in allowing you to respond. In the spinal cord, signals are soon initiated in motor neurons (also highly elongated) that control muscles in your arm, and those signals travel equally quickly from the spinal cord to the muscles, which contract and pull your finger away from the danger.

The speed of signal transmission by neurons is evident in many ordinary actions. During banjo plucking, each pluck of a string takes place because of neuronal signals that originate in the brain and travel to the player's fingers (**FIGURE 31.4**). These signals must be generated by brain neurons—and then move down the spinal cord and out to fingers—at enormous speeds for notes to burst from the instrument as fast as they do.

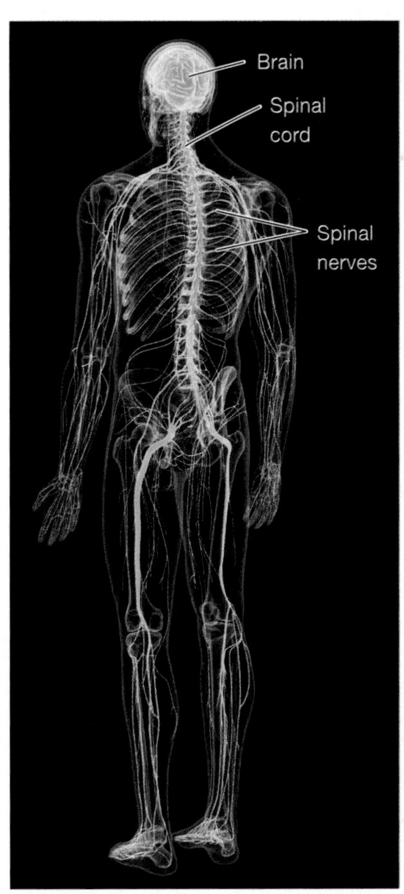

FIGURE 31.1 The Human Nervous System Structures shown in yellow are parts of the nervous system.

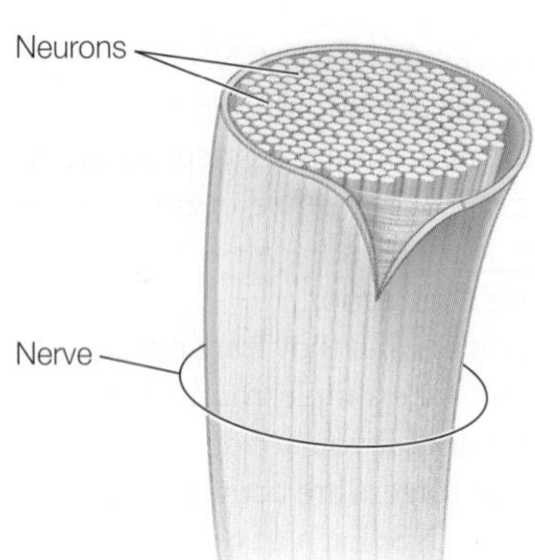

FIGURE 31.2 A Nerve Consists of a Bundle of Neurons within a Sheath

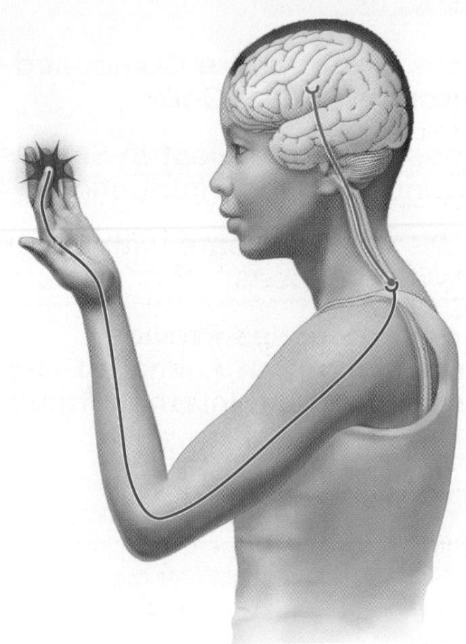

FIGURE 31.3 Neurons Involved in Sensing Pain This diagram shows two neurons. The red neuron has a sensory ending in the index finger and carries pain signals all the way to the spinal cord near the base of the neck. As we will discuss in Key Concept 31.5, these signals can be processed in a reflex in the spinal cord, giving rise directly to outgoing signals in motor neurons that activate finger and arm muscles to withdraw the finger from the pain stimulus. The sensory neuron (red) may also activate a neuron (blue) that transmits signals to the brain, where conscious awareness of pain can take place and more complex responses can be initiated.

Banjo plucking highlights an additional fundamental attribute of neuronal function, namely that neuronal signals are *addressed*, meaning the signals are delivered only to defined locations. A neuron typically must make physical contact with a muscle cell or other target cell for signals in the neuron to affect the target cell directly. Conversely, only cells contacted by a neuron typically receive the neuron's signals. Each motor neuron involved in banjo plucking extends from the spinal cord to a specific muscle in a specific finger. Because of the specificity of this anatomical connection, signals in the neuron effectively are addressed to just the relevant muscle. Thus different fingers can be controlled independently and specifically, permitting music to be made. Summarizing the points we've made thus far, signal transmission by neurons is *fast* and *addressed*.

Neurons are cells specialized to produce electric signals

A neuron is a type of **excitable cell**, meaning that in general its cell membrane can generate and conduct *impulses*, or *action potentials*. Most types of cells in an animal's body are not excitable. The two principal types of excitable cells are neurons and muscle cells (see Chapter 33). The excitable nature of neurons is one of their key specializations because it permits neurons to carry signals over long distances. Up to now we have spoken of "signals" that travel rapidly along the lengths of neurons. Each of these signals is an impulse, or action potential.

What is the nature of an impulse, or action potential? Recall the electrical polarity of cell membranes; ordinarily the outside of a cell membrane is more positive than the inside (see Figure 28.1A). An **action potential** is a state of reversed polarity of the cell membrane. In a region of the cell membrane where an action potential is present, the cell membrane is said to be **depolarized** because its electrical polarity is outside-negative instead of outside-positive. In an excitable cell, by definition, after an action potential is generated at one point in the cell membrane, it propagates (spreads) over the whole membrane. During this process, the region of depolarization moves along the cell membrane, and the membrane is said to "conduct" or "propagate" the impulse (**FIGURE 31.5**). In a banjo player, this conduction, or propagation, is what happens when a neuron carries a signal from the CNS to a finger, telling the finger to pluck a string.

FIGURE 31.4 Banjo Plucking Dramatizes the Speed of Neuronal Signaling To initiate each finger pluck, nerve impulses are generated in the brain and travel down the spinal cord and out to the finger that plucks the string. The entire process occurs several times per second (up to 10 times per second in championship playing—check out YouTube).

We will discuss action potentials in more detail in Key Concept 31.2.

Neurons are very diverse in structure, although they have several features in common. **FIGURE 31.6** illustrates the structure of a generalized neuron. As we discuss the structure of neurons, a good starting point is to mention their contacts with other cells.

The places where neurons make functionally relevant contact with other cells are called synapses. More specifically, a **synapse** is a cell-to-cell contact point that is specialized for signal transmission from one cell to another. Where a neuron makes a synapse with another cell, the latter cell may be another neuron, or it may be some other type of cell such as a muscle cell. Signal transmission at a synapse is usually one-way. The cell that conducts signals into the synapse is called the **presynaptic cell**, and the cell that conducts signals away from the synapse is the **postsynaptic cell**.

Most neurons have *four anatomical regions*:

1. A set of dendrites
2. A cell body
3. An axon
4. A set of presynaptic axon terminals

We can also think of a neuron as having *four functional regions*, which generally correspond to the four anatomical regions (see Figure 31.6).

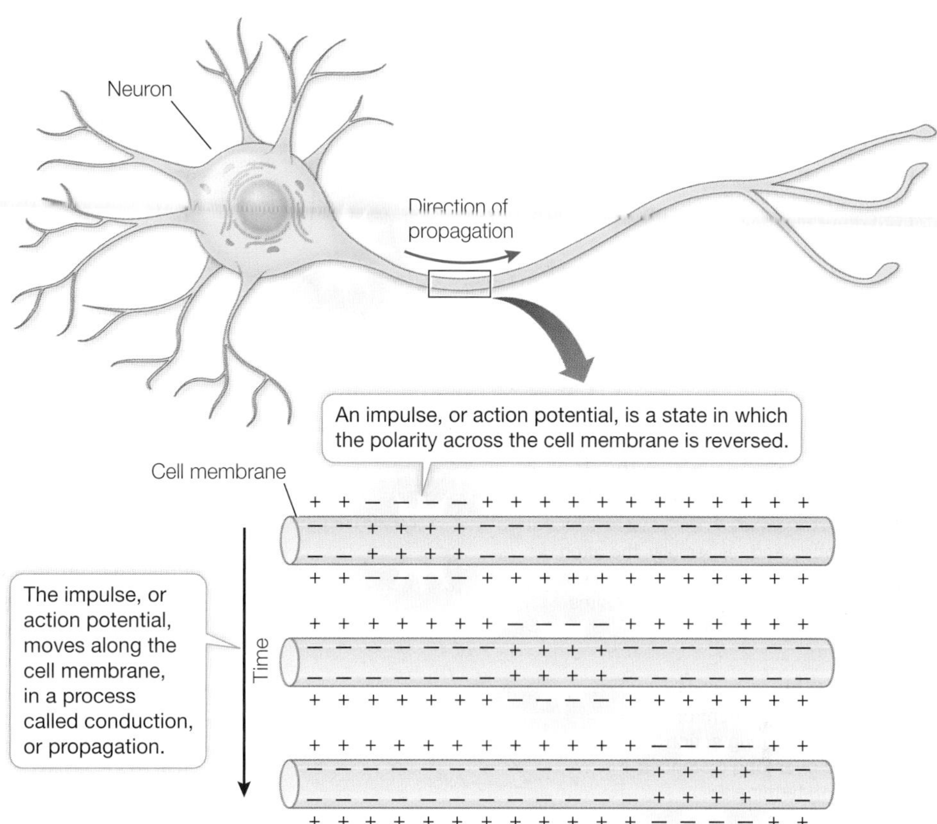

FIGURE 31.5 Propagation of a Nerve Impulse The neuron at the top is propagating (conducting) an impulse, or action potential, from left to right. The three diagrams below show (from top to bottom) electric charges present in a part of the neuron at three successive, closely spaced times. The yellow region shows the position of the impulse, where the cell membrane is depolarized (outside-negative). Note how the impulse moves along the length of the cell from left to right as time passes.

The **dendrites** (from the Greek *dendron*, "tree") are relatively short cell processes (cell extensions) that tend to branch from the cell body like twigs on a shrub. The dendrites are typically the principal sites where incoming signals arrive from other cells. Hundreds or thousands of other cells—neurons or sensory cells—may bring signals to a single neuron! These cells typically make synaptic contact with the neuron's dendrites or its cell body.

The **cell body** contains the neuron's nucleus and most of its other organelles. From the perspective of signaling, the most important function of the cell body is to *integrate* incoming signals. Earlier we stressed that neurons achieve fast and addressed signal transmission. Of equal importance, neurons integrate signals. This means that a neuron can receive many inputs and combine them to produce a single output, and it can do this in ways that are adaptive and help promote harmonious function of the animal. The **axon hillock**—the part of a neuron where the cell body transitions to the axon—is particularly engaged in integration. Later (see Figure 31.21) we will further discuss the mechanism of integration by a neuron.

The **axon** is a particularly long cell process that is anatomically specialized for long-distance signal conduction. Usually the signals (action potentials) conducted by the axon originate in or near the cell body and travel outward from the cell body along the axon's entire length. As we have stressed, neurons are often highly elongated cells. This is true because the axon may extend a great distance, such as the entire distance from your spinal cord to your fingers or toes (see Figure 31.3).

At its end, an axon branches, and each branch terminates in a small swelling called a **presynaptic axon terminal**. These terminals make synaptic contact with other cells, enabling signals generated by one neuron to initiate signals in other neurons or in muscle cells. The presynaptic axon terminals of a neuron are said to **innervate** the cells with which they make synaptic contact.

Glial cells work with neurons and help guide nervous system development

Nervous systems are composed mostly of two types of cells, neurons and **glial cells**, also called glia or neuroglia. Unlike neurons, glial cells typically are not excitable and do not conduct action

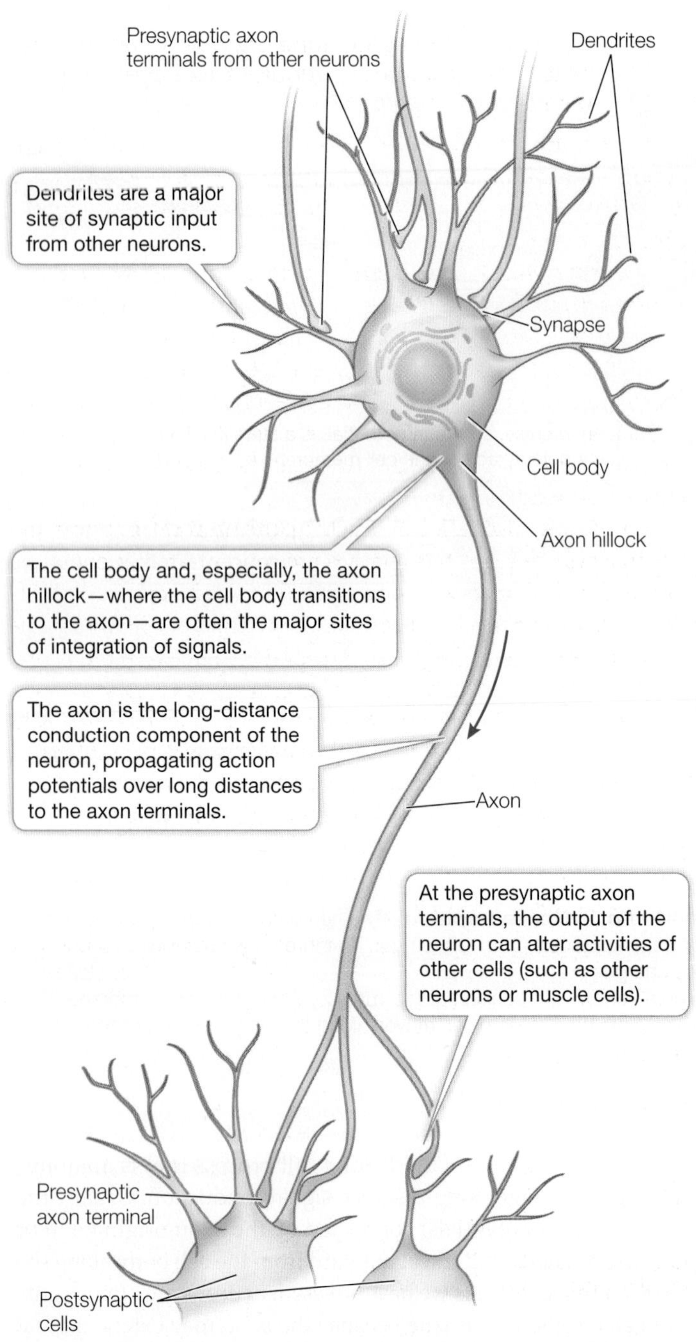

FIGURE 31.6 A Generalized Neuron The balloons identify the four major *anatomical* regions of a neuron, corresponding approximately to four *functional* regions. The drawing shows the features typical of most neurons (red arrow denotes direction of impulse propagation). Details, such as the length of the axon and the density and branching pattern of the dendrites, vary greatly among the many different types of neurons in an animal's body.

potentials. Nonetheless, they are known to play many vital roles, and knowledge of their functions continues to expand. Recent research indicates that glial cells are approximately as numerous as neurons in the mammalian brain (some older research suggests they are much more numerous than neurons).

Several types of glial cells are present in the vertebrate central nervous system, among them astrocytes, oligodendrocytes, and microglia. Glial cells in the brain and spinal cord provide metabolic support for neurons, help regulate the composition of the extracellular fluids bathing the neurons, and perform immune functions. Glial cells also help regulate signal transmission across synapses. During development, glial cells play many key roles, such as helping guide growing neurons in the central nervous system to synaptic contacts and removing unneeded neurons or neuron parts.

In vertebrates, but generally not in invertebrates, certain glial cells wrap around axons, forming concentric layers of cell membrane called **myelin**. The multilayered wrap of cell membrane forms a lipid-rich, electrically insulating sheath around the axon. **Oligodendrocytes** perform this function in the brain and spinal cord, and other glial cells, called **Schwann cells**, do so in other parts of the nervous system (**FIGURE 31.7**). Parts of the nervous system that consist mostly of myelinated axons have a glistening white appearance and are sometimes called **white matter**. Not all vertebrate axons are myelinated, but those that are can conduct action potentials more rapidly than unmyelinated axons, as we will discuss in Key Concept 31.2.

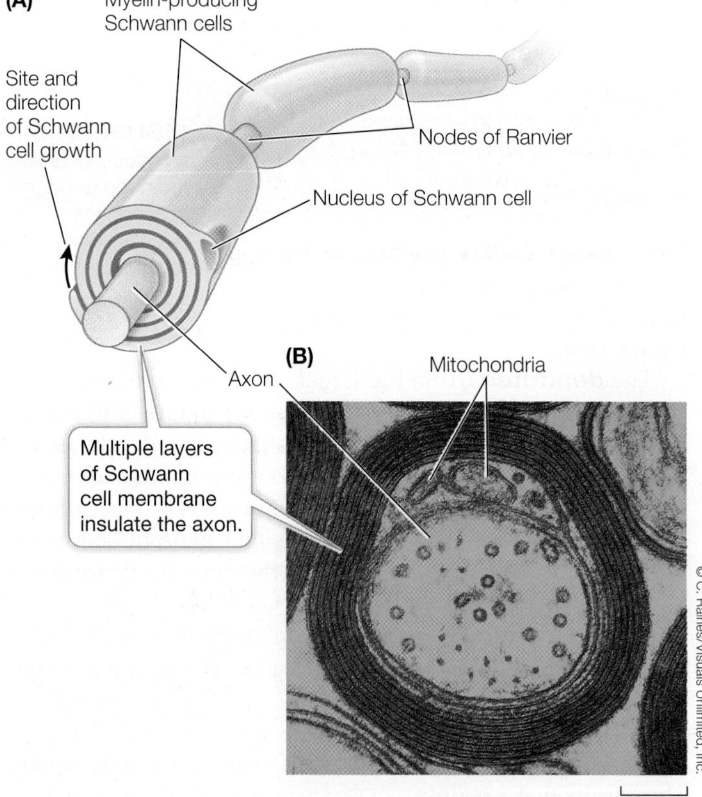

FIGURE 31.7 Electrically Insulating an Axon (A) Schwann cells wrap around an axon, enclosing the axon in many layers of lipid-rich Schwann cell membrane, called myelin, without intervening cytoplasm. Myelin provides electrical insulation to the axon. At the intervals between Schwann cells—called the nodes of Ranvier—the axon is exposed. **(B)** A myelinated axon, seen in cross section at high magnification by use of an electron microscope.

REVIEW & APPLY | 31.1

R Neurons are one of the two major types of excitable cells (muscle cells being the other type). Being excitable means that a cell can produce and propagate electric impulses termed action potentials. Neurons are often highly elongated. A neuron receives action potentials from large numbers of other neurons at synapses, usually located on the dendrites or cell body of the neuron. As the neuron processes these signals, it integrates them to initiate its own action potentials, which travel outward along its axon. Multiple presynaptic axon terminals at the end of the axon make synaptic contact with other neurons or muscle cells. Glial cells, which typically are not excitable, are the second major class of cells in the nervous system. Glial cells provide electrical insulation for neurons, support synaptic transmission of signals between neurons, and guide neurons toward establishing correct synaptic contacts during development.

1. What is the fundamental difference between neurons and glial cells?
2. Use a tape measure to measure the distance from your brain to the part of your spine at shoulder level, then measure the distance from that part of your spine to the tip of your index finger. Calculate a rough estimate of the speed (meters per second) at which action potentials would have to travel for you to pluck four guitar or banjo strings per second with your index finger. For simplification, assume that the action potentials merely need to travel the distance you've measured when your brain orders your finger to contract.
3. What is meant when we say signals transmitted by a neuron are "addressed"?

We've now seen that nervous systems are composed of neurons and glial cells. Next we'll focus on how the neurons generate and transmit action potentials.

31.2 Neurons Generate Electric Signals by Controlling Ion Distributions

LEARNING OBJECTIVES

By the end of this key concept you should be able to:

31.2.1 **Explain how the membrane potential of a cell can change by a *large* amount because of a *small* movement of ions across the cell membrane.**

31.2.2 **Explain how gating of ion channels can produce an electric signal.**

31.2.3 **Explain why an all-or-none change in membrane potential is critical for impulses to travel long distances in neurons.**

The one feature common to all nervous systems is that they encode and transmit information in the form of action potentials. Some basic electrical principles help us understand how neurons produce action potentials and other electric signals.

A **current** is a flow of electric charges from place to place. In a wire, electrons flow, and current is based on the flow of electrons. However, in the cells of organisms, current is based on the flow of ions in solution (see Key Concept 4.2), such as sodium ions (Na^+), potassium ions (K^+), or chloride ions (Cl^-). Each Na^+ ion, for example, bears a single positive charge. When Na^+ ions move from one place to another, positive charges flow, and this flow of charges is an electric current.

A **voltage**, also called an electric **potential difference**, exists in a system if positive charges are concentrated in one place and negative charges are concentrated in a different place. Voltages *produce currents* because opposite charges attract and move toward one another. Suppose, for example, that two regions of a solution differ, such that positive charges are concentrated in one region and negative charges in the other. Positive ions dissolved in the solution, such as Na^+ ions, will be attracted to and tend to flow toward the negative area, giving rise to a current.

As we start now to understand how neurons produce action potentials and other electric signals, let's look at **FIGURE 31.8A**, which shows a part of a resting axon in a familiar, standard way. An important fact to recognize is that diagrams of this type are not to scale because the cytoplasm is actually far larger than shown. A

(A) Part of an axon

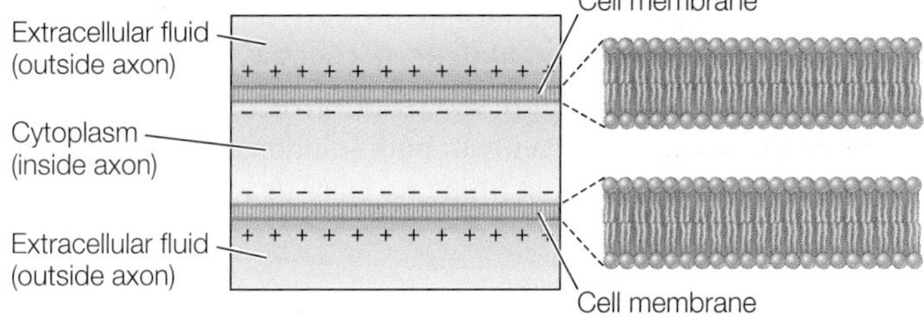

(B) An axon cell membrane with channel proteins

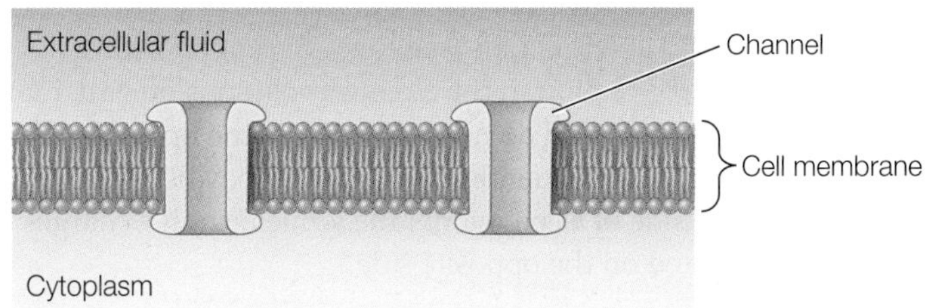

FIGURE 31.8 Basic Features of an Axon **(A)** Diagram of a part of an axon, with magnified drawings of the cell membrane. As discussed in the text, a diagram like this is not to scale because the cytoplasm is actually far larger relative to the membranes than shown. The electric charges shown are typical of a resting axon. **(B)** An axon cell membrane, showing channel proteins embedded in the lipid bilayer.

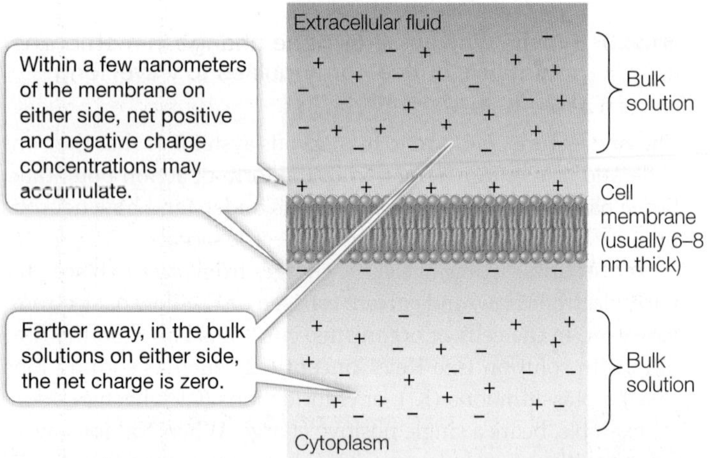

FIGURE 31.9 Distributions of Electric Charges in an Axon The charge difference across the cell membrane is called the membrane potential. In a resting neuron, as shown here, the inside of the membrane is negative relative to the outside.

cell membrane is about 7 nm thick on average, and many axons are about 1 μm (1,000 nm) in diameter. This means that, with the two cell membranes drawn as they are in Figure 31.8A, the space between them would need to be about 20 times greater than shown to represent the physical dimensions of an axon accurately. There is a lot more cytoplasm separating the two cell membranes than the familiar, standard diagram suggests!

When thinking about voltages there are two critically important principles to consider:

1. *No voltages (potential differences) can exist from one place to another within an open—or bulk—solution.* When we speak of an open or bulk solution, we refer to a solution not interrupted by physical objects or by other impediments to free diffusion. In **FIGURE 31.9**, the extracellular fluid outside the membrane is a bulk solution. Similarly, the cytoplasm inside is a bulk solution. As the figure shows, positive and negative ions distribute themselves evenly in bulk solutions. Therefore no potential differences exist from one place to another within either of the solutions.

2. *Voltages (potential differences) can be established across membranes,* such as the cell membrane in Figure 31.9. A cell membrane is composed primarily of lipids. Because ions have an electric charge, they tend to associate with water (a polar molecule) and not with lipids (see Key Concept 2.5). Ions therefore tend to remain in the aqueous solutions on either side of a membrane and do not readily cross the lipid core of the membrane. Consequently, positive ions may be concentrated on one side of a cell membrane while negative charges are concentrated on the opposite side.

In summary, a bulk solution is electrically neutral and does not show charge differences from place to place. Charge differences can, however, occur across a membrane, caused by ion imbalances in the thin layers of water immediately next to the membrane surfaces. A potential difference (voltage) across a membrane is called the **membrane potential**.

Let's now talk more specifically about the potential differences that exist in a neuron. A neuron is described as "resting" when it is not carrying a signal. A resting neuron, like other cells, has a potential difference across its cell membrane (that is, a membrane potential). This potential difference is outside-positive, because there are more positive ions than negative ions on the outside surface of the membrane—and more negative ions than positive ions on the inside surface (see Figure 31.9). The membrane potential of a resting neuron is referred to as the **resting potential** and is typically between –60 and –70 millivolts (mV). The minus sign means that the inside of the cell is electrically negative relative to the outside. Membrane potentials are always written in this way: the sign (negative or positive) always refers to the charge inside relative to the charge outside.

As already emphasized, ions are essentially unable to pass directly through the lipid core—the lipid bilayer—of a cell membrane. This circumstance permits a charge difference to exist across a cell membrane, and if the membrane consisted only of the molecules shown in Figure 31.8A, the charge difference would persist over a long period.

As shown in **FIGURE 31.8B**, however, neuron cell membranes contain **channel proteins** that serve as ion channels, and **carrier proteins** that serve as transporters (see Figures 4.5 and 4.6). Although ions cannot pass through the membrane lipid bilayer, they can pass through the channels and carriers. The channel proteins and carrier proteins in a cell membrane determine the magnitudes and directions of the voltages and currents across the membrane.

Only small shifts of ions are required for rapid changes in membrane potential

As we go forward in our discussion of neuron function, we will often encounter cases in which rapid changes of the membrane potential are of critical importance. We need, therefore, to focus briefly on some key properties of these changes.

The membrane potential changes when ions are redistributed. During the rapid events of routine electrical signaling such as action potentials, ion redistribution occurs through the membrane channel proteins—usually called simply channels (see Figure 31.8B). Channels permit ions to cross a membrane by simple diffusion, as discussed in Key Concept 4.2.

Suppose that a membrane undergoes a dramatic shift from being negative on the inside to being positive on the inside. Suppose also that this change occurs because channels in the membrane allow a current of positive charges (ions) to flow inward across the membrane.

A critically important point is that only relatively small numbers of positive charges need to move through the membrane for this dramatic change of membrane potential to occur. This is true because *only the charges near the membrane surfaces are changing.* To quantify this concept, let's ask: Of all the ions within 1,000 nanometers of a membrane surface, what proportion participate in changing the membrane potential? Researchers have carried out this calculation, and the answer is fewer than 100 out of every 1 million ions. Just a relatively few ions—positioned within a few nanometers of the membrane surfaces—are involved in rapid changes of membrane potential. The other ions, located in the bulk solutions on either side of the membrane, are not directly involved. The *current* that passes through the membrane channels during a rapid change of membrane potential is small because it is not changing the compositions of the bulk solutions. A small current suffices to redistribute ions on the inner and outer membrane surfaces and lead to a large change in membrane potential—a phenomenon that is fundamental to how neurons generate signals.

Animation 31.1 The Resting Membrane Potential

The sodium–potassium pump sets up concentration gradients of Na^+ and K^+

One of the key membrane proteins is the carrier protein **Na^+-K^+ ATPase**, also called the **sodium–potassium pump**, which carries out active transport of Na^+ and K^+. (Recall from Key Concept 4.2 that a carrier protein engaged in active transport differs from a channel protein in that a carrier protein uses ATP-bond energy, whereas a channel protein simply permits diffusion.) The sodium–potassium pump goes through repeating cycles in which it actively transports ions. During each cycle, this pump uses the energy of an ATP molecule to expel three Na^+ ions from the inside of the neuron to the outside—and simultaneously to pump two K^+ ions into the neuron from the outside. Na^+ and K^+ are the principal positive ions in both the intracellular and extracellular fluids.

LINK Active transport moves a substance across a membrane by use of energy. Because energy is used, the direction of transport may be opposite from the direction of diffusion. For more on the sodium–potassium pump, see **Key Concept 4.2**.

The sodium–potassium pump creates steady ion concentration differences between the intracellular and extracellular fluids. Because of the action of the pump, the concentration of Na^+ is higher in the bulk solution outside the neuron than inside, and the K^+ concentration is higher in the bulk solution inside than outside.

The resting potential is mainly a consequence of K^+ leak channels

What is the cause of the resting potential? This potential difference is principally the result of K^+ channels that are open all the time. As we have seen, K^+ is more concentrated inside a neuron than outside. Some (fewer than half) of the K^+ channels—channels that specifically permit K^+ to diffuse through the cell membrane—are always open. They are like little holes in the walls of a water bottle. Much as holes let water leak out of the bottle, these open channels let K^+ leak out of a neuron. K^+ diffuses out through these so-called leak channels because of the K^+ concentration gradient (a higher K^+ concentration inside than outside).

This diffusion of K^+ (positive charges) out of a neuron leaves behind unbalanced negative charges inside the neuron. These negative charges tend to pull K^+ back into the cell because of the attraction of positive and negative charges.

The charges we are discussing are positioned on the *membrane surfaces*. As K^+ leaks out, positive K^+ charges accumulate on the outer surface of the cell membrane. At the same time, unbalanced negative charges accumulate on the inner surface. In this way, the outward leak of K^+ creates an electric potential difference (voltage)—outside-positive—across the cell membrane of a resting neuron.

What is the magnitude of this potential difference? It can be *predicted* by a key relationship called the Nernst equation, as we will see below. It can also be *measured*, by placing electrodes on either side of the membrane. The Nernst equation predicts a potential difference in a resting neuron of approximately –85 mV. The measured potential difference in a resting neuron is similar, approximately –65 mV, as we said before.

The Nernst equation predicts an ion's equilibrium potential

To carry out a theoretical analysis, let's consider a cell membrane across which only one type of ion can diffuse because the only channels open are specific for one type of ion. For example, only K^+ might be diffusing across the membrane because only K^+ channels are open. We would then say that the membrane is *permeable* only to K^+, and that K^+ is the only *permeating* ion (see Key Concept 4.2).

If the concentration of the permeating ion is different on one side of the membrane than the other, and if one side of the membrane is more positive than the other, there will be two effects on ion diffusion: (1) The *concentration effect* will cause the ion to diffuse across the membrane from the side of high concentration to the side of low concentration. (2) The *electrical effect* will cause the ion to diffuse across the membrane because of electrical attraction, moving toward the side where the charge on the membrane is opposite to the ion's charge. As the ion diffuses across the membrane, its diffusion will modify both effects by changing the electric charge difference and concentration difference across the membrane.

If allowed to continue without interference, ion diffusion across the membrane will continue until an equilibrium is reached in which the concentration effect and the electrical effect become equal and opposite. **FIGURE 31.10** illustrates this point using K^+ as an example. K^+ is at higher concentration inside a neuron than outside. As K^+ diffuses outward, leaving unbalanced negative

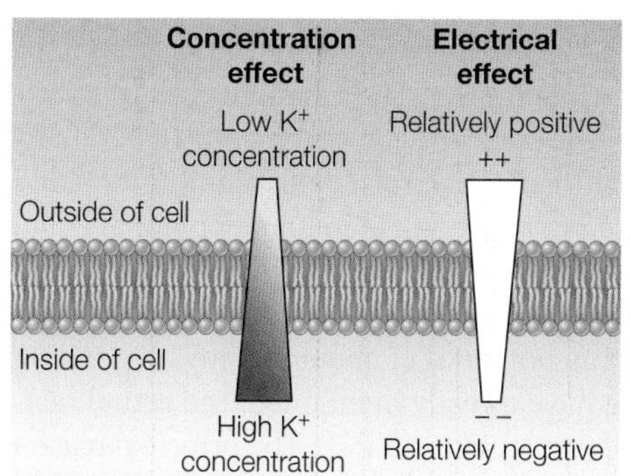

FIGURE 31.10 Both Concentration and Charge Differences Affect the Movement of an Ion through the Cell Membrane of a Resting Neuron This simplified illustration shows the two effects on the movement of K^+ through a membrane. The arrow-shaped icons point in the direction of K^+ movement caused by each effect. At electrochemical equilibrium, the two effects are equal and opposite, and no net K^+ movement occurs across the membrane.

charges behind, the inside becomes relatively negative. These two effects have opposite effects on K^+ diffusion: the concentration effect drives K^+ out, whereas the electrical effect drives K^+ in. K^+ will diffuse until the two effects are equal and opposite, at which point a state of **electrochemical equilibrium** will have been achieved. When an ion has reached electrochemical equilibrium, there is no net diffusion of the ion across the membrane in either direction.

When only one ion is diffusing across the cell membrane of a neuron, the electric potential difference at electrochemical equilibrium is called the **equilibrium potential of the ion** and is symbolized E_{ion}. The **Nernst equation** calculates the equilibrium potential as follows:

Expressions with brackets represent concentrations. Thus "[ion]" represents the concentration of the ion of interest.

$$\text{Potential difference at electrochemical equilibrium} = E_{ion} = 2.3\frac{RT}{zF}\log\frac{[\text{ion}]_{outside}}{[\text{ion}]_{inside}}$$

In this equation, z is the number of charges per ion, T is the absolute temperature, R is a constant (the universal gas constant), and F is another constant (the Faraday constant). (Both constants can be looked up in reference books or online.)

Activity 31.1 The Nernst Equation

Although we have used K^+ as an example, the Nernst equation can be applied to any ion (all that's needed is to plug the prevailing concentrations of the ion into the equation). Strictly speaking, the membrane potential will equal the calculated equilibrium potential only if just the single ion of interest is able to diffuse across the membrane (and equilibrium has been attained). However, in cases of multiple permeating ions, the Nernst equation often makes useful predictions of the existing membrane potential if only one ion accounts for most of the diffusion across the membrane and the other ions are diffusing to just a minor extent. The Nernst equation helps explain each of the major phases of an action potential, as we will soon see.

Let's now apply the Nernst equation to a resting neuron. As we have said, some of the K^+ channels—termed the K^+ leak channels—are open in a resting neuron. The channels for other ions are mostly closed. With the open channels being mostly K^+ channels, K^+ is expected to diffuse to electrochemical equilibrium, and this diffusion is expected to produce a membrane potential equal to the equilibrium potential of K^+, E_K. Plugging the inside and outside concentrations of K^+ into the Nernst equation, the equation predicts a membrane potential of about –85 mV, depending on temperature. As we have already mentioned, the actual resting potential is a little lower (about –65 mV). The principal reason the equation's prediction is a little off is that resting neurons are slightly permeable to other ions besides K^+.

Gated ion channels alter the membrane potential

Unlike the K^+ leak channels, most ion channels are gated: they are open under some conditions and closed under others (see Key Concept 4.2). Even most K^+ channels are gated. Most gated channels are closed in a resting neuron, explaining why the K^+ leak channels determine the resting membrane potential.

How does gating occur? Protein molecules are noted for changing their three-dimensional molecular conformation (shape) in response to changes in their molecular environment. For example, a protein molecule may assume a different conformation at a relatively low pH (when H^+ is relatively abundant in the fluid bathing the molecule) than at a higher pH. Gating results from conformation changes in channel proteins. When a channel protein changes its molecular conformation, not only does the physical shape of the molecule change, but the positions of the electrically charged parts of the molecule also shift. These physical alterations affect the ease with which ions (e.g.,

(A) Voltage-gated channel

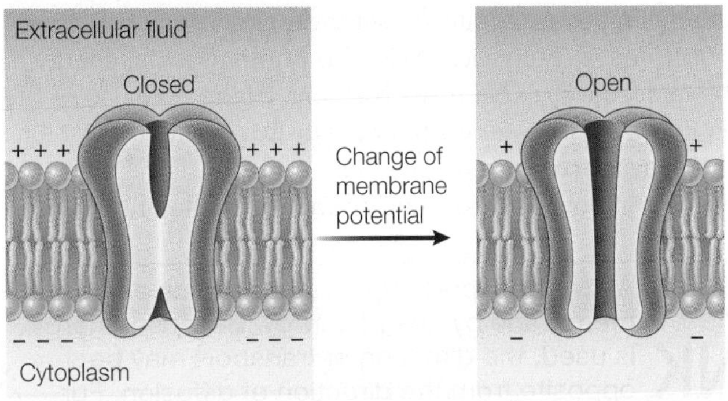

(B) Stretch-gated (tension-gated) channel

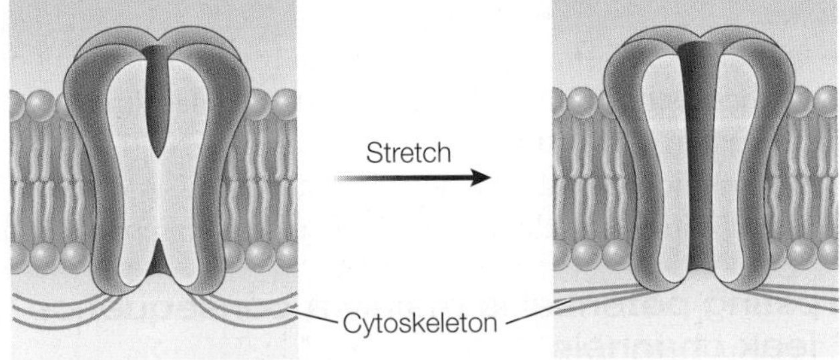

(C) Ligand-gated channel

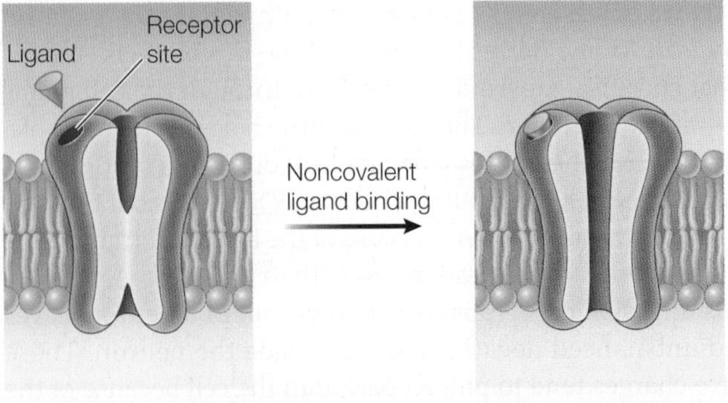

FIGURE 31.11 Three Types of Gated Ion Channels (After E. R. Kandel. 2013. *Principles of Neural Science*, 5th ed. McGraw-Hill Companies, Inc.)

K^+) diffuse through the channel protein. There are not literally gates (swinging doors) in these proteins, but they behave as if there were gates.

Gated channels are categorized on the basis of the agent that causes them to open or close (**FIGURE 31.11**):

- **Voltage-gated channels** open or close in response to local changes in the membrane potential.
- **Stretch-gated channels** open or close in response to stretch or tension applied to the cell membrane.
- **Ligand-gated channels** have binding sites where they bind noncovalently with specific chemical compounds that control them. A compound that controls a channel is termed a ligand of that channel. A ligand-gated channel opens and closes depending on whether it is bound to its ligand. Although Figure 31.11C shows a ligand-binding site on the outside of the cell membrane, binding sites (e.g., for second messengers) can also be on the inside.

Openings and closings of gated channels alter the membrane potential. For example, let's consider what happens in a resting neuron if some of the Na^+ channels in the cell membrane open (Na^+ channels are ordinarily closed in a resting neuron). If Na^+ channels open, Na^+ ions diffuse into the neuron from the outside because Na^+ is far from electrochemical equilibrium in a resting neuron. Na^+ is far more concentrated outside the cell than inside, and this concentration difference tends to drive diffusion into the cell. The electrical difference also tends to drive Na^+ diffusion into the cell because the cell membrane is negative on the inside in a resting neuron, attracting positive ions. The inward diffusion of Na^+—caused by both concentration and electrical effects—represents an inward flow of positive charges which makes the inside of the neuron cell membrane become less negative. In other words, the voltage difference across the membrane (the membrane potential)—which is ordinarily near –65 mV in a resting neuron—becomes smaller. Perhaps it becomes –58 mV instead.

Depolarization is said to occur whenever the charge on the inside of the cell membrane becomes less negative—relative to the charge on the outside—than it is at rest. In Key Concept 31.1 we implied that depolarization is either present or absent. Actually, in functioning neurons, many different degrees of depolarization are possible. **Hyperpolarization** occurs when the charge on the inside becomes more negative than it is at rest. Opening and closing of gated ion channels can cause either depolarization or hyperpolarization, depending on which ions are affected.

Changes in membrane potential can be graded or all-or-none, depending on whether a threshold is crossed

Two types of changes in membrane potential occur: graded and all-or-none. The type of change that occurs depends on whether a voltage threshold of about –50 mV is crossed. We will start by discussing graded changes.

FIGURE 31.12A shows variations in membrane potential over time in a neuron in which the voltage threshold is not crossed. Note that the membrane potential can be high, low, or

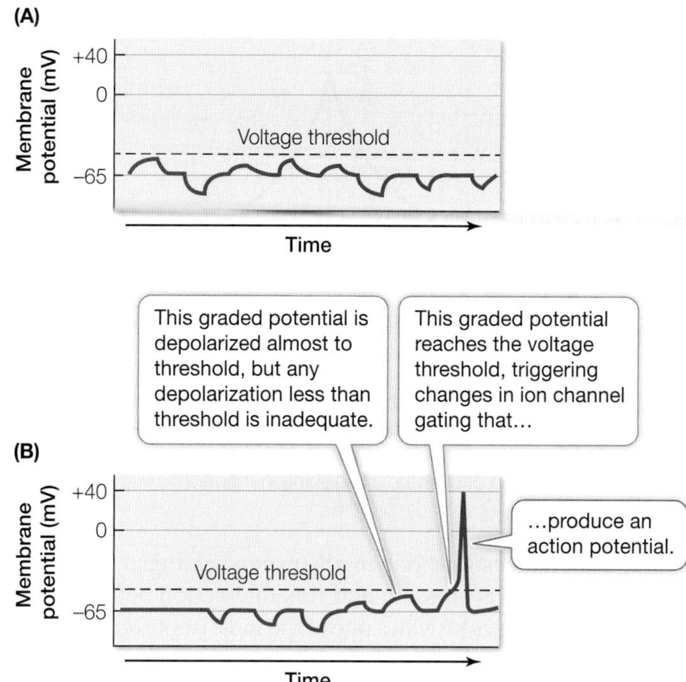

FIGURE 31.12 Graded and All-or-None Changes in Membrane Potential If you place an electrode on the inside of a neuron's cell membrane and another electrode on the outside, you can measure the membrane potential continuously as time passes. **(A)** During graded changes, the membrane potential goes up and down as time passes, but it never crosses the voltage threshold. **(B)** When a graded change in membrane potential reaches the voltage threshold, the membrane generates a large, all-or-none change in membrane potential—an action potential.

in between. These changes are described as **graded** because any value of the membrane potential is possible. The changes are caused by various gated ion channels opening or closing—some causing the membrane potential to shift in one direction, some causing it to shift in the other direction.

Graded changes of the membrane potential spread only a short distance from the locations of the ion channels that cause them. If a set of Na^+ channels opens, for example, the change in membrane potential will be greatest at the location along the cell membrane where the channels are found. The change will spread on either side. But it will become smaller and smaller at greater and greater distances from the channels (**FIGURE 31.13**).

All-or-none changes in the membrane potential are very different. As **FIGURE 31.12B** shows, if a neuron is undergoing graded changes and, in the course of these changes, depolarizes sufficiently for the membrane potential to cross the threshold, a massive, all-or-none event is triggered: an action potential. An action potential is described as all-or-none because, when it occurs, it is always the same size (i.e., it is not graded). Moreover, as we'll soon discuss, an action potential does not become smaller and smaller as it moves away from the location on a cell membrane where it originates. It stays the same in size as it propagates along the cell membrane.

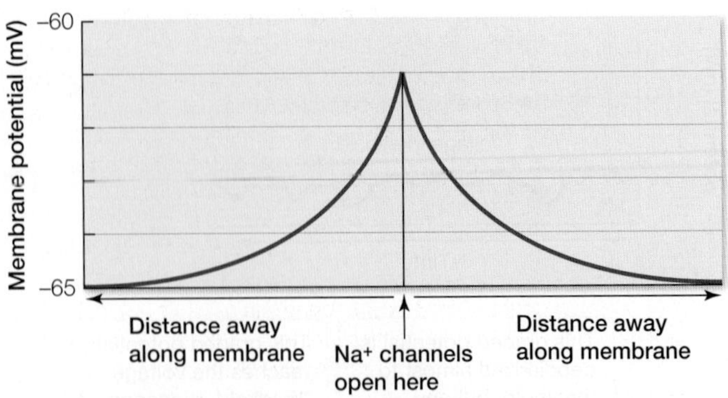

FIGURE 31.13 The Spread of a Graded Potential along the Length of a Neuron Membrane When a graded potential is induced at a point on the cell membrane of a neuron, it spreads away from that point in both directions, diminishing in magnitude with distance.

Graded changes give rise to all-or-none changes when the voltage threshold is crossed, and this interaction between the two types of changes provides one of the most important mechanisms of *integration*: the process by which each *output* of a neuron can be determined by *multiple inputs*. To illustrate, suppose a neuron simultaneously receives four signals—A, B, C, and D—that affect different types and numbers of gated ion channels in the neuron cell membrane, producing graded changes in the membrane potential. Suppose that at a particular location on the cell membrane, signal A causes a 2.5-mV depolarization, B causes a 1.5-mV depolarization, C causes a 1-mV hyperpolarization, and D causes a 2-mV depolarization. These graded changes will sum together, causing this part of the cell membrane to be depolarized by 5 mV. The membrane potential will be –60 mV instead of –65 mV as at rest.

The location of greatest importance for this type of integration in a neuron is the axon hillock, where the axon arises from the cell body (see Figure 31.6). A neuron may be receiving hundreds or thousands of signals at its dendrites and cell body. These signals bring about graded changes in membrane potential that spread to the axon hillock, where all the depolarizations and hyperpolarizations sum.

If the summation of all signals at the axon hillock fails to raise the membrane potential above the –50 mV threshold, as in Figure 31.12A, no action potentials are produced. Accordingly, no electric signals reach the presynaptic axon terminals because the graded potentials disappear over the distance between the axon hillock and the axon terminals (see Figure 31.13). In sharp contrast, if the summation of graded potentials causes the membrane potential at the axon hillock to cross the threshold (as in Figure 31.12B), an action potential will be initiated. The action potential will then travel along the axon without loss of magnitude, all the way to the axon terminals. In turn, the axon terminals will send signals to the cells they innervate.

The processes described here are the major processes by which a single neuron integrates its inputs to determine its outputs. In short, all the inputs create small, graded changes of membrane potential that sum together as they spread. If their integrated effect is a membrane potential that is not as depolarized as about –50 mV, the neuron produces no action potentials. But if their integrated effect is a membrane potential that is depolarized to at least –50 mV, the neuron generates one or more action potentials that propagate long distances down the axon.

An action potential is a large depolarization that propagates with no loss of size

We have seen that an action potential is a large, brief, localized change in the membrane potential—so large that the polarity reverses from inside-negative to inside-positive. During an action potential, the membrane depolarizes from –65 mV at rest to about +40 mV: a change of more than 100 mV. This event is extremely brief; it lasts only about 1 millisecond. It is also very localized, being restricted to just a small region of the cell membrane.

An action potential propagates with no loss of magnitude, however. Propagation occurs because an action potential at one location in the cell membrane of a neuron causes currents to flow that depolarize neighboring regions to the voltage threshold. For example, when an action potential occurs in the axon hillock, currents flow into the neighboring region of the axon, causing a graded depolarization there that exceeds the threshold. That region then produces an action potential. Currents from that action potential then spread into the axon region still closer to the axon terminals. That region then produces an action potential. Each region of the axon stimulates the next region to produce an action potential until the action potential reaches the axon terminals. This process is what is meant by "propagation" or "conduction" in discussions of neuron function. Recall that action potentials are all-or-none and always the same size. Thus there is no loss of magnitude as an action potential propagates along the length of an axon.

Why does the cell membrane depolarize so profoundly during an action potential? And why does this depolarization end so quickly—within a millisecond? The properties of voltage-gated K^+ and Na^+ channels in the neuron cell membrane are chiefly responsible for the behavior of the action potential in its three successive phases: the initial resting phase, the rising or depolarization phase, and the falling or repolarization phase. As we describe the events in these three phases, **FIGURE 31.14** will provide useful additional information.

When the membrane potential in any local region of an axon reaches threshold, massive numbers of voltage-gated Na^+ channels in that region are triggered to open (compare Figure 31.14A and B). Suddenly, therefore, the principal ion that can cross the membrane becomes Na^+, instead of K^+. This results in a sudden, rapid influx of Na^+ ions into that part of the axon because, as we discussed earlier, both the Na^+ concentration difference and the electrical difference across the cell membrane favor inward Na^+ diffusion. If Na^+ were the only ion that could cross the membrane at this stage and if enough time were available to reach electrochemical equilibrium, the membrane potential would become E_{Na}—the equilibrium potential for Na^+ (about +50 mV according to the Nernst equation). The membrane potential in fact almost reaches E_{Na}: it typically becomes about +40 mV. Quickly, however (compare Figure 31.14B and C), the Na^+ channels close, and voltage-gated K^+ channels, which respond in a delayed fashion to the local membrane potential, open in massive numbers. With

(A) Resting membrane potential

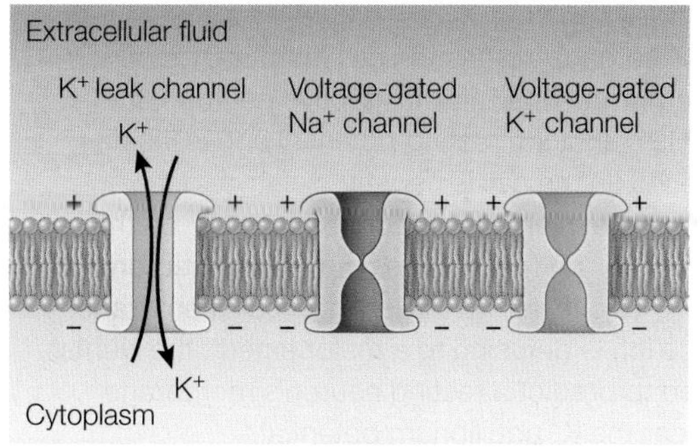

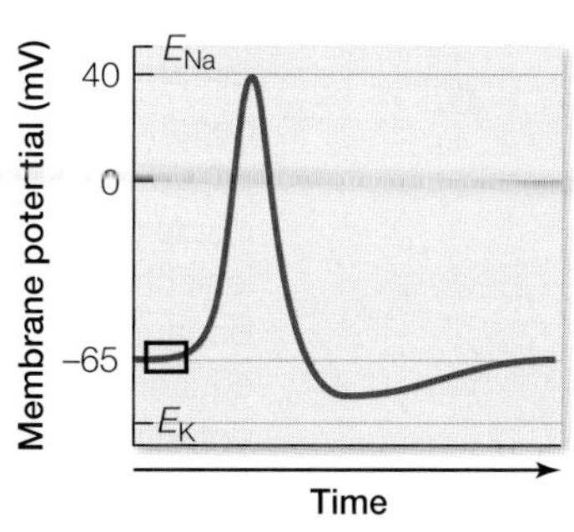

(B) Rising phase (depolarization) of action potential

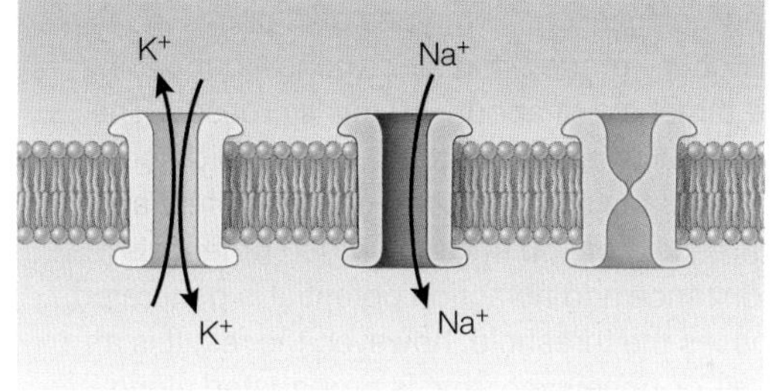

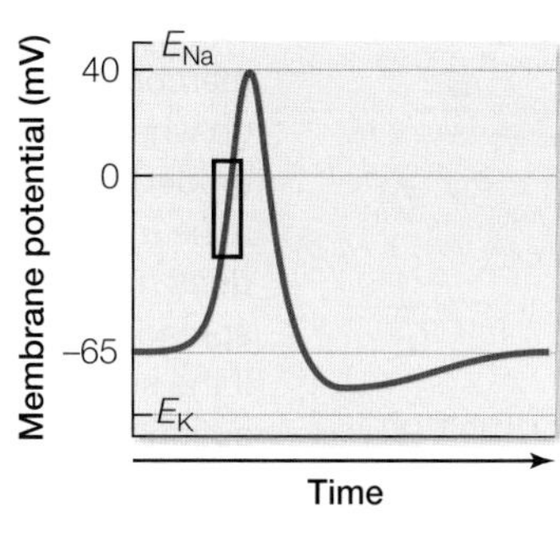

(C) Falling phase (repolarization) of action potential

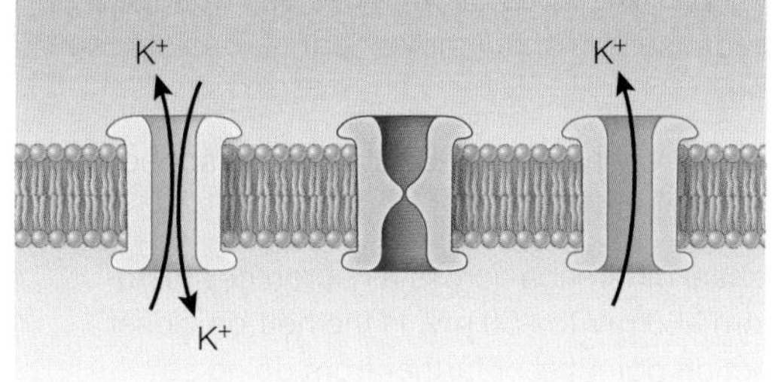

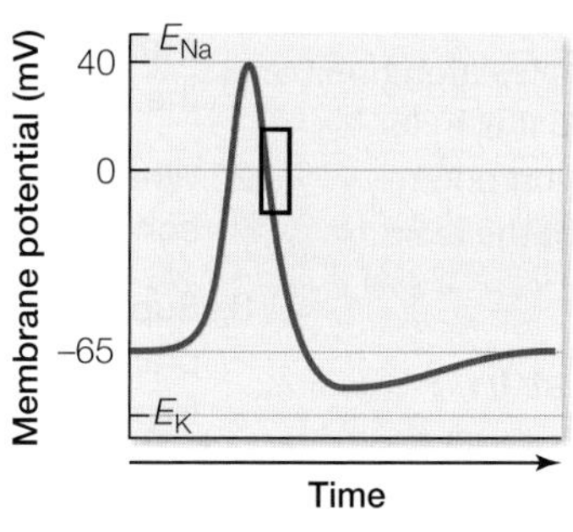

FIGURE 31.14 Production of an Action Potential The diagrams to the left show changes in the voltage-gated Na^+ and K^+ channels in the cell membrane (here diagrammed stylistically), and the curves on the right show changes in membrane potential during rest **(A)**, the rising phase of an action potential **(B)**, and the falling phase **(C)**. The black rectangles on the membrane potential curves indicate the part of the curve during which the events diagrammed at left occur. The closing of the Na^+ channels in (C) is of a specialized type (called inactivation) that—for a brief period—cannot be reversed (i.e., the channels temporarily cannot reopen). K^+ leak channels are always open. (After M. F. Bear et al. 2001. *Neuroscience: Exploring the Brain*. Lippincott: Baltimore.)

Animation 31.2 The Action Potential

the Na^+ channels closed and the voltage-gated K^+ channels open, the principal ion that can cross the membrane becomes K^+. K^+ diffuses out of the cell because K^+ is more concentrated inside than outside and also because, at this moment (with the membrane potential at +40 mV), the outside of the cell membrane is relatively negative, attracting K^+. The movement of K^+ out of the cell drives the membrane potential back toward –65 mV, which is close to E_K, the K^+ equilibrium potential.

Positive feedback (see Key Concept 28.3) plays a major role in the rising (depolarization) phase of an action potential (see Figure 31.14B). Positive feedback (**FIGURE 31.15**) occurs because the Na^+ channels are voltage-gated and the voltage across the neuron cell membrane is rapidly changing. When the cell membrane becomes partly depolarized, some voltage-gated Na^+ channels open. Na^+ ions then cross the cell membrane faster (diffusing from outside to inside) and depolarize the membrane even more. This greater depolarization causes more voltage-gated Na^+ channels to open. This positive-feedback loop continues until all the voltage-gated Na^+ channels open and maximum depolarization (approaching E_{Na}) occurs.

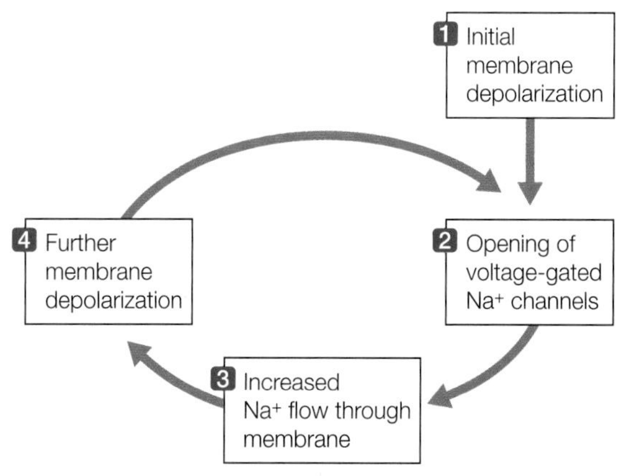

FIGURE 31.15 Positive Feedback Plays a Key Role in the Generation of an Action Potential During the rising phase of an action potential, positive feedback progressively drives all voltage-gated Na^+ channels to open.

What is the definition of positive feedback?

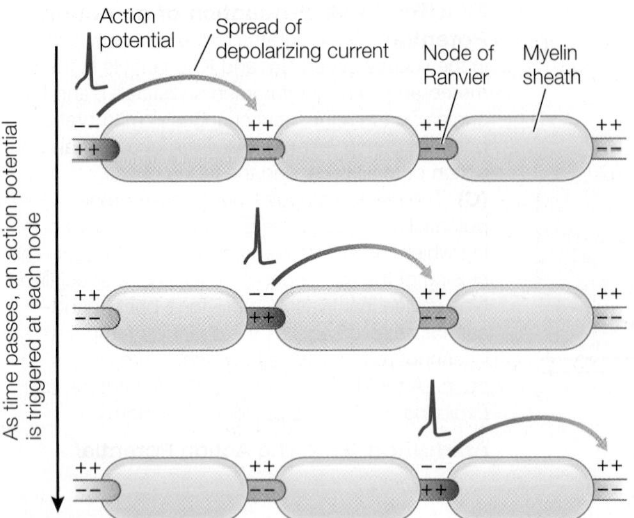

FIGURE 31.16 Saltatory Conduction along a Myelinated Axon

Why does an action potential travel in just one direction along an axon? After any particular local part of the axon membrane has produced an action potential, it is unable to produce another action potential for a short interval of time called the *refractory period*. Thus as an action potential propagates along an axon, the region of the cell membrane just behind it is in its refractory period. Currents are able to induce an action potential only in the neighboring region in the direction of the axon terminals.

Action potentials travel particularly fast in large axons and in myelinated axons

Action potentials do not travel along all axons at the same speed. Because of physical laws, they tend to travel faster in large-diameter than in small-diameter axons. This principle has been particularly important in the evolution of invertebrate nervous systems. Many invertebrates have evolved large-diameter axons that increase the speed of conduction of action potentials to muscles on which the animals depend for rapid escape from danger. Squid, for example, have "giant" axons—1 millimeter in diameter—that carry impulses to their escape muscles. Many of the basic principles of neuron function were first discovered by studying these extraordinarily large cells.

Vertebrates have evolved an additional means of increasing the speed of conduction: myelination. When glial cells wrap around axons, covering them with myelin, they leave uncovered spaces, called **nodes of Ranvier**, in between (see Figure 31.7). In a myelinated axon, action potentials are generated only at these nodes, because voltage-gated ion channels are clustered at the nodes and also because the myelin-covered axon segments are too well insulated electrically to permit rapid ion movements. An action potential jumps from one node of Ranvier to the next, a process called saltatory ("jumping") conduction (**FIGURE 31.16**). In this way the action potential travels far faster than it would by continuously traveling the entire length of the axon.

REVIEW & APPLY | 31.2

R The electric charge difference across the cell membranes of neurons is called the membrane potential. In inactive neurons, the membrane potential is called the resting potential and is inside-negative (the charge inside the cell is negative relative to the outside). The sodium–potassium pump concentrates K^+ on the inside of a neuron and Na^+ on the outside. In a resting neuron, K^+ leak channels allow K^+ to diffuse out through the cell membrane. Because only the one ion is defusing to a major extent, the Nernst equation can be used to predict a resting neuron's membrane potential, which is near the K^+ equilibrium potential.

Openings and closings of voltage-gated ion channels in the cell membrane alter the membrane potential by causing either depolarization or hyperpolarization. Two types of changes in membrane potential occur—graded and all-or-none—depending on whether a voltage threshold of about –50 mV is crossed. The graded potentials from multiple sources are summed to determine a neuron's outputs. If the summation of graded potentials does not reach the threshold of –50 mV, the graded potentials spread only a short distance and no action potential is generated. If the summation crosses the threshold, however, the result is an action potential—an all-or-none event that is propagated along the entire length of a neuron's axon without loss of magnitude.

A
1. Explain how the membrane potential of a cell can change by a *large* amount because of a *small* movement of ions across the cell membrane.
2. What are two different evolutionary adaptations to speed the conduction of action potentials?
3. Suppose the potential difference across a neuron cell membrane changes from –60 mV to –50 mV. Is the neuron closer to producing an action potential or farther from doing so? Explain.
4. Why is an all-or-none change in membrane potential critical for impulses to travel long distances in neurons?

Having described how an action potential is generated and propagated, let's consider what happens when the action potential arrives at a presynaptic axon terminal. How is the information communicated to the next cell?

31.3 | Neurons Communicate with Other Cells at Synapses

LEARNING OBJECTIVES

By the end of this key concept you should be able to:

31.3.1 Explain the roles of neurotransmitters in the function of nervous systems.

31.3.2 Distinguish between chemical and electrical synapses.

31.3.3 Outline ways that synaptic plasticity can function as a mechanism of learning and memory.

Recall that synapses are cell-to-cell contact points specialized for signal transmission from one cell to another. Synapses play extremely important roles in information processing and integration. They also play central roles in learning and memory. A great deal of modern research on the function of the nervous system focuses on synapses.

Chemical synapses are most common, but electrical synapses also exist

The most common type of synapse is a **chemical synapse**, so called because the signal passed by the presynaptic cell to the next neuron or target cell is chemical in nature. At a chemical synapse, the cell membranes of the pre- and postsynaptic cells do not make direct contact, but instead are separated by an extremely narrow space, about 20–30 nanometers wide, termed the **synaptic cleft**. Action potentials cannot cross this space. As shown in **FIGURE 31.17**, when an action potential arrives at the end of the presynaptic cell, that cell releases molecules of a chemical compound called a **neurotransmitter** into the synaptic cleft. The neurotransmitter molecules diffuse across the narrow synaptic cleft rapidly, then bind to receptor proteins in the cell membrane of the postsynaptic cell. These receptor molecules are specialized to generate an electrical change (or some other sort of change) in the postsynaptic cell when they bind to neurotransmitter molecules—completing signal transmission across the synapse. In many cases, the entire process of synaptic signal transmission takes just a few milliseconds.

Neurotransmitter molecules are rapidly removed from the synaptic cleft. Removal terminates the signal and is essential for the proper function of a chemical synapse. Removal may occur by two principal mechanisms: (1) neurotransmitter molecules may be broken down enzymatically, or (2) they may be taken up by nearby neurons or glial cells.

In an **electrical synapse**, signal transmission is "all electrical" instead of employing a neurotransmitter. At an electrical synapse, the cell membranes of the pre- and postsynaptic cells are joined by gap junctions at which the cytoplasms of the two cells merge and are continuous (see Figure 4.19). Thus when an action potential arrives in the presynaptic cell, the action potential can spread directly to the postsynaptic cell. Signals cross electrical synapses with essentially no delay.

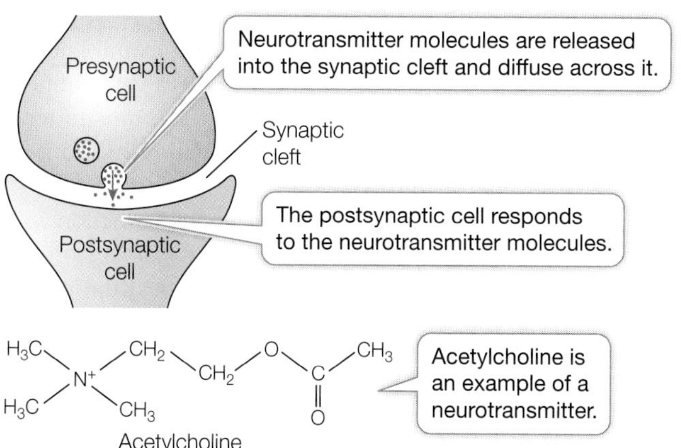

FIGURE 31.17 A Chemical Synapse

Electrical synapses have evolved principally where very fast, invariant signal transmission is advantageous. They are found in neurons that control escape swimming in some fishes and crustaceans. They are also found in some tissues where great numbers of cells must be stimulated synchronously to act together, such as fish electric organs (see Key Concept 33.3).

The vertebrate neuromuscular junction is a model chemical synapse

In a vertebrate skeletal muscle such as our biceps (upper-arm muscle), each muscle cell (muscle fiber) is typically innervated by one neuron, which as we have noted before is called a *motor neuron*, because it carries impulses to a muscle cell. There are many muscle cells in a muscle; our biceps muscle contains thousands of cells, for example. Each is typically innervated by one neuron and will contract only if stimulated to do so by that neuron. A **neuromuscular junction**—a type of chemical synapse—is a synapse between a motor neuron and a skeletal muscle cell (**FIGURE 31.18A**). This type of synapse, sometimes called a motor end-plate, has been studied intensively to understand how chemical synapses work. It is not entirely representative of all chemical synapses but provides a lot of insight into them.

At a neuromuscular junction, the axon of the presynaptic neuron branches close to the muscle cell, creating several buttonlike axon terminals (called boutons) that are in synaptic contact with the muscle cell (see Figure 31.18A). Each of these axon terminals contains spherical synaptic vesicles filled with molecules of acetylcholine (ACh), the neurotransmitter used by all vertebrate neuromuscular junctions (**FIGURE 31.18B**).

When an action potential arrives at an axon terminal, the terminal releases ACh into the synaptic cleft by a mechanism that depends on voltage-gated calcium channels. The action potential causes Ca^{2+} channels in the presynaptic membrane to open (see Figure 31.18B, step 1). Ca^{2+} diffuses into the axon terminal through the open channels because the concentration of Ca^{2+} is greater outside the cell than inside (step 2). Increased Ca^{2+} inside the terminal induces synaptic vesicles to fuse with the presynaptic membrane, releasing ACh into the synaptic cleft by exocytosis (step 3).

The response of the muscle cell depends on ligand-gated receptor proteins (receptors) in the postsynaptic cell membrane (i.e., the cell membrane of the muscle cell). These receptors, which are gated by ACh, act as ion channels when they are open. When ACh is released by the presynaptic cell, it diffuses across the synaptic cleft (step 4) and binds to the ACh-gated receptors. The receptors open, allowing Na^+ and K^+ to diffuse through, with influx of Na^+ being dominant (step 5). The rapid influx of Na^+ causes a large, graded depolarization of the postsynaptic membrane, which induces graded ionic currents that spread to neighboring regions of the postsynaptic cell membrane (step 6) where there are voltage-gated Na^+ channels (and K^+ channels). The membrane depolarization in these neighboring regions is great enough for the voltage-gated Na^+ channels to open, initiating an action potential (step 7). This action potential is propagated throughout the cell membrane of the muscle cell and triggers muscle contraction, as we will discuss in Chapter 33.

(A) Overall structure of a neuromuscular junction

Motor neuron
Myelinated axon
Neuromuscular junction
Muscle cell (muscle fiber)
Myelin
Axon
Presynaptic terminal boutons

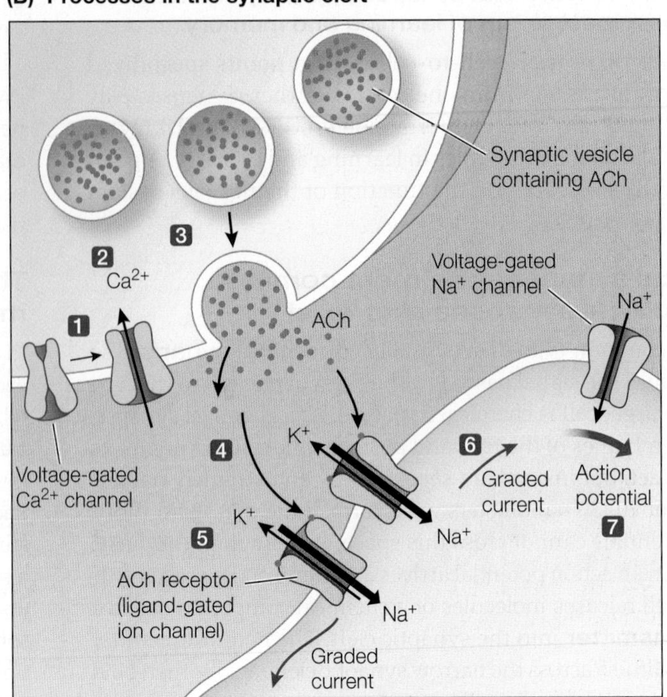

FIGURE 31.18 Synaptic Transmission at a Vertebrate Neuromuscular Junction **(A)** A motor neuron communicates with a skeletal muscle cell at a type of synapse called a neuromuscular junction. **(B)** The motor neuron releases acetylcholine (ACh) into the synaptic cleft, stimulating an action potential in the muscle cell. STEPS 1–3: The action potential stimulates influx of Ca^{2+}, which in turn causes vesicles filled with ACh to fuse with the presynaptic membrane. STEPS 4–5: ACh diffuses across the synaptic cleft and binds to ACh-gated ion channels in the postsynaptic membrane, causing them to open. STEPS 6–7: The resulting voltage changes trigger an action potential in the postsynaptic cell.

Animation 31.3 Synaptic Transmission
Activity 31.2 Neurons and Synapses Simulation
Media Clip 31.1 Put Some ACh into It!

Signals cross chemical synapses in just one direction. They cannot cross in the "wrong" direction because the pre- and postsynaptic cell membranes are specialized in different ways. The postsynaptic membrane in a neuromuscular synapse, for example, cannot produce a neurotransmitter.

Many different neurotransmitters have been identified

Several dozen neurotransmitters are now known in chemical synapses, and more are being discovered all the time. They fall into three major chemical categories:

1. *Amino acids* such as glutamate, glycine, and γ-aminobutyric acid (GABA; γ = gamma)
2. *Biogenic amines* such as acetylcholine, dopamine, norepinephrine, and serotonin
3. *Peptides* (strings of amino acids)

Each presynaptic neuron produces specific neurotransmitters. Most produce only one neurotransmitter, according to present knowledge. However, some presynaptic neurons produce two or more neurotransmitters. Adjectives ending in "ergic" are commonly used to specify the neurotransmitter secreted by a presynaptic neuron (and likewise the neurotransmitter to which receptors in the postsynaptic neuron respond). For example, "cholinergic" refers to use of acetylcholine, "dopaminergic" refers to use of dopamine, and so forth. **INVESTIGATION FIGURE 31.19** describes the discovery of a key role for neurons in the mammalian spinal cord that employ the amino acid glutamate as their neurotransmitter.

Within the nervous system, a postsynaptic cell may have synapses with hundreds of presynaptic cells

The neuromuscular junction is an informative example for understanding the basics of synapse function. It is a very simple example, however, because it employs just one neurotransmitter and one type of receptor, and the postsynaptic cell receives input from just a single presynaptic cell.

As brain scientists seek to understand synaptic function within the nervous system—where neurons make synapses on other neurons—they face a massively more complex situation. This helps explain the challenge of understanding the function of the nervous system. Neurons typically have great numbers of synapses that function with a variety of neurotransmitters, and often with two or more receptor proteins for any one neurotransmitter.

The stunning image in **FIGURE 31.20** shows the cell body and dendrites of a postsynaptic neuron in a mouse brain, labeled

INVESTIGATION

FIGURE 31.19 Revealing the Mechanisms That Initiate Walking Motions When mammals walk, their leg muscles alternately contract and relax to make the legs swing forward and backward. Motor neurons to these muscles originate in the lower part of the spinal cord. Within the spinal cord there are sets of neurons that interact to function as *central pattern generators* to initiate impulses in the motor neurons and thereby control the alternating rhythm of muscle activity required for walking. But what activates these neurons? Must they interact with the brain to be effective? To answer these questions, Ole Kiehn, Martin Hägglund, and their colleagues have studied the spinal cords of nestling mice. The results of such research could be of huge importance for the medical treatment of people who have suffered spinal injuries.[a]

HYPOTHESIS

The neurotransmitter glutamate activates neurons in the central pattern generators that control walking.

METHOD

The researchers used molecular genetic techniques—termed optogenetic techniques—to create a strain of mouse that expresses a gated Na^+ channel that opens when it is exposed to blue light. (The gene for this channel came from an alga; mammalian Na^+ channels are not light-activated.) Because of the technique the investigators used, the light-gated channel was expressed only in glutamate-releasing neurons in the central nervous system (i.e., neurons that employ glutamate as a neurotransmitter). The cell membranes of these neurons became populated with the light-gated Na^+ channel; opening of the channel could therefore alter the membrane potential of these neurons, leading to glutamate release. Using narrow light beams, the investigators illuminated small, highly specific sets of neurons in the abdominal spinal cord with blue light. At the same time, they measured the electrical activity in nearby nerves leading to the leg muscles involved in walking. The photograph shows a rat in an optogenetic study of brain neurons; in the mice discussed here, spinal-cord neurons were activated with blue light.

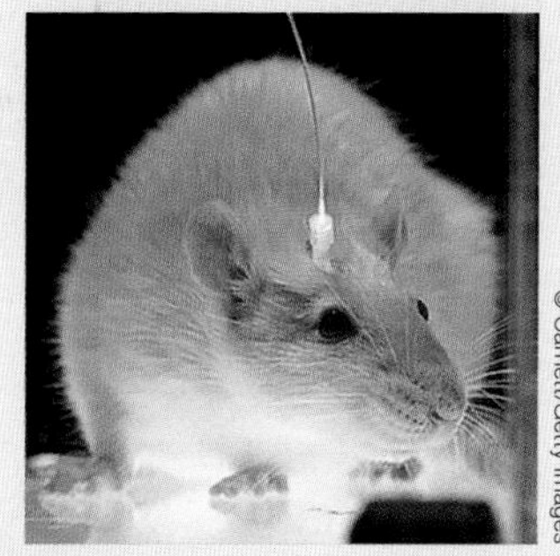

RESULTS

At the instant the blue light was turned on, the investigators saw rhythmic voltage signals in nerves to the walking muscles in the right and left hind legs. (The graph at right shows signals from whole nerves, reflecting the summed activity of many neurons.)

CONCLUSION

Glutamate released by specific neurons in the spinal cord plays a key role in activating the central pattern generators that control walking movements.

ANALYZE THE DATA

The figure shows the voltages recorded in the nerves leading to the right and left hind legs when the spinal cord neurons were being illuminated with blue light.

1. Why do these results support the conclusion that glutamate is a key activator of the central pattern generators?
2. Compare the recordings for the left and right nerves, in particular the alignment of their peaks and valleys. Is this what you would expect for nerves controlling leg muscles during walking? Explain.
3. Eighty years ago, when investigators wanted to see if a chemical compound such as glutamate acted as an activator for walking, they would simply have poured a solution containing glutamate on the spinal cord and observed what happened. Why is Kiehn's and Hägglund's discovery new? Why might early investigators not have discovered the action of glutamate by the method they used, and how might the new method be superior?

[a]M. Hägglund et al. 2010. *Nat Neurosci* 13: 246–252.

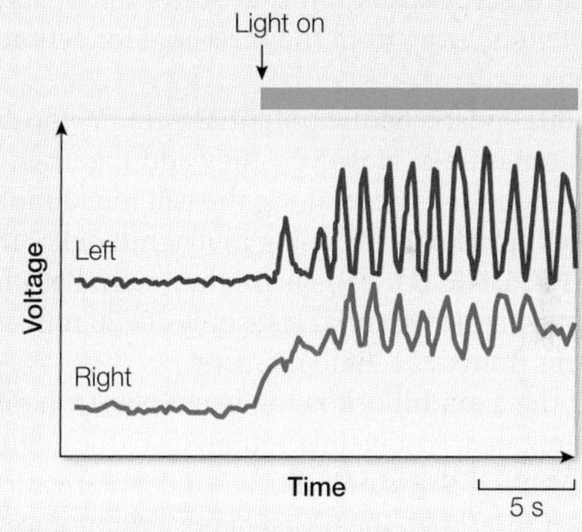

A similar **Analyze the Data** exercise may be assigned in **Achieve**.

so that the synapses on the neuron are yellow. This neuron is typical in that there are hundreds or thousands of synapses, meaning a great many presynaptic neurons make synaptic contact with this single postsynaptic neuron. Each presynaptic neuron will typically use one or two neurotransmitters. Collectively, however, all the presynaptic neurons may use a wide diversity of neurotransmitters. Moreover, at any particular synapse, the postsynaptic cell may express two or more types of receptor proteins—having different actions—that respond to a single neurotransmitter released by the presynaptic cell.

Another highly consequential difference exists between neuron-to-neuron synapses and the neuromuscular junction. At most neuron-to-neuron synapses the response of the postsynaptic cell to a single stimulation by a presynaptic cell is typically just a small, graded, and short-lived change in membrane potential. That is, neuron-to-neuron synapses are specialized so that each presynaptic impulse produces just a small (usually less than 1 mV) and short-lived (lasting ~10–20 milliseconds) postsynaptic effect. These small, graded changes in the postsynaptic membrane potential are termed either **excitatory**

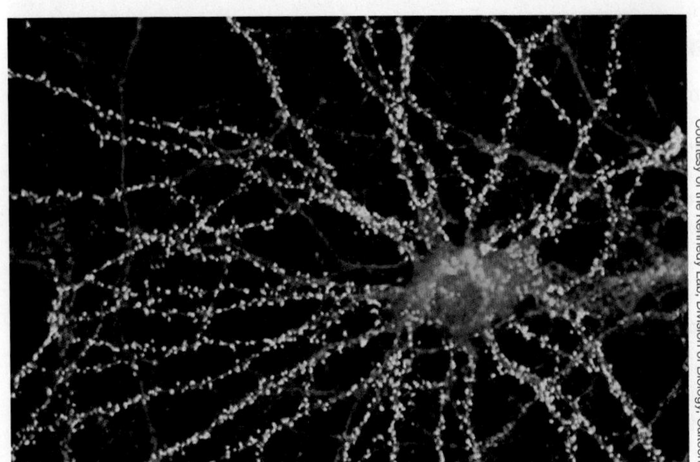

FIGURE 31.20 Synapses on a Single Postsynaptic Neuron in the Brain of a Mouse This microscopic image shows the cell body and many dendrites of a single neuron, labeled green. Synapses on the neuron are yellow. The synapses were labeled red, but the combination of green and red produces yellow at most synapses.

postsynaptic potentials (EPSPs) or **inhibitory postsynaptic potentials (IPSPs)**. An *excitatory* postsynaptic potential is caused by depolarization of the postsynaptic membrane, which shifts the membrane potential toward the threshold for action potential production. An *inhibitory* postsynaptic potential is caused by a hyperpolarization that shifts the postsynaptic membrane potential away from the threshold for action potential production.

The individual postsynaptic potentials are far too small to initiate action potentials. However, they add together. All the postsynaptic potentials spread along the cell membrane of the postsynaptic cell (gradually declining in magnitude) to the cell's axon hillock (**FIGURE 31.21**), where they sum. Because each individual postsynaptic potential lasts only 10–20 milliseconds, the sum changes from moment to moment.

Recall that the axon hillock is the usual place where action potentials are initiated. At any time, if the sum of the graded postsynaptic potentials at the axon hillock is a depolarization great enough to reach threshold, the postsynaptic cell produces an action potential (see the graph in Figure 31.21). If the depolarization remains greater than threshold after the refractory period, the cell produces another action potential. In this way, a long period of depolarization above threshold results in the production of a long sequence of closely spaced action potentials by the postsynaptic cell. Here we see a concrete example of *integration*. The postsynaptic neuron produces action potentials (or fails to) based on information received from hundreds or thousands of other neurons and based on the timing of this information (e.g., whether inputs are received synchronously or not).

Synaptic plasticity is a mechanism of learning and memory

The synapses in an individual animal's nervous system are continuously remodeled over the animal's lifetime, a phenomenon called **synaptic plasticity**. In this process the synapses undergo long-term changes in their functional properties and sometimes even their physical shape. Such changes are currently hypothesized to be one of the major mechanisms of learning and memory, explaining why synaptic plasticity is a principal focus of modern synapse research.

Learning refers to the ability of an individual animal to modify its behavior as a consequence of individual experiences that took place at earlier times in its life. For instance, suppose that Julie is accustomed to nibbling mushrooms while walking in the woods but one day becomes terribly ill after doing so. She will change her behavior in the future: seeing mushrooms, she will no longer grab one for a snack—an example of learning. **Memory** refers to a record in the nervous system of past experience and is closely related to learning. In Julie's case, she would have a recollection—a memory—of her illness that led her to learn and adopt her new behavior.

By what mechanism does learning occur? Synaptic plasticity is one hypothesis. According to this hypothesis, an individual's experiences at one time in its life can produce long-term changes in the

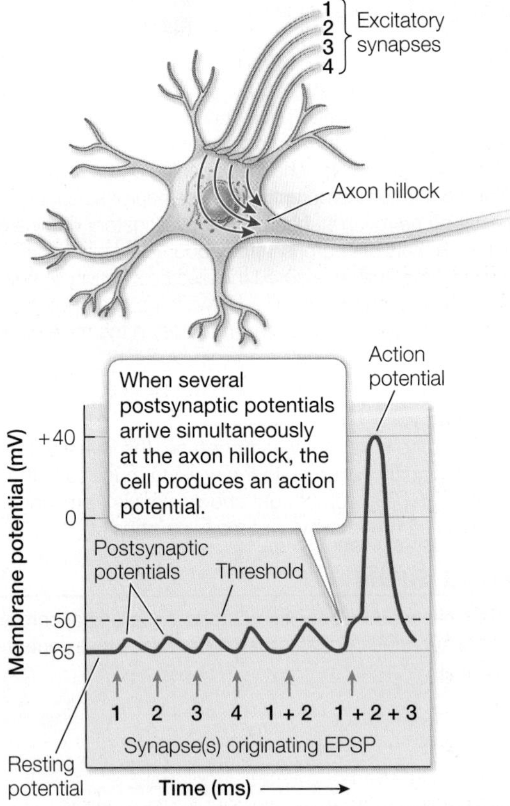

FIGURE 31.21 Signal Integration Based on Postsynaptic Potentials An individual neuron sums postsynaptic potentials received at spatially separate synapses. The drawing shows that the postsynaptic potentials spread to the axon hillock, where the summation occurs. The graph shows the membrane potential at the axon hillock under six different scenarios: input only from synapse 1, only from synapse 2, only from synapse 3, only from synapse 4, simultaneous input from synapses 1 and 2, and simultaneous input from synapses 1, 2, and 3. When the sum of the potentials present at one time depolarizes the axon hillock to threshold, the neuron generates an action potential, which will propagate down its axon. EPSP, excitatory postsynaptic potential.

(A) Sea hare (*Aplysia*)

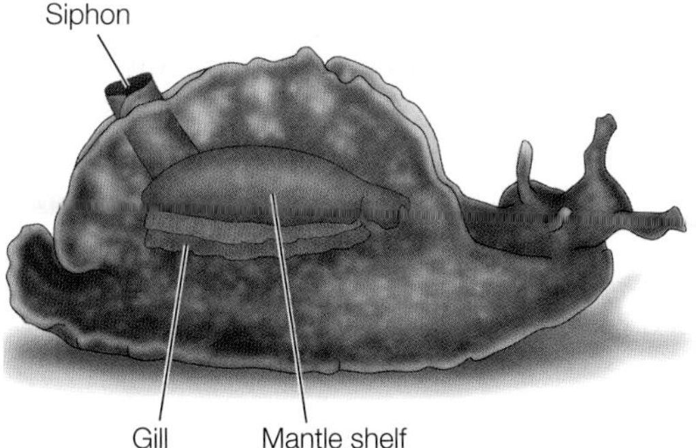

(B) The human brain, showing the position of the hippocampus

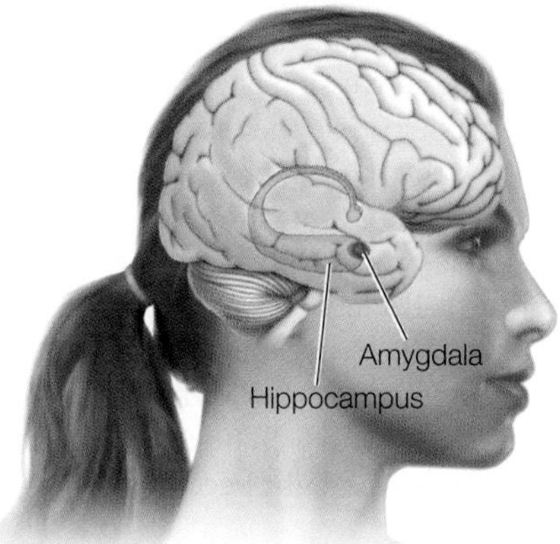

FIGURE 31.22 Two Cases Where Researchers Have Documented Synaptic Plasticity (A after E. R. Kandel. 1976. *Cellular Basis of Behavior*. Freeman: San Francisco.)

individual's synapses, so that future experiences are then processed by the nervous system in altered ways, leading to altered behaviors.

LINK Some of the many ways animal behavior is modified by learning and early experiences are described in **Key Concept 37.2**.

Two model examples of synaptic plasticity illustrate its mechanisms and importance. Sea hares (marine mollusks in the genus *Aplysia*) pull their gills inside when certain parts of their body are touched (**FIGURE 31.22A**). The sea hares learn from past experience how vigorously to do this. They withdraw their gills more vigorously if they have previously been exposed to a very noxious agent—a type of learning called sensitization. Researchers have shown that sensitization occurs because of synaptic plasticity in synapses between the sensory neurons that detect touching and the motor neurons that stimulate the gill-withdrawal muscles. When a sea hare is subjected to a noxious stimulus, these synapses become altered so that the amount of neurotransmitter released per presynaptic signal is increased. After this change, the postsynaptic cell in a synapse is excited to a greater degree by an identical presynaptic signal.

Studies of the mammalian hippocampus (**FIGURE 31.22B**) provide a vivid example of synaptic plasticity in which synapses change not only in function but also in physical characteristics. The hippocampus, positioned deep within the brain, is involved in spatial learning and memory formation. Studies have been carried out on living hippocampal tissue slices isolated from fully mature mice. Certain neuronal circuits in these slices can be stimulated while neighboring circuits are not. Investigators have shown that when a particular circuit is repeatedly stimulated, certain of the postsynaptic structures in that circuit physically grow. Along with this physical response, the postsynaptic structures also change functionally: the number of active receptor molecules in each postsynaptic cell increases. Consequently, the postsynaptic cell responds more strongly when it receives synaptic input. If you imagine some information being processed by circuits that are identical except that some have received a lot of prior stimulation and others have not, it is clear that the previously stimulated circuits will process the information differently, providing a basis for both learning and memory.

REVIEW & APPLY | 31.3

R A synapse is a cell-to-cell contact point specialized for signal transmission from one cell to another. Most synapses are chemical synapses (with neurotransmitters). Some are electrical synapses. At a chemical synapse, when an action potential reaches the axon terminal of the presynaptic neuron, neurotransmitter is released, diffuses across the synaptic cleft, and binds to receptors in the cell membrane of the postsynaptic cell, causing changes in the postsynaptic membrane potential. Within the nervous system, individual neurons may receive synaptic inputs from hundreds or thousands of other neurons. Each incoming signal briefly changes the membrane potential of the postsynaptic neuron by just a small amount in a way that is either excitatory or inhibitory. The postsynaptic neuron integrates information by summing graded postsynaptic potentials, producing action potentials based on the integrated effect (sum) of all the graded potentials.

Synaptic plasticity is the process by which synapses in the nervous system of an individual animal undergo long-term changes in their functional properties. Such changes are important in learning and memory.

A

1. At a vertebrate neuromuscular junction, what would happen if a poison blocked the mechanism that removes acetylcholine from the synapse after it has had its usual effect?
2. Consider the arrival of a single excitatory presynaptic signal at a single chemical synapse. What is the difference between the magnitude of the postsynaptic graded potential at a vertebrate neuromuscular junction and at a neuron-to-neuron synapse, and why does this difference matter?
3. What is meant by synaptic plasticity, and speaking generally, how can it be a mechanism of learning?

We have seen how neurons use ion pumps and ion channels to generate action potentials, and how action potentials trigger neurotransmitter release, allowing neurons to communicate with each other and with other cells. Next we'll look at how systems of neurons receive sensory information.

31.4 Sensory Processes Provide Information on an Animal's External Environment and Internal Status

LEARNING OBJECTIVES

By the end of this key concept you should be able to:

31.4.1 **State the meaning and significance of the point that "sensory receptor cells transform stimuli into electric signals."**

31.4.2 **Explain how hearing is achieved by use of mechanoreceptors.**

31.4.3 **Order the steps in signal processing within the vertebrate eye and explain how this affects the signals the eye sends to the brain.**

Animals require information on their external environments to function successfully. For example, they need information to locate food, find mates, and avoid danger. Animals have evolved an enormous variety of sensory processes that enable them to gather information on their external environments. Imagine, for example, a male moth, an insect-eating bat, and a migrating bird moving across the landscape on a moonless, pitch-black night in late summer. The moth might be able to detect a female of his species, hundreds of meters away, by using sense organs that are extremely sensitive to odors that females release into the air. The bat might try to zero in on him as he flies toward the female by listening to sounds (echoes) reflected off his body. Meanwhile the bird migrating overhead might use its eyes to locate a key star, which it can use as a navigational aid to fly south (see Key Concept 37.2).

Sensory processes are equally important for providing animals with information on their internal status. For instance, internal sensory processes provide animals with information on the level of O_2 in their blood and the amount of tension produced by their muscles during contraction. The information on muscle tension can be used to adjust stimulation of the muscles by the nervous system so that the tension is neither too high nor too low.

As we stressed in our initial overview of the nervous system (see Figure 31.1), sensory information is carried to the central nervous system by action potentials traveling in sensory neurons. After processing of this information by the central nervous system, responses are activated by action potentials transmitted via other neurons to muscles or other tissues. An informative example is provided by our response to a sudden pain in one of our fingers (see Figure 31.3).

Sensory receptor cells transform stimuli into electric signals

Sensory processes depend on **sensory receptor cells**, which are cells, typically neurons, specialized to transform the energy of a stimulus—called *stimulus energy*—into an electric signal. In physics, **transduction** refers to the conversion of energy from one form to another. Biologists, by analogy, use the term *transduction* (or *sensory transduction*) to refer to the process by which a sensory receptor cell produces an electric signal from stimulus energy.

The initial electric signal that is produced by a sensory receptor cell is a graded change in the cell's membrane potential. This phenomenon is illustrated by rod cells—specialized neurons that detect light—in the retina of the vertebrate eye. When light strikes a rod cell, the membrane potential of the cell changes in a graded way: the change becomes greater as the light becomes brighter (**FIGURE 31.23**).

The next step in signal transduction is that the graded response in the sensory receptor cell—directly or indirectly—causes action

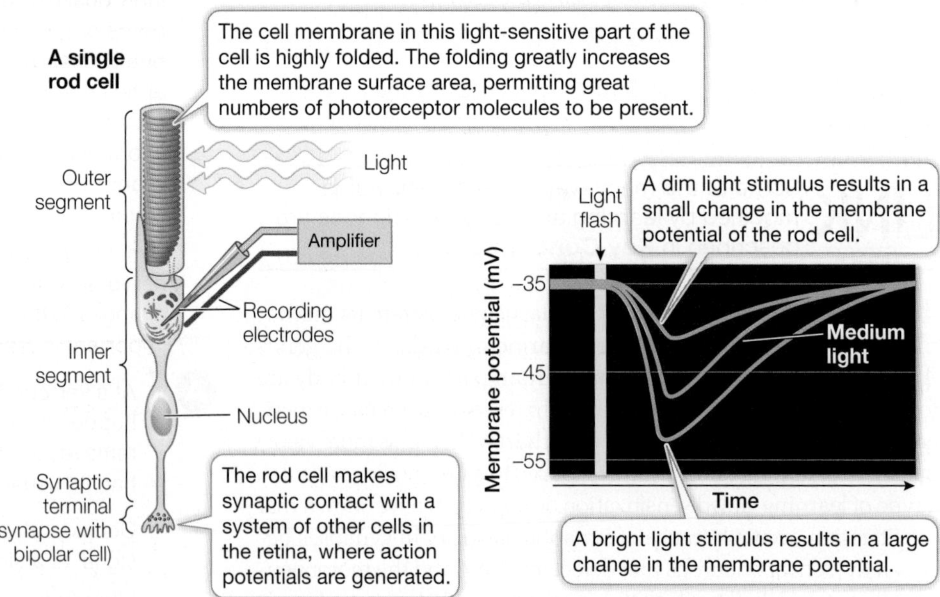

FIGURE 31.23 Signal Transduction in a Rod Cell in the Vertebrate Eye A rod cell produces a graded change in membrane potential in response to light. Elsewhere in the retina, action potentials are produced as a consequence, and the action potentials propagate to the brain, providing for sensory perception of light in the environment. Although the graded change in membrane potential in a rod cell is a hyperpolarization, the graded changes in membrane potential produced by most sensory receptor cells are depolarizations. (After D. A. Baylor et al. 1979. *J Physiol* 288: 589–611.)

potentials to be generated in a neuron or neurons. The action potentials—which encode the initial stimulus—convey the sensory information to the brain or other integrative parts of the nervous system for processing and interpretation. In this way, the animal acquires a sensory perception of its external environment and its internal status.

What do we mean when we speak of "the energy of a stimulus" or "stimulus energy"? Some stimuli are simply forms of energy. Light, heat, and mechanical motion are examples. But what do we mean when we speak of stimulus energy when the stimulus is a chemical compound (e.g., a sweet-tasting molecule)? In this case, the stimulus energy is not the bond energy of the molecule (the energy that would be released by burning or respiration). Instead, in the case of a chemical stimulus, when a stimulus molecule binds with a protein during sensory reception, the protein's conformation (molecular shape) changes, and the energy involved in this process represents the stimulus energy.

Sensory receptor cells depend on specific receptor proteins

A **sensory receptor protein** is a membrane protein in a sensory receptor cell that initially detects a stimulus and directly or indirectly produces a graded change in the receptor cell's membrane potential. This graded change in membrane potential (such as that seen in Figure 31.23) is termed a **receptor potential**. There are many types of sensory receptor proteins, and they respond to different kinds of stimuli. For example, some sensory receptor proteins respond to light, as in rod cells. Others respond to particular chemicals. Still others respond to stretch. A receptor cell is specific in the sensory receptor protein it expresses, helping explain why receptor cells are highly specific in the stimuli to which they respond.

Receptor cells often have modified cell membranes with exceptionally large surface areas, as the rod cell (see Figure 31.23) illustrates. The cell membrane may be greatly enlarged by the presence of microvilli, cilia, or membrane foldings. The enlarged cell membrane enables a receptor cell to have great numbers of molecules of the cell's receptor protein, making the cell highly sensitive to stimuli.

Sensory receptor cells may be *ionotropic* or *metabotropic*. Although we did not call attention to this distinction earlier, it is also true that in a chemical synapse, the neurotransmitter receptor protein in the postsynaptic cell can be either ionotropic or metabotropic. Here we are speaking just about sensory receptor cells.

An **ionotropic receptor cell** typically has a receptor protein that is a stimulus-gated Na^+ channel (**FIGURE 31.24A**). What happens during stimulus reception? First, when the stimulus acts on the receptor protein, the protein changes conformation so that it operates as an open Na^+ channel. Second, Na^+ diffuses in through the open channel, creating a receptor potential: a graded change in the cell membrane potential. Third, the graded change of membrane potential directly or indirectly results in action potentials. These steps can occur very rapidly. Na^+ plays a key role in this rapid response because (as we earlier discussed) it is far out of electrochemical equilibrium across cell membranes, not only being more concentrated outside than inside but also being attracted to the negatively charged inner side of the membranes.

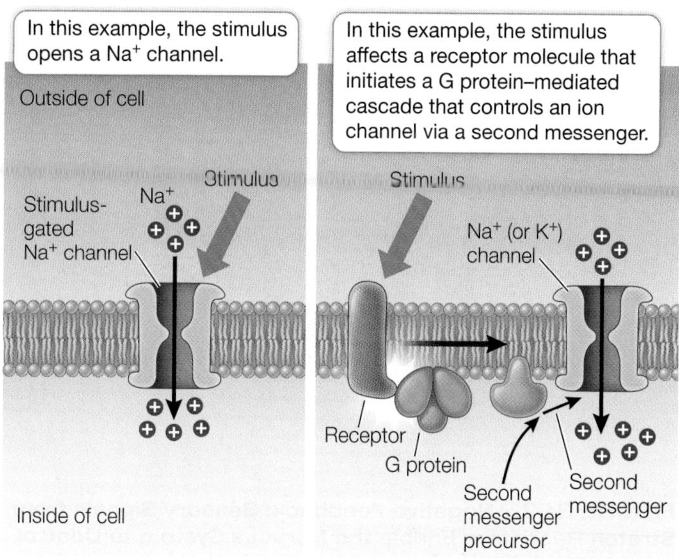

FIGURE 31.24 Sensory Receptor Cells Are Ionotropic or Metabotropic Membrane events in an ionotropic receptor cell **(A)** and a metabotropic receptor cell **(B)**. A metabotropic cell that affects an ion channel is illustrated here, but metabotropic cells do not always work in this way.

A **metabotropic receptor cell** (such as a rod cell) typically has a receptor protein that activates a G protein when exposed to the stimulus (**FIGURE 31.24B**). In most cases, the receptor is itself a G protein-coupled receptor. Activation of the G protein typically leads to a change in the concentration of a second messenger in the receptor cell. This change can directly or indirectly alter ion fluxes across the cell membrane, producing a receptor potential. The receptor cell itself may then generate action potentials, or it may secrete a neurotransmitter that affects another cell that produces action potentials.

The structure and properties of G protein-coupled receptors and the cellular responses they activate are described in **Key Concepts 6.2** and **6.3**.

Sensation depends on which neurons in the brain receive action potentials from sensory cells

All sensory systems convey information to the brain or other integrative parts of the nervous system in the form of action potentials. Accordingly, the brain receives action potentials from all types of sensory systems. How, then, do we perceive some of these action potentials as representing light, others as representing smells, and still others as representing sounds? How do we decode the action potentials?

Our perception of different action potentials depends in part on where they are delivered in the brain. In a highly specific way, axons from different sensory systems travel to different brain regions. In vertebrates, for example, axons from the eyes travel to a specific brain region called the visual cortex. In turn, action potentials

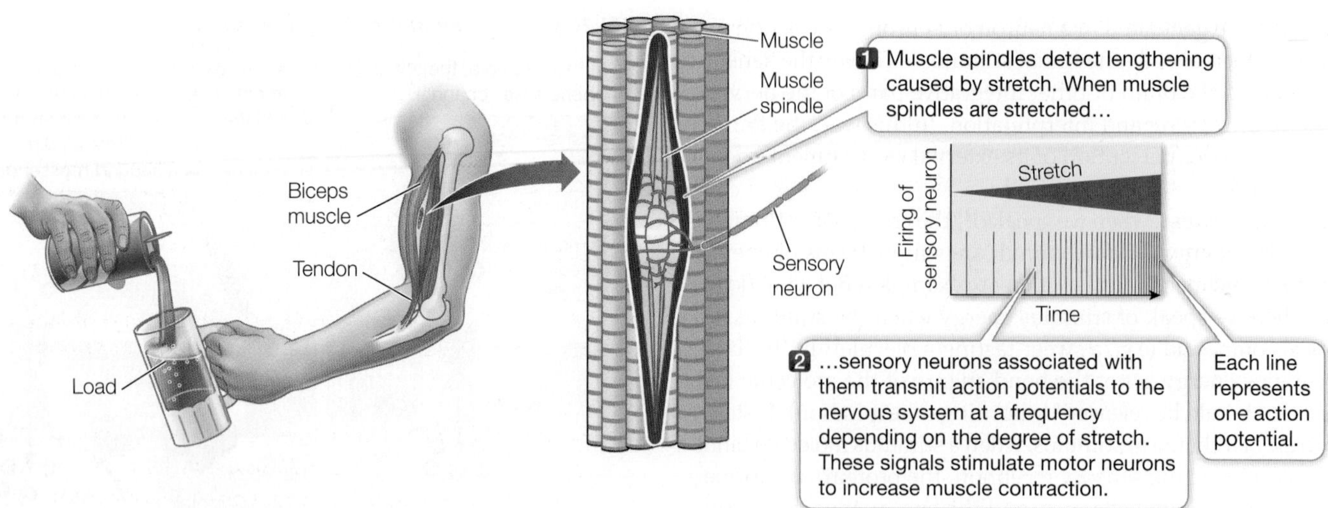

FIGURE 31.25 Negative Feedback: Sensory Signals from Stretch Receptors Enable the Nervous System to Control Muscle Contraction In vertebrates, tiny sensory organs called muscle spindles send information to the nervous system about how much a muscle is being lengthened by stretching. This information is used here to keep limb position stable by increasing the contractile force as the downward force increases. The process represents negative feedback because if the arm starts to be pulled down by the load, the process opposes such movement and keeps the arm in the desired position.

arriving in the visual cortex are interpreted as representing light or other visual sensations. Axons from the inner ear travel to another brain region called the auditory cortex, where arriving action potentials are interpreted as sounds. The intensity of the sensations (e.g., the brightness of a light or loudness of a sound) is coded by the frequencies of the action potentials (number per millisecond).

The importance of these principles is demonstrated when action potentials are generated by the wrong stimulus. People sometimes see flashes of light when they are poked in the eye. A strong poke can produce action potentials in axons from the retina that go to the visual cortex. Because the action potentials arrive in the visual cortex, the brain interprets them as representing light.

The parts of the brain that receive particular types of sensory information are determined during development, at least in part by interaction with the axons conveying the information. In a person blind from birth, for example, the visual cortex does not simply occupy space in the brain, unused, for life. Instead, much of the brain region is taken over to serve other sensory functions.

Sensations of stretch and smell exemplify ionotropic and metabotropic reception

Animals commonly have stretch receptor cells associated with their muscles. These receptor cells are types of **mechanoreceptors**: cells that respond specifically to mechanical distortion of their cell membrane. When a muscle contracts, its stretching forces are applied to its stretch receptor cells, which provide the nervous system with information on the effectiveness of the contraction. Mechanoreceptors are typically ionotropic, as are **thermoreceptors**, which detect heat and cold. Certain receptors involved in taste are also ionotropic.

In some cases, single, specialized neurons act as stretch receptors. A stretch receptor cell of a crayfish is of this type and provides a model of how an ionotropic receptor cell generates action potentials. Dendrites of the receptor cell attach to a structure, such as a muscle cell, where stretch is measured. When the dendrites are stretched, stretch-gated Na^+ channels in the cell membrane (see Figure 31.11B) open, producing a graded, depolarizing receptor potential, which spreads, helping trigger action potentials at the axon hillock. The action potentials then propagate along the axon of the cell to the brain.

Activity 31.3 Mechanoreceptor Simulation

Some stretch receptor cells are parts of tiny organs called *muscle spindles*, located within skeletal muscles. Stretch receptors of this type in our biceps help control the muscle's strength of contraction to match the load the muscle must sustain (**FIGURE 31.25**). For example, suppose someone pours a drink into a glass you're holding in your hand. Your biceps muscle tends to be stretched to a longer length as the weight of the glass increases. Receptors in the muscle detect this effect and encode it in the frequency of action potentials. Your nervous system then responds to the signals by stimulating stronger contraction as the glass fills, so that the glass does not pull your arm downward.

Now let's look at an example of metabotropic reception, our sense of smell. Chemicals detected by smell are called *odorants*, and the sense of smell is called **olfaction**. The sensory cells for odorants, called olfactory receptor cells, are specialized neurons. They exemplify **chemoreceptors**: cells that respond to the presence or absence of specific chemicals. Mammalian olfactory receptor cells illustrate the workings of metabotropic sensory cells (see Figure 31.24B). They also illustrate in a spectacular way the importance of receptor protein diversity and specificity.

The olfactory receptor cells of a mammal are found in the olfactory epithelium, a part of the lining of the internal nasal cavity. One end of each receptor cell projects into the mucus that covers the surface of this epithelium. The other end of the cell is an axon that extends into a part of the brain called the olfactory bulb (see Figure 31.38). Odorants must dissolve in the liquid of the mucus to be detected.

The receptor protein of a mammalian olfactory receptor cell is a G protein-coupled receptor found in the part of the cell that projects into the mucus. When an odorant molecule binds to the receptor, the receptor activates a G protein. The G protein in turn activates an enzyme that causes an increase of a second messenger in the cytoplasm. The second messenger opens ion channels in the cell membrane, causing a graded depolarization, which in turn causes action potentials to be generated in the cell. The action potentials propagate into the brain by way of the cell axon.

One of the amazing aspects of olfactory receptor cells is the number of different G protein-coupled receptor proteins (GPCRs) they can express. Each cell expresses just one type of GPCR and can detect only the odorant molecules that bind to that GPCR. However, the genome codes for many options. About 1,000 different GPCRs are found in mice, for example. Collectively, the tens of thousands of olfactory receptor cells—each with one type of GPCR—can detect a very wide range of odorants. Each specific type of GPCR typically can bind to a range of odorants. However, different types of GPCRs have different sets of odorants they bind. The combination of olfactory cells that are stimulated by any particular odorant is unique to that odorant. Consequently, the brain can interpret the pattern of signals from the olfactory cells as pointing to a particular smell.

Some animals have olfactory receptor cells of extraordinary specificity. Because these cells effectively respond to only one ordinary odorant, they can be extremely sensitive to it. The feathery antennae of some male moths (**FIGURE 31.26**) have huge numbers of odorant receptor cells that respond in a highly specific way to pheromone molecules emitted by the females of their species (see Key Concept 37.3). These moths can detect a female when her pheromone molecules represent only 0.000000000000001 percent of the molecules in the air!

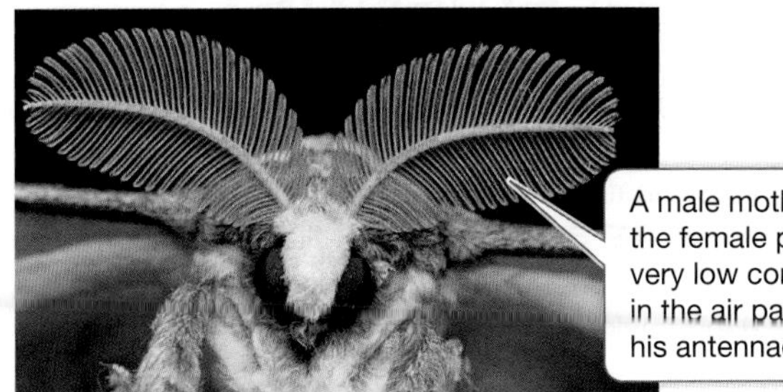

FIGURE 31.26 A Male Moth Detects a Female of His Species Using Chemoreceptor Cells in His Antennae

Auditory systems use mechanoreceptors to sense sound pressure waves

Sounds consist of waves of compression of the air. An easy way to understand the nature of sound is to envision a loudspeaker. A close look reveals that the membrane of a loudspeaker moves back and forth. When the membrane moves toward you, it compresses the air, creating a region of relatively high pressure that moves toward you like a wave (at the speed of sound). When the speaker membrane moves away from you, it creates a region of relatively low pressure that also moves toward you between the regions of high pressure. You thus experience a *sound pressure wave* of alternating high and low pressure. The frequency of these alternations accounts for tone or pitch. We perceive high frequencies as high-pitch (treble) tones and low frequencies as low-pitch (bass) tones.

The hearing organs of many different types of animals are based on a single principle. The hearing organ contains a thin membrane that is free to move when a sound pressure wave strikes it—moving in when high pressure strikes and out when low pressure strikes. Specialized mechanoreceptors are linked to the membrane and detect the movements.

The hearing organs of moths (and other insects) provide graphic examples of this basic principle. A moth has two hearing organs, positioned on opposite sides of its body. Each hearing organ consists of a thin membrane to which two mechanoreceptor neurons are attached. The neurons produce action potentials, which propagate to the brain. As simple as these hearing organs may be, they can save the lives of some moths that are being hunted by insect-eating bats, because the bats—hunting in the dark of night—emit ultrasonic calls for echolocation (see the opening photo in Chapter 13). Some species of moths can hear the ultrasonic calls, and having thus been forewarned of attack, undertake evasive maneuvers. Recent research reveals that the responses of certain species of moths are even more elaborate (**FIGURE 31.27**).

The hearing organs of terrestrial vertebrates are far more complex and provide a greatly enhanced ability to detect and discriminate many different frequencies. Here we discuss the ears of mammals, using the human ear as our example. When we listen to someone playing the piano, our brain receives specific information on each frequency in the music, allowing us to perceive its full tonal complexity.

In the ear of a mammal, the membrane that initially responds to sound pressure waves is the tympanic membrane (eardrum). On the inside of the tympanic membrane is an air-filled cavity called the middle ear (**FIGURE 31.28A**). Because the tympanic membrane has air on both sides, it is free to vibrate when a sound pressure wave strikes it. Three interconnected, delicate bones—called ossicles and named the malleus, incus, and stapes—are found in the middle ear (**FIGURE 31.28B**). They act as a lever system that turns vibrations of the tympanic membrane into more forceful vibrations of another flexible membrane called the oval window.

The oval window is part of the inner ear. Specifically, it is part of the **cochlea**, a coiled, fluid-filled tube where sound energy is transduced into electric signals. A cross section of the cochlea reveals that it is composed of three parallel canals separated by two membranes. Of these the **basilar membrane** is of particular importance (**FIGURE 31.28C**).

The basilar membrane gradually changes in width and stiffness from one end to the other. At the end nearest the oval window, it is only about 0.04 millimeter wide in the adult human ear and relatively stiff. It becomes progressively wider and more flexible until, at its far end, it is 0.5 millimeter wide and not as stiff.

When the sound pressure wave is transmitted into the cochlear fluid at the oval window, different frequencies in the sound

THINK LIKE A SCIENTIST

Do some moths jam bats' echolocation mechanism?

Insect-eating bats hunt moths using echolocation (**FIGURE 31.27**). As a bat flies, it emits pulses of ultrasonic sound, which can strike a flying moth and echo back to the bat, informing the bat of the moth's location—reminiscent of the use of sonar to find submarines during naval warfare. Just as in submarine warfare, the potential victim can use the sound pulses as a warning. Some moths can hear a bat's sound pulses and quickly change their flight path to evade capture.

The tiger moth *Bertholdia trigona*—an abundant nocturnal species in the American West—is distinctive in that it is believed to employ "jamming" to interfere with bats' echolocation mechanism. That is, when a moth of this species hears the sounds of an echolocating bat, it is believed to avoid capture in two ways. First, it evades the bat by changing its flight path. Second, it is believed to jam the bat's echolocation system.

Regarding the jamming concept, researchers in the 1990s discovered that *Bertholdia trigona* can produce bursts of ultrasonic sound at relatively high intensity. Moreover, laboratory measurements and theoretical analyses indicated that this species might use these sound bursts to disrupt echolocation. Specifically, researchers postulated that if a moth is being pursued by a bat and if the moth emits ultrasonic pulses so they arrive in the bat's ears in synchrony with the bat's echoes, the moth's sounds might prevent the bat from carrying out an accurate calculation of the moth's position.

In the process of science, however, laboratory measurements and theoretical analyses lead to *hypotheses*, not to *conclusions*, regarding the question of how animals function in their natural environments. Hypotheses must be tested before they can be accepted.

FIGURE 31.27 A Cave Myotis Bat (*Myotis velifer*) Hunting a Tiger Moth (*Bertholdia trigona*)

Aaron Corcoran and William Conner wanted to conduct a direct test of the jamming hypothesis on moths and bats flying freely in their natural habitat. Making full use of modern technology, they could directly observe nighttime encounters between bats and moths, and they carried out a manipulative experiment by testing two categories of *Bertholdia trigona* moths: (1) *normal moths* with their normal ability to produce sound and (2) *soundless moths* that could not produce sound because a hole had been made in a key part of each of the two sound-producing organs.

Corcoran and Conner observed 44 cases in which a bat approached a normal moth. In only 3 cases (see the table), however, did the bat make physical contact with the moth. In the other 41 cases, the bat "missed": it made no physical contact.

Corcoran and Conner observed 38 cases in which a bat approached a soundless moth. Of these moths, 31 were contacted by the bat and a majority of those were captured and eaten. Only 7 of the soundless moths avoided contact with the bat.

Type of moth	Number approached by a bat	Number with which the bat made contact (meaning moth might be eaten)	Number not contacted by the bat (meaning moth escaped)
Normal	44	3	41
Soundless	38	31	7

Source: A. J. Corcoran and W. E. Conner. 2012. *J Exp Biol* 215: 4278–4287.

1. A manipulative experiment is one in which investigators deliberately modify a feature of the system they are studying, rather than simply describing it. What insight did Corcoran and Conner obtain that they would not have obtained without manipulation?
2. Design a manipulative experiment of your own, either on the bat–moth system or on another system of your choosing. State why it is manipulative.
3. A common type of manipulative experiment is to eliminate or block a function and observe the consequences. Speaking in principle, how might this type of experiment introduce confounding factors?
4. Is there a statistically significant difference between the normal and soundless moths in Corcoran and Conner's experiment? Use a chi-square test or G-test to answer.
5. Consider the two mechanisms used by the moths to avoid capture: evasion by flight and jamming of the bat's echolocation system. Based on the data, which mechanism seems to be more effective? Explain.

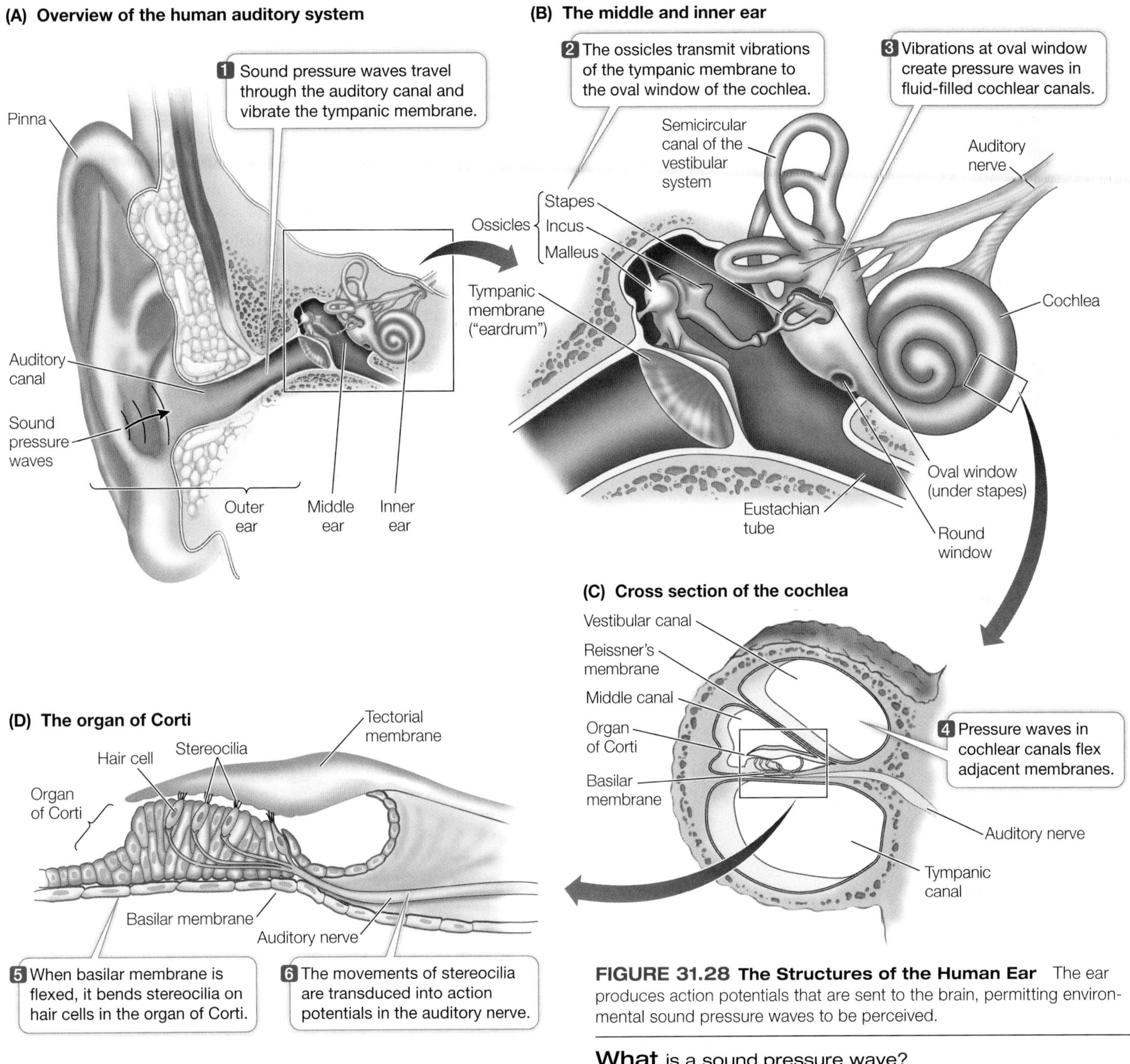

FIGURE 31.28 The Structures of the Human Ear The ear produces action potentials that are sent to the brain, permitting environmental sound pressure waves to be perceived.

What is a sound pressure wave?

Activity 31.4 Structures of the Human Ear

pressure wave cause different, local regions of the basilar membrane to oscillate. Different regions move in response to different frequencies because of the gradations of membrane width and stiffness. Low-frequency vibrations (representing low pitches) have their greatest effect on the wide end of the membrane farthest from the oval window, setting that part into motion to a far greater degree than other parts (**FIGURE 31.29A**). Other frequencies affect other parts of the membrane (**FIGURE 31.29B**).

How is information on these oscillations sent to the brain? Sitting on the basilar membrane is a structure, the organ of Corti, that houses distinctive mechanoreceptor cells called hair cells (**FIGURE 31.28D**). Each hair cell bears extensions called stereocilia (which are actually microvilli, not cilia) that project into a semirigid membrane, the tectorial membrane. When a particular part of the basilar membrane oscillates, the stereocilia of the hair cells in that region are bent because they are effectively anchored at one end in the tectorial membrane but moving at the other end.

The hair cells transduce the bending motions of their stereocilia into electric signals. They do not produce action potentials. Instead, they synapse with neurons, and when they are stimulated by the

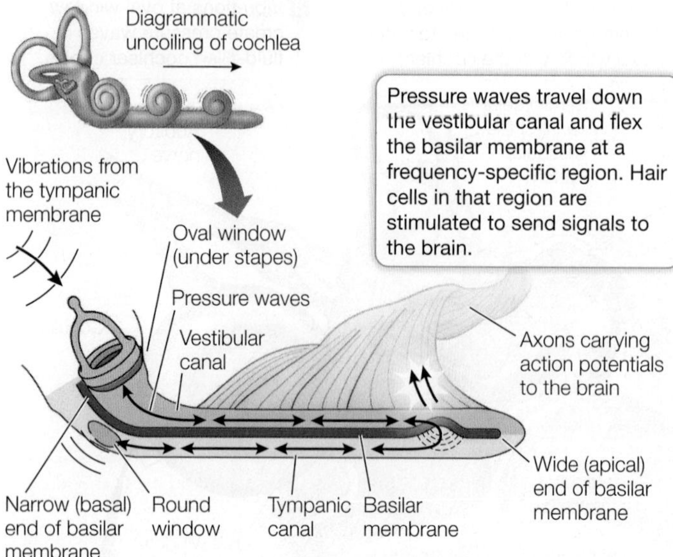

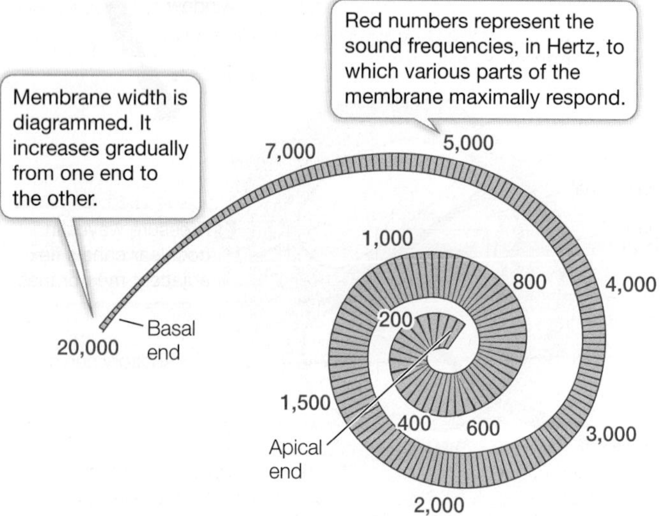

FIGURE 31.29 Frequencies in Sound Pressure Waves Are Transduced into Signals from Distinctive Sets of Hair Cells **(A)** Each frequency causes oscillations at a particular region of the basilar membrane. Here low-frequency sound pressure waves are shown flexing a region of the basilar membrane far from the oval window. **(B)** Different regions of the basilar membrane move in response to different frequencies because of the gradations in the membrane's width and stiffness. Hair cells positioned along the basilar membrane produce signals that lead to generation of action potentials that encode the regions that are oscillating.

Animation 31.4 Sound Transduction in the Human Ear

bending of their stereocilia, they release a neurotransmitter that excites the neurons to produce action potentials, which travel to the brain. In this way, the frequency composition of a sound pressure wave is encoded by the particular hair cells that trigger action potentials as the wave exerts its effect in the inner ear (see Figure 31.29A).

The photoreceptors involved in vision detect light using opsin molecules such as rhodopsin

Both simple and complex animals have visual systems by which they sense and respond to light. The simplest visual systems give animals the ability distinguish light from dark. Complex visual systems provide animals with detailed images of their environment. Remarkably, the basic process of detecting light in visual systems has been conserved across the entire animal kingdom. **Photoreceptors** are sensory receptor cells that are sensitive to light. In all photoreceptors employed in vision, the sensory receptor protein is a member of a family of closely similar membrane pigments, sometimes called collectively the *visual opsins*. In the rod cells of vertebrate eyes (which we introduced above and will discuss in more detail shortly), the specific visual opsin present is **rhodopsin**. (Sometimes "rhodopsin" is used as a general term that refers to all visual opsins.) A visual opsin molecule consists of two parts: (1) a protein (opsin) conjugated with (2) a nonprotein structure, 11-*cis*-retinal, that absorbs light. The retinal part of each molecule is derived from vitamin A, explaining why sufficient vitamin A is essential for proper vision (see Key Concept 28.1).

Detection of light depends on the ability of visual opsins to absorb photons of light and undergo a change in molecular conformation (shape) as a consequence. Exposed to light, 11-*cis*-retinal undergoes a change in shape (**FIGURE 31.30**), placing strains on bonds in the opsin part of the molecule. In turn the conformation of the opsin changes, activating a G protein-mediated signal transduction cascade in which a second messenger exerts effects on Na^+ channels in the photoreceptor cell membrane. This metabotropic control has a huge potential advantage in that it can greatly amplify signals. With large numbers of visual opsin molecules and strong amplification, some photoreceptor cells undergo

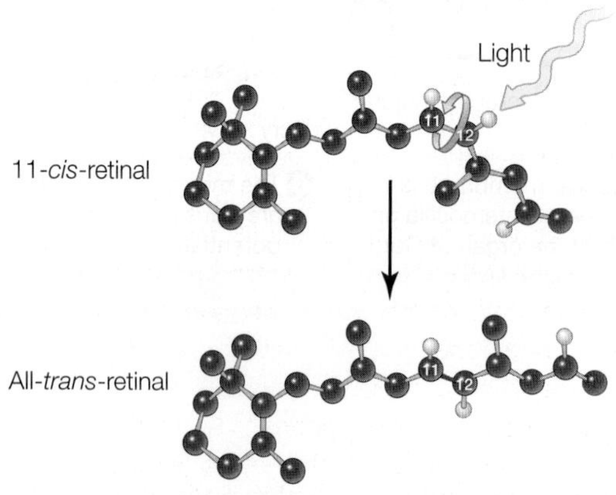

FIGURE 31.30 The Molecular Conformation of 11-*cis*-Retinal Changes When It Absorbs Light Rhodopsin and all the other visual opsins contain 11-*cis*-retinal as a molecular subpart (they differ in their protein—opsin—subparts). When the 11-*cis*-retinal in a molecule of visual opsin absorbs light, it undergoes a twist around one of its bonds, so its atoms become reoriented in space to the all-*trans*-retinal isomer. This process changes the molecular conformation of the entire visual opsin molecule.

Animation 31.5 Photosensitivity

a measurable change in membrane potential in response to just a single photon of light!

The vertebrate retina is a developmental outgrowth of the brain and consists of specialized neurons

Vertebrates have image-forming eyes. Like a camera, an image-forming eye has a lens that focuses images on an internal surface that is sensitive to light (**FIGURE 31.31A**). The eye is a fluid-filled structure enclosed in a tough layer of connective tissue. At the front, the connective tissue forms the transparent **cornea**. Light enters the light-sensing part of the eye through the **pupil**, the central opening of a circle of colored tissue, the iris. The iris is equipped with small muscles that enlarge the diameter of the pupil in dim light and reduce it in bright light. Behind the iris is the **lens**, made of crystal-clear proteins called crystallins. In mammals, muscle cells pulling on the lens produce fine adjustments in its shape, focusing images on the photosensitive layer—the **retina**—at the back of the eye.

Activity 31.5 Structure of the Human Eye

The retina is a developmental outgrowth of the brain and consists of specialized neurons of several types. Two types, the **rod** cells and **cone** cells, are the photoreceptor cells responsible for vision (**FIGURES 31.31B** and **31.32A**). Both rods and cones are specialized to have a large membrane area and large numbers of rhodopsin molecules (rods) or other visual opsin molecules (cones). The protein (opsin) subpart varies from one specific type of visual opsin to another, affecting wavelength sensitivity. Rods outnumber cones in the human retina: 100 million rods versus 5 million cones. The rods are far more sensitive to light than the cones and are especially important for vision in dim light. Cones provide high-acuity color vision. Humans and some other primates have three types of cones, which are distinguished by having three different molecular forms of visual opsin. These forms differ from one another in their sensitivities to various wavelengths of light, and all three differ in this respect from the rhodopsin in rod cells (**FIGURE 31.32B**). As a group, the cone cells are able to detect and encode color because each wavelength of light has a unique collective effect on the three types.

Rod and cone cells do not produce action potentials. They only produce graded membrane potentials (see Figure 31.23). Compared with other sensory receptor cells, they are unusual in that they hyperpolarize in response to their stimulus. How does this work? When in darkness, they have *open* Na^+ channels in their cell membranes. Na^+ ions entering by way of these channels depolarize the cells to some extent. When the cells are stimulated by light, the second messenger produced under G protein control *closes* Na^+ channels. This closing lowers the Na^+ current flowing into the cell, permitting the charge on the inside of the cell membrane to become more negative relative to the charge on the outside. This effect is graded, depending on the light intensity (again see Figure 31.23). The rod and cone cells synapse with other neurons in the retina and relay signals to them by way of graded neurotransmitter release.

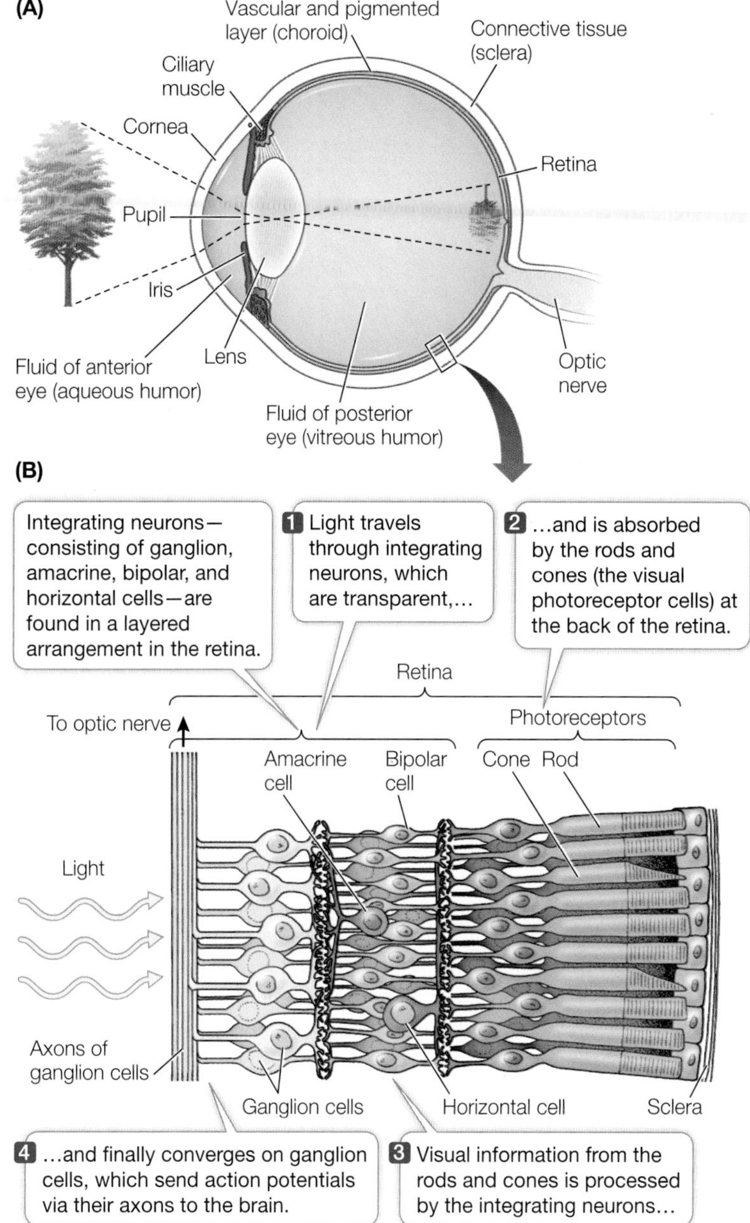

FIGURE 31.31 The Human Eye and Retina **(A)** A sagittal (side view) section of the human eye. **(B)** The retina is found at the back of the eye. It consists of the visual photoreceptor cells (rods and cones) and four types of integrating neurons arranged in layers.

Why are the rods and cones positioned at the back of the retina, away from the light entering the eye?

Activity 31.6 Structure of the Human Retina

Media Clip 31.2 Into the Eye

An enormous amount of signal processing occurs in the vertebrate retina before visual signals are relayed to the brain by way of axons in the optic nerve. In addition to the visual photoreceptor cells (rods and cones), four types of visual integrating neurons—called bipolar cells, horizontal cells, amacrine cells, and ganglion

(A) Rods and cones in the retina

(B) The absorption spectra of rod cells and three types of cone cells

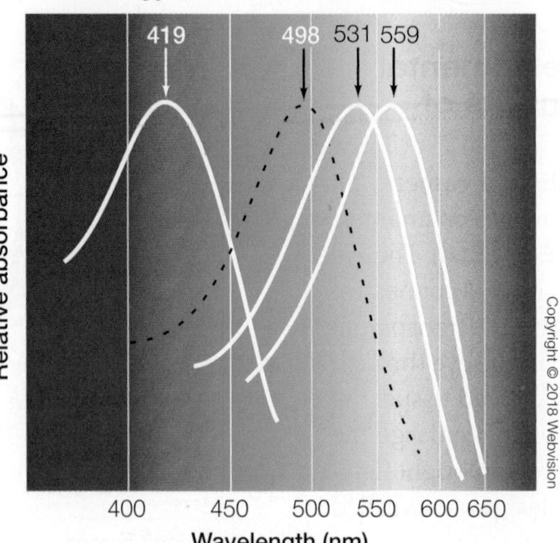

FIGURE 31.32 Rods and Cones **(A)** A scanning electron micrograph of photoreceptor cells in the retina of a mud puppy (an amphibian) showing cylindrical rods and tapered cones. The cells are artificially colored to distinguish them. **(B)** The color spectrum of light, with the wavelengths plotted along the *x* axis. Each of the lines superimposed on the spectrum shows, in relative terms, how the absorption of light by a particular pigment varies with the wavelength of light. The dashed line shows the absorption spectrum of human rod cells. The three solid lines show the absorption spectra of the three types of cone cells in the human retina. The numbers at the top are the wavelengths (nanometers) at which the four pigments exhibit maximum absorption.

cells—are found in the retina in a layered arrangement (see Figure 31.31B; see also Figure 28.1B). A striking feature of the arrangement of these cells is that light must pass through the integrating cells (which are transparent) to stimulate the rods and cones! This unexpected arrangement is a by-product of the embryonic developmental path by which the eyes form from the brain.

The photoreceptor cells (rods and cones) synapse with bipolar cells, which in turn synapse with ganglion cells (see Figure 31.31B). In contrast to the cells in these straight-through connections from the back to the front of the retina, the horizontal cells and amacrine cells communicate laterally within the retina and in this way are able to extract information on relative signal strengths in neighboring photoreceptor cells and neighboring bipolar cells. Horizontal cells help sharpen the perception of contrast between light and dark. Amacrine cells have several functions, including detecting motion. Ultimately, all of this information converges onto the ganglion cells, which are the only cells that produce action potentials. Axons of these cells converge to form the optic nerve and carry the action potentials to the visual cortex in the brain.

As a consequence of the great amount of signal processing within the retina, the ganglion cells send a great deal more information to the brain than simply the raw outputs of the rods and cones. The ganglion cells provide the brain with information about *patterns* present in the field of vision, such as spots, edges, and areas of contrast.

The huge numbers of cells involved in vision say a lot about the amount of processing of visual images that takes place in the retina and brain. Each human eye sends a million axons to the brain. The visual cortex consists of hundreds of millions of neurons that receive and process the information from those million axons.

Some retinal ganglion cells are photoreceptive and interact with the circadian clock

One of the most revolutionary recent discoveries about the mammalian eye is that 1–2 percent of the retinal ganglion cells are photoreceptive. Most ganglion cells (the other 98–99 percent) cannot directly detect light and function only to help integrate visual information that originates in the rods and cones.

The photoreceptive ganglion cells are not employed in vision. Instead, these ganglion cells provide information simply on whether light is present and how bright it is. This information is used in two ways. It is relayed to the suprachiasmatic nuclei, where it is used to entrain the master circadian biological clock to the environmental rhythm of day and night (see Key Concept 29.4). The information is also employed in regulating the size of the pupils of the eyes, so the pupils are large in dim light and small in bright light.

Arthropods have compound eyes

Invertebrates have evolved a variety of types of eyes that differ from vertebrate eyes. Even the eyes of octopuses and squid, which look very similar to vertebrate eyes on the outside, have very different properties than vertebrate eyes.

The eyes of arthropods (insects, crayfishes, crabs, and lobsters) are called **compound eyes** because each eye consists of many optical units, called **ommatidia** (singular *ommatidium*), each with its own lens and photoreceptor cells (containing visual opsins) that send axons to the brain. Light entering an ommatidium through its lens strikes the ommatidium's photoreceptor cells, which generate action potentials (**FIGURE 31.33**). The individual ommatidia in a compound eye are directed at slightly different parts of the visual world. The brain compiles an image of the world using information from all the ommatidia. The more ommatidia in the eye, the higher the resolution of the image, rather like the effect of pixel density in a computer monitor. The number of ommatidia in a compound eye varies from only a few in some ants to 800 in fruit flies and to 30,000 in some fast-flying predators such as dragonflies.

Animals have evolved a remarkable diversity of sensory abilities

Some animals have evolved distinctive sensory abilities that let them make use of information sources that we humans cannot use. The philosopher Bertrand Russell titled one of his most famous books *Our Knowledge of the External World.* In the book (first published in 1914) he stressed that, as humans, we can only directly

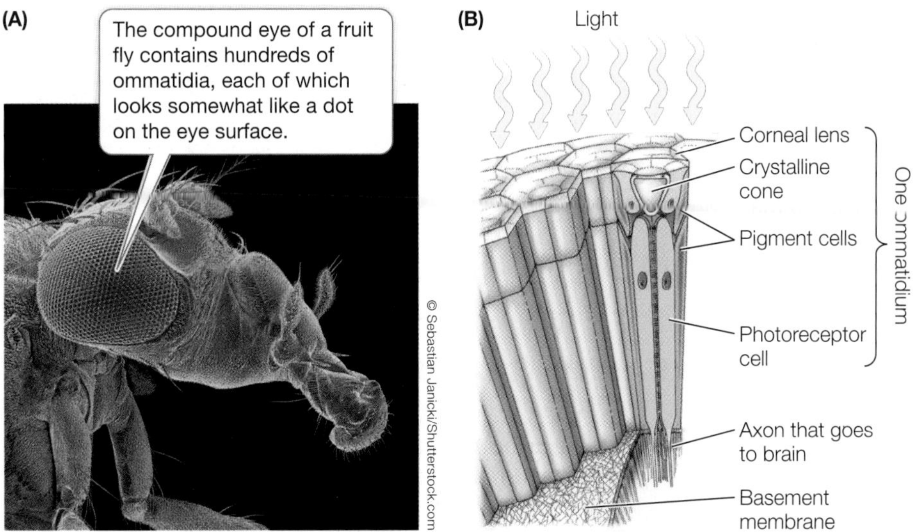

FIGURE 31.33 An Insect Eye and Ommatidia **(A)** A scanning electron micrograph of the compound eye of a fruit fly. **(B)** Several ommatidia in a compound eye. Light enters each ommatidium independently and is detected by visual-opsin-containing photoreceptor cells in the ommatidium. (B after M. Friedrich. 2006. *Arthropod Struct Dev* 35: 357–378.)

know about those aspects of our external environment that our human sensory systems detect. With this fact in mind, we realize that some animals that live around us perceive our external environment in very different ways than we do. Some can see electromagnetic wavelengths to which we are blind, or (like the bats and moths we have discussed) hear sound frequencies to which we are deaf.

The electromagnetic spectrum consists of electromagnetic energy at a broad range of wavelengths from 0 to over 100 micrometers (μm) (see Figure 5.16). Our eyes can see the wavelengths between 0.4 and 0.7 μm (400–700 nm). We have names for the parts of the spectrum we cannot see. The wavelengths that are too long for our eyes to see (i.e., longer than 0.7 μm) are called **infrared wavelengths**. The wavelengths too short for our eyes to see (i.e., shorter than 0.4 μm) are called **ultraviolet wavelengths**.

Some snakes obtain visionlike images of their surroundings using infrared wavelengths. They are therefore able to see patterns of warmth and coolness in darkness—an ability that helps them find prey. The best-understood snakes with this ability are the pit vipers, such as rattlesnakes, which have specialized sense organs in pits near their eyes (**FIGURE 31.34A**). The organ in each pit focuses infrared electromagnetic energy onto a membrane that can detect it, much like our eyes focus visible light onto our retina. The mechanism that pit vipers use to detect infrared wavelengths does not employ visual opsins and is not entirely understood.

At the other end of the electromagnetic spectrum, many animals have eyes that can see ultraviolet wavelengths to which we are blind. Honey bees, for example, see flowers in different ways than we see them because they see in the ultraviolet. Many flowers that look uniform in color to our eyes have bold color patterns at the ultraviolet wavelengths that bees see (**FIGURE 31.34B**). Many species of birds also see ultraviolet wavelengths, using specialized cone cells. Cedar Waxwings (*Bombycilla cedrorum*), found almost everywhere in North America, illustrate one potential advantage of ultraviolet vision. Male and female waxwings look the same to us. But we see only at 0.4–0.7 μm. When we use instruments to learn what the waxwings look like at ultraviolet wavelengths, we find that males and females look different. Thus the birds themselves can tell males and females apart by sight even though we cannot.

Some animals can sense electric fields—fields of voltage and current—enabling them to perceive their surroundings using

(A) Pit vipers can detect infrared wavelengths

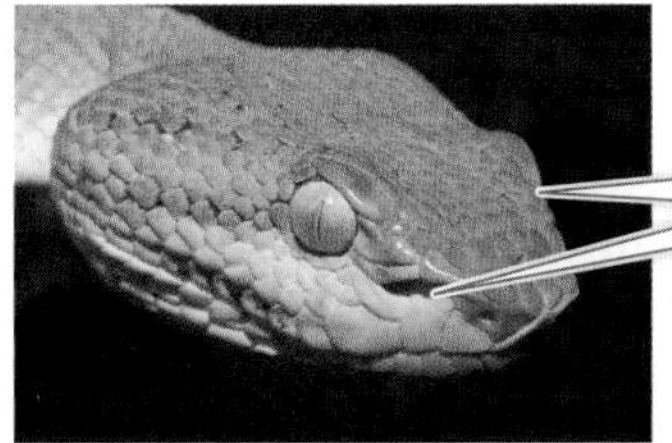

(B) Bees can detect ultraviolet light

(C) Electric fishes use electric fields to detect objects

FIGURE 31.34 Diversity in Sensory Abilities Three examples of distinctive sensory abilities in animals.

information that's entirely foreign to us. The weakly electric fishes provide examples. They emit weak electric fields into their surroundings by use of modified muscles (see Key Concept 33.3). The electric currents in these self-produced environmental fields pass through the outer skin layers of the fishes, where sensory receptor cells detect them. Objects in the environment distort the electric fields and currents, and the fishes can detect these distortions. They thus can perceive objects even in complete darkness. Fishes in the genus *Gnathonemus* use this sensory ability to locate and capture stream insects at night (**FIGURE 31.34C**).

REVIEW & APPLY | 31.4

R Sensory receptor cells transduce information about an animal's external and internal environment, directly or indirectly, into action potentials. The cells have sensory receptor proteins that respond to sensory input, causing a graded change of membrane potential called a receptor potential. An ionotropic receptor cell typically has a receptor protein that is a stimulus-gated Na^+ channel. A metabotropic receptor cell typically has a receptor protein that activates a G protein. Action potentials sent to the brain from sensory receptor cells or organs are interpreted as particular sensations based on which neurons in the brain receive them.

Chemoreceptors respond to the presence or absence of specific chemicals. Mechanoreceptors respond to mechanical distortion of their cell membrane. Hearing relies on mechanoreceptors in the hearing organs that respond to sound pressure waves. In the mammalian auditory system, sound pressure waves are relayed to the fluid-filled cochlea, where the basilar membrane vibrates—producing action potentials in hair cells—at specific locations that depend on the wave frequency. Photoreceptors are sensory receptor cells that contain visual opsins (e.g., rhodopsin) that are sensitive to light. In vertebrates, extensive processing of visual information occurs in the retina of the eye before action potentials are sent to the brain. Arthropods have compound eyes consisting of many optical units called ommatidia, each with its own lens. Collectively, animals have an enormous diversity of sensory receptors, enabling some types of animals to perceive the external environment in very different ways than humans do.

A

1. Suppose that as a mosquito lands on your arm, a cell in the mosquito's leg responds to altered stretch within the leg. Another cell in its antennae detects the CO_2 you exhale. What types of sensory receptor proteins do these cells probably contain?
2. Explain how the structure of the vertebrate retina makes it possible for visual signals to be processed extensively in the retina itself.
3. How is the structure of the vertebrate cochlea related to the range of sound frequencies that it processes?

Having discussed the cellular processes of neurons—from action potentials to receptor function—let's now return to our examination of how neurons are organized to form the nervous systems of animals.

31.5 | Neurons Are Organized into Nervous Systems

LEARNING OBJECTIVES

By the end of this key concept you should be able to:

31.5.1 **Describe how nervous systems have often been highly influenced in the course of evolution by trends toward centralization and cephalization.**

31.5.2 **Differentiate between the autonomic nervous system and the rest of the nervous system.**

31.5.3 **Outline what happens after a physician taps the front of your knee with a rubber hammer.**

A nervous system consists of many neurons, interacting with each other by way of synapses. When we say "many," we mean a huge number. The human brain is currently estimated to contain about 86 billion neurons!

The simplest nervous systems in animals alive today are seen in anemones, corals, jellyfishes, and other cnidarians. In these animals the individual neurons are widely and relatively randomly dispersed throughout the body. Together they form a *nerve net* (**FIGURE 31.35A**). Biologists generally assume that when nervous systems first evolved they were similar to nerve nets.

As nervous systems evolved further, they followed two major trends, *centralization* and *cephalization*. Some neurons became specialized to integrate the signals of other neurons. Most of these integrating neurons became clustered together in centralized integrating organs, rather than being distributed randomly throughout the body. **Centralization** refers to this evolutionary trend toward clustering. At the same time, the major integrating areas became concentrated toward the anterior end of the animal's body, a trend called **cephalization** (from a Greek word for "head"). Mammals vividly exemplify the outcomes of these evolutionary trends. Most of the integrating neurons of a mammal are centralized in the central nervous system—the brain and spinal cord (see Figure 31.1)—and the major integrating organ—the brain—is at the anterior end of the body in the head. Why has cephalization occurred? Most kinds of animals have particularly important sets of sense organs—such as eyes, hearing organs, and sensory tentacles—at their anterior end. The evolutionary advantage of a centralized anterior processing center is easy to understand when you consider that as an animal moves forward in its environment, its anterior end is the first to make contact with new dangers or opportunities. Concentration of the integrating parts of the nervous system at the anterior end facilitates the processing of signals from the concentration of sense organs there.

Here we focus on the groups of animals with high degrees of centralization and cephalization. In all such animals, the nervous system consists of central and peripheral systems. We have already noted the existence of these two parts in humans (see Figure 31.1). However, we need general definitions of the two parts that we can apply to all types of animals, not just humans or other vertebrates, because arthropods, mollusks, and other invertebrates also have central and peripheral nervous systems. Speaking generally, the central nervous system (CNS) consists of relatively large structures (containing great numbers of neurons) that are composed principally of integrating neurons and associated glial cells. The **brain** is the largest part of the CNS—the part containing the greatest number of neurons. The peripheral nervous system (PNS) consists of all the neurons and subparts of neurons (e.g., axons) that are located outside the CNS. In this definition, we need to mention "subparts of neurons" because there are individual neurons that are partly positioned in the CNS and partly in the PNS, such as motor neurons with their cell bodies in the CNS but with axons extending into the PNS.

The CNS functions as a control system, and like all control systems it needs to interact with sensors and effectors to function (see Figure 28.18. The *sensors* are the sensory cells and sense organs, just discussed, that provide the CNS with information on the animal's external environment and internal status. The *effectors* are cells or tissues that "carry out orders" issued by the control system. Muscle cells, for example, are effectors that produce motion under orders from the CNS. Glands are effectors that secrete materials in response to signals from the nervous system. Other effectors include specialized structures such as fish electric organs, which in some species produce large voltages or currents used in the capture of prey.

Neurons are classified as interneurons, sensory neurons, and efferent (including motor) neurons. Again, although we have already discussed this topic in a limited way, we need a completely general overview that applies to all types of animals and neurons, not just the cases discussed earlier in this chapter:

- **Interneurons** are neurons that are confined to the CNS—they do not physically extend into the PNS. For the most part, they engage in integrative, command, and information-storage functions. Most neurons in the CNS are interneurons.
- **Sensory neurons** (also called afferent neurons) carry signals to the CNS from sensory cells or organs.
- **Efferent neurons** convey signals from the CNS to effectors, including muscles, glands, and others. As we've noted before, those that convey signals to the skeletal muscles are called motor neurons.

The cellular components of nervous systems—the neurons, synapses, and sensory cells—are remarkably similar in the fundamental ways they function in all groups of animals. However, different groups of animals can differ substantially in the ways their neurons are organized to form their nervous systems. Let's briefly compare vertebrates and arthropods to see this point.

In vertebrates, the central nervous system is positioned in the dorsal part of the body (**FIGURE 31.35B**). Moreover, both major parts of the CNS, the brain and spinal cord, are to some degree hollow inside. In short, the vertebrate CNS is dorsal and hollow. The neurons and glial cells in the spinal cord form a continuous column of tissue, in the sense that the spinal cord does not exhibit sharply differentiated subparts along its length.

(A) Cnidarians have a nerve net

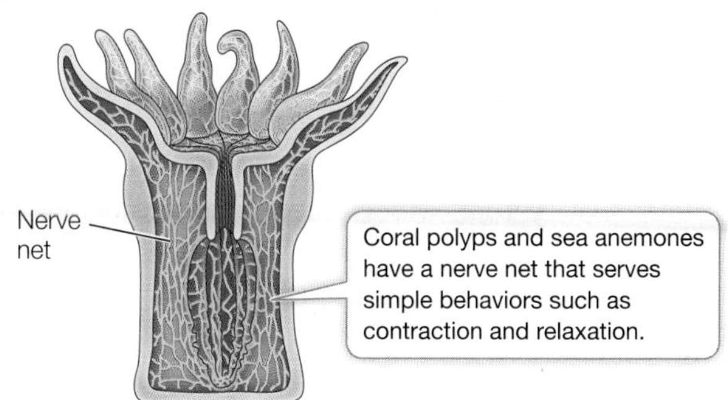

(B) In vertebrates the central nervous system is dorsal

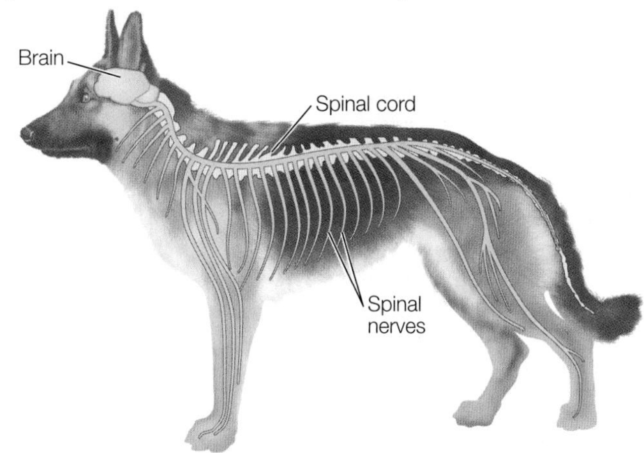

(C) In arthropods the central nervous system is mostly ventral

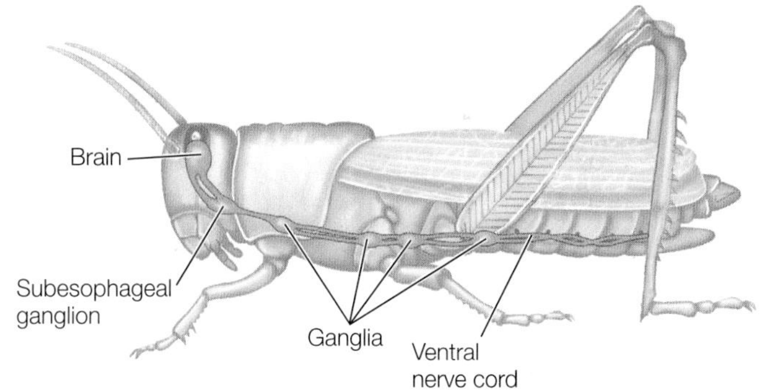

FIGURE 31.35 The Organization of Three Nervous Systems Only a few of the ganglia are shown in (C). (B after W. B. Thomas. 2018. *Merck Veterinary Manual*. Merck & Co., Inc., Kenilworth, NJ.)

In arthropods, such as insects, lobsters, and crayfishes, the CNS is mostly positioned in the ventral part of the body (**FIGURE 31.35C**; see also Figure 30.22). It consists of two major parts: the *brain*, and a *ventral nerve cord* that extends for most of the length of the body behind the brain. These structures are not hollow. In short, the arthropod CNS is mostly ventral and solid. Moreover, instead of being a continuous column of tissue like a vertebrate spinal cord, the ventral nerve cord is composed of a series of alternating subparts. The bodies of arthropods are visibly segmented (see Key Concept 22.1), and in the ventral nerve cord

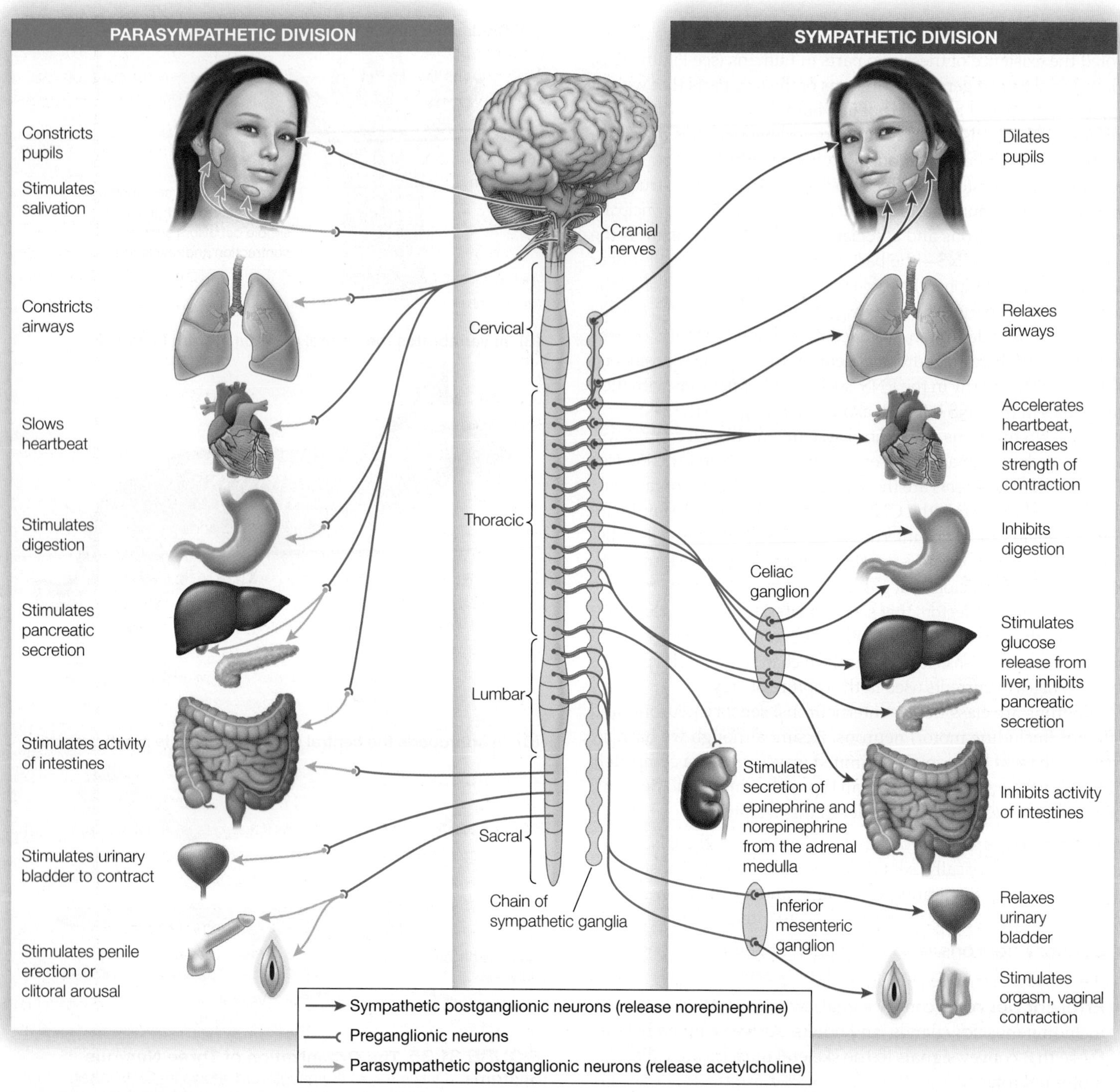

FIGURE 31.36 The Autonomic Nervous System The sympathetic and parasympathetic divisions typically work in opposition to each other in their effects on most organs. Whereas one division stimulates an increase in activity, the other division causes a decrease in activity. The preganglionic neurons of both divisions mostly release acetylcholine.

there is an enlarged structure called a *ganglion* (plural ganglia) in each segment of the animal's body. Two strands of nervous tissue, called connectives and each containing many neurons, run from each ganglion to the next. (Important note: the word "ganglion" is also used in describing vertebrate nervous systems, but it has very different meanings in discussing vertebrates and arthropods, as we'll soon see.)

The autonomic nervous system controls involuntary functions

The nervous system as a whole controls several types of effectors: skeletal muscles (muscles attached to the skeleton), smooth muscles (muscles not associated with the skeleton), glands, and so forth. The **autonomic nervous system** (**ANS**) is the part of the nervous system (both CNS and PNS) that controls effectors other than the skeletal muscles. The ANS exists in all groups of vertebrates. Some invertebrates are also considered to have an ANS.

To see more clearly what the ANS does, let's consider ourselves and other mammals. The ANS of a human or other mammal controls a wide variety of smooth muscles, such as those in the gut, blood vessels, bladder, eyes (e.g., iris muscles), airways of the lungs, and reproductive organs (e.g., during orgasm) (**FIGURE 31.36**). The ANS also controls a variety of exocrine glands (glands that secrete their products elsewhere than into the blood) such as the tear and sweat glands. It also controls a few endocrine glands (glands that secrete hormones into the blood), most notably the adrenal medullary glands. And the ANS controls the pacemaker and other parts of the heart, the acid-secreting cells of the stomach, and several other effectors.

Collectively, the effectors controlled by the ANS are called the **autonomic** (or internal) **effectors**. We often are unaware of their actions. We also have little or no ability to influence them by conscious, voluntary effort. ("Autonomic" means "involuntary" or "unconscious.")

In vertebrates, the ANS has three divisions. The **enteric division** is composed of the nerve cells internal to the gut wall (see Figure 28.20). Here we focus on the **sympathetic** and **parasympathetic divisions** (see Figure 31.36).

We can think of the sympathetic and parasympathetic divisions as consisting mainly of efferent neurons that carry orders from the CNS to autonomic effectors. To understand the neuronal structure of these two divisions, we need to take account of the fact that synapses in a ganglion are always involved. (As we discuss this topic, remember that "ganglion" has different meanings in vertebrates and arthropods.) A vertebrate **ganglion** is a discrete, anatomically clustered set of neuron cell bodies in the peripheral nervous system. Neurons of the ANS exit the CNS and travel to a ganglion. There they synapse directly with neurons that travel from the ganglion to the effector being controlled. The neurons traveling from CNS to ganglion are termed *preganglionic*, whereas those traveling from ganglion to effector are called *postganglionic*. The preganglionic neurons of the *parasympathetic* division exit the CNS from two regions (see Figure 31.36): a cranial region where the brain and spinal cord connect and a sacral region, the most posterior region of the spinal cord. The axons of these preganglionic neurons travel almost to their target cells before synapsing with the postganglionic neurons.

In contrast, the preganglionic neurons of the *sympathetic* division exit the CNS from the thoracic and lumbar (abdominal) regions of the spinal cord (see Figure 31.36). Most of the ganglia of the sympathetic division lie next to the spinal cord, where they are lined up in two chains of ganglia (the "sympathetic chains") on either side. Accordingly, the postganglionic neurons need to cover most of the distance to the target cells.

The most important neuronal difference between the sympathetic and parasympathetic divisions is that their postganglionic neurons employ different neurotransmitters where they synapse with target cells. The sympathetic neurons release norepinephrine as their neurotransmitter. The parasympathetic neurons release acetylcholine (ACh). In organs that receive both sympathetic and parasympathetic input, the target cells typically respond in opposite ways to norepinephrine versus ACh.

Thus the sympathetic and parasympathetic divisions often work in opposition to each other, as illustrated in Figure 31.36. In this way, the two divisions—acting together—can control autonomic effectors in fine-tuned ways, adjusting their functions up or down as needed. For example, both divisions innervate the pacemaker of the heart. Sympathetic stimulation causes the heart rate to increase, and parasympathetic stimulation causes it to decrease.

The *fight-or-flight response* is the most famous effect of the sympathetic division. When the sympathetic division is strongly activated, it greatly increases the heart rate, force of heart contraction, and cardiac output. It dilates the passageways of the lungs and increases release of glucose into the blood from the liver. In all these ways, it prepares the body to meet an emergency. At the same time, it reduces less urgent activities such as digestion.

Spinal reflexes represent a simple type of skeletal muscle control

Having discussed the control of the autonomic effectors, we now need to turn to the skeletal muscles. These are the muscles associated with the skeleton that are responsible for walking, swimming, flying, chewing, and all the other actions that animals undertake that involve movements of the limbs and other parts of the skeleton (see Chapter 33). The skeletal muscles are not controlled by the ANS but instead are under the control of other parts of the CNS and PNS (parts sometimes collectively called the somatic nervous system).

As we all know, we can consciously (voluntarily) control our skeletal muscles. This is possible because neurons in our brain can control signals in the motor neurons to the muscles. Some signals to the skeletal muscles, however, originate by way of **spinal reflexes**, neuron-mediated responses that do not involve participation of the brain. In a spinal reflex all the neuron interactions occur in the spinal cord, requiring no conscious attention. Because spinal reflex systems are relatively simple, they are among the best-known vertebrate neural circuits.

We are all familiar with the spinal reflex known as the knee-jerk reflex, in which a physician or nurse taps the front of your knee with a little hammer and—if your nervous system is healthy—your lower leg jerks forward (**FIGURE 31.37**). The tap stretches the patellar tendon that connects your quadriceps muscle—the muscle on the forward-facing side of your thigh—to your tibia bone in your lower leg. The stretch caused by the tap stimulates stretch receptors in the muscle. These receptors initiate action potentials in sensory neurons, which propagate these action potentials along their axons to the spinal cord via a spinal nerve (see Figures 31.1 and 31.35B for the anatomy of spinal nerves).

In the spinal cord, each sensory axon synapses with a motor neuron that serves the same muscle, the quadriceps. This synapse is excitatory and causes the motor neuron to generate action potentials. These signals travel in the axon of the motor neuron to the quadriceps via the spinal nerve (spinal nerves are composed of both sensory and motor neurons). When the action potentials arrive in the quadriceps, they excite the muscle at neuromuscular junctions (see Figure 31.18). The muscle contracts, and your lower leg kicks forward. This all happens without any sort of conscious control because your brain is not involved.

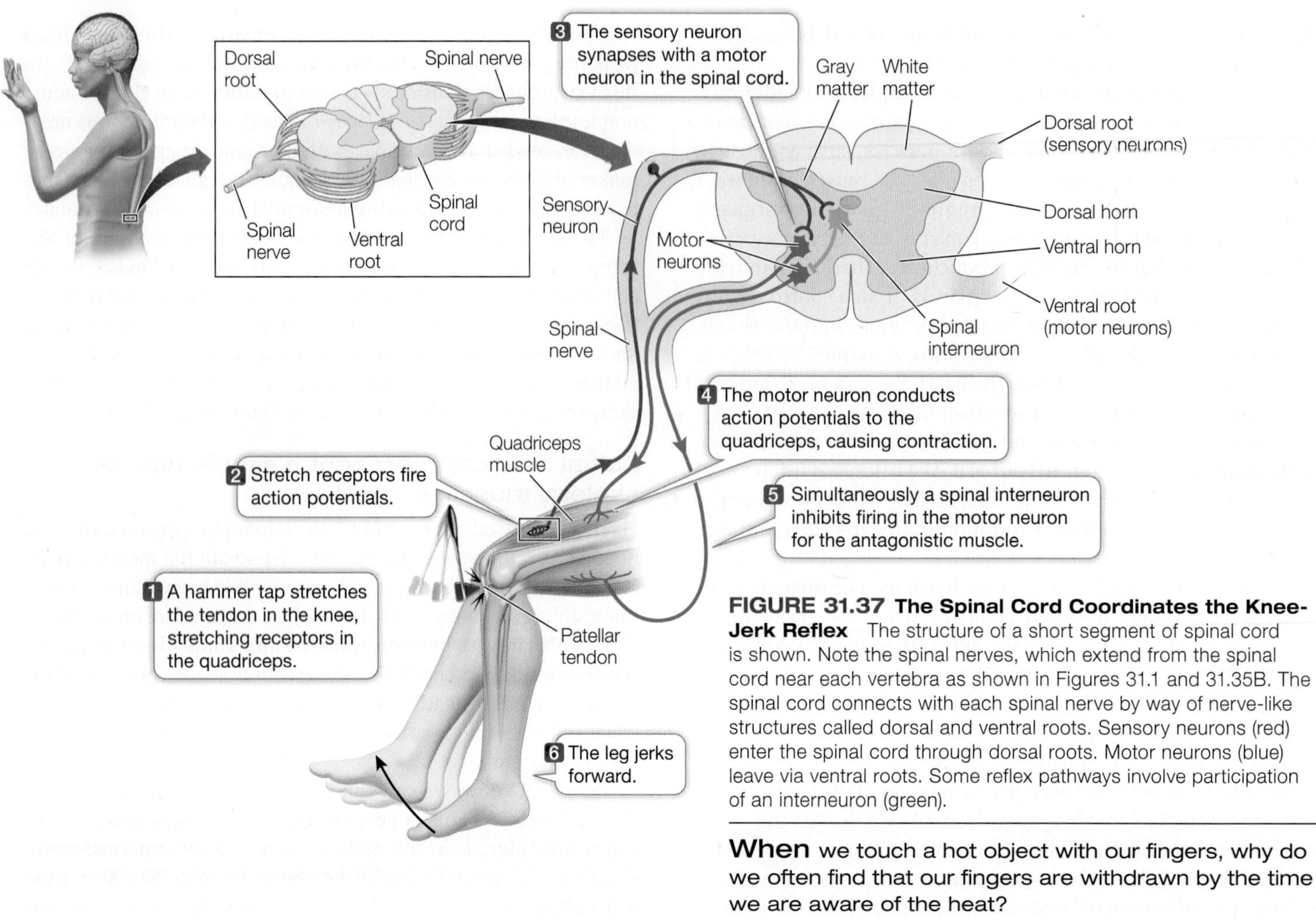

FIGURE 31.37 The Spinal Cord Coordinates the Knee-Jerk Reflex The structure of a short segment of spinal cord is shown. Note the spinal nerves, which extend from the spinal cord near each vertebra as shown in Figures 31.1 and 31.35B. The spinal cord connects with each spinal nerve by way of nerve-like structures called dorsal and ventral roots. Sensory neurons (red) enter the spinal cord through dorsal roots. Motor neurons (blue) leave via ventral roots. Some reflex pathways involve participation of an interneuron (green).

When we touch a hot object with our fingers, why do we often find that our fingers are withdrawn by the time we are aware of the heat?

Animation 31.6 Information Processing at the Spinal Cord

There's more to this story. Limb muscles are organized into opposing (antagonistic) pairs (see Key Concept 33.2). For contraction of the quadriceps to move your lower leg forward easily, the opposing muscle on the back of your thigh needs to relax. These two opposing muscle responses are coordinated by the nervous system. In addition to synapsing on the motor neurons to the quadriceps, the sensory neurons coming from the stretch receptors also synapse onto interneurons in the spinal cord. Those interneurons make inhibitory synapses onto the motor neurons that control the opposing muscle (see Figure 31.37, step 5). The motor neurons are inhibited from stimulating that muscle to contract, allowing your lower leg to move forward when the quadriceps contracts.

Earlier we discussed the rapid way in which we pull our fingers away from a painful stimulus (see Figure 31.3). This process also is controlled by a spinal reflex, explaining why our fingers have already been withdrawn by the time sensory information reaches our brain and we consciously realize our fingers are in pain.

The interactions among cells in a spinal reflex are unusually simple, but they nicely illustrate the interaction of sensory and motor neurons, interneurons, and the PNS and CNS in controlling contractions of skeletal muscles.

The most dramatic changes in vertebrate brain evolution have been in the forebrain

All vertebrate brains consist of three major regions: the forebrain (the most anterior part), midbrain, and hindbrain (the most posterior part). The most posterior part of the hindbrain is the medulla oblongata, and all information traveling between the brain and spinal cord must pass through this structure. During the course of vertebrate brain evolution, the overall shape and relative size of the medulla oblongata have changed little (**FIGURE 31.38**).

In contrast, the cerebral hemispheres, which are a major component of the forebrain, have undergone dramatic changes during vertebrate evolution. Both mammals and birds have evolved especially large cerebral hemispheres (see Figure 31.38). This is noteworthy because the cerebral hemispheres are particularly important in carrying out high-order sensory, motor, and integrative functions. In humans, for example, the cerebral hemispheres are responsible for language and reasoning.

Mammals and birds also have evolved brains that are about ten times larger than the brains of lizards or other non-avian reptiles that are the same in body size. A mouse, for example, has a brain about ten times larger than that of a mouse-sized lizard.

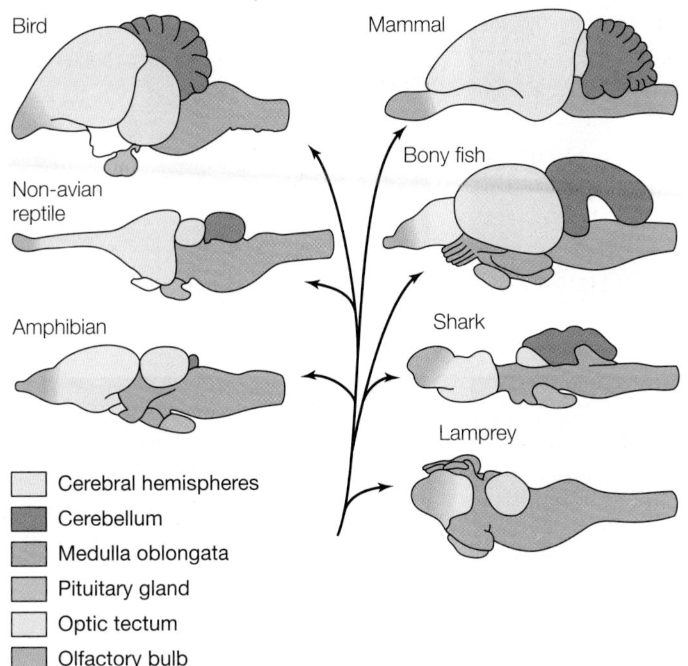

FIGURE 31.38 Evolution of the Vertebrate Brain A reasonably representative brain of a modern species in each group is shown. The optic tectum is involved in processing visual information. "Medulla oblongata," a term used with several variations in exact meaning, is here used in its most general sense. In the mammal, the large cerebral hemispheres keep us from seeing the pituitary, which is found in a position similar to that in other groups. (After C. B. Braun and R. G. Northcutt. 1999. In R. Singer [ed.], *Encyclopedia of Paleontology*, pp. 185–192. Fitzroy Dearborn Publishers: Chicago.)

Evidently the evolution of enhanced functionality in mammals and birds has gone hand in hand with large increases in the numbers of neurons.

We need to be careful, however, not to overextrapolate the significance of this broad trend in brain size. Brain size matters in a general sense, but it does not matter when we look at all particulars. Some animals with small brains exhibit stunning behavioral capacities, as exemplified by certain ants that have sophisticated navigational abilities (see Figure 37.10). Moreover, among human beings—where the subject has been studied for well over a century—all available evidence indicates that individual intelligence is not correlated with individual brain size. We know too little about brain function to explain how all these pieces of the puzzle fit together.

Location specificity is an important property of the mammalian cerebral hemispheres

In humans, the cerebral hemispheres are so enlarged that they cover most other parts of the brain (**FIGURE 31.39A**). The outermost layer of the cerebral hemispheres is the **cerebral cortex**, a thin, important layer rich in cell bodies. The cerebral cortex is only about 4 millimeters thick, but it is folded into ridges (convolutions) that increase its size, as Figure 31.39 shows. The cerebral hemispheres play major roles in sensory perception, learning, memory, and conscious behavior (**FIGURE 31.39B**). Indeed, they provide the neural resources for the great intellectual capacity of humans.

A curious feature of the mammalian nervous system is that the left side of the body is served (in both sensory and motor aspects) mostly by the right side of the brain, and the right side of the body is served by the left side of the brain. Thus sensory input from your right hand goes to your left cerebral hemisphere, and sensory input

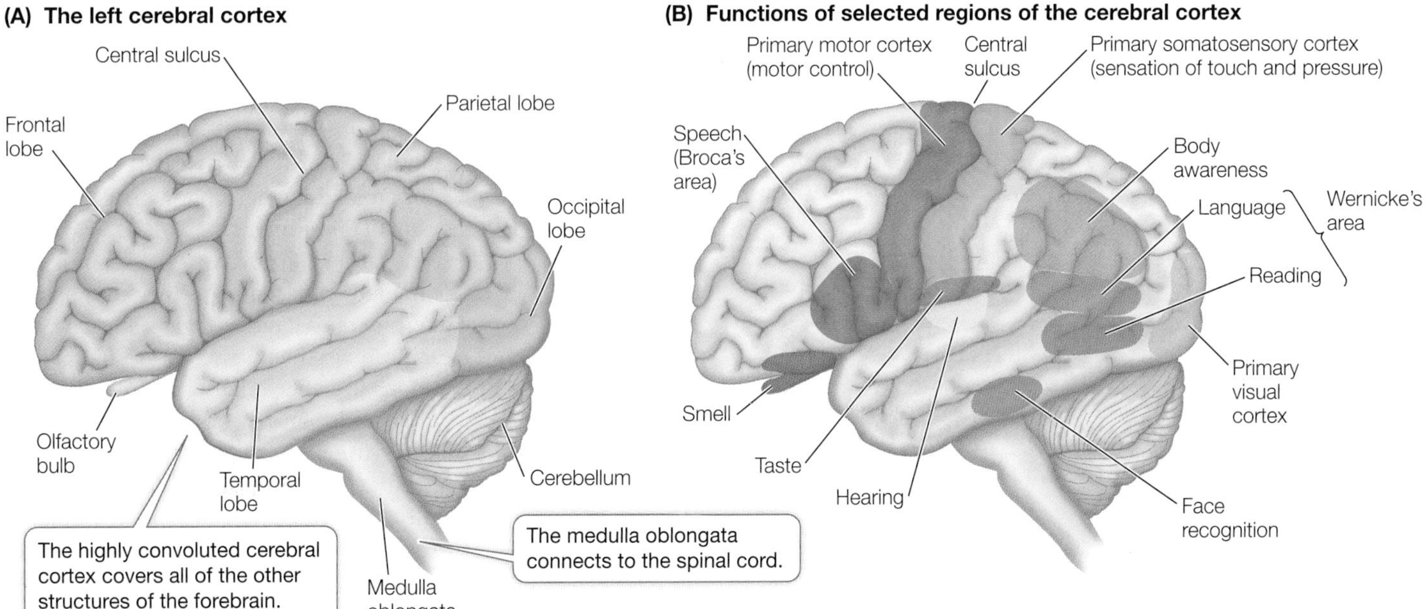

FIGURE 31.39 The Human Left Cerebral Hemisphere **(A)** The major anatomical regions of the cerebral cortex, the outermost layer of the cerebral hemisphere. Almost all the parts shown belong to the cerebral cortex. **(B)** Different areas of the cerebral cortex have specific, localized functions.

Activity 31.7 **Structures of the Human Brain**

FIGURE 31.40 Imaging Techniques Reveal Active Parts of the Brain Positron emission tomography (PET) scanning reveals the brain regions activated by different aspects of language use. Radioactively labeled glucose is given to the subject. Brain areas take up radioactivity in proportion to their metabolic use of glucose. The PET scan visualizes levels of radioactivity in specific brain regions when a particular activity is performed. The orange, yellow, and white areas are the most active. Blue and magenta areas are regions with lower-level activity. (From M. I. Posner and M. E. Raichle. 1994. *Images of Mind*. New York: W. H. Freeman: New York.)

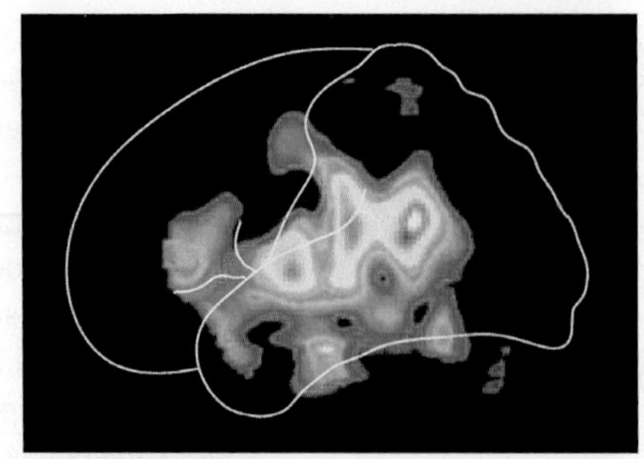

Listening to words

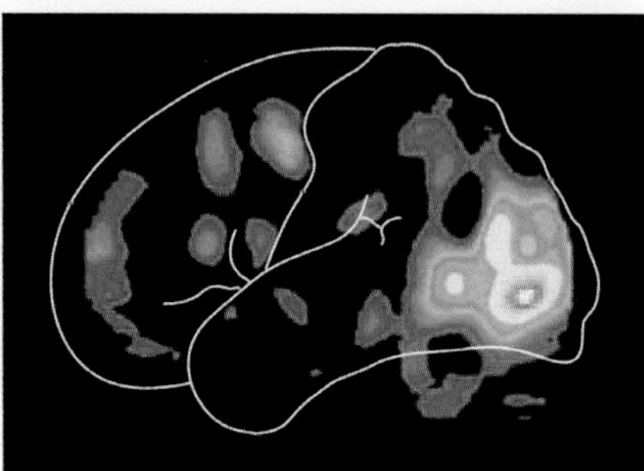

Passively viewing words

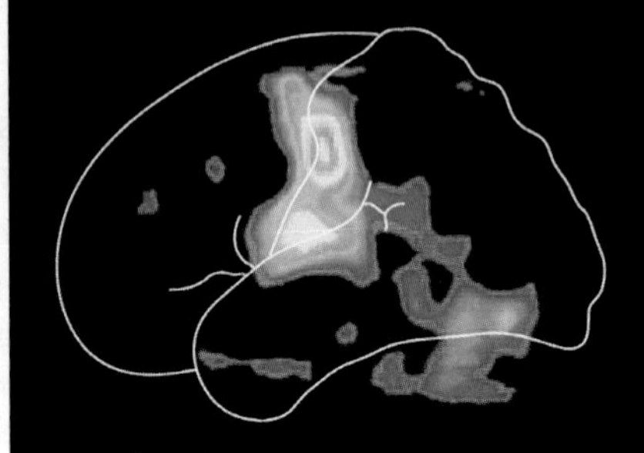

Speaking words

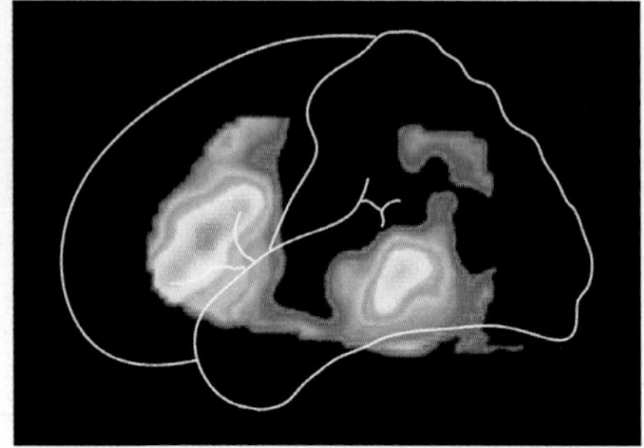

Generating word associations

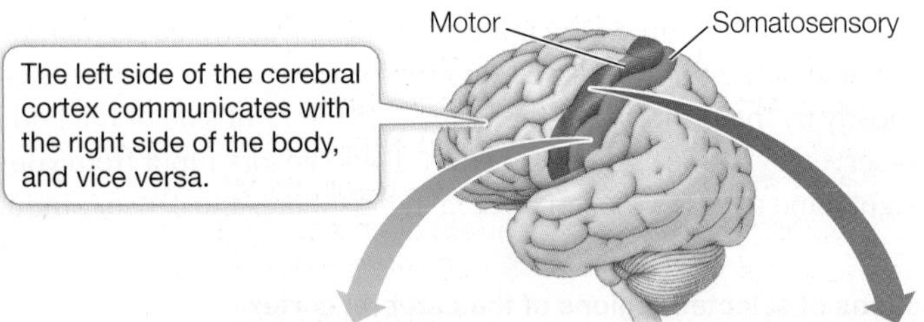

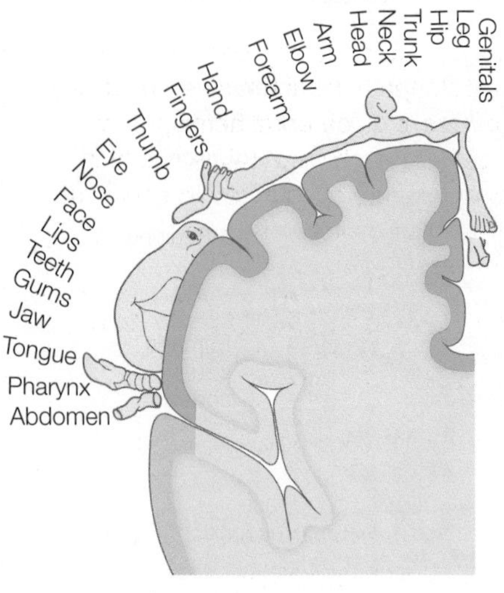

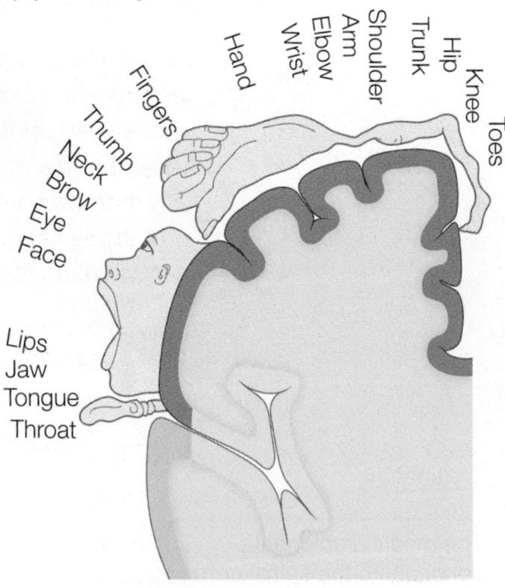

FIGURE 31.41 The Body Is Represented in Maplike Ways in the Primary Somatosensory and Primary Motor Parts of the Cerebral Cortex The amount of cortical tissue devoted to various body parts is not even. More cortical tissue is devoted to some body parts than others, as reflected in the sizes of the parts in the drawings. **(A)** The primary somatosensory cortex. **(B)** The primary motor cortex. (After W. Penfield and T. Rasmussen. 1950. *The Cerebral Cortex of Man: A Clinical Study of Localization of Function*. The Macmillan Company: New York.)

from your left hand goes to your right cerebral hemisphere.

Earlier we emphasized that all action potentials arriving in the brain are the same, even though some represent visual images, others represent sounds, and still others represent odors. Similarly, some action potentials from touch receptors originate in the fingers, others in the feet or abdomen. How does the brain process a simple, invariant electrical event—an action potential—and develop an accurate perception of the type of stimulus (sight, sound, or odor) that caused it, or an accurate perception of the place of origin of the stimulus?

Location specificity in the brain is one of the major mechanisms the brain uses to interpret the action potentials it receives. In each cerebral hemisphere, various specific regions are specialized in adults to carry out specific sensory and motor functions (see Figure 31.39B). Axons from the retina of the eye connect to the specific region called the primary visual cortex. Similarly, axons from the cochlea in the ear connect to a specific brain region responsible for hearing, and axons from the odorant receptor cells in the nasal epithelium connect to the region responsible for smell. Because of these physical connections, action potentials are effectively labeled in terms of what they represent. For example, the brain "knows" that action potentials arriving in the visual cortex provide information on vision because the axons carrying those action potentials come from the eye.

Motor functions are also localized in the brain. Moreover, combined sensory and motor functions often occur in certain, relatively localized brain areas. An explosive area of research at present is the exploration of sensory, motor, and combined sensory and motor localization by use of imaging methods that allow us to visualize where neurons exhibit evidence of increased electrical activity. For instance, the positron emission tomography (PET) images in **FIGURE 31.40** show brain regions where there is evidence of increased neuronal activity when several language functions are carried out. Functional magnetic resonance imaging (fMRI) is another technique used for similar research.

A dramatic and informative example of combined sensory and motor localization is provided by the somatosensory ("body sensing") region of the cerebral cortex and a neighboring motor region (**FIGURE 31.41**). The somatosensory region (blue) processes sensory information from touch and pressure receptors on the body surface. As Figure 31.41A shows, different parts of the body surface are sensed in different parts of the cortex. The relative size of each body part in the drawing reflects the amount of cortical area devoted to that part. The large size of the lips, for example, reflects the fact that a large cortical area is devoted to processing sensory signals from the lips. As Figure 31.41B shows, the parts of the cortex that control motor signals to the body surface (orange) closely parallel the somatosensory parts (blue) in their physical locations.

Localization of function is not limited to the cerebral hemispheres. **FIGURE 31.42**, for example, is an fMRI image of the brain of a person experiencing fear. The image indicates increased neuronal activity in the amygdala, a part of the forebrain (outside the cerebral hemispheres) that is well known to participate in fear responses. The hippocampus (see Figure 31.22B), another part of the forebrain, is critical for spatial learning and memory. Damage to the hippocampus renders mice and rats unable to solve maze problems that depend on learning the relationships of objects in space.

Not all parts of the brain fit easily into the concepts of localized function and mapped brain regions. Thus we need to recognize that these concepts are both important and limited in their ability to fully clarify brain function.

A curious aspect of language processing in humans is that it resides principally in one cerebral hemisphere—which in about 85 percent of people is the left hemisphere. The two hemispheres share responsibilities for other functions. Individuals with damage to the left hemisphere frequently experience deficits in their ability to use or understand words. The deficit depends on the specific part of the left hemisphere that is damaged. Damage to Broca's area, for example, interferes with speech and writing, whereas damage to Wernicke's area interferes with understanding language (see Figure 31.39B).

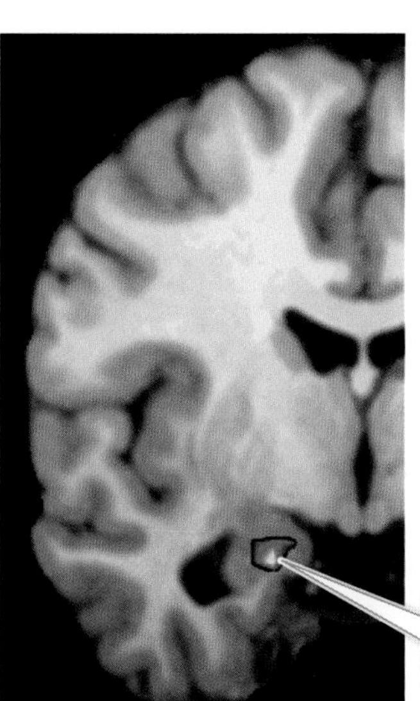

Courtesy of Dr. Kevin LaBar, Duke University

FIGURE 31.42 Source of the Fear Response Frightening situations—or even memories of such situations—activate the amygdala, as shown in this functional magnetic resonance image (fMRI) of the left side of the brain of a person experiencing fear. Figure 31.22B shows the location of the amygdala.

REVIEW & APPLY | 31.5

R Nerve nets (in cnidarians) are the simplest nervous systems. Nervous system evolution followed two major trends: centralization—the clustering of neurons into centralized integrating organs—and cephalization—the concentration of major integrating centers at the anterior end of the animal's body. The brain and spinal cord (in vertebrates) or the brain and ventral nerve cord (in arthropods) make up the central nervous system (CNS). Neurons that extend or reside outside the CNS make up the peripheral nervous system (PNS). In vertebrates, the CNS is positioned in the dorsal part of the body. In arthropods, the CNS is primarily positioned in the ventral part of the body. The autonomic nervous system (ANS) is the part of the nervous system (both CNS and PNS) that controls involuntary functions. In vertebrates the ANS consists of enteric, sympathetic, and parasympathetic divisions, which differ in anatomy, neurotransmitters, and effects on target tissues. The sympathetic and parasympathetic divisions usually exert opposite effects on an organ.

During the course of vertebrate evolution, some parts of the brain (e.g., medulla oblongata) have remained relatively unchanged, whereas others (e.g., cerebral hemispheres) have changed dramatically. In each cerebral hemisphere of a mammal, specific regions are specialized to carry out specific sensory and motor functions—for example, language functions and fear.

A

1. Give two specific examples in which the sympathetic and parasympathetic divisions exert opposite effects on the function of an organ.
2. Why is localization of function in the brain critical for sensory perception?
3. What is cephalization of the nervous system, and why has it evolved in so many animals? Try to think this through using the knowledge you have gained about the relationships between the brain and sense organs. As part of your answer, describe a sketch of a lobster or a rattlesnake that you could make and that would show the anatomical locations of sense organs that sense the outside environment.

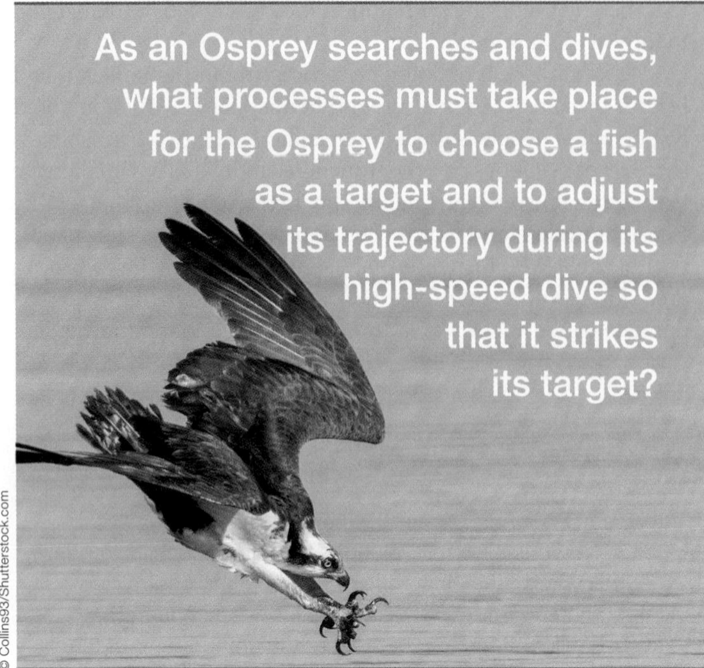

Answer: Being a visual predator, the Osprey uses light rays reflected off the water surface and off objects below the surface to gain information on its environment. When the reflected light rays strike the retinas of the Osprey's eyes, nerve impulses that encode the visual information must travel in neurons from the retinas to the brain, where sets of neurons in the brain must carry out calculations to determine the suitability of each potential target observed and ultimately decide on a dive directed at a particular target. As soon as the calculations are complete, the brain must initiate nerve impulses in neurons that will carry the impulses to all the flight muscles involved in pulling in the wings and angling the Osprey's body downward for a headfirst free fall. When these nerve impulses arrive in the muscles, they must activate muscle contractions, which actually start the dive. During the Osprey's rapid descent, the brain—using visual information—must initiate course-correcting nerve impulses that travel to the muscles during the dive, and as the Osprey nears the target the brain must send impulses to the legs, where leg muscles must contract or relax in ways to project the open talons directly toward the fish to be captured.

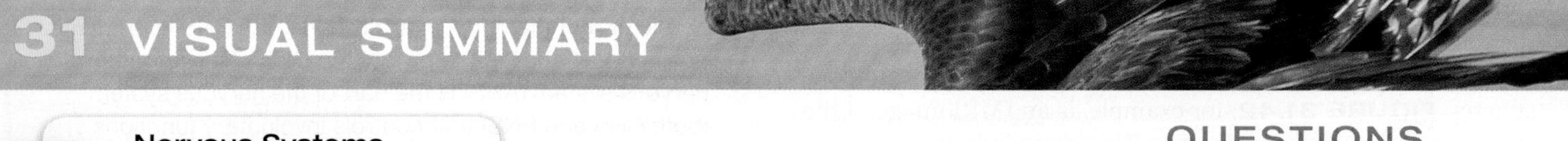

31 VISUAL SUMMARY

31.1 Nervous Systems Are Composed of Neurons and Glial Cells

FIGURE 31.6

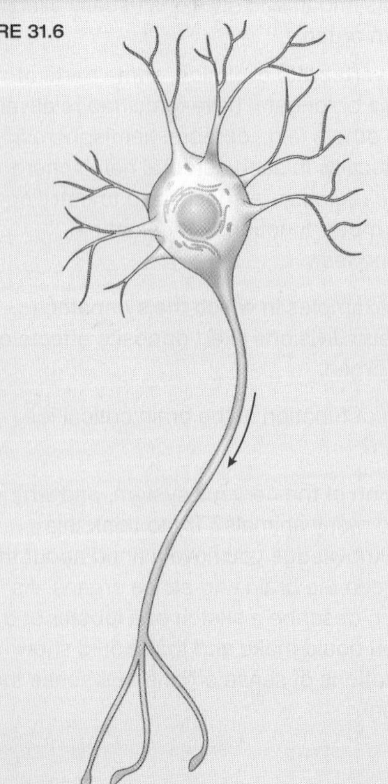

- A **NEURON** is a type of **EXCITABLE CELL**, meaning its cell membrane can generate and conduct **ACTION POTENTIALS**.
- A neuron typically consists of **DENDRITES** (where other neurons make **SYNAPSES**), a **CELL BODY** (where the cell nucleus is located and which also receives synapses), an **AXON HILLOCK** (part of the cell body where action potentials originate), an **AXON** (which conducts action potentials over long distances), and **AXON TERMINALS** (which initiate synaptic transmission) (**FIGURE 31.6**).
- **GLIAL CELLS**, which typically are not excitable, play roles in nervous system development and in the function of synapses and axons.

QUESTIONS

1. Approximately how long does it take an action potential to cover 1 meter as it moves along an axon?
2. Besides neurons, what is the other major type of excitable cell?
3. What is the difference between a neuron and a nerve?

31.2 Neurons Generate Electric Signals by Controlling Ion Distributions

Go to ANIMATIONS 31.1 and 31.2 and ACTIVITY 31.1

FIGURE 31.14

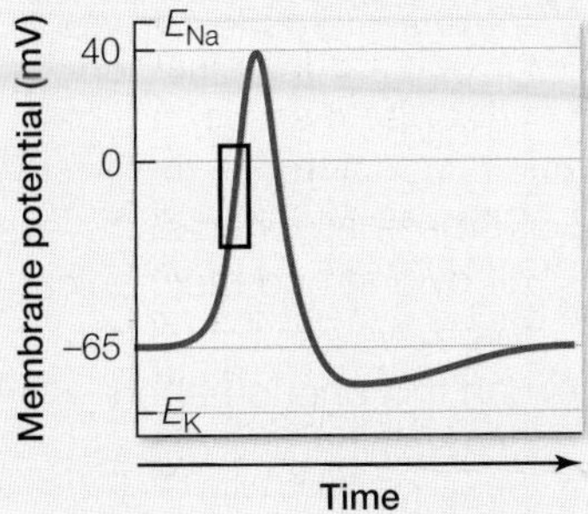

- To change the **MEMBRANE POTENTIAL**, only relatively small numbers of ions need cross the cell membrane of a neuron because the potential is determined by ions within nanometers of the membrane surfaces.
- Gated ion channels enable a rapid and major change in membrane potential resulting in either **DEPOLARIZATION** (**FIGURE 31.14**) or **HYPERPOLARIZATION**.
- An action potential propagates without loss of magnitude.

QUESTIONS

1. The membrane potential is close to the K^+ equilibrium potential in a neuron at rest. In the initial stage of an action potential, to what potential does the membrane potential switch?
2. How does myelination of an axon speed propagation of action potentials?
3. Describe the function of a stretch-gated channel.

31.3 Neurons Communicate with Other Cells at Synapses

Go to ANIMATION 31.3 and ACTIVITY 31.2

- Where signals pass between neurons, the neurons typically do not touch, and action potentials cannot cross the microscopic distance between them.
- At a **CHEMICAL SYNAPSE**, action potentials in the presynaptic cell cause the release of **NEUROTRANSMITTER** into the **SYNAPTIC CLEFT**. The neurotransmitter activates receptor molecules in the postsynaptic cell, leading to initiation of action potentials (**FIGURE 31.18**).
- A neuron often receives signals from hundreds of other neurons and integrates these signals by summation to determine its production of action potentials.

FIGURE 31.18

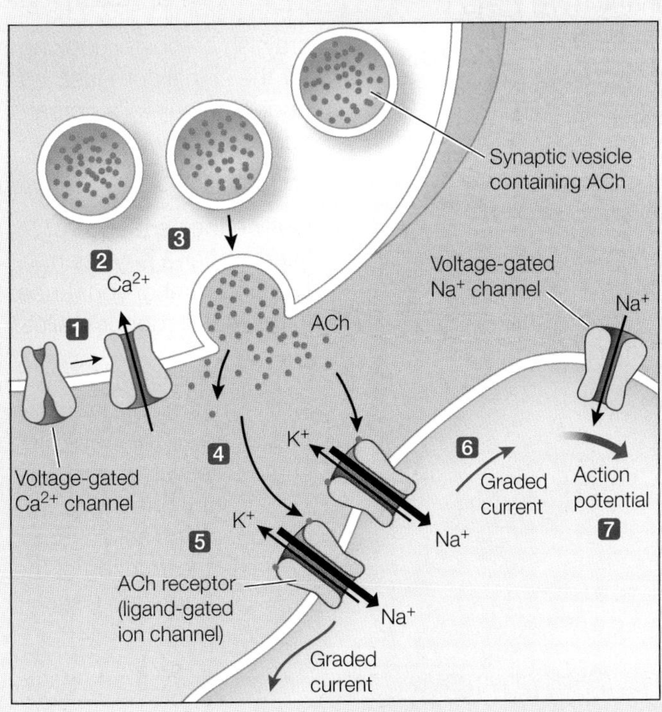

QUESTIONS

1. How do sea hares illustrate synaptic plasticity?
2. How does an electrical synapse work?
3. Acetylcholinesterase is an enzyme that breaks down acetylcholine. If you read labels on insecticides, you'll notice that many are anti-acetylcholinesterases. How do they work to kill?

31 VISUAL SUMMARY

31.4 Sensory Processes Provide Information on an Animal's External Environment and Internal Status

Go to ANIMATIONS 31.4, 31.5 and ACTIVITIES 31.3–31.6

- **SENSORY RECEPTOR CELLS** provide information on an animal's external environment and internal status.
- Stimulus energy (e.g., light or pressure) is transformed into electric signals in the process called **TRANSDUCTION**.
- The vertebrate **RETINA** (**FIGURE 31.31**) is among the sense organs that send highly synthesized information to the central nervous system on the nature of stimuli received.

FIGURE 31.31

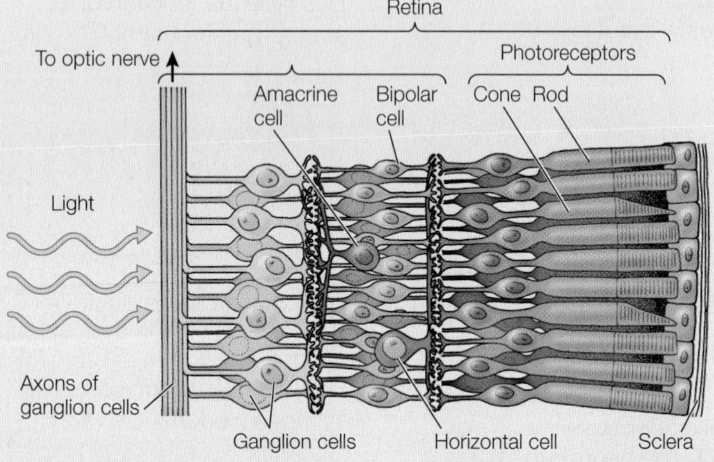

QUESTIONS

1. Describe a specific example of how the central nervous system needs information on internal status to carry out proper coordination of body function.
2. Sometimes when our eye is hit by a strong mechanical force, we see light. How can this phenomenon be explained? What fundamental principle of sensory biology does it illustrate?
3. How does an ionotropic receptor cell transduce sensory information?

31.5 Neurons Are Organized into Nervous Systems

Go to ANIMATION 31.6 and ACTIVITY 31.7

FIGURE 31.40

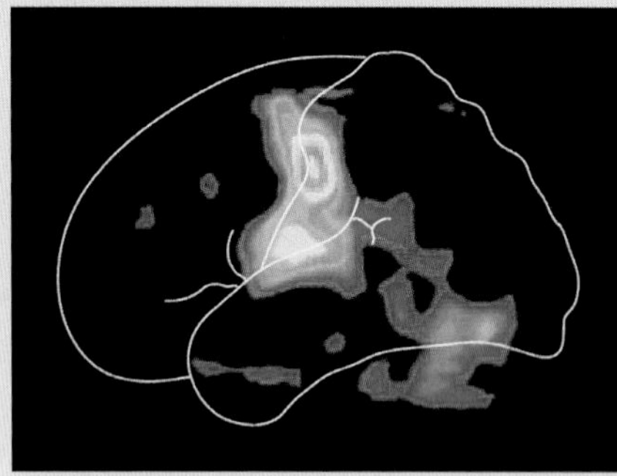

- **CENTRALIZATION** and **CEPHALIZATION** are hallmarks of the evolution of animal nervous systems.
- The **AUTONOMIC NERVOUS SYSTEM** (**ANS**) controls effectors other than skeletal muscles.
- Location specificity is what enables the **BRAIN** to differentiate among action potentials from various sensory neurons and is also important in originating motor signals (**FIGURE 31.40**).

QUESTIONS

1. If you carefully dissect a crayfish or lobster, looking for the spinal cord just under the animal's dorsal surface, you will find nothing. Why?
2. Outline the sequence of events after a physician taps the front of your knee with a small rubber mallet.
3. Suppose a person has a stroke that might damage their brain. In the days immediately afterward, explain why one of the first questions asked is, "What is the location of the possible damage?"

32 Control by the Endocrine and Nervous Systems

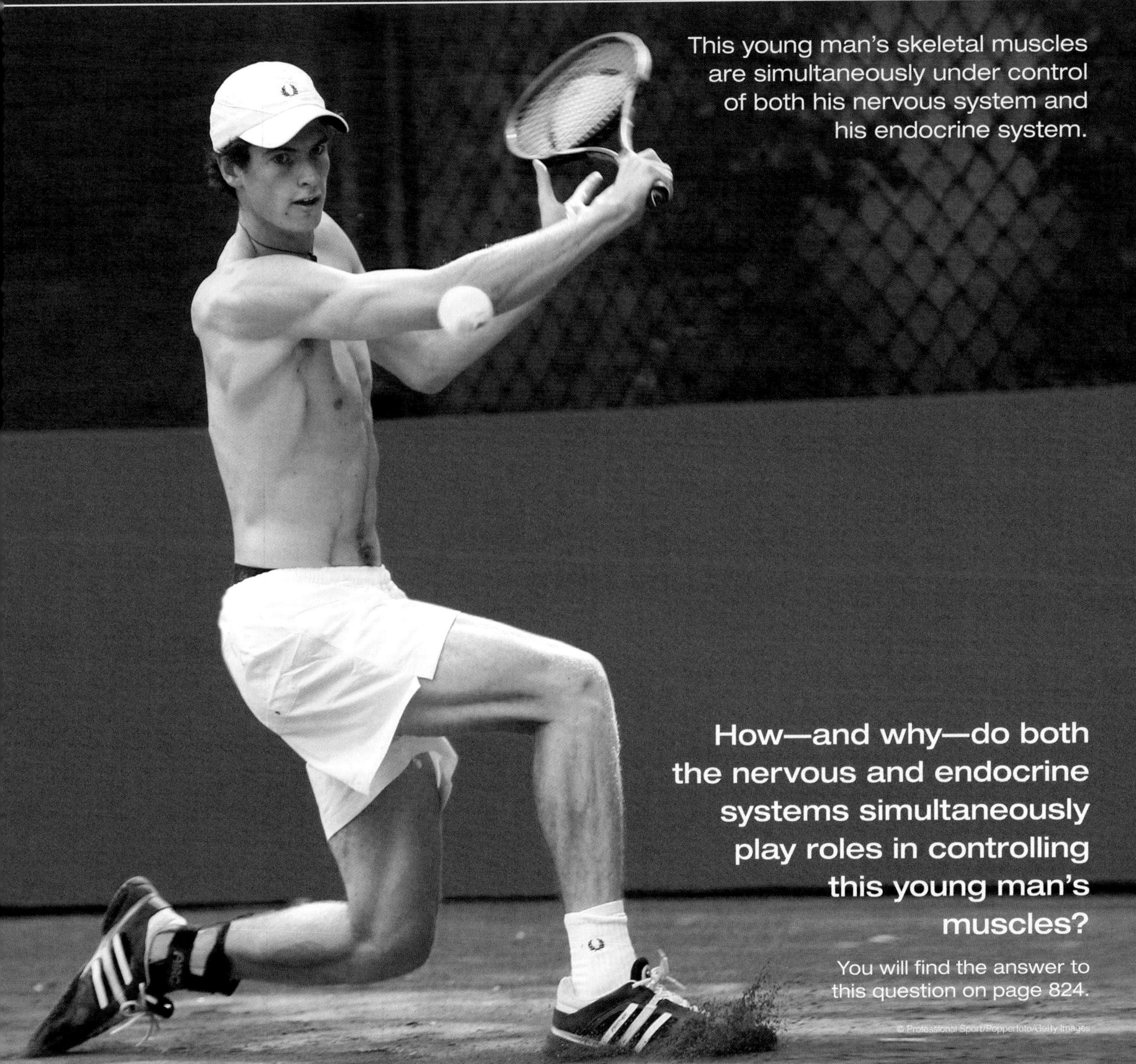

This young man's skeletal muscles are simultaneously under control of both his nervous system and his endocrine system.

How—and why—do both the nervous and endocrine systems simultaneously play roles in controlling this young man's muscles?

You will find the answer to this question on page 824.

KEY CONCEPTS

32.1 The Endocrine and Nervous Systems Play Distinct, Interacting Roles

LEARNING OBJECTIVES

By the end of this key concept you should be able to:

32.1.1 **Postulate why animals have evolved both a nervous system and an endocrine system.**

32.1.2 **Illustrate how an animal's nervous system and endocrine system work together.**

32.1.3 **Describe the full range of spatial scales over which chemical control is exerted, distinguishing among autocrine, paracrine, endocrine, and pheromone functions.**

All the cells in an animal's body need to act together harmoniously. One of the keys to harmonious function is that certain types of cells—notably the nerve and endocrine cells—have evolved the specialized role of controlling all the other types of cells in coordinated ways. We discussed nerve cells in Chapter 31. **Endocrine cells**—our chief focus in this chapter—are defined by the fact that they secrete hormones into the blood. Nerve and endocrine cells work together to ensure that an animal functions in ways that are harmonious and coordinated, rather than clashing and disjointed.

Nerve and endocrine cells need to communicate with other cells to carry out their functions of control and coordination. Most such cell-to-cell communication takes place by means of chemical signals—signaling molecules—that are released from a nerve cell or an endocrine cell and travel to another cell, called the **target cell** (**FIGURE 32.1**). As discussed in Chapter 6, signaling molecules bind to protein receptor molecules on or in the target cell, triggering a target-cell response that may affect the animal's physiology, anatomy, or behavior.

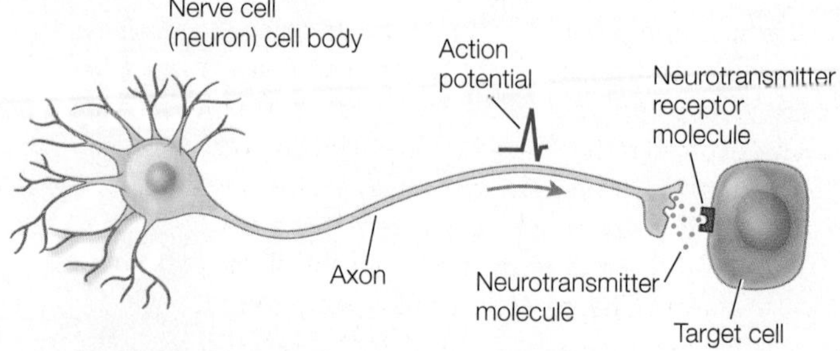

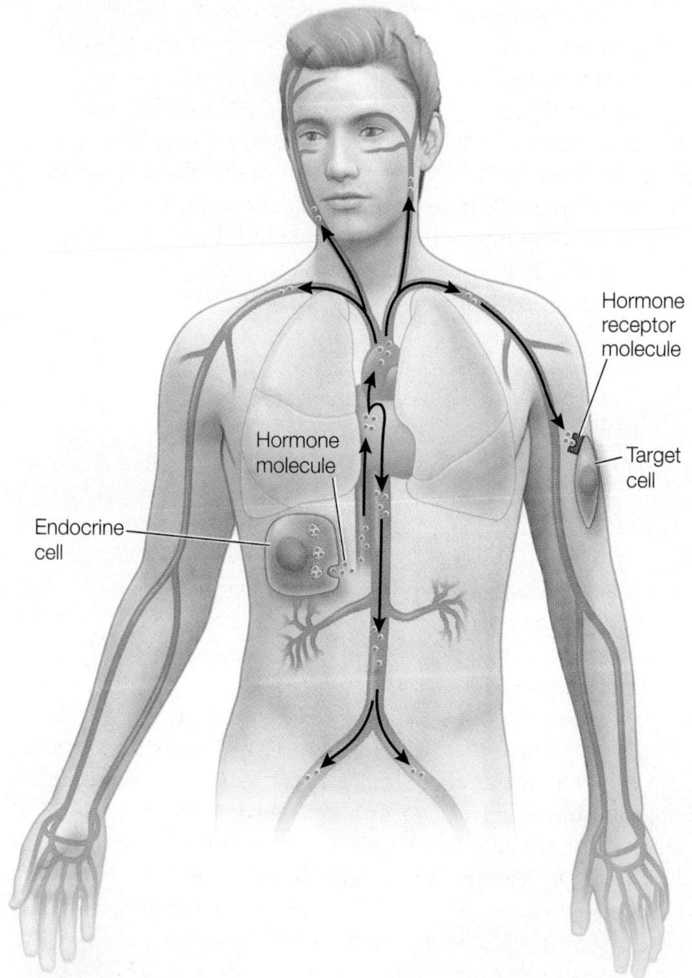

FIGURE 32.1 Chemical Signaling by Nerve and Endocrine Cells Both nerve cells (neurons) and endocrine cells release chemical signaling molecules (blue dots) that affect target cells. **(A)** A nerve cell makes synaptic contact with a target cell at the end of its axon. Action potentials in the axon trigger release of neurotransmitter molecules at the synapse. These molecules then diffuse a short distance to bind with neurotransmitter receptor molecules in the postsynaptic cell, initiating a response in the postsynaptic cell. **(B)** An endocrine cell releases hormone molecules into the blood. These molecules flow with the blood, circulating throughout the body, and bind with hormone receptor molecules on target cells, initiating responses. The process is illustrated here with a pancreatic endocrine cell secreting a hormone that affects a muscle cell.

The nervous and endocrine systems work in different ways

Most tissues and organs in an animal's body are affected by both the nervous and endocrine systems. The heart, for example, is subject to both nervous and endocrine controls. Why have two systems evolved rather than just one? The answer is undoubtedly complicated. One important reason is that the nervous and endocrine systems work in different ways. The two systems are specialized to carry out different types of control and coordination.

As we saw in Chapter 31, nervous system control is fast and addressed. Neuronal signals are fast in that they travel at high speeds along axons (see Figure 32.1A), and they begin and end abruptly. They are addressed—like a letter or an e-mail—because they are delivered to the specific target cells with which a neuron makes synaptic contact, and only those cells are affected by the neuron.

In contrast, endocrine control is slow and broadcast. As shown in Figure 32.1B, endocrine cells release their chemical signals—the hormones—into the blood, and the signals are then carried throughout the body by the circulation of the blood, potentially reaching all the cells in most or all tissues and organs.

Individual hormonal signals are relatively slow because a hormone, once it is released into the blood, must circulate to target tissues and diffuse to effective concentrations in the tissues before it can elicit a response. Thus the initiation of a hormonal effect requires at least several seconds or minutes—in sharp contrast to neuronal signaling, which can initiate an effect in 10–20 milliseconds. Hormonal signaling is also slow because after a hormone has entered the blood, it may act on target cells for a substantial amount of time—minutes, hours, or even days—before its blood concentration is reduced to ineffective levels by metabolic destruction or excretion. One more reason that hormonal signaling can be slow is that hormones sometimes exert their effects on target cells by altering gene transcription and protein synthesis—processes that require at least many minutes to have effects on cell function.

Endocrine control is said to be broadcast because after a hormone is released into the blood, all cells in the body are potentially exposed to it. Those that respond are the ones that express a receptor protein for the hormone. That is, hormone action has specificity, but its specificity depends not on addressed delivery of the chemical signal but on which cells have receptor molecules for the signal. In any one tissue, it is common for all cells to be similar in the receptor molecules they express, meaning all cells in the tissue respond similarly to a single release of a hormone. Moreover, cells in more than one tissue may respond, perhaps with different cell types responding in different ways. In principle, hormones may exert either limited or widespread effects, but in practice they commonly affect at least an entire tissue, and often multiple tissues.

Nervous systems and endocrine systems tend to control different processes

Lines of communication in the nervous system are capable of much finer control—in both time and space—than is possible in the endocrine system. Not surprisingly, the two systems tend to be used to control different functions in the body. The nervous system predominantly controls the fine, rapid movements of skeletal muscles. The endocrine system typically controls more widespread, prolonged processes such as development, reproductive cycles, or changes in metabolism that are sustained for many minutes, hours, or days.

Consider, for example, the young tennis player in our opening photo. Tennis requires rapid body movements that anticipate ball movements and an opponent's possible responses—processes that entail specific control of multiple skeletal muscles in split-second time. These functions could only be coordinated by the nervous system. In contrast, the coordination of adolescent development in a young person requires adjusting the activities of many tissues over a prolonged period. In principle, the nervous system *could* carry out a coordination task of this sort. To do so, however, the nervous system would require tens of thousands of discrete axons running between integrating centers (such as the brain) and target cells, and would need to send trains of impulses along all these axons for years. In contrast, endocrine glands can accomplish this task with greater economy, by secreting a relatively small number of long-lasting chemicals into the blood.

The dual controls of the nervous and endocrine systems are evident in our tennis player's skeletal muscles. A single muscle, such as the biceps, typically contains thousands of muscle cells innervated by more than 100 motor neurons. The nervous system can selectively activate a few, many, or all of the motor neurons to control rapidly and precisely the amount of force the player's biceps generates. And it can separately control each of the other skeletal muscles in his body. Simultaneously—as days and months go by—testosterone and other hormones secreted steadily by endocrine glands during adolescence facilitate protracted, widespread muscle growth and development.

The nervous and endocrine systems work together

In spite of the differences between the nervous and endocrine systems, the two systems do not operate in mutually exclusive ways. Instead, they work together closely, and each system can directly affect the other.

The nervous system exerts control over the endocrine system in many ways. The brain initiates endocrine development of the gonads during puberty in mammals, for example. Moreover, throughout life the brain helps control the secretion of reproductive hormones, thyroid hormones, and other hormones. We will discuss these controls in Key Concepts 32.3 and 32.4.

Conversely, the endocrine system sometimes exerts control over the nervous system. Sex hormones from the testes and ovaries affect brain development during mammalian puberty, for example. Studies on rats, hamsters, and other research animals indicate that the sex hormones have an organizing effect on the development of certain neural circuits in the brain so that the circuits develop different specific properties in females and males.

Chemical signaling operates over a broad range of distances

Animals use chemical signaling over a very broad range of spatial scales (**FIGURE 32.2**). We have already discussed the two

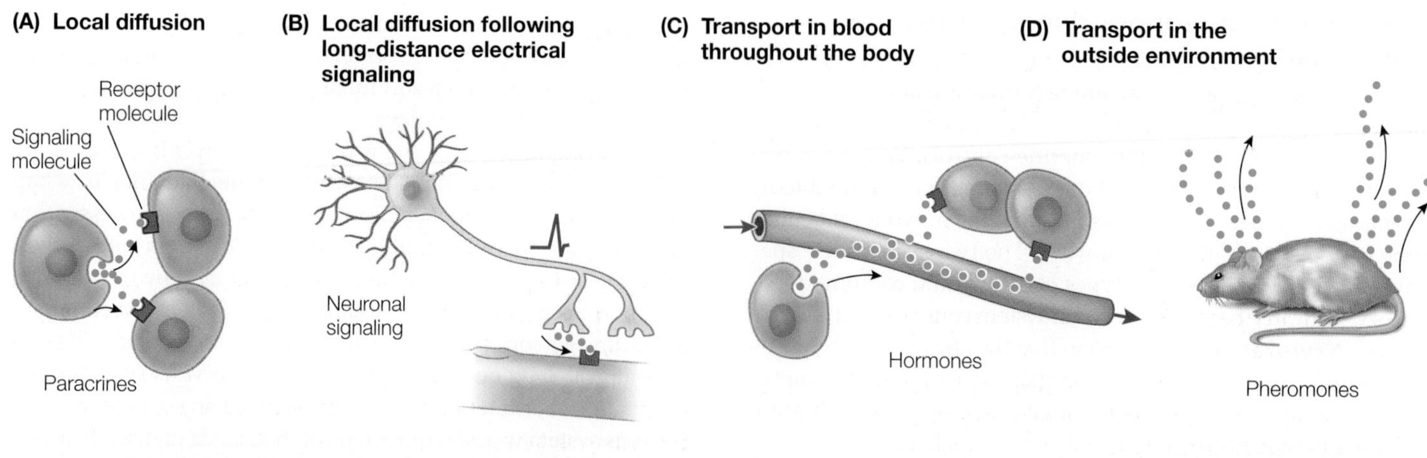

FIGURE 32.2 Chemical Signaling Occurs over a Broad Range of Spatial Scales **(A)** Cells communicate with neighboring cells in a tissue using molecules called paracrines. **(B)** Neurons have long processes (axons), at the ends of which they release molecules called neurotransmitters to communicate across synapses with target cells. **(C)** Endocrine cells secrete molecules called hormones that can travel long distances by circulating in the blood. **(D)** Animals give off molecules called pheromones that can communicate specific information (such as their sex) over very long distances to other members of their species.

different scales of distance seen in neuronal signaling and endocrine signaling. However, chemical signaling also takes place on even shorter and even longer scales of distance.

In many types of chemical signaling, the signaling molecules diffuse from cell to cell in a tissue without entering the blood. Paracrines and autocrines are two categories of molecules of this type that operate over particularly short distances. **Paracrines** are secreted by one cell and affect the functions of neighboring cells by binding to receptors on or in the neighboring cells (see Figure 32.2A). **Autocrines** are secreted by a cell into the surrounding tissue fluids (intercellular fluids) and then diffuse to receptors on that same cell and affect its function (see Figure 6.2).

Neurotransmitters and hormones work at intermediate distances. Neurotransmitters resemble paracrines in that they diffuse just a short distance as they cross the synaptic cleft that separates a presynaptic neuron from its target cell. However, the secretion of neurotransmitters is controlled from farther away by electric signals that travel the length of the presynaptic cell to reach the synapse (see Figure 32.2B). Hormones are carried to the farthest reaches of an animal's body by the circulation of the blood (see Figure 32.2C).

Pheromones are chemical signaling molecules that an individual animal releases into its external environment and that exert specific effects (e.g., behavioral effects) on other individuals of the same species (see Figure 32.2D). Some pheromones—such as male-attracting pheromones released into the air by the females of some species of moths—can travel hundreds of meters before another individual detects them and is influenced by them.

REVIEW & APPLY | 32.1

R Nerve and endocrine cells control and coordinate the functions of the body by releasing chemical signals that travel to another cell, called the target cell. Neuronal signals are fast and addressed, whereas endocrine signals are slow and broadcast. The nervous system controls predominantly the fine, rapid movements of discrete skeletal muscles. The endocrine system typically controls more widespread, prolonged activities such as developmental or metabolic changes. Animals use chemical signaling over a broad range of spatial scales. Autocrines and paracrines act, respectively, on the cells producing the signals and on their immediate neighbors. Neurons employing neurotransmitters work at intermediate distances, as do hormones. Pheromones are released into the environment and can affect targets hundreds of meters away.

A

1. Why is endocrine control described as slow and broadcast?
2. Why is it correct to say that both nervous and endocrine control depend on chemical signaling?
3. If all the cells in an animal's body are bathed by a hormone, why are some cells affected by the hormone while others are not?

Now that we've compared and contrasted the nervous and endocrine systems, let's turn our spotlight on the endocrine system for the remainder of this chapter.

32.2 Hormones Are Chemical Messengers Distributed by the Blood

LEARNING OBJECTIVES

By the end of this key concept you should be able to:

32.2.1 **Differentiate between endocrine glands and exocrine glands.**

32.2.2 **Explain how neuroendocrine cells serve as a direct interface between the nervous system and endocrine system.**

32.2.3 **Explain how it is possible for one type of hormone molecule to have many different effects throughout an animal's body.**

As we begin our study of the endocrine glands, it is important to recognize that there are two types of glands—exocrine and endocrine—and that only the endocrine glands secrete hormones and play major roles in coordination within the body. Both types of glands produce and secrete materials. However, the **exocrine glands**—exemplified by the mammary, salivary, sweat, and tear glands—typically have outflow tubes, called ducts, and their secretions flow out by way of these tubular ducts, often exiting the body (e.g., milk, sweat, and tears). The prostate gland, which contributes fluid to the semen, is another example of an exocrine gland. In general, the exocrine glands do not have a signaling function.

The **endocrine glands**—our focus in this chapter—do not have outflow ducts. Instead, endocrine cells, by definition, secrete their products into the blood flowing through nearby blood capillaries or other blood passages (keep in mind that in this book we use the term "blood" to refer to the fluid circulating in the circulatory system in all animals; see Key Concept 30.4). Because endocrine cells secrete into the blood, endocrine glands require only a blood supply, not ducts, for their secretions to be carried away.

In some cases, endocrine cells are found as single cells scattered within a tissue composed mostly of other types of cells. For example, in the epithelium lining the midgut (see Figure 28.20), scattered among the predominant digestive-absorptive cells are endocrine cells that emit signals concerning the state of food digestion. In other cases, large numbers of endocrine cells of a particular kind are grouped together to form a tissue or organ called an endocrine gland; the thyroid gland and adrenal glands are examples.

A **hormone** is defined to be a chemical substance that is secreted into the blood by endocrine cells and that regulates the function of other cells it reaches by blood circulation. An additional defining feature of hormones is that they act at very low blood concentrations: sometimes as low as 10^{-12} molar (moles per liter). Still another defining feature of hormones is that they initiate their effects on target cells by noncovalent binding of the hormone molecules to receptor protein molecules synthesized by the target cells. This feature determines which cells in the body respond to a hormone. Only the cells that synthesize receptor molecules respond (and thereby qualify as target cells).

The specific action of a hormone on a target cell depends on properties of the target cell. It depends on the receptor proteins that the target cell synthesizes and on the biochemical pathways the receptor proteins activate in the cell.

Growth, development, reproductive cycles, water balance, nutrient homeostasis, and long-term stress responses are some of the processes that typically are under primarily hormonal control. This list reflects the fact that, as we discussed in Key Concept 32.1, hormones are well suited to control processes that involve many tissues and that occur on time scales of hours, days, months, or years.

Endocrine cells are neurosecretory or non-neural

There are two broad classes of endocrine cells: neurosecretory cells and non-neural endocrine cells. **Neurosecretory cells**, also called neuroendocrine cells, resemble neurons in that they are excitable cells that propagate action potentials. **Non-neural endocrine cells**, also called epithelial endocrine cells, are not excitable.

The cell bodies of neurosecretory cells are located in the central nervous system (CNS). These cells have axons, which typically extend outside the CNS. Instead of releasing neurotransmitters at synapses as most neurons do, neurosecretory cells release hormones into the blood at their axon terminals (**FIGURE 32.3A**). Structures called **neurohemal organs** (neuro, "related to neurons"; hemal, "related to blood") are commonly associated with neurosecretory cells. At a neurohemal organ, the axon terminals of groups of neurosecretory cells are positioned in a specialized part of the circulatory system where the terminals are closely juxtaposed with the flowing blood.

Neurosecretory cells provide a direct interface between the nervous and endocrine systems. To see this, consider the signal inputs and outputs of a neurosecretory cell (see Figure 32.3A). Ordinary neurons typically make synaptic contact with the dendrites and cell body of a neurosecretory cell in the CNS. The cell thus receives ordinary synaptic signals from many neurons and can integrate these signals in the usual ways (see Figure 31.21). Then, as its output, the neurosecretory cell secretes a hormone (which some scientists call a *neurohormone* or *neurosecretion*) into the blood. The typical order of events is that the neurosecretory cell initiates action potentials in response to the neuronal signals it receives. These action potentials propagate down the cell's axon and activate hormone release into the blood at the axon terminals.

As we've noted, non-neural endocrine cells do not employ action potentials. Non-neural endocrine cells are typically stimulated to secrete their hormones by other hormones (**FIGURE 32.3B**). A non-neural endocrine cell expresses receptor proteins for the hormones that control it, and the typical order of events is that when these hormones bind to the receptor proteins, the non-neural endocrine cell is stimulated to secrete its own hormone into the blood. Additionally, some non-neural endocrine cells release hormones in direct response to changes in their cellular environment. For example, pancreatic beta cells (see Figure 28.22) increase or decrease secretion of insulin in direct response to changes in blood glucose concentration.

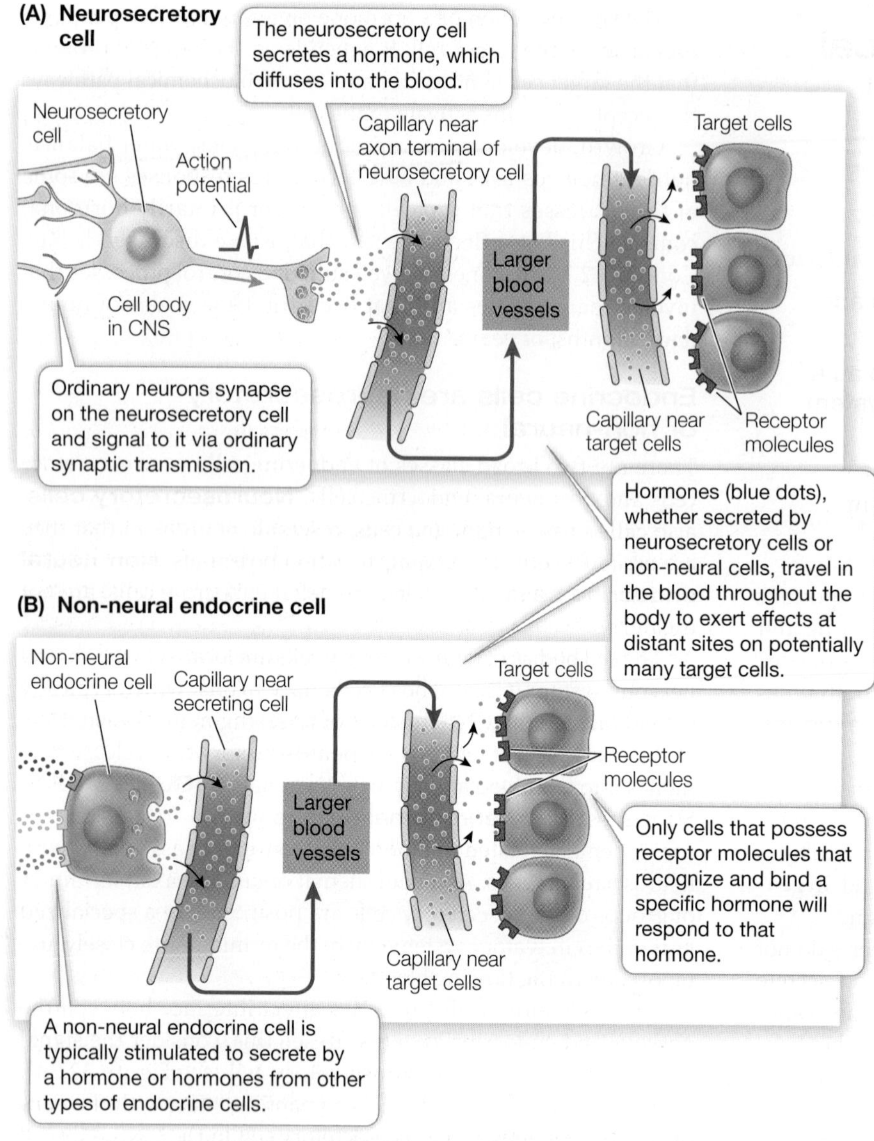

FIGURE 32.3 Two Classes of Endocrine Cells **(A)** A neurosecretory cell. Release of hormone into the blood often occurs in a specialized neurohemal organ, not shown in this diagram. **(B)** A non-neural endocrine cell.

LINK The crucial role of insulin in regulating glucose metabolism is detailed in **Key Concept 28.4**.

Most hormones belong to one of three chemical groups

There is enormous diversity in the chemical structure of hormones, but we can classify most hormones into one of three groups:

- *Peptide hormones and protein hormones* are molecules composed of chains of amino acids. They are water-soluble and therefore easily transported in the blood. Most hormones are in this group. Some are large molecules that may contain nearly 200 amino acid units (**FIGURE 32.4A**). Others are small molecules containing as few as three amino acid units. For simplicity, these hormones are often all referred to as peptide hormones (disregarding the fact that the larger ones are technically proteins), and we follow that practice here. During their synthesis, the peptide hormones are packaged in vesicles within the cells that make them. To be secreted, they are released from the cells by exocytosis (see Key Concept 4.3). The receptor proteins for peptide hormones are on the exterior surfaces of target cells (i.e., they are cell surface receptors). This makes sense because the hormones, being water-soluble, cannot easily cross the cell membrane to enter a cell.
- *Steroid hormones* are derived from cholesterol and—in common with cholesterol—all share a similar structure of four interlinked rings (**FIGURE 32.4B**). Steroid hormones are lipid-soluble and readily pass through cell membranes, but they do not dissolve well in blood plasma. Thus they are usually bound to carrier proteins for transport in the blood. Most receptors for these hormones are inside target cells, in the cytoplasm or nucleus. When a steroid hormone reaches a target cell, it typically diffuses through the cell membrane and binds internally to receptors. Some steroid hormones also bind to cell surface receptors, in addition to binding to receptors inside the cell.
- *Amine hormones* are small molecules that are modified amino acids. For example, both the thyroid hormone thyroxine and the adrenal hormone epinephrine are made from the amino acid tyrosine (**FIGURE 32.4C**). Depending on whether the modified amino acid is polar or nonpolar, an amine hormone may be water-soluble (such as epinephrine) or lipid-soluble (such as thyroxine).

Receptor proteins can be on the cell surface or inside a cell

Some receptor proteins are positioned in the cell membrane of a target cell. As we have just said, for example, the receptors for peptide hormones are cell surface receptors. Certain amine hormones, such as epinephrine, also have cell surface receptors. The hormone binds to a part of the receptor protein that projects outside the cell membrane, and then the receptor molecule initiates processes that alter cell function. Many receptors of this type are G protein-coupled receptors (discussed at length in Key Concepts 6.2 and 6.3) that, after binding to a hormone, initiate second messenger cascades inside the target cell. These cascades activate or inactivate enzymes in the cell cytoplasm, leading to the cell's response (see Figures 6.8 and 6.10).

(A) Peptide hormones

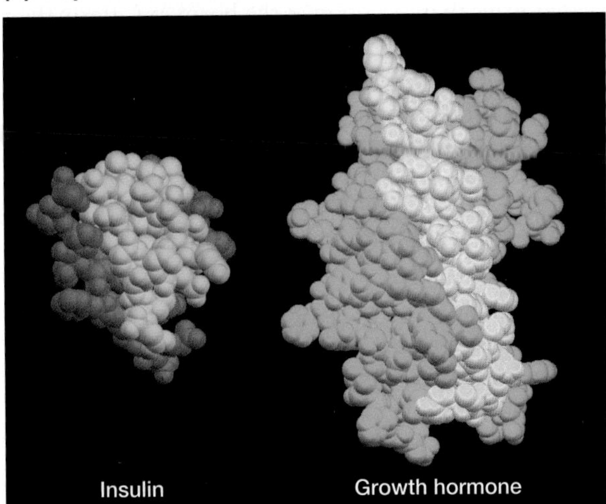

(B) Steroid hormones

Sterol backbone

Cholesterol

Corticosteroids | **Sex steroids**

Aldosterone

Cortisol

Testosterone

Estradiol (an estrogen)

(C) Amine hormones

Tyrosine

Epinephrine

Thyroxine

FIGURE 32.4 The Three Principal Chemical Classes of Hormones **(A)** Peptide hormones are composed of strings of amino acids. They include proteins and small polypeptide molecules. Shown here are molecules of two protein hormones, insulin and growth hormone. Each of these proteins consists of two subunits, which are shown in different colors. **(B)** Steroid hormones are synthesized from cholesterol, and all include a characteristic set of four ring structures, here called the sterol backbone and colored green. They include the corticosteroids produced by the adrenal cortex and the sex steroids produced primarily by the gonads. **(C)** Amine hormones are small molecules synthesized from single amino acid molecules. Thyroxine and epinephrine are both made from the amino acid tyrosine. The structures in B and C are in abbreviated form; not all carbon (C) atoms are identified. (A, left after data from Q. X. Hua et al. 1995. *Nat Struct Biol* 2: 129–138; right after data from L. Chantalat et al. 1995. *Protein Pept Lett* 2: 333–340.)

Other receptor proteins called **intracellular receptors** are located *inside* the target cell, in the cytoplasm or nucleus. Most receptors for steroid hormones are intracellular receptors. After a steroid hormone (being lipid-soluble) diffuses into a target cell, it binds to its receptor, forming a hormone–receptor complex that then alters gene expression (transcription and translation), resulting in synthesis of new proteins (**FIGURE 32.5**). For example, when testosterone binds to receptors in skeletal muscle cells, testosterone–receptor complexes activate transcription of several genes. These genes include those that code for synthesis of the contractile proteins actin and myosin, which constitute much of the substance of muscle tissue. In this way, testosterone tends to bring about an increase in muscle mass.

A target cell is not limited to expressing only a single kind of hormone receptor protein. Individual cells often express two or more receptor proteins, and therefore are target cells for two or more hormones.

In a tissue where the cells have receptors for a particular hormone (and therefore the tissue responds to the hormone), tissue sensitivity to the hormone is not necessarily constant. Instead, tissue sensitivity might be high at one time but low at a later time. Changes in hormone sensitivity are caused by many mechanisms, of which the simplest to understand is a change in the number of receptor molecules per cell; a decrease in receptor number tends to reduce cell sensitivity to a hormone.

Hormone action depends on the nature of the target cells

The effect of a hormone on a target cell depends on characteristics of the target cell—in particular, the cellular processes that the cell activates in response to the hormone. For this reason, the action of a hormone is not determined by the hormone itself, and one hormone can have dramatically different effects on different cells.

In the course of evolution, many hormones have been conserved for long periods of evolutionary time, but their functions have changed as various types of animals have evolved. This pattern—in which the functions of a single hormone molecule evolve—arises because the actions of a hormone depend on cellular responses, which can change. The peptide hormone prolactin provides a dramatic example.

Prolactin was first discovered in mammals, where it stimulates milk synthesis by the mammary glands (breasts). Prolactin has since been found in all other vertebrate groups—from fishes to

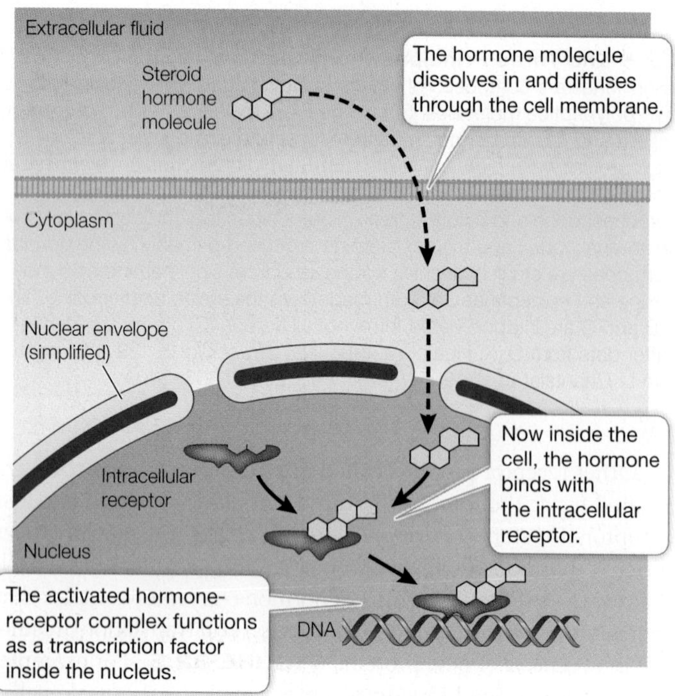

FIGURE 32.5 A Steroid Hormone Interacts with an Intracellular Receptor Lipid-soluble hormones such as steroids can diffuse into the cell to interact with intracellular receptors. By contrast, peptide hormones and some amine hormones (e.g., epinephrine) interact with receptors on the cell surface (see Figure 6.12).

birds—even though these other groups of vertebrates have neither mammary glands nor milk. In salmon, prolactin is involved in stabilizing blood ion composition as the fishes migrate between seawater and fresh water. In birds, prolactin can help control reproductive behavior and egg production. In one molecular form or another, prolactin has existed in vertebrates throughout their evolutionary history. It has taken on radically different functions, however, in various vertebrate groups by evolution of the target cells it affects.

Within one individual animal, some target cells for a hormone can respond in very different ways from others. An excellent example is provided by human responses to epinephrine (also called adrenaline), which is secreted by the adrenal medullary glands. Each adrenal gland is actually composed of two glands: the adrenal medulla in the center and the adrenal cortex surrounding it (**FIGURE 32.6**). The adrenal medulla secretes epinephrine, which plays a key role in one of the most famous and easily experienced of all hormonally controlled processes: the "fight or flight" response. When we suddenly experience a crisis, our adrenal medullary glands (activated by the sympathetic nervous system) secrete epinephrine, which—within seconds—is circulating throughout our body in the blood, helping contribute to our sense of internal upheaval. The epinephrine affects many tissues. In the heart, it stimulates the muscle cells to increase their strength of contraction. In the liver, it stimulates the liver cells to break down glycogen, yielding glucose which floods into the blood (see Figure 6.12). Epinephrine decreases blood flow to the midgut, thereby suppressing digestive activity (which is not immediately needed in a crisis). This single hormone elicits these and many other diverse responses in cells throughout the body. The reason the responses can be so diverse is that different target cells differ dramatically in the particular ways that reception of the hormone affects their cellular processes.

LINK Signal transduction pathways—the cellular processes activated by receptors—are discussed in **Key Concept 6.3**.

A hormonal signal is initiated, has its effect, and is terminated

When endocrine cells are activated to secrete a hormone, they vary in how quickly they start hormone release. Because peptide and amine hormones (including epinephrine) are usually synthesized and stored in vesicles prior to use, they can be secreted very rapidly. In contrast, steroid hormones are typically synthesized on demand. Initiation of secretion of steroid hormones is relatively slow because of the need to synthesize the hormones before they can be released.

Just as there are processes for secreting a hormone into the blood, there are mechanisms to remove it. Removal takes place in several ways. Target cells sometimes enzymatically degrade hormones. Certain organs, such as the liver and kidneys in vertebrates, also enzymatically degrade hormones, removing them from the blood. Hormones may also be excreted.

The removal processes ultimately terminate a hormonal signal. The *half-life* of a hormone in the blood is the time required for half of a group of simultaneously secreted hormone molecules to be removed from the blood. Some hormones have half-lives measured in minutes, and in these cases an individual hormonal signal does not last long. Epinephrine, for example, has a half-life of 1–2 minutes, explaining why the jolt of a sudden fight-or-flight response disappears quickly. Other hormones have far longer half-lives. In the human bloodstream, for example, the hormones antidiuretic

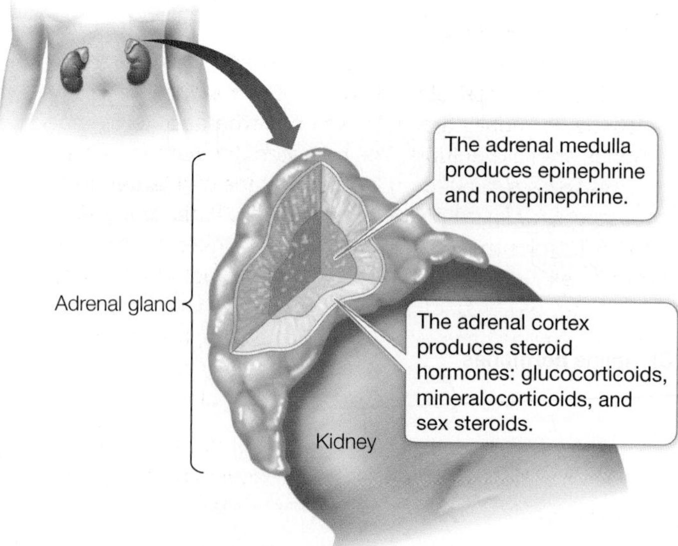

FIGURE 32.6 The Adrenal Gland Consists of Two Distinct Parts That Secrete Different Hormones An adrenal gland sits above each kidney. The gland consists of two parts, each of which is itself a gland: the outer cortex and the inner medulla, which produce different hormones.

hormone (ADH, involved in water conservation), cortisol (involved in long-term responses to stress), and thyroxine (involved in cellular metabolism) display average half-lives of about 15 minutes, 1 hour, and nearly 1 week, respectively. They may therefore have prolonged effects on target tissues.

When a hormone is secreted steadily over a period of time, the blood hormone concentration is determined by an interaction between the rate of secretion (adding hormone to the blood) and the rate of removal. Removal of any particular hormone tends to take place at a fairly steady rate. Thus changes in the blood concentration depend principally on changes in the rate of secretion.

Some hormones are bound to carrier proteins when in the blood. For example, steroid hormones must be bound with water-soluble proteins to be in solution in the blood. Binding to carrier proteins tends to lengthen hormone half-life.

Some hormones are converted to more active forms after they are secreted, a process called *peripheral activation*. Thyroid hormones provide a classic example. The thyroid gland secretes a molecule—thyroxine (sometimes called T_4 or tetraiodothyronine)—that contains four iodine atoms (see Figure 32.4C). The hormone then undergoes activation when peripheral tissues enzymatically remove one of the iodine atoms, forming T_3 (triiodothyronine).

REVIEW & APPLY | 32.2

R Animals have two types of secretory glands: exocrine glands, which have ducts to carry away their secretions; and endocrine glands, which lack ducts and secrete chemical signals into the blood. Hormones secreted into the blood by endocrine cells regulate the function of other cells they reach by blood circulation. Neurosecretory endocrine cells propagate action potentials and secrete hormones from their axon terminals. Other endocrine cells are nonneural (and not excitable).

Most hormones fall into one of three chemical groups: peptide hormones, steroid hormones, or amine hormones. Peptide hormones are water-soluble, whereas steroid hormones are lipid-soluble. The receptors for water-soluble hormones are located on the cell surface of a target cell; receptors for lipid-soluble hormones are typically inside the cell. Hormones cause different responses in different target cells, depending on the processes activated inside the target cell by binding of the hormone to its receptor. Each hormone has a characteristic half-life, the time required for half of a group of simultaneously secreted hormone molecules to be removed from the blood.

A

1. Growth hormone is a large, water-soluble peptide. Given these characteristics, how would you expect this hormone to interact with its target cells?
2. Are sweat glands correctly described as endocrine glands? Explain your answer.
3. Name one major similarity and one major difference between an ordinary neuron and a neurosecretory cell.

For understanding the vertebrate endocrine system, one of the most important themes is that the brain often controls the secretion of hormones by the endocrine system. We will start with that theme as we now turn our attention to the vertebrate endocrine system.

32.3 | The Vertebrate Hypothalamus and Pituitary Gland Link the Nervous and Endocrine Systems

LEARNING OBJECTIVES

By the end of this key concept you should be able to:

32.3.1 Explain how the posterior pituitary gland is a neurohemal organ.

32.3.2 List the steps in a typical endocrine control axis mediated by the anterior pituitary gland.

32.3.3 Explain how negative feedback is often operative in endocrine control mediated by the anterior pituitary gland.

The **pituitary gland**, found at the base of the brain (**FIGURE 32.7A**; see also Figure 31.38), is sometimes called the "master gland" because it secretes hormones that control the functions of many other glands. It is attached by a stalk to a part of the brain called the **hypothalamus**. The pituitary has two parts—the **anterior pituitary gland** (adenohypophysis) and **posterior pituitary gland** (neurohypophysis). The two have different developmental origins, which helps account for marked differences in the way they function. Both parts of the pituitary gland have close functional links with the brain.

Hypothalamic neurosecretory cells produce the posterior pituitary hormones

The posterior pituitary is basically an extension of the brain (**FIGURE 32.7B**). Its hormones are synthesized and secreted by brain neurosecretory cells. The cell bodies of these cells are in the hypothalamus, where they receive synapses from other, ordinary brain cells (neurons). The axons of the neurosecretory cells extend into the posterior pituitary, where they terminate in a rich blood capillary network, forming a neurohemal organ where the hormones of the neurosecretory cells are secreted into the blood.

An important general principle is that different sets of neurosecretory cells secrete different hormones. We can apply this principle as we focus on the hypothalamus and posterior pituitary. Most mammals and many other vertebrates have two distinct sets of neurosecretory cells in the posterior pituitary system, and the two sets secrete two different hormones. Both hormones are small peptide molecules containing nine amino acid units. In mammals these two hormones are **antidiuretic hormone** (**ADH**; also called vasopressin) and **oxytocin**. ADH controls the permeability to water of key membranes in the kidneys and thereby helps determine how much water is excreted in the urine.

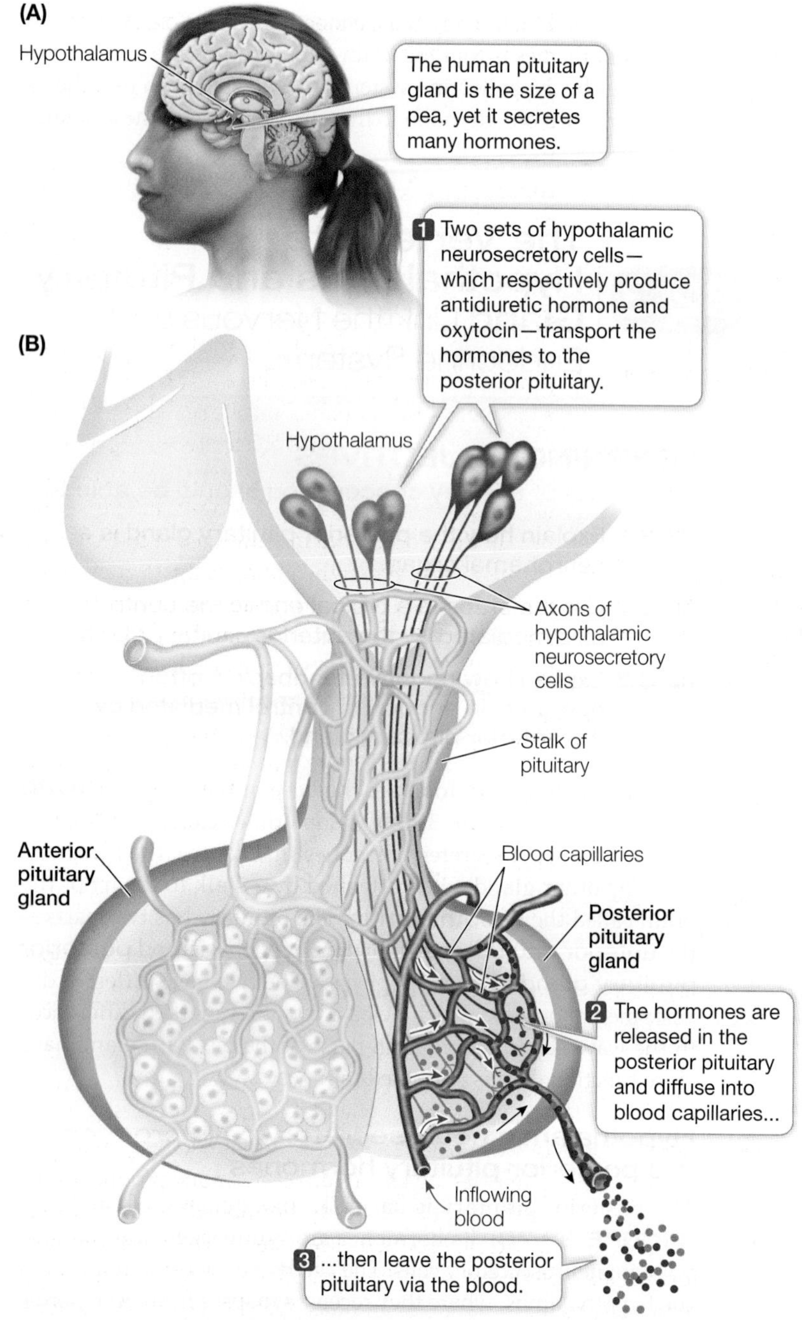

FIGURE 32.7 The Posterior Pituitary Gland **(A)** The human pituitary gland is located just below the hypothalamus of the brain. **(B)** The posterior pituitary is a neurohemal organ for neurosecretory cells in the brain. Two sets of neurosecretory cells in the hypothalamus produce two peptide hormones that are released in the posterior pituitary.

Oxytocin controls milk ejection from the mammary glands and helps control contractions of the uterus during the birth process (see Figure 34.14).

Signals produced by the brain control secretion by the neurosecretory cells in the posterior pituitary system. In this way, the enormous computational and integrative powers of the brain can be marshaled to control secretion of the posterior pituitary hormones. When the neurosecretory cells are activated, they generate action potentials that travel to their axon terminals, leading to hormone release into the blood.

In short, in the posterior pituitary system, *brain neurons control the secretion of hormones into the general circulation by neurosecretory cells*. This is one type of interface between the nervous and endocrine systems—a type of interface commonly found in both vertebrates and invertebrates.

Secretion of anterior pituitary hormones is controlled by hormones from hypothalamic neurosecretory cells

Unlike the posterior pituitary (which is a neurohemal organ), the anterior pituitary is an endocrine gland. The anterior pituitary hormones are synthesized in the anterior pituitary gland itself by non-neural endocrine cells (**FIGURE 32.8**). All these hormones are peptides or closely related to peptides.

Four of the anterior pituitary hormones are called **tropins**, or **tropic hormones**—terms meaning that their principal functions are to control the activities of other endocrine glands. They are **adrenocorticotropic hormone** (**ACTH**), **follicle-stimulating hormone** (**FSH**), **luteinizing hormone** (**LH**), and **thyroid-stimulating hormone** (**TSH**). A different set of pituitary cells produces each tropic hormone. ACTH controls the secretion of hormones by the adrenal cortex (see Figure 32.6). FSH and LH, which together are called **gonadotropins**, control the gonads: the testes in males and the ovaries in females. TSH controls the thyroid gland.

The anterior pituitary also produces several other hormones. These hormones act on non-endocrine targets and therefore are not tropic hormones. One is **growth hormone** (**GH**; see Figure 32.4A), which acts on a wide variety of tissues to promote growth. Others include prolactin (which we've already mentioned) and melanocyte-stimulating hormone (which is named for its action on skin-color cells in amphibians).

How is secretion of the anterior pituitary hormones controlled? To understand this important topic, we first need to consider a specialized network of tiny blood vessels, the **hypothalamo–hypophysial portal system** ("hypophysis" is an alternative name for pituitary). This vascular system consists of two capillary beds that are connected so that blood flows through them sequentially. Blood flows first through a capillary bed in the hypothalamus. Neurosecretory cells in the hypothalamus secrete hormones into the blood as it flows through this capillary bed. The blood then travels a short distance to the second capillary bed, which is in the anterior pituitary (see Figure 32.8). This sequential arrangement of capillary beds enables the hormones secreted by the hypothalamic neurosecretory cells to travel to the pituitary non-neural endocrine cells, where the hypothalamic hormones control secretion by the pituitary cells. A typical sequence of action begins when ordinary brain neurons that synapse on hypothalamic neurosecretory cells stimulate those cells to secrete hormones. Those

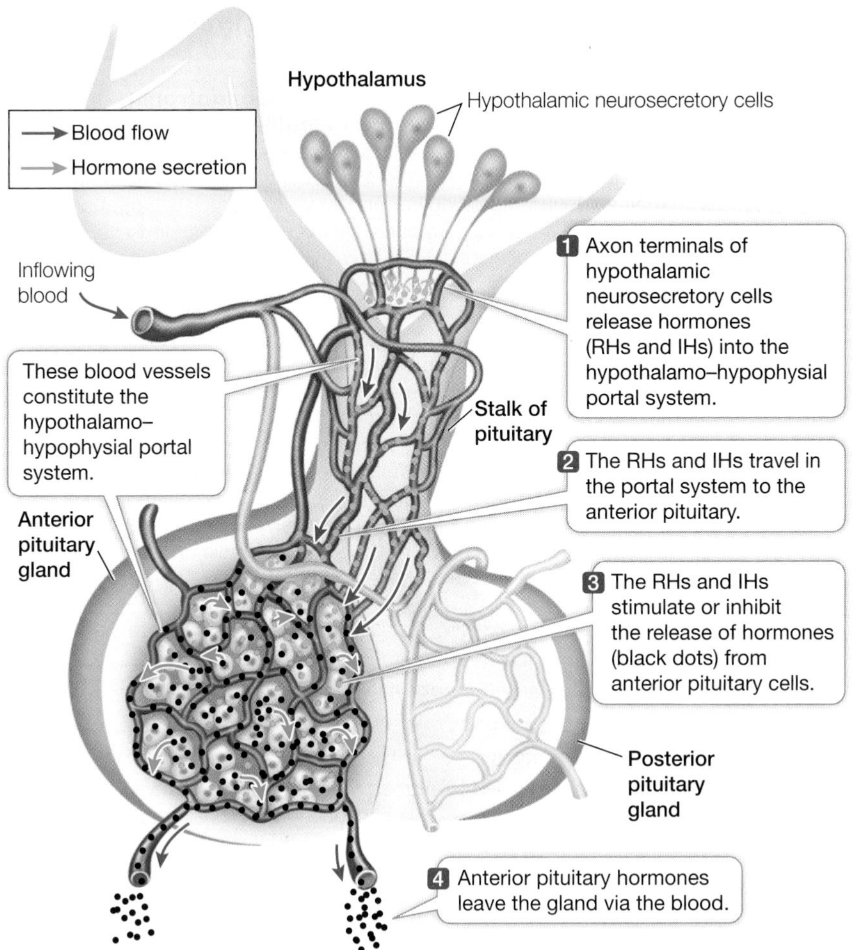

FIGURE 32.8 The Anterior Pituitary Gland Cells of the anterior pituitary gland produce four tropic hormones—which control other endocrine glands in the body—and several other peptide hormones. The anterior pituitary cells are controlled by releasing hormones (RHs) and inhibiting hormones (IHs) produced by neurosecretory cells in the hypothalamus. The RHs and IHs are delivered to the cells in the anterior pituitary by blood flowing through portal blood vessels that run between the hypothalamus and the anterior pituitary through the pituitary stalk. The hypothalamo-hypophysial system is sometimes called the hypothalamo-hypohyseal system.

hormones then travel to the anterior pituitary and control secretion by the anterior pituitary non-neural endocrine cells.

The hormones added to the blood by the neurosecretory cells in the hypothalamus are called **releasing hormones** (**RHs**) and **inhibiting hormones** (**IHs**; also called release-inhibiting hormones). Each RH or IH is specific in its actions. For example, there are separate RHs for growth hormone and thyroid-stimulating hormone. When the RH for growth hormone is secreted in the hypothalamus, it affects the specific cells in the anterior pituitary that secrete growth hormone, stimulating them to release (secrete) their hormone into the general circulation. The RH for thyroid-stimulating hormone, in contrast, stimulates the specific pituitary cells that secrete TSH.

In summary, in the anterior pituitary system, *neurosecretory cells in the brain control the secretion of hormones into the general circulation by other endocrine cells*. This is a second type of interface between the nervous and endocrine systems. It too is a type of interface that is common in both vertebrates and invertebrates.

Endocrine cells are organized into control axes

A typical sequence in vertebrates is for a hormone secreted by brain neurosecretory cells to control the secretion of a tropic hormone by anterior pituitary cells, and for the tropic hormone then to control hormone secretion by the cells in a gland elsewhere in the body. A system of this type—in which endocrine cells act on each other in sequence—is called an **axis**.

A good example of the operation of an axis is provided by the HPA axis, where "HPA" refers to the sequence hypothalamus–pituitary gland–adrenal cortex. This axis controls the secretion of adrenal cortex hormones—termed glucocorticoids—that are involved in responding to long-term stress. The primary glucocorticoid in mammals is cortisol (see Figure 32.4B). When the brain, using its great computational capacity to integrate inputs, detects that an animal is under stress, some of the neurosecretory cells in the hypothalamus secrete a releasing hormone (RH) called corticotropin-releasing hormone. This RH travels in the portal system to cells in the anterior pituitary that respond by increasing secretion of the tropic hormone adrenocorticotropic hormone (ACTH). ACTH, transported by blood flow in the general circulation, then travels to the adrenal cortex, which responds by increasing secretion of glucocorticoids into the blood.

Negative feedback is often a principle of enormous importance in the operation of an endocrine axis. To explain, let's consider an axis such as the HPA axis to consist of three steps, with the hypothalamus being step 1, the pituitary step 2, and a third gland step 3, as shown in the systems diagram in **FIGURE 32.9**. Hormones secreted in step 3 often "feed back" to affect hormone production in steps 1 and 2. Similarly, hormones secreted in step 2 often feed back to affect step 1. The most common form of such feedback is for the hormones secreted in later steps to inhibit production of the hormones in earlier steps. It's because of this inhibition that the feedback is a form of negative feedback.

LINK As discussed in **Key Concepts 1.2** and **28.3**, a negative feedback process is called "negative" because the control mechanism activates effectors to reduce the magnitude of variations (i.e., negate variations) in the controlled variable, thereby tending to stabilize the controlled variable.

Negative feedback of this sort prevents hormone concentrations from shooting up to very high levels or plunging to very low levels. It tends to stabilize hormone levels. If the hormones being produced in step 3 start to accumulate in the blood to excessively highly concentrations, the negative feedback that these hormones exert at steps 1 and 2 acts to reduce the stimulus for secretion at step 3.

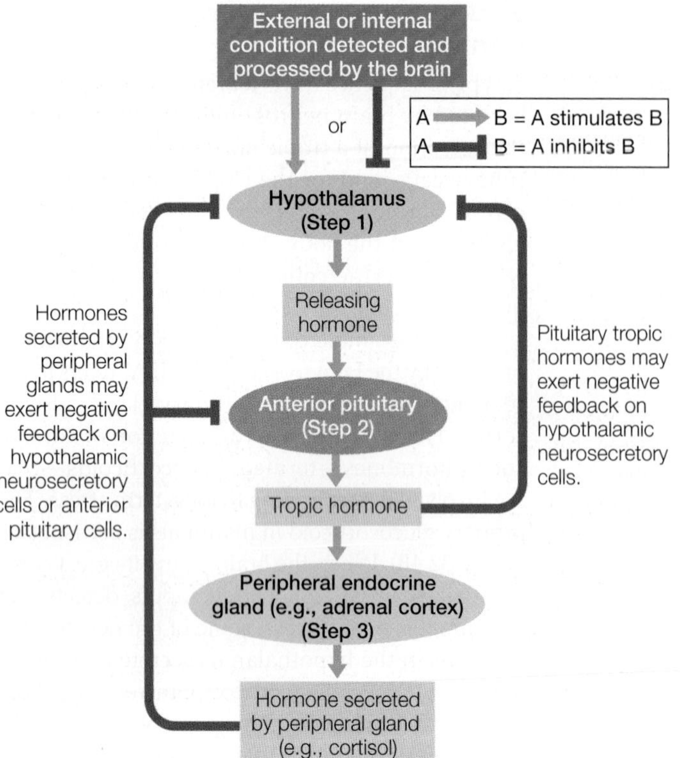

FIGURE 32.9 Multiple Feedback Loops Control Hormone Secretion Multiple negative-feedback loops regulate the chain of command from hypothalamus to anterior pituitary to peripheral endocrine glands. For purposes of this diagram, the glands controlled by anterior pituitary tropic hormones are termed peripheral glands.

Why are the negative-feedback loops important for stability of blood hormone concentrations?

The HPA axis exemplifies the general principles diagrammed in Figure 32.9. When the blood concentration of glucocorticoid (e.g., cortisol) increases following stimulation of the adrenal cortex (step 3), blood glucocorticoid tends to cause cells in the hypothalamus (step 1) to reduce release of corticotropin-releasing hormone. It also tends to cause cells in the anterior pituitary (step 2) to reduce secretion of ACTH. This negative feedback helps stabilize the blood glucocorticoid concentration at a relatively constant level.

Animation 32.1 The Hypothalamus and Negative Feedback

Hypothalamic and anterior pituitary hormones are often released in pulses

In many cases the brain stimulates the neurosecretory cells in the hypothalamus to secrete in pulses instead of continuously. Corticotropin-releasing hormone, for example, is typically secreted in mammals in two or three pulses per hour. Pulsed secretion of RHs or IHs often proves to be necessary for these hormones to act, as discussed in **INVESTIGATION FIGURE 32.10**. Continuous secretion can lead target cells in the anterior pituitary to become less responsive to the RHs or IHs. Pulsed secretion is hypothesized to prevent this loss of responsiveness. At the next step in the axis, the anterior pituitary hormones are often also secreted in pulses.

INVESTIGATION

FIGURE 32.10 Effectiveness of Gonadotropin-Releasing Hormone during Pulsatile and Continuous Infusion Gonadotropin-releasing hormone (GnRH) is the hypothalamic releasing hormone that controls pituitary secretion of the two gonad-regulating hormones, follicle-stimulating hormone (FSH) and luteinizing hormone (LH). In adult rhesus monkeys, the hypothalamus secretes GnRH in relatively short pulses, about once every 1–3 hours. A monkey can be rendered incapable of secreting GnRH by procedures that make the relevant part of its hypothalamus nonfunctional. When a monkey stops secreting GnRH, its anterior pituitary no longer secretes FSH or LH, and blood concentrations of these gonad-regulating hormones decline to low levels. In one of the most cited papers in the history of endocrinology, investigators at the University of Pittsburgh studied rhesus monkeys that could not make GnRH.[a] These monkeys were tested in two ways: either GnRH was infused into their blood in pulses, or GnRH was infused continuously. The objective was to determine if the *timing* of GnRH delivery (pulsatile or continuous) had any significant effect on the *actions* of GnRH.

HYPOTHESIS

The effectiveness of GnRH depends on the timing of its delivery.

METHOD

1. Administer GnRH to a group of 7 female rhesus monkeys unable to produce GnRH because the GnRH-secreting part of their hypothalamus has been rendered nonfunctional. Infuse the GnRH into the circulating blood either (A) in hourly pulses over a period of 20 days or (B) continuously for 10 days at one of five different rates. During pulsatile infusion, one 6-minute pulse (at a rate of 1 microgram [μg] per minute) was infused each hour—matching the timing of pulses seen naturally.
2. Measure the LH and FSH blood concentrations attained either daily (for pulsatile infusions) or toward the end of the 10-day infusion period (for continuous infusions).

RESULTS

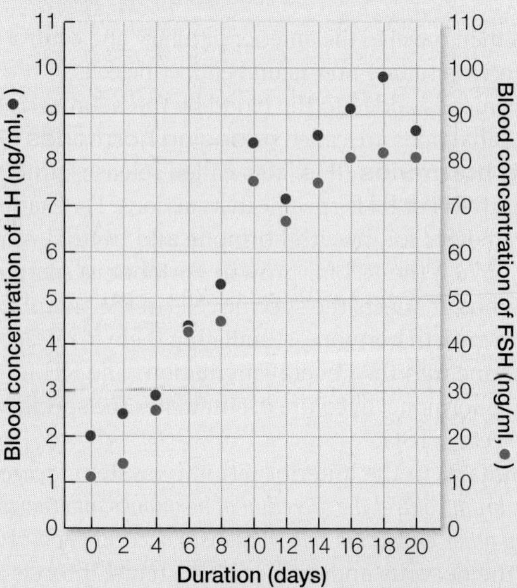

(B) Continuous infusion after 10 days

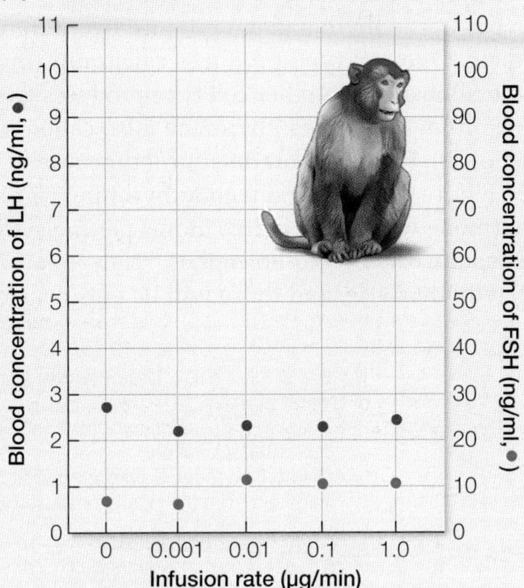

The graphs show average blood concentrations of luteinizing hormone (LH; *y* axis on the left) and follicle-stimulating hormone (FSH; *y* axis on the right) under both conditions. At 10 days, the LH and FSH levels are much higher in animals receiving pulsatile GnRH delivery than in those receiving continuous infusion at any of the five rates.

CONCLUSION

The response to GnRH differs depending on the timing of administration.

ANALYZE THE DATA

1. What do you conclude from the data presented on pulsatile delivery of GnRH? Explain, being sure to note both starting and ending hormone levels.
2. To focus on *timing* by comparing pulsatile and continuous delivery, why might you want to keep the dosage (total amount of GnRH infused) the same in the two delivery methods? Explain.
3. What continuous rate of GnRH infusion provides the same total daily dose of GnRH as the pulsatile delivery? To analyze the data with a focus on timing, what specific data should be compared in the results for pulsatile and continuous infusion? What do you conclude? Explain.
4. Notice that in the experiments using continuous infusion one of the infusion rates is 0. Why did the investigators measure LH and FSH during infusion without GnRH? Focusing on only the data in part B of the figure, what do you conclude about continuous infusion?
5. What is your final conclusion about control of the anterior pituitary by the hypothalamus?

[a]P. E. Belchetz et al. 1978. *Science* 202: 631–633.

A similar **Analyze the Data** exercise may be assigned in **Achieve**.

REVIEW & APPLY | 32.3

R The pituitary gland has two parts—the anterior pituitary and posterior pituitary—which have different developmental origins and function in different ways. Both parts have close functional links with the brain. The posterior pituitary is a neurohemal organ where hormones produced by neurosecretory cells in the hypothalamus of the brain are released into the blood. In mammals the posterior pituitary secretes two peptide hormones: antidiuretic hormone (ADH) and oxytocin. The anterior pituitary is a non-neural endocrine gland that secretes four tropic hormones as well as growth hormone (GH), prolactin, and a few other hormones. The anterior pituitary is controlled by releasing hormones (RHs) and inhibiting hormones (IHs) secreted by neurosecretory cells in the hypothalamus, often in pulses. Pulsed release is hypothesized to prevent loss of sensitivity in the target cells in the pituitary.

Endocrine cells often act on each other in sequence, a system known as an axis. For example, the hypothalamus–pituitary gland–adrenal cortex (HPA) axis controls the secretion of glucocorticoids (e.g., cortisol) by the adrenal cortex. Negative feedback often controls rates of hormone secretion within an axis.

1. Considering the hierarchy of control, why might it be misleading to call the pituitary gland the "master gland"?
2. Why are many anterior pituitary hormones called tropic hormones?
3. Neurosecretory cells in the hypothalamus and cells in the anterior pituitary gland both have cortisol receptors. If they were to lose their cortisol receptors, these cells would immediately begin secreting unusually large amounts of corticotropin-releasing hormone and ACTH. Why?

The hypothalamus and anterior pituitary gland control the secretion of hormones from several other endocrine glands. We will now look at the hormones that some of these other glands secrete, and their functions.

32.4 Hormones Regulate Mammalian Physiological Systems

LEARNING OBJECTIVES

By the end of this key concept you should be able to:

32.4.1 **State the physiological reason why iodized salt is beneficial to human health.**

32.4.2 **Explain how, during intrauterine development, the external sex organs of boys and girls can be identical at first but be different at birth.**

32.4.3 Specify where endocrine cells are located in the human body.

Hormones are involved in controlling and coordinating a wide range of mammalian physiological systems, and these regulatory processes in mammals often have parallels in other vertebrates. Here we take a brief look at the some of the major roles of hormones in the lives of mammals.

FIGURE 32.11 shows the locations of the principal mammalian endocrine glands and summarizes their major hormones and actions. In addition to the principal glands, many organs—such as the heart and stomach—contain endocrine cells even though they are not named as endocrine glands (the box in the figure lists some of these). **FIGURE 32.12** identifies several endocrine control systems discussed in other chapters.

The thyroid gland is essential for normal development and provides examples of hormone deficiency disease

The thyroid gland is found in the neck, wrapped around the front of the trachea (windpipe) (see Figure 32.11). Two different cell types in the thyroid gland produce two different types of hormones. One cell type produces calcitonin (involved in calcium metabolism). The other cell type produces two iodine-containing amine hormones: thyroxine (also called tetraiodothyronine or T_4)—which contains four iodine atoms per molecule (see Figure 32.4C)—and triiodothyronine (T_3), a closely similar hormone with three iodine atoms per molecule. When biologists speak of "thyroid hormones," they mean T_4 and T_3 unless stated otherwise, and those will be our focus here. The

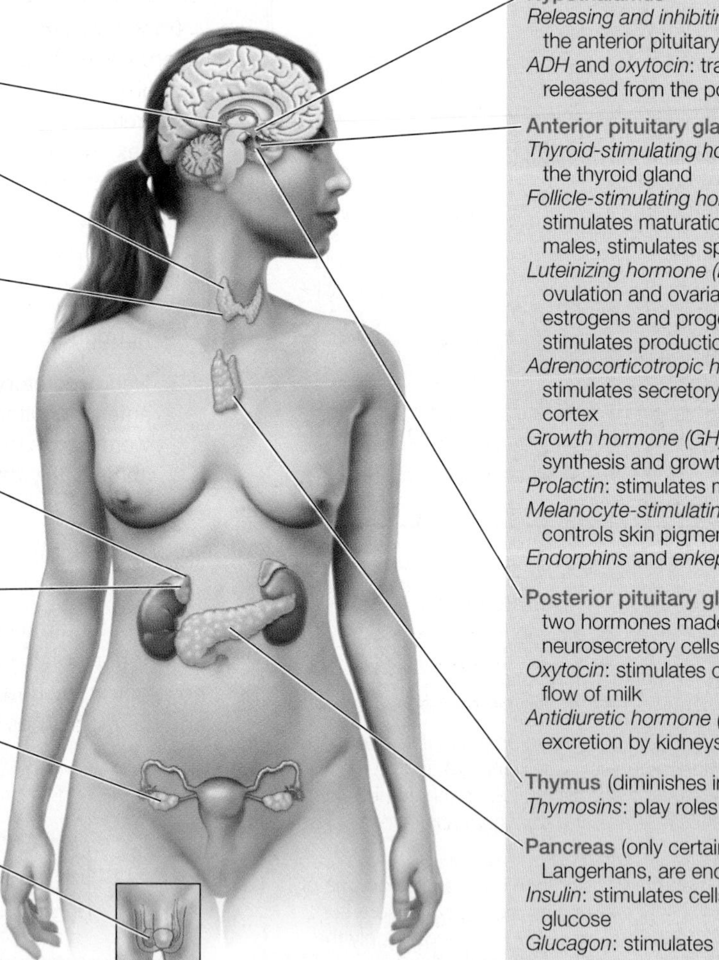

Other organs include cells that produce and secrete hormones:

Organ	Hormone
White adipose tissue	Leptin
Heart	Atrial natriuretic peptide
Kidney	Erythropoietin
Stomach	Gastrin, ghrelin
Intestine	Secretin, cholecystokinin
Liver	Insulin-like growth factors

FIGURE 32.11 The Human Endocrine System

Activity 32.1 The Human Endocrine Glands

Chapter 28
Transformations of Energy and Matter: Nutrition, Temperature, and Homeostasis

Insulin and glucagon (both secreted by endocrine cells in the pancreas) regulate processing of absorbed food materials from one meal to the next meal (see Key Concept 28.4).

Chapter 34
Animal Reproduction

Follicle-stimulating hormone and luteinizing hormone (from the anterior pituitary) help control the menstrual and estrous cycles and are essential for male fertility (see Key Concept 34.2; Figure 34.13).

Estrogens (from the ovaries) help control the menstrual and estrous cycles (see Key Concept 34.2; Figure 34.13).

Progesterone and estrogens (from the ovaries and placenta) help control uterine development that is required for pregnancy (see Key Concept 34.2; Figure 34.13).

Oxytocin (from the posterior pituitary) plays a crucial role in controlling milk output from the mammary glands and in controlling uterine muscle contractions essential for the birth process (see Figure 34.14).

FIGURE 32.12 Some Endocrine Control Systems Are Discussed in Other Chapters

presence of iodine in these hormones is the reason for most of our dietary iodine requirement. The thyroid secretes T_4 and a relatively small amount of T_3. Of the two, T_3 is more active in affecting target cells, and after T_4 has been secreted into the blood, peripheral tissues convert it to T_3, as noted earlier.

The thyroid hormones are vital during development and growth. They promote cellular amino acid uptake and protein synthesis. They enter cells and bind to an intracellular receptor that promotes transcription of numerous genes. Insufficiency of the thyroid hormones in a human fetus or growing child can greatly retard mental and physical development. The thyroid hormones also are noted for elevating metabolic rate in mammals and birds.

Soils deficient in iodine occur in many parts of the world. Plants that grow in these soils tend to be low in iodine, and people who depend on local foods in these areas often suffer from iodine deficiency. Iodized table salt (salt containing traces of iodine) was invented to solve this problem. If people use iodized salt in a customary way to salt their food, they get enough iodine regardless of iodine levels in their local soils. Impaired childhood mental development caused by deficiency of thyroid hormones—resulting from a dietary shortage of iodine—remains a massive worldwide problem because in many poverty-stricken regions, iodized salt is not readily available. Global human mental performance would be raised in a generation if iodine deficiency during pregnancy and childhood could be eliminated.

Goiter—a type of enlarged thyroid gland (**FIGURE 32.13**)—is another hormone deficiency disease that (in its most common form) occurs in adults who receive inadequate iodine in their diet. It was common in many parts of the United States until use of iodized salt became the norm. Goiter illustrates that the proper operation of feedbacks in endocrine control systems can be critical for health. The thyroid hormones, when present in sufficient amounts in the blood, exert negative feedback on the production of thyroid-stimulating hormone (TSH) by the anterior pituitary. This negative feedback does not take place adequately in people suffering from iodine deficiency who produce only small amounts of thyroid hormones. With inadequate negative feedback, the pituitary steadily produces TSH at high levels, stimulating the thyroid gland to grow to excessive size.

Sex steroids control reproductive development

Many hormones, including the thyroid hormones, play critical roles in mammalian development. In certain ways, however, the hormones of greatest importance for development are the sex steroids and the hypothalamic and pituitary hormones that control them.

The gonads—the testes of males and the ovaries of females—produce steroid hormones as well as gametes (sperm and ova). Many of these hormones are described as androgens or estrogens. Masculinizing steroid hormones are called **androgens** (Greek, "male-makers"). **Testosterone** (see Figure 32.4B) is an androgen and is the principal hormone produced by the testes. Feminizing steroid hormones are called **estrogens**. Two or more types of estrogen molecules are typically secreted, one often being estradiol (see Figure 32.4B). The ovaries produce estrogens and another female sex steroid, **progesterone**, which

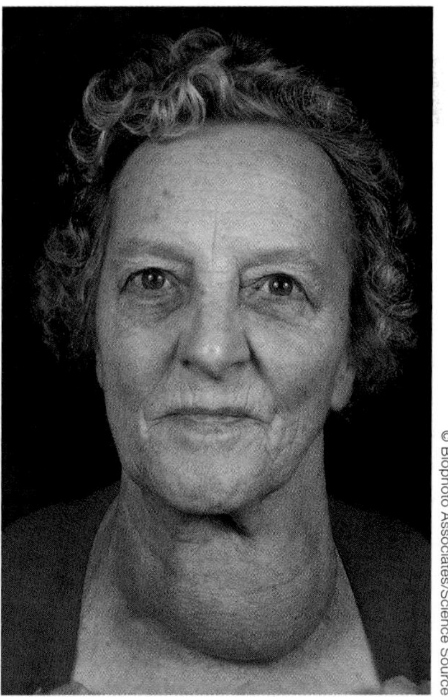

FIGURE 32.13 Goiter Enlargement of the thyroid gland can occur when a person has insufficient iodine. This iodine deficiency results in insufficient secretion of thyroid hormones and heightened stimulation of the gland by thyroid-stimulating hormone (TSH).

THINK LIKE A SCIENTIST

Commercialization of hormones: New choices for people to make

The marketplace is flooded with testosterone products, and if you choose to pay attention, you will be inundated with confident claims that the life of every man will be extraordinarily improved by the simple act of taking testosterone pills, using testosterone patches, or injecting testosterone. In some of the advertising there is no mention of possible downsides. A scientist would use the process of science to examine the claims critically, asking for objective evidence and applying hard-nosed logic.

1. Sperm production depends on the anterior pituitary hormones FSH and LH, which act on the testes, as we will discuss in detail in Chapter 34. Clinical evidence shows that men who take testosterone supplements often develop low sperm counts and become infertile, a downside for men who want to have children. What might be the cause of this outcome? Base your answer on the topics we have covered in this chapter.
2. Do an internet search for "testosterone supplements" or some other similar words of your choosing. Make a list of the first 25 (or first 50) "hits" you get in the order they are presented. Then categorize the hits according to the degree to which they reflect scientific thinking. For example, to categorize a hit, you might look at the source of the information presented (e.g., an article in a peer-reviewed journal, the National Institutes of Health, a health supplements marketplace, etc.). Then write a short essay on the conclusions you reach, taking into account the order of the hits and the methods you used to categorize them. Thinking about the task of communicating science to the general public, what advice would you offer if you were to have the opportunity to speak to a random group of young adults?

is involved principally in coordinating processes associated with pregnancy (see Chapter 34). During pregnancy, the placenta also secretes estrogens and progesterone.

Androgens are not exclusive to males, and estrogens are not exclusive to females. Both sexes make both androgens and estrogens. However, the relative blood concentrations differ in the two sexes. In some cases androgens are simply precursors that are converted to estrogens by a process called aromatization.

During human embryonic development, the sex steroids play a key role in controlling whether an embryo develops phenotypically into a male or female. Androgens are required for male phenotypic development. In their absence an embryo develops as a female.

In the early weeks of embryonic development, the gonads and external genital structures are the same whether an individual is male or female (**FIGURE 32.14**). The sex of an individual human or other mammal is determined genetically at the time of conception. Females inherit two X chromosomes, whereas males inherit an X and a Y chromosome (see Key Concept 34.1). Early in development, a protein encoded by a gene on the male's Y chromosome, the *SRY* gene, causes the gonads to differentiate as testes (otherwise they become ovaries). As the differentiation into testes proceeds, the gonads of a male embryo begin to produce testosterone. In humans, starting 2–3 months after conception, testosterone increases from a low blood concentration to a high blood

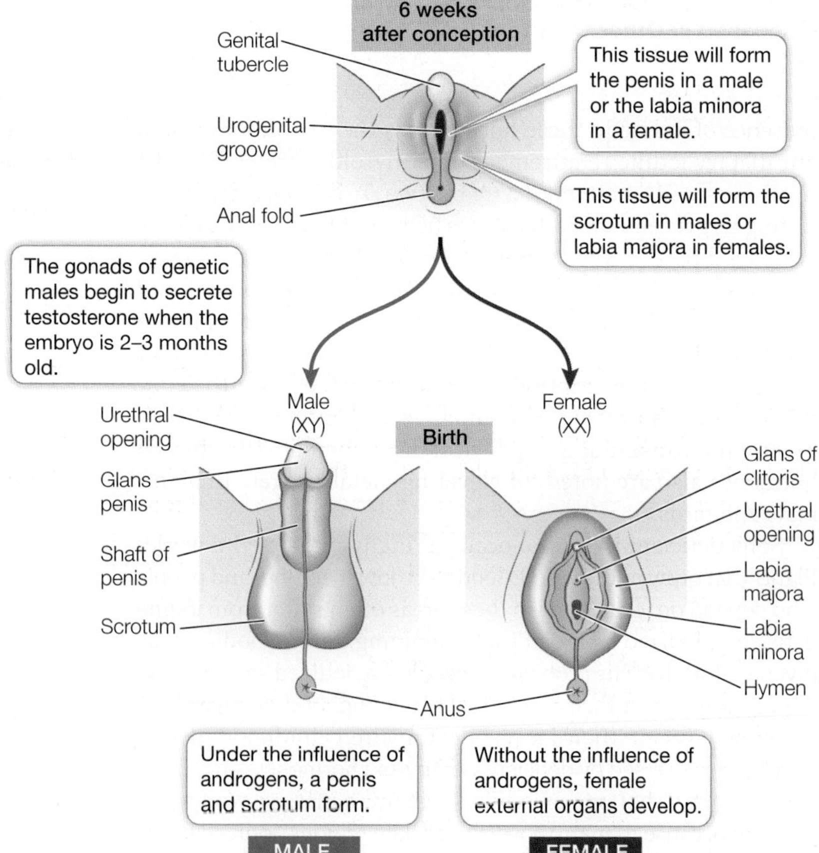

FIGURE 32.14 Sex Steroids Direct the Prenatal Development of Human Sex Organs The external sex organs of early human embryos are undifferentiated. Testosterone promotes the development of male external sex organs. In its absence, female sex organs form.

concentration, remains high for about 2 months, and then falls to a low concentration again because the gonads stop producing it rapidly. This period of high blood testosterone causes phenotypic differentiation of the external genital structures into the male form. Without testosterone, the structures differentiate into female form. The scrotum of males and certain labial folds of females develop from the same early structures (see Figure 32.14). The head (glans) of the penis and the head of the clitoris also have the same embryonic origins. The gonads are initially located in the abdomen in both sexes. In males, the testes migrate out of the abdomen into the scrotum.

Media Clip 32.1 The Testosterone Factor

LINK As discussed in **Key Concept 8.2**, sex chromosomes do not always occur in homologous pairs. Human females have a homologous pair of large X chromosomes (and are thus designated XX). The sex chromosomes of the human male, by contrast, consist of a large X and a very small Y chromosome (designated XY).

Just as hormones play a crucial role in embryonic development, they again play a crucial role at puberty. Recall that luteinizing hormone (LH) and follicle-stimulating hormone (FSH)—both tropic hormones that act on the gonads—are secreted by the anterior pituitary gland, and their secretion is controlled by gonadotropin-releasing hormone (GnRH) secreted by neurosecretory cells in the hypothalamus. When a person is between about 1 year of age and the start of puberty, the GnRH neurosecretory cells in the hypothalamus are quiet, producing a GnRH pulse only once every few hours. Puberty begins when these cells shift to producing GnRH pulses at far higher rates. Despite many theories, researchers remain uncertain about the immediate causes of this change in activity of the GnRH neurosecretory cells. The anterior pituitary responds to the increase in the pulse rate of GnRH secretion by greatly increasing secretion of FSH and LH. In males, the elevated blood concentration of LH stimulates cells in the testes to produce testosterone vigorously. A large increase occurs in the blood concentration of testosterone, which (along with other androgens) brings about many changes. Testosterone enters cells, binds to intracellular receptors, and alters gene expression throughout the body. As a consequence, a boy's voice deepens, hair begins to grow on his face and body, his skeletal muscles increase in mass, and his testes and penis grow larger. The increased blood concentration of FSH stimulates the production of sperm.

In females, increased blood levels of LH stimulate the ovaries to increase production of estrogens. Concentrations of estrogens in the blood rise and initiate development of many traits: a girl's breasts, vagina, and uterus become larger; her hips broaden; she develops increased subcutaneous fat; and her menstrual cycles begin. Increased FSH stimulates the maturation of ovarian follicles, which are necessary for production of mature eggs (see Key Concept 34.1).

REVIEW & APPLY | 32.4

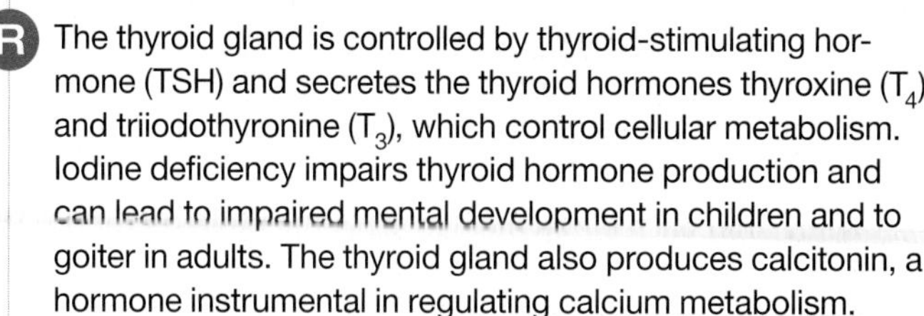

R The thyroid gland is controlled by thyroid-stimulating hormone (TSH) and secretes the thyroid hormones thyroxine (T_4) and triiodothyronine (T_3), which control cellular metabolism. Iodine deficiency impairs thyroid hormone production and can lead to impaired mental development in children and to goiter in adults. The thyroid gland also produces calcitonin, a hormone instrumental in regulating calcium metabolism.

Sex steroids (predominantly androgens in males, estrogens and progesterone in females) are produced by the gonads under control of tropic hormones, the gonadotropins, secreted by the anterior pituitary. In both sexes, the two principal gonadotropins are follicle-stimulating hormone (FSH) and luteinizing hormone (LH). Pituitary secretion of the gonadotropins is controlled by gonadotropin-releasing hormone (GnRH) secreted by cells in the hypothalamus. The axis consisting of GnRH, the gonadotropins, and the sex steroids controls prenatal sexual development, puberty, and adult reproductive functions.

A
1. Boys are occasionally born with a testicle missing from the scrotum. The missing testicle is nearly always in the abdomen, and its position can be corrected surgically. But why might a testicle be in the abdomen at birth?
2. Explain why the thyroid gland can grow to great size when it secretes inadequate amounts of thyroid hormones because of iodine deficiency.
3. What endocrine event in the hypothalamus initiates puberty?

Now that we've examined vertebrate endocrine systems, let's conclude with a look at endocrine systems in invertebrates. Of all the invertebrates, the insects are the best understood in this regard. A striking parallel is that endocrine controls are of enormous importance in insect development, just as they are in vertebrate development.

32.5 | The Insect Endocrine System Is Crucial for Development

LEARNING OBJECTIVES

By the end of this key concept you should be able to:

32.5.1 **Explain why the size and shape of an insect's body change in stages (rather than gradually) during the insect's development into an adult.**

32.5.2 **Describe how juvenile hormone determines the timing of when an insect develops into an adult.**

Insects and other arthropods, such as crayfishes and crabs, have elaborate endocrine systems. For example, many insects have diuretic and antidiuretic hormones that control excretion of water by the insect organs that serve kidney functions. Diuretic hormones promote excretion of a high volume of water. Some of the blood-sucking insects secrete diuretic hormones immediately after each blood meal to promote rapid excretion of much of the water in the meal, thereby reducing weight and concentrating the nutritious part of the meal (e.g., blood proteins) in the gut.

The best-understood endocrine systems in insects control growth and development. A key feature of insects (and other arthropods) is that they have a hard exterior skeleton, the exoskeleton (see Key Concept 22.3). For an insect to grow, it must periodically shed its exoskeleton and produce a larger one—a process called *molting* or *ecdysis*. The life of an insect consists of a series of discrete steps: growing followed by molting, then more growing followed by molting, and so on. Insects typically undergo changes in their external shape as they develop from young individuals to adults. However, the rigid exoskeleton cannot change shape between molts. Thus all the changes of external shape in the life of an insect occur during molting. In this way, the molting process is intimately connected with both growth and the development of body form.

Although many hormones are involved in controlling molting and development, three are of principal importance (**FIGURE 32.15**): *prothoracicotropic hormone (PTTH)*, a peptide tropic hormone secreted by brain neurosecretory cells; *ecdysone*; and *juvenile hormone (JH)*. Both ecdysone and JH are secreted by non-neural endocrine cells—ecdysone by cells in the prothoracic glands, JH by cells in the corpora allata (singular corpus allatum) (see Figure 32.15). Whereas ecdysone is a steroid, JH is a terpene.

In the regulation of molting, the brain neurosecretory hormone PTTH controls the activity of cells in the prothoracic glands that secrete ecdysone into the blood. That is—as occurs in some cases in the interface between the vertebrate nervous and endocrine systems—*neurosecretory cells in the insect brain control secretion of a hormone into the general circulation by other endocrine cells.* Invertebrates and vertebrates, we see, sometimes exhibit similar major organizational features (see page 815).

The insect brain uses its integrative and computational powers to determine when each molt will occur. The neurosecretory cells that secrete PTTH receive synaptic inputs from other, ordinary brain neurons that determine the times when the neurosecretory cells secrete PTTH. In an arrangement reminiscent of that in the posterior pituitary gland in vertebrates, the axons of the brain neurosecretory cells extend out of the brain, and their axon terminals are positioned in the corpora allata (see Figure 32.15). The corpora allata (among other attributes) are neurohemal organs, where PTTH is released into the blood.

During each episode of PTTH secretion, the flow of blood carries PTTH to the prothoracic glands, where it stimulates ecdysone secretion. However, ecdysone is not the molecule that directly activates molting. Instead, as the ecdysone circulates, it undergoes peripheral activation to form *20-hydroxyecdysone*, a hormone

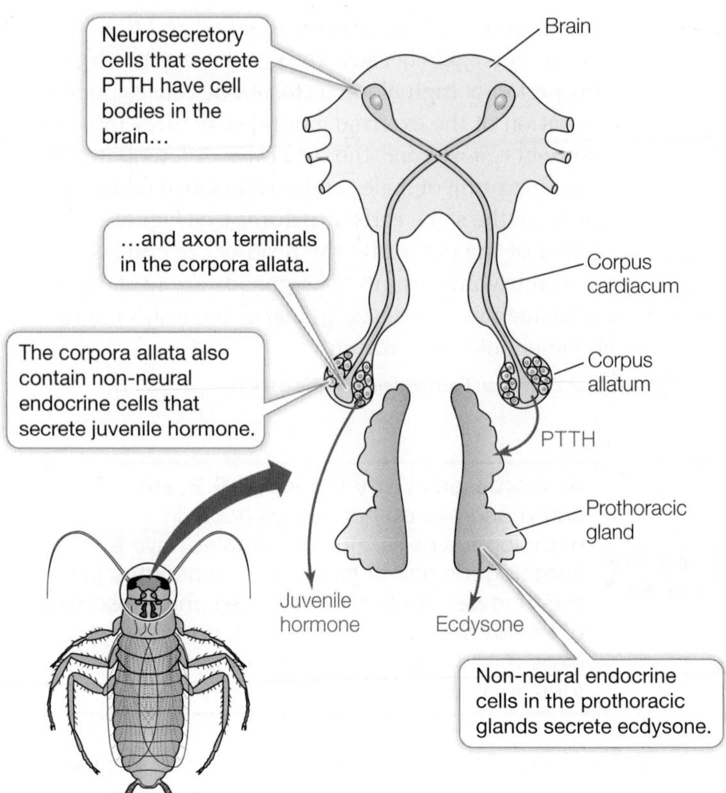

FIGURE 32.15 Key Structures and Hormones in the Control of Insect Molting and Development Hormones are in red. Ecdysone is peripherally activated to 20-hydroxyecdysone, which is the active form of the hormone. PTTH, prothoracicotropic hormone.

What is another example of peripheral activation we have recognized in this chapter?

(a hydroxylated steroid sometimes called *molting hormone*) that primarily activates molting. One site of action is the insect's epidermis, the superficial layer of living tissue that produces the exoskeleton. Stimulated by 20-hydroxyecdysone, the epidermal cells secrete enzymes that loosen their connection with the old exoskeleton, allowing the old exoskeleton to be shed. Then the epidermal cells produce a new, larger exoskeleton.

Juvenile hormone (JH) affects an insect's *developmental stage* as it molts. Some insects, such as moths and butterflies, undergo a complete metamorphosis during their development. Each individual starts life as a larva (caterpillar), retains its larval form for several molts, and then metamorphoses into an adult of radically different form (see Key Concept 22.4). Here we discuss the role of JH in such insects.

JH is lipid-soluble. Thus when it is secreted into the blood and circulates throughout the body, it enters target cells. Many types of cells in an insect's body have intracellular receptors for JH and respond to it by undergoing changes in gene transcription.

When, during molting, a larva's blood concentration of 20-hydroxyecdysone is high, a key determinant of the outcome of the molt is whether the concentration of JH is simultaneously high.

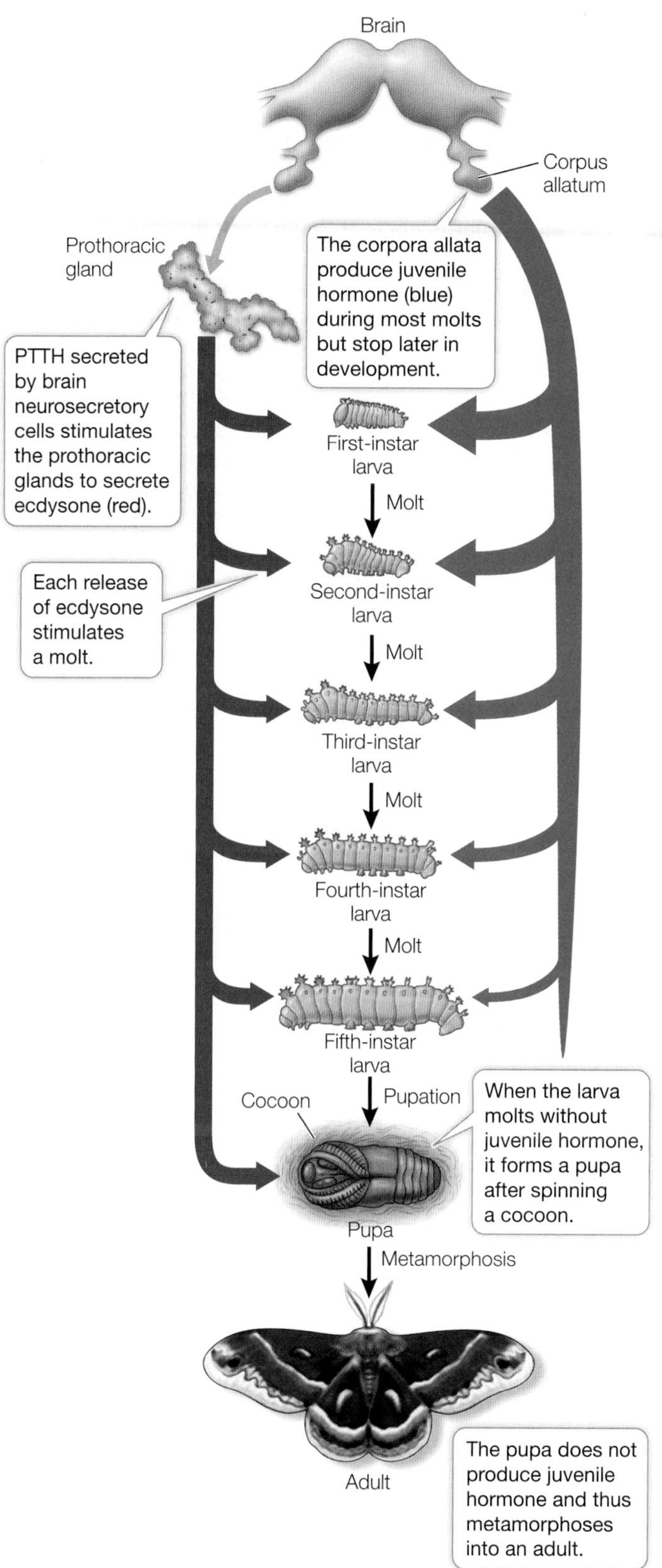

FIGURE 32.16 Hormonal Control of Molting and Metamorphosis in a Moth The silkworm moth *Hyalophora cecropia* is illustrated. An instar is a stage between molts, thus the expressions "first-instar larva," "second-instar larva," and so forth. (After C. M. Williams. 1952. *Biol Bull* 103: 120–138.)

Animation 32.2 Complete Metamorphosis

If it is, the insect molts into a larger larva. Throughout an insect's early development, the non-neural endocrine cells in the corpora allata maintain a high blood concentration of JH. Thus each successive molt results in an ever-larger larval form (**FIGURE 32.16**). However, the endocrine cells in the corpora allata become inactive during the final larval stage. At that point, when the concentration of 20-hydroxecdysone rises and molting occurs (now in the absence of JH), the larval form is not retained. Instead, the insect enters a resting stage of distinctive body form, called a pupa, during which the insect's body is extensively remodeled inside the pupal exoskeleton. Then, when the insect molts yet again without a high concentration of JH in the blood, the individual emerges as an adult.

REVIEW & APPLY | 32.5

R Molting in insects is controlled by two hormones. One is prothoracicotropic hormone (PTTH), a tropic hormone secreted by brain neurosecretory cells. The second is ecdysone, which is secreted by the prothoracic glands and undergoes peripheral activation to 20-hydroxyecdysone. A third hormone, juvenile hormone (JH)—secreted by cells in the corpora allata—prevents maturation. When an insect stops producing JH, it molts into an adult.

A
1. Why is juvenile hormone called "juvenile" hormone?
2. What two actions does ecdysone have on insect epithelial cells?
3. In what way is there a fundamental similarity between (1) the control of ecdysone secretion by PTTH in insects and (2) the control of FSH secretion by GnRH in mammals?

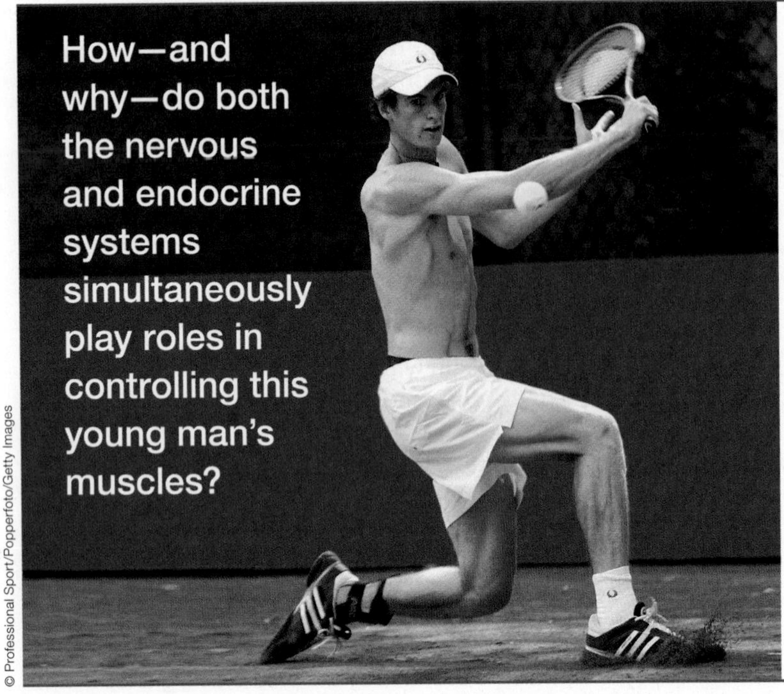

Answer: Two types of control of skeletal muscle are evident in this scene. First, the young man's skeletal muscles are engaged in the quick, moment-by-moment moves of the game. To return the ball with a backhand shot, he runs to a good position to execute his return, plants his feet, pulls back the racket, focuses his eyes on the ball, and swings—all in a few seconds. Fast and addressed signal transmission by the nervous system (see Chapter 31) is required to provide moment-by-moment control of the numerous muscles involved. Second, as the young man goes through adolescence, his muscles gradually become more highly developed, larger, and stronger—a process starting at puberty and requiring many years to be completed. His endocrine system controls these long-term muscle changes. As discussed in this chapter, the endocrine system is more suited than the nervous system for controlling long-term processes.

32 VISUAL SUMMARY

32.1 The Endocrine and Nervous Systems Play Distinct, Interacting Roles

- Signaling by the nervous system is fast and addressed, whereas signaling by the endocrine system is slow and broadcast.
- Mechanisms of chemical signaling specific to many different spatial scales have evolved, from **PARACRINES** to **PHEROMONES**. Neuronal and endocrine signaling are both types of chemical signaling, operative at intermediate scales (**FIGURE 32.2**).
- The nervous and endocrine systems exert effects on one another.

FIGURE 32.2

Local diffusion following long-distance electrical signaling

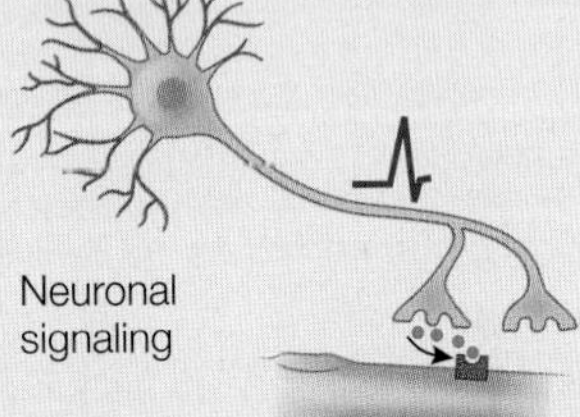

Transport in blood throughout the body

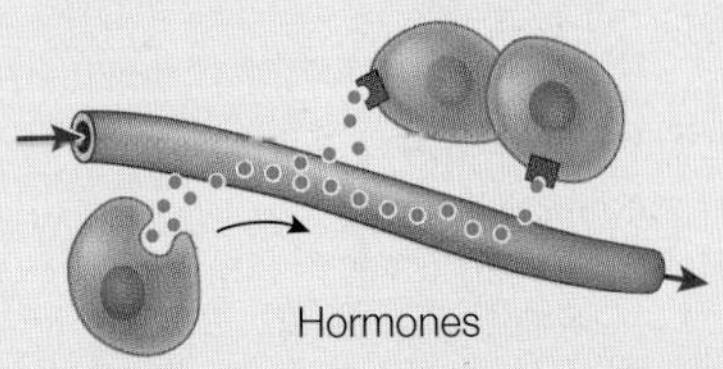

QUESTIONS

1. Define a hormone.
2. Why are growth and development typically under endocrine control more than nervous control?
3. Describe a process that is controlled by both the nervous and endocrine systems.

32.2 Hormones Are Chemical Messengers Distributed by the Blood

FIGURE 32.3

Neurosecretory cell

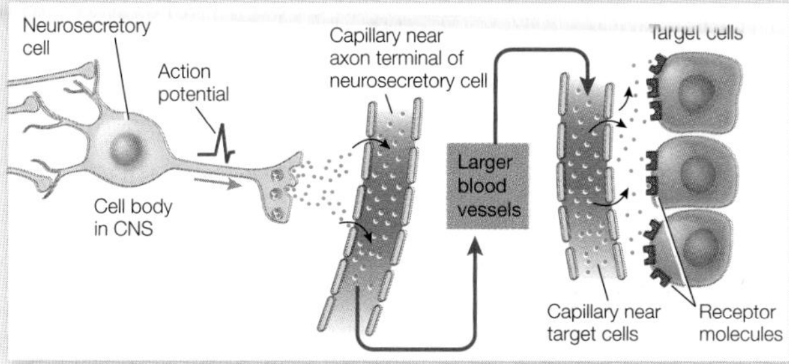

Non-neural endocrine cell

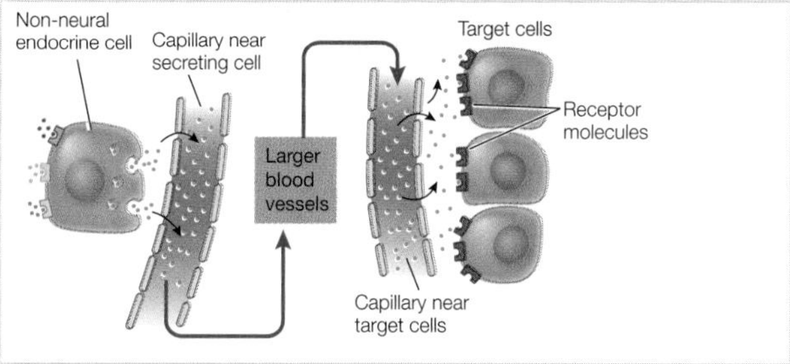

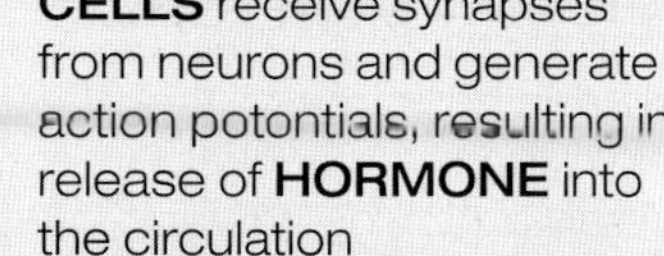

- **NEUROSECRETORY CELLS** receive synapses from neurons and generate action potentials, resulting in release of **HORMONE** into the circulation (**FIGURE 32.3**).
- Target cells for a hormone express a receptor protein for the hormone. A target cell's specific response to a hormone depends on the target cell, not on the hormone itself.
- Receptors for water-soluble hormones are typically in the cell membrane. Lipid-soluble hormones can pass through the cell membrane to reach **INTRACELLULAR RECEPTORS** inside a cell, where they can directly affect gene transcription.

QUESTIONS

1. What are the chemical classes of hormones?
2. What is a neurohemal organ, and why are neurorhemal organs critical for integration of the nervous and endocrine systems?
3. How is it possible for one hormone to have exceedingly different effects in different groups of evolutionarily related animals?

32.3 The Vertebrate Hypothalamus and Pituitary Gland Link the Nervous and Endocrine Systems

FIGURE 32.8

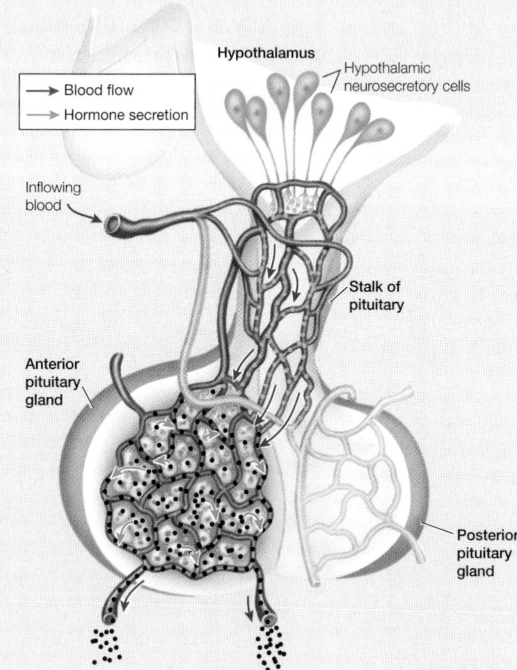

Go to ANIMATION 32.1

- Secretion of two key hormones in the **POSTERIOR PITUITARY GLAND—ANTIDIURETIC HORMONE (ADH)** and **OXYTOCIN**—is carried out by neurosecretory cells with cell bodies in the **HYPOTHALAMUS**.
- Hormones secreted by the **ANTERIOR PITUITARY GLAND** are produced by non-neural endocrine cells. These cells are controlled by brain neurosecretory cells that secrete **RELEASING HORMONES** and **INHIBITING HORMONES** into the **HYPOTHALAMO–HYPOPHYSIAL PORTAL SYSTEM** (**FIGURE 32.8**).
- Endocrine cells often act on each other in a sequence termed an **AXIS**. For example, hypothalamic neurosecretory cells secrete hormones that control anterior pituitary cells that secrete **TROPIC HORMONES** that control peripheral endocrine cells (e.g., in the thyroid or gonad).

1. Explain how negative feedback in the HPA axis helps prevent large increases or decreases in the blood concentration of cortisol.
2. What does it mean to say that a hormone is secreted in "pulses"? (Avoid the word "pulse" in your answer). Why might secretion in pulses have evolved?
3. Why might it be true to say that the brain controls secretion of antidiuretic hormone, oxytocin, and glucocorticoids?

32 VISUAL SUMMARY

32.4 Hormones Regulate Mammalian Physiological Systems

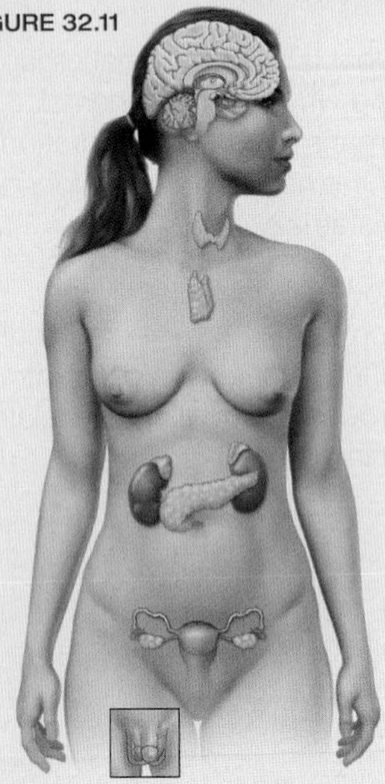

Go to ACTIVITY 32.1

FIGURE 32.11

- There are more than ten endocrine glands in the human body (**FIGURE 32.11**).
- The thyroid hormones T_4 and T_3 affect cellular protein metabolism, development, and (in mammals and birds) metabolic rate. The thyroid gland is part of an axis in which hypothalamic neurosecretory cells control anterior pituitary cells that secrete thyroid-stimulating hormone (TSH), and TSH directly controls the thyroid.
- Sex hormones in mammals—**ANDROGENS** and **ESTROGENS**—control whether the embryonic sex organs differentiate into male or female sex organs during intrauterine development.
Another key role for sex hormones is in the initiation and control of puberty.

QUESTIONS

1. What is a hormone deficiency disease? Give an example.
2. What does it mean to say, "A person's sex is determined genetically, but the form of their sex organs is determined hormonally."
3. Iodized salt is cheap. How might "global human mental performance…be raised in a generation" (quoting the text) by making iodized salt available to everyone on Earth?

32.5 The Insect Endocrine System Is Crucial for Development

Go to ANIMATION 32.2

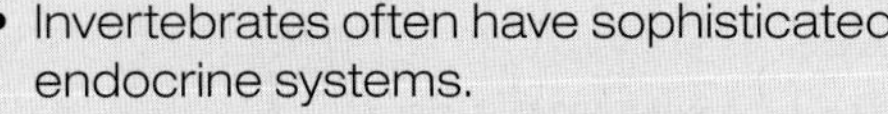

FIGURE 32.16

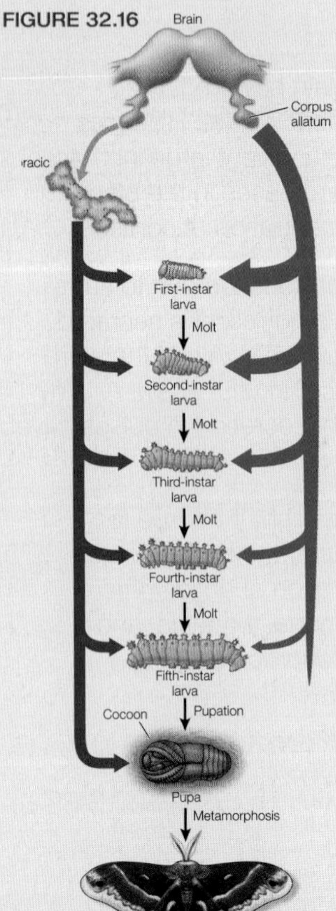

- Invertebrates often have sophisticated endocrine systems.
- In an insect, a molt is initiated when brain neurosecretory cells secrete prothoracicotropic hormone (PTTH), which stimulates ecdysone secretion (**FIGURE 32.16**).
- Juvenile hormone (JH) controls the timing of insect metamorphosis. While JH is present in the blood, the insect remains juvenile. Transformation to the adult form occurs when JH production is curtailed.

1. In what way is the control of ecdysone secretion in an insect closely similar to the control of thyroid hormone secretion in a vertebrate?
2. What is peripheral activation, and why is it important in understanding molting in insects?

Activity 32.2 **Concept Matching: Vertebrate Hormones**

Go to **Achieve** for the e-book, animations, activities, and additional resources and assignments.

Muscle and Movement 33

By having muscles, animals are able to move—one of their hallmark characteristics. Christmas Island red crabs (*Gecarcoidea natalis*) are descended from ocean crabs and must return to the ocean to breed. Yet they spend most of their time—and collect most of their food—in highly productive, inland rainforests. Here they walk toward the sea.

How do muscles make possible a life history in which ocean-reproducing animals use rainforests as their principal source of food? Thinking as broadly as possible, how has the evolution of muscles affected the life histories and survival of diverse animal species?

You will find the answer to these questions on page 846.

KEY CONCEPTS

33.1 Muscle Cells Develop Forces by Means of Cycles of Protein-Protein Interaction

33.2 The Function of Skeletal Muscle Depends on Interaction with the Skeleton and on ATP Supply, Cell Type, and Training

33.3 Many Distinctive Types of Muscle Have Evolved

33.1 Muscle Cells Develop Forces by Means of Cycles of Protein–Protein Interaction

LEARNING OBJECTIVES

By the end of this key concept you should be able to:

33.1.1 Explain how proteins interact to develop mechanical forces.

33.1.2 Describe the role of energy in muscular contraction and the mechanism by which energy is supplied to the contractile apparatus.

33.1.3 Distinguish between excitation and contraction and explain how they are linked.

Muscle cells—one of the defining properties of animals—develop forces by means of interactions among fiberlike, contractile proteins within the cells. These forces typically set matter in motion. As we learn in physics, energy is required for matter to be set in motion. The contractile proteins draw the necessary energy from adenosine triphosphate (ATP) made by the mitochondria in the muscle cells.

Muscle makes up a large portion of body mass in vertebrates and many other animals. In humans, for example, muscles account for almost half of body mass (**FIGURE 33.1**). Muscles are as important functionally as they are anatomically. They are the basis for virtually all behavior. They permit animals to walk, run, fly, or swim to acquire food or escape dangers. In humans and some other mammals, muscles of the face account for changing facial expressions. Besides behavior, many other physiological actions also depend on muscle contraction, such as the circulation of blood by the beating of the heart.

There are several types of muscle tissue in the animal kingdom. The contractile mechanism is basically similar in all muscle types. We begin our study of muscle function by focusing on the contractile mechanism in vertebrate **skeletal muscle**—the type of muscle that is attached to the bones of the skeleton and that provides the power for walking, flying, swimming, and other forms of locomotion.

Contraction occurs by a sliding-filament mechanism

In the scientific study of muscle, **contraction** refers to the *development of a force*. A muscle does not necessarily *shorten* when it undergoes contraction. To see the distinction between contraction and shortening, imagine grabbing a steel handrail attached to a concrete wall with your hand and pulling as hard as you can with your arm muscles. Your muscles will develop great forces, but they will not shorten because the handrail is immovable. For a muscle to shorten during contraction, the force it exerts must be greater than opposing forces. When you use your arm to pull on a drawer, your arm muscles shorten during contraction because the force they exert exceeds the opposing forces, and the drawer slides toward you.

The current theory of how muscle contracts is called the **sliding-filament theory**. This name describes what the process *looks* like and is useful in that way. When muscle cells contract in circumstances in which they shorten, tiny filaments in the cells look like they slide past each other during contraction. However, the name "sliding" can be misleading, because it suggests effortless motion driven by outside forces, like a sled sliding down a snow-covered hill under force of gravity. In fact, the sequence of

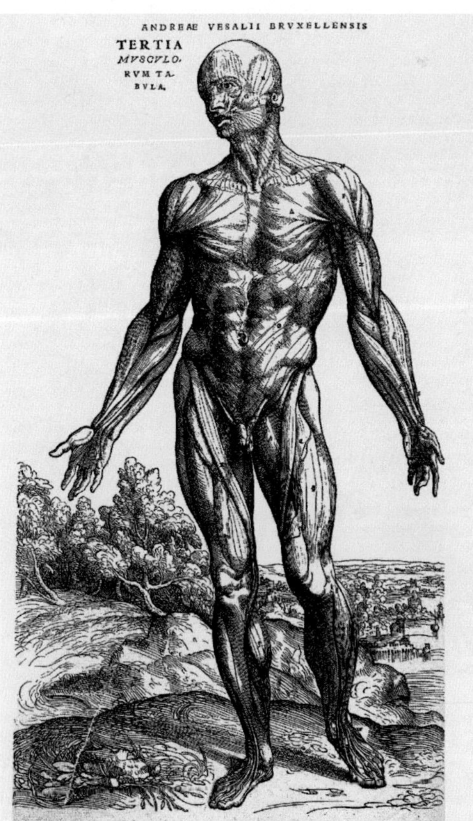

FIGURE 33.1 Muscles Account for a Large Portion of Body Mass Andrea Vesalius produced this wood block print, the first accurate depiction of the skeletal muscles of the human body, in 1543. He was noted for posing his subjects as they might have stood in life. (From A. Vesalius. 1543. *De humani corporis fabrica libri septem.* Ex officina Ioannis Oporini: Basel, Switzerland.)

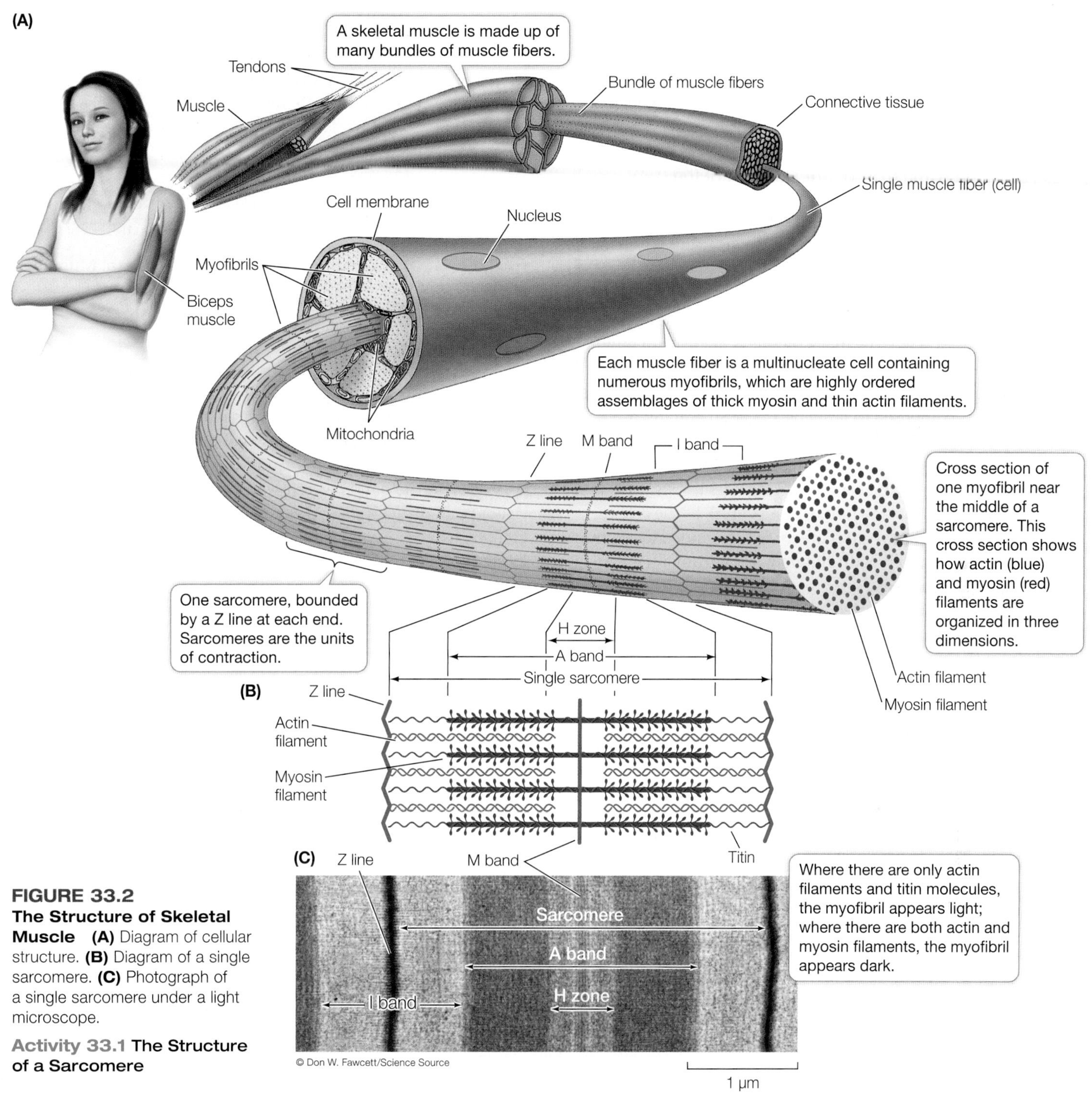

FIGURE 33.2
The Structure of Skeletal Muscle **(A)** Diagram of cellular structure. **(B)** Diagram of a single sarcomere. **(C)** Photograph of a single sarcomere under a light microscope.

Activity 33.1 The Structure of a Sarcomere

events is the opposite during contraction. The filaments start by generating mechanical forces, and those self-generated forces pull certain of the filaments past others so their relative positions change.

Actin and myosin filaments slide in relation to each other during muscle contraction

The cells of skeletal muscle are sometimes called "cells" and sometimes "fibers." These two terms are synonyms: a **muscle fiber** is a muscle cell. The cells in vertebrate skeletal muscle are large and have many nuclei, making them multinucleate. During development they form through the fusion of many individual embryonic muscle cells. A muscle such as your biceps (the muscle that pulls your forearm up toward your upper arm) is composed of hundreds or thousands of muscle fibers that are organized in bundles surrounded by connective tissue (**FIGURE 33.2A**).

Muscle contraction results from the interaction of two types of proteins in each muscle fiber: **actin** and **myosin** (technically *muscle myosin*, because there are many types of myosins). Molecules of actin are organized into **actin filaments**, also called *thin filaments*. Molecules of myosin are organized into **myosin filaments**, or *thick filaments*. These two types of filaments interdigitate in parallel arrays as diagrammed in **FIGURE 33.2B**.

LINK Actin and myosin are also part of the cytoskeleton, which is responsible for other types of movement in cells; see **Key Concept 4.4**.

Let's now focus in more detail on the anatomical features of skeletal muscle. If you look at Figure 33.2A and scan the structures from left to right, you will see that a muscle is made up of bundles of long, parallel muscle fibers (i.e., muscle cells). Similarly, each muscle fiber is composed in large part of long, parallel **myofibrils**, each of which is composed in large part of actin and myosin filaments oriented parallel to the long axis of the myofibril. At the far right of Figure 33.2A, a myofibril is shown in cross section so you can see the three-dimensional arrangement of its actin and myosin filaments. In most regions of a myofibril, as the cross section shows, each thick myosin filament is surrounded by six thin actin filaments, and each actin filament sits within a triangle of three myosin filaments.

Understanding the detailed structure of a myofibril is critical. In the foreground of Figure 33.2A, you can see a myofibril viewed from the side. Notice that, along its length, the myofibril consists of repeating units called **sarcomeres**. Sarcomeres are the units of contraction. Each sarcomere starts and ends with a Z line, a protein structure called a "line" for historical reasons. Figure 33.2B shows the interdigitating filaments and other aspects of the internal structure of one sarcomere.

In side view, the structures we have been discussing give skeletal muscle a distinctly banded appearance when the muscle is examined under a light microscope (**FIGURE 33.2C**). Individual actin and myosin filaments are far too thin to be seen with a light microscope. Thus the image in Figure 33.2C does not show individual filaments, but instead shows patterns that result from the positions of large sets of the individually invisible filaments—much as an image on a computer screen shows patterns that result from pixels even though the individual pixels are not evident. If you compare Figure 33.2B and C, you will see how the bands seen with the light microscope line up with the molecular structures in a sarcomere.

In **FIGURE 33.3** we see what skeletal muscle looks like when it is viewed from the side at a far lower magnification, about 1/40th as great as in Figure 33.2C. The banding pattern is very apparent in this view. Because of this banding pattern, skeletal muscle is called *striated muscle*.

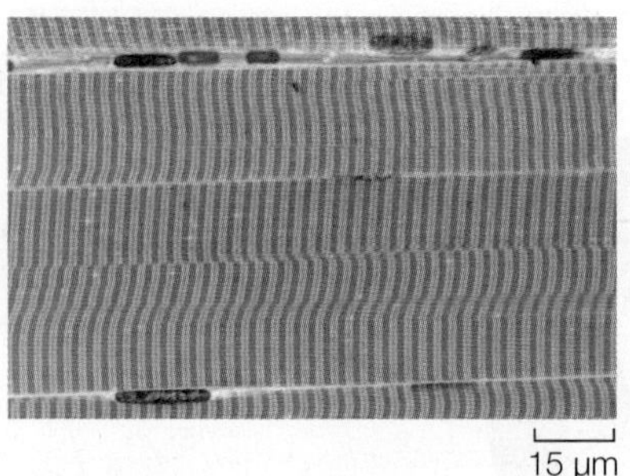

FIGURE 33.3 A Longitudinal Section of Skeletal Muscle at Relatively Low Magnification (From M. H. Ross et al. 2009. *Atlas of Descriptive Histology*, Oxford University Press/Sinauer Associates: Sunderland, MA.)

Before the molecular nature of the muscle banding pattern was known, the bands were given names that are still used today (see Figure 33.2B and C). Each sarcomere is bounded by Z lines, as already mentioned, which represent protein structures to which the actin filaments are attached. At the center of each sarcomere is a structure called the M band (for "middle"). The M band is composed of proteins that help hold the myosin filaments in position. Each myosin filament extends out symmetrically on either side of the M band. Collectively, the myosin filaments in each sarcomere give rise to a dark region—the A band—centered on the M band.

Finally, in each sarcomere there are three relatively light bands: an H zone centered on the M band, and one half of an I band at each end of the sarcomere. These bands appear light because in a partially or completely relaxed myofibril there are regions where

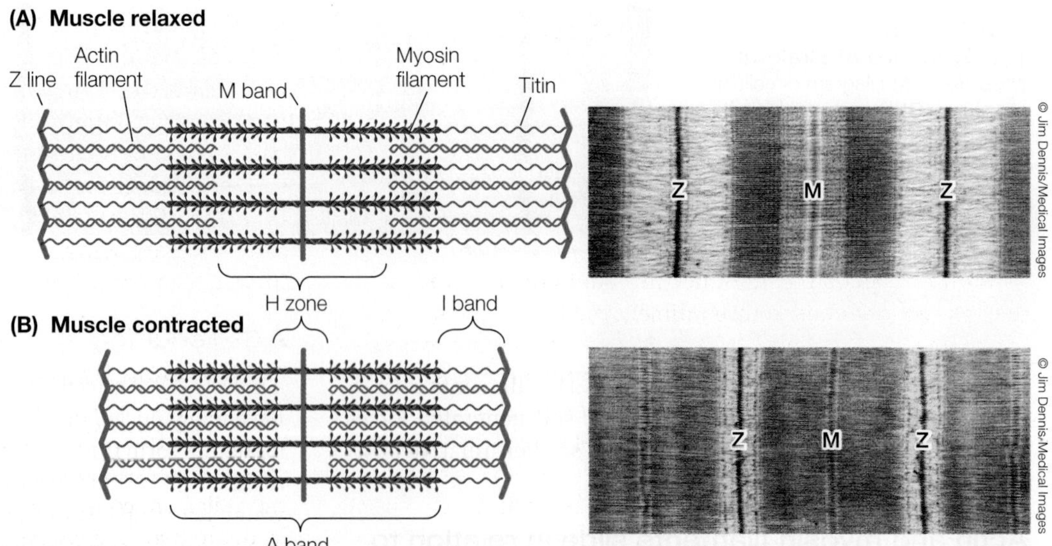

FIGURE 33.4 Sliding Filaments Diagrams and light-microscope images of **(A)** a relaxed sarcomere and **(B)** the same sarcomere shortened during contraction. The banding pattern of the sarcomere changes as it shortens. This shortening results from the sliding-filament contractile mechanism of muscle contraction.

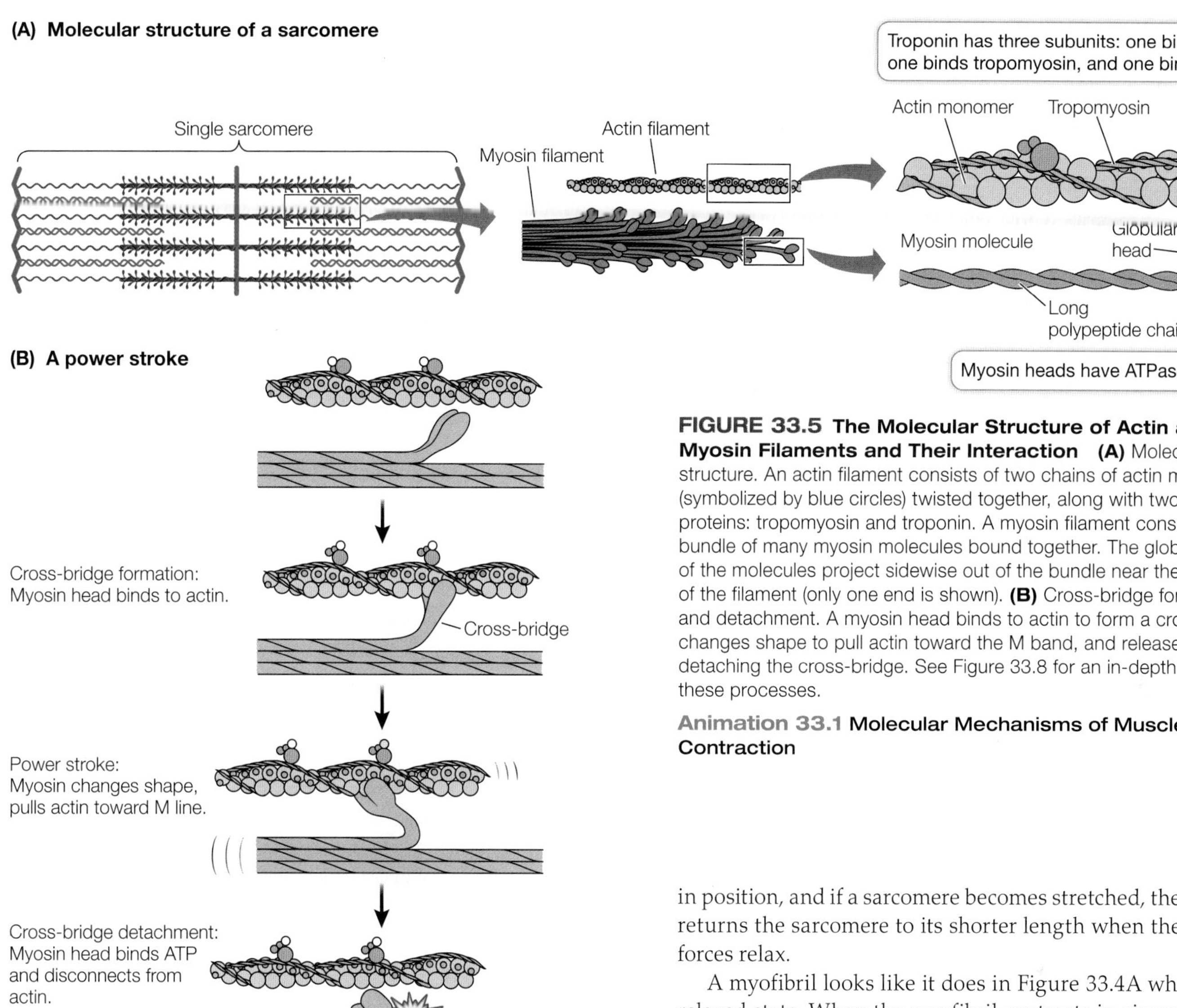

FIGURE 33.5 The Molecular Structure of Actin and Myosin Filaments and Their Interaction (A) Molecular structure. An actin filament consists of two chains of actin monomers (symbolized by blue circles) twisted together, along with two other proteins: tropomyosin and troponin. A myosin filament consists of a bundle of many myosin molecules bound together. The globular heads of the molecules project sidewise out of the bundle near the two ends of the filament (only one end is shown). **(B)** Cross-bridge formation and detachment. A myosin head binds to actin to form a cross-bridge, changes shape to pull actin toward the M band, and releases actin, detaching the cross-bridge. See Figure 33.8 for an in-depth view of these processes.

Animation 33.1 Molecular Mechanisms of Muscle Contraction

the actin and myosin filaments do not overlap: only myosin filaments are found in the H zone, and only actin filaments are found in the I bands (see Figure 33.2B and C).

Within a sarcomere, the thick myosin filaments are held in their correct positions in three-dimensional space by molecules of a protein called **titin** (**FIGURE 33.4A**). This giant molecule is the largest known protein. Each myosin filament—consisting of many myosin molecules bound together—is associated with a single titin molecule on each side of the M band. The titin molecule runs the entire distance between the M band and Z line. Where the titin molecule runs through the myosin filament, it is inelastic and integrated with the myosin molecules. Between the end of the filament closest to the Z line and the Z line itself, the titin molecule is elastic, like a bungee cord. Collectively, the titin molecules do two things. They hold the myosin filaments in position, and if a sarcomere becomes stretched, their elasticity returns the sarcomere to its shorter length when the stretching forces relax.

A myofibril looks like it does in Figure 33.4A when it is in a relaxed state. When the myofibril contracts in circumstances in which it can shorten, its sarcomeres shorten (the Z lines move closer together) and the banding pattern changes to look like it does in **FIGURE 33.4B**. The light-colored H zone and I bands become much narrower. Why do these changes occur? At a molecular level, during contraction the actin and myosin filaments slide past each other, so that they overlap to an increased degree and shorten the distance between Z lines. More specifically, the actin filaments slide toward the M band, entering the A band more fully and narrowing the H zone (the region occupied only by myosin filaments).

ATP-requiring actin–myosin interactions are responsible for contraction

To understand how a sarcomere develops mechanical forces during contraction, the structures of actin and myosin filaments must be understood at a molecular level (**FIGURE 33.5A**). An actin filament consists of many small actin monomers polymerized into two long chains, somewhat analogous to two strands of pearls twisted together. Twisting around each actin chain is another protein, **tropomyosin**, and attached to tropomyosin at intervals are molecules of still another protein, **troponin**. These molecules control contraction and relaxation, as we will see in the next section.

An individual myosin molecule (see Figure 33.5A) consists of two long polypeptide chains coiled together, each ending in a large globular head. A myosin filament is made up of about 200–400 such myosin molecules bound together, with their heads projecting sideways near each of the two ends of the filament.

How do myosin and actin interact? The globular heads of myosin can bind to specific sites on actin, forming bridgelike links, called **cross-bridges**, between the myosin and actin filaments (**FIGURE 33.5B**). Two functional properties of the myosin heads are exceedingly important. First, when a myosin head binds to an actin filament, the head's molecular shape (conformation) undergoes a change—called the *power stroke*—in which the head bends and exerts a tiny force on the actin filament. This force pulls the actin filament a distance of about 10 nanometers toward the M band along the myosin filament. Second, when a myosin head is bound to actin, it has ATP-binding activity. The myosin head binds to an ATP molecule, with the immediate result that the head changes shape again—this time in a way that disconnects it from actin. The myosin head then exhibits ATPase activity. It hydrolyzes the ATP molecule, releasing energy. The energy released is essentially stored in the myosin molecule briefly until the myosin head binds again to actin, at which time the energy powers another power stroke that again exerts a tiny force on the actin filament, pulling the actin filament another 10 nanometers toward the M band along the myosin filament.

Together, these details explain how the actin and myosin filaments in a sarcomere interact to develop mechanical forces and pull the actin filaments past the myosin filaments during contraction, so the two types of filaments "slide" relative to each other. The details also explain rigor mortis—the stiffening of muscles soon after death. For actin–myosin bonds to be broken, ATP is required. When ATP production ceases with death, actin–myosin bonds cannot be broken and muscles stiffen.

To fully understand contraction, remember that each myosin filament has many myosin heads near both of its ends (see Figure 33.5A) and is surrounded by six actin filaments (see Figure 33.2A). As contraction takes place, numerous myosin heads go through cycles of interaction with the actin filaments. Their cycles of binding and release are not exactly synchronous, however. Accordingly, when a single myosin head disconnects from actin, other cross-bridges are still connected and undergoing power strokes. As a result, the actin filaments do not slip backward.

Excitation leads to contraction, mediated by calcium ions

Muscle cells—like nerve cells—are *excitable*. As we discussed in Key Concept 31.1, this means that the cell membrane can generate and conduct (propagate) action potentials. In an excitable cell, after an action potential—a region of reversed polarity, or depolarization—is initiated at one point in the cell membrane, it travels without loss of magnitude along the full length of the cell membrane (see Figure 31.5).

What triggers a muscle cell to contract? In skeletal muscle, contraction is triggered by nerve impulses (i.e., action potentials in neurons). Each muscle fiber is in synaptic contact with the axon terminals of a neuron called a *motor neuron*. The motor neuron is said to **innervate** the muscle fiber. The synapse where the motor neuron makes contact with the muscle fiber is a specialized type of chemical synapse known as a **neuromuscular junction** or *motor end-plate* (**FIGURE 33.6**). When a nerve impulse arrives at the neuromuscular junction, it is said to *excite* the muscle fiber, and the event is called **excitation**. During excitation, an action potential is initiated in the cell membrane of the muscle fiber, and that action potential then propagates throughout the cell membrane (see Figure 31.18).

Excitation-contraction coupling is the process by which the *excitation* of a muscle fiber leads to *contraction* of the muscle fiber. Several decades ago, when excitation–contraction coupling was first being investigated, a stunning experiment on muscle

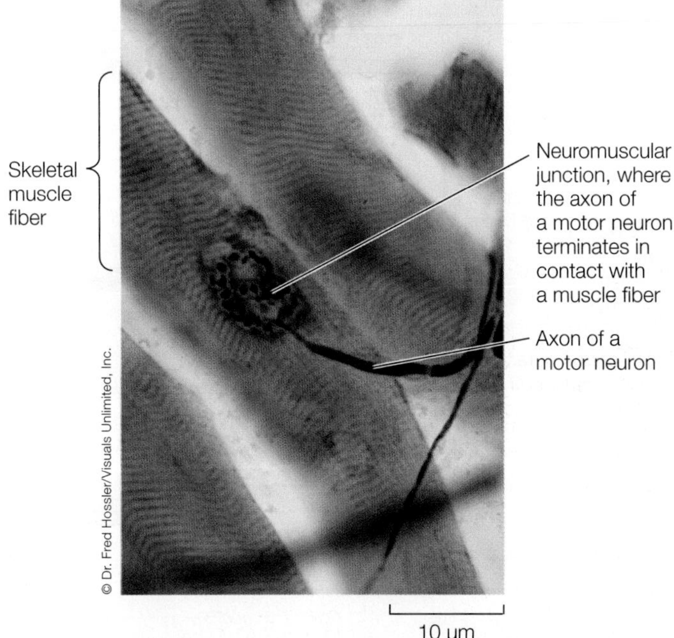

FIGURE 33.6 Innervation of a Skeletal Muscle Fiber
In this view of several skeletal muscle fibers, we see the axon of a motor neuron that controls the contraction of one of the muscle fibers. The cell body of this neuron is far away in the spinal cord. The point of contact of the axon with the muscle fiber is a neuromuscular junction. When the neuron is stimulated, an action potential travels along its axon to the neuromuscular junction and excites the muscle fiber (see Figure 31.18 for details of synaptic transmission). It is common for one motor neuron to send an axon process to more than one muscle fiber. In such cases, all the muscle fibers that the neuron innervates are simultaneously excited, and contract together, when the neuron is stimulated.

Synaptic transmission at a vertebrate neuromuscular junction is distinctive, in comparison with transmission at neuron-to-neuron synapses. Basing your answer on what we learned in Chapter 31, how do they differ?

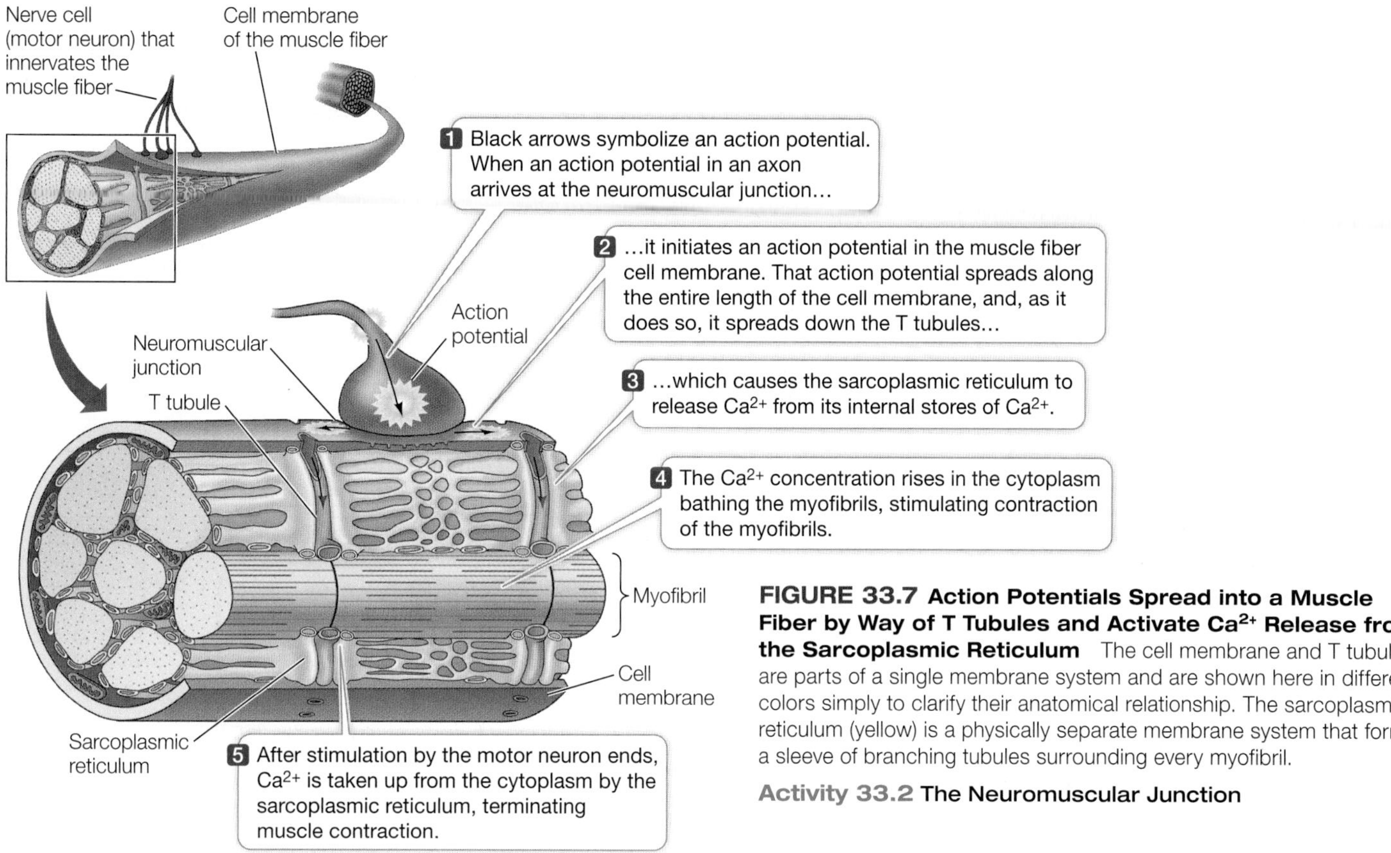

FIGURE 33.7 Action Potentials Spread into a Muscle Fiber by Way of T Tubules and Activate Ca^{2+} Release from the Sarcoplasmic Reticulum The cell membrane and T tubules are parts of a single membrane system and are shown here in different colors simply to clarify their anatomical relationship. The sarcoplasmic reticulum (yellow) is a physically separate membrane system that forms a sleeve of branching tubules surrounding every myofibril.

Activity 33.2 The Neuromuscular Junction

fibers in frog legs (see *Think Like a Scientist*, p. 835) revealed that Ca^{2+} plays a key role in excitation–contraction coupling. After a muscle fiber is excited, Ca^{2+} is released into the general cytoplasm of the fiber from internal Ca^{2+}-storage locations, and the Ca^{2+} activates contraction.

Where is the Ca^{2+} stored in a muscle fiber, and how is it released? A distinctive feature of muscle cells is that each cell has many spots where its cell membrane is indented inward into the cell cytoplasm to form a **transverse tubule**, or **T tubule** (**FIGURE 33.7**). The membrane that forms the walls of each T tubule is continuous with the outer cell membrane. Thus when an action potential travels over the outer cell membrane, it also travels inward along each T tubule. The action potential thereby spreads into the interior of the cell at multiple places.

The T tubules of a muscle cell are physically very close to the cell's endoplasmic reticulum. In a muscle cell, the endoplasmic reticulum is called the **sarcoplasmic reticulum** (*sarco*, "flesh"). The sarcoplasmic reticulum is a closed, membrane-bounded compartment that surrounds every myofibril in the cell with a sleeve of branching tubules (see Figure 33.7). Calcium pumps in the membrane of the sarcoplasmic reticulum take up Ca^{2+} from the cytoplasm and pump the Ca^{2+} to the interior. Thus when a muscle cell is at rest there is a high concentration of Ca^{2+} inside the sarcoplasmic reticulum and a low concentration in the cytoplasm. Essentially, Ca^{2+} is stored inside the sarcoplasmic reticulum when a muscle cell is resting.

Spanning the space between the membranes of the T tubules and the membranes of the sarcoplasmic reticulum are two physically connected proteins. One of these, the dihydropyridine (DHP) receptor, is located in the T tubule membrane. The other protein, the ryanodine receptor, is a gated Ca^{2+} channel located in the sarcoplasmic reticulum membrane. These two proteins act to release Ca^{2+} into the cytoplasm when a muscle fiber is excited. When the DHP receptor is activated by an action potential traveling down a T tubule, both the DHP and ryanodine receptor proteins change conformation (molecular shape). As a consequence of the shape change, the ryanodine receptor allows Ca^{2+} to leave the sarcoplasmic reticulum. Ca^{2+} then floods into the cytoplasm surrounding the actin and myosin filaments (**FIGURE 33.8**, step 1).

The Ca^{2+} ions in the cytoplasm trigger actin and myosin to interact so that the muscle fiber contracts. How do the Ca^{2+} ions do this? Recall that strands of the protein tropomyosin are twisted around the chainlike actin polymers (see Figure 33.5A), and molecules of the globular protein troponin occur at regular intervals. Each troponin molecule has three subunits: one binds actin, one binds tropomyosin, and one binds Ca^{2+}.

When a muscle fiber is at rest, the tropomyosin strands are positioned so they block the sites on actin filaments where myosin heads can bind. When Ca^{2+} is released into the cytoplasm, it binds to troponin, changing troponin's conformation. Because

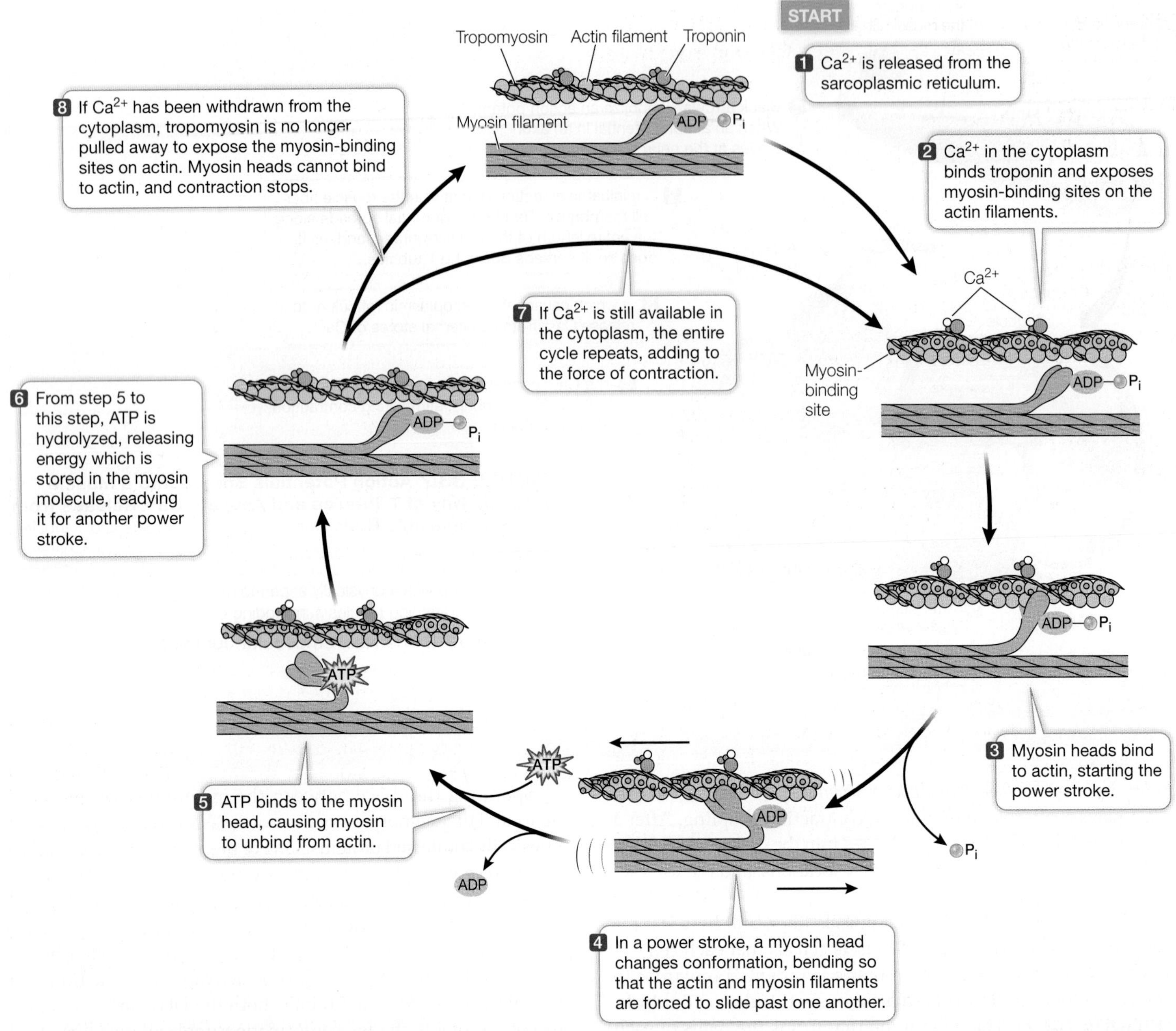

FIGURE 33.8 Ca^{2+} Initiates Muscle Contraction When Ca^{2+} binds to troponin, it causes a change in molecular conformation that exposes the myosin-binding sites on actin. As long as the binding sites are exposed and ATP is available, actin and myosin repeatedly go through the cycle of interactions shown here, using ATP-bond energy to pull the actin and myosin filaments past each other, thereby developing mechanical forces. Adenosine diphosphate (ADP) and inorganic phosphate (P_i) are shown because their bonding and unbonding help control steps in the cycle.

troponin is bound to tropomyosin, this troponin change affects the conformation of tropomyosin, which twists enough to expose the actin–myosin binding sites (see Figure 33.8, step 2). Myosin heads can then form cross-bridges with actin (step 3). When a myosin head starts to form a cross-bridge, it is in a conformation that poises it to undergo its power stroke. Binding activates the power stroke (step 4), pulling the actin and myosin filaments past each other. ATP then binds to the myosin head, causing myosin and actin to dissociate (step 5). Hydrolysis of the ATP (going from step 5 to step 6) then occurs, releasing energy and leading to another conformational change in myosin, poising it again for a power stroke (step 6). As long as Ca^{2+} remains available, the cycle repeats (step 7).

When excitation ends, the calcium pumps in the sarcoplasmic reticulum remove Ca^{2+} from the cytoplasm. The conformations of troponin and tropomyosin revert to their original state, tropomyosin blocks the binding of myosin heads to actin, and contraction stops (step 8).

THINK LIKE A SCIENTIST

Process of Science

Interdisciplinary

Communicate & Collaborate

From the shores of ancient Rome to flashing muscle fibers: progress in a stunning collaboration across generations of scientists, disciplines, animals, and tissues

The discovery of bioluminescence—biological production of light—in jellyfish, and the subsequent investigations into its source, provide a beautiful example of how the process of science involves collaboration across multiple disciplines.

The story starts in ancient Rome, where the famous naturalist Pliny the Elder (killed in the eruption of Vesuvius in CE 79) recorded that he had observed bioluminescence by a jellyfish. By 1885, scientists had started to concern themselves with molecules, and a Frenchman named Raphael Dubois searched for the mechanism by which insects such as fireflies can produce light. Dubois discovered that two proteins from the insects—luciferin and luciferase—needed to be mixed together in the presence of O_2, and then—almost miraculously—a “cold” light was emitted (“cold” because no flame or heat was needed). There followed the discovery of many other luciferin–luciferase systems in other animals.

Meanwhile, new information on jellyfishes had emerged in 1810 as, over the centuries, naturalists poked and prodded a great diversity of light-producing organisms. In England, Mr. J. Macartney, a naturalist and marine biologist, reported placing bioluminescent jellyfishes in a vacuum. The jellyfishes continued to produce light. Thus when Raphael Dubois discovered the luciferin–luciferase mechanism in 1885, those who were keeping their facts straight realized that the jellyfishes were unlikely to use that mechanism, since the luciferin–luciferase mechanism requires O_2.

The jellyfish mystery was resolved at a marine biology laboratory in Washington State in 1962 by a Japanese biochemist (and future Nobel laureate)—Osamu Shimomura—who had traveled to the laboratory to study a jellyfish named *Aequorea*. He and his colleagues would walk out of their biochemistry lab to the seashore often, wade into the ocean, and use hand nets to catch jellyfishes gently, then return indoors to do their studies—perhaps often thinking of Pliny on the Italian coast in the first century CE.

Shimomura discovered in *Aequorea* a totally different system for making light than the luciferin–luciferase system (**FIGURE 33.9**). Just as Macartney had concluded a century and a half earlier, this system operates without any immediate need for O_2. Moreover, the system does not require two parts. It is based on a single macromolecule, which Shimomura named *aequorin*. In the system discovered by Shimomura, Ca^{2+} activates light production: one needs only add Ca^{2+} at a very low concentration to aequorin, and blue light is emitted.

Biologists were already beginning to realize that tissues occasionally use Ca^{2+} as a signaling molecule. Thus soon after Shimomura discovered aequorin, two animal physiologists, a Briton and American, published the idea that aequorin (purified from jellyfishes) could be used to *detect* Ca^{2+}, and they demonstrated this use in the muscle fibers of an invertebrate (a barnacle).

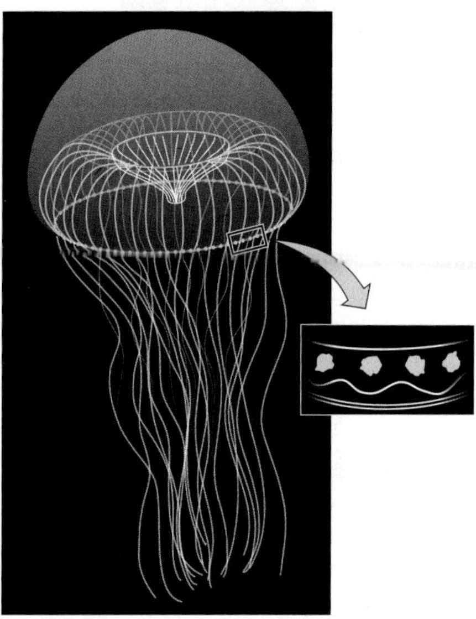

FIGURE 33.9 The Bioluminescent Hydromedusan Jellyfish *Aequorea victoria* The inset shows the clusters of light-producing cells. Although aequorin—the bioluminescent macromolecule in this jellyfish—emits blue light, the intact cells in the jellyfish emit green light because of a fluorescent molecule (green fluorescent protein), synthesized by the jellyfish, that changes the wavelength of the light.

The culmination of this story came in the mid-1970s at the Mayo Medical School in Minnesota. It had been hypothesized that Ca^{2+} might be the immediate stimulus for contraction in vertebrate skeletal muscle. Thus one day three medical physiologists stood together in a totally dark room looking in the direction of a muscle fiber that they had dissected out of frog muscle. They had used a micro-injection apparatus to inject purified jellyfish aequorin into the cytoplasm of the frog muscle fiber. They had also arranged to excite the fiber to contract with mild electric shocks.

The physiologists delivered a shock, and—in the total darkness—the muscle put out a flash of blue light, which the researchers detected with a light-amplifying device. Another shock, another flash.

And thus the modern theory of excitation–contraction coupling was provided with its most dramatic initial confirmation.

1. How did the experiment with frog muscle fibers support the theory of excitation–contraction coupling that we have discussed in this chapter?
2. Would you agree or disagree that the final, famous study on frog muscle by medical physiologists could not have been accomplished without contributions from many different scientists? Explain. What thought processes characterized the scientists in this story? (If you want to delve further into this question, read the two scientific papers cited below.)
3. In the 1970s, aequorin was extracted from jellyfish for use in Ca^{2+} detection in living cells. Injection put a tiny hole in the cell membrane and introduced a foreign solution into the cell cytoplasm. How might modern methods of genetic engineering be used to circumvent these deficiencies in the aequorin method?

For more on this topic, see J. R. Blinks et al. 1978. *J Physiol* 277: 291–323 and O. Shimomura et al. 1962. *J Cell Comp Physiol* 59: 223–239.

REVIEW & APPLY | 33.1

R The fibers (cells) of vertebrate skeletal muscle consist of myofibrils, which contain bundles of partially overlapping actin and myosin filaments arranged in sarcomeres, giving the muscle a striated appearance. Contraction is the development of force by a muscle. The molecular mechanism of contraction is described by the sliding-filament theory and involves the binding of the globular heads of myosin molecules to actin molecules to form cross-bridges. Upon binding to actin, the conformation of a myosin head changes, pulling the myosin filament past neighboring actin filaments. Release of the myosin head from actin requires ATP.

Motor neurons synapse with muscle fibers at neuromuscular junctions, with each muscle fiber typically being controlled by one motor neuron. Excitation causes an action potential to spread over a muscle fiber cell membrane and into the fiber by way of T tubules, causing Ca^{2+} to be released into the cytoplasm from the sarcoplasmic reticulum, initiating the process of excitation–contraction coupling.

A

1. Imagine planting your feet and trying to push through a concrete wall that's far too heavy to move. As you push, would you describe the associated muscles in your back, arms, and legs as contracting, shortening, lengthening, or a combination of these words? Explain.
2. In a muscle fiber, how is force development aided by the overlapping arrangement of actin and myosin filaments?
3. Describe how the concentration of Ca^{2+} in the sarcoplasmic reticulum of a muscle cell changes before, during, and after the cell is excited by an action potential.

Now that we have seen how the contractile apparatus of a muscle generates forces, let's look at how these forces are put to use.

33.2 The Function of Skeletal Muscle Depends on Interaction with the Skeleton and on ATP Supply, Cell Type, and Training

LEARNING OBJECTIVES

By the end of this key concept you should be able to:

33.2.1 Present and explain an example of how skeletal muscles work together to control an animal's movements.

33.2.2 Explain how skeletal muscle performance depends on the biochemistry of ATP production.

33.2.3 Distinguish between slow oxidative and fast glycolytic muscle fibers.

33.2.4 Describe how the effects of skeletal muscle training depend on the type of exercise performed during training.

The skeletal system consists of rigid structural elements against which the skeletal muscles can pull to create directed movements. Two major types of skeletal systems exist in animals. In vertebrates, the skeleton is surrounded by the other tissues of the body (**FIGURE 33.10**) and is called an **endoskeleton** (*endo*, "inside") because it is inside the body. In arthropods, such as crabs and insects, the skeleton encases the rest of the body and is an **exoskeleton** (*exo*, "outside").

In vertebrates, muscles pull on the bones of the endoskeleton

The endoskeleton consists mainly of bone in most adult vertebrates. **Bone** is composed principally of an extracellular matrix of collagen protein fibers with insoluble calcium phosphate crystals deposited among them, giving the bone hardness and rigidity. Bone also contains several types of living cells, the activities of which make bone a dynamic tissue throughout life. Because of the activities of these cells, bone is repaired if it is damaged, and it can be remodeled; for example, in response to forces placed on it, a bone routinely undergoes long-term, modest changes in shape or structure. Bone also serves as a reservoir of calcium, and bone Ca^{2+} is in dynamic exchange with Ca^{2+} elsewhere in the body. This exchange is under the control of hormones, such as parathyroid hormone (see Figure 32.11), and vitamin D. The endoskeleton also includes cartilage in its structure. **Cartilage** is a flexible, rather than hard, form of skeletal tissue that plays several roles, such as imparting flexibility to skeletal elements (e.g., the rib cage).

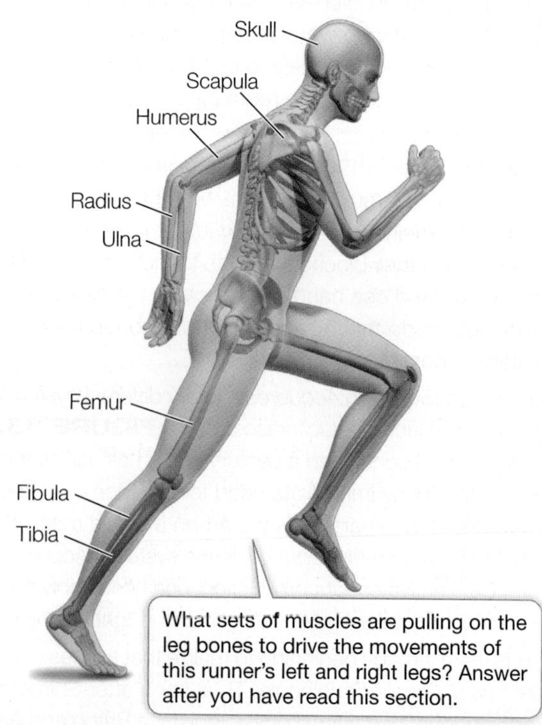

FIGURE 33.10 The Human Skeletal System For running, the skeleton is as important as the muscles because the muscles pull on bones.

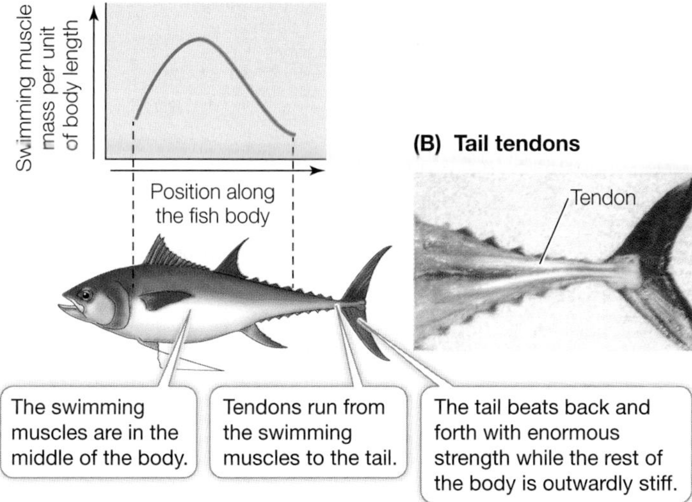

FIGURE 33.11 Tuna Swimming **(A)** The blue line is a graph of swimming muscle mass. Where the line is highest, the muscle mass is greatest. **(B)** Tendons transfer force from the swimming muscles to the tail. (B from R. E. Shadwick et al. 2002. *Comp Biochem Physiol A* 133: 1109–1125.)

When tunas are used for food, they are famed for the thick "steaks" of meat cut from the middle of the body behind the head. It seems safe to say that the tissue in these steaks is present for the benefit of the tunas, not the benefit of a person eating the tunas. Why is this tissue present?

To produce movements, skeletal muscles and bones work together around **joints**, the structures where two or more bones come together. Skeletal muscles attach to bones by bands of flexible connective tissue called **tendons**. Tendons often extend *across* joints. In such cases, muscles on one side of a joint exert forces on bones on the opposite side of the joint. Tunas, which are among the most powerful swimmers of all fishes, provide dramatic examples. A tuna swims by beating its tail fin, but there are no muscles of any size in its tail. The massive swimming muscles of a tuna are in the middle of its body (**FIGURE 33.11A**). Forces developed by these muscles are transmitted to the tail by tendons that run from the swimming muscles to the tail fin across the joint at the base of the tail (**FIGURE 33.11B**).

A skeletal muscle can exert force in only one direction. Thus for a bone at a joint to move in two opposite directions, two muscles are required. These muscles work as an **antagonistic pair** in which the two muscles have opposite actions: when either one contracts, the other relaxes. You can see this action in the motions of your forearm (**FIGURE 33.12A**). When your biceps muscle contracts, your triceps muscle relaxes, and your forearm (as it is viewed in the figure) moves up because the tendons of the biceps cross the elbow joint to attach to the forearm. When your triceps contracts, your biceps relaxes, and your forearm moves down.

Different sets of skeletal muscles, coordinated by the nervous system, typically must work together to control complex sets of movements. For example, consider the large muscle on the back of your calf (lower leg), the gastrocnemius muscle. When it contracts, the movement it creates depends on the state of contraction of the muscle on the front of your thigh, the quadriceps femoris (**FIGURE 33.12B**). If your quadriceps is not contracting strongly, contraction of your gastrocnemius bends your knee joint and lifts your calf as in walking or running (see Figure 33.12B, left). If your quadriceps is contracting strongly—an action that keeps your knee joint from bending—contraction of your gastrocnemius extends your foot as in jumping (Figure 33.12B, right).

In arthropods, muscles pull on interior extensions of the exoskeleton

Crabs, lobsters, insects, and other arthropods have an exoskeleton that covers every part of their body. It is composed of the

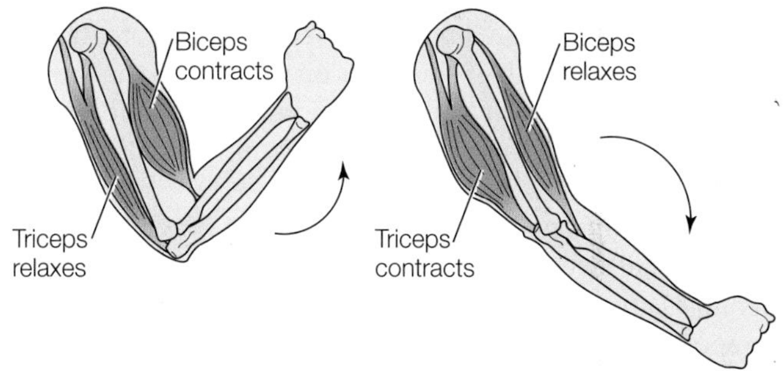

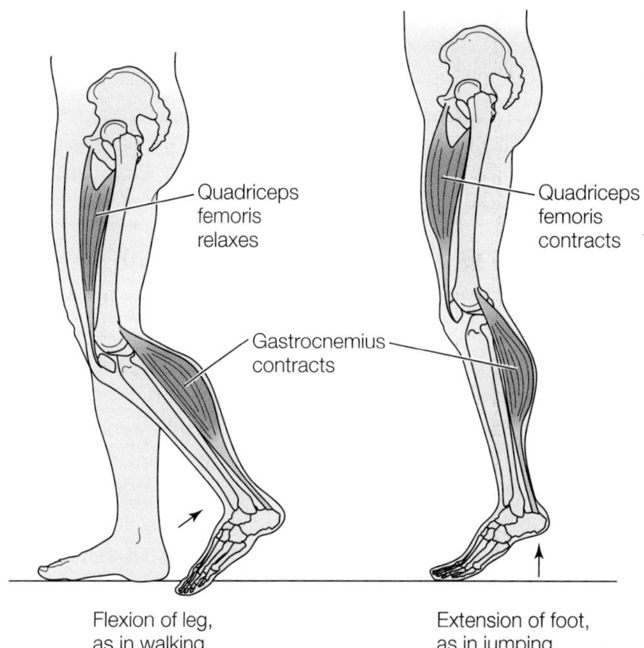

FIGURE 33.12 Two or More Muscles Are Coordinated during Common Movements **(A)** Antagonistic muscles are coordinated to allow the forearm to move in two opposite directions at a single joint. **(B)** The same muscle sets are used both when a person lifts their calf (left) and when a person jumps (right), but the muscles are coordinated in different ways in the two activities. The antagonistic muscles (not shown) in the upper and lower leg are coordinated as well.

polymerized carbohydrate chitin, and in crabs and lobsters it is extensively hardened by deposition of calcium-containing minerals. An exoskeleton functions like armor, protecting the soft tissues of the body inside. However, the exoskeleton of an arthropod cannot grow after it is formed. Thus as an arthropod grows, it must replace its exoskeleton periodically with a new, larger version.

LINK The process of shedding an exoskeleton—called molting—is controlled by hormones, as described in **Key Concept 32.5**.

In an arthropod, the skeletal muscles have a different relationship to the skeleton than they do in animals with endoskeletons. Where a joint exists, a part of the exoskeleton projects inside the body as a structure called an *apodeme*. Muscles inside the body attach to the apodeme. Thus the muscles pull on the exoskeleton when they contract—by pulling on the apodeme that is attached to the exoskeleton. This arrangement is illustrated in **FIGURE 33.13** by the muscles that operate the large, pinching claw of a crab. If you focus on the claw in the drawing, you'll see that the lower part has no joint and therefore cannot swivel. When the claw opens and closes, only the upper part—the *dactyl*—moves.

The dactyl has two major apodemes, each of which is a long, hard, internal structure. The opener and closer muscles—which operate as an antagonistic pair—connect, respectively, to these two apodemes. When the closer muscle contracts and pulls on one of the apodemes, the dactyl swivels to close the claw. The closer muscle is powerful, and it pulls the claw closed with great force, leading to a strong pinching movement (and a loud scream if your finger is caught!). The opener muscle (which is weaker) pulls on the other apodeme, opening the claw when it contracts.

LINK The diverse structures and compositions of exoskeletons are detailed in **Key Concepts 22.3** and **22.4**.

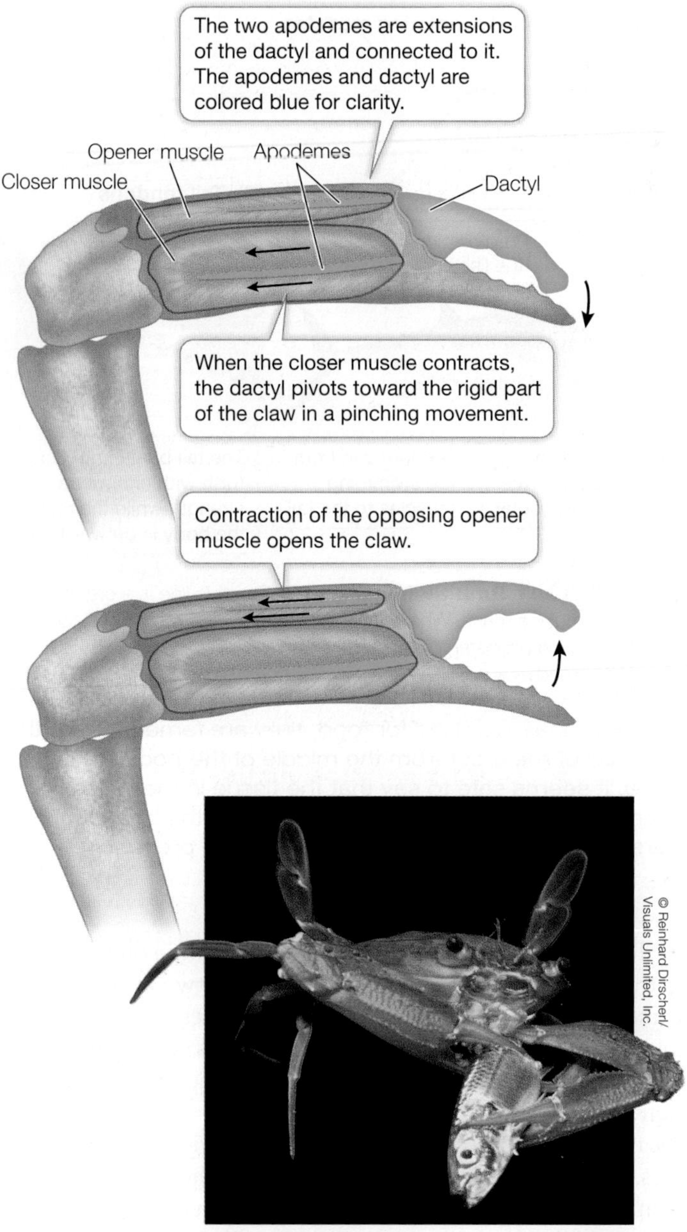

FIGURE 33.13 In Animals with an Exoskeleton, Muscles Pull on Interior Projections of the Exoskeleton This principle is illustrated here by the muscles that operate the claw of a crab. The arrows indicate the directions of movement of the apodemes and dactyl. (After A. G. Vidal-Gadea and J. H. Belanger. 2009. *Arthropod Struc Dev* 38: 179–194.)

Hydrostatic skeletons have important relationships with muscle

An animal is said to have a **hydrostatic skeleton** if its body or a part of its body becomes stiff and skeleton-like because of a high fluid pressure inside. The most familiar hydrostatic skeleton is the human erected penis; blood pressure, produced by the beating of the heart muscle, gives the penis its rigid, skeleton-like character (see Key Concept 34.2).

An earthworm digs through soil by using muscles in its body wall to create a hydrostatic skeleton (**FIGURE 33.14**). Each time the worm moves forward (red arrows in the figure), contractions of certain muscles raise the pressure inside the worm to a high enough level that the front end of the worm's body expands forward through the soil, much as a long, skinny balloon lengthens when inflated. The worm then anchors its front end and pulls its back end toward its front end (blue arrows in the figure) before repeating the process.

Clams do much the same thing to dig into sand. Through muscular contractions, they alternately inflate their foot under pressure, so that the foot shoves deeper into the sand, and then pull the rest of their body down toward the foot.

A muscle's power output depends on the current rate at which it is making ATP

Muscles perform work. This statement is accurate in the strict sense used in the study of physics. Muscles apply forces to bones or other structures—forces that often make the structures move. In the study of mechanics in physics, work is done when a force causes a mass to move over a distance.

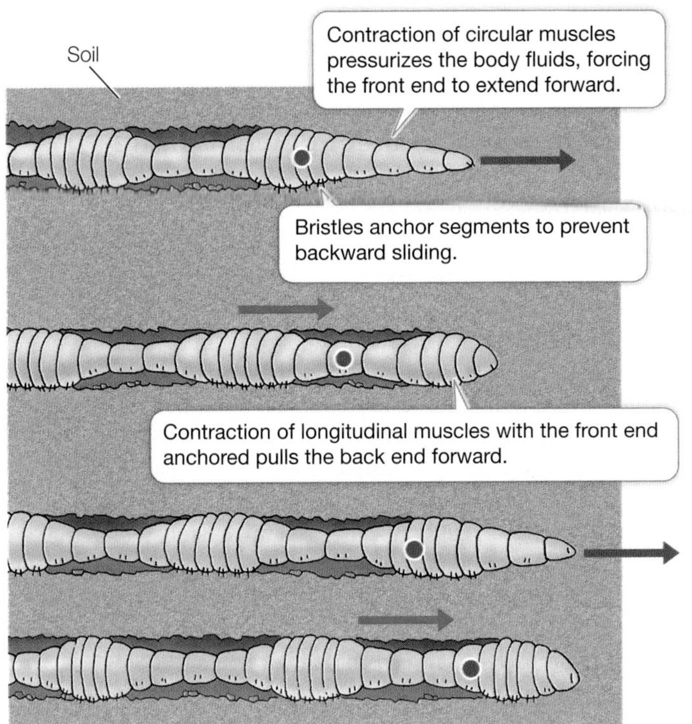

FIGURE 33.14 Use of a Hydrostatic Skeleton Red arrows show forward extension of the earthworm's body when circular muscles in the body wall contract and pressurize the body fluids inside the worm, and the interior pressure forces the anterior end to extend forward. Blue arrows show trailing parts of the body being pulled forward by contraction of longitudinal muscles while the anterior is anchored. The red dot marks the same body segment in all four drawings.

For understanding the performance of work by a skeletal muscle cell, the cell's *fuel supply* for work is often as important as the properties of the contractile apparatus that we have emphasized up to here in this chapter. Muscle cells depend on ATP as their source of energy to do work. You may know from physics that **power** is the *rate* at which work is done. A muscle cell's power output depends on the rate at which ATP is made available for use in contraction.

Now recall also one of the major points emphasized in Chapter 28: ATP is not transported from cell to cell. Instead, each cell must make its own ATP (see Key Concept 28.4). Thus the rate at which a muscle cell can do work depends on the rate at which that very cell can produce ATP.

Skeletal muscle cells have three biochemical systems that supply ATP for contraction:

- The **oxidative system** is very familiar, consisting of the citric acid cycle and electron transport (see Key Concept 5.2). The oxidative system synthesizes ATP by means of the aerobic breakdown of food molecules, principally carbohydrates and lipids, producing fully oxidized products, CO_2 and H_2O. The system is termed "oxidative" (or "aerobic") because it requires O_2 to function.
- The **glycolytic system** employs anaerobic glycolysis to produce ATP (see Key Concept 5.2) and has no requirement for O_2. It synthesizes ATP anaerobically (i.e., without using O_2) by breaking down carbohydrates. In vertebrates and crustaceans (and some other animals), this system produces lactic acid as well as ATP.
- The **immediate system** uses preformed ATP that is already present in a cell when contraction begins. Some preformed ATP is present literally as ATP, but muscle cells can also store small amounts of preformed ATP in chemically combined form, such as in creatine phosphate, a compound found in vertebrate skeletal muscle.

These three systems vary greatly in how *rapidly* they can supply ATP and in how *long* they can continue to supply ATP at a high rate. Accordingly, the *power output* and *endurance* of a working muscle cell vary depending on which system the cell is using to make ATP. Let's explore this important point by looking at power output and endurance during human exercise when the ATP for exercise is supplied by each of the three systems (**FIGURE 33.15**).

When a muscle cell starts to do work, the immediate system can accelerate supply of ATP extremely rapidly, enabling the cell to achieve a very high power output almost instantly. The immediate system can do this because it uses ATP made at an earlier time and already present in the cell when work begins. Only a small amount of preformed ATP can be present in a cell, however. Thus the immediate system can continue to supply ATP at a high rate for only a brief period of time. As seen in Figure 33.15, during exercise powered by the immediate system,

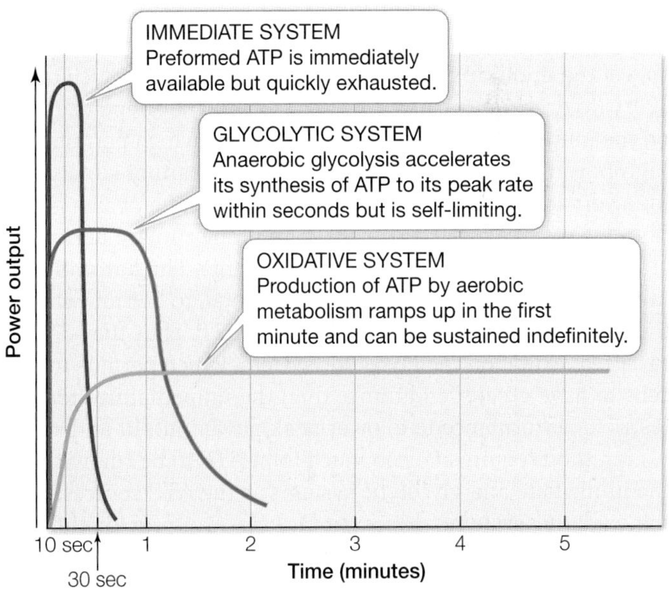

FIGURE 33.15 Power Output of a Skeletal Muscle Cell during All-Out Exertion versus the Length of Time the Cell Can Maintain All-Out Exertion Each line shows what happens when all ATP comes from a single source: the immediate system, the glycolytic system, or the oxidative system. Values shown are for human performance. In actual exercise, all three sources of ATP may be used simultaneously.

power output rises quickly to a high level, but the exercise must soon end.

The glycolytic system can accelerate production of ATP rapidly and then synthesize ATP at a very high rate, permitting high power output—but not as high a rate as that of the immediate system (see Figure 33.15). Although the glycolytic system has greater endurance than the immediate system, the glycolytic system cannot make ATP indefinitely. Its mechanism of synthesizing ATP is self-limiting, and within minutes the glycolytic system stops and fatigue sets in. Decades ago, the accumulation of lactic acid was blamed for this rapid fatigue, but modern research points to other immediate causes. After the glycolytic system has made as much ATP as it can at a particular time, it must rest for a long time (tens of minutes or hours) before it can be used again to supply ATP.

The oxidative system accelerates more slowly and synthesizes ATP at a lower peak rate—meaning lower peak power output—than the other two systems (see Figure 33.15). However, its products—besides ATP—are simply H_2O and CO_2, which healthy animals can deal with effectively. Thus the oxidative system is not self-limiting and can continue making ATP for a very long period of time if a cell has a sufficient supply of O_2 and also a sufficient supply of fuel molecules for the citric acid cycle (e.g., fatty acids).

Reflecting on Figure 33.15, we see that *if we focus on power output*, the three systems of ATP supply rank as follows: the immediate system permits a muscle cell to reach highest power output, the glycolytic system ranks next, and the oxidative system ranks third. However, *if we focus on endurance*, the three systems rank in the opposite order: the immediate system permits the least endurance, the glycolytic system ranks next, and the oxidative system permits the greatest endurance.

These principles explain why, in human competitive running, speed and endurance are inversely related. Sprints such as the 100-yard dash are run at a very high speed. They depend greatly on the immediate system for supply of ATP, which permits very high power output. The immediate system cannot supply ATP for a very long time, however, so sprints must be of short duration. Events like the half-mile run and mile run are performed at lower speeds and last longer than sprints. The half-mile and mile runs require far too much ATP, in total, to be run with just the immediate system as their source of ATP. Thus they depend to a great extent on the glycolytic system, which permits muscle cells to have greater endurance than the immediate system but restricts performance to a lower peak power output and speed. A marathon requires far too much total ATP to be run with just the immediate and glycolytic systems as the ATP sources. Only the oxidative system—based on O_2 use—can provide the total amount of ATP a marathon requires. The speed of running in a marathon must therefore be lower than in races fueled by the immediate and glycolytic systems, although the pace can be sustained for hours.

Long-endurance casual sports, such as jogging and swimming laps, depend almost entirely on the oxidative system. This is true because the oxidative system can accelerate its ATP production as rapidly as desired at the start, and then it can meet all the ATP needs of these forms of exercise on a minute-by-minute basis, indefinitely. This is why these sports are described as being *aerobic exercise*.

Muscle cell types affect power output and endurance

All skeletal muscle cells are not equal in their ability to use the glycolytic and oxidative systems to synthesize ATP. Some cells principally use the glycolytic system, whereas others principally use the oxidative system. There are no cells, however, that depend principally on the immediate system, which can supply only small amounts of ATP.

Individual animals typically have multiple types of muscle cells in their skeletal muscles. Two major skeletal muscle cell types are recognized: **slow oxidative cells** (sometimes called slow-twitch cells) and **fast glycolytic cells** (fast-twitch cells). In vertebrates, these cell types are also called, respectively, "red" and "white." The two types of muscle cells typically occur mixed together in the skeletal muscles of mammals. In some other animals, however, certain skeletal muscles are composed almost entirely of one muscle cell type or the other. For example, many fishes have distinct masses of swimming muscle that are visibly red or white (**FIGURE 33.16**). The red muscles are composed mostly of slow oxidative cells. The white muscles consist mostly of fast glycolytic cells.

Slow oxidative and fast glycolytic muscle cells differ in three main ways:

1. Their principal mechanisms for making ATP
2. Their mechanisms for taking up O_2 from the blood
3. Their molecular forms of muscle myosin; the two types of cells express different genes for myosin

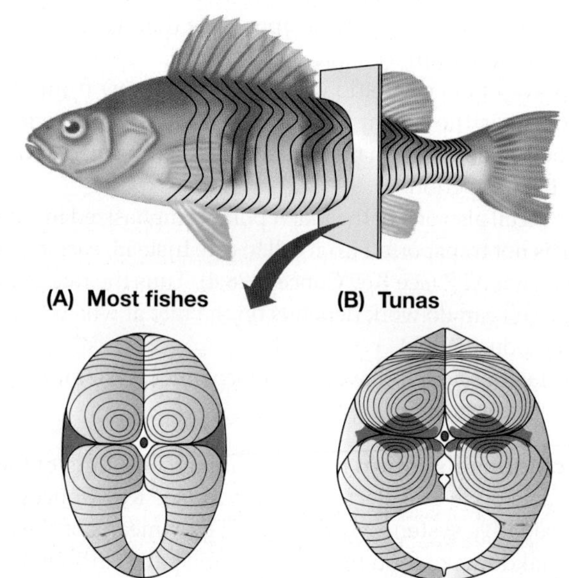

FIGURE 33.16 Red and White Swimming Muscles in Fishes Cross sections were made where shown on the drawing of a whole fish (perch). The cross sections show the red (dark) and white (light) swimming muscles in **(A)** most fishes and **(B)** tunas. (A, B after R. E. Shadwick. 2005. *Am Sci* 93: 524–531.)

Slow oxidative cells are poised to make most of their ATP aerobically using the oxidative system of ATP synthesis. Slow oxidative cells need lots of O_2. This requirement for O_2 explains why, in vertebrates, the cells are red: they have evolved high levels of the red, hemoglobin-like compound **myoglobin** in their cytoplasm. The myoglobin speeds entry of O_2 into the cells from the cell surfaces. Slow oxidative cells are poised biochemically for aerobic ATP production because they have large numbers of mitochondria and are rich in mitochondrial enzymes, notably the enzymes of the citric acid cycle and electron transport system (see Key Concept 5.2). Slow oxidative cells are not, however, well endowed with the enzymes required for the glycolytic system to make ATP anaerobically. Being predominantly aerobic, the cells can sustain ATP production for long, continuous periods of time, but their peak rates of ATP production are only moderately high (see Figure 33.15). Slow oxidative cells have molecular forms of myosin with relatively low ATPase activity, meaning that actin–myosin cross-bridges are made and broken at relatively low rates. All things considered, slow oxidative cells tend to contract and develop tension relatively slowly (accounting for the word "slow" in their name). They are well suited for maintaining relatively low power outputs for sustained periods.

Fast glycolytic cells contrast with slow oxidative cells in all the ways we've mentioned. They have high levels of the enzymes needed for ATP production by the glycolytic system. They also have molecular forms of myosin with relatively high ATPase activity, meaning that actin–myosin cross-bridges are made and broken rapidly. Because of these properties, fast glycolytic cells tend to contract and develop tension relatively rapidly (thus they are "fast"). They attain high peak power output, but they fatigue rapidly (see Figure 33.15). They have relatively few mitochondria, and they lack myoglobin, so they are not red.

Fishes use principally their red muscles, composed mainly of slow oxidative cells, during routine cruising, when they move slowly for long periods. They use principally their white muscles, composed mainly of fast glycolytic cells, during burst swimming, when they swim at high speeds to evade danger or catch prey. Their white muscles enable them to achieve very high power outputs for brief periods.

Individual people vary greatly in the proportions of red and white cells in their major skeletal muscles. These differences generally do not matter in casual sports. However, researchers know from many studies that, in competition at the national or international level, champion long-distance runners, swimmers, and cyclists—who require endurance—tend to differ from champion weight-lifters, wrestlers, and sprinters—who require high power outputs for brief periods. The long-distance runners, swimmers, and cyclists—the endurance athletes—tend to have leg and arm muscles consisting mostly of slow oxidative muscle cells, whereas the other athletes tend to have muscles with a high percent of fast glycolytic cells (**FIGURE 33.17**).

(A) Cross sections of leg muscles viewed under a light microscope

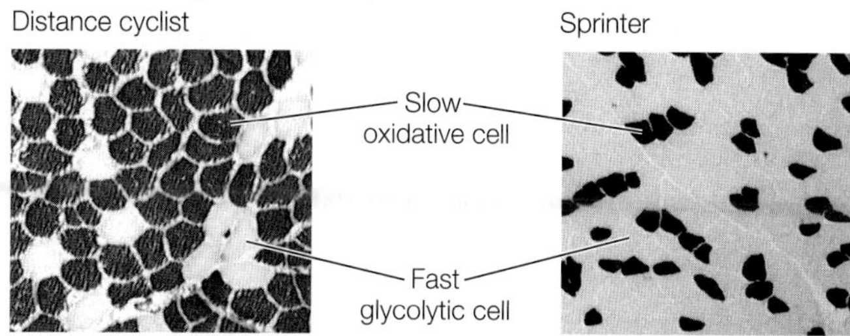

(B) Individual differences in skeletal muscle composition

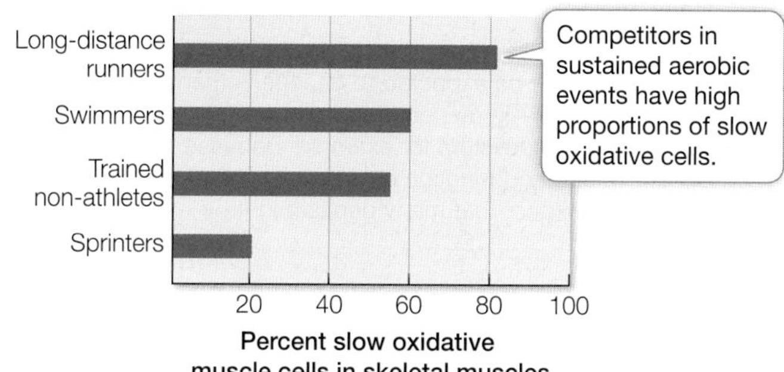

FIGURE 33.17 Skeletal Muscle Fiber Types Vary by Sport in Competitive Athletes **(A)** The cellular composition of the skeletal muscles of competitive athletes varies depending on the nature of their sport. In these microscopic images, cells were stained with a reagent that shows slow oxidative cells as dark and fast glycolytic cells as light. **(B)** At the level of international or national competition, athletes successful in sustained aerobic events tend to have a high percent of slow oxidative cells, whereas successful sprinters tend to have a high percent of fast glycolytic cells. As with all correlations, identification of cause and effect can be a complex undertaking; heredity, early development, and training are involved, and moreover, athletes choose sports in which they can succeed. When people engage in casual sports for fun, their individual muscle fiber types are generally not relevant for their performance or enjoyment. (A courtesy of Jesper L. Andersen; B adapted from graphic by J. Johansen in J. L. Andersen et al. 2000. Muscle, genes and athletic performance. *Sci Am* 283: 48–55.)

Training modifies muscle performance

Skeletal muscle has a striking ability to change its properties with use. In this way, skeletal muscle provides dramatic examples of phenotypic plasticity (see Key Concept 29.3). When a muscle changes in response to use, the change is often termed a "training effect." We observe a familiar training effect when we lift weights or do other similar forms of exercise. Our muscles gradually become visibly larger. The number of muscle cells in a muscle does not change under these conditions. Instead, already existing cells increase in size by increasing their content of actin and myosin.

To fully understand training effects, two types of exercise need to be distinguished. **Endurance exercise** (endurance training) refers to steady, long-duration exercise such as cross-country running or long-distance cycling. **Resistance exercise** (resistance training) refers to exercise that generates particularly large forces, typically in relatively few repetitions of movement spread over a short period of time. Weight-lifting and wrestling are forms of resistance exercise.

During endurance training, skeletal muscle cells increase their numbers of mitochondria and their amounts of mitochondrial enzymes. Endurance training also stimulates growth of additional blood capillaries in skeletal muscles—an effect that reduces the average distance between each muscle cell and its nearest blood capillary (source of O_2). These training effects enhance the muscles' capability to carry out steady, sustained forms of exercise. Recent studies show that when people or laboratory animals engage in endurance training after a period of sedentary existence, the expression of more than 100 genes is dramatically increased in their skeletal muscle cells (**INVESTIGATION FIGURE 33.18**). These changes in gene expression help control the changes in muscle properties that occur (with hormonal and metabolic control mechanisms also playing roles).

One of the principal effects of resistance training, as we have already seen, is that skeletal muscle cells build a larger contractile apparatus. This response enhances the ability of the cells to develop the forces required for weight-lifting or other resistance activities.

INVESTIGATION

FIGURE 33.18 Exercise and Muscle Gene Expression Skeletal muscles can be expected to change in many ways—both known and unknown—in response to endurance training. We know that more capillaries develop, for example. Because many changes in muscle structure and function require the synthesis of new proteins, one might expect that many genes might be expressed at higher levels after a bout of endurance training. Techniques for measuring changes in gene expression have undergone revolutionary advances in recent decades, so that today thousands of genes can be assessed simultaneously. Using one of these techniques, termed microarrays (see Figure 12.18), investigators measured muscle gene expression before and after a single bout of endurance training.[a]

HYPOTHESIS

Endurance exercise causes changes in muscle gene expression.

METHOD

© iStock.com/dardespot

1. Recruit healthy and fit young men (university students)—who have been leading relatively sedentary lives—to exercise intensely on a bicycle for about 75 minutes.
2. Prior to exercise, remove a tiny tissue sample (150 mg) from each individual by biopsy from the vastus lateralis muscle, a thigh muscle that is a subpart of the quadriceps femoris (see Figure 33.12B).
3. Three hours after the exercise ends, remove a second tissue sample. Extract messenger RNA (mRNA) from the tissue samples, and use microarrays and sophisticated statistical methods to measure changes in the expression of about 7,000 genes between the first and second samples.

RESULTS

The investigators found that 117 genes exhibited increased expression following the single bout of exercise. On average, these genes showed a 3.7-fold increase in expression (i.e., a 3.7-fold increase in mRNA transcription) after exercise, compared with before exercise. By contrast, the investigators found that only 8 genes (out of the set of 7,000) exhibited decreased expression after exercise.

CONCLUSION

About 2 percent (117 out of 7,000) of the studied genes showed increased expression after a single bout of exercise; relatively few genes showed decreased expression.

ANALYZE THE DATA

The table lists common names for ten of the genes that showed increased expression after exercise. (The names given to genes are not highly standardized. Thus each of these genes has other names, which are often called "aliases.")

CEBPD	*MT2A*	*PGC1A*
FOXO1A	*MYC*	*PPARG*
MRPL2	*PDK4*	*SERCA3*
MT1B		

1. Why might these genes be increased in expression during exercise? How might the increased expression be affecting muscle function? One way to make progress on these questions is to look up what is known about the roles of the genes. The National Center for Biotechnology Information (NCBI) website hosts the most authoritative tool for this approach. Go to its homepage (www.ncbi.nlm.nih.gov), click on the "All Databases" dropdown menu, and choose "Gene." Now you can enter gene names in the search box and execute searches. To get started, enter *SERCA3*. You'll see that information is provided from studies of many species. Choose "human" or "*Homo sapiens*" for your first pass. Based on what you've learned in this chapter, why might increased expression of *SERCA3* be of advantage during exercise? Now do the same search for *PGC1A* and *MT2A*. Based on what you find, why might *PGC1A* and *MT2A* expression be increased?
2. What is the advantage of measuring the expression of 7,000 random genes?
3. Is it correct to assume that all changes of gene expression are beneficial? Do the data prove that they are beneficial? Why might we expect changes to be beneficial?
4. Suppose you are giving a talk to class of eighth graders on whether they should exercise regularly. Might the data we are discussing be relevant? Why or why not?

[a]D. J. Mahoney et al. 2005. *FASEB J* 19: 1498–1500.

A similar **Analyze the Data** exercise may be assigned in **Achieve**.

REVIEW & APPLY | 33.2

R Vertebrates have an endoskeleton, and the skeletal muscles and bones work together around joints to produce movements. Arthropods have an exoskeleton, with the skeletal muscles attaching to apodemes at the joints of the exoskeleton. A hydrostatic skeleton exists if an animal's body or a part of its body becomes stiff and skeleton-like because of high internal fluid pressure produced by muscle contraction. A muscle cell's power output depends on the rate at which the cell can supply the contractile apparatus with ATP. For producing ATP, skeletal muscle cells have three systems: immediate, glycolytic, and oxidative. The glycolytic system can produce ATP rapidly but is self-limiting. The oxidative system produces ATP more slowly but can continue for a long time. Individual animals typically have more than one type of skeletal muscle cell, with the two major types in vertebrates being slow oxidative (red) cells and fast glycolytic (white) cells. Skeletal muscle displays a wide range of phenotypic plasticity, as illustrated by training effects during endurance and resistance training.

A

1. Can components of the endoskeleton continue to change shape and structure after an animal is done growing? Explain.
2. Are all muscles that function in locomotion necessarily attached to hard skeletal elements?
3. Consider a skeletal muscle cell that contains a type of myosin specialized for rapidly making and breaking actin–myosin cross-bridges. Is the cell likely to synthesize most of its ATP by using the glycolytic system or the oxidative system?
4. Weekly measurements of an athlete's gastrocnemius (calf) muscle reveal that the density of capillaries in the muscle and the number of mitochondria per muscle cell are increasing. Would you predict that the athlete has been training for a marathon or lifting weights at the gym? Explain.

Skeletal muscles, as we have seen, are enormously important. We breathe using skeletal muscles, and the skeletal muscles are responsible for most behaviors, including athletic performance. Skeletal muscle is not the only type of muscle, however, and next we look at other types.

33.3 | Many Distinctive Types of Muscle Have Evolved

LEARNING OBJECTIVES

By the end of this key concept you should be able to:

33.3.1 **Explain how the cells of our heart muscle are able to contract nearly simultaneously.**

33.3.2 **Describe three types of muscle contraction that occur in the animal kingdom, including the specialized excitation–contraction coupling seen in most insect species.**

33.3.3 **Explain the relationship between the transmembrane voltage that exists in all muscle cells and the huge electric discharges put out by electric fishes.**

Besides skeletal muscle, several other distinctive types of muscle have evolved. Vertebrate **cardiac muscle**—heart muscle—is one type. During each of our lives, our cardiac muscle contracts and relaxes regularly from the time of our early embryonic development until the time we die.

Vertebrate cardiac muscle is both similar to and different from skeletal muscle

Like skeletal muscle, cardiac muscle appears striated when viewed under a light microscope. This is because its actin and myosin filaments are arranged in regular patterns in sarcomeres. In other ways, however, cardiac muscle differs morphologically from skeletal muscle. Cardiac muscle cells are much smaller than skeletal muscle cells, and unlike skeletal muscle cells, each has only one nucleus.

A key way that cardiac muscle differs from skeletal muscle is in the relationship of adjacent cells. In cardiac muscle, adjacent cells are joined by structures called *intercalated discs*, where the cells are electrically coupled by gap junctions (**FIGURE 33.19A**). At a gap junction (see Figure 4.19 and Key Concept 4.4), tiny pores (lined with proteins termed connexins) in the cell membrane of one cell line up with similar pores in the other cell, establishing cytoplasmic continuity: the cytoplasm of one cell is continuous with the cytoplasm of the other. Because of this continuity, if one of the two cells is excited and an action potential spreads over its cell membrane, the action potential spreads directly, without delay, to the second cell.

Accordingly, when an action potential is initiated at any one point in a mass of heart muscle, it spreads rapidly throughout the muscle by moving directly from one cell to another. This process ensures that all the cells within a large mass of muscle are excited at approximately the same time. Thus all the cells contract almost simultaneously, raising the blood pressure in the heart chambers to a high level and propelling the blood out of the heart and through the blood vessels.

In a vertebrate heart, certain of the muscle cells in the heart muscle, called *pacemaker cells*, generate the heartbeat rhythm by endogenously initiating action potentials in a rhythm corresponding to the heart rate (see Key Concept 30.5 and Figure 30.26). The heart rhythm is thus very different from the breathing rhythm. The breathing muscles of vertebrates are skeletal muscles, and like all skeletal muscles, they contract only if stimulated by neurons. If a person suffers a neck fracture, they will immediately stop breathing if the neurons responsible for stimulating the breathing muscles are damaged. The person's heart will keep beating, however, because the heartbeat rhythm is endogenously initiated in the heart muscle.

Media Clip 33.1 Be Still My Beating Stem Cell Heart

Vertebrate smooth muscle powers slow contractions of many internal organs

In addition to skeletal and cardiac muscle, vertebrates have a third type of muscle, **smooth muscle**. Smooth muscle cells—which are smaller than skeletal muscle cells—are usually spindle-shaped, and each has a single nucleus (**FIGURE 33.19B**). Smooth muscle is termed "smooth" because the tissue does not have a striated appearance when viewed under a light microscope. It lacks striation because its actin and myosin filaments are not as regularly arranged as those in skeletal and cardiac muscle.

Smooth muscles provide the contractile forces for movements in most of the internal organs, under the control of the nervous system (especially the autonomic nervous system). Smooth muscles in the walls of the digestive tract move food through the digestive tract. Smooth muscles in the walls of small arteries and arterioles adjust the diameters of these blood vessels, helping control blood flow through them (see Figure 30.21). During sexual climax, smooth muscles in the ejaculatory passages of the male propel the ejaculation of semen, and in the female smooth muscles contribute to the rhythmic contractions of orgasm. Smooth muscles in the walls of the bladder produce forces for bladder emptying.

Some smooth muscle tissue, such as that in the walls of the digestive tract (see Figure 28.20), consists of sheets of cells in which adjacent cells have cytoplasmic continuity because of gap junctions. As a result, excitation of one cell can spread directly to adjacent cells, helping coordinate the contractions of the cells. This occurs during **peristalsis**, a pattern of contraction in which a wave of contraction moves along the gut tube from one end to the other.

LINK The structure and function of the mammalian gut are detailed in **Key Concept 28.4**.

In both cardiac muscle and smooth muscle, excitation–contraction coupling depends on Ca^{2+} influx into the cytoplasm, as it does in skeletal muscle. Details of excitation–contraction coupling differ among the three types of vertebrate muscle, however.

Animation 33.2 Smooth Muscle Action

(A) Cardiac muscle

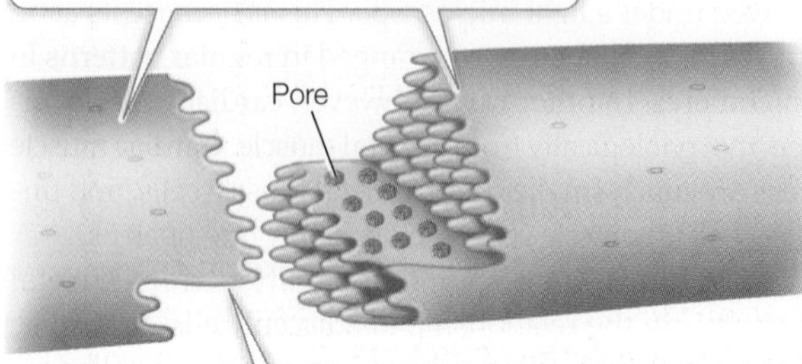

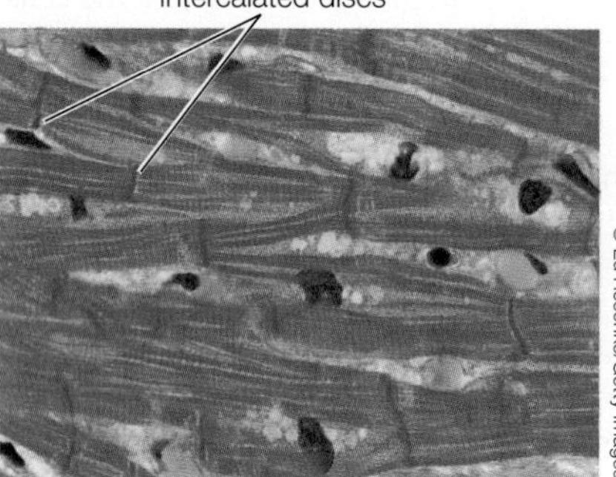

(B) Smooth muscle

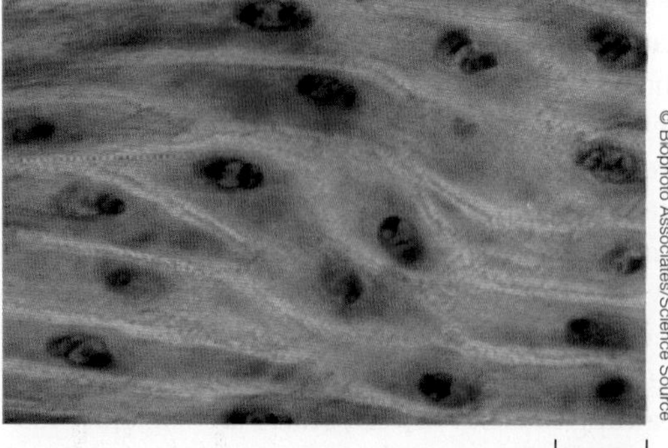

FIGURE 33.19 Vertebrate Cardiac and Smooth Muscle
(A) Cardiac muscle. The diagram at the top shows an enlarged view of two cardiac muscle cells at an intercalated disc. The photograph below shows a section of cardiac muscle seen under a light microscope.
(B) Smooth muscle. The photograph shows a section of smooth muscle seen under a light microscope.

Some insect flight muscle has evolved unique excitation–contraction coupling

Invertebrates have evolved a variety of specialized types of muscle. Here we discuss just two of them, plus a fascinating specialized type of vertebrate muscle. The two invertebrate muscle types, asynchronous flight muscle and catch muscle, are specialized to contract in unusual ways that are highly adaptive for the animals in which they occur. The specialized type of vertebrate muscle does not contract at all but instead has become modified to produce electricity!

The muscles that drive the wings during insect flight are of two types in terms of their excitation–contraction coupling. In about 25 percent of insect species, each contraction requires excitation, and the ratio of excitations to contractions is therefore 1:1. For flight muscle of this type to contract at a higher and higher frequency, all the processes of excitation–contraction coupling must take place faster and faster. At high frequencies, contraction becomes severely compromised in several ways as a result. Both muscular efficiency and power output diminish at high frequencies.

A different and unique form of muscle that has evolved only in insects solves these problems. It is termed **asynchronous**

(A) Asynchronous flight muscle

(B) Catch muscle

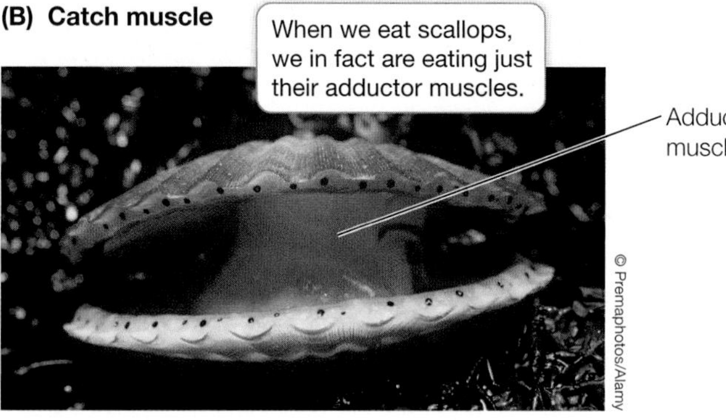

FIGURE 33.20 Two Distinctive Types of Invertebrate Muscle **(A)** Bees are one of several insect groups that have asynchronous fight muscle, meaning the flight muscles contract multiple times in response to each single neuronal excitation. **(B)** When a scallop gapes as in this photo, its adductor muscle is easily seen. In times of danger, the adductor muscle contracts, pulling the two half-shells tightly together. The muscle enters a state of catch in which it can stay contracted for long periods with little ATP use.

flight muscle because each excitation results in many contractions. At high frequencies of contraction, asynchronous flight muscle is believed to maintain higher efficiency and greater power output than synchronous flight muscle. Asynchronous flight muscle has evolved independently in insects about ten times and is found today in about 75 percent of insect species, including bees, beetles, and flies (**FIGURE 33.20A**).

Catch muscle in clams and scallops stays contracted with little ATP use

A scallop or clam has a shell composed of two half-shells. One or two powerful *adductor muscles* connect these half-shells (**FIGURE 33.20B**). Contraction of the adductor muscles pulls the two half-shells tightly together so that the animal becomes fully enclosed in its shell—a form of armor.

When predators are nearby, a scallop's or clam's adductor muscles may need to remain contracted for long periods of time. We think this is why the adductor muscles have evolved the ability to enter a specialized state of contraction termed **catch**. During catch, the muscles maintain high contractile forces continuously for tens of minutes—or even for hours—with almost no use of ATP. The way catch works is not well understood. However, measurements show that catch muscle maintains contraction at a far lower ATP cost than most muscles. Catch is a highly controlled state that can be quickly ended by signals from the nervous system. Its energy-saving feature is adaptive for animals that possess body armor that depends on muscular contraction to be effective.

Fish electric organs are composed of modified muscle

Electric eels (*Electrophorus*), which live in tropical rivers, can produce external voltage pulses as large as 700 volts. Electric rays (*Torpedo*) in the ocean also produce large external electric pulses, although they produce high electric currents (up to 20 amperes) rather than high voltages. Fishes such as these use their electric pulses to stun prey. Many other fishes produce small external electric pulses that they use to detect their surroundings for orientation (see Figure 31.34C).

The electric organs of nearly all electric fishes have evolved from skeletal muscle and consist of modified muscle cells. These cells cannot contract because their contractile apparatus (i.e., actin and myosin) is poorly developed. However, the cell membranes of these muscle cells have a unique property: during excitation, the small voltages that exist across the cell membranes of many cells all add together (**FIGURE 33.21**). The voltage across each cell membrane is about 0.1 volts (a very ordinary voltage for an animal cell). In the electric organ, however, the voltages of many membranes add together, in the same way that the voltages of batteries add up when the batteries are connected in series. In this way, an electric organ with 1,000 cell membranes can produce 100 volts (0.1 volt × 1000 = 100 volt).

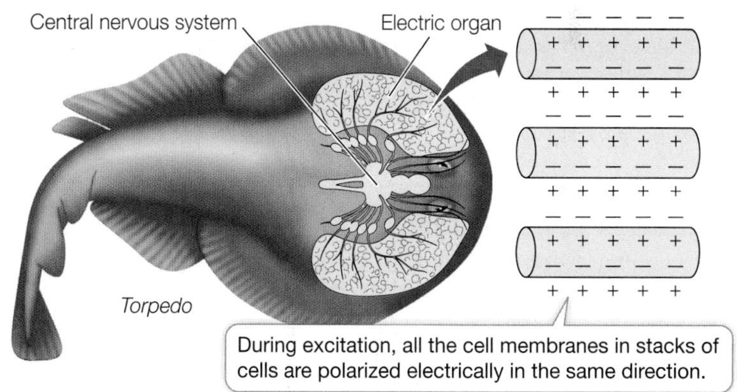

FIGURE 33.21 The Electric Organs in an Electric Ray (*Torpedo*) Are Composed of Modified Muscle Cells Each organ consists of many cells. Three cells are shown diagrammatically at the right, illustrating how the membrane potentials of adjacent cells can add together, in a way similar to the way that the voltages of batteries in a flashlight add together. Ancient fisherman reported that they could tell if they'd caught a *Torpedo* in their hand nets when they were fishing in the ocean because a sense of numbness spread over their hands. (Ray after C. P. Hickman, Sr. et al. 1979. *Integrated Principles of Zoology* 6e. C. V. Mosby Company: St. Louis, MO.)

REVIEW & APPLY | 33.3

R Vertebrate cardiac muscle cells are electrically connected by gap junctions, so that action potentials spread rapidly throughout large masses of cardiac muscle and excite synchronized contractions. In a vertebrate heart, modified muscle cells serve as the pacemaker cells, meaning neuronal stimulation is not required for the heart rhythm. Smooth muscle provides contractile force for internal organs such as the gut, blood vessels, and reproductive ducts. In a majority of insects, flight is powered by asynchronous flight muscle, which unlike most muscle undergoes multiple contractions with each excitation. Catch muscle, such as the adductor muscles of clams and scallops, can sustain strong contractions for long periods with little ATP. The electric organs of nearly all electric fishes evolved from skeletal muscle and consist of modified, noncontractile muscle cells.

A

1. Why does the heart keep beating in victims of neck fractures, although the victims stop breathing?
2. What type of muscle would be used in each of the following activities: the extension of a lizard's forelimbs during walking; the pumping of O_2-poor blood from a person's heart to their lungs; the expulsion of food during vomiting; the holding shut of a clam's shell?
3. If needed, could the muscle cells in an electric fish's electric organs be used to generate forces for swimming? Why or why not?

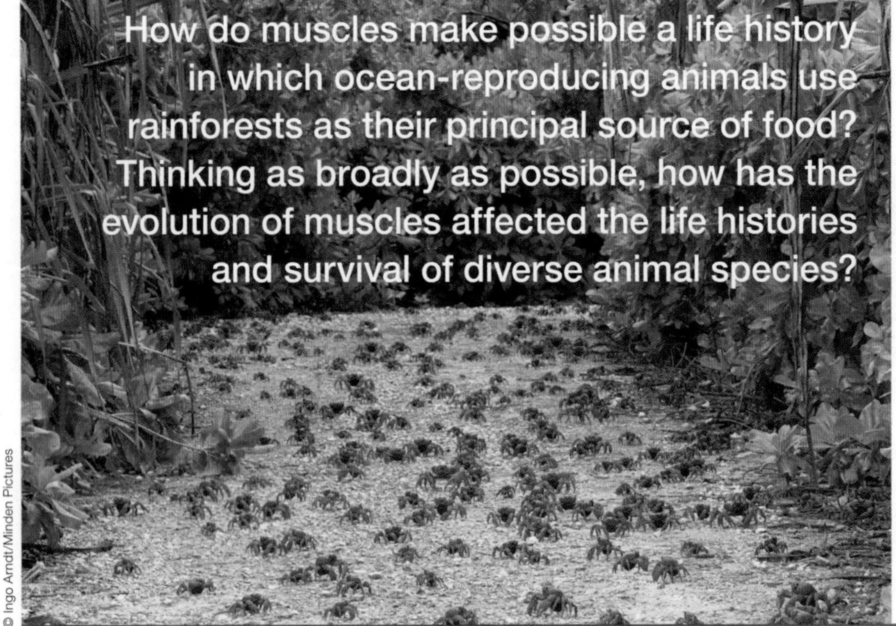

Answer: With the ability to move, animals do not need to collect most of their food and reproduce in the same place. They can exploit different habitats for feeding and reproduction because they can travel between habitats in a way that's dependable and cost effective. From the viewpoint of evolutionary theory, the evolution of muscle and movement permitted animal species a large degree of independence in the choice of feeding habitats and reproductive habitats. The feeding habitat is not strongly constrained by reproductive needs, and vice versa. Emperor Penguins (*Aptenodytes forsteri*) do all their feeding in the ocean but reproduce inland, 100 miles (160 km) away. Eastern Pacific gray whales (*Eschrichtius robustus*) do much of their feeding in sub-Arctic waters but reproduce thousands of miles to the south in subtropical lagoons. There are many other examples.

33.1 Muscle Cells Develop Forces by Means of Cycles of Protein–Protein Interaction

FIGURE 33.7

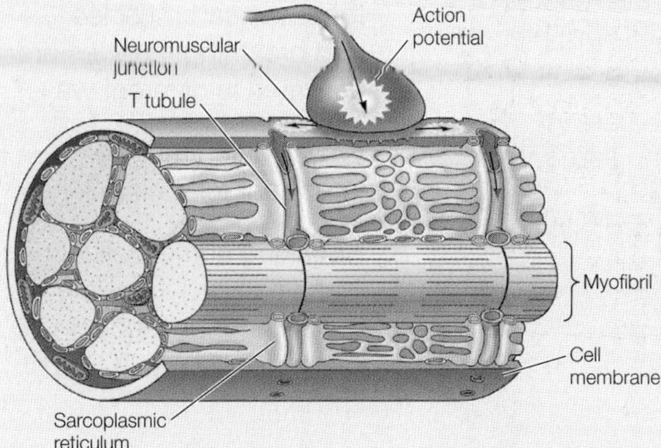

Go to ACTIVITIES 33.1 and 33.2 and ANIMATION 33.1

- The contractile proteins **ACTIN** and **MYOSIN** develop forces by increasing their extent of overlap in the **MYOFIBRILS**—a process achieved by flexing of myosin heads bonded to actin.
- Bonding with ATP permits a myosin head to detach from actin. ATP hydrolysis releases energy that is stored in myosin to drive the next power stroke.
- Excitation of a **MUSCLE FIBER** spreads to the cell interior in **TRANSVERSE**, or **T**, **TUBULES**, triggering the release of Ca^{2+} from the **SARCOPLASMIC RETICULUM** in the process of **EXCITATION–CONTRACTION COUPLING** (FIGURE 33.7).

QUESTIONS

1. What is the role of titin?
2. Why does rigor mortis occur?
3. What are the roles of troponin and tropomyosin?

33.2 The Function of Skeletal Muscle Depends on Interaction with the Skeleton and on ATP Supply, Cell Type, and Training

FIGURE 33.12

Coordination of antagonistic muscles acting at one joint

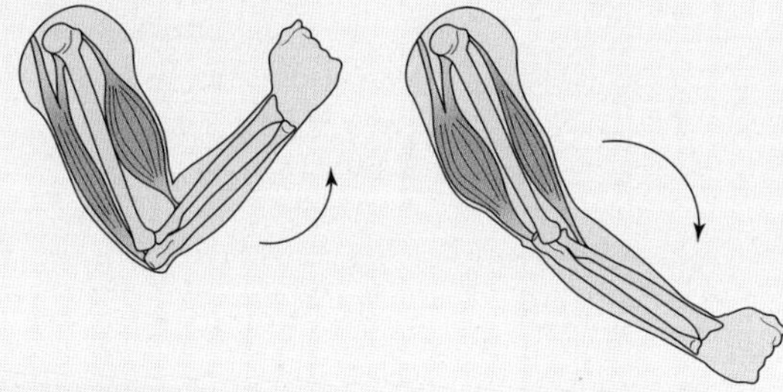

Coordination of different muscle sets acting on one limb

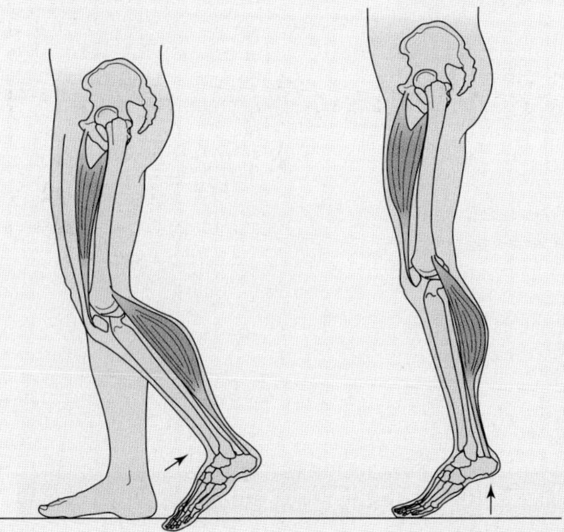

- For skeletal muscles to power locomotion, they pull on elements of the skeleton, whether it is an **ENDOSKELETON** or **EXOSKELETON**. **ANTAGONISTIC PAIRS** of muscles produce opposing movements (**FIGURE 33.12**).
- The **IMMEDIATE**, **GLYCOLYTIC**, and **OXIDATIVE SYSTEMS** vary dramatically in how rapidly and for how long they can produce ATP for muscle contraction.
- Two major types of vertebrate skeletal muscle cells are the **FAST GLYCOLYTIC** (white) **CELLS** and **SLOW OXIDATIVE** (red) **CELLS**.

1. What different motions can result when a person's gastrocnemius muscle contracts, and why?
2. How do worms and clams employ a hydrostatic skeleton to move through soil or sand?
3. How and why are people and fishes similar in the types of exercise they power primarily with the glycolytic system of ATP production?

33.3 Many Distinctive Types of Muscle Have Evolved

Go to ANIMATION 33.2

- Vertebrates have three major types of muscle: skeletal muscle, **CARDIAC MUSCLE**, and **SMOOTH MUSCLE** (**FIGURE 33.19**). Two distinctive types of invertebrate muscle are the **ASYNCHRONOUS FLIGHT MUSCLE** found in the majority of insects and the **CATCH** muscle found in clams and scallops.
- The electric organs of most electric fishes consist of sets of modified muscle cells in which all the cell membranes are capable of assuming electrical polarity in the same direction, permitting addition of all the membrane potentials.

FIGURE 33.19

Cardiac muscle

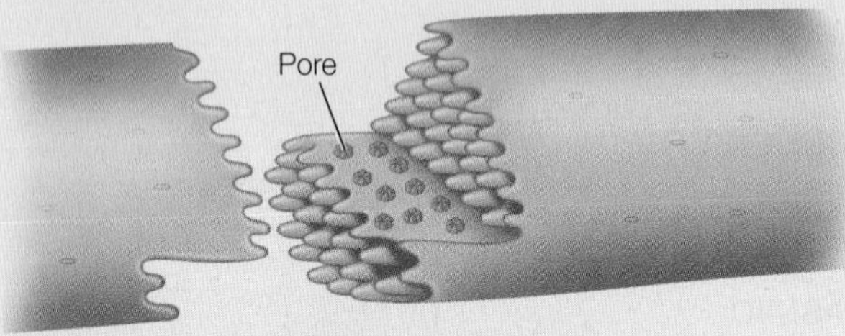

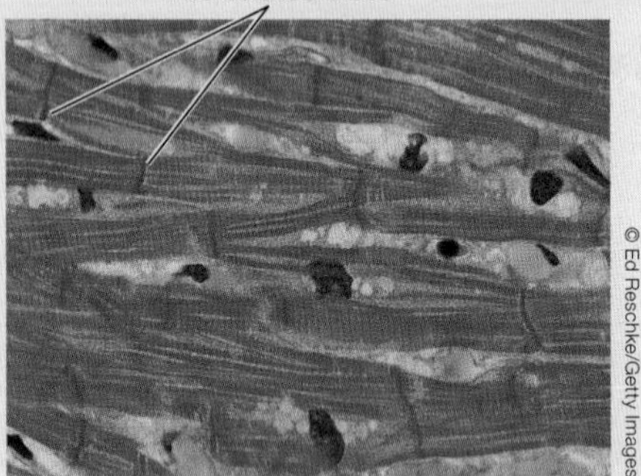

© Ed Reschke/Getty Images

Smooth muscle

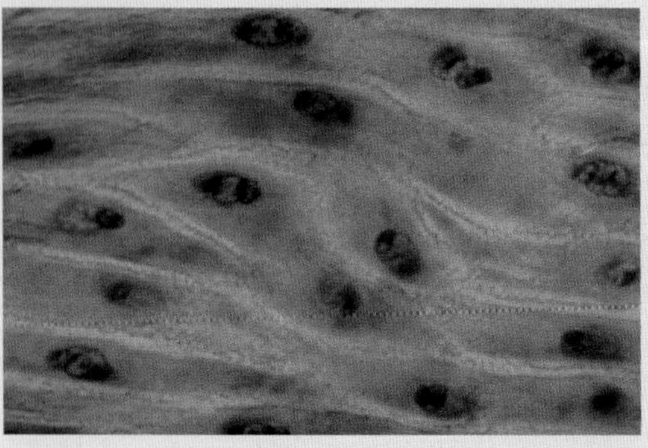

© Biophoto Associates/Science Source

QUESTIONS

1. In aiding people who experience severe trauma, why is it important to know that the heart muscles and breathing muscles are activated in different ways?
2. Why is catch muscle important for a clam?
3. Speaking numerically (quantitatively), how is it possible for an electric fish to produce a 700-volt shock?

Animal Reproduction 34

Rabbits are famed for their reproductive potential. Some of this fame is of a shocking kind. European rabbits (*Oryctolagus cuniculus*) were introduced into Australia around 1860 so they could be hunted for sport. Populations quickly grew into the millions, and today rabbits are blamed for dramatic habitat destruction in much of the country.

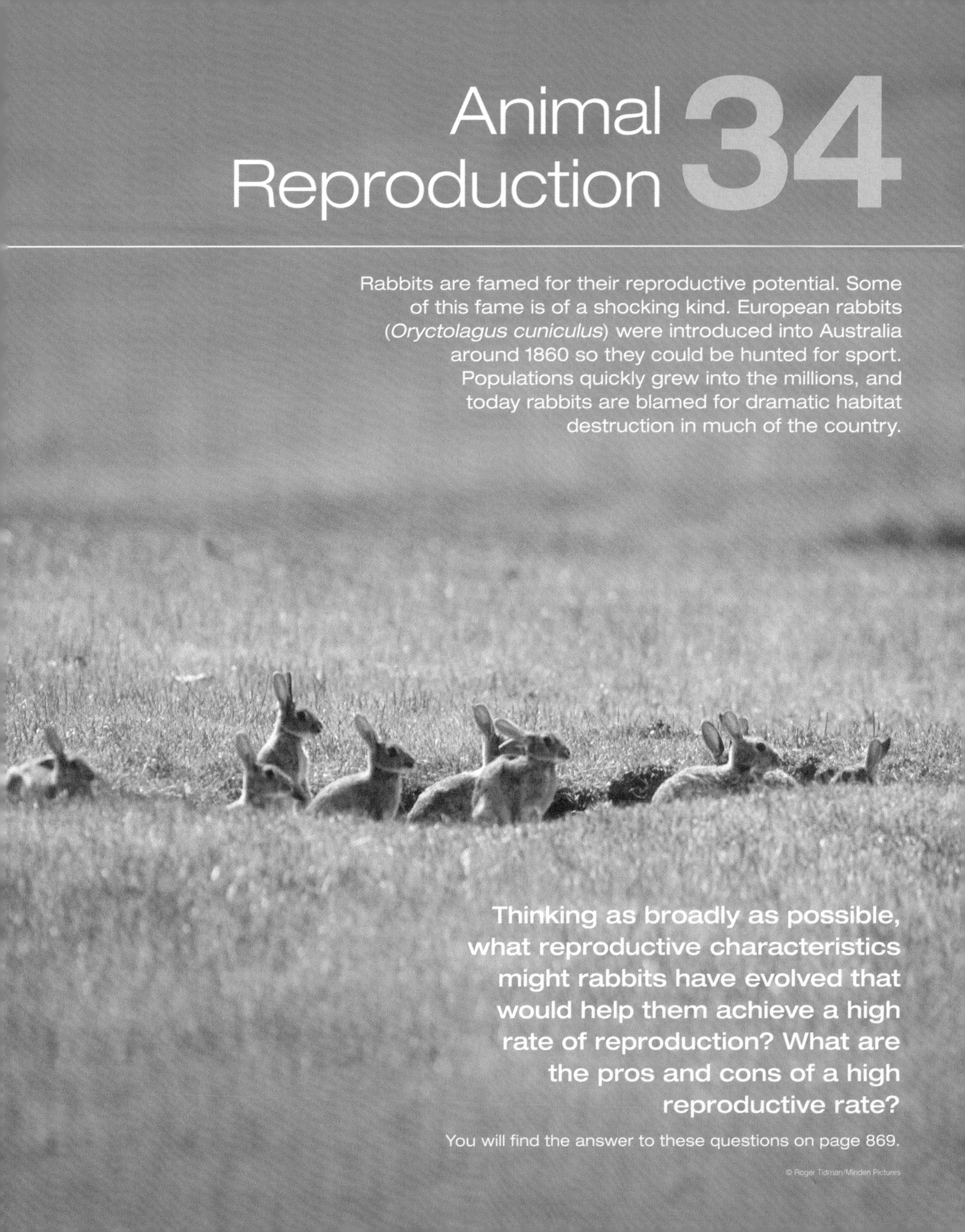

Thinking as broadly as possible, what reproductive characteristics might rabbits have evolved that would help them achieve a high rate of reproduction? What are the pros and cons of a high reproductive rate?

You will find the answer to these questions on page 869.

KEY CONCEPTS

34.1 Sexual Reproduction Depends on Gamete Formation and Fertilization

34.2 The Mammalian Reproductive System Is Hormonally Controlled

34.3 Reproduction Is Integrated with the Life Cycle

34.1 Sexual Reproduction Depends on Gamete Formation and Fertilization

LEARNING OBJECTIVES

By the end of this key concept you should be able to:

34.1.1 **Outline differences in gametogenesis between males and females.**

34.1.2 **Outline the mechanisms of sex determination in animals.**

34.1.3 **Specify potential advantages and disadvantages of internal and external fertilization.**

In casual conversation, "having sex" refers to human sexual activity. Biologists, however, mean something far broader when speaking of sex. As applied to animals, the process of **sex** is a mechanism by which the genes of two individuals are combined to produce offspring. In animal **sexual reproduction**, each parent produces specialized reproductive cells called **gametes**, which differ in important ways from the other cells—termed the **somatic cells**—in the body. The gametes are produced by the process of meiosis, which has two important consequences. First, each gamete cell has only half as many chromosomes as a somatic cell (see Figure 7.14), and second, the gametes produced by an individual animal vary greatly in their genetic composition because of the processes of independent assortment and crossing over that occur during meiosis (see Key Concept 7.3). The gametes of the female parent are relatively large, nonmotile cells called **eggs** or **ova** (singular, *ovum*). The gametes of the male, called **spermatozoa** or **sperm**, are small cells that swim, typically using flagella (**FIGURE 34.1A**). A sperm and egg fuse to produce a single-celled zygote that has a full set of chromosomes and develops into a sexually produced offspring (**FIGURE 34.1B**; see Chapter 35). In this way, chromosomes and genes from two parents are combined to produce each offspring, and the new individual is not genetically identical to either parent.

LINK Sexual reproduction also occurs in plants (see **Key Concept 26.2**) and in fungi (see **Key Concept 21.3**). The genetic aspects of sexual reproduction and the diversity it generates are discussed in **Key Concepts 7.1** and **7.3**; the evolutionary consequences are discussed in **Key Concept 15.2**.

During **asexual reproduction**, by contrast, offspring are produced by just a single parent, usually by mitosis: parental cells with a full set of chromosomes divide to produce offspring cells that each have the same full set of chromosomes. Offspring are thus genetically identical to their single parent. A variety of animals, mostly invertebrates, reproduce asexually. Usually they can also reproduce sexually.

Budding and fission are the two most common types of asexual reproduction. In **budding**, a new individual forms as a mitotically produced outgrowth (bud) of the parental individual. In **fission**, the parental individual splits into two or more pieces that grow into new individuals. Reef corals undergo both of these types of asexual reproduction. Corals occasionally reproduce sexually with eggs and sperm. However, the large reef structures

(A) A sperm and egg

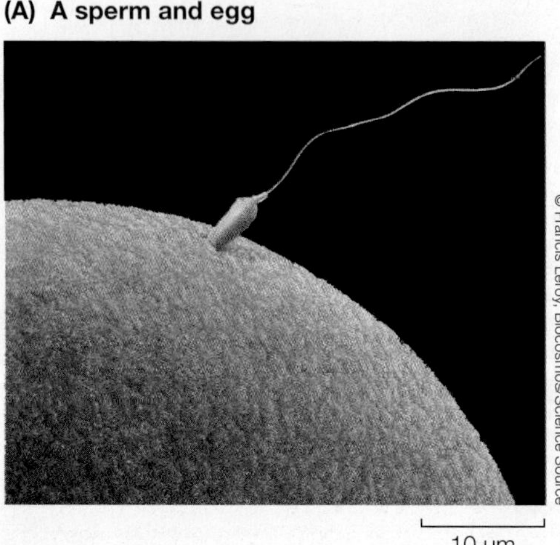

(B) Overview of animal life cycle

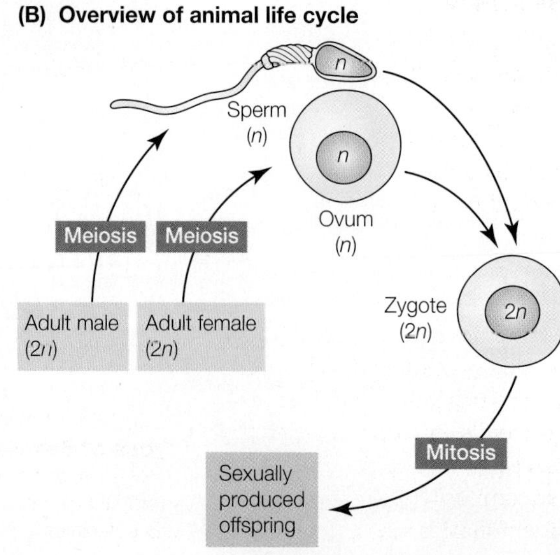

FIGURE 34.1 Sexual Reproduction in Animals **(A)** A human sperm and egg. **(B)** The essence of animal sexual reproduction is the combination of genes from two individuals. The somatic cells in the body of an animal have two copies of each chromosome and thus are diploid (2*n*). The gametes, by contrast, have only a single copy of each chromosome and thus are haploid (*n*). When a sperm and egg combine, they form a single-celled zygote that is 2*n* (diploid), and the zygote subsequently multiplies by mitosis to develop into the adult. The life cycle of an animal is categorized as diplontic (see Figure 7.3).

FIGURE 34.2 Asexual Reproduction in Animals Sexual reproduction in reef corals produces a single polyp **(A)**, which then reproduces asexually by budding to produce a colony **(B, C)**. Each polyp secretes skeletal material. The skeletal material of all the polyps in a colony fuses into a single structure, creating the massive skeletal structures that typify coral reefs.

with which we are familiar are produced principally by budding. Sexual reproduction results in a single, small individual called a polyp (**FIGURE 34.2A**). Then that initial polyp produces other polyps by budding (**FIGURE 34.2B**), resulting in a large colony of hundreds or thousands of genetically identical polyps (**FIGURE 34.2C**). During violent storms, a large colony may break into two or more pieces that currents carry to separate locations—leading to asexual reproduction by fission because each piece can subsequently grow into a new large colony.

Asexual reproduction has the advantage that it maintains favorable combinations of genes. An individual with a genotype that is highly suited to its environment can rapidly produce offspring with the same advantageous genotype by budding or other asexual means.

Most animals reproduce sexually

Most vertebrates reproduce only sexually, and the same can be said of many invertebrates. Almost all animals have the capacity to reproduce sexually even if, like corals, they also reproduce asexually. We will focus on sexual reproduction for the rest of this chapter.

Sexual reproduction can seem to us to be the obvious best option for producing offspring. Why would reproduction occur by any other means? However, sexual reproduction can have disadvantages in comparison with asexual reproduction. For example, during sexual reproduction potential mates have to find each other and succeed in mating despite the presence of competitors and predators. Possibly the greatest downside of sexual reproduction is that it breaks up favorable combinations of genes. An individual with a genotype that is virtually perfect for its environment cannot simply pass that genotype along to its offspring. Biologists have advanced many theories on why sex is so common despite these limitations. One often-stated hypothesis emphasizes the changeable nature of environments. Environments, it is argued, change sufficiently—from time to time and from place to place—that no single genotype is likely to be highly successful for very long. Sexual reproduction has an enormous potential for producing genetic diversity in offspring, maximizing the odds that at least some offspring will succeed in a changing environment.

In a species of sexually reproducing animal, individuals of the two sexes, male and female, differ in three ways:

1. Primary reproductive organs (primary sex organs)
2. Accessory reproductive organs (accessory sex organs)
3. Secondary sexual characteristics

The **primary reproductive organs** are the organs that produce the gametes: the **gonads**. These are termed the **ovaries** in females and the **testes** (singular *testis*)—or sometimes testicles—in males. The **accessory reproductive organs** are the reproductive organs other than the gonads. In human males, for example, the penis is an accessory reproductive organ, as are the vas deferens and the other ducts that carry sperm out of the testes to the urethra in the penis (**FIGURE 34.3**). The females of some animals, such

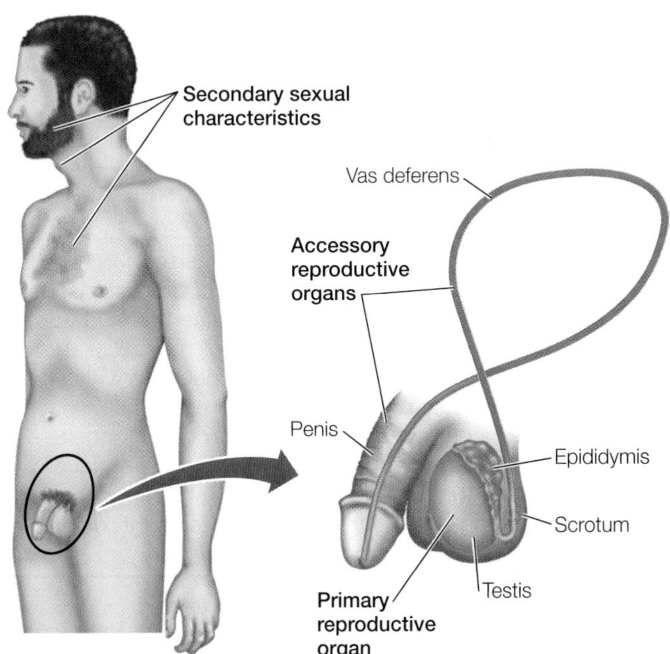

FIGURE 34.3 Sexually Reproducing Animals Have Three Types of Sex-Specific Characteristics Individuals of each sex have sex-specific primary reproductive organs (gonads) and accessory reproductive organs. They may also have sex-specific secondary sexual characteristics.

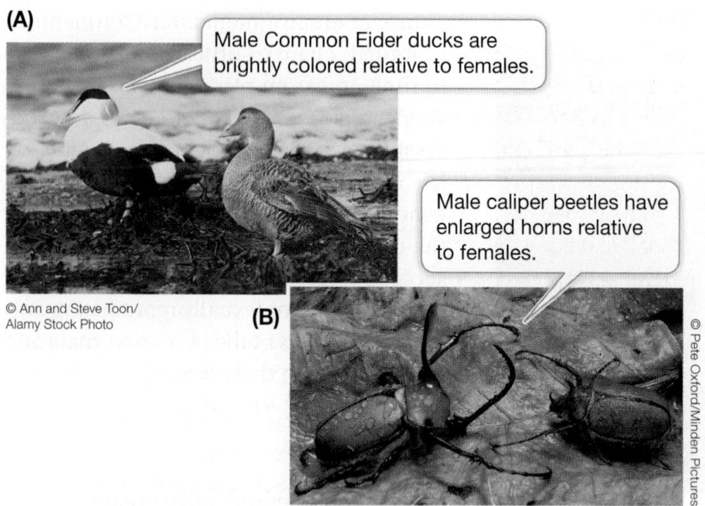

FIGURE 34.4 Secondary Sexual Characteristics The coloration of male and female Common Eider ducks (*Somateria mollissima*) **(A)** and the size of horns in male and female caliper beetles (*Golofa porteri*) **(B)** are examples of secondary sexual characteristics.

as mammals and many sharks, nurture offspring internally for a long time before giving birth. In these cases, the uterus and other structures responsible for nurturing the young are among the accessory reproductive organs. In other animals, such as birds, the females lay eggs, and their accessory organs include the glands that add a shell or capsule to each egg.

The **secondary sexual characteristics** are properties of nonreproductive tissues and organs that are distinctive in each sex. These characteristics often result from the actions of sex hormones on nonreproductive structures. In humans, the secondary sexual characteristics of males include facial hair, chest hair, and a deep voice (see Figure 34.3), and those of females include broad hips and a higher voice. The bright feather colors of many male birds and the enlarged horns of male caliper beetles (*Golofa porteri*) are also secondary sexual characteristics (**FIGURE 34.4**).

LINK As **Key Concept 13.2** emphasizes, conspicuous secondary sexual characteristics (e.g., the long tails of male African widowbirds discussed there) play a key role in *sexual selection*, the evolutionary process in which individuals of one sex mate preferentially with particular types of individuals of the opposite sex rather than at random. Sexual selection favors traits that enhance an individual's chances of mating—even when these traits might reduce the individual's chances of survival.

Gametogenesis in the gonads produces the haploid gametes

Gametogenesis is the process by which gametes are produced. More specifically, it is called *spermatogenesis* in males and *oogenesis* in females. The process begins with diploid germ cells (diploid, denoted $2n$, means two copies of each chromosome per cell; see Figure 34.1B). The **germ cells** in the gonads, by definition, are capable of undergoing meiosis to produce gametes. The diploid germ cells first divide by mitosis—producing increased numbers of diploid germ cells—until the cells reach a stage at which they switch to meiosis, ultimately resulting in haploid gametes (haploid, $1n$, means one copy of each chromosome per cell). Recall that meiosis consists of two cell divisions, meiosis I and meiosis II, during which the DNA is replicated only once (see Figure 7.11).

In adult males, sperm are typically produced continuously during the time of year when reproduction occurs (i.e., all year in human males). During this time, the diploid germ cells in the testes multiply continuously, producing cells that go through the two meiotic divisions, resulting in four equal-sized haploid cells from each diploid cell (**FIGURE 34.5A**). The haploid cells then mature into fully differentiated sperm.

Gametogenesis in females differs in two ways from that in males. First, meiosis in females typically produces only one gamete from each germ cell because the cytoplasm of the germ cell is divided in highly unequal ways during each of the two meiotic divisions (**FIGURE 34.5B**). One of the cells produced by meiosis is a large cell, the ovum. The others are tiny cells, called **polar bodies**, that degenerate.

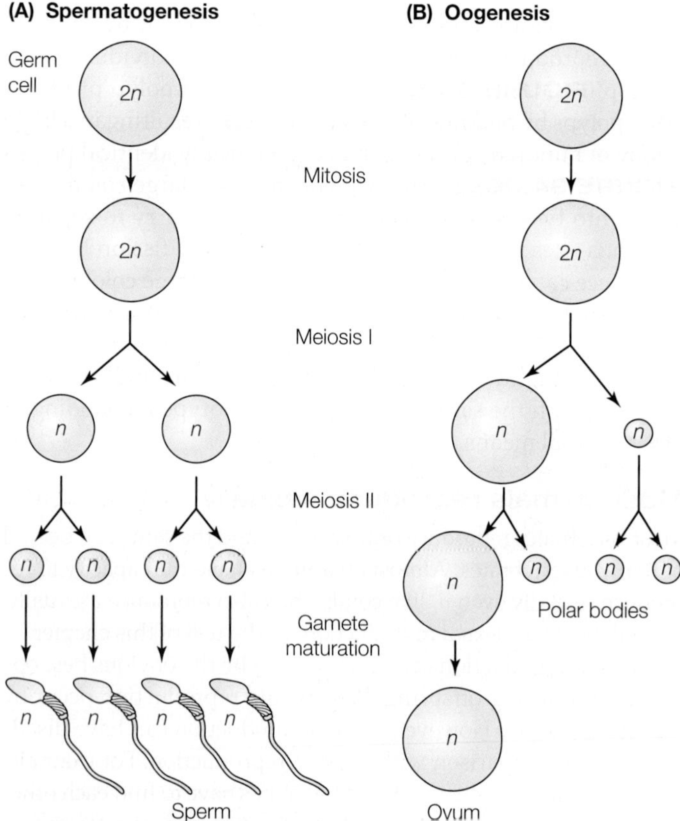

FIGURE 34.5 Gametogenesis Male and female germ cells proliferate by mitosis to produce an abundant supply of diploid cells that initiate and complete meiosis to produce sperm or ova. **(A)** Spermatogenesis. **(B)** Oogenesis. During oogenesis in many types of animals, the first polar body to be produced does not divide, so only two polar bodies are produced in total. $2n$ is the diploid number of chromosomes per cell; n is the haploid number. See Key Concept 7.3 for details of meiosis.

The second difference that often occurs during gametogenesis in females is that cells enter a stage of developmental arrest as they go through meiosis. During such arrest, a cell can continue to accumulate resources and differentiate, but its meiosis temporarily stops. Mammals exemplify developmental arrest of this type. According to the prevailing view of most researchers, all the germ cells of a young female placental mammal form and enter developmental arrest early in her life. Specifically, while she is a fetus in the uterus, or shortly thereafter, her germ cells (having been produced by mitosis) start meiosis, undergo the first step in meiosis I (prophase I; see Figure 7.11), and go into developmental arrest. The cells then remain in arrest for months or years until the female matures to reproductive age. Then some of the arrested cells resume meiosis to become ova that are released from the ovaries. In humans, one of the arrested cells typically develops fully and is released from the ovaries during each month between the time a girl enters puberty (near age 10) and the time she goes through menopause (near age 50).

The gonads typically consist of more than one cell type: they contain somatic cells that contribute to reproduction, in addition to the germ cells and gametes. Certain somatic cells in the gonads provide metabolic support to the germ cells during gametogenesis. The gonads also often contain somatic cells that secrete reproductive hormones. The mammalian testis illustrates these points.

There are three principal cell types in the testis:

1. Germ cells that produce the sperm
2. Somatic **Sertoli cells** that assist sperm production
3. Somatic **interstitial cells (Leydig cells)** that secrete the steroid sex hormone **testosterone** (see Key Concept 32.4)

The testis is principally composed of long, coiling tubules, the seminiferous tubules, where sperm are produced (**FIGURE 34.6A**). As shown in **FIGURE 34.6B** and **C**, the lining of each seminiferous tubule consists of germ cells—undergoing all the stages of gametogenesis—and Sertoli cells. The Sertoli cells envelop the developing germ cells, providing them with nutrients and helping regulate sperm production. The germ cells undergo meiosis and develop into sperm as they move from the periphery of a seminiferous tubule to the lumen (the open central cavity), and the sperm are shed into the lumen. The Sertoli cells are so important for gametogenesis that collectively they constitute well over half the volume of each

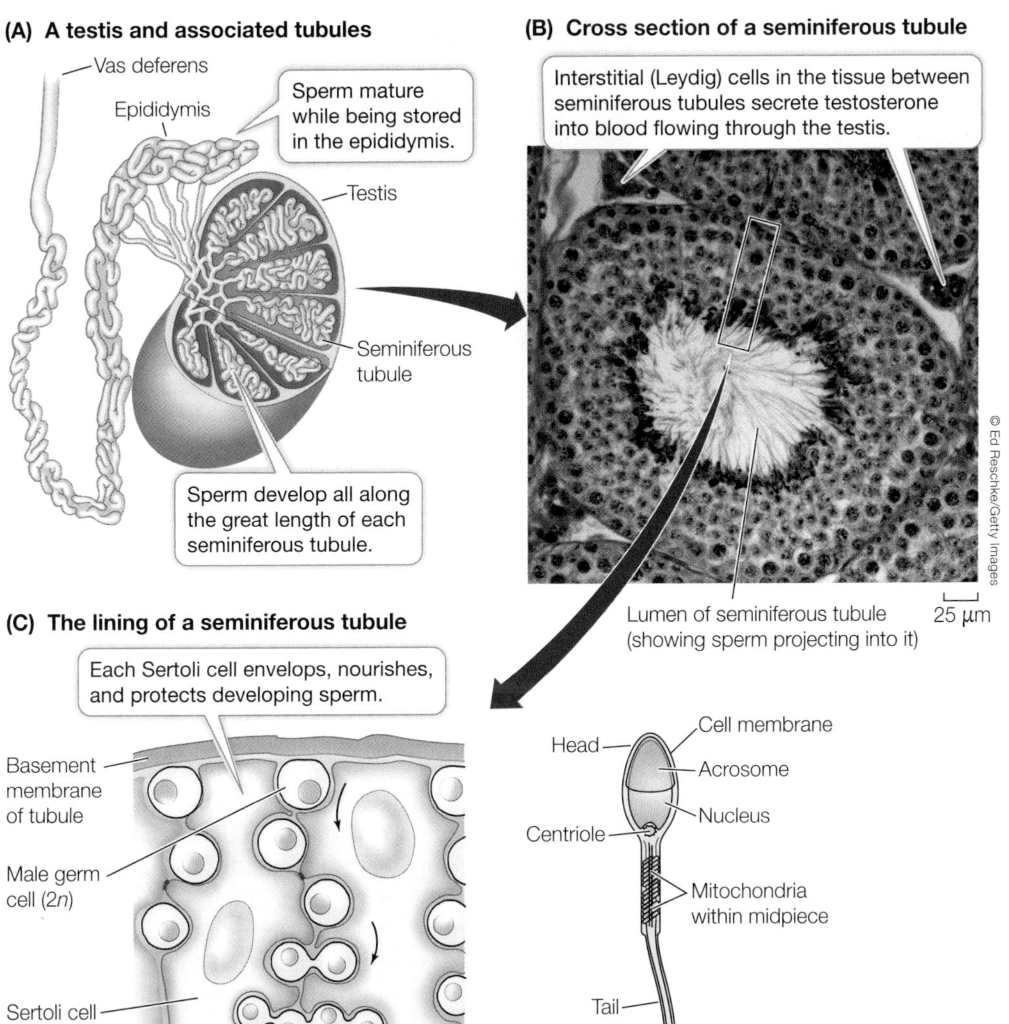

FIGURE 34.6 A Mammalian Testis **(A)** Seminiferous tubules fill the testis and continuously produce sperm at a high rate during the reproductive season. **(B)** This cross section of a testis—greatly magnified by comparison with (A)—shows both a seminiferous tubule and groups of interstitial (Leydig) cells in the spaces between this tubule and neighboring tubules. **(C)** This diagram of the lining of a seminiferous tubule shows the relationship between Sertoli cells and germ cells. As germ cells undergo gametogenesis, the cells move from the outer region of a seminiferous tubule toward its center (shown by arrows; compare with Figure 34.5A).

Activity 34.1 Spermatogenesis

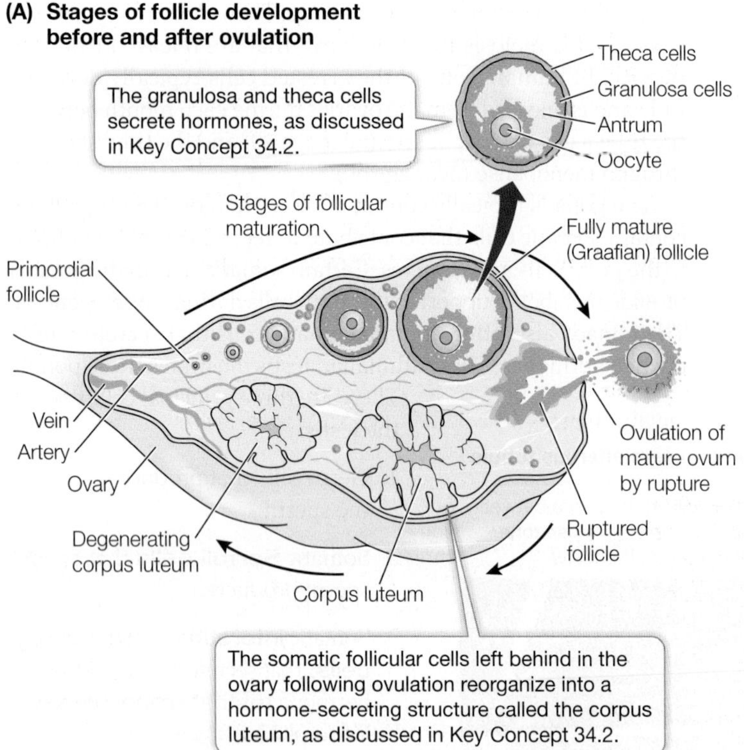

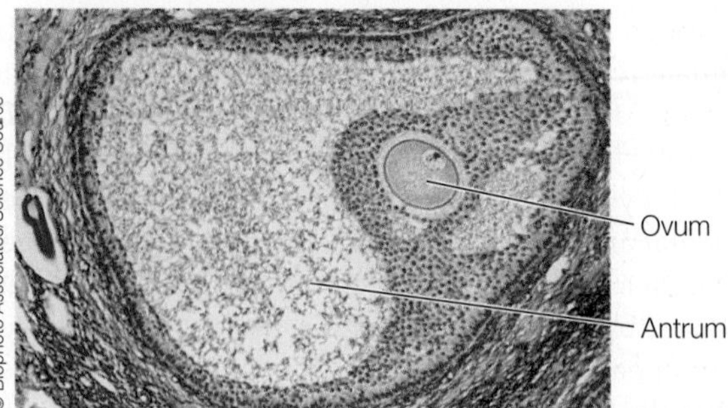

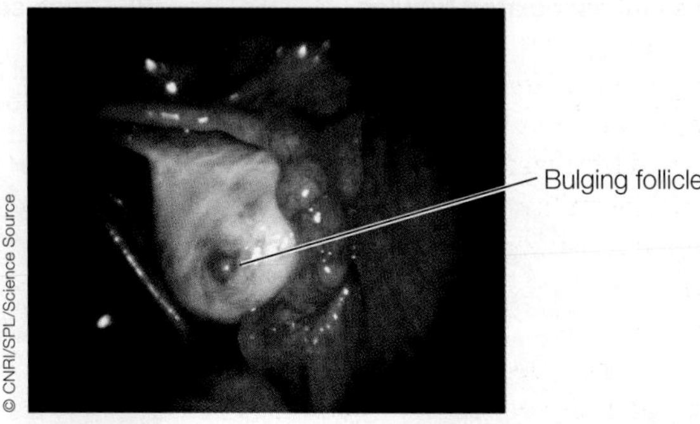

FIGURE 34.7 A Human Ovary Showing Stages of Follicle Development **(A)** The stages of follicle development are shown in a clockwise series, starting at the upper left with a primordial follicle. The primordial follicle progresses through all these stages in the course of approximately 28 days (one menstrual cycle). A new primordial follicle starts the next month. The clockwise arrangement is for learning purposes only. Follicles do not actually progress clockwise through an ovary. **(B)** The structure of a mature follicle as seen under a light microscope. **(C)** A mature follicle seen from outside the ovary, showing how the follicle is positioned at the ovary surface. Rupture of the follicle releases the ovum into the body fluids surrounding the ovary, and the ovum is then drawn into an oviduct for transit to the uterus, as discussed in Key Concept 34.2.

Media Clip 34.1 Human Ovulation

testis in a human or other mammal. The interstitial (or Leydig) cells, endocrine cells that secrete testosterone into the blood flowing through the testis, are located in the spaces between adjacent seminiferous tubules (see Figure 34.6B).

The sperm that are shed into the lumens of all the seminiferous tubules in a testis make their way to the *epididymis*, a highly convoluted tubular structure attached to the outside of the testis, where the sperm undergo maturation. The epididymis connects to the *vas deferens*, a tube that carries sperm out of the testis during ejaculation (see Figure 34.3). Most of the fluid that is ejaculated is added to the sperm by glands outside the testes—notably the *prostate gland*—during the actual process of sexual climax and ejaculation.

Somatic cells typically also play essential roles in the ovaries of females, just as they do in the testes. In vertebrates and also in some invertebrates (e.g., insects), each developing ovum is associated with a specific set of somatic support cells during its development in the ovary. The developing ovum and its support cells are together called an **ovarian follicle**. To exemplify this, let's focus on humans. Early in its development (i.e., during its developmental arrest), an ovum becomes surrounded by somatic cells termed *granulosa* and *theca* cells, forming a primordial follicle (**FIGURE 34.7A**). These somatic cells play two major roles that are reminiscent of the roles played by somatic cells in the testis: the somatic cells provide metabolic support to the developing ovum, and they secrete reproductive hormones into the blood flowing through the ovary. As we have said, once a human female has reached puberty, one ovum emerges from developmental arrest and completes its development each month. As that ovum matures, it enlarges and its support cells multiply—with the consequence that the primordial follicle transforms into a mature follicle (often called a Graafian follicle), as shown in Figure 34.7A. Along the way, a fluid-filled cavity (antrum) opens up within the follicle (**FIGURE 34.7B**). Moreover, the follicle migrates to be at the very surface of the ovary, where at full maturity it bulges out like a large blister, 1.5–2 centimeters in diameter (**FIGURE 34.7C**). The release of an ovum from the ovary is called **ovulation**. In vertebrates, including humans, it occurs by rupture of the mature follicle (see Figure 34.7A).

Fertilization may be external or internal

Fertilization is the fusion of sperm and ovum. It can be external or internal. The product is a single diploid cell, the zygote, which develops into an embryo (see Chapter 35).

To achieve fertilization, most aquatic animals simply release their gametes into the water, a process termed **spawning**. Fertilization then occurs in the water outside the female's body and is described as **external fertilization**. The zygotes typically do not receive parental care. They develop into feeding larvae that must find food before the yolk or other nutrients provided in the ovum run out. Mass deaths of commercially important fish larvae sometimes occur when the larvae are only 2–3 weeks old because at that time the larvae exhaust the yolk provided in the eggs but fail to find food. With the larvae dead, adults are not produced.

Internal fertilization occurs when a male inserts sperm inside the reproductive tract of a female, permitting fertilization to occur inside the female's body. Most terrestrial animals employ internal fertilization—not surprising, because it ensures that the gametes and zygotes are kept moist. Some aquatic animals also employ internal fertilization.

Why has internal fertilization evolved? In addition to the simple fact that it keeps gametes and zygotes moist, internal fertilization has three potential advantages. First, it permits an egg to be ensconced in a protective shell or capsule before the egg is laid (**FIGURE 34.8**). Because a shell blocks sperm access, sperm must reach the ovum inside the body, before the shell gland adds the shell. Second, internal fertilization permits an especially effective form of parental protection: the young remain inside their mother's body for part or all of their development. Third, in some cases—notably mammals and certain sharks (e.g., requiem and hammerhead sharks)—nutrients are steadily provided to the young developing inside the mother's body by way of a placenta or placenta-like structure, meaning the young can become large and developmentally advanced before they are born.

What are some of the ways that sperm are introduced into the female reproductive tract for internal fertilization? Mammals have a penis that is stiffened by an internal bone or by blood pressure (see Key Concept 34.2). Insects and some snails also have penises. Sharks have a specialized, penis-shaped fin element (clasper) with a groove. The clasper is inserted in the female genital opening, and sperm are washed down the groove into the female reproductive tract. Crabs assemble sperm into packets called spermatophores that they grasp with modified legs and insert. Birds achieve sperm transfer without a penis or similar organ, merely by the act of the male and female touching their cloacal openings together (a "cloacal kiss").

Whether fertilization is external or internal, a key challenge that must be met is for sperm and eggs of the same species to be brought together. This process often begins behaviorally through processes by which males and females of a species recognize each other and associate (see Key Concept 37.4). Behavior often also plays a role in synchronization. For example, the females of many species of mammals enter a sexually receptive behavioral condition called **estrus**, or *heat*, only at times when there is a high likelihood that they are releasing ova from their ovaries (see Key Concept 34.2). Otherwise they ignore or repel males.

Important mechanisms also exist at the cellular level for ensuring that eggs are fertilized by sperm of the same species. Eggs released into the environment often attract sperm with species-specific chemical attractants. After sperm and ova meet, cellular mechanisms continue to play key roles. For example, there are cellular mechanisms that prevent fertilization by sperm of the wrong species. Such cellular mechanisms are particularly important in aquatic species that employ external fertilization because the sperm and ova of many different species may be mixed together in the water. There are also cellular mechanisms—termed **blocks to polyspermy**—that ensure

(A) Bird eggs

© Wut_Moppie/Shutterstock.com

© Gertjan Hooijer/Shutterstock.com

(B) The egg capsule (case) of a skate

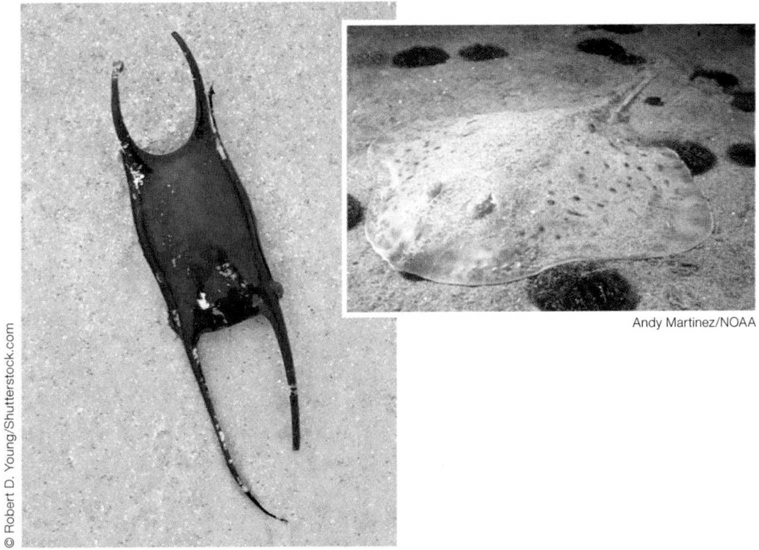

© Robert D. Young/Shutterstock.com

Andy Martinez/NOAA

FIGURE 34.8 Eggs Enclosed in a Shell or Capsule After the egg of a bird is fertilized internally, a shell gland in the mother's reproductive tract adds a hard, calcified shell before the egg is laid **(A)**. In skates (and some sharks), after an egg is fertilized internally, it is enclosed in a tough but flexible, proteinaceous capsule (case) before being laid **(B)**. In both cases, after the fertilized egg has developed into a young bird or skate, the youngster breaks or cuts its way out of its protective covering. The discarded capsules of skates are often found on ocean beaches.

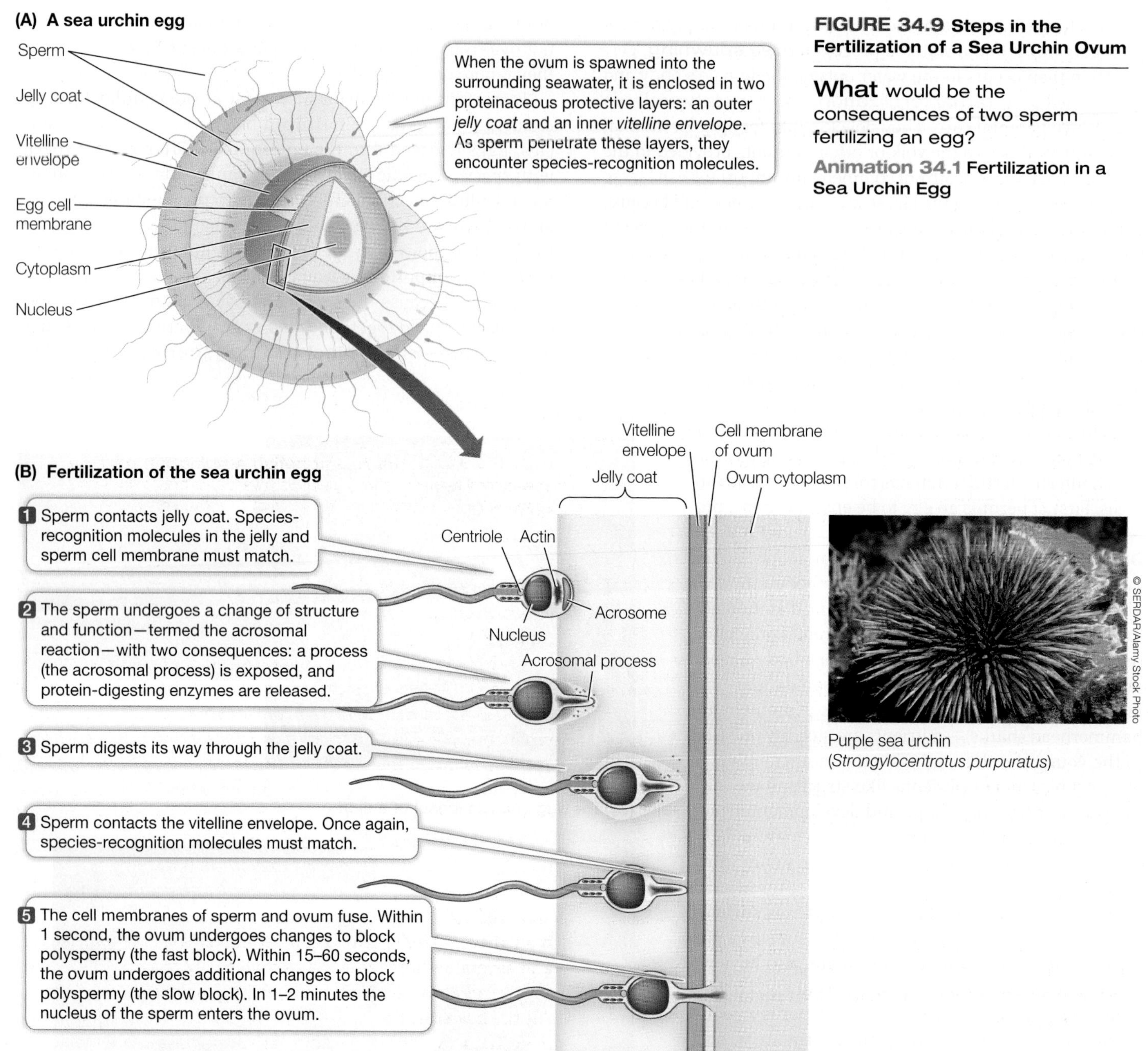

Purple sea urchin (*Strongylocentrotus purpuratus*)

FIGURE 34.9 Steps in the Fertilization of a Sea Urchin Ovum

What would be the consequences of two sperm fertilizing an egg?

Animation 34.1 Fertilization in a Sea Urchin Egg

that only a single sperm can transfer its nucleus to an ovum, regardless of how many sperm may be trying. Sea urchins have been used as model organisms for the study of many of these processes. **FIGURE 34.9** summarizes the successive mechanisms in a sea urchin that permit sperm to reach the egg (by penetrating the egg's outer coverings), ensure that sperm and egg are of the same species, and block polyspermy.

The sex of an offspring is sometimes determined at fertilization

Sex determination is the process by which the sex of an individual becomes fixed. Sex determination is not understood in all animals, but among the animals in which it is understood, a wide variety of mechanisms are known.

Sex is determined at fertilization in humans and other placental mammals. This was first analyzed through studies of chromosomes. Human diploid cells have 23 pairs of chromosomes. The two chromosomes in 22 of these pairs are similar to each other in their appearance under a light microscope. However, the two chromosomes in the 23rd pair (termed the *sex chromosomes*) differ in appearance in males—one being called the X chromosome and the other the Y chromosome—whereas in females both of the sex chromosomes are X chromosomes. Sex is determined at fertilization by the sex chromosomes a zygote receives. All ova have an

X chromosome because adult females have two X chromosomes at the 23rd position and meiosis therefore always provides an X chromosome to each ovum that is produced. By contrast, in the male parent, meiosis provides an X chromosome to half the sperm and a Y chromosome to the other half. When a sperm fuses with an ovum and fertilization occurs, some zygotes receive an X chromosome from the father and therefore (being XX) develop into females. Other zygotes receive a Y chromosome from the father and (being XY) develop into males.

Long after this chromosomal explanation of sex determination was discovered, researchers identified a particularly critical gene that they named *SRY* (*s*ex-determining *r*egion of the *Y* chromosome). This one gene exerts exceptional influence on sexual development, although many other genes are involved. The gene is nearly always found on the Y chromosome. An XY zygote thus has the gene, which directs the primordial gonads (the gonads early in development) to become testes and secrete testosterone. An XX zygote lacks the *SRY* gene and develops into a female.

LINK **Key Concept 32.4** and **Figure 32.14** discuss how testosterone secretion in XY embryos leads to phenotypic sex differentiation.

Sex is also determined at fertilization in many other types of animals, although the exact mechanisms vary widely. Sex is determined at fertilization in fruit flies (*Drosophila*), for example, but the mechanism is only superficially similar to that in mammals. Although fruit flies have X and Y chromosomes (see Key Concept 8.2), genes on the Y chromosome are not involved in sex determination.

In a minority of animals, sex is not determined at fertilization. Of these, crocodilians and turtles are of particular interest because many species exhibit **environmental sex determination**: the sex of each individual is determined by the temperature it experiences during its embryonic development. American alligators, for example, become males if they develop at 32°C–33°C but develop into females at 35°C. Great worry exists that populations of such animals will experience maladaptive changes in their sex ratios if global warming causes changes in the temperatures at which individuals develop.

Some animals undergo sex change during their adult lives

Some animals do not have a fixed sex. Instead, individuals are **sequential hermaphrodites**, meaning they can be males at one time and females at another during their adult lives. Hermaphrodites are animals that can produce both sperm and ova. In sequential hermaphrodites, sperm and ova are not produced simultaneously but instead are produced at different times ("sequentially") in an individual's life. Sex change in these animals does not entail anything as dramatic as exchanging a penis for a vagina. The animals are aquatic and release their gametes into the water by spawning. Their accessory reproductive organs are suited to carrying either sperm or eggs to the outside, and their sex change occurs because of a switch in the type of gametes they produce. Great numbers of fish species, including many coral-reef species, undergo sex change. The phenomenon also occurs in certain aquatic annelid worms and in certain snails and other gastropod mollusks.

Amphiprion percula

FIGURE 34.10 The Anemonefish *Amphiprion percula* These coral-reef fish, often called clownfish, gain protection from enemies by living among the tentacles of a sea anemone.

Biologists hypothesize that sex change has evolved in species in which an individual's circumstances can change dramatically during its life and the individual can achieve a greater lifetime reproductive output by functioning as a male in some circumstances and a female in others. What circumstances might change? An individual's size and social position are often important, or the density of the population in which it lives may be important.

The anemonefish *Amphiprion percula*, which lives among the tentacles of sea anemones, provides a well-studied example of sex change (**FIGURE 34.10**). Under many circumstances, the adult fish living on a single sea anemone consist of one breeding male that spawns sperm, one breeding female that spawns ova, and several nonbreeding males that do not spawn. The breeding female is largest, and the breeding male is second largest. All the fish start life as males. If the female dies, the breeding male becomes the largest fish. It then turns into a female, and one of the nonbreeding males becomes the new breeding male. According to the prevailing hypothesis of sex change, the reason the breeding male changes to a breeding female is that—under the new circumstances that prevail—he can have more offspring by being female. To see why, consider that egg output depends on body size in females (the larger a female, the more eggs she makes), and remember that there are only two breeding fish on a sea anemone. A fish of *any* size can produce huge numbers of sperm. If the biggest fish remained male, his reproductive output would be limited by the egg output of a smaller fish. By switching to producing eggs, the biggest fish increases the number of eggs that get fertilized and thus its reproductive output. In some species that undergo sex change, progress has been made in understanding the hormonal mechanisms involved (**INVESTIGATION FIGURE 34.11**).

INVESTIGATION

FIGURE 34.11 Investigating a Mechanism of Sex Change Blackeye gobies (*Rhinogobiops nicholsii*) live under different circumstances than anemonefishes, and in contrast to anemonefishes, adult female gobies sometimes change into males. Investigators have studied the mechanism by which this occurs. By simply describing blood hormones, the investigators found that the female-to-male transition is correlated with a decline in the blood concentration of the female sex steroid 17β-estradiol and an increase in the concentration of the male sex steroid 11-ketotestosterone (11-KT). But do these changes in hormones *cause* the sex change? Are they *sufficient* to make a female change into a male? To find out, the investigators artificially manipulated the hormone levels.[a]

HYPOTHESIS

In blackeye gobies, an increase in the whole-body concentration of 11-KT is sufficient to cause a female to change into a male.

METHOD

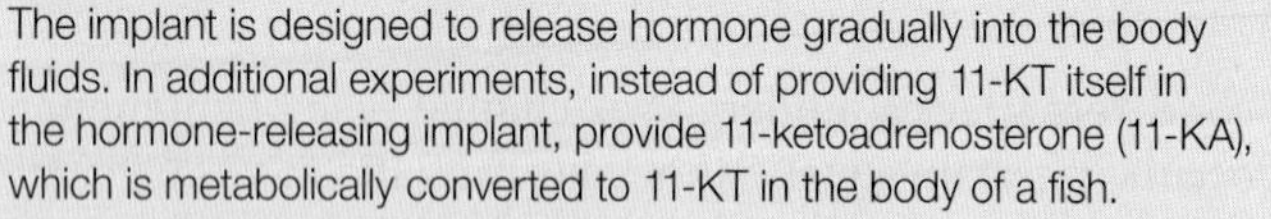

© Amar and Isabelle Guillen/Guillen Photo LLC/Alamy Stock Photo

1. Capture female blackeye gobies in the wild and hold in individual tanks.
2. Insert in each fish a hormone-releasing implant containing 11-KT or an implant with no hormone (control treatment). The implant is designed to release hormone gradually into the body fluids. In additional experiments, instead of providing 11-KT itself in the hormone-releasing implant, provide 11-ketoadrenosterone (11-KA), which is metabolically converted to 11-KT in the body of a fish.
3. After 43 days of treatment, examine the gonads of the fish to determine the sex.

RESULTS

Hormone in implant	Sex at end of treatment		
	Female	Transitional	Male
None (control)	13	0	0
11-KT or 11-KA	0	1	13

CONCLUSION

Because the body concentration of 11-KT is increased by administering either 11-KT or 11-KA, the results support the hypothesis that an increase in the body concentration of 11-KT is sufficient to cause female blackeye gobies to become male. (Results with 11-KA were closely similar to those with 11-KT.)

ANALYZE THE DATA

In a related experiment, 8 female blackeye gobies were treated with fadrozole, a chemical that inhibits the synthesis of 17β-estradiol and thus is expected to lower the concentration of this female sex hormone. At the end of treatment, 1 fish was still female, 5 fish were transitional, and 2 fish were male.

1. For causing females to become males, was this treatment as effective as elevating the blood 11-KT concentration (by 11-KT or 11-KA administration)? What data support your conclusion?
2. The researchers hypothesized that a decrease in 17β-estradiol causes sex change. Do the data support the hypothesis? Why or why not? What additional information would be helpful in deciding whether the hypothesis is supported?

[a]F. J. Kroon and N. R. Liley. 2000. *Gen Comp Endocrinol* 118: 273–283.

A similar **Analyze the Data** exercise may be assigned in **Achieve**.

REVIEW & APPLY | 34.1

R In sexual reproduction the parents produce—through meiosis—gametes that have only half as many chromosomes as the other cells in the body. The female and male gametes fuse to produce a single-celled zygote that has a full set of chromosomes. In asexual reproduction offspring are produced by mitosis. Whereas sexual reproduction produces genetic diversity, asexual reproduction maintains favorable combinations of genes.

Gametogenesis occurs in the gonads (testes or ovaries), where germ cells proliferate mitotically before undergoing meiosis. In vertebrate testes, Sertoli cells assist sperm production and interstitial (Leydig) cells produce testosterone. In vertebrate ovaries, somatic cells in the ovarian follicles aid ova development and produce reproductive hormones. External fertilization is common among aquatic animals, most of which spawn gametes into the water. Internal fertilization occurs when a male inserts sperm into the reproductive tract of a female, permitting encapsulation of the fertilized ovum in a shell, parental protection, or placental development.

Sex determination occurs at fertilization in many animals. Exceptions include animals capable of environmental sex determination or sequential hermaphroditism.

A

1. What are some of the advantages and disadvantages of internal and external fertilization?
2. Vasectomy is a birth control procedure in which the principal ducts carrying sperm out of the testes are tied off. Basing your answer on the cellular structure of the human testis, explain why a man's testosterone level is not affected by vasectomy.
3. Rarely, men are found who have two X chromosomes. The *SRY* gene is responsible for this syndrome. What type of unusual gene placement could account for the syndrome? Explain.

Having surveyed the general aspects of animal gametogenesis and fertilization, we will next consider the male and female reproductive systems in mammals, using our own species as the primary example.

34.2 The Mammalian Reproductive System Is Hormonally Controlled

LEARNING OBJECTIVES

By the end of this key concept you should be able to:

34.2.1 Give two examples of hormones that play essential roles in both male and female reproduction and specify their roles.

34.2.2 Explain the differences between estrous cycles and menstrual cycles, and between spontaneous and induced ovulation.

34.2.3 Explain how the birth process in mammals depends on positive feedback.

Hormones play critical roles in almost every aspect of mammalian sexual reproduction. Follicle-stimulating hormone (FSH) and luteinizing hormone (LH), secreted by the anterior pituitary gland, are key players in both sexes (see Key Concepts 32.3 and 32.4). These hormones are named for their roles in females, but the same hormones are secreted in males and exert major effects on tissues of the male reproductive system. Secretion of FSH and LH is controlled by neurosecretory cells in the hypothalamus of the brain. These cells secrete gonadotropin-releasing hormone (GnRH) into blood that travels to the anterior pituitary via the hypothalamo–hypophysial portal system (see Figure 32.8). In the pituitary, GnRH regulates the cells that secrete FSH and LH.

Somatic cells in the gonads of both sexes produce steroid sex hormones. Although we often say that "the ovaries" or "the testes" secrete hormones, it's important to recognize that only certain cells in the gonads are responsible. The principal sex steroid in males is testosterone, which we've already seen is secreted by the interstitial (Leydig) cells in the testes. In females, the somatic cells of the ovarian follicles secrete feminizing steroids called **estrogens**. Typically, two or more specific types of estrogen molecules are secreted, but here we refer to them simply as "estrogens." After a follicle has ruptured during ovulation, the somatic cells remaining in the ovary reorganize to form a structure called the corpus luteum (see Figure 34.7A) which is principally an endocrine structure secreting the steroid **progesterone** as well as estrogens. Progesterone is involved principally in preparing for and coordinating pregnancy.

LINK As discussed in more detail in **Key Concept 32.4**, the hormones of greatest importance for reproductive development are the sex steroids and the hypothalamic and pituitary hormones that control them. In mammals, androgens are required for male phenotypic development. In their absence an embryo develops as a female.

Ova mature in the ovaries and move to the uterus

The ovaries are positioned in the lower abdomen near the uterus (**FIGURE 34.12**). Ova released from the ovaries make their way to the uterus by way of tubes called the **oviducts** or fallopian tubes. Each ovary is positioned near the funnel-shaped opening of an oviduct. When an ovum is discharged from a follicle at the surface of the ovary during ovulation, it is drawn into the oviduct, and then it is gradually propelled toward the uterus by beating cilia and contractions of smooth muscles.

Sperm are deposited in the vagina during copulation (sexual intercourse). Fertilization of an ovum, if it occurs, takes place in the upper third of the oviduct. To reach the ovum and achieve fertilization, sperm in the vagina enter the uterus at the cervix, travel through the uterus, and then travel up the oviduct. Sperm are often in the upper oviduct within 30 minutes after ejaculation, indicating that they are probably transported not only by their own swimming powers but also by uterine contractions.

Passage of an ovum or zygote down the oviduct is relatively slow, requiring about 4 days in humans. If fertilization has occurred, the zygote undergoes its early development during this time and arrives in the uterus as an early embryo called a blastocyst (see Chapter 35). The uterus is a thick-walled organ consisting of two major parts: an outer layer of smooth muscle, the **myometrium**, and an inner, nurturing layer, the **endometrium** (see Figure 34.12). The endometrium is highly vascular and composed of nutrient-providing cells, and it is where the zygote will develop further. After the blastocyst arrives in the uterus, it remains in the lumen of the uterus for a few days and then buries itself in the endometrium by means of enzyme-catalyzed processes. The entry of the blastocyst into the tissue of the endometrium is termed **implantation**.

Ovulation is either spontaneous or induced

The immediate stimulus for ovulation is a large increase—a surge—in the blood concentration of LH. However, the cause of this LH surge differs between two categories of mammalian species. The species in one category are said to undergo spontaneous ovulation, whereas those in the other undergo induced ovulation.

Most mammals, including humans and other primates, are spontaneous ovulators. In a species with **spontaneous ovulation**, the timing of the LH surge—and thus the timing of ovulation—is controlled by endogenous hormonal cycles in the female. "Endogenous" refers to processes that originate within an animal. The hormonal cycles in a spontaneous ovulator occur on their own rhythm, more or less independently of whether copulation takes place.

Induced ovulation, by contrast, is ovulation directly triggered by copulation. Rabbits, featured in the chapter's opening photo, are examples of induced ovulators, as are certain cats, shrews, and camelids. During copulation in an induced ovulator, the copulatory act stimulates key sensory neurons in the cervix. These neurons send action potentials to GnRH-secreting neurosecretory cells in the hypothalamus of the brain, stimulating GnRH secretion that causes a sharp increase in LH secretion by the anterior pituitary gland into the general circulation. In rabbits that have been studied, the blood LH concentration increases within 1–2 hours after copulation to more than five times its usual level, and this LH surge causes ovulation.

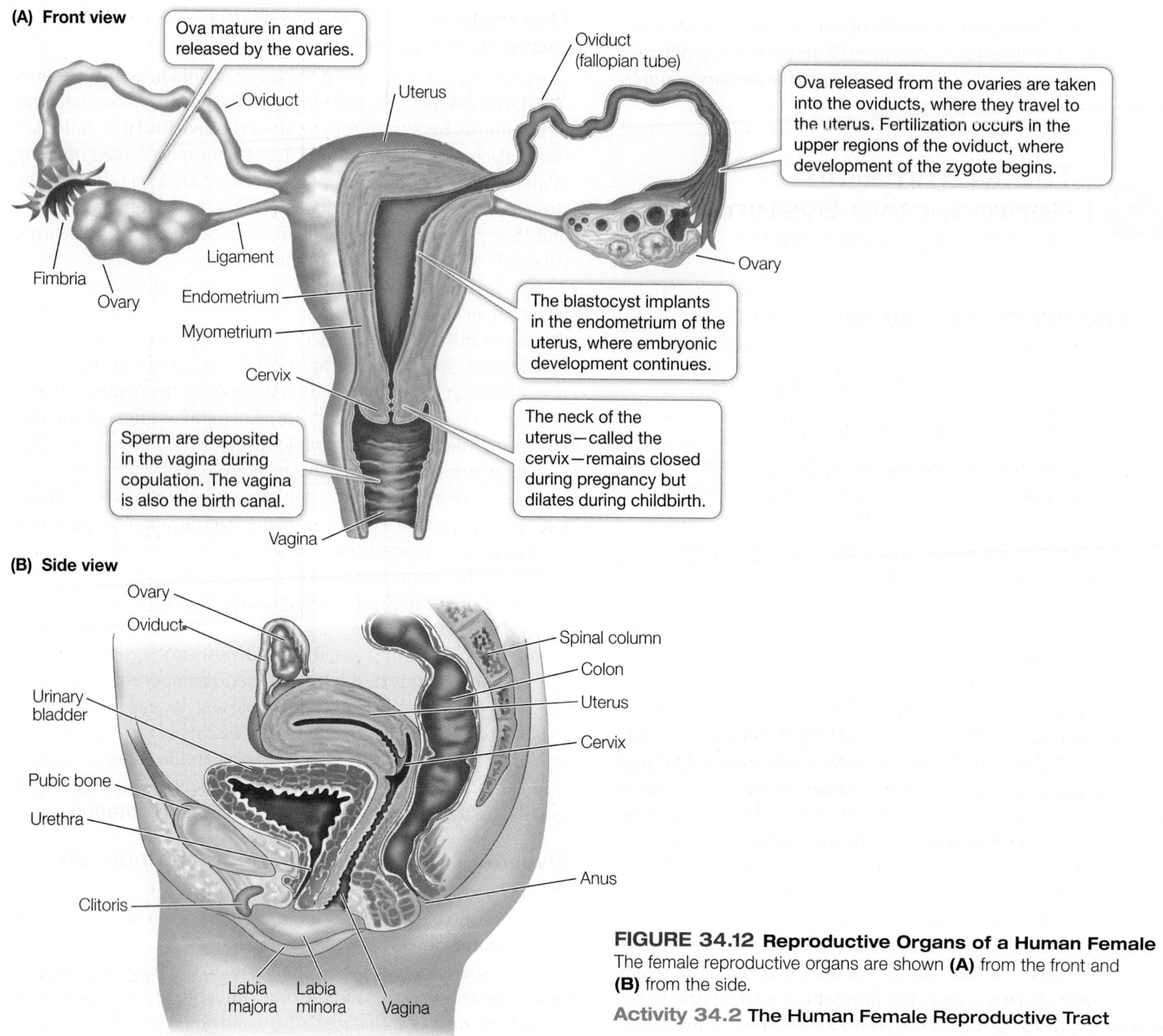

FIGURE 34.12 Reproductive Organs of a Human Female The female reproductive organs are shown **(A)** from the front and **(B)** from the side.
Activity 34.2 The Human Female Reproductive Tract

In a spontaneous ovulator, there is a large element of chance in whether pregnancy occurs because copulation does not control ovulation: a female might mate and get lots of sperm, yet not become pregnant because her ovaries happen to release eggs at a different time. In a species with induced ovulation, by contrast, the act of mating directly causes ovulation to occur—guaranteeing that eggs will be present for fertilization in a female's reproductive tract at the same time that sperm are there.

As we have said, most mammals are spontaneous ovulators. With this is mind, we will now focus on spontaneous ovulation for the rest of this key concept.

In mammals that display spontaneous ovulation, females typically ovulate in endogenously timed cycles. The duration of each cycle varies among species. For instance, cycles in rats and mice are typically 4–6 days long (i.e., females ovulate every 4–6 days), whereas the human cycle averages 28 days. The cycles repeat, over and over, during the reproductive season (i.e., all year in humans) unless pregnancy occurs. If pregnancy occurs, the cycles stop.

Primates menstruate in each cycle that does not result in pregnancy. **Menstruation** is a discharge of bloody material from the uterus via the vagina. During each cycle, the endometrium undergoes preparation for pregnancy. It develops glands and an abundant proliferation of blood vessels, and it thickens (in humans, by a factor of three to five). If pregnancy fails to occur, much of the extra growth is sloughed off, creating the bloody discharge. The cycles of the reproductive system are called **menstrual cycles** because of this highly visible aspect.

Mammals other than primates do not menstruate or have menstrual cycles. Among the species that exhibit spontaneous ovulation, however, females often undergo dramatic cycles in behavior. Synchronized with their times of ovulation, they cyclically enter estrus (or "heat"), a behavioral state in which they express readiness to mate (whereas otherwise they avoid or repel males). These cycles of behavior are the most highly visible aspect of their reproductive cycles, which accordingly are called **estrous cycles**. (Note that the cycles are "estrous" cycles, whereas the period of sexual receptivity is "estrus.") The endometrium prepares for pregnancy during each cycle and then regresses if pregnancy fails to occur, but the regression occurs by reabsorption, not sloughing off, of tissue.

In species with estrous cycles, a female stops exhibiting cyclic ovulation and estrus when she becomes pregnant. In some cases the estrous cycle is not resumed until the offspring are weaned. In other cases, however, an estrus—termed *postpartum estrus*—occurs immediately after birth ("postpartum" means "following birth"). Some induced ovulators, such as rabbits, also display postpartum estrus. In species that undergo postpartum estrus, a female is often simultaneously both nursing a set of recently born young and pregnant with a new set of young that were conceived by mating during her postpartum estrus. Postpartum estrus is a mechanism that shortens the gap between one pregnancy and the next, enhancing the rate at which offspring can be produced.

To see how cycling occurs in a spontaneous ovulator, let's look at a specific example: the human female reproductive cycle. Hormones secreted by the brain, anterior pituitary gland, and ovaries are involved, and the cycles occur because the actions of these hormones on tissues interact and affect each other. Hormone receptor molecules are as important as the hormones. Cellular responses to hormones depend on the numbers of hormone receptors present, and hormones can cause cells to modify their receptor numbers.

By convention, in the human female reproductive cycle, each cycle is considered to start on the day menstruation begins, and that day is numbered day "0." The follicles that will mature in the cycle start their growth and development then, and they undergo ovulation (follicular rupture) about 14 days later. Accordingly days 0–14 are called the *follicular phase* of the cycle. What happens over those 14 days? The answer is complex, but by mastering it you will be rewarded with an understanding of one of the most fascinating and important processes in the human body. Follow the steps listed here by referring to the circled numbers in **FIGURE 34.13A–D**:

(A) FSH and LH secreted by the anterior pituitary gland

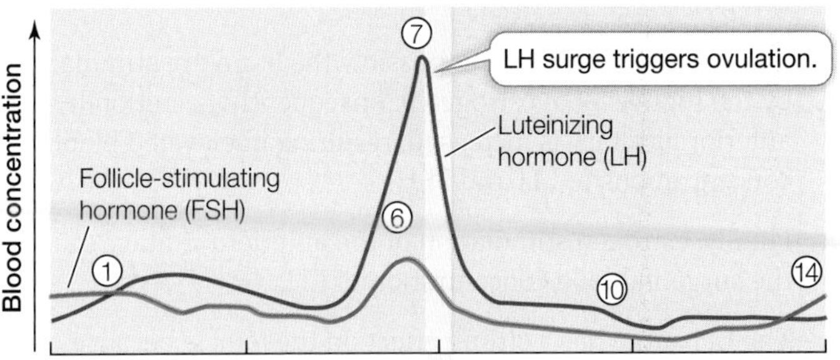

(B) Events in the ovary

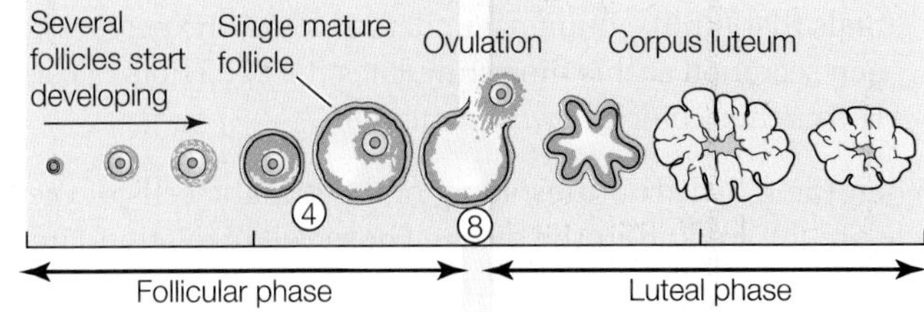

(C) Ovarian hormones secreted by the somatic endocrine cells in the ovary

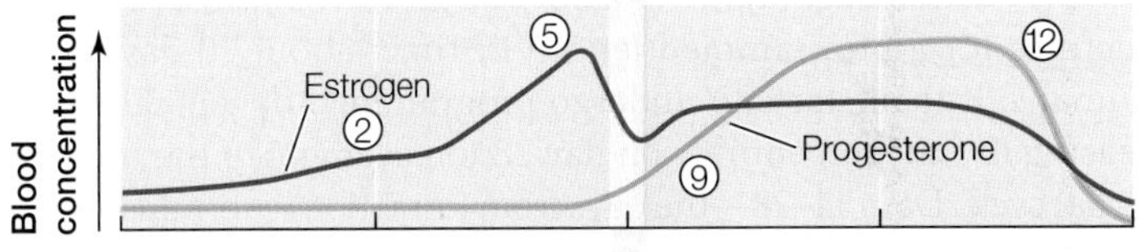

(D) Events in the endometrium of the uterus

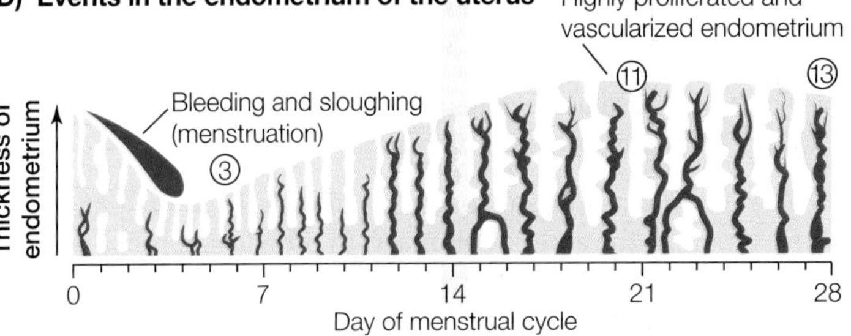

FIGURE 34.13 Events in the Human Female Reproductive Cycle Circled numbers correlate with the sequence of events described in the text. Yellow bar marks day 14.

Hormonal contraceptive pills of some types contain estrogen and progesterone. Which portion of the ovarian cycle do they simulate and how do they prevent ovulation?

Animation 34.2 The Menstrual Cycle

1. At the start, FSH and LH are secreted by the anterior pituitary gland at fairly high rates compared with the rates seen following ovulation during the previous cycle (compare with step 10). The FSH and LH stimulate development of follicles, including the one follicle that will mature fully and release its ovum. Specifically, FSH and LH stimulate the somatic cells of that follicle—the granulosa and theca cells—to secrete estrogen.

2. Estrogen secretion gradually increases as the follicle develops.

3. The increasing estrogen secretion affects the endometrium, stimulating endometrial development that prepares the uterus to accept a blastocyst if fertilization occurs. Estrogen also stimulates the endometrial cells to express progesterone receptors, preparing the cells to respond to progesterone later on.

4. In the second half of the follicular phase, the granulosa cells of the maturing follicle increase their numbers of LH receptors.

5. As day 14 approaches, the blood estrogen concentration is high.

6. The anterior pituitary and possibly the brain are stimulated by the high level of estrogen. Consequently, secretion by the anterior pituitary is increased, resulting in elevated blood concentrations of LH and FSH.

7. Finally, estrogen stimulation of the anterior pituitary causes the surge in blood concentration of LH.

8. The LH surge has a strong effect on the follicle because of the abundance of LH receptors in the granulosa cells (recall step 4). The follicle ruptures, and ovulation occurs. In mammals that (unlike humans) undergo estrus, the peak estrogen secretion at this time stimulates the behaviors of estrus. Estrus thus coincides with ovulation.

When the follicle ruptures, most of the somatic cells of the follicle are left behind in the ovary. Those cells are stimulated by LH (which, you'll recall, is *luteinizing* hormone) to reorganize to form a new hormone-producing structure, the **corpus luteum** (see Figure 34.13B). If pregnancy has not occurred, the corpus luteum—at this stage known as a **corpus luteum of the cycle**—secretes for a programmed length of time: about 10 days in humans. It then starts to undergo programmed degeneration, ceasing to function entirely on day 28 (on average) of the menstrual cycle. Days 14–28—the days from formation to termination of the corpus luteum—are termed the *luteal phase* of the cycle. What happens during it? Again, follow the steps by referring to the circled numbers in Figure 34.13A–D:

9. The corpus luteum secretes progesterone (its principal secretion), estrogen, and a third hormone, inhibin.

10. These hormones, acting together, suppress FSH and LH secretion by the anterior pituitary. In primates, this suppression is so strong that it prevents new follicles from starting to develop at this stage.

11. Progesterone secretion from the corpus luteum has a major effect on the endometrium. Exocrine glands in the endometrium are stimulated to secrete nutrient materials, and the endometrium becomes more vascularized. These changes fully prepare the endometrium to accept a blastocyst.

12. If fertilization has not occurred, the corpus luteum starts to undergo its programmed degeneration after about 10 days, leading to a steep decline in progesterone and estrogen secretion.

13. Deprived of the hormones secreted by the corpus luteum, the endometrium cannot maintain its highly developed state. It starts to regress, with menstruation beginning on day 0 of the next cycle.

14. The disappearance of luteal hormones (hormones secreted by the corpus luteum) also sets the stage for a new cycle to begin because the luteal hormones no longer act on the pituitary or brain to suppress FSH and LH secretion. FSH and LH increase, returning the cycle to step 1, in which development of a new follicle is stimulated.

Pregnancy is a specialized hormonal state

If fertilization takes place, the blastocyst burrows into the tissue of the endometrium within a few days after arriving in the uterus, and there is an urgent need to maintain the endometrium in its highly developed state. In humans and many other mammals, the endometrium is prevented from deteriorating at this point by *rescue of the corpus luteum*. A signal of the existence of pregnancy is sent to the corpus luteum. In response, the corpus luteum does not undergo programmed degeneration. Instead it develops into a **corpus luteum of pregnancy**, which continues to grow and secrete hormones. The signal in primates is a hormone, chorionic gonadotropin (CG), secreted by the blastocyst. In other words, the embryo emits a signal that tells the mother's tissues to maintain the uterus in a state suitable for pregnancy. The concentration of CG rises high enough in the mother's blood and urine to be easily detected, and many pregnancy tests work by detecting it.

The uterus requires progesterone to remain in a state suitable for pregnancy. In humans, after the corpus luteum is rescued, it provides adequate progesterone for 7–10 weeks. By then, the placenta—which also secretes progesterone—has become sufficiently well developed to take over.

The **placenta** is a structure in which large arrays of maternal blood vessels and embryonic (or fetal) blood vessels are closely juxtaposed (see Figure 35.25). The blood of the mother and that of the embryo do not mix in the placenta. However, nutrients and O_2 move across vessel walls from the maternal blood into the blood of the embryo in the placenta, and CO_2 and other wastes move from the embryo into the maternal blood.

Birth depends on a hormonally mediated positive-feedback loop

Another hormone, **oxytocin**, plays essential roles at the end of pregnancy. It is synthesized by neurosecretory cells in the hypothalamus and secreted into the blood in the posterior pituitary gland (see Key Concept 32.3 and Figure 32.7).

As we discussed in Key Concept 28.3, evolution has produced systems in which positive feedback occurs, even though we often think of feedback as being entirely negative. The birth process in mammals is an outstanding example, in which oxytocin is intimately involved. During the birth process, oxytocin stimulates waves of contraction of the mother's uterine muscle (myometrium)—waves that tend to push the fetus outward via the cervix. A control mechanism in the hypothalamus receives information from sensors in the cervix on the force of the uterine contractions. When this control mechanism detects an increased force of contraction, it activates *increased* hypothalamic oxytocin secretion. The heightened oxytocin secretion further increases the force of contraction in a self-reinforcing, positive-feedback loop (**FIGURE 34.14**). Ultimately the force becomes great enough to push the fetus out of the uterus into the outside world, ending further stimulus for oxytocin production and uterine contraction.

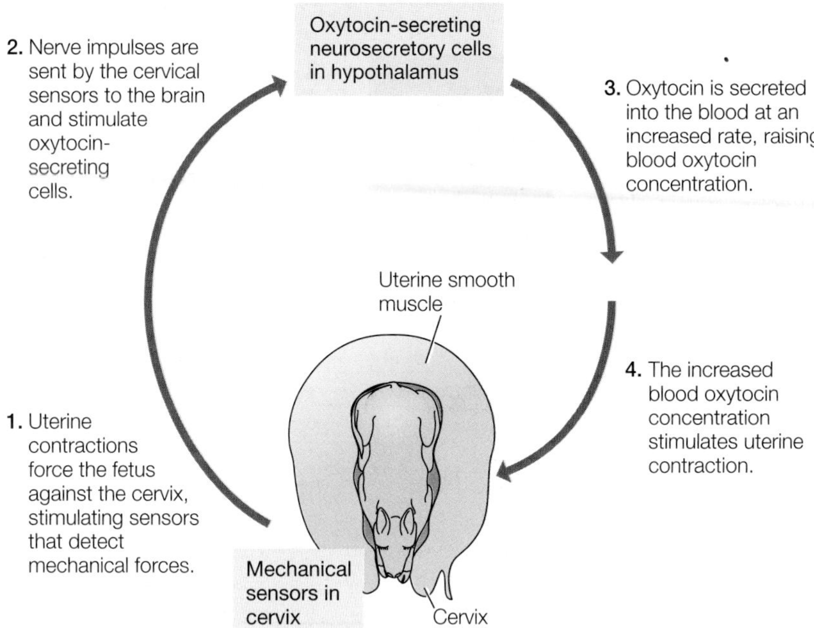

FIGURE 34.14 Positive Feedback: The Development of the Force Required for Birth to Occur in Mammals This system produces an escalating force by positive-feedback control of uterine muscle contractions.

Oxytocin also plays a role in providing milk for the newborn youngster. Following birth, oxytocin and the anterior pituitary hormone prolactin stimulate the mammary glands (breasts) to produce and eject milk.

Male sex organs produce and deliver semen

The testes of most mammals are located outside the body cavity in a sac of skin called the **scrotum** (see Figure 34.3). During early development they are inside the abdomen in the same positions as the ovaries, but they migrate into the scrotum as development proceeds. In human males, the testes are normally in the scrotum all the time. In many other species, such as certain mice, the testes are in the scrotum only during the reproductive season, being withdrawn into the abdomen at other times. The testes of most mammals need to be about 2 degrees Celsius cooler than ordinary abdominal temperature (37°C) for normal production of viable sperm, and being in the scrotum permits this.

After sperm are produced in a testis, they are stored in the epididymis (see Figure 34.6A) and vas deferens. The vas deferens is the tube that leads from the epididymis to the urethra, the passage in the penis through which urine and semen flow to the outside (see Figure 34.3). The vas deferens and urethra are endowed with smooth muscle, which contracts to propel semen during ejaculation.

Semen, the fluid expelled from the penis during ejaculation, consists of a mix of sperm and fluids secreted by male accessory reproductive glands. In humans, the glands that provide most of the fluids in semen are the paired seminal vesicles and the single prostate gland (**FIGURE 34.15**). Sperm are mixed with these fluids as sexual climax and ejaculation occur. The fluids provide a supportive medium for the sperm, including fructose and other energy sources. The paired bulbourethral glands secrete a clear mucus prior to sexual climax. This fluid often emerges from the penis as climax approaches and may contain sperm even though it is not semen.

The shaft of the human penis is filled almost entirely by three compartments of spongy tissue (see Figure 34.15) that are expanded with blood during erection under control of nitric oxide (NO). When NO is released from parasympathetic nerve endings in the penis, it leads to dilation of blood vessels that permit blood to fill the spongy tissues at high pressure, stiffening the penis. NO is short-lived. It acts by increasing synthesis of cyclic GMP (cyclic guanosine monophosphate), which in turn acts as a second messenger (see Key Concept 6.3). Medications for erectile dysfunction (impotence) inhibit an enzyme that breaks down cyclic GMP. This permits the second messenger to act longer.

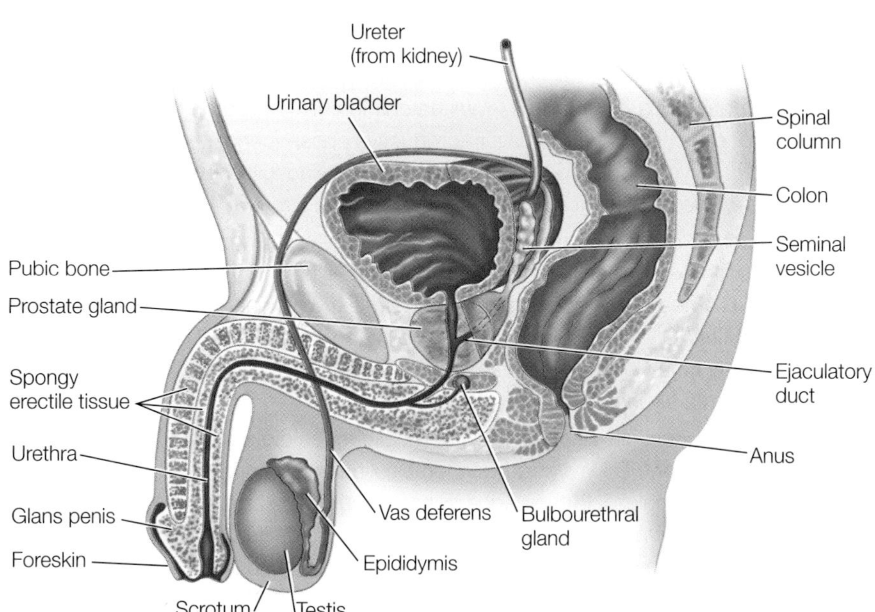

FIGURE 34.15 Reproductive Organs of the Human Male The two vas deferens, from the right and left testes, join at a point upstream from the single prostate gland. The foreskin, the skin covering the glans of the penis, is removed during circumcision.

Activity 34.3 The Human Male Reproductive Tract

THINK LIKE A SCIENTIST

The value of manipulative experiments

When we observe a strong correlation between two traits, we commonly wonder whether one trait actually *causes* the other or whether the two traits occur together for merely incidental reasons. One of the best ways to determine whether *x* causes *y* is to set up a *manipulative* experiment in which trait *x* (the hypothesized explanatory variable) is altered deliberately to see if trait *y* (the hypothesized response variable) then also changes. Elegant manipulative experiments have been used to clarify aspects of reproductive biology.

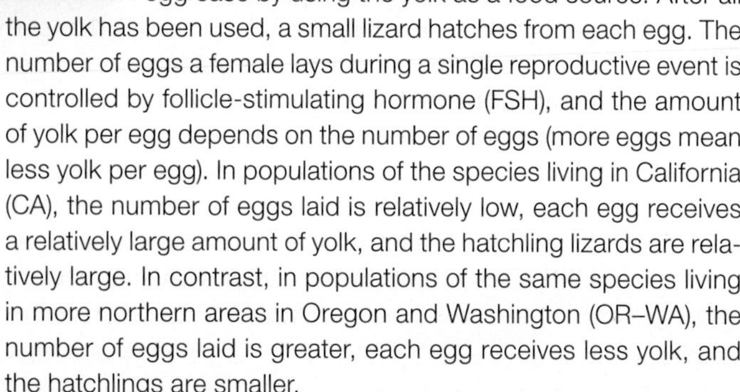

Female western fence lizards (*Sceloporus occidentalis*) reproduce by laying eggs containing a lot of yolk (in many ways similar to bird eggs). Each egg contains one embryo, which grows and develops inside the egg case by using the yolk as a food source. After all the yolk has been used, a small lizard hatches from each egg. The number of eggs a female lays during a single reproductive event is controlled by follicle-stimulating hormone (FSH), and the amount of yolk per egg depends on the number of eggs (more eggs mean less yolk per egg). In populations of the species living in California (CA), the number of eggs laid is relatively low, each egg receives a relatively large amount of yolk, and the hatchling lizards are relatively large. In contrast, in populations of the same species living in more northern areas in Oregon and Washington (OR–WA), the number of eggs laid is greater, each egg receives less yolk, and the hatchlings are smaller.

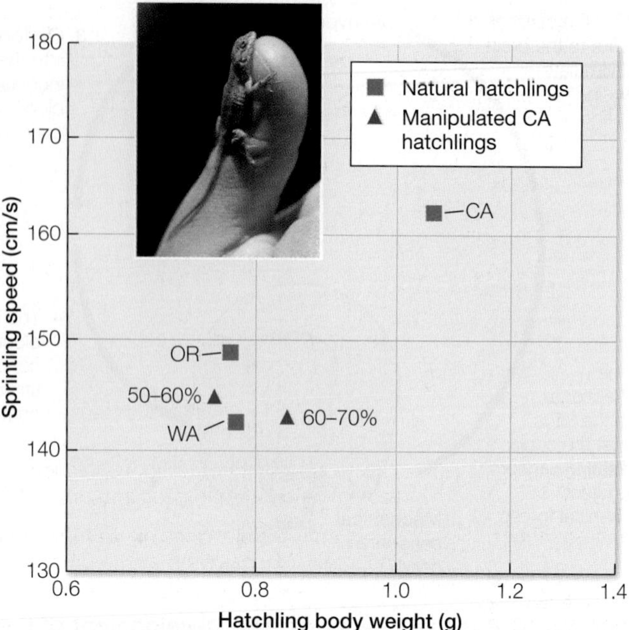

FIGURE 34.16 Average Sprinting Speed and Body Weight in Groups of Hatchling Western Fence Lizards Sprinting speed and body weight were measured at 3 weeks of age in five groups of hatchlings. Blue symbols show data for three natural populations: one in California (CA), one in Oregon (OR), and one in Washington (WA). Red symbols show results for two groups of CA hatchlings from manipulated eggs subjected, respectively, to 50–60 percent or 60–70 percent yolk removal soon after eggs were laid. Both axes are logarithmic. (After B. Sinervo. 1993. *BioScience* 43: 210–218.)

An intriguing question arose when hatchlings were tested (at age 3 weeks) to see how fast they could sprint away from danger. CA hatchings can run faster than OR–WA hatchlings. This greater sprinting speed is highly likely to have reproductive consequences. Hatchings can live to adulthood only if they avoid being killed by enemies, and the CA hatchlings are faster sprinters.

Why are the CA hatchlings faster sprinters? Possibly their large body size is the explanation (CA hatchlings are bigger than OR–WA hatchlings). But does large body size actually *cause* their enhanced sprinting ability? Or is their high sprinting ability an independent specialization in the CA population so that fast sprinting and large body size are merely incidentally correlated?

To answer this question, Barry Sinervo and colleagues did a manipulative experiment. They worked with females in the CA population. Soon after females laid their eggs, Sinervo removed some of the yolk from each egg with a sterile hypodermic needle. He then permitted the eggs to develop and hatch under the same conditions as usual. In the case of CA eggs from which he removed 50–70 percent of the yolk, the CA offspring hatched at about the same body weight as normal OR–WA hatchlings (**FIGURE 34.16**). Sinervo then measured and compared the sprinting speeds of CA hatchlings versus OR–WA hatchlings. He found that the manipulated CA hatchlings had slower sprinting speeds than unmanipulated CA hatchlings. He also found that the sprinting speeds of the manipulated CA hatchings were similar to those of the normal OR–WA hatchlings (see Figure 34.16). Sinervo therefore concluded that low sprinting speed is caused by small body size. In other words, the normal (unmanipulated) hatchlings in the CA population exhibit high sprinting speeds because they are large. Sinervo further concluded that as the OR–WA populations evolved smaller egg size (which allowed them to lay greater numbers of eggs), they also evolved—as a direct consequence—a lower ability to sprint away from danger.

1. Why is Sinervo's experiment classified as a manipulative experiment? Are there other types of experiments? Do you agree that a manipulative experiment can help distinguish between (a) mere correlation and (b) cause and effect? Explain.
2. Suppose you wonder if the simple act of inserting a hypodermic needle could affect hatchling body size or sprinting speed. What experiment could you do to gain insight into this concern?
3. An alternative way to manipulate egg number and yolk size in fence lizards would be to manipulate the blood FSH concentration in reproducing females. What would be the pros and cons of that method compared with the yolk manipulations?

TABLE **34.1** | Failure Rates of Some Major Methods of Routine Contraception

Method	How is the method used?	Annual failure rate (pregnancies/100 women) Typical use[a]	Perfect use[b]
Unprotected	No form of birth control	85	85
Fertility awareness methods (e.g., rhythm method)	Engaging in intercourse only on days of the ovarian cycle when fertility is judged to be low based on cervical mucus or standard cycle properties	24	3–5
Withdrawal (pulling out, coitus interruptus)	The man withdraws his penis prior to ejaculating	22	4
Male condom	A sheath of impermeable material (usually latex) is fitted over the erect penis	18	2
Female condom; diaphragm	A physical device is inserted in the vagina to block entry of sperm into the cervix and uterus	12–21	5–6
Oral hormones ("the pill")	A hormonal pill that prevents ovulation is taken daily; contains either a combination of estrogens and progestin (progesterone) or just progestin	9	0.3
Nonoral hormones	Based on the same principle as the pill, hormones are instead delivered nonorally (e.g., by long-acting injection or skin patch).	6–9	0.3
Intrauterine devices (IUD)	A small plastic or metal device is inserted semipermanently into the uterus. The device impairs sperm or prevents implantation.	0.2–0.8	0.2–0.6

Source: J. Trussell. 2011. *Contraception* 83: 397–404.

[a]"Typical use" failure rates are based on real-world statistics for people who sometimes apply methods incorrectly or skip using them.

[b]"Perfect use" refers to using methods according to directions without fail, every time intercourse takes place.

Many mammals have a bone in the penis that helps stiffen it for copulation. Included are dogs, seals, rodents, and some primates (but not humans).

Hormones play as crucial a role in male reproductive function as in female. Male reproductive processes and the controlling hormones that induce them do not, however, cycle in the dramatic ways seen in females.

Testosterone is required for spermatogenesis. In human males, the interstitial cells secrete an abundance of testosterone during a period of fetal life (when sexual differentiation is occurring) and again during the first year after birth. After that, testosterone levels remain low until puberty begins. At puberty, the blood testosterone level increases dramatically, and thereafter it stays high for the rest of life, although declining gradually after middle age. Spermatogenesis begins at puberty and continues without interruption throughout life. In contrast to humans, many nonhuman mammals exhibit strongly seasonal reproduction. In these animals, it is common for testosterone secretion and spermatogenesis to occur only during the defined reproductive season.

The pituitary hormones FSH and LH play central roles in males. LH stimulates testosterone secretion by the interstitial cells. In this way it plays a mandatory role in spermatogenesis. FSH and testosterone, acting together, stimulate the Sertoli cells to support spermatogenesis.

Many contraceptive methods are available

Useful contraceptive methods are based on an accurate understanding of the biology of reproduction. The simplest method to understand is abstinence. Without sperm in the female reproductive tract, pregnancy cannot occur. Several methods are based on physically blocking sperm from entering the uterus during and after sexual intercourse. This is the rationale for condoms (male and female) and diaphragms. Some methods, such as birth control pills, manipulate a woman's hormone cycle so that ovulation does not occur.

TABLE 34.1 presents contraceptive failure rates in the United States for a short list of methods, according to a recent assessment (the original publication can be consulted for a much more comprehensive list). In many ways, the most informative aspect of the table is the comparison of failure rates for "typical" and "perfect" applications of the methods. For example, latex male condoms are estimated have an annual failure rate of 2 pregnancies per 100 women (whose partners use condoms) when used in a "perfect" way (i.e., used exactly according to directions). However, among people in the general population who say they use male condoms as their method of contraception, the failure rate is nine times higher. This difference reflects deficiencies in how and when condoms are put on and removed, and it also reflects instances when avowed condom users skip using them. The message is clear: many methods are not very effective unless users follow directions in detail and without fail.

REVIEW & APPLY | 34.2

R In mammals, ova mature within ovarian follicles in the ovaries, and after ovulation enter the oviducts. Sperm move through the cervix and uterus into the oviducts, where fertilization occurs. The resulting zygote undergoes preliminary development and arrives in the uterus as a blastocyst, which implants itself in the endometrium. Ovulation—always stimulated by a surge of luteinizing hormone (LH)—is either induced (triggered by copulation) or spontaneous (under control of an endogenous hormonal cycle).

In most species of mammals that display spontaneous ovulation, a female ovulates in cycles. In primates these cycles, named menstrual cycles, entail menstruation (sloughing off of part of the endometrium). Other female mammals ovulate in estrous cycles that entail dramatic changes in behavior—notably estrus (heat). The cycles of all mammals are under control of hormones secreted by the hypothalamus (GnRH), anterior pituitary gland (FSH and LH), ovarian follicle somatic cells (estrogen), and corpus luteum (progesterone and estrogen). The uterus requires progesterone to support pregnancy. An important part of the birth process is a positive-feedback loop entailing uterine muscle contractions stimulated by oxytocin.

The male reproductive organs produce and deliver semen containing sperm. Spermatogenesis depends on testosterone, which the interstitial (Leydig) cells of the testes secrete under control of LH. FSH and testosterone support Sertoli cell function.

A

1. What is the role of chorionic gonadotropin (CG) in human reproduction, and why is it a target for pregnancy tests?
2. Where do the fluids come from that are expelled from a man's penis before and during ejaculation?
3. Explain how FSH is involved in orchestrating ovulation over a series of menstrual cycles in a woman. Also explain why FSH is required for reproduction in a man.
4. What differences and similarities exist between mammals that exhibit induced ovulation and spontaneous ovulation?

In countless ways, reproduction must be integrated with the rest of an animal's life cycle. If reproduction requires nutrients, for example, it must be integrated with processes that provide nutrients. Thus a final aspect of reproduction is how it relates to other aspects of the life cycles of animals.

34.3 Reproduction Is Integrated with the Life Cycle

LEARNING OBJECTIVES

By the end of this key concept you should be able to:

34.3.1 Explain why controlled delays between steps in the reproductive process can sometimes aid reproductive success.

34.3.2 Specify two distinctly different ways in which animals with an annual reproductive season can time their reproduction to occur in their reproductive season.

Reproduction is an essential part of an animal's life cycle. Because of this, a full understanding of reproduction often involves asking questions about the life cycle. For example, does reproduction place limits on the rest of the life cycle? Does the life cycle compel reproduction to take place in certain ways?

Animals often gain flexibility by having mechanisms to decouple the steps in reproduction

In humans, the individual steps in the reproductive process are rigidly linked. Mating leads promptly to fertilization, fertilization leads promptly to embryonic development, and embryonic development adheres to a relatively rigid schedule, culminating in birth at a relatively fixed time after fertilization. Some other animals also show this sort of rigid sequencing. In cases like this, the individual steps in the reproductive process cannot be separately coordinated with conditions in the environment.

Many animal species, however, have evolved mechanisms of decoupling successive steps in the reproductive process, so that the time that elapses between one step and the next is flexible. Such mechanisms increase options for certain steps to be coordinated with environmental conditions independently of other steps.

Sperm storage in the female reproductive tract is a common mechanism that provides for flexible timing between copulation and fertilization. When a female stores sperm for a long time after mating, fertilization can occur long after copulation.

Female blue crabs (*Callinectes sapidus*) illustrate the advantages of sperm storage. A female crab typically copulates during only one short period (lasting less than a day) in her life. However, she stores the sperm she acquires. Thus she can produce new masses of fertilized eggs for over a year. Queen honey bees store sperm for multiple years. Sperm storage also occurs in certain other crustaceans and insects, and in certain bats, sharks, and reptiles (including birds).

Embryonic diapause—a programmed state of arrested or profoundly slowed embryonic development—occurs in many animals. In cases of embryonic diapause, embryos start to develop but then stop for a while before continuing. This permits adjustment of the time between fertilization and completion of embryonic development.

Silkworm moths (*Bombyx*), for example, use embryonic diapause to ensure that their offspring do not hatch out of their eggs in winter. In autumn months, the adult females copulate and lay fertilized eggs. However, these eggs do not immediately develop into hatchlings. Instead the autumn-laid eggs are programmed to undergo a complete arrest of development (cessation of mitosis) early in embryonic development. After the eggs have entered this arrest, they must be exposed to a low temperature (5°C or lower) for about 2 months before they can end the arrested state and

Estrus and conception
Birth
DELAYED IMPLANTATION (diapause)
Implantation
Placental development starts
Placental development requires 250 days
Birth
SPRING
SUMMER
AUTUMN
WINTER
SPRING

FIGURE 34.17 The Role of Delayed Implantation in the Life Cycle of the Antarctic Fur Seal

complete their development. These interacting processes—the programmed appearance of diapause and the need for extended cold exposure to terminate it—ensure that eggs laid in the autumn do not hatch into hungry caterpillars during winter. The eggs need winter cold to emerge from developmental arrest. They accordingly hatch in the spring.

> **LINK** Plants have mechanisms to ensure that their seeds germinate when conditions are favorable for seedling growth. For example, a photoreceptor sensitive to hours of sunlight promotes seed germination in plants, as discussed in **Key Concept 25.3**.

Embryonic diapause in placental mammals is usually called **delayed implantation**. As its name implies, delayed implantation involves postponing the implantation of the blastocyst into the endometrium. Antarctic fur seals (*Arctocephalus gazella*) provide an astounding example. They use delayed implantation to achieve an exact match between the length of their reproductive cycle and the length of the calendar year. The challenge these seals face is that they have only a narrow window of time, early in the Antarctic summer, to give birth. If they give birth much earlier, their offspring can be killed by late winter storms. If they give birth much later, their offspring cannot grow as much as they need to before the next winter starts.

These fur seals have evolved a reproductive cycle in which the time between copulation and birth is almost exactly 365 days. Thus if a female copulates in the early summer one year, she gives birth to the offspring from that copulation 365 days later in the early summer of the next year, at which time she copulates again, and so forth.

However, the placental development of a young Antarctic fur seal (measured from implantation to birth) lasts only 250 days. If these seals were like people—with all the steps rigidly linked—a youngster conceived in the early summer would be born in the cold of the Antarctic winter, but delayed implantation solves this problem (**FIGURE 34.17**). After fertilization, the blastocyst enters the uterus, but it does not implant. Instead it goes into arrest for more than 3 months and then implants. At that point, 250 days remain before the proper birthing moment in early summer.

Some animals can reproduce only once, but most can reproduce more than once

One of the most important reproductive characteristics of a species is the number of times that individuals are physiologically capable of reproducing during their lives. **Semelparous** animals are physiologically programmed to reproduce only once. Octopuses, some fishes, many insects, and many marine worms are semelparous.

Octopuses and Pacific salmon provide two dramatic examples. A species of octopus that is particularly well studied is *Octopus vulgaris*. Many other octopus species are believed to have a similar reproductive strategy. Males place *spermatophores* (packets of sperm) in the reproductive tracts of females during mating, and the females soon lay fertilized eggs. At that point, a female stops eating and guards her eggs for 1–2 months (**FIGURE 34.18A**).

(A) A female *Octopus vulgaris* guards her eggs until they hatch and then she dies

(B) Pacific salmon die after spawning

FIGURE 34.18 Semelparity (A) In many species of octopuses, a female produces eggs just once in her life and guards them. She does not eat while guarding her eggs, and she dies after they hatch. **(B)** Sockeye salmon (*Oncorhynchus nerka*) expend enormous resources to reach freshwater spawning areas, then die after spawning.

Guarding her eggs is her final achievement, because she dies soon after her eggs hatch. Males also die at about the same time. Both the females and males reproduce only once in their lives.

Like the octopus, Pacific salmon such as sockeye salmon (*Oncorhynchus nerka*) also die soon after reproducing. After sockeye salmon grow to adulthood over a period of 1–4 years in the Pacific Ocean (see Key Concept 29.2), they migrate up rivers—sometimes hundreds of miles—before spawning. As soon as they enter the rivers from the sea, both sexes stop eating. Accordingly, they power their upriver migration entirely by breaking down their own body substance. After reaching a suitable spot to spawn, they release their gametes and die (**FIGURE 34.18B**).

In species that are semelparous, parents often sacrifice their own well-being to an extraordinary extent to provide resources to their offspring. This is understandable because the parents have no future reproductive potential after their single period of reproduction. By the time sockeye salmon spawn, they are often capable of little else but spawning because they have so thoroughly depleted their bodies.

Most animal species are **iteroparous**, meaning that individuals are physiologically capable of two or more separate periods of reproductive activity during their lives (i.e., multiple "iterations" of reproduction). In many species, for example, individuals reproduce at least once per year for as many years as they live.

The question of parental investment in offspring is far more complex in iteroparous animals than in semelparous ones. This is because when iteroparous parents produce offspring at any one time, they remain able to produce more offspring at future times. To gain the advantages of future reproduction, they must survive: as they reproduce, they must keep enough resources for themselves to preserve their own health. Their use of resources over their lives thus follows a different pattern than in semelparous species.

Seasonal reproductive cycles are common

In iteroparous animals that live in environments with regular seasonal cycles, the reproductive cycle is nearly always timed to coordinate with the environmental seasonal cycle. We humans are unusual in not showing such seasonality.

The most common pattern is for animals to reproduce in months when temperature, food supply, and other conditions are favorable—such as spring and summer—and suspend reproduction during unfavorable months such as those of winter. Some species of rabbits and mice, for example, shut down their gonads during winter. The testes of the males stop making sperm, become small, and sometimes are withdrawn from the scrotum into the abdomen. The testes then regrow, move into the scrotum, and resume sperm production in the spring. Females also suspend reproductive function and then restore it.

To describe life cycles like this, biologists often speak of animals as having a reproductive season and a nonreproductive season. In this context, "season" does not refer to the defined four seasons of the year. Instead, it refers to the months in which the animals are reproductive and those in which they are nonreproductive.

Some animals breed only once in each reproductive season. A female red fox (*Vulpes vulpes*), for example, enters estrus for a single 1- to 6-day period—and ovulates once—each year during her reproductive season. Some other species of foxes and wolves are similar. If a female fails to mate at this time or loses her offspring, she does not enter estrus again until a full year later. Many birds also breed once per year.

Mammals more commonly superimpose an estrous cycle on their seasonal cycle. For example, during each month of her reproductive season each year, a female rat or mouse cycles in and out of estrus every 4–6 days unless she gets pregnant, as we've mentioned. Males produce sperm continuously during the months of the reproductive season.

Seasonal breeders require mechanisms to time their seasonal cycle. Often they employ environmental cues. Of these, the most important, in both invertebrates and vertebrates, is the amount of daylight per 24-hour day—termed the **photoperiod**. For example, many mice enter reproductive readiness when the photoperiod is increasing or long (i.e., during the "long days" of late spring and early summer). In this case, photoperiod is being used as a signal that the time of year is spring or summer. Because photoperiod varies in a mathematically exacting way with the time of year, it provides unambiguous information.

Circannual biological clocks (see Key Concept 29.4) are a second important mechanism for timing seasonal cycles. Sheep have internal timing mechanisms that time events for an entire year. Some other animals do as well. They use these circannual clocks to control their seasonal reproductive cycles.

REVIEW & APPLY | 34.3

R Many animal species have evolved mechanisms that decouple successive steps in the reproductive process and that coordinate steps with environmental conditions. Such mechanisms include sperm storage and embryonic diapause (called delayed implantation in mammals). Semelparous animals are physiologically programmed to reproduce only once. Most animal species, however, are iteroparous—physiologically capable of two or more separate periods of reproductive activity during their lives. In environments with regular seasonal cycles, the reproductive cycle of iteroparous animals is nearly always timed to coordinate with the environmental seasonal cycle. Coordination is most often achieved by monitoring the photoperiod in the environment, but some animals time their annual reproductive cycles with circannual biological clocks.

A

1. If red foxes (*Vulpes vulpes*) enter estrus and ovulate only once a year, why are they not classified as semelparous?
2. How does sperm storage provide flexibility in timing the parts of the reproductive cycle? Use a specific example to explain your answer.
3. Why is embryonic diapause in mammals called delayed implantation?

Thinking as broadly as possible, what reproductive characteristics might rabbits have evolved that would help them achieve a high rate of reproduction? What are the pros and cons of a high reproductive rate?

Answer: Several characteristics could contribute to the high reproductive rate of rabbits. The most obvious is that they have several offspring in each litter, and they can produce more than one litter per year. There also are less obvious characteristics that we discuss in this chapter. Rabbits exhibit postpartum estrus; thus as a female nurses a litter, she often is simultaneously pregnant and nurturing a new litter in her uterus. Rabbits also exhibit induced ovulation, which synchronizes mating and ovulation, so eggs are ovulated when sperm have just been provided. One pair of European rabbits (*Oryctolagus cuniculus*) has the biological potential to produce 15 pairs in a year and (by the principles of exponential growth) more than 200 pairs in 2 years.

A species with a high reproductive rate can succeed by being able to rapidly fill (populate) any habitat that has lots of unused resources. A high reproductive rate is a mechanism of competing successfully in such habitats. However, a downside of a high reproductive rate is the possibility of needless expenditure of parental resources in habitats that are already relatively fully occupied; in a steady-state context, only a small percentage of young rabbits will live, meaning a high percentage can represent wasted effort. In habitats that are already relatively fully occupied, parents might better succeed if they were to invest their resources in just a few offspring, giving those offspring as great an abundance of resources as possible.

VISUAL SUMMARY 34

34.1 Sexual Reproduction Depends on Gamete Formation and Fertilization

Go to ACTIVITY 34.1 and ANIMATION 34.1

FIGURE 34.7

Stages of follicle development before and after ovulation

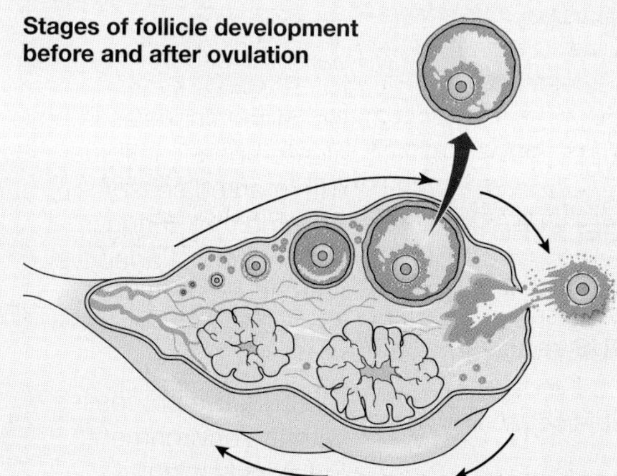

- **GERM CELLS** replicate by mitosis until they undergo meiosis, producing highly diverse haploid **GAMETES**: **SPERM** in males, **OVA** in females. In female **GAMETOGENESIS** only one ovum results from each germ cell, and cellular development arrest is common.
- In males, gonadal **SERTOLI CELLS** support sperm development, and **INTERSTITIAL (LEYDIG) CELLS** secret testosterone. In females, somatic cells envelope the ovum in an **OVARIAN FOLLICLE**, supporting ovum development and secreting estrogens (**FIGURE 34.7**).
- **FERTILIZATION** can be external or internal. Cellular mechanisms help ensure that sperm and ovum are of the same species and that polyspermy is prevented. **SEX DETERMINATION** can occur at fertilization, or in some cases, during embryonic development because of environmental effects.

QUESTIONS

1. What is the difference between accessory reproductive organs and secondary sexual characteristics?
2. How does sexual reproduction produce offspring with diverse characteristics? Give one reason why this property of sexual reproduction can be adaptive.
3. Describe the morphological development of an ovarian follicle during one reproductive (menstrual) cycle in a woman.

34 VISUAL SUMMARY

34.2 The Mammalian Reproductive System Is Hormonally Controlled

Go to ACTIVITIES 34.2 and 34.3 and ANIMATION 34.2

- The human female reproductive cycle is orchestrated by multiple hormones that exert interacting effects on key tissues. For example, **PROGESTERONE** and **ESTROGEN** secreted by a **CORPUS LUTEUM** can inhibit secretion of follicle-stimulating hormone (FSH) by the anterior pituitary gland strongly enough to temporarily block maturation of new ovarian follicles.
- Follicle-stimulating hormone (FSH) and luteinizing hormone (LH), secreted by the anterior pituitary gland under control of hypothalamic gonadotropin-releasing hormone (GnRH), play key roles in regulating reproductive processes in both males and females (**FIGURE 34.13**). Other endocrine structures include the ovarian follicles (secrete estrogen), corpus luteum (progesterone and estrogen), **PLACENTA** (progesterone), posterior pituitary gland (**OXYTOCIN**), and in males, the testis **INTERSTITIAL CELLS** (**TESTOSTERONE**).
- After it plays an early role in the differentiation of male organs during fetal development, testosterone is at low blood concentration until its secretion escalates at the onset of puberty.

FIGURE 34.13 **FSH and LH secreted by the anterior pituitary gland**

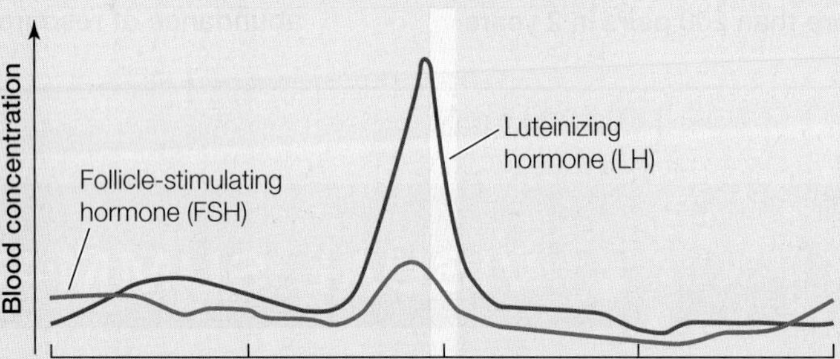

QUESTIONS

1. What critical role is played by nitric oxide (NO) in male reproductive function?
2. Why is oxytocin said to mediate a positive-feedback loop during the mammalian birth process?
3. Describe processes in a woman's uterus during a single menstrual cycle, assuming pregnancy does not occur.

34.3 Reproduction Is Integrated with the Life Cycle

- Reproduction must be integrated with the entire life cycle of an animal. Animal reproductive processes are exceedingly diverse, reflecting the diversity of life cycles and environmental contexts.
- Animals have often evolved mechanisms of decoupling successive steps in the reproductive process—mechanisms that increase options for certain steps to be coordinated with environmental conditions independently of other steps. These mechanisms include sperm storage and **EMBRYONIC DIAPAUSE** (termed **DELAYED IMPLANTATION** in mammals) (**FIGURE 34.17**).
- **ITEROPAROUS** species typically have annual reproductive seasons timed by external cues (notably **PHOTOPERIOD**) or circannual biological clocks. **SEMELPAROUS** species reproduce just once in their lifetime.

FIGURE 34.17

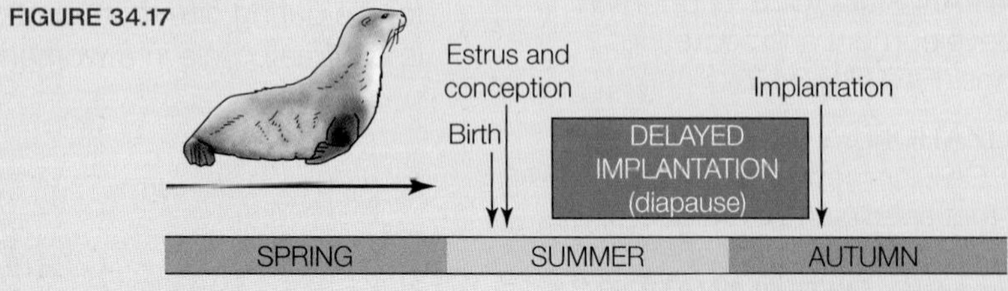

1. What are the potential implications of semelparity for use of parental resources?
2. Consider a species living at a temperate latitude with marked seasons. Assume that spring is the best (most adaptive) time of year for the species to reproduce. For coordinating reproductive processes with the environment, individuals might monitor temperature or photoperiod in the environment. Why might photoperiod be a more reliable cue for timing reproduction?

Go to **Achieve** for the e-book, animations, activities, and additional resources and assignments.

Animal Development 35

These *Xenopus* (frog) eggs, if fertilized, will develop into tadpoles that feed and metamorphose into the familiar froglets. Developmental biologists have long used frog eggs to study development because they are readily available and easy to manipulate. Before fertilization, each unfertilized egg is a single huge cell. Thus, scientists can use unfertilized amphibian eggs as factories for the production of molecules they wish to investigate. These eggs are especially helpful for studies of cell membrane proteins. In nature, *Xenopus* eggs have few ion channels of their own, and yet these eggs are favorite materials of scientists working on ion channels in cell membranes.

Why is the *Xenopus* egg an ideal expression system, and what steps would a scientist take to evoke ion channels for study in these eggs?

You will find the answer to this question on page 897.

This chapter is contributed by Dr. Susan Douglas Hill, Michigan State University

KEY CONCEPTS

35.1 Fertilization Activates Development

35.2 Cleavage Creates Building Blocks to Make an Embryo

35.3 Gastrulation Sets the Stage for Morphogenesis

35.4 Extraembryonic Membranes Protect and Nourish the Embryo

35.5 Development Continues throughout Life

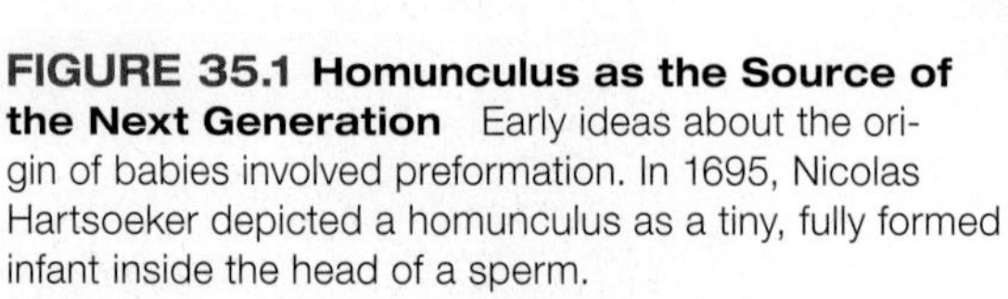

FIGURE 35.1 Homunculus as the Source of the Next Generation Early ideas about the origin of babies involved preformation. In 1695, Nicolas Hartsoeker depicted a homunculus as a tiny, fully formed infant inside the head of a sperm.

What logical argument can be made against this idea?

35.1 Fertilization Activates Development

LEARNING OBJECTIVES

By the end of this key concept you should be able to:

35.1.1 **Outline the different contributions of egg and sperm.**

35.1.2 **Describe the effects of fertilization.**

35.1.3 **Describe some of the ways that polarity is established in the fertilized animal egg.**

The development of an organism is one of the most breathtaking processes you will encounter in your study of biology. In sexually reproducing animals, one cell, the fertilized egg, gives rise to an entire organism. Early investigators assumed that a tiny person was encapsulated in a human egg or sperm and that this "homunculus" required only the appropriate stimulus and nutrients for growth (**FIGURE 35.1**). As microscopes improved, scientists were able to see that no tiny animal resides in either sperm or egg. That left the mystery—how does an organism achieve its final form?

Scientists obtain much information by studying "model" organisms. In developmental biology certain animals are used as models not because they represent the majority of the group under investigation, but because of particular characteristics that make them desirable for research, such as short developmental times and accessibility of embryos. Remember that there are probably millions of animal species—many not yet discovered. Each has tweaked its development to adapt to its environment. The doors are open for further investigations in developmental biology.

Development is a continuous process; even in adulthood animals continue to develop. In this chapter we focus on the events of **embryonic development**—the stages by which a fertilized egg is transformed into an embryo. During development, cells become **differentiated**, or different from each other; that is, they come to display unique characteristics that enable their functioning. For example, differentiated muscle fibers contain large amounts of actin and myosin and are very different from red blood cells (cells that have lost their nuclei and most organelles to become sacs of hemoglobin). Even before cells are recognizable as being differentiated, they become committed to a specific fate and to that fate only. At that point, we say they are **determined**. A cell that is determined to be a nerve cell will differentiate into a neuron even in isolation or if transplanted to an environment unfavorable for neuronal development. Determination and differentiation are key developmental events. Two others are **morphogenesis** (the processes of cell and tissue growth, migration, and death by which the animal body acquires its final organized form) and **growth** (an energy-requiring increase in size that can result from an increase in cell number, cell size, production of extracellular material, or a combination). Despite great variation in body plans, the stages of embryonic development are strikingly similar across a wide range of animal species. As this chapter unfolds you will learn how these key developmental events take place.

Egg and sperm make different contributions to the zygote

Animal development usually begins with the union of egg and sperm to form a fertilized egg, or **zygote**. Egg cells are very large compared with other cells—and compared with sperm, they are gigantic. Even the non-yolky eggs of sea urchins and mammals are large compared with other cells, because they contain the materials and information needed to initiate and maintain early development. The cytoplasm of an egg contains most of the molecular machinery for cell division, energy production, and growth (ribosomes, tRNAs, mitochondria, yolk, etc.), as well as informational molecules such as transcription factors and messenger RNAs (mRNAs) transcribed from the mother's DNA. All of the elements in the egg are crucial, providing resources and starting the signaling cascades that orchestrate the major processes of development: determination, differentiation, morphogenesis, and growth.

The union of egg and sperm is complex. Many sperm are attracted to an egg. However, one sperm, and only one, must convey its nuclear material to the egg cytoplasm (**FIGURE 35.2**). All of the egg cell membrane may be available for sperm entry, as in sea urchins, or entry may be restricted to certain parts. In amphibians, for example, sperm can penetrate only half the egg surface. In *Drosophila,* sperm must enter through a tunnel, the position of which is fixed prior to fertilization.

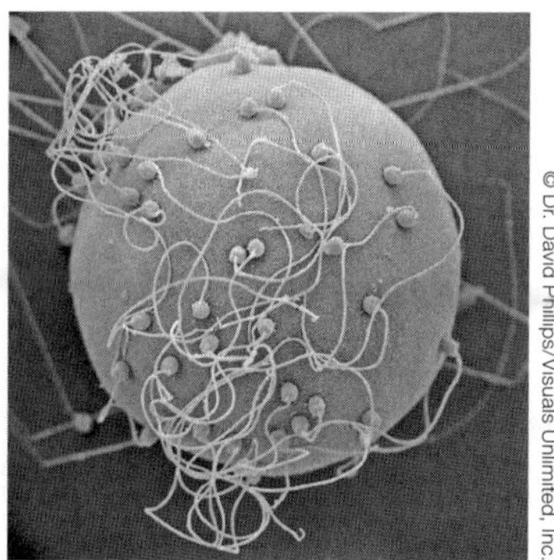

FIGURE 35.2 A Sea Urchin Egg Surrounded by Sperm One sperm, and only one, must fertilize the egg.

© Dr. David Phillips/Visuals Unlimited, Inc.

Activation of an egg by a sperm brings about immediate changes in the egg cell membrane and the release of ions from stores within the egg cytoplasm. These ion changes trigger a series of crucial events. Entry of additional sperm is blocked, metabolic rate rises dramatically, protein synthesis is initiated, and cytoplasm is often reorganized. In other words, the processes of development begin.

LINK Eggs and sperm have ways to recognize each other and to block fertilization by more than one sperm (polyspermy); see **Key Concept 34.1**.

Most of the sperm cytoplasm is discarded when the egg and sperm membranes fuse, but the nuclear material and usually the centriole (the microtubule organizing center of an animal cell) are transferred from the sperm to the egg. The centriole then plays a major role in forming the zygote centrosome and organizing the zygote's microtubules. When the haploid egg and sperm nuclei join, the fertilized egg becomes a single-celled, diploid zygote ready for its first mitotic division.

Polarity is established early in development

Most animals have a defined head and tail, back and belly, and right and left sides. These differences, known as polarities, are usually established early in development. In most animals, polarities are established during early cell divisions as materials deposited in the egg cytoplasm by the mother are redistributed and segregated. In some animals, however, polarities are set in place before or at fertilization.

In the fruit fly *Drosophila melanogaster*, a favorite model organism, the position of the egg in the ovary determines both anterior–posterior (head–tail) and dorsal–ventral (back–belly) polarity. Cells that surround the egg, called nurse cells or follicle cells, pass mRNAs from the mother to specific regions of the egg cytoplasm. After the egg is fertilized and development is underway, these mRNAs encode regulatory proteins that determine polarity.

In amphibians, anterior–posterior and dorsal–ventral polarities are determined at fertilization by the position of the yolk and the point of sperm penetration. Cytoplasmic rearrangement following fertilization is easily observed in some frog species. In the unfertilized eggs, dense yolk granules are concentrated by gravity in the lower half of each egg, the **vegetal hemisphere**. This yolky vegetal hemisphere is unpigmented. In the top half of each egg, or **animal hemisphere**, the outermost (cortical) cytoplasm is heavily pigmented, while the pigment in the underlying cytoplasm is more diffuse. Sperm-binding sites occur only on the animal hemisphere. When a sperm enters the egg, the cortical cytoplasm rotates toward the site of sperm entry, producing a band of lightly pigmented cytoplasm called the **gray crescent** on the opposite side (**FIGURE 35.3**). The anus later forms within the gray

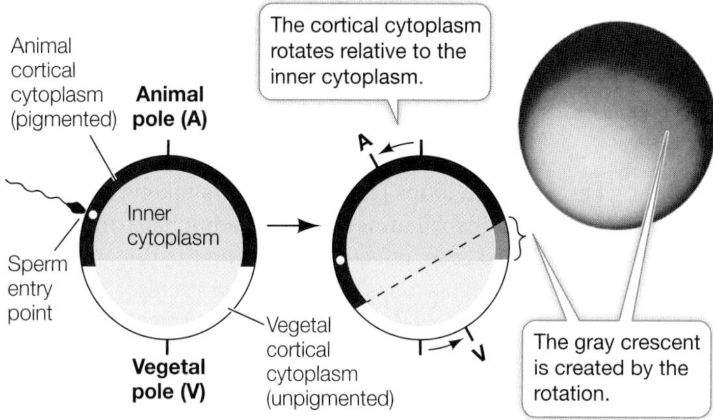

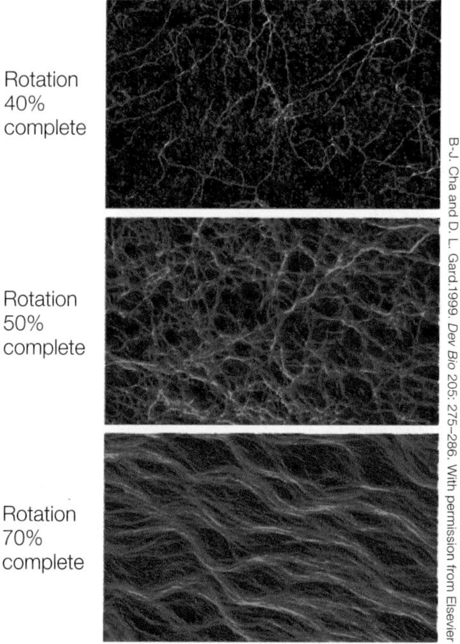

B.-J. Cha and D. L. Gard.1999. *Dev Bio* 205: 275–286. With permission from Elsevier

FIGURE 35.3 The Gray Crescent In some frog eggs, rearrangement of the cytoplasm after fertilization creates the gray crescent and imposes anterior–posterior and dorsal–ventral axes on the egg. **(A)** The sperm centriole acts as the microtubule organizing center, and the zygote cytoplasm reorganizes. Cortical cytoplasm rotates toward the sperm entry point, and a crescent of lightly pigmented cortical cytoplasm is revealed directly opposite. **(B)** Microtubules that polymerize and align during first cleavage bring about the cytoplasmic rotation. Small numbers indicate the percentage completion of the first cleavage (0.4 NT = 40%, 0.5 NT = 50%, and 0.7 NT = 70%). (A after R. P. Elinson and B. Rowning. 1988. *Dev Biol* 128: 185–197.)

crescent. The point of sperm entry will become the ventral (belly) part of the embryo, and the area 180 degrees removed will become the dorsal (spinal) part. Thus the zygote already has an anterior–posterior axis for a future head and tail, as well as a dorsal–ventral axis, both determined by the point of sperm entry. The cytoplasmic rearrangement is initiated by the sperm centriole, which organizes microtubules in the egg cytoplasm in a parallel array that extends through the vegetal hemisphere. This cytoskeleton guides the shift of the cortical cytoplasm that results in the gray crescent.

Much of our understanding of developmental mechanisms builds on the insightful experiments of scientists in the late 1800s and early 1900s who performed simple but elegant experiments that were very revealing. In the early 1900s, beautiful work by Hans Spemann and colleagues who were studying the development of salamander eggs showed that all parts of even the uncleaved zygote are not identical. With great patience and dexterity, Spemann formed loops from single hairs taken from his baby daughter's head (baby hair is finer than adult hair) and tied them around fertilized eggs, effectively dividing the eggs in half (**FIGURE 35.4**). When his loops bisected the gray crescent, both halves of the zygote developed into complete embryos. When he tied the loops so the gray crescent was on only one side of the constriction, only the half with the gray crescent developed normally. The half lacking gray crescent material became an unorganized clump of cells. Spemann concluded that important factors in the cytoplasm—now called **cytoplasmic determinants**—that establish polarity and are necessary for normal development reside in the gray crescent.

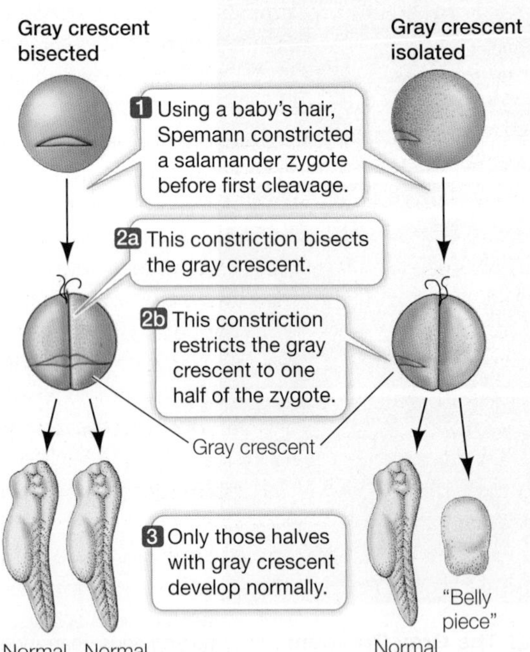

FIGURE 35.4 Fertilization Results in Major Cytoplasmic Rearrangements As the cytoplasm is reorganized, molecules necessary to initiate cleavage are activated, in part, by their new location. Spemann's research revealed the importance of the gray crescent for normal salamander development. (After H. Spemann. 1938. *Embryonic Development and Induction*. Yale University Press: New Haven.)

How did Spemann interpret his results?

Experiments using amphibian eggs have contributed much to our understanding of development. Keep in mind, though, that although amphibians are useful models, many animals develop differently, and in many, polarities are not established until cell divisions are underway.

REVIEW & APPLY | 35.1

R Fertilization is the mechanism by which the haploid nuclei of egg and sperm fuse to form the diploid zygote nucleus. The formation of the zygote is the first step in creating a new individual. The process of sperm penetration activates the egg and initiates development. Egg and sperm make different contributions to the zygote. The egg contributes a haploid nucleus, most of the molecular machinery for cell division, energy production, and growth (yolk, mitochondria, ribosomes, etc.), and informational molecules such as transcription factors and mRNAs for early protein synthesis. The sperm contributes its nuclear material and usually its centriole, the major microtubule organizing center. Upon fertilization of the egg, metabolism increases, protein synthesis begins, and the cytoplasm is rearranged. Often, polarities are established. In amphibians, rearrangement of the cytoplasm establishes the major axes of the future embryo and produces the gray crescent.

A

1. In addition to nuclear material, what does the sperm contribute to the zygote? Why is this important?
2. Why do scientists sometimes use unfertilized frog eggs to study expression of exogenous proteins (proteins that are not normally present in the egg)?
3. What usually determines initial polarity in developing embryos?
4. How are anterior–posterior and dorsal–ventral polarities established in fruit flies? In frogs?

Animation 35.1 Early Asymmetry in the Embryo

We have now seen how the zygote forms and how polarities may be established. We will turn next to the early series of cell divisions that produce a multicellular embryo.

35.2 | Cleavage Creates Building Blocks to Make an Embryo

LEARNING OBJECTIVES

By the end of this key concept you should be able to:

35.2.1 Explain the importance of cleavage in the redistribution of cytoplasmic determinants such as transcription factors.

35.2.2 Explain the two basic cleavage patterns seen in non-yolky eggs of protostomes and deuterostomes.

35.2.3 Describe the typical blastula stage of an embryo derived from an egg with little yolk, such as a sea urchin.

35.2.4 Describe complete cleavage in the yolky eggs of amphibians and incomplete cleavage in the very yolky eggs of birds, and compare them with cleavage in mammals.

Cleavage is the initial process that creates cellular building blocks to form the new organism by producing smaller and smaller cells from the fertilized egg. At some point, as a result of cleavage and perhaps after only one cleavage, these smaller cells will differ in their cytoplasmic determinants. During cleavage, no growth, at the level of the cell or the whole organism, occurs. These are specialized, rapid mitotic divisions during which there is no "gap" or growth phase, so no new cytoplasm is added. There is also no new energy source ("food"), even from the placenta in mammals or the yolk in bird eggs at this point. Thus as cells increase in number, they become smaller and smaller. While a frog egg stays the same size, it becomes divided into tens of thousands of cells. The cells produced are called **blastomeres**, and the zygote becomes a **blastula**.

Early investigators thought that these cell divisions divided genes among blastomeres so that, later, each could form only its specific tissue or organ. We now know that this is not the case. Each blastomere receives the entire genome of the animal. How, then, can cells with the same genetic information give rise to all the different structures of the adult?

During very early stages of development, events are controlled by specific information passed on by the mother. Cytoplasmic determinants, consisting of proteins such as transcription factors (see Key Concept 11.2) and mRNAs, are deposited in the cytoplasm of the egg as it matures. Early cleavages occur rapidly, and mRNA already placed by the mother in the egg is used to direct protein synthesis. In most animals, it is not until the mid- to late blastula stage that the genome of the developing embryo takes over. At that time old mRNA from the mother in the egg cytoplasm is broken down and information in the zygote nucleus controls further development. This **maternal-to-zygotic transformation** marks the first real step toward independence for the developing embryo.

As we saw in the formation of the gray crescent, the cytoskeleton may redistribute cytoplasmic determinants from one point to another. Also, as mitotic divisions continue, new cell membranes separate cytoplasmic determinants so they become distributed differentially among blastomeres. These processes, plus cell–cell communication, play a role in the determination of blastomere fates. All of these processes are involved in the development of most animals, but the stage at which each process comes into play differs from group to group. However, at some point the fate of a blastomere becomes restricted, or *determined*, and it can no longer form an entire embryo.

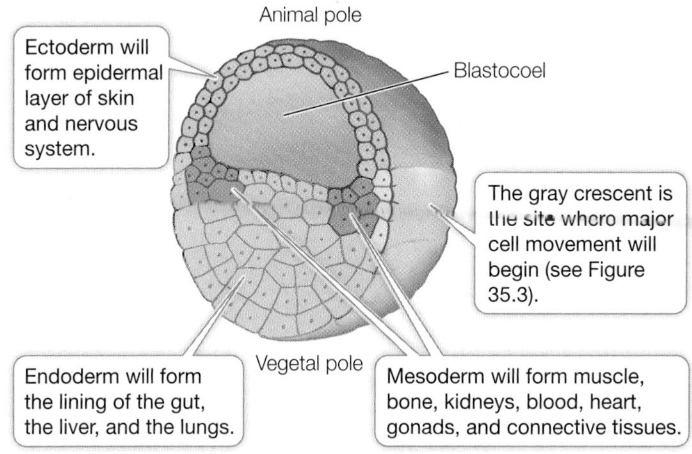

FIGURE 35.5 Fate Map of a Frog Blastula Colors indicate the portions of the early-stage frog embryo (blastula) that will form the three germ layers—ectoderm, mesoderm, and endoderm—and subsequently the frog's tissues and organs.

Scientists have learned much about the developmental potential of blastomeres by separating the blastomeres at different times of development. In such experiments, blastomeres are isolated to learn how they will develop on their own. It is also possible to leave embryos intact, label a specific cell (e.g., with a dye), and then identify the tissues and organs that form from that cell. New techniques (e.g., light sheet microscopy and multiphoton microscopy) allow individual cells to be followed through many cleavages without the addition of a dye or tracer. Experiments of all types—isolation experiments and tracing experiments—have been used to generate fate maps for many species (**FIGURE 35.5**). **Fate maps** show from which portion of the embryo various adult or larval structures arise. To understand the developmental potential of blastomeres, and to follow cell lineages, we need to explore different cleavage patterns.

Cleavage is easiest to understand in animals that produce eggs with little yolk, such as sea urchins and some mollusks. If the egg cytoplasm is divided equally among blastomeres, the blastomeres will be the same size. If it is divided unequally, large blastomeres and small blastomeres will result. In both cases, as cleavage continues, the zygote becomes divided into smaller and smaller blastomeres (**FIGURE 35.6A**). If the cytoplasm is completely divided by each cell division, cleavage is termed **complete cleavage**. If it is not, as you will see is the case in reptiles (including birds), insects, and some other animals, it is termed **incomplete cleavage**.

Two different patterns of cleavage that seem to have deep evolutionary roots are **radial cleavage** and **spiral cleavage**. These patterns are clearly seen in eggs that have little yolk and undergo complete cleavage, such as those of snails and sea urchins. In all types of cleavage, the plane of cleavage (shown by the position of the new cell membrane that divides the cells) is determined by the orientation of the mitotic spindle. Differences in radial and spiral cleavage begin with the first cleavage but are seen most readily after the third cleavage, at the eight-cell stage. At this time, in radial cleavage the four daughter cells lie directly above the four parent cells (**FIGURE 35.6B**). During spiral cleavage, mitotic spindles align so that, after the third and following cleavages, the

(A) The first three cleavages

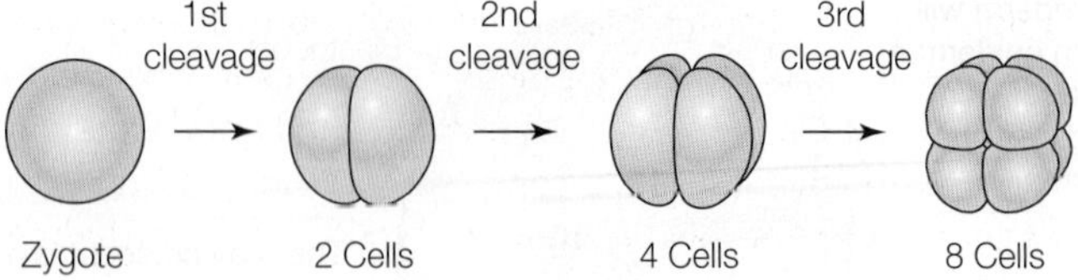

(B) Radial cleavage

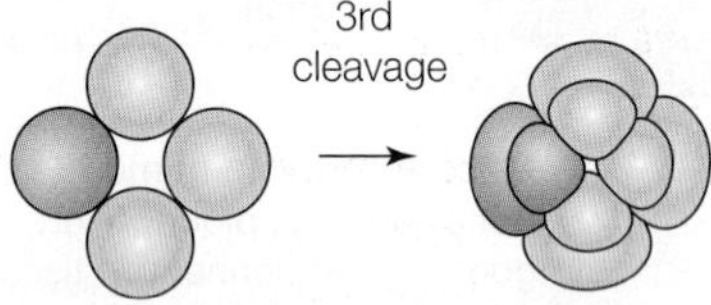

(C) Spiral cleavage

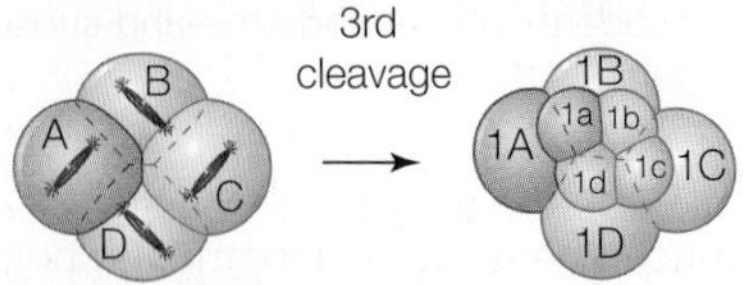

FIGURE 35.6 Cleavage: One Cell Becomes Many
The plane of cleavage furrows determines different cleavage patterns. **(A)** Complete cleavage showing division of the cytoplasm into equal-sized blastomeres, as seen from the side. The pattern of cleavage shown in (A) is radial cleavage. **(B, C)** Radial and spiral cleavage, seen from the top. Cleavage is unequal, resulting in four larger and four smaller blastomeres. (B) Radial cleavage. The four smaller blastomeres lie directly above the larger four. (C) Spiral cleavage. The four smaller blastomeres lie in cleavage furrows produced by previous divisions. For clarity, one cell lineage is highlighted in purple.

four daughter cells lie in the grooves between the four parent cells, creating the appearance of a spiral (**FIGURE 35.6C**). A spirally cleaving egg has the most thermodynamically stable shape possible. Appropriately sized soap bubbles assume the same configuration as a spirally cleaving egg!

You are probably most familiar with deuterostome animals, the large clade (monophyletic group) that includes the vertebrates and all other chordates, echinoderms, and a few other groups. Echinoderms and several other deuterostomes display radial cleavage. Radial cleavage is considered to be the basic deuterostome pattern, although it is often highly modified in the presence of yolk. The other large group of animals, the protostomes, includes most of the invertebrates. Protostomes do not have radial cleavage. In the lophotrochozoans, a clade that includes about half of all animal phyla, including the annelids and mollusks, spiral cleavage predominates.

LINK **Key Concept 22.1** shows how early development determines body plans that reflect phylogenetic relationships among the animals; **Key Concept 22.3** discusses the protostomes, explaining that there are two major clades; and **Key Concept 22.5** discusses the deuterostomes and some of their developmental characteristics.

Specific blastomeres generate specific tissues and organs

Blastomeres become determined—irreversibly committed to specific fates—at different times in different species. In many protostomes with spirally cleaving eggs and in a few deuterostomes, blastomere fates are determined as early as the two-cell stage. If one of these blastomeres is experimentally removed, a particular portion of the embryo will not form (**FIGURE 35.7**). Experiments like the one shown in Figure 35.7B clearly demonstrate one way in which information for further development is distributed to specific blastomeres. This type of development has been called **mosaic development** because each blastomere appears to contribute a specific set of "tiles" to the final "mosaic" that is the adult animal.

In contrast to what happens in mosaic development, in **regulative development** the loss of some cells during cleavage does not affect the developing embryo. Instead, the remaining cells compensate for the loss. If blastomeres of a sea urchin egg are separated at the two-cell stage, two half-sized, but complete, larvae form. Even at the four-cell stage, separation produces four small larvae (**FIGURE 35.8**). We now know that cytoplasmic determinants are found in the cytoplasm of eggs showing both mosaic and regulative development. In animals that display regulative

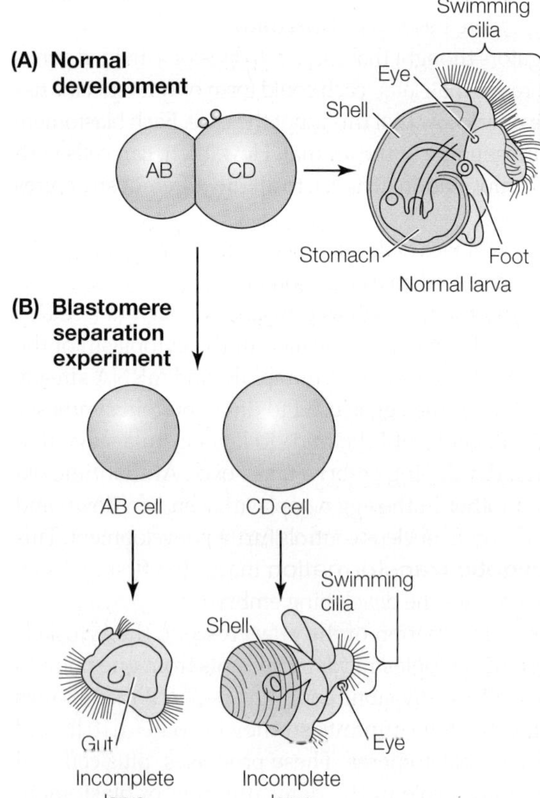

FIGURE 35.7 Mosaic Development of a Snail Egg
(A) Normal Development. **(B)** Separation of the two blastomeres results in two abnormal larvae, incomplete in different ways. Determinants in the CD blastomere lead to a more complete larva, but critical organs and features are missing.

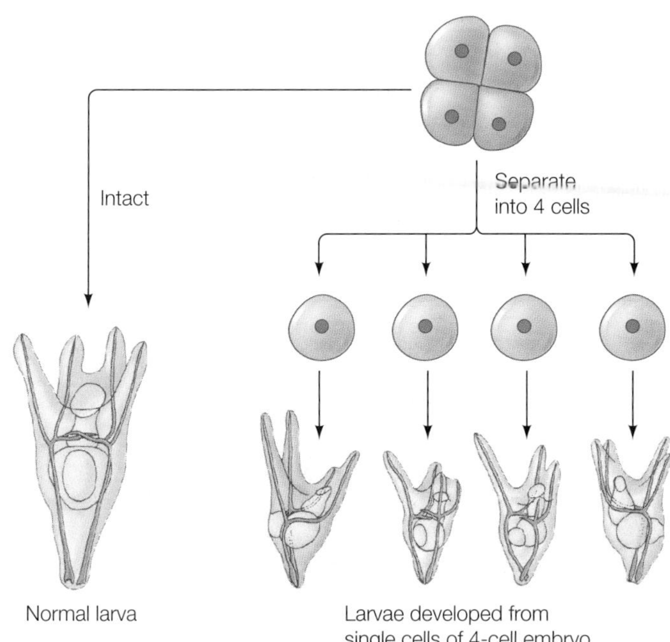

FIGURE 35.8 Regulative Development of a Sea Urchin Egg Separation of four blastomeres results in four small but almost normal larvae. (After S. Hörstadius and A. Wolsky. 1936. *Archiv für Entwicklungs-mechanik* 135: 69–113.)

development, however, the distribution of these determinants happens later as cell divisions continue. Regulative development is typical of deuterostomes, including humans and other vertebrates.

As noted earlier, cleavage leads to the production of a ball of cells termed a blastula. Blastulas produced by radially cleaving eggs, and by eggs in which cleavage is derived from this basic pattern, have an inner fluid-filled cavity called a **blastocoel** (see Figure 35.5). The blastula of a sea urchin, for example, is a hollow sphere. Blastulas produced by spirally cleaving eggs may have a blastocoel, but often they are a solid ball of cells, without the fluid-filled space.

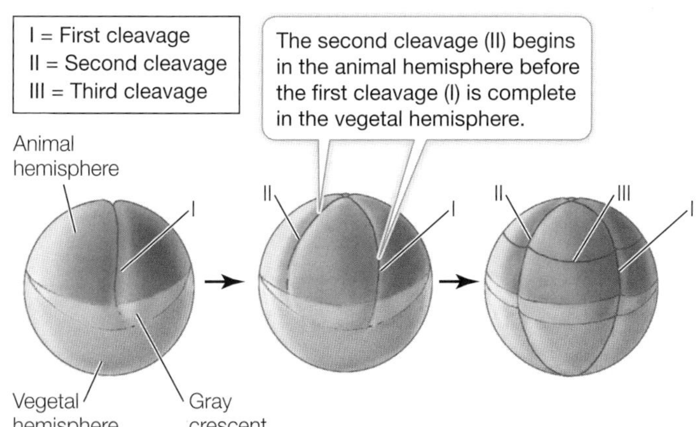

FIGURE 35.9 Complete Cleavage in a Frog Egg Cleavage proceeds more rapidly in the animal hemisphere than in the vegetal hemisphere. (After B. M. Carlson, 1981. *Patten's Foundations of Embryology*, [4th ed.] McGraw-Hill: New York.)

The amount of yolk affects cleavage

The amount of yolk has a profound effect on both cleavage patterns and *gastrulation*, the developmental stage that follows cleavage. Some eggs, such as those of sea urchins, snails, and placental mammals, have little or no yolk. Some, such as those of frogs and salamanders, have a moderate amount of yolk. Others, including those of birds and egg-laying mammals, are extremely yolky. As you will see, it is as if the presence of yolk slows down or impedes cell divisions. This means that the eggs of many animals do not follow the radial or spiral cleavage patterns that we consider basic in deuterostomes and protostomes.

In eggs with little yolk, like those of sea urchins and snails, cleavage usually proceeds as described above and is complete. In moderately yolky eggs such as those of frogs, cleavage is complete and cell membranes form around every nucleus. However, cells divide much more slowly in the yolk-filled vegetal hemisphere than in the animal hemisphere (**FIGURES 35.9** and **35.10A**). In the vegetal hemisphere, the yolk retards the process of division so that the second round of cleavage has begun in the animal hemisphere before the first round is completed in the vegetal hemisphere, and this continues, over and over. In the example shown in Figure 35.9, the third cleavage is horizontal and occurs in the animal hemisphere. Because cells divide faster in the animal hemisphere, more cells are produced in the dorsal portion of the egg than in the ventral, resulting in the animal hemisphere consisting of many small blastomeres and the vegetal hemisphere consisting of fewer, larger blastomeres. The blastocoel forms within the small blastomeres of the animal hemisphere (see Figure 35.5).

In very yolky eggs such as those of a squid (a molluscan protostome), fish, bird, or egg-laying mammal (e.g., the platypus), the yolk remains undivided and the embryo forms as a small **blastodisc** of cleaving cells on the yolk surface (**FIGURE 35.10B**). The yolk never divides, and so the cleavage is incomplete. The disc of cells on top of the yolk enlarges and becomes several cells thick. For example, the blastula of a bird, or of the zebrafish shown in Figure 35.10B, is a flattened structure that sits atop the large undivided yolk and consists of an upper layer, the epiblast, which will give rise to the embryo proper, and a lower hypoblast, with a fluid-filled space between the two. (Epiblast and hypoblast in the bird egg are shown in Figure 35.15.)

A different type of incomplete cleavage is seen in insects. In *Drosophila* eggs, for example, the original zygote nucleus undergoes a series of divisions unaccompanied by divisions of the cytoplasm. This early embryo is a **syncytium**: a mass of cytoplasm in which several nuclei are surrounded by the same cell membrane. As development proceeds, nuclei migrate to the periphery of the egg, and cell membranes form (**FIGURE 35.10C**).

Cleavage in placental mammals is unique

Placental mammals have non-yolky eggs (which undergo development within the mother's body after fertilization), but their cleavage pattern differs from that of other non-yolky deuterostome eggs. You might expect that placental mammals' eggs would show radial cleavage like sea urchin eggs, but they do not. Remember that mammals evolved from a reptilian ancestor. As a result, several aspects of their development—such as the formation of

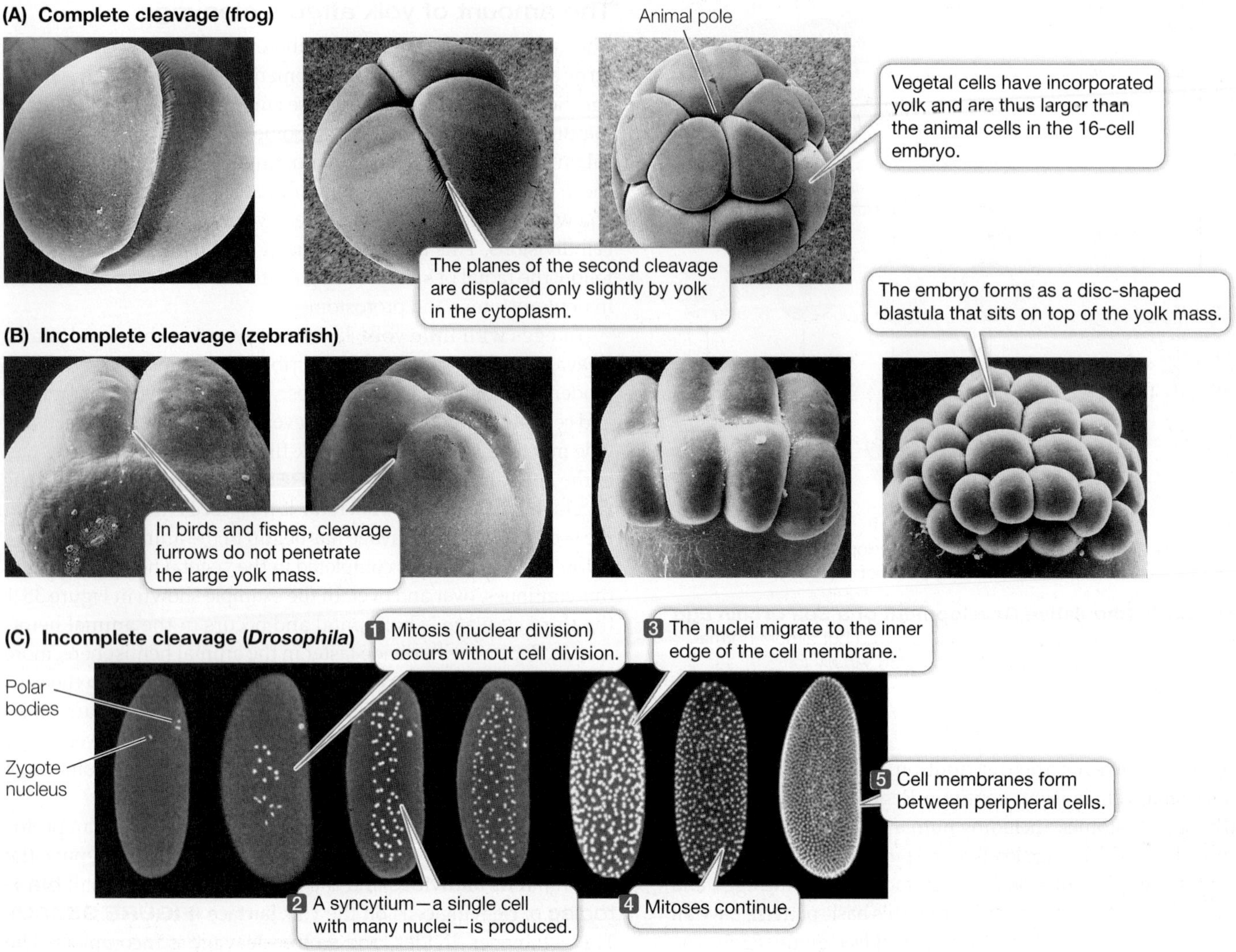

(A) Complete cleavage (frog)

Animal pole

Vegetal cells have incorporated yolk and are thus larger than the animal cells in the 16-cell embryo.

The planes of the second cleavage are displaced only slightly by yolk in the cytoplasm.

(B) Incomplete cleavage (zebrafish)

The embryo forms as a disc-shaped blastula that sits on top of the yolk mass.

In birds and fishes, cleavage furrows do not penetrate the large yolk mass.

(C) Incomplete cleavage (*Drosophila*)

1 Mitosis (nuclear division) occurs without cell division.

3 The nuclei migrate to the inner edge of the cell membrane.

Polar bodies

Zygote nucleus

5 Cell membranes form between peripheral cells.

2 A syncytium—a single cell with many nuclei—is produced.

4 Mitoses continue.

FIGURE 35.10 Some Patterns of Cleavage Differences in patterns of early embryonic development reflect differences in egg cytoplasm. **(A)** The frog is a model organism representing complete cleavage. **(B)** Zebrafish embryos illustrate incomplete cleavage, in which the large yolk mass limits the planes of cleavage. **(C)** Fruit flies have another type of incomplete cleavage. Nuclear staining reveals the syncytial nuclei characteristic of their early development. These nuclei migrate to the periphery of the egg. Cleavage furrows then move inward to separate the nuclei into individual cells. (A [left], B from H. W. Beams and R. G. Kessel. 1976. *Am Sci* 64: 279–290; A [middle, right] © Dr. Lloyd M. Beidler/Science Source; C from S. Kotadia et al. 2010. *eLS*, [ed.]. doi:10.1002/9780470015902.a0001071.pub2.)

epiblast and hypoblast, as well as of the yolk sac and allantois (which we will discuss in Key Concept 35.4)—were derived from processes in yolky eggs of reptiles. Other features of mammalian development, however, are unique. The two groups of placental mammals, the marsupials and the eutherians, have evolved slightly different developmental steps. In this chapter, we will now consider development in only eutherian mammals unless specified otherwise. As stated above, cell divisions in eutherians are not radial. In many eutherians, at the second cleavage division, the two blastomeres divide in different planes, so that one blastomere cleaves vertically and the other horizontally, in a pattern called **rotational cleavage** (**FIGURE 35.11**). Blastomeres within the

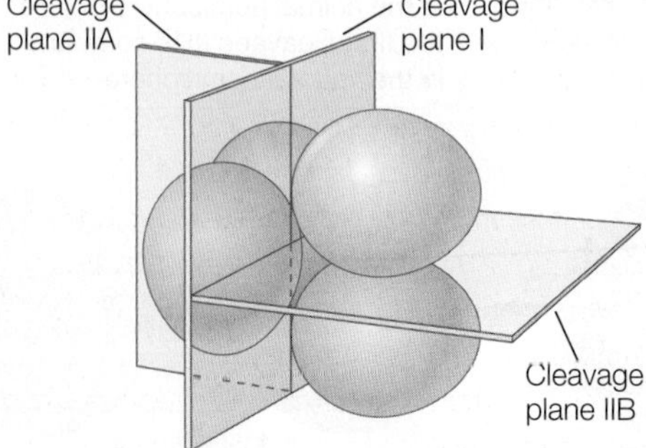

FIGURE 35.11 Rotational Cleavage Is Characteristic of Eutherian Mammals Cleavage planes and division rates differ from those of other non-yolky deuterostomes such as sea urchins. At second cleavage, the divisions of the two blastomeres are at right angles to each other. (After B. J. Gulyas. 1975. *J Exp Zool* 193: 235–247. doi:10.1002/jez.1401930212.)

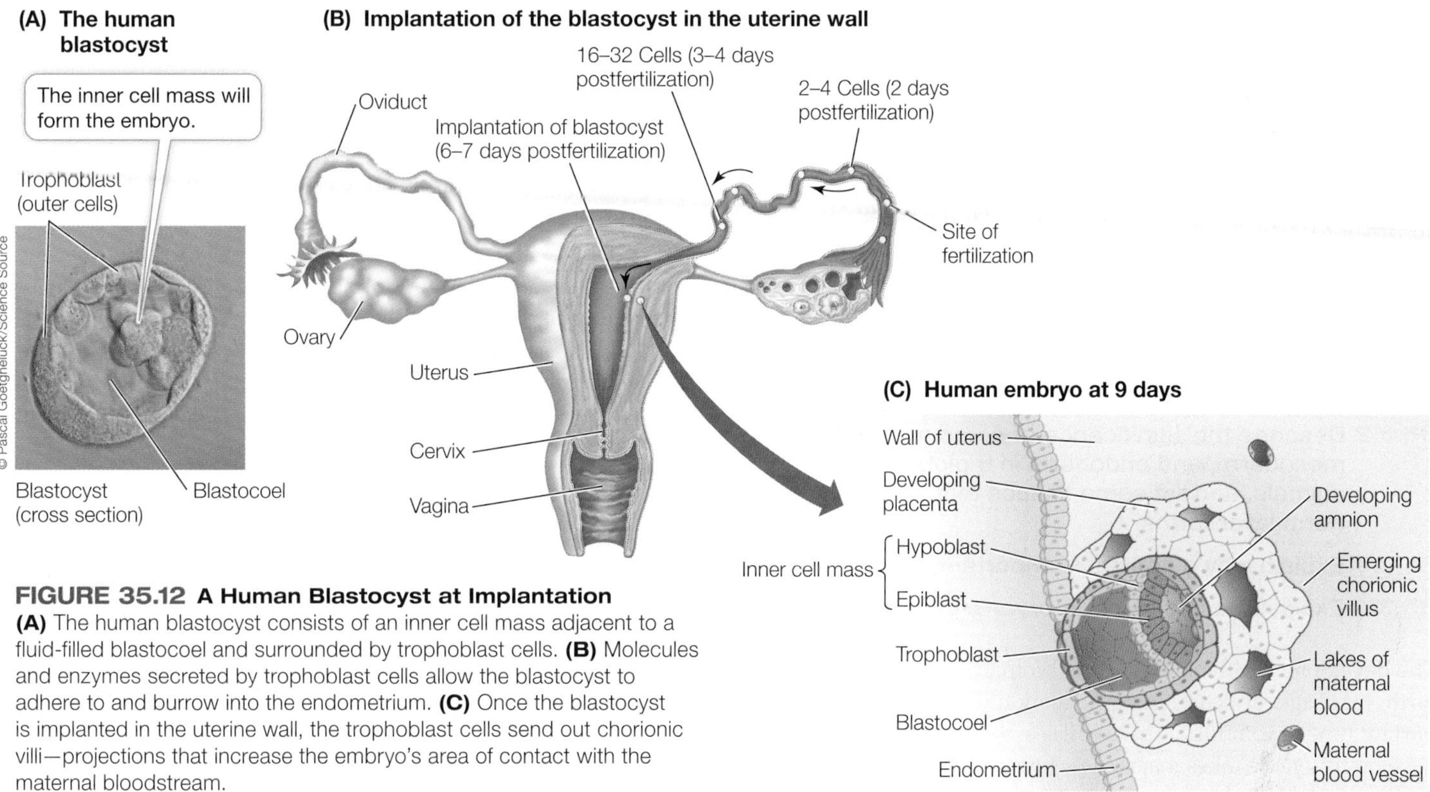

FIGURE 35.12 A Human Blastocyst at Implantation
(A) The human blastocyst consists of an inner cell mass adjacent to a fluid-filled blastocoel and surrounded by trophoblast cells. **(B)** Molecules and enzymes secreted by trophoblast cells allow the blastocyst to adhere to and burrow into the endometrium. **(C)** Once the blastocyst is implanted in the uterine wall, the trophoblast cells send out chorionic villi—projections that increase the embryo's area of contact with the maternal bloodstream.

same egg divide at different times, so the eutherian mammal egg may contain an odd number of cells. This is not usually the case during early cleavage in other groups. The mammalian genome is activated very early, so cleavage and development are governed by the new nuclei of the early blastula. In most deuterostomes, the maternal-to-zygotic transformation happens considerably later, around the mid- to late-blastula stage. In humans, it begins at the eight-cell stage; in mice, at the two-cell stage.

Another unique feature of mammalian development occurs during the fourth cleavage, when the cells separate into two groups (**FIGURE 35.12A**). The **inner cell mass** will become the embryo, while the surrounding outer cells become an encompassing sac called the **trophoblast**. Trophoblast cells secrete fluid, creating a blastocoel with the inner cell mass at one end. At this stage the mammalian embryo is called a **blastocyst**. Later, trophoblast cells will contribute to the formation of a unique mammalian organ, the placenta, which we will discuss in Key Concept 35.4.

In a mammal, fertilization occurs in the upper reaches of the mother's oviduct, and cleavage takes place as the zygote travels down the oviduct. When the blastocyst arrives in the uterus, it adheres to the lining of the uterus, the endometrium, and begins burrowing into the uterine wall. In humans, this process of **implantation** begins about 6 days after fertilization (**FIGURE 35.12B** and **C**).

REVIEW & APPLY | 35.2

R Cleavage is a period of rapid cell division without cell growth. Cleavage can be complete or incomplete, often correlating with the amount of yolk, and results in a ball or mass of cells called a blastula. Cell membranes that form during cleavage segregate cytoplasmic determinants into different blastomeres. Some species undergo mosaic development, in which the fate of each cell is already determined at the first division. Other species, including vertebrates, undergo regulative development, in which cell fates become determined later. Usually, little if any gene expression occurs until the mid-blastula stage, although in mammals it happens much earlier. Early cell divisions in mammals are unique. These cell divisions produce a blastocyst composed of an inner cell mass that will become the embryo and an outer layer of cells called the trophoblast that will form part of the placenta.

A

1. What two major functions does cleavage perform?
2. Compare radial and spiral cleavage. What determines the direction of the cleavage plane?
3. Describe complete and incomplete cleavage and explain where mammals fit into this scheme.
4. What type of development must a species have for identical twins to occur? What do you think leads to the development of identical twins?
5. A cat has three kittens—one orange, one gray, and one black. What can you conclude about the kittens?

After formation of the blastula, the next stage of development is gastrulation—the process we'll look at next.

35.3 | Gastrulation Sets the Stage for Morphogenesis

LEARNING OBJECTIVES

By the end of this key concept you should be able to:

35.3.1 Explain the significance of gastrulation.

35.3.2 Describe the significance of ectoderm, mesoderm, and endoderm in triploblastic animals, and list some tissues that each germ layer forms.

35.3.3 Explain why a coelom is important.

35.3.4 Explain the basics of embryonic induction.

The developmental biologist Lewis Wolpert once said, "It is not birth, marriage, or death, but gastrulation which is the most important time in your life." As we will see, this is the time when the developing zygote takes major steps toward independence.

During **gastrulation**, many cells move from the external surface to the interior of the embryo. These cell movements will create a three-layered body plan and position cells so they have new neighbors. The new associations allow the cellular interactions required for development. Moreover, the zygote genome now controls protein synthesis, and the processes of differentiation and morphogenesis are initiated.

Triploblastic animals (see Key Concept 22.1) create all of their organs and tissues from three basic germ layers—ectoderm, endoderm, and mesoderm. At the end of gastrulation, a gastrula with ectoderm on the outside, endoderm on the inside, and mesoderm in between has formed.

Gastrulation is easiest to understand in radially cleaving eggs with little yolk, such as the much-studied eggs of sea urchins (**FIGURE 35.13**). One wall of the blastula (the vegetal pole) bulges inward, or invaginates. Imagine poking a finger into a soft, hollow ball. The first "hollow" is the blastocoel. The new space that forms as a result of the "poke" is the archenteron, or "first gut." The opening into the archenteron is the blastopore. Recall that in deuterostomes the blastopore becomes the anus, and the mouth (the second opening) forms opposite it. The outer layer of cells is now ectoderm, and the wall of the archenteron is endoderm and future mesoderm. The archenteron elongates, assisted by contractions of cells that leave the single layer and become wandering cells called **mesenchyme** cells. The archenteron will eventually squeeze the blastocoel out of existence.

Yolk affects gastrulation

In eggs that contain a moderate amount of yolk, such as those of amphibians, cleavage continues to produce many ever-smaller cells in the animal hemisphere and larger cells in the vegetal hemisphere. To initiate gastrulation, some cells in the gray crescent invaginate to form a slitlike blastopore (**FIGURE 35.14**). Small blastomeres of the animal pole then begin to roll like a sheet over the dorsal lip of the blastopore and push into the blastocoel. At the same time, cell divisions continue to increase the number of cells. Because the small cells of the animal hemisphere continue to divide more rapidly than the larger cells of the vegetal hemisphere, these smaller cells grow over and surround the larger cells of the vegetal hemisphere. As these smaller cells reach the blastopore, they too move inward so that the inward migration occurs all around the blastopore. These cells will form much of the endoderm and mesoderm of the developing embryo. The blastopore becomes the anus.

In very yolky eggs such as those of birds, in which the blastodisc forms on top of the yolk, the blastopore forms as a groove in the **primitive streak**, a collection of cells marked at the anterior end by **Hensen's node** (**FIGURE 35.15**). Cells that

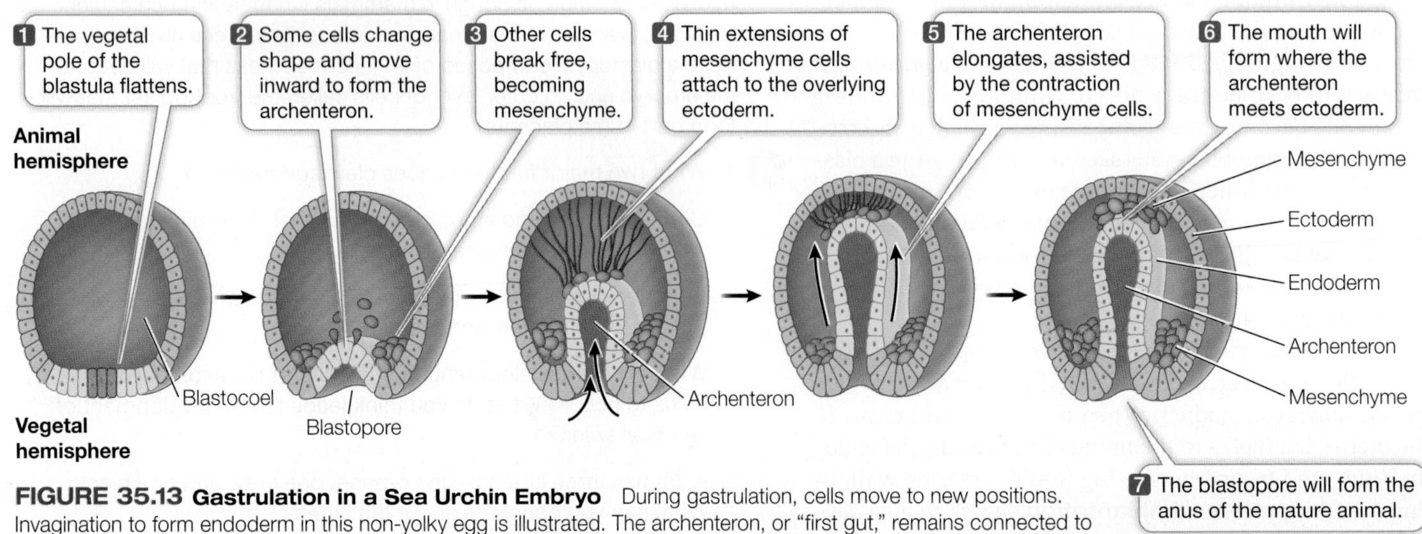

FIGURE 35.13 Gastrulation in a Sea Urchin Embryo During gastrulation, cells move to new positions. Invagination to form endoderm in this non-yolky egg is illustrated. The archenteron, or "first gut," remains connected to the exterior via the blastopore. Mesenchyme cells assist with gastrulation and will also form the temporary larval skeleton.

1 Gastrulation begins when cells in the region of the gray crescent move inward, forming the dorsal lip of the future blastopore.
2 Cells of the animal pole spread out, pushing surface cells below them toward and across the dorsal lip. These cells move into the interior of the embryo, where they form the endoderm and mesoderm.
3 The archenteron expands, eliminating the blastocoel. The blastopore lip forms a circle, with cells moving to the interior all around the blastopore; the yolk plug is visible through the blastopore.
Animal pole
Blastocoel
Vegetal pole
Dorsal lip of blastopore
Archenteron begins to form
Blastocoel displaced
Archenteron
Mesoderm
Dorsal lip of blastopore
Endoderm
Archenteron (future digestive tract)
Ectoderm
Mesoderm (notochord)
Dorsal lip of blastopore
Yolk plug
Ventral lip of blastopore

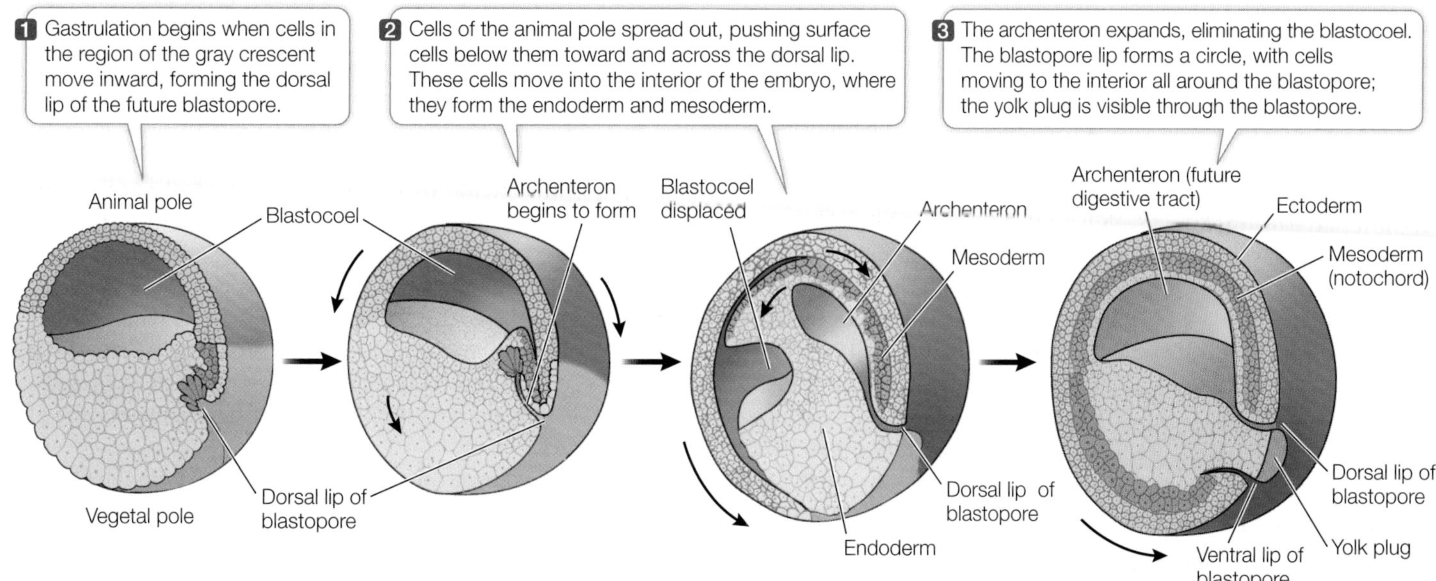

FIGURE 35.14 Gastrulation in a Frog Embryo The colors in this diagram are matched to those in Figure 35.5, the frog fate map. (After R. E. Keller. 1986. In *Developmental Biology: A Comprehensive Synthesis*, Vol. 2, L. Browder [ed.], pp. 241–327. Springer: New York.)

Animation 35.2 Gastrulation in a Frog Embryo
Media Clip 35.1 Frog Gastrulation Time-Lapse

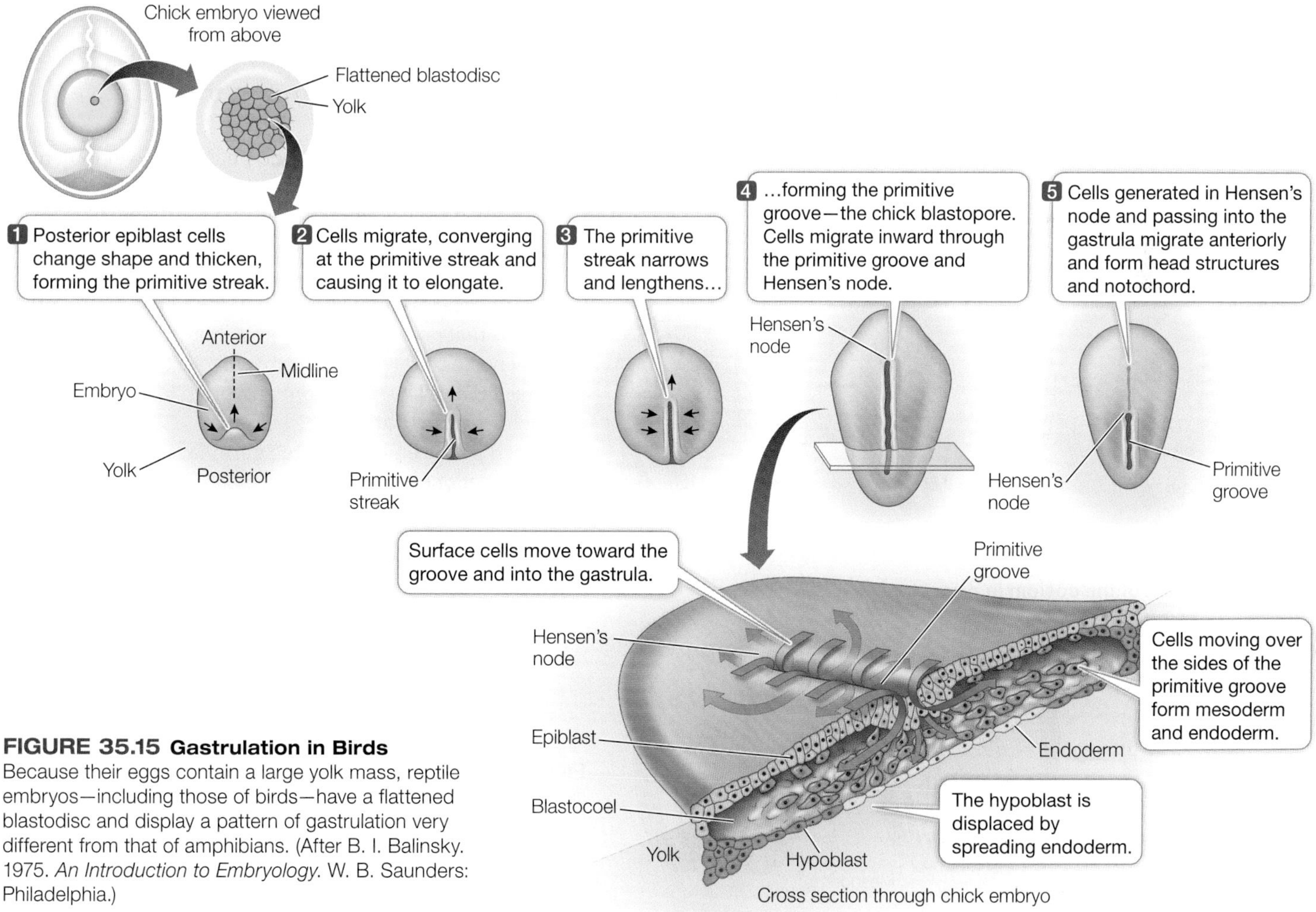

Chick embryo viewed from above
Flattened blastodisc
Yolk
1 Posterior epiblast cells change shape and thicken, forming the primitive streak.
2 Cells migrate, converging at the primitive streak and causing it to elongate.
3 The primitive streak narrows and lengthens…
4 …forming the primitive groove—the chick blastopore. Cells migrate inward through the primitive groove and Hensen's node.
5 Cells generated in Hensen's node and passing into the gastrula migrate anteriorly and form head structures and notochord.
Anterior
Midline
Embryo
Yolk
Posterior
Primitive streak
Hensen's node
Hensen's node
Primitive groove
Surface cells move toward the groove and into the gastrula.
Primitive groove
Hensen's node
Epiblast
Blastocoel
Yolk
Hypoblast
Endoderm
Cells moving over the sides of the primitive groove form mesoderm and endoderm.
The hypoblast is displaced by spreading endoderm.
Cross section through chick embryo

FIGURE 35.15 Gastrulation in Birds
Because their eggs contain a large yolk mass, reptile embryos—including those of birds—have a flattened blastodisc and display a pattern of gastrulation very different from that of amphibians. (After B. I. Balinsky. 1975. *An Introduction to Embryology*. W. B. Saunders: Philadelphia.)

will become mesoderm and endoderm migrate inward through Hensen's node and all along the groove of the primitive streak.

Placental mammals retain the avian/reptilian gastrulation pattern, even though they lack yolk. Mammals and reptiles, including birds, are all amniotes (i.e., they produce eggs that develop extraembryonic membranes), so it is not surprising that they share certain patterns of early development. In Key Concept 35.2 we discussed the development of the mammalian inner cell mass and the outer trophoblast. As in avian development, the inner cell mass of placental mammals splits into an upper layer called the epiblast and a lower layer called the hypoblast (see Figure 35.12C). The embryo forms from the epiblast, while the hypoblast contributes to the extraembryonic membranes that will encase the developing embryo and help form the **placenta**, the temporary organ that nourishes most mammalian embryos. Gastrulation occurs in the mammalian epiblast as it does in the avian epiblast. A primitive groove forms, and epiblast cells migrate through the groove to become layers of endoderm and mesoderm.

All three types of gastrulation described above can be found among the protostomes as well as deuterostomes, but recall that there is a major difference in the fate of the blastopore. In protostomes, the blastopore (the first opening) becomes the mouth and the anus forms at some distance away; in deuterostomes, the blastopore becomes the anus, and the mouth breaks through secondarily. This is why animals are called protostomes ("mouth first") or deuterostomes ("mouth second").

LINK For more on the differences between protostomes and deuterostomes, see **Key Concept 22.1**.

(A) Development of the coelom in deuterostomes

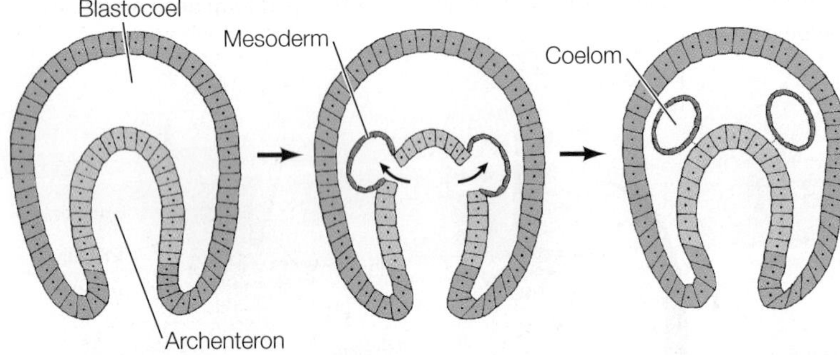

(B) Development of the coelom in protostomes

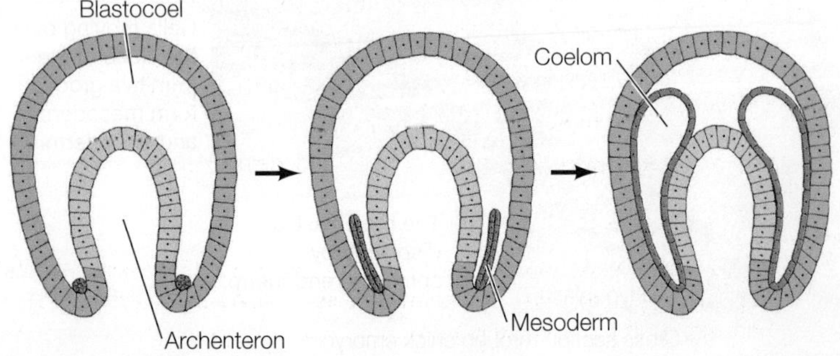

FIGURE 35.16 Formation of the Coelom Non-yolky eggs of deuterostomes and protostomes form their coeloms in different ways. In deuterostomes, the coelom often forms as an outpouching of the archenteron **(A)**. In protostomes, the coelom forms as a split in the mesoderm **(B)**. In all cases, the coelom displaces the blastocoel, which disappears as development proceeds.

A new body cavity, the coelom, forms

Gastrulation sets the stage for the formation of the **coelom** (pronounced see' loam), a space completely enclosed in mesoderm. Your body (or peritoneal) cavity—the space between the gut and body wall—is a coelom (see Key Concept 22.1). Your coelom allows muscles in your organs, like your heart and gut, to contract independently of your outer body wall. To understand the formation of the coelom, it is best to consider the mesoderm and coelom together.

In deuterostomes, the mesoderm often forms from the archenteron wall, and the coelom develops within in it, sometimes as a direct out-pouching of the archenteron (**FIGURE 35.16A**). This type of mesoderm and coelom formation is believed to be characteristic of deuterostomes, although it may be highly modified by yolk and other factors.

Protostomes show a different pattern that fits with the fact that specific blastomeres usually give rise to each tissue and organ (mosaic development). For example, in the snail *Ilyanassa*, and in many invertebrates, one recognizable small blastomere, 4d (the 4th small descendent of D; see Figure 35.6C), undergoes many divisions and forms most of the mesoderm. The coelom forms as a split in the prospective mesoderm cells (**FIGURE 35.16B**).

Whatever the mechanism of gastrulation, the process produces an embryo with three germ layers—ectoderm, endoderm, and mesoderm—that are positioned to influence one another through tissue interactions. The three germ layers and their interactions are responsible for morphogenesis, the process by which an animal achieves its final form or shape.

Organs develop from the three germ layers

Organogenesis refers to the phase in development in which all of the organs and organ systems form. We will discuss organs sequentially, but keep in mind that in an actual embryo the processes of organogenesis occur simultaneously. Here we focus on organogenesis in the chordates, the large clade that includes the vertebrates and all other animals that possess a notochord. Chordates also possess a dorsal nerve cord with a central cavity—a defining chordate characteristic. The nerve cord and brain form by a process known as **neurulation** (see below).

Each of the embryo's three germ layers gives rise to a defined set of tissues and organs:

- **Ectoderm**, the outer germ layer, forms the epidermis of the skin and structures such as hair, claws, and sweat glands, the brain and all of the nervous system, and pigment cells that provide color to skin, hair, feathers, and more.

- **Endoderm**, the innermost germ layer, forms the lining of the digestive tract and the organs that arise from it, among them the lungs, liver, gall bladder, and pancreas. Endoderm is responsible for the uptake of oxygen and nutrients into the body.
- **Mesoderm**, the middle layer that surrounds the coelom, gives rise to the notochord, heart, blood and blood vessels, urogenital system, muscles (including smooth muscles of the gut), many bones, and the dermis (inner layer of the skin). In vertebrates the dermis interacts with the overlying ectoderm (the epidermis) to produce many structures, including scales, teeth, hair, feathers, horns, sweat glands, and mammary glands.

Taking a big-picture view, the basic body plan can be seen as a tube within a tube. The outer tube, arising from ectoderm and mesoderm, is the body wall. The inner tube, arising from endoderm and mesoderm, is the gut. In between is the coelom. The major organs and organ systems develop as elaborations of this basic plan.

As development proceeds, cells become recognizable as a particular type (neurons, blood cells, liver cells, and so on). When a cell displays its final specialized characteristics, it is said to be differentiated. Differentiation takes place gradually throughout development. Before cells begin to express the traits characteristic of their differentiated state, they are often said to be determined—that is, the fate of the cell is now set, even though we do not recognize it.

The notochord induces formation of the neural tube

During gastrulation, cells that move to the interior form the endoderm and mesoderm of the embryo. In many chordates, some of the cells that move through the blastopore early in gastrulation separate from the roof of the archenteron (**FIGURE 35.17**) and form the notochord, a mesodermal structure. In amphibians, for example, these are cells from the dorsal blastopore lip. In addition to providing structural support, the notochord plays a critical role during development because it induces the overlying ectoderm to form the dorsal nerve cord—a process known as neurulation. **Induction** refers to the ability of one tissue to direct or lead another along a specific pathway of development. Inducers—signals that pass from cells in one tissue to those in another tissue—can determine the fate of the recipient cells. This is clearly seen during neurulation where signaling pathways have been revealed. Induction is one of the most important mechanisms in biology. It probably takes place in all animals and plays a role in the formation of most organ systems.

Under the influence of the developing notochord, a portion of the dorsal ectoderm flattens to form the neural plate. The edges

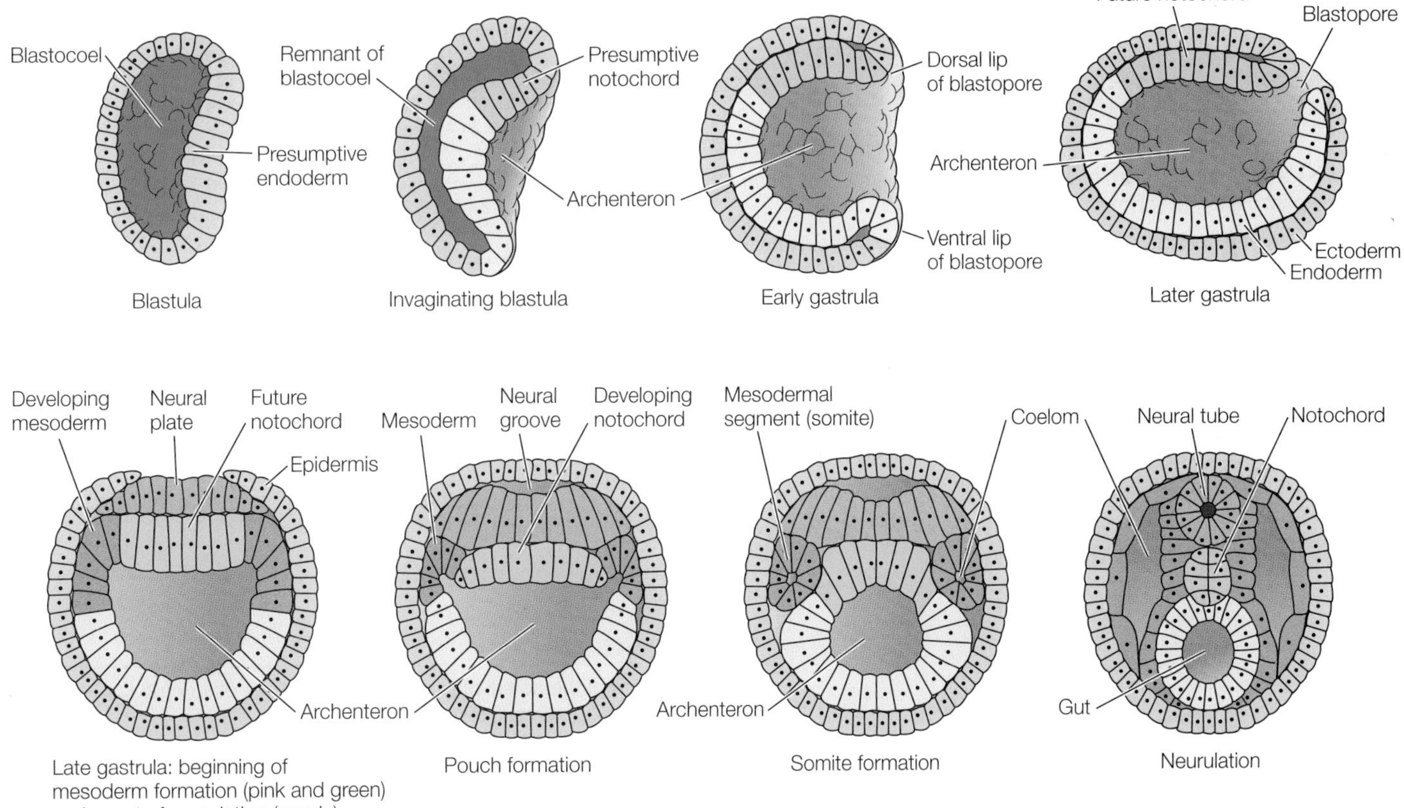

FIGURE 35.17 Gastrulation, Mesoderm Formation, and Neurulation in the Lancelet Amphioxus, a Non-Vertebrate Chordate In these classic descriptions (by Edwin Grant Conklin and Berthold Hatschek), the formation of tissue layers in a non-yolky chordate egg is clearly seen. The bottom row depicts sections of embryos cut at right angles to those in the top row.

of the flattened area begin to roll up so that a trough-like neural groove develops. The walls of the groove continue to expand dorsally and fuse when the edges reach each other, forming the neural tube. Ectoderm closes over them (see Figure 35.17). At its anterior end, the vertebrate neural tube forms three swellings, which will become the major divisions of the adult brain: the **hindbrain**, **midbrain**, and **forebrain**. The rest of the neural tube will become the spinal cord. If the neural tube does not close properly, serious problems result. Examples in humans include spina bifida (an opening in the spine that allows the spinal cord to protrude) and anencephaly (absence of a major portion of the brain).

Much of our understanding of induction comes from work that began on amphibian embryos in the early twentieth century and has continued since. When Hans Spemann and his student Hilde Mangold were studying the dorsal lip of the blastopore in salamander eggs, they transplanted this region to another embryo at the same stage of development. The results were momentous. This small piece of tissue stimulated a second site of gastrulation, and a second complete embryo formed (**INVESTIGATION FIGURE 35.18**). Because the dorsal blastopore lip of amphibians was apparently capable of inducing host tissue to form an entire embryo, Spemann and Mangold dubbed it the **organizer**. For more than 90 years, the organizer has been an active area of research. Hensen's node acts as an organizer in birds, and a homologous region does so in mammals.

We now know that the protein β-catenin plays a key role in determining which cells become the organizer. It does this by triggering a complex series of interactions between various transcription factors and growth factors that control gene expression. Several of these transcription factors have been identified.

Now, modern techniques are enabling scientists to understand more about inductive interactions. One of the best-studied examples is the induction of the dorsal nerve cord in amphibians. Much of development is controlled by molecular switches that

INVESTIGATION

FIGURE 35.18 The Dorsal Lip Induces Embryonic Organization In a classic experiment, Hans Spemann and Hilde Mangold transplanted the dorsal blastopore lip mesoderm of an early-gastrula-stage salamander embryo. The results showed that the cells of this embryonic region, which they dubbed "the organizer," could direct the formation of an entire embryo.[a]

HYPOTHESIS

Cytoplasmic factors in the early dorsal blastopore lip organize cell differentiation in amphibian embryos.

METHOD

1. Excise a patch of tissue from above the dorsal blastopore lip of an early-gastrula-stage salamander embryo (the donor).
2. Transplant the donor tissue onto a recipient embryo at the same stage. The donor tissue is transplanted onto a region of ectoderm that should become epidermis (skin).

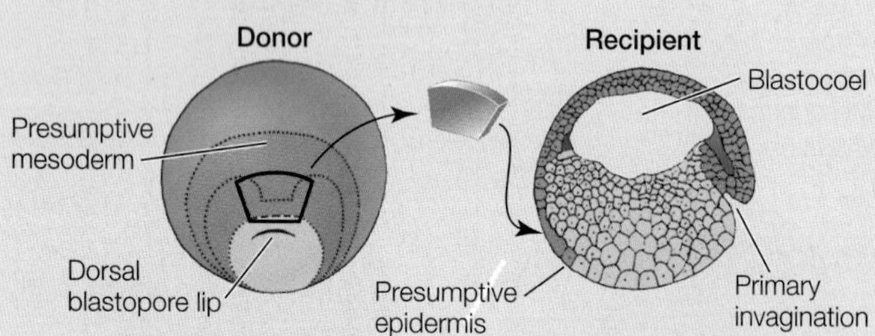

RESULTS

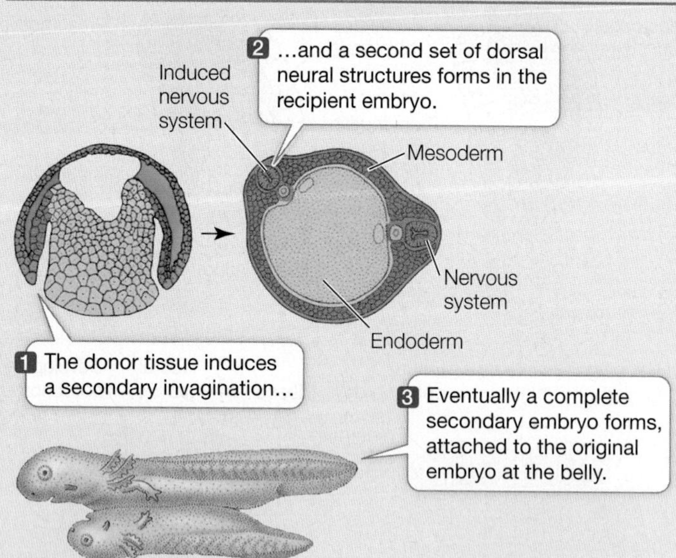

CONCLUSION

The cells of the dorsal blastopore lip can induce other cells to change their developmental fates.

Activity 35.1 Cell Fates Simulation

Animation 35.3 Tissue Transplants Reveal the Process of Determination

[a]H. Spemann and H. Mangold. 1924. *Roux' Arch Entw Mech* 100: 599–638. Viktor Hamburger's translation appeared in *Foundations of Experimental Embryology*, 1964 (B. H. Willier and J. M. Oppenheimer, eds.), pp. 146–184.

A similar **Analyze the Data** exercise may be assigned in **Achieve**.

THINK LIKE A SCIENTIST

Differentiation can be due to inhibition of transcription factors

Spemann and Mangold demonstrated that cells from the dorsal lip of the blastopore are able to induce the formation of dorsal neural structures. When these organizer cells roll under the dorsal ectoderm along the embryo midline, becoming mesoderm, the overlying ectoderm becomes neural tissue rather than skin (epidermis). Other experiments have demonstrated that when animal cap cells (presumptive ectoderm excised from the "cap" of a blastula) are isolated and maintained in a dispersed state, they become neural tissue, not skin. So do the organizer cells *cause* dorsal ectoderm to become neural tissue, or do they *prevent* this ectoderm from becoming skin? Think of how you might use the process of science to test these options. There are different ways, but in the experiment shown in **FIGURE 35.19**, epidermal caps from amphibian eggs were isolated and exposed to a growth factor (BMP4) that is released by the mesoderm of the forming notochord.

Use the gel results shown in step 3 of Figure 35.19 to answer the questions.

1. Does BMP4 induce expression of neural-specific or epidermal-specific genes?
2. Does BMP4 block any gene expression in the ectodermal cells?
3. What is the evidence that BMP4 has an inductive and/or an inhibitory effect on gene expression in ectodermal cells?
4. Do these results support the hypothesis that the default state of amphibian dorsal ectoderm is neural?

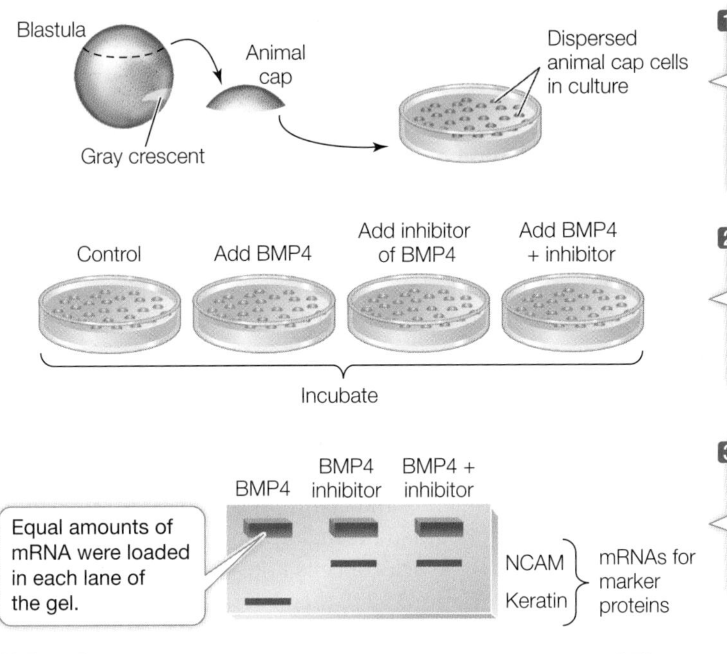

FIGURE 35.19 The Role of Induction in the Differentiation of Tissues
(After P. A. Wilson and A. Hemmati-Brivanlou. 1995. *Nature* 376: 331–333.)

allow a cell to proceed down one of two alternative paths. Will a cell become epidermis or nerve cell? The development of the central nervous system depends on the activation of a specific genetic cascade within certain ectodermal cells that will become neuronal instead of epidermal. Work on amphibian embryos has shown that specific polypeptides released from the developing notochord act as inducers that control this activation. Two of these polypeptides that induce the formation of the neural tube are appropriately named Noggin and Chordin. They act not by direct activation of genes, but by inhibiting the activity of another substance, the TGF-β bone morphogenetic protein BMP4. Among its many actions, BMP4 induces ectoderm to form epidermis. In the absence of BMP4, ectoderm forms neural tissue. By blocking the effect of BMP4, Chordin and Noggin allow the ectoderm to carry out its "default" differentiation of becoming neural.

Another important factor controlling differentiation of the neural tube is Sonic hedgehog, symbolized by Shh. Shh is released by the notochord and diffuses into the ventral region of the neural

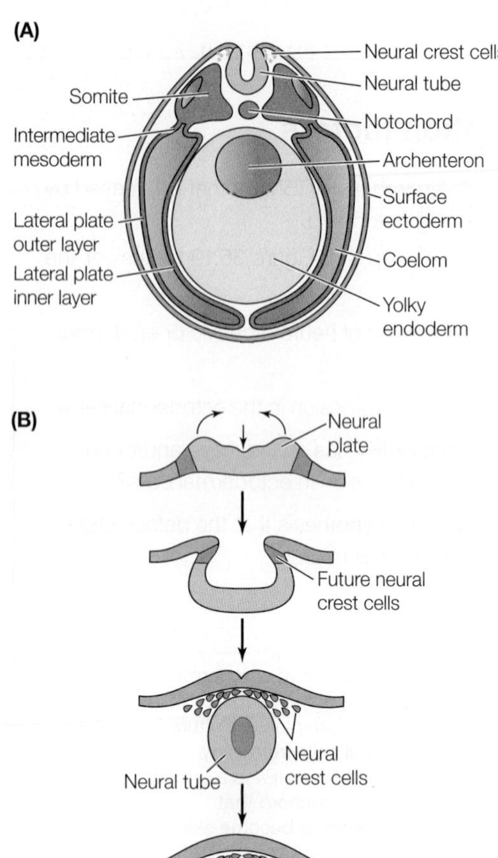

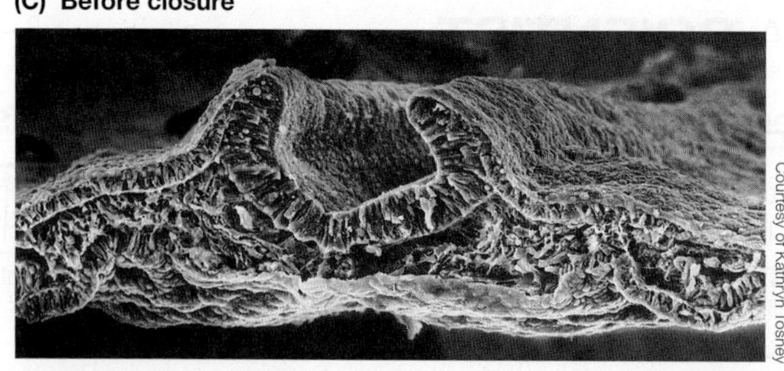

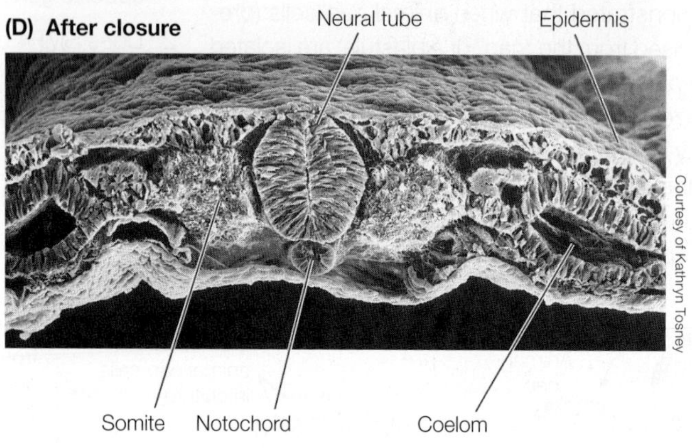

FIGURE 35.20 Neurulation and Differentiation of Mesoderm in Vertebrates **(A)** This illustration of a frog egg shows how the neural folds draw together and close, releasing neural crest cells. The mesoderm extends between the ectoderm and endoderm. The three divisions of trunk mesoderm—somite, intermediate mesoderm, and lateral plate mesoderm—are apparent. **(B)** At the start of vertebrate neurulation, the ectoderm of the neural plate is flat. The neural plate invaginates and folds, forming a tube. Neural crest cells are released. **(C, D)** Scanning electron micrograph of the neural tube in a chick, before and after closure.

tube, where it directs the development of this region into the ventral structures and circuits of the spinal cord. These circuits will carry the major output from the motor pathways of the nervous system. Shh also plays an important role in the development of vertebrate limbs, as we will see shortly.

As the neural tube is closing, certain cells dissociate from it and come to lie between the neural tube and the overlying epidermis. Known as **neural crest cells** (**FIGURE 35.20**), these star-shaped mesenchymal ("wandering") cells are termed "multipotent" because they can differentiate in many different ways. Neural crest cells migrate extensively and have amazing developmental capacities. Not only do they form sensory neurons and major parts of the autonomic nervous system, but they are also responsible for most of the skull bones, the pigment cells, and many other structures. Without neural crest cells, vertebrates would not have heads as we recognize them.

Mesoderm forms tissues of the middle layer

On either side of the forming notochord, three regions of mesoderm can be recognized—**somites**, **intermediate mesoderm**, and **lateral plate mesoderm**. The vertebrate body plan, like that of annelids and arthropods, consists of repeating segments that are modified during development. These segments are first seen as the blocks of cells known as somites (**FIGURE 35.21**; also see Figure 35.20), then later as the repeating patterns of vertebrae, ribs, nerves, and muscles along the anterior–posterior axis as differentiation proceeds.

Somites form bones, cartilage, skeletal muscle, and the dermis of the skin. Intermediate mesoderm forms much of our urinary and reproductive system. Lateral plate mesoderm surrounds the coelom and lines the vertebrate body cavity (the peritoneal cavity). The peritoneal cavity makes up much of the coelom. The thin tissue layer that surrounds the peritoneal cavity is the **peritoneum**. (You may have heard of peritonitis, a severe inflammation of the covering of the internal organs. A major cause is the entry of bacteria into the peritoneal cavity, as occurs when an appendix ruptures.) The inner layer of lateral plate mesoderm lies close to the endoderm. It forms part of the peritoneum, the muscles of the digestive tract, and most of the circulatory system, including the heart. The outer layer of lateral plate mesoderm lies adjacent to the ectoderm and forms part of the peritoneum and some of the muscle of the body wall and limbs.

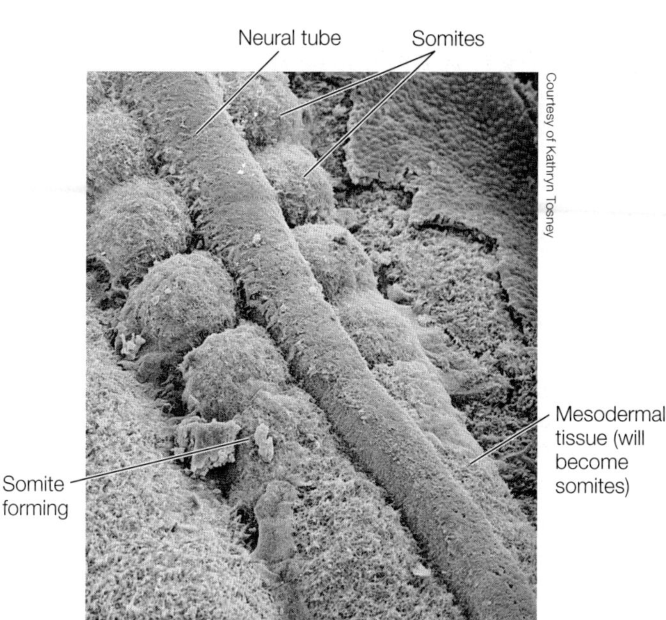

FIGURE 35.21 Body Segmentation In this scanning electron micrograph of somite formation in a chick embryo, the overlying ectoderm has been removed and the neural tube and somites are seen from above. Skeletal muscle, cartilage, bone, and the lower layer of the skin (the dermis) form from the somites.

Positional information guides vertebrate limb formation

During development, the key cellular question "What am I (or what will I be)?" is often answered in part by "Where am I?" This spatial "sense" is called positional information. Positional information often comes in the form of an inducer called a **morphogen**, which diffuses from one group of cells to surrounding cells, setting up a concentration gradient. There are two requirements for a signal to be considered a morphogen:

1. It must specifically affect target cells.
2. Different concentrations of the signal must cause different effects.

An example can be seen in the development of vertebrate limbs.

The vertebrate limb develops from a paddle-shaped limb bud (**FIGURE 35.22A**). The cells that develop into different digits must receive positional information; if they do not, the limb will not be organized properly (imagine a hand with only thumbs or only little fingers). How do the cells know where they are? A group of cells at the posterior base of the limb bud, just where it joins the body wall, is called the **zone of polarizing activity** (**ZPA**). The cells of the ZPA secrete Sonic hedgehog (Shh), which forms a gradient that determines the posterior–anterior (little finger to thumb) axis of the developing limb. The cells getting the highest dose of Shh form the little finger; those getting the lowest dose develop into the thumb.

You might expect morphogenesis to involve a lot of cell division, followed by differentiation—and it does. But what you might not expect is the amount of programmed cell death—**apoptosis**—that occurs during morphogenesis. For example, in an early human embryo, the hands and feet look like tiny paddles: the tissues that will become fingers and toes are linked by connective tissue. Between days 41 and 56 of development, the cells between the digits die, freeing the individual fingers and toes (**FIGURE 35.22B**). (In the webbed feet of ducks, this programmed cell death does not occur.) Many cells and structures form and then disappear or are reshaped during development, in processes involving apoptosis.

(A) A morphogen gradient determines the anterior–posterior axis in hands

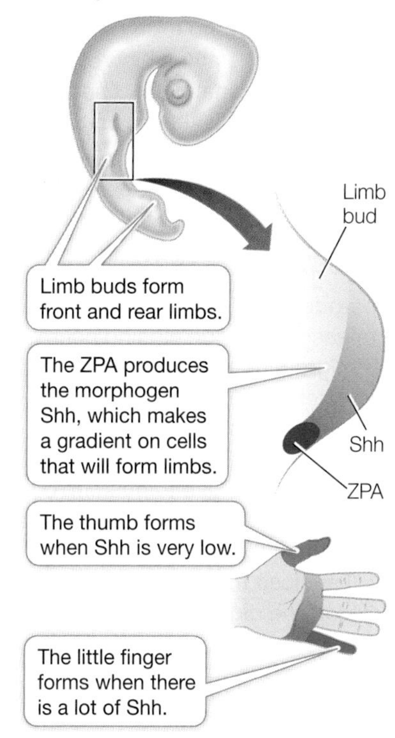

(B) Digits are formed by localized apoptosis

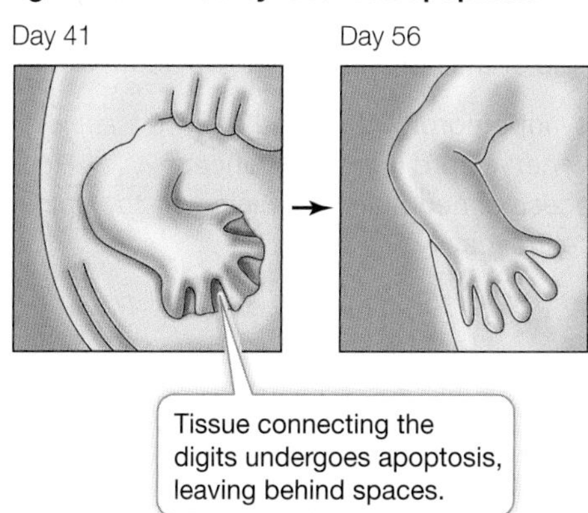

FIGURE 35.22 Vertebrate Limb Development Limbs appear as small buds and develop into paddle-shaped structures. **(A)** A gradient of Shh produced by the zone of polarizing activity (ZPA) determines the anterior–posterior axis. **(B)** Programmed cell death (apoptosis) plays a role in shaping the digits.

REVIEW & APPLY | 35.3

R Gastrulation involves massive cell movements during which cells move from the outside to the inside of the embryo. These movements result in three germ layers and place cells from various regions of the blastula in new associations with one another. Cells become differentiated by responding to signals from within their own cytoplasm that activate specific parts of the genome and by responding to signals received from other cells—a process known as induction. Both types of signaling processes probably occur and are important in all animals. These signals set off genetic cascades that lead cells down specific developmental pathways so that they become different from one another. Organogenesis is the process whereby tissues interact to form organs and organ systems. The dorsal lip of the amphibian blastopore is a critical site for cell determination and has been called the organizer. Hensen's node is the organizer during avian development. In the formation of the chordate nervous system, one group of cells that migrates through the blastopore becomes mesodermal tissue and forms the notochord. Ectoderm overlying this mesoderm becomes either epidermis or neural tissue. The notochord induces the overlying ectoderm to form the neural tube. Neural crest cells are released from the closing neural tube. Neural crest cells migrate to different regions and form many parts of the vertebrate body—the sensory nervous system, many bones of the head, pigment cells, and other structures. Mesoderm on either side of the notochord forms somites that will give rise to muscle, bone, dermis, and other tissues; intermediate mesoderm that will form urinary and reproductive tissues; and lateral plate mesoderm that will enclose the coelom and give rise to the heart, muscles of the digestive system, and other tissues.

A

1. What is the blastopore, and what does it become in protostomes? What does it become in deuterostomes?
2. Why is the primitive streak in the chick embryo the equivalent of the blastopore?
3. What is the role of the notochord in neurulation?
4. A chordate synapomorphy (a shared derived trait) is a dorsal hollow nerve cord (see Key Concept 22.5). How does the process of neurulation explain why the spinal cord has a central cavity?

We've seen now how the three germ layers form the animal. Now we will focus on how also the germ layers protect and nourish the embryo in many of the animals with which you are most familiar.

Activity 35.2 Stages of Animal Development

35.4 | Extraembryonic Membranes Protect and Nourish the Embryo

LEARNING OBJECTIVES

By the end of this key concept you should be able to:

35.4.1 Explain the evolutionary significance of the amniotic egg.

35.4.2 Describe the formation and function of the four extraembryonic membranes in reptiles (including birds) and mammals.

35.4.3 Explain the significance of the four extraembryonic membranes in mammalian development.

35.4.4 Compare the formation of a yolk sac in a mammal and a fish.

A major step in the evolution of tetrapods (four-limbed vertebrates) was the formation of the amniotic egg, seen today in reptiles, including birds, and mammals. These animals are amniotes, and it is a result of the amniotic egg that they are able to live on land so successfully without returning to the water to reproduce. As we saw in Key Concept 22.6, the amniotic egg provides the embryo with a contained aqueous environment, freeing the processes of reproduction and development from dependence on an external water supply. The three germ layers that form the embryo also create the membranes that provide this protective, self-contained environment.

Extraembryonic membranes form with contributions from all germ layers

Whether they develop inside or outside the mother's body, embryos of amniotes are assisted by four **extraembryonic membranes** that function in protection, nutrition, gas exchange, and waste removal. The chicken embryo provides a good example, but the process is similar in other reptiles and in egg-laying mammals. (We discuss placental mammals below.) The four membranes are the yolk sac, allantois, amnion, and chorion (**FIGURE 35.23**).

1. The **yolk sac** forms first, by extension of the endoderm and the inner layer of lateral plate mesoderm (see Figure 35.23). These two layers grow around the yolk and enclose it. The inner layer of lateral plate mesoderm is special because it can form blood cells and blood vessels. The first blood cells of the embryo and the vessels that carry them—the vitelline artery and vein and their branches—form in the mesoderm of the yolk sac. Stored yolk is not transferred directly to the gut of the embryo through the yolk stalk. Rather, it is digested by the endoderm of the yolk sac, and the products are transported to the embryo by the developing blood and blood vessels in the yolk sac wall (**FIGURE 35.24**).
2. The **allantois**, a sac for storage of metabolic wastes, is also an outgrowth of endoderm and the inner layer of lateral plate

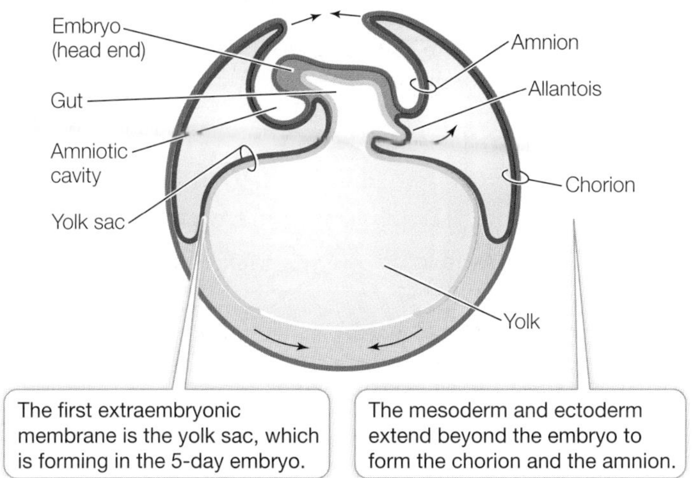

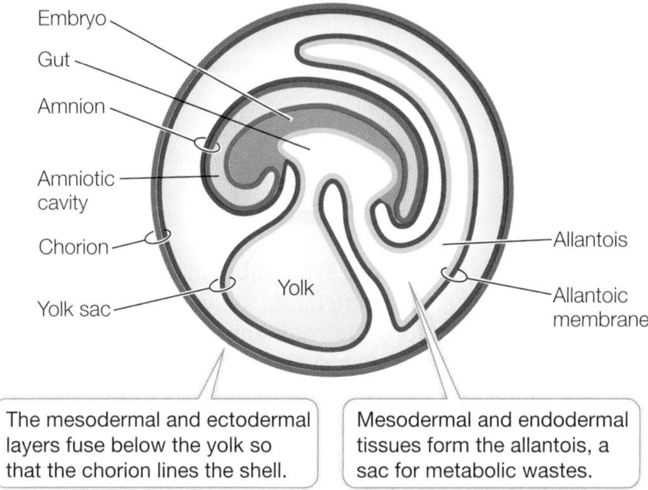

FIGURE 35.23 The Extraembryonic Membranes of Amniotes In birds, other reptiles, and mammals, the embryo constructs four extraembryonic membranes. **(A)** Developing amnion and chorion form from ectoderm and mesoderm, as seen in this 5-day chick embryo. The membranes will fuse when they meet dorsally (top arrows), enclosing the embryo in two envelopes. The yolk sac and allantois form from endoderm and mesoderm. The yolk sac grows around the yolk (arrows), and the allantois forms a pouch (single arrow at right). **(B)** Membranes surround the embryo in this 9-day embryo. Some will continue to expand as the embryo grows, but the yolk sac shrinks as yolk is used by the embryo. Fluids secreted by the amnion fill the amniotic cavity, providing an aqueous environment for the embryo. The allantois stores the embryo's waste products. The allantoic membrane will become pressed against the chorion and shell and functions in gas exchange between the embryo and its environment.

Activity 35.3 Extraembryonic Membranes

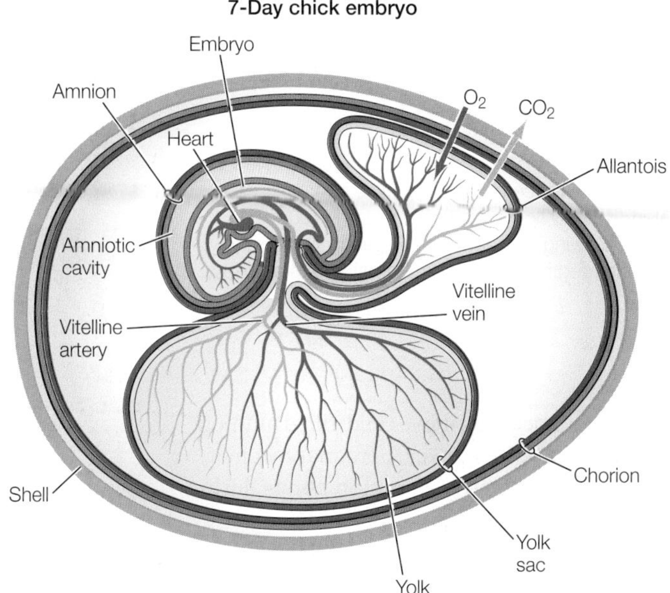

FIGURE 35.24 Development of the Circulatory System in a 7-Day Chick Developing chick in shell showing development of the circulatory system, which is well underway by day 7. Only the yolk sac and the allantois are capable of making blood vessels. As a result, they are involved in nutrient transport and gas exchange.

mesoderm (see Figures 35.23 and 35.24). Because the yolk sac and the allantois are the extraembryonic membranes that contain the inner layer lateral plate mesoderm, they are the only two that develop blood vessels (see Figure 35.24).

3. The **amnion** surrounds the embryo and secretes fluid into the enclosed cavity, providing a protective, shock-absorbing environment. Ectoderm and the outer layer of the lateral plate mesoderm grow out from the embryo and surround it (see Figure 35.23A). Where they meet, they fuse, forming two membranes, an inner amnion and outer chorion (see Figure 35.23B).
4. The **chorion**, located just beneath the egg's shell, forms a continuous membrane that limits water loss. The chorion cannot make blood vessels, but it fuses with the blood vessel–forming allantoic membrane, forming the chorioallantoic membrane that provides for exchange of O_2 and CO_2 between the embryo and the outside world (see Figure 35.24).

Extraembryonic membranes in mammals form the placenta

In eutherian mammals, the entire trophoblast becomes embedded in the endometrium of the uterus (see Figure 35.12) and forms the chorion. Hypoblast cells proliferate to form what in birds would be the yolk sac, even though there is little or no yolk in the eggs of placental mammals. Cells that would form the allantois in birds join with the chorion to form the embryonic portion of the placenta. The chorion and allantoic cells interact with maternal tissue of the uterine wall (the endometrium) to produce the placenta. Note that the placenta is a unique organ because it is composed of tissues from two organisms—the mother and her offspring. Interestingly, the yolk sac in mammals continues to be an early site of blood cell production as it is in birds, even though no or very little yolk is present. Meanwhile, the amnion grows around

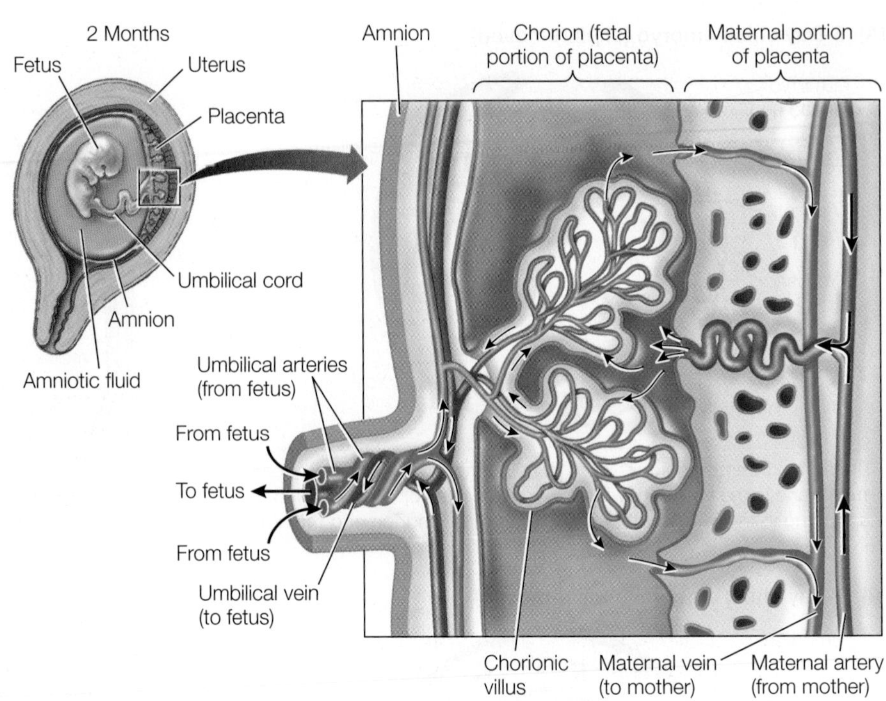

FIGURE 35.25 The Mammalian Placenta In humans and other placental mammals, nutrients and wastes are exchanged between maternal and fetal blood in the placenta. This organ forms from both fetal tissue and tissues of the mother's uterine wall. The embryo is attached to the placenta by the umbilical cord. Embryonic blood vessels invade the placental tissue to form fingerlike chorionic villi. Maternal blood flows into the spaces surrounding the villi, allowing nutrients and respiratory gases to be exchanged between the maternal and fetal blood.

the embryo, enclosing it in a fluid-filled amniotic cavity (**FIGURE 35.25**). (A pregnant woman's "water breaks" when the amnion bursts during labor and releases the amniotic fluid.)

Human gestation is traditionally divided into three periods of roughly 12 weeks each, called trimesters. The first trimester is a time of rapid cell division and tissue differentiation, and the embryo is very sensitive during this time to damage from radiation, drugs, chemicals, and pathogens that can cause birth defects. By the end of the first trimester, most organs have started to form, and the embryo is about 8 centimeters long. At about this time, the human embryo is medically and legally referred to as a **fetus**.

Fishes also make yolk sacs

Many fishes have a very yolky egg (see Figure 35.10B). Fishes are not amniotes, however, so how do they obtain nutrients from this stored material? Embryonic fishes produce a yolk sac, but it differs from that of amniotes. As the fish embryo forms, all three germ layers—ectoderm, mesoderm, and endoderm—grow around the yolk. The yolk sac becomes vascularized as in amniotes, and materials are carried in the blood vessels to the embryo (**FIGURE 35.26**).

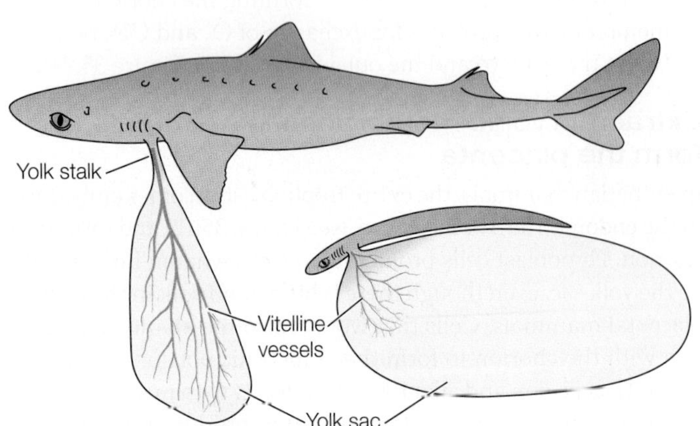

FIGURE 35.26 Fish Yolk Sac The yolk sac in this developing shark is trilaminar because it is composed of all three tissue layers. Contrast this with the yolk sac of mammals, which forms from just two layers and thus is bilaminar.

REVIEW & APPLY | 35.4

R Amniotic eggs contain four extraembryonic membranes. In reptiles, including birds, the yolk sac surrounds the yolk and provides nutrients to the embryo. The chorion lines the eggshell and participates with the allantois in gas exchange. The amnion surrounds the embryo and encloses it in an aqueous environment. The allantois is a storage sac for metabolic wastes and participates with the chorion in gas exchange (as part of the chorioallantoic membrane). Only the yolk sac and the allantois can develop blood vessels, and the first blood cells and blood vessels appear in the yolk sac. In eutherian mammals, the chorion interacts with maternal uterine tissues to form a placenta, which provides the embryo with nutrients and gas exchange. The amnion surrounds the embryo, bathing it in amniotic fluid and a yolk sac forms.

A

1. Would you expect a frog embryo to have the same extraembryonic membranes as a lizard? Why or why not?
2. What is the function of the blood vessels that form in the yolk sac of birds?
3. The yolk sac of fishes is referred to as being trilaminar. Why is this term appropriate? How does a fish yolk sac differ from a chick yolk sac?
4. What outstanding feature of the placenta makes it a unique organ?

We have now seen how a single-celled zygote becomes a complex organism, but development does not stop at birth or hatching. Animals must grow. For some, growth stops when they reach adulthood. Others grow throughout their lives. In either case, this growth is part of development. We will consider post-embryonic growth next.

35.5 Development Continues throughout Life

LEARNING OBJECTIVES

By the end of this key concept you should be able to:

35.5.1 Describe how growth can lead to a change in body shape and proportions.

35.5.2 Describe the difference between direct and indirect development and explain the two major functions that a larval stage serves.

35.5.3 Explain the need for continued cell divisions even in organisms in which body size is not enlarging.

35.5.4 Explain why changes in gene expression underlie cell fate determination and differentiation.

35.5.5 Appreciate the potential contributions and the concerns regarding cloning and stem cell research.

No sexually produced animal is as large as its parent at birth or hatching. All must obtain nourishment and grow. Some animals immediately look like their parents and undergo **direct development** to reach full size. Others appear very different and spend part of their lives as one or more different forms, as they undergo **indirect development**. All must continue to make new cells throughout their lives.

Offspring of some animals undergo direct development

Some animals appear very similar to their parents when they are born or emerge from eggs. Such animals are said to undergo direct development. In contrast, animals that undergo a larval stage and/or metamorphosis are said to undergo indirect development. **Allometry**—a change in the proportions of various parts of an organism relative to one another as a consequence of unequal growth—is often seen in direct developers. Isometric growth is a 1:1 size increase; allometric growth is any change from this. For example, at birth a human baby is not a miniature adult. Not only are many of its systems immature, but its overall morphology (form and shape) is quite different than that of an adult. One difference is the ratio of head to trunk size. A baby's head is much larger in comparison with the rest of its body than an adult's head is, and the head stops growing much earlier in life than the rest of the body (**FIGURE 35.27**). It has been suggested that the large head of infant mammals, especially humans, corresponds to the large brain and development of the cerebral cortex. Many examples of allometry exist. Indeed, nearly all direct developers experience allometric growth of some sort.

Offspring of other animals undergo indirect development

Indirect development involves changes in morphology that can be dramatic. Many animals, both protostome and deuterostome, have one or more larval stages that may be quite unlike the adult. A familiar example is the monarch butterfly, which has a caterpillar that feeds on the milkweed plant on which its egg was placed, and a

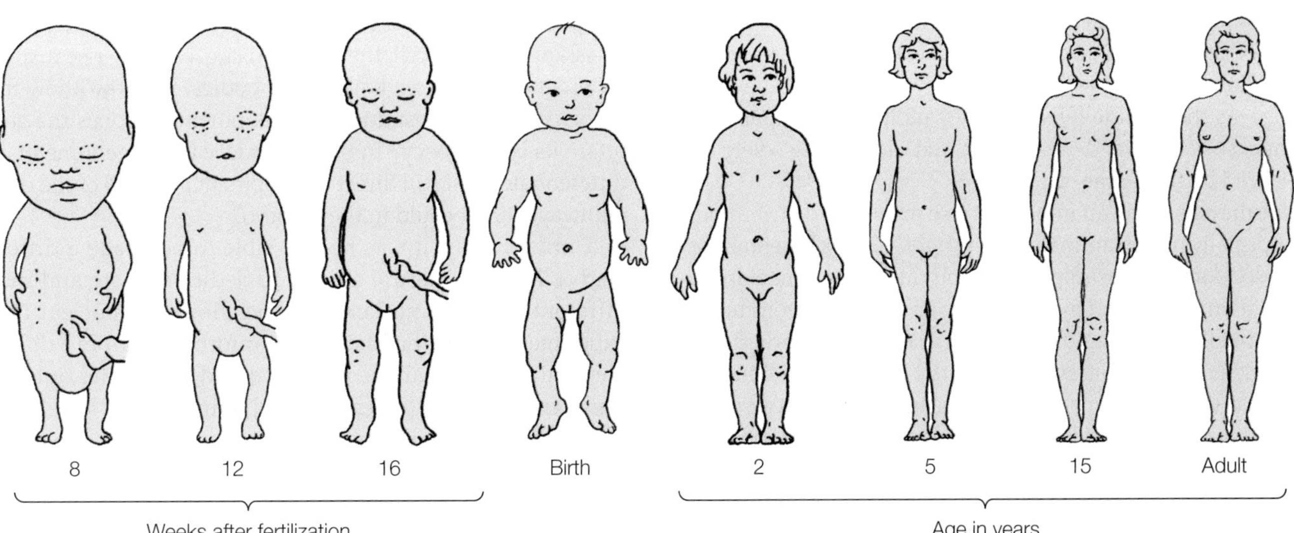

FIGURE 35.27 Allometric Growth of a Human Proportions of body parts change as an individual develops from fetus to adult.

(A) Shore crab (*Carcinus maenas*)

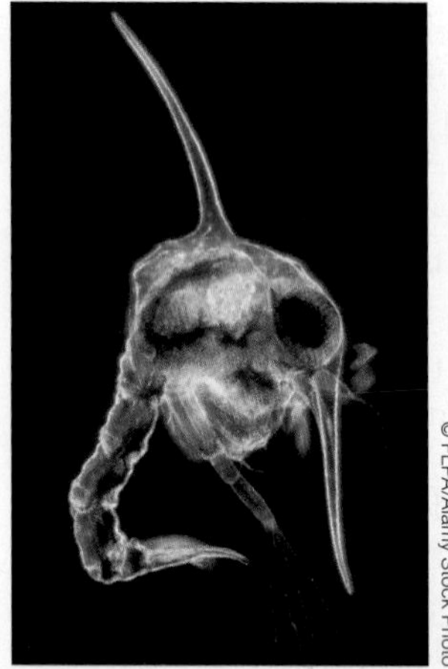

© FLPA/Alamy Stock Photo

© Sabena Jane Blackbird/Alamy Stock Photo

(B) Sea star (*Asterias rubens*)

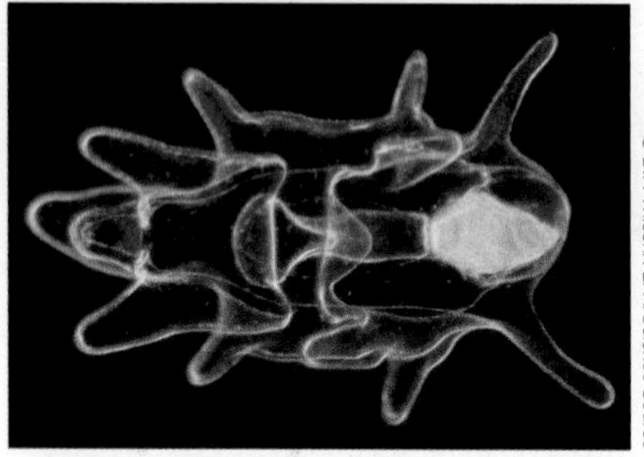

© D.P. Wilson/FLPA/Science Source

© Arterra Picture Library/Alamy Stock Photo

FIGURE 35.28 Larval Forms Are Often Quite Unlike Their Parents Here we show larval (left) and adult (right) forms of **(A)** a crab (a protostome) and **(B)** a sea star (a deuterostome).

winged adult that can migrate from Canada to Mexico. **FIGURE 35.28** shows two larvae with which you might not be familiar.

A larval stage usually performs one of two important roles—or both. Like a caterpillar, it may be primarily a feeding stage. This is seen in animals that provision their eggs with little or moderate yolk resources. Insects, fishes, and amphibians are prime examples of this lifestyle. Some adult insects do not feed at all. Ephemeridae—mayflies—live only 1 day as adults. (They are ephemeral, as their name tells us.) They molt from the pupal to adult stage, seek a mate, deposit eggs, and die, relying solely on energy stored by the larvae.

Larvae may also play an important role in species distribution. This is especially true in animals in which the adult is sedentary or relatively so. Corals, for example, can reproduce by fragmentation, but new sites are colonized by swimming larvae. Some polychaete worms (capitellids) are rapid colonizers of freshly disturbed, organically rich areas (substrates of polluted harbors and beneath fish farms, oil spills, and whale falls—decaying whale carcasses on the ocean floor) that are widely distributed around the world. The nonfeeding larvae, swimming only with ciliary bands and using only energy reserves deposited in eggs by their mother, are carried for miles by currents and then settle in these areas.

When a larva transforms into an adult at metamorphosis, changes can be dramatic, as in the monarch and other butterflies and moths. Major portions of the body may be remodeled or even discarded (see Key Concept 22.4). Many frogs lay their eggs in spring, and after quite rapid development, small tadpoles emerge. A tadpole feeds, and its size increases. At metamorphosis, frog tadpoles show changes in virtually every system. Tail and gills are resorbed, and limbs grow. In many species, the tadpole is herbivorous whereas the adult is carnivorous. Accompanying this change is an alteration of gut length. The gut of a tadpole is actually several times longer than the gut of the small froglet the tadpole will become! The tadpole of a sea squirt (a tunicate; see Figure 22.41) settles and rapidly resorbs its tail, including the notochord and nerve cord. Since the three major chordate characteristics (notochord, hollow dorsal nerve chord, and post-anal tail) are no longer present, the adult tunicate does not resemble a chordate! The larva of the lamprey is so different from the adult that it was once considered a different species. It is still called by the earlier name of "ammocoetes."

Determination precedes differentiation

When a cell is recognizable as a skin cell, a liver cell, a red blood cell, and so on, we say that it is differentiated. Usually, differentiated cells no longer undergo mitotic divisions. Even before the cell is recognizable as a certain cell type, it becomes committed to follow a specific developmental pathway—it is said to be determined. This can be easily shown by transplanting the cell (or cells) to a different location in the body. Will the cell's "old address" or its new one control its differentiation? If the cell differentiates as it would have before transplantation, the cell was already determined prior to transplantation. How fixed are determination and differentiation? Are the fates of determined or differentiated cells completely locked in, or can they be altered? We began the chapter by asking how one cell, the fertilized egg, can give rise to all of the different cell types we have in our bodies. We now know that no chromosomes are lost during development, but does the genome retain its equivalency? In other words, can the genome of a fully differentiated cell still direct development along all of the different pathways that it could in the embryo?

Until recently, it was not possible to persuade a differentiated cell from a mature animal to dedifferentiate and then redifferentiate into another cell type. However, nuclear transfer experiments show that the genetic information from a differentiated animal cell remains unaltered and, under the right conditions, can be used to create cloned animals. You have probably heard of Dolly the sheep, the first mammal to be experimentally cloned. To clone animals, the nucleus from an unfertilized egg is removed and discarded, leaving an enucleated "host" egg (**FIGURE 35.29**). A donor nucleus from a somatic cell of the animal to be cloned is then introduced into the "empty" egg. Remember, this is a diploid nucleus, so no sperm nuclear material is necessary. However, the egg must still be activated, a process normally

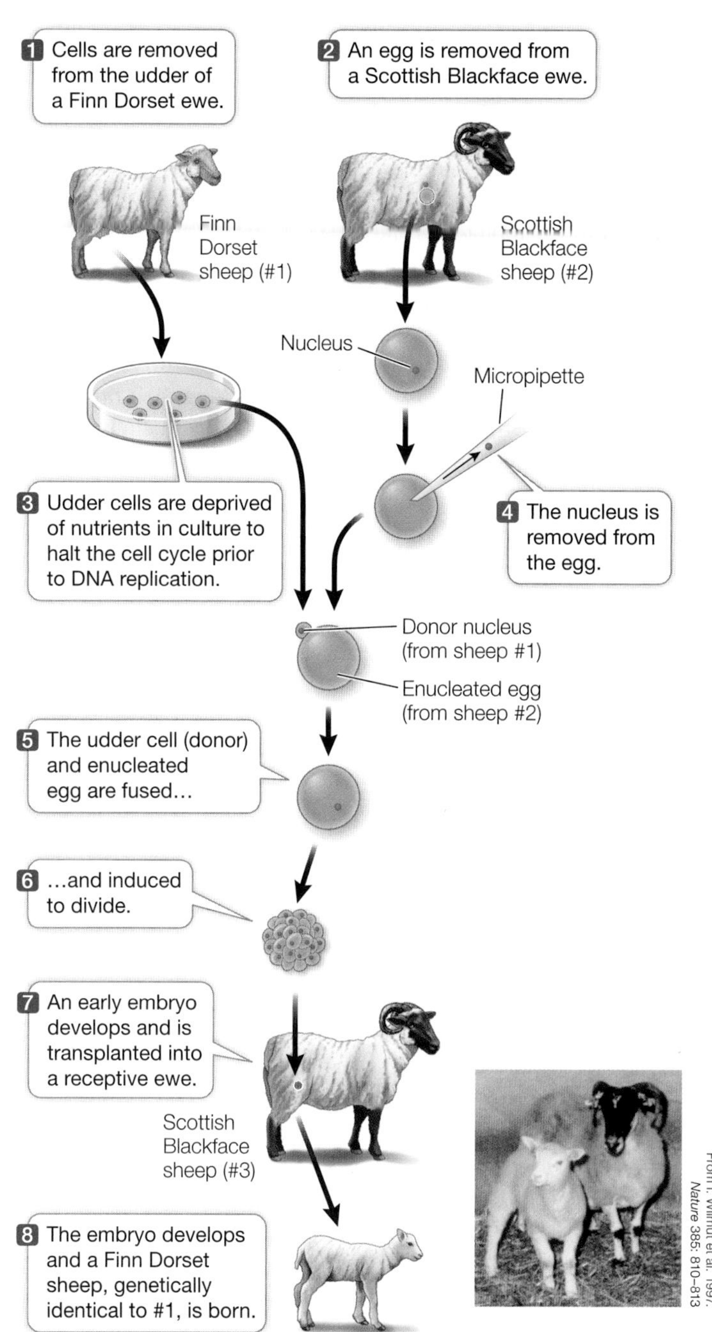

FIGURE 35.29 Cloning a Mammal The experimental procedure described here produced the first cloned mammal, a Finn Dorset lamb named Dolly (shown on the left in the photo with her black-faced surrogate mother). As an adult, Dolly mated and subsequently gave birth to a normal offspring, thus proving the genetic viability of cloned mammals. (From I. Wilmut et al. 1997. *Nature* 385: 810–813.)

accomplished by sperm. (In the egg that produced Dolly, this activation was accomplished by electric shock, stimulating the egg to divide.) The activated egg is then transplanted to the uterus of a surrogate mother. If the process is successful, the resulting embryo will develop into an adult with a genetic composition identical to that of its nuclear donor.

Dolly, born in 1996, was cloned from an epithelial cell taken from the udder of a white-faced Finn Dorset ewe. Her surrogate mother was a Scottish Blackface ewe. Dolly grew up and had lambs of her own. Dolly's cloning has shown conclusively that cells retain genomic equivalence as they differentiate. There are many differences between Scottish Blackface sheep and Finn Dorset sheep in addition to face color. There is no doubt that Dolly's genetic makeup came from the nucleus of the differentiated udder cell of her Finn Dorset donor mother. Dolly was a normal sheep, able to bear lambs of her own. Thus the nucleus used to clone her retained its full potential.

Other animal species, including cats, dogs, horses, pigs, rabbits, and mice, have since been cloned by nuclear transfer. The first primate clones (two macaque monkeys) were born in 2018. To produce these monkey clones, researchers in China modified the techniques used to clone Dolly. They were unsuccessful using cell nuclei from adult monkeys, but when they used nuclei from embryos and modified other techniques, two live clones were born.

Cloning of animals has some practical applications. For example, the number of valuable animals with particularly useful traits could be increased for agriculture and for scientific and medical purposes. A cow that was genetically engineered to produce human growth hormone (HGH) in its milk has been successfully cloned. It is estimated that 15 such cattle could provide enough HGH to supply the world's need for this protein, which is used to treat people with growth hormone deficiency.

Cloning might also help preserve some endangered species. For example, the banteng, a relative of the cow, was the first endangered animal to be cloned and survive. The cloned banteng was made by transplanting the nucleus from a banteng cell into the enucleated egg from a cow and implanting the zygote into a cow surrogate mother. Cloning may be the only way to save endangered species with low rates of natural reproduction.

Serious issues and problems are associated with cloning, however. To date, the success rate of cloning is extremely low, and failed attempts to clone result in many abnormalities. Of 277 nuclear transplants of epithelial cell nuclei to enucleated sheep eggs, only 29 developed successfully enough to be transferred to surrogate mothers, and of these, only one lamb, Dolly, resulted. The two healthy macaque monkey babies resulted from egg transfers to more than 60 surrogate mothers.

The cloning of monkeys has been especially controversial, raising ethical as well as practical questions concerning not only our nonhuman primate cousins but the looming issue of human cloning. No country in the world is known to be attempting to clone a human at this time, but it is no longer a subject in the realm of science fiction. Pet cloning is an increasingly offered service. Are humans next? In addition to the conundrum of playing God, there are overwhelming practical objections. Who would determine who has the right to be cloned? Would it be a perk for only the extremely wealthy and powerful? Who would regulate this "industry" and how? What would be the fate of abnormal human fetuses and babies? All epigenetic and environmental factors that affect gene expression cannot be controlled, so an individual would not be perfectly duplicated even if the genome were shared. (Note how different identical twins are, even though they are clones, beginning with an identical genetic makeup and sharing the same uterus. No one suggests that identical twins are really the same person duplicated!)

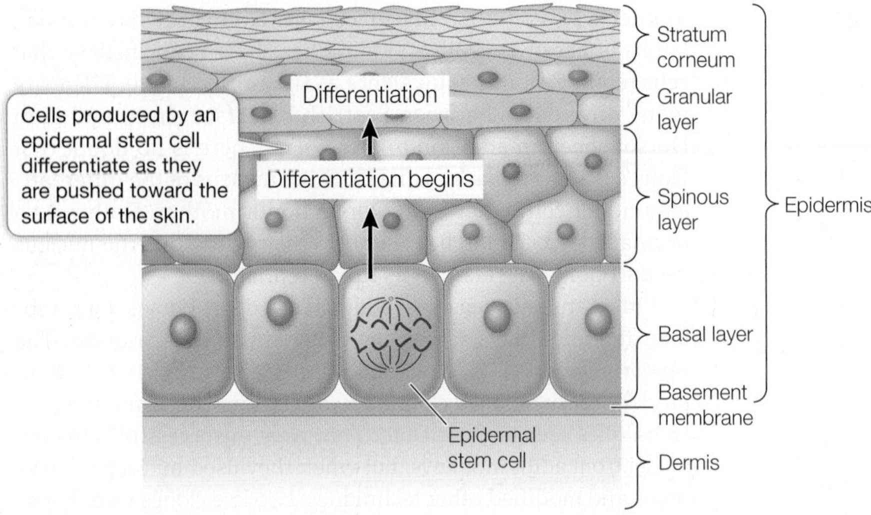

FIGURE 35.30 Differentiation of Epidermal Epithelial Cells in Mammalian Skin When a cell in the basal layer undergoes mitosis, one daughter cell remains on the basement membrane; the other daughter cell is pushed toward the surface of the skin. Cells differentiate as they approach the surface, synthesizing keratin, eliminating the nucleus, and becoming flattened sacs of keratin. (After Y. C. Hsu et al. 2014. *Nat Med* 20: 847–856.)

Owners of cloned pets are sometimes disappointed to learn that a clone of "Fifi" will not be Fifi. These and many other questions arise as we consider the implications of cloning. So far, it is a process fraught with difficulties. One of the scientists responsible for cloning Dolly now feels that the most important aspect of the project was not the demonstration that cloning is possible but the confirmation that cells retain genomic equivalence as they differentiate. This is opening doors for new investigations and techniques.

Stem cells provide new cells for growth and maintenance

As an animal increases in size, more cells are needed. Even after growth ceases, in most cases our individual cells do not last throughout our lifetime. They wear out and die and must be replaced. One estimate suggests that the skin cells that cover your body today (the outer layer of epidermis) will be completely replaced in a month. The old sloughed-off cells are the major component of the house dust we battle. Approximately 2 million of your red blood cells are replaced every second!!

We have seen that as an embryo develops, its cells become committed to develop into specific cell types. Often, as with red blood cells or mature skin cells, differentiated cells no longer divide. So where do new red blood cells and skin cells come from? In many tissues, cells called **stem cells** remain undifferentiated and act as a reserve population, able to undergo mitosis and produce the next round of differentiated cells. A stem cell is often defined as one that, on division, produces another stem cell and a cell that is closer to the differentiated state. In the case of the epidermis, a layer of epidermal stem cells sits atop the basement membrane that separates the dermis and the epidermis (**FIGURE 35.30**). When one of these cells divides, it gives rise to another stem cell and to a daughter cell that is a bit closer to the differentiated state and does not divide. As these daughter cells are pushed closer to the body surface (pushed by more divisions of the underlying stem cells), they start to synthesize keratin, the major protein of our skin, hair, and nails. By the time the daughter cells reach the body surface, they have lost their nuclei and are little more than flat sacs of keratin that protect our bodies. Similarly, mammalian red blood cells are produced by stem cells in bone marrow. Some progeny of these **hematopoietic** cells (stem cells that produce blood cells) differentiate to become cells that are highly specialized to carry hemoglobin. As they mature, they begin to synthesize hemoglobin, losing their nuclei and other organelles. When they begin circulating in the blood stream as red blood cells, they are tiny sacs filled with hemoglobin. Other progeny of the hematopoietic stem cells become different types of blood cells (e.g., T and B lymphocytes, macrophages, neutrophils, platelets) (**FIGURE 35.31**). Because hematopoietic stem cells can differentiate into several types of cells, they are called

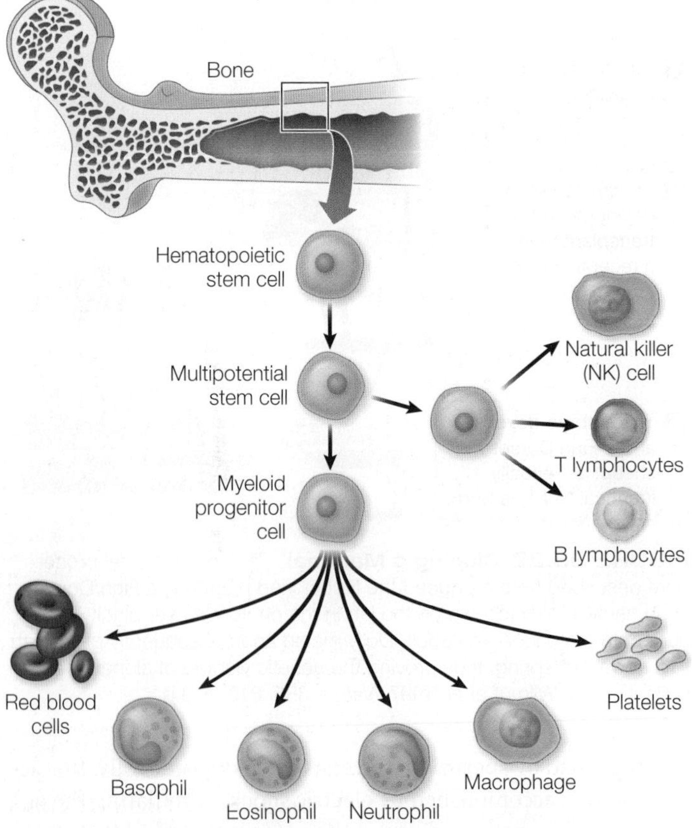

FIGURE 35.31 Hematopoietic Stem Cells Reside in Bone Marrow Stem cells that give rise to all of our different types of blood cells lie in marrow in our bones. They divide to form more stem cells and cells that will differentiate, for example, into red blood cells, lymphocytes, granulocytes, or platelets. Because hematopoietic stem cells are capable of producing different kinds of blood cells, they are multipotent.

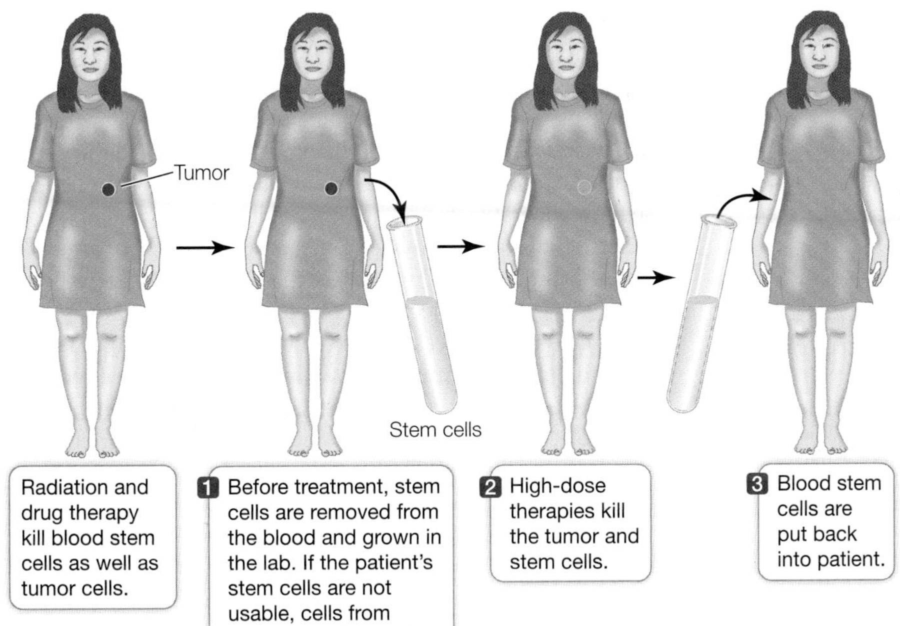

FIGURE 35.32 Multipotent Stem Cells In hematopoietic stem cell transplantation, blood stem cells are used to replace stem cells destroyed by cancer therapy. The stem cells are harvested from bone marrow or from circulating blood as shown here. Prior to cell harvesting, the patient or donor is given daily injections of a cytokine (a protein signal molecule) that, in this case, promotes rapid growth of hematopoietic cells and mobilizes their release into the peripheral blood stream. These multipotent stem cells repopulate the bone marrow of the patient.

Animation 35.4 **Embryonic Stem Cells**

multipotent stem cells. (Note that their pathways are quite restricted, however. They become blood cells.)

The potential uses of stem cells are of extreme interest to researchers in developmental biology, regenerative medicine, and many other fields. Since the 1950s, bone marrow transplants have been used following some disease treatments. For example, some treatments that kill cancer cells also kill other dividing cells, including the hematopoietic cells of patients exposed to these treatments. Now the isolation of stem cells and the ability to maintain them is the basis of an important cancer therapy called **hematopoietic stem cell transplantation** (**HSCT**; **FIGURE 35.32**). Before treatment, stem cells are harvested from the patient (or from a cancer-free donor). The stem cells are stored under appropriate conditions and retain their ability to differentiate. When reintroduced back into the patient, they are able to differentiate into different types of blood cells. By allowing the use of high doses of chemical or radiation treatment to kill cancer cells, HSCT saves thousands of lives each year.

Stem cells can be isolated from embryos or induced from differentiated cells

The term "stem cell" is currently used in different contexts, so it is important to understand exactly what is meant by the various usages. Especially confusing are the prefixes: "toti-" (meaning "all" or "everything"), "multi-" ("many"), and "pluri-" (also "many"). Each has a different meaning.

Cells in an early mammalian embryo behave as stem cells during normal development. (Think of how identical twins are produced.) There are two types of **embryonic stem cells** (**ESCs**). In mammals, **totipotent stem cells**—cells that can individually give rise to an entire organism—are found only in very early embryos, in the very short time before the formation of the blastocyst. Once the blastocyst has formed, cells of the inner cell mass are no longer totipotent because they cannot form the trophoblast, a part of the embryo that participates in placenta formation. However, they are capable of forming all the other cell types, as they would in the intact embryo, and as such they are called **pluripotent stem cells**. Timing is very specific, and cells are only pluripotent before the blastocyst implants (this is the last stage before they begin to differentiate).

In 1998, two years after the cloning of Dolly, researchers found that pluripotent stem cells can be grown and maintained in culture in an undifferentiated state almost indefinitely, if provided with the right conditions. Then, when presented with the appropriate signals, pluripotent stem cells are capable of differentiating into cells of all three germ layers. For example, treatment of mouse ESCs with a derivative of vitamin A causes them to form neurons (nerve cells), whereas other growth factors induce the ESCs to form blood cells. Such experiments demonstrate both the developmental potential of ESCs and the roles of environmental signals. This finding raises the possibility of using ESC cultures as sources of differentiated cells to repair specific tissues, such as a damaged pancreas in diabetics, or a brain that malfunctions in Parkinson's disease. However, problems, both technical and ethical, surround the use of ESCs. Think about what these problems are.

In a major breakthrough in 2006, researchers in Japan were able to bring about the dedifferentiation of differentiated mouse fibroblast cells (loose, mesodermal cells) and produce pluripotent stem cells from them (**FIGURE 35.33**). Shinya Yamanaka and coworkers at Kyoto University began by comparing the expression patterns of genes in ESCs and in differentiated cells. They first selected genes that were uniquely expressed at high levels during early development in ESCs. Using retroviruses (see Key Concept 18.4) to incorporate these candidate active genes, in a variety of combinations, into the genomes of differentiated cells, they identified four genes that were involved in maintaining a stem

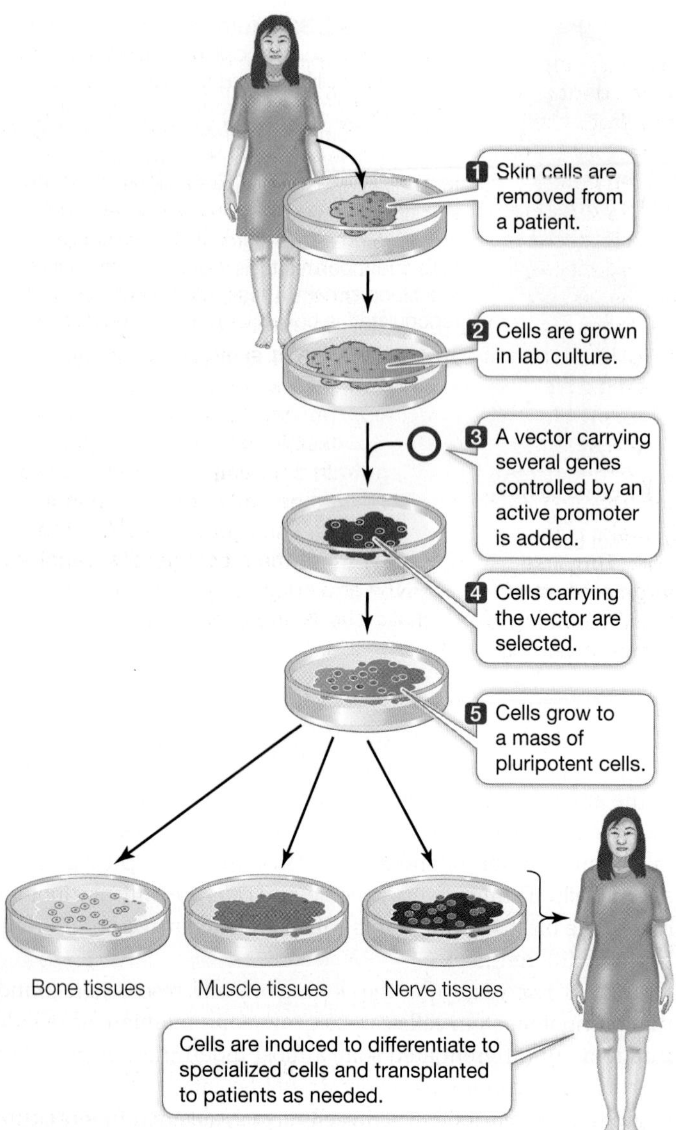

FIGURE 35.33 Induced Pluripotent Stem Cells Pluripotent stem cells can be obtained by adding genes that are highly expressed in stem cells to skin or other differentiated cells to transform them into stem cells.

cell–like state. These four genes coded for key transcription factors normally implicated in keeping mouse ESCs in an immature state (Sox2, Oct3/4, cMyc, and Klf4—now called Yamanaka factors). When activated copies of these four genes were inserted into a differentiated cell, the cell lost its differentiated characteristics and regained the ability to divide and to differentiate (following appropriate signaling) into different cell types.

Yamanaka's work has now been repeated in many laboratories using many differentiated tissues as starting points. The term **induced pluripotent stem cells (iPSCs)** is used to identify stem cells that have been induced back into a pluripotent state from a differentiated state. (Note that although the prefixes "multi" and "pluri" both mean "many," they have specific meanings, as described above, when referring to stem cells.) Shinya Yamanaka and John B. Gurdon were awarded the Nobel Prize in Physiology or Medicine in 2012 "for the discovery that mature cells can be reprogrammed to become pluripotent."

The use of iPSCs is of real value in disease modeling, drug discovery and development, and helping us understand the plasticity of the differentiated state. Yamanaka's technique holds much promise for regenerative medicine because it destroys no embryos, and because iPSCs can be made from skin cells or other types of cells from the individual being treated, there is no immune response to the cells (i.e., tissue rejection) as there may be with ESCs. The hope is that cells could be easily harvested from a patient (perhaps from skin) and reprogrammed to grow into the type of cell required—neurons for Parkinson's disease, pancreatic cells for the treatment of diabetes. The time and cost to grow such cells, as well as safety issues, remain hurdles. Early work using retroviruses as vectors to introduce the Yamanaka factor genes often led to cancer. Adenoviruses and plasmids can be used to introduce the genetic material but are less efficient. Some of the transcription factors can be replaced by treatment with a set of chemical compounds. The search for a way to generate iPSCs by treating somatic cells with a set of compounds that safely promote reprogramming continues.

REVIEW & APPLY | 35.5

R Development continues throughout life. All sexually reproducing organisms increase in size between birth (or hatching) and sexual maturity. Some animals have very different morphologies at different times in their life cycles. Even when growth ceases, individual cells must be replaced and tissues must be repaired. Cells become differentiated, but probably all animals retain some cells in an undifferentiated state (stem cells) that are capable of generating both more stem cells to replace themselves and differentiated cells. These cells are used to treat some illnesses, and their plasticity is being explored. It has been found that differentiated cells retain their full genomic capacity and can be forced to dedifferentiate and redifferentiate along many different pathways. The potential of these induced pluripotent stem cells (iPSCs) is an area of active research.

A

1. Prior to the Renaissance, painters often depicted infants as miniature adults. Why do these babies not appear natural?
2. On a walk through a field, you spot a grasshopper that looks unlike the ones you have seen before. It has very short wings, its body seems too short, and its head seems too big. What would you conclude?
3. When a tadpole metamorphoses into a froglet, what two major developmental processes take place?
4. What is probably the greatest significance of the cloning success of Dolly?
5. Generating iPSCs is an ongoing challenge in biology and medicine. What problems, seen with the use of ESCs, do iPSCs circumvent?

Why is the *Xenopus* egg an ideal expression system, and what steps would a scientist take to evoke ion channels for study in these eggs?

Answer: The answer lies in the machinery and molecules deposited in the egg by the female frog. The oocyte is essentially a factory primed for protein synthesis. Even before fertilization, the egg is fully equipped to initiate protein synthesis, and is provided with all of the organelles, nutrients, and enzymes necessary for that task. Transcription factors and RNA based on the maternal DNA are already in the oocyte. After investigators choose a protein or a complex (such as an ion channel) of interest, they isolate the coding gene or genes and, using polymerase chain reaction (PCR; see Key Concept 9.2), make many, many copies of the DNA. They use the DNA to create a large amount of complementary RNA (cRNA), which is then injected into an unfertilized frog egg. The injection of the cRNA activates the egg (as a sperm would), and the cRNA becomes the blueprint for protein synthesis. The egg uses the materials already inside to make multiple copies of the protein or structure of interest coded by the cRNA. In the example above, in about 24 hours, many, many ion channels would be expressed in the cell membrane of the unfertilized egg. The egg is a large, single cell, and the channels are positioned on its surface, making them readily available and accessible for study. Note that the protein or channel of interest does not have to be one normally found in the oocyte. The injected cRNA is the blueprint for the message; the transfer RNAs and enzymes just follow the instructions and protein synthesis proceeds.

VISUAL SUMMARY 35

35.1 Fertilization Activates Development

FIGURE 35.2

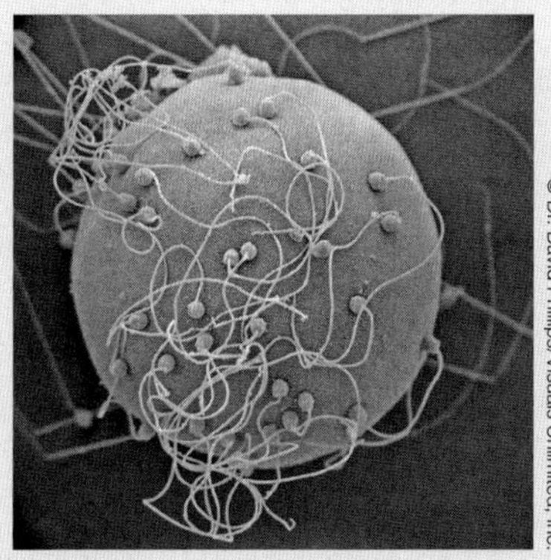

© Dr. David Phillips/Visuals Unlimited, Inc.

Go to ANIMATION 35.1

- Fertilization activates development and initiates the beginning of a new organism.
- One, and only one, sperm must fertilize the egg (**FIGURE 35.2**).
- Often, the cytoplasm is reorganized.
- The $2n$ cell formed by the union of the sperm and egg is the **ZYGOTE**.

QUESTIONS

1. What is the function of the change in the egg cell membrane immediately following fertilization?
2. What do the egg and sperm each contribute?
3. What are the major results of fertilization?

35 VISUAL SUMMARY

35.2 Cleavage Creates Building Blocks to Make an Embryo

FIGURE 35.5

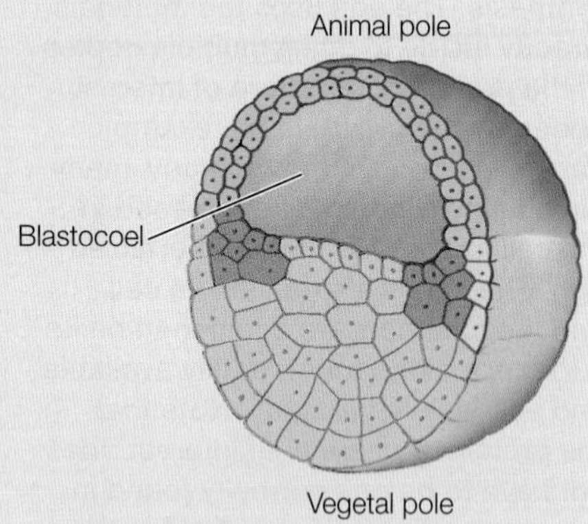

- **CLEAVAGE** leads to the production of the **BLASTULA**, which is composed of many smaller cells called **BLASTOMERES** that often surround a fluid-filled **BLASTOCOEL** (**FIGURE 35.5**).
- No **GROWTH** occurs during cleavage, but many cell divisions occur.
- In most animals, the genome of the embryo is not transcribed until about the mid-blastula stage. The mother's genome, through mRNAs, tRNAs, and transcription factors deposited in the egg cytoplasm, usually controls most of the events of cleavage. The transition is much earlier in mammals.
- Cleavage is affected both by the amount of yolk in the egg and the evolutionary history of the group.

QUESTIONS

1. What major functions does cleavage serve?
2. What is unique about the mitotic divisions during cleavage?
3. What is the effect of yolk on cleavage?
4. How do radial and spiral cleavage differ?
5. What is special about cleavage in mammals?

35.3 Gastrulation Sets the Stage for Morphogenesis

Go to ANIMATIONS 35.2 and 35.3, and ACTIVITIES 35.1 and 35.2

- **GASTRULATION** creates a three-layered (ectoderm, endoderm, and mesoderm) organism and sets the stage for **ORGANOGENESIS**.
- Deuterostomes and protostomes follow different developmental patterns to form the three germ layers and the **COELOM** (**FIGURE 35.16**).
- Like cleavage, gastrulation is affected by the amount of yolk in the egg.
- The rearrangement of cells allows new cell interactions to occur. For example, one group of cells may induce another group to follow a certain path toward differentiation.

FIGURE 35.16

Development of the coelom in deuterostomes

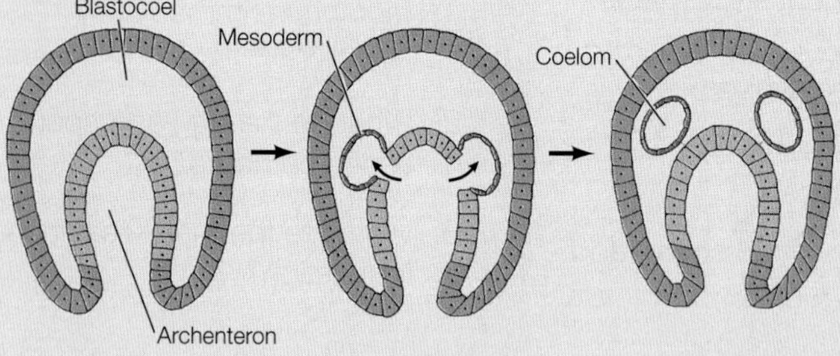

Development of the coelom in protostomes

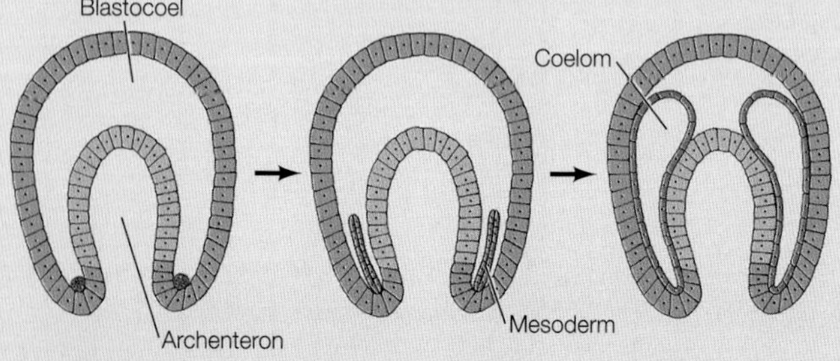

QUESTIONS

1. Describe the early gastrula of a sea urchin.
2. What are the major similarities and differences between sea urchin gastrulation and frog gastrulation?
3. Compare mesoderm and coelom formation in the non-yolky eggs of a deuterostome such as a sea urchin and in a protostome such as a snail.
4. What two major mechanisms are responsible for determining the differentiated state of a cell?
5. What major tissues does ectoderm form? What three major mesodermal regions form in chordates, and what does each form?
6. What is the significance of endoderm?

35.4 Extraembryonic Membranes Protect and Nourish the Embryo

FIGURE 35.24

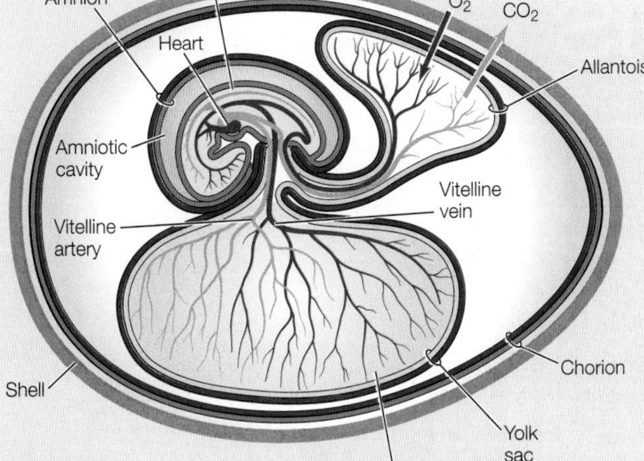

Go to ACTIVITY 35.3

- The amniotic egg allows animals to live on land without returning to water to breed.
- Four **EXTRAEMBRYONIC MEMBRANES** that develop from endoderm, mesoderm, and ectoderm form the membranes of the amniotic egg (**FIGURE 35.24**). These membranes serve essential functions in birds, and also in mammals such as humans even though the embryo is retained in the mother's body.

QUESTIONS

1. What are the four extraembryonic membranes of a chick embryo? Describe their functions.
2. What germ layers form the amnion?
3. Which two membranes are formed from endoderm and mesoderm?
4. What function does the yolk sac have in egg-laying and in placental mammals?

35.5 Development Continues throughout Life

FIGURE 35.29

From I. Wilmut et al. 1997. *Nature* 385: 810–813

Go to ANIMATION 35.4

- All sexually reproducing organisms grow between birth and sexual maturity. Some animals have very different appearances and lifestyles at different parts of their life cycles. Even when growth ceases, individual cells must be replaced and tissues must be repaired.
- **STEM CELLS** are undifferentiated cells that are capable of generating more stem cells and differentiated cells. Stem cells play a role throughout our lives replacing cells. Stem cells are also used to treat some illnesses, and their plasticity is being explored.
- Completely **DIFFERENTIATED** cells retain their full genetic capacity (**FIGURE 35.29**). They can be forced to dedifferentiate and then redifferentiate along the same or a different pathway. The potential of these **INDUCED PLURIPOTENT STEM CELLS** (**iPSCs**) for medicine and research is being actively investigated.

1. Give an example, in addition to head size, of a feature that illustrates allometric growth in humans.
2. What are two major functions of larvae?
3. What are embryonic stem cells (ESCs)?

Go to **Achieve** for the e-book, animations, activities, and additional resources and assignments.

36 Immunology: Animal Defense Systems

Smallpox, caused by a viral infection, has been a scourge of humanity for centuries. Indeed, more than 300 million people died from smallpox in the twentieth century. But smallpox made its last appearance in 1978, thanks to vaccination, an application of knowledge of the immune system.

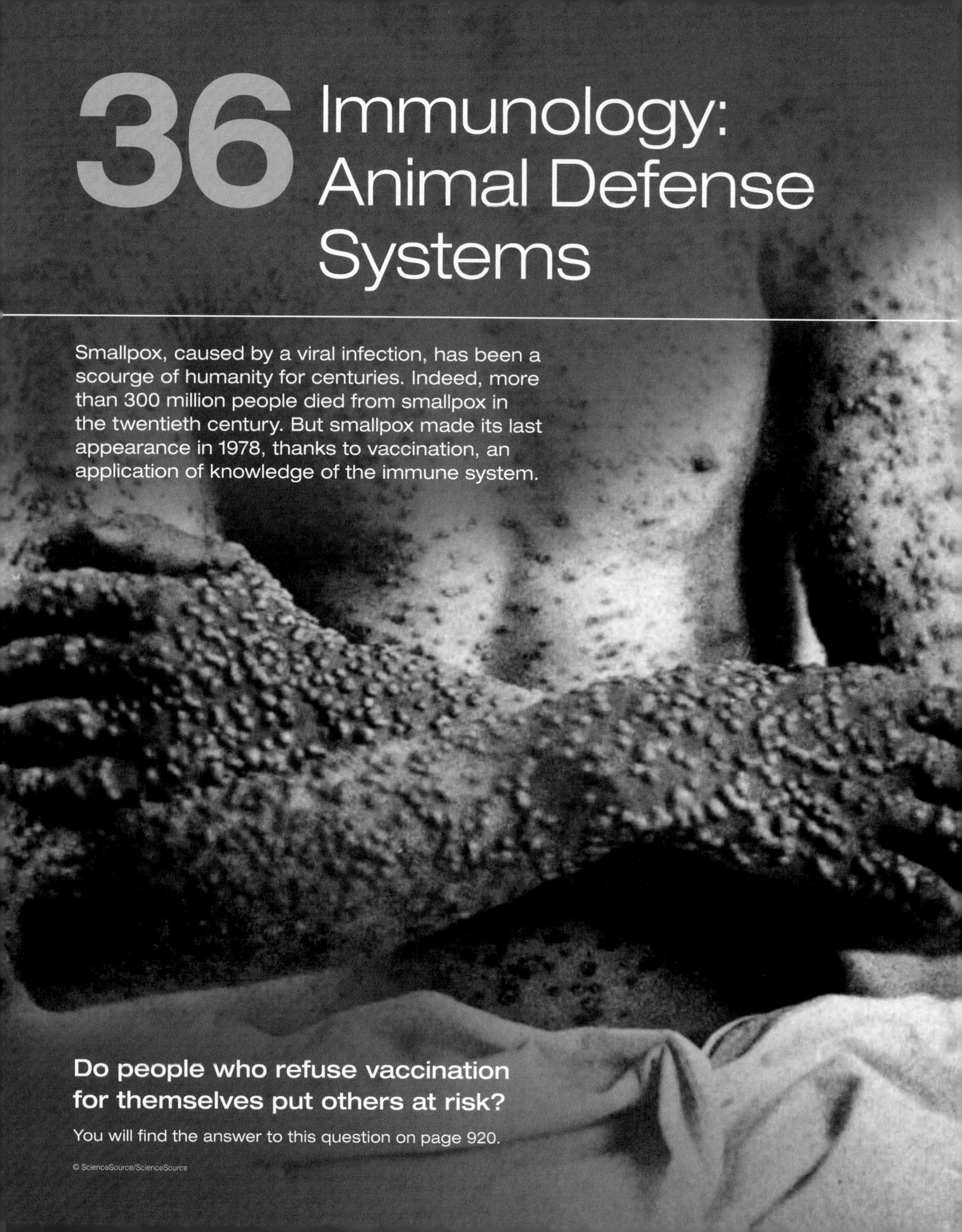

Do people who refuse vaccination for themselves put others at risk?

You will find the answer to this question on page 920.

KEY CONCEPTS

36.1 Animals Use Innate and Adaptive Mechanisms to Defend Themselves against Pathogens

LEARNING OBJECTIVES

By the end of this key concept you should be able to:

36.1.1 **Distinguish between innate and adaptive defense mechanisms.**

36.1.2 **Describe the role of the Toll-like receptor (TLR) pathway in innate immunity.**

36.1.3 **List types of cells produced in mammalian blood and lymphoid tissues that function in immune responses.**

Animals have several ways of defending themselves against **pathogens**—harmful organisms and viruses that can cause disease. These defense systems are based on the distinction between self—the animal's own molecules—and nonself, or foreign, molecules. Some defensive mechanisms are present all the time. For example, the skin is always present to protect a mammal from invaders. Other defenses are activated in response to invaders. Typically the defense response involves three phases:

1. *Recognition phase*: The organism must be able to recognize pathogens and discriminate between self and nonself.
2. *Activation phase*: The recognition event leads to a mobilization of cells and molecules to fight the invader.
3. *Effector phase*: The mobilized cells and molecules destroy the invader.

There are two general types of defense mechanisms:

1. **Innate defenses**, or nonspecific defenses, are inherited mechanisms that provide the first line of defense against pathogens. Innate defenses include physical barriers such as the skin, molecules that are toxic to invaders, and phagocytic cells that ingest invaders. This system recognizes broad classes of organisms or molecules and gives a quick response, within minutes or hours. Some innate defenses are present all the time, whereas others are rapidly activated in response to an injury or invasion by a pathogen. All animals have innate defenses (as do plants, where they are called constitutive defenses; see Chapter 27).
2. **Adaptive defenses** are aimed at specific pathogens and are activated by the innate immune system. For example, cells in the adaptive defense system can make **antibodies**—proteins that will recognize, bind to, and aid in the destruction of specific pathogens, if the pathogens enter the body. Adaptive defenses are typically slow to develop and long-lasting. Adaptive defenses evolved in vertebrate animals.

Immunity occurs when an organism has sufficient defenses to successfully avoid biological invasion by a pathogen.

LINK Phagocytosis is a form of endocytosis in which a cell engulfs a large particle or another cell; see **Key Concept 4.3**.

Innate defenses evolved before adaptive defenses

All animals have innate defenses against their enemies. For example, the crustacean *Tachypleus tridentatus*—the Japanese horseshoe crab—first appeared in the fossil record about 400 million years ago. It relies only on innate defenses. These defenses include barriers, defensive cells, and defensive molecules.

- Barriers include physical, chemical, and biological mechanisms for resisting infections (more on this in Key Concept 36.2). For example, the Atlantic horseshoe crab (*Limulus polyphemus*) has the hard exoskeleton that is characteristic of arthropods. This shell acts to protect the crab from invasion by pathogens (**FIGURE 36.1A**).
- Cells involved in innate defenses include phagocytes that bind to microbial pathogens, ingest them by endocytosis, and destroy their large molecules by hydrolysis. Amebocytes in the blood of the horseshoe crab fulfill this defensive role by ingesting bacteria that secrete dangerous endotoxins (**FIGURE 36.1B**).
- Molecules that are toxic to invading pathogens are important in innate defense. The horseshoe crab has a wide array of such molecules that are released from cells in its blood. These molecules include peptides that disrupt the bacterial cell membrane, rendering it permeable; and peptides that bind to bacterial surfaces and cross-link them.

Studies of innate immunity, along with genome sequencing, have revealed that the recognition and activation phases of innate immunity evolved very early in animals. For example, animals as diverse as humans and fruit flies share a class of receptors, called **Toll-like receptors** (**TLRs**), that participate in innate defense responses. In vertebrates, each Toll-like receptor recognizes and

(A) An Atlantic horseshoe crab

(B) Amebocytes in horseshoe crab blood

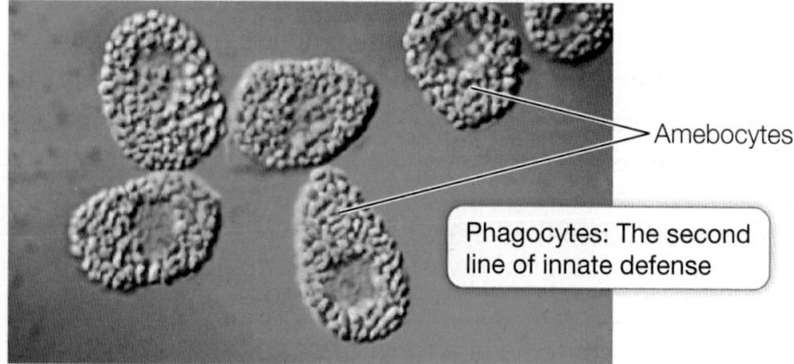

FIGURE 36.1 Innate Defenses in the Atlantic Horseshoe Crab **(A)** A hard, impermeable exoskeleton provides defense against large pathogens. **(B)** Phagocytes in blood release a protein that coagulates endotoxins made on bacterial pathogens. The phagocytes then ingest and hydrolyze the pathogens.

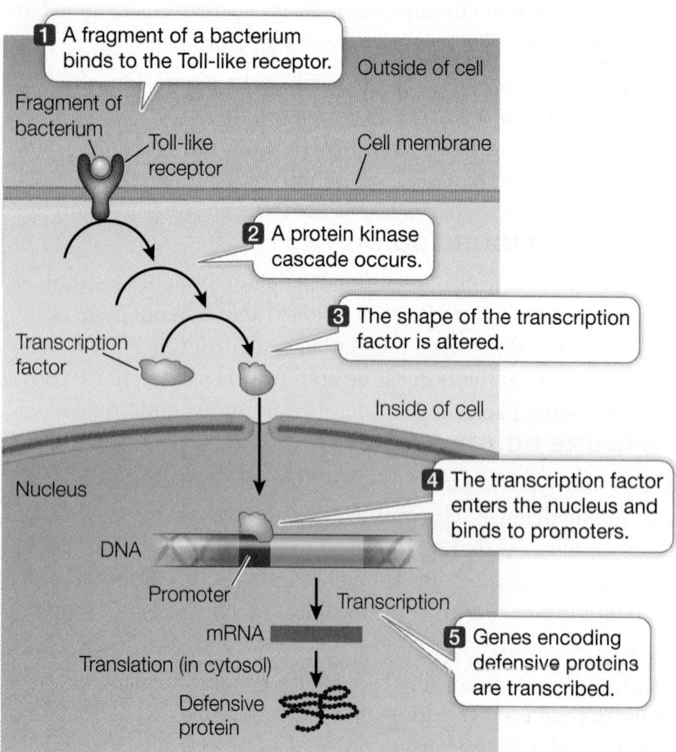

FIGURE 36.2 Cell Signaling and Innate Defense Binding of a pathogen molecule to a Toll-like receptor initiates a signal transduction pathway that results in the expression of genes involved in innate defense.

binds to certain conserved structures on microbes called **pathogen-associated molecular patterns**, or **PAMPs** (described in more detail in Key Concept 36.2). Binding sets off a signal transduction pathway that ends with the expression of genes for anti-pathogen molecules (**FIGURE 36.2**). This pathway exists in some form in many animal groups, including humans. A similar mechanism occurs in plants (see Key Concept 27.4).

Mammals have both innate and adaptive defenses

In mammals and other vertebrates, the innate and adaptive defenses operate together, usually in sequence, as a coordinated defense system. We will focus on these defenses for the rest of the chapter. **TABLE 36.1** gives an overview of innate and adaptive defenses during the course of an infection. Innate defenses are the body's first line of defense; adaptive defenses often require days or even weeks to become effective.

The major players in adaptive immunity are specific cells and proteins. These are produced in the blood and lymphoid tissues and are circulated throughout the body, where they interact with almost all the other tissues and organs.

One milliliter of human blood typically contains about *5 billion* red blood cells and *7 million* white blood cells. Whereas the main function of red blood cells is to carry oxygen throughout the body, **white blood cells** (also called leukocytes) are specialized for various functions in the immune system (**FIGURE 36.3**). There are two major groups of white blood cells: phagocytes and lymphocytes. **Phagocytes** are large cells that engulf pathogens and other substances by phagocytosis. Some phagocytes are involved in both innate and adaptive immunity. In particular, macrophages and dendritic cells play key roles in communicating between the innate and adaptive immune systems. Targeting invariant

TABLE **36.1** | Innate and Adaptive Immune Defenses

Response (time after infection by a pathogen)	System	Mechanisms
Early (0–4 hr)	Innate, nonspecific (first line)	Barrier (skin and lining of organs)
		Dryness, low pH
		Mucus
		Lysozyme, defensins
Middle (4–96 hr)	Innate, nonspecific (second line)	Inflammation
		Phagocytosis
		Natural killer cells
		Complement system
		Interferons
Late (>96 hr)	Adaptive, specific	Humoral immunity (B cells, antibodies)
		Cellular immunity (T cells)

TYPE OF CELL	FUNCTION
Basophils (I)	Release histamine and other molecules involved in inflammation
Eosinophils (A)	Kill antibody-coated parasites
Phagocytes:	
Neutrophils (I)	Stimulate inflammation
Mast cells (I)	Release histamine
Monocytes (I, A)	Develop into macrophages and dendritic cells
Macrophages (I, A)	Antigen presentation
Dendritic cells (I, A)	Present antigens to T cells
Lymphocytes:	
B lymphocytes (A)	Differentiate to form antibody-producing cells and memory cells
T lymphocytes (A)	Kill pathogen-infected cells; regulate activities of other white blood cells
Natural killer cells (I, A)	Attack and lyse virus-infected or cancerous body cells

FIGURE 36.3 White Blood Cells White blood cells have key roles in both innate (I) and adaptive (A) immunity.

Animation 36.1 Cells of the Immune System

molecules of pathogens works reasonably well but is insufficient to generate immunological memory (see Key Concept 36.3). **Lymphocytes** include B cells and T cells, which are involved in adaptive immunity; and natural killer cells, which are involved in both innate and adaptive immunity.

As you will see in Key Concept 36.3, important features of the adaptive immune system include specificity, the ability to distinguish self from nonself, immunological memory, and the ability to respond to an enormous diversity of nonself molecules. In humans, for example, lymphocytes such as B cells can produce in excess of 10^{12} different potential receptors for pathogens!

Blood and lymph tissues play important roles in defense

The components of the mammalian defense system are dispersed throughout the body and interact with almost all of its other tissues and organs. The lymphoid tissues, which include the thymus (an organ in the upper part of the chest), bone marrow, spleen, and lymph nodes, are essential parts of the defense system (**FIGURE 36.4**). The blood and lymph have central roles in defense.

The blood and lymph both consist of liquids in which cells are suspended:

- **Blood plasma** is a yellowish solution containing ions, small molecule solutes, and soluble proteins. Suspended in the blood plasma are red blood cells, white blood cells, and platelets (cell fragments essential to blood clotting). Whereas red blood cells are normally confined to the closed circulatory system (the heart, arteries, capillaries, and veins), white blood cells and platelets are also found in the lymph.
- **Lymph** is a fluid that is derived from the blood (but lacking red blood cells) and other tissues and accumulates in intercellular spaces throughout the body. From these spaces, the lymph moves slowly into the vessels of the lymphatic system. Tiny lymph capillaries conduct this fluid to larger ducts that eventually join together, forming two large vessels, the thoracic duct and right lymphatic duct, which join a major vein near the heart. By this system of vessels, the lymph is eventually returned to the blood and the circulatory system.

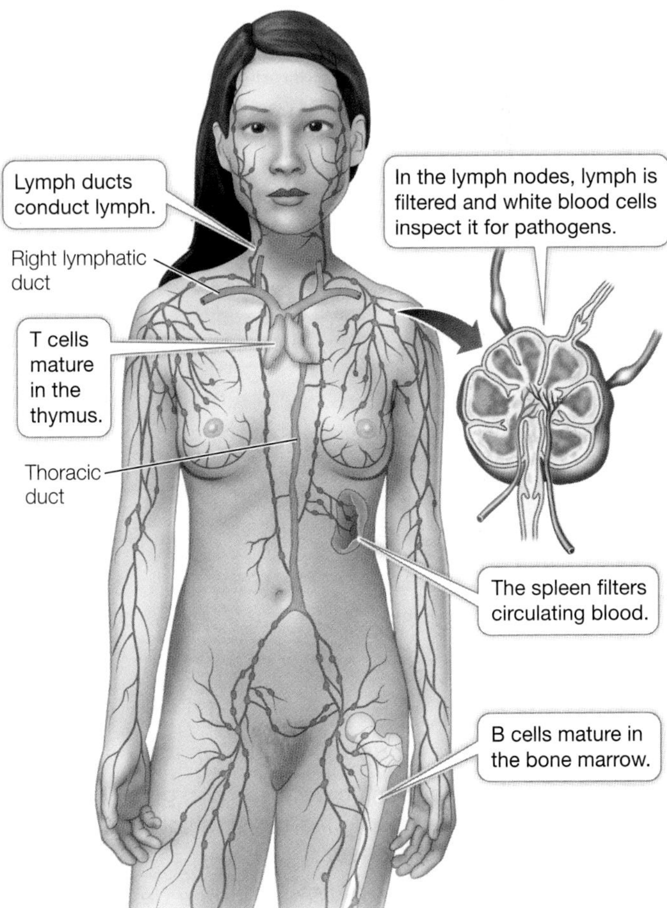

FIGURE 36.4 The Human Lymphatic System A network of ducts and vessels collects lymph from body tissues and carries it toward the heart, where it mixes with blood to be pumped back to the tissues. Other lymphoid tissues, including the thymus, spleen, and bone marrow, are also essential to the body's defense system.

Activity 36.1 The Human Defense System

At many sites along the lymph vessels there are small, roundish structures called **lymph nodes**, which contain lymphocytes. As lymph passes through a lymph node, the lymphocytes encounter foreign cells and molecules that have entered the body, and if they are recognized as nonself, an immune response is initiated.

REVIEW & APPLY | 36.1

R All animals have innate defenses against pathogens, and vertebrates have both innate and adaptive defenses. Both kinds of mechanisms are based on the ability to differentiate self from nonself, which is aided by pattern-recognizing receptors such as Toll-like receptors (TLRs). Innate defenses include barriers, defensive cells, and defensive molecules targeting nonspecific molecules and organisms, whereas adaptive defenses target specific pathogens. Both innate and adaptive defenses involve specialized white blood cells.

A
1. What are the characteristics that differentiate innate immunity and adaptive immunity?
2. Some people have a rare genetic disease that results in defective signaling in the TLR pathway. What effect does this have on the phenotype of such individuals?
3. Differentiate between phagocytes and lymphocytes, and name at least two examples of each.

The outcome of an infectious disease depends on the success of both rapid innate responses and longer-lasting adaptive responses to invading pathogens. We'll now consider in more detail some of the innate defenses that mammals use to keep invading organisms at bay.

36.2 | Innate Defenses Are Nonspecific

LEARNING OBJECTIVES

By the end of this key concept you should be able to:

36.2.1 **Describe nonspecific defense and give examples of nonspecific physical, chemical, and cellular defenses.**

36.2.2 **Explain the role of PAMPs in innate defenses.**

36.2.3 **List the steps in the inflammatory response.**

Innate, nonspecific defenses are general protection mechanisms that attempt to either stop pathogens from invading the body or quickly eliminate pathogens that do manage to invade. They are genetically programmed (innate) and "ready to go," in contrast to adaptive, specific defenses, which take time to develop after a pathogen or toxin has been recognized as nonself. In mammals, innate defenses include physical barriers as well as cellular and chemical defenses (**FIGURE 36.5**).

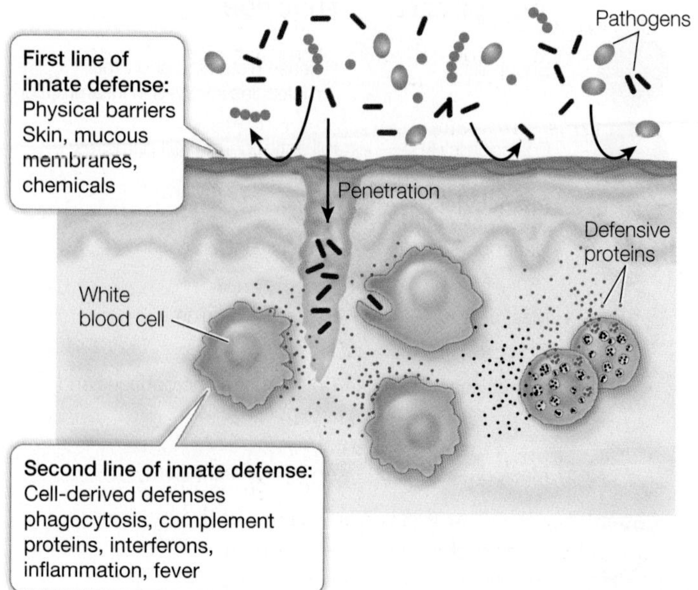

FIGURE 36.5 Innate Defenses Physical barriers, cells, and proteins (complement and interferons) provide nonspecific defenses against invading pathogens.

Media Clip 36.1 The Chase Is On: Phagocyte versus Bacteria

Barriers and local agents defend the body against invaders

The first line of innate defense is encountered by a potential pathogen as soon as it lands on the surface of the animal. Consider a pathogenic bacterium that lands on human skin. The challenges (physical, chemical, and biological) faced by the bacterium just to invade the body are formidable:

- *The physical barrier of the skin*: Bacteria rarely penetrate intact skin (which explains why broken skin increases the risk of infection).
- *The saltiness of the skin*: This condition is usually not hospitable to the growth of the bacterium.
- *The presence of normal flora*: Bacteria and fungi that normally live in great numbers on body surfaces without causing disease will compete with potential pathogens for space and nutrients.

If a pathogen lands inside the nose or another internal organ, it faces other innate defenses:

- **Mucus** is a slippery secretion produced by mucous membranes, which line various body cavities that are exposed to the external environment. Mucus traps microorganisms so they can be removed by the beating of cilia (see Key Concept 4.4), which continuously move the mucus and its trapped debris toward the outside of the body.

- **Lysozyme**, an enzyme made by mucous membranes, cleaves bonds in the cell walls of many bacteria, causing them to lyse (burst open).
- **Defensins**, also made by mucous membranes, are peptides of 18–45 amino acids. They contain hydrophobic domains and are toxic to a wide range of pathogens, including bacteria, microbial eukaryotes, and enveloped (membrane-enclosed) viruses. Defensins insert themselves into the cell membranes of these pathogens and make the membranes freely permeable to water and all solutes, thus killing the invaders. Defensins are also produced inside phagocytes, where they kill pathogens trapped by phagocytosis. Plants also produce molecules that kill pathogens (see Key Concept 27.4).
- Harsh conditions in the internal environment can kill pathogens. For example, the gastric juice in the stomach is a deadly environment for many bacteria because of the hydrochloric acid and proteases that are secreted into it.

Cell signaling pathways stimulate additional innate defenses

Pathogens that are able to penetrate the body's outer and inner surfaces encounter a second line of innate defenses. These include the activation of defensive cells. In some animals and plants, this activation involves recognition of nonself molecules by a class of receptors called **pattern recognition receptors** (**PRRs**). In mammals, these receptors are mainly present on phagocytes and natural killer cells. The Toll-like receptor shown in Figure 36.2 is an example of a PRR. As we mentioned in Key Concept 36.1, the molecules recognized by PRRs are called pathogen-associated molecular patterns (PAMPs). PAMPs consist of unique constituents of microbes such as bacterial flagellin and fungal chitin. Other well-known PAMPs include bacterial lipopolysaccharide, found in bacterial cell membranes, and nucleic acid variants that are unique to viruses, such as double-stranded RNA. Binding of PAMPs to PRRs stimulates signal transduction pathways that lead to a variety of responses. One response is the production of antimicrobial peptides such as defensins (see above). Other responses include phagocytosis of invading organisms, activation of natural killer cells, activation of the complement system, and production of cytokines.

PHAGOCYTOSIS Pathogenic cells, viruses, and virus-infected cells can be recognized by phagocytes, which then ingest the invaders by phagocytosis (**FIGURE 36.6**). Once inside the phagocyte, the invader is killed either by hydrolysis within the lysosomes or by defensins.

NATURAL KILLER CELLS One class of lymphocytes, known as **natural killer cells**, can distinguish between healthy body cells and those that are infected by viruses or have become cancerous. In the latter cases they initiate apoptosis (programmed cell death; see Key Concept 7.5) in the target cells. In addition to this nonspecific action, natural killer cells interact with the adaptive defense mechanisms by lysing target cells that have been tagged with antibodies (as we will discuss in Key Concept 36.4).

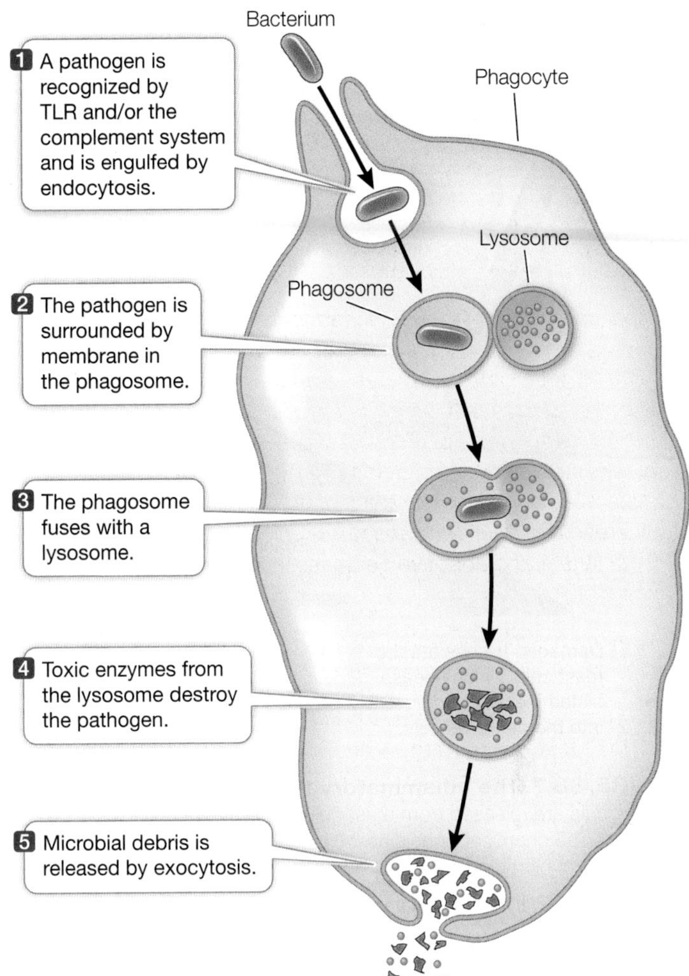

FIGURE 36.6 Phagocytosis

COMPLEMENT PROTEINS Vertebrate blood contains more than 20 different proteins that make up the antimicrobial **complement system**. Once the complement system has been activated, the proteins function in a characteristic sequence, or cascade, much like signal transduction cascades (see Key Concept 6.3), with each protein activating the next:

1. One complement protein binds to components on the surface of an invading cell. This binding helps phagocytes recognize and destroy the invading cell.
2. Another protein activates the inflammation response (see below) and attracts phagocytes to the site of infection.
3. Finally, other proteins insert into cell membranes of the invading cell, making pores that cause leakage from the cytosol and eventually, lysis.

The complement system can be activated by various mechanisms, including the innate and adaptive immune systems. In the latter case, the complement cascade is initiated by the binding of a complement protein to an antibody bound to the surface of the invading cell.

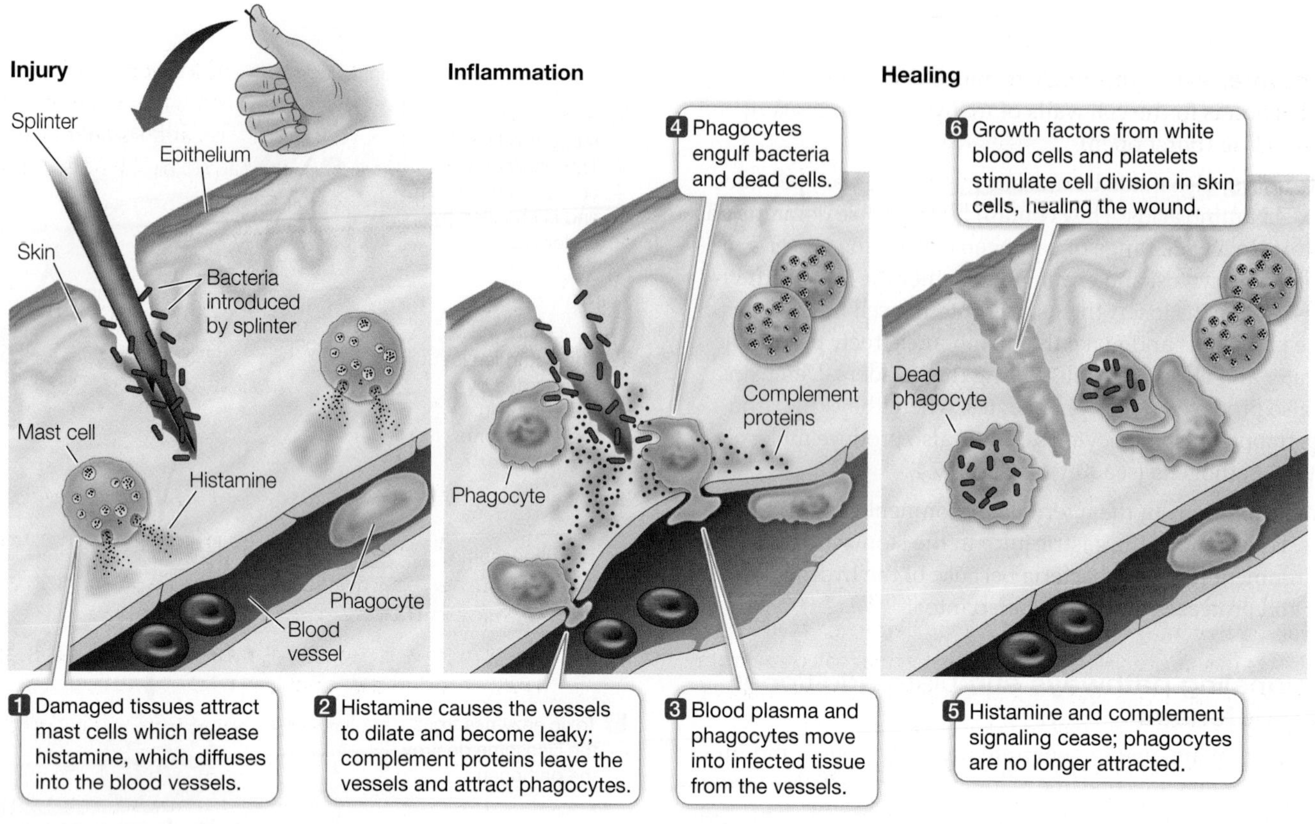

FIGURE 36.7 The Inflammatory Response Histamine and other signals are released from mast cells to initiate the inflammatory response. The chemical signals associated with inflammation attract phagocytes, which digest the pathogens and damaged cells.

Activity 36.2 Inflammatory Response

Antihistamines are a class of drugs widely used to treat people with inflammation. How do you think they work?

CYTOKINES **Cytokines** are signaling proteins that are released by many cell types. Cytokines that are produced in response to PRR activation include inflammatory cytokines (some interleukins and tumor necrosis factor; see below) and **interferons**, a class of proteins that help increase the resistance of neighboring cells to infection. Interferons are found in many vertebrates, and their production is induced by a variety of molecules. One such inducer is double-stranded viral RNA; thus interferons are particularly important in defense against viruses. Interferons bind to receptors on the cell membranes of uninfected cells, stimulating a signaling pathway that inhibits viral reproduction if the cells are subsequently infected. In addition, interferons stimulate the cells to hydrolyze bacterial or viral proteins to peptides, an initial step in adaptive immunity.

Inflammation is a coordinated response to infection or injury

When a tissue is damaged because of infection or injury, the body responds with **inflammation**: redness, swelling, and heat near the damaged site, which can then become painful. This response can happen almost anywhere in the body, internally as well as on the outer surface. Inflammation is an important phenomenon: it isolates the area to stop the spread of the damage; it recruits cells and molecules to the damaged location to kill any pathogens that might be present; and it promotes healing.

Among the first responders to tissue damage are **mast cells**, which adhere to the skin and the linings of organs and release numerous chemical signals, including:

- **Tumor necrosis factor**, a cytokine protein that kills target cells and activates immune cells
- **Prostaglandins**, fatty acid derivatives that play roles in various responses, including the initiation of inflammation in nearby tissues
- **Histamine**, an amino acid derivative that increases the permeability of blood vessels to white blood cells and molecules so they can act in nearby tissues

The redness and heat of inflammation result from the dilation of blood vessels in the infected or injured area (**FIGURE 36.7**). Phagocytes enter the inflamed area, where they engulf pathogens and dead tissue cells. Phagocytes are responsible for most of the healing associated with inflammation. They produce several cytokines, which (among other functions) can signal the brain to produce a fever. This rise in body temperature accelerates lymphocyte production and phagocytosis, thereby speeding the immune response. In some cases pathogens are temperature-sensitive, and their growth is inhibited by the fever. The pain of inflammation results from increased pressure due to swelling, the action of leaked enzymes on nerve endings, and the action of prostaglandins, which increase the sensitivity of the nerve endings to pain. Aspirin alleviates pain associated with inflammation by blocking the synthesis of prostaglandins.

Following inflammation, pus may accumulate. Pus is a mixture of leaked fluid and dead cells: bacteria, white blood cells, and damaged body cells. Pus is a normal result of inflammation, and is gradually consumed and further digested by macrophages. If there is a wound, platelets—small, irregularly shaped cell fragments that are present in the blood—aggregate at the wound site. The platelets produce growth factors that stimulate nearby skin cells to divide and heal the wound.

Inflammation can cause medical problems

Although inflammation is generally a good thing, sometimes the inflammatory response is inappropriately strong:

- In an **allergic reaction**, a nonself molecule that is normally harmless binds to mast cells, causing the release of histamine and subsequent inflammation (along with itchy, watery eyes, and rashes in some cases). The nonself molecule may come from food or from the environment—for example, from the surface of a plant pollen grain (as in hay fever) or a particle in dust.
- In **autoimmune diseases** such as rheumatoid arthritis, the immune system fails to distinguish between self and nonself, and attacks tissues in the organism's own body. In these cases the inflammation can be local, attacking an organ, or general, affecting tissues throughout the body.
- In **sepsis**, the inflammation that is due to a bacterial infection does not remain local. Instead it becomes widespread, with the dilation of blood vessels throughout the body. The resulting drop in blood pressure is a medical emergency and can be lethal.

REVIEW & APPLY | 36.2

R Innate immunity is the first line of defense against nonspecific pathogens, and includes physical barriers such as the skin and mucus, and cellular responses involving the recognition of self and nonself molecules. In some animals and plants, recognition of pathogens and/or pathogen constituents (PAMPs) is initiated by pattern recognition receptors (PRRs) such as the Toll-like receptor. Recognition of nonself molecules by white blood cells leads to coordinated responses such as phagocytosis, activation of natural killer cells, and the production of defensive proteins and inflammation.

1. A pathogenic bacterium lands on your skin. Outline the innate defenses that come into play.
2. When a splinter of wood penetrates the skin, the skin swells up. Outline the events by which this occurs.
3. Defensins made by mucous membranes are amphipathic, with polar and nonpolar regions. How does this relate to their function?

In most instances innate immunity is sufficient to block a pathogen from affecting the body. But many pathogens are present in huge numbers (think of a viral infection), and some may escape the innate defenses and begin to proliferate in the body. In these cases, adaptive immunity takes over.

36.3 | The Adaptive Immune Response Is Specific

LEARNING OBJECTIVES

By the end of this key concept you should be able to:

36.3.1 **Describe four key features of adaptive immunity.**

36.3.2 **Describe the interactions between humoral and cellular immunity.**

36.3.3 **Outline how plasma cells and memory cells are generated in the primary immune response.**

36.3.4 **Differentiate between B cells, cytotoxic T cells, and T-helper cells.**

Long before the twentieth century, scientists suspected that blood was somehow involved in immunity against pathogens, but they did not have definitive experimental evidence for this. More than a century ago, Emil von Behring and Shibasaburo Kitasato at the University of Marburg in Germany performed a key experiment that suggested that blood contained important factors for adaptive immunity (**INVESTIGATION FIGURE 36.8**). They showed that the blood serum (the noncellular fluid that remains after blood is clotted) of guinea pigs injected with a sublethal dose of diphtheria toxin developed a factor that, when injected into other guinea pigs, was able to protect them from a lethal dose of the same toxin. In other words, the serum recipients had somehow acquired immunity. The response of the donor guinea pigs to the diphtheria toxin is an example of adaptive immunity: after exposure to the toxin, they made a protective factor that was not present before their exposure. Moreover, the immunity was *specific*: the factor made by the first guinea pig protected the others only against the specific toxin produced by the strain of bacteria with which the first guinea pig had been injected.

Based on the animal model, Behring realized that serum protection might work for human diseases as well. It did, and he won a Nobel Prize for his efforts in protecting children against diphtheria. Later, the agent of this immunity was identified as an antibody protein, and the process of acquiring immunity from antibodies received from another individual was called **passive immunity**.

The injection of antibodies to produce a passive immune response is an artificial system that allows an organism to mount a defense against a single pathogenic strain. Vertebrates have evolved systems of adaptive immunity that have the potential to recognize and eliminate virtually any pathogen that may infect the body. We will now turn to some key features of adaptive immunity.

INVESTIGATION

FIGURE 36.8 The Discovery of Specific Immunity Until the twentieth century, most people did not survive an attack of diphtheria, but a few did. Emil von Behring and Shibasaburo Kitasato performed a key experiment using an animal model. They demonstrated that the factor(s) responsible for immunity against diphtheria was in the blood serum.[a]

HYPOTHESIS

Serum from guinea pigs injected with a sublethal dose of diphtheria toxin protects other guinea pigs that are exposed to a lethal dose of the same toxin.

METHOD

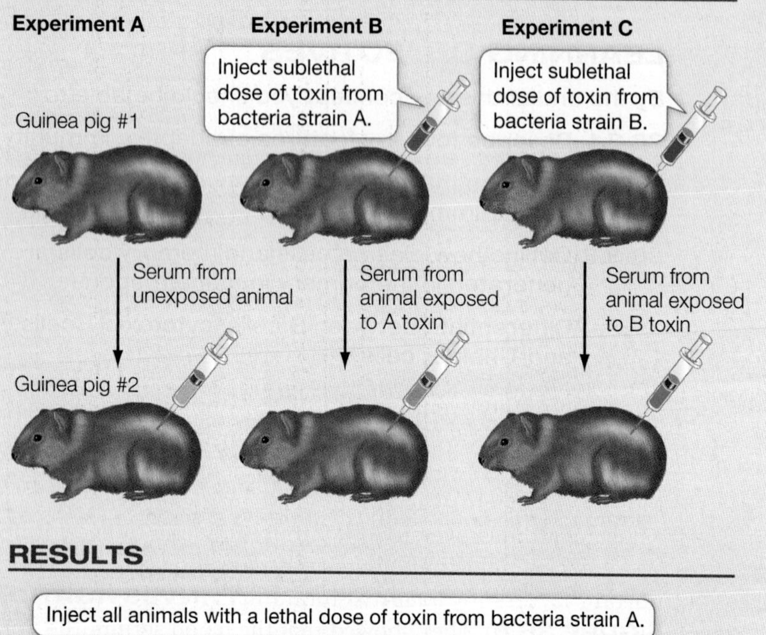

RESULTS

Inject all animals with a lethal dose of toxin from bacteria strain A.

Diphtheria | No diphtheria | Diphtheria

CONCLUSION

Serum of toxin-exposed guinea pigs is protective against later exposure to a lethal dose of toxin from the same genetic strain of bacteria, but not a different strain.

ANALYZE THE DATA

In their experiments, Behring and Kitasato used different doses of toxin to test the immunity of guinea pig #2 in each experiment. The results are shown in the table.

	Symptoms	
Experiment	0.5-ml dose	10-ml dose
A	Diphtheria	Diphtheria
B	No diphtheria	No diphtheria
C	Diphtheria	Diphtheria

1. Explain these data in terms of the level of protection afforded by the serum.
2. These experiments could be performed with either intact bacteria that cause diphtheria, or with a bacteria-free filtrate of a 10-day-old culture of the bacteria. Explain.

[a]E. von Behring and S. Kitasato. 1890. *Dtsch Med Wochenschrift* 49: 1113–1114.

A similar **Analyze the Data** exercise may be assigned in **Achieve**.

Adaptive immunity has four key features

Four important features of the adaptive immune system are: specificity; the ability to distinguish self from nonself; the ability to respond to an enormous diversity of nonself molecules; and immunological memory.

SPECIFICITY B cells and T cells are crucial for the specificity of adaptive immune responses. T cell receptors and the antibodies produced by B cells recognize and bind to specific nonself substances called antigens, and this interaction initiates a specific immune response. An **antigen** is any molecule that will elicit the synthesis of antibodies that specifically bind the antigen. Each T cell and each antibody-producing B cell is specific for a single antigen. Antigens are usually proteins or polysaccharides, and there can be multiple antigens on a single invading bacterium.

The sites on antigens that the immune system recognizes are called **antigenic determinants** or **epitopes** (**FIGURE 36.9**). An antigenic determinant is a specific portion of an antigen, such as a certain sequence of amino acids. A single antigenic molecule can have multiple different antigenic determinants. For the remainder of the chapter, we will often refer to antigenic determinants simply as "antigens."

DISTINGUISHING SELF FROM NONSELF As we saw in Key Concept 36.2, the innate immune system can recognize certain nonself molecules (PAMPs), and mount defensive responses to

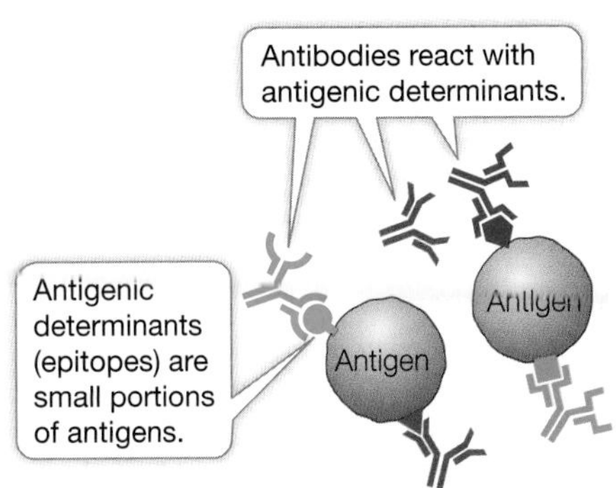

FIGURE 36.9 Antigenic Determinants An antigenic determinant (also called an epitope) is a specific region of an antigen to which the immune system generates a response. B cells produce antibodies (and T cells produce T cell receptors) that are specific for individual antigens.

the pathogens from which these nonself molecules were derived. The adaptive immune system recognizes both self and nonself molecules. The human body contains tens of thousands of different molecules, each with a specific three-dimensional structure capable of generating an immune response. Thus every cell in the body bears a tremendous number of antigens. A crucial requirement of an individual's immune system is that it recognize the body's own antigens and not attack them.

Normally the body is tolerant of its own molecules—the same molecules that would generate an immune response in another individual. This occurs primarily during the early differentiation of T and B cells, when they first encounter self antigens. Any immature B or T cell that shows the potential to mount a strong immune response against self antigens undergoes apoptosis within a short time. The elimination of potentially self-harming T and B cells (immature T and B cells expressing receptors for self antigens) is called **clonal deletion**.

A failure of clonal deletion can lead to an immune response to self antigens, a condition called **autoimmunity**. While the causes of autoimmunity are unclear, there appears to be a strong genetic component. The most common autoimmune diseases are:

- Systemic lupus erythematosus (SLE—often called lupus): People develop antibodies to cellular components, such as proteins in the cell nucleus. Indeed, antinuclear antibodies are a hallmark of this disease. These antibodies can cause serious damage when they bind to normal tissue antigens and form large circulating antigen–antibody complexes that can become stuck in tissues and provoke inflammation.
- Hashimoto's thyroiditis: This is the most common autoimmune disease in women over age 50, with 1 person in 20 afflicted at some point in life. Immune system cells attack and ultimately destroy thyroid gland tissue, resulting in fatigue, depression, weight gain, and other symptoms associated with impaired thyroid function.
- Type 1 diabetes: Immune system cells attack and destroy the cells in the pancreas that normally make insulin. As noted in Chapter 6, a lack of insulin results in excess blood glucose levels, with significant clinical effects.

DIVERSITY Pathogens take many forms: viruses, bacteria, protists, fungi, and multicellular parasites. Furthermore, each pathogenic species usually exists as one of many subtly different genetic strains, and each strain possesses multiple surface features. Estimates vary, but a reasonable guess is that humans can respond specifically to more than 10 million different antigens. Upon recognizing an antigen, the immune system responds by activating lymphocytes of the appropriate specificity.

To have the ability to respond to the large number of potential pathogens, the body needs to generate a vast diversity of lymphocytes, each of which is specific for a unique antigen. These lymphocytes represent a pool from which specific cells are selected when needed.

- *Diversity is generated primarily by DNA changes*—chromosomal rearrangements and other mutations—that occur just after the B and T cells are formed in the bone marrow. Each B cell is able to produce only one kind of antibody; thus there are millions of different B cells. Similarly, there are millions of different T cells with specific T cell receptors. The adaptive immune system is "predeveloped"—*all of the machinery available to respond to an immense diversity of antigens is already there, even before the antigens are encountered.*
- *Antigen binding "selects" a particular B or T cell for proliferation.* For example, when an antigen fits the surface receptor on a B cell and binds to it, that B cell is activated. It divides to form a clone of cells (a genetically identical group derived from a single cell), all of which produce and/or secrete antibodies with the same specificity as the receptor (**FIGURE 36.10**). Binding, activation, and proliferation also apply to T cells. A particular lymphocyte is selected via binding and activation, and then it proliferates to generate a clone—hence the name **clonal selection** for this mechanism of producing an immune response.

IMMUNOLOGICAL MEMORY After responding to a particular antigen on a particular type of pathogen once, the immune system "remembers" that pathogen and can usually respond more rapidly and powerfully to the same threat in the future. **Immunological memory** usually saves us from repeats of childhood infectious diseases.

Immunological memory is a product of the adaptive immune system and develops in two phases called, respectively, the primary immune response and the secondary immune response. Several days after the first time a vertebrate animal is exposed to a particular antigen, the **primary immune response** produces antibodies and T cells specific for that antigen. After a primary immune response to a particular antigen, subsequent encounters with the same antigen will trigger a much more rapid and powerful **secondary immune response**. Immunological memory arises from the fact that activated lymphocytes divide and differentiate to produce *two types* of daughter cells: effector cells and memory cells (see Figure 36.10).

- An immune response generates **effector cells**, which carry out the attack on the antigen. Effector B cells, called **plasma cells**, secrete antibodies. Effector T cells

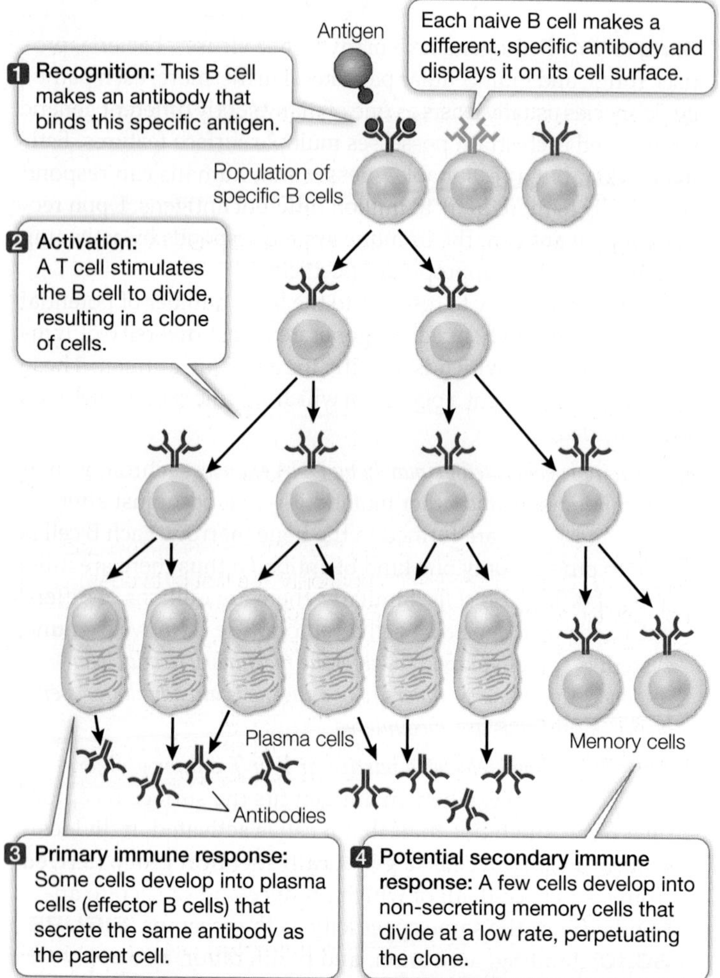

FIGURE 36.10 Clonal Selection in B Cells The binding of an antigen to a specific receptor on the surface of a B cell stimulates that cell to divide, producing a clone of genetically identical cells to fight that invader.

Assuming you never have been exposed to the virus that causes smallpox, do you have cells that make antibodies against this virus?

release cytokines and other molecules that initiate reactions that destroy nonself or altered cells. Effector cells live for only a few days.

- The primary immune response also generates **memory cells**, long-lived cells that retain the ability to start dividing on short notice to produce more effector and more memory cells. Memory B and T cells may survive in the body for decades, rarely dividing.

In a secondary immune response, the memory cells that bind with that antigen proliferate, launching a huge "army" of plasma cells and effector T cells. The principle behind vaccination is to trigger a primary immune response that prepares the body to mount a stronger, quicker secondary response if it encounters the actual pathogen. Today vaccines are routinely administered to prevent disease caused by many highly infectious and often devastating microbial and viral pathogens. Probably you were vaccinated as an infant against hepatitis B, measles, mumps, chicken pox, diphtheria, tetanus, whooping cough, bacterial and viral influenza, and polio. In most regions of the developed world, all children receive these vaccines and it is usually required for starting school. When about 90 percent of a population is vaccinated, **herd immunity** occurs, meaning that there are not enough unvaccinated, susceptible individuals for the pathogen to infect, and so it is lost from the population.

Macrophages and dendritic cells play a key role in activating the adaptive immune system

We have just seen how the adaptive immune system uses antigens to distinguish between self and nonself, and how it uses clonal selection to produce large numbers of specific B and T cells. In Key Concepts 36.4 and 36.5 we will focus on how these cells function to eliminate specific pathogens from the body. But how is the adaptive immune system activated in the first place?

One mechanism for eliminating pathogens in the innate immune system is phagocytosis (see Figure 36.6). After ingestion of a pathogenic organism or infected host cell, phagocytic cells display fragments of the pathogen on their cell surfaces. These fragments function as antigens, and **antigen presentation** is one way that components of the innate immune system communicate with the adaptive immune system. Macrophages and dendritic cells play a key role in activating the adaptive immune system. After engulfing pathogens or infected host cells, these cells migrate to lymph nodes, where they present antigen to immature T cells. In addition, the antigen-presenting cells secrete cytokines and other signals that stimulate the activation and differentiation of the T cells.

Two types of adaptive immune responses interact

The adaptive immune system mounts two types of responses against invaders: the **humoral immune response** and the **cellular immune response**. ("Humor" is an old term for a body fluid.) These two responses work simultaneously and cooperatively, sharing many mechanisms. We will use the example of a viral infection in an overview of these two types of responses (**FIGURE 36.11**). These responses also occur when bacteria or other pathogens infect and grow inside host cells.

B cells that make antibodies are the workhorses of the humoral immune response, and **cytotoxic T (T_C) cells** are the workhorses of the cellular immune response. There are three phases in both types of immune response:

- *Recognition phase.* In both cellular and humoral immunity, recognition occurs when an antigen is inserted into the cell membrane of an antigen-presenting cell, with the unique antigen structure protruding from the cell membrane. In addition to dendritic cells and macrophages, developing B cells can also perform phagocytosis and function as antigen-presenting cells. Figure 36.11 shows a bacteria-infected host cell being engulfed by an antigen-presenting cell; in other cases, the free viral particles may be engulfed. The antigen on the

THINK LIKE A SCIENTIST

Science & Society

Avoiding incompatibilities in blood type: the immune response to the Rh factor

Many scientific insights have benefited society through the medical advances that have followed. Our current understanding of the Rh factor provides a prime example.

In humans, the dominant gene *D* codes for the RhD protein (Rhesus—so named because its discoverers thought, incorrectly, that the human version also occurs in monkeys) on the surfaces of red blood cells. The recessive allele ***d*** does not encode a cell surface protein. If cells from an individual who is *DD* or *Dd* (Rh^+) enter the bloodstream of an individual who is *dd* (Rh^-), the Rh^- individual, encountering a nonself antigen, activates B cells that make antibodies that react with the RhD protein.

During pregnancy, blood vessels from the developing fetus and mother come near one another in an organ called the placenta. As you learned in Chapter 35, exchange of gases and nutrients between maternal and fetal blood vessels is important. But there can also be some leakage of cells across the thin capillaries. If a mother who is Rh^- (about 15% of the U.S. population) encounters fetal red blood cells that are Rh^+, the mother mounts an immune response. The effects of maternal anti-protein D antibodies on fetal red blood cells can be dangerous, including severe anemia (low red blood cell amounts) in the newborn.

Mother		Fetus		
Genotype	Rh phenotype	Genotype	Rh phenotype	Result
Dd		*Dd*		
dd		*Dd*		

1. Hemolytic disease of the newborn (sometimes resulting in stillbirth) can occur when antibodies from the mother pass into the fetal circulatory system. Fill in the table above, which predicts RhD incompatibility.
2. During birth there is mixing of the blood supplies of mother and fetus. In a situation of RhD incompatibility, the first birth does not result in a clinical problem. But a second RhD incompatible birth with this mother results in hemolytic disease. Explain how this can happen (*Hint*: immunological memory).
3. Today RhD incompatibility does not result in stillbirths. The mother is treated with serum containing anti-RhD antibodies late in the first and subsequent pregnancies. How can this prevent the immunological reactions?

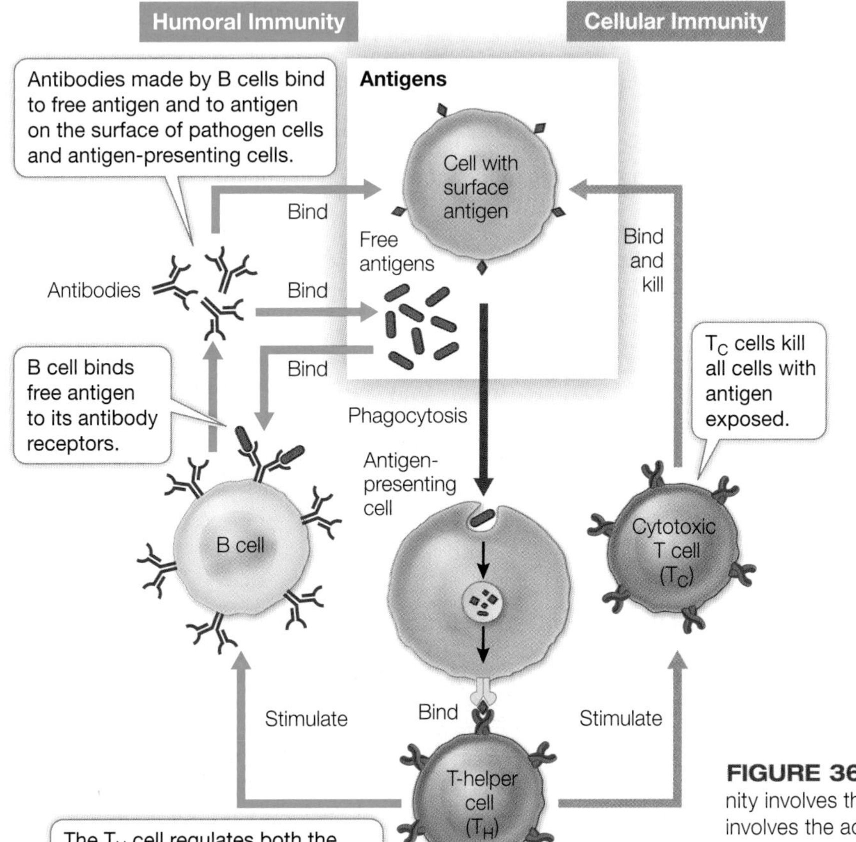

FIGURE 36.11 The Adaptive Immune System Humoral immunity involves the production of antibodies by B cells. Cellular immunity involves the activation of cytotoxic T cells that bind to and destroy self cells that are mutated or infected by pathogens. Details of these processes are described later in this chapter.

surface of the antigen-presenting cell is recognized by a **T-helper (T_H) cell** bearing a T cell receptor protein that is specific for the antigen. In both humoral and cellular immunity, binding initiates the activation phase.

- *Activation phase.* When the T_H cell recognizes an antigen on an antigen-presenting cell, it propagates and releases cytokines that stimulate B cells and T_C cells bearing receptors to the same antigen to divide. The results are a clone of B cells that function in humoral immunity and a clone of T_C cells that function in cellular immunity.
- *Effector phase.* In the humoral immune response, cells of the B clone produce antibodies that bind to bacterial components and/or virus-infected cells. The bound antibodies attract phagocytes and complement proteins that engulf and destroy bacteria (or the virus-infected cells, as the case may be). In cellular immunity, cells of the T_C clone bind to cells with antigen exposed and destroy them.

REVIEW & APPLY | 36.3

R The adaptive immune system reacts against nonself molecules called antigens. The system generates amazing diversity in both antibodies (produced by B cells) and in T cell receptors. In the primary immune response, B cells and T cells that recognized a particular antigen proliferate by clonal selection. Immunological memory prepares the body for a much stronger secondary immune response. The principle behind vaccination is to trigger a primary immune response that prepares the body to mount a stronger, quicker secondary response if it encounters the actual pathogen.

A
1. List the four key features of the adaptive immune system.
2. Describe the process of clonal selection in B cells, and the functions of the cells generated.
3. Distinguish between humoral and cellular immunity, discussing the functions of B cells, T_H cells, and T_C cells.

With this overview of the basic characteristics of adaptive immunity and the events of the immune response, we will now examine some of the cellular and molecular characteristics of the humoral response, particularly how antibody diversity arises.

36.4 The Adaptive Humoral Immune Response Involves Specific Antibodies

LEARNING OBJECTIVES

By the end of this key concept you should be able to:

36.4.1 Sketch the structure of an antibody.

36.4.2 Describe how genomic rearrangements make antibody diversity possible.

36.4.3 Describe how classes of antibodies differ, and identify which class is most common and why.

Every day in the human body, billions of B cells survive the test of clonal deletion and are released from the bone marrow into the circulatory system. B cells are the basis for the humoral immune response.

Plasma cells produce antibodies that share a common overall structure

A B cell begins as a "naive" B cell with a receptor protein on its cell surface that is specific for a particular antigen. The B cell is activated by antigen binding to this receptor, and after stimulation by a T_H cell, it gives rise to a clone of plasma cells that make antibodies as well as to a smaller number of memory cells (see Figure 36.10). The stimulation occurs after the B cell presents antigen to a T_H cell with a receptor that can recognize the antigen. The T_H cell then secretes cytokines that stimulate the B cell to divide. We'll describe this process in more detail in Key Concept 36.5.

All the plasma cells arising from a given B cell produce antibodies that are specific for the antigen that originally bound to the parent B cell. Thus antibody specificity is maintained as B cells proliferate.

Animation 36.2 Humoral Immune Response

There are several classes of antibodies—also called **immunoglobulins**—and all contain a minimum of two antigen-binding sites. An antibody is composed of four polypeptide chains: two identical longer **heavy chains** that anchor the antigen receptor to the B cell, and two identical shorter **light chains**, flanking the heavy chains to form the antigen-binding site (**FIGURE 36.12**).

(A) General structure of an immunoglobulin

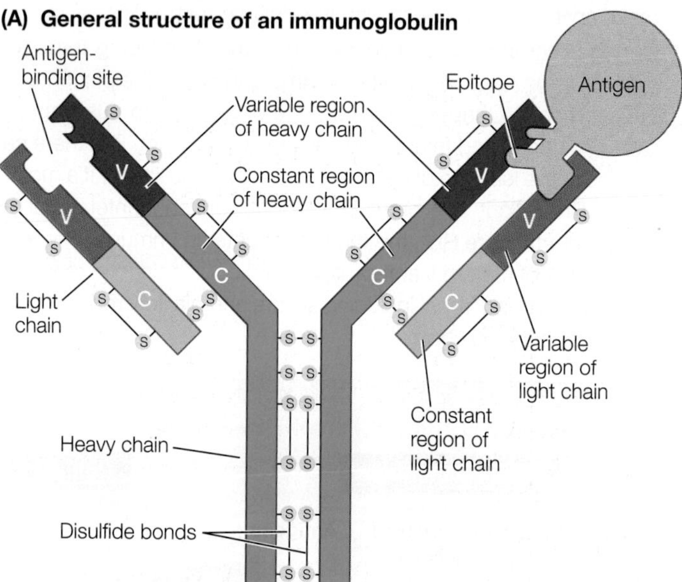

(B) Space-filling model of an immunoglobulin

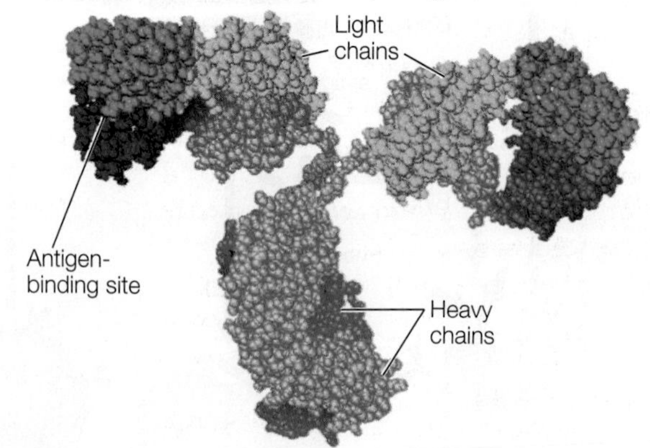

FIGURE 36.12 The Structure of an Immunoglobulin Four polypeptide chains (two light, two heavy) make up an immunoglobulin molecule. Here we show both diagrammatic **(A)** and space-filling **(B)** representations of an immunoglobulin. The variable regions (V) of the antigen-binding site distinguish an antibody from other antibodies and allow it to bind to a specific antigen. The constant regions (C) on light and heavy chains are so called because their amino acid sequences vary little among different antibodies.

Activity 36.3 Immunoglobulin Structure

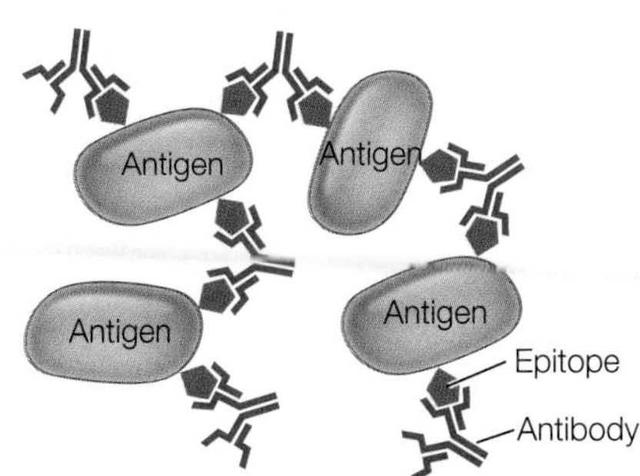

FIGURE 36.13 An Antigen-Antibody Complex Because each antibody is bivalent, antibodies can form large complexes with antigens that are easy targets for ingestion and breakdown by phagocytes.

The four chains combine to form a Y-shaped structure, held together by disulfide bridges. Each of the light and heavy chains is subdivided into a constant region and a variable region:

- The amino acid sequence of the **constant region** determines the general structure and function (the class) of an immunoglobulin. All immunoglobulins in a particular class have an identical constant region. When an antibody acts as a B cell receptor, the constant region is inserted into the B cell membrane, anchoring the antibody to the cell surface.
- The amino acid sequence of the **variable region** is different for each specific immunoglobulin. The three-dimensional shape of the variable region's antigen-binding site is responsible for antibody specificity.

The two antigen-binding sites on each immunoglobulin molecule are identical, making the antibody bivalent (*bi*, "two," + *valent*, "binding"). The ability to bind two antigen molecules at once, along with the existence of multiple epitopes (antigenic determinants) on each antigen, permits antibodies to form large complexes with the antigens. For example, one antibody might bind two molecules of an antigen. Another antibody might bind the same antigen at a different epitope or to an antigen that is already bound to the first antibody, along with a third antigen molecule (**FIGURE 36.13**). The binding of multiple antigens and multiple antibodies—a process called **agglutination**—clumps the antigens together into complexes that are later endocytosed by phagocytes and then hydrolyzed within their lysosomes.

There are five classes of immunoglobulins (Ig), which are defined by five different types of heavy chains distinguished by characteristics of their constant regions. Each class differs in function:

1. IgG constitutes about 80 percent of circulating antibodies.
2. IgD is a cell surface receptor on a B cell.
3. IgM is the initial surface and circulating antibody released by a B cell.
4. IgA protects mucous membranes exposed to the environment.
5. IgE binds to mast cells and is involved with inflammation.

Differences in the constant regions endow different classes of immunoglobulins with different capabilities. Some, for example, can bind complement proteins; some cannot. A key property of IgG antibodies is their ability to bind and **opsonize** microbes in blood and tissue fluids—a process by which microbes are "tagged" for destruction by phagocytes (**FIGURE 36.14**). We will focus on IgG antibodies in this chapter.

Antibody diversity results from DNA rearrangements and other mutations

Each mature B cell makes many identical antibodies targeted to only a single antigenic determinant. And there are millions of possible antigenic determinants to which a human can be exposed. A single complex molecule or virus might have many different epitopes, each provoking an immune response. Because each B cell type producing a specific antibody is a clone of identical cells, the response to a complex antigen is called **polyclonal**, in contrast to the **monoclonal** response to a single epitope.

There are millions of epitopes on antigens that the immune system recognizes as nonself. This means that there are millions of individual B cells, each encoding an antibody with a unique amino acid sequence that allows it to bind to a particular epitope. How can the genome encode enough different antibodies to protect the body against all the possible pathogens expressing unique epitopes? There are two possibilities:

1. Each person inherits millions of different genes encoding immunoglobulins. Simply put, this does not happen. Indeed, there are about 20,000 protein-coding genes in the human genome (see Key Concept 12.4), less than a few hundred of them encoding immunoglobulin light and heavy chains.

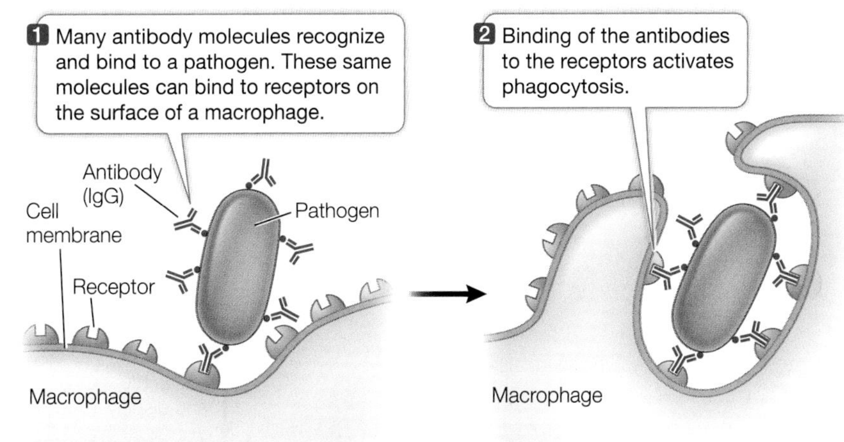

FIGURE 36.14 Opsonization by IgG Antibodies Binding of IgG antibodies to the antigen either directly stimulates phagocytosis, or initiates the complement system, which then stimulates phagocytosis and clearance of the foreign material.

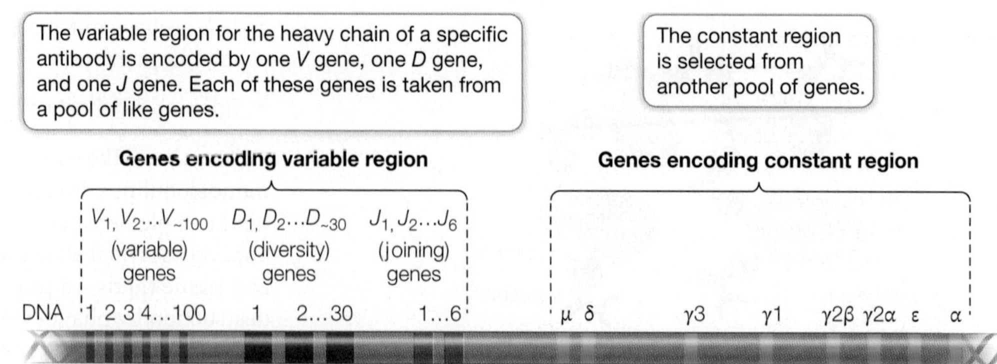

FIGURE 36.15 Supergenes Mouse immunoglobulin heavy chains have four domains, each of which is coded for by a gene selected from a cluster of similar genes. The immunoglobulin protein has one domain from each cluster.

2. Each person inherits a moderate number of genes that are rearranged randomly to form the genes encoding the light and heavy chains (or parts thereof). The randomness generates diversity through mixing and matching of the genes. This is what occurs, as we explain below.

Each gene encoding an immunoglobulin chain is in reality a "supergene" assembled by means of genetic recombination from several clusters of smaller genes scattered along part of a chromosome (**FIGURE 36.15**). Every cell in the body has hundreds of immunoglobulin genes located in separate clusters that are potentially capable of participating in the synthesis of both the variable and constant regions of immunoglobulin chains. In most body cells and tissues, these genes remain intact and separated from one another. But during B cell development, these genes are cut out, rearranged, and joined together in DNA recombination events. One gene from each cluster is chosen randomly for joining, and the others are deleted. In the case of one multigene set, the *J* genes, some of the extra sequences are removed by RNA splicing (**FIGURE 36.16**).

In this manner, a unique immunoglobulin supergene is assembled from randomly selected "parts." Each B cell precursor

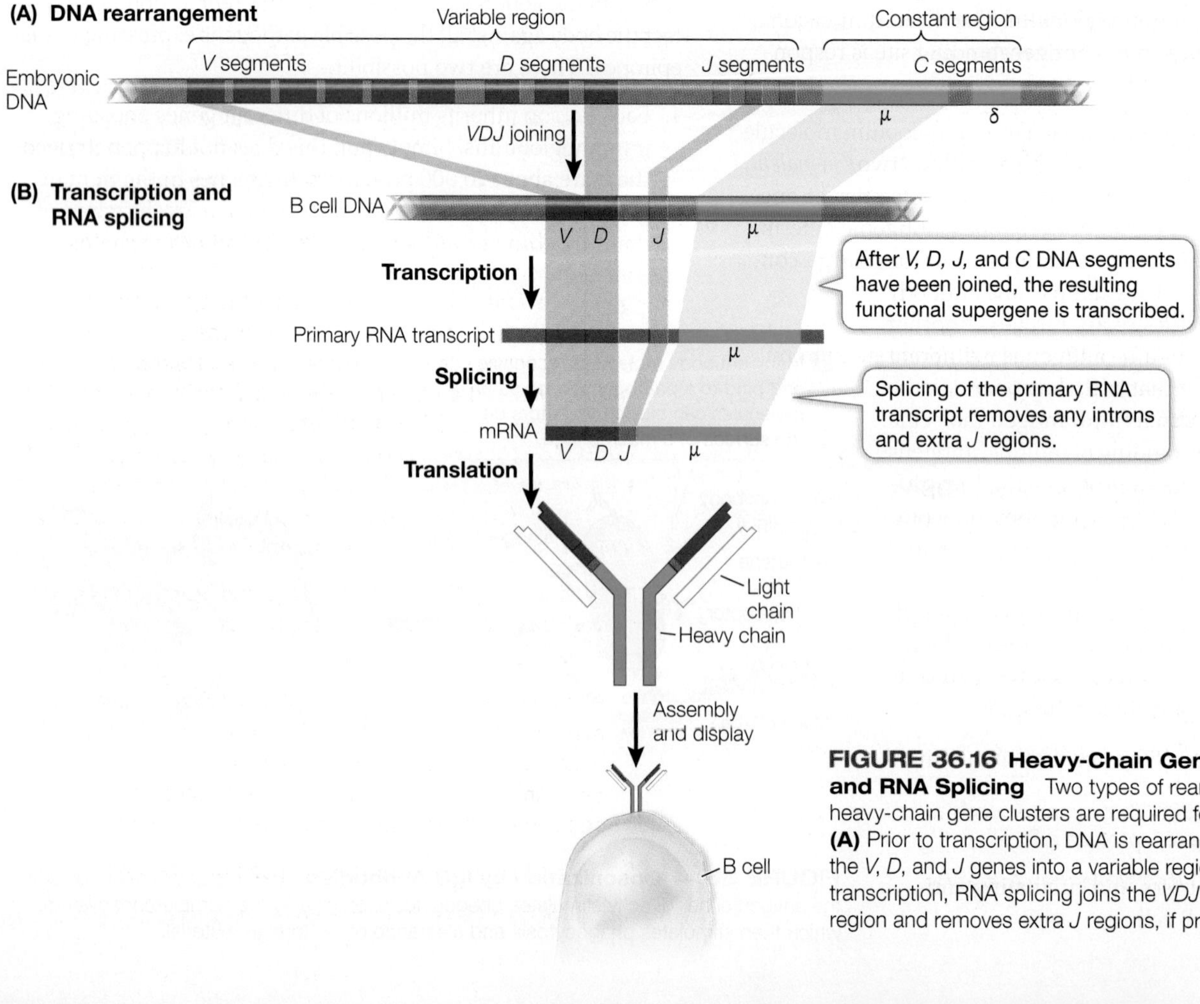

FIGURE 36.16 Heavy-Chain Gene Recombination and RNA Splicing Two types of rearrangements in the heavy-chain gene clusters are required for antibody formation. **(A)** Prior to transcription, DNA is rearranged to join one each of the *V*, *D*, and *J* genes into a variable region supergene. **(B)** After transcription, RNA splicing joins the *VDJ* region to the constant region and removes extra *J* regions, if present.

assembles two supergenes, one for a specific heavy chain and the other, assembled independently, for a specific light chain. This remarkable example of irreversible cell differentiation generates an enormous diversity of immunoglobulins from the same genome. It is a major exception to the generalization that all somatic cells derived from the fertilized egg have identical DNA.

Figure 36.15 illustrates the gene families that encode the constant and variable regions of the heavy chain in mice. Multiple genes encode each of the three parts of the variable region: 100 *V*, 30 *D*, and 6 *J* genes. Each B cell randomly selects one gene from each of these clusters to make the final coding sequence (*VDJ*) of the heavy-chain variable region. So the number of different heavy chains that can be made through this random recombination process is quite large:

$$100\ V \times 30\ D \times 6\ J = 18{,}000 \text{ possible combinations}$$

Now consider that the light chains are similarly constructed, with a similar amount of diversity made possible by random recombination. If we assume that the degree of potential light-chain diversity is the same as that for heavy-chain diversity, the number of possible combinations of light- and heavy-chain variable regions is:

$$18{,}000 \text{ different light chains} \times 18{,}000 \text{ different heavy chains} = 324 \text{ million possibilities!}$$

Other mechanisms generate even more diversity:

- When the DNA sequences that encode the *V*, *D*, and *J* regions are rearranged so that they are next to one another, the recombination event is not precise, and errors occur at the junctions. This imprecise recombination can create frame-shift mutations, generating new codons at the junctions, with resulting amino acid changes.
- After the DNA sequences are cut and before they are rejoined, the enzyme terminal transferase often adds some nucleotides to the free ends of the DNA pieces. These additional bases create insertion mutations.
- There is a relatively high spontaneous mutation rate in immunoglobulin genes. Once again, this process creates many new alleles and adds to antibody diversity.

LINK You can review the effects of DNA mutations on the amino acid sequences of proteins in **Key Concept 10.3**.

When we include these possibilities with the millions of combinations that can be made by random DNA rearrangements, it is not surprising that the immune system can mount a response to almost any natural or artificial substance.

Once the DNA rearrangements are completed, each supergene is transcribed and then translated to produce an immunoglobulin light chain or heavy chain. These chains combine to form an active immunoglobulin protein.

Animation 36.3 A B Cell Builds an Antibody

Antibodies bind to antigens and activate defense mechanisms

Recall that antibodies have two roles in B cells after they undergo DNA rearrangements and RNA splicing. First, by being expressed on the cell surface, a unique antibody can act as a receptor for an antigen in the recognition phase of the humoral response (see Figures 36.10 and 36.11). Second, in the effector phase of the humoral response, specific antibodies are produced in large amounts by a clone of B cells. These antibodies are secreted from the B cells and enter the bloodstream, where they act in either of two ways:

- *Activation of the complement system*. When antibodies bind to antigens that are present on the surfaces of pathogens, they attract components of the complement system (see Key Concept 36.2). Some complement proteins attack and lyse the pathogenic cells. In addition, the binding of antibody and complement proteins stimulates phagocytes (monocytes, macrophages, or dendritic cells) to ingest the pathogens.
- *Activation of effector cells*. Because of its cross-linking function, antibody binding can result in large, insoluble antibody–pathogen or antibody–antigen complexes. These attract effector cells such as phagocytes and natural killer cells.

REVIEW & APPLY | 36.4

R Naive B cells are activated by binding of the antigen and by stimulation from T_H cells with the same specificity, and then form plasma cells. These cells synthesize and secrete specific antibodies (immunoglobulins). The basic unit of an immunoglobulin is a tetramer of four polypeptides: two identical light chains and two identical heavy chains, each consisting of a constant region and a variable region. The variable regions determine the specificity of an immunoglobulin, and the constant regions of the heavy chain determine its class. There are five classes of immunoglobulins, differing in function and in the constant region of the heavy chain. B cell genomes undergo recombination events in which the genes that encode specific domains of the immunoglobulin variable regions are randomly selected from large clusters of genes. This DNA rearrangement yields millions of different immunoglobulin proteins.

1. Sketch an IgG antibody, the light and heavy chains, variable and constant regions, and antigen binding sites.
2. The bacterium that causes diphtheria (see Figure 36.8) synthesizes a toxic protein. You have probably not been exposed to this bacterium or its toxin. At the present time, are you making B cells and antibodies that bind specifically to diphtheria toxin? Explain your answer.
3. Why is IgG the most common class of antibody?

The B cells that make antibodies during the humoral immune response work in concert with the T cells activated by the cellular immune response. In the next key concept we'll look at several types of T cells and their roles in the immune response.

36.5 The Adaptive Cellular Immune Response Involves T Cells and Their Receptors

LEARNING OBJECTIVES

By the end of this key concept you should be able to:

36.5.1 Summarize the role of MHC proteins in the humoral immune response.

36.5.2 Describe the roles of T cells and MHC proteins in transplantation of organs from one unrelated individual to another.

36.5.3 Predict consequences of inappropriate levels of Treg activity.

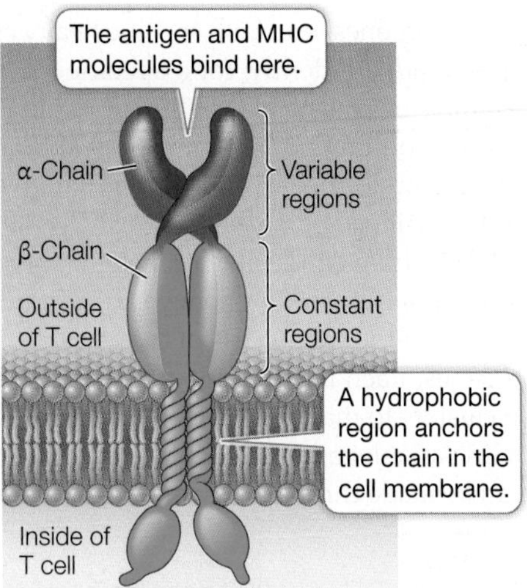

FIGURE 36.17 A T Cell Receptor The receptors on T lymphocytes are smaller than those on B lymphocytes. The T cell receptor is formed by two polypeptide chains containing both variable and constant regions. As with the B cell receptors, the constant regions fix the receptor in the cell membrane, while the variable regions establish the specificity for binding to antigen.

Two types of effector T cells, T-helper cells and cytotoxic T cells, are involved in the cellular immune response, along with proteins of the major histocompatibility complex (MHC). MHC proteins (called HLA, or human leukocyte antigens, in humans) present antigens on the surfaces of cells and contribute to the immune system's tolerance for the body's own cells.

Animation 36.4 Cellular Immune Response

T cell receptors specifically bind to antigens on cell surfaces

Like B cells, T cells possess specific receptors protruding from their cell membranes. The T cell receptor is not an immunoglobulin, however, but a glycoprotein made up of two polypeptide chains, each encoded by a separate gene (**FIGURE 36.17**). Like immunoglobulins, T cell receptors have distinct regions with constant and variable amino acid sequences, and the latter provide the site for specific binding to antigens. However, there is an important difference between immunoglobulin binding and T cell receptor binding: whereas an immunoglobulin can bind to an antigen whether it is present on the surface of a cell or in soluble form, unattached to a cell, a T cell receptor binds only to an antigen displayed along with an MHC protein on the surface of an antigen-presenting cell. This is a key difference between humoral and cellular immunity.

MHC proteins present antigens to T cells and result in recognition

T cells are so called because they mature within the thymus. So far we have described two types of T cells: T-helper (T_H) and cytotoxic T (T_C) cells (see Figure 36.11). Both cell types express T cell receptors that bind to antigen expressed on the cell surface. But the response of each T cell type to binding to a presented antigen is quite different:

- T_H binding results in activation of the adaptive immune response.
- T_C binding results in death of the cell displaying the antigen.

The key to which type of T cell binds to an antigen-presenting cell is the kind of MHC protein used to present the antigen. **Class I MHC** proteins are present on the surface of every nucleated cell in the mammalian body. MHC I proteins present antigens to T_C cells. These antigens can be fragments of virus proteins in virus-infected cells or abnormal proteins made by cancer cells as a result of mutations. **Class II MHC** proteins are on the surfaces of macrophages, B cells, and dendritic cells. MHC II proteins present antigens to T_H cells. The macrophages, B cells, and dendritic cells ingest antigens and break them down; one of the fragments then binds to MHC II for presentation (**FIGURE 36.18**).

To accomplish its role in antigen presentation, an MHC protein has an antigen-binding site that can hold a peptide of about 10–20 amino acids. In humans there are six genetic loci for class I MHC proteins, with more than 1,000 alleles, and six genetic loci for class II MHC proteins, with about 500 alleles. Thus when an antigen-presenting cell uses an MHC molecule to present an antigen to a T cell, the T cell receptor in the latter recognizes two structures: the unique epitope of the antigen and the particular three-dimensional structure of the MHC molecule. Normally the antigen-presenting cell has the same MHC genes as the T cell; after all, both are from the same individual.

But what happens if the presenting cell is from a different person? In this case, with all the genetic diversity noted, the MHC molecule on the presenting cell will be recognized as nonself by the recipient's T cells. The latter will signal an immune response, and ultimately T_c cells will destroy the nonself cells. This is what happens in rejection of tissue transplants. One way around this

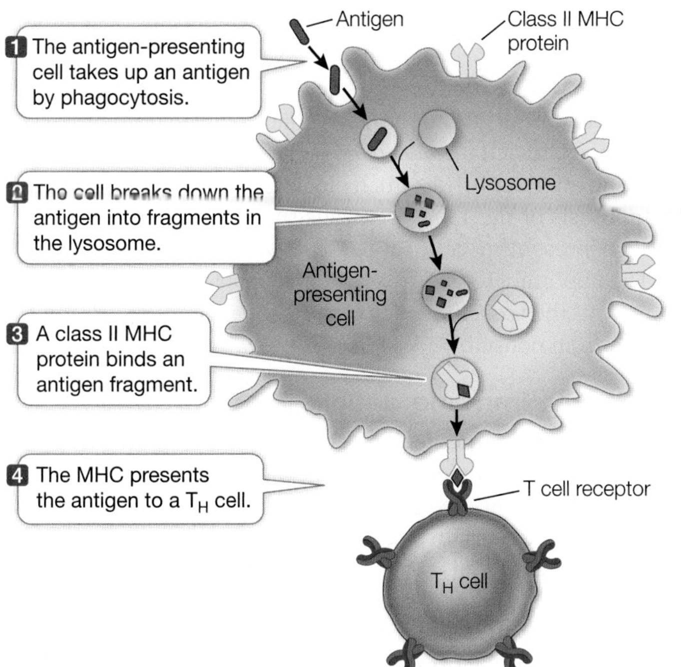

FIGURE 36.18 Macrophages Are Antigen-Presenting Cells A fragment of an antigen is displayed by MHC II protein on the surface of a macrophage. T cell receptors on a specific T-helper cell can then bind to and interact further with the antigen–MHC II protein complex.

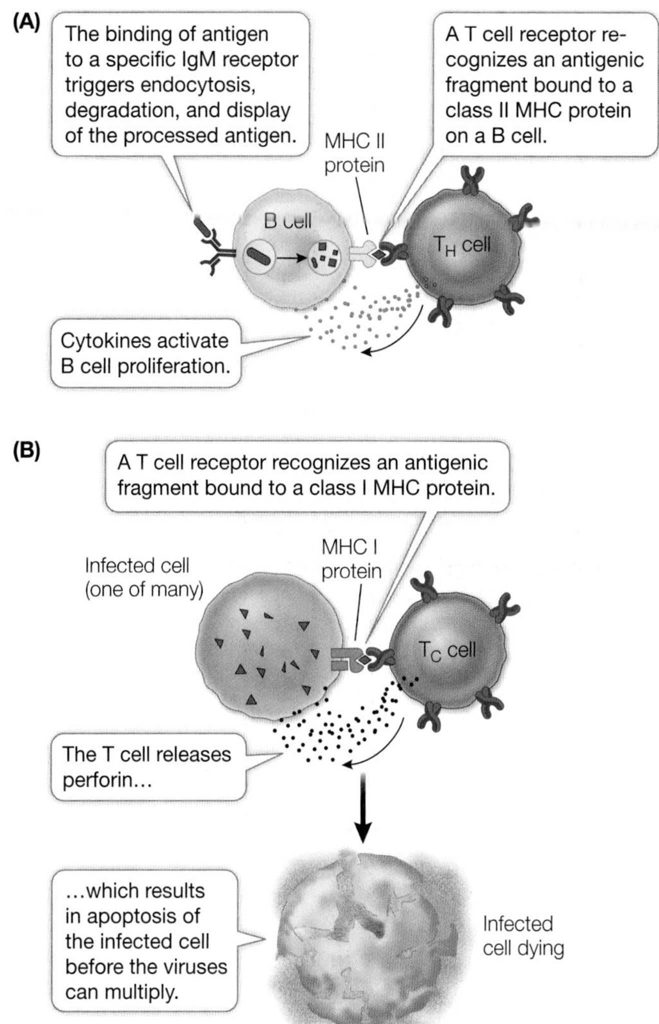

FIGURE 36.19 The Effector Phase in the Adaptive Immune Response **(A)** B cells acting as antigen-presenting cells display antigenic fragments on class II MHC proteins that are recognized by a receptor on a T_H cell. The subsequent release of cytokines causes the B cell to produce a clone of plasma cells and memory cells. The plasma cells will release antibody to complete the effector phase of the humoral immune response. **(B)** When a cytotoxic T cell recognizes and binds to an antigen–MHC I complex, it releases perforin, which destroys the infected cell.

problem is to use donor tissue from the same or a closely related person so that the MHC molecules on the donor tissue are the same as, or close to, those of the recipient, and the latter's T cell receptors do not recognize the donor tissue as nonself. Rapid genetic tests are available to identify a person's MHC genes. There is a "bank" of donors with identified MHC genotype. When a person needs a transplant, the genotypes of millions of people are scanned in the database for a possible genetic match.

Information on MHC proteins, types of antigens, and T lymphocytes is summarized in **TABLE 36.2**.

T_H cells contribute to the humoral and cellular immune responses

A T_H cell is activated when it has receptors that can bind a specific antigen on the surface of an antigen-presenting cell. This results in the rapid clonal propagation of the T_H cell. The T_H cells then release cytokines that stimulate other immune cells, including T_C cells, to propagate.

A T_H cell can also directly interact with a B cell that recognizes the same antigen. Like dendritic cells and macrophages, B cells are antigen-presenting cells. B cells take up antigens bound to their surface immunoglobulin receptors by endocytosis, break them down, and display antigenic fragments on class II MHC proteins (**FIGURE 36.19A**). When a T_H cell binds to the displayed antigen–MHC II complex, it stimulates the B cell to propagate and to secrete antibodies.

TABLE **36.2** | The Interaction between T Cells and Antigen-Presenting Cells

Presenting cell type	Antigen presented	MHC class	T cell type
Any nucleated cell	Intracellular protein fragment	Class I	Cytotoxic T cell (T_C)
Macrophages, dendritic cells, and B cells	Fragments from extracellular proteins	Class II	T-helper cell (T_H)

Activation of the cellular response results in death of the targeted cell

T_C cells are activated when they bind to cells carrying an antigen–MHC I protein complex, such as a virus-infected cell (**FIGURE 36.19B**). This activation, along with signals from T_H cells, results in the production of a clone of T_C cells with the same specific receptor. When bound, the T_C cells do two things to eliminate the antigen-carrying cell:

1. They produce **perforin**, an enzyme that lyses the bound target cell.
2. They stimulate apoptosis (programmed cell death) in the target cell.

Regulatory T cells suppress the humoral and cellular immune responses

A third class of T cells called **regulatory T cells** (**Tregs**) ensures that the immune response does not spiral out of control.

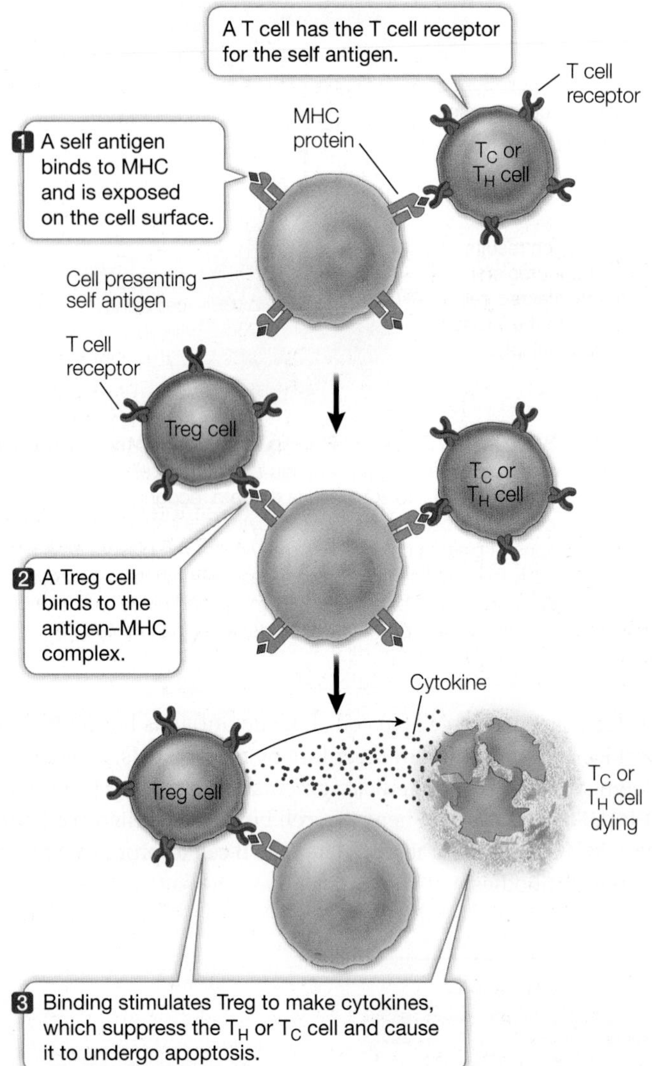

FIGURE 36.20 Tregs and Tolerance Regulatory T cells (Tregs) inhibit the activation of the immune system in response to self antigens.

Like T_H and T_C cells, Tregs are made in the thymus, express the T cell receptor, and become activated if they bind to antigen–MHC protein complexes. But Tregs are different in one important way: the antigens that Tregs recognize are self antigens. The activation of Tregs causes them to secrete the cytokine interleukin-10, which blocks T cell activation and leads to apoptosis of the T_C and T_H cells that are bound to the same antigen-presenting cell (**FIGURE 36.20**).

The important role of Tregs is to mediate tolerance to self antigens. Thus they constitute one of the mechanisms for distinguishing self from nonself.

AIDS is an immune deficiency disorder

There are several inherited and acquired immune deficiency disorders. In some individuals, T or B cells never form; in others, B cells lose the ability to differentiate into a clone of plasma cells and make antibodies. In either case, the affected individual is unable to mount an immune response and thus lacks a major line of defense against pathogens. A more common disease is **acquired immune deficiency syndrome** (**AIDS**), which results from infection of T_H cells by **human immunodeficiency virus** (**HIV**).

The course of HIV infection provides a good review of adaptive immunity (see Figure 36.11). HIV is transmitted from person to person sexually in semen or vaginal fluid, or rarely, in breast milk or through a blood transfusion; the recipient tissue is usually a mucous membrane lining an organ (vagina or anus). HIV initially infects macrophages, T_H cells, and antigen-presenting dendritic cells. At first there is an immune response to the viral infection, and some T_H cells are activated. But because HIV infects the T_H cells, they are killed both by HIV itself and by T_C cells that lyse infected T_H cells. Consequently, T_H cell numbers decline after the first month or so of infection. Meanwhile, the extensive production of HIV by infected cells activates the humoral immune system. Antibodies bind to HIV and the complexes are removed by phagocytes. The HIV level in blood goes down. There is still a low level of infection, however, mostly because of the depletion of T_H cells (**FIGURE 36.21**).

During this dormant period, people carrying HIV generally feel fine, and their T_H cell levels are adequate for them to mount immune responses against other infections. Eventually, however, the virus destroys the T_H cells, and their numbers fall to the point where the infected person is susceptible to infections that the T_H cells would normally eliminate. These serious and often lethal infections can result in a skin tumor and a white blood cell tumor caused by different viral infections, and pneumonia caused by a bacterium. There are about 1.2 million deaths per year worldwide from AIDS.

The molecular biology of HIV and its life cycle have been intensively studied. Drug treatments focused on inhibiting processes in the life cycle are quite successful at prolonging the lives of those infected. Since AIDS is caused by a viral infection, a huge international effort has been mounted to develop a vaccine.

LINK The life cycle of HIV is described in **Key Concept 11.3**.

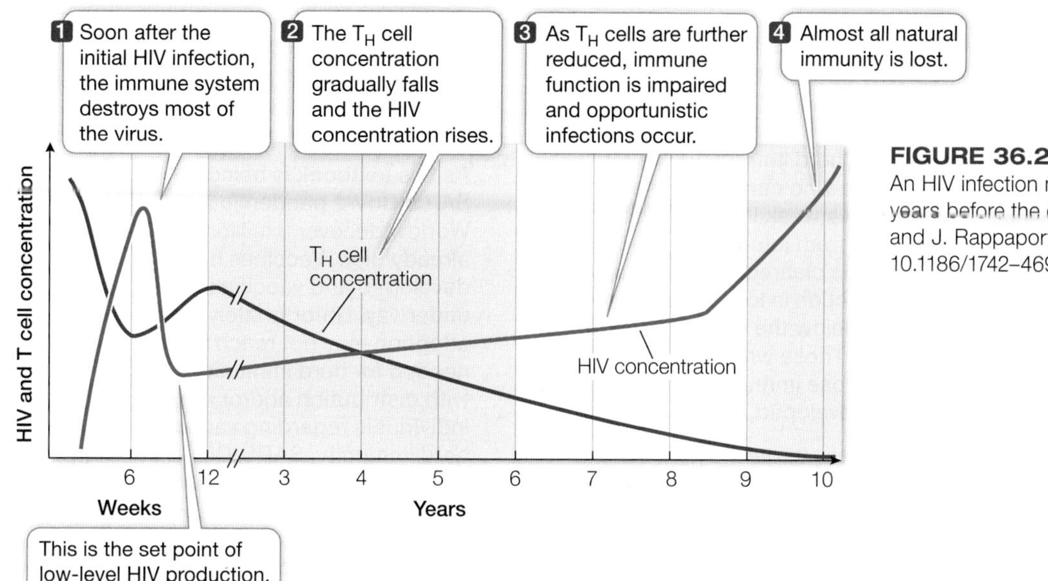

FIGURE 36.21 The Course of an HIV Infection An HIV infection may be carried, unsuspected, for many years before the onset of symptoms. (After M. Kogan and J. Rappaport. 2001. *Retrovirology* 8: 25. doi: 10.1186/1742–4690–8–25.)

Vaccination induces long-lasting immunity

We have seen how activated B cells and T cells produce agents that fight infection in the primary immune response. Small numbers of these cells remain as memory cells after an initial infection. A subsequent infection with the same pathogen activates these memory cells to mount a rapid, massive secondary immune response (see Figure 36.10). This sequence of primary and secondary responses has been exploited for the development of vaccines.

Instead of using an active pathogen as antigen, a **vaccine** uses either an inactive form of the pathogen that is still antigenic (e.g., a heat-killed bacterium or chemically treated virus), or an antigenic component of the pathogen (e.g., an inactive bacterial toxin or virus surface protein, or even an mRNA molecule that is translated by host cells to produce viral proteins, or protein epitopes). These antigens provoke a primary immune response, generating memory cells without causing illness. Immunity can be boosted by repeated exposure to antigen, so some vaccination strategies employ multiple inoculations over time. Later, if an intact pathogen carrying the same antigen attacks, the already formed memory cells recognize the antigen and quickly mount a secondary immune response (**FIGURE 36.22**).

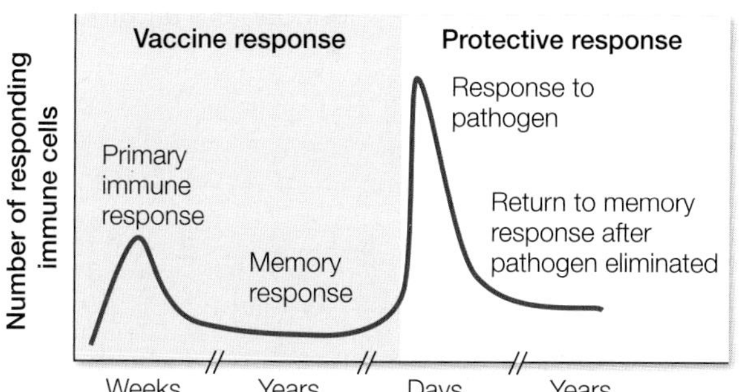

FIGURE 36.22 Vaccination Immunological memory from exposure to an antigen that does not cause disease can result in a massive response to the disease agent when it appears later.

The use of vaccines, especially in children, has been a great success in public health worldwide. Widespread vaccination has completely or almost completely wiped out some deadly diseases, including smallpox, diphtheria, and polio.

REVIEW & APPLY | 36.5

R The cellular immune response acts against virus-infected or mutated body cells. Specific receptors on T cells bind to antigen–MHC complexes displayed on cell surfaces. T-helper cells bind antigens on MHC II proteins and contribute to the humoral immune response. Cytotoxic T cells bind antigens on MHC I proteins and contribute to the cellular immune response. Tregs suppress immune response to self antigen. Vaccinations induce adaptive immune responses to specific antigens, responses that result in the production of memory cells that fight future infections by those antigens.

A

1. What are the similarities and differences in function between class I and II MHC proteins?
2. MHC proteins almost always differ among unrelated people, so organ transplants usually provoke a cellular immune response by the host, causing organ rejection. People receiving transplants are given a drug called cyclosporine that inhibits T cell development. How do you think cyclosporine works?
3. Cancer cells can express proteins that the body has never seen, so they are targeted by the cellular adaptive immune system. In some cancers, however, Tregs are increased in numbers and activity. Why would this indicate a poor prognosis?

Answer: Vaccination has been a spectacular public health success. Yet in the United States, more than one-half of adults do not get the flu vaccine. Some parents even refuse vaccination for their children.

Unfortunately, those who refuse vaccination may harm people other than themselves. A vaccination program succeeds only if a high percentage (typically above 90 percent) of people are vaccinated, thus disrupting the chain of infection from person to person. This results in "herd immunity," a state in which even those who cannot be vaccinated or who have weak immune systems are protected from infection, because the pathogen is cleared from the population. This protection is lost if the vaccination rate falls below the level needed for herd immunity. Those who are old or sick, and infants whose immune systems have not yet fully developed, are most at risk.

The best way to develop herd immunity in a population is compulsory vaccination. For example, during the periodic smallpox epidemics during the twentieth century in the United States, doctors accompanied by police would go into neighborhoods where the disease raged, vaccinating all those who were uninfected and removing infected people to quarantine. Henning Jacobson was arrested when he refused vaccination during a smallpox epidemic. He took his case to the U.S. Supreme Court, which ruled in 1905 that each state is entitled to protect its citizens. This provided a legal framework for compulsory vaccination that continues to this day. But opposition and court challenges continue.

As this textbook is being updated, the SARS-CoV-2 pandemic is ongoing. Worldwide, over 3 million people have already died. Vaccines have been developed and vaccinations are currently underway. Unfortunately, vaccine adoption may not reach the 90 percent needed for herd immunity due to difficulty with distribution and/or apprehension by individuals regarding vaccines. Without herd immunity, SARS-COV-2 will likely remain in the human population long-term. It will continue to produce new variants by mutation, some of which may have different enough antigens that they evade immune memory caused by vaccination or prior infection. Such variants will be able to infect individuals who are immune to the current strains. Annual, or more frequent, SARS-COV-2 vaccinations against the currently circulating strain(s) may be needed, similar to the annual vaccines required for the influenza virus.

36 VISUAL SUMMARY

36.1 Animals Use Innate and Adaptive Mechanisms to Defend Themselves against Pathogens

FIGURE 36.2

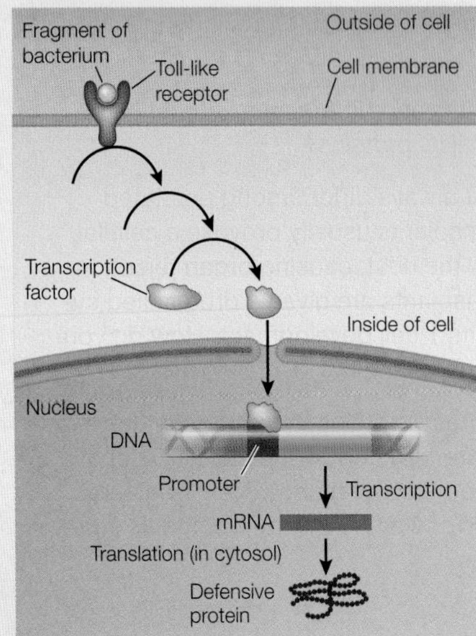

Go to ACTIVITY 36.1 and ANIMATION 36.2

- Animal defenses against **PATHOGENS** begin by distinguishing between self and nonself; **TOLL-LIKE RECEPTORS** (**TLRs**) recognize pathogenic patterns (**FIGURE 36.2**).
- **INNATE DEFENSES** are nonspecific and typically act rapidly. **ADAPTIVE DEFENSES** develop in response to specific pathogens and are long-lasting.

QUESTIONS

1. List three kinds of lymphocytes and three kinds of phagocytes, along with their functions.
2. Describe what happens when a Toll-like receptor encounters a pathogen

36.2 Innate Defenses Are Nonspecific

Go to ACTIVITY 36.2

FIGURE 36.6

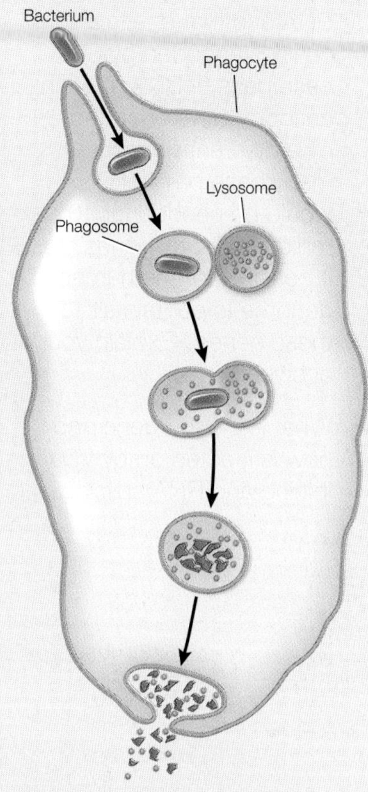

- **INNATE DEFENSES** include physical barriers such as the skin, mucous membranes, and competition with normal flora.
- **PATHOGEN-ASSOCIATED MOLECULAR PATTERNS (PAMPs)** are targeted for destruction by **PHAGOCYTES (FIGURE 36.6)**.
- In **INFLAMMATION**, **MAST CELLS** release **HISTAMINE**, which increases the permeability of blood vessels.

QUESTIONS

1. What is the complement system, and what is the result of its activation?
2. Describe three chemical factors that are released in the inflammation response.

36.3 The Adaptive Immune Response Is Specific

FIGURE 36.10

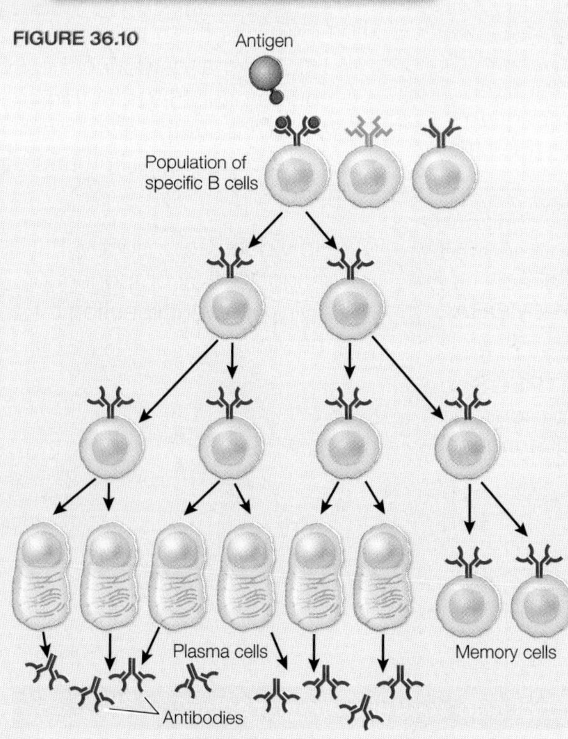

- The adaptive immune system recognizes an enormous diversity of **ANTIGENIC DETERMINANTS**.
- **CLONAL SELECTION** accounts for the specificity and diversity of the immune response and for **IMMUNOLOGICAL MEMORY (FIGURE 36.10)**.
- The **HUMORAL IMMUNE RESPONSE** involves **ANTIBODY** production by B cells, and the **CELLULAR IMMUNE RESPONSE** is mediated by T cells.

1. Describe the phases of the adaptive immune response and how the T-helper cell regulates both the humoral and cellular systems.
2. Clonal expansion of activated B cells results in the production of what two types of cells? What are these cells' functions?

36.4 The Adaptive Humoral Immune Response Involves Specific Antibodies

FIGURE 36.12

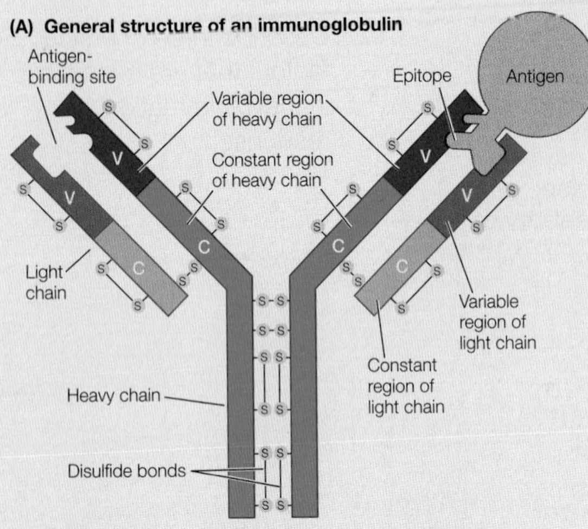

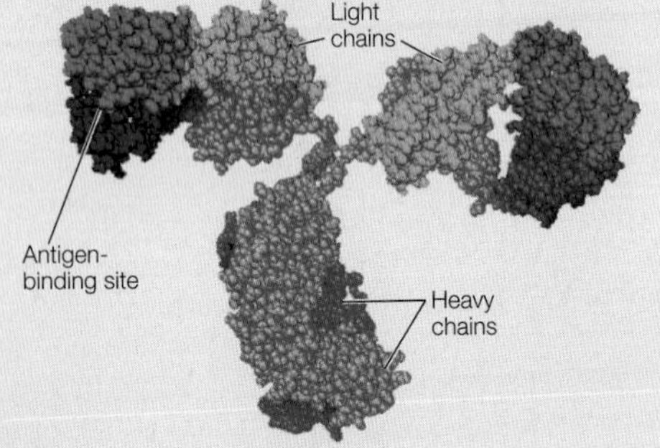

Go to ANIMATIONS 36.2 and 36.3 and ACTIVITY 36.3

- An antibody is composed of two **LIGHT CHAINS** and two **HEAVY CHAINS**, each having a **CONSTANT REGION** and a **VARIABLE REGION** (**FIGURE 36.12**).
- B cell genomes undergo DNA recombination events that yield millions of different **IMMUNOGLOBULIN** proteins.

QUESTIONS

1. Assume that one IgG heavy chain is encoded by 1,500 base pairs of DNA and that one IgG light chain is encoded by 600 base pairs of DNA. If each antibody were encoded by a different gene, how many base pairs of DNA would be needed for 10 million antibody genes? Is it possible for the human genome to contain all these genes? Explain your answer.
2. What is a supergene and how is it acted on by *VDJ* joining and RNA splicing?

36.5 The Adaptive Cellular Immune Response Involves T Cells and Their Receptors

FIGURE 36.18

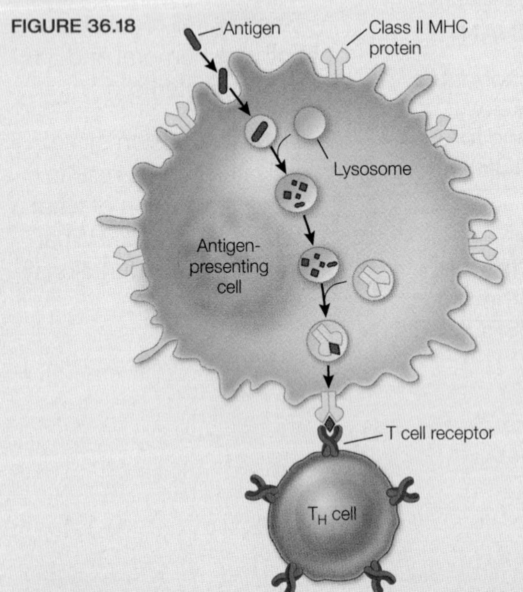

Go to ANIMATION 36.4

- T cell receptors bind only to antigen displayed on an antigen-presenting cell such as an MHC protein (**FIGURE 36.18**).
- **CYTOTOXIC T** (T_C) cells recognize and kill virus-infected cells or mutated cells. **T-HELPER** (T_H) cells direct both the **CELLULAR** and **HUMORAL IMMUNE RESPONSES**.
- **REGULATORY T CELLS** (**TREGS**) inhibit the other T cells from mounting an immune response to self antigens.

1. How does the number of T_H cells change during the course of an HIV infection, and what are the effects of these changes on immune function?
2. How does an antibody that recognizes a component of a virus prevent infection by the virus?

Go to **Achieve** for the e-book, animations, activities, and additional resources and assignments.

Animal Behavior 37

Coral reefs are hotbeds of fish diversification. More than 4,000 fish species—many of them dramatically colored—are found nowhere else. The fishes on a reef are not randomly distributed in a state of chaos. Instead, the behaviors of the fishes impart a great deal of regularity and structure to these fish communities.

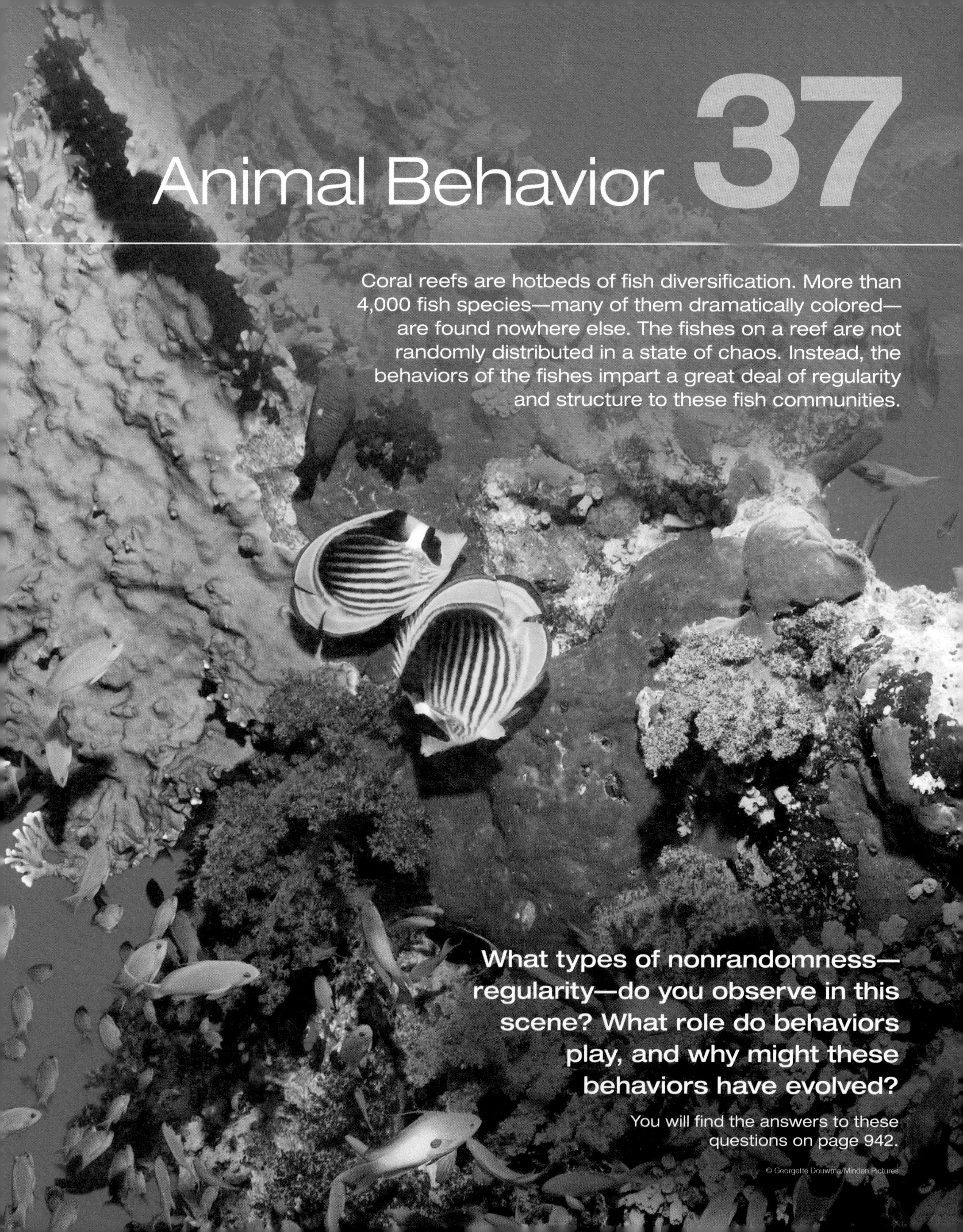

What types of nonrandomness—regularity—do you observe in this scene? What role do behaviors play, and why might these behaviors have evolved?

You will find the answers to these questions on page 942.

KEY CONCEPTS

37.1 Behavior Is Controlled by the Nervous System and Integrated with the Rest of Function

37.2 Behavior Is Influenced by Learning and Early Experience

37.3 Moving through Space Presents Distinctive Challenges

37.4 Behavior Structures Social Groups and Plays Key Ecological Roles

37.1 Behavior Is Controlled by the Nervous System and Integrated with the Rest of Function

LEARNING OBJECTIVES

By the end of this key concept you should be able to:

37.1.1 Explain the difference between proximate and ultimate causes.

37.1.2 Provide evidence that behavior can undergo organic evolution from generation to generation.

37.1.3 Explain and provide examples of how behavior can be constrained by other attributes of an animal.

Animals are distinctive in that behaviors play central roles in their interactions with each other and with their physical environments. They can rapidly approach each other, flee, vocalize, seek shelter, or migrate—all by using their muscles under control of their nervous system. Individuals use behaviors to relate to other individuals, such as by seeking mates or avoiding enemies. They also use behaviors to position themselves in favorable physical environments—as illustrated by North American and European animals that migrate to warm regions of the globe during winter, such as monarch butterflies and many birds.

Because behavior can rapidly enable a wide range of responses to conditions in the environment, it is one of the major mechanisms by which animals are responding to climate change. For example, during recent decades, Pied Flycatchers (*Ficedula hypoleuca*)—migratory birds that breed and spend their summer in the Netherlands and migrate to Africa each winter—have been modifying the timing of their migration as spring temperatures in the Netherlands have risen because of global warming. During their spring migration from Africa, the flycatchers arrive in the Netherlands in April. As April temperatures have risen, the birds have adjusted the timing of their annual arrival to an earlier and earlier date in the month (**FIGURE 37.1**), a response that helps ensure coincidence of their arrival with a spring peak in the abundance of insects, their primary food. Of course, Pied Flycatchers are just one species, and we must wonder whether other species exhibit similar behavioral responses to changes in local temperature. Recent surveys of hundreds of species highlight the enormous role that behavior often plays in responding to global warming (see Think Like a Scientist on next page).

The effects of climate change on ecosystems and communities are complex, and changes in the behavior of a species may not be sufficient to ensure its survival; see **Key Concept 42.5**.

The distinction between proximate and ultimate causes is a fundamental concept for all studies of behavior

Throughout our study of behavior, it will be important to distinguish between proximate and ultimate causes. This distinction is important in all aspects of biology. The study of behavior is one of the branches of biology in which it is of particularly great significance.

Whenever an animal initiates a process, biologists ask "What are the immediate, mechanistic causes of the process?" These are the **proximate causes**. Biologists also ask "Why did natural selection favor the evolution of the process? What were the causes that led the process to evolve?" These are the **ultimate causes**. To illustrate, let's consider an example that we will shortly discuss again as we consider the control of behavior by the nervous system. Adult Herring Gulls (*Larus argentatus*)—and the adults

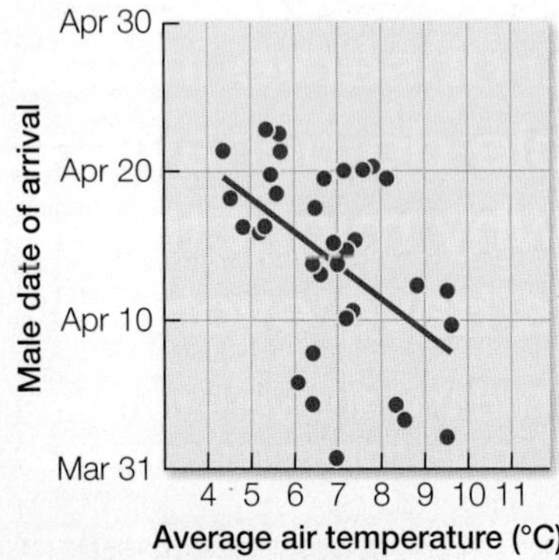

A male Pied Flycatcher

FIGURE 37.1 Global Warming Is Influencing Migratory Behaviors Pied Flycatchers (*Ficedula hypoleuca*) migrate from the Netherlands to Africa each winter, then return to the Netherlands in the spring to breed. This graph summarizes data on the spring migration. Each dot on the graph corresponds to a particular year during a 34-year study (i.e., there are 34 dots) carried out between 1969 and 2003. The dot shows, on the *y* axis, the mean date that males arrived in the Netherlands in the spring and, on the *x* axis, the average air temperature in the Netherlands in March–April. The graph shows that there has been a statistically significant correlation between arrival date and air temperature: as the air temperature in March–April has increased, male flycatchers have arrived earlier. (After C. Both et al. 2005. *J Avian Biol* 36: 368–373.)

THINK LIKE A SCIENTIST

How are animals reacting to global warming?

In theory, behavioral shifts of populations into new, cooler areas may represent many species' greatest chance of surviving global warming. But are animal populations actually shifting in this way?

Suppose a population of terrestrial animals is living in a range subjected to global warming. (The *range* is the sum total of all the areas where the population lives.) If rising temperatures are detrimental, we can hypothesize that the population will behaviorally shift its range in latitude, to be closer to the North Pole (in the Northern Hemisphere) or to the South Pole (in the Southern Hemisphere). Alternatively, if the population is living in a mountainous area, we can hypothesize that the population will behaviorally shift its range to a higher, cooler elevation. By means of such shifts, the population might keep its habitat temperature constant even as its original habitat heats up.

Great concept! But do populations actually shift their ranges as hypothesized? If they do, can we assume that these shifts fully "solve" the problems that global warming poses?

In 2011, Chris Thomas and colleagues published a comprehensive summary (a meta-analysis; see *Think Like a Scientist* in Chapter 30, p. 745) of all the data they could find in the scientific literature on possible range shifts in natural populations of terrestrial animals. Their study of latitudinal range shifts (i.e., shifts toward the poles) incorporated data on more than 700 species of animals (including birds, mammals, spiders, and many subgroups of insects), and they also studied elevational shifts, with data on more than 800 species. To illustrate, **FIGURE 37.2** shows two subsets of data included in their overall study of latitudinal shifts: data for spiders (85 species) and butterflies (29 species). The graphs show the number of kilometers of range shift observed from the beginning to the end of a 25-year study period (ca. 1975–2000). The amount of variation among species is typical.

Thomas and his colleagues reached a conclusion that took account of all the hundreds of species included in their analysis. They concluded that natural populations of terrestrial animals have shown a median range shift of about 17 kilometers per decade toward the poles (i.e., away from the Equator), and in mountainous regions, terrestrial animals have shown a median range shift of about 11 meters per decade toward higher elevations.

1. Thinking about real animals and real animal life spans, how do these range shifts likely occur, and what roles are likely played by behavior? Do individual animals present at the start of a decade simply shift where they live by the end of a decade?
2. When we think about the hundreds of diverse species studied by Thomas and his colleagues, we realize that many different life histories were involved and many different methods were used to get the data. Thus the study involved a lot of "lumping apples and oranges" (i.e., combining dissimilar things in a potentially fuzzy way). With the same effort, Thomas and his colleagues could have done an extremely careful study of a single model species: a study that would leave no doubt about that species' range shift. Why conduct the study as Thomas and his colleagues did? In answering, take explicit account of the actual data sets presented in Figure 37.2.
3. We started by saying "behavioral shifts of populations into new, cooler areas may represent many species' greatest chance of surviving global warming." Think, however, about human land-use patterns as they exist today. Taking account of human land-use patterns, do behavioral shifts provide a sure solution for the problems that animals face because of global warming? Include in your answer a consideration of animal populations that live in national parks or other protected areas.

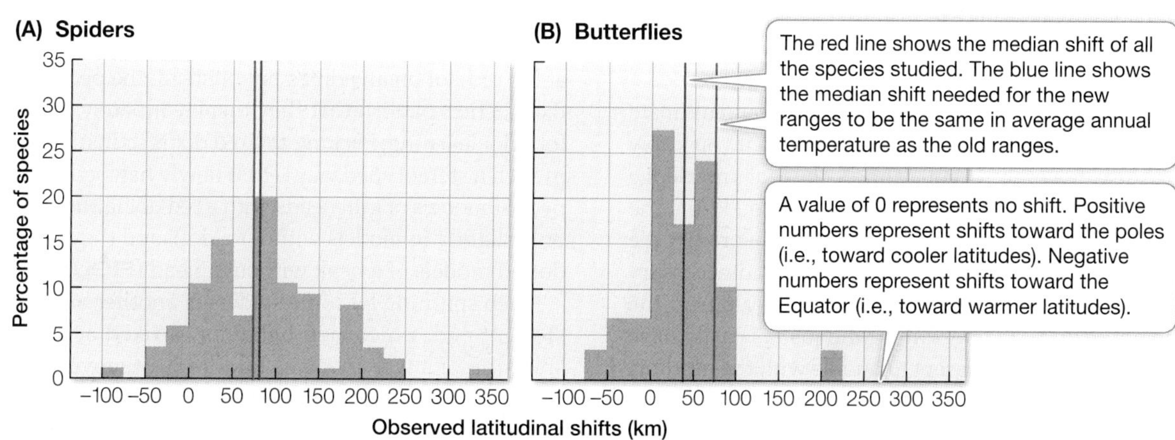

FIGURE 37.2 Latitudinal Range Shifts in 85 Species of Spiders and 29 Species of Butterflies over a 25-Year Period of Global Warming These two data sets illustrate the type of analysis used by Thomas and colleagues; their overall study included many additional data sets on other animals. In the case of spiders **(A)** and butterflies **(B)**, the populations were monitored for 25 years, and the *x* axis shows how much the range shifted from the beginning to the end of the 25-year study period (ca. 1975–2000). (After I.-C. Chen et al. 2011. *Science* 333: 1024–1026.)

FIGURE 37.3 The Proximate Cause of Nestling-Feeding Behavior in Adult Herring Gulls Adult Herring Gulls (*Larus argentatus*) have a red dot on their bill. Nestlings peck at the dot to stimulate their parents to feed them.

of some other gull species—have a red dot on their bill (**FIGURE 37.3**). When a parent returns to its nest from a trip to forage for food, young nestlings peck at this dot. Their pecking stimulates the adult to regurgitate some of the food it collected so the nestlings can eat the food. Let's ask "What is the cause of the adult's nestling-feeding behavior?" In fact, there are two causes, one proximate, one ultimate. The proximate cause—the immediate, mechanistic cause—is that the nestlings peck at the red dot and the adult responds to this pecking with feeding behavior. The ultimate cause, we surmise, is that the process ensures and coordinates the provision of food to the offspring. The nestlings require food, and—until they are old enough to forage on their own—they must obtain the food from their parents. Thus we surmise that natural selection favored the evolution of the process because it enables parents to provide food in a coordinated way to their offspring.

An important point to recognize is that proximate and ultimate causes cannot be predicted from each other. That is, if you know the proximate cause of a process, you cannot—by that knowledge alone—predict the ultimate cause. Conversely, if you know the ultimate cause, you cannot—from that knowledge—predict the proximate cause. To illustrate, notice that there is no necessary reason why pecking at a dot on a parent's bill should lead the parent to feed a nestling. The pecking could just as easily have evolved to be a stimulus for the parent to huddle with the nestling. Conversely, although it is indeed essential that parents feed their offspring, there is no necessary reason why bill pecking should be the stimulus. An alternative proximate stimulus could have evolved instead, such as a unique vocalization by the nestlings.

These considerations explain why biologists must directly study both proximate and ultimate causes. If we know of a proximate cause, we still must independently study the ultimate cause in order to know the nature of the ultimate cause, and vice versa.

Proximate and ultimate causation are relevant to all aspects of behavior, and they are highlighted repeatedly in this chapter. For now, however, let's focus on another of the most fundamental aspects of behavior, its relation to the nervous system.

Behavior has a neural basis

Biologists today broadly agree that when an animal engages in behaviors, its nervous system activates and coordinates the behaviors. When an animal runs, for instance, neurons stimulate the leg motions. Similarly, when people engage in an enthusiastic conversation, their brains activate their hand movements and facial expressions. According to this viewpoint, although behaviors—such as leg motions and smiles—are fleeting and not in themselves material objects, they have a material basis: they arise from brain tissue, neurons, muscle cells, and the movements of ions that give rise to nerve impulses and activate muscle contractions.

There are many types of evidence for the neural basis of behavior. Studies of the human brain provide one type of evidence, namely that particular types of behavior depend on the function of particular brain regions. For example, if the part of a person's brain known as Broca's area (see Figure 31.39B) is badly damaged in an accident, the person may be perfectly fine in almost all ways but will have extreme difficulty with talking or writing. This sort of "negative" evidence (what goes wrong when a brain region is missing?) is complemented by positive evidence from brain-imaging studies in normal people. As seen in Figure 31.40, when a person speaks, specific brain regions show evidence of increased metabolism. These types of evidence point to a neural basis for complex behaviors involved in human communication.

Additional evidence for the neural basis of behavior comes from a very different source: the study of certain highly stereotyped animal behaviors, often called **fixed action patterns**, which are expressed by animals without prior learning and are often resistant to modification by learning. These behaviors point to control by the nervous system because they depend solely on the presence of a healthy nervous system to be performed. A classic example is provided by the bill-pecking behavior of gull nestlings we have just discussed. The nestlings do not learn to peck at the red dot on an adult's bill. Instead, the pecking behavior is a fixed action pattern that they display innately, without any need for prior learning. Pecking at a red dot is so instinctual—and done in such a stereotyped way—that newly hatched, naive gull chicks peck at models of gull heads with a red dot (although they respond only slightly to models without a dot), and they even peck at red-dotted models of a beak without a head (**FIGURE 37.4**).

Web spinning by some spiders is another example of a highly stereotyped, unlearned behavior: a fixed action pattern. The spinning of a web involves hundreds of movements performed in sequence, and each time a spider builds a web, the sequence is largely the same. Yet in many species there is no opportunity for a spider to learn how to build a web from another spider because the adults die before their eggs hatch.

Behaviors evolve

Recognizing that neural tissue gives rise to behaviors, we can expect behaviors to undergo organic evolution. The properties of

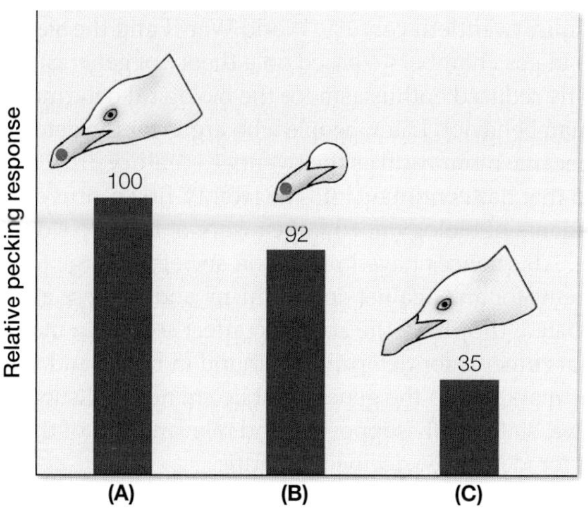

FIGURE 37.4 Expression of a Fixed Action Pattern
(A) Herring Gull (*Larus argentatus*) chicks instinctively peck at a cardboard model of an adult gull head with a red dot on the bill and **(B)** even at a model of just a bill with a red dot. **(C)** If presented with a model of a gull head without a red dot, the chicks respond to only a small extent. (After N. Tinbergen and A. C. Perdeck. 1950. *Behavior* 3: 1–39.)

The adults of most bird species do not have a red dot on their bill and the nestlings of most species do not peck at a red dot. How can the lack of the red dot be explained in view of the fact that the nestlings of most bird species need to get food from their parents?

tissues—including neural tissue—are in part encoded by genes, which are mutable and passed from generation to generation. In the case of neural tissue, if certain alleles (gene variants) produce more adaptive behaviors than others, natural selection can favor those alleles. In this way the frequencies of pertinent alleles in a population can change so that the properties of the neural tissue—and the behaviors elicited by the tissue—adaptively evolve.

Many studies establish that genes exert important effects on behavior. Perhaps the most graphic evidence comes from studies of mutations in fruit flies (*Drosophila melanogaster*). For example, there are specific gene mutations that reduce the spontaneous activity of the flies, making the flies with the mutations more subdued and passive than flies without them. One of the most important examples involves the "clock" mutations discussed in Investigation Figure 29.21. Recall that around 1970, Ronald Konopka and Seymour Benzer discovered mutations in genes that radically altered the flies' daily rhythm of activity under control of their circadian clock. These mutations, discovered because they modified fly behavior, ultimately allowed biologists to clarify the biochemical mechanisms of the biological clock.

Current laboratory studies employing artificial selection demonstrate that behavior can evolve remarkably rapidly. In artificial selection (as opposed to natural selection), human researchers choose which combinations of animals have the most offspring, mimicking natural selection but in a highly controlled way. In one such experiment, researchers gave each mouse in a large population a running wheel and then monitored the number of revolutions each mouse voluntarily ran per day. Males that ran the highest numbers of revolutions were then mated with females that ran the highest numbers to produce a new generation of mice. When those mice grew to adulthood, they were put through the same process. This selection procedure was repeated generation after generation. After only 13 generations, the selected mice on average ran 12.1 kilometers per day, more than twice as far as control mice (5.5 km per day). Primarily, the difference reflected faster running by the selected mice, rather than a longer time spent running. Studies revealed critical changes in the brains of the selected mice, indicating that a difference had evolved in the neural control of running behavior.

House mice (*Mus musculus*) in North America provide an elegant illustration of adaptive evolution of behavior in the wild. House mice did not live in North America prior to European colonization. They arrived as stowaways on European ships, presumably in random groups. They then established wild populations throughout the continent. Carol Lynch trapped wild house mice at five locations, from warm Florida to cold Maine. Within the set of mice from each location, she then paired males and females in her lab to obtain lab-born, lab-reared offspring, which she studied. All the mice studied had been born and grew up in exactly the same lab environment. When Lynch measured the nests they made, she found a trend: under identical conditions, the mice from progressively more northern populations tended to build progressively bigger (and better insulating) nests (**FIGURE 37.5**). This result points to the evolution by natural selection of a genetically

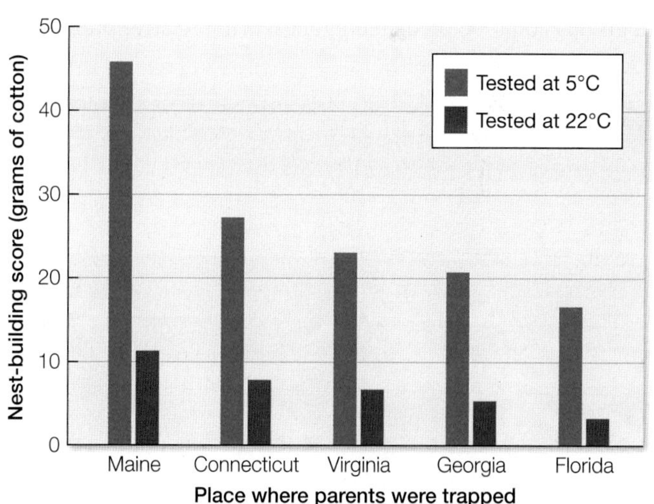

FIGURE 37.5 Nest Building by House Mice from Five Locations All mice were born in the lab and tested under identical conditions. Their parents, however, were trapped from wild populations living at the five locations shown. Mice were tested at two different temperatures. Nest building by each mouse was measured by summing the weight of cotton the mouse used to build four nests. This sum is termed the "nest-building score." The graph shows the average score of mice tested at each temperature. (After C. B. Lynch. 1992. *The Am Nat* 139: 1219–1236.)

controlled, behavioral propensity to build bigger nests in wild mouse populations that settled in locations where big nests are particularly advantageous.

Despite its neural basis, behavior is not necessarily simplistically deterministic

Earlier we noted that behaviors are often fleeting and are not in themselves material objects, but they have a material basis because they arise from neural tissue. Regarding human behavior, this dichotomy—that nonmaterial behavior arises from material substance—attracts the attention of virtually all branches of scholarship. Biologists, psychologists, anthropologists, sociologists, theologians, and criminologists are all concerned with the dichotomy and ponder it.

A long-standing argument in the modern study of biology focuses on **biological determinism**, the concept that an individual's attributes are determined by their particular genetic endowment. The concept of biological determinism, as applied to behavior, suggests that the behaviors of animals are hardwired by genetics. In this view, an individual's genes orchestrate their neural properties in fixed ways that in turn affect their behavior in fixed ways, leading to predictable behaviors.

Almost all biologists are convinced that some simple animals exhibit determinism. Clams, for example, are predictable and inflexible in many of their responses to their environment. Here, however, we focus on animals with advanced nervous systems. In particular, we focus on people.

Belief in human biological determinism flourished in the nineteenth and early twentieth centuries. For example, at that time some prominent biologists believed (incorrectly) that a person's intelligence is determined by their inherited brain size. According to this discarded view, the relative mental capacities of racial groups could be predicted by measuring relative brain sizes. In the mid-twentieth century, World War II and the slaughter of people in gas chambers—based on a theory of genetic inferiority—greatly reduced enthusiasm for the biological determinist view of human behavior. Later, people who argue for the deterministic view became more vocal as the twentieth century drew to a close, a trend that has continued into the twenty-first century.

Students of biology need to be aware of the two sides in this debate. All sciences have impacts on society at large. There are arguments for and against determinism, and as these arguments are debated, they have the ability to affect society in many ways. Many arguments for determinism found in books and magazine articles marketed to the general public are not well supported by objective, statistically supported, and relevant data of the sort required for ideas to be deemed scientific.

In what ways could the behavior of animals with advanced nervous systems not be simplistically deterministic despite its neural origins? Behavior is dramatically more flexible than any other biological trait. This is true in part because learning modifies behavior: two individuals with the same genes can differ in behavior because of differences in what they have learned. In addition, many types of early life experiences can affect the later behavior of individuals. And lately, biologists have become ever more aware of important epigenetic effects on behavior: non-genomic effects that can exert lifelong influences and in some cases may be transmissible from one generation to the next (see Key Concept 11.4). As animals develop and live their lives, each individual can interact in unique ways with its environment, and—because of learning, early life experiences, and epigenetic effects (discussed in more detail in Key Concept 37.2)—these interactions can result in dramatic behavioral differences among genetically similar individuals.

The greatest unresolved question in all of biology is probably the question of how the human mind emerges from the human brain. The unresolved nature of this question, despite its profound importance, emphasizes that scientists have much to learn about mental phenomena and the behaviors they unleash.

Behavior is integrated with the rest of function

Over evolutionary time, all attributes of a species evolve together in an integrated way. Behaviors therefore do not evolve in isolation. Instead, the evolution of behaviors often depends on evolutionary developments in other functional systems. For example, navigation by use of polarized light—a behavior seen in many insects—cannot evolve unless eyes sensitive to polarized light evolve.

Pronghorn (*Antilocapra americana*), seen in **FIGURE 37.6**, strikingly illustrate how an animal's behavior often depends on—and is integrated with—the animal's other characteristics. Although cheetahs (*Acinonyx jubatus*) achieve the highest speeds of all running animals (about 110 km per hour), they can sustain those speeds for only 1–2 minutes. Pronghorn, in contrast, achieve the highest sustained speeds of all running

Mark Gocke/USDA

FIGURE 37.6 Pronghorn Run at the Fastest Sustained Speeds of Any Running Animals Their running behavior is possible only because they have also evolved systems of O_2 transport, ATP production, and muscular contraction that are up to the task. Pronghorn occur in grasslands from northern Mexico to southern Alberta and Saskatchewan.

animals: they can run at more than 50 kilometers per hour for tens of minutes, meaning they can keep up with a car traveling 30 miles per hour for a long distance. For pronghorn to behave as they do, their running muscles must be able to produce enormous forces over long periods. To do so, the muscles require high, sustained rates of O_2 delivery, and the muscle cells must rapidly use the O_2 for aerobic ATP production (see Key Concept 33.2). Research has revealed that almost all attributes of pronghorn are specialized for sustained production of high forces during running. For example, compared with other mammals of their size, pronghorn have exceptionally large lungs and skeletal muscles, and their muscle cells are exceptionally tightly packed with mitochondria.

The integration of behavior with other functional attributes is also nicely illustrated by certain toads and frogs, which have evolved contrasting behavioral specializations that depend on contrasting biochemical mechanisms of ATP synthesis. When toads such as the western toad (*Anaxyrus boreas*) are chased, they hop away at modest speeds that they can maintain for many minutes. In contrast, when leopard frogs (*Rana pipiens*) are chased, they leap away at lightning speeds—far faster than toads—but they become fatigued and stop leaping after a minute or two. These contrasting escape behaviors in the toads and frogs reflect differences in the biochemical mechanisms used by their leg muscles to make ATP for escape running. The usual rule, when comparing related animals, is that aerobic ATP production is relatively slow and resistant to fatigue, whereas anaerobic ATP production is relatively fast and subject to rapid fatigue (see Key Concept 33.2). Enzyme studies show that the toads have evolved high levels of enzymes required for aerobic ATP production in their leg muscles. The frogs, by contrast, have an enzyme profile that emphasizes high rates of anaerobic ATP production. These evolved enzyme differences help explain the differences in their hopping behavior.

Behaviors are often integrated with body size and growth

Body size and growth are additional animal attributes with which behaviors must be integrated. Body size, for instance, is sometimes a factor affecting the sound frequencies used by animals for communication. This is true in many insects—such as crickets and katydids—that sing to communicate with other members of their species. The tonal frequencies of the songs of these insects tend to vary regularly with body size. As in a pipe organ, in which the large pipes produce the lowest frequencies of sound (the deepest tones), insect species of large body size tend to produce songs of lower frequency than species of small size. This same trend is observed in singing frogs.

Growth also can have pivotal impacts on behavior. In many species, for example, individuals must grow to a large adult size before they can successfully consummate reproductive behaviors. This is true of male elk (*Cervus elaphus*) in Scotland, which rarely father offspring before they are 5 years old, even though they are reproductively mature at 2 years of age. Being reproductively mature (e.g., being capable of sperm production) is not sufficient for successful reproductive behavior. A male must also be big enough and experienced enough to dominate other males.

When spotted hyenas (*Crocuta crocuta*) capture a zebra or antelope by hunting as a group in a pack, they are famous for eating everything, including the skeleton, within minutes in a feeding frenzy. For young hyenas to get food during this group feeding behavior, they need to be as effective as possible in grabbing bites to eat. Being able to bite through bone is a great advantage because it permits much faster feeding than stripping flesh off bone. A young hyena's feeding behavior is transformed—in ways that raise its odds of survival—as its jaws, jaw muscles, and teeth mature, permitting faster and faster bone consumption (**FIGURE 37.7**).

FIGURE 37.7 Young Hyenas Must Mature to Feed on Bone Rapidly Spotted hyenas (*Crocuta crocuta*) of various ages were given a standardized "bone-crunching test." The amount of bone they could consume in 15 minutes increased dramatically as they aged and their jaws and jaw muscles became stronger. Young hyenas are limited in their ability to compete with adults during group feeding behavior at a kill. The line was fitted by least-squares linear regression (see Appendix B). (Data from W. Binder and B. Van Valkenburgh. 2000. *J Zool* 252: 273–283.)

REVIEW & APPLY | 37.1

R In studying behaviors, a distinction must often be made between proximate and ultimate causes. Changes in behavior are an important way for animals to respond to changes in their environment. An animal's behaviors are activated and coordinated by the animal's nervous system. Although some behaviors, called fixed action patterns, are highly stereotyped, behavior is more often flexible and modified by an animal's experiences. Animal behavior has a genetic basis and is subject to natural and artificial selection. Behavior and other aspects of animal function are interdependent; the behavior of an animal is constrained by its physiological capacity to perform the required actions. Body structure, size, and growth rate can influence behavior, as can cellular biochemistry and metabolism.

A

1. What is the physical basis for animal behaviors? How are fixed action patterns passed from one generation of a species to the next?
2. How do experiments in which mice are artificially selected for wheel running provide insight into the rate of evolution of running behavior?
3. Mice do not build especially large nests in warm environments, but the mice collect more nesting material and build bigger nests in cold environments. What are the likely proximate and ultimate causes of the nest building observed in cold environments?
4. Simply because of the laws of physics, if an animal falls from a substantial height, its risk of injury from whole-body impact on the ground depends on its body size. For a mouse-sized animal, the odds of being hurt by the whole-body impact of a fall are near zero. The risk increases with body size; for a horse, a substantial fall causes death. How would this size-dependent trend affect the evolution of behaviors that involve leaping from branch to branch within trees?

We have already noted that development and learning often affect an animal's behavior. In the next key concept we'll focus on these important processes.

37.2 Behavior Is Influenced by Learning and Early Experience

LEARNING OBJECTIVES

By the end of this key concept you should be able to:

37.2.1 Explain the adaptive value of learning.

37.2.2 Explain how behavioral imprinting and other types of early learning are often of critical importance.

37.2.3 Summarize how early experience can affect an individual's lifetime behavior and physiology through epigenetic tagging.

The particulars of an individual animal's life—such as the exact place it lives and the exact appearances of its parents—are typically not predictable. Information about these particulars therefore cannot be inherited. Biologists think this is the fundamental reason why animals have evolved the ability to **learn**—by which we mean the ability of an individual animal to modify its behaviors as a consequence of individual experiences earlier in its life. Suppose that a mouse living in a forest is less likely to be caught by a predator if it can rapidly return to its burrow. Prior to the mouse's birth, there would be no way to predict the location of that mouse's burrow or landmarks useful for finding the burrow. For this reason, natural selection could not provide inherited information on these particulars. However, natural selection has favored the evolution of learning abilities. A mouse inherits mechanisms by which it can learn locations. Then, during its life, it uses those mechanisms to incorporate specific information on its actual burrow location and useful landmarks into its escape behavior.

Lee Metzgar did experiments on white-footed mice (*Peromyscus leucopus*) to test the hypothesis that animals learn about their surroundings in ways that aid their survival. An important background point for understanding Metzgar's experiments is that when these mice move around on the forest floor, they commonly run next to logs or sticks. An adult mouse was placed for several days in a seminatural arena with a simulated forest floor, consisting of small logs, sticks, and dead leaves. There was no burrow in this arena, but there were logs big enough to provide protection from aerial attack. Then the mouse was temporarily removed from the arena and a hungry screech-owl was introduced. Soon afterward, the mouse was returned to the arena, along with a new mouse that lacked prior experience with the arena. The arena was then observed until the screech-owl caught one of the mice (**FIGURE 37.8**). To a strongly significant extent, the new mouse was the one more likely to be caught. However, if the logs, sticks, and leaves were replaced with new but similar items in different locations immediately before the test with the owl, the two mice in the arena were equally likely to be caught. These results indicate that the experienced mouse learned the locations of logs and sticks during the days it lived in the arena. It knew escape routes and places where it was safe, provided the logs and sticks were where the mouse had learned them to be.

Specific information of critical survival value is often learned during early postnatal development

Konrad Lorenz, one of the pioneers of the modern study of behavior, discovered that if he associated with young Graylag Geese (*Anser anser*) soon after they hatched, the geese thereafter treated him as if he were their parent. They associated with him instead of other geese and followed him everywhere. This attachment bordered on impossible to change. Lorenz named this phenomenon **behavioral imprinting** (not to be confused with genomic imprinting; see Key Concept 11.4). Behavioral imprinting also occurs in the young of many other species of birds and some other animals. Today behavioral imprinting is considered a distinctive type

© Darin Vance/Alamy Stock Photo

FIGURE 37.8 Screech-Owls Prey on Mice, Which Employ Learning to Increase Their Chances of Evading Capture Screech-owls search for mice on the ground from perches above the forest floor. When an owl sees or hears a mouse, it drops from its perch and grasps the mouse with its talons unless the mouse behaviorally evades capture. Mice learn details of their habitat that help them evade detection and capture. The bird shown here is an Eastern Screech-Owl (*Megascops asio*).

of learning: distinctive because the learning takes place within a relatively narrow window of time early in postnatal life and, after that, is inflexible. Imprinting can have lifelong consequences. When male geese imprint on a person at the start of their lives, they later, in adulthood, prefer trying to mate with people. During normal development, geese imprint on—and establish a strong attraction to—their true parents and species.

The importance of early learning is illustrated in a different way by Indigo Buntings (*Passerina cyanea*) and some other bird species that navigate by the stars. Adult buntings migrate at night, using the stars to determine the direction in which they fly. After this amazing phenomenon was first discovered, Stephen Emlen studied young buntings in a planetarium to learn more about the mechanism they use. Emlen established that the birds need to know the North Star to navigate correctly. But how do they know which star is the North Star? His experiments revealed that each young bird has to learn the correct star during its first few weeks of life following hatching. In the planetarium, Emlen could make the artificial sky rotate around any star he chose. Young birds learned to identify that star—the one the sky rotated around—and later, when they were under a natural sky, they treated that star as the North Star for the rest of their lives. When buntings develop outdoors in natural environments in the Northern Hemisphere, the sky appears to rotate around the true North Star (**FIGURE 37.9**). In this way, buntings in nature learn the true North Star.

The learning of species-specific songs by young birds—a process seen in many species—is another example of the critical importance of early learning. The birds known as Darwin's finches in the Galápagos archipelago (see Key Concept 16.3) provide well-studied examples. In adulthood, the males of each finch species sing a species-specific song that attracts females (which do not sing) in a species-specific way. The songs are not inherited. Instead, each male must learn the song he will sing for the rest of his life during a period of about 30 days, 10–40 days after he hatches. During that period a young male is in the nest built by his parents, and his father sings nearby. The young male learns his song by an imprinting-like process, by listening to (and later duplicating) his father's song. Particular brain regions, larger in males than in females, are required for this learning.

One piece of evidence for song learning comes from studies of young male finches whose fathers die early in their development. In this circumstance, a young male cannot hear his father sing and may learn the song of a different species by hearing an adult male of a different finch species sing during his critical learning period. Young males subjected to this experience later sing an incorrect song and attract females of the incorrect species.

© Malcolm Park sciences/Alamy Stock Photo

FIGURE 37.9 The Stars in the Northern Hemisphere Seem to Make Circles around the North Star Because of Earth's rotation around its axis, the stars seem to move in circles around the northern end of the axis—the position of the North Star. This apparent movement of the stars is evident in a long exposure of the northern sky, as in this photograph. The camera shutter was kept open for several hours so that as stars moved on their apparent paths through the sky, their paths were recorded. The North Star is in the center.

Early experience also has other, more global effects on an individual's behavior

A large body of knowledge demonstrates that early experience often has multiple lifelong effects on an individual's behavior. These effects are specific to each individual, depending on the individual's particular experiences.

An ongoing series of experiments on lab rats, for example, is bringing to light new revelations about the relationship between adult behavior and early development. You may recall from Chapter 11 that rat mothers vary in their parenting behaviors. Two categories of mothers are recognized. Here we will call them "high-caring" and "low-caring." Mothers in both categories are statistically the same in how often they nurse their young and in the absolute amount of contact they have with their suckling offspring. They differ, however, in how much they lick and groom their young during nursing and in how much they adopt a favorable posture for easy suckling. Mothers in the high-caring category lick and groom a lot and adopt a favorable posture. Low-caring mothers lick and groom to a lesser extent and adopt a less favorable posture.

Offspring of the two categories of mothers have been raised to adulthood and then studied in two novel, threatening environments. In both environments, adult offspring of low-caring mothers are far more likely to exhibit fear than adult offspring of high-caring mothers. For example, after adult offspring have been deprived of food for 24 hours, they have been placed, one at a time, in a novel, brightly lit arena where food is provided at only one place in the very center. This environment is intimidating because rats prefer dim light and prefer not being out in the open. The behavioral differences between the two groups of adult offspring are dramatic. The hungry adult offspring of low-caring mothers wait much longer than the hungry adult offspring of high-caring mothers to go to the food and start eating, and they spend far less time eating.

How is it possible for a young animal's early experiences to change multiple aspects of its behavior throughout its life? In the case we are discussing, research has demonstrated that epigenetic effects are in part responsible and play a pivotal role (see Key Concept 11.4). Key regulatory genes in the stress-response pathways (both biochemical and hormonal pathways) of rats are subjected to epigenetic "tagging" or "marking" during nestling life—meaning that DNA methylation changes or histone changes occur in ways that affect the expression of the genes. These changes, termed epigenetic "tags" or "marks," are then maintained throughout the life of the individual, and (although the tags, by definition, do not alter the base sequence of the DNA) they affect the extent to which various genes are expressed throughout life. The tagging of these genes is different in the offspring of low- and high-caring mothers. This difference in epigenetic tagging permanently alters the stress responses of individuals in the two groups, giving rise to the differences observed between the adult offspring.

Malnutrition in early life is also known to affect an individual rat's epigenetic tags. Abandonment does so as well. Epigenetic tags from these early experiences persist into adulthood, altering an individual's gene expression and behavior throughout life.

Migratory locusts (*Locusta*) and other similar locust species provide perhaps the most dramatic example of a change in individual phenotype caused by individual experience. In Africa, massive swarms of millions of migratory locusts sometimes form and fly across the landscape, devastating crops wherever they land to eat. An amazing feature of this phenomenon is that each individual locust is capable of an entirely different type of behavior. In fact, each individual locust can display either of two radically different behavioral phenotypes. When the individual expresses one phenotype, it avoids other individuals. A population of individuals expressing this phenotype is spread out and inconspicuous. Alternatively, each individual may express a highly gregarious phenotype—meaning the population forms a swarm. Individual experience determines each individual's phenotype. Individuals that are never forced to live closely together retain the solitary phenotype. The same individuals become a swarm if forced into close contact—for instance, if they are forced to feed next to each other, in the same places, because of food shortage. Recently, researchers have found evidence in locusts that epigenetic tags play key roles in controlling individual phenotype (solitary versus gregarious).

REVIEW & APPLY | 37.2

R Animals in general are able to learn—meaning that individuals can alter their behaviors on the basis of previous individual experiences. Learning helps give behavior specificity appropriate to the environments and challenges of particular individuals. Behavioral imprinting is a specialized type of learning in which an individual learns key details during a limited period early in postnatal life, after which certain behavioral responses are fixed. Early experiences such as maternal care and malnutrition can affect the behavior of an animal for the remainder of its life via epigenetic effects on gene expression.

A
1. In captive breeding programs for Whooping Cranes (*Grus americana*), an endangered species, the young birds must be prevented from seeing people because otherwise they will not mate with cranes. How does this behavior arise?
2. In the lives of mice in a woodlot and of finches in the Galápagos archipelago, explain how learning helps give specificity to an individual's behavior that aids the individual's ecological success.
3. Do epigenetic tags alter the sequence of base pairs in an animal's genome?

Let's now turn our attention to the crucial topic of how animals get from one place to another. Most animals move from place to place over Earth's surface during their lives. Travel behavior of this sort can help animals find things they need, but also entails risks, including the risk of getting lost. Travel behavior is a particularly fascinating and important aspect of animal behavior.

37.3 Moving through Space Presents Distinctive Challenges

LEARNING OBJECTIVES

By the end of this key concept you should be able to:

37.3.1 **Explain the difference between trail following and path integration.**

37.3.2 **Explain how the waggle dance of honey bees translates navigation information from a horizontal plane to a vertical plane.**

37.3.3 **Explain and provide examples of the concept of redundancy in direction finding.**

For most animals, movement through space is critical for success. The distances covered may be short, as when animals search for food each day. At the other extreme, some animals travel almost halfway around the globe each year during seasonal migrations. Moving through space requires muscle function. It also requires that animals meet the challenges of navigation and orientation, ensuring that they can find their goals and return home. **Navigation** refers to the act of moving toward a particular destination or along a particular course. **Orientation** refers to adopting a position, or a path of locomotion, relative to an environmental cue such as the Sun.

Trail following and path integration are two mechanisms of navigation

To navigate, animals sometimes simply follow trails. Foraging workers in many species of ants use this method in a specialized way, by following trails they create by use of a pheromone. A **pheromone** is a chemical compound or mix of compounds that is emitted into the outside environment by individuals of a species and that elicits specific behavioral responses from other members of the same species. One sex sometimes uses a pheromone to attract the other sex. In the case of the ants, as worker ants return home from a profitable food source, they deposit minute quantities of their species-specific trail pheromone along the way. In this manner they leave behind a fully marked trail between the colony nest and the food source. Other workers then detect the pheromone and find the same food source by following the trail. That is, they use the presence of the pheromone as a proximate cue to find the food source. The ultimate cause of the behavior is that it aids food collection for the colony.

Rather than employing trail following, ants of the genus *Cataglyphis* use path integration to navigate during foraging. These ants are specialized to live in hot, dry deserts, notably the Sahara. Colonies of *Cataglyphis* live underground, where temperatures are mild. Worker ants emerge to collect food during the daytime, when the desert surface is fiercely hot. Their principal food is the dead bodies of other desert insects killed by overheating. Why do the workers forage during the heat of the day? The answer is hypothesized to be that this behavior enables them to find heat-killed insects promptly after death, before the bodies of the dead insects dry out. In this way, when the workers acquire a food item, they obtain water—which is extremely valuable in the desert—as well as food. These ants must be able to get back underground quickly at any moment, so they can save their own lives if they start to overheat. How do they do this?

Experiments have shown that a *Cataglyphis* worker that is searching far and wide for a dead body always knows how to run straight back to its underground nest. A worker does not retrace its path to get home. Instead, if it finds a dead body or starts to overheat, it returns home by the shortest possible, straight path.

Years of research have revealed how a *Cataglyphis* worker knows how to find its home nest without retracing its path. To understand the mechanism, let's consider the worker's outbound path to consist of a series of segments (**FIGURE 37.10**, black line). The worker monitors the length and the compass direction of each segment. Then, by the process of *path integration*, it integrates, or puts together, all this information on segment lengths and directions to know where it is relative to its nest. Thus it can return directly (see Figure 37.10, red line) at any moment. We know this sort of process can work because it closely resembles the process (called dead reckoning) that sailors often used 150 years ago to know where they were at sea on an hour-by-hour basis.

How does a *Cataglyphis* worker measure the length and direction of each segment of its foraging path? It measures the length of each segment by counting steps (**INVESTIGATION FIGURE 37.11**). It can determine the direction by several means, as we will soon discuss.

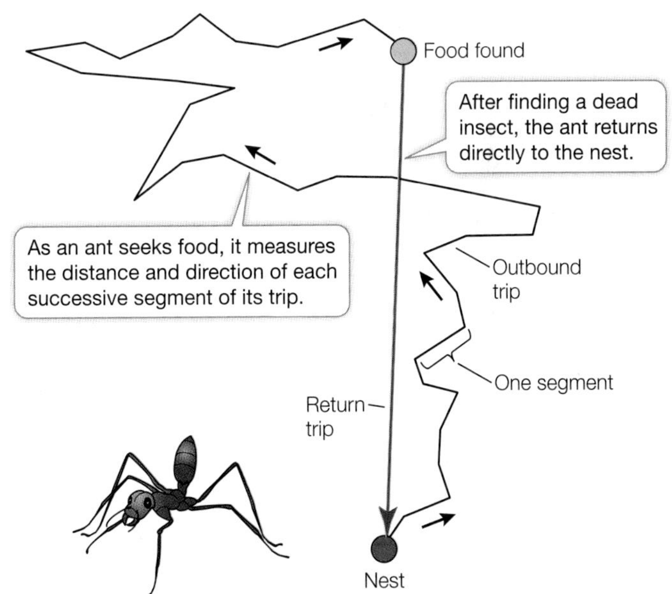

FIGURE 37.10 A Foraging Trip by a Worker Ant of a Desert Ant Species (*Cataglyphis*) The worker ant uses path integration to find the quickest route home. Its outbound trip (black line) consists of a series of segments oriented in various directions as the ant wanders the desert seeking heat-killed insects. At any point, however, the ant can return to its home nest quickly in a straight line (red line). *Cataglyphis* ants accomplish this task with a brain that weighs 0.1 milligram! (After R. Wehner and S. Wehner. 1990. *Ethol Ecol Evol* 2: 27–48.)

INVESTIGATION

FIGURE 37.11 An Insect Odometer Desert ants navigate by path integration, which requires information about both direction and distance. How do they measure distance? Matthias Wittlinger and his colleagues investigated the possibility that Saharan desert ants (*Cataglyphis fortis*) "count" steps to keep track of how far they've walked.[a] Because every stride is a certain length, the number of steps can theoretically measure the distance traveled.

HYPOTHESIS

Desert ants measure distance by keeping track of the number of steps they take.

METHOD

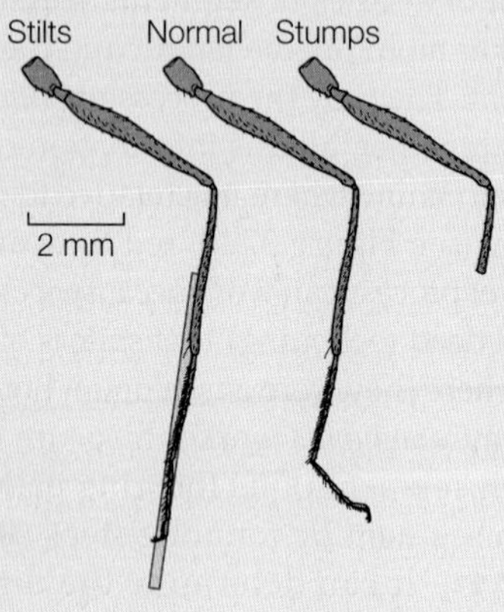

1. Train ants to walk from their nest to a feeder 10 meters away and then back to their nest.
2. Capture trained ants that have walked from the nest to the feeder and modify some of them before they start back to their nest. Separate them into three groups, with 25 ants in each group. Shorten the legs of one group ("Stumps") by removing their lower leg segments. Lengthen the legs of another group ("Stilts") by gluing pig bristle extensions to their legs. Leave the legs of the third group ("Normal") unchanged as a control.
3. Release the ants at the feeder, so that they can return to their nest. Measure how far they walk before they stop to search for the nest entrance (indicating that they've arrived where they expect the nest to be).

A similar **Analyze the Data** exercise may be assigned in **Achieve**.

RESULTS[b]

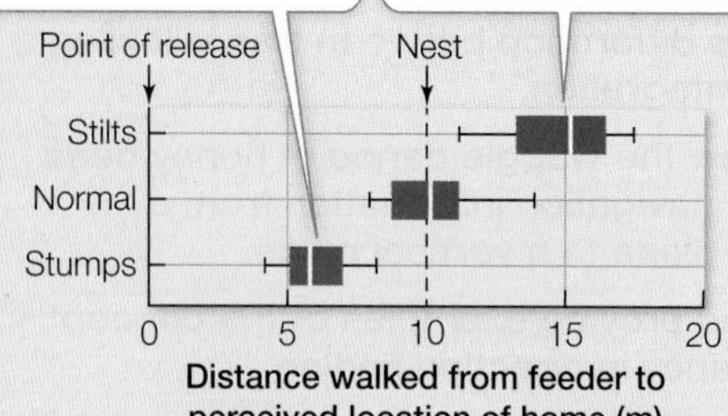

CONCLUSION

Desert ants behave as if they keep track of how far they travel by "counting" their steps.

ANALYZE THE DATA

In a second experiment, the same ant groups (two of which had altered legs) were allowed to walk from their nest to the feeder and then return to their nest. The graph shows the distances the three groups of ants walked to get back to their nest.

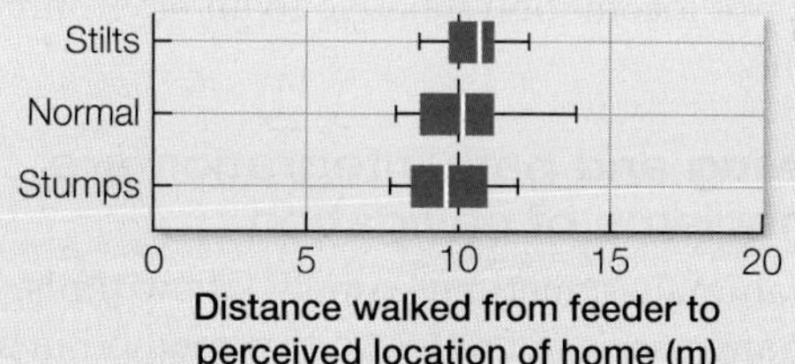

1. How do these results differ from those obtained in the first experiment? Explain why they differ.
2. Why is this experiment an important control?

[a]M. Wittlinger et al. 2006. *Science* 312: 1965–1967.

[b]The data are shown as a box plot. The vertical line inside the box is the median, the box indicates the middle 50% of the data, and the horizontal line that extends outside the box indicates the middle 90% of the data.

Animals have evolved multiple mechanisms for determining direction

As we noted in Chapter 29, some of the earliest research on animal orientation focused on homing pigeons, amazing birds that are able to fly back to their home nests even after they have been transported tens of kilometers away. During both World Wars, the pigeons carried messages from soldiers on the battlefield to their home nests next to command centers, where the messages could be retrieved by strategic and tactical commanders.

In the years after World War II, researchers put a high priority on understanding the mechanisms by which homing pigeons determine direction so they can fly home. The researchers discovered that the pigeons have a **sun compass**, meaning that, in essence, they can determine where north, south, east, and west are by observing the Sun. As we discussed in Key Concept 29.4, animals require knowledge of the time of day to determine compass directions from the position of the Sun. The experiments described in Figure 29.24 demonstrated convincingly that pigeons use their endogenous circadian clock to provide the time information. Thus, for example, if a pigeon observes the Sun near the horizon, it uses its circadian clock to determine if the position of the Sun corresponds to east or west (east if the time is morning according to the clock, west if it is evening). Then all the other compass directions can be deduced from the position of the Sun.

By now, a wide variety of animals have been shown to have a sun compass. In most cases, their circadian clocks play an essential role.

In addition to elucidating the sun compass, the pioneering studies of homing pigeons also led to another major discovery of enduring importance: the discovery of *redundancy* in the mechanisms of orientation. Investigators found that instead of having a single orientation mechanism, the pigeons have multiple mechanisms. We now know that this sort of redundancy often occurs in the animal kingdom.

Redundancy makes it possible for homing pigeons to find home on both sunny and cloudy days. On cloudy days when they cannot see the Sun and thus cannot use their sun compass, a key orientation method they use is to detect Earth's magnetic field and orient to it. They also sometimes use landmarks such as hills to orient, and they may use odors, low-frequency environmental sounds, and learning from other pigeons.

Activity 37.1 Homing Simulation

Many insects and birds can determine compass directions by detecting patterns of polarized light in the sky. This mechanism is particularly common in insects, where it depends on specialized photoreceptors. The atmosphere is filled with tiny particles, such as dust, water droplets, and ice crystals. When sunlight strikes such particles, the reflected light rays become polarized—aligned parallel to one another to some degree—giving rise to polarization patterns in the sky. Although our eyes cannot detect these patterns, the eyes of insects and birds can. The patterns are present even in small patches of clear sky. For this reason, the "polarization compass" is often functional on cloudy days.

We discussed earlier how *Cataglyphis* ants in the desert employ path integration by measuring the distances (see Investigation Figure 37.11) and directions of all the segments of a trip they make away from their nest. How do they measure directions? If a prominent landmark is present, the ants determine their direction of travel relative to the landmark. However, they function perfectly well without any landmarks. Under those circumstances the principal mechanism they use to determine direction is a compass based on polarized light detection. *Cataglyphis* ants also have a sun compass.

Birds that migrate at night can use the stars to determine their directions of travel (see Figure 37.9). A great many animals besides birds are now known to be able to orient by using Earth's magnetic field.

Honey bee workers communicate distance and direction by a waggle dance

Honey bees (*Apis mellifera*), like ants, live in colonies in which worker individuals perform services for the colony as a whole. When a worker bee goes on a foraging flight to gather food and finds flowers containing pollen and nectar, she returns to the hive and communicates her discovery to other workers by a specialized behavior called the waggle dance. During her foraging flight, she measures the distance and direction to the flowers. Her dance communicates both of those pieces of information to the other workers so that they also can fly to the flowers.

A bee that discovers a new patch of flowers measures the distance to the patch by visually monitoring the rate at which she flies past local landmarks. To measure the direction to the flower patch, she monitors the angle of her flight relative to the compass position of the Sun. Of course, as she measures these critical pieces of information, she is flying *horizontally* outdoors. After she returns to the hive, she performs her waggle dance on *vertical* honeycomb surfaces inside the hive, where it is dark and the Sun cannot be seen. The marvel of the waggle dance is that it communicates the information on horizontal distance and angle to the Sun while the bee is running ("dancing") on vertical surfaces in the dark.

During her waggle dance, the bee moves in a repeating figure-eight pattern. She alternates half-circles to the left and right with a straight run between the half-circles (**FIGURE 37.12**). During each straight run, she waggles her abdomen. The duration of the straight, waggle run communicates the distance to the food source. The angle of the straight run relative to vertical communicates the

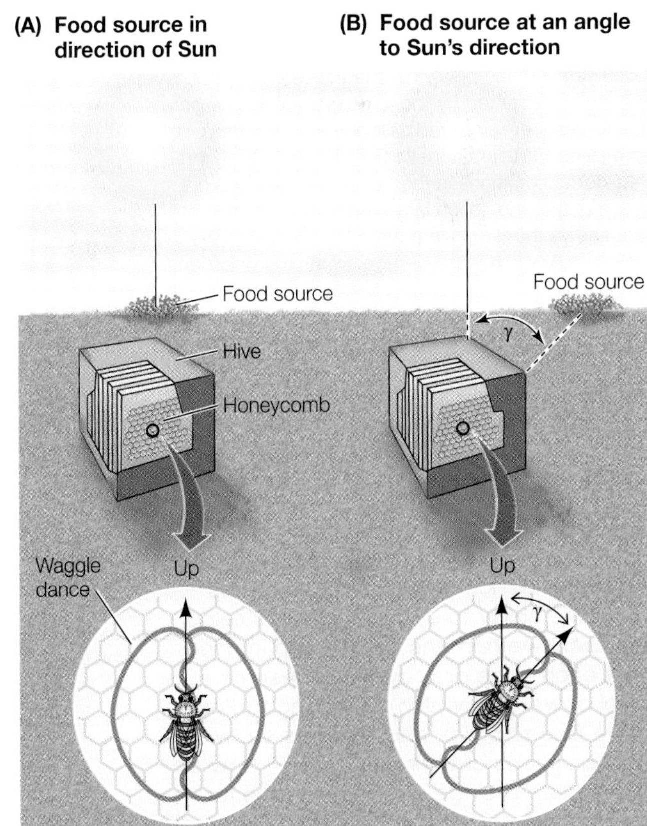

FIGURE 37.12 The Honey Bee Waggle Dance A honey bee (*Apis mellifera*) worker performs a waggle dance in her hive to communicate the location of a food source to other workers. The direction of the middle segment of the dance communicates the direction to fly to get to the food source. The direction to fly is communicated relative to the compass position of the Sun, the compass position being the spot on the horizon immediately below the Sun's position in the sky. In each part of this diagram, a vertical line under the Sun identifies the compass position of the Sun. **(A)** If the food source is in the same direction as the Sun, the bee runs straight up on the vertical surface of the honeycomb in the dark interior of the hive each time she executes the middle segment of her waggle dance. **(B)** If the food source is at an angle γ to the right of the Sun's compass position, the bee runs up at angle γ to the right of vertical. (After T. D. Seeley. 1985. *Honeybee Ecology: A Study of Adaptation in Social Life*. Princeton University Press: Princeton, NJ.)

Activity 37.2 Honey Bee Dance Communication

FIGURE 37.13 A Bar-tailed Godwit (*Limosa lapponica*) in Flight The godwits feed principally on insects and other small invertebrates, some of which they collect by probing in mud with their long bill.

direction to the food source. That angle matches the angle a bee needs to fly relative to the Sun's compass position to find the food source. What happens if the Sun is hidden by clouds? Honey bees can infer the position of the Sun from the polarization patterns in the sky.

Many animals undertake migrations

Today biologists are better able to track moving animals than ever before, because of technological innovations. Consequently, a revolution is underway in our understanding of migrations. For example, natural historians have long known that Bar-tailed Godwits (*Limosa lapponica*; **FIGURE 37.13**) migrate between Alaska and New Zealand, spending the Northern Hemisphere summer in Alaska and the winter in New Zealand. Recently, godwits were outfitted with miniaturized electronics that permitted global tracking of their flights, and for the first time, mind-boggling details of their migrations were revealed. We now know that when the birds leave Alaska, they fly *nonstop* to New Zealand over the open water of the central Pacific Ocean—a trip of more than 10,000 kilometers. They need 6–9 days to complete the trip, and during that time they cannot eat or drink. They presumably do not sleep. Their flight muscles must engage in powering or controlling flight continuously day and night—not to mention the necessity of navigating accurately over endless expanses of open water. How is this performance possible? As often happens in the practice of science, the tracking studies of godwits seem to raise more questions than they answer.

"Migration" is defined in various ways by various specialists. A useful general definition is that animals exhibit **migration** when they move from one location to a different location, remain for a

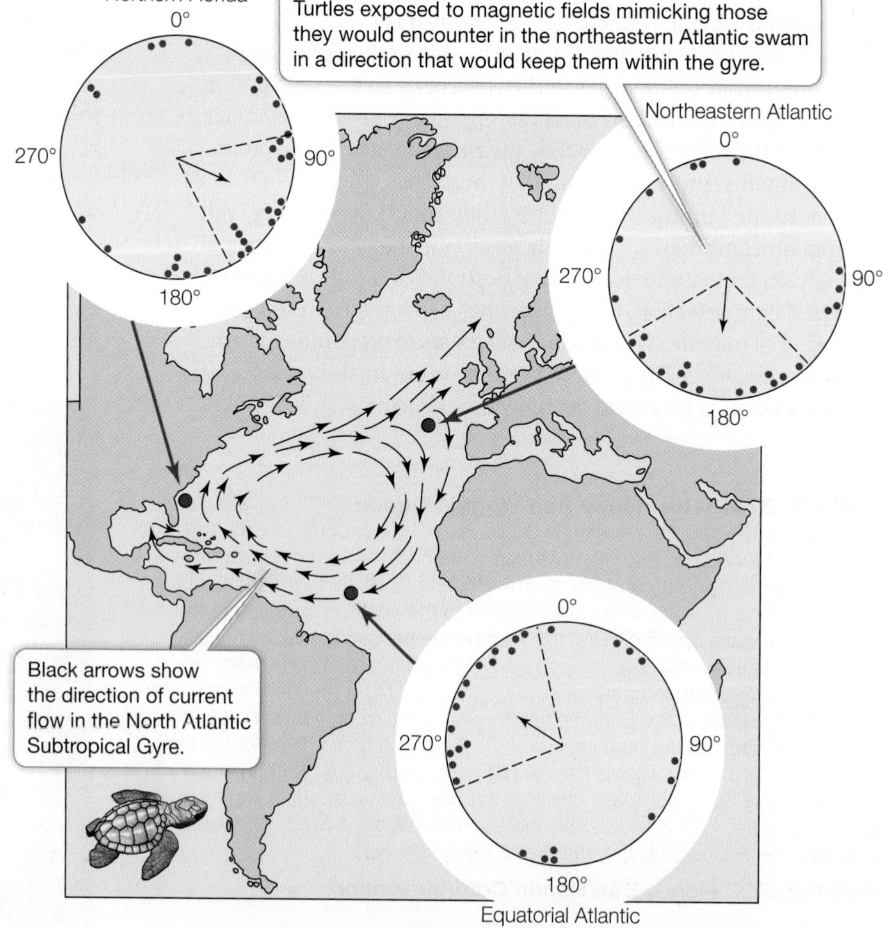

FIGURE 37.14 Young Loggerhead Sea Turtles Orient to Magnetic Fields as They Migrate During their migration, the young sea turtles travel in a giant closed circle across the Atlantic Ocean and back. To better understand the migration, biologists studied newly hatched turtles in a pool of water—represented by the large gray circles—where the turtles were exposed to magnetic fields that mimicked Earth's magnetic field at each of the three red-dotted locations in the ocean. Within each diagram of the pool (each large gray circle), each small dot shows the direction a turtle tried to swim, the solid arrow shows the average swimming direction, and the dashed lines show the 95 percent confidence limits for swimming direction. The solid arrows (showing average swimming direction) indicate that as the sea turtles travel in the currents of the North Atlantic Subtropical Gyre, they orient relative to local magnetic fields to swim in directions that keep them within the closed circle of current flow. (After K. J. Lohmann et al. 2001. *Science* 294: 364–366, which is based on misc. sources.)

The results gathered in this study provide strong evidence that young sea turtles have the ability to detect magnetic fields: an ability we humans lack. Keep in mind that each large gray circle represents a pool of water in which the young turtles were free to swim, and the dots show the directions of their swimming. How do these data provide strong evidence for magnetic-field detection?

substantial length of time, and later return—often in a periodic (repeating) way year after year.

One of the best-understood long-distance migrations is that of young loggerhead sea turtles (*Caretta caretta*). Several decades ago, Archie Carr coined the term "the missing year" in describing these turtles, which hatch out of eggs on beaches in Florida (among other places). For a while, the little turtles can be found in the ocean along the Florida coast, but then they disappear. The next time they are seen in Florida, about a year later, they are much bigger. Where have they been during their missing year, and how did they find their way back?

Genetic studies have been crucial in determining where the turtles go. The allele frequencies of turtles that hatch in a particular region of Florida provide a distinctive genetic signature. Researchers discovered that turtles in the ocean west of Africa have the same signature. This evidence (which is reinforced by other evidence) indicates that young loggerhead sea turtles that hatch in Florida migrate across the Atlantic Ocean to African waters. Then, growing as they go, they migrate back to the waters off Florida. The turtles do not need to power their migration entirely by swimming because a large circular water current, the North Atlantic Subtropical Gyre (shown by black arrows in **FIGURE 37.14**) aids their travel all along the way. The turtles use detection of Earth's magnetic field to help them stay within the gyre and ensure that they do not get lost at sea. The large gray circles in the figure show the results of experiments demonstrating that the turtles orient their direction of swimming in relation to magnetic fields so that they stay in the circular water current of the gyre.

REVIEW & APPLY | 37.3

R Animals find their way in the environment by navigation and orientation. Some animals navigate by following trails (marked with pheromones in some cases), whereas other animals navigate by path integration. Animals may determine direction by using a sun compass, magnetic fields, landmarks, and atmospheric polarization patterns. Animals often have multiple, redundant mechanisms of direction determination. Honey bees use a waggle dance to communicate the distance and direction of a food source relative to the hive. Some animals display remarkable feats of navigation when they migrate, traveling as far as thousands of kilometers between locations.

A
1. In some species of ants, when workers employ a trail pheromone to mark a trail to a food source, each worker that goes to the food source adds more pheromone to the trail on its way home if it readily found food at the source. How could this process act as a positive-feedback loop to draw large numbers of workers to a distinctively valuable food source?
2. Why might natural selection favor the emergence of redundant methods of direction determination?
3. How does the waggle dance encode the angle that worker bees should fly to find a food source?

Now we turn to one of the themes with which we began this chapter, namely that animals often do not function alone. When animals interact with each other, behavior is the paramount way in which they do so.

37.4 Behavior Structures Social Groups and Plays Key Ecological Roles

LEARNING OBJECTIVES

By the end of this key concept you should be able to:

37.4.1 **Identify documented advantages and disadvantages of group living.**

37.4.2 **Distinguish social dominance and eusociality.**

37.4.3 **Specify ways that ecological communities are structured by behavior.**

A moment's reflection reveals that many animal species live in groups. Flocks of birds, schools of fishes, and herds of antelopes are just a few examples. Many types of groups are considered societies. Specialists debate the exact meaning of a "society," but a broad, commonly used definition is that a **society** is a group of individuals of a single species, organized to some degree in a cooperative manner. The subject of **social behavior** includes both the behaviors of individuals that integrate the individuals into societies and the group behaviors of entire societies.

Living in a group has downsides. Grouping tends to make animals more visible. Diseases can spread more easily within a group, and a group may rapidly deplete food or other resources in the place where it is living.

Biologists presume that group living has evolved only in circumstances where it provides advantages that exceed its disadvantages. Potential advantages are thus of great interest, and in much of the research being done on societies today, emphasis is placed on testing hypotheses regarding how individuals in a society benefit from living together in a group. Physiological advantages prevail in some cases. Penguins, for example, reduce their thermoregulatory costs in the cold of winter by huddling together. As we discuss societies here, our focus will be on potential behavioral advantages of group living.

Some societies consist of individuals of equal status

Sometimes the adult individuals in a flock, school, or herd seem to play similar, approximately equal roles in the function of the group. In a group of this sort, one advantage hypothesized for group living is that a group may enjoy enhanced sensory awareness of its environment because the members of the group can pool their individual sensory capabilities.

A group of 50 vertebrate animals has 100 eyes instead of just 2. Groups may therefore have greater awareness of danger than do individuals. Robert Kenward found that the success of

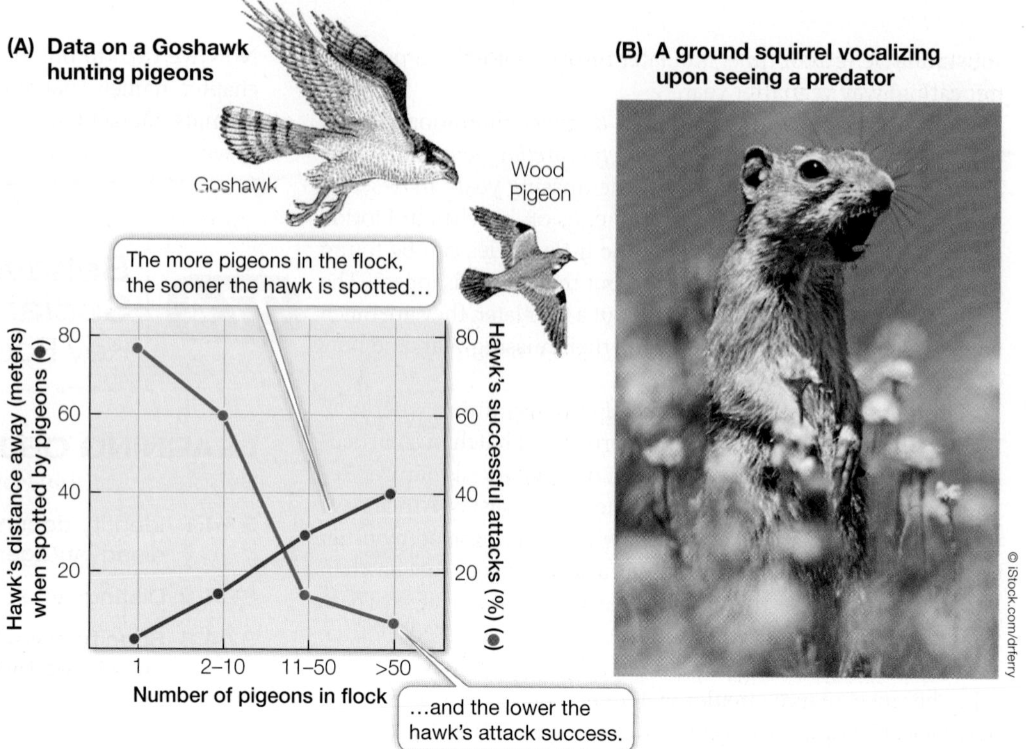

FIGURE 37.15 Group Living Provides Protection from Predators Animals that live in groups can spread the cost of detecting predators. **(A)** The larger the number of Wood Pigeons (*Columba palumbus*) in a flock, the greater the chances that one of the pigeons will spot a Goshawk (*Accipiter gentilis*) before it attacks, and the lower the chances that the Goshawk will capture one of the pigeons. **(B)** A young male Belding's ground squirrel (*Spermophilus beldingi*) gives an alarm call upon spotting a predator. (A after R. E. Kenward. 1978. *J Anim Ecol* 47: 449–460.)

a trained Goshawk (*Accipiter gentilis*) in capturing a Wood Pigeon (*Columba palumbus*) in a flock decreased as the number of pigeons in the flock increased (**FIGURE 37.15A**). The larger the flock, the sooner one of the individuals in the flock spotted the Goshawk and started to fly away, stimulating the other individuals to take flight as well. Belding's ground squirrels (*Spermophilus beldingi*), which live in large colonies in open meadows, use alarm calls to reduce predation risk. When an aerial predator such as a hawk arrives overhead, the first squirrel to spot the hawk whistles loudly to warn others (**FIGURE 37.15B**). Immediately, other individuals also whistle, and the animals run to their burrows or other shelter. Aerial predators almost never capture a ground squirrel after the alarm whistles have begun.

Animals in groups may also discover preferred environments more efficiently. In a recent experiment, investigators studied a species of fish that prefers to be in areas where light levels are low. The investigators studied schools of the fish of various sizes in a large aquarium that had both darkened and well-lit areas of water. The darkened areas moved from place to place, so the fish could not simply find a dark spot and stay there. The efficiency with which a school could find dark areas increased dramatically as the number of fish in the school increased (**FIGURE 37.16**). By analogy, we could expect that large schools find food-rich regions in the water relatively quickly.

(A) Successive images of a fish school finding a dark patch

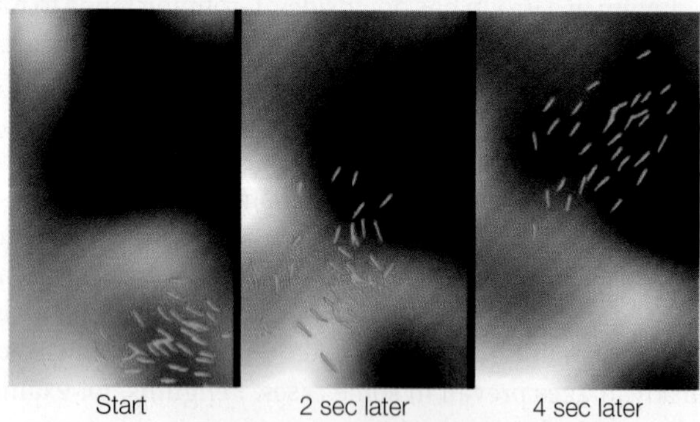

(B) Patch-finding success in relation to school size

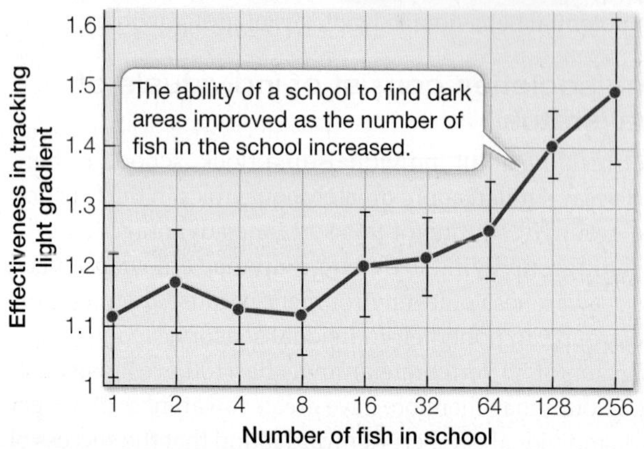

FIGURE 37.16 Preferred Habitat Is Discovered More Quickly When Many Seek It Juvenile golden shiners (*Notemigonus*) prefer areas in the water where the light intensity is low. Schools were placed in an aquarium where dark areas were present but moved from place to place. **(A)** Three successive images of a school, taken 2 seconds apart, as the school finds and enters a dark patch. **(B)** An index of a school's effectiveness in finding dark patches. The index is plotted against the number of fish in the school. Success in locating dark patches increases with the number of fish in the school. Vertical bars are 95 percent confidence intervals (see Appendix B). (From Berdahl et al. 2013. *Science* 339: 574–576.)

Some societies are composed of individuals of differing status

The individuals in a group of vervet monkeys (*Chlorocebus*) are far from equal. A group typically consists of multiple subadult and adult males (**FIGURE 37.17**) plus females and youngsters. One of the adult males typically dominates all the other adult and subadult males in the group, meaning it "wins" during one-on-one behavioral contests with each of the others. This dominant male has by far the greatest chances of mating with the adult females in the group.

Dominance by a particular male is even more evident in impala antelopes (*Aepyceros melampus*). If you ever go on safari in Africa and encounter a group of impala during the breeding season, you'll notice the group consists of two spatially separate subgroups. One subgroup consists entirely of males. By marking the animals in this type of subgroup, researchers established that these males do not get to mate with females. The second subgroup consists of a single male and a large number of females. This single male does essentially all of the mating! As you might guess, the dominant male—the single male in the group of females—must expend extraordinary effort to repel intruding males and intercept females that try to leave. This activity is exhausting, and dominant males tend to retain their dominance for only a few months.

Dominance behaviors may result in sexual selection, which would tend to favor increases in these behaviors in subsequent generations; see **Key Concept 13.2**.

Within groups in which certain individuals have distinctive status—such as groups of vervet monkeys or impala with dominant males—individuals are able to benefit from group living in the same ways we discussed in the last section. Biologists hypothesize, however, that there are additional benefits, which explain the evolution of social systems based on distinctive status. In cases of dominant males, researchers currently hypothesize that the process of becoming dominant may often serve as a test of a male's strength, endurance, and other properties critical for success. Females that mate with the dominant male may thereby ensure that their offspring are genetically well endowed. In this way, both the individual males and the individual females that mate in such systems of social behavior may realize advantages.

Eusociality represents an extreme type of differing status in a society

In most animals, all the adult members of a social group are physiologically capable of reproducing, even though some may be behaviorally excluded from reproducing at certain times. An entirely different situation exists in many social insects, such as honey bees.

In a honey bee colony, a single female—the queen—is reproductive and lays eggs. Occasionally she produces a few male offspring (drones) that are fertile, producing sperm. However, most of the thousands of other individuals in the colony—all of which are her offspring—are sterile female worker bees that never have offspring of their own and instead help rear the offspring of the queen. Similar colony structures occur in many other types of bees, wasps, ants, and termites.

FIGURE 37.17 An Adult Male Vervet Monkey (*Chlorocebus*) Note his colorful genitalia, which signal his adult status.

Eusociality refers to a social structure in which some adult members of a social group are nonreproductive and assist the reproduction of fertile members of the group, typically their mother. It represents an extreme form of differing status within a social group. Nearly all known cases of eusociality occur within the insects, although eusociality is also documented in naked mole-rats (*Heterocephalus glaber*) and certain coral-reef shrimps.

All cases of structured, unequal reproductive status among adults in social groups raise the question "Why has evolution produced systems in which reproductive inequality exists?" This question reaches its extreme manifestation in the eusocial animals. As the question is debated, the concept of altruism arises. In discussions of evolution, **altruism** refers to any characteristic of an individual that imposes a cost on that individual while aiding another individual. Workers in eusocial insect colonies exemplify altruism because, whereas they themselves are unable to breed, they help another individual breed. Altruism also occurs if an individual—when seeing a predator—gives warning calls that draw the predator's attention to itself (endangering its life) while giving neighboring individuals a better chance to hide.

The reasons why altruism exists have been hotly debated in evolutionary biology for more than a half century. Kin selection is an often-mentioned hypothesis. **Kin selection** stresses that when an individual acts altruistically, the individuals benefitted may be kin (relatives) of the altruist and thus likely to possess similar genetic alleles; in that case, even though the copies of alleles in the altruist may be disadvantaged (or even extinguished if the altruist dies), the other copies in kin that are aided by altruism may be advantaged to an even greater extent. In this way alleles favoring altruism could be favored by natural selection. Although kin selection is a well-known hypothesis for the evolution of altruism, not all specialists are convinced that it is an accurate hypothesis.

Media Clip 37.1 **Eusocial Animals**

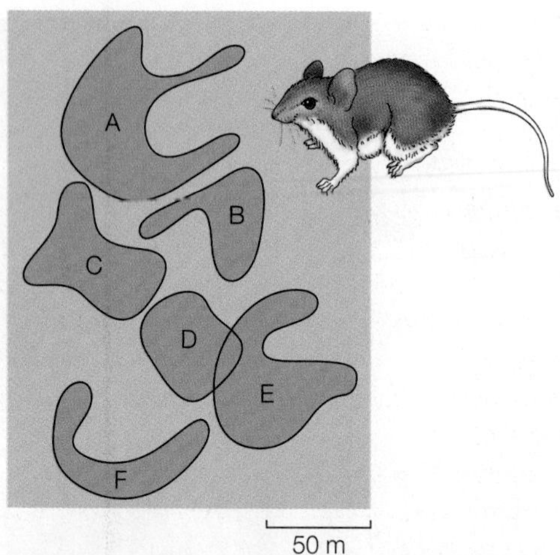

FIGURE 37.18 Home Ranges of Male White-footed Mice (*Peromyscus leucopus*) in a Michigan Woodland The drawing shows a view from above of the forest floor. At the time of study, six adult males, designated A–F, were living in this approximately 4-hectare area within a large woodland. Each mouse moved around just within the circumscribed home range shown. (After L. H. Metzgar and R. W. Hill. 1971. *J Mammal* 52: 12–20.)

Animals often behaviorally partition space into territories or home ranges

Now that we have discussed behavioral relationships among individuals in single-species social groups, let's turn to a more ecological perspective. As you might guess, there are no strict distinctions between the social and ecological perspectives on animal behavior. A topic of importance in both contexts is the use of space.

When we consider animals living in their natural habitats, we find that individuals in a population often restrict their movements to limited areas within the space the population occupies. This common phenomenon—which has the effect of structuring space—is seen in animals as diverse as fishes, dragonflies, birds, and mammals. For example, **FIGURE 37.18** shows the use of space by the six adult male white-footed mice that lived within a 4-hectare region of a forest. Each mouse roamed within a defined area on the forest floor, with almost no overlap among the areas occupied by the different mice.

The region occupied by an individual is known as its **territory** if the individual actively keeps out other individuals of the same species. The region is known as an individual's **home range** if other individuals are not excluded. Home ranges often overlap considerably. Both territories and home ranges provide familiarity. An individual occupying a territory or home range learns all the locations within its living area where it can escape attacks or find resources or shelter. When an individual defends a territory, it enjoys the additional benefit of having sole use of any resources present.

Animation 37.1 The Costs of Defending a Territory

Behavior helps structure ecological communities

Within an ecological community consisting of a diversity of species living together, behaviors often provide critically important structure to the community. As we have just seen, behaviors structure the use of space. Behaviors also structure the use of time, and they structure interrelationships among species. We discuss these points in this and the next two sections.

To illustrate the structuring of time and its importance, consider the structuring of the 24-hour day into activity periods. Some animals in a community are active only during daylight hours. Others are active only at night. These behavioral differences can determine whether two species encounter each other and directly interact. For example, because insect-eating bats are typically nocturnal (spending their days in secluded places such as caves and attics), diurnal insects are not threatened by the bats even when the two live in the same ecological community.

Behavior often maintains species distinctions

In natural environments, behavior can be critical for preventing hybridization among species, thereby maintaining the distinctions among species that otherwise might blend. For example, in much of North America the principal native mice are members of the genus *Peromyscus*, and it is common for two species in this genus to be found living near each other in one general region. White-footed mice (*P. leucopus*) and cotton mice (*P. gossypinus*) are one example. When males and females of these two species are housed together in cages, they breed readily and produce fertile hybrid offspring. Hybrids, however, are rarely found in nature because of behaviors that keep the species distinct. The two species prefer different types of woodlands. Individuals of the two species thus live in different places and probably rarely meet. Moreover, we can hypothesize that if they were to meet, they would detect oddities in each other's interactive behavior and separate before mating (whereas caged individuals cannot separate).

Many additional cases are known in which coexisting species capable of hybridizing remain distinct in undisturbed natural habitats because of behavior. One case of particular interest is provided by the species of Darwin's finches in the Galápagos archipelago. This case is of special note because the effectiveness of behavior has been quantified. Peter and Rosemary Grant have established by analyses of DNA that—in free-living, wild populations of the several species of ground-living Darwin's finches—about 1 percent of individuals that grow to adulthood are fertile hybrids of two distinct species. The hybridization that is observed demonstrates that the species of Darwin's finches are not genetically walled off from each other. The very low likelihood of hybridization shows, however, that behavior is ordinarily very effective in preventing hybridization and keeping the species distinct. Males ordinarily learn their songs from their fathers (see Key Concept 37.2), and a female ordinarily will not mate with a male singing a song other than the correct one for her species. These behaviors prevent most crossbreeding.

Behaviors that prevent crossbreeding are an important mechanism of reproductive isolation and speciation; see **Key Concepts 16.1** and **16.4**.

FIGURE 37.19 A Bumblebee Landing on a Flower to Collect Nectar When foraging in cold weather, bumblebees must focus on species of flowers that provide sufficient energy reward to meet the bees' foraging costs at low air temperatures.

Behavior helps structure ecological relationships among species

For understanding behavioral relationships among species in ecosystems, ecologists often use a cost–benefit approach. The **cost-benefit approach** assumes that an individual animal has only a limited amount of time and energy, and therefore—in its relationships with individuals of other species—cannot afford to engage in behaviors that cost more to perform than they bring in benefits. In cost–benefit analysis, ecologists quantify both costs and benefits. Then, by calculating relative costs and benefits, they make—and test—predictions about which behaviors to expect.

Bernd Heinrich took a cost–benefit approach in his detailed study of flower choice in the foraging ecology of bumblebees (*Bombus*). As a bumblebee forages in a flower patch, it flies from flower to flower, landing on flowers to collect nectar (**FIGURE 37.19**). During this process, the bee must keep its flight muscles warm so it will be able to fly. Keeping them warm during flight is no problem because the working muscles produce heat at a high rate. Keeping them warm when the bee has landed on a flower requires extra heat production when the air is cold, for the same reason a house furnace needs to make heat at an increased rate in cold weather. Adding all these considerations together, Heinrich calculated the energy cost per minute of foraging and determined that the cost is considerably greater in cold air than in warm air.

Heinrich also calculated the benefit of foraging. To do this, he measured how much nectar could be obtained from each flower and the sugar concentration of the nectar. Then he calculated the energy a bee could obtain from each flower by metabolically oxidizing the sugar in the flower's nectar.

One common flowering plant in Heinrich's study area was a rhododendron with large flowers. A bumblebee obtained 1.7 joules (J) of energy by taking the nectar from each rhododendron flower. In cold air at 0°C, a bee's cost of foraging was about 12.5 J per minute. Thus a bumblebee needed to harvest the nectar from only about 7 flowers each minute for its foraging benefit to equal its foraging cost. Actually, a bee could harvest 20 flowers per minute, receiving a benefit that far exceeded its cost—and enabling the bee to take energy back to its hive. Heinrich predicted that bumblebees would forage on rhododendron flowers in both cold and warm weather, and this is what the bees do.

Another common plant was wild cherry. Heinrich noticed that bumblebees foraged on cherry flowers only in warm weather. By applying his cost–benefit analysis, he discovered why. Cherry flowers are smaller than rhododendron flowers and provide only 0.2 J of energy each. Accordingly, at 0°C a bee would need to harvest more than 60 cherry flowers per minute to get enough nectar to equal its cost of foraging. Bumblebees cannot harvest that many flowers per minute. Thus they would have to go into an energy deficit—spending more energy than they would receive—to forage on cherry flowers in 0°C air.

With this study, Heinrich showed that bumblebee foraging behavior is structured in ways that keep a bee in energy balance. This fact, in turn, structures the relationship between bumblebees and the various particular plant species that flower in their ecosystems.

Animation 37.2 **Foraging Behavior**

LINK A cost–benefit approach can also be used to analyze life history trade-offs during evolution; see **Key Concept 39.3**.

REVIEW & APPLY | 37.4

R A society is a group of individuals of a single species that exhibits some degree of cooperative organization. Group living confers benefits (e.g., greater foraging efficiency), but it also has costs (e.g., increased disease transmission). Individuals in a society can be of equal status—as in a school of fishes—or of unequal status—as in a group of impala antelopes. An extreme form of unequal status is seen in eusociality (e.g., in social insects), in which some individuals are infertile and assist the reproduction of fertile individuals rather than reproducing themselves. An individual within a population often restricts its movements to a limited area—termed a territory if the individual actively keeps out other individuals of its species or a home range if other individuals are not excluded. Behavior plays important roles in maintaining species distinctions and structuring ecological communities.

A

1. Summarize advantages and disadvantages of group living.
2. What is altruism, and why might a biologist have difficulty reconciling it with the theory of evolution by natural selection?
3. How might Lee Metzgar's study (see Key Concept 37.2, Figure 37.8) of the survival of white-footed mice in the presence of screech-owls help explain the existence of home range behavior?
4. Imagine that on a cold day a certain bumblebee requires 15 J of energy per minute to forage continuously. If the bee can harvest 20 flowers of a certain species every minute, and gets 0.8 J of energy from each flower it visits, would it be to the bee's energetic benefit to forage on those flowers that day? Explain.

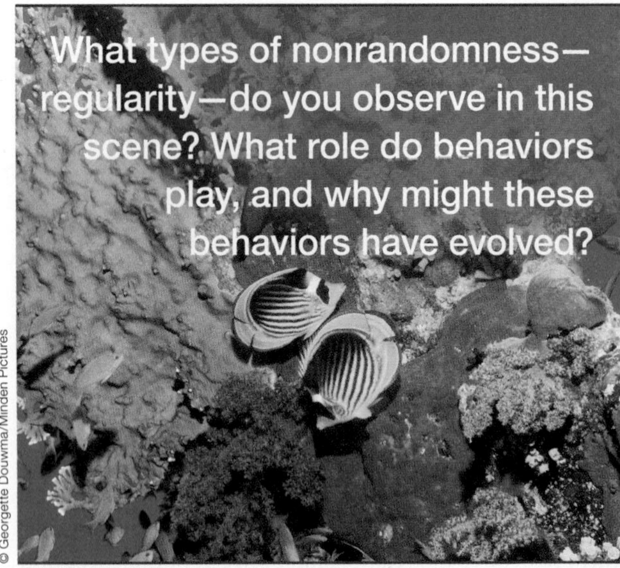

© Georgette Douwma/Minden Pictures

Answer: The small orange fish, which are a type of anthias fish, are highly nonrandom in their distribution. They are organized into a loose school in the lower left. The two large yellow and black fish, which are diagonal butterflyfish (*Chaetodon fasciatus*), seem to be traveling together. In fact they are a male and female, and it is common in butterflyfish for pairs to travel together like this. The fish community is not a random assemblage of all the species that might be present. It is structured by the behaviors of the fishes. Schooling behavior can be advantageous in several ways. For example, a school has more sense organs than a single individual. For this reason, a school may detect predators sooner than an individual so that evasive action can be taken. Similarly, a school may be able to find valuable resources in the environment more rapidly or efficiently. Pairing behavior could have several advantages—one being that in male–female pairs, each individual is assured of a mate.

37 VISUAL SUMMARY

37.1 Behavior Is Controlled by the Nervous System and Integrated with the Rest of Function

- To understand a behavior, both **PROXIMATE** and **ULTIMATE CAUSES** must be considered (**FIGURE 37.3**).
- Because it is controlled by the nervous system, behavior is subject to genetic mutation and evolution by natural selection (**FIGURE 37.6**).
- Although behavior has a genetic basis, there is often not an inflexible relation between genes and behavior because behavior is flexible and affected by learning, early life experiences, and **EPIGENETICS**.

FIGURE 37.3

© Johner Images/Alamy Stock Photo

FIGURE 37.6

Mark Gocke/USDA

QUESTIONS

1. What is a fixed action pattern?
2. How might behavior be related to the biochemistry of ATP production?
3. What is biological determinism, and is it a fact or a hypothesis? Explain.

37.2 Behavior Is Influenced by Learning and Early Experience

- **LEARNING** is the modification of an individual's behavior by the individual's prior experiences and observations (**FIGURE 37.9**).
- **BEHAVIORAL IMPRINTING** is a specialized type of learning that occurs during a narrow window of time early in life and that resists modification later in life.
- Early life experiences such as malnutrition can lead to patterns of epigenetic tagging that affect an individual for the rest of life.

FIGURE 37.9

QUESTIONS

1. What must star-navigating birds learn early in life, and what is the evidence?
2. What do mice in a woodlot learn that aids their survival, and what is the evidence?
3. What is the genetic relationship between solitary-living locusts and the highly destructive swarms of migratory locusts?

37.3 Moving through Space Presents Distinctive Challenges

Go to ACTIVITIES 37.1 and 37.2

FIGURE 37.12

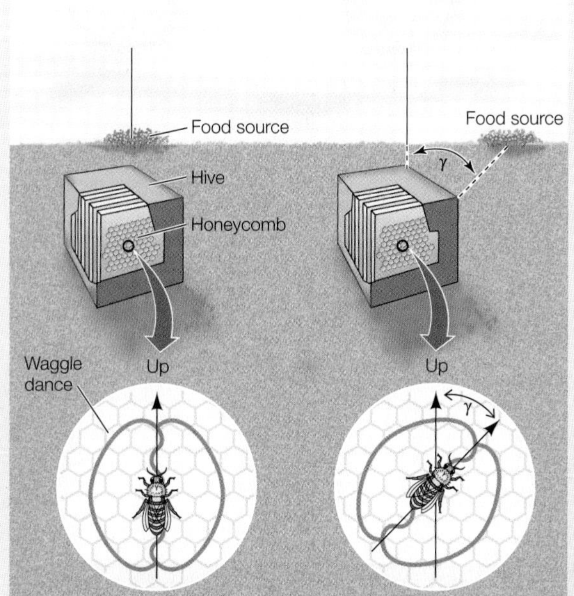

- Animals can determine direction using the Sun, stars, Earth's magnetic field, landmarks, and patterns of light polarization in the sky (**FIGURE 37.12**).
- Animals often employ redundancy in their direction-determining capabilities to ensure success.
- Animals are said to migrate when they move from one location to another, remain, and later return—often on an annual basis. **NAVIGATION** is as important as muscular performance during **MIGRATION**.

1. How do migrating juvenile loggerhead sea turtles use Earth's magnetic field to guide their migration?
2. What is the role of the circadian clock in a sun compass?
3. What information does a foraging *Cataglyphis* ant use to calculate the way back to its home nest? Give a general answer, and add specific mechanisms if possible.

 Behavior Structures Social Groups and Plays Key Ecological Roles

Go to ANIMATIONS 37.1 and 37.2

- Living in a **SOCIETY** can increase the abilities of animals to detect dangers and find resources (**FIGURE 37.16**).
- In their natural habitats animals often live in **HOME RANGES** or **TERRITORIES**, making it possible for them to be familiar with their surroundings.
- Behavior helps maintain species distinctions and helps structure ecological communities both spatially and temporally.

FIGURE 37.16

Successive images of a fish school finding a dark patch

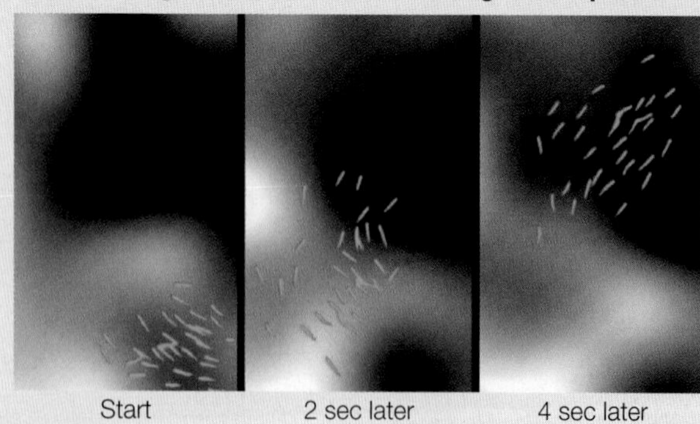

Patch-finding success in relation to school size

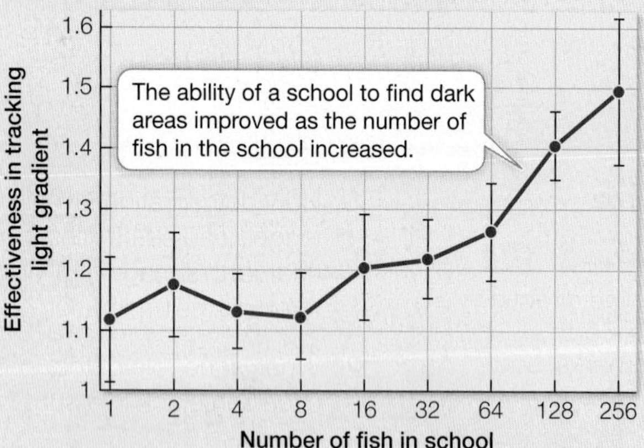

Activity 37.3 Concept Matching: Animal Behavior

QUESTIONS

1. What is eusociality, and how might kin selection explain it?
2. Animals may be diurnal or nocturnal. How does this behavioral property determine the effective structure of their ecological communities?
3. Focusing on the ground-living finches of the Galápagos archipelago, what do we learn from the fact that about 1 percent of the fertile adults are hybrids of two species?

Part 7
Ecology

John Cancalosi/Alamy Stock Photo

38 Ecological Systems in Time and Space

In the 1880s the U.S.–Mexico Borderlands of southeastern Arizona were clothed in lush grasslands that supported vast cattle herds. A decade later, overgrazing favored mesquite and other prickly shrubs over grass. Removing cattle failed to restore the grasslands, and today the Borderlands still sustain fewer cattle than they once could.

How can an understanding of interactions between physical and biological factors help us explain why removing cattle did not reverse this transition from grassland to shrubland?

You will find the answer to this question on page 966.

KEY CONCEPTS

38.1 Ecological Systems Vary over Space and Time

LEARNING OBJECTIVES

By the end of this key concept you should be able to:

38.1.1 **Describe the basic components of ecological systems at each level in the hierarchy of biological organization.**

38.1.2 **Give an example of variation in an ecological system over space and time.**

38.1.3 **Distinguish three tools for understanding ecological systems: descriptive observation, experiment, and systems modeling.**

During the eighteenth and nineteenth centuries, the prospect of visiting unknown lands filled with exotic plants and animals lured many European scientists—including Charles Darwin—to travel the world. They studied and described a tremendous variety of environments and organisms on their journeys, thus initiating the study of two new topics: **physical geography**, the spatial distributions of Earth's climates and surface features; and **biogeography**, the spatial distributions of species. It soon became apparent that the distributions of species and environments are linked.

Organisms plus their environments are dynamic ecological systems

The link between physical geography and biogeography makes sense when we think of organisms as components of **ecological systems** (often shortened to **ecosystems**). The "state" of any **system** (how it appears at any time) and its "dynamics" (how its state changes through time) are shaped by its components and their interactions (see Key Concept 1.2). Organisms interact with other organisms—the other **biotic** (living) components of their ecosystem. But they also interact with, and must be able to tolerate, the **abiotic** (nonliving) components of their systems. The types of organisms found in any location on Earth therefore reflect that location's physical environment.

The term "ecological" derives from **ecology**, a word that the German biologist Ernst Haeckel constructed in 1866 from the Greek roots *oikos*, "household," and *logia*, "study of." Haeckel meant this word to describe "the entire science of the relations of the organism to its external environment." This was not a new area of knowledge—even the earliest humans understood much about the plants and animals they depended on for food, the animals that competed with them for food, and the organisms and physical circumstances that could harm them. Indeed, the earliest available religious texts and scholarly writings are full of ecological topics, and indigenous cultures harbor a wealth of knowledge about their "households." But by coining a formal term, Haeckel helped establish the study of "all those complicated interrelations described by Darwin as the conditions of the struggle for existence" as a formal science, and emphasized ecology's relevance to Darwin's theory of evolution by natural selection.

Ecological systems occur on a hierarchy of levels

The components of ecological systems can be identified at several levels in the hierarchy of biological organization (see Figure 1.5B). The lowest level is the individual **organism** and its immediate environment. Individuals interact with their environment when they take in or put out materials and energy, or when their activities alter its physical structure. The individual organism is part of a **population**: a group of individuals of the same species that live in a particular location. The populations of different species that occur together in a location and interact with one another form an ecological **community**. Sets of neighboring communities that inhabit larger areas can be thought of as ecological **landscapes**. At the highest level, all the landscapes of Earth make up our planet's **biosphere**.

Ecological systems at higher levels of the hierarchy tend to be more complex than those at lower levels, both because they contain more components and because their components interact over a greater span of space and time. Different landscapes within a geographic area, for example, may be linked not only by rapid and localized exchanges of individual organisms and by shared local climate, but also by slower and more extensive migrations of organisms, great cycles of air and water movement, and the much slower geological churning of Earth's crust.

Ecologists choose a level of hierarchy and a scale of time and space for their research that are appropriate for the questions they wish to answer. They commonly use "systems thinking"—they identify the interacting components, measure the types and strengths of interactions, and develop systems models to understand the properties of the study system.

Ecological systems can be small

Ecological systems at any given level of biological organization can vary in their physical scale, and some are surprisingly small. Consider an ecosystem that is found within each of us—the human gut (**FIGURE 38.1A**). Our gut is one of the most densely populated ecosystems on Earth, hosting hundreds of species of bacteria, archaea, and yeasts, and on the order of 3.8×10^{13} individual

(A) Microbes in the human gut

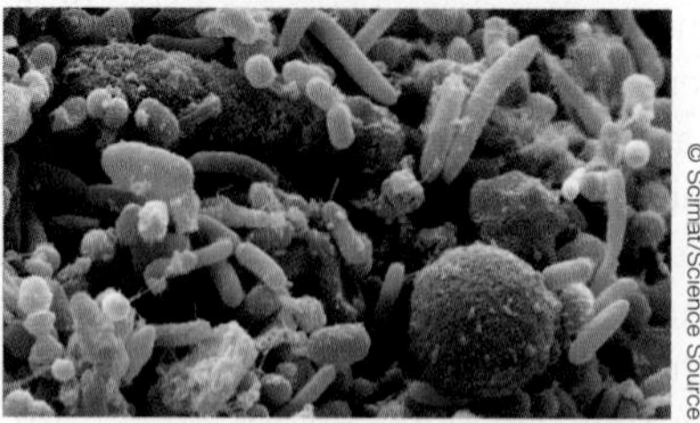

(B) Genetic influence on gut communities

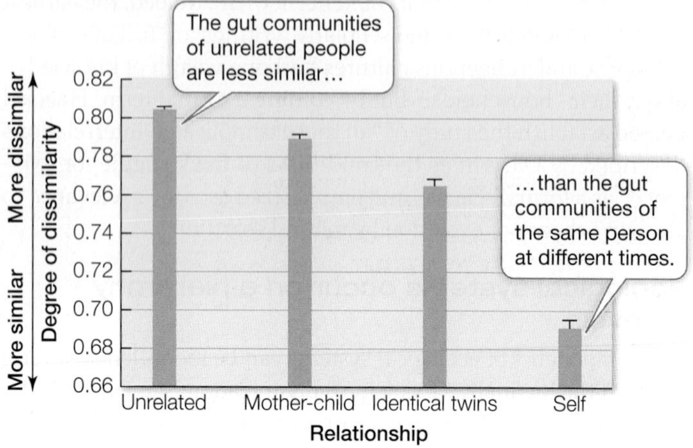

(C) Comparison of gut communities

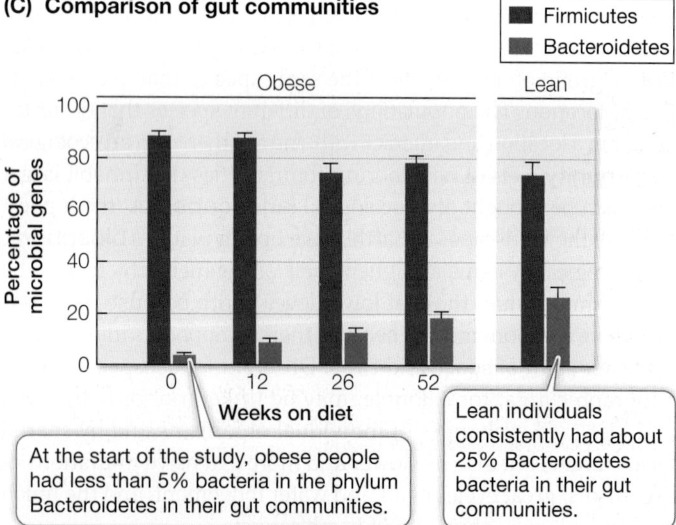

FIGURE 38.1 Host Genetics and Diet Affect the Composition of the Gut Microbial Community (A) This colorized electron micrograph reveals a diverse community of microorganisms in a human intestine. **(B)** The closer the genetic relationship between two people, the more similar their gut microbial communities are. **(C)** Obese people have proportionally more bacteria in the phylum Firmicutes, and proportionally fewer bacteria in the phylum Bacteroidetes, than do lean people. However, the microbial community of obese people approaches that of lean people as they lose weight on a calorie-restricted diet. Error bars in B and C indicate 1 standard error of the mean (see Appendix B). (B after P. J. Turnbaugh. 2009. *Nature* 457: 480–484; C after R. E. Ley et al. 2006. *Nature* 444: 1022–1023.)

Why are the gut microbial communities of identical twins, who share the same genotype, more dissimilar than those of the same person taken at two different times?

microbes. The microbial cells in the biofilms that coat the lining of the gut outnumber our own trillion (10^{12}) or so nucleated human cells, and their combined metabolism rivals that of an organ such as the liver.

It's no surprise that microbes thrive in the mammalian gut. Mammals regulate the gut's physical environment within narrow limits of temperature and chemistry, and they supply it with a steady stream of ingested nutrients. Gut microbes metabolize food materials, including some that the host cannot digest, and excrete waste products that provide nutrition to the host or to other members of the microbial community.

As with any system, these interactions determine the properties of the whole and its components: microbe–microbe interactions affect which species make up the microbial community, which in turn affects host nutrition and immune function; and the host's genotype and diet in turn affect the microbial community. Several human diseases—including infection by pathogenic gut microbes, obesity, autoimmune diseases such as irritable bowel syndrome, and wasting diseases such as kwashiorkor—have been linked to disruption of the gut's microbial community.

LINK Human health is linked to the health of our microbial communities. Review the description of the communities of microbes associated with our bodies in **Key Concept 18.3**.

Ecological systems vary, but in ways that can be understood with scientific methods

The biotic and abiotic components of ecosystems are distributed unevenly in space. Individual ecosystems also change over time, as cattle ranchers in the Borderlands (mentioned at the opening of the chapter) discovered to their dismay. Such spatial and temporal variations may be bewildering at first, but further observation often reveals patterns that provide clues to their possible causes—hypotheses that, when refined by further observation and systems modeling, can be tested experimentally.

Recent research on the human gut ecosystem illustrates how scientists have gone about understanding how and why gut microbial communities vary spatially and temporally, and what the variation means for human nutrition and health. Modern molecular genetics methods—such as metagenomics, which isolates and analyzes the DNA found in a sample (see Figure 12.9)—have made it possible to describe the phylogenetic and functional properties of entire microbial communities contained in gut samples. The first metagenomic surveys revealed astonishing variation: each person, it turns out, hosts a unique gut community, but ongoing research demonstrates that there is order to this variation. The gut communities of genetic relatives, for instance, are more similar than are those of unrelated people (**FIGURE 38.1B**), suggesting that a host's genotype favors some microbes over others. In addition, the gut communities of obese and lean people differ in their ratios of two bacterial phyla. As obese people lose weight on experimental calorie-restricted diets, the ratios of these two bacterial phyla become more like those of lean people (**FIGURE 38.1C**), suggesting that diet also affects the gut community. The relationship between gut community and obesity appears to be a two-way street, because systems models of nutrient processing by different microbial groups indicate that obesity might be related to how much energy the community extracts from food and passes on to the host. This conclusion is supported by transplant experiments done with mice, whose gut ecology is similar to ours (**INVESTIGATION FIGURE 38.2**). Exciting results such as these open the door to new therapies that use ecological principles to restore human health by manipulating our "ecosystem within."

INVESTIGATION

FIGURE 38.2 The Microbial Communities of Genetically Obese Mice Contribute to Their Obesity Mice (*Mus musculus*) that are homozygous (*ob/ob*) for a particular gene, *Lep*^{ob}, are fatter and eat more chow than their *ob*/+ and +/+ littermates. Like obese humans, *ob/ob* mice host a gut microbial community that is relatively rich in bacteria of the phylum Firmicutes and poor in the phylum Bacteroidetes. Does this gut community contribute to the obesity of *ob/ob* mice, or does it simply reflect their greater food consumption?[a]

HYPOTHESIS

The gut microbial community of obese mice contributes to their obesity.

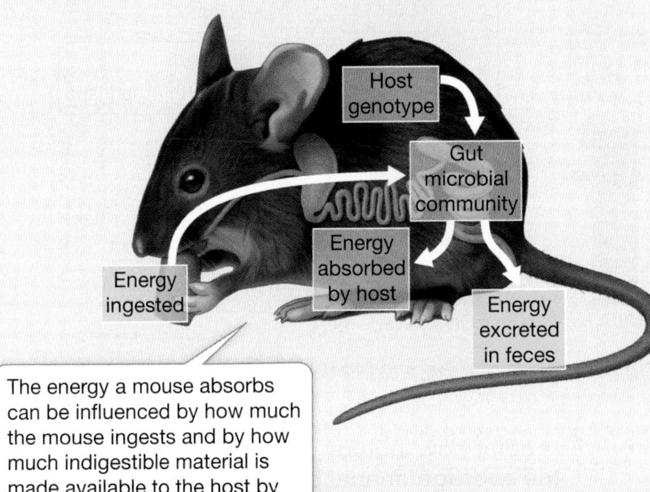

METHOD

1. Rear +/+ mice to adulthood in a germ-free environment so that their guts contain no microbes.
2. Select 9 mice at random to have their guts inoculated with fecal material obtained from obese *ob/ob* donors and select 10 mice at random to be inoculated with material from lean +/+ donors.
3. Verify that the recipient mice have developed microbial communities similar to those of the donors.
4. Feed both groups of recipient mice on mouse chow containing 3.7 kcal/g. Measure the food intake of the mice over the next 2 weeks.
5. Use a noninvasive body-scanning method to measure changes in the body fat content of the two groups of recipient mice over the same 2-week period.

RESULTS

Both groups of recipients became fatter during the 2 weeks, but those inoculated with gut material from an *ob/ob* donor gained proportionally more fat than mice that received +/+ gut material, even though they ate the same amount of food. Error bars indicate 1 standard error of the mean; bars with different letters are statistically different from one another at the 5% level.

Increase in body fat (%) (0, 20, 40, 60); bars: +/+ (a), *ob/ob* (b); Donor genotype

CONCLUSION

The Firmicutes-dominated gut microbial community of obese mice contributes to their obesity.

ANALYZE THE DATA

Gut bacteria break down indigestible plant polysaccharides into short-chain fatty acids that the mammalian gut can absorb, but phyla vary in how efficiently they do this. Firmicutes produce more lipid- and carbohydrate-metabolizing enzymes than do Bacteroidetes. The researchers therefore hypothesized that the reason a Firmicutes-dominated gut community contributes to obesity is that it extracts more energy from food that the host can use. To evaluate this hypothesis, the researchers collected all feces produced in 24 hours by 13 *ob/ob* mice and 9 +/+ or *ob*/+ littermates, then burned the feces in a closed chamber and used the temperature change to measure the energy content of each gram of feces. Error bars indicate 1 standard error of the mean; bars with different letters are statistically different from one another.

Fecal energy content (kcal/g) (2.0, 2.5, 3.0, 3.5, 4.0); bars: Lean (a), Obese (b)

1. Do these results support the hypothesis that the gut communities of obese mice allow them to harvest more energy from ingested food? Explain.
2. Does it matter to your conclusion that *ob/ob* mice typically eat more chow than +/+ or *ob*/+ mice?
3. Assume that mice store all the excess energy they obtain in the form of body fat, and that mouse fat contains 9.3 kcal/g. If the *ob/ob* recipient mice added 0.44 g more body fat than the +/+ recipients in 2 weeks, how much more energy did they obtain from their food in this time period?
4. If both groups of recipients consumed 55 g of chow, what percentage of the energy consumed in chow does this extra energy gain represent? (*Hint*: Recall that the energy content of chow is 3.7 kcal/g.)

[a]P. J. Turnbaugh et al. 2006. *Nature* 444: 1027–1031.

A similar **Analyze the Data** exercise may be assigned in **Achieve**.

REVIEW & APPLY | 38.1

R Ecology is the study of ecological systems, which consist of one or more organisms interacting with the biotic and abiotic components of their environment. Ecological systems exist on levels ranging from a single organism and its immediate environment up to the entire biosphere. System properties, which are determined by the system's component parts and interactions, vary over space and time. Ecologists deploy the standard scientific tools of descriptive natural history observation, experiments, and modeling to understand these complex systems.

A Researchers sampled the stools (feces) of 35 Hadza hunter–gatherers in Tanzania during the 2013–2014 wet and dry seasons and those of 32 North Americans. They compared samples with respect to microbial genes that code for enzymes involved in various carbohydrate-processing pathways (called CAZymes). The three panels below show the relative number of distinct CAZymes involved in breaking down carbohydrates from animal, vegetable, and fruit sources in the Hadza during the dry and wet season (red bars) and in North Americans (blue bars). Bar height indicates the median; error bars indicate 1 standard deviation (see Appendix B, Figure B.6). Bars with different letters are statistically different from one another at the 5 percent level. (*Figure source*: S. A. Smits et al. 2017. *Science* 357: 802–806.)

1. Identify and describe one comparison among bars that illustrates temporal variation in the microbial community of the human gut, and one comparison that illustrates spatial variation.
2. Assuming that the microbial community of the human gut responds to the diet of its host, predict how the diversity of animal matter, vegetables, and fruit in the Hadza diet differs (a) between wet and dry seasons and (b) from the diet of North Americans.
3. Design a study to test your predictions and identify which scientific tool (descriptive observation, experiment, modeling) your study uses.

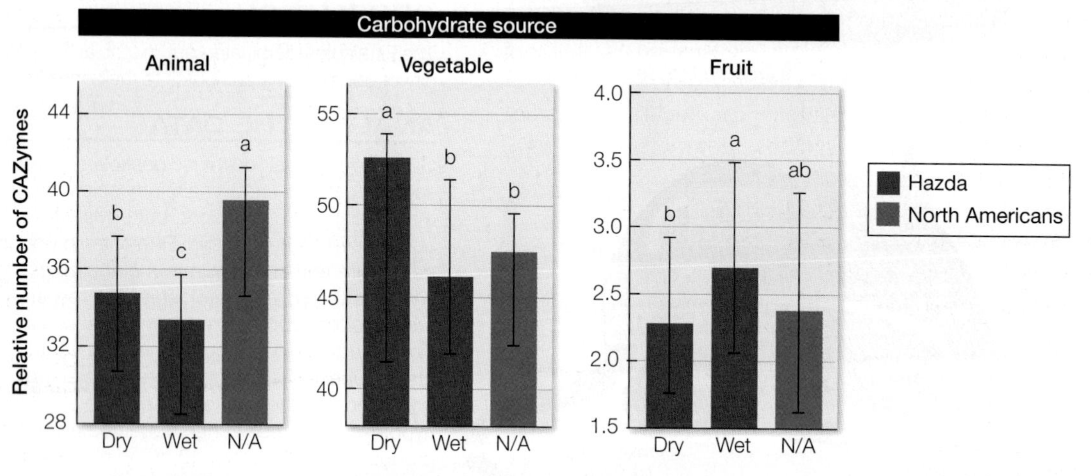

Variation over space and time can be seen in ecosystems at all levels, from the human gut to the entire biosphere. We will now turn to some of the physical factors that underlie this heterogeneity at the biosphere level: Earth's climates and topography.

38.2 Solar Energy Input and Topography Shape Earth's Physical Environments

LEARNING OBJECTIVES

By the end of this key concept you should be able to:

38.2.1 Predict how latitude will affect average and seasonal variation in temperature of a location on Earth.

38.2.2 Predict how latitude and elevation will affect the average annual precipitation of a location on Earth.

38.2.3 Illustrate, using Hadley and polar cell circulation, why hot deserts are found at latitudes 30°N and 30°S.

38.2.4 Use the climate diagram of a place on Earth to infer its location and growing season.

Earth's physical environments vary enormously—from hot to cold, wet to dry, aquatic to terrestrial, salt water to fresh water. To this variation across space we must add variation over time, on scales from the very slow comings and goings of oceans, mountains, or ice ages to the very fast shifts in temperature as clouds pass overhead. On the short time scales that we experience, physical conditions depend largely on solar energy input, which drives the circulation of the atmosphere and the oceans. Patterns of input and circulation determine whether a particular place at a particular time of year tends to be hot or cold, wet or dry—and as a consequence, the types of organisms found in that place at that time of year.

LINK Review the effects that changes in Earth's physical environment—such as continental drift, climate shifts, changes in sea level, and changing atmospheric concentrations of O_2 and CO_2—have had on the evolution of life in **Key Concept 17.2**.

Variation in solar energy input drives patterns of weather and climate

All of us are familiar with **weather**, the atmospheric conditions—temperature, humidity, precipitation (rainfall or snowfall), and wind direction and speed—that we experience at a particular place and time. Averaged over longer periods, the state of these atmospheric conditions and their pattern of variation over time constitute the **climate** of a place. In other words, climate is what you expect; weather is what you get. We and other organisms respond to both weather and climate. Responses to weather are usually short-term: animals may seek shelter from a sudden rainstorm (human animals open their umbrellas), and plants may close their stomata (small openings in the leaf epidermis) on a hot summer day to reduce water loss from their tissues. Responses to climate, by contrast, tend to involve adaptations that prepare organisms for expected conditions. Organisms in hot climates can tolerate heat; those that inhabit seasonal climates use cues such as day length to prepare for an impending hot, cold, or dry season.

Some global patterns of climate stem from variation in solar energy input over space. Because Earth is spherical, sunlight strikes the planet at a shallower angle near the poles than near the equator (**FIGURE 38.3**). This difference in incident angle results in less solar energy input at high latitudes (near the poles) than at low latitudes (near the equator), for two basic reasons. First, the energy of sunlight is spread over a larger surface area at high latitudes, so less is delivered per unit of area. Second, sunlight passes through more atmosphere when it strikes at a shallow angle, and more photons are therefore absorbed or reflected back into space by atmospheric gas molecules before they reach Earth's surface. A consequence of this latitudinal gradient in energy input is that air temperatures at Earth's surface decrease from low latitudes to high latitudes. Averaged over the course of a year, air temperature drops by about 0.8°C with every degree of latitude (about 110 km) north or south of the equator.

At any given place, the Sun's input also varies through the year, causing seasonal variation in climate. This **seasonality** is a consequence of the 23.5° tilt of Earth's axis of rotation relative to the plane of its yearly orbit around the Sun. The tilt causes different latitudes to receive their greatest solar energy input at different times of the year (**FIGURE 38.4**). At the solstice in late June, the Sun is directly above the Tropic of Cancer (latitude 23.5°N): the Northern Hemisphere experiences summer, with its long days and warm temperatures. At the solstice in late December, the Sun is directly above the Tropic of Capricorn (latitude 23.5°S), and the climate pattern is reversed: it is summer in the Southern Hemisphere and winter in the Northern Hemisphere. At the equinoxes in late March and late September, the Sun is directly above the equator (90° above the equatorial horizon), and both hemispheres experience equal lengths of day and night. The yearly seasonal variation in solar input and in day length is small at the equator and becomes pronounced at the poles, which cycle between periods of no sunlight at all and of continuous sunlight.

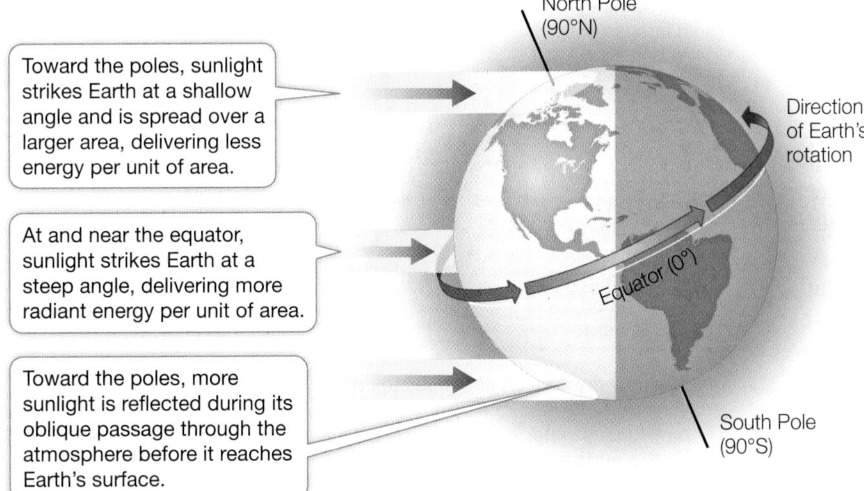

FIGURE 38.3 Solar Energy Input Varies with Latitude The angle of incoming sunlight affects the amount of solar energy that reaches a given area of Earth's surface. On average, sunlight strikes Earth's surface at a steeper angle at low than at high latitudes, resulting in greater input of energy per unit of area at low (i.e., equatorial) latitudes because the energy is spread over a smaller area and because less sunlight is reflected back into space during its passage through the atmosphere. (After C. D. Ahrens. 2005. *Essentials of Meteorology: An Invitation to the Atmosphere*. Brooks/Cole, Cengage Learning. USA.)

The circulation of Earth's atmosphere redistributes heat energy

The yearly input of solar energy is greatest in the **tropics** (the latitudes between the Tropics of Cancer and Capricorn) and decreases poleward. This latitudinal gradient in solar input drives the vast circulation patterns of Earth's gaseous atmosphere.

As the Sun beats down on the tropics, the air there heats up—its molecules move faster (temperature is a measure of disordered molecular motion), colliding and pushing one another farther apart (see Key Concept 2.4). The heated air expands, becoming less dense and more buoyant, and rises. As it rises, it cools **adiabatically** (this is the principle on which an air conditioner works: when compressed gas expands, it cools). The air rises until it reaches a level at which the surrounding air has its same temperature and density. Because tropical air gets so warm at Earth's surface, it ascends to great altitudes before it stops rising.

Rising tropical air creates a vacuum that pulls in cooler, denser surface air from the north and south. As that air in turn heats, expands, and

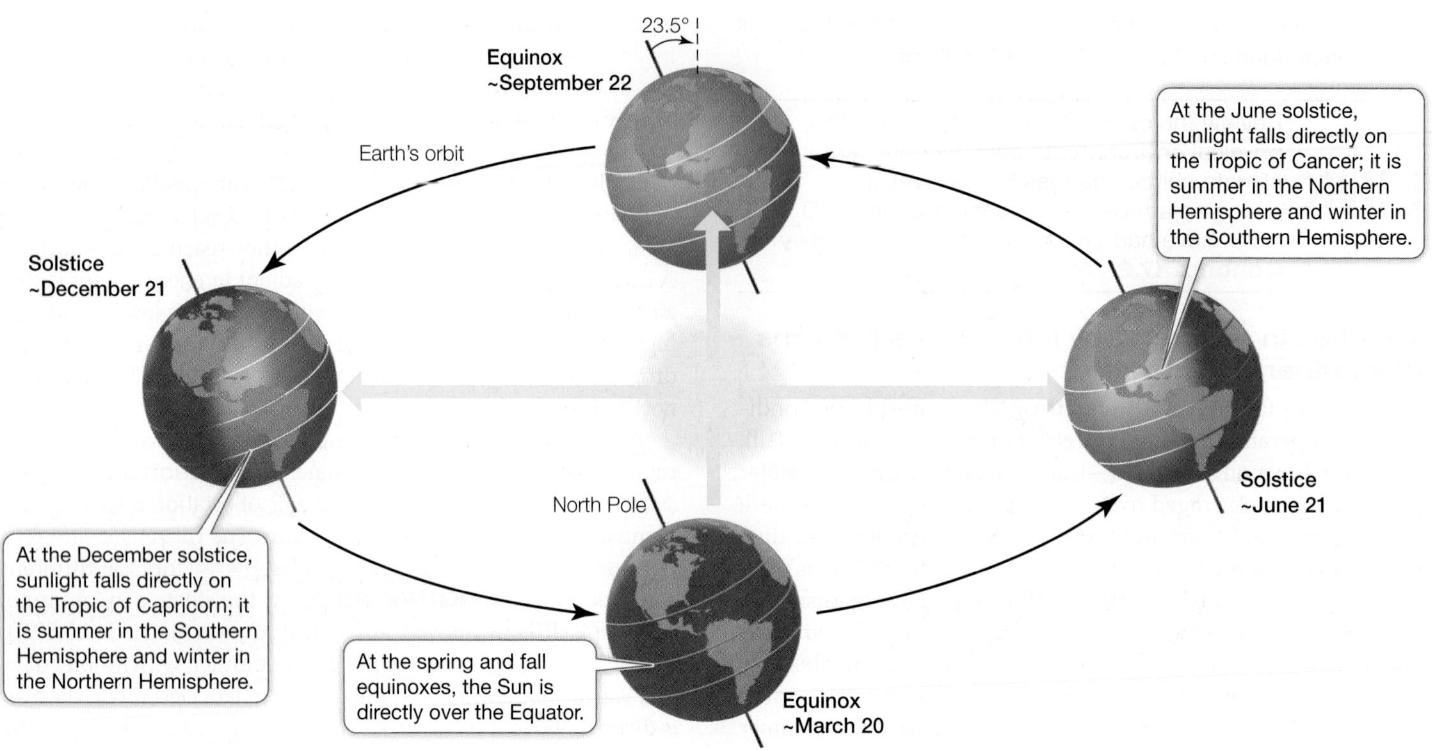

FIGURE 38.4 The Tilt of Earth's Axis of Rotation Causes the Seasons As Earth orbits the Sun, the North Pole alternately tilts toward and away from the Sun. This tilt causes the latitude of greatest solar energy input to cycle between the Tropics of Cancer (23.5°N) and Capricorn (23.5°S) over the course of a year. (After C. D. Ahrens. 2005. *Essentials of Meteorology: An Invitation to the Atmosphere*. Brooks/Cole, Cengage Learning: USA.)

If you were at the equator, how high in the sky would the Sun appear at its highest point during the day (solar noon) at the March equinox, the September equinox, the June solstice, and the December solstice? Express your answer in degrees N or S from straight overhead.

rises, it pushes the cool air above it to the side. As this air aloft moves away from the tropics, it radiates heat to outer space, becoming denser and less buoyant. The cool air sinks back toward Earth, is compressed by the overlying atmosphere, and warms adiabatically (the reverse of adiabatic cooling). It reaches Earth's surface at latitudes of roughly 30°N and 30°S, at which point some of it flows back toward the equator to replace air that is rising in the tropics. Thus it forms two cycles of vertical atmospheric circulation—one north and one south of the equator—called **Hadley cells** (**FIGURE 38.5**).

The Hadley cell circulation produces strong latitudinal patterns of precipitation. Water molecules at Earth's surface in the tropics absorb the intense sunlight and evaporate (change from a liquid to a gaseous state), so the air that rises in the tropics is rich in water vapor. As this warm, moist air rises and cools, fewer water molecules are moving fast enough to bounce apart after they collide. Instead, they change state from gaseous to liquid or solid. The resulting water droplets or ice crystals grow and—when massive enough—fall to the ground as precipitation. Thus the warm, rising air of the Hadley cells drops most of its moisture near the equator as rain. The high-altitude air that eventually descends near 30°N and 30°S is largely depleted of water. Moreover, because it warms as it descends and is compressed, the little water

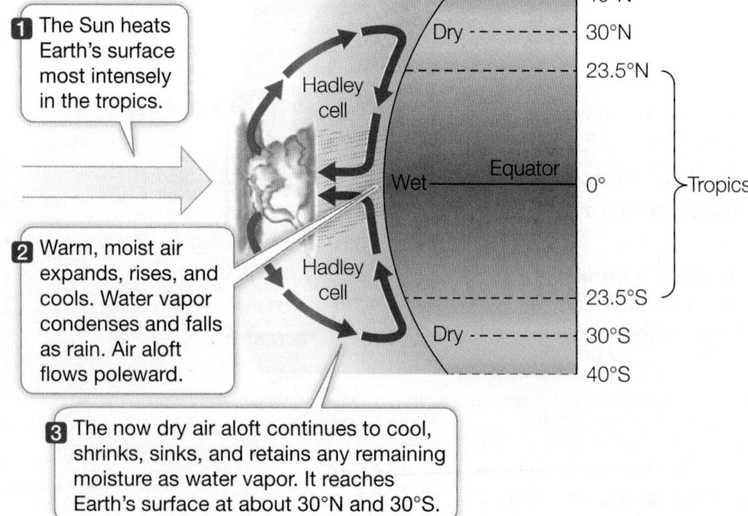

FIGURE 38.5 Tropical Solar Energy Input Sets the Atmosphere in Motion Intense sunlight in the tropics drives a vertical pattern of atmospheric circulation that reaches the subtropics and that results in high precipitation near the equator. This illustration of the circulation of Hadley cells is simplified: they are never exactly symmetrical around the equator. In the Northern Hemisphere winter, the rising portion of the northern Hadley cell actually occurs south of the equator and the northern cell is larger than the southern cell; the reverse occurs in the Southern Hemisphere winter.

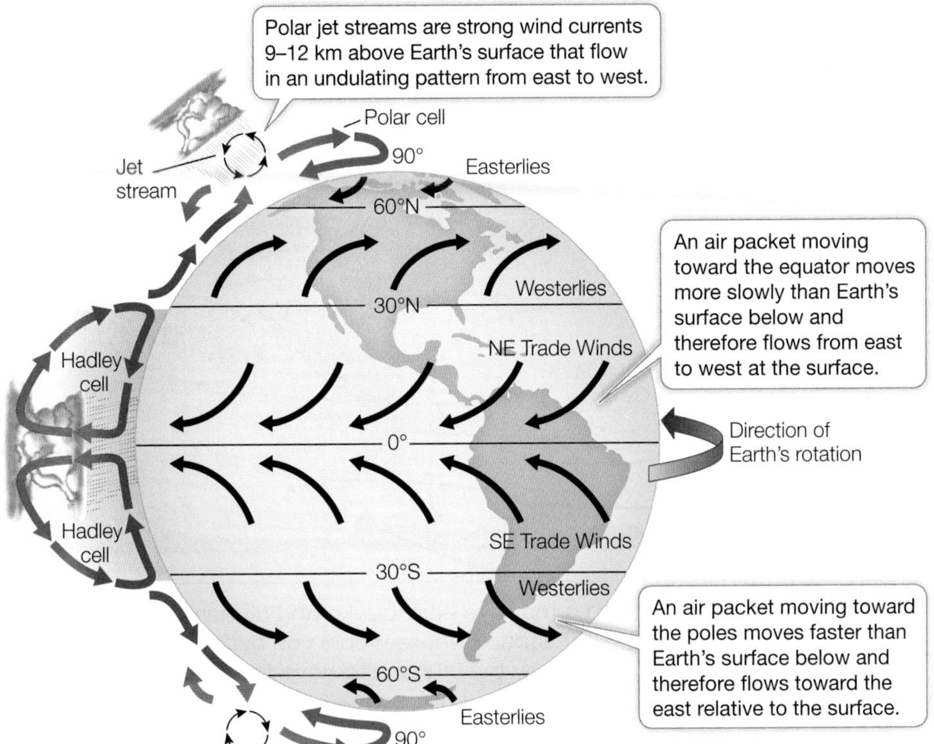

FIGURE 38.6 Global Atmospheric Circulation and Prevailing Winds Atmospheric circulation in the middle latitudes is complex. Some of the equatorial air descending from the Hadley cells travels poleward, encountering polar air. At this "polar front" the warm air over-rides the cold, dense polar air. As it rises and cools, water vapor condenses, forming a temperate-zone rain belt as well as strong, high-altitude winds that travel in a meandering path from west to east in the middle latitudes. These polar jet streams, and the spinning turbulent storms they transport, return air both northward and southward. Prevailing surface winds blow as they do, and storms rotate, because of the Coriolis effect: Earth's spherical shape causes its surface rotational speed to decrease with latitude. Consequently, air flowing toward the equator lags behind, and air flowing toward the poles moves faster than, the surface beneath it.

that remains stays in a gaseous state. Earth's great hot deserts are located around these latitudes (see Figure 38.5).

Not all of the descending air in the Hadley cells returns immediately to the tropics. Some of it continues to flow toward the poles, where it encounters frigid, dense polar air that is flowing toward the equator as part of the circulation of polar cells that are smaller and weaker than the Hadley cells. The interaction of these warm and cold air masses generates uplift and the turbulent, spinning storms that sweep sinuously from west to east through the middle latitudes, eventually returning polar air to the tropics (see Figure 38.6).

To understand these storms and other patterns of horizontal air movement, we must consider one more thing—Earth's rotation on its axis from west to east (see Figure 38.3). We have already seen that the latitudinal gradient in solar input creates a latitudinal, or north–south, movement of air. The planet's rotation adds an east–west component to air movement. Because Earth is a sphere, the rotation of its surface is fastest at the equator, where its circumference is greatest, and slower toward the poles. Surface air that is not moving north or south is carried eastward at the same speed as the surface beneath it. But as soon as air moves toward the equator, it lags behind the faster-moving surface below and therefore has a westward-movement component relative to the surface. Conversely, air that is moving poleward races ahead of the surface below it and appears to be flowing eastward. This **Coriolis effect** causes prevailing surface winds to blow from the northeast or southeast in the tropics (the northeast or southeast trade winds, so called because of their importance during the era of trade by sailing ships), from the northwest or southwest in the middle latitudes (the westerlies), and from northeast or southeast again above latitudes 60°N and 60°S (the easterlies) (**FIGURE 38.6**). The Coriolis effect also leads to the rotation of air associated with storms.

Latitudinal patterns of atmospheric circulation have a powerful effect on climate patterns by transferring an immense amount of heat energy from the hot tropics to the cold poles. Without this transfer, the poles would sink toward absolute zero in winter, and the equator would reach fantastically high temperatures throughout the year.

Ocean circulation also influences climate

The prevailing surface winds and Coriolis effect drive massive circulation patterns in the waters of the oceans, called **currents**, which carry materials, organisms, and heat with them. In the northern tropics, the trade winds drag surface water toward the west. When this water nears the western edge of an ocean basin, much of it flows northward, only to be turned to the east by the Coriolis effect and by drag from the westerlies of the northern middle latitudes. This eastward flow is deflected southward along the western edge of the continents and then westward again, completing a clockwise circuit, or **gyre**. The pattern in the Southern Hemisphere is essentially a mirror image of that just described, with surface circulation moving in counterclockwise rather than clockwise rotation (**FIGURE 38.7**). The westward-flowing tropical water that is not caught up in these gyres flows back to the east in an **equatorial countercurrent**.

As they move poleward, tropical surface waters transfer heat from low to high latitudes, adding to the transfer that accompanies atmospheric circulation. The Gulf Stream and North Atlantic Drift, for example, bring warm water from the tropical Atlantic Ocean and the Gulf of Mexico north and east across the Atlantic, which warms the air above the North Atlantic. Westerlies then carry this warm air across northern Europe, moderating temperatures there. The effect of ocean currents on organisms is dramatic: polar bears roam the shores of Hudson Bay in Canada but are missing from comparable latitudes in Scotland and Sweden.

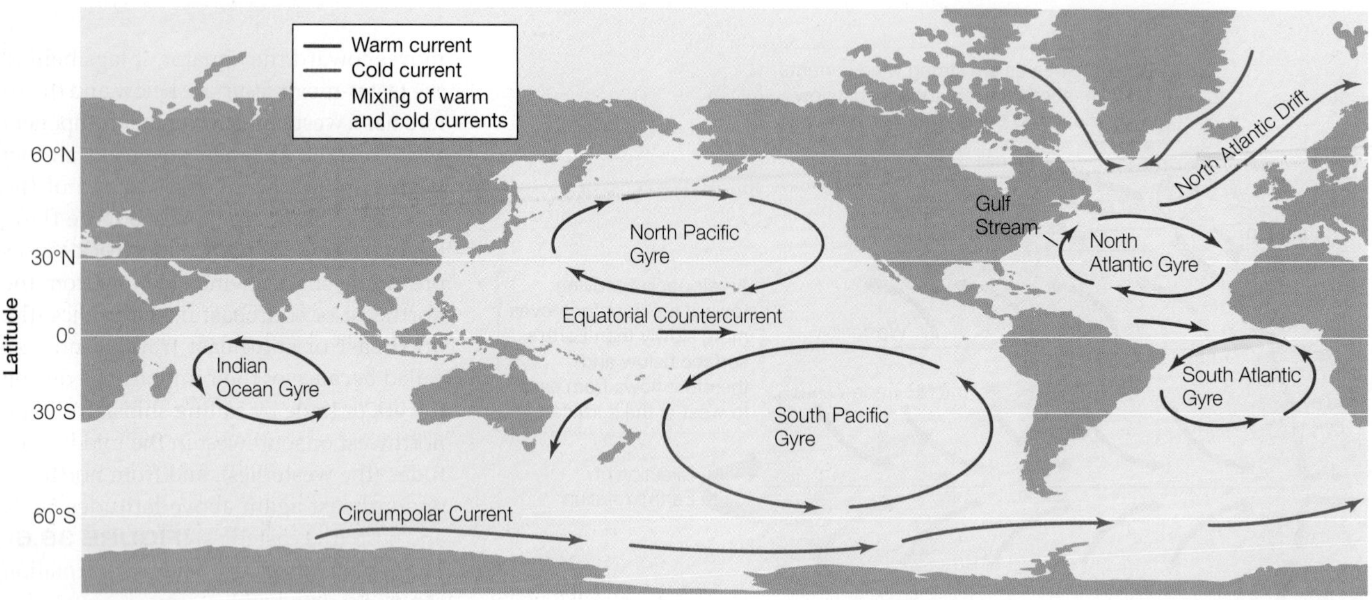

FIGURE 38.7 Ocean Surface Currents Ocean surface currents are characterized by giant patterns (gyres) of clockwise rotation in the Northern Hemisphere and counterclockwise rotation in the Southern Hemisphere, caused by frictional drag from prevailing winds (see Figure 38.6), the Coriolis effect, and deflection at the edges of continents.

In March 2011 a massive earthquake and tsunami struck the eastern coast of Japan. Where would you expect floating debris carried out to sea by the tsunami to have first reached beaches?

Media Clip 38.1 Perpetual Ocean Currents

Surface currents are only part of the circulation of ocean waters. Gradients in water density caused by variation in temperature and salinity cause surface waters to sink in certain regions, forming deep currents. These deep currents return to the surface in areas of **upwelling**, completing a vertical oceanic circulation. Oceanic circulation, like atmospheric circulation, is three-dimensional.

Oceans and large lakes moderate Earth's terrestrial climates because water has a high heat capacity, as we saw in Key Concept 2.5. This means that the temperature of water changes relatively slowly as it exchanges heat with the air it contacts and as it absorbs or emits radiant energy. As a result, water temperatures fluctuate less with the seasons and with the day–night cycle than do land temperatures, and the air over land close to oceans or lakes also shows less seasonal and daily temperature fluctuation.

Topography contributes to environmental heterogeneity

Earth's surface is not flat; it is crumpled into mountains, valleys, and ocean basins. This variation in the elevation of Earth's surface, known as **topography**, affects the physical environment.

Mountains offer a good example of how topography affects weather. Because rising air expands and cools adiabatically, temperatures become cooler with increases in elevation (distance above sea level). If you climb a mountain, you will find that the temperature drops by about 1°C for each 220 meters of elevation gained. Where prevailing winds encounter a mountain range, air flows up the windward side (the side facing into the wind). As the air cools, water vapor condenses, falling as rain or snow on higher reaches of the mountain. There is relatively little precipitation on the leeward side of a mountain range (the side away from the wind) because air descending on that side warms, keeping any remaining water within it in a gaseous state. Thus the mountain range is said to form a **rain shadow** (**FIGURE 38.8**); greater precipitation near the mountaintop fosters more luxuriant growth of plants on the windward than on the leeward side.

Topography also influences physical conditions in aquatic environments. The velocity of water flow, for example, is dictated by topography: water flows rapidly down steep slopes, more slowly down

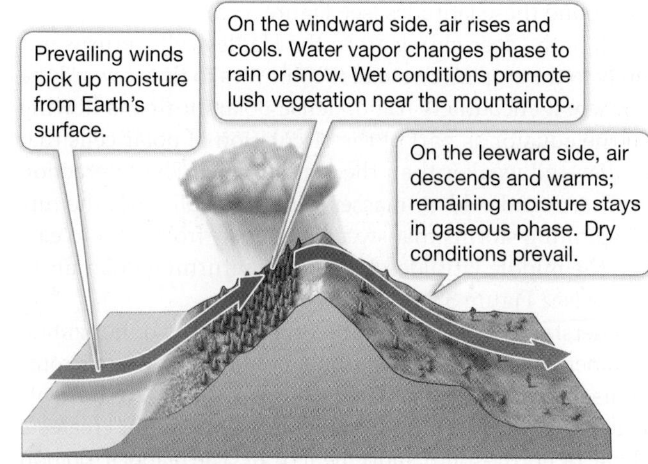

FIGURE 38.8 Mountains Are "Weather Formers" That Cast Rain Shadows Precipitation often is higher at higher elevations on the windward side of a mountain range than on the leeward side.

Animation 38.1 Rain Shadow

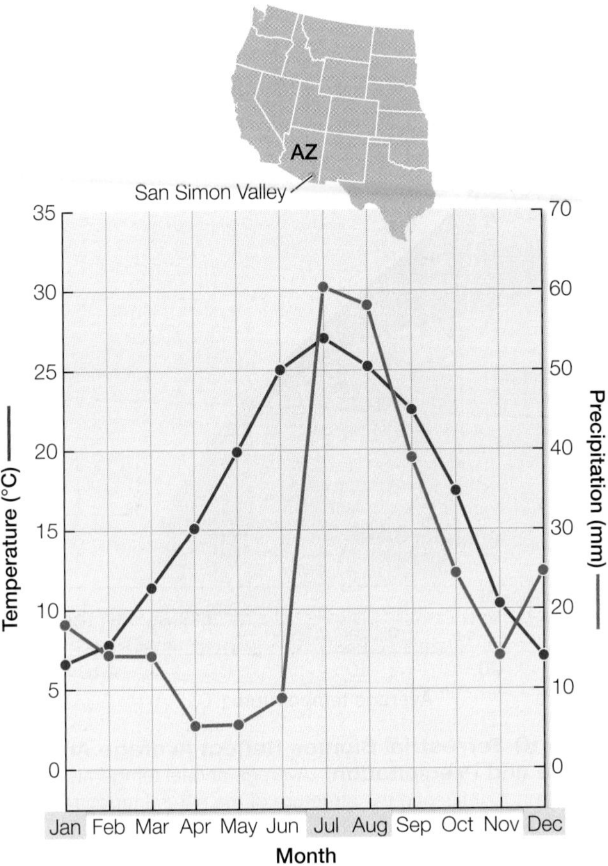

FIGURE 38.9 Climate Diagrams Summarize Climate in an Ecologically Relevant Way Climate diagrams traditionally combine graphs of average temperature (left axis) and average precipitation (right axis) throughout the year, as shown here for San Simon Valley near the Arizona–New Mexico border. The summer solstice at the location in question (late June in the Northern Hemisphere, late December in the Southern Hemisphere) is placed at the center of the diagram. The axes are scaled such that precipitation favors plant growth when the precipitation line is above the temperature line. The months highlighted in green are typically favorable for plant growth.

gradual slopes, and pools as lakes, ponds, or oceans in depressions. The depth of a water-filled depression determines the gradients of many abiotic factors, including temperature, pressure, light penetration, and water movement, as we will discuss in Key Concept 38.3.

Climate diagrams summarize climates in an ecologically relevant way

The climate at any location can be summarized in a **climate diagram**. Recognizing that ecological processes are strongly influenced by temperature and moisture, German biogeographer Heinrich Walter superimposed graphs of average monthly temperature and average monthly precipitation through the year. He scaled the axes of the two graphs to incorporate the rule of thumb that plant growth requires at least 20 millimeters of precipitation per 10°C of temperature above freezing (0°C). This scaling makes it easy to see when conditions allow terrestrial plant growth—when temperature is greater than 0°C and the precipitation line is above the temperature line, a period known as the growing season (**FIGURE 38.9**).

REVIEW & APPLY | 38.2

R Climate is a central attribute of a terrestrial ecosystem—the average weather it experiences over a yearly cycle. Earth's spherical shape, the tilt of its axis of rotation relative to its yearly orbit around the Sun, and its surface topography cause latitudinal and elevational gradients in temperature, precipitation, and seasonality. Latitudinal gradients in solar energy input, coupled with Earth's rotation and the distribution of continents, drive three-dimensional circulation in the atmosphere and the oceans that redistributes heat energy around the planet.

A Shown below are climate diagrams for three different localities (*figure source*: Wikimedia Commons/Hedwig in Washington. CC BY-SA 2.5). (Note the discontinuity in the precipitation axis in the bottom graph.) Using the information in these diagrams and what you've learned about factors that determine climate, answer the following questions and justify your answers.

1. Which locality is nearest the equator?
2. Which locality has the greatest seasonality in temperature?
3. Which locality has the longest growing season?
4. Which locality is in the Northern Hemisphere?

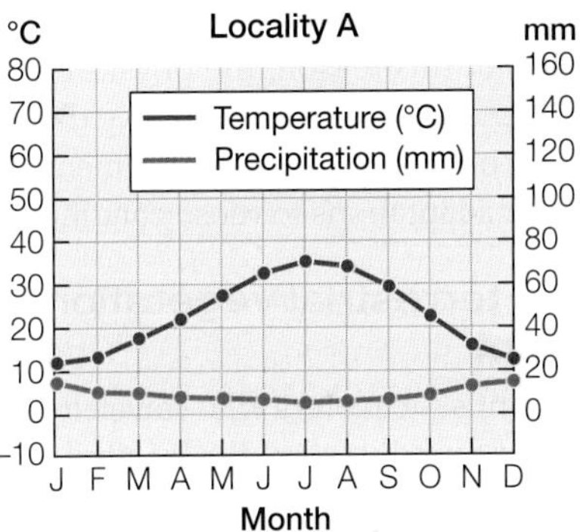

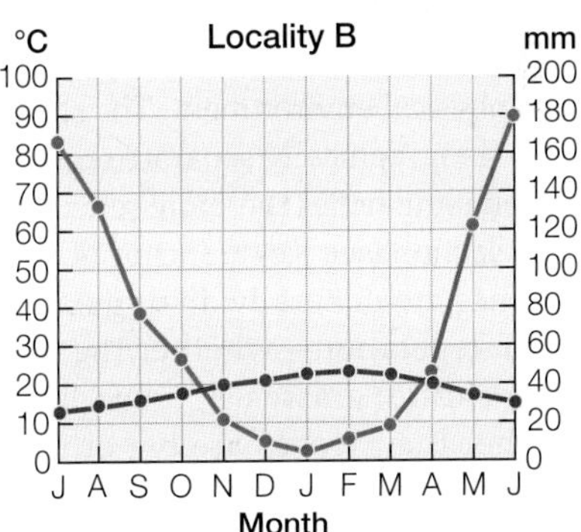

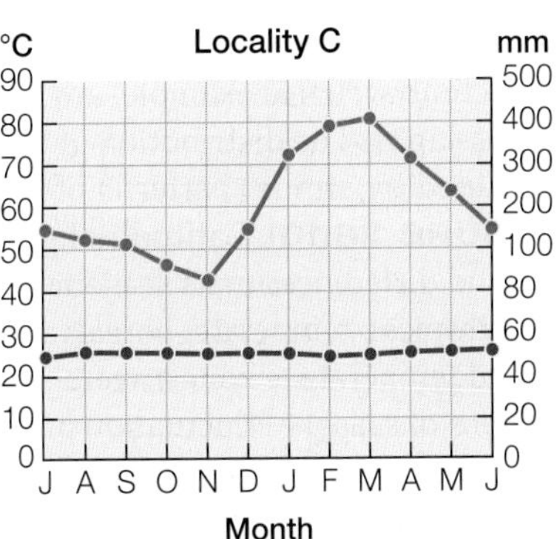

The patterns in the physical world that we have just discussed limit where organisms can live. For this reason, they strongly influence biogeography—the spatial distributions of species.

38.3 | Biogeography Reflects Physical Geography

LEARNING OBJECTIVES

By the end of this key concept you should be able to:

38.3.1 Define "biome" and provide two examples of how terrestrial biomes reflect climate.

38.3.2 List two characteristics of the physical environment of a terrestrial location, other than climate, that can affect its biome.

38.3.3 List two aspects of the physical environment that distinguish aquatic biomes.

An organism's physiology, morphology, and behavior—all aspects of its phenotype—affect how well it can tolerate a particular physical environment. For this reason, the physical environment of a place greatly influences what species can live there. Moreover, we expect species that occur in similar environments to have evolved similar phenotypic adaptations to those conditions.

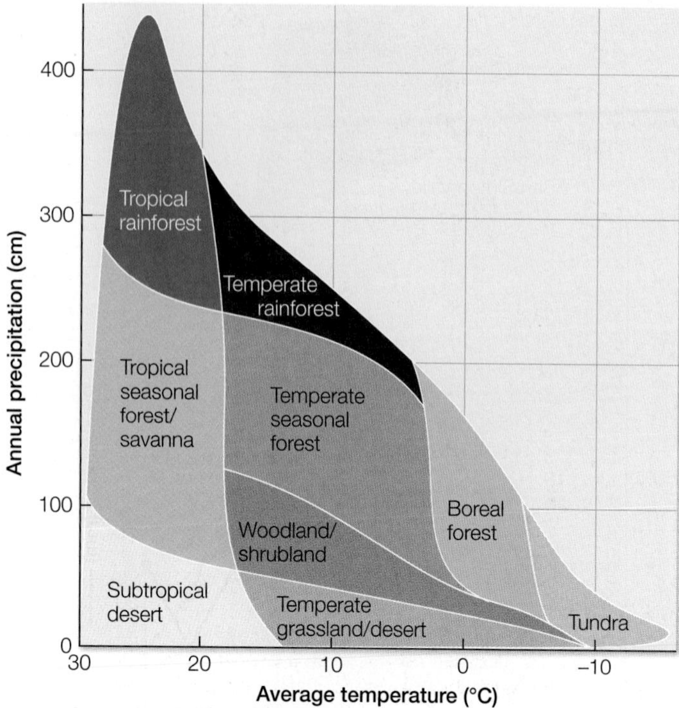

FIGURE 38.10 Terrestrial Biomes Reflect Average Annual Temperature and Precipitation Average annual temperature and precipitation strongly influence the structure of the natural terrestrial vegetation of a location. (After Jwratner1/Wikimedia CC0 1.0.)

Similarities in terrestrial vegetation led to the biome concept

The European scientist–explorers who traveled the globe noticed that the vegetation found in climatically similar regions on different continents was strikingly convergent. This observation led to the concept of a **biome**: an ecosystem with a distinct physical environment and ecologically similar organisms with similar adaptations to the physical environment. The species that occupy the same biome in geographically separate regions are often not closely related. Their similarities therefore reflect convergent evolution by natural selection (see Key Concept 14.1).

Terrestrial biomes are generally distinguished by the characteristics of their vegetation—whether the dominant plants are woody trees or shrubs, grasses, or broad-leaved herbaceous plants, and whether their leaves are big or small, deciduous (dropped seasonally) or evergreen. Characteristics of organisms other than plants also distinguish the biomes, since they represent adaptations to the vegetation as well as the climate. Animals that can climb well, for example, are more common in forests than in grasslands. The distribution of terrestrial biomes is broadly determined by annual patterns of temperature and precipitation (**FIGURE 38.10**). Latitudinal gradients in average temperature and differences in its seasonal fluctuation separate tropical, subtropical, temperate, boreal, and polar biomes. Within those broad temperature zones, average annual precipitation and its pattern of seasonal fluctuation predict what biome will be found. In tropical biomes, for example, the length of the dry season determines whether tropical rainforest, tropical seasonal forest, or savanna develops. Elevation-associated gradients in temperature also determine the distribution of biomes; for example, tundra can be found on mountaintops, even in the tropics (**FIGURE 38.11**).

LINK Repeated exposure to the cold temperatures of the tundra induces cold-hardening in some plant species, a process that helps cell membranes retain their fluidity by altering lipid proportions. This and other examples of plant adaptations to environmental stresses are explored in **Key Concept 27.3**.

Climate is not the only factor that molds terrestrial biomes

Many attributes of the physical environment interact with climate to influence the character of terrestrial vegetation. Consider southwestern Australia, for example, which has a Mediterranean climate with hot, dry summers and cool, moist winters. The vegetation of southwestern Australia (woodland and shrubland) is not, however, altogether similar to that of the Mediterranean. Succulent plants, which are well adapted to summer drought, are common in some Mediterranean climates, but not in Australia, which has nutrient-poor soils. A lack of nitrogen in soil makes it expensive for plants to construct the nitrogen-rich photosynthetic machinery of leaves. As a result, many plants in Australia's shrublands and woodlands produce long-lived leaves that are defended against herbivores in ways that do not

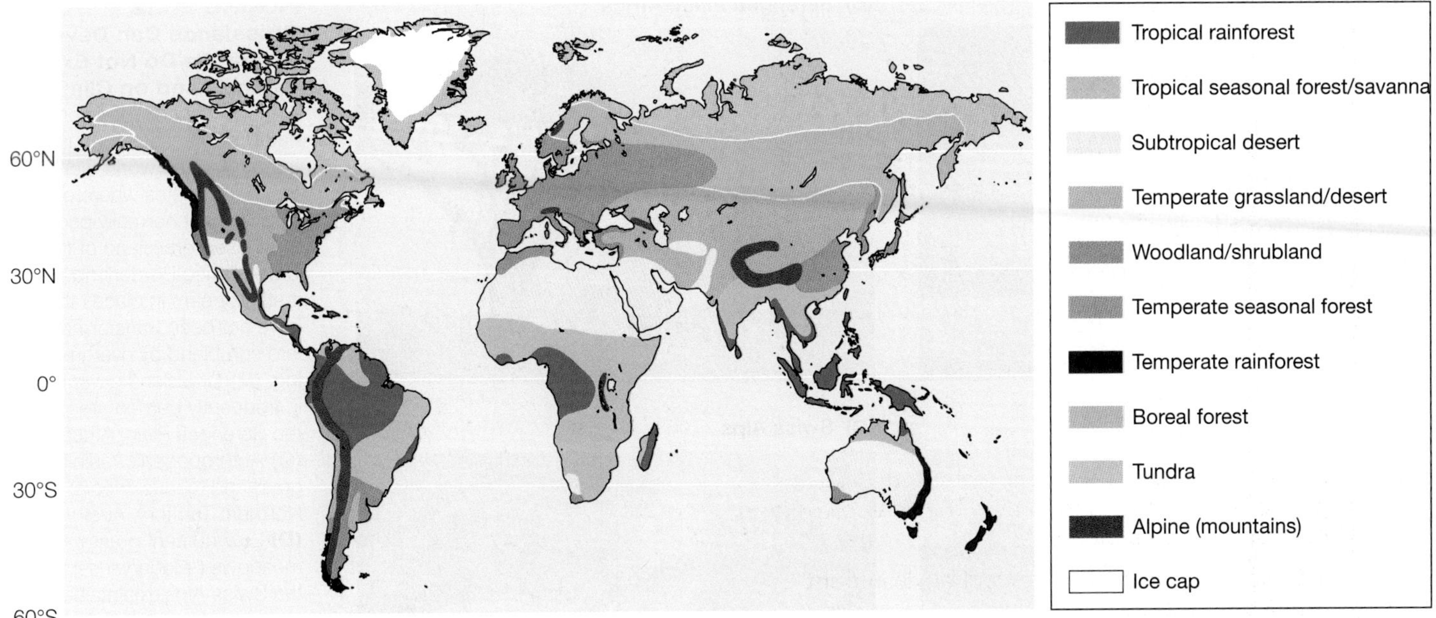

FIGURE 38.11 Global Terrestrial Biomes The global distribution of terrestrial biomes reflects latitudinal and elevational gradients in temperature and precipitation. The same biomes can be found in widely separated locations. (Note that tundra occurs on the tops of high mountains, even in the tropics, but in patches too small to see on this map.)

Why does temperate seasonal forest occur farther north in western Europe than it does in eastern North America? (*Hint*: See Figure 38.7.)

Activity 38.1 **Terrestrial Biomes**

require nitrogen (see Key Concept 27.2). These defenses include tough, indigestible cell walls, compounds that reduce the nutritional quality of leaves, and toxic oils and resins—features that also make the vegetation highly flammable. Plant material produced during the winter rains dries out during the summer and can fuel intense fires that periodically sweep across the landscape in areas prone to summer lightning storms. Fires easily decimate succulent plants, which is why they are rare in southwestern Australia.

The biomes of southwestern Australia are not, of course, the only regions that lack soil nutrients and are subject to periodic fires. We find similar conditions in the desert grasslands of the U.S.–Mexico Borderlands as well (**FIGURE 38.12A**). Grasslands—areas dominated by grasses and herbs rather than trees or shrubs—normally occur where precipitation is too scanty, or too sporadic, to support forests but is more plentiful than is typical of deserts. Fully one-fourth of Earth's land surface experiences such conditions, and much of this area supports grasslands. But grasslands are also found in some places where we do not expect them based on climate alone. Their presence in these places demonstrates that biome boundaries are fuzzy and are not perfectly predicted by temperature and precipitation because other factors also affect vegetation.

One such factor is fire, which rarely kills grasses—indeed, grasses propagate fire and are adapted to survive it—but often kills young shrubs and trees, which lack fire adaptations. Before humans arrived, the U.S.–Mexico Borderlands experienced periodic natural lightning-ignited wildfires that promoted desert grassland. Humans themselves have contributed to the establishment of grasslands in many areas. Fires lit by humans in the Serengeti Plain of northern Tanzania, Africa, may have helped convert the landscape from tropical seasonal forest to savanna, a biome in which grasses dominate and trees are widely scattered (**FIGURE 38.12B**). Such human effects are not necessarily recent. Humans have inhabited Australia for at least 45,000 years, for instance, and their fires appear to have transformed some of the original woodlands in the arid "red center" of the continent into grasslands dominated by "spinifex" grasses (**FIGURE 38.12C**). Ecosystems in which grasses dominate are also found in other places where we might expect to find forests. When humans brought domesticated livestock into temperate regions such as the European Alps, they often converted forests into grassy meadows to feed their animals (**FIGURE 38.12D**).

The properties of soils interact with climate, fire, or grazing to affect the character of vegetation. Trees and shrubs grow particularly slowly in nutrient-poor soils. Lack of sufficient nutrients might prevent a tree or shrub seedling from sending its roots deep enough into the soil to survive the next dry season, or from growing large enough to survive the next wildfire, or from being able to recover from being eaten. If so, then the availability of soil nutrients might tip the balance between woody vegetation and grassland in seasonally dry climates, fire-prone regions, or grazing lands. Among the four grassland examples shown in Figure 38.12, the alpine meadows and Australian spinifex grow on relatively nutrient-poor soils, whereas the other two occupy richer volcanic soils.

(A) U.S.–Mexico Borderlands

Sunpix Travel/Alamy Stock Photo

(B) Serengeti Plain, Africa

Ulrich Doering/Alamy Stock Photo

(C) Northern Territory, Australia

© Paul Mayall/Imagebroker/AGE Fotostock

(D) Swiss Alps

Niels van Kampenhout/Alamy Stock Photo

FIGURE 38.12 Grasslands Can Develop Where We Do Not Expect Them Based on Climate Alone Landscapes dominated by grasses are found throughout the world, sometimes in climates where grasslands do not normally occur. **(A)** Desert grassland of the U.S.–Mexico Borderlands, as it still appears in places that have not been transformed into shrubland by overgrazing. **(B)** Grassland savanna in frequently burned areas of the Serengeti Plain, Africa. **(C)** Anthropogenic "spinifex" grassland (genus *Triodia*) in Northern Territory, Australia. **(D)** Low-nutrient grassy meadows ("Magerwiesen") in the Swiss Alps, maintained by grazing, where temperate forest normally occurs.

Media Clip 38.2 Grasslands and Fire

TABLE **38.1** | Major Aquatic Biomes

Biome[a]	Description
FRESH WATER	
Rivers and streams	Flowing water. Many small, fast-flowing source streams form on high ground, feeding into networks of ever-larger, slower-flowing streams and rivers. Biota adapted to constantly moving water.
Wetlands	Glades, swamps, and marshes. Biota adapted to water-saturated soil or standing fresh water.
Ponds and lakes	Significant bodies of standing fresh water. Ponds are smaller and shallower, subject to drying. Biotic zones determined by distance from shore and light penetration (see Figure 38.13A).
ESTUARINE	
Salt marshes	Cool-temperate stands of salt-tolerant grasses, herbaceous plants, and low-growing shrubs. Crucial to nutrient cycling and coastal protection. Habitat supports diverse aquatic and terrestrial life.
Mangrove forests	Tropical and warm subtropical coasts and river deltas. Dominated by mangrove trees with aerial roots. Rich in animal life; protect against coastal erosion.
MARINE	
Intertidal	Sandy or rocky coastlines subject to rising and falling tides; organisms adapted to withstand both submerged and dry conditions, as well as the force of waves and moving water.
Kelp forests	Found in shallow coastal waters of temperate and cold regions. Dominated by large, leaflike brown algae (kelp) that support a wide variety of marine life.
Seagrass beds	"Meadows" of grasses (see Figure 20.31C) found in shallow, light-filled temperate and tropical waters.
Coral reefs	Species-rich, highly endangered ecosystems of shallow tropical waters. Dependent on corals (see Figure 22.13C) and their photosynthetic endosymbionts (see Key Concept 19.4).
Open ocean	The pelagic zone (see Figure 38.13B) hosts photosynthetic planktonic (free-floating) organisms that support a rich array of marine animals. At the greatest depths, below the level of light penetration, the abyssal zone supports a fauna largely dependent on detritus that sinks down from pelagic regions.
Hydrothermal vents	Abyssal ecosystems warmed by volcanic emissions. Chemoautotrophic prokaryotes (see Key Concept 18.3) nourish large annelid worms (see Figure 22.21) and other invertebrates.

[a]A benthic region—a bottom zone containing silt, sand, or some other substrate and the organisms that occur there—is part of all three biome types.

The biome concept can be extended to aquatic environments

The biome concept can be applied to aquatic environments as well as terrestrial ones (**TABLE 38.1**), but there are some differences. Unlike the biomes of terrestrial environments, no single group of organisms is useful for distinguishing aquatic biomes and climate (atmospheric conditions of temperature, precipitation, and the like) is relatively unimportant in determining what organisms grow in an aquatic biome. Aquatic biomes instead are categorized by factors such as water depth and movement, temperature, pressure, chemistry, oxygen content, and characteristics of the substrate.

Activity 38.2 Aquatic Biomes

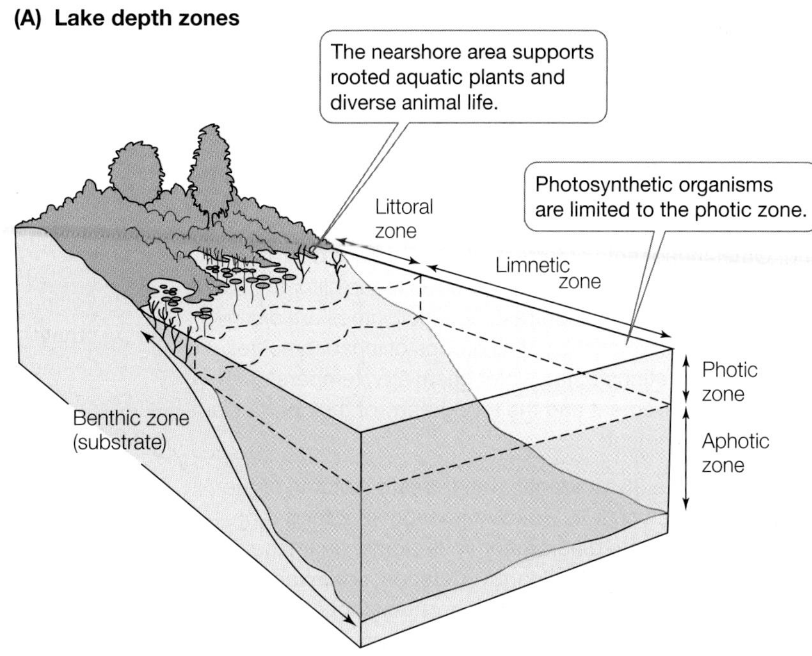

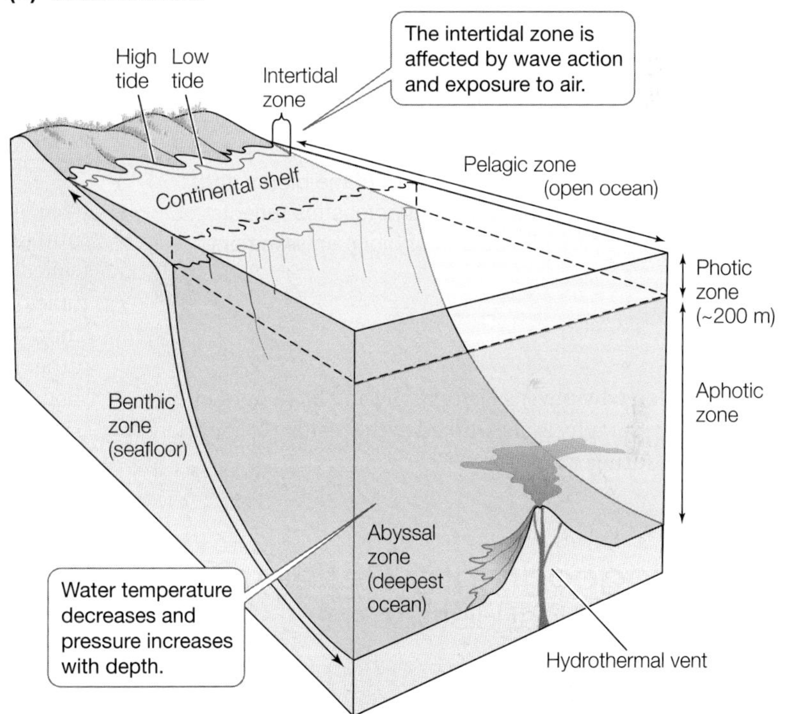

FIGURE 38.13 Water-Depth Zones Freshwater **(A)** and marine **(B)** environments can be divided into water-depth zones.

Salinity—one aspect of water chemistry—distinguishes freshwater biomes, estuarine biomes (biomes at river mouths where fresh and salt water mix), and marine biomes. Salt concentration strongly affects the ability of aquatic organisms to regulate the water content of their bodies (see Key Concept 29.2) and thus influences which species are found where.

Water movement also distinguishes aquatic biomes because moving water exerts force. Stream velocity determines the force that organisms must resist to avoid being swept downstream. Coastal currents and waves that break on the shores of oceans and lakes exert forces that can physically damage or dislodge resident organisms. Water flow also determines the presence or absence of sediment. Rapid flow scours away sediment, exposing rocky surfaces to which organisms can cling. As water slows, it deposits sediment, forming a soft bottom in which organisms can burrow.

Aquatic biomes such as lakes and oceans can be divided into water-depth zones (**FIGURE 38.13**). The nearshore regions of lakes (**littoral zone**) and oceans (littoral or **intertidal zone**) are shallow, affected by wave action, and periodically exposed to air by fluctuations in water level. Here, different species live in different depth zones according to their tolerance of heat, desiccation, and wave energy.

Because only the surface water is in contact with air, dissolved oxygen concentrations are highest in surface waters. Light, too, penetrates only a short distance into water. Photosynthetic organisms are confined to this zone of light penetration, called the **photic zone**. Rooted aquatic vascular plants and their multicellular algal equivalents (as well as corals with their photosynthetic algal symbionts) grow where the photic zone extends to the bottom, sometimes forming communities with a complex vertical structure much like that of terrestrial forests. In the open-water **limnetic zone** of lakes and the **pelagic zone** of oceans beyond the continental shelf, the prominent photosynthesizers are **phytoplankton** (free-floating photosynthetic organisms).

No photosynthetic organisms inhabit the **aphotic zone**, below the reach of light, which as a consequence is sparsely populated. The lake bottom or ocean floor is called the **benthic zone**. Water is heavy, so pressure increases with water depth. Temperature generally decreases with depth because cold water is denser than warm water and sinks. Organisms that dwell in the deepest **abyssal zone** of the oceans experience very high pressures and (except near hydrothermal vents) cold temperatures.

REVIEW & APPLY | 38.3

R Biomes are ecosystems consisting of a distinct physical environment and ecologically similar organisms with similar adaptations. Terrestrial biomes generally are distinguished by characteristics of dominant plants, which reflect climate, soil, and frequency of disturbances such as wildfire or grazing. Other inhabitants of terrestrial biomes have characteristic adaptations to the vegetation as well as to the climate and disturbance regime. In contrast, aquatic biomes are distinguished by characteristics of all groups of inhabitants, reflecting their adaptations to the water chemistry, temperature, depth, and movement and the topography of their particular aquatic environment.

A
1. Using Figure 38.11, identify the terrestrial biome of the place where you live. Go to Animation 38.2 for a representative climate diagram for your biome, an image and description of the dominant vegetation, and examples of animal inhabitants. Write a short description of your biome's climate that includes average annual temperature and precipitation, degree of seasonality in temperature and precipitation, and length of the growing season.
2. Write a short description of the characteristics of the dominant plants of your biome and how at least one plant feature represents an adaptation to an aspect of the climate.
3. Pick one of the representative animals of the biome and describe how at least one of its characteristics reflects an adaptation to either its abiotic or biotic environment.

We have just seen why physical geography and biogeography are linked. But biogeography is also linked to the history of continental movements during Earth's history.

38.4 | Biogeography Reflects Geological History

LEARNING OBJECTIVES

By the end of this key concept you should be able to:

38.4.1 Describe the difference between a biome and a biogeographic region.

38.4.2 Relate continental movements and environmental barriers to the current distribution of biogeographic regions.

During the scientific expeditions of the eighteenth and nineteenth centuries it became clear that spatial patterns in the physical environment were not sufficient by themselves to explain the distributions of organisms. The observations of one scientist in particular suggested an important role for geological history.

Barriers to dispersal affect biogeography

Alfred Russel Wallace—who, along with Charles Darwin, advanced the idea that natural selection could account for the evolution of life's diversity—first noticed an odd pattern of species distributions during an exploration of the Malay Archipelago. He observed that dramatically different bird faunas inhabited two neighboring islands, Bali and Lombok. These differences could not be explained by climate or by soil characteristics, because in those respects the two islands are similar. Instead, Wallace suggested that the Malay Archipelago was divided into two distinct parts by a line (now known as Wallace's Line; see Figure 38.14) that follows a deep-water channel separating Bali and Lombok. This channel is so deep that it would have remained full of water—and thus would have been a barrier to the movement of terrestrial animals, including birds—even during the glaciations of the Pleistocene era, when sea level dropped more than 100 meters and Bali and the islands to its west were connected to the Asian mainland. As a consequence, the terrestrial faunas on either side of Wallace's Line evolved mostly in isolation over a long period.

LINK As discussed in **Key Concept 13.1**, Wallace sent Darwin a manuscript in 1858 offering an explanation of evolution by natural selection that was similar to Darwin's. Later that year, both men were recognized for the theory of natural selection.

Wallace's observations of animal distributions worldwide led him to divide Earth's land masses into six continental-scale areas called **biogeographic regions**. Each region encompasses multiple biomes and contains a distinct assemblage of species that are more closely related phylogenetically than they are to species in other biogeographic regions. Although Wallace based his regions' boundaries on animal distributions, schemes based on plants have similar boundaries, and the seven regions that we recognize today, with our improved understanding of phylogenetic relationships, have changed little from what Wallace proposed (**FIGURE 38.14**).

Many boundaries between biogeographic regions correspond to geographic barriers to movement of organisms. These barriers include bodies of water (for terrestrial organisms), land (for aquatic organisms), and areas with extreme physical environments, such as deserts or high mountain ranges. At Wallace's Line, for example, a deep-water channel separates the Oriental and Australasian biogeographic regions.

Continental drift creates barriers to dispersal

Not all biogeographic regions are bounded by obvious physical barriers, and some regions span several continents. It's not at all clear at first why there are major biogeographic boundaries in Mexico, northern Africa, and Asia; or why the Antarctic region includes the southern tips of South America and Africa as well as New Zealand. These biogeographic patterns remained a mystery until the second half of the twentieth century, when the theory of continental drift became widely accepted (see Key Concept 17.2). Until then, biogeographers could only propose unsatisfactory explanations for the distributions of many organisms, such as southern beeches (genus

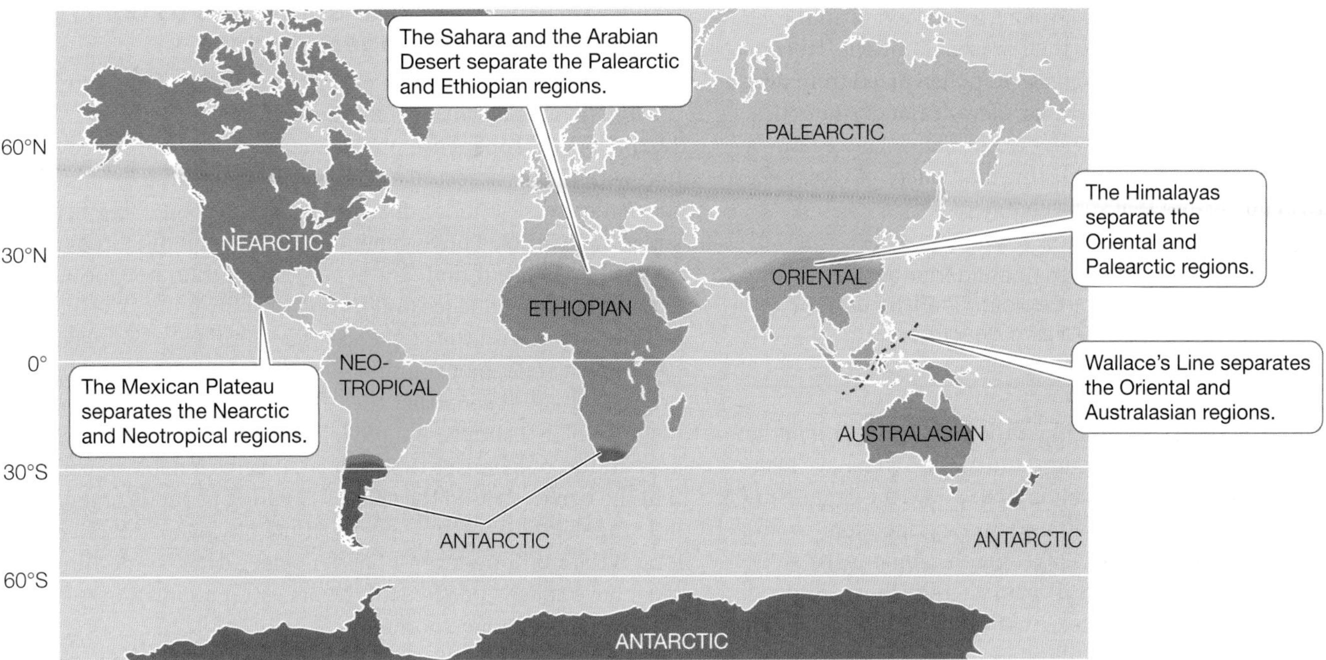

FIGURE 38.14 Earth's Terrestrial Biogeographic Regions The terrestrial biogeographic regions we recognize today are essentially the six regions Wallace proposed based on their distinctive assemblages of animals; modern biogeographers add an Antarctic region. Today we know that the boundaries between these regions are largely the result of climatic, topographic, and aquatic barriers to dispersal caused by the drift of continents carried on Earth's major lithospheric plates (see Key Concept 17.2). (After H. Kreft and W. Jetz. 2010. *J Biogeogr* 37: 2029–2053.)

Activity 38.3 Major Biogeographic Regions

Nothofagus). These trees are found in South America, New Zealand, Australia, and on certain islands of the southern Pacific Ocean (**FIGURE 38.15**). How could their distribution be explained, if not by several unlikely transoceanic colonization events?

Once the movements of the continents were understood, it became clear that they have shaped the distributions of organisms by creating and disrupting dispersal routes. The genus *Nothofagus* is now thought to have originated on the southern supercontinent

Nothofagus sp.

FIGURE 38.15 Distribution of *Nothofagus* The modern distribution of southern beeches is best explained by their origin on Gondwana during the late Jurassic. The breakup of Gondwana and subsequent continental drift resulted in the modern distribution of *Nothofagus* in South America, Australia, New Zealand, and some islands of the South Pacific. mya, millions of years ago. (After Swenson et al. 2001. *Cladistics* 17: 28–47; C. R. Scotese. 2001. Atlas of Earth History, vol. 1, *Paleogeography*. PALEOMAP project, Arlington, TX.)

Gondwana and to have been carried along when Gondwana broke apart and its fragments drifted to their present locations. Paleontologists have found fossilized *Nothofagus* pollen dated from over 80 million years ago (mya) in Australia, New Zealand, Antarctica, and South America. This fossil evidence pre-dates the breakup of those portions of Gondwana, suggesting that *Nothofagus* was once widely distributed across that supercontinent.

LINK The movements of the continents over geological time are depicted in **Figure 17.12** and **Animation 17.1** and described in **Table 17.1**.

Our understanding of continental drift makes it clear that the major biogeographic regions occupy land masses that have been isolated from one another long enough to allow the organisms present at the time of isolation to undergo independent evolutionary radiations. The northern biogeographic regions (Nearctic and Palearctic) became isolated from the southern regions (Neotropical, Ethiopian, Oriental, Australasian, and Antarctic) during the Jurassic, about 150 million years ago, when Pangaea broke into two great land masses, Laurasia to the north and Gondwana to the south (see Figure 38.15). Southern biogeographic regions subsequently became isolated from one another when Gondwana began to break apart, about 130 million years ago. The Neotropical, Ethiopian, Oriental, and Australasian regions were separated approximately 110 million years ago, and Antarctica became isolated from Australia about 60 million years ago. At about the same time, the Nearctic and Palearctic regions became isolated by the opening of the Atlantic Ocean, except for a periodic connection via the Bering land bridge between Asia and North America. Thus the biotas of the seven biogeographic regions developed largely in isolation throughout the Tertiary Period (about 65.5–2.6 mya), when extensive evolutionary radiations of flowering plants and vertebrates took place.

More recently, continued movements of continents have eliminated some barriers to dispersal and have caused mixing of species that previously were isolated, referred to as **biotic interchange**. The collision of India with Asia, for example, allowed mixing of Gondwanan and Laurasian lineages to form the Oriental biogeographic region and at the same time caused uplift of the Himalayas, which eventually limited further north–south dispersal. These high mountains separate the Oriental and Palearctic regions. Exchange of species between the Ethiopian and Palearctic regions became possible when Africa collided with Eurasia, but the Sahara and the Arabian Desert present a climate-based barrier to exchange. Similarly, mixing of species between the Neotropical and Nearctic regions began when a land bridge formed between South America and North America. During this Great American Interchange, many North American species displaced South American lineages and drove them to extinction. Currently the Mexican Plateau provides a climate-based barrier to the northward dispersal of tropical species that slows their mixing with the communities to the north.

Phylogenetic methods contribute to our understanding of biogeography

Taxonomists have developed powerful methods of reconstructing phylogenetic relationships among organisms (see Key Concept 14.2). Biogeographers use phylogenetic information, in conjunction with the fossil record and geological history, to explore how the modern distributions of organisms came about. They begin by superimposing on a phylogenetic tree the geographic areas where fossils have been found, and where each group now occurs. By comparing the sequence and timing of splits (nodes) in the phylogenetic tree with the sequence of separation or connection of geographic areas, biogeographers can determine where a lineage originated and reconstruct the history of its diversification and dispersal. For example, if a phylogenetic split coincides with the appearance of a barrier to dispersal, it is reasonable to conclude that the barrier caused the split by isolating portions of the original range of the ancestral species. If the phylogenetic split comes long before or long after a barrier arises, then something else must have caused portions of the ancestral population to evolve in isolation from one another.

REVIEW & APPLY | 38.4

R Concordant boundaries in the geographic distributions of species led to the concept of biogeographic regions—continental-scale areas, often containing multiple biomes, that contain distinctive assemblages of phylogenetically related species. The boundaries of biogeographic regions correspond to physical or climatic barriers to the dispersal of organisms that resulted from the movements of lithospheric plates by tectonic drift. Biogeographers reconstruct how the current distribution of organisms came about by matching geological events with evidence of evolutionary splits revealed by phylogenetic analysis and the fossil record.

A

1. Most boundaries between terrestrial biogeographic regions coincide with the margins of continents that became separated by oceans when larger land masses split apart. Other boundaries, such as that between the Oriental and Palearctic regions, coincide with high mountain ranges that present obvious climatic barriers to the movement of organisms. The Antarctic–Neotropical and Antarctic–Ethiopian boundaries are exceptions: they do not coincide with a body of water or with a high mountain range. With what barrier do they coincide? To answer this question, consult Figures 38.14, 38.11, and 38.6 to correlate the distributions of biogeographic region, biomes, and climate, as follows:
   a. In Figure 38.14, identify the approximate latitudes of these boundaries.
   b. In Figure 38.11, list biomes that have boundaries around those latitudes.
   c. In Figure 38.6, identify a feature of atmospheric circulation that occurs around those latitudes.
2. Propose a hypothesis to explain the location of these biogeographic region boundaries.

THINK LIKE A SCIENTIST

Phylogenetic methods contribute to our understanding of biogeography

FIGURE 38.16 shows a phylogenetic tree of placental mammals based on molecular data, accompanied by the original and current continental distribution of each lineage. Underneath the tree is the time course of major continental drift events from 150 million years ago (mya). Use your understanding of the process of science and interdisciplinary comparison to infer which phylogenetic splits were caused by a continental separation event, and which current patterns of distribution reflect dispersal following removal of a barrier to dispersal. You may wish to refer to Figure 38.14 and Figure 17.12 to visualize continental drift events.

1. Which one node (split) in the tree most clearly corresponds to an event of continental separation? Explain your reasoning.
2. List two groups of mammals that appear to have dispersed to new areas following removal of a barrier to dispersal by continental drift. Explain your reasoning.
3. List two distinct scientific disciplines that have contributed to our understanding of boundaries between biogeographic regions and describe their contributions in a few words.

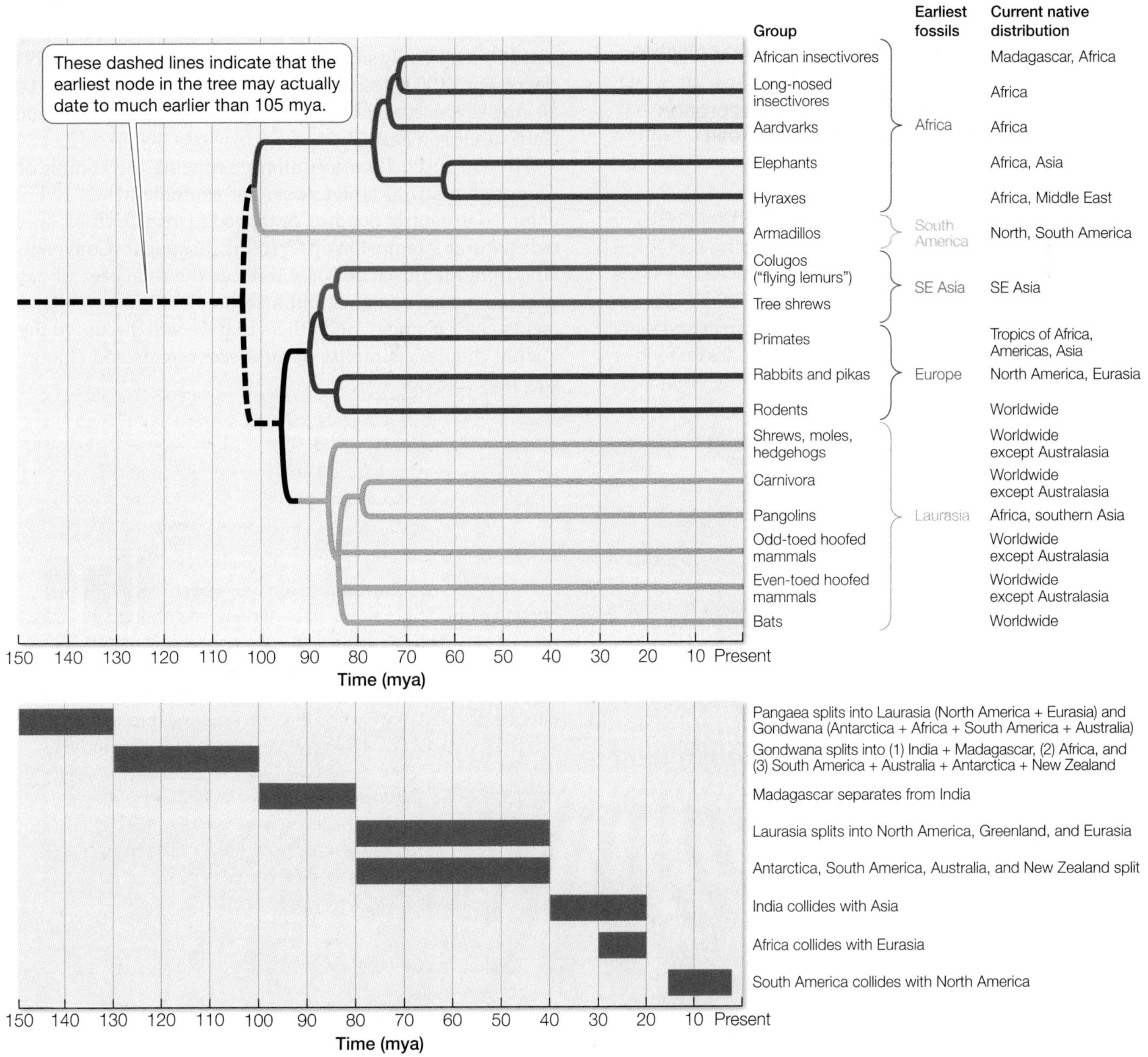

FIGURE 38.16 Relationships between Phylogeny and Biogeography of Major Mammal Groups and Continental Drift Events from Cretaceous to Present Periods

(Top after J. W. Murphy et al. 2007. *Genome Res* 17: 413–421; bottom after I. Sanmartín and F. Ronquist. 2004. *Syst Bio* 53: 216–243.)

As we have seen, the makeup of ecological systems on Earth is enormously variable over time and space. Recently, however, this variation has been greatly influenced by the activities of one species: *Homo sapiens*. How are our activities affecting ecological systems?

38.5 Human Activities Influence Ecological Systems on a Global Scale

LEARNING OBJECTIVES

By the end of this key concept you should be able to:

38.5.1 List at least two ways in which human activities are influencing ecological systems.

38.5.2 Define conservation ecology and restoration ecology and contrast the goals of these subdisciplines of ecology.

We are entering a new geological period, the Anthropocene, or Age of Humans, because our activities are altering ecological systems on a global scale. This new age also has been called the Homogenocene, or Homogeneous Age, because the net effect of our activities is to make ecological systems less complex and more uniform. We affect ecological systems by using species for food or timber or fiber, by altering landscapes, by spreading organisms across the globe without regard for natural barriers to dispersal, and as Key Concept 42.4 will describe, by changing Earth's climates. All of these anthropogenic effects on ecological systems change their properties, as we will see in the chapters to come.

We are altering natural ecosystems as we use their resources

When we extract natural resources—for example, through hunting, fishing, grazing, or logging—we remove individuals of particular species and thereby change their abundances. If we remove too many, we can even cause some species to disappear entirely—to go extinct. The resulting shifts in the relative abundances of components of natural ecosystems change the patterns of interaction among species and thereby change how entire ecosystems function, as we will see in the following chapters. The decreased abundance of grasses relative to shrubs in the U.S.–Mexico Borderlands, for example, has reduced the ability of these lands to support grass-eating animals, whether domesticated or wild.

We are converting natural ecosystems to human-modified ecosystems

Analysis of satellite imagery indicates that by 2007 humans had converted over half of Earth's ice-free land area of 130.1 million square kilometers to ecosystems heavily affected by human activities: infrastructure such as cities and roads occupies approximately 6.8 percent, while land devoted to agriculture and forestry occupies another 50 percent or so.

Human-dominated ecosystems contain fewer interacting species than do natural ecosystems and are therefore less complex. Pastures or heavily grazed rangelands, for example, often contain fewer species than do natural grasslands. And when an area is converted to agriculture, monocultures (plantings of single crops) often replace species-rich natural plant communities (**FIGURE 38.17**). Furthermore, agricultural systems worldwide are dominated by a relatively small number of plant species. Some 19 species comprise 95 percent of total global crop production, and overall crop diversity is a tiny fraction of the nearly 400,000 named plant species on Earth.

Human activities are similarly reducing the complexity of seemingly natural landscapes. For example, when rivers are dammed to control flooding or tapped to irrigate crops, species-rich natural streamside ecosystems disappear. Conversion of land to human uses not only reduces the total area of natural vegetation that remains, but also breaks it into smaller fragments. Such changes, and others that we will discuss in the following chapters, simplify natural ecosystems and change the way they function.

FIGURE 38.17 Human Agricultural Practices Produce a Uniform Landscape The cultivation of crop plants for food and other uses diminishes the diversity of ecosystems. This monoculture is a field of corn plants (*Zea mays*) in Minnesota.

We are blurring biogeographic boundaries and changing communities

Humans are also moving organisms around on a global scale, sometimes deliberately—as when Spanish settlers brought domesticated cattle to the U.S.–Mexico Borderlands—and sometimes inadvertently. Only about 1 percent of inadvertent introductions result in self-sustaining populations in the new locality, but their cumulative effect on geographic distributions is huge. Of the insects found in both Europe and North America, for example, half have been transported between the two continents by humans. Similarly, more than half of the plant species on many oceanic islands are not native, and in many continental areas the figure is 20 percent or more. The pace of introductions is astonishing: for example, in California's San Francisco Bay, one new species on average became established every 12 weeks during the 1990s. When new species establish in a community, they add novel interactions to the original system that may precipitate further changes in species abundances—even the complete loss of some species from the system.

Humans have promoted unprecedented biotic interchanges that are homogenizing the biota of the planet, blurring the differences among communities that evolved during long periods of continental isolation. One wonders how Alfred Wallace would draw the boundaries of biogeographic regions if he were using present-day data!

Science provides tools for conserving and restoring ecological systems

With the dawning of the Anthropocene, ecologists are called on as never before to understand how humans are influencing Earth's ecological systems and how we can preserve their ability to sustain life on our planet. The goal of **conservation ecology** is to explain the process of extinction and devise ways to prevent the extinction of vulnerable species. Often this requires that we protect the ecosystems of which these species are a part. The goal of **restoration ecology** is to restore the health of damaged ecosystems. These two fields are related because natural systems are sometimes damaged so badly that extinction will occur unless the systems are restored.

To achieve their goals, conservation and restoration ecologists deploy the same tools of scientific inquiry that all other biologists (and scientists) use: observation, questioning, logic, and experimentation (see Key Concept 1.5). They often begin with **natural history**—the observation of nature outside of a formal, hypothesis-testing investigation—because descriptive knowledge of nature is fundamental to all stages of ecological inquiry. Natural history observations are the source of new questions and hypotheses and are critical for the design of good experiments—experiments based on natural conditions and actual ecological systems. Without knowledge of natural history, our ability to answer important questions is often limited. For example, we cannot fully answer the question "Has the introduction of honey bees caused declines in native bee species?" without knowing what bee species were present before honey bees were introduced and how abundant they were. Similarly, knowledge of the identities and metabolic capabilities of gut microbes is essential to understanding how the gut microbiome affects human health.

As we saw in Chapter 1, scientists test hypotheses by seeing if their predictions are true. Sometimes we can use simple rules of logic to develop predictions. But with systems as complex as ecological systems, we often need to use a mathematical systems model, or perhaps even a computer simulation. What is critical is that we base the model on the natural history of the system. In the degraded rangeland system of the Borderlands, for example, identifying a viable strategy for restoring desirable grasses requires a complex model of the many factors that affect the growth of grasses relative to shrubs in that area (see the answer to the opening question on the next page).

Conservation and restoration ecology are only two examples of the important practical applications of ecology in the Anthropocene. The chapters that follow will highlight numerous additional examples.

REVIEW & APPLY | 38.5

Humans are influencing natural ecosystems on a global scale by extracting organisms from them through such activities as hunting and logging, by introducing nonnative species to them, and by converting them for agriculture, livestock grazing, or urban uses. These activities alter how ecosystems function worldwide because they simplify them through the loss of species. Conservation ecology seeks to prevent the extinction of species. Restoration ecology seeks to restore the health and functioning of damaged ecosystems. Ecologists must rely especially heavily on two tools of scientific inquiry—natural history observation and modeling—to understand the properties of complex ecological systems.

1. Explain why both "Age of Humans" and "Homogeneous Age" describe the current period of Earth's history.
2. Do medical researchers who seek ways of treating human digestive disorders have anything in common with restoration ecologists who seek to restore degraded grasslands in the Borderlands? Justify your answer.

Now that we have explored some of the properties of ecological systems, we are ready to take a closer look at them, beginning with the smallest levels in the ecological hierarchy: individual organisms and populations.

How can an understanding of interactions between physical and biological factors help us explain why removing cattle did not reverse this transition from grassland to shrubland?

© iStock.com/eyecrave

ANSWER: Ecologists have come to realize that the biosphere, much like traditional Russian "matryoshka" nested wooden dolls, is made up of systems within systems within systems. At each level, system properties depend on component parts and interactions among those parts. The conversion of grassland to shrubland in the southwestern Borderlands ecosystem was precipitated by a change in one part—the addition of too many grass-eating cattle. But studies of grassland biomes the world over have shown that reduction of grass cover by cattle grazing has multiple consequences for physical components of ecosystems, which in turn can permanently tip the balance in favor of shrubs over grasses. Physical changes include erosion of nutrient-rich topsoil and reduced water infiltration into the soil, which inhibits establishment of grass seedlings, and fewer wildfires, which kill young shrubs but not grasses. Increased shrub cover in turns shades out and inhibits grasses, completing a "vicious cycle." Ecologists are now building conceptual models of the Borderlands ecosystem, such as the one shown in **FIGURE 38.18**, that include all of these interacting components. Such systems models constitute hypotheses that can be tested with controlled field experiments designed to determine what interventions will restore grasses to the Borderlands ecosystem. This model suggests that a combination of erosion control, reduction of shrub cover, grazing management, seeding in native grasses, and controlled burns may all be necessary to break the vicious cycle.

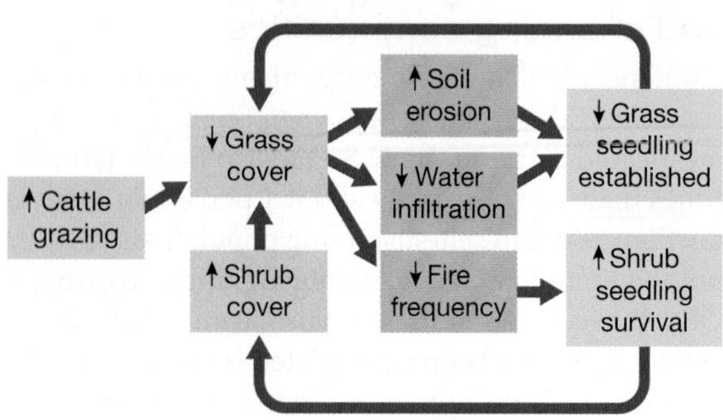

FIGURE 38.18 A Conceptual Model of the Borderlands Ecosystem

38 VISUAL SUMMARY

38.1 Ecological Systems Vary over Space and Time

FIGURE 38.1

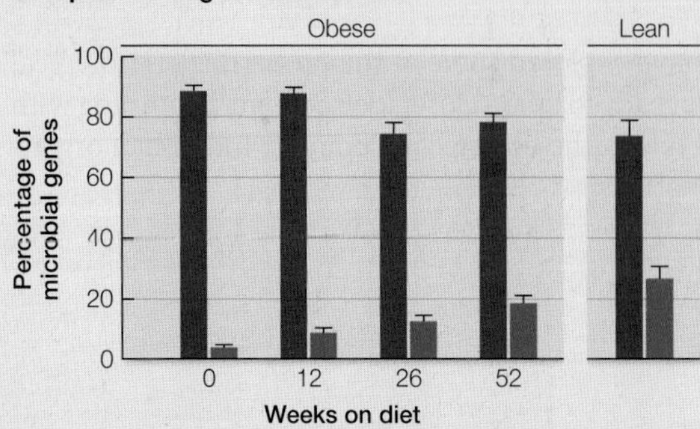

- An **ECOLOGICAL SYSTEM** consists of one or more organisms and the **BIOTIC** and **ABIOTIC** components of the environment with which they interact.
- Ecological systems can be studied at levels ranging from an individual **ORGANISM** to **POPULATIONS**, **COMMUNITIES**, **LANDSCAPES**, and the entire **BIOSPHERE**.
- Ecological systems are complex and vary over space and time, but their properties can be understood with scientific methods (**FIGURE 38.1**).

QUESTIONS

1. Explain why ecological systems at a higher level in the hierarchy of biological organization tend to be more complex.
2. People sometimes experience digestive and other health problems after taking antibiotics to treat a bacterial infection. Discuss how a restoration ecologist would approach the challenge of restoring the health of the gut ecosystem.

38.2 Solar Energy Input and Topography Shape Earth's Physical Environments

Go to ANIMATION 38.1

FIGURE 38.6

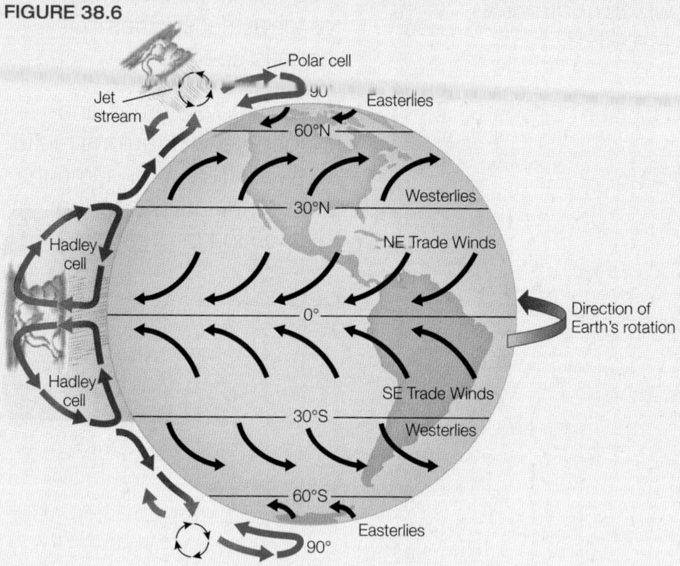

- Variation in solar input and circulation of the atmosphere and oceans drive patterns in Earth's **CLIMATES**.
- When prevailing winds arising from north–south atmospheric circulation interact with Earth's rotation, they gain an east-west component, a phenomenon known as the **CORIOLIS EFFECT** (**FIGURE 38.6**).
- Variation in Earth's surface **TOPOGRAPHY** influences the physical environment in both terrestrial and aquatic biomes.
- **CLIMATE DIAGRAMS** are useful for discerning when conditions are suitable for terrestrial plant growth.

QUESTIONS

1. Predict where in Peru you would find the rain shadow of the Andes, a north–south range of high mountains that extends from the equator to about latitude 18°S.
2. Evaluate how Earth's climates would differ if the axis of Earth's rotation were not tilted relative to the plane of Earth's orbit around the Sun.

38.3 Biogeography Reflects Physical Geography

Go to ACTIVITIES 38.1 and 38.2

FIGURE 38.10

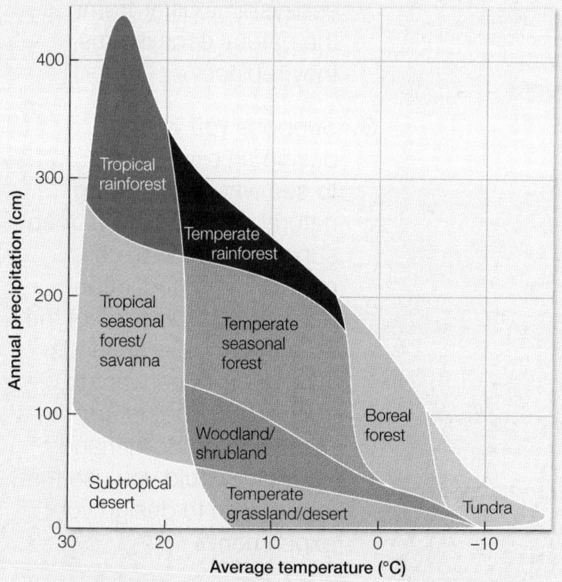

- A **BIOME** is a type of ecosystem recognized by characteristic adaptations of its species to the distinct physical environment they inhabit.
- The distribution of terrestrial biomes is broadly determined by annual patterns of temperature and precipitation, but other factors—such as soil characteristics and fire—also influence terrestrial vegetation (**FIGURE 38.10**).
- Aquatic biomes are defined by water depth and flow, temperature, pressure, water chemistry, and characteristics of the substrate.

1. Defend the proposition that the human gut can be considered a biome.
2. Contrast the primary factors that limit the distribution of photosynthetic organisms in aquatic biomes versus terrestrial biomes.

38.4 Biogeography Reflects Geological History

Go to ACTIVITY 38.3

- Earth can be divided into **BIOGEOGRAPHIC REGIONS**, each of which contains a distinct assemblage of species, many of which are phylogenetically related (**FIGURE 38.14**).
- The boundaries of biogeographic regions generally correspond to present or past barriers to dispersal and can be explained by continental drift.

FIGURE 38.14

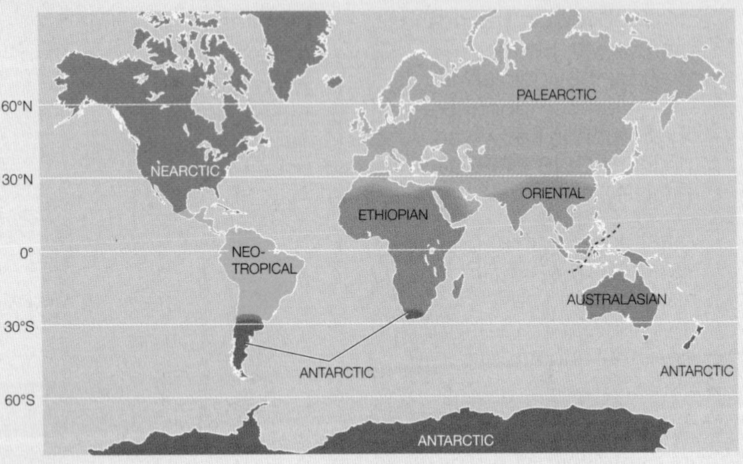

QUESTIONS

1. What kinds of barriers might limit the dispersal of terrestrial species? Of aquatic species?
2. Explain how biomes and biogeographic regions relate to two different fundamental factors that influence the distribution of organisms.

38.5 Human Activities Influence Ecological Systems on a Global Scale

- Humans have converted much of Earth's land area into human-dominated ecosystems, which are less diverse than natural ecosystems. Many natural ecosystems are now fragmented (**FIGURE 38.17**).
- Humans move species around the globe without regard for natural barriers. These movements are homogenizing Earth's biota.
- **CONSERVATION ECOLOGY** and **RESTORATION ECOLOGY** strive to prevent extinction of species and to restore damaged ecosystems, respectively, and are dependent on knowledge of **NATURAL HISTORY**.

FIGURE 38.17

1. Identify two human activities that alter the distribution of species.
2. Explain how agricultural ecosystems differ from the natural ecosystems they replace.
3. Suppose you wish to design an experiment to see whether allowing natural wildfires, rather than suppressing them, can restore grassland in the degraded shrublands of the U.S.–Mexico Borderlands. What would you want to know about the natural fire regime in the Borderlands, and how would you use that information to design your experiment?

Populations 39

Lyme disease is caused by a spirochete bacterium in the genus *Borrelia* that can be transmitted to people when they are bitten by a black-legged tick (*Ixodes scapularis*) that is seeking a blood meal. Ecologists are playing a central role in devising strategies to slow the incidence of Lyme disease. They are doing so by studying the life cycle of ticks and the factors, such as abundance of nonhuman tick hosts (various vertebrates), that influence the number of infected ticks.

How has an understanding of the ecology of tick populations helped us combat Lyme disease?

You will find the answer to this question on page 987.

Lezh/E+/Getty Images

KEY CONCEPTS

39.1 Populations Are Patchy in Space and Dynamic over Time

39.2 Births Increase and Deaths Decrease Population Size

39.3 Life Histories Determine Population Growth Rates

39.4 Populations Grow Multiplicatively, but the Multiplier Can Change

39.5 Immigration and Emigration Can Influence Population Dynamics

39.6 Ecology Provides Tools for Conserving and Managing Populations

39.1 Populations Are Patchy in Space and Dynamic over Time

LEARNING OBJECTIVES

By the end of this key concept you should be able to:

39.1.1 **Explain the difference between population size and population density.**

39.1.2 **Calculate population density and population size from census data and knowledge of the spatial extent of a population.**

In Chapter 38 we saw that individual organisms are the fundamental units of all larger ecological systems. Many important insights are gained by focusing on individuals, but additional insights arise by studying **populations**—groups of individuals of the same species that occur together at a single place and time. Populations have properties that individuals do not have, and we cannot understand larger ecological systems, such as communities, without understanding their component populations. One important property of a population is its **population size**—the number of individuals it contains—because a population's aggregate role in the larger ecological system depends not only on properties of the average individual, but also on how many individuals there are.

Humans have long had a practical interest in understanding what determines population size because we want to manage our ecological interactions with other species. We want to know how to increase the population sizes of species that provide us with resources such as food or fiber, how to conserve beneficial wild organisms such as pollinators, and how to decrease the population sizes of undesirable species such as crop pests, weeds, and pathogens.

Population size is usually estimated from population density and spatial extent

Measuring population size can be challenging. It would be difficult, for instance, to count every single Edith's checkerspot butterfly (*Euphydryas editha*) in any of the individual populations found in the San Francisco Bay area of California (see Figure 39.1). For most species, therefore, an ecologist will derive an estimate of the size of a population by taking several samples from within the population. To obtain these samples the ecologist first identifies a place of interest, which might be a special area set aside to protect an endangered species or simply a representative place where a certain species can be studied. Within this study place the ecologist then chooses several smaller sample sites that represent all the types of sites that the population occupies in the study place, and takes a census (a count of individuals) within each sample site. Dividing each census count by the size of the site it came from gives a measure of **population density**—the number of individuals per unit of two-dimensional area (for surface-dwelling organisms) or three-dimensional volume (for organisms that live in air, soil, or water). Average density across the replicate sample sites can then be multiplied by the total area or volume that the population occupies to estimate the total population size in that place.

Counting individuals within each sample site can also be challenging. If individuals do not move (for example, plants that are rooted in fixed locations), or if they move but each one has a distinctive feature (for example, African wild dogs, *Lycaon pictus*, which have unique spot patterns on their fur), each can be counted only once. Otherwise, individuals must be caught and marked to avoid double-counting.

Population size varies in space and time

Most species are made up of many populations, and as we saw in Chapter 38, these populations usually are found only within a particular geographic region of Earth—called the species' **geographic range**. Within the geographic range, a species may be restricted to particular kinds of environments, called **habitats**, which are often patchily distributed. For example, Edith's checkerspot butterfly is found from southern British Columbia and Alberta to Baja California, Nevada, Utah, and Colorado (**FIGURE 39.1A**). Within this geographic range, it occurs in a series of distinct populations in sunny habitats (coastal chaparral, meadows, grasslands, or open woodlands) where the plants grow that its caterpillars eat. In the San Francisco Bay area, suitable caterpillar food plants occur in soils derived from a chemically unique rock called *serpentine* (**FIGURE 39.1B**). Outcrops of serpentine rock constitute habitat "islands" for the butterflies—that is, patches of suitable habitat separated by areas of unsuitable habitat. Habitat patchiness has consequences for populations that we will explore in Key Concept 39.5.

In any given locality, population sizes are **dynamic**—they change over time. For example, the black-legged ticks mentioned in the opener to this chapter live in eastern forests that also are home to rodents from which the ticks obtain blood meals. Population densities—and therefore population sizes—vary greatly from year to year for the ticks, the rodents, and the acorns that the rodents eat (**FIGURE 39.2**).

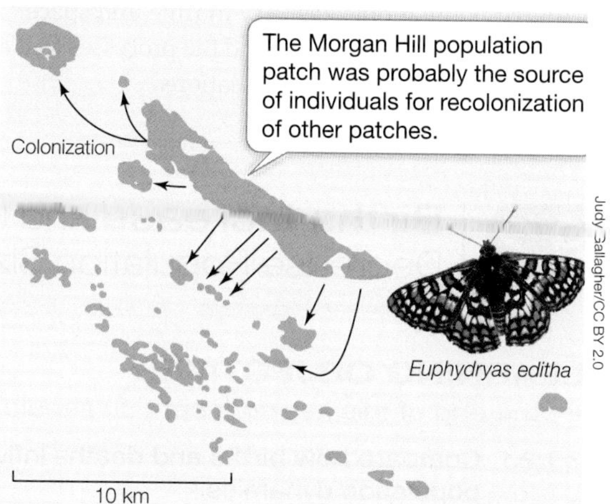

FIGURE 39.1 Species Are Patchily Distributed on Several Spatial Scales **(A)** The geographic range of Edith's checkerspot butterfly (*Euphydryas editha*) extends from British Columbia and Alberta to Baja California. **(B)** Within the geographic range, populations of the butterflies are found in patches of suitable habitat, which in the San Francisco Bay area are on serpentine soils. Arrows indicate recolonization events that are discussed in Key Concept 39.5. (A after R. E. Stanford and P. A. Opler 1993. *Atlas of Western Butterflies, Including Adjacent Parts of Canada and Mexico.* Published by authors. Denver, CO.; K. Lotts and T. Naberhaus, coordinators. 2017. *Butterflies and Moths of North America*. http://www.butterfliesandmoths.org; B after S. Harrison et al. 1988. *Am Nat* 132: 360–382.)

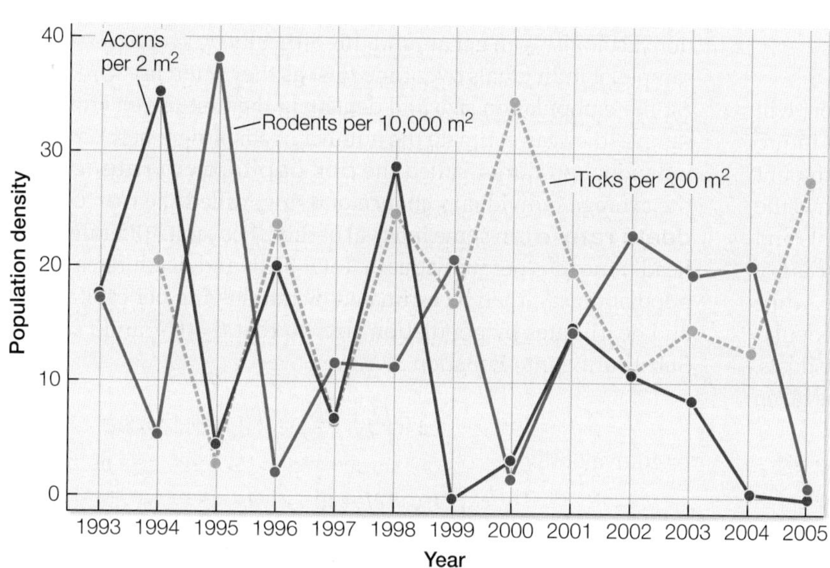

FIGURE 39.2 Populations Are Dynamic Densities of acorns, rodents, and nymphal black-legged ticks varied over time in an oak forest near Millbrook, New York. (Notice that the population density of each organism is measured on a spatial scale appropriate for that organism's average density.) (After R. S. Ostfeld et al. 2006. *PLoS Biol* 4: e145.)

In what year were ticks most abundant, and in what year were they least abundant? Were rodents ever more numerous than acorns or ticks?

REVIEW & APPLY | 39.1

R Populations are groups of individual organisms of the same species that occur together in time and space. One important attribute of a population is its size—the total number of individuals it contains. Ecologists estimate population size by multiplying an estimate of population density (averaged across replicate small sample sites) by the overall space occupied by the population. Population densities vary in space; each species usually occurs only in specific habitats within its larger geographic range. Population sizes fluctuate through time even within suitable habitats.

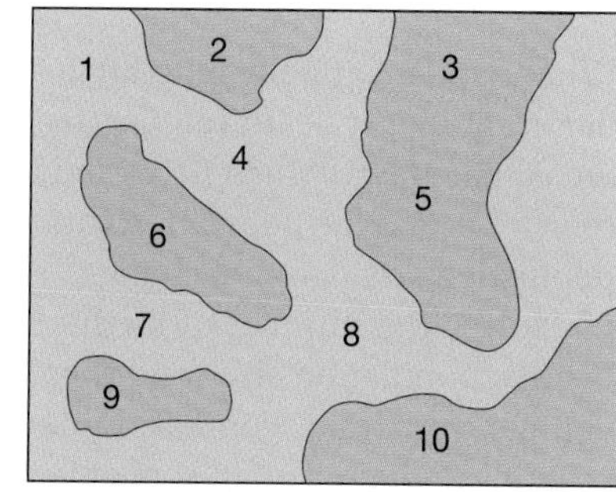

A The map shows a hypothetical nature reserve with two distinct habitat types indicated in green and blue. Ecologists establish 10 sampling sites in the reserve, each 100 m^2 in area, at the locations indicated by the numbers on the map. The numbers of oak trees found in each of the 10 sampling sites are given in the table.

Census site	Population size
1	0
2	4
3	6
4	0
5	5
6	7
7	0
8	0
9	6
10	8

1. Are both habitats suitable for oak trees?
2. What is the average density of oaks per 1 m^2 in suitable habitat within the reserve?
3. Describe how you would estimate the total population size of oaks in the nature reserve.

Why do population sizes vary in time and space? To answer this question we need to understand the processes that add individuals to or subtract them from populations.

39.2 Births Increase and Deaths Decrease Population Size

LEARNING OBJECTIVES

By the end of this key concept you should be able to:

39.2.1 Compare how births and deaths influence population dynamics.

39.2.2 Explain the relationship between per capita and total population birth and death rates.

39.2.3 Use birth and death data to calculate changes in population size.

The size of anything, whether a bank account or a population, will change over time if there are additions and subtractions. The future size of your bank account, for example, will be equal to the amount of money you have now, plus the amount that is added, minus the amount that is taken out. What differs between bank accounts and populations are the processes that add or subtract. In the case of your bank account, deposits and interest add to your balance, while withdrawals and bank fees subtract from it. In the case of populations, births (or the production of seeds by plants) add individuals, while deaths remove individuals. If no other factors were to be taken into consideration, we can express this dynamic as follows:

the number of individuals in a population at some future time = the number now + the number that are born – the number that die

This word equation can be converted into a basic algebraic model of population size:

$$N_{t+1} = N_t + B - D \qquad \text{Equation 39.1}$$

where N_t is the population size at time t, B is the number of births in the time interval from time t to time $t + 1$, and D is the number of deaths in that same interval.

Recall that any **computational model**—whether it is expressed in words, graphs, or mathematical symbols—is a quantitative hypothesis about the important components of a system and their interactions that can be used to predict the properties of the system. The model can be tested by comparing model predictions with observations of real populations; if observations deviate in important ways from predictions, then the model needs to be modified. The "birth–death" model, or **BD model**, of Equation 39.1 works well for species whose individuals rarely move among populations. For example, it is useful for understanding the population dynamics of species—such as the marine copepod *Tigriopus californicus*, which occupies intertidal rock pools along the Pacific coast of North America. Individuals of this species so rarely move among habitat patches that nearby populations are genetically distinguishable.

Equation 39.1 tells us that the size of a population is likely to change over time—to grow or shrink. Ecologists call the amount of change over a specific amount of time the **growth rate** of the population (a confusing term until we realize that when a population shrinks, its growth rate is negative). The growth rate can be obtained from Equation 39.1 by subtracting N_t (the present population size) from both sides of the equation, which shows that the change in population size, ΔN, is simply $B - D$. This change occurs over the time interval from t to $t + 1$, which we symbolize as ΔT. Putting this together we have:

$$\frac{\Delta N}{\Delta T} = B - D \qquad \text{Equation 39.2}$$

The model indicates, logically enough, that the change in population size over one time interval is the number of births minus the number of deaths during that interval.

Change in the size of a population can be measured directly only for very small, countable populations, such as a zoo population. Ecologists can get around this difficulty by keeping track of a sample of individuals over time (just as they often use samples to estimate population size and density in the first place). From this sample, they can estimate the number of offspring that the average individual produces, called the **per capita birth rate (*b*)**, and the average individual's chance of dying, called the **per capita death rate (*d*)**, in some interval of time. Per capita (literally, "per head") means "per individual." Total birth and death rates in the population can then be estimated by multiplying per capita birth and death rates by population size, so that $B = bN_t$ and $D = dN_t$. Substituting into Equation 39.1, we have

$$N_{t+1} = N_t + bN_t - dN_t$$

or equivalently,

$$N_{t+1} = N_t + (b - d)N_t$$

The value $(b - d)$—the difference between per capita birth rate and per capita death rate—represents the average individual's contribution to total population growth rate. This value is the **per capita growth rate**, which we symbolize as ***r***. Substituting r for $(b - d)$, we have

$$N_{t+1} = N_t + rN_t \qquad \text{Equation 39.3}$$

By converting Equation 39.3 into an equation of growth rate analogous to Equation 39.2, we get

$$\frac{\Delta N}{\Delta T} = rN \qquad \text{Equation 39.4}$$

What happens if per capita birth rate is greater than per capita death rate, or if death rate is greater than birth rate, or if they are equal? If $b > d$, then $r > 0$, and the population grows. If $b < d$, then $r < 0$, and the population shrinks. If $b = d$, then $r = 0$, and the population size does not change.

Unless the number of births exactly equals the number of deaths, population size (and therefore population density) will change over time. Population densities of a species also vary spatially and are zero in some places. We have already seen (see Key Concept 38.4) that a species may be absent from a location if barriers to dispersal have prevented individuals from reaching that place. But negative population growth can also exclude species from some places. A species will not persist in a given location if its per capita growth rate, r, is negative there, because any population that decreases steadily in size over time eventually reaches a density of zero—that is, it goes extinct.

REVIEW & APPLY | 39.2

R Births of new individuals add to the size of biological populations, whereas deaths subtract. For species whose individuals rarely if ever move between different populations, we can model the size of any one population at some future time if we know the numbers of births and deaths in the population over that time period. It often is convenient to estimate total births and deaths by first obtaining values for birth and death rates of a sample of representative individuals, and then multiplying these average per capita values by the total number of individuals in the population. Whether the population will grow, shrink, or stay the same size depends on whether the per capita growth rate, r, is positive, negative, or zero.

A Jane is a rancher who is considering shifting her operation from cattle to American bison (*Bison bison*), to take advantage of the demand for lean, healthy bison meat. To decide, she needs to know how well bison will do on her ranch. She buys 50 female bison that are already inseminated and places 10 of them, picked at random, into their own pasture. These 10 females serve as a sample from which Jane collects birth and death data over 1 year, shown in the table below.

Female number	Alive at end of year?	Number of offspring
1	Yes	1
2	Yes	0
3	Yes	1
4	Yes	0
5	No	0
6	Yes	1
7	Yes	1
8	No	0
9	Yes	1
10	Yes	0

1. How many total births and deaths occurred in the sample population of bison?
2. Based on the sample, what are the estimated per capita birth and death rates (b and d) and per capita growth rate (r) for the entire bison herd?
3. Based on the sample, how many bison does Jane expect to have in her entire herd at the end of the year? (*Hint*: Use Equation 39.3.)

We have seen that the dynamics of populations depend on average birth and death rates. But as rancher Jane's bison example illustrates, individuals within populations vary around those averages—some females produced a calf, and some did not. Furthermore, the average birth and death rates within a single population can change through time and can differ among populations. What causes this within- and between-population variation in birth and death rates?

39.3 | Life Histories Determine Population Growth Rates

LEARNING OBJECTIVES

By the end of this key concept you should be able to:

39.3.1 Identify the primary components of a quantitative life history.

39.3.2 Explain how life histories determine per capita population growth rates.

39.3.3 Describe how the principle of allocation leads to life history trade-offs.

39.3.4 Outline how environments shape life histories.

To understand variation in birth and death rates and the consequences for population growth, we turn to the field of **demography** (from the Greek *demos*, "population," + *graphia*, "description"), which arose from interest in practical issues related to humans—such as life expectancy (of interest to the life insurance industry), or forecasting future demand for elementary schools or maternity wards. Recognizing that an individual's probability of reproducing or dying changes during its lifetime, demographers begin by characterizing a species' **life history**: the sequence and duration of stages an organism passes through from the beginning to the end of its life. Let's consider the life history of the black-legged tick. The tick's life begins as one of the thousands of fertilized eggs laid by an adult female in spring (**FIGURE 39.3**). If the egg survives, it hatches into a larva in midsummer. If the larva can obtain a blood meal from a mammal, bird, or lizard, it molts into a nymph and goes dormant for the winter. If it survives the winter, the nymph becomes active again the following summer and seeks another blood meal from a vertebrate host. If it is successful, the nymph molts into an adult in the fall and seeks a final host, such as a deer, raccoon, opossum—or human. Only a few ticks survive to adulthood, find a host, and successfully mate on that host. If the tick is male, it dies in the fall; if the tick is female, it goes dormant for another winter, lays eggs in the spring, and dies. Demographers usually focus on females, whose reproductive success determines the birth rate.

Media Clip 39.1 **Dangerous Deer Ticks**

A *qualitative* description of a tick's life history like the one we've just given provides important insights into its ecology. But to obtain a deeper understanding, we need *quantitative* information about the duration of life stages, survival probabilities, and

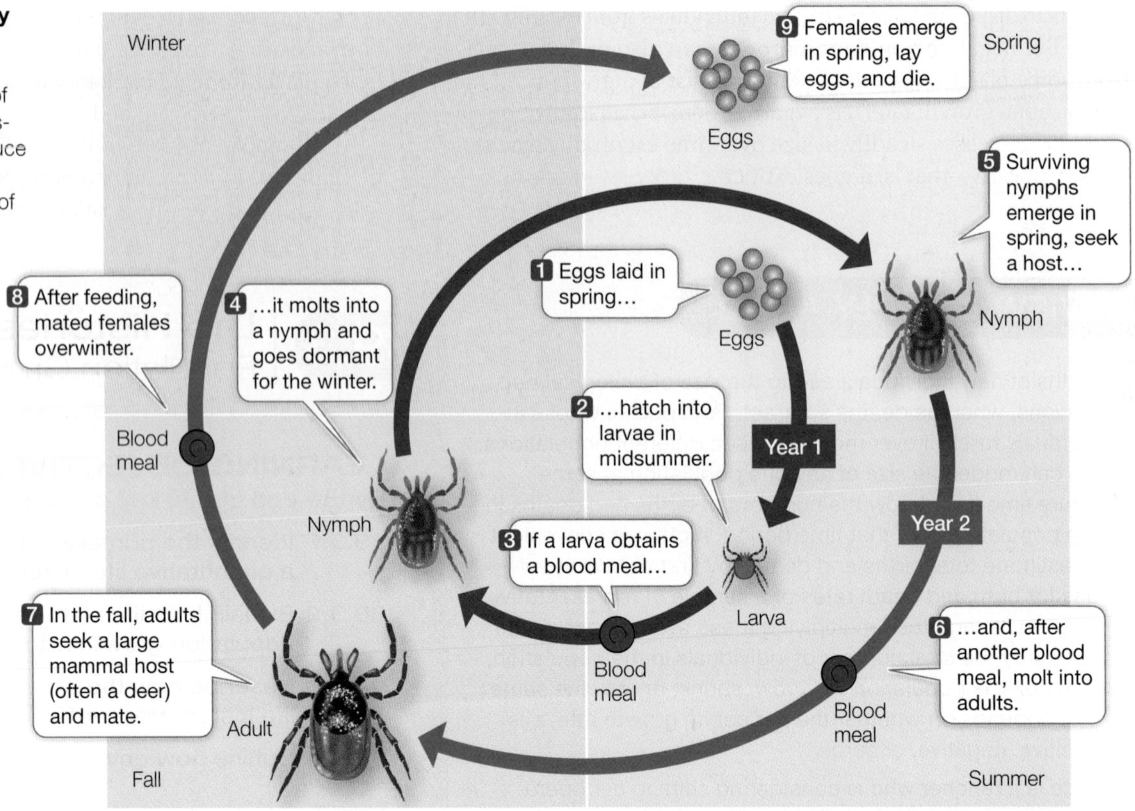

FIGURE 39.3 Life History of the Black-Legged Tick The seasonality of the tick's environment and the difficulty of obtaining the blood meal necessary to grow, molt, and reproduce have led to a 2-year life history that includes two long periods of winter dormancy.

the number of offspring produced by survivors during each stage. Obtaining such information is not easy, and ecologists still don't have all of it for the black-legged tick. We do have quantitative life histories for other species, however, such as the Cactus Finch (*Geospiza scandens*) in the Galápagos archipelago. **TABLE 39.1** shows the quantitative life history of a sample of 90 female Cactus Finches that successfully fledged (left the nest) on Isla Daphne in 1978. The table quantifies two important aspects of any life history: (1) **survivorship**, or l_x, the proportion of individuals that survive from birth to life stage x (which are ages in the case of the finches), and (2) **fecundity**, or m_x, the average number of daughters that each surviving female produces during life stage x. Table 39.1 also shows the proportion of females that became reproductively mature (bred for the first time) in each year.

Factors that influence birth and death rates can be inferred from quantitative life histories. For example, we can see from Table 39.1 that age is one such factor: birds spent on average more than a year as juveniles before they started breeding. We can see that environmental conditions also affect fecundity: once females were reproductively mature, fecundity was high in wet years and low in dry years. From quantitative life histories we also can derive useful statistics such as **life expectancy** (the average age at death), the average age of the mother when her daughters are produced (also called the **generation time**), and average lifetime reproductive success, or R_0. The average female Cactus Finch described in Table 39.1 lived 2.19 years and produced 2.87 daughters in her lifetime; the females that lived to breed were 4.5 years old when their average baby fledged.

Quantitative life histories allow us to estimate per capita growth rate and, together with information about the starting numbers of individuals of various ages in the population, to project the population's growth into the future. Quantitative life histories also can be used to ask "what if" questions about how a change in survivorship, or fecundity, or timing of reproductive maturity will affect per capita growth rate if other variables are held constant. The conclusion from such theoretical studies is that higher fecundity, higher survivorship, and earlier reproduction all increase r. We conclude that per capita growth rates vary because life histories vary.

Activity 39.1 Quantitative Life Histories and Per Capita Rate of Population Growth

Life histories are diverse

Species vary considerably in how many and what types of life stages they go through, how old they are when they begin to reproduce, how often they reproduce, how many offspring they produce, and how long they live. Most black-legged ticks spin out their lives over 2 years, whereas the life history of a periodical cicada spans nearly 2 decades. Other arthropods have life spans of days or weeks. Some species, like the tick, go through a discrete series of life stages of variable duration, each separated by a molt. Other species, such as Cactus Finches and humans, develop continuously as they age. Plants known as "annuals" germinate, grow, flower, produce seeds, and die all in one growing season (see Key Concept 27.2). Other plants live for centuries. Some species spend long periods in a dormant state, others are continuously active. Some organisms, including the tick, reproduce only once and then die; others, such as finches and humans, can reproduce multiple times.

TABLE **39.1** | Quantitative Life History for 90 Female Cactus Finches Fledged in 1978 on Isla Daphne

Calendar year	Rain (mm)[a]	Bird age (yr)	Survivorship[b]	First breeding[c]	Fecundity[d]
1978	130	0	1.000	0	0.000
1979	60	1	0.434	0.062	0.051
1980	50	2	0.367	0.875	0.667
1981	60	3	0.333	0.063	1.500
1982	40	4	0.322	0	0.655
1983	1360	5	0.300	0	5.500
1984	50	6	0.178	0	0.687
1985	T	7	0.089	0	0
1986	40	8	0.056	0	0
1987	610	9	0.056	0	2.200
1988	0	10	0.033	0	0
1989	T	11	0.022	0	0
1990	195	12	0.000	0	0

Source: P. R. Grant and B. R. Grant. 1992. *Ecology* 73: 766–784.
[a]T = trace.
[b]Survivorship = l_x = the proportion of the 90 birds that survived from birth to a given age.
[c]First breeding = the proportion of the 90 birds that reproduced for the first time.
[d]Fecundity = m_x = the average number of daughters produced (fledged) per female of a given age.

What was the average age at which finches first bred?

© Tui De Roy/Minden Pictures

Media Clip 39.2 The Finches of the Galápagos

Life histories vary not only among species but also among populations within species—including humans. For example, the Aeta people of the Philippines, who hunt and gather wild food, live for only 16.5 years on average. Many of the children die young, but girls who survive childhood and reach puberty can expect to live for about a decade beyond that. The Turkana, herders of East Africa, have a life expectancy of 47.5 years, whereas life expectancy in the United States before COVID-19 was 78.5 years. We do not yet know the reasons for these differences in life expectancy among human populations. But the Aeta do experience periodic food shortages that leave them vulnerable to diseases, whereas cattle provide the Turkana with a reliable food supply that may increase their disease resistance. And the low mortality in the United States clearly results from access to abundant food and medical care. These patterns suggest why life histories are diverse: environments vary.

Resources and physical conditions shape life histories

Individual organisms must acquire materials and energy to maintain themselves and to fuel metabolism, growth, activity, defense, and reproduction. Materials and energy—and the time available to acquire them—constitute **resources** for organisms. Organisms also need tolerable physical conditions. The primary distinction between resources and conditions is that resources (such as mineral nutrients, photons of light, time, space) are consumed by organisms, whereas conditions are experienced, not consumed.

The rate at which an organism can acquire resources increases with resource availability in the environment, up to the point at which the organism's capacity to take in and process resources is saturated. A plant's net photosynthetic rate, for example, increases with sunlight intensity (**FIGURE 39.4A**), and an animal's rate of food intake increases as food becomes more abundant (**FIGURE 39.4B**)—but in both cases intake rates level off at high resource availability.

Organisms use the resources they obtain for various life functions such as maintaining their bodies, growing, defending themselves, reproducing, and obtaining more resources (**FIGURE 39.5**). In general, resources can't be used for two functions simultaneously—this is called the **principle of allocation**—so they must be divided up among life functions.

An organism's first priority usually is to maintain the integrity of its body. The investment required for maintenance depends on how stressful the environmental conditions are. In stressful environments, more resources must be allocated to maintenance, leaving fewer resources available for other life functions (compare Figure 39.5A and C). For example, the hotter their environment, the more time day-active lizards must spend in a cool retreat to keep their body temperature below a lethal level, and thus the less time they have to look for food (see Figure 39.6). Once an organism obtains more resources than it needs for maintenance, it can allocate the excess to other functions, such as growth, defense, or reproduction. In general, as the average individual in a population acquires more resources, average fecundity and survivorship—and *r*—increase (see Figure 39.5A and B).

As discussed in **Key Concept 29.3** and **Figure 29.3**, poikilotherms such as lizards often use behavior to regulate their body temperature.

Life history variation among species and populations often reflects the principle of allocation. An individual that invests heavily

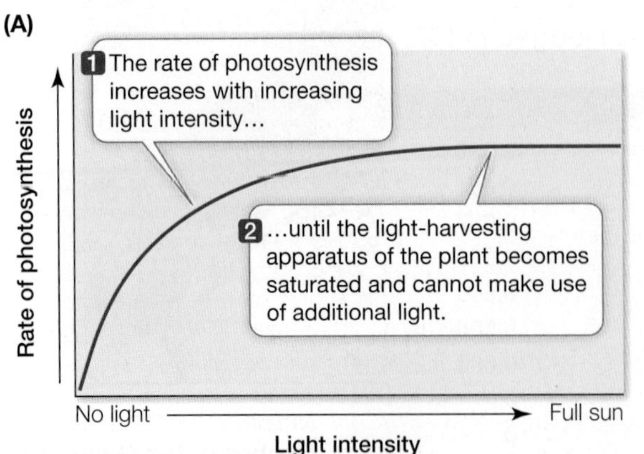

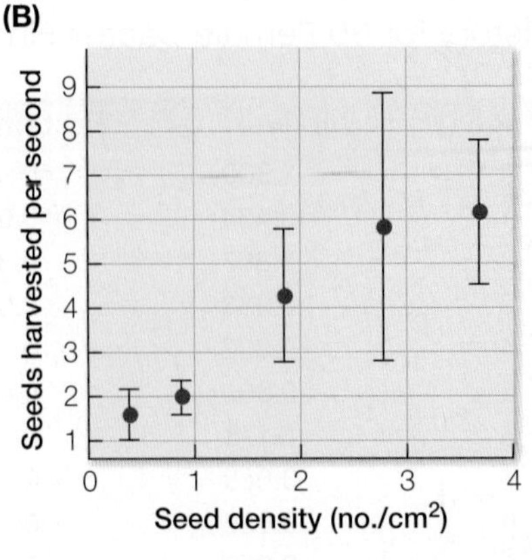

Dipodomys sp.

FIGURE 39.4 Resource Acquisition Increases with Resource Availability—Up to a Point **(A)** A plant's photosynthetic rate increases with light intensity. **(B)** Kangaroo rats (*Dipodomys* spp.) are able to harvest seeds faster when seeds are abundant. (Note the cheek pouches bulging with seeds.) Error bars indicate 95 percent confidence intervals around the mean. See Appendix B for discussion of statistical concepts. (B graph after M. V. Price and K. M. Heinz. 1984. *Oecologia* 61: 420–425.)

in growth early in life, for example, cannot simultaneously invest heavily in defense (e.g., protective structures or chemical defenses). As a consequence, it may reach adult size quickly, but at the cost of lower survivorship than an individual that invests more in defense. An individual that starts investing in reproduction early in life stops growing sooner and matures at a smaller adult size than one that waits to reproduce. Individuals that invest heavily in reproduction often do so at the expense of adult survivorship: they have high fecundity but short life spans. Such negative relationships among growth, reproduction, and survival are called **life history trade-offs**.

The environment shapes much of this life history variation because, together with a species' way of life, the environment determines the relative costs and benefits, in terms of survivorship and fecundity, of any particular allocation pattern. In most cases a single allocation pattern yields the highest per capita growth rate in a given environment. This means that individuals that deviate from the best allocation strategy contribute fewer offspring—and genes—to future generations. An individual's per capita growth rate is therefore a measure of its **fitness**, and allocation patterns can evolve by natural selection to the extent that they are inherited.

The Aeta people mentioned earlier, for example, live in an environment that imposes higher mortality than is the case for the Turkana. High mortality means that for the Aeta, the benefit of an early start to reproduction (more individuals survive to reproduce at least once) outweighs the fecundity benefit of delaying puberty and growing to a larger adult size (in general, larger human females produce more offspring). On average, Aeta women reach puberty and stop growing at 12 years of age and attain an adult height of 140 centimeters. For the Turkana, lower mortality shifts the balance—the advantage of early reproduction is outweighed by the fecundity advantage of attaining a larger adult size. On average, Turkana women mature at age 15 and at a height of

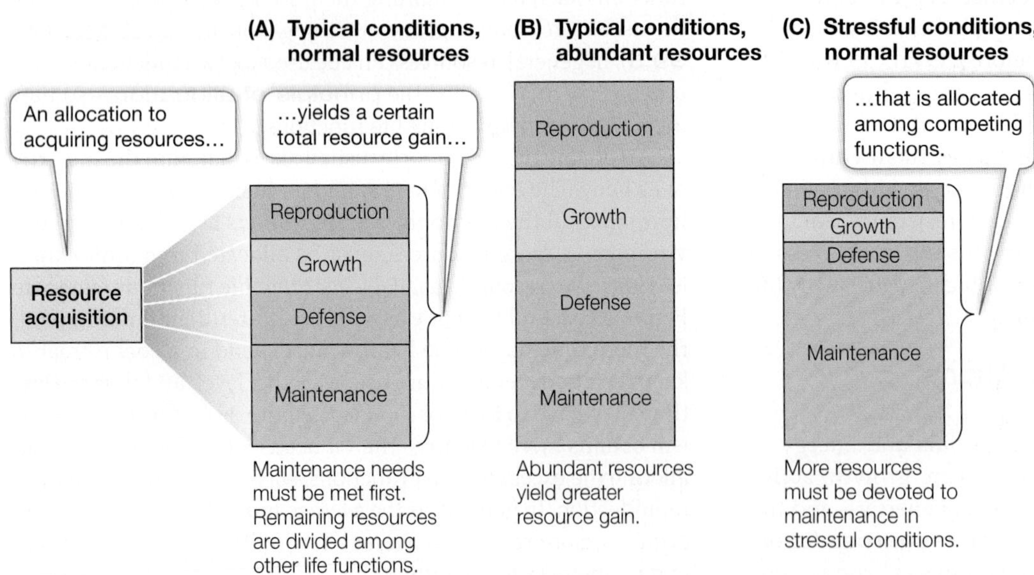

FIGURE 39.5 The Principle of Allocation Organisms use resources such as energy and time to acquire a certain amount of additional resources (indicated here by the height of each box). Acquired resources are then partitioned among competing life functions. In general, an organism's first priority is to maintain the integrity of its body. This figure assumes that the same amount of energy or time is allocated to resource acquisition in three environments that vary either in resource availability or physical conditions.

Redraw the column of stacked boxes for B to show what the allocation pattern would look like for this environment if the organism were to invest more in reproduction relative to other life functions than is shown in the figure.

166 centimeters. In the United States, well-nourished girls can grow fast *and* mature early, reaching puberty at age 12 on average, at a size equivalent to that of the Turkana.

In a similar way, the short summers in New York State make it advantageous for black-legged tick nymphs to invest in overwinter survival, even though it means that most individuals do not reproduce until their second year (see Figure 39.3), and a few take even longer. Ticks find their hosts by waiting: they climb up on vegetation, extend their forelegs, and climb aboard when a host bumps into them (see the image that opens this chapter). Finding a host is so rare that a tick can expect at most one meal during the short growing season. As a result, nymphs do not allocate resources to further host-seeking behavior during their first season of life, but instead allocate resources to overwinter survival. The fact that ticks take 2 years to complete their life history is thus a consequence of the difficulty of encountering a host and the seasonality of the environment.

Species' distributions reflect the effects of environment on per capita growth rates

A population cannot persist in an environment where its per capita growth rate is negative, because its size will inevitably shrink to zero—it will go extinct. We can predict where populations are likely to be found if we know how resource availability and physical conditions influence survivorship and fecundity. Obtaining complete knowledge of life histories is often impossible for species in the wild, but even incomplete knowledge helps us understand species' distributions. **INVESTIGATION FIGURE 39.6** illustrates this point for the lizard *Sceloporus serrifer*. By measuring conditions in the lizard's natural environment and studying its physiology and behavior, researchers were able to draw conclusions about how climate change is influencing the lizard's life history and distribution—with important implications for its conservation.

Sometimes it is possible to explore the links between environmental conditions, life histories, and species distributions with the help of laboratory experiments. Charles Birch used this approach to understand why two species of beetles that infest stored grain are serious pests in some parts of Australia but not in others. Birch grew populations of the rice weevil (*Sitophilus oryzae*) and the lesser grain borer (*Rhyzopertha dominica*) under various conditions of temperature and humidity. He quantified beetle life histories for each set of conditions and used them to calculate per capita growth rates for each species under those conditions. The results, summarized in **FIGURE 39.7**, explain the more tropical geographic range of *R. dominica*. In addition, both species had negative per capita growth rates in cool, dry environments. This finding suggested a strategy for managing stores of grain to minimize losses: keep the grain cool and dry.

INVESTIGATION

FIGURE 39.6 Climate Warming Stresses Spiny Lizards
Barry Sinervo and colleagues wanted to know whether climate warming will stress Mexican blue spiny lizards (*Sceloporus serrifer*) by reducing the number of hours they can remain outside their burrows without overheating. These lizards feed only during daytime and retreat to their cool burrows when their body temperature gets too high (see Figure 29.8). In 2008 the researchers conducted experiments in four locations in Mexico where these lizards had been found in 1975. Climate records indicated that the four locations did not warm at the same rate between 1975 and 2008, perhaps because they differ in such things as proximity to the ocean, elevation above sea level, and latitude.[a]

HYPOTHESIS

Spiny lizards can't forage for as many hours on hotter days.

METHOD

1. Construct model "lizards" that have the same thermal properties as real lizards.
2. At four sites on the Yucatán Peninsula where lizards lived in 1975, place replicate model lizards in various perches that real lizards use when they forage. Monitor body temperature T_b of the models each hour and record maximum daily air temperatures during the breeding season (March and April).

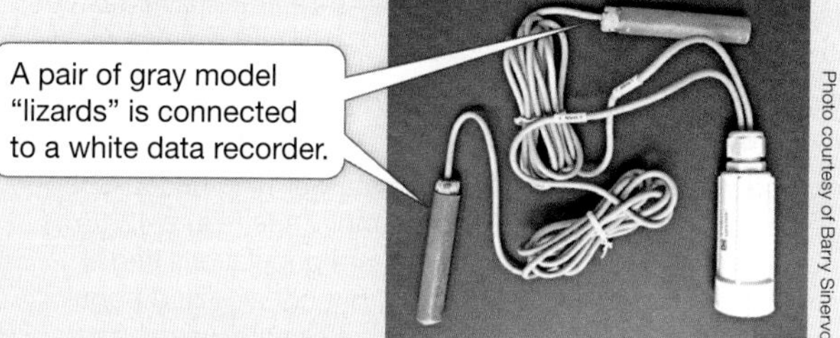

3. For each day, calculate the number of hours that the body temperature T_b of the model lizards exceeded 31°C—the temperature at which *S. serrifer* are known to stop foraging and retreat to their burrows—to arrive at a predicted number of inactive hours for real lizards.
4. Determine whether the predicted number of inactive hours increased with maximum daily air temperature.

RESULTS

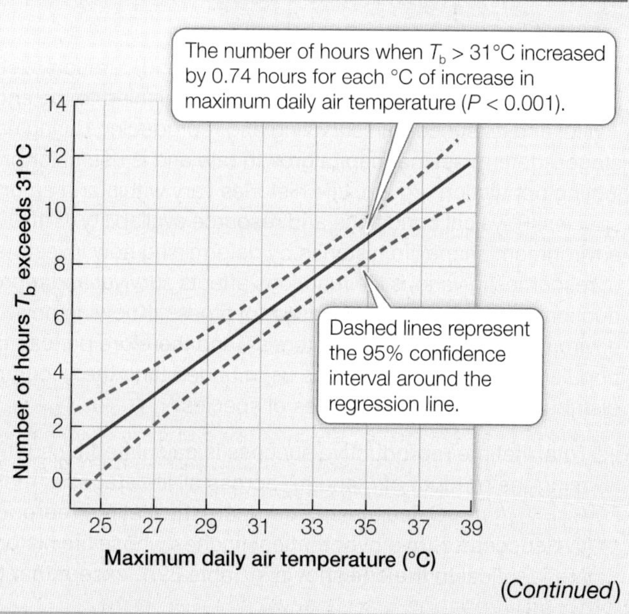

(Continued)

INVESTIGATION (continued)

CONCLUSION

The hypothesis is supported by the data: *S. serrifer* can forage without overheating for fewer hours on days when maximum air temperature is higher.

ANALYZE THE DATA

Between 1975 and 2008, *S. serrifer* populations went extinct at two of the four sites monitored by Sinervo and colleagues. The graph shows March and April averages in 2008 for the number of hours (in some cases including hours after sundown!) when $T_b > 31°C$ at the two "extinct" sites and at the two "persistent" sites. Error bars are 1 standard error of the mean (see Appendix B); sample size for each point is 240 observations (2 sites × 30 days × 4 lizard models). Use this graph to answer the questions below.

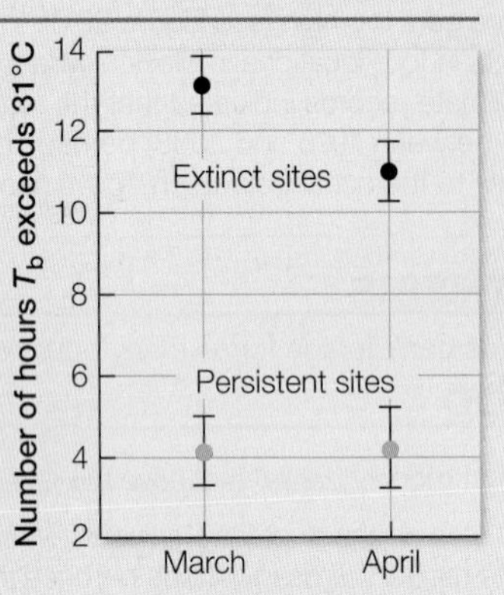

1. How did the hours available for foraging differ between the "extinct" sites (black symbols) and the "persistent" sites (green symbols)?
2. How might the availability of foraging time influence lizard fecundity, survivorship, and per capita growth rates?
3. Sinervo and colleagues knew that climate warming had taken place between 1975 and 2008 (for example, see Figure 42.13), the period during which some of the lizard populations had gone extinct. Is the extinction of some, but not all, lizard populations consistent with the conclusion that climate warming was one of the causes of extinction? Why or why not?

[a]B. Sinervo et al. 2010. *Science* 328: 894–899.

A similar **Analyze the Data** exercise may be assigned in **Achieve**.

Sitophilus oryzae (rice weevil)

Rhyzopertha dominica (lesser grain borer)

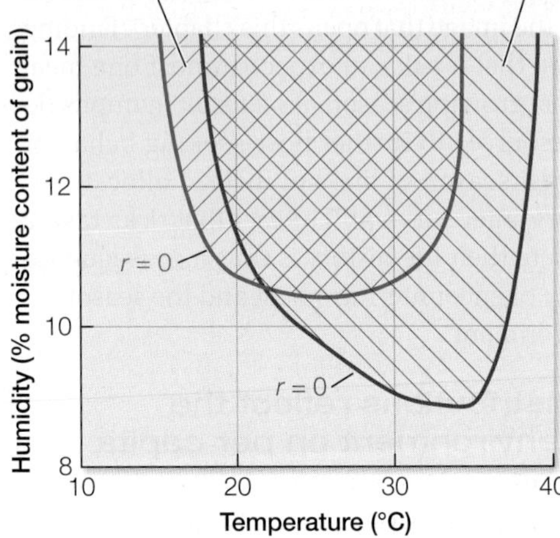

FIGURE 39.7 Environmental Conditions Influence Per Capita Growth Rates and Species Distributions Charles Birch's laboratory research revealed that *R. dominica* populations performed better under warmer conditions than did *S. oryzae* populations. The lines marked $r = 0$ indicate the limit of the conditions under which populations of each species can grow; within the hatched areas, per capita growth rates are positive for the species, and outside the hatched areas, r is negative for the species. (After L. C. Birch. 1953. *Ecology* 34: 698–711.)

Are the results in this figure consistent with the lesser grain borer having a more tropical geographic range than the rice weevil? (*Hint*: See Key Concept 38.2.)

REVIEW & APPLY | 39.3

R The life history of a species—the sequence and duration of stages an individual passes through from the beginning to the end of its life, and its success in surviving and reproducing during each stage—determines per capita growth rate and is useful for projecting population growth. Life histories vary within and among species. Physical conditions and resource availability in the environment shape life histories by determining how investment of resources in various life functions affects survival and reproduction, and therefore r, a measure of fitness. Knowing how environments influence life histories—and therefore per capita population growth rates—helps us to understand the geographic distribution and habitat affinities of species.

A

1. Total lifetime reproductive success is calculated by summing the product of l_x and m_x across all life stages x; that is, $R_0 = \Sigma_x l_x m_x$. Calculate and compare the lifetime reproductive success of two hypothetical finches whose life histories are identical to the one shown in Table 39.1, except that the first hypothetical finch produced one additional fledgling when it was 1 year old (substitute $m_1 = 1.051$ for 0.051), and the second produced one additional fledgling when it was 2 years old (substitute $m_2 = 1.667$ for 0.667). Which hypothetical finch has higher fitness? Explain why.
2. Female hummingbirds in the Rocky Mountains live long enough on average to reproduce during several summer seasons. If they still have young in the nest when their food supply (nectar produced by flowers) disappears at the end of a summer, they abandon the nest and the young. Explain this behavior in terms of the principle of allocation.
3. In an experiment, ecologists added predatory fish to some streams, leaving other streams predator-free as controls. How would you expect size and age of first reproduction of prey fish to evolve in the presence of predators? (*Hint*: Consider the example of the Aeta and Turkana.)

Knowing whether population growth rates are negative or positive in particular environments tells us why species may be absent or present in those environments. But this knowledge does not help us understand why population densities vary among different places where population growth rates are all positive. To do so, we must take a closer look at the dynamics of population growth.

39.4 Populations Grow Multiplicatively, but the Multiplier Can Change

LEARNING OBJECTIVES

By the end of this key concept you should be able to:

39.4.1 **Distinguish between multiplicative and additive growth processes.**

39.4.2 **Explain why populations do not grow indefinitely and why environmental conditions can cause carrying capacity to vary through time and space.**

39.4.3 **Explain why the human population has not stopped growing.**

We saw in Key Concept 39.2 that a population will grow as long as the per capita growth rate, r, is greater than zero. Recall also that r refers to a specific period of time (it might be a day, a week, or a year, depending on the species) over which we have estimated per capita birth and death rates. During this period a population with a positive value of r will increase by a number of individuals that is precisely r times its initial size. In other words, growth is *multiplicative*. We saw this in Equation 39.3, which states that $N_{t+1} = N_t + rN_t$. What may not be obvious at first is that this pattern continues in the next time period: the population will again increase by a multiple r of its size at the start of that period, since $N_{t+2} = N_{t+1} + rN_{t+1}$. Because N_{t+1} was already larger than the initial population N_t, this means that the population grows by an ever-larger number of individuals in each successive period.

Multiplicative growth with a constant and positive r differs dramatically from **additive growth**, in which a *number* (rather than a *multiple*) is added in each period (**FIGURE 39.8**). Many things in our everyday experience grow in a constant additive manner—for example, the distance traveled as we drive at a constant speed, the total volume of water that moves past a point on the bank of a steadily flowing stream, and the number of years added to our age through time. But not populations: with rare exceptions, populations grow multiplicatively.

Multiplicative growth with constant r can generate large numbers very quickly

We stress the distinction between multiplicative and additive growth because we tend to think in additive terms and often fail to appreciate how multiplicative growth can generate phenomenally large numbers in a short time. For example, in 1911 L. O. Howard, then chief entomologist (insect scientist) of the U.S. Department of Agriculture, estimated that a pair of flies reproducing for the first time on April 15 would give rise to a population of 5,598,720,000,000 adults by September 10 if all the offspring survived and reproduced. Other entomologists disagreed with Howard's calculation—they pegged the number much *higher*!

Charles Darwin was well aware of the power of multiplicative growth.

> Every organic being naturally increases at so high a rate, that, if not destroyed, the earth would soon be covered by the progeny of a single pair. ...Hence, as more individuals are produced than can possibly survive, there must in every case be a struggle for existence... . (C. Darwin. 1859.)

This ecological struggle for existence, fueled by population growth, underlies natural selection and adaptation.

LINK As discussed in **Key Concept 13.1**, Darwin's theory of evolution by natural selection follows from the observation that individuals of the same species differ in their success at surviving and reproducing—that is, in their per capita contributions to overall population growth rate, or fitness. As long as these differences arise because successful individuals possess traits that adapt them better to their environments, and as long as successful individuals pass their traits to their offspring, adaptive traits will increase in frequency within the population over time. This connection shows how ecology and evolution are intimately related.

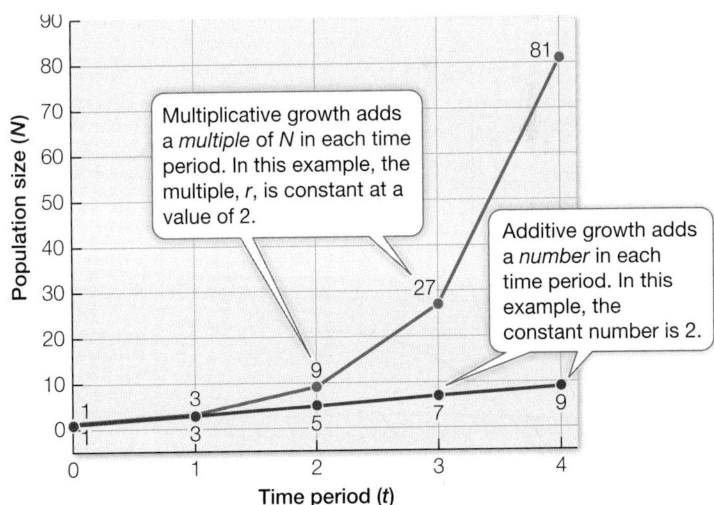

FIGURE 39.8 Comparing Multiplicative and Additive Growth Processes With multiplicative growth (blue line), in each time period a population adds a *multiple*, r, of its size (N) at the beginning of the period. As the population grows, more individuals are added in each successive time period. It is important to see how different this is from additive growth (red line), in which a population adds a certain *number* in each period. In this example the multiple remains constant at a value of 2, so a number equal to two times the population size at the beginning of each period is added by the end of that period; in the additive growth example, the number added is constant at 2 individuals.

Populations that grow multiplicatively with constant *r* have a constant doubling time

Although some things in everyday experience change additively, many other things change multiplicatively—for example, the money in interest-bearing bank accounts, the human consumption of fossil fuels, and the decay of radioactive elements (see Figure 17.1).

As long as *r* does not change, multiplicative growth has a very striking property: a constant **doubling time**. A population growing multiplicatively with a positive and constant *r* adds an ever-increasing number of individuals as time goes on. At some specific time, the number added exactly equals the initial population, and the population has doubled. The time the population takes to double will also remain constant. The doubling time depends on *r*—it gets shorter as *r* increases.

Activity 39.2 Multiplicative Population Growth Simulation

It is easy to see the implications of multiplicative growth with a constant, positive *r* if we think in terms of doubling times. For example, by doing so a municipality would realize that even a seemingly small but constant per capita growth rate of 3 percent per year would mean that its human population would double every 23 years—and that its sewage treatment capacity and other municipal services would need to do the same.

Density dependence prevents populations from growing indefinitely

If growing populations grow multiplicatively, why isn't Earth covered with flies, or any other organism? It turns out that growth in real populations deviates from the J-shaped curve characteristic of multiplicative growth with constant *r*. Initially the growth curves are similar. But as population size increases, it becomes apparent that growth of real populations slows and eventually stops when the population reaches a particular size (*K* in **FIGURE 39.9**).

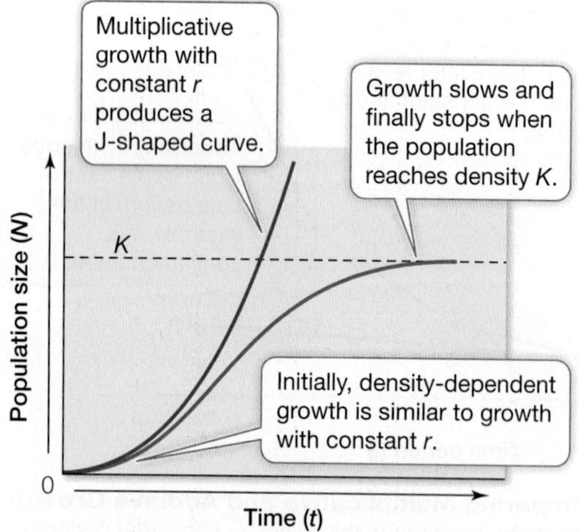

FIGURE 39.9 Density-Dependent Population Growth Initially, density-dependent growth is similar to multiplicative growth with constant *r*, but growth then slows and eventually stops when the population reaches density *K*, the population's "carrying capacity."

Activity 39.3 Population Growth

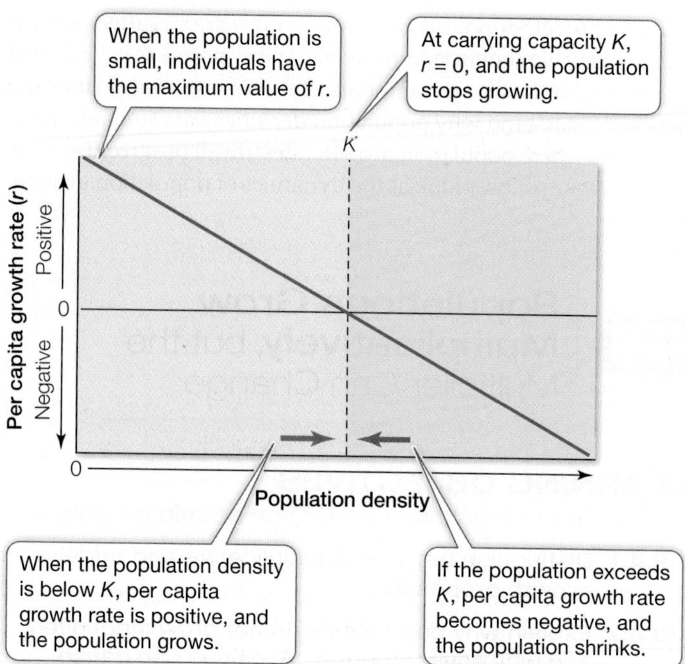

FIGURE 39.10 Per Capita Growth Rate Decreases with Population Density A schematic showing the simplest form of density-dependent growth, in which *r* declines linearly with population density.

Why do populations stop growing? The answer comes from observing what happens to *r* as the population grows. Instead of staying constant, *r* is **density-dependent**—it changes as the population grows and becomes denser (we assume that the area or volume that the population occupies stays constant). As a result, population growth rate (the slope of the growth trajectory, which equals rN_t) decreases steadily as the population grows relative to multiplicative growth with constant *r* (Figure 39.9).

We can model density-dependent population growth by replacing the constant *r* in Equation 39.4 with an expression that is a function of *N*. The simplest function is a linear decrease in *r* as *N* increases, which gives us the "logistic" model of density dependence (**FIGURE 39.10**). The logistic captures several essential features of growth in real populations, although linear density dependence is rare. Figure 39.10 shows schematically that when *r* decreases with density, the per capita growth rate is highest at very low population density. As the population grows, *r* decreases and population growth slows. Eventually, at a certain density called the **carrying capacity**, or ***K***, *r* is zero and the population stops growing—it reaches a dynamic equilibrium in which birth and death are balanced. If the population exceeds *K*, growth is negative and the population shrinks back to *K*. The blue line in Figure 39.9 illustrates the time trajectory of logistic growth relative to that of density-independent growth. Note that at low density the blue and red lines are similar, but the trajectories diverge as time goes on and the populations grow. Eventually, the density-dependent trajectory reaches a plateau when $N = K$. More realistic models that assume nonlinear density dependence share these essential properties.

Why does r decrease with density? Unless the production of resources keeps pace with increased consumption as the population grows, each individual will experience lower resource availability and therefore a lower rate of resource acquisition (see Figure 39.4). And if maintenance requirements don't change with density, individuals will have fewer resources to allocate to survival and reproduction (see Figure 39.5), and r will decrease.

Activity 39.4 Density-Dependent Population Growth Simulation

Key Concept 40.2 describes how density dependence can result from intraspecific competition—competition among members of the same species for the same limiting resources.

Now we are prepared to return to the question posed at the end of Key Concept 39.1: Why do species population sizes vary in time and space?

Changing environmental conditions cause the carrying capacity to change

We have learned that resource availability in the environment influences an individual's success in acquiring resources (see Figure 39.4), and that physical conditions influence the costs of maintenance (see Figure 39.5). If these factors vary over space, then the carrying capacity will vary over space as well.

Similarly, if resource availability or environmental conditions are good in one year but bad in another, the population will find itself alternately above and below the current carrying capacity. As a result, populations will fluctuate around an average K over time. Figure 39.2 illustrates this dynamic for rodents and ticks near Millbrook, New York. Rodent densities tend to increase following years of high acorn production because the rodents have enough food to produce many offspring, and they tend to decrease following poor acorn years. Similarly, nymphal tick densities go up and down following high and low rodent years, respectively, because host abundance determines the success of larval ticks in obtaining a blood meal. These increases and decreases are not immediate—there is a delay while acorns are converted into newborn rodents, and while newly fed larval ticks molt into nymphs.

Technology has increased Earth's carrying capacity for humans

The human population is unique among populations of large animals. Not only has it continued to grow for a long time, but it has grown at an ever-faster rate over the last 500 years or so, as is indicated by steadily decreasing doubling times (**FIGURE 39.11**). In the last half century, however, the population multiplier, or per capita growth rate (r), has decreased, to less than half the peak of 2.2 percent per year that was reached in 1963. The explanation for this decrease appears to be a *demographic transition*—as wealth and education rise worldwide, birth rates have decreased faster than death rates. Even if r continues to decrease,

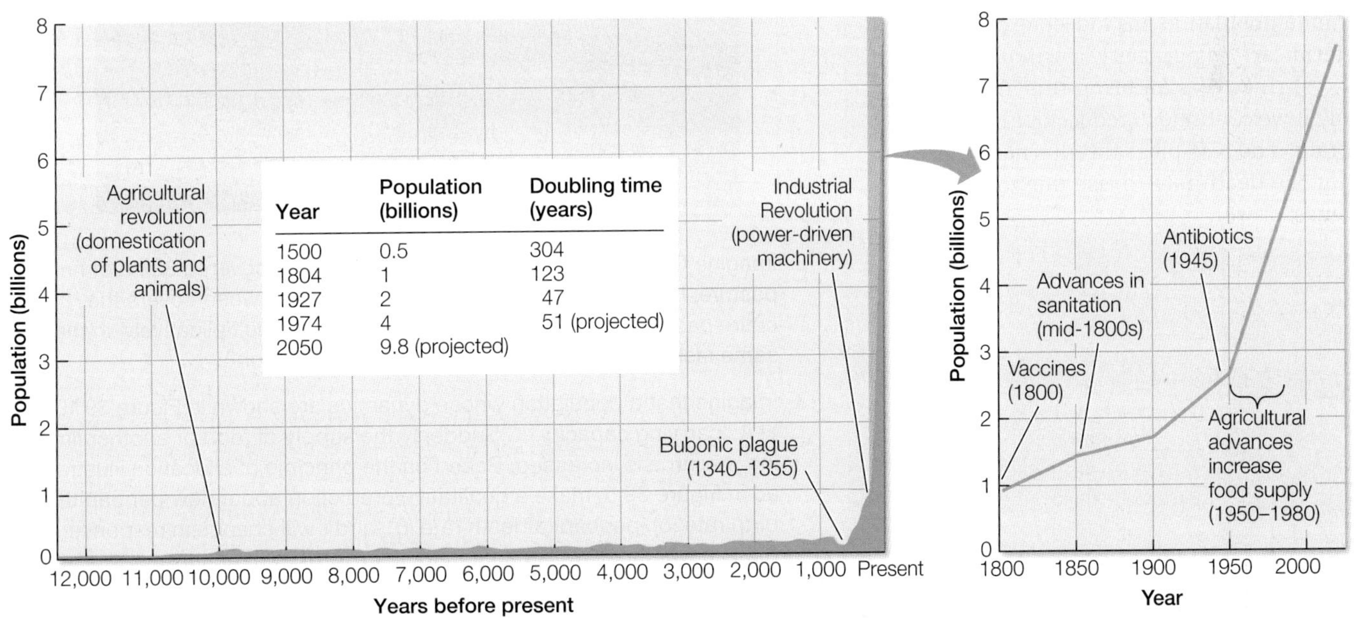

Year	Population (billions)	Doubling time (years)
1500	0.5	304
1804	1	123
1927	2	47
1974	4	51 (projected)
2050	9.8 (projected)	

FIGURE 39.11 Human Population Growth Advances in technology have allowed the human population to grow at an ever-faster rate. (Data from various sources, including Wikimedia using US Census Bureau data [www.census.gov/ipc/www/worldhis.html] and "World Population over 12000 years," including Medium Projection – UN Population [2015 revision], published online at OurWorldInData.org.)

however, the human population is projected by the United Nations to reach almost 10 billion by about 2050.

The human population growth rate has jumped with every technological advance that has raised carrying capacity by increasing food production (thereby increasing per capita birth rate, *b*) or by improving health (thereby reducing per capita death rate, *d*). The first jump occurred with the rise of agriculture (about 10,000 years ago). In the nineteenth and twentieth centuries, advances in sanitation and the development of vaccines and antibiotics decreased death rates to an unprecedented low level (see Figure 39.11). Finally, the "green revolution" in agriculture more than doubled world grain production between 1950 and 1980. These increases in *r* are reflected in shorter and shorter doubling times (see the table in Figure 39.11).

Many ecologists believe that the human population has overshot its long-term carrying capacity. One concern they raise is human dependence on fossil fuels, which are not renewable. For example, agricultural productivity depends on fertilizers derived from natural gas, pesticides derived from oil, and petroleum-fueled engines that pump irrigation water, till the soil, and process and transport food. Another concern is the climate change and degradation of Earth's ecosystems that have been a consequence of the Industrial Revolution and the recent growth of the human population. We will consider these topics further in Chapter 42. If the human population has indeed exceeded its long-term carrying capacity, it will ultimately have to shrink. We can bring this about voluntarily if we continue to reduce our per capita birth rate. The less pleasant alternative is to wait for the death rate to rise due to famine, pestilence, or war.

REVIEW & APPLY | 39.4

R Biological populations grow by adding a multiple of their current size in each time period. Such multiplicative growth would lead to infinite population sizes very quickly if essential resources were unlimited. Instead, individual survival and reproduction become more limited by resource availability as population density increases. Populations reach an equilibrium size, known as the carrying capacity, or *K*, at which births are balanced by deaths and *r* is zero. The carrying capacity changes as resource availabilities vary over time and space. Human ingenuity has increased the carrying capacity for our species, but the human population cannot continue to grow indefinitely.

A

1. Yellow star-thistle (*Centaurea solstitialis*) is an invasive plant unpalatable to livestock, including bison. When rancher Jane discovers that this spiny weed has invaded her 128-ha pasture, she establishes replicate plots and measures how much ground is covered by star-thistle plants in each plot. She estimates that, on average, star-thistles cover a total area of 1 ha in her pasture. When she repeats her measurements a year later, the weed population has grown, and now covers an area of 2 ha. If star-thistle populations grow additively, how many hectares do you predict that the plant will cover in years 3, 4, and 5? What do you predict if you instead assume that star-thistle populations grow multiplicatively with a constant *r*?

Courtesy of Steve Dewey, Utah State University, Bugwood.org; inset Courtesy of Peggy Greb, USDA Agricultural Research Service, Bugwood.org

2. Imagine that Jane discovers that star-thistle has covered 32 ha of her pasture; how many years does she have until the weed covers the entire pasture if it is growing (a) additively or (b) multiplicatively, in the same fashion as described in the previous question?
3. Imagine that a population whose dynamics are shown in Figure 39.10 is at its carrying capacity, *K*. Suddenly the supply of food or another limiting resource is increased. Based on the principle of allocation illustrated in Figure 39.5, make a qualitative prediction about how per capita birth rate (*b*), per capita death rate (*d*), and *r* will change in response. Explain whether *K* will change and, if so, in what direction.
4. If a population has a per capita growth rate (*r*) that is negative, will the population have a constant halving time or "half-life" (time over which its size becomes half as large as in the previous time period), and why?

So far we have considered the effects of two universal demographic processes—birth and death—on populations. But the dynamics of many populations are also influenced by the movement of individuals among populations. Such "open" populations grow when immigrants enter from the outside world and shrink when emigrants leave. Immigration and emigration influence populations in important ways.

39.5 Immigration and Emigration Can Influence Population Dynamics

LEARNING OBJECTIVES

By the end of this key concept you should be able to:

39.5.1 Outline how the BD model for population dynamics is modified to incorporate immigration and emigration.

39.5.2 Explain how population size influences risk of extinction for an individual population.

39.5.3 Outline how population size, number of populations, and dispersal among populations influence extinction risk for a metapopulation.

Populations of many species have a patchy distribution within the geographic range, occurring in suitable habitat patches that are separated by unsuitable environments where the species cannot persist (**FIGURE 39.12**; see also Figure 39.1). Each habitat patch that is occupied by the species represents a separate population, and the set of populations in a region is known as a **metapopulation** (meaning a "population of populations"). The BD model (see Equation 39.1) describes growth of separate populations if they are so isolated from one another that individuals never move between them. But if between-population movement, or *dispersal*, occurs, then any given population can change not only through births (B) and deaths (D) but also with the added input of individuals that move *into* the population (immigrants) and loss of individuals that move *away* (emigrants). Adding the number of immigrants (I) and emigrants (E) to the BD growth model, we get the **BIDE model**:

the number of individuals in a population at some future time = the number now + the number that are born + the number that immigrate − the number that die − the number that emigrate

In the BD model, populations are considered **closed systems** because there is negligible immigration or emigration—few if any individuals move into and out of them. In the BIDE model, populations are considered **open systems** because individuals can move in and out. This difference has a profound effect on population dynamics. In the BD model, a population that goes extinct stays extinct: once the population reaches zero, there are no individuals left to reproduce. With the BIDE model, however, immigration can bring an extinct population "back to life."

You may think that extinction of a population is a rare event, but it is fairly common in small habitat patches. When populations are very small, there is a chance that no female will give birth and that all individuals will die, even during a time period when environmental conditions are favorable—in which case the population will be extinct. Small populations are also more likely than large ones to go extinct following environmental disturbances, such as a hurricane.

Once a population goes extinct, it will remain at zero density—and its patch of habitat will remain unoccupied—unless individuals move into, and thus recolonize, the patch. Thus if there is no dispersal among patches, the populations of a metapopulation will "wink off" one by one. The time to extinction of the entire metapopulation system is longer the larger each population is, and the more populations there are, but extinction of the entire metapopulation will occur eventually. If populations are connected by dispersal, however, the metapopulation as a whole will persist much longer, for two reasons. First, immigration can repopulate unoccupied patches, thus "rescuing" extinct populations. The repopulated patches can, in turn, act as sources of colonists for other unoccupied patches, causing them to "wink on." Second, immigrants contribute to the genetic diversity within populations by bringing new alleles with them, which counteracts the erosion of genetic diversity through genetic drift that can reduce r and compromise a species' evolutionary potential.

FIGURE 39.12 A Metapopulation Consists of Many Populations The coast of Scandinavia is characterized by natural freshwater rock pools in which species of *Daphnia* thrive. Each pool supports a population within a metapopulation.

Activity 39.5 Metapopulation Simulation

LINK As discussed in **Key Concept 13.2**, the movement of individuals between populations also represents movement of genes. This gene flow can change allele frequencies both in the population that individuals leave and the one into which they migrate.

We can see the recolonization process at work in the metapopulation of Edith's checkerspot butterfly that occupies the San Francisco Bay area (see Figure 39.1). Many populations of this butterfly in the Bay Area went extinct during a severe drought that gripped California between 1975 and 1977. The only population that did not go extinct was the largest population (Morgan Hill). The extinct populations remained at zero density until 1986, when nine suitable habitat patches were recolonized from the Morgan Hill population. Patches closest to the Morgan Hill source were most likely to be recolonized because adult checkerspot butterflies do not fly very far.

REVIEW & APPLY | 39.5

R Populations of many species have a patchy distribution—individuals live in patches of suitable habitat that are separated by unsuitable habitat. If individuals move among habitat patches, then immigration and emigration, as well as birth and death, will affect the dynamics not only of the population that inhabits each patch, but also of the entire assemblage of populations—the metapopulation. Population size affects the risk of extinction. Dispersal of individuals among populations greatly reduces the risk of extinction for a metapopulation.

A Refer to Figure 39.12. Swedish ecologist Jan Bengtsson established artificial rock pools containing either 12, 50, or 300 liters (L) of water, introduced the same number of *Daphnia* into each one, and covered the pools with mesh to prevent dispersal between pools. Over the next 4 years he monitored population sizes. The table shows average sizes of populations that went extinct (measured *before* they disappeared) and of those that persisted.

Pool size (L)	Number of replicate populations	Percentage of populations going extinct	Average size of extinct populations	Average size of persisting populations
12	27	18.5	65	1,500
50	25	16	530	2,705
300	11	9.1	1,800	9,690

Source: J. Bengtsson. 1989. *Nature* 340: 713–715.

1. Graph the relationship between pool size and the percentage of populations that went extinct, using pool size as the *x* axis. What does your graph show you?
2. Within each pool size, which populations were at greatest risk of extinction?
3. How are your answers to Questions 1 and 2 related to each other?
4. What might have happened if Bengtsson had not placed mesh over his experimental pools?

Chapter 38 described ways in which an understanding of ecology influences applied fields of science, such as medicine and agriculture. In the next key concept we will explore how our understanding of population dynamics can be used to influence whether populations and metapopulations persist or decline.

39.6 Ecology Provides Tools for Conserving and Managing Populations

LEARNING OBJECTIVES

By the end of this key concept you should be able to:

39.6.1 Provide one example of how knowledge of life histories helps design a strategy for managing a natural population.

39.6.2 Explain how an understanding of the effects of dispersal on metapopulation dynamics is useful for conservation.

For millennia, humans have tried to maintain or increase populations of desirable or useful species and reduce populations of species they consider undesirable. Such efforts are most successful if they are based on knowledge of how those populations grow and what determines their densities.

Knowledge of life histories helps us manage populations

Knowing the life history of a species helps us identify those life stages that are most important for the species' reproduction and survival, and hence for its population growth rate. Here are a few examples of how this knowledge can be applied.

CONSERVING ENDANGERED SPECIES Populations of Edith's checkerspot butterfly in the San Francisco Bay area are threatened with extinction. Larval survival is critical to the population dynamics of these butterflies (see Figure 39.1). Temporal and spatial variation in butterfly density is tied to the availability of two larval food plants, California plantain (*Plantago erecta*) and purple owl's clover (*Castilleja densiflora*), which are found only on serpentine soils. Maintaining healthy populations of these plants is critical for conservation of the butterflies, but the low-growing *Plantago* and *Castilleja* are suppressed by non-native grasses. Grazing by cattle can control the invasive grasses, and grazing is an important strategy for conserving *Euphydryas editha*. This example shows that achieving conservation goals is often compatible with human use of the environment.

MANAGING FISHERIES The black rockfish (*Sebastes melanops*) is an important game fish that lives off the Pacific coast of North America. Rockfish grow continually throughout their lives. As is true of many animals, larger females produce more eggs than smaller females. Larger females also provision their eggs with larger oil droplets. Larger droplets provide more energy to the newly hatched larvae, allowing them to grow faster and survive better than larvae with smaller oil droplets.

THINK LIKE A SCIENTIST

Quantitative Reasoning

Process of Science

Dispersal corridors can "rescue" fragmented populations from extinction

Andrew Gonzalez and Enrique Chaneton devised an experiment to test the effect of dispersal corridors on persistence of species. They selected seven replicate moss-covered boulders that are habitat for tiny arthropods. On each boulder they scraped off moss to create a cluster of 13 moss patches with different properties (**FIGURE 39.13A**). In the center they left a large 50- × 50-cm moss "mainland" (M); this was surrounded by 12 small "island" patches of moss, each 10 cm^2 in area. Three treatments were applied to the 12 patches and compared. In the "isolated" treatment (I), the small patches were surrounded by bare rock (which is inhospitable to the small arthropods and is thus a barrier to dispersal). In the "corridor" treatment (C), the small patches were connected to the mainland by a 7- × 2-cm strip of live moss. In the "broken-corridor" treatment (B), the configuration was the same as in the "corridor" treatment, except that the moss corridor was interrupted in the middle by a 2-cm strip of bare rock. After 6 months, the scientists determined the number of small arthropod species present in each island patch. The results are shown in **FIGURE 39.13B**.

Did the presence of a corridor help "rescue" species from extinction? Here you will use quantitative reasoning and the process of science to answer new questions.

1. The experimental patches of moss varied somewhat in the area of continuous habitat they contained, because there was moss habitat in the corridors as well as in the 10-cm^2 area of the island patches. Calculate the habitat area of the four patch types (M, I, C, and B treatments) and order them by the number of species you would expect them to retain if population size (which is proportional to habitat area) alone influenced extinction risk.
2. Figure 39.13B shows the average number of species retained by the four types of patches across the seven replicate boulders, with error bars that indicate the variation among replicates. Discuss what the results indicate about the relative effect of patch area versus dispersal on extinction risk in the moss metapopulation system.
3. Design and describe a follow-up experiment to test the hypothesis that the width of a dispersal corridor influences population persistence. Present hypothetical results of your experiment in the same graphical form as the figure, under the assumption that the hypothesis is upheld.

(A)

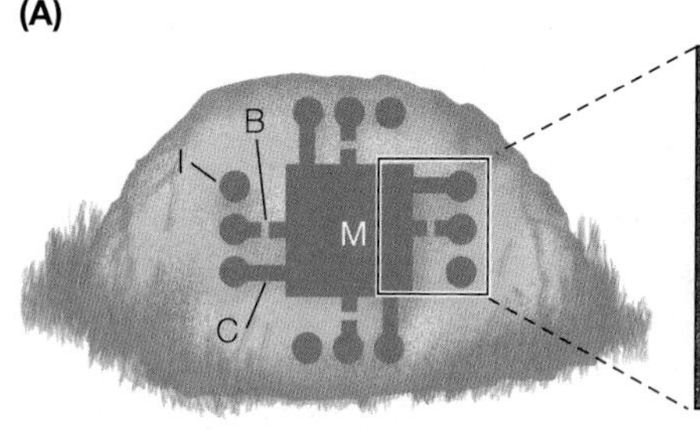

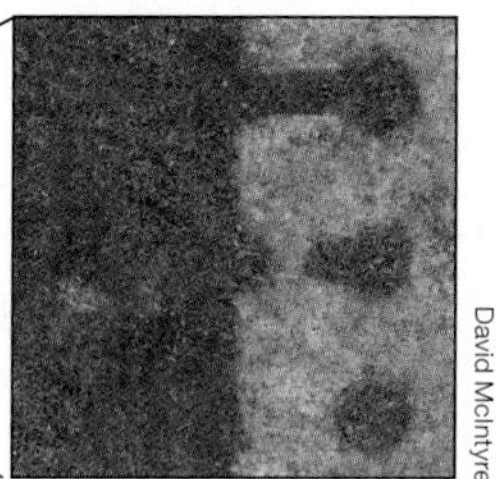

(B)

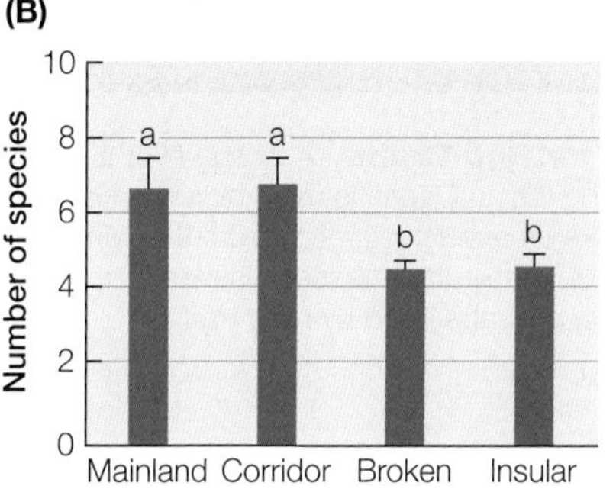

FIGURE 39.13 An Experimental System to Study Dispersal Corridors (A) Moss-covered boulders and the tiny arthropods that inhabit the moss plants served as an excellent system for manipulating the dispersal connections between a habitat "mainland" and habitat "islands." **(B)** Error bars indicate 1 standard error of the mean; bars with different letters are statistically different from one another. (After A. Gonzalez and E. J. Chaneton. 2002. *J Anim Ecol* 71: 594–602.)

Animation 39.1 Dispersal Corridors

These life history characteristics have important implications for rockfish populations. Because fishermen prefer to catch big fish, intensive fishing off the Oregon coast between 1996 and 1999 reduced the average age of female rockfish from 9.5 to 6.5 years. Thus in 1999, females were, on average, smaller than in 1996. This change decreased the average number of eggs produced by females in the population and reduced the average growth rate of larvae by about 50 percent, causing a rapid decline in population size. Because a relatively small number of large females can produce enough eggs to maintain the population, one strategy for maintaining rockfish without shutting fishing down completely is to set aside a few no-fishing areas where some females are protected and can grow to large sizes.

REDUCING DISEASE RISK Because adult black-legged ticks feed and mate primarily on large mammals—white-tailed deer, for example—it seems logical that controlling deer densities would reduce the abundance of reproducing females and thus of new tick nymphs. But experimental reductions in deer density have had little effect on nymph abundance. Studies of the tick's life history (see Figure 39.3) indicate that the success of larval ticks in obtaining a blood meal, not the number of them that hatch from

Photo by Yvonne McKenzie © Wondrous World Images 2018

FIGURE 39.14 A Corridor for Red Crabs A bridge over a highway on Christmas Island in the Indian Ocean is used by red crabs (*Gecarcoidea natalis*) as they make an annual migration from the rainforest to the beaches. The bridge (like many underpasses elsewhere) was constructed to reduce the number of crabs killed by road traffic.

eggs, has the greatest effect on the subsequent abundance of surviving nymphs. As is true of rockfish, a few adult female ticks can produce enough eggs to maintain populations. Hence controlling the abundance of larval hosts (small mammals) is a more effective strategy for reducing tick populations and reducing the risk of Lyme disease than is controlling the abundance of adult hosts (deer and other large mammals).

Knowledge of metapopulation dynamics helps us conserve species and control epidemics

Conservation biology strives to avoid the extinction of species. From studies of extinction rates, we know that the risk of extinction for a species with a metapopulation structure is influenced by the number and average size of its populations, and by rates of dispersal among them. Conservation planners therefore begin with an inventory of remaining areas of natural suitable habitat and an evaluation of the risks to populations in those areas. They then devise ways to protect as many habitat patches as possible, giving priority to those with the largest area because large patches potentially support the largest and most genetically diverse populations. Planners also evaluate the quality of the patches, as measured by their carrying capacity for the species (often estimated from population densities), and develop ways to maintain their quality.

Finally, planners must consider opportunities for individuals to move among populations. For some species (such as Edith's checkerspot butterfly, which can fly through unsuitable habitat to reach a new serpentine outcrop), simple proximity of patches determines rates of dispersal and recolonization. For those species, patches that are distant from one another can be connected by a series of intervening patches that serve as "stepping stones." For other species, however, a continuous area of habitat through which individuals can move, called a **dispersal corridor**, is needed to connect populations. Dispersal corridors can sometimes be created by maintaining vegetation along roadsides, fence lines, or streams, or by creating bridges or underpasses that allow individuals to circumvent roads or other barriers to movement (**FIGURE 39.14**). Public health programs use metapopulation models to devise strategies to control spread of infectious diseases within a host population.

Activity 39.6 COVID-19 Simulation

LINK Strategies for conserving communities of interacting species are discussed in **Key Concept 41.4**.

REVIEW & APPLY | 39.6

Population ecology provides useful tools that we can use to limit populations of undesirable species and to conserve or augment populations of desirable species. Detailed knowledge of the life history of a species may indicate life stages where we can best intervene to influence population growth. The recognition that most species have a metapopulation structure shows us that the fate of individual populations, and of metapopulations, depends on how large and interconnected populations are.

1. Propose at least two principles you would use to design a plan to conserve a patchily distributed endangered species.
2. Provide an example of how knowledge about the life history of a species suggests a strategy for managing its abundance.
3. Effective conservation plans for marine species such as rockfish often lack a corridor component. Why?

This chapter has focused primarily on single-species populations, but it has also touched on various interactions among species—rodents interacting with acorns and ticks, ticks with hosts and bacteria, butterflies with food plants and invasive plants. These ubiquitous interactions among species, which play important roles in their population dynamics, will be the subject of the next two chapters.

Answer: How frequently people come into contact with a dangerous pathogen depends on its abundance and distribution and on those of any disease vectors (carriers) that transmit it to humans. Population ecology helps us understand what determines the abundance and distribution of pathogens and their vectors, and therefore allows us to devise ways of controlling their abundance or avoiding contact with them. By studying the survivorship and fecundity of the black-legged tick throughout its life history (see Key Concept 39.3), scientists learned that the abundance of hosts for larvae and nymphs determines tick abundance, rather than weather or the abundance of hosts for adult ticks. Furthermore, the researchers discovered that ticks become infected with *Borrelia* when they feed on infected hosts, and that small mammals are the most effective reservoirs for *Borrelia*. Long-term studies of rodent population dynamics had already shown that rodent abundance is positively correlated with acorn availability (see Key Concept 39.1). This knowledge suggests a way to reduce the probability of contracting Lyme disease (see Key Concept 39.6). Acorn production can be used to predict areas that are likely to become infested with disease-carrying ticks, and various measures can be taken to reduce the risk of human contact with those ticks. These measures include boosting public education for several years after a big acorn crop, posting warning signs in high-risk areas, and alerting health-care specialists whose patients live or work in high-risk areas. A longer-term possibility that has yet to be assessed is to reduce the fraction of ticks that carry *Borrelia* by augmenting populations of hosts that do not harbor *Borrelia* (such as lizards and birds) in the host community. Ticks that bite these hosts will not become infected with *Borrelia*.

VISUAL SUMMARY 39

39.1 Populations Are Patchy in Space and Dynamic over Time

- Ecological **POPULATIONS** are groups of individuals of the same species. Important population properties are **POPULATION SIZE**—the total number of individuals in a population—and **POPULATION DENSITY**—the number of individuals per unit of area or volume.
- The region in which a species lives is its **GEOGRAPHIC RANGE**. Within that range, the species may be restricted to particular **HABITATS**, which may occur in patches separated by areas of unsuitable habitat (**FIGURE 39.1**).
- Population sizes vary over time and over space.

FIGURE 39.1

Geographic range of *Euphydryas editha*

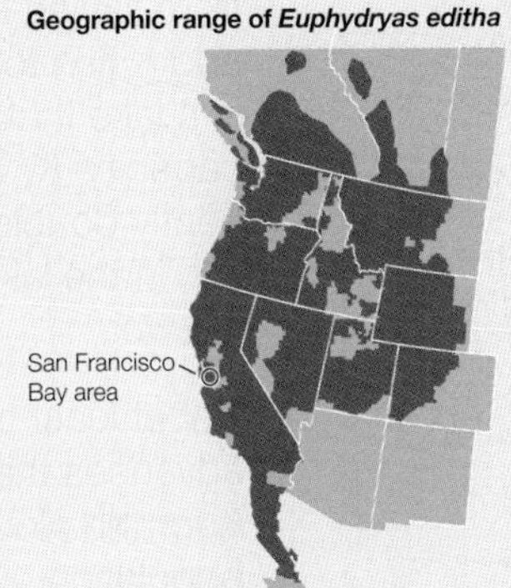

Serpentine outcrops (potential butterfly habitat) in Santa Clara Valley

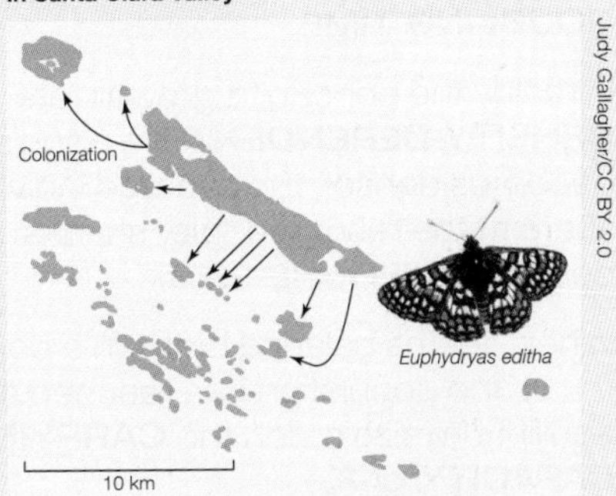

QUESTIONS

1. Define population size and population density, and explain their units of measure.
2. Barriers to dispersal (see Key Concepts 39.5 and 38.4) and tolerance of the environment both influence where a species is found. Discuss the relative importance of these two factors in determining the geographic range and the habitat distribution, respectively, of a species.

39 VISUAL SUMMARY

39.2 Births Increase and Deaths Decrease Population Size

- The **BD MODEL** is useful for understanding how births and deaths affect population dynamics.
- Knowing the average **PER CAPITA BIRTH RATE** (***b***) and **PER CAPITA DEATH RATE** (***d***) for a representative sample of individuals in the population allows ecologists to estimate the **PER CAPITA GROWTH RATE** (*r*) (**EQUATION 39.3**).
- The size of a population changes through time unless the per capita growth rate is zero. If *r* is positive ($b > d$), the population grows; if *r* is negative ($b < d$), the population shrinks.

EQUATION 39.3 $N_{t+1} = N_t + rN_t$

QUESTIONS

1. What is the relationship between per capita birth rate and the total number of births in a population during a time period?
2. What do we need in addition to per capita birth and death rates to predict population size at some future time?
3. Describe two basic ways in which rancher Jane (see Review & Apply 39.2) could expand her bison herd more quickly.

39.3 Life Histories Determine Population Growth Rates

FIGURE 39.3

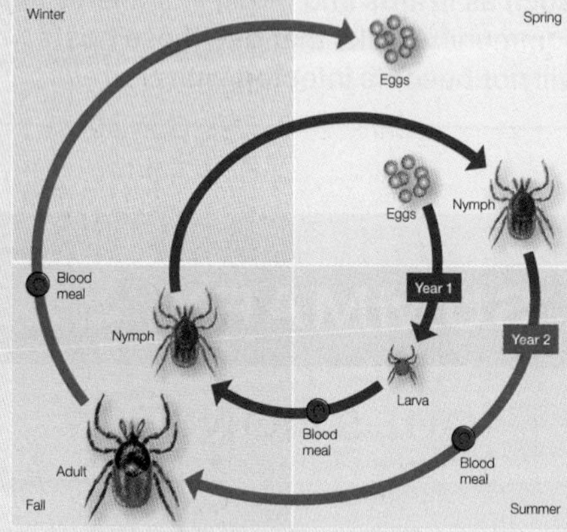

Go to ACTIVITY 39.1

- A species' **LIFE HISTORY** describes the sequence and timing of key events during an average individual's life, such as growth and development, reproduction, and death (**FIGURE 39.3**).
- **SURVIVORSHIP** (the proportion of individuals that survive to particular life stages) and **FECUNDITY** (the average number of daughters that each surviving female produces at those stages) determine the per capita growth rate, *r*.
- The **PRINCIPLE OF ALLOCATION** states that resources devoted to one life function cannot simultaneously be used for another function. This leads to **LIFE HISTORY TRADE-OFFS** among competing life functions.
- Species can persist only in environments where they have positive per capita growth rate.

QUESTIONS

1. What information needs to be added to the tick life history diagram in Figure 39.3 to make it a quantitative life history?
2. What are four basic life functions to which an organism can allocate resources?
3. In the United States, girls now reach puberty earlier than they did 100 years ago. Describe how this affects life history and explain what effect, if any, this change will have on per capita population growth rate.

39.4 Populations Grow Multiplicatively, but the Multiplier Can Change

FIGURE 39.9

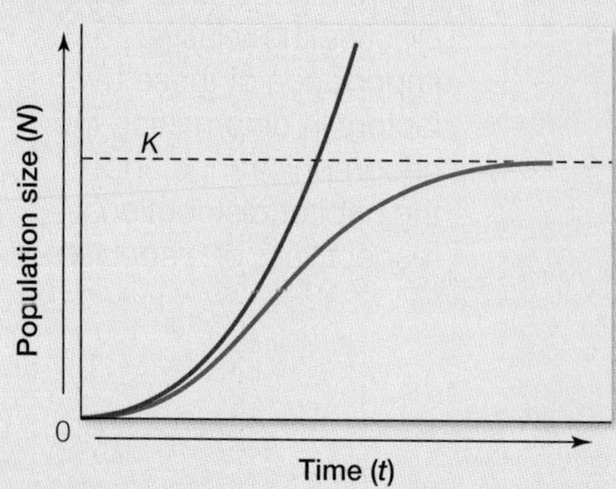

Go to ACTIVITIES 39.2–39.4

- Populations grow **MULTIPLICATIVELY**, changing by a multiple *r* of their current size in each time period. If *r* is constant and positive, populations have a constant **DOUBLING TIME**.
- In reality the per capita growth rate *r* is **DENSITY-DEPENDENT**: if a population becomes denser, the average individual's share of the resource "pie" shrinks, and *r* decreases (**FIGURE 39.9**).
- When deaths balance births in a population ($r = 0$), the population has reached an equilibrium size called the **CARRYING CAPACITY**, or ***K***.

QUESTIONS

1. The State of Colorado wants to increase trout populations in mountain lakes to promote tourism and generate revenue from fishing licenses. If they add *K* fish to the population in a lake that is at its carrying capacity, *K*, will they permanently double the number of fish in the lake? Explain your answer.
2. List at least two technological advances that have affected *K* for the global human population.

39.5 Immigration and Emigration Can Influence Population Dynamics

Go to ACTIVITY 39.5

FIGURE 39.12

© DGP_travel/Alamy Stock Photo; inset © Nancy Nehring/iStock.com

- The distribution of a species in a region often takes the form of a **METAPOPULATION**, a cluster of distinct populations in separate habitat patches (**FIGURE 39.12**).
- The **BIDE MODEL** adds immigration and emigration to the BD model.
- Immigration can repopulate a site where a population has gone locally extinct.

QUESTIONS

1. Describe the fundamental difference between the BD model (see Key Concept 39.2) and the BIDE model of population growth.
2. Human activities often break up a large area of continuous habitat into smaller habitat patches separated by a different habitat. How does this habitat fragmentation affect the risk of extinction of populations that occupied the original habitat area?

39.6 Ecology Provides Tools for Conserving and Managing Populations

Go to ANIMATION 39.1

FIGURE 39.14

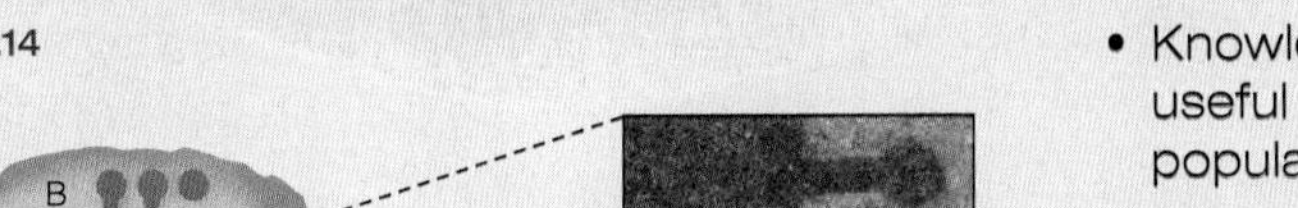

David McIntyre

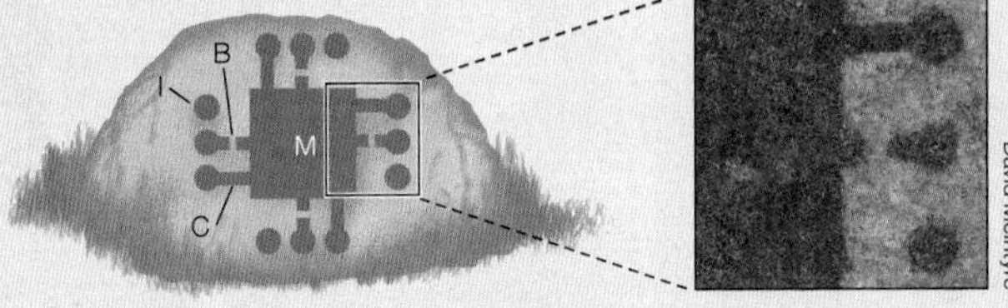

- Knowledge of life histories is useful for managing individual populations.
- One way to reduce extinction risk of species with a patchy population structure is to maximize the size of habitat patches, because larger patches support larger populations with lower extinction risk.
- Another strategy for conserving metapopulations is to provide **DISPERSAL CORRIDORS** that facilitate dispersal and gene flow among populations (**FIGURE 39.14**).

QUESTIONS

1. Big-leaf mahogany (*Swietenia macrophylla*) is a valuable timber tree native to South America. In Brazil, habitat destruction and illegal logging threaten its populations. In the Philippines, mahogany is grown in commercial plantations but is invading undisturbed rainforest and threatening native species. What goals would you set for managing wild mahogany populations in these two areas?
2. Identify a strategy to conserve mahogany in its native range, and a strategy to limit its abundance in forests outside its native range.

40 Interactions within and among Species

Leaf-cutter ants are so numerous in the New World tropics that they denude entire trees and are serious agricultural pests. But the ants don't eat the leaf fragments that they cut and collect. Instead, they grow a fungus on the leaves and eat specialized "food bodies" that the fungus produces—the ants are farmers! These social insects evolved agriculture more than 20 million years ago and now are completely dependent on the fungus, just as the fungus is completely dependent on the ants—it grows in leaf-cutter nests and nowhere else.

How might the intricate ecological interaction between leaf-cutter ants and fungus have evolved?

You will find the answer to this question on page 1005.

KEY CONCEPTS

40.1 Interactions between Species May Increase, Decrease, or Have No Effect on Fitness

40.2 Species Interactions Influence Population Dynamics and Distributions

40.3 Species Are Embedded in Complex Interaction Webs

40.4 Species Interactions Can Result in Evolution

40.1 Interactions between Species May Increase, Decrease, or Have No Effect on Fitness

LEARNING OBJECTIVES

By the end of this key concept you should be able to:

- 40.1.1 **Explain how per capita growth rate is a useful measure of fitness.**
- 40.1.2 **Distinguish between the five types of interspecific ecological interactions and describe the effect of each interaction on the fitness of participants.**
- 40.1.3 **Describe an example of an interspecific interaction whose fitness effects vary depending on environmental conditions.**

The relationship between leaf-cutter ants and the fungus they "farm" is just one of many interactions these ants have with other species. Another involves the green mold *Escovopsis,* which can invade nests of some leaf-cutter species and destroy their fungal gardens, and the bacterium *Pseudonocardia,* which lives on the ants' exoskeletons and produces antibiotics that suppress growth of the mold without affecting the ants or the fungus. Leaf-cutters interact with many plant species in their environment by harvesting their leaves. Anteaters, armadillos, parasitic flies, beetles, and pathogenic fungi kill and consume leaf-cutters. The bodies of dead ants are discarded on the colony's refuse dump, where they become food for fungi and microbes.

Leaf-cutters exemplify one of life's certainties: at some point between birth and death, every organism will encounter and interact with individuals of other species. **Interspecific** (Latin, "between species") **interactions** affect each individual's life history, and thereby its success in surviving and reproducing. Individual differences in birth and death rates define both an individual's contribution to total population growth rate—and its **fitness**. By affecting survival and reproduction, interspecific interactions influence the dynamics and densities of populations, alter the distributions of species, and lead to evolutionary change in one or both of the interacting species.

As noted in **Key Concept 13.4**, fitness is the reproductive contribution of an inherited phenotype to subsequent generations, relative to other phenotypes. Variation in fitness among phenotypes within populations leads by natural selection to evolutionary change in phenotype (and genotype) frequencies.

Interspecific interactions are classified by their effects on fitness

The details of interspecific interactions are bewildering, but we can cut through those details by asking, "Does the interaction have a beneficial (+), detrimental (–), or neutral (0) effect on the fitness of individuals of each interacting species?" Given the possible outcomes for fitness, five broad categories of interspecific interactions can be defined: competition, consumption, mutualism, commensalism, and amensalism (**FIGURE 40.1A**).

INTERSPECIFIC COMPETITION (–/–) **Interspecific competition** refers to interactions in which individuals of two different species decrease one another's fitness (–/–) because they require some of the same resources. Recall from Key Concept 39.4 that when an organism uses a resource, that resource becomes unavailable to other organisms, potentially reducing their survival and reproduction. We say "potentially" because not all resources are in such short supply as to affect fitness. At any given time, however, there generally is some **limiting resource** that is in shortest supply relative to demand. Species that share limiting resources at the same place and time will compete.

Competition may occur among organisms that consume the same food, whether they are African lions and spotted hyenas that attack wildebeest, or aphids and grasshoppers that eat the same plant. Even the interaction between the mold *Escovopsis* and leaf-cutter ants can be considered competitive, because both derive energy and nutrients from the fungus that the ants cultivate. For plants, the limiting resource may be sunlight, water, or inorganic nutrients in the soil (**FIGURE 40.1B**). But sources of energy or materials are not the only limiting resources. To give a few examples, some organisms are limited by space in which to grow (think of barnacles encrusting a rocky shore; see Figure 40.5) or in which to build a nest, and some plants may compete for the attention of animal partners that pollinate flowers or disperse seeds and thereby contribute to the plants' reproduction.

CONSUMPTION INTERACTIONS (+/–) Consumption interactions are those in which organisms obtain nutrition by consuming or harming other living organisms or are themselves consumed or harmed. It's obvious why these are +/– interactions: the consumer gains fitness while the consumed or harmed organism loses fitness. Consumption interactions include **predation**, in which an individual of one species (the predator) kills and eats individuals of another species (the prey); **herbivory**, in which an herbivore eats part or all of a plant, to the plant's detriment (**FIGURE 40.1C**); and

(A)

Major Types of Species Interactions

Type of interaction	Effect on species 1	Effect on species 2
Competition	–	–
Consumption: Predation, herbivory, parasitism	+	–
Mutualism	+	+
Commensalism	+	0
Amensalism	–	0

(C)

(B)

(D)

Mutualism
The fleshy berries of holly plants entice Eastern Bluebirds to disperse their seeds.

© Gay Bumgarner/Alamy Stock Photo

FIGURE 40.1 Types of Interspecific Interactions
(A) Interactions between species can be classified based on whether their influence on fitness is positive (+), negative (–), or neutral (0). **(B)** Competition for light shapes the characteristics of canopy (treetop) and understory (low-growing) vegetation in this forest in Olympic National Park, Washington, USA. **(C)** Commensalisms between large grazing mammals such as this American bison (*Bison bison*) and insect-eating birds such as these Brown-headed Cowbirds (*Molothrus ater*) are found in grassland environments around the world. **(D)** The interaction between berry-producing plants and berry-eating birds is a mutualism. Here an Eastern Bluebird (*Sialia sialis*) gains nutrition by eating holly berries (*Ilex opaca*). The holly's seeds pass through the bird's gut unharmed and are deposited in nutrient-rich feces away from the parent plant.

Activity 40.1 Ecological Interactions

parasitism, in which a parasite consumes part of a host individual but usually does not kill it. These categories blur together to some degree. For example, herbivorous caterpillars that consume some of a plant's leaves (see Figure 27.19) are functionally similar to parasitic black-legged ticks (see Chapter 39 opening photo) that consume some of a rodent's blood. And some parasites and pathogens are functionally similar to predators because they consume so much of the host's body, or otherwise harm it so severely, that the host dies.

MUTUALISM (+/+) A **mutualism** is an interaction that benefits both species (+/+). The interaction between leaf-cutter ants and the fungus they cultivate is mutualistic: the ants feed, cultivate, and disperse the fungus; the fungus, in turn, converts inedible leaf fragments into special fungal structures that the ants can eat. The interactions between plants and pollinating or seed-dispersing animals (**FIGURE 40.1D**), or between humans and some of their gut microbes (see Key Concept 38.1), similarly enhance the fitness of both partners. Mutualisms take many forms and involve many kinds of organisms; mutualisms also vary in the degree to which they are essential—or simply beneficial but not essential—to one or the other interacting partner.

LINK We have seen several examples of mutualisms in this book, including interactions between mycorrhizal fungi and plants (see **Key Concepts 21.2 and 25.2**); among fungi, algae, and cyanobacteria in lichens (see **Key Concept 21.2**); and between corals and dinoflagellates (see **Key Concept 20.4**).

COMMENSALISM (+/0) Interspecific competition, consumption interactions, and mutualism all influence the fitness of both participants. **Commensalism** influences only one participant, which benefits, while the other participant is unaffected (+/0). Some examples involve species whose feeding behavior makes food more accessible for another. For example, the Brown-headed

Cowbird owes its common name to its habit of following grazing ungulates (hoofed mammals such as cattle or bison) and feeding on insects flushed as they move and feed (see Figure 40.1C). The ungulate–cowbird interaction is commensal because the birds have no effect on ungulate fitness. In other cases, the feeding of one species benefits a second species by converting food into a form it can use. Cattle, for example, convert plants into dung, which certain beetles feed on; these "dung beetles" have no immediate direct effect on the cattle. Dung beetles, in turn, are on the giving end of another commensalism: dung-dwelling organisms that cannot fly—such as mites, nematodes, and fungi—attach themselves to the bodies of the beetles, which not only can fly but are very good at locating fresh dung. The hitchhikers have no apparent effect on dung beetle survival or reproduction.

AMENSALISM (–/0) Like commensalism, **amensalism** affects only one party in the interaction, which is harmed (–/0). Elephants moving through a forest or bison grazing the plains crush insects and plants with each step (see Figure 40.1C), but the large mammals are themselves unaffected by this carnage. Amensal interactions tend to be more unintentional than the other interactions, as these examples suggest.

The effects of interactions can vary

Ecologists find it useful to distinguish the five categories of interactions just described, but it's important to keep in mind that these categories, just like many others devised by humans, are only starting points. Some interactions are more complex than the categories suggest. For example, the commensalism we described between cattle and dung beetles may actually tend toward mutualism when viewed over a longer time period, since the activity of the beetles can fertilize the soil and lead to better growth of grasses that cattle graze.

Furthermore, many interactions have both beneficial and harmful aspects, and the net effect of one species on another can be positive or negative depending on environmental conditions. Sea anemones, for example, sting and consume small animals, including fishes. But a few fish species (mostly anemonefishes; genus *Amphiprion*) are protected by a special mucous coat and thus can live among the anemones' stinging tentacles (**FIGURE 40.2**). Although the benefit of this association to the anemonefishes is clear—they escape their own predators by hiding behind the anemones' nematocysts and can scavenge food caught by the anemones—more research is needed to clarify the consequences for the anemones. Are they nourished by the fishes' nitrogen-rich feces? Do they suffer because the fish steal some of their prey? The net effect of the interaction for anemones may change from beneficial when prey are scarce to detrimental when prey are abundant.

Amphiprion ocellaris

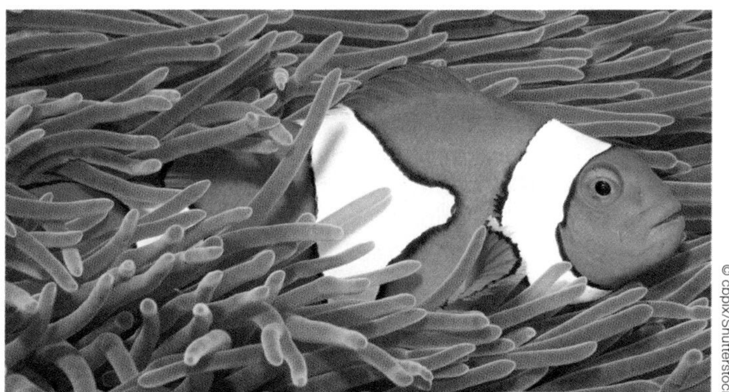

FIGURE 40.2 Interactions between Species Are Not Always Clear-Cut Ecologists long believed that the relationship between sea anemones and anemonefishes was a commensalism: that the fishes, by living among the anemones' stinging tentacles, gained protection from predators. But could it also be considered a mutualism—if the fishes' feces provide nutrients that are important for the anemones, or competition—if the fishes occasionally steal the anemones' prey?

REVIEW & APPLY | 40.1

R Interactions between two species can have negative, positive, or neutral effects on the fitness of the interacting individual organisms. Drawing on this observation, ecologists distinguish five basic types of interspecific interactions: competition, consumption, mutualism, commensalism, and amensalism. Nature is not, however, easily categorized—immediate and longer-term fitness effects of interactions may differ, and interactions sometimes switch from one category to another depending on environmental conditions.

A Kangaroo rats (*Dipodomys* spp.) (see Figure 39.4B) harvest seeds of desert plants and bury the seeds in shallow "caches." In dry years with low seed production, the animals return to eat most of their cached seeds. In wet years, however, the kangaroo rats store more seeds than they can eat, and many caches remain in the soil. These "planted" seeds germinate and survive well. Researchers in Arizona removed kangaroo rats from half of a series of fenced plots in 1931. They measured the density of native grasses relative to 1931 levels in plots with and without kangaroo rats over the next 15 years, as shown in the graph below (*figure source*: H. G. Reynolds. 1950. *Ecology* 31: 456–463).

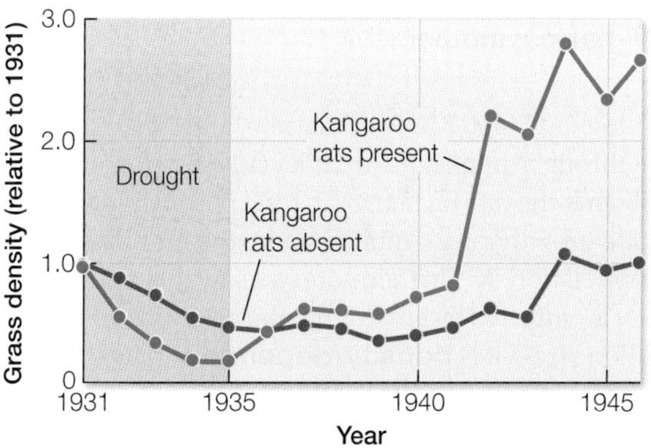

1. From the graph, determine whether kangaroo rats had a positive, negative, or neutral effect on the fitness of native grasses (a) during the 1931–1935 period of drought, and (b) during the wetter years after 1935, which included the unusually wet year of 1941.

(Continued)

REVIEW & APPLY | 40.1 *(continued)*

2. Drawing on the data and the summary of kangaroo rat biology, identify aspects of kangaroo rat behavior that are detrimental to grass fitness, and aspects that can benefit grasses.
3. Imagine a hypothetical kangaroo rat species that finds and eats 50 percent of the seeds in its environment without storing any seeds. Draw a new line on the graph that shows the probable effect of this hypothetical species on grass density, and explain how you derived this line.

Interspecific interactions influence population dynamics because they change the contributions of individuals to population growth rate. How exactly are population dynamics affected?

40.2 Species Interactions Influence Population Dynamics and Distributions

LEARNING OBJECTIVES

By the end of this key concept you should be able to:

40.2.1 **Describe how an increase in the population size of one species will influence population dynamics of the other species, for each of the five categories of interspecific interactions.**

40.2.2 **Describe a scenario in which a given species can persist in an environment only if another species is present, and another in which a given species can persist only if another species is absent.**

40.2.3 **Design an experiment that would allow you to determine whether one species benefits or harms another.**

As we saw in Chapter 39, populations are dynamic—their sizes change through time—and are unevenly distributed in space. We also saw that the rate of change in total population size equals the per capita growth rate, *r*, multiplied by the number of individuals in the population, *N*. Furthermore, we saw that *r* is not a constant. Not only is *r* affected by environmental conditions, it is sensitive to population size—it is **density-dependent**. The decrease in *r* as a population grows represents **intraspecific** (Latin, "within species") **competition**—competition among members of the same species for the same limiting resources.

Interspecific interactions modify per capita population growth rates

Just as *r* is sensitive to intraspecific competition, it is sensitive to interspecific competition and other interspecific interactions. To see how, let's write a word equation for growth of a population of one species (species A) that includes intraspecific competition and interspecific competition with a second species (species B):

per capita growth rate (*r*) of species A =
(maximum possible *r* for species A in uncrowded conditions)
– (an amount that is a function of A's own population density)
– (an amount that is a function of the population density of competing species B)

To fully describe interspecific competition, we also would need a second equation parallel to this one for species B, the other species involved in the interaction.

Interspecific interactions other than competition can be described in a similar way. Depending on the type of interaction, we must either subtract or add the effect of the other species (see Figure 40.1A). In consumption interactions, we subtract the effect of the consumer species (e.g., the predator) in the per capita equation for the consumed species (e.g., the prey), since the consumer increases the mortality or decreases the reproduction of the consumed species. Conversely, we add an effect of the consumed species in the equation for the consumer, since the consumer benefits. In mutualistic interactions, we add an effect of each mutualist to the equation for the other. In commensal interactions, we add a term in the equation of the species that benefits, and in amensal interactions, we subtract a term in the equation of the species that is harmed. (In these last two types of interactions, equations for the partner species contain no reciprocal terms, because, by definition, the partner is unaffected by the interaction.)

Activity 40.2 Species Interactions and Population Dynamics

Interspecific interactions affect population dynamics and can lead to local extinction

Because interspecific interactions modify per capita growth rates, populations show different dynamics in the presence and absence of other species. In a series of classic studies, the Russian ecologist G. F. Gause explored the dynamics of competition and consumption interactions between species of unicellular aquatic protists (see Figure 19.6). Gause's experimental results confirm general predictions that ecologists have derived from mathematical models of interspecific interactions. In the case of interspecific competition, all the models give the following predictions:

- The presence of a competitor reduces population growth rate. In **FIGURE 40.3A**, the population growth curves for *Paramecium caudatum* and *P. bursaria* rise more slowly from low densities when the two species are together.
- When two competing species coexist—when both persist in the system—each achieves a lower equilibrium population density than it would in the absence of the other. In Figure 40.3A, populations of the two *Paramecium* species stop growing at lower densities when the two species are together.
- Some species cannot coexist—one persists, and the other is excluded from the system. In **FIGURE 40.3B**, the population of *P. caudatum* is driven extinct by *P. aurelia*, even though *P. caudatum* could coexist with *P. bursaria*.

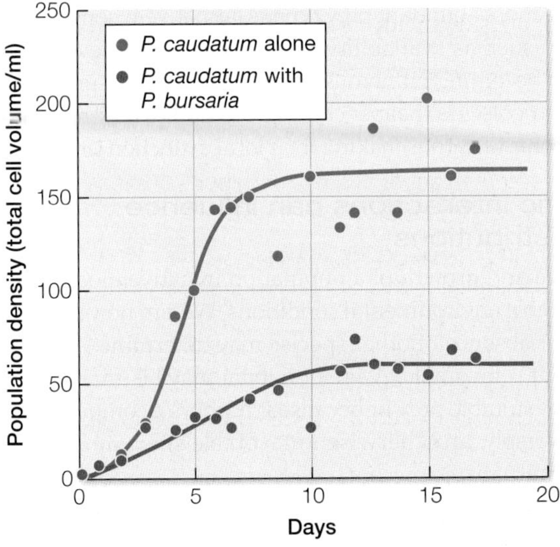

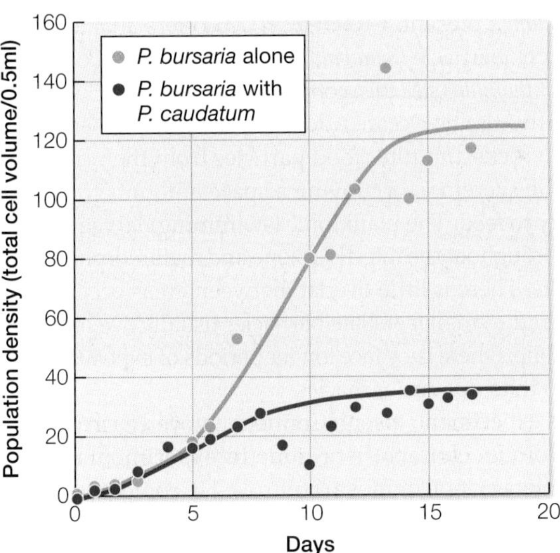

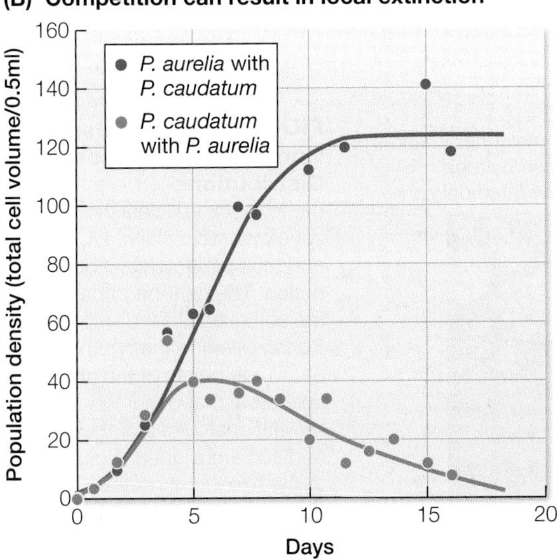

FIGURE 40.3 Interspecific Competition Affects Population Growth **(A)** The unicellular protists *Paramecium caudatum* and *P. bursaria* are able to coexist, but their populations grow more slowly, and reach lower equilibrium densities, when the competitor is present. **(B)** Competition between some *Paramecium* species results in local extinction of one species. In this case *P. caudatum* is excluded by *P. aurelia*. (A after G. F. Gause. 1935. *Vérifications Expérimentales de la Théorie Mathématique de la Lutte pour la Vie*. Hermann et Cie: Paris; B after G. F. Gause. 1934. *The Struggle for Existence*. Williams and Wilkins: Baltimoro.)

What effect of *P. bursaria* on *P. caudatum* population dynamics is evident up to about day 8, and what is evident after about day 10?

Gause's experiments, and similar ones conducted by others with a variety of organisms, confirm predictions about other types of interspecific interactions as well:

- In all interspecific interactions, the per capita growth rate of at least one species is modified by the presence of the other, sometimes positively and sometimes negatively.
- In all interspecific interactions, the average population density of at least one species differs in the presence or absence of the other species. Population densities are increased in species for which the interaction has a positive effect and are decreased in species for which the interaction has a negative effect.
- Local extinction of one or both of the interacting species is a possible outcome of interactions that have negative effects—that is, consumption interactions, amensalism, and competition.

Rarity advantage promotes species coexistence

Mathematical models have been useful not only in helping us understand how interspecific interactions affect population dynamics, but also by giving us insights into the conditions that allow species to coexist. For example, models of interspecific competition indicate that two species will coexist when individuals of each species suppress their own per capita growth rate more than the per capita growth rate of their competitor. In other words, intraspecific competition must be stronger than interspecific competition: *you harm yourself more than you harm your competitor*. When this is the case, a species gains a growth advantage as it becomes rare—that is, when it is at a low density and its competitor is at a high density—and this **rarity advantage** prevents the population from decreasing to zero. The result is coexistence.

Resource partitioning—differences in the ways in which organisms living in the same locale use a limiting resource—can cause intraspecific competition to be stronger than interspecific competition. Individuals of the same species, which compete for very similar resources, will suppress their own per capita resource availability more than they suppress that of the competitor, unless the competitor uses identical resources. When differences in resource use are sufficiently large, competing species can coexist (**FIGURE 40.4**).

Gause's experiments provide a concrete example. The coexistence of *Paramecium caudatum* and *P. bursaria* (see Figure 40.3A) is possible because *P. bursaria* can feed on microbes in the low-oxygen sediment layer at the bottom of a pond, whereas *P. caudatum* cannot tolerate this habitat. This partitioning of habitat

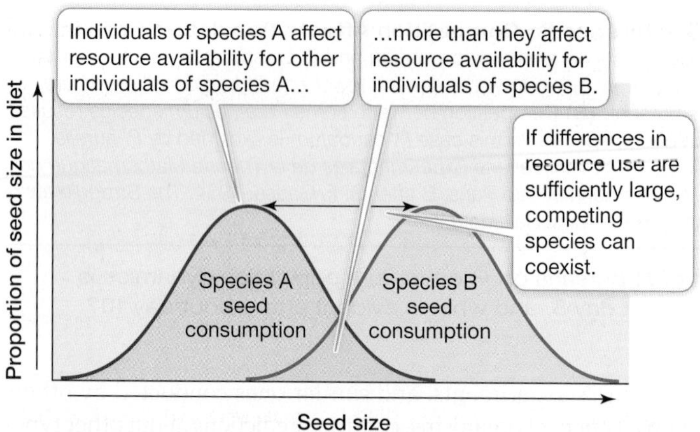

FIGURE 40.4 Resource Partitioning Can Cause Intraspecific Competition to Be Greater Than Interspecific Competition When two species differ in their use of resources, individuals will have a greater effect on the availability of resources to individuals of their own species than to individuals of the other species. In this hypothetical scenario, birds of species A eat smaller seeds on average than birds of species B. Individuals of species A affect the food available to their own species (smaller seeds) more than they affect the food available to species B (larger seeds), and vice versa. In other words, the two species partition, or divide up, the seed resource.

is possible because *P. bursaria* harbors symbiotic algae that provide it with additional oxygen as a by-product of algal photosynthesis. *P. caudatum* outcompetes *P. bursaria* when there is no sediment. *P. caudatum* cannot coexist with *P. aurelia* (see Figure 40.3B) because both feed on microbes in the water column; in this case, water chemistry determines which species persists.

In the case of predators and prey, several processes can result in a rarity advantage that protects prey from being driven extinct by their predators. For example, prey may become harder to find as they become rarer because they can all hide in the best refuges, and because they may be able to invest more in defending themselves against predators when they are at low density and have more resources available per capita. Predators may also stop looking for rare prey and switch to a more abundant prey type. Alternatively, some other limiting factor (such as availability of nest sites) may prevent the predators from becoming numerous enough to eat all the prey. The effect of such processes is that per capita growth rates may rebound when prey become rare, thus making their local extinction less likely.

Interspecific interactions can influence species distributions

The persistence and growth of a population in a given locale depends on favorable environmental conditions. We can now see that the presence or absence of other species may determine whether an environment is favorable. A site or habitat may be unfavorable because it lacks suitable prey or because it harbors a competitor or predator; conversely, an otherwise unfavorable site may become favorable if a mutualist is present.

An example of how competition can influence distributions is the restriction of *Paramecium bursaria* to the bottom sediment of a pond when *P. caudatum* is present. **FIGURE 40.5** offers an additional example. The rock barnacle (*Semibalanus balanoides*) and Poll's stellate barnacle (*Chthamalus stellatus*) compete for space on rocky shorelines of the North Atlantic Ocean. Adult barnacles attach themselves permanently to rocks and filter food particles from the water that flows past them; therefore, occupying a space amounts to having the opportunity to feed. The planktonic (swimming) larvae of both species settle throughout the intertidal zone and metamorphose into immobile adults. There is little overlap between areas occupied by adults, however: the smaller stellate barnacles usually live higher in the intertidal zone, where they face longer periods of exposure to air and desiccation than do rock barnacles.

In a famous experiment, Joseph Connell removed each of these two species from its characteristic zone in experimental plots, leaving other plots as non-removal controls. He compared population growth of each species in experimental and control plots. Stellate barnacle larvae settled in large numbers throughout much

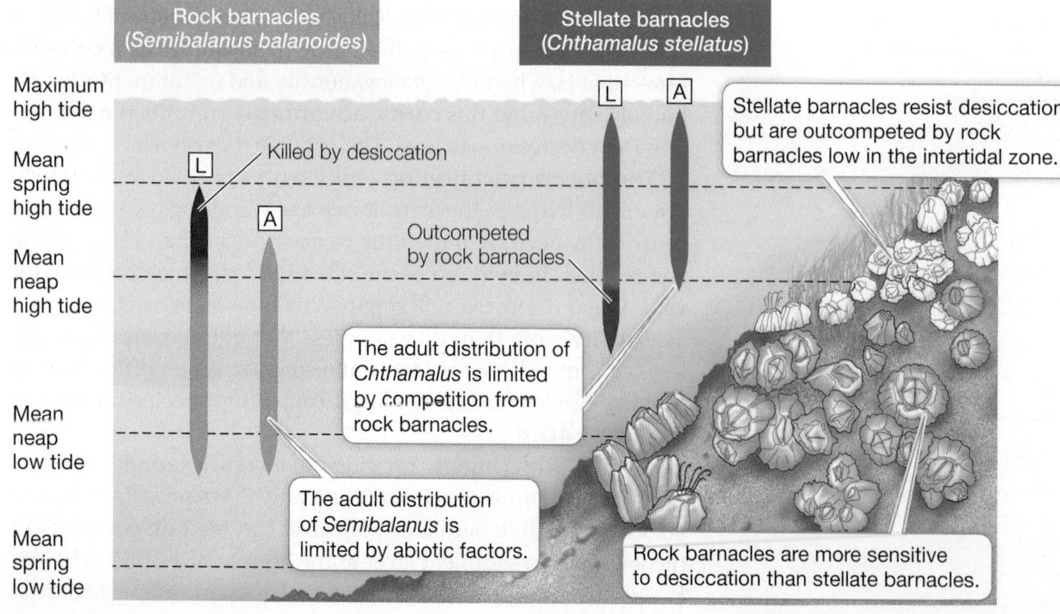

FIGURE 40.5 Interspecific Competition Can Restrict Distributions The left-hand bar for each species shows the zone over which larvae (L) settle, and thus the potential adult distribution. The right-hand bar shows the actual adult (A) distribution. Spring refers to the two high-amplitude points of the monthly tidal cycle that occur when gravitational forces of the moon and sun are aligned. Neap refers to the two low-amplitude points of the monthly tidal cycle that occur when gravitational forces of the moon and sun are at right angles to one another. (After J. H. Connell. 1961. *Ecology* 42: 710–723.)

of the intertidal zone, including the lower levels where rock barnacles are normally found, but they thrived at those levels only when rock barnacles were absent; the rock barnacles grew so fast when present that they smothered, crushed, or undercut the stellate barnacles. By outcompeting them for space, rock barnacles made the lower intertidal zone unsuitable for stellate barnacles. In contrast, removing stellate barnacles from higher in the intertidal zone did not lead to their replacement by rock barnacles because the rock barnacles were less tolerant of desiccation and failed to thrive there even when stellate barnacles were absent.

REVIEW & APPLY | 40.2

R Interspecific interactions influence per capita growth rates of interacting species, increasing population growth rates and equilibrium population densities if the interaction has a positive effect on fitness, and decreasing them if the effect is negative. Negative interactions can cause local extinction of one species if processes that increase per capita growth rate at low density are too weak to allow coexistence. Interspecific interactions thus influence where any given species can persist.

A A leaf-cutter ant nest is a multispecies system in which species interact with one another. Major components of the system are shown in the flowchart below, in which arrows indicate the effect of each component species on others.

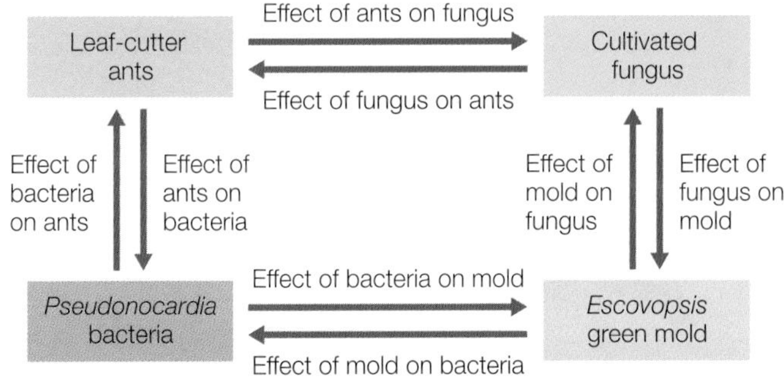

Review the description of interactions in leaf-cutter nests given in the chapter opening and first paragraph of Key Concept 40.1 (p. 991), and then:

1. Label each arrow with a "+," "–," or "0" to indicate whether the effect of each species on the other species is positive, negative, or neutral, respectively.
2. Categorize each of the four two-species interactions and describe how they culminate in a positive, negative, or neutral effect on each species.
3. Predict how an increase in the ability of *Escovopsis* to tolerate dry conditions would influence the distribution of leaf-cutter ant nests along an environmental gradient that extends from wet to dry habitats.

The two-species interspecific interactions that we have discussed do not occur in isolation. As may already be apparent, they form parts of a larger web of interactions that connects each species to many others living in the same area.

40.3 | Species Are Embedded in Complex Interaction Webs

LEARNING OBJECTIVES

By the end of this key concept you should be able to:

40.3.1 **Draw a trophic diagram from information about food habits of species in a community.**

40.3.2 **Define a trophic cascade and give a specific example.**

40.3.3 **Predict, from the way an invasive plant or animal makes a living, how such a species will interact with other species in a community.**

Some species are specialized in their interspecific interactions, competing with, consuming, or being consumed by few other species. Examples include the *Escovopsis* mold that infects only the fungus cultivated by leaf-cutter ants, yucca moths, which pollinate and lay eggs only in flowers of particular *Yucca* species (their larvae eat some of the developing seeds); and many parasitic insects that attack only a single host species. However, many species are generalized, engaging in diverse interactions with a variety of other species. For example, most animal-pollinated flowering plants are pollinated by more than one animal species, and few pollinators visit only one plant species. Furthermore, each species in a web of pollination interactions is connected to a much larger set of species that act as its competitors, hosts, consumers, and so forth. As a case in point, the mesquite shrubs and trees (*Prosopis* spp.) of the U.S.–Mexico Borderlands (see the opener to Chapter 38) are pollinated by many species of native bees and other insects, and their flowers and leaves are eaten, or harvested, by dozens of species of herbivorous insects, including leaf-cutter ants. Their seeds are eaten or dispersed by rodents, beetles, coyotes, foxes, and hoofed mammals, including cattle. The roots of a *Prosopis* shrub interact with an entire community of soil fungi and bacteria as well as with roots of other plants. Each of the species that interacts with *Prosopis* in turn interacts with a multitude of other species.

Consumption interactions form the core of interaction webs

Extensive webs of interspecific interactions may at first seem bewildering, but ultimately they boil down to consumption. Interaction webs are based on the reality that everybody has to eat—or (to be more accurate) every organism needs energy and nutrients. Every species participates in consumption interactions as the consumer, the consumed, or both. Interaction webs also reflect competition that commonly occurs between consumers that use the same sources of energy and nutrients—for example, sunlight, chemical nutrients, or food—or (in the case of immobile

TABLE 40.1 | The Major Trophic Levels

Trophic level	Source of energy	Examples
Primary producers (photosynthesizers, chemoautotrophs; see Table 18.2)	Solar energy, inorganic chemicals	Green plants, protists, bacteria, archaea
Primary consumers (herbivores)	Tissues of primary producers	Elk, grasshoppers, gypsy moth larvae, pollinating bees, geese
Secondary consumers (carnivores)	Tissues of herbivores	Spiders, Great Tits, cheetahs, parasites
Tertiary consumers (carnivores)	Tissues of carnivores	Tuna, killer whales, parasites
Omnivores	Several trophic levels	Coyotes, opossums, crabs, robins, white-footed mice
Decomposers	Dead bodies and waste products of other organisms	Fungi, many bacteria, earthworms, dung beetles

Activity 40.3 The Major Trophic Levels **Activity 40.4 Predator-Prey System Simulation**

organisms such as barnacles or plants) the space that gives them access to these resources. Even mutualisms usually involve consumers and consumed species. For example, leaf-cutter ants both feed and eat their fungus; algae that live inside the cells of sea anemones (see Figure 40.2) photosynthesize, producing carbohydrates that the animals consume, and the anemones provide the algae with nitrogen. To be sure, energy and nutrients are not the only types of resources involved in interspecific interactions. Bird species, for example, compete for nest sites that are safe for their offspring. Nevertheless feeding, or **trophic** (Greek *trophes*, "nourishment"), interactions form the core of interaction webs.

Trophic interactions determine the flow of resources through communities. Nutrients and energy enter communities via **primary producers**, organisms that obtain the *materials* that form their bodies from inorganic sources and the *energy* that fuels their metabolism from solar radiation or inorganic chemicals. Primary producers are so named because they convert energy and inorganic materials into organic compounds that can be used by the rest of the community.

Primary producers are also known as **autotrophs** (Greek, "self-feeders") because they create their own "food" from inorganic sources (see Key Concept 5.5). Species that obtain energy by breaking apart organic compounds that have been assembled by other organisms are called **heterotrophs** (Greek, "other-feeders"). Heterotrophs that dine on primary producers are called **primary consumers**, or herbivores (Latin, "plant eaters"). Those that eat herbivores are called **secondary consumers**, or carnivores (Latin, "flesh eaters"); those that eat the flesh of secondary consumers are called **tertiary consumers**, and so on. Organisms that feed from multiple trophic levels are called **omnivores**; they are especially common among

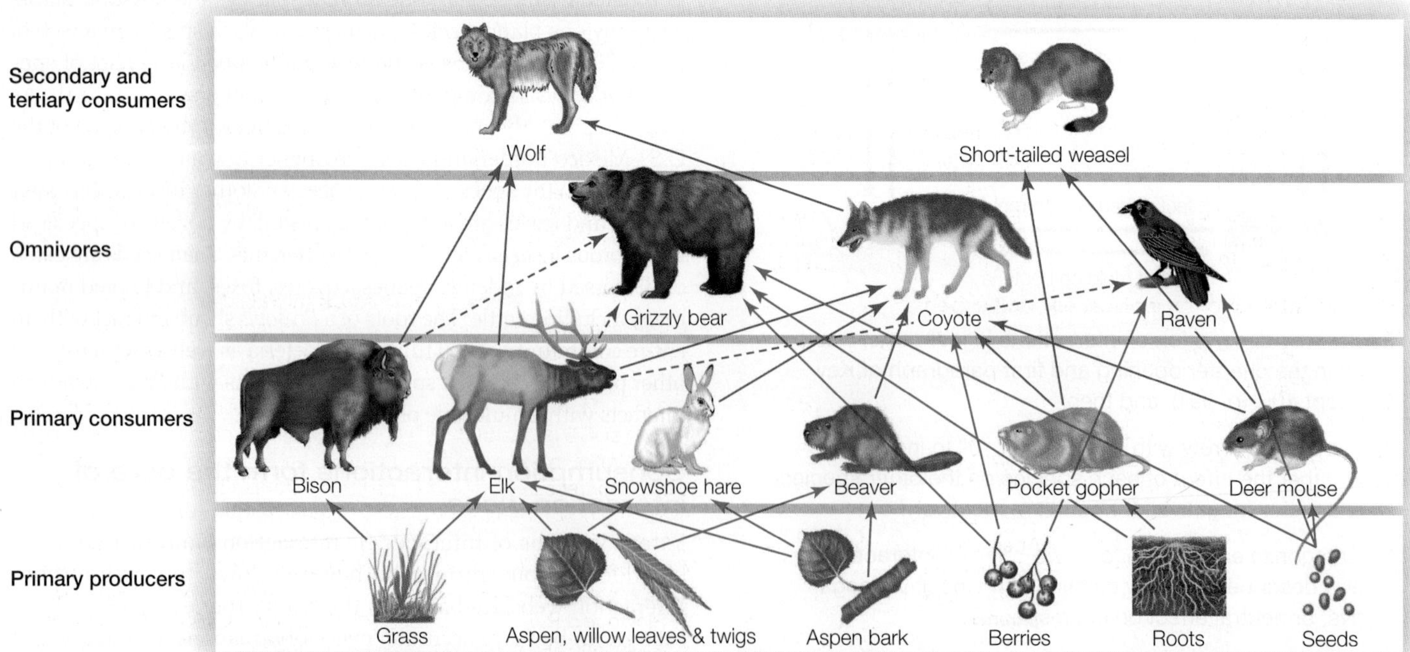

FIGURE 40.6 A Food Web in the Grasslands of Yellowstone National Park The arrows point from consumed to consumer species. Even this highly simplified food web is complex. Primary consumers eat plant material (blue arrows). Secondary and tertiary consumers are carnivores that kill and eat live animals (red arrows). Omnivores such as grizzly bears, coyotes, and ravens eat both plant and animal tissues; ravens and grizzlies also eat carrion (flesh of dead animals that they did not kill; dashed red arrows).

What trophic level is missing from this figure?

filter feeders in marine environments. **Decomposers** feed on waste products or dead bodies of organisms. These feeding positions are called **trophic levels** (**TABLE 40.1**).

Trophic interactions can be diagrammed as a **food web** that shows who eats whom. Arrows from consumed to consumer species show how energy and materials flow, and species are arranged vertically according to trophic level (**FIGURE 40.6**). Food web diagrams depict many interspecific interactions within the system, and some others can be inferred from them. Interspecific competition, for example, is a possibility (although not a certainty) when multiple consumers use the same resource. In the Yellowstone grasslands (see Figure 40.6), elk (*Cervus canadensis*) and American bison (*Bison bison*) compete for grass, and plants compete for sunlight and soil nutrients.

Effects of species loss or addition can cascade through communities

As the naturalist and conservationist John Muir noted in his 1911 book *My First Summer in the Sierra*, "When we try to pick out anything by itself, we find it hitched to everything else in the Universe." This poetic statement encapsulates an ecological fact: adding, removing, or changing the abundance of any species has effects that reverberate throughout an interaction web.

An example of what we might call the "Muir Effect" comes from Yellowstone National Park in Wyoming, USA. Wolves (*Canis lupus*) in the park prey on large mammals, particularly elk (see Figure 40.6). Unrestricted hunting eliminated wolves from the park by 1926, and they were reintroduced in 1995. This shift in trophic interactions has had cascading effects on the entire ecological system.

Park managers initiated annual elk counts in 1920. Concerned by the rapid increase in elk numbers after wolves disappeared, they deliberately kept elk numbers in check for many years. After the culling of elk was curtailed in 1968, elk densities increased rapidly (**FIGURE 40.7A**). In the absence of wolves, populations of aspen trees (*Populus tremuloides*) declined near the grasslands where elk spend the winter (**FIGURE 40.7B**). An elk exclusion experiment (**FIGURE 40.7C**) showed that elk were the culprits: after 1960 their browsing was so intense that it prevented any young trees from surviving to maturity. The elk also browsed streamside willows (*Salix* spp.), with the result that beavers (*Castor canadensis*), which depend on willows for food, nearly went extinct in the park.

After park managers reintroduced wolves in 1995, elk populations decreased. As a consequence, young aspen now survive to become trees, willows regrow, and species associated with these trees (beavers as well as birds, insects, and many others) have increased. Beaver dams have slowed streams, allowing water to infiltrate the soil rather than run off downstream. The result has been replenished groundwater reservoirs, stream flows, and aquatic systems. This history offers an example of how removal and reintroduction of species can initiate a **trophic cascade**—a cascade of effects across trophic levels. What happens to predators is felt not only by their prey, but also by the prey of their prey, and by many other species

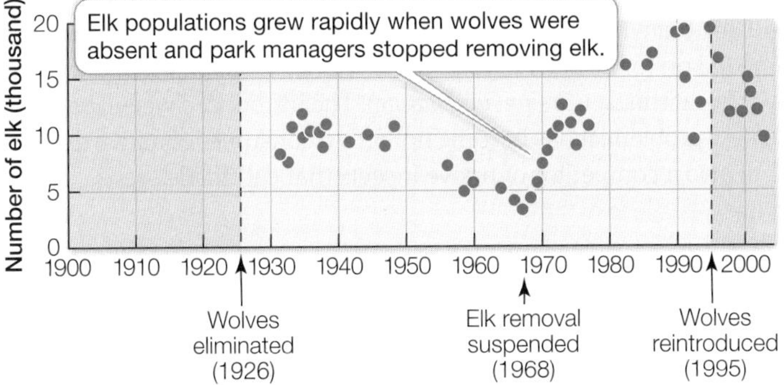

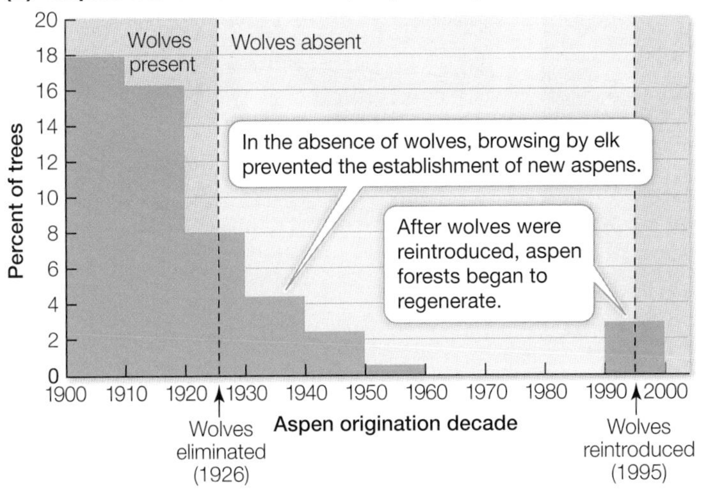

R. L. Beschta and W. J. Ripple. 2009. *Biol Conserv* 142: 2401–2414. With permission from Elsevier

FIGURE 40.7 Removing Wolves Initiated a Trophic Cascade **(A)** The elk population in Yellowstone National Park increased rapidly after wolves were eliminated and the park stopped removing elk. Elk numbers decreased after wolves were reintroduced. **(B)** In 2004, researchers measured the trunk diameters at breast height (1.3 m) of 476 aspen trees in the Park and used knowledge of aspen growth rates to estimate when each tree had originated. They found almost no trees in their sample that had sprouted between 1950 and 1990 and concluded that few young aspens had survived to maturity when wolves were absent and elk were abundant. **(C)** Researchers constructed an "elk exclusion" fence to demonstrate that aspen forests regenerated only inside the fence, where elk were absent. (A after W. J. Ripple and R. L. Beschta. 2004. *BioScience* 54: 755–766; B after J. Halofsky and W. Ripple. 2008. *Forestry* 81: 195–207.)

Was the per capita growth rate of the aspen population between 1960 and 1990 positive, zero, or negative? (*Hint*: Recall that $r = b - d$.)

connected with all of them. Conversely, removal or introduction at a low trophic level cascades up through higher levels.

Media Clip 40.1 Wolf Cascade

The web of species interactions has implications for conservation

The Yellowstone example illustrates how ecological understanding can guide interventions to restore damaged ecosystems and conserve threatened species. Reintroducing predators that humans once eliminated is a bold attempt to restore ancestral ecosystems. It also illustrates a bigger point: if we wish to restore or conserve ecological systems, we must consider the entire web of ecological interactions. And if we strive to conserve a particular species that is in danger of extinction, we must consider the web from the standpoint of that species, looking outward at all of its ecological interactions.

Nothing illustrates these points better than the issue of species introductions. As discussed in Key Concept 38.5, humans are introducing many species to new places, both intentionally and unintentionally. Such introductions tend to blur biogeographic boundaries and also have the potential to alter interactions among native species. Many species that are introduced into a new region cannot persist and disappear, but a fraction (roughly 10%) persist, and a fraction of those that persist (again roughly 10%) spread aggressively, with dramatic ecological—and economic—impacts. These non-native **invasive species** may be successful because they have left their enemies behind in their homeland—their predators and parasites have "missed the boat." Invasive species often harm native species that lack adaptations to compete with or defend themselves against the newcomers. However, some invasive species benefit natives by providing a resource that was not available previously. In all cases, however, non-native species—and invasive ones in particular—alter interaction webs, as we will see in Chapter 41.

Invasive species can harm native species in several ways (**FIGURE 40.8**). Introduced yellow star-thistle (*Centaurea solstitialis*), for example, grows rapidly in grasslands and crowds out native species, covering entire areas with its spiny flowering heads (see p. 982). Introduced flowering plants can alter relationships between native plants and their mutualists. Purple loosestrife (*Lythrum salicaria*) (see Figure 40.8A), introduced into North America in the early 1800s, not only crowds out natives but, adding insult to injury, also competes with them for the attention of insect pollinators. The native species *Lythrum alatum*, for example, receives fewer visits from bumblebees, and produces fewer seeds as a result, when purple loosestrife is present.

Some invaders alter ecological interactions by causing local extinctions of native species. Key Concept 21.3 described the case of an introduced sac fungus, a pathogen that drove the once-dominant American chestnut (*Castanea dentata*) so close to extinction that it no longer plays any real ecological role. The chestnut was replaced by oaks throughout the forests of the northeastern United States. The production of oak acorns, unlike that of chestnut nut crops, varies greatly from year to year, which causes year-to-year fluctuations in rodents, ticks, and Lyme disease (see Figure 39.2).

Sometimes non-native species are introduced deliberately, with the intention of improving an ecological system. When the hoped-for results are achieved, they can be dramatic, as exemplified by successful biocontrol (short for "biological control") of pests. But the result may be unexpected and undesirable. *Rhinocyllus conicus*, for example, is a Eurasian weevil whose larvae eat developing seeds (see Figure 40.8C). This insect was introduced into North America in 1968 to control invasive musk thistle (*Carduus nutans*) (see Figure 40.8B), but when the abundance of musk thistle declined, the weevil moved on to the native Platte thistle (*Cirsium canescens*) and then to wavyleaf thistle (*Cirsium undulatum*). The introduced weevil became a new problem: it has become not only a consumer of native thistles but also a competitor of native insects that eat thistle seeds.

(A) Purple loosestrife

David McIntyre

(B) Musk thistle

Lazaregagnidze/CC BY-SA 3.0

(C) Rhinocyllus

© natalean/Shutterstock.com

FIGURE 40.8 Some Non-Native Species Become Harmful Invasives **(A)** Purple loosestrife, *Lythrum salicaria*, is a wetland plant species that competes with native species for space and with some of them for the attention of pollinators. **(B)** Throughout North American grassland habitats, the introduced musk thistle, *Carduus nutans*, also crowds out native plants. **(C)** *Rhinocyllus conicus* was imported to control musk thistle, but this weevil unfortunately also eats native thistles.

REVIEW & APPLY | 40.3

R Some species are specialized and interact with only a few other species, but most interact with many others in interaction webs. These webs are organized around feeding (trophic) relationships—who eats whom. Adding or removing species can have effects that cascade throughout an interaction web. Understanding interaction webs is crucial for devising strategies to restore damaged ecosystems or conserve threatened species.

A

1. Anthropological research indicates that prehistoric people in the Yellowstone area gathered seeds, berries, and plant tubers; hunted bison, elk, and smaller mammals; and were themselves rarely taken as prey by wolves or bears. Did the trophic position of the prehistoric hunter–gatherers resemble that of any species in the present-day Yellowstone web (see Figure 40.6), or did they instead occupy a unique trophic position that is now vacant?
2. European immigrants to North America hunted many bison populations to extinction during the 1800s. Use Figure 40.6 to predict how loss of bison from the Yellowstone area would affect population densities of (a) wolves, (b) elk, (c) aspen and willows, (d) snowshoe hares, and (e) beavers. Compare predicted effects of bison removal to known effects of wolf removal as described in Key Concept 40.3.
3. The figure below shows the thistle food web, described above, that includes the invasive musk thistle and the Eurasian weevil introduced to control it (shaded boxes), two native thistle species, and the native community of thistle-feeding insects. Modify the figure to replace the Eurasian weevil with a hypothetical weevil species that would control musk thistle without affecting native Platte and Wavyleaf species. Explain what is different between the innocuous hypothetical weevil and the Eurasian weevil.

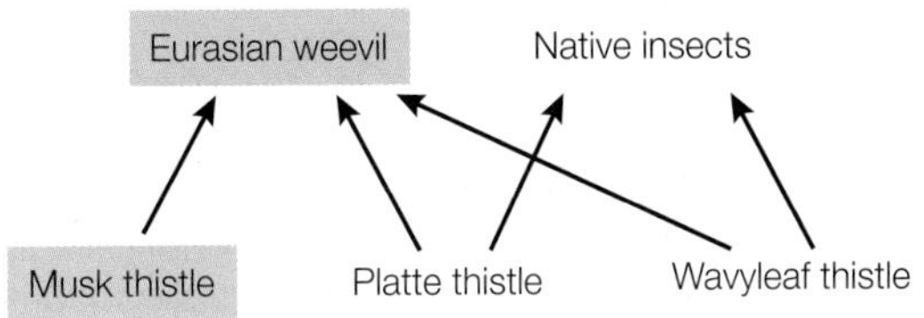

The effects of interspecific and intraspecific interactions on individual members of a population depend on those individuals' characteristics. For instance, if individuals vary in inherited traits that determine how much a competitor affects their access to resources, or that determine their risk of being eaten by a predator, then the interaction will result in evolution by natural selection.

40.4 | Species Interactions Can Result in Evolution

LEARNING OBJECTIVES

By the end of this key concept you should be able to:

- 40.4.1 **Explain how competition within and between species leads to evolution by natural selection.**
- 40.4.2 **Explain, using an example, how consumption interactions can lead to an evolutionary arms race.**
- 40.4.3 **Explain why mutualists do not evolve traits in order to benefit one another.**

We saw in Key Concept 40.1 that species interactions can influence the success of individuals in surviving and reproducing, that is, their fitness. Inherited traits that lead to greater benefit from positive interactions, or less harm from negative interactions, will become more common in the population.

Species evolve in response to both intraspecific and interspecific competition

Competition within a single species (intraspecific competition) will reduce the per capita population growth rate when individuals reduce each other's supply of limiting resources such as food (see Key Concept 40.2). This effect on fitness will not be the same for every individual because individuals vary in phenotypic traits that influence how they use resources. Traits that allow better access to resources, or more efficient use of them, will increase in frequency by natural selection. A counterintuitive consequence of such evolution will be that the carrying capacity—the number of individuals that can be supported by available resources—will increase as individuals that convert resources into offspring most efficiently become more and more common in the population.

The seed-eating finches of the Galápagos archipelago off the west coast of Ecuador (see Figure 16.8) illustrate how species evolve in response to intraspecific competition. Peter and Rosemary Grant, and their colleagues, have studied these finches for many years, characterizing their distributions among islands, morphologies, energy requirements, diets, food supply, population dynamics, genetics, and life histories. They have documented that the average beak size for any given finch species varies from island to island, and that the beak sizes present on each island are close to the best size for processing the types of seeds available there. Such patterns suggest that beak size has evolved in response to intraspecific competition for available food.

To visualize how evolution of beak size may occur, consider the following scenario. Imagine a finch-less island that supports an abundant plant species that produces seeds of a certain size and hardness. Next, suppose that several individuals of one finch species colonize the island. The beak sizes of the new colonists will vary somewhat because their body sizes will vary: bigger individuals will have bigger, stronger beaks. Within this range of beak sizes, one beak (and body) size will be best. Relative to this "optimal" size, smaller birds will have more difficulty cracking open the seeds of the abundant plant species (and so will have a lower rate of food intake), whereas larger birds will be able to

Geospiza fuliginosa

© Tui De Roy/Minden Pictures

Nectar Use and Size of *G. Fuliginosa*

Island	Time spent feeding on flower nectar (%)	Mean size (wingspan, mm)
BEES ABSENT		
Pinta	10	59.8
Marchena	28	58.2
BEES PRESENT		
Fernandina	1	64.8
Santa Cruz	14	64.0
San Salvador	0	63.8
Española	0	64.7
Isabela	7	64.5

Xylocopa darwinii

Avalon/Picture Nature/Alamy Stock Photo

FIGURE 40.9 Finch Morphology Evolves in Response to Competition with Carpenter Bees On those Galápagos islands where Small Ground-Finches are the sole pollinators of small flowers, the birds are small (as measured by their short wingspans). Small size lowers their energy requirements and may increase their ability to negotiate the flowers. On islands where carpenter bees compete with these finches for nectar, the birds are larger and have bigger, stronger beaks (which allows them to include seeds in their diet). (After D. Schluter. 1986. *Am Nat* 127: 95–102.)

crack the seeds easily but will need to use more energy to maintain their larger bodies, leaving less energy for reproduction (see Key Concept 39.3). Individuals with beaks closest to the optimal size will leave the most offspring, and average beak size will evolve toward the optimum by natural selection.

Interspecific competition, like intraspecific competition, can lead to evolution, as illustrated by another interaction in the Galápagos archipelago. The Small Ground-Finch (*Geospiza fuliginosa*) consumes flower nectar as well as seeds. Small individuals are better at obtaining nectar from the islands' small flowers, but large individuals have larger beaks that are better at cracking seeds. Carpenter bees (*Xylocopa darwinii*) compete with *G. fuliginosa* for nectar on some islands but are absent from others. Small finches are affected more strongly by competition from bees than are larger individuals, because the bees rely more heavily on nectar. As a consequence, the finches are larger in size, and drink less nectar, on islands where carpenter bees are present (**FIGURE 40.9**). This evolutionary shift means that resource use by *G. fuliginosa* has diverged from that of its bee competitor on islands where the two species coexist—an example of resource partitioning (see Figure 40.4) that resulted from natural selection.

Consumption interactions can lead to an evolutionary arms race

Predators, parasites, and herbivores benefit if they can acquire nutritious food quickly and with minimal energy expenditure, whereas the organisms that they consume or harm benefit if they can evade or deter these natural enemies at little cost. The interests of consumer and consumed are obviously at odds. These opposing interests can lead to an "evolutionary arms race," in which prey continually evolve better defenses, predators continually evolve better offenses, and neither gains any lasting advantage over the other (**FIGURE 40.10**). The Red Queen summed up this situation in Lewis Carroll's *Through the Looking-Glass* (1871) with the words "It takes all the running you can do, to keep in the same place."

Various defense strategies are available to consumed species. Some mobile animals use speed, size, or weapons to thwart predators. Others hide or use camouflage to avoid being detected, or to mimic unpalatable species. Immobile organisms have other tricks up their sleeves; for example, they may have evolved thick armor, or they may be poisonous or not nutritious. In response to these strategies, selection in populations of consumers favors greater speed, size, or strength; keen senses; armor-piercing or crushing tools; or means of detoxifying poisons.

Anodorhynchus hyacinthinus

© Pete Oxford/Minden Pictures

Activity 40.5 Evolutionary Arms Race Simulation

FIGURE 40.10 An Evolutionary Arms Race Brazil nuts (the seeds of *Bertholletia excelsa*) are protected by extremely hard shells that most birds cannot penetrate. The evolution of beak strength in the Hyacinth Macaw (*Anodorhynchus hyacinthinus*), however, has kept pace with the nut's hardness.

THINK LIKE A SCIENTIST

Intra- and interspecific competition influence the morphology of coexisting species

In the Galápagos archipelago, seed-eating finches (*Geospiza* spp.) use their beaks to collect and crack open the hard outer shell of seeds, which are often in short supply. The seeds of any given species can be opened only by birds with a beak larger than a threshold size; the harder a seed is, the larger the threshold. Dolph Schluter and Peter Grant determined which species of seeds can be eaten by *Geospiza* species with different beak sizes, as well as each species' body size and food requirements. They found that birds with bigger beaks could eat harder seeds, but had larger bodies and metabolic needs. They then measured the hardness and abundance of edible seeds produced on each island and calculated the availability of food for finches of different beak sizes. From this and from knowledge of metabolic needs, they calculated, for each island, the density of finches with a given beak size that could be supported if no other finch species were present (this is an estimate of carrying capacity). They made this calculation for hypothetical finches with beak sizes that span the possible range for seed-eating *Geospiza* species. **FIGURE 40.11** displays the results of their calculations for six islands. To make the graphs easier to read, beak size has been expressed as the logarithm to the base *e* of beak depth in millimeters, and carrying capacity has been scaled relative to the largest value. A peak in the graph predicts a high carrying capacity for a finch species with that beak (and body) size. A low point means the opposite—a species with that beak size is predicted to achieve only a low equilibrium density. The carrying capacity lines are jagged because each island supported at most eight plant species that produce edible seeds, and their seeds fall into only four hardness categories. Finally, Schluter and Grant placed dots on the graph for each of the six islands to show the average beak size of each *Geospiza* species actually present on the island. The authors concluded that the availability of seeds of different hardness on each island determines how many finch species occur there and what their beak sizes are.

Use your understanding of the process of science and quantitative reasoning to answer the following questions.

1. Identify and describe two aspects of the results that support the authors' conclusion.
2. Suppose the *G. fuliginosa* population on Marchena goes extinct, and *G. fuliginosa* individuals from Rábida colonize Marchena. Do you expect the average beak size of the colonizing population to increase or decrease through time? Justify your answer.
3. Propose an explanation for why the carrying capacity peaks rise so abruptly as beak size increases, and then decline more gradually.
4. Propose an explanation for the observation that peaks in carrying capacity have no more than one associated finch species.

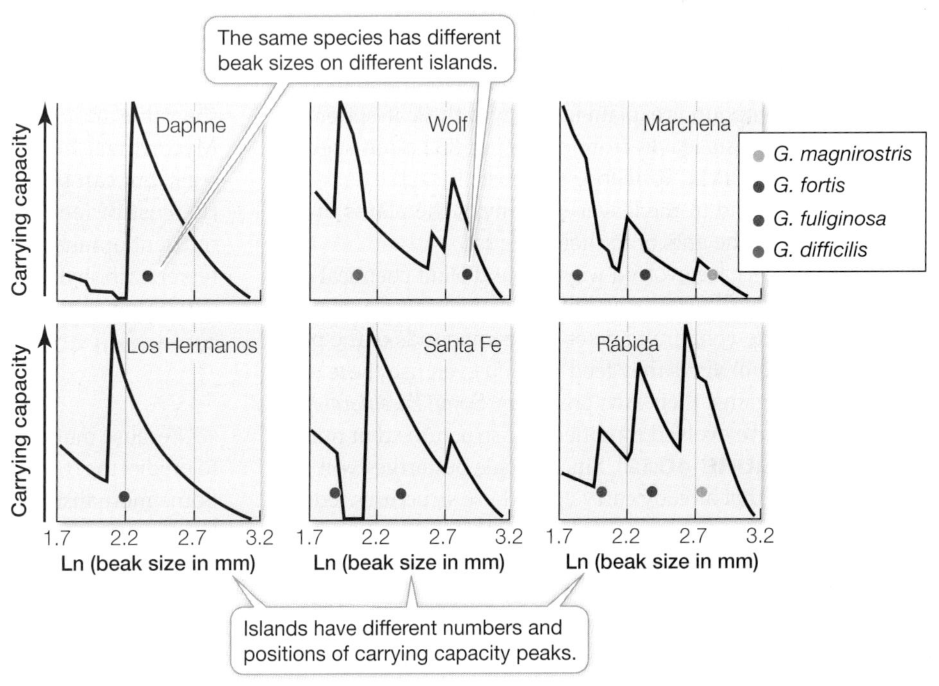

FIGURE 40.11 Predicted and Observed Beak Sizes of Finch Species on Six Islands in the Galápagos Archipelago From measurements of the abundance of seeds of different sizes on each island and the food requirements of seed-eating finches (*Geospiza* spp.) with different beak sizes, scientists estimated the carrying capacities for finches as a function of beak size (jagged black lines). They also determined the average beak sizes of the finch species actually found on each island (colored dots). (After D. Schluter and P. R. Grant. 1984. *Am Nat* 123: 175–196.)

FIGURE 40.12 Using Mimicry to Avoid Being Eaten The raised yellow bumps on leaves of passionflower (*Passiflora reticulata*) resemble the eggs of the plant's principal herbivores, zebra butterflies (*Heliconius cydno*). Because female butterflies will not lay their eggs on leaves that already carry eggs, these "false eggs" protect the plant from being eaten by caterpillars.

Plants produce a great variety of defensive chemicals, as we saw in Chapter 27. Used in small quantities, many of these chemicals, including the caffeine of coffee and tea, the mustard oils of cabbages, the capsaicins of chilis, and the piperine of black pepper, spice up our lives. But the production of such compounds evolved because they are toxic or repellant to some herbivores and pathogens. Defensive compounds explain in part why the leaf-cutter ant *Atta cephalotes* harvests leaves from only 17 of 332 plant species that grow in the Florencia Norte forest of Costa Rica: the terpene-like compounds found in the leaves of many of the plants make them toxic either to the ants or to their fungus.

Many herbivores have evolved ways around plant chemical defenses. The caterpillars of *Heliconius* butterflies, for example, store or detoxify the cyanide-containing defensive compounds of the passionflower (*Passiflora*) vines they feed on, and even use these poisons as defenses against their own predators. Some *Passiflora* species, conversely, have evolved modified leaf structures that mimic butterfly eggs (**FIGURE 40.12**). Since female butterflies will not lay eggs on leaves that already carry eggs, these structures reduce the plant's probability of being eaten by *Heliconius* caterpillars.

Mutualisms involve self interest

Mutualisms are often misunderstood. We sometimes hear that "bees visit flowers to pollinate them," or that "flowers provide food for their pollinators," but such statements imply an impossible evolutionary process in which species exhibit traits that evolve to benefit another species. As Charles Darwin said about his theory of natural selection:

> If it could be proved that any part of the structure of any one species had been formed for the exclusive good of another species, it would annihilate my theory, for such could not have been produced through natural selection.
> (C. Darwin. 1859. *On the Origin of Species by Means of Natural Selection*. J. Murray: London. Chapter 6, p. 180.)

In truth, most pollinators visit flowers to get food (see Figure 37.19) and just happen to transfer pollen among flowers in the process; flowers provide food (usually as little as possible) to lure animals, not to fatten them up; birds and bees respond to any conspicuous color that signals the presence of food; and most flowers are visited by any animal that gains economically by doing so, whether it pollinates the plant or not. In the rough-and-tumble world of ecological interactions, species benefit other species not out of **altruism** (when one individual benefits another but at a cost to its own fitness), but because acting in their own self-interest happens to benefit others.

All mutualisms involve "biological barter": the exchange of resources and services. Plants exchange food (usually nectar or pollen) for pollen transport by pollinators. Plants also exchange food (fruit pulp) for seed dispersal by frugivores (fruit eaters). Some mutualists (such as ants and plants) exchange defense or parasite and pathogen control for food or shelter. Others, such as plants and mycorrhizae (see Figure 24.12), or mammals and gut bacteria (see Chapter 38), exchange one kind of nutrient for another. In all mutualisms, natural selection favors individuals that obtain the greatest benefit for the least expenditure of energy and materials. As a consequence, the net effect of the interaction on fitness may depend on environmental conditions that influence the value and cost of the goods and services that are exchanged. Mycorrhizal fungi, for example, benefit plants in nutrient-poor soils, but can be a liability to plants in nutrient-rich soils, where the cost of feeding the mycorrhizae outweighs their value in nutrient uptake. Accordingly, plants in nutrient-rich soils excrete fewer carbohydrate-rich compounds from their roots, thus discouraging colonization by mycorrhizae.

Animation 40.1 Mutualism

Because mutualisms are diverse and often complex, it is difficult to predict the consequences for the evolution of the species involved. Some mutualisms evolve into one-on-one relationships with ever-tighter reciprocal adaptation—leaf-cutter ants and their fungus are a case in point. Others evolve into multispecies relationships that rely on generalized adaptations; many mutualisms between plants and the animals that disperse their seeds fall into this category. Some mutualisms are unstable and become consumption relationships. Understanding the conditions under which mutualisms follow different evolutionary trajectories is an active area of current research.

REVIEW & APPLY | 40.4

R When interactions within and among species affect per capita growth rates, they also affect individual fitness. If individuals vary in heritable traits that reduce the cost or increase the benefit of an interaction, then interactions can cause evolution by natural selection. Examples include the evolution of higher carrying capacity, resource partitioning between competitors, arms races between the consumers and the consumed, and variable degrees of dependence between mutualists.

A Blue mussels (*Mytilus edulis*) throughout the Atlantic coast of North America have been preyed on for 200 years by introduced European green crabs (*Carcinus maenas*). Asian shore crabs (*Hemigrapsus sanguineus*) also eat mussels but arrived only 25 years ago and have not yet spread to northern Maine. Some mussels have inherited traits that allow them to detect crabs and grow a thicker shell in response. Researchers collected young mussels from northern and southern populations, grew them with no crabs present or with a caged green or shore crab nearby, and measured their shell thickness after 3 months. The graph shows shell thicknesses of mussels from northern versus southern populations that were exposed to the three treatments (*figure source*: A. S. Freeman and J. E. Byers. 2006. *Science* 313: 831–833). Error bars indicate one standard error of the mean; bars with different letters are statistically different from one another ($P < 0.05$). See Appendix B for discussion of statistical concepts.

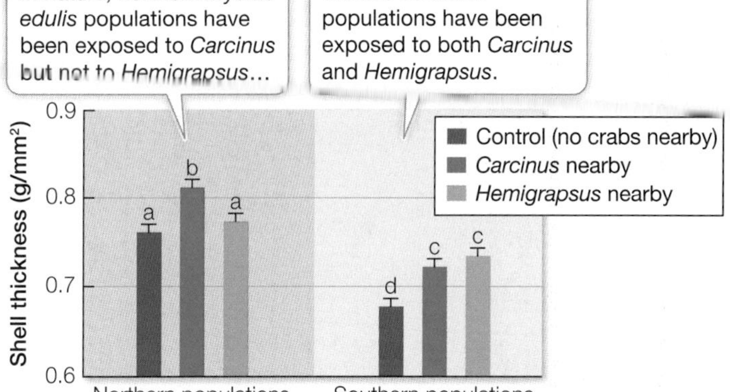

1. Identify and describe results that support the hypothesis that mussel populations have evolved the ability to detect nonnative crabs and respond to them by thickening their shells.
2. Explain what the results tell us about how fast mussels can evolve the ability to detect and respond to a new predator.
3. Predict how the strength of crab claws may respond to increasing mussel shell thickness.

Answer: As is true of many ants, the ancestors of modern leaf-cutters disposed of their organic refuse in garbage dumps outside the nest, which would have supported the growth of wild fungi. At some point ancestral leaf-cutters must have started to feed on some of these fungi. Once fungal feeding began, natural selection would have favored colonies that provided better growing conditions for the most nutritious fungus by, for example, excavating special chambers inside the ant nest and providing leaf mulch that the fungus could grow on (**FIGURE 40.13**). Colonies with the best leaf-collecting and garden-tending behaviors would have had high fitness because their fungal gardens would produce more food, and so the behaviors would have spread by natural selection. Selection would also have favored fungal traits that provided a more complete diet to the ant "farmers," since this would allow ants to form more and larger colonies that cultivate more fungus.

(A)

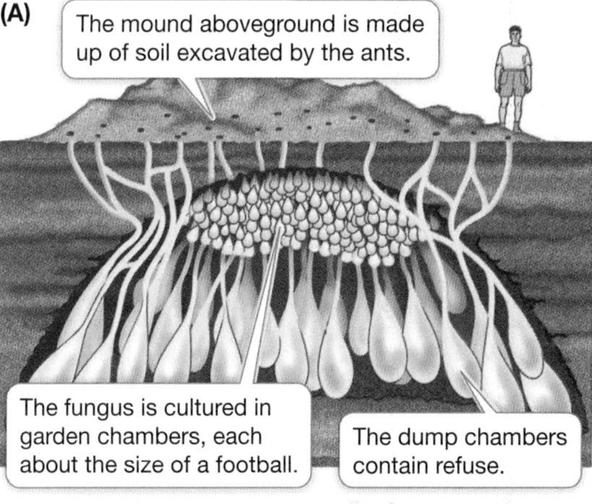

(B)

FIGURE 40.13 A Fungal Garden **(A)** Diagram of the nest chamber of a large *Atta columbica* leaf-cutter ant colony. **(B)** Leaf cutter ant workers tending their fungal garden. (A after B. Hölldobler and E. O. Wilson. 1990. *The Ants*. Belknap, Harvard: Cambridge, MA, modified from J. C. M. Jonkman, in Weber 1979.)

40 VISUAL SUMMARY

40.1 Interactions between Species May Increase, Decrease, or Have No Effect on Fitness

Go to ACTIVITY 40.1

FIGURE 40.1

Major Types of Species Interactions

Type of interaction	Effect on species 1	Effect on species 2
Competition	–	–
Consumption: Predation, herbivory, parasitism	+	–
Mutualism	+	+
Commensalism	+	0
Amensalism	–	0

- **INTERSPECIFIC INTERACTIONS** can have positive (+), negative (–), or neutral (0) effects on the **FITNESS** of interacting individuals. Interactions are categorized depending on their fitness effects (**FIGURE 40.1**).
- The net effect of an interaction can vary depending on environmental conditions.

QUESTIONS

1. Explain how interspecific interactions are classified and give an example of each of the five types of interactions.
2. Classify the type of interspecific interaction that occurs between humans and crop plants and defend your answer.
3. Design an experiment that would allow you to determine whether the effect of anemonefishes on sea anemones is positive, negative, or neutral.

40.2 Species Interactions Influence Population Dynamics and Distributions

Go to ACTIVITY 40.2

- The equation for **DENSITY-DEPENDENT** population growth, which describes **INTRASPECIFIC COMPETITION**, can be extended to include the effects of interspecific competition or other interspecific interactions.
- The per capita growth rate and the average population size of each interacting species are modified by the presence of the other in a way that depends on the type of interaction (**FIGURE 40.3**).
- Coexistence of species competing for resources, and persistence of prey species and their predators, is possible if each species has a **RARITY ADVANTAGE**.

FIGURE 40.3

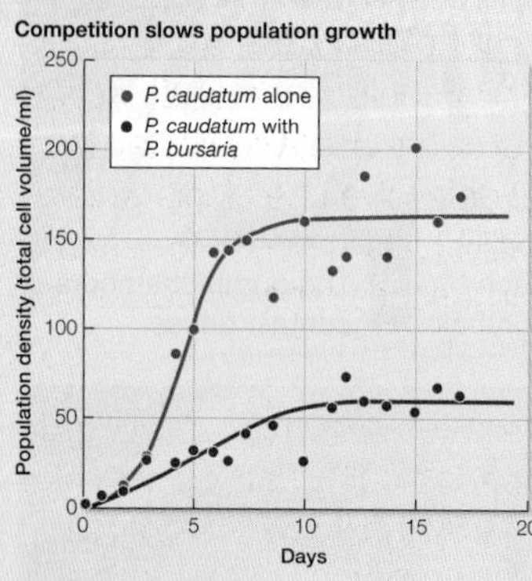

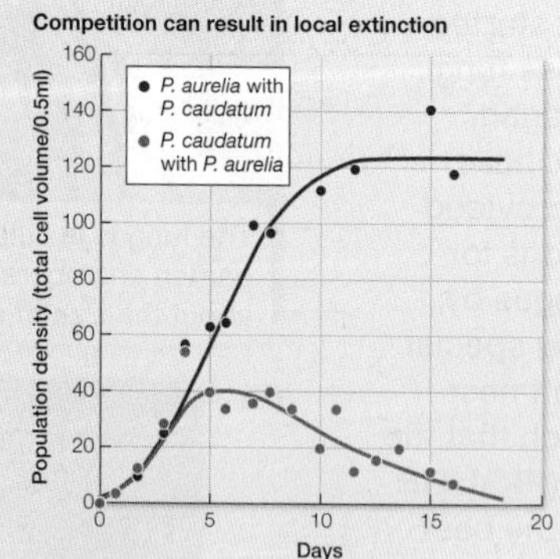

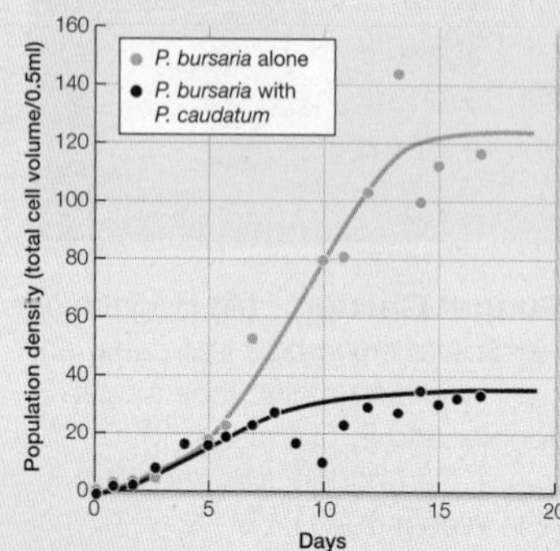

1. Write two word equations that describe the interaction between a plant (species A) and one of its pollinators (species B). (*Hint*: Modify the word equation on p. 994.)
2. Explain, in terms of resource availability, why the equilibrium population size of one species is reduced by the presence of a competing species.
3. What are the possible outcomes of a predator–prey interaction in terms of the eventual population sizes of the two interacting species?

40.3 Species Are Embedded in Complex Interaction Webs

Go to ACTIVITIES 40.3 and 40.4

- Most species are involved in interaction webs with many other species.
- **TROPHIC** interactions form the core of interaction webs. They can be diagrammed as **FOOD WEBS** (**FIGURE 40.6**).
- Changes in the abundance or identities of species in a food web can cause a **TROPHIC CASCADE** of additional changes across trophic levels.

FIGURE 40.6

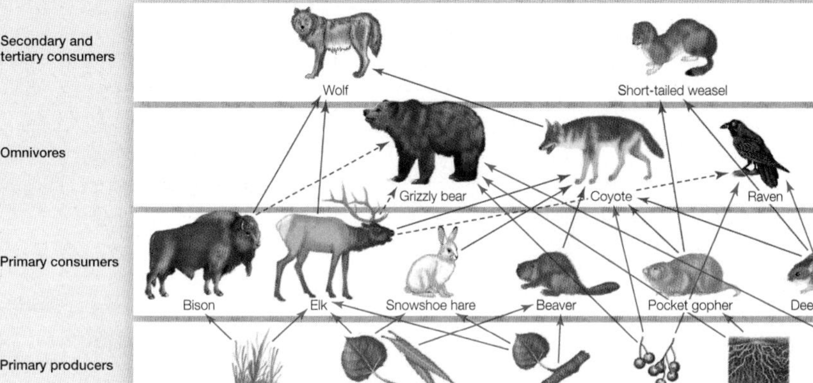

QUESTIONS

1. Describe a terrestrial landscape whose food webs contain no decomposers.
2. What type of interspecific interaction best describes the relationship between (a) coyotes and snowshoe hares and (b) bison and elk?
3. Discuss whether loss of a species of primary producer can cause a trophic cascade.

40.4 Species Interactions Can Result in Evolution

Go to ANIMATION 40.1 and ACTIVITY 40.5

- Both intraspecific and interspecific competition can lead to evolution of more efficient resource use within a species; this can lead to resource partitioning in the case of interspecific competition.
- The opposing interests of consumer and consumed species can lead to an "evolutionary arms race" in which consumed species continually evolve better defenses and consumer species evolve better offenses (**FIGURE 40.12**).
- Mutualisms involve mutual exploitation and the exchange of goods and services.

FIGURE 40.12

Courtesy of Lawrence E. Gilbert

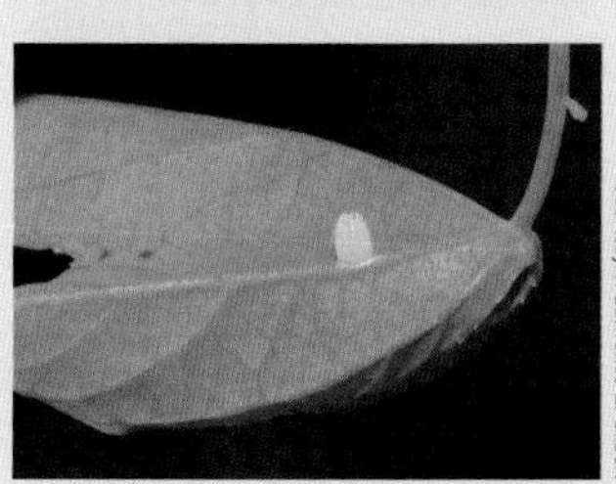

Courtesy of Lawrence E. Gilbert

Courtesy of Lawrence E. Gilbert

1. What goods or services are exchanged in the "biological barter" between leaf-cutter ants and their fungus?
2. Why might individual plants in populations within the geographic range of an insect herbivore contain high levels of defensive chemicals while individuals in populations outside the insect's range produce no defensive chemicals?
3. Females of some yucca moth species collect pollen, carry it to the next flower they visit, and deliberately place it on the stigma (the female part of the flower that receives pollen) before depositing eggs into the flower's ovary (where their larval offspring eat some of the developing seeds). Does this behavior, which benefits the plant, "annihilate" Darwin's theory of natural selection? Explain why or why not.

41 Ecological Communities

Over the past 1,500 years the cultivation of coffee (genus *Coffea*) has spread from Ethiopia to Arabia to Europe and finally to Asia and the Americas. This important crop is grown mostly in the tropics using methods that range from high-intensity sun cultivation (shown here) to cultivation in the shade of undisturbed forests. High-intensity methods give higher yields but require costly fertilizer, herbicides, pesticides, and labor.

Can we use principles of community ecology to improve methods of coffee cultivation?

You will find the answer to this question on page 1027.

KEY CONCEPTS

41.1 Communities Contain Species That Colonize and Persist

LEARNING OBJECTIVES

By the end of this key concept you should be able to:

41.1.1 **Define "ecological community" and explain how ecologists choose to study communities.**

41.1.2 **Identify three elements of community structure.**

41.1.3 **Outline how communities are assembled.**

We saw in Chapter 40 that the abundances, distributions, and phenotypes of species are affected by their interactions with other species. These interactions occur in the context of an ecological **community**—the set of species that occur together at the same time and place. Humans have practical reasons for studying how communities are put together and how they function, because we depend on them in many ways. As with any other dynamic system, a community's properties depend on its components—its species—and their interactions.

Ecological communities may have a simple definition, but studying them is not simple, for two reasons. First, communities in nature typically contain many interacting species. It is usually not possible to study every aspect of such a complex community, so ecologists often focus on a subset of the species or interactions—either a particular taxonomic group, or the species that occur in a particular habitat or location, or the species involved in particular types of interaction or that make their living in a particular way. Hence ecologists may speak of a bird community, the *Daphnia* community of rock pools (see Figure 39.12), all the aquatic species in a pond, the microbial community of the human gut, the plant–pollinator community of a meadow, or the community of grazing herbivores in the marine rocky intertidal zone.

Second, communities often occupy a larger area than can be studied easily. In these cases, ecologists can choose to study a representative portion of the area occupied by the community, such as a 10- × 10-meter area of forest. Fortunately, the scientific understanding that is gained by studying portions of communities provides useful insights into properties of the larger systems.

Communities vary in their structure

Ecologists characterize the structure of communities in terms of aspects of their component species: the *identities* of the species they contain, *how many* species they contain, and *how abundant* species are relative to one another. Communities vary substantially in these three aspects of their structure: they vary in the particular species they contain, in how many species live together (from few to many), and in the distribution of abundances of different species (from a few abundant and many rare species, to relatively equal abundances of species). Ecologists usually describe communities in terms of their **species composition**—the mix of species in the community. The other two aspects of community structure relate to **species diversity**, a topic which we will develop in Key Concept 41.3.

Communities are assembled via gains and losses of species

The mix of species in a community is determined by the same factors that explain the distributions of individual species—a species will not be present in a community at a particular time if it never reached that location, or if it did arrive but did not persist (see Key Concept 39.1). We can therefore think of communities as being assembled by a dynamic process in which previously resident species go extinct and new species arrive, either by colonizing from elsewhere or, over much longer spans of time, by evolving in place (see Key Concept 16.2).

Species may fail to colonize a community or be lost from it for several reasons. Some species may never reach a location, even if it is suitable—consider how few dispersers from a mainland are likely to reach a remote island before they perish! If colonists do arrive, they may be unable to establish a population, for example if a resource or an essential mutualistic partner is missing from the community, or if the physical environment is unsuitable. Or a population may become established and grow for a while, only to be excluded eventually by competitors, predators, or pathogens within the community. Finally, some species may go locally extinct because their populations are so small that, just by chance, all the individuals die at some point without reproducing (see Key Concept 39.5).

The abundances of the species that persist in a community are determined by population-level processes discussed in **Chapters 39** and **40**.

The processes involved in community assembly are well illustrated by Krakatau, a small (17 km^2) volcanic island in the Sunda Strait of Indonesia (see the map in Figure 41.10). The volcano exploded in 1883, sterilizing what was left of the island and covering it with a thick layer of ash. Scientists quickly mounted expeditions to observe the return of life to Krakatau, and they have surveyed it periodically ever since. By 1886,

(A)

(B)

FIGURE 41.1 Vegetation Recolonized Krakatau **(A)** Beach-adapted plants such as beach morning glory (*Ipomoea pes-caprae*) were among the first organisms to take hold on the denuded island. **(B)** One century after a massive eruption left Krakatau an ash-covered shell, vegetation again covered much of the island.

10 plant species that grow on nearby tropical beaches had established themselves on Krakatau after their seeds floated to the island (**FIGURE 41.1A**), and wind had brought seeds or spores of 14 additional species of grasses and ferns. These first, or "pioneering," plant species had adaptations that allowed them to thrive in the intense sunlight and ashy soil of the barren, post-eruption environment, but they could not tolerate shade. Many of them disappeared from Krakatau by the 1920s after tree seeds had arrived and the forest canopy had closed in. Once forests developed, forest-dwelling birds and bats began to be attracted to the island, bringing new animal-dispersed seeds with them. In the 100 years after the volcano exploded, 400 plant species colonized and were recorded on the island, and 267 were still present during the 1980s in the lush tropical communities that once more covered the island (**FIGURE 41.1B**).

REVIEW & APPLY | 41.1

R Ecological communities are groups of species that live together at the same place and time. Ecologists often study subsets of a community because it is impractical to include all species and their interactions and all physical locations that the community occupies. Communities vary in all three aspects of their structure—the particular species they contain, the number of species, and the relative abundances of the species. Communities are dynamic—new species colonize, populations interact, and some species persist for shorter or longer periods, but then disappear from the community.

A

1. Researchers decide to study a coral-reef community in the Pacific Ocean. The community is complex and occupies a large area, so it is not practical to study all of it. Describe at least two ways in which the researchers might choose a manageable part of the community to study.
2. The table below shows the number of species in various taxonomic groups that were recorded on Krakatau during biological surveys of the island done in 1883 (immediately after the volcano erupted), in 1908, and in the 1980s.

	Number of species present		
Taxonomic group	**1883**	**1908**	**1980s**
Butterflies	0	5	52
Land birds	0	13	22
Ferns and lycophytes	0	12	65
Seed plants	0	88	202

   a. Visualize these data by drawing a graph with year on the *x* axis and number of species on the *y* axis; include five species-number curves, one for each taxonomic group and another that shows the total number of species.
   b. Use your graph to identify and describe aspects of community structure that changed over time on Krakatau.
   c. Discuss whether additional data are needed to draw any conclusions about whether other aspects of community structure also changed over time.

Krakatau provides a dramatic illustration of a general pattern: the species composition of communities changes over time. Species composition also changes over space. Are there general rules that govern which species we find where and when?

41.2 Communities Change over Space and Time

LEARNING OBJECTIVES

By the end of this key concept you should be able to:

41.2.1 **Identify one environmental factor that is more likely to cause spatial turnover in the species composition of an animal than of a plant community, and one that is likely to be more important for plants than for animals.**

41.2.2 **Distinguish three processes that cause communities to change over time.**

41.2.3 **Contrast two distinct patterns of community change that occur after a sudden disturbance.**

Ecologists have noted repeated patterns of spatial and temporal change, or **turnover**, in the species composition of communities. Let's take a closer look at these patterns and what they suggest about the processes involved.

Species composition varies along environmental gradients

We saw in Chapter 38 that physical conditions on Earth change over space. These spatial changes, called environmental gradients, can span large or small distances. Species vary in their environmental tolerances, so as we move along environmental gradients we encounter constantly changing communities, whether they are human-managed or natural. In tropical regions, for example, we find coffee plantations around 1,400 meters in elevation, where coffee grows best—reflecting its evolutionary origin in African highlands—and we find other crops at higher or lower elevations. Natural communities also change along environmental gradients. Krakatau presents, for example, elevational gradients from sea level to the 800-meter-high mountain top, gradients from wave- and wind-buffeted beaches to sheltered inland areas and from gentle ash-covered slopes to steep rocky cliffs. The species composition of Krakatau's plant communities reflects these gradients in environmental conditions, just as the vegetation of terrestrial biomes reflects climate (see Key Concept 38.3). Sandy beaches are dominated by low-growing herbs and vines that can withstand wind or wave disturbance and salt spray. Farther inland, outside the reach of waves, a fringe of wind-tolerant short trees and shrubs takes hold. Wind-sheltered inland slopes are dominated by taller trees festooned with vines, and an understory of shade-tolerant shrubs, herbs, and ferns. Krakatau's damp, chilly summit hosts a thick growth of dwarf trees covered with mosses—a "cloud forest." These spatial changes in community structure are easy to see from afar because plant species that are adapted to different environments tend to look different—they vary in growth form (whether short or tall, woody or herbaceous, climbing or self-supporting) or in leaf size, shape, or color (compare Figure 41.1A and B).

Spatial turnover in the species composition of communities usually indicates an underlying environmental gradient (**FIGURE 41.2**). In many cases (such as elevation), the gradient is obvious,

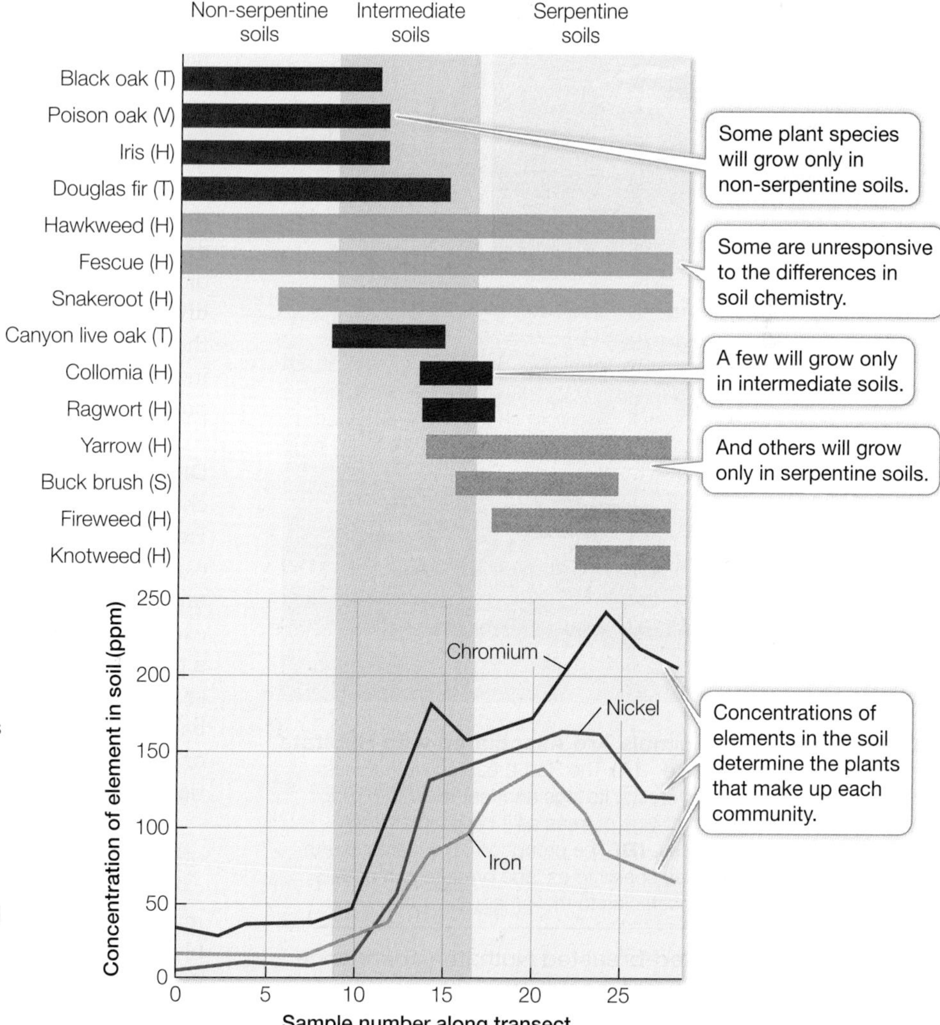

FIGURE 41.2 Change in Species Composition along an Environmental Gradient Ecologists sampled plant species along a transect running from non-serpentine to serpentine soils. They found that as concentrations of heavy metals in the soil changed, some species dropped out of the plant community and new ones appeared. Bars indicate samples in which a given species was present. The growth form is indicated for each species. (T = tree, S = woody shrub, V = woody vine, H = herbaceous.) (After R. E. Ricklefs. 1993. *The Economy of Nature*, [3rd ed.] New York: W. H. Freeman. Data from C. D. White. 1971. *Vegetation–Soil Chemistry in Serpentine Ecosystems*. Ph.D dissertation, University of Oregon, Eugene.)

but in others it can be hard to detect. The rock outcrops that provide habitat for Edith's checkerspot butterfly near San Francisco (see Figure 39.1), for example, contain soils derived from serpentine rock, which is unusually high in heavy metals such as chromium and nickel. If we measure off a transect—a straight line used for ecological surveys—running from a non-serpentine to a serpentine area and identify the plants that occur in evenly spaced samples along it, we will find that several species drop out and new ones appear partway along the transect, where concentrations of heavy metals jump.

When plant communities change along environmental gradients, animal communities do too. It is obvious that the distributions of herbivores such as Edith's checkerspot butterfly must overlap the distributions of the plant species they eat. But plants do more than provide food. They also modify physical conditions such as light, temperature, or humidity, and thereby make a location more or less favorable for animal species, which vary in their physical tolerances. Finally, the plant community determines **habitat structure**—the three-dimensional distribution of objects and characteristics of their surfaces in the habitats where animals obtain food, avoid predators, communicate, rest, and reproduce. Forests, for example, contain trees whose trunks, branches, and twigs provide surfaces along which animals can climb or crawl, crevices in which insects can hide, perches of varying diameters for birds, and nesting sites. Woody vegetation also presents obstacles to movement that may affect an animal's ability to catch prey or escape pursuing predators. Grassland or open desert vegetation presents fewer obstacles to movement, but also fewer sturdy structures for climbing, perching, or hiding. An animals' morphological, physiological, and behavioral traits shape how it makes a living—how it moves, obtains food, avoids predators, nests, and so on (see Key Concept 37.4). And since habitat structure shapes the effectiveness of alternative ways of making a living, animal species tend to be associated with particular plant communities (**FIGURE 41.3**).

(A) ***Sitta canadensis***

© Stuart Clarke /Alamy Stock Photo

(B) ***Antilocapra americana***

© Mcmaster Studio/Alamy Stock Photo

FIGURE 41.3 Many Animals Are Associated with Habitats of a Particular Structure **(A)** The Red-breasted Nuthatch is most often found in conifer forests. Its legs and feet allow it to cling to the trunks, branches, and twigs of trees as it picks small arthropods and seeds from crevices. **(B)** The pronghorn, the fastest North American land mammal, uses speed to escape predators. It prefers open, flat habitats without obstacles to movement.

Would you expect Red-breasted Nuthatch abundance to change along the environmental gradient shown in Figure 41.2?

Several processes cause communities to change over time

Communities change not only over space, but also over time. Krakatau is not unique in this regard—all communities, not just those in the early stages of assembly, are dynamic over both short and long time periods. Three major processes contribute to this dynamism: colonization and extinction, disturbance, and global change.

COLONIZATION AND EXTINCTION Ecological communities are likely to change even in a constant environment because new species arrive from time to time, and resident species may be excluded by the newcomers or may go locally extinct by chance (see Key Concept 39.5). Dispersal is an ongoing process that delivers a constant influx of new individuals to all but the most isolated locations. When individuals arrive at a location that already contains a population of their species, they add to its size (and genetic diversity); when those individuals are members of a novel species, they may establish a new population and add another species to the community.

DISTURBANCE An event that causes sudden environmental change—referred to as a **disturbance**—can alter communities. Examples of such events are volcanic eruptions, wildfires, hurricanes, landslides, floods, and activities of animals that alter habitat structure. Disturbance can occur on a small spatial scale—for example, when a falling tree creates a gap in the forest canopy, a pocket gopher creates areas of disturbed soil as it burrows in search of plant roots, or a large mammal deposits dung. Disturbance can also occur on a large scale, as when a volcanic eruption or forest fire—or a human farmer— influences existing communities over large areas.

As communities develop following a disturbance, their structure often undergoes a more or less predictable sequence of changes called **succession**. The sequence on Krakatau from a simple community consisting of a few pioneering species, to the gradual assembly of a species-rich and structurally complex community similar to the undisturbed tropical forest communities of the region, is typical of successional sequences following severe disturbances.

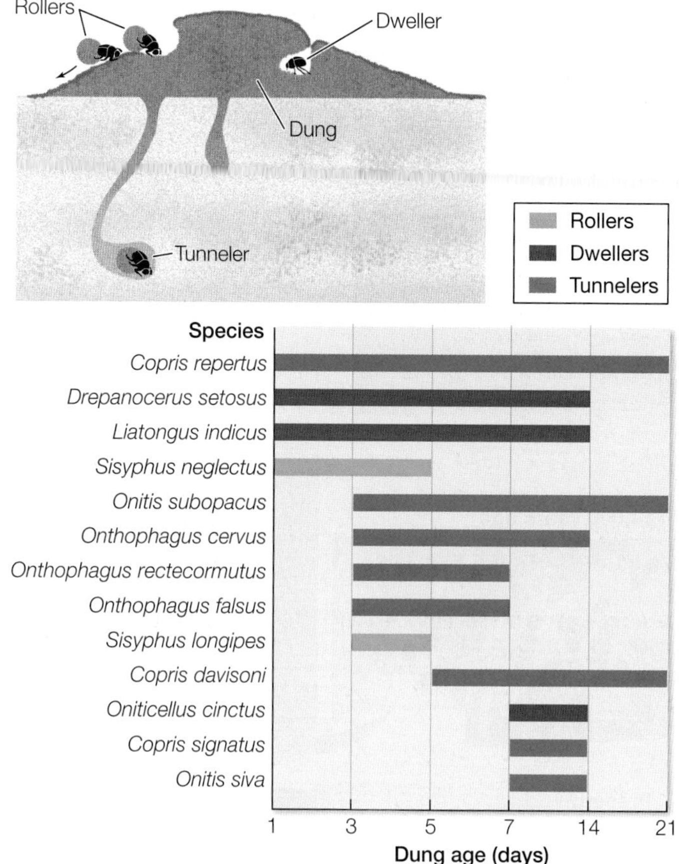

FIGURE 41.4 Dung Beetle Species Composition Changes over Time The Indian dung beetles represented in this diagram all consume elephant dung. Each bar indicates the time period when a species is present. Whereas *Copris repertus* (top) colonizes newly produced dung and is still present 3 weeks later, other species enter and drop out of the community at various times, depending on their adaptations for finding dung and for using dung of different ages. Thus the composition of the beetle community changes over time. Notice the different ways in which dung is used: some species roll balls of dung away and then bury them, some dwell in the dung, and some take bits of dung into chambers they dig under the dung pile. (After T. K. Sabu et al. 2006. *J Insect Sci* 6: 17; CC BY 2.5.)

Relatively predictable changes can occur in animal communities as well as plant communities. A fresh pile of elephant dung, for example, is colonized by a sequence of dung beetle species (family Scarabaeidae). These animals feed on the dung and lay their eggs in it. Some species typically arrive early, when the dung is wet, and are replaced by later arrivals as the dung dries and becomes aerated through activities of pioneering microbes and beetles (**FIGURE 41.4**).

Media Clip 41.1 **Dung Beetles**

Several factors are responsible for predictable features of post-disturbance community changes. Colonization, the process by which new species enter communities, has an element of chance—a population can't take hold until dispersing individuals by chance encounter the disturbed area and manage to survive and reproduce. But as time goes on, more and more successful colonization events occur, and the number of species predictably increases. Species also tend to colonize in a relatively predictable sequence because they vary in their dispersal ability. Pioneering plant species on Krakatau were those that produce abundant seeds that are readily dispersed by water or wind. Early-arriving dung beetles tend to be strong fliers with an excellent sense of smell, or are "hitchhikers" that ride on dung-producing animals and colonize dung as it is deposited.

A second factor that contributes to successional patterns is predictable post-disturbance environmental change. Physical processes, and the presence and activities of early-arriving species, modify environmental conditions in the disturbed site over time. As a result, pioneering and late-arriving species experience different environments, and the characteristics that make a species good at colonizing newly disturbed sites often decrease its ability to succeed later on. Pioneering plant species, for example, typically produce many small seeds that are easily carried long distances by the wind. These small seeds contain little stored food that the seedlings can use early in their growth. As a result, the seedlings are small and are easily outcompeted by larger-seeded competitors. Small-seeded species typically succeed only in uncrowded, open environments and can be maintained in communities over the long term only if there is ongoing, small-scale disturbance.

Animation 41.1 **Succession after Glacial Retreat**

On Krakatau, the tropical forests destroyed by the volcanic eruption eventually came back; in elephant dung, beetles and other decomposers eventually consume the dung, disappear, and the plant community where the dung was deposited returns to something resembling the pre-dung community. But a return to the original community is far from guaranteed. Instead, disturbance may push the system past a tipping point, causing a **state transition** (or "regime change") to a distinctly different community. The system may then stay in this alternative state and not return to its original state, even when the disturbance factor is removed. For example, intensive cattle grazing helped to foster the conversion of grasslands into shrublands in the U.S.–Mexico Borderlands (see opener to Chapter 38). Grasslands have not returned, even in places where cattle have been removed, because the initial overgrazing precipitated long-term changes in the soils and wildfire regimes of the Borderlands.

GLOBAL CHANGE Earth's physical environments are never static. Instead they have changed due to natural factors over large areas and over long spans of time, causing species to arise, evolve, go extinct, and shift their geographic distributions (see Key Concept 17.3). Any change in the physical environment of a location, or in the regional pool of species that can colonize it, can alter the structure of the resident community. We can often deduce the history of community change by uncovering and dating the fossilized remains of plants and animals present at a site. Pollen preserved in lake sediments of known age provides a record of the plants that occurred around wetlands in the past. And in dry uplands, fossils can be obtained from the middens

FIGURE 41.5 Species Composition Changes as the Climate Changes **(A)** A packrat midden. **(B)** Fossilized plant remains extracted from ancient packrat middens indicate that 14,000 years ago, when the climate was moist, valleys in the U.S.–Mexico Borderlands held lakes whose shores were clothed in piñon–juniper woodland. **(C)** In today's more arid climate, the lakes have dried up and the woodland has given way to desert grassland—and more recently to shrubland, following human disturbance (see the opener of Chapter 38). (After T. W. Swetnam et al. 1999. *Ecol Appl* 9: 1189–1206. Data from T. R. Van Devender et al. 1987; R. S. Thompson et al. 1993; J. L. Betancourt et al. 1990.)

(A) Packrat midden

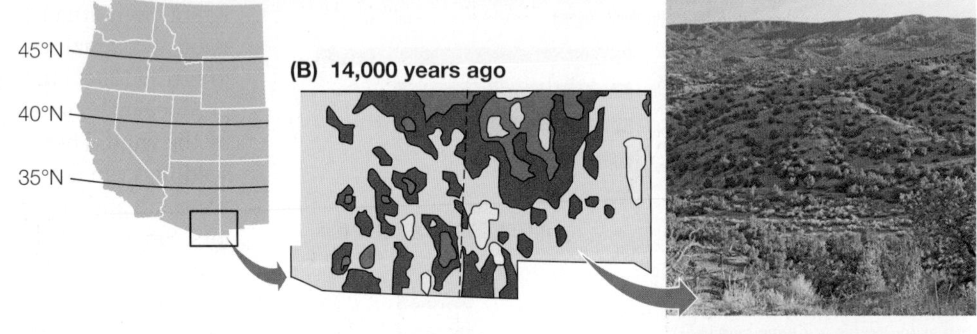

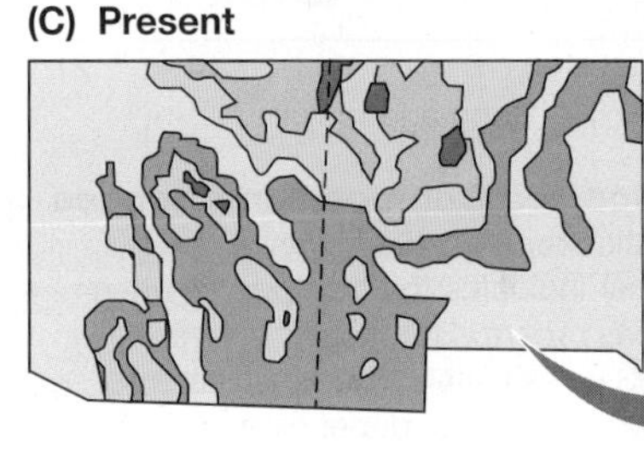

(refuse piles) left by rodents such as the packrats (*Neotoma* spp.) of North America, which have the peculiar behavior of hauling plant and animal material into rock shelters to make their nests (**FIGURE 41.5A**). Carbon-14 dating of packrat-midden fossils in the U.S.–Mexico Borderlands indicates, for example, that plant communities characteristic of desert grassland biomes now occupy sites that once supported shrubland or woodland communities (**FIGURE 41.5B** and **C**). From current biome–climate associations (see Figure 38.10), we can conclude that the climate of the Borderlands region became drier after the last ice age, causing species associated with moister conditions to retreat to higher elevations.

REVIEW & APPLY | 41.2

R The species composition of communities changes through space along environmental gradients because species have different tolerances for physical conditions, or require different habitats or resources that also change along environmental gradients. Ongoing processes of extinction and colonization cause the species composition of communities to change through time. Disturbance can cause abrupt changes in the physical environment and the community. After a disturbance, community structure sometimes changes in a relatively predictable sequence that eventually leads to a community that resembles the pre disturbance community. This process of succession is not universal, however, and disturbance sometimes leads to an alternative stable community very different from the original community. Global change in environmental conditions also leads to change in community structure.

A

1. Give an example of an environmental gradient that you have observed yourself. Explain what physical factors changed along it and what visible changes you saw in the species present.
2. Figure 41.5 shows that the vegetation of the U.S.–Mexico Borderlands shifted over the past 14,000 years from a piñon–juniper woodland to a desert grassland and shrubland biome. Use Figure 38.10 to estimate how the climate changed over that time period.
3. The gut microbiome of individual Hadza hunter–gatherers in Tanzania cycles back and forth during the year as these people shift their diet from more animal and vegetable matter in the tropical dry season to more berries and fruits in the wet season (see Key Concept 38.1). In what respect does this cycle resemble an ecological state transition, and in what respect does it resemble ecological succession?

We saw in Key Concept 41.1 that species composition, species number, and species abundances all are aspects of community structure. How do these attributes affect the way a community functions?

41.3 Community Structure Affects Community Function

LEARNING OBJECTIVES

By the end of this key concept you should be able to:

41.3.1 **Characterize the source of energy for primary producers, primary consumers, and higher-level consumers, respectively, in a typical terrestrial community.**

41.3.2 **Explain why net primary productivity (NPP) is less than gross primary productivity (GPP), and why the transfer of energy from one trophic level to the next is less than 100 percent.**

41.3.3 **Describe two different interpretations of the ecological niche.**

41.3.4 **Identify the two components of species diversity and describe a reason that diversity can increase community function.**

It is useful to think of an ecological community as a dynamic system (see Key Concept 1.2) similar to a human-made machine, with inputs, "internal workings," and outputs. The inputs to a community are energy and materials from the surrounding abiotic environment. The "internal workings" of the community—the metabolism and activities of its individuals, the dynamics of its populations, and the interactions among its species—produce outputs of transformed energy and materials, or a changed physical environment.

If we think of a community as a machine, we can measure **community function** as we would the function of a machine, in terms of its inputs and outputs. Community function can be measured by the amount of energy or matter that moves into and out of the community per unit of time and by the community's impacts on physical processes. The **flux**, or flow rate, of energy and materials through the community, and its impacts on physical processes, correspond in many cases to "goods" and "services" for humans, as we will discuss in Key Concept 41.5.

Looking inside the community "machine," we see that its components also have inputs and outputs. Wolves that feed on elk, for example, convert an input of elk biomass into an output of new wolf biomass. The output of seeds and fruits by many flowering plants, including coffee, requires a different type of input: visits by insect pollinators.

Energy flux is a critical aspect of community function

To develop the idea of community function, let's focus on how energy moves through communities. (The flux of matter through communities is coupled with that of energy, as will become more apparent in Chapter 42.) **FIGURE 41.6** shows, in simplified

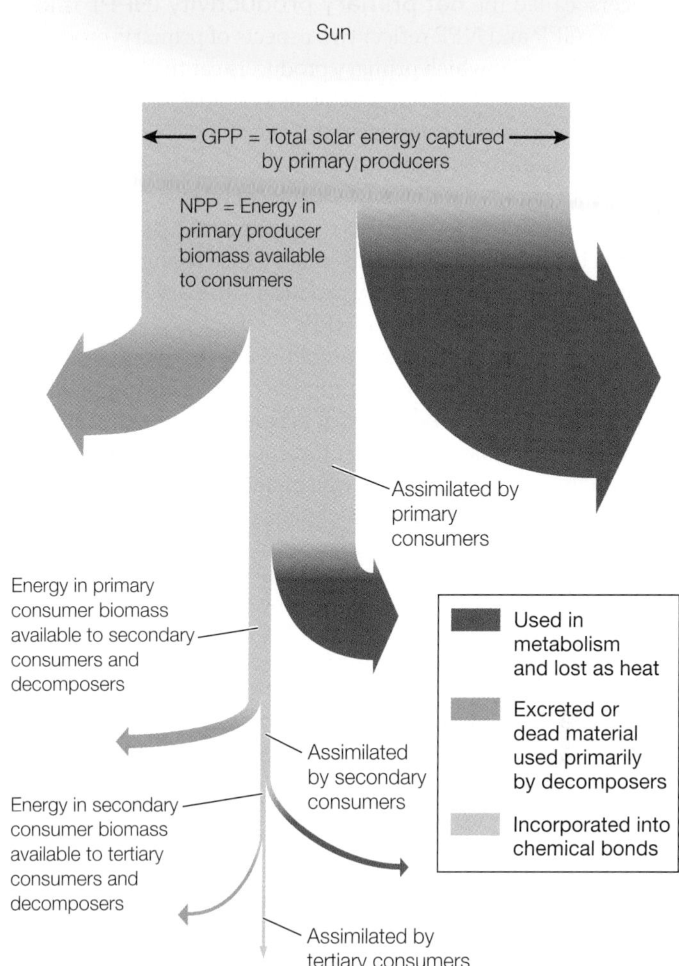

FIGURE 41.6 Energy Flow through Ecological Communities On average, the total amount of energy incorporated into the biomass of a trophic level per unit of time is about one-tenth that of the level it consumes. Of the 90 percent of energy *not* available to the next trophic level, some is waste or dead material used by decomposers, and some is lost during metabolism as heat.

Activity 41.1 Energy Flow through an Ecological Community

form, the flux of energy through the most common type of food web (for example, the Yellowstone food web shown in Figure 40.6), in which the primary producers are photosynthetic organisms that use solar energy. The pattern of energy flux is similar in those rare communities (such as deep-sea hydrothermal vents; see Chapter 18) in which primary producers are chemoautotrophs.

Energy enters communities through primary producers; the total amount of energy that they capture per unit of time and convert to chemical-bond energy (often simply called chemical energy) is a rate termed the **gross primary productivity** (**GPP**). Primary producers feed the carbohydrate products of photosynthesis into catabolic pathways that break down the carbohydrates to supply the energy (and materials) needed to fuel active transport and the construction of new macromolecules. Energy is lost as heat with every step in these metabolic pathways. As a consequence, the rate at which chemical-bond energy is stored in the tissues of primary producers, called the **net primary productivity** (**NPP**), is less than GPP. GPP and NPP reflect two aspects of primary-producer function: the rate at which primary producers capture energy, and the rate at which they convert it to forms available to other trophic levels. GPP can be measured from the rate at which primary producers take up carbon dioxide during photosynthesis, and NPP from the rate of CO_2 uptake minus its rate of release during respiration. New technologies allow us to measure such gas exchanges in nature, but it is often more convenient to use change per unit of time in the **biomass** of primary producers (the dry mass of their tissues) as an approximation for NPP.

The percentage of the energy stored in the dry biomass of one trophic level that is incorporated into the dry biomass of the next trophic level is called the **ecological efficiency** of energy transfer between those levels. Ecological efficiency is only about 10 percent—roughly 1 kilogram of new wolf biomass, for example, is produced for every 10 kilograms of elk biomass that wolves consume.

LINK The structure of ecological communities is related to their food webs (the webs of connections that describe which species consume which other species) and trophic structure (the positions of species in terms of the flow of energy through communities); food webs and trophic structure are explained in **Key Concept 40.3**.

Just as NPP measures primary-producer function, the biomass produced by primary consumers measures how well they function to convert primary-producer tissue into their own tissue, and so on for secondary and tertiary consumers. The low efficiency of these energy transfers limits the number of trophic levels found in a community—few communities contain more than four—because fewer and fewer individuals can be supported as one moves to higher trophic levels. (This is why we sometimes are urged to "eat low on the food chain": more humans can be fed if we act as primary rather than secondary consumers.)

There are two reasons for low ecological efficiency:

1. Not all the biomass at one trophic level is assimilated (ingested and digested) by the next one. Primary consumers do not ingest hard-to-get, toxic, or low-nutrient plant material; higher-level consumers do not eat hard-to-catch, toxic, or dangerous prey; and decomposers do not break down all nonliving organic material, as the existence of fossil fuels shows. Furthermore, some ingested matter is not digested and is excreted as organic waste. Plant cell walls, for example, contain cellulose, which most herbivores cannot digest.
2. Consumers use much of the chemical-bond energy they take in to fuel aspects of their own metabolism that do not add to their biomass through growth or reproduction. Through catabolic pathways, consumers convert energy in the biomass they have eaten into forms that do the work of assembling and disassembling their own biological molecules, maintaining their internal physiological state, moving their bodies, and so on. With each metabolic step, some energy is lost as heat.

Community function is affected by species diversity

The structure of a community influences how much energy enters it and how much is transferred through its food web, because each species has a unique **niche** that determines its function in the community. The niche is a central ecological concept with two important meanings. The first meaning refers to the physical and biological environments in which a species has a positive per capita growth rate—these are the environments where we might find that species. Figure 39.7 shows, for example, that the niche of the beetle *Rhyzopertha dominica* includes drier and warmer conditions than that of *Sitophilus oryzae*. Physical tolerances do not completely define a species' niche, however, because the biological environment—the presence of competitors, predators, mutualists, and so on—also affects population growth. Figure 40.5 shows, for example, that low

FIGURE 41.7 Species Richness and Species Evenness Contribute to Diversity These hypothetical communities of fungi (mushrooms) are all the same size (12 individuals), but they differ in species richness (3 versus 4 species) and species evenness, both of which affect diversity.

Activity 41.2 Measures of Species Diversity

intertidal zones are included in the niche of stellate barnacles only where rock barnacles are absent.

A species' niche also refers to what the species does in the community—its functional role, or "profession." This role is largely defined by how the species affects other species—what resources it uses and makes unavailable to other species, when and where it obtains them, how efficiently it does so, what it produces from them that other species can use, and whether it affects the physical environment. Many of these effects, but not all of them, are determined by the species' trophic interactions.

Once we think of functional roles, we can begin to see that community structure affects community function. Loss of wolves from Yellowstone National Park in the 1920s, for example, decreased the flux of energy from elk to secondary consumers and increased energy flux from aspen and willow to primary consumers (see Figures 40.6 and 40.7). But how do we go from specific examples like this to a more general understanding of how structure affects function? Ecologists have begun to tackle this problem by looking for broad patterns in structure–function relationships across communities. They have discovered that one important aspect of community structure—species diversity—influences community function.

Species diversity has two components, illustrated in **FIGURE 41.7**. One is simply the number of species in the community, called **species richness**. The other is how similar the species are in their abundances, a property called **species evenness**. A community that has four equally abundant species is more diverse than one with three equally abundant species, and also more diverse than one with four species in which most individuals belong to one of the species and few belong to the other three species. This second point may not be intuitive until we consider how species evenness influences communities: the properties of a community with a few abundant species will be defined mostly by those species, whereas the properties of a community with equally abundant species will reflect the influences of all of them. Ecologists sometimes use species richness as a simple measure of diversity, but often prefer a mathematical diversity index that incorporates both richness and evenness.

Community outputs often vary with species diversity. In a long-term study of prairie plant communities, for instance, David Tilman and his colleagues at the University of Minnesota cleared outdoor plots into which they planted seeds from 0 up to 32 native perennial prairie species. (The "zero" treatment controlled for unplanned seed dispersal into experimental plots.) After 3 years they harvested the plants and measured their aboveground biomass. They found that this measure of NPP—an aspect of community function—increased as the species richness of a plot increased (**FIGURE 41.8A**).

A positive relationship between species diversity and community function might be explained by "niche complementarity." Consider that species can be grouped into different "functional groups" that have somewhat different niches; that is, they use somewhat different resources or thrive under somewhat different conditions. Examples of different functional plant groups are shrubs and trees that tap nutrients and water from deeper soil layers than low-growing herbaceous plants do, and C_4 plants that grow better than C_3 plants in warm conditions but worse in cool conditions (see Key Concept 5.5). Species-rich communities may be more likely than species-poor communities to contain representatives of several such complementary functional groups that together make more complete use of available resources. Tilman and colleagues could test the niche complementarity hypothesis because they planted some plots with species belonging to a single functional group, and others with species belonging to up to five functional groups. They found that NPP increased with the number of functional groups in a plot (**FIGURE 41.8B**).

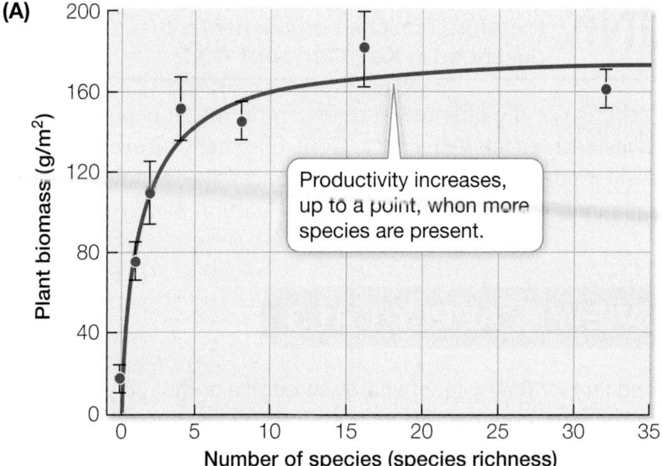

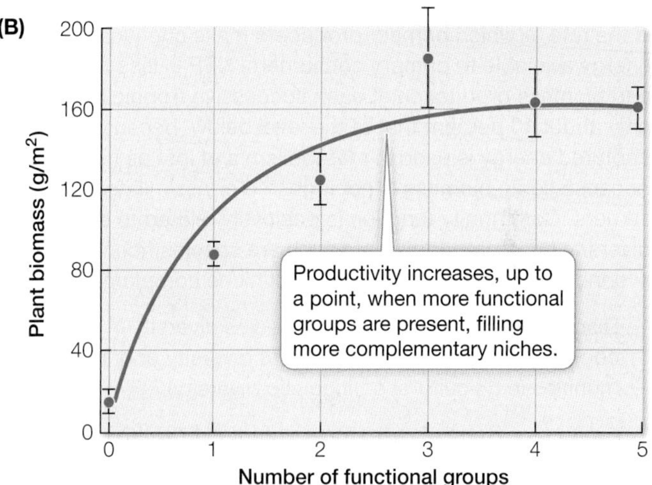

FIGURE 41.8 Species Richness and Number of Functional Groups Affect Primary Productivity **(A)** In 289 experimental prairie plots, NPP (measured as aboveground plant biomass produced per square meter after 3 years) increased with species richness. **(B)** The researchers found that NPP increased with the number of different functional groups of species. These results suggest that primary productivity is enhanced by a greater variety of species with complementary niches. The plots with zero species or functional groups were unplanted controls. Error bars indicate ± 1 standard error of the mean. See Appendix B for review of statistical concepts. (After D. Tilman et al. 1997. *Science* 277: 1300–1302.)

Use the unplanted controls (0 species) in (A) to estimate how much of the plant biomass in each treatment was due to uncontrolled seed input to plots.

LINK The evolution of resource partitioning—and therefore of niche complementarity—is described in **Key Concept 40.2**.

NPP is not the only measure of community function that increases with species diversity. Control of agricultural pests by natural predators and parasites is enhanced when multiple crops are planted together or when a diverse natural plant community borders agricultural fields. Diverse plant communities reduce soil erosion by maintaining more aboveground biomass through cycles of wet and dry years. And two recent studies of coffee pollination found that production was higher, and pollination rates were less variable between years, in plantations that were close to natural tropical forests, which harbor a great diversity of native coffee-visiting bee species.

REVIEW & APPLY | 41.3

R The flux, or flow rate, of energy through a community—one measure of community function—can be measured by the rate of biomass production at each successive trophic level. Gross primary productivity (GPP) measures the rate at which primary producers capture energy, and net primary productivity (NPP) is the rate at which primary producers make chemical-bond energy available to primary consumers. NPP is less than GPP, and biomass production at each successive trophic level is only about 10 percent that of the level below, because some captured energy is used in metabolism and lost as heat, and because some biomass is not ingested or assimilated by consumers. Community function is positively related to species diversity (the components of which are species richness and evenness) at least in part because of niche complementarity.

A 1. Discuss whether measures of species diversity in ecological communities can be applied to the diversity of human communities (e.g., cultural or linguistic diversity).

2. A group of German researchers studied how the diversity of bee species (both native species and the domesticated honey bee) visiting individual flowers in 24 Indonesian coffee plantations influenced fruit set (the percent of flowers that produced coffee beans). Their results are shown in the top graph at right (the best-fit regression line [see Appendix B] is $y = 58.56 + 1.81x$; the probability $r = 0$ is less than 0.001). (*Figure source*: A.-M. Klein et al. 2003. *Proc R Soc Lond B Biol Sci* 270: 955–961.) An American researcher then studied how, in a large Costa Rican coffee plantation, the number of native bee species visiting coffee bushes was influenced by distance to the nearest natural tropical forest. His results are shown in the bottom graph (bars with different letters are significantly different [$P < 0.05$]). (*Figure source*: T. Ricketts. 2004. *Conserv Biol* 18: 1262–1271.) Use these graphs to answer the following questions:

   a. Describe, using one sentence each, the primary conclusion you would draw from each graph.

   b. Imagine that an indigenous community in Costa Rica owns a piece of land that includes a square patch of remnant tropical forest that covers 100 ha (1 ha = an area of 100 × 100 m). Around this patch of forest is open land that was cleared for cattle pasture. The community wishes to plant coffee in the open land, and can afford to plant 21 ha. The community cannot import honey bees as pollinators and will rely on native bees instead. The grid to the right represents the 100-ha area of forest (green squares; all squares in the grid = 1 ha) and some of the surrounding open land. On the grid, shade in which areas, for a total of 21 ha, the community would do best to plant its coffee, and explain your choice in a few words.

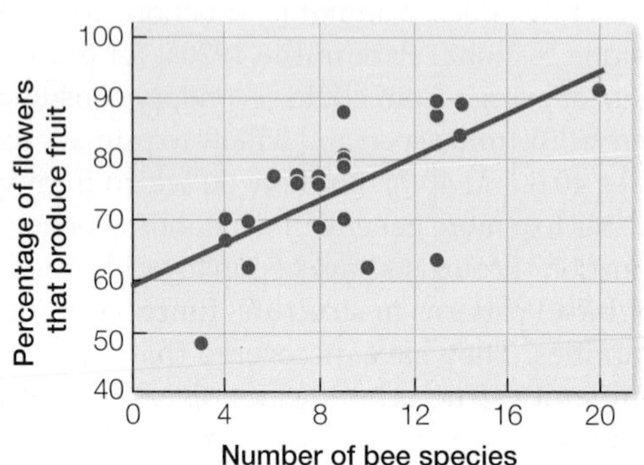

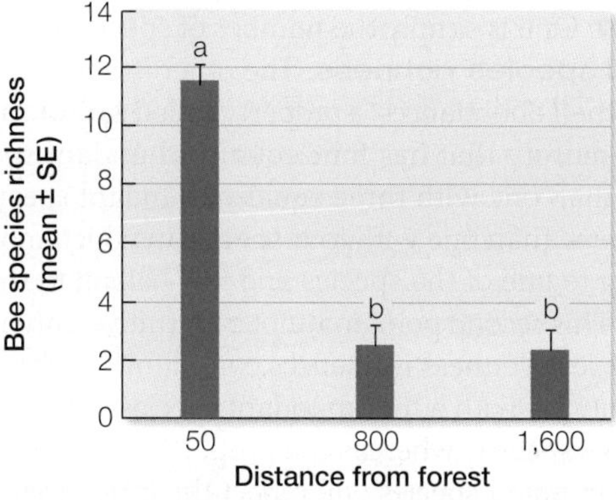

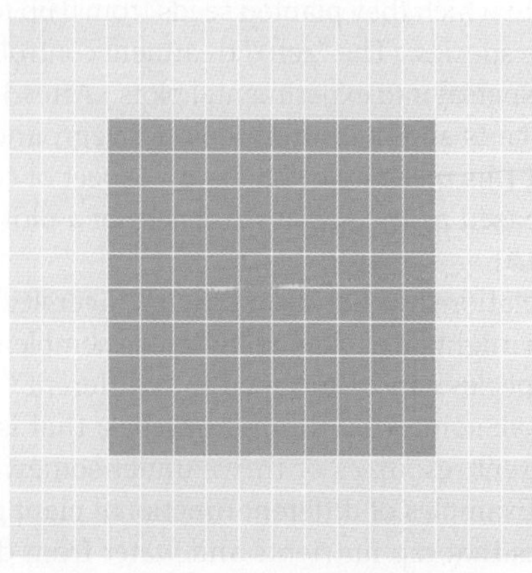

Species diversity is an important characteristic of communities because it affects how those communities function as ecological systems. But what factors determine species diversity in a particular community?

41.4 Diversity Patterns Provide Clues to What Determines Diversity

LEARNING OBJECTIVES

By the end of this key concept you should be able to:

41.4.1 Describe global patterns of variation in species diversity.

41.4.2 Use island biogeography theory to explain how species richness varies with the size and isolation of islands.

Species diversity varies dramatically among locations on Earth. Geographic patterns in diversity shed light on several factors that influence species diversity by modifying the balance between gain and loss of species from communities.

Species richness varies with latitude

When the German naturalist Alexander von Humboldt traveled around Central and South America two centuries ago, he noted that "the nearer we approach the tropics, the greater the ... variety of structure ... of organic life." This recognition that species richness increases toward the equator has been amply confirmed, with just one example shown in **FIGURE 41.9**. A 2-hectare plot of tropical rainforest in Malaysia can contain 227 tree species, whereas an equivalent plot of temperate forest in Michigan contains only 10 to 15. There are 60 bird species that breed in Greenland, 244 in New York State, 501 in Guatemala, and 1,746 in Colombia. And although some taxa appear not to follow the pattern—of some 20,000 known bee species on Earth, for instance, about half are found in temperate regions (especially in deserts) and half in the tropics—most taxa do, regardless of whether they are aquatic or terrestrial, what their trophic level is, and other aspects of their biology.

What is it about the tropics that supports such diversity? This question has intrigued generations of ecologists. A definitive answer is still in the future, but ecologists have identified several factors that could contribute to the pattern.

First, climate conditions over long time spans have been more stable in the tropics than in temperate regions. Tropical communities are certainly dynamic—they experience local disturbances (such as the fall of individual rainforest trees, which causes small disturbances)—but major disturbances at large spatial scales have been rare since the meteorite impact at the end of the Cretaceous (see Key Concept 17.2). Whereas higher latitudes have been subjected to eight glacial cycles over the last 740,000 years, the tropics were spared these "ice ages," which caused major shifts in the geographic ranges of species, and much extinction. Given relatively stable climates over millions of years, tropical species likely experienced fewer extinctions and more speciation events, leading to high diversity today.

Second, latitude influences climate. The tropics receive abundant solar energy input, which makes them warm and wet (see Key Concept 38.2). These conditions promote rapid growth of primary producers, so NPP, like diversity, increases toward the equator. High NPP can enhance diversity if greater energy input into food webs allows species to maintain larger population sizes and thereby have a lower risk of extinction. Greater energy flow through tropical communities could also allow species to be more specialized in their use of resources or habitats and therefore facilitate the coexistence of a larger number of species.

Third, animal diversity is higher in more structurally complex habitats. Variation in habitat structure appears to interact with greater niche specialization to amplify tropical diversity. Bird species diversity, for example, is low in forests where foliage occurs at only one height above the ground and high in forests where foliage occurs at many heights. If tropical species have more specialized habitat requirements, and if tropical vegetation is more structurally complex by virtue of greater plant species richness, then complex habitat structure and productivity could interact to increase tropical diversity.

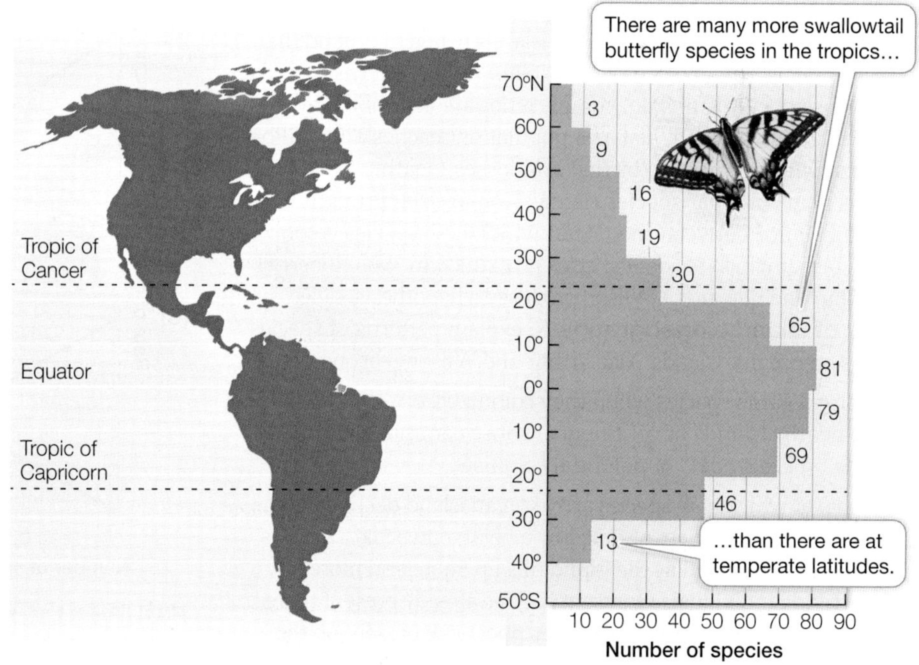

FIGURE 41.9 Species Richness Increases toward the Equator Among swallowtail butterflies (Papilionidae), species richness decreases with latitude both north and south of the equator. Similar latitudinal gradients of species richness have been observed in most other taxa. (After J. M. Scriber. 1973. *Psyche* 80: 355–373.)

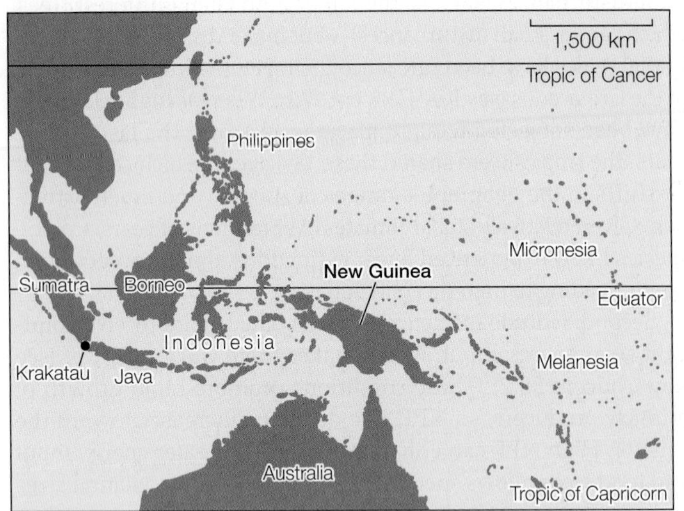

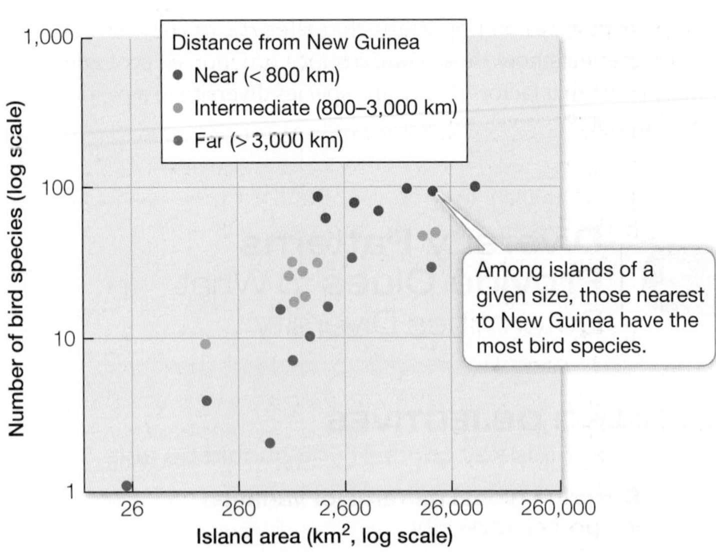

FIGURE 41.10 Area and Isolation Influence Species Richness on Islands On islands in the tropical eastern Pacific (left), species richness increases with island area and decreases with distance from New Guinea, the major source of colonists (right). (After R. H. MacArthur and E. O. Wilson. 1963. *Evolution* 17: 373–387.)

Ecologists still have much to learn about the relative contributions of these various factors to the puzzle of tropical diversity, but the research stimulated by that puzzle has demonstrated that disturbance patterns, primary productivity, and habitat structure all affect species diversity.

Species richness varies with the size and isolation of islands

Two regular patterns in the number of species found on oceanic islands have long fascinated ecologists (**FIGURE 41.10**). First, species richness is greater on large than on small islands. And second, species richness is greater on islands near a mainland than on more distant islands. What can explain these patterns?

As we saw with Krakatau, communities are assembled by colonization and extinction. Just as a population reaches an equilibrium size when additions (births) and losses (deaths) are equal, it seems plausible that species richness on an island might represent a balance between the rate at which new species colonize and the rate at which resident species go extinct. In 1963 Robert H. MacArthur and Edward O. Wilson proposed a hypothesis, called the **theory of island biogeography**, to explain patterns of species richness on oceanic islands. MacArthur and Wilson reasoned that:

- An island gains species when they colonize from elsewhere—from nearby islands or continents, whose communities make up the "species pool" of potential colonists.
- The rate at which new species arrive on an island declines as the island fills with species. At first, any arrival represents a species new to the island, but as the island fills up, more and more of the arriving individuals represent species already present there. If all species in the mainland species pool were present on the island, the rate of colonization by new species would be zero.
- The overall rate at which resident species are lost from the island—the extinction rate—increases as the island fills with species. In any time period, each species has some chance of

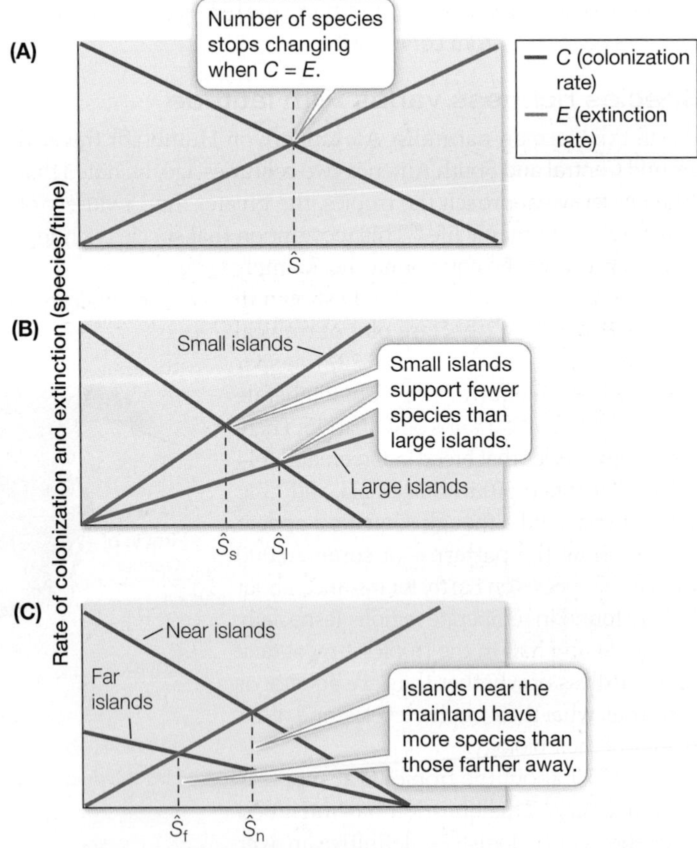

FIGURE 41.11 MacArthur and Wilson's Theory of Island Biogeography **(A)** Species richness reaches a stable equilibrium ($\hat{S}$) when the colonization rate equals the extinction rate. **(B)** Smaller islands have larger per-species extinction rates, and hence lower equilibrium species richness. **(C)** Islands more distant from the mainland have lower colonization rates, and hence lower equilibrium species richness. (After R. H. MacArthur and E. O. Wilson. 1963. *Evolution* 17: 373–387.)

going extinct from various causes. Therefore as the number of species increases, so must the total number that go extinct per unit of time.

- The number of species on an island—its species richness—stops changing when the colonization rate equals the extinction rate. When we plot colonization and extinction rates on a graph, this dynamic equilibrium species richness, $\hat{S}$, occurs where the two lines cross (**FIGURE 41.11A**).
- The equilibrium species richness should be greater on large than on small islands. As island area decreases, so does the size of each resident species population. Since small populations are more at risk of extinction than large ones are, the extinction rate for small islands should rise more steeply than that of large islands as a function of total species on the islands, and thus should intersect the colonization curve at a lower species richness (**FIGURE 41.11B**).
- The equilibrium species richness should be greater on close than on distant islands. Fewer wind- and water-borne seeds, and fewer dispersing animals, will reach a distant island than one closer to the mainland. Thus the colonization rate for close islands should start higher and decline more steeply than that of distant islands as a function of total species on the islands, and thus should intersect the extinction curve at a greater species richness (**FIGURE 41.11C**).

The theory of island biogeography successfully explains the patterns of species richness found not only on oceanic islands, but also on "islands" of one habitat type surrounded by a "sea" of a different habitat type. It has proved to be one of the most successful explanatory theories in ecology (**INVESTIGATION FIGURE 41.12**).

Activity 41.3 **Island Biogeography Simulation**

Animation 41.2 **Island Biogeography in the Florida Keys**

INVESTIGATION

FIGURE 41.12 The Theory of Island Biogeography Can Be Tested D. S. Simberloff and E. O. Wilson removed all the arthropods from (defaunated) four small mangrove islands of equal size but at different distances from the mainland. They then observed the process of community reassembly and compared the results with two predictions of island biogeography theory.[a]

HYPOTHESIS

The number of species on an island before it is experimentally defaunated will decrease with distance from a mainland source of colonists, and this number will be reestablished after the island has been recolonized.

METHOD

1. Census the terrestrial arthropod species on four small mangrove islands of equal size (11–12 m diameter) but at different distances from a mainland source of colonists.
2. Erect scaffolding and tent the islands. Fumigate with methyl bromide (a chemical that kills arthropods but does not kill the mangroves or other plants).

Courtesy of E. O. Wilson

3. Remove tenting. Monitor reassembly for the following 2 years by periodically censusing arthropod species.

RESULTS

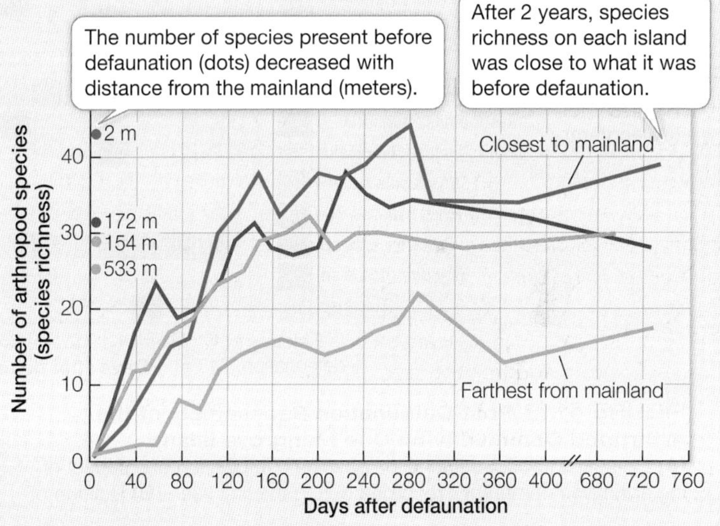

The number of species present before defaunation decreased with distance from the mainland. Two years after defaunation, each island had about the same number of species it had before the experiment.

CONCLUSION

The results agree qualitatively with two predictions of island biogeography theory.

ANALYZE THE DATA

1. Immediately after defaunation, when there were no species present on the mangrove islands, the rate of community reassembly was dictated by colonization rate (see Figure 41.11). (a) What does the theory of island biogeography predict about the effect of island distance on the initial rate of community reassembly? (b) Are the results consistent with this prediction? (c) How confident would you be in this test of the theory?
2. Discuss an additional prediction of island biogeography theory about equilibrium species diversity that is not tested by this experiment, and outline how you would test it.

[a]D. S. Simberloff and E. O. Wilson. 1969. *Ecology* 51: 934–937.

A similar **Analyze the Data** exercise may be assigned in **Achieve**.

THINK LIKE A SCIENTIST

Additional predictions of the MacArthur-Wilson theory can be tested

Before MacArthur and Wilson proposed their model of island biogeography, the prevailing hypothesis to explain variation in the communities found on different islands was the "niche hypothesis"—each island has a particular mix of niches that supports a specific community. In Simberloff and Wilson's mangrove island experiment, the niche hypothesis predicts that each island will reassemble its original arthropod community, and that the number of species will stop changing when the full complement of species is restored. Use your understanding of the process of science and your quantitative reasoning to answer the following questions.

1. **FIGURE 41.13A** shows the original community and post-defaunation census results for a portion of the arthropod community on one mangrove island in the Simberloff and Wilson experiment. In the empty right-hand column, labeled "After reassembly," fill in the boxes of species that the niche hypothesis predicts would be present on the island once the community was completely reassembled.
2. Compare the actual observed results of reassembly one year after defaunation (the "364" column) with your predictions in the "After reassembly" column. Assuming that reassembly is complete after one year, does the species composition of the island at that point support the niche hypothesis?
3. **FIGURE 41.13B** shows the accumulation of arthropod species in three functional groups—herbivores, carnivores, and parasites—for the same island. Summarize the pattern of species accumulation and discuss what might explain it.

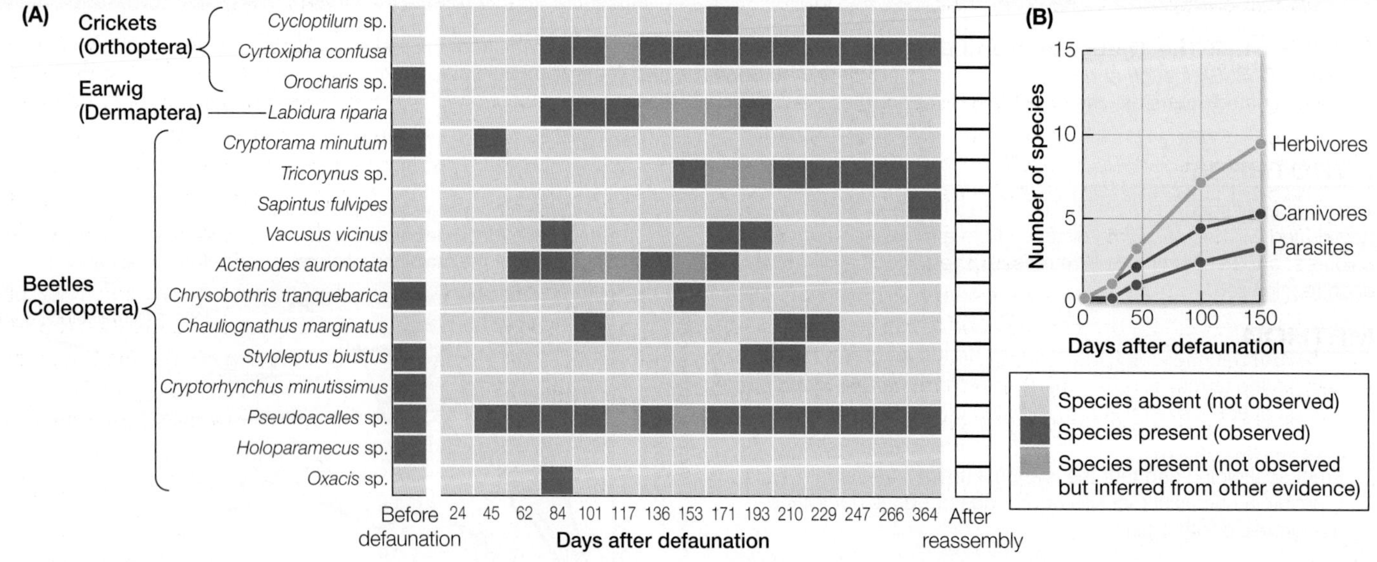

FIGURE 41.13 Post-Defaunation Reassembly of the Arthropod Community on One Mangrove Island **(A)** Original and post-defaunation species composition of a subset of the arthropod community recorded during the first year. **(B)** Buildup of arthropod species richness after defaunation. The insect species have been assigned to three functional groups—herbivores, carnivores, and parasites. (After D. S. Simberloff and E. O. Wilson. 1969. *Ecology* 50: 278–296; J. W. Glasser. 1982. *Am Nat* 119: 375–390.)

REVIEW & APPLY | 41.4

R The striking trend toward higher species diversity as one approaches the tropics could result from climate stability over long time periods; warm, wet climate conditions that enhance primary productivity; and habitat complexity. Patterns of increasing diversity with the size of oceanic islands and decreasing diversity with isolation of oceanic islands are successfully predicted by the hypothesis that species richness reflects a dynamic equilibrium in which rates of colonization by new species are balanced by rates of extinction of resident species. The same processes affect species diversity in habitat patches embedded in a "sea" of a different habitat.

A

1. As is true of terrestrial tropical rainforests, tropical coral reefs are among the most species-rich communities on Earth. Identify at least one factor that might explain the high diversity of coral reefs.
2. Analyze what Figure 41.11B and C indicate about the predicted rate of change in species composition (species turnover rate) at equilibrium on small versus large islands and on islands distant from versus close to a mainland.

We will now consider how our knowledge of community structure and function can be applied to conserving and restoring ecological communities.

41.5 Community Ecology Suggests Strategies for Conserving Community Function

LEARNING OBJECTIVES

By the end of this key concept you should be able to:

41.5.1 Define and provide at least one example each of a "good" and a "service" provided by ecological communities.

41.5.2 Explain why species diversity and abundance of critical species are often used to evaluate the effectiveness of interventions to restore community function.

41.5.3 Apply island biogeography principles to explain why a system of natural areas designed to conserve species diversity should contain large habitat patches that are clustered or connected by corridors.

The benefits that we derive from natural ecological communities and human-managed ones (such as agricultural communities) are commonly called their "goods and services." The goods include resources such as food, clean water, clean air, building materials, and fuel; the services include beneficial processes such as flood control, soil stabilization, pest control, pollination of plants, and climate regulation. The ability of an ecological community to provide critical goods and services often depends on its diversity. How can we use this knowledge to conserve community function?

Ecological communities provide humans with goods and services

Humans, like all species, depend on the interactions that occur among species within communities, and between communities and their physical environments. A short list of some of the goods and services provided by communities—also known as **ecosystem services**—is given in **TABLE 41.1**.

Goods that are of direct biological origin are obvious: everybody knows that we get food from plants and animals, wood from trees, wool from sheep, and aesthetic enjoyment from flowers and birds. The roles of properly functioning communities in providing ecosystem services are less obvious, however, and we often take them for granted—until something goes wrong.

European settlers of Australia learned the hard way about the importance of the services provided by decomposers. They brought cattle along when they arrived in 1788. From the beginning, however, the Australian cattle industry ran into problems: native dung beetles, adapted to the dry, fibrous dung of forest marsupials, shunned the wet manure produced by cattle in open pastures. As a consequence, manure piled up higher and deeper. The accumulated dung not only reduced pasture productivity (because the nitrogen in the dung was not being recycled, and because cattle do not eat fouled plant material), but also provided a perfect nursery for native bush flies and blood-sucking buffalo flies. Populations of these pests exploded.

In 1964, Australia's national scientific organization, the Commonwealth Scientific and Industrial Research Organisation (CSIRO), finally tackled the problem. Insect biologists traveled the world to find dung beetles that could process cattle dung in Australian environments without disrupting native beetle communities—an important precaution, since many well-intended introductions have had unforeseen, detrimental impacts (see Key Concept 40.3).

TABLE **41.1** | Some Major Ecosystem Goods and Services

Good or service	Community process	Examples
Food production	Trophic interactions	Production of wild food; production of crops and livestock
Materials production	Trophic interactions	Production of lumber, fuel, fiber
Pollination and seed dispersal	Plant–pollinator and plant–disperser interactions	Plant reproduction and dispersal
Maintenance of fertile soil	Decomposition; composition of vegetation; plant– and animal–microbe mutualisms	Nitrogen fixation, nutrient recycling, erosion control
Waste treatment	Decomposition	Breakdown of toxins and wastes
Pest control	Predator–prey, host–parasite, and competitive interactions	Decreased pest abundance
Water supply	Regulation of water infiltration and runoff by vegetation structure and animal activity	Retention of water in watersheds, reservoirs, and aquifers
Climate regulation	Metabolic gas exchange	Regulation of greenhouse gases and cloud formation
Disturbance control	Composition and biomass of vegetation	Damping of storm winds and wave surges; flood control
Recreational and cultural opportunities	Ecological processes; species diversity	Ecotourism, outdoor recreation; aesthetic, educational, spiritual, scientific values

Between 1968 and 1984, CSIRO introduced more than 50 species of beetles. The project was successful: soil nitrogen content improved, pasture productivity increased, fly numbers plummeted, cattle became healthier, and native dung beetles were unharmed.

Ecosystem services have economic value

Ecosystem services have economic value, as the Australian Dung Beetle Project shows. Some services related to food production are relatively straightforward to value because they are included in the price of the food. In the United States, for instance, wild pollinating insects (including native bees but not domesticated honey bees) contribute $4 billion annually to crop production. The values of many other services—such as regulation of the composition of Earth's atmosphere (which we will discuss in Chapter 42)—are more difficult to estimate because they are not included in commercial markets, but this does not mean they lack economic value.

Consider the case of New York City's water, which comes from reservoirs in the Catskill Mountains. Natural ecosystems in the Catskills provided the city with pure water for many years. But conversion of forests into agricultural fields and housing developments eventually began to decrease absorption of rain and snow by the soil, which increased the discharge of sewage, fertilizers, pesticides, and sediment into streams. New York was faced with the prospect of building a new $6–$8 billion water treatment facility. When the city considered alternatives, it realized that it could meet water quality standards by investing $1.5 billion in land protection and better sewage treatment in the Catskills. Although expensive, this natural alternative cost a lot less than the technological alternative of a new treatment plant. New York's Long-Term Watershed Protection Plan, completed in 2006, has been a complete success in providing water of high quality at low cost to consumers.

The New York example is not unique. Many natural communities can provide more goods and services, at lower cost, than human-engineered substitutes can. But as we have seen, the community functions that deliver those goods and services depend on species diversity. Ecological communities can best meet our needs if we do not manage them so intensively, or disturb them so severely, that we lower their diversity.

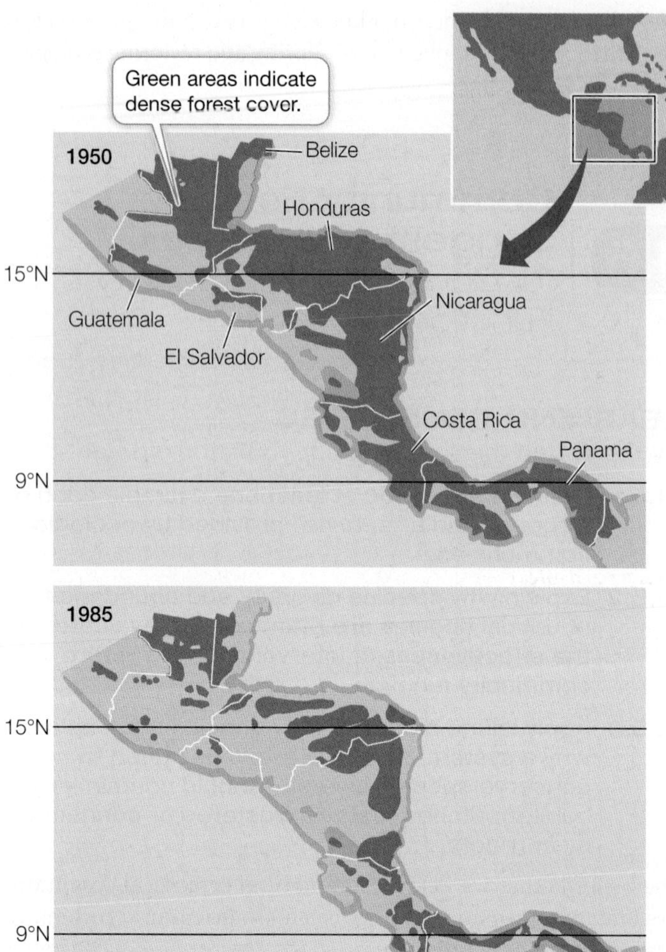

FIGURE 41.14 Habitat Fragmentation in Tropical Forests Less than half of the moist tropical forest that existed in Central America in 1950 (top) remained by 1985 (bottom), most of it in small patches. Although the rate of deforestation slowed somewhat after 1985, an additional 19 percent of the forest had disappeared by 2005, and deforestation continues today. (Maps after P. Utting. 1993. *Trees, People and Power: Social Dimensions of Deforestation and Forest Protection in Central America*. UNRISD and Earthscan. London.)

Island biogeography suggests strategies for conserving community diversity

Key Concept 38.5 described how humans are converting natural communities into less-diverse, human-managed communities such as croplands, pastures, and urban settlements. As a result, once-continuous areas of natural habitat are being reduced to scattered fragments—habitat islands embedded in human-modified landscapes (**FIGURE 41.14**). We expect such **fragmentation** of habitat to cause losses of species from communities because it reduces the total amount of habitat and the size of habitat patches, and it increases inter-patch distances. Given what you have learned about metapopulations (see Key Concept 39.5) and island biogeography, you now can predict the results. Populations become smaller as habitat area shrinks and are therefore more prone to extinction. The human-dominated landscape surrounding the fragments may serve as a barrier to dispersal, decreasing colonization rates. The net consequence is lower species richness in the fragments than in the original continuous habitat.

The insight that species richness reflects a balance between colonization and extinction suggests that the detrimental effects of fragmentation can be minimized by enhancing colonization and reducing extinction. Colonization rates can be increased by clustering habitat fragments close to one another or to a large "mainland" patch, or by connecting fragments with dispersal corridors (see Key Concept 39.6). Extinction rates can be minimized by retaining at least some large patches of the original habitat and by preserving the capacity of remaining patches to support healthy populations.

Several of these predictions have been tested in a remarkable large-scale experiment. Landowners in Brazil agreed to preserve tropical forest patches of certain sizes and configurations laid out by biologists (**FIGURE 41.15**), and they allowed the biologists to

FIGURE 41.15 A Large-Scale Study of Habitat Fragmentation Biologists studied patches of tropical rainforest near Manaus, Brazil, before and after the parcels were isolated by forest clearing. The results of the study demonstrated that small, isolated habitat fragments support fewer species than larger or less isolated patches of the same habitat. (After T. E. Bierregaard, Jr. et al. 1992. *BioScience* 42: 859–866.)

Animation 41.3 Fragmentation Effects

survey the parcels before and after the intervening forest was cleared for pasture. Species soon began to disappear from the patches. Monkeys that travel over large areas of forest were lost quickly, along with army ants and birds that follow the ant columns. Species were lost more rapidly in isolated fragments than in fragments of equal area that were connected to nearby unfragmented forest, and small isolated fragments lost species more rapidly than larger isolates.

Results of studies such as these now guide land-use planning worldwide. They suggest that dispersal corridors and large habitat patches are necessary to maintain healthy, diverse communities.

Trophic cascades suggest the importance of conserving certain species

It is fine to plan for dispersal corridors and large patches of habitat, but species differ considerably in the area of habitat they need and in the kinds of dispersal corridors they will use. How big should natural areas be? What constitutes an effective corridor? These decisions can often be facilitated if they focus on species that play particularly important roles in communities.

The wolves of Yellowstone National Park provide an example of a species whose presence has a major effect on community structure and function (see Key Concept 40.3). Wolves are crucial for maintaining healthy aspen forests and watersheds via trophic cascades that also involve elk, aspen and willow, and beavers. Conservation planning is now targeting such critical species. The Yellowstone to Yukon Conservation Initiative (**FIGURE 41.16**), for example, has the goal of maintaining a continuous corridor of wolf habitat between Yellowstone National Park and areas of similar habitat to the north, which should help maintain the Yellowstone wolf population and hence a healthy Yellowstone ecosystem.

The relationship of diversity to community function suggests strategies for restoring degraded habitats

We have seen that even highly disturbed communities, such as those on Krakatau, can recover. This insight from community ecology suggests that degraded ecosystems can often be restored—the goal of restoration ecology (see Key Concepts 38.5 and 40.3). The relationship between species diversity and community function suggests that one of the best ways to restore degraded communities is to restore the diversity of functional groups (**INVESTIGATION FIGURE 41.17**).

As we have also seen, however, disturbance sometimes results in an ecological state transition to a community that is very different from the pre-disturbance community. In such cases it may be difficult to reverse the transition and restore the function of the original community. This possibility is a sobering reminder that restoration is challenging and requires deep understanding of individual communities and their physical environments.

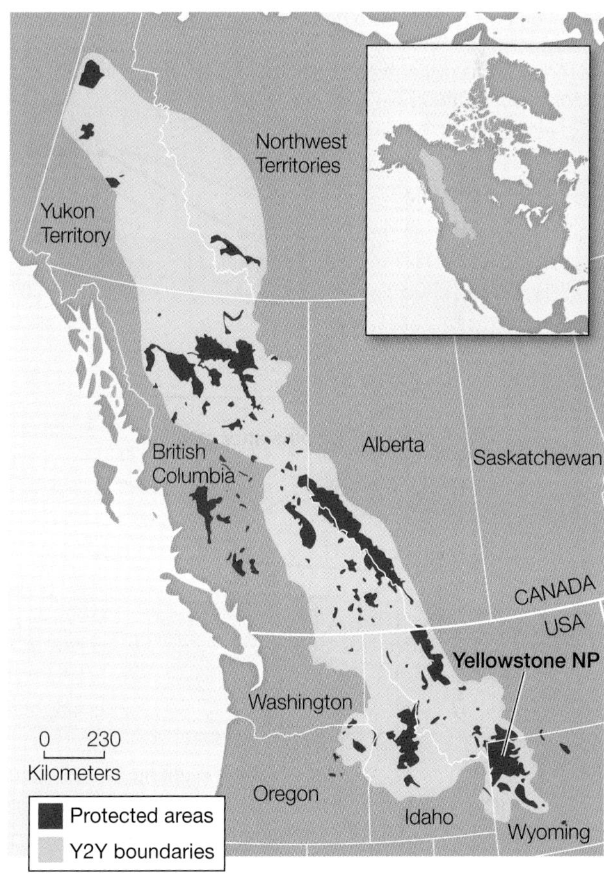

FIGURE 41.16 The Yellowstone to Yukon (Y2Y) Conservation Initiative Y2Y is a nonprofit organization that works with governments, non-government agencies, businesses, Native peoples, and other land managers to connect the mountain ecosystems between Yellowstone National Park in the USA and the Yukon Territory in Canada with corridors through which wildlife can move freely. This map outlines the territory identified on scientific grounds as Y2Y's target for conservation (yellow), and also shows areas that are protected at this time (purple). (After J. L. Pearce et al. 2008. *Biol Conserv* 41: 908–924.)

INVESTIGATION

FIGURE 41.17 Species Richness Can Enhance Wetland Restoration In one large-scale field experiment, ecologists compared different methods for restoring denuded areas of the Tijuana Estuary, a wetlands environment near San Diego, California. They found that several measures of community function improved more rapidly in species-rich than in species-poor plantings.[a]

HYPOTHESIS

Faster progress toward restoring the community's original function will be made by planting more rather than fewer species.

METHOD

1. In an area of wetland denuded of vegetation, mark off replicate small experimental plots, all of the same size.
2. Choose 8 native species typical of the region. Plant some plots with only one of the 8 species, others with different 3-species subsets, and others with different 6-species subsets.
3. Plant the same total number of seedlings per plot. Leave some plots unplanted as controls.
4. Return over the next 18 months to measure the vegetation and ecosystem nitrogen levels.

RESULTS

Vegetation covered the bare ground more quickly in plots planted with more species (A). Species-rich plots developed complex vertical structure more quickly (B) and accumulated more nitrogen in plant roots per square meter of area (C).

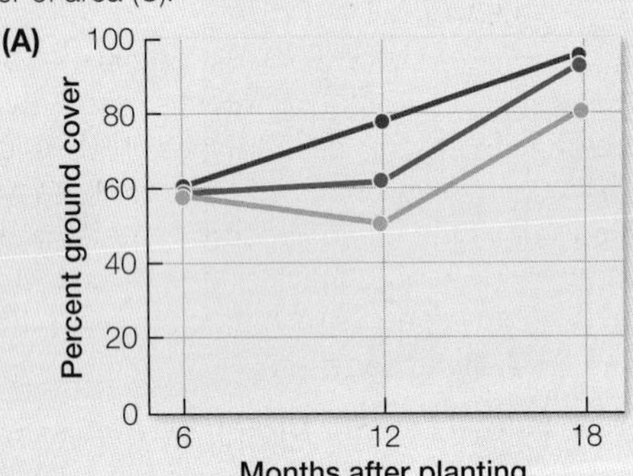

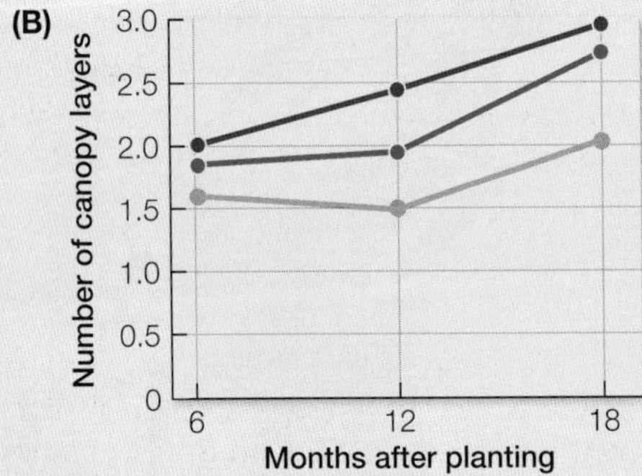

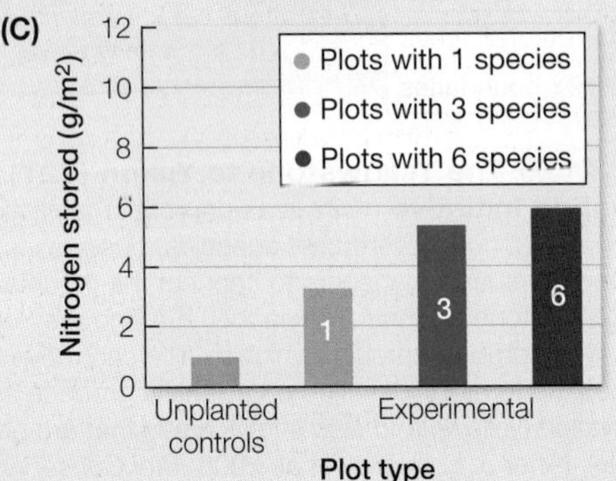

CONCLUSION

Planting more species into a community leads to more rapid restoration of wetland primary productivity.

ANALYZE THE DATA

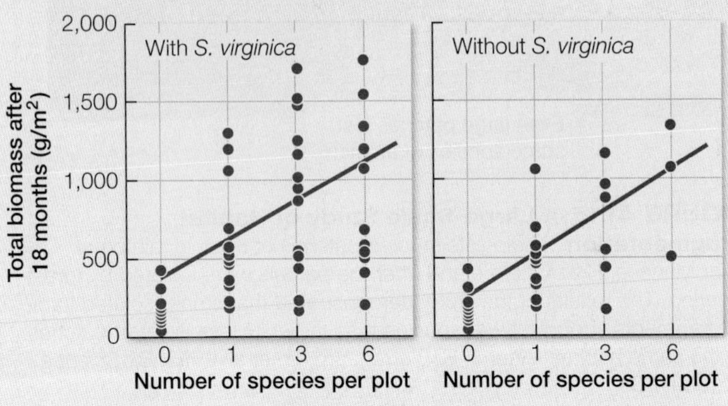

1. The unplanted control plots (graph C) showed nonzero values of stored nitrogen after 18 months. Is this a surprise? Why or why not?
2. In addition to using the niche complementarity hypothesis (see Key Concept 41.3) to explain a positive effect of species diversity on community function, ecologists have proposed an alternative "superspecies" hypothesis. Communities with more species may be more likely, just by chance, to include some "superspecies" that are exceptionally good at predation, or at photosynthesizing, or at whatever they do, and so have a strong influence on total community output. One of the eight species in the Tijuana Estuary study, *Salicornia virginica*, contributed a disproportionate share of the biomass, cover, and nitrogen to plots in which it was planted. In the two graphs above, each circle represents the performance of a particular plot; the lines indicate best-fit regressions (see statistics primer in Appendix B). Based on these two graphs, is the greater chance of including *S. virginica* in species-rich plantings the likely explanation for the positive diversity–function relationship?

[a]G. H. Keer and J. B. Zedler. 2002. *Ecol Appl* 12: 456–473; J. C. Callaway et al. 2003. *Ecol Appl* 13: 1626–1639.

A similar **Analyze the Data** exercise may be assigned in **Achieve**.

REVIEW & APPLY | 41.5

R Natural and human-managed communities provide many essential goods and services to humanity that often cost less and are of higher quality than what our technology can produce. Consequently, it pays to conserve and restore natural communities and their functions. Knowledge of how species diversity and food web structure affect community function, and of how colonization and extinction dynamics affect community structure, has suggested an effective strategy for maintaining and restoring community diversity and function: focus on conserving critical components of food webs, facilitate colonization rates, and reduce extinction rates.

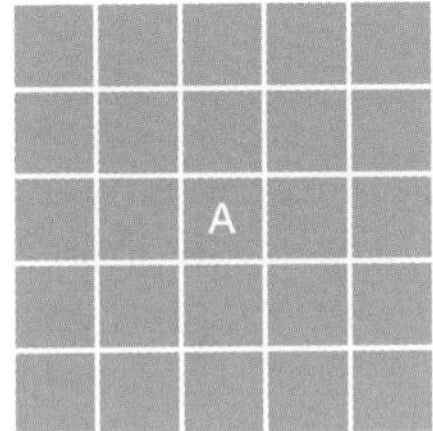

Unfragmented

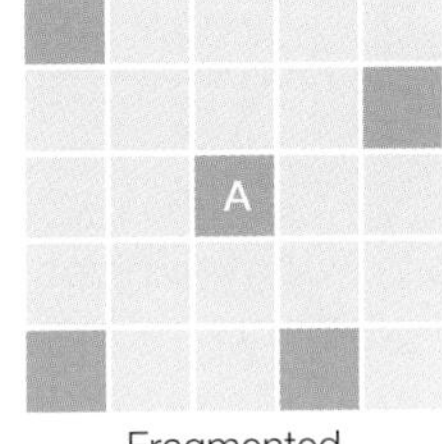

Fragmented

A 1. The figure on the left shows a large area of continuous unfragmented forest (indicated by green squares). The figure on the right shows the same area after much of the forest has been converted to pasture (indicated by yellow squares). Draw a diagram showing the rates of species colonization and extinction for the central forest patch ("A") as a function of the number of species in this patch before forest fragmentation. Now add colonization and extinction curves after forest fragmentation. Use your diagram to predict the effect of fragmentation on equilibrium species richness in patch "A." (*Hint*: Recall Figure 41.11.)

2. Since 1970, coffee cultivation in the Neotropics has shifted away from the tradition of planting in the shade of diverse tropical forest vegetation, toward practices intended to increase yields by reducing shade (for example, by planting underneath a monoculture of timber trees), using modern high-yield cultivars planted at high density, pruning the coffee bushes to stimulate flower production, and applying manufactured fertilizers and herbicides. The table below summarizes information on the economics of traditional coffee farms, semi-traditional farms that use some modern cultivation methods, and farms that use the full range of modern production technology.

Cultivation methods	Coffee yield (kg/ha)	Production cost (U.S.$/ha)	Cost/kg
Traditional	317	$269.47	
Semi-traditional	953	$1,092.00	
Modern	1,397	$1,738.94	

Source: I. Perfecto et al. 1996. *BioScience* 46: 598–608.

a. Calculate values for the Cost/kg column by dividing the values in the third column by those in the second column. Which cultivation method is the most economical (lowest cost per kilogram)?

b. We saw in Review & Apply to Key Concept 41.3 that traditional planting in tropical forest exposes coffee to higher bee diversity, which increases yields of coffee. Yet the modern farms in the table above have the highest yields per ha. Identify two aspects of modern cultivation methods that more than compensate for the loss of natural pollination services and discuss how they affect per-kilogram production costs.

In this chapter we have discussed factors that influence the structure of communities and the interactions that shape the flow of energy through them. We have seen how community structure affects a community's function, including its outputs of goods and services. But outputs are limited by inputs, and to understand what determines inputs to communities, we need to consider their abiotic environments. The next chapter will apply this ecosystem perspective and consider how humans are influencing the global ecosystem.

Answer: Croplands are human-managed ecological communities that function in ways similar to natural communities. Extrapolating from our knowledge of natural communities, we can predict that coffee will produce good crops with little human intervention when it is grown in the shade of natural forests, where diverse species provide pest control, pollination, and other ecosystem services. This prediction is borne out: in recent economic comparisons from southern Mexico, sun cultivation produced high yields but at high monetary cost for labor, pesticides, and fertilizers, whereas low-intensity cultivation avoided most of these costs, was more profitable overall, and caused less pollution from chemicals.

41 VISUAL SUMMARY

41.1 Communities Contain Species That Colonize and Persist

FIGURE 41.1

Courtesy of Robert Whittaker

- A **COMMUNITY** is a group of species that occur together at the same time and place. Aspects of community structure include **SPECIES COMPOSITION**, species number, and species relative abundances (**FIGURE 41.1**).
- A community contains species that have colonized a location and persisted there.

QUESTIONS

1. Describe the processes involved in community assembly.
2. Explain when local extinction of species occurs during the process of community assembly.

41.2 Communities Change over Space and Time

FIGURE 41.4

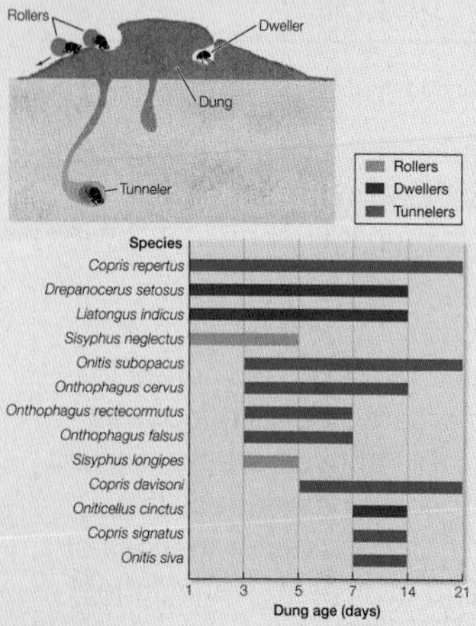

Rollers
Dweller
Dung
Tunneler
Rollers
Dwellers
Tunnelers
Species
Copris repertus
Drepanocerus setosus
Liatongus indicus
Sisyphus neglectus
Onitis subopacus
Onthophagus cervus
Onthophagus rectecornutus
Onthophagus falsus
Sisyphus longipes
Copris davisoni
Oniticellus cinctus
Copris signatus
Onitis siva
1 3 5 7 14 21
Dung age (days)

Go to ANIMATION 41.1

- **TURNOVER** in species composition of communities in space occurs along environmental gradients.
- Turnover also occurs through time, even in constant environments, due to ongoing extinction and colonization.
- After **DISTURBANCE**, the structure of a community sometimes changes in a predictable sequence called **SUCCESSION** that eventually reestablishes a community similar to the pre-disturbance one (**FIGURE 41.4**). Disturbance also can precipitate an ecological **STATE TRANSITION** in which the pre-disturbance community is never recreated.
- Global change in environmental conditions also causes temporal change in community structure.

1. Explain why animal species often are associated with particular plant communities.
2. Compare and contrast ecological succession and temporal species turnover.

41.3 Community Structure Affects Community Function

FIGURE 41.6

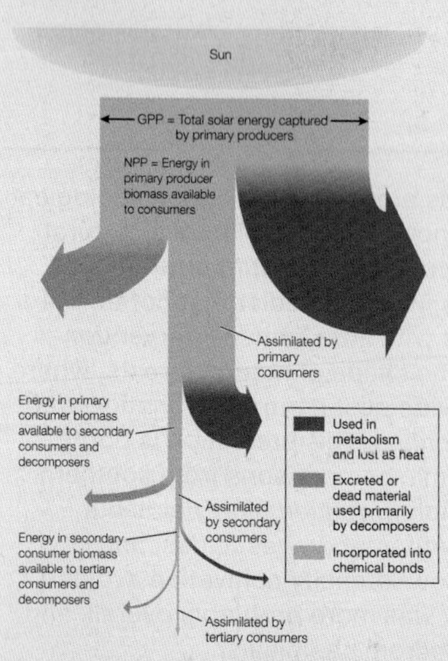

Go to ACTIVITIES 41.1 and 41.2

- An ecological community is a dynamic system with inputs, internal workings, and outputs that are aspects of **COMMUNITY FUNCTION**. Critical inputs and outputs involve **FLUXES** of energy and matter and altered physical environments (**FIGURE 41.6**).
- Primary producers generate chemical-bond energy at a rate termed the **GROSS PRIMARY PRODUCTIVITY** (**GPP**). Some of this energy is used to fuel metabolism, and some is retained in primary-producer **BIOMASS**, which is produced at a rate termed the **NET PRIMARY PRODUCTIVITY** (**NPP**). NPP is smaller than GPP because energy is lost as heat during each step in metabolism.
- Community function increases with **SPECIES DIVERSITY**, the two components of which are **SPECIES RICHNESS** and **SPECIES EVENNESS**.
- One reason community function increases with species diversity is that diverse communities contain species with different **NICHES** that fill complementary functional roles.

1. What is the major source of energy for Earth's communities, and how does it enter them?
2. Describe two uses of the term "ecological niche" and discuss whether they are aspects of the same thing.
3. We have seen that GPP in any community is greater than NPP, and that ecological efficiency—the transfer of energy from one trophic level to the next higher level—is low. Describe the shared mechanism that underlies these two phenomena.

QUESTIONS

41.4 Diversity Patterns Provide Clues to What Determines Diversity

Go to ANIMATION 41.2 and ACTIVITY 41.3

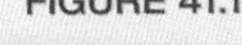

FIGURE 41.11

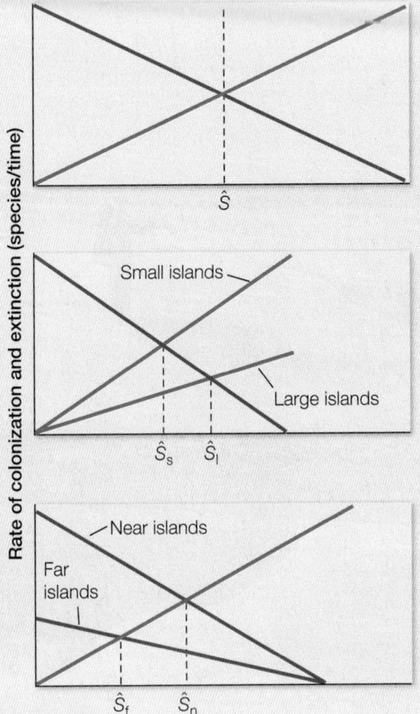

- Species richness is greatest in the tropics and decreases with increasing latitude. Possible causes of this pattern include climate stability, greater energy input, and greater habitat complexity at low latitudes.
- Species richness also varies among oceanic islands. Large islands and those close to a mainland support more species than do small islands and those far from a mainland.
- The **THEORY OF ISLAND BIOGEOGRAPHY** proposes that species richness on islands reflects a dynamic equilibrium between the rate at which new species colonize and the rate at which resident species go extinct (**FIGURE 41.11**).

1. Describe three factors that may contribute to the high species richness of tropical rainforests.

41.5 Community Ecology Suggests Strategies for Conserving Community Function

Go to ANIMATION 41.3

FIGURE 41.15

Richard Bierregaard, courtesy of the Smithsonian Institution, Office of Environmental Awareness

- Communities provide humans with a variety of goods and services. These **ECOSYSTEM SERVICES** have economic value.
- Ecosystem services are threatened by human activities, including habitat **FRAGMENTATION**, that reduce species diversity. Detrimental effects of fragmentation can be reduced by maintaining large habitat fragments, which minimizes extinction, and by clustering or connecting fragments to maximize dispersal (**FIGURE 41.15**).
- Ecosystem function can be preserved or restored by conserving species that play critical roles in the community and by maintaining a diversity of species functional groups.

1. Give two examples each of goods and services provided to humans by natural ecosystems and by ecosystems constructed and managed by humans.
2. Relate the idea of trophic cascades (see Key Concept 40.3) to the proposal that a conservation or restoration plan might focus on a particular species in a community.

42 The Global Ecosystem

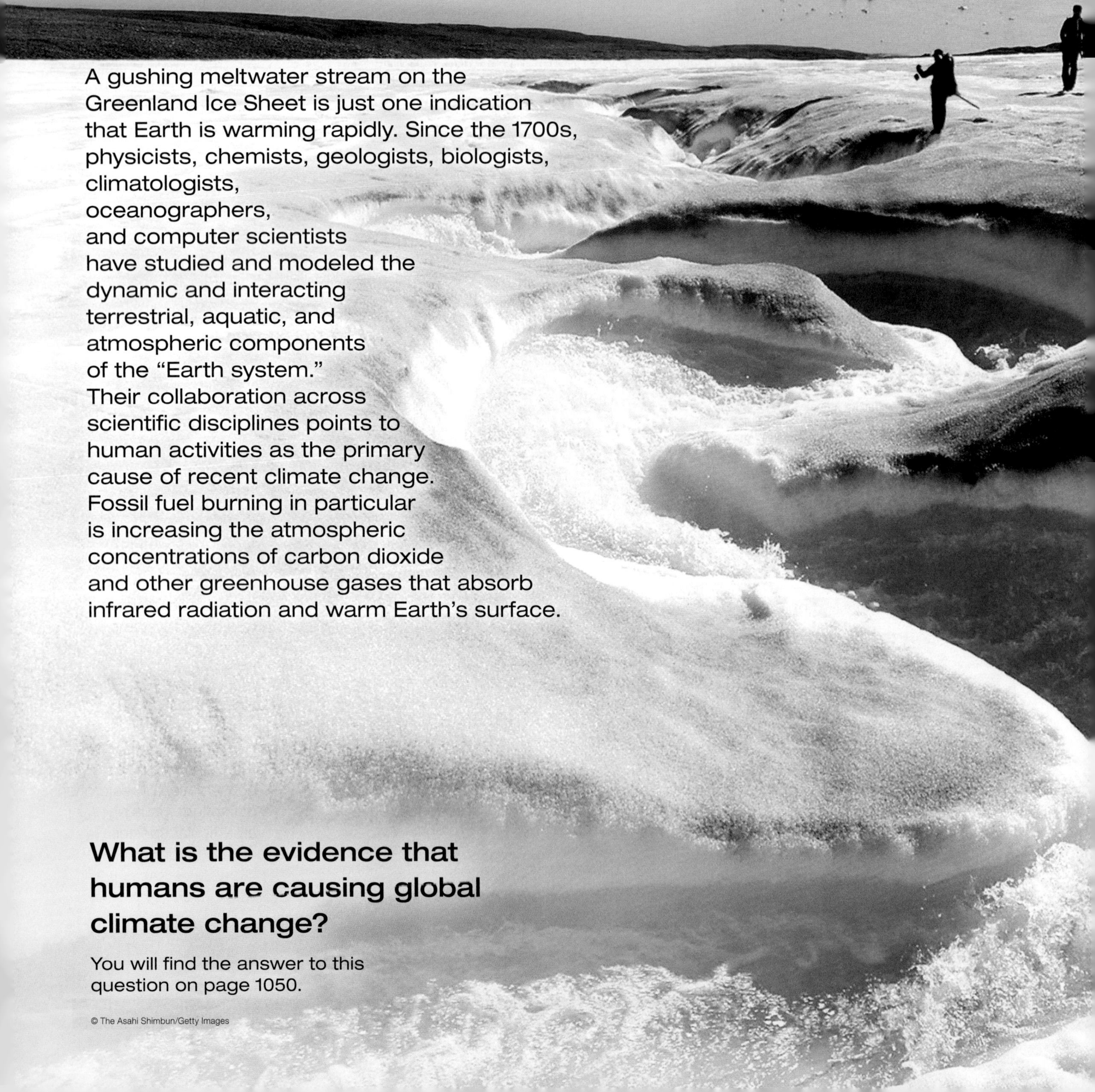

A gushing meltwater stream on the Greenland Ice Sheet is just one indication that Earth is warming rapidly. Since the 1700s, physicists, chemists, geologists, biologists, climatologists, oceanographers, and computer scientists have studied and modeled the dynamic and interacting terrestrial, aquatic, and atmospheric components of the "Earth system." Their collaboration across scientific disciplines points to human activities as the primary cause of recent climate change. Fossil fuel burning in particular is increasing the atmospheric concentrations of carbon dioxide and other greenhouse gases that absorb infrared radiation and warm Earth's surface.

What is the evidence that humans are causing global climate change?

You will find the answer to this question on page 1050.

© The Asahi Shimbun/Getty Images

KEY CONCEPTS

42.1 Climate and Nutrients Influence Ecosystem Function

42.2 Biological, Geological, and Chemical Processes Move Materials through Ecosystems

42.3 Certain Biogeochemical Cycles Are Especially Critical for Ecosystems

42.4 Biogeochemical Cycles Influence Global Climate

42.5 Rapid Climate Change Influences Species and Communities

42.6 Ecological Challenges Can Be Addressed through Science and International Cooperation

42.1 Climate and Nutrients Influence Ecosystem Function

LEARNING OBJECTIVES

By the end of this key concept you should be able to:

42.1.1 **Explain why net primary productivity (NPP) is a useful measure of the flow of materials as well as energy through the living component of an ecosystem.**

42.1.2 **Describe global geographic patterns in terrestrial and marine NPP.**

42.1.3 **Identify the most important factors that influence NPP in terrestrial and aquatic ecosystems.**

The interactions within and between biotic and abiotic components of ecosystems involve exchanges and transformations of energy and materials. Most of these interactions happen at a small spatial scale. A plant, for example, can absorb only those photons of solar radiation that are intercepted by a leaf, or only those nitrate molecules that are in physical contact with its root hairs; an elk in Canada cannot fall prey to a wolf living in Wyoming's Yellowstone National Park; and a raindrop interacts directly only with the soil particles it contacts. For this reason, most studies of ecosystems, even at the community level, are done at relatively small spatial scales, such as a single river drainage, lake, or patch of forest.

Such local ecosystems are linked, however, by slower exchanges of organisms, energy, and materials that occur on spatial scales ranging from tens of kilometers within regional landscapes to thousands of kilometers around the planet. These long-distance exchanges occur when organisms move among communities and when energy and materials are transported by the physical circulation of the atmosphere and the oceans, by flowing water, and by geological processes that occur below Earth's crust. Links over large scales make it impossible to understand a local ecosystem completely without considering it in the context of the larger systems of which it is a part. In this chapter we consider community-level ecosystems from both local and global perspectives.

NPP is a measure of ecosystem function

Like other biological systems, ecosystems can be characterized by their components and by patterns of interaction among those components. The community as a whole is often treated as one component, and distinct portions of the abiotic environment, such as soil, atmosphere, or water, are treated as other components. The dynamic processes by which these components exchange and transform energy and materials are aspects of **ecosystem function**—an expansion of the concept of community function (see Key Concept 41.3), to include the abiotic context of the community.

Ultimately, movements of materials between the abiotic and biotic components of ecosystems are tied to carbon. Primary producers (both photosynthetic organisms and chemoautotrophs) combine carbon, hydrogen, and oxygen from inorganic sources to make organic carbohydrates. Energy stored in the chemical bonds of carbohydrates then fuels the synthesis of all the biomass in the bodies of organisms at every trophic level. Most of the elemental constituents of that biomass enter communities through the primary producers as well. Calcium in the bones of a wolf, for example, came from the elk it ingested; the elk obtained calcium from the plants it ate; and the plants used energy from the breakdown of carbohydrates to extract inorganic calcium from the soil. Heterotrophs do exchange some materials directly with the abiotic environment—elk and wolves can take in or excrete water and salts directly to maintain homeostasis—but such exchanges are fueled by energy obtained from the breakdown of carbohydrates and are therefore also coupled to the carbon contained in primary producer biomass. This linkage between materials and energy means that net primary productivity (NPP)—the rate at which primary producers generate dry biomass (see Key Concept 41.3)—is a measure of the rate at which materials, as well as energy, flow between abiotic and biotic components of ecosystems.

International groups of scientists interested in the global ecosystem have teamed up to study the global distribution of NPP and the factors that influence it. For terrestrial systems, the annual production of plant biomass has been measured repeatedly at a network of permanent study sites in biomes across the planet. In order to compare NPP across vegetation types that vary in the carbon content of plant tissue, NPP is often expressed as grams of carbon instead of dry biomass. From these data, global maps of NPP can then be drawn, based on statistical relationships between NPP and variables such as temperature, precipitation, soils, solar radiant input, and vegetation structure. Scientists now can supplement direct measures of NPP from biomass samples with instruments on orbiting satellites that measure the reflection of different wavelengths of light by Earth's surface. This "view from space" allows them to calculate how much incident sunlight is absorbed by chlorophyll, and thus to map photosynthetic biomass. When coupled with information on how rates of photosynthesis and respiration vary with environmental conditions, photosynthetic biomass can predict NPP. Such remote-sensing methods are

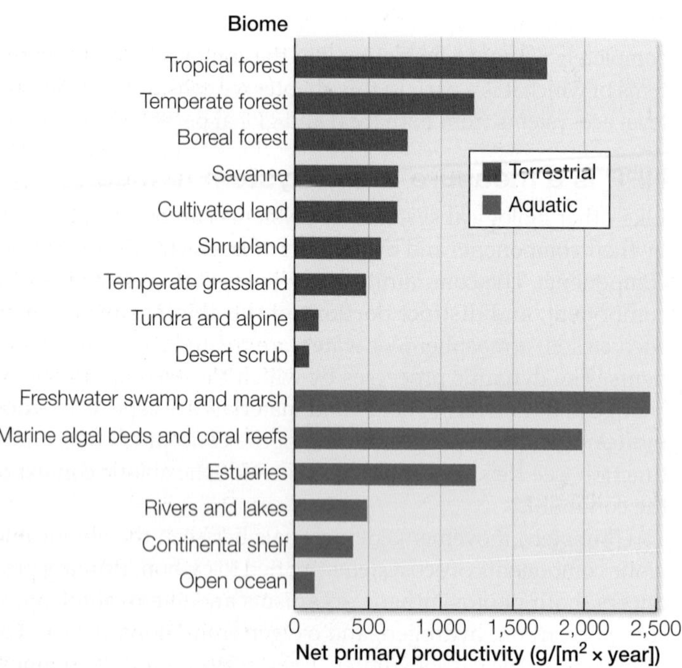

FIGURE 42.1 NPP Varies among Biomes Net primary productivity is expressed here as grams of dry biomass produced per square meter of area per year. (Data from R. H. Whittaker and G. E. Likens. 1973. *Hum Ecol* 1: 357–369.)

particularly important for the open ocean, where annual production of primary-producer biomass is difficult to measure directly.

As we will see next, patterns of variation in NPP point to factors that determine this aspect of ecosystem function.

NPP varies predictably with temperature, precipitation, and nutrients

Net primary productivity varies considerably among ecosystems. Of terrestrial biomes, tropical forests and temperate forests are the most productive per unit of area, whereas tundra and deserts are the least productive. Among aquatic biomes, freshwater swamps and marshes, marine algal beds, and coral reefs are the most productive, followed by estuarine habitats; open ocean is the least productive (**FIGURE 42.1**). Much of this pattern derives from variation in climate and nutrient availability.

Satellite images of terrestrial ecosystems show that NPP varies with latitude: highly productive areas are concentrated in the tropics, and higher latitudes are less productive (**FIGURE 42.2**). This pattern parallels latitudinal gradients in solar energy input (see Figure 38.3) and temperature, suggesting that NPP increases with temperature. Indeed, terrestrial NPP generally does increase with temperature (**FIGURE 42.3A**), which makes sense when we consider that the activity of photosynthetic enzymes, like that of other enzymes, increases up to the critical temperature at which enzymes denature.

The satellite images also show areas of low terrestrial NPP at 30°N and S latitude (see Figure 42.2), where deserts occur. This pattern parallels gradients in precipitation, suggesting that moisture also affects NPP. Terrestrial NPP does increase with precipitation up to about 240 centimeters per year, then decreases at higher levels (**FIGURE 42.3B**). The initial increase with moisture makes physiological sense: moist soil conditions facilitate water and nutrient uptake by roots and allow plants to maintain pressure in their cells without closing their stomata (leaf pores), thus facilitating CO_2 uptake (see Key Concept 24.3). The decrease in NPP in extremely moist climates may result from increased cloud cover and lower solar input or from a lack of oxygen in water-saturated soils.

Finally, we also know that elements such as nitrogen, phosphorus, and potassium, which are critical for plant growth, can be in short enough supply to limit NPP in both natural and agricultural ecosystems. Soils in southwestern Australia, for example, are nutrient-poor because most nutrients have been washed (leached) out of the soil over hundreds of millions of years, during which time there has been no nutrient-renewing geological uplift. As we saw in Chapter 38, many plants in this region cannot afford to allocate their scarce nitrogen to defensive compounds; instead, they allocate it to proteins that are essential for metabolism and growth.

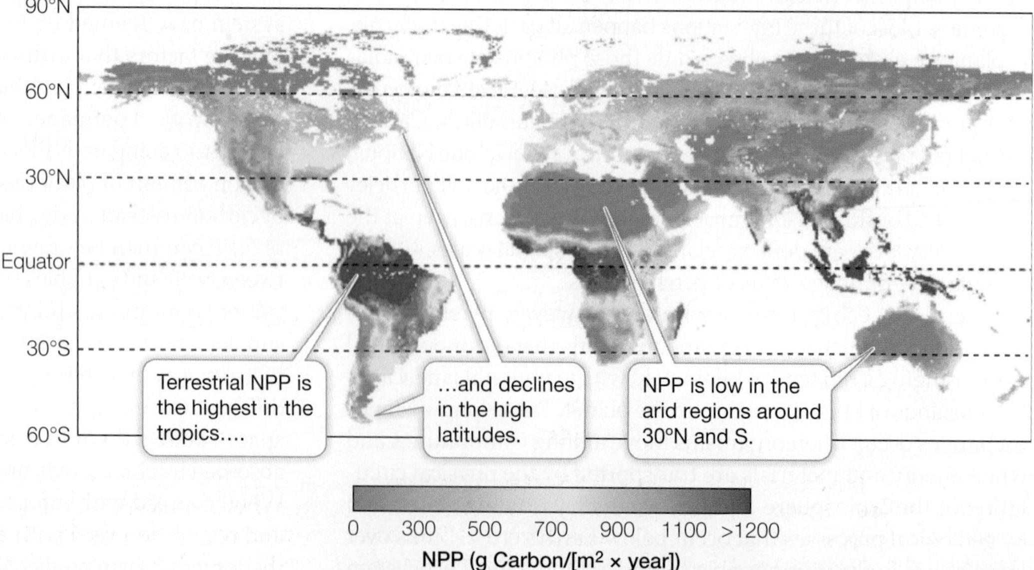

FIGURE 42.2 Terrestrial NPP Corresponds to Climate Values are mean terrestrial NPP from 1961 to 2010, expressed as grams of carbon contained in the biomass produced per square meter of area per year. They are based on projections from regression of observed NPP as a function of climate and vegetation variables, measured at permanent study sites located in all major terrestrial biomes. (From P. Li et al. 2017. *Ecol Indic* 76: 245–255.)

If carbon constitutes 50% of the dry biomass of plants in the most productive biomes, what would the maximum scale value be if NPP in this figure were expressed as the annual production of dry biomass per square meter?

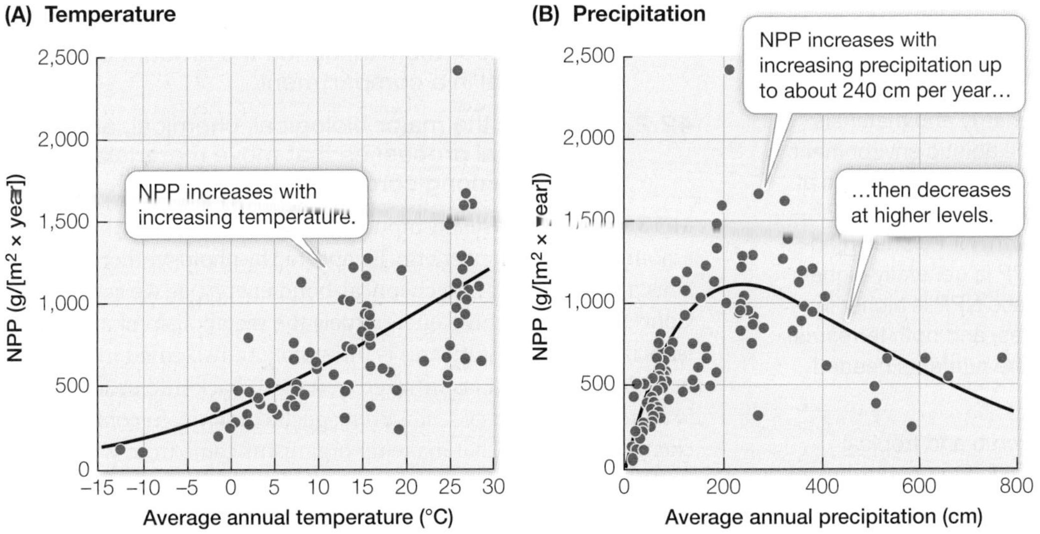

FIGURE 42.3 Terrestrial NPP Varies with Temperature and Precipitation NPP is given in grams of carbon produced per square meter of area per year. (After E. A. G. Schuur. 2003. *Ecology* 84: 1165–1170.)

less variable than air temperature. Light penetrates only the top layers of lakes and oceans (the photic zone; see Figure 38.13), so photosynthesis is restricted to those layers. As a result, NPP is generally higher in surface than in deep waters; the only productive deep-water areas are hydrothermal vents, where chemoautotrophs use chemical-bond energy (often called chemical energy) rather than sunlight to generate new biomass. But not all surface waters are productive. The open ocean, for example (see Figure 42.1), is unproductive because dissolved nutrients are often limited there—phytoplankton absorb surface nutrients quickly and move them to deep waters when they (or the consumers that eat them) die and sink. Aquatic NPP tends to be highest where nutrients are abundant—where rivers and streams discharge nutrients leached from terrestrial ecosystems into coastal marine areas, the near-shore zones of lakes, and in shallow freshwater wetlands (**FIGURE 42.4**). NPP is also high in areas of upwelling, where deep water from the bottom of an ocean or lake rises to the surface, bringing with it nutrients from benthic sediments.

LINK The principle of allocation, discussed in **Key Concept 39.3**, states that resources devoted to one life function (maintenance, growth, defense, or reproduction) cannot be used simultaneously for another function, thereby leading to trade-offs among competing life functions.

In aquatic ecosystems, NPP is affected more strongly by light and nutrients than by water temperature, which is far

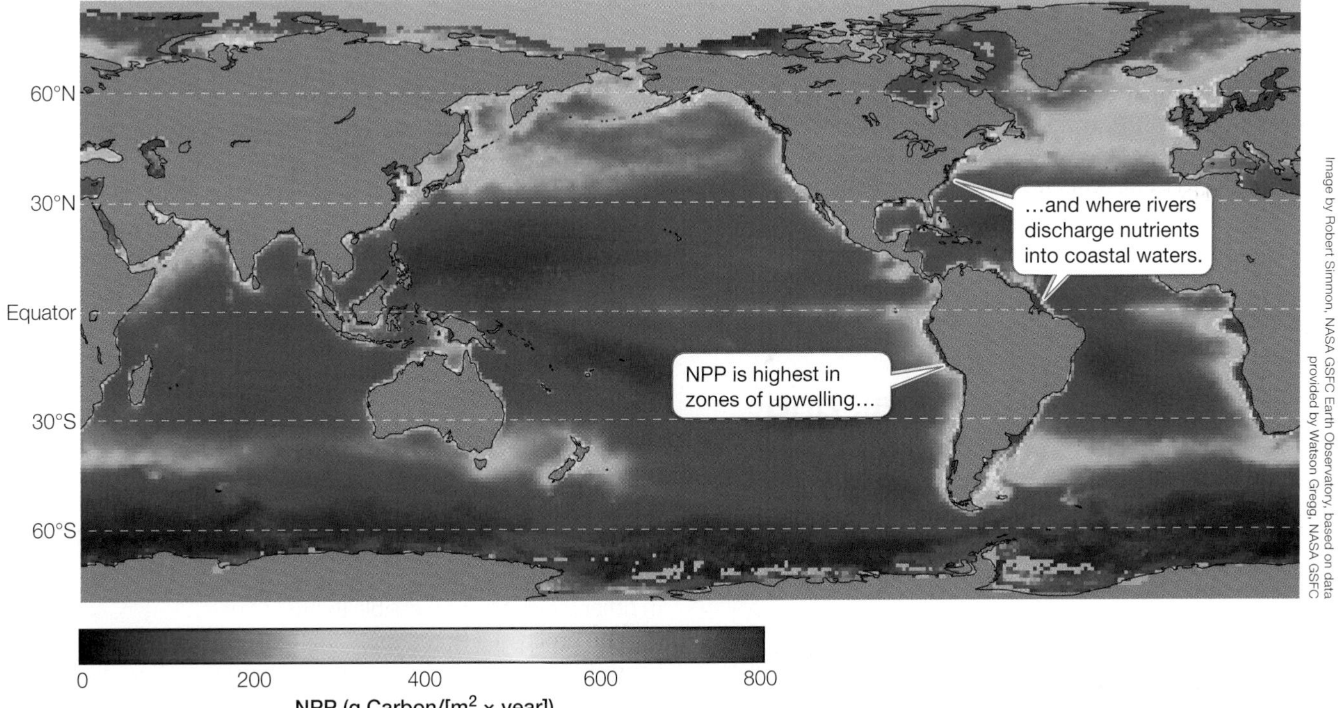

FIGURE 42.4 Marine NPP Is Highest in Surface Waters Near Coastlines Sensors on orbiting satellites have been used to estimate the chlorophyll content of the ocean's photic zone by measuring the wavelengths of reflected light. Between 1997 and 2002, the highest average NPP (expressed here as grams of carbon per square meter per year) occurred where runoff from land brings nutrients into shallow coastal waters, and where upwelling of deep waters brings nutrients from the benthic zone to the surface.

REVIEW & APPLY | 42.1

R In any ecosystem the exchanges of energy and materials between the biotic community and the abiotic environment are tied to the movement of carbon. This makes the rate at which primary producers accumulate dry biomass—NPP—a useful measure of ecosystem function. NPP varies globally in predictable ways: terrestrial NPP is generally higher in warmer and moister climates, aquatic NPP is higher in surface waters that sunlight penetrates, and both terrestrial NPP and aquatic NPP are higher where nutrients needed by primary producers are abundant.

A 1. a. Predict the NPP of the desert scrub and tropical forest ecosystem types (biomes) in Figure 42.1 based on the temperature and precipitation profiles in Figure 38.10, and on the relationships between NPP and these environmental factors in Figure 42.3A and B. (Assume that the desert scrub and tropical forest biomes correspond, respectively, to the subtropical desert and the midpoint of the tropical rainforest and tropical seasonal forest biomes in Figure 38.10.)

   b. Compare your estimate with what is reported in Figure 42.1.

2. Suggest why NPP in human-cultivated agricultural ecosystems is lower than that in several natural ecosystem types, such as temperate and tropical forest (see Figure 42.1). (*Hint*: See Key Concept 41.3.)

3. Design an experiment that would allow you to determine whether NPP in a grassland ecosystem is limited by the availability of water in the soil.

We now have an overview of global patterns of ecosystem function, as measured by NPP, and some of the factors that cause them. We have seen that nutrient availability plays an important role in these patterns, which makes sense because nutrients are chemical elements that organisms require for their metabolism and as materials to build their bodies. What processes affect the availability of nutrients?

42.2 Biological, Geological, and Chemical Processes Move Materials through Ecosystems

LEARNING OBJECTIVES

By the end of this key concept you should be able to:

42.2.1 **Explain why Earth is considered an open system with respect to energy but a closed system with respect to matter.**

42.2.2 **Identify the four major compartments for materials in the Earth system and how fluxes in and out of them influence the amount of a material in a compartment.**

42.2.3 **Diagram the major biological, chemical, and geological processes that move materials and energy among compartments.**

The Sun provides a steady input of radiant energy to Earth. A small amount of this radiant energy is captured by photosynthetic organisms and converted into chemical-bond energy, as we saw in Key Concept 41.3. Captured energy fuels the metabolism of all organisms in the community and is ultimately transformed into waste heat that is emitted (as infrared radiation) back into space. Earth is therefore an *open system* with respect to energy. In contrast, the chemical elements that make up organisms come from within the Earth system itself. There is essentially a fixed amount of each element because so little matter escapes Earth's gravity or enters the Earth system from space. Earth is therefore a *closed system* with respect to matter. The distribution of matter within the Earth system is not static, however: energy from the Sun and heat from Earth's interior drive biological, geological (physical), and chemical processes that transform matter and move it around the global ecosystem.

The forms and locations of elements determine their accessibility to organisms

Imagine an atom. At any time, this atom occurs in a particular molecular form and occupies a particular location in Earth's closed system. It might be part of a living organism or of dead organic matter. It might be part of an inorganic molecule in the atmosphere, in a body of fresh or salt water, or in soil, sediment, or rock. If in rock, it might be exposed at the surface of Earth or buried more deeply. These alternative categories of form and location in which matter can occur are commonly called **compartments**. Atoms cycle repeatedly among compartments as they are chemically transformed or physically moved. The atoms in Earth's crust, for example, have gone through repeated cycles of being buried, compressed, heated, uplifted, chemically altered, eroded, and deposited again in sediments (see Figure 17.2). The compartments of the global ecosystem, and the processes that cycle materials among them, can be depicted in a box-and-arrow **systems diagram** (**FIGURE 42.5**).

The chemical and physical form in which an atom exists determines whether organisms can use it. With rare exceptions, autotrophs obtain carbon as CO_2 (or its aquatic form, HCO_3^-) and absorb other nutrients in the form of dissolved ions when they take up liquid water (see Key Concept 25.1). These same atoms cannot be taken up by autotrophs in other forms. Heterotrophs (including decomposers) extract most of these essential materials from living or dead organic matter. But even when an atom is in a usable form, organisms cannot access it if it is in the wrong place. Atoms that are buried too deeply or occur in places that are too hot, cold, or hostile in other ways, are inaccessible to life.

The processes that supply matter to organisms occur in the **biosphere**, the thin layer at Earth's surface where atmosphere, land, and water are in contact with one another and where organisms live. The biosphere is about 23 kilometers thick at its thickest, extending from the deepest abyssal zones of the oceans to the stratosphere.

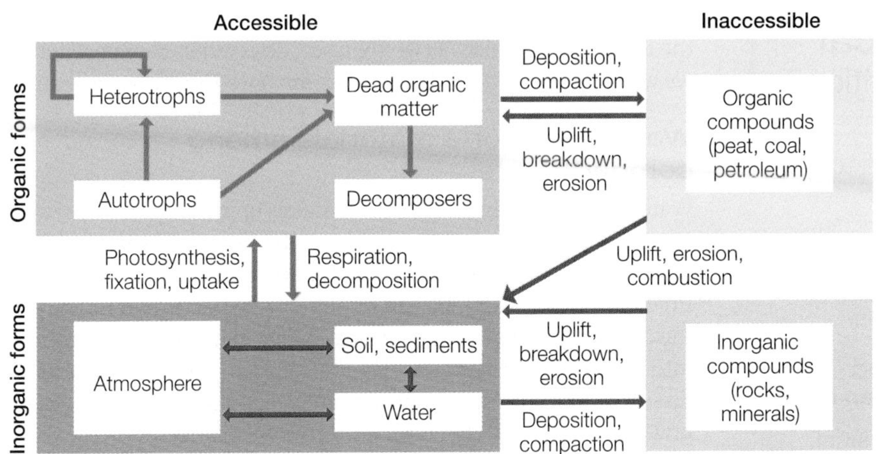

FIGURE 42.5 Chemical Elements and Compounds Cycle among Compartments of the Global Ecosystem The different chemical forms and locations of elements determine whether or not they can be used and are accessible to living organisms. These different categories of location and form of matter represent compartments (boxes) in the diagram shown here. Biological processes (green arrows) and geological and chemical processes (brown arrows) lead to fluxes (movements) of matter among the compartments.

Movement of matter is driven by biogeochemical processes

Matter is cycled among the compartments of the global ecosystem by processes that are biological, geological (physical), and chemical; thus these movements are called **biogeochemical cycles**. Biological and abiotic chemical reactions combine elements with other elements and compounds, converting them between inorganic and organic forms or between oxidized and reduced states. Many of these reactions involve the abundant elements oxygen and hydrogen. Chemical reactions are also involved in the breakdown of rock during soil formation. Physical processes—such as the circulation of air and water and the convective flows within Earth's interior—move matter within the Earth system, alternately exposing it to air, water, and solar radiation or to cold, heat, or pressure.

The total amount, or **pool**, of an element or molecule in a given compartment at a particular time depends on its rate of movement into and out of that compartment. The rate of movement, or **flux**, from one compartment to another is measured in amount per unit of time. If the fluxes into and out of a compartment are not identical, the pool will grow or shrink. Increased oxidation of organic matter by respiration, wildfires, and fossil fuel combustion, for example, will increase the atmospheric pool of carbon unless photosynthesis and other biogeochemical processes remove carbon from the atmosphere at an equal or greater rate.

All materials in the bodies of living organisms are ultimately derived from abiotic sources—that is, all biomass has its origin in either atmospheric gases or materials dissolved in water. Biomass would not exist were it not for the primary producers that take up materials from these inorganic compartments and transform them into the molecular building blocks on which living things depend. Through trophic interactions (see Key Concept 40.3), those chemical elements are passed on to heterotrophs, which also take in other inorganic materials such as oxygen and water as part of their metabolism. Death eventually returns all of these materials to the compartment of nonliving organic matter. As decomposers break down the dead organic matter into simpler chemical compounds and elements that can be taken up by primary producers, the cycle continues. Much of this recycling of materials occurs locally, but some occurs on a larger scale when materials diffuse away in gaseous form or are carried out of the local ecosystem by wind or water.

REVIEW & APPLY | 42.2

R Earth is an open system with respect to energy because it receives a steady input of solar radiation and steadily outputs energy back to space as infrared (heat) radiation. In contrast, Earth is a closed system with respect to matter because so few atoms are exchanged between Earth and space. Within the Earth system, biological, geological, and chemical processes transform matter between alternative chemical and physical forms that are usable or unusable to organisms and move it among locations that are accessible or inaccessible to life. Compartments are categories of form and location of matter. The amounts of matter in different compartments are pools, and the rates of movement of matter between compartments are fluxes.

A

1. Write an equation that predicts what the pool of a substance in a compartment will be at time $t + 1$ (P_{t+1}) as a function of the current pool (P_t) and the fluxes into and out of the compartment. (*Hint*: See Equation 39.1.)
2. The accessible, organic compartment of the biosphere in Figure 42.5 is shown below; arrows indicate fluxes of matter. Add and label arrows to this diagram to show how, and in what forms, energy from sunlight enters, flows through, and leaves this compartment. (*Hint*: See Figure 41.6.)

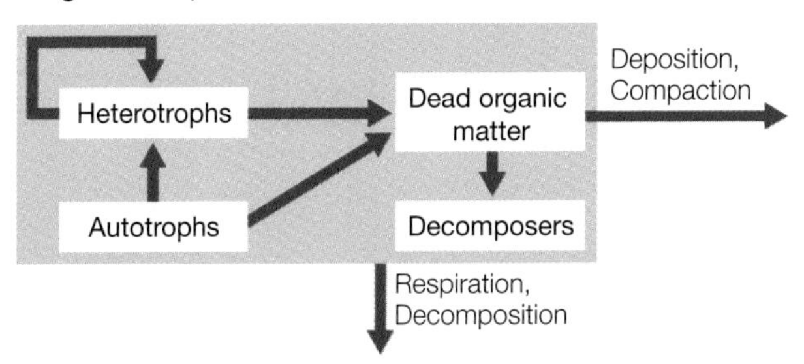

We have now seen that the materials that make up living tissue—chemical elements and compounds—cycle through ecosystems. Let's next consider the cycles of three materials that are especially critical for ecosystem function.

42.3 Certain Biogeochemical Cycles Are Especially Critical for Ecosystems

LEARNING OBJECTIVES

By the end of this key concept you should be able to:

42.3.1 Compare the relative magnitudes of fluxes in the water cycle.

42.3.2 Calculate the relative sizes of usable and unusable pools of nitrogen in the nitrogen cycle.

42.3.3 Assign fluxes in the carbon cycle to physical, biological, or inorganic-chemical processes. Explain how human activities have altered these fluxes.

42.3.4 Describe two examples in which the biogeochemical cycles of different materials are linked.

All biogeochemical cycles can be described in terms of compartments containing the material involved, pools of the material contained in each compartment, and fluxes of the material into and out of each compartment. The cycles of different materials vary, however, in their spatial scales and in the processes that drive them. We will illustrate this variation by tracing the cycles for three materials that are essential for life: water, nitrogen, and carbon. We also will point out how each cycle affects the global ecosystem, particularly the atmosphere, which, as we shall see, determines climate. Finally, we will stress how each cycle is being influenced by human activities.

Water transports materials among compartments

Water is a remarkable molecule. It makes up more than 70 percent of living tissue. It is the medium for metabolism and the solvent in which biologically usable forms of many nutrients are dissolved. Precipitation of water as rain or snow transports materials from the atmosphere to Earth's surface. Water helps break down rock into soil. And as the primary agent of erosion and transport of dissolved ions and sediment, water is responsible for much of the physical movement of materials around the planet. By virtue of its high heat capacity and its ability to change between solid, liquid, and gaseous states at normal Earth temperatures, water redistributes heat around the planet as it circulates through the oceans and atmosphere. And as we will soon see, water in the atmosphere plays an important role in the global radiation balance.

The global ecosystem contains almost 1.4 billion cubic kilometers of water. About 96.5 percent of this water is in the oceans, which cover 71 percent of the planet's surface—making it obvious why Earth is sometimes called the "blue planet." Other compartments of water include ice and snow (1.76%), groundwater (1.70%), the atmosphere (0.001%), and fresh water in lakes and rivers (surface fresh water; 0.013%). Even though living tissue is mostly water, only about 0.0001 percent of Earth's water is found in living organisms.

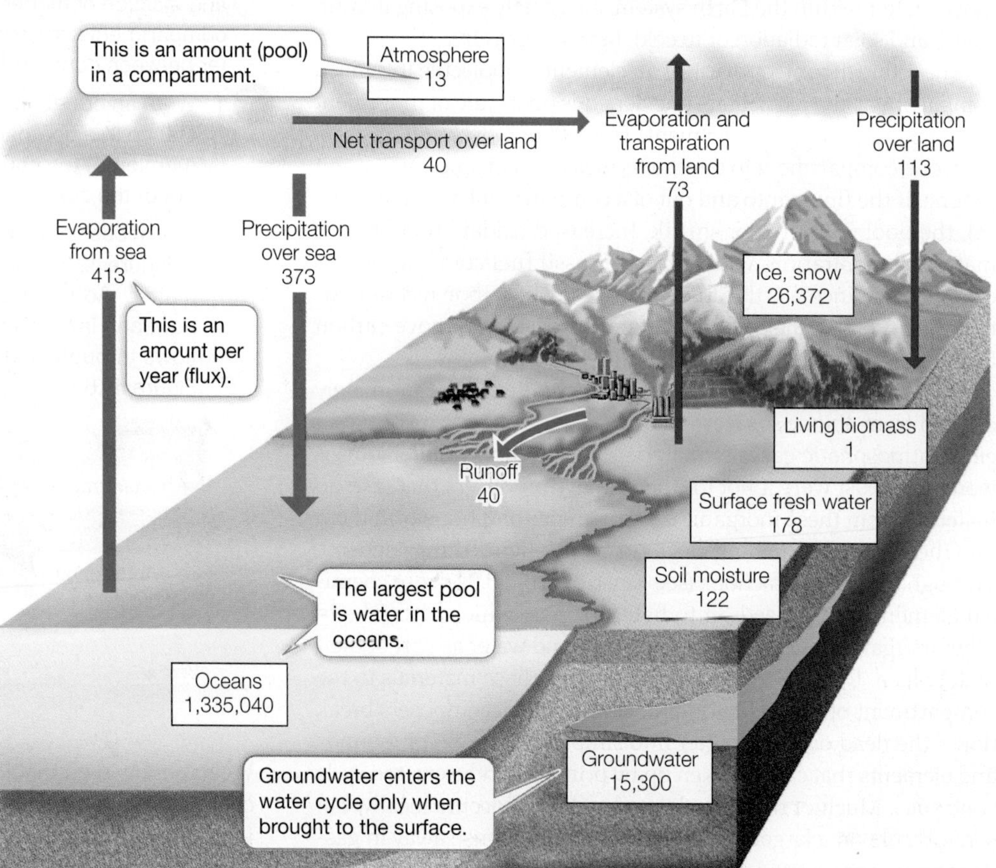

FIGURE 42.6 The Global Water Cycle Estimated pools in major compartments (boxes) and per-year fluxes between compartments (arrows) are expressed in units of 10^{18} grams. (After K. E. Trenberth et al. 2007. *J Hydrometeor* 8: 758–769.)

Is the total atmospheric pool of water increasing or decreasing?

Animation 42.1 The Global Water Cycle

Some fluxes among these water compartments (**FIGURE 42.6**) involve gravity-driven flows of water from the atmosphere to Earth's surface (precipitation) and subsurface (infiltration), and from land to the oceans and other bodies of water (runoff). Other fluxes, however, are associated with changes in the physical state of water—from liquid to gas (evaporation), from gas to liquid (condensation), between liquid and solid (freezing and melting), and between solid and gas (sublimation and deposition).

The driver of the water cycle is solar-powered evaporation, which transforms water from liquid to gas and, in the process, moves it into the atmosphere from oceans and other bodies of water, from the soil, and from plants and other organisms. The evaporation of 1 gram of water takes considerable energy—2.24 kilojoules, about the energy stored in one AA battery. Fully one-third of the solar energy that reaches Earth's surface is involved in evaporating water; this energy is released again as heat when the water vapor changes state to rain or snow. Fluxes of water from Earth's surface to the atmosphere and from the atmosphere to the surface are equivalent on a global basis, although precipitation exceeds evaporation over land, and evaporation exceeds precipitation over the oceans.

Humans affect the global water cycle when they change the way land is used. Removing vegetation reduces the amount of precipitation that is retained by soil and recycled locally, thereby increasing the amount that leaves the local ecosystem as surface runoff. As a result, deforestation, grazing, and cultivation of most crops all tend to dry out local ecosystems. Pumping moves groundwater to the surface, where much of it either evaporates or runs off into streams and, eventually, into oceans.

The water cycle also influences and is influenced by the global climate. Global warming is melting polar ice caps and glaciers (see the chapter opener photo), increasing the total amount of water in the oceans and causing sea level to rise. With more liquid water comes more evaporation, and thus more water entering the atmosphere, which causes more precipitation globally. Water vapor absorbs infrared radiation, but it also forms clouds, which reflect incoming sunlight back into space. Scientists are working to predict the net effects on climate of these counterbalancing changes in the water cycle.

Within-ecosystem recycling dominates the global nitrogen cycle

Unlike water—whose availability to organisms depends only on where it is found and its physical state—the availability of nitrogen to organisms depends on its chemical form in addition to where it is found. As a result, the nitrogen cycle, unlike the water cycle, is dominated by chemical transformations (**FIGURE 42.7**).

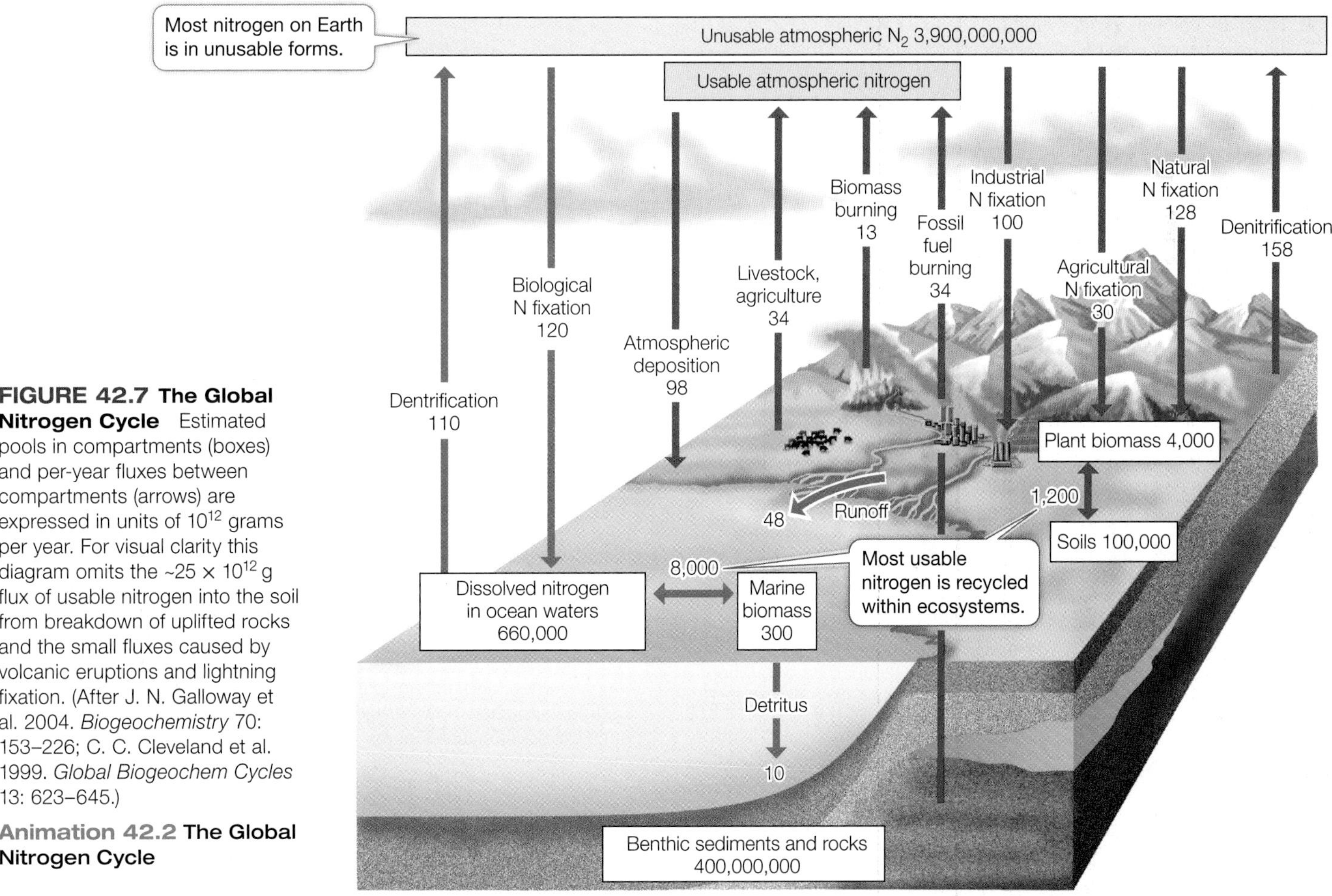

FIGURE 42.7 The Global Nitrogen Cycle Estimated pools in compartments (boxes) and per-year fluxes between compartments (arrows) are expressed in units of 10^{12} grams per year. For visual clarity this diagram omits the ~25×10^{12} g flux of usable nitrogen into the soil from breakdown of uplifted rocks and the small fluxes caused by volcanic eruptions and lightning fixation. (After J. N. Galloway et al. 2004. *Biogeochemistry* 70: 153–226; C. C. Cleveland et al. 1999. *Global Biogeochem Cycles* 13: 623–645.)

Animation 42.2 The Global Nitrogen Cycle

Nitrogen is an abundant element on Earth. For example, the gas N_2 (N≡N) constitutes about 78 percent of the molecules in Earth's atmosphere. But most organisms cannot use nitrogen in this form because they cannot break the strong triple bond between the two nitrogen atoms. The exceptions are certain microbes that are capable of **nitrogen fixation**—a metabolic process that breaks the triple bond and attaches hydrogen to make ammonium (NH_4^+), a form in which the nitrogen is usable by other organisms. In terrestrial ecosystems, nitrogen fixation is carried out mostly by free-living bacteria in the soil and by symbiotic bacteria associated with plant roots, but it also occurs in other places, such as the fungal gardens of leaf-cutter ants described in the opening of Chapter 40 (**INVESTIGATION FIGURE 42.8**). Analogous biochemical processes in aquatic ecosystems are carried out by free-living microorganisms and by microbial symbionts of phytoplankton and other aquatic

INVESTIGATION

FIGURE 42.8 Where Does the Extra Nitrogen Come From? The nitrogen content of leaf-cutter ants is higher than that of the fresh leaves that they supply to their fungal gardens. Perhaps the fungus concentrates nitrogen from the leaves into the special fungal structures that the ants eat. If so, then spent leaf material in the ants' refuse dumps (see Figure 40.13) should be lower in nitrogen than fresh leaf material—but the opposite turns out to be true. How might nitrogen enrichment of the discarded leaf material be explained? One possible answer is that ants eat protein-rich insects as well as the fungus and fertilize the fungal garden with their feces. Another possibility is that the fungus absorbs nitrogen from the soil. A third possibility is that the fungus absorbs atmospheric nitrogen that has been fixed by microorganisms in the nest. Researcher Pinto-Tomás and colleagues tested this third hypothesis.[a]

HYPOTHESIS

Nitrogen-fixing microorganisms in leaf-cutter ant nests supply nitrogen to the ants.

METHOD

1. Bring ant colonies into the laboratory and allow them to live in an environment with no other insects and no soil, so that the only non-atmospheric source of N is fresh leaves supplied by the researchers.
2. Measure the nitrogen content of fresh leaves, fungus garden, ants, and leaf refuse.

RESULTS

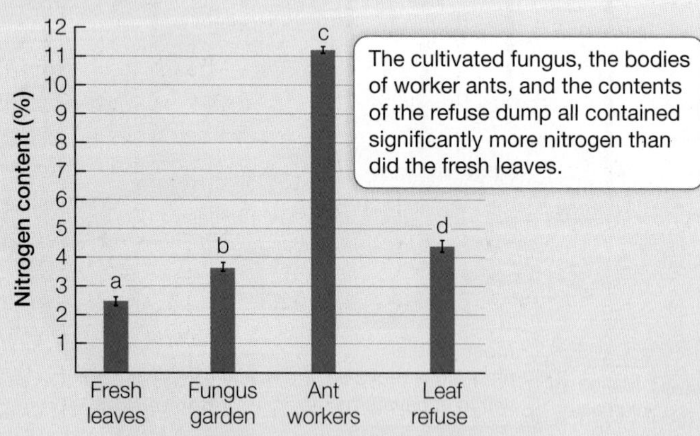

Error bars indicate ± 1 standard error from the mean; bars with different letters are statistically different from one another. (See Appendix B for discussion of statistical concepts.)

CONCLUSION

Atmospheric nitrogen is being fixed in the ant nest.

ANALYZE THE DATA

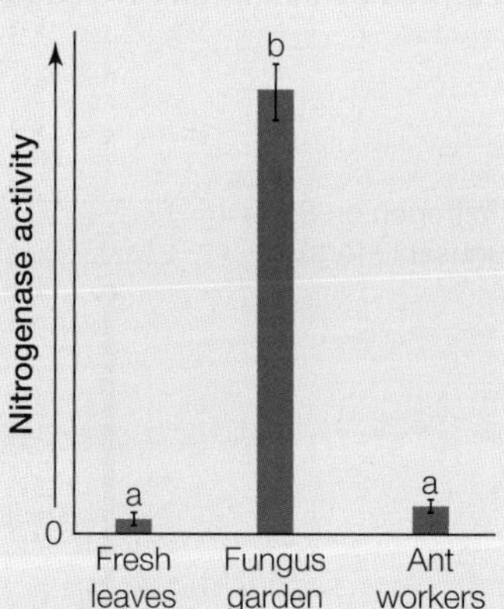

The researchers measured activity of nitrogenase, an enzyme in the bacterial metabolic pathway that fixes atmospheric nitrogen. The results are shown in the figure above. Error bars indicate ± 1 standard error from the mean; different letters indicate significant differences in nitrogenase activity.

1. Where in the nest is nitrogen fixation occurring?
2. Can the researchers now conclude that the extra nitrogen comes from nitrogen-fixing bacteria living in the fungus gardens?
3. Notice in the figure to the left that the ant workers are richer in nitrogen than the fungus garden, where nitrogen fixation takes place. The explanation for this surprising finding is that chitin, the polymer from which the durable ant exoskeleton is constructed during development, is rich in nitrogen. Draw a systems diagram of an ant nest with one box each for the nitrogen in the atmosphere, fresh leaves, fungal garden, and ant compartments; and arrows between compartments to show how nitrogen ends up in the chitin of the adult ant's exoskeleton (see Figures 1.6 and 42.5).

[a] A. A. Pinto-Tomás et al. 2009. *Science* 326: 1120–1123.

A similar **Analyze the Data** exercise may be assigned in **Achieve**.

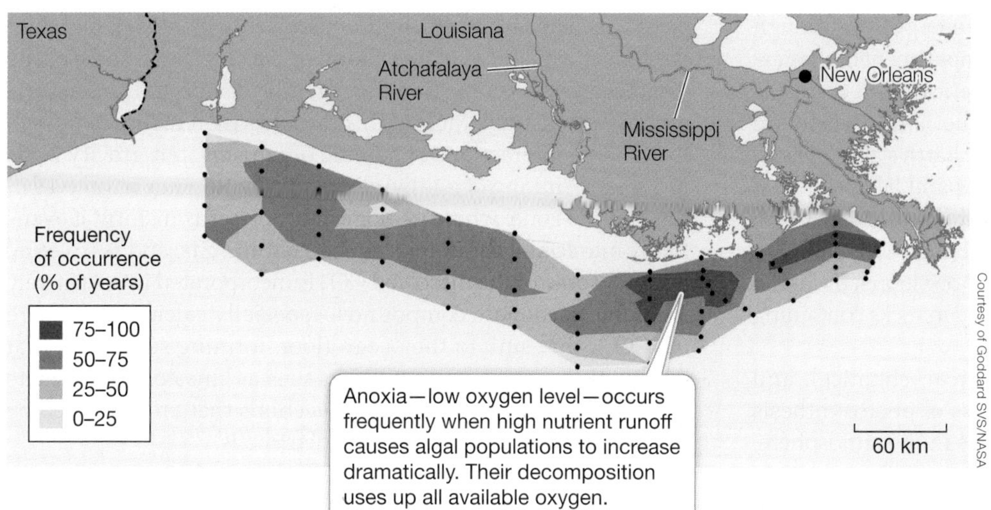

FIGURE 42.9 High Nutrient Input Creates Dead Zones Runoff from agricultural lands in the midwestern United States carries dissolved nitrogen and phosphorus down the Mississippi and Atchafalaya Rivers and into the Gulf of Mexico. The resulting eutrophication frequently creates an oxygen-poor "dead zone" that few aquatic organisms other than microbes can tolerate, causing economic losses to regional fisheries. (After R. E. Turner and N. N. Rabalais. 2018. *2018 Forecast: Summer Hypoxic Zone Size Northern Gulf of Mexico.* Louisiana State University. Data provided by N. N. Rabalais and R. E. Turner. https://gulfhypoxia.net/research/shelfwide-cruise.)

organisms. A small amount of nitrogen is fixed in the atmosphere by lightning. Diverse other microbial species in terrestrial and aquatic ecosystems convert ammonium into other usable forms of nitrogen such as nitrate (NO_3^-). Nitrogen fixation is reversed by another group of microbes in a process called **denitrification**, which completes the cycle by returning N_2 gas to the atmosphere.

LINK The metabolic processes by which microbes transform nitrogen are described in **Key Concepts 18.3 and 24.2**.

Autotrophs can use ammonium and nitrate because these forms of nitrogen dissolve in water and thus are easily taken up through cell membranes. Heterotrophs obtain nitrogen by breaking proteins down into amino acids, which can also be transported through cell membranes. Much of the nitrogen in terrestrial ecosystems is recycled locally because decomposition of dead organic matter releases usable nitrogen into the soil. Some nitrogen, however, is lost to the atmosphere when biomass burns or volcanoes erupt, or is leached from soils and transported by streams into lakes and oceans. In aquatic systems, primary production occurs in surface waters, and some nitrogen is recycled there. However, much of it is recycled at greater depths in the water column as organisms in deep waters intercept and consume sinking organic matter. The nitrogen in this organic matter eventually accumulates in sediments. Some of the deep nitrogen is returned to the surface by upwelling; much more of it is incorporated over geological time into a large pool of fossil organic nitrogen in sedimentary rock. Geological uplift and weathering of sedimentary rocks are now thought to provide a substantial input of usable fixed nitrogen to the biosphere.

Human activities are affecting nitrogen fluxes and pools. Livestock raising and burning of plant material, coal, and petroleum release oxides of nitrogen and ammonia into the atmosphere. These molecules contribute to atmospheric smog and acid rain, and one of the molecules, nitrous oxide (N_2O), absorbs infrared radiation. When these molecules are eventually deposited on Earth's surface, sometimes far from their source, they can add as much nitrogen "fertilizer" as farmers place on their crops. Humans also alter the atmospheric pool of N_2 by fixing nitrogen industrially to manufacture fertilizer, explosives, and other nitrogen-containing compounds. The rate of industrial fixation plus the rate of fixation by legume crops such as soybeans now exceeds the rate of natural terrestrial nitrogen fixation (see Figure 42.7). Transport of topsoil and dissolved nitrogen from fertilized croplands and deforested areas by wind or water runoff is increasingly exporting nitrogen from terrestrial to aquatic ecosystems.

Increased nitrogen inputs can increase primary production, but too much of a good thing can be bad. In aquatic systems, nutrient-stimulated increases in primary productivity—a phenomenon called **eutrophication**—can result in rapid growth of phytoplankton (including cyanobacteria) and algae. Respiration by these primary producers and by the decomposers that process their dead bodies can reduce oxygen levels below the tolerances of many aquatic organisms, including fishes and crustaceans. Eutrophication thus can lead to "dead zones" devoid of oxygen-requiring aquatic life, like that found where the Mississippi River discharges nutrient-rich agricultural runoff into the Gulf of Mexico (**FIGURE 42.9**). In nutrient-poor terrestrial ecosystems, excess nitrogen can change the species composition of plant communities. Species that are adapted to low nutrient levels grow slowly, even when fertilized, and thus are displaced by faster-growing species that can take advantage of the additional nutrient supply. In the Netherlands, nitrogen deposition from upwind industrial and agricultural regions has caused species-rich plant communities to give way to species-poor communities, contributing measurably to a recent loss of plant species diversity there.

Media Clip 42.1 Tracking Dead Zones from Space

Movement of carbon is linked to energy flow through ecosystems

All the macromolecules that make up living organisms contain carbon, and much of the energy that organisms use to fuel their metabolic activities comes from the oxidation of organic carbon compounds. The movement of carbon into, through, and out of communities is therefore intimately linked to energy flow through ecosystems, and biomass is an important compartment of the carbon cycle (**FIGURE 42.10**). Most atmospheric carbon occurs as the gases CO_2 (carbon dioxide) and CH_4 (methane), which are mixed on a global scale by the circulation of Earth's atmosphere. Some atmospheric CO_2 diffuses into oceans and lakes where it combines with water molecules to form carbonic acid (H_2CO_3); this molecule easily dissociates to yield bicarbonate ions (HCO_3^-) that are the carbon source for aquatic primary producers. By far the largest pools of carbon occur in fossil fuels and in rocks containing carbonate compounds.

Fluxes of carbon are driven by biological, chemical, and physical processes. The biochemical process of photosynthesis moves carbon from inorganic compartments in the atmosphere, oceans, and lakes into the organic compartment; cellular respiration reverses this flux. Carbon dioxide and methane move from their atmospheric compartments into the much larger compartments in the oceans when they dissolve, and they are returned to the atmosphere when they outgass (leave the water surface in gaseous form); both of these movements are purely physical processes. The rate at which CO_2 dissolves is slightly greater than the rate at which it outgasses, for two reasons. First, some of the dissolved CO_2 is converted into organic compounds by marine primary production. Most of this carbon is then recycled through the trophic interactions of aquatic organisms, but gravity leads to a steady sinking of nonliving organic matter into sediments in the benthic zone, where geological processes transform it eventually into **fossil fuels** such as coal, natural gas, and petroleum. Second, some of the dissolved CO_2 is incorporated into relatively insoluble carbonate compounds, especially calcium carbonate ($CaCO_3$), that sink to the ocean floor, forming sediments that eventually become carbonate rocks such as limestone. Carbonate precipitation involves inorganic reactions that are catalyzed by marine organisms as they make their shells.

Human activities influence the carbon cycle in several ways. Anything that changes primary productivity, such as nitrogen deposition or altered land use, changes the movement of carbon between inorganic and organic compartments. Any activity that influences water runoff, such as deforestation or alteration and impoundment of river flows, modifies the movement of carbon

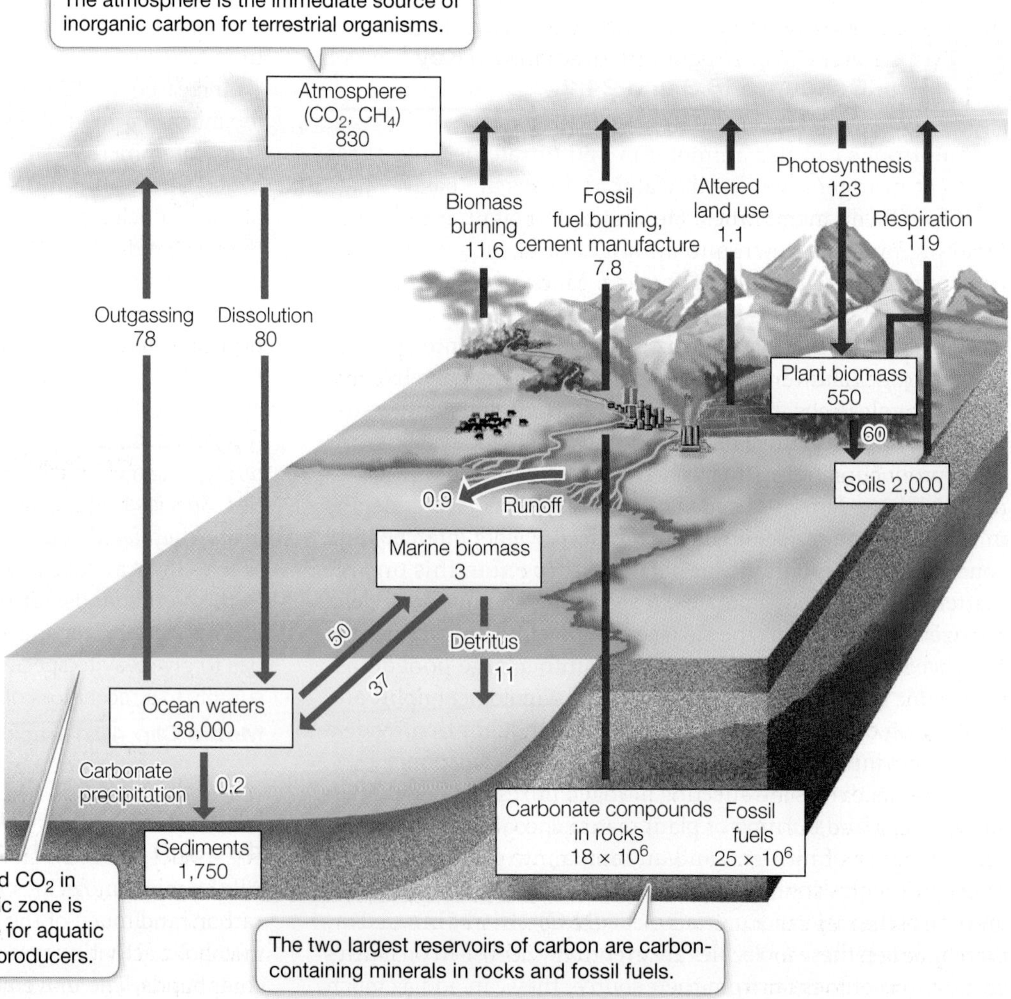

FIGURE 42.10 The Global Carbon Cycle Estimated pools in compartments (boxes) and per-year fluxes between compartments (arrows) are expressed in units of 10^{15} grams. Not shown here is the enormous pool of inaccessible carbon ($100,000 \times 10^{15}$ g) in the deeper layers of Earth's mantle that contributes to volcanic emissions of carbon. (After P. Ciais et al. 2013. In *Climate Change 2013: The Physical Science Basis. Contribution of Working Group I to the Fifth Assessment Report of the Intergovernmental Panel on Climate Change*, T. F. Stocker et al. [eds.]. Cambridge University Press: Cambridge, UK and New York, NY, USA; W. H. Schlesinger and E. S. Bernhardt. 2013. In *Biogeochemistry: An Analysis of Global Change* 3rd ed. W. H. Schlesinger and E. S. Bernhardt [eds.], pp. 445–467. Academic Press: Cambridge, MA.)

What percentage of the flux of carbon from terrestrial ecosystems into the atmosphere is due to human activities?

Animation 42.3 The Global Carbon Cycle

between the terrestrial and aquatic compartments; deforestation and burning or decomposition of biomass also release carbon to the atmosphere that had been stored in living plants. Fossil fuel burning increases the atmospheric pool of CO_2, as does the heating of $CaCO_3$ to manufacture cement (CaO; the chemical equation is $CaCO_3 + \text{heat} \rightarrow CaO + CO_2$). The atmospheric pool of CH_4 is increased by livestock production, rice cultivation, and water storage in reservoirs, because microbes in the guts of cattle and in waterlogged sediments break down organic compounds anaerobically to produce methane gas. Atmospheric pools of CH_4 and CO_2 affect the global climate because both gases absorb infrared radiation, as we will soon see.

Biogeochemical cycles are diverse and are linked

The water, nitrogen, and carbon cycles illustrate the diversity of processes, relative sizes of pools and fluxes, and spatial and temporal scales that are involved in the cycling of materials within the Earth system. As you can imagine, there is even more variation if we consider other materials with different chemical and physical properties. Few forms of phosphorus are gases, for example, and the phosphorus cycle differs from the water, nitrogen, and carbon cycles because it lacks a significant atmospheric component. Volcanic eruptions are a particularly prominent feature of the sulfur cycle because they release large amounts of this element into the atmosphere, where it forms tiny, airborne particles (aerosols) that affect Earth's radiation balance (as we will discuss in Key Concept 42.4).

Activity 42.1 The Phosphorus and Sulfur Cycles

Biogeochemical cycles are not independent of one another. The fluxes of different materials are linked, for example, if they share common underlying physical or chemical processes. Many elements, including carbon, nitrogen, phosphorus, and sulfur, have water-soluble forms that are transported together in precipitation and runoff. If runoff increases, the fluxes of all water-soluble materials from terrestrial to aquatic compartments will increase.

The fluxes of different materials through communities are also linked because the functional macromolecules in living tissue—proteins, nucleic acids, polysaccharides, and so forth—have a precise structure with fixed proportions of elemental building blocks. When biomass is produced, there is coordinated movement of these building blocks into living organic compartments; and when biomass decomposes or burns, they all move back into inorganic compartments.

Finally, the cycles of different materials can interact because of idiosyncratic chemical or physical properties. An increasing anthropogenic (human-generated) flux of CO_2 to the atmosphere, for example, can increase the flux of water from soil to the atmosphere. Most terrestrial plants obtain the CO_2 they need for photosynthesis by opening their stomata, which increases the evaporation of water from their tissues. In a high-CO_2 environment, these plants can leave their stomata open for a shorter time, which reduces the movement of water from soil to roots to leaves and ultimately to the atmosphere.

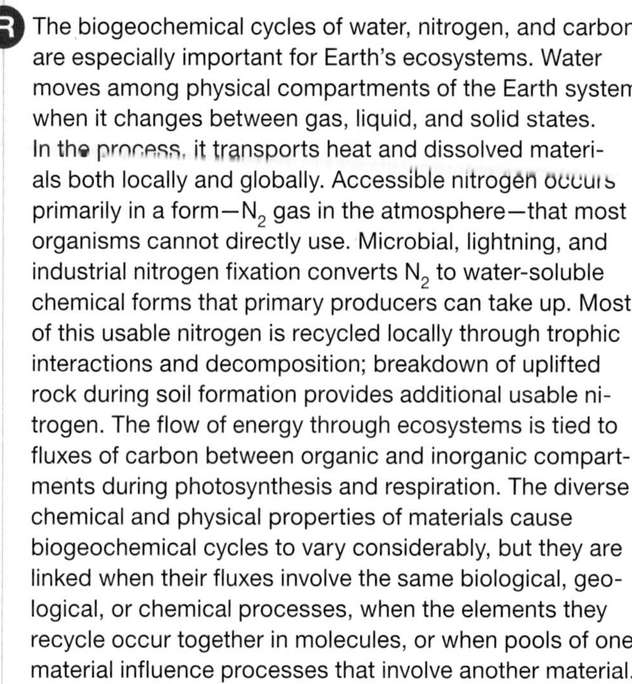

REVIEW & APPLY | 42.3

R The biogeochemical cycles of water, nitrogen, and carbon are especially important for Earth's ecosystems. Water moves among physical compartments of the Earth system when it changes between gas, liquid, and solid states. In the process, it transports heat and dissolved materials both locally and globally. Accessible nitrogen occurs primarily in a form—N_2 gas in the atmosphere—that most organisms cannot directly use. Microbial, lightning, and industrial nitrogen fixation converts N_2 to water-soluble chemical forms that primary producers can take up. Most of this usable nitrogen is recycled locally through trophic interactions and decomposition; breakdown of uplifted rock during soil formation provides additional usable nitrogen. The flow of energy through ecosystems is tied to fluxes of carbon between organic and inorganic compartments during photosynthesis and respiration. The diverse chemical and physical properties of materials cause biogeochemical cycles to vary considerably, but they are linked when their fluxes involve the same biological, geological, or chemical processes, when the elements they recycle occur together in molecules, or when pools of one material influence processes that involve another material.

A
1. Describe ways in which human activities directly affect fluxes and their associated pools in the carbon cycle (see Figure 42.10).
2. List two biogeochemical cycles, other than the three described in this chapter, that are likely to have major impacts on ecosystem function, and justify your choice. (*Hint*: Consider the major chemical constituents of living tissue, discussed in Key Concept 2.1.)

Biogeochemical cycles with significant atmospheric pools can be especially important for Earth's climate, as we will see next.

42.4 Biogeochemical Cycles Influence Global Climate

LEARNING OBJECTIVES

By the end of this key concept you should be able to:

42.4.1 **Explain what is meant by the "greenhouse effect" and how greenhouse gases cause it.**

42.4.2 **Describe two scientific observations that support the proposition that humans are influencing average global temperature.**

Key Concept 38.2 discussed how latitudinal and seasonal variation in solar energy input influences the geographic distribution of Earth's climates. Climate is also influenced by the chemical makeup of Earth's atmosphere. As mentioned at the opening of this chapter, increasing atmospheric concentrations of gases that absorb infrared radiation and warm Earth's surface can cause global **climate change**.

Earth's surface is warm because of its atmosphere

All objects that are warmer than absolute zero emit energy in the form of electromagnetic radiation (see Figure 5.16). The energy and the wavelengths emitted depend on the object's temperature: the rate of energy emission goes up exponentially with temperature, and hotter objects emit absolutely and proportionally more short-wavelength, high-energy photons. Because the surface of the Sun is so hot—about 6,000°C—its radiant output peaks at wavelengths visible to the human eye. Earth's surface, being much cooler than the Sun—about 16°C—emits infrared photons (low-energy photons with longer wavelengths than the red extreme of the visible spectrum) that we perceive as heat rather than light.

What happens when photons emitted by the Sun reach Earth? Approximately 22 percent of them are reflected back into space when they hit tiny airborne particles, water droplets or ice crystals in clouds, or gas molecules in the atmosphere. Another 7 percent are reflected back into space when they hit Earth's surface. About 23 percent are absorbed, rather than reflected, by materials in the atmosphere. Finally, about 48 percent of incoming solar energy (163 watts per square meter, averaged across day and night, all seasons, and all locations) is absorbed by Earth's surface (**FIGURE 42.11**).

What happens to the infrared photons emitted by Earth's surface? Some of them escape immediately into space. But others are absorbed by, and warm, gas molecules in the atmosphere, which then reradiate infrared photons in all directions, some of which are absorbed again by the surface. The ensuing warming of Earth's lower atmosphere and surface is called the **greenhouse effect**.

Water vapor, carbon dioxide, methane, nitrous oxide, and other molecules that absorb infrared radiation are referred to as **greenhouse gases**. The pools of these molecules in the atmosphere influence how much energy is retained in the lower atmosphere. Without its atmosphere, Earth's average surface temperature would be a frigid −18°C, about 34 degrees Celsius colder than it

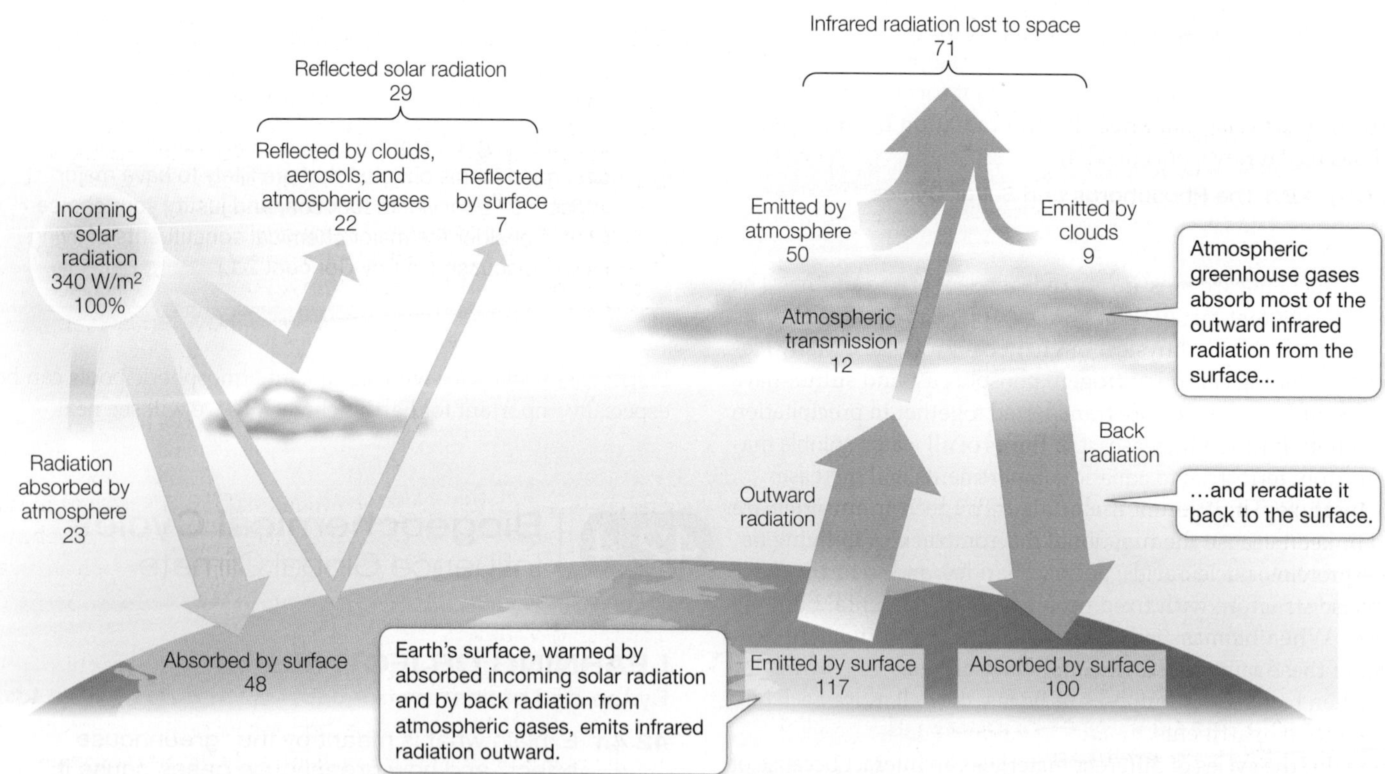

FIGURE 42.11 Earth's Radiation Budget An average of 340 watts/m² of energy from short-wave solar radiation arrives at the top of Earth's atmosphere each year. Seventy-one percent of that is absorbed by the atmosphere or surface; the rest is reflected back into space (left-hand portion of figure). This average annual energy gain from absorbed short-wave solar radiation (left-hand portion of figure) is balanced by the average annual loss to outer space in outgoing long-wavelength (infrared) radiation (right-hand portion of figure). Greenhouse gases in the atmosphere absorb some of the infrared radiation emitted by Earth's surface and reradiate it to the surface, adding to the warming caused by absorbed solar radiation. This "greenhouse effect" causes the total yearly infrared energy emitted by Earth's surface to exceed the average incoming solar energy at the top of the atmosphere. Energy fluxes (arrows) are given as percentages of incoming solar energy. (After K. E. Trenberth et al. 2009. *Bull Am Meteorol Soc* 90: 311–323; R. Lindsey. 2009. *Climate and Earth's energy budget*. Earth Observatory, EOS Project Science Office, NASA Goddard Space Flight Center: Greenbelt, MD.)

Animation 42.4 Earth's Radiation Budget

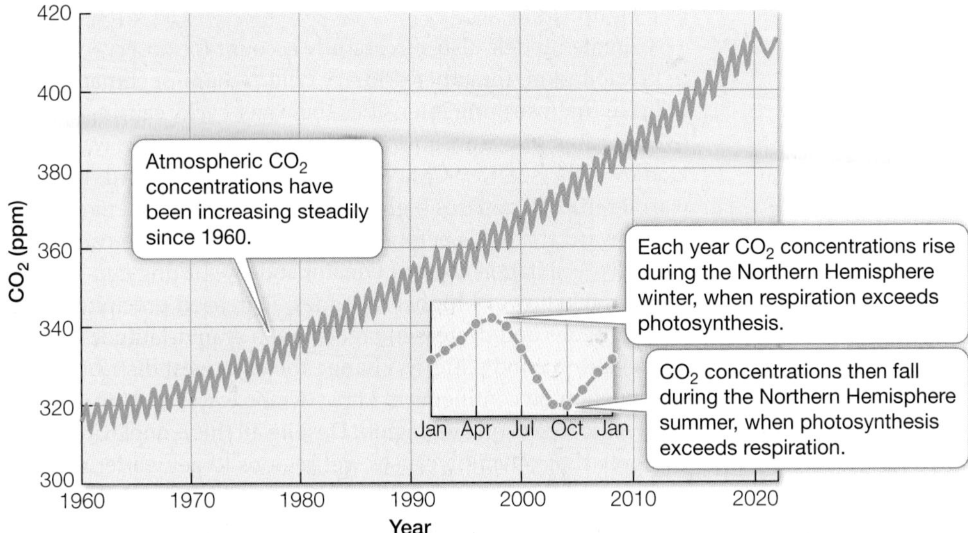

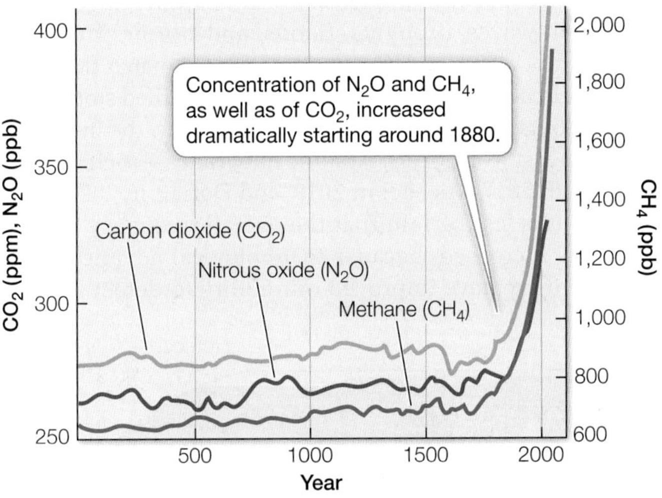

FIGURE 42.12 Atmospheric Greenhouse Gas Concentrations Are Increasing **(A)** Atmospheric CO_2 concentrations atop Mauna Loa, Hawaii, expressed as parts per million (ppm) by volume of dry air, vary seasonally but have been increasing steadily since 1960. **(B)** Measurements of air trapped in glacial ice have allowed researchers to extend our record of greenhouse gas concentrations back 2,000 years. Recent increases began during the Industrial Revolution. Concentrations are expressed as parts per million (ppm) for CO_2 and parts per billion (ppb) for nitrous oxide (N_2O) and methane (CH_4). (A after C. D. Keeling et al. 2001. *I. Global aspects, SIO Reference Series, No. 01-06. Scripps Institution of Oceanography.* San Diego; B after P. Forster et al. 2007. In *Climate Change 2007: The Physical Science Basis. Contribution of Working Group I to the Fourth Assessment Report of the Intergovernmental Panel on Climate Change*, S. Solomon et al. [eds.]. Cambridge University Press: Cambridge, UK and New York, NY, USA. Updates from NOAA.)

Media Clip 42.1 Time-Lapse of Global CO_2 Emissions

is at present. The greenhouse effect influences any planet with an atmosphere, but its magnitude depends on the amounts and types of photon-absorbing materials in the atmosphere. The thin atmosphere of Mars warms its surface by only 3 degrees Celsius, whereas the thick atmosphere of Venus warms that planet's surface by a whopping 468 degrees Celsius.

Recent increases in greenhouse gases are warming Earth's surface

Improved methods for measuring the chemical content of air samples show us that the composition of the atmosphere has changed over Earth's history and continues to change. The CO_2 content of air sampled at the Mauna Loa Observatory in Hawaii has been increasing steadily since measurements began in 1957 (**FIGURE 42.12A**). Analyses of fossil air trapped in polar ice sheets (see the chapter opener image), which allow scientists to extend this record back in time, indicate that CO_2 and other greenhouse gases began their recent increase in the 1800s, when the coal- and petroleum-fueled Industrial Revolution was in full swing (**FIGURE 42.12B**). As expected, given how atmospheres influence the temperature of planets in general, average annual global temperatures have followed suit (**FIGURE 42.13**)—there has been **global warming**.

The parallel trends between greenhouse gas concentrations and average global temperature caused scientists to ask whether CO_2, the most abundant greenhouse gas, could be responsible for the warming. Many scientists initially thought this was unlikely, since greenhouse gases make up such a small fraction (approximately 0.04%) of the atmosphere, but others decided to test the hypothesis by developing quantitative models of how the atmosphere influences the radiation budget—and therefore the climate—of the Earth system. Their first models were simple, containing just solar inputs, the atmosphere, and Earth's surface. But as knowledge of the interactions among and within terrestrial, aquatic, atmospheric, and biological components grew, and as computing technology improved, the models became more

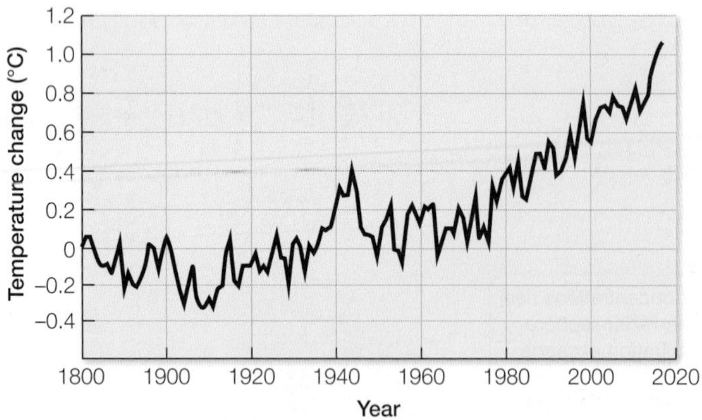

FIGURE 42.13 Global Average Temperature Has Increased in Concert with Rising Atmospheric Greenhouse Gas Concentrations "Temperature change" means the difference, in degrees Celsius, from the average global temperature for the period 1901–1960 (indicated by 0). (After T. Knutson et al. 2017. In *Climate Science Special Report: Fourth National Climate Assessment, Volume I*, D. J. Wuebbles et al. [eds.], pp. 114–132. U.S. Global Change Research Program: Washington, DC.)

complex and realistic. Today's models, which must be run on supercomputers, divide Earth's surface into grid cells that include vertical sections through layers of atmosphere or ocean. They calculate, from a set of starting conditions and assumptions about biogeochemical cycles and solar input, how materials and energy move among layers of each cell and between cells through time, and they output climate variables such as temperature and precipitation by grid cell. Models typically are run multiple times with different sets of input variables to simulate known variation and uncertainties about biogeochemical fluxes or solar inputs, and the outputs are averaged across these replicate simulations. Climate models are tested by comparing their outputs with historical climate records in a process called **hindcasting**. If a model does well in hindcasting, we gain confidence that it can also forecast the future.

Today's models have been tested successfully in this way. Even the first-generation models successfully hindcast the upward trend in temperature during the twentieth century (see Figure 42.13). Most locations on Earth have been getting warmer, as these averages indicate. But some locations have warmed more than others (**FIGURE 42.14**). High latitudes, for example, have warmed more than low latitudes—as models predict—and the North Atlantic Ocean has cooled slightly, perhaps because increased freshwater runoff from a melting Greenland Ice Sheet (see Chapter Opener) has slowed the Gulf Stream current (see Figure 38.7).

Climate models also successfully account for observed trends in precipitation, the other primary determinant of climate. They indicate that warming intensifies the water cycle, causing greater overall rates of evaporation and precipitation. With warming, Hadley cells (see Key Concept 38.2) expand poleward, because warmer tropical air rises higher in the atmosphere and moves farther toward the poles before sinking. This expansion causes increased precipitation near the equator and, by shifting mid-latitude storm tracks toward higher latitudes, increased precipitation at high latitudes and decreased precipitation at mid-latitudes. Shifts in prevailing wind patterns change the east–west distribution of precipitation on continents in idiosyncratic ways, causing shifts to be specific to season and region. Despite all this complexity, models agree that warming causes wet regions to get wetter and dry regions to get drier. Precipitation trends over the twentieth century agree with these patterns in model outputs (**FIGURE 42.15**).

Finally, climate models indicate that global warming increases the frequency and intensity of extreme weather events such as heat waves, droughts, floods, and storms. Trends in extreme temperature and precipitation events match these expectations. In addition, satellite images and detailed storm data point to stronger hurricanes worldwide. Recently, the incidence of the strongest (category 5) Atlantic hurricanes—such as Irma and Maria in 2017, Michael in 2019, and Dorian in 2020—has increased. It is less certain that the overall frequency of hurricanes has increased, because technological advances over the last century have improved our ability to detect storms.

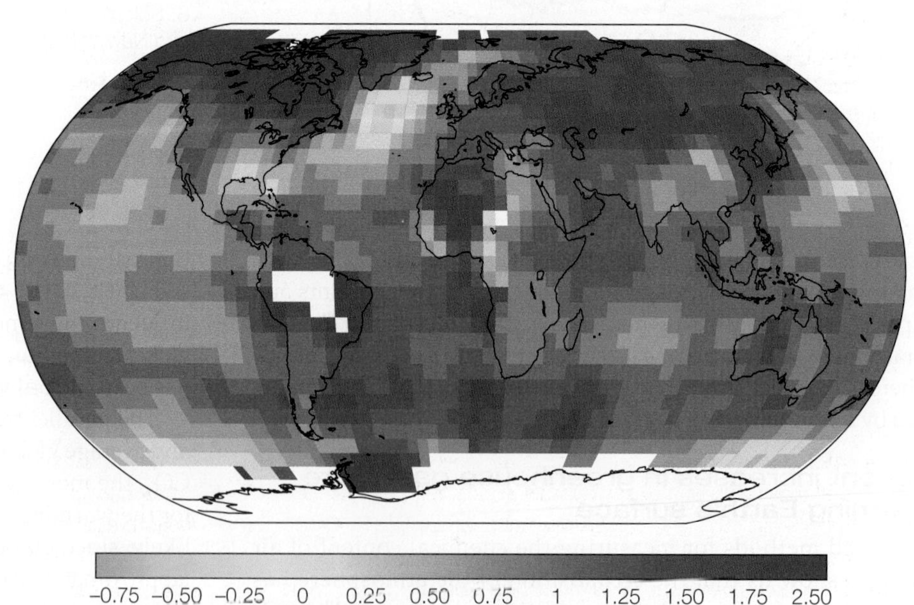

FIGURE 42.14 Temperatures Rose Faster in Some Locations than in Others from 1901 to 2012 The color scale indicates the change in average annual temperature (in degrees Celsius) from 1901 to 2012. White cells indicate missing data. (After D. L. Hartmann et al. 2013. In *Climate Change 2013: The Physical Science Basis. Contribution of Working Group I to the Fifth Assessment Report of the Intergovernmental Panel on Climate Change*, T. F. Stocker et al. [eds.]. Cambridge University Press: Cambridge, UK and New York, NY, USA.)

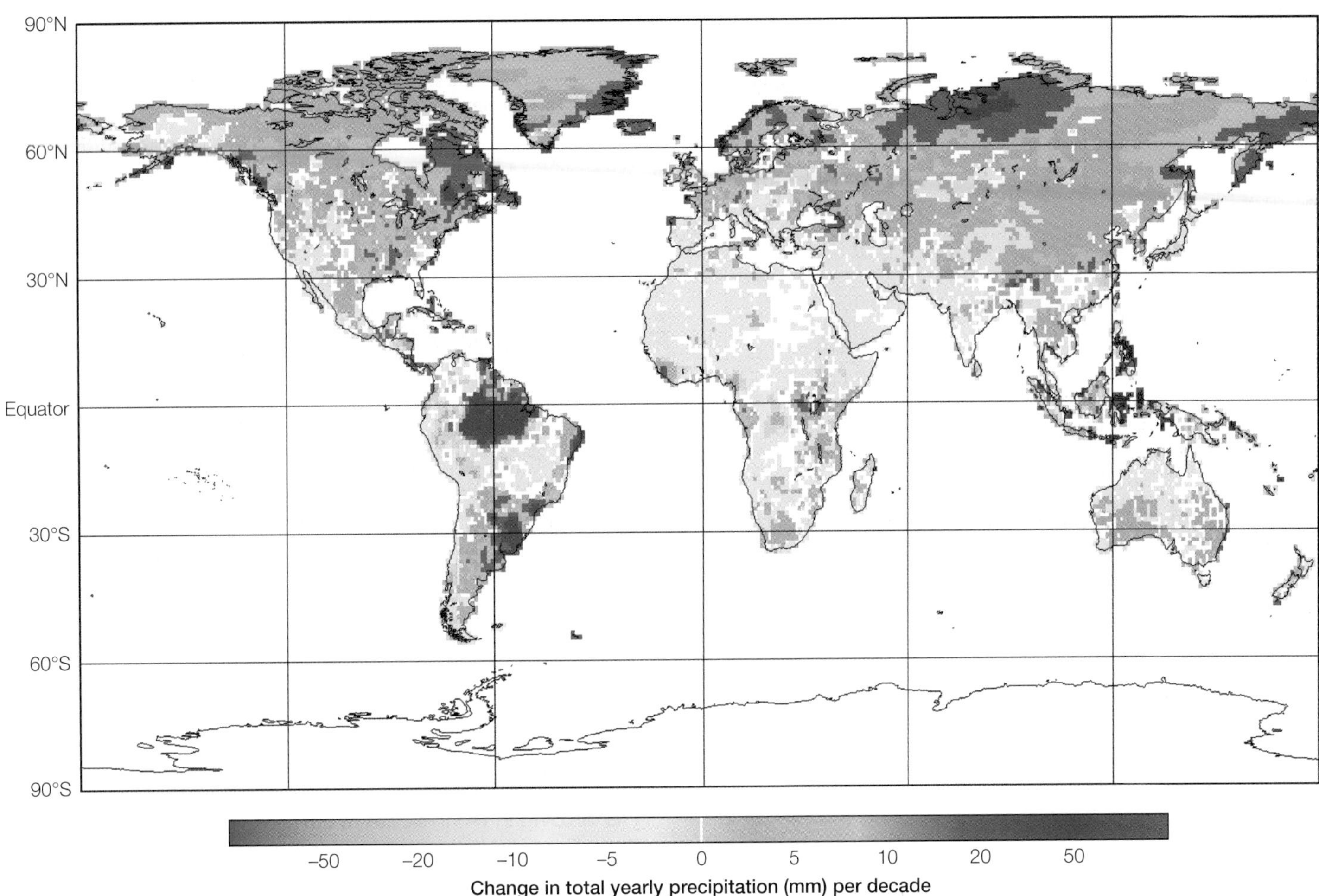

FIGURE 42.15 During the Twentieth Century Wet Regions on Land Became Wetter and Dry Regions Became Drier Annual total precipitation increased especially in already-wet high latitudes and moist tropical regions, but decreased in some already-dry mid- and low-latitude arid zones. Values indicate the rate of change in total annual precipitation (mm/decade) observed at land-based weather stations over the period 1901–2010. (After A. Becker et al. 2013. *Earth Syst Sci Data* 5: 71–99; CC BY 3.0.)

Media Clip 42.2 Time-Lapse History of Global CO_2 Emissions

Time will tell if the planet is experiencing an increased frequency of "megastorms" as the planet warms.

Human activities are contributing to changes in Earth's radiation budget

We've just discussed evidence that humans are affecting Earth's biogeochemical cycles. Fossil fuel burning and many other human activities are adding the greenhouse gases CO_2, CH_4, and N_2O to the atmosphere. Other human activities are also influencing Earth's radiation budget. Expansion of livestock grazing, agriculture, and off-road vehicle traffic in arid regions are increasing airborne dust. Dust and dark-colored soot particles ("black carbon") from fossil fuel burning that are deposited on snow and ice increase the absorption of incoming solar energy. The result is faster melting of persistent ice fields such as Arctic sea ice and the Greenland Ice Sheet of the chapter opener, as well as an earlier date of spring snowmelt in temperate regions. However, some human activities have the potential to counteract these warming effects. When we increase particulates in the atmosphere by generating dust or clearing land, for example, we increase the amount of incoming solar radiation that is reflected back into space, which has the effect of decreasing the temperature of the lower atmosphere and Earth's surface.

What is the net effect of all of these human activities? Are we the cause of the warming that Earth has experienced over the past century? To answer this question, climate scientists have run computer simulation models with and without anthropogenic factors, while holding constant natural factors such as volcanic eruptions (which spew all sorts of materials into the atmosphere). When climate scientists incorporate only natural factors into their models, the models fail to accurately hindcast observed warming since 1880 (**FIGURE 42.16A**). By contrast, the models accurately hindcast temperature patterns when anthropogenic factors and natural factors are included (**FIGURE 42.16B**). This is the reason there is such strong consensus among scientists that human activities are responsible for recent climate warming. A sobering realization, it nonetheless tells us that we have the power to change future climate by our actions.

(A) Natural factors

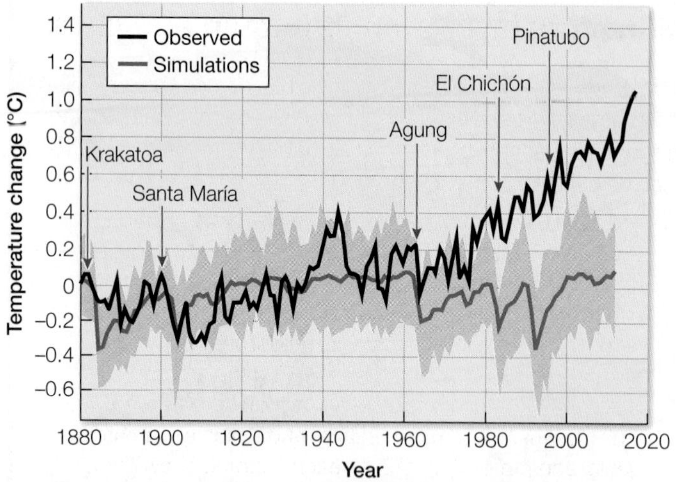

(B) Natural plus anthropogenic factors

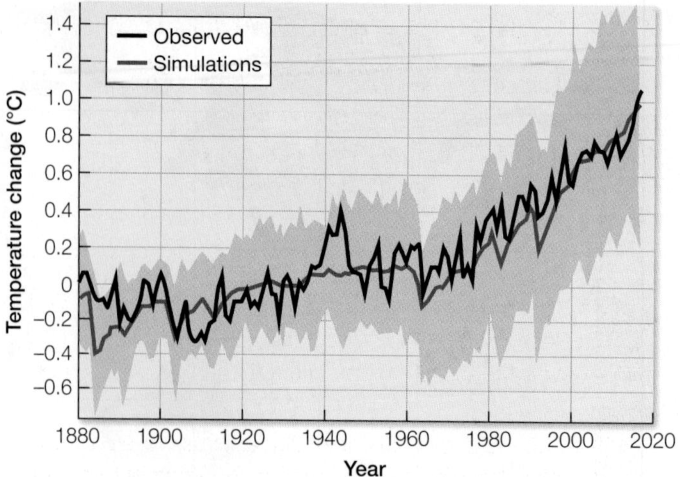

FIGURE 42.16 Human Activities Have Caused Recent Global Warming Climate scientists used computer models to see if they could successfully hindcast the global warming observed between 1880 and 2016. Dark black lines indicate observed average global surface temperatures. Dark blue or red lines indicate means of replicate simulations. Pale blue- or pink-shaded envelopes encompass the range of outputs from replicate computer simulations. Observed and simulation temperatures are expressed as deviations from the average observed global temperature over the period 1901–1960 (indicated by the dashed horizontal line). Results of two models are shown. **(A)** This model considers only natural factors, such as solar radiation input (which varies with Earth's orbit and sunspot cycles), volcanic eruptions (which eject massive amounts of fine particles into the atmosphere), and oscillations in three-dimensional oceanic circulation (which influences the spatial distribution of warm surface waters). **(B)** This model takes into account both natural and anthropogenic factors. Anthropogenic factors include activities such as land clearing (which decreases solar energy absorption by Earth's surface, a cooling effect), soot deposition (which increases solar energy absorption by ice and snow, a warming effect), and emission of greenhouse gases (which absorb infrared radiation, a warming effect). Simulations without anthropogenic factors did not successfully predict the observed upward temperature trend in the twentieth century. This indicates that human activities are primarily responsible for recent global warming. (After T. Knutson et al. 2017. In *Climate Science Special Report: Fourth National Climate Assessment, Volume I*, D. J. Wuebbles et al. [eds.], pp. 114–132. U.S. Global Change Research Program: Washington, DC.)

The arrows in (A) indicate five major volcanic eruptions. What effect did these natural events have on average global temperature, and why did they have this effect?

REVIEW & APPLY | 42.4

R Greenhouse gases in the atmosphere—primarily CO_2, CH_4, and N_2O—absorb infrared (heat) radiation emitted from Earth's surface and thereby retain energy in the lower atmosphere, which warms the planet. Atmospheric concentrations of greenhouse gases have been increasing for more than a century, due to human activities. Average global temperatures have increased in parallel, although this global warming has been uneven across Earth's surface. Continued warming is predicted to change patterns of precipitation worldwide and to increase the frequency of extreme weather events.

A

1. Figure 42.12B shows that atmospheric concentrations of CO_2, CH_4, and N_2O started to increase rapidly in the late 1800s. How has human population growth contributed to this increase? (*Hint*: Recall the trajectory of human population growth illustrated in Figure 39.11.)
2. Why is global warming linked to greater storm intensity? (*Hint*: See Key Concept 38.2 and consider what happens when a pot of water is heated.)

Now that we have seen how human activities are changing Earth's radiation budget and climate, we can consider how ecological systems are responding.

42.5 Rapid Climate Change Influences Species and Communities

LEARNING OBJECTIVES

By the end of this key concept you should be able to:

42.5.1 Discuss how climate change influences environmental cues used by plants and animals

42.5.2 Discuss how altered cues can lead to changes in species interactions

42.5.3 Give examples of mechanisms by which climate change can alter the composition of ecological communities

THINK LIKE A SCIENTIST

Computer models of Earth's climate link global warming to human activities

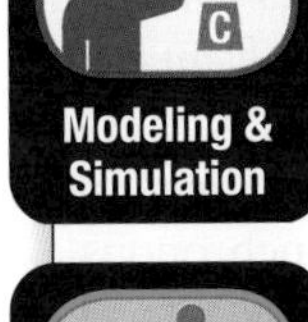

The ability of the latest generation of climate models to successfully hindcast observed global temperature trends since 1880 (see Figure 42.16) illustrates how quantitative reasoning, modeling and simulation, and interdisciplinary communication and collaboration have contributed to progress in our understanding of global climate. The following exercises require you to use data presented in this chapter to explore how different disciplines have contributed to climate science, and the critical role of interaction between science and society for solving global problems.

Computer simulations of global climate are so complex that they must be run on supercomputers. But the basic approach used by computer modelers (or modelers of any sort, for that matter) is not hard to understand. They start by identifying the important components of the system and the interactions among them that influence the property of the system that is of interest. Then they decide what information to input and how to incorporate variability, for example by varying the starting conditions and time-sequence of inputs for replicate simulations. Finally, they give the computer instructions on what calculations to perform and what variables to output.

1. Figure 42.16 shows the outputs from two sets of computer simulations designed to answer the question "Can recent trends in average global temperatures be explained by natural factors alone?" One set (see Figure 42.16A) included only natural factors that affect average global temperatures. The other set (see Figure 42.16B) included anthropogenic factors as well as natural factors. **FIGURE 42.17** shows the basic components of these climate models of Earth's energy budget. Solar energy enters an Earth system that consists of three compartments—atmosphere, land, and ocean. Properties of these compartments (which are determined by anthropogenic as well as natural factors) influence energy fluxes among them as well as rates of loss to space, and together determine the output variable we're interested in—global temperature at Earth's surface.
   a. Populate each of the three empty (blue) boxes of the Earth system in Figure 42.17 with at least one natural factor and one anthropogenic factor that drive global temperature either by directly influencing the amount of short-wave solar energy that is reflected versus absorbed by the Earth system (mechanism 1), or by modifying the greenhouse gas content of the atmosphere (mechanism 2). Be sure to indicate which factors are natural and which are anthropogenic, and put a (1) or a (2) by each factor to show the primary mechanism by which it influences global temperature. See the Figure 42.16 legend for examples, but feel free to include other factors you can justify.
   b. Describe a computer study that would allow you to determine how much of an effect a particular anthropogenic factor has on global temperature.
2. The 1957–1958 International Geophysical Year was the first international collaborative effort to fund research on the Earth system. Since then, nations have continued to pool their resources to support research that helps them plan for future environmental challenges such as flooding, drought, severe storms, pest and pathogen outbreaks, and weather extremes that affect human health.
   a. List one major hazard that global warming is likely to present to a community that you care about.
   b. Imagine you are tasked with devising a plan to minimize the negative impacts of this hazard on your community. Describe what kinds of expertise you would want on your planning team.
3. As the chapter opener indicates, our current understanding of Earth's changing climate and what causes climate change has relied on centuries of study by, and communication between, experts in such fields as physics, chemistry, geology, biology, climatology, oceanography, and computer science. Pick three insights that have improved our understanding of climate change and identify the scientific disciplines that produced them.
4. Imagine that you have agreed to participate in a formal debate on the topic "Are human activities causing recent global warming?" You have been assigned the position "Yes" in the debate. Outline the debating points you would use.

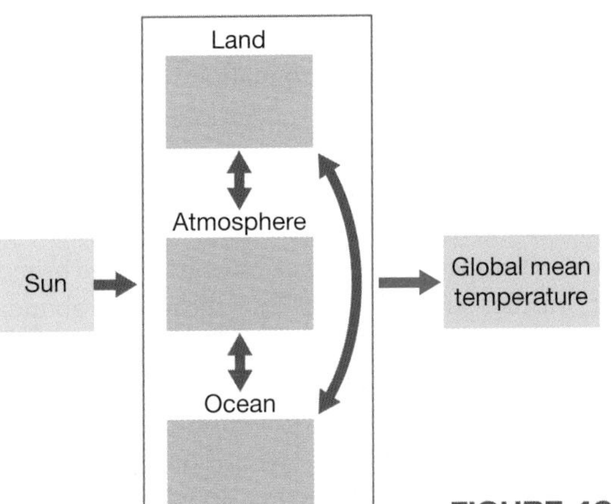

FIGURE 42.17 Systems Diagram

FIGURE 42.18 Climate Change Affects Life Histories
A long-term study in the Netherlands shows that warming temperatures have affected the seasonal cycle of the winter moth, whose caterpillars are now emerging before the leaves on which they feed have expanded, causing more caterpillars to starve. Great Tits are not laying eggs earlier. They now have difficulty finding enough caterpillars to feed their nestlings because they have missed the time of peak caterpillar abundance, and there are fewer caterpillars overall.

Recent climate change has been far more rapid than the species that are alive today have experienced in their evolutionary histories. How will they respond?

Rapid climate change presents ecological challenges

We have seen many examples of how the environment shapes life histories. Organisms are adapted to environmental conditions experienced over past annual cycles, including adaptations that allow crucial life events to coincide with favorable periods. Alpine plants, for example, start growing when the snow melts; seeds of desert plants germinate in response to certain combinations of soil temperature and moisture; seasonal changes in day length often trigger plant flowering, bird migration, or the onset or cessation of insect dormancy. Rapid anthropogenic climate change not only alters conditions that were once beneficial to the organism but also affects the reliability of current environmental cues that in the past heralded predictable seasonal changes. In a warmer world where snow melts faster, for example, snowmelt may no longer signal the last of the late-season killing frosts. Species can evolve by natural selection in response to a changing environment, but the available genetic variation may be too little to allow species to keep up with an environment that changes rapidly, and in the same direction, for too long.

We have ample evidence that the seasonal timing of life events is changing for many species. In temperate ecosystems, for example, plants are leafing out and flowering earlier in spring—the first flowers of a sample of 385 British plant species have appeared, on average, 4.5 days earlier per 1-degree-Celsius increase in average temperature. But not all species have responded in the same way—some have not shifted their time of first flowering, and some now flower later in the spring rather than earlier.

Changes in seasonal timing can disrupt interspecific interactions

Species in a community can interact only if they are biologically active at the same time. If species vary in how they respond to environmental cues, as appears to be the case for the onset of flowering among British plants, then climate change may well cause temporal "mismatches" among interacting species—and thus alter their interactions.

Such timing mismatches are now becoming apparent in nature. In the Netherlands, for example, the leaves of oak trees (*Quercus robur*) are fed on by caterpillars of winter moths (*Operophtera brumata*), and these caterpillars are a main source of food for nestlings of Great Tits (*Parus major*). The emergence of leaves, hatching of moth eggs, and nesting of the birds used to be synchronous. However, each species uses different environmental cues to time springtime events: oak leaves emerge when warm days follow a certain number of cool days; moth eggs hatch when temperatures warm after a fixed number of frosts, and Great Tits use increasing day length to time their nesting. With global warming, oak leaves are appearing earlier in the spring, but moth eggs have been hatching even earlier, so that many emerging caterpillars have no food and starve. Because day length does not change with global warming, Great Tits are not nesting earlier, their nestlings have less food, and nestling survival is declining (**FIGURE 42.18**).

Climate change can alter community composition by several mechanisms

If local populations cannot accommodate changing environmental conditions, they will go extinct, and their extinctions will change the species composition of their communities. As we saw in Figure 39.6, *Sceloporus* lizards on the Yucatán Peninsula of Mexico are experiencing hotter temperatures during the spring breeding season, and this warming restricts the time during which they can forage. The lizards have gone locally extinct where they have insufficient feeding time to support reproduction. Extrapolating to lizards worldwide, researchers predict that continued global warming will put most lizard populations at risk of extinction by the year 2080. Losses of lizards may have cascading effects on populations of their insect prey, the plants those insects feed on, and other animals that also eat the insects. And as we emphasized in Key Concept 40.3, loss of diversity in communities can lead to overall loss of community function.

Altered relationships among environmental cues can also change communities even if species do not go locally extinct. In southwestern North America, the winter rainy season now arrives later in the year, after soil temperatures have cooled. This shift favors plant species whose seeds use cool-season moisture as the cue to germinate, at the expense of species whose germination is cued by moisture in the earlier, warmer part of the fall. Thus the relative abundance of cool-season species is increasing in these communities.

Climate-caused shifts in environmental gradients can also lead to the assembly of novel communities (see Key Concept 41.2). Distributions of some species have moved up mountains as low-elevation habitats become so warm that populations die out and dispersers succeed in now-favorable higher-elevation habitats. Others have moved toward higher latitudes for similar reasons; such poleward shifts have been documented for numerous taxonomic groups. Different species shift to different degrees or not at all, so the mixture of species at a particular place changes. These distributional shifts also influence species interactions. Warmer temperatures in western North America, for example, are speeding up the life history of the mountain bark beetle (*Dendroctonus ponderosae*) and decreasing its winter mortality rate. As a result, the beetle's geographic range has expanded and its abundance has increased, causing widespread tree mortality in pine and spruce forests from New Mexico to Alaska. Loss of the conifers has cascading effects on the consumers that depend on them, as well as on water runoff, wildfire frequency, and NPP.

Extreme weather events have a lasting impact

One of the predictions of climate models is that the frequency of extreme weather events will increase. Severe storms, flooding, droughts, heat waves, and other such events can cause sudden shifts in species distributions and in community composition. For example, a brief but intense drought in the 1950s caused the boundary between ponderosa pine forest (*Pinus ponderosa*) and drought-adapted piñon-juniper woodland in northern New Mexico to shift by more than 2 kilometers in fewer than 5 years. The new community boundary remained stable even after the drought ended. Although in this case the piñons (*Pinus edulis*) and junipers (*Juniperus* sp.) were able to fill in the gap left by the loss of ponderosa pines, other species did not follow suit. This suggests that many species will be "left behind" by ongoing shifts in environmental conditions, and the number of species, their relative abundances, and their interactions will be altered in communities worldwide.

REVIEW & APPLY | 42.5

R Recent climate change presents unique challenges to Earth's biota. Evolution by natural selection may not be rapid enough for species—and their interactions with each other—to track changing environmental conditions and seasonal cues, or to tolerate increasingly common periods of extreme weather. The anticipated results are species extinctions and shifts in the structure of ecological communities, with consequent changes in species interactions and community function.

A

1. Extrapolating from the British plant data presented in this key concept, calculate how much the average first flowering date will change by the middle of this century, when atmospheric CO_2 is predicted to reach double its preindustrial concentration and the average temperature in Britain is expected to rise an additional 3 degrees Celsius.
2. Discuss how changes in the seasonal timing of some events (such as the onset of warm weather at the end of winter) are likely to affect mutualistic interactions between plants and insect pollinators.

The changes that human activities are causing in Earth's biogeochemical cycles and in its climate have serious ecological implications. What can we do to address them?

42.6 | Ecological Challenges Can Be Addressed through Science and International Cooperation

LEARNING OBJECTIVE

By the end of this key concept you should be able to:

42.6.1 **Identify elements of human behavior that facilitate or impede resolution of global ecological challenges.**

The current period of directional climate change is not unprecedented. Throughout the history of life on Earth, wobble in the planet's orbit around the Sun, continental drift, volcanic activity, sunspots, and even asteroid impacts have caused Earth's climates to change, precipitating five major episodes of mass extinction (see Figure 17.12). There is even precedent for organism-caused change in the atmosphere. The first photosynthetic microbes increased atmospheric oxygen concentrations to a level that was toxic to the anaerobic prokaryotes that inhabited Earth at that time, and the first land plants raised oxygen concentrations even more about 250 million years ago (see Key Concept 17.2). What is unprecedented about the present climate change is that it has been precipitated by a single species: *Homo sapiens*.

It is sobering to realize that one species can wield such power. However, as Steven Pinker concludes in *Enlightenment Now: The Case for Reason, Science, Humanism, and Progress* (**FIGURE 42.19**), it is encouraging to contemplate that, of all species on Earth, humans are uniquely able to address the problems they have caused, not only because science equips us to understand the natural world and to solve problems, but also because *Homo sapiens* has a remarkable capacity for cooperative action. We certainly are capable of selfishness and conflict too, but cooperative behaviors are central features of all successful human societies. Other group-living animals cooperate as well, but scientific studies suggest that cooperation with unrelated individuals is especially highly developed in humans—even as infants we routinely offer help to another individual in need. We support cooperative societal ventures of all sorts, we police "bad" behaviors that can undermine cooperation—and we are unique, as far as we know, in empathizing with other creatures and caring about their welfare.

The scale of cooperative human groups has gradually expanded through human history from prehistoric hunter–gatherer family groups to the United Nations, which spans the planet. It is encouraging that governments of separate nations have been cooperating to support global-scale scientific initiatives to understand complex environmental issues, such as the United Nations' Intergovernmental Panel on Climate Change. Earth's nations have also negotiated international agreements to achieve environmental goals. These include the 1963 Convention on International Trade in Endangered Species of Wild Fauna and Flora (CITES), which seeks to conserve species by eliminating economic incentives to exploit them, the 1987 Montreal Protocol to prevent depletion of UV-absorbing atmospheric ozone, the 1992 United Nations Framework Convention on Climate Change to reduce emissions of greenhouse gases, and the 2015 Paris Agreement to adopt a specific plan to keep global temperature rise in this century below 2 degrees Celsius.

Activity 42.2 The Benefits of Cooperation

Nonetheless, humans face huge challenges to effective cooperation on a global scale. A major challenge is that the economic policies of virtually every nation are structured to achieve continual economic growth—ever-increasing production and consumption of goods and services—despite the fact that Earth has a finite capacity to provide those goods and services. We will need to make the transition to sustainable steady-state economies, and that will require an overhaul of economic models and institutions. Another, related, challenge is the continued multiplicative growth of the human population. As the planet becomes more crowded and competition within and among societies for limited resources intensifies, cooperation becomes more difficult. Addressing both challenges will require us to devise international systems for establishing—and enforcing—rules of mutually beneficial behavior within and among nations.

We are born into a pitiless universe, facing steep odds against life-enabling order and in constant jeopardy of falling apart. We were shaped by a force that is ruthlessly competitive. We are made from crooked timber, vulnerable to illusions, self-centeredness, and at times astounding stupidity.

Yet human nature has also been blessed with resources that open a space for a kind of redemption. We are endowed with the power to combine ideas recursively, to have thoughts about our thoughts. We have an instinct for language, allowing us to share the fruits of our experience and ingenuity. We are deepened with the capacity for sympathy—for pity, imagination, compassion, commiseration.

.... From a few oases, the territories with peace and prosperity are growing, and could some day encompass the globe. Much suffering remains, and tremendous peril. But ideas on how to reduce them have been voiced, and an infinite number of others are yet to be conceived.

(S. Pinker. 2018. *Enlightenment Now: The Case for Reason, Science, Humanism and Progress*. Viking: New York: Image provided by: NASA)

FIGURE 42.19 Earth—The Blue Planet This lovely image of verdant Earth rising above the barren surface of the moon, captured on December 24, 1968 by Bill Anders of the Apollo 8 moon mission, forever changed our view of Earth and its place in the universe.

Answer: Precise measurements of the chemical content of Earth's atmosphere and of fossil air trapped in permanent ice fields confirm that greenhouse gas concentrations in Earth's atmosphere began to increase rapidly in the late 1800s and continue to increase. The concentration of CO_2, the most abundant greenhouse gas, was relatively stable at approximately 275 parts per million (ppm) for thousands of years before 1800 and rose from 315 ppm in 1958 and surpassed 410 ppm in 2019. These observed increases are reproduced in outputs of computer models of the global carbon cycle that include natural processes plus estimates of the rate at which human activities such as fossil fuel burning, water impoundment, and livestock production are generating greenhouse gases. Supercomputer models of the global climate that include human influences on climate also reproduce the rising global temperature that has been measured over the last century, whereas models that do not include human influences do not. Thus all the scientific "detective work" points to human activities as the cause of recent directional climate change.

42.1 Climate and Nutrients Influence Ecosystem Function

- Interactions within ecosystems involve exchanges of energy and materials on both local and larger spatial scales.
- Net primary productivity (NPP) reflects the rate at which a community exchanges energy and materials with its abiotic surroundings and is a measure of **ECOSYSTEM FUNCTION** (**FIGURE 42.2**).
- Patterns of geographic variation in terrestrial NPP show that it is influenced by temperature, precipitation, and soil nutrients; aquatic NPP varies with availability of light and nutrients.

FIGURE 42.2

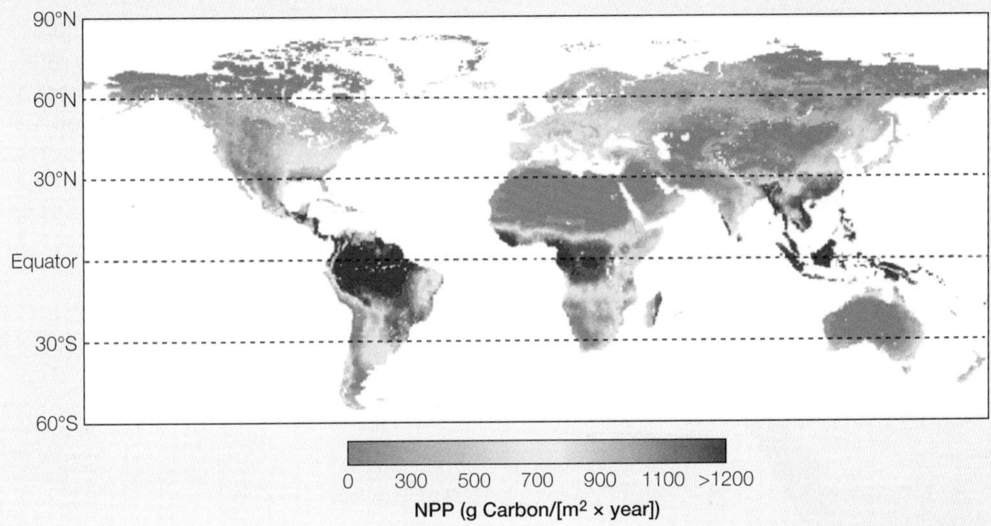

QUESTIONS

1. Discuss how NPP serves to measure the flux of materials as well as energy through an ecosystem. (*Hint*: See also Key Concept 41.3.)
2. Explain why NPP is generally higher in freshwater wetlands (swamps and marshes) than in shallow marine waters (algal beds, coral reefs, and estuaries; see Figure 42.1).

42.2 Biological, Geological, and Chemical Processes Move Materials through Ecosystems

- The chemical and physical form in which an element exists and where it occurs on Earth determine whether the element is usable and accessible to organisms. Alternative forms and locations of matter can be thought of as **COMPARTMENTS** of the global ecosystem (**FIGURE 42.5**).
- The chemical transformations and physical movements of matter among compartments are called **BIOGEOCHEMICAL CYCLES**.
- The **POOL**, or amount, of an element or molecule in a compartment depends on its rate of movement, or **FLUX**, into and out of that compartment.

FIGURE 42.5

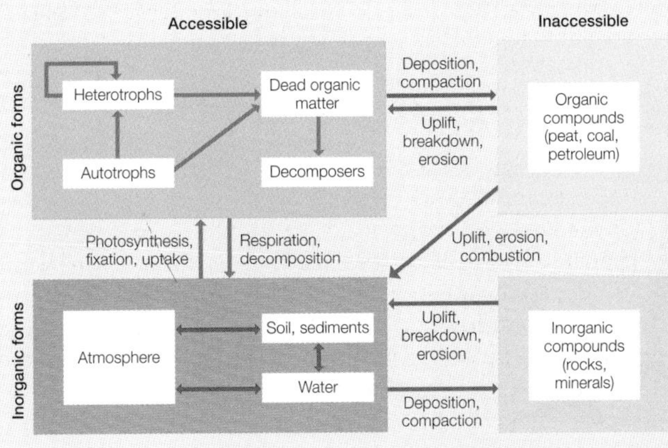

QUESTIONS

1. Distinguish between the concepts of materials being *accessible* to versus *usable* by organisms.
2. List at least one example each of a biological, geological, and chemical process that moves matter among compartments of the global ecosystem.

42 VISUAL SUMMARY

42.3 Certain Biogeochemical Cycles Are Especially Critical for Ecosystems

Go to ACTIVITY 42.1 and ANIMATIONS 42.1, 42.2, and 42.3

- Incoming solar radiation drives the water cycle, transforming liquid water to a gas that rises into the atmosphere and then cools, condenses, and falls to Earth's surface as rain or snow (**FIGURE 42.6**).
- Microbial **NITROGEN FIXATION** and other biochemical processes convert gaseous N_2 into water-soluble forms that other organisms can use.
- Photosynthesis and respiration move carbon between inorganic and organic compartments; thus energy flow and the flux of carbon are linked. Various human activities, such as **FOSSIL FUEL** burning, are increasing the atmospheric carbon pool.

FIGURE 42.6

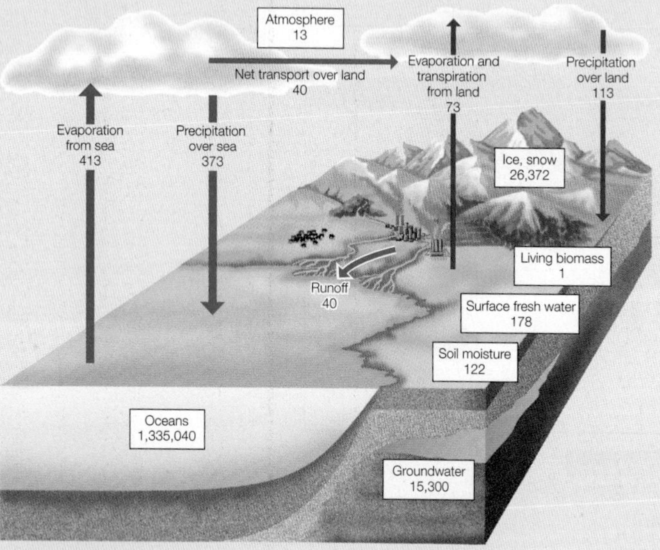

42.4 Biogeochemical Cycles Influence Global Climate

Go to ANIMATION 42.4

- Earth absorbs incoming solar radiation and re-emits it at infrared wavelengths. Some of this infrared radiation is absorbed by **GREENHOUSE GASES**, causing a **GREENHOUSE EFFECT** (**FIGURE 42.11**).
- Increasing pools of greenhouse gases are causing **GLOBAL WARMING** and changing Earth's climates. Computer models of the Earth system show that human activities are responsible for recent increases in greenhouse gases.

FIGURE 42.11

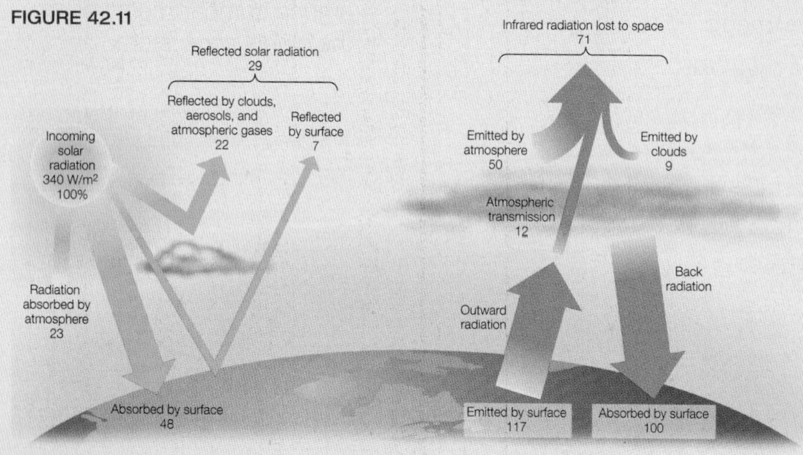

QUESTIONS

42.3

1. Identify two ways in which the water cycle is linked to cycles of other materials.
2. Summarize why nitrogen often is a limiting nutrient for plant growth, even though it is the most abundant element in Earth's atmosphere.
3. Explain why the rate at which CO_2 outgasses from bodies of still water is slightly lower than the rate at which it dissolves into those waters (see Figure 42.10).

42.4

1. Interpret the following statement: the greenhouse effect occurs without humans, but also is influenced by humans.

42.5 Rapid Climate Change Influences Species and Communities

- Climate change is causing mismatches between environmental cues that predict changing seasons and when the seasonal events actually occur. Seasonal timing of organisms' life history events and interactions with other organisms can be disrupted as a result (**FIGURE 42.18**).
- Climate change is altering the distributions and abundances of species, resulting in the assembly of novel communities.
- Climate change is associated with increased probability of extreme weather events, which can cause sudden shifts in species distributions and in community composition.

FIGURE 42.18

© blickwinkel/Alamy Stock Photo

42.6 Ecological Challenges Can Be Addressed through Science and International Cooperation

Go to ACTIVITY 42.2

- Human nature has also been blessed with resources that open a space for a kind of redemption.

FIGURE 42.19

Image provided by NASA

Activity 42.3
Concept Matching: Global Ecosystem

QUESTIONS

1. Key Concept 42.5 points out that species may not be able to adapt to rapid directional climate change. Explain what factors can limit the rate of evolution by natural selection. (*Hint*: Review the mechanisms of evolution by natural selection outlined in Key Concept 15.2.)

Appendix A | The Tree of Life

Phylogeny is the organizing principle of modern biological taxonomy. A guiding principle of modern phylogeny is monophyly. A monophyletic group is considered to be one that contains an ancestral lineage and all of its descendants. Any such group can be extracted from a phylogenetic tree with a single cut.

The trees shown here provide a guide to the relationships among the major groups of extant (living) organisms in the tree of life as we have presented them throughout this book. The position of the branching "splits" indicates the relative branching order of the lineages of life, but the time scale is not meant to be uniform. In addition, the groups appearing at the branch tips do not necessarily carry equal phylogenetic "weight." For example, the ginkgo [78] is indeed at the apex of its lineage; this gymnosperm group consists of a single living species. In contrast, a phylogeny of the eudicots [86] could continue on from this point to fill many more trees the size of this one.

The glossary entries that follow are informal descriptions of some major features of the organisms described in Part Four of this book. We show greater detail about the relationships and content of some groups than we have presented in this book so that this appendix may serve as a reference. Each entry gives the group's common name, followed by the formal scientific name of the group (in parentheses). Numbers in square brackets reference the location of the respective groups on the tree.

It is sometimes convenient to use an informal name to refer to a collection of organisms that are not monophyletic but nonetheless all share (or all lack) some common attribute. We call these "convenience terms"; such groups are indicated in these entries by quotation marks, and we do not give them formal scientific names. Examples include "prokaryotes," "protists," and "algae." Note that these groups cannot be removed with a single cut; they represent a collection of distantly related groups that appear in different parts of the tree. We also use quotation marks here to designate two groups of fungi that are not believed to be monophyletic.

Go to **Achieve** for interactive versions of these trees, with links to photos, distribution maps, species lists, and identification keys.

A

acorn worms (*Enteropneusta*) Benthic marine hemichordates [124] with an acorn-shaped proboscis, a short collar (neck), and a long trunk.

actinobacteria (*Actinobacteria*) [7] High-GC Gram-positive bacteria [2] with genomes that have a high ratio of G-C to A-T nucleotide base pairs.

"algae" A term of convenience encompassing various distantly related groups of aquatic, photosynthetic eukaryotes [17].

alveolates (*Alveolata*) [18] Unicellular eukaryotes with a layer of flattened vesicles (alveoli) supporting the cell membrane. Major groups include the dinoflagellates [54], apicomplexans [53], and ciliates [52].

amborella (*Amborella*) [81] An understory shrub or small tree found only on the South Pacific island of New Caledonia. Thought to be the sister group of the remaining living angiosperms [28].

ambulacrarians (*Ambulacraria*) [41] The echinoderms [123] and hemichordates [124].

amniotes (*Amniota*) [49] Mammals, reptiles, and their extinct close relatives. Characterized by many adaptations to terrestrial life, including an amniote egg (with a unique set of membranes—the yolk sac, amnion, chorion, and allantois), a water-repellant epidermis (with epidermal scales, hair, or feathers), and in males, a penis that allows internal fertilization.

amoebozoans (*Amoebozoa*) [87] A group of eukaryotes [17] that use lobe-shaped pseudopods for locomotion and to engulf food. Major amoebozoan groups include the loboseans, plasmodial slime molds, and cellular slime molds.

Figure A1

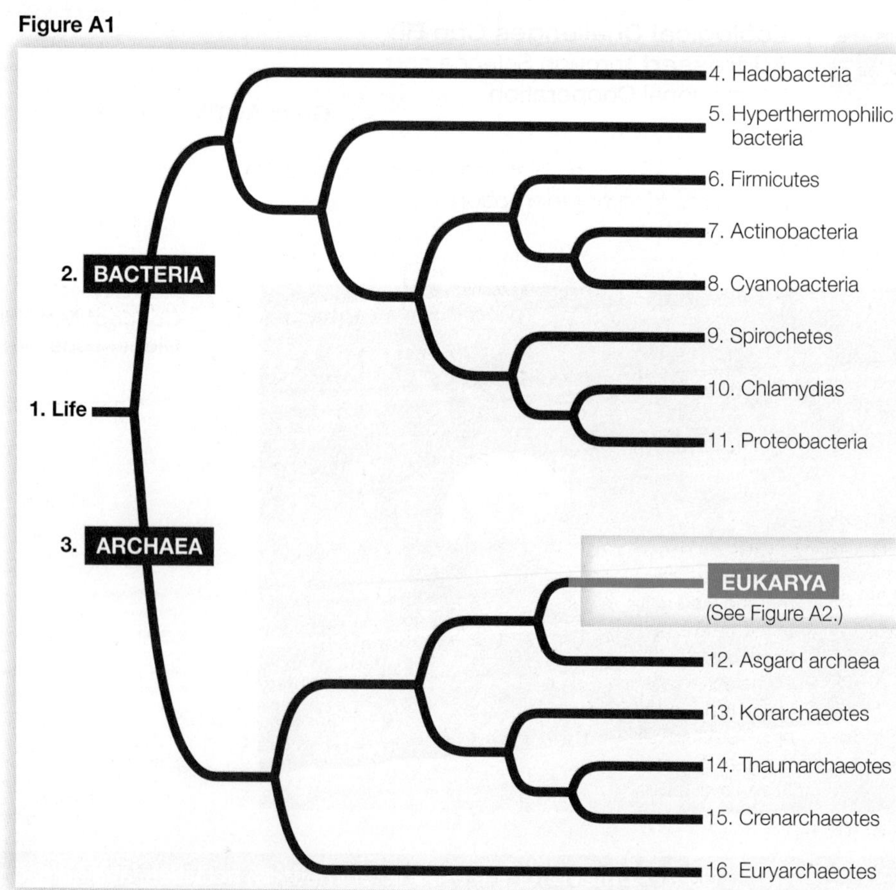

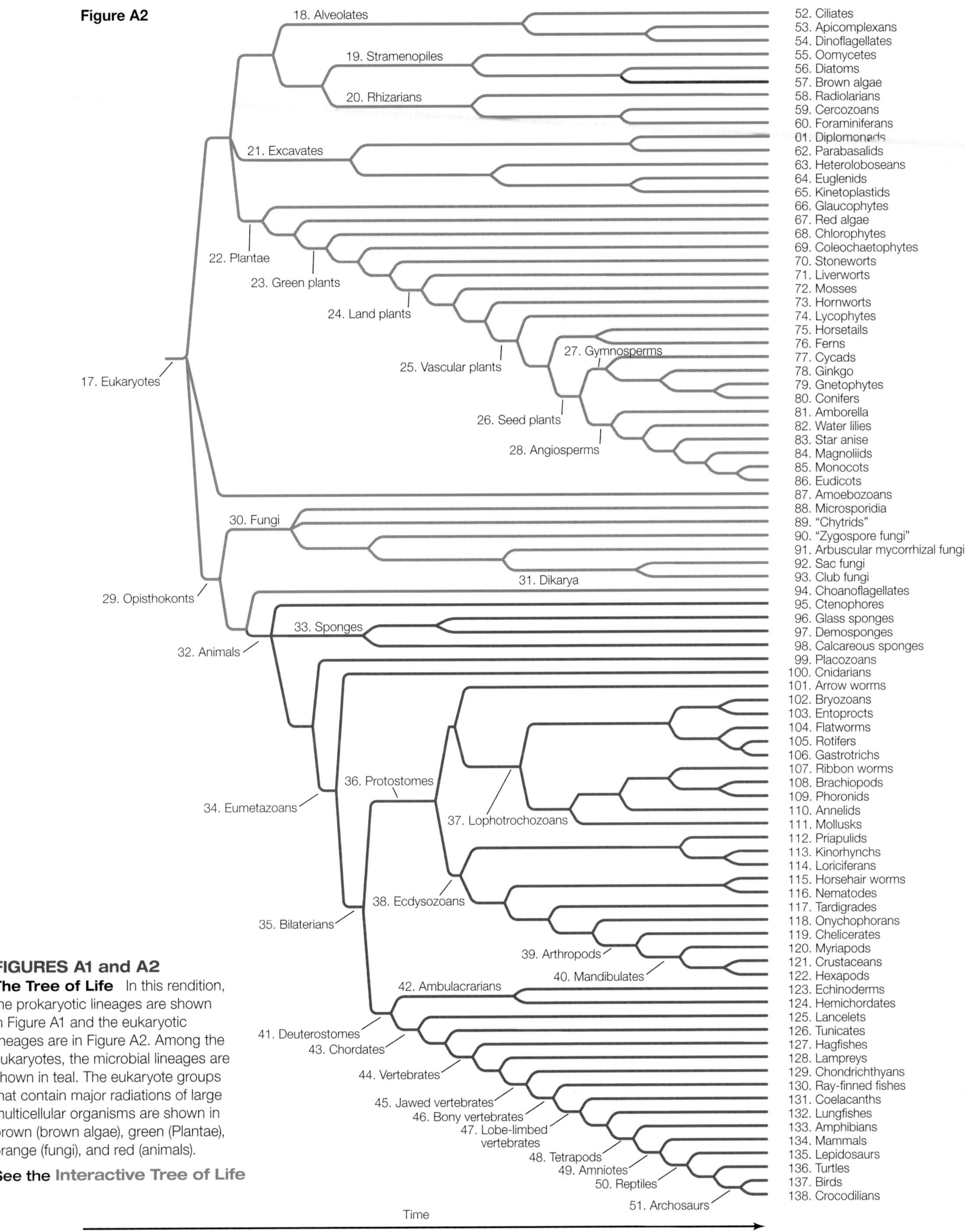

FIGURES A1 and A2

The Tree of Life In this rendition, the prokaryotic lineages are shown in Figure A1 and the eukaryotic lineages are in Figure A2. Among the eukaryotes, the microbial lineages are shown in teal. The eukaryote groups that contain major radiations of large multicellular organisms are shown in brown (brown algae), green (Plantae), orange (fungi), and red (animals).

See the Interactive Tree of Life

amphibians (*Amphibia*) [133] Tetrapods [48] with glandular skin that lacks epidermal scales, feathers, or hair. Many amphibian species undergo a complete metamorphosis from an aquatic larval form to a terrestrial adult form, although direct development is also common. Major amphibian groups include frogs and toads (anurans), salamanders, and caecilians.

amphipods (*Amphipoda*) Small crustaceans [121] that are abundant in many marine and freshwater habitats. They are important herbivores, scavengers, and micropredators, and are an important food source for many aquatic organisms.

angiosperms (*Anthophyta* or *Magnoliophyta*) [28] The flowering plants; major groups include the monocots [85], eudicots [86], and magnoliids [84].

animals (*Animalia* or *Metazoa*) [32] Multicellular heterotrophic eukaryotes. The majority of animals are bilaterians [35]. Other groups of animals include the sponges [33], ctenophores [95], placozoans [99], and cnidarians [100]. The closest living relatives of the animals are the choanoflagellates [94].

annelids (*Annelida*) [110] Segmented worms, including earthworms, leeches, and polychaetes. One of the major groups of lophotrochozoans [37].

anthozoans (*Anthozoa*) One of the major groups of cnidarians [100]. Includes the sea anemones, sea pens, and corals.

anurans (*Anura*) Comprising the frogs and toads, this is the largest group of living amphibians [133]. They are tail-less, with a shortened vertebral column and elongate hind legs modified for jumping. Many species have an aquatic larval form known as a tadpole.

apicomplexans (*Apicomplexa*) [53] Parasitic alveolates [18] characterized by the possession of an apical complex at some stage in the life cycle.

arachnids (*Arachnida*) Chelicerates [119] with a body divided into two parts: a cephalothorax that bears six pairs of appendages (four pairs of which are usually used as legs) and an abdomen that bears the genital opening. Familiar arachnids include spiders, scorpions, mites and ticks, and harvestmen.

arbuscular mycorrhizal fungi (*Glomeromycotina*) [91] A group of fungi [30] that associate with plant roots in a close symbiotic relationship.

archaeans (*Archaea*) [3] One of the two primary divisions of life [1]. The term is often used only for the prokaryotic archaeans, which are unicellular organisms that lack a nucleus and lack peptidoglycan in the cell wall. Recent studies indicate that eukaryotes [17] are most closely related to the lokiarchaeotes [12] among the prokaryotic archaeans.

archosaurs (*Archosauria*) [51] A group of reptiles [50] that includes dinosaurs and crocodilians [138]. Most dinosaur groups became extinct at the end of the Cretaceous; birds [137] are the only surviving dinosaurs.

arrow worms (*Chaetognatha*) [101] Small planktonic or benthic predatory marine worms with fins and a pair of hooked, prey-grasping spines on each side of the head.

arthropods (*Arthropoda*) The largest group of ecdysozoans [38]. Arthropods are characterized by a stiff exoskeleton, segmented body, and jointed appendages. Includes the chelicerates [119], myriapods [120], crustaceans [121], and hexapods (insects and their relatives) [122].

Asgard archaea [12] A group of recently discovered archaeal lineages (including Lokiarchaeotes) that are the closest known prokaryotic relatives of eukaryotes.

ascidians (*Ascidiacea*) "Sea squirts"; the largest group of tunicates [126]. They are sessile (as adults), marine, saclike filter feeders.

B

bacteria (*Bacteria* or *Eubacteria*) [2] Unicellular organisms lacking a nucleus, possessing distinctive ribosomes and initiator tRNA, and generally containing peptidoglycan in the cell wall. Different bacterial groups are distinguished primarily on nucleotide sequence data. One of the two primary divisions of life [1].

barnacles (*Cirripedia*) Crustaceans [121] that undergo two metamorphoses—first from a feeding planktonic larva to a nonfeeding swimming larva, and then to a sessile adult that forms a "shell" composed of four to eight plates cemented to a hard substrate.

bilaterians (*Bilateria*) [35] Those animal groups characterized by bilateral symmetry and three distinct tissue types (endoderm, ectoderm, and mesoderm). Includes the protostomes [36] and deuterostomes [41].

birds (*Aves*) [137] Feathered, flying (or secondarily flightless) tetrapods [48].

bivalves (*Bivalvia*) A major mollusk [111] group; clams, oysters, and mussels. Bivalves typically have two similar hinged shells that are each asymmetrical across the midline.

bony vertebrates (*Osteichthyes*) [46] Vertebrates [44] in which the skeleton is usually ossified to form bone. Includes the ray-finned fishes [130], coelacanths [131], lungfishes [132], and tetrapods [48].

brachiopods (*Brachiopoda*) [108] Lophotrochozoans [37] with two similar hinged shells that are each symmetrical across the midline. Superficially resemble bivalve mollusks, except for the shell symmetry.

brittle stars (*Ophiuroidea*) Echinoderms [123] with five long, whip-like arms radiating from a distinct central disk that contains the reproductive and digestive organs.

brown algae (*Phaeophyta*) [57] Multicellular, almost exclusively marine stramenopiles [19] generally containing the pigment fucoxanthin as well as chlorophylls *a* and *c* in their chloroplasts.

bryozoans (*Ectoprocta* or *Bryozoa*) [102] A group of marine and freshwater lophotrochozoans [37] that live in colonies attached to substrates; also known as ectoprocts or moss animals.

C

caecilians (*Gymnophiona*) A group of burrowing or aquatic amphibians [133]. They are elongate, legless, with a short tail (or none at all), reduced eyes covered with skin or bone, and a pair of sensory tentacles on the head.

calcareous sponges (*Calcarea*) [98] Filter-feeding marine sponges with spicules composed of calcium carbonate.

cellular slime molds (*Dictyostelida*) Amoebozoans [87] in which individual amoebas aggregate under stress to form a multicellular pseudoplasmodium.

cephalochordates (*Cephalochordata*) [125] *See* lancelets.

cephalopods (*Cephalopoda*) Active, predatory mollusks [111] in which the molluscan foot has been modified into muscular hydrostatic arms or tentacles. Includes octopuses, squids, and nautiluses.

cercozoans (*Cercozoa*) [59] Unicellular eukaryotes [17] that feed by means of threadlike pseudopods. Group together with foraminiferans [60] and radiolarians [58] to comprise the rhizarians [20].

charophytes (*Charales*) [70] *See* stoneworts.

chelicerates (*Chelicerata*) [119] A major group of arthropods [39] with pointed appendages (chelicerae) used to grasp food (as opposed to the chewing mandibles of most other arthropods). Includes the arachnids, horseshoe crabs, pycnogonids, and extinct sea scorpions.

chimaeras (*Holocephali*) A group of bottom-dwelling, marine, scaleless chondrichthyan fishes [129] with large, permanent, grinding tooth plates (rather than the replaceable teeth found in other chondrichthyans).

chitons (*Polyplacophora*) Flattened, slow-moving mollusks [111] with a dorsal protective calcareous covering made up of eight articulating plates.

chlamydias (*Chlamydiae*) [10] A group of very small Gram-negative bacteria; they live as intracellular parasites of other organisms.

chlorophytes (*Chlorophyta*) [68] The most abundant and diverse group of green algae, including freshwater, marine, and terrestrial forms; some are unicellular, others colonial, and still others multicellular. Chlorophytes use chlorophylls *a* and *c* in their photosynthesis.

choanoflagellates (*Choanozoa*) [94] Unicellular eukaryotes [17] with a single flagellum surrounded by a collar. Most are sessile, some are colonial. The closest living relatives of the animals [32].

chondrichthyans (*Chondrichthyes*) [129] One of the two main groups of jawed vertebrates [45]; includes sharks, rays, and chimaeras. They have cartilaginous skeletons and paired fins.

chordates (*Chordata*) [43] One of the three major groups of deuterostomes [41], characterized by the presence (at some point in development) of a notochord, a hollow dorsal nerve cord, and a post-anal tail. Includes the lancelets [125], tunicates [126], and vertebrates [44].

"chytrids" [89] A term of convenience used for a paraphyletic group of mostly aquatic, microscopic fungi [30] with flagellated gametes. Some exhibit alternation of generations.

ciliates (*Ciliophora*) [52] Alveolates [18] with numerous cilia and two types of nuclei (micronuclei and macronuclei).

clitellates (*Clitellata*) Annelids [110] with gonads contained in a swelling (called a clitellum) toward the head of the animal. Includes earthworms (oligochaetes) and leeches.

club fungi (*Basidiomycota*) [93] Fungi [30] that, if multicellular, bear the products of meiosis on club-shaped basidia and possess a long-lasting dikaryotic stage. Some are unicellular.

club mosses (*Lycopodiophyta*) [74] Vascular plants [25] characterized by microphylls. *See* lycophytes.

cnidarians (*Cnidaria*) [100] Aquatic, mostly marine eumetazoans [34] with specialized stinging organelles (nematocysts) used for prey capture and defense, and a blind gastrovascular cavity. The sister group of bilaterians [35].

coelacanths (*Actinista*) [131] A group of marine lobe-limbed vertebrates [47] that was diverse from the Middle Devonian to the Cretaceous but is now known from just two living species. The pectoral and anal fins are on fleshy stalks supported by skeletal elements, so they are also called lobe-finned fishes.

coleochaetophytes (*Coleochaetales*) [69] Multicellular green algae characterized by flattened growth form composed of thin-walled cells. Thought to be the sister group to the stoneworts [70] plus land plants [24].

conifers (*Pinophyta* or *Coniferophyta*) [80] Cone-bearing, woody seed plants [26].

copepods (*Copepoda*) Small, abundant crustaceans [121] found in marine, freshwater, or wet terrestrial habitats. They have a single eye, long antennae, and a body shaped like a teardrop.

craniates (*Craniata*) Some biologist exclude the hagfishes [127] from the vertebrates [44], and use the term craniates to refer to the two groups combined.

crenarchaeotes (*Crenarchaeota*) [15] A major and diverse group of prokaryotic archaeans [3], defined on the basis of rRNA base sequences. Many are extremophiles (inhabit extreme environments), but the group may also be the most abundant archaeans in the marine environment.

crinoids (*Crinoidea*) Echinoderms [123] with a mouth surrounded by feeding arms, and a U-shaped gut with the mouth next to the anus. They attach to the substratum by a stalk or are free-swimming. Crinoids were abundant in the middle and late Paleozoic, but only a few hundred species have survived to the present. Includes the sea lilies and feather stars.

crocodilians (*Crocodylia*) [138] A group of large, predatory, aquatic archosaurs [51]. The closest living relatives of birds [137]. Includes alligators, caimans, crocodiles, and gharials.

crustaceans (*Crustacea*) [121] A major group of marine, freshwater, and terrestrial arthropods [39] with a head, thorax, and abdomen (although the head and thorax may be fused), covered with a thick exoskeleton, and with two-part appendages. Crustaceans undergo metamorphosis from a larva. Includes decapods, isopods, krill, barnacles, amphipods, copepods, and ostracods.

ctenophores (*Ctenophora*) [95] Radially symmetrical, gelatinous marine animals [32], with eight rows of fused plates of cilia (called ctenes).

cyanobacteria (*Cyanobacteria*) [8] A group of unicellular, colonial, or filamentous bacteria that conduct photosynthesis using chlorophyll *a*.

cycads (*Cycadophyta*) [77] Palmlike gymnosperms with large, compound leaves.

cyclostomes (*Cyclostomata*) This term refers to the possibly monophyletic group of lampreys [128] and hagfishes [127]. Molecular data support this group, but morphological data suggest that lampreys are more closely related to jawed vertebrates [45] than to hagfishes.

D

decapods (*Decapoda*) A group of marine, freshwater, and semiterrestrial crustaceans [121] in which five of the eight pairs of thoracic appendages function as legs (the other three pairs, called maxillipeds, function as mouthparts). Includes crabs, lobsters, crayfishes, and shrimps.

demosponges (*Demospongiae*) [97] The largest of the three groups of sponges [33], accounting for 90 percent of all sponge species. Demosponges have spicules made of silica, spongin fiber (a protein), or both.

deuterostomes (*Deuterostomia*) [41] One of the two major groups of bilaterians [35], in which the mouth forms at the opposite end of the embryo from the blastopore in early development (contrast with protostomes). Includes the ambulacrarians [42] and chordates [43].

diatoms (*Bacillariophyta*) [56] Unicellular, photosynthetic stramenopiles [19] with glassy cell walls in two parts.

Dikarya [31] A group of fungi [30] in which two genetically different haploid nuclei coexist and divide within the same hypha; includes club fungi [93] and sac fungi [92].

dinoflagellates (*Dinoflagellata*) [54] A group of alveolates [18] usually possessing two flagella, one in an equatorial groove and the other in a longitudinal groove; many are photosynthetic.

dinosaurs (*Dinosauria*) A group of archosaurs [51] that includes birds [137] as well as many extinct groups from the Mesozoic. Extinct Mesozoic dinosaurs included some of the largest terrestrial vertebrates that have ever lived. Informally, many people use the term to refer only to the extinct Mesozoic species.

diplomonads (*Diplomonadida*) [61] A group of eukaryotes [17] lacking mitochondria; most have two nuclei, each with four associated flagella.

dry-nosed primates (*Haplorhini*) A group of primate mammals [134] that includes apes, monkeys, and tarsiers.

E

ecdysozoans (*Ecdysozoa*) [38] One of the two major groups of protostomes [36], characterized by periodic molting of their exoskeletons. Nematodes [116] and arthropods [39] are the largest ecdysozoan groups.

echinoderms (*Echinodermata*) [123] A major group of marine deuterostomes [41] with five-fold radial symmetry (at some stage of life) and an endoskeleton made of calcified plates and spines. Includes sea stars, crinoids, sea urchins, sea cucumbers, and brittle stars.

elasmobranchs (*Elasmobranchii*) The largest group of chondrichthyan fishes [129]. Includes sharks, skates, and rays. In contrast to the other group of living chondrichthyans (the chimaeras), they have replaceable teeth.

embryophytes *See* land plants [24].

entoprocts (*Entoprocta*) [103] A group of marine and freshwater lophotrochozoans [37] that live as single individuals or in colonies attached to substrates. They are the sister group of bryozoans, from which they differ in having both their mouth and anus inside the lophophore (the anus is outside the lophophore in bryozoans).

eudicots (*Eudicotyledones*) [86] A group of angiosperms [28] with pollen grains possessing three openings. Typically with two cotyledons, net-veined leaves, taproots, and floral organs typically in multiples of four or five.

euglenids (*Euglenida*) [64] Flagellated excavates [21] characterized by a pellicle composed of spiraling strips of protein under the cell membrane; the mitochondria have disc-shaped cristae. Some are photosynthetic.

eukaryotes (*Eukarya*) [17] Organisms made up of one or more complex cells in which the genetic material is contained in nuclei. Contrast with archaeans [3] and bacteria [2].

eumetazoans (*Eumetazoa*) [34] A group of animals [32] characterized by body symmetry, a gut, a nervous system, specialized types of cell junctions, and well-organized tissues in distinct cell layers.

euphyllophytes (*Euphyllophyta*) The group of vascular plants [25] that is sister to the lycophytes [74] and which includes all plants with megaphylls.

euryarchaeotes (*Euryachaeota*) [16] A major group of prokaryotic archaeans [3], diagnosed on the basis of rRNA sequences. Includes many methanogens, extreme halophiles, and thermophiles.

eutherians (*Eutheria*) A group of viviparous mammals [134], eutherians are well developed at birth (contrast with prototherians and marsupials, the other two groups of mammals). Most familiar mammals outside the Australian and South American regions are eutherians (see Table 22.3).

excavates (*Excavata*) [21] Diverse group of unicellular, flagellate eukaryotes, many of which possess a feeding groove; some lack mitochondria.

F

ferns (*Pteridopsida* or *Polypodiopsida*) [76] Vascular plants [25] usually possessing large, frond-like leaves and possessing thin-walled sporangia.

firmicutes (*Firmicutes*) [6] Mostly low-GC Gram-positive bacteria [2] with genomes that have a low ratio of G-C to A-T nucleotide base pairs.

flatworms (*Platyhelminthes*) [104] A group of dorsoventrally flattened and generally elongate soft-bodied lophotrochozoans [37]. May be free-living or parasitic, found in marine, freshwater, or damp terrestrial environments. Major flatworm groups include the tapeworms, flukes, monogeneans, and turbellarians.

flowering plants *See* angiosperms [28].

flukes (*Trematoda*) A group of wormlike parasitic flatworms [104] with complex life cycles that involve several different host species. May be paraphyletic with respect to tapeworms.

foraminiferans (*Foraminifera*) [60] Amoeboid organisms with fine, branched pseudopods that form a food-trapping net. Most produce external shells of calcium carbonate.

fungi (*Fungi*) [30] Eukaryotic heterotrophs with absorptive nutrition based on extracellular digestion; cell walls contain chitin. Major fungal groups include the microsporidia [88], "chytrids" [89], "zygospore fungi" [90], arbuscular mycorrhizal fungi [91], sac fungi [92], and club fungi [93].

G

gastropods (*Gastropoda*) The largest group of mollusks [111]. Gastropods possess a well-defined head with two or four sensory tentacles (often terminating in eyes) and a ventral foot. Most species have a single coiled or spiraled shell. Common in marine, freshwater, and terrestrial environments.

gastrotrichs (*Gastrotricha*) [106] Tiny (0.06–3.0 mm), elongate acoelomate lophotrochozoans [37] that are covered in cilia. They live in marine, freshwater, and wet terrestrial habitats. They are simultaneous hermaphrodites.

ginkgo (*Ginkgophyta*) [78] A gymnosperm [27] group with only one living species. The ginkgo seed is surrounded by a fleshy tissue not derived from an ovary wall and hence not a fruit.

glass sponges (*Hexactinellida*) [96] Sponges [33] with a skeleton composed of four- and/or six-pointed spicules made of silica.

glaucophytes (*Glaucophyta*) [66] Unicellular freshwater algae with chloroplasts containing traces of peptidoglycan, the characteristic cell wall material of bacteria. Thought to be the sister group to the rest of Plantae.

gnathostomes (*Gnathostomata*) *See* jawed vertebrates [45].

gnetophytes (*Gnetophyta*) [79] A gymnosperm [27] group with three very different lineages; all have wood with vessels, unlike other gymnosperms.

green plants (*Viridiplantae*) [23] Organisms with chlorophylls *a* and *b*, cellulose-containing cell walls, starch as a carbohydrate storage product, and chloroplasts surrounded by two membranes.

gymnosperms (*Gymnospermae*) [27] Seed plants [26] with seeds "naked" (i.e., not enclosed in carpels). Probably monophyletic, but status still in doubt. Includes the conifers [80], gnetophytes [79], ginkgo [78], and cycads [77].

H

hadobacteria (*Hadobacteria*) [4] A group of extremophilic bacteria [2] that includes the genera *Deinococcus* and *Thermus*.

hagfishes (*Myxini*) [127] Elongate, slimy-skinned vertebrates [44] with three small accessory hearts, a partial cranium, and no stomach or paired fins. *See also* craniata; cyclostomes.

hemichordates (*Hemichordata*) [124] One of the two primary groups of ambulacrarians [42]; marine wormlike organisms with a three-part body plan.

heteroloboseans (*Heterolobosea*) [63] Colorless excavates [21] that can transform among amoeboid, flagellated, and encysted stages.

hexapods (*Hexapoda*) [122] Major group of arthropods [39] characterized by a reduction (from the ancestral arthropod condition) to six walking appendages, and the consolidation of three body segments to form a thorax. Includes insects and their relatives (see Table 22.2).

hornworts (*Anthocerophyta*) [73] Nonvascular land plants with sporophytes that grow from the base. Cells contain a single large, platelike chloroplast.

horsehair worms (*Nematomorpha*) [115] A group of very thin, elongate, wormlike freshwater ecdysozoans [38]. Largely nonfeeding as adults, they are parasites of insects and crayfish as larvae.

horseshoe crabs (*Xiphosura*) Marine chelicerates [119] with a large outer shell in three parts: a carapace, an abdomen, and a tail-like telson. There are only five living species, but many additional species are known from fossils.

horsetails (*Sphenophyta* or *Equisetophyta*) [75] Vascular plants [25] with reduced megaphylls in whorls.

hydrozoans (*Hydrozoa*) A group of cnidarians [100]. Most species go through both polyp and medusa stages, although one stage or the other is eliminated in some species.

hyperthermophilic bacteria [5] A group of thermophilic bacteria [2] that live in volcanic vents, hot springs, and in underground oil reservoirs; includes the genera *Aquifex* and *Thermotoga*.

I

insects (*Insecta*) The largest group within the hexapods [122]. Insects are characterized by exposed mouthparts and one pair of antennae containing a sensory receptor called a Johnston's organ. Most have two pairs of wings as adults. There are more described species of insects than all other groups of life [1] combined, and many species remain to be discovered. The major insect groups are described in Table 23.2.

"invertebrates" Convenience term encompassing any animal [32] that is not a vertebrate [44].

isopods (*Isopoda*) Crustaceans [121] characterized by a compact head, unstalked compound eyes, and mouthparts consisting of four pairs of appendages. Isopods are abundant and widespread in salt, fresh, and brackish water, although some species (the sow bugs) are terrestrial.

J

jawed vertebrates (*Gnathostomata*) [45] A major group of vertebrates [44] with jawed mouths. Includes chondrichthyans [129], ray-finned fishes [130], and lobe-limbed vertebrates [47].

K

kinetoplastids (*Kinetoplastida*) [65] Unicellular, flagellated organisms characterized by the presence in their single mitochondrion of a kinetoplast (a structure containing multiple, circular DNA molecules).

kinorhynchs (*Kinorhyncha*) [113] Small (< 1 mm) marine ecdysozoans [38] with bodies in 13 segments and a retractable proboscis.

korarchaeotes (*Korarchaeota*) A group of prokaryotic archaeans [3] known only by evidence from nucleic acids derived from hot springs.

krill (*Euphausiacea*) A group of shrimplike marine crustaceans [121] that are important components of the zooplankton.

L

lampreys (*Petromyzontiformes*) [128] Elongate, eel-like vertebrates [44] that often have rasping and sucking disks for mouths.

lancelets (*Cephalochordata*) [125] A group of weakly swimming, eel-like benthic marine chordates [43].

land plants (*Embryophyta*) [24] Plants with embryos that develop within protective structures; also called embryophytes. Both sporophytes and gametophytes of land plants are multicellular. Land plants possess a cuticle. Major groups are the liverworts [71], mosses [72], hornworts [73], and vascular plants [25].

larvaceans (*Larvacea*) Solitary, planktonic tunicates [126] that retain both notochords and nerve cords throughout their lives.

lepidosaurs (*Lepidosauria*) [135] Reptiles [50] with overlapping scales. Includes tuataras and squamates (lizards, snakes, and amphisbaenians).

life (*Life*) [1] The monophyletic group that includes all known living organisms. Characterized by a nucleic-acid based genetic system (DNA or RNA), metabolism, and cellular structure. Some parasitic forms, such as viruses, have secondarily lost some of these features and rely on the cellular environment of their host.

liverworts (*Hepatophyta*) [71] Nonvascular land plants lacking stomata; stalk of sporophyte elongates along its entire length.

lobe-limbed vertebrates (*Sarcopterygii*) [47] One of the two major groups of bony vertebrates [46], characterized by jointed appendages (paired fins or limbs).

loboseans (*Lobosea*) A group of unicellular amoebozoans [87]; includes the most familiar amoebas (e.g., *Amoeba proteus*).

lokiarchaeotes (*Lokiarchaeota*) [12] A group of prokaryotic archaeans [3] known from deep-sea hydrothermal vents. Among the known prokaryotic archaeans, they are among the closest relatives of the eukaryotes [17]. *See* Asgard archaea.

"lophophorates" Convenience term used to describe several groups of lophotrochozoans [37] that have a feeding structure called a lophophore (a circular or U-shaped ridge around the mouth that bears one or two rows of ciliated, hollow tentacles). Not a monophyletic group.

lophotrochozoans (*Lophotrochozoa*) [37] One of the two main groups of protostomes [36]. This group is morphologically diverse and is supported primarily by information from gene sequences. Includes bryozoans [102], entoprocts [103], flatworms [104], rotifers [105], gastrotrichs [106], ribbon worms [107], brachiopods [108], phoronids [109], annelids [110], and mollusks [111].

loriciferans (*Loricifera*) [114] Small (< 1 mm) ecdysozoans [38] with bodies in four parts, covered with six plates.

lungfishes (*Dipnoi*) [132] A group of aquatic lobe-limbed vertebrates [47] that are the closest living relatives of the tetrapods [48]. They have a modified swim bladder used to absorb oxygen from air, so some species can survive the temporary drying of their habitat.

lycophytes (*Lycopodiophyta*) [74] Vascular plants [25] characterized by microphylls; includes club mosses, spike mosses, and quillworts.

M

magnoliids (*Magnoliidae*) [84] A major group of angiosperms [28] possessing two cotyledons and pollen grains with a single opening. The group is defined primarily by nucleotide sequence data; it is more closely related to the eudicots and monocots than to three other small angiosperm groups.

mammals (*Mammalia*) [134] A group of tetrapods [48] with hair covering all or part of their skin; females produce milk to feed their developing young. Includes the prototherians, marsupials, and eutherians.

mandibulates (*Mandibulata*) [40] Arthropods [39] that include mandibles as mouth parts. Includes myriapods [120], crustaceans [121], and hexapods [122].

marsupials (*Marsupialia*) Mammals [134] in which the female typically has a marsupium (a pouch for rearing young, which are born at an extremely early stage in development). Includes such familiar mammals as opossums, koalas, and kangaroos.

metazoans (*Metazoa*) *See* animals [32].

microbial eukaryotes *See* "protists."

microsporidia (*Microsporidia*) [88] A group of parasitic unicellular relatives of fungi [30] that lack mitochondria and have walls that contain chitin.

mollusks (*Mollusca*) [111] One of the major groups of lophotrochozoans [37], mollusks have bodies composed of a foot, a mantle (which often secretes a hard, calcareous shell), and a visceral mass. Includes monoplacophorans, chitons, bivalves, gastropods, and cephalopods.

monilophytes (*Monilophyta*) A group of vascular plants [25], sister to the seed plants [26], characterized by overtopping and possession of megaphylls; includes the horsetails [75] and ferns [76].

monocots (*Monocotyledones*) [85] Angiosperms [28] characterized by possession of a single cotyledon, usually parallel leaf veins, a fibrous root system, pollen grains with a single opening, and floral organs usually in multiples of three.

monogeneans (*Monogenea*) A group of ectoparasitic flatworms [104].

monoplacophorans (*Monoplacophora*) Mollusks [111] with segmented body parts and a single, thin, flat, rounded, bilateral shell.

mosses (*Bryophyta*) [72] Nonvascular land plants with true stomata and erect, "leafy" gametophytes; sporophytes elongate by apical cell division.

moss animals *See* bryozoans [102].

mycoplasmas Firmicutes [6] that lack cell walls and are among the smallest known cellular organisms.

myriapods (*Myriapoda*) [120] Arthropods [39] characterized by an elongated, segmented trunk with many legs. Includes centipedes and millipedes.

N

nanoarchaeotes (*Nanoarchaeota*) A group of extremely small, thermophilic, prokaryotic archaeans [3] with a much-reduced genome. The only described example can survive only when attached to a host organism.

nematodes (*Nematoda*) [116] A very large group of elongated, unsegmented ecdysozoans [38] with thick, multilayer cuticles. They are among the most abundant and diverse animals, although most species have not yet been described. Include free-living predators and scavengers, as well as parasites of most species of land plants [24] and animals [32].

neognaths (*Neognathae*) The main group of birds [137], including all living species except the ostrich, emu, rheas, kiwis, cassowaries, and tinamous. *See* palaeognaths.

neopterans (*Neoptera*) The largest group of flying (winged) hexapods [122], recognizable by the fact that they can fold their wings close to their body upon landing.

O

oligochaetes (*Oligochaeta*) Annelid [110] group whose members lack parapodia, eyes, and anterior tentacles, and have few setae. Earthworms are the most familiar oligochaetes.

onychophorans (*Onychophora*) [118] Elongated, segmented ecdysozoans [38] with many pairs of soft, unjointed, claw-bearing legs. Also known as velvet worms.

oomycetes (*Oomycota*) [55] Water molds and relatives; absorptive heterotrophs with nutrient-absorbing, filamentous hyphae.

opisthokonts (*Opisthokonta*) [29] A group of eukaryotes [17] in which the flagellum on motile cells, if present, is posterior. The opisthokonts include the fungi [30], animals [32], and choanoflagellates [94].

ostracods (*Ostracoda*) Marine and freshwater crustaceans [121] that are laterally compressed and protected by two clamlike calcareous or chitinous shells.

P

palaeognaths (*Palaeognathae*) A group of secondarily flightless or weakly flying birds [137]. Includes the flightless ostrich, emu, rheas, kiwis, and cassowaries, and the weakly flying tinamous.

parabasalids (*Parabasalia*) [62] A group of unicellular eukaryotes [17] that lack mitochondria; they possess flagella in clusters near the anterior of the cell.

phoronids (*Phoronida*) [109] A small group of sessile, wormlike marine lophotrochozoans [37] that secrete chitinous tubes and feed using a lophophore.

placoderms (*Placodermi*) An extinct group of jawed vertebrates [45] that lacked teeth. Placoderms were the dominant predators in Devonian oceans.

placozoans (*Placozoa*) [99] A poorly known group of structurally simple, asymmetrical, flattened, transparent animals found in coastal marine tropical and subtropical seas. Most evidence suggests that placozoans are secondarily simplified eumetazoans [34].

Plantae (*Plantae*) [22] The most broadly defined plant group, corresponding to the group that descended from the primary endosymbiotic event that gave rise to chloroplasts. Includes green plants [23] as well as glaucophytes [66] and red algae [67]. In most parts of this book, we use the word **plant** as synonymous with land plant [24], a more restrictive definition.

plasmodial slime molds (*Myxogastrida*) Amoebozoans [87] that in their feeding stage consist of a coenocyte called a plasmodium.

pogonophorans (*Pogonophora*) Deep-sea annelids [110] that lack a mouth or digestive tract; they feed by taking up dissolved organic matter, facilitated by endosymbiotic bacteria in a specialized organ (the trophosome).

polychaetes (*Polychaeta*) A group of mostly marine annelids [110] with one or more pairs of eyes and one or more pairs of feeding tentacles; parapodia and setae extend from most body segments. May be paraphyletic with respect to the clitellates.

priapulids (*Priapulida*) [112] A small group of cylindrical, unsegmented, wormlike marine ecdysozoans [38] that takes its name from its phallic appearance.

primates (*Primates*) A group of mammals [134] that includes apes, monkeys, tarsiers, lemurs, and lorises.

"prokaryotes" Unicellular organisms that do not have nuclei. Not a monophyletic group; as commonly used, includes the bacteria [2] and prokaryotic archaeans [3]. A term of convenience encompassing all cellular organisms that are not eukaryotes. (Contrast with eukaryotes.)

progymnosperms (*Progymnospermophyta*) An extinct group of seedless, vascular plants [25] that were among the first large, woody trees; lived in the Middle Devonian and Early Carboniferous periods.

proteobacteria (*Proteobacteria*) [11] A large and extremely diverse group of Gram-negative bacteria that includes many pathogens, nitrogen fixers, and photosynthesizers. Includes the alpha, beta, gamma, delta, and epsilon proteobacteria.

"protists" A term of convenience is used to encompass a large number of distinct and distantly related groups of eukaryotes, many but far from all of which are microbial and unicellular. Essentially a "catch-all" term for any eukaryotic group not contained within the land plants [24], fungi [30], or animals [32].

protostomes (*Protostomia*) [36] One of the two major groups of bilaterians [35]. In protostomes, the mouth typically forms from the blastopore (if present) in early development (contrast with deuterostomes). The major protostome groups are the lophotrochozoans [37] and ecdysozoans [38].

prototherians (*Prototheria*) A mostly extinct group of mammals [134], common during the Cretaceous and early Cenozoic. The five living species—four echidnas and the duck-billed platypus—are the only extant egg-laying mammals.

pterobranchs (*Pterobranchia*) A small group of sedentary marine hemichordates [124] that live in tubes secreted by the proboscis. They have one to nine pairs of arms, each bearing long tentacles that capture prey and function in gas exchange.

pterygotes (*Pterygota*) Flying (winged) insects, the largest group of hexapods [122].

pycnogonids (*Pycnogonida*) Treated in this book as a group of chelicerates [119], but sometimes considered an independent group of arthropods [39]. Pycnogonids have reduced bodies and very long, slender legs. Also called sea spiders.

R

radiolarians (*Radiolaria*) [58] Amoeboid organisms with needlelike pseudopods supported by microtubules. Most have glassy internal skeletons.

ray-finned fishes (*Actinopterygii*) [130] A highly diverse group of freshwater and marine bony vertebrates [46]. They have reduced swim bladders that often function as hydrostatic organs and fins supported by soft rays (lepidotrichia). Includes most familiar fishes.

red algae (*Rhodophyta*) [67] Mostly multicellular, marine and freshwater algae characterized by the presence of phycoerythrin in their chloroplasts.

reptiles (*Reptilia*) [50] One of the two major groups of extant amniotes [49], supported on the basis of similar skull structure and gene sequences. The term "reptiles" traditionally excluded birds [137], but the resulting group is then clearly paraphyletic. As used in this book, living reptiles include turtles [136], lepidosaurs [135], birds [137], and crocodilians [138].

rhizarians (*Rhizaria*) [20] Mostly amoeboid unicellular eukaryotes with pseudopods, many with external or internal shells. Includes the foraminiferans [60], cercozoans [59], and radiolarians [58].

rhyniophytes (*Rhyniophyta*) A group of early vascular plants [25] that appeared in the Silurian and became extinct in the Devonian. Possessed dichotomously branching stems with terminal sporangia but no true leaves or roots.

ribbon worms (*Nemertea*) [107] A group of unsegmented lophotrochozoans [37] with an eversible proboscis used to capture prey. Mostly marine, but some species live in fresh water or on land.

rotifers (*Rotifera*) [105] Tiny (< 0.5 mm) lophotrochozoans [37] with a pseudocoelomic body cavity that functions as a hydrostatic organ, and a ciliated feeding organ called the corona that surrounds the head. Rotifers live in freshwater and wet terrestrial habitats.

roundworms (*Nematoda*) [116] *See* nematodes.

S

sac fungi (*Ascomycota*) [92] Fungi that bear the products of meiosis within sacs (asci) if the organism is multicellular. Some are unicellular.

salamanders (*Caudata*) A group of amphibians [133] with distinct tails in both larvae and adults and limbs set at right angles to the body.

salps *See* thaliaceans.

sarcopterygians (*Sarcopterygii*) [47] *See* lobe-limbed vertebrates.

scyphozoans (*Scyphozoa*) Marine cnidarians [100] in which the medusa stage dominates the life cycle. Commonly known as jellyfish.

sea cucumbers (*Holothuroidea*) Echinoderms [123] with an elongate, cucumber-shaped body and leathery skin. They are scavengers on the ocean floor.

sea spiders *See* pycnogonids.

sea squirts *See* ascidians.

sea stars (*Asteroidea*) Echinoderms [123] with five (or more) fleshy "arms" radiating from an indistinct central disk. Also called starfishes.

sea urchins (*Echinoidea*) Echinoderms [123] with a test (shell) that is covered in spines. Most are globular in shape, although some groups (such as the sand dollars) are flattened.

"seed ferns" A paraphyletic group of loosely related, extinct seed plants that flourished in the Devonian and Carboniferous. Characterized by large, frondlike leaves that bore seeds.

seed plants (*Spermatophyta*) [26] Heterosporous vascular plants [25] that produce seeds; most produce wood; branching is axillary (not dichotomous). The major seed plant groups are gymnosperms [27] and angiosperms [28].

simians (*Simiiformes*) A group of primate mammals [134] that consists of apes and monkeys.

sow bugs *See* isopods.

spirochetes (*Spirochaetes*) [9] Motile, Gram-negative bacteria with a helically coiled structure and characterized by axial filaments.

sponges (*Porifera*) [33] A group of relatively asymmetrical, filter-feeding animals that lack a gut or nervous system and generally lack differentiated tissues. Includes glass sponges [96], demosponges [97], and calcareous sponges [98].

springtails (*Collembola*) Wingless hexapods [122] with springing structures on the third and fourth segments of their bodies. Springtails are extremely abundant in some environments (especially in soil, leaf litter, and vegetation).

squamates (*Squamata*) The major group of lepidosaurs [135], characterized by the possession of movable quadrate bones (which allow the upper jaw to move independently of the rest of the skull) and hemipenes (a paired set of eversible penises, or penes) in males. Includes the lizards (a paraphyletic group), snakes, and amphisbaenians.

staphylococci (*Staphylococcus*) Firmicutes [6] that are abundant on the surface of the human body; can cause skin diseases and respiratory, intestinal, and wound infections.

star anise (*Austrobaileyales*) [83] A group of woody angiosperms [28] thought to be the sister-group of the clade of flowering plants that includes eudicots [86], monocots [85], and magnoliids [84].

starfish (*Asteroidea*) *See* sea stars.

stoneworts (*Charales*) [70] Multicellular green algae with branching, apical growth and plasmodesmata between adjacent cells. The closest living relatives of the land plants [24], they retain the egg in the parent organism.

stramenopiles (*Heterokonta* or *Stramenopila*) [19] Organisms having, at some stage in their life cycle, two unequal flagella, the longer possessing rows of tubular hairs. Chloroplasts, when present, surrounded by four membranes. Major stramenopile groups include the brown algae [57], diatoms [56], and oomycetes [55].

streptophytes (*Streptophyta*) All green plants [23] other than chlorophytes [68].

T

tapeworms (*Cestoda*) Parasitic flatworms [104] that live in the digestive tracts of vertebrates as adults and usually in various other species of animals as juveniles.

tardigrades (*Tardigrada*) [117] Small (< 1.5 mm) ecdysozoans [38] with fleshy, unjointed legs and no circulatory or gas exchange organs. They live in marine sands, in temporary freshwater pools, and on the water films of plants. Also called water bears.

tetrapods (*Tetrapoda*) [48] The major group of lobe-limbed vertebrates [47]; includes the amphibians [133] and the amniotes [49]. Named for the presence of four jointed limbs (although limbs have been secondarily reduced or lost completely in several tetrapod groups).

thaliaceans (*Thaliacea*) A group of solitary or colonial planktonic marine tunicates [126]. Also called salps.

thaumarchaeotes (*Thaumarchaeota*) [14] A group of prokaryotic archaeans [3] known from hot environments and marine waters; they oxidize ammonia and are thought to play an important role in the nitrogen cycle.

therians (*Theria*) Mammals [134] characterized by viviparity (live birth). Includes eutherians and marsupials.

theropods (*Theropoda*) Archosaurs [51] with bipedal stance, hollow bones, a furcula ("wishbone"), elongated metatarsals with three-fingered feet, and a pelvis that points backward. Includes many well-known extinct dinosaurs (such as *Tyrannosaurus rex*), as well as the living birds [137].

tracheophytes *See* vascular plants [25].

trilobites (*Trilobita*) An extinct group of arthropods [39] related to the chelicerates [119]. Trilobites flourished from the Cambrian through the Permian.

tuataras (*Rhyncocephalia*) A group of lepidosaurs [135] known mostly from fossils; there are only two living tuatara species. The quadrate bone of the upper jaw is fixed firmly to the skull. The sister group of the squamates.

tunicates (*Tunicata*) [126] A group of chordates [43] that are mostly saclike filter feeders as adults, with motile larval stages that resemble tadpoles.

turbellarians (*Turbellaria*) A group of free-living, generally carnivorous flatworms [104]. Their monophyly is questionable.

turtles (*Testudines*) [136] A group of reptiles [50] with a bony carapace (upper shell) and plastron (lower shell) that encase the body in a fashion unique among the vertebrates.

U

urochordates (*Tunicata*) [126] *See* tunicates.

V

vascular plants (*Tracheophyta*) [25] Plants with xylem and phloem. Major groups include the lycophytes [74] and euphyllophytes.

vertebrates (*Vertebrata*) [44] The largest group of chordates [43], characterized by a rigid endoskeleton supported by the vertebral column and an anterior skull encasing a brain. Includes hagfishes [127], lampreys [128], and the jawed vertebrates [45], although some biologists exclude the hagfishes from this group. *See also* craniates.

W

water bears *See* tardigrades.

water lilies (*Nymphaeaceae*) [82] A group of aquatic, freshwater angiosperms [28] that are rooted in soil in shallow water, with round floating leaves and flowers that extend above the water's surface. They are the sister-group to most of the remaining flowering plants, with the exception of the genus *Amborella* [81].

wet-nosed primates (*Strepsirrhini*) A group of primate mammals [134] that includes lemurs and lorises.

X

xenoturbellids (*Xenoturbellida*) Wormlike organisms that feed on or parasitize marine mollusks. Relationships uncertain; thought to be a sister group of either ambulacrarians [42] or bilaterians [35].

Y

"yeasts" A term of convenience for several distantly related groups of unicellular fungi [30].

Z

"zygospore fungi" (*Zoopagomycota*, if monophyletic) [90] A term of convenience for a probably paraphyletic group of fungi [30] in which hyphae of differing mating types conjugate to form a zygosporangium.

Appendix B | Making Sense of Data: A Statistics Primer

This appendix is designed help you understand the analysis of biological data in a statistical context. We outline major concepts involved in the collection and analysis of data, and we present basic statistical analyses that will help you interpret, understand, and complete the problems throughout this book as well as understand and interpret scientific studies in general. We present formulas for some common statistical tests as examples, but the main purpose of this appendix is to help you understand the purpose and reasoning behind statistical tests. Once you understand the basis of an analysis, you may wish to use one of many free, online websites for conducting the tests (such as http://vassarstats.net).

Why Do We Need Statistical Analysis?

ALMOST EVERYTHING VARIES We live in a variable world, but within the variation we see among biological organisms there are predictable patterns. We use concepts and methods from the field of statistics to find and analyze these patterns. Consider any group of common things in nature—all women aged 22, all the cells in your liver, or all the blades of grass in your yard. Although they will have many similar characteristics, they will also have important differences. Men aged 22 tend to be taller than women aged 22, but of course not every man will be taller than every woman in this age group.

Natural variation can make it difficult to find general patterns. For example, scientists have determined that smoking increases the risk of getting lung cancer. But we know that not all smokers will develop lung cancer and not all nonsmokers will remain cancer-free. If we compare just one smoker with just one nonsmoker, we may end up drawing the wrong conclusion. So how did scientists discover this general pattern? How many smokers and nonsmokers did they examine before they felt confident about the risk of smoking?

Statistics helps us detect and describe general patterns in nature, and draw conclusions from those patterns.

AVOIDING FALSE POSITIVES AND FALSE NEGATIVES When a woman takes a pregnancy test, there is some chance that it will be positive even if she is not pregnant, and there is some chance that it will be negative even if she is pregnant. We call these kinds of mistakes "false positives" and "false negatives."

Doing science is a bit like taking a medical test. We observe patterns in the world, and we try to draw conclusions about how the world works from those observations. Sometimes our observations lead us to draw the wrong conclusions. We might conclude that a phenomenon occurs, when it actually does not; or we might conclude that a phenomenon does not occur, when it actually does.

For example, planet Earth has been warming over the past century (see Key Concept 42.4). Ecologists are interested in whether plant and animal populations have been affected by global warming. If we have long-term information about the locations of species and about temperatures in certain areas, we can determine whether shifts in species distributions coincide with temperature changes. Such information, however, can be very complicated. Without proper statistical methods, one may not be able to detect the true impact of temperature, or instead may think a pattern exists when it does not.

Statistics helps us avoid drawing the wrong conclusions.

How Does Statistical Analysis Help Us Understand the Natural World?

Statistical analysis is essential to scientific discovery. Most biological studies involve five basic steps, each of which involves aspects of statistical analysis.

- **Step 1: Experimental Design**
  Clearly define the scientific question and the methods necessary to tackle the question.
- **Step 2: Data Collection**
  Gather information about the natural world through observations and experiments.
- **Step 3: Organize and Visualize the Data**
  Use tables, graphs, and other useful representations to gain intuition about patterns in the data.
- **Step 4: Summarize the Data**
  Summarize the data with a few key statistical calculations.
- **Step 5: Inferential Statistics**
  Use statistical methods to draw general conclusions from the data about the world and the ways it works.

Step 1: Experimental Design

We make observations and conduct experiments to gain knowledge about the world. Scientists come up with scientific ideas based on prior research and their own observations. These ideas start as a question such as "Does smoking cause cancer?" From prior experience, scientists then propose possible answers to the question in the form of hypotheses such as "Smoking increases the risk of cancer." Experimental design then involves devising comparisons that can test predictions of the hypotheses. For example, if we are interested in evaluating the hypothesis that smoking increases cancer risk, we might decide to compare the incidence of cancer in nonsmokers and smokers. Our hypothesis would predict that the cancer incidence is higher in smokers than in nonsmokers. There are various ways we could make that comparison. We could compare the smoking history of people newly diagnosed with cancer with that of people whose tests came out negative for cancer. Alternatively, we could assess the current smoking habits of a sample of people and see whether those who smoke more heavily are more likely to develop cancer in, say, 5 years. Statistics provides us with tools for assessing which approach will provide us with an answer with smaller costs in time and effort in data collection and analysis.

We use statistics to guide us in planning exactly how to make our comparisons—what kinds of data and how many observations we will need to collect.

Step 2: Data Collection

TAKING SAMPLES When biologists gather information about the natural world, they typically collect representative pieces of information, called **data** or observations. For example, when evaluating the efficacy of a candidate drug for medulloblastoma brain cancer, scientists may test the drug on tens or hundreds of patients, and then draw conclusions about its efficacy for all patients with these tumors. Similarly, scientists studying the relationship between body weight and clutch size (number of eggs) for female spiders of a particular species may examine tens to hundreds of spiders to make their conclusions.

We use the expression "sampling from a population" to describe this general method of taking representative pieces of information from the system under investigation (**FIGURE B1**). The pieces of information together make up a **sample** of the larger system, or **population**. In the cancer therapy example, each observation was the change in a patient's tumor size 6 months after initiating treatment, and the population of interest was all individuals with medulloblastoma tumors. In the spider example, each observation was a pair of measurements—body weight and clutch size—for a single female spider, and the population of interest was all female spiders of this species.

Sampling is a matter of necessity, not laziness. We cannot hope (and would not want) to collect and weigh *all* of the female spiders of the species of interest on Earth! Instead, we use statistics to determine how many spiders we must collect to confidently infer something about the general population and then use statistics again to make such inferences.

DATA COME IN ALL SHAPES AND SIZES In statistics, we use the word "variable" to mean a measurable characteristic of an individual component of a system. Some variables are on a numerical scale, such as the daily high temperature or the clutch size of a spider. We call these **quantitative variables**. Quantitative variables that take on only whole number values (such as spider clutch size) are called **discrete variables**, whereas variables that can also take on a fractional value (such as temperature) are called **continuous variables**.

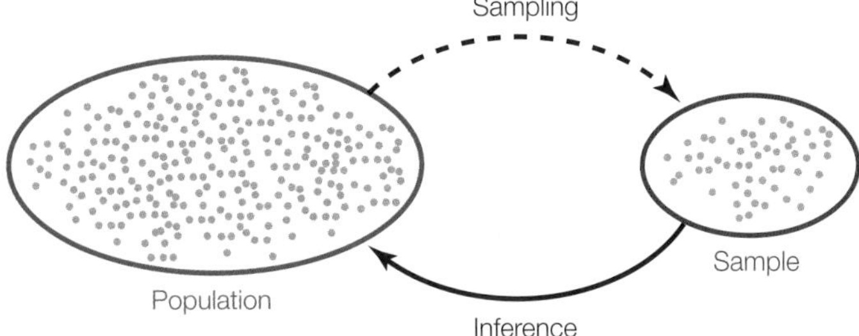

FIGURE B1 Sampling from a Population Biologists take representative samples from a population, use descriptive statistics to characterize their samples, and then use inferential statistics to draw conclusions about the original population.

Other variables take categories as values, such as a human blood type (A, B, AB, or O) or an ant caste (queen, worker, or male). We call these **categorical variables**. Categorical variables with a natural ordering, such as a final grade in Introductory Biology (A, B, C, D, or F), are called **ordinal variables**.

Each class of variables comes with its own set of statistical methods. We will introduce a few common methods in this appendix that will help you work on the problems presented in this book, but you should consult a biostatistics textbook for more advanced tests and analyses for other data sets and problems.

Step 3: Organize and Visualize the Data

Your data consist of a series of values of the variable or variables of interest, each from a separate observation. For example, **TABLE B1**

TABLE **B1** | Weight and Length Measurements for a Sample of *Abramis brama* from Lake Laengelmavesi

Individual number	Fish weight (g)	Fish length (cm)
1	242	30.0
2	290	31.2
3	340	31.1
4	363	33.5
5	390	35.0
6	430	34.0
7	450	34.7
8	450	35.1
9	475	36.2
10	500	34.5
11	500	36.2
12	500	36.2
13	500	36.4
14	575	38.7
15	600	37.3
16	600	37.2
17	610	38.5
18	620	39.2
19	650	38.6
20	680	39.7
21	685	39.5
22	700	37.2
23	700	38.3
24	700	40.6
25	714	40.6
26	720	40.9
27	725	40.5
28	850	41.5
29	920	42.6
30	925	44.0
31	950	45.9
32	955	44.1
33	975	45.3
34	1,000	41.6

Source: fishcatch.txt from http://jse.amstat.org/jse_data_@archive.htm

TABLE **B2** | Summary of Fish Weights of *Abramis brama* from Lake Laengelmavesi

Weight (g)	Frequency	Relative frequency
201–300	2	0.06
301–400	3	0.09
401–500	8	0.24
501–600	3	0.09
601–700	8	0.24
701–800	3	0.09
801–900	1	0.03
901–1,000	6	0.18
Total	34	1.0

shows the weight and length of 34 fish (*Abramis brama*) from Lake Laengelmavesi in Finland. From this list of numbers, it's hard to get a sense of how big the fish in the lake are, or how variable they are in size. It's much easier to gain intuition about your data if you organize them. One way to do this is to group (or bin) your data into **classes**, and count up the number of observations that fall into each class. The result is a **frequency distribution**. **TABLE B2** shows the fish weight data as a frequency distribution. For each 100-gram weight class, Table B2 shows the number, or frequency, of observations in that weight class, as well as the relative frequency (proportion of the total) of observations in that weight class. Notice that the data take up much less space when organized in this fashion. Also notice that we can now see that most of the fish fall in the middle of the weight range, with relatively few very small or very large fish.

Media Clip B1 Interpreting Frequency Distributions

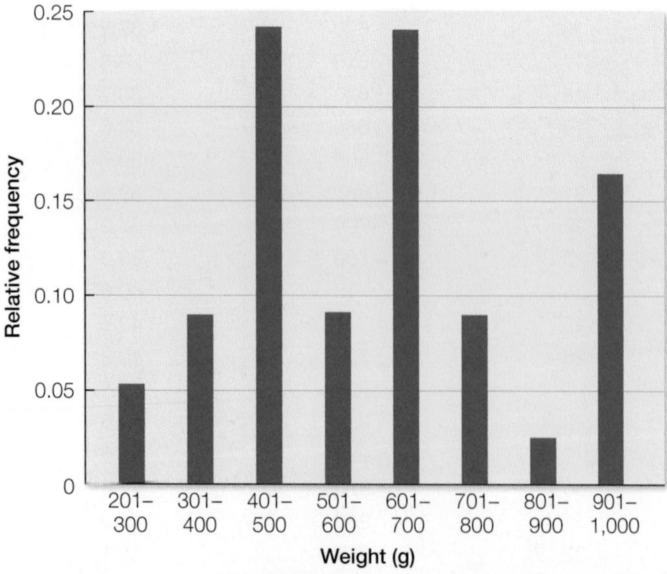

FIGURE B2 Histograms Depict Frequency Distributions of Quantitative Data This histogram shows the relative frequency of different weight classes of fish (*Abramis brama*).

TABLE **B3** | Poinsettia Colors

Color	Frequency	Proportion
Red	108	0.59
Pink	34	0.19
White	40	0.22
Total	182	1.0

It is even easier to visualize the frequency distribution of fish weights if we graph them in the form of a **histogram** such as the one in **FIGURE B2**. When grouping quantitative data, it is necessary to decide how many classes to include. It is often useful to look at multiple histograms before deciding which grouping offers the best representation of the data.

Frequency distributions are also useful ways of summarizing categorical data. **TABLE B3** shows a frequency distribution of the colors of 182 poinsettia plants (red, pink, or white) resulting from an experimental cross between two parent plants. Notice that, as with the fish example, the table is a much more compact way to present the data than a list of 182 color observations would be. For categorical data, the possible values of the variable are the categories themselves, and the frequencies are the number of observations in each category. We can visualize frequency distributions of categorical data like this by constructing a **bar chart**. The heights of the bars indicate the number of observations in each category (**FIGURE B3**). Another way to display the same data is in a **pie chart**, which shows the proportion of each category represented like pieces of a pie (**FIGURE B4**).

Sometimes we wish to compare two quantitative variables. For example, the researchers at Lake Laengelmavesi investigated the relationship between fish weight and length, from the data presented in Table B1. We can visualize this relationship using a **scatter plot** in which the weight and length of each fish is represented as a single point (**FIGURE B5**). These two variables have a **positive relationship** since the slope of a line drawn through

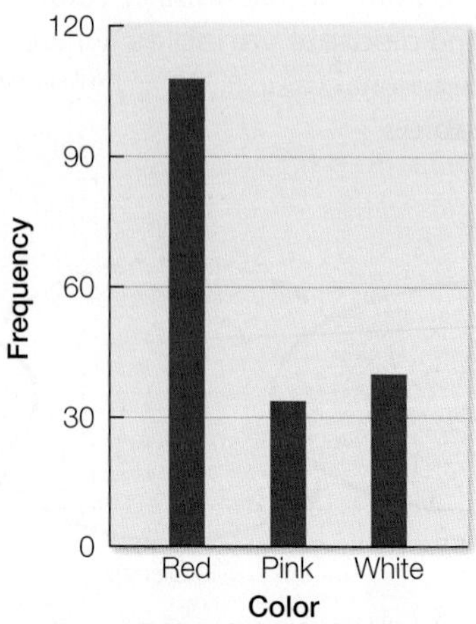

FIGURE B3 Bar Charts Compare Categorical Data This bar chart shows the frequency of three poinsettia colors that result from an experimental cross.

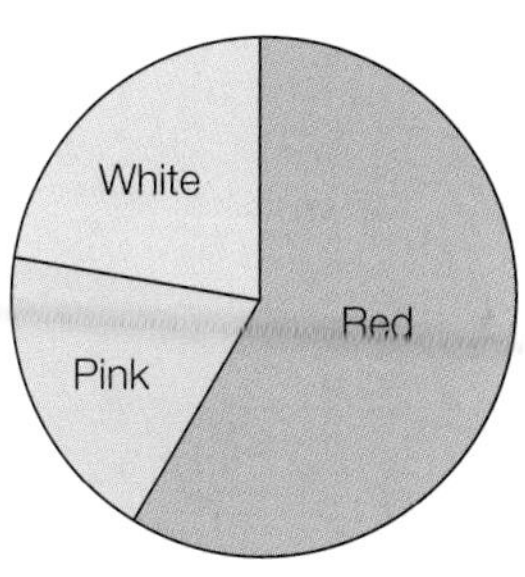

FIGURE B4 Pie Charts Show Proportions of Categories
This pie chart shows the proportions of the three poinsettia colors presented in Table B3.

the points is positive. As the length of a fish increases, its weight tends to increase in an approximately linear manner.

Tables and graphs are critical to interpreting and communicating data, and thus should be as self-contained and understandable as possible. Their content should be easily understood simply by looking at them. Axes, captions, and units should be clearly labeled, statistical terms should be defined, and appropriate groupings should be used when tabulating or graphing quantitative data.

Step 4: Summarize the Data

A **statistic** is a numerical quantity calculated from data, whereas **descriptive statistics** are quantities that describe general patterns in data. Descriptive statistics allow us to make straightforward comparisons among different data sets and concisely communicate basic features of our data.

DESCRIBING CATEGORICAL DATA For categorical variables, we typically use proportions to describe our data. That is, we construct tables containing the proportions of observations in each category. For example, the third column in Table B3 provides the proportion of poinsettia plants in each color category, and the pie chart in Figure B4 provides a visual representation of those proportions.

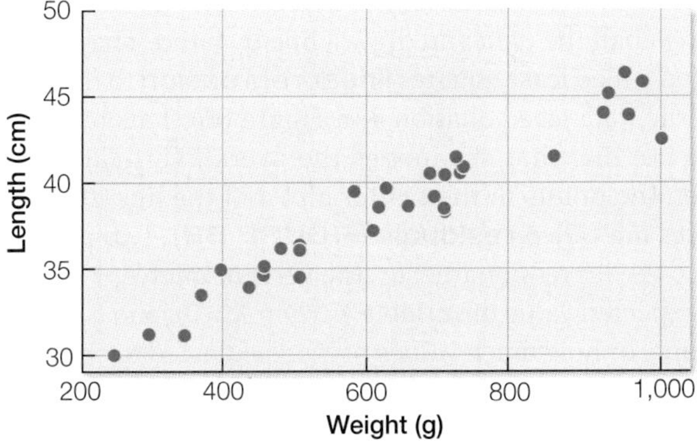

FIGURE B5 Scatter Plots Contrast Two Variables
Scatter plot of *Abramis brama* weights and lengths (measured from nose to end of tail). These two variables have a positive relationship since the slope of a line that could be drawn through the points is positive.

DESCRIBING QUANTITATIVE DATA For quantitative data, we often start by calculating **measures of center**, quantities that roughly tell us where the center of our data lies. There are three commonly used measures of center:

- The **mean**, or average value, of our sample is simply the sum of all the values in the sample divided by the number of observations in our sample (**FIGURE B6**).
- The **median** is the value at which there are equal numbers of smaller and larger observations.
- The **mode** is the most frequent value in the sample.

It is often just as important to quantify the variation in the data as it is to calculate its center. There are several statistics that tell us how much the values differ from one another. We call these **measures of dispersion**. The easiest one to understand and calculate is the **range**, which is simply the largest value in the sample minus the smallest value. The most commonly used measure of dispersion is the **standard deviation**, which calculates the extent to which the data are spread out from the mean. A

RESEARCH TOOLS

FIGURE B6 Descriptive Statistics for Quantitative Data

Below are the equations used to calculate the descriptive statistics we discuss in this appendix. You can calculate these statistics yourself, or use free internet resources to help you make your calculations.

Notation:

$x_1, x_2, x_3, \ldots x_n$ are the n observations of variable X in the sample.

$\sum_{i=1}^{n} x_i = x_1 + x_2 + x_3, \ldots + x_n$ is the sum of all of the observations. (The Greek letter sigma, Σ, is used to denote "sum of.")

In regression, the independent variable is X, and the dependent variable is Y. b_0 is the vertical intercept of a regression line. b_1 is the slope of a regression line.

Equations

1. Mean: $\bar{x} = \dfrac{\sum_{i=1}^{n} x_i}{n}$

2. Standard deviation: $s = \sqrt{\dfrac{\Sigma(x_i - \bar{x})^2}{n-1}}$

3. Correlation coefficient: $r = \dfrac{\Sigma(x_i - \bar{x})(y_i - \bar{y})}{\sqrt{\Sigma(x_i - \bar{x})^2 (y_i - \bar{y})^2}}$

4. Least-squares regression line: $Y = b_0 + b_1 X$
   where $b_1 = \dfrac{\Sigma(x_i - \bar{x})(y_i - \bar{y})}{\Sigma(x_i - \bar{x})^2}$ and $b_0 = \bar{y} - b_1\bar{x}$

5. Standard error of the mean: $SE_{\bar{x}} = \dfrac{s}{\sqrt{n}}$

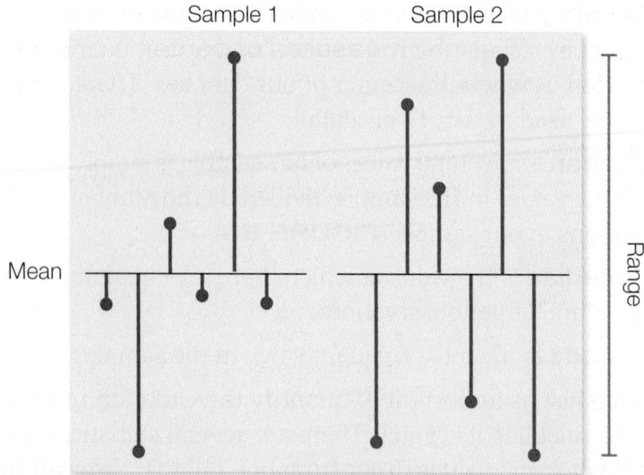

FIGURE B7 Measures of Dispersion Two samples with the same mean (black horizontal lines) and range (blue vertical line). Red lines show the deviations of each observation from the mean. Samples with large deviations have large standard deviations. Sample 1 has a smaller standard deviation than sample 2.

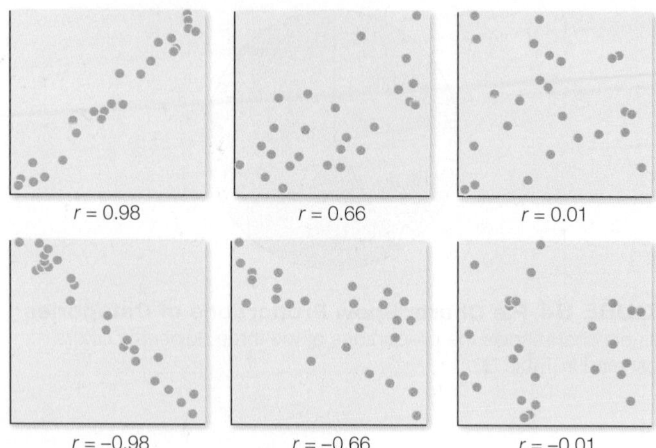

FIGURE B8 Correlation Coefficients The correlation coefficient (*r*) indicates both the strength and the direction of the relationship.

deviation is the difference between an observation and the mean of the sample. The standard deviation is a measure of how far the average observation in the sample is from the sample mean. Two samples can have the same range, but very different standard deviations if observations in one are clustered closer to the mean than in the other. In **FIGURE B7**, for example, sample 1 has a smaller standard deviation ($s = 2.6$) than does sample 2 ($s = 3.6$), even though the two samples have the same means and ranges.

To demonstrate these descriptive statistics, let's return to the Lake Laengelmavesi study (see the data in Table B1). The mean weight of the 34 fish (see equation 1 in Figure B6) is:

$$\bar{x} = \frac{\text{sum of the weight of all fish in sample}}{\text{number of fish in sample}} = \frac{21{,}284 \text{ g}}{34} = 626 \text{ g}$$

Since there is an even number of observations in the sample, then the median weight is the value halfway between the two middle values:

$$\frac{610 \text{ g} + 620 \text{ g}}{2} = 615 \text{ g}$$

The mode of the sample is 500 g, which appears four times. The standard deviation (see equation 2 in Figure B6) is:

$$s = \sqrt{\frac{\sum (x_i - \bar{x})^2}{n-1}} = 206.6 \text{ g}$$

and the range is 1,000 g – 242 g = 758 g.

DESCRIBING THE RELATIONSHIP BETWEEN TWO QUANTITATIVE VARIABLES Biologists often are interested in understanding the relationship between two different quantitative variables: How does the height of an organism relate to its weight? How does air pollution relate to the prevalence of asthma? How does lichen abundance relate to levels of air pollution? Recall that scatter plots visually represent such relationships.

We can quantify the strength of the relationship between two quantitative variables using a single value called the Pearson product–moment **correlation coefficient** (see equation 3 in Figure B6). This statistic ranges between –1 and 1 and tells us how closely the points in a scatter plot conform to a straight line. A negative correlation coefficient indicates that one variable decreases as the other increases; a positive correlation coefficient indicates that the two variables increase together; and a correlation coefficient of zero indicates that there is no linear relationship between the two variables (**FIGURE B8**).

One must always keep in mind that *correlation does not mean causation*. Two variables can be closely related without one causing the other. For example, the number of cavities in a child's mouth correlates positively with the size of his or her feet. Clearly cavities do not enhance foot growth; nor does foot growth cause tooth decay. Instead the correlation exists because both quantities tend to increase with age.

Intuitively, the straight line that tracks the cluster of points on a scatter plot tells us something about the typical relationship between the two variables. Statisticians do not, however, simply eyeball the data and draw a line by hand. They often use a method called least-squares **linear regression** to fit a straight line to the data (see equation 4 in Figure B6). This method calculates the line that minimizes the overall vertical distances between the points in the scatter plot and the line itself. These distances are called **residuals** (**FIGURE B9**). Two parameters describe the regression line: b_0 (the vertical intercept of the line, or the expected value of variable Y when $X = 0$), and b_1 (the slope of the line, or how much values of Y are expected to change with changes in values of X).

Step 5: Inferential Statistics

Data analysis often culminates with statistical inference—an attempt to draw general conclusions about the system under investigation—the larger population—from observations of

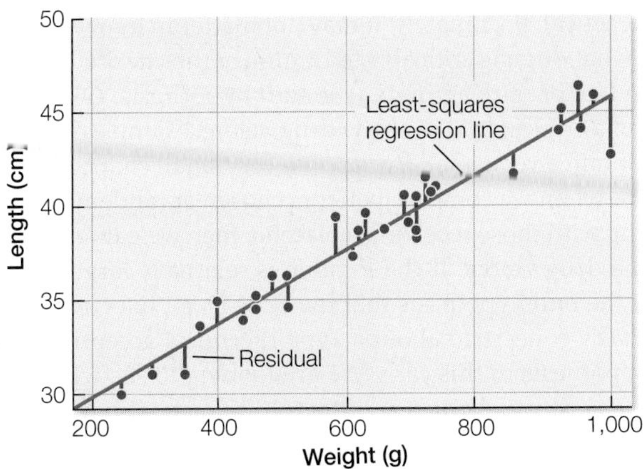

FIGURE B9 Linear Regression Estimates the Typical Relationship between Two Variables Least-squares linear regression line for *Abramis brama* weights and lengths (measured from nose to end of tail). The regression line (blue line) is given by the equation $Y = 26.1 + 0.02X$. It is the line that minimizes the sum of the squares of the residuals (red lines).

a representative sample (see Figure B1). When we test a new medulloblastoma brain cancer drug on ten patients, we do not simply want to know the fate of those ten individuals; rather, we hope to predict the drug's efficacy on the much larger population of all medulloblastoma patients.

STATISTICAL HYPOTHESES Once we have collected our data, we want to evaluate whether or not they fit the predictions of our hypothesis. For example, we want to know whether or not cancer incidence is greater in our sample of smokers than in our sample of nonsmokers, whether or not clutch size increases with spider body weight in our sample of spiders, or whether or not growth of a sample of fertilized plants is greater than that of a sample of unfertilized plants.

Before making statistical inferences from data, we must translate our "whether or not" question into a pair of opposing statistical hypotheses. Often our "whether" hypothesis is difficult to translate because it does not make specific predictions such as *how much* greater the cancer incidence should be in smokers compared with nonsmokers, or *how much* clutch size will increase with body weight, or *how big* the growth advantage of fertilized plants will be. For this reason, scientists often evaluate the "whether" hypothesis indirectly by calculating the fit of the data to an opposing "or not" hypothesis that does make a specific prediction. This "or not" hypothesis is often called the **null hypothesis** (denoted H_0) because it proposes that there is *no difference* between sample means, or *no correlation* between variables in the sample, or *no difference* between our sample and an expected frequency distribution. The statistical **alternative hypothesis** (denoted H_A) is often our biological hypothesis: that there *is* a difference in means, or that there *is* a correlation between variables, or that the sample distribution *does* differ from the expected one. We base our conclusion about our biological hypothesis on the fit of our data to the null hypothesis because it is so difficult to calculate the fit to the many alternative hypotheses.

Suppose, for example, we would like to know whether or not a new vaccine is more effective than an existing vaccine at immunizing children against influenza. We have measured flu incidence in a group of children who received the new vaccine and want to compare it with flu incidence in a group of children who received the old vaccine. Our statistical hypotheses would be as follows:

H_0: *Flu incidence was the same in both groups of children.*

H_A: *Flu incidence was different between the two groups of children.*

In the next few sections we will discuss how we decide when to reject the null hypothesis in favor of the alternative hypothesis.

JUMPING TO THE WRONG CONCLUSIONS There are two ways that we can draw incorrect conclusions from a statistical test (**FIGURE B10**). We can reject the null hypothesis when it is actually true (**Type I error**), or we can accept the null hypothesis when it is actually false (**Type II error**). These kinds of errors are analogous to false positives and false negatives in medical testing, respectively. If we mistakenly reject the null hypothesis when it is actually true, then we falsely endorse the incorrect alternative hypothesis. If we mistakenly accept the null hypothesis when it is actually false, then we falsely reject the correct alternative hypothesis.

Fortunately, statisticians have developed statistical tests to minimize the chance of committing these kinds of errors and to increase our confidence in the reliability of our conclusions. The methods are based on calculating the probabilities of obtaining different possible data values. Although you may have heard or even used the word "probability" on multiple occasions, it is important that you understand its mathematical meaning. A **probability** is a numerical quantity that expresses the chance of making a particular observation. It ranges between 0 and 1; 0 means that the observation never occurs, and 1 that the observation always occurs. This makes sense only if there is an element of chance—that is, if alternative observations are possible. For example, in a coin flip, heads and tails are possible observations. Similarly, when we select an individual from a human population, we may select a female or a male, or when we select an *Abramis brama* from Lake Laengelmavesi, its length can take on any number of values.

Probability is at the core of statistics. To draw conclusions about the real world (the population) from our sample, we first

Our conclusion	The real world: Null hypothesis true (*not more females*)	The real world: Null hypothesis false (*more females*)
Null hypothesis true (*not more females*)	√	Type II error (*false negative*)
Null hypothesis false (*more females*)	Type I error (*false positive*)	√

FIGURE B10 Two Types of Errors Possible outcomes of a statistical test. Statistical inference can result in correct and incorrect conclusions about the population of interest.

calculate the *probability of obtaining the data in our sample if the null hypothesis is true*. We refer to this probability as the ***P*-value** of our test. Specifically, we ask "What is the probability of observing what we see or measure in our random sample from the total population if the null hypothesis is true?" If our sample is highly improbable under the null hypothesis, then we gain confidence in the alternative hypothesis.

To take an example, suppose we would like to know whether there are equal numbers of females and males in a population of 10,000 students. To determine the makeup of the population, we choose 20 individuals at random and record their sex. Our null hypothesis is that the sex ratio in the population is 50:50, and our alternative hypothesis is that the sex ratio is not 50:50. Suppose that our sample contains 5 women and 15 men. The probability by chance alone of a sex ratio that skewed or more, if the sex ratio is really 50:50, turns out to be 0.02 (that is, 2%). (We will explain below how to arrive at this *P*-value.)

What sense can we make of this result and its *P*-value? We are justified in being skeptical that the population from which we drew the sample really has a 50:50 sex ratio. Yes, it's possible to randomly choose 5 women and 15 men if the null hypothesis is true, but we would expect this outcome in only 2% of all samples of 20 students that we could take. Said another way, if we are wrong and the sex ratio is actually 50:50, we have made a Type I error, but the chance of doing that is only 2%. We will likely go ahead to reject the null hypothesis and instead conclude that the sex ratio is not 50:50 in the population of 10,000 students.

Now, suppose instead that our sample of 20 students contains 8 women and 12 men. Here $P = 0.44$ (that is, 44%). What can we conclude from this result? We are much less certain about rejecting the null hypothesis that the sex ratio is 50:50. The *P*-value means that 44% of samples would be this skewed, or more, if the sex ratio were 50:50.

With a result like this, we are tempted to conclude that the sex ratio indeed is 50:50, and it is a common practice to accept the null hypothesis when the *P*-value is moderate. But this action might very well represent a Type II error (accepting a false hypothesis). Note that a probability of 44% of obtaining our result if the sex ratio is 50:50 is not very strong support for that specific null hypothesis. Other hypotheses, such as a hypothesis that the student population actually contains 40% women and 60% men (the observed proportions) fit the data even better than the null hypothesis of a 50:50 sex ratio.

For this reason, it is better to conclude simply that we cannot reject the null hypothesis when we obtain a moderate *P*-value. How moderate does the *P*-value have to be for us to conclude this? It has become conventional to use $P = 0.05$ (5%) as a cutoff, and to label a result with $P < 0.05$ as "significant" and one with $P > 0.05$ as "not significant." You should keep in mind, however, that $P = 0.05$ is not some magic number that determines whether one or another hypothesis is absolutely "proved" or "falsified."

What can we do to go further, with the example of $P = 0.44$ for our sample sex ratio? If possible, we could obtain additional samples (for example, of 20 individuals) from the overall population, to gather more information on the underlying sex ratio. But this sort of replication of sampling may often be impossible or impractical. Fortunately, it may not matter in many cases, because what often is critical is to minimize the chance of making a Type I error of rejecting a true null hypothesis. Once again, think of a clinical trial of a new drug, using a sample of patients. If we obtain a very small *P*-value—a very small chance of a Type I error—when we compare patients chosen at random to receive the drug with those receiving a placebo, then we gain confidence that the drug works. If the *P*-value is relatively large, then we accept the null hypothesis that the drug is not effective, and we aren't very concerned about a Type II error of accepting a false null hypothesis in this case. We are looking to see whether the drug has a strong beneficial effect. If it does not have a strong effect, we are not so interested in finding out whether its effect is exactly zero or just close to zero.

This agrees with our intuition. When choosing 20 people randomly from an evenly divided population, we would be surprised if almost all of them were female, but we would not be surprised at all if we ended up with a few more females than males (or a few more males than females). Exactly how many females do we need in our sample before we can confidently infer that they make up the majority of the population? And how confident are we when we reach that conclusion? Statistics allows us to answer these questions precisely.

STATISTICAL SIGNIFICANCE: AVOIDING FALSE POSITIVES

Whenever we test hypotheses, we calculate the probability just discussed, and refer to this value as the *P*-value of our test. Specifically, the *P*-value is the probability of getting data as extreme as our data (just by chance) if the null hypothesis is in fact true. In other words, it is the probability that chance alone would produce data that differ from the null hypothesis as much as our data differ from the null hypothesis. How we measure the fit of our data to the null hypothesis depends on the kind of data in our sample (categorical or quantitative) and the nature of the null hypothesis (assertions about proportions, single variables, multiple variables, differences between variables, correlations between variables, etc.).

For many statistical tests, *P*-values can be calculated mathematically. One option is to quantify the extent to which the data depart from the null hypothesis and then use look-up tables (available in most statistics textbooks or on the internet) to find the probability that chance alone would produce a difference of that magnitude. Most scientists, however, find *P*-values primarily by using statistical software rather than hand calculations combined with look-up tables. Regardless of the technology, the most important steps of the statistical analysis are still left to the researcher: constructing appropriate null and alternative hypotheses, choosing the correct statistical test, and drawing correct conclusions.

After we calculate a *P*-value from our data, we have to decide whether it is small enough to conclude that our data are inconsistent with the null hypothesis. This is decided by comparing the *P*-value to a threshold called the **significance level**, which is often chosen even before making any calculations. We reject the null hypothesis only when the *P*-value is less than or equal to the significance level, denoted α. This ensures that, if the null hypothesis is true, we have at most a probability α of rejecting it when it is true. Therefore the lower the value of α, the less likely you are to make a Type I error (see the lower left cell of Figure B10). The most commonly used significance level is $\alpha = 0.05$, which limits the probability of a Type I error to 5%.

If our statistical test yields a *P*-value that is less than our significance level α, then we conclude that the effect described by our alternative hypothesis is statistically significant at the level α and we reject the null hypothesis. If our *P*-value is greater than α, then we conclude that we are unable to reject the null hypothesis.

POWER: AVOIDING FALSE NEGATIVES The **power** of a statistical test is the probability that we will correctly reject the null hypothesis when it is false (see the lower right cell of Figure B10). Therefore the higher the power of the test, the less likely we are to make a Type II error (see the upper right cell of Figure B10). The power of a test can be calculated, and such calculations can be used to improve your methodology. Generally, there are several steps that can be taken to increase power and thereby avoid false negatives:

- **Increase the value of α.** The lower the value of α (our significance level), the harder it is to reject the null hypothesis, even if it is actually false.
- **Increase the sample size.** The more data one has, the more likely one is to find evidence against the null hypothesis, if it is actually false.
- **Decrease variability in the sample.** The more variation there is in the sample, the harder it is to discern a clear effect (the alternative hypothesis) when it actually exists.

It is always a good idea to design your experiment to reduce any variability that may obscure the pattern you seek to detect. For example, it is possible that the chance of a child contracting influenza varies depending on whether he or she lives in a crowded (e.g., urban) environment or one that is less so (e.g., rural). To reduce variability, a scientist might choose to test a new influenza vaccine only on children from one environment or the other. After you have minimized such extraneous variation, you can use power calculations to choose the right combination of α and sample size to reduce the risks of Type I and Type II errors to desirable levels.

There is a trade-off between Type I and Type II errors: as α decreases, the risk of a Type I error decreases but the risk of a Type II error increases. As discussed above, scientists tend to be more concerned about Type I than Type II errors. However, there are many real-world scenarios in which it would be worse to make a Type II error than a Type I error. For example, suppose a new cold medication is being tested for dangerous (life-threatening) side effects. The null hypothesis is that there are no such side effects. A Type II error might lead regulatory agencies to approve a harmful medication that could cost human lives. In contrast, a Type I error would simply mean one less cold medication among the many that already line pharmacy shelves. In such cases, policymakers take steps to avoid a Type II error, even if, in doing so, they increase the risk of a Type I error.

STATISTICAL INFERENCE WITH QUANTITATIVE DATA Statistics that describe patterns in our samples are used to estimate properties of the larger population. Earlier we calculated the mean weight of a sample of *Abramis brama* in Lake Laengelmavesi, which provided us with an estimate of the mean weight of all the *Abramis brama* in the lake. But how close is our estimate to the true value in the larger population? Our estimate from the sample is unlikely to exactly equal the true population value. For example, our sample of *Abramis brama* may, by chance, have included an excess of large individuals. In this case, our sample would overestimate the true mean weight of fish in the population.

The **standard error** of a sample statistic (such as the mean) is a measure of how close it is likely to be to the true population value. The **standard error of the mean**, for example, provides an estimate of how far we might expect a sample mean to deviate from the true population mean. It is a function of how much individual observations vary within samples (the standard deviation) and the size of the sample (n). Standard errors increase with the degree of variation within samples, and they decrease with the number of observations in a sample (because large samples provide better estimates about the underlying population than do small samples). For our sample of 34 *Abramis brama*, we would calculate the standard error of the mean using equation 5 in Figure B6:

$$SE_{\bar{x}} = \frac{s}{\sqrt{n}} = \frac{206.6\text{ g}}{\sqrt{34}} = 35.4\text{ g}$$

The standard error of a statistic is related to the **confidence interval**—a range around the sample statistic that has a specified probability of including the true population value. The formula we use to calculate confidence intervals depends on the characteristics of the data, and the particular statistic. For many types of continuous data, the bounds of the **95% confidence interval** of the mean can be calculated by taking the mean and adding and subtracting 1.96 times the standard error of the mean. Consider our sample of 34 *Abramis brama*, for example. We would say that the mean of 626 grams has a 95% confidence interval from 556.6 grams to 695.4 grams (626 ± 69.4 grams). If all our assumptions about our data have been met, we would expect the true population mean to fall in this confidence interval 95% of the time.

Researchers typically use graphs or tables to report sample statistics (such as the mean) as well as some measure of confidence in them (such as their standard error or a confidence interval). This book is full of examples. If you understand the concepts of sample statistics, standard errors, and confidence intervals, you can see for yourself the major patterns in the data, without waiting for the authors to tell you what they are. For example, if samples from two groups have 95% confidence intervals of the mean that do not overlap, then you can conclude that it is unlikely that the groups have the same true mean.

Activity B1 Standard Deviations, Standard Errors, and Confidence Intervals Simulation

Biologists conduct statistical tests to obtain more precise estimates of the probability of observing a given difference between samples if the null hypothesis that there is no difference in the populations is true. The appropriate test depends on the nature of the data and the experimental design. For example, we might want to calculate the probability that the mean weights of two different

RESEARCH TOOLS

FIGURE B11 The *t*-test

What is the *t*-test? It is a standard method for assessing whether the means of two groups are statistically different from each other.

Step 1: State the null and alternative *hypotheses*:

H_0: The two populations have the same mean.

H_A: The two populations have different means.

Step 2: Choose a significance level, α, to limit the risk of a Type I error.

Step 3: Calculate the *test statistic*:

$$t_s = \frac{\bar{y}_1 - \bar{y}_2}{\sqrt{\frac{s_1^2}{n_1} + \frac{s_2^2}{n_2}}}$$

Notation: $\bar{y}_1$ and $\bar{y}_2$ are the sample means; s_1 and s_2 are the sample standard deviations; and n_1 and n_2 are the sample sizes.

Step 4: Use the test statistic to assess whether the data are consistent with the null hypothesis:

Calculate the *P-value* (*P*) using statistical software or by hand using statistical tables.

Step 5: Draw conclusions from the test:

If $P \leq \alpha$, then reject H_0, and conclude that the population distribution is significantly different.

If $P > \alpha$, then we do not have sufficient evidence to conclude that the means differ.

fish species in Lake Laengelmavesi, *Abramis brama* and *Leusiscus idus*, are the same. A simple method for comparing the means of two groups is the *t*-test, described in **FIGURE B11**. We looked earlier at data for *Abramis brama*; the researchers who collected these data also collected weights for six individuals of *Leusiscus idus*: 270, 270, 306, 540, 800, and 1,000 grams. We begin by stating our hypotheses and choosing a significance level:

H_0: *Abramis brama* and *Leusiscus idus* have the same mean weight.

H_A: *Abramis brama* and *Leusiscus idus* have different mean weights.

$\alpha = 0.05$

The test statistic is calculated using the means, standard deviations, and sizes of the two samples:

$$t_s = \frac{626 - 531}{\sqrt{\frac{207^2}{34} + \frac{310^2}{6}}} = 0.724$$

We can use statistical software or one of the free statistical sites on the internet to find that the *P*-value for this result is $P = 0.497$. Since *P* is considerably greater than α, we fail to reject the null hypothesis and conclude that our study does not provide evidence that the two species have different mean weights.

RESEARCH TOOLS

FIGURE B12 The Chi-Square Goodness-of-Fit Test

What is the chi-square goodness-of-fit test? It is a standard method for assessing whether a sample came from a population with a specific distribution.

Step 1: State the null and alternative *hypotheses*:

H_0: The population has the specified distribution.

H_A: The population does not have the specified distribution.

Step 2: Choose a significance level, α, to limit the risk of a Type I error.

Step 3: Determine the *observed frequency* and *expected frequency* for each category:

The observed frequency of a category is simply the number of observations in the sample of that type.

The expected frequency of a category is the probability of the category specified in H_0 multiplied by the overall sample size.

Step 4: Calculate the *test statistic*:

$$\chi_s^2 = \sum_{i=1}^{C} \frac{(O_i - E_i)^2}{E_i}$$

Notation: *C* is the total number of categories, O_i is the observed frequency of category *i*, and E_i is the expected frequency of category *i*.

Step 5: Use the test statistic to assess whether the data are consistent with the null hypothesis:

Calculate the *P-value* (*P*) using statistical software or by hand using statistical tables.

Step 6: Draw conclusions from the test:

If $P \leq \alpha$, then reject H_0, and conclude that the population distribution is significantly different than the distribution specified by H_0.

If $P > \alpha$, then we do not have sufficient evidence to conclude that population has a different distribution.

You may want to consult an introductory statistics textbook to learn more about confidence intervals, *t*-tests, and other basic statistical tests for quantitative data.

STATISTICAL INFERENCE WITH CATEGORICAL DATA With categorical data, we often wish to ask whether the frequencies of observations in different categories are consistent with a hypothesized frequency distribution. We can use a **chi-square goodness-of-fit** test to answer this question.

FIGURE B12 outlines the steps of a chi-square goodness-of-fit-test. As an example, consider the data described in Table B3. Many plant species have simple Mendelian genetic systems in which parent plants produce progeny with three different colors of flowers in a ratio of 2:1:1. However, a botanist believes that these particular poinsettia plants have a different genetic system that does not produce a 2:1:1 ratio of red, pink, and white plants. A chi-square goodness-of-fit can be used to assess whether or not the data are consistent with this ratio, and thus whether or not this simple genetic explanation is valid. We start by stating our hypotheses and significance level:

H_0: The progeny of this type of cross have the following probabilities of each flower color:

Pr{Red} = .50, Pr{Pink} = .25, Pr{White} = .25

H_A: At least one of the probabilities of H_0 is incorrect.

$\alpha = 0.05$

We next use the probabilities in H_0 and the sample size to calculate the expected frequencies:

	Red	Pink	White
Observed	108	34	40
Expected	(.50)(182) = 91	(.25)(182) = 45.5	(.25)(182) = 45.5

Based on these quantities, we calculate the chi-square test statistic:

$$\chi_s^2 = \sum_{i=1}^{C} \frac{(O_i - E_i)^2}{E_i} = \frac{(108-91)^2}{91} + \frac{(34-45.5)^2}{45.5} + \frac{(40-45.5)^2}{45.5} = 6.747$$

We find the P-value for this result to be $P = 0.034$ using statistical software. Since P is less than α, we reject the null hypothesis and conclude that the botanist is likely correct: the plant color patterns are not explained by the simple Mendelian genetic model under consideration.

This appendix is meant to provide only a brief introduction to the concepts of statistical analysis, with a few example tests. **FIGURE B13** summarizes some of the commonly used statistical tests that you may encounter in biological studies.

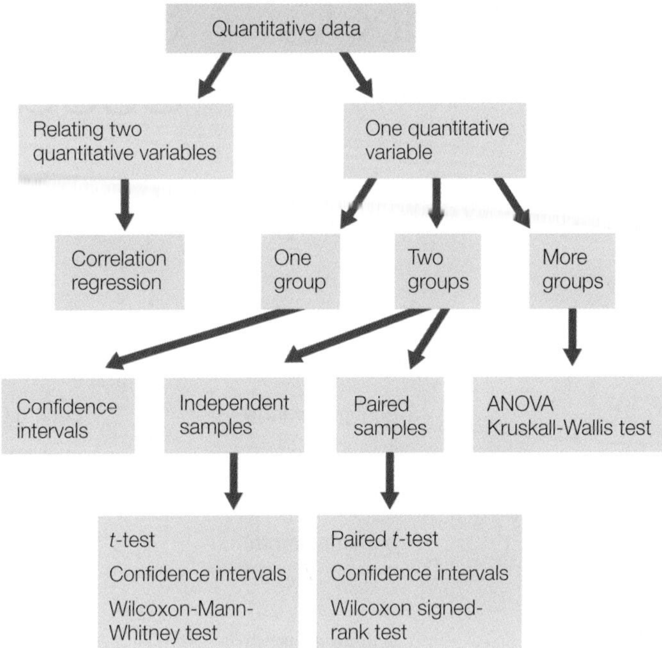

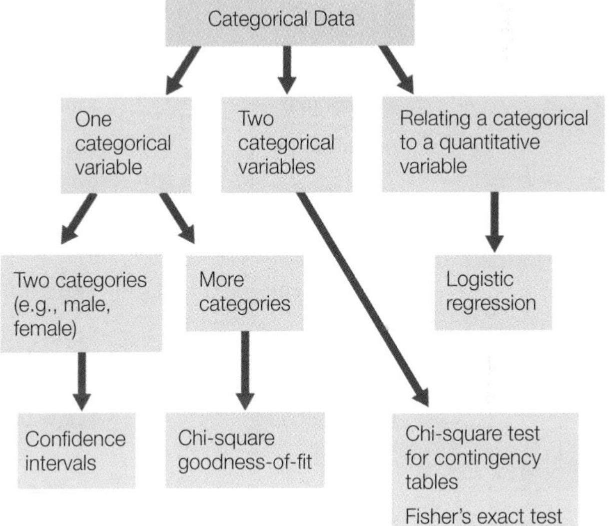

FIGURE B13 Some Common Methods of Statistical Inference This flowchart shows some of the commonly used methods of statistical inference for different combinations of data. Detailed descriptions of these methods can be found in most introductory biostatistics textbooks.

Appendix C | Some Measurements Used in Biology

Measures of	Unit	Equivalents	Metric → English conversion
LENGTH			
	meter (m)	base unit	1 m = 39.37 inches = 3.28 feet = 1.196 yards
	kilometer (km)	1 km = 1,000 (10^3) m	1 km = 0.62 miles
	centimeter (cm)	1 cm = 0.01 (10^{-2}) m	1 cm = 0.39 inches
	millimeter (mm)	1 mm = 0.1 cm = 10^{-3} m	1 mm = 0.039 inches
	micrometer (µm)	1 µm = 0.001 mm = 10^{-6} m	
	nanometer (nm)	1 nm = 0.001 µm = 10^{-9} m	
AREA			
	square meter (m^2)	base unit	1 m^2 = 1.196 square yards
	hectare (ha)	1 ha = 10,000 m^2	1 ha = 2.47 acres
VOLUME			
	liter (L)	base unit	1 L = 1.06 quarts
	milliliter (ml)	1 ml = 0.001 L = 10^{-3} L	1 ml = 0.034 fluid ounces
	microliter (µl)	1 µl = 0.001 ml = 10^{-6} L	
MASS			
	gram (g)	base unit	1 g = 0.035 ounces
	kilogram (kg)	1 kg = 1,000 g	1 kg = 2.20 pounds
	metric ton (mt)	1 mt = 1,000 kg	1 mt = 2,200 pounds = 1.10 ton
	milligram (mg)	1 mg = 0.001 g = 10^{-3} g	
	microgram (µg)	1 µg = 0.001 mg = 10^{-6} g	
TEMPERATURE			
	degree Celsius (°C)	base unit	°C = (°F – 32)/1.8
			0°C = 32°F (water freezes)
			100°C = 212°F (water boils)
			20°C = 68°F ("room temperature")
			37°C = 98.6°F (human internal body temperature)
	Kelvin (K)*	K = °C – 273	0 K = –460°F
ENERGY			
	joule (J)		1 J ≈ 0.24 calorie = 0.00024 kilocalorie†

*0 K (–273°C) is "absolute zero," a temperature at which molecular oscillations approach 0—that is, the point at which motion all but stops.
†A *calorie* is the amount of heat necessary to raise the temperature of 1 gram of water 1°C. The *kilocalorie*, or nutritionist's calorie, is what we commonly think of as a calorie in terms of food.

Appendix D | Working with DNA

Since the demonstration that DNA is the genetic material and the discovery of its structure (see Chapter 9), many techniques have been developed to work with DNA. We describe some of these techniques elsewhere in the book, such as PCR (see Chapter 9) and DNA sequencing (see Chapter 12). In this appendix we briefly discuss some of the other major techniques used in labs that work with DNA, highlighting some of the applications of this technology.

Restriction enzymes cleave DNA at specific sequences

There are many methods for efficiently isolating DNA from cells. However, DNA molecules—especially genomic DNA molecules—are generally very long and must be broken into smaller pieces before they can be studied. One way to do this is to use mechanical methods to shear the DNA at random locations. Alternatively, and more usefully, DNA can be cut at particular sites along the molecule using **restriction enzymes**. These enzymes are produced by numerous species of bacteria as a defense against viruses.

There are many different restriction enzymes, each of which cleaves DNA at a specific sequence of nucleotides called a **restriction site** or recognition sequence. Hundreds of these enzymes have been purified from various microorganisms and can be used to cut DNA in the laboratory (by setting up a "restriction digest"). Restriction sites are usually 4 to 6 base pairs (bp) long and are palindromes, meaning they have the same sequence when read from the 5′ end in either direction. A restriction enzyme breaks the ester bond between the 3′ hydroxyl group of one nucleotide and the 5′ phosphate group of the next nucleotide at a particular location in the restriction site. The enzyme cuts both DNA strands, resulting in a double-stranded break.

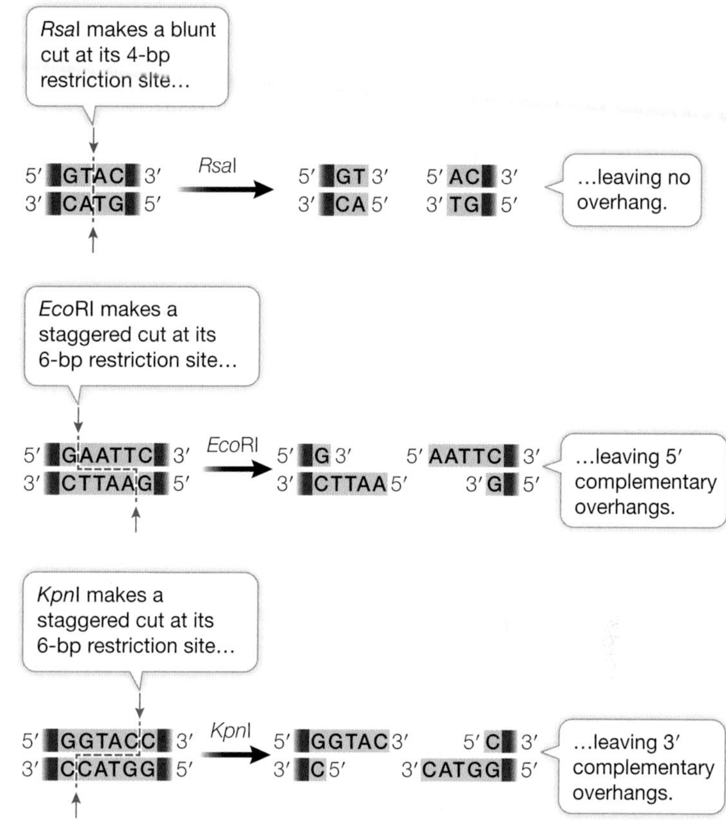

FIGURE D1 Restriction Enzymes Cut at Specific Sites Restriction enzymes make blunt or staggered double-stranded cuts—indicated by the red arrows—at particular DNA sequences that are usually 4- to 6-bp palindromes, meaning they have the same sequence in both directions. Staggered cuts can leave overhanging 3′ or 5′ single-stranded "sticky ends."

FIGURE D1 shows the restriction sites for three restriction enzymes. The names of these enzymes reflect their organism of origin. *Eco*RI, for example, was found in *E. coli*. Some enzymes cut DNA at sites that are directly opposite one another, resulting in blunt-ended fragments. Other enzymes make cuts that are offset from one another, which results in staggered ends that have a 5′ or 3′ overhang. These are referred to as "sticky ends" because they are able to form hydrogen bonds with (that is, "stick to") complementary sequences on other DNA molecules.

On average, an enzyme with a 6-bp restriction site is expected to cut DNA about once in every 4,500 bp in a genome that has equal frequencies of the four nucleotides. Of course, "on average" does not mean that the enzyme cuts all stretches of DNA at regular intervals. For example, the *Eco*RI restriction site does not occur even once in the 40,000 bp of the T7 phage genome—a fact that is crucial to the survival of this virus, since its host is *E. coli*. Fortunately for *E. coli*, the *Eco*RI restriction site does appear in the DNA of other bacteriophages.

Why don't restriction enzymes cut the DNA of the organism that produces them? For each restriction enzyme, there is a methyltransferase that recognizes and methylates the restriction site for that enzyme. An organism that produces a restriction enzyme also produces the corresponding methyltransferase. Methylation of the organism's DNA prevents the restriction enzyme from cutting at its restriction sites, thereby protecting the organism's own genome.

Gel electrophoresis separates DNA fragments

After a sample of DNA has been cut with a restriction enzyme, the fragments can be separated from each other to determine the number of fragments and their sizes (in base pairs). In this way, an individual fragment can be identified, and it can then be purified for further analysis or for use in an experiment.

A convenient way to separate or purify DNA fragments is by **gel electrophoresis**. Samples containing the fragments are placed in wells at one end of a semisolid gel (usually made of agarose or polyacrylamide polymers), and an electric field is applied to the gel (**FIGURE D2**). Because of its phosphate groups, DNA is negatively charged at neutral pH. Therefore, because opposite charges attract, the DNA fragments move through the gel toward the positive end of the electric field. Because the spaces between

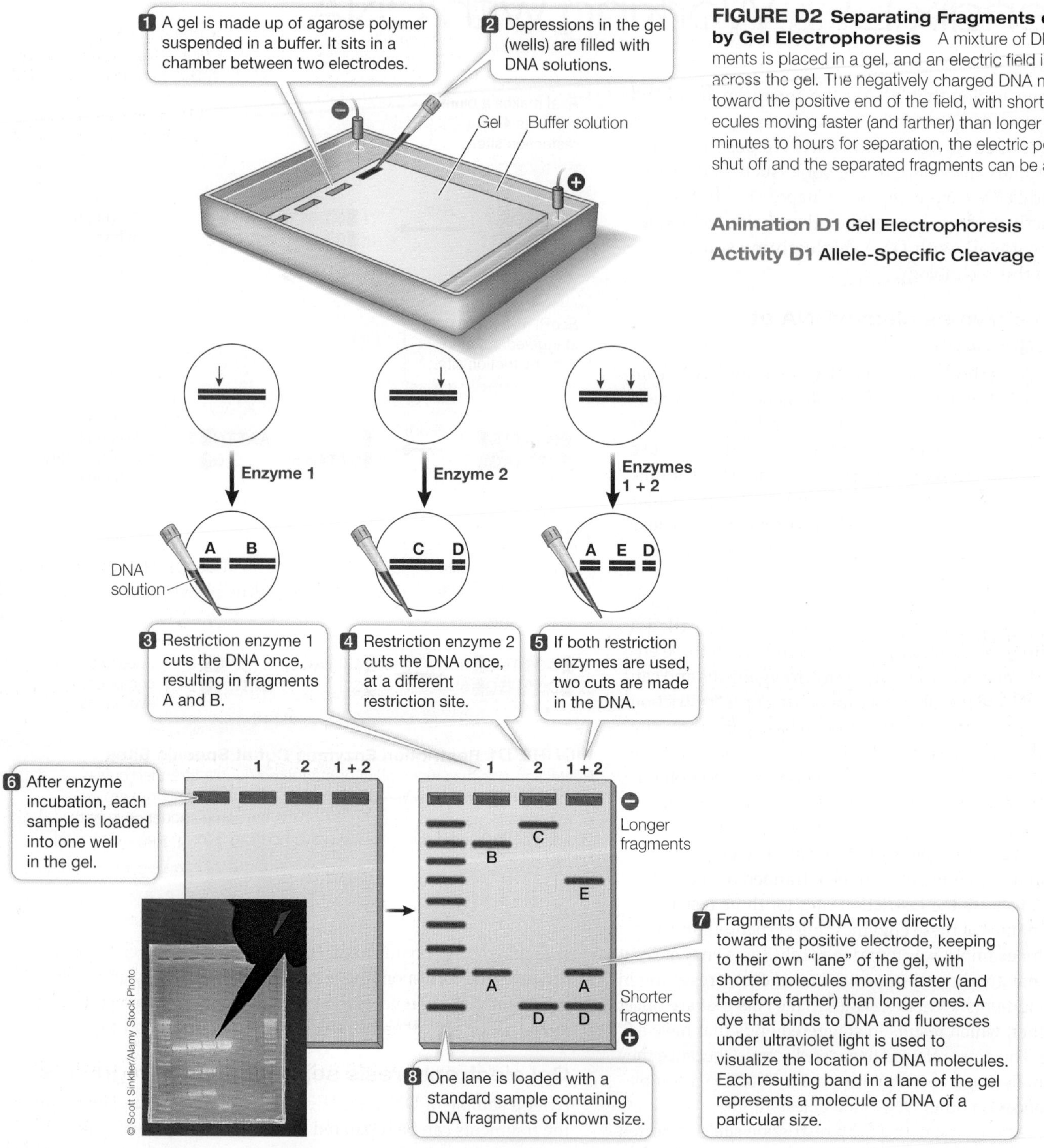

FIGURE D2 Separating Fragments of DNA by Gel Electrophoresis A mixture of DNA fragments is placed in a gel, and an electric field is applied across the gel. The negatively charged DNA moves toward the positive end of the field, with shorter molecules moving faster (and farther) than longer ones. After minutes to hours for separation, the electric power is shut off and the separated fragments can be analyzed.

Animation D1 Gel Electrophoresis

Activity D1 Allele-Specific Cleavage

the polymers of the gel are small, shorter DNA molecules can move through the gel faster than longer ones. Thus DNA fragments of different sizes separate from one another and can be detected with a dye that binds to DNA. This gives us three types of information about a DNA sample:

- *The number of fragments.* The number of fragments produced by digestion of a DNA sample with a given restriction enzyme depends on how many times that enzyme's restriction site occurs in the sample. Thus gel electrophoresis can provide some information about the presence of specific DNA sequences in the DNA sample.
- *The sizes of the fragments.* DNA fragments of known size are often placed in one well of the gel to provide a standard for comparison. This tells us how large the DNA fragments in the

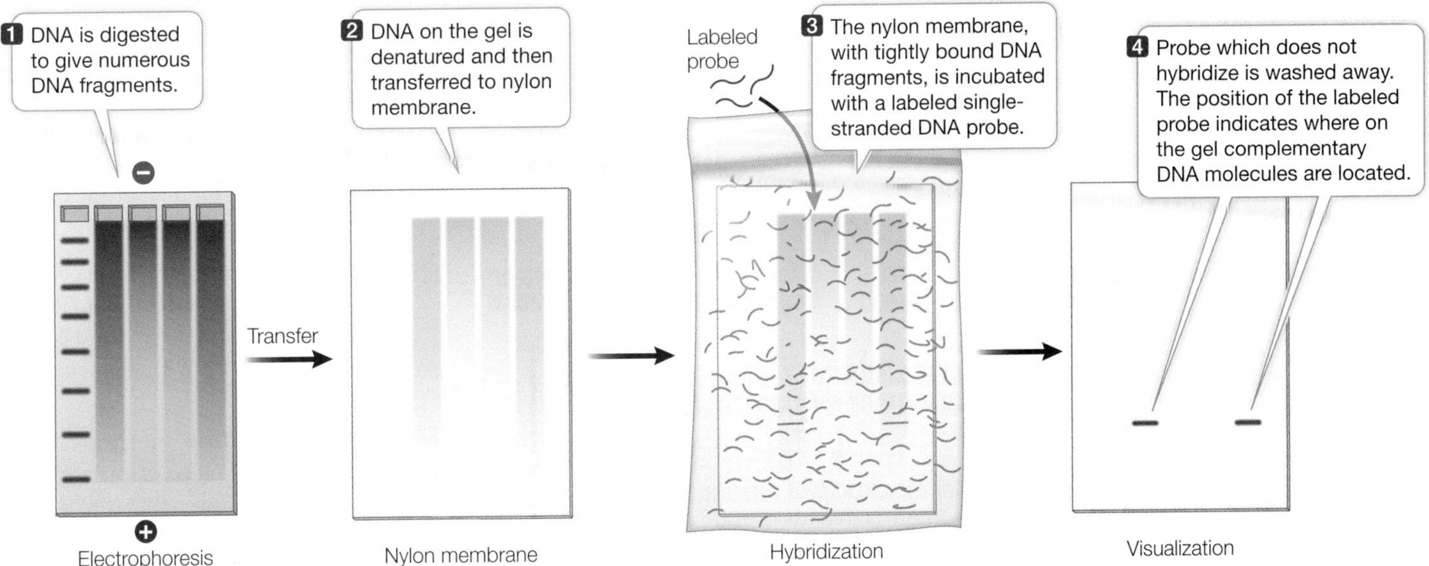

FIGURE D3 Southern Blotting A particular DNA sequence can be located on a gel by hybridizing it to a labeled single-stranded DNA molecule.

other wells are. By comparing the fragment sizes obtained with two or more restriction enzymes, the locations of their restriction sites relative to one another can be worked out (mapped).

- *The relative abundance of a fragment.* In many experiments, the investigator is interested in how much DNA is present. The relative intensity of a band produced by a specific fragment can indicate the amount of that fragment.

After separation on the gel, a slice of gel containing the desired DNA fragment (identified by its size) can be cut out and the DNA fragment purified by one of a variety of methods. This fragment can then be analyzed to determine its sequence (see Figure 12.1) or used to make recombinant DNA.

DNA fragments of known sequence can be located using hybridization

A large DNA molecule that is sheared, or digested with a restriction enzyme, may produce thousands of DNA fragments. When these fragments are subjected to gel electrophoresis, there are so many bands that the DNA looks like a continuous smear. Often a scientist is interested in DNA fragments containing a particular DNA sequence. How can those fragments be distinguished from the many other fragments present on the gel? The answer is to use the **Southern blotting** technique, in which a **probe**—a short piece of single-stranded DNA—is used to locate complementary DNA in the gel. The probe is labeled—chemically tagged—so that it and the target DNA can be visualized at the end of the process. In brief, the DNA fragments in the gel are chemically treated to separate the complementary strands, and then transferred to a sheet of nylon membrane (or nitrocellulose paper), while maintaining the fragments' relative positions. The paper is bathed in a solution containing the probe, and the probe hybridizes to (hydrogen-bonds with) complementary DNA sequences. The locations where the probe hybridized are then visualized using the label (**FIGURE D3**).

Southern blotting is named after the scientist, Edwin Southern, who invented it. Other techniques have been developed for visualizing particular RNA sequences on gels (Northern blotting) using single-stranded DNA probes, or particular protein sequences (Western blotting) using antibodies (see Chapter 36). The names Northern and Western follow from Southern and represent molecular biologists' attempts at humor!

Recombinant DNA can be made from DNA fragments

Using restriction enzymes and DNA ligase, which can seal nicks in DNA strands (see Figure 9.12), scientists are able to cut DNA into fragments and then join the fragments together in new combinations as **recombinant DNA**. As shown in **FIGURE D4**, two fragments with complementary sticky ends first join by hydrogen bonding, and then DNA ligase covalently bonds each strand, making a single intact DNA molecule. DNA from any sources can be used to make recombinant DNA molecules: bacterial DNA can be joined to human DNA, for example.

DNA can transform cells and be cloned

In Chapter 9 we described experiments with S and R bacterial strains that demonstrated the alteration of a bacterium's genotype by DNA taken up from the environment. This discovery, along with that of plasmids—small, autonomously replicating DNA molecules that often carry antibiotic resistance genes—allowed scientists to develop techniques for **cloning**—the production of many copies of a particular sequence of DNA.

We have seen the term "clone" used in the context of whole cells or organisms that are genetically identical to one another (see Key Concept 7.1). A gene can be cloned by inserting it into a plasmid and then transforming a bacterial cell such as *E. coli* with the plasmid. The bacterium is allowed to reproduce and multiply into millions of identical cells, all carrying several copies of the

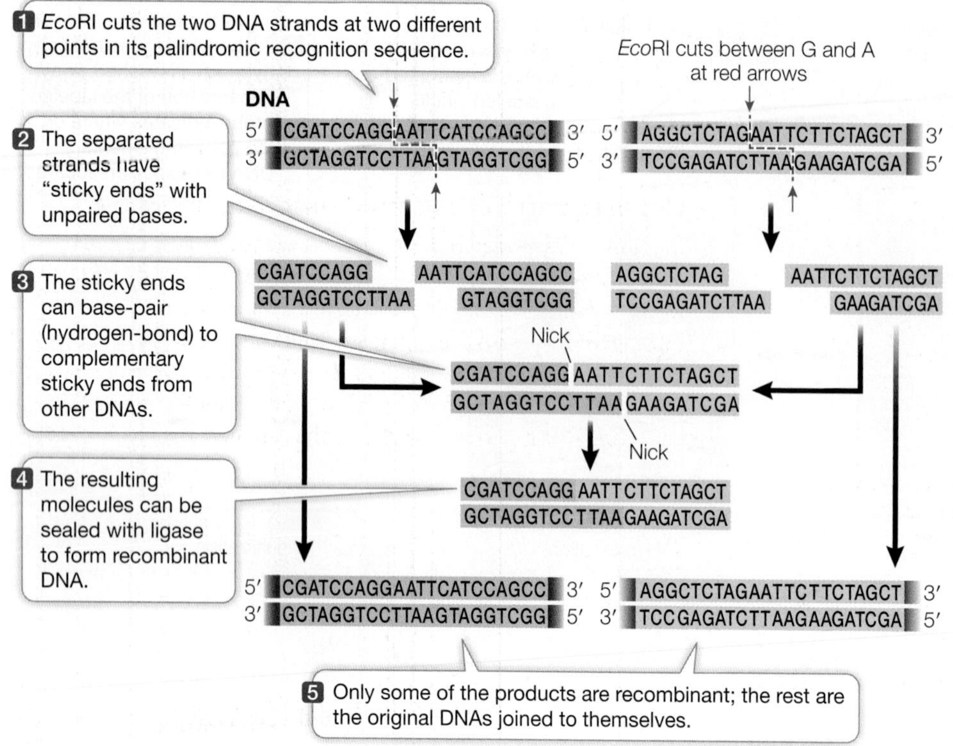

FIGURE D4 Recombinant DNA Cutting two different DNA molecules with a restriction enzyme that makes staggered cuts gives identical sticky ends (complementary overhangs). These ends can hydrogen-bond with one another, and ligase can then be used to join the molecules together.

Media Clip D1 Striking Views of Recombinant DNA Being Made

plasmid, resulting in several million identical copies of the gene. Cloning might be done to produce DNA for sequence analysis, to produce a protein product in quantity (see Figure 10.15A), or as a step toward creating an organism with a new phenotype (see Figure D8).

The process of inserting recombinant DNA into a host cell is called **transformation**, or **transfection** if the host cell is derived from an animal. A host cell or organism that has undergone this process is described as **transgenic**. Methods for getting DNA into host cells vary. In some cases cells will simply take up the DNA from the environment. However, this ability varies enormously across cells, so other methods are often needed. Cells may be chemically treated to make their outer membranes more permeable, or a short electric shock may be used to create temporary pores in the membranes through which the DNA can enter. Viruses can be altered so that they carry or insert recombinant DNA into cells. A common method for transforming plants involves a specific bacterium that inserts DNA into plant cells. Transgenic animals can be produced by injecting recombinant DNA into the nuclei of fertilized eggs. There are even "gene guns," which "shoot" the host cells with tiny particles coated with recombinant DNA.

Once the new DNA has entered a host cell, it must replicate as the cell divides. There are two general ways in which the newly introduced DNA can be replicated within the host cell:

1. It may be inserted into a host chromosome. Although the site of insertion is usually random (but see below), this is nevertheless a common method of integrating new genes into host cells.
2. It can be on a cloning vector (see below) that has its own origin of DNA replication. A bacterial plasmid is an example of a vector with its own *ori* sequence.

Many transformation methods use a **cloning vector**, a piece of carrier DNA such as a bacterial plasmid that can be replicated in the host. The cloning vector usually encodes a **selectable marker** gene that can be used to distinguish host cells with the vector from host cells lacking the vector. The first, and still most widely used, selectable markers are antibiotic resistance genes.

Some cloning vectors also contain one or more **reporter genes**—genes that encode proteins whose expression can be easily detected visually. Two commonly used reporter genes are the β-galactosidase (*lacZ*) gene from the *lac* operon (see Key Concept 11.2) and the gene for green fluorescent protein (GFP). Strains of bacteria growing on a medium containing the chemical X-Gal will be blue if they produce a functional β-galactosidase protein and white if they do not—an easy assay for expression of functional β-galactosidase. GFP, which normally occurs in the jellyfish *Aequorea victoria* (see Figure 33.9), emits green light when exposed to ultraviolet light—again, an easy assay for expression of a functional gene product.

To illustrate how to clone a gene, consider the steps for inserting foreign DNA into a plasmid (**FIGURE D5**). A plasmid carrying an antibiotic resistance gene (in this case, for ampicillin) is cut with a restriction enzyme that makes a single cut and leaves sticky ends (steps 1 and 2). The restriction site for the enzyme is often in the coding sequence of a reporter gene such as *lacZ*. The foreign DNA is cut with the same restriction enzyme, or one that leaves the same sticky ends (step 3). Plasmid and target DNA are mixed, the sticky ends are allowed to anneal (hydrogen-bond), and then DNA ligase is added to seal the nicks in the molecules (step 4). The DNA mixture is then added to a bacterial strain that is sensitive to the antibiotic corresponding to the resistance gene on the plasmid and the bacteria are then grown on medium containing that antibiotic (in this case, ampicillin; step 5). Successful transformation results in cells that can grow (form colonies) in the presence of the antibiotic, which selects for cells carrying the plasmid (steps 6 and 7).

A final step is to confirm that the antibiotic-resistant cells contain recombinant plasmids—that is, that their plasmids carry the foreign DNA. This can be done by isolating the plasmid DNA from

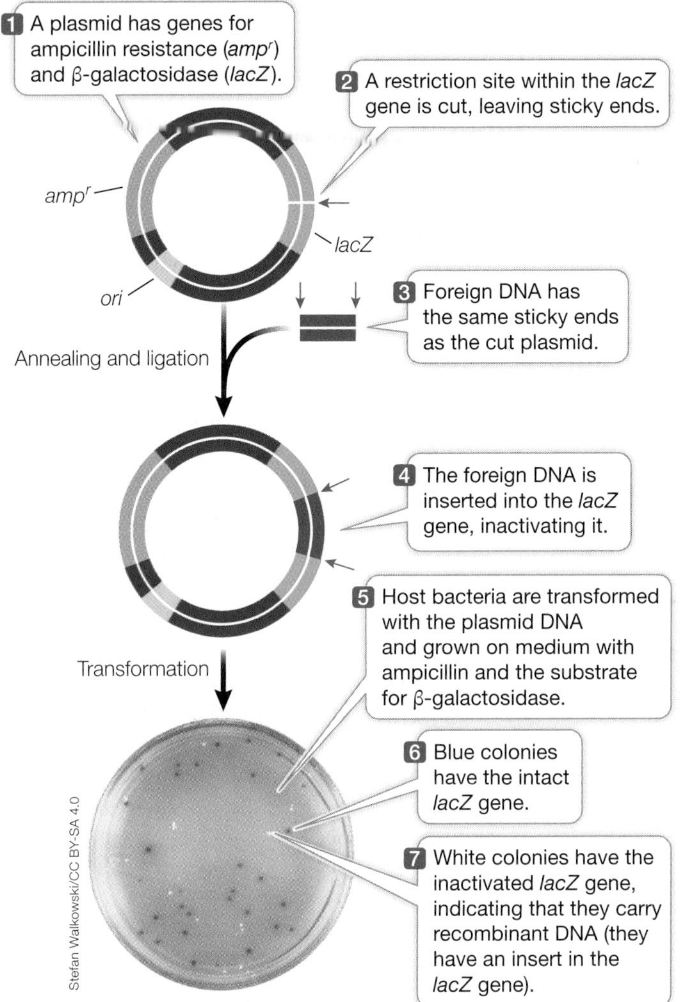

FIGURE D5 Cloning a Gene by Inserting It into a Bacterial Plasmid Vector Selectable marker genes are used by scientists to select for bacteria that have taken up a plasmid. In this example a second gene, *lacZ*, is used to screen for an insert of cloned DNA: in the absence of an insert, the *lacZ* gene product, β-galactosidase, is functional and colonies turn blue. An *ori* sequence is included so the plasmid can replicate.

the cells and checking its size (a plasmid containing inserted DNA is larger than the original plasmid) and the number of restriction sites it has for the restriction enzyme used in the original digests (a plasmid containing an insert has two sites instead of one). In the case of the cloning vector shown in Figure D5, bacteria carrying recombinant plasmids form white colonies because insertion of foreign DNA into the plasmid's restriction site inactivates the *lacZ* reporter gene (step 7). Once a bacterial strain is determined to carry a recombinant plasmid, the identity of the insert can be confirmed using Southern hybridization (see Figure D3) or sequencing. Transformed cells can be stored indefinitely in a freezer until needed.

Reporter genes have many other uses besides being an aid to cloning. The protein-coding sequence of a reporter gene can be attached to promoters or other regulatory sequences from other genes to study how these sequences regulate gene expression under different conditions or in different tissues of a transgenic multicellular organism (**FIGURE D6A**). The coding sequence of a reporter gene can also be added to the amino or carboxyl end of another protein-coding gene to "tag" the protein produced by that gene. For example, GFP can be used to study how and where a protein is localized in eukaryotic cells (**FIGURE D6B**).

Cloning can be used to generate libraries

In many cases, researchers want to clone a single gene of interest. In other cases, however, the goal is to clone all the genes or even all the DNA in a genome. Cloning an entire genome might seem counterintuitive at first since the organism itself can already make copies of its own genome. However, cloning a genome has two main advantages over using the organism itself. First, it allows the genome to be broken down into smaller pieces that can be more easily studied.

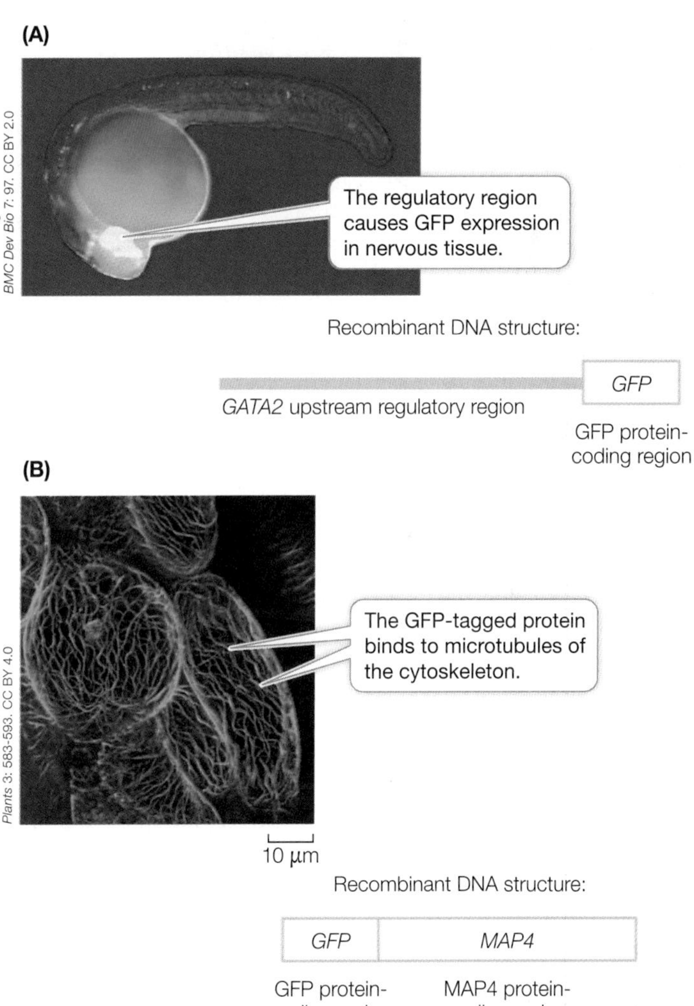

FIGURE D6 GFP as a Reporter Gene (A) Expression of GFP protein in a transgenic zebrafish 24 hours after fertilization. The recombinant DNA consists of 7.3 kilo-base pairs (kbp) of the upstream region of a transcription factor gene (*GATA2*) joined to the coding sequence of GFP. The GFP fluorescence is strongest in neuronal cells (in the brain) of the embryo, indicating that this is where the GATA2 transcription factor is normally expressed. **(B)** GFP-tagged MAP4 protein expressed in *Arabidopsis thaliana* shoot tissue. Because MAP4 binds to microtubules, the GFP fluorescence shows the structure of the cytoskeleton in the cells.

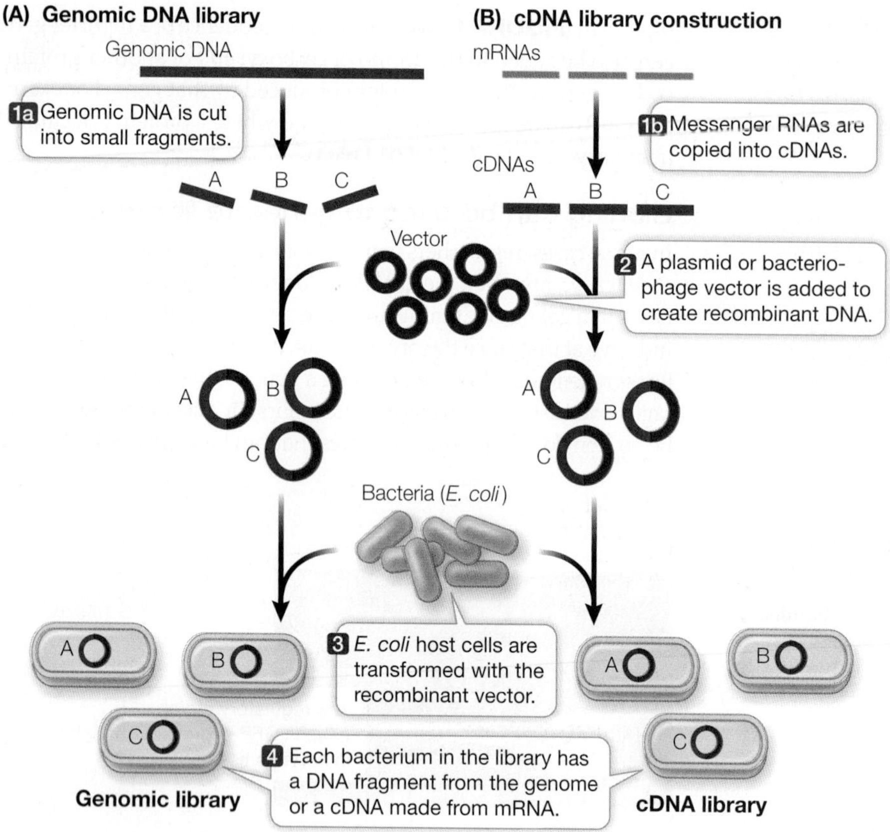

FIGURE D7 Constructing Libraries **(A)** A genomic library can be made by breaking the DNA into small fragments, incorporating the fragments into a vector, and then transforming host cells with the recombinant vectors. Each colony of cells contains many copies of a small part of the genome. **(B)** Similarly, cDNA libraries can be made from the mRNAs present in a sample. mRNAs are first copied into cDNAs and a library made from them. Because cDNA libraries only include transcribed genes after processing, they are often substantially smaller than genomic libraries.

Second, it means the genomic DNA can be studied in an organism that is more amenable to lab methods (the cloning host) than the organism from which the genome was obtained—elephants and whales are harder to work with than bacteria!

Cloning an entire genome requires the creation of a **genomic library**: a collection of cloned DNA fragments that together make up the entire genome of an organism. This is the starting point for some methods of genome sequencing. The genome is broken into small pieces and then each fragment is inserted into a vector and transformed into a host (**FIGURE D7A**). The resulting colonies contain vectors with inserts that (it is hoped) represent the entire genome. Making a complete genomic library for some organisms can be a daunting challenge. Plasmids can be used to clone pieces of DNA up to about 15,000 bp (15 kilo-base pairs, or kbp). Cloning a bacterial genome that is 4.6 million bp thus requires about 300 clones in the library. Cloning a human genome that is 3.1 billion bp into a plasmid library would require more than 206,000 clones. These are the minimum numbers assuming that non-overlapping pieces representing the entire genome could be chosen. In reality, this is not possible and the number of clones needed is several times larger. Some vectors allow larger pieces of DNA to be cloned, which reduces the number of clones needed to cover the entire genome. For example, yeast artificial chromosomes can be used to clone pieces of foreign DNA of up to about 1 million bp.

A much smaller DNA library—one that includes only the genes transcribed in a particular tissue—can be made from **complementary DNA**, or **cDNA**, which is obtained by reverse transcribing mRNA into DNA. The viral enzyme reverse transcriptase catalyzes this reaction (see Figure 11.14). This collection of cDNAs can be cloned to create a **cDNA library**, which is a "snapshot" of the transcription pattern of the cells in the sample (**FIGURE D7B**). cDNA libraries are much smaller than genomic libraries, and vary across tissues, development times, environments, and genotypes. cDNA libraries have been invaluable for comparing gene expression (mRNA levels) in different tissues. For example, if cDNAs derived from developing red blood cells are examined, the globin sequences (encoding the subunits of hemoglobin) are prominent. But a cDNA library derived from hair follicles does not contain those sequences.

DNA sequences can be manipulated to study cause-and-effect relationships

Mutations that occur in nature have been important in demonstrating cause-and-effect relationships in biology. However, mutations in nature are rare events. Recombinant DNA technology allows us to ask "what if" questions by creating artificial genes. Because synthetic DNA can be made with any desired sequence, it can be manipulated to create specific mutations or combinations of sequences in a gene. The phenotypes caused by these manipulations can be observed when the recombinant DNA is expressed in host cells.

Such techniques have revealed thousands of cause-and-effect relationships. One example involves the auxin response element, a short sequence of DNA that binds a specific transcription factor in plants. This element is found in the promoters of genes that are switched on in the presence of the plant hormone auxin (see Key Concept 25.2). To study the role of the auxin response element in plants, scientists made an artificial promoter containing many copies of the auxin response element and ligated the artificial promoter to a reporter gene. The recombinant DNA was used to transform *Arabidopsis thaliana* plants. When the plants were treated with auxin, the reporter gene was switched on at very high levels (higher than those produced by a wild-type auxin-responsive promoter). This experiment helped show that the presence of the auxin response element (the "cause") results in gene expression in response to auxin (the "effect").

Genes can be intentionally altered in cells

Another way to understand a gene's function is to inactivate it so it is not transcribed and translated into a protein. An example of this approach is the use of transposon mutagenesis in experiments designed to describe the minimal genome in a bacterium (see Figure 12.11). In animals, these "knockout" experiments often involve the use of CRISPR-Cas9 to generate double-stranded breaks in target genes (see Figure 12.12). These breaks can be repaired either by a process that simply joins the broken pieces back together (nonhomologous end joining; see Key Concept 9.3), which is error-prone and thus likely to inactivate the gene. Alternatively, repair can involve crossing over with a piece of DNA supplied by the researcher. As we saw in Chapter 8, a key feature of crossing over is that it involves an exchange of DNA between molecules with identical, or nearly identical, sequences. The supplied DNA contains sequences homologous to the gene of interest, so crossing over will occur, and often a reporter gene, so that transformation and knockout of the gene can be confirmed. The supplied DNA can also include a functional version of the gene if the cells to be transformed have an inactive version. This is the basis for **gene therapy**, a method of repairing damaged genes to correct genetic disorders.

A major hurdle to knocking out or adding a gene in a multicellular organism is the number of cells involved. How can transformation of a single cell (the usual method) lead to transformation of an entire organism? This is especially difficult in animals because differentiated animal cells cannot generate an entire new organism. One solution is to transform a newly fertilized egg, but getting enough eggs to successfully perform transformation can be challenging. Instead, transformation is usually performed on stem cells (cells that have the potential to develop into any other kind of cell) in cell culture. An essentially unlimited number of stem cells are available, and once a transformant is obtained, it can be grown into a large number of transformed cells. These transformed cells are then injected into an embryo during an early stage of development. If the procedure works, some of the injected cells give rise to germ line cells in the adult animal, in which case the recombinant DNA can be passed down to offspring, resulting in a transgenic animal (**FIGURE D8**).

The knockout technique in mice has been important in assessing the roles of many genes and is especially valuable in studying human genetic diseases. Many such diseases, including phenylketonuria (see Key Concept 10.1) and cystic fibrosis (see Figure 8.17), have knockout mouse models: mouse strains with the same genes knocked out by the process outlined in Figure D8. These mice have similar phenotypes to affected humans and are used as models to study the diseases and to test potential treatments.

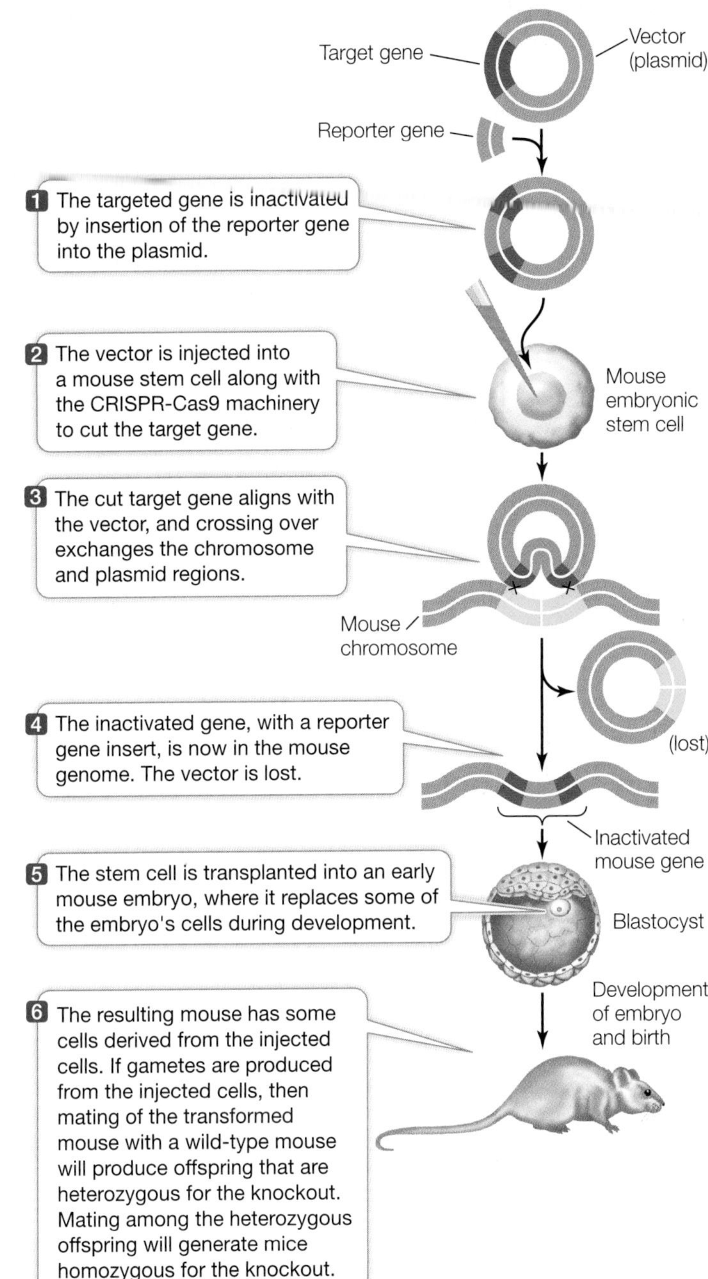

FIGURE D8 Making a Knockout Mouse Animals carrying mutations are rare. Crossing over is used to replace a normal mouse gene with an inactivated copy of that gene, thus "knocking out" the gene. Discovering what happens to a mouse with an inactive gene tells us much about the normal role of that gene.

Transgenic organisms can express proteins from other species

As mentioned above, a transgenic organism is one that contains genes, or some other DNA sequence, from a different species. While many transgenic organisms are used for cloning DNA sequences, others are engineered to express proteins from another species.

In order for a gene from another species to be appropriately expressed, several hurdles must be overcome. A special kind of cloning vector, called an **expression vector**, is used to express foreign genes in an organism. In the case of a eukaryotic gene in a prokaryote, key prokaryotic DNA sequences must be included with the gene. These sequences include the bacterial promoter, a signal for transcription termination, and a special sequence that is necessary for ribosome binding on the mRNA (see Chapter 10). In addition, the eukaryotic coding region must be made using cDNA so that it has no introns, since these cannot be removed in prokaryotes.

Activity D2 Expression Vectors

In addition to appropriate signals for transcription and translation, foreign genes expressed in multicellular eukaryotes often have: an *inducible promoter*, which responds to a specific signal, so that the transgene will be expressed at high levels only when the inducer is added; a *tissue-specific promoter*, which is expressed only in a certain tissue at a certain time; and *signal sequences* added to the protein-coding region to make sure that the gene product is directed to an appropriate destination.

Transgenic technology has many applications in medicine and agriculture. We give a few examples here.

SYNTHESIS OF THERAPEUTIC PROTEINS Many human diseases are caused by missing or dysfunctional proteins. For example, people suffering from type I diabetes make insufficient levels of the insulin protein, which regulates sugar metabolism. Human insulin is now expressed in bacteria for use in humans (see Figure 10.15A). Human growth hormone, which is expressed in cows and secreted in milk, is used to treat children suffering from a type of dwarfism (see Key Concept 35.5).

ADVANCING CROP YIELD AND NUTRITION CONTENT For more than 10,000 years, people have been selecting plants and animals for desirable characteristics such as large seeds, high fat content in milk, or resistance to disease. While traditional breeding has made large gains, recombinant DNA technology is much faster and allows traits from other species to be introduced into a target species.

Plants have been engineered to make their own insecticides, so that they are resistant to insect attack and thus give higher yield. A gene that makes crop plants resistant to the herbicide glyphosate (Roundup) allows fields of so-called Roundup Ready crops to be sprayed with the herbicide and survive, while adjacent, competing weeds are killed, again increasing yield. A strain of rice (Golden Rice) has been developed that contains a bacterial gene and a daffodil gene that together allow production of β-carotene—a precursor of vitamin A that is often deficient in the diets of children in some parts of the world—in rice seeds. The next few decades will see a continued emergence of such genetically engineered crops.

Glossary

5′ cap A chemically modified GTP added to the 5′ end of mRNA; facilitates binding of mRNA to ribosome and prevents mRNA breakdown. Also called G cap.

A

abiotic (a′ bye ah tick) [Gk. *a*: not + *bios*: life] Nonliving. In ecology, the nonliving (physical and chemical) components of the environment of organisms. (Contrast with biotic.)

abscisic acid (ABA) (ab sighs′ ik) A plant growth substance with growth-inhibiting action. Causes stomata to close; involved in a plant's response to salt and drought stress.

absorption (1) Of light: retention (in contrast to reflection or transmission). (2) In the study of nutrition, the process of transporting molecules from the gut lumen into the blood.

absorption spectrum A graph of light absorption versus wavelength of light; shows how much light is absorbed at each wavelength.

absorptive heterotroph An organism (usually a fungus) that obtains its food by **absorptive heterotrophy**, i.e., by secreting digestive enzymes into the environment to break down large food molecules, then absorbing the breakdown products.

abyssal zone (uh biss′ ul) [Gk. *abyssos*: bottomless] The deepest portions of the oceans; characterized by extremely high pressures, low dissolved oxygen levels, and cold temperatures.

accessory fruit In a flowering plant, a fruit in which some of the flesh is derived from tissues other than the ovary; examples include apples and pears.

accessory reproductive organs Reproductive organs other than the gonads.

acclimation Refers to increased tolerance for environmental extremes (e.g., extreme cold) after prior exposure to them.

acclimatization The change in an individual's phenotype that results from long-term exposure to a new environment. A form of phenotypic plasticity. Sometimes used only to describe responses in a natural (non-laboratory) setting. (*See also* acclimation.)

acetyl CoA A compound that reacts with oxaloacetate to produce citrate at the beginning of the citric acid cycle; a key metabolic intermediate in the formation of many compounds.

acid [L. *acidus*: sharp, sour] A substance that can release a proton in solution. (Contrast with base.)

acoelomate An animal that does not have a coelom.

acquired immune deficiency syndrome (AIDS) A condition caused by human immunodeficiency virus (HIV) in which the body's T-helper cells are reduced, leaving the infected individual subject to opportunistic diseases.

actin [Gk. *aktis*: ray] A protein that makes up the cytoskeletal microfilaments in eukaryotic cells and is one of the two contractile proteins in muscle. (*See also* myosin.)

actin filaments Chains of linked actin molecules. Also called thin filaments. (*See also* myosin filaments.)

actinobacteria *See* Appendix A.

action potential An impulse in a neuron; that is, a brief, local, high-amplitude depolarization of the cell membrane that propagates over the whole membrane without loss of amplitude. (Contrast with resting potential.)

action spectrum A graph of a biological process plotted against wavelength of light; shows which wavelengths are involved in the process.

activation energy (E_a) The total amount of energy that must be supplied for a reaction to begin.

active site The region on the surface of an enzyme or ribozyme where the substrate binds, and where catalysis occurs.

active transport The energy-dependent transport of a substance across a biological membrane against a concentration gradient—that is, from a region of low concentration (of that substance) to one of high concentration. (*See also* primary active transport, secondary active transport; contrast with facilitated diffusion, passive transport.)

adaptation (1) In evolutionary biology, a particular structure, physiological process, or behavior that makes an organism better able to survive and reproduce. Also, the evolutionary process that leads to the development or persistence of such a trait. (2) In sensory neurophysiology, a sensory cell's loss of sensitivity as a result of repeated stimulation.

adaptive defenses In animals, one of two general types of defenses against pathogens. Involves antibody proteins and other proteins that recognize, bind to, and aid in the destruction of specific viruses and bacteria. Present only in vertebrate animals. (Contrast with innate defenses.)

adaptive radiation A series of evolutionary events that result in an array (radiation) of related species that live in a variety of environments, differing in the characteristics each uses to exploit those environments.

additive growth A form of growth in which a certain number is added during each successive time interval. (Contrast with multiplicative growth.)

adenosine triphosphate *See* ATP.

adhesion Chemical attraction of one molecule, substance or structure to another, as in a cell binding to its extracellular matrix.

adiabatic (ad′ e uh bat′ ik) [Gk. *adiabatos*: impassable] Referring to a system that exchanges no heat with its surroundings. When such a "heat-insulated" system—for example, a parcel of air—contracts or expands, its temperature increases or decreases, respectively. This thermodynamic principle applies to the functioning of air conditioners and refrigerators.

adrenocorticotropic hormone (ACTH) A tropic hormone produced by the anterior pituitary gland that stimulates release of cortisol and certain other hormones from the adrenal cortex.

adventitious roots (ad ven ti′ shus) [L. *adventitius*: arriving from outside] Roots originating from the stem at ground level or below; typical of the fibrous root system of monocots.

aerobic (air oh′ bic) [Gk. *aer*: air + *bios*: life] In the presence of oxygen; requiring or using oxygen (as in **aerobic metabolism**). (Contrast with anaerobic.)

aerobic respiration A metabolic pathway in which glucose is fully catabolized in the presence of oxygen such that oxygen is the final electron acceptor in electron transport. (Contrast with anaerobic respiration.)

aerotolerant anaerobes Organisms that tolerate the presence of oxygen but do not use it for metabolism.

agglutination The binding of multiple antigens and multiple antibodies together into complexes that are later endocytosed by phagocytes and then hydrolyzed within the phagocytes' lysosomes.

aggregate fruit In a flowering plant, a fruit that develops from the merger of several different ovaries that were separate in the flower.

alcoholic fermentation *See* fermentation.

algae *See* Appendix A.

allantois (al′ lun toh is) [Gk. *allant*: sausage] An extraembryonic membrane sac that stores the embryo's nitrogenous wastes; because it contains blood vessels, it provides the means for gas exchange. In eutherian mammals, forms part of the placenta.

allele (a leel′) [Gk. *allos*: other] A specific form of a gene at a given locus on a chromosome, among multiple possible forms.

allele frequency The proportion of each allele in a gene pool.

allergic reaction [Ger. *allergie*: altered] An overreaction of the immune system to amounts of an antigen that do not affect most people; often involves IgE antibodies.

allometry (1) In studies of animals of different sizes, nonproportional variation in relation to body size. (2) In development, a change in the proportions of various parts of an organism relative to one another as a consequence of unequal growth.

allopatric speciation (al′ lo pat′ rick) [Gk. *allos*: other + *patria*: homeland] Speciation that results when reproductive isolation occurs because of the interposition of (or crossing of) a physical geographic barrier such as a river. (Contrast with sympatric speciation.)

allopolyploidy The possession of more than two entire chromosome sets that are derived from more than one species.

allosteric regulation (al lo steer′ ik) [Gk. *allos*: other + *stereos*: structure] Regulation of the activity of a protein (usually an enzyme) by the binding of an effector molecule to a site other than the active site.

allosteric site A site on a protein (often on a polypeptide that is distinct from the one containing the active site) that can bind a substance that alters the conformation of the protein, changing its functionality.

alpha (α) cells Pancreatic cells that secrete the hormone glucagon.

α (alpha) helix A prevalent type of protein secondary structure; a right-handed spiral.

alternation of generations The succession of multicellular haploid and diploid phases in some sexually reproducing organisms, notably plants.

alternative hypothesis In statistical inference, the hypothesis that contrasts with the null hypothesis; typically the hypothesis of primary interest.

alternative splicing A process for generating different mature mRNAs from a single gene by splicing together different sets of exons during RNA processing.

altruism Behavior that benefits another individual at a cost of reduced fitness to the individual performing the behavior.

alveolar sac A blind sac found at the end of a branch of the lung airways in a mammal; consists of many alveoli.

alveolates *See* Appendix A.

alveolus (al ve′ o lus) (plural: alveoli) [L. *alveus*: cavity] A semispherical outpocketing of the wall of an alveolar sac in the mammalian lung. The alveoli are the location of gas exchange between air and blood.

ambulacrarians *See* Appendix A.

amensalism (a men′ sul ism) A"–/0" interaction between individuals of two different species that harms one individual (i.e., it suffers reduced fitness) but does not affect the other. (Contrast with commensalism, mutualism.)

amino acid An organic compound containing both NH_2 and COOH groups, and one of 20 different side chains (in common amino acids). Peptides are polymers of amino acids.

amnion (am′ nee on) The fluid-filled sac within which the embryos of reptiles (including birds) and mammals develop.

amniote egg A shelled egg surrounding four extraembryonic membranes and embryo-nourishing yolk. This evolutionary adaptation permitted mammals and reptiles to live and reproduce in drier environments than can most amphibians.

amniotes *See* Appendix A.

amoebozoans *See* Appendix A.

amphibians *See* Appendix A.

amphipathic (am′ fi path′ ic) [Gk. *amphi*: both + *pathos*: emotion] Of a molecule, having both hydrophilic and hydrophobic regions.

anabolic Referring to a synthetic reaction in which simple molecules are linked to form more complex ones; requires an input of energy and captures it in the chemical bonds that are formed. (Contrast with catabolic.)

anabolism The synthesis of complex molecules from simpler ones; requires an input of energy and captures it in the chemical bonds that are formed. (Contrast with catabolism.)

anaerobic (an ur row′ bic) [Gk. *an*: not + *aer*: air + *bios*: life] Occurring without the use of molecular oxygen, O_2 (as in **anaerobic metabolism**). (Contrast with aerobic.)

anaerobic respiration A metabolic pathway in which glucose is fully catabolized in the absence of oxygen, such that a molecule other than oxygen is used as the final electron acceptor in electron transport. (Contrast with aerobic respiration.)

analogous In biology, refers to similar structures found in two or more groups of organisms, but with separate, independent evolutionary origins. An example is the wings of bats and the wings of birds, which evolved independently as wings from wingless ancestors.

anaphase (an′ a phase) [Gk. *ana*: upward] The stage in cell nuclear division at which the first separation of sister chromatids (or, in the first meiotic division, of paired homologs) occurs.

ancestral trait The trait originally present in the ancestor of a given group; may be retained or changed in the descendants of that ancestor.

anchored membrane proteins Proteins that are associated with the membrane by covalent attachments to lipids. (Contrast with integral membrane proteins, peripheral membrane proteins.)

androgen (an′ dro jen) [Gk. *andros*: man + *genein*: to produce] Any of the several masculinizing steroid hormones (most notably testosterone).

aneuploidy (an′ you ploy dee) A condition in which the number of one or more chromosomes is either lacking or in excess.

angiosperms *See* Appendix A.

animal hemisphere The non-yolky upper portion of some animal eggs. (Contrast with vegetal hemisphere.)

anion (an′ eye on) [Gk. *ana*: upward] A negatively charged ion. (Contrast with cation.)

annelids *See* Appendix A.

antagonistic muscle pair Two muscles, or groups of muscles, that perform coordinated, opposing actions, thus allowing the bones at a joint to move in opposite directions. When one muscle contracts (to close the angle of a joint), its antagonist relaxes. To open the joint, the latter muscle contracts and the former relaxes.

antenna complex In photosynthesis, composed of several light-harvesting complexes that surround the reaction center to make up a photosystem in thylakoid membranes of chloroplasts.

anterior Toward or pertaining to the tip or headward region of the body axis. (Contrast with posterior.)

anterior pituitary gland The anterior portion of the vertebrate pituitary gland that is composed of non-neural endocrine cells and produces several hormones, including tropic hormones. Also called the adenohypophysis. (*See also* pituitary gland, posterior pituitary gland.)

anther (an′ thur) [Gk. *anthos*: flower] A pollen-bearing portion of the stamen of a flower.

antheridium (an′ thur id′ ee um) [Gk. *antheros*: blooming] The multicellular structure that produces the sperm in nonvascular land plants and ferns.

antibody One of the myriad proteins produced by the immune system that specifically binds to a foreign substance in blood or other tissue fluids and initiates its removal from the body.

anticodon The three nucleotides in transfer RNA that pair with a complementary triplet (a codon) in messenger RNA.

antidiuretic hormone (ADH) A hormone that promotes water reabsorption by the kidney. ADH is produced by neurons in the hypothalamus and released from nerve terminals in the posterior pituitary gland. Also called vasopressin in mammals.

antigen (an′ ti jun) Any substance that stimulates the production of an antibody or antibodies in the body of a vertebrate.

antigen presentation In cellular immunity, the process in which a cell ingests and digests an antigen, and then exposes fragments of that antigen, bound to a major histocompatibility complex (MHC) molecule, to the outside of the cell, where the antigen can be recognized by T cells.

antigenic determinant The specific region of an antigen that is recognized and bound by a specific antibody. Also called an epitope.

antiparallel Pertaining to molecular orientation in which a molecule or parts of a molecule have opposing directions.

aorta (a or′ tah) [Gk. *aorte*: aorta] The main trunk of the arteries leading to the systemic (as opposed to the breathing-organ) circulation.

APETALA 1 (AP1) A gene in *Arabidopsis thaliana* that encodes the transcription factor AP1; one of the class A floral identity genes.

aphotic zone The part of lakes and oceans below the photic zone, where no sunlight penetrates.

apical meristem The meristem at the tip of a shoot or root; responsible for a plant's primary growth.

apicomplexans *See* Appendix A.

apomixis (ap oh mix′ is) [Gk. *apo*: away from + *mixis*: sexual intercourse] The asexual production of seeds.

apoplast (ap′ oh plast) In plants, the continuous meshwork of cell walls and extracellular spaces through which material can pass without crossing a plasma membrane. (Contrast with symplast.)

apoptosis (a pop toe′ sis) Programmed cell death. A controlled cell death important in morphogenesis during animal development and growth (e.g., removal of tissue between digits, metamorphosis of frog tadpole tail and gut).

aquaporin [L. *aqua*: water + *porus*: pore] A transport protein in plant and animal cell membranes through which water passes by osmosis.

arbuscular mycorrhizae Mutualistic fungi that enter the root and penetrate the cell walls of the root cells, forming arbuscular (treelike) structures inside the cell wall but outside the cell membrane. (Contrast with ectomycorrhizae.)

arbuscular mycorrhizal fungi *See* Appendix A.

archaeans *See* Appendix A.

archegonium (ar′ ke go′ nee um) The multicellular structure that produces eggs in nonvascular land plants, ferns, and gymnosperms.

archosaurs *See* Appendix A.

arrow worms *See* Appendix A.

arteriole A microscopically small blood vessel with muscular walls that carries blood from a tiny artery into a capillary bed. (Contrast with venule.)

artery A muscular blood vessel carrying blood away from the heart to other parts of the body. (Contrast with vein.)

arthropods *See* Appendix A.

artificial selection Selective breeding of organisms, commonly practiced by animal and plant breeders, to increase the frequency of a favored trait from one generation to the next. (Contrast with natural selection.)

ascospores Haploid spores contained within the sexual reproductive structure (the ascus) of sac fungi.

ascus (ass′ cus) (plural: asci) [Gk. *askos*: bladder] In sac fungi, the club-shaped sporangium within which spores (ascospores) are produced by meiosis.

asexual reproduction The formation of new individuals without the union of genetic material from two different parents. Often takes place by budding or fragmentation. The offspring are genetically identical to their parent.

Asgard archaea A group of archaeal lineages that includes the closest known prokaryotic relatives of eukaryotes.

asynchronous flight muscle A flight muscle—known to occur in about 75 percent of insects—in which each excitation results in multiple contractions.

atom [Gk. *atomos*: indivisible] The smallest unit of a chemical element. Consists of a nucleus containing protons and (except for hydrogen) neutrons, and one or more electrons in orbitals.

atomic mass Equal to the mass of protons plus the mass of neutrons (electrons have so little mass they are usually ignored).

atomic number The number of protons in the nucleus of an atom; also equals the number of electrons around the neutral atom. Determines the chemical properties of the atom.

atomic orbital The defined area of space within which the electrons in each electron shell move; also called orbital or electron orbital.

ATP (adenosine triphosphate) A component of nucleic acids and an energy-storage compound containing adenine, ribose, and three phosphate groups. When it is formed from ADP, useful energy is stored; when it is broken down (to ADP or AMP), energy is released to drive endergonic reactions.

ATP synthase An integral membrane protein that couples the transport of protons with the formation of ATP.

atrium (a′ tree um) [L. *atrium*: central hall] In the hearts of vertebrates, a relatively thin-walled muscular chamber(s) entered by blood on its way to the ventricle(s).

autocrine A molecule that a cell secretes and that binds to and affects the cell that makes it. (Contrast with paracrine.)

autocrine signaling A type of cell signaling in which the chemical signal binds to and affects the cell that makes it. (Contrast with juxtacrine signaling, paracrine signaling.)

autoimmune diseases Diseases (e.g., rheumatoid arthritis) that result from failure of the immune system to distinguish between self and nonself, causing it to attack tissues in the organism's own body.

autoimmunity An immune response by an organism to its own molecules or cells.

autonomic effector An effector other than skeletal muscle; includes smooth and cardiac muscles, exocrine glands, and endocrine glands. Also called an internal effector.

autonomic nervous system (ANS) The portion of the nervous system that controls autonomic effectors (effectors other than the skeletal muscles). Also called the involuntary nervous system.

autophosphorylation A process in which a protein catalyzes the addition of one or more phosphate groups to itself.

autopolyploidy The possession of more than two entire chromosomes sets that are derived from a single species.

autosome Any chromosome (in a eukaryote) other than a sex chromosome.

autotroph (au′ tow trowf′) [Gk. *autos:* self + *trophe*: food] An organism that is capable of living exclusively on inorganic materials, water, and an energy source other than the chemical bonds of organic compounds. Some autotrophs (photoautotrophs) use sunlight as their energy source. Others (chemoautotrophs) use oxidation of inorganic compounds. (Contrast with heterotroph.)

auxin (awk′ sin) [Gk. *auxein*: to grow] In plants, a substance (the most common being indoleacetic acid) that regulates growth and various aspects of development.

avirulence (Avr) factors Factors produced by a pathogen that are directly or indirectly recognized by plant resistance (R) proteins.

axil In plants, the space formed where a leaf (or petiole) meets a stem.

axillary bud A bud that forms in the axil (the space where a leaf meets a stem).

axis In the study of endocrinology, a system in which two or more types of endocrine cells act on each other in sequence.

axon [Gk. *axle*] The part of a neuron—often highly elongated—that conducts action potentials away from the cell body.

axon hillock The junction between an axon and the neuron's cell body; where action potentials are typically generated.

B

bacillus (bah sil′ us) (plural *bacilli*) [L: little rod] Any of various rod-shaped bacteria.

bacteria *See* Appendix A.

bacterial chromosome A single, circular DNA molecule containing the vast majoity of the genes present in a bacterial cell.

bacterial conjugation *See* conjugation.

bacteriophage (bak teer′ ee o fayj) [Gk. *bakterion*: little rod + *phagein*: to eat] Any of a group of viruses that infect bacteria. Also called phage.

bacteroids Nitrogen-fixing organelles that develop from endosymbiotic bacteria.

bar chart A figure that displays frequency distributions of categorical data using bar lengths to represent relative frequency.

basal metabolic rate (BMR) The minimum rate of energy use in an awake (but resting) bird or mammal. By definition, BMR is measured when the animal is resting, fasting, and in its thermoneutral zone (i.e., its metabolic rate is not elevated for thermoregulation).

basal transcription apparatus The RNA polymerase and general transcription factors bound to a promoter.

base (1) A substance that can accept a hydrogen ion in solution. (Contrast with acid.) (2) The purine or pyrimidine attached to the sugar in a nucleoside.

base-pair substitution A change of a single base pair in a nucleotide sequence (e.g., AT to GC).

basidioma (plural: basidiomata) A fruiting structure produced by club fungi.

basidium (bass id′ ee yum) In club fungi, the characteristic sporangium in which four **basidiospores** are formed by meiosis and then borne externally before being shed.

basilar membrane A membrane in the human inner ear that flexes in response to sound waves and thereby activates hair cells; flexes at different locations in response to different pitches of sound.

BD model A mathematical model that describes how birth and death rates affect the size and change in size (dynamics) of ecological populations.

behavioral imprinting In animal behavior, a rapid form of learning in which an individual animal learns—during a brief critical period in its life—to make a particular response, which is maintained for life, to some object or other individual.

benthic zone [Gk. *benthos*: bottom] The substrate at the bottom of an ocean, lake, stream, or other body of water.

beta (β) cells Pancreatic cells that secrete the hormone insulin.

β-oxidation In catabolism, the oxidation of fatty acids to produce acetyl CoA that can then be oxidized in the citric acid cycle.

β (beta) pleated sheet A type of protein secondary structure; results from hydrogen bonding between polypeptide regions lying next to one other.

BIDE model A mathematical model that describes how birth, immigration (movement of individuals into populations), death, and emigration (movement of individuals out of populations) affect the size and change in size (dynamics) of ecological populations.

bilateral symmetry The condition in which only the right and left sides of an organism, divided by a single plane through the midline, are mirror images of each other.

bilaterians *See* Appendix A.

bilayer A structure that is two layers in thickness. In biology, most often refers to the phospholipid bilayer of membranes. (*See* phospholipid bilayer.)

binary fission Reproduction of a prokaryote by division of a cell into two comparable progeny cells.

binding affinity The strength of the interaction between a ligand and the protein molecule to which it binds.

binomial nomenclature A taxonomic naming system in which each species is given two names (a genus name followed by a species name).

biofilm A community of microorganisms embedded in a polysaccharide matrix, forming a highly resistant coating on almost any moist surface.

biogeochemical cycle A description of the cyclical pathways by which biological, geological, and chemical processes move a chemical element or molecule through biotic and abiotic compartments of planet Earth.

biogeographic region A large, continental-scale region of Earth, encompassing multiple biomes, that contains a distinct assemblage of species, many of which are phylogenetically related. (Contrast with biome.)

biogeography The spatial distribution of Earth's species.

biological clock A self-contained, metabolic mechanism that keeps track of time.

biological determinism As applied to behavior, the belief that genes completely determine an organism's behavior.

biological species concept The definition of a species as a group of actually or potentially interbreeding natural populations that are reproductively isolated from other such groups. (Contrast with lineage species concept, morphological species concept.)

biomass The mass of a sample of biological tissue, living or dead. The term usually refers to the total mass of a designated group of organisms (or of a certain portion of their tissues), and may be measured as the mass of their intact bodies or of their dried bodies.

biome (bye′ ome) An ecosystem characterized by a distinct physical environment and ecologically similar organisms that have similar, convergent adaptations to the physical environment (Contrast with biogeographic region.)

biosphere (bye′ oh sphere) The region that supports living organisms on Earth, extending about 23 km from the abyssal depths of the oceans into the stratosphere (an upper portion of the atmosphere).

biota (bye oh′ tah) All of the organisms—animals, plants, fungi, and microorganisms—found in a given area. (Contrast with flora, fauna.)

biotic interchange The mixing of distinct groups of species that were isolated by physical or climatic barriers, after the barrier is removed.

biotic (bye ah′ tick) [Gk. *bios*: life] Living. In ecology, the living components of the environment of organisms. (Contrast with abiotic.)

bipedal locomotion The act of walking or running on two appendages.

birds *See* Appendix A.

bivalves *See* Appendix A.

blastocoel (blass' toe seal) [Gk. *blastos*: sprout + *koilos*: hollow] The central, hollow cavity of a blastula.

blastocyst (blass' toe cist) A mammalian blastula. The sphere of cells formed by early cell divisions and composed of an inner cell mass that becomes the embryo and an outer layer of cells called the trophoblast.

blastodisc (blass' toe disk) An embryo that forms as a disc of cells on the surface of a large yolk mass; comparable to a blastula, but occurring in animals such as birds and other reptiles, in which the massive yolk restricts complete cleavage.

blastomere Any of the cells produced by the early divisions of a fertilized animal egg.

blastula (blass' chu luh) An early stage of the animal embryo; in many species, a hollow sphere of cells surrounding a central cavity, the blastocoel. (Compare with blastodisc.)

blocks to polyspermy Any of several responses to entry of a sperm into an egg that prevent more than one sperm from entering the egg.

blood plasma The watery solution in which red blood cells and/or other blood cells are suspended. The part of the blood other than cells.

blood pressure The extent to which the pressure of the blood exceeds the pressure in an animal's external environment.

blood–brain barrier The selective impermeability of blood capillaries in the brain that prevents most dissolved molecules from freely diffusing between the blood plasma and the brain tissue fluids. The impermeability arises from peculiarities of the capillary walls and from the structural relationship between capillaries and glial cells.

body cavity An internal, fluid-filled space, enclosed by mesoderm in animals.

body plan The general structure of an animal, the arrangement of its organ systems, and the integrated functioning of its parts.

bone A rigid component of vertebrate skeletal systems that consists mostly of an extracellular matrix of collagen fibers with insoluble calcium phosphate crystals deposited among them. Also contains living cells that, for example, can remodel bone throughout life.

brachiopods *See* Appendix A.

bracts In plants, modified leaf-like organs.

brain The part of the central nervous system of greatest size.

branch point A conserved sequence (A followed by several pyrimidines) in the interior of an intron that is used during intron splicing to attach the 5′ end of the intron.

breathing The exchange of O_2 and CO_2 between the environmental medium next to the gas exchange membranes and the outside world; often occurs by ventilation, but can occur entirely by diffusion. Also called external respiration.

breathing-organ circuit In a circulatory system, the blood vessels that carry blood to and from, and through, the breathing organs. Also called the gill circuit or lung circuit. (Contrast with systemic circuit.)

brown adipose tissue (BAT) A specialized fatty tissue, known only in mammals, that serves as the primary site of nonshivering thermogenesis.

brown algae *See* Appendix A.

brown fat In mammals, fat tissue that is specialized to produce heat. It has many mitochondria and capillaries, and a protein that uncouples oxidative phosphorylation.

bryozoans *See* Appendix A.

budding Asexual reproduction in which a more or less complete new organism grows from the body of the parent organism, eventually detaching itself.

buffer A substance that can transiently accept or release hydrogen ions and thereby resist changes in pH.

bulk flow Mass movement of a fluid (gas or liquid) driven by a difference of pressure, from a region of high pressure to a region of low pressure.

C

callus In plant cell cultures, a group of dividing cells that have lost the differentiated state they had when part of the parent plant.

calorie (cal) [L. *calor*: heat] A unit of energy equal to the amount of heat required to raise the temperature of 1 gram of water by 1°C. Physiologists commonly use the kilocalorie (kcal) as a unit of energy (1 kcal = 1,000 calories). Nutritionists also use the kilocalorie, but refer to it as the *Calorie* (capital C).

Calvin cycle The stage of photosynthesis in which CO_2 reacts with RuBP to form 3PG, 3PG is reduced to a sugar, and RuBP is regenerated, while other products are released to the rest of the plant. Also known as the Calvin–Benson cycle.

Cambrian explosion The rapid diversification of multicellular life that took place during the Cambrian period.

capillary [L. *capillaris*: hair] In the circulatory system, a microscopic blood vessel with walls composed of just a simple epithelium (called the capillary endothelium), one cell layer thick. Capillaries are the principal sites of exchange between blood and other tissues in animals with a closed circulatory system.

carbohydrates Organic compounds containing carbon, hydrogen, and oxygen in the ratio 1:2:1 (i.e., with the general formula $C_m(H_2O)_n$. Common examples are sugars, starch, and cellulose.

carbon-fixation reactions The phase of photosynthesis in which chemical energy captured in the light reactions is used to drive the reduction of CO_2 to form carbohydrates.

cardiac cycle Alternating contraction and relaxation of the heart. In a mammal, contraction of the two atria of the heart, followed by contraction of the two ventricles and then by relaxation until the next contraction begins.

cardiac muscle A type of muscle tissue that makes up the wall of the heart. In vertebrates characterized by branching cells with single nuclei and a striated (striped) appearance, electrically coupled at intercalated discs. (Contrast with smooth muscle, skeletal muscle.)

cardiac output The volume of blood the heart pumps per minute. Calculated by multiplying stroke volume by heart rate (the number of beats per minute).

cardiovascular system [Gk. *kardia*: heart + L. *vasculum*: small vessel] The heart and vessels of the circulatory system. Also called circulatory system.

carotenoids Lipids that can absorb energy from particular wavelengths of light.

carpel (kar' pel) [Gk. *karpos*: fruit] The organ of the flower that contains one or more ovules.

carrier protein An integral protein in a cell membrane that noncovalently binds specific molecules and transports them through the membrane. Many are capable of moving molecules against their electrochemical gradients by direct or indirect use of ATP-bond energy.

carrying capacity (*K*) The maximum population size or density of a given species that a given unchanging environment can support indefinitely.

cartilage In vertebrates, a tough but flexible connective tissue found in joints, the outer ear, part of the nose, and elsewhere. Forms the entire skeleton in some animal groups, notably sharks and their relatives.

Casparian strip A band of cell wall containing suberin and lignin, found in the endodermis. Restricts the movement of water across the endodermis.

catabolic (kat uh bah′ lik) [Gk. *kata*: to break down + *ballein*: to throw] Referring to a synthetic reaction in which complex molecules are broken down into simpler ones and energy is released. (Contrast with anabolic.)

catabolism The breakdown of complex molecules into simpler ones; releases energy. (Contrast with anabolism.)

catalyze The ability to increase the rate of a reaction without undergoing a chemical change.

catch A state in which some invertebrate muscles maintain a state of contraction over extended periods of time with minimal energy expenditure.

categorical variables Variables that take on qualitative categories as values, such as human blood types (A, AB, B, or O).

cation (cat′ eye on) An ion with one or more positive charges. (Contrast with anion.)

cation exchange The process in soil by which cations bound to soil particles are released from the particles and made available to plant roots in soil water through the binding of hydrogen ions to the soil particles.

cDNA *See* complementary DNA.

cDNA library A collection of complementary DNAs derived from mRNAs of a particular tissue at a particular time in the life cycle of an organism.

cell The simplest structural unit of a living organism. In multicellular organisms, many individual cells serve as the building blocks of tissues and organs.

cell body In a neuron, the portion of the neuron that contains the cell nucleus and most of the neuron's other organelles.

cell cycle The stages through which a cell passes between one division and the next. Includes all stages of interphase and mitosis.

cell cycle checkpoints Points of transition between different phases of the cell cycle, which are regulated by cyclins and cyclin-dependent kinases (CDKs).

cell division The reproduction of a cell to produce two new cells. In eukaryotes, involves nuclear division (mitosis) and cytoplasmic division (cytokinesis).

cell division signals The one or more signals required to initiate cell division. The signals may originate from either inside or outside the cell.

cell fate The type of cell that an undifferentiated cell in an embryo will become in the adult.

cell junctions Specialized structures associated with the cell membrane that join cells in multicellular organisms. Some contribute to cell adhesion, others to intercellular communication.

cell membrane A lipid bilayer also containing proteins and other molecules that encloses the cytoplasm of the cell and separates it from the surrounding environment.

cell theory States that cells are the basic structural and physiological units of all living organisms, and that all cells come from preexisting cells.

cell wall A relatively rigid structure that encloses cells of plants, fungi, many protists, and most prokaryotes, and which gives these cells their shape and limits their expansion in hypotonic media.

cellular immune response The immune system response mediated by T cells and directed against parasites, fungi, intracellular viruses, and foreign tissues (grafts). (Contrast with humoral immune response.)

cellular respiration The catabolic pathways by which electrons are removed from various molecules and passed through intermediate electron carriers to O_2, generating H_2O and releasing energy.

cellular slime molds *See* Appendix A.

cellular specialization In multicellular organisms, the division of labor such that different cell types become responsible for different functions (e.g., reproduction or digestion) within the organism.

cellulose microfibrils Long filaments of cellulose that are hydrogen bonded to one another, forming a rigid structure that strengthens plant cell walls. A single filament of cellulose is a polymer composed of repeating glucose units joined in a straight chain.

cellulose synthases Protein complexes that sit in the cell membrane and synthesize cellulose.

central dogma of molecular biology The premise that information flows from DNA to RNA to polypeptide (protein).

central nervous system (CNS) Large structures composed principally of integrating neurons and associated glial cells, forming the site of most information processing, storage, and retrieval; in vertebrates, the brain and spinal cord. (Contrast with peripheral nervous system.)

centralization (of the nervous system) Over the course of evolution, the tendency of neurons to become clustered into centralized, integrating organs.

centriole (sen′ tree ole) A paired organelle that helps organize the microtubules in animal and protist cells during nuclear division.

centromere (sen′ tro meer) [Gk. *centron*: center + *meros*: part] The region where sister chromatids join.

centrosome (sen′ tro soam) The major microtubule organizing center of an animal cell.

cephalization (sef ah luh zay′ shun) [Gk. *kephale*: head] Over the course of evolution, the trend toward increasing concentration of nervous tissue (e.g., the brain) and sensory organs at the anterior end of the animal.

cephalopods *See* Appendix A.

cercozoans *See* Appendix A.

cerebral cortex The outer layer of the cerebral hemispheres in the vertebrate brain. Plays key roles in sensory perception, learning, memory, and motor control.

channel protein An integral protein in a cell membrane that forms an aqueous passageway across the membrane in which it is inserted and through which specific solutes (e.g., specific ions) may pass by diffusion in the direction dictated by their electrochemical gradient.

character In genetics, an observable feature, such as eye color. (Contrast with trait.)

chelicerates *See* Appendix A.

chemical bond An attractive force stably linking two atoms.

chemical reaction The change in the composition or distribution of atoms of a substance with consequent alterations in properties.

chemical synapse A point of contact between two neurons at which the neurons are separated by only a very narrow space and neurotransmitter molecules released from the presynaptic cell induce changes in ion flux and membrane potential in the postsynaptic cell. Also refers to similar points of contact between neurons and other types of cells such as muscle cells. (Contrast with electrical synapse.)

chemiosmosis The formation of ATP in mitochondria and chloroplasts, resulting from a pumping of protons across a membrane (against a gradient of electric charge and of pH), followed by the return of the protons through a protein channel with ATP synthase activity.

chemoautotroph An organism that uses carbon dioxide as a carbon source and obtains energy by oxidizing inorganic substances from its environment. (Contrast with chemoheterotroph, photoautotroph, photoheterotroph.)

chemoheterotroph An organism that must obtain both carbon and energy from organic substances. (Contrast with chemoautotroph, photoautotroph, photoheterotroph.) Most known bacteria and prokaryotic archaea are chemoheterotrophs, as are all animals and fungi and many protists.

chemoreceptor A sensory receptor cell that senses specific molecules (such as CO_2 or O_2 internally or environmental odorant molecules externally).

chi-square goodness-of-fit A statistical test used to assess whether the frequencies of observations in different categories are consistent with a hypothesized frequency distribution.

chiasma (kie az′ muh) (plural: chiasmata) [Gk. cross] An X-shaped connection between paired homologous chromosomes in prophase I of meiosis. A chiasma is the visible manifestation of crossing over between homologous chromosomes.

chitin (kye′ tin) [Gk. *kiton*: tunic] The characteristic tough but flexible organic component of the exoskeleton of arthropods, consisting of a complex, nitrogen-containing polysaccharide. Also found in cell walls of fungi.

chitons *See* Appendix A.

chlamydias *See* Appendix A.

chlorophytes *See* Appendix A.

chloroplast [Gk. *kloros*: green + *plast*: a particle] An organelle bounded by a double membrane containing the enzymes and pigments that perform photosynthesis. Chloroplasts occur only in eukaryotes.

choanocytes (ko′ an uh sites) The collared, flagellated feeding cells of sponges.

chondrichthyans *See* Appendix A.

chordates *See* Appendix A.

chorion (kor′ ee on) [Gk. *khorion*: afterbirth] The outermost of the membranes protecting mammal and reptile (including bird) embryos; in mammals it forms part of the placenta.

chromatin The nucleic acid–protein complex that makes up eukaryotic chromosomes.

chromophore [Gk. *chromo*: color] In plant photoreceptors, the part of the molecule that absorbs visible light.

chromosome arm In linear, eukaryotic chromosomes, the part between a centromere and a telomere.

chytrids *See* Appendix A.

ciliates *See* Appendix A.

cilium (sil′ ee um) (plural: cilia) [L. eyelash] A hairlike organelle used for locomotion or signaling. Generally shorter than a flagellum. (*See* primary cilium, motile cilia.)

circadian clock A biological clock that has a free-running cycle of about 24 hours.

circulatory system A system consisting of a muscular pump (heart), a fluid (blood), and a series of conduits (e.g., blood vessels) that transports materials around the body by mass flow. Often called cardiovascular system.

clade [Gk. *klados*: branch] A monophyletic group made up of an ancestor and all of its descendants.

class (1) In biological classification, a group of related orders. (2) Generally, a group of things defined by particular attributes.

class I MHC proteins Cell surface proteins that are present on the surface of all nucleated cells. They present antigens to cytotoxic T (T_C) cells.

class II MHC proteins Cell surface proteins that are present on the surface of macrophages, B cells, and dendritic cells. They present antigens to T-helper (T_H) cells.

cleavage Early cell divisions of an animal zygote. Cleavage divides the embryo without increasing its mass. (*See* complete cleavage, incomplete cleavage.)

climate The long-term average state of the weather at a given place.

climate change Long-term change in climatic conditions.

climate diagram A way of summarizing the climate of a given location by superimposing graphs of average monthly temperature and precipitation through a year in a way that indicates periods that are favorable or unfavorable for plant growth.

clinal variation [Gk. *klinein*: to lean] Gradual change in the phenotype of a species over a geographic gradient.

clonal analysis A technique that allows one to trace cell lineages—to determine what parts of a plant descended from a parent cell. A genetic mutation that occurs in a single cell results in production of descendent cells that can be identified because they are clones of the original cell—that is, they share the same mutation.

clonal deletion Inactivation or destruction of lymphocyte clones that would produce immune reactions against the animal's own body.

clonal lineages Asexually reproduced groups of nearly identical organisms.

clonal selection A mechanism by which exposure to antigen results in the activation of selected T or B cell clones, resulting in an immune response.

clone [Gk. *klon*: twig, shoot] (1) Genetically identical cells or organisms produced from a common ancestor by asexual means. (2) To produce many identical copies of a DNA sequence by its introduction into, and subsequent asexual reproduction of, a cell or organism.

cloning vector A piece of carrier DNA such as a bacterial plasmid that can be replicated in the host.

closed circulatory system A circulatory system in which the circulating fluid is contained within a continuous system of vessels. (Contrast with open circulatory system.)

closed system A system that is isolated from, and does not interact with, its surroundings. (Contrast with open system.)

club fungi *See* Appendix A.

cnidarians *See* Appendix A.

co-repressor A molecule that binds to a repressor protein, altering its shape so that it binds to DNA and reduces transcription.

co-transport *See* secondary active transport.

coccus (kock′ us) (plural: cocci) [Gk. *kokkos*: berry, pit] Any of various spherical or spheroidal bacteria.

cochlea (kock′ lee uh) [Gk. *kokhlos*: snail] A spiral tube in the inner ear of vertebrates; it contains the mechanosensory cells involved in hearing.

cocurrent gas exchange The exchange of dissolved gases between blood and water in a **cocurrent system** (a system in which the blood and water flow in the same direction along the gas exchange membrane of the animal; sometimes called a concurrent system). Cocurrent exchange can also occur in air-breathers, in which case the exchange is between blood and air.

coding region The nucleotide sequences in a gene that directly specify amino acids in a protein.

coding strand One of the two strands of DNA that for a particular gene specifies the amino acids in a protein. Same base sequence as transcribed RNA, but with Ts instead of Us (compare template strand).

codominance A condition in which two alleles at a locus produce different phenotypic effects and both effects appear in heterozygotes.

codon Three nucleotides in messenger RNA that direct the placement of a particular amino acid into a polypeptide chain. (Contrast with anticodon.)

coelom (see′ loam) [Gk. *koiloma*: cavity] An animal body cavity completely enclosed by mesoderm.

coelomate (see′ lo mate) An animal that contains a body cavity, the coelom, that develops within the mesoderm.

coenocyte (see′ no site) An organism that contains many nuclei within each cell membrane.

coenocytic (seen′ a sit ik) [Gk. *koinos*: common + *kytos*: container] Referring to the condition of "cells" containing many nuclei but enclosed by a single cell membrane. Results from nuclear division without cytokinesis.

cohesion The chemical attraction of the same molecules or substances to one another.

cold-hardening A process by which plants can acclimate to cooler temperatures; requires repeated exposure to cool temperatures over many days.

coleochaetophytes *See* Appendix A.

collagen [Gk. *kolla*: glue] A fibrous protein found extensively in bone and connective tissue.

collenchyma (cull en' kyma) [Gk. *kolla*: glue + *enchyma*: infusion] A type of plant cell, living at functional maturity, which lends flexible support by virtue of primary cell walls thickened at the corners. (Contrast with parenchyma, sclerenchyma.)

commensalism [L. *com*: together + *mensa*: table] A "+/0" interaction between individuals of two different species that benefits one individual (i.e., increases its fitness) but does not affect the other. (Contrast with amensalism, mutualism.)

community The assemblage of interacting individuals of different species within a particular geographic area.

community function Any aspect of the interactions that occur within an ecological community that affects the rate of input of energy and matter into the community and its output of transformed matter and energy.

companion cell In angiosperms, a specialized cell found adjacent to a sieve tube element.

comparative experiment Experimental design in which data from various unmanipulated samples or populations are compared, but in which variables are not controlled or even necessarily identified. (Contrast with controlled experiment.)

comparative genomics Computer-aided comparison of DNA sequences between different organisms to reveal genes with related functions.

compartment One of the interacting components of a system. Applied to the global ecosystem, the different forms and physical locations in which chemical elements or molecules may occur.

complement system A group of eleven proteins that play a role in some reactions of the immune system. The complement proteins are not immunoglobulins.

complementary base pairing The AT (or AU), TA (or UA), CG, and GC pairing of bases in double-stranded DNA, in transcription, and between tRNA and mRNA.

complementary DNA (cDNA) DNA formed by reverse transcriptase acting with an RNA template; an essential intermediate in the reproduction of retroviruses; used as a tool in recombinant DNA technology; lacks introns.

complete cleavage A pattern of cleavage that occurs in eggs that have little or moderate yolk. Early cleavage furrows divide the egg completely. (*See also* radial cleavage, spiral cleavage; contrast with incomplete cleavage.)

complete metamorphosis A change of state during the life cycle of an organism in which the body is almost completely rebuilt to produce an individual with a very different body form. Characteristic of insects such as butterflies, moths, beetles, ants, wasps, and flies.

complex life cycle In reference to parasitic species, a life cycle that requires more than one host to complete.

components The interacting parts of a biological system.

compound A molecule formed by at least two different elements.

compound eye An eye characteristic of arthropods, composed of many individual optical units called ommatidia.

compound leaves Leaves in which several leaflets (repeating leaf–petiole units) are attached to a central stalk. (Contrast with simple leaves.)

computational model A description of a system in which the interactions among components are expressed as mathematical functions that can be used to predict and understand the observable characteristics of the system.

condensation reaction A chemical reaction in which two molecules become connected by a covalent bond and a molecule of water is released ($AH + BOH \rightarrow AB + H_2O$). (Contrast with hydrolysis reaction.)

conditional mutation A mutation that results in a characteristic phenotype only under certain environmental conditions.

conducting airways In the lungs of animals, the airways that do not participate in the exchange of respiratory gases between air and blood, but rather simply conduct air in and out of the lungs. (Contrast with respiratory airways.)

conducting system In the vertebrate heart, a system of specialized muscle cells by which depolarization rapidly spreads from the atrial to the ventricular heart muscle.

cone (1) In conifers, a reproductive structure consisting of spore-bearing scales extending from a central axis. (2) In the vertebrate retina, a type of photoreceptor cell responsible for color vision.

confidence interval A numerical interval that is calculated to contain the true value of some parameter of interest at a stated probability level. For example, a **95% confidence interval** contains the true value of some parameter of interest in 95 percent of cases.

conformational change Change in the shape of a protein.

conformer An animal in which an internal condition (e.g., body temperature) adjusts to match an external condition (e.g., environmental temperature). (Contrast with regulator.)

conformity The state in which an internal condition of an animal (e.g., body temperature) changes to match an external condition (e.g., environmental temperature). (Contrast with regulation.)

conidium (ko nid' ee um) (plural: conidia) [Gk. *konis*: dust] A type of haploid fungal spore borne at the tips of hyphae, not enclosed in sporangia.

conifers *See* Appendix A.

conjugation (kon ju gay' shun) [L. *conjugare*: yoke together] (1) A process by which DNA is passed from one cell to another through a conjugation tube, as in bacteria. (2) A nonreproductive sexual process by which *Paramecium* and other ciliates exchange genetic material.

consensus sequences Short stretches of DNA that appear, with little variation, in many different genes.

conservation ecology The subdiscipline of ecology whose goals are to understand the processes of extinction of populations and species and to devise ways to conserve species and the ecological systems of which they are a part.

***CONSTANS* (*CO*)** A gene in *Arabidopsis thaliana* that codes for a transcription factor (CO) that activates the synthesis of florigen (FT); involved in the induction of flowering.

constant region The portion of an immunoglobulin molecule whose amino acid composition determines its class and does not vary among immunoglobulins in that class. (Contrast with variable region.)

constitutive genes Genes that are expressed all the time. (Contrast with inducible and repressible genes.)

consumption interactions A "+/–" interaction between individuals of two different species in which one individual is harmed by being partly or completely consumed and the other benefits from the nutrition it gains.

continental drift The gradual movements of the world's continents that have occurred over billions of years.

continuous variables Variables that can take on a continuous range of values.

contractile vacuole A specialized vacuole that collects excess water taken in by osmosis, then contracts to expel the water from the cell.

contraction The condition in which a muscle produces force; not necessarily accompanied by shortening.

control mechanism One of four essential elements of a control system, modifies a controlled variable by using information from sensors to determine which effectors to activate and how intense the activation needs to be.

control system A set of cells and organs that work together to determine the value of *x*, where *x* is some attribute of an animal such as body temperature. Composed of four elements: the controlled variable, sensors, effectors, and a control mechanism.

controlled experiment An experiment in which a sample is divided into groups whereby experimental groups are exposed to manipulations of an independent variable while one group serves as an untreated control. The data from the various groups are then compared to see if there are changes in a dependent variable as a result of the experimental manipulation. (Contrast with comparative experiment.)

controlled variable One of four essential elements of a control system; the characteristic of a biochemical system, cell, or animal that is being controlled.

convergent evolution Independent changes to the same state or trait in two or more groups of organisms.

coral bleaching A phenomenon in which corals lose their endosymbionts and become whitened as a result; leads to weakening, and sometimes death, of the coral.

core genome The part of a genome found in all individuals (or strains) within a species.

Coriolis effect The apparent deflection of a moving body as viewed from a rotating frame of reference. In the system of the rotating Earth, an object (such as a packet of atmospheric gases) moving toward the equator will appear to be deflected toward the west from the perspective of an observer on the surface; an object moving toward a pole will appear to be deflected toward the east.

cork cambium [L. *cambiare*: to exchange] In plants, a lateral meristem that produces secondary growth, mainly in the form of waxy-walled protective cells, including some of the cells that become bark.

cornea In a vertebrate, the clear, transparent tissue that covers the forward-facing surface of the eye and allows light to pass through to the retina.

corolla (ko role' lah) [L. *corolla*: a small crown] All of the petals of a flower, collectively.

coronary circulatory system A system of blood vessels that carries blood to and from the heart muscle.

corpus luteum (kor' pus loo' tee um) (plural: corpora lutea) [L. yellow body] A hormone-producing (endocrine) structure formed from reorganization of the cells of an ovarian follicle that has undergone ovulation. In mammals, it secretes progesterone, estrogen, and inhibin. If fertilization occurs, it remains active during pregnancy; if not, it degenerates.

corpus luteum of pregnancy A corpus luteum that functions for an extended period of time during pregnancy. Secretes hormones that are important for maintenance of pregnancy.

corpus luteum of the cycle A corpus luteum that functions for a short time (e.g., about 10 days in humans) and then undergoes internally programmed degeneration.

correlation coefficient A measure of the strength of relationship between two quantitative variables, ranging from –1 (a perfect negative relationship) to 1 (a perfect positive relationship).

cortical microtubules Cytoskeletal proteins (microtubules) that are located just inside the cell membrane.

cost–benefit approach An approach to ecological and evolutionary analysis that posits that an animal has a limited amount of time and energy to devote to its activities, and accordingly activities will occur only if benefits exceed costs. (*See also* trade-off.)

cotyledon (kot' ul lee' dun) [Gk. *kotyledon*: hollow space] A "seed leaf." An embryonic organ that stores and digests reserve materials; may expand when the seed germinates.

countercurrent gas exchange The exchange of dissolved gases between blood and water in a **countercurrent system** (a system in which the blood and water flow in opposite directions along the gas exchange membrane of the animal). Countercurrent exchange can also occur in air-breathers, in which case the exchange is between blood and air.

countercurrent heat exchange The exchange of heat that occurs between two closely juxtaposed fluid streams that are flowing in opposite directions (e.g., blood flowing in opposite directions in two juxtaposed blood vessels).

coupled transport *See* secondary active transport.

covalent bond A chemical bond based on the sharing of electrons between two atoms.

CpG islands DNA regions rich in C nucleotides adjacent to G nucleotides. Especially abundant in promoters, these regions are where methylation of cytosine usually occurs.

crenarchaeotes *See* Appendix A.

CRISPR-Cas9 gene editing A technique that leverages an antiviral system found in prokaryotes. A Cas9 nuclease is used to cut DNA at particular sequences (chosen by the scientist) that are complementary to a guide RNA, which has a similar structure to RNAs produced at the CRISPR (clustered regularly interspaced short palindromic repeats) gene.

crocodilians *See* Appendix A.

cross-bridges During muscle contraction, the links formed when the globular heads of myosin filaments bind to specific sites on actin filaments. In this state, the myosin heads heads flex to produce contraction.

cross-fertilization The mating of one parent with another. (Contrast self-fertilization.)

crossing over The mechanism by which linked genes undergo recombination. In general, the term refers to the reciprocal exchange of corresponding segments between two homologous chromatids.

crustaceans *See* Appendix A.

crustose A growth form of organisms, such as lichens, in which the organism forms a thin, close, tight bond with the surface of a rock, tree, or other object.

cryptic species Species that are morphologically indistinguishable but do not interbreed.

ctene (teen) [Gk. *cteis*: comb] In ctenophores, a comblike row of cilia-bearing plates. Ctenophores move by beating the cilia on their eight ctenes.

ctenophores *See* Appendix A.

current (1) Flow of a fluid (gas or liquid), for example, in ocean waters or in the atmosphere. (2) The flow of electrical charge.

cuticle (1) In plants, a waxy layer on the outer body surface that retards water loss. (2) In ecdysozoans, an outer body covering that provides protection and support and is periodically molted.

cyanobacteria *See* Appendix A.

cycads *See* Appendix A.

cyclin A protein that activates a cyclin-dependent kinase, bringing about transitions in the cell cycle.

cyclin-dependent kinase (CDK) A protein kinase whose target proteins are involved in transitions in the cell cycle and which is active only when complexed with additional protein subunits, called cyclins.

cytokine A regulatory protein made by immune system cells that affects other target cells in the immune system.

cytokinesis (sy′ toe kine ee′ sis) [Gk. *kytos*: container + *kinein*: to move] The division of the cytoplasm of a dividing cell. (Contrast with mitosis.)

cytoplasm The contents of the cell, excluding the nucleus.

cytoplasmic determinates Materials in the cytoplasm, the spatial distribution of which may determine such things as embryonic axes and cell determination.

cytoplasmic streaming A form of amoeboid movement in which a fluid cytoplasm moves and stretches the organism's body in new directions.

cytoskeleton The network of microtubules, intermediate filaments, and microfilaments that gives a eukaryotic cell its shape and its capacity to arrange its organelles and to move.

cytosol The fluid portion of the cytoplasm, excluding organelles and other solids.

cytotoxic T (T_C) cells Cells of the cellular immune system that recognize and directly eliminate virus-infected cells. (Contrast with T-helper cells.)

D

Dalton (Da) A measure of mass for atoms and atomic particles equal to the atomic mass unit.

data Quantified observations about a system under study.

decomposer An organism that obtains the energy and materials it needs to survive, grow, and reproduce from the waste products or dead bodies of other organisms.

defensin A type of protein made by phagocytes that kills bacteria and enveloped viruses by insertion into their cell membranes.

defensive protein A protein that recognizes and responds to substances or particles that invade an organism from the environment.

delayed implantation Embryonic diapause in placental mammals. A state of arrested embryonic development that occurs after an embryo has arrived in the uterus but before it implants in the uterine wall. Occurs only in certain species (not humans).

deletion A mutation resulting from the loss of a continuous segment of a gene or chromosome. Such mutations almost never revert to wild type. (Contrast with duplication, point mutation.)

demethylase An enzyme that catalyzes the removal of the methyl group from cytosine, reversing DNA methylation.

demography [Gk. *demos*: population + *graphia*: description] The quantitative study of the processes that determine the structure of populations (e.g., the relative proportions of individuals of different sizes, ages, stages, and genders), the size of populations, and the changes in these attributes.

denaturation Loss of activity of an enzyme or a nucleic acid molecule as a result of structural changes induced by heat or other means.

dendrite [Gk. *dendron*: tree] The receptive element of most neurons, which receives synaptic input from other neurons. Usually much branched and relatively short compared with the axon.

denitrification The chemical process by which "fixed" (biologically useable) nitrogen is converted back into N_2 gas. (Contrast with nitrogen fixation.)

denitrifiers Bacteria that release nitrogen to the atmosphere as nitrogen gas (N_2).

density-dependent Pertaining to any factor or interaction that varies with density. For example, individual survival and reproductive success often change as population density changes.

deoxygenation The release of O_2 from a combined state with a respiratory pigment such as hemoglobin.

deoxyribonucleic acid *See* DNA.

depolarization A change in the membrane potential across a cell membrane so that the inside of the membrane becomes less negative, or even positive, compared with the outside of the membrane. When this type of change in membrane potential occurs, the membrane is said to be depolarized. (Contrast with hyperpolarization.)

derived trait A trait that differs from the ancestral trait. (Contrast with synapomorphy.)

dermal tissue system The outer covering of a plant, consisting of epidermis in the young plant and periderm in a plant with extensive secondary growth. (Contrast with ground tissue system, vascular tissue system.)

descent with modification Darwin's premise that all species share a common ancestor and have diverged from one another gradually over time.

descriptive statistics Quantitative measures that describe general patterns in data.

desmosome (dez′ mo sowm) [Gk. *desmos*: bond + *soma*: body] An adhering junction between animal cells.

determinate Having a fixed number or size; often used in reference to the growth of an organism in which the number and approximate size of the parts, such as arms or leaves, can be predicted before birth or germination. (Contrast with indeterminate.)

determination In development, the process whereby the fate of an embryonic cell or group of cells (e.g., to become epidermal cells or neurons) is set ("determined").

deuterostomes *See* Appendix A.

development The process by which a multicellular organism, beginning with a single cell, goes through a series of changes, taking on the successive forms that characterize its life cycle.

diaphragm (dye′ uh fram) [Gk. *diaphrassein*: barricade] (1) A sheet of skeletal muscle that separates the thoracic and abdominal cavities, found only in mammals; helps drive air flow during breathing. (2) A birth control device in which a sheet of rubber is fitted over the woman's cervix, blocking the entry of sperm.

diastole (dye ass′ toll ee) [Gk. dilation] The portion of the cardiac cycle when the heart muscle relaxes. (Contrast with systole.)

diatoms *See* Appendix A.

dichotomous (dye cot′ oh mus) [Gk. *dichot*: split in two + *tomia*: removed] A branching pattern in which the shoot divides at the apex, producing two equivalent branches that subsequently never overlap.

differentiation The process whereby originally similar cells follow different developmental pathways (become "differentiated"); the actual expression of determination.

diffusion Random movement of molecules or other particles, resulting in even distribution of the particles when no barriers are present.

digestion The breakdown (by hydrolysis) of ingested food molecules into smaller chemical components that an animal is capable of absorbing and distributing to the tissues of its body.

digestive vacuole In protists, an organelle specialized for digesting food ingested by endocytosis.

dihybrid cross A mating in which the parents differ with respect to the alleles of two loci of interest.

Dikarya *See* Appendix A.

dikaryon (di care′ ee ahn) [Gk. *di*: two + *karyon*: kernel] A cell or organism carrying two genetically distinguishable nuclei. Common in fungi.

dinoflagellates *See* Appendix A.

dinosaurs *See* Appendix A.

dioecious (die eesh′ us) [Gk. *di*: two + *oikos*: house] Pertaining to organisms in which the two sexes are "housed" in two different individuals, so that eggs and sperm are not produced in the same individuals.

Examples include humans, fruit flies, date palms. (Contrast with monoecious.)

diploblastic Having two cell layers. (Contrast with monoblastic, triploblastic.)

diploid (dip′ loid) [Gk. *diplos*: double] Having a chromosome complement consisting of two copies (homologs) of each chromosome. Designated 2*n*. (Contrast with haploid.)

diplomonads *See* Appendix A.

diplontic Referring to a life cycle in which the organism is diploid and the gametes are the only haploid stage.

dipole A separation of opposite electric charges, as in a polar bond.

dipole–dipole interactions An attraction or repulsion of similar or opposite electric charges respectively between two dipoles.

direct development A pattern of development in which the young resemble the adults. (Contrast with indirect development.)

directional selection Selection in which phenotypes at one extreme of the population distribution are favored. (Contrast with disruptive selection, stabilizing selection.)

disaccharide A carbohydrate made up of two monosaccharides (simple sugars).

discrete traits Traits that can take only a few values and often determined by one or a few genes. (Contrast with quantitative traits.)

discrete variables Quantitative variables that take on only whole number values.

dispersal corridor A continuous strip of habitat that connects larger pieces of habitat, thus facilitating movement of organisms between them.

disruptive selection Selection in which phenotypes at both extremes of the population distribution are favored. (Contrast with directional selection, stabilizing selection.)

disturbance Events that cause sudden change in the environment and in ecological systems as a result. Examples include fire, landslides, floods, strong storms, and volcanic eruptions, as well as less obvious events such as the death and fall of a large tree that opens up the forest canopy.

DNA (deoxyribonucleic acid) The fundamental hereditary material of all living organisms. In eukaryotes, stored primarily in the cell nucleus. A nucleic acid using deoxyribose rather than ribose.

DNA binding domain The region in a protein, especially transcription factors, that is able to bind to a particular sequence of a DNA molecule.

DNA helicase An enzyme that catalyzes the unwinding of a nucleic acid double helix.

DNA ligase An enzyme that unites broken DNA strands during replication and recombination.

DNA methyltransferase An enzyme that catalyzes the methylation of DNA.

DNA microarray A small glass or plastic square onto which thousands of single-stranded DNA sequences are fixed so that hybridization of cell-derived RNA or DNA to the target sequences can be performed.

DNA replication The creation of a new strand of DNA in which DNA polymerase catalyzes the exact reproduction of an existing (template) strand of DNA.

DNA segregation The separation of two DNAs formed by replication into two new cells during cell division.

DNA transposons Mobile genetic elements that move without making an RNA intermediate. (Contrast with retrotransposons.)

domain (1) An independent structural element within a protein. Encoded by recognizable nucleotide sequences, a domain often folds separately from the rest of the protein. Similar domains can appear in a variety of different proteins across phylogenetic groups (e.g., homeobox domain; calcium-binding domain). (2) In phylogenetics, the three monophyletic branches of life (Bacteria, Archaea, and Eukarya).

dominant In genetics, an allele that determines the phenotype in an individual who is heterozygous for it and a different (recessive) allele. (Contrast with recessive.)

dorsal [L. *dorsum*: back] Toward or pertaining to the back or upper surface. (Contrast with ventral.)

double fertilization In angiosperms, a process in which the nuclei of two sperm fertilize one egg. One sperm's nucleus combines with the egg nucleus to produce a zygote, while the other combines with the same egg's two polar nuclei to produce the first cell of the triploid endosperm (the tissue that will nourish the growing plant embryo).

doubling time The time over which an ecological population doubles in size (number of individuals). Populations growing multiplicatively with a constant value of *r* have a constant doubling time, no matter what their current size is.

dry-nosed primates *See* Appendix A.

duplication A mutation in which a segment of a chromosome is duplicated, often by the attachment of a segment lost from its homolog. (Contrast with deletion.)

dynamic [Gk. *dunamikos*: powerful] Characterized by activity or change.

dynamic instability The co-occurrence of addition (assembly) and removal (disassembly) of subunits from microfilaments and microtubules.

E

ecdysozoans *See* Appendix A.

echinoderms *See* Appendix A.

ecological efficiency The transfer of energy from one trophic level to the next, often estimated as the ratio of dry biomass at the higher level to dry biomass at the lower level. Note that this use of "efficiency" as the ratio of energy output to input is similar to the measure of efficiency used for many human-made machines.

ecological system One or more organisms and the abiotic and biotic environment with which they interact.

ecology [Gk. *oikos*: household + *logia*: study of] Defined by the German biologist Ernst Haeckel in 1866 as "the entire science of the relations of the organism to its external environment," both its biotic and abiotic components.

ecosystem In general, shorthand for "ecological system." Often used more specifically to refer to an ecological community and its abiotic environmental context.

ecosystem function Any aspect of the interactions that occur within an ecosystem that affects the rate of input of energy and matter into the ecosystem and its output of transformed matter and energy.

ecosystem services Ecosystem outputs that benefit humans.

ectoderm [Gk. *ektos*: outside + *derma*: skin] The outermost of the three embryonic germ layers first delineated during gastrulation. Gives rise to the skin, sense organs, and nervous system.

ectomycorrhizae Mutualistic fungi that cover the roots of plants and assist in the uptake of water and minerals from the soil by the plant. (Contrast with arbuscular mycorrhizae.)

ectotherm [Gk. *ektos*: outside + *thermos*: heat] *See* poikilotherm.

effector One of four essential elements of a control system. A tissue, organ, or cell that performs actions that modify the level of the controlled variable ("carries out orders"), under the direction of the control mechanism.

effector cell In cellular immunity, B cells and T cells that attack an antigen, either by secreting antibodies that bind to the antigen or by releasing molecules that destroy any cell bearing the antigen.

efferent neurons In a nervous system, nerve cells that carry commands from the central nervous system to physiological and behavioral effectors such as muscles and glands.

egg In all sexually reproducing organisms, the female gamete (also called an ovum). In reptiles (including birds) and some other animals, it also refers to a structure within which the early embryonic development of a fertilized ovum occurs. (*See also* amniote egg.)

egg cell The female gamete. Can be fertilized by a sperm to form a zygote.

electrical synapse A type of synapse at which there is direct continuity between the cytoplasms of the presynaptic and postsynaptic neurons, meaning action potentials spread directly from presynaptic cell to postsynaptic cell. (Contrast with chemical synapse.)

electrocardiogram (ECG or **EKG)** A graphic recording of differences of electrical potential that are measured on the body surface but originate from the heart.

electrochemical equilibrium When ions diffuse, their diffusion is driven simultaneously by both concentration differences and electric-charge differences, i.e., by chemical and electrical differences. The equilibrium state in such diffusion is termed an electrochemical equilibrium to recognize this dual nature of the forces driving diffusion.

electron One of the three fundamental subatomic particles of matter (along with neutrons and protons), outside the atomic nucleus carrying a negative charge and very little mass.

electron shell The region surrounding the atomic nucleus at a fixed energy level in which electrons orbit.

electron transport The passage of electrons through a series of proteins with a release of energy which may be captured in a concentration gradient or in chemical form such as NADH or ATP.

electronegativity The tendency of an atom to attract electrons when it occurs as part of a compound.

element A substance consisting of only one kind of atom that cannot be converted to another substance by ordinary chemical means.

elongation (1) In molecular biology, the addition of monomers to make a longer DNA, RNA or protein during replication, transcription or translation. (2) Growth of a plant axis or cell primarily in the longitudinal direction.

embryo [Gk. *en*: within + *bryein*: to grow] A young animal, or young plant sporophyte, while it is still contained within a protective structure such as a seed, egg, or uterus.

embryo sac In angiosperms, the female gametophyte. Found within the ovule, it consists of eight or fewer cells, membrane bounded, but without cellulose walls between them.

embryonic development The stages by which a fertilized egg is transformed into an embryo and its subsequent development.

embryonic diapause A programmed state of arrested or profoundly slowed embryonic development. (*See also* delayed implantation.)

embryonic stem cell (ESC) A totipotent cell from a very early cleavage stage embryo or a pluripotent cell from the inner cell mass of a blastocyst.

embryophytes *See* Appendix A.

endergonic Referring to a chemical reaction in which the free energy is positive such that the products have higher energy than the reactants, thereby requiring energy input to occur. (Contrast with exergonic.)

endocrine cells Cells that secrete hormones into the blood. (*See also* endocrine gland.)

endocrine gland (en' doh krin) [Gk. *endo*: within + *krinein*: to separate] A gland or tissue, lacking ducts, in which endocrine cells secrete hormones into the blood. (Contrast with exocrine gland.)

endocrine signaling Cell-to-cell signaling involving a chemical signal (hormone) produced by specialized cells that is carried by bulk flow to target cells.

endocytosis A process by which liquids or solid particles are taken up by a cell through invagination of the cell membrane. (Contrast with exocytosis.)

endoderm [Gk. *endo*: within + *derma*: skin] The innermost of the three embryonic germ layers delineated during gastrulation. Gives rise to the digestive and respiratory tracts and structures associated with them.

endodermis In plants, a specialized cell layer in roots and some stems that lies exterior to the pericycle (outside the stele). Frequently a barrier to free diffusion of solutes.

endogenous retrovirus A DNA sequence derived from a retrovirus that persists in and is transmitted as part of the genome of a nonviral organism. Some endogenous retroviruses can become active and produce new viral infections.

endomembrane system A system of intracellular membranes that exchange material with one another, consisting of the nuclear envelope, Golgi apparatus, endoplasmic reticulum, and lysosomes when present.

endometrium The nurturing inner tissue layer of the uterus of a mammal. Embryos implant in the endometrium, which later helps form the placenta.

endophytic fungi Fungi that live within the aboveground portions of plants without causing obvious harm to the host plant.

endoplasmic reticulum (ER) [Gk. *endo*: within + L. *reticulum*: net] A system of membranous tubes and flattened sacs found in the cytoplasm of eukaryotes. Exists in two forms: rough ER, studded with ribosomes; and smooth ER, lacking ribosomes.

endoskeleton [Gk. *endo*: within + *skleros*: hard] An internal skeleton covered by other, soft body tissues. (Contrast with exoskeleton.)

endosperm [Gk. *endo*: within + *sperma*: seed] A specialized triploid seed tissue found only in angiosperms; contains stored nutrients for the developing embryo.

endospore [Gk. *endo*: within + *spora*: to sow] In some bacteria, a resting structure that can survive harsh environmental conditions.

endosymbiosis [Gk. *endo*: within + *sym*: together + *bios*: life] The engulfing of one cell by another.

endotoxin A lipopolysaccharide that forms part of the outer membrane of certain Gram-negative bacteria that is released when the bacteria grow or lyse. (Contrast with exotoxin.)

endurance exercise Exercise that consists of many repetitions of relatively low-intensity muscular actions over long periods of time. Also termed endurance training. (Contrast with resistance exercise.)

energy The capacity to create and maintain organization. More narrowly defined, the capacity do work or move matter against an opposing force.

enhancers Regulatory DNA sequences that bind transcription factors that either activate or increase the rate of transcription.

enteric division One of three divisions of the vertebrate autonomic nervous system; composed of neurons located in the smooth muscle layers of the gut.

entrainment The process of synchronizing an organism's internal circadian clock with a 24-hour rhythm in the organism's environment, usually the 24-hour light–dark cycle.

entropy (*S*) (en' tro pee) [Gk. *tropein*: to change] A measure of the degree of disorder in any system. Spontaneous reactions in a closed system are always accompanied by an increase in entropy.

environmental genomics Sequencing technique used when biologists are unable to work with the whole genome of a prokaryote species and instead examine individual genes collected from a random sample of the organism's environment.

environmental sex determination A mechanism—seen, for example, in certain nonavian reptiles—in which the sex of an individual is determined by the temperature it experiences during embryonic development.

enzyme (en′ zime) [Gk. *zyme*: to leaven (as in yeast bread)] A catalytic protein that speeds up a biochemical reaction.

enzyme–substrate complex (ES) An intermediate in an enzyme-catalyzed reaction; consists of the enzyme bound to its substrate(s).

eons The primary division of geological time. The history of Earth is divided into four eons: Hadeon, Archean, Proterozoic, and Phanerozoic.

epigenetics The scientific study of changes in the expression of a gene or set of genes that occur without change in the DNA sequence.

epiphytes [Gk. *Epi-*: upon] A type of plant that grows on a tree or other plant. Being higher up in a forest canopy can be advantageous, but these plants remain photosynthetic and are not otherwise parasitic.

epistasis Interaction between genes in which the presence of a particular allele of one gene determines how another gene will be expressed.

epithelium (plural: epithelia) A sheet of cells that lines a body cavity or covers an organ or body surface; one of the major tissue types in multicellular animals. Termed a simple epithelium when the sheet of cells is only one cell thick.

epitope *See* antigenic determinant.

epoch A subdivision of geological time that divides geological periods.

equatorial countercurrent West-to-East-flowing ocean currents at the Equator between the flanking east-to-west-flowing North and South Equatorial currents in the Pacific, Atlantic, and Indian Oceans.

equilibrium potential of an ion (E_{ion}) The membrane potential at which an ion is at electrochemical equilibrium, i.e., at which the concentration and electrical effects on diffusion of the ion are balanced so there is no net diffusion of the ion across the membrane.

equilibrium The state of a system in which there is no net change through time.

erythrocyte (ur rith′ row site) [Gk. *erythros*: red + *kytos*: container] *See* red blood cell.

essential amino acids Amino acids that an animal cannot synthesize for itself and must obtain from its food.

essential element A mineral nutrient required for normal growth and reproduction in plants and animals.

essential fatty acids Fatty acids that an animal cannot synthesize for itself and must obtain from its food.

essential mineral *See* essential nutrient.

essential nutrient A nutrient (e.g., amino acid, fatty acid, vitamin, mineral) that an animal cannot synthesize for itself and must obtain from its food.

ester linkage A condensation (water-releasing) reaction in which the carboxyl group of a fatty acid reacts with the hydroxyl group of an alcohol. Lipids, including most membrane lipids, are formed in this way. (Contrast with ether linkage.)

estrogen Any of several feminizing steroid sex hormones; produced chiefly by the ovaries (specifically, somatic cells of ovarian follicles) in mammals, and by the placenta during pregnancy.

estrous cycle A cycle of behavioral readiness to copulate (called estrus and correlated with ovulation) in most female mammals.

estrus (es′ trus) [L. *oestrus*: frenzy] The period of heat, or maximum sexual receptivity, observed in most female mammals at times correlated with ovulation.

ether linkage The linkage of two hydrocarbons by an oxygen atom (HC—O—CH). Ether linkages are characteristic of the membrane lipids of the Archaea. (Contrast with ester linkage.)

ethylene One of the plant growth hormones, the gas $H_2C{=}CH_2$. Involved in fruit ripening and other growth and developmental responses.

euchromatin Diffuse, uncondensed chromatin. Contains active genes that will be transcribed into mRNA. (Contrast with heterochromatin.)

eudicots *See* Appendix A.

euglenids *See* Appendix A.

eukaryotes *See* Appendix A.

eukaryotic cells Cells that contain (or used to contain) a nucleus.

euphyllophytes *See* Appendix A.

euryarchaeotes *See* Appendix A.

eusociality A type of social group in which some adults are infertile and aid the reproduction of other, fertile adults.

eutherians *See* Appendix A.

eutrophication (yoo trofe′ ik ay′ shun) [Gk. *eu*: truly + *trephein*: to flourish] Responses of aquatic ecosystems to excessive nutrient input, which include greatly increased primary productivity and resulting depletion of dissolved oxygen.

evolution Any gradual change. Most often refers to organic or Darwinian evolution, which is the genetic and resulting phenotypic change in populations of organisms from generation to generation. (Contrast with speciation.)

evolutionary radiation The rapid proliferation of many species within a single evolutionary lineage.

evolutionary reversal The reappearance of an ancestral trait in a group that had previously acquired a derived trait.

evolutionary theory The understanding and application of the mechanisms of evolutionary change to biological problems.

excavates *See* Appendix A.

excitable cell A cell that can generate and propagate action potentials because its cell membrane contains voltage-gated ion channels. Neurons and muscle cells are the two types of excitable cells in animals.

excitation In the study of muscle, the event in which a nerve impulse arrives at a neuromuscular junction and initiates an action potential that propagates over the cell membrane (and into the transverse tubules) of a muscle cell.

excitation–contraction coupling In a muscle cell, the process by which electrical excitation of the cell membrane leads to contractile activity by the contractile proteins (actin and myosin) in the cell. Key events occur at the transverse tubules and sarcoplasmic reticulum, and result in release of Ca^{2+} into the cytoplasm.

excitatory postsynaptic potential (EPSP) At a chemical synapse (specifically, an excitatory synapse), a graded membrane depolarization produced in the postsynaptic cell by arrival of an impulse in the presynaptic cell. (Contrast with inhibitory postsynaptic potential.)

exergonic Referring to a chemical reaction in which the free energy is negative, such that the products of the reaction have lower energy than the reactants, thereby releasing energy. (Contrast with endergonic.)

exocrine gland (eks′ oh krin) [Gk. *exo*: outside + *krinein*: to separate] A gland that has outflow ducts and that produces secretions that exit by these ducts, such as tear glands, sweat glands, mammary glands (breasts), and the prostate gland in males. (Contrast with endocrine gland.)

exocytosis A process by which a vesicle within a cell fuses with the cell membrane and releases its contents to the outside. (Contrast with endocytosis.)

exon A portion of a DNA molecule, in eukaryotes, that is present in the mature mRNA and codes for part of a polypeptide. (Contrast with intron.)

exoskeleton [Gk. *exos*: outside + *skleros*: hard] A hard covering on the outside of the body to which muscles are attached. (Contrast with endoskeleton.)

exotoxin A highly toxic, usually soluble protein released by living, multiplying bacteria. (Contrast with endotoxin.)

expression vector A DNA vector, such as a plasmid, that carries a DNA sequence for the expression of an inserted gene into mRNA and protein in a host cell.

external environment The conditions of the surroundings outside an animal's body.

external fertilization Fertilization that occurs after release of sperm and eggs into the external environment; typical of aquatic animals. Also called spawning. (Contrast with internal fertilization.)

external respiration *See* breathing.

extracellular fluid The aqueous solution (body fluid) outside cells (in the extracellular compartment). In animals with a closed circulatory system, subdivided into blood plasma and interstitial fluid (tissue fluid).

extracellular matrix A material of heterogeneous composition surrounding cells and performing many functions, including adhesion of cells.

extraembryonic membranes Four membranes that support but are not part of the developing embryos of reptiles (including birds) and mammals, defining these groups phylogenetically as amniotes. (*See* amnion, allantois, chorion, and yolk sac.)

extreme halophiles A group of euryarchaeotes that live exclusively in very salty environments.

extremophiles Archaea and bacteria that live and thrive under conditions (e.g., extremely high temperatures) that would kill most organisms.

exudates In plants, any compounds that are exuded (secreted) by the plant. For example, compounds secreted into the soil environment by plant roots that influence the type and abundance of organisms living near the roots.

F

F_1 The first filial generation; the immediate progeny of a parental (P) mating.

F_2 The second filial generation; the immediate progeny of a mating between members of the F_1 generation.

facilitated diffusion Passive movement through a membrane involving a specific integral membrane protein; does not proceed against a concentration gradient. (Contrast with active transport, diffusion.)

facultative anaerobe A prokaryote that can shift its metabolism between anaerobic and aerobic modes depending on the presence or absence of O_2.

facultative parasite An organism that can parasitize other living organisms but is also capable of growing independently.

fasciated Term applied to a plant stem whose shape is unusually wide and thin (ribbon-like) rather than cylindrical; also used to describe individual flowers or roots that have unusually elongated shapes.

fast glycolytic cells Skeletal muscle cells that are highly dependent on ATP production by anaerobic glycolysis and that contract and develop high tension rapidly, but fatigue quickly. Also called fast-twitch cells or white muscle cells. (Contrast with slow oxidative cells.)

fate map A diagram of the blastula showing which cells (blastomeres) are "fated" to contribute to specific tissues and organs in the mature body.

fatty acid A molecule made up of a long nonpolar hydrocarbon chain and a polar carboxyl group. Found in many lipids.

fauna All the animals found in a given area or time. (Contrast with flora.)

fecundity The average number of female offspring produced by an individual female during a particular life stage, age, or period of time.

feedback In regulatory systems, information about the relationship between the set point of the system and its current state. (Contrast with feedforward information.)

feedback inhibition A mechanism for regulating a metabolic pathway in which the end product of the pathway can bind to and inhibit the enzyme that catalyzes the first committed step in the pathway. Also called end-product inhibition.

female gametophyte A haploid organism that divides by mitosis to give rise to several types of cells including the egg cell.

fermentation [L. *fermentum*: yeast] (1) Speaking specifically about energy metabolism, the partial catabolism of glucose when oxygen is unavailable, resulting in the production of lactic acid or alcohol, with no involvement of electron transport. (2) Speaking generally, metabolic processes that occur in the absence of O_2.

ferns *See* Appendix A.

fertilization The union of gametes.

fetus A medical and legal term for the stages of a developing human embryo from about the eighth week of pregnancy (the point at which all major organ systems have formed) to the moment of birth.

fiber In angiosperms, an elongated, tapering sclerenchyma cell, usually with a thick cell wall, that serves as a support function in xylem. (*See also* muscle fiber.)

Fick's law An equation that describes the factors that determine the rate of diffusion of molecules from place to place.

filament In flowers, the part of a stamen that supports the anther.

firmicutes *See* Appendix A.

first filial generation *See* F_1.

first law of thermodynamics The principle that energy can be neither created or destroyed.

fission (1) A type of asexual reproduction in animals, in which an individual breaks into two or more pieces and each piece then grows into a new individual. *See* asexual reproduction. (2) Reproduction of a prokaryote by division of a cell into two comparable progeny cells.

fitness An individual's contribution of genes to the next generation, as a consequence of its success in surviving and reproducing.

fixed The condition in which a particular allele is the only allele present at a genetic locus within a population (i.e., there is no genetic variation at the locus).

fixed action pattern In animal behavior, a genetically determined behavior that is performed without prior learning in a stereotyped way, and that is resistant to modification by learning.

flagellum (fla jell' um) (plural: flagella) [L. *flagellum*: whip] A long, whiplike appendage that propels cells. Prokaryotic flagella differ sharply from those found in eukaryotes.

flatworms *See* Appendix A.

flavin adenine dinucleotide (FAD/FADH2) A compound used in redox reactions to transport high-energy electrons. In carbohydrate catabolism, electrons are picked up by FAD^+ in the citric acid cycle to give $FADH_2$, and transported to the electron transport chain.

flora All of the plants found in a given area or time. (Contrast with fauna.)

floral meristem In angiosperms, a meristem that forms the floral organs (sepals, petals, stamens, and carpels).

florigen A plant hormone involved in the conversion of a vegetative shoot apex to a flower. Also called FLOWERING LOCUS T (FT).

flower The sexual structure of an angiosperm.

FLOWERING LOCUS T (FT) A gene in *Arabidopsis thaliana* that codes for the protein FT which acts as a plant hormone involved in the conversion of a vegetative shoot apex to a flower. FT is also called florigen.

fluid mosaic model A molecular model for the structure of biological membranes consisting of a fluid phospholipid bilayer in which suspended proteins are free to move in the plane of the bilayer.

flux The rate of movement of energy, or of matter (an element or a molecular compound), between two compartments of an ecosystem, or between the Earth system and space.

foliose Having a leafy growth form.

follicle-stimulating hormone (FSH) A hormone, produced by the anterior pituitary gland, that helps control secretion of hormones by the gonads; a gonadotropin. Also involved in supporting gamete production in both sexes.

food web A diagrammatic representation of which organisms consume which other organisms in an ecological community, i.e., a diagram of trophic relationships.

foot One of the three main body parts of a mollusk.

foraminiferans *See* Appendix A.

forebrain The region of the vertebrate brain that comprises the cerebrum, thalamus, and hypothalamus.

forward genetics An approach that begins with an interesting phenotype, finds the gene(s) underlying it, and then determines as much as possible about those genes.

fossil fuels Combustible materials, such as coal, natural gas, petroleum, and peat, that are formed by geological processes over long periods of time from the remains of living organisms.

founder effect Random changes in allele frequencies resulting from establishment of a population by a very small number of individuals.

fragmentation In ecology, the subdivision of a larger piece of habitat into smaller pieces that are isolated from one another by a different habitat.

frame-shift mutation The addition or deletion of a single or two adjacent nucleotides in a gene's sequence. Results in the misreading of mRNA during translation and the production of a nonfunctional protein. (Contrast with missense mutation, nonsense mutation, loss-of-stop, silent mutation.)

free energy (or Gibbs free energy) is the change in the total amount of energy that is available to perform work in a chemical reaction. (*See* endergonic and exergonic.)

free-running A descriptor referring to an endogenous biological clock that is not synchronized with (entrained by) an environmental cue.

frequency distribution A figure that displays the frequency of different classes of data.

frequency-dependent selection Selection that changes in intensity with the proportion of individuals in a population having the trait.

fruit In angiosperms, a ripened and mature ovary (or group of ovaries) containing the seeds. Sometimes applied to reproductive structures of other groups of plants.

fruticose Having a shrubby growth form.

functional genomics The assignment of functional roles to the proteins encoded by genes identified by sequencing entire genomes.

functional group A characteristic combination of atoms that contributes specific properties when attached to larger molecules.

G

G protein-coupled receptors (GPCRs) A class of receptors that change configuration upon ligand binding such that a G protein binding site is exposed on the cytosolic region of the receptor, initiating a signal transduction pathway.

G0 A resting phase in which a cell is not preparing for the cell division cycle.

G1 phase In the cell cycle, the gap between the end of mitosis and the onset of S phase.

G2 phase In the cell cycle, the gap between S phase and the onset of mitosis.

gain-of-function mutation A mutation that results in a protein with a new function. (Contrast with loss-of-function mutation.)

gametangium (gam uh tan′ gee um) (plural: gametangia) [Gk. *gamos*: marriage + *angeion*: vessel] Any plant or fungal structure within which a gamete is formed.

gamete (gam′ eet) [Gk. *gamete/gametes*: wife, husband] The mature sexual reproductive cell: the egg or the sperm.

gametogenesis (ga meet′ oh jen′ e sis) The specialized series of cellular divisions, including meiosis, that leads to the formation of gametes.

gametophyte (ga meet′ oh fyte) In plants and photosynthetic protists with alternation of generations, the multicellular haploid phase that produces the gametes. (Contrast with sporophyte.)

ganglion (gang′ glee un) (plural: ganglia) [Gk. tumor] (1) In a vertebrate, a discrete cluster of neuron cell bodies in the peripheral nervous system. (2) In an arthropod, an enlarged part of the ventral nerve cord found in each body segment.

gap genes In *Drosophila* (fruit fly) development, segmentation genes that define broad areas along the anterior–posterior axis of the early embryo. Part of a developmental cascade that includes maternal effect genes, pair rule genes, segment polarity genes, and Hox genes.

gap junction A 2.7-nanometer gap between cell membranes of two animal cells, spanned by protein channels. Gap junctions allow chemical substances or electric signals to pass from cell to cell.

gas exchange membranes Thin sheets of tissue (typically one or two cells thick) where respiratory gases move between the internal tissues of an animal and the animal's environmental medium.

gastropods *See* Appendix A.

gastrovascular cavity A body cavity that serves for both digestion ("gastro") and circulation ("vascular"); in particular, the central cavity of the body of jellyfishes and other cnidarians.

gastrulation The process by which blastomeres move relative to one another, and from the exterior to the interior, resulting in the three germ layers of the embryo.

gel electrophoresis (e lek′ tro fo ree′ sis) [L. *electrum*: amber + Gk. *phorein*: to bear] A technique for separating molecules (such as DNA fragments) from one another on the basis of their electric charges and molecular weights by applying an electric field to a gel.

gene [Gk. *genes*: to produce] A unit of heredity. Used here as the unit of genetic function which carries the information for a polypeptide or RNA.

gene duplication A way that genomes can acquire new functions.

gene family A set of similar genes derived from a single parent gene; need not be on the same chromosomes. The vertebrate globin genes constitute a classic example of a gene family.

gene flow Exchange of genes between populations through migration of individuals or movements of gametes.

gene pool All of the different alleles of all of the genes existing in all individuals of a population.

gene therapy Treatment of a genetic disease by providing patients with cells containing functioning alleles of the genes that are nonfunctional in their bodies.

gene tree A graphic representation of the evolutionary relationships of a single gene in different species or of the members of a gene family.

gene-for-gene resistance In plants, a mechanism of resistance to pathogens in which resistance is triggered by the specific interaction of the products of a pathogen's *Avr* genes and a plant's *R* genes.

general transcription factors In eukaryotes, transcription factors that bind to the promoters of most protein-coding genes and are required for their expression. Distinct from transcription factors that have specific regulatory effects only at certain promoters or classes of promoters.

generation time The average age of a mother when her daughters are produced.

generative cell In flowering plants, one of the cells in the male gametophyte. Divides by mitosis to form the two sperm cells.

genetic code The set of instructions, in the form of nucleotide triplets, that translates a linear sequence of nucleotides in mRNA into a linear sequence of amino acids in a protein.

genetic drift Changes in gene frequencies from generation to generation as a result of random (chance) processes.

genetic map distance The distance between genes on a chromosome in terms of the sum of the frequencies of recombinant offspring produced in crosses between intervening genes.

genetic structure The frequencies of the different alleles at each locus and the frequencies of the different genotypes in a population.

genetic switches Mechanisms that control how the genetic toolkit is used, such as promoters and the transcription factors that bind them. The signal cascades that converge on and operate these switches determine when and where genes will be turned on and off.

genetic toolkit A set of developmental genes and proteins that is common to most animals and is hypothesized to be responsible for the evolution of their differing developmental pathways.

genome (jee′ nome) The complete DNA (or in the case of some viruses, RNA) sequence for a particular organism or individual.

genomic imprinting A developmental process whereby the form of a gene's expression is determined by parental source (i.e., whether the gene is inherited from the male or female parent).

genomic library All of the cloned DNA fragments generated by the breakdown of genomic DNA into smaller segments.

genotype (jean′ oh type) [Gk. *gen*: to produce + *typos*: impression] An exact description of the genetic constitution of an individual, either with respect to a single trait or with respect to a larger set of traits. (Contrast with phenotype.)

genotype frequency The proportion of each genotype among individuals in a population.

genus (jean′ us) (plural: genera) [Gk. *genos*: stock, kind] A group of related, similar species recognized by taxonomists with a distinct name used in binomial nomenclature.

geographic range The entire region within which a species occurs.

geological time scale The standardized scale accepted by geologists to mark the major divisions in the history of Earth.

germ cells [L. germen: to beget] Diploid cells in the gonads that multiply mitotically and are capable of undergoing meiosis to produce gametes. (Contrast with somatic cells.)

germline A lineage of cells that gives rise to the gametes.

gibberellins (jib er el′ lins) A class of plant growth hormones playing roles in stem elongation, seed germination, flowering of certain plants, etc.

gill A breathing organ that is evaginated (folded outward from the body) and surrounded by the environmental medium. The type of breathing organ found most commonly in water breathers. (Contrast with lung.)

ginkgo *See* Appendix A.

glaucophytes *See* Appendix A.

glial cells One of the two major types of cells in the nervous system (the other being neurons); unlike neurons, glial cells are typically not excitable and do not conduct action potentials. Also called glia or neuroglia.

global warming An increase through time in Earth's average annual surface temperature.

glucagon A hormone produced by endocrine cells in the pancreas, specifically alpha cells of the pancreatic islets of Langerhans. Glucagon stimulates the liver to break down glycogen and release glucose into the blood.

gluconeogenesis The biochemical synthesis of glucose from other substances, such as amino acids, lactate, and glycerol.

glucose transporter A carrier protein used to facilitate diffusion of the monosaccharide glucose.

glyceraldehyde 3-phosphate (G3P) A phosphorylated three-carbon sugar; an intermediate in glycolysis and photosynthetic carbon fixation.

glycerol (gliss′ er ole) A three-carbon alcohol with three hydroxyl groups; a component of phospholipids and triglycerides.

glycolipid A lipid to which sugars are attached.

glycolytic system A shorthand expression for discussing sources of ATP during muscle contraction, referring to production of ATP by anaerobic glycolysis. (Contrast with immediate system, oxidative system.)

glycoprotein A protein to which sugars are attached.

glycosidic bond Bond between carbohydrate (sugar) molecules through an intervening oxygen atom (—O—). Also called glycosidic linkage.

glyoxysome (gly ox′ ee soam) An organelle found in plants, in which stored lipids are converted to carbohydrates.

gnathostomes *See* Appendix A.

gnetophytes *See* Appendix A.

Golgi apparatus (goal′ jee) A system of concentrically folded membranes found in the cytoplasm of eukaryotic cells; functions in secretion from the cell by exocytosis.

gonad (go′ nad) [Gk. *gone*: seed] An organ that produces gametes in animals: either an ovary (female gonad) or testis (male gonad). Certain of the tissues in a gonad secrete hormones.

gonadotropin A type of tropic hormone that controls hormone secretion by gonadal endocrine cells, and that supports and maintains gonadal tissue, including gamete production. The two gonadotropins in vertebrates are follicle-stimulating hormone and luteinizing hormone, both secreted by the anterior pituitary gland.

Gondwana The southernmost of the two large continents produced by the breakup of Pangaea. Present-day remnants are South America, Africa, India, Australia, and Antarctica.

graded membrane potentials Small local changes in membrane potential caused by opening or closing of ion channels. They can be of any size (thus "graded") and—instead of propagating like an action potential—decrease in magnitude as they spread from their points of initiation.

grafting Artificial transplantation of tissue from one organism to another. In horticulture, the transfer of a bud or stem segment from one plant onto the root of another as a form of asexual reproduction.

Gram stain A differential purple stain useful in characterizing bacteria. The peptidoglycan-rich cell walls of Gram-positive bacteria stain blue to purple; cell walls of Gram-negative bacteria generally stain pink to red.

Gram-negative bacteria Bacteria that appear red when stained using the Gram-staining technique. These bacteria have an outer membrane outside the relatively thin peptidoglycan layer of the cell wall.

Gram-positive bacteria Bacteria that appear blue to purple when stained using the Gram-staining technique. These bacteria have an outer cell wall consisting of a thick layer of peptidoglycan.

gray crescent In frog development, a band of diffusely pigmented cytoplasm on the side of the egg opposite the site of sperm entry. Arises as a result of cytoplasmic rearrangements that establish the anterior–posterior axis of the zygote. Gastrulation starts in this location.

green plants *See* Appendix A.

greenhouse effect The retention of heat energy near Earth's surface as a consequence of absorption and reradiation of infrared radiation by certain gases (such as carbon dioxide and methane) in the atmosphere.

greenhouse gases Gases in the atmosphere, such as carbon dioxide and methane, that absorb infrared radiation and thereby retain heat energy within Earth's lower atmosphere.

gross primary productivity (GPP) The rate at which primary producers in a given area (or volume) capture energy from sunlight or other abiotic sources and convert it to chemical energy.

ground tissue system Those parts of the plant body not included in the dermal or vascular tissue systems. Ground tissues function in storage, photosynthesis, and support. (Contrast with dermal tissue system, vascular tissue system.)

growth An increase in the size of the body and/or its organs by cell division, cell expansion, and the production of extracellular materials. In ecology, change in size of the population of a species.

growth factor A chemical signal that stimulates cells to divide.

growth hormone (GH) A peptide hormone released by the anterior pituitary gland that stimulates many processes involved in tissue formation and growth.

growth rate The change in a quantity, such as the size of an organism or a population, per unit of time.

guard cells In plants, specialized, paired epidermal cells that surround and control the opening of a stoma (pore).

gut An animal's digestive tract.

gymnosperms *See* Appendix A.

gyre Large-scale circular ocean currents caused by prevailing winds and Earth's rotation.

H

habitat The characteristic environment(s) occupied by a species.

habitat structure (1) The physical structure of a habitat, for example, the horizontal and vertical distribution of objects (such as the stems and leaves of plants) in a habitat. (2) The three-dimensional distribution of objects and characteristics of their surfaces in the environments typically occupied by an animal species.

Hadley cells Large-scale vertical patterns of atmospheric circulation in which warm air rises near the equator, moves toward the poles, and cools and sinks around latitudes 30°N and S.

hadobacteria *See* Appendix A.

hagfishes *See* Appendix A.

half-life The time required for half of a sample of a radioactive isotope to decay to its stable, nonradioactive form, or for a drug or other substance to reach half its initial dosage.

halophyte (hal′ oh fyte) [Gk. *halos*: salt + *phyton*: plant] A plant that grows in a saline (salty) environment.

halotropism Growth of roots away from regions of high salt concentration.

haplo-diplontic Referring to a life cycle in which the organism spends significant time in both the haploid and diploid stages.

haploid (hap′ loid) [Gk. *haploeides*: single] Having a chromosome complement consisting of just one copy of each chromosome; designated $1n$ or n. (Contrast with diploid.)

haplontic Referring to a life cycle in which the mature organism is haploid and the zygote is the only diploid stage.

Hardy–Weinberg equilibrium The expected frequencies of alleles and genotypes in a population under random mating, in the absence of natural selection, mutation, migration, and genetic drift. The expectations of Hardy–Weinberg equilibrium can be used to calculate the expected frequencies of genotypes given the frequency of alleles, under the stated assumptions.

haustorium (haw stor′ ee um) (plural: haustoria) [L. *haustus*: draw up] A specialized hypha or other structure by which fungi and some parasitic plants draw nutrients from a host plant.

Hayflick limit The maximum number of cell divisions for a normal (human) cell before apoptosis occurs (due to telomere shortening).

heart In a circulatory system, a muscular pump that drives flow of blood around the body.

heat of vaporization The energy that must be supplied to convert a molecule from a liquid to a gas at its boiling point.

heat shock proteins Chaperone proteins expressed in cells exposed to high or low temperatures or other forms of environmental stress.

heavy chains The longer two of the four chains of an antibody molecule (immunoglobulin). The other two chains are the light chains.

hematopoietic (heem a′ toe poe et′ ick) Capable of generating blood cells.

hematopoietic stem cell transplantation (HSCT) A medical procedure in which cells that are capable generating blood cells are harvested from an individual and injected into a patient following loss of their ability to replace blood cells.

hemichordates *See* Appendix A.

hemizygous (hem′ ee zie′ gus) [Gk. *hemi*: half + *zygotos*: joined] In a diploid organism, having only one allele for a given trait, typically the case for X-linked genes in male mammals and Z-linked genes in female birds. (Contrast with homozygous, heterozygous.)

hemocyanin A type of respiratory pigment found in arthropods and mollusks, consisting of a copper-based protein. Undergoes reversible combination with O_2 at copper-containing loci.

hemoglobin (hee′ mo glow bin) [Gk. *heaema*: blood + L. *globus*: globe] A type of respiratory pigment consisting of an iron-containing protein. Undergoes reversible combination with O_2 at iron-containing loci called heme sites.

Hensen's node In avian embryos, a structure at the anterior end of the primitive groove; determines the fates of cells passing over it during gastrulation.

herbivore (ur′ bi vore) [L. *herba*: plant + *vorare*: to devour] An animal that consumes plant tissues. *See* primary consumer.

herbivory An interaction between individuals of two different species in which one individual, the herbivore, consumes all or part of the other individual, which is a plant.

herd immunity The situation in which a population's rate of vaccination against a disease is high enough that an unvaccinated person will not get the disease because the probability of contacting an infected person is low.

heritability The degree to which phenotypes are determined by individual genes transmitted from parents, rather than by interactions between different genes or between alleles of the same genes.

heterochromatin Densely packed, dark-staining chromatin; any genes it contains are usually not transcribed. (Contrast with euchromatin.)

heterochrony [Gk. different time] Alteration in the timing of developmental events, leading to different results in the adult organism. (Contrast with heterometry, heterotopy.)

heterocyst A large, thick-walled cell type in the filaments of certain cyanobacteria that performs nitrogen fixation.

heteroloboseans *See* Appendix A.

heterometry [Gk. different measure] Alteration in the level of gene expression, and thus in the amount of protein produced, during development, contributing to the evolution of different phenotypes in the adult. (Contrast with heterochrony, heterotopy.)

heteromorphic (het′ er oh more′ fik) [Gk. *heteros*: different + *morphe*: form] Having a different form or appearance, as two heteromorphic life stages of a plant. (Contrast with isomorphic.)

heterosporous (het′ er os′ por us) Producing two types of spores, one of which gives rise to a female megaspore and the other to a male microspore. (Contrast with homosporous.)

heterotopy [Gk. different place] Spatial differences in gene expression during development, controlled by developmental regulatory genes and contributing to the evolution of distinctive adult phenotypes. (Contrast with heterochrony, heterometry.)

heterotroph (het′ er oh trof) [Gk. *heteros*: different + *trophe*: feed] An organism that requires preformed organic molecules as sources of energy and chemical building blocks. (Contrast with autotroph.)

heterozygous (het′ er oh zie′ gus) [Gk. *heteros*: different + *zygotos*: joined] In diploid organisms, having different alleles of a given gene on the pair of homologs carrying that gene. (Contrast with homozygous.)

hexapods *See* Appendix A.

hibernation [L. *hibernum*: winter] In mammals, the state of inactivity of some species during winter, during which body temperature typically falls low enough to match ambient temperature and metabolic rate is profoundly depressed.

high-throughput sequencing Rapid DNA sequencing on a micro scale in which many fragments of DNA are sequenced in parallel.

hindbrain The region of the developing vertebrate brain that gives rise to the medulla, pons, and cerebellum.

hindcasting A method for testing a mathematical model by comparing its outputs to known past events.

histamine (hiss′ tah meen) A substance released by damaged tissue, or by mast cells in response to allergens. Histamine increases vascular permeability, leading to edema (swelling).

histogram A figure that displays frequencies of classes of quantitative data binned by ranges of a particular variable.

histone acetyltransferases In chromatin remodeling, enzymes that add acetyl groups to the tail regions of histone proteins. (Contrast with histone deacetylase.)

histone deacetylases In chromatin remodeling, enzymes that remove acetyl groups from the tail regions of histone proteins. (Contrast with histone acetyltransferases.)

holometabolous Undergoing complete metamorphosis.

home range A region inhabited by an individual in which other individuals are not excluded. (Contrast with territory.)

homeobox A 180-base-pair segment of DNA found in certain homeotic genes; regulates the expression of other genes and thus controls large-scale developmental processes.

homeodomain A 60-amino acid sequence within the homeobox that regulates the expression of other genes and through this regulation controls large-scale developmental processes.

homeostasis (home′ ee o sta′ sis) [Gk. *homos*: same + *stasis*: position] Stability of the internal environment of an individual, such as a constant body temperature, and the physiological or behavioral feedback responses that maintain that stability. In the words of Walter Cannon, who coined the term,"the coordinated physiological processes which maintain most of the [constant] states in the organism."

homeotherm An animal that maintains a relatively constant body temperature by physiological means (e.g., changes in metabolic rate) rather than simply by behavior.

homeothermy The condition of being a homeotherm. *See* homeotherm.

homeotic mutation Mutation in a homeotic gene that results in the formation of a different organ than that normally made by a region of the embryo.

hominids The great apes—gorillas, chimpanzees, orangutans, and humans—and their extinct relatives. (Contrast with hominins.)

hominins Modern humans (*Homo sapiens*) and their extinct close relatives (e.g., Australopithecines, *Homo erectus*). (Contrast with hominids.)

homolog (1) In cytogenetics, one of a pair (or larger set) of chromosomes having the same overall genetic composition and sequence. In diploid organisms, each chromosome inherited from one parent is matched by an identical (except for mutational differences) chromosome—its homolog—from the other parent. (2) In evolutionary biology, one of two or more features in different species that are similar by reason of descent from a common ancestor.

homologous pair A pair of matching chromosomes made up of a chromosome from each of the two sets of chromosomes in a diploid organism.

homology (ho mol′ o jee) [Gk. *homologia*: of one mind, agreement] A similarity between two or more features that is due to inheritance from a common ancestor. The structures are said to be **homologous**, and each is a **homolog** of the others.

homoplasy (home′ uh play zee) [Gk. *homos*: same + *plastikos*: shape, mold] The presence in multiple groups of a trait that is not inherited from the common ancestor of those groups. Can result from convergent evolution, evolutionary reversal, or parallel evolution.

homosporous Producing a single type of spore that gives rise to a single type of gametophyte, bearing both female and male reproductive organs. (Contrast with heterosporous.)

homozygous (home′ oh zie′ gus) [Gk. *homos*: same + *zygotos*: joined] In diploid organisms, having identical alleles of a given gene on both homologous chromosomes. An individual may be a homozygote with respect to one gene and a heterozygote with respect to another. (Contrast with heterozygous.)

horizons The horizontal layers of a soil profile, including the topsoil (A horizon), subsoil (B horizon), and parent rock or bedrock (C horizon).

horizontal gene transfer *See* lateral gene transfer.

hormone [Gk. *hormon*: to excite, stimulate] A chemical substance secreted by endocrine cells into the blood at low concentrations and transported in the blood to distant target cells, where it exerts regulatory influences on target-cell function.

hornworts *See* Appendix A.

horsehair worms *See* Appendix A.

horsetails *See* Appendix A.

Hox genes Conserved homeotic genes found in vertebrates, *Drosophila*, and other animal groups. Hox genes contain the homeobox and specify pattern and axis formation in these animals.

Human Genome Project A publicly and privately funded research effort, successfully completed in 2003, to produce a complete DNA sequence for the entire human genome.

human immunodeficiency virus (HIV) The retrovirus that causes acquired immune deficiency syndrome (AIDS).

humoral immune response The response of the immune system mediated by B cells that produces circulating antibodies active against extracellular bacterial and viral infections. (Contrast with cellular immune response.)

humus (hew′ mus) The partly decomposed remains of plants and animals on the surface of a soil.

hybrid zone A region of overlap in the ranges of two closely related species where the species hybridize.

hydration shell The cluster of water molecules that forms around a polar or charged solute.

hydrogen bond A weak electrostatic bond that arises from the attraction between the slight positive charge on a hydrogen atom and a slight negative charge on a nearby oxygen or nitrogen atom.

hydrolysis reaction (high drol′ uh sis) [Gk. *hydro*: water + *lysis*: break apart] A chemical reaction that breaks a bond by inserting the components of water (AB + H_2O → AH + BOH). (Contrast with condensation reaction.)

hydrophilic (high dro fill′ ik) [Gk. *hydro*: water + *philia*: love] Having an affinity for water. (Contrast with hydrophobic.)

hydrophobic (high dro foe′ bik) [Gk. *hydro*: water + *phobia*: fear] Having no affinity for water. Uncharged and nonpolar groups of atoms are hydrophobic. (Contrast with hydrophilic.)

hydroponic Pertaining to a method of growing plants with their roots suspended in nutrient solutions instead of soil.

hydrostatic skeleton Stiffness imparted to the body or part of the body of an animal by high blood pressure.

hyperaccumulators Species of plants that store large quantities of heavy metals such as arsenic, cadmium, nickel, aluminum, and zinc.

hyperosmotic In comparing two solutions, refers to the one having a higher osmotic pressure. The term is relative and has meaning only in comparing two specified solutions. (Contrast with hyposmotic, isosmotic.)

hyperosmotic regulator An aquatic animal that maintains a stable blood osmotic pressure higher than the osmotic pressure of the water in which it lives.

hyperpolarization A change in the membrane potential across a cell membrane so that the inside of the membrane becomes more negative compared with the outside of the membrane. (Contrast with depolarization.)

hypersensitive response A defensive response of plants to microbial infection in which phytoalexins and pathogenesis-related proteins are produced and the infected tissue undergoes apoptosis to isolate the pathogen from the rest of the plant.

hypertension High blood pressure.

hyperthermophilic bacteria *See* Appendix A.

hypertonic Having a higher concentration of membrane-impermeable solutes; said of one solution compared with another. (Contrast with hypotonic, isotonic.)

hypha (high′ fuh) (plural: hyphae) [Gk. *hyphe*: web] In the fungi and oomycetes, any single filament.

hyposmotic In comparing two solutions, refers to the one having a lower osmotic pressure. The term is relative and has meaning only in comparing two specified solutions. (Contrast with hyperosmotic, isosmotic.)

hyposmotic regulator An aquatic animal that maintains a stable blood osmotic pressure lower than the osmotic pressure of the water in which it lives.

hypothalamo–hypophysial portal system In vertebrates, a system of blood vessels that connects capillaries in the hypothalamus to capillaries in the anterior pituitary gland; provides a direct pathway by which hypothalamic releasing and inhibiting hormones can reach specific populations of cells in the anterior pituitary gland.

hypothalamus A part of the forebrain lying below the thalamus and relatively near the base of the brain. It plays roles in learning, memory, spatial orientation, and control or production of many hormones involved, for example, in control of water balance and reproduction.

hypothesis A tentative answer to a question, from which testable predictions can be generated. (Contrast with theory.)

hypotonic Having a lower concentration of membrane-impermeable solutes; said of one solution compared with another. (Contrast with hypertonic, isotonic.)

I

igneous rock Rock formed from the cooling and hardening of molten magma.

immediate system A shorthand expression for discussing sources of ATP during muscle contraction, referring to ATP that is already present in a muscle cell when a contraction begins and thus available for immediate use. (Contrast with glycolytic system, oxidative system.)

immunity In animals, the ability to avoid disease when invaded by a pathogen by deploying various defense mechanisms.

immunoglobulins A class of proteins containing a tetramer consisting of four polypeptide chains—two identical light chains and two identical heavy chains—held together by disulfide bonds; active as receptors and effectors in the immune system.

immunological memory The capacity to more rapidly and massively respond to a second exposure to an antigen than occurred on the first exposure.

imperfect flower A flower lacking either functional stamens or functional carpels. (Contrast with perfect flower.)

implantation The process by which the early mammalian embryo becomes attached to and embedded in the endometrium (lining of the uterus).

imprinting (behavioral) *See* behavioral imprinting.

in vitro evolution A method based on natural molecular evolution that uses artificial selection in the laboratory to rapidly produce molecules with novel enzymatic and binding functions.

incomplete cleavage A pattern of cleavage that occurs in many very yolky eggs, in which the cleavage furrows do not penetrate and divide all of the cytoplasm. (Contrast with complete cleavage.)

incomplete dominance A condition in which the heterozygous phenotype is intermediate between the two homozygous phenotypes.

incomplete metamorphosis Insect development in which changes between instars are gradual.

incorporation error rate The probability that an incorrect base will be inserted into a new strand during DNA replication.

independent assortment During meiosis, the random separation of genes carried on nonhomologous chromosomes into gametes so that inheritance of these genes is random. This principle was articulated by Mendel as his second law.

indeterminate In plants, refers to the growth of an organism in which the number or size of organs is unspecified (variable). (Contrast with determinate.)

indirect development Development in which the young are morphologically very different than the adult (e.g., larval stage and adult). (Contrast with direct development.)

induced mutation Changes in the sequence of DNA caused by a mutagen. (Compare with spontaneous mutation.)

induced ovulation Ovulation (release of an egg from the ovaries of a female) triggered by the act of copulation. (Contrast with spontaneous ovulation.)

induced pluripotent stem cells (iPSCs) Pluripotent animal cells derived from differentiated cells.

inducer (1) A compound that stimulates the synthesis of a protein. (2) In embryonic development, a substance that causes a group of target cells to differentiate in a particular way.

inducible enzyme An enzyme that is present in variable amounts, depending on the conditions to which a cell, tissue, or organism is exposed. Expression of an inducible enzyme depends on levels of specific inducing agents.

inducible genes Genes that are expressed only when their products are needed. (Contrast with constitutive and repressible genes.)

induction In embryonic development, the process by which one cell population influences the fate of another cell population.

infection thread A tube-like structure produced during the process of nodule formation whose presence provides bacteria with a route by which they can enter the root.

inflammation A nonspecific defense against pathogens; characterized by redness, swelling, pain, and increased temperature.

inflorescence A structure composed of several to many flowers.

inflorescence meristem A meristem that produces floral meristems as well as other small leafy structures (bracts).

infrared wavelengths Wavelengths in the electromagnetic spectrum that are too long to be seen by the human eye (i.e., longer than 0.7 μm).

ingroup In a phylogenetic study, the group of organisms of primary interest. (Contrast with outgroup.)

inhibiting hormone (IH) A hormone, released by neurosecretory cells in the hypothalamus of a vertebrate, that travels to the anterior pituitary gland through the hypothalamo–hypophysial portal system and inhibits the secretion of a hormone by a specific population of anterior pituitary endocrine cells. Also called a release-inhibiting hormone. (Contrast with releasing hormone.)

inhibitors Regulatory molecules that bind to a protein (usually an enzyme) and prevent it from functioning (often by preventing binding of the substrate(s) of the enzyme).

inhibitory postsynaptic potential (IPSP) At a chemical synapse (specifically, an inhibitory synapse), a graded membrane hyperpolarization produced in the postsynaptic cell by arrival of an impulse in the presynaptic cell. (Contrast with excitatory postsynaptic potential.)

initiation In molecular biology, the beginning of replication, transcription or translation.

initiation complex In protein translation, a combination of a small ribosomal subunit, an mRNA molecule, and the tRNA charged with the first amino acid coded for by the mRNA; formed at the onset of translation.

initiation factors The proteins involved in helping to assemble the translation initiation complex.

innate defenses In animals, one of two general types of defenses against pathogens. Nonspecific and present in most animals. (Contrast with adaptive defenses.)

inner cell mass A group of internal cells in the mammalian blastula (blastocyst) that gives rise to the embryo proper (via the epiblast) and to the yolk sac, allantois, and amnion (via the hypoblast). (Contrast with trophoblast.)

innervate To provide neural input.

instar (in′ star) An immature stage of an insect between molts.

insulin (in′ su lin) [L. *insula*: island] A hormone secreted by the endocrine beta cells of the pancreas that promotes use of glucose by cells and the conversion of glucose to the storage material glycogen in the liver.

integral membrane proteins Proteins that are at least partially embedded in the cell membrane. (Contrast with anchored membrane proteins, peripheral membrane proteins.)

integument [L. *integumentum*: covering] A protective surface structure. In gymnosperms and angiosperms, a layer of tissue around the ovule that will become the seed coat.

interferon A glycoprotein produced by virus-infected animal cells; increases the resistance of neighboring cells to the virus.

intermediate filaments Cytoskeleton components with a diameter that falls between those of the larger microtubules and those of the smaller microfilaments.

intermediate mesoderm The mesoderm that forms urinary and reproductive tissues.

internal environment In multicellular organisms, the properties of the extracellular body fluids (interstitial fluids, or tissue fluids), which bathe the cells of the body and therefore constitute the immediate environment of the cells.

internal fertilization Fertilization inside the body of the female that occurs because sperm are deposited into the female reproductive tract and encounter ova there. (Contrast with external fertilization.)

interneuron A neuron that communicates information between two other neurons.

internode The region between two nodes of a plant stem.

interphase In the cell cycle, the period between successive nuclear divisions during which the chromosomes are diffuse and the nuclear envelope is intact. During interphase the cell is most active in transcribing and translating genetic information.

interspecific competition A mutually detrimental (–/–) interaction between individuals of two different species that causes both to suffer reduced fitness. (Contrast with intraspecific competition.)

interspecific interaction An interaction between individuals of different species that affects the per capita birth and/or death rate (or equivalently the fitness) of at least some of the individuals involved.

interstitial cells Testosterone-secreting cells located between the seminiferous tubules of the vertebrate testis. Also called Leydig cells.

interstitial fluid Extracellular fluid that is found bathing the cells in tissues throughout the body. Also called tissue fluid.

intertidal zone The nearshore region of oceans that lies between the level of the high and low tides, so that it is alternately inundated with water and exposed to air.

intracellular fluid The aqueous solution inside a cell.

intracellular receptor A receptor protein that binds its ligand inside the cell. An example is the estrogen receptor in mammals. The ligand enters the cell and then binds to the intracellular receptor.

intraspecific competition A mutually detrimental (–/–) interaction between individuals of the same species that causes both to suffer reduced fitness. (Contrast with interspecific competition.)

intron A portion of a gene within the coding region that is transcribed into pre-mRNA but is spliced out prior to translation. (Contrast with exon.)

invasive species A species that increases in abundance and spreads widely, often to the detriment of other species, when it is introduced to a new location.

invasiveness (1) The ability of a pathogen to multiply in a host's body. (2) The ability of a species to be invasive. (*See* invasive species.)

inversion A rare 180° reversal of the order of genes within a segment of a chromosome.

ion [Gk. *ion*: wanderer] An atom that loses or gains one or more electrons.

ion channel A transmembrane protein that allows specific ions to diffuse across the membrane in which it is embedded.

ionic bond An electrostatic attraction between positively and negatively ions.

ionotropic receptor cell A sensory receptor cell that typically has a receptor protein that is a stimulus-gated Na^+ channel, the effect of which activates the cellular response. The receptor protein is termed an ionotropic receptor; for more information *see* ionotropic receptor. (Contrast with metabotropic receptor cell.)

irreversible inhibition A type of inhibition in which an inhibitor permanently binds to the protein, rendering it nonfunctional.

isomorphic (eye so more' fik) [Gk. *isos*: equal + *morphe*: form] Having the same form or appearance, as when the haploid and diploid life stages of an organism appear identical. (Contrast with heteromorphic.)

isosmotic Having the same osmotic pressure; said of a solution in comparison to another solution. The term is relative and has meaning only in comparing two specified solutions. (Contrast with hyperosmotic, hyposmotic.)

isotonic Having the same concentration of membrane-impermeable solutes; said of two solutions. (Contrast with hypertonic, hypotonic.)

isotope (eye' so tope) [Gk. *isos*: equal + *topos*: place] Isotopes of a given chemical element have the same number of protons in their nuclei (and thus are in the same position on the periodic table) but differ in the number of neutrons.

iteroparity In describing the reproductive life cycle of a species, a physiological ability to reproduce more than once in a lifetime.

iteroparous Characterized by iteroparity.

J

jasmonate A plant hormone involved in triggering responses to pathogen attack as well as other processes. Also called jasmonic acid.

jawed vertebrates *See* Appendix A.

joint In skeletal systems, a junction between two or more bones.

joule (J) A metric unit for measuring energy: 1 J = 0.24 calorie; 1 calorie = 4.2 joule.

juxtacrine signaling A type of cell signaling that requires that the signaling and target cells be in direct contact. Usually involves interaction between signaling molecules bound to the surfaces of the two cells. (Contrast with autocrine signaling, paracrine signaling.)

K

karyogamy The fusion of nuclei of two cells. (Contrast with plasmogamy.)

karyotype The number, forms, and types of chromosomes in a cell.

kilocalorie (kcal) *See* calorie.

kin selection During evolution, positive selection on individuals that results from their helping the survival of relatives (kin) that possess the same gene alleles by descent from a common ancestor. (Contrast with direct fitness.)

kinases A class of enzymes that catalyze the addition of phosphate groups to proteins.

kinetic energy (kuh-net' ik) [Gk. *kinetos*: moving] The energy associated with movement. (Contrast with potential energy.)

kinetochore (kuh net' oh core) A specialized structure on a centromere to which microtubules attach.

kinetoplastids *See* Appendix A.

kinorhynchs *See* Appendix A.

Koch's postulates A set of rules for establishing that a particular microorganism causes a particular disease.

korarchaeotes *See* Appendix A.

L

lactic acid fermentation Anaerobic series of reactions that convert glucose to lactic acid, in some bacteria and animal cells.

lagging strand In DNA replication, the daughter strand that is synthesized in discontinuous stretches. (Contrast with leading strand.) *See* Okazaki fragments.

lancelets *See* Appendix A.

land plants *See* Appendix A.

landscape An ecological system consisting of multiple ecological communities within a geographic area larger than the area occupied by a single community.

lateral gene transfer The transfer of individual genes, organelles, or fragments of genomes from one species to another, common among bacteria and archaea. Also called horizontal gene transfer.

lateral plate mesoderm One of three regions of mesoderm on either side of the notochord; encloses the coelom and gives rise to the heart and circulatory system, muscles of the digestive system, many bones, muscles of the outer body wall, and other tissues.

lateral root A root extending outward from the taproot in a taproot system; typical of eudicots.

Laurasia The northernmost of the two large continents produced by the breakup of Pangaea.

law of independent assortment During meiosis, the random separation of genes carried on nonhomologous chromosomes into gametes so that inheritance of these genes is random. This principle was articulated by Mendel as his second law.

law of segregation In genetics, the separation of alleles, or of homologous chromosomes, from each other during meiosis so that each of the haploid daughter nuclei produced contains one or the other member of the pair found in the diploid parent cell, but never both. This principle was articulated by Mendel as his first law.

leaching In soils, a process by which mineral nutrients in upper soil horizons are dissolved in water and carried to deeper horizons, where they are unavailable to plant roots.

leading strand In DNA replication, the daughter strand that is synthesized continuously. (Contrast with lagging strand.)

learning The process by which an individual modifies its behavior as a consequence of its prior experiences.

leaves In plants, the chief organ of photosynthesis.

leghemoglobin In nitrogen-fixing plants, an oxygen-carrying protein in the cytoplasm of nodule cells that transports enough oxygen to the nitrogen-fixing bacteria to support their respiration, while keeping free oxygen concentrations low enough to protect nitrogenase.

lens In an eye, a structure composed of transparent proteins that focuses images on the retina or other light-sensing structures.

lepidosaurs *See* Appendix A.

leukocytes *See* white blood cells.

Leydig cells *See* interstitial cells.

lichen (lie' kun) An organism resulting from the symbiotic association of a fungus and either a cyanobacterium or a unicellular alga.

life expectancy The average time from birth to death of an individual organism.

life history The sequence and timing of key events, such as growth and development, reproduction, and death, that occur during the life of the average individual of a given species.

life history trade-offs Negative relationships among growth and development, reproduction, and survival that occur because an individual cannot simultaneously allocate resources to such competing life functions as resource acquisition, maintenance, growth, defense, or reproduction.

ligand (lig' and) Any molecule that binds to a receptor site of another (usually larger) molecule.

ligand-gated channel A channel protein that can open and close and that opens to allow diffusion of a solute as a result of binding by a specific signaling molecule (e.g., neurotransmitter) to a receptor site on the channel protein.

light chains The shorter two of the four chains of an antibody molecule (immunoglobulin). The other two chains are the heavy chains.

light reactions The initial phase of photosynthesis, in which light energy is converted into chemical energy.

light-harvesting complex In photosynthesis, a group of proteins and pigment molecules in thylakoid membranes that function to absorb light energy and transfer it to a reaction center.

lignin A complex, hydrophobic polyphenolic polymer in plant cell walls that crosslinks other wall polymers, strengthening the walls, especially in wood.

limiting resource A resource that constrains the fitness of individuals at a given time and place because it is in short supply relative to demand.

limnetic zone [Gk. *limné*: pool] The "open-water" zone of freshwater lakes.

lineage A series of populations, species, or genes descended from a single ancestor over evolutionary time.

lineage species concept The definition of a species as a branch on the tree of life, which has a history that starts at a speciation event and ends either at extinction or at another speciation event. (Contrast with biological species concept, morphological species concept.)

linear regression A statistical method of fitting a straight line to describe the relationship between two variables in a scatter plot.

linkage In a cross, the observation that parental gametes/offspring are more frequent than recombinant gametes/offspring, indicating genes are on the same chromosome.

lipids [Gk. *lipos*: fat] Nonpolar, hydrophobic molecules that include fats, oils, waxes, steroids, and the phospholipids that make up biological membranes.

lipoproteins Lipids packaged inside a covering of protein so that they can be circulated in the blood.

littoral zone The nearshore regions of lakes and oceans. In oceans, this includes the intertidal zone.

liverworts *See* Appendix A.

loam A type of soil consisting of a mixture of sand, silt, clay, and organic matter. One of the best soil types for agriculture.

lobe-limbed vertebrates *See* Appendix A.

loboseans *See* Appendix A.

locus (low' kus) (plural: loci, low' sigh) In genetics, a specific location on a chromosome. For locations encoding functional RNAs or proteins, may be considered synonymous with *gene*.

lokiarchaeotes *See* Appendix A.

long-day plant (LDP) A plant that requires long days (actually, short nights) in order to flower. (Compare with short-day plant.)

lophophore A U-shaped fold of the body wall with hollow, ciliated tentacles that encircle the mouth of animals in several different groups. Used for filtering prey from the surrounding water.

lophotrochozoans *See* Appendix A.

loriciferans *See* Appendix A.

loss-of-function mutation A mutation that results in the loss of a functional protein. (Contrast with gain-of-function mutation.)

loss-of-stop mutation A mutation that results in a change from a stop (nonsense) codon to a sense codon, causing additional amino acids to be added to the end of the protein. (Contrast with frame-shift mutation, missense mutation, nonsense mutation, silent mutation.)

lumen (loo'men) [L. *lumen*: light] The open cavity inside any tubular organ or structure, such as the gut, a renal tubule, or a blood vessel.

lung A breathing organ that is invaginated (folded inward) into the body and contains the environmental medium. The most common type of breathing organ in terrestrial animals. (Contrast with gill.)

lung surfactant A thin layer of lipids and proteins that coats the inside of a lung; decreases surface tension at the surfaces of the lung epithelium, thereby stabilizing lung structure and reducing the amount of work necessary to inflate the lung.

lungfishes *See* Appendix A.

luteinizing hormone (LH) A hormone, produced by the anterior pituitary gland, that helps control secretion of hormones by the gonads; a gonadotropin. Also involved in ovulation, which occurs in response to a surge in the blood concentration of LH.

lycophytes *See* Appendix A.

lymph [L. *lympha*: liquid] A fluid derived from blood and other tissues that accumulates in intercellular spaces throughout the body and is returned to the blood by the lymphatic system.

lymph node A specialized structure in the vessels of the lymphatic system. Lymph nodes contain lymphocytes, which encounter and respond to foreign cells and molecules in the lymph as it passes through the vessels.

lymphatic system An elaborate system of vessels, separate from the blood vascular system, that picks up excess interstitial fluid (lymph) and returns it to the blood.

lymphocyte One of the two major classes of white blood cells; includes T cells, B cells, and other cell types important in the immune system.

lysogeny A form of viral replication in which the virus becomes incorporated into the host chromosome and remains inactive. Also called a lysogenic cycle. (Contrast with lytic cycle.)

lysozyme (lie' so zyme) An enzyme in saliva, tears, and nasal secretions that hydrolyzes bacterial cell walls.

lytic cycle A viral reproductive cycle in which the virus takes over a host cell's synthetic machinery to replicate itself, then bursts (lyses) the host cell, releasing the new viruses. (Contrast with lysogeny.)

M

M phase The portion of the cell cycle in which mitosis takes place.

macromolecule A giant (molecular weight > 1,000) polymeric molecule. The macromolecules are the proteins, polysaccharides, nucleic acids, and lipids. Note that lipids are included here because they often occur as large aggregates of smaller molecules.

macronutrient A mineral element required in large amounts. (Contrast with micronutrient.)

magnoliids See Appendix A.

maintenance methyltransferase An enzyme that transfers methyl groups to DNA after DNA replication.

male gametophyte A haploid organism that divides by mitosis to give rise to several types of cells including sperm.

mammals *See* Appendix A.

mandibulates *See* Appendix A.

mantle (1) In mollusks, a fold of tissue that covers the organs of the visceral mass and secretes the hard shell that is typical of many mollusks. (2) In geology, Earth's crust below the solid lithospheric plates.

map units A measure of map distance that is equivalent to 1 percent recombination. Also called centimorgans (cM).

marsupials *See* Appendix A.

mass extinction A period of evolutionary history during which rates of extinction are much higher than during intervening times.

mast cells Cells, typically found in connective tissue, that release histamine in response to tissue damage.

master regulators Transcription factors (proteins) at the head of a cascade, these control the expression of other transcription factors, often used to describe factors that are key to establishing the identity of an organ.

maternal effect genes Genes coding for morphogens that determine the polarity of the egg and larva in fruit flies. Part of a developmental cascade that includes gap genes, pair rule genes, segment polarity genes, and Hox genes.

maternal-to-zygotic transformation The transition that occurs, often at the mid-blastula stage, during which the zygote genome is activated and assumes genetic control of development.

mating type A particular strain of a species that is incapable of sexual reproduction with another member of the same strain but capable of sexual reproduction with members of other strains of the same species.

maximum likelihood A statistical method of determining which of two or more hypotheses (such as phylogenetic trees) best fit the observed data, given an explicit model of how the data were generated.

mean The sum of all values in a sample divided by the number of observations in the sample.

measures of center Quantities that describe various aspects of the center of a group of observations.

measures of dispersion Measures that quantify the dispersion of observations in a sample of observations.

mechanoreceptor A sensory receptor cell that is sensitive to physical movement or physical distortion and that directly or indirectly generates action potentials in response.

median The value at which there are equal numbers of larger and smaller observations in a sample.

mediator A regulatory protein composed of multiple subunits that interacts with the general transcription factors and RNA polymerase of the basal transcription apparatus and with specific transcription factors.

medusa (plural: medusae) In cnidarians, a free-swimming, sexual life cycle stage shaped like a bell or an umbrella.

megagametophyte In heterosporous plants, the female gametophyte; produces eggs. (Contrast with microgametophyte.)

megaphyll The generally large leaf of a fern, horsetail, or seed plant, with several to many veins. (Contrast with microphyll.)

megasporangia The structures on a heterosporous plant that produce a few large megaspores (which develop into female gametophytes).

megaspore [Gk. *megas*: large + *spora*: to sow] In plants, a haploid spore that produces a female gametophyte.

megastrobilus (plural: megastrobili) In conifers, the female (seed-bearing) cone. (Contrast with microstrobilus.)

meiosis (my oh' sis) [Gk. *meiosis*: diminution] Division of a diploid nucleus to produce four haploid daughter cells. The process consists of two successive nuclear divisions with only one cycle of chromosome replication. In **meiosis I**, homologous chromosomes separate but retain their chromatids. The second division **meiosis II**, is similar to mitosis, in which chromatids separate.

membrane potential The difference in electric charge (voltage) between the inside surface and the outside surface of a cell membrane, typically expressed in millivolts. The name derives from the fact that potential difference is a synonym for voltage.

membrane receptor A transmembrane protein with an extracellular, ligand-binding region.

memory A record in the nervous system of past experience.

memory cells In the immune system, long-lived lymphocytes produced after exposure to antigen. They persist in the body and are able to mount a rapid response to subsequent exposures to the antigen.

menstrual cycle The cycle in which oocytes periodically mature and are ovulated—and the reproductive tract prepares for possible pregnancy—in females of primates, including humans; one phase of each cycle is characterized by menstruation, the shedding of the endometrial uterine lining in a blood-tinged discharge from the vagina.

menstruation In humans and other primates, a process by which the lining of the uterus (the endometrium) breaks down (after a failure of pregnancy to occur), and the sloughed-off tissue, including blood, flows from the body.

meristem [Gk. *meristos*: divided] Plant tissue made up of undifferentiated actively dividing cells.

meristem culture A method for the asexual propagation of plants, in which pieces of shoot apical meristem are cultured to produce plantlets.

meristematic zone In plant roots, the region near the tip of a root that contains the meristem; the source of all cells of the root's primary tissues.

meristemoid A type of cell that occurs in leaves, a precursor to the formation of stomatal guard cells.

mesenchyme (mez' en kyme) [Gk. *mesos*: middle + *enchyma*: infusion] Embryonic or unspecialized cells, often migratory; often, but not always, derived from the mesoderm. *See* neural crest cells.

mesoderm [Gk. *mesos*: middle + *derma*: skin] The middle of the three embryonic germ layers first delineated during gastrulation. Gives rise to the skeleton, circulatory system, muscles, excretory system, and most of the reproductive system.

mesoglea (mez' uh glee uh) [Gk. *mesos*: middle + *gloia*, glue] A thick, gelatinous noncellular layer that separates the two cellular tissue layers of ctenophores, cnidarians, and scyphozoans.

mesophyll (mez' uh fill) [Gk. *mesos*: middle + *phyllon*: leaf] Chloroplast-containing, photosynthetic cells in the interior of leaves.

messenger RNAs (mRNAs) The processed transcript from the template strand of a protein-coding gene; carries information (as a sequence of codons) for the synthesis of one or more proteins.

meta-analysis A statistical approach that assembles a large data set by seeking out and combining data from many smaller data sets.

metabolic pathway A series of enzyme-catalyzed reactions so arranged that the product of one reaction is the substrate of the next.

metabolic rate (MR) An animal's rate of energy consumption; the rate at which it converts chemical-bond energy to heat and external work.

metabolome The quantitative description of all the small molecules in a cell, tissue, or organism.

metabolomics The study of the metabolome as it relates to the physiological state of a cell, tissue, or organism.

metabotropic receptor cell A sensory receptor cell in which the sensory receptor protein is typically a G protein-coupled receptor and initiates production of a second messenger in the cell. (Contrast with ionotropic receptor cell.)

metagenomics The practice of analyzing DNA from environmental samples without isolating intact organisms.

metamorphosis [Gk. *meta*: between + *morphe*: form, shape] A change occurring between one developmental stage and another, as, for example, from a tadpole to a frog. (*See* complete metamorphosis, incomplete metamorphosis.)

metaphase (met' a phase) The stage in nuclear division at which the centromeres of the highly supercoiled chromosomes are all lying on a plane (the metaphase plate) perpendicular to a line connecting the division poles.

metapopulation A "population of populations"—the group of spatially separated populations that occurs within a defined geographic area.

methanogens Methane-producing microbes.

microbial eukaryotes *See* Appendix A.

microbiomes The diverse communities of bacteria that live on or within the body and are essential to bodily function.

microcirculation The part of a closed circulatory system that consists of the smallest diameter blood vessels (i.e., arterioles, capillaries, and venules).

microclimate The set of climate conditions (temperature, humidity, wind speed, etc.) prevailing in a microhabitat.

microenvironment *See* microhabitat.

microfilament In eukaryotic cells, a fibrous structure made up of actin monomers. Microfilaments play roles in the cytoskeleton, in cell movement, and in muscle contraction.

microgametophyte In heterosporous plants, the male gametophyte; produces sperm. (Contrast with megagametophyte.)

microhabitat A place within a habitat that potentially differs from the statistically average conditions in the habitat at large. Also called a microenvironment.

micronutrient In plants, a mineral element required in concentrations of less than 100 micrograms per gram of plant dry matter; in animals, a mineral element required in concentrations of less than 100 micrograms per day. (Contrast with macronutrient.)

microphyll A small leaf with a single vein, found in club mosses and their relatives. (Contrast with megaphyll.)

micropyle (mike' roh pile) [Gk. *mikros*: small + *pylon*: gate] Opening in the integument(s) of a seed plant ovule through which pollen grows to reach the female gametophyte within.

microRNA (miRNA) A small, noncoding RNA molecule, typically about 21 bases long, that binds to mRNA to inhibit its translation.

microsatellites Simple 1–5 base pair DNA repeats, present in multiple tandem copies.

microsporangia The structures on a heterosporous plant that produce many small microspores (which develop into male gametophytes).

microspore [Gk. *mikros*: small + *spora*: to sow] In plants, a haploid spore that produces a male gametophyte.

microsporidia *See* Appendix A.

microstrobilus (plural: microstrobili) In conifers, male pollen-bearing cone. (Contrast with megastrobilus.)

microtubule A tubular structure composed of tubulin dimers found in the centrioles, spindle apparatus, cilia, flagella, and cytoskeleton of eukaryotic cells. Microtubules play roles in the motion and maintenance of shape of eukaryotic cells.

midbrain One of the three regions of the vertebrate brain. Part of the brainstem, it serves as a relay station for sensory signals sent to the cerebral hemispheres, in addition to other functions.

migration Movement by an animal that entails traveling to a different location, remaining there for a substantial length of time, and then returning—often carried out in a periodic (repeating) way year after year.

mismatch A base pair in which the two bases are not complementary (e.g., AC or AG instead of AT).

mismatch repair A mechanism that scans DNA after it has been replicated and corrects any base-pairing mismatches.

missense mutations A change in a gene's sequence that changes the amino acid at that site in the encoded protein. (Contrast with frame-shift mutation, nonsense mutation, loss-of-stop, silent mutation.)

missense substitution *See* nonsynonymous substitution.

mitochondrion (my' toe kon' dree un) (plural: mitochondria) [Gk. *mitos*: thread + *chondros*: grain] An organelle in eukaryotic cells that contains the enzymes of the citric acid cycle, the respiratory chain, and oxidative phosphorylation.

mitosis (my toe' sis) [Gk. *mitos*: thread] Nuclear division in eukaryotes leading to the formation of two daughter nuclei, each with a chromosome complement identical to that of the original nucleus.

mitosomes Reduced structures derived from mitochondria found in some organisms.

mode The most frequent value in a sample of observations.

model organisms Also known as **model systems**, these include the small group of species that are the subject of extensive research. They are organisms that adapt well to laboratory situations, and findings from experiments on them can apply across a broad range of species. Classic examples include white mice and the fruit fly *Drosophila*.

molds Sac fungi composed of filamentous hyphae that do not form large ascomata.

mole A quantity of a compound whose weight in grams is numerically equal to its molecular weight expressed in atomic mass units. Avogadro's number of molecules: 6.023×10^{23} molecules.

molecular clock The approximately constant rate of divergence of macromolecules from one another over evolutionary time; used to date past events in evolutionary history.

molecule A chemical substance made up of two or more atoms joined by covalent bonds or ionic bonds.

mollusks *See* Appendix A.

molting The process of shedding part or all of an outer covering, as the shedding of feathers by birds or of the entire exoskeleton by arthropods.

monilophytes *See* Appendix A.

monoblastic Having a body derived from a single embryonic cell layer. (Contrast with diploblastic, triploblastic.)

monoclonal response An immune response against a single epitope, involving a single clone of B cells.

monocots *See* Appendix A.

monoecious (mo nee' shus) [Gk. *mono*: one + *oikos*: house] Pertaining to organisms in which both sexes are "housed" in a single individual that produces both eggs and sperm. (In some plants, these are found in different flowers within the same plant.) Examples include corn, peas, earthworms, hydras. (Contrast with dioecious.)

monohybrid cross A mating in which the parents differ with respect to the alleles of only one locus/gene of interest.

monomer The minimal repeating subunit of a macromolecule (amino acids in proteins, nucleotides in nucleic acids, monosaccharides in polysaccharides).

monophyletic (mon' oh fih leht' ik) [Gk. *mono*: one + *phylon*: tribe] Pertaining to a group that consists of an ancestor and all of its descendants. (Contrast with paraphyletic, polyphyletic.)

monosaccharide A simple sugar. Oligosaccharides and polysaccharides are made up of monosaccharides.

morphogen A diffusible substance whose concentration gradient determines a developmental pattern in animals and plants.

morphogenesis (more' fo jen' e sis) [Gk. *morphe*: form + *genesis*: origin] The development of form; the overall consequence of determination, differentiation, and growth.

morphological species concept The definition of a species as a group of individuals that look alike. (Contrast with biological species concept, lineage species concept.)

morphology (more fol' o jee) [Gk. *morphe*: form + *logos*: study, discourse] The scientific study of organic form, including both its development and function.

mosaic development A pattern of animal embryonic development in which each blastomere contributes a specific part of the adult body. (Contrast with regulative development.)

mosses *See* Appendix A.

motile cilia A hairlike organelle used for locomotion by many unicellular organisms and for moving water and mucus by many multicellular organisms.

motor neuron A neuron that carries signals from the central nervous system to a skeletal muscle, stimulating the muscle to contract. A type of efferent neuron.

motor proteins Specialized proteins that use energy to change shape and move cells or structures within cells.

mucus A slippery substance secreted by mucous membranes (e.g., mucosal epithelium). A barrier defense against pathogens in innate immunity in animals.

Muller's ratchet The accumulation—"ratcheting up"—of deleterious mutations in the nonrecombining genomes of asexual species.

multi-organ system A system in which multiple organs work together.

multiple fruit A fruit derived from carpels of several flowers. An example is a pineapple.

multiplicative growth A form of growth in which a certain multiple of a current quantity is added during each successive time interval. The multiple of growth may be a constant, or it may change. (Contrast with additive growth.)

multipotent stem cells Stem cells that have the ability to differentiate into several, but limited, cell types. (Contrast with pluripotent stem cells, totipotent stem cells.)

muscle fiber A single muscle cell. In the case of skeletal muscle, a multinucleate cell.

mutagen (mute' ah jen) [L. *mutare*: change + Gk. *genesis*: source] Any agent (e.g., a chemical, radiation) that increases the mutation rate.

mutation A change in the genetic material that is not caused by recombination.

mutualism A mutually beneficial (+/+) interaction between individuals of two different species that causes both to benefit, in terms of increased fitness. (Contrast with amensalism, commensalism.)

mutualistic A pair of symbiotic organisms in which both partners benefit from the interaction.

mycelium (my seel' ee yum) (plural: mycelia) [Gk. *mykes*: fungus] In the fungi, a mass of hyphae.

mycologist A biologist who studies fungi.

mycoplasmas *See* Appendix A.

mycorrhiza (my' ko rye' za) (plural: mycorrhizae) [Gk. *mykes*: fungus + *rhiza*: root] An association of the root of a plant with the mycelium of a fungus.

myelin (my' a lin) Concentric layers of glial cell membrane wrapped around some axons, forming an electrically insulating sheath around the axon segments where it occurs; myelin increases the rate of propagation of action potentials. (*See also* node of Ranvier.)

myocardium The muscle tissue of a heart.

myofibril (my' oh fy' bril) [Gk. *mys*: muscle + L. *fibrilla*: small fiber] A long, longitudinally oriented, internal component of a muscle cell that consists of a series of sarcomeres and extends the length of the cell. A myofibril is composed principally of actin and myosin filaments. In cross section, a muscle cell consists of multiple myofibrils.

myogenic heart A heart in which the electrical impulse to contract during each beating cycle originates in muscle cells or modified muscle cells. (Contrast with neurogenic heart.)

myoglobin (my' oh globe' in) [Gk. *mys*: muscle + L. *globus*: sphere] A molecule (pigment) that can undergo reversible combination with O_2, found in red muscle cells. A type of hemoglobin, consisting of a single heme unit and a single globin chain per molecule.

myometrium An outer layer of smooth muscle of the uterus.

myosin filaments Bundles of linked myosin molecules. Also called thick filaments. (*See also* actin filaments.)

myosin One of the two contractile proteins of muscle. (*See also* actin.)

myriapods *See* Appendix A.

N

Na^+–K^+ ATPase *See* sodium–potassium pump.

natural history The observation of nature outside a formal hypothesis-testing context, which provides the basis for further investigation.

natural history (1) The study of nature outside of a formal hypothesis-testing investigation. (2) The body of knowledge so obtained.

natural killer cell A type of lymphocyte that attacks virus-infected cells and some tumor cells as well as antibody-labeled target cells.

natural selection The differential contribution of offspring to the next generation by various genetic types belonging to the same population. The mechanism of evolution proposed by Charles Darwin.

navigation The act of moving on a particular course or toward a particular destination.

necrosis (nec roh' sis) [Gk. *nekros*: death] Premature cell death caused by external agents such as toxins.

negative feedback In a regulatory system, a type of control that acts to reduce differences that arise between the level of a controlled variable and its set-point level. It tends to stabilize the controlled variable at a level close to the set-point level. (Contrast with positive feedback.)

negative regulation Regulation in which binding of a regulatory protein reduces the rate of transcription. (Contrast with positive regulation.)

negative-sense RNA RNA that is complementary to mRNA. Before it can be translated, it must be converted to positive-sense RNA by an RNA polymerase.

nematocyst (ne mat' o sist) [Gk. *nema*: thread + *kystis*: cell] An elaborate, threadlike structure produced by cells of jellyfishes and other cnidarians, used chiefly to paralyze and capture prey.

nematodes *See* Appendix A.

neopterans *See* Appendix A.

neoteny (nee ot' enny) [Gk. *neo*: new, recent; *tenein*: to extend] The retention of juvenile or larval traits by the fully developed adult organism.

Nernst equation A mathematical equation that calculates the electrical potential difference required for electrochemical equilibrium across a membrane that is permeable to a single type of ion, when the ion differs in concentration on the two sides of the membrane.

nerve A structure consisting of many neuronal axons bound together like wires in a telephone cable.

nerve cell *See* neuron.

net primary productivity (NPP) The rate at which primary producers in a given area (or volume) convert energy they have captured from sunlight or other abiotic sources into energy stored in the molecules that make up their tissues, and which therefore is available for use by primary consumers. NPP is often estimated as the gain in dry biomass of primary producers in a unit of time.

neural crest cells During vertebrate neurulation, cells that migrate outward from the neural plate; they are multipotent mesenchymal cells that form the sensory nervous system, autonomic nervous system, many bones of the skull, pigment cells, and other tissues.

neuroendocrine cells *See* neurosecretory cells.

neurogenic heart A heart in which the electrical impulse to contract during each beating cycle originates in neurons. (Contrast with myogenic heart.)

neurohemal organ A specialized organ in the circulatory system that is made up of axon terminals of neurosecretory cells in association with a well-developed network of capillaries.

neuromuscular junction A synapse where a motor neuron axon makes synaptic contact with a muscle fiber (muscle cell).

neuron (noor′ on) [Gk. *neuron*: nerve] A nervous system cell that can generate and conduct action potentials along an axon to a synapse with another cell. Also called nerve cell.

neurosecretory cells Neurons that synthesize and release hormones into the blood. Also called neuroendocrine cells.

neurotransmitter At a chemical synapse, a molecule produced and released by the presynaptic cell that diffuses across the synapse and excites or inhibits the postsynaptic cell.

neurulation The stage in vertebrate development during which the nervous system begins to form.

neutral theory A view of molecular evolution that postulates that most mutations do not affect the amino acid being coded for, and that such mutations accumulate in a population at rates driven by genetic drift and mutation rates.

neutron (new′ tron) One of the three fundamental subatomic particles of matter (along with protons and electrons), part of the atomic nucleus, with mass slightly larger than that of a proton and no electrical charge.

niche (nitch) [L. *nidus*: nest] In ecology, the abiotic and biotic conditions under which a given species can persist, and the functional role of the species in its community.

nicotinamide adenine dinucleotide (NAD/NADH) A compound used in redox reactions to transport high-energy electrons. In carbohydrate catabolism, electrons are picked up by NAD^+ in the citric acid cycle and in glycolysis to give NADH, and transported to the electron transport chain.

nicotinamide adenine dinucleotide phosphate ($NADP^+$/NADPH) A compound used in redox reactions to transport high-energy electrons. In photosynthesis, electrons are picked up by $NADP^+$ from electron transport to give NADPH and used during the Calvin Cycle to synthesize carbohydrates.

nitrifiers Chemoautotrophic bacteria that oxidize ammonia to nitrate in soil and in seawater.

nitrogen fixation Conversion of atmospheric nitrogen gas (N_2) to a water-soluble, biologically usable form (usually ammonium, NH_4^+), in nature by free-living or symbiotic microbes, but also by industrial processes. (Compare with denitrification.)

nitrogen fixers Organisms that convert atmospheric nitrogen gas into a chemical form (ammonia) that is usable by the nitrogen fixers themselves as well as by other organisms.

nitrogenase An enzyme complex found in nitrogen-fixing bacteria that mediates the stepwise reduction of atmospheric N_2 to ammonia and which is strongly inhibited by oxygen.

noble gases Elements in the last column of the periodic table that don't react with other elements (because their *p*- and *s*-orbitals are full).

node [L. *nodus*: knob, knot] (1) In plants, a (sometimes enlarged) point on a stem where a leaf is or was attached. (2) In phylogenetics, a split in a phylogenetic tree when one lineage diverges into two.

node of Ranvier In a myelinated axon, a gap in the myelin sheath where the axon cell membrane is exposed (not covered with myelin).

nodule Specialized structure produced by a plant in response to the presence of nitrogen-fixing bacteria that provides an environment in which nitrogen fixation can occur.

non-neural endocrine cells Cells that secrete hormones into the blood and that are not neurons or derived from neurons. Also called epithelial endocrine cells. (*See also* neurosecretory cells.)

nondisjunction Failure of sister chromatids to separate in meiosis II or mitosis, or failure of homologous chromosomes to separate in meiosis I. Results in aneuploidy.

nonhomologous end joining A repair pathway that functions to join together (ligate) the two ends of a broken chromosome.

nonpolar Having electric charges that are evenly balanced from one end to the other. (Contrast with polar.)

nonsense codon *See* stop codon.

nonsense mutation Change in a gene's sequence that prematurely terminates translation by changing one of its codons to a stop codon. Also called stop mutation. (Contrast with frame-shift mutation, missense mutation, loss-of-stop, silent mutation.)

nonsense substitution A nucleotide substitution in a protein-coding sequence that results in a stop codon, thus terminating protein translation.

nonshivering thermogenesis (NST) In mammals and some birds, the elevation of heat production for thermoregulation by means other than shivering.

nonsynonymous substitution A change in a gene from one nucleotide to another that changes the amino acid specified by the corresponding codon (i.e., AGC → AGA, or serine → arginine). Also called **missense substitution**. (Contrast with synonymous substitution.)

nonvascular land plants Land plants that lack specialized vascular tissues for the conduction of water or nutrients through the plant body. There are three living groups of nonvascular land plants: the liverworts, hornworts, and mosses.

notochord (no′ tow kord) [Gk. *notos*: back + *chorde*: string] A flexible rod of gelatinous material serving as a support in the embryos of all chordates and in the adults of tunicates and lancelets.

nuclear envelope A double membrane that encloses the cell nucleus.

nuclear pores Perforations in the nuclear envelope through which the regulated passage of large molecules occurs.

nucleic acid (new klay′ ik) A polymer made up of nucleotides, specialized for the storage, transmission, and expression of genetic information. DNA and RNA are nucleic acids.

nucleic acid hybridization A technique in which a single-stranded nucleic acid probe is made that is complementary to, and binds to, a target sequence, either DNA or RNA. The resulting double-stranded molecule is a hybrid.

nucleoid (new′ klee oid) The region that harbors the chromosomes of a prokaryotic cell. Unlike the eukaryotic nucleus, it is not enclosed by a membrane.

nucleolus (new klee′ oh lus) A small, generally spherical body found within the nucleus of eukaryotic cells. The site of synthesis of ribosomal RNA.

nucleoside A nucleotide without the phosphate group; a nitrogenous base attached to a sugar.

nucleosome A portion of a eukaryotic chromosome, consisting of part of the DNA molecule wrapped around a group of histone molecules. A chromosome is made up of many nucleosomes.

nucleotide A nucleoside containing 1-3 phosphate groups. Building blocks of nucleic acids and important co-enzymes (especially ATP).

nucleus [L. *nux*: kernel or nut] (1) In cells, the centrally located compartment of eukaryotic cells that is enclosed by a double membrane and contains the chromosomes.

(2) In the brain, an identifiable group of neurons that share common characteristics or functions. (3) In atoms, the central region composed of one or more protons and (except in hydrogen) one or more neutrons.

null hypothesis In statistics, the premise that any differences observed in an experiment are simply the result of random differences that arise from drawing two finite samples from the same population.

nutrition The study of how organisms obtain the chemical materials they require to meet their needs for chemical building blocks and energy.

O

obligate aerobe An organism that requires oxygen for metabolism.

obligate anaerobe An anaerobic prokaryote that cannot survive exposure to O_2.

obligate parasite An organism that can only survive and grow in or on other living organisms, to the detriment of the host.

octet rule The description of processes that atoms undergo whereby they obtain, give up, or share electrons such that the *p*- and *s*- orbitals in their outer (valence) shell contains eight electrons.

Okazaki fragments Newly formed DNA making up the lagging strand in DNA replication. DNA ligase links Okazaki fragments together to give a continuous strand.

olfaction [L. *olfacere*: to smell] The sense of smell.

oligochaetes *See* Appendix A.

oligodendrocyte A type of glial cell that forms a myelin sheath on axons in the central nervous system.

oligonucleotide A nucleic acid made up of 20 or fewer monomers.

oligopeptide A peptide made up of fewer than 20 amino acids.

oligosaccharide A polymer containing a small number of monosaccharides.

ommatidium (plural: ommatidia) [Gk. *omma*: eye] A single visual unit in the compound eye of an arthropod.

omnivore [L. *omnis*: everything + *vorare*: to devour] An organism that obtains the energy and materials it needs to survive, grow, and reproduce from a variety of trophic levels.

one gene–one polypeptide The idea, since shown to be an oversimplification, that each gene in the genome encodes only a single polypeptide—that there is a one-to-one correspondence between genes and polypeptides.

oomycetes *See* Appendix A.

open circulatory system A circulatory system in which the blood leaves blood vessels and travels through spaces (sinuses and lacunae) bounded by ordinary tissue cells as it flows through the body. (Contrast with closed circulatory system.)

open reading frames Sequences of DNA within genes that begin with a start codon and end with a stop codon.

open system A system that is not isolated from, and interacts with, its surroundings. (Contrast with closed system.)

operator The region of an operon that acts as the binding site for the repressor.

operon In prokaryotes, a genetic unit of transcription, typically consisting of several structural genes that are transcribed together; the operon contains at least two control regions: the promoter and the operator.

opisthokonts *See* Appendix A.

opsonization A process by which microbes in blood and tissue fluids bind IgG antibodies and thereby are "tagged" for destruction by phagocytes.

ordinal variables Categorical variables with a natural ordering, such as the grades A, B, C, D, and F.

organ [Gk. *organon*: tool] A body part, such as the heart, liver, brain, root, or leaf, that is composed of two or more tissues integrated to perform a distinct function.

organ identity genes In angiosperms, genes that specify the different organs of the flower.

organ system An interrelated and integrated group of tissues and organs that work together in a physiological function.

organelle (or gan el′) Any of the membrane-enclosed structures within a eukaryotic cell. Examples include the nucleus, endoplasmic reticulum, and mitochondria.

organic chemistry The study of the properties of and ways in which carbon compounds interact.

organism An individual living entity that grows and reproduces as an independent unit.

organizer A region of the early amphibian embryo (the dorsal lip of the blastopore) that develops into the notochord and thus establishes the basic body plan. Also known as the primary embryonic inducer.

organogenesis The formation of organs and organ systems during development.

orientation A behavioral process in which an animal positions itself or moves in relation to an environmental cue such as a landmark or the Sun.

origin of replication (ori) A DNA sequence at which helicase unwinds the DNA double helix and DNA polymerase binds to initiate DNA replication.

orthologs Homologous genes whose divergence can be traced to speciation events. (Contrast with paralogs.)

osmolarity The total number of solute particles per liter of water.

osmosis [Gk. *osmos*: to push] Movement of water across a selectively permeable membrane, from a hypotonic solution to a hypertonic solution.

osmotic conformer An aquatic animal in which the osmotic pressure of the blood plasma and other extracellular fluids matches the osmotic pressure of the external environment, and thus varies as the external osmotic pressure varies. Also called osmoconformer. (Contrast with osmotic regulator.)

osmotic pressure The property of aqueous solutions that determines the direction of osmotic water movement when water can move from one solution to another. When water moves by osmosis between two solutions, it always moves from the lower to the higher osmotic pressure. Each individual dissolved entity (regardless of chemical nature) contributes approximately equally to the osmotic pressure of a solution.

osmotic regulator An aquatic animal that maintains a relatively constant osmotic pressure in its blood plasma and other extracellular fluids regardless of changes in the osmotic pressure of its external environment. Also called osmoregulator. (Contrast with osmotic conformer.)

outcrossing In plants, fertilization of an egg with pollen from another individual, especially one that is genetically distinct. (Contrast with self-pollination.)

outgroup In phylogenetics, a group of organisms used as a point of reference for comparison with the groups of primary interest (the ingroup).

ovarian follicle During gamete development in the ovary of a female animal, an oocyte (developing ovum) together with somatic support cells that are associated with the oocyte (theca and granulosa cells in vertebrates).

ovary [L. *ovum*: egg] A female organ, in plants or animals, that produces ova (eggs).

overtopping Plant growth pattern in which one branch differentiates from and grows beyond the others.

oviduct In female mammals, one of the two tubes that serve to transport eggs from the ovaries to the uterus. Also called a fallopian tube.

ovule (oh′ vule) In plants, a structure comprising the megasporangium and the integument, which develops into a seed after fertilization.

ovum (plural: ova) *See* egg.

oxidative phosphorylation ATP formation in the mitochondrion using a proton gradient set up during electron transport through the respiratory chain.

oxidative system A shorthand expression for discussing sources of ATP during muscle contraction, referring to ATP produced by the citric acid cycle and electron transport, requiring O_2. (Contrast with glycolytic system, immediate system.)

oxygen equilibrium curve A graph of the amount of O_2 per unit of blood volume as a function of the O_2 partial pressure of the blood.

oxygenation (1) Referring to water, the dissolution of O_2 in that water. (2) The reversible combination of a respiratory pigment with O_2; not equivalent to oxidation.

oxytocin In mammals, a hormone synthesized in the hypothalamus and released in the posterior pituitary gland; its major functions are to stimulate contraction of the uterus during birth and the flow of milk from mammary glands.

ozone layer A layer of Earth's atmosphere, 20–30 km above the surface and containing a high concentration of ozone (O_3). Absorbs much of the Sun's damaging ultraviolet radiation and prevents it from reaching Earth's surface.

P

***P*-value** The calculated probability of observing a given result by chance sampling, given the null hypothesis is true.

pacemaker cells In a heart, the cells that initiate depolarization and contraction of the heart muscle during each heart beat. These are modified muscle cells in vertebrate hearts.

pair rule genes In *Drosophila* (fruit fly) development, segmentation genes that divide the early embryo into units of two segments each. Part of a developmental cascade that includes maternal effect genes, gap genes, segment polarity genes, and Hox genes.

paleomagnetic dating A method for determining the age of rocks based on properties relating to changes in the patterns of Earth's magnetism over time.

palisade cells Cells that lie just below the epidermis in a plant leaf and that typically contain many chloroplasts and capture light.

pan genome The entirety of genome sequence found across all individuals (or strains) within a species.

Pangaea (pan jee′ uh) [Gk. *pan*: all, every] The single land mass formed when all the continents came together in the Permian Period.

parabasalids *See* Appendix A.

paracrine [Gk. *para*: near] A molecule that is secreted by a cell and that binds to and affects the function of neighboring cells. (Contrast with autocrine.)

paracrine signaling [Gk. *para*: near] A type of cell signaling in which the chemical signal acts locally, near the site of its secretion. (Contrast with autocrine signaling, juxtacrine signaling.)

paralogs Homologous genes whose divergence can be traced to gene duplication events. (Contrast with orthologs.)

paraphyletic (par′ a fih leht′ ik) [Gk. *para*: beside + *phylon*: tribe] Pertaining to a group that consists of an ancestor and some, but not all, of its descendants. (Contrast with monophyletic, polyphyletic.)

parasitism A "+/–" interaction between individuals of two different species that causes one individual (the parasite) to benefit (gain fitness) from consuming part or all of the other (the host), which suffers reduced fitness as a consequence.

parasympathetic division A division of the vertebrate autonomic nervous system that is connected to the central nervous system via cranial and sacral nerves; the parasympathetic and sympathetic divisions tend to exert opposing controls on autonomic effectors. (Contrast with sympathetic division.)

parenchyma (pair eng′ kyma) A plant tissue composed of relatively unspecialized cells without secondary walls. (Contrast with collenchyma, sclerenchyma.)

parent rock The soil horizon consisting of the rock that is breaking down to form the soil. Also called bedrock or the C horizon.

parental (P) generation The individuals that mate in a genetic cross. Their offspring are the first filial (F_1) generation.

parental gametes The two genotypes of gametes produced by an individual that are identical to the two gametes that fused to make the diploid cell undergoing meiosis.

parsimony principle The principle that states that the preferred explanation of observed data is the simplest explanation.

partial pressure A measure of the chemical activity of a gas that is essential for understanding gas diffusion because a gas always diffuses from where its partial pressure is high to where its partial pressure is low. When discussing a gas in a gas phase (e.g., the atmosphere), the partial pressure of the gas is the part of the total pressure exerted by the gas. When discussing a gas dissolved in a liquid phase, the meaning of partial pressure is more complex.

passive immunity The process of acquiring immunity from antibodies received from another individual.

passive transport Diffusion across a membrane; may or may not require an integral membrane protein. (Contrast with active transport.)

pathogen (path′ o jen) [Gk. *pathos*: suffering + *genesis*: source] An organism that causes disease.

pathogen-associated molecular patterns (PAMPs) The molecules recognized by pattern recognition receptors.

pattern recognition receptors (PRRs) Proteins made by cells that recognize molecular patterns on pathogens; part of innate immunity.

pedigree The pattern of transmission of a genetic trait within a family.

pelagic zone [Gk. *pelagos*: open sea] The "open-water" zone of oceans that lies beyond the continental shelves.

pentaradial symmetry Symmetry in five or multiples of five; a feature of adult echinoderms.

peptide A molecule containing two or more amino acids.

peptide bond The bond between amino acids in a protein; formed between a carboxyl group and amino group (—CO—NH—) with the loss of water molecules. Also called peptide linkage.

peptidoglycan The cell wall material of many bacteria, consisting of a single enormous molecule that surrounds the entire cell.

per capita birth rate (*b*) The number of offspring that the average individual produces in some specified interval of time.

per capita death rate (*d*) The average individual's chance of dying in some specified interval of time.

per capita growth rate (*r*) The average individual's contribution to total population growth rate in some specified interval of time, which equals per capita birth rate (*b*) minus per capita death rate (*d*).

perfect flower A flower with both stamens and carpels; a hermaphroditic flower. (Contrast with imperfect flower.)

perforin An enzyme produced by cytotoxic T cells that lyses target cells (e.g., pathogens) bound to the T cells.

perfusion The flow of blood through a tissue or organ.

pericycle [Gk. *peri*: around + *kyklos*: ring or circle] In plant roots, tissue just within the endodermis, but outside the root vascular tissue. Meristematic activity of pericycle cells produces lateral root primordia.

periodic table The arrangement of elements in order of increasing atomic number, with vertical groupings of similar valence electrons.

peripheral membrane proteins Proteins associated with but not embedded within the cell membrane. (Contrast with anchored membrane proteins, integral membrane proteins.)

peripheral nervous system (PNS) The portion of the nervous system other than the central nervous system (CNS). It consists of neurons and parts of neurons located outside the CNS. (Contrast with central nervous system.)

peristalsis (pair' i stall' sis) Wavelike muscular contractions proceeding along a tubular organ, propelling the contents along the tube.

peritoneum The mesodermal lining of the body cavity in coelomate animals.

peroxisome An organelle that houses reactions in which toxic peroxides are formed and then converted to water.

personal genomics The use of an individual's genome sequence to inform ancestry determination, risks of genetic disease and response to drugs.

petal [Gk. *petalon*: spread out] In an angiosperm flower, a sterile modified leaf, nonphotosynthetic, frequently brightly colored, and often serving to attract pollinating insects.

petiole (pet' ee ole) [L. *petiolus*: small foot] The stalk of a leaf.

P_{fr} *See* phytochrome.

pH The negative logarithm of the hydrogen ion concentration; a measure of the acidity of a solution. A solution with pH = 7 is said to be neutral; pH values higher than 7 characterize basic solutions, while acidic solutions have pH values less than 7.

phage therapy A strategy for treating bacterial infections using bacteriophage.

phagocyte [Gk. *phagein*: to eat + *kystos*: sac] One of two major classes of white blood cells; one of the nonspecific defenses of animals; ingests invading microorganisms by phagocytosis.

phagocytosis Endocytosis by a cell of another cell or large particle.

phagosome A membrane-enclosed vesicle inside a cell that results from infolding of the cell membrane and enclosing a particle to be taken into the cell.

pharmacogenomics The study of how an individual's genetic makeup affects their response to drugs or other agents, with the goal of predicting the effectiveness of different treatment options.

phenotype (fee' no type) [Gk. *phanein*: to show] The observable properties of an individual resulting from both genetic and environmental factors. (Contrast with genotype.)

phenotypic plasticity The ability of an individual animal to express two or more phenotypes during its life. (*See also* acclimation.)

pheromone (feer' o mone) [Gk. *pheros*: carry + *hormon*: excite, arouse] A type of molecule or mixture of molecules that is released into the external environment by an individual and that brings about specific behavioral responses in other individuals of the same species.

phloem (flo' um) [Gk. *phloos*: bark] In vascular plants, the vascular tissue that transports the products of photosynthesis from sites where they are produced or released to sites where they are used or stored.

phoronids *See* Appendix A.

phosphatase An enzyme that removes phosphate groups from proteins.

phosphodiester bond The connection in a nucleic acid strand, formed by linking two nucleotides. Also called phosphodiester linkage.

phospholipid A lipid containing a phosphate group; an important constituent of cellular membranes. (*See* lipids.)

phospholipid bilayer The basic structural unit of biological membranes; a sheet of phospholipids two molecules thick in which the phospholipids are lined up with their hydrophobic "tails" packed tightly together and their hydrophilic, phosphate-containing "heads" facing outward. Also called lipid bilayer.

photic zone The surface zone of lakes or oceans that is penetrated by sunlight.

photoautotroph An organism that obtains energy from light and carbon from carbon dioxide. (Contrast with chemoautotroph, chemoheterotroph, photoheterotroph.)

photoheterotroph An organism that obtains energy from light but must obtain its carbon from organic compounds. (Contrast with chemoautotroph, chemoheterotroph, photoautotroph.)

photomorphogenesis In plants, a process by which physiological and developmental events are controlled by light.

photon (foe' ton) [Gk. *photos*: light] A quantum of visible radiation; a "packet" of light energy.

photoperiod Day length; the number of hours of daylight in a 24-hour day.

photophosphorylation A mechanism for ATP formation in chloroplasts in which electron transport is coupled to the transport of hydrogen ions (protons, H^+) across the thylakoid membrane. (Contrast with chemiosmosis.)

photoreceptor (1) In plants, a pigment that triggers a physiological response when it absorbs a photon. (2) In animals, a sensory receptor cell that senses and responds to light energy.

photorespiration Light-driven uptake of oxygen and release of carbon dioxide, the carbon being derived from the early reactions of photosynthesis.

photosynthate Carbohydrate products of photosynthesis.

photosynthesis [literally, "synthesis from light"] Metabolic processes carried out by green plants and cyanobacteria, by which visible light is trapped and the energy used to convert CO_2 into organic compounds.

photosynthetic lamellae Elaborate internal membrane systems found in cyanobacteria that are used for photosynthesis.

photosystem [Gk. *phos*: light + *systema*: assembly] A reaction center plus antenna complex in the chloroplast thylakoid.

photosystem I In photosynthesis, the photosystem whose reaction center absorbs light at 700 nm, passing electrons to ferrodoxin and thence to NADPH.

photosystem II In photosynthesis, the photosystem whose reaction center absorbs light at 680 nm, passing electrons to the electron transport chain in the chloroplast.

phototropins A class of blue light receptors that mediate phototropism and other plant responses.

phototropism [Gk. *photos*: light + *trope*: turning] A directed plant growth response to light.

phycoerythrin A red accessory photosynthetic pigment found in red algae.

phylogenetic tree A graphic representation of the lines of descent among organisms or their genes.

phylogeny (fy loj' e nee) [Gk. *phylon*: tribe, race + *genesis*: source] The evolutionary history of a particular group of organisms or their genes.

physical geography The spatial distribution of Earth's climates and physical features.

phytoalexins Substances toxic to pathogens, produced by plants in response to fungal or bacterial infection.

phytochrome (fy' tow krome) [Gk. *phyton*: plant + *chroma*: color] A plant pigment regulating a large number of developmental and other phenomena in plants. It has two isomers: P_r, which absorbs red light, and P_{fr}, which absorbs far red light. P_{fr} is the active form.

phytomers In plants, the repeating modules that comprise a shoot, each consisting of one or more leaves, attached to the stem at a node; an internode; and one or more axillary buds.

phytoplankton Photosynthetic plankton.

phytoremediation A form of bioremediation that uses plants to clean up environmental pollution.

phytosterol A plant compound with structural similarity to cholesterol. Affects membrane fluidity.

pie chart A circular figure that displays proportions of different classes of data in an observed sample.

pigment A substance that absorbs visible light.

pinocytosis Endocytosis by a cell of liquid containing dissolved substances.

pistil [L. *pistillum*: pestle] The structure of an angiosperm flower within which the ovules are borne. May consist of a single carpel or of several carpels fused into a single structure. Usually differentiated into ovary, style, and stigma.

pituitary gland A small endocrine gland attached to the base of the brain in vertebrates. Many of its hormones control the activities of other endocrine glands. Also known as the hypophysis. (*See also* posterior pituitary gland, anterior pituitary gland.)

placenta (pla sen′ ta) The organ in female placental mammals that is made up of maternal and fetal tissues and is the interface between the maternal and fetal circulatory systems. It is thus responsible for meeting nutritional, respiratory, excretory, hormonal and immunological requirements of the fetus and for providing protection.

placozoans *See* Appendix A.

plankton Aquatic organisms that drift with the current. Photosynthetic members of the plankton are referred to as phytoplankton.

Plantae *See* Appendix A.

planula (plan′ yew la) [L. *planum*: flat] A free-swimming, ciliated larval form typical of the cnidarians.

plasma cell An antibody-secreting cell that develops from a B cell; the effector cell of the humoral immune system.

plasmid A circular DNA molecule distinct from the chromosome(s); that is, an extrachromosomal element; found in many bacteria. May replicate independently of the chromosome.

plasmodesma (plural: plasmodesmata) [Gk. *plassein*: to mold + *desmos*: band] A cytoplasmic strand connecting two adjacent plant cells.

plasmodial slime molds *See* Appendix A.

plasmogamy The fusion of the cytoplasm of two cells. (Contrast with karyogamy.)

plate tectonics [Gk. *tekton*: builder] The scientific study of the structure and movements of Earth's lithospheric plates, which are the cause of continental drift.

platelet In blood, a membrane-enclosed body without a nucleus, arising as a fragment of a cell in the bone marrow of mammals. Important to blood-clotting action.

pleiotropic The influence of a single gene on more than one character.

pluripotent stem cells Stem cells that have the ability to form all of the cells of the developing embryo (except trophoblast cells) and so all the cells in the body. (Contrast with multipotent stem cells, totipotent stem cells.)

poikilotherm An animal in which the body temperature matches the temperature of the external environment and varies as the external temperature varies. Also called an ectotherm.

point mutation A mutation that results from the gain, loss, or substitution of a single nucleotide.

polar Having separate and opposite electric charges at two ends, or poles; the water molecule (H_2O) is the most prevalent example. (Contrast with nonpolar.)

polar auxin transport Transport of the plant hormone auxin in a particular direction (polarity). For example, localization of auxin efflux proteins (PIN proteins) in the basal side of cells in the stem can result in auxin transport being directed toward the roots.

polar bodies During female meiosis in some animals, tiny cells that are produced directly or indirectly by unequal division of the cytoplasm during cell division. They are reproductively nonfunctional and degenerate.

polar nuclei In angiosperms, the two nuclei in the central cell of the megagametophyte; following fertilization they give rise to the endosperm.

polarity (1) In chemistry, the property of unequal electron sharing in a covalent bond that defines a polar molecule. (2) In development, the difference between one end of an organism or structure and the other.

pollen [L. *pollin*: fine flour] In seed plants, microscopic grains that contain the male gametophyte (microgametophyte) and gamete (microspore).

pollen grains *See* pollen.

pollen tube A structure that develops from a pollen grain through which sperm are released into the megagametophyte.

pollination The process of transferring pollen from an anther to the stigma of a pistil in an angiosperm or from a strobilus to an ovule in a gymnosperm.

poly A tail A long sequence of adenine nucleotides (50–250) added after transcription to the 3′ end of most eukaryotic mRNAs.

polyclonal response An immune response against multiple epitopes, involving multiple clones of B cells.

polymer [Gk. *poly*: many + *meros*: unit] A large molecule made up of similar or identical subunits called monomers.

polymerase chain reaction (PCR) An enzymatic technique for the rapid production of millions of copies of a particular stretch of DNA where only a small amount of the parent molecule is available.

polymorphic (pol′ lee mor′ fik) [Gk. *poly*: many + *morphe*: form, shape] Referring to the coexistence in a population of two or more distinct traits.

polynucleotide A nucleic acid strand consisting of more than 20 nucleotides.

polyp (pah′ lip) [Gk. *poly*: many + *pous:* foot] In cnidarians, a sessile, asexual life cycle stage.

polypeptide A large molecule made up of many amino acids joined by peptide linkages. Large polypeptides are called proteins.

polyphyletic (pol′ lee fih leht′ ik) [Gk. *poly*: many + *phylon*: tribe] Pertaining to a group that consists of multiple distantly related organisms, and does not include the common ancestor of the group. (Contrast with monophyletic, paraphyletic.)

polyploidy (pol′ lee ploid ee) The possession of more than two entire sets of chromosomes.

polysaccharide A macromolecule composed of many monosaccharides (simple sugars). Common examples are cellulose and starch.

polyspermy The fertilization of a single female gamete by more than one male gamete.

pool The total amount of an element of molecule in a compartment of the Earth system at a particular time.

population (1) In ecology, a group of individuals of the same species that live, interact, and reproduce together in a particular geographic area. (2) In statistics, a set of similar items.

population bottleneck A period during which only a few individuals of a normally large population survive.

population density The number of individuals in a population per unit of area (for organisms that live in two-dimensional habitats such as the land surface) or volume (for organisms that live in three-dimensional habitats such as air, soil, or water).

population size The total number of individuals in a population.

positional information In development, the basis of the spatial sense that induces cells to differentiate as appropriate for their location within the developing organism; often comes in the form of a morphogen gradient.

positive feedback In a regulatory system, a type of control that acts to increase differences that arise between the level of a controlled variable and its set-point level. The period of amplifying deviation is followed by a period in which stabilization is restored in most biological systems. (Contrast with negative feedback.)

positive regulation Regulation in which binding of a regulatory protein increases the rate of transcription. (Contrast with negative regulation.)

positive relationship A relationship in which two variables tend to vary among observations in the same direction.

positive selection Natural selection that acts to establish a trait that enhances survival in a population. (Contrast with purifying selection.)

posterior Toward or pertaining to the rear. (Contrast with anterior.)

posterior pituitary gland A portion of the pituitary gland that is derived from neural tissue. In mammals, it is involved in the release of antidiuretic hormone and oxytocin, both of which are synthesized by hypothalamic neurosecretory cells. Also called neurohypophysis. (*See also* pituitary gland, anterior pituitary gland.)

postsynaptic cell A neuron or effector cell that receives a signal (chemical or electrical) from a presynaptic cell at a synapse.

postzygotic isolating mechanisms Barriers to the reproductive process that occur after the union of the nuclei of two gametes. (Contrast with prezygotic isolating mechanisms.)

potential difference *See* voltage.

potential energy Stored energy, such as the energy stored in chemical bonds. (Contrast with kinetic energy.)

power (1) In reference to statistical tests, the probability of correctly rejecting a null hypothesis when it is false. (2) The *rate* at which work is done.

Pr *See* phytochrome.

Precambrian The first three eons of geological time, during which life originated.

precursor RNA (pre-mRNA) The initial gene transcript before it is modified to produce functional mRNA. Also known as the primary transcript.

predation A "+/–" interaction between individuals of two different species in which one individual (the predator) gains fitness by killing and consuming the other individual (the prey), which loses fitness as a consequence.

prepenetration apparatus (PPA) In plants forming mycorrhizal associations, a sub-cellular structure that spans a root cell creating a pathway through which the fungus can pass through the cell.

pressure flow model An effective model for phloem transport in angiosperms. It holds that sieve element transport is driven by an osmotically generated pressure gradient between source and sink.

pressure potential (Ψ_p) The hydrostatic pressure of an enclosed solution in excess of the surrounding atmospheric pressure. (Contrast with turgor pressure, solute potential, water potential.)

presynaptic axon terminal One of the end processes of an axon, where the axon terminates at a synapse.

presynaptic cell A neuron that transmits a signal to a postsynaptic cell at a synapse. (Contrast with postsynaptic cell.)

prezygotic isolating mechanisms Barriers to the reproductive process that occur before the union of the nuclei of two gametes. (Contrast with postzygotic isolating mechanisms.)

priapulids *See* Appendix A.

primary active transport Active transport in which ATP is hydrolyzed, yielding the energy required to transport an ion or molecule against its concentration gradient. (Contrast with secondary active transport.)

primary cell wall In plant cells, a structure made up of cellulose microfibrils, hemicelluloses, and pectins that forms after cytokinesis and is located exterior to the cell membrane. (Contrast with secondary cell wall.)

primary cilium A nonmotile hairlike organelle used for signaling in vertebrate cells.

primary consumer An organism that consumes primary producers.

primary endosymbiosis The engulfment of a cyanobacterium by a larger eukaryotic cell that gave rise to the first photosynthetic eukaryotes with chloroplasts.

primary immune response The first response of the immune system to an antigen, involving recognition by lymphocytes and the production of effector cells and memory cells. (Contrast with secondary immune response.)

primary lysosome A lysosome before fusion with an endosome. (Contrast with secondary lysosome.)

primary meristem Meristem that produces the tissues of the primary plant body.

primary phloem Phloem tissue produced independent of the vascular cambium (generally, before the vascular cambium forms).

primary producer A photosynthetic or chemosynthetic organism that synthesizes complex organic molecules from simple inorganic ones.

primary reproductive organs The gonads: the ovaries in females and the testes (or testicles) in males.

primary structure The specific sequence of amino acids in a protein. (Contrast with secondary, tertiary, quaternary structure.)

primary xylem Xylem tissue produced independent of the vascular cambium (generally, before the vascular cambium forms).

primase An enzyme that catalyzes the synthesis of a primer for DNA replication.

primates *See* Appendix A.

primer A strand of nucleic acid, usually RNA, that is the necessary starting material for the synthesis of a new DNA strand, which is synthesized from the 3′ end of the primer.

primitive streak A midline ridge that forms in the avian blastula that will become the blastopore.

primordium An organ or tissue in its earliest recognizable stage of development.

principle of allocation The principle that a unit of resource (such as food) cannot be used simultaneously for multiple functions (such as growth and reproduction), but instead must be allocated to one or another function.

probability A numerical quantity that expresses the likelihood of an event occurring on a scale from 0 (no chance of the event) to 1 (certainty of the event).

probe A segment of single-stranded nucleic acid used to identify DNA molecules containing the complementary sequence.

proboscis A hollow, muscular feeding organ.

processes The ways in which the components of a biological system interact (e.g., protein synthesis, nutrient metabolism, grazing).

products The molecules that result from the completion of a chemical reaction.

progesterone [L. *pro*: favoring + *gestare*: to bear] A female steroid sex hormone that maintains pregnancy. Secreted by the corpus luteum and placenta.

prokaryotes *See* Appendix A.

prokaryotic cells Cells that do not contain (and have never contained) a nucleus.

prometaphase The phase of nuclear division that begins with the disintegration of the nuclear envelope.

promoter A DNA sequence to which RNA polymerase binds to initiate transcription.

proofreading During DNA replication, a mechanism that excises a base that is incorrectly inserted according to the template (e.g., an A opposite a template G) and inserts the correct base (e.g., C opposite template G).

prophase (pro′ phase) The first stage of nuclear division, during which chromosomes condense from diffuse, threadlike material to discrete, compact bodies.

prostaglandin Any one of a group of specialized lipids with hormone-like functions. It is not clear that they act at any considerable distance from the site of their production.

proteasome In the eukaryotic cytoplasm, a huge protein structure that binds to and digests cellular proteins that have been tagged by ubiquitin.

protein [Gk. *protos*: first] A compound consisting of one, or more polypeptides. Occurs with its polypeptide chains extended in fibrous proteins, or coiled into a compact macromolecule in enzymes and other globular proteins.

proteobacteria *See* Appendix A.

proteoglycan A glycoprotein containing a protein core with attached long, linear carbohydrate side chains.

proteolysis [protein + Gk. *lysis*: break apart] An enzymatic digestion of a protein or polypeptide.

proteome The set of proteins that can be made by an organism. Because of alternative splicing of pre-mRNA, the number of proteins that can be made is usually much larger than the number of protein-coding genes present in the organism's genome.

proteomics The study of the proteome—the complete complement of proteins produced by an organism.

protists *See* Appendix A.

proton (pro′ ton) [Gk. *protos*: first, before] (1) One of the three fundamental subatomic particles of matter (along with neutrons and electrons), part of the atomic nucleus, with a single positive charge. The number of protons in the nucleus of an atom determines its element. (2) A hydrogen ion, H^+.

proton pump An active transport system that uses ATP energy to move hydrogen ions across a membrane, generating an electric potential.

protostomes *See* Appendix A.

prototherians *See* Appendix A.

provirus Double-strand DNA made by a virus that is integrated into the host's chromosome and contains promoters that are recognized by the host cell's transcription apparatus.

proximate causes (1) In the study of behavior, the immediate mechanistic causes of an individual's behavior. (2) In the study of evolution, the immediate genetic, physiological, neurological, and developmental explanations for the advantages of an adaptation. (Contrast with ultimate causes.)

proximate explanations The immediate genetic, physiological, neurological, and developmental explanations for the advantages of an adaptation. (Contrast with ultimate explanations.)

pseudocoelomate (soo′ do see′ low mate) [Gk. *pseudes*: false + *koiloma*: cavity] An animal that has a body cavity, called a pseudocoel, consisting of a fluid-filled space in which many of the internal organs are suspended, but which is enclosed by mesoderm only on its outside.

pseudogene [Gk. *pseudes*: false] A DNA segment that is homologous to a functional gene but is not expressed because of changes to its sequence or changes to its location in the genome.

pseudoplasmodium An aggregate of individuals in cellular slime molds that act in a coordinated fashion to form a fruiting structure.

pterygotes *See* Appendix A.

Punnett square A method of predicting the results of a genetic cross by arranging the gametes of each parent at the edges of a square.

pupil The opening in the vertebrate eye through which light passes from the outside to the retina.

purifying selection The elimination by natural selection of detrimental characters from a population. (Contrast with positive selection.)

purine (pure′ een) One of the two types of double-ring nitrogenous bases in nucleic acids. Each of the purines—adenine and guanine—pairs with a specific pyrimidine.

pyrimidine (per im′ a deen) One of the two types of single-ring nitrogenous bases in nucleic acids. Each of the pyrimidines—cytosine, thymine, and uracil—pairs with a specific purine.

Q

qualitative traits Traits that differ from one another by discrete qualities (e.g., black vs. white or smooth vs. wrinkled).

quantify To turn an observations into explicit counts or measures that can be used and verified by others.

quantitative trait Traits that show *continuous* quantitative variation rather than *discrete* qualitative variation.

quantitative variables Variables that can take on values along a numerical scale.

quaternary structure The specific three-dimensional arrangement of polypeptide subunits in a protein composed of multiple polypeptides. (Contrast with primary, secondary, tertiary structure.)

quorum sensing The use of chemical communication signals to trigger density-linked activities such as biofilm formation in prokaryotes.

R

R group The distinguishing group of atoms of a particular amino acid. Also known as a side chain.

R_0 The average number of offspring produced by an individual during its lifetime.

radial cleavage Embryonic development in some deuterostomes in which the planes of cell division are parallel and perpendicular to the animal–vegetal axis of the embryo. Considered the basic deuterostome cleavage pattern. (Contrast with spiral cleavage.)

radial symmetry The condition in which any two halves of a body are mirror images of each other, providing the cut passes through the center; a cylinder cut lengthwise down its center displays this form of symmetry.

radioactive decay Spontaneous disintegration of a substance with the emission of ionizing radiation.

radiolarians *See* Appendix A.

radiometric dating A method for determining the age of objects such as fossils and rocks based on the decay rates of radioactive isotopes.

rain shadow The relatively dry area on the downwind side of a mountain range.

range The largest minus the smallest observed value for a variable in a sample.

rarity advantage Any situation in which individuals of a species lose less fitness from interspecific competition when they are rare than when they are common.

RAS A G protein involved in activating cell division.

ray-finned fishes *See* Appendix A.

reactant A chemical substance that enters into a chemical reaction with another substance.

reaction center A group of electron transfer proteins that receive energy from light-absorbing pigments and convert it to chemical energy by redox reactions.

reaction rate In a chemical reaction, a measure of how fast products are made per unit of time.

receptor potential The change in the membrane potential of a sensory cell when it is stimulated.

receptor protein A protein that can bind to a specific molecule (ligand), or detect a specific stimulus, within the cell or in the cell's external environment.

receptor-mediated endocytosis Endocytosis initiated by macromolecular binding to a specific membrane receptor.

recessive In genetics, an allele that does not determine phenotype in an individual who is heterozygous for it and a different (dominant) allele. (Contrast with dominance.)

reciprocal cross A cross in which the genotypes of the male and female parent are switched.

recombinant chromatid In meiosis, a chromatid after crossing over that contains part of a sister chromatid, typically having exchanged that part with the sister chromatid.

recombinant DNA A DNA molecule made in the laboratory that is derived from two or more genetic sources.

recombinant gametes Any genotype of gametes produced by an individual that is different from the two gametes that fused to make the diploid cell undergoing meiosis.

recombination frequency The proportion of offspring of a genetic cross that received gametes that are different from the gametes that fused to make the diploid cell undergoing meiosis due to crossing over between linked genes and independent assortment.

red algae *See* Appendix A.

red blood cell (RBC) A cell in the blood of an animal that contains hemoglobin and transports O_2. Also called an erythrocyte.

redox reaction A chemical reaction in which one reactant becomes oxidized and the other becomes reduced. Short for reduction–oxidation reaction.

reduction–oxidation reaction *See* redox reaction.

regional hypothermia A phenomenon in which appendage tissues are cooler than body core tissues.

regulation The maintenance of internal conditions at an approximately constant level while external conditions vary. (Contrast with conformity.)

regulative development A pattern of animal embryonic development in which the fates of the first blastomeres are not fixed. (Contrast with mosaic development.)

regulator An animal that maintains relatively constant internal conditions in the presence of changing external conditions (i.e., the animal exhibits regulation.) (Contrast with conformer.)

regulatory genes Genes that encode proteins that controls the rates of biological processes.

regulatory network A collection of interacting regulators that determine the expression levels of a gene.

regulatory protein A protein that controls the rate of a biological process.

regulatory system A system that uses feedback information to maintain a physiological function or parameter at an optimal level. (Contrast with controlled system.)

regulatory T cells (Tregs) The class of T cells that mediates tolerance to self antigens.

reinforcement The evolution of enhanced reproductive isolation between populations due to natural selection for greater isolation.

releasing hormone (RH) A hormone secreted by neuroendocrine cells in the hypothalamus of a vertebrate that travels to the anterior pituitary gland through the hypothalamo–hypophysial portal system and stimulates the secretion of a hormone by a specific population of anterior pituitary endocrine cells. (Contrast with inhibiting hormone.)

replication fork A point at which a DNA molecule is replicating. The fork forms by the unwinding of the parent molecule.

reporter gene A genetic marker included in recombinant DNA to indicate the presence of the recombinant DNA in a host cell.

repressible genes Genes that are usually expressed but can be shut off when their products are not needed. (Contrast with constitutive and inducible genes.)

repressor A protein encoded by a regulatory gene that can bind to a promoter and prevent transcription of the associated gene. (Contrast with activator.)

reproductive growth In a plant, a period of growth during which the plant produces flowers and seeds.

reproductive isolation Condition in which two divergent populations are no longer exchanging genes. Can lead to speciation.

reptiles *See* Appendix A.

residuals The deviations, along the y axis from the linear regression line, of individual observations in a bivariate scatter plot.

residue A subunit of a macromolecule.

resistance (R) proteins Plant proteins that allow a pathogen to be recognized by directly or indirectly serving as a receptor for a pathogen-produced factor.

resistance exercise Exercise that consists of relatively short periods of high-intensity muscular actions against a large load, often repeated with intervening interruptions. Also called resistance training. (Contrast with endurance exercise.)

resource partitioning A situation in which competing organisms (usually of different species) differ in their use of resources.

resources Materials and energy for organisms.

respiratory airways Airways in the lungs where O_2 and CO_2 are exchanged between the air and blood. (Contrast with conducting airways.)

respiratory gases Oxygen (O_2) and carbon dioxide (CO_2).

respiratory minute volume In an animal with lungs, the total volume of air inhaled and exhaled per minute.

respiratory pigments Any of the blood pigments that undergo reversible combination with O_2 and thus are able to pick up O_2 in certain places in an animal's body (e.g., the breathing organs) and release it in other places (e.g., systemic tissues). Respiratory pigments include hemoglobin and hemocyanin.

resting potential The potential difference (voltage) across the cell membrane of a living cell at rest. In cells at rest, the interior is negative to the exterior. (Contrast with action potential.)

restoration ecology The applied subdiscipline of ecology whose goal is to restore the function of damaged ecosystems.

restriction enzyme Any of a type of enzyme that cleaves double-stranded DNA at specific sites; extensively used in recombinant DNA technology. Also called a restriction endonuclease.

restriction point (R) The specific time during G1 of the cell cycle at which the cell becomes committed to undergo the rest of the cell cycle.

restriction site A specific DNA base sequence that is recognized and acted on by a restriction enzyme.

retina (ret′ in uh) [L. *rete*: net] The light-sensitive layer of cells in the vertebrate eye and some invertebrate eyes. In the vertebrate eye, it consists of photoreceptor cells and cells that process signals from the photoreceptor cells.

retrotransposons Mobile genetic elements that are reverse transcribed into RNA as part of their transfer mechanism. (Contrast with DNA transposons.)

retrovirus An RNA virus that contains reverse transcriptase. Its RNA serves as a template for cDNA production, and the cDNA is integrated into a chromosome of the host cell.

reverse genetics An approach that begins with a gene and attempts to determine its function, often by examining what happens when the gene is knocked out.

reverse transcriptase An enzyme that catalyzes the production of DNA (cDNA), using RNA as a template; essential to the reproduction of retroviruses.

rhizarians *See* Appendix A.

rhizoids (rye′ zoids) [Gk. root] Hairlike extensions of cells in mosses, liverworts, and a few vascular plants that serve the same function as roots and root hairs in vascular plants. The term is also applied to branched, rootlike extensions of some fungi and algae.

rhizome (rye′ zome) An underground stem (as opposed to a root) that runs horizontally beneath the ground.

rhizosphere In plants, the region of soil close enough to a root to be affected by the presence of the root. The composition of this region changes over time depending on the environment and state of the plant.

rhodopsin A vertebrate visual pigment (a type of visual opsin) involved in transducing photons of light into changes in the membrane potential of certain photoreceptor cells, the rod cells. Sometimes in the past used as a general term that refers to all visual opsins.

rhyniophytes *See* Appendix A.

ribbon worms *See* Appendix A.

ribosomal RNA (rRNA) Several species of RNA that are incorporated into the ribosome. Involved in peptide bond formation.

ribosome A structure in the cell that is the site of protein synthesis.

ribozyme An RNA molecule with catalytic activity.

ribulose bisphosphate carboxylase/oxygenase *See* rubisco.

RNA (ribonucleic acid) An often single-stranded nucleic acid whose nucleotides use ribose rather than deoxyribose and in which the base uracil replaces thymine found in DNA. Serves as genome from some viruses. (*See* ribosomal RNA, transfer RNA, messenger RNA, ribozyme.)

RNA interference (RNAi) A mechanism for reducing mRNA translation whereby a double-strand RNA, made by the cell or synthetically, is processed into a small, single-strand RNA, whose binding to a target mRNA results in the latter's breakdown.

RNA sequencing (RNA-seq) The sequencing of all of the complementary DNA in a sample, obtained from the RNA by reverse transcription.

RNA splicing The last stage of RNA processing in eukaryotes, in which the transcripts of introns are excised through the action of small nuclear ribonucleoprotein particles (snRNP).

rod cells Light-sensitive cells in the vertebrate retina that are sensitive to extremely low-intensity light and are responsible for black-and-white vision and vision in dim light.

root (1) In reference to phylogenetic trees: the base (oldest) part of the tree. (2) In reference to plants: the organ responsible for anchoring the plant in the soil, absorbing water and minerals, and producing certain hormones. Some roots are storage organs.

root apical meristem Undifferentiated tissue at the apex of the root that gives rise to the organs of the root.

root cap A thimble-shaped mass of cells, produced by the root apical meristem, that protects the meristem; the organ that perceives the gravitational stimulus in root gravitropism.

root hair A long, thin process from a root epidermal cell that absorbs water and minerals from the soil solution.

root meristem Undifferentiated tissue at the apex of a primary or lateral root that gives rise to the organs of the root. Can be used interchangeably with root apical meristem but preferred by some as it avoids the confusion that can arise from the root apical meristem being located at the base of the plant.

root system The organ system that anchors a plant in place, absorbs water and dissolved minerals, and may store products of photosynthesis from the shoot system.

rotational cleavage A cleavage pattern typical of eutherian mammals in which the first cleavage is meridional whereas in the second cleavage, one blastomere divides meridionally and the other equatorially.

rotifers *See* Appendix A.

rough endoplasmic reticulum (RER) The portion of the endoplasmic reticulum in which the outer surface has attached ribosomes. (Contrast with smooth endoplasmic reticulum.)

rubisco Contraction of ribulose bisphosphate carboxylase/oxygenase, the enzyme that combines CO_2 or O_2 with ribulose bisphosphate to catalyze the first step of photosynthetic carbon fixation or photorespiration, respectively.

S

S phase In the cell cycle, the stage of interphase during which DNA is replicated. (Contrast with G0, G1, G2, M phase.)

sac fungi *See* Appendix A.

salt glands (1) In plants, glands on the leaves of some halophytic (salt-loving) plants that secrete salts onto the outer plant surfaces, thereby ridding the tissue fluids of the plants of excess salts. (2) In animals, organs other than kidneys that excrete concentrated salt solutions, thereby ridding the body fluids of excess salts.

sample A set of observations made from a population.

saprobe [Gk. *sapros*: rotten] An organism (usually a bacterium or fungus) that obtains its carbon and energy by absorbing nutrients from dead organic matter.

saprobic Feeding on dead organic matter.

sarcomere (sark′ o meer) [Gk. *sark*: flesh + *meros*: unit] The contractile unit of a muscle cell. Each myofibril in a muscle cell consists of a series of sarcomeres, each composed of actin and myosin filaments.

sarcoplasmic reticulum The endoplasmic reticulum of a muscle cell, used for Ca^{2+} storage and release.

saturated fatty acid A fatty acid in which all the bonds between carbon atoms in the hydrocarbon chain are single bonds—that is, all the bonds are saturated with hydrogen atoms. (Contrast with unsaturated fatty acid.)

scaling relationships Relationships between physiological (or morphological) characteristics and body size within sets of phylogenetically related species (e.g., metabolism–weight relationships among mammals of different sizes).

scatter plot A figure that displays the values of observations for two variables along perpendicular axes.

Schwann cell A type of glial cell that forms a myelin sheath on axons in the peripheral nervous system.

scion In horticulture, the bud or stem from one plant that is grafted to a root or root-bearing stem of another plant (the stock).

sclereid One of the principle types of cells in sclerenchyma.

sclerenchyma (skler eng′ kyma) [Gk. *skleros*: hard + *kymus*: juice] A plant tissue composed of cells with heavily thickened cell walls. The cells are dead at functional maturity. The principal types of sclerenchyma cells are fibers and sclereids. (Contrast with parenchyma, collenchyma.)

scrotum In the males of most mammals, a pouch outside the abdomen that contains the testes (testicles), providing them with a lower temperature than prevails in the abdominal cavity.

seasonality Predictable variation in the climate of a given place during the course of an annual cycle.

second filial generation *See* F_2.

second law of thermodynamics The principle that when energy is converted from one form to another, some of that energy becomes unavailable for doing work.

second messenger A compound, such as cAMP, that is released within a target cell after the first messenger has bound to a surface receptor on a cell; the second messenger triggers further reactions in the cell.

secondary active transport A form of active transport that does not use ATP as an energy source; rather, transport is coupled to ion diffusion down a concentration gradient established by primary active transport. Also called coupled transport or co-transport. (Contrast with primary active transport.)

secondary cell wall A thick, cellulosic structure internal to the primary cell wall formed in some plant cells after cell expansion stops. (Contrast with primary cell wall.)

secondary consumer An organism that consumes primary consumers.

secondary endosymbiosis The engulfment of a photosynthetic eukaryote by another eukaryotic cell that gave rise to certain groups of photosynthetic eukaryotes (e.g., euglenids).

secondary growth In plants, growth that contributes to an increase in girth. (Contrast with primary growth.)

secondary immune response A rapid and intense response to a second or subsequent exposure to an antigen, initiated by memory cells. (Contrast with primary immune response.)

secondary lysosome A membrane-enclosed organelle formed by the fusion of a primary lysosome with a phagosome, in which macromolecules taken up by phagocytosis are hydrolyzed into their monomers. (Contrast with primary lysosome.)

secondary meristem A lateral meristem such as the vascular cambium, generally adds to the width of the plant. (Contrast with apical meristem.)

secondary metabolite A compound synthesized by a plant that is not needed for basic cellular metabolism. Typically has an antiherbivore or antiparasite function.

secondary phloem The inner bark of a woody plant; this part of the phloem is produced by the vascular cambium.

secondary plant body The portion of the plant body produced by the secondary meristems.

secondary sexual characteristics Sex-specific properties of nonreproductive tissues and organs.

secondary structure In reference to a protein, localized regularities of structure, such as the α helix and the β pleated sheet. (Contrast with primary, tertiary, quaternary structure.)

secondary xylem Also called wood, this part of the xylem is produced by the vascular cambium.

secrete To discharge a substance from a cell or gland.

sedimentary rock Rock formed by the accumulation of sediment grains on the bottom of a body of water.

seed A fertilized, ripened ovule of a gymnosperm or angiosperm. Consists of the embryo, nutritive tissue, and a seed coat.

seed coat A covering over the seed.

segment polarity genes In *Drosophila* (fruit fly) development, segmentation genes that determine the boundaries and anterior–posterior organization of individual segments. Part of a developmental cascade that includes maternal effect genes, gap genes, pair rule genes, and Hox genes.

segmentation The division of an animal body into segments.

segmentation genes Genes that determine the number and polarity of body segments.

selectable marker A gene, such as one encoding resistance to an antibiotic, that can be used to identify (select) cells that contain recombinant DNA from among a large population of untransformed cells.

selective permeability In reference to membranes, the characteristic of allowing certain substances to pass through while other substances are excluded.

self-fertilization A cross in which the male and female parent are the same individual.

self-incompatibility In plants, the possession of mechanisms that prevent self-fertilization.

self-pollination Fertilization of an egg by pollen from the same individual. (Contrast with outcrossing.)

selfing *See* self-fertilization.

semelparity A type of reproductive life history in which individuals are physiologically capable of reproducing only one time during their lives. In **semelparous** species, individuals are often programmed to die after reproducing once. (Contrast with iteroparity.)

semen (see′ men) [L. *semin*: seed] The fluid ejaculated by the male during sexual intercourse, consisting of sperm suspended in secretions of exocrine reproductive glands, such as the seminal vesicles and prostate gland.

sense codons A sequence of three nucleotides in an mRNA that encodes a particular amino acid.

sensor One of four essential elements of a control system; an organ or cell that detects the current level of the controlled variable.

sensory neuron A neuron that functions as a sensory receptor cell or that carries signals from sensory cells or organs to the central nervous system. Also called an afferent neuron.

sensory receptor cell A cell, typically a neuron, that is specialized to transform the energy of a stimulus into an electric signal (change of membrane potential). Each type is highly specific in the stimuli to which it ordinarily responds.

sensory receptor protein A protein in a sensory receptor cell that responds to sensory input, causing a receptor potential (a graded change of membrane potential) in the receptor cell.

sepal (see′ pul) [L. *sepalum*: covering] One of the outermost structures of the flower, usually protective in function and enclosing the rest of the flower in the bud stage.

sepsis Generalized inflammation caused by bacterial infection. Can cause a dangerous drop in blood pressure.

septate [L. wall] Divided, as by walls or partitions.

septum (plural: septa) (1) A partition or cross-wall appearing in the hyphae of some fungi. (2) The bony structure dividing the nasal passages.

sequential hermaphrodite An individual that is male at some times and female at other times during its lifespan.

Sertoli cells Cells in the seminiferous tubules that nurture the developing sperm. These cells account for over half of testis volume in mammals.

sex *See* sexual reproduction.

sex chromosome In organisms with a chromosomal mechanism of sex determination, one of the chromosomes involved in sex determination.

sex determination The process by which the sex (male or female) of an individual becomes established during the conception and development of the individual.

sex pilus (plural: pili) A thin projection of a prokaryotic cell that can attach to other cells, triggering the formation of a mating pilus.

sex-linked inheritance Inheritance of a gene that is carried on a sex (usually the X) chromosome. Also called sex linkage.

sexual reproduction Reproduction in which the genes of two individuals are combined to produce offspring, typically involving the union of male and female gametes.

sexual selection Selection by one sex of characteristics in individuals of the opposite sex. Also, the favoring of characteristics in one sex as a result of competition among individuals of that sex for mates.

shiver (1) In a mammal or bird, the subtle contraction and relaxation of skeletal muscles that produces heat rather than motion as the primary product. (2) In an insect, contraction of the flight muscles in a nonflying mode to generate heat rather than flight.

shoot apical meristem Undifferentiated tissue at the apex of the shoot that gives rise to the organs of the shoot.

shoot system In plants, the organ system consisting of the leaves, stem(s), and flowers.

short-day plant (SDP) A plant that flowers when nights are longer than a critical length specific for that plant's species. (Compare with long-day plant.)

side chain *See* R group.

sieve plate The end wall of a sieve tube element that contains a set of pores.

sieve tube element The characteristic cell of the phloem in angiosperms, which contains cytoplasm but relatively few organelles, and whose end walls (sieve plates) contain pores that form connections with neighboring cells.

sigma factor In prokaryotes, a protein that binds to RNA polymerase, allowing the complex to bind to and stimulate the transcription of a specific class of genes (e.g., those involved in sporulation).

signal protein A protein that is used to communicate with other cells to elicit a response.

signal sequence The sequence within a protein that directs the protein to a particular organelle.

signal transduction cascade A multistep signal transduction pathway.

signal transduction The series of events that occur between the binding of a signal to its receptor and the cellular response.

significance level A particular threshold of making a Type I error (incorrectly rejecting a true null hypothesis) in a statistical test, selected a priori by the investigator.

silencer A gene sequence binding transcription factors that repress transcription. (Contrast with promoter.)

silent mutation A change in a gene's sequence that has no effect on the amino acid sequence of a protein because it occurs in noncoding DNA or because it does not change the amino acid specified by the corresponding codon. (Contrast with frame-shift mutation, missense mutation, nonsense mutation.)

silent substitution *See* synonymous substitution.

simians *See* Appendix A.

simple diffusion Diffusion that doesn't involve a direct input of energy or assistance by integral membrane proteins.

simple epithelium An epithelium (sheet of cells that lines a body cavity or covers an organ or body surface) consisting of a single cell layer.

simple fruit A fruit derived from a single ovary; examples include grapes and tomatoes.

simple leaves Leaves that have a single blade and not more than one petiole. (Contrast with compound leaves.)

single nucleotide polymorphisms (SNPs) Inherited variations in a single nucleotide base in DNA that differ between individuals.

sink In plants, any organ that imports the products of photosynthesis, such as roots, developing fruits, and immature leaves. (Contrast with source.)

sister chromatid Each of a pair of newly replicated chromatids.

sister clades Two phylogenetic groups that are each other's closest relatives.

sister species Two species that are each other's closest relatives.

skeletal muscle A type of muscle tissue, often connected to bones, characterized by multinucleated cells containing highly ordered arrangements of actin and myosin filaments that give the tissue a striated appearance under a light microscope. A type of striated muscle. (Contrast with cardiac muscle, smooth muscle.)

sliding-filament theory The theory that the mechanism of muscle contraction (use of ATP to produce contractile forces) is based on the formation and flexing of cross-bridges between actin and myosin filaments, causing the filaments to slide past each other.

slow oxidative cells Skeletal muscle cells that are highly dependent on aerobic ATP production by the citric acid cycle and electron transport, and that are specialized for sustained, relatively low-intensity work; contain myoglobin and abundant mitochondria. Also called slow-twitch cells or red cells. (Contrast with fast glycolytic cells.)

small interfering RNAs (siRNAs) Short, double-strand RNA molecules used in RNA interference.

small nuclear ribonucleoprotein particle (snRNP) A complex of an enzyme and a small nuclear RNA molecule, functioning in RNA splicing.

smooth endoplasmic reticulum (SER) The portion of the endoplasmic reticulum that lacks ribosomes and has a tubular appearance. (Contrast with rough endoplasmic reticulum.)

smooth muscle Nonstriated muscle tissue consisting of small, mononucleated muscle cells innervated by the autonomic nervous system. Found, for example, in the walls of the gut, blood vessels, and reproductive organs. (Contrast with cardiac muscle, skeletal muscle.)

social behavior Behavior that helps integrate individuals into social groups (societies), and the group behaviors of social groups. (*See also* society.)

society A group of individuals of a single species that exhibits some degree of cooperative action.

sodium–potassium (Na^+–K^+) pump An integral membrane protein (anti-porter) that carries out primary active transport of ions; it uses energy from ATP to pump sodium ions out of a cell and potassium ions into the cell. Also called a sodium–potassium ATPase.

solute potential (Ψ_s) A property of any solution, resulting from its solute contents; it may be zero or have a negative value. The more negative the solute potential, the greater the tendency of the solution to take up water through a differentially permeable membrane. (Contrast with pressure potential, water potential.)

solvent Liquid in which a substance (solute) is dissolved to form a solution.

somatic cells [Gk. *soma*: body] All the cells of the body that are not specialized for reproduction. (Contrast with germ cells.)

somite (so' might) A segmental block of mesoderm; differentiates to form dermis, axial skeleton (vertebrae, ribs), and skeletal muscle.

soredia (sing.: soredium) Propagules of lichens consisting of one or a few photosynthetic cells bound by fungal hyphae.

sori (sing.: sorus) Clusters of stalked sporangia that occur on the underside of fern fronds.

source In plants, any organ that exports the products of photosynthesis in excess of its own needs, such as a mature leaf or storage organ. (Contrast with sink.)

Southern blotting Named after the scientist, Edwin Southern. A technique for identifying a specific sequence of DNA.

spawning *See* external fertilization.

speciation The process of splitting one biological lineage into two biological lineages that evolve independently from one another.

species [L. kind] The base unit of taxonomic classification, consisting of an ancestor–descendant lineage of evolutionarily closely related, similar organisms that share a suite of genetic and morphological attributes, and are reproductively isolated from other such groups.

species composition The identities of all the species that make up a given ecological community.

species concepts Ways that biologists think about the existence of species.

species diversity In ecology, a measure of the chance that two individuals in a community are of different species. Measures of species diversity incorporate the number of distinct species in a community (species richness) and the relative abundances of different species (species evenness).

species evenness A measure of the similarity in the abundances of species in a community. If species are equally abundant, evenness is at a maximum.

species richness The total number of species in an ecological community.

specific heat The amount of energy that must be absorbed by a gram of a substance to raise its temperature by 1 degree Celsius. By convention, water is assigned a specific heat of one.

sperm [Gk. *sperma*: seed] *See* spermatozoa.

spermatozoa Haploid gametes that are capable of swimming (usually by use of flagella), produced by spermatogenesis in the testis of a male animal. Also called sperm.

spicule [L. arrowhead] A hard, calcareous skeletal element typical of sponges.

spinal nerves Nerves that extend laterally from the spinal column. One spinal nerve goes to the left and another to the right in association with each vertebra. Composed of both sensory and motor neurons.

spinal reflex In a vertebrate, the initiation of motor neuronal signals in response to sensory neuronal signals in the spinal cord without participation of the brain.

spindle Array of microtubules emanating from both poles of a dividing cell during mitosis and playing a role in the movement of chromosomes at nuclear division. Named for its shape.

spiral cleavage Embryonic development in many protostomes in which the plane of cell division is diagonal to the vertical axis of the embryo. (Contrast with radial cleavage.)

spirillum (plural *spirilla*) A bacterium that is shaped like a corkscrew.

spirochetes *See* Appendix A.

spliceosome An RNA–protein complex that splices out introns from eukaryotic pre-mRNAs.

sponges *See* Appendix A.

spongy mesophyll Chlorophyll-containing cells in a leaf, generally on the bottom side, that are lobed and separated by air spaces.

spontaneous mutation A genetic change caused by internal cellular mechanisms, such as an error in DNA replication. (Contrast with induced mutation.)

spontaneous ovulation Ovulation (release of an egg from the ovaries of a female) that results from endogenous processes within the female, more or less independent of copulation. (Contrast with induced ovulation.)

sporangiophore A stalked reproductive structure produced by zygospore fungi that extends from a hypha and bears one or many sporangia.

sporangium (spor an′ gee um) (plural: sporangia) [Gk. *spora*: seed + *angeion*: vessel or reservoir] In plants and fungi, any specialized structure within which one or more spores are formed.

spore [Gk. *spora*: seed] (1) Any asexual reproductive cell capable of developing into an adult organism without gametic fusion. In plants, haploid spores develop into gametophytes, diploid spores into sporophytes. (2) In prokaryotes, a resistant cell capable of surviving unfavorable periods.

sporocyte Specialized cells of the diploid sporophyte that will divide by meiosis to produce four haploid spores. Germination of these spores produces the haploid gametophyte.

sporophyte (spor′ o fyte) [Gk. *spora*: seed + *phyton*: plant] In plants and protists with alternation of generations, the diploid phase that produces the spores. (Contrast with gametophyte.)

sporulation The formation of specialized cells (spores) that are capable of developing into new individuals.

squamates *See* Appendix A.

stabilizing selection Selection against the extreme phenotypes in a population, so that the intermediate types are favored. (Contrast with directional selection, disruptive selection.)

stamen (stay′ men) [L. *stamen*: thread] A male (pollen-producing) unit of a flower, usually composed of an anther, which bears the pollen, and a filament, which is a stalk supporting the anther.

standard amino acid One of the 20 amino acids that animals use to synthesize proteins.

standard deviation A measure of the spread of observations in a sample. *See* Appendix B for the mathematical formula.

standard error A measure of how close a sample statistic (such as the mean) is likely to be to the true population value.

standard error of the mean A measure of how close a sample mean is likely to be to the true population value. Calculated by dividing the standard deviation of a sample by the square root of the sample size.

staphylococci *See* Appendix A.

starch A polymer of glucose; used by plants to store energy.

start codon The mRNA triplet (AUG) that acts as a signal for the beginning of translation at the ribosome. (Contrast with stop codon.)

state transition A shift in the composition of an ecological community following disturbance to a stable alternative community that is distinctly different from the pre-disturbance community.

statistic A numerical quantity calculated from data.

stem In plants, the organ that holds leaves or flowers and transports and distributes materials among the other organs of the plant.

stem cell An undifferentiated cell that is capable of continuous proliferation. A stem cell generates more stem cells and a large clone of differentiated progeny cells. (*See also* embryonic stem cell.)

stereoisomers Molecules made up of the same kinds and numbers of atoms, in which the atoms have the same bonds, but different three-dimensional geometry.

steroids A four-ringed lipid molecule (examples include cholesterol in membranes and steroid hormones).

stigma [L. *stigma*: mark, brand] The part of the pistil at the apex of the style that is receptive to pollen, and on which pollen germinates.

stock In horticulture, the root or root-bearing stem to which a bud or piece of stem from another plant (the scion) is grafted.

stolon A stem that grows horizontally and forms roots at intervals. Also called a runner.

stoma (plural: stomata) [Gk. *stoma*: mouth, opening] Small opening in the plant epidermis that permits gas exchange; bounded by a pair of guard cells, the osmotic status of which regulates the size of the opening.

stomatal compromise Amount of stomatal opening that effectively balances the requirements for CO_2 entry and water retention.

stomatal guard cells In plants, specialized, paired epidermal cells that surround and control the opening of a stoma (pore).

stoneworts *See* Appendix A.

stop codon Any of the three mRNA codons that signal the end of protein translation at the ribosome: UAG, UGA, UAA. Also called nonsense codon. (Contrast with start codon.)

stop mutation *See* nonsense mutation.

storage protein A protein that functions to store amino acids for protein synthesis or energy.

stramenopiles *See* Appendix A.

stratigraphy The study of geological strata, or layers.

stratum (plural: strata) [L. *stratos*: layer] A layer of sedimentary rock laid down at a particular time in the past.

streptophytes *See* Appendix A.

stress-relaxation cycle The process by which plant cells increase in size. Changes in the structure of the cell wall result in a decrease in stress within the wall, thereby lowering turgor pressure in the cell and allowing more water to enter.

stretch-gated channel A membrane ion channel that opens and closes in response to local stretching or pulling forces applied to the cell membrane.

strigolactones Signaling molecules produced by plant roots that attract the hyphae of mycorrhizal fungi.

strobilus (plural: strobili) One of several conelike structures in various groups of plants (including club mosses, horsetails, and conifers) associated with the production and dispersal of reproductive products. Contrast with cone.

stroke volume The volume of blood the heart pumps per beat.

stroma The aqueous fluid inside of the chloroplast surrounding the thylakoid membranes. The location of carbohydrate synthesis during photosynthesis.

structural gene A gene that encodes the primary structure of a protein not involved in the regulation of gene expression.

structural isomers Molecules made up of the same kinds and numbers of atoms, in which the atoms are bonded differently.

structural protein A protein that is involved in physical stability or movement of a cell or organism.

style [Gk. *stylos*: pillar or column] In the angiosperm flower, a column of tissue extending from the tip of the ovary, and bearing the stigma or receptive surface for pollen at its apex.

subduction In plate tectonics, the movement of one plate under another.

suberin A waxy substance present in the walls of some plant cells; helps create the water-impermeable nature of endodermal cell walls that have a Casparian strip.

subsoil The soil horizon lying below the topsoil and above the parent rock (bedrock); the zone of infiltration and accumulation of materials leached from the topsoil. Also called the B horizon.

substrate (1) The molecule or molecules on which an enzyme exerts catalytic action. (2) The base material on which a sessile organism lives.

substrate-level phosphorylation The transfer of a phosphate group from one molecule to another.

succession A sequence of relative predictable changes in the composition of an ecological community following disturbance that eventually restores a community resembling the pre-disturbance community.

succulence In plants, possession of fleshy, water-storing leaves or stems; an adaptation to dry environments.

sugar–phosphate backbone The repeating deoxyribose then phosphate groups that form a strand of a nucleic acid.

sun compass A mechanism by which an animal uses the position of the Sun and an internal clock to determine compass direction in navigation.

surface area-to-volume ratio For any cell, organism, or geometrical solid, the ratio of surface area to volume; this is an important factor in setting an upper limit on the size a cell or organism can attain.

surface tension The attractive intermolecular forces at the surface of liquid; an especially important property of water.

survivorship The average individual's chance of surviving over a specified set of life stages, ages, or period of time—often from birth to a given stage or age.

suspension feeding In aquatic animals, a method of obtaining food by collecting small food particles suspended in the water. Sometimes called filter feeding.

suspensor In the embryos of seed plants, the stalk of cells that pushes the embryo into the endosperm and is a source of nutrient transport to the embryo.

swim bladders Organs used to regulate buoyancy in fish.

symbiotic A relationship in which two or more organisms live in close association with one another.

symmetry Pertaining to an attribute of an animal body in which at least one plane can divide the body into similar, mirror-image halves. (*See* bilateral symmetry, radial symmetry.)

sympathetic division A division of the vertebrate autonomic nervous system that is connected to the central nervous system via thoracic and lumbar spinal nerves; the sympathetic and parasympathetic divisions tend to exert opposing controls over autonomic effectors. (Contrast with parasympathetic division.)

sympatric speciation (sim pat' rik) [Gk. *sym*: same + *patria*: homeland] Speciation due to reproductive isolation without any physical separation of the population. (Contrast with allopatric speciation.)

symplast The continuous meshwork of the interiors of living cells in the plant body, resulting from the presence of plasmodesmata. (Contrast with apoplast.)

synapomorphy A trait that arose in the ancestor of a phylogenetic group and is present (sometimes in modified form) in all of its members, thus helping delimit and identify that group. Also called a shared derived trait. (Contrast with derived trait.)

synapse (sin' aps) [Gk. *syn*: together + *haptein*: to fasten] A specialized type of junction where a neuron communicates with its target cell (which can be another neuron or some other type of cell). Can be chemical or electrical. *See* chemical synapse, electrical synapse.

synaptic cleft The space between the presynaptic cell and the postsynaptic cell in a chemical synapse.

synaptic plasticity The process by which synapses in the nervous system of an individual animal can undergo long-term changes in their functional properties. Believed to play roles in learning and memory.

syncytium Many nuclei surrounded by cytoplasm and enclosed within one cell membrane, as, for example, in superficial cleavage in insects.

synergid In flowering plants, a type of cell in the female gametophyte that is located next to the egg cell, produces a chemical signal that attracts the pollen tube.

synonymous substitution A change of one nucleotide in a sequence to another when that change does not affect the amino acid specified (i.e., UUA → UUG, both specifying leucine). Also called silent substitution. (Contrast with nonsynonymous substitution, nonsense mutation.)

system A set of interacting parts in which neither the parts nor the whole can be understood without taking into account the interactions among the parts.

systematics The scientific study of the diversity and relationships among organisms.

systemic circuit In a circulatory system, the blood vessels that take blood to and from, and through, the systemic tissues. (Contrast with breathing-organ circuit.)

systemic tissues All tissues other than those of the breathing organs (lungs or gills).

systems analysis A process in which the parts or components of a biological system are identified and the processes by which the components interact are specified.

systems diagram A visual representation of a system that shows its components and their interactions.

systole (sis' tuh lee) [Gk. *systole*: contraction] During each repetition of the cardiac cycle, the period of contraction of the chambers of the heart. (Contrast with diastole.)

T

T tubules *See* transverse tubules.

T-helper (T_H) cell A type of T cell that stimulates events in both the cellular and humoral immune responses by binding to the antigen on an antigen-presenting cell; target of the HIV virus, the agent of AIDS. (Contrast with cytotoxic T cell.)

tardigrades *See* Appendix A.

target cell A cell with the appropriate receptor proteins to bind and respond to a particular hormone or other chemical mediator (e.g., neurotransmitter).

TATA box An eight-base-pair sequence, found about 25 base pairs before the starting point for transcription in many eukaryotic promoters, that binds a general transcription factor and helps initiate transcription.

taxon (plural: taxa) [Gk. *taxis*: put in order] A biological group (typically a species or clade) that is given a name.

telomerase An enzyme that catalyzes the addition of telomeric sequences lost from chromosomes during DNA replication.

telomeres (tee' lo merz) [Gk. *telos*: end + *meros*: units, segments] Repeated DNA sequences at the ends of eukaryotic chromosomes.

telophase (tee' lo phase) [Gk. *telos*: end] The final phase of mitosis or meiosis during which chromosomes become diffuse, nuclear envelopes re-form, and nucleoli begin to reappear in the daughter nuclei.

template A molecule or surface on which another molecule is synthesized in complementary fashion, as in the replication of DNA.

template strand In double-stranded DNA, the strand that is transcribed to create an RNA transcript. Also refers to a strand of RNA that is used to create a complementary RNA. (Compare coding strand.)

tendon A collagen-containing band of connective tissue that connects a muscle with a bone.

termination In molecular biology, the end of transcription or translation.

territory A region that an individual occupies and defends against occupation by other individuals of the same species. (Contrast with home range.)

tertiary consumer An organism that consumes secondary consumers.

tertiary endosymbiosis The mechanism by which some eukaryotes acquired the capacity for photosynthesis; for example, a dinoflagellate that apparently lost its chloroplast became photosynthetic by engulfing another protist that had acquired a chloroplast through secondary endosymbiosis.

tertiary structure In reference to a protein, the relative locations in three-dimensional space of all the atoms in the molecule. The overall shape of a protein. (Contrast with primary, secondary, quaternary structures.)

test cross Mating of a dominant-phenotype individual (who may be either heterozygous or homozygous) with a homozygous-recessive individual.

testis (tes' tis) (plural: testes) [L. *testis*: witness] The male gonad; the organ that produces the male gametes. Also sometimes called a testicle.

testosterone The principal androgen (masculinizing steroid hormone) in vertebrates, produced by the interstitial cells (Leydig cells) of the testes.

tetraploid Having a chromosome complement consisting of three copies (homologs) of each chromosome. Designated 4*n*.

tetrapods *See* Appendix A.

thallus The vegetative body of a lichen.

thaumarchaeotes *See* Appendix A.

theory [Gk. *theoria*: analysis of facts] A far-reaching explanation of observed facts that is supported by such a wide body of evidence, with no significant contradictory evidence, that it is scientifically accepted as a factual framework. Examples are Newton's theory of gravity and Darwin's theory of evolution. (Contrast with hypothesis.)

theory of island biogeography A theory proposing that the number of species on an island (or in another geographically defined and isolated area) represents a balance, or equilibrium, between the rate at which species immigrate to the island and the rate at which resident species go extinct.

therians *See* Appendix A.

thermoneutral zone (TNZ) [Gk. *thermos*: temperature] In a homeotherm, the range of ambient temperatures over which the animal is under neither heat nor cold stress and the metabolic rate is constant.

thermoreceptor A sensory receptor cell that is sensitive to temperature and generates action potentials in response.

thermoregulation The maintenance of a relatively constant body temperature.

theropods *See* Appendix A.

thylakoid (thigh la koid) [Gk. *thylakos*: sac or pouch] A flattened sac within a chloroplast. Thylakoid membranes contain all of the chlorophyll in a plant, in addition to the electron carriers of photophosphorylation. Thylakoids stack to form grana.

thyroid-stimulating hormone (TSH) A hormone, released by the anterior pituitary gland, that signals the thyroid gland to secrete its hormones.

tidal ventilation Bidirectional pumping of air in and out of lungs, during which air enters and leaves the lungs by the same route. (Contrast with unidirectional ventilation.)

tidal volume The volume of air breathed in and out of the lungs during each breath.

tight junction A junction between epithelial cells in which there is no gap between adjacent cells.

tissue A group of similar cells organized into a functional unit; usually integrated with other tissues to form part of an organ.

tissue fluid *See* interstitial fluid (which is a synonym).

titin A giant, elongated, partially elastic protein molecule that in a muscle cell spans an entire half-sarcomere from the M band to the Z line, serving to stabilize the position of a myosin filament and help restore muscle length after stretching.

Toll-like receptors (TLRs) Membrane proteins of macrophages and dendritic cells with extracellular domains that recognize molecules from microbial pathogens.

tonicity A measure of the relative concentration of membrane-impermeable solutes on either side of a membrane. Determines the direction and extent of net water movement across a membrane.

topography Variation in the elevation of Earth's surface, in the form of mountains, valleys, ocean basins, and so on.

topsoil The uppermost soil horizon; contains most of the organic matter of soil, but may be depleted of most mineral nutrients by leaching. Also called the A horizon.

totipotent [L. *toto*: whole, entire + *potens*: powerful] Possessing all the genetic information and other capacities necessary to form an entire individual. (Contrast with multipotent, pluripotent.)

totipotent stem cells Stem cells possessing all the genetic information and other capacities necessary to form an entire individual. (Contrast with multipotent stem cells, pluripotent stem cells.)

trachea (tray' kee ah) [Gk. *trakhoia*: tube] (1) In a vertebrate lung, the principal tube leading into the lungs from the mouth cavity. (2) One of the airways in the tracheal breathing system of an insect.

tracheal breathing system The type of breathing system found in insects, in which airways (called tracheae and tracheoles) branch throughout the animal's body so that airways are nearby all cells.

tracheid (tray' kee id) A type of tracheary element found in the xylem of nearly all vascular plants, characterized by tapering ends and walls that are pitted but not perforated. (Contrast with vessel element.)

tracheoles In the tracheal breathing system of an insect, the finest gas-filled tubules.

tracheophytes *See* Appendix A.

trait In genetics, a specific form of a character: eye color is a character; brown eyes and blue eyes are traits. (Contrast with character.)

transcription factors Proteins that assemble on a eukaryotic chromosome, allowing RNA polymerase II to perform transcription.

transcription The synthesis of RNA using one strand of DNA as a template.

transcriptome The subset of the genome that is expressed as RNA in a particular cell or tissue at a particular time.

transduction In physics, the conversion of energy from one form to another. In the study of sensory receptor cells, the conversion of stimulus energy to a cellular electric signal.

transfection Insertion of recombinant DNA into animal cells.

transfer RNA (tRNA) A family of folded RNA molecules. Each tRNA carries a specific amino acid and anticodon that will pair with the complementary codon in mRNA during translation.

transformation (1) A mechanism for transfer of genetic information in bacteria in which pure DNA from a bacterium of one genotype is taken in through the cell surface of a bacterium of a different genotype and incorporated into the chromosome of the recipient cell. (2) Insertion of recombinant DNA into a host cell.

transgenic Containing recombinant DNA incorporated into the genetic material.

transgenic organism An organism engineered to contain, and usually express, a gene from another organism.

translation The synthesis of a protein (polypeptide). Takes place on ribosomes, using the information encoded in messenger RNA.

translational repressor A protein that binds to an mRNA and interferes with translation.

translocation (1) In genetics, a rare mutational event that moves a portion of a chromosome to a new location, generally on a nonhomologous chromosome. (2) In vascular plants, the movement of solutes in the phloem.

transmembrane protein An integral membrane protein that spans the phospholipid bilayer.

transpiration [L. *spirare*: to breathe] The evaporation of water from plant leaves and stems, driven by heat from the Sun, and providing the motive force to raise water (plus mineral nutrients) from the roots.

transpiration–cohesion–tension mechanism The theoretical basis for water movement in plants: evaporation of water from cells in leaves (transpiration) causes an increase in surface tension, pulling water up through the xylem. Cohesion of water occurs because of hydrogen bonding.

transport proteins Proteins that carry substances within an organism and across biological membranes.

transposons Mobile DNA segments that can insert into a chromosome and cause genetic change.

transverse tubules (T tubules) Indentations of the cell membrane at regular intervals over the entire surface of a muscle cell. Transverse tubules conduct electrical excitation into the interior of the cell and thus play a key role in linking electrical excitation with contraction.

tree of life A term that encompasses the evolutionary history of all life, or a graphic representation of that history.

triglyceride A simple lipid in which three fatty acids are combined with one molecule of glycerol.

triploblastic Having three cell layers. (Contrast with diploblastic, monoblastic.)

triploid Having a chromosome complement consisting of three copies (homologs) of each chromosome. Designated $3n$.

trochophore (troke' o fore) [Gk. *trochos*: wheel + *phoreus*: bearer] A radially symmetrical larval form typical of annelids and mollusks, distinguished by a wheel-like band of cilia around the middle.

trophic [Gk *trophes*: nourishment] Referring to feeding.

trophic cascade Effects of abundances of species at one trophic level on species at other trophic levels. For example, introduction of a tertiary consumer to a community may reduce the population sizes of secondary consumers that it eats, which in turn may increase the population sizes of primary consumers that the secondary consumers eat, which may reduce the population sizes of primary producers that the primary consumers eat.

trophic level Hierarchical feeding positions in a food web. Levels are defined in terms of the number of feeding links away from the primary source of energy for the web (usually the Sun).

trophoblast [Gk *trophes*: nourishment + *blastos*: sprout] In mammalian development, the outer group of cells of the blastocyst; will form the chorion and thus become part of the placenta, thereby helping nourish the growing embryo. (Contrast with inner cell mass.)

tropic hormone A hormone produced by one endocrine gland that controls the secretion of hormones by another endocrine gland. Also called a tropin.

tropics The latitudes between the Tropic of Cancer and the Tropic of Capricorn (23.5°N and S, respectively), defined by the region of the planet in which the Sun is directly overhead at least once during each annual cycle.

tropin *See* tropic hormone.

tropomyosin [troe poe my' oh sin] A muscle protein that—by blocking or unblocking myosin-binding sites on actin filaments—controls the interactions of actin and myosin necessary for muscle contraction.

troponin A muscle protein that—by helping to control the position of tropomyosin—plays a key role in controlling the interactions of actin and myosin necessary for muscle contraction.

tuataras *See* Appendix A.

tube cell In flowering plants, one of the cells in the male gametophyte. Produces the pollen tube that contains the generative cell and, eventually, the sperm.

tube feet A unique feature of echinoderms; extensions of the water vascular system, which functions in gas exchange, locomotion, and feeding.

tuber In potato, an enlarged portion of the underground stem. More generally, a portion of a plant such as a root or stem that is unusually large, often serving as a site where carbohydrate reserves are stored.

tubulin A protein that polymerizes to form microtubules.

tumor necrosis factor A family of cytokines (growth factors) that causes cell death and is involved in inflammation.

tunicates *See* Appendix A.

turgid Swollen or distended. In plants, a cell is turgid when its water potential is lower than that of the surrounding environment.

turgor pressure [L. *turgidus*: swollen] The force exerted by a plant cell against the cell wall when it has taken up water by osmosis. (Contrast with pressure potential.)

turnover In ecology, the change through time or over space in the composition of an ecological community.

turtles *See* Appendix A.

Type I error The incorrect rejection of a true null hypothesis.

Type II error The incorrect acceptance of a false null hypothesis.

U

ubiquitin A small protein that is covalently linked to other cellular proteins identified for breakdown by the proteosome.

ultimate causes (1) In the study of behavior, the causes that led to the evolution of a behavior over evolutionary time. (2) In the study of evolution, historical explanations of the processes that led, over evolutionary time, to the evolution of an adaptation. (Contrast with proximate causes.)

ultimate explanations The historical explanations of the processes that led to the evolution of an adaptation. (Contrast with proximate explanations.)

ultraviolet wavelengths Wavelengths in the electromagnetic spectrum that are too short to be seen by the human eye (i.e., shorter than 0.4 μm).

unidirectional ventilation Forced flow (of water or air) over or through a breathing organ in a single direction from the point of intake to the point of outflow. (Contrast with tidal ventilation.)

unsaturated fatty acid A fatty acid whose hydrocarbon chain contains one or more double bonds. (Contrast with saturated fatty acid.)

upwelling A process by which offshore winds in combination with the Coriolis effect push warmer surface water away from the shore, allowing cold, nutrient-rich water from deeper waters to rise to the surface.

V

vaccine An inactive form of a pathogen that is still antigenic (e.g., a heat-killed bacterium or chemically treated virus), or an antigenic component of the pathogen (e.g., an inactive bacterial toxin or virus surface protein) administered to trigger a primary immune response that prepares the body to mount a stronger, quicker secondary response if it encounters the actual pathogen.

vacuole (vac′ yew ole) A membrane-enclosed organelle in plant cells that can function for storage, water concentration for turgor, or hydrolysis of stored macromolecules.

valence shell The outermost energy shell of an atom, containing the valence electrons involved in chemical reactivity.

van der Waals interactions Weak electrostatic attractions between atoms resulting from temporary dipoles that spontaneously occur or are induced in the atoms. This type of attraction is about one-fourth as strong as a hydrogen bond.

variable A factor or variable that is compared to a constant.

variable region The portion of an immunoglobulin molecule or T cell receptor that includes the antigen-binding site and is responsible for its specificity. (Contrast with constant region.)

vascular cambium (kam′ bee um) [L. *cambiare*: to exchange] In plants, a lateral meristem that gives rise to secondary xylem and phloem.

vascular endothelium In the walls of all blood vessels, the layer of cells—a simple epithelium—that immediately surrounds the blood.

vascular plants *See* Appendix A.

vascular tissue system The transport system of a vascular plant, consisting primarily of xylem and phloem. (Contrast with dermal tissue system, ground tissue system.)

vasomotor control Related to changes in the inside (luminal) diameters of blood vessels mediated by contraction and relaxation of smooth muscles in the blood vessel walls. Changes in inside diameter have large effects on rate of blood flow through a vessel.

vegetal hemisphere The yolky portion of some animal eggs. (Contrast with animal hemisphere.)

vegetative cells (1) Generally: non-reproductive cells. (2) The photosynthesizing cells of a cyanobacterial colony.

vegetative growth In a plant, a phase of development between germination and reproduction during which the plant produces leaves.

vegetative reproduction Asexual reproduction through the modification of stems, leaves, or roots.

vein [L. *vena*: channel] A blood vessel that returns blood to the heart. (Contrast with artery.)

velvet worm Elongated, segmented ecdysozoan with many pairs of soft, unjointed, claw-bearing legs.

ventilation Forced flow of air or water over or through structures used for breathing (external respiration).

ventral [L. *venter*: belly, womb] Toward or pertaining to the belly or lower side. (Contrast with dorsal.)

ventricle In reference to a heart, the principal pumping chamber. The walls of the ventricle are highly muscular, composed of cardiac muscle.

venule A microscopically small blood vessel draining a capillary bed that joins others of its kind to form a vein. (Contrast with arteriole.)

vernalization [L. *vernalis*: spring] Events occurring during a required chilling period, leading eventually to flowering.

vertebrates *See* Appendix A.

vesicle Within the cytoplasm, a membrane-enclosed compartment that is associated with other organelles; the Golgi apparatus is one example.

vessel element A type of tracheary element with perforated end walls; found only in angiosperms. (Contrast with tracheid.)

visceral mass A centralized, internal portion of a mollusk's body that contains the heart as well as digestive, excretory, and reproductive organs.

vitamin [L. *vita*: life] A carbon compound that an organism cannot synthesize but nevertheless requires in small quantities for normal growth and health.

voltage A measure of the difference in electric charge between two points. Also called potential difference.

voltage-gated channel A type of gated channel protein, located in a cell membrane, that opens or closes in response to changes in the local membrane potential.

W

water potential In osmosis, the tendency for a system (a cell or solution) to take up water from pure water through a differentially permeable membrane. Water flows toward the system with a more negative water potential. (Contrast with solute potential, pressure potential.)

water vascular system In echinoderms, a network of water-filled canals that functions in gas exchange, locomotion, and feeding.

weather The atmospheric conditions of temperature, humidity, precipitation, and wind speed and direction experienced at a given place and time. (Contrast with climate.)

wet-nosed primates *See* Appendix A.

white blood cells Cells in the blood plasma that play defensive roles in the immune system. Also called leukocytes.

white fat In mammals, fat tissue that stores energy and provides thermal insulation that helps regulate body temperature.

white matter In the central nervous system, tissue that is rich in myelinated axons and therefore has a glistening white appearance.

whorl In a flower, the floral organs arranged in a circle that are inserted at a single level, for example, all of the sepals.

wild type Geneticists' term for standard or reference type. Deviants from this standard, even if the deviants are found in the wild, are usually referred to as mutant. (Note that this terminology is not usually applied to human genes.)

wood Secondary xylem tissue.

X

xerophyte (zee' row fyte) [Gk. *xerox*: dry + *phyton*: plant] A plant adapted to an environment with limited water supply.

xylem (zy' lum) [Gk. *xylon*: wood] In vascular plants, the tissue that conducts water and minerals; xylem consists, in various plants, of tracheids, vessel elements, fibers, and other highly specialized cells.

Y

yeasts *See* Appendix A.

yolk sac In reptiles (including birds) and mammals, the extraembryonic membrane that forms from the endoderm and lateral plate mesoderm and surrounds the yolk; the first source of blood and blood vessels. During development, the yolk sac digests the yolk, and its blood vessels carry products to the embryo.

Z

Z-ring Protein fibers on the inside of a prokaryotic cell membrane in the middle of the cell that tightens, causing cytokinesis.

zone of cell differentiation The part of a plant root, shootward of the zone of cell elongation, where cells are differentiating. Also called the maturation zone.

zone of cell elongation The part of a plant root, generally above the zone of cell division, where cells are expanding (growing), primarily in the longitudinal direction.

zone of polarizing activity (ZPA) Tissue near the base of the developing limb bud that determines the anterior–posterior axis of the developing limb.

zygospore A multinucleate, diploid cell that is a resting stage in the life cycle of zygospore fungi.

zygospore fungi *See* Appendix A.

zygote (zye' gote) [Gk. *zygotos*: yoked] The fertilized egg. The cell created by the union of two gametes, in which the gamete nuclei are fused. The earliest stage of the diploid generation.

Index

Numbers in *italic* indicate that the information will be found in an illustration or table.

A

D

I

M

Q

R